A TREATISE ON GENERAL PROPERTIES OF MATTER

Recommended as a textbook for Honours Course by the University of Calcutta
[vide Notification No. T/6/81 dated 19.06.1981]

A TREATISE ON GENERAL PROPERTIES OF MATTER

[For Degree Honours Students]

Hemendralal Chatterjee MSc
Principal and Professor of Physics [Retd], City College, Kolkata

Ratneswar Sengupta MSc
Ex-Head of the Department of Physics, Anandamohan College, Kolkata
Part-time Lecturer, Rammohan College, Kolkata
Formerly Professor of Physics, City College, and
Barisha Vivekananda College, Kolkata

New Central Book Agency (P) Ltd

Co-Published by

Bijay Publications
E-mail : bijaypublications@gmail.com

Exclusively Distributed by

Bijay Pustak Bhandar (P) Ltd
90/6A M. G. Road, Bhawani Dutta Lane, Kolkata-700007
✆ : (033) 4605-7856, 8777898576, 8274916104
E-mail : bijaypustakbhandar@gmail.com

NCBA
REGD OFFICE
8/1 Chintamoni Das Lane, Kolkata 700 009, India
email: ncbakolkata@gmail.com

CORPORATE OFFICE
501A/1 Right Side [Upper Ground Floor]
Govind Puri, Kalkaji, New Delhi 110 019
email: delhincba@gmail.com

BIJAY PUBLICATIONS
90/6A M. G. Road, Bhawani Dutta Lane, Kolkata-7
E-mail : bijaypublications@gmail.com

BIJAY PUSTAK BHANDAR (P) LTD
90/6A M. G. Road, Bhawani Dutta Lane, Kolkata-7
E-mail : bijaypustakbhandar@gmail.com

A TREATISE ON GENERAL PROPERTIES OF MATTER • Hemendralal Chatterjee, Rajneswar Sengupta

First Published: September 1981
Seventh Edition [Thoroughly Revised, Updated and Enlarged]: 2010
Reprinted: 2011, 2012, 2013, 2015, 2016, 2017, 2018, 2019, 2021, 2022, 2025

PUBLISHER • TYPESETTER • PRINTER
New Central Book Agency (P) Ltd
8/1 Chintamoni Das Lane, Kolkata 700 009

CO-PUBLISHED BY
Bijay Publications
90/6A M G Road, Bhawani Dutta Lane, Kolkata 700007

PROJECT TEAM
Pradip Biswas and Prabhat Jas

ISBN: 978 81 7381 174 6

Price ₹ 1250.00

3.39 Newton's Equation of Motion in Moving Frames of Reference 185
3.40 Rotating Frames of Reference 186
3.41 Velocity and Acceleration of a Particle Relative to a Rotating Frame 187
3.42 Centrifugal Acceleration due to the Earth's Rotation 188
3.43 Coriolis Force 189
3.44 Foucault's Pendulum 191
3.45 Motion of Projectiles 196
3.46 Collision of Particles—Direct and oblique impact of two bodies 200
3.47 Application of the principles of linear and angular momentum conservation in the case of a moving bullet striking with a large wooden block: The Ballistic pendulum 208

4 Dynamics of a Rigid Body—Moment of Inertia 237
4.1 Rotational Motion of a Rigid Body—Calculation of K.E. 237
4.2 Moment of Inertia and Moment of Momentum or Angular Momentum 237
4.3 Moment of Inertia and Torque 238
4.4 Radius of Gyration 238
4.5 Physical Significance of Moment of Inertia 238
4.6 General Theorems on Moment of Inertia 239
4.6.1 Theorem of perpendicular axes 239
4.6.2 Theorem of parallel axes 240
4.7 Calculation of Moment of Inertia for Different Cases 242
4.8 Momental Ellipsoid or Ellipsoid of Inertia 262
4.9 Plane Lamina 264
4.9.1 Discussion on Principal Axes and Principal Moments of Inertia 265
4.10 Kinetic Energy of a Rigid Body in Terms of Angular Velocity and Principal Moments of Inertia 269
4.11 Angular Momentum of a Rigid Body 270
4.12 Tensor of Inertia 271
4.13 Moments and products of inertia and hence ellipsoids of inertia of some symmetrical bodies 276
4.14 Routh's Rule 283
4.15 Experimental Determination of Moment of Inertia 284

5 Simple Harmonic Motion 297
5.1 Periodic Motion 297
5.2 Vibratory Motion 297
5.3 Simple Harmonic Motion 297
5.4 Time-period 298
5.5 Frequency 298
5.6 Amplitude 298
5.7 Phase 298
5.8 Calculation of Displacement of a Particle Executing S.H.M. 299
5.9 Calculation of Velocity 299
5.10 Calculation of Acceleration and Equation of Motion of S.H.M. 299
5.11 Energy of a Particle Executing S.H.M. 301
5.12 Potential Energy 301
5.13 Graphical Representation of S.H.M. 302
5.14 Examples of S.H.M. 302
5.15 Damped Harmonic Vibration 313
5.16 Forced Vibration 316
5.17 Composition of Simple Harmonic Motions 325

5.18 Composition of Two Rectangular Simple Harmonic Motions differing in Amplitude, Phase and Frequency—Frequencies being in the Ratio 1 : 2 329
5.19 Two Rectangular S.H. Motions differing in Amplitude, Phase and Frequency—Frequencies being in the Ratio 1 : 3 330
5.20 Graphical Treatment 331
5.20.1 Composition of two collinear simple harmonic motions of same period but differing in amplitude and phase 331
5.20.2 Composition of two rectangular simple harmonic motions 332
5.21 Composition of a Uniform Circular Motion from Two Simple Harmonic Motions at Right Angles 336
5.22 Composition of Two Uniform Circular Motions of same radius and periodic Time, but described in opposite directions 336
5.23 Lissajous' Figures—uses 337
5.24 Experimental Methods for Producing Lissajous' Figures 338

6 The Acceleration Due to Gravity 345
6.1 The Acceleration due to Gravity 345
6.2 Simple Pendulum 345
6.3 Laws of Simple Pendulum 347
6.4 Seconds Pendulum 348
6.5 Gain or Loss of Time by Pendulum Clock 348
6.6 Finite Arc Correction of a Pendulum: Anharmonic Oscillator 350
6.7 Simple Pendulum with a Movable Support 353
6.8 Conical Pendulum 355
6.9 Demerits of a Simple Pendulum 356
6.10 Borda's Pendulum 356
6.11 Compound Pendulum 357
6.12 Bar Pendulum 360
6.13 Kater's Pendulum 360
6.14 Bessel's Computed Time-Period 361
6.15 Different Corrections of Compound Pendulum 366
6.15.1 Finite Arc Correction 366
6.15.2 Air Correction 367
6.15.3 Viscous Drag of Air 369
6.15.4 Knife-edge Correction 371
6.15.5 Correction for Yielding of Support 372
6.16 Summary of Corrections in the use of a Compound Pendulum 374
6.17 Variation of g 375
6.17.1 Effect of altitude 375
6.17.2 Effect of Depth 376
6.17.3 Effect of Latitude 377
6.18 Different Methods for the Determination of g 379
6.18.1 Atwood's Machine 379
6.18.2 Inclined Plane 380
6.18.3 Bifilar Suspension Method 381
6.18.4 Body oscillating on a concave surface 384
6.19 Gravity Survey 386
6.19.1 Invariable Pendulum 386
6.19.2 Gravity Meter 387
6.19.3 Eötvös Balance 389
6.20 The Gravity Gradient and Horizontal Directive Tendency 393
6.20.1 The Gravity Gradient 393
6.20.2 Horizontal Directive Tendency 394

6.21	Geophysical Prospecting	394
6.22	Change in the Direction of Force of Gravity with Time— The Horizontal Pendulum	395
6.23	Measurement of g at Sea	396
Gravitation		**401**
7.1	Newton's Law of Gravitation	401
	7.1.1 Dimension of G	401
	7.1.2 Astronomical Units	402
7.2	Universality of the Law of Gravitation	402
7.3	Newton's Law of Gravitation and Einstein's Theory of Relativity	403
7.4	Gravitational Field or Attraction or Intensity of the Field	407
7.5	Attraction of a System of Particles	407
7.6	Gravitational Potential	408
7.7	Potential due to a System of Attracting Particles	409
7.8	Physical Interpretation of Potential	410
7.9	Equipotential Surface	410
7.10	Lines of Force	411
7.11	Calculation of Potential—Spherical Bodies	411
	7.11.1 Thin Spherical Shell	411
	7.11.2 Uniform Solid Sphere	413
	7.11.3 Thick Spherical Shell	415
7.12	Calculation of Attraction: Spherical Bodies	416
	7.12.1 Thin Spherical Shell	416
	7.12.2 Uniform Solid Sphere	418
	7.12.3 Thick Spherical Shell	419
7.13	Graphical representation of potential and attraction	420
7.14	Attraction between two portions of a solid sphere divided by a plane	422
7.15	Motion of a body in a tunnel bored through the earth	424
7.16	Uniform shell bounded by non-intersecting eccentric spheres	428
	7.16.1 Calculation of Potential	428
	7.16.2 Calculation of Intensity	428
7.17	Calculation of potential and intensity in some special cases	429
	7.17.1 Uniform Circular Disc or Plate	429
	7.17.2 Discussions	430
	7.17.3 Uniform ring or Circular hoop—potential and attraction	431
	7.17.4 Homogeneous Solid Cylinder	433
	7.17.5 Thin uniform straight rod	435
	7.17.6 Uniform Rod bent into the form of a circular arc	436
	7.17.7 Attraction at a point P due to a thin uniform straight Rod	437
7.18	Self-attracting System	441
7.19	Laplace's Equation for Potential	444
7.20	Poisson's Equation for Potential	445
7.21	The flux of gravitational intensity or attraction	446
	7.21.1 Gauss's theorem	446
7.22	Kepler's laws	449
	7.22.1 Deduction of Kepler's laws from Newton's law of gravitation	449
7.23	Deduction of Newton's Inverse Square Law of Gravitation from Kepler's Laws	459
7.24	Velocity of Escape	463
7.25	Artificial Satellites	464
	7.25.1 Condition to put a satellite into an orbit	465
	7.25.2 Period of a satellite revolving round the earth	466
	7.25.3 Geo-stationary Satellite	467

7.25.4 System of variable mass—motion of a rocket 470
7.26 Mass and Mean Density of the Earth 488
7.27 Airy's Mine Experiment 488
7.28 Determination of 'G'—Mountain Experiments 490
7.28.1 Bouger's Experiment 490
7.28.2 Maskelyne's Experiment 491
7.29 Determination of G—Laboratory Methods 491
7.29.1 Cavendish's Experiment 492
7.29.2 Boys' Modification of Cavendish's Experiment 493
7.29.3 Poynting's Method 495
7.29.4 Heyl's Method 497
7.30 Variation in G 500
7.30.1 Selectivity 500
7.30.2 Temperature 500
7.30.3 Directivity 500
7.30.4 Permeability 500

8 Gyroscopic Motion **514**
8.1 Precessional Torque 514
8.2 Gyroscope 515
8.3 Euler's Equation of Motion 515
8.4 Kinetic Energy of Force-free Rotation of a Rigid Body 517
8.5 Force-free Rotation of a Symmetrical Rigid Body 518
8.6 Precession of the Earth's Axis of Rotation 519
8.7 Top Motion: Theory of the Steady Precession of a Top 520
8.8 General Motion of a Symmetrical Top 523
8.9 Symmetrical Top (Alternative Treatment)—Eulerian Angles 527
8.10 Gyroscopic Applications 530
8.10.1 The Gyroscopic Compass 530
8.10.2 Pendulum Gyrocompass 530
8.10.3 Other Gyroscopic Applications 530

9 Elasticity **532**
9.1 Elastic Property 532
9.2 Strain 532
9.3 Stress 532
9.4 Characteristics of Homogeneous Strain 534
9.5 Principal Axes of Strain 535
9.6 Calculation of Volume Strain in terms of the three Principal Strains 536
9.7 Hooke's Law and Elastic Constants 536
9.7.1 Poisson's Ratio 538
9.8 Elastic Properties of Matter 540
9.9 Viscosity of Metals and Elastic Fatigue 541
9.10 A Shear is Equivalent to a Compression and an equal Extension at Right Angles to each other 541
9.11 Relations connecting Y, k, n and σ and to show that σ lies between $+0.5$ and -1 544
9.12 Relation between Axial Modulus and other Elastic Constants 545
9.12.1 Relations connecting Y, k, n and σ (Alternative Treatment) 546
9.13 Torsion of a Cylinder or Wire 552
9.13.1 A Hollow shaft is stronger than a Solid one of the same mass 553
9.14 Torsional Rigidities of Bars of Non-circular Cross-sections—St. Venant's Approximate Formula 554

9.15	Elasticities of Liquids and Gases	557
9.16	Isothermal and Adiabatic Elasticities	557
9.17	Bending of Beams	558
	9.17.1 Internal bending moment	558
9.18	Cantilever	559
9.19	Reciprocal Theorem	565
9.20	A weightless beam supported near the ends on two knife edges held in the same horizontal plane and a concentrated load W is applied at the mid-point	566
	9.20.1 Beam supported near the ends on two knife edges held in the same horizontal plane, and a concentrated load W is applied at the mid-point, weight of the beam being taken into consideration	568
9.21	Uniform bending of a beam equally loaded at two free ends, and supported on two knife edges at equal distances from the two ends	569
9.22	Upward displacement at any point of a beam bent in the form of a circular arc, convex outward by suspending two equal weights at equal distances from the two knife edges	570
9.23	General method of determining deflection due to bending of a beam—application to calculate deflections in different cases	571
9.24	Transverse Vibration of a Loaded Cantilever	582
9.25	Bending of Beams with Considerable Curvature	583
9.26	Rods under Thrust—Struts	585
9.27	Strain Energy	588
	9.27.1 Bending of a beam: Strain Energy	589
	9.27.2 Twisting of a rod of circular cross-section: Strain Energy	591
9.28	Flat Spiral Springs	591
	9.28.1 Angular oscillation	594
9.29	Non-flat Spiral Spring	594
9.30	Experimental Determination of Elastic Constants	596
	9.30.1 Young's modulus	596
	9.30.2 Modulus of rigidity	598
	9.30.3 Searle's method for the determination of Young's modulus and modulus of rigidity	599
	9.30.4 Determination of Poisson's ratio	601
9.31	Variation of Elasticity with Temperature	603
10	**Surface Tension**	**615**
10.1	Introduction	615
	10.1.1 Alternative definition of Surface Tension—Surface Energy and Surface Tension	616
	10.1.2 Case 1. Spraying of a Liquid into Droplets	619
	10.1.3 Case 2. Inverse Phenomenon of Case 1	619
10.2	Angle of Contact	623
10.3	General consideration of interfacial Surface Tension	624
10.4	Neumann's Triangle	625
10.5	Measurement of Angle of Contact	626
10.6	Shape of Liquid Meniscus in a Capillary Tube	628
10.7	Capillarity—Elevation or Depression of a liquid in a Capillary Tube	630
	10.7.1 Rise of a liquid in a Capillary Tube	630
	10.7.2 Capillary tube of insufficient length	631
	10.7.3 Energy changes associated with capillary rise	632
	10.7.4 Expression for the capillary rise	632
10.8	Pressure on a Curved Membrane of Uniform Tension	634
	10.8.1 Rise of a Liquid in a Conical Capillary tube	635

10.9	Shape of Films	636
10.10	Instability of Cylindrical Films	640
10.11	Shape of an interfacial surface	646
10.12	Rise of a liquid along an inclined solid surface	647
	10.12.1 Rise of a liquid along the side of a plane plate dipped vertically	650
	10.12.2 Sessile drop—Theory of Quincke's method	652
10.13	Condition of equilibrium of a floating needle	654
10.14	Rise of a liquid between two vertical plane plates making a small angle with each other	656
	10.14.1 Rise of a liquid between two vertical parallel plates	657
	10.14.2 Rise of water between two vertical parallel plates close to each other	658
10.15	Force between two parallel plates separated by a thin layer of liquid, due to surface tension	659
10.16	Force to pull a horizontal plate upwards from a liquid surface	660
10.17	Liquid pulled upwards by a horizontal flat ring	662
10.18	Surface Tension and Vapour Pressure	662
	10.18.1 Surface Tension and Evaporation	664
	10.18.2 Alternative Treatment	665
10.19	Surface Tension and Boiling	666
10.20	Explanation of Surface Tension: Laplace's theory of capillary action	667
	10.20.1 Estimation of the range of molecular action (C)	670
	10.20.2 Intrinsic pressure (K) and its estimation	670
10.21	Dependence of Surface Tension on Temperature	673
10.22	Experimental Determination of Surface Tension	675
	10.22.1 Capillary rise method (Laboratory method)	675
	10.22.2 Balance method	676
	10.22.3 Measurement of surface tension and angle of contact by Quincke's method	677
	10.22.4 Jaeger's method	678
	10.22.5 Sugden's theory and experiment	679
	10.22.6 *U*-tube arrangement for measuring surface tension of a liquid at different temperatures	683
	10.22.7 Drop-weight method	684
	10.22.8 Rayleigh's jet method	686
	10.22.9 Rayleigh's ripple method	688
10.23	Some practical aspects of surface tension	693
11	**Viscosity**	**705**
11.1	Introduction	705
11.2	Coefficient of viscosity	705
	11.2.1 Dimension of η	706
11.3	Fugitive elasticity	706
11.4	Streamline and Turbulent motion—Critical velocity	707
11.5	Experimental demonstration of streamline and turbulent motion	708
11.6	Poiseuille's formula for flow of an incompressible liquid through a horizontal narrow tube	708
	11.6.1 Alternative treatment	710
11.7	Flow of an incompressible liquid through the annular space between two concentric cylinders	711
11.8	Corrections to Poiseuille's formula	712
11.9	Downward flow of a liquid through a vertical narrow tube	713
11.10	Flow of a liquid through an inclined narrow tube	715
11.11	Flow of a liquid through capillaries in series and in parallel	716

11.12 Torque acting on a cylinder immersed in a rotating fluid 722
11.13 Viscosity of high viscous liquids 723
11.13.1 Discussions 724
11.14 Motion of a body falling through a viscous medium: Terminal Velocity 725
11.15 Flow of a gas through a narrow tube 728
11.16 Dependence of viscosity on temperature 730
11.16.1 Liquids 730
11.16.2 Gases 731
11.17 Effect of pressure on the viscosity 731
11.17.1 Liquids 731
11.17.2 Gases 731
11.18 Experimental determination of the viscosity of liquids 732
11.18.1 Viscosity of water 732
11.18.2 Viscosity of a liquid by Couette's method 732
11.18.3 Measurement of variation of viscosity with temperature 733
11.18.4 Viscometer for comparison of viscosities 733
11.18.5 Measurement of viscosity of a liquid by Searle's viscometer (Co-axial cylinder method) 735
11.18.6 Alternative arrangement for co-axial cylinder method 735
11.18.7 Determination of the viscosity of a transparent viscous liquid by Stokes' method 737
11.19 Experimental determination of viscosity of gases 738
11.20 Non-Newtonian liquids 743

12 Fluid Mechanics **753**
12.1 Introduction 753
12.2 Fluid statics 753
12.3 Condition of equilibrium of a fluid 754
12.4 Fluid dynamics 756
12.5 General equations of fluid motion—Euler's equation of motion 757
12.6 Equation of continuity 759
12.7 Bernoulli's equation and Conservation of energy 760
12.8 Bernoulli's theorem 761
12.9 Experimental arrangement for the illustration of Bernoulli's theorem 766
12.10 Applications of Bernoulli's theorem 767
12.10.1 Venturimeter 767
12.10.2 Velocity of efflux of a liquid—Torricelli's Theorem 768
12.10.3 Pitot tube 769
12.11 Bernoulli's equation for gases 771
12.12 Efflux of gases 772
12.13 Derivation (alternative), from the principle of conservation of energy, of Bernoulli's equation for variation of pressure along a streamline 773

13 Osmosis and Diffusion **779**
13.1 Osmosis 779
13.2 Abbe Nollet's Experiment—Demonstration of osmotic pressure 779
13.3 Van't Hoff's Equation and the laws of osmotic pressure 780
13.3.1 Conclusion—Solution behaves like a gas 781
13.4 Determination of osmotic pressure 781
13.4.1 Pfeffer's method 781
13.4.2 Berkeley and Hartley's apparatus 782
13.4.3 Measurement of osmotic pressure 782
13.4.4 Verification of the laws of osmotic pressure 782

13.5	Theories of osmotic pressure and semi-permeability	783
	13.5.1 Theory of osmotic pressure	783
	13.5.2 Theory of semi-permeability	783
13.6	Relations between the osmotic pressure and the other properties of a solution	783
	13.6.1 Relation between osmotic pressure and vapour pressure	783
	13.6.2 Relation between osmotic pressure and boiling point	784
	13.6.3 Relation between osmotic pressure and freezing point	785
	13.6.4 Relation between osmotic pressure and association or dissociation of molecules	785
	13.6.5 Illustrative examples	785
13.7	Diffusion	786
13.8	Fick's Law	787
13.9	Differential equation for diffusion	787
13.10	Determination of the coefficient of diffusion	788
	13.10.1 Inter-diffusion of Solids	791
14	**Kinetic Theory of Matter**	**792**
14.1	Introduction	792
14.2	Kinetic Theory of Gases—Pressure exerted by a Perfect Gas	792
14.3	Expression for Pressure of a Perfect Gas	793
14.4	Deductions from the Pressure Expression	795
	14.4.1 Boyle's law	795
	14.4.2 Charle's law	796
	14.4.3 Avogadro's law	796
	14.4.4 Dalton's law	796
14.5	Maxwell's Law of Distribution of Molecular Velocities	797
14.6	Calculation of Average Velocity, Mean Square Velocity and the Most Probable Velocity on the basis of Maxwell's Velocity Distribution Law	799
	14.6.1 Average velocity	799
	14.6.2 Mean square velocity	800
	14.6.3 The value of C, for which dn_C is maximum, is the most probable velocity C_p	800
14.7	Evaluation of the Constant λ and Discussion of the Different Results	800
14.8	Energy Distribution of Molecules	802
14.9	Temperature Dependence and Reduced form of Velocity Distribution	803
	14.9.1 Reduced Form of Maxwell's law	803
14.10	Experimental Verification of Maxwell's Law of Distribution of Velocities	803
14.11	Law of Equipartition of Energy	804
14.12	Degrees of Freedom and Ratio of Specific Heats	806
14.13	Mean Free Path—Clausius's and Maxwell's mean-free path	807
14.14	Probability of no collision and the Law of Distribution of Free paths—Expression for mean free path from it	809
14.15	Deduction of the Expression for the mean free path from Maxwell's law of the Distribution of velocities	811
14.16	Experimental determination of the mean free path	812
14.17	Transport phenomena in gases	814
	14.17.1 Viscosity from Kinetic theory	814
	14.17.2 Thermal conductivity from kinetic theory	816
	14.17.3 Diffusion in gases	817
14.18	General treatment for the transport phenomena	819
14.19	Brownian Movement	822
14.20	Brownian Movements and the Law of Atmospheres	823

14.21	The displacement of the particles undergoing Brownian Movements and verification of Einstein's equation	825
14.22	Einstein's theory of translational Brownian Movement	827
14.23	Translational Brownian Movement in gases	828
14.24	Rotational Brownian Movement	829
14.25	Van der Waals' equation of state	830
14.26	Deduction of Van der Waals' equation from Virial Theorem	831
15	**Exhaust Pumps and Gauges**	**847**
15.1	Production and measurement of low pressures	847
15.2	The Speed of a pump	847
15.3	Conductance of a tube	848
15.4	Water-jet pumps	849
15.5	Sprengel pump	849
15.6	Geissler pump	849
15.7	Rotary oil pumps	850
	15.7.1 Gaede's rotary oil pump	850
	15.7.2 Cenco rotary oil pump	851
15.8	Molecular pump	851
15.9	Diffusion-Condensation pumps	851
	15.9.1 Gaede's diffusion pump	851
	15.9.2 Langmuir's pump	852
15.10	Pressure Gauges	853
	15.10.1 McLeod gauge	853
	15.10.2 Pirani gauge	854
	15.10.3 Ionisation gauge	854
	15.10.4 Knudsen gauge	855
16	**Fourier Analysis**	**858**
16.1	Fourier's Theorem	858
16.2	Fourier Series	858
16.3	Evaluation of the constants a_0 and a_r	859
16.4	Evaluation of the constants A_r and B_r	860
16.5	Fourier Series involving phase angles	861
16.6	Fourier Representation of Even and Odd Functions	862
	16.6.1 Evaluation of Riemann Zeta Function by the method of Fourier Expansion	863
16.7	Examples illustrating 'Analysis and Synthesis' by Fourier's Theorem	865
16.8	Expansion of a periodic function $f(x)$	868
16.9	Extension of the Range	869
16.10	Half-Range Series	870
16.11	Advantages of Fourier Series	870
16.12	Importance of Fourier's Theorem	871
17	**Differential Equations**	**886**
17.1	Classification of differential equations	886
17.2	Linear Differential Equations of the First order	888
17.3	Exact and Inexact differential equations: Use of suitable integrating factors for reducing inexact equations to exact ones	891
17.4	Solution of second order differential equations	893
	17.4.1 Second Order Linear Differential Equations with Constant Coefficients	893
	17.4.2 Determination of the Complementary Function: Application of differential operator cum trial solution method	895

17.5 Determination of the Particular Integral 898
17.5.1 Method of undetermined coefficients 898
17.5.2 Operator method 900
17.6 Applications of Linear Equations to problems in Physics 907
17.7 Power Series Solutions of Differential Equations 920
17.8 Gamma function 923
17.9 Frobenius' Method for Power Series Solutions of Differential Equations 924
17.10 Bessel's differential equation and its solution 926
17.10.1 Generating Function for $J_n(x)$ 929
17.10.2 Recurrence formula for $J_n(x)$ 930
17.11 Legendre's differential equation and its solution 932
17.11.1 Generating Function for Legendre's Polynomial $P_n(x)$ 935
17.11.2 Rodrigues' formula for the Legendre's Polynomials 937
17.11.3 Properties of Legendre's polynomial 938
17.11.4 Associated Legendre's equation 939
17.11.5 Recurrence formula for $P_n(x)$ 940
17.12 Hermite's Differential Equation and its solution 940
17.13 Hermite's Polynomials and their generating function 943
17.13.1 Recurrence Formulae for $H_n(x)$ and to show that $H_n(x)$ is a solution of Hermite's equation 945
17.13.2 Orthogonal property of $H_n(x)$ 946

18 Some Important Types of Partial Differential Equations occurring in Physics and their Solutions **966**
18.1 Introduction 966
18.1.1 Three Dimensional Laplace's equation in Cartesian Co-ordinates 966
18.1.2 The electric potential in the steady flow of electric current 967
18.1.3 The potential flow of a fluid 968
18.2 Method of separation of variables 968
18.3 Solution of three-dimensional Laplace's Equation in Cartesian Co-ordinates 969
18.3.1 Three-dimensional Laplace's Equation in Cylindrical Co-ordinates 970
18.3.2 Solution of three-dimensional Laplace's Equation in Cylindrical Co-ordinates 971
18.4 Three-Dimensional Laplace's Equation in Spherical Polar Co-ordinates 974
18.4.1 Solution of three-dimensional Laplace's Equation in Spherical Polar Co-ordinates 976
18.5 Solution of two-dimensional Laplace's Equation in Cartesian Co-ordinates—Problem of Steady-State flow of heat 979
18.6 Application of Partial Derivatives to Thermodynamical problems 984
18.6.1 Wave equation 985
18.6.2 Fourier Equation for Heat Conduction 986
18.7 Solution of one-dimensional wave equation by D'Alembert's method 986
18.7.1 Solution by the method of separation of variables or the Fourier method 987
18.8 Solution of two-dimensional wave equation in rectangular and polar co-ordinates 990
18.8.1 Transverse vibrations of a rectangular membrane (rectangular co-ordinates) 990
18.8.2 Vibrations of a circular membrane (Polar co-ordinates) 993
18.9 Solution of one-dimensional Fourier equation for heat conduction by the method of separation of variables 994
18.10 Solution of three-dimensional Wave Equation 996
18.10.1 Cartesian Co-ordinates 996
18.10.2 Spherical Polar Co-ordinates 998
18.10.3 Cylindrical Polar Co-ordinates 999

19 Matrices **1013**
19.1 Introduction 1013
19.1.1 Linear equations and matrix notation 1013
19.2 Definitions 1014
19.3 Matrix addition 1015
19.4 Scalar multiplication 1015
19.5 Subtraction of matrices 1016
19.6 Null Matrix 1016
19.7 Laws of Matrix Algebra 1016
19.7.1 Commutative law of addition 1016
19.7.2 Associative law of addition 1016
19.7.3 Commutative law of scalar multiplication 1016
19.7.4 Associative law of scalar multiplication 1017
19.7.5 Distributive law for scalar multiplication 1017
19.8 Matrix multiplication 1017
19.9 Laws of Matrix multiplication 1018
19.9.1 Commutative law 1018
19.9.2 Associative law 1018
19.9.3 Distributive law 1019
19.10 Transpose of a matrix 1019
19.11 Properties 1020
19.12 Diagonal matrix 1021
19.12.1 Scalar matrix 1021
19.12.2 Unit matrix or Identity matrix 1021
19.13 Adjoint matrix 1022
19.13.1 Reciprocal matrix 1022
19.14 Symmetric matrix 1023
19.14.1 Skew (or anti-symmetric) matrix 1023
19.14.2 Complex conjugate matrix 1023
19.14.3 Hermitian conjugate 1024
19.14.4 Hermitian matrix 1024
19.14.5 Unitary matrix 1024
19.14.6 Co-factor matrix 1025
19.15 Orthogonal matrix 1025
19.15.1 Properties of an Orthogonal matrix 1025
19.16 Matrix division—Inverse of a matrix 1026
19.17 Trace of a matrix 1028
19.17.1 The derivative of a matrix 1028
19.17.2 The integral of a matrix 1029
19.17.3 Minor and Rank 1029
19.18 Applications of the matrix method 1029
19.18.1 Solution of problems involving matrix algebra 1029
19.18.2 Solution of a set of linear equations by matrix method 1033
19.18.3 Solution of the problem of linear inversion 1034
19.18.4 Diagonalisation of a matrix 1036
19.18.5 Diagonalisation Technique 1037
19.18.6 The Process of Diagonalisation of a matrix 1042
19.18.7 Cayley-Hamilton Theorem 1048

20 Infinite Sequences and Infinite Series **1065**
20.1 Introduction 1065
20.2 Different Types of Sequences 1066

20.3	Monotone Sequence	1066
20.4	Series—Infinite Series	1067
20.5	Different Types of Infinite Series	1067
20.6	Geometric Series	1068
20.7	Nature of Infinite Series	1069
20.8	General Principle of Convergence for Series	1070
20.9	The Comparison Test	1071
20.10	Cauchy's Integral Test	1072
20.11	D'Alembert's Ratio Test	1073
20.12	Cauchy's Ratio Test	1073
20.13	Cauchy's Root Test	1077
20.14	Raabe's Test	1077
20.15	Cauchy's Logarithmic Test	1078
20.16	Cauchy's Condensation Test	1078
20.17	Alternating Series	1079
20.18	Absolute and Conditional Convergence	1080
20.19	Power Series	1081
20.20	Theorems regarding Power Series	1082
20.21	Use of Series for the Computation of Functions	1087
21	**Complex-Valued Functions—Complex Variables**	**1102**
21.1	Introduction	1102
21.2	Geometrical interpretation of the imaginary number j	1103
21.3	Graphical representation of complex numbers—Argand Diagram	1103
21.4	Functions of a complex variable	1104
21.5	Analytic Function	1107
21.6	Necessary and sufficient conditions for $f(z)$ to be analytic	1108
21.7	Polar form of Cauchy-Riemann conditions	1111
21.8	Laplace's equations	1112
21.9	Zeros of an analytic complex function	1112
21.10	Singularities of an analytic function	1113
21.11	The point at infinity	1116
21.12	Evaluation of residues	1117
21.13	Methods of finding the residue at a finite pole	1119
22	**Functions of Several Real Variables**	**1132**
22.1	Real-valued Function of a Real Variable	1132
22.2	Functions in Higher Dimensional Space, i.e., Functions of more than One Variable	1132
22.3	Limit of a Function of Two Variables: Simultaneous Limit or Double Limit	1134
22.4	Repeated or Iterated Limits	1137
22.5	Continuity in R^2 for a function f of two variables x and y at a point (a, b) in the domain D of definition of the function f	1138
22.6	Directional Derivative: Partial Derivative—Differentiability and Differential	1140
22.7	Higher Order Partial Derivatives	1143
22.8	On the Equality of f_{yx} and f_{xy} at $(a, b) \in D$	1146
	22.8.1 Theorem of Schwarz	1147
	22.8.2 Theorem of Young	1147
22.9	Function of Three Variables	1149
22.10	Homogeneous Function	1151
	22.10.1 Euler's Theorem on Homogeneous Function	1151
	22.10.2 Converse of Euler's Theorem	1151
22.11	Jacobians	1154

22.11.1 Chain Rule for Jacobians 1154
22.11.2 Jacobian of Implicit Functions 1155
22.11.3 Some Important Cases 1159
22.11.4 Deductions of Equations (22.11.2) and (22.11.3) in Some Special Cases 1159
22.12 Functional Relationship 1162
22.12.1 The Condition is Necessary 1163
22.12.2 The Sufficiency of the Condition 1163
22.13 Implicit Function: Existence Theorem 1165
22.13.1 Existence Theorem (for Two Variables) 1165

23 Partial Differentiation 1174
23.1 Introduction 1174
23.2 Partial Derivatives 1174
23.2.1 Continuity 1174
23.2.2 Order of differentiation 1175
23.2.3 Conditions for equality of F_{xy} and F_{yx} 1176
23.3 Total Differential 1177
23.4 Taylor's Theorem for Functions of Two Variables 1180
23.4.1 Symbolic Form of Taylor's Series Expansion of Functions of Two or More Variables 1182
23.5 Differentiation of Composite Functions: Chain Rule 1184
23.5.1 Second and Higher Derivatives 1185
23.5.2 Change of Variables 1186
23.6 Alternative Treatment of Total Differential 1186
23.7 Differential of Implicit Functions 1189
23.7.1 One equation, more than two variables 1190
23.7.2 Chain Rule—Alternative Treatment 1191
23.8 Maxima and Minima 1192
23.8.1 Maxima and Minima of Functions of Two Variables 1193
23.9 Lagrange's Method of Undetermined Multipliers 1196
23.10 Differentiation of a Definite Integral 1200
23.11 Integration under the Integral Sign—Multiple Integrations 1202
23.12 Evaluation of Certain Definite Integrals 1203
23.13 Leibnitz's Rule 1206
23.14 Double Integrals 1208
23.15 Change of the Order of Integration 1209
23.16 Triple Integrals 1211
23.17 Errors and Approximations 1212

24 Statistics and Probability 1219
24.1 Introduction 1219
24.2 Definitions of Probability: A Priori Probability or Mathematical Probability 1220
24.2.1 A Posteriori Probability or the Statistical Definition of Probability 1221
24.3 Fundamental Laws of Probability 1223
24.3.1 Mutually Exclusive Events 1224
24.3.2 Statistically Independent Events 1224
24.4 Relations among Conditional Probabilities: Bayes's Theorem 1225
24.5 Repeated Trials 1226
24.6 Sample Space 1227
24.7 Random Variables 1229
24.8 The Statistical Expectation Value 1231
24.8.1 Expectation values of the sum and product of two random variables 1231
24.9 Statistical Distributions 1233

24.9.1 Frequency 1233
24.9.2 The Mean of a Distribution 1233
24.10 Second Moments and Standard Deviation 1234
24.10.1 The Second Moment 1234
24.10.2 The Standard Deviation 1235
24.11 Continuous Distributions 1235
24.12 Expectation, Moments and Standard Deviation 1236
24.12.1 An Important Statistical Inequality—Chebyshev's Inequality 1238
24.13 The Binomial Distribution—Characteristics 1239
24.13.1 The Mean of the Binomial Distribution 1240
24.13.2 The Standard Deviation of the Binomial Distribution 1241
24.14 The Poisson Distribution—Examples of Poisson variates 1241
24.14.1 The Mean of the Poisson Distribution 1243
24.14.2 The Standard Deviation of the Poisson Distribution 1243
24.15 The Normal or Gaussian Distribution—Properties of the normal curve 1243
24.15.1 The Mean Value of the Normal Distribution 1245
24.15.2 The Standard Deviation of the Normal Distribution 1246
24.15.3 The Area under Some Part of the Gaussian Distribution Curve 1247
24.16 Distribution of a Sum of Normal Variates 1248
24.17 Applications to Experimental Measurements 1249
24.18 The Standard Deviation of the Mean 1250
24.19 Stirling's Approximation for the Factorial 1251

Appendix

A Vectors and Tensors **1262**
A.1 Introduction 1262
A.2 Symmetric and Antisymmetric Tensors 1263
A.3 Scalars, Vectors and Tensors 1263
A.4 The Origin and Concept of Tensor Calculus 1263
A.5 Definitions—Contravariant and Covariant Vectors 1264

B Mechanics and Central Force-field Problems **1266**
B.1 Transformation of Co-ordinates through matrix operation—Rotation of co-ordinates in two dimensions 1266
B.2 Calculation of the centre of mass (C.M.) in some important cases 1269
B.3 Lagrangian formulation of conservation theorems 1272
B.4 Equation of Central Orbit in Pedal Form 1276
B.4.1 Vectorial Treatment for the Central Orbit 1278
B.4.2 Motion of an electron in an atom 1279
B.4.3 Scattering in a central force field—Rutherford's scattering of α-particles 1279

C Stess and Strain Tensors **1285**

D Introduction to Fourier Transforms **1286**
D.1 Integral Transforms 1286
D.2 The Fourier Transform 1287
D.3 Some Important Properties of the Fourier Transform 1288
D.3.1 Some Important Properties of Sine and Cosine Fourier Transforms 1290
D.4 Applications of Fourier Transforms 1294
D.5 Application of Fourier Transforms to Solve Differential Equations 1296
D.6 The Dirac-delta Function—its important properties—its Fourier Transform and some Simple illustrations 1301

Bibliography 1309

Index 1311

Questions of C.U. Physics (Honours) Paper I and Paper II (New Syllabus) 2009—relevant to different chapters of this book—with answers 1–7

Bibliography 1309

Index 1311

Questions of C.U. Physics (Honours) Paper I and Paper II (New Syllabus) 2009—relevant to different chapters of this book, with answers 1-7

Preface to the 7th Edition

It is always a pleasure to welcome a new edition of a book, but the publication of the 7th edition in a thoroughly revised, enlarged and updated form has given me much pleasure from that point of view in my mind that the book will be of great help for those for whom it is intended.

Each and every chapter has been replenished with different useful items, especially with the solution of current problems and those set in the recent examinations of different universities. Exercises of each and every chapter have also been updated.

I hope that this edition will also be equally appreciated by the students as well as the teachers. The author is inviting constructive criticisms and suggestions from students and teachers for further improvement of the book.

In spite of our best efforts, some errors might have crept in. The authors will be grateful to any one who points out any such error.

I shall be failing in my duty if I do not express my sincere gratefulness to Late Prof. Madhusudan Dutta of City College and Late (Prof.) Dr. Murari Mohan Som of the Federal University of Rio-de-Jeneiro of Brasil. Both of them have helped me much in solving different problems, and with suggestions for further improvement of the book.

In conclusion, the author expresses his deep sense of gratitude to Late Sri J. N. Sen, Sri Amitabha Sen and Sri Abhijit Sen of New Central Book Agency (P) Ltd. for their ungrudging cooperation at every stage of the publication of the book.

Rathajatra, the 24th June, 2009 H. L. Chatterjee

Preface to the 6th Edition

Inclusion of several important topics for BSc Hons course under the revised syllabus of the Calcutta University, which will come into effect from the academic session 2003-2004, has necessitated the publication of the sixth edition of the book in a thoroughly revised, updated and enlarged form.

The preliminary topics under Mathematical Methods of Physics have been given in the chapters 20, 21, 22, 23 and 24 of the book. These, include—Infinite Sequences and Series—Convergence and Divergence, Conditional and Absolute Convergence, Ratio Test for Convergence—Complex-Valued Functions—Analytic Functions defined in terms of Taylor Series Expansion—Functions of Several Real Variables—Partial Differentiation—Taylor's Series, Multiple Integrals—Random Variables and Probabilities—Statistical Expectation Value, Variance—Binomial Distribution, Gaussian Distribution and Poisson Distribution—Simple Examples.

We have given in the Appendix the following topics which are also included in the revised syllabus. These include—Concept of Tensors—Introduction to Fourier transforms: the Dirac-delta function and its Fourier transform—other simple examples—Rutherford scattering as a central orbit problem.

The exercises of different chapters have also been updated with important and current problems—which will also be helpful for the students for whom the book is meant.

We hope that this edition will also be equally appreciated by the students as well as the teachers. The author is inviting constructive criticisms and suggestions from students and teachers for further improvement of the book.

In spite of our best efforts, some errors might have crept in. I shall be grateful to anyone who points out any such error.

The author takes this opportunity to express his deep sense of gratitude to Sri J. N. Sen, Sri Amitabha Sen, Sri Abhijit Sen and Sri Subir Pal of New Central Book Agency (P) Ltd., for their undergrudging cooperation at every stage of the publication of the book.

Dipawali, the 24th October, 2003 H. L. Chatterjee

Preface to the 5th Edition

We take much pleasure in presenting the fifth edition of the book in a thoroughly revised, updated and enlarged form. Almost all the chapters have been updated in accordance with the syllabi introduced recently in different universities of West Bengal and other important Indian universities to meet the requirements of the students qualifying for Degree Part I and II Hons. examinations. Various problems including those set at recent university examinations—of mainly the University of Calcutta—have also been given in the exercises of different chapters. Some topics have been included in the Appendix of the book to supplement some chapters of the book.

We will consider our efforts to be amply rewarded if the students for whom the book is intended as well as the teachers find it helpful. The author will be highly indebted to those who will offer constructive suggestions for further improvement of the book.

The author takes this opportunity to express his deep sense of gratitude to Sri J. N. Sen, Sri Amitabha Sen, Sri Abhijit Sen and Sri Subir Pal of New Central Book Agency (P) Ltd. for their ungrudging cooperation at every stage of publication of the fifth edition of the book.

Rathajatra, 1st July 2001 H. L. Chatterjee

Preface to the 1st Edition

The first edition of A TREATISE ON GENERAL PROPERTIES OF MATTER is an attempt to present different topics of the subject in a systematic and comprehensive manner. The students qualifying for Degree Physics Honours Examinations of different Indian Universities find it difficult to have a grasp of different topics of Honours standard in one single book. This book is an answer to that. The authors, during their long experience of teaching in the Honours classes are well aware of the needs of the students, and with that end in view, have tried to present the subject matter. The chapter on "Vector Analysis and its Applications" has been included in order that the students may follow the subsequent chapters in which many problems have been solved by vector treatment. One of the attractive features of this book is that the students will find many illustrative examples worked out.

The authors are indebted to their numerous friends for their valuable suggestions, and will feel obliged if they receive suggestions for further improvement of the book. The authors will deem their labour amply rewarded if the book can cater to the needs of those for whom it is intended.

It would be worthwhile to mention here that in writing the book, the authors were inspired by their revered teacher late Prof. Jnanendranath Sen of City College. The authors are also grateful to late Dr M. M. Ghosh, Ex-Principal, City College, for his helpful suggestions. We thank Prof. R. Ghosh of Barisha Vivekananda College for his helpful suggestions.

We are really grateful to Prof. M. K. Dasgupta of the Institute of Radio Physics and Electronics, Calcutta University, and also to Prof. D. L. Bhattacharyya, of the Pure Physics Department, University of Calcutta, who have highly appreciated the book and helped the authors with their valuable suggestions.

It will be worthwhile to mention in this connection that the book has been recommended by the University of Calcutta as a Textbook for undergraduate Honours students.

In conclusion, the authors express their indebtedness to Sri J. N. Sen of New Central Book Agency (P) Ltd. for his ungrudging cooperation at every stage of publication of the present volume.

R. Sengupta
H. L. Chatterjee

Chapter 1

Basic Principles

1.1 Measurement

The study of science involves observations and measurements. Physics is concerned with precise measurements of physical quantities such as length, time, weight, velocity, density, energy, temperature and many other quantities. Physicists are to measure quantities as small as the diameter of the atomic nucleus and as large as the distances of stars and nebulae. Advancing science and technology have devised instruments for precise measurements of physical quantities for which we require well-defined units of measurement. In mechanics and general properties of matter all physical quantities are expressed in terms of three units—length, mass and time. These units are called fundamental units. The units of physical quantities obtained from these three fundamental units are known as derived units.

Different systems of fundamental units

The following three systems of fundamental units are in use :

(i) The C.G.S. System (Metric System)

(ii) The F.P.S. System (British System)

(iii) The M.K.S. System.

The C.G.S. or the metric system is widely used and has been adopted commercially by all except the English-speaking nations. In this system C denotes centimetre (cm) as the unit of length, G gramme (gm) as the unit of mass and S second (sec) as the unit of time. In the F.P.S. or the British system of units F denotes foot (ft) as the unit of length, P pound (lb) as the unit of mass and S second (sec) as the unit of time.

In the M.K.S. system M denotes metre (m) as the unit of length, K kilogramme (kg) as the unit of mass and S second (sec) as the unit of time. In electrical engineering this system of units is commonly used. The great advantage of this unit is that it is of the same magnitude as the practical units such as volt, ampere and ohm. In mechanics also the M.K.S. units are in use.

In the international system of units (abbreviated as S.I.) three additional dimensional quantities are used so as to cover all fields of Physics.

They are :

(a) In the field of electricity the basic quantity is current intensity, the unit being the Ampere (A).

(b) In the field of thermodynamics the basic quantity is thermodynamic temperature, the unit being the degree Kelvin (°K).

(c) In the field of photometry the basic quantity is luminous intensity, the unit being the candela (cd).

This international system of units was recommended in the year 1960 by the International Conference of Weights and Measures. The S.I. system coincides with the M.K.S. system in its mechanical quantities and with the M.K.S.A. system in its electromagnetic quantities. M.K.S.A. means length in metre, mass in kg, time in sec and current in ampere.

1.2 Dimensions of Physical Quantities

The value of a derived unit depends on the values of the fundamental units from which it is derived. The units of length, mass and time are denoted by $[L]$, $[M]$ and $[T]$ respectively. Thus the unit of volume, which is a product of three units of lengths is represented by $[L^3]$ and is said to have three dimensions in length. As it does not depend on the units of mass and time, its dimensions are zero in each of them and so the unit of volume is represented by $[M^0L^3T^0]$. The dimension of a derived unit may be defined as the power to which the fundamental units of mass, length and time must be raised to represent it.

We give below the dimensions of different physical quantities deduced either directly from their definitions or from any relation that may exist between them and quantities of known dimensions.

Density : The density D of a substance is defined as its mass per unit volume and hence its dimension is given by

$$[D] = [M]/[L^3] = [ML^{-3}].$$

Specific gravity : The dimension of specific gravity $[S] = [M]/[M]$ = a number. Hence specific gravity has no dimension.

Velocity : Velocity of a particle is defined as the distance traversed in unit time and hence its dimension is

$$[V] = [L]/[T] = [LT^{-1}].$$

Acceleration : As the acceleration denotes change of velocity in unit time, its dimension is

$$[f] = [LT^{-1}]/[T] = [LT^{-2}].$$

Force : The force on a body is the product of its mass and the acceleration produced. Hence the dimension of force is

$$[P] = [M][LT^{-2}] = [MLT^{-2}].$$

Work : Work or energy is defined as the product of force and displacement. Hence the dimension of work is

$$[W] = [MLT^{-2}][L] = [ML^2T^{-2}].$$

Pressure : Pressure is defined as the force acting on unit area. Hence its dimension is

$$[p] = [MLT^{-2}]/[L^2] = [ML^{-1}T^{-2}].$$

The following table gives the dimensional formulae for some physical quantities.

Quantity	Dimensional formulae
1. Length	L
2. Mass	M
3. Time	T
4. Area	L^2
5. Volume	L^3
6. Velocity	LT^{-1}
7. Acceleration	LT^{-2}
8. Angular speed	T^{-1}
9. Angular acceleration	T^{-2}
10. Density	ML^{-3}
11. Moment of inertia	ML^2
12. Energy of work	ML^2T^{-2}
13. Torque	ML^2T^{-2}
14. Gravitational constant	$M^{-1}L^3T^{-2}$
15. Young's modulus of elasticity	$ML^{-1}T^{-2}$
16. Surface tension	ML^{-2}
17. Coefficient of viscosity	$ML^{-1}T^{-1}$

1.3 Dimensional Equations

We have already seen that the dimension of volume is $[L^3]$. Hence, we may write $[V] = [L^3]$, where $[V]$ stands for the dimension of volume. The above equation is called dimensional equation. *The dimensional equation is thus an expression which shows how and which of the fundamental units enter into the unit of a physical quantity.*

The main uses of dimensional equations are :

(i) **Checking of physical equations.** The dimensions of all terms on the two sides of an equation must be the same. This is called the *principle of homogeneity.* Hence, by substituting the dimensions on both sides of the equation, we can check any physical equation. For example, let us take the well-known equation $W =$ mg. The dimension of the L.H.S. is the dimension of force which is equal to $[MLT^{-2}]$. The dimension of the R.H.S. is $[M][LT^{-2}] = [MLT^{-2}]$.

The dimensions on the two sides of the equation being equal, the correctness of the equation is established.

(ii) **Recapitulation of equations.** In case there is any doubt about the correctness of a formula, dimensional analysis helps us to recapitulate the exact formula. For example, let us consider the equation for an accelerated linear motion given by $v = u + ft$. Here v has the dimension $[LT^{-1}]$ and, u and ft have also the same dimension. If we write the equation as $v = u + ft^2$, the dimensions of L.H.S. and R.H.S of the equation will not be the same which clearly indicates wrong inclusion of ft^2 instead of ft in the equation.

(iii) **Change of unit.** The dimensional equations are very helpful in changing one system of units to some other system of units. For example, let us express a velocity of 10 cms/sec in metres/hour.

$$\text{The velocity in metre/hour} = 10 \times (\text{cm/metre}) \times (\text{hour/sec})$$
$$= (10/100) \times 3600 = \mathbf{360}.$$

The same value may be obtained by dimensional method. Considering the required velocity to be x metres/hour, we may write dimensionally $x[L']/[T'] = 10[L]/[T]$, where $[L], [T]$ stand for unit of length in cm and unit of time in second respectively, and $[L'], [T']$ stand for unit of length in metre and unit of time in hour respectively.

$$\therefore\ x = 10[L]/[L'] \cdot [T']/[T] = 10 \times (\text{cm/metre}) \times (\text{hour/sec}) = \mathbf{360.}$$

So the required velocity is 360 metres/hour.

(iv) **Derivation of physical equations.** Many physical equations have been derived by the method of dimensions. The following examples will show how the physical equations have been deduced by dimensional method.

(a) Time period of a simple pendulum

Let t be the time period of oscillation of a simple pendulum which, we assume, depends on :

(i) length l of the pendulum,

(ii) mass m of the bob,

(iii) acceleration g due to gravity.

Hence, we may write $t = \phi(l, m, g)$, where ϕ is the function to be determined. The above expression may also be written in the form $t = Kl^x m^y g^z$, where K is a constant having no dimension. Transforming this relation to dimensional equation, we have

$$[T] = K[L]^x[M]^y[LT^{-2}]^z$$
$$= K[L^{x+z}][M^y][T^{-2z}].$$

Equating the powers of L, M and T on both sides, we have

$$x + z = 0$$

$$y = 0$$

$$-2z = 1$$

$$\therefore \quad z = -\frac{1}{2} \text{ and } x = \frac{1}{2}$$

$$\therefore \quad t = K\sqrt{\frac{l}{g}}.$$

Thus t is independent of the mass of the bob. The constant K may be experimentally determined and has been found to be equal to 2π.

$$\therefore t = 2\pi\sqrt{\frac{l}{g}}. \tag{1.3.1}$$

(b) Period of vibration of a tuning fork

Let t be the period of vibration of a tuning fork. Assuming that the period of vibration of a fork depends on the length of the prongs of the fork, the density and Young's modulus of the material of the fork, we may write

$$t = Kl^x \rho^y Y^z,$$

where l = length of the prongs,

ρ = density of the material,

Y = Young's modulus of the material.

Transforming the above relation into dimensional equation we have

$$[T] = K[L]^x[ML^{-3}]^y[ML^{-1}T^{-2}]^z \tag{1.3.2}$$

$$\left\{ \because Y = \frac{\text{stress}}{\text{strain}} = \frac{\text{force/area}}{\text{change in length/original length}} \right.$$

$$\left. \therefore \text{ The dimension of } Y = \frac{[MLT^{-2}]/[L^2]}{[L]/[L]} = [ML^{-1}T^{-2}] \right\}$$

$$\therefore \quad [T] = K[L^{x-3y-z}][M^{y+z}][T^{-2z}].$$

Hence, equating the powers of L, M and T on both sides, we have

$$x - 3y - z = 0$$

$$y + z = 0$$

$$-2z = 1$$

$$\therefore z = -\frac{1}{2} \text{ and } y = \frac{1}{2}, \text{ whence}$$

$$x = \frac{3}{2} - \frac{1}{2} = 1$$

$$\therefore t = Kl\sqrt{\frac{\rho}{Y}} \tag{1.3.3}$$

(c) Velocity of sound in a gas

The velocity of sound V in a gas depends on its pressure P and its density D.

Hence we may write

$$V = KP^x D^y$$

The dimensional equation of the above relation is

$$\begin{aligned}[LT^{-1}] &= K[ML^{-1}T^{-2}]^x[ML^{-3}]^y \\ &= K[M^{x+y}][L^{-x-3y}][T^{-2x}].\end{aligned}$$

Hence equating the powers of L, M and T on both sides,

we have

$$\begin{aligned}x + y &= 0 \\ -(x + 3y) &= 1 \\ -2x &= -1\end{aligned}$$

whence $x = \frac{1}{2}$ and $y = -\frac{1}{2}$

$$\therefore V = K\sqrt{\frac{P}{D}}. \tag{1.3.4}$$

1.4 Limitations of the Method of Dimensions

At first sight it appears that the method of dimensions is free from all troubles, but actually this method is attended with many restrictions which are discussed below :

(a) In any case where a physical quantity depends on more than three other physical quantities, the method of dimensions fails to find out a relation between them. This is because by dimensional method we get three equations with the powers of $[L]$, $[M]$ and $[T]$ and so from the three equations we cannot find out more than three quantities. However, the difficulty of handling more than three physical quantities may be overcome by grouping two or more quantities so as to reduce the number of quantities to three.

(b) The method of dimensions may lead to wrong solution if we insert wrong physical quantity in the equation.

(c) If any equation contains a dimensionless factor or a pure number, then the factor will not appear in the equation when it is deduced dimensionally. For example, the dimensional method cannot give the value of K in equation (1.3.3).

(d) In this method we have assumed the expression as power function as in equation (1.3.2). This is not always true as the expression may be an exponential or trigonometrical function or there may be some added terms of different functions but of same dimensions.

(e) Many physical quantities of different nature have got the same dimension. As for example, the moment of a force has the same dimension as that of work or energy though these two physical quantities are not identical.

Despite these limitations, the dimensional method may be considered as a most powerful tool in attacking many complicated problems.

1.5 Equations with Non-dimensional Variables

Let us consider a body of dimension l which moves with a velocity v through a liquid of density ρ and viscosity η. Then the retarding force acting against the motion of the body is given by $F = \phi(l, v, \eta, \rho)$, where ϕ is the function to be determined. The above expression may also be written in the form, $F = Kl^p v^q \eta^r \rho^s$, where K is a constant. Transforming this relation to dimensional one, we have

$$
\begin{aligned}
[MLT^{-2}] &= K[L]^p[LT^{-1}]^q[ML^{-1}T^{-1}]^r[ML^{-3}]^s \\
&= KL^{p+q-r-3s}M^{r+s}T^{-q-r} \\
\therefore\ r+s &= 1, p+q-r-3s = 1 \quad \text{and} \quad q+r = 2 \\
\text{whence } p &= 1+s, r = -s+1 \quad \text{and} \quad q = 1+s \\
\therefore\ F &= Kl^{1+s}v^{1+s}\eta^{-s+1}\rho^s \\
\text{or } F/(\rho v^2 l^2) &= Kl^{s-1}v^{s-1}\eta^{-(s-1)}\rho^{s-1} \\
\text{or } F/(\rho v^2 l^2) &= K(lv\rho/\eta)^{s-1} = \phi(vl\rho/\eta) \qquad (1.5.1)
\end{aligned}
$$

$\phi(vl\rho/\eta)$ is non-dimensional. Its value depends on geometrical conditions

$$[(vl\rho/\eta) = \{(LT^{-1})[L][ML^{-3}]\}/[ML^{-1}T^{-1}] = 1.$$

Thus, $\phi(vl\rho/\eta)$ is dimensionless.

The non-dimensional number $\frac{vl\rho}{\eta}$ is known as *Reynold's number.*

Let there be n number of quantities on which the physical relation depends. According to the π-theorem of dimensional analysis when the relation is expressed by 3 fundamental units it is expressed in terms of the product of $n-3$ non-dimensional quantities, each of which being called π-quantity. In equation (1.5.1), there are five physical quantities such as F, l, v, η, ρ and hence the number of non-dimensional quantities $= 5-3 = 2$. Now the equation (1.5.1) may be written in the form $F/\rho v^2 l^2 \times \phi^{-1}(vl\rho/\eta) = 1$. Thus the relation is expressed as the product of two non-dimensional quantities.

Illustrative Examples

1. Express the viscosity of air, 0.000170 C.G.S. units, in terms of corresponding F.P.S. units if 1 lb = 453.6 gm and 1 ft = 30.48 cm.

$0.000170[M_1L_1^{-1}T_1^{-1}] = x[M_2L_2^{-1}T_2^{-1}]$, where x is the viscosity in F.P.S. units.

The above expression may be written in the form $x = 0.00017 \times [M_1L_1^{-1}]/[M_2L_2^{-1}]$

[$\because$ the time in both units is expressed in secs]

$$
\begin{aligned}
\therefore\ x &= 0.00017 \times \{(1/453.6) \times 30.48\} \\
&= 1.142 \times 10^{-5} \text{ F.P.S. units.}
\end{aligned}
$$

2. Find the number of dynes in the force which, acting on 1 cwt for one minute, produces a velocity of 1 mile per hour. Given 1 ft = 30.5 cm and 1 lb = 453 gm.

Let the number of dynes in the force = x

$$\therefore\ x \text{ dyne} = 1 \text{ cwt} \frac{\text{mile}}{\text{hour}} \cdot \frac{1}{\text{min}}.$$

Hence, x gm cm sec^{-2} = 112 lb$\frac{1 \text{ mile}}{\text{hour}} \cdot \frac{1}{\text{min}}$.

$$\text{or } x = 112\left(\frac{\text{lb}}{\text{gm}}\right)\cdot\left(\frac{\text{mile}}{\text{cm}}\right)\cdot\left(\frac{\text{sec}}{\text{hour}}\right)\cdot\left(\frac{\text{sec}}{\text{min}}\right)$$

$$= 112 \times 453 \times 1760 \times 3 \times 30.5 \times \frac{1}{60 \times 60} \times \frac{1}{60}$$

$$= 3.78 \times 10^4.$$

So the number of dynes in the force = 3.78×10^4.

The required force = 3.78×10^4 dynes.

3. The frequency n of vibration of a stretched string is a function of its tension F, its length l and its mass per unit length m. Show that $n \propto \{(1/l)\sqrt{(F/m)}\}$, making use of the knowledge of dimensions.

Let $n = KF^x l^y m^z$, where K has no dimension. It is a number. Transforming this relation into dimensional equation, we have,

$$[T^{-1}] = K[MLT^{-2}]^x[L]^y[ML^{-1}]^z$$

$$= KM^{x+z}L^{x+y-z}T^{-2x}.$$

Hence, equating the powers of $[L]$, $[M]$ and $[T]$ on both sides, we have

$$x + y - z = 0$$

$$x + z = 0$$

$$-2x = -1$$

$\therefore\ x = \frac{1}{2}, z = -\frac{1}{2}$ and $y = -1$.

Hence

$$n = K\left\{\frac{1}{l}\sqrt{\frac{F}{m}}\right\}.$$

4. The viscosity η of a gas is determined by its density ρ, average molecular velocity C, and its mean free path l. Show that $\eta = K\rho Cl$.

Let $\eta = K\rho^x C^y l^z$

$$\therefore\ [ML^{-1}T^{-1}] = K[ML^{-3}]^x[LT^{-1}]^y[L]^z$$

$$= KM^xL^{-3x+y+z}T^{-y}$$

$$\therefore\ x = 1, -3x + y + z = -1 \text{ and } y = 1$$

$$\therefore\ z = 1, \quad \therefore\ \eta = K\rho Cl.$$

5. In a rotating cylinder method of measuring the viscosity of a liquid there is critical angular velocity ω above which turbulence occurs. Find how ω is related to the viscosity η, the separation distance x between the cylinders, and the density ρ of the liquid.

Let $\omega = K\eta^\alpha \rho^\beta x^\gamma$

$$\therefore\ [T^{-1}] = K[ML^{-1}T^{-1}]^\alpha [ML^{-3}]^\beta [L]^\gamma$$
$$= KM^{\alpha+\beta} L^{-\alpha-3\beta+\gamma} T^{-\alpha}.$$

Hence, $\alpha + \beta = 0$, $-\alpha - 3\beta + \gamma = 0$ and $\alpha = 1$

$$\therefore\ \beta = -1 \quad \text{and} \quad \gamma = -2.$$

So,

$$\omega = K\eta\rho^{-1}x^{-2} = K\eta/(\rho x^2).$$

6. Keeping the unit of time 1 sec, units of length and mass are so changed that the values of C, the velocity of light in vacuo and G, Newton's gravitational constant both become unity. Find out the values of the new unit of length and unit of mass in terms of cm and gm ($C = 3 \times 10^{10}$ cm s^{-1}, $G = 6.7 \times 10^{-8}$ cm^3 s^{-2} gm^{-1}). **[C.U. (Hons.) 1977]**

In the new unit of length, the velocity of light in vacuo = 1 new unit of length per sec

or 3×10^{10} cm per sec = 1 new unit of length per sec.

$\therefore$ 1 new unit of length = 3×10^{10} cm.

The value of G in the new units of length and mass = 1 (new unit of length)^3 s^{-2} (new unit of mass)$^{-1}$.

$\therefore$ 1 (new unit of length)^3 s^{-2} (new unit of mass)$^{-1}$ = 6.7×10^{-8} cm^3 s^{-2} gm^{-1}

or $(3 \times 10^{10})^3$ cm$^3 \cdot 1$ (new unit of mass)$^{-1}$ = 6.7×10^{-8} cm^3 gm^{-1}.

$\therefore$ 1 new unit of mass = $\frac{27}{6.7} \times 10^{38}$ gm = 4.03×10^{38} gm.

1.6 Mass and Weight

The mass of a body usually denotes the quantity of matter contained in the body. If the mass of a body is small, a given force applied on the body will produce large acceleration. The same force when applied on a larger body will produce smaller acceleration. Thus the mass and acceleration are inversely proportional to each other. So it appears that mass also denotes the property of the body which makes it resist an acceleration. This property of the body is called *inertia*. Mass may thus be measured by inertia. Hence, it should be noted that in defining the mass, we should take into account two things—quantity of matter contained in the body and the inertia possessed by the body.

The weight is attributable to the attraction which the earth exercises on bodies near its surface. The weight is thus dependent on the existence of earth and the body, whereas the mass depends only on the body. We can measure mass conveniently and precisely with the help of a common balance. The weight of a body may be determined by using a spring balance.

1.7 Newton's Laws of Motion

The three fundamental laws of motion were enunciated by Sir Isaac Newton. These laws which constitute the foundation of dynamics may be stated in the following form.

I. *Every body continues in its state of rest or of uniform motion in a straight line except in so far as it is compelled to change that state by forces impressed on it from outside.*

II. *The rate of change of momentum is proportional to the impressed force and takes place in the direction in which the force acts.*

III. *To every action there is an equal and opposite reaction or the mutual actions of two bodies on each other are always equal and oppositely directed.*

The first law may be called the law of inertia and provides us with the definition of force. From this law it follows that the force is that agency which changes or tends to change the state of motion or of rest of a body.

According to Newton's second law of motion

$$F \propto \frac{d}{dt}(mv),$$

where m is the mass of the body, v its velocity and F the impressed force. The above relation may be written in the form

$$F = K\frac{d}{dt}(mv),$$

where K is a constant of proportionality. As the mass of a body is usually constant, we can write

$$F = Km\frac{dv}{dt} = Kmf,$$

where f denotes the acceleration of the body.

If we choose unit of force such that it acting on a unit mass produces unit acceleration, i.e., $m = 1$, $f = 1$, when $F = 1$ and K will then be equal to unity. Hence

$$F = mf. \tag{1.7.1}$$

Thus the second law of motion gives us the quantitative measure of a force. The absolute units of force in C.G.S. and F.P.S. systems are dyne and poundal respectively. In M.K.S. unit, the unit of force is newton [1 newton $= 10^5$ dynes].

The force which makes a body accelerate towards the centre of the earth is a measure of its weight and is given by $W = mg$, where m is the mass of the body and g the acceleration due to gravity.

In gravitational unit we compare forces with weights. Thus the magnitude of some force is said to be the same as the weight of some mass. The C.G.S. gravitational unit of force is gm-wt. 1 gm-wt means the weight of 1 gm which is equal to g dynes ($g = 980\,\text{cm}/\text{sec}^2$). The F.P.S. gravitational unit of force is pound-wt. 1 pound-wt means the weight of 1 lb which is equal to g poundals [$g = 32.2\,\text{ft}/\text{sec}^2$].

1.8 Accelerated Linear Motion

If a force acts on a body of mass m then according to Newton's 2nd law of motion, the body will move with acceleration given by $f = F/m$.

The distance s traversed by a body in a given time 't', the initial velocity u, final velocity v and the uniform acceleration f of the body are related to each other by the following equations :

$$v = u + ft \tag{1.8.1}$$

$$s = ut + \frac{1}{2}ft^2 \tag{1.8.2}$$

$$v^2 = u^2 + 2fs \tag{1.8.3}$$

Falling bodies

One of the most interesting examples of uniformly accelerated rectilinear motion is that of a body freely falling under gravity. The above equations may be used in case of bodies falling under gravity. The acceleration to be used is that due to gravity.

1.9 Graphical Treatment of Linear Motion

If the distance s traversed by a body is plotted in rectangular co-ordinates against time t taken by the body to cover the distance, the curve obtained is called space-time curve of the motion. The slope of the tangent drawn at any point of the curve gives the speed at that instant. Plotting the values of speed thus obtained against time, we get a curve from which in a similar way the acceleration of the body may be obtained. The distance traversed by the body in any time may be obtained by the area under corresponding portion of the time-speed curve. This will also be true if the acceleration varies. Thus the graphical method is the simple method for treating cases of non-uniform acceleration.

Exercises

1. What are the different systems of fundamental units? What do you mean by derived units?
2. Explain what you mean by the dimension of a physical quantity. Calculate the dimensions of the following quantities : (a) Density, (b) Acceleration, (c) Work, (d) Young's modulus.
3. Explain the principle of homogeneity of dimensions in a physical equation. What are M.K.S. and S.I. systems of units? Explain clearly what is meant by M.K.S.A. system.
4. Explain what you mean by dimensional equations. What are the main uses of dimensional equations?
5. Using the method of dimensions obtain an expression for the period of vibration of a tuning fork.
6. Explain clearly the limitations of the method of dimensions.

7. The time period of oscillation of a small drop of liquid under surface-tension depends on the density ρ, radius r and surface tension S. Prove dimensionally that the time period of oscillation $T \propto \left(\frac{r^3 \rho}{S}\right)^{1/2}$.

8. Find the number of dynes in a force which acting on 15 kilograms for one minute produces a velocity 4.6 kilometres per sec. **[Ans.** 11.5×10^7 dynes] **[C.U. 1948]**

9. Convert 1 HP into watts, given 1 inch = 2.54 cm, 1 lb = 454 gm. **[Ans.** 742.4 Watts]

10. If instead of the dyne, the unit force is defined as the gravitational force between two particles each of mass 1 gm placed 1 cm apart, what will be the new form of the equation $P = mf$?

[Ans. $P = Gmf$] **[C.U. (Hons.) 1977]**

11. State the principle of homogeneity of dimensions. Hence show that in $f(x) = e^x$, x is dimensionless. **[C.U. (Hons.) 1997]**

Answer to last part

$$f(x) = e^x = 1 + \frac{x}{\underline{|1}} + \frac{x^2}{\underline{|2}} + \cdots + \frac{x^n}{\underline{|n}}$$

From the above, it is evident that if x has got some dimensions, the above equation violates the principle of homogeneity of dimensions. So x should be dimensionless.

12. Use dimensional analysis to show that pressure is physically equal to energy per unit volume.

[Hints. $\frac{\text{Energy}}{\text{Volume}} = \frac{\text{Work done}}{\text{Volume}} = \frac{\text{Force} \times \text{distance}}{\text{Area} \times \text{distance}}$ = Pressure] **[C.U. (Hons.) 2002]**

Chapter 2

Vector Analysis and its Applications

2.1 Vector Analysis

Vector analysis forms an essential part of the mathematical treatment of physics. The study of physical phenomena involves both scalar and vector quantities.

2.2 Scalar and Vector Quantities

Quantities such as displacement, velocity, force, acceleration, momentum etc., which have got both magnitude and direction are called *vector* quantities. We represent graphically a vector by an arrow, say, AB (Fig. 2.2.1) giving the direction, the magnitude being represented by the length of the arrow. Analytically a vector is represented by a letter with an arrow mark over it, as $\vec{V}$ in Fig. 2.2.1. The *scalar* magnitude of the vector is represented as $|\vec{V}|$ or V. A vector is also represented by Clarendon letter $\mathbf{V}$ and its magnitude by corresponding italic letter V. Quantities such as length, mass, time, volume, density, temperature, electric potential, etc., which have got only magnitude but no directions are called scalar quantities.

Fig. 2.2.1

2.3 Operations with Scalars and Vectors

Operations with scalars follow the rules as in the elementary algebra. For operations with vectors the following laws are fundamental:

(i) If two vectors $\vec{A}$ and $\vec{B}$ have the same magnitude and direction, and whatever may be their positions they are *equal*, i.e.,

$$\vec{A} = \vec{B} \quad \text{[Fig. 2.3.1]}.$$

Reversing the arrow-head the sign of a vector is changed. Here the two vectors $\vec{C}$ and $\vec{D}$ are of equal size but are oppositely directed

$$\therefore\ \vec{C} = -\vec{D}.$$

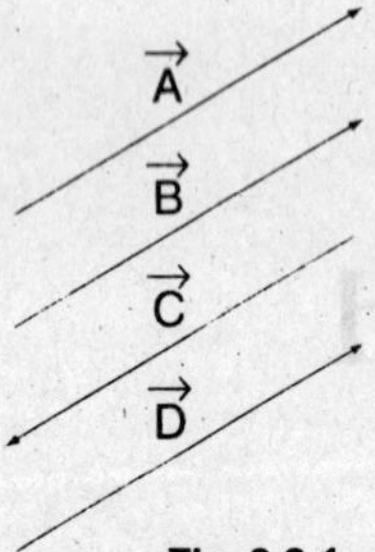

Fig. 2.3.1

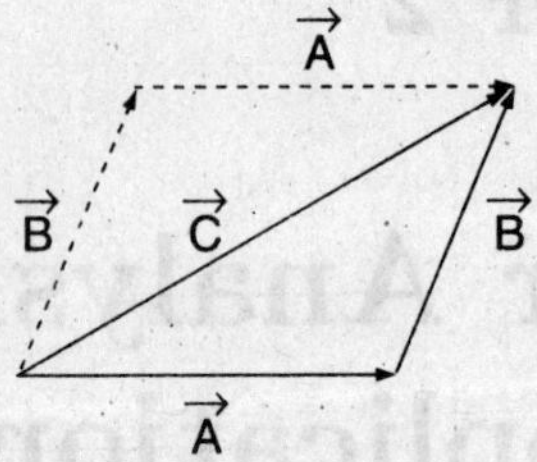

Fig. 2.3.2

(ii) Let us consider two vectors $\vec{A}$ and $\vec{B}$. Their resultant $\vec{C}$ is obtained by placing the vector $\vec{B}$ at the end of $\vec{A}$ and drawing the vector $\vec{C}$ joining the initial point of $\vec{A}$ to the terminal point of $\vec{B}$ (Fig. 2.3.2). The same result is obtained by starting with $\vec{B}$ and then joining $\vec{A}$ to it in a similar way as shown by the dotted lines. Then $\vec{C} = \vec{A} + \vec{B} = \vec{B} + \vec{A}$. Thus we see that the sum of two vectors is the diagonal of the parallelogram with sides formed by the two vectors.

(iii) If we want to subtract a vector $\vec{B}$ from $\vec{A}$, the vector $\vec{B}$ is to be reversed and then the rule for addition is to be applied, i.e., $\vec{C} = \vec{A} + (-\vec{B}) = \vec{A} - \vec{B}$. If $\vec{A} = \vec{B}$, then $\vec{A} - \vec{B}$ is called the *null* or zero vector and is represented by the symbol $\vec{0}$. A null vector has zero magnitude and specific direction. Vectors other than null vector, are called proper vectors.

(iv) The product of a vector $\vec{A}$ with a scalar K is a vector $K\vec{A}$ with magnitude $|K|$ times the magnitude of $\vec{A}$. The direction of the vector $K\vec{A}$ may be same or opposite to that of $\vec{A}$ according as K is positive or negative.

In handling vector quantities the following laws of Vector Algebra are to be noted.

1. $\vec{A} + \vec{B} = \vec{B} + \vec{A}$—Commutative law for addition.
2. $K\vec{A} = \vec{A}K$—Commutative law for multiplication.
3. $\vec{A} + (\vec{B} + \vec{C}) = (\vec{A} + \vec{B}) + \vec{C}$—Associative law for addition.
4. $K(l\vec{A}) = (Kl)\vec{A}$—Associative law for multiplication.
5. $\vec{A}(K + l) = K\vec{A} + l\vec{A}$—Distributive law.
6. $K(\vec{A} + \vec{B}) = K\vec{A} + K\vec{B}$—Distributive law.

2.4 Unit Vectors and Rectangular Unit Vectors

A vector having unit magnitude is called a unit vector. If any vector $\vec{A}$ is of magnitude $A (A \neq 0)$ then $(\vec{A}/A)$ is a measure of the unit vector having the same direction as the vector $\vec{A}$.

The rectangular unit vectors are usually represented by $\hat{\imath}$, $\hat{\jmath}$ and $\hat{k}$ and denote vectors of unit magnitude along the three mutually perpendicular axes, X, Y and Z. Any vector $\vec{A}$ can be resolved into three component vectors $\vec{A}_x$, $\vec{A}_y$ and $\vec{A}_z$ of magnitudes A_x, A_y and A_z parallel to OX, OY and OZ respectively. Then $\vec{A}_x = A_x\hat{\imath}$, $\vec{A}_y = A_y\hat{\jmath}$ and $\vec{A}_z = A_z\hat{k}$ (Fig. 2.4.1)

$$\therefore\ \vec{A} = A_x\hat{\imath} + A_y\hat{\jmath} + A_z\hat{k}. \tag{2.4.1}$$

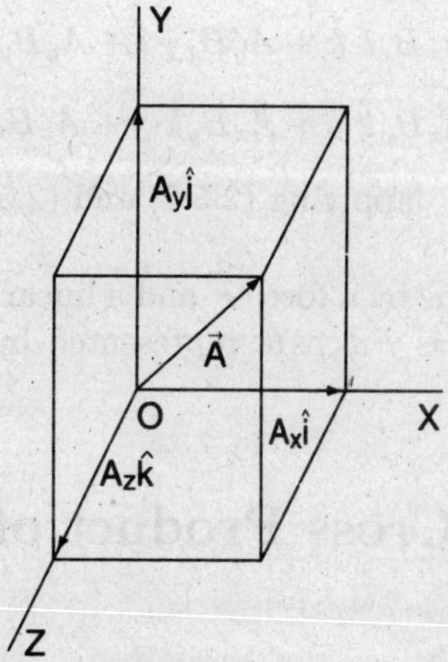

Fig. 2.4.1

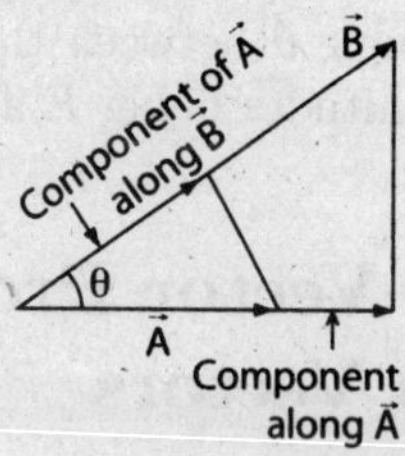

Fig. 2.5.1

2.5 The Scalar Product or Dot Product of Two Vectors

The scalar product or the dot product of two vectors is defined as the product of the magnitudes of the two vectors and the cosine of the angle between the two vectors. The scalar or the dot product of two vectors $\vec{A}$ and $\vec{B}$ having angle θ between them is given by

$$\vec{A}\cdot\vec{B} = AB\cos\theta. \quad \text{(See Fig. 2.5.1)} \tag{2.5.1}$$

It is to be noted that $\vec{A}\cdot\vec{B}$ is a scalar quantity.

The scalar product of two vectors may also be defined as the product of the magnitude of one of the vectors and the component of the other in the direction of the first. The commutative law holds for scalar product, i.e., $\vec{A}\cdot\vec{B} = \vec{B}\cdot\vec{A}$ [since $\cos\theta = \cos(-\theta)$].

If the two vectors $\vec{A}$ and $\vec{B}$ are perpendicular to each other, then $\cos\theta = 0$.

$$\therefore\ \vec{A}\cdot\vec{B} = 0.$$

In case of unit vectors $\hat{\imath}$, $\hat{\jmath}$ and $\hat{k}$, we have, therefore,

$$\hat{\imath}\cdot\hat{\jmath} = \hat{\jmath}\cdot\hat{k} = \hat{k}\cdot\hat{\imath} = 0. \tag{2.5.2}$$

If the two vectors are parallel to each other, we have then $\cos\theta = 1$.

$$\therefore\ \vec{A}\cdot\vec{B} = AB.$$

In case of unit vectors, we have, then

$$\hat{\imath}\cdot\hat{\imath} = \hat{\jmath}\cdot\hat{\jmath} = \hat{k}\cdot\hat{k} = 1. \tag{2.5.3}$$

From equation (2.4.1) the scalar product of two vectors $\vec{A}$ and $\vec{B}$ may be written in the form

$$\begin{aligned}\vec{A}\cdot\vec{B} &= \left(A_x\hat{\imath}+A_y\hat{\jmath}+A_z\hat{k}\right)\cdot\left(B_x\hat{\imath}+B_y\hat{\jmath}+B_z\hat{k}\right)\\ &= A_xB_x\hat{\imath}\cdot\hat{\imath}+A_xB_y\hat{\imath}\cdot\hat{\jmath}+A_xB_z\hat{\imath}\cdot\hat{k}+A_yB_x\hat{\jmath}\cdot\hat{\imath}+A_yB_y\hat{\jmath}\cdot\hat{\jmath}\\ &\quad +A_yB_z\hat{\jmath}\cdot\hat{k}+A_zB_x\hat{k}\cdot\hat{\imath}+A_zB_y\hat{k}\cdot\hat{\jmath}+A_zB_z\hat{k}\cdot\hat{k}\\ &= A_xB_x+A_yB_y+A_zB_z \quad \text{[applying (2.5.2) and (2.5.3)].} \end{aligned} \tag{2.5.4}$$

As a simple example in Physics, the work done by a force $\vec{F}$ and a linear displacement vector $\vec{d}$, having an angle θ between their directions, is $Fd\cos\theta$, represented by the scalar product $\vec{F}\cdot\vec{d}$, their magnitudes being F and d.

2.6 The Vector Product or Cross Product of Two Vectors

The vector product or the cross product of two vectors is a vector having magnitude equal to the product of the magnitudes of the two vectors and the sine of the angle between their directions, and a direction perpendicular to the plane containing the vectors. Thus the vector product of two vectors $\vec{A}$ and $\vec{B}$ having angle θ between them is another vector $\vec{C}$ which is given by

$$\vec{C}=\vec{A}\times\vec{B}=AB\sin\theta\hat{n}, \ \{\text{the cross product } \vec{A}\times\vec{B} \text{ may also be written as } [\vec{A}\vec{B}]\} \tag{2.6.1}$$

where $\hat{n}$ is a unit vector in the direction of $\vec{C}$ which is in a direction perpendicular to the plane containing $\vec{A}$ and $\vec{B}$ (Fig. 2.6.1).

The sense of the product along this perpendicular is defined by a right-handed system, i.e., a right-threaded screw rotated through an angle less than 180° from $\vec{A}$ to $\vec{B}$ will advance in the direction of $\vec{C}$ as shown in Figs. (2.6.1) and (2.6.2).

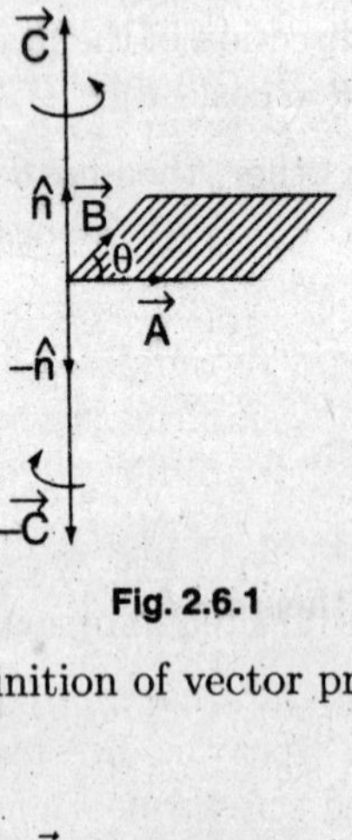

Fig. 2.6.1

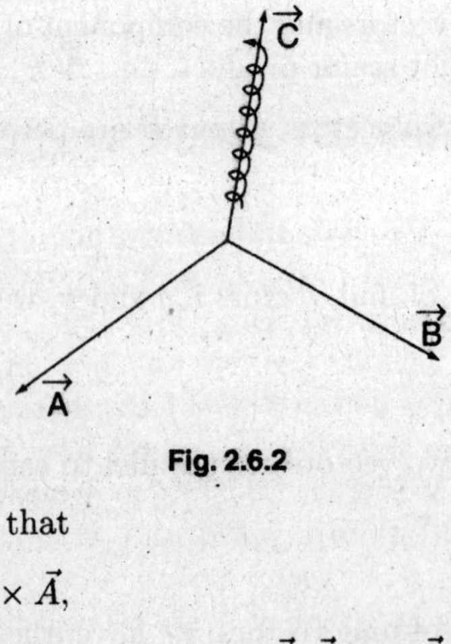

Fig. 2.6.2

From the definition of vector product, it follows that

$$\vec{A}\times\vec{B}=-\vec{B}\times\vec{A},$$

because $\vec{A}\times\vec{B}=\vec{C}$ has magnitude $AB\sin\theta$ and direction such that $\vec{A}, \vec{B}$ and $\vec{C}$ form a right-handed system [Fig. 2.6.3(a)], and $\vec{B}\times\vec{A}=\vec{C}'$ has magnitude $BA\sin\theta$ and direction such that

$\vec{B}, \vec{A}$ and $\vec{C}'$ form a right-handed system [Fig. 2.6.3(b)]. Thus $\vec{C}'$ has the same magnitude as $\vec{C}$ but is opposite in direction.

$$\therefore\ \vec{A} \times \vec{B} = -\vec{B} \times \vec{A}. \tag{2.6.2}$$

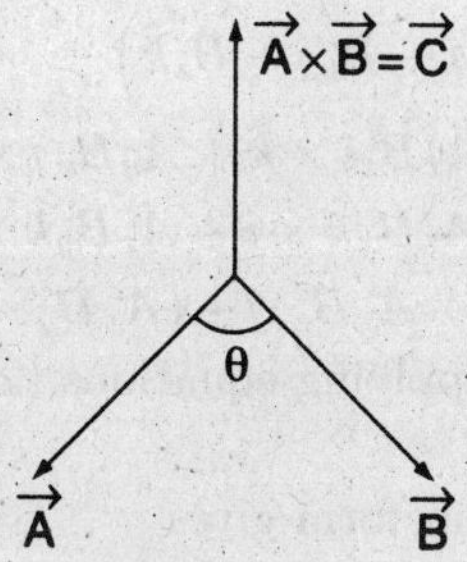

Fig. 2.6.3(a)

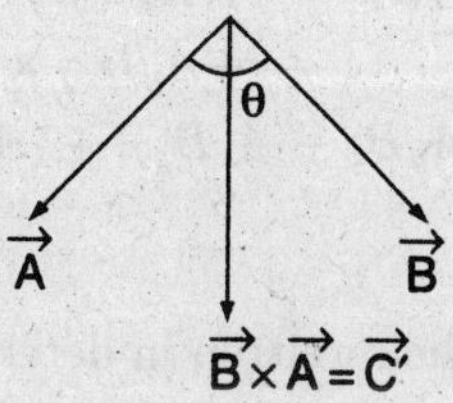

Fig. 2.6.3(b)

Thus the commutative law for vector product does not hold good. When the two vectors are parallel $\sin\theta = 0 \therefore\ \vec{A} \times \vec{B} = 0$ and when they are perpendicular $\sin\theta = 1$

$$\therefore\ \vec{A} \times \vec{B} = AB\hat{n}.$$

Hence, for rectangular unit vectors, we have

$$\hat{\imath} \times \hat{\imath} = \hat{\jmath} \times \hat{\jmath} = \hat{k} \times \hat{k} = 0 \tag{2.6.3}$$

$$\left.\begin{aligned} \text{and}\quad \hat{\imath} \times \hat{\jmath} &= -\hat{\jmath} \times \hat{\imath} = \hat{k}, \hat{\jmath} \times \hat{k} = -\hat{k} \times \hat{\jmath} = \hat{\imath}, \\ \hat{k} \times \hat{\imath} &= -\hat{\imath} \times \hat{k} = \hat{\jmath}. \end{aligned}\right\} \tag{2.6.4}$$

As a simple example in Physics, the couple exerted by a force vector $\vec{F}$ and a linear displacement vector $\vec{d}$, having an angle θ between their directions is $Fd\sin\theta$, represented by the vector product $\vec{F} \times \vec{d}$ and acts about an axis perpendicular to the plane containing $\vec{F}$ and $\vec{d}$.

2.7 Vector Area

We have seen that the magnitude of the vector product of two vectors $\vec{A}$ and $\vec{B}$ is $AB\sin\theta$ which is equal to the area of the parallelogram with sides A and B, and θ the angle between them (Fig. 2.6.1). Hence, any plane area of magnitude $AB\sin\theta$ with its positive unit normal $\hat{n}$ represents a vector product. Thus a plane area a as in Fig. 2.7.1 possesses both magnitude and direction. The sign to be assigned to a vector area is defined with reference to the order in which it is traced out when viewed from an external point such as O. If the direction of tracing is counter clockwise as viewed from O, the positive direction of the vector area is along the unit normal $\hat{n}$. Here the direction of tracing and that of $\hat{n}$ are related by the right-handed screw rule. Then $\vec{a} = a\hat{n}$ is called a vector area. Like other vector quantities we can resolve or add vector areas.

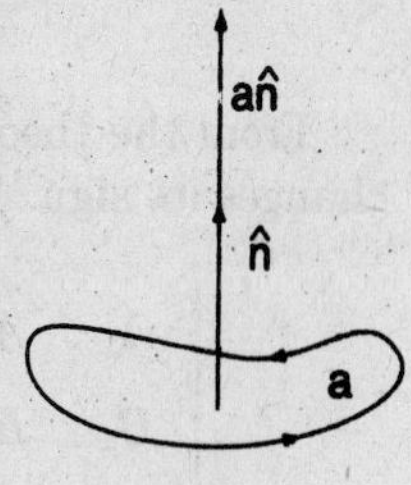

Fig. 2.7.1

2.8 Vector Product in Rectangular Co-ordinates

We can express vector product of two vectors in terms of rectangular components. Using equation (2.4.1), we have,

$$\begin{aligned}\vec{A}\times\vec{B} &= \left(A_x\hat{\imath}+A_y\hat{\jmath}+A_z\hat{k}\right)\times\left(B_x\hat{\imath}+B_y\hat{\jmath}+B_z\hat{k}\right)\\ &= A_xB_x\hat{\imath}\times\hat{\imath}+A_xB_y\hat{\imath}\times\hat{\jmath}+A_xB_z\hat{\imath}\times\hat{k}+A_yB_x\hat{\jmath}\times\hat{\imath}+A_yB_y\hat{\jmath}\times\hat{\jmath}\\ &\qquad +A_yB_z\hat{\jmath}\times\hat{k}+A_zB_x\hat{k}\times\hat{\imath}+A_zB_y\hat{k}\times\hat{\jmath}+A_zB_z\hat{k}\times\hat{k}\\ &= (A_yB_z-A_zB_y)\hat{\imath}+(A_zB_x-A_xB_z)\hat{\jmath}+(A_xB_y-A_yB_x)\hat{k}\end{aligned}$$

[applying equations (2.6.3) and (2.6.4)].

This expression when written in determinant form gives

$$\vec{A}\times\vec{B}=\begin{vmatrix}\hat{\imath} & \hat{\jmath} & \hat{k}\\ A_x & A_y & A_z\\ B_x & B_y & B_z\end{vmatrix} \tag{2.8.1}$$

2.9 Triple Products

The dot product and cross product of three vectors $\vec{A}$, $\vec{B}$ and $\vec{C}$ produce meaningful products of the forms $\vec{A}\cdot(\vec{B}\times\vec{C})$ and $\vec{A}\times(\vec{B}\times\vec{C})$.

These forms frequently occur in physical applications. Of the two forms the first may be written as

$$\begin{aligned}\vec{A}\cdot(\vec{B}\times\vec{C}) &= \left(A_x\hat{\imath}+A_y\hat{\jmath}+A_z\hat{k}\right)\cdot\begin{vmatrix}\hat{\imath} & \hat{\jmath} & \hat{k}\\ B_x & B_y & B_z\\ C_x & C_y & C_z\end{vmatrix}\\ &= \left(A_x\hat{\imath}+A_y\hat{\jmath}+A_z\hat{k}\right)\cdot\left[(B_yC_z-B_zC_y)\hat{\imath}+(B_zC_x-B_xC_z)\hat{\jmath}+(B_xC_y-B_yC_x)\hat{k}\right]\\ &= A_x(B_yC_z-B_zC_y)+A_y(B_zC_x-B_xC_z)+A_z(B_xC_y-B_yC_x)\\ &= \begin{vmatrix}A_x & A_y & A_z\\ B_x & B_y & B_z\\ C_x & C_y & C_z\end{vmatrix}\end{aligned} \tag{2.9.1}$$

From the theorem of determinants it follows that interchange of two rows of a determinant changes its sign. Hence we have,

$$\begin{vmatrix}A_x & A_y & A_z\\ B_x & B_y & B_z\\ C_x & C_y & C_z\end{vmatrix} = -\begin{vmatrix}B_x & B_y & B_z\\ A_x & A_y & A_z\\ C_x & C_y & C_z\end{vmatrix} = \begin{vmatrix}B_x & B_y & B_z\\ C_x & C_y & C_z\\ A_x & A_y & A_z\end{vmatrix} = \vec{B}\cdot(\vec{C}\times\vec{A}).$$

Similarly it can be shown that

$$\begin{vmatrix} A_x & A_y & A_z \\ B_x & B_y & B_z \\ C_x & C_y & C_z \end{vmatrix} = \vec{C}\cdot(\vec{A}\times\vec{B}).$$

Thus we have

$$\vec{A}\cdot(\vec{B}\times\vec{C}) = \vec{B}\cdot(\vec{C}\times\vec{A}) = \vec{C}\cdot(\vec{A}\times\vec{B}). \tag{2.9.2}$$

The second form may be written as

$$\begin{aligned}
\vec{A}\times(\vec{B}\times\vec{C}) &= (A_x\hat{\imath}+A_y\hat{\jmath}+A_z\hat{k})\times \begin{vmatrix} \hat{\imath} & \hat{\jmath} & \hat{k} \\ B_x & B_y & B_z \\ C_x & C_y & C_z \end{vmatrix} \\
&= (A_x\hat{\imath}+A_y\hat{\jmath}+A_z\hat{k})\times \\
&\qquad \left[(B_yC_z-B_zC_y)\hat{\imath}+(B_zC_x-B_xC_z)\hat{\jmath}+(B_xC_y-B_yC_x)\hat{k}\right] \\
&= \begin{vmatrix} \hat{\imath} & \hat{\jmath} & \hat{k} \\ A_x & A_y & A_z \\ B_yC_z-B_zC_y & B_zC_x-B_xC_z & B_xC_y-B_yC_x \end{vmatrix} \\
&= \hat{\imath}(A_yB_xC_y - A_yB_yC_x - A_zB_zC_x + A_zB_xC_z) \\
&\quad +\hat{\jmath}(A_zB_yC_z - A_zB_zC_y - A_xB_xC_y + A_xB_yC_x) \\
&\quad +\hat{k}(A_xB_zC_x - A_xB_xC_z - A_yB_yC_z + A_yB_zC_y)
\end{aligned}$$

Now,

$$\begin{aligned}
\vec{B}(\vec{A}\cdot\vec{C}) - \vec{C}(\vec{A}\cdot\vec{B}) &= (B_x\hat{\imath}+B_y\hat{\jmath}+B_z\hat{k})(A_xC_x+A_yC_y+A_zC_z) \\
&\quad -(C_x\hat{\imath}+C_y\hat{\jmath}+C_z\hat{k})(A_xB_x+A_yB_y+A_zB_z) \\
&= \hat{\imath}(A_yB_xC_y - A_yB_yC_x - A_zB_zC_x + A_zB_xC_z) \\
&\quad +\hat{\jmath}(A_zB_yC_z - A_zB_zC_y - A_xB_xC_y + A_xB_yC_x) \\
&\quad +\hat{k}(A_xB_zC_x - A_xB_xC_z - A_yB_yC_z + A_yB_zC_y)
\end{aligned}$$

$$\therefore\ \vec{A}\times(\vec{B}\times\vec{C}) = \vec{B}(\vec{A}\cdot\vec{C}) - \vec{C}(\vec{A}\cdot\vec{B}). \tag{2.9.3}$$

Similarly, it can be shown that

$$\vec{B}\times(\vec{C}\times\vec{A}) = \vec{C}(\vec{B}\cdot\vec{A}) - \vec{A}(\vec{B}\cdot\vec{C}) \tag{2.9.4}$$

$$\text{and}\quad \vec{C}\times(\vec{A}\times\vec{B}) = \vec{A}(\vec{C}\cdot\vec{B}) - \vec{B}(\vec{C}\cdot\vec{A}). \tag{2.9.5}$$

2.10 Vector Differentiation

Let $\vec{V}(u)$ be a vector depending on a single scalar variable u. Then

$$\frac{\delta\vec{V}}{\delta u} = \frac{\vec{V}(u+\delta u) - \vec{V}(u)}{\delta u},$$

where $\delta\vec{V}$ denotes small increment in $\vec{V}$ and δu a small increment in u [Fig. (2.10.1)].

The ratio $\dfrac{\delta\vec{V}}{\delta u}$ denotes the average rate of change of $\vec{V}$ with u. As δu becomes vanishingly small the ratio attains a limiting value which gives the derivative of $\vec{V}$ with respect to the scalar variable u. Then

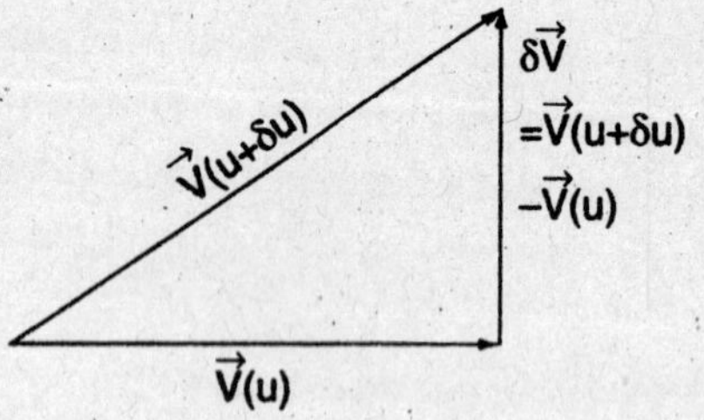

Fig. 2.10.1

$$\frac{d\vec{V}}{du} = \lim_{\delta u\to 0}\frac{\delta\vec{V}}{\delta u} = \lim_{\delta u\to 0}\frac{\vec{V}(u+\delta u) - \vec{V}(u)}{\delta u}.$$

As $\dfrac{d\vec{V}}{du}$ is a vector depending on u, its first, second, third and higher derivatives are similarly obtained. If the second derivative exists it is denoted by $\dfrac{d^2\vec{V}}{du^2}$.

Let us consider a physical example where $\vec{V}$ denotes a vector of displacement and u is time. The extremity of the vector then gives the path of a moving body and the derivatives $\dfrac{d\vec{V}}{du}$ and $\dfrac{d^2\vec{V}}{du^2}$ denote respectively the velocity and acceleration along the path at any instant.

2.11 Differentiation of Sums and Products

If $\vec{A}$, $\vec{B}$ and $\vec{C}$ are differentiable vector functions of a scalar u, then

(i) $\dfrac{d}{du}\left(\vec{A}\pm\vec{B}\right) = \dfrac{d\vec{A}}{du}\pm\dfrac{d\vec{B}}{du}.$

(ii) $\dfrac{d}{du}\left(\vec{A}\cdot\vec{B}\right) = \dfrac{d\vec{A}}{du}\cdot\vec{B} + A\cdot\dfrac{d\vec{B}}{du}.$

(iii) $\dfrac{d}{du}\left(\vec{A}\times\vec{B}\right) = \vec{A}\times\dfrac{d\vec{B}}{du} + \dfrac{d\vec{A}}{du}\times\vec{B}.$

(iv) $\dfrac{d}{du}\left\{\vec{A}\cdot\left(\vec{B}\times\vec{C}\right)\right\} = \vec{A}\cdot\left(\vec{B}\times\dfrac{d\vec{C}}{du}\right) + \vec{A}\cdot\left(\dfrac{d\vec{B}}{du}\times\vec{C}\right) + \dfrac{d\vec{A}}{du}\cdot\left(\vec{B}\times\vec{C}\right).$

(v) $\dfrac{d}{du}\left\{\vec{A}\times\left(\vec{B}\times\vec{C}\right)\right\} = \vec{A}\times\left(\vec{B}\times\dfrac{d\vec{C}}{du}\right) + \vec{A}\times\left(\dfrac{d\vec{B}}{du}\times\vec{C}\right) + \dfrac{d\vec{A}}{du}\times\left(\vec{B}\times\vec{C}\right).$

These simple properties of differentiation as applied to vectors can be extended to partial derivatives when a vector is a function of more than one scalar independent variables.

2.12 Vector Integration

For integrating a vector with respect to a scalar, let us take $\frac{d\vec{A}}{du} = \vec{B}(u)$. We may write on integration

$$\vec{A}(u) = \int_{u_0}^{u} \vec{B}(u)du + C,$$

where C is a constant of integration which gives the value of $\vec{A}$ when $u = u_0$. If $\vec{B}(u)$ stands for the velocity $\vec{V}(t)$ of a point, its position vector $\vec{r}(t)$ may be substituted in place of $\vec{A}(u)$ which is given by

$$\vec{r}(t) = \int_{t_0}^{t} \vec{V}(t)dt + \vec{r}_0,$$

where $\vec{r}_0$ is the position vector at time $t = t_0$.

2.13 Integration of Sums and Products

(i) $$\int \left(\vec{A}\cdot\frac{d\vec{B}}{du} + \vec{B}\cdot\frac{d\vec{A}}{du}\right) = \int \frac{d}{du}\left(\vec{A}\cdot\vec{B}\right) du = \vec{A}\cdot\vec{B} + C,$$

where C is the constant of integration.

(ii) $$\int \left(2\vec{A}\cdot\frac{d\vec{A}}{du}\right) du = \int \frac{d}{du}\left(\vec{A}\cdot\vec{A}\right) du = \vec{A}\cdot\vec{A} + C = A^2 + C.$$

(iii) $$\int \left(2\frac{d\vec{A}}{du}\cdot\frac{d^2\vec{A}}{du^2}\right) du = \int \frac{d}{du}\left(\frac{d\vec{A}}{du}\cdot\frac{d\vec{A}}{du}\right) du = \frac{d\vec{A}}{du}\cdot\frac{d\vec{A}}{du} + C = \left(\frac{dA}{du}\right)^2 + C.$$

(iv) $$\int \left(\vec{A}\times\frac{d^2\vec{A}}{du^2}\right) du = \int \frac{d}{du}\left(\vec{A}\times\frac{d\vec{A}}{du}\right) du = \vec{A}\times\frac{d\vec{A}}{du} + C.$$

(v) $$\int \left(2\vec{A}\times\frac{d\vec{A}}{du}\right) du = \int \frac{d}{du}\left(\vec{A}\times\vec{A}\right) du = \vec{A}\times\vec{A} + C = C \quad [\because\ \vec{A}\times\vec{A} = 0].$$

Illustrative Examples

1. Find the angle between $\vec{r}_1 = 2\hat{\imath} + \hat{\jmath} + 2\hat{k}$ and $\vec{r}_2 = 4\hat{\imath} - 3\hat{\jmath}$.

$$\vec{r}_1\cdot\vec{r}_2 = \left(2\hat{\imath} + \hat{\jmath} + 2\hat{k}\right)\cdot(4\hat{\imath} - 3\hat{\jmath}) = 8 - 3 + 0 = 5$$

$$r_1^2 = 4 + 1 + 4 = 9 \quad \text{or} \quad r_1 = 3$$

$$r_2^2 = 16 + 9 = 25 \quad \text{or} \quad r_2 = 5$$

$$\therefore\ \cos\theta = \frac{\vec{r}_1\cdot\vec{r}_2}{r_1 r_2} = \frac{5}{3\times 5} = \frac{1}{3} \quad \text{or} \quad \theta = 70°32'.$$

2. Determine the value of x so that $\vec{a} = \hat{\imath} + x\hat{\jmath} + \hat{k}$ and $\vec{b} = 3i - 2\hat{\jmath} - 2\hat{k}$ are perpendicular.

When $\vec{a}$ and $\vec{b}$ are perpendicular, $\vec{a}\cdot\vec{b} = 0$.

$$\therefore \left(\hat{\imath} + x\hat{\jmath} + \hat{k}\right)\cdot\left(3\hat{\imath} - 2\hat{\jmath} - 2\hat{k}\right) = 3 - 2x - 2 = 0 \quad \text{or} \quad 2x = 1 \quad \text{or} \quad x = \frac{1}{2}.$$

3. Find the work done in moving a body along a vector $\vec{r} = 3\hat{\imath} - 6\hat{\jmath} + 2\hat{k}$ if the force applied is given by $\vec{F} = 2\hat{\imath} - \hat{\jmath} + \hat{k}$.

The work done in moving the body = (magnitude of the force along the vector $\vec{r}$) (distance moved)

$$= (F\cos\theta)r = \vec{F}\cdot\vec{r} = \left(2\hat{\imath} - \hat{\jmath} + \hat{k}\right)\cdot\left(3\hat{\imath} - 6\hat{\jmath} + 2\hat{k}\right) \quad \text{[Fig. 1]}$$

$$= 6 + 6 + 2 = 14.$$

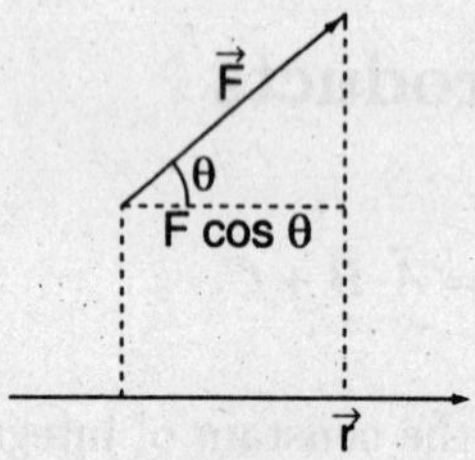

Fig. 1

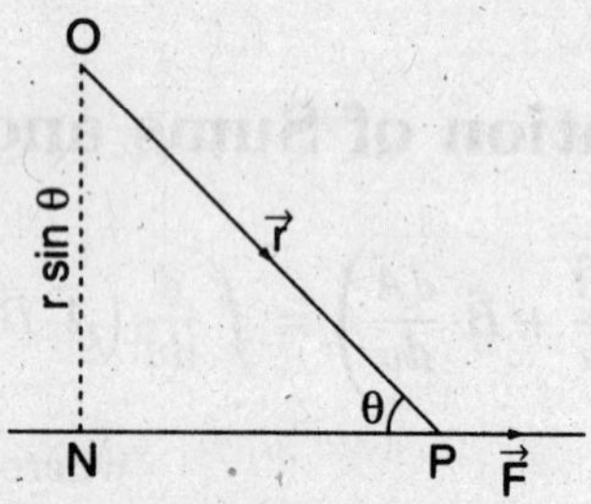

Fig. 2

4. Obtain an expression for the moment of a force about a point.

In Fig. 2, $\vec{F}$ denotes a force, the moment of which about the point O is to be determined. The moment of the force is a vector and is given by $\vec{M}$ = (magnitude of force) (perpendicular distance ON from the point O to the line of action of the force) $= F(r\sin\theta) = \vec{r}\times\vec{F}$, where $\vec{r}$ is the vector from O to the initial point P of the force. The direction of the vector $\vec{M}$ is perpendicular to the plane of $\vec{r}$ and $\vec{F}$ and the vectors $\vec{r}, \vec{F}$ and $\vec{M}$ form a right-handed system.

5. A rigid body is rotating about an axis with angular velocity $\vec{\omega}$. Prove that the linear velocity $\vec{v}$ of a point of the body with position vector $\vec{r}$ relative to a fixed point on the axis is given by $\vec{v} = \vec{\omega}\times\vec{r}$.

In Fig. 3, the point P of the body describes a circle of radius $r\sin\theta$, where r denotes the distance of the point P from the point O on the axis and $\angle PON = \theta$.

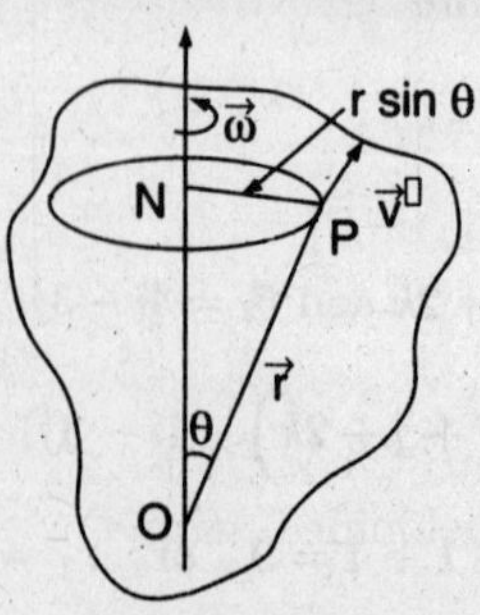

Fig. 3

$\therefore$ linear velocity $\vec{v} = (r\sin\theta)\omega = \vec{\omega}\times\vec{r}$; $\vec{v}$ is perpendicular to both $\vec{\omega}$ and $\vec{r}$ and the vectors $\vec{r}, \omega$ and $\vec{v}$ form a right-handed system.

6. If $\vec{A} = 5u^2\hat{\imath} + u\hat{\jmath} - u^3\hat{k}$ and $\vec{B} = \sin u\hat{\imath} - \cos u\hat{\jmath}$, find $\frac{d}{du}(\vec{A}\cdot\vec{B})$ and $\frac{d}{du}(\vec{A}\times\vec{B})$.

$$\frac{d}{du}(\vec{A}\cdot\vec{B}) = \vec{A}\cdot\frac{d\vec{B}}{du} + \frac{d\vec{A}}{du}\cdot\vec{B}$$

$$= (5u^2\hat{\imath} + u\hat{\jmath} - u^3\hat{k})\cdot(\cos u\hat{\imath} + \sin u\hat{\jmath}) + (10u\hat{\imath} + \hat{\jmath} - 3u^2\hat{k})\cdot(\sin u\hat{\imath} - \cos u\hat{\jmath})$$

$$= (5u^2\cos u + u\sin u) + (10u\sin u - \cos u)$$

$$= \cos u(5u^2 - 1) + 11u\sin u$$

$$\frac{d}{du}(\vec{A}\times\vec{B}) = \vec{A}\times\frac{d\vec{B}}{du} + \frac{d\vec{A}}{du}\times\vec{B}$$

$$= (5u^2\hat{\imath} + u\hat{\jmath} - u^3\hat{k})\times(\cos u\hat{\imath} + \sin u\hat{\jmath}) + (10u\hat{\imath} + \hat{\jmath} - 3u^2\hat{k})\times(\sin u\hat{\imath} - \cos u\hat{\jmath})$$

$$= \begin{vmatrix} \hat{\imath} & \hat{\jmath} & \hat{k} \\ 5u^2 & u & -u^3 \\ \cos u & \sin u & 0 \end{vmatrix} + \begin{vmatrix} \hat{\imath} & \hat{\jmath} & \hat{k} \\ 10u & 1 & -3u^2 \\ \sin u & -\cos u & 0 \end{vmatrix}$$

$$= \hat{\imath}(u^3\sin u) - \hat{\jmath}(u^3\cos u) + \hat{k}(5u^2\sin u - u\cos u) + \hat{\imath}(-3u^2\cos u)$$
$$-\hat{\jmath}(3u^2\sin u) + \hat{k}(-10u\cos u - \sin u)$$

$$= \hat{\imath}(u^3\sin u - 3u^2\cos u) - \hat{\jmath}(u^3\cos u + 3u^2\sin u)$$
$$+\hat{k}(5u^2\sin u - 11u\cos u - \sin u).$$

7. Prove that if $\vec{A}, \vec{B}$ and $\vec{C}$ are three non-collinear vectors then the triple scalar product $\vec{A}\cdot(\vec{B}\times\vec{C})$ is equal to the volume of the parallelopiped with $\vec{A}, \vec{B}$ and $\vec{C}$ as its edges. Hence show that $\vec{A}\cdot(\vec{B}\times\vec{C}) = \vec{B}\cdot(\vec{C}\times\vec{A}) = \vec{C}\cdot(\vec{A}\times\vec{B})$. **[C.U. (Hons.) 1980]**

We consider a parallelopiped with edges $\vec{OP} = \vec{A}$, $\vec{OQ} = \vec{B}$ and $\vec{OR} = \vec{C}$ as shown in Fig. 4.

Let $\vec{A}\times\vec{C} = a\hat{n}$ and its direction ON is perpendicular to the plane $OPTR$. From the definition of vector area it follows that the magnitude of $a\hat{n}$ gives the area of the parallelogram $OPTR$. Hence the volume of the parallelopiped

$$= (\text{magnitude of } a\hat{n})(OQ\cos\theta)$$

$$= aB\cos\theta \quad [B = \text{magnitude of } \vec{B}]$$

$$= \vec{B}\cdot a\hat{n} = \vec{B}\cdot(\vec{A}\times\vec{C}).$$

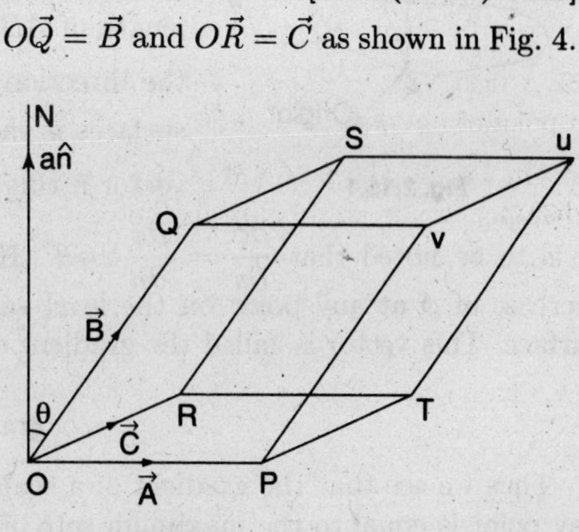

Fig. 4

Thus the triple scalar product $\vec{B}\cdot(\vec{A}\times\vec{C})$ is equal to the volume of the parallelopiped with edges $\vec{A}, \vec{B}$ and $\vec{C}$.

Similarly it can be shown that the volume of the parallelopiped $= \vec{A}\cdot(\vec{B}\times\vec{C}) = \vec{C}\cdot(\vec{A}\times\vec{B})$.

Hence

$$\vec{A}\cdot(\vec{B}\times\vec{C}) = \vec{B}\cdot(\vec{A}\times\vec{C}) = \vec{C}\cdot(\vec{A}\times\vec{B}).$$

2.14 The Vector Differential Operator $\vec{\nabla}$

The differential operator $\vec{\nabla}$ is defined in Cartesian notation by

$$\vec{\nabla} = \hat{\imath}\frac{\partial}{\partial x} + \hat{\jmath}\frac{\partial}{\partial y} + \hat{k}\frac{\partial}{\partial z}.$$

This vector operator may be applied as a directive differentiator to a scalar or to a vector function of space. As the symbol $\vec{\nabla}$ denotes an invariant vector differentiator which requires an operand whether scalar or vector on which it can operate, its expression in vectorial form should only be considered as symbolic or formal. In vector analysis there are three fundamental operations with $\vec{\nabla}$ which are of physical interest. If ϕ is a scalar function and $\vec{V}$ a vector function of space these operations are as follows: (i) $\vec{\nabla}\phi$, (ii) $\vec{\nabla}\cdot\vec{V}$ and (iii) $\vec{\nabla}\times\vec{V}$.

2.15 The Gradient of a Scalar Field

Certain physical quantities such as density, temperature, electric potential, etc. or any such non-directed quantity can be represented from point to point in space by a scalar point-function ϕ of the co-ordinates.

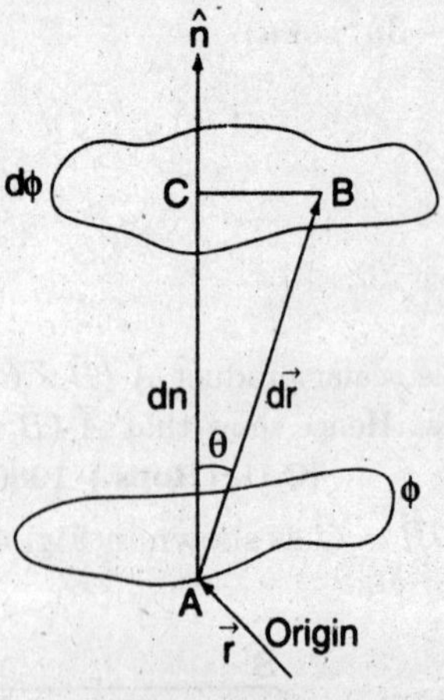

Fig. 2.15.1

The entire scalar field can be mapped out by level surfaces, the scalar function on each of which remains constant. We consider two such surfaces very close to each other. Let the two surfaces be characterised by the constant values ϕ and $\phi+d\phi$ of the scalar function. We shall examine a small portion of the two surfaces near a given point A on the surface denoted by the function ϕ [Fig. 2.15.1]. Let $\vec{r}$ and $\vec{r}+d\vec{r}$ be the radius vectors from the origin to the points A and B respectively. From the figure it follows that AC is the least distance between the two surfaces in the direction of normal unit vector $\hat{n}$ at A which is of length dn. If dr be the length of AB, the magnitude of the rate of increase of ϕ at A in the direction of AB is $\dfrac{\partial\phi}{\partial r}$, when the distance between the two surfaces is vanishingly small. In the direction of the unit normal vector $\hat{n}$ this rate of increase is maximum, i.e., $\dfrac{\partial\phi}{\partial n}$ is maximum. It is to be noted that $\dfrac{\partial\phi}{\partial r} = \dfrac{\partial\phi}{\partial n}\cos\theta$. Hence, the vector $\hat{n}\dfrac{\partial\phi}{\partial n}$ gives the maximum rate of increase of ϕ at any point on the level surface, $\hat{n}$ being the unit vector normal to the level surface. This vector is called the gradient of ϕ at the point and is written as

$$\text{grad}\,\phi = \hat{n}\frac{\partial\phi}{\partial n}. \tag{2.15.1}$$

Thus we see that the gradient of a scalar field is a vector field, the magnitude of which at any point is equal to the maximum rate of increase of ϕ and the direction is perpendicular to the level surface at that point.

A simple illustrative example in this case is an equipotential surface. The electric intensity X at any point is in the direction of the maximum rate of decrease of potential ϕ, i.e., normal to the equipotential surface, and has a magnitude equal to the rate of decrease of potential. Hence, the electric intensity is given by $X = -\,\text{grad}\,\phi$.

We consider the vector $\vec{\nabla}\phi$ given by

$$\vec{\nabla}\phi = \hat{\imath}\frac{\partial\phi}{\partial x} + \hat{\jmath}\frac{\partial\phi}{\partial y} + \hat{k}\frac{\partial\phi}{\partial z}$$

$\hat{\imath}\frac{\partial\phi}{\partial x}$, $\hat{\jmath}\frac{\partial\phi}{\partial y}$ and $\hat{k}\frac{\partial\phi}{\partial z}$ are the vector rates of increase of ϕ in the directions of the axes of X, Y and Z. We will now show that the sum of the above vector rates is equal to the gradient of ϕ. Taking scalar product on the two sides of equation (2.15.1) with an element of radius vector $d\vec{r}$, we have

$$(\text{grad}\,\phi)\cdot d\vec{r} = \frac{\partial\phi}{\partial n}\hat{n}\cdot d\vec{r} = \frac{\partial\phi}{\partial n}dr\cos\theta = \frac{\partial\phi}{\partial n}dn = d\phi$$

$\frac{\partial\phi}{\partial n}$ is the total normal rate of change of ϕ.

Now,

$$d\phi = \frac{\partial\phi}{\partial x}dx + \frac{\partial\phi}{\partial y}dy + \frac{\partial\phi}{\partial z}dz.$$

$$\therefore\ (\text{grad}\,\phi)\cdot d\vec{r} = \frac{\partial\phi}{\partial x}dx + \frac{\partial\phi}{\partial y}dy + \frac{\partial\phi}{\partial z}dz$$

$$= \left(\hat{\imath}\frac{\partial\phi}{\partial x} + \hat{\jmath}\frac{\partial\phi}{\partial y} + \hat{k}\frac{\partial\phi}{\partial z}\right)\cdot\left(\hat{\imath}dx + \hat{\jmath}dy + \hat{k}dz\right)$$

$$= \left(\vec{\nabla}\phi\right)\cdot d\vec{r}$$

$$\therefore\ \vec{\nabla}\phi = \text{grad}\,\phi = \hat{\imath}\frac{\partial\phi}{\partial x} + \hat{\jmath}\frac{\partial\phi}{\partial y} + \hat{k}\frac{\partial\phi}{\partial z} \qquad (1.15.2)$$

1.16 The Divergence of a Vector

The divergence of a vector over the surface of a unit volume element surrounding a point P in a vector field is given by

$$\text{div}\,\vec{A} = \lim_{V\to 0}\frac{\iint_S \vec{A}\cdot d\vec{S}}{V} = \lim_{V\to 0}\frac{\iint_S \vec{A}\cdot\hat{n}d\vec{S}}{V}$$

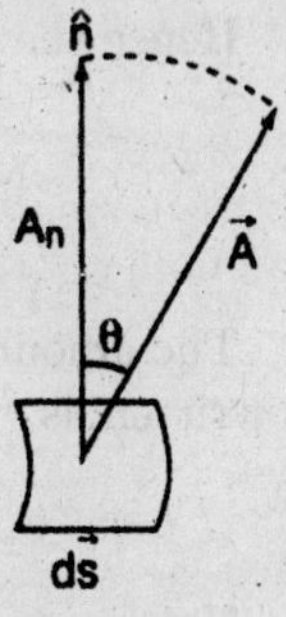

Fig. 2.16.1

where S is a closed surface surrounding the point P and encloses the volume V of the element.

Now, we know that $\vec{A}\cdot d\vec{S} = \vec{A}\cdot\hat{n}ds = A\ dS\cos\theta$, where θ is the angle between the vectors $\vec{A}$ and $d\vec{S}$ (Fig. 2.16.1). The vector $d\vec{S}$ represents the area of magnitude dS and having direction normal to the area dS. Since $A\cos\theta$ is the component of $\vec{A}$ along outward normal to dS we may put A_n for $A\cos\theta$. Hence,

$$\text{div}\,\vec{A} = \lim_{V\to 0}\frac{\iint_S A_n\,dS}{V} \qquad (1.16.1)$$

The expression $\iint_S \vec{A}\cdot\hat{n}d\vec{S} = \iint_S A_n dS$ is the surface integral of A_n over the surface area S.

This gives the total flux of $\vec{A}$ through the entire surface S. The divergence of a vector is thus concerned with the net outflow of some physical entity, such as heat or electric flux, through the surface area of a unit volume element in a vector field.

2.17 Analytical Expression for the Divergence of a Vector in Terms of Rectangular Co-ordinates

We consider a vector $\vec{A}$ at the midpoint of an infinitesimal element of volume with sides dx, dy and dz parallel to the axes OX, OY and OZ. The vector $\vec{A}$ has components of magnitudes A_x, A_y and A_z parallel to OX, OY and OZ respectively. The value of x-component of the vector at the midpoint of the left-hand face of the volume element is $A_x - \frac{1}{2}\frac{\partial A_x}{\partial x}dx$. When the volume element is infinitesimally small this component of vector may be considered all over the face. Similarly, on the right-hand face the x-component of the vector is $A_x + \frac{1}{2}\frac{\partial A_x}{\partial x}dx$.

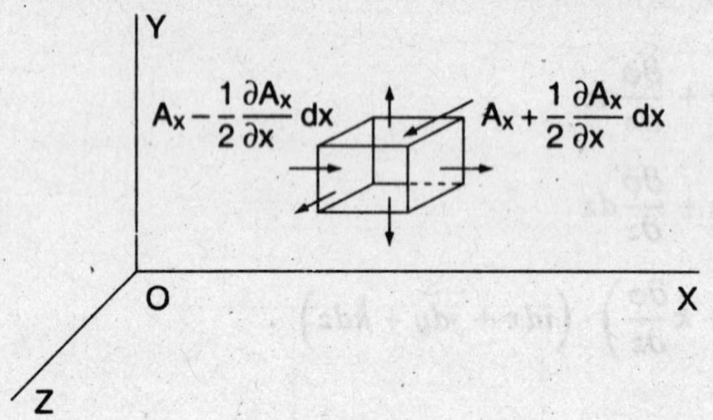

Fig. 2.17.1

Then the excess of flux leaving the element over that entering it parallel to OX is

$$\left(A_x + \frac{1}{2}\frac{\partial A_x}{\partial x}dx\right) dy\, dz - \left(A_x - \frac{1}{2}\frac{\partial A_x}{\partial x}dx\right) dy\, dz = \frac{\partial A_x}{\partial x}dx\, dy\, dz.$$

By similar reasons the contributions to the flux parallel to OY and OZ are

$$\frac{\partial A_y}{\partial y}dx\, dy\, dz \text{ and } \frac{\partial A_z}{\partial z}dx\, dy\, dz.$$

Hence the net flux diverging from the element is

$$\left(\frac{\partial A_x}{\partial x} + \frac{\partial A_y}{\partial y} + \frac{\partial A_z}{\partial z}\right) dx\, dy\, dz.$$

The amount of this flux per unit volume is defined as the divergence of the vector $\vec{A}$ which is written as

$$\text{div } \vec{A} = \frac{\partial A_x}{\partial x} + \frac{\partial A_y}{\partial y} + \frac{\partial A_z}{\partial z}. \tag{2.17.1}$$

Now,

$$\frac{\partial A_x}{\partial x} + \frac{\partial A_y}{\partial y} + \frac{\partial A_z}{\partial z} = \left(\hat{\imath}\frac{\partial}{\partial x} + \hat{\jmath}\frac{\partial}{\partial y} + \hat{k}\frac{\partial}{\partial z}\right)\cdot\left(\hat{\imath}A_x + \hat{\jmath}A_y + \hat{k}A_z\right) = \vec{\nabla}\cdot\vec{A}$$

$$\therefore \text{ div } \vec{A} = \vec{\nabla}\cdot\vec{A}. \tag{2.17.2}$$

When the flux entering the volume element is equal to that leaving it, we have,

$$\text{div}\vec{A} = 0 \quad \text{or} \quad \vec{\nabla} \cdot \vec{A} = 0.$$

In case of fluid motion this means that there is no accumulation of fluid anywhere. This is called the continuity equation for an incompressible fluid. As the fluid is neither created nor destroyed at any point it is said to have no sources or sinks. The lines of flow of the vector $\vec{A}$, in this case, either form closed curves or terminate upon bounding surfaces or extend to infinity. A vector satisfying the above conditions is called *solenoidal.*

2.18 Gauss's Divergence Theorem

We consider an elementary cube of volume dV within the surface S as shown in Fig. (2.18.1). Instead of taking an elementary cube we can also take an elementary volume of any shape. If $\vec{A}$ be the vector at the midpoint of the element, then the total flux diverging from the elementary volume is $(\text{div}\vec{A})dV$ [See equation 2.16.1]. Now, the outward flux of the vector $\vec{A}$ from the surface $ABCD$ of the cube is positive. The flux from the same surface $ABCD$ of an adjacent elementary cube $ABCC'D'A'$ is negative. Hence the fluxes from the common surface of the two elementary cubes cancel. Adding a number of cubical elements until one face of an element is on the surface S, it can be easily shown by similar arguments that this face (marked shaded) only makes contribution to the normal flux through S. In a similar way if we apply this treatment to the surfaces of elements of volume throughout the space within the surface S, we have the total flux

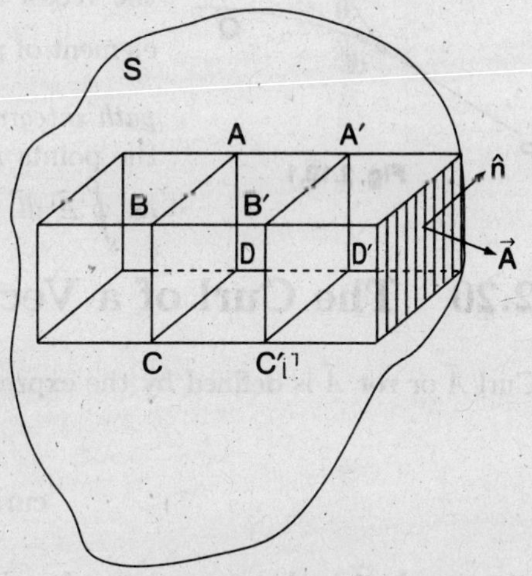

Fig. 2.18.1

$$\iint_S \vec{A}\cdot d\vec{S}.$$

Again integrating $\text{div}\vec{A}dV$ throughout the enclosed volume V we have the total flux

$$\iiint_V \text{div}\ \vec{A}dV.$$

Hence

$$\iint_S \vec{A}\cdot d\vec{S} = \iiint_V (\text{div}\ \vec{A})dV = \iiint_V \vec{\nabla}\cdot \vec{A}dV.$$

Now $d\vec{S} = \hat{n}dS$, $\hat{n}$ being the outward normal unit vector

$$\therefore \quad \iint_S \vec{A}\cdot d\vec{S} = \iint_S \vec{A}\cdot \hat{n}dS.$$

Hence we may write

$$\iint_S \vec{A}\cdot \hat{n} dS = \iiint_V (\text{div } \vec{A}) dV.$$

This is Gauss's divergence theorem, i.e., the volume integral of the divergence of a vector $\vec{A}$ in a vector field, taken through a given volume, is equal to the surface integral of the normal component of $\vec{A}$ over the surface which encloses the volume.

2.19 The Line Integral or the Path Integral

We consider a vector $\vec{A}$ in a region where a vector field exists [Fig. 2.19.1]. In the figure $d\vec{l}$ denotes an element of a path between the two points P and Q in the vector field. The scalar product $\vec{A}\cdot d\vec{l} = Adl \cos\theta$, where θ denotes the angle between the vector at any point on the curve PQ and the direction of the element of path $d\vec{l}$. Then $\int_{PQ} \vec{A}\cdot d\vec{l}$ denotes the line integral or the *path integral* of the vector $\vec{A}$ along the curve obtained by joining the points P and Q. For a closed path the line integral is written as $\oint \vec{A}\cdot d\vec{l}$.

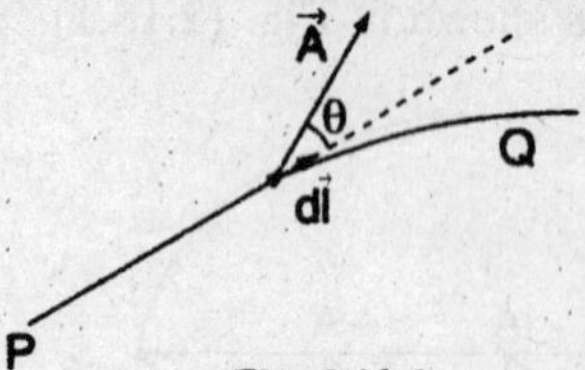

Fig. 2.19.1

2.20 The Curl of a Vector

Curl $\vec{A}$ or rot $\vec{A}$ is defined by the expression

$$\text{curl}_n \vec{A} = \lim_{S\to 0} \frac{\oint \vec{A}\cdot d\vec{l}}{S}$$

where $\text{curl}_n \vec{A}$ is the component of curl $\vec{A}$ in the direction of the normal unit vector $\hat{n}$ at any point which is in the plane of a closed curve surrounding the point. The closed curve encloses the area S. The direction assigned to the vector $\hat{n}$ is considered positive if the rotation is anticlockwise, i.e., right-handed. The integration along the closed curve is also carried out in this direction.

2.21 Analytical Expression for the Curl of a Vector in Terms of Rectangular Co-ordinates

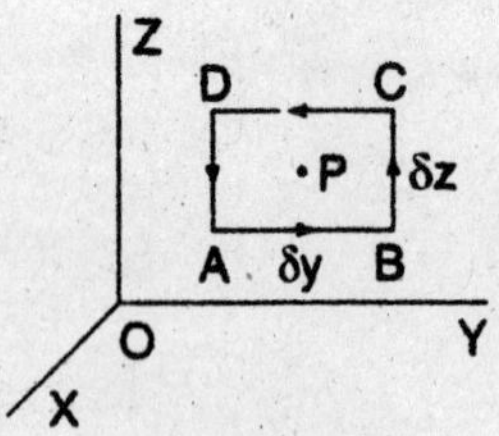

Fig. 2.21.1

We consider a point $P(x, y, z)$ in a right-handed set of cartesian axes. Let a rectangle $ABCD$ be drawn symmetrically surrounding the point P in the plane YZ Fig. (2.21.1). Let the length of the side of the rectangle parallel to Y-axis be δy and that parallel to Z-axis be δz. Let A_x, A_y, A_z be the components of $\vec{A}$ at P.

$$\therefore \int_{AB} \vec{A}\cdot d\vec{l} = \int_{-\delta y/2}^{+\delta y/2} A_y\left(x, y, z - \frac{\delta z}{2}\right) dy$$

where $A_y\left(x, y, z-\frac{\delta z}{2}\right)$ denotes the component of $\vec{A}$ at the midpoint of AB in the direction of Y-axis. The value of A_y at the midpoint of AB, i.e., at $\left(x, y, z-\frac{\delta z}{2}\right)$ is $\left(A_y-\frac{\partial A_y}{\partial z}\frac{\delta z}{2}\right)$

$$\therefore \int_{AB} \vec{A}\cdot d\vec{l} = \int_{-\delta y/2}^{+\delta y/2}\left(A_y-\frac{\partial A_y}{\partial z}\frac{\delta z}{2}\right)dy = \left(A_y-\frac{\partial A_y}{\partial z}\frac{\delta z}{2}\right)\delta y.$$

Similarly it can be shown

$$\int_{CD} \vec{A}\cdot d\vec{l} = -\left(A_y+\frac{\partial A_y}{\partial z}\frac{\delta z}{2}\right)\delta y$$

and $$\int_{BC} \vec{A}\cdot d\vec{l} = \left(A_z+\frac{\partial A_z}{\partial y}\frac{\delta y}{2}\right)\delta z$$

and $$\int_{DA} \vec{A}\cdot d\vec{l} = -\left(A_z-\frac{\partial A_z}{\partial y}\frac{\delta y}{2}\right)\delta z$$

$$\begin{aligned}\therefore \int_{ABCDA} \vec{A}\cdot d\vec{l} &= \left(A_y-\frac{\partial A_y}{\partial z}\frac{\delta z}{2}\right)\delta y - \left(A_y+\frac{\partial A_y}{\partial z}\frac{\delta z}{2}\right)\delta y \\ &\quad + \left(A_z+\frac{\partial A_z}{\partial y}\frac{\delta y}{2}\right)\delta z - \left(A_z-\frac{\partial A_z}{\partial y}\frac{\delta y}{2}\right)\delta z \\ &= -\frac{\partial A_y}{\partial z}\delta y\delta z + \frac{\partial A_z}{\partial y}\delta y\delta z \\ &= \left(\frac{\partial A_z}{\partial y}-\frac{\partial A_y}{\partial z}\right)\delta y\delta z\end{aligned}$$

$$\therefore \text{curl}_x\vec{A} = \lim_{S\to 0}\frac{\oint \vec{A}\cdot d\vec{l}}{S} = \left(\frac{\partial A_z}{\partial y}-\frac{\partial A_y}{\partial z}\right) \quad [\because S=\delta y\delta z.] \qquad (2.21.1)$$

Similarly it can be shown that

$$\text{curl}_y\vec{A} = \left(\frac{\partial A_x}{\partial z}-\frac{\partial A_z}{\partial x}\right) \qquad (2.21.2)$$

and $$\text{curl}_z\vec{A} = \left(\frac{\partial A_y}{\partial x}-\frac{\partial A_x}{\partial y}\right) \qquad (2.21.3)$$

$$\therefore \text{curl}\,\vec{A} = \hat{\imath}\left(\frac{\partial A_z}{\partial y}-\frac{\partial A_y}{\partial z}\right)+\hat{\jmath}\left(\frac{\partial A_x}{\partial z}-\frac{\partial A_z}{\partial x}\right)+\hat{k}\left(\frac{\partial A_y}{\partial x}-\frac{\partial A_x}{\partial y}\right). \qquad (2.21.4)$$

The above expression may be written in the form

$$\text{curl}\,\vec{A} = \begin{vmatrix}\hat{\imath} & \hat{\jmath} & \hat{k}\\ \frac{\partial}{\partial x} & \frac{\partial}{\partial y} & \frac{\partial}{\partial z}\\ A_x & A_y & A_z\end{vmatrix}.$$

Now,

$$\vec{\nabla} \times \vec{A} = \left(\hat{\imath}\frac{\partial}{\partial x} + \hat{\jmath}\frac{\partial}{\partial y} + \hat{k}\frac{\partial}{\partial z}\right) \times \left(\hat{\imath}A_x + \hat{\jmath}A_y + \hat{k}A_z\right)$$

$$= \hat{\imath}\left(\frac{\partial A_z}{\partial y} - \frac{\partial A_y}{\partial z}\right) + \hat{\jmath}\left(\frac{\partial A_x}{\partial z} - \frac{\partial A_z}{\partial x}\right) + \hat{k}\left(\frac{\partial A_y}{\partial x} - \frac{\partial A_x}{\partial y}\right) \tag{2.21.5}$$

From (2.21.4) and (2.21.5) we have

$$\text{curl}\,\vec{A} = \vec{\nabla} \times \vec{A}. \tag{2.21.6}$$

For an example of curl, see the worked out example on page 44.

2.22 Stokes's Theorem

Let us divide a surface S by means of a system of intersecting lines drawn in the surface into a number of infinitesimal surface elements. We consider the shaded element of vector area $\hat{n}dS$. Here $\hat{n}$ is the positive unit normal on the area. Let us now consider the line integral of a vector $\vec{A}$ around each of the paths which are the perimeters of the surface elements. The sum of these line integrals, taking in the same direction (anticlockwise direction) will be equal to the line integral around perimeter L enclosing the area S since all parts of the line integral round the surface elements cancel out except the parts which are along the perimeter L. For each surface element of area such as δS_i having perimeter L_i, we may write

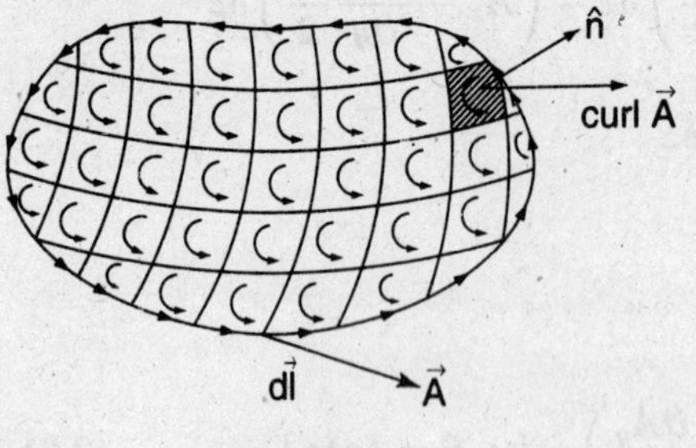

Fig. 2.22.1

$$\oint_{L_i} \vec{A}\cdot d\vec{l} = (\text{curl}_n \vec{A})\delta S_i.$$

Adding up these equations for as such surface elements, we have,

$$\oint_L \vec{A}\cdot d\vec{l} = \iint_S (\text{curl}_n \vec{A})dS.$$

Again

$$(\text{curl}_n \vec{A})dS = \text{curl}\,\vec{A}\cdot \hat{n}dS = \text{curl}\,\vec{A}\cdot d\vec{S},$$

where $d\vec{S}$ is the vector area perpendicular to the area dS, i.e., along the unit normal vector $\hat{n}$.

$$\therefore \oint_L \vec{A}\cdot d\vec{l} = \iint_S \text{curl}\,\vec{A}\cdot d\vec{S}.$$

This is Stokes's theorem. It can be expressed in the following way:

The line integral of the tangential component of the vector $\vec{A}$ round any closed path is equal to the normal surface integral of the vector curl $\vec{A}$ *over the surface having the path as its boundary.*

2.23 Green's Theorem

We know that a volume integral may be converted into a surface integral by Gauss's theorem which is given by

$$\iint_S \vec{A}\cdot d\vec{s} = \iiint_V (\text{div } \vec{A}) dV.$$

Let $\vec{A}$ be the product of a scalar φ and a second vector $\vec{F}$, then $\vec{A} = \varphi\vec{F}$

$$\therefore \text{ div } \vec{A} = \vec{\nabla}\cdot(\varphi\vec{F}) = \vec{\nabla}\cdot(\varphi\hat{\imath}F_x + \varphi\hat{\jmath}F_y + \varphi\hat{k}F_z)$$

$$= \frac{\partial}{\partial x}(\varphi F_x) + \frac{\partial}{\partial y}(\varphi F_y) + \frac{\partial}{\partial z}(\varphi F_z)$$

$$= \left(\frac{\partial\varphi}{\partial x}F_x + \frac{\partial\varphi}{\partial y}F_y + \frac{\partial\varphi}{\partial z}F_z\right) + \varphi\left(\frac{\partial F_x}{\partial x} + \frac{\partial F_y}{\partial y} + \frac{\partial F_z}{\partial z}\right)$$

$$= \left(\hat{\imath}\frac{\partial\varphi}{\partial x} + \hat{\jmath}\frac{\partial\varphi}{\partial y} + \hat{k}\frac{\partial\varphi}{\partial z}\right)\cdot\left(\hat{\imath}F_x + \hat{\jmath}F_y + \hat{k}F_z\right) + \varphi\left(\frac{\partial F_x}{\partial x} + \frac{\partial F_y}{\partial y} + \frac{\partial F_z}{\partial z}\right)$$

$$= \left(\vec{\nabla}\varphi\right)\cdot\vec{F} + \varphi\vec{\nabla}\cdot\vec{F}.$$

Hence Gauss's theorem may be written as

$$\iint_S \varphi\vec{F}\cdot d\vec{s} = \iiint_V \left\{\left(\vec{\nabla}\varphi\right)\cdot\vec{F} + \varphi\vec{\nabla}\cdot\vec{F}\right\} dV. \quad (2.23.1)$$

If the vector $\vec{F}$ be represented as the gradient of a scalar ψ, then $\vec{F} = \text{grad}\,\psi = \vec{\nabla}\psi$;

$$\therefore \vec{\nabla}\cdot\vec{F} = \nabla\cdot\left(\vec{\nabla}\psi\right) = \nabla^2\psi \quad \text{[see Art. 2.24 (vi)]}. \quad (2.23.2)$$

$\therefore$ From (2.23.1) and (2.23.2) we have,

$$\iint_S \left(\varphi\vec{\nabla}\psi\right)\cdot d\vec{s} = \iiint_V \left\{\left(\vec{\nabla}\varphi\right)\cdot\left(\vec{\nabla}\psi\right) + \varphi\nabla^2\psi\right\} dV. \quad (2.23.3)$$

If from (2.23.3) we subtract the equation obtained by interchanging φ and ψ, we have,

$$\iint_S \left(\varphi\vec{\nabla}\psi - \psi\vec{\nabla}\varphi\right)\cdot d\vec{s} = \iiint_V \left\{\varphi\nabla^2\psi - \psi\nabla^2\varphi\right\} dV. \quad (2.23.4)$$

Equations (2.23.3) and (2.23.4) are referred to as Green's theorem.

2.24 Relations involving $\vec{\nabla}$

We shall derive here the following important relations involving $\vec{\nabla}$:

(i) $\vec{\nabla}(\varphi + \psi) = \vec{\nabla}\varphi + \vec{\nabla}\psi$;

(ii) $\vec{\nabla}(\varphi\psi) = \varphi\vec{\nabla}\psi + \psi\vec{\nabla}\varphi$;

(iii) $\vec{\nabla}(\vec{A}\cdot\vec{B}) = (B\cdot\vec{\nabla})\vec{A} + (\vec{A}\cdot\vec{\nabla})\vec{B} + \vec{B}\times(\vec{\nabla}\times\vec{A}) + \vec{A}\times\vec{\nabla}\times\vec{B}$;

(iv) $\vec{\nabla}\cdot(\vec{A}+\vec{B}) = \vec{\nabla}\cdot A + \vec{\nabla}\cdot B$;

(v) $\vec{\nabla}\cdot(\varphi\vec{A}) = (\vec{\nabla}\varphi)\cdot\vec{A} + \varphi(\vec{\nabla}\cdot\vec{A})$;

(vi) $\vec{\nabla}\cdot(\vec{\nabla}\varphi) = \nabla^2\varphi$;

(vii) $\vec{\nabla}\times(\vec{A}+\vec{B}) = \vec{\nabla}\times\vec{A} + \vec{\nabla}\times\vec{B}$;

(viii) $\vec{\nabla}\times(\varphi\vec{A}) = (\vec{\nabla}\varphi)\times\vec{A} + \varphi(\vec{\nabla}\times\vec{A})$;

(ix) $\vec{\nabla}\times(\vec{\nabla}\varphi) = 0$;

(x) $\vec{\nabla}\cdot(\vec{\nabla}\times\vec{A}) = 0$;

(xi) $\vec{\nabla}\times(\vec{\nabla}\times\vec{A}) = \vec{\nabla}(\vec{\nabla}\cdot\vec{A}) - \nabla^2\vec{A}$.

In these relations $\vec{A}$ and $\vec{B}$ denote differentiable vector functions, and φ and ψ denote the scalar functions.

Solution.

(i)
$$\vec{\nabla}(\varphi+\psi) = \hat{\imath}\frac{\partial}{\partial x}(\varphi+\psi) + \hat{\jmath}\frac{\partial}{\partial y}(\varphi+\psi) + \hat{k}\frac{\partial}{\partial z}(\varphi+\psi)$$
$$= \left(\hat{\imath}\frac{\partial\varphi}{\partial x} + \hat{\jmath}\frac{\partial\varphi}{\partial y} + \hat{k}\frac{\partial\varphi}{\partial z}\right) + \left(\hat{\imath}\frac{\partial\psi}{\partial x} + \hat{\jmath}\frac{\partial\psi}{\partial y} + \hat{k}\frac{\partial\psi}{\partial z}\right)$$
$$= \vec{\nabla}\varphi + \vec{\nabla}\psi \quad \text{or} \quad \mathrm{grad}(\varphi+\psi) = \mathrm{grad}\,\varphi + \mathrm{grad}\,\psi.$$

(ii)
$$\vec{\nabla}(\varphi\psi) = \hat{\imath}\frac{\partial}{\partial x}(\varphi\psi) + \hat{\jmath}\frac{\partial}{\partial y}(\varphi\psi) + \hat{k}\frac{\partial}{\partial z}(\varphi\psi)$$
$$= \psi\left(\hat{\imath}\frac{\partial\varphi}{\partial x} + \hat{\jmath}\frac{\partial\varphi}{\partial y} + \hat{k}\frac{\partial\varphi}{\partial z}\right) + \varphi\left(\hat{\imath}\frac{\partial\psi}{\partial x} + \hat{\jmath}\frac{\partial\psi}{\partial y} + \hat{k}\frac{\partial\psi}{\partial z}\right)$$
$$= \psi\vec{\nabla}\varphi + \varphi\vec{\nabla}\psi \quad \text{or} \quad \mathrm{grad}(\varphi\psi) = \psi\,\mathrm{grad}\,\varphi + \varphi\,\mathrm{grad}\,\psi.$$

(iii)
$$(\vec{A}\cdot\vec{\nabla})\vec{B} = \left(A_x\frac{\partial}{\partial x} + A_y\frac{\partial}{\partial y} + A_z\frac{\partial}{\partial z}\right)\vec{B} = A_x\frac{\partial\vec{B}}{\partial x} + A_y\frac{\partial\vec{B}}{\partial y} + A_z\frac{\partial\vec{B}}{\partial z}$$
$$= \hat{\imath}\left(A_x\frac{\partial B_x}{\partial x} + A_y\frac{\partial B_x}{\partial y} + A_z\frac{\partial B_x}{\partial z}\right) + \hat{\jmath}\left(A_x\frac{\partial B_y}{\partial x} + A_y\frac{\partial B_y}{\partial y} + A_z\frac{\partial B_y}{\partial z}\right)$$
$$+\hat{k}\left(A_x\frac{\partial B_z}{\partial x} + A_y\frac{\partial B_z}{\partial y} + A_z\frac{\partial B_z}{\partial z}\right). \tag{2.24.1}$$

Similarly

$$(\vec{B}\cdot\vec{\nabla})\vec{A} = \hat{\imath}\left(B_x\frac{\partial A_x}{\partial x} + B_y\frac{\partial A_x}{\partial y} + B_z\frac{\partial A_x}{\partial z}\right) + \hat{\jmath}\left(B_x\frac{\partial A_y}{\partial x} + B_y\frac{\partial A_y}{\partial y} + B_z\frac{\partial A_y}{\partial z}\right)$$
$$+\hat{k}\left(B_x\frac{\partial A_z}{\partial x} + B_y\frac{\partial A_z}{\partial y} + B_z\frac{\partial A_z}{\partial z}\right). \tag{2.24.2}$$

Also

$$\vec{B} \times (\vec{\nabla} \times \vec{A}) = (\hat{\imath}B_x + \hat{\jmath}B_y + \hat{k}B_z) \times \begin{vmatrix} \hat{\imath} & \hat{\jmath} & \hat{k} \\ \frac{\partial}{\partial x} & \frac{\partial}{\partial y} & \frac{\partial}{\partial z} \\ A_x & A_y & A_z \end{vmatrix}$$

$$= (\hat{\imath}B_x + \hat{\jmath}B_y + \hat{k}B_z) \times \left\{\hat{\imath}\left(\frac{\partial A_z}{\partial y} - \frac{\partial A_y}{\partial z}\right) + \hat{\jmath}\left(\frac{\partial A_x}{\partial z} - \frac{\partial A_z}{\partial x}\right) + \hat{k}\left(\frac{\partial A_y}{\partial x} - \frac{\partial A_x}{\partial y}\right)\right\}$$

$$= \begin{vmatrix} \hat{\imath} & \hat{\jmath} & \hat{k} \\ B_x & B_y & B_z \\ \left(\frac{\partial A_z}{\partial y} - \frac{\partial A_y}{\partial z}\right) & \left(\frac{\partial A_x}{\partial z} - \frac{\partial A_z}{\partial x}\right) & \left(\frac{\partial A_y}{\partial x} - \frac{\partial A_x}{\partial y}\right) \end{vmatrix}$$

$$= \hat{\imath}\left\{B_y\left(\frac{\partial A_y}{\partial x} - \frac{\partial A_x}{\partial y}\right) - B_z\left(\frac{\partial A_x}{\partial z} - \frac{\partial A_z}{\partial x}\right)\right\}$$
$$+\hat{\jmath}\left\{B_z\left(\frac{\partial A_z}{\partial y} - \frac{\partial A_y}{\partial z}\right) - B_x\left(\frac{\partial A_y}{\partial x} - \frac{\partial A_x}{\partial y}\right)\right\}$$
$$+\hat{k}\left\{B_x\left(\frac{\partial A_x}{\partial z} - \frac{\partial A_z}{\partial x}\right) - B_y\left(\frac{\partial A_z}{\partial y} - \frac{\partial A_y}{\partial z}\right)\right\} \qquad (2.24.3)$$

Similarly

$$\vec{A} \times (\vec{\nabla} \times \vec{B}) = \hat{\imath}\left\{A_y\left(\frac{\partial B_y}{\partial x} - \frac{\partial B_x}{\partial y}\right) - A_z\left(\frac{\partial B_x}{\partial z} - \frac{\partial B_z}{\partial x}\right)\right\}$$
$$+\hat{\jmath}\left\{A_z\left(\frac{\partial B_z}{\partial y} - \frac{\partial B_y}{\partial z}\right) - A_x\left(\frac{\partial B_y}{\partial x} - \frac{\partial B_x}{\partial y}\right)\right\}$$
$$+\hat{k}\left\{A_x\left(\frac{\partial B_x}{\partial z} - \frac{\partial B_z}{\partial x}\right) - A_y\left(\frac{\partial B_z}{\partial y} - \frac{\partial B_y}{\partial z}\right)\right\} \qquad (2.24.4)$$

Adding (2.24.1), (2.24.2), (2.24.3) and (2.24.4), we have the RHS of the expression (iii)

$$= \hat{\imath}\frac{\partial}{\partial x}(A_xB_x + A_yB_y + A_zB_z) + \hat{\jmath}\frac{\partial}{\partial y}(A_xB_x + A_yB_y + A_zB_z) + \hat{k}\frac{\partial}{\partial z}(A_xB_x + A_yB_y + A_zB_z)$$

$$= \left(\hat{\imath}\frac{\partial}{\partial x} + \hat{\jmath}\frac{\partial}{\partial y} + \hat{k}\frac{\partial}{\partial z}\right)(A_xB_x + A_yB_y + A_zB_z)$$

$$= \left(\hat{\imath}\frac{\partial}{\partial x} + \hat{\jmath}\frac{\partial}{\partial y} + \hat{k}\frac{\partial}{\partial z}\right)\left\{(\hat{\imath}A_x + \hat{\jmath}A_y + \hat{k}A_z)\cdot(\hat{\imath}B_x + \hat{\jmath}B_y + \hat{k}B_z)\right\} = \vec{\nabla}(\vec{A}\cdot\vec{B}).$$

Thus we have

$$\vec{\nabla}(\vec{A}\cdot\vec{B}) = (\vec{B}\cdot\vec{\nabla})\vec{A} + (\vec{A}\cdot\vec{\nabla})\vec{B} + \vec{B}\times(\vec{\nabla}\times\vec{A}) + \vec{A}\times(\vec{\nabla}\times\vec{B})$$

$$\text{or}\quad \text{grad}(\vec{A}\cdot\vec{B}) = (\vec{B}\cdot\vec{\nabla})\vec{A} + (\vec{A}\cdot\vec{\nabla})\vec{B} + \vec{B}\times\text{curl}\,\vec{A} + \vec{A}\times\text{curl}\,\vec{B}.$$

(iv) $$\vec{\nabla}\cdot(\vec{A}+\vec{B}) = \left(\hat{\imath}\frac{\partial}{\partial x} + \hat{\jmath}\frac{\partial}{\partial y} + \hat{k}\frac{\partial}{\partial z}\right)\cdot\left\{\hat{\imath}(A_x+B_x) + \hat{\jmath}(A_y+B_y) + \hat{k}(A_z+B_z)\right\}$$

$$= \frac{\partial}{\partial x}(A_x+B_x) + \frac{\partial}{\partial y}(A_y+B_y) + \frac{\partial}{\partial z}(A_z+B_z)$$

$$= \vec{\nabla}\cdot\vec{A} + \vec{\nabla}\cdot\vec{B} \quad\text{or}\quad \text{div}(\vec{A}+\vec{B}) = \text{div}\vec{A} + \text{div}\vec{B}.$$

(v) $$\vec{\nabla}\cdot(\varphi\vec{A}) = \vec{\nabla}\cdot\left\{\varphi(\hat{\imath}A_x + \hat{\jmath}A_y + \hat{k}A_z)\right\}$$

$$= \left(\hat{\imath}\frac{\partial}{\partial x} + \hat{\jmath}\frac{\partial}{\partial y} + \hat{k}\frac{\partial}{\partial z}\right)\cdot\varphi\left\{(\hat{\imath}A_x + \hat{\jmath}A_y + \hat{k}A_z)\right\}$$

$$= \frac{\partial}{\partial x}(\varphi A_x) + \frac{\partial}{\partial y}(\varphi A_y) + \frac{\partial}{\partial z}(\varphi A_z)$$

$$= \varphi\left(\frac{\partial A_x}{\partial x} + \frac{\partial A_y}{\partial y} + \frac{\partial A_z}{\partial z}\right) + \left(A_x\frac{\partial\varphi}{\partial x} + A_y\frac{\partial\varphi}{\partial y} + A_z\frac{\partial\varphi}{\partial z}\right)$$

$$= \varphi\left(\frac{\partial A_x}{\partial x} + \frac{\partial A_y}{\partial y} + \frac{\partial A_z}{\partial z}\right) + \left(\hat{\imath}\frac{\partial\varphi}{\partial x} + \hat{\jmath}\frac{\partial\varphi}{\partial y} + \hat{k}\frac{\partial\varphi}{\partial z}\right)\cdot(\hat{\imath}A_x + \hat{\jmath}A_y + \hat{k}A_z)$$

$$= \varphi(\vec{\nabla}\cdot\vec{A}) + (\vec{\nabla}\varphi)\cdot\vec{A} \quad\text{or}\quad \text{div}\varphi\vec{A} = \varphi\,\text{div}\vec{A} + (\text{grad}\,\varphi)\cdot\vec{A}.$$

(vi) $$\vec{\nabla}\cdot(\vec{\nabla}\varphi) = \left(\hat{\imath}\frac{\partial}{\partial x} + \hat{\jmath}\frac{\partial}{\partial y} + \hat{k}\frac{\partial}{\partial z}\right)\cdot\left(\hat{\imath}\frac{\partial\varphi}{\partial x} + \hat{\jmath}\frac{\partial\varphi}{\partial y} + \hat{k}\frac{\partial\varphi}{\partial z}\right)$$

$$= \frac{\partial^2\varphi}{\partial x^2} + \frac{\partial^2\varphi}{\partial y^2} + \frac{\partial^2\varphi}{\partial z^2} = \vec{\nabla}^2\varphi \quad\text{or}\quad \text{div grad}\,\varphi = \vec{\nabla}^2\varphi = \text{a scalar field.}$$

(vii) $$\vec{\nabla}\times(\vec{A}+\vec{B}) = \hat{\imath}\left\{\frac{\partial}{\partial y}(A_z+B_z) - \frac{\partial}{\partial z}(A_y+B_y)\right\}$$

$$+\hat{\jmath}\left\{\frac{\partial}{\partial z}(A_x+B_x) - \frac{\partial}{\partial x}(A_z+B_z)\right\}$$

$$+\hat{k}\left\{\frac{\partial}{\partial x}(A_y+B_y) - \frac{\partial}{\partial y}(A_x+B_x)\right\}$$

$$= \hat{\imath}\left(\frac{\partial A_z}{\partial y} - \frac{\partial A_y}{\partial z}\right) + \hat{\jmath}\left(\frac{\partial A_x}{\partial z} - \frac{\partial A_z}{\partial x}\right) + \hat{k}\left(\frac{\partial A_y}{\partial x} - \frac{\partial A_x}{\partial y}\right)$$

$$+\hat{\imath}\left(\frac{\partial B_z}{\partial y} - \frac{\partial B_y}{\partial z}\right) + \hat{\jmath}\left(\frac{\partial B_x}{\partial z} - \frac{\partial B_z}{\partial x}\right) + \hat{k}\left(\frac{\partial B_y}{\partial x} - \frac{\partial B_x}{\partial y}\right)$$

$$= \vec{\nabla}\times\vec{A} + \vec{\nabla}\times\vec{B} \quad\text{or}\quad \text{curl}(\vec{A}+\vec{B}) = \text{curl}\,\vec{A} + \text{curl}\,\vec{B}.$$

(viii) $$\vec{\nabla}\times(\varphi\vec{A}) = \hat{\imath}\left\{\frac{\partial}{\partial y}(\varphi A_z) - \frac{\partial}{\partial z}(\varphi A_y)\right\} + \hat{\jmath}\left\{\frac{\partial}{\partial z}(\varphi A_x) - \frac{\partial}{\partial x}(\varphi A_z)\right\} + \hat{k}\left\{\frac{\partial}{\partial x}(\varphi A_y) - \frac{\partial}{\partial y}(\varphi A_x)\right\}$$

$$= \varphi\left\{\hat{\imath}\left(\frac{\partial A_z}{\partial y} - \frac{\partial A_y}{\partial z}\right) + \hat{\jmath}\left(\frac{\partial A_x}{\partial z} - \frac{\partial A_z}{\partial x}\right) + \hat{k}\left(\frac{\partial A_y}{\partial x} - \frac{\partial A_x}{\partial y}\right)\right\}$$

$$+\hat{\imath}\left(\frac{\partial\varphi}{\partial y}A_z - \frac{\partial\varphi}{\partial z}A_y\right) + \hat{\jmath}\left(\frac{\partial\varphi}{\partial z}A_x - \frac{\partial\varphi}{\partial x}A_z\right) + \hat{k}\left(\frac{\partial\varphi}{\partial x}A_y - \frac{\partial\varphi}{\partial y}A_x\right)$$

$$= \varphi(\vec{\nabla}\times\vec{A}) + (\vec{\nabla}\varphi)\times\vec{A} \quad \text{or} \quad \text{curl}\,\varphi\vec{A} = \varphi\,\text{curl}\,\vec{A} + \text{grad}\,\varphi\times\vec{A}.$$

(ix) $$\vec{\nabla}\times(\vec{\nabla}\varphi) = \hat{\imath}\left\{\frac{\partial}{\partial y}\left(\frac{\partial\varphi}{\partial z}\right) - \frac{\partial}{\partial z}\left(\frac{\partial\varphi}{\partial y}\right)\right\} + \hat{\jmath}\left\{\frac{\partial}{\partial z}\left(\frac{\partial\varphi}{\partial x}\right) - \frac{\partial}{\partial x}\left(\frac{\partial\varphi}{\partial z}\right)\right\} + \hat{k}\left\{\frac{\partial}{\partial x}\left(\frac{\partial\varphi}{\partial y}\right) - \frac{\partial}{\partial y}\left(\frac{\partial\varphi}{\partial x}\right)\right\} = 0$$

$$\text{or} \quad \text{curl grad}\,\varphi = 0.$$

(x) $$\vec{\nabla}\cdot(\vec{\nabla}\times\vec{A}) = \vec{\nabla}\cdot\left\{\hat{\imath}\left(\frac{\partial A_z}{\partial y} - \frac{\partial A_y}{\partial z}\right) + \hat{\jmath}\left(\frac{\partial A_x}{\partial z} - \frac{\partial A_z}{\partial x}\right) + \hat{k}\left(\frac{\partial A_y}{\partial x} - \frac{\partial A_x}{\partial y}\right)\right\}$$

$$= \left(\hat{\imath}\frac{\partial}{\partial x} + \hat{\jmath}\frac{\partial}{\partial y} + \hat{k}\frac{\partial}{\partial z}\right)\cdot\left\{\hat{\imath}\left(\frac{\partial A_z}{\partial y} - \frac{\partial A_y}{\partial z}\right) + \hat{\jmath}\left(\frac{\partial A_x}{\partial z} - \frac{\partial A_z}{\partial x}\right) + \hat{k}\left(\frac{\partial A_y}{\partial x} - \frac{\partial A_x}{\partial y}\right)\right\}$$

$$= \frac{\partial}{\partial x}\left(\frac{\partial A_z}{\partial y} - \frac{\partial A_y}{\partial z}\right) + \frac{\partial}{\partial y}\left(\frac{\partial A_x}{\partial z} - \frac{\partial A_z}{\partial x}\right) + \frac{\partial}{\partial z}\left(\frac{\partial A_y}{\partial x} - \frac{\partial A_x}{\partial y}\right) = 0$$

$$\text{or} \quad \text{div curl}\,\vec{A} = 0.$$

(xi) $$\vec{\nabla}\times(\vec{\nabla}\times\vec{A}) = \left(\hat{\imath}\frac{\partial}{\partial x} + \hat{\jmath}\frac{\partial}{\partial y} + \hat{k}\frac{\partial}{\partial z}\right)\times\left\{\hat{\imath}\left(\frac{\partial A_z}{\partial y} - \frac{\partial A_y}{\partial z}\right) + \hat{\jmath}\left(\frac{\partial A_x}{\partial z} - \frac{\partial A_z}{\partial x}\right) + \hat{k}\left(\frac{\partial A_y}{\partial x} - \frac{\partial A_x}{\partial y}\right)\right\}$$

$$= \hat{\imath}\left\{\frac{\partial}{\partial y}\left(\frac{\partial A_y}{\partial x} - \frac{\partial A_x}{\partial y}\right) - \frac{\partial}{\partial z}\left(\frac{\partial A_x}{\partial z} - \frac{\partial A_z}{\partial x}\right)\right\}$$

$$+\hat{\jmath}\left\{\frac{\partial}{\partial z}\left(\frac{\partial A_z}{\partial y} - \frac{\partial A_y}{\partial z}\right) - \frac{\partial}{\partial x}\left(\frac{\partial A_y}{\partial x} - \frac{\partial A_x}{\partial y}\right)\right\}$$

$$+\hat{k}\left\{\frac{\partial}{\partial x}\left(\frac{\partial A_x}{\partial z} - \frac{\partial A_z}{\partial x}\right) + \frac{\partial}{\partial y}\left(\frac{\partial A_z}{\partial y} - \frac{\partial A_y}{\partial z}\right)\right\}.$$

On simplification we have,

$$\vec{\nabla} \times (\vec{\nabla} \times \vec{A}) = -\left(\frac{\partial^2}{\partial x^2} + \frac{\partial^2}{\partial y^2} + \frac{\partial^2}{\partial z^2}\right)\left(\hat{\imath}A_x + \hat{\jmath}A_y + \hat{k}A_z\right)$$

$$+\hat{\imath}\frac{\partial}{\partial x}\left(\frac{\partial A_x}{\partial x} + \frac{\partial A_y}{\partial y} + \frac{\partial A_z}{\partial z}\right) + \hat{\jmath}\frac{\partial}{\partial y}\left(\frac{\partial A_x}{\partial x} + \frac{\partial A_y}{\partial y} + \frac{\partial A_z}{\partial z}\right)$$

$$+\hat{k}\frac{\partial}{\partial z}\left(\frac{\partial A_x}{\partial x} + \frac{\partial A_y}{\partial y} + \frac{\partial A_z}{\partial z}\right)$$

$$= -\vec{\nabla}^2\vec{A} + \nabla(\vec{\nabla}\cdot\vec{A}) \quad \text{or} \quad \text{curl curl}\, A = -\vec{\nabla}^2\vec{A} + \text{grad div}\,\vec{A}.$$

2.25 Irrotational and Solenoidal Vectors

We know that a vector $\vec{V}$ is *solenoidal*, if div$\vec{V} = 0$, but the divergence of the curl of a vector is zero, i.e., $\vec{\nabla}\cdot\vec{\nabla}\times\vec{A} = 0$. Hence, the solenoidal vector $\vec{V}$ can always be expressed as the curl of a vector function $\vec{A}$. This latter vector function is called a *vector potential*. For example, for magnetic induction $\vec{B}$, we have $\vec{\nabla}\cdot\vec{B} = 0$. Hence $\vec{B} = \vec{\nabla}\times\vec{A}$ where $\vec{A}$ is called the *magnetic vector potential*.

$\vec{\nabla}\cdot\vec{B} = 0$ signifies that the fluxes or the magnetic lines of force entering any arbitrary closed surface must exactly be the same as those leaving it. It immediately follows that the lines of flow of the vector $\vec{B}$ must either form closed curves (cf. the magnetic field of a current), or terminate upon the bounding surfaces (cf. the electric field in a condenser) or extend to infinity.

We also know that a vector $\vec{V}$ is *irrotational* if curl$\vec{V} = \vec{\nabla}\times\vec{V} = 0$, but curl of the gradient of a scalar function is always zero, i.e., $\vec{\nabla}\times\vec{\nabla}\phi = 0$. Hence an irrotational vector $\vec{V}$ can always be expressed as the gradient of some scalar function. Thus we may write $\vec{V} = \vec{\nabla}\phi = \text{grad}\,\phi$. This scalar function ϕ is called a *scalar potential*. In the case of gravitational or electrostatic field, we may write

$$\vec{\nabla}\times\vec{E} = 0 \quad \text{or} \quad E = -\vec{\nabla}\phi.$$

The negative sign indicates that the vector $\vec{E}$ is directed along which ϕ decreases.

2.26 Reciprocal System of Vectors

We consider a set of three *non-co-planar* vectors $(\vec{a}, \vec{b}, \vec{c})$ so that $\vec{a}\cdot(\vec{b}\times\vec{c}) \neq 0$. We next consider another set of vectors $(\vec{a}', \vec{b}', \vec{c}')$, such that

$$\left.\begin{array}{lll} \vec{a}\cdot\vec{a}' = 1, & \vec{a}\cdot\vec{b}' = 0, & \vec{a}\cdot\vec{c}' = 0, \\ \vec{b}\cdot\vec{a}' = 0, & \vec{b}\cdot\vec{b}' = 1, & \vec{b}\cdot\vec{c}' = 0, \\ \vec{c}\cdot\vec{a}' = 0, & \vec{c}\cdot\vec{b}' = 0, & \vec{c}\cdot\vec{c}' = 1, \end{array}\right\} \tag{2.26.1}$$

We then say that the set $(\vec{a}', \vec{b}', \vec{c}')$ is *reciprocal* to the set $(\vec{a}, \vec{b}, \vec{c})$.

Expressions for $\vec{a}', \vec{b}', \vec{c}'$ in terms of $\vec{a}, \vec{b}, \vec{c}$

We consider the three relations in the first column of (2.26.1). The relations $\vec{b}\cdot\vec{a}' = \vec{c}\cdot\vec{a}' = 0$ imply that $\vec{a}'$ is perpendicular to $\vec{b}$ as well as $\vec{c}$ and hence parallel to $\vec{b}\times\vec{c}$. In other words $\vec{a}' = k(\vec{b}\times\vec{c})$, where k is a constant to be determined. Again putting this value of $\vec{a}'$ in $\vec{a}\cdot\vec{a}' = 1$, we immediately obtain $k = \frac{1}{\vec{a}\cdot(\vec{b}\times\vec{c})}$, so that we may write

$$\vec{a}' = \frac{\vec{b}\times\vec{c}}{\vec{a}\cdot(\vec{b}\times\vec{c})},\ \vec{b}' = \frac{\vec{c}\times\vec{a}}{\vec{a}\cdot(\vec{b}\times\vec{c})},\ \vec{c}' = \frac{\vec{a}\times\vec{b}}{\vec{a}\cdot(\vec{b}\times\vec{c})} \qquad (2.26.2)$$

$\vec{b}'$, and $\vec{c}'$ being derived in a similar manner from the relations of second and third columns respectively of (2.26.1).

The symmetry of the relations in (2.26.2) also give

$$\vec{a} = \frac{\vec{b}'\times\vec{c}'}{\vec{a}'\cdot(\vec{b}'\times\vec{c}')},\ \vec{b} = \frac{\vec{c}'\times\vec{a}'}{\vec{a}'\cdot(\vec{b}'\times\vec{c}')},\ \vec{c} = \frac{\vec{a}'\times\vec{b}'}{\vec{a}'\cdot(\vec{b}'\times\vec{c}')}. \qquad (2.26.3)$$

We may thus call the set $(\vec{a}, \vec{b}, \vec{c})$ reciprocal to $(\vec{a}', \vec{b}', \vec{c}')$. Relations (2.26.1) may be derived from relations (2.26.2) in the following way. We suppose that $\vec{a}', \vec{b}', \vec{c}'$ are defined in terms of $\vec{a}, \vec{b}, \vec{c}$ by the relations (2.26.2). We observe that

$$\vec{a}\cdot\vec{a}' = \vec{a}\cdot\frac{\vec{b}\times\vec{c}}{\vec{a}\cdot(\vec{b}\times\vec{c})} = 1 \quad \text{and} \quad \vec{a}\cdot\vec{b}' = \vec{a}\cdot\frac{\vec{c}\times\vec{a}}{\vec{a}\cdot(\vec{b}\times\vec{c})} = 0.$$

Similarly other relations of (2.26.1) can be derived.

Illustrative Examples

1. $\vec{a}_1, \vec{a}_2, \vec{a}_3$ are three non-co-planar vectors. Three other vectors $\vec{b}_1, \vec{b}_2, \vec{b}_3$ are related to the former vectors by $\vec{a}_i\cdot\vec{b}_j = \delta_{ij}$, where δ is known as Kroneĕcker delta. Show that $\vec{b}_1 = \frac{\vec{a}_2\times\vec{a}_3}{\vec{a}_1\cdot(\vec{a}_2\times\vec{a}_3)}$ and $(ij = 1, 2, 3)$. **[C.U. (Hons.) 1990]**

The notation $\vec{a}_i\cdot\vec{b}_j = \delta_{ij}$, has the meaning that $\delta_{ij} = 1$ for $i = j$ and $\delta_{ij} = 0$ for $i \neq j$. So we may write

$$\vec{a}_1\cdot\vec{b}_1 = \vec{a}_2\cdot\vec{b}_2 = \vec{a}_3\cdot\vec{b}_3 = 1$$
$$\text{and} \quad \vec{a}_1\cdot\vec{b}_2 = \vec{a}_1\cdot\vec{b}_3 = \vec{a}_2\cdot\vec{b}_1 = 0 \quad \text{etc.}$$

Thus the two vector sets are reciprocal. Taking the relations $\vec{b}_1\cdot\vec{a}_2 = \vec{b}_1\cdot\vec{a}_3 = 0$ and $\vec{a}_1\cdot\vec{b}_1 = 1$ and proceeding in the same way as above, the required relation may be obtained.

2. If $(\vec{a}, \vec{b}, \vec{c})$ and $(\vec{a}', \vec{b}', \vec{c}')$ are reciprocal vector sets, then prove that $\vec{a}\times\vec{a}' + \vec{b}\times\vec{b}' + \vec{c}\times\vec{c}' = 0$.

We know that

$$\vec{a}\times(\vec{b}\times\vec{c}) = (\vec{a}\cdot\vec{c})\vec{b} - (\vec{a}\cdot\vec{b})\vec{c}$$
$$\vec{b}\times(\vec{c}\times\vec{a}) = (\vec{b}\cdot\vec{a})\vec{c} - (\vec{b}\cdot\vec{c})\vec{a}$$
$$\vec{c}\times(\vec{a}\times\vec{b}) = (\vec{c}\cdot\vec{b})\vec{a} - (\vec{c}\cdot\vec{a})\vec{b}$$

Adding we get

$$\vec{a} \times (\vec{b} \times \vec{c}) + \vec{b} \times (\vec{c} \times \vec{a}) + \vec{c} \times (\vec{a} \times \vec{b}) = 0.$$

Dividing both sides by $\vec{a} \cdot (\vec{b} \times \vec{c})$, we have

$$\vec{a} \times \frac{\vec{b} \times \vec{c}}{\vec{a} \cdot (\vec{b} \times \vec{c})} + \vec{b} \times \frac{\vec{c} \times \vec{a}}{\vec{a} \cdot (\vec{b} \times \vec{c})} + \vec{c} \times \frac{\vec{a} \times \vec{b}}{\vec{a} \cdot (\vec{b} \times \vec{c})} = 0$$

or $\vec{a} \times \vec{a}' + \vec{b} \times \vec{b}' + \vec{c} \times \vec{c}' = 0$ [using relations (2.26.2)].

2.27 Polar and Axial Vectors

Vectors are generally divided into two classes. The first class includes quantities such as, displacement, velocity, acceleration, force etc. in which the vector is drawn in the direction of the quantity concerned; mere linear action in a particular direction is involved and the vector is called a *polar* vector. A polar vector remains invariant on mirror reflection, and so its direction does not depend on the reference frame.

The second class includes quantities such as angular velocity, angular acceleration etc. in which rotatory action of some kind takes place about an axis. Here the vector is drawn parallel to the axis about which the quantity acts. The length of the vector gives the magnitude of the quantity. The direction of the vector is conventionally fixed by the rule of the right-handed screw, i.e., when sighting along the direction of the vector shown by an arrow-head, rotation is taken as clock-wise. Such vectors are known as *axial* vectors. An axial vector therefore reverses its direction on mirror reflection, and so sense of direction of such a vector depends on the reference frame.

2.28 Curvilinear Co-ordinates

Let the rectangular co-ordinates (x, y, z) of any point be expressed as functions of u_1, u_2, u_3) so that

$$x = x(u_1, u_2, u_3), y = y(u_1, u_2, u_3), z = z(u_1, u_2, u_3). \tag{2.28.1}$$

We assume that (2.28.1) can be solved for u_1, u_2, u_3 in terms of x, y, z, i.e.,

$$u_1 = u_1(x, y, z), u_2 = u_2(x, y, z), u_3 = u_3(x, y, z). \tag{2.28.2}$$

The functions in (2.28.1) and (2.28.2) are assumed to be single valued and to have continuous derivatives, so that the correspondence between (x, y, z) and (u_1, u_2, u_3) is unique.

Given a point P with rectangular co-ordinates (x, y, z), we can from (2.28.2) associate a unique set of co-ordinates (u_1, u_2, u_3) called the *curvilinear co-ordinates* of P.

Orthogonal curvilinear co-ordinates

The surfaces $u_1 = c_1$, $u_2 = c_2$, $u_3 = c_3$, where c_1, c_2, c_3 are constants, are called *co-ordinate surfaces* and each pair of these surfaces intersects in curves called *co-ordinate curves or lines* as shown in the attached figure (2.28.1). If the co-ordinate surfaces intersect at right angles, the curvilinear co-ordinate system is called *orthogonal*. The u_1, u_2 and u_3 co-ordinate curves of a curvilinear system are analogous to the X, Y and Z co-ordinate axes of a rectangular system.

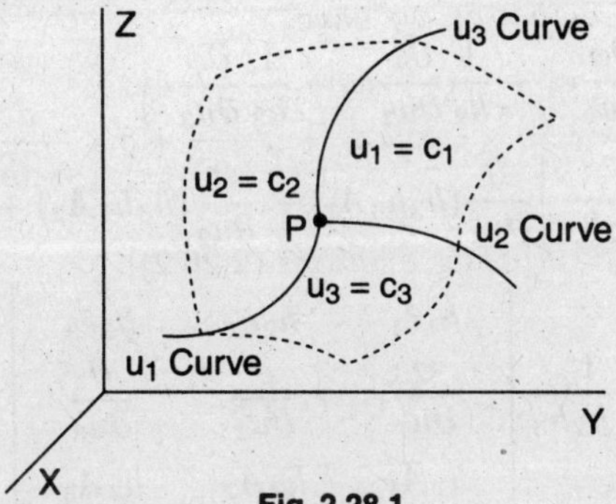

Fig. 2.28.1

Unit vectors in curvilinear systems

Let $\vec{r} = x\hat{\imath} + y\hat{\jmath} + z\hat{k}$ be the position vector of a point P. Then equations (2.28.1) can be written as $\vec{r} = \vec{r}(u_1, u_2, u_3)$. A tangent vector to u_1 curve at P (for which u_2 and u_3 are constants) is $\dfrac{\partial \vec{r}}{\partial u_1}$. Then a unit tangent vector in this direction is $\hat{e}_1 = \dfrac{\partial \vec{r}}{\partial u_1} \Big/ \left|\dfrac{\partial \vec{r}}{\partial u_1}\right|$, so that $\dfrac{\partial \vec{r}}{\partial u_1} = h_1 \hat{e}_1$ where $h_1 = \left|\dfrac{\partial \vec{r}}{\partial u_1}\right|$. Similarly if $\hat{e}_2$ and $\hat{e}_3$ are unit tangent vectors to the u_2 and u_3 curves at P respectively, then $\dfrac{\partial \vec{r}}{\partial u_2} = h_2 \hat{e}_2$ and $\dfrac{\partial \vec{r}}{\partial u_3} = h_3 \hat{e}_3$ where $h_2 = \left|\dfrac{\partial \vec{r}}{\partial u_2}\right|$ and $h_3 = \left|\dfrac{\partial \vec{r}}{\partial u_3}\right|$. The quantities h_1, h_2, h_3 are called *scale factors*. The unit vectors $\hat{e}_1, \hat{e}_2, \hat{e}_3$ are in the directions of increasing u_1, u_2, u_3 respectively.

Since $\vec{\nabla} u_1$ is a vector at P normal to the surface $u_1 = c_1$, a unit vector in this direction is given by $\hat{E}_1 = \vec{\nabla} u_1 / |\vec{\nabla} u_1|$. Similarly the unit vectors $\hat{E}_2 = \vec{\nabla} u_2 / |\vec{\nabla} u_2|$ and $\hat{E}_3 = \vec{\nabla} u_3 / |\vec{\nabla} u_3|$ at P are normal to the surfaces $u_2 = c_2$ and $u_3 = c_3$ respectively.

Thus at each point P of a curvilinear system, there exist, in general, two sets of unit vectors, $\hat{e}_1, \hat{e}_2, \hat{e}_3$ — tangent to the co-ordinate curves and $\hat{E}_1, \hat{E}_2, \hat{E}_3$ — normal to the co-ordinate surfaces as shown in Fig. (2.28.2). The sets become identical, if and only if the curvilinear co-ordinate system is orthogonal. Both sets are analogous to the $\hat{\imath}, \hat{\jmath}, \hat{k}$ unit vectors in rectangular co-ordinates, but are unlike them in that they may change directions from point to point. It can be shown that the sets $\dfrac{\partial \vec{r}}{\partial u_1}, \dfrac{\partial \vec{r}}{\partial u_2}, \dfrac{\partial \vec{r}}{\partial u_3}$ and $\vec{\nabla} u_1, \vec{\nabla} u_2, \vec{\nabla} u_3$ are reciprocal to each other.

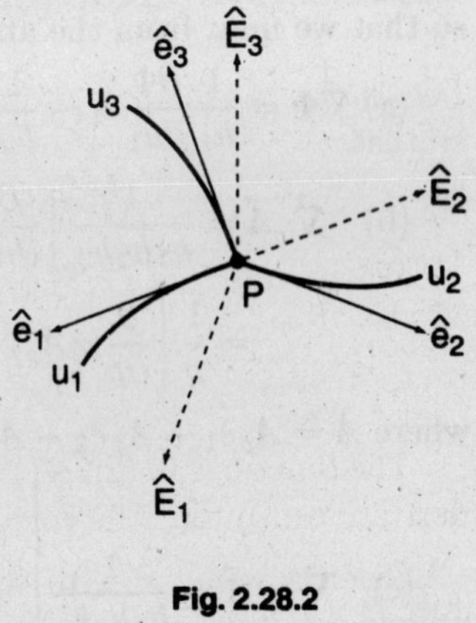

Fig. 2.28.2

A vector $\vec{A}$ can be represented in terms of the unit base vectors $\hat{e}_1, \hat{e}_2, \hat{e}_3$ or $\hat{E}_1, \hat{E}_2, \hat{E}_3$ in the form

$$\vec{A} = A_1 \hat{e}_1 + A_2 \hat{e}_2 + A_3 \hat{e}_3 = a_1 \hat{E}_1 + a_2 \hat{E}_2 + a_3 \hat{E}_3$$

where A_1, A_2, A_3 and a_1, a_2, a_3 are the respective components of $\vec{A}$ in each system.

The *Gradient, Divergence* and *Curl* can be expressed in terms of curvilinear co-ordinates. If ϕ be a scalar function and $\vec{A} = A_1\hat{e}_1 + A_2\hat{e}_2 + A_3\hat{e}_3$, a vector function of orthogonal curvilinear co-ordinates u_1, u_2, u_3, then the following results can be obtained.

1. $\vec{\nabla}\phi = \text{grad}\,\phi = \dfrac{1}{h_1}\dfrac{\partial\phi}{\partial u_1}\hat{e}_1 + \dfrac{1}{h_2}\dfrac{\partial\phi}{\partial u_2}\hat{e}_2 + \dfrac{1}{h_3}\dfrac{\partial\phi}{\partial u_3}\hat{e}_3.$

2. $\vec{\nabla}\cdot\vec{A} = \text{div}\,\vec{A} = \dfrac{1}{h_1h_2h_3}\left[\dfrac{\partial}{\partial u_1}(h_2h_3A_1) + \dfrac{\partial}{\partial u_2}(h_3h_1A_2) + \dfrac{\partial}{\partial u_3}(h_1h_2A_3)\right].$

3. $\vec{\nabla}\times\vec{A} = \text{curl}\,\vec{A} = \dfrac{1}{h_1h_2h_3}\begin{vmatrix} h_1\hat{e}_1 & h_2\hat{e}_2 & h_3\hat{e}_3 \\ \dfrac{\partial}{\partial u_1} & \dfrac{\partial}{\partial u_2} & \dfrac{\partial}{\partial u_3} \\ h_1A_1 & h_2A_2 & h_3A_3 \end{vmatrix}.$

4. $\vec{\nabla}^2\phi$ = Laplacian of ϕ

$$= \frac{1}{h_1h_2h_3}\left[\frac{\partial}{\partial u_1}\left(\frac{h_2h_3}{h_1}\frac{\partial\phi}{\partial u_1}\right) + \frac{\partial}{\partial u_2}\left(\frac{h_3h_1}{h_2}\frac{\partial\phi}{\partial u_2}\right) + \frac{\partial}{\partial u_3}\left(\frac{h_1h_2}{h_3}\frac{\partial\phi}{\partial u_3}\right)\right].$$

If $h_1 = h_2 = h_3 = 1$ and $\hat{e}_1, \hat{e}_2, \hat{e}_3$ are replaced by $\hat{\imath}, \hat{\jmath}, \hat{k}$, these reduce to the usual expressions in rectangular co-ordinates, where (u_1, u_2, u_3) is replaced by (x, y, z).

Expressions for $\vec{\nabla}\Phi$, $\vec{\nabla}\cdot\vec{A}$, $\vec{\nabla}\times\vec{A}$ and $\vec{\nabla}^2\Phi$ in cylindrical polar co-ordinates

For cylindrical polar co-ordinates (ρ, ϕ, z),

$$u_1 = \rho, u_2 = \phi, u_3 = z; \quad \hat{e}_1 = \hat{e}_\rho, \hat{e}_2 = \hat{e}_\phi, \hat{e}_3 = \hat{e}_z$$

$$\text{and} \quad h_1 = h_\rho = 1, \quad h_2 = h_\phi = \rho, \quad h_3 = h_z = 1,$$

so that we may, from the above expressions, write

(a) $\vec{\nabla}\Phi = \dfrac{1}{h_1}\dfrac{\partial\Phi}{\partial u_1}\hat{e}_1 + \dfrac{1}{h_2}\dfrac{\partial\Phi}{\partial u_2}\hat{e}_2 + \dfrac{1}{h_3}\dfrac{\partial\Phi}{\partial u_3}\hat{e}_3 = \dfrac{\partial\Phi}{\partial\rho}\hat{e}_\rho + \dfrac{1}{\rho}\dfrac{\partial\Phi}{\partial\phi}\hat{e}_\phi + \dfrac{\partial\Phi}{\partial z}\hat{e}_z.$

(b) $\vec{\nabla}\cdot\vec{A} = \dfrac{1}{h_1h_2h_3}\left[\dfrac{\partial}{\partial u_1}(h_2h_3A_1) + \dfrac{\partial}{\partial u_2}(h_3h_1A_2) + \dfrac{\partial}{\partial u_3}(h_1h_2A_3)\right]$

$$= \frac{1}{\rho}\left[\frac{\partial}{\partial\rho}(\rho A_\rho) + \frac{\partial}{\partial\phi}A_\phi + \frac{\partial}{\partial z}(\rho A_z)\right]$$

where $\vec{A} = A_\rho\hat{e}_1 + A_\phi\hat{e}_2 + A_z\hat{e}_3$, i.e., $A_1 = A_\rho$, $A_2 = A_\phi$, $A_3 = A_z$.

(c) $\vec{\nabla}\times\vec{A} = \dfrac{1}{h_1h_2h_3}\begin{vmatrix} h_1\hat{e}_1 & h_2\hat{e}_2 & h_3\hat{e}_3 \\ \dfrac{\partial}{\partial u_1} & \dfrac{\partial}{\partial u_2} & \dfrac{\partial}{\partial u_3} \\ h_1A_1 & h_2A_2 & h_3A_3 \end{vmatrix} = \dfrac{1}{\rho}\begin{vmatrix} \hat{e}_\rho & \rho\hat{e}_\phi & \hat{e}_z \\ \dfrac{\partial}{\partial\rho} & \dfrac{\partial}{\partial\phi} & \dfrac{\partial}{\partial z} \\ A_\rho & \rho A_\phi & A_z \end{vmatrix}$

$$= \frac{1}{\rho}\left[\left(\frac{\partial A_z}{\partial\phi} - \frac{\partial}{\partial z}(\rho A_\phi)\right)\hat{e}_\rho + \left(\rho\frac{\partial A_\rho}{\partial z} - \rho\frac{\partial A_z}{\partial\rho}\right)\hat{e}_\phi + \left(\frac{\partial}{\partial\rho}(\rho A_\phi) - \frac{\partial A_\rho}{\partial\phi}\right)\hat{e}_z\right]$$

(d) $$\vec{\nabla}^2\Phi = \frac{1}{h_1h_2h_3}\left[\frac{\partial}{\partial u_1}\left(\frac{h_2h_3}{h_1}\frac{\partial\Phi}{\partial u_1}\right)+\frac{\partial}{\partial u_2}\left(\frac{h_3h_1}{h_2}\frac{\partial\Phi}{\partial u_2}\right)+\frac{\partial}{\partial u_3}\left(\frac{h_1h_2}{h_3}\frac{\partial\Phi}{\partial u_3}\right)\right]$$

$$= \frac{1}{\rho}\left[\frac{\partial}{\partial\rho}\left(\rho\frac{\partial\Phi}{\partial\rho}\right)+\frac{\partial}{\partial\phi}\left(\frac{1}{\rho}\frac{\partial\Phi}{\partial\phi}\right)+\frac{\partial}{\partial z}\left(\rho\frac{\partial\Phi}{\partial z}\right)\right]$$

$$= \frac{1}{\rho}\frac{\partial}{\partial\rho}\left(\rho\frac{\partial\Phi}{\partial\rho}\right)+\frac{1}{\rho^2}\frac{\partial^2\Phi}{\partial\phi^2}+\frac{\partial^2\Phi}{\partial z^2}.$$

Expressions for $\vec{\nabla}\psi$, $\vec{\nabla}\cdot\vec{A}$, $\vec{\nabla}\times\vec{A}$ and $\vec{\nabla}^2\psi$ in spherical polar co-ordinates

For spherical polar co-ordinates, $u_1 = r$, $u_2 = \theta$, $u_3 = \phi$; $\hat{e}_1 = \hat{e}_r$, $\hat{e}_2 = \hat{e}_\theta$, $\hat{e}_3 = \hat{e}_\phi$; and $h_1 = h_r = 1$, $h_2 = h_\theta = r$, $h_3 = h_\phi = r\sin\theta$.

(a) $$\vec{\nabla}\psi = \frac{1}{h_1}\frac{\partial\psi}{\partial u_1}\hat{e}_1+\frac{1}{h_2}\frac{\partial\psi}{\partial u_2}\hat{e}_2+\frac{1}{h_3}\frac{\partial\psi}{\partial u_3}\hat{e}_3 = \hat{e}_r\frac{\partial\psi}{\partial r}+\frac{1}{r}\frac{\partial\psi}{\partial\theta}\hat{e}_\theta+\frac{1}{r\sin\theta}\frac{\partial\psi}{\partial\phi}\hat{e}_\phi$$

(b) $$\vec{\nabla}\cdot\vec{A} = \frac{1}{h_1h_2h_3}\left[\frac{\partial}{\partial u_1}(h_2h_3A_1)+\frac{\partial}{\partial u_2}(h_3h_1A_2)+\frac{\partial}{\partial u_3}(h_1h_2A_3)\right]$$

$$= \frac{1}{r^2\sin\theta}\left[\sin\theta\frac{\partial}{\partial r}(r^2A_r)+r\frac{\partial}{\partial\theta}(A_\theta\sin\theta)+r\frac{\partial A_\phi}{\partial\phi}\right]$$

$$= \frac{1}{r^2}\frac{\partial}{\partial r}(r^2A_r)+\frac{1}{r\sin\theta}\frac{\partial}{\partial\theta}(\sin\theta A_\theta)+\frac{1}{r\sin\theta}\frac{\partial A_\phi}{\partial\phi}.$$

where $\vec{A} = A_r\hat{e}_1 + A_\theta\hat{e}_2 + A_\phi\hat{e}_3$, i.e., $A_1 = A_r, A_z = A_\theta, A_3 = A_\phi$.

(c) $$\vec{\nabla}\times\vec{A} = \frac{1}{h_1h_2h_3}\begin{vmatrix} h_1\hat{e}_1 & h_2\hat{e}_2 & h_3\hat{e}_3 \\ \frac{\partial}{\partial u_1} & \frac{\partial}{\partial u_2} & \frac{\partial}{\partial u_3} \\ h_1A_1 & h_2A_2 & h_3A_3 \end{vmatrix} = \frac{1}{r^2\sin\theta}\begin{vmatrix} \hat{e}_r & r\hat{e}_\theta & r\sin\theta\hat{e}_\phi \\ \frac{\partial}{\partial r} & \frac{\partial}{\partial\theta} & \frac{\partial}{\partial\phi} \\ A_r & rA_\theta & r\sin\theta A_\phi \end{vmatrix}$$

$$= \frac{1}{r^2\sin\theta}\left[\left\{\frac{\partial}{\partial\theta}(r\sin\theta A_\phi)-\frac{\partial}{\partial\phi}(rA_\theta)\right\}\hat{e}_r+\left\{\frac{\partial A_r}{\partial\phi}-\frac{\partial}{\partial r}(r\sin\theta A_\phi)\right\}r\hat{e}_\theta\right.$$
$$\left.+\left\{\frac{\partial}{\partial r}(rA_\theta)-\frac{\partial A_r}{\partial\theta}\right\}r\sin\theta\hat{e}_\phi\right]$$

(d) $$\vec{\nabla}^2\psi = \frac{1}{h_1h_2h_3}\left[\frac{\partial}{\partial u_1}\left(\frac{h_2h_3}{h_1}\frac{\partial\psi}{\partial u_1}\right)+\frac{\partial}{\partial u_2}\left(\frac{h_3h_1}{h_2}\frac{\partial\psi}{\partial u_2}\right)+\frac{\partial}{\partial u_3}\left(\frac{h_1h_2}{h_3}\frac{\partial\psi}{\partial u_3}\right)\right]$$

$$= \frac{1}{r^2\sin\theta}\left[\sin\theta\frac{\partial}{\partial r}\left(r^2\frac{\partial\psi}{\partial r}\right)+\frac{\partial}{\partial\theta}\left(\sin\theta\frac{\partial\psi}{\partial\theta}\right)+\frac{1}{\sin\theta}\frac{\partial^2\psi}{\partial\phi^2}\right]$$

$$= \frac{1}{r^2}\frac{\partial}{\partial r}\left(r^2\frac{\partial\psi}{\partial r}\right)+\frac{1}{r^2\sin\theta}\frac{\partial}{\partial\theta}\left(\sin\theta\frac{\partial\psi}{\partial\theta}\right)+\frac{1}{r^2\sin^2\theta}\frac{\partial^2\psi}{\partial\phi^2}.$$

Illustrative Examples

1. Calculate the gradient of a function, $f(r)$, that depends only on the absolute value of the radius vector $\vec{r}$

$$\text{grad}\, f(r) = \hat{\imath}\frac{\partial f}{\partial r}\cdot\frac{\partial r}{\partial x} + \hat{\jmath}\frac{\partial f}{\partial r}\cdot\frac{\partial r}{\partial y} + \hat{k}\frac{\partial f}{\partial r}\cdot\frac{\partial r}{\partial z}.$$

Now,

$$r = \sqrt{x^2+y^2+z^2}$$

$$\therefore\ \frac{\partial r}{\partial x} = \frac{x}{\sqrt{x^2+y^2+z^2}} = \frac{x}{r},\ \frac{\partial r}{\partial y} = \frac{y}{r} \text{ and } \frac{\partial r}{\partial z} = \frac{z}{r}$$

$$\therefore\ \text{grad}\, f(r) = \frac{\partial f}{\partial r}\frac{(\hat{\imath}x+\hat{\jmath}y+\hat{k}z)}{r} = \frac{\partial f}{\partial r}\frac{\vec{r}}{r}.$$

2. Calculate div $\dfrac{\vec{r}}{r^3}$.

From the relation (v) of Art. 2.24, we have

$$\text{div}\,(\varphi\vec{A}) = (\vec{\nabla}\varphi)\cdot\vec{A} + \varphi(\vec{\nabla}\cdot\vec{A}).$$

Taking $\varphi = \frac{1}{r^3}$ and $\vec{A} = \vec{r}$, we have

$$\text{div}\left(\frac{\vec{r}}{r^3}\right) = \left\{\vec{\nabla}\left(\frac{1}{r^3}\right)\right\}\cdot\vec{r} + \frac{1}{r^3}(\vec{\nabla}\cdot\vec{r}).$$

Now,

$$\vec{\nabla}\left(\frac{1}{r^3}\right) = -\frac{3\hat{\imath}}{r^4}\frac{\partial r}{\partial x} - \frac{3\hat{\jmath}}{r^4}\frac{\partial r}{\partial y} - \frac{3\hat{k}}{r^4}\frac{\partial r}{\partial z} = -\frac{3}{r^4}\left(\frac{\hat{\imath}x+\hat{\jmath}y+\hat{k}z}{r}\right) = -\frac{3\vec{r}}{r^5},$$

again

$$\vec{\nabla}\cdot\vec{r} = \frac{\partial x}{\partial x} + \frac{\partial y}{\partial y} + \frac{\partial z}{\partial z} = 3$$

$$\text{div}\left(\frac{\vec{r}}{r^3}\right) = -\frac{3}{r^5}(\vec{r}\cdot\vec{r}) + \frac{3}{r^3} = -\frac{3}{r^3} + \frac{3}{r^3} = 0.$$

3. Calculate curl $\vec{r}$.

$$\text{curl}\,\vec{r} = \vec{\nabla}\times\vec{r} = \left(\hat{\imath}\frac{\partial}{\partial x} + \hat{\jmath}\frac{\partial}{\partial y} + \hat{k}\frac{\partial}{\partial z}\right)\times(\hat{\imath}x+\hat{\jmath}y+\hat{k}z)$$

$$= \begin{vmatrix} \hat{\imath} & \hat{\jmath} & \hat{k} \\ \dfrac{\partial}{\partial x} & \dfrac{\partial}{\partial y} & \dfrac{\partial}{\partial z} \\ x & y & z \end{vmatrix} = 0.$$

4. The electrostatic force acting between two point charges q and q' at a distance r apart is

$$\vec{F} = \frac{qq'}{r^2}\hat{r}_0$$

where $\hat{r}_0$ is a unit vector along r. Show that this force is conservative. **[C.U. (Hons.) 1979]**

The above expression may be written in the form

$$\vec{F} = \frac{qq'}{r^3}\vec{r} = f(r)\vec{r} \quad \text{where } f(r) = \frac{qq'}{r^3}.$$

Now

$$\text{curl}\,\vec{F} = \text{curl}\, f(r)\vec{r} = \vec{\nabla} \times \left[f(r)x\hat{\imath} + f(r)y\hat{\jmath} + f(r)z\hat{k}\right]$$

$$= \begin{vmatrix} \hat{\imath} & \hat{\jmath} & \hat{k} \\ \dfrac{\partial}{\partial x} & \dfrac{\partial}{\partial y} & \dfrac{\partial}{\partial z} \\ f(r)x & f(r)y & f(r)z \end{vmatrix}$$

$$= \hat{\imath}\left(z\frac{\partial f(r)}{\partial y} - y\frac{\partial f(r)}{\partial z}\right) + \hat{\jmath}\left(x\frac{\partial f(r)}{\partial z} - z\frac{\partial f(r)}{\partial x}\right) + \hat{k}\left(y\frac{\partial f(r)}{\partial x} - x\frac{\partial f(r)}{\partial y}\right) \tag{1}$$

Now

$$\frac{\partial f(r)}{\partial x} = \frac{\partial f(r)}{\partial r}\cdot\frac{\partial r}{\partial x} = -\frac{3qq'}{r^4}\cdot\frac{x}{r} = -\frac{3qq'x}{r^5}.$$

Similarly

$$\frac{\partial f(r)}{\partial y} = -\frac{3qq'y}{r^5} \quad \text{and} \quad \frac{\partial f(r)}{\partial z} = -\frac{3qq'z}{r^5}.$$

Substituting these values in expression (1), we have,

$$\text{curl}\,\vec{F} = 0.$$

This shows that $\vec{F}$ is conservative.

5. If $\vec{\nabla}\cdot\vec{E} = 0, \vec{\nabla}\cdot\vec{H} = 0, \vec{\nabla}\times\vec{E} = -\dfrac{\partial\vec{H}}{\partial t}$ and $\vec{\nabla}\times\vec{H} = \dfrac{\partial E}{\partial t}$, show that $\nabla^2\vec{E} = \dfrac{\partial^2\vec{E}}{\partial t^2}$ and $\nabla^2\vec{H} = \dfrac{\partial^2\vec{H}}{\partial t^2}$.

We know from the relation (xi) of Art. 2.24

$$\vec{\nabla}\times(\vec{\nabla}\times\vec{E}) = \vec{\nabla}(\vec{\nabla}\cdot\vec{E}) - \nabla^2\vec{E} = -\nabla^2\vec{E} \quad (\because\ \vec{\nabla}\cdot\vec{E} = 0) \tag{1}$$

Also

$$\vec{\nabla}\times(\vec{\nabla}\times\vec{E}) = \vec{\nabla}\times\left(-\frac{\partial\vec{H}}{\partial t}\right) = -\frac{\partial}{\partial t}(\vec{\nabla}\times\vec{H}) = -\frac{\partial^2\vec{E}}{\partial t^2} \tag{2}$$

From (1) and (2) we have,

$$\nabla^2\vec{E} = \frac{\partial^2\vec{E}}{\partial t^2}.$$

Similarly

$$\vec{\nabla}\times(\vec{\nabla}\times\vec{H}) = -\nabla^2\vec{H}; \quad \text{but} \quad \vec{\nabla}\times(\vec{\nabla}\times\vec{H}) = \vec{\nabla}\times\left(\frac{\partial\vec{E}}{\partial t}\right) = \frac{\partial}{\partial t}(\vec{\nabla}\times\vec{E}) = -\frac{\partial^2\vec{H}}{\partial t^2}.$$

Hence

$$\nabla^2 \vec{H} = \frac{\partial^2 \vec{H}}{\partial t^2}.$$

6. If $\vec{V} = \vec{\omega} \times \vec{r}$, prove that $\vec{\omega} = \frac{1}{2}\,\mathrm{curl}\,\vec{V}$, where $\vec{\omega}$ is a constant vector.

(From the above expression, it follows that if a motion is such that the velocity $\vec{V}$ has a curl, the ultimate particles of the body rotate with instantaneous angular velocity $\vec{\omega}$. This motion is called rotational or vortical. On the other hand when the particles do not rotate, curl $\vec{V} = 0$ and the motion is called irrotational or non-vortical.)

$$\frac{1}{2}\,\mathrm{curl}\,\vec{V} = \frac{1}{2}(\vec{\nabla} \times \vec{V}) = \frac{1}{2}(\vec{\nabla} \times \vec{\omega} \times \vec{r})$$

$$= \frac{1}{2}\left[\left(\hat{\imath}\frac{\partial}{\partial x} + \hat{\jmath}\frac{\partial}{\partial y} + \hat{k}\frac{\partial}{\partial z}\right) \times \begin{vmatrix} \hat{\imath} & \hat{\jmath} & \hat{k} \\ \omega_1 & \omega_2 & \omega_3 \\ x & y & z \end{vmatrix}\right]$$

$$= \frac{1}{2}\left[\left(\hat{\imath}\frac{\partial}{\partial x} + \hat{\jmath}\frac{\partial}{\partial y} + \hat{k}\frac{\partial}{\partial z}\right) \times \left\{\hat{\imath}(\omega_2 z - \omega_3 y) + \hat{\jmath}(\omega_3 x - \omega_1 z) + \hat{k}(\omega_1 y - \omega_2 x)\right\}\right]$$

$$= \frac{1}{2}\begin{vmatrix} \hat{\imath} & \hat{\jmath} & \hat{k} \\ \frac{\partial}{\partial x} & \frac{\partial}{\partial y} & \frac{\partial}{\partial z} \\ (\omega_2 z - \omega_3 y) & (\omega_3 x - \omega_1 z) & (\omega_1 y - \omega_2 x) \end{vmatrix}$$

$$= \hat{\imath}\omega_1 + \hat{\jmath}\omega_2 + \hat{k}\omega_3 = \vec{\omega}.$$

[Thus the curl of the linear velocity of a particle of a rigid body is equal to twice the angular velocity. Curl $\vec{V}$, therefore, describes the rotation of a vector field $\vec{V}$ and hence the name *curl* or *rotation* is justified.]

7. Show that curl of $\dfrac{x\hat{\imath} + y\hat{\jmath}}{x + y} = \dfrac{x - y}{(x + y)^2}\hat{k}$. **[C.U. (Hons.) 1974]**

$$\mathrm{curl}\left(\frac{x\hat{\imath} + y\hat{\jmath}}{x + y}\right) = \vec{\nabla} \times \left(\frac{x\hat{\imath} + y\hat{\jmath}}{x + y}\right) = \begin{vmatrix} \hat{\imath} & \hat{\jmath} & \hat{k} \\ \frac{\partial}{\partial x} & \frac{\partial}{\partial y} & \frac{\partial}{\partial z} \\ \frac{x}{x + y} & \frac{y}{x + y} & 0 \end{vmatrix}$$

$$= \hat{k}\left[\frac{\partial}{\partial x}\left(\frac{y}{x + y}\right) - \frac{\partial}{\partial y}\left(\frac{x}{x + y}\right)\right]$$

$$= \hat{k}\left[-\frac{y}{(x + y)^2} + \frac{x}{(x + y)^2}\right]$$

$$= \frac{(x - y)}{(x + y)^2}\hat{k}.$$

8. **(a)** Find the unit vector perpendicular to the plane of the vectors $\vec{A} = 3\hat{\imath} - 2\hat{\jmath} + 4\hat{k}$ and $\vec{B} = \hat{\imath} + \hat{\jmath} - 2\hat{k}$.

(b) A force given by $\vec{F} = 3\hat{\imath} - 2\hat{\jmath} - 4\hat{k}$ is applied at the point $(1, -1, 2)$. Find the moment of $\vec{F}$ about the point $(2, -1, 3)$.

(c) If $\vec{A}+\vec{B}+\vec{C}=0$, prove that $\vec{A}\times\vec{B}=\vec{B}\times\vec{C}=\vec{C}\times\vec{A}$. Is the converse true?

(d) A particle moves along the curve $x=2t^2, y=t^2-4t, z=3t-5$, where t is the time. Find the components of velocity and acceleration at time $t=1$ in the direction $\hat{\imath}-3\hat{\jmath}+2\hat{k}$.

[C.U. (Hons.) 1994, 2008]

(a) The resulting perpendicular vector is given by

$$\vec{C}=\pm(3\hat{\imath}-2\hat{\jmath}+4\hat{k})\times(\hat{\imath}+\hat{\jmath}-2\hat{k})=\pm(0\hat{\imath}+10\hat{\jmath}+5\hat{k}).$$

$\therefore$ The unit vector in the direction of $\vec{C}=\dfrac{\pm(10\hat{\jmath}+5\hat{k})}{\sqrt{10^2+5^2}}=\pm\dfrac{1}{\sqrt{5}}(2\hat{\jmath}+\hat{k}).$

(b) The position vector of the particle at $(1,-1,2)$ with respect to $(2,-1,3)$ is given by

$$\vec{r}=(1-2)\hat{\imath}+(-1+1)\hat{\jmath}+(2-3)\hat{k}=-\hat{\imath}-\hat{k}$$

$\therefore$ Moment of the force $\vec{F}$ about the given point is

$$\vec{M}=\vec{r}\times\vec{F}=(-\hat{\imath}-\hat{k})\times(3\hat{\imath}-2\hat{\jmath}-4\hat{k})=-2\hat{\imath}-7\hat{\jmath}+2\hat{k}$$

$\therefore$ Magnitude of the moment $=\sqrt{2^2+7^2+2^2}=\sqrt{57}.$

(c) $\because\ \vec{A}+\vec{B}+\vec{C}=0$, we may write

$$\vec{A}\times(\vec{A}+\vec{B}+\vec{C})=0 \quad\text{or}\quad \vec{A}\times\vec{B}+\vec{A}\times\vec{C}=0 \quad \therefore\ \vec{A}\times\vec{B}=\vec{C}\times\vec{A}$$

Similarly multiplying by $\vec{B}$, we can show that $\vec{B}\times\vec{C}=\vec{A}\times\vec{B}$

$$\therefore\ \vec{A}\times\vec{B}=\vec{B}\times\vec{C}=\vec{C}\times\vec{A}.$$

Converse part: If $\vec{B}\times\vec{C}=\vec{A}\times\vec{B}$, then

$$\vec{B}\times\vec{C}+\vec{B}\times\vec{A}=0 \quad\text{or}\ \vec{B}\times(\vec{A}+\vec{C})=0 \quad\text{or}\ \vec{B}\times(\vec{A}+\vec{B}+\vec{C})=0$$

$\therefore$ either $\vec{B}=0$ or $\vec{A}+\vec{B}+\vec{C}=0$.

Similarly from $\vec{B}\times\vec{C}=\vec{C}\times\vec{A}$, we can show that either $\vec{C}=0$ or $\vec{A}+\vec{B}+\vec{C}=0$ and also from $\vec{A}\times\vec{B}=\vec{C}\times\vec{A}$, we can show that either $\vec{A}=0$ or $\vec{A}+\vec{B}+\vec{C}=0$.

Thus we can conclude that either $\vec{A},\vec{B}$ and $\vec{C}$ are individually equal to zero or the sum of the vectors $(\vec{A}+\vec{B}+\vec{C})$ is a null vector.

(d) Here $\vec{r}=2t^2\hat{\imath}+(t^2-4t)\hat{\jmath}+(3t-5)\hat{k}$

$$\therefore\ \frac{d\vec{r}}{dt}=4t\hat{\imath}+(2t-4)\hat{\jmath}+3\hat{k} \quad\text{or}\quad \left.\frac{d\vec{r}}{dt}\right]_{t=1}=\vec{v}\Big]_{t=1}=4\hat{\imath}-2\hat{\jmath}+3\hat{k}$$

$$\text{and}\quad \frac{d^2\vec{r}}{dt^2}=4\hat{\imath}+2\hat{\jmath} \quad\text{or}\quad \left.\frac{d^2\vec{r}}{dt^2}\right]_{t=1}=\vec{f}\Big]_{t=1}=4\hat{\imath}+2\hat{\jmath}.$$

The components of $\vec{v}$ and $\vec{f}$ at $t=1$ along the direction $(\hat{\imath}-3\hat{\jmath}+2\hat{k})$ are given by the scalar products of $\vec{v}$ and $\vec{f}$ with the unit vector along $(\hat{\imath}-3\hat{\jmath}+2\hat{k})$.

Thus the velocity component $= (4\hat{\imath} - 2\hat{\jmath} + 3\hat{k})\cdot\dfrac{(\hat{\imath} - 3\hat{\jmath} + 2\hat{k})}{\sqrt{1^2+3^2+2^2}} = \dfrac{16}{\sqrt{14}}$ and the acceleration component $= (4\hat{\imath} + 2\hat{\jmath})\cdot\dfrac{(\hat{\imath} - 3\hat{\jmath} + 2\hat{k})}{\sqrt{1^2+3^2+2^2}} = -\dfrac{2}{\sqrt{14}}$.

9. A force is defined as $\vec{F} = (-\hat{\imath}x + \hat{\jmath}y)/(x^2+y^2)$. Evaluate $\int \vec{F}\cdot d\vec{r}$ along the following paths connecting the points $(-1,0)$ and $(1,0)$:

(i) along a semi-circular path with the origin as centre;

(ii) along the straight lines joining $(-1,0),(0,1)$ and $(0,1),(1,0)$. Comparing the two values, comment on the nature of the force. **[C.U. (Hons.) 1990]**

Here $\vec{F}\cdot d\vec{r} = \dfrac{-xdx + ydy}{x^2+y^2}$.

(i) Along a semi-circular path as given in the problem:

Let us put

$$x = a\cos\theta \quad \therefore\ dx = -a\sin\theta d\theta$$

$$\text{and} \quad y = a\sin\theta \quad \therefore\ dy = a\cos\theta d\theta$$

$$\therefore \quad \int \vec{F}\cdot d\vec{r} = \int_0^{\pi} \sin 2\theta d\theta = 0$$

(ii) (a) Along a straight line joining $(-1,0)(0,1)$: The equation of the straight line joining $(-1,0),(0,1)$ is

$$x + 1 = y \quad \text{or} \quad dx = dy$$

$$\therefore\ \vec{F}\cdot d\vec{r} = \frac{dx}{2x^2+2x+1}$$

$$\therefore \int \vec{F}\cdot d\vec{r} = \frac{1}{2}\int_{-1}^{0} \frac{dx}{\left(x+\frac{1}{2}\right)^2 + \left(\frac{1}{2}\right)^2} = \frac{1}{2}\cdot 2\left[\tan^{-1}\frac{x+\frac{1}{2}}{\frac{1}{2}}\right]_{-1}^{0} = \frac{\pi}{2}$$

(b) Along the straight line joining $(0,1),(1,0)$:

The equation of the straight line is given by

$$y = 1 - x \quad \text{or} \quad dy = -dx$$

$$\therefore\ \vec{F}\cdot d\vec{r} = \frac{-dx}{2x^2-2x+1}$$

$$\therefore \int \vec{F}\cdot d\vec{r} = -\frac{1}{2}\int_{0}^{1} \frac{dx}{\left(x-\frac{1}{2}\right)^2 + \left(\frac{1}{2}\right)^2} = -\frac{1}{2}\cdot 2\left[\tan^{-1}\frac{x-\frac{1}{2}}{\frac{1}{2}}\right]_{0}^{1} = -\frac{\pi}{2}$$

$\therefore \int \vec{F}\cdot d\vec{r}$ joining the two points $(-1,0)$ and $(1,0)$ via the above straight lines $= \dfrac{\pi}{2} - \dfrac{\pi}{2} = 0$

$\therefore$ The force $\vec{F}$ is conservative.

10. Show that the line integral of $t = -\hat{\imath}y + \hat{\jmath}x$ around a continuous closed curve in the xy plane is twice the area enclosed by the curve. **[C.U. (Hons.) 1986, 2003]**

Let

$$\vec{r} = \hat{i}x + \hat{j}y \quad \therefore \; d\vec{r} = \hat{i}dx + \hat{j}dy$$

$$\text{Also let } x = r\cos\theta \quad \therefore \; dx = -r\sin\theta d\theta$$

$$\text{and } y = r\sin\theta \quad \therefore \; dy = r\cos\theta d\theta.$$

So the required line integral around a closed path is given by

$$\int_0^{2\pi} \vec{t}\cdot d\vec{r} = \int_0^{2\pi} (-\hat{i}y + \hat{j}x)\cdot(\hat{i}dx + \hat{j}dy)$$

$$= \int_0^{2\pi} -ydx + xdy = \int_0^{2\pi} (r^2\sin^2\theta d\theta + r^2\cos^2\theta d\theta)$$

$$= r^2\int_0^{2\pi} d\theta = 2\pi r^2 = 2\times \text{ area enclosed by the simplest closed curve.}$$

11. Given $\vec{V} = \hat{i}x + \hat{j}y + \hat{k}z$, evaluate the surface integral $\int_S \vec{V}\cdot\hat{n}ds$ over the entire surface of a cylinder (top and bottom surfaces included) of radius R and length h, where $\hat{n}$ is the unit vector normal to the surface. The centre of the bottom surface is at the origin of the co-ordinate system and the axis of the cylinder is along the z-axis. **[C.U. (Hons.) 1984]**

The unit normal vector on the curved surface is given by

$$\hat{n} = \frac{\hat{i}x + \hat{j}y}{\sqrt{x^2+y^2}}$$

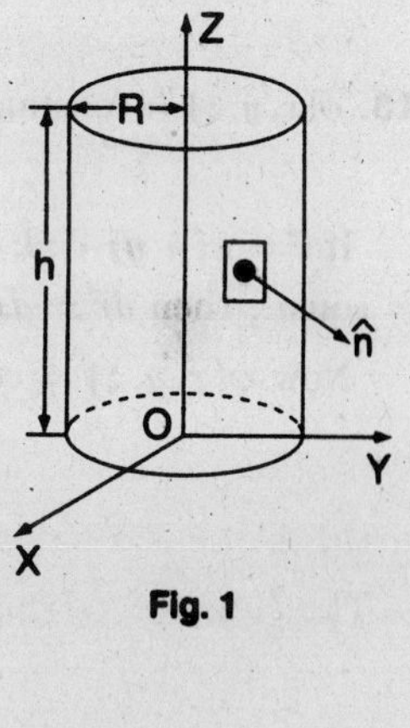

Fig. 1

$\therefore\; \vec{V}\cdot\hat{n} = \dfrac{x^2+y^2}{\sqrt{x^2+y^2}} = R$, so that the surface integral over the curved surface is

$$\int_S \vec{V}\cdot\hat{n}ds = R\int_S dS = 2\pi R^2 h$$

Now for the bottom surface, $\hat{n} = -\hat{k}$ and $z = 0$, so that

$$\int_S \vec{V}\cdot\hat{n}ds = -\int_S (\hat{i}x + \hat{j}y)\cdot\hat{k}ds = 0$$

for the bottom surface.

Again for the top surface $\hat{n} = \hat{k}$ and $z = h$, so that

$$\int_S \vec{V}\cdot\hat{n}ds = \int_S \vec{V}\cdot\hat{k}ds = h\int_S ds = \pi R^2 h$$

for the top surface. Hence $\int_S \vec{V}\cdot\hat{n}ds = 3\pi R^2 h$ over the entire surface of the cylinder.

12. If $\vec{r}$ be the position vector of a point on a closed contour C, prove that the line integral

$$\oint_C \vec{r}\cdot d\vec{r} = 0.$$

[C.U. (Hons.) 1981, '92]

Considering the vector $\vec{r}$ to be the gradient of a scalar U, we have $\vec{r} = \vec{\nabla}U$.

The closed contour C consists of two paths as shown in the figure 2. If $\vec{r}$ makes an angle θ with an element of a path length $d\vec{r}$ on the path 1, then

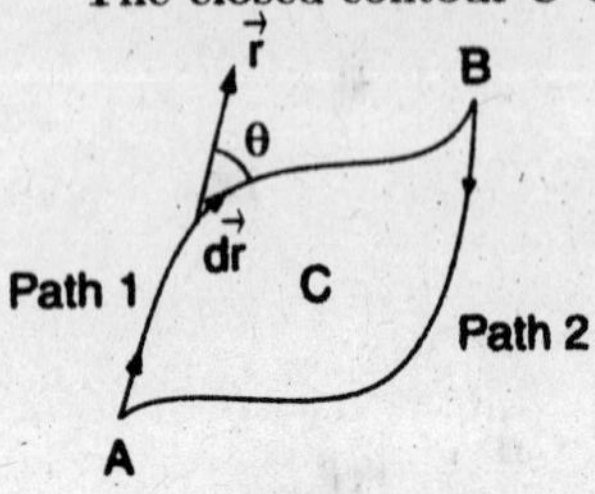

Fig. 2

$$\vec{r} \cdot d\vec{r} = \vec{\nabla}U \cdot d\vec{r} = dU$$

where dU denotes the variation of the scalar function U over the path length $d\vec{r}$.

$\therefore$ The line integral of $\vec{r}$ from A to B

$$= \int_A^B \vec{r} \cdot d\vec{r} = \int_A^B dU = U_B - U_A$$

where U_B and U_A are the values of the scalar at the two ends of path 1.

Similarly the line integral from B to A along path 2

$$= \int_B^A \vec{r} \cdot d\vec{r} = \int_B^A dU = U_A - U_B$$

$\therefore$ The line integral over the closed contour C is given by

$$\oint_C \vec{r} \cdot d\vec{r} = (U_B - U_A) + (U_A - U_B) = 0.$$

13. $\phi(x, y, z) =$ **constant is an equipotential surface. Show that $\vec{\nabla}\phi$ is perpendicular to it.**

[C.U. (Hons.) 1983, '88, '90, '92, 2005, 2008]

If $\vec{r} = x\hat{\imath} + y\hat{\jmath} + z\hat{k}$ be the position vector of any point on the equipotential surface $\phi(x, y, z) =$ const., then $d\vec{r} = dx\hat{\imath} + dy\hat{\jmath} + dz\hat{k}$ also lies in the tangent plane to the surface at that point.

Now $\phi(x, y, z) =$ const.

$$\therefore \quad d\phi = \frac{\partial \phi}{\partial x}dx + \frac{\partial \phi}{\partial y}dy + \frac{\partial \phi}{\partial z}dz = 0$$

$$\text{or} \quad \left(\frac{\partial \phi}{\partial x}\hat{\imath} + \frac{\partial \phi}{\partial y}\hat{\jmath} + \frac{\partial \phi}{\partial z}\hat{k}\right) \cdot (dx\hat{\imath} + dy\hat{\jmath} + dz\hat{k}) = 0$$

$$\text{or} \quad \vec{\nabla}\phi \cdot d\vec{r} = 0.$$

So $\vec{\nabla}\phi$ is perpendicular to $d\vec{r}$ and therefore to the surface.

14. For a given vector $\vec{A}$, prove by the application of Gauss's divergence theorem and Stokes' theorem that div curl $\vec{A} = 0$. **[C.U. (Hons.) 1980, '97]**

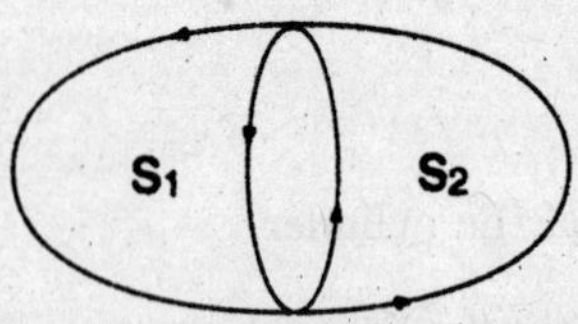

Fig. 3

Let a surface S enclose a volume V in a vector field. Let a closed line on the surface divide the volume into two parts as shown in the figure 3. S_1 is the surface on the left hand side and S_2 that on the right hand side of the line.

By Gauss's theorem, over the closed volume,

$$\iiint \text{div curl } \vec{A} dV = \iint \text{curl } \vec{A} \cdot d\vec{S} = \iint \text{curl } \vec{A} \cdot d\vec{S}_1 + \iint \text{curl } \vec{A} \cdot d\vec{S}_2$$

By Stokes' theorem if

$$\iint \text{curl}\,\vec{A}\cdot d\vec{S}_1 = \oint \vec{A}\cdot d\vec{l},$$

$$\text{then} \quad \iint \text{curl}\,\vec{A}\cdot d\vec{S}_2 = -\oint \vec{A}\cdot d\vec{l},$$

so that div curl $\vec{A} = 0$.

15. Evaluate $\frac{1}{3}\int_s \vec{r}\cdot\hat{n}ds$ over the unit cube defined by the point $(0,0,0)$ and the unit intercepts on the positive x, y, and z axes. **[C.U. (Hons.) 1986,'90,'96]**

$$\int_S \vec{r}\cdot\hat{n}ds = \iiint_V (\text{div}\,\vec{r})dV \quad \text{(from Divergence theorem)}$$

$$= \iiint_V (\vec{\nabla}\cdot\vec{r})dV = \iiint \left(\hat{\imath}\frac{\partial}{\partial x} + \hat{\jmath}\frac{\partial}{\partial y} + \hat{k}\frac{\partial}{\partial z}\right)\cdot(\hat{\imath}x + \hat{\jmath}y + \hat{k}z)dx\,dy\,dz$$

$$= \int_{x=0}^{1}\int_{y=0}^{1}\int_{z=0}^{1}\left(\frac{\partial x}{\partial x} + \frac{\partial y}{\partial y} + \frac{\partial z}{\partial z}\right)dx\,dy\,dz$$

$$= 3\int_0^1 dx \int_0^1 dy \int_0^1 dz = 3 \quad \therefore \frac{1}{3}\int_S \vec{r}\cdot\hat{n}ds = 1.$$

16. Verify divergence theorem for a vector $\vec{V} = r^2\vec{r}$ over the region bounded by the surface S of a sphere of radius a whose centre is at the origin. **[C.U. (Hons.) 1984, '91]**

From Gauss's theorem

$$\iiint \vec{\nabla}\cdot\vec{V}dV = \iint \vec{V}\cdot d\vec{S} = \iint \vec{V}\cdot\hat{n}dS.$$

We have to prove

$$\iiint \vec{\nabla}\cdot(r^2\vec{r})dV = \iint (r^2\vec{r})\cdot\hat{n}dS.$$

$$\vec{\nabla}\cdot(r^2\vec{r}) = \left(\hat{\imath}\frac{\partial}{\partial x} + \hat{\jmath}\frac{\partial}{\partial y} + \hat{k}\frac{\partial}{\partial z}\right)\cdot r^2\left(\hat{\imath}x + \hat{\jmath}y + \hat{k}z\right) = \frac{\partial(r^2x)}{\partial x} + \frac{\partial(r^2y)}{\partial y} + \frac{\partial(r^2z)}{\partial z}.$$

Now

$$\frac{\partial(r^2x)}{\partial x} = (x^2+y^2+z^2) + x\cdot 2x$$

$$\frac{\partial(r^2y)}{\partial y} = (x^2+y^2+z^2) + y\cdot 2y$$

$$\frac{\partial(r^2z)}{\partial z} = (x^2+y^2+z^2) + z\cdot 2z.$$

Adding we get

$$\vec{\nabla}\cdot(r^2\vec{r}) = 5(x^2+y^2+z^2) = 5r^2$$

$$\therefore \iiint \vec{\nabla}\cdot(r^2\vec{r})dV = \iiint 5r^2 dV$$

$$= 5\iiint r^2\cdot r^2 \sin\theta d\theta d\phi dr \quad \text{(transforming } dV \text{ into spherical polar co-ordinates)}$$

$$= 5\int_0^a r^4 dr \int_0^\pi \sin\theta d\theta \int_0^{2\pi} d\phi = 4\pi a^5$$

Again

$$\iint (r^2\vec{r})\cdot \hat{n}ds = \iint (x^2+y^2+z^2)(\hat{\imath}x+\hat{\jmath}y+\hat{k}z)\cdot\left(\frac{x}{r}\hat{\imath}+\frac{y}{r}\hat{\jmath}+\frac{z}{r}\hat{k}\right)dS$$

$$\left[\because\ \hat{n} = \frac{(x^2+y^2+z^2)x\hat{\imath}+(x^2+y^2+z^2)y\hat{\jmath}+(x^2+y^2+z^2)z\hat{k}}{\sqrt{r^4(x^2+y^2+z^2)}}\right.$$

$$\left.= \frac{r^2(x\hat{\imath}+y\hat{\jmath}+z\hat{k})}{r^3} = \frac{x}{r}\hat{\imath}+\frac{y}{r}\hat{\jmath}+\frac{z}{r}\hat{k}\right]$$

$$\therefore \iint (r^2\vec{r})\cdot\hat{n}dS = \iint (x^2+y^2+z^2)\frac{1}{r}(x^2+y^2+z^2)dS$$

$$= \iint r^3 dS = \iint r^3 r^2 \sin\theta d\theta d\phi \quad \text{(transforming } dS \text{ into spherical polar co-ordinates)}$$

$$= 4\pi a^5 \quad \text{(putting } r = a \text{ and proper limits for } \theta \text{ and } \phi)$$

So the divergence theorem is verified for the given vector.

17. If $\vec{A} = 4xz\hat{\imath} - y^2\hat{\jmath} + yz\hat{k}$, evaluate $\iint\limits_S \vec{A}\cdot\hat{n}dS$, where S is the surface of a unit cube with two opposite corners at $(0,0,0)$ and $(1,1,1)$ respectively. **[C.U. (Hons.) 1988,'92]**

According to divergence theorem,

$$\iint\limits_S \vec{A}\cdot\hat{n}dS = \iiint\limits_V \vec{\nabla}\cdot\vec{A}dV = \iiint\limits_V \left(\hat{\imath}\frac{\partial}{\partial x}+\hat{\jmath}\frac{\partial}{\partial y}+\hat{k}\frac{\partial}{\partial z}\right)\cdot(4xz\hat{\imath}-y^2\hat{\jmath}+yz\hat{k})dV$$

$$= \iiint\limits_V \left[\frac{\partial}{\partial x}(4xz)+\frac{\partial}{\partial y}(-y^2)+\frac{\partial}{\partial z}(yz)\right]dV$$

$$= \iiint\limits_V (4z-2y+y)dV = \int_{x=0}^1\int_{y=0}^1\int_{z=0}^1 (4z-y)dx\,dy\,dz$$

$$= \int_{x=0}^1\int_{y=0}^1 \Big[(2z^2-yz)\Big]_{z=0}^1 dx\,dy = \int_{x=0}^1\int_{y=0}^1 (2-y)dx\,dy$$

$$= \int_{x=0}^1 \left(2y-\frac{y^2}{2}\right)_{y=0}^1 dx = \int_{x=0}^1\left(2-\frac{1}{2}\right)dx = \left(\frac{3}{2}x\right)_{x=0}^1 = \frac{3}{2}.$$

18. **(i)** Let $\vec{r}(t)$ be a vector of fixed magnitude. Show that $\dfrac{d\vec{r}(t)}{dt}$ is perpendicular to $\vec{r}(t)$.

(ii) If $\vec{V}(t)$ is a time dependent vector, show that $\vec{V}\cdot\dfrac{d\vec{V}}{dt} = v\dfrac{dv}{dt}$ where v is the magnitude of $\vec{V}$.

[C.U. (Hons.) 1985, '91, 2007]

(i) Since $\vec{r}(t)$ is a vector of fixed magnitude, we may write $\vec{r}(t)\cdot\vec{r}(t) = \text{constant}$

$$\therefore\ \frac{d}{dt}[\vec{r}(t)\cdot\vec{r}(t)] = 0 \quad \text{or} \quad \vec{r}(t)\cdot\frac{d\vec{r}(t)}{dt} + \frac{d\vec{r}(t)}{dt}\cdot\vec{r}(t) = 0 \quad \text{or} \quad 2\frac{d\vec{r}(t)}{dt}\cdot\vec{r}(t) = 0$$

$$\text{or} \quad \frac{d\vec{r}(t)}{dt}\cdot\vec{r}(t) = 0 \quad \text{i.e.,} \quad \frac{d\vec{r}(t)}{dt} \text{ is perpendicular to } \vec{r}(t).$$

(ii) We may write $\vec{V}\cdot\vec{V} = v^2$ since v is the magnitude of $\vec{V}$

$$\text{or} \quad \frac{d}{dt}(\vec{V}\cdot\vec{V}) = \frac{d}{dt}(v^2) \quad \text{or} \quad 2\vec{V}\cdot\frac{d\vec{V}}{dt} = 2v\frac{dv}{dt} \quad \therefore\ \vec{V}\cdot\frac{d\vec{V}}{dt} = v\frac{dv}{dt}.$$

If now $\vec{V}(t)$ is a vector of constant magnitude then $\frac{d\vec{V}}{dt} = \frac{dv}{dt} = 0$

$$\therefore\ \vec{V}\cdot\frac{d\vec{V}}{dt} = 0 \quad \text{i.e.,} \quad \frac{d\vec{V}}{dt} \text{ is perpendicular to } \vec{V}.$$

This relation is very much relevant in connection with the motion of a particle in a circular orbit. If a particle moves under an acceleration such that the magnitude of its velocity remains constant, the acceleration is then at right angles to the velocity which is exactly the case of the motion of a particle in a circular orbit.

19. Prove that $\vec{\nabla}\cdot(\vec{A}\times\vec{B}) = \vec{B}\cdot(\vec{\nabla}\times\vec{A}) - \vec{A}\cdot(\vec{\nabla}\times\vec{B})$. If $\vec{A}$ and $\vec{B}$ are irrotational, show that $\vec{A}\times\vec{B}$ is solenoidal.

[C.U. (Hons.) 1991]

Let $\vec{V} = \vec{A}\times\vec{B} = \begin{vmatrix} \hat{\imath} & \hat{\jmath} & \hat{k} \\ A_x & A_y & A_z \\ B_x & B_y & B_z \end{vmatrix}$, so that we may write

$$V_x\hat{\imath} + V_y\hat{\jmath} + V_z\hat{k} = (A_yB_z - A_zB_y)\hat{\imath} + (A_zB_x - A_xB_z)\hat{\jmath} + (A_xB_y - A_yB_x)\hat{k}$$

Now

$$\begin{aligned}
\vec{\nabla}\cdot(\vec{A}\times\vec{B}) &= \vec{\nabla}\cdot\vec{V} = \frac{\partial}{\partial x}V_x + \frac{\partial}{\partial y}V_y + \frac{\partial}{\partial z}V_z \\
&= \frac{\partial}{\partial x}(A_yB_z - A_zB_y) + \frac{\partial}{\partial y}(A_zB_x - A_xB_z) + \frac{\partial}{\partial z}(A_xB_y - A_yB_x) \\
&= B_x\left(\frac{\partial}{\partial y}A_z - \frac{\partial}{\partial z}A_y\right) + B_y\left(\frac{\partial}{\partial z}A_x - \frac{\partial}{\partial x}A_z\right) + B_z\left(\frac{\partial}{\partial x}A_y - \frac{\partial}{\partial y}A_x\right) \\
&\quad - A_x\left(\frac{\partial}{\partial y}B_z - \frac{\partial}{\partial z}B_y\right) - A_y\left(\frac{\partial}{\partial z}B_x - \frac{\partial}{\partial x}B_z\right) - A_z\left(\frac{\partial}{\partial x}B_y - \frac{\partial}{\partial y}B_x\right)
\end{aligned}$$

$$= \left(B_x\hat{i} + B_y\hat{j} + B_z\hat{k}\right) \cdot \begin{vmatrix} \hat{i} & \hat{j} & \hat{k} \\ \dfrac{\partial}{\partial x} & \dfrac{\partial}{\partial y} & \dfrac{\partial}{\partial z} \\ A_x & A_y & A_z \end{vmatrix} - (A_x\hat{i} + A_y\hat{j} + A_z\hat{k}) \cdot \begin{vmatrix} \hat{i} & \hat{j} & \hat{k} \\ \dfrac{\partial}{\partial x} & \dfrac{\partial}{\partial y} & \dfrac{\partial}{\partial z} \\ B_x & B_y & B_z \end{vmatrix}$$

$$= \vec{B} \cdot (\vec{\nabla} \times \vec{A}) - \vec{A} \cdot (\vec{\nabla} \times \vec{B}).$$

If now $\vec{A}$ and $\vec{B}$ are irrotational, then $\vec{\nabla} \times \vec{A} = 0$ and $\vec{\nabla} \times \vec{B} = 0$, so that from the above relation, we have $\vec{\nabla} \cdot (\vec{A} \times \vec{B}) = 0$. Hence $(\vec{A} \times \vec{B})$ is solenoidal.

20. Prove that $(\vec{A} \cdot \vec{\nabla})\vec{A} = \frac{1}{2}\vec{\nabla}A^2 - \vec{A} \times (\vec{\nabla} \times \vec{A})$, i.e., $\vec{A} \times (\vec{\nabla} \times \vec{A}) = \frac{1}{2}\vec{\nabla}A^2 - (\vec{A} \cdot \vec{\nabla})\vec{A}$.

[C.U. (Hons.) 1993]

$$\vec{A} \times (\vec{\nabla} \times \vec{A}) = \vec{A} \times \begin{vmatrix} \hat{i} & \hat{j} & \hat{k} \\ \dfrac{\partial}{\partial x} & \dfrac{\partial}{\partial y} & \dfrac{\partial}{\partial z} \\ A_x & A_y & A_z \end{vmatrix}$$

$$= \left(A_x\hat{i} + A_y\hat{j} + A_z\hat{k}\right) \times \left[\hat{i}\left(\frac{\partial}{\partial y}A_z - \frac{\partial}{\partial z}A_y\right) + \hat{j}\left(\frac{\partial}{\partial z}A_x - \frac{\partial}{\partial x}A_z\right) + \hat{k}\left(\frac{\partial}{\partial x}A_y - \frac{\partial}{\partial y}A_x\right)\right]$$

$$= \begin{vmatrix} \hat{i} & \hat{j} & \hat{k} \\ A_x & A_y & A_z \\ \left(\dfrac{\partial}{\partial y}A_z - \dfrac{\partial}{\partial z}A_y\right) & \left(\dfrac{\partial}{\partial z}A_x - \dfrac{\partial}{\partial x}A_z\right) & \left(\dfrac{\partial}{\partial x}A_y - \dfrac{\partial}{\partial y}A_x\right) \end{vmatrix}$$

$$= \hat{i}\left(A_y\frac{\partial A_y}{\partial x} - A_y\frac{\partial A_x}{\partial y} - A_z\frac{\partial A_x}{\partial z} + A_z\frac{\partial A_z}{\partial x}\right)$$
$$+\hat{j}\left(A_z\frac{\partial A_z}{\partial y} - A_z\frac{\partial A_y}{\partial z} - A_x\frac{\partial A_y}{\partial x} + A_x\frac{\partial A_x}{\partial y}\right)$$
$$+\hat{k}\left(A_x\frac{\partial A_x}{\partial z} - A_x\frac{\partial A_z}{\partial x} - A_y\frac{\partial A_z}{\partial y} + A_y\frac{\partial A_y}{\partial z}\right)$$

$$= \hat{i}\left[\frac{1}{2}\frac{\partial}{\partial x}(A_x^2 + A_y^2 + A_z^2) - A_x\frac{\partial A_x}{\partial x} - A_y\frac{\partial A_x}{\partial y} - A_z\frac{\partial A_x}{\partial z}\right]$$
$$+\hat{j}\left[\frac{1}{2}\frac{\partial}{\partial y}(A_x^2 + A_y^2 + A_z^2) - A_x\frac{\partial A_y}{\partial x} - A_y\frac{\partial A_y}{\partial y} - A_z\frac{\partial A_y}{\partial z}\right]$$
$$+\hat{k}\left[\frac{1}{2}\frac{\partial}{\partial z}(A_x^2 + A_y^2 + A_z^2) - A_x\frac{\partial A_z}{\partial x} - A_y\frac{\partial A_z}{\partial y} - A_z\frac{\partial A_z}{\partial z}\right]$$

$$= \frac{1}{2}\left(\hat{\imath}\frac{\partial}{\partial x} + \hat{\jmath}\frac{\partial}{\partial y} + \hat{k}\frac{\partial}{\partial z}\right)(A_x^2 + A_y^2 + A_z^2)$$

$$- \left(A_x\frac{\partial}{\partial x} + A_y\frac{\partial}{\partial y} + A_z\frac{\partial}{\partial z}\right)\left(A_x\hat{\imath} + A_y\hat{\jmath} + A_z\hat{k}\right)$$

$$= \frac{1}{2}\vec{\nabla}A^2 - \left[\left(A_x\hat{\imath} + A_y\hat{\jmath} + A_z\hat{k}\right)\cdot\left(\hat{\imath}\frac{\partial}{\partial x} + \hat{\jmath}\frac{\partial}{\partial y} + \hat{k}\frac{\partial}{\partial z}\right)\right]\left(A_x\hat{\imath} + A_y\hat{\jmath} + A_z\hat{k}\right)$$

$$= \frac{1}{2}\vec{\nabla}A^2 - \left(\vec{A}\cdot\vec{\nabla}\right)\vec{A}.$$

21. From Gauss's divergence theorem, show that if $\vec{B} = \vec{\nabla} \times \vec{A}$, then $\int_S \vec{B}\cdot \hat{n} dS = 0$ for any closed surface S, n being the out-ward drawn unit normal vector to S. **[C.U. (Hons.) 1987,'90]**

Applying Gauss's divergence theorem, we may write

$$\int_S \vec{B}\cdot \hat{n} dS = \int_V \vec{\nabla}\cdot \vec{B} dV = \int_V \vec{\nabla}\cdot(\vec{\nabla} \times \vec{A}) dV = 0$$

since $\vec{B} = \vec{\nabla} \times \vec{A}$ (given) and $\vec{\nabla}\cdot(\vec{\nabla} \times \vec{A})$, i.e., div curl $\vec{A} = 0$.

22. Find the line integral of the vector $\vec{F}$ along the line segment joining the points $(0,0,0)$ and $(2,3,4)$; $\vec{F} = x^2\hat{\imath} + y^2\hat{\jmath} + (xz - y)\hat{k}$. **[C.U. (Hons.) 1999, 2007]**

Here $\vec{F} = x^2\hat{\imath} + y^2\hat{\jmath} + (xz - y)\hat{k}$ and $d\vec{r} = dx\hat{\imath} + dy\hat{\jmath} + dz\hat{k}$.

$$\therefore \quad \vec{F}\cdot d\vec{r} = x^2\,dx + y^2\,dy + (xz - y)dz$$

The equation of the line segment joining $(0,0,0)$ and $(2,3,4)$ is given in parametric form by

$$\frac{x-0}{2-0} = \frac{y-0}{3-0} = \frac{z-0}{4-0} = t \quad \text{(say)}$$

$\therefore$ $x = 2t$ or $dx = 2dt$; $y = 3t$ or $dy = 3dt$; $z = 4t$ or $dz = 4dt$, and when $t = 1$, $x = 2$, $y = 3$ and $z = 4$.

$$\therefore \quad \int_L \vec{F}\cdot d\vec{r} = \int_L \left[x^2\,dx + y^2\,dy + (xz - y)dz\right]$$

$$= \int_{t=0}^{t=1} (8t^2 + 27t^2 + 32t^2 - 12t)dt \qquad \text{[Putting } x,\ y,\ z,\ dx,\ dy,\ dz \text{ in terms of } t\text{]}$$

$$= \int_0^1 (67t^2 - 12t)dt = \left(\frac{67}{3}t^3 - 12\frac{t^2}{2}\right)_0^1 = \frac{49}{3}$$

23. *ABCD* is a square in the $z = 3$ plane. The x, y coordinates of A, B, C and D are $(0,0)$, $(0,a)$, (a,a), and $(a,0)$ respectively. Show that the line integral of $\vec{F} = (2xy + z^3)\hat{\imath} + x^2\hat{\jmath} + 3xz^2\hat{k}$ is zero over *ABCDA*. What is the corresponding scalar potential? **[C.U. (Hons.) 2001]**

Here

$$\int \vec{F}\cdot d\vec{r} = \int (2xy + z^3)\,dx + x^2\,dy + 3xz^2\,dz$$

Now along the straight line joining A and B, $x = 0, z = 3$.

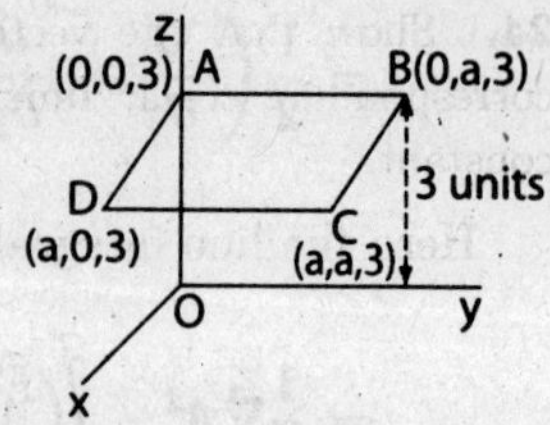

$$\therefore \quad dx = dz = 0$$

and

$$\int_{y=0\;AB}^{a} \vec{F} \cdot d\vec{r} = \int_{0}^{a} 0 \, dy = 0 \qquad (1)$$

Along the straight line joining B and C, $y = a$, $z = 3$.

$$\therefore \quad dy = dz = 0$$

and

$$\int_{x=0\;BC}^{a} \vec{F} \cdot d\vec{r} = \int_{0}^{a} (2xa + 27)dx = \left[2a\frac{x^2}{2} + 27x\right]_0^a = a^3 + 27a \qquad (2)$$

Along the straight line joining C and D, $x = a$, $z = 3$.

$$\therefore \quad dx = dz = 0$$

and

$$\int_{y=a\;CD}^{0} \vec{F} \cdot d\vec{r} = \int_{a}^{0} x^2 dy = \int_{a}^{0} a^2 dy = -a^3 \qquad (3)$$

Along along the straight line joining D and A, $y = 0$, $z = 3$.

$$\therefore \quad dy = dz = 0$$

and

$$\int_{x=a\;DA}^{0} \vec{F} \cdot d\vec{r} = \int_{a}^{0} (2x \cdot 0 + 27)dx = -27a \qquad (4)$$

Adding (1), (2), (3) and (4)

$$\int_{ABCDA} \vec{F} \cdot d\vec{r} = 0 + (a^3 + 27a) - a^3 - 27a = 0$$

To find the scalar potential, we know that $\vec{F} = \vec{\nabla}\phi$

$$\text{or} \quad \vec{F} \cdot d\vec{r} = \vec{\nabla}\phi \cdot d\vec{r} = \left(\hat{i}\frac{\partial \phi}{\partial x} + \hat{j}\frac{\partial \phi}{\partial y} + \hat{k}\frac{\partial \phi}{\partial z}\right) \cdot \left(\hat{i}dx + \hat{j}dy + \hat{k}dz\right)$$

$$= \frac{\partial \phi}{\partial x}dx + \frac{\partial \phi}{\partial y}dy + \frac{\partial \phi}{\partial z}dz = d\phi$$

$$\therefore \quad d\phi = \vec{F} \cdot d\vec{r} = (2xy + z^3)dx + x^2 dy + 3xz^2 dz$$

$$= (2xydx + x^2 dy) + (z^3 dx + 3xz^2 dz)$$

$$= d(x^2 y) + d(z^3 x) = d(x^2 y + z^3 x)$$

$$\therefore \quad \phi = x^2 y + z^3 x + C$$

24. Show that the vector $\vec{F} = yz\hat{i} + zx\hat{j} + xy\hat{k}$ has path independent line integral. Find the corresponding scalar function ϕ from which $\vec{F}$ may be obtained. Take due care in obtaining the constant. **[C.U. (Hons.) 1991]**

Here the line integral is given by

$$\int \vec{F} \cdot d\vec{r} = \int \left(yz\vec{i} + zx\vec{j} + xy\vec{k}\right) \cdot \left(\hat{i}dx + \hat{j}dy + \hat{k}dz\right)$$
$$= \int (yzdx + zxdy + xydz) = \int d(xyz)$$

Let us consider that the particle is moving from $P_1(x_1, y_1, z_1)$ to $P_2(x_2, y_2, z_2)$, so that

$$\int_{P_1}^{P_2} \vec{F} \cdot d\vec{r} = \int_{P_1}^{P_2} d(xyz) = P_2(x_2, y_2, z_2) - P_1(x_1, y_1, z_1).$$

So the line integral depends only on the end points P_1 and P_2, but not on the path joining them—so it is path independent.

To find the corresponding scalar function ϕ from which $\vec{F}$ may be obtained, we know that $\vec{F} = \vec{\nabla}\phi$.

$\vec{F} \cdot d\vec{r} = \vec{\nabla}\phi \cdot d\vec{r} = d\phi$ as shown in example 23.

$$\therefore \quad d\phi = \vec{F} \cdot d\vec{r} = (yz\hat{i} + zx\hat{j} + xy\hat{k}) \cdot (\hat{i}dx + \hat{j}dy + \hat{k}dz)$$
$$= yzdx + zxdy + xydz = d(xyz)$$
$$\therefore \quad \phi = xyz + C$$

Alternative method. We know that the line integral of the vector $\vec{F}$ evaluated between two points will be independent of the path connecting the two points if the vector field is conservative, i.e., when $\vec{\nabla} \times \vec{F} = 0$.

For the problem

$$\vec{\nabla} \times \vec{F} = \begin{vmatrix} \hat{i} & \hat{i} & \hat{k} \\ \dfrac{\partial}{\partial x} & \dfrac{\partial}{\partial y} & \dfrac{\partial}{\partial z} \\ yx & zx & xy \end{vmatrix} = 0$$

As $\vec{F}$ is conservative, $\vec{F} = \vec{\nabla}\phi$.

To calculate ϕ, proceed as in example 4 of page 83 on Mechanics.

25. From Gauss's divergence theorem, prove that for two scalar functions u and v,

$\int u\vec{\nabla}v \cdot d\vec{s} = \int u\vec{\nabla} \cdot (\nabla v)dV \int \vec{\nabla}u \cdot \vec{\nabla}vdV.$ **[C.U. (Hons.) 1998]**

If $\vec{A}$ be any vector field through a volume V which is enclosed by the surface S, then according to Gauss's theorem,

$$\int_V \vec{\nabla} \cdot \vec{A}dV = \int_S \vec{A} \cdot d\vec{S}$$

Now the two scalar functions given are u and v.

So clearly $(u\vec{\nabla}v)$ will be regarded as a vector function on the volume V. Then applying Gauss's theorem, we may write

$$\int_S u\vec{\nabla}v \cdot d\vec{S} = \int_V \vec{\nabla} \cdot (u\vec{\nabla}v)dV \tag{1}$$

Now if we take $u = \phi$ and $\vec{\nabla}v = \vec{A}$, then

$$\vec{\nabla} \cdot (u\vec{\nabla}v) = \vec{\nabla} \cdot (\phi A) = (\vec{\nabla}\phi) \cdot \vec{A} + \phi(\vec{\nabla} \cdot \vec{A}),$$

so that the r.h.s. of equation (1) becomes

$$\vec{\nabla} \cdot (u\vec{\nabla}v) = (\vec{\nabla}u) \cdot \vec{\nabla}v + u(\vec{\nabla} \cdot \vec{\nabla}v)$$

which proves the identity.

26. Two vector fields $\vec{A}_1$ and $\vec{A}_2$ have the same divergence and curl at every point in volume V and have the same normal component at every point on the surface S enclosing V. Prove that $\vec{A}_1$ and $\vec{A}_2$ are equal everywhere. **[C.U. (Hons.) 1998]**

From the first condition, $\vec{\nabla} \cdot \vec{A}_1 = \vec{\nabla} \cdot \vec{A}_2$

$$\text{or} \quad \vec{\nabla} \cdot (\vec{A}_1 - \vec{A}_2) = 0 \quad \text{or} \quad \vec{\nabla} \cdot \vec{B} = 0 \quad (\text{putting } \vec{A}_1 - \vec{A}_2 = \vec{B}) \tag{1}$$

From the second condition, $\vec{\nabla} \times \vec{A}_1 = \vec{\nabla} \times \vec{A}_2$

$$\text{or} \quad \vec{\nabla} \times (\vec{A}_1 - \vec{A}_2) = 0 \quad \text{or} \quad \vec{\nabla} \times \vec{B} = 0 \tag{2}$$

If ϕ be a scalar differential function,

$$\vec{B} = -\vec{\nabla}\phi \tag{3}$$

using (3), in (1), we obtain

$$\nabla^2\phi = 0 \tag{4}$$

From Gauss's divergence theorem,

$$\int_\tau \vec{\nabla} \cdot (\phi\vec{\nabla}\phi)d\tau = \int_S (\phi\vec{\nabla}\phi) \cdot d\vec{S}$$

$$\text{or} \quad \int_\tau (\vec{\nabla}\phi \cdot \vec{\nabla}\phi + \phi\nabla^2\phi)d\tau = \int_S (\phi\vec{\nabla}\phi) \cdot d\vec{S}$$

$$\text{or} \quad \int_\tau \left[|\vec{\nabla}\phi|^2 + \phi\nabla^2\phi\right] d\tau = \int_S \left(\phi\vec{\nabla}\phi\right) \cdot d\vec{S}$$

$$\text{or} \quad \int_\tau |\vec{\nabla}\phi|^2 d\tau = \int_S (\phi\vec{\nabla}\phi) \cdot d\vec{S} \quad (\because \quad \nabla^2\phi = 0) \tag{5}$$

Now

$$\int_S \phi\vec{\nabla}\phi \cdot d\vec{S} = -\int_S \phi\vec{B} \cdot d\vec{S} \quad \text{from (3)}$$

$$= -\int_S \phi(\vec{A}_1 - \vec{A}_2) \cdot d\vec{S}$$

$$= -\left[\int_S \phi\vec{A}_1 \cdot d\vec{S} - \int_S \phi\vec{A}_2 \cdot d\vec{S}\right] = 0 \quad \left(\because \quad \vec{A}_1 \cdot d\vec{S} = \vec{A}_2 \cdot d\vec{S}\right) \tag{6}$$

Using (6) in (5), we obtain

$$\int_\tau |\vec{\nabla}\phi|^2 d\tau = 0 \quad \therefore \quad \vec{\nabla}\phi = 0 \quad \text{or} \quad -\vec{B} = 0 \quad \text{i.e.,} \quad \vec{A}_1 - \vec{A}_2 = 0 \quad \text{or} \quad \vec{A}_1 = \vec{A}_2,$$

so that $\vec{A}_1$ and $\vec{A}_2$ are equal everywhere.

27. Show that $f(r) = \frac{1}{r}$ is a solution of the equation $\nabla^2 f(r) = 0$.

[C.U. (Hons.) 1973, 2001, '05, '07]

Let us evaluate $\nabla^2\left(\frac{1}{r}\right)$, where r is the magnitude of the position vector of any point (x, y, z), so that $r = (x^2 + y^2 + z^2)^{1/2}$, and

$$\nabla^2\left(\frac{1}{r}\right) = \left(\frac{\partial^2}{\partial x^2} + \frac{\partial^2}{\partial y^2} + \frac{\partial^2}{\partial z^2}\right)(x^2 + y^2 + z^2)^{-1/2}$$

Now

$$\frac{\partial}{\partial x}(x^2 + y^2 + z^2)^{-1/2} = -\frac{1}{2}(x^2 + y^2 + z^2)^{-3/2}(2x),$$

and therefore we may write

$$\begin{aligned}\frac{\partial^2}{\partial x^2}(x^2 + y^2 + z^2)^{-1/2} &= \frac{\partial}{\partial x}\left[-x(x^2 + y^2 + z^2)^{-3/2}\right]\\ &= 3x^2(x^2 + y^2 + z^2)^{-5/2} - (x^2 + y^2 + z^2)^{-3/2}\\ &= \frac{2x^2 - y^2 - z^2}{(x^2 + y^2 + z^2)^{5/2}}\end{aligned}$$

Similarly

$$\frac{\partial^2}{\partial y^2}(x^2 + y^2 + z^2)^{-1/2} = \frac{2y^2 - z^2 - x^2}{(x^2 + y^2 + z^2)^{5/2}}$$

$$\text{and} \quad \frac{\partial^2}{\partial z^2}(x^2 + y^2 + z^2)^{-1/2} = \frac{2z^2 - x^2 - y^2}{(x^2 + y^2 + z^2)^{5/2}}$$

By addition, we obtain

$$\nabla^2\left(\frac{1}{r}\right) = \frac{(2x^2 - y^2 - z^2) + (2y^2 - z^2 - x^2) + (2z^2 - x^2 - y^2)}{(x^2 + y^2 + z^2)^{5/2}} = 0$$

It therefore follows that $f(r) = \frac{1}{r}$ is a solution of $\nabla^2\left(\frac{1}{r}\right) = 0$ which is Laplace's equation.

28. (a) Prove $\nabla^2 f(r) = \dfrac{d^2 f}{dr^2} + \dfrac{2}{r}\dfrac{df}{dr}$ and hence (b) find $f(r)$ such that $\nabla^2 f(r) = 0$.

[C.U. (Hons.) 1991]

(a) Here

$$\nabla^2 f(r) = \frac{\partial^2}{\partial x^2} f(r) + \frac{\partial^2}{\partial y^2} f(r) + \frac{\partial^2}{\partial z^2} f(r)$$

Now

$$\frac{\partial}{\partial x} f(r) = \frac{\partial}{\partial r} f(r) \frac{\partial r}{\partial x} = f'(r) \frac{\partial}{\partial x}(x^2 + y^2 + z^2)^{1/2} = f'(r) \frac{x}{(x^2 + y^2 + z^2)^{1/2}},$$

so that we have

$$\frac{\partial^2}{\partial x^2}f(r) = \frac{\partial}{\partial x}\left[f'(r)\frac{x}{(x^2+y^2+z^2)^{1/2}}\right]$$

$$= \frac{x}{(x^2+y^2+z^2)^{1/2}}\cdot\frac{\partial}{\partial x}f'(r) + f'(r)\frac{1}{(x^2+y^2+z^2)^{1/2}} - \frac{f'(r)\cdot x\cdot\frac{1}{2}\cdot 2x}{(x^2+y^2+z^2)^{3/2}}$$

$$= \frac{x}{(x^2+y^2+z^2)^{1/2}}\cdot\frac{\partial}{\partial r}f'(r)\cdot\frac{\partial r}{\partial x} + \frac{f'(r)}{(x^2+y^2+z^2)^{1/2}} - \frac{f'(r)x^2}{(x^2+y^2+z^2)^{3/2}}$$

$$= \frac{x}{(x^2+y^2+z^2)^{1/2}}\cdot f''(r)\cdot\frac{1}{2}(x^2+y^2+z^2)^{-1/2}\cdot 2x + \frac{f'(r)}{(x^2+y^2+z^2)^{1/2}} - \frac{f'(r)x^2}{(x^2+y^2+z^2)^{3/2}}$$

$$= \frac{x^2 f''(r)}{(x^2+y^2+z^2)} + \frac{f'(r)}{(x^2+y^2+z^2)^{1/2}} - \frac{f'(r)x^2}{(x^2+y^2+z^2)^{3/2}} \tag{1}$$

Similarly

$$\frac{\partial^2}{\partial y^2}f(r) = \frac{y^2 f''(r)}{(x^2+y^2+z^2)} + \frac{f'(r)}{(x^2+y^2+z^2)^{1/2}} - \frac{f'(r)y^2}{(x^2+y^2+z^2)^{3/2}} \tag{2}$$

and

$$\frac{\partial^2}{\partial z^2}f(r) = \frac{z^2 f''(r)}{(x^2+y^2+z^2)} + \frac{f'(r)}{(x^2+y^2+z^2)^{1/2}} - \frac{f'(r)z^2}{(x^2+y^2+z^2)^{3/2}} \tag{3}$$

Adding (1), (2) and (3), we obtain

$$\nabla^2 f(r) = f''(r) + \frac{3f'(r)}{(x^2+y^2+z^2)^{1/2}} - \frac{f'(r)}{(x^2+y^2+z^2)^{1/2}}$$

$$= \frac{d^2 f}{dr^2} + \frac{3}{r}\frac{df}{dr} - \frac{1}{r}\frac{df}{dr}$$

$$= \frac{d^2 f}{dr^2} + \frac{2}{r}\frac{df}{dr}$$

(b) According to the condition of the problem, we have

$$\frac{d^2 f}{dr^2} + \frac{2}{r}\frac{df}{dr} = 0 \quad\text{or}\quad r^2\frac{d^2 f}{dr^2} + 2r\frac{df}{dr} = 0 \quad \text{(multiplying both sides by } r^2\text{)}$$

$$\text{or}\quad \frac{d}{dr}\left(r^2\frac{df}{dr}\right) = 0 \quad\text{or}\quad d\left(r^2\frac{df}{dr}\right) = 0 \quad\text{or}\quad r^2\frac{df}{dr} = C \quad \text{(Integration constant} = C\text{)}$$

$$\text{or}\quad df = C\frac{dr}{r^2} \quad\text{or}\quad f(r) = -\frac{C}{r} + A \quad \text{(integrating and putting } A = \text{ another integration constant)}$$

$$= A + \frac{B}{r} \quad \text{(putting } B = -C\text{)}$$

29. For an arbitrary scalar function ϕ, prove that

$$\oint_S \phi d\vec{S} = \int_V \vec{\nabla}\phi dV,$$

where V is the volume enclosed by the surface S. **[C.U. (Hons.) 1999, 2007]**

From the Divergence theorem, we known that

$$\int_V \vec{\nabla}\cdot\vec{A}dV = \int_S \vec{A}\cdot d\vec{S}, \tag{1}$$

where V is the volume enclosed by the closed surface S and $\vec{A}$ is an arbitrary vector.

Now let us take $\vec{A} = \phi\vec{C}$, $\vec{C}$ being a constant vector.

$\therefore$ from (1), we have

$$\int_V \vec{\nabla}\cdot(\phi\vec{C})dV = \int_S \phi\vec{C}\cdot d\vec{S} \tag{2}$$

Now

$$\vec{\nabla}\cdot(\phi\vec{C}) = (\vec{\nabla}\phi)\cdot\vec{C} = \vec{C}\cdot\vec{\nabla}\phi,$$

so that (2) can be written as

$$\int_S \phi\vec{C}\cdot d\vec{S} = \int_V \vec{\nabla}\cdot(\phi\vec{C})dV \text{ or } \vec{C}\cdot\int_S \phi d\vec{S} = C\cdot\int_V \vec{\nabla}\phi dV \text{ or } \int_S \phi d\vec{S} = \int_V \vec{\nabla}\phi dV$$

$\because \vec{C}$ is any arbitrary constant vector.

30. Prove that when $\vec{B}$ is solenoidal, a vector $\vec{A}$ exists such that $\vec{B} = \vec{\nabla}\times\vec{A}$.

[cf. C.U. (Hons.) 2006]

Here we have

$$\vec{B} = \vec{\nabla}\times\vec{A} \tag{1}$$

Let $\vec{B} = \hat{\imath}B_x + \hat{\jmath}B_y + \hat{k}B_z$ and the unknown vector $\vec{A} = \hat{\imath}A_x + \hat{\jmath}A_y + \hat{k}A_z$; then from (1) we may write

$$\frac{\partial A_z}{\partial y} - \frac{\partial A_y}{\partial z} = B_x \tag{2a}$$

$$\frac{\partial A_x}{\partial z} - \frac{\partial A_z}{\partial x} = B_y \tag{2b}$$

$$\frac{\partial A_y}{\partial x} - \frac{\partial A_x}{\partial y} = B_z \tag{2c}$$

Let us now choose $\vec{A}$ in the yz plane, i.e., $A_x = 0$, then

$$B_y = -\frac{\partial A_z}{\partial x} \quad \text{and} \quad B_z = \frac{\partial A_y}{\partial x}$$

which on integration gives

$$A_y = \int_0^x B_z dx + f_2(y,z) \tag{3}$$

$$\text{and} \quad A_z = -\int_{x_0}^x B_y dx + f_3(y,z) \tag{4}$$

where f_2 and f_3 are arbitrary functions of y and z, but are not functions of x. Then equation (2a) gives

$$\frac{\partial A_z}{\partial y}-\frac{\partial A_y}{\partial z}=-\int_{x_0}^{x}\left(\frac{\partial B_y}{\partial y}+\frac{\partial B_z}{\partial z}\right)dx+\frac{\partial f_3}{\partial y}-\frac{\partial f_2}{\partial z}$$

$$=\int_{x_0}^{x}\frac{\partial B_x}{\partial x}+\frac{\partial f_3}{\partial y}-\frac{\partial f_2}{\partial z}\quad(\text{using } \vec{\nabla}\cdot\vec{B}=0) \tag{5}$$

$$=B_x(x,y,z)-B_x(x_0,y,z)+\frac{\partial f_3}{\partial y}-\frac{\partial f_2}{\partial z} \tag{6}$$

Let us now choose

$$f_2=0,\quad \text{and}\quad f_3=\int_{y_0}^{y}B_x(x_0,y,z)dy,$$

so that right-hand side of (6) reduces to $B_x(x,y,z)$ in agreement with equation (2a). With this choice of f_2 and f_3, we can construct A as follows

$$\vec{A}=\hat{j}\int_{x_0}^{x}B_z(x,y,z)dx+\hat{k}\left[\int_{y_0}^{y}B_x(x,y,z)dy-\int_{x_0}^{x}B_y(x,y,z)dx\right]$$

and hence $\vec{A}$ exists.

31. Prove that $\int_C \vec{r}\times d\vec{r}=2\int_S d\vec{S}$, where S is a surface enclosed by a line C.

If $\vec{b}$ be a constant vector, then application of Stoke's theorem to the function $\vec{b}\times\vec{r}$ gives

$$\int_C \vec{b}\times\vec{r}\cdot d\vec{r}=\int_S \operatorname{curl}(\vec{b}\times\vec{r})\cdot d\vec{S}=\int\vec{\Delta}\times(\vec{b}\times\vec{r})\cdot d\vec{S} \tag{1}$$

Here if $\vec{b}=\hat{i}b_1+\hat{j}b_2+\hat{k}b_3$ and $\vec{r}=\hat{i}x+\hat{j}y+\hat{k}z$, then

$$\vec{\nabla}\times(\vec{b}\times\vec{r})=\left(\hat{i}\frac{\partial}{\partial x}+\hat{i}\frac{\partial}{\partial y}+\hat{i}\frac{\partial}{\partial z}\right)\times\begin{vmatrix}\hat{i}&\hat{j}&\hat{k}\\ b_1&b_2&b_3\\ x&y&z\end{vmatrix}$$

$$=\left(\hat{i}\frac{\partial}{\partial x}+\hat{i}\frac{\partial}{\partial y}+\hat{i}\frac{\partial}{\partial z}\right)\times\hat{i}(b_2z-b_3y)+\hat{j}(b_3x-b_1z)+\hat{k}(b_1y-b_2x)$$

$$=\begin{vmatrix}\hat{i}&\hat{j}&\hat{k}\\ \dfrac{\partial}{\partial x}&\dfrac{\partial}{\partial y}&\dfrac{\partial}{\partial z}\\ (b_2z-b_3y)&(b_3x-b_1z)&(b_1y-b_2x)\end{vmatrix}$$

$$=2(b_1\hat{i}+b_2\hat{j}+b_3\hat{k})=2\vec{b}$$

Substituting this value in (1), we have

$$\int_C \vec{b}\times\vec{r}\cdot d\vec{r}=2\int_S \vec{b}\cdot d\vec{S}\quad\text{or}\quad\int_C \vec{b}\cdot\vec{r}\times d\vec{r}=2\int_S \vec{b}\cdot d\vec{S}$$

Since $\vec{b}$ is a constant vector, we can write

$$\vec{b} \cdot \int_C \vec{r} \times d\vec{r} = 2\vec{b} \cdot \int_S d\vec{S} \quad \text{or} \quad \vec{b} \cdot \left[\int_C \vec{r} \times d\vec{r} - 2 \int_S d\vec{S} \right] = 0$$

Since $\vec{b} \neq 0$, we may write

$$\int_C \vec{r} \times d\vec{r} - 2 \int_S d\vec{S} = 0, \quad \text{i.e.,} \quad \int_C \vec{r} \times d\vec{r} = 2 \int_S d\vec{S}$$

2.29 Hamilton's Action Integral

The time integral of the Lagrangian function is called Hamilton's action integral and may be written as:

$$S = \int_{t_1}^{t_2} Ldt,$$

where t_1 denotes the initial time in which the system occupies a position denoted by the set of values of generalised co-ordinates $q_{i1} = (q_{11}, q_{21}, q_{31}, \cdots)$ and generalised velocities $\dot{q}_{i1} = (\dot{q}_{11}, \dot{q}_{21}, \dot{q}_{31}, \cdots)$ and t_2 denotes the time at which the position of the system is determined by the set of generalised co-ordinates $q_{i2} = (q_{12}, q_{22}, q_{32}, \cdots)$ and generalised velocities $\dot{q}_{i2} = (\dot{q}_{12}, \dot{q}_{22}, \dot{q}_{32}, \cdots)$.

2.30 Hamilton's Principle

Let us now suppose that the system starts from the position $P(q_{i1})$ at time t_1 and reaches the position $Q(q_{i2})$ at time t_2 by the path PRQ (Fig. 2.30.1). If now the system S follows a virtually displaced path (shown by the dotted line), the initial and the final time remaining the same, then according to Hamilton's principle the virtual change (δ) of S is zero. Thus we can summarise Hamilton's principle by saying that the motion of the system from time t_1 to t_2 is such that the action integral is an extremum for the path of motion. Hence

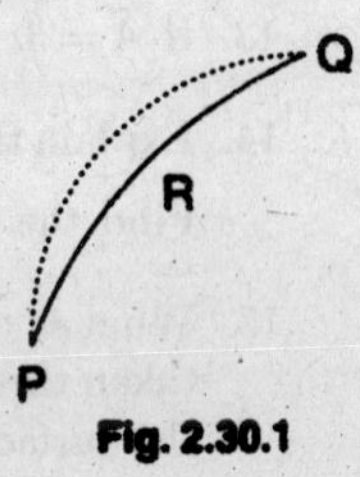

Fig. 2.30.1

$$\delta S = \delta \int_{t_1}^{t_2} Ldt = 0.$$

From this principle Lagrange's equations of motion can be derived. For a conservative system $\int_{t_1}^{t_2} Ldt$ is minimum and its value is

$$\int_{t_1}^{t_2} 2Tdt.$$

Thus for a conservative system we may write

$$\delta \int_{t_1}^{t_2} 2Tdt = 0.$$

This is called the principle of least action.

Exercises

1. State which of the following are scalars and which are vectors: (a) temperature, (b) specific heat, (c) speed, (d) momentum, (e) energy, (f) magnetic field.

2. Show that addition of vectors is commutative.

3. Define unit vectors. Given three unit vectors $\hat{\imath}, \hat{\jmath}$ and $\hat{k}$, find an expression for any vector $\vec{r}$ in three dimensional space.

4. Prove $\vec{A} \cdot \vec{B} = \vec{B} \cdot \vec{A}$.

5. If $\vec{A} = A_1\hat{\imath} + A_2\hat{\jmath} + A_3\hat{k}$ and $\vec{B} = B_1\hat{\imath} + B_2\hat{\jmath} + B_3\hat{k}$, prove that $\vec{A} \cdot \vec{B} = A_1B_1 + A_2B_2 + A_3B_3$.

6. Show that $\hat{\imath} \cdot \hat{\jmath} = \hat{\jmath} \cdot \hat{k} = \hat{k} \cdot \hat{\imath} = 0$ and $\hat{\imath} \cdot \hat{\imath} = \hat{\jmath} \cdot \hat{\jmath} = \hat{k} \cdot \hat{k} = 1$.

7. If $\vec{A} \cdot \vec{B} = 0$, show that $\vec{A}$ and $\vec{B}$ are perpendicular to each other, provided $\vec{A}$ and $\vec{B}$ are not zero.

8. Prove that $\vec{A} \times \vec{B} = -\vec{B} \times \vec{A}$.

9. If $\vec{A} \times \vec{B} = 0$ and if $\vec{A}$ and $\vec{B}$ are not zero, prove that $\vec{A}$ is parallel to $\vec{B}$.

10. Show that $\hat{\imath} \times \hat{\imath} = \hat{\jmath} \times \hat{\jmath} = \hat{k} \times \hat{k} = 0$ and $\hat{\imath} \times \hat{\jmath} = -\hat{\jmath} \times \hat{\imath} = \hat{k}$.

11. If $\vec{A} = A_1\hat{\imath} + A_2\hat{\jmath} + A_3\hat{k}$ and $B = B_1\hat{\imath} + B_2\hat{\jmath} + B_3\hat{k}$ prove that

$$\vec{A} \times \vec{B} = \begin{vmatrix} \hat{\imath} & \hat{\jmath} & \hat{k} \\ A_1 & A_2 & A_3 \\ B_1 & B_2 & B_3 \end{vmatrix}$$

12. Show that $\vec{A} \times (\vec{B} \times \vec{C}) = \vec{B}(\vec{A} \cdot \vec{C}) - \vec{C}(\vec{A} \cdot \vec{B})$.

13. If $\vec{A} = 3\hat{\imath} - \hat{\jmath} + 2\hat{k}, \vec{B} = 2\hat{\imath} + \hat{\jmath} - \hat{k}$, show that $\vec{A} \times \vec{B} = -\hat{\imath} + 7\hat{\jmath} + 5\hat{k}$.

14. Explain the following terms used in vector algebra: (a) dot product (b) cross product (c) gradient. Find the values of $\vec{A} \cdot (\vec{B} \times \vec{C})$ and $\vec{A} \times (\vec{B} \times \vec{C})$, where A, B and C are vectors.

15. What is divergence of a vector? Prove that the volume integral of a vector $\vec{A}$ in a vector field, taken through a given volume, is equal to the surface integral of the normal component of $\vec{A}$ over the surface enclosing the volume.

    Prove that $\nabla^2(1/r) = 0$, where $r = (x^2 + y^2 + z^2)^{1/2}$. The symbols have usual meanings.

    [C.U. (Hons.) 1973]

16. State Stokes's theorem.

17. If $\vec{r} = x\hat{\imath} + y\hat{\jmath} + z\hat{k}$, prove that div $\vec{r} = 3$.

18. Prove that curl$(\vec{a} \times \vec{r}) = 2\vec{a}$.

19. Show that the force on a charge q moving with velocity $\vec{v}$ is

$$\vec{F} = q\left[\vec{E} + \frac{\vec{v} \times \vec{B}}{c}\right],$$

$\vec{E}$ and $\vec{B}$ being the electric and the magnetic fields and c denotes the velocity of light in vacuum.

20. Find the value of p which will make the following three vectors co-planar $\vec{A} = 3\hat{\imath} + 2\hat{\jmath} + \hat{k}$, $\vec{B} = 3\hat{\imath} + 4\hat{\jmath} + 5\hat{k}$, $C = \hat{\imath} + \hat{\jmath} - p\hat{k}$. **[C.U. (Hons.) 1980]**

[**Hints.** To make the three vectors co-planar, put $\vec{A}\cdot(\vec{B} \times \vec{C}) = 0$.] [**Ans.** $p = -1$]

21. State what is meant by an irrotational vector field and explain how it can be related to a scalar field. Derive the equation satisfied by such a scalar field, when the parent vector field is also solenoidal. What is the name of this partial differential equation? Give two examples from Physics where this equation is satisfied. **[C.U. (Hons.) 1980]**

[**Ans.** See the chapter on Partial Differential Equations.]

22. (a) The directions of two unit vectors $\hat{n}_1$ and $\hat{n}_2$ are given by spherical polar angles (θ_1, ϕ_1) and (θ_2, ϕ_2) respectively. Using scalar product of $\hat{n}_1$ and $\hat{n}_2$, show that the angle γ between these unit vectors is given by $\cos\gamma = \cos\theta_1 \cos\theta_2 + \sin\theta_1 \sin\theta_2 \cos(\phi_1 - \phi_2)$. **[C.U. (Hons.) 1984]**

[**Hints.** $\hat{n}_1 = \hat{\imath}\sin\theta_1\cos\phi_1 + \hat{\jmath}\sin\theta_1\sin\phi_1 + \hat{k}\cos\theta_1$;
$\hat{n}_2 = \hat{\imath}\sin\theta_2\cos\phi_2 + \hat{\jmath}\sin\theta_2\sin\phi_2 + \hat{k}\cos\theta_2$.
$\therefore\ \hat{n}_1\cdot\hat{n}_2 = |n_1||n_2|\cos\gamma$. $\quad\therefore\ \cos\gamma = \hat{n}_1\cdot\hat{n}_2$.]

(b) Let $\hat{r}$ be the unit vector along the radial direction. Given a vector $\vec{A}$, write down the radial $\vec{A}_r$ and tangential $\vec{A}_t$ vectors in terms of scalar and vector products of $\vec{A}$ and $\hat{r}$. **[C.U. (Hons.) 1984]**

23. Establish the vector identity $\vec{\nabla} \times (\vec{A} \times \vec{B}) = (\vec{B}\cdot\vec{\nabla})\vec{A} - (\vec{A}\cdot\vec{\nabla})\vec{B} - \vec{B}(\vec{\nabla}\cdot\vec{A}) + \vec{A}(\vec{\nabla}\cdot\vec{B})$.

Utilise the above identity to show that the curl of the linear velocity of a mass point of a rotating rigid body is twice the angular velocity. **[C.U. (Hons.) 1986]**

24. (a) If $\vec{R}(t)$ be a vector from the origin to the instantaneous position of a particle moving along a curve C, show that $\vec{R} \times d\vec{R}$ is equal to twice the area of the sector defined by the two vectors $\vec{R}(t), \vec{R}(t + dt) = \vec{R} + d\vec{R}$ and the arc of C, which they intercept.
(For answer see Art. 3.14 and Art. 7.22.1)

(b) Hence show that

i. if $\vec{R} \times d\vec{R} = 0$, the vector $\vec{R}$ has a constant direction.
[**Hints**: Take $\vec{R} = \hat{\imath}\cos\omega t + \hat{\jmath}\sin\omega t$; find $\frac{d\vec{R}}{dt}$ and then show that $\vec{R} \times \frac{d\vec{R}}{dt} = \omega\hat{k}$, a constant vector.]

ii. if $\vec{R} \times \frac{d^2\vec{R}}{dt^2} = 0$, the particle moves so that its radius vector sweeps out equal areas in equal times. **[C.U. (Hons.) 1995]**
(For answer See Art. 3.14)

(c) If $\phi(x, y, z) = xy^2 + yz^3$ be a scalar function, find the component of grad ϕ at the point $(2, -1, 1)$ in the direction of the vector $\hat{\imath} - 2\hat{\jmath} + 2\hat{k}$. **[C.U. (Hons.) 1995]**

25. (a) Show that the area of a triangle ABC formed by the vectors $\overrightarrow{AB}$ and $\overrightarrow{AC}$ is given by $\Delta ABC = \frac{1}{2}(\overrightarrow{AB} \times \overrightarrow{AC})$.

(b) The co-ordinates of the three vertices of a triangle are $(2, 1, 5), (5, 2, 8)$ and $(4, 8, 2)$. Compute the area.

(c) Explain clearly the difference between a *polar* and an *axial* vector, giving examples. **[C.U. (Hons.) 1999]**

(d) Write down expressions for div $\vec{A}$ and curl $\vec{A}$ in terms of spherical polar co-ordinates. A central field $\vec{A}$ is defined by $\vec{A} = \vec{e}_r f(r)$ where $\vec{e}_r$ is the unit vector in the direction of $\vec{r}$. Determine $f(r)$ so that $\vec{A}$ may be irrotational or solenoidal. **[C.U. (Hons.) 1996]**
(For answer see Art. 3.14 and worked out examples)

26. (a) Find a unit vector along a certain direction in terms of its direction cosines (l, m, n).

[**Ans.** $\vec{A} = A_x\hat{\imath} + A_y\hat{\jmath} + A_z\hat{k} = A\cos\alpha\hat{\imath} + A\cos\beta\hat{\jmath} + A\cos\gamma\hat{k} = Al\hat{\imath} + Am\hat{\jmath} + An\hat{k}$ or $\frac{\vec{A}}{|\vec{A}|} = l\hat{\imath} + m\hat{\jmath} + n\hat{k}$]

(b) Two lines have direction cosines (l_1, m_1, n_1) and (l_2, m_2, n_2). Show that the angle θ between the lines is given by $\cos\theta = l_1l_2 + m_1m_2 + n_1n_2$. **[C.U. (Hons.) 1997, 2000]**

[**Ans.** The unit vectors along the two lines are

$$\frac{\vec{r}_1}{|r_1|} = \hat{r}_1 = l_1\hat{\imath} + m_1\hat{\jmath} + n_1\hat{k} \quad \text{and} \quad \hat{r}_2 = l_2\hat{\imath} + m_2\hat{\jmath} + n_2\hat{k} = \frac{\vec{r}_2}{|r_2|}$$

$\therefore\ \vec{r}_1\cdot\vec{r}_2 = |r_1||r_2|\cos\theta$ or $\cos\theta = (l_1\hat{\imath} + m_1\hat{\jmath} + n_1\hat{k})\cdot(l_2\hat{\imath} + m_2\hat{\jmath} + n_2\hat{k}) = l_1l_2 + m_1m_2 + n_1n_2$]

(c) It is known that $\vec{\nabla}\cdot\vec{B} = 0$. where $\vec{B}$ is the magnetic induction. Interpret the result.

[**Ans.** See Art 2.25]

(d) Show that $\vec{r}\cdot d\vec{r} = rdr$. **[C.U. (Hons.) 1997]**

[**Ans.** $\vec{r}\cdot d\vec{r} = xdx + ydy + zdz = \frac{1}{2}(2xdx + 2ydy + 2zdz) = \frac{1}{2}d(r^2) = rdr$.]

27. (a) For an arbitrary scalar function ϕ and vector function $\vec{A}$, prove that $\vec{\nabla}\cdot(\phi\vec{A}) = (\vec{\nabla}\phi)\cdot\vec{A} + \phi(\vec{\nabla}\cdot\vec{A})$. [**Ans.** See Art. 2.24(v)]

(b) From Gauss's divergence theorem, prove that for two scalar functions u and v,

$$\int u\vec{\nabla}v\cdot d\vec{S} = \int u\vec{\nabla}\cdot(\vec{\nabla}v)dV + \int \vec{\nabla}u\cdot\vec{\nabla}v dV$$

where the volume V is enclosed by surface S. [**Ans.** See worked out examples]

(c) Two vector fields $\vec{A}_1$ and $\vec{A}_2$ have the same divergence and curl at every point in volume V and have the same normal component at every point on the surface S enclosing V. Prove that $\vec{A}_1$ and $\vec{A}_2$ are equal every where. **[C.U. (Hons.) 1998]**

(See worked out examples.)

28. (a) $\vec{a}_1$, $\vec{a}_2$, $\vec{a}_3$ are three non-coplanar vectors. Three other vectors $\vec{b}_1$, $\vec{b}_2$, $\vec{b}_3$ are related to the former vectors by $\vec{a}_i\cdot\vec{b}_j = \delta_{ij}(ij = 1, 2, 3)$ where δ is known as Kronečker delta. Show that

$$\vec{b}_1 = \frac{\vec{a}_2\times\vec{a}_3}{\vec{a}_1\cdot(\vec{a}_2\times\vec{a}_3)}$$ **[C.U. (Hons.) 1990]**

[**Ans.** See article 2.26.]

(b) If $(\vec{a}_1, \vec{b}_1, \vec{c}_1)$ and $(\vec{a}_2, \vec{b}_2, \vec{c}_2)$ are reciprocal to each other, then prove that

$$\vec{a}_1\times\vec{a}_2 + \vec{b}_1\times\vec{b}_2 + \vec{c}_1\times\vec{c}_2 = 0.$$ [**Ans.** See Article 2.26.]

29. (a) Find out a unit vector perpendicular to the two vectors $-\hat{\imath} + \hat{\jmath} + \hat{k}$ and $2\hat{\imath} - 3\hat{\jmath} + 2\hat{k}$.

(b) If $\vec{r}$ be the co-ordinate vector, prove that for a central force $\vec{F} = f(r)\vec{r}$, $\vec{\nabla}\times\vec{F} = 0$.

(c) For an arbitrary scalar function ϕ, prove that

$$\oint_s \phi d\vec{S} = \int_V \vec{\nabla}\phi dV,$$

where V is the volume enclosed by the surface S.

(See worked out examples.)

(d) Find the line integral of the vector $\vec{F}$ along the line segment joining the points (0, 0, 0) and (2, 3, 4) — $\vec{F} = x^2\hat{\imath} + y^2\hat{\jmath} + (xz - y)\hat{k}$. **[C.U. (Hons.) 1999, 2007]**

(See worked out examples.)

30. (a) Prove the identity $\vec{\nabla} \times (\vec{\nabla} \times \vec{A}) = \vec{\nabla}(\vec{\nabla} \cdot \vec{A}) - \vec{\nabla}^2 A$. **[C.U. (Hons.) 2000]**

(b) State Stokes' theorem. Using this theorem prove that

$$\oint_C \phi d\vec{r} = \iint_S d\vec{S} \times \vec{\nabla}\phi,$$

where ϕ is a scalar function of $\vec{r}$, S is an open surface bounded by a closed curve. **[C.U. (Hons.) 1993, 2000]**

Answer of (b) We consider $\vec{F} = \vec{a}\phi$, where $\vec{a}$ is a constant vector and ϕ is a differentiable scalar point function.

Then

$$\int_C (\vec{a}\phi) \cdot d\vec{r} = \iint_S \text{curl}(\vec{a}\phi) \cdot \hat{n} dS \text{ (using Stokes' theorem)}$$

$$= \iint_S (\phi \, \text{curl}\, \vec{a} - \vec{a} \times \text{grad}\, \phi) \cdot \hat{n} dS = \iint_S (\vec{\nabla}\phi \times \vec{a}) \cdot \hat{n} dS$$

$$= -\iint_S \vec{a} \cdot (\vec{\nabla}\phi \times \hat{n}) dS = -\vec{a} \cdot \iint_S (\vec{\nabla}\phi \times \hat{n}) dS$$

$$\text{or} \quad \vec{a} \cdot \int_C \phi d\vec{r} = \vec{a} \cdot \iint_S (\hat{n} \times \vec{\nabla}\phi) dS$$

$$\text{or} \quad \int_C \phi dr = \iint_S d\vec{S} \times \vec{\nabla}\phi \quad \text{(since } \vec{a} \text{ is arbitrary)}$$

31. Prove the identity $(\vec{A} \cdot \vec{\nabla})\vec{A} = \frac{1}{2}\vec{\nabla}A^2 - \vec{A} \times (\vec{\nabla} \times \vec{A})$. **[C.U. (Hons.) 1993]**

[**Ans.** See worked out examples.]

32. Verify divergence theorem for a vector $\vec{A} = \hat{\imath}\frac{x}{r} + \hat{\jmath}\frac{y}{r} + \hat{k}\frac{z}{r} = \frac{\vec{r}}{r}$, where $\hat{\imath}, \hat{\jmath}, \hat{k}$ are the unit vectors along X, Y, Z — the region of integration being a sphere of radius a with centre at the origin. (For answer see worked out examples) **[C.U. (Hons.) 1980]**

33. Show that each of the three equations

$$\vec{F} = \vec{\nabla}\phi, \quad \vec{\nabla} \times \vec{F} = 0, \quad \oint \vec{F} \cdot d\vec{r} = 0,$$

implies the other two.

34. Prove that when $\vec{B}$ is solenoidal, a vector $\vec{A}$ exists such that $\vec{B} = \vec{\nabla} \times \vec{A}$. (For answer see worked out examples)

35. If $\vec{F} = 3xy\hat{\imath} - y^2\hat{\jmath}$, evaluate $\int_C \vec{F} \cdot d\vec{r}$ where C is the curve in the $x - y$ plane, $y = 2x^2$, from $(0,0)$ to $(1,2)$. [**Ans.** $-\frac{7}{6}$.]

36. Verify the divergence theorem for $\vec{A} = 4x\hat{\imath} - 2y^2\hat{\jmath} + z^2\hat{k}$ taken over the region bounded by $x^2 + y^2 = 4$, $z = 0$, $z = 3$.

37. Prove Stokes' theorem for the vector $\vec{V}(x + y,\ 2x - z,\ y + z)$ taken over the triangle ABC cut from the plane $3x + 2y + z = 6$ by the co-ordinate axes.

38. Verify divergence theorem for the vector $\vec{V} = x^2\hat{\imath} + y^2\hat{\jmath} + z^2\hat{k}$ taken over the cube $0 \le x,\ y,\ z \le 1$.

39. Verify Stokes' theorem for $\vec{A} = (2x - y)\hat{\imath} - yz^2\hat{\jmath} - y^2z\hat{k}$, where S is the upper half surface of the sphere $x^2 + y^2 + z^2 = 1$ and C is the boundary.

40. Evaluate $\int_S \vec{A}\cdot d\vec{S}$ where $\vec{A} = \hat{\imath}x + \hat{\jmath}y + \hat{k}z^2$ and S is the closed surface bounded by $x^2 + y^2 = z^2$ and the plane $z = 4$. [**Ans.** $\frac{128\pi}{3}$.]

41. Given $\vec{V} = 4y\hat{\imath} + x\hat{\jmath} + 2z\hat{k}$, find $\iint_S (\vec{\nabla} \times \vec{V})\cdot d\vec{S}$ over the hemisphere $x^2 + y^2 + z^2 = a^2, z \geq 0$.

[**Ans.** $-3\pi a^2$.]

42. Evaluate $\iiint_V \vec{F}dV$, where $\vec{F} = 2xz\hat{\imath} - x\hat{\jmath} + y^2\hat{k}$ and V is the region bounded by the surfaces $x = 0$, $y = 6$, $z = x^2$ and $z = 4$. [**Ans.** $128\hat{\imath} - 24\hat{\jmath} - 384\hat{k}$]

43. (a) Prove the following relations

(i) $\vec{\nabla}r^n = nr^{n-2}\vec{r}$, (ii) $\vec{\nabla}^2r^n = n(n+1)r^{n-2}$, (iii) $\vec{\nabla}\cdot\frac{\vec{r}}{r^3} = 0$, (iv) $\vec{\nabla}\times\vec{r} = 0$.

[C.U. (Hons.) 1991]

(b) A vector $\vec{a} = \vec{a}(t)$ has a constant magnitude (t = time). Show that $\vec{a}\cdot\frac{d\vec{a}}{dt} = 0$. Indicate the relevance of this relation in case of the motion of a particle in a circular orbit.

[**Ans.** See worked examles] **[C.U. (Hons.) 1985, 1991, 2007]**

44. (a) Show that the vector $\vec{F} = yz\hat{\imath} + zx\hat{\jmath} + xy\hat{k}$ has path independent line integral. Find the corresponding scalar function ϕ from which $\vec{F}$ may be obtained. Take due care in obtaining the constant.

(For answer see worked out examples)

(b) Calculate the amount of work done in moving a particle around a closed path in a force field given by $\vec{F} = f(r)\vec{r}$, where $f(r) = K/r^3$ (K = constant).

[**Ans.** See Art. 3.14] **[C.U. (Hons.) 1991]**

45. (a) Find the unit vector perpendicular to $\vec{A} = 4\hat{\imath} - \hat{\jmath} + 3\hat{k}$ and $\vec{B} = -2\hat{\imath} + \hat{\jmath} - 2\hat{k}$.

(b) Show that $f(r) = \frac{1}{r}$ is a solution of the equation $\nabla^2 f(r) = 0$.

(c) If $\vec{F}$ is a conservative field, prove that $\vec{F}$ can be expressed as the gradient of a scalar potential using Stokes' theorem. (See Art 3.13)

(d) $ABCD$ is a square in the $z = 3$ plane. The x, y co-ordinates of A, B, C and D are $(0,0)$, $(0,a)$, (a,a) and $(a,0)$ respectively. Show that the line integral of $\vec{F} = (2xy + z^3)\hat{\imath} + x^2\hat{\jmath} + 3xz^2\hat{k}$ is zero over $ABCDA$. What is the corresponding scalar potential?

(See worked-out problems) **[C.U. (Hons.) 2001]**

46. (a) State Stoke's theorem. **[C.U. (Hons.) 2002]**

(b) In a vector field $\vec{A} = \vec{A}(x,y,z)$, a vector $\vec{B}$ is defined as $\vec{B} = \vec{\nabla}\times\vec{A}$. Consider any two surfaces S_1 and S_2 that have same contour L generated in the same direction (fig). Show that the flux of $\vec{B}$ over the two surfaces are same.

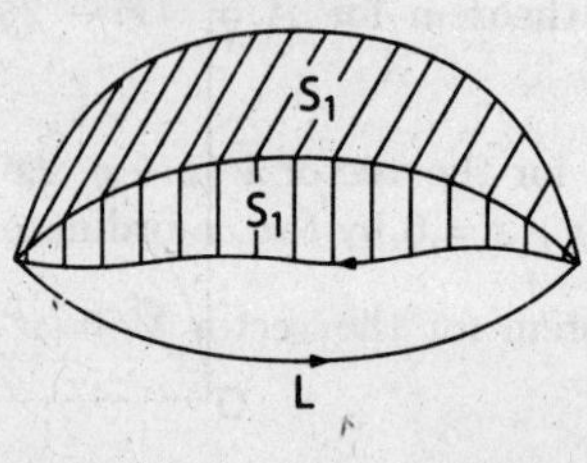

[**Ans.** If the flux through the surface S_1 and S_2 be ϕ_1 and ϕ_2 respectively, we may write

$$\phi_1 = \int_{S_1} \vec{B}\cdot\hat{n}dS_1 \quad \text{and} \quad \phi_2 = \int_{S_2} \vec{B}\cdot\hat{n}dS_2$$

Using $\vec{B} = \vec{\nabla}\times\vec{A}$ and applying Stoke's theorem, we have

$$\phi_1 = \int_{S_1}\left(\vec{\nabla}\times\vec{A}\right)\cdot\hat{n}dS_1 = \oint_L \vec{A}\cdot d\vec{r} \quad \text{and} \quad \phi_2 = \int_{S_2}\left(\vec{\nabla}\times\vec{A}\right)\cdot\hat{n}dS_2 = \oint_L \vec{A}\cdot d\vec{r}$$

$\therefore \phi_1 = \phi_2$.]

(c) $\oint_S d\vec{S} = 0$, where S is a closed surface.

[**Ans.**

$$\oint d\vec{S} = \int_{S_1} d\vec{S} + \int_{S_2} d\vec{S}$$ where S_1 and S_2 are the two parts of the closed surface as shown in fig 3 of example 14 on page 48

$$= \int_{S_1}\hat{n}dS + \int_{S_2} -\hat{n}dS.$$

So making $S_1 \to S_2$, $\oint d\vec{S} = 0$.

(d) What is the difference between $\vec{\nabla}\cdot\vec{u}$ and $\vec{u}\cdot\vec{\nabla}$ where $\vec{u}$ is a vector point function?

[**Ans.**

$$\vec{\nabla}\cdot\vec{u} = \text{div}\,\vec{u} = \frac{\partial u_x}{\partial x} + \frac{\partial u_y}{\partial y} + \frac{\partial u_z}{\partial z}$$

which is a scalar quantity.

Again

$$\vec{u}\cdot\vec{\nabla} = \left(\hat{i}u_x + \hat{j}u_y + \hat{k}u_z +\right)\cdot\left(\hat{i}\frac{\partial}{\partial x} + \hat{j}\frac{\partial}{\partial y} + \hat{k}\frac{\partial}{\partial z}\right) = \frac{\partial u_x}{\partial x} + \frac{\partial u_y}{\partial y} + \frac{\partial u_z}{\partial z}$$

(using commutative rule) $\to$ which is same as $\vec{\nabla}\cdot\vec{u}$ and a scalar quality.

(e) Evaluate $\int \vec{A}\cdot d\vec{r}$ around the closed path shown in the figure below where

$$\vec{A} = (x+y)\hat{i} + (x-y)\hat{j}.$$

Here $\vec{A}\cdot d\vec{r} = (x+y)dx + (x-y)dy$ $\left[\vec{r} = \hat{i}x + \hat{j}y \text{ or } d\vec{r} = \hat{i}dx = \hat{j}dy\right]$

Along path 1

$$\int_0^1 \vec{A}\cdot d\vec{r} = \int_0^1 (y^2+y)2ydy + \int_0^1 (y^2-y)dy \quad [\because x = y^2 \text{ or } dx = 2ydy]$$

$$= \int_0^1 2y^3dy + \int_0^1 2y^2dy + \int_0^1 y^2dy - \int_0^1 ydy$$

$$= \left(\frac{y^4}{2}\right)_0^1 + \frac{2}{3}(y^3)_0^1 + \left(\frac{y^3}{3}\right)_0^1 - \left(\frac{y^2}{2}\right)_0^1 = +1$$

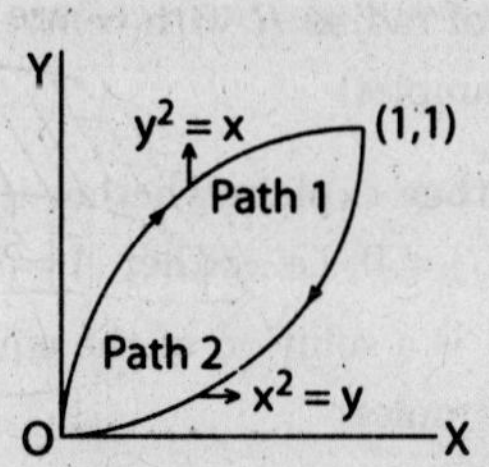

Similarly along path 2,

$$\int_1^0 \vec{A}\cdot d\vec{r} = \int_1^0 (x+x^2)dx + \int_1^0 (x-x^2)2xdx \quad [\because x^2 = y \text{ or } 2xdx = dy]$$

$$= \int_1^0 x\,dx + \int_1^0 x^2\,dx + 2\int_1^0 x^2 dx - 2\int_1^0 x^3 dx = -1, \text{ so that } \oint \vec{A}\cdot d\vec{r} = 1 - 1 = 0$$

which is therefore a path independent line integral.

(f) Show that $\hat{e}_r \cdot d\vec{r} = dr$ where the notations have usual significance.

Let $\vec{r} = \hat{i}x + \hat{j}y + \hat{k}z$ and a unit vector along $\vec{r}$ be $\hat{e}_r$ i.e. $\hat{r}$;

$$\hat{e}_r \text{ i.e. } \hat{r} = \frac{\vec{r}}{r} = \hat{i}\frac{x}{r} + \hat{j}\frac{y}{r} + \hat{k}\frac{z}{r} \text{ where } r = \sqrt{x^2+y^2+z^2}$$

$$\hat{e}_r \cdot d\vec{r} = \left(\hat{i}\frac{x}{r} + \hat{j}\frac{y}{r} + \hat{k}\frac{z}{r}\right)\cdot\left(\hat{i}dx + \hat{j}dy + \hat{k}dz\right) = \frac{xdx + ydy + zdz}{r}$$

Again $r = (x^2+y^2+z^2)^{1/2}$.

$$\therefore \quad dr = \frac{1}{2}\cdot\frac{2(xdx+ydy+zdz)}{(x^2+y^2+z^2)^{1/2}} = \frac{xdx+ydy+zdz}{r}$$

$\therefore \hat{e}_r \cdot d\vec{r} = dr.$

47. (a) i. Let $\vec{r}$ be the position vector and $\vec{a}$ be a constant vector. Find $\vec{\nabla}\cdot\vec{r}$ and $\vec{\nabla}(\vec{a}\cdot\vec{r})$.

ii. What is the curl of $\hat{i}x^2 + \hat{j}y^2 + \hat{k}(x^2 - y^2)$?

iii. A solid thin disc is rotating with a constant angular velocity $\vec{\omega}$ about the symmetry axis passing through the centre of the disc and perpendicular to the plane of the disc. Write an expression for the velocity field of the points on the disc. What is the magnitude and direction of the curl of the velocity field?
(See worked out examples)

(b) Use Stoke's theorem to show that the line integral of $\vec{F} = -\hat{i}y + \hat{j}x$ around a closed curve in the xy-plane is twice the area enclosed by the curve.
(See worked out examples) **[C.U. (Hons.) 2003]**

48. (a) Find the unit vector along a line with direction cosines (α, β, γ).

(b) For a radius vector $\vec{r}$ prove that for the vector field $\vec{F} = f(r)\hat{e}_r$, $\vec{\nabla}\times\vec{F} = 0$, where $\hat{e}_r$ is a unit vector along $\vec{r}$.
(See Art 3.14)

(c) It is known that $\vec{\nabla}\cdot\vec{B} = 0$ always, where $\vec{B}$ is the magnetic induction. Explain the nature of $\vec{B}$.
(See Art. 2.25)

(d) State and explain Gauss's divergence theorem for a vector field $\vec{A} = \vec{A}(x,y,z)$, having continuous first derivatives.

(e) Verify Gauss' theorem for the vector field $\vec{r} = \hat{i}x + \hat{j}y + \hat{k}z$ over a surface enclosing the volume of a sphere of radius R with centre at $O(0,0,0)$. **[C.U. (Hons.) 2004]**
(See worked out examples)

49. (a) If $\vec{A}\times\vec{B} = \vec{A}\times\vec{C}$, then explain whether it always implies $\vec{B} = \vec{C}$.
[**Hints.** $\vec{A}\times(\vec{B}-\vec{C}) = 0$, i.e., either $\vec{A} = 0$ or $\vec{B} = \vec{C}$]

(b) Show that $f(r) = \frac{1}{r}$ is a solution of the equation $\nabla^2 f(r) = 0$.
(See worked out examples)

(c) State Stoke's theorem. Give an example where this theorem is not valid.

[**Ans.** For an open surface i.e. a surface not bounded by a closed line]

(d) For an arbitrary scalar function ϕ and vector function $\vec{A}$, prove that

$$\vec{\nabla}\cdot(\phi\vec{A}) = (\vec{\nabla}\phi\cdot\vec{A}) + \phi(\vec{\nabla}\cdot\vec{A}) \quad \text{(See Art. 2.24)}$$

(e) If $\phi(\vec{r})$ is constant on a surface, prove that $\vec{\nabla}\phi$ is perpendicular to that surface at every point. **[C.U. (Hons.) 2005]**
(See worked out examples)

50. (a) Two particles 1 and 2 travel along x and y axes with respective velocities $\vec{v}_1 = 2\hat{x}$ and $\vec{v}_2 = 3\hat{y}$. At time $t = 0$, $x_1 = -3$, $y_1 = 0$ and $x_2 = 0$, $y_2 = -3$. Show that the vector $\vec{r} = (3-2t)\hat{x} + (3t-3)\hat{y}$ represents the position of particle 2 relative to 1. At what time are the particles closest?

(b) If $\vec{F}$ is a conservative force, prove that $\vec{F}$ can be expressed as the gradient of a scalar potential. Is the potential unique?

(c) If $\vec{A} = 2xz\hat{\imath} - y^2\hat{\jmath} + 2yz\hat{k}$, evaluate $\int_C \vec{A}\cdot d\vec{l}$, where C is the contour of a square with one pair of opposite corners at $(0,0)$ and $(1,1)$ lying in the $z = 3$ plane. What will be the value of $\int_S (\vec{\nabla}\times\vec{A})\cdot d\vec{S}$ where S is the surface of this square? **[C.U. (Hons.) 2005 (old syll.)]**
(See worked out examples)

51. (a) i. In a vector field $\vec{A} = \vec{A}(x,y,z)$, define div $\vec{A}$ at a point (x,y,z) and show that div$\vec{A} = \vec{\nabla}\cdot\vec{A}$.

ii. For magnetic flux density vector $\vec{B}$, $\vec{\nabla}\cdot\vec{B} = 0$ everywhere. What is the significance of this result?

(b) Given that $\vec{B} = \hat{\imath}(x^2+2yz) + \hat{\jmath}2yz - \hat{k}(z^2+2zx)$, find a vector $\vec{A}$ such that $\vec{\nabla}\times\vec{A} = \vec{B}$. Is the solution unique?
(See worked out examples)

(c) Prove $\oint \vec{r}\times d\vec{S} = 0$, where notations have usual significance

[**Ans.** $\oint \vec{r}\times d\vec{S} = \oint \vec{r}\times\hat{n}dS \quad \because \oint d\vec{S} = 0, \quad \oint \vec{r}\times d\vec{S} = 0$]

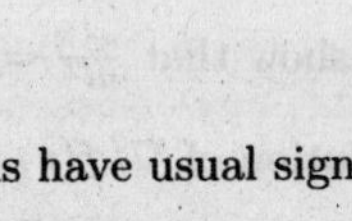

(d) How are covariant and contra-variant vectors defined? (See Appendix)

(e) Show that $d\vec{r}$ transforms like a contra-variant vector, but $\vec{\nabla}\phi$ transforms like a covariant vector. (See Appendix) **[C.U. (Hons.) 2006]**

52. (a) Find the unit vectors which are perpendicular to the vector $2\hat{\imath}+\hat{\jmath}-3\hat{k}$ and lie in the plane of the vectors $7\hat{\imath}-\hat{\jmath}-\hat{k}$ and $\hat{\imath}+5\hat{\jmath}-3\hat{k}$. (See worked out examples)

(b) Using Stoke's theorem, prove that $\oint_C \phi d\vec{r} = \int_S d\vec{S}\times\vec{\nabla}\phi$ where ϕ is a scalar function of $\vec{r}$ and S is an open surface bounded by a closed curve C. (See Q. 30 of Exercises)

(c) Classify the vectors $\vec{r}$ and $\vec{L}$ as axial or polar vector where $\vec{r}$ is the position vector of a particle and $\vec{L}$ is its angular momentum.

(d) If $\vec{r}$ be the radius vector joining a particle of mass m with centre of force and $\vec{A}$ the area swept by the radius vector, show that

$$d\vec{A} = \frac{1}{2}\vec{r} \times d\vec{r} \quad \text{and} \quad \frac{d\vec{A}}{dt} = \frac{\vec{L}}{2m} \quad \text{(See Art. 3.14 and Art. 7.22.1)}$$

where $\vec{L}$ is the angular momentum of the particle about the centre of force.

[C.U. (Hons.) 2006 (old syll.)]

53. (a) If the magnitude of a vector $\vec{A}(t)$ is constant w.r.t. time, show that $\frac{d\vec{A}}{dt}$ is perpendicular to $\vec{A}$. (See worked out examples)

(b) If $\vec{A}$ is irrotational, then show that $\vec{A} \times \vec{r}$ is solenoidal.
[**Hints.** Given $\vec{\nabla} \times \vec{A} = 0$; now $\vec{\nabla} \cdot (\vec{A} \times \vec{r}) = \vec{r} \cdot (\vec{\nabla} \times \vec{A}) - \vec{A} \cdot (\vec{\nabla} \times \vec{r}) = 0$ provided $\vec{\nabla} \times \vec{r} = 0$ i.e. $\vec{r}$ is also irrotational.]

(c) Prove that $\int_V \vec{\nabla}\phi dV = \oint_S \phi d\vec{S}$, where $\phi(\vec{r})$ denotes a well-behaved scalar field and S is a closed surface enclosing volume V.
(See worked out examples)

(d) For an arbitrary scalar function ϕ and vector function $\vec{A}$, prove that

$$\vec{\nabla} \cdot (\phi\vec{A}) = (\vec{\nabla}\phi) \cdot \vec{A} + \phi(\vec{\nabla} \cdot \vec{A}) \quad \text{(See worked out examples)}$$

(e) The force on a particle is given by $\vec{F} = k(-y\hat{i} + x\hat{j})$, where k is a constant. Calculate the work done by the particle as it moves from $(a, 0)$ to $(0, a)$ going round the arc of a circle of radius a.
[**Hints.** Line integral to be calculated with proper limits—See worked out examples]

(f) A particle of mass m moves in the XY-plane so that its position vector is $\vec{r} = a\cos\omega t\hat{i} + b\sin\omega t\hat{j}$, where a, b, ω are constants. Show that the force acting on the particle is always directed towards the origin. **[C.U. (Hons.) 2007]**
[**Hints.** Calculate $\frac{d^2\vec{r}}{dt^2}$ and show that $\frac{d^2\vec{r}}{dt^2} = -\omega^2\vec{r}$]

54. (a) Show that $f(r) = \frac{1}{r}$ is a solution of $\nabla^2 f(r) = 0$ (See worked out examples)

(b) If $\vec{F}$ is a conservative field, prove that $\vec{F}$ can be expressed as the gradient of a scalar potential. (See Art. 3.13)

(c) State Gauss's divergence theorem.

(d) Find the line integral of the vector $\vec{F} = x^2\hat{i} + y^2\hat{j} + (xy - y)\hat{k}$ along the straight line joining the point $(0, 0, 0)$ to $(2, 3, 4)$. (See worked out examples)

(e) Prove the identity $\vec{\nabla} \times (\vec{\nabla} \times \vec{A}) = \vec{\nabla}(\vec{\nabla} \cdot \vec{A}) - \nabla^2\vec{A}$.

(f) What do you mean by polar vectors and axial vectors? Find a unit vector normal to the surface $x^2y + 2xz = 4$ at the point $(2, -2, 3)$. (See worked out examples)

(g) Evaluate $\int_S \vec{F} \cdot d\vec{S}$ where $\vec{F} = 4xz\hat{i} - y^2\hat{j} + yz\hat{k}$ and S is the surface of a cube bounded by $x = 0$, $x = 1$, $y = 0$, $y = 1$, $z = 0$, $z = 1$. (See worked out examples)

(h) Write down expressions for div $\vec{A}$ and curl $\vec{A}$ in terms of spherical polar co-ordinates. A central field $\vec{A}$ is defined by $\vec{A} = \hat{e}_r f(r)$ where $\hat{e}_r$ is the unit vector along $\vec{r}$. Determine $f(r)$ so that $\vec{A}$ may be irrotational or solenoidal. (See Art. 3.14 and worked out examples)

[C.U. (Hons) 2007 (old syll.)]

55. (a) Show that $\vec{\nabla} \cdot (\vec{\nabla} \times \vec{A}) = 0$ for any vector $\vec{A}$.

(b) A particle moves along a curve $x = 2t^2$, $y = t^2 - 4t$, and $z = 3t - 5$, where t is the time. Find its component of velocity at time $t = 1$ in the direction $\hat{\imath} - 2\hat{\jmath} + 2\hat{k}$. (See worked out examples)

(c) A surface in three dimensions is described by $\phi(x, y, z) = c$, where c is a constant. Show that $\vec{\nabla}\phi = 0$ on this surface (See worked out examples)

(d) If $\vec{A} = ax\hat{\imath} + by\hat{\jmath} + cz\hat{k}$, where a, b, and c are constants, evaluate $\int \vec{A} \cdot d\vec{S}$, where S is the surface of a unit sphere. (See worked out examples)

(e) Express the divergence operator in spherical polar coordinates starting from the expression of the same in Cartesian coordinates.

(f) State Green's theorem in a plane.

(g) Show that the area bounded by a simple closed curve C in a plane is given by $\frac{1}{2}\oint_C (xdy - ydx)$.

[**Hints.** Take $\vec{A} = -\hat{\imath}y + \hat{\jmath}x$, find $\oint_C \vec{A} \cdot d\vec{r}$ and then apply Stoke's theorem.]

(h) Find the directional derivative of $\phi = x^3 + y^3 + z^3$ at the point $(2, 1, -1)$ in the direction of the vector $\hat{\imath} + 3\hat{\jmath} + \hat{k}$. (See worked out examples)

(i) In the xy-plane, a force has Cartesian components $F_x = x^3 + xy^2$, $F_y = y^3 + 3x^2y$. Compute the work done in a displacement from the point $(1, 1)$ to $(2, 2)$ where the path is

i. a straight line from $(1, 1)$ to $(2, 2)$;

ii. a straight from $(1, 1)$ to $(1, 2)$, followed by another straight line from $(1, 2)$ to $(2, 2)$.

Is this force conservative? (See worked out examples) **[C.U. (Hons.) 2008]**

56. A Cartesian coordinate axes system is rotated about Z-axis in anticlockwise direction through an angle θ such that a point $P(x, y, z)$ changes to (x', y', z') and $\begin{pmatrix} x' \\ y' \\ z' \end{pmatrix} = R_z(\theta) \begin{pmatrix} x \\ y \\ z \end{pmatrix}$. Find $R_z(\theta)$. (See Appendix B.1) **[C.U. (Hons.) 2006]**

57. Consider an orthogonal transformation of coordinate system $OXYZ$ to $OX'Y'Z'$ with unit vectors $\hat{\imath}, \hat{\jmath}, \hat{k} \rightarrow \hat{\imath}', \hat{\jmath}', \hat{k}'$. A position vector $\vec{r}$ in these systems can be expressed as $\vec{r} = \hat{\imath}x + \hat{\jmath}y + \hat{k}z = \hat{\imath}'x' + \hat{\jmath}'y' + \hat{k}'z'$.

Show that $x' = (\hat{\imath}' \cdot \hat{\imath})x + (\hat{\imath}' \cdot \hat{\jmath})y + (\hat{\imath}' \cdot \hat{k})z$.

Hence establish

$$\begin{bmatrix} x' \\ y' \\ z' \end{bmatrix} = \begin{bmatrix} a_{11} & a_{12} & a_{13} \\ a_{21} & a_{22} & a_{23} \\ a_{31} & a_{32} & a_{33} \end{bmatrix} \begin{bmatrix} x \\ y \\ z \end{bmatrix}$$

where $a_{11} = \hat{\imath}' \cdot \hat{\imath}, a_{12} = \hat{\imath}' \cdot \hat{\jmath}$, etc. **[C.U. (Hons.) 1997]**

(See Appendix B.1)

58. A coordinate system XOY is rotated about Z-axis in anticlockwise sense through θ, so that it takes a point (x, y) to (x', y'). Establish the transformation relation $\begin{pmatrix} x' \\ y' \end{pmatrix} = \begin{pmatrix} \cos\theta & \sin\theta \\ -\sin\theta & \cos\theta \end{pmatrix} \begin{pmatrix} x \\ y \end{pmatrix}$. (See Appendix B.1) **[C.U. (Hons.) 2002]**

Chapter 3

Mechanics

3.1 Introduction

The subject of mechanics deals with motion and the causes of motion. The cause of motion is force. Mechanics also deals with forces which acting on a rigid body mutually balance each other. This part of the subject which deals with forces mutually balancing each other and causing no motion is called *Statics.* The other part which deals with forces causing motion is called *Dynamics.* The branch of dynamics which deals with motion alone without considering its cause is called *Kinematics.*

3.2 Kinematics

Here we shall discuss the path described by a particle and determine the velocity and acceleration at any point along its path. Let the position of a particle P at time t be specified by the vector $\vec{r}(t)$ relative to the fixed origin O (Fig. 3.2.1). $\vec{r}(t)$ is called position vector and is given by

$$\vec{r}(t) = \hat{\imath}x(t) + \hat{\jmath}y(t) + \hat{k}z(t),$$

where x, y, z are the co-ordinates of the point P, and $\hat{\imath}, \hat{\jmath}, \hat{k}$ denote the rectangular unit vectors.

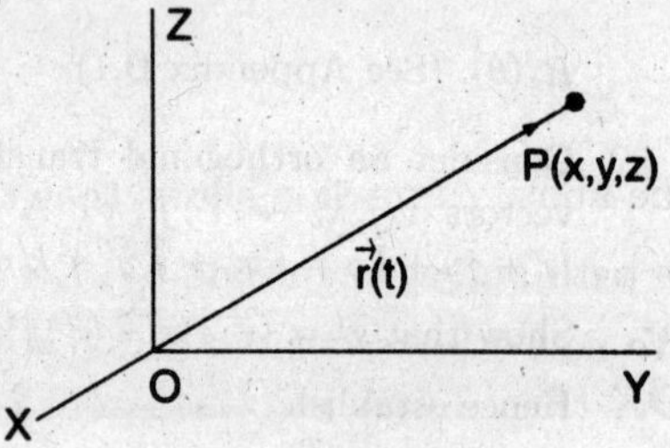

Fig. 3.2.1

3.3 Velocity and Acceleration

From the above expression we shall calculate the velocity in rectangular co-ordinates. The velocity $\vec{v}$ of a particle is a vector which is measured by the time rate of change of the position of the particle.

$$\therefore \ \vec{v} = \frac{d\vec{r}}{dt} = \hat{\imath}\frac{dx}{dt} + \hat{\jmath}\frac{dy}{dt} + \hat{k}\frac{dz}{dt} = \hat{\imath}v_x + \hat{\jmath}v_y + \hat{k}v_z, \tag{1}$$

where v_x, v_y, v_z denote the three components of velocity.

The acceleration $\vec{f}$ is also a vector which is the time derivative of the velocity and is given by

$$\vec{f} = \frac{d\vec{v}}{dt} = \frac{d^2\vec{r}}{dt^2} = \hat{\imath}\frac{d^2x}{dt^2} + \hat{\jmath}\frac{d^2y}{dt^2} + \hat{k}\frac{d^2z}{dt^2} = \hat{\imath}f_x + \hat{\jmath}f_y + \hat{k}f_z, \tag{2}$$

where f_x, f_y, f_z are the three components of acceleration.

3.4 Tangential and Normal Components of Velocity and Acceleration

We consider a particle at P in a plane in which the axes OX and OY are fixed (Fig. 3.4.1). Let the particle move along the curve AB from P to P' during the interval of time from t to $t + \delta t$. The position vectors of the points P and P' are respectively $\vec{r}$ and $\vec{r} + \delta\vec{r}$. The velocity $\vec{v}$ of the particle at P is $\frac{\delta\vec{r}}{\delta t}$.

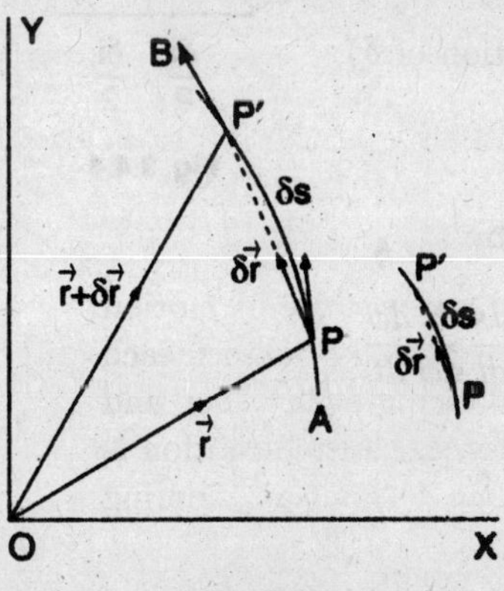

Fig. 3.4.1

Now, $\frac{\delta\vec{r}}{\delta t} = \frac{\delta\vec{r}}{\delta s}\frac{\delta s}{\delta t}$, where δs is an element of length on the path AB and $\delta\vec{r}$ denotes the displacement of the particle in time δt along the dotted line.

Now

$$\vec{v} = \lim_{\delta t \to 0} \frac{\delta\vec{r}}{\delta t} = \frac{d\vec{r}}{dt} = \frac{d\vec{r}}{ds}\frac{ds}{dt}.$$

From the figure it follows that as P approaches P', then in the limit $\frac{d\vec{r}}{ds} = \hat{\imath}$, where $\hat{\imath}$ is the unit vector along the tangent at the point P.

$$\therefore\ \vec{v} = \hat{\imath}\frac{ds}{dt} = \hat{\imath}v \quad \left[\because\ v = \frac{ds}{dt}\right]. \tag{1}$$

The above expression shows that the velocity of a moving particle has a direction tangent to the path and a magnitude $\frac{ds}{dt}$. Let $\hat{\imath} + \delta\hat{\imath}$ be the unit vector along the tangent at the point P'. We now suppose that ϕ and $\phi + \delta\phi$ are the inclinations of the vectors $\hat{\imath}$ and $\hat{\imath} + \delta\hat{\imath}$ to the axis OX (Fig. 3.4.2). In this figure $\hat{\jmath}$ is the unit normal vector which is drawn from the point

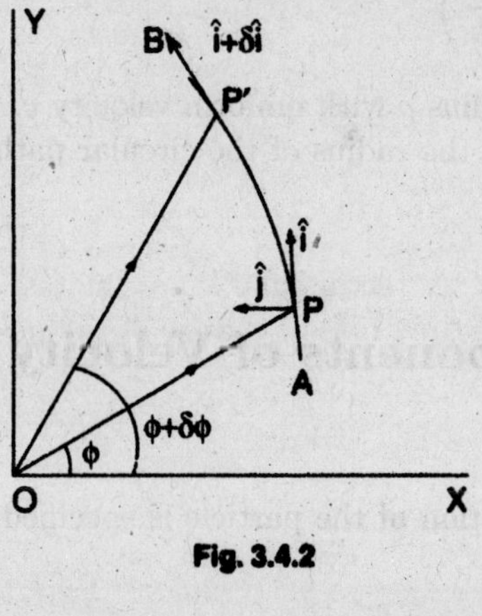

Fig. 3.4.2

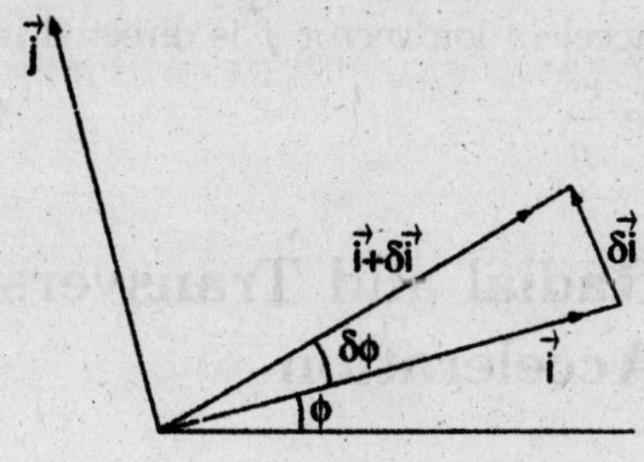

Fig. 3.4.3

P perpendicular to the vector $\hat{\imath}$. Now the vectors $\hat{\imath}$ and $\hat{\imath}+\delta\hat{\imath}$ are transferred to a common origin as shown in Fig. 3.4.3. Since both $\hat{\imath}$ and $\hat{\imath}+\delta\hat{\imath}$ are unit vectors, the triangle formed by the vectors $\hat{\imath}, \hat{\imath}+\delta\hat{\imath}$ and $\delta\hat{\imath}$ are isosceles. The magnitude of $\delta\hat{\imath}$ is given by

$$\frac{\delta i}{2}=\sin\frac{\delta\phi}{2} \quad \text{or} \quad \delta i = 2\sin\frac{\delta\phi}{2} \text{ (Fig. 3.4.4).} \quad \therefore \frac{\delta\hat{\imath}}{\delta\phi}=\frac{\sin\frac{\delta\phi}{2}}{\frac{\delta\phi}{2}}.$$

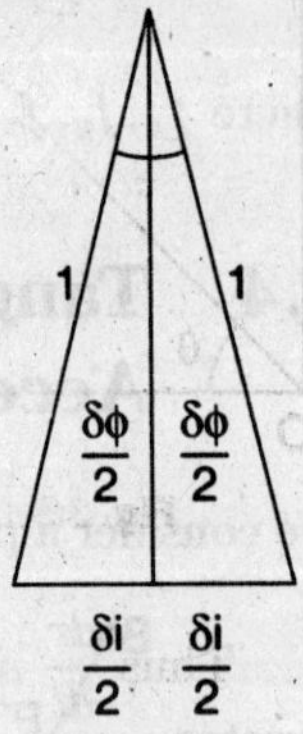

Fig. 3.4.4

Hence the limit of the magnitude of $\dfrac{\delta\hat{\imath}}{\delta\phi}$ is unity. Thus $\dfrac{d\hat{\imath}}{d\phi}$ is a unit vector pointing in the limiting direction defined by δi as $\delta\phi$ tends to zero. The direction is therefore given by the direction of $\hat{\jmath}$. $\therefore \dfrac{d\hat{\imath}}{d\phi}=\hat{\jmath}$. In a similar way we see that $\dfrac{d\hat{\jmath}}{d\phi}$ is perpendicular to $\hat{\jmath}$. Now since the limiting direction of $\delta\hat{\jmath}$ is that of $-\hat{\imath}$, we may write $\dfrac{d\hat{\jmath}}{d\phi}=-\hat{\imath}$.

Differentiating the expression (3.4.1) we get acceleration given by

$$\vec{f}=\frac{d\vec{v}}{dt}=\frac{d}{dt}\left(\hat{\imath}\frac{ds}{dt}\right)=\hat{\imath}\frac{d^2s}{dt^2}+\frac{ds}{dt}\frac{d\hat{\imath}}{dt}=\hat{\imath}\frac{d^2s}{dt^2}+\frac{ds}{dt}\frac{d\hat{\imath}}{d\phi}\frac{d\phi}{dt}$$

$$=\hat{\imath}\frac{d^2s}{dt^2}+\frac{ds}{dt}\hat{\jmath}\frac{d\phi}{ds}\frac{ds}{dt}=\hat{\imath}\frac{d^2s}{dt^2}+\hat{\jmath}\frac{d\phi}{ds}\left(\frac{ds}{dt}\right)^2.$$

The radius of curvature of the path AB at P is given by $\rho=\dfrac{ds}{d\phi}$.

$$\therefore \vec{f}=\hat{\imath}\frac{d^2s}{dt^2}+\frac{\hat{\jmath}}{\rho}\left(\frac{ds}{dt}\right)^2=\hat{\imath}\frac{dv}{dt}+\frac{\hat{\jmath}}{\rho}v^2. \tag{2}$$

Thus we see that the acceleration of a moving particle has a component $\dfrac{dv}{dt}$ along the tangent and a component $\dfrac{v^2}{\rho}$ along the normal to the path. The magnitude of the acceleration is given by

$$f=\sqrt{\left(\frac{dv}{dt}\right)^2+\left(\frac{v^2}{\rho}\right)^2}.$$

In case of a particle moving in a circular path of radius ρ with uniform velocity v, it follows that the acceleration vector $\vec{f}$ is directed inward along the radius of the circular path and has magnitude $\dfrac{v^2}{\rho}$.

3.5 Radial and Transverse Components of Velocity and Acceleration

We consider a particle P moving in a plane. The position of the particle is specified by polar co-ordinates r, θ (Fig. 3.5.1).

Here we take $\hat{\imath}$ and $\hat{\jmath}$ as the unit vectors along and perpendicular to OP. The directions of the unit vectors are as shown in Fig. 3.5.1. As in the previous article, we may write here

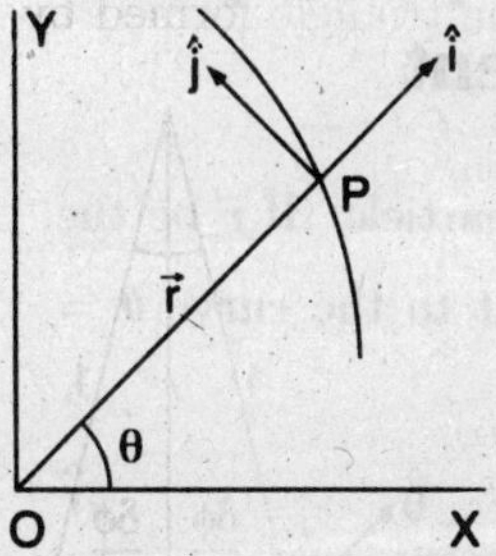

Fig. 3.5.1

$$\frac{d\hat{\imath}}{d\theta} = \hat{\jmath}, \quad \frac{d\hat{\jmath}}{d\theta} = -\hat{\imath}.$$

The position vector of the point P is

$$\vec{r} = r\hat{\imath}.$$

Hence,

$$\vec{v} = \frac{dr}{dt}\hat{\imath} + r\frac{d\hat{\imath}}{dt} = \frac{dr}{dt}\hat{\imath} + r\frac{d\hat{\imath}}{d\theta}\frac{d\theta}{dt} = \frac{dr}{dt}\hat{\imath} + r\frac{d\theta}{dt}\hat{\jmath}. \quad (1)$$

Thus $\frac{dr}{dt}$ and $r\frac{d\theta}{dt}$ denote the components of velocity along and perpendicular to the radius vector.

Differentiating the above expression with respect to t, we get the acceleration given by

$$\begin{aligned}
\vec{f} = \frac{d\vec{v}}{dt} &= \frac{d^2r}{dt^2}\hat{\imath} + \frac{dr}{dt}\frac{d\hat{\imath}}{dt} + r\frac{d\theta}{dt}\frac{d\hat{\jmath}}{dt} + \frac{d}{dt}\left(r\frac{d\theta}{dt}\right)\hat{\jmath} \\
&= \frac{d^2r}{dt^2}\hat{\imath} + \frac{dr}{dt}\frac{d\hat{\imath}}{d\theta}\frac{d\theta}{dt} + r\frac{d\theta}{dt}\frac{d\hat{\jmath}}{d\theta}\frac{d\theta}{dt} + \frac{d}{dt}\left(r\frac{d\theta}{dt}\right)\hat{\jmath} \\
&= \frac{d^2r}{dt^2}\hat{\imath} + \frac{dr}{dt}\frac{d\theta}{dt}\hat{\jmath} - r\left(\frac{d\theta}{dt}\right)^2\hat{\imath} + \frac{d}{dt}\left(r\frac{d\theta}{dt}\right)\hat{\jmath} \\
&= \left[\frac{d^2r}{dt^2} - r\left(\frac{d\theta}{dt}\right)^2\right]\hat{\imath} + \frac{1}{r}\frac{d}{dt}\left(r^2\frac{d\theta}{dt}\right)\hat{\jmath}.
\end{aligned}$$

Thus the components of acceleration along and perpendicular to the radius vector, i.e., the radial and transverse accelerations are,

$$\frac{d^2r}{dt^2} - r\left(\frac{d\theta}{dt}\right)^2 \quad \text{and} \quad \frac{1}{r}\frac{d}{dt}\left(r^2\frac{d\theta}{dt}\right).$$

Calling the radial and transverse accelerations f_r and f_θ respectively, the magnitude of the acceleration is given by

$$f^2 = f_r^2 + f_\theta^2. \quad (2)$$

In case the particle moves along the radius, θ remains constant.

$$\therefore\ f_r = \frac{d^2r}{dt^2} \quad \text{and} \quad f_\theta = 0;$$

when the particle moves along a circular path of radius r, r remains constant, but θ changes.

$$\therefore\ f_r = -r\left(\frac{d\theta}{dt}\right)^2 = -\frac{r^2\left(\frac{d\theta}{dt}\right)^2}{r} = -\frac{v^2}{r}$$ [where $v = r\frac{d\theta}{dt}$ = tangential component of velocity.]

This acceleration f_r is towards the centre of the circular path and is called centripetal acceleration.

3.6 Velocity and Acceleration of a particle in plane polar Co-ordinate system: Alternative Treatment

Let (r, θ) represent the polar co-ordinates describing the position P of a particle. If $\vec{r}$ be the position vector of the particle at any time t, then $\frac{\partial \vec{r}}{\partial r}$ is a vector tangent to the curve $\theta =$ constant i.e., a vector in the direction of $\vec{r}$ increasing. A unit vector in this direction is given by

$$\hat{r} = \frac{\partial \vec{r}}{\partial r} \Big/ \left|\frac{\partial \vec{r}}{\partial r}\right|. \tag{1}$$

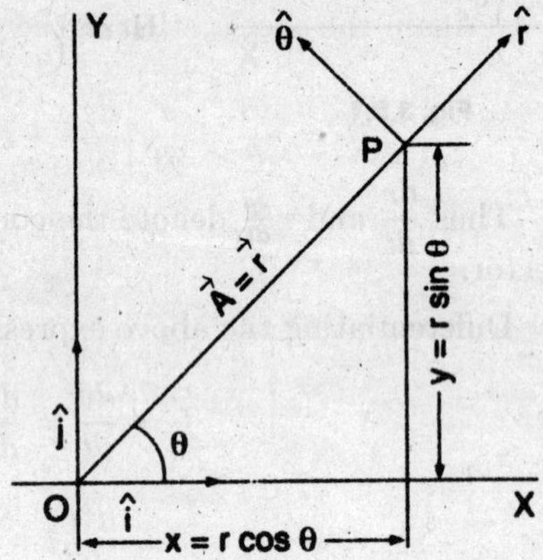

Fig. 3.6.1

Since

$$\vec{r} = x\hat{\imath} + y\hat{\jmath} = r\cos\theta\hat{\imath} + r\sin\theta\hat{\jmath} \tag{2}$$

as evident from Fig. 3.6.1, we may write

$$\frac{\partial \vec{r}}{\partial r} = \cos\theta\hat{\imath} + \sin\theta\hat{\jmath} \text{ and } \left|\frac{\partial \vec{r}}{\partial r}\right| = 1,$$

so that

$$\hat{r} = \cos\theta\hat{\imath} + \sin\theta\hat{\jmath}. \tag{3}$$

Similarly, $\frac{\partial \vec{r}}{\partial \theta}$ is a vector tangent to the curve $r =$ constant, i.e., a vector in the direction of θ increasing in the anti-clockwise sense and normal to $\vec{r}$. A unit vector in this direction is given by

$$\hat{\theta} = \frac{\partial \vec{r}}{\partial \theta} \Big/ \left|\frac{\partial \vec{r}}{\partial \theta}\right|. \tag{4}$$

From (3.6.2), we may write

$$\frac{\partial \vec{r}}{\partial \theta} = -r\sin\theta\hat{\imath} + r\cos\theta\hat{\jmath} \text{ and } \left|\frac{\partial \vec{r}}{\partial \theta}\right| = r,$$

so that

$$\hat{\theta} = -\sin\theta\hat{\imath} + \cos\theta\hat{\jmath} \tag{5}$$

Solving the simultaneous equations (3.6.3) and (3.6.5) for $\hat{\imath}$ and $\hat{\jmath}$ i.e., multiplying (3.6.3) on both sides by $\cos\theta$ and (3.6.5) on both sides by $\sin\theta$ and then subtracting, we get $\hat{\imath} = \cos\theta\hat{r} - \sin\theta\hat{\theta}$.

Also multiplying (3.6.3) by $\sin\theta$ and (3.6.5) by $\cos\theta$ on both sides and then adding we get

$$\hat{\jmath} = \sin\theta\hat{r} + \cos\theta\hat{\theta}.$$

On inspection of (3.6.3) and (3.6.5), it is evident that

$$\frac{d\hat{r}}{d\theta} = \hat{\theta} \text{ and } \frac{d\hat{\theta}}{d\theta} = -\hat{r}. \tag{6}$$

Also we may write, $\vec{r} = r\hat{r}$, so that the velocity of the particle is given by

$$\vec{v} = \frac{d}{dt}(r\hat{r}) = \frac{dr}{dt}\hat{r} + r\frac{d\hat{r}}{dt} = \frac{dr}{dt}\hat{r} + r\frac{d\hat{r}}{d\theta}\frac{d\theta}{dt} = \frac{dr}{dt}\hat{r} + r\frac{d\theta}{dt}\hat{\theta} \text{ from (3.6.6)}$$

$$= \dot{r}\hat{r} + r\dot{\theta}\hat{\theta}$$

$$= \text{radial component} + \text{transverse or cross-radial component.}$$

Again the acceleration of the particle is given by

$$\vec{f} = \frac{d\vec{v}}{dt} = \frac{d}{dt}\left[\frac{dr}{dt}\hat{r} + r\frac{d\theta}{dt}\hat{\theta}\right] = \frac{d^2r}{dt^2}\hat{r} + \frac{dr}{dt}\frac{d\hat{r}}{dt} + \frac{dr}{dt}\frac{d\theta}{dt}\hat{\theta} + r\frac{d^2\theta}{dt^2}\hat{\theta} + r\frac{d\theta}{dt}\frac{d\hat{\theta}}{dt}$$

$$= \frac{d^2r}{dt^2}\hat{r} + \frac{dr}{dt}\left(\frac{d\theta}{dt}\hat{\theta}\right) + \frac{dr}{dt}\frac{d\theta}{dt}\hat{\theta} + r\frac{d^2\theta}{dt^2}\hat{\theta} + r\frac{d\theta}{dt}\left(-\frac{d\theta}{dt}\hat{r}\right)$$

$$= \left[\frac{d^2r}{dt^2} - r\left(\frac{d\theta}{dt}\right)^2\right]\hat{r} + \left[r\frac{d^2\theta}{dt^2} + 2\frac{dr}{dt}\frac{d\theta}{dt}\right]\hat{\theta}$$

$$= [\ddot{r} - r\dot{\theta}^2]\hat{r} + [r\ddot{\theta} + 2\dot{r}\dot{\theta}]\hat{\theta}$$

$$= \text{radial component} + \text{cross-radial component.}$$

$$\left.\begin{aligned} \because \quad \frac{d\hat{r}}{dt} &= \frac{d\hat{r}}{d\theta}\frac{d\theta}{dt} = \frac{d\theta}{dt}\hat{\theta} \\ \text{and} \quad \frac{d\hat{\theta}}{dt} &= \frac{d\hat{\theta}}{d\theta}\frac{d\theta}{dt} = -\frac{d\theta}{dt}\hat{r} \end{aligned}\right\} \text{from (3.6.6)}$$

3.7 Velocity and Acceleration of a Particle in Cylindrical Polar Co-ordinates

We consider a vector $\vec{A}$ which is a function of parameter t. We shall show in the following way that

$$\frac{d\vec{A}}{dt} = \left(\frac{dA_\rho}{dt} - A_\phi\frac{d\phi}{dt}\right)\hat{\rho} + \left(\frac{dA_\phi}{dt} + A_\rho\frac{d\phi}{dt}\right)\hat{\phi} + \frac{dA_z}{dt}\hat{z},$$

where A_ρ, A_ϕ and A_z are the components of $\vec{A}$ in cylindrical polar co-ordinates.

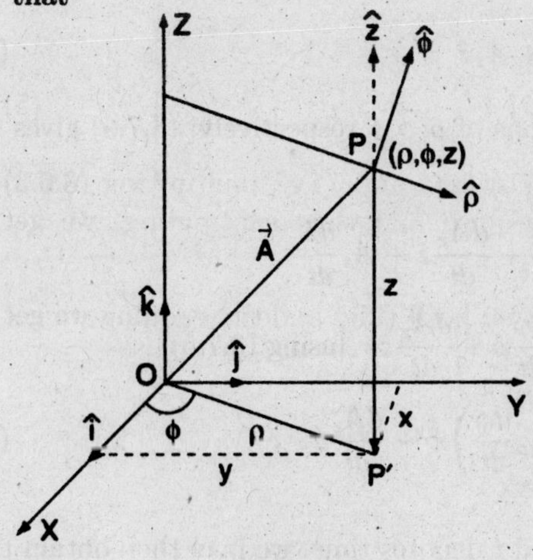

Fig. 3.7.1

In cylindrical co-ordinate system, the cartesian co-ordinates (x, y, z) of a point P are replaced by (ρ, ϕ, z), where ρ represents OP' which is the projection of P in the $X - Y$ plane as shown in the fig. 3.7.1. ϕ is the angle between OX and OP'. We may therefore write

$$\vec{A} = \hat{\imath}x + \hat{\jmath}y + \hat{k}z = \hat{\imath}\rho\cos\phi + \hat{\jmath}\rho\sin\phi + \hat{k}z \qquad (1)$$

where $x = \rho\cos\phi$, $y = \rho\sin\phi$, $z = z$ as evident from the fig. 3.7.1.

From (3.7.1) we have

$$\frac{\partial \vec{A}}{\partial \rho} = \hat{\imath}\cos\phi + \hat{\jmath}\sin\phi, \quad \frac{\partial \vec{A}}{\partial \phi} = -\hat{\imath}\rho\sin\phi + \hat{\jmath}\rho\cos\phi \text{ and } \frac{d\vec{A}}{dz} = \hat{k},$$

so that a unit vector in the direction of ρ increasing is given by

$$\hat{\rho} = \frac{\partial \vec{A}}{\partial \rho} \Big/ \left|\frac{\partial \vec{A}}{\partial \rho}\right| = \hat{\imath}\cos\phi + \hat{\jmath}\sin\phi. \tag{2}$$

Similarly, unit vectors in the directions of ϕ increasing and z increasing are respectively given by

$$\hat{\phi} = \frac{\partial \vec{A}}{\partial \phi} \Big/ \left|\frac{\partial \vec{A}}{\partial \phi}\right| = \frac{-\hat{\imath}\rho\sin\phi + \hat{\jmath}\rho\cos\phi}{\rho} = -\hat{\imath}\sin\phi + \hat{\jmath}\cos\phi \tag{3}$$

$$\hat{z} = \hat{k}. \tag{4}$$

Also we may write,

$$\left.\begin{aligned}
\frac{d\hat{\rho}}{dt} &= \frac{d\hat{\rho}}{d\phi}\frac{d\phi}{dt} = (-\hat{\imath}\sin\phi + \hat{\jmath}\cos\phi)\frac{d\phi}{dt} \\
&= \frac{d\phi}{dt}\hat{\phi} \text{ from (3.7.2) and (3.7.3)} \\
\text{Similarly, } \frac{d\hat{\phi}}{dt} &= \frac{d\hat{\phi}}{d\phi}\frac{d\phi}{dt} = -(\hat{\imath}\cos\phi + \hat{\jmath}\sin\phi)\frac{d\phi}{dt} = -\frac{d\phi}{dt}\hat{\rho} \\
\text{and } \frac{d\hat{z}}{dt} &= 0
\end{aligned}\right\} \tag{5}$$

Now the vector $\vec{A}$ may be written as

$$\vec{A} = A_\rho\hat{\rho} + A_\phi\hat{\phi} + A_z\hat{z} \tag{6}$$

A_ρ, A_ϕ, A_z being the components of A in the directions of $\hat{\rho}, \hat{\phi}, \hat{z}$ respectively; (3.7.6) gives on differentiation

$$\begin{aligned}
\frac{d\vec{A}}{dt} &= \frac{dA_\rho}{dt}\hat{\rho} + A_\rho\frac{d\hat{\rho}}{dt} + \frac{dA_\phi}{dt}\hat{\phi} + A_\phi\frac{d\hat{\phi}}{dt} + \frac{dA_z}{dt}\hat{z} + A_z\frac{d\hat{z}}{dt} \\
&= \frac{dA_\rho}{dt}\hat{\rho} + A_\rho\frac{d\phi}{dt}\hat{\phi} + \frac{dA_\phi}{dt}\hat{\phi} - A_\phi\frac{d\phi}{dt}\hat{\rho} + \frac{dA_z}{dt}\hat{z} \quad \text{[using (3.7.5)]} \\
&= \left(\frac{dA_\rho}{dt} - A_\phi\frac{d\phi}{dt}\right)\hat{\rho} + \left(\frac{dA_\phi}{dt} + A_\rho\frac{d\phi}{dt}\right)\hat{\phi} + \frac{dA_z}{dt}\hat{z}.
\end{aligned} \tag{7}$$

If $\vec{A}$ in the above case be the position vector $\vec{r}$ and t denotes time, we may then obtain the corresponding expression for $\frac{d\vec{r}}{dt}$ in the following way.

As $\vec{A} = \vec{r}$, $\dfrac{d\vec{r}}{dt}$ denotes the velocity vector.

Solving the equations (3.7.2) and (3.7.3) for $\hat{\imath}$ and $\hat{\jmath}$, i.e., multiplying (3.7.2) on both sides by $\cos\phi$ and (3.7.3) by $\sin\phi$ and then subtracting, we have

$$\hat{\imath} = \hat{\rho}\cos\phi - \hat{\phi}\sin\phi. \tag{8}$$

Again multiplying (3.7.2) by $\sin\phi$ on both sides and (3.7.3) by $\cos\phi$ and then adding, we have

$$\hat{\jmath} = \hat{\rho}\sin\phi + \hat{\phi}\cos\phi. \tag{9}$$

Now,

$$\vec{A} = \vec{r} = \hat{\imath}x + \hat{\jmath}y + \hat{k}z$$

$$= (\hat{\rho}\cos\phi - \hat{\phi}\sin\phi)\rho\cos\phi + (\hat{\rho}\sin\phi + \hat{\phi}\cos\phi)\rho\sin\phi + z\hat{z}$$

[using (3.7.4), (3.7.8) and (3.7.9)]

$$= (\rho\cos^2\phi + \rho\sin^2\phi)\hat{\rho} + (-\rho\cos\phi\sin\phi + \rho\cos\phi\sin\phi)\hat{\phi} + z\hat{z}$$

$$= \rho\hat{\rho} + z\hat{z}. \tag{10}$$

But

$$\vec{A} = \vec{r} = A_\rho\hat{\rho} + A_\phi\hat{\phi} + A_z\hat{z}. \tag{11}$$

(according to the problem)

Hence comparing (3.7.10) and (3.7.11), we get $A_\rho = \rho$, $A_\phi = 0$ and $A_z = z$, so that

$$\frac{dA_\rho}{dt} = \frac{d\rho}{dt}, \quad \frac{dA_\phi}{dt} = 0 \text{ and } \frac{dA_z}{dt} = \frac{dz}{dt}.$$

Putting these values in (3.7.7), we get

$$\vec{v} = \frac{d\vec{r}}{dt} = \left(\frac{d\rho}{dt} - 0\cdot\frac{d\phi}{dt}\right)\hat{\rho} + \left(0 + \rho\frac{d\phi}{dt}\right)\hat{\phi} + \frac{dz}{dt}\hat{z}$$

$$= \frac{d\rho}{dt}\hat{\rho} + \rho\frac{d\phi}{dt}\hat{\phi} + \frac{dz}{dt}\hat{z} \tag{12}$$

$$= \dot{\rho}\hat{\rho} + \rho\dot{\phi}\hat{\phi} + \dot{z}\hat{z} = \text{ velocity at any instant.}$$

Expression for acceleration

Differentiating equation (3.7.12) w.r.t time, we get

$$\vec{f} = \frac{d\vec{v}}{dt} = \frac{d}{dt}\left[\frac{d\rho}{dt}\hat{\rho} + \rho\frac{d\phi}{dt}\hat{\phi} + \frac{dz}{dt}\hat{z}\right]$$

$$= \frac{d^2\rho}{dt^2}\hat{\rho} + \frac{d\rho}{dt}\frac{d\hat{\rho}}{dt} + \frac{d\rho}{dt}\frac{d\phi}{dt}\hat{\phi} + \rho\frac{d^2\phi}{dt^2}\hat{\phi} + \rho\frac{d\phi}{dt}\frac{d\hat{\phi}}{dt} + \frac{d^2z}{dt^2}\hat{z} + \frac{dz}{dt}\frac{d\hat{z}}{dt}$$

$$= \frac{d^2\rho}{dt^2}\hat{\rho} + \frac{d\rho}{dt}\frac{d\phi}{dt}\hat{\phi} + \frac{d\rho}{dt}\frac{d\phi}{dt}\hat{\phi} + \rho\frac{d^2\phi}{dt^2}\hat{\phi} - \rho\frac{d\phi}{dt}\frac{d\phi}{dt}\hat{\rho} + \frac{d^2z}{dt^2}\hat{z}$$

(using the results of 3.7.5)

$$= \left[\frac{d^2\rho}{dt^2} - \rho\left(\frac{d\phi}{dt}\right)^2\right]\hat{\rho} + \left[\rho\frac{d^2\phi}{dt^2} + 2\frac{d\rho}{dt}\frac{d\phi}{dt}\right]\hat{\phi} + \frac{d^2z}{dt^2}\hat{z}$$

$$= \left[\ddot{\rho} - \rho\dot{\phi}^2\right]\hat{\rho} + \left[\rho\ddot{\phi} + 2\dot{\rho}\dot{\phi}\right]\hat{\phi} + \ddot{z}\hat{z}$$

= acceleration at any instant.

Important Applications

(a) A force $\vec{F}$ is given in cartesian co-ordinates as $\vec{F} = \hat{\imath}yB - \hat{\jmath}Bx$, where B is a constant. We are to find F_ϕ component.

We have

$$\vec{F} = \hat{\imath}yB - \hat{\jmath}Bx \quad (B = \text{ constant})$$

$$= (\hat{\rho}\cos\phi - \hat{\phi}\sin\phi)B\rho\sin\phi - (\hat{\rho}\sin\phi + \hat{\phi}\cos\phi)B\rho\cos\phi$$

(putting the values of $\hat{\imath}, \hat{\jmath}, x$ and y)

$$= (B\rho\cos\phi\sin\phi - B\rho\cos\phi\sin\phi)\hat{\rho} - B\rho(\sin^2\phi + \cos^2\phi)\hat{\phi} = -B\rho\hat{\phi}.$$

Again we may write $\vec{F} = F_\rho\hat{\rho} + F_\phi\hat{\phi}$.

Comparing the above two expressions, we have $F_\phi = -B\rho$.

(b) A particle of charge q and mass m moves in a uniform magnetic field $\vec{B} = B\hat{z}$ and an electric field $\vec{E} = \frac{a}{\rho}\hat{\rho}$, where the constants a and B may be either positive or negative. We are to set up the equations of motion in cylindrical co-ordinates and show that $m\rho^2\dot{\phi} + \frac{qB}{2c}\rho^2$ is a constant of motion, assuming that the force due to electric field is $q\vec{E}$ and that due to the magnetic field is $\frac{q}{c}(\vec{v} \times \vec{B})$.

We know that the velocity and acceleration of a particle in cylindrical co-ordinates are given by

$$\left.\begin{aligned} \vec{v} &= \dot{\rho}\hat{\rho} + \rho\dot{\phi}\hat{\phi} + \dot{z}\hat{z} \\ \text{and } \vec{f} &= (\ddot{\rho} - \rho\dot{\phi}^2)\hat{\rho} + (\rho\ddot{\phi} + 2\dot{\rho}\dot{\phi})\hat{\phi} + \ddot{z}\hat{z} \end{aligned}\right\} \tag{a}$$

According to the problem, we have

$$\vec{F} = m\vec{f} = q\vec{E} + \frac{q}{c}(\vec{v} \times \vec{B})$$

$$\text{or } m[(\ddot{\rho} - \rho\dot{\phi}^2)\hat{\rho} + (\rho\ddot{\phi} + 2\dot{\rho}\dot{\phi})\hat{\phi} + \ddot{z}\hat{z}] = q\frac{a}{\rho}\hat{\rho} + \frac{q}{c}\left[\left(\dot{\rho}\hat{\rho} + \rho\dot{\phi}\hat{\phi} + \dot{z}\hat{z}\right) \times B\hat{z}\right] \quad \text{[using (a)]}$$

$$= q\frac{a}{\rho}\hat{\rho} - \frac{q}{c}\dot{\rho}B\hat{\phi} + q\frac{\rho}{c}\dot{\phi}B\hat{\rho}$$

$$= \left(q\frac{a}{\rho} + q\frac{\rho}{c}B\dot{\phi}\right)\hat{\rho} - \frac{q}{c}B\dot{\rho}\hat{\phi}$$

so that the equations of motion required to be set up are given by

$$\left.\begin{aligned} m(\ddot{\rho} - \rho\dot{\phi}^2)\hat{\rho} &= q\left(\frac{a}{\rho} + \frac{\rho B}{c}\dot{\phi}\right) \\ m(\rho\ddot{\phi} + 2\dot{\rho}\dot{\phi}) &= -\frac{qB}{c}\dot{\rho} \text{ and } m\ddot{z} = 0 \end{aligned}\right\}. \tag{b}$$

Multiplying the second equation of (b) by ρ, we may write

$$m\rho^2\ddot{\phi} + 2m\rho\dot{\rho}\dot{\phi} + \frac{qB}{c}\rho\dot{\rho} = 0 \text{ or } \frac{d}{dt}\left(m\rho^2\dot{\phi} + \frac{qB}{2c}\rho^2\right) = 0$$

$$\therefore\ m\rho^2\dot{\phi} + \frac{qB}{2c}\rho^2 = \text{a constant of motion.}$$

3.8 Velocity and Acceleration of a particle in spherical polar Co-ordinates

In spherical polar co-ordinate system, the cartesian co-ordinates (x, y, z) of a point P are replaced by (r, θ, ϕ), the relations connecting them being given by

$$x = r\sin\theta\cos\phi, y = r\sin\theta\sin\phi \text{ and } z = r\cos\theta$$

as shown in fig. 3.8.1, so that any vector $\vec{A}$, which is a function of parameter t, may be written as

$$\vec{A} = \hat{\imath}x + \hat{\jmath}y + \hat{k}z = \hat{\imath}r\sin\theta\cos\phi + \hat{\jmath}r\sin\theta\sin\phi + \hat{k}r\cos\theta. \tag{3.8.1}$$

From (3.8.1), a unit vector in the direction of $\vec{r}$ increasing is given by

$$\hat{r} = \frac{\partial\vec{A}}{\partial r}\Bigg/\left|\frac{\partial\vec{A}}{\partial r}\right|$$

$$= \hat{\imath}\sin\theta\cos\phi + \hat{\jmath}\sin\theta\sin\phi + \hat{k}\cos\theta. \tag{3.8.2}$$

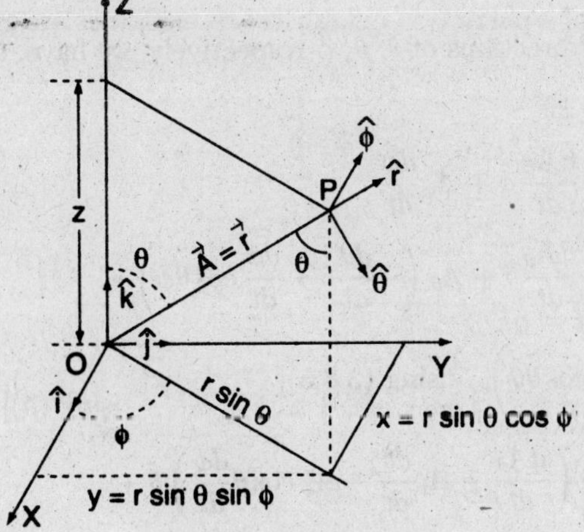

Fig. 3.8.1

Similarly, a unit vector in the direction of θ increasing i.e., perpendicular to $\hat{r}$ but in the same plane as $\hat{r}$ is given by

$$\hat{\theta} = \frac{\partial\vec{A}}{\partial\theta}\Bigg/\left|\frac{\partial\vec{A}}{\partial\theta}\right|$$

$$= \hat{\imath}\cos\theta\cos\phi + \hat{\jmath}\cos\theta\sin\phi - \hat{k}\sin\theta. \tag{3.8.3}$$

Also a unit vector in the direction of ϕ increasing i.e., perpendicular to both $\hat{r}$ and $\hat{\theta}$ and lying in the $X-Y$ plane is given by

$$\hat{\phi} = \frac{\partial \vec{A}}{\partial \phi} \Big/ \left|\frac{\partial \vec{A}}{\partial \phi}\right| = -\hat{\imath}\sin\phi + \hat{\jmath}\cos\phi. \tag{3.8.4}$$

Now we may write

$$\begin{aligned}
\frac{d\hat{r}}{dt} &= \left(\cos\theta\frac{d\theta}{dt}\cos\phi - \sin\theta\sin\phi\frac{d\phi}{dt}\right)\hat{\imath} + \\
&\quad \left(\cos\theta\frac{d\theta}{dt}\sin\phi + \sin\theta\cos\phi\frac{d\phi}{dt}\right)\hat{\jmath} - \sin\theta\frac{d\theta}{dt}\hat{k} \\
&= \frac{d\theta}{dt}(\hat{\imath}\cos\theta\cos\phi + \hat{\jmath}\cos\theta\sin\phi - \hat{k}\sin\theta) + \frac{d\phi}{dt}\sin\theta(-\hat{\imath}\sin\phi + \hat{\jmath}\cos\phi) \\
&= \frac{d\theta}{dt}\hat{\theta} + \frac{d\phi}{dt}\sin\theta\hat{\phi} \quad \text{[using (3.8.3) and (3.8.4)]}
\end{aligned}$$

Similarly,

$$\frac{d\hat{\theta}}{dt} = -\frac{d\theta}{dt}\hat{r} + \frac{d\phi}{dt}\cos\theta\hat{\phi} \tag{3.8.5}$$

$$\text{and} \quad \frac{d\hat{\phi}}{dt} = -\frac{d\phi}{dt}(\hat{\imath}\cos\phi + \hat{\jmath}\sin\phi) = -\frac{d\phi}{dt}(\sin\theta\hat{r} + \cos\theta\hat{\theta})$$

since we can show

$$\sin\theta\hat{r} + \cos\theta\hat{\theta} = \hat{\imath}\cos\phi + \hat{\jmath}\sin\phi.$$

Now writing the vector $\vec{A}$ as

$$\vec{A} = A_r\hat{r} + A_\theta\hat{\theta} + A_\phi\hat{\phi}. \tag{3.8.6}$$

A_r, A_θ, A_ϕ being the components of A in the directions of $\hat{r}$, $\hat{\theta}$, $\hat{\phi}$ respectively, we have, on differentiation of (3.8.6).

$$\begin{aligned}
\frac{d\vec{A}}{dt} &= \frac{dA_r}{dt}\hat{r} + A_r\frac{d\hat{r}}{dt} + \frac{dA_\theta}{dt}\hat{\theta} + A_\theta\frac{d\hat{\theta}}{dt} + \frac{dA_\phi}{dt}\hat{\phi} + A_\phi\frac{d\hat{\phi}}{dt} \\
&= \frac{dA_r}{dt}\hat{r} + A_r\left(\frac{d\theta}{dt}\hat{\theta} + \frac{d\phi}{dt}\sin\theta\hat{\phi}\right) + \frac{dA_\theta}{dt}\hat{\theta} + A_\theta\left(-\frac{d\theta}{dt}\hat{r} + \frac{d\phi}{dt}\cos\theta\hat{\phi}\right) + \\
&\quad \frac{dA_\phi}{dt}\hat{\phi} + A_\phi\left(-\frac{d\phi}{dt}\sin\theta\hat{r} - \frac{d\phi}{dt}\cos\theta\hat{\theta}\right) \quad \text{[using (3.8.5)]} \\
&= \left(\frac{dA_r}{dt} - A_\theta\frac{d\theta}{dt} - A_\phi\sin\theta\frac{d\phi}{dt}\right)\hat{r} + \left(\frac{dA_\theta}{dt} + A_r\frac{d\theta}{dt} - A_\phi\cos\theta\frac{d\phi}{dt}\right)\hat{\theta} + \\
&\quad \left(\frac{dA_\phi}{dt} + A_r\sin\theta\frac{d\phi}{dt} + A_\theta\cos\theta\frac{d\phi}{dt}\right)\hat{\phi}.
\end{aligned} \tag{3.8.7}$$

If in the above case $\vec{A} = \vec{r}$, the position vector, then $\frac{d\vec{r}}{dt}$ which denotes the velocity vector, may be obtained in the following way. We may write

$$\vec{A} = \vec{r} = \hat{\imath}x + \hat{\jmath}y + \hat{k}z;$$

using the values of $\hat{\imath}, \hat{\jmath}$ and $\hat{k}$ obtained by solving the simultaneous equations (3.8.2), (3.8.3) and (3.8.4) in the above expression we may show that

$$\vec{A} = \vec{r} = r\hat{r}. \tag{3.8.8}$$

Again

$$\vec{A} = \vec{r} = A_r\hat{r} + A_\theta\hat{\theta} + A_\phi\hat{\phi}. \tag{3.8.9}$$

Comparing (3.8.8) and (3.8.9), we get $A_r = r, A_\theta = 0, A_\phi = 0$, so that

$$\frac{dA_r}{dt} = \frac{dr}{dt}, \frac{dA_\theta}{dt} = 0 \text{ and } \frac{dA_\phi}{dt} = 0.$$

Putting these values in (3.8.7), we get

$$\vec{v} = \frac{d\vec{r}}{dt} = \frac{dr}{dt}\hat{r} + r\frac{d\theta}{dt}\hat{\theta} + r\sin\theta\frac{d\phi}{dt}\hat{\phi} = \dot{r}\hat{r} + r\dot{\theta}\hat{\theta} + r\sin\theta\dot{\phi}\hat{\phi} \tag{3.8.10}$$

$$= \text{velocity at any instant.}$$

The above expression for velocity may also be obtained by writing the position vector as $\vec{r} = r\hat{r}$ and then differentiating w.r.t time and putting the value of $\frac{d\hat{r}}{dt}$.

Expression for Acceleration

Differentiating equation (3.8.10) w.r.t time, we get

$$\vec{f} = \frac{d\vec{v}}{dt} = \frac{d}{dt}\left[\frac{dr}{dt}\hat{r} + r\frac{d\theta}{dt}\hat{\theta} + r\sin\theta\frac{d\phi}{dt}\hat{\phi}\right]$$

$$= \left(\frac{d^2r}{dt^2}\hat{r} + \frac{dr}{dt}\frac{d\hat{r}}{dt}\right) + \left(\frac{dr}{dt}\frac{d\theta}{dt}\hat{\theta} + r\frac{d^2\theta}{dt^2}\hat{\theta} + r\frac{d\theta}{dt}\frac{d\hat{\theta}}{dt}\right)$$

$$+ \left(\frac{dr}{dt}\sin\theta\frac{d\phi}{dt}\hat{\phi} + r\cos\theta\frac{d\theta}{dt}\frac{d\phi}{dt}\hat{\phi} + r\sin\theta\frac{d^2\phi}{dt^2}\hat{\phi} + r\sin\theta\frac{d\phi}{dt}\frac{d\hat{\phi}}{dt}\right)$$

$$= \left[\frac{d^2r}{dt^2}\hat{r} + \frac{dr}{dt}\left(\frac{d\theta}{dt}\hat{\theta} + \frac{d\phi}{dt}\sin\theta\hat{\phi}\right)\right] + \left[\frac{dr}{dt}\frac{d\theta}{dt}\hat{\theta} + r\frac{d^2\theta}{dt^2}\hat{\theta} + r\frac{d\theta}{dt}\left(-\frac{d\theta}{dt}\hat{r} + \frac{d\phi}{dt}\cos\theta\hat{\phi}\right)\right]$$

$$+ \left[\frac{dr}{dt}\sin\theta\frac{d\phi}{dt}\hat{\phi} + r\cos\theta\frac{d\theta}{dt}\frac{d\phi}{dt}\hat{\phi} + r\sin\theta\frac{d^2\phi}{dt^2}\hat{\phi}\right.$$

$$\left. + r\sin\theta\frac{d\phi}{dt}\left(-\frac{d\phi}{dt}\sin\theta\hat{r} - \frac{d\phi}{dt}\cos\theta\hat{\theta}\right)\right] \quad \text{[using (3.8.5)]}$$

$$= \left[\frac{d^2r}{dt^2} - r\left(\frac{d\theta}{dt}\right)^2 - r\sin^2\theta\left(\frac{d\phi}{dt}\right)^2\right]\hat{r} + \left[r\frac{d^2\theta}{dt^2} + 2\frac{dr}{dt}\frac{d\theta}{dt} - r\sin\theta\cos\theta\left(\frac{d\phi}{dt}\right)^2\right]\hat{\theta}$$

$$+ \left[r\sin\theta\frac{d^2\phi}{dt^2} + 2\frac{dr}{dt}\frac{d\phi}{dt}\sin\theta + 2r\frac{d\theta}{dt}\frac{d\phi}{dt}\cos\theta\right]\hat{\phi}$$

$$= (\ddot{r} - r\dot{\theta}^2 - r\sin^2\theta\dot{\phi}^2)\hat{r} + (r\ddot{\theta} + 2\dot{r}\dot{\theta} - r\sin\theta\cos\theta\dot{\phi}^2)\hat{\theta}$$

$$+ (r\sin\theta\ddot{\phi} + 2\dot{r}\dot{\phi}\sin\theta + 2r\dot{\theta}\dot{\phi}\cos\theta)\hat{\phi}$$

= Acceleration at any instant.

3.9 Expressions for velocity and acceleration of a particle in plane polar co-ordinates obtained in the same way (general treatment) as in 3.7 or 3.8

Any vector $\vec{A}$, which is a function of parameter t, may be written as, as shown in fig. 3.6.1,

$$\vec{A} = \hat{\imath}x + \hat{\jmath}y = \hat{\imath}r\cos\theta + \hat{\jmath}r\sin\theta \tag{3.9.1}$$

whence a unit vector in the direction of $\vec{r}$ increasing is

$$\hat{r} = \frac{\partial\vec{A}}{\partial r}\bigg/\left|\frac{\partial\vec{A}}{\partial r}\right| = \hat{\imath}\cos\theta + \hat{\jmath}\sin\theta \tag{3.9.2}$$

and a unit vector in the direction of θ increasing is

$$\hat{\theta} = \frac{\partial\vec{A}}{\partial\theta}\bigg/\left|\frac{\partial\vec{A}}{\partial\theta}\right| = -\hat{\imath}\sin\theta + \hat{\jmath}\cos\theta \tag{3.9.3}$$

From (3.9.2),

$$\left.\begin{aligned} \frac{d\hat{r}}{dt} &= \frac{d\hat{r}}{d\theta}\frac{d\theta}{dt} = \frac{d\theta}{dt}\hat{\theta} \\ \text{and from (3.9.3)}\quad \frac{d\hat{\theta}}{dt} &= \frac{d\hat{\theta}}{d\theta}\frac{d\theta}{dt} = -\frac{d\theta}{dt}\hat{r} \end{aligned}\right\}. \tag{3.9.4}$$

Now writing $\vec{A}$ as

$$\vec{A} = A_r\hat{r} + A_\theta\hat{\theta}. \tag{3.9.5}$$

A_r, A_θ being the components of A along $\hat{r}$, $\hat{\theta}$ respectively, we have, taking time derivative of (3.9.5),

$$\frac{d\vec{A}}{dt} = \frac{dA_r}{dt}\hat{r} + A_r\frac{d\hat{r}}{dt} + \frac{dA_\theta}{dt}\hat{\theta} + A_\theta\frac{d\hat{\theta}}{dt}$$

$$= \frac{dA_r}{dt}\hat{r} + A_r\frac{d\theta}{dt}\hat{\theta} + \frac{dA_\theta}{dt}\hat{\theta} - A_\theta\frac{d\theta}{dt}\hat{r} \quad \text{(using 3.9.4)}$$

$$= \left(\frac{dA_r}{dt} - A_\theta\frac{d\theta}{dt}\right)\hat{r} + \left(A_r\frac{d\theta}{dt} + \frac{dA_\theta}{dt}\right)\hat{\theta} \tag{3.9.6}$$

If in the above case $\vec{A} = \vec{r}$, the position vector, then $\frac{d\vec{r}}{dt}$, which denotes the velocity vector, may be obtained in the following way.

We may write

$$\vec{A} = \vec{r} = \hat{\imath}x + \hat{\jmath}y.$$

Solving the simultaneous equations (3.9.2) and (3.9.3) for $\hat{\imath}$ and $\hat{\jmath}$, we have

$$\left.\begin{aligned}\hat{\imath} &= \cos\theta\hat{r} - \sin\theta\hat{\theta}\\ \hat{\jmath} &= \sin\theta\hat{r} + \cos\theta\hat{\theta}\end{aligned}\right\} \tag{3.9.7}$$

so that

$$\vec{A} = \vec{r} = \hat{\imath}x + \hat{\jmath}y = (\cos\theta\hat{r} - \sin\theta\hat{\theta})r\cos\theta + (\sin\theta\hat{r} + \cos\theta\hat{\theta})r\sin\theta$$

[using (3.9.7) and putting the values of x and y]

$$= r\hat{r}. \tag{3.9.8}$$

Comparing (3.9.5) and (3.9.8), we get

$$A_r = r, \quad A_\theta = 0,$$

so that

$$\frac{dA_r}{dt} = \frac{dr}{dt}, \quad \frac{dA_\theta}{dt} = 0.$$

Putting these values in (3.9.6), we get

$$\vec{v} = \frac{d\vec{r}}{dt} = \frac{dr}{dt}\hat{r} + r\frac{d\theta}{dt}\hat{\theta} = \dot{r}\hat{r} + r\dot{\theta}\hat{\theta} = \text{ velocity at any instant}$$

which gives on differentiation

$$\vec{f} = \frac{d\vec{v}}{dt} = (\ddot{r} - r\dot{\theta}^2)\hat{r} + (r\ddot{\theta} + 2\dot{r}\dot{\theta})\hat{\theta} = \text{Acceleration at any instant.}$$

3.10 Impulsive Force or Time Integral of Force

From Newton's second law of motion we know that the time derivative of the momentum of a particle is equal to the force acting on it.

Thus

$$\frac{d}{dt}(m\vec{v}) = \vec{F}. \tag{3.10.1}$$

The vector $\vec{F}dt$ is called an *impulse of force* $\vec{F}$ acting in time dt. The impulse of force acting in the finite interval of time from t_1 to t_2 is

$$\int_{t_1}^{t_2} \vec{F}dt = \int_{v_1}^{v_2} d(m\vec{v}) = m\vec{v_2} - m\vec{v_1}. \tag{3.10.2}$$

Here $\vec{v_1}$ and $\vec{v_2}$ are the velocities of the particle at time t_1 and t_2 respectively, i.e., the impulse of a force is equal to the change in momentum. The expression (3.10.2) may be written in the form

$$\vec{F}_{av}\int_{t_1}^{t_2} dt = m\vec{v_2} - m\vec{v_1} \quad \text{or} \quad \vec{F}_{av}(t_2 - t_1) = m\vec{v_2} - m\vec{v_1}. \tag{3.10.3}$$

$$\vec{F}_{av} = \frac{\int_{t_1}^{t_2} \vec{F} dt}{t_2 - t_1}$$

is the average of the force in the time interval from t_1 to t_2. If the force $\vec{F}$ is constant, its impulse during the time interval $(t_2 - t_1)$ is $\vec{F}(t_2 - t_1)$. It should be noted that an impulsive force is a large force which acts on a body for a very short time. The displacement of the body during this small interval is negligible. The impulse of force though it acts for a very short time, may be measured by the change in momentum.

Illustrative Example

The polar co-ordinates of a particle moving in a plane are given by $r = a\sin w_1 t$ and $\theta = w_2 t$. Obtain expressions for the polar components of the velocity and acceleration of the particle.

[C.U. (Hons.) 1980]

$\because$ $r = a\sin w_1 t$ and $\theta = w_2 t$, we have

$$\frac{dr}{dt} = aw_1\cos w_1 t, \quad \frac{d^2r}{dt^2} = -aw_1^2 \sin w_1 t \quad \text{and} \quad \frac{d\theta}{dt} = w_2.$$

Radial component of velocity $= \dfrac{dr}{dt} = aw_1\cos w_1 t$.

Transverse component of velocity $= r\dfrac{d\theta}{dt} = rw_2 = aw_2 \sin w_1 t$.

Radial component of acceleration

$$f_r = \frac{d^2r}{dt^2} - r\left(\frac{d\theta}{dt}\right)^2 = -w_1^2 r - rw_2^2 = -r(w_1^2 + w_2^2)$$
$$= -a\sin w_1 t \times (w_1^2 + w_2^2).$$

Transverse component of acceleration

$$f_\theta = \frac{1}{r}\frac{d}{dt}\left(r^2\frac{d\theta}{dt}\right) = \frac{1}{r}\frac{d}{dt}(a^2\sin^2 w_1 t \times w_2)$$
$$= \frac{2a^2}{r} w_1 w_2 \sin w_1 t \cos w_1 t$$
$$= \frac{2a^2}{a\sin w_1 t} w_1 w_2 \sin w_1 t \cos w_1 t$$
$$= 2aw_1 w_2 \cos w_1 t.$$

3.11 Work

The work done by the force $\vec{F}$ causing displacement $d\vec{r}$ of a particle is given by

$$\delta W = \vec{F} \cdot d\vec{r} = F ds \cos\theta,$$

where θ is the angle between the vectors $\vec{F}$ and $d\vec{r}$ and ds is the magnitude of $d\vec{r}$; ds also denotes an element of length along the path of the particle and $F\cos\theta$ denotes the projection of $\vec{F}$ on the tangent to the path. It is to be noted that the element of work is denoted by δW and not dW since it is not a total differential in the general case. This means that the line integral of δW along an arbitrary closed path of the point of application of the force is not equal to zero. For example, the work done by frictional forces along a closed path may not be zero. The work done by the force $\vec{F}$ along a finite length s of the path is given by

$$W = \int_0^S F ds \cos\theta$$

F may not constant and may depend on the co-ordinates of the particle.

If the displacement is measured in centimeters and force in dynes, work will be obtained in ergs. Work will be obtained in foot-poundals, if the displacement is measured in feet and the force is poundals. As erg is a very small unit, it is convenient to define a larger unit for practical purposes. The practical unit of work in C.G.S. system is Joule which is equal to 10^7 ergs.

The M.K.S. unit of work will be in newton-metre where the displacement is measured in metre and the force in newton.

[1 newton-metre = 10^5 dynes $\times$ 100 cm = 10^7 ergs = 1 Joule.]

The gravitational unit of work in C.G.S. system is gram-centimetre which is equal to 981 ergs and that in F.P.S. system is foot-pound which is equal to 32.2 ft-poundals.

3.12 Power or Activity

The power of a force $\vec{F}$ denotes the time rate at which work is done by the force and is given by

$$P = \frac{dW}{dt} = \vec{F} \cdot \frac{d\vec{r}}{dt} = \vec{F} \cdot \vec{v}.$$

Thus the power is equal to the scalar product of the force and the velocity of displacement of the point of application of the force.

The C.G.S. unit of power is erg per second and its F.P.S. unit is foot-poundal per second. The C.G.S. practical units of power are watt and kilowatt.

1 watt = 1 Joule per second = 10^7 ergs/sec.

1 kilowatt – 1,000 watts.

The British practical unit used by engineers is called horse-power which is equal to 33,000 ft-lbs. per min. or 550 ft-lbs. per second.

[1 H.P. = 746 watts.]

3.13 Conservative Forces

Forces acting on a body are said to be conservative if the work done by the forces in causing displacement of the body depends only on the initial and final positions of the body and not on the path traversed from the initial to the final point. Hence the work done by a conservative force $\vec{F}$ along an arbitrary closed path of its point of application is zero. Thus

$$\oint \vec{F} \cdot d\vec{r} = 0 [\oint \text{ denotes the integration over a closed path}]. \tag{3.13.1}$$

If friction or other dissipative forces are present, the forces cannot be said to be conservative. When the friction is present $\vec{F} \cdot d\vec{r}$ is positive and hence the above integral is not zero. The Equation (3.13.1) may be written in the form

$$\text{curl}\, \vec{F} = \vec{\nabla} \times \vec{F} = 0 \quad [\text{by Stokes's theorem}] \tag{3.13.2}$$

Now, since the curl of a gradient always vanishes, $\vec{F}$ must be gradient of a scalar. Hence

$$\vec{F} = -\vec{\nabla} V = -\,\text{grad}\, V \tag{3.13.3}$$

V is called the potential energy, also called the scalar potential function. The existence of V may be established in the following way. If equation (3.13.1) holds, the work must be independent of the path of integration between the initial and final positions.

Hence it is possible to express work as the change in a quantity which depends only on the initial and final positions. This quantity may be represented by $-V$. So for differential path in case of conservative force, $\vec{F} \cdot d\vec{r} = -dV$.

3.14 Central Forces

A central force is a vector which is always directed towards a fixed point. The magnitude of this force solely depends on the distance r of the particle on which the force acts from the fixed point. Let us suppose that the force on a particle of mass m is $\vec{F} = m\vec{f}(r)$. This shows that the force is of magnitude $mf(r)$ and is radial. Now, the moment of this force about the fixed point is zero. Hence we may write

$$\vec{r} \times \vec{F} = 0 \quad \text{or} \quad \vec{r} \times \vec{f}(r) = 0 \quad \text{or} \quad \vec{r} \times \frac{d^2\vec{r}}{dt^2} = 0. \tag{3.14.1}$$

$\because \; \vec{f}(r) = \frac{d^2\vec{r}}{dt^2}$ = acceleration directed towards the fixed point.

Now,

$$\frac{d}{dt}\left(\vec{r} \times \frac{d\vec{r}}{dt}\right) = \left(\frac{d\vec{r}}{dt} \times \frac{d\vec{r}}{dt}\right) + \left(\vec{r} \times \frac{d^2\vec{r}}{dt^2}\right) = 0 + \vec{r} \times \frac{d^2\vec{r}}{dt^2}.$$

$$\therefore \; \vec{r} \times \frac{d^2\vec{r}}{dt^2} = \frac{d}{dt}\left(\vec{r} \times \frac{d\vec{r}}{dt}\right).$$

Hence from equation (3.14.1) we have

$$\frac{d}{dt}\left(\vec{r} \times \frac{d\vec{r}}{dt}\right) = 0.$$

Integrating, we have, $\left(\vec{r} \times \frac{d\vec{r}}{dt}\right) = h$, where h is a constant.

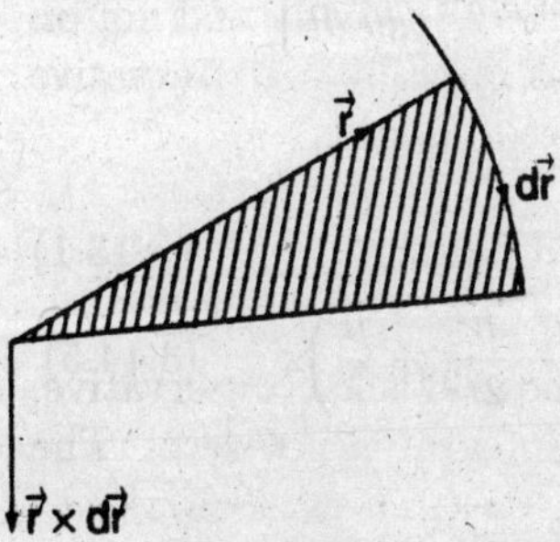

Fig. 3.14.1

The magnitude of $\left(\vec{r} \times \frac{d\vec{r}}{dt}\right)$ is equal to twice the area swept out by the radius vector in unit time (Fig. 3.14.1). This shows that the areal velocity is constant for motion under a central force. This will be discussed in the chapter on Gravitation.

It should also be noted that the vector product $\left(\vec{r} \times \frac{d\vec{r}}{dt}\right)$ is a vector normal to the plane determined by the fixed directions of $\vec{r}$ and $d\vec{r}$. The vector $m\left(\vec{r} \times \frac{d\vec{r}}{dt}\right)$ is called the moment of momentum or the angular momentum about the fixed point. Thus we may conclude that for motion of a particle under a central force field, the angular momentum vector remains constant, and the orbit of the particle is such that for any value of $\vec{r}$ and $\frac{d\vec{r}}{dt}$ on it, there is a constant vector associated, and hence the orbit described will lie in a plane.

It is to be remembered that in case of gravitation

$$f(r) = -\frac{\mu}{r^2} = -\frac{\mu}{r^3}\vec{r},$$

where μ is a constant.

Now, the potential energy per unit mass is given by

$$V(r) = -\int_{\infty}^{r} f(r)\,dr = \int_{\infty}^{r} \frac{\mu}{r^2}\,dr = -\frac{\mu}{r}.$$

The kinetic energy per unit mass $= \frac{1}{2}\dot{r}^2$.

Hence the total energy $E = \frac{1}{2}\dot{r}^2 - \frac{\mu}{r}$.

If the motion is in $(r - \theta)$ plane, the kinetic energy per unit mass $= \frac{1}{2}(\dot{r}^2 + r^2\dot{\theta}^2)$.

Hence the total energy

$$E = \frac{1}{2}(\dot{r}^2 + r^2\dot{\theta}^2) - \frac{\mu}{r}. \tag{3.14.2}$$

Again the areal velocity is

$$\frac{1}{2}r^2\dot{\theta} = \frac{h}{2} = \text{constant}. \tag{3.14.3}$$

$$\therefore\ r^2\dot{\theta}^2 = \frac{h^2}{r^2}. \tag{3.14.4}$$

Hence equation (3.14.2) gives

$$E = \frac{1}{2}\left(\dot{r}^2 + \frac{h^2}{r^2}\right) - \frac{\mu}{r} = \frac{1}{2}\left[\left(\frac{dr}{d\theta}\right)^2 \dot{\theta}^2 + \frac{h^2}{r^2}\right] - \frac{\mu}{r} \quad \left[\because \dot{r} = \frac{dr}{d\theta}\frac{d\theta}{dt}\right]$$

$$= \frac{1}{2}\left[\frac{h^2}{r^4}\left(\frac{dr}{d\theta}\right)^2 + \frac{h^2}{r^2}\right] - \frac{\mu}{r} \quad \text{[From (3.14.4)]}$$

$$\text{or} \quad 2E = \frac{h^2}{r^4}\left(\frac{dr}{d\theta}\right)^2 + \frac{h^2}{r^2} - \frac{2\mu}{r} \quad \text{or} \quad \left(\frac{dr}{d\theta}\right)^2 = \frac{2r^4}{h^2}\left(E - \frac{h^2}{2r^2} + \frac{\mu}{r}\right). \tag{3.14.5}$$

From this we get the polar equation of Conic. [See Chapter VII.]

It can be shown that the *Central force is conservative.* If the central force is conservative,

$$\text{curl}\,\vec{F} = 0; \quad \text{or,} \quad \text{curl}\,\vec{f}(r) = 0,$$

$$\text{or,} \quad \text{curl}\, f(r)\hat{r} = 0, \quad \text{where } \hat{r} \text{ is the unit vector along } \vec{r},$$

$$\text{i.e.,} \quad \vec{\nabla} \times [f(r)\hat{r}] = 0,$$

$$\text{or,} \quad \vec{\nabla} \times [f(r)\vec{r}/r] = 0.$$

Now

$$\vec{\nabla} \times \left[f(r)\frac{\vec{r}}{r}\right] = \vec{\nabla} \times \left[f(r)\frac{x\hat{\imath}}{r} + f(r)\frac{y\hat{\jmath}}{r} + f(r)\frac{z\hat{k}}{r}\right]$$

$$= \begin{vmatrix} \hat{\imath} & \hat{\jmath} & \hat{k} \\ \dfrac{\partial}{\partial x} & \dfrac{\partial}{\partial y} & \dfrac{\partial}{\partial z} \\ \dfrac{x}{r}f(r) & \dfrac{y}{r}f(r) & \dfrac{z}{r}f(r) \end{vmatrix}$$

$$= \hat{\imath}\left(\frac{z}{r}\frac{\partial f}{\partial y} - \frac{y}{r}\frac{\partial f}{\partial z}\right) + \hat{\jmath}\left(\frac{x}{r}\frac{\partial f}{\partial z} - \frac{z}{r}\frac{\partial f}{\partial x}\right) + \hat{k}\left(\frac{y}{r}\frac{\partial f}{\partial x} - \frac{x}{r}\frac{\partial f}{\partial y}\right).$$

Now

$$\frac{\partial f}{\partial x} = \frac{\partial f}{\partial r}\frac{\partial r}{\partial x} = \frac{\partial f}{\partial r}\frac{x}{r} \quad [\because r^2 = x^2 + y^2 + z^2]$$

Similarly,

$$\frac{\partial f}{\partial y} = \frac{\partial f}{\partial r}\frac{y}{r} \quad \text{and} \quad \frac{\partial f}{\partial z} = \frac{\partial f}{\partial r}\frac{z}{r}.$$

Hence

$$\vec{\nabla} \times \left[f(r)\frac{\vec{r}}{r}\right] = 0.$$

This shows that the central force is conservative.

3.15 Conservation of energy for Central force field

If $F(r)$ be the central force, the equation of motion of a body of mass m under central force is given by

$$m(\ddot{r} - r\dot{\theta}^2) = F(r). \tag{3.15.1}$$

Here $(\ddot{r} - r\dot{\theta}^2)$ denotes the radial acceleration. [See Art. 3.5 or 3.6]

From equation (3.14.3) we have

$$m\dot{\theta} = \frac{mh}{r^2} = \frac{H}{r^2} \; (H = \text{a constant}) \tag{3.15.2}$$

$\therefore$ Substituting the value of $\dot{\theta}$ in equation (3.15.1), we have

$$m\left(\ddot{r} - \frac{H^2}{m^2r^3}\right) = F(r) \quad \text{or} \quad m\ddot{r} = \frac{H^2}{mr^3} + F(r).$$

Since a central force is a conservative field it can be derived from a potential V which is a function of r alone.

$$\therefore\; m\ddot{r} = \frac{H^2}{mr^3} - \frac{\partial V}{\partial r} = -\frac{1}{2}\left(\frac{\partial}{\partial r}\right)\left(\frac{H^2}{mr^2}\right) - \frac{\partial V}{\partial r}$$

$$= -\frac{\partial}{\partial r}\left[\frac{1}{2}\left(\frac{H^2}{mr^2}\right) + V\right].$$

Multiplying both sides of the equation by $\dot{r}$, we have

$$m\dot{r}\ddot{r} = -\dot{r}\left(\frac{\partial}{\partial r}\right)\left[\frac{1}{2}\left(\frac{H^2}{mr^2}\right) + V\right]$$

$$\text{or} \quad \frac{d}{dt}\left(\frac{1}{2}m\dot{r}^2\right) = -\frac{d}{dt}\left[\frac{1}{2}\left(\frac{H^2}{mr^2}\right) + V\right]$$

$$\text{or} \quad \frac{1}{2}m\dot{r}^2 + \frac{1}{2}\left(\frac{H^2}{mr^2}\right) + V = \text{Constant}. \tag{3.15.3}$$

Now, the Kinetic energy

$$T = \frac{1}{2}m(\dot{r}^2 + r^2\dot{\theta}^2) = \frac{1}{2}m\dot{r}^2 + \frac{1}{2}mr^2\dot{\theta}^2$$

$$= \frac{1}{2}m\dot{r}^2 + \frac{1}{2}mr^2\frac{H^2}{r^4m^2} = \frac{1}{2}m\dot{r}^2 + \frac{1}{2}\left(\frac{H^2}{mr^2}\right). \tag{3.15.4}$$

$\therefore$ Total energy E = Kinetic energy + potential energy

$$= \frac{1}{2}m\dot{r}^2 + \frac{1}{2}\left(\frac{H^2}{mr^2}\right) + V.$$

Hence from the equation (3.15.3), we have E = constant.

This proves the conservation of energy for a central force field.

3.16 Potential Energy

The potential energy is the energy which a body possesses by virtue of its position or configuration and is measured by the work done by the body in passing from its present position or configuration to some standard position or configuration. The potential energy in the standard position is taken to be zero.

When a body is lifted, it will gain potential energy because we do work and expend energy in lifting the body. When the body is allowed to fall from this elevated position, it loses potential energy which is transformed into kinetic energy. By stretching a spring, we do work which is stored up in the spring as potential energy.

We consider a body of mass m raised to a vertical height h above the earth's surface. The potential energy of the body = work done in raising the body against the force due to gravity from the surface of the earth or the ground level which is usually considered as the standard position = mgh. In case of an extended body, it is to be noted that the height h is the vertical distance between the centres of gravity at the two positions.

In a conservative field, the potential energy $V(r)$ at any point is equal to the amount of work done in taking any particle from the point to any other fixed point. [The fixed point is usually taken to be at infinity.] Hence

$$V(r) = \int_{\vec{r_1}}^{\vec{r_2}} \vec{F}\cdot d\vec{r},$$

where $\vec{F}$ denotes the force acting on the particle whose position vector is $\vec{r_1}$; $\vec{r_2}$ denotes the position vector of the fixed point.

The above expression may be written in the form

$$V(\vec{r}) = -\int_{\vec{r_2}}^{\vec{r_1}} \vec{F}\cdot d\vec{r} = -\int_{P_2}^{P_1} (\hat{\imath}F_x + \hat{\jmath}F_y + \hat{k}F_z)\cdot(\hat{\imath}dx + \hat{\jmath}dy + \hat{k}dz)$$

$$= -\int_{P_2}^{P_1} (F_x dx + F_y dy + F_z dz). \tag{3.16.1}$$

The right-hand side denotes the work done against the force in transferring the particle from the fixed position P_2 to the position P_1 in question.

Now, $V = f(x, y, z)$, where x, y, z denote the co-ordinates of the particle.

$$\therefore\ dV = \frac{\partial V}{\partial x}dx + \frac{\partial V}{\partial y}dy + \frac{\partial V}{\partial z}dz. \tag{3.16.2}$$

Also from equation (3.16.1), we have

$$dV = -(F_x dx + F_y dy + F_z dz). \tag{3.16.3}$$

As dx, dy, dz are arbitrary, we have comparing equations (3.16.2) and (3.16.3)

$$F_x = -\frac{\partial V}{\partial x}, \quad F_y = -\frac{\partial V}{\partial y}, \quad F_z = -\frac{\partial V}{\partial z}. \tag{3.16.4}$$

Now

$$\vec{F} = \hat{\imath}F_x + \hat{\jmath}F_y + \hat{\jmath}F_z = -\left(\hat{\imath}\frac{\partial V}{\partial x} + \hat{\jmath}\frac{\partial V}{\partial y} + \hat{k}\frac{\partial V}{\partial z}\right) = -\,\text{grad}\, V.$$

This is same as equation (3.13.3).

3.17 Conservation of Energy

We know that

$$dV = -(F_x dx + F_y dy + F_z dz) = -\vec{F}\cdot d\vec{r}$$

$$\text{or} \quad \int_{\vec{r_1}}^{\vec{r_2}} \vec{F}\cdot d\vec{r} = -\int_{V_1}^{V_2} dV = V_1 - V_2.$$

Also,

$$\int_{\vec{r_1}}^{\vec{r_2}} \vec{F}\cdot d\vec{r} = \int_{\vec{r_1}}^{\vec{r_2}} m\vec{f}\cdot d\vec{r} = \int m\frac{d\vec{v}}{dt}\cdot\frac{d\vec{r}}{dt}\,dt = \int_{\vec{v_1}}^{\vec{v_2}} m\frac{d\vec{v}}{dt}\cdot\vec{v}dt$$

$$= m\int_{v_1}^{v_2} v\,dv = \frac{m}{2}(v_2^2 - v_1^2).$$

$$\therefore\ V_1 - V_2 = \frac{m}{2}(v_2^2 - v_1^2).$$

$$\therefore\ \frac{m}{2}v_1^2 + V_1 = \frac{m}{2}v_2^2 + V_2. \tag{3.17.1}$$

This shows that in a conservative field the sum of the kinetic energy and potential energy of a particle at any point remains constant. This is called conservation of energy.

3.17.1 Torque Acting on a Particle

A force is required to produce linear acceleration in a particle. In a similar way, a torque (or moment of a force) is required to produce angular acceleration in a particle about some axis. Like force, the torque is also a vector quantity, and is defined in the following way.

Let a force $\vec{F}$ act on a particle at P whose position vector is $\vec{r}$ w.r.t. the origin O of an inertial reference frame (Fig 3.17.1). Then the torque $\vec{N}$ acting on the particle about the origin O is defined as the *vector product* of $\vec{r}$ and $\vec{F}$. That is

$$\vec{N} = \vec{r}\times\vec{F} \tag{3.17.2}$$

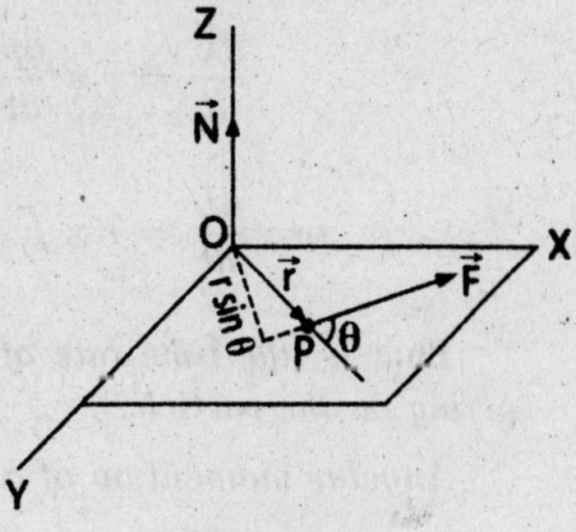

Fig. 3.17.1

Its scalar magnitude is $N = rF\sin\theta$, where θ is the angle between $\vec{r}$ and $\vec{F}$ and is along a direction perpendicular to the plane containing $\vec{r}$ and $\vec{F}$. In the figure, $\vec{r}$ and $\vec{F}$ lie in the X-Y plane, and the torque $\vec{N}$ is along the *positive* z-axis according to right hand rule (If the fingers of the right-hand be curled in the direction of turning from $\vec{r}$ to $\vec{F}$ through the smaller angle between them, the thumb will point in the direction of $\vec{N}$).

It is evident that the torque is a maximum when the force is applied at right angles to $\vec{r}$ (i.e. when $\theta = 90°$). This is why in opening or closing a heavy revolving door, the force is applied (by hand) at right angles to the door at its outer edge. Besides this, the torque depends also on the position of the point relative to origin at which the force is applied. A force applied at the origin O (i.e. $\vec{r}$ is zero) produces zero torque about O. This is why we cannot open or close a door by applying force at the hinge.

Angular momentum of a particle: We consider a particle at P of mass m, whose position vector is $\vec{r}$ w.r.t. the origin O of an inertial frame (Fig. 3.17.2). Let $\vec{p}(= m\vec{v})$ be the linear momentum of the particle. Then the angular momentum $\vec{L}$ of the particle about the origin O is defined as the *vector product* of $\vec{r}$ and $\vec{p}$. That is

$$\vec{L} = \vec{r} \times \vec{p} \qquad (3.17.3)$$

$\vec{L}$ is a vector quantity. Its scalar magnitude is $L = rp\sin\theta$, when θ is the angle between $\vec{r}$ and $\vec{p}$, and is along a direction perpendicular to the plane containing $\vec{r}$ and $\vec{p}$. In the Fig 3.17.2, $\vec{r}$ and $\vec{p}$ lie in the X-Y plane, and the angular momentum vector $\vec{L}$ is along the *positive* z-axis (according to right hand rule). It is evident that the angular momentum about O is zero, if the line of action of $\vec{p}$ passes through O (i.e. $\vec{r} = 0$) or if the angle between $\vec{r}$ and $\vec{p}$ is 0° or 180°.

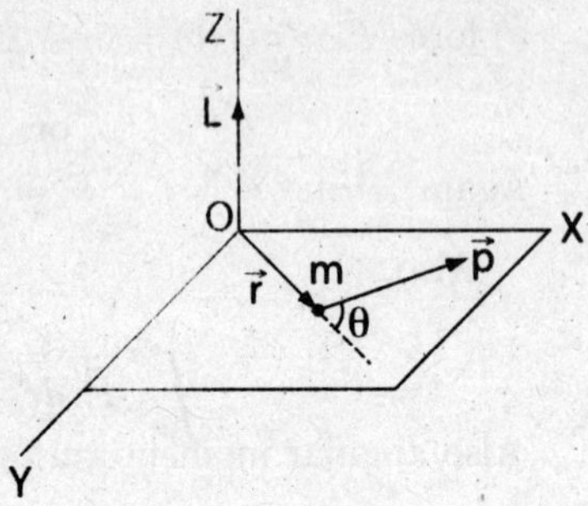

Fig. 3.17.2

Relation between Torque and Angular Momentum: Differentiating (3.17.3) w.r.t. time, we obtain

$$\frac{d\vec{L}}{dt} = \frac{d}{dt}(\vec{r} \times \vec{p}) = \left(\frac{d\vec{r}}{dt} \times \vec{p}\right) + \left(\vec{r} \times \frac{d\vec{p}}{dt}\right)$$

$$= (\vec{v} \times m\vec{v}) + \left(\vec{r} \times \frac{d\vec{p}}{dt}\right)$$

But $\vec{v} \times m\vec{v}$ is the *vector product* of two parallel vectors, and is therefore zero. Thus

$$\frac{d\vec{L}}{dt} = \vec{r} \times \frac{d\vec{p}}{dt} = \vec{r} \times \vec{F}$$ [$\because$ force $\vec{F} = \frac{d\vec{p}}{dt}$, the rate of change of momentum, from Newton's second law]

or $$\frac{d\vec{L}}{dt} = \vec{r} \times \vec{F} = \vec{N} \quad \text{[from eqn (3.17A.1)]} \qquad (3.17.4)$$

That is, *the time rate of change of angular momentum of a particle is equal to the torque acting on the particle.*

Angular momentum of a particle with constant velocity: We have shown that

$$\frac{d\vec{L}}{dt} = \vec{r} \times m\frac{d\vec{v}}{dt}$$

If $\vec{v}$ is constant, then $\frac{d\vec{v}}{dt} = 0$, so that $\frac{d\vec{L}}{dt} = 0$.

$\therefore$ $\vec{L}$ = constant—which shows that in absence of external torque $\vec{N}$, angular momentum remains conserved which is the law of conservation of angular momentum.

Illustrative Examples

1. A particle of unit mass moves according to the equation $\vec{r} = \hat{\imath}(2 + 3t^2) + \hat{\jmath}5t^2 + \hat{k}t$. Find the force $\vec{F}$ acting on it, torque $\vec{N}$ and angular momentum $\vec{L}$ about the origin. Hence verify $\vec{N} = \frac{d\vec{L}}{dt}$.

[C.U. (Hons.) 2002]

Here $\vec{r} = \hat{\imath}(2 + 3t^2) + \hat{\jmath}5t^2 + \hat{k}t$.

$\therefore$ velocity $\vec{v} = \frac{d\vec{r}}{dt} = \hat{\imath}6t + \hat{\jmath}10t + \hat{k}$ and acceleration $\vec{f} = \frac{d^2\vec{r}}{dt^2} = \hat{\imath}6 + \hat{\jmath}10$.

$\therefore$ force $\vec{F} = m\vec{f}(t) = 6\hat{\imath} + 10\hat{\jmath}$ (for $m = 1$).

$$\text{Again torque } \vec{N} = \vec{r} \times \vec{F} = \begin{vmatrix} \hat{\imath} & \hat{\jmath} & \hat{k} \\ 2+3t^2 & 5t^2 & t \\ 6 & 10 & 0 \end{vmatrix}$$

$$= \hat{\imath}(-10t) - \hat{\jmath}(-6t) + \hat{k}(20 + 30t^2 - 30t^2) = -10t\hat{\imath} + 6t\hat{\jmath} + 20\hat{k}$$

Also angular momentum for unit mass $\vec{L} = \vec{r} \times \vec{v}$

$$= \begin{vmatrix} \hat{\imath} & \hat{\jmath} & \hat{k} \\ 2+3t^2 & 5t^2 & t \\ 6t & 10t & 1 \end{vmatrix}$$

$$= \hat{\imath}(5t^2 - 10t^2) - \hat{\jmath}(2 + 3t^2 - 6t^2) + \hat{k}(20t + 30t^3 - 30t^3)$$

$$= -5t^2\hat{\imath} - (2 - 3t^2)\hat{\jmath} + 20t\hat{k} = -5t^2\hat{\imath} + (3t^2 - 2)\hat{\jmath} + 20t\hat{k}$$

$$\therefore \quad \frac{d\vec{L}}{dt} = -10t\hat{\imath} + 6t\hat{\jmath} + 20\hat{k} = \hat{N} \quad \text{(verified)}$$

2. The trajectory of a particle of unit mass is given by the radius vector $\vec{r} = \hat{\imath}a\cos\omega t + \hat{\jmath}b\sin\omega t$, where a, b are constants. Calculate angular momentum of the particle about the origin. Show that it is constant and along $\hat{k}$.

[C.U. (Hons.) 1997]

Here $\vec{r} = \hat{\imath}a\cos\omega t + \hat{\jmath}b\sin\omega t$.

$$\therefore \quad \vec{v} = \frac{d\vec{r}}{dt} = -\hat{\imath}a\omega\sin\omega t + \hat{\jmath}b\omega\cos\omega t$$

Angular momentum of the particle of unit mass is given by

$$\vec{L} = \vec{r} \times \vec{v} = \begin{vmatrix} \hat{\imath} & \hat{\jmath} & \hat{k} \\ a\cos\omega t & b\sin\omega t & 0 \\ -a\omega\sin\omega t & b\omega\cos\omega t & 0 \end{vmatrix} = ab\omega\hat{k}$$

which is constant and along $\hat{k}$.

3. The trajectory of a particle of mass m moving on a plane is given by $x = x_0 \cos\omega_1 t$, $y = y_0 \sin\omega_2 t$. Find the force $\vec{F}$, torque $\vec{N}$ about the origin acting on the particle. Find also the angular momentum $\vec{L}$ about the origin.

Find expressions for P.E. and K.E. of the particle at any instant of time t. Is the total energy conserved? Under what condition is the force central? What is $\vec{N}$ when the force is central?

[C.U. (Hons.) 2003]

Here $\vec{r} = \hat{i}x + \hat{j}y = \hat{i}x_0 \cos\omega_1 t + \hat{j}y_0 \sin\omega_2 t$.

Velocity $\vec{v} = \dfrac{d\vec{r}}{dt} = -\hat{i}x_0\omega_1 \sin\omega_1 t + \hat{j}y_0\omega_2 \cos\omega_2 t$.

$\therefore$ force acting on the particle

$$\vec{F} = m\frac{d^2\vec{r}}{dt^2} = m\left(-\hat{i}x_0\omega_1^2 \cos\omega_1 t - \hat{j}y_0\omega_2^2 \sin\omega_2 t\right) = -m\left(\hat{i}\omega_1^2 x + \hat{j}\omega_2^2 y\right)$$

Again, torque acting is given by

$$\vec{N} = \vec{r}\times\vec{F}\begin{vmatrix} \hat{i} & \hat{j} & \hat{k} \\ x_0\cos\omega_1 t & y_0 \sin\omega_2 t & 0 \\ -x_0\omega_1^2\cos\omega_1 t & -y_0\omega_2^2 \sin\omega_2 t & 0 \end{vmatrix}$$

$$= \hat{k}\left(-x_0y_0\omega_2^2 \cos\omega_1 t \sin\omega_2 t + x_0y_0\omega_1^2 \cos\omega_1 t\sin\omega_2 t\right)$$

$$= \hat{k}(\omega_1^2 - \omega_2^2)xy \quad \text{which is along } \hat{k}$$

Angular momentum

$$\vec{L} = \vec{r}\times m\vec{v}\begin{vmatrix} \hat{i} & \hat{j} & \hat{k} \\ x_0\cos\omega_1 t & y_0 \sin\omega_2 t & 0 \\ -mx_0\omega_1\sin\omega_1 t & my_0\omega_2 \cos\omega_2 t & 0 \end{vmatrix}$$

$$= \hat{k}(mx_0y_0\omega_2 \cos\omega_1 t\cos\omega_2 t + mx_0y_0\omega_1 \sin\omega_1 t\sin\omega_2 t)$$

$$= \hat{k}(m\omega_1 x_0 y \sin\omega_1 t + m\omega_2 y_0 x\cos\omega_2 t)$$

which is also directed along $\hat{k}$.

To find K.E., we see that

$$x = x_0\cos\omega_1 t \quad \text{or} \quad \dot{x} = -x_0\omega_1 \sin\omega_1 t$$

$$\text{and} \quad y = y_0 \sin\omega_2 t \quad \text{or} \quad \dot{y} = y_0\omega_2\cos\omega_2 t$$

$\therefore$ K.E. at any instant $= \frac{1}{2}m\dot{x}^2 + \frac{1}{2}m\dot{y}^2 = \frac{1}{2}mx_0^2\omega_1^2 \sin^2\omega_1 t + \frac{1}{2}my_0^2\omega_2^2\cos^2\omega_2 t$.

Again P.E. at any instant

$$= -\int \vec{F}\cdot d\vec{r} = -\int_0^t \vec{F}\cdot\frac{d\vec{r}}{dt}dt$$

$$= -\int_0^t -m(\hat{i}x_0\omega_1^2\cos\omega_1 t + \hat{j}y_0\omega_2^2\sin\omega_2 t)\cdot(-\hat{i}x_0\omega_1\sin\omega_1 t + \hat{j}y_0\omega_2\cos\omega_2 t)dt$$

$$= \int_0^t m(-x_0^2\omega_1^3\sin\omega_1 t\cos\omega_1 t + y_0^2\omega_2^3\sin\omega_2 t\cos\omega_2 t)dt$$

$$= -mx_0^2\omega_1^3\int_0^t \frac{\sin 2\omega_1 t}{2}dt + my_0^2\omega_2^3\int_0^t \frac{\sin 2\omega_2 t}{2}dt$$

$$= -\frac{mx_0^2\omega_1^3}{2}\left[\frac{-\cos 2\omega_1 t}{2\omega_1}\right]_0^t + \frac{my_0^2\omega_2^3}{2}\left[\frac{-\cos 2\omega_2 t}{2\omega_2}\right]_0^t$$

$$= -\frac{mx_0^2\omega_1^2}{4}\left(\cos 2\omega_1 t\right)_t^{\circ} + \frac{my_0^2\omega_2^2}{4}\left(\cos 2\omega_2 t\right)_t^{\circ}$$

$$= -\frac{mx_0^2\omega_1^2}{4}(1-\cos 2\omega_1 t) + \frac{my_0^2\omega_2^2}{4}(1-\cos 2\omega_2 t)$$

$$= -\frac{mx_0^2\omega_1^2}{2}\sin^2\omega_1 t + \frac{my_0^2\omega_2^2}{2}\sin^2\omega_2 t$$

$\therefore$ T.E. = K.E. + P.E. = $\frac{my_0^2\omega_2^2}{2}$ which is constant.

$\therefore$ T.E. is conserved.

The force will be central if the angular momentum remains constant, i.e. conserved when $\vec{L}$ = constant.

Now $\vec{N} = \frac{d\vec{L}}{dt} = 0$ i.e. $\vec{N}$ will be zero if the force is central; or in other words, in absence of external torque angular momentum remains constant i.e. conserved.

4. Estimate whether the following forces are conservative:

(i) $\vec{F} = \left(\frac{\alpha}{r^4}x,\ \frac{\alpha}{r^4}y,\ \frac{\alpha}{r^4}z\right)$; (ii) $\vec{F} = (3z+5y)\hat{\imath} + (5x+2z)\hat{k} + (2y+3x+4z)\hat{l}$,

where α is a constant and $\hat{\imath}, \hat{k}$ and $\hat{l}$ are the unit vectors in the x, y and z, directions respectively.

[C.U. (Hons.) 1977]

(i) $\vec{F} = \frac{\alpha}{r^4}x\hat{\imath} + \frac{\alpha}{r^4}y\hat{k} + \frac{\alpha}{r^4}z\vec{l}.$

If $\vec{F}$ is conservative $\vec{\nabla}\times\vec{F}$ must be equal to zero. Now $\vec{\nabla}\times\vec{F}$

$$= \begin{vmatrix} \hat{\imath} & \hat{k} & \vec{l} \\ \frac{\partial}{\partial x} & \frac{\partial}{\partial y} & \frac{\partial}{\partial z} \\ \frac{\alpha}{r^4}x & \frac{\alpha}{r^4}y & \frac{\alpha}{r^4}z \end{vmatrix}$$

$$= \hat{\imath}\left[\frac{\partial}{\partial y}\left(\frac{\alpha}{r^4}z\right) - \frac{\partial}{\partial z}\left(\frac{\alpha}{r^4}y\right)\right] + \hat{k}\left[\frac{\partial}{\partial z}\left(\frac{\alpha}{r^4}x\right) - \frac{\partial}{\partial x}\left(\frac{\alpha}{r^4}z\right)\right] + \vec{l}\left[\frac{\partial}{\partial x}\left(\frac{\alpha}{r^4}y\right) - \frac{\partial}{\partial y}\left(\frac{\alpha}{r^4}x\right)\right]$$

$$= \hat{\imath}\left[-\frac{4\alpha z}{r^5}\frac{\partial r}{\partial y} + \frac{4\alpha y}{r^5}\frac{\partial r}{\partial z}\right] + \hat{k}\left[-\frac{4\alpha x}{r^5}\frac{\partial r}{\partial z} + \frac{4\alpha z}{r^5}\frac{\partial r}{\partial x}\right] + \vec{l}\left[-\frac{4\alpha y}{r^5}\frac{\partial r}{\partial x} + \frac{4\alpha x}{r^5}\frac{\partial r}{\partial y}\right]$$

$$= \hat{\imath}\left[-\frac{4\alpha z}{r^5}\frac{y}{r} + \frac{4\alpha y}{r^5}\frac{z}{r}\right] + \hat{k}\left[-\frac{4\alpha x}{r^5}\frac{z}{r} + \frac{4\alpha z}{r^5}\frac{x}{r}\right] + \vec{l}\left[-\frac{4\alpha y}{r^5}\frac{x}{r} + \frac{4\alpha x}{r^5}\frac{y}{r}\right]$$

$$= 0 \quad \left[\because x^2+y^2+z^2=r^2 \quad \therefore \frac{\partial r}{\partial x} = \frac{x}{r}, \frac{\partial r}{\partial y} = \frac{y}{r}, \text{ and } \frac{\partial r}{\partial z} = \frac{z}{r}\right].$$

$\therefore$ **F is conservative.**

(ii) Here,

$$\vec{\nabla} \times \vec{F} = \begin{vmatrix} \hat{\imath} & \hat{k} & \hat{l} \\ \frac{\partial}{\partial x} & \frac{\partial}{\partial y} & \frac{\partial}{\partial z} \\ (3z + 5y) & (5x + 2z) & (2y + 3x + 4z) \end{vmatrix}$$

$$= \hat{\imath}\left[\frac{\partial}{\partial y}(2y + 3x + 4z) - \frac{\partial}{\partial z}(5x + 2z)\right] + \hat{k}\left[\frac{\partial}{\partial z}(3z + 5y) - \frac{\partial}{\partial x}(2y + 3x + 4z)\right]$$

$$+ \hat{l}\left[\frac{\partial}{\partial x}(5x + 2z) - \frac{\partial}{\partial y}(3z + 5y)\right]$$

$$= \hat{\imath}(2 - 2) + \hat{k}(3 - 3) + \hat{l}(5 - 5)$$

$$= 0.$$

Hence F is conservative.

5. Prove that the total energy of a particle of mass m acted upon by a central force is given by

$$E = \frac{h^2}{2m}\left[u^2 + \left(\frac{du}{d\theta}\right)^2\right] + V(r)$$

where $V(r)$ is the potential energy and h is the angular momentum of the body; $u = \frac{1}{r}$, r and θ being the polar co-ordinates of the particle. [C.U. (Hons.) 1980, '85, '87, 2006 (old syll)]

Kinetic energy in $(r - \theta)$ plane $= \frac{1}{2}m(\dot{r}^2 + r^2\dot{\theta}^2)$.

Hence the total energy is given by

$$E = \frac{1}{2}m(\dot{r}^2 + r^2\dot{\theta}^2) + V(r),$$

where $V(r)$ is the potential energy. Now the angular momentum is $mr^2\dot{\theta} = h$.

$$\therefore\ r^2\dot{\theta}^2 = \frac{h^2}{m^2r^2} = \frac{h^2u^2}{m^2} \qquad \left[\text{Putting } \frac{1}{r^2} = u^2\right]$$

Again

$$\dot{r} = \frac{d}{dt}\left(\frac{1}{u}\right) = -\frac{1}{u^2}\frac{du}{d\theta}\frac{d\theta}{dt} = -\frac{1}{u^2}\frac{du}{d\theta}\frac{h}{m}u^2 = -\frac{h}{m}\left(\frac{du}{d\theta}\right).$$

Substituting the values of $r^2\dot{\theta}^2$ and $\dot{r}^2$ in the expression for total energy we have

$$E = \frac{1}{2}m\left[\frac{h^2}{m^2}\left(\frac{du}{d\theta}\right)^2 + \frac{h^2}{m^2}u^2\right] + V(r) = \frac{h^2}{2m}\left[u^2 + \left(\frac{du}{d\theta}\right)^2\right] + V(r).$$

6. Show that the gravitational force is conservative.

The gravitational force is given by

$$\vec{F} = -\frac{Gm_1m_2}{r^3}\vec{r} = -\frac{K}{r^3}\vec{r}.$$

If the gravitational force is conservative,

$$\vec{\nabla} \times \vec{F} = 0 \text{ or } \vec{\nabla} \times \left(-\frac{K}{r^3}\vec{r}\right) = 0.$$

Now

$$\vec{\nabla} \times \left(-\frac{K}{r^3}\vec{r}\right) = \vec{\nabla}\left(-\frac{K}{r^3}\right) \times \vec{r} - \frac{K}{r^3}(\vec{\nabla} \times \vec{r}).$$

Again

$$\because \quad \vec{\nabla} \times r = \begin{vmatrix} \hat{\imath} & \hat{\jmath} & \hat{k} \\ \frac{\partial}{\partial x} & \frac{\partial}{\partial y} & \frac{\partial}{\partial z} \\ x & y & z \end{vmatrix} = 0,$$

$$\therefore \vec{\nabla} \times \left(-\frac{K}{r^3}\vec{r}\right) - \vec{\nabla}\left(-\frac{K}{r^3}\right) \times \vec{r} = \frac{3K}{r^4}\hat{r} \times \vec{r} = 0 \quad [\because \hat{r} \times \vec{r} = 0].$$

$\therefore\ \vec{\nabla} \times \vec{F} = 0$. Thus the gravitational force is conservative.

7. A particle moves in a field of force given by

$$F_x = yz(1 - 2xyz),$$

$$F_y = zx(1 - 2xyz),$$

$$\text{and} \quad F_z = xy(1 - 2xyz).$$

Verify that the force is conservative and find the potential function from which it is derivable.

[C.U. (Hons.) 1980, cf. 1998]

Here

$$\vec{F} = \hat{\imath}F_x + \hat{\jmath}F_y + \hat{k}F_z = \hat{\imath}yz(1 - 2xyz) + \hat{\jmath}zx(1 - 2xyz) + \hat{k}xy(1 - 2xyz)$$

and $\vec{\nabla} \times \vec{F} = 0$—so the force is conservative and the field of force is irrotational.

If now ϕ be the potential function from which the force is derivable, then $\vec{F} = \vec{\nabla}\phi$

$$\text{or} \quad \hat{\imath}yz(1 - 2xyz) + \hat{\jmath}zx(1 - 2xyz) + \hat{k}xy(1 - 2xyz) = \hat{\imath}\frac{\partial\phi}{\partial x} + \hat{\jmath}\frac{\partial\phi}{\partial y} + \hat{k}\frac{\partial\phi}{\partial z}.$$

$$\therefore \text{(i) } \frac{\partial\phi}{\partial x} = yz(1 - 2xyz); \text{ (ii) } \frac{\partial\psi}{\partial y} = zx(1 - 2xyz); \text{ (iii) } \frac{\partial\phi}{\partial z} = xy(1 - 2xyz),$$

whence on integration of (i) w.r.t. x, keeping y and z constant, we get $\phi = xyz(1 - xyz) + g_1(y,z)$ where $g_1(y,z) = a$ constant being a function of y and z. Similarly from (ii) $\phi = xyz(1 - xyz) + g_2(z,x)$ and from (iii) $\phi = xyz(1 - xyz) + g_3(x,y)$. Now, choosing

$g_1(y, z) = g_2(z, x) = g_3(x, y) = c$, c being an arbitrary constant, the above three expressions give a common potential $\phi = xyz(1 - xyz) + c$.

8. Two bodies of mass m and M are held at a distance h apart. Only force to be considered is a force of mutual attraction, which has a constant value P, independent of their separation. Show, from momentum and energy consideration, that the body of mass m has acquired just before collision a velocity given by $v^2 = 2Ph/m\left(1 + \frac{m}{M}\right)$. **[C.U. (Hons.) 1984]**

Let the velocities of masses m and M be v and V respectively, so that their K.E. is given by,

$$E = \frac{1}{2}MV^2 + \frac{1}{2}mv^2.$$

Since they move as a result of the force of their mutual attraction, we may write,

$$E = \int_0^h P dr = Ph,$$

dr being an infinitesimal displacement against P.

$$\therefore\ \frac{1}{2}MV^2 + \frac{1}{2}mv^2 = Ph \tag{1}$$

Again from momentum consideration,

$$MV = mv \quad \text{or} \quad V = \frac{mv}{M} \tag{2}$$

Putting the value of V from (2) in (1), we have

$$\frac{1}{2}M\frac{m^2v^2}{M^2} + \frac{1}{2}mv^2 = Ph \quad \text{or} \quad \frac{1}{2}mv^2\left(1 + \frac{m}{M}\right) = Ph$$

$$\therefore\ v^2 = 2Ph/m\left(1 + \frac{m}{M}\right)$$

9. (a) Show that for a single particle with constant mass, the equation of motion can be put in the form $\frac{dT}{dt} = \vec{F}\cdot\vec{V}$, where T is the K.E., F is the force applied and V, the velocity.

If the mass varies with time, show that the corresponding form is

$$\frac{d}{dt}(mT) = \vec{F}\cdot\vec{p},$$

where p is the momentum.

(b) A uniform chain of total length a is placed on a horizontal frictionless table, so that a length b of the chain dangles over the side. Show that the time t_0 required for the chain to slide off the table is give by

$$t_0 = \sqrt{\frac{a}{g}}\log_e\left(\frac{a + \sqrt{a^2 - b^2}}{b}\right).$$ **[C.U. (Hons.) 1986]**

(a) According to Newton's second law, $\vec{F} = m\frac{d\vec{V}}{dt}$.

$$\therefore\ \vec{F}\cdot\vec{V} = m\frac{d\vec{V}}{dt}\cdot\vec{V} = \frac{d}{dt}\left(\frac{1}{2}mV^2\right) = \frac{dT}{dt}.$$

Again

$$\frac{d}{dt}(mT) = \frac{d}{dt}\left(m\frac{1}{2}mV^2\right) = \frac{d}{dt}\left(\frac{1}{2}m^2V^2\right) = \frac{d}{dt}\left(\frac{1}{2}p^2\right)$$

$$= \frac{d\vec{p}}{dt}\cdot\vec{p} = \vec{F}\cdot\vec{p}.$$

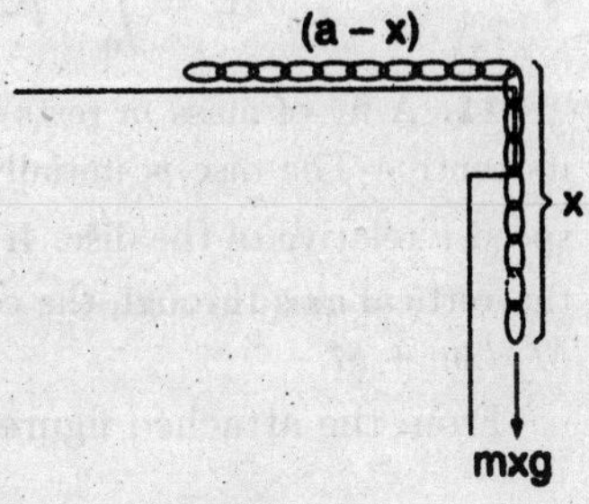

Fig. 1

(b) Since the chain is uniform, we assume m to be its mass per unit length, so that the total mass of the chain is am. Let at any instant t, the length x of the chain dangles over the side as shown in the figure, so that the down-ward force acting on the chain is mxg, g being the acceleration due to gravity. Since the table is frictionless, this force produces acceleration of the chain. So the equation of motion is

$$am\frac{d^2x}{dt^2} = mxg \quad \text{or} \quad \frac{d^2x}{dt^2} = \frac{g}{a}x.$$

To solve this equation, let us multiply both sides by $2\frac{dx}{dt}$, so that

$$2\frac{dx}{dt}\frac{d^2x}{dt^2} = 2\frac{g}{a}x\frac{dx}{dt} \quad \text{or} \quad \frac{d}{dt}\left(\frac{dx}{dt}\right)^2 = \frac{g}{a}\frac{d}{dt}(x^2)$$

Integrating we get

$$\left(\frac{dx}{dt}\right)^2 = \frac{g}{a}x^2 + c,$$

c being the constant of integration.

According to the problem, at $t = 0$, $x = b$ and $\frac{dx}{dt} = 0$ which gives $c = -\frac{gb^2}{a}$.

$$\therefore \frac{dx}{dt} = \sqrt{\frac{g}{a}}\sqrt{x^2 - b^2} \quad \text{or} \quad \frac{dx}{\sqrt{x^2 - b^2}} = \sqrt{\frac{g}{a}}\,dt.$$

Let at $t = t_0$, $x = a$ i.e., the chain dangles over fully. So integrating the above expression between the limits $x = b$ and $x = a$ for $t = 0$ and $t = t_0$, we have

$$\int_b^a \frac{dx}{\sqrt{x^2 - b^2}} = \sqrt{\frac{g}{a}}\int_0^{t_0} dt$$

$$\text{or} \quad t_0\sqrt{\frac{g}{a}} = \left[\log_e\left(x + \sqrt{x^2 - b^2}\right)\right]_b^a = \log_e\frac{a + \sqrt{a^2 - b^2}}{b}$$

$$\therefore \quad t_0 = \sqrt{\frac{a}{g}}\log_e\frac{a + \sqrt{a^2 - b^2}}{b}$$

10. A particle moves around a semi-circle of radius R from one end A of a diameter to the other B. It is attracted towards its starting point A by a force proportional to its distance from A. It is also given that when the particle is at B, the force towards A is F_0. Show that the work done against this force when the particle moves around the semi-circle from A to B is F_0R. **[C.U. (Hons.) 1986]**

From the condition of the problem we have $F \propto x$ or $F = \mu x$, x being the distance of the particle from A.

When $x = 2R, F = F_0$. $\therefore F_0 = \mu 2R$.

$\therefore$ required work done

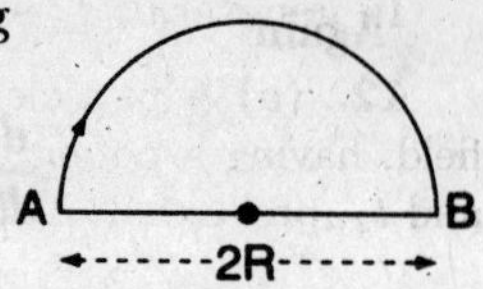

Fig. 2

$$W = \int_0^{2R} \mu x \, dx = \mu 2R^2 = \mu.2R.R = F_0 R.$$

11. A fly of mass m rests on a horizontal disc that is freely pivoted about a vertical axis through its centre. The disc is initially stationary when the fly starts to walk round the circumference at a speed v relative to the disc. If the disc has radius r, mass M and moment of inertia $\frac{1}{2}Mr^2$ about the the vertical axis through the centre, show that the speed of the fly relative to a stationary observer is $Mv/2m + M$. **[C.U. (Hons.) 1986]**

From the attached figure, velocity of the fly relative to the disc is

$$\vec{v_{fd}} = \vec{v_f} - \vec{v_d} \quad \text{or} \quad v_{fd} = v = v_f + v_d.$$

$$\therefore v_d = (v - v_f) \qquad (1)$$

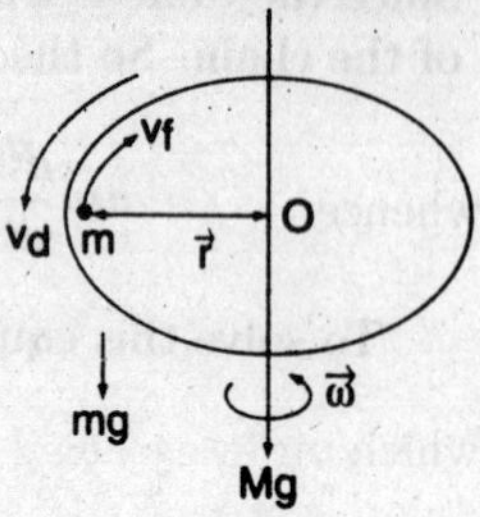

Fig. 3

From conservation of angular momentum, we may write

$$mv_f r - \frac{1}{2}Mr^2 \frac{v_d}{r} = 0$$

(Since the total torque on the disc-fly system about the vertical axis through O is zero, the angular momentum of the system about this axis is conserved, and it was initially zero w.r.t. a stationary observer at O).
Or $2mv_f - M(v - v_f) = 0$ from (1) whence $v_f = Mv/2m + M$.

The above treatment, although very simple, is not at all rigorous, for which we should consider the disc-fly system as a rotating frame of reference w.r.t. a stationary observer, at O, and the angular momentum conservation equation should be written as

$$I\vec{\omega} - \vec{r} \times m\vec{v_f} = 0 \quad \text{or} \quad \vec{\omega} = \frac{\vec{r} \times m\vec{v_f}}{I} \qquad (2)$$

which implies that as the fly moves clock-wise, the disc should rotate anti-clock-wise. Here $\vec{r}$ is the position vector of the fly w.r.t. O, $\vec{\omega} \rightarrow$ the angular velocity of the disc about the vertical axis, and $v_f \rightarrow$ the linear speed of the fly w.r.t. O.

From (3.40.1)

$$\frac{d\vec{r}}{dt} = \frac{\delta \vec{r}}{\delta t} + \vec{\omega} \times \vec{r} \quad \text{or} \quad \vec{v} = \vec{v_f} + \vec{\omega} \times \vec{r}. \qquad (3)$$

Now using (2) in (3), we have

$$\vec{v_f} = \vec{v} - \left(\frac{\vec{r} \times m\vec{v_f}}{I}\right) \times \vec{r} = \vec{v} + \vec{r} \times \frac{(\vec{r} \times m\vec{v_f})}{I}$$

$$\therefore \quad = \vec{v} + \frac{(\vec{r} \cdot m\vec{v_f})\vec{r} - r^2 m\vec{v_f}}{\frac{1}{2}Mr^2}$$

$$= \vec{v} - \frac{2m}{M}\vec{v_f} \quad \because \vec{r} \cdot m\vec{v_f} = 0.$$

$$\therefore \vec{v_f} = \frac{M\vec{v}}{2m + M}.$$

In general $\vec{v_f} \leq \vec{v}$ and both are in the same direction and $\vec{v_f} = \vec{v}$ when $m \ll M$.

12. (a) A particle of mass m moves along the X-axis under the influence of a conservative force field, having a potential $V(x)$. If the particle is located at positions x_1 and x_2 at respective times t_1 and t_2, prove that, if E be the total energy,

$$t_2 - t_1 = \sqrt{\frac{m}{2}} \int_{x_1}^{x_2} \frac{dx}{\sqrt{E - V(x)}}$$

(b) If in the above problem $V = \frac{1}{2}kx^2$ and the particle starts from rest at $x = a$, find the equation of motion in terms of x and t. **[C.U. (Hons.) 1995, 2006 (old syll)]**

(a) From conservation of energy, K.E. + P.E. $= E$

$$\text{or} \quad \frac{1}{2}m\left(\frac{dx}{dt}\right)^2 + V(x) = E \quad \text{or} \quad \left(\frac{dx}{dt}\right)^2 = \frac{2}{m}[E - V(x)] \tag{1}$$

whence

$$dt = \sqrt{\frac{m}{2}} \frac{dx}{\sqrt{E - V(x)}} \quad \text{(taking only the positive root)}$$

which on integration gives

$$\int_{t_1}^{t_2} dt = t_2 - t_1 = \sqrt{\frac{m}{2}} \int_{x_1}^{x_2} \frac{dx}{\sqrt{E - V(x)}}$$

(b) From (1) we have

$$\left(\frac{dx}{dt}\right)^2 = \frac{2}{m}\left(E - \frac{1}{2}kx^2\right)$$

$\because$ at $x = a$, $\frac{dx}{dt} = 0$, we have $E = \frac{1}{2}ka^2$, so that we may write

$$\left(\frac{dx}{dt}\right)^2 = \frac{k}{m}(a^2 - x^2) \quad \text{or} \quad \frac{dx}{\sqrt{a^2 - x^2}} = \pm\sqrt{\frac{k}{m}}\, dt,$$

which, on integration gives

$$\sin^{-1}\frac{x}{a} = \pm\sqrt{\frac{k}{m}}t + c$$

Putting $x = a$ at $t = 0$, we get $c = \frac{\pi}{2}$, so that

$$x = a \sin\left(\frac{\pi}{2} \pm \sqrt{\frac{k}{m}}t\right) = a \cos\sqrt{\frac{k}{m}}t$$

which is the required equation of motion. It follows from this equation that the particle oscillates back and forth along the X-axis from $x = a$ to $x = -a$ and the period of oscillation is given by $2\pi\sqrt{\frac{m}{k}}$.

13. A spherical rain-drop falling under constant gravity grows by absorption of moisture from its surroundings. The rate of increase of its mass is proportional to the instantaneous surface area. Assuming that the drop starts with an infinitely small radius, determine its acceleration at a time t.

[C.U. (Hons.) 1998, 2007]

Let V = volume of the spherical rain-drop at any instant of time; r = corresponding radius; m = corresponding mass of the rain drop, g = acceleration due to gravity.

If S be the instantaneous surface area of the spherical rain-drop, then according to the problem

$$\frac{dm}{dt} \propto S \text{ or } \frac{d}{dt}\left(\frac{4}{3}\pi r^3 \rho\right) \propto 4\pi r^2$$

$$\text{or } \frac{d}{dt}\left(\frac{4}{3}\pi r^3 \rho\right) = k4\pi r^2 \qquad \begin{matrix}(\rho = \text{density,} \\ k = \text{proportionality constant})\end{matrix} \tag{1}$$

$$\text{or } \frac{d}{dt}\left(\frac{1}{3}r^3 \rho\right) = kr^2$$

$$\text{or } \frac{d}{dt}\left(\frac{1}{3}r^3\right) = \frac{k}{\rho}r^2$$

$$\text{or } \frac{1}{3}3r^2\frac{dr}{dt} = \frac{k}{\rho}r^2$$

$$\text{or } \frac{dr}{dt} = \frac{k}{\rho} = \text{constant.} \tag{2}$$

Since the rain drop is falling under constant gravity, its acceleration is independent of its variable mass. We can show it from the principle of conservation of energy in the following way. If the instantaneous vertical position and velocity of the rain drop be $x(t)$ and $v(t)$ respectively, the equation of conservation of energy is given by

$$\frac{1}{2}mv^2 - mgx = \text{constant} \tag{3}$$

which on differentiation yields

$$\frac{dm}{dt}\left(\frac{1}{2}v^2 - gx\right) + mv\left(\frac{dv}{dt} - g\right) = 0 \tag{4}$$

Substitution of (1) and expression for mass m in (4) gives

$$k4\pi r^2\left(\frac{1}{2}v^2 gx\right) + \frac{4}{3}\pi r^3 \rho v\left(\frac{dv}{dt} - g\right) = 0 \tag{5}$$

$$\text{or } k\left(\frac{1}{2}v^2 - gx\right) + \frac{1}{3}r\rho v\left(\frac{dv}{dt} - g\right) = 0$$

$$\text{or } k\left(\frac{1}{2}v^2 - gx\right) = -\frac{1}{3}r\rho v\left(\frac{dv}{dt} - g\right) = 0$$

$$\text{or } -\frac{1}{3}\frac{r\rho}{k}\left[\frac{d\left(\frac{1}{2}v^2-gx\right)}{\left(\frac{1}{2}v^2-gx\right)}\right]=0 \quad \left[\because \frac{d}{dt}\left(\frac{1}{2}v^2-gx\right)=v\left(\frac{dv}{dt}-g\right)\right]$$

$$\text{or } -\frac{1}{\frac{d(r^3)}{r^3}}\left[\frac{d\left(\frac{1}{2}v^2-gx\right)}{\left(\frac{1}{2}v^2-gx\right)}\right]=0$$

$$\left[\because \frac{d(r^3)}{dt}=3r^2\frac{dr}{dt}=\frac{3r^2\cdot r}{r}\frac{k}{\rho} \text{ from (2) } \therefore \frac{d(r^3)}{r^3}=\frac{3k}{\rho r}\right]$$

On integration, we have

$$\ln r^3 \ln\left(\frac{1}{2}v^2-gx\right)=\text{constant} \quad \text{or} \quad \left(\frac{1}{2}v^2-gx\right)r^3=C \tag{6}$$

From initial conditions, we have at $t=0$, $v=0$, $x=x(0)$ and $r(0)=$ infinitesimal, so that $[r(0)]^3=0$. So at $t=0$, we have from (6) $C=0$—which implies that for any value of r at any instant t

$$\frac{1}{2}v^2-gx=0 \quad \text{or} \quad \frac{d}{dt}(v)=g$$

14. A particle of mass m at rest at $(a,0,0)$ is subjected to a force

$$\vec{F}=-\left(\frac{\alpha}{x^3}\right)\hat{\imath} \quad (\alpha=\text{ constant}).$$

Find the time taken by the particle to reach the origin $(a>0, \alpha>0)$. **[C.U. (Hons.) 1998]**

From Newton's second law, we have

$$m\frac{d^2x}{dt^2}=-\frac{\alpha}{x^3} \text{ or } 2\frac{dx}{dt}m\frac{d^2x}{dt^2}=-\frac{2\alpha}{x^3}\frac{dx}{dt} \quad \left(\text{Multiplying both sides by } 2\tfrac{dx}{dt}\right)$$

$$\text{or } m\frac{d}{dt}\left(\frac{dx}{dt}\right)^2=\alpha\frac{d}{dt}\left(\frac{1}{x^2}\right)$$

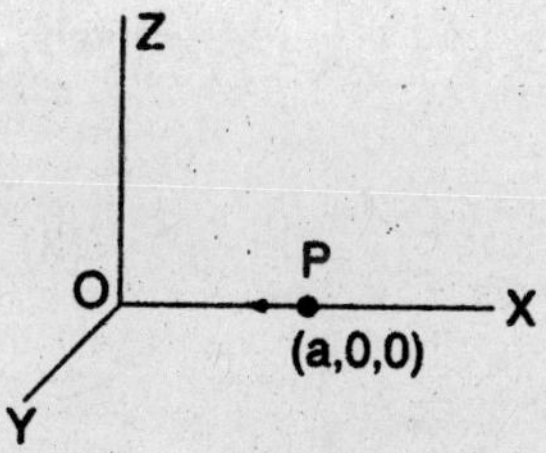

Fig. 4

Integrating we have

$$m\left(\frac{dx}{dt}\right)^2=\alpha\frac{1}{x^2}+C \quad (C=\text{ Integration constant})$$

Now at $x=a$, $\frac{dx}{dt}=0$ which gives $C=-\frac{\alpha}{a^2}$, so that we may write

$$m\left(\frac{dx}{dt}\right)^2=\alpha\left(\frac{1}{x^2}-\frac{1}{a^2}\right)=\alpha\frac{(a^2-x^2)}{a^2x^2}.$$

$$\therefore \quad \frac{dx}{dt}=\sqrt{\frac{\alpha}{m}}\frac{\sqrt{a^2-x^2}}{ax} \quad \text{(taking only the positive root)}$$

$$\text{or} \quad dt=\sqrt{\frac{m}{\alpha}}\frac{ax\,dx}{\sqrt{a^2-x^2}}=\sqrt{\frac{m}{\alpha}}\frac{aa\sin\theta a\cos\theta\,d\theta}{a\cos\theta}=\sqrt{\frac{m}{\alpha}}a^2\sin\theta d\theta.$$

(Putting $x = a\sin\theta$, we have $dx = a\cos\theta\ d\theta$ and when $x = 0$, $\theta = 0$ and when $x = a$, $\theta = \pi/2$.)

Integrating between proper limits, we have

$$\int_0^t dt = \sqrt{\frac{m}{\alpha}a^2}\int_0^{\pi/2} \sin\theta\ d\theta = \sqrt{\frac{m}{\alpha}a^2}\Big[-\cos\theta\Big]_0^{\pi/2}$$

$$\text{or} \quad t = \sqrt{\frac{m}{\alpha}a^2}\Big[\cos\theta\Big]_{\pi/2}^{0} = \sqrt{\frac{m}{\alpha}a^2}$$

which is the time required by the particle to reach the origin.

15. A particle moves under the influence of a force field given by $\vec{F} = a(\hat{\imath}\sin\omega t + \hat{\jmath}\cos\omega t)$. If the particle be initially at rest, prove that at an instant t, work done on the particle is given by $\frac{a^2}{m\omega^2}$ $(1-\cos\omega t)$. Do not ignore the initial condition. **[C.U. (Hons.) 2002]**

From Newton's second law, time rate of change of momentum = applied force

$$\therefore\ \frac{d}{dt}(mv) = \vec{F} = a(\hat{\imath}\sin\omega t + \hat{\jmath}\cos\omega t), \quad v \text{ being the final velocity at time } t$$

$$\text{or } m\frac{dv}{dt} = a(\hat{\imath}\sin\omega t + \hat{\jmath}\cos\omega t)$$

$$\text{or} \quad \int dv = \frac{a}{m}\int(\hat{\imath}\sin\omega t + \hat{\jmath}\cos\omega t)dt$$

$$\text{or } v = \frac{a}{mw}(-\hat{\imath}\cos\omega t + \hat{\jmath}\sin\omega t) + C \quad (C = \text{Integration constant})$$

Now the initial condition at $t = 0$, $v = 0$, gives $C = \frac{a\hat{\imath}}{mw}$

$$\therefore \quad v = \frac{a}{mw}\Big[(1-\cos\omega t)\hat{\imath} + \hat{\jmath}\sin\omega t\Big]$$

Now the work done

$$W = \int_0^t \vec{F}\cdot d\vec{r} \quad \text{(during finite time interval from 0 to } t\text{)}$$

$$= \int_0^t \vec{F}\cdot\frac{d\vec{r}}{dt}dt = \int_0^t \vec{F}\cdot\vec{v}dt = \int_0^t \frac{a^2}{m\omega}\{\sin\omega t(1-\cos\omega t) + \sin\omega t\cos\omega t\}dt$$

$$= \frac{a^2}{m\omega}\int_0^t \sin\omega t dt = \frac{a^2}{m\omega^2}(1-\cos\omega t)$$

16. A particle is subjected to a frictional force so that its velocity decreases with time according to $v = c(t-t_1)^2$ where t_1 is the time when it stops and c a constant. Find the force $F(v)$ and the distance travelled before it stops. **[C.U. (Hons.) 2001]**

Here $v = c(t-t_1)^2$. $\quad \therefore \quad (t-t_1) = \sqrt{\dfrac{v}{c}}$

Again

$$\frac{dx}{dt} = c(t-t_1)^2 \quad \text{(At } t = t_1, v = 0)$$

$$\text{or} \int_0^x dx = \int_0^{t_1} c(t-t_1)^2 dt = \int_0^{t_1} c(t^2 - 2tt_1 + t_1^2)dt$$

$$\text{or } x = \frac{ct_1^3}{3}$$

Again

$$F = m\frac{dv}{dt} = 2mc(t-t_1) \quad \text{or} \quad F(v) = 2mc\sqrt{\frac{v}{c}}$$

17. A particle is moving under the action of a force $\hat{k} \times \vec{v}$ where $\hat{k}$ is a unit vector along z-axis and v is the velocity of the particle. Show that K.E. of the particle remains constant.

[C.U. (Hons.) 2000]

Let $\vec{r}$ be the position vector of the particle, then we have $\vec{r} = \hat{i}x + \hat{j}y + \hat{k}z$ or $d\vec{r} = \hat{i}dx + \hat{j}dy + \hat{k}dz$ and

$$\vec{v} = \frac{d\vec{r}}{dt} = \hat{i}\frac{dx}{dt} + \hat{j}\frac{dy}{dt} + \hat{k}\frac{dz}{dt};$$

Now according to the problem,

$$\vec{F} = \hat{k} \times \vec{v} = \begin{vmatrix} \hat{i} & \hat{j} & \hat{k} \\ 0 & 0 & 1 \\ \frac{dx}{dt} & \frac{dy}{dt} & \frac{dz}{dt} \end{vmatrix} = -\frac{dy}{dt}\hat{i} + \frac{dx}{dt}\hat{j}$$

So work done

$$= \vec{F} \cdot d\vec{r} = \left(-\frac{dy}{dt}\hat{i} + \frac{dx}{dt}\hat{j}\right) \cdot (\hat{i}dx + \hat{j}dy + \hat{k}dz)$$

$$= -\frac{dy}{dt}dx + \frac{dx}{dt}dy = -\frac{dy}{dt}\frac{dx}{dt}dt + \frac{dx}{dt}dy$$

$$= -\frac{dx}{dt}(dy - dy) = 0$$

[Also since $dy/dt = dx/dt$ from the given condition $\vec{F} = \hat{k} \times \vec{v}$]

Again

$$\vec{F} \cdot d\vec{r} = \frac{d\vec{p}}{dt} \cdot d\vec{r} \quad \text{(where } \vec{p} \text{ is the momentum)}$$

$$= \frac{d}{dt}(m\vec{v}) \cdot d\vec{r} = m\frac{d\vec{v}}{dt} \cdot \frac{d\vec{r}}{dt}dt = m\vec{v} \cdot d\vec{v} = mvdv = md\left(\frac{1}{2}v^2\right) = d\left(\frac{1}{2}mv^2\right)$$

$\because \vec{F} \cdot d\vec{r} = 0$, $d\left(\frac{1}{2}mv^2\right) = 0$ or $\frac{1}{2}mv^2 =$ constant i.e. K.E. = constant.

18. Set up the one-dimensional equation of motion of a particle of mass m subjected to a force such that its potential energy is $\frac{1}{2}kx^2$. If a damping force $F = bv$ is also present, where v is the velocity and b a constant, find out the resultant motion. Discuss the case when $\frac{b}{2m} = \sqrt{\frac{k}{m}}$. Compute the total (kinetic + potential) energy. Is it a constant? Justify your answer. **[C.U. (Hons.) 2001, 2005]**

(a) From the definition of P.E.,

$$V(x) = -\int_0^x kx dx = -k\frac{x^2}{2}$$

so that the force is given by $-kx$ or $F = -kx$ or $F \propto -x$ i.e. F is a restoring force.

Thus

$$F = -kx \quad \text{or} \quad m\frac{d^2x}{dt^2} + kx = 0 \quad \text{or} \quad \frac{d^2x}{dt^2} + \frac{k}{m}x = 0$$

$$\therefore \frac{d^2x}{dt^2} + \omega_0^2 x = 0 \quad \text{where } \omega_0^2 = \frac{k}{m}$$

which is the required one-dimensional equation of motion.

(b) When restoring as well as damping or resisting force (caused by friction) is present, the damping force $F_b \propto v$, instantaneous velocity of the particle, i.e. $F_b \propto -\frac{dx}{dt}$ (negative sign indicates that F_b resists motion).

Here net force $= F + F_b$, so that the equation of motion is written as

$$m\frac{d^2x}{dt^2} = -kx - b\frac{dx}{dt} \quad (\because F_b = bv, \text{ according to the question})$$

$$\text{or} \quad m\frac{d^2x}{dt^2} + b\frac{dx}{dt} + kx = 0 \quad \text{or} \quad \frac{d^2x}{dt^2} + 2p\frac{dx}{dt} + \omega_0^2 x = 0$$

where $\frac{k}{m} = \omega_0^2$ = restoring per unit mass per unit displacement, and $\frac{b}{m} = 2p$ = resisting force per unit mass per unit velocity.

To find out the resultant motion, we solve the above equation by taking

$$x = e^{\alpha t} \quad \text{or} \quad \frac{dx}{dt} = \alpha x \quad \text{and} \quad \frac{d^2x}{dt^2} = \alpha^2 x$$

Putting these values, we have $(\alpha^2 + 2p\alpha + \omega_0^2)x = 0$

$$\text{or } \alpha^2 + 2p\alpha + \omega_0^2 = 0 \quad \because x \neq 0$$

$$\text{or } \alpha = -p \pm \sqrt{p^2 - \omega_0^2}$$

so that the general solution of the above equation is

$$x = A_1 e^{(-p+\sqrt{p^2-\omega_0^2})t} + A_2 e^{(-p-\sqrt{p^2-\omega_0^2})t}$$

$$= e^{-pt}\left(A_1 e^{\sqrt{p^2-\omega_0^2}\,t} + e^{-\sqrt{p^2-\omega_0^2}\,t}\right)$$

where A_1 and A_2 are arbitrary constants.

(i) For $p > \omega_0$, the motion is over-damped.

(ii) For $p < \omega_0$, the motion is damped oscillatory.

When $\frac{b}{2m} = \sqrt{\frac{k}{m}}$ i.e. $p = \omega_0$, the equation of motion becomes

$$\frac{d^2w}{dt^2} + 2p\frac{dx}{dt} + px = 0$$

and the motion is critically damped.

(c) Because of the presence of damping force, total energy is not conserved. Now loss of energy (per unit time) at any instant due to damping force is given by $\frac{dE}{dt}$ = damping force × velocity $= bv \cdot v = 2pmv^2 = 2pm\dot{x}^2$.

Equation of motion under damping force is

$$\ddot{x} + 2p\dot{x} + \omega_0^2 x = 0$$

$$\text{or } m\dot{x}\ddot{x} + 2pm\dot{x}^2 + m\dot{x}x\omega_0^2 = 0 \quad (\text{multiplying both sides by } m\dot{x})$$

$$\begin{aligned}\therefore \quad 2pm\dot{x}^2 &= -\left(m\dot{x}\ddot{x} + m\omega_0^2 x\dot{x}\right)\\ &= -\left[\frac{m}{2}\frac{d}{dt}(\dot{x}^2) + \frac{m}{2}\omega_0^2\frac{d}{dt}(x^2)\right]\\ &= -\left[\frac{d}{dt}\left(\frac{1}{2}m\dot{x}^2 + \frac{1}{2}m\omega_0^2x^2\right)\right] = -\frac{dE}{dt}\end{aligned}$$

$\because \frac{dE}{dt} \neq 0$, $E \neq$ conserved.

19. A particle of mass m moves in the X-Y plane and its position vector is given by $\vec{r} = \hat{i}a\cos\omega t + \hat{j}b\sin\omega t$, where a, b, ω are constants. Show that the force acting on the particle is always directed towards the origin and the force field is conservative. Find the work done by the force field in moving the particle from a point $(a, 0)$ to $(0, b)$. **[C.U. (Hons.) 2004]**

(a) Here $\vec{r} = \hat{i}a\cos\omega t + \hat{j}b\sin\omega t$.

$$\therefore \quad \vec{v} = -\hat{i}a\omega\sin\omega t + \hat{j}b\omega\cos\omega t$$

$$\text{and } \vec{f} = \frac{d\vec{v}}{dt} = \frac{d^2\vec{r}}{dt^2} = -\hat{i}a\omega^2\cos\omega t - \hat{j}b\omega^2\sin\omega t$$

$\therefore \vec{F} = m\vec{f} = -m\omega^2(\hat{i}a\cos\omega t + \hat{j}b\sin\omega t) = -m\omega^2\vec{r}$.

$\therefore \vec{F} \propto -\vec{r}$ which shows that the force is directed towards the origin.

(b) To prove that the force field is conservative, we have to show $\vec{\nabla} \times \vec{F} = 0$.

Now

$$\vec{\nabla} \times \vec{F} = \begin{vmatrix} \hat{i} & \hat{j} & \hat{k} \\ \frac{\partial}{\partial x} & \frac{\partial}{\partial y} & \frac{\partial}{\partial z} \\ -m\omega^2 a\cos\omega t & -m\omega^2 b\sin\omega t & 0 \end{vmatrix} = 0 \quad \because m, \omega, a, b \text{ are all constants}$$

$\therefore \vec{F}$ is conservative.

(c) $\vec{r} = x\hat{i} + y\hat{j} = \hat{i}a\cos\omega t + \hat{j}b\sin\omega t$.

$\therefore x = a\cos\omega t, y = b\sin\omega t$ which are the parametric equations of an ellipse having a and b as semi-major and semi-minor axes respectively as shown in the attached fig.

$\therefore$ The required work done

$$= \int_a^b \vec{F}\cdot d\vec{r} = \int_a^b F\cdot\frac{d\vec{r}}{dt}dt$$

$$= \int_a^b \vec{F}\cdot\vec{v}dt = -m\omega^3\int_0^{\pi/2}\left(\hat{i}a\cos\omega t + \hat{j}b\sin\omega t\right)\cdot\left(-\hat{i}a\sin\omega t + \hat{j}b\cos\omega t\right)dt$$

[$\because$ from $\vec{r} = \hat{i}a\cos\omega t + \hat{j}b\sin\omega t$, when $\omega t = 0, r = a$ and when $\omega t = \frac{\pi}{2}, r = b$]

$$= -m\omega^3\int_0^{\pi/2}(-a^2\sin\omega t\cos\omega t + b^2\sin\omega t\cos\omega t)dt$$

$$= m\omega^3(a^2 - b^2)\int_0^{\pi/2}\frac{\sin 2\omega t}{2}dt$$

$$= \frac{m\omega^3(a^2-b^2)}{2}\left[-\frac{\cos 2\omega t}{2\omega}\right]_0^{\pi/2}$$

$$= \frac{m\omega^2(a^2-b^2)}{4}\Big[\cos 2\omega t\Big]_{\pi/2}^{0} = \frac{m\omega^2(a^2-b^2)}{2}$$

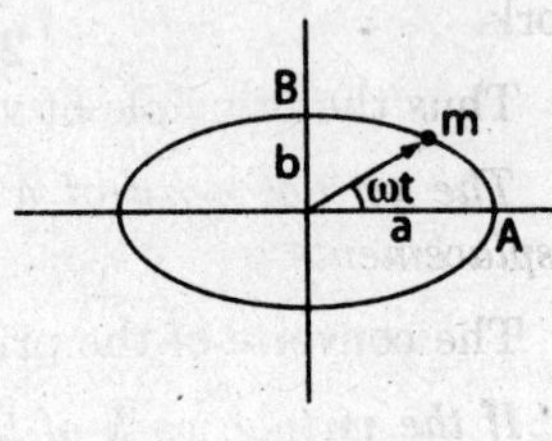

3.18 D'Alembert's Principle

If a number of forces acting on a particle of mass m produces an acceleration $\vec{f}$, the equation of motion of the particle is given by $\Sigma\vec{F} = m\vec{f}$, where $\Sigma\vec{F}$ denotes the sum of all the forces acting on the particle. The above expression may be written in the form

$$\Sigma\vec{F} - m\vec{f} = 0. \tag{3.18.1}$$

The quantity $m\vec{f}$ is called the effective force acting on the particle. From the above equation, it follows that the sum of all the forces *minus* the effective force is zero. In other words, the sum of all the forces including the reversed effective force $-m\vec{f}$ is zero, i.e., the forces are in equilibrium. This principle is called D'Alembert's principle.

This principle enables us to apply statical rules to problems in Dynamics. As an example, let us consider a particle moving along a circular path of radius r with uniform velocity $\vec{v}$. We know that the particle has an acceleration $\frac{\vec{v}^2}{r}$ directed towards the centre of the circle. By D'Alembert's principle the forces $\Sigma\vec{F}$ acting on the particle together with the reversed effective force $-\frac{m\vec{v}^2}{r}$ are in equilibrium, i.e.,

$$\Sigma\vec{F} - \frac{m\vec{v}^2}{r} = 0.$$

It is to be noted that the forces $\Sigma\vec{F}$ are directed towards the centre of the circular path and effective force $\frac{m\vec{v}^2}{r}$ is directed away from the centre of the circle. This force is called the centrifugal force.

Here centrifugal force is not an actual force but it is a force devised to tackle a statical problem having great resemblance to a dynamical one.

3.19 Principle of Virtual Work

If we imagine a number of forces acting on a particle, then for a slight displacement of the particle, the work done by the forces is equal to the work done by the resultant of the forces. When the forces are in equilibrium, i.e., the resultant force is zero, the work done for such a slight displacement is zero. This is true whether actually the displacement occurs or not. In mechanics we use in many problems such work which is done on a body producing an imaginary small displacement. This imaginary displacement which never really occurs is called virtual displacement. The work done by any force for such virtual displacement is called virtual work.

Thus the principle of virtual work may be stated as follows:

The virtual work of a system of forces in equilibrium acting on a particle is zero for all displacements.

The converse of the principle of virtual work is also true which is as follows:

If the virtual work of the forces acting on a particle is zero for all displacements, then the forces are in equilibrium.

We consider a force $\vec{F_i}$ which acting on ith particle produces slight virtual displacement $\delta\vec{S_i}$. Then the work done by the force is $\vec{F_i}\cdot\delta\vec{S_i}$. For a number of particles in equilibrium we have, $\Sigma\vec{F_i}\cdot\delta\vec{S_i} = 0$, since the total force on each particle in the condition of equilibrium is zero.

We shall now consider an illustrative example in support of the principle of virtual work.

We consider two heavy particles of weights ω_1 and ω_2. The particles are connected by a light inextensible string and suspended over a fixed smooth circular cylinder of radius r, the axis of the cylinder being horizontal (Fig. 3.19.1). We want to find the condition of equilibrium of the system.

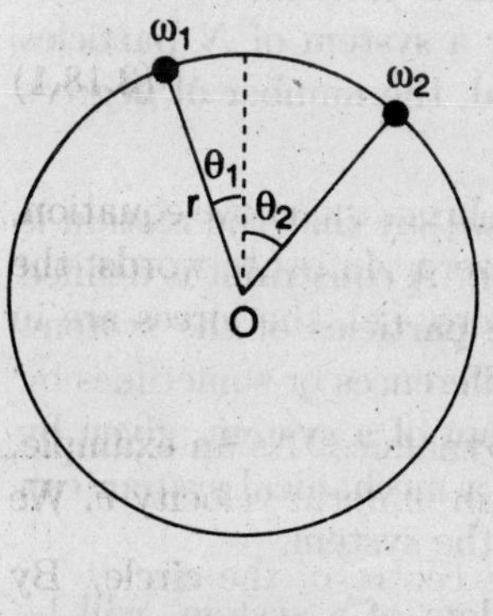

Fig. 3.19.1

If we give a virtual displacement satisfying the constraints, only the force due to gravity works.

Here the system is conservative and the potential energy is given by

$$V = \omega_1 r\cos\theta_1 + \omega_2 r\cos\theta_2,$$

where θ_1, θ_2 are the angles of inclination to the vertical of the radii meeting the particles. For an infinitesimal displacement,

$$\delta V = -\omega_1 r\sin\theta_1\delta\theta_1 - \omega_2 r\sin\theta_2\delta\theta_2.$$

Now, $\theta_1 + \theta_2 =$ constant, since the length of the string remains unaltered as it is inextensible;

$$\therefore\ \delta\theta_1 = -\delta\theta_2.$$

Hence

$$\delta V = r\delta\theta_1(\omega_2 \sin\theta_2 - \omega_1 \sin\theta_1)$$

When the system is in equilibrium, the work done is zero, i.e., the change in potential energy $\delta V = 0$.

$$\therefore\ \omega_2 \sin\theta_2 = \omega_1 \sin\theta_1 \quad \text{or} \quad \frac{\omega_2}{\omega_1} = \frac{\sin\theta_1}{\sin\theta_2}.$$

Thus from the principle of virtual work, we get the condition of equilibrium of a system.

3.20 Degrees of Freedom of a Mechanical System

The total number of independent parameters defining the position of a mechanical system in space is called the number of degrees of freedom of the system. When a material particle is constrained to move along a straight line, it is said to have one degree of freedom. If it is allowed to move in a plane, it has then two degrees of freedom, but when the particle moves in space it has three degrees of freedom, i.e., the position of the particle in space is defined by three independent parameters, say x, y, z in Cartesian co-ordinates. If we consider a system having N particles, then the position of the system is fully known by $3N$ independent parameters. If the distribution of points of the system is fixed in any case, then the number of degrees of freedom may be less than $3N$. As for example, let us consider two points of co-ordinates (x_1, y_1, z_1) and (x_2, y_2, z_2) rigidly constrained, then we have

$$(x_2 - x_1)^2 + (y_2 - y_1)^2 + (z_2 - z_1)^2 = R^2, \tag{3.20.1}$$

where R denotes the distance between the points. It should be noted that only five of the six variables $x_1, x_2, y_1, y_2, z_1, z_2$ are independent. Hence the degrees of freedom of a system of two particles separated by a fixed distance is five. It can be shown in a similar way that for a system of three particles rigidly fixed on the three vertices of a triangle, we get three equations including equation (3.20.1) and so the number of degrees of freedom is six, since six of the nine quantities $(x_1, x_2, x_3, y_1, y_2, y_3, z_1, z_2, z_3)$ are independent. If for a system of N particles there exist constraints, expressed in K equations, we have, in general, the number of degrees of freedom given by $n = 3N - K$.

When the motion of a system is subjected to certain limitations, we say that the motion is *constrained,* and the limitations are known as the ***constraints of motion.*** A constraint is defined to be some geometrical restriction on the freedom of motion of the particles of the system, which might have to be satisfied by the co-ordinates or co-ordinate differences or sometimes by velocities rather than co-ordinates. The number of degrees of freedom of a system, given by $n = 3N - k$, means the total number of independent ways in which a mechanical system can move, without violating any constraints which might be imposed on the system.

That constraints actually reduce the number of degrees of freedom of a system, will be evident when we consider the motion of a single mass point. Without any restriction, the particle will possess 3 degrees of freedom in the 3-dimensional space. If, however, the particle is constrained to move on a plane only, the number of degrees of freedom is reduced to *two.* A further restriction that the particle moves along a straight line, reduces this number by *one* more, and the particle is then left with only one degree of freedom.

Types of constraints

When the constraints imposed on the system are expressible in the form of an equation,

$$f(\vec{r_1}, \vec{r_2}, \cdots, \vec{r_N}, t), \tag{3.20.2}$$

connecting position vectors of the particles and the time, the constraint is called a *holonomic* constraint, whether it be due to the properties of the system itself (as in the case of a rigid body) or due to the interaction of the system with the surroundings (for example a bead moving along a wire.)

Examples of holonomic constraints

The simplest example of holonomic constraint is the motion of a simple pendulum bob which is constrained to move in a circle of radius equal to the length of the string from the point of suspension (position vector $\vec{a}$). If $\vec{r}$ be the position vector of the bob, and l, the length of the string, the equation of constraint may be written as

$$(\vec{r} - \vec{a})^2 - l^2 = 0. \tag{3.20.3}$$

Another very good example is the motion of a rigid body in which the body is constrained to move such that the distances between the various particles of the body remain fixed.

In a holonomic constraint, a restriction may be imposed on the velocities instead of co-ordinates, provided the constraint equation in terms of velocities is integrable to a form containing co-ordinates. An example is the motion of a cylinder of radius 'a' rolling and sliding down on an inclined plane with its axis always horizontal.

Constraints which cannot be expressed in the form of an equation containing position vectors and time only, i.e., in the form of equation (3.20.2), are called *non-holonomic* constraints. For example, the motion of a particle on the surface of a sphere under the action of gravity, expressed by the equation

$$(\vec{r} - \vec{r_0})^2 \leq a^2, \tag{3.20.4}$$

where $\vec{r_0}$ is the position vector of the centre of the sphere and $\vec{r}$ is the position vector of the particle. This equation expresses that the distance between the centre of the sphere and the particle cannot be less than the radius a of the sphere. If however $(\vec{r} - \vec{r_0})^2 = a^2$, then it will reduce to an equation of holonomic constraint.

Other types of constraints

Constraint equations which do not contain time explicitly, i.e., which are independent of time, are called *Scleronomic* constraints. Again, constraint equations in which time does enter explicitly are called *Rheonomic* constraints. In other words, a scleronomic system is one which has only *fixed* constraints, whereas a rheonomic system has *moving* constraints. For example, a pendulum with a fixed support is scleronomic, while a pendulum for which the point of support is given an assigned motion is rheonomic.

We are mostly concerned with holonomic constraints. Constraints introduce some difficulties in the solution of mechanical problems. For example, all the co-ordinates r_i are not independent, but are governed by the conditions of the constraints, and it may not be obvious as to which of the co-ordinates are to be regarded as independent variables. This difficulty is got rid of by using generalised co-ordinates.

Illustrative Examples

1. Classify each of the following according as they are: (i) scleronomic or rheonomic, (ii) holonomic or non-holonomic and (iii) conservative or non-conservative.

(a) A small sphere rolling down from the top of a fixed large sphere.

(b) A cylinder rolling without slipping down a rough inclined plane of angle α.

(c) A particle sliding down the inner surface, with coefficient of friction μ, of a paraboloid of revolution having its axis vertical and vertex downward.

(d) A particle moving on a very long frictionless wire which rotates with constant angular speed about a horizontal axes.

(*a*)	Scleronomic	Since equations of constraint do not involve time t explicitly;
	Non-holonomic	Since the small rolling sphere leaves the fixed large sphere at some point;
	Conservative	Gravitational force acting is derivable from a potential.
(*b*)	Scleronomic	
	Holonomic	Since equation of constraint is that of a line or plane;
	Conservative	
(*c*)	Scleronomic	
	Holonomic	
	Non-conservative	Since force due to friction is not derivable from a potential;
(*d*)	Rheonomic	Since constraint involves time t explicitly—constraint equation being $\theta = \omega t$ i.e. $\dot{\theta} = \omega$;
	Holonomic	Since equation of constraint is that of a line which involves t explicitly;
	Conservative	

3.21 Generalised Co-ordinates

The elimination of the dependent variables can be expressed by introducing a new, $3N - K$, independent variables $q_1, q_2, q_3, \cdots, q_{3N-K}$ in terms of which the cartesian co-ordinates $x_1, x_2, x_3, \cdots, x_n$ are expressed in the form

$$x_i = x_i(q_1, q_2, q_3, \cdots, q_{3N-K}) \tag{3.21.1}$$

These new independent variables $q_1, q_2, q_3, \cdots, q_{3N-K}$ are called generalised co-ordinates.

They may be space co-ordinates, angular co-ordinates or any other kind of variables or the functions of these variables. In order to choose a suitable set of generalised co-ordinates for a given system, we should be guided by the following three principles.

(1) Their values determine the configuration of the system.

(2) They may be varied arbitrarily and independently of each other without violating the constraints on the system.

(3) There is no uniqueness in the choice of generalised co-ordinates. Our choice should then fall on a set of co-ordinates that will give us a reasonable mathematical simplification of the problem.

The number of co-ordinates satisfying (1) and (2) is the number of degrees of freedom. Principle (3) then implies that after ascertaining the number of degrees of freedom, we look for a suitable system of co-ordinates to achieve mathematical simplicity in setting up the equations of motion and then solving them.

Some examples of mechanical systems and the generalised co-ordinates to be associated with them are given below.

System	Generalised Co-ordinates
1. Simple pendulum	θ; the angle which the pendulum makes with the vertical line through the point of suspension.
2. Fly-wheel	θ; the angle between a definite radius of the fly-wheel and a fixed line perpendicular to the axis.
3. A particle moving on the surface of a sphere	θ, ϕ; the usual polar angles of a point on the sphere.
4. A particle moving on the inside surface of a cone	r, θ; r the radius vector drawn from the vertex as origin to the position of the particle, and θ, the angle which the radius vector makes with a fixed slant-edge of the cone.

Generalised velocities

If $q_i(i = 1, 2, 3, \cdots, 3N - K)$ represent generalised co-ordinates, $\dfrac{dq_i}{dt} = \dot{q}_i$ etc. are called generalised velocities.

It should be kept in mind that each of the generalised co-ordinates q_i is a function of the cartesian co-ordinates of all the particles of the system, i.e.,

$$q_i = q_i(x_1, y_1, z_1, x_2, y_2, z_2, \cdots)$$

With the help of the above equation, a relation between the generalised velocities $\dot{q}_i$ and cartesian velocities $\dot{x}_1$, $\dot{y}_1$, $\dot{z}_1$, etc. can be found out.

Generalised momenta

The K.E. T of a system of N free particles in terms of cartesian co-ordinates is

$$T = \sum_{i=1}^{N} \frac{1}{2} m_i \dot{r}_i^2.$$

From the above expression we see that the linear momentum associated with the linear velocity $\dot{x}_i$ is $m_i \dot{x}_i$ given by

$$p_{x_i} = \frac{\partial T}{\partial \dot{x}_i} = m_i \dot{x}_i.$$

Then momentum associated with generalised co-ordinate q_i is similarly defined and is called the generalised momentum p_i associated with a co-ordinate q_i and is written as

$$p_i = \frac{\partial T}{\partial \dot{q}_i}$$

p_i need not always have dimensions of linear momentum, for if q_i happens to be an angular co-ordinate, p_i is the corresponding angular momentum.

3.22 Lagrange's Equations for Generalised Co-ordinates

Applying the principle of virtual work to the case of dynamical equilibrium (D'Alembert's equation) we have

$$\Sigma_i(F_i - m_i f_i)\delta x_i = 0$$

$$\text{or} \quad \Sigma_i(F_i - m_i \ddot{x}_i)\delta x_i = 0 \tag{3.22.1}$$

Now

$$\delta x_i = \frac{\partial x_i}{\partial q_1}\delta q_1 + \frac{\partial x_i}{\partial q_2}\delta q_2 + \cdots + \frac{\partial x_i}{\partial q_l}\delta q_l \tag{3.22.2}$$

where $l = 3N - K$.

Similarly,

$$\dot{x}_i = \frac{\partial x_i}{\partial q_1}\dot{q}_1 + \frac{\partial x_i}{\partial q_2}\dot{q}_2 + \cdots + \frac{\partial x_i}{\partial q_l}\dot{q}_l \tag{3.22.3}$$

Now,

$$\frac{d}{dt}\left(\dot{x}_i\frac{\partial x_i}{\partial q_k}\right) = \ddot{x}_i\frac{\partial x_i}{\partial q_k} + \dot{x}_i\frac{d}{dt}\left(\frac{\partial x_i}{\partial q_k}\right)$$

$$\text{or} \quad \ddot{x}_i\frac{\partial x_i}{\partial q_k} = \frac{d}{dt}\left(\dot{x}_i\frac{\partial x_i}{\partial q_k}\right) - \dot{x}_i\frac{d}{dt}\left(\frac{\partial x_i}{\partial q_k}\right) \tag{3.22.4}$$

Now, the equation (3.22.3) gives

$$\frac{\partial \dot{x}_i}{\partial \dot{q}_k} = \frac{\partial x_i}{\partial q_k}$$

Then the first term on the right-hand side of the equation (3.22.4) becomes

$$\frac{d}{dt}\left(\dot{x}_i\frac{\partial x_i}{\partial q_k}\right) = \frac{d}{dt}\left(\dot{x}_i\frac{\partial \dot{x}_i}{\partial \dot{q}_k}\right) = \frac{d}{dt}\left(\frac{1}{2}\frac{\partial \dot{x}_i^2}{\partial \dot{q}_k}\right)$$

Now, from (3.22.3), we have

$$\frac{d}{dt}\left(\frac{\partial x_i}{\partial q_k}\right) = \frac{\partial^2 x_i}{\partial q_1 \partial q_k}\dot{q}_1 + \frac{\partial^2 x_i}{\partial q_2 \partial q_k}\dot{q}_2 + \cdots + \frac{\partial^2 x_i}{\partial q_l \partial q_k}\dot{q}_l$$

$$= \frac{\partial}{\partial q_k}\left(\frac{\partial x_i}{\partial q_1}\dot{q}_1 + \frac{\partial x_i}{\partial q_2}\dot{q}_2 + \cdots + \frac{\partial x_i}{\partial q_l}\dot{q}_l\right) = \frac{\partial}{\partial q_k}\dot{x}_i$$

Hence

$$\dot{x}_i\frac{d}{dt}\left(\frac{\partial x_i}{\partial q_k}\right) = \frac{\partial}{\partial q_k}\left(\frac{1}{2}\dot{x}_i^2\right)$$

This gives the second term of equation (3.22.4). Hence we may write equation (3.22.4) in the form

$$\ddot{x}_i \frac{\partial x_i}{\partial q_k} = \frac{d}{dt}\left(\frac{1}{2}\frac{\partial \dot{x}_i^2}{\partial \dot{q}_k}\right) - \frac{\partial}{\partial q_k}\left(\frac{1}{2}\dot{x}_i^2\right).$$

From equation (3.22.2) the above expression may be written in the form

$$\ddot{x}_i \delta x_i = \sum_k \left[\frac{d}{dt}\frac{\partial}{\partial \dot{q}_k}\left(\frac{1}{2}\dot{x}_i^2\right) - \frac{\partial}{\partial q_k}\left(\frac{1}{2}\dot{x}_i^2\right)\right]\delta q_k$$

$$\text{or} \quad m_i \ddot{x}_i \delta x_i = \sum_k \left[\frac{d}{dt}\frac{\partial}{\partial \dot{q}_k}\left(\frac{1}{2}m_i\dot{x}_i^2\right) - \frac{\partial}{\partial q_k}\left(\frac{1}{2}m_i\dot{x}_i^2\right)\right]\delta q_k \tag{3.22.5}$$

where m_i = mass of the ith particle.

Considering that the acting forces are derivable from potentials U_i we may write $F_i = -\dfrac{\delta U_i}{\delta x_i}$.

Let us assume that the potentials are functions of q_k. Hence we have

$$F_i \delta x_i = -\delta U_i = -\left(\frac{\partial U_i}{\partial q_1}\delta q_1 + \frac{\partial U_i}{\partial q_2}\delta q_2 + \cdots + \frac{\partial U_i}{\partial q_l}\delta q_l\right) \tag{3.22.6}$$

Now substituting (3.22.5) and (3.22.6) in equation (3.22.1), we have,

$$\sum_k \left[\frac{d}{dt}\frac{\partial}{\partial \dot{q}_k}\left(\sum_i \frac{1}{2}m_i\dot{x}_i^2\right) - \frac{\partial}{\partial q_k}\left(\sum_i \frac{1}{2}m_i\dot{x}_i^2\right) + \sum_i \frac{\partial U_i}{\partial q_k}\right]\delta q_k = 0.$$

Now since U_i depends only upon the position co-ordinate q_k and not on $\dot{q}_k$, the above equation may be written in the form

$$\frac{d}{dt}\frac{\partial (T-U)}{\partial \dot{q}_k} - \frac{\partial}{\partial q_k}(T-U) = 0$$

[$\because$ the displacements δq_k are independent.]

Putting $T - U = L$, this expression may be written as

$$\frac{d}{dt}\left(\frac{\partial L}{\partial \dot{q}_k}\right) - \frac{\partial L}{\partial q_k} = 0 \tag{3.22.7}$$

The difference between the kinetic and potential energies L is called the Lagrangian function and the equation (3.22.7) is known as Lagrange's equation. This equation is true for conservative forces.

3.23 Hamilton's Equations

Since U is independent of $\dot{q}_k$ the Lagrange's equation may be expressed in generalised momentum co-ordinate p_k, as

$$\frac{dp_k}{dt} = \frac{\partial L(q_k, \dot{q}_k)}{\partial q_k},$$

where L is function of q_k and $\dot{q}_k$. We now introduce a new function H of q_k and p_k such that

$$H(p_k, q_k) = \sum_k p_k \dot{q}_k - L(q_k, \dot{q}_k) \tag{3.23.1}$$

Hence

$$\begin{aligned} dH &= \sum_k \left(\frac{\partial H}{\partial p_k} dp_k + \frac{\partial H}{\partial q_k} dq_k \right) \\ &= \sum_k p_k d\dot{q}_k + \sum_k \dot{q}_k dp_k - \sum_k \frac{\partial L}{\partial \dot{q}_k} d\dot{q}_k - \sum_k \frac{\partial L}{\partial q_k} dq_k \\ &= \sum_k \dot{q}_k dp_k - \sum_k \frac{\partial L}{\partial q_k} dq_k \quad \left[\text{since } \frac{\partial L}{\partial \dot{q}_k} = \frac{\partial T(q_k, \dot{q}_k)}{\partial \dot{q}_k} = p_k \right] \end{aligned}$$

Now, equating the co-efficients of dp_k and dq_k, we have

$$\dot{q}_k = \frac{\partial H}{\partial p_k} \tag{3.23.2}$$

$$\text{and} \quad \frac{\partial L}{\partial q_k} = -\frac{\partial H}{\partial q_k} \quad \text{or} \quad \dot{p}_k = -\frac{\partial H}{\partial q_k} \tag{3.23.3}$$

H is called the Hamiltonian function.

Now,

$$H(p_k, q_k) = \sum_k p_k \dot{q}_k - L(q_k, \dot{q}_k) = 2T^* - (T - U) = T + U = E.$$

Thus the Hamiltonian function is equal to the sum of kinetic and potential energies. The relations (3.23.2) and (3.23.3) are called Canonical equations of motion.

If the Lagrangian function of a system is not the function of co-ordinate q_k, the co-ordinate q_k is called cyclic. Hence if the generalised co-ordinate is cyclic we have from equation (3.22.7)

$$\frac{d}{dt}\left(\frac{\partial L}{\partial \dot{q}_k}\right) = 0 \text{ or } \frac{\partial L}{\partial \dot{q}_k} = \text{constant}$$

$$\text{or } \frac{\partial (T - U)}{\partial \dot{q}_k} = \text{constant}$$

$$\text{or } \frac{\partial T}{\partial \dot{q}_k} = \text{constant}$$

i.e., the generalised momentum p_k = constant.

Thus the generalised momentum is conserved if the corresponding generalised co-ordinate is cyclic.

---

*The kinetic energy T is a homogeneous quadratic function of $\dot{q}_k$. According to Euler's theorem, if f is a homogeneous function, of order n of a set of variables q_k, then

$$\sum_k q_k \frac{\partial f}{\partial q_k} = nf.$$

Here n is 2, so that

$$\sum_k \dot{q}_k \frac{\partial T}{\partial \dot{q}_k} = \sum p_k \dot{q}_k = 2T.$$

3.24 Stable and Unstable Equilibrium

While dealing with the small oscillations executed by various systems, we shall frequently come across the conditions of equilibrium. We shall give, in the following discussion, a brief account of the equilibrium state.

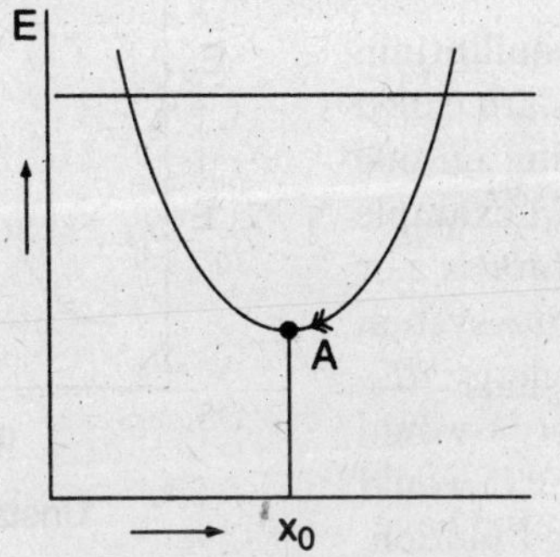

Point of stable equilibrium, $\frac{d^2V}{dx^2}$ = + ve at a neighbouring point A.

Fig. 3.24.1

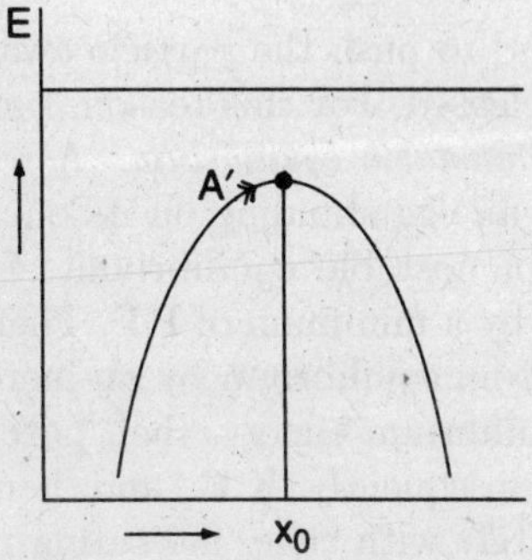

Point of unstable equilibrium, $\frac{d^2V}{dx^2}$ = – ve at a neighbouring point A′.

Fig. 3.24.2

In Figs. (3.24.1 and 3.24.2) potential energy as a function of distance x (say), for a particle executing bounded and unbounded motion respectively has been shown. At x_0, slope of the $P.E$ curve dV/dx is zero. Consequently force is zero. That is $F(x) = -dV/dx = 0$, and a particle placed at such points with zero velocity will continue to remain at rest. When the force acting on a particle vanishes, the particle is said to be in *equilibrium.* Point x_0 in Fig. 3.24.1 is at the minimum of potential energy, and represents a state of equilibrium. Now force also vanishes at points, like x_0 in Fig. 3.24.2, at which potential energy is equal to the total energy and is maximum (because at these points $E = V = \frac{1}{2}m\dot{x}^2$ = max, i.e., constant—that means force $m\ddot{x} = 0$). So we should distinguish between the equilibrium points at which potential energy has a minimum value and those at which it has a maximum value.

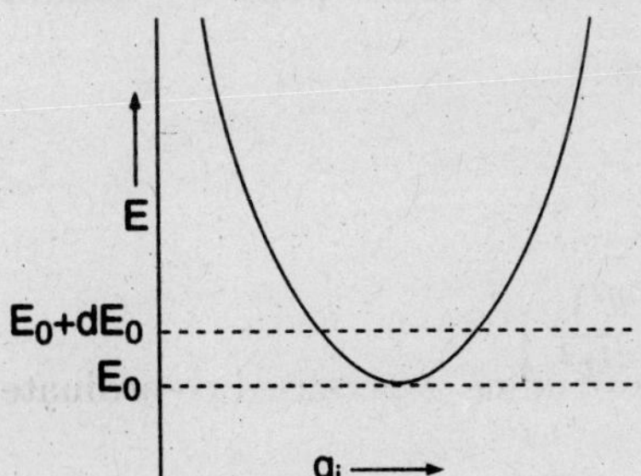

Stable equilibrium

Fig. 3.24.3

At the former point x_0 (in Fig. 3.24.1), at which V is minimum, if the particle is displaced slightly from the position of equilibrium, then the force $F(x) = -dV/dx$ will tend to restore it and it will oscillate about the equilibrium point, performing bounded motion. These points are called points of *stable equilibrium.* A bar pendulum at rest, a suspension galvanometer at its zero position etc. are examples of state of stable equilibrium. We suppose that such a system is disturbed from equilibrium position by an increase in energy dE_0 above the equilibrium energy. Since the potential energy is minimum at equilibrium, any deviation from the equilibrium position will result in an increase in potential energy, and the K.E. would then decrease (as the total energy is conserved). Due to the decrease in K.E., the velocity would also decrease and approach zero in due course of time—the motion is thus bounded. Such a conclusion can be drawn from the shape of the curve of Fig. 3.24.3.

On the other hand, if the particle is displaced slightly from the point of equilibrium x_0 in Fig. 3.24.2, then it will be acted upon by a force

$$F(x) = -\frac{(-dV)}{dx} = \frac{dV}{dx}$$

which will tend to push the particle away from equilibrium point when released. For this reason, such points are called the points of *unstable equilibrium*. A rod standing on one of its ends or an egg standing on its one end is an example of the state of unstable equilibrium. Such a state is not characterised by a minimum of P.E. Therefore if the system is disturbed from equilibrium by an increase in energy dE_0 above the equilibrium energy, then potential energy would decrease. Consequently K.E. and hence velocity would increase infinitely with time, indicating unbounded motion. Such a conclusion can easily be drawn from the shape of the curve of Fig. 3.24.4.

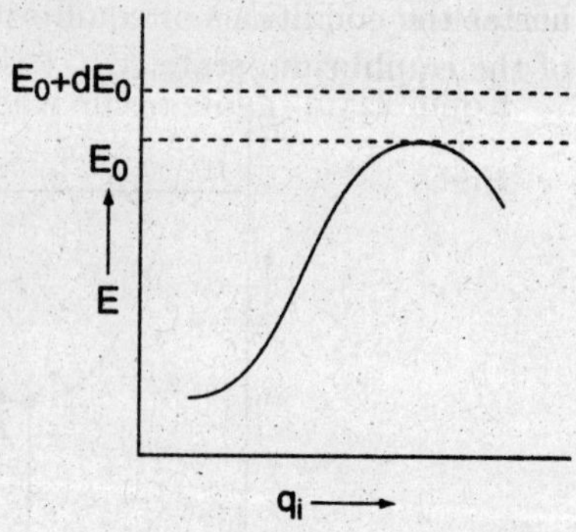

Unstable equilibrium

Fig. 3.24.4

Potential energy about a point of stable equilibrium can be calculated in the following way.

Let the particle be slightly displaced from the point of stable equilibrium, executing small oscillations. Then the potential energy function can be expressed in the form of Taylor-series expansion, as

$$V(x) = V(x_0) + (x - x_0)\left(\frac{dV}{dx}\right)_{x_0} + \frac{1}{2}(x - x_0)^2\left(\frac{d^2V}{dx^2}\right)_{x_0} + \cdots$$

$$= V(x_0) + \frac{1}{2}(x - x_0)^2\left(\frac{d^2V}{dx^2}\right)_{x_0} + \cdots$$

$$\left(\because \left(\frac{dV}{dx}\right)_{x_0} = 0 \text{ at a point } x_0 \text{ of stable equilibrium}\right).$$

Since zero level of potential energy is arbitrary, we can choose it at the point x_0, making $V(x_0) = 0$. Then to a first approximation, we can write

$$V(x) = \frac{1}{2}\left(\frac{d^2V}{dx^2}\right)_{x_0}(x - x_0)^2$$

$$= \frac{1}{2}k(x - x_0)^2, \quad \text{where } k = \left(\frac{d^2V}{dx^2}\right)_{x_0}$$

To find out the points of stable and unstable equilibrium, we are to put $\frac{dV}{dx} = 0$ and find the values of x. Then putting these values of x—one by one in $\frac{d^2V}{dx^2}$, if we see that $\frac{d^2V}{dx^2} = +ve$ for some value, then that point will be a stable equilibrium; if however $\frac{d^2V}{dx^2} = -ve$ for some value, then that point will be an unstable equilibrium as will be evident from the following illustrative example.

Illustrative Example

The potential energy of a particle is given by $V = 3x^4 - 8x^3 - 6x^2 + 24x$. Find the points of stable and unstable equilibrium. [C.U. (Hons.) 1998, cf. 2003]

Equilibrium points occur when $\dfrac{dV}{dx} = 0$.

Here

$$\frac{dV}{dx} = 12x^3 - 24x^2 - 12x + 24 = 0 \quad \text{or} \quad 12(x^3 - 2x^2 - x + 2) = 0$$

i.e., $x^3 - 2x^2 - x + 2 = 0$ (1)

or $(x^2 - 1)(x - 2) = 0$ or $x = +1, -1$ and 2.

Thus there are three equilibrium points at $x = +1, -1$ and 2.

Now calculating $\dfrac{d^2V}{dx^2}$ from (1), we see that

$$\frac{d^2V}{dx^2} = 3x^2 - 4x - 1 = -2 = -ve \text{ when } x = +1.$$

Again

$$\frac{d^2V}{dx^2} = 6 = +ve \text{ when } x = -1.$$

Also

$$\frac{d^2V}{dx^2} = 3 = +ve \text{ when } x = 2.$$

Thus V is minimum when $x = -1$ and 2, which therefore give the points of stable equilibrium. Again V is maximum when $x = +1$, which therefore gives the point of unstable equilibrium.

3.25 Small oscillations

(i) Here we shall consider a special form of motion known as small oscillations. In the case of a simple pendulum oscillating free from any friction with deflection θ the equation relating the deflection to the time is given by

$$t = \sqrt{l/2g} \int_\theta^\alpha \frac{d\theta}{\sqrt{\cos\theta - \cos\alpha}} \quad \text{[See Art. 6.6]}$$

This integral is known as elliptical integral and cannot be evaluated in terms of elementary function. The problem is simplified if we take deflection θ and the amplitude α to be very small. Under this condition,

$$\cos\theta = 1 - \frac{\theta^2}{2} \quad \text{and} \quad \cos\alpha = 1 - \frac{\alpha^2}{2}.$$

Hence

$$t = -\sqrt{\frac{l}{g}} \int_{\alpha}^{\theta} \frac{d\theta}{\sqrt{\alpha^2 - \theta^2}} \quad \text{or} \quad -\int_{\alpha}^{\theta} \frac{d\theta}{\sqrt{\alpha^2 - \theta^2}} = \sqrt{\frac{g}{l}}\, t.$$

∴ Integrating we have,

$$\cos^{-1} \frac{\theta}{\alpha} = \sqrt{\frac{g}{l}}\, t \quad \text{or} \quad \theta = \alpha \cos \sqrt{\frac{g}{l}}\, t.$$

Thus we see that the angle θ is a periodic function of time. The angular frequency is given by $w = \sqrt{g/l}$ and the periodic time $T = 2\pi\sqrt{l/g}$. This gives the period of small oscillation of the pendulum. It should be noted here that the frequency and period of small oscillations do not depend on the amplitude of oscillation α.

(ii) General Problem of Small oscillations

In tackling the general problem of small oscillations we perform an appropriate simplification of Lagrangian method. We know that the oscillations whether large or small take place about a position of equilibrium. Thus a pendulum oscillates about a vertical which is its stable equilibrium position. When the pendulum is deflected from the stable position a restoring force acts on the pendulum in a direction opposite to the direction of deflection. In the equilibrium position this force is zero.

We now suppose that the force is acting on a system which is wholly conservative and hence it is derivable from a potential. In the equilibrium condition of the system the generalised force Q_k will be equal to zero.

The potential energy is a function of position only i.e., the generalised co-ordinates $q_1, q_2, q_3, \cdots, q_n$. Hence

$$\left(\frac{\partial V}{\partial q_k}\right) = 0. \tag{3.25.1}$$

Let us denote the deviation of the generalised co-ordinates from equilibrium by β_k.

$$\therefore\ q_k = q_{0k} + \beta_k, \tag{3.25.2}$$

q_{0k} denotes the equilibrium position of the co-ordinate system. Expanding the potential $V(q_1, q_2, \cdots, q_n)$ in a Taylor's series about q_{0k} we may write

$$V(q_1, q_2, \cdots, q_n) = V(q_{01}, q_{02}, \cdots, q_{0n}) + \sum_k \left(\frac{\partial V}{\partial q_k}\right)_0 \beta_k + \frac{1}{2}\sum_{k,j}\left(\frac{\partial^2 V}{\partial q_k \partial_{q_j}}\right)_0 \beta_k\beta_j + \cdots,$$

The first term on the right hand side of the equation is constant and hence may be taken to be equal to zero and the second term vanishes due to equation (3.25.1). So to the first approximation we may write

$$V(q_1, q_2, \cdots, q_n) = \frac{1}{2}\sum_{k,j}\left(\frac{\partial^2 V}{\partial q_k \partial q_j}\right)_0 \beta_k\beta_j.$$

Writing η_{kj} for $\left(\frac{\partial^2 V}{\partial q_k \partial q_j}\right)_0$ we obtain

$$V(q_1, q_2, \cdots, q_n) = \frac{1}{2}\sum_{k,j} \eta_{kj}\beta_k\beta_j. \tag{3.25.3}$$

We shall now examine the expression for kinetic energy of the system of particles which is given by

$$T = \sum_{k,j} \frac{1}{2} m_{kj} \dot{q}_k \dot{q}_j.$$

The coefficients m_{kj} are in general functions of the co-ordinates q_k.

Here $m_{kj} = m_{jk}$, i.e., m_{kj} form a symmetric matrix. From Taylor's series we may write

$$m_{kj}(q_1, q_2, \cdots, q_n) = m_{kj}(q_{01}, q_{02}, \cdots, q_{0n}) + \sum_i \left(\frac{\partial m_{kj}}{\partial q_i}\right)_0 \beta_i + \cdots$$

Hence the kinetic energy

$$T = \frac{1}{2}\left\{ m_{kj}\left(q_{01}, q_{02}, \cdots, q_{0n}\right) + \sum_i \left(\frac{\partial m_{kj}}{\partial q_i}\right)_0 \beta_i + \cdots \right\} \dot{\beta}_k \dot{\beta}_j.$$

This equation is quadratic in $\dot{\beta}_k$'s and the lowest non-vanishing approximation to T is obtained by dropping all the terms on the right hand side of the above equation except the first term.

$$\therefore \quad T = \frac{1}{2}\sum_{k,j} m_{kj}(q_{01}, q_{02}, \cdots, q_{0n})\dot{\beta}_k\dot{\beta}_j = \frac{1}{2}\sum_{k,j} \alpha_{kj}\dot{\beta}_k\dot{\beta}_j. \tag{3.25.4}$$

Here the constant value of m_{kj} at equilibrium is denoted by α_{kj}.

Hence the Lagrangian function

$$L = T - V = \frac{1}{2}\sum_{kj}\left(\alpha_{kj}\dot{\beta}_k\dot{\beta}_j - \eta_{kj}\beta_k\beta_j\right).$$

Thus the n-equations of motion derived from function L is

$$\frac{d}{dt}\left(\frac{\partial L}{\partial \dot{\beta}_k}\right) - \left(\frac{\partial L}{\partial \beta_k}\right) = 0 \quad (k = 1, 2, 3, \cdots, n)$$

$$\text{or} \quad \sum_j \alpha_{kj}\ddot{\beta}_j + \sum_j \eta_{kj}\beta_j = 0. \tag{3.25.5}$$

In the above equation both α_{kj} and η_{kj} have symmetric properties.

We can write this equation in the following matrix form

$$\alpha\ddot{\beta} + \eta\beta = 0 \tag{3.25.6}$$

where $\alpha = \begin{pmatrix} \alpha_{11} & \alpha_{12} & \cdots & \alpha_{1n} \\ \alpha_{21} & \alpha_{22} & \cdots & \alpha_{2n} \\ \cdots & \cdots & \cdots & \cdots \\ \alpha_{n1} & \alpha_{n2} & \cdots & \alpha_{nr} \end{pmatrix}$; $\eta = \begin{pmatrix} \eta_{11} & \eta_{12} & \cdots & \eta_{1n} \\ \eta_{21} & \eta_{22} & \cdots & \eta_{2n} \\ \cdots & \cdots & \cdots & \cdots \\ \eta_{n1} & \eta_{n2} & \cdots & \eta_{nn} \end{pmatrix}$; $\beta = \begin{pmatrix} \beta_1 \\ \beta_2 \\ \vdots \\ \beta_n \end{pmatrix}$.

Here α denotes the *inertia coefficient matrix* and η the *stiffness coefficient matrix.* η plays the role of restoring force which tends to bring the system to the position of equilibrium which may be called restoring tensor. We now consider the solution of the equation (3.25.6) taking the motion to be purely harmonic. Let the solution be of the form

$$\beta = A\sin(wt+\delta), \tag{3.25.7}$$

where A is a column matrix of constant amplitude. From (3.25.6) and (3.25.7), we have

$$-\alpha A w^2 + \eta A = 0 \text{ or } (\eta - w^2\alpha)A = 0 \tag{3.25.8}$$

$$\text{or } \eta^{-1}(\eta - w^2\alpha)A = 0$$

$$\text{or } (I - w^2\eta^{-1}\alpha)A = 0$$

$$(\eta^{-1}\eta = I, \text{ being an } n\times n \text{ unit matrix})$$

$$\text{or } (I - w^2U)A = 0[\eta^{-1}\alpha = U = \text{ a dynamical matrix}].$$

Putting $w^2 = 1/Z$ in the above equation we have,

$$(ZI - U)A = 0 \tag{3.25.9}$$

The above equation constitutes n linear homogeneous equations and can have non-trivial solution (i.e., $A \neq 0$) only when

$$|ZI - U| = \begin{vmatrix} Z-U_{11} & -U_{12} & \cdots & U_{1n} \\ -U_{21} & Z-U_{22} & \cdots & U_{2n} \\ \cdots & \cdots & \cdots & \cdots \\ -U_{n1} & -U_{n2} & \cdots & Z-U_{nn} \end{vmatrix} = 0 \tag{3.25.10}$$

since for unit matrix each diagonal element is equal to 1 and each non-diagonal element is equal to zero.

Let the roots of the equation (3.25.10) be $Z_1, Z_2, \cdots, Z_n$. If we consider ω_r corresponding to each Z_r, then

$$w_r = \frac{1}{\sqrt{Z_r}}, \quad \text{where } r = 1, 2, \cdots, n.$$

Thus the solution may be written in the form

$$\beta = \sum_{r=1}^{n} A_r \sin(w_r t + \delta_r). \tag{3.25.11}$$

The column matrices A_r are the *eigen vectors* corresponding to *eigen values* Z_r and are called *modal columns.* Every oscillation represented by each modal column is called a ***principal mode*** of oscillation of the system. The number of principal modes of oscillation is called number of degrees of freedom of the system and the lowest of these is called fundamental frequency. Each oscillation with a definite frequency is called *eigen vibration* or *normal mode* of the system.

From the above solution (3.25.11), we may write

$$\left.\begin{aligned}\beta_1 &= A_1 \sin(w_1 t + \delta_1),\\ \beta_2 &= A_2 \sin(w_2 t + \delta_2),\\ &\cdots\cdots\cdots\cdots\cdots\\ \beta_n &= A_n \sin(w_n t + \delta_n),\end{aligned}\right\} \tag{3.25.12}$$

Normal Co-ordinates

These generalised co-ordinates β_1, β_2, β_3 etc. are known as *normal co-ordinates.* These are such that each represents oscillation with a single normal frequency. These functions are orthogonal and kinetic and potential energies of a system in terms of these co-ordinates become diagonal. It is important to note that the off-diagonal elements in the K.E. and P.E. matrices phyically represent the coupling between the particles of the system. So, with the use of normal co-ordinates, the co-ordinates become uncoupled and the problem of coupled systems become completely separable into independent motions with normal co-ordinates.

3.26 Small oscillation with one degree of freedom

In this case the kinetic and the potential energies may be written in the form

$$T = \frac{1}{2}\alpha(q_0)\dot{\beta}^2; \quad V = \frac{1}{2}\eta\beta^2$$

[from equations (3.25.4) and (3.25.3)]

If $q_0 = 0$, i.e., q is measured from the equilibrium position, we have

$$L = \frac{1}{2}\alpha(0)\dot{q}^2 - \frac{1}{2}\eta q^2.$$

Hence the Lagrange's equation is

$$\alpha(0)\ddot{q} + \eta q = 0 \quad \text{or} \quad \ddot{q} + \frac{\eta}{\alpha(0)}q = 0.$$

Putting

$$\frac{\eta}{\alpha(0)} = w^2, \quad \text{i.e.,} \quad \frac{\left(\dfrac{\partial^2 V}{\partial q^2}\right)_0}{\alpha(0)} = w^2,$$

we obtain

$$\ddot{q} + w^2 q = 0. \tag{3.26.1}$$

The general solution of the equation may be written in the form

$$q = A_1 \sin wt + A_2 \cos wt$$

$$\text{or} \quad q = A\cos(wt + \delta).$$

A system described by the equation (3.26.1) is called a linear harmonic oscillator.

3.27 Small oscillation with two degrees of freedom

In this case

$$L = \frac{1}{2}(\alpha_{11}\dot{\beta}_1^2 + 2\alpha_{12}\dot{\beta}_1\dot{\beta}_2 + \alpha_{22}\dot{\beta}_2^2) - \frac{1}{2}(\eta_{11}\beta_1^2 + 2\eta_{12}\beta_1\beta_2 + \eta_{22}\beta_2^2).$$

If $q_0 = 0$, i.e., q is measured from the equilibrium position, we have,

$$L = \frac{1}{2}(\alpha_{11}\dot{q}_1^2 + 2\alpha_{12}\dot{q}_1\dot{q}_2 + \alpha_{22}\dot{q}_2^2) - \frac{1}{2}(\eta_{11}q_1^2 + 2\eta_{12}q_1q_2 + \eta_{22}q_2^2). \quad (3.27.1)$$

(a) Problem of double pendulum

It is a case of small oscillation with two degrees of freedom. In the double pendulum, in mass m_1 of a pendulum there is a hinge from which another pendulum of mass m_2 is suspended (Fig. 3.27.1).

The second pendulum is forced to oscillate. Let the masses and lengths of the two pendulums be m_1, l_1 and m_2, l_2. Let the angle of deflection of the pendulum of mass m_1 be ϕ and that of the pendulum of mass m_2 be ψ.

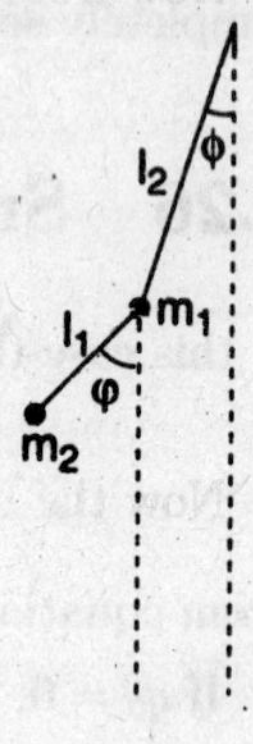

Fig. 3.27.1

The co-ordinates of the mass m_2 are

$$x_2 = l_1 \sin\phi + l_2 \sin\psi$$

$$z_2 = l_1(1 - \cos\phi) + l_2(1 - \cos\psi),$$

whence we get the velocity components

$$\dot{x}_2 = l_1 \cos\phi\dot{\phi} + l_2 \cos\psi\dot{\psi}$$

$$\dot{z}_2 = l_1 \sin\phi\dot{\phi} + l_2 \sin\psi\dot{\psi}.$$

Squaring and adding we obtain the kinetic energy of the body of mass m_2 given by

$$T_2 = \frac{m_2}{2}\left[l_1^2\dot{\phi}^2 + l_2^2\dot{\psi}^2 + 2l_1l_2\cos(\phi - \psi)\dot{\phi}\dot{\psi}\right].$$

The potential energy

$$V_2 = m_2 g\,[l_1(1 - \cos\phi) + l_2(1 - \cos\psi)]$$

For a single pendulum of mass m_1, kinetic energy $T_1 = \frac{m_1}{2}l_1^2\dot{\phi}^2$ and potential energy $V_1 = m_1gl_1(1 - \cos\phi)$.

$\therefore$ The Lagrangian function for a double pendulum is

$$L = \left(\frac{m_1 + m_2}{2}\right) l_1^2\dot{\phi}^2 + \frac{m_2}{2}l_2^2\dot{\psi}^2 + m_2l_1l_2\cos(\phi - \psi)\dot{\phi}\dot{\psi}$$
$$-(m_1 + m_2)gl_1(1 - \cos\phi) - m_2gl_2(1 - \cos\psi).$$

Considering ϕ and ψ close to zero we must write $\cos(\phi - \psi) = 1$ and $\cos\phi = 1 - \phi^2/2$ and $\cos\psi = 1 - \psi^2/2$.

Hence

$$L = \frac{m_1 + m_2}{2} l_1^2 \dot{\phi}^2 + \frac{m_2}{2} l_2 \dot{\psi}^2 + m_2 l_1 l_2 \dot{\phi} \dot{\psi} - \left(\frac{m_1 + m_2}{2} \right) g l_1 \phi^2 - \frac{m_2 g l_2}{2} \psi^2. \quad (3.27.2)$$

Comparing (3.27.1) and (3.27.2) we have for the problem of the double pendulum,

$$a_{11} = (m_1 + m_2) l_1^2, \quad a_{12} = m_2 l_1 l_2, \quad a_{22} = m_2 l_2^2$$

$$\eta_{11} = (m_1 + m_2) l_1 g, \quad \eta_{12} = 0, \quad \eta_{22} = m_2 l_2 g$$

η_{11}, η_{12} and η_{22} are expressed by the equations

$$\eta_{11} = \left(\frac{\partial^2 V}{\partial q_1^2} \right)_0, \quad \eta_{12} = \left(\frac{\partial^2 V}{\partial q_1 \partial q_2} \right)_0, \quad \eta_{22} = \left(\frac{\partial^2 V}{\partial q_2^2} \right)_0$$

Now from equation (3.27.1), we have

$$\frac{\partial L}{\partial \dot{q}_1} = \alpha_{11} \dot{q}_1 + \alpha_{12} \dot{q}_2; \qquad -\frac{\partial L}{\partial q_1} = \eta_{11} q_1 + \eta_{12} q_2.$$

$$\frac{\partial L}{\partial \dot{q}_2} = \alpha_{12} \dot{q}_1 + \alpha_{22} \dot{q}_2; \qquad -\frac{\partial L}{\partial q_2} = \eta_{12} q_1 + \eta_{22} q_2.$$

Now the Lagrange's equation is given by

$$\frac{d}{dt} \left(\frac{\partial L}{\partial \dot{q}_k} \right) - \frac{\partial L}{\partial q_k} = 0.$$

$$\left. \begin{aligned} \therefore \alpha_{11} \ddot{q}_1 + \alpha_{12} \ddot{q}_2 + \eta_{11} q_1 + \eta_{12} q_2 = 0 \\ \alpha_{12} \ddot{q}_1 + \alpha_{22} \ddot{q}_2 + \eta_{12} q_1 + \eta_{22} q_2 = 0 \end{aligned} \right\} \quad (3.27.3)$$

The solutions of the above equations must be written in the form

$$\left. \begin{aligned} q_1 &= B_1 e^{iwt} \\ q_2 &= B_2 e^{iwt} \end{aligned} \right\}. \quad (3.27.4)$$

The real part of the solution must be taken. Substituting (3.27.4) in (3.27.3) we have,

$$(\eta_{11} - \alpha_{11} w^2) B_1 + (\eta_{12} - \alpha_{12} w^2) B_2 = 0$$

$$(\eta_{12} - \alpha_{12} w^2) B_1 + (\eta_{22} - \alpha_{22} w^2) B_2 = 0.$$

Hence we obtain

$$\frac{\eta_{11} - \alpha_{11} w^2}{\eta_{12} - \alpha_{12} w^2} = \frac{\eta_{12} - \alpha_{12} w^2}{\eta_{22} - \alpha_{22} w^2}$$

whence we get

$$(\alpha_{11} \alpha_{22} - \alpha_{12}^2) w^4 - (\eta_{11} \alpha_{22} + \eta_{22} \alpha_{11} - 2 \alpha_{12} \eta_{12}) w^2 + (\eta_{11} \eta_{22} - \eta_{12}^2) = 0 \quad (3.27.5)$$

or substituting the values of α_{11}, α_{22}, α_{12}, η_{11}, η_{12} and η_{22} in the above equation we obtain the following equation for the frequencies of a double pendulum.

$$m_1m_2l_1^2l_2^2w^4 - (m_1+m_2)m_2l_1l_2(l_1+l_2)gw^2 + (m_1+m_2)m_2l_1l_2g^2 = 0 \quad (3.27.6)$$

Putting $\frac{l_2}{l_1} = \lambda$ and $\frac{m_2}{m_1} = m$ we have

$$w^2 = \frac{g}{2l_1}\left[(1+m)(1+\lambda) \pm \sqrt{(1+m)^2(1+\lambda)^2 - 4\lambda(1+m)}\right]$$

It is easy to see that the above expression gives only the real values of the frequencies. 'w' has four values, both pairs of which are equal in absolute value. It we take the solution of the form (3.27.4) we may take only the positive value of w.

(b) Problem of Coupled oscillation

In this case the kinetic energy and the potential energy for simple motion in q_1, q_2 directions of a coupled system are given by

$$T = \frac{1}{2}(m_1\dot{q}_1^2 + m_2\dot{q}_2^2)$$

$$V = \frac{1}{2}(s_1q_1^2 + s_2q_2^2 + 2s_1s_2q_1q_2),$$

where s_1, s_2 are the restoring forces per unit displacement along q_1 and q_2 directions and k denotes the coupling constant between the oscillating masses m_1 and m_2.

$$\therefore\ L = \frac{1}{2}\left[(m_1\dot{q}_1^2 + m_2\dot{q}_2^2) - (s_1q_1^2 + s_2q_2^2 + 2kq_1q_2)\right].$$

Comparing with equation (3.27.1), we have

$$\alpha_{11} = m_1$$

$$\alpha_{22} = m_2$$

$$\eta_{11} = s_1$$

$$\eta_{22} = s_2$$

$$\eta_{12} = \eta_{21} = k.$$

From equation (3.25.8), we have the eigen frequencies given by

$$|\eta - w^2\alpha| = 0,$$

i.e., $\begin{vmatrix} \eta_{11} - w^2\alpha_{11} & \eta_{12} - w^2\alpha_{12} \\ \eta_{21} - w^2\alpha_{21} & \eta_{22} - w^2\alpha_{22} \end{vmatrix} = 0$ or $\begin{vmatrix} s_1 - w^2m_1 & k \\ k & s_2 - w^2m_2 \end{vmatrix} = 0$

or $(m_1w^2 - s_1)(m_2w^2 - s_2) - k^2 = 0$; when $m_1 = m_2 = m$ and $s_1 = s_2 = s$,

$$w^2 = \frac{s}{m} \pm \frac{k}{m}.$$

Thus there are two distinct frequencies in the coupled system.

The eigen vectors A_{11}, A_{12} corresponding to frequency $w^2 = \frac{s}{m} + \frac{k}{m}$ are obtained from

$$\begin{bmatrix} s - \left(\frac{s}{m} + \frac{k}{m}\right) m & k \\ k & s - \left(\frac{s}{m} + \frac{k}{m}\right) m \end{bmatrix} \begin{bmatrix} A_{11} \\ A_{12} \end{bmatrix} = 0$$

$$\text{or} \quad \begin{bmatrix} -k & k \\ k & -k \end{bmatrix} \begin{bmatrix} A_{11} \\ A_{12} \end{bmatrix} = 0$$

$$\text{or} \quad -A_{11} + A_{12} = 0 \quad \text{or} \quad A_{11} = A_{12};$$

when $w^2 = \frac{s}{m} - \frac{k}{m}$, we get

$$\begin{bmatrix} k & k \\ k & k \end{bmatrix} \begin{bmatrix} A_{11} \\ A_{12} \end{bmatrix} = 0 \quad \text{or} \quad A_{11} + A_{12} = 0.$$

$$\therefore A_{11} = -A_{12}.$$

Thus when $w^2 = \frac{s}{m} + \frac{k}{m}$ the oscillations are in phase and when $w^2 = \frac{s}{m} - \frac{k}{m}$ the oscillations are in opposite phase.

3.28 Oscillation of a linear triatomic molecule

We consider a linear triatomic molecule in which two atoms each of mass m_p are placed at same distances on either side of the third atom of mass m_Q.

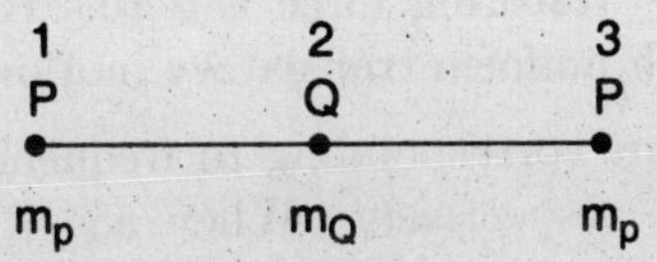

Fig. 3.28.1

We assume that the oscillation is along the line joining the atoms and there is no interaction between the atoms of mass m_p.

Let q_1, q_2, q_3 be the displacements of each atom from the equilibrium position. Then the kinetic and potential energies are given by

$$T = \frac{1}{2} m_p(\dot{q}_1^2 + \dot{q}_3^2) + \frac{1}{2} m_Q \dot{q}_2^2$$

$$V = \frac{1}{2} s(q_2 - q_1)^2 + \frac{1}{2} s(q_2 - q_3)^2 = \frac{s}{2}(q_1^2 + 2q_2^2 + q_3^2 - 2q_1q_2 - 2q_2q_3).$$

$$\therefore L = \frac{1}{2}\left[m_p(\dot{q}_1^2 + \dot{q}_3^2) + m_Q \dot{q}_2^2 - s(q_1^2 + 2q_2^2 + q_3^2 - 2q_1q_2 - 2q_2q_3)\right].$$

Here

$$\begin{aligned} \alpha_{11} &= m_p & \eta_{11} &= s \\ \alpha_{22} &= m_Q & \eta_{12} &= \eta_{21} = -s \\ \alpha_{33} &= m_p & \eta_{22} &= 2s \\ & & \eta_{23} &= \eta_{32} = -s \\ & & \eta_{33} &= s. \end{aligned}$$

Now the eigen frequencies are given by

$$|\eta - w^2\alpha| = 0, \quad \text{i.e.,} \quad \begin{vmatrix} s - w^2 m_p & -s & 0 \\ -s & 2s - w^2 m_Q & -s \\ 0 & -s & s - w^2 m_p \end{vmatrix} = 0.$$

Evaluation of the determinant leads to the following equation;

$$(s - w^2 m_p)(w^4 m_Q m_p - s w^2 m_Q - 2 s w^2 m_p) = 0$$

$$\text{or} \quad w^2 (s - w^2 m_p)\left[w^2 m_Q m_p - s(m_Q + 2m_p)\right] = 0.$$

The roots are

$$w_1 = 0; \quad w_2 = \sqrt{\frac{s}{m_p}}$$

$$w_3 = \sqrt{\frac{s(m_Q + 2m_p)}{m_Q m_p}} = \sqrt{\frac{s}{m_p}\left(1 + \frac{2m_p}{m_Q}\right)}.$$

The frequency $w_1 = 0$ corresponds to the case where the restoring force is zero, i.e., the acceleration acting against the restoring force is zero. Hence the motion is uniform. The molecule as a whole moves with uniform translatory motion.

Case I. Let the eigen vectors corresponding to frequencies w_1, w_2, w_3 be (A_{11}, A_{12}, A_{13}), (A_{21}, A_{22}, A_{23}), (A_{31}, A_{32}, A_{33}) respectively. When $w_1 = 0$ there is no change in potential energy and so we have

$$\begin{bmatrix} s & -s & 0 \\ -s & 2s & -s \\ 0 & -s & s \end{bmatrix} \begin{bmatrix} A_{11} \\ A_{12} \\ A_{13} \end{bmatrix} = 0$$

whence we may write

$$sA_{11} - sA_{12} = 0 \quad \text{or} \quad A_{11} - A_{12} = 0 \tag{3.28.1}$$

$$-sA_{11} + 2sA_{12} - sA_{13} = 0$$

$$\text{or} \quad -A_{11} + 2A_{12} - A_{13} = 0 \tag{3.28.2}$$

$$\text{and} \quad -sA_{12} + sA_{13} = 0$$

$$\text{or} \quad -A_{12} + A_{13} = 0. \tag{3.28.3}$$

From (3.28.1) and (3.28.3) we have

$$A_{11} = A_{12} = A_{13}.$$

This shows that each atom of the triatomic molecule is equally displaced in the same direction. The eigen vectors are always taken with unit magnitude and hence we apply the normalisation condition given by

$$m_p(A_{1j}^2 + A_{3j}^2) + m_Q A_{2j}^2 = 1$$

whence we have

$$m_p(A_{11}^2 + A_{13}^2) + m_Q A_{12}^2 = 1$$

$$\text{or } 2m_p A_{11}^2 + m_Q A_{11}^2 = 1 \quad [\because A_{11} = A_{12} = A_{13}]$$

$$\text{or } A_{11} = \frac{1}{\sqrt{2m_p + m_Q}}$$

$$\text{or } A_{12} = \frac{1}{\sqrt{2m_p + m_Q}}$$

$$\text{or } A_{13} = \frac{1}{\sqrt{2m_p + m_Q}}.$$

Thus the normalisation condition for $\omega_1 = 0$ fixes the values of A_{11}, A_{12} and A_{13}. Each atom is equally displaced.

Case II. When $w_2 = \sqrt{\dfrac{s}{m_p}}$, we have

$$\begin{bmatrix} s - w_2^2 m_p & -s & 0 \\ -s & 2s - w_2^2 m_Q & -s \\ 0 & -s & s - w_2^2 m_p \end{bmatrix} \begin{bmatrix} A_{21} \\ A_{22} \\ A_{23} \end{bmatrix} = 0$$

$$\text{or} \quad \begin{bmatrix} 0 & -s & 0 \\ -s & 2s - s\left(\dfrac{m_Q}{m_p}\right) & -s \\ 0 & -s & 0 \end{bmatrix} \begin{bmatrix} A_{21} \\ A_{22} \\ A_{23} \end{bmatrix} = 0$$

whence we get

$$-sA_{22} = 0$$

$$\text{or } A_{22} = 0;$$

$$-sA_{21} + \left[2s - s\left(\frac{m_Q}{m_p}\right)\right] A_{22} - sA_{23} = 0$$

$$\text{or } -A_{21} + \left(2 - \frac{m_Q}{m_p}\right) A_{22} - A_{23} = 0$$

$$\text{or } A_{21} + A_{23} = 0 \quad (\because A_{22} = 0)$$

$$\therefore A_{21} = -A_{23}.$$

Since $A_{22} = 0$, the central atom Q remains at rest. The end atoms are equally displaced in opposite directions as $A_{21} = -A_{23}$. The normalisation condition gives

$$m_p(A_{21}^2 + A_{23}^2) = 1 \quad [A_{22} = 0]$$

$$\text{or } A_{21} = \frac{1}{\sqrt{2m_p}} \quad (\because A_{23} = -A_{21})$$

$$\therefore A_{23} = -\frac{1}{\sqrt{2m_p}}.$$

Case III. Also when $w_3 = \sqrt{\frac{s}{m_p}\left(1 + \frac{2m_p}{m_Q}\right)}$ we have

$$\begin{bmatrix} s - w_3^2 m_p & -s & 0 \\ -s & 2s - w_3^2 m_Q & -s \\ 0 & -s & s - w_3^2 m_p \end{bmatrix} \begin{bmatrix} A_{31} \\ A_{32} \\ A_{33} \end{bmatrix} = 0$$

$$\text{or} \begin{bmatrix} -2s\frac{m_p}{m_Q} & -s & 0 \\ -s & -s\frac{m_Q}{m_p} & -s \\ 0 & -s & -\frac{2sm_p}{m_Q} \end{bmatrix} \begin{bmatrix} A_{31} \\ A_{32} \\ A_{33} \end{bmatrix} = 0$$

whence we have

$$-2s\frac{m_p}{m_Q}A_{31} - sA_{32} = 0 \quad \text{or} \quad 2\frac{m_p}{m_Q}A_{31} + A_{32} = 0 \tag{3.28.4}$$

$$-sA_{31} - s\frac{m_Q}{m_p}A_{32} - sA_{33} = 0 \quad \text{or} \quad A_{31} + \frac{m_Q}{m_p}A_{32} + A_{33} = 0 \tag{3.28.5}$$

$$\text{and} \quad -sA_{32} - 2s\frac{m_p}{m_Q}A_{33} = 0 \quad \text{or} \quad A_{32} + 2\frac{m_p}{m_Q}A_{33} = 0. \tag{3.28.6}$$

From (3.28.4) and (3.28.6), we have

$$A_{31} = A_{33} \tag{3.28.7}$$

$\therefore$ From (3.28.6) we get

$$A_{32} = -2\frac{m_p}{m_Q}A_{31} \tag{3.28.8}$$

The relation (3.28.7) shows that the displacements of the end atoms are equal in the same direction and from the relation (3.28.8) we see that the displacement of the central atom is not equal to that of either of end atoms and moreover they are not in phase.

The normalisation condition gives

$$m_p(A_{31}^2 + A_{33}^2) + m_Q A_{32}^2 = 1 \text{ or } 2m_p A_{31}^2 + m_Q\left(\frac{2m_p}{m_Q}\right)^2 A_{31}^2 = 1$$

$$\text{or } A_{31} = \frac{1}{\sqrt{2m_p\left(1 + \frac{2m_p}{m_Q}\right)}}.$$

$$\therefore\ A_{33} = \frac{1}{\sqrt{2m_p\left(1+\dfrac{2m_p}{m_Q}\right)}}$$

$$\text{and } A_{32} = -\frac{2m_p}{m_Q}\frac{1}{\sqrt{2m_p\left(1+\dfrac{2m_p}{m_Q}\right)}} = \frac{-2}{\sqrt{2m_Q\left(2+\dfrac{m_Q}{m_p}\right)}}.$$

Case I	m_p $\bullet\!\rightarrow$ 1	m_Q $\bullet\!\rightarrow$ 2	m_p $\bullet\!\rightarrow$ 3
Case II	m_p $\leftarrow\!\bullet$ 1	m_Q $\bullet$ 2	m_p $\bullet\!\rightarrow$ 3
Case III	m_p $\bullet\!\rightarrow$ 1	m_Q $\leftarrow\!\bullet$ 2	m_p $\bullet\!\rightarrow$ 3

Illustrative Examples

1. Using Lagrange's equation of motion find the motion of a simple pendulum without damping and calculate its time period oscillation.

In this case the kinetic and the potential energies are

$$T = \frac{1}{2}ml^2\dot{\theta}^2$$

$$V = mgl(1-\cos\theta),$$

where l = length of the pendulum,

θ = angular displacement,

m = mass of the bob of the simple pendulum.

Hence from Lagrange's equation we have

$$\frac{d}{dt}\left(\frac{\partial L}{\partial\dot{\theta}}\right) - \frac{\partial L}{\partial\theta} = 0. \tag{1}$$

Now

$$T - V = L = \frac{1}{2}ml^2\dot{\theta}^2 - mgl + mgl\cos\theta.$$

$$\therefore\ \frac{\partial L}{\partial\dot{\theta}} = ml^2\dot{\theta} \quad\text{and}\quad \frac{\partial L}{\partial\theta} = -mgl\sin\theta.$$

$\therefore$ From equation (1), we have

$$\frac{d}{dt}(ml^2\dot{\theta}) + mgl\sin\theta = 0 \quad\text{or}\quad \ddot{\theta} + \frac{g}{l}\sin\theta = 0.$$

For small angular displacement θ, the above equation may be written in the form

$$\ddot{\theta} + \frac{g}{l}\theta = 0 \quad\text{or}\quad \ddot{\theta} + \omega^2\theta = 0 \quad \left[\text{where } \omega^2 = \frac{g}{l}\right].$$

Hence the time period of oscillation is given by

$$T = \frac{2\pi}{\omega} = 2\pi\sqrt{\frac{l}{g}}.$$

2. Deduce Kepler's laws of motion from the Canonical equations. **[C.U. (Hons.) '93, '96]**

Considering the motion to be confined in (r, θ) plane, we have

$$T = \frac{1}{2}m(\dot{r}^2 + r^2\dot{\theta}^2) \quad \text{(See Chap. VII)}$$

$$V = -\frac{GMm}{r}.$$

Now, from the Canonical equations, we have

$$\dot{p}_r = -\frac{\partial H}{\partial r} \quad \text{and} \quad \dot{p}_\theta = -\frac{\partial H}{\partial \theta}.$$

But

$$H = \frac{1}{2}m(\dot{r}^2 + r^2\dot{\theta}^2) - \frac{GMm}{r}$$

$$= \frac{1}{2m}\left(p_r^2 + \frac{p_\theta^2}{r^2}\right) - \frac{GMm}{r} \quad \left[\because p_r = \frac{\partial T}{\partial \dot{r}} = m\dot{r} \text{ and } p_\theta = \frac{\partial T}{\partial \dot{\theta}} = mr^2\dot{\theta}\right].$$

$$\therefore \dot{p}_\theta = -\frac{\partial H}{\partial \theta} = 0 \quad \text{or} \quad p_\theta = mr^2\dot{\theta} = \text{ constant.}$$

Hence the areal velocity $= \frac{1}{2}r^2\dot{\theta} =$ constant. This proves Kepler's second law of motion.

Again,

$$\dot{p}_r = -\frac{\partial H}{\partial r} = -\left(-\frac{p_\theta^2}{mr^3} + \frac{GMm}{r^2}\right) = \frac{p_\theta^2}{mr^3} - \frac{GMm}{r^2}$$

$$\text{or } \frac{d}{dt}(m\dot{r}) = \frac{(mr^2\dot{\theta})^2}{mr^3} - \frac{GMm}{r^2} = mr\dot{\theta}^2 - \frac{GMm}{r^2}$$

$$\text{or } \ddot{r} = r\dot{\theta}^2 - \frac{GM}{r^2}$$

$$\text{or } \ddot{r} - r\dot{\theta}^2 = -\frac{GM}{r^2}.$$

From this expression other laws can be deduced.

3. (i) Write down the Lagrangian for a one dimensional simple harmonic oscillator and obtain the equation of motion.

(ii) Derive the Hamiltonian of the oscillator from the Lagrangian function. **[C.U. (Hons.) '81, '88]**

(i) For a one dimensional simple harmonic oscillator

$$T = \frac{1}{2}m\dot{x}^2 \quad \text{and} \quad V = -\int F dx, \quad F = -kx$$

being the restoring force proportional to displacement x from the mean position and directed towards the mean position.

$$\therefore\ V = \int_0^x kx\,dx = \frac{1}{2}kx^2.$$

$\therefore$ Lagrangian $L = T - V = \frac{1}{2}m\dot{x}^2 - \frac{1}{2}kx^2$ and the Lagrangian equation for a simple harmonic oscillator corresponding to generalised co-ordinates x and $\dot{x}$ is

$$\frac{d}{dt}\left(\frac{\partial L}{\partial \dot{x}}\right) - \frac{\partial L}{\partial x} = 0$$

$$\text{or } \frac{d}{dt}(m\dot{x}) + kx = 0 \quad \because \frac{\partial L}{\partial \dot{x}} = m\dot{x} \text{ and } \frac{\partial L}{\partial x} = -kx$$

$$\text{or } \ddot{x} + \frac{k}{m}x = 0$$

which is the required equation of motion.

(ii) The Hamiltonian of the oscillator corresponding to generalised co-ordinates x and $\dot{x}$ is

$$H = p_x\dot{x} - L \tag{1}$$

p_x being the momentum of the oscillator.

Here

$$L = \frac{1}{2}m\dot{x}^2 - \frac{1}{2}kx^2, \quad p_x = m\dot{x} = \frac{\partial L}{\partial \dot{x}} \quad \text{or} \quad \dot{x} = \frac{p_x}{m}.$$

Putting these values in (1), we have

$$H = p_x\frac{p_x}{m} - \frac{1}{2}m\left(\frac{p_x}{m}\right)^2 + \frac{1}{2}kx^2 = \frac{1}{2}\frac{p_x^2}{m} + \frac{1}{2}kx^2.$$

The Hamiltonian equations are

$$\dot{x} = \frac{\partial H}{\partial p_x} \quad \text{or} \quad \dot{x} = \frac{\partial}{\partial p_x}\left[\frac{1}{2}\frac{p_x^2}{m} + \frac{kx^2}{2}\right] = \frac{p_x}{m} \quad \text{or} \quad p_x = m\dot{x}.$$

$$\therefore\ \dot{p}_x = m\ddot{x}. \tag{2}$$

$$\text{and } \dot{p}_x = -\frac{\partial H}{\partial x} \quad \text{or} \quad \dot{p}_x = -\frac{\partial}{\partial x}\left[\frac{p_x^2}{2m} + \frac{kx^2}{2}\right] = -kx. \tag{3}$$

From (2) and (3), we have

$$m\ddot{x} = -kx \quad \text{or} \quad \ddot{x} + \frac{k}{m}x = 0$$

which is the equation of a one-dimensional simple harmonic oscillator from the Hamiltonian function.

4. (i) A particle moves under the influence of a central field of force which depends only on its distance from the origin. Show that the angular momentum is conserved during its motion.

(ii) Set up the Hamiltonian for the system in (i). [C.U. (Hons.) '82, '87, 2005, 2007]

Since the force is directed opposite to $\vec{r}$, we may take it as $F = -\frac{k}{r^2}$, k being the force constant.

$$\therefore \ V = -\int F dr = \int \frac{k}{r^2}\, dr = -\frac{k}{r} + c.$$

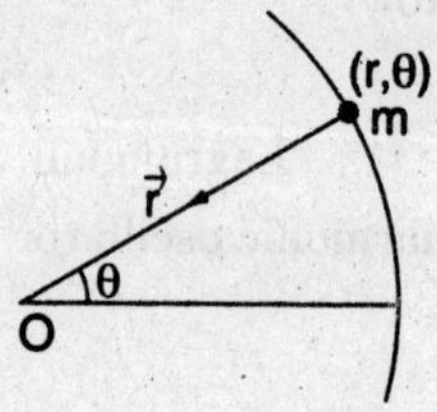

Now at $r = \infty, V = 0$.

$\therefore \ c = 0$, or, $V = -\frac{k}{r}$ and $T = \frac{1}{2}m(\dot{r}^2 + r^2\dot{\theta}^2)$, so that the Lagrangian is given by

$$L = T - V = \frac{1}{2}m(\dot{r}^2 + r^2\dot{\theta}^2) + \frac{k}{r}.$$

So the Lagrangian equations for a conservative system of generalised co-ordinates r and θ are

$$\frac{d}{dt}\left(\frac{\partial L}{\partial \dot{r}}\right) - \frac{\partial L}{\partial r} = 0 \qquad (1)$$

$$\text{and} \quad \frac{d}{dt}\left(\frac{\partial L}{\partial \dot{\theta}}\right) - \frac{\partial L}{\partial \theta} = 0. \qquad (2)$$

Now

$$\frac{\partial L}{\partial \dot{r}} = m\dot{r} \quad \text{and} \quad \frac{\partial L}{\partial r} = mr\dot{\theta}^2 - \frac{k}{r^2}.$$

Also

$$\frac{\partial L}{\partial \dot{\theta}} = mr^2\dot{\theta} \quad \text{and} \quad \frac{\partial L}{\partial \theta} = 0$$

suggest that the acceleration has only radial component.

With these values, the Lagrangian equations become

$$\frac{d}{dt}(m\dot{r}) - mr\dot{\theta}^2 + \frac{k}{r^2} = 0 \quad \text{from (1)}$$

$$\text{or} \quad m\ddot{r} - mr\dot{\theta}^2 + \frac{k}{r^2} = 0 \quad \text{or} \quad \ddot{r} - r\dot{\theta}^2 + \frac{k}{m}\frac{1}{r^2} = 0 \qquad (3)$$

$$\text{and} \quad \frac{d}{dt}(mr^2\dot{\theta}) = 0 \quad \text{from (2)}$$

$$\text{or} \quad mr^2\ddot{\theta} + 2mr\dot{r}\dot{\theta} = 0 \qquad (4)$$

which are the Lagrange's equations of motion for a particle moving in a central force field.

Also from above, we have

$$\frac{d}{dt}(mr^2\dot{\theta}) = 0 \quad \text{or} \quad mr^2\dot{\theta} = \text{constant},$$

θ being a cyclic co-ordinate in this case, angular momentum is conserved during its motion.

(ii) Hamiltonian of a conservative system having generalised co-ordinate q_k is

$$H = \sum_k p_k\dot{q}_k - L,$$

where p_k is the generalised momentum.

For a conservative system having generalised co-ordinates r, θ, the above expression for Hamiltonian becomes $H = p_r\dot{r} + p_\theta\dot{\theta} - L$.

Now

$$L = T - V = \frac{1}{2}m(\dot{r}^2 + r^2\dot{\theta}^2) + \frac{k}{r}.$$

$$\therefore\ p_r = \frac{\partial L}{\partial \dot{r}} = \frac{\partial T}{\partial \dot{r}} = m\dot{r} \quad \text{or} \quad \dot{r} = \frac{p_r}{m}$$

$$\text{and}\ p_\theta = \frac{\partial L}{\partial \dot{\theta}} = \frac{\partial T}{\partial \dot{\theta}} = mr^2\dot{\theta} \quad \text{or} \quad \dot{\theta} = \frac{p_\theta}{mr^2},$$

so that the Hamiltonian for the system may be written as

$$H = \frac{p_r^2}{m} + \frac{p_\theta^2}{mr^2} - \frac{1}{2}m\left(\frac{p_r^2}{m^2} + r^2\frac{p_\theta^2}{m^2r^4}\right) + \frac{k}{r} = \frac{p_r^2}{2m} + \frac{p_\theta^2}{2mr^2} + \frac{k}{r}.$$

The Hamiltonian equations for the system are

$$\dot{\theta} = \frac{\partial H}{\partial p_\theta}, \quad \dot{p}_\theta = -\frac{\partial H}{\partial \theta}, \quad \text{i.e.,} \quad \dot{\theta} = \frac{p_\theta}{mr^2} \quad \text{or} \quad p_\theta = mr^2\dot{\theta} \quad \text{and} \quad \dot{p}_\theta = 0, \tag{5}$$

θ being a cyclic co-ordinate in this case.

$$\dot{r} = \frac{\partial H}{\partial p_r}, \quad \dot{p}_r = -\frac{\partial H}{\partial r}, \quad \text{i.e.,} \quad \dot{r} = \frac{p_r}{m} \quad \text{or} \quad p_r = m\dot{r} \quad \text{and} \quad \dot{p}_r = \frac{p_\theta^2}{mr^3} + \frac{k}{r^2} \tag{6}$$

From (5) and (6), we can write

$$\frac{d}{dt}(p_\theta) = 0 \quad \text{or} \quad \frac{d}{dt}(mr^2\dot{\theta}) = 0.$$

$\therefore\ mr^2\dot{\theta} =$ constant i.e., angular momentum is conserved and $m\ddot{r} = \dfrac{p_\theta^2}{mr^3} + \dfrac{k}{r^2}$.

5. An electric dipole with opposite charges of masses m_1 and m_2 separated by a distance l is placed in an external homogeneous electric field. Write down the Lagrangian of the dipole.

[C.U. (Hons.) '83]

Here $AB = l, P =$ centre of mass of the system. So

$$m_1AP = m_2BP \quad \text{or} \quad \frac{m_1}{m_2} = \frac{BP}{AP} \quad \text{or} \quad \frac{m_1}{m_1 + m_2} = \frac{BP}{l}.$$

$$\therefore\ BP = \frac{m_1 l}{m_1 + m_2} \quad \text{and} \quad AP = \frac{m_2 l}{m_1 + m_2}$$

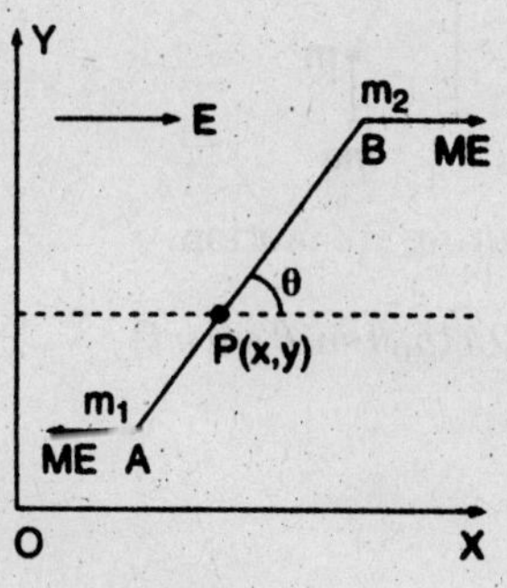

Degrees of freedom for the system are 3, which are x, y, θ and so the Lagrange's equations for the system are

$$\frac{d}{dt}\left(\frac{\partial L}{\partial \dot{x}}\right) - \frac{\partial L}{\partial x} = 0, \frac{d}{dt}\left(\frac{\partial L}{\partial \dot{y}}\right) - \frac{\partial L}{\partial y} = 0 \text{ and } \frac{d}{dt}\left(\frac{\partial L}{\partial \dot{\theta}}\right) - \frac{\partial L}{\partial \theta} = 0.$$

Now

$$\text{K.E.}_{\text{Translational}} = \frac{1}{2}(m_1 + m_2)(\dot{x}^2 + \dot{y}^2)$$

$$\text{and K.E.}_{\text{Rotational}} = \frac{1}{2}I\dot{\theta}^2.$$

Again

$$I = m_1 AP^2 + m_2 BP^2 = m_1 \left(\frac{m_2 l}{m_1 + m_2}\right)^2 + m_2 \left(\frac{m_1 l}{m_1 + m_2}\right)^2 = \frac{m_1 m_2}{m_1 + m_2} l^2 = \mu l^2$$

where $\mu = \frac{m_1 m_2}{m_1+m_2}$ = reduced mass. (Since the masses are unequal, the dipole will execute rotational motion about the CM of the system with the above reduced mass. See Art. 3.31)

$$\therefore \quad T = \tfrac{1}{2}(m_1 + m_2)(\dot{x}^2 + \dot{y}^2) + \tfrac{1}{2}\mu l^2 \dot{\theta}^2 \quad \text{and} \quad V = ME(1 - \cos\theta),$$

M being the moment of the dipole.

$$\therefore \quad L = T - V = \tfrac{1}{2}(m_1 + m_2)(\dot{x}^2 + \dot{y}^2) + \tfrac{1}{2}\mu l^2 \dot{\theta}^2 - ME(1 - \cos\theta),$$

so that

$$\frac{\partial L}{\partial \dot{x}} = (m_1 + m_2)\dot{x}, \quad \frac{\partial L}{\partial x} = 0,$$

$$\frac{\partial L}{\partial \dot{y}} = (m_1 + m_2)\dot{y}, \quad \frac{\partial L}{\partial y} = 0,$$

$$\frac{\partial L}{\partial \dot{\theta}} = \mu l^2 \dot{\theta} \quad \text{and} \quad \frac{\partial L}{\partial \theta} = -ME\sin\theta.$$

Therefore Lagrange's equations reduce to

$$\frac{d}{dt}\left(\frac{\partial L}{\partial \dot{x}}\right) - \frac{\partial L}{\partial x} = 0 \quad \text{or} \quad (m_1 + m_2)\ddot{x} = 0;$$

$$\frac{d}{dt}\left(\frac{\partial L}{\partial \dot{y}}\right) - \frac{\partial L}{\partial y} = 0 \quad \text{or} \quad (m_1 + m_2)\ddot{y} = 0$$

and

$$\frac{d}{dt}\left(\frac{\partial L}{\partial \dot{\theta}}\right) - \frac{\partial L}{\partial \theta} = 0 \quad \text{or} \quad \mu l^2 \ddot{\theta} + ME\sin\theta = 0.$$

6. The point of suspension of a simple pendulum moves harmonically along the vertical line. Find the Lagrangian and the equation of motion for small oscillations of the pendulum.

[C.U. (Hons.) '83, 2000, 2005]

The point of suspension O of a simple pendulum of length l and having a bob of mass m, moves harmonically along OO'—the vertical line between OX and $O'X'$. If θ be the angular displacement of the bob from its equilibrium position at any instant, the co-ordinates (x, y) of the bob are given by

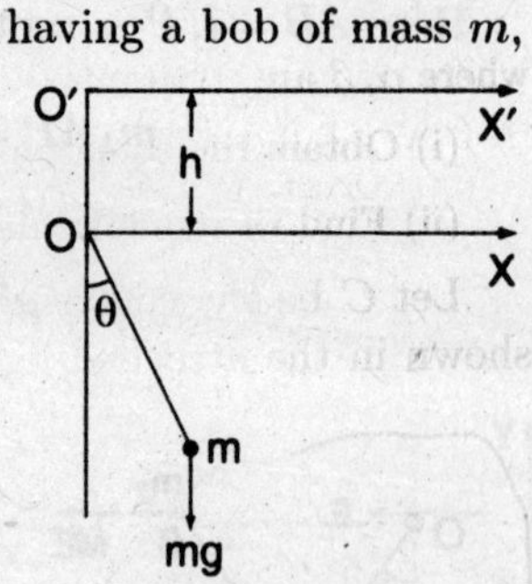

$$x = l\sin\theta$$

$$\text{and} \quad y = l\cos\theta + h = l\cos\theta + h_0 \cos wt,$$

(h_0 being the amplitude of oscillation of the point of suspension which moves harmonically).

Now K.E. of the system

$$T = \tfrac{1}{2}m(\dot{x}^2 + \dot{y}^2) = \tfrac{1}{2}m(l^2\cos^2\theta\dot{\theta}^2 + l^2\sin^2\theta\dot{\theta}^2 + w^2h_0^2\sin^2 wt + 2wlh_0\dot{\theta}\sin\theta\sin wt)$$

$$= \tfrac{1}{2}m(l^2\dot{\theta}^2 + w^2h_0^2\sin^2 wt + 2wlh_0\dot{\theta}\sin\theta\sin wt)$$

and P.E. of the system $V = mgy = mg(l\cos\theta + h_0\cos wt)$.

$\therefore$ Lagrangian of the system is given by

$$L = T - V = \tfrac{1}{2}m\left[l^2\dot{\theta}^2 + w^2h_0^2\sin^2 wt + 2wlh_0\dot{\theta}\sin\theta\sin wt\right] - mg(l\cos\theta + h_0\cos wt).$$

Lagrangian equation of the system is

$$\frac{d}{dt}\left(\frac{\partial L}{\partial\dot{\theta}}\right) - \frac{\partial L}{\partial\theta} = 0 \tag{1}$$

Now
$$\frac{\partial L}{\partial\dot{\theta}} = ml^2\dot{\theta} + mwlh_0\sin\theta\sin wt.$$

$$\therefore\quad \frac{d}{dt}\left(\frac{\partial L}{\partial\dot{\theta}}\right) = ml^2\ddot{\theta} + mw^2h_0l\sin\theta\cos wt + mwh_0l\sin wt\cos\theta\dot{\theta}$$

and
$$\frac{\partial L}{\partial\theta} = mwh_0l\dot{\theta}\cos\theta\sin wt + mgl\sin\theta$$

Putting these values in equation (1), we get

$$ml^2\ddot{\theta} + mw^2h_0l\sin\theta\cos wt + mwh_0l\dot{\theta}\cos\theta\sin wt - mwh_0l\dot{\theta}\cos\theta\sin wt - mgl\sin\theta = 0$$

or $\quad l\ddot{\theta} + w^2h_0\sin\theta\cos wt - g\sin\theta = 0\quad$ or $\quad \ddot{\theta} + \dfrac{(w^2h_0\cos wt - g)}{l}\sin\theta = 0$

For small oscillations, $\sin\theta \simeq \theta$, and the above equation reduces to

$$\ddot{\theta} + \frac{(w^2h_0\cos wt - g)}{l}\theta = 0$$

which is the required equation of the system.

7. (a) Set up the Lagrangian for a compound pendulum oscillating in a vertical plane about a fixed horizontal axis and obtain the equation of motion.

(b) A Lagrangian is given in the form

$$L = \tfrac{1}{2}\alpha\dot{q}^2 - \tfrac{1}{2}\beta q^2,$$

where α, β are constants.

(i) Obtain the equation of motion.

(ii) Find the Hamiltonian of the system. **[C.U. (Hons.) '84]**

Let C be the centre of mass of the compound pendulum, oscillating in the $X - Y$ plane, as shown in the attached figure, where O is its point of intersection with the axis of rotation. If θ be the instantaneous angle which OC makes with the vertical axis through O, then

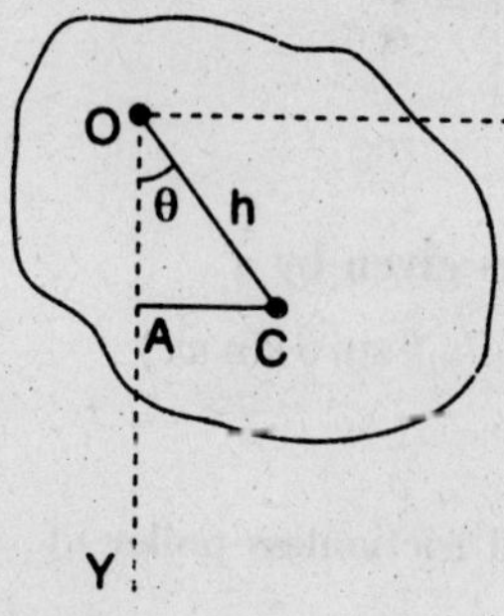

$$\text{K.E.} = T = \tfrac{1}{2}I\dot{\theta}^2 = \tfrac{1}{2}mk^2\dot{\theta}^2,$$

where $I = mk^2$ is the moment of inertia of the body about the axis of rotation, k being the radius of gyration. The P.E. of the system relative to the horizontal plane through O is

$$V = -mgOA$$
$$= -mgh\cos\theta, \quad \text{where } h = OC.$$

The Lagrangian of the system is then written as

$$L = T - V = \tfrac{1}{2}mk^2\dot{\theta}^2 + mgh\cos\theta,$$

which gives

$$\frac{\partial L}{\partial \theta} = -mgh\sin\theta \quad \text{and} \quad \frac{\partial L}{\partial \dot{\theta}} = mk^2\dot{\theta},$$

so that the Lagrangian equation in the generalised co-ordinates $\theta, \dot{\theta}$ is

$$\frac{d}{dt}\left(\frac{\partial L}{\partial \dot{\theta}}\right) - \frac{\partial L}{\partial \theta} = 0 \quad \text{or} \quad mk^2\ddot{\theta} + mgh\sin\theta = 0 \quad \text{or} \quad \ddot{\theta} + \frac{gh}{k^2}\sin\theta = 0$$

which is the equation of motion of the compound pendulum.

If the amplitude of oscillation is small, then the above equation reduces to

$$\ddot{\theta} + \frac{gh}{k^2}\theta = 0,$$

so that the period of oscillation

$$T = \frac{2\pi}{w} = 2\pi\sqrt{\frac{k^2}{gh}} \quad \text{where } w^2 = \frac{gh}{k^2}.$$

N.B. In the above treatment I should be written as $(mk^2 + mh^2)$, applying the theorem of parallel axes, where mk^2 is the moment of inertia of the body about an axis through C and parallel to axis of rotation, so that the equation of motion for small amplitude of oscillation is given by

$$\ddot{\theta} + \frac{gh}{k^2 + h^2}\theta = 0 \quad \text{and} \quad T = 2\pi\sqrt{\frac{k^2 + h^2}{gh}}.$$

(b) (i) Here

$$L = \tfrac{1}{2}\alpha\dot{q}^2 - \tfrac{1}{2}\beta q^2. \quad \therefore \quad \frac{\partial L}{\partial \dot{q}} = \alpha\dot{q} \quad \text{and} \quad \frac{\partial L}{\partial q} = -\beta q.$$

∴ Lagrange's equation or the equation of motion is

$$\frac{d}{dt}\left(\frac{\partial L}{\partial \dot{q}}\right) - \frac{\partial L}{\partial q} = 0 \text{ or } \frac{d}{dt}(\alpha\dot{q}) - (-\beta q) = 0 \text{ or } \alpha\ddot{q} + \beta q = 0$$

$$\text{or } \ddot{q} + \frac{\beta}{\alpha}q = 0 \text{ or } \ddot{q} + w^2 q = 0 \text{ where } w^2 = \frac{\beta}{\alpha}.$$

(ii) The Hamiltonian of the system is $H = p_k\dot{q}_k - L$.

Here momenta $p_k = \dfrac{\partial L}{\partial \dot{q}} = \alpha\dot{q}$, so that the Hamiltonian of the system is given by

$$H = p_k\dot{q}_k - L = \alpha\dot{q}^2 - \left(\tfrac{1}{2}\alpha\dot{q}^2 - \tfrac{1}{2}\beta q^2\right) = \tfrac{1}{2}\alpha\dot{q}^2 + \tfrac{1}{2}\beta q^2.$$

8. A uniform string of length l and mass ρ per unit length passes over a small frictionless pulley of negligible mass. Two masses m_1 and m_2 are tied at its two ends:

(i) Write down the Lagrangian of the system;

(ii) Write down the Lagrange's equation of motion and hence find the acceleration of the mass m_1.

[C.U. (Hons.) '85, 91]

Let $PQ = l$; then if $PA = x$, $QA = (l - x)$. x is the only independent co-ordinate in this case. K.E. of mass $m_1 = \frac{1}{2}m_1\dot{x}^2$; K.E. of mass $m_2 = \frac{1}{2}m_2\dot{x}^2$; K.E. of the portion PA of the string $= \frac{1}{2}x\rho\dot{x}^2$ and K.E. of the portion QA of the string $= \frac{1}{2}(l - x)\rho\dot{x}^2$.

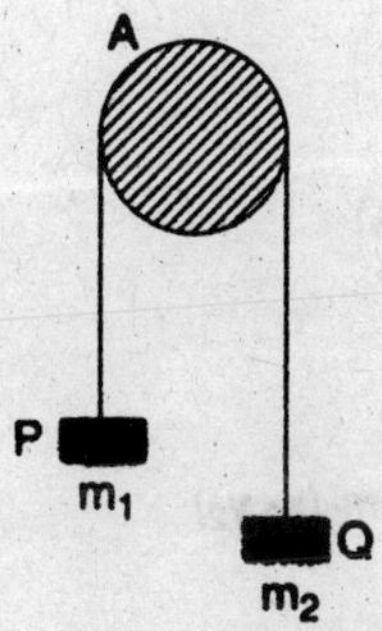

$$\therefore \quad T = \tfrac{1}{2}(m_1 + m_2)\dot{x}^2 + \tfrac{1}{2}(x\rho + l\rho - x\rho)\dot{x}^2 = \tfrac{1}{2}(m_1 + m_2 + l\rho)\dot{x}^2.$$

Considering the horizontal plane through A as the reference level, P.E. of mass $m_1 = -m_1gx$; P.E. of mass $m_2 = -m_2g(l - x)$; P.E. of the portion

$$PA = -\rho x\frac{x}{2}g = -\rho\frac{x^2}{2}g;$$

P.E. of the portion

$$QA = -\rho(l - x)\frac{(l - x)}{2}g = -\rho\frac{(l - x)^2}{2}g$$

$$\therefore \quad V = -m_1gx - m_2g(l - x) - \frac{\rho g}{2}x^2 - \frac{\rho g}{2}(l - x)^2.$$

The Lagrangian of the system is given by

$$L = T - V = \frac{1}{2}(m_1 + m_2 + l\rho)\dot{x}^2 + m_1gx + m_2g(l - x) + \frac{\rho g}{2}x^2 + \frac{\rho g}{2}(l - x)^2$$

$$= \frac{1}{2}(m_1 + m_2 + l\rho)\dot{x}^2 + m_1gx + m_2g(l - x) + \frac{\rho g}{2}l^2 - \rho glx + \rho gx^2$$

so that

$$\frac{\partial L}{\partial \dot{x}} = (m_1 + m_2 + l\rho)\dot{x}$$

$$\text{and} \quad \frac{\partial L}{\partial x} = (m_1 - m_2)g - \rho lg + 2\rho gx = (m_1 - m_2 - \rho l)g + 2\rho gx.$$

$\therefore$ Lagrange's equation of motion of the system is given by

$$\frac{d}{dt}\left(\frac{\partial L}{\partial \dot{x}}\right) - \frac{\partial L}{\partial x} = 0 \text{ or } (m_1 + m_2 + l\rho)\ddot{x} - (m_1 - m_2 - l\rho)g - 2\rho gx = 0$$

$$\text{or } \ddot{x} = \frac{(m_1 - m_2 - l\rho)}{(m_1 + m_2 + l\rho)}g + \frac{2\rho g}{(m_1 + m_2 + l\rho)}x.$$

9. A double pendulum consisting of two masses m_1 and m_2 oscillates in a vertical plane through small angles. m_1 is suspended from a fixed point by a light inextensible string of length l_1 and m_2 is suspended from m_1 by a similar string of length l_2. Find the equations of motion by using Lagrange's formulation.

[C.U. (Hons.) '86, 89]

If θ_1 and θ_2 be the deflections of the pendulums from the vertical and (x_1, y_1) and (x_2, y_2) the rectangular co-ordinates of m_1 and m_2 respectively at any instant, then it is obvious from the attached figure that

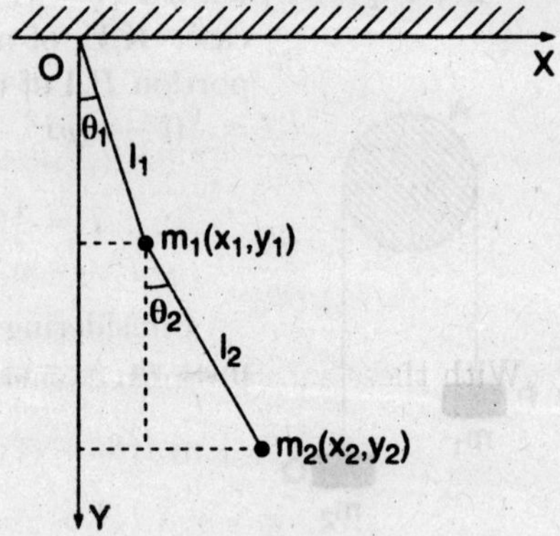

$$x_1 = l_1 \sin\theta_1,$$
$$y_1 = l_1 \cos\theta_1,$$
$$x_2 = l_1 \sin\theta_1 + l_2 \sin\theta_2$$
and $$y_2 = l_1 \cos\theta_1 + l_2 \cos\theta_2,$$

whence we have

$$\left.\begin{aligned} \dot{x}_1 &= l_1 \cos\theta_1 \dot{\theta}_1, \\ \dot{y}_1 &= -l_1 \sin\theta_1 \dot{\theta}_1, \\ \dot{x}_2 &= l_1 \cos\theta_1 \dot{\theta}_1 + l_2 \cos\theta_2 \dot{\theta}_2 \\ \text{and} \quad \dot{y}_2 &= -l_1 \sin\theta_1 \dot{\theta}_1 - l_2 \sin\theta_2 \dot{\theta}_2 \end{aligned}\right\} . \quad (1)$$

∴ K.E. of the system,

$$\begin{aligned} T &= \frac{1}{2} m_1 \left(\dot{x}_1^2 + \dot{y}_1^2\right) + \frac{1}{2} m_2 \left(\dot{x}_2^2 + \dot{y}_2^2\right) \\ &= \frac{1}{2} m_1 \left[(l_1 \cos\theta_1 \dot{\theta}_1)^2 + (-l_1 \sin\theta_1 \dot{\theta}_1)^2\right] \\ &\quad + \frac{1}{2} m_2 \left[(l_1 \cos\theta_1 \dot{\theta}_1 + l_2 \cos\theta_2 \dot{\theta}_2)^2 + (-l_1 \sin\theta_1 \dot{\theta}_1 - l_2 \sin\theta_2 \dot{\theta}_2)^2\right] \\ &= \frac{1}{2} m_1 l_1^2 \dot{\theta}_1^2 + \frac{1}{2} m_2 \left[l_1^2 \dot{\theta}_1^2 + l_2^2 \dot{\theta}_2^2 + 2 l_1 l_2 \dot{\theta}_1 \dot{\theta}_2 \cos(\theta_1 - \theta_2)\right] \end{aligned}$$

Considering the horizontal plane at a distance $(l_1 + l_2)$ below the point of suspension O as the reference level, the P.E. of the system,

$$V = m_1 g(l_1 + l_2 - l_1 \cos\theta_1) + m_2 g[l_1 + l_2 - (l_1 \cos\theta_1 + l_2 \cos\theta_2)],$$

so that the Lagrangian L is

$$\begin{aligned} L = T - V &= \frac{1}{2} m_1 l_1^2 \dot{\theta}_1^2 + \frac{1}{2} m_2 \left[l_1^2 \dot{\theta}_1^2 + l_2^2 \dot{\theta}_2^2 + 2 l_1 l_2 \dot{\theta}_1 \dot{\theta}_2 \cos(\theta_1 - \theta_2)\right] \\ &\quad - m_1 g(l_1 + l_2 - l_1 \cos\theta_1) - m_2 g(l_1 + l_2 - l_1 \cos\theta_1 - l_2 \cos\theta_2). \quad (2) \end{aligned}$$

The Lagrange's equations associated with θ_1 and θ_2 are

$$\frac{d}{dt}\left(\frac{\partial L}{\partial \dot{\theta}_1}\right) - \frac{\partial L}{\partial \theta_1} = 0 \quad \text{and} \quad \frac{d}{dt}\left(\frac{\partial L}{\partial \dot{\theta}_2}\right) - \frac{\partial L}{\partial \theta_2} = 0 \quad (3)$$

From (2)

$$\frac{\partial L}{\partial \theta_1} = -m_2 l_1 l_2 \dot{\theta}_1 \dot{\theta}_2 \sin(\theta_1 - \theta_2) - m_1 g l_1 \sin\theta_1 - m_2 g l_1 \sin\theta_1$$

$$\frac{\partial L}{\partial \dot{\theta}_1} = m_1 l_1^2 \dot{\theta}_1 + m_2 l_1^2 \dot{\theta}_1 + m_2 l_1 l_2 \dot{\theta}_2 \cos(\theta_1 - \theta_2)$$

$$\frac{\partial L}{\partial \theta_2} = m_2 l_1 l_2 \dot{\theta}_1 \dot{\theta}_2 \sin(\theta_1 - \theta_2) - m_2 g l_2 \sin\theta_2$$

$$\frac{\partial L}{\partial \dot{\theta}_2} = m_2 l_2^2 \dot{\theta}_2 + m_2 l_1 l_2 \dot{\theta}_1 \cos(\theta_1 - \theta_2).$$

With these substitutions, equations (3) become,

$$m_1 l_1^2 \ddot{\theta}_1 + m_2 l_1^2 \ddot{\theta}_1 + m_2 l_1 l_2 \ddot{\theta}_2 \cos(\theta_1 - \theta_2) - m_2 l_1 l_2 \dot{\theta}_2 (\dot{\theta}_1 - \dot{\theta}_2) \sin(\theta_1 - \theta_2)$$
$$= -m_2 l_1 l_2 \dot{\theta}_1 \dot{\theta}_2 \sin(\theta_1 - \theta_2) - m_1 g l_1 \sin\theta_1 - m_2 g l_1 \sin\theta_1$$

and $m_2 l_2^2 \ddot{\theta}_2 + m_2 l_1 l_2 \ddot{\theta}_1 \cos(\theta_1 - \theta_2) - m_2 l_1 l_2 \dot{\theta}_1 (\dot{\theta}_1 - \dot{\theta}_2) \sin(\theta_1 - \theta_2)$

$$= m_2 l_1 l_2 \dot{\theta}_1 \dot{\theta}_2 \sin(\theta_1 - \theta_2) - m_2 g l_2 \sin\theta_2$$

which reduce respectively to

$$(m_1 + m_2) l_1^2 \ddot{\theta}_1 + m_2 l_1 l_2 \ddot{\theta}_2 \cos(\theta_1 - \theta_2) + m_2 l_1 l_2 \dot{\theta}_2^2 \sin(\theta_1 - \theta_2)$$
$$= -(m_1 + m_2) g l_1 \sin\theta_1 \qquad (4)$$

and $m_2 l_2^2 \ddot{\theta}_2 + m_2 l_1 l_2 \ddot{\theta}_1 \cos(\theta_1 - \theta_2) - m_2 l_1 l_2 \dot{\theta}_1^2 \sin(\theta_1 - \theta_2)$

$$= -m_2 g l_2 \sin\theta_2. \qquad (5)$$

Equations (4) and (5) are the required equations.

If $m_1 = m_2, l_1 = l_2$, equations (4) and (5) assume the form

$$2l\ddot{\theta}_1 + l\cos(\theta_1 - \theta_2)\ddot{\theta}_2 + l\dot{\theta}_2^2 \sin(\theta_1 - \theta_2) = -2g\sin\theta_1 \qquad (6)$$

and $l\cos(\theta_1 - \theta_2)\ddot{\theta}_1 + l\ddot{\theta}_2 - l\dot{\theta}_1^2 \sin(\theta_1 - \theta_2) = -g\sin\theta_2.$ (7)

If we use the approximation that $\sin\theta = \theta$ and $\cos\theta = 1$ and neglect the terms involving $\dot{\theta}^2$, equations (6) and (7) reduce to $2l\ddot{\theta}_1 + l\ddot{\theta}_2 = -2g\theta_1$ and $l\ddot{\theta}_1 + l\ddot{\theta}_2 = -g\theta_2$.

10. A bead of mass m slides along a frictionless straight rod of negligible mass. If the rod is uniformly rotating in a force-free space at a constant angular speed ω in a horizontal plane, set up the Lagrangian and the equation of motion and also find the radial position (r) of the bead as a function of time, assuming that $r = R, \dot{r} = v$ at $t = 0$. **[C.U. (Hons.) '95, 98; cf. C.U. (Hons.) 2003]**

In this case, the constraint is time-dependent given by the relation $\dot{\theta} = w$, w being the angular velocity of rotation. Considering the bead as a particle of mass m, the relations between cartesian (x, y) and polar (r, θ) co-ordinates of the particle at any instant t (polar co-ordinates r, θ being the appropriate generalised co-ordinates) must contain time explicitly and can be expressed as

$$\left.\begin{aligned} x &= r\cos\theta = r\cos wt \\ y &= r\sin\theta = r\sin wt \end{aligned}\right\} \quad (\because \theta = wt) \qquad (1)$$

$$\therefore\ T = \tfrac{1}{2}m\,(\dot{x}^2 + \dot{y}^2)$$
$$= \tfrac{1}{2}m\left[(\dot{r}\cos wt - rw\sin wt)^2 + (\dot{r}\sin wt + rw\cos wt)^2\right] \quad \text{[From (1)}$$
$$= \tfrac{1}{2}m(\dot{r}^2 + r^2w^2),$$

so that

$$\frac{\partial L}{\partial r} = \frac{\partial T}{\partial r} = mrw^2 \quad \text{and} \quad \frac{\partial L}{\partial \dot{r}} = \frac{\partial T}{\partial \dot{r}} = m\dot{r} \tag{2}$$

($\because\ L = T - V = T$, V being zero in the problem).

In this case Lagrange's equation in r is given by

$$\frac{d}{dt}\left(\frac{\partial L}{\partial \dot{r}}\right) - \frac{\partial L}{\partial r} = 0$$

[we need not consider the Lagrange's equation in θ, since the wire is rotating uniformly in a force-free space]

or $m\ddot{r} - mrw^2 = 0$

or $\ddot{r} = rw^2$ (using (2) and since the bead is sliding in a force-free space)

which shows that the bead moves out-ward due to centri-petal acceleration.

To find the solution, let $r = e^{\alpha t}$ $\therefore \alpha^2 = \omega^2$ or $\alpha = \pm\omega$.

$\therefore$ the general solution is $r = C_1e^{\omega t} + C_2e^{-\omega t}$.

Now the condition that at $t = 0$, $r = R$, gives $R = C_1 + C_2$.

Again we have $\dot{r} = C_1\omega e^{\omega t} - C_2\omega e^{-\omega t}$ and the condition that at $t = 0$, $\dot{r} = v$, gives $v = C_1\omega - C_2\omega = \omega(C_1 - C_2)$.

$$\therefore \left.\begin{aligned} C_1 + C_2 &= R \\ C_1 - C_2 &= \tfrac{v}{\omega} \end{aligned}\right\} \quad \text{whence } C_1 = \frac{1}{2}\left(R + \frac{v}{\omega}\right) \quad \text{and} \quad C_2 = \frac{1}{2}\left(R - \frac{v}{\omega}\right)$$

$$\therefore\quad r = \frac{1}{2}\left(R + \frac{v}{\omega}\right)e^{\omega t} + \frac{1}{2}\left(R - \frac{v}{\omega}\right)e^{-\omega t}$$

11. A nucleus of charge Ze and mass M and an electron of charge $-e$ and mass m, situated at a distance r from each other, are rotating in circles about their common centre of mass. Writing down the Lagrangian of the system, work out the equation of motion with the help of Lagrange's equation.

[C.U. (Hons.) '90]

Applying the condition of rotation about the common centre of mass C of the system, we have

$$Mr_1 = mr_2 \quad \text{or} \quad M(r - r_2) = mr_2. \quad \therefore\ r_2 = \frac{M}{m + M}r$$

Similarly $r_1 = \frac{m}{m+M} r$, so that K.E. of the system

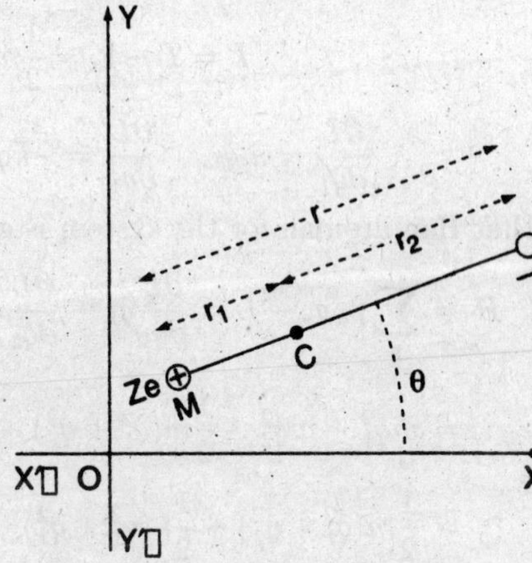

$$T = \frac{1}{2}m(r_2\dot{\theta})^2 + \frac{1}{2}M(r_1\dot{\theta})^2$$

$$= \frac{1}{2}m\left(\frac{M}{m+M}r\dot{\theta}\right)^2 + \frac{1}{2}M\left(\frac{m}{m+M}r\dot{\theta}\right)^2$$

$$= \frac{1}{2}mM\dot{\theta}^2 r^2\left[\frac{M}{(m+M)^2} + \frac{m}{(m+M)^2}\right]$$

$$= \frac{1}{2}\frac{mM}{m+M}r^2\dot{\theta}^2 = \frac{1}{2}\mu r^2\dot{\theta}^2,$$

where $\mu = \frac{mM}{m+M}$ = reduced mass.

Again, P.E. of the system, $V = -\frac{Ze^2}{r}$, so that Lagrangian of the system is given by

$$L = T - V = \frac{1}{2}\mu r^2\dot{\theta}^2 + \frac{Ze^2}{r} \tag{1}$$

Now Lagrange's equation of motion of the system is

$$\frac{d}{dt}\left(\frac{\partial L}{\partial \dot{\theta}}\right) - \frac{\partial L}{\partial \theta} = 0 \tag{2}$$

From (1), we have

$$\frac{\partial L}{\partial \dot{\theta}} = \mu r^2\dot{\theta} \quad \text{and} \quad \frac{\partial L}{\partial \theta} = 0.$$

$$\therefore \quad \frac{d}{dt}(\mu r^2\dot{\theta}) = 0 \quad \text{from (2)} \quad \text{or} \quad \mu r^2\dot{\theta} = \text{constant},$$

i.e., angular momentum of the system is conserved.

12. **Consider two identical simple harmonic oscillators which are coupled together. Their kinetic and potential energies are given by**

$$T = \tfrac{1}{2}m(\dot{q}_1^2 + \dot{q}_2^2) \quad \text{and} \quad V = \tfrac{1}{2}k(q_1^2 + q_2^2) - hq_1q_2.$$

(a) Find the Hamiltonian of the system;

(b) Find the frequencies for normal modes and the ratios of the amplitudes of each of the co-ordinates q_1 and q_2 in those modes;

(c) Introduce normal co-ordinates and set up the corresponding Lagrange's equations.

[C.U. (Hons.) '86]

(a) K.E. and P.E. of the two coupled simple harmonic oscillators are given by

$$T = \tfrac{1}{2}m(\dot{q}_1^2 + \dot{q}_2^2) \tag{1}$$

$$\text{and} \quad V = \tfrac{1}{2}k(q_1^2 + q_2^2) - hq_1q_2 \tag{2}$$

where q_1 and q_2 are the generalised co-ordinates for the system.

$\therefore$ Lagrangian for the system

$$L = T - V = \frac{1}{2}m(\dot{q}_1^2 + \dot{q}_2^2) - \frac{1}{2}k(q_1^2 + q_2^2) + hq_1q_2. \tag{3}$$

$$\therefore \quad \frac{\partial L}{\partial \dot{q}_1} = m\dot{q}_1, \quad \frac{\partial L}{\partial q_1} = -kq_1 + hq_2, \quad \frac{\partial L}{\partial \dot{q}_2} = m\dot{q}_2, \quad \frac{\partial L}{\partial q_2} = -kq_2 + hq_1,$$

so that Hamiltonian for the system is given by

$$H = \sum_k p_k\dot{q}_k - L = \frac{\partial L}{\partial \dot{q}_1}\dot{q}_1 + \frac{\partial L}{\partial \dot{q}_2}\dot{q}_2 - L \quad [\because p_k = \text{ Generalised momentum } = \frac{\partial L}{\partial \dot{q}_k}]$$

$$= m\dot{q}_1^2 + m\dot{q}_2^2 - \frac{1}{2}m(\dot{q}_1^2 + \dot{q}_2^2) + \frac{1}{2}k(q_1^2 + q_2^2) - hq_1q_2$$

$$= \frac{1}{2}m(\dot{q}_1^2 + \dot{q}_2^2) + \frac{1}{2}k(q_1^2 + q_2^2) - hq_1q_2 \tag{4}$$

(b) We know that the determinant for normal modes is given by

$$\begin{vmatrix} V_{11} - w^2T_{11} & V_{12} - w^2T_{12} \\ V_{21} - w^2T_{21} & V_{22} - w^2T_{22} \end{vmatrix} = 0$$

where

$$V_{11} = \left(\frac{\partial^2 V}{\partial q_1^2}\right)_0, \ V_{12} = \left(\frac{\partial^2 V}{\partial q_1 \partial q_2}\right)_0,$$

$$V_{21} = \left(\frac{\partial^2 V}{\partial q_2 \partial q_1}\right)_0, \ V_{22} = \left(\frac{\partial^2 V}{\partial q_2^2}\right)_0, \ T_{11} = (m_{11})_0,$$

$T_{12} = T_{21} = (m_{12})_0, T_{22} = (m_{22})_0$ and w is the normal mode frequency.

Now from (1) and (2), for the given problem, we can write,

$$V_{11} = k, \ V_{12} = -h, \ V_{21} = -h, \ V_{22} = k,$$

$$\text{and } \ T_{11} = m, \ T_{12} = 0, \ T_{21} = 0, \ T_{22} = m$$

Substituting these values in the determinant, we get

$$\begin{vmatrix} k - mw^2 & -h \\ -h & k - mw^2 \end{vmatrix} = 0 \quad \text{or} \quad (k - mw^2)^2 - h^2 = 0$$

$$\text{or} \quad k - mw^2 = \pm h$$

$$\text{or} \quad mw^2 = k \mp h$$

$$\text{or} \quad w^2 = \frac{k}{m} \mp \frac{h}{m}.$$

Thus we see that w has two values w_1 and w_2 given by

$$\left.\begin{aligned} w_1^2 &= \frac{k}{m} - \frac{h}{m} \\ \text{and} \quad w_2^2 &= \frac{k}{m} + \frac{h}{m} \end{aligned}\right\} \tag{5}$$

and these are the frequencies for normal modes.

Now Lagrange's equation of motion for generalised co-ordinate q_1 is given by

$$\frac{d}{dt}\left(\frac{\partial L}{\partial \dot{q}_1}\right) - \frac{\partial L}{\partial q_1} = 0$$

Substituting the values of $\frac{\partial L}{\partial \dot{q}_1}$ and $\frac{\partial L}{\partial q_1}$ from (3) in the above equation, we have

$$m\ddot{q}_1 + kq_1 - hq_2 = 0 \tag{6}$$

Similarly Lagrange's equation of motion for generalised co-ordinate q_2 is given by

$$m\ddot{q}_2 + kq_2 - hq_1 = 0 \tag{7}$$

(6) and (7) are the equations of motion for the generalised co-ordinates, and let the solutions of these equations be given by

$$q_1 = Ae^{jwt} \tag{8}$$

$$q_2 = Be^{jwt} \tag{9}$$

where w is the normal mode frequency and A, B are the amplitudes of oscillation in the normal modes (corresponding to the above generalised co-ordinates), so that

$$\ddot{q}_1 = -Aw^2 e^{jwt}$$

$$\text{and} \quad \ddot{q}_2 = -Bw^2 e^{jwt}$$

Putting these values in (6), we get,

$$(-mAw^2 + kA - hB)e^{jwt} = 0$$

$$\text{or} \quad -mAw^2 + kA - hB = 0 \quad \because e^{jwt} \neq 0$$

$$\text{or} \quad (k - mw^2)A = hB$$

$$\text{or} \quad \frac{A}{B} = \frac{h}{k - mw^2}. \tag{10}$$

Similarly from (7) we have,

$$\frac{A}{B} = \frac{k - mw^2}{h}. \tag{11}$$

Now putting $w^2 = w_1^2 = \left(\frac{k}{m} - \frac{h}{m}\right)$ in (10), we get

$$\frac{A}{B} = \frac{h}{k - m\left(\frac{k}{m} - \frac{h}{m}\right)} = \frac{h}{k - k + h} = 1$$

So when $w = w_1$, the two oscillators are in the same phase (i.e., oscillating in the same direction and sense) having amplitudes of equal magnitude.

Again putting $w^2 = w_2^2 = \left(\frac{k}{m} + \frac{h}{m}\right)$ in (11), we get

$$\frac{A}{B} = \frac{k - m\left(\frac{k}{m} + \frac{h}{m}\right)}{h} = -\frac{h}{h} = -1$$

In this case, i.e., when $w = w_2$, the two oscillators are oscillating in opposite phases, having amplitudes of equal magnitude.

(c) If we now introduce *two new co-ordinates* in place of the generalised co-ordinates, such that each new co-ordinate has one single frequency which is one of the two normal mode frequencies, then *these new co-ordinates* are known as *normal co-ordinates.*

To do this, we may proceed as follows.

The general solution given in equation (8) can be written as

$$q_1 = A_1 e^{jw_1 t} + A_2 e^{jw_2 t}$$

$$\text{or} \quad q_1 = P\cos w_1 t + Q\sin w_1 t + R\cos w_2 t + S\sin w_2 t, \tag{12}$$

where P, Q, R and S are constants which can be evaluated from the initial conditions.

Similarly from (9), we can write

$$q_2 = B_1 e^{jw_1 t} + B_2 e^{jw_2 t}$$

$$\text{or} \quad q_2 = -P\cos w_1 t - Q\sin w_1 t + R\cos w_2 t + S\sin w_2 t \tag{13}$$

so that

$$q_1 - q_2 = X_1 = 2(P\cos w_1 t + Q\sin w_1 t)$$

$$\text{and} \quad q_1 + q_2 = X_2 = 2(R\cos w_2 t + S\sin w_2 t)$$

Since each of the new co-ordinates X_1 and X_2 contains a single frequency, these may be the normal co-ordinates. To prove it we proceed as follows.

Since $q_1 - q_2 = X_1$ and $q_1 + q_2 = X_2$, we can write

$$q_1 = \frac{X_1 + X_2}{2} \quad \text{and} \quad q_2 = \frac{X_2 - X_1}{2}$$

Substituting these values in the expression of Lagrangian [i.e., in equation (3)], we get on simplification,

$$L = \frac{1}{2}m\left[\frac{\dot{X}_1^2 + \dot{X}_2^2}{2}\right] - \frac{1}{2}k\left[\frac{X_1^2 + X_2^2}{2}\right] + h\frac{(X_1 + X_2)(X_2 - X_1)}{4}$$

$$= \frac{1}{4}m(\dot{X}_1^2 + \dot{X}_2^2) - \frac{1}{4}k(X_1^2 + X_2^2) + h\frac{(X_2^2 - X_1^2)}{4} \tag{14}$$

$\therefore$ We have

$$\frac{\partial L}{\partial \dot{X}_1} = \frac{m}{2}\dot{X}_1, \quad \frac{\partial L}{\partial X_1} = -\frac{k}{2}X_1 - \frac{h}{2}X_1, \quad \frac{\partial L}{\partial \dot{X}_2} = \frac{m}{2}\dot{X}_2, \quad \frac{\partial L}{\partial X_2} = -\frac{k}{2}X_2 + \frac{h}{2}X_2$$

so that the new Lagrange's equation for co-ordinate X_1 is

$$\frac{d}{dt}\left(\frac{\partial L}{\partial \dot{X}_1}\right) - \frac{\partial L}{\partial X_1} = 0 \quad \text{or} \quad \frac{m}{2}\ddot{X}_1 + \left(\frac{k}{2} + \frac{h}{2}\right) X_1 = 0$$

$$\text{or} \quad m\ddot{X}_1 + m\left(\frac{k}{m} + \frac{h}{m}\right) X_1 = 0$$

$$\text{or} \quad \ddot{X}_1 + w_2^2 X_1 = 0 \tag{15}$$

putting $w_2^2 = \left(\frac{k}{m} + \frac{h}{m}\right)$.

Similarly the Lagrange's equation for co-ordinate X_2 can be shown as

$$\ddot{X}_2 + w_1^2 X_2 = 0 \tag{16}$$

where $w_1^2 = \left(\frac{k}{m} - \frac{h}{m}\right)$.

Thus we see that each of the co-ordinates X_1 and X_2 executes simple harmonic motion with arbitrary amplitude and phase, but with a normal mode frequency. X_1 and X_2 are, therefore, the normal co-ordinates and the corresponding Lagrange's equations are given by (15) and (16).

13. Two identical simple harmonic oscillators, described by the potential functions $V_1 = \frac{1}{2}kr_1^2$ and $V_2 = \frac{1}{2}kr_2^2$, are coupled to each other through an interaction $V' = -k'r_1r_2 (k' > 0)$.

(i) Set up the Lagrangian of the system.

(ii) Write down Lagrange's equations of motion.

(iii) Solve the equations to obtain the eigen frequencies and eigen-functions.

(iv) What is the physical interpretation of the fact that the out-of-phase motion has a frequency higher than that of the in-phase motion? **[C.U. (Hons.) '92]**

The total P.E. of the system is

$$V = \frac{1}{2}kr_1^2 + \frac{1}{2}kr_2^2 - k'r_1r_2$$

$$\text{or} \quad 2V = kr_1^2 + kr_2^2 - 2k'r_1r_2 = \{r_1 \quad r_2\} \begin{bmatrix} k & -k' \\ -k' & k \end{bmatrix} \begin{Bmatrix} r_1 \\ r_2 \end{Bmatrix}$$

[Any quadratic expresion can be put into matrix form.

Considering a quadratic in q_1, q_2 as $V = aq_1^2 + 2bq_1q_2 + cq_2^2$, we may write

$$V = \underset{\substack{\downarrow \\ \text{Transpose} \\ \text{matrix } \{A\}^T}}{\{q_1 \quad q_2\}} \quad \underset{\substack{\downarrow \\ \text{Square} \\ \text{matrix}}}{\begin{bmatrix} a & b \\ b & c \end{bmatrix}} \quad \underset{\substack{\downarrow \\ \text{Column} \\ \text{matrix } \{A\}}}{\begin{Bmatrix} q_1 \\ q_2 \end{Bmatrix}}$$

The right hand side of the above being a quadratic expression has been put into matrix form.]

Again K.E. of the system is given by

$$T = \frac{1}{2}m\dot{r}_1^2 + \frac{1}{2}m\dot{r}_2^2 \quad \text{or} \quad 2T = m\dot{r}_1^2 + m\dot{r}_2^2 = \{\dot{r}_1 \quad \dot{r}_2\}\begin{bmatrix} m & 0 \\ 0 & m \end{bmatrix}\begin{Bmatrix} \dot{r}_1 \\ \dot{r}_2 \end{Bmatrix}$$

(Here also the right hand side has been expressed in matrix notation.)

Hence Lagrangian of the above system is

$$L = T - V = \frac{1}{2}\{\dot{r}_1 \quad \dot{r}_2\}\begin{bmatrix} m & 0 \\ 0 & m \end{bmatrix}\begin{Bmatrix} \dot{r}_1 \\ \dot{r}_2 \end{Bmatrix} - \frac{1}{2}\{r_1 \quad r_2\}\begin{bmatrix} k & -k' \\ -k' & k \end{bmatrix}\begin{Bmatrix} r_1 \\ r_2 \end{Bmatrix}$$

$$= \frac{1}{2}\{\dot{r}_1 \quad \dot{r}_2\}[M]\begin{Bmatrix} \dot{r}_1 \\ \dot{r}_2 \end{Bmatrix} - \frac{1}{2}\{r_1 \quad r_2\}[K^{11}]\begin{Bmatrix} r_1 \\ r_2 \end{Bmatrix} \tag{1}$$

where $[M]$ is the inertia coefficient matrix and $[K^{11}]$ is the Stiffness coefficient matrix.

Hence Lagrange's equation of motion for r_1 is written as

$$\frac{d}{dt}\left(\frac{\partial L}{\partial \dot{r}_1}\right) - \frac{\partial L}{\partial r_1} = 0$$

Now from (1), we have

$$\frac{\partial L}{\partial \dot{r}_1} = \frac{1}{2}\{1 \quad 0\}[M]\begin{Bmatrix} \dot{r}_1 \\ \dot{r}_2 \end{Bmatrix} + \frac{1}{2}\begin{Bmatrix} \dot{r}_1 \\ \dot{r}_2 \end{Bmatrix}[M]\begin{Bmatrix} 1 \\ 0 \end{Bmatrix}.$$

$$\therefore \frac{d}{dt}\left(\frac{\partial L}{\partial \dot{r}_1}\right) = \frac{1}{2}\{0 \quad 0\}[M]\begin{Bmatrix} \dot{r}_1 \\ \dot{r}_2 \end{Bmatrix} + \frac{1}{2}\{1 \quad 0\}[M]\begin{Bmatrix} \ddot{r}_1 \\ \ddot{r}_2 \end{Bmatrix} + \frac{1}{2}\begin{Bmatrix} \ddot{r}_1 \\ \ddot{r}_2 \end{Bmatrix}[M]\begin{Bmatrix} 1 \\ 0 \end{Bmatrix}$$

$$+\frac{1}{2}\begin{Bmatrix} \dot{r}_1 \\ \dot{r}_2 \end{Bmatrix}[M]\begin{Bmatrix} 0 \\ 0 \end{Bmatrix}$$

$$= \{1 \quad 0\}[M]\begin{Bmatrix} \ddot{r}_1 \\ \ddot{r}_2 \end{Bmatrix}$$ [The first term and the last term on the r.h.s. being zero, and the two middle terms being equal]

and $$\frac{\partial L}{\partial r_1} = -\frac{1}{2}\{1 \quad 0\}[K^{11}]\begin{Bmatrix} r_1 \\ vecr_2 \end{Bmatrix} - \frac{1}{2}\{r_1 \quad r_2\}[K^{11}]\begin{Bmatrix} 1 \\ 0 \end{Bmatrix}$$

[These two terms on the r.h.s. are also equal.]

Equation of motion is therefore given by

$$\{1 \quad 0\}[M]\begin{Bmatrix} \ddot{r}_1 \\ \ddot{r}_2 \end{Bmatrix} + \{1 \quad 0\}[K^{11}]\begin{Bmatrix} r_1 \\ r_2 \end{Bmatrix} = 0$$

which can be written as

$$[M]\{\ddot{r}\} + [K^{11}]\{r\} = 0$$

Now the secular equation of the above matrix relation is given by

$$\left|[K] - w^2[M]\right| = 0 \quad \text{or} \quad \begin{vmatrix} k - w^2 m & -k' \\ -k' & k - w^2 m \end{vmatrix} = 0 \tag{2}$$

Solving we get $w^2 = \frac{k'}{m} \pm \frac{k}{m}$. If w_1 and w_2 be the eigen frequencies, then we have

$$w_1 = \frac{k'}{m} + \frac{k}{m} \quad \text{and} \quad w_2 = \frac{k'}{m} - \frac{k}{m}$$

If η_1 and η_2 be the normal co-ordinates, then the equations of motion corresponding to normal co-ordinates are given by

$$\ddot{\eta}_1 + w_1^2\eta_1 = 0 \quad \text{and} \quad \ddot{\eta}_2 + w_2^2\eta_2 = 0$$

To express the above notation is generalised co-ordinates, we may write

$$\begin{pmatrix} r_1 \\ r_2 \end{pmatrix} = \begin{pmatrix} A_{11} & A_{12} \\ A_{21} & A_{22} \end{pmatrix} \begin{pmatrix} \eta_1 \\ \eta_2 \end{pmatrix}$$

where A_{11}, A_{21} are the eigen vectors of A_1 and A_{12}, A_{22} are the eigen vectors of A_2.

If $w_1^2 = w^2 = \left(\frac{k'}{m} + \frac{k}{m}\right)$, then from (2)

$$\begin{pmatrix} -k' & -k' \\ -k' & -k' \end{pmatrix} \begin{pmatrix} A_{11} \\ A_{21} \end{pmatrix} = 0 \quad \text{or} \quad A_{11}(-k') + A_{21}(-k') = 0$$

$$\text{or} \quad (-k')A_{11} + (-k')A_{21} = 0$$

$$\therefore \quad A_{11} = -A_{21} = \alpha \quad \text{(say)}.$$

If $w_2^2 = w^2 = \left(\frac{k'}{m} - \frac{k}{m}\right)$, then from (2)

$$\begin{pmatrix} k' & -k' \\ -k' & k' \end{pmatrix} \begin{pmatrix} A_{12} \\ A_{22} \end{pmatrix} = 0 \quad \text{or} \quad A_{12}(k') - A_{22}(k') = 0$$

$$\therefore \quad A_{12} = A_{22} = \beta \quad \text{(say)}.$$

$$\therefore \; A = \begin{pmatrix} \alpha & \beta \\ -\alpha & \beta \end{pmatrix} \quad \text{and} \quad A^T = \begin{pmatrix} \alpha & -\alpha \\ \beta & \beta \end{pmatrix} \quad (A^T = \text{Transpose matrix.})$$

$$\therefore \; A^T T A = 1 \quad \text{or} \quad \begin{pmatrix} \alpha & -\alpha \\ \beta & \beta \end{pmatrix} \begin{pmatrix} m & 0 \\ 0 & m \end{pmatrix} \begin{pmatrix} \alpha & \beta \\ -\alpha & \beta \end{pmatrix} = \begin{pmatrix} \delta_{11} & \delta_{12} \\ \delta_{21} & \delta_{22} \end{pmatrix} \quad (T = \text{K.E.})$$

$$\text{or} \quad \begin{pmatrix} \alpha & -\alpha \\ \beta & \beta \end{pmatrix} \begin{pmatrix} m\alpha & m\beta \\ -m\alpha & m\beta \end{pmatrix} = \begin{pmatrix} 1 & 0 \\ 0 & 1 \end{pmatrix}$$

$$\text{or} \quad \begin{pmatrix} 2m\alpha^2 & 0 \\ 0 & 2m\beta^2 \end{pmatrix} = \begin{pmatrix} 1 & 0 \\ 0 & 1 \end{pmatrix}$$

$$\therefore \; 2m\alpha^2 = 1 \quad \text{or} \quad \alpha^2 = \frac{1}{2m} \quad \text{or} \quad \alpha = \frac{1}{\sqrt{2m}}$$

Again $2m\beta^2 = 1$ or $\beta^2 = \frac{1}{2m}$. $\therefore\ \beta = \frac{1}{\sqrt{2m}}$.

$$\therefore \quad A_{11} = -A_{21} = \frac{1}{\sqrt{2m}} \quad \text{and} \quad A_{12} = A_{22} = \frac{1}{\sqrt{2m}}.$$

Thus when $w^2 = \frac{k}{m} - \frac{k'}{m}$, the oscillations are in phase and when $w^2 = \frac{k}{m} + \frac{k'}{m}$, the oscillations are in opposite phases.

In the out of phase motion, the displacements of the particles are in opposite directions and in the case of in-phase motion, the displacements of the particles are in the same direction. Hence the particles in the out-of-phase motion have a frequency higher than that of the particles in the in-phase motion.

14. A particle rotating with angular velocity $\vec{w}$ around the Z-axis is given a linear velocity $\vec{u} = \dot{\vec{r}}$ w.r.t. a fixed origin situated on the Z-axis. Assuming a potential energy $V(\vec{r})$, write down the Lagrangian in component form. Hence obtain Lagrange's equation of motion, treating $\vec{r}$ as a generalised co-ordinate. Interpret the various terms in the equation. **[C.U. (Hons.) '83]**

With reference to Art. 3.40, we consider the frame S' rotating with angular velocity $\vec{w}$ relative to the point O fixed in the Newtonian frame S. The velocity of a particle in S' relative to S is given by

$$\frac{d\vec{r}}{dt} = \frac{\delta\vec{r}}{\delta t} + \vec{w} \times \vec{r} \quad \text{from (3.40.1)}$$

$$\text{or} \quad \vec{v} = \vec{u} + \vec{w} \times \vec{r} \tag{1}$$

Here $\vec{u} = \hat{\imath}\dot{x} + \hat{\jmath}\dot{y} + \hat{k}\dot{z}$, $\vec{r} = \hat{\imath}x + \hat{\jmath}y + \hat{k}z$, and $\vec{w}$ is along the Z-axis;

$$\therefore \quad \vec{w} \times \vec{r} = \begin{vmatrix} \hat{\imath} & \hat{\jmath} & \hat{k} \\ 0 & 0 & w \\ x & y & z \end{vmatrix} = -\hat{\imath}wy + \hat{\jmath}wx.$$

$\therefore$ from (1),

$$\begin{aligned} \vec{v} &= \hat{\imath}\dot{x} + \hat{\jmath}\dot{y} + \hat{k}\dot{z} - \hat{\imath}wy + \hat{\jmath}wx \\ &= \hat{\imath}(\dot{x} - wy) + \hat{\jmath}(\dot{y} + wx) + \hat{k}\dot{z}. \\ \therefore \quad v^2 &= (\dot{x} - wy)^2 + (\dot{y} + wx)^2 + \dot{z}^2. \end{aligned}$$

So, the K.E. of the particle relative to S is given by

$$T = \frac{1}{2}mv^2 = \frac{1}{2}m\left[(\dot{x} - wy)^2 + (\dot{y} + wx)^2 + \dot{z}^2\right]$$

and P.E. of the system is $V = V(r)$, so that the Lagrangian of the system is

$$L = T - V = \frac{1}{2}m\left[(\dot{x} - wy)^2 + (\dot{y} + wx)^2 + \dot{z}^2\right] - V(r)$$

$$\therefore \quad \frac{\partial L}{\partial \dot{x}} = m(\dot{x} - wy), \quad \frac{\partial L}{\partial x} = mw(\dot{y} + wx) - \frac{\delta V(r)}{\delta x}$$

$$\frac{\partial L}{\partial \dot{y}} = m(\dot{y} + wx), \quad \frac{\partial L}{\partial y} = -mw(\dot{x} - wy) - \frac{\delta V(r)}{\delta y}$$

$$\frac{\partial L}{\partial \dot{z}} = m\dot{z}, \quad \frac{\partial L}{\partial z} = -\frac{\partial V(r)}{\partial z}$$

Lagrange's equations in component form may therefore be represented as

$$\frac{d}{dt}\left(\frac{\partial L}{\partial \dot{x}}\right) - \frac{\partial L}{\partial x} = 0 \tag{2}$$

$$\frac{d}{dt}\left(\frac{\partial L}{\partial \dot{y}}\right) - \frac{\partial L}{\partial y} = 0 \tag{3}$$

and
$$\frac{d}{dt}\left(\frac{\partial L}{\partial \dot{z}}\right) - \frac{\partial L}{\partial z} = 0 \tag{4}$$

From (2), we get

$$m(\ddot{x} - w\dot{y}) - m\dot{w}y - mw(\dot{y} + wx) + \frac{\partial V(r)}{\partial x} = 0$$

or
$$m\ddot{x} = 2mw\dot{y} + mw^2x + m\dot{w}y - \frac{\partial V(r)}{\partial x} \tag{5}$$

From (3), we get

$$m(\ddot{y} + w\dot{x}) + m\dot{w}x + mw(\dot{x} - wy) + \frac{\partial V(r)}{\partial y} = 0$$

or
$$m\ddot{y} = -2mw\dot{x} + mw^2y - m\dot{w}x - \frac{\partial V(r)}{\partial y} \tag{6}$$

From (3), we get

$$m\ddot{z} + \frac{\partial V(r)}{\partial z} = 0 \quad \text{or} \quad m\ddot{z} = -\frac{\partial V(r)}{\partial z} \tag{7}$$

From (5), (6) and (7), we can write

$$m(\hat{i}\ddot{x} + \hat{j}\ddot{y} + \hat{k}\ddot{z}) = m(\hat{i}\dot{w}y - \hat{j}\dot{w}x) + 2m(\hat{i}w\dot{y} - \hat{j}w\dot{x}) + mw^2(\hat{i}x + \hat{j}y) - \frac{\partial V(\vec{r})}{\partial \vec{r}}$$

$$\text{or} \quad m\ddot{\vec{r}} = -m(-\hat{i}\dot{w}y + \hat{j}\dot{w}x) - 2m(-\hat{i}w\dot{y} + \hat{j}w\dot{x}) + mw^2(\hat{i}x + \hat{j}y) - \frac{\partial V(\vec{r})}{\partial \vec{r}}$$

$$= m(\vec{r} \times \dot{\vec{w}}) + 2m(\vec{v} \times \vec{w}) + m\left[\vec{w} \times (\vec{w} \times \vec{r})\right] - \frac{\partial V(\vec{r})}{\partial \vec{r}}.$$

If the rigid body rotates with constant angular velocity, then $\dot{\vec{w}} = 0$, and the above equation reduces to

$$m\ddot{\vec{r}} = 2m(\vec{v} \times \vec{w}) - \frac{\partial V(\vec{r})}{\partial \vec{r}} + m[\vec{w} \times (\vec{w} \times \vec{r})]$$

For interpretation of different terms, see Art. 3.41.

N.B. the term $\frac{\partial V(\vec{r})}{\partial \vec{r}}$ should however be written as $\vec{\nabla}V(\vec{r})$ which denotes a force derivable from potential function $V(\vec{r})$ in a conservative field.

15. For a free particle moving in a straight line with constant velocity v in an inertial co-ordinate system, show by physical reasonings, that the Lagrangian is a scalar quantity. Hence show that if the Lagrangian is invariant w.r.t. Galilean transformations of co-ordinate systems, it can be written as $L(v^2) = \frac{1}{2}mv^2$, where $\frac{1}{2}m$ is a constant. (Assume the relative velocity of the co-ordinate systems to be small compared with the particle velocity.) **[C.U. (Hons.) '83]**

We consider a particle moving freely in an inertial frame of reference. Let its cartesian position vector and velocity vector at a given instant of time t be $\vec{r}$ and $\vec{v}$ respectively. Then as the particle is far removed from the influence of all other particles in the universe, it follows that the particle continues to move with velocity $\vec{v}$. This is the well known law of inertia. Now we look at the same example in the language of a Lagrangian. For the given motion, the number of degrees of freedom is three and the Lagrangian $L = L(\vec{r}, \vec{v}, t)$. In an inertial frame, space is homogeneous, and isotropic and time is homogeneous; hence it follows that L cannot contain explicitly either $\vec{r}$ or t and should be independent of the orientation of the co-ordinate axes as well as the choice of origin. For the Lagrangian to be independent of the orientation of the co-ordinate axes, it must be a scalar quantity.

To summarise, then, the Lagrangian is a scalar that depends only on the velocity of the free particle relative to the given inertial system. The only scalar quantity which can be formed from a vector is the absolute value of the vector. Therefore

$$L = L(v^2) \tag{1}$$

From Lagrange's equation

$$\frac{d}{dt}\left(\frac{\partial L}{\partial \vec{v}}\right) - \frac{\partial L}{\partial \vec{r}} = 0,$$

we obtain $\frac{d}{dt}\left(\frac{\partial L}{\partial \vec{v}}\right) = 0$ since here $\frac{\partial L}{\partial \vec{r}} = 0$.

Thus $\frac{\partial L}{\partial \vec{v}}$ = a constant. Since $\frac{\partial L}{\partial \vec{v}}$ is a function of velocity only, it follows that $\vec{v}$ is a constant vector.

The form of the function in (1) can be found from the relativity principle, according to which the Lagrangian must not change with the transformation from one inertial system to another. In Newtonian mechanics, this transformation is effected with the aid of Galilean transformation equations. The Galilean transformations lead to the law of addition of velocities,

$$\vec{v} = \vec{v'} + \vec{V},$$

where $\vec{V}$ is the relative velocity of the inertial systems.

Since the Lagrangian is determined by a total derivative, it is sufficient (for its invariance w.r.t. Galilean transformations) that the following equality should be satisfied

$$L = L(v^2) = L[(\vec{v'} + \vec{V})^2] = L(v'^2) + \frac{df}{dt} \tag{2}$$

where the functions $L(v^2)$ and $L(v'^2)$ have the same form is accordance with the principle of relativity.

Now if the relative velocity of the inertial systems $\vec{V}$ is very small compared with the particle velocity $\vec{v}$, then we can write

$$(\vec{v'}+\vec{V})^2 = v'^2 + 2\vec{v'}\vec{V} \quad \text{(neglecting } V^2\text{)}$$

Expanding $L[(\vec{v'}+\vec{V})^2]$ in a series, and with the same approximation, we may write

$$L[(\vec{v'}+\vec{V})^2] = L(v'^2) + \frac{\partial L}{\partial(v'^2)} 2\vec{v'}\vec{V}$$

Comparing with (2), we find

$$\frac{\partial L}{\partial(v'^2)} 2\vec{v'}\vec{V} = \frac{\partial L}{\partial(v'^2)} 2\vec{V}\frac{d\vec{r'}}{dt} = \frac{df(\vec{r'})}{dt}$$

However the expression on the left hand side of the equation can be a total derivative of the function of a co-ordinates only if $\dfrac{\partial L}{\partial(v'^2)}$ is independent of velocity.

Writing $\dfrac{\partial L}{\partial(v'^2)} = \dfrac{m}{2}$ = constant, we obtain

$$\frac{df(\vec{r'})}{dt} = \frac{d}{dt}(m\vec{V}\vec{r'}) \quad \because \frac{d\vec{V}}{dt} = 0$$

Thus the Lagrangian for a free particle is $L(v^2) = \dfrac{m}{2}v^2$.

16. (a) Two undamped one-dimensional oscillators of different natural frequencies are coupled together. Prove that the general oscillation is a sum of two vibrations of different frequencies. Investigate the frequencies and the ratio of amplitudes when the coupling is loose.

(b) Introduce normal co-ordinates for the problem in (a) and prove, using Lagrange's equation, that each normal co-ordinate executes a simple harmonic motion with arbitrary amplitude and phase, but with frequency given in (a).

(c) Discuss the formation of beats in (a) when both the vibrations are simultaneously excited.

[C.U. (Hons.) '81]

Let us consider two identical simple pendulums, each having length l and a bob of mass m coupled to each other by means of a horizontal spring as shown in the attached figure. The mass of the spring is assumed to be negligible. Let the force constant of the spring be k. Let the spring be unstretched when the two bobs are in equilibrium positions. At any instant t, let x_1 and x_2 represent the displacements of the two bobs from their equilibrium positions. Then the spring is stretched by an amount $(x_2 - x_1)$ and supplies the force $k(x_2 - x_1)$. We assume that the motion is undamped and the displacements are small. Now the restoring force due to mg may be written as mgx/l as in the case of a simple pendulum.

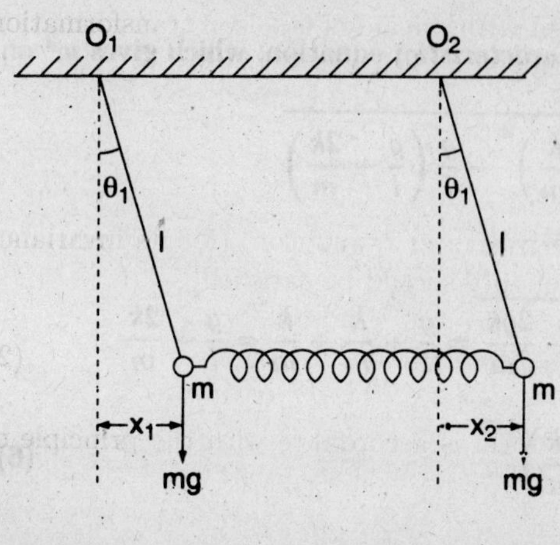

The equations of motion of the two pendulums are, therefore, written as

$$m\ddot{x}_1 = -\frac{mgx_1}{l} + k(x_2 - x_1) \quad (1)$$

$$m\ddot{x}_2 = -\frac{mgx_2}{l} - k(x_2 - x_1) \quad (2)$$

Assuming the solutions as $x_1 = A_1e^{jwt}$ and $x_2 = A_2e^{jwt}$, (w being the angular frequency of each of the two systems), we have

$$\ddot{x}_1 = -A_1w^2e^{jwt} \quad \text{and} \quad \ddot{x}_2 = -A_2w^2e^{jwt}$$

$\therefore$ from (1), with these substitutions, we have

$$-mA_1w^2e^{jwt} + \frac{mg}{l}A_1e^{jwt} - k(A_2 - A_1)e^{jwt} = 0$$

$$\text{or} \quad \left(-mw^2 + \frac{mg}{l} + k\right)A_1 = kA_2 \quad \because e^{jwt} \neq 0$$

$$\therefore \quad \frac{A_1}{A_2} = \frac{k}{\left(-mw^2 + \frac{mg}{l} + k\right)} \tag{3}$$

Again from (2), we may similarly have,

$$\left(-mw^2 + \frac{mg}{l} + k\right)A_2 = kA_1,$$

so that

$$\frac{A_1}{A_2} = \frac{\left(-mw^2 + \frac{mg}{l} + k\right)}{k} \tag{4}$$

From (3) and (4) we obtain

$$\frac{k}{\left(-mw^2 + \frac{mg}{l} + k\right)} = \frac{\left(-mw^2 + \frac{mg}{l} + k\right)}{k}$$

$$\text{or} \quad m^2w^4 - 2\left(\frac{mg}{l} + k\right)w^2m + \left(\frac{mg}{l} + k\right)^2 = k^2$$

$$\text{or} \quad w^4 - 2\left(\frac{g}{l} + \frac{k}{m}\right)w^2 + \left(\frac{g}{l} + \frac{k}{m}\right)^2 = \frac{k^2}{m^2}$$

$$\text{or} \quad w^4 - 2\left(\frac{g}{l} + \frac{k}{m}\right)w^2 + \frac{g}{l}\left(\frac{g}{l} + \frac{2k}{m}\right) = 0 \tag{5}$$

The above equation is known as the *secular* (or *characteristic*) equation, which gives w^2 on simplification.

$$\therefore \quad w^2 = \left(\frac{g}{l} + \frac{k}{m}\right) \pm \sqrt{\left(\frac{g}{l} + \frac{k}{m}\right)^2 - \frac{g}{l}\left(\frac{g}{l} + \frac{2k}{m}\right)}$$

"w^2" will thus have two values given by

$$w_1^2 = \left(\frac{g}{l} + \frac{k}{m}\right) + \sqrt{\frac{g^2}{l^2} + \frac{2gk}{lm} + \frac{k^2}{m^2} - \frac{g^2}{l^2} - \frac{2gk}{lm}} = \frac{g}{l} + \frac{k}{m} + \frac{k}{m} = \frac{g}{l} + \frac{2k}{m}$$

$$\therefore \quad w_1 = \left(\frac{g}{l} + \frac{2k}{m}\right)^{1/2} \tag{6}$$

and

$$w_2^2 = \frac{g}{l} + \frac{k}{m} - \frac{k}{m} = \frac{g}{l}. \quad \therefore \quad w_2 = \sqrt{\frac{g}{l}} \tag{7}$$

With the above values of w_1 and w_2, we can now find out A_1/A_2, i.e., the ratio of the amplitudes.

From (3), we have, on putting $w = w_1 = \left(\frac{g}{l} + \frac{2k}{m}\right)^{1/2}$ obtained from (6),

$$\frac{A_1}{A_2} = \frac{k}{-m\left(\frac{g}{l} + \frac{2k}{m}\right) + \frac{mg}{l} + k} = -1, \quad \text{i.e.,} \quad A_1 = -A_2 \quad \text{and} \quad k > 0,$$

so that the two pendulums oscillate in opposite directions with a frequency larger than the uncoupled one.

Similarly by putting $w = w_2$, we get $\frac{A_1}{A_2} = +1$, i.e., $A_1 = A_2$ and $k > 0$, so that the two pendulums oscillate together with a frequency equal to that for a simple pendulum.

The above two normal modes of oscillation are shown in the figures given below. In the first mode, there is a phase difference of 180° in the oscillations of the two pendulums and the spring will exert a force leading to a higher frequency. In the second mode, the pendulums oscillate in phase and there will be no force in the spring, so that the frequency of oscillation is same as that for a simple pendulum.

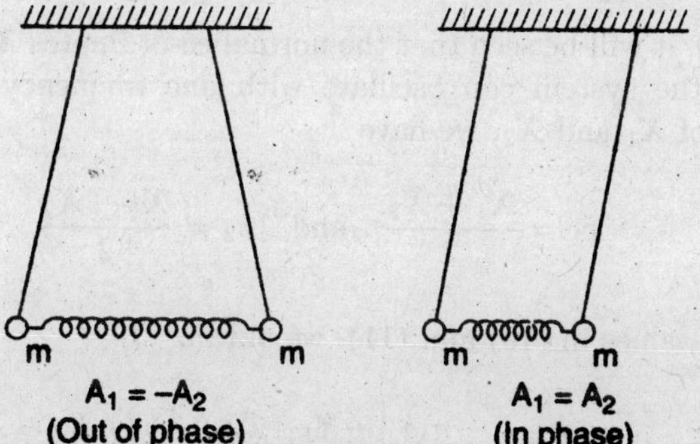

Normal modes of the coupled oscillators

It can be shown in the following way that the general solution is a linear superposition of the two modes of oscillations. The general solutions can be written as

$$\begin{aligned} x_1 &= A_1 e^{jwt} = A_{1+}e^{jw_1t} + A_{1-}e^{jw_2t} \\ &= A_{1+}\cos w_1t + A'_{1+}\sin w_1t + A_{1-}\cos w_2t + A'_{1-}\sin w_2t \\ &= P\cos w_1t + Q\sin w_1t + R\cos w_2t + S\sin w_2t \end{aligned} \tag{8}$$

$$\begin{aligned} \text{and} \quad x_2 &= A_{2+}e^{jw_1t} + A_{2-}e^{jw_2t} \\ &= A_{2+}\cos w_1t + A'_{2+}\sin w_1t + A_{2-}\cos w_2t + A'_{2-}\sin w_2t \\ &= -P\cos w_1t - Q\sin w_1t + R\cos w_2t + S\sin w_2t. \end{aligned} \tag{9}$$

The constants can be found from the initial conditions (i.e., positions and velocities) of the two pendulums. We have thus shown that the general oscillation will be the superposition of the two normal modes of oscillation.

(b) The total energy of the coupled pendulum system is partly (i) potential and partly (ii) kinetic.

Now the potential energy of the system is due to (i) the extension of the spring; and (ii) the raising of the pendulum bobs against gravity.

$$\therefore \quad V = \text{P.E. of the system} = \int_0^{(x_2-x_1)} kx dx + \left[\int_0^{x_1} \frac{mgx}{l}\, dx + \int_0^{x_2} \frac{mgx}{l}\, dx\right]$$

$$= \frac{1}{2}k(x_2 - x_1)^2 + \frac{mg}{2l}(x_1^2 + x_2^2) \tag{10}$$

$$T = \text{K.E. of the system} = \frac{1}{2}m(\dot{x}_1^2 + \dot{x}_2^2). \tag{11}$$

From (8) and (9), we may write

$$x_1 + x_2 = X_1 = 2(R\cos w_2 t + S\sin w_2 t) \tag{12}$$

$$\text{and} \quad x_1 - x_2 = X_2 = 2(P\cos w_1 t + Q\sin w_1 t) \tag{13}$$

The modes of oscillation given by (12) and (13) are called the *normal modes* of oscillation for the reason that X_1 or X_2 involves only one frequency w_2 or w_1.

From (12) and (13), it will be seen that the normal co-ordinates X_1 and X_2 are independent of each other, since the system can oscillate with one frequency only, while the other is suppressed. In terms of X_1 and X_2, we have

$$x_1 = \frac{X_1 + X_2}{2} \quad \text{and} \quad x_2 = \frac{X_1 - X_2}{2}$$

Substituting these values in (10) and (11), we obtain

$$V = \frac{k}{2}X_2^2 + \frac{mg}{2l}\left[\left(\frac{X_1 + X_2}{2}\right)^2 + \left(\frac{X_1 - X_2}{2}\right)^2\right]$$

$$= \frac{mg}{4l}X_1^2 + \left(\frac{mg}{4l} + \frac{k}{2}\right)X_2^2 \tag{14}$$

$$\text{and} \quad T = \frac{1}{4}m(\dot{X}_1^2 + \dot{X}_2^2) \tag{15}$$

Hence the Lagrangian of the system can be written as

$$L = T - V = \frac{1}{4}m(\dot{X}_1^2 + \dot{X}_2^2) - \frac{mg}{4l}X_1^2 - \left(\frac{mg}{4l} + \frac{k}{2}\right)X_2^2 \tag{16}$$

The Lagrange's equations for a conservative system in two generalised co-ordinates X_1 and

X_2 can be written as

$$\left.\begin{aligned} &\frac{d}{dt}\left(\frac{\partial L}{\partial \dot{X}_1}\right) - \frac{\partial L}{\partial X_1} = 0 \qquad \ldots \qquad (a) \\ \text{and} \quad &\frac{d}{dt}\left(\frac{\partial L}{\partial \dot{X}_2}\right) - \frac{\partial L}{\partial X_2} = 0 \qquad \ldots \qquad (b) \end{aligned}\right\} \tag{17}$$

From (16), we have

$$\frac{\partial L}{\partial \dot{X}_1} = \frac{1}{2} m \dot{X}_1, \quad \frac{\partial L}{\partial X_1} = -\frac{mg}{2l} X_1,$$

$$\frac{\partial L}{\partial \dot{X}_2} = \frac{1}{2} m \dot{X}_2, \quad \frac{\partial L}{\partial X_2} = -\left(\frac{mg}{2l} + k\right) X_2$$

From (17a), we get,

$$\frac{m}{2}\ddot{X}_1 + \frac{mg}{2l} X_1 = 0 \quad \text{or} \quad \ddot{X}_1 + \frac{g}{l} X_1 = 0$$

$$\text{or} \quad \ddot{X}_1 + w_2^2 X_1 = 0 \tag{18}$$

Again from (17b), we get

$$\frac{m}{2}\ddot{X}_2 + \left(\frac{mg}{2l} + k\right) X_2 = 0 \quad \text{or} \quad \ddot{X}_2 + \left(\frac{g}{l} + \frac{2k}{m}\right) X_2 = 0$$

$$\text{or} \quad \ddot{X}_2 + w_1^2 X_2 = 0 \tag{19}$$

Equations (18) and (19) show that each normal co-ordinate executes a simple harmonic motion with arbitrary amplitude and phase, but with frequency obtained in answer to part (a) of the question.

(c) In such a system, there is resonance between the two parts of the system. We suppose that the pendulum A oscillates first with pendulum B at rest. The energy from A gets transferred to B and it alternates between the pendulums A and B. When pendulum A has maximum amplitude, pendulum B has zero amplitude and vice-versa.

Again

$$w_1 = \left(\frac{g}{l} + \frac{2k}{m}\right)^{1/2} = \sqrt{\frac{g}{l}}\left(1 + \frac{2kl}{gm}\right)^{1/2} = \sqrt{\frac{g}{l}}\left(1 + \frac{kl}{gm}\right).$$

(expanding binomially and neglecting smaller terms)

$$\therefore \quad w_1 - \sqrt{\frac{g}{l}} + \sqrt{\frac{l}{g}\frac{k}{m}} = \sqrt{\frac{g}{l}} + n \quad \text{where } n = \sqrt{\frac{l}{g}\frac{k}{m}} = \text{a small quantity}$$

$$\text{and} \quad w_2 = \sqrt{\frac{g}{l}}.$$

Since $(w_1 - w_2) = n =$ very small, beats will be formed.

17. A particle of mass m moves in a conservative force field. Find (a) the Lagrangian and (b) the equations of motion in cylindrical polar co-ordinates. **[C.U. (Hons.) '98]**

(a) A particle moving freely in space has three degrees of freedom. For the independent generalised co-ordinates, we may take cartesian or some other co-ordinates giving the position of the particle. For the case when cylindrical co-ordinates (r, θ, z) are chosen as generalised co-ordinates (q_1, q_2, q_3), from Art. 3.21 and the attached figure, we may write $x = r\cos\theta$, $y = r\sin\theta$, $z = z$ (which are the relations connecting the cartesian (x, y, z) and cylindrical co-ordinates (r, θ, z) of the particle of mass m at P at any instant), whence

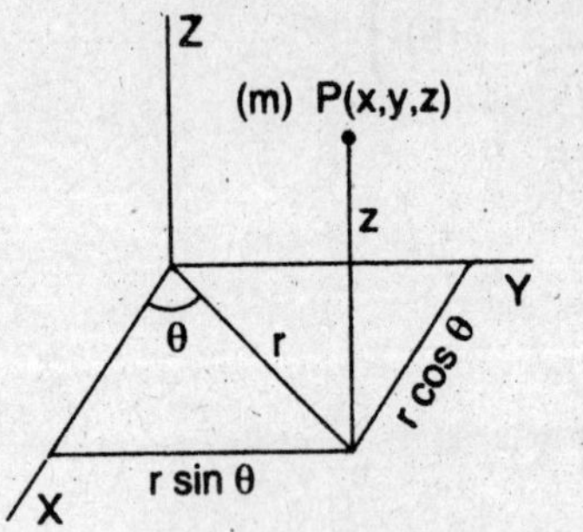

$$\left.\begin{aligned} \dot{x} &= \dot{r}\cos\theta - r\sin\theta\dot{\theta} \\ \dot{y} &= \dot{r}\sin\theta + r\cos\theta\dot{\theta} \\ \dot{z} &= \dot{z} \end{aligned}\right\} \tag{1}$$

$$\therefore \quad T = \frac{1}{2}m(\dot{x}^2 + \dot{y}^2 + \dot{z}^2) = \frac{1}{2}m\left[(\dot{r}\cos\theta - r\sin\theta\dot{\theta})^2 + (\dot{r}\sin\theta + r\cos\theta\dot{\theta})^2 + \dot{z}^2\right]$$

$$= \frac{1}{2}m(\dot{r}^2 + r^2\dot{\theta}^2 + \dot{z}^2)$$

and $\quad V = V(r, \theta, z)$,

so that the Lagrangian L is given by

$$L = T - V = \frac{1}{2}m(\dot{r}^2 + r^2\dot{\theta}^2 + \dot{z}^2) - V(r, \theta, z) \tag{2}$$

(b) From (2), we have

$$\frac{\partial L}{\partial \dot{r}} = m\dot{r}, \quad \frac{\partial L}{\partial r} = mr\dot{\theta}^2 - \frac{\partial V}{\partial r}$$

$$\frac{\partial L}{\partial \dot{\theta}} = mr^2\dot{\theta}, \quad \frac{\partial L}{\partial \theta} = -\frac{\partial V}{\partial \theta}$$

$$\frac{\partial L}{\partial \dot{z}} = m\dot{z}, \quad \frac{\partial L}{\partial z} = -\frac{\partial V}{\partial z}$$

With there results, Lagrange's equations in cylindrical co-ordinates are written as

(i) $\dfrac{d}{dt}\left(\dfrac{\partial L}{\partial \dot{r}}\right) - \dfrac{\partial L}{\partial r} = 0$ or $m(\ddot{r} - r\dot{\theta}^2) = -\dfrac{\partial V}{\partial r}$;

(ii) $\dfrac{d}{dt}\left(\dfrac{\partial L}{\partial \dot{\theta}}\right) - \dfrac{\partial L}{\partial \theta} = 0$ or $mr^2\ddot{\theta} + 2mr\dot{r}\dot{\theta} + \frac{\partial V}{\partial \theta} = 0$

(iii) $\dfrac{d}{dt}\left(\dfrac{\partial L}{\partial \dot{z}}\right) - \dfrac{\partial L}{\partial z} = 0$ or $m\ddot{z} + \dfrac{\partial V}{\partial z} = 0$

18. Set up Lagrange's equations of a particle moving on the surface of the earth, using spherical polar co-ordinates.

With the centre of the earth as origin, we consider a particle moving on its surface. If (r, θ, ϕ) be the spherical polar co-ordinates of the particle at P at any instant as shown in the attached figure, then we have

$$\left.\begin{aligned} x &= r\sin\theta\cos\phi \\ y &= r\sin\theta\sin\phi \\ z &= r\cos\theta \end{aligned}\right\}$$ (x, y, z) being the cartesian co-ordinates at the instant; here r is constant being the radius of the earth.

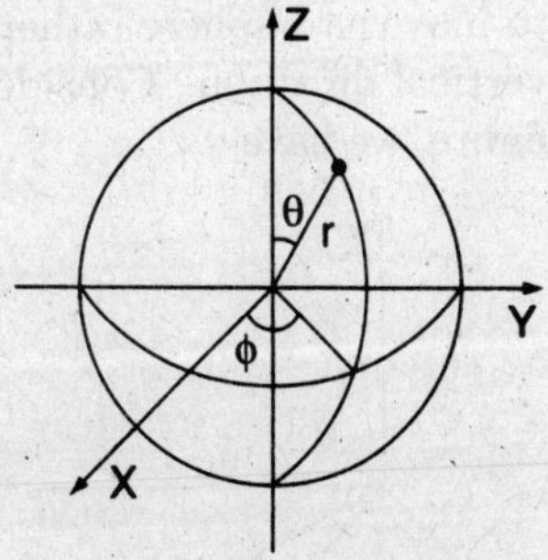

whence $\dot{x} = r\cos\phi\cos\theta\dot{\theta} - r\sin\theta\sin\phi\dot{\phi}$

$\dot{y} = r\sin\theta\cos\phi\dot{\phi} + r\sin\phi\cos\theta\dot{\theta}$

$\dot{z} = -r\sin\theta\dot{\theta}$

$$\therefore\ T = \frac{1}{2}m\left[(r\cos\phi\cos\theta\dot{\theta} - r\sin\theta\sin\phi\dot{\phi})^2 + (r\sin\theta\cos\phi\dot{\phi} + r\sin\phi\cos\theta\dot{\theta})^2 + (-r\sin\theta\dot{\theta})^2\right]$$

$$= \frac{1}{2}mr^2(\dot{\theta}^2 + \sin^2\theta\dot{\phi}^2).$$

Let the P.E. of the particle be $V(r, \theta, \phi)$, so that the Lagrangian is

$$L = T - V = \frac{1}{2}mr^2(\dot{\theta}^2 + \sin^2\theta\dot{\phi}^2) - V(r, \theta, \phi) \qquad (1)$$

In this case, r being constant, there are two generalised co-ordinates θ and ϕ, and so there will be two Lagrangian equations. Now Lagrange's equation for a conservative system in variable θ is

$$\frac{d}{dt}\left(\frac{\partial L}{\partial\dot{\theta}}\right) - \frac{\partial L}{\partial\theta} = 0 \qquad (2)$$

From (1) we have

$$\frac{\partial L}{\partial\dot{\theta}} = mr^2\dot{\theta} \quad \text{and} \quad \frac{\partial L}{\partial\theta} = mr^2\sin\theta\cos\theta\dot{\phi}^2 - \frac{\partial V}{\partial\theta},$$

so that (2) reduces to

$$mr^2\ddot{\theta} - mr^2\sin\theta\cos\theta\dot{\phi}^2 + \frac{\partial V}{\partial\theta} = 0 \qquad (3)$$

Again Lagrange's equation for a conservative system in variable ϕ is given by

$$\frac{d}{dt}\left(\frac{\partial L}{\partial\dot{\phi}}\right) - \frac{\partial L}{\partial\phi} = 0 \qquad (4)$$

From (1) we have

$$\frac{\partial L}{\partial\dot{\phi}} = mr^2\sin^2\theta\dot{\phi} \quad \text{and} \quad \frac{\partial L}{\partial\phi} = -\frac{\partial V}{\partial\phi},$$

so that (4) becomes

$$mr^2\sin^2\theta\ddot{\phi} + 2mr^2\sin\theta\cos\theta\dot{\phi}\dot{\theta} + \frac{\partial V}{\partial\phi} = 0 \qquad (5)$$

(3) and (5) are the required Lagrange's equations.

19. Find the Lagrangian and hence the equations of motion of a spherical pendulum.

A spherical pendulum is nothing but a plane pendulum whose bob of mass m is constrained to move on a sphere rather than on a circle. Here the force is not central, but is constant in the vertical direction. Considering the co-ordinates of the bob as r, θ, ϕ as shown in the attached figure, we have

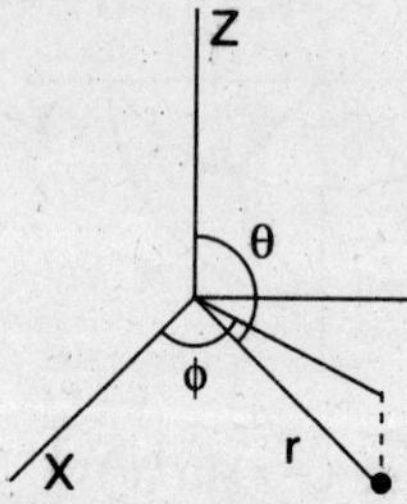

$$\text{K.E.} = T = \frac{1}{2}mr^2(\dot{\theta}^2 + \sin^2\theta\dot{\phi}^2)$$

(r being the radius of the sphere which is constant here)

and P.E. of the system due to gravity relative to the horizontal plane, $V = mgr\cos\theta$, so that the Lagrangian

$$L = T - V = \frac{1}{2}mr^2(\dot{\theta}^2 + \sin^2\theta\dot{\phi}^2) - mgr\cos\theta \qquad (1)$$

which gives

$$\left.\begin{aligned} \frac{\partial L}{\partial \dot{\theta}} &= mr^2\dot{\theta}, & \frac{\partial L}{\partial \theta} &= mr^2\sin\theta\cos\theta\dot{\phi}^2 + mgr\sin\theta \\ \frac{\partial L}{\partial \dot{\phi}} &= mr^2\sin^2\theta\dot{\phi}, & \frac{\partial L}{\partial \phi} &= 0 \end{aligned}\right\} \qquad (2)$$

The equations of motion are

$$\frac{d}{dt}\left(\frac{\partial L}{\partial \dot{\theta}}\right) - \frac{\partial L}{\partial \theta} = 0$$

$$\text{or} \quad mr^2\ddot{\theta} - mr^2\sin\theta\cos\theta\dot{\phi}^2 - mgr\sin\theta = 0 \qquad (3)$$

using (2)

and

$$\frac{d}{dt}\left(\frac{\partial L}{\partial \dot{\phi}}\right) - \frac{\partial L}{\partial \phi} = 0$$

$$\text{or} \quad mr^2\sin^2\theta\ddot{\phi} + 2mr^2\sin\theta\cos\theta\dot{\theta}\dot{\phi} = 0 \qquad (4)$$

using (2)

(3) and (4) are the required Lagrange's equations.

20. Three equal masses are so connected by springs that they move along a circle. The point A is fixed (see attached figure). (i) Write down the Lagrangian, (ii) Obtain the equations of motion, (iii) Find the eigen frequencies and normal co-ordinates of the system. **[C.U. (Hons.) '99]**

Let k be the spring constant and x_i (i = 1, 2, 3) be the displacement of the ith mass from equilibrium along the circle. The Lagrangian of the system is given by

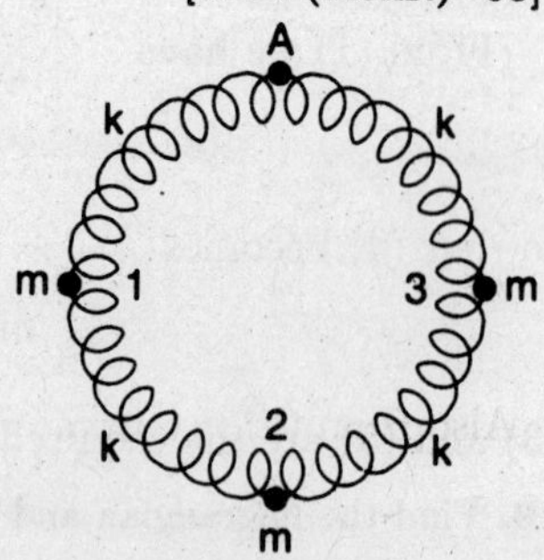

$$\begin{aligned} L &= T - V \\ &= \frac{1}{2}m(\dot{x}_1^2 + \dot{x}_2^2 + \dot{x}_3^2) - \frac{1}{2}k\left[x_1^2 + (x_2 - x_1)^2 + (x_3 - x_2)^2 + x_3^2\right] \\ &= \frac{1}{2}m(\dot{x}_1^2 + \dot{x}_2^2 + \dot{x}_3^2) - kx_1^2 - kx_2^2 - kx_3^2 + kx_1x_2 + kx_2x_3 \qquad (1) \end{aligned}$$

whence
$$\frac{\partial L}{\partial \dot{x}_1} = m\dot{x}_1, \quad \frac{\partial L}{\partial x_1} = -2kx_1 + kx_2 = -k(2x_1 - x_2)$$
$$\frac{\partial L}{\partial \dot{x}_2} = m\dot{x}_2, \quad \frac{\partial L}{\partial x_2} = -2kx_2 + kx_1 + kx_3 = -k(2x_2 - x_3 - x_1)$$
$$\frac{\partial L}{\partial \dot{x}_3} = m\dot{x}_3, \quad \frac{\partial L}{\partial x_3} = -2kx_3 + kx_2 = -k(2x_3 - x_2),$$

so that the three Lagrange's equations of motion are written as

$$\frac{d}{dt}\left(\frac{\partial L}{\partial \dot{x}_1}\right) - \frac{\partial L}{\partial x_1} = 0 \quad \text{or} \quad m\ddot{x}_1 + k(2x_1 - x_2) = 0 \tag{2}$$
$$\frac{d}{dt}\left(\frac{\partial L}{\partial \dot{x}_2}\right) - \frac{\partial L}{\partial x_2} = 0 \quad \text{or} \quad m\ddot{x}_2 + k(2x_2 - x_3 - x_1) = 0 \tag{3}$$
and
$$\frac{d}{dt}\left(\frac{\partial L}{\partial \dot{x}_3}\right) - \frac{\partial L}{\partial x_3} = 0 \quad \text{or} \quad m\ddot{x}_3 + k(2x_3 - x_2) = 0 \tag{4}$$

We assume real solutions of the form

$$x_i = A_i \cos(wt + \phi), \quad i = 1, 2, 3$$

$$\therefore\ x_1 = A_1 \cos(wt + \phi), \quad x_2 = A_2 \cos(wt + \phi), \quad x_3 = A_3 \cos(wt + \phi)$$

whence $\ddot{x}_1 = -A_1 w^2 \cos(wt + \phi)$, $\ddot{x}_2 = -A_2 w^2 \cos(wt + \phi)$, $\ddot{x}_3 = -A_3 w^2 \cos(wt + \phi)$.

Substituting these values in (2), (3) and (4) and knowing that $A_i \cos(wt + \phi) \neq 0$, we may write, after re-arranging the terms

$$(2k - mw^2)A_1 - kA_2 + 0A_3 = 0 \tag{5}$$
$$-kA_1 + (2k - mw^2)A_2 - kA_3 = 0 \tag{6}$$
$$0A_1 - kA_2 + (2k - mw^2)A_3 = 0 \tag{7}$$

which are the equations to be satisfied.

From (5)

$$A_1 = \frac{k}{(2k - mw^2)} A_2$$

Putting this value of A_1 in (6), we have

$$\left[(2k - mw^2) - \frac{k^2}{(2k - mw^2)}\right] A_2 - kA_3 = 0$$

$$\therefore\ \frac{A_2}{A_3} = \frac{k}{(2k - mw^2) - \dfrac{k^2}{(2k - mw^2)}} \tag{8}$$

Also from (7), we have

$$\frac{A_2}{A_3} = \frac{(2k - mw^2)}{k} \tag{9}$$

Equating (8) and (9) and on simplification, the *characteristic equation* is

$$(2k - mw^2)^3 - 2k^2(2k - mw^2) = 0$$

$$\text{or} \quad (2k - mw^2)(2k - mw^2 + \sqrt{2}\,k)(2k - mw^2 - \sqrt{2}\,k) = 0 \tag{10}$$

From (10), the eigen frequencies may be written as

$$w_1^2 = (2 - \sqrt{2})\frac{k}{m}, \quad w_3^2 = (2 + \sqrt{2})\frac{k}{m}, \quad w_2^2 = 2\frac{k}{m}$$

Of the three algebraic equations for the A's, only two are independent. Taking $w = w_1$, we obtain, from (5), (6) and (7)

$$\sqrt{2}\,kA_1 - kA_2 = 0 \tag{11}$$

$$-kA_1 + \sqrt{2}\,kA_2 - kA_3 = 0 \tag{12}$$

$$-kA_2 + \sqrt{2}\,kA_3 = 0 \tag{13}$$

These equations give

$$A_1 = \tfrac{1}{\sqrt{2}}A_2 = A_3 = A$$

Similarly when $w = w_3$, we obtain

$$A_1 = -\tfrac{1}{\sqrt{2}}A_2 = A_3 = C$$

Also $w = w_2$ gives $A_1 = -A_3 = B$ and $A_2 = 0$

Thus the general solutions are

$$\left.\begin{aligned} x_1 &= A\cos(w_1 t + \phi_1) + B\cos(w_2 t + \phi_2) + C\cos(w_3 t + \phi_3) \\ x_2 &= \sqrt{2}\,A\cos(w_1 t + \phi_1) - \sqrt{2}\,C\cos(w_3 t + \phi_3) \\ x_3 &= A\cos(w_1 t + \phi_1) - B\cos(w_2 t + \phi_2) + C\cos(w_3 t + \phi_3) \end{aligned}\right\} \tag{14}$$

The six constants A, B, C, ϕ_1, ϕ_2, ϕ_3 are to be determined from the initial conditions.

Finally, proceeding exactly in the same way as in problem (16), we may obtain the normal co-ordinates given by

$$X_1 = 2A\cos(w_1 t + \phi_1) = \tfrac{1}{2}\left(x_1 + \sqrt{2}\,x_2 + x_3\right)$$

$$X_2 = \sqrt{2}\,B\cos(w_2 t + \phi_2) = \tfrac{1}{\sqrt{2}}(x_1 - x_3)$$

$$\text{and} \quad X_3 = 2C\cos(w_3 t + \phi_3) = \tfrac{1}{2}\left(x_1 - \sqrt{2}\,x_2 + x_3\right)$$

In terms of these normal co-ordinates, the Lagrangian becomes

$$L = \tfrac{1}{2}m\left[\dot{X}_1^2 - w_1^2X_1^2 + \dot{X}_2^2 - w_2^2X_2^2 + \dot{X}_3^2 - w_3^2X_3^2\right]$$

whence the equations of motion may easily be written.

21. A bead of mass m is sliding freely on a smooth circular wire of radius 'b', which rotates in a horizontal plane about one of its points with a constant angular frequency ω. Show that the motion of the bead is that of a pendulum of length $a = g/\omega^2$.

Here

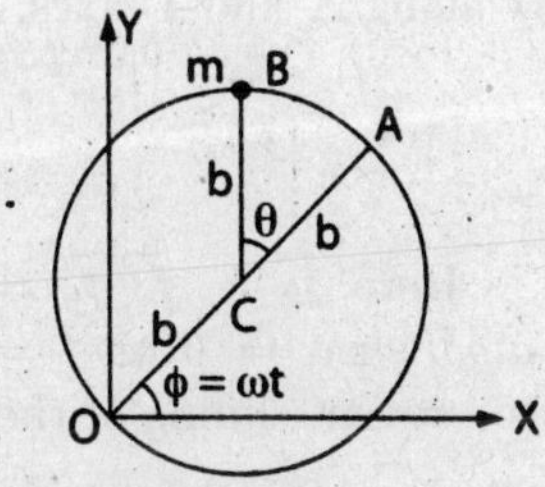

$$x = b\cos\omega t + b\cos(\theta + \omega t)$$

$$y = b\sin\omega t + b\sin(\theta + \omega t)$$

$$\dot{x} = -b\omega\sin\omega t - b(\dot{\theta} + \omega)\sin(\theta + \omega t)$$

$$\dot{y} = b\omega\cos\omega t + b(\dot{\theta} + \omega)\cos(\theta + \omega t) \quad \text{and } V = 0$$

$$T = \tfrac{1}{2}m(\dot{x}^2 + \dot{y}^2) = \tfrac{1}{2}mb^2\left[\omega^2 + (\dot{\theta} + \omega)^2 + 2\omega(\dot{\theta} + \omega)\cos\theta\right],$$

so that $L = T - V = T$ and

$$\frac{\partial L}{\partial \dot{\theta}} = \frac{\partial T}{\partial \dot{\theta}} = mb^2(\dot{\theta} + \omega + \omega\cos\theta)$$

$$\frac{\partial L}{\partial \theta} = \frac{\partial T}{\sqrt{\theta}} = -mb^2\omega(\dot{\theta} + \omega)\sin\theta$$

$\therefore$ Lagrange's equation is

$$mb^2(\ddot{\theta} - \omega\dot{\theta}\sin\theta) + mb^2\omega(\dot{\theta} + \omega)\sin\theta = 0$$

$$\text{or } \ddot{\theta} + \omega^2\sin\theta = 0 \quad \text{or} \quad \ddot{\theta} + \omega^2\theta = 0 \quad (\text{when } \theta \text{ is very small})$$

which represents a simple harmonic motion of period $T = \frac{2\pi}{\omega} = 2\pi\sqrt{\frac{a}{g}}$ where a is the length of the pendulum given by $a = g/\omega^2$.

22. A frictionless block of mass 'm' is placed on an incline making an angle α with the horizontal. The incline is now given a horizontal acceleration in the vertical plane of the incline such that the block can not slide. Use D'Alembert's principle to find the acceleration given. **[C.U. (Hons.) 2003]**

In the adjoining figure, we consider a frictionless block of mass 'm' at P on AB, the smooth plane inclined to the horizontal BC at angle α. In applying D'Alembert's principle, we consider the impressed force $\vec{F_i} = m\vec{g}$ due to gravity and the effective force $m\vec{a}$ only. So by D'Alembert's principle, $(m\vec{g} - m\vec{a}) \cdot \delta\vec{r} = 0$, where $\delta\vec{r}$ is taken down the plane, the constraint forces appearing through it. Hence $m\vec{g} \cdot \delta\vec{r} = m\vec{a} \cdot \delta\vec{r}$.

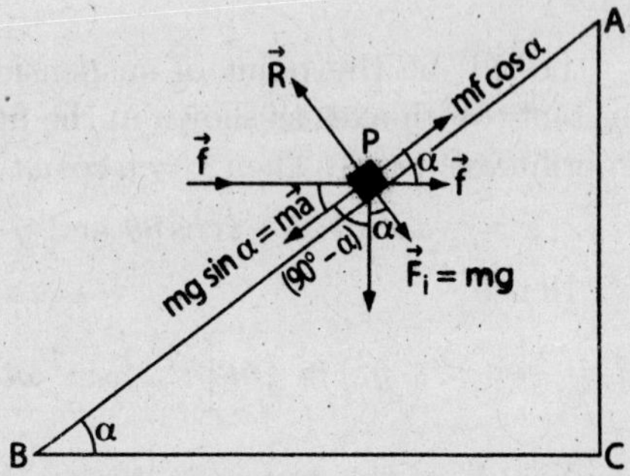

But $m\vec{g} \cdot \delta\vec{r} = mg\sin\alpha\delta r$ and $m\vec{a} \cdot \delta\vec{r} = ma\delta r$.

$\therefore \quad ma\delta r = mg\sin\alpha\delta r$ or $a = g\sin\alpha$.

If now the incline is given a horizontal acceleration $\vec{f}$ in its vertical plane such that the block stops sliding then

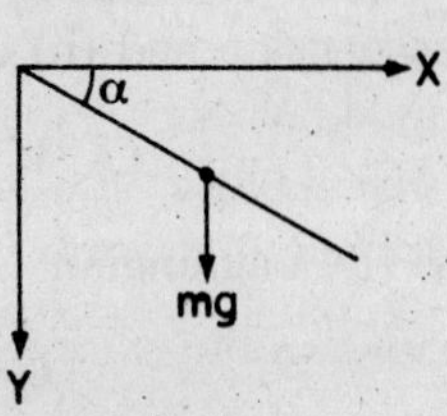

$$m\vec{a} = m \times \text{component of } \vec{f} \text{ up along the incline}$$

$$= m\vec{f}\cos\alpha$$

$$\text{or} \quad mg\sin\alpha = m\vec{f}\cos\alpha \quad \text{or} \quad \vec{f} = g\tan\alpha$$

The above result can also be alternatively obtained as follows.

If $\delta\vec{r}$ represents the possible virtual displacement along the incline, then $\delta\vec{x}$ = component along $X = \delta\vec{r}\cos\alpha$ and $\delta\vec{y}$ = component along $y = \delta\vec{r}\sin\alpha$. The forces acting on the block of mass m are $\vec{F}_x = 0$ and $\vec{F}_y = mg$. If $\vec{f}_x$ and $\vec{f}_y$ be the components of the applied acceleration $\vec{f}$ along X and Y axes, then using D'Alembert's principle $(\vec{F}_f - m\ddot{\vec{r}})\cdot\delta\vec{r} = 0$, we may write

$$\left(\vec{F}_x - m\vec{f}_x\right)\cdot\delta\vec{x} + \left(\vec{F}_y - m\vec{f}_y\right)\cdot\delta\vec{y} = 0$$

Here $\vec{f}_x = \vec{f}$, $\vec{f}_y = 0$, $\vec{F}_x = 0$ so that the above equation reduces to $-mf\delta r\cos\alpha + mg\delta r\sin\alpha = 0$ or $f = g\tan\alpha$ which is the required horizontal acceleration of the incline to prevent sliding of the block on the incline.

23. Establish the equations of motion of a homogeneous sphere rolling on a plane under an external force $\vec{F}$ and torque $\vec{T}$.

[C.U. (Hons.) 1997]

The sphere rolling on a plane executes simultaneously a translational and a rotational motion. Using D'Alembert's principle, the equation of motion corresponding to virtual displacement $\delta\vec{r}$ is written as

$$(\vec{F} - m\ddot{\vec{r}})\cdot\delta\vec{r} = 0.$$

$\because$ $\delta\vec{r}$ is arbitrary, we have $\vec{F} - m\ddot{\vec{r}} = 0$.

Again the equation of motion due to rotation may be written as, using D'Alembert's principle,

$$(\vec{T} - I\ddot{\vec{\theta}})\cdot\delta\vec{\theta} = 0$$

whence $\vec{T} - I\ddot{\vec{\theta}} = 0$, $\delta\vec{\theta}$ being arbitrary.

Here I = the moment of inertia of the sphere about its axis of rotation $= \frac{2}{5}ma^2$, 'a' being the radius of the sphere, and $\delta r = a\delta\theta$ is the constraint equation connecting the two coordinates r and θ.

24. The point of support of a simple pendulum (length $= l$, mass of the bob $= m$) is moving on a horizontal line according to the law $x = a\cos\omega t$, (a and w are constants). Find the Lagrangian and hence show that the equation of motion for small angular displacement θ is $\ddot{\theta} + \frac{g}{l}\theta = \frac{a\omega^2}{l}\cos\omega t$.

[C.U. (Hons.) 1995]

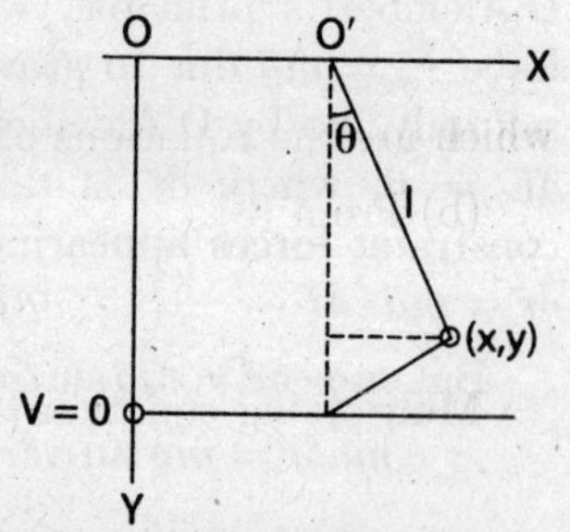

Let O' be the point of suspension of the pendulum at any instant. With axes as shown in the figure, the pendulum bob has coordinates (x, y). Then $x = a\cos\omega t + l\sin\theta$, and $y = l\cos\theta$.

$\therefore \dot{x} = -a\omega\sin\omega t + l\cos\theta\dot{\theta}$ and $\dot{y} = -l\sin\theta\dot{\theta}$.

Hence

$$T = \tfrac{1}{2}m(\dot{x}^2 + \dot{y}^2) = \tfrac{1}{2}m\left[a^2\omega^2\sin^2\omega t + l^2\dot{\theta}^2 - 2a\omega l\dot{\theta}\sin\omega t\cos\theta\right]$$

Now V of the system is due to (i) oscillation of the point of suspension and (ii) oscillation of the bob

$$\therefore \quad V = \tfrac{1}{2}kx^2 + mgl(1-\cos\theta) = \tfrac{1}{2}m\omega^2a^2\sin^2\omega t + mgl(1-\cos\theta);$$

so the Lagrangian

$$L = T - V = \tfrac{1}{2}m[l^2\dot{\theta}^2 - 2a\omega l\dot{\theta}\sin\omega t\cos\theta - mgl(1-\cos\theta)]$$

and the equation of motion is

$$\frac{d}{dt}\left(\frac{\partial L}{\partial \dot{\theta}}\right)-\frac{\partial L}{\partial \theta}=0$$

i.e. $$\frac{d}{dt}\left[\tfrac{1}{2}m\left(l^2 2\dot{\theta}-2a\omega l\sin\omega t\cos\theta\right)\right]-ma\omega l\dot{\theta}\sin\omega t\sin\theta+mgl\sin\theta=0$$

or $$l^2\ddot{\theta}-a\omega^2 l\cos\omega t\cos\theta+a\omega l\sin\omega t\sin\theta\dot{\theta}-a\omega l\dot{\theta}\sin\omega t\sin\theta+gl\sin\theta=0$$

or $$\ddot{\theta}+\frac{g}{l}\sin\theta=\frac{a\omega^2}{l}\cos\omega t\cos\theta.$$

For small angular displacement θ, the above equation reduces to

$$\ddot{\theta}+\frac{g}{l}\theta=\frac{a\omega^2}{l}cos\omega t.$$

25. (a) Using plane polar coordinates, find the Lagrangian of a particle moving under a central force; hence obtain the equation of motion.

(b) Show that this motion of the particle is equivalent to that in one dimension with a fictitious potential energy $V'=V+\dfrac{l^2}{2mr^2}$, where 'V' is the original central force potential, 'l' is the angular momentum and 'm' is the mass of the particle. **[C.U. (Hons.) 1999]**

(a) K.E. of the particle $T=\frac{1}{2}m\left(\dot{r}^2+r^2\dot{\theta}^2\right)$ and P.E. $=V(r)$, so that the Lagrangian is given by

$$L=T-V=\tfrac{1}{2}m(\dot{r}^2+r^2\dot{\theta}^2)-V(r)\quad\text{using } r,\theta \text{ as generalised coordinates} \tag{1}$$

Lagrange's equations of motion are written as

$$\frac{d}{dt}\left(\frac{\partial L}{\partial \dot{r}}\right)-\frac{\partial L}{\partial r}=0\quad\text{and}\quad\frac{d}{dt}\left(\frac{\partial L}{\partial \dot{\theta}}\right)-\frac{\partial L}{\partial \theta}=0$$

whence we have

$$\frac{d}{dt}(m\dot{r})-mr\dot{\theta}^2+\frac{\partial V}{\partial r}=0\quad\text{or}\quad m\ddot{r}-mr\dot{\theta}^2=-\frac{\partial V}{\partial r} \tag{2}$$

and

$$\frac{d}{dt}(mr^2\dot{\theta})=0\quad\text{or}\quad mr^2\dot{\theta}=l=\text{constant} \tag{3}$$

which are the equations of motion.

(b) From (3)

$$mr^2\dot{\theta}=l\text{ (angular momentum)}\quad\text{or}\quad\dot{\theta}=\frac{l}{mr^2} \tag{4}$$

Multiplying equation (2) by $\dot{r}$, we have

$$m\dot{r}\ddot{r}-mr\dot{r}\dot{\theta}^2=-\dot{r}\frac{\partial V}{\partial r}\quad\text{or}\quad m\dot{r}\ddot{r}-mr\dot{r}\frac{l^2}{m^2r^4}=-\dot{r}\frac{\partial V}{\partial r}\quad\text{(using 4)}$$

or $$\frac{d}{dt}\left(\frac{1}{2}m\dot{r}^2\right)+\frac{d}{dt}\left(\frac{l^2}{2mr^2}\right)=-\frac{dV}{dt}\quad\left[\because\frac{dV}{dt}=\frac{dV}{dr}\frac{dr}{dt}=\dot{r}\frac{dV}{dr}\right]$$

or $$\frac{d}{dt}\left(\frac{1}{2}m\dot{r}^2+\frac{l^2}{2mr^2}+V\right)=0$$

which shows that the motion of the particle is equivalent to that in one dimension with a fictitious P.E. $V' = V + \frac{l^2}{2mr^2}$.

26. A particle of mass m moves on a plane in the field of force given by $\vec{F} = -\hat{r}kr\cos\theta$ where k is a constant and $\hat{r}$ is the radial unit vector. Find the Lagrange's equations of motion.

[C.U. (Hons.) 2002]

Here P.E. $= V = -\int \vec{F}\cdot d\vec{r} = +\int_{\infty}^{r} \hat{r}kr\cos\theta \cdot d\vec{r} = \frac{1}{2}kr^2\cos\theta.$

K.E. $= T = \frac{1}{2}m(\dot{r}^2 + r^2\dot{\theta}^2)$.

$$\therefore \quad L = T - V = \frac{1}{2}m(\dot{r}^2 + r^2\dot{\theta}^2) - \frac{1}{2}kr^2\cos\theta,$$

so that Lagrange's equations *for a conservative system in terms of generalised coordinates r and θ* are written as

$$\frac{d}{dt}\left(\frac{\partial L}{\partial \dot{r}}\right) - \frac{\partial L}{\partial r} = 0 \tag{1}$$

and

$$\frac{d}{dt}\left(\frac{\partial L}{\partial \dot{\theta}}\right) - \frac{\partial L}{\partial \theta} = 0 \tag{2}$$

Now

$$\frac{\partial L}{\partial \dot{r}} = m\dot{r} \quad \text{and} \quad \frac{\partial L}{\partial r} = mr\dot{\theta}^2 - kr\cos\theta,$$

so that from (1) we may write

$$m\ddot{r} - mr\dot{\theta}^2 + kr\cos\theta = 0 \quad \text{or} \quad \ddot{r} - r\dot{\theta}^2 + \frac{k}{m}r\cos\theta = 0 \tag{3}$$

Again

$$\frac{\partial L}{\partial \dot{\theta}} = mr^2\dot{\theta} \quad \text{and} \quad \frac{\partial L}{\partial \theta} = \frac{kr^2}{2}\sin\theta,$$

so that from (2) we may write

$$mr^2\ddot{\theta} - \frac{kr^2}{2}\sin\theta = 0 \quad \text{or} \quad \ddot{\theta} - \frac{k}{2m}\sin\theta = 0 \tag{4}$$

Equations (3) and (4) are the Lagrange's equations for the given field, provided it is conservative.

However, *if the force field is a non-conservative one,* Lagrange's equations of motion in terms of r and θ are

$$\frac{d}{dt}\left(\frac{\partial T}{\partial \dot{r}}\right) - \frac{\partial T}{\partial r} = Q_r \quad \text{and} \quad \frac{d}{dt}\left(\frac{\partial T}{\partial \dot{\theta}}\right) - \frac{\partial T}{\partial \theta} = Q_\theta$$

($\because L = T - V = T$ as $V = 0$ in the given field)

Here $\vec{F} = -\hat{r}kr\cos\theta$.

$\therefore Q_r = -kr\cos\theta$ and $Q_\theta = 0$, so that Lagrange's equations of motion are given by

$$\frac{d}{dt}(m\dot{r}) - mr\dot{\theta}^2 = -kr\cos\theta \quad \text{and} \quad \frac{d}{dt}(mr^2\dot{\theta}) = 0$$

$$\text{or } m\ddot{r} - mr\dot{\theta}^2 + kr\cos\theta = 0 \quad \text{and} \quad \frac{d}{dt}(mr^2\dot{\theta}) = 0 \quad \text{i.e.} \quad mr^2\dot{\theta} = \text{constant}$$

27. A particle of mass m is constrained to move on the surface of a cylinder. The particle is subjected to a force directed towards the origin (mid-point of the cylinder) and is proportional to the distance of the particle from the origin. Construct the Lagrangian, Hamiltonian and Hamilton's equations of motion.

Force acting on the particle at P is $\vec{F} = -kr\hat{r}$, so that P.E. $= \frac{1}{2}kr^2 = \frac{1}{2}k(x^2+y^2+z^2)$.

In cylindrical coordinates $x = R\cos\phi$, $y = R\sin\phi$, ($R =$ radius of the cylinder).

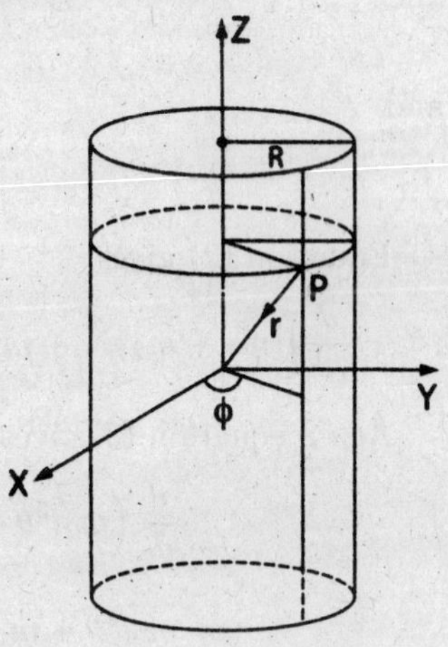

$\therefore x^2 + y^2 = R^2$, so that

$$\text{P.E.} = V = \tfrac{1}{2}k(R^2+z^2) \quad \text{and} \quad \text{K.E.} = T = \tfrac{1}{2}m(\dot{x}^2+\dot{y}^2+\dot{z}^2).$$

Now $\dot{x} = -R\sin\phi\dot{\phi}$, $\dot{y} = R\cos\phi\dot{\phi}$. $\therefore \dot{x}^2 + \dot{y}^2 = R^2\dot{\phi}^2$

$$\therefore T = \tfrac{1}{2}m\left(R^2\dot{\phi}^2 + \dot{z}^2\right) \qquad \therefore L = \tfrac{1}{2}m\left(R^2\dot{\phi}^2 + \dot{z}^2\right) - \tfrac{1}{2}k(R^2+z^2)$$

Now the Hamiltonian H is defined as

$$H = \sum_k p_k\dot{q}_k - L = p_\phi\dot{\phi} + p_z\dot{z} - L$$

But

$$p_\phi = \frac{\partial L}{\partial\dot{\phi}} = mR^2\dot{\phi} \quad \text{and} \quad p_z = \frac{\partial L}{\partial\dot{z}} = m\dot{z},$$

so that we can write

$$H = mR^2\dot{\phi}^2 + m\dot{z}^2 - \frac{1}{2}m\left(R^2\dot{\phi}^2 + \dot{z}^2\right) + \frac{1}{2}k(R^2+z^2)$$

$$= \frac{1}{2}mR^2\dot{\phi}^2 + \frac{1}{2}m\dot{z}^2 + \frac{1}{2}k(R^2+z^2) = \frac{1}{2}\frac{p_\phi^2}{mR^2} + \frac{p_z^2}{2m} + \frac{1}{2}k(R^2+z^2) \tag{1}$$

and Hamilton's equations of motion

$$\dot{\phi} = \frac{\partial H}{\partial p_\phi}, \quad \dot{p}_\phi = -\frac{\partial H}{\partial\phi}, \quad \dot{z} = \frac{\partial H}{\partial p_z}, \quad \dot{p}_z = -\frac{\partial H}{\partial z}$$

which can be calculated from (1) and may be put in simplified forms.

28. A simple pendulum consisting of mass m_2 at one end and with a mass m_1 at the point of support, moves in the vertical plane under gravity. Mass m_1 can move on a horizontal line lying in the plane in which m_2 moves. Construct the Lagrangian and obtain the equation of motion of the pendulum.

[C.U. (Hons.) 2004; cf. 2001]

If, at any instant, (x_2, y_2) be the coordinates of the mass m_2 and $(x_1, 0)$ that of m_1, then from the attached figure $x_2 = x_1 + l\sin\theta$, $y_2 = l\cos\theta$;

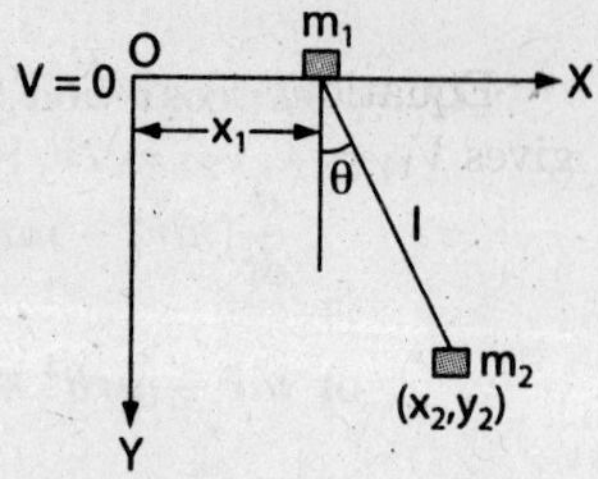

$\therefore\ \dot{x}_2 = \dot{x}_1 + l\cos\theta\dot{\theta}$ and $\dot{y}_2 = -l\sin\theta\dot{\theta}$, so that the K.E. of the system

$$T = \tfrac{1}{2}m_1\dot{x}_1^2 + \tfrac{1}{2}m_2(\dot{x}_2^2 + \dot{y}_2^2)$$
$$= \tfrac{1}{2}m_1\dot{x}_1^2 + \tfrac{1}{2}m_2(\dot{x}_1^2 + l^2\dot{\theta}^2 + 2\dot{x}_1\dot{\theta}l\cos\theta)$$
$$= \tfrac{1}{2}(m_1+m_2)\dot{x}_1^2 + \tfrac{1}{2}m_2l^2\dot{\theta}^2 + m_2\dot{x}_1\dot{\theta}l\cos\theta$$

Again P.E. $= V = -m_2gl\cos\theta$.

$$\therefore\quad L = \tfrac{1}{2}(m_1+m_2)\dot{x}_1^2 + \tfrac{1}{2}m_2l^2\dot{\theta}^2 + m_2\dot{x}_1\dot{\theta}l\cos\theta + m_2gl\cos\theta \qquad (1)$$

Here the generalised coordinates are x_1 and θ and so the Lagrange's equations of motion are

$$\frac{d}{dt}\left(\frac{\partial L}{\partial \dot{x}_1}\right) - \frac{\partial L}{\partial x_1} = 0 \qquad (2)$$

and

$$\frac{d}{dt}\left(\frac{\partial L}{\partial \dot{\theta}}\right) - \frac{\partial L}{\partial \theta} = 0 \qquad (3)$$

Equation (2) gives

$$\frac{d}{dt}\Big[(m_1+m_2)\dot{x}_1 + m_2\dot{\theta}l\cos\theta)\Big] = 0 \quad \text{or} \quad (m_1+m_2)\ddot{x}_1 + m_2l\ddot{\theta} = 0 \quad (\text{for small } \theta) \qquad (4)$$

Again equation (3) gives

$$\frac{d}{dt}\left(m_2l^2\dot{\theta} + m_2\dot{x}_1l\cos\theta\right) + m_2gl\sin\theta = 0$$

$$\text{or } m_2l^2\ddot{\theta} + m_2\ddot{x}_1l + m_2gl\theta = 0 \ (\text{for small } \theta) \quad \text{or} \quad \ddot{\theta} + \frac{\ddot{x}_1}{l} + \frac{g}{l}\theta = 0 \qquad (5)$$

From (4) we have

$$\ddot{x}_1 = -\frac{m_2l}{(m_1+m_2)}\ddot{\theta}$$

so that the equation (5) reduces to

$$\ddot{\theta} - \frac{m_2}{(m_1+m_2)}\ddot{\theta} + \frac{g}{l}\theta = 0 \quad \text{or} \quad \frac{m_1}{(m_1+m_2)}\ddot{\theta} + \frac{g}{l}\theta = 0 \quad \text{or} \quad \ddot{\theta} + \frac{(m_1+m_2)}{m_1}\frac{g}{l}\theta = 0$$

which is the required equation of motion.

29. Three particles of equal mass m move without friction in one dimension. Two of the particles are each connected to the third by a massless spring of spring constant k. Find the eigen frequencies and solutions for each mode of oscillation. Interpret the solution for each mode. **[C.U. (Hons.) 2004]**

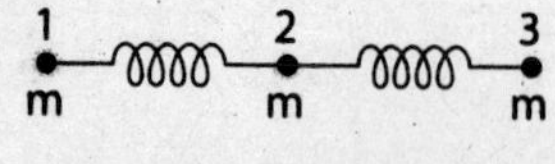

If, at any instant, q_1, q_2 and q_3 be the displacements of the 1st, 2nd and 3rd particles respectively, then

$$\text{K.E. of system} = T = \tfrac{1}{2}m\left(\dot{q}_1^2 + \dot{q}_2^2 + \dot{q}_3^2\right) \qquad (1)$$

and

$$\text{P.E.} = V = \tfrac{1}{2}k(q_1-q_2)^2 + \tfrac{1}{2}k(q_2-q_3)^2$$
$$= \tfrac{1}{2}k\left[q_1^2 + 2q_2^2 + q_3^2 - 2q_1q_2 - 2q_2q_3\right] \qquad (2)$$

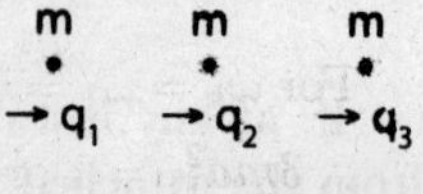

Equation (1) gives $T_{ij} = 0$ for $i \neq j$ i.e. $T_{ii} = T_{11} = T_{22} = T_{33} = m$. Again equation (2) gives $V_{11} = k, V_{22} = 2k, V_{33} = k, V_{12} = V_{21} = -k, V_{23} = V_{32} = -k$ i.e.

$$T = \begin{bmatrix} m & 0 & 0 \\ 0 & m & 0 \\ 0 & 0 & m \end{bmatrix} \quad \text{and} \quad V = \begin{bmatrix} k & -k & 0 \\ -k & 2k & -k \\ 0 & -k & k \end{bmatrix}$$

Secular determinant $|V_{ij} - \omega^2 T_{ij}| = 0$ gives

$$\begin{vmatrix} k-\omega^2 m & -k & 0 \\ -k & 2k-\omega^2 m & -k \\ 0 & -k & k-\omega^2 m \end{vmatrix} = 0$$

$$\text{or } (k-\omega^2 m)\left[(k-\omega^2 m)(2k-\omega^2 m) - k^2\right] - k.k(k-\omega^2 m) = 0$$

$$\text{or } (k-\omega^2 m)\left[(k-\omega^2 m)(2k-\omega^2 m) - 2k^2\right] = 0$$

$$\text{or } (k-\omega^2 m)\left[2k^2 - mk\omega^2 - 2mk\omega^2 + m^2\omega^4 - 2k^2\right] = 0$$

$$\text{or } (k-\omega^2 m)\omega^2(m^2\omega^2 - 3mk) = 0$$

whence we have $\omega_1 = 0, \omega_2 = \sqrt{\frac{k}{m}}, \omega_3 = \sqrt{\frac{3k}{m}}$

For frequency ω_k, the eigen value equation is

$$\begin{bmatrix} k-\omega_k^2 m & -k & 0 \\ -k & 2k-m\omega_k^2 & -k \\ 0 & -k & k-mk_k^2 \end{bmatrix} \begin{bmatrix} a_{1k} \\ a_{2k} \\ a_{3k} \end{bmatrix} = 0$$

For $\omega_k = \omega_1 = 0$ we may write

$$\begin{bmatrix} k & -k & 0 \\ -k & 2k & -k \\ 0 & -k & k \end{bmatrix} \begin{bmatrix} a_{1k} \\ a_{2k} \\ a_{3k} \end{bmatrix} = 0$$

which gives $a_{11} = a_{21} = a_{31}$

Normalisation condition is $\tilde{a}Ta = 1$ i.e.

$$\sum_{ij} T_{ij} a_{il} a_{jk} = \delta_{lk}$$

Here T_{ij} exists for $i = j$ and since $\delta_{lk} = 0$ for $l \neq k$, the normalisation condition reduces to

$$\sum T_{ii} a_{ik}^2 = 1 \quad \text{i.e.} \quad m\left(a_{1k}^2 + a_{2k}^2 + a_{3k}^2\right) = 1$$

For $\omega_k = \omega_1 = 0$ (i.e. $k = 1$), $a_{11} = a_{21} = a_{31}$.

$\therefore 3ma_{11}^2 = 1$ or $a_{11} = \frac{1}{\sqrt{3m}}$ i.e. $a_{11} = a_{21} = a_{31} = \frac{1}{\sqrt{3m}}$.

For $\omega_k = \omega_2 = \sqrt{\frac{k}{m}}$, the eigen value equation becomes

$$\begin{bmatrix} 0 & -k & 0 \\ -k & k & -k \\ 0 & -k & 0 \end{bmatrix} \begin{bmatrix} a_{12} \\ a_{22} \\ a_{32} \end{bmatrix} = 0$$

which gives $a_{22} = 0$ and $a_{12} = -a_{32}$.

So the normalisation condition gives

$$\sum T_{ii} a_{ik}^2 = 1 \quad \text{i.e.} \quad m(a_{12}^2 + a_{32}^2) = 1 \quad \text{or} \quad 2ma_{12}^2 = 1 \quad \text{i.e.} \quad a_{12} = \frac{1}{\sqrt{2m}}$$

$$\therefore \quad a_{22} = 0 \quad \text{and} \quad a_{12} = -a_{32} = \frac{1}{\sqrt{2m}}$$

Again for $\omega_k = \omega_3 = \sqrt{\frac{3k}{m}}$, the eigen value equation becomes

$$\begin{bmatrix} -2k & -k & 0 \\ -k & -k & -k \\ 0 & -k & -2k \end{bmatrix} \begin{bmatrix} a_{13} \\ a_{23} \\ a_{33} \end{bmatrix} = 0$$

whence we have $2a_{13} + a_{23} = 0$ and $a_{23} + 2a_{33} = 0$ i.e. $a_{23} = -2a_{13}$ and $a_{33} = -\frac{a_{23}}{2} = a_{13}$.

So normalisation condition gives

$$m\left(a_{13}^2 + a_{23}^2 + a_{33}^2\right) = 1 \quad \text{or} \quad m\left(a_{13}^2 + 4a_{13}^2 + a_{13}^2\right) = 1 \quad \text{or} \quad a_{13} = \frac{1}{\sqrt{6m}}$$

$$\therefore a_{13} = a_{33} = \frac{1}{\sqrt{6m}} \quad \text{and} \quad a_{23} = -\frac{2}{\sqrt{6m}}$$

Thus for $\omega = 0$, •→ •→ •→, all the three particles move in the same phase with same amplitude.

For $\omega = \sqrt{\frac{k}{m}}$, ←• • •→, the middle one is at rest and the two end particles move in opposite phase with equal amplitude.

For $\omega = \sqrt{\frac{3K}{m}}$, •→←••→, the two end particles are in phase with same amplitude, while the middle one is in opposite phase with double amplitude.

30. Write down the Lagrangian of a coplanar double pendulum with equal masses and equal lengths of strings. Expand it to quadratic order in the coordinates neglecting the second order of their difference. Find the frequencies of the normal modes of small oscillations of the system. **[C.U. (Hons.) 2006]**

From the figure

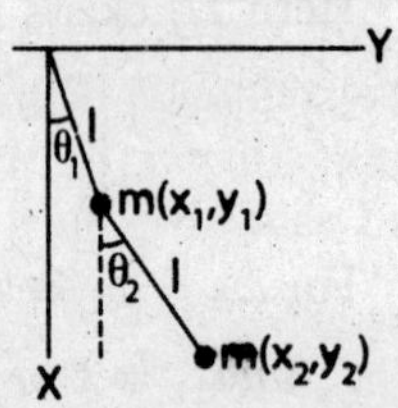

$$x_1 = l\cos\theta_1, \; y_1 = l\sin\theta_1; \qquad \therefore \dot{x}_1 = -l\sin\theta_1\dot{\theta}_1, \; \dot{y}_1 = l\cos\theta_1\dot{\theta}_1$$

$$x_2 = l\cos\theta_1 + l\cos\theta_2 \qquad \therefore \dot{x}_2 = -l\sin\theta_1\dot{\theta}_1 - l\sin\theta_2\dot{\theta}_2$$

$$y_2 = l\sin\theta_1 + l\sin\theta_2 \qquad \dot{y}_2 = l\cos\theta_1\dot{\theta}_1 + l\cos\theta_2\dot{\theta}_2$$

$$\text{So K.E.} = T = \tfrac{1}{2}m\left(\dot{x}_1^2 + \dot{y}_1^2\right) + \tfrac{1}{2}m\left(\dot{x}_2^2 + \dot{y}_2^2\right)$$

$$= \tfrac{1}{2}ml^2\dot{\theta}_1^2 + \tfrac{1}{2}m\left[l^2\dot{\theta}_1^2 + l^2\dot{\theta}_2^2 + 2l^2\dot{\theta}_1\dot{\theta}_2\cos(\theta_1 \sim \theta_2)\right]$$

$$= \tfrac{1}{2}ml^2\left[2\dot{\theta}_1^2 + \dot{\theta}_2^2 + 2\dot{\theta}_1\dot{\theta}_2\cos(\theta_1 \sim \theta_2)\right]$$

and P.E. $= V = -mgl\cos\theta_1 - mgl(\cos\theta_1 + \cos\theta_2)$.

$\therefore$ Lagrangian $= L = ml^2\dot{\theta}_1^2 + \tfrac{1}{2}ml^2\dot{\theta}_2^2 + ml^2\dot{\theta}_1\dot{\theta}_2\cos(\theta_1 \sim \theta_2) + mgl(2\cos\theta_1 + \cos\theta_2)$.

For small θ_1 and θ_2, expanding $\cos\theta_1, \cos\theta_2$ and $\cos(\theta_1 \sim \theta_2)$ to quadratic order of magnitude and neglecting the second order term involving $(\theta_1 \sim \theta_2)$, we have

$$\cos\theta_1 = 1 - \frac{\theta_1^2}{2}, \quad \cos\theta_2 = 1 - \frac{\theta_2^2}{2} \quad \text{and} \quad \cos(\theta_1 \sim \theta_2) \simeq 1$$

$$\therefore \quad L = ml^2\dot{\theta}_1^2 + \tfrac{1}{2}ml^2\dot{\theta}_2^2 + ml^2\dot{\theta}_1\dot{\theta}_2 + mgl\left(3 - \theta_1^2 - \frac{\theta_2^2}{2}\right)$$

Since the system has two (2) degrees of freedom corresponding to the coordinates θ_1 and θ_2, Lagrange's equations are written as

$$\frac{d}{dt}\left(\frac{\partial L}{\partial \dot{\theta}_1}\right) - \frac{\partial L}{\partial \theta_1} = 0 \tag{1}$$

$$\text{and} \quad \frac{d}{dt}\left(\frac{\partial L}{\partial \dot{\theta}_2}\right) - \frac{\partial L}{\partial \theta_2} = 0 \tag{2}$$

From (1)

$$\frac{d}{dt}\left[2ml^2\dot{\theta}_1 + ml^2\dot{\theta}_2\right] + mgl2\theta_1 = 0$$

$$\text{or } 2ml^2\ddot{\theta}_1 + ml^2\ddot{\theta}_2 + mgl2\theta_1 = 0$$

$$\text{or } 2l\ddot{\theta}_1 + l\ddot{\theta}_2 + 2g\theta_1 = 0 \tag{3}$$

From (2)

$$\frac{d}{dt}\left[ml^2\dot{\theta}_2 + ml^2\dot{\theta}_1\right] + mgl\theta_2 = 0 \quad \text{or} \quad l\ddot{\theta}_2 + l\ddot{\theta}_1 + g\theta_2 = 0 \tag{4}$$

Taking $\theta_k = a_k e^{i\omega t}$ as the trial solution, we have from (3)

$$-2l\omega^2 a_1 - l\omega^2 a_2 + 2ga_1 = 0 \quad \text{or} \quad 2(g - l\omega^2)a_1 - l\omega^2 a_2 = 0 \tag{5}$$

and from (4)

$$-l\omega^2 a_2 - l\omega^2 a_1 + ga_2 = 0 \quad \text{or} \quad -l\omega^2 a_1 + (g - l\omega^2)a_2 = 0 \tag{6}$$

For non-trivial solution

$$\begin{vmatrix} 2(g - l\omega^2) & -l\omega^2 \\ -l\omega^2 & (g - l\omega^2) \end{vmatrix} = 0 \quad \text{or} \quad 2(g - l\omega^2)^2 - l^2\omega^4 = 0$$

$$\text{or} \quad 2(g^2 + l^2\omega^4 - 2gl\omega^2) - l^2\omega^4 = 0$$

$$\text{or} \quad l^2\omega^4 - 4gl\omega^2 + 2g^2 = 0$$

whence we have

$$\omega^2 = \frac{4gl \pm \sqrt{16g^2l^2 - 8g^2l^2}}{2l^2} = \frac{4gl \pm \sqrt{8g^2l^2}}{2l^2} = (2 \pm \sqrt{2})\frac{g}{l}$$

The normal frequencies are

$$f_1 = \frac{\omega_1}{2\pi} = \frac{1}{2\pi}\sqrt{\frac{(2+\sqrt{2})g}{l}} \quad \text{and} \quad f_2 = \frac{\omega_2}{2\pi} = \frac{1}{2\pi}\sqrt{\frac{(2-\sqrt{2})g}{l}}$$

3.29 Dynamics of a System of Particles

Here we shall consider the motion of a system of particles.

(a) Centre of Mass

We consider a system of n particles of masses $m_1, m_2, m_3, \cdots, m_n$, situated at points P_1, P_2, $P_3, \cdots, P_n$. Let the position vectors of these points relative to some assigned point O be $\vec{r_1}$, $\vec{r_2}, \vec{r_3}, \cdots, \vec{r_n}$.

$\therefore$ the linear moment of the system is

$$\sum_{i=1}^{n} m_i\vec{r_i},$$

where m_i is the mass of the ith particle and $\vec{r_i}$ is the position vector of the ith particle relative to the point O. The centre of mass of the system of particles is defined as that point with respect to which the linear moment of the system vanishes. To establish this, we consider a point C, the position vector of which relative to a fixed point O is given by

$$\overrightarrow{OC} = \frac{\sum_{i=1}^{n} m_i\vec{r_i}}{\sum_{i=1}^{n} m_i} \tag{3.29.1}$$

N.B. The linear moment defined above is also known as the *first moment of mass.*

$\therefore$ the position vector of the ith particle P_i relative to C is $\vec{r_i} - \overrightarrow{OC}$ (Fig. 3.29.1). Hence the linear moment of the system of particles relative to C is

$$\sum_{i=1}^{n} m_i(\vec{r_i} - \overrightarrow{OC}) = \sum_{i=1}^{n} m_i\vec{r_i} - \sum_{i=1}^{n} m_i\overrightarrow{OC}$$

$$= \sum_{i=1}^{n} m_i\vec{r_i} - \sum_{i=1}^{n} m_i\vec{r_i} = 0 \quad \text{[from equation (3.29.1)]}$$

Thus C denotes the mass centre and equation (3.29.1) gives the position vector of the mass centre relative to an arbitrary origin.

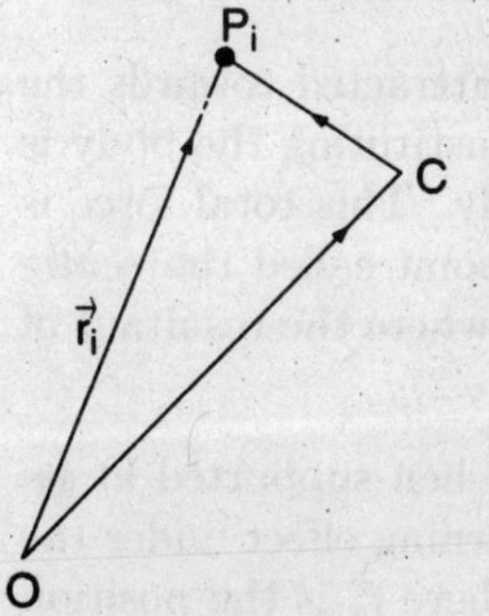

Fig. 3.29.1

We are still left to establish the uniqueness of the centre of mass, i.e., we shall prove that there exists only one centre of mass. Let us now assume that there are two centres of mass C and C', and let the position vectors of the particles relative to these centres of mass be $\vec{r_1}$, $\vec{r_2}, \cdots, \vec{r_n}$ and $\vec{r'_1}, \vec{r'_2}, \cdots, \vec{r'_n}$.

$$\therefore \sum_{i=1}^{n} m_i \vec{r_i} = 0 \quad \text{and} \quad \sum_{i=1}^{n} m_i \vec{r'_i} = 0 \tag{3.29.2}$$

But $\vec{r_i} = \vec{r'_i} + \overrightarrow{CC'}$.

When this value of r_i is substituted in equation (3.29.2) we get

$$\sum_{i=1}^{n} m_i \overrightarrow{CC'} = 0. \quad \therefore \overrightarrow{CC'} = 0$$

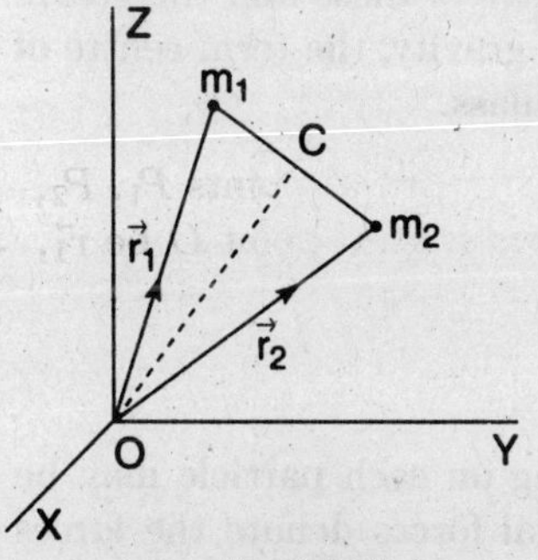

Fig. 3.29.2

This shows that C and C' coincide. Thus the uniqueness of the centre of mass is established.

For the system having only two particles with masses m_1 and m_2, the centre of mass lies on the line joining them and divides the line in the ratio $m_2 : m_1$ (Fig. 3.29.2). The position of the mass centre from the origin in this case is given by

$$\overrightarrow{OC} = \frac{m_1 \vec{r_1} + m_2 \vec{r_2}}{m_1 + m_2} \tag{3.29.3}$$

We can express equation (3.29.3) in scalar form. The co-ordinates of the centre of mass of a system of particles in Cartesian system expressed in scalar form are given by

$$\bar{x} = \frac{\sum_{i=1}^{n} m_i x_i}{\sum_{i=1}^{n} m_i}, \quad \bar{y} = \frac{\sum_{i=1}^{n} m_i y_i}{\sum_{i=1}^{n} m_i}, \quad \bar{z} = \frac{\sum_{i=1}^{n} m_i z_i}{\sum_{i=1}^{n} m_i}$$

where x_i, y_i, z_i are the three components of r_i and $\bar{x}$, $\bar{y}$, $\bar{z}$ are the components of OC.

For a continuous distribution of matter instead of a system of particles, we may write

$$\bar{x} = \frac{\int x \, dm}{M}, \quad \bar{y} = \frac{\int y \, dm}{M}, \quad \bar{z} = \frac{\int z \, dm}{M},$$

where dm is an elementary mass; the position of the mass is specified by the co-ordinates x, y, z. M denotes the total mass.

Writing $\overrightarrow{OC} = \vec{R}$, the velocity $\vec{V}$ of the centre of mass is obtained from (3.29.3) and is given by

$$\vec{V} = \dot{\vec{R}} = \frac{m_1 \dot{\vec{r}}_1 + m_2 \dot{\vec{r}}_2}{m_1 + m_2} = \frac{m_1 \vec{V_1} + m_2 \vec{V_2}}{m_1 + m_2} \tag{3.29.4}$$

Again the velocity of the particle 1 w.r.t. the particle 2 is given by

$$\vec{V} = \dot{\vec{r}} = \dot{\vec{r}}_1 - \dot{\vec{r}}_2 = \vec{V_1} - \vec{V_2} \tag{3.29.5}$$

(b) Centre of Gravity

When a body is near the earth's surface, each particle of the body is attracted towards the centre of the earth. The sum of the attractive forces on the particles constituting the body is equal to the total attractive force which the earth exercises on the body. This total force is equal to the weight of the body and it acts vertically downwards at a point called the *centre of gravity* of the body. Thus the *centre of gravity* is defined as the point where the resultant of the weights of all particles constituting the body acts.

It follows from the definition of the centre of gravity that a body when supported at its centre of gravity remains in equilibrium, i.e., the body experiences no turning effect under the action of the force due to gravity. Hence we may write $\Sigma m_i g_i \vec{r_i} = 0$, where $\vec{r_i}$ is the position vector of the mass relative to the centre of gravity. If the value of acceleration due to gravity g_i remains constant, the above expression may be written in the form $\Sigma m_i \vec{r_i} = 0$. This equation is same as the equation (3.29.2) obtained by considering the position vector of the mass m_i relative to the centre of mass showing that the positions of the centre of mass and the centre of gravity are the same. In free space where there is no influence of gravity, the term centre of gravity carries no sense at all and hence we use the term centre of mass.

3.30 Motion of the Centre of mass

As in Art. 3.29 we consider a system of particles. The force acting on each particle may be divided into two groups—external and internal forces. The internal forces denote the forces exerted by other parts of the same system on the particle; the external forces denote the forces on the particle exerted by the bodies which are not parts of the system. Let $\vec{F_i}$ be the resultant of the external forces acting on the ith particle. The internal force acting on ith particle due to another particle, say jth particle is denoted by $\vec{F_{ij}}$ (Fig. 3.30.1). We can apply Newton's second law for each particle. Hence we may write for ith particle

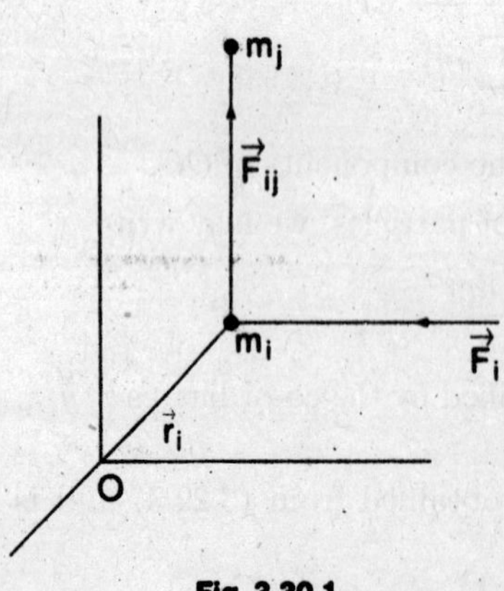

Fig. 3.30.1

$$\vec{F_i} + \sum_{\substack{j=1 \\ j \neq i}}^{n} \vec{F_{ij}} = m_i \frac{d^2\vec{r_i}}{dt^2} \tag{3.30.1}$$

From Newton's third law of motion, it is easily seen that $\vec{F_{ij}} = -\vec{F_{ji}}$ because each of the forces $\vec{F_{ij}}$ is equal and opposite to the force $\vec{F_{ji}}$. Considering n number of particles we get

$$\sum_{i=1}^{n} \vec{F_i} + \sum_{i=1}^{n} \sum_{\substack{j=1 \\ j \neq i}}^{n} \vec{F_{ij}} = \sum_{i=1}^{n} m_i \frac{d^2\vec{r_i}}{dt^2} \tag{3.30.2}$$

The first term of the L.H.S. of the above expression gives the resultant external force acting on the system of particles. Let this force be denoted by $\vec{F}$. The second term of L.H.S. gives the resultant of internal forces which vanishes as the sum of the pair of forces $\vec{F_{ij}}$ and $\vec{F_{ji}}$ is zero.

$\therefore$ The above expression may be written in the form,

$$\vec{F} = \sum_{i=1}^{n} m_i \frac{d^2\vec{r_i}}{dt^2} = \frac{d^2}{dt^2}\sum_{i=1}^{n} m_i\vec{r_i} \tag{3.30.3}$$

$$\text{or} \quad \vec{F} = \sum_{i=1}^{n} m_i \frac{d^2\overrightarrow{OC}}{dt^2} \quad \text{[from the equation (3.29.1)]}$$

$$= M\frac{d^2\overrightarrow{OC}}{dt^2}, \tag{3.30.4}$$

where $\sum_{i=1}^{n} m_i = M$, the total mass of the system of particles.

Thus the motion of the centre of mass C of a system of particles relative to an inertial frame of reference is the same as that of a single particle of mass equal to the total mass of the system under the action of the external force $\vec{F}$ which is equal to the vector sum of the external forces acting on the system.

3.31 The two body problem

We consider two particles of masses m_1 and m_2 separated by a distance r and acted on by external forces F_1, F_2 and internal forces F_{12}, F_{21} respectively. The internal forces satisfy Newton's third law given by

$$\vec{F}_{12} = -\vec{F}_{21}$$

The equations of motion for the two particles are

$$m_1\frac{d^2\vec{r_1}}{dt^2} = \vec{F}_{12} + \vec{F}_1 \tag{3.31.1}$$

$$m_2\frac{d^2\vec{r_2}}{dt^2} = \vec{F}_{21} + \vec{F}_2 \tag{3.31.2}$$

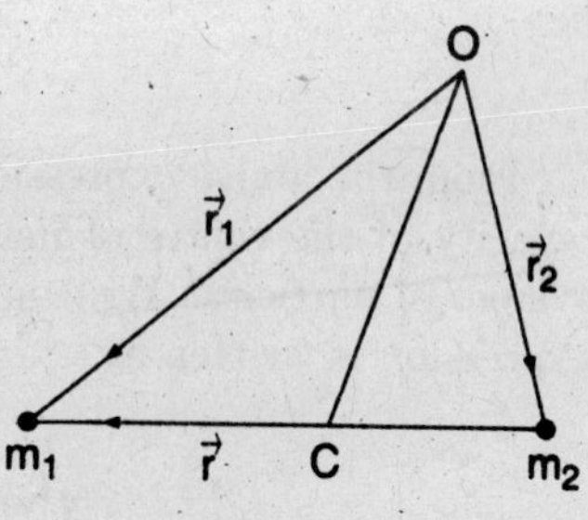

Fig. 3.31.1

Multiplying (3.31.1) by m_2 and (3.31.2) by m_1 and then subtracting we get

$$m_1m_2\frac{d^2}{dt^2}(\vec{r_1}-\vec{r_2}) = (m_2\vec{F}_{12} - m_1\vec{F}_{21}) + (m_2\vec{F}_1 - m_1\vec{F}_2)$$

$$\text{or} \quad m_1m_2\frac{d^2}{dt^2}(\vec{r_1}-\vec{r_2}) = (m_1+m_2)\vec{F}_{12} + (m_2\vec{F}_1 - m_1\vec{F}_2)$$

$$\text{or} \quad m_1m_2\frac{d^2}{dt^2}(\vec{r_1}-\vec{r_2}) = (m_1+m_2)\vec{F}_{12} + m_1m_2\left(\frac{\vec{F}_1}{m_1} - \frac{\vec{F}_2}{m_2}\right)$$

$$\text{or} \quad \mu\frac{d^2}{dt^2}(\vec{r_1}-\vec{r_2}) = \vec{F}_{12} + \mu\left(\frac{\vec{F}_1}{m_1} - \frac{\vec{F}_2}{m_2}\right) \quad \text{where } \mu = \frac{m_1m_2}{m_1+m_2}$$

If the motion of the two particles is only due to internal forces arising from the interaction between the two particles the equation is reduced to

$$\mu \frac{d^2}{dt^2}\vec{r} = \vec{F}_{12} \quad [\because \vec{r_1} - \vec{r_2} = \vec{r}] \tag{3.31.3}$$

Here $\vec{r}$ denotes the relative co-ordinate of m_1 with respect to m_2. Equation (3.31.3) gives the equation of motion of a particle of mass μ acted on by the internal force $\vec{F}_{12}$ which the particle of mass m_2 exerts on the particle of mass m_1. Thus the motion of the particle of mass m_1, when viewed from particle of mass m_2 is the same as if the particle of mass m_2 is fixed and the particle in motion has a mass μ. μ is called the *reduced mass.* In case of the motion of a planet round the sun also, we can treat it as a two-body problem. This two body problem can be considered as a single body problem by assuming the sun to be fixed and the planet in motion having a reduced mass.

3.32 Linear Momentum of a System of Particles

From equation (3.29.1) we may write

$$\overrightarrow{OC} = \frac{\sum_{i=1}^{n} m_i \vec{r_i}}{M}$$

Hence the velocity of the centre of mass is given by

$$\vec{V} = \frac{\sum_{i=1}^{n} m_i \vec{v_i}}{M} \quad \text{or} \quad M\vec{V} = \sum_{i=1}^{n} m_i \vec{v_i} = \sum_{i=1}^{n} \vec{p_i}$$

From the above expression it follows that if a fictitious particle of mass M moves with the velocity of the centre of mass, its linear momentum is equal to the linear momentum of the system of particles. Writing $\vec{P}$ for the linear momentum of the system of particles, the above expression is written as

$$\vec{P} = M\vec{V} \quad \text{or} \quad \frac{d\vec{P}}{dt} = M\frac{d\vec{V}}{dt} = M\frac{d^2\overrightarrow{OC}}{dt^2} = \vec{F}$$

Thus the rate of change of momentum of the system of particles is equal to the resultant of the external forces acting on the system.

3.33 Angular Momentum of a System of Particles

The angular momentum $\vec{l}$ of the ith particle of a system of particles relative to an arbitrary fixed point O as origin is defined to be equal to $\vec{r_i} \times \vec{p_i}$ (Fig. 3.33.1).

Hence the total angular momentum of a system of particles referred to the origin O is given by

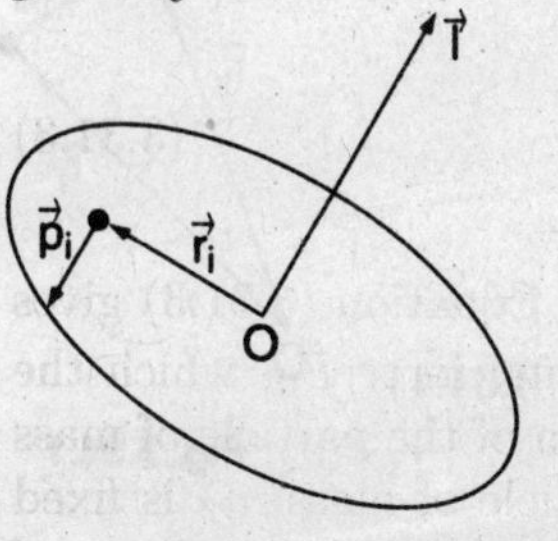

$$\vec{L} = \sum_{i=1}^{n} \vec{r_i} \times \vec{p_i} \tag{3.33.1}$$

The above expression may be written in the form

$$\vec{L} = \sum_{i=1}^{n} (\vec{r_i} - \overrightarrow{OC}) \times \vec{p_i} + \sum_{i=1}^{n} \overrightarrow{OC} \times \vec{p_i},$$

where $\overrightarrow{OC}$ denotes the position vector of the mass centre relative to the origin O.

Fig. 3.33.1

The above expression may also be written in the form

$$\vec{L} = \sum_{i=1}^{n} \vec{r'}_i \times \vec{p_i} + \sum \overrightarrow{OC} \times \vec{p_i} = \vec{L'} + \overrightarrow{OC} \times \vec{P} \tag{3.33.2}$$

where $\vec{L'}$ = total angular momentum about the centre of mass [$\vec{r'}_i$ denotes the position vector of ith particle relative to centre of mass].

And $\vec{P} = \Sigma \vec{p_i}$ is the total linear momentum of the system.

$\overrightarrow{OC} \times \vec{P}$ is the angular momentum of the centre of mass about the origin O. It is to be noted that the term $\overrightarrow{OC} \times \vec{P}$, depends on the choice of the origin whereas the term $\vec{L'}$ does not.

3.34 External and Internal Torques

Differentiating the expression (3.33.1), we have

$$\begin{aligned}\frac{d\vec{L}}{dt} &= \frac{d}{dt}\sum_{i=1}^{n} \vec{r_i} \times \vec{p_i} = \sum_{i=1}^{n}\left(\frac{d\vec{r_i}}{dt} \times \vec{p_i} + \vec{r_i} \times \frac{d\vec{p_i}}{dt}\right) \\ &= \sum_{i=1}^{n}\left(\vec{v_i} \times m_i\vec{v_i} + \vec{r_i} \times \frac{d\vec{p_i}}{dt}\right) = \sum_{i=1}^{n} \vec{r_i} \times \frac{d\vec{p_i}}{dt} \qquad \left[\because \sum \vec{v_i} \times m_i\vec{v_i} = 0\right] \\ &= \sum_{i=1}^{n} \vec{r_i} \times \vec{F_i} = \vec{N}_T \end{aligned} \tag{3.34.1}$$

where $\vec{N}_T$ denotes the total torque about O.

Now the internal forces $= \sum_{\substack{j=1 \\ i \neq j}}^{n} \vec{F}_{ij}$, gives the sum of the forces on the particle i from all other particles j.

Calling $\vec{N}$ and $\vec{N'}$ the external and internal torques respectively, the internal torque may be written as

$$\vec{N'} = \sum_{i=1}^{n} \sum_{\substack{j=1 \\ i \neq j}}^{n} \vec{r_i} \times \vec{F}_{ij}$$

In the right-hand side of the above expression, we get pair of terms like $\left(\vec{r_k} \times \vec{F_{kl}} + \vec{r_l} \times \vec{F_{lk}}\right)$ which may also be written in the form

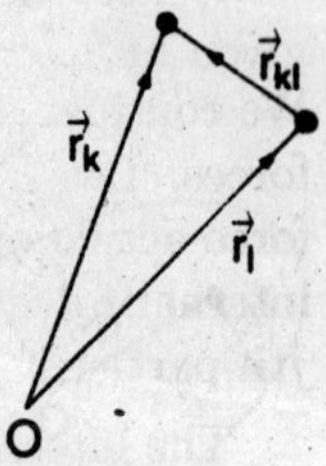

Fig. 3.34.1

$$\left(\vec{r_k} \times \vec{F_{kl}} - \vec{r_l} \times \vec{F_{kl}}\right) = (\vec{r_k} - \vec{r_l}) \times \vec{F_{kl}} \quad \left[\because \vec{F_{kl}} = -\vec{F_{lk}}\right]$$

$$= \vec{r_{kl}} \times \vec{F_{kl}},$$

where $\vec{r_{kl}}$ denotes the distance between the kth and lth particles (Fig. 3.34.1). As $\vec{r_{kl}}$ and $\vec{F_{kl}}$ are in the same direction $\vec{r_{kl}} \times \vec{F_{kl}} = 0$. $\quad \therefore \vec{N}' = 0$.

Hence from relation (3.34.1), we have

$$\frac{d\vec{L}}{dt} = \vec{N} \tag{3.34.2}$$

Thus from the above relation it follows that the time rate of change of angular momentum relative to an arbitrary fixed point is equal to the external torque or the moment of the external force on the system about the same point.

If the external torque $\vec{N} = 0$, then $\dfrac{d\vec{L}}{dt} = 0$ or $\vec{L}$ is constant, i.e., the angular momentum of the system of particles remains constant in absence of external torque. This is called the Law of Conservation of Angular Momentum.

3.35 Kinetic Energy of a System of Particles

According to the definition the kinetic energy of the system of particles is given by

$$T = \frac{1}{2}\sum_{i=1}^{n} m_i \vec{v_i}^2$$

Now in Fig. 3.35.1 $GP = \vec{v_i'} = \vec{v_i} - \vec{V}$, where $\vec{V}$ is the velocity of the centre of mass C relative to a fixed point O; $\vec{v_i'}$ is the velocity of the ith particle P relative to mass centre and $\vec{v_i}$ is the velocity of the ith particle relative to the fixed point O.

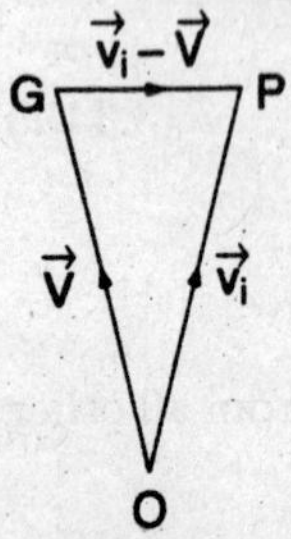

Fig. 3.35.1

$$T = \frac{1}{2}\sum_{i=1}^{n} m_i \left(\vec{v_i'} + \vec{V}\right)^2 = \frac{1}{2}\sum_{i=1}^{n} m_i \left(\vec{v_i}'^2 + \vec{V}^2 + 2\vec{v_i'}\vec{V}\right)$$

Hence

$$T = \frac{1}{2}M\vec{V}^2 + \frac{1}{2}\sum_{i=1}^{n} m_i \vec{v_i}'^2 \quad \left[\because \sum m_i \vec{v'}_i \vec{V} = \vec{V}\frac{d}{dt}\sum m_i \vec{r_i'} = 0\right]$$

In the right-hand side of the above expression the first term indicates the kinetic energy of the mass centre and the second term the kinetic energy of all particles in the system relative to the mass centre.

3.36 Potential energy of a system of particles

We consider a system of particles. The forces acting on each particle are external and internal forces. Let $\vec{F}_i$ be the resultant of the external forces acting on the ith particle. The internal force acting on the ith particle due to another particle, say jth particle is denoted by $\vec{F}_{ij}$. The internal force between the ith and jth particles acting along the line joining the ith and the jth particles is given by $\vec{F}_{ij} = F_{ij}\vec{r}_{ij}$.

The internal force is a central force as it acts along the line joining the ith and jth particles, $\vec{r_{ij}}$ being the distance between the two particles. As the internal force is a central force and it is derivable from an internal potential V_{ij},

$$\therefore \quad \vec{F}_{ij} = -\vec{\nabla}_i V_{ij}$$

$$\vec{F}_{ji} = -\vec{\nabla}_j V_{ij} = -\vec{F}_{ij}$$

$$\therefore \quad \vec{F}_{ij} - (-\vec{F}_{ij}) = -(\vec{\nabla}_i - \vec{\nabla}_j)V_{ij}$$

$$\text{or} \quad \vec{F}_{ij} = -\frac{1}{2}\vec{\nabla}_{ij} V_{ij} \tag{3.36.1}$$

$\vec{\nabla}_{ij}$ denotes the gradient with respect to $\vec{r}_{ij}$.

We shall now calculate the work done due to both external and internal forces due to the displacement of the system from an initial configuration 1 to a final configuration 2.

Work done due to external forces acting on the system of particles is given by

$$W_1 = \sum_i \int_1^2 \vec{F}_i \cdot d\vec{r}_i = \sum_i \int_1^2 \vec{\nabla}_i V_i' d\vec{r}_i = -\left[\sum_i V_i'\right]_1^2.$$

Here V_i' is the potential derived from the external force $\vec{F}_i$.

Considering the sum of internal forces acting over all pairs of particles, the work done

$$W_2 = \sum_{\substack{i,j \\ i\neq j}} \int_1^2 \vec{F}_{ij} \cdot d\vec{r}_i$$

This expression may be written as the sum over the pairs of particles. Considering pairs of particles we may write the work done given by

$$W_2 = \sum_{\substack{i,j \\ i\neq j}} \int_1^2 \left(\vec{F}_{ij} \cdot d\vec{r}_i + \vec{F}_{ji} \cdot d\vec{r}_j\right) = \sum_{\substack{i,j \\ i\neq j}} \int_1^2 \left(\vec{F}_{ij} \cdot d\vec{r}_i - \vec{F}_{ij} \cdot d\vec{r}_j\right)$$

$$= \sum_{\substack{i,j \\ i\neq j}} \int_1^2 \vec{F}_{ij} \cdot (d\vec{r}_i - d\vec{r}_j) = -\frac{1}{2}\sum_{\substack{i,j \\ i\neq j}} \int_1^2 \vec{\nabla}_{ij} V_{ij} \cdot d\vec{r}_{ij}$$

[From equation 3.36.1]

$$= -\frac{1}{2}\left[\sum_{\substack{i,j \\ i\neq j}} V_{ij}\right]_1^2$$

Hence the total work done or the total potential energy function for the whole system is given by

$$W_1 + W_2 = -\left[\sum_i V_i' + \frac{1}{2}\sum_{\substack{i,j\\ i\neq j}} V_{ij}\right]_1^2 = -\Big[V\Big]_1^2 = -[V_2 - V_1],$$

where V is the total potential energy function of the whole system and is given by

$$V = \left[\sum_i V_i' + \frac{1}{2}\sum_{\substack{i,j\\ i\neq j}} V_{ij}\right].$$

As in the case of a single particle, we calculate the work done by all forces in moving the system from an initial configuration 1, to a final configuration 2.

$$\text{The work done} = \sum_i \int_1^2 \vec{F_i}\cdot d\vec{r_i} = \sum_i \int_1^2 m_i\vec{v_i}\cdot \dot{\vec{v_i}}dt$$

$$= \sum_i \int_1^2 \frac{d}{dt}\left(\frac{1}{2}m_i\vec{v}_i^2\right) dt = T_2 - T_1, \text{ where } T = \frac{1}{2}\sum m_i\vec{v}_i^2$$

$$= \text{Kinetic energy of the system.}$$

$$\therefore\ -(V_2 - V_1) = T_2 - T_1 \quad \text{or} \quad V_1 + T_1 = V_2 + T_2.$$

Thus the total energy $T + V$ i.e., the sum of kinetic and potential energies is conserved provided both the internal and external forces are conservative.

3.37 Inertial and Non-inertial frames of Reference

Motion of a body has no meaning unless it is described with respect to some well-defined co-ordinates system known as a *frame of reference.* There are two types of reference frames.

(i) The frames with respect to which a body, not acted upon by an external force, is un-accelerated, i.e., is at rest or moving with constant linear velocity are called *inertial* frames.

In absence of any external force acting on a body having position co-ordinates (x, y, z) with reference to any co-ordinate system, we may write

$$m\frac{d^2x}{dt^2} = 0, \quad m\frac{d^2y}{dt^2} = 0, \quad m\frac{d^2z}{dt^2} = 0,$$

$$\text{or} \quad \frac{d^2x}{dt^2} = 0, \quad \frac{d^2y}{dt^2} = 0, \quad \frac{d^2z}{dt^2} = 0,$$

whence we have $\frac{dx}{dt} = u_x = \text{constant}$, $\frac{dy}{dt} = u_y = \text{constant}$, $\frac{dz}{dt} = u_z = \text{constant}$. From above we see that the components of velocity are constant, i.e., we may say that without the application of an external force, a body in motion continues its motion with uniform velocity in a straight line which is Newton's first law. We may therefore say that an inertial frame is one in which law of inertia or Newton's first law is valid. Any frame of reference moving with constant velocity relative to an inertial frame, is also inertial, since acceleration of a body in both frames is zero, while velocity of the body is different but uniform (See Art. 3.38).

(ii) The frames w.r.t. which a body, not acted upon by an external force, is accelerated are called *non-inertial* frames. In a non-inertial frame of reference which is accelerated w.r.t. an inertial frame, acceleration of a particle is partly due to the external force and partly due to the acceleration of the frame itself. Its acceleration may be translational or rotational or both. Forces in a non-inertial frame which arise out of the acceleration of the frame itself, are often called *fictitious* forces or *pseudo* forces.

From experiments, we know that a frame of reference fixed in stars is an inertial frame. A co-ordinate system fixed in earth is not an inertial frame, since earth rotates about its axis and also about the sun. In fact every rotating system is intrinsically non-inertial.

To describe the motion of a body, we should always choose inertial frames, for if we choose non-inertial frames, these involve different forces and hence it is very complicated to formulate the laws common to all observers. When mechanical laws are stated w.r.t. unaccelerated or inertial frames, they preserve the same form, whatever may be the relative motion between the observers.

3.37.1 Galilean transformation—Invariance of Newton's laws

In classical physics, the motion of a body or any physical system is governed by Newton's laws of motion. We consider the simplest mechanical system i.e. a particle of mass m acted upon by a force $\vec{F}$. Then by Newton's second law, we have

$$m\ddot{x} = F_x, \quad m\ddot{y} = F_y, \quad m\ddot{z} = F_z \tag{3.37.1}$$

where F_x, F_y, F_z are the x, y and z components of the force vector $\vec{F}$ and $\ddot{x}, \ddot{y}, \ddot{z}$ are the accelerations of the particle along x, y and z axes respectively.

The above equations are valid only if the frame of reference S, defined by the coordinates x, y, z, t i.e. $S(x, y, z, t)$ is an *inertial frame.*

We consider now another frame $S'(x', y', z', t')$ whose axes are parallel to the corresponding axes of the old frame $S(x, y, z, t)$. The frame S' moves to the right with constant velocity v along the x-axis relative to S—the corresponding axes of the two frames being parallel to one another (Fig. 3.37.1). Here t and t' are the times measured in the two frames and defined such that $t = t' = 0$ when the two frames coincide. This also follows from our classical concept according to which time-measurement is to be regarded as same in both the frames. The relations between the two sets of coordinates (x, y, z, t) and (x', y', z', t') are written as

$$\left.\begin{aligned} x' &= x - vt \\ y' &= y \\ z' &= z \\ t' &= t \end{aligned}\right\} \tag{3.37.2}$$

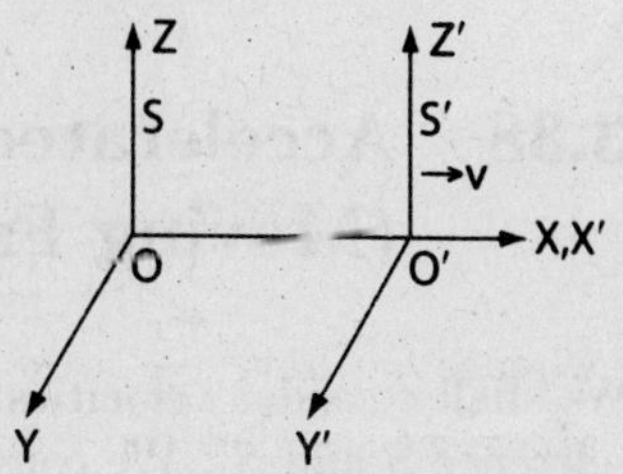

Fig. 3.37.1

These equations are known as the *Galilean transformation equations* which make it possible to transform the four coordinates (x, y, z, t) of S frame into the corresponding ones (x', y', z', t') of S'-frame.

The conjugate set of transformation equations, (that is transformation from S'-system to S-system) is

$$\left.\begin{aligned} x &= x' + vt \\ y &= y' \\ z &= z' \\ t &= t' \end{aligned}\right\} \tag{3.37.3}$$

From the first equation of the set (3.37.2), we obtain by differentiation

$$\dot{x}' = \dot{x} - v \quad (\because v = \text{constant}) \tag{3.37.4}$$

which may be considered as the transformation equation for velocity. Differentiating (3.37.4)

$$\left.\begin{aligned} &\text{we have} & \ddot{x}' &= \ddot{x} \\ &\text{and evidently from the 2nd} & \ddot{y}' &= \ddot{y} \\ &\text{and 3rd equations of set 3.37.2} & \ddot{z}' &= \ddot{z} \end{aligned}\right\} \tag{3.37.5}$$

So in the reference frame $S'(x', y', z', t')$, Newton's second law can be written as

$$m\ddot{x}' = F_x, \quad m\ddot{y}' = F_y, \quad m\ddot{z}' = F_z \tag{3.37.6}$$

These equations have exactly the same mathematical form as equations (3.37.1). So Newton's laws, which govern the behaviour of the system, do not change when we make a Galilean transformation. *Newton's law are therefore said to remain invariant in Galilean transformation* i.e. *they are covariant in inertial frames.* The behaviour of all mechanical systems will thus be identical in all inertial frames in uniform translation (or at rest) w.r.t. each other. This implies that an observer at rest in a particular frame will not able to decide, by performing any mechanical experiment, if his frame is at rest or in uniform rectilinear motion. This equivalence of all inertial frames with regard to the laws of classical mechanics is known as *Newtonian relativity.*

3.38 Accelerated Frames of Reference (Moving Frames of Reference)

We shall consider velocities and accelerations of a particle relative to two different frames of reference S and S'. We consider here that there is relative motion between the frames, but no relative rotation of the two frames. Let O and O' be the origins for the frames of reference S and S' respectively (Fig. 3.38.1).

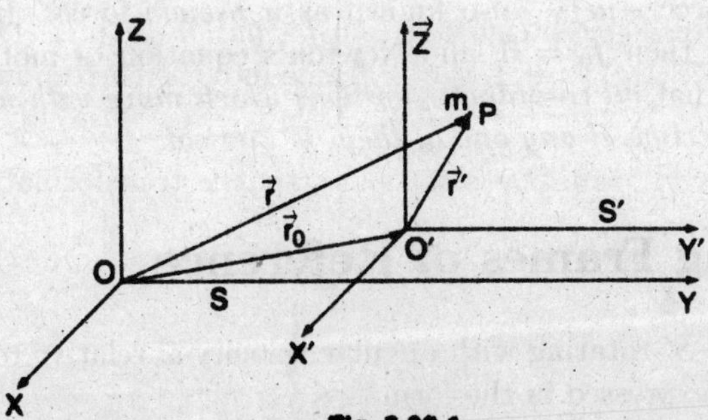

Fig. 3.38.1

The position vector of the particle P relative the origin O is $\vec{r} = \overrightarrow{OP}$ and the position vector relative to O' is $\vec{r'} = \overrightarrow{O'P}$.

$$\therefore\ \vec{r} = \overrightarrow{OO'} + \overrightarrow{O'P} = \vec{r_0} + \vec{r'} \quad \text{where, } \vec{r_0} = \overrightarrow{OO'}. \tag{3.38.1}$$

Differentiating the above expression we have,

$$\frac{d\vec{r}}{dt} = \frac{d\vec{r_0}}{dt} + \frac{d\vec{r'}}{dt} \tag{3.38.2}$$

$$\text{or, } \quad \vec{v} = \vec{v_0} + \vec{v'}, \tag{3.38.3}$$

where $\vec{v}$ = velocity of the particle relative to the frame S,

$\vec{v_0}$ = velocity of the frame S' relative to S,

$\vec{v'}$ = velocity of the particle relative to the frame S'.

The expression (3.38.3) gives the relation between the velocities of a particle relative to two different frames of reference.

Differentiating the equation (3.38.2), we have

$$\frac{d^2\vec{r}}{dt^2} = \frac{d^2\vec{r_0}}{dt^2} + \frac{d^2\vec{r'}}{dt^2} \quad \text{or} \quad \vec{f} = \vec{f_0} + \vec{f'}.$$

This gives the relation between the accelerations of a particle relative to two different frames of reference.

3.39 Newton's Equation of Motion in Moving Frames of Reference

We consider S to be a Newtonian frame of reference and S' a frame of reference having only a translational motion relative to S.

Hence Newton's second law of motion may be written as

$$\vec{F} = m\vec{f} = m(\vec{f_0} + \vec{f'}) \quad \text{or} \quad m\vec{f'} = \vec{F} - m\vec{f_0}. \tag{3.39.1}$$

Here $\vec{f'}$ denotes the acceleration in S' and $\vec{f_0}$ the acceleration of S' relative to S. The equation (3.39.1) shows that we can consider S' as a Newtonian frame if we add to the actual force $\vec{F}$ a fictitious force $-m\vec{f}_0$—also known as a *pseudo* force. If S' moves with constant velocity relative to S, then $\vec{f}_0 = 0$. and Newton's equation of motion will hold in both the systems. This proves that *all co-ordinate systems which move with uniform velocity relative to one another will be inertial, if any one of them is inertial.*

3.40 Rotating Frames of Reference

We consider the frame S' rotating with angular velocity $\vec{\omega}$ relative to the Newtonian frame S. Any vector $\vec{A}$ may be expressed in the form

$$\vec{A} = A_x\hat{\imath} + A_y\hat{\jmath} + A_z\hat{k},$$

where $\hat{\imath}, \hat{\jmath}, \hat{k}$ are the unit vectors in the frame of reference S'.

We shall now consider the rate of change of $\vec{A}$ as observed by an observer in the frame S. Remembering that A_x, A_y, A_z and $\hat{\imath}, \hat{\jmath}, \hat{k}$ vary with time we have

$$\frac{d\vec{A}}{dt} = \frac{dA_x}{dt}\hat{\imath} + \frac{dA_y}{dt}\hat{\jmath} + \frac{dA_z}{dt}\hat{k} + A_x\frac{d\hat{\imath}}{dt} + A_y\frac{d\hat{\jmath}}{dt} + A_z\frac{d\hat{k}}{dt}.$$

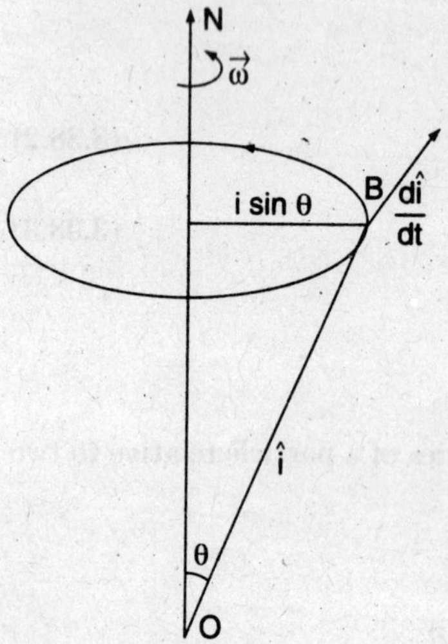

Fig. 3.40.1

Now $\hat{\imath}$ is a vector fixed in a rigid body (system S') which rotates with angular velocity $\vec{\omega}$. We may consider $\hat{\imath}$ as the position vector of a particle B of this rigid body relative to the origin O. Then $\frac{d\hat{\imath}}{dt}$ is the velocity of B relative to O. Let the rigid body rotate with angular velocity $\vec{\omega}$ about the axis ON (Fig. 3.40.1). Then the point B on the rigid body moves in a circular path about ON with tangential linear velocity $wi\sin\theta$. This tangential velocity in vector notation may be written in the form $\vec{\omega} \times \hat{\imath}$. Hence

$$\frac{d\hat{\imath}}{dt} = \vec{\omega} \times \hat{\imath}.$$

Similarly,

$$\frac{d\hat{\jmath}}{dt} = \vec{\omega} \times \hat{\jmath} \quad \text{and} \quad \frac{d\hat{k}}{dt} = \vec{\omega} \times \hat{k}.$$

$$\therefore \quad \frac{d\vec{A}}{dt} = \frac{dA_x}{dt}\hat{\imath} + \frac{dA_y}{dt}\hat{\jmath} + \frac{dA_z}{dt}\hat{k} + A_x(\vec{\omega} \times \hat{\imath}) + A_y(\vec{\omega} \times \hat{\jmath}) + A_z(\vec{\omega} \times \hat{k})$$

$$= \frac{\delta\vec{A}}{\delta t} + \vec{\omega} \times \vec{A}, \tag{3.40.1}$$

where

$$\frac{\delta\vec{A}}{\delta t} = \frac{dA_x}{dt}\hat{\imath} + \frac{dA_y}{dt}\hat{\jmath} + \frac{dA_z}{dt}\hat{k}$$

$\frac{\delta}{\delta t}$ denotes partial differentiation in which $\hat{\imath}, \hat{\jmath}, \hat{k}$ are kept fixed.

Thus we see that the right-hand side of the expression (3.40.1) consists of two terms. $\frac{\delta\vec{A}}{\delta t}$ denotes the time rate of change of $\vec{A}$ for an observer moving with S' and the second part arises due to rotation of the unit vectors $\hat{\imath}, \hat{\jmath}, \hat{k}$.

3.41 Velocity and Acceleration of a Particle Relative to a Rotating Frame

We consider a frame S' rotating with angular velocity $\vec{\omega}$ about a point O fixed in a Newtonian frame S. Hence the velocity of a moving particle relative to S is

$$\vec{v} = \frac{d\vec{r}}{dt} = \frac{\delta\vec{r}}{\delta t} + \vec{\omega}\times\vec{r} \quad \text{[from (3.40.1)]} \tag{3.41.1}$$

and the acceleration is given by

$$\vec{f} = \frac{\delta\vec{v}}{\delta t} + \vec{\omega}\times\vec{v}. \tag{3.41.2}$$

From the expressions (3.41.1) and (3.41.2) we have

$$\vec{f} = \frac{\delta^2\vec{r}}{\delta t^2} + \frac{\delta\vec{\omega}}{\delta t}\times\vec{r} + \vec{\omega}\times\frac{\delta\vec{r}}{\delta t} + \vec{\omega}\times\left(\frac{\delta\vec{r}}{\delta t} + \vec{\omega}\times\vec{r}\right)$$

$$= \frac{\delta^2\vec{r}}{\delta t^2} + \frac{\delta\vec{\omega}}{\delta t}\times\vec{r} + \vec{\omega}\times(\vec{\omega}\times\vec{r}) + 2\vec{\omega}\times\frac{\delta\vec{r}}{\delta t}. \tag{3.41.3}$$

In this expression $\dfrac{\delta^2\vec{r}}{\delta t^2}$ denotes the acceleration of the particle relative to S' and the term $2\vec{\omega}\times\dfrac{\delta\vec{r}}{\delta t}$ is called Coriolis acceleration.

If the angular velocity $\vec{\omega}$ of the rotating frame S' is constant, then the term $\dfrac{\delta\vec{\omega}}{\delta t}\times\vec{r}$ vanishes. The term $\vec{\omega}\times(\vec{\omega}\times\vec{r})$ may be calculated for constant angular velocity in the following way.

Since $\vec{\omega}$ is a constant vector, it determines a fixed axis of rotation through O as shown in Fig. 3.41.1. In the figure, AN is the perpendicular drawn from the position of the particle A to the axis ON.

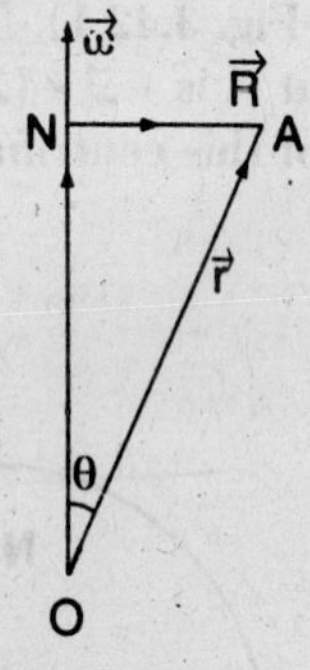

Fig. 3.41.1

$$\therefore \quad \vec{\omega}\times(\vec{\omega}\times\vec{r}) = \vec{\omega}(\vec{\omega}\cdot\vec{r}) - \vec{r}\omega^2 = \vec{\omega}(\omega r\cos\theta) - \vec{r}\omega^2$$

$$= \overrightarrow{ON}\omega^2 - \vec{r}\omega^2 = \overrightarrow{ON}\omega^2 - (\overrightarrow{ON} + \overrightarrow{NA})\omega^2$$

$$= -\overrightarrow{NA}\vec{\omega}^2 = -\vec{R}\omega^2 \quad [\text{where } \overrightarrow{NA} = \vec{R}]$$

$-\vec{R}w^2$ is the centrifugal acceleration.

Thus when the frame S' rotates with constant angular velocity, we may write expression (3.41.3) in the form

$$\vec{f} = \frac{\delta^2\vec{r}}{\delta t^2} - \vec{R}\omega^2 + 2\vec{\omega}\times\frac{\partial\vec{r}}{\delta t}. \tag{3.41.4}$$

For a particle of mass m acted on by a force F the law of motion in the system S is given by

$$m\vec{f} = \vec{F} = m\left\{\frac{\delta^2\vec{r}}{\delta t^2} + \frac{\delta\vec{\omega}}{\delta t}\times\vec{r} + \vec{\omega}\times(\vec{\omega}\times\vec{r}) + 2\vec{\omega}\times\frac{\delta\vec{r}}{\delta t}\right\}.$$

In S' system, the law of motion is given by

$$m\vec{f'} = m\frac{\delta^2 \vec{r'}}{\delta t^2} = \vec{F} - m\left\{\frac{\delta\vec{\omega}}{\delta t}\times\vec{r} + \vec{\omega}\times(\vec{\omega}\times\vec{r}) + 2\vec{\omega}\times\frac{\delta\vec{r}}{\delta t}\right\}. \quad (3.41.5)$$

From the equation (3.41.5) it follows that we may consider S' as a Newtonian frame provided we add to the applied force the fictitious forces such as

$$-m\frac{\delta\vec{\omega}}{\delta t}\times\vec{r},\quad -m\vec{\omega}\times(\vec{\omega}\times\vec{r})\quad\text{and}\quad -2m\vec{\omega}\times\frac{\delta\vec{r}}{\delta t}$$

of which the last one is called Coriolis force.

When the frame S' rotates with constant angular velocity the equation (3.41.5) becomes

$$m\vec{f'} = m\frac{\delta^2 \vec{r'}}{\delta t^2} = \vec{F} - m\left\{-\vec{R}\omega^2 + 2\vec{\omega}\times\frac{\delta\vec{r}}{\delta t}\right\}.$$

Here $m(-\vec{R}\omega^2)$ is the centrifugal force.

3.42 Centrifugal Acceleration due to the Earth's Rotation

We suppose that the earth rotates about the axis OZ with constant angular velocity $\vec{\omega}$ (Fig. 3.42.1). Then the centrifugal acceleration experienced by a particle on the earth's surface at A is $-\vec{\omega}\times(\vec{\omega}\times\vec{r})$ directed along NA which is normal to the axis of rotation. The magnitude of this centrifugal acceleration is

$$\omega^2(NA) = \omega^2 r\sin\theta = \omega^2 r\cos\lambda,$$

where λ is the latitude of the point A.

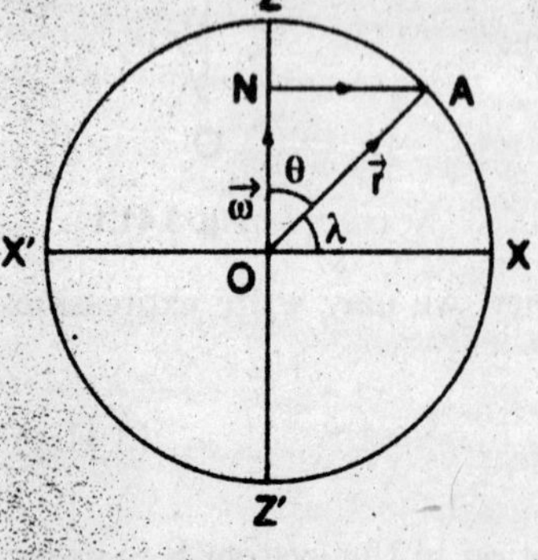

Fig. 3.42.1

Now,

$$\omega = \frac{2\pi}{24\times 3600} = 7.29\times 10^{-5}\ \text{rad/sec}.$$

At the equator the magnitude of the centrifugal acceleration is

$$\omega^2 r = (7.29\times 10^{-5})^2\times 6.4\times 10^8 = 3.38\ \text{cm/sec.}^2$$

[$\because$ r = radius of the earth = 6.4×10^8 cm and $\lambda = 0$ at the equator.]

Thus at the equator the centrifugal acceleration is maximum and is only about 3.38 cm/sec^2 or 0.3% of the acceleration due to gravity.

Hence at the equator the effective acceleration is given by $\vec{g}_e = \vec{g} - \omega^2\vec{r}$, where $\vec{g}$ is the acceleration due to the earth's attraction. $\vec{g}_e$ is directed towards the centre of the earth. However, at other latitudes, the effective acceleration is not directed towards the centre of the earth. Now, the equilibrium position of a plumbline is the resultant of the force of gravity and the centrifugal force. Thus except at the equator, the plumbline is deviated slightly from the direction of the radius vector $\vec{r}$. This is illustrated in Fig. 3.42.2. In the figure the effective acceleration $\vec{g}_e$ is the resultant of the acceleration due to gravity $\vec{g}$ and the centrifugal acceleration $\omega^2\vec{r}\cos\lambda$. The resultant acceleration is not directed along the radius vector $\vec{r}$, but deviated as shown in the figure.

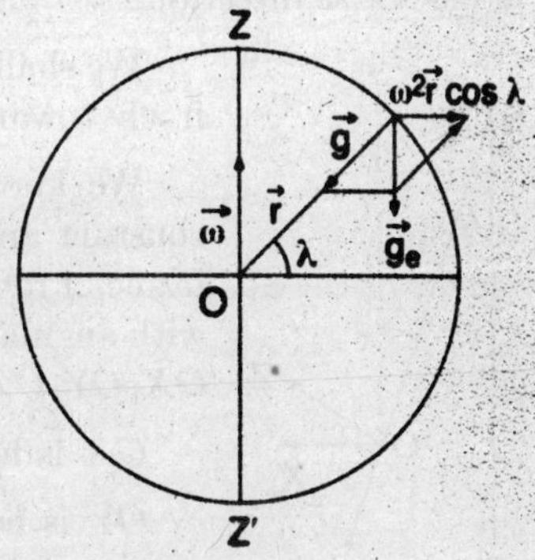

Fig. 3.42.2

Here we have neglected the centrifugal acceleration due to revolution of the earth about the sun which is negligibly small.

3.43 Coriolis Force

When a body is moving in a rotating system, Coriolis force comes into play. This force

$$= -2m\vec{\omega} \times \frac{\delta\vec{r}}{\delta t} = 2m\frac{\delta\vec{r}}{\delta t} \times \vec{\omega} = 2m\vec{v} \times \vec{\omega}.$$

From cross-product term in Coriolis force it is clear that Coriolis force on a moving body is perpendicular to both $\vec{\omega}$ and $\vec{v}$. In the Northern Hemisphere the Coriolis force tends to deflect a projectile shot along the earth's surface to the right of its direction of motion (Fig. 3.43.1). The deflection of the projectile shot in the Southern Hemisphere will be in reverse direction.

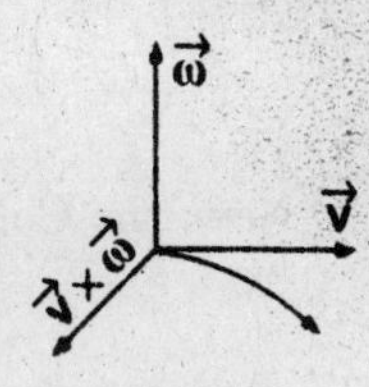

Fig. 3.43.1

If θ be the angle between the vectors $\vec{v}$ and $\vec{\omega}$ then the magnitude of the Coriolis force is $2mv\omega\sin\theta$. Hence the maximum value of the Coriolis acceleration $= 2\omega v = 2 \times 7.29 \times 10^{-5}v \simeq 1.5 \times 10^{-4}v$.

Thus Coriolis acceleration is negligible for a body moving with ordinary velocity, but there are instances where the Coriolis acceleration plays an important role. For example, when a projectile is shot horizontally at the north pole, the Coriolis acceleration on the projectile is $2\omega v$. Hence the linear deflection after time $t = \frac{1}{2}(2\omega v)t^2 = \omega vt^2$. Thus the deflection is proportional to the square of the time of flight of the projectile. The angular deflection is therefore $\dfrac{\omega vt^2}{vt} = \omega t =$ the angle, the earth describes in time t. Taking the time of flight $t =$ 100 secs [this may be the time of flight of large projectiles] we have the angular deflection $= 7.29 \times 10^{-5} \times 100 = 7.29 \times 10^{-3}$ radian $\simeq 0.4°$. This value is not negligible.

In case of guided missiles the time of flight is much longer and hence the angular deflection of the guided missiles is considerably large.

In meteorological problem of wind circulation the Coriolis force plays a very important role. The effective time of flight for wind circulation is much larger than that for projectiles. In absence of the Coriolis force the motion of wind takes place from high to low pressure, but the Coriolis force deflects the wind to the right of its direction of motion in the Northern Hemisphere. A region of low pressure with nearly concentric isobars is technically called cyclone. In the Northern Hemisphere the Coriolis forces make the winds circulate round the cyclone in a counter clockwise direction. In the Southern Hemisphere the winds circulate round the cyclone in clockwise direction.

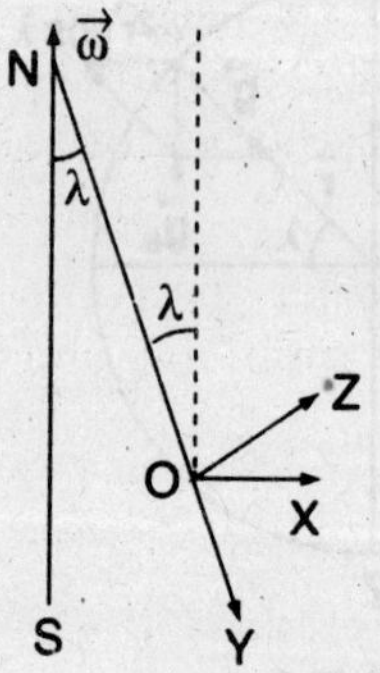

Fig. 3.43.2

We shall now consider the deflection of a body due to Coriolis force falling freely towards the earth.

We know that the earth is turning about its axis of symmetry with constant angular velocity. The axis is supposed to be fixed in a Newtonian frame of reference. In Fig. 3.43.2 we consider SN as the earth's axis rotating with angular velocity $\vec{\omega}$. O is a point taken on or near the earth's surface. Let OX, OY, OZ be the axes fixed relative to the earth and directed as follows:

OX is horizontal and points east;

OY is horizontal and points south;

OZ is vertical and points upward;

λ is the latitude of the point O, which is the angle of elevation of the earth's axis above the horizontal plane.

Let the body fall freely vertically along ZO and reach the earth's surface in time t. Then in absence of Coriolis force the vertical velocity on reaching the earth's surface is given by $v = gt$. If the body falls through a height z, then

$$z = \frac{1}{2}gt^2 \quad \text{or} \quad t = \sqrt{\frac{2z}{g}}. \tag{3.43.1}$$

Now, the Coriolis acceleration is

$$-2\vec{\omega} \times \vec{v} = 2v\omega \sin(90 - \lambda) = 2v\omega \cos\lambda = 2\omega gt \cos\lambda.$$

Considering x to be the deflection towards east in the horizontal plane due to Coriolis force, we have

$$\frac{d^2x}{dt^2} = 2\omega gt \cos\lambda.$$

Integrating the above expression twice w.r.t. t, and making the integration constants zero with suitable initial conditions, we have

$$x = \omega\frac{gt^3}{3}\cos\lambda = \frac{\omega g}{3}\left(\frac{2z}{g}\right)^{3/2}\cos\lambda \quad \text{[from equation (3.43.1)]}$$

$$= \frac{2\sqrt{2}}{3}\omega\sqrt{\frac{z^3}{g}}\cos\lambda. \tag{3.43.2}$$

At the equator the deflection of a body falling freely is maximum and is given by

$$x_{\max} = \frac{2\sqrt{2}}{3}\omega\sqrt{\frac{z^3}{g}}. \tag{3.43.3}$$

If a body falls through a height of 100 ft. at the equator, then the deflection of the body towards east will be nearly 0.15 inch.

In actual experiment it is very difficult to measure such small deflection due to disturbing effect of wind currents, viscosity, etc.

The effect of Coriolis force may easily be observed by Foucault's pendulum. If the pendulum is made to oscillate at the north pole in a given plane, then to an observer on the earth the plane of oscillation appears to rotate once a day. The rotation of the plane of oscillation takes place due to Coriolis force.

Coriolis force also has its use in atomic physics. Perturbation in molecular spectra is explained by Coriolis force coming to play due to interactions between the rotational and vibrational levels of the molecule.

3.44 Foucault's Pendulum

It is a simple pendulum having a heavy bob suspended by a long string from a frictionless support. If the pendulum is made to oscillate, its plane of oscillation appears to rotate with respect to the floor. If we look down at the earth's surface in the northern hemisphere, the precession of the plane of oscillation will be in the clockwise direction, while the precession in the southern hemisphere will be in the anticlockwise direction. It was Foucault who for the first time used this type of pendulum to demonstrate the rotation of the earth. The heavy bob (28 Kg) of the pendulum is suspended by a long string of length 67 metres to make the time period of oscillation of the pendulum large.

The equation of motion of the bob of the pendulum relative to a fixed system, i.e., relative to Newtonian frame of reference subjected to gravitational force $m\vec{g}$ and another force $\vec{F}$ is given by

$$m\vec{f} = \vec{F} + m\vec{g}.$$

The equation of motion of the bob relative to a moving frame of reference rotating with constant angular velocity $\vec{\omega}$ is given by

$$m\vec{f'} = (\vec{F} + m\vec{g}) - m\left\{\vec{\omega} \times (\vec{\omega} \times \vec{r}) + 2\vec{\omega} \times \frac{\delta\vec{r}}{\delta t}\right\} \quad \text{[See equation (3.41.5)]}.$$

If $\vec{T}$ be the tension of the string the above equation may be written as

$$m\vec{f'} = (\vec{T} + m\vec{g}) - m\left\{\vec{\omega} \times (\vec{\omega} \times \vec{r}) + 2\vec{\omega} \times \frac{\delta\vec{r}}{\delta t}\right\}.$$

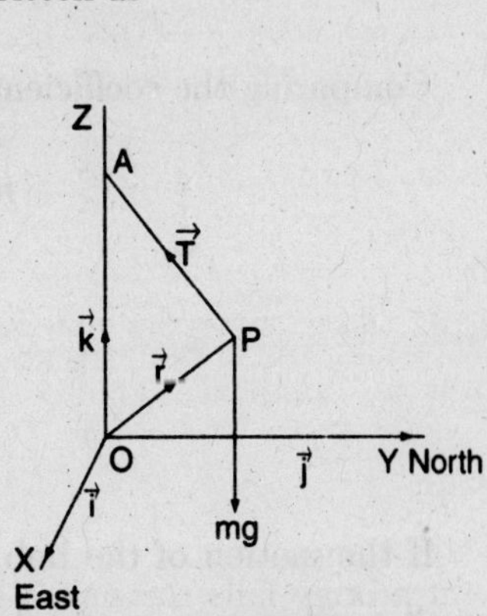

Fig. 3.44.1

Neglecting the second term in the R.H.S. of the above equation, we have,

$$m\vec{f'} = (\vec{T} + m\vec{g}) - 2m\left(\vec{\omega} \times \frac{\delta\vec{r}}{\delta t}\right). \qquad (3.44.1)$$

In the figure 3.44.1 OX, OY and OZ denote three co-ordinate axes and O the equilibrium position of the bob P of the pendulum. The three directions OX, OY and OZ denote east, north and vertical directions respectively.

Let $\hat{\imath}, \hat{\jmath}$ and $\hat{k}$ be the unit vectors along OX, OY and OZ axes respectively. Then

$$\vec{T} = (\vec{T}.\hat{\imath})\hat{\imath} + (\vec{T}.\hat{\jmath})\hat{\jmath} + (\vec{T}.\hat{k})\hat{k}$$

$$= T\cos\alpha\hat{\imath} + T\cos\beta\hat{\jmath} + T\cos\gamma\hat{k}.$$

But,

$$\cos\alpha = -\frac{x}{l}, \quad \cos\beta = -\frac{y}{l}, \quad \cos\gamma = \frac{l-z}{l}.$$

$\therefore$ the equation (3.44.1) becomes

$$m\vec{f}' = \left\{-T\frac{x}{l}\hat{\imath} - T\frac{y}{l}\hat{\jmath} + T\left(\frac{l-z}{l}\right)\hat{k}\right\} - mg\hat{k} - 2m\left(\vec{\omega} \times \frac{\delta\vec{r}}{\delta t}\right\}, \tag{3.44.2}$$

$$\text{or} \quad m\left(\ddot{x}\hat{\imath} + \ddot{y}\hat{\jmath} + \ddot{z}\hat{k}\right) = \left\{-T\frac{x}{l}\hat{\imath} - T\frac{y}{l}\hat{\jmath} + T\frac{(l-z)}{l}\hat{k}\right\} - mg\hat{k} - 2m\left[\vec{\omega} \times \frac{\delta\vec{r}}{\delta t}\right]. \tag{3.44.3}$$

Now $\vec{\omega} = \omega\cos\lambda\hat{\jmath} + \omega\sin\lambda\hat{k}$, where λ is the latitude.

$$\therefore \vec{\omega} \times \frac{\delta\vec{r}}{\delta t} = (\omega\cos\lambda\hat{\jmath} + \omega\sin\lambda\hat{k}) \times (\hat{\imath}\dot{x} + \hat{\jmath}\dot{y} + \hat{k}\dot{z})$$

$$= \begin{vmatrix} \hat{\imath} & \hat{\jmath} & \hat{k} \\ 0 & \omega\cos\lambda & \omega\sin\lambda \\ \dot{x} & \dot{y} & \dot{z} \end{vmatrix}$$

$$= \hat{\imath}(\omega\cos\lambda\dot{z} - \omega\sin\lambda\dot{y}) + \hat{\jmath}\omega\sin\lambda\dot{x} - \hat{k}\omega\cos\lambda\dot{x}.$$

Hence the equation (3.44.3) may be written as

$$m\left(\ddot{x}\hat{\imath} + \ddot{y}\hat{\jmath} + \ddot{z}\hat{k}\right) = \left\{-T\frac{x}{l}\hat{\imath} - T\frac{y}{l}\hat{\jmath} + T\frac{(l-z)}{l}\hat{k}\right\} - mg\hat{k} - 2m\hat{\imath}(\omega\cos\lambda\dot{z} - \omega\sin\lambda\dot{y})$$

$$- 2m\hat{\jmath}\omega\sin\lambda\dot{x} + 2m\hat{k}\omega\cos\lambda\dot{x}.$$

Comparing the coefficients of $\hat{\imath}, \hat{\jmath}$ and $\hat{k}$ on both sides we have,

$$m\ddot{x} = -T\left(\frac{x}{l}\right) - 2m\omega(\cos\lambda\dot{z} - \sin\lambda\dot{y}). \tag{3.44.4}$$

$$m\ddot{y} = -T\left(\frac{y}{l}\right) - 2m\omega\sin\lambda\dot{x}. \tag{3.44.5}$$

$$m\ddot{z} = T\left(\frac{l-z}{l}\right) - mg + 2m\omega\cos\lambda\dot{x}. \tag{3.44.6}$$

If the motion of the bob takes place in a horizontal plane we can assume $\ddot{z} = \dot{z} = 0$. For small oscillation we may take $\dfrac{l-z}{l} \to 1$.

∴ From the equation (3.44.6) we have

$$O = T - mg + 2m\omega\cos\lambda\dot{x}.$$

$$\therefore\ T = mg - 2m\omega\cos\lambda\dot{x}.$$

∴ From (3.44.4), we have

$$m\ddot{x} = -\frac{mgx}{l} + 2m\omega\cos\lambda\dot{x}\left(\frac{x}{l}\right) + 2m\omega\sin\lambda\dot{y}$$

$$\text{or}\quad \ddot{x} = -\frac{gx}{l} + \frac{2\omega\cos\lambda\dot{x}x}{l} + 2\omega\sin\lambda\dot{y} \tag{3.44.7}$$

and similarly from (3.44.5),

$$\ddot{y} = -\frac{gy}{l} + \frac{2\omega\cos\lambda\dot{x}y}{l} - 2\omega\sin\lambda\dot{x}. \tag{3.44.8}$$

Since $\dot{x}x$ and $\dot{x}y$ are small, equation (3.44.7) and (3.44.8) may be written as

$$\ddot{x} = -\frac{gx}{l} + 2\omega\sin\lambda\dot{y} \tag{3.44.9}$$

$$\text{and}\quad \ddot{y} = -\frac{gy}{l} - 2\omega\sin\lambda\dot{x}. \tag{3.44.10}$$

According to the boundary condition, at $t = 0$

$$x = 0,\ \dot{x} = 0$$

$$y = a,\ \dot{y} = 0$$

a is the displacement along Y-axis at $t = 0$.

Putting $k^2 = g/l$ and $\phi = \omega\sin\lambda$ equations (3.44.9) and (3.44.10) become

$$\ddot{x} = -k^2x + 2\phi\dot{y}$$

$$\ddot{y} = -k^2y - 2\phi\dot{x}.$$

Applying the boundary conditions the solutions of the above equations are obtained and are given by

$$x = a\cos kt\sin\phi t$$

$$y = a\cos kt\cos\phi t.$$

Substituting the values of k and ϕ in the above solutions we may write

$$\left.\begin{aligned} x &= a\cos\sqrt{\frac{g}{l}}t\sin(\omega\sin\lambda t) \\ y &= a\cos\sqrt{\frac{g}{l}}t\cos(\omega\sin\lambda t) \end{aligned}\right\}. \tag{3.44.11}$$

In vector form the equation (3.44.11) may be written as

$$\vec{r} = \hat{\imath}x + \hat{\jmath}y = a\cos\sqrt{\frac{g}{l}}t[\hat{\imath}\sin(\omega\sin\lambda t) + \hat{\jmath}\cos(\omega\sin\lambda t)].$$

From the above equation it follows that the period of rotation of the plane of oscillation of Foucault's pendulum is given by

$$T = \frac{2\pi}{\omega\sin\lambda} = \frac{24}{\sin\lambda}\ \text{hours.} \tag{3.44.12}$$

From this expression for the time period we see that it depends on the latitude λ.

Illustrative Examples

1. Three particles of masses 4 gm., 3 gm. and 2 gm. are at the points $(2, 0, -1)$, $(1, 1, 3)$ and $(3, -1, 0)$ respectively. Find the co-ordinates of the centre of mass.

From equation (3.29.1) the position vector of the centre of mass may be written as

$$\overrightarrow{OC} = \frac{m_1\vec{r_1} + m_2\vec{r_2} + m_3\vec{r_3}}{m_1 + m_2 + m_3}$$

here

$$\begin{aligned}\vec{r_1} &= 2\hat{\imath} - \hat{k}\\ \vec{r_2} &= \hat{\imath} + \hat{\jmath} + 3\hat{k}\\ \vec{r_3} &= 3\hat{\imath} - \hat{\jmath}\end{aligned}$$

and

$$\begin{aligned}m_1 &= 4\ \text{gm.}\\ m_2 &= 3\ \text{gm.}\\ m_3 &= 2\ \text{gm.}\end{aligned}$$

$$\therefore\ \overrightarrow{OC} = \frac{4(2\hat{\imath} - \hat{k}) + 3(\hat{\imath} + \hat{\jmath} + 3\hat{k}) + 2(3\hat{\imath} - \hat{\jmath})}{4 + 3 + 2} = \frac{17\hat{\imath} + \hat{\jmath} + 5\hat{k}}{9}.$$

Hence the co-ordinates of the centre of mass $= \left(\frac{17}{9}, \frac{1}{9}, \frac{5}{9}\right)$.

2. Calculate the deflection of a freely falling body from a height of 100 metres at latitude 30^0N due to coriolis acceleration. **[cf. C.U. (Hons.) 2001]**

The deflection of a freely falling body from the vertical at latitude λ is given by

$$x = \frac{2\sqrt{2}}{3}\omega\sqrt{\frac{z^3}{g}}\cos\lambda \quad \text{[From (3.43.2)]}$$

Now

$$w = \frac{2\pi}{24 \times 60 \times 60} \text{ radian/sec.}; \quad z = 100 \text{ metres} = 10^4 \text{ cm.}$$

$$\lambda = 30^0, \quad g = 981 \text{ cm/sec}^2.$$

$$\therefore x = \frac{2\sqrt{2}}{3} \times \frac{2\pi}{24 \times 60 \times 60} \times \frac{\sqrt{3}}{2}\sqrt{\frac{10^{12}}{981}} = 1.895 \text{ cm.}$$

The deflection is towards the east.

3. Calculate the rate at which the plane of oscillation of a Foucault's pendulum rotates at a latitude 45^0N.

The rate at which the plane of oscillation of the pendulum rotates is the angle traced per hour due to rotation. Hence the rate of rotation at latitude λ

$$= \frac{360^0}{24} \sin\lambda \quad \text{[From equation 3.43.12]}$$

$$= \frac{360^0}{24} \cdot \frac{1}{\sqrt{2}} \quad \left[\because \sin 45^0 = \frac{1}{\sqrt{2}}\right] = 10^0.63 \text{ per hour.}$$

4. A particle is thrown vertically upwards with a velocity 70 m/sec. at a place with latitude 60^0. Find how far from the original position it will land. **[C.U. (Hons.) 1991, cf. '03]**

We know that the coriolis acceleration at a place having latitude λ and in a co-ordinate system with X-axis in the east-ward direction and Z-axis vertically upwards is given by

$$\ddot{x} = -2wv_z \cos\lambda = -2w(v_0 - gt)\cos\lambda,$$

where v_0 = velocity of projection.

Integrating w.r.t. time, we have

$$\dot{x} = -2w\cos\lambda\left(v_0 t - \frac{1}{2}gt^2\right) + c, \tag{1}$$

c being the constant of integration.

From the initial conditions, $\dot{x} = 0$ at $t = 0$, we have $c = 0$

$$\therefore \dot{x} = -2w\cos\lambda\left(v_0 t - \frac{1}{2}gt^2\right). \tag{2}$$

Integrating equation (2) for the total time of flight $\left(t = \frac{2v_0}{g}\right)$, we have the total shift given by

$$x = -2w\cos\lambda\int_0^{2v_0/g}\left(v_0 t - \frac{1}{2}gt^2\right)dt = -2w\cos\lambda\left[v_0\frac{t^2}{2} - \frac{1}{2}g\frac{t^3}{3}\right]_0^{2v_0/g}$$

$$= -2w\cos\lambda\left[\frac{v_0}{2}\left(\frac{2v_0}{g}\right)^2 - \frac{1}{6}g\left(\frac{2v_0}{g}\right)^3\right] = -\frac{4wv_0^3}{3g^2}\cos\lambda$$

$$= -\frac{4}{3}\omega\cos\lambda\frac{(2gh)^{3/2}}{g^2} = -\frac{4}{3}\omega\cos\lambda\left(\frac{8h^3}{g}\right)^{1/2} \quad \left[\because v_0 = (2gh)^{1/2}\right]$$

which is obviously west-ward.

In the given problem, all data on the right hand side being given, x can be calculated.

5. A river of width D flows southward with a speed v_0 at a co-latitude λ in the northern hemi-sphere. Prove that the difference H between water-levels of the two banks due to coriolis force is given by $H = 2Dwv_0 \cos\lambda/g$. **[C.U. (Hons.) '98]**

The coriolis force on a mass $m = 2m\vec{v_0} \times \vec{w}$, and it is to the right of $\vec{v_0}$ in the northern hemisphere and hence in the direction of the west-bank. Thus the water level of the west-bank will be higher than that of the east-bank as shown in the fig. (a).

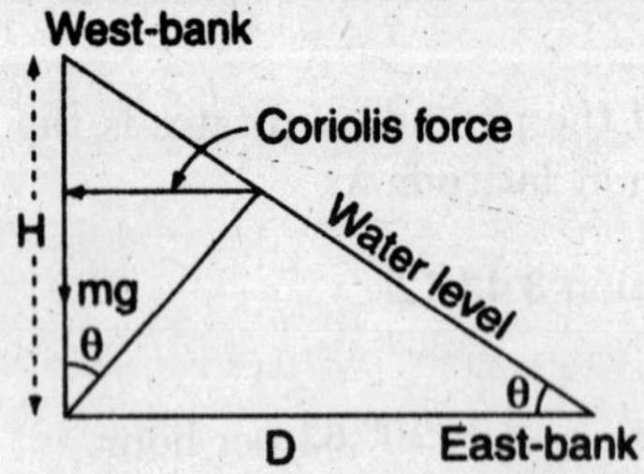

Fig. (a)

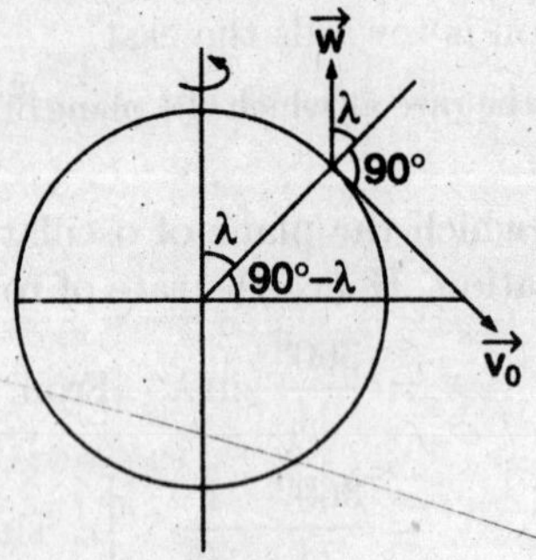

Fig. (b)

From the fig. (a), the coriolis force on a mass m of water towards the west $= 2mwv_0 \cos\lambda$, [$\because$ $\vec{v_0} \times \vec{w} = v_0 w \sin(90^0 + \lambda) = v_0 w \cos\lambda$ as evident from fig. (b)] and also the force due to gravity $= mg$.

$$\therefore\ \tan\theta = \frac{2mv_0 w\cos\lambda}{mg} = \frac{2wv_0\cos\lambda}{g}.$$

The difference between the water levels of the two banks is given by

$$H = D\tan\theta = \frac{2Dwv_0\cos\lambda}{g}.$$

3.45 Motion of Projectiles

A projectile is launched with initial speed v_0 at an angle α with the horizontal. We are to find (a) the position vector at any time, (b) the time to reach the highest point, (c) the maximum height reached, (d) the time of flight back to earth, and (e) the range.

(a) Let $\vec{r}$ be the position vector of the projectile and $\vec{v}$, the velocity at any time t. Then by Newton's law, we may write

$$m\frac{d^2\vec{r}}{dt^2} = -mg\hat{k} \quad \text{or} \quad \frac{d^2\vec{r}}{dt^2} = -g\hat{k} \quad \text{or} \quad \frac{d\vec{v}}{dt} = -g\hat{k}, \tag{3.45.1}$$

whence on integration, we have

$$\vec{v} = -gt\hat{k} + c, \tag{3.45.2}$$

where c = the integration constant.

As shown in Fig. 3.45.1, the initial velocity of the projectile is in the $Y - Z$ plane, so that the initial velocity is written as

$$\vec{v_0} = v_0\cos\alpha\hat{j} + v_0\sin\alpha\hat{k} \quad (\text{where } v_0 = \text{magnitude of } \vec{v_0}). \tag{3.45.3}$$

Now since at $t = 0$, $\vec{v} = \vec{v_0}$, we have from (3.45.2), $c = \vec{v_0}$. Putting this value of c in (3.45.2) and then writing the value of $\vec{v_0}$ from (3.45.3), we get

$$\vec{v} = v_0 \cos\alpha \hat{\jmath} + (v_0 \sin\alpha - gt)\hat{k} \qquad (3.45.4)$$

$$\text{or} \quad \frac{d\vec{r}}{dt} = v_0 \cos\alpha \hat{\jmath} + (v_0 \sin\alpha - gt)\hat{k}.$$

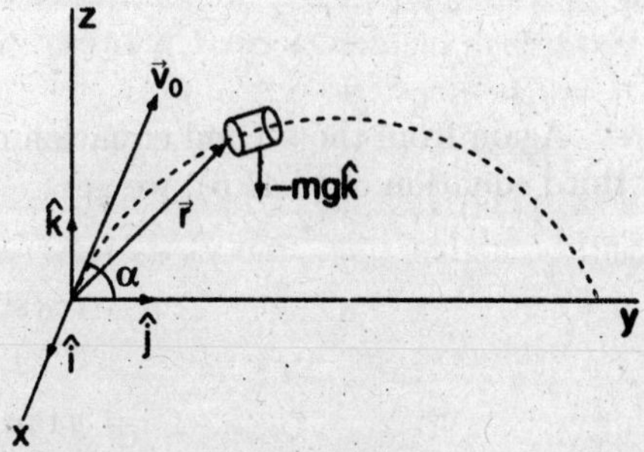

Fig. 3.45.1

On integration,

$$\vec{r} = (v_0 \cos\alpha)t\hat{\jmath} + \left\{(v_0 \sin\alpha)t - \frac{1}{2}gt^2\right\}\hat{k} + c_1$$

c_1 = Integration constant).

Now at $t = 0$, $\vec{r} = 0$, which gives $c_1 = 0$, so that we may write

$$\vec{r} = (v_0 \cos\alpha)t\hat{\jmath} + \left\{(v_0 \sin\alpha)t - \frac{1}{2}gt^2\right\}\hat{k}. \qquad (3.45.5)$$

That is, we have equivalently,

$$x = 0, \quad y = (v_0 \cos\alpha)t, \quad z = (v_0 \sin\alpha)t - \frac{1}{2}gt^2. \qquad (3.45.6)$$

From (3.45.6), we may conclude that the projectile remains in the $Y - Z$ plane.

(b) At the highest point of the path, the component of velocity $\vec{v}$ in the Z-direction is zero, so that from (3.45.4)

$$v_0 \sin\alpha - gt = 0 \quad \text{or} \quad t = \frac{v_0 \sin\alpha}{g} \qquad (3.45.7)$$

is the required time.

(c) Using the value of t from (3.45.7), we have from (3.45.6), maximum height attained given by

$$(v_0 \sin\alpha)\left(\frac{v_0 \sin\alpha}{g}\right) - \frac{1}{2}g\left(\frac{v_0 \sin\alpha}{g}\right)^2 = \frac{v_0^2 \sin^2\alpha}{2g}. \qquad (3.45.8)$$

(d) The time of flight back to earth is the time when $z = 0$ i.e., when

$$(v_0 \sin\alpha)t - \frac{1}{2}gt^2 = t\left[v_0 \sin\alpha - \frac{1}{2}gt\right] = 0.$$

But $t \neq 0$,

$$\therefore\ t = \frac{2v_0 \sin\alpha}{g} \qquad (3.45.9)$$

This time is twice the time obtained in (b).

(e) The range is the value of y in (3.45.6) at the time given by (3.45.9), so that we may write

$$\text{Range } R = (v_0 \cos\alpha)\left(\frac{2v_0 \sin\alpha}{g}\right) = \frac{2v_0^2 \sin\alpha \cos\alpha}{g} = \frac{v_0^2 \sin 2\alpha}{g}, \qquad (3.45.10)$$

which will have a maximum value, when $\sin 2\alpha = 1$, i.e., $2\alpha = 90°$ or $\alpha = 45°$ or when the launching angle α is 45°,

$$\text{or} \quad R_{\max} = \frac{v_0^2}{g}. \tag{3.45.11}$$

Again from the second equation of (3.45.6), we have $t = y/v_0 \cos\alpha$, and putting this in the third equation of (3.45.6), we get

$$\begin{aligned} z &= (v_0 \sin\alpha)\frac{y}{v_0 \cos\alpha} - \frac{1}{2}g\left(\frac{y}{v_0 \cos\alpha}\right)^2 \\ &= y\tan\alpha - \frac{g}{2v_0^2}y^2 \sec^2\alpha, \end{aligned}$$

which is the equation of a parabola in the $Y - Z$ plane. Thus the path of a projectile is a parabola.

Illustrative Examples

1. A cannon has its maximum range given by $R_{\max}$. Prove that (a) the height reached in such case is $\frac{1}{4}R_{\max}$, (b) the time to reach the maximum height is $\sqrt{R_{\max}/2g}$ and (c) the time of flight is $\sqrt{2R_{\max}/g}$.

[C.U. (Hons.) '95]

(a) Using equation (3.45.8) and putting $\alpha = 45°$, we have the height required in the problem $= \dfrac{v_0^2}{4g} = \dfrac{1}{4}R_{\max}$, using equation (3.45.11).

(b) From (3.45.11), we have

$$R_{\max} = \frac{v_0^2}{g} \quad \text{or} \quad \frac{R_{\max}}{2g} = \frac{v_0^2}{2g^2} = \left(\frac{v_0}{\sqrt{2}g}\right)^2$$

$$\text{or} \quad \sqrt{\frac{R_{\max}}{2g}} = \frac{v_0}{\sqrt{2}g} = \text{the time to reach the maximum height, using (3.45.7) and putting } \alpha = 45°.$$

(c) We know

$$R_{\max} = \frac{v_0^2}{g} \quad \text{or} \quad \frac{2R_{\max}}{g} = \frac{2v_0^2}{g^2} = \left(\frac{\sqrt{2}v_0}{g}\right)^2$$

$$\text{or} \quad \sqrt{\frac{2R_{\max}}{g}} = \sqrt{2}\frac{v_0}{g} = \text{the time of flight, using (3.45.9) and putting } \alpha = 45°.$$

2. An inclined plane (Fig. 4) makes an angle α with the horizontal. A projectile is launched from the bottom A of the incline with speed v_0 in a direction making an angle β with the horizontal.

(a) Prove that the range R up the incline is given by

$$R = \frac{2v_0^2 \sin(\beta - \alpha)\cos\beta}{g\cos^2\alpha}.$$

(b) Prove that the maximum range up the incline is given by

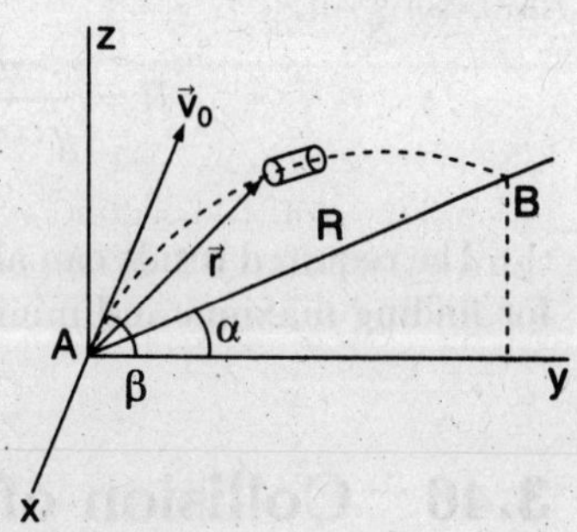

$$R_{\max} = \frac{v_0^2}{g(1+\sin\alpha)}$$

and is achieved when $\beta = \dfrac{\pi}{4} + \dfrac{\alpha}{2}$.

(a) As shown in Art. 3.45, the position vector of the projectile at any time t is given by equation (3.45.5) and may be written as

$$\vec{r} = (v_0\cos\beta)t\hat{\jmath} + \left\{(v_0\sin\beta)t - \frac{1}{2}gt^2\right\}\hat{k} \qquad (1)$$

$$\text{or} \quad y = (v_0\cos\beta)t, \quad z = (v_0\sin\beta)t - \frac{1}{2}gt^2. \qquad (2)$$

The equation of the incline, (which is a line in the $Y - Z$ plane) is given by

$$z = y\tan\alpha. \qquad (3)$$

Using (2) in (3), we see that the projectile's path and the incline intersect for those values of t, satisfying

$$(v_0\sin\beta)t - \frac{1}{2}gt^2 = \{(v_0\cos\beta)t\}\tan\alpha$$

$$\text{or} \quad (2v_0\sin\beta\cos\alpha)t - gt^2\cos\alpha - (2v_0\cos\beta\sin\alpha)t = 0$$

$$\text{or} \quad t(2v_0\sin\beta\cos\alpha - 2v_0\cos\beta\sin\alpha - gt\cos\alpha) = 0,$$

which gives $t = 0$ and $t = \dfrac{2v_0\sin(\beta-\alpha)}{g\cos\alpha}$.

The value $t = 0$ gives the intersection point A, and the second value of t gives the point B, which is the required point. Using this second value of t in the first equation of (2), we have the required range R up the incline given by

$$R = y\sec\alpha = (v_0\cos\beta)\left\{\frac{2v_0\sin(\beta-\alpha)}{g\cos\alpha}\right\}\sec\alpha \qquad \left[\because \sec\alpha = \frac{R}{y}\right]$$

$$= \frac{2v_0^2\sin(\beta-\alpha)\cos\beta}{g\cos^2\alpha}. \qquad (4)$$

(b) Equation (4) may be written as

$$R = \frac{v_0^2}{g\cos^2\alpha}\{\sin(2\beta-\alpha) - \sin\alpha\}$$

{using the trigonometric identity,

$$\cos A\sin B = \frac{1}{2}\{\sin(A+B) - \sin(A-B)\}$$

which will assume a maximum value when

$$\sin(2\beta-\alpha) = 1, \quad \text{i.e.,} \quad 2\beta - \alpha = \frac{\pi}{2} \quad \text{or} \quad \beta = \frac{\pi}{4} + \frac{\alpha}{2}$$

and the value of this maximum is given by

$$R = \frac{v_0^2}{g\cos^2\alpha}(1-\sin\alpha) = \frac{v_0^2(1-\sin\alpha)}{g(1-\sin^2\alpha)} = \frac{v_0^2}{g(1+\sin\alpha)}.$$

The required result can also be alternatively obtained by the methods of differential calculus for finding maxima and minima.

3.46 Collision of Particles

When two elastic bodies collide or impinge with each other, the collision is called *impact.* The time during which such objects are in contact is composed of a *compression time,* during which slight deformation may take place, and *restitution time* during which the shape is restored. We assume that the colliding objects, such as spheres, are smooth so that forces are exerted along the *common normal* to the spheres at their point of contact—known as *the line of impact.* In case of two impinging spheres, the common normal (i.e., the line of impact) is the line joining their centres. A collision can be *direct* or *oblique.* In a *direct collision,* the direction of motion of both the colliding bodies *is along the line of impact both before and after collision,* while in an *oblique collision,* the direction of motion of either or both *is not along the line of impact.*

In the general case, both the particles move in such directions that they come closer to each other and collision takes place [Fig. 3.46.1(a)]. Experiments are often carried out in which one of the particles is at rest in the laboratory and the other approaches it and collision takes place [Fig. 3.46.1(b)]. Such a set-up in which a particle collides with another particle at rest is called the laboratory frame of reference. Again the discussion of collision between two bodies becomes very simple and symmetric, if we describe the effects of the collision in a co-ordinate system in which the centre of mass is at rest. The system is referred to as the centre of mass system or in short, as the C.M. system. In this system, the collision between two particles is treated as if they have equal and opposite momenta initially [Fig. 3.46.1(c)].

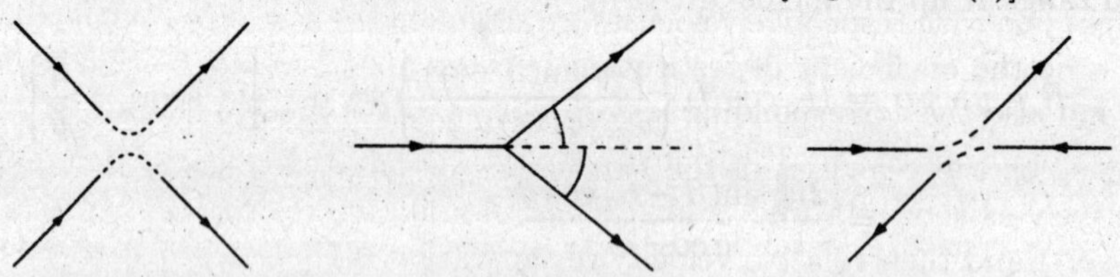

(a) General (b) Laboratory frame (c) Centre of mass frame

Fig. 3.46.1

Fundamental in collision problems is the following principle called *Newton's Collision rule,* which is based on experimental evidence. We shall take it as a postulate.

Newton's Collision Rule: If two bodies impinge on each other, the relative velocity of separation of the two bodies immediately after impact measured along the line of impact (i.e., along the common normal at the point of contact), bears a constant ratio to their relative velocity of approach along the same direction just before impact.

If u_1, u_2 be the components of velocities of two colliding bodies along the line of impact immediately before impact, and $v_1; v_2$ their corresponding component velocities just after impact along the same line and in the same sense, then

$$\frac{v_2 - v_1}{u_1 - u_2} = e \quad \text{or} \quad v_2 - v_1 = e(u_1 - u_2).$$

The constant e, known as the *coefficient of restitution* (or *resilience*), depends on the materials of which the bodies are made, and is independent of the masses of the colliding bodies, is always positive, and lies between 0 and 1. If $e = 0$, the collision is called *perfectly inelastic* or briefly *in-elastic,* and in this case, the two bodies do not separate after impact. If $e = 1$, the collision is called *perfectly elastic* or briefly *elastic,* and in this case, the velocity of separation just after impact is equal to the velocity of approach just before impact, and therefore the total K.E. before and after impact is the same.

Elastic and In-elastic Collisions: If the forces of inter-action between the colliding particles are conservative, the K.E. remains conserved in the collision process and the collision is said to be *elastic.* In such a process, therefore, the internal nature or the structure of the particles does not change and we can apply the law of conservation of energy without taking into consideration the internal energies at all. Collisions between atomic, nuclear and fundamental particles are usually elastic. Collisions between ivory and glass balls are approximately elastic.

When, the K.E. is changed in the collision process, the collision is said to be *inelastic* (the momentum as well as the total energy is still conserved). Most of the collisions which occur in nature are in-elastic. When two large bodies collide, some amount of heat and sound is produced. Thus, some part of the initial K.E. is lost in the process. This loss must be taken into account, while using the law of conservation of energy. When two bodies stick together after collision, the collision is said to be completely in-elastic. When a bullet hitting a target remains embedded in the target, the collision is completely in-elastic. In a completely in-elastic collision the whole of initial K.E. is not lost, but the loss is as great as is consistent with momentum conservation. The difference is converted into heat or to potential energy of deformation in the collision. The final K.E. may also exceed the initial value when potential energy is released in the collision process. Changes are known to take place in the internal structure of the particles during an in-elastic collision process.

Case 1. Direct impact of two bodies: Two bodies of masses m_1 and m_2 travelling in the same straight line with respective velocities u_1 and u_2 in the same direction impinge directly (head-on). If e be the coefficient of restitution between them, we are to find their velocities after impact, and also the corresponding loss of K.E., and the mutual transfer of momentum.

Let v_1 and v_2 be the velocities of the two bodies just after impact in the direction of u_1 and u_2 respectively as shown in Fig. 3.46.2. Then the relative velocity of approach of the two bodies is $(u_1 - u_2)$ and their relative velocity of separation is $(v_2 - v_1)$.

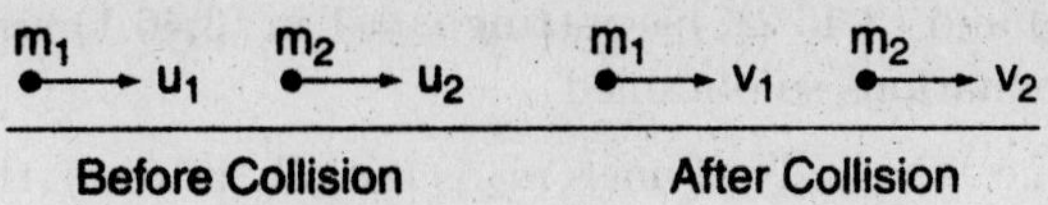

Fig. 3.46.2

By Newton's collision rule, we have

$$v_2 - v_1 = e(u_1 - u_2). \tag{3.46.1}$$

Again since the impulsive action as well as reaction is along the same line, we have by the principle of conservation of momentum

$$m_1 v_1 + m_2 v_2 = m_1 u_1 + m_2 u_2, \tag{3.46.2}$$

i.e., the total momentum along the line of impact, before and after impact, remains unaltered.

Multiplying (3.46.1) by m_2 and then subtracting from (3.46.2), we get

$$m_1v_1 + m_2v_1 = m_1u_1 + m_2u_2 - em_2u_1 + em_2u_2,$$

$$\text{or,} \quad (m_1 + m_2)v_1 = u_1(m_1 - em_2) + m_2u_2(1 + e)$$

$$\text{or,} \quad v_1 = \frac{(m_1 - em_2)}{(m_1 + m_2)}u_1 + \frac{m_2(1 + e)}{(m_1 + m_2)}u_2. \tag{3.46.3}$$

Again multiplying (3.46.1) by m_1 and adding with (3.46.2), we obtain

$$(m_1 + m_2)v_2 = m_1(1 + e)u_1 + (m_2 - em_1)u_2$$

$$\text{or,} \quad v_2 = \frac{m_1(1 + e)}{(m_1 + m_2)}u_1 + \frac{(m_2 - em_1)}{(m_1 + m_2)}u_2. \tag{3.46.4}$$

Equations (3.46.3) and (3.46.4) give the velocities v_1 and v_2 of the two bodies after impact.

(a) Now for *a perfectly inelastic collision,* we put $e = 0$ in (3.46.3) and (3.46.4) above and obtain

$$v_1 = \frac{m_1u_1 + m_2u_2}{m_1 + m_2} \quad v_2 = \frac{m_1u_1 + m_2u_2}{m_1 + m_2}.$$

Thus after collision, the two particles move with the same velocity, i.e., they move as if they were stuck together as a single particle.

(b) Again for *a perfectly elastic collision,* we put $e = 1$ in (3.46.3) and (3.46.4) above, and obtain

$$v_1 = \frac{(m_1 - m_2)}{(m_1 + m_2)}u_1 + \frac{2m_2}{(m_1 + m_2)}u_2 \tag{3.46.5}$$

$$\text{and} \quad v_2 = \frac{2m_1}{(m_1 + m_2)}u_1 + \frac{(m_2 - m_1)}{(m_1 + m_2)}u_2, \tag{3.46.6}$$

which shows that after collision, the two particles move with different velocities.

Special cases: (1) When $m_1 = m_2$, then from (3.46.5) $v_1 = u_2$ and from (3.46.6), $v_2 = u_1$ which shows that *in a one-dimensional elastic collision of two bodies of equal masses, the bodies simply exchange or interchange their velocities as a result of collision.* This result may also be obtained from (3.46.1) and (3.46.2), by putting $e = 1$ in (3.46.1) and $m_1 = m_2$ in (3.46.2) and then solving the two equations so obtained.

(2) When $u_2 = 0$ (i.e., the body of mass m_2 is initially at rest), the final velocities are given, from (3.46.5) and (3.46.6), by

$$v_1 = \left(\frac{m_1 - m_2}{m_1 + m_2}\right)u_1 = \frac{\left(1 - \dfrac{m_2}{m_1}\right)}{\left(1 + \dfrac{m_2}{m_1}\right)}u_1 \quad \text{and} \quad v_2 = \frac{2m_1}{m_1 + m_2}u_1 = \frac{2u_1}{1 + \dfrac{m_2}{m_1}}. \tag{3.46.7}$$

Now here we may consider three situations:

(i) If $m_1 = m_2$, then $v_1 = 0$ and $v_2 = u_1$ as expected. The first body is *stopped cold* and the second one *takes off,* with the velocity which the first one originally had. Both the momentum and the K.E. of the first are completely transferred to the second.

(ii) If $m_2 \gg m_1$, then $v_1 \simeq -u_1$ and $v_2 \simeq 0$. That is, when a light body collides with a much heavier body at rest, the velocity of the light body is approximately reversed and the heavier body remains approximately at rest. A ball dropped on earth rebounds with reversed velocity attaining approximately the same height from which it falls, provided the collision is elastic.

(iii) If $m_2 \ll m_1$, then $v_1 \simeq u_1$, and $v_2 \simeq 2u_1$. That is, when a heavy body collides with a much lighter body at rest, the velocity of the heavy body remains practically unchanged, but the light body rebounds with approximately twice the velocity of the heavy body.

The above considerations show that in order to "*slow down*" the fast neutrons in a nuclear reactor, they must be made to collide with the stationary targets (nuclei) of nearly the same mass as that of the neutrons themselves ($m_1 = m_2$). That is why paraffin (which is rich in hydrogen whose nucleus "*proton*" has nearly the same mass as that of a neutron) is a very good moderator.

Loss of K.E. due to impact: Loss of K.E. due to impact = K.E. before impact − K.E. after impact

$$
\begin{aligned}
&= \left(\frac{1}{2}m_1u_1^2 + \frac{1}{2}m_2u_2^2\right) - \left(\frac{1}{2}m_1v_1^2 + \frac{1}{2}m_2v_2^2\right) \\
&= \frac{1}{2}\{(m_1u_1^2 + m_2u_2^2) - (m_1v_1^2 + m_2v_2^2)\} \\
&= \frac{1}{2(m_1+m_2)}\{(m_1+m_2)(m_1u_1^2 + m_2u_2^2) - (m_1+m_2)(m_1v_1^2 + m_2v_2^2)\} \\
&= \frac{1}{2(m_1+m_2)}\Big[\{m_1^2u_1^2 + m_2^2u_2^2 + m_1m_2(u_1^2 + u_2^2)\} \\
&\qquad - \{m_1^2v_1^2 + m_2^2v_2^2 + m_1m_2(v_1^2 + v_2^2)\}\Big] \\
&= \frac{1}{2(m_1+m_2)}\Big[\{(m_1u_1 + m_2u_2)^2 + m_1m_2(u_1 - u_2)^2\} \\
&\qquad - \{(m_1v_1 + m_2v_2)^2 + m_1m_2(v_2 - v_1)^2\}\Big] \\
&= \frac{1}{2(m_1+m_2)}\left[m_1m_2(u_1-u_2)^2 - m_1m_2(u_1-u_2)^2e^2\right] \quad \text{using (3.46.1) and (3.46.2)} \\
&= \frac{m_1m_2}{2(m_1+m_2)}(1-e^2)(u_1-u_2)^2.
\end{aligned}
$$

Now for a *perfectly inelastic collision,* $e = 0$ and the loss of

$$
\text{K.E.} = \frac{m_1m_2}{2(m_1+m_2)}(u_1-u_2)^2.
$$

If, however, the collision be *perfectly elastic,* $e = 1$ and the loss of K.E. = 0, i.e., the total K.E. remains unchanged after impact.

Impulsive Action: Impulsive action on m_2 due to impact

$$\begin{aligned}
&= \text{the change of momentum of } m_2 \text{ due to impact}\\
&= m_2v_2 - m_2u_2 = m_2(v_2 - u_2)\\
&= \frac{m_2}{(m_1+m_2)}[(m_1+m_2)v_2 - (m_1+m_2)u_2]\\
&= \frac{m_2}{(m_1+m_2)}[m_1u_1(1+e) + u_2(m_2 - em_1) - (m_1+m_2)u_2] \quad \text{using (3.46.4)}\\
&= \frac{m_2}{(m_1+m_2)}[m_1u_1(1+e) + u_2(m_2 - em_1 - m_1 - m_2)]\\
&= \frac{m_2}{(m_1+m_2)}[m_1u_1(1+e) - m_1u_2(1+e)]\\
&= \frac{m_1m_2}{(m_1+m_2)}(1+e)(u_1 - u_2),
\end{aligned}$$

which is the momentum transferred from the first body to the second. Similarly we can find the impulsive action on m_1 due to impact, and it will be equal and opposite to that on m_2. By putting $e = 0$ and $e = 1$, we can get the expression of the momenta transferred (due to impact) for an *inelastic* as well as for an *elastic* collision.

Maximum Energy Transfer in a Head-on Elastic Collision: Let a ball of mass m_1 travelling with a velocity u_1 collide with a ball of mass m_2 at rest. Let v_1 be the final velocity of the first ball. Then the initial K.E. of the ball is $k_i = \frac{1}{2}m_1u_1^2$ and the final K.E. is $k_f = \frac{1}{2}m_1v_1^2$, so that the fractional decrease in K.E. is

$$\frac{k_i - k_f}{k_i} = \frac{u_1^2 - v_1^2}{u_1^2} = 1 - \frac{v_1^2}{u_1^2}.$$

But for such a collision, we may write, using (3.46.7)

$$v_1 = \left(\frac{m_1 - m_2}{m_1 + m_2}\right)u_1 \quad \text{or} \quad \frac{v_1^2}{u_1^2} = \left(\frac{m_1 - m_2}{m_1 + m_2}\right)^2.$$

Therefore

$$\frac{k_i - k_f}{k_i} = 1 - \left(\frac{m_1 - m_2}{m_1 + m_2}\right)^2 = \frac{4m_1m_2}{(m_1+m_2)^2} = \frac{4(m_2/m_1)}{\left(1 + \dfrac{m_2}{m_1}\right)^2}. \tag{3.46.8}$$

Let us now put $m_1 = m$ and $m_2 = nm$, then

$$\frac{k_i - k_f}{k_i} = \frac{4n}{(1+n)^2}.$$

The transfer of energy will be maximum when $k_f = 0$, so that

$$\frac{4n}{(1+n)^2} = 1 \quad \text{or} \quad n = 1.$$

Thus when the mass ratio is unity, the "*whole*" of the K.E. of the moving ball is transferred to the ball initially at rest.

Special cases: (i) When $\frac{m_2}{m_1} = \frac{\text{mass of hydrogen atom}}{\text{mass of an electron}} = 1840$,

$$\frac{k_i - k_f}{k_i} = \frac{4 \times 1840}{(1841)^2} = 0.00217 \text{ or } 0.217\%.$$

(ii) When $\frac{m_2}{m_1} = \frac{\text{mass of hydrogen nucleus}}{\text{mass of a neutron}} = 1$, $\frac{k_i - k_f}{k_i} = 1$ or 100%.

(iii) When $\frac{m_2}{m_1} = \frac{\text{mass of lead nucleus}}{\text{mass of a neutron}} = 206$, $\frac{k_i - k_f}{k_i} = 0.02$ or 2%.

Case 2. Collision of particles in the C.M. frame: Discussion of problems on collision between two particles becomes simple, symmetric and easy if we describe the effects of the collision in C.M. (C) co-ordinate system instead of treating the same in the laboratory (L) system. The velocity $\vec{V}$ of the C.M. of the two particles of masses m_1 and m_2 moving with velocities $\vec{V}_1$ and $\vec{V}_2$ respectively is given by [from 3.29.4]

$$\vec{V} = \frac{m_1\vec{V}_1 + m_2\vec{V}_2}{m_1 + m_2} \qquad (3.46.9)$$

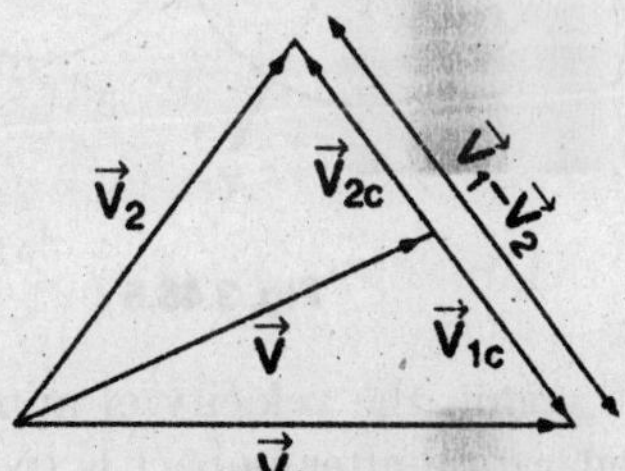

Fig. 3.46.3

The velocities of the particles in the C-system (Fig. 3.46.3) are

$$\vec{V}_{1c} = \vec{V}_1 - \vec{V} = \frac{m_2}{m_1 + m_2}(\vec{V}_1 - \vec{V}_2). \qquad (3.46.10)$$

$$\vec{V}_{2c} = \vec{V}_2 - \vec{V} = -\frac{m_1}{m_1 + m_2}(\vec{V}_1 - \vec{V}_2). \qquad (3.46.11)$$

Thus $\vec{V}_{1c}$ and $\vec{V}_{2c}$ are oppositely directed along the relative velocity vector $\vec{V} = \vec{V}_1 - \vec{V}_2$. Now the linear momenta of the two particles in the C-system are $m_1\vec{V}_{1c} = \mu\vec{V}$ and $m_2\vec{V}_{2c} = -\mu\vec{V}$, (where $\mu = m_1m_2/(m_1 + m_2)$ = reduced mass of the system—which is the natural unit of mass in a two-particle system), so that the total linear momentum in the C-system is zero, and as such $\vec{V}_{1c}$ and $\vec{V}_{2c}$ are back to back before collision and so also $\vec{V}'_{1c}$ and $\vec{V}'_{2c}$ after collision as shown in Fig. 3.46.4. The initial and final velocities in the C-system determine a plane known as *the plane of scattering,* where each particle is deflected through the same scattering angle θ. Interaction force is directly related to this scattering angle θ. The situation is further simplified if the collision is elastic. From conservation of K.E.

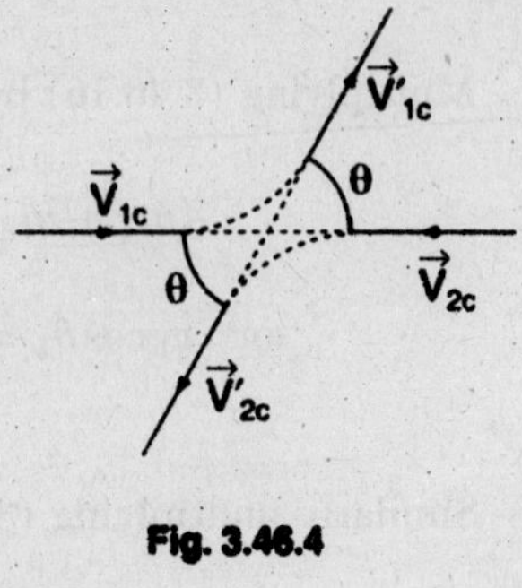

Fig. 3.46.4

$$\frac{1}{2}m_1v_{1c}^2 + \frac{1}{2}m_2v_{2c}^2 = \frac{1}{2}m_1v'^2_{1c} + \frac{1}{2}m_2v'^2_{2c} \qquad (3.46.12)$$

and from conservation of linear momentum, which is zero in the C-system

$$m_1v_{1c} + m_2v_{2c} = 0 = m_1v'_{1c} + m_2v'_{2c} \qquad (3.46.13)$$

(3.46.12) and (3.46.13) give

$$v_{1c} = v'_{1c} \quad \text{and} \quad v_{2c} = v'_{2c}.$$

Thus in an elastic collision, speed of each particle remains unchanged in the C-system. Therefore the velocity vectors simply rotate in the scattering plane—thereby giving all permissible values of θ.

Case 3. Oblique Impact of two Spheres: Two smooth spheres of masses m_1 and m_2 moving with respective velocities u_1 and u_2 at angles α_1 and α_2 with their line of centres, impinge obliquely. If e be the coefficient of restitution, we are to find the magnitudes and directions of the velocities just after impact, loss of K.E., and mutual transfer of momentum due to impact.

Let v_1 and v_2 be the velocities of the two spheres just after impact in the directions, making angles θ_1 and θ_2 with the line of impact C_1C_2. Since the spheres are smooth, the impulsive action and re-action will be along the line of centres C_1C_2, and therefore, there will no change in the velocity components in the direction perpendicular to C_1C_2.

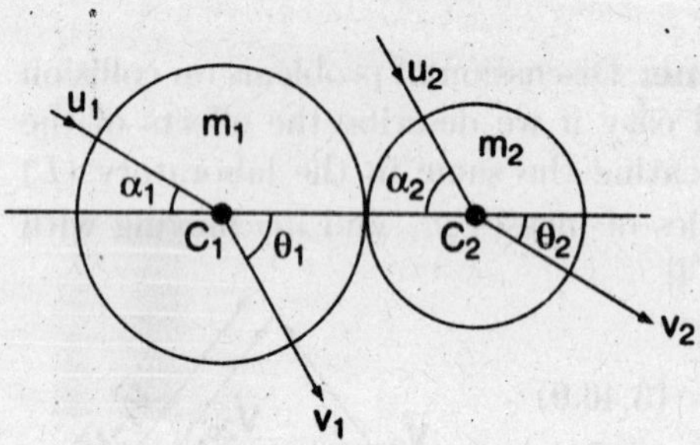

Fig. 3.46.5

$\therefore$ resolving the velocities perpendicular to C_1C_2, we get

$$v_1 \sin\theta_1 = u_1 \sin\alpha_1 \tag{3.46.14}$$

and $$v_2 \sin\theta_2 = u_2 \sin\alpha_2. \tag{3.46.15}$$

Again the velocity of approach before impact is $(u_1 \cos\alpha_1 - u_2 \cos\alpha_2)$ and the velocity of separation after impact is $(v_2 \cos\theta_2 - v_1 \cos\theta_1)$, so that by Newton's Collision rule, we get

$$(v_2 \cos\theta_2 - v_1 \cos\theta_1) = e(u_1 \cos\alpha_1 - u_2 \cos\alpha_2). \tag{3.46.16}$$

Also, by the principle of conservation of linear momentum, along the line of impact, we have

$$m_1 v_1 \cos\theta_1 + m_2 v_2 \cos\theta_2 = m_1 u_1 \cos\alpha_1 + m_2 u_2 \cos\alpha_2. \tag{3.46.17}$$

Multiplying (3.46.16) by m_2 and then subtracting from (3.46.17), we get

$$(m_1 + m_2) v_1 \cos\theta_1 = (m_1 - em_2) u_1 \cos\alpha_1 + m_2(1 + e) u_2 \cos\alpha_2$$

or $$v_1 \cos\theta_1 = \frac{(m_1 - em_2) u_1 \cos\alpha_1 + m_2(1 + e) u_2 \cos\alpha_2}{(m_1 + m_2)}. \tag{3.46.18}$$

Similarly multiplying (3.46.16) by m_1 and then adding with (3.46.17), we have

$$v_2 \cos\theta_2 = \frac{m_1(1 + e) u_1 \cos\alpha_1 + (m_2 - em_1) u_2 \cos\alpha_2}{(m_1 + m_2)}. \tag{3.46.19}$$

Squaring (3.46.14) and (3.46.18) and then adding, we get v_1 and dividing (3.46.14) by (3.46.18), we obtain θ_1. Similarly, v_2 and θ_2 can be obtained from (3.46.15) and (3.46.19).

Loss of K.E. due to oblique impact: Loss of K.E. due to impact

$$= \text{K.E. before impact} - \text{K.E. after impact}$$

$$= \frac{1}{2}\left[(m_1u_1^2 + m_2u_2^2) - (m_1v_1^2 + m_2v_2^2)\right]$$

$$= \frac{1}{2}\Big[m_1u_1^2(\cos^2\alpha_1 + \sin^2\alpha_1) + m_2u_2^2(\cos^2\alpha_2 + \sin^2\alpha_2) - \left\{m_1v_1^2(\cos^2\theta_1 + \sin^2\theta_1) + m_2v_2^2(\cos^2\theta_2 + \sin^2\theta_2)\right\}\Big]$$

$$= \frac{1}{2}[m_1u_1^2\cos^2\alpha_1 + m_2u_2^2\cos^2\alpha_2 - (m_1v_1^2\cos^2\theta_1 + m_2v_2^2\cos^2\theta_2)]$$

using (3.46.14) and (3.46.15)

$$= \frac{1}{2(m_1+m_2)}[m_1^2u_1^2\cos^2\alpha_1 + m_2^2u_2^2\cos^2\alpha_2 + m_1m_2(u_1^2\cos^2\alpha_1 + u_2^2\cos^2\alpha_2) - \{m_1^2v_1^2\cos^2\theta_1 + m_2^2v_2^2\cos^2\theta_2 + m_1m_2(v_1^2\cos^2\theta_1 + v_2^2\cos^2\theta_2)\}]$$

$$= \frac{1}{2(m_1+m_2)}[(m_1u_1\cos\alpha_1 + m_2u_2\cos\alpha_2)^2 + m_1m_2(u_1\cos\alpha_1 - u_2\cos\alpha_2)^2 - \{(m_1v_1\cos\theta_1 + m_2v_2\cos\theta_2)^2 + m_1m_2(v_2\cos\theta_2 - v_1\cos\theta_1)^2\}]$$

$$= \frac{1}{2(m_1+m_2)}[m_1m_2(u_1\cos\alpha_1 - u_2\cos\alpha_2)^2 - m_1m_2e^2(u_1\cos\alpha_1 - u_2\cos\alpha_2)^2]$$

using (3.46.16) and (3.46.17)

$$= \frac{1}{2}\frac{m_1m_2}{(m_1+m_2)}(1-e^2)(u_1\cos\alpha_1 - u_2\cos\alpha_2)^2.$$

If $e = 0$, i.e., if the collision is perfectly in-elastic, then the loss of K.E. is

$$\frac{1}{2}\frac{m_1m_2}{(m_1+m_2)}(u_1\cos\alpha_1 - u_2\cos\alpha_2)^2.$$

If however $e = 1$, i.e., if the collision is perfectly elastic, then the loss of K.E. = 0, i.e., the total K.E. remains unchanged after oblique impact.

Impulsive Action: Impulsive action on m_2 due to oblique impact

= the change of momentum of m_2 due to oblique impact

$$= m_2v_2\cos\theta_2 - m_2u_2\cos\alpha_2 = m_2(v_2\cos\theta_2 - u_2\cos\alpha_2)$$

$$= \frac{m_2}{(m_1+m_2)}[(m_1+m_2)v_2\cos\theta_2 - (m_1+m_2)u_2\cos\alpha_2]$$

$$= \frac{m_2}{(m_1+m_2)}[m_1(1+e)u_1\cos\alpha_1 + (m_2 - em_1)u_2\cos\alpha_2 - m_1u_2\cos\alpha_2 - m_2u_2\cos\alpha_2]$$

using (3.46.19)

$$= \frac{m_2}{(m_1+m_2)}[m_1(1+e)u_1\cos\alpha_1 - em_1u_2\cos\alpha_2 - m_1u_2\cos\alpha_2]$$

$$= \frac{m_2}{(m_1+m_2)}[m_1(1+e)u_1\cos\alpha_1 - m_1u_2\cos\alpha_2(1+e)]$$

$$= \frac{m_1m_2}{m_1+m_2}(1+e)(u_1\cos\alpha_1 - u_2\cos\alpha_2),$$

which is the momentum transferred from the first sphere to the second. Similarly we can find the impulsive action on m_1 due to oblique impact, and it will be equal and opposite to that on m_2. By putting $e = 0$ and $e = 1$, we can get the expressions of the momenta transferred (due to oblique impact) for an *in-elastic* as well as for an *elastic* collision.

3.47 Application of the principles of linear and angular momentum conservation in the case of a moving bullet striking with a large wooden block : The Ballistic pendulum

We consider a rigid body, say a large wooden block, hanging in equilibrium from a horizontal axis O (Fig. 3.47.1). A bullet, travelling horizontally, strikes the body at A, and becomes embedded in it. As a result of the impact, the body swings as a compound pendulum, the centre of mass G rising to G' through an angular displacement θ before coming to rest. On account of its importance in ballistics, the apparatus is called a ***ballistic pendulum***. From the angle θ, and the constants of the system, we can calculate the velocity of the bullet as shown below.

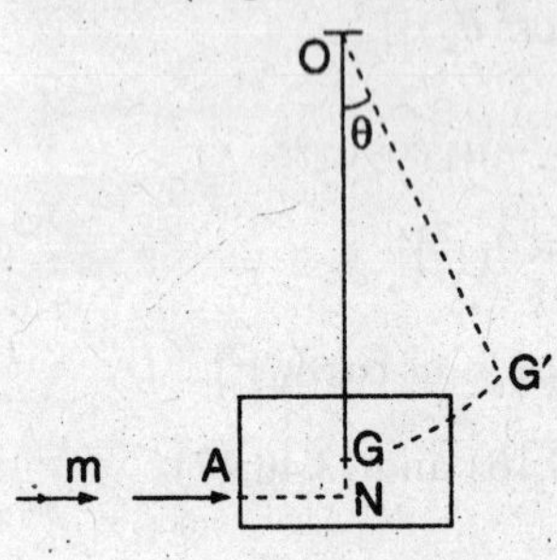

Fig. 3.47.1

Let us take the body and the bullet as a dynamical system. Then the forces between the body and the bullet are internal. During the brief interval of impact, the only external forces acting are: (i) gravity and (ii) the reaction at O. The force of gravity is a finite force and so contributes no impulsive action. Since the reaction at O has no moment about O, it is evident that the principle of angular momentum enables us to state that—angular momentum of the system about O before impact = angular momentum of the system about O after impact.

Let ON be the vertical through O, AN being horizontal. Let m be the mass of bullet, v its speed, I the moment of inertia of the body about O, and w its angular velocity immediately after impact. Then the angular momentum of the system about O is $mv\ ON$ before impact and is $[m(AO)^2 + I]w$ after impact.

If the mass of the bullet be very small in comparison with that of the body, we may neglect $m(AO)^2$ in comparison with I, so that we may write

$$mvl = Iw, \tag{3.47.1}$$

where $l = ON$. Although we cannot apply the principle of energy conservation during impact, we can apply it in the subsequent motion. Thus, again neglecting the mass of the bullet in comparison with that of the body, we have

$$\frac{1}{2}Iw^2 = Mgh(1 - \cos\theta), \tag{3.47.2}$$

where M is the mass of the body and h the distance of the centre of mass G from O. From (3.47.1) and (3.47.2), we get

$$v = \frac{\sqrt{2IMgh(1 - \cos\theta)}}{ml}, \tag{3.47.3}$$

which gives the speed of the bullet in terms of θ and constants of the system.

Illustrative Examples

1. A ball which is dropped from a height H onto a floor rebounds to a height $h < H$. Determine the coefficient of restitution.

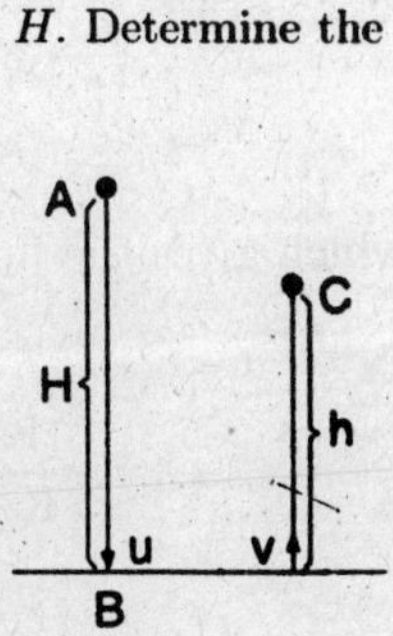

Let u be the velocity acquired by the ball on reaching the fixed horizontal floor at B, when it is dropped from a height H above the floor. Then

$$u^2 = 0 + 2gH = 2gH \quad \text{or} \quad u = \sqrt{2gH}.$$

Here u is the velocity of approach just before impact. Let v be the upward velocity of the ball at B immediately after impact. Both u and v are along the common normal at the point of contact of the ball and the floor.

$\therefore$ By Newton's collision rule

$$v = eu \quad \text{or} \quad v^2 = e^2u^2 = e^2 2gH.$$

If h be the height to which the ball rebounds and $BC = h$, then the ball comes to rest at C at a height h above the floor and we may write

$$0 = v^2 - 2gh \quad \text{or} \quad h = \frac{v^2}{2g} = \frac{e^2 2gH}{2g} = e^2H. \quad \therefore\ e = \sqrt{h/H}.$$

2. A gun fires a bullet of mass m with horizontal velocity $\vec{v}$ into a block of wood of mass M which rests on a horizontal frictionless plane. If the bullet becomes embedded in the wood, (a) determine the subsequent velocity of the system, and (b) find the loss in K.E.

There is no horizontal force before collision, and also during collision, if the collision time (the time taken by the bullet to come to rest w.r.t. the block) is very small compared to the time of swing of the block, and then the linear momentum is conserved in such a collision in the horizontal direction. So we may write $m\vec{v} = (M+m)\vec{V}$, where $\vec{V}$ is the required horizontal speed of the bullet-block system immediately after collision.

$$\therefore\ \vec{V} = \frac{m\vec{v}}{(M+m)}$$

and the loss in K.E.

$$= \frac{1}{2}mv^2 - \frac{1}{2}(M+m)\vec{V}^2 \quad (v \rightarrow \text{magnitude of } \vec{v})$$

$$= \frac{1}{2}mv^2 - \frac{1}{2}(M+m)\frac{m^2v^2}{(M+m)^2} \quad (\text{putting the value of } \vec{V})$$

$$= \frac{1}{2}mv^2 - \frac{1}{2}\frac{m^2v^2}{(M+m)} = \frac{mMv^2}{2(M+m)}.$$

N.B. After the collision is over, the bullet-block system moves in the conservative gravitational field. Therefore, the conservation of mechanical energy will be valid for this part of the motion, so that we may write

$$\frac{1}{2}(M+m)V^2 = (M+m)gh \quad \text{or} \quad V = \sqrt{2gh},$$

where h is the maximum height to which the bullet-block system swings if the block is suspended by a cord from a support, and V = magnitude of $\vec{V}$.

$$\therefore\ v = \frac{(M+m)}{m}\sqrt{2gh},$$

which gives the initial speed of the bullet in terms of M, m and h.

Also

$$\frac{\text{K.E. of the system after collision}}{\text{K.E. of the system before collision}} = \frac{\frac{1}{2}(M+m)V^2}{\frac{1}{2}mv^2} = \frac{m}{M+m}.$$

As $m \ll M$, a very small fraction of the original K.E. remains, most of the energy being converted into heat etc.

3. A bullet of mass 5 gm. is shot through a 1 kg. wood-block suspended by a string 2 m. long. The C.G. of the block rises a distance of 0.50 cm. Find the speed of the bullet as it emerges from the block if the initial speed is 300 m/sec.

If v be the velocity of the emerging bullet and V the velocity acquired by the block just after collision, then from the conservation of momentum, we have, Initial momentum of the bullet = Final momentum of the bullet + momentum of the block

$$\text{or} \quad (0.005\text{ kg})(300\text{ m/sec}) = (0.005\text{ kg})\ v + (1\text{ kg})V. \tag{1}$$

Now K.E. $\frac{1}{2}MV^2$ acquired by the block is converted into gravitational P.E. Mgh, so that

$$V = \sqrt{2gh} = \sqrt{2 \times (9.8\,m/s^2) \times (0.50 \times 10^{-2}\,m)} = 0.3125\text{ m/s}.$$

Putting this value of V in (1), we get

$$1.5 = 0.005v + 0.3125 \quad \text{or} \quad v = \frac{1.5 - 0.3125}{0.005} = 237\text{ m/s}.$$

4. Consider two particles of masses m_1 and m_2 which collide and stick together on collision. If m_2 be at the origin and at rest, and m_1 is moving with velocity $\vec{V_1}$ along the $+ve$ direction of X-axis before collision, then find (i) the velocity of the coalesced particles after collision, (ii) the ratio of the K.E. after collision to the initial K.E., (iii) Discuss the motion of the system before and after the collision in the C.M. frame.

or

Discuss an in-elastic collision between two particles in the laboratory frame and also in the C.M. frame.

When the colliding particles stick together after the collision and move as a single mass, the collision is completely an inelastic collision.

(i) In the laboratory frame of reference, the velocity of m_1 is $\vec{V_1}$ and m_2 is at rest before collision. If $\vec{V}$ be the velocity of the coalesced particle of mass $(m_1 + m_2)$ after the collision, then from conservation of linear momentum, we get

$$m_1\vec{V_1} + m_2(0) = (m_1 + m_2)\vec{V} \quad \text{or} \quad \vec{V} = \frac{m_1\vec{V_1}}{m_1 + m_2}, \tag{1}$$

i.e., both particles move with the same velocity $\vec{V}$ after the collision.

(ii) K.E. before collision $T_1 = \frac{1}{2}m_1v_1^2$ (v_1 is the magnitude of $\vec{V_1}$) and K.E. after collision,

$$T_2 = \frac{1}{2}(m_1+m_2)v^2 = \frac{1}{2}(m_1+m_2)\frac{m_1^2v_1^2}{(m_1+m_2)^2} = \frac{m_1^2v_1^2}{2(m_1+m_2)} \quad \text{using (1).}$$

$\therefore \ \frac{T_2}{T_1} = \frac{m_1}{m_1+m_2}$, which shows that $T_2 < T_1$, implying that in an in-elastic collision a part of the K.E. is lost. This lost energy appears as heat energy, and the lost fraction

$$1 - \frac{T_2}{T_1} = \frac{m_2}{m_1+m_2}.$$

(iii) If x be the distance traversed by the centre of mass in time t, then $x = \vec{V}t$, since we know that the velocity of the C.M. of the coalesced particles is the same at every instant in absence of external forces. So from (1) we get

$$x = \frac{m_1}{m_1+m_2}\vec{V_1}t,$$

which on differentiation w.r.t. t, gives the velocity of the C.M. of the system, before collision. So we have

$$\vec{V}_{CM} = \frac{m_1}{m_1+m_2}\vec{V_1}.$$

In the CM reference frame, the velocity of the CM is assumed to be zero. Therefore, the initial velocity of m_1 relative to CM frame is given by

$$\vec{V}_{1c} = \vec{V_1} - \vec{V}_{CM} = \vec{V_1} - \frac{m_1\vec{V_1}}{m_1+m_2} = \frac{m_2\vec{V_1}}{m_1+m_2}$$

and that of m_2 is given by

$$\vec{V}_{2c} = 0 - \vec{V}_{CM} = -\vec{V}_{CM} = -\frac{m_1\vec{V_1}}{m_1+m_2}.$$

Thus before collision, both the particles appear to approach the observer sitting at the centre of mass.

After collision, the velocity $\vec{V}'$ of the coalesced particle in the CM frame is, from the conservation of linear momentum, given by

$$(m_1+m_2)\vec{V}' = m_1\vec{V}_{1c} + m_2\vec{V}_{2c} = \frac{m_1m_2\vec{V_1}}{m_1+m_2} - \frac{m_1m_2\vec{V_1}}{m_1+m_2} = 0.$$

Thus $\vec{V}' = 0$. That is in the CM frame, the system (coalesced particle) is at rest. This is obvious, since the CM coincides with the coalesced particle. So the K.E. after collision is zero, indicating that the initial K.E. is entirely converted into heat.

5. Two particles of masses m_1 and m_2 moving with velocities $\vec{u}_1$ and $\vec{u}_2$ respectively collide; find (a) the velocities of the particles after collision; (b) the ratio of the final K.E. to the initial K.E; and (c) discuss the motion of the system before and after collision, as viewed from the CM frame. The collision may be assumed to be elastic.

(a) If the velocities of the particles after collision be $\vec{V}_1$ and $\vec{V}_2$ respectively, then, from the principle of conservation of linear momentum, we have

$$m_1\vec{u}_1 + m_2\vec{u}_2 = m_1\vec{V}_1 + m_2\vec{V}_2. \tag{1}$$

Also from conservation of energy, we get

$$\frac{1}{2}m_1u_1^2 + \frac{1}{2}m_2u_2^2 = \frac{1}{2}m_1v_1^2 + \frac{1}{2}m_2v_2^2. \tag{2}$$

$u_1, u_2 \rightarrow$ magnitudes of $\vec{u}_1, \vec{u}_2$

$v_1, v_2 \rightarrow$ magnitudes of $\vec{V}_1, \vec{V}_2$.

Solving (1) and (2), we get

$$\left.\begin{aligned} \vec{V}_1 &= \frac{(m_1 - m_2)\vec{u}_1 + 2m_2\vec{u}_2}{m_1 + m_2} \\ \vec{V}_2 &= \frac{2m_1\vec{u}_1 + (m_2 - m_1)\vec{u}_2}{m_1 + m_2} \end{aligned}\right\}, \tag{3}$$

(b) $$\frac{\text{Final K.E.}}{\text{Initial K.E.}} = \frac{T_2}{T_1} = \frac{\frac{1}{2}m_1v_1^2 + \frac{1}{2}m_2v_2^2}{\frac{1}{2}m_1u_1^2 + \frac{1}{2}m_2u_2^2} = 1 \tag{4}$$

(c) If $\vec{r_1}$ and $\vec{r_2}$ be the position vectors of the two particles of masses m_1 and m_2, then the position vector of the CM is given by

$$\vec{r} = \frac{m_1\vec{r_1} + m_2\vec{r_2}}{m_1 + m_2}$$

and the velocity of the CM is given by

$$\vec{V} = \dot{\vec{r}} = \frac{m_1\dot{\vec{r}}_1 + m_2\dot{\vec{r}}_2}{m_1 + m_2} = \frac{m_1\vec{u}_1 + m_2\vec{u}_2}{m_1 + m_2}. \tag{5}$$

The velocity of the particle of mass m_1 before collision relative to CM frame is given by

$$\vec{V}_{1c} = \vec{u}_1 - \vec{V} = \vec{u}_1 - \frac{(m_1\vec{u}_1 + m_2\vec{u}_2)}{m_1 + m_2} \quad \text{from (5)}$$

$$= \frac{m_2(\vec{u}_1 - \vec{u}_2)}{m_1 + m_2}. \tag{6}$$

The velocity of the particle of mass m_2 before collision relative to CM frame is

$$\vec{V}_{2c} = \vec{u}_2 - \vec{V} = \vec{u}_2 - \frac{(m_1\vec{u}_1 + m_2\vec{u}_2)}{m_1 + m_2} = \frac{m_1(\vec{u}_2 - \vec{u}_1)}{m_1 + m_2}. \tag{7}$$

The velocity of the particle of mass m_1 after collision relative to CM frame is

$$\vec{V}'_{1c} = \vec{V}_1 - \vec{V} = \frac{(m_1 - m_2)\vec{u}_1 + 2m_2\vec{u}_2}{m_1 + m_2} - \frac{m_1\vec{u}_1 + m_2\vec{u}_2}{m_1 + m_2}$$

$$= -\frac{m_2(\vec{u}_1 - \vec{u}_2)}{m_1 + m_2}. \tag{8}$$

The velocity of the particle of mass m_2 after collision relative to CM frame is

$$\vec{V}'_{2c} = \vec{V}_2 - \vec{V} = \frac{2m_1\vec{u}_1 + (m_2 - m_1)\vec{u}_2}{m_1 + m_2} - \frac{m_1\vec{u}_1 + m_2\vec{u}_2}{m_1 + m_2}$$

$$= -\frac{m_1(\vec{u}_2 - \vec{u}_1)}{m_1 + m_2}. \tag{9}$$

From (6), (7), (8) and (9) we conclude that in an elastic collision, speed of each particle remains unchanged in the CM frame. The above results were also obtained in Art. 3.46. (Case 2), where we have shown, with the help of vector diagrams (see Figs. 3.46.3 and 3.46.4), that the velocity vectors rotate in the scattering plane.

6. A nucleus at rest undergoes spontaneous radio-active decay by emitting an electron of momentum 1.73 MeV/c and at right angles to the direction of the electron, a neutrino with momentum 1.00 MeV/c. Find the direction in which the nucleus recoils. What is its momentum? Here c is the velocity of light.

[C.U. (Hons.) 2000]

Let the electron move along OA with momentum p_e and the neutrino move along OB with momentum p_n. The resultant motion is along OC. From the principle of conservation of linear momentum, the nucleus at O should recoil along OD. Resolving along OD and applying the principle of conservation of linear momentum, we may write

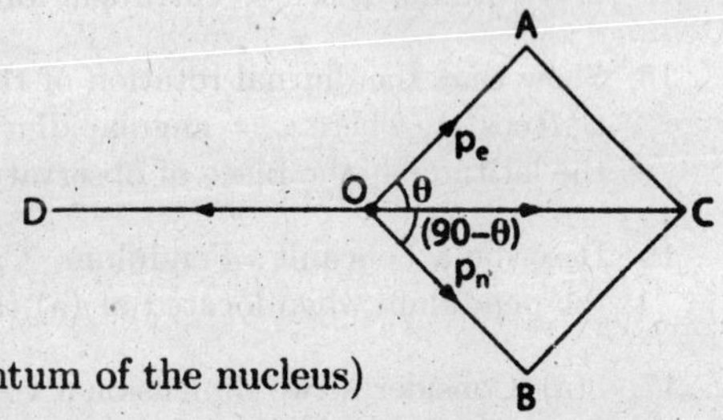

$$p_N = p_n \cos\theta + p_e \sin\theta \quad (p_N = \text{momentum of the nucleus})$$

$$= 1 \times \cos 30^\circ + 1.73 \sin 30^\circ$$

$$= 1 \times \frac{\sqrt{3}}{2} + 1.73 \times \frac{1}{2}$$

$$= \frac{1.73}{2} + \frac{1.73}{2}$$

$$= 1.73\,\text{MeV/c}$$

From the figure

$$\tan\theta = \frac{p_n}{p_e} = \frac{1}{1.73} = \frac{1}{\sqrt{3}}$$

whence $\theta = 30^\circ$

Exercises

1. Calculate the tangential and normal components of acceleration of a particle moving along a curve. What is the magnitude of the resultant acceleration? Hence show that for a particle moving in a circular path of radius ρ with uniform velocity the acceleration is directed inward along the radius of the circular path and has magnitude v^2/ρ.

2. Calculate the radial and transverse accelerations of a particle moving in a plane.

3. What do you mean by conservative forces? Show that all central forces are conservative. Show that in a conservative field the sum of the kinetic energy and potential energy of a particle at any point is constant. **[C.U. (Hons.) 2004, 2008]**

4. What is D'Alembert's principle? Explain the use of this principle.

5. What is meant by the principle of virtual work? Explain the principle by an illustrative example.

6. What is centre of mass? Show that there exists only one centre of mass in a system of particles. Explain the motion of the centre of mass.

7. Show that the rate of change of momentum of a system of particles is equal to the resultant of the external forces acting on the system.

8. Calculate the angular momentum of a system of particles.

9. Calculate the kinetic energy and potential energy of a system of particles.

10. Show that the time rate of change of angular momentum relative to an arbitrary fixed point is equal to the external torque.

11. Describe a double pendulum with necessary theory.

12. Explain what you understand by Coriolis force. Explain the action of Coriolis force on a body freely falling towards the earth.

13. Discuss the free vibrations of a linear triatomic molecule.

14. Show that a rotating frame of reference gives rise to forces which are due only to the acceleration of the frame. Discuss centrifugal force in this context.

15. Show that the diurnal rotation of the earth has the effect of reducing gravity at the surface by $\omega^2 R\cos^2\lambda$, where ω = angular diurnal speed of rotation of the earth, R = its radius and λ = the latitude at the place of observation. **[C.U. (Hons.) 1972]**

16. Describe a Foucault's Pendulum. Calculate the period of rotation of the plane of oscillation of the pendulum when located at (a) the north pole, and (b) latitude 45°N.

17. (a) Consider a two dimensional vector $\vec{A}$ which is a function of time. Show that

$$\frac{d\vec{A}}{dt} = \left(\frac{dA_r}{dt} - A_\theta\frac{d\theta}{dt}\right)\hat{r} + \left(A_r\frac{d\theta}{dt} + \frac{dA_\theta}{dt}\right)\hat{\theta},$$

where $\hat{r}$ and $\hat{\theta}$ are the unit vectors in plane polar co-ordinates, and A_r and A_θ are the magnitudes of the corresponding components of $\vec{A}$.

(b) If $\vec{A}$ in the above case is the position vector $\vec{r}$, obtain the corresponding expression for $\dfrac{d\vec{A}}{dt}$.

[**Ans.** See Art. 3.9] **[C.U. (Hons.) 1991]**

18. (a) Consider a vector $\vec{A}$ which is a function of a parameter t. Show that

$$\frac{d\vec{A}}{dt} = \left(\frac{dA_\rho}{dt} - A_\phi\frac{d\phi}{dt}\right)\hat{\rho} + \left(\frac{dA_\phi}{dt} - A_\rho\frac{d\phi}{dt}\right)\hat{\phi} + \frac{dA_z}{dt}\hat{z},$$

where A_ρ, A_ϕ and A_z are the components of $\vec{A}$ in cylindrical co-ordinates.

(b) If $\vec{A}$ in the above case be the position vector $\vec{r}$, obtain the corresponding expression for $\dfrac{d\vec{r}}{dt}$.

(c) A force $\vec{F}$ is given in cartesian co-ordinates as $\vec{F} = \hat{\imath}yB - \hat{\jmath}Bx$ where B is a constant. Find F_ϕ component. **[C.U. (Hons.) 1985]**

[**Ans.** See Art. 3.7]

19. A particle of charge q and mass m moves in a uniform magnetic field $\vec{B} = B\hat{z}$ and an electric field $\vec{E} = \dfrac{a}{\rho}\hat{\rho}$, where the constants a and B may be either positive or negative.

(a) Set up the equations of motion in cylindrical co-ordinates.

(b) Show that $m\rho^2\dot{\phi} + \frac{qB}{2c}\rho^2$ is a constant of motion.

[Assume that the force due to electric field is $q\vec{E}$ and that due to the magnetic field is $\frac{q}{c}(\vec{v} \times \vec{B})$]. **[C.U. (Hons.) 1984]**

[**Ans.** See Art. 3.7]

20. A particle is attracted towards the Z-axis by a force $\vec{F}$ proportional to the square of its distance from the $X - Y$ plane and inversely proportional to its distances from the Z-axis. Add an additional force perpendicular to $\vec{F}$ such that the total force is conservative. Find also the potential energy corresponding to this total force. **[C.U. (Hons.) 1986]**

21. (a) Find an expression for the maximum range of a projectile near the surface of the earth (neglecting the frictional resistance of air).

(b) A cannon has its maximum range given by R_{max}. Prove that when the maximum range is actually realised, (i) the height reached in such a case is $\frac{1}{4}R_{max}$ and (ii) the total time of flight is $\sqrt{\frac{R_{max}}{2g}}$. **[C.U. (Hons.) 1995]**

[**Ans.** See Art. 3.45]

22. (a) Consider a particle freely falling towards the earth from a height h at a place, with latitude λ. Calculate the deflection of the particle caused by coriolis force when it is dropped from rest.

(b) A particle is thrown vertically upwards with a velocity 70 m/sec at a place with latitude 60^0. Find how far from the original position it will land. **[C.U. (Hons.) 1991]**

(For Answer see Art. 3.42 and worked out examples.)

23. (a) Define generalised momentum for a conservative system and show from Lagrange's equations that it is conserved if the corresponding generalised co-ordinate is cyclic.

(b) Prove that if the Lagrangian of a conservative system does not contain time explicitly, the total energy is conserved.

(c) Write down the Lagrangian for a one-dimensional simple harmonic oscillator and obtain the equation of motion. Derive the Hamiltonian of the oscillator from the Lagrangian function.

[**Hints.** See worked out example.] **[C.U. (Hons.) 1981, '88]**

24. (a) Two undamped one-dimensional oscillators of different natural frequencies are coupled together. Prove that the general oscillation is a sum of two vibrations of different frequencies. Investigate the frequencies and the ratio of amplitudes, when the coupling is loose.

(b) Introduce normal co-ordinates for the problem in (a), and prove, using Lagrange's equation, that each normal co-ordinate executes a simple harmonic motion with arbitrary amplitude and phase but with frequency given in (a).

(c) Discuss the formation of beats in (a), when both the vibrations are simultaneously excited.

[**Hints.** See worked out examples.] **[C.U. (Hons.) 1981]**

25. (a) Prove from Lagrange's equation that if the algebraic sum of the total moments of external forces acting on a system of particles vanishes, then the total angular momentum of the system is conserved.

(b) A particle moves under the influence of a central field of force which depends only on its distance from the origin. Show that the angular momentum is conserved during its motion.

(c) Set up the Hamiltonian for the system in (b). **[C.U. (Hons.) 1982, '87, '93, '96]**

[**Ans.** See worked out examples.]

26. (a) For a free particle moving in a straight line with constant velocity v in an inertial co-ordinate system, show, by physical reasonings that the Lagrangian is a scalar quantity. Hence show that if the Lagrangian is invariant w.r.t. Galilean transformations of co-ordinate systems, it can be written as $L(v^2) = \frac{1}{2}mv^2$, where $\frac{1}{2}m$ is a constant.

(Assume the relative velocity of the co-ordinate systems to be small compared with the particle velocity.)

(b) An electric dipole with opposite charges, of masses m_1 and m_2 separated by a distance l is placed in an external, homogeneous electric field. Write down the Lagrangian of the dipole.

[**Ans.** See worked out examples.] **[C.U. (Hons.) 1983]**

27. (a) The point of suspension of a simple pendulum moves harmonically along the vertical line. Find the Lagrangian and the equation of motion for small oscillations of the pendulum.

(b) A particle rotating with angular velocity $\vec{w}$ around the Z-axis is given a linear velocity $\vec{u} = \dot{\vec{r}}$ w.r.t. a fixed origin situated on the Z-axis. Assuming a potential energy $V(\vec{r})$, write down the Lagrangian in component form. Hence obtain Lagrange's equation of motion treating $\vec{r}$ as a generalised co-ordinate. Interpret the various terms in the equation.

[**Ans.** See worked out examples.] **[C.U. (Hons.) 1983]**

28. (a) Set up the Lagrangian for a compound pendulum oscillating in a vertical plane about a fixed horizontal axis and obtain the equation of motion.

(b) A Lagrangian is given in the form $L = \frac{1}{2}\alpha\dot{q}^2 - \frac{1}{2}\beta q^2$, where α, β are constants;

i. Obtain the equation of motion,

ii. Find the Hamiltonian of the system. **[C.U. (Hons.) 1984, '94]**

[**Ans.** See worked out examples.]

29. A uniform string of length l and of mass ρ per unit length passes over a small frictionless pulley of negligible mass. Two masses m_1 and m_2 are tied at its two ends.

(a) Write down the Lagrangian of the system.

(b) How many degrees of freedom does it possess?

(c) Write down the Lagrange's equation of motion and hence find the acceleration of the mass m_1.

[**Ans.** See worked out examples.] **[C.U. (Hons.) 1985, 91]**

30. Consider the following double pendulum.

A mass m_1 is suspended from a fixed point by a light inextensible string of length L_1. A second mass m_2 is suspended from m_1 by a similar string of length L_2. m_1 and m_2 perform small oscillations in a vertical plane;

(a) What is the Lagrangian of the system?

(b) How many degrees of freedom does the system possess?

(c) Write down the Lagrange's equations of motion for the system. **[C.U. (Hons.) 1986, '89]**

[**Ans.** See worked out examples.]

31. Consider two identical simple harmonic oscillators which are coupled together. Their kinetic and potential energies are given by

$$T = \frac{1}{2}m\left(\dot{q}_1^2 + \dot{q}_1^2\right) \quad \text{and} \quad V = \frac{1}{2}k\left(q_1^2 + q_2^2\right) - hq_1q_2;$$

(a) Find the Hamiltonian of the system.

(b) Find the frequencies for normal modes and the ratios of the amplitudes of each of the co-ordinates q_1 and q_2 in those modes.

(c) Introduce normal co-ordinates and set up the corresponding Lagrange's equations.

[**Ans.** See worked out examples.] **[C.U. (Hons.) 1986, '92]**

32. A nucleus of charge Ze and mass M and an electron of charge $-e$ and mass m, situated at a distance r from each other are rotating in circles about their common centre of mass. Writing down the Lagrangian of the system, work out the equations of motion with the help of Lagrange's equation. Is there any cyclic co-ordinate in the equation? Explain your answer.

[**Ans.** See worked out examples.] **[C.U. (Hons.) 1990]**

33. The point of support of a simple pendulum (length l, mass of the bob $= m$) is moving on a horizontal line according to the law $x = a\cos wt$ (a and w are constants). Find the Lagrangian and hence show that the equation of motion for small angular displacement θ is

$$\ddot{\theta} + \frac{g}{l}\theta = \frac{aw^2}{l}\cos wt.$$ **[C.U. (Hons.) 1995, 2001]**

[**Ans.** See Art. 6.7 and worked out examples]

34. (a) Explain holonomic and non-holonomic constraints. State D'Alembert's principle. Extablish Lagrange's equation for conservative holonomic system using D'Alembert's principle.

(b) Establish the equations of motion of a homogeneous sphere rolling on a plane under an external force $\vec{F}$ and torque $\vec{T}$. **[C.U. (Hons.) 1997]**

[**Ans.** See worked out examples]

35. Set up the Lagrangian of a heavy symmetrical top capable of rotation about its axis in the gravitational field of the earth. Find the constants of motion.

What is precessional motion?

Find the frequencies of precessional motion. **[C.U. (Hons.) 1998]**

[**Ans.** See Art. 8.8 and 8.9.]

36. A bead of mass m is sliding along a uniformly rotating wire with angular velocity w in a force-free space. Using the Lagrangian or the Hamiltonian of the system, find the centri-petal acceleration.

[**Ans.** See worked out examples.] **[C.U. (Hons.) 1994, '95, 98, 2003]**

37. A river of width D flows north-ward with a speed v_0 at a co-latitude λ in the northern hemisphere. Prove that the difference H between water-levels of the two banks due to coriolis force is given by $H = 2Dwv_0 \cos\lambda/g$. **[C.U. (Hons.) 1998]**

[**Ans.** See worked out examples.]

38. (a) Derive the eigenvalue equation of motion of a system undergoing free small oscillations in three dimensional space.

(b) Find the linear harmonic oscillator approximation for the potential energy function $V(x) = ax^{-2} + bx^2$.

(c) Consider a symmetrical triatomic molecule O-C-O. Neglect the interaction between like atoms. Determine the frequencies of vibrations assuming that the potential energy of the molecule depends only on the distances OC and CO and the angle OCO.

[**Ans.** See arts 3.25, 3.27 and 3.28.] **[C.U. (Hons.) 1997]**

39. (a) Examine whether the cylindrical polar co-ordinates (ρ, ϕ, z) represent a set of generalised co-ordinates.

[**Ans.** See worked out Examples.]

(b) Two identical particles each of mass m are fixed to the trisections of a horizontal string of length l stretched with a tension T as shown in the figure. The two particles are performing transverse oscillation. Assuming small amplitude of oscillation, find the following quantities, (i) eigen frequencies, (ii) eigen vectors, (iii) normal co-ordinates.

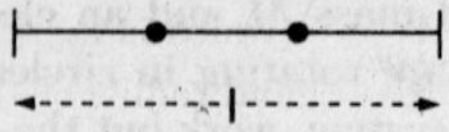

[**Ans.** See worked out examples]

(c) The potential energy of a particle is given by $V = x^4 - 8x^3 - 6x^2 + 24x$. Find the points of stable and unstable equilibrium. **[C.U. (Hons.) 1998, cf. 2003]**

[**Ans.** See Art. 3.24.]

40. Three equal masses are so connected by a spring that they move along a circle. The point A is fixed.

(a) Write down the Lagrangian. (b) Obtain the equations of motion. (c) Find the eigen frequencies and normal co-ordinates of the system.

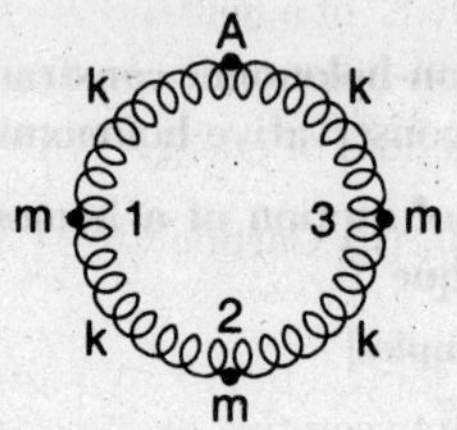

[**Ans.** See worked out examples.] **[C.U. (Hons.) 1999]**

41. A $yo - yo$ (see fig.) consists of a cylinder of mass 90 gm. around which a mass-less string of length 30 cm. is wound. If one end of the string is kept fixed and the $yo - yo$ is allowed to fall vertically starting from rest, find its speed when it reaches the end of the string. Show that the speed is independent of the mass and radius of the $yo - yo$. ($g = 980$ cm/sec^2).

[C.U. (Hons.) 1997]

[**Ans.** For translational motion of the cylinder along the string as shown in the attached figure, we have

$$Mg - F = Ma, \tag{1}$$

where $F \rightarrow$ frictional force along the string, M = mass of $yo - yo$, $a \rightarrow$ linear acceleration of the cylinder. Again for rotational motion about CM, we have

$$RF = I\alpha, \tag{2}$$

where $R \rightarrow$ radius of the cylinder, $I \rightarrow$ moment of inertia of the cylinder about the axis of rotation through the CM, $\alpha \rightarrow$ angular acceleration.

Now as the $yo - yo$ is rotating without slipping, we have

$$\therefore \quad a = R\alpha. \tag{3}$$

From (2) and (3),

$$F = \frac{I\alpha}{R} = \frac{Ia}{R^2},$$

so that from (1) we have

$$Mg - \frac{Ia}{R^2} = Ma \quad \text{or} \quad Mg - \frac{\frac{1}{2}MR^2a}{R^2} = Ma \quad \text{since for a cylinder } I = \frac{1}{2}MR^2.$$

$$\therefore \quad g - \frac{a}{2} = a \quad \text{or} \quad a = \frac{2}{3}g.$$

Now equation of motion

$$v^2 = u^2 + 2as = 2as \quad \because u = 0$$

$$= 2.\frac{2}{3}gl \quad \because s = l = \text{ the length of the string and } a = \frac{2}{3}g$$

$$= \frac{4}{3}gl \quad \text{(which is independent of mass and radius of the } yo - yo\text{)}$$

$$= \frac{4}{3} \cdot 980.30 = 4.98.10^2.$$

$$\therefore \quad v = 140\sqrt{2} = 198 \text{ cm/sec.}$$

42. (a) What do you mean by elastic and inelastic collisions? Two masses m_1 and m_2 travelling along the same straight line collide; find the velocities of the particles after collision (both for a perfectly elastic and an elastic collision).

(b) Show that for a perfectly elastic collision in the above case, the total K.E. before collision equals the total K.E. after collision.

(c) By what fraction is the K.E. of a neutron of mass m_1 decreased in a head-on elastic collision with an atomic nucleus of mass m_2 initially at rest?

(d) Calculate the percentage loss of energy when the nucleus is that of

(i) hydrogen $\left(\frac{m_2}{m_1} = 1\right)$, (ii) Carbon $\left(\frac{m_2}{m_1} = 12\right)$, (iii) lead $\left(\frac{m_2}{m_1} = 206\right)$.

[**Ans.** See Art. 3.46.] **[C.U. (Hons.) 1996]**

43. Consider an elastic collision between two particles of masses m_1 and m_2 the in CM frame (fig 1)

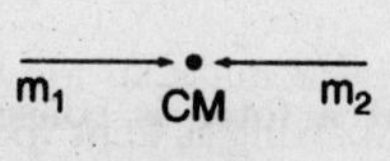

Fig. 1

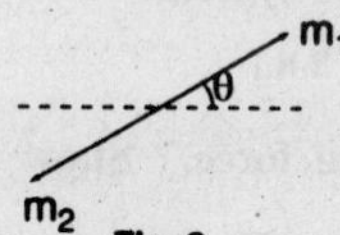

Fig. 2

Show that after collision m_1 and m_2 move off in opposite directions with equal linear momentum and all values of θ (Fig. 2) are permissible. **[C.U. (Hons.) 1997]**

[**Ans.** See Art. 3.46.]

44. A block of wood of mass M is suspended by a string of length L. Initially the string is vertical and the mass is at rest. A bullet of mass m is fired into the block and becomes embedded in it. The block then swings upward and rises until the cord makes an angle θ with the vertical. Find the initial speed of the bullet. **[C.U. (Hons.) 1998, 2005]**

[**Ans.** See Art. 3.47.]

45. A spherical rain-drop falling under constant gravity grows by absorption of moisture from its surroundings. The rate of increase of its mass is proportional to the instantaneous surface area. Assuming that the drop starts with an infinitely small radius, determine its acceleration at a time t. **[C.U. (Hons.) 1998, 2007]**

[**Ans.** See worked out examples.]

46. Define centre of mass for a system of particles. Show that the K.E. of the system is equal to the K.E. of a single particle of total mass M situated at the centre of mass, together with the K.E. of the system of particles with their motion relative to the centre of mass.

[**Ans.** See Art. 3.35.] **[C.U. (Hons.) 1997, '99, 2005]**

47. Prove the angular momentum theorem $\left(\vec{N} = \frac{d\vec{L}}{dt}\right)$ for a system of particles, mentioning the condition for its validity. **[C.U. (Hons.) 1998, '99, 2005]**

[**Ans.** See Art. 3.34.]

48. Two particles of masses m_1 and m_2 are travelling in the same line with velocities $\vec{v_1}$ and $\vec{v_2}$ respectively. Find out the loss in total K.E. if they collide perfectly inelastically.

[**Ans.** See Art. 3.46.] **[C.U. (Hons.) 1999, 2004]**

49. A body is falling from rest under uniform gravity in a medium whose resistance is proportional to velocity. Find out the velocity and distance travelled after time t. Draw v-t curve. What is the limiting velocity of the body? **[C.U. (Hons.) 1994, '99, 2004]**

[**Ans.** See Art. 11.14.]

50. Find out the centre of mass of a uniform solid hemisphere. **[C.U. (Hons.) 1999]**

[**Ans.** From symmetry we see that CM lies on the Z-axis. The hemisphere is made up of solid circular plates; let one such circular plate of radius r, thickness dz, be considered with its centre C at a distance z from O, the centre of the hemi-sphere; then $r^2 + z^2 = a^2$, and the mass of the elementary plate $= \pi r^2 dz\rho = \pi(a^2 - z^2)\,dz\rho$, ρ being the mass per unit area of the hemisphere. So from the definition of CM, we have

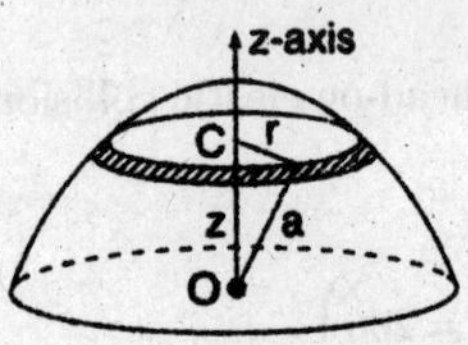

$$\bar{z} = \frac{\int_{z=0}^{a} \pi(a^2 - z^2)z\rho dz}{\int_{z=0}^{a} \pi(a^2 - z^2)dz\rho} = \frac{3}{8}a.$$

51. Draw a right handed Cartesian co-ordinate system XYZ and take a point $O(x, y, z)$. Show in the diagram the cylindrical polar and spherical polar co-ordinates of P. Express x, y, z in terms of these two co-ordinate system. **[C.U. (Hons.) 2000, 2005]**

[**Ans.** See Arts. 3.7 and 3.8.]

52. (a) Defie conservative force. Show that a force is conservative when there exists a scalar potential fuction. **[C.U. (Hons.) 1998, 2000, 2007]**

[See Art. 3.16 and worked out examples.]

(b) Two particles move under a force of mutual interaction. Obtain the equation of motion for the reduced mass. **[C.U. (Hons.) 2000, 2005]**

[See Art. 3.31.]

(c) A particle is moving under the action of a force $\hat{k} \times \vec{v}$, where $\hat{k}$ is the unit vector along Z-axis, and $\vec{v}$ is the velocity of the particle. Show that the K.E. of the particle remains constant. **[C.U. (Hons.) 2000]**

[See worked out examples.]

(d) A nucleus at rest undergoes spontaneous radio-active decay by emitting an electron of momentum 1.73 MeV/c and at right angles to the direction of the electron, a neutrino with momentum 1.00 MeV/c. Find the direction in which the nucleus recoils. What is its momentum? Here c is the velocity of light. **[C.U. (Hons.) 2000]**

[See Art. 3.46 and worked out examples.]

53. A boat was travelling with velocity u when the boatman stopped rowing. Assuming the retarding force due to water to be $a\exp(bv)$, at velocity v, find the time required by the boat to stop. a, b are constants. Assume the water to be stationary. **[C.U. (Hons.) 2000, 2005]**

[**Hints.** $m\frac{du}{dt} = -ae^{bu}$ or $e^{-bu}du = -\frac{a}{m}dt$ or $\int_u^0 e^{-bu}du = -\frac{a}{m}\int_0^t dt = -\frac{a}{m}t$ whence t can be found out.]

54. Consider a plate in the form of an isosceles right angled triangle. The length of the hypotenuse is 2 cm and surface density at a point is $\rho = ah$ where $a = 2gm/cm^3$ and h is the distance of the point from the hypotenuse. Locate the centre of mass of the plate. **[C.U.(Hons.) 2000]**

[**Ans.** By symmetry, the CM lies on the median AD, i.e., along Z-axis. From the geometry of the figure $BC = 2$ cm, $AB = AC = \sqrt{2}$ cm, $BD = DC = 1$ cm and $AD = 1$ cm $= z$ (say)

Considering the elementary rectangular slice of length x and thickness dh at a distance h from the hypotenuse, we have the co-ordinate of the CM given by

$$\bar{z} = \frac{\int_0^z xdh\rho h}{\int_0^z xdh\rho} = \frac{\int_0^z xdh(ah)h}{\int_0^z xdh(ah)} = \frac{2a\int_0^1 h^2(1-h)dh}{2a\int_0^1 h(1-h)dh}$$

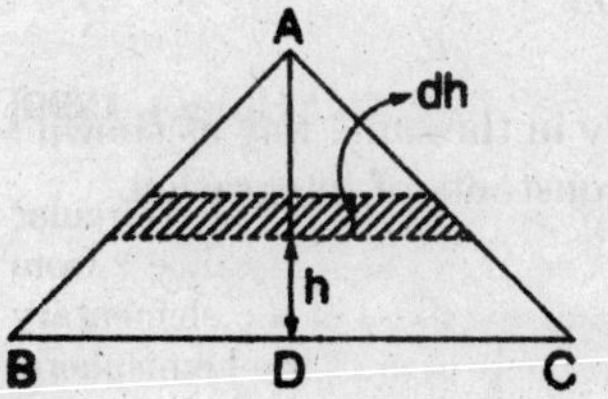

[x being variable, we may write from the geometry of the fig.]

$$\frac{x}{2} = \frac{(1-h)}{1} \text{ or } x = 2(1-h)$$

$$= \frac{\int_0^1 (h^2-h^3)dh}{\int_0^1 (h-h^2)dh} = \frac{\frac{1}{3}-\frac{1}{4}}{\frac{1}{2}-\frac{1}{3}} = \frac{1}{2} \text{ cm on the line } DA \text{ from } D.$$

55. (a) Explain the term centre of mass (CM) frame. Show that the total linear momentum is zero in the centre of mass frame.

(b) Derive the conditions for which the angular momentum $\vec{L}$ is conserved. What kind of motion is expected for $\vec{L} = 0$. **[C.U. (Hons.) 2001]**

56. (a) A particle moves in a plane. Find expressions for radial and transverse velocity and acceleration of the particle. **[C.U. (Hons.) 1997, 2003, 2005]**

(b) If the particle moves with uniform angular velocity along a circle of radius a, use the previous results to find velocity and acceleration of the particle. How are they related? What is the acceleration called?

(c) Given that $\vec{F} = r^2\vec{r}$ is conservative, find the scalar potential. **[C.U. (Hons.) 1997, 2007]**

[**Hints.** $\vec{\nabla} \times \vec{F} = 0$ or $\vec{\nabla} \times (\vec{\nabla}\phi) = 0$ where ϕ is the scalar potential.

$$\therefore \quad \vec{F} = -\vec{\nabla}\phi \quad \text{or} \quad -\vec{\nabla}\phi = -\left(\frac{\partial\phi}{\partial x}\hat{i} + \frac{\partial\phi}{\partial y}\hat{j} + \frac{\partial\phi}{\partial z}\hat{k}\right) = \vec{F} = r^2\left(\hat{i}x + \hat{j}y + \hat{k}z\right)$$

$$= (x^2+y^2+z^2)(\hat{i}x + \hat{j}y + \hat{k}z)$$

$$\therefore \quad \frac{\partial\phi}{\partial x} = -(x^2+y^2+z^2)x \quad \text{or} \quad d\phi = -(x^2+y^2+z^2)xdx$$

which on integration gives

$$\phi = -\frac{x^4}{4} - \frac{x^2y^2}{2} - \frac{z^2x^2}{2} + c_1(y,z);$$

similarly from $\frac{\partial\phi}{\partial y}$ and $\frac{\partial\phi}{\partial z}$, we may obtain ϕ with proper choice of constants of integration.]

57. (a) Establish the angular momentum theorem $\frac{d\vec{L}}{dt} = \vec{N}$ for a particle of mass m. $\vec{L}$ and $\vec{N}$ are respectively the angular momentum and torque on the particle about some point. In which frame is this equation valid? **[C.U. (Hons.) 1997, 2004]**

(b) Trajectory of a particle of unit mass is given by the radius vector $\vec{r} = \hat{i}a\cos\omega t + \hat{j}b\sin\omega t$, where a, b are constants. Calculate angular momentum of the particle about the origin. Show that it is constant and along $\hat{k}$. **[C.U. (Hons.) 1997]**

[**Ans.** See worked out examples.]

58. (a) Prove that the force field $\vec{F} = (yz - y)\hat{i} + (xz - x - 1)\hat{j} + (xy - 2z)\hat{k}$ is conservative and find the corresponding potential function. Is the angular momentum (about the origin) conserved for this force?

[**Hints.** (a) Here

$$\vec{\nabla} \times \vec{F} = \begin{vmatrix} \hat{i} & \hat{j} & \hat{k} \\ \dfrac{\partial}{\partial x} & \dfrac{\partial}{\partial y} & \dfrac{\partial}{\partial z} \\ (yz - y) & (zx - x - 1) & (xy - 2z) \end{vmatrix} = 0$$

$\therefore \vec{F}$ is conservative. Potential function ϕ can be obtained exactly in the same way as shown in problem 56(c) (of the exercise)—taking due care w.r.t. the constants of integration.

(b) Again here

$$\frac{d\vec{L}}{dt} = \vec{N} = \vec{r} \times \vec{F} \begin{vmatrix} \hat{i} & \hat{j} & \hat{k} \\ x & y & z \\ (yz - y) & (zx - x - 1) & (xy - 2z) \end{vmatrix} \neq 0,$$

so that $\vec{L}$ is not conserved for this force field.

(c) A particle (mass $= m$) at rest at $(a, 0, 0)$ is subjected to a force

$$\vec{F} = -\left(\frac{\alpha}{x^3}\right)\hat{i} \quad (\alpha = \text{constant}).$$

Find the time taken by the particle to reach that origin $(a > 0, \alpha > 0)$.

[**Ans.** See worked out examples] **[C.U. (Hons.) 1998]**

59. Prove that for a system of particles, the total external torque is equal to the rate of change of angular momentum of the system provided that the internal force between the particles is central. **[C.U. (Hons.) 1999, 2005, 2007]**

60. (a) i. Set up the one-dimensional equation of motion of a particle of mass m subject to a force such that its potential energy $\frac{1}{2}kx^2$. If a damping force $F = bv$ is also present, where v is the velocity and $b =$ a constant, find out the resultant motion. Discuss the case when $\frac{b}{2m} = \sqrt{\frac{k}{m}}$.

ii. Compute the total (kinetic + potential) energy. Is it a constant? Justify your answer. **[C.U. (Hons.) 2001, 2005]**

[**Ans.** See worked out examples]

(b) A particle is subjected to a frictional force so that its velocity decreases with time according to $v = c(t - t_1)^2$, where t_1 is the time when it stops and c a constant. Find the force $F(v)$ and the distance travelled before it stops. **[C.U. (Hons.) 2001]**

[**Ans.** See worked out examples.]

61. (a) Show that the acceleration $\vec{a}$ of a particle which travels along a space curve with velocity $\vec{v}$ is given by

$$\vec{a} = \frac{dv}{dt}\hat{T} + \frac{v^2}{\rho}\hat{N}$$

where $\hat{T}$ is the unit-tangent vector to the space curve, $\hat{N}$ is the unit principal normal vector and ρ is the radius of curvature. **[C.U. (Hons.) 2002]**

[**Ans.** Velocity $\vec{v}$ = magnitude of $\vec{v}$ multiplied by the unit tangent vector $\hat{T}$, i.e. $\vec{v} = v\hat{T}$ which on differentiation gives

$$\vec{a} = \frac{d\vec{v}}{dt} = \frac{d}{dt}(v\hat{T}) = \frac{dv}{dt}\hat{T} + v\frac{d\hat{T}}{dt}$$

But

$$\frac{d\hat{T}}{dt} = \frac{d\hat{T}}{ds}\frac{ds}{dt} = k\hat{N}\frac{ds}{dt} = k\hat{N}v = \frac{v\hat{N}}{\rho}$$

$$\therefore \quad \vec{a} = \frac{dv}{dt}\hat{T} + v\left(\frac{v\hat{N}}{\rho}\right) = \frac{dv}{dt}\hat{T} + \frac{v^2}{\rho}\hat{N}$$

$\because \hat{N}$ is a unit vector in the direction of $\frac{d\hat{T}}{ds}$, we may write $\frac{d\hat{T}}{ds} = k\hat{N} - \hat{N}$ is called the unit *principal normal vector* to the space curve and the scalar $k = \left|\frac{d\hat{T}}{ds}\right|$ is called the *curvature*, while $\rho = \frac{1}{k}$ is the radius of curvature.

The component $\frac{dv}{dt}$ is in a direction tangent to the space curve and $\frac{v^2}{\rho}$ in the direction of the principal normal is often called the *centripetal acceleration*.]

(b) S and S' are two inertial frames. At $t = t' = 0$, axes of the two frames coincide with each other. Taking S as fixed frame, S' is moving with velocity $\vec{u}$ along x-axis. Write down the Galilean transformation relations relating the event points (x, y, z, t) and (x', y', z', t') in the two frames.

(c) Show that Newton's second law remains invariant under Galilean transformation. **[C.U. (Hons.) 2002, 2006]**

[**Ans.** See Art. 3.37.1.]

(d) A particle of mass m_1 moving with velocity u, undergoes head-on elastic collision with another particle of mass $m_2 = km_1$, initially at rest. Show that after collision K.E. transferred to m_2 is

$$\Delta E = \frac{1}{2}m_1u^2\frac{4k}{(1+k)^2}$$

[**Ans.** See Art. 3.46.] **[C.U. (Hons.) 2002]**

(e) Two particles of equal masses undergo oblique elastic collision. Initially one particle was at rest. Apply momentum and energy conservation laws and show that after collision the particles are scattered at 90° to each other. **[C.U. (Hons.) 2002]**

[**Ans.** Here $e = 1, m_1 = m_2$. If u_1, u_2 and v_1, v_2 be the velocities of the two particles before and after impact respectively, then $u_2 = 0$, and so from (3.46.15) of Art. 3.46, we have $v_2 \sin\theta_2 = u_2 \sin\alpha_2 = 0 \times \sin\alpha_2 = 0. \therefore \theta_2 = 0$.

Again from (3.46.18) we have

$$v_1 \cos\theta_1 = \frac{(m_1 - em_2)u_1 \cos\alpha_1 + m_2(1+e)u_2 \cos\alpha_2}{m_1 + m_2}$$

$$= \frac{(m_1 - m_2)u_1 \cos\alpha_1 + (1+1)m_2 \times 0 \times \cos\alpha_2}{m_1 + m_2} = 0$$

$\therefore \cos\theta_1 = 0$ or $\theta_1 = 90°$.

Since $\theta_1 = 90°, \theta_2 = 0$, their paths after impact will be at right angles to each other.]

62. (a) Show that for a conservative force field, provided, (i) K.E. is a quadratic function of velocity and (ii) P.E. does not contain time explicitly, the total energy is conserved. **[C.U. (Hons.) 2002]**

[**Ans.** K.E. of a particle of mass m and velocity $\vec{v}$ is

$$T = \frac{1}{2}m\vec{v}^2 \quad \text{or} \quad \frac{dT}{dt} = m\vec{v}\frac{d\vec{v}}{dt} = m\frac{d\vec{v}}{dt}\frac{d\vec{r}}{dt} = \vec{F}\frac{d\vec{r}}{dt} \tag{1}$$

Again if V be the P.E., we may write

$$dV = \left(\hat{i}\frac{\partial V}{\partial x} + \hat{j}\frac{\partial V}{\partial y} + \hat{k}\frac{\partial V}{\partial z}\right) d\vec{r} + \frac{\partial \vec{r}}{\partial t}dt \quad \text{or} \quad \frac{dV}{dt} = \vec{\nabla}V\frac{d\vec{r}}{dt} + \frac{\partial \vec{r}}{\partial t} \tag{2}$$

Now

$$E = T + V \quad \text{or} \quad \frac{dE}{dt} = \frac{dT}{dt} + \frac{dV}{dt} = \left(\vec{F} + \vec{\nabla}V\right)\frac{d\vec{r}}{dt} + \frac{\partial \vec{r}}{\partial t} \quad \text{[from (1) and (2)]} \tag{3}$$

Again $\vec{F}$ being conservative, $\vec{F} = -\vec{\nabla}V$ whence from (3) we may write $\frac{dE}{dt} = \frac{\partial \vec{r}}{\partial t}$; also since P.E. does not contain time explicitly, we have

$$\frac{\partial \vec{r}}{\partial t} = 0 \quad \therefore \frac{dE}{dt} = 0 \quad \text{or} \quad E = \text{constant.]}$$

(b) A particle of unit mass moves according to the equation $\vec{r} = \hat{i}(2+3t^2) + \hat{j}5t^2 + \hat{k}t$. Find the force $\vec{F}$ acting on it, torque $\vec{N}$ and angular momentum $\vec{L}$ about the origin. Hence verify $\vec{N} = \frac{d\vec{L}}{dt}$.

[**Ans.** See worked out examples] **[C.U. (Hons.) 2002]**

(c) A particle moves under the influence of a force-field given by $\vec{F} = a(\hat{i}\sin\omega t + \hat{j}\cos\omega t)$. If the particle be initially at rest, prove that at an instant t, work done on the particle is given by $\frac{a^2}{m\omega^2}(1-\cos\omega t)$. Do not ignore the initial condition. **[C.U. (Hons.) 2002]**

[**Ans.** See worked out examples]

63. (a) For a particle moving under a central force field, show that

i. torque about the force centre is zero;

ii. the central orbit is a planar path. **[C.U. (Hons.) 2002, 2005]**

(b) For a system of particle, show that the total instantaneous angular momentum of the system is equal to the angular momentum of a single particle of total mass M situated at the CM together with the angular momentum of the system of particles with their motion relative to the CM. **[C.U. (Hons.) 2002, 2004]**

64. A bead slides without friction along a straight rod of negligible mass as the rod rotates with constant angular velocity $\vec{\omega}$ about a horizontal axis through the mid-point O of the rod. The bead is initially at rest at O. Set up the equation of motion and find the distance of the bead from O as a function of time. **[C.U. (Hons.) 2003]**

[**Ans.** See worked out examples]

65. (a) Consider a system of particles which interact with one another through pairwise central forces. Show, by using Newton's second and third laws, that the total angular momentum of the system is a constant. **[C.U. (Hons.) 2003]**

(b) Show that for a system of particles, the total linear momentum of the particles is equal to that of a single particle of mass equal to the total mass of the particles placed at their CM. **[C.U. (Hons.) 2003]**

Ans. We consider a system of n particles in a coordinate system with O as origin. Position vector of ith particle $= \vec{r}_i$ w.r.t. O; position vector CM(C) w.r.t. $O = \vec{R}$; position vector of ith particle w.r.t. $C = \vec{r}_i'$.

$\therefore$ from the attached figure $\vec{r}_i = \vec{R} + \vec{r}_i{}'$ or $\dot{\vec{r}}_i = \dot{\vec{R}} + \dot{\vec{r}}_i{}'$.

Linear momentum of ith particle of mass m_i is $\vec{p}_i = m_i\dot{\vec{r}}_i = m_i\left(\dot{\vec{R}} + \dot{\vec{r}}_i{}'\right)$.

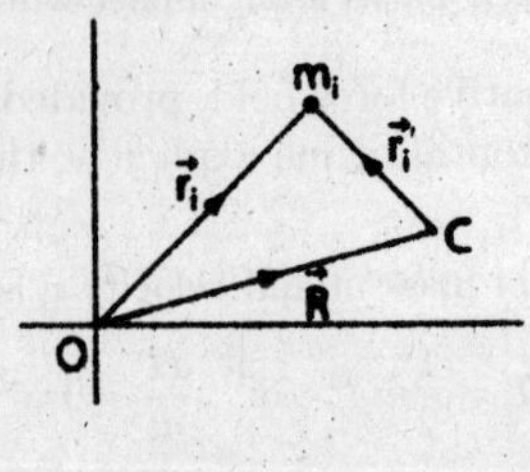

$\therefore$ total linear momentum of the system is

$$\vec{p} = \sum_{i=1}^{n} \vec{p}_i = \sum_{i=1}^{n} m_i \left(\dot{\vec{R}} + \dot{\vec{r}}_i{}'\right) = \dot{\vec{R}} \sum_{i=1}^{n} m_i + \sum_{i=1}^{n} m_i \dot{\vec{r}}_i{}'$$

Now $\sum_{i=1}^{n} m_i = M$ = the mass of system and we know that $m_i \vec{r}_i{}' = 0$.

$$\therefore \quad \sum_{i=1}^{n} m_i \dot{\vec{r}}_i{}' = 0$$

$\therefore \vec{p} = M\dot{\vec{R}}$ which is to be proved.

(c) Consider the motion of two mass points m_1 and m_2. Show that the K.E. of the particles in the CM frame of reference is equal to $\frac{1}{2}\mu\dot{\vec{r}}^2$, where $\vec{r}$ is the relative position vector of the particles and μ is the reduced mass given by $\frac{1}{\mu} = \frac{1}{m_1} + \frac{1}{m_2}$. **[C.U. (Hons.) 2003]**

Ans. We consider two mass points m_1 and m_2 in the CM(C) frame of reference.

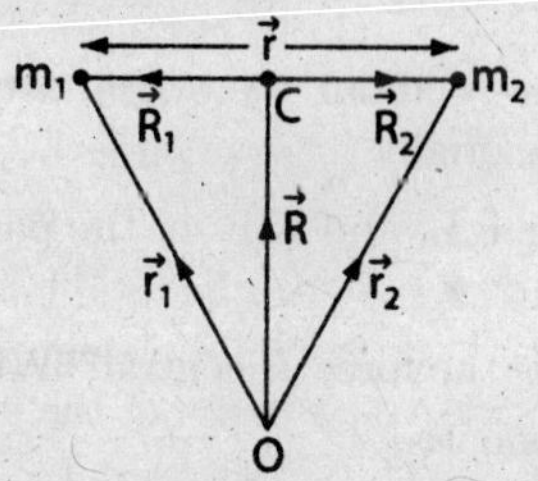

Let $\vec{r}_1$ = position vector of m_1 w.r.t. origin O

$\vec{r}_2$ = ,, ,, ,, m_2 ,, ,, ,,

$\vec{R}$ = ,, ,, ,, CM(C) ,, ,, ,,

$\vec{R}_1$ = ,, ,, ,, m_1 ,, ,, C

$\vec{R}_2$ = ,, ,, ,, m_2 ,, ,, C

$\vec{r}$ is the relative position vector of the particles.

$\therefore$ from the figure,

$$\vec{r}_1 = \vec{R} + \vec{R}_1 \quad \text{or} \quad \vec{R}_1 = \vec{r}_1 - \vec{R} \tag{1}$$

$$\vec{r}_2 = \vec{R} + \vec{R}_2 \quad \text{or} \quad \vec{R}_2 = \vec{r}_2 - \vec{R} \tag{2}$$

$$\text{and} \quad \vec{r} = \vec{r}_1 - \vec{r}_1 \tag{3}$$

Now we know from definition of CM

$$\vec{R} = \frac{m_1\vec{r}_1 + m_2\vec{r}_2}{m_1 + m_2}$$

From (1) we may write

$$\vec{R}_1 = \vec{r}_1 - \frac{m_1\vec{r}_1 + m_2\vec{r}_2}{m_1 + m_2} = \frac{m_2(\vec{r}_1 - \vec{r}_2)}{m_1 + m_2} = \frac{\mu}{m_1}\vec{r}$$

$$\therefore \quad \dot{\vec{R}}_1 = \frac{\mu}{m_1}\dot{\vec{r}}$$

Again from (2)

$$\vec{R}_2 = \vec{r}_2 - \frac{m_1\vec{r}_1 + m_2\vec{r}_2}{m_1 + m_2} = \frac{m_1(\vec{r}_2 - \vec{r}_1)}{m_1 + m_2} = -\frac{\mu}{m_2}\vec{r}$$

$$\therefore \quad \dot{\vec{R}}_2 = \frac{\mu}{m_2}\dot{\vec{r}}$$

$\therefore$ K.E. of the system in the CM frame of reference

$$= \frac{1}{2}m_1\dot{\vec{R}}_1^2 + \frac{1}{2}m_2\dot{\vec{R}}_2^2 = \frac{1}{2}\mu^2\dot{\vec{r}}^2\left(\frac{1}{m_1} + \frac{1}{m_2}\right) = \frac{1}{2}\mu\dot{\vec{r}}^2$$

putting the values of $\dot{\vec{R}}_1$ and $\dot{\vec{R}}_2$ and

$$\because \frac{1}{m_1} + \frac{1}{m_2} = \frac{1}{\mu}$$

which is identical with the motion of a single particle of mass μ.

66. The trajectory of a particle of mass m moving on a plane is given by $x = x_0 \cos\omega_1 t$ and $y = y_0 \sin\omega_2 t$.

(a) Find the force $\vec{F}$, torque $\vec{N}$ about the origin, acting on the particle. Find also the angular momentum $\vec{L}$ about the origin.

(b) Find the expressions for P.E. and K.E. of the particle at any instant of time t. Is the total energy conserved?

(c) Under what condition is the force $\vec{F}$ central? What is $\vec{N}$ when the force is central?

[**Ans.** See worked out examples.] **[C.U. (Hons.) 2003]**

67. (a) The position of a particle moving in a plane is given by $x = \sin\omega t, y = \cos\alpha\omega t$. Show that the trajectory repeats itself only if α is a rational number. **[C.U. (Hons.) 2004]**

[**Ans.** Here $x = \sin\omega t, y = \cos\alpha\omega t$ i.e. $x^2 + y^2 = \sin^2\omega t + \cos^2\alpha\omega t = 1$ if $\alpha = 1$ i.e. the equation represents a circle of radius 1 provided $\alpha = 1$. So we may say that the trajectory repeats itself only if α is a rational number.]

(b) For a particle of mass m moving in a conservative force field, establish that the total mechanical energy of the system remains conserved. State and explain the assumptions clearly.

[**Ans.** See Q.62(a) above.] **[C.U. (Hons.) 2004, 2008]**

(c) To the above system we add an additional resistive force $\vec{F}$. Establish that $\frac{dE}{dt} = -\vec{F}\cdot\vec{v}$, where E is the total mechanical energy and $\vec{v}$ is the velocity. **[C.U. (Hons.) 2004]**

[**Ans.** If a resistive force $\vec{F}$ is added to the system, there will be dissipation of total energy (the force $\vec{F}$ being dissipative in nature), so that the work done by the force will result in a loss of energy. Therefore, we may write

$$-dE = \vec{F}\cdot d\vec{r} \quad \text{or} \quad -\frac{dE}{dt} = \vec{F}\cdot\vec{v}. \quad \therefore \frac{dE}{dt} = -\vec{F}\cdot\vec{v}.]$$

68. A particle of mass m moves in the $(x - y)$ plane and its position vector is given by $\vec{r} = \hat{i}a\cos\omega t + \hat{j}b\sin\omega t$, where a, b, ω are constants. Show that

(a) the force acting on the particle is always directed towards the origin and the force field is conservative; and

(b) find the work done by the force field in moving the particle from the point $(a, 0)$ to $(0, b)$.

[**Ans.** See worked out examples] **[C.U. (Hons.) 2004, 2007]**

69. A radio-active particle decays exponentially in time with half-life τ. Find out its equation of motion in a gravitational field. Let the initial mass of the particle be m_0. **[C.U. (Hons.) 2005]**

[**Hints.** Let the particle fall through a height x in time t—then its equation of motion is $m\frac{d^2x}{dt^2} = mg$ (where $m = m_0e^{-\lambda t}$, λ being the decay constant for the particle) $\cdot \frac{d^2y}{dt^2} = g$.]

70. Explain the term 'centre of mass frame'. In an elastic collision between two particles, what are the conditions to be satisfied? **[C.U. (Hons.) 2005]**

71. (a) A particle of mass m is connected to a light spring of spring constant k (fig). The table is horizontal and frictionless. At $t = 0$, the block is displaced to the right by x_0 from equilibrium position and released. Find the instantaneous displacement x_0 of the block as a function of m and k. **[C.U. (Hons.) 2006]**

[**Hints.** If x be the displacement at time t, then the equation of motion is

$$m\frac{d^2x}{dt^2} = -kx \quad \text{or} \quad \frac{d^2x}{dt^2} + \omega^2 x \quad (\text{where } \omega^2 = k/m)$$

which represents a simple harmonic motion of time period

$$T = \frac{2\pi}{\omega} = 2\pi\sqrt{\frac{m}{k}}.$$

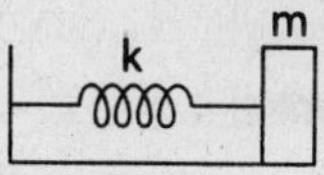

The solution is $x = x_0 \cos\left(\sqrt{\frac{k}{m}}t - \delta\right)$, x_0 being the amplitude and δ being a constant.]

(b) Show that the locus of the above particle in $(x - p_x)$ plane (p_x is the linear momentum along the line of motion and x is the displacement) is an ellipse. **[C.U. (Hons.) 2006]**

[**Hints.** Since the particle is executing s.h.m., its total energy is

$$E = \frac{p_x^2}{2m} + \frac{1}{2}kx^2$$

which may be written as

$$\frac{p_x^2}{2Em} + \frac{x^2}{2E/k} = 1$$

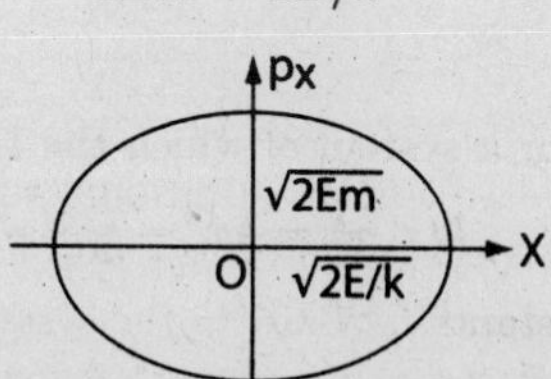

This represents an ellipse with semi-axes equal to $\sqrt{2E/K}$ and $\sqrt{2Em}$. The trajectory of the particle in the phase space will therefore be an ellipse. Any point on the ellipse represents the phase (x, p_x) of the particle at sometime; and the ellipse is described over and over again forever, since the particle is vibrating periodically in a straight line, repeating the same positions and momenta forever.]

72. (a) A particle at rest is acted on by a force $\vec{F} = \hat{i}\pi \sin 2\pi t$ Newton. Calculate the linear momentum of the particle after $t = 0.5$ secs. **[C.U. (Hons.) 2006]**

[**Hints.** $\frac{d\vec{p}}{dt} = \vec{F} = \pi \sin 2\pi t$ or $\int_0^{\vec{p}} d\vec{p} = \int_0^{1/2} \pi \sin 2\pi t \, dt$ or $\vec{p} = \pi\left[-\frac{\cos 2\pi t}{2\pi}\right]_0^{1/2} = \frac{1}{2}(\cos 2\pi t)_{1/2}^{0} = 1$ kg-m/sec.]

(b) Establish that the force field given by $\vec{F} = \frac{k}{r^2}\hat{e}_r$ is a conservative one. Calculate the corresponding potential energy at r. (Notations have usual significance).

[**Ans.** See worked out examples] **[C.U. (Hons.) 2006]**

(c) A gun of mass 2 kg fires a bullet of mass 20 gm. Energy of explosion is 1.6×10^3 Joule. Calculate the nozzle velocity of the bullet. **[C.U. (Hons.) 2006]**

[**Hints.** $\frac{1}{2}mv^2 = 1.6 \times 10^3 \times 10^7$ ergs; putting $m = 20$ gm, get the value of v in cm/sec.]

73. (a) A frame of reference rotates with angular velocity $\vec{\omega}$. For this frame establish the identity $\frac{d}{dt} = \frac{d'}{dt} + \vec{\omega}\times$. **[C.U. (Hons.) 2006, 2007]**

(b) Explain the effect of Coriolis force on the formation of cyclone in the northern hemisphere. **[C.U. (Hons.) 2006]**

74. Write down the expression for a volume element in spherical polar coordinates. Hence find the volume of a hemi-spherical shell bounded by hemispherical surfaces of radii r_1 and r_2 $(r_1 > r_2)$, using this relation. **[C.U. (Hons.) 2007]**

75. (a) Prove that if the total momentum of a system is conserved, then the CM is either at rest or in uniform motion. **[C.U. (Hons.) 2007]**

[**Hints.** See Art. 3.32.]

(b) The force on a particle is given by $\vec{F} = k(-y\hat{\imath} + x\hat{\jmath})$ where k is a constant. Calculate the work done by the particle as it moves from $(a, 0)$ to $(0, a)$ going round the arc of a circle of radius a. **[C.U. (Hons.) 2007]**

[**Ans.** See worked out examples]

76. In the x-y plane, a force has Cartesian components $F_x = x^3 + xy^2$, $F_y = y^3 + 3x^2y$. Compute the work done in a displacement from the point $(1, 1)$ to $(2, 2)$, where the path is

(a) a straight line from $(1, 1)$ to $(2, 2)$;

(b) a straight line from $(1, 1)$ to $(1, 2)$, followed by another straight line from $(1, 2)$ to $(2, 2)$.

Is this force conservative? **[C.U. (Hons.) 2008]**

[**Ans.** See worked out problems]

77. Fine particles of sand are being dropped continuously from a fixed container on a moving belt. Find out the force necessary to have the belt moving at *constant speed,* and show that the power supplied by this force is twice the rate of increase of kinetic energy of the system.

[C.U. (Hons.) 2006 (Old Syllabus)]

78. Set up the Hamiltonian for a system of which the Lagrangian is given as

$$L(x, \dot{x}) = \tfrac{1}{2}\dot{x}^2 - \tfrac{1}{2}\omega^2 x^2 - \alpha x^3 + \beta x\dot{x}^2$$

where α, β and ω are constants. **[C.U. (Hons.) 1995]**

The Hamiltonian of a system is given by

$$H = \sum_k p_k \dot{q}_k - L = \sum_k \frac{\partial L}{\partial \dot{q}_k}\dot{q}_k - L$$

Here

$$\begin{aligned} H &= \frac{\partial L}{\partial \dot{x}}\dot{x} - L = \dot{x}(\dot{x} + 2\beta x\dot{x}) - \tfrac{1}{2}\dot{x}^2 + \tfrac{1}{2}\omega^2 x^2 + \alpha x^3 - \beta x\dot{x}^2 \\ &= \frac{\dot{x}^2}{2} + 2\beta x\dot{x}^2 + \tfrac{1}{2}\omega^2 x^2 + \alpha x^3 - \beta x\dot{x}^2 \\ &= \tfrac{1}{2}\dot{x}^2 + \beta x\dot{x}^2 + \tfrac{1}{2}\omega^2 x^2 + \alpha x^3 \end{aligned}$$

79. Classify each of the following according as they are: (a) Scleronomic or rheonomic, (b) holonomic or non-holonomic and (c) conservative or non-conservative.

(a) A small sphere rolling down from the top of a fixed large sphere.

(b) A particle moving on a very long frictionless wire which rotates with a constant angular speed about a horizontal axis. **[C.U. (Hons.) 1994]**

[**Ans.** See worked out examples]

80. What do you mean by *degrees of freedom* and *generalized coordinates*? Find the number of degrees of freedom in the following cases, given reasons in brief: **[C.U. (Hons.) 1994, 1999]**

(a) Two mass points connected by a rigid massless rod or by a massless spring:

[**Ans.** Here the number of degrees of freedom $= n = 3N - K = (3 \times 2 - 1) = 5 \because N = 2$ and $K = 1$.]

(b) Three mass points connected by three rigid massless rods.

[**Ans.** Here $n = 3N - K = 3 \times 3 - 3 = 6$. $\because$ number of constraints for 3 particles, $K = 3$.]

(c) A rigid body fixed at one point.

[**Ans.** Since the rigid body is fixed at one point, it has only one degree of freedom.]

(d) A rigid body rotating about a fixed axis in space.

[**Ans.** Here the motion is completely specified if we know the coordinates of two points say $(x_1 y_1 z_1)$ and $(x_2 y_2 z_2)$, where the fixed point is taken at the origin of the coordinate system. But since the body is rigid, we must have $x_1^2+y_1^2+z_1^2 = \text{constant} = x_2^2+y_2^2+z_2^2$ and $(x_1-x_2)^2+(y_1-y_2)^2+(z_1-z_2)^2 = \text{constant}$—from which 3 coordinates can be found in terms of the remaining 3. Thus there are 3 degrees of freedom.]

(e) A particle moving along a specified curve in three-dimensional space.

[**Ans.** Here $n = 3N - K = 3 \times 1 - 1 = 2 \because K = 1$, as the particle moves along a specified curve which can be represented by one equation.]

(f) A bead sliding on a smooth horizontal rod which is rotating about a vertical axis with a constant angular velocity.

[**Ans.** Here $n = 3N - K = 3 \times 1 - 1 = 2$ as $K = 1$.]

81. Two particles A and B having masses m_1 and m_2 respectively are moving in a two-dimensional plane such that the distance r between them remains constant.

(a) What is the number of degrees of freedom of the system?

(b) Find the coordinates of the CM and the reduced mass.

(c) Choose an appropriate set of generalized coordinates for describing the dynamics of the two particle system.

(d) Set up Lagrange's equations of motion for these coordinates in absence of any external applied field.

(e) Obtain solutions of these equations.

(f) Give physical interpretation for each solution.

(g) If the angular velocity of the two-particle-system is 5 rad/sec, find how far the line joining A and B rotates in degrees after 5 seconds. **[C.U. (Hons.) 1992]**

[**Ans.** (a) Since the two particles are moving in a two-dimensional plane, they would have 4 degrees of freedom, if they were free. However, since there is 1 constraint acting between them, they will have 3 degrees of freedom.

(b) See the answer of Q.65(c) above.

(c) In order to solve this problem for the motion of two interacting particles, we choose a coordinate system with origin at CM (called the CM coordinates in the CM frame) because this choice reduces the two body problem to a one body problem.

(d) The Lagrangian of such a system is

$$L = \tfrac{1}{2}m_1\dot{\vec{R}}_1^2 + \tfrac{1}{2}m_2\dot{\vec{R}}_2^2 - U(\vec{R}_1 - \vec{R}_2) = \tfrac{1}{2}\mu\dot{\vec{r}}^2 - U(\vec{r})$$

Thus the problem of motion of two interacting particles is equivalent to that of the motion of one particle of mass μ in a given external field $U(\vec{r})$. In absence of any external field (i.e. $U(\vec{r}) = 0$), $L = \frac{1}{2}\mu\dot{\vec{r}}^2$ and hence Lagrange's equation of motion is written as

$$\frac{d}{dt}\left(\frac{\partial L}{\partial \dot{\vec{r}}}\right) - \frac{\partial L}{\partial r} = 0 \quad \text{or} \quad \frac{d}{dt}(\mu\dot{\vec{r}}) = 0 \quad \text{or} \quad \mu\ddot{\vec{r}} = 0$$

(e) The solution of the above equation is obtained as follows

$$\mu\ddot{\vec{r}} = 0 \quad \text{or} \quad \frac{d\dot{\vec{r}}}{dt} = 0 \quad \text{or} \quad \dot{\vec{r}} = a$$

(on integration, a being integration constant).

Again on integration $\vec{r} = at + b$ (b being another constant) Putting initial conditions at $t = 0, \vec{r} = 0$, we have

$$\vec{r} = at = \vec{r}(t)]$$

(f) From this solution, the distances $\vec{R}_1 = \vec{R}_1(t)$ and $\vec{R}_2 = \vec{R}_2(t)$ of the two particles separately relative to their common CM are obtained from the relations

$$\vec{R}_1 = \frac{\mu}{m_1}\vec{r} = \frac{m_2}{m_1 + m_2}\vec{r} \quad \text{and} \quad \vec{R}_2 = -\frac{\mu}{m_2}\vec{r} = -\frac{m_1}{m_1 + m_2}\vec{r}$$

as deduced in (b).

(g) The angular velocity of the line joining A and B being 5 rad/sec., the angle by which AB will rotate in 5 secs is given by 5×5 rad $= 25$ rad $= 25 \times \frac{180}{\pi}$ degree $= 1433°$.

82. State with reasons if the following constraints are holonomic:

(a) Two mass points connected by a massless rigid rod.

(b) A gas molecule in an enclosed volume.

(c) A disc rolling on a plane without slipping. **[C.U. (Hons.) 2001]**

[**Ans.** (a) The two mass points are rigidly constrained by the equation $(x_2 - x_1)^2 + (y_2 - y_1)^2 + (z_2 - z_1)^2 = R^2$, where (x_1, y_1, z_1), (x_2, y_2, z_3) are the coordinates of the points and R is the distance between them. As the constraint can be expressed by an equation, it is a holonomic one.

(b) A gas molecule in an enclosed volume is a non-holonomic constraint, because, as it collides with other gas molecules, it constantly changes its direction every moment, and there is no certain direction of motion of the molecule, and therefore it cannot be expressed by any single equation.

(c) The disc rolling on a plane without slipping is an example of non-holonomic constraint, since it cannot be expressed by any single equation. This is due to the fact that during rolling without slipping, it constantly changes its direction.]

83. Show that, for conservative systems, $\frac{\partial T}{\partial p_i} = \dot{q}_i$, the symbols having their usual meanings.

[C.U. (Hons.) 2000]

From Art. 3.23, we can show that

$$\dot{q}_i = \frac{\partial H}{\partial p_i} = \frac{\partial}{\partial p_i}(T + V) = \frac{\partial T}{\partial p_i} \quad \because \frac{\partial V}{\partial p_i} = 0$$

for conservative systems.

84. Write down the Lagrangian of a point mass thrown at an angle α with the horizontal. Obtain the Hamiltonian and the Hamilton's equations of motion.

[**Ans.** Let $P(x, y)$ be the position of the particle of mass m at any instant t in the X-Y plane with X-axis as horizontal. The K.E. of the particle is $T = \frac{1}{2}m(\dot{x}^2 + \dot{y}^2)$ and the P.E. is (taking X-axis as the reference level) $V = mgy$, so that the Lagrangian

$$L = T - V = \tfrac{1}{2}m(\dot{x}^2 + \dot{y}^2) - mgy \tag{1}$$

whence we may write

$$\left.\begin{array}{ll} \dfrac{\partial L}{\partial \dot{x}} = m\dot{x} \quad \text{or} \quad \dot{x} = \dfrac{p_x}{m} & \text{Also} \quad \dfrac{\partial L}{\partial \dot{y}} = m\dot{y} \quad \text{or} \quad \dot{y} = \dfrac{p_y}{m} \\ \text{and} \quad \dfrac{\partial L}{\partial x} = 0 & \text{and} \quad \dfrac{\partial L}{\partial y} = -mg \end{array}\right\} \tag{2}$$

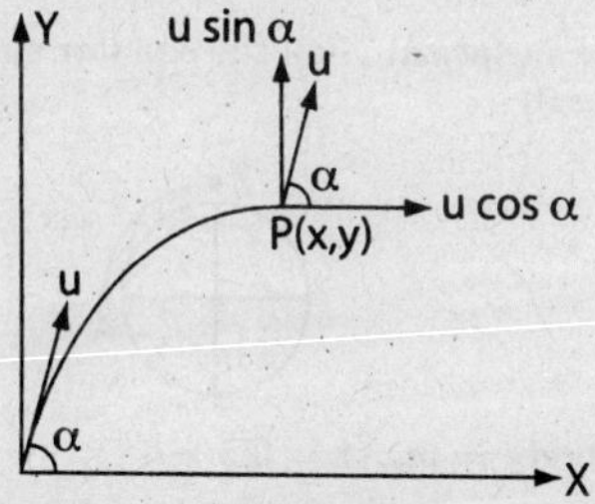

Again Hamiltonian of the system is given by

$$H = \sum p_k \dot{q}_k - L = p_x\dot{x} + p_y\dot{y} - L = \frac{p_x^2}{m} + \frac{p_y^2}{m} - \frac{1}{2}m\left(\frac{p_x^2}{m^2} + \frac{p_y^2}{m^2}\right) - mgy \quad \text{[using (1) and (2)]}$$

$$= \frac{1}{2}\left(\frac{p_x^2}{m} + \frac{p_y^2}{m}\right) - mgy \tag{3}$$

To obtain Hamilton's equations for the system we proceed as follows.

For one generalised coordinate x, we have

$$\dot{x} = \frac{\partial H}{\partial p_x}; \quad \dot{p}_x = -\frac{\partial H}{\partial x} \tag{4}$$

$$\frac{\partial H}{\partial p_x} = \frac{\partial}{\partial p_x}\left[\frac{1}{2}\left(\frac{p_x^2}{m} + \frac{p_y^2}{m}\right) - mgy\right] \quad \text{or} \quad \dot{x} = \frac{p_x}{m} \quad \text{and} \quad \dot{p}_x = 0 \quad \text{[using (4)]} \tag{5}$$

For another generalized coordinate y, we have

$$\dot{y} = \frac{\partial H}{\partial p_y}; \quad \dot{p}_y = -\frac{\partial H}{\partial y} \tag{6}$$

$$\therefore \quad \frac{\partial H}{\partial p_y} = \frac{p_y}{m} \quad \therefore \quad \dot{y} = \frac{p_y}{m} \quad \text{and} \quad \dot{p}_y = mg \quad \text{[using (5) and (6)]}$$

So the Hamilton's equations of motion are

$$\left.\begin{array}{l|l} \dot{x} = \dfrac{\partial H}{\partial p_x} \quad \text{or} \quad \dfrac{\partial H}{\partial p_x} = u\cos\alpha & \dfrac{\partial H}{\partial x} = -\dot{p}_x \quad \text{or} \quad \dfrac{\partial H}{\partial x} = 0 \\ \dot{y} = \dfrac{\partial H}{\partial p_y} \quad \text{or} \quad \dfrac{\partial H}{\partial p_y} = u\sin\alpha & \dfrac{\partial H}{\partial y} = -\dot{p}_y = -mg \end{array}\right\}$$

85. An object is dropped from a height of 500 m in the northern hemisphere at a latitude 30°. Find out the deflection of the body from the point vertically below the starting point.

[C.U. (Hons.) 2001]

[**Ans.** Apply equation (3.43.2)

$$x = \frac{2\sqrt{2}}{3}\omega\sqrt{\frac{z^3}{g}}\cos\lambda$$

Here $z = 500$ m, $\omega = 7.29 \times 10^{-5}$ rad/sec, $\lambda = 30°$ and $g = 9.81$ m/sec^2

$$x = \frac{2\sqrt{2}}{3} \times 7.29 \times 10^{-5} \times \sqrt{\frac{(500)^3}{9.81}} \times \frac{\sqrt{3}}{2}\ \text{m} = 0.212\ \text{m} = 21.2\ \text{cm towards east.}]$$

86. A body of mass 0.5 kg starts sliding down from the top of a smooth sphere of radius $R = 1$ m. The sphere rotates with constant angular velocity $\omega = 6$ rad/sec about a vertical axis passing through its centre. Find the centrifugal force and the coriolis force at the moment when the body breaks off the surface. **[C.U. (Hons.) 2002]**

[**Ans.** Let the body break off the surface at A on the equator since the centrifugal acceleration $\omega^2 r\cos\lambda$ = maximum at the equator ($\because \lambda = 0$).

So the centrifugal force on the body $= m\omega^2 r = 0.5 \times 6^2 \times 1 = 18N$.

Again the coriolis force on the body at $A = 2mv\cos\lambda = 2mv$ ($\because \lambda = 0$)

$$= 2mgt = 2mg\sqrt{\frac{2z}{g}} = 2m\sqrt{2gz} = 2 \times 0.5 \times \sqrt{2 \times 9.8 \times 1}$$

$$= 2 \times 0.5 \times 2 \times 0.7 = 1.4N.]$$

87. A block of mass M can move in a horizontal line. A simple pendulum of length l and bob-mass m is suspended from the block. The pendulum with the block moves in the same plane. Set up the Lagrangian and obtain the equations of motion. **[C.U. (Hons.) 2001, cf. 2004]**

[**Ans.** See worked out examples.]

88. (a) What do you mean by holonomic and scleronomic system? Given one exmple. Show that for such a system K.E. is a quadratic function of generalized velocities. If the system is conservative, prove also that the Hamiltonian of the system is equal to its total energy. **[C.U. (Hons.) 2005]**

For answer to 1st and 3rd parts see Arts. 3.20 and 3.23.

Answer to 2nd part. K.E. T is given, in general, by

$$T = \sum_i \frac{1}{2} m_i \dot{\vec{r}}_i^{\,2} = \sum_i \frac{1}{2} m_i \left(\sum_j \frac{\partial \vec{r}_i}{\partial q_j}\dot{q}_j\right)\left(\sum_k \frac{\partial \vec{r}_i}{\partial q_k}\dot{q}_k\right)$$

because from transformation equation

$$\vec{r}_i = \vec{r}_i(q_1, q_2, \ldots)$$

we may write

$$\dot{\vec{r}}_i = \sum_j \frac{\partial \vec{r}_i}{\partial q_j}\dot{q}_j$$

Thus

$$T = \sum_j \sum_k \left(\sum_i \frac{1}{2} m_i \frac{\partial \vec{r}_i}{\partial q_j}\frac{\partial \vec{r}_i}{\partial q_k}\right)\dot{q}_j\dot{q}_k = \sum_{jk} a_{jk}\dot{q}_j\dot{q}_k \tag{1}$$

only if transformation equations do not contain time explicitly i.e. *constraints are independent of time.* In equation (1)

$$a_{jk} = a_{jk}(q_1, q_2, \ldots, q_n) = \sum_i \frac{1}{2} m_i \frac{\partial \vec{r}_i}{\partial q_j} \frac{\partial \vec{r}_i}{\partial q_k}$$

$$= \sum_i \frac{1}{2} m_i \left(\frac{\partial x_i}{\partial q_j} \frac{\partial x_i}{\partial q_k} + \frac{\partial y_i}{\partial q_j} \frac{\partial y_i}{\partial q_k} + \frac{\partial z_i}{\partial q_j} \frac{\partial z_i}{\partial q_k} \right)$$

Equation (1) shows that for holonomic scleronomous systems, K.E. is always a homogeneous quadratic function of generalized velocities i.e. $\dot{q}_j, \dot{q}_k$.

(b) State whether the constraint given by $x\dot{y} - y\dot{x} =$ constant is holonomic one. **[C.U. (Hons.) 2005]**

[**Ans.** The given constraint equation can be written as $x\frac{dy}{dt} - y\frac{dx}{dt} =$ constant. Multiplying throughout by dt and dividing by xy, we obtain

$$\frac{dy}{y} - \frac{dx}{x} = \text{constant} \times \frac{dt}{xy}$$

Since this is non-integrable, the relation between x and y is unknown.

So the constraint is *non-holonomic*.]

(c) For the Hamiltonian $H = p_1 q_1 - p_2 q_2 - a q_1^2 - b q_2^2$, solve Hamilton's equation of motion and prove that $(p_2 - b q_2)/q_1 =$ constant and $q_1 q_2 =$ constant; a, b are constants, and p_1, p_2, q_1, q_2 are generalized coordinates. **[C.U. (Hons.) 2005]**

[**Ans.** We have $H = p_1 q_1 - p_2 q_2 - a q_1^2 - b q_2^2$ whence we may write

$$\dot{q}_1 = \frac{\partial H}{\partial p_1} = q_1; \quad \dot{q}_2 = \frac{\partial H}{\partial p_2} = q_2 \tag{1}$$

and

$$\left.\begin{aligned} \dot{p}_1 &= -\frac{\partial H}{\partial q_1} = -p_1 + 2a q_1 = -(p_1 - 2a q_1) \\ \text{and} \quad \dot{p}_2 &= -\frac{\partial H}{\partial q_2} = p_2 + 2b q_2 \end{aligned}\right\} \tag{2}$$

From (1) $\frac{dq_1}{dt} = q_1$ or $\frac{dq_1}{q_1} = dt$ or $\ln q_1 = t + \ln c_1$ i.e. $q_1 = c_1 e^t$ and similarly we have $q_2 = c_2 e^{-t}$ so that $q_1 q_2 = c_1 c_2 =$ constant.

Now

$$\frac{d}{dt}\left(\frac{p_2 - b q_2}{q_1}\right) = \frac{q_1(\dot{p}_2 - b\dot{q}_2) - \dot{q}_1(p_2 - b q_2)}{q_1^2}$$

$$= \frac{C_1 e^t [p_2 + 2b q_2 + b q_2] - C_1 e^t [p_2 - b q_2]}{C_1^2 e^{2t}} \quad \text{[using the values of } q_1, q_2, \dot{q}_1, \dot{q}_2 \text{ and } \dot{p}_2]$$

$\neq 0$ so the given problem is not correct.

However, if $H = p_1 q_1 - p_2 q_2 - a q_1^2 + b q_2^2$, then

$$\frac{d}{dt}\left(\frac{p_2 - b q_2}{q_1}\right) = 0 \quad \text{and hence} \quad \frac{p_2 - b q_2}{q_1} = \text{constant of motion.}]$$

89. A particle is moving under a potential $V(r, \theta, \phi)$ where (r, θ, ϕ) are spherical polar coordinates of the particle at any instant. Obtain Hamilton's equations of motion. **[C.U. (Hons.) 2003]**

[**Ans.** In spherical polar coordinate system, K.E. of the particle is given by

$$T = \frac{1}{2} m \left[\dot{r}^2 + r^2 \dot{\theta}^2 + r^2 \sin^2 \theta \dot{\phi}^2\right]$$

$$\therefore \quad p_r = \frac{\partial T}{\partial \dot{r}} = m\dot{r}, \quad p_\theta = \frac{\partial T}{\partial \dot{\theta}} = m r^2 \dot{\theta} \quad \text{and} \quad p_\phi = \frac{\partial T}{\partial \dot{\phi}} = m r^2 \sin^2 \theta \dot{\phi}$$

So in terms of p_r, p_θ and p_ϕ, K.E. T can be written as

$$T = \frac{m}{2}\left[\frac{p_r^2}{m^2} + r^2\frac{p_\theta^2}{m^2r^4} + r^2\sin^2\theta\frac{p_\phi^2}{m^2r^4\sin^4\theta}\right] = \frac{1}{2m}\left[p_r^2 + \frac{p_\theta^2}{r^2} + \frac{p_\phi^2}{r^2\sin^2\theta}\right]$$

$\therefore$ Hamiltonian of the particle

$$H = T + V = \frac{1}{2m}\left(p_r^2 + \frac{p_\theta^2}{r^2} + \frac{p_\phi^2}{r^2\sin^2\theta}\right) + V(r,\theta,\phi)$$

Hamilton's equations of motion are then written as

$$\dot{p}_r = -\frac{\partial H}{\partial r} = \frac{1}{2m}\left(\frac{2p_\theta^2}{r^3} + \frac{2}{r^3}\frac{p_\phi^2}{\sin^2\theta}\right) - \frac{\partial V}{\partial r}; \quad \left[\because \dot{p}_k = -\frac{\partial H}{\partial q_k} \text{ and } \dot{q}_k = \frac{\partial H}{\partial p_k}\right]$$

$$\dot{r} = -\frac{\partial H}{\partial p_r} = \frac{1}{2m}2p_r = \frac{p_r}{m}$$

$$\dot{p}_\theta = -\frac{\partial H}{\partial \theta} = -\frac{p_\phi^2}{2mr^2}\frac{\partial}{\partial\theta}\left(\frac{1}{\sin^2\theta}\right) - \frac{\partial V}{\partial\theta} = \frac{p_\phi^2}{2mr^2}\frac{2\cos\theta}{\sin^3\theta} - \frac{\partial V}{\partial\theta} = \frac{p_\phi^2\cos\theta}{mr^2\sin^3\theta} - \frac{\partial V}{\partial\theta};$$

$$\dot{\theta} = \frac{\partial H}{\partial p_\theta} = \frac{p_\theta}{mr^2}; \quad \dot{p}_\phi = -\frac{\partial H}{\partial\phi} = -\frac{\partial V}{\partial\phi}; \quad \dot{\phi} = \frac{\partial H}{\partial p_\phi} = \frac{p_\phi}{mr^2\sin^2\theta}$$

90. Consider a simple pendulum of mass m whose point of support moves uniformly on a vertical circle with constant frequency ω. Obtain expression for the Lagrangian of the system and hence obtain the equation of motion for small θ.

Ans. With the position of the point of support as shown in the figure, the coordinates (x, y) of the pendulum bob are given by $x = a\cos\omega t + l\sin\theta$ and $y = a\sin\omega t + l\cos\theta$.

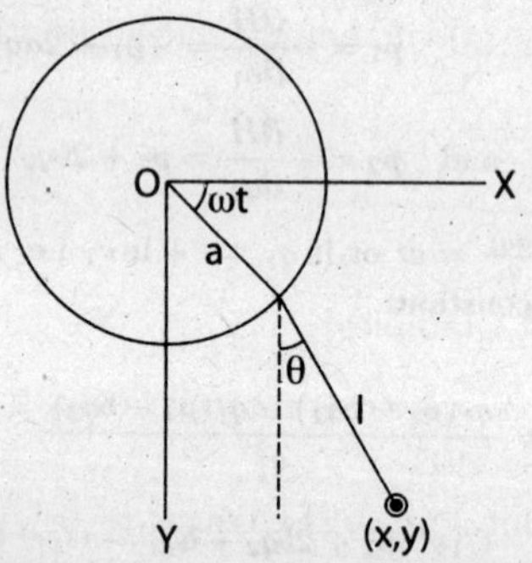

$\therefore \dot{x} = -a\omega\sin\omega t + l\dot{\theta}\cos\theta$ and $\dot{y} = a\omega\cos\omega t - l\dot{\theta}\sin\omega t$, so that

$$\begin{aligned}
\text{K.E.} = T &= \tfrac{1}{2}m(\dot{x}^2 + \dot{y}^2) \\
&= \tfrac{1}{2}m\left[a^2\omega^2 + l^2\dot{\theta}^2 - 2a\omega l\dot{\theta}(\sin\omega t\cos\theta + \cos\omega t\sin\theta)\right] \\
&= \tfrac{1}{2}m\left[a^2\omega^2 + l^2\dot{\theta}^2 - 2a\omega l\dot{\theta}\sin(\omega t + \theta)\right]
\end{aligned}$$

P.E. of the bob $= V = -mgy = -mg(a\sin\omega t + l\cos\theta)$.

Hence the Lagrangian of the system $= L = T - V = \frac{1}{2}m\left[a^2\omega^2 + l^2\dot{\theta}^2 - 2a\omega l\dot{\theta}\sin(\omega t + \theta)\right] + mg(a\sin\omega t + l\cos\theta)$.

$$\therefore \quad \frac{\partial L}{\partial\dot{\theta}} = \frac{1}{2}m\left[l^2 2\dot{\theta} - 2a\omega l\sin(\omega t + \theta)\right] \quad \text{and} \quad \frac{\partial L}{\partial\theta} = -ma\omega l\dot{\theta}\cos(\omega t + \theta) - mgl\sin\theta$$

So the equation of motion is written as

$$\frac{d}{dt}\left(\frac{\partial L}{\partial \dot{\theta}}\right) - \frac{\partial L}{\partial \theta} = 0$$

$$\text{or } \frac{d}{dt}[ml^2\dot{\theta} - ma\omega l \sin(\omega t + \theta)] + ma\omega l\dot{\theta}\cos(\omega t + \theta) + mgl\sin\theta = 0$$

$$\text{or } l\ddot{\theta} - a\omega\cos(\omega t + \theta)(\omega + \dot{\theta}) + a\omega\dot{\theta}\cos(\omega t + \theta) + g\sin\theta = 0$$

$$\text{or } l\ddot{\theta} - a\omega^2\cos(\omega t + \theta) - a\omega\dot{\theta}\cos(\omega t + \theta) + a\omega\dot{\theta}\cos(\omega t + \theta) + g\sin\theta = 0$$

$$\text{or } \ddot{\theta} + \frac{g}{l}\sin\theta = \frac{a\omega^2}{l}\cos(\omega t + \theta)$$

which, for small θ, reduces to $\ddot{\theta} + \frac{g}{l}\theta = \frac{a\omega^2}{l}\cos\omega t$.

91. An artificial earth satellite is revolving round the earth. It is in the form of a toroid (as shown in the figure given below) of radius 2.5 m. Calculate the angular velocity $\vec{\omega}$ of the satellite about its axis in order to generate artificial gravity (g = 10 m/sec^2) inside it. **[C.U. (Hons.) 2006]**

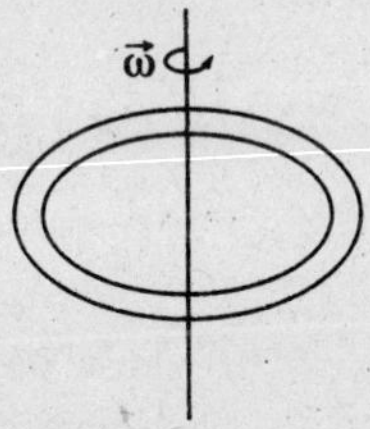

For answer see Q.62 in the Exercises of the Chapter on Gravitation, i.e. Chapter 7.

92. A spherical rain drop of radius a cms falls from rest through a vertical height h, receiving throughout the motion an accumulation of condensed vapour at the rate of k gms per sq cm per second, no vertical force but gravity acting; show that when it reaches the ground, its radius will be

$$k\sqrt{\frac{2h}{g}}\left[1 + \sqrt{\left(1 + \frac{ga^2}{2hk^2}\right)}\right]$$

[cf. C.U. (Hons.) 1998]

(For answer see worked out examples)

93. A Cartesian coordinate axes system is rotated about Z-axis in anticlockwise direction through an angle θ such that a point $P(x, y, z)$ changes to (x', y', z') and $\begin{pmatrix} x' \\ y' \\ z' \end{pmatrix} = R_z(\theta)\begin{pmatrix} x \\ y \\ z \end{pmatrix}$. Find $R_z(\theta)$. (See Appendix B.1) **[C.U. (Hons.) 2006]**

94. Consider an orthogonal transformation of coordinate system $OXYZ$ to $OX'Y'Z'$ with unit vectors $\hat{\imath}, \hat{\jmath}, \hat{k} \rightarrow \hat{\imath}', \hat{\jmath}', \hat{k}'$. A position vector $\vec{r}$ in these systems can be expressed as $\vec{r} = \hat{\imath}x + \hat{\jmath}y + \hat{k}z = \hat{\imath}'x' + \hat{\jmath}'y' + \hat{k}'z'$.

Show that $x' = (\hat{\imath}' \cdot \hat{\imath})x + (\hat{\imath}' \cdot \hat{\jmath})y + (\hat{\imath}' \cdot \hat{k})z$.

Hence establish

$$\begin{bmatrix} x' \\ y' \\ z' \end{bmatrix} = \begin{bmatrix} a_{11} & a_{12} & a_{13} \\ a_{21} & a_{22} & a_{23} \\ a_{31} & a_{32} & a_{33} \end{bmatrix}\begin{bmatrix} x \\ y \\ z \end{bmatrix}$$

where $a_{11} = \hat{\imath}' \cdot \hat{\imath}, a_{12} = \hat{\imath}' \cdot \hat{\jmath}$, etc. **[C.U. (Hons.) 1997]**

(See Appendix B.1)

95. A coordinate system XOY is rotated about Z-axis in anticlockwise sense through θ, so that it takes a point (x, y) to (x', y'). Establish the transformation relation $\begin{pmatrix} x' \\ y' \end{pmatrix} = \begin{pmatrix} \cos\theta & \sin\theta \\ -\sin\theta & \cos\theta \end{pmatrix} \begin{pmatrix} x \\ y \end{pmatrix}$. (See Appendix B.1) **[C.U. (Hons.) 2002]**

Chapter 4

Dynamics of a Rigid Body—Moment of Inertia

4.1 Rotational Motion of a Rigid Body—Calculation of K.E.

Let us consider a rigid body rotating about a fixed axis AB with uniform angular velocity ω. We imagine the body to be made up of a large number of particles of masses m_1, m_2, m_3, etc. situated at distances r_1, r_2, r_3, etc. from the axis of rotation (Fig. 4.1.1). The angular velocity of different particles is same as that of the rigid body but their linear velocities are different due to their different positions with respect to the axis. Hence the kinetic energy of the whole body is given by,

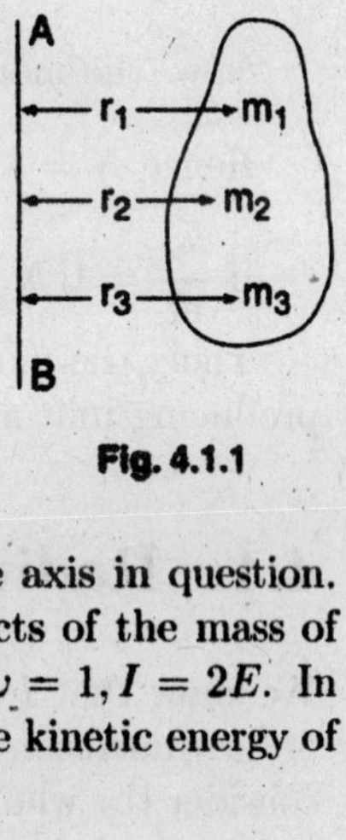

Fig. 4.1.1

$$E = \frac{1}{2}m_1\omega^2r_1^2 + \frac{1}{2}m_2\omega^2r_2^2 + \cdots = \frac{1}{2}\omega^2(m_1r_1^2 + m_2r_2^2 + \cdots)$$

$$= \frac{1}{2}\omega^2\Sigma mr^2 = \frac{1}{2}I\omega^2 \qquad (1)$$

where $I = \Sigma mr^2$. I is called the moment of inertia of the body about the axis in question. Hence moment of inertia of a body about an axis is the sum of the products of the mass of each particle and the square of its distance from the axis. From (4.1.1) if $\omega = 1, I = 2E$. In other words, moment of inertia of a body about any axis is equal to twice the kinetic energy of a body rotating with unit angular velocity about the same axis.

4.2 Moment of Inertia and Moment of Momentum or Angular Momentum

Referring to Fig. 4.1.1, we consider the particle of mass m_1, at a distance r_1 from the axis rotating with angular velocity ω. Linear momentum of the particle is $m_1r_1\omega$. Therefore, moment of momentum about the axis under consideration is $m_1r_1\omega \times r_1 = m_1\omega r_1^2$.

Considering the rigid body to be made up of all such particles of masses m_1, m_2, etc. at distances r_1, r_2, etc. from the axis, the moment of momentum of the whole body is

$$L = m_1\omega r_1^2 + m_2\omega r_2^2 + \cdots = \omega(m_1r_1^2 + m_2r_2^2 + \cdots) = \omega\Sigma mr^2 = I\omega.$$

If $\omega = 1$, moment of momentum $L = I$.

Thus moment of inertia of the body about the axis is numerically equal to the moment of momentum of the body rotating with unit angular velocity. The moment of momentum of a body about an axis is also known as 'Angular Momentum'.

4.3 Moment of Inertia and Torque

Considering the same rigid body rotating with a uniform angular acceleration $\frac{d\omega}{dt}$, the particle of mass m_1 of the body at a distance r_1 from the axis has a linear acceleration $r_1\frac{d\omega}{dt}$. Hence the force acting on the particle is $m_1r_1\frac{d\omega}{dt}$. The moment of this force about the axis in question is $m_1r_1\frac{d\omega}{dt}\cdot r_1 = m_1r_1^2\frac{d\omega}{dt}$.

Therefore, the moment of this force acting on the entire rigid body about the axis is

$$m_1r_1^2\frac{d\omega}{dt} + m_2r_2^2\frac{d\omega}{dt} + \cdots = \frac{d\omega}{dt}\left(m_1r_1^2 + m_2r_2^2 + \cdots\right) = I\frac{d\omega}{dt}.$$

Now, the moment of this force is commonly known as torque (N).

Hence $N = I\frac{d\omega}{dt}$.

If $\frac{d\omega}{dt} = 1, N = I$.

Thus, the moment of inertia of the body about an axis is numerically equal to the torque producing unit angular acceleration about the same axis.

4.4 Radius of Gyration

We know that $I = \Sigma mr^2 = Mk^2$ (say), where M denotes the mass of the whole body and k is of the dimension of length and is called the radius of gyration of the body about the axis. If we consider the whole mass of the body to be concentrated at a single point such that the moment of inertia of this concentrated point mass is same as the moment of inertia of the whole body about the axis, the distance of that single point from the axis is called the radius of gyration of the body about the axis.

It is sometimes called the 'Swing Radius'.

4.5 Physical Significance of Moment of Inertia

We know that the kinetic energy of a body in translational motion is $\frac{1}{2}Mv^2$, but the kinetic energy in rotational motion is $\frac{1}{2}I\omega^2$. Comparing these two expressions for kinetic energy, we see that since ω is the rotational analogue of v, I corresponds to M and represents the effect of the mass and its space distribution in rotational motion.

From Newton's first law of motion, it follows that a body continues in its state of rest or of uniform motion in a straight line due to inertia. Greater the mass of the body, greater will be the force required to change its state. The mass of the body is thus taken to be a measure of its inertia for translatory motion. In case of rotational motion, the body will continue in its state of uniform rotational motion due to its 'rotational inertia'. Now, greater the moment of inertia of the body, greater will be the torque required to change its state of uniform rotational motion. The moment of inertia is thus taken to be a measure of its inertia for rotational motion.

4.6 General Theorems on Moment of Inertia

There are two general theorems of importance on moment of inertia. They help us in determining the moment of inertia of a body about an axis, if its moments of inertia about some other axes be known.

4.6.1 Theorem of perpendicular axes

(i) *For a plane laminar body :*

Let us consider a lamina and two axes OX and OY at right angles to each other in the plane of the lamina, meeting at O (Fig. 4.6.1). According to this theorem, the moment of inertia of the lamina about an axis OZ, passing through O and perpendicular to the plane of the lamina, is equal to the sum of the moments of inertia of the lamina about the axes OX and OY. Representing the moments of inertia of the laminar body about OX, OY and OZ, by I_x, I_y and I_z respectively, the theorem states that $I_z = I_x + I_y$.

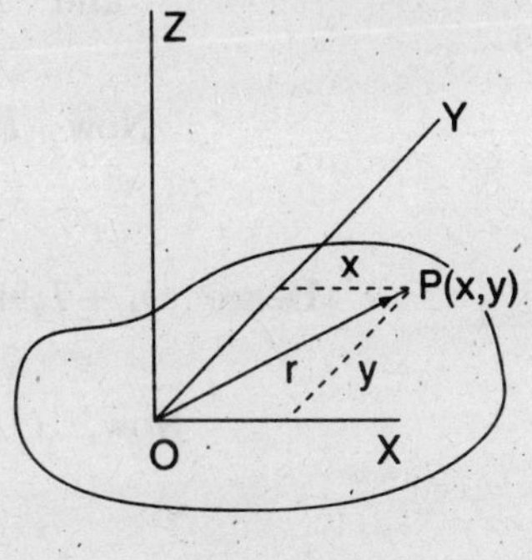

Fig. 4.6.1

Proof : We consider an elementary particle of mass m at P having co-ordinates x and y. Then the distances of P from the axes OX and OY are y and x respectively. Hence

$$I_x = \Sigma my^2 \quad \text{and} \quad I_y = \Sigma mx^2$$

$$\therefore \quad I_x + I_y = \Sigma m(y^2 + x^2) = \Sigma mr^2 = I_z \quad [\because x^2 + y^2 = r^2]$$

This proves the theorem.

(ii) *For a three-dimensional body :*

If the moments of inertia of a three-dimensional body about the three mutually perpendicular axes OX, OY and OZ be I_x, I_y and I_z respectively then according to this theorem, $I_x + I_y + I_z = 2\Sigma mOP^2$, where m denotes an elementary mass of the body and OP the distance of the elementary mass from the origin O of the co-ordinate axes OX, OY and OZ (Fig. 4.6.2).

Proof : We consider an elementary mass m of the three dimensional body at a point P having co-ordinates x, y and z with respect to the co-ordinate axes OX, OY and OZ. We now drop a perpendicular PL from P on the plane XY. From L, LQ and LN are drawn parallel to axes

OX and OY respectively. Now OP, QP and NP are joined and PR is drawn parallel to LO (Fig. 4.6.2).

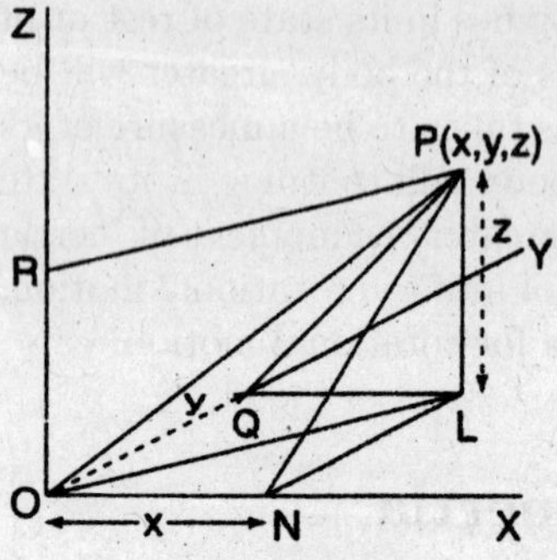

Fig. 4.6.2

Then $I_x = \Sigma mPN^2$

[$\because$ PN is normal to the axis OX.]

Again,

$$PN^2 = PL^2 + LN^2 \quad [\because \angle PLN \text{ is a right angle.}]$$
$$= PL^2 + OQ^2 = z^2 + y^2$$
$$\therefore I_x = \Sigma m(y^2 + z^2).$$

Also,

$$I_y = \Sigma mPQ^2 \quad [\because PQ \text{ is normal to the axis } OY.]$$

Again, $$PQ^2 = PL^2 + LQ^2 \quad [\because \angle PLQ \text{ is a right angle.}]$$
$$= PL^2 + ON^2 = z^2 + x^2. \quad \therefore I_y = \Sigma m(z^2 + x^2)$$

and $$I_z = \Sigma mPR^2 \quad [\because PR \text{ is normal to the axis } OZ.]$$

Now $$PR^2 = OL^2 = ON^2 + LN^2 \quad [\because \angle LNO \text{ is a right angle.}]$$
$$= ON^2 + OQ^2 = x^2 + y^2. \quad \therefore I_z = \Sigma m(x^2 + y^2).$$

Hence $$I_x + I_y + I_z = 2\Sigma m(x^2 + y^2 + z^2).$$

Now, $$OP^2 = OL^2 + PL^2 \quad [\because \angle PLO \text{ is a right angle.}]$$
$$= ON^2 + NL^2 + PL^2 = ON^2 + OQ^2 + PL^2$$
$$= x^2 + y^2 + z^2. \quad \therefore I_x + I_y + I_z = 2\Sigma mOP^2.$$

This proves the theorem.

4.6.2 Theorem of parallel axes

(i) *For a plane laminar body :*

Let us consider a lamina and two axes AB and CD in the plane of the lamina. The two axes are parallel to each other of which the axis CD passes through the centre of mass G of the lamina. According to this theorem, the moment of inertia of the lamina about the axis AB is equal to its moment of inertia about the axis CD, plus the product of the mass of the lamina and the square of the distance between the two axes. Representing the moments of inertia about AB and CD by I and I_G respectively, we may write according to this theorem,

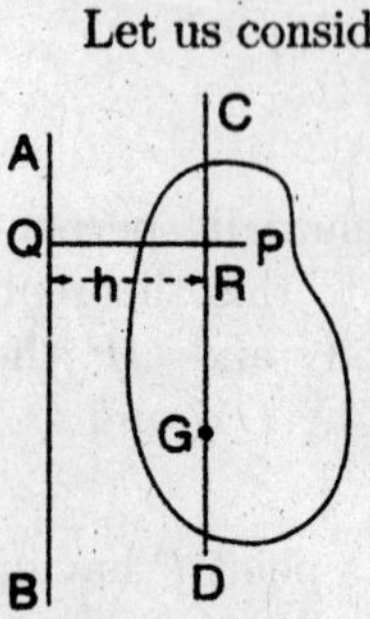

Fig. 4.6.3

$$I = I_G + Mh^2$$

where M is the mass of the lamina and h the distance between the two parallel axes AB and CD.

Proof : We consider an elementary particle of the lamina of mass m at P. From P, a perpendicular PRQ is drawn meeting CD and AB in R and Q respectively. Let $RQ = h$.

$$\therefore\ I = \Sigma mPQ^2 = \Sigma m(PR + RQ)^2 = \Sigma m(PR + h)^2$$

$$= \Sigma mPR^2 + \Sigma mh^2 + \Sigma 2mPR\cdot h$$

$$= \Sigma mPR^2 + h^2\Sigma m + 2h\Sigma mPR \quad [\because h \text{ is a constant.}]$$

The first term ΣmPR^2 is the moment of inertia of the lamina about the axis CD. $\therefore\ \Sigma mPR^2 = I_G$. The second term $h^2\Sigma m = Mh^2$ where M is the mass of the lamina.

Now since a body always balances about an axis passing through its centre of mass, it is obvious that the algebraic sum of the moments of the weights of its individual particles about the centre of mass must be equal to zero. Hence, here $\Sigma mg\cdot PR$ (algebraic sum of such moments about an axis through G) is zero and, therefore the expression $\Sigma m\cdot PR$ is also equal to zero, g being constant at a given place. Consequently the third term $2h\Sigma m\cdot PR = 0$, so that,

$$I = I_G + Mh^2.$$

(ii) *For a three-dimensional body :*

We consider three axes GX, GY, GZ through the centre of mass G of a three-dimensional body and OX', OY', OZ' parallel axes through any point O (Fig. 4.6.4). Let the co-ordinates of any element m of the body be (x, y, z) referred to the first set of axes and (x', y', z') the co-ordinates of the same element referred to the second set of axes. If (a, b, c) be the co-ordinates of G referred to the second set of axes we may write

$$x' = x + a, y' = y + b \text{ and } z' = z + c.$$

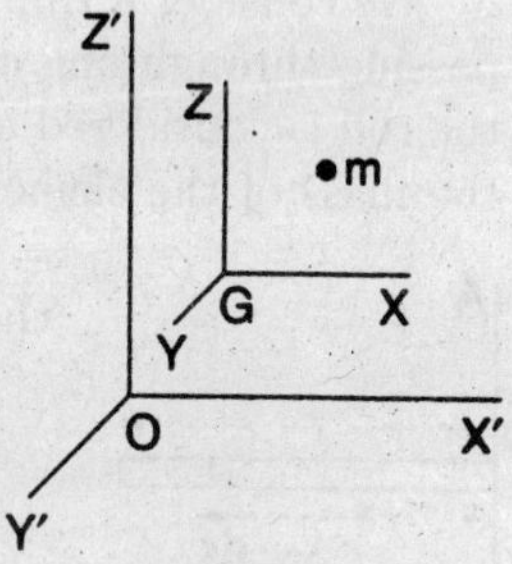

Fig. 4.6.4

Hence the moment of inertia of the body about OX'

$$= \Sigma m(y'^2 + z'^2) = \Sigma m\left[(y + b)^2 + (z + c)^2\right]$$

$$= \Sigma m\left[y^2 + b^2 + 2yb + z^2 + c^2 + 2zc\right]$$

$$= \Sigma m(y^2 + z^2) + \Sigma m(b^2 + c^2) + 2\Sigma myb + 2\Sigma mzc.$$

The last two terms are zero, because $\dfrac{\Sigma my}{\Sigma m}, \dfrac{\Sigma mz}{\Sigma m}$ denote the co-ordinates of the mass centre G which is taken as origin (see Art. 3.29).

$\therefore$ The moment of inertia about OX'

$$= \Sigma m(y^2 + z^2) + \Sigma m(b^2 + c^2)$$

$$= \Sigma m(y^2 + z^2) + M(b^2 + c^2)$$

= the moment of inertia of the body about GX + moment of inertia of mass M supposed to be concentrated at G, i.e., the moment of inertia of the body about any axis OX' = the moment of inertia of the body about an axis passing through the centre of mass and parallel to OX' + mass of the body × the square of the distance between the two parallel axes GX and OX'. This proves the theorem of parallel axes.

In case of product* of inertia the theorem also applies.

The product of inertia about OY' and $OZ' = \Sigma m(y'z') = \Sigma m[(y+b)(z+c)]$

$$= \Sigma m[yz + yc + bz + bc]$$

$$= \Sigma myz + \Sigma myc + \Sigma mbz + \Sigma mbc.$$

Now $\Sigma myc = \Sigma mzb = 0$.

$\therefore$ The product of inertia about OY' and OZ' $= \Sigma myz + \Sigma mbc = \Sigma myz + Mbc =$ the product of inertia of the body about GY and GZ + the product of inertia of the mass of the body concentrated at G about the axes OY' and OZ'.

4.7 Calculation of Moment of Inertia for Different Cases

I. Thin Uniform Rod

(i) *About an axis passing through one end of the rod and perpendicular to its length :*

We consider a thin uniform rod of length l and an axis AB perpendicular to its length and passing through one end of the rod as shown in the figure 4.7.1. An element of length dx of the rod is considered at a distance x from the axis. If ρ be the mass per unit length of the rod, the mass of the element $= \rho dx$ and its moment of inertia about the axis under consideration $= \rho dx \cdot x^2$. Integrating this expression between the limits $x = 0$ and $x = l$, the moment of inertia of the rod will given by

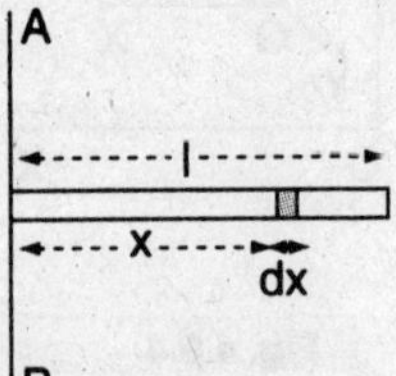

Fig. 4.7.1

$$I = \int_0^l \rho x^2 dx = \rho \left[\frac{x^3}{3}\right]_0^l = \frac{1}{3}\rho l^3 = \frac{1}{3}Ml^2$$

where $M = \rho l$, the mass of the rod. The radius of gyration about the axis $= \dfrac{l}{\sqrt{3}}$.

(ii) *About an axis through its centre and perpendicular to its length :*

The treatment in this case is same as above, except that, since the axis AB passes through the centre of the rod (Fig. 4.7.2), the expression for the moment of inertia of the element dx of the rod is now to be integrated between the limits $x = -l/2$ and $x = +l/2$. Thus moment of inertia of the rod about the axis in question is given by,

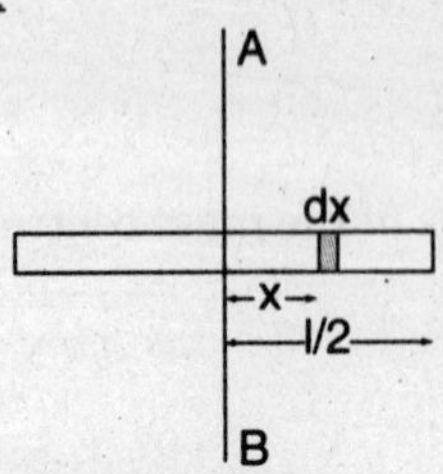

Fig. 4.7.2

$$I = \int_{-l/2}^{+l/2} \rho x^2 dx = \rho \left[\frac{x^3}{3}\right]_{-l/2}^{+l/2} = \frac{\rho l^3}{12} = \frac{Ml^2}{12}$$

where $M = \rho l$, the mass of the rod. The radius of gyration about the axis $= \dfrac{l}{2\sqrt{3}}$.

Illustrative Examples

1. A uniform rod 2 ft. long weighing 5 lbs. is revolving 60 times a minute about one end. Calculate its moment of inertia and kinetic energy.

---

*See article 4.8

Here $I = \frac{Ml^2}{3}$, $M = 5$ lbs., $l = 2$ ft. $\therefore I = \frac{5 \times 4}{3}$ lb.ft.^{2} = 6.66 lb.ft.^{2}.

Again,

$$\text{K.E.} = \frac{1}{2}I\omega^2 = \frac{1}{2}I(2\pi n)^2 \quad [\because \omega = 2\pi n, n \text{ being the frequency of revolution.}]$$

Here, $n = \frac{60}{60} = 1$ rev./sec.

$\therefore$ K.E. $= \pi^2 I n^2 = 2 \times (3.14)^2 \times 6.66 \times 1 = 131.31$ ft. poundals = 4.07 ft. lbs.

2. Show that the kinetic energy of a thin rod of length l and m mass per unit length rotating about an axis through the middle point and perpendicular to its length with angular velocity ω is $\frac{1}{24}m\omega^2 l^3$.

[C.U. 1948]

$$\text{Here, } I = \frac{Ml^2}{12} = ml\cdot\frac{l^2}{12} \quad [\because M = ml]$$

$$= \frac{ml^3}{12}.$$

$$\text{Now, K.E.} = \frac{1}{2}I\omega^2 = \frac{1}{2}\frac{ml^3}{12}\omega^2 = \frac{ml^3}{24}\omega^2.$$

II. Rectangular Lamina

(i) *About an axis parallel to its breadth and passing through its centre of mass—the axis lying in the plane of the lamina :*

Let us consider a rectangular lamina of length a and an axis passing through the centre of mass O of the lamina (Fig. 4.7.3). The axis YOY' is parallel to its breadth and lies in the plane of the lamina. We divide the lamina into a number of elementary plane strips parallel to its length and each strip may be looked upon as a thin uniform rod. If m be the mass of an elementary rod, its moment of inertia about the axis under consideration $= \frac{ma^2}{12}$.

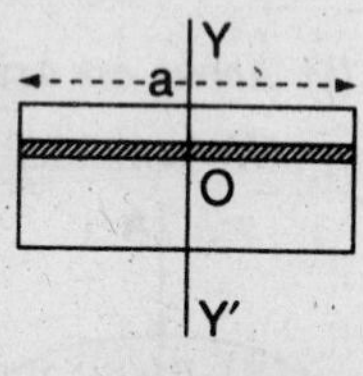

Fig. 4.7.3

Therefore, the moment of inertia of the entire lamina about the axis is given by

$$I_y = \Sigma\frac{ma^2}{12} = \frac{a^2}{12}\Sigma m = \frac{Ma^2}{12}$$

where M = mass of the lamina.

(ii) *About an axis parallel to its length and passing through its centre of mass—the axis lying in the plane of the lamina :*

In this case, the rectangular lamina of breadth b is divided into a number of elementary plane strips parallel to its breadth, and each such strip may be looked upon as a thin uniform rod (Fig. 4.7.4). If m' be the mass of such an elementary rod, its moment of inertia about the axis $XOX' = \frac{m'b^2}{12}$.

$\therefore$ The moment of inertia of the entire lamina about the axis in question is given by

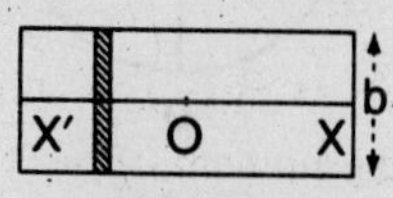

Fig. 4.7.4

$$I_x = \Sigma\frac{m'b^2}{12} = \frac{b^2}{12}\Sigma m' = \frac{Mb^2}{12}.$$

(iii) *About an axis passing through its centre of mass and perpendicular to the plane of the lamina* :

By the theorem of perpendicular axes the moment of inertia of the rectangular lamina about the axis under consideration is given by

$$I_z = I_x + I_y = \frac{Mb^2}{12} + \frac{Ma^2}{12} = \frac{M(a^2+b^2)}{12}.$$

The radius of gyration about the axis $= \sqrt{\dfrac{a^2+b^2}{12}}$.

Illustrative Example

1. Calculate the moment of inertia of a rectangular plate about an axis passing through one corner and perpendicular to its plane. **[C.U. (Hons.) 1967]**

Moment of inertia of the rectangular plate about an axis perpendicular to its plane and passing through any corner = moment of inertia of the plate about an axis perpendicular to its plane and passing through its centre O + mass of the plate × (distance of the corner from O)^2 [applying the theorem of parallel axes] (Fig.1)

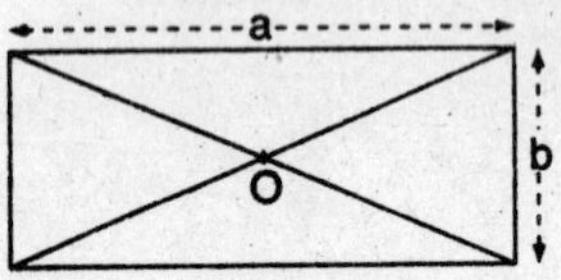

Fig. 1

$$= M\frac{(a^2+b^2)}{12} + M\frac{(a^2+b^2)}{4} = M\frac{(a^2+b^2)}{3}.$$

III. Thin Uniform Circular Ring or Hoop

(i) *About an axis passing through its centre and perpendicular to the plane of the ring* :

We consider a thin uniform ring of radius a and an axis AB passing through its centre and perpendicular to its plane (Fig. 4.7.5).

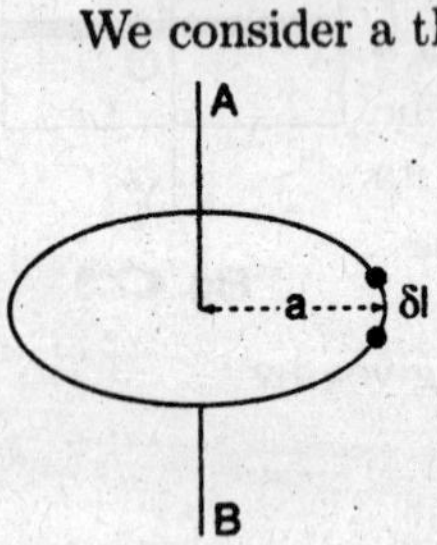

Fig. 4.7.5

Let us consider an element of length δl of the ring and let ρ be the mass per unit length of the ring. Moment of inertia of the element δl about the axis $= \rho \delta l a^2$.

∴ Moment of inertia of the ring about the axis under consideration is given by

$$I = \Sigma \rho \delta l a^2 = \rho a^2 \Sigma \delta l = \rho a^2 \times 2\pi a = Ma^2$$

where $M = 2\pi a \rho$, the mass of the ring.

(ii) *About an axis passing through its centre and lying in the plane of the ring, i.e., about a diameter* :

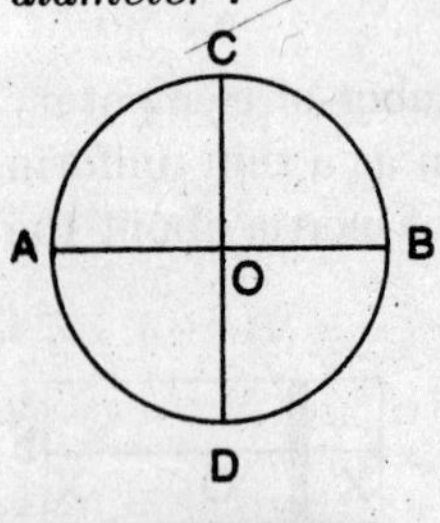

Fig. 4.7.6

If I_d be the moment of inertia of the ring about the diameter AB, it will, also be its moment of inertia about the diameter CD which is perpendicular to AB (Fig. 4.7.6). Then by the theorem of perpendicular axes, its moment of inertia about an axis passing through the centre O and perpendicular to its plane, is equal to the sum of its moments of inertia about the axes AB and CD in its own plane. Therefore we have,

$$I_d + I_d = Ma^2 \text{ or } I_d = \frac{Ma^2}{2}.$$

Alternative Treatment

The above result can also be obtained in the following way.

We consider a circular ring of radius a and mass per unit length ρ (Fig. 4.7.7.).

Let us consider an element PP' of the ring subtending an angle $d\theta$ at the centre O. The mass of the element $= ad\theta\rho$ and its distance from the axis $AB = a\sin\theta$, so that its moment of inertia about the axis in question $= ad\theta\rho(a\sin\theta)^2 = \rho a^3 \sin^2\theta d\theta$.

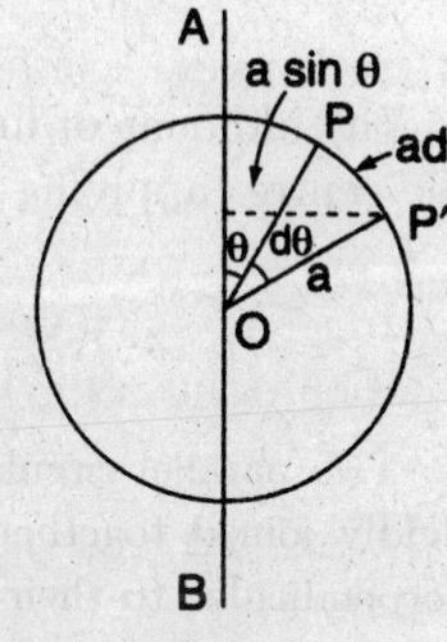

Fig. 4.7.7

Integrating this expression between the limits $\theta = 0$ and $\theta = \pi$ and multiplying the result by 2, to include both halves of the ring, we have the moment of inertia of the ring about the aforesaid axis given by

$$I = 2\int_0^{\pi} \rho a^3 \sin^2\theta d\theta = \rho a^3 \int_0^{\pi} 2\sin^2\theta d\theta = \rho a^3 \int_0^{\pi} (1-\cos 2\theta)d\theta$$

$$= \rho a^3 \left[\theta - \frac{\sin 2\theta}{2}\right]_0^{\pi} = \rho a^3 \pi = 2\pi a\rho \cdot \frac{a^2}{2} = \frac{Ma^2}{2},$$

where $M - 2\pi a\rho$, the mass of the ring.

(iii) *About a tangent (in the plane of the ring)* :

If I_T be the required moment of inertia, by the theorem of parallel axes, we have (Fig. 4.7.8).

$$I_T = \frac{Ma^2}{2} + Ma^2 = \frac{3}{2}Ma^2.$$

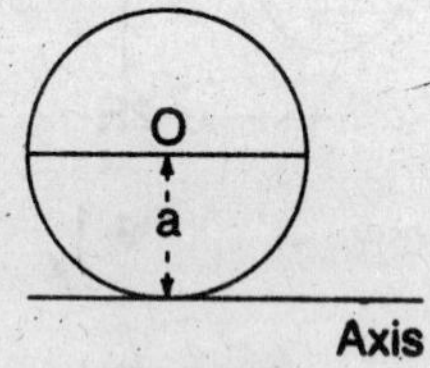

Fig. 4.7.8

Illustrative Examples

1. A hoop of mass 2 kgms and radius 20 cms rolls along the ground at the rate of 10 metres per second. Calculate its kinetic energy.

Here, $I = Ma^2$ and

$$\text{K.E.} = \frac{1}{2}I\omega^2 = \frac{1}{2}Ma^2\omega^2 = \frac{1}{2}Ma^2 \cdot \frac{v^2}{a^2} = \frac{1}{2}Mv^2, \quad \text{where } v \text{ is the linear speed.}$$

Here, $M = 2$ kgms $= 2000$ gms, $v = 10$ metres/sec. $= 1000$ cms/sec.

$\therefore$ K.E. $= \frac{1}{2} \times 2000 \times (1000)^2 = 10^9$ ergs.

2. Calculate the moment of inertia for a thin circular ring of mass 250 gms and radius 15 cms about an axis in the plane of the ring—

(i) when the axis passes through the centre in the plane of the circle;

(ii) when the axis is 10 cms from the centre in the plane of the circle. **[C.U. (Hons.) 1959]**

(i) Moment of inertia of the ring about an axis passing through the centre and in the plane of the circle, i.e., the moment of inertia of the ring about is diameter $= \frac{ma^2}{2}$, where m = mass of the ring and a = radius of the ring.

∴ The moment of inertia about the diameter

$$\frac{ma^2}{2} = \frac{250 \times 15^2}{2} = 125 \times 225 = 28125 \text{ gm.cm}^2.$$

(ii) Moment of inertia of the ring about an axis in its plane at a distance of 10 cms from the centre (applying the theorem of parallel axes).

$$= \frac{ma^2}{2} + m \times (10)^2 = 28125 + 250 \times 100 = 53125 \text{ gm.cm}^2$$

3. Two parallel circular wheels, each of mass M and radius R and fitted with weightless spokes are rigidly joined together by a weightless rod of length $2R$ passing through the centre of the wheels perpendicular to their planes. Calculate the moment of inertia of the wheels about an axis passing perpendicularly through the centre of the rod. **[C.U. (Hons.) 1980]**

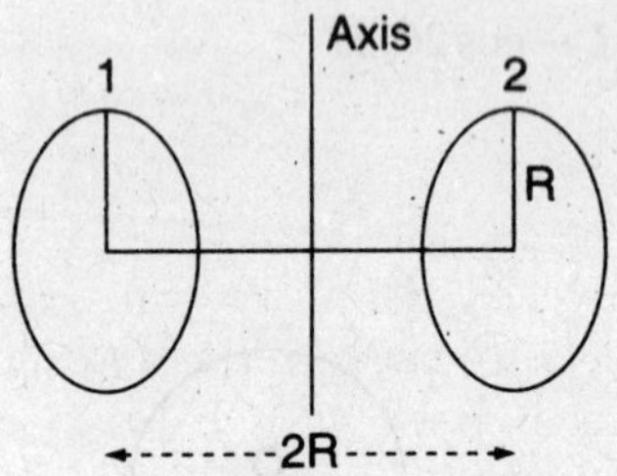

Fig. 1

Applying the theorem of parallel axes the moment of inertia of wheel No. 1 or 2 about an axis passing perpendicularly through the centre of the rod as shown in Fig. 1.

$$= \frac{MR^2}{2} + MR^2 = \frac{3}{2}MR^2.$$

Hence the moment of inertia of both the wheels about the required axis

$$= 2 \times \frac{3}{2}MR^2 = 3MR^2.$$

IV. Circular Lamina or Disc

(i) *About an axis through its centre and perpendicular to its plane :*

We consider a disc of raidus a and mass per unit area ρ. Let us consider a narrow concentric circular ring of radius x and width dx as shown in the figure 4.7.9. Evidently the area of the elementary ring

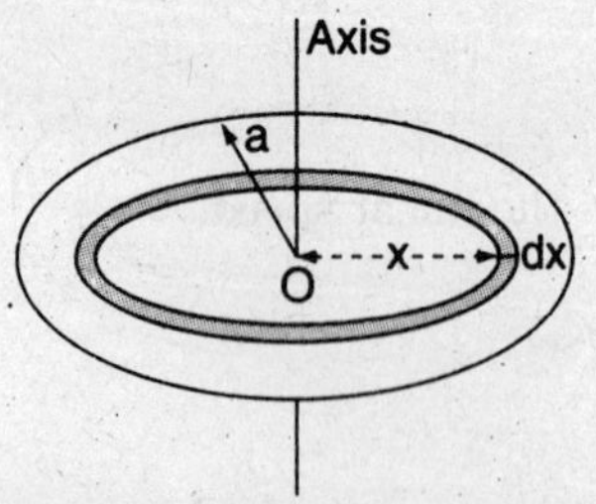

Fig. 4.7.9

$$= \pi(x + dx)^2 - \pi x^2 = 2\pi x dx$$

[neglecting square of dx which is negligible].

Hence the moment of inertia of this ring about an axis through O and perpendicular to its plane

$$= 2\pi x dx \rho \cdot x^2 = 2\pi \rho x^3 dx.$$

Integrating the above expression between the limits $x = 0$ and $x = a$, we have the moment of inertia of the disc given by

$$I = 2\pi\rho \int_0^a x^3 dx = 2\pi\rho \cdot \frac{a^4}{4} = \frac{\pi a^4 \rho}{2} = \pi a^2 \rho \times \frac{a^2}{2} = \frac{Ma^2}{2}$$

where $M = \pi a^2 \rho$, the mass of the disc.

(ii) *About a diameter :*

Let I_d be the moment of inertia of the disc about a diameter AB. Then I_d is also its moment of inertia about a perpendicular diameter CD (Fig. 4.7.10). Now, since AB and CD both lie in the plane of the disc, by the theorem of perpendicular axes, moment of inertia of the disc about $AB+$ its moment of inertia about CD = its moment of inertia about an axis through O and perpendicular to its plane

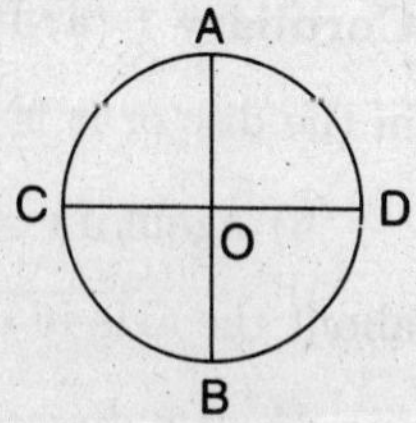

Fig. 4.7.10

$$\text{or} \quad I_d + I_d = \frac{Ma^2}{2} \quad \therefore\ I_d = \frac{Ma^2}{4}.$$

(iii) *About a tangent parallel to one diameter :*

Let EF be the tangent about which the moment of inertia is to be determined (Fig. 4.7.11). Let CD be a diameter parallel to EF, so that by the theorem of parallel axes, we have moment of inertia of the disc about the tangent EF = its moment of inertia about

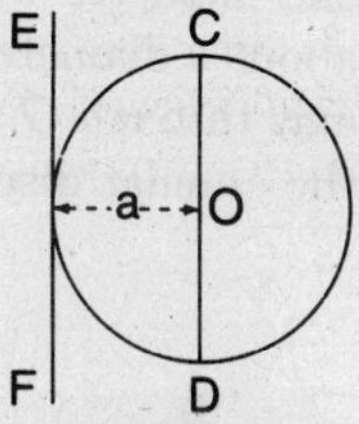

Fig. 4.7.11

$$CD + Ma^2 = \frac{Ma^2}{4} + Ma^2 \quad \text{or} \quad I_T = \frac{5}{4}Ma^2.$$

(iv) *About an axis tangential to the disc and perpendicular to its plane :*

Let I_T be the required moment of inertia. This tangent is evidently parallel to the axis through the centre of the disc and perpendicular to its plane, the distance between them being equal to the radius of the disc. Therefore, by the theorem of parallel axes, we have,

$$I_T = \text{Moment of inertia about the perpendicular axis through the centre} + Ma^2$$

$$= \frac{Ma^2}{2} + Ma^2 = \frac{3}{2}Ma^2.$$

V. Annular Disc or Ring

(i) *About an axis passing through its centre and perpendicular to its plane :*

An annular ring consists of an ordinary circular disc from which the central portion has been removed so that there is a concentric circular hole in it. Let the internal and external radii of the ring be r and a respectively, and let ρ be the mass per unit area of the ring.

We imagine the annular disc to be made up of a number of thin concentric circular rings, and let us consider one such ring of radius x and width dx (Fig. 4.7.12).

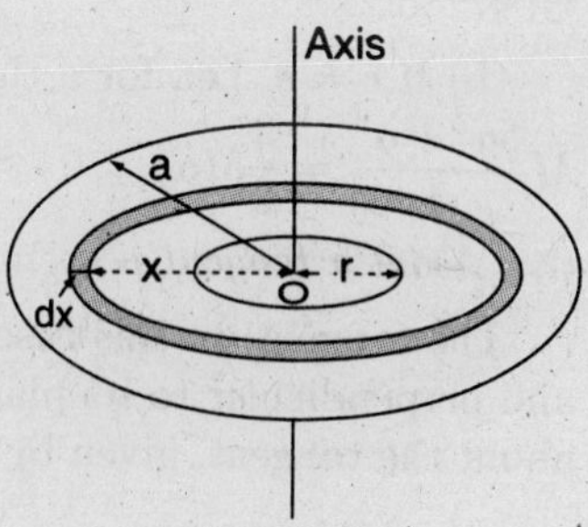

Fig. 4.7.12

The moment of inertia of this elementary ring about the axis through the centre O and perpendicular to its plane = $2\pi x dx \rho x^2 = 2\pi\rho x^3 dx$.

By integrating the above expression between the limits $x = r$ and $x = a$, the moment of inertia of the whole annular disc is given by

$$I = 2\pi\rho \int_r^a x^3 dx = 2\pi\rho \cdot \frac{1}{4}(a^4 - r^4) = \frac{1}{2}\pi\rho(a^2 - r^2)(a^2 + r^2) = M\frac{(a^2 + r^2)}{2}$$

where $M = \pi\rho(a^2 - r^2)$, the mass of the annular disc.

Corollary : (a) It follows at once from the above result that if $r = 0$, i.e., if there is no hole in the disc or it is just a plane disc, its moment of inertia $= \dfrac{Ma^2}{2}$.

(b) Again if $r = a$, i.e., if we have a hoop or a circular ring of radius a, its moment of inertia about the axis in question $= M\dfrac{(a^2 + a^2)}{2} = Ma^2$.

(ii) *About a diameter :*

Obviously, due to the symmetrical shape, the moment of inertia of the annular disc about one diameter will be the same as that about any other diameter. Let the moment of inertia about a diameter be I_d. By the theorem of perpendicular axes, its moment of inertia about an axis through O and perpendicular to its plane is equal to the sum of the moments of inertia of the annular disc about two perpendicular diameters.

$$\text{or } I_d + I_d = M\frac{(a^2 + r^2)}{2} \quad \text{whence} \quad I_d = M\frac{(a^2 + r^2)}{4}.$$

Corollary : (a) If $r = 0$, i.e., the disc is a plane one, we have its moment of inertia about a diameter $= \dfrac{Ma^2}{4}$.

(b) If $r = a$, we have a thin circular ring whose moment of inertia about a diameter $= \dfrac{Ma^2}{2}$.

(iii) *About a tangent in its own plane :*

Here the tangent being parallel to a diameter of the ring and at a distance a from it, we have, by the theorem of parallel axes, the moment of inertia about a tangent given by

$$I_T = M\frac{(a^2 + r^2)}{4} + Ma^2 = M\frac{(5a^2 + r^2)}{4}.$$

Corollary : (a) If $r = 0$, i.e., for a plane circular disc of radius a, moment of inertia about a tangent in its own plane $= \dfrac{5}{4}Ma^2$.

(b) If $r = a$, i.e., for a thin circular ring, moment of inertia about a tangent in its plane $= M\dfrac{5a^2 + a^2}{4} = \dfrac{3}{2}Ma^2$.

(iv) *About a tangent perpendicular to its own plane :*

The tangent, in this case, being parallel to the axis through the centre of the annular ring and perpendicular to its plane, we have, by the theorem of parallel axes, its moment of inertia about the tangent, given by

$$I_T = M\frac{a^2 + r^2}{2} + Ma^2 = M\frac{3a^2 + r^2}{2}.$$

Corollary : (a) If $r = 0$, i.e., for a plane circular disc of radius a, its moment of inertia about the required tangent $= \dfrac{3Ma^2}{2}$.

(b) If $r = a$, i.e., for a circular ring, its moment of inertia about the required tangent $= M\dfrac{3a^2 + a^2}{2} = 2Ma^2$.

VI. (a) Solid Circular Cylinder

(i) *About its own axis :*

A solid cylinder is nothing but a number of thin circular discs piled one above the other perpendicular to its axis and hence its moment of inertia about its own axis will be the same as the sum of the moments of inertia of all the discs about the same axis—the axis passing through centres of all such discs and being perpendicular to their planes.

The moment of inertia of an elementary disc of mass m amd radius a about the axis under consideration $= \frac{1}{2}ma^2$.

$\therefore$ The moment of inertia of the whole solid cylinder about the same axis

$$= \Sigma\frac{1}{2}ma^2 = \frac{1}{2}a^2\Sigma m = \frac{1}{2}Ma^2$$

where $M = \Sigma m$, the mass of the solid cylinder.

(ii) *About an axis passing through the centre of gravity of the cylinder and perpendicular to its length :*

Let us consider a cylinder of radius a and mass per unit volume ρ. Let l be the length of the cylinder. We imagine the cylinder to be made up of a number of thin discs, and consider one such disc of thickness dx, at a distance x from the axis AB as shown in figure 4.7.13.

Mass of the elementary disc $= \pi a^2 dx\rho$, so that its moment of inertia about a diameter parallel to the axis in question $= \pi a^2 dx\rho\frac{a^2}{4}$ and by the theorem of parallel axes, its moment of inertia about the axis under consideration $= \pi a^2 dx\rho\frac{a^2}{4} + \pi a^2 dx\rho\cdot x^2$.

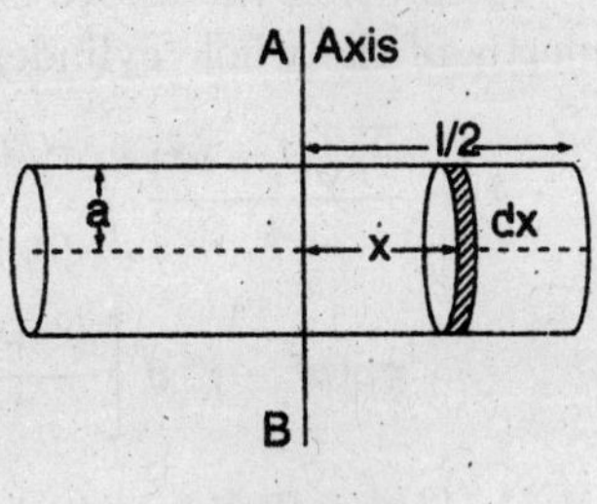

Fig. 4.7.13

Integrating the above expression between the limits $x = -\frac{l}{2}$ and $x = +\frac{l}{2}$, the moment of inertia of the whole cylinder about the axis in question is given by

$$I = \frac{\pi\rho a^4}{4}\int_{-l/2}^{+l/2} dx + \pi\rho a^2\int_{-l/2}^{+l/2} x^2 dx = \frac{\pi\rho a^4}{4}\cdot l + \pi\rho a^2\cdot\frac{l^3}{12} = \pi a^2 l\rho\left[\frac{a^2}{4}+\frac{l^2}{12}\right] = M\left[\frac{a^2}{4}+\frac{l^2}{12}\right]$$

where $M = \pi a^2 l\rho$, the mass of the cylinder.

(b) Hollow Circular Cylinder

(i) *About its own axis :*

The hollow cylinder may be supposed to consist of a large number of annular discs of internal and external radii r and a respectively, placed one above the other—the axis of the cylinder passing through their centres and being perpendicular to their planes.

Let us consider one such elementary annular disc and let its mass be m. Its moment of inertia about the axis under consideration $= m\frac{a^2+r^2}{2}$.

Hence the moment of inertia of the whole hollow cylinder about its own axis is given by

$$I = \Sigma m\frac{a^2+r^2}{2} = \frac{a^2+r^2}{2}\Sigma m = M\frac{a^2+r^2}{2}$$

where $M = \Sigma m$, the mass of the hollow cylinder. Now, if the cylinder is considered very thin, then $a = r$, and $I = Ma^2$.

(ii) *About an axis passing through the centre of gravity of the cylinder, and perpendicular to its length :*

As mentioned already, we consider an annular disc of internal and external radii r and a respectively, at a distance x from the axis AOB and of thickness dx as shown in figure 4.7.14.

Mass of the elementary annular disc $= \pi(a^2 - r^2)dx\rho$, (where ρ is the mass per unit volume of the hollow cylinder), so that its moment of inertia about a diameter parallel to the axis under consideration.

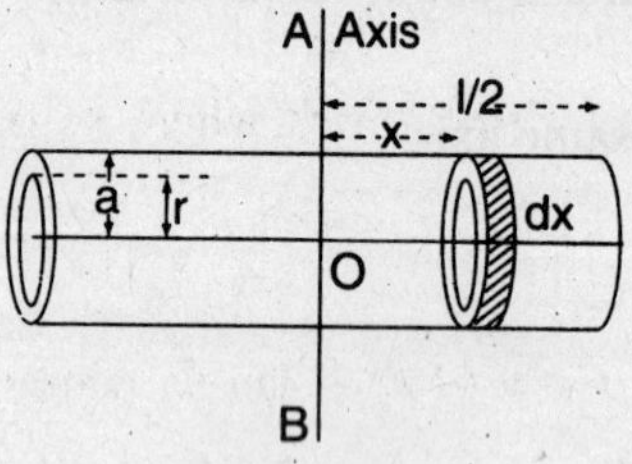

Fig. 4.7.14

$$= \pi(a^2 - r^2)dx\rho\cdot\frac{a^2 + r^2}{4}$$

and by the theorem of parallel axes, its moment of inertia about the axis in question

$$= \frac{\pi\rho(a^2 - r^2)(a^2 + r^2)}{4}dx + \pi\rho(a^2 - r^2)x^2 dx.$$

Integrating the above expression between the limits $x = -\frac{l}{2}$ and $x = +\frac{l}{2}$ the moment of inertia of the whole cylinder about the axis AB is given by

$$I = \frac{\pi\rho(a^4 - r^4)}{4}\int_{-l/2}^{+l/2} dx + \pi\rho(a^2 - r^2)\int_{-l/2}^{+l/2} x^2 dx = \frac{\pi\rho(a^4 - r^4)}{4}l + \pi\rho(a^2 - r^2)\frac{l^3}{12}$$

$$= \pi\rho(a^2 - r^2)l\left[\frac{a^2 + r^2}{4} + \frac{l^2}{12}\right] = M\left[\frac{a^2 + r^2}{4} + \frac{l^2}{12}\right]$$

where $M = \pi\rho(a^2 - r^2)l$, the mass of the hollow cylinder.

Illustrative Examples

1. A circular disc of mass M and radius r is set rolling on a table. If ω be the angular velocity, show that its total energy is given by $\frac{3}{4}Mr^2\omega^2$. **[P.U. 1950]**

Total K.E. $= \frac{1}{2}I\omega^2 + \frac{1}{2}Mv^2$.

Here $\frac{1}{2}I\omega^2$ is kinetic energy due to rotational motion and $\frac{1}{2}Mv^2$ is that due to translational motion.

Now $I = \dfrac{Mr^2}{2}$.

$$\therefore \text{ K.E.} = \frac{1}{2}\frac{Mr^2}{2}\cdot\omega^2 + \frac{1}{2}Mr^2\omega^2 \quad [\because v = r\omega] = \frac{1}{4}Mr^2\omega^2 + \frac{1}{2}Mr^2\omega^2 = \frac{3}{4}Mr^2\omega^2.$$

2. A cylinder has a mass M, length l and radius a. Find the ratio of l to a if the moment of inertia about an axis through the centre and perpendicular to the length is a minimum.

Here $I = M\left[\dfrac{a^2}{4} + \dfrac{l^2}{12}\right]$

Now $M = \pi a^2 l\rho$ or $a^2 l$ = constant $= k$ (say) $\quad\therefore\ I = M\left[\dfrac{k}{4l} + \dfrac{l^2}{12}\right]$

For I to be a minimum, we must have $\frac{dI}{dl} = 0$, i.e. $M\left[-\frac{k}{4l^2} + \frac{l}{6}\right] = 0$

$$\text{or } \frac{k}{4l^2} = \frac{l}{6} \quad \text{or} \quad \frac{a^2 l}{4l^2} = \frac{l}{6} \quad \text{or} \quad \frac{l}{a} = \frac{\sqrt{3}}{\sqrt{2}}$$

3. A thin hollow cylinder, open at both ends and weighing 96 lbs., (a) slides with a speed of 10 ft. per second, without rotating, (b) rolls with same speed without slipping. Compare the kinetic energies of the cylinder in the two cases.

(a) When the cylinder slides without rotating, its kinetic energy $E_1 = \frac{1}{2}mv^2$, where m is the mass of the hollow cylinder, and v its linear speed.

$$\therefore E_1 = \tfrac{1}{2} \times 96 \times (10)^2 = 4800 \text{ ft. poundals.}$$

(b) When the cylinder rolls without slipping, its kinetic energy $E_2 = \frac{1}{2}I\omega^2 + \frac{1}{2}mv^2$, where I is the moment of inertia of the cylinder and ω is its angular speed.

$\because$ $I = mr^2$, where r is the radius of the cylinder,

$$\therefore E_2 = \frac{1}{2}mr^2 \cdot \frac{v^2}{r^2} + \frac{1}{2}mv^2 \quad \left[\because \frac{v}{r} = \omega\right]$$

$$\text{or } E_2 = mv^2 = 96 \times (10)^2 = 9600 \text{ ft. poundals.} \quad \therefore \frac{E_2}{E_1} = \frac{9600}{4800} = \frac{2}{1}.$$

4. A solid cylinder is rotating about its symmetry axis aligned in the vertical direction. Measuring z from the bottom of the cylinder along the axis, the density of the body is given by

$$\rho(z) = \rho_0\left(1 - \tfrac{1}{2}\tfrac{z}{l}\right)$$

where ρ_0 is a constant, and l is the length of the cylinder. Calculate the moment of inertia of the cylinder, its radius being R. [C.U. (Hons.) 1982]

The moment of inertia of the cylinder about its axis of symmetry is given by

$$I = \int_0^l \pi R^2 dz \rho(z) \cdot R^2/2 \quad \left[\begin{array}{l}dz \text{ being the width of an elementary circular disc} \\ \text{at a distance } z \text{ from the bottom of the cylinder as} \\ \text{shown in the attached figure 1}\end{array}\right]$$

$$= \frac{\pi R^4}{2}\int_0^l \rho_0\left(1 - \frac{1}{2}\cdot\frac{z}{l}\right)dz = \frac{\pi R^4 \rho_0}{2}\int_0^l \left(dz - \frac{1}{2l}z\,dz\right)$$

$$= \frac{\pi R^4 \rho_0}{2}\left(l - \frac{1}{2l}\cdot\frac{l^2}{2}\right) = \frac{\pi R^4 \rho_0 l}{2}\left(1 - \frac{1}{4}\right) = \frac{3}{8}\pi R^4 \rho_0 l$$

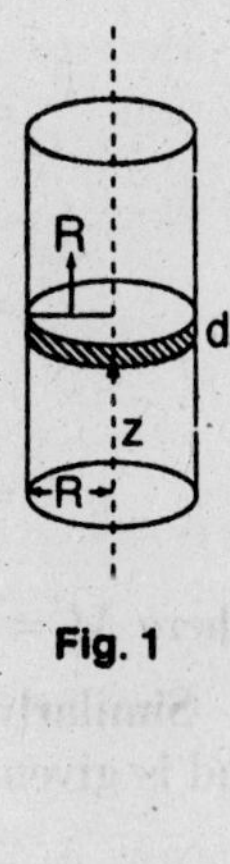

Fig. 1

Now mass of the cylinder is given by

$$M = \int_0^l \pi R^2 dz \rho(z) = \pi R^2 \int_0^l \rho_0\left(1 - \frac{1}{2}\cdot\frac{z}{l}\right)dz = \pi R^2 \rho_0 l \cdot \frac{3}{4} = \frac{3}{4}\pi R^2 \rho_0 l$$

$$\therefore I = \frac{MR^2}{2}.$$

VII. Elliptic Disc

(i) *About any of its principal axes :*

We consider a thin elliptic disc having XX' and YY' as major and minor axes respectively of lengths $2a$ and $2b$ respectively (Fig. 4.7.15). Let us calculate the moment of inertia of the disc about YY'. We consider a strip $PQRS$ of the disc parallel to YY', at a distance x from the axis, and of width dx. The mass of the strip is given by $2ydx\rho$, where $2y$ is the length of the strip and ρ, the mass per unit area of the disc. The moment of inertia of the elementary strip about YY'

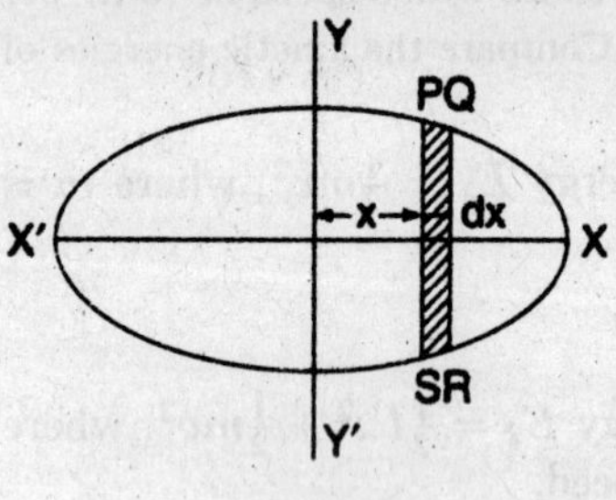

Fig. 4.7.15

$$= 2ydx\rho \cdot x^2$$

$\therefore$ The moment of inertia of the entire disc about YY' is given by

$$I_y = 2\rho \int_{-a}^{+a} yx^2 dx.$$

Now from the equation of ellipse with its centre as the origin and with the co-ordinate axes coinciding with the major and minor axes, we have,

$$\frac{x^2}{a^2} + \frac{y^2}{b^2} = 1 \quad \text{or} \quad y^2 = b^2\frac{(a^2 - x^2)}{a^2} \quad \text{or} \quad y = \frac{b}{a}\sqrt{a^2 - x^2}.$$

Hence the expression for moment of inertia may be written as,

$$I_y = \frac{2\rho b}{a} \int_{-a}^{+a} \sqrt{a^2 - x^2} \cdot x^2 dx.$$

Let us put, $x = a \sin\phi \quad \therefore\ dx = a\cos\phi d\phi$.

Also when $\phi = \frac{\pi}{2}, x = a$ and when $\phi = -\frac{\pi}{2}, x = -a$. Hence with these substitutions the expression for moment of inertia may be written as

$$I_y = \frac{2\rho b}{a} \int_{-\pi/2}^{\pi/2} a\sqrt{1 - \sin^2\phi} \cdot a^2 \sin^2\phi \cdot a\cos\phi d\phi = 2\rho ba^3 \int_{-\pi/2}^{\pi/2} \sin^2\phi \cos^2\phi d\phi$$

$$= 2\rho ba^3 \int_{-\pi/2}^{+\pi/2} \frac{(1 - \cos 2\phi)}{2} \cdot \frac{(1 + \cos 2\phi)}{2} d\phi = \frac{\rho ba^3}{2} \int_{-\pi/2}^{+\pi/2} (1 - \cos^2 2\phi) d\phi$$

$$= \frac{\rho ba^3}{2} \left[\int_{-\pi/2}^{+\pi/2} d\phi - \int_{-\pi/2}^{+\pi/2} \frac{(1 + \cos 4\phi)}{2} d\phi\right] = \frac{\rho ba^3}{2}\left[\pi - \frac{\pi}{2}\right]$$

$$= \frac{\rho ba^3}{2}\left[\frac{\pi}{2}\right] = \frac{\pi ab\rho}{4} \cdot a^2 = \frac{Ma^2}{4}$$

where $M = \pi ab\rho$, the mass of the elliptic disc.

Similarly, the moment of inertia of the disc about XX', the major axis, may be calculated, and is given by

$$I_x = \frac{Mb^2}{4}.$$

(ii) *About an axis passing through the centre of the disc and perpendicular to its plane :*

By the theorem of perpendicular axes, the moment of inertia about this axis is

$$I = I_x + I_y = \frac{Mb^2}{4} + \frac{Ma^2}{4} = \frac{M}{4}(a^2 + b^2).$$

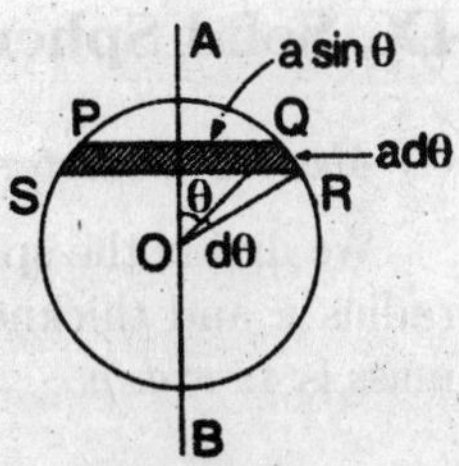

Fig. 4.7.16

VIII. Spherical Shell

(i) *About a diameter as axis :*

We consider a section of a spherical shell of radius a through its centre O and take a thin slice $PQRS$ of the shell perpendicular to the diameter AB. The slice in question is obviously a circular ring of diameter PQ, and width QR. From the figure (4.7.16), it follows that, radius of the ring $= a\sin\theta$ and width of the ring $= ad\theta$.

Thus the mass of the ring $= 2\pi a \sin\theta \cdot ad\theta \cdot \sigma$, where σ = mass per unit area of the ring.

$\therefore$ Moment of inertia of the elementary ring about AB

$$= 2\pi a \sin\theta \cdot ad\theta \cdot \sigma \cdot (a\sin\theta)^2$$

and the moment of inertia of the entire spherical shell about the diameter AB is given by

$$I = 2\int_0^{\pi/2} 2\pi a^4 \sigma \sin^3\theta d\theta.$$

Now, $\sin 3\theta = 3\sin\theta - 4\sin^3\theta$ or $\sin^3\theta = \frac{1}{4}(3\sin\theta - \sin 3\theta)$

$$\therefore\ I = 4\pi a^4\sigma \int_0^{\pi/2} \sin^3\theta d\theta = 4\pi a^4\sigma \cdot \frac{1}{4}\int_0^{\pi/2} (3\sin\theta - \sin 3\theta)d\theta$$

$$= \pi a^4\sigma\left(3 - \frac{1}{3}\right) = \frac{8}{3}\pi a^4\sigma = 4\pi a^2\sigma \times \frac{2}{3}a^2 = \frac{2}{3}Ma^2$$

where $M = 4\pi a^2\sigma$, the mass of the shell.

Alternative Treatment

From the theorem of perpendicular axes for a three dimensional body, we know that the sum of the moments of inertia of the shell about its three mutually perpendicular diameters must be equal to twice the summation Σma^2 about the centre of the shell. Thus,

$$I_x + I_y + I_z = 2\Sigma ma^2 = 2Ma^2.$$

Here $I_x = I_y = I_z = I$ (say) $\quad\therefore\ 3I = 2Ma^2$ or $I = \frac{2}{3}Ma^2$.

(ii) *About a tangent :*

By the theorem of parallel axes, the moment of inertia of the shell about a tangent parallel to its diameter, is given by

$$I_T = \frac{2}{3}Ma^2 + Ma^2 = \frac{5}{3}Ma^2.$$

IX. Solid Sphere

(i) *About a diameter as axis:*

We divide the sphere into a number of concentric sphereical shells. Let one such shell have radius x and thickness dx (Fig. 4.7.17). If ρ be the mass per unit volume of the sphere, its mass is $4\pi x^2 dx\rho$.

Hence moment of inertia of the spherical shell about AB

$$4\pi x^2 dx\rho \times \frac{2}{3}x^2 = \frac{8}{3}\pi\rho x^4 dx.$$

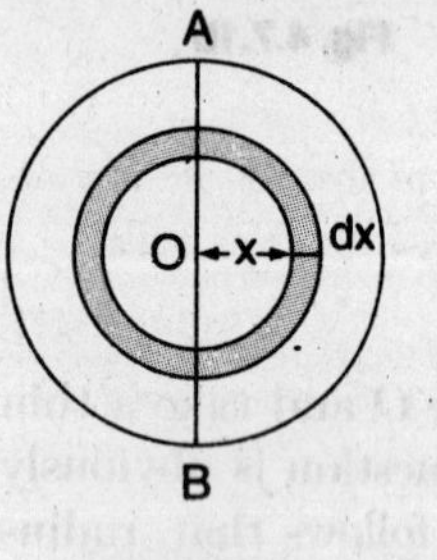

Fig. 4.7.17

$\therefore$ the moment of inertia of the whole sphere of radius a about its diameter AB as shown in the figure is given by

$$= \frac{8}{3}\pi\rho\int_0^a x^4 dx = \frac{8}{3}\pi\rho\cdot\frac{a^5}{5} = \frac{4}{3}\pi a^3\rho\cdot\frac{2}{5}a^2 = \frac{2}{5}Ma^2,$$

where M = mass of the sphere $= \dfrac{4}{3}\pi a^3\rho$.

Alternative Treatment

We consider a circular slice $PQRS$ of the sphere at a distance x from the centre O and of thickness dx (Fig. 4.7.18). The slice is perpendicular to the diameter AB about which the moment of inertia is to be determined.

The moment of inertia of the slice about

$$AB = \pi(a^2 - x^2)dx\rho\cdot(a^2 - x^2)/2,$$

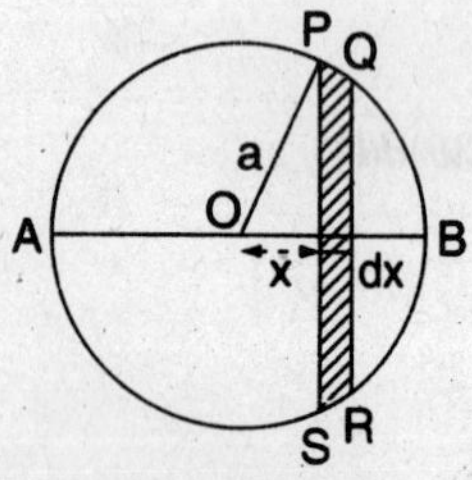

Fig. 4.7.18

where $\sqrt{a^2 - x^2}$ is the radius of the slice.

Hence for the whole sphere, the moment of inertia about AB

$$= 2\int_0^a \frac{\pi(a^2 - x^2)^2}{2}\rho dx = \pi\rho\left[\int_0^a a^4 dx - \int_0^a 2a^2x^2 dx + \int_0^a x^4 dx\right]$$

$$= \pi\rho\left[a^5 - \frac{2}{3}a^5 + \frac{a^5}{5}\right] = \pi\rho\frac{8}{15}a^5 = \frac{4}{3}\pi a^3\rho\cdot\frac{2}{5}a^2 = \frac{2}{5}Ma^2.$$

(ii) *About a tangent :*

By the theorem of parallel axes, the moment of inertia of the solid sphere about a tangent parallel to its diameter is given by

$$I_T = \frac{2}{5}Ma^2 + Ma^2 = \frac{7}{5}Ma^2.$$

Illustrative Examples

1. A sphere of mass 50 gm and diameter 2 cm. rolls without slipping with a velocity 5 cm sec. Calculate the total kinetic energy in ergs. **[P.U. 1943]**

Here total kinetic energy of the sphere which rolls without slipping

$$= \frac{1}{2}I\omega^2 + \frac{1}{2}mv^2 = \frac{1}{2}\cdot\frac{2}{5}mr^2\frac{v^2}{r^2} + \frac{1}{2}mv^2$$

$$= \frac{1}{5}mv^2 + \frac{1}{2}mv^2 = \frac{7}{10}mv^2$$

$$= \frac{7}{10}\times 50 \times 25 = 875 \text{ ergs.}$$

[where I is the moment of inertia of the sphere about its diameter and is given by $\frac{2}{5}mr^2$, m being the mass of the sphere sphere and r its radius.]

2. You are given two spheres of the same mass and size and appearance, but one of them is hollow at the centre and the other is solid throughout. How will you find which is hollow and which is solid?

[Delhi 1937]

We know that the moment of inertia of a solid sphere about its diameter is given by $I_S = \frac{2}{5}mr_1^2$ where m and r_1 are respectively the mass and radius of the solid sphere.

Again the moment of inertia of a hollow sphere of internal radius r_2 and external radius r_1 (size of the solid as well as the hollow sphere being the same) is given by

$$I_H = \frac{2}{5}m_1r_1^2 - \frac{2}{5}m_2r_2^2 = \frac{2}{5}\cdot\frac{4}{3}\pi r_1^3\rho\cdot r_1^2 - \frac{2}{5}\cdot\frac{4}{3}\pi r_2^3\rho\cdot r_2^2 = \frac{2}{5}\cdot\frac{4}{3}\pi\rho\left[r_1^5 - r_2^5\right]$$

where m_1 and m_2 are respectively the masses of the spheres of radii r_1 and r_2, and ρ is the density of the material.

Now, the mass of the hollow sphere is the same as that of the solid sphere and is given by

$$m = \frac{4}{3}\pi r_1^3\rho - \frac{4}{3}\pi r_2^3\rho = \frac{4}{3}\pi\rho(r_1^3 - r_2^3). \quad \therefore \frac{4}{3}\pi\rho = \frac{m}{(r_1^3 - r_2^3)}$$

$$\therefore I_H = \frac{2}{5}\cdot\frac{m(r_1^5 - r_2^5)}{(r_1^3 - r_2^3)} = \frac{\frac{2}{5}mr_1^5\left(1 - \frac{r_2^5}{r_1^5}\right)}{r_1^3\left(1 - \frac{r_2^3}{r_1^3}\right)} = \frac{\frac{2}{5}mr_1^2\left(1 - \frac{r_2^5}{r_1^5}\right)}{\left(1 - \frac{r_2^3}{r_1^3}\right)} = \frac{I_S\left(1 - \frac{r_2^5}{r_1^5}\right)}{\left(1 - \frac{r_2^3}{r_1^3}\right)}$$

Now $r_2 < r_1$.

$$\therefore \frac{r_2^5}{r_1^5} \ll \frac{r_2^3}{r_1^3} \quad \text{i.e.,} \quad \frac{\left(1 - \frac{r_2^5}{r_1^5}\right)}{\left(1 - \frac{r_2^3}{r_1^3}\right)} > 1 \quad \therefore I_H > I_S.$$

Now both the spheres have got the same mass;

$$\therefore \quad mK_H^2 > mK_S^2 \quad \text{or} \quad K_H > K_S$$

[K_H and K_S being the radii of gyration for the hollow and solid sphere respectively.]

Now both the spheres are allowed to roll down an inclined plane. The hollow sphere having greater radius of gyration will roll down with acceleration less than that of a solid sphere [See the case of "rolling of a body down an inclined plane"].

3. A sphere of radius $2r$ and density d has an internal spherical cavity, of radius r, the diameter of which is the radius of the sphere. Find the moment of inertia, I, of the body about an axis perpendicular to the common diameter of sphere and cavity and at a distance x from the centre of the sphere.

Moment of inertia of a solid sphere having centre O and of radius $2r$ about the axis AB or $A'B'$ is given by

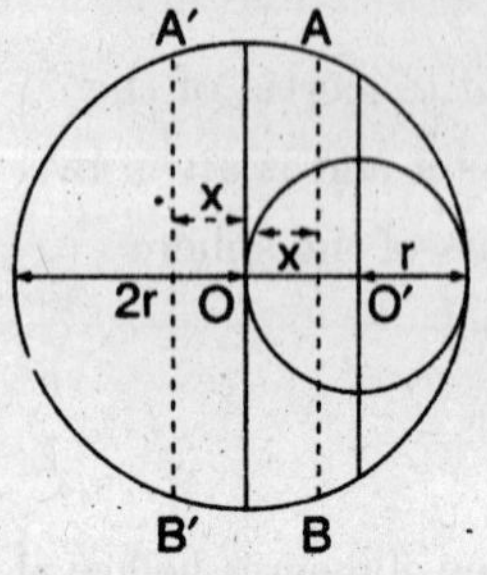

Fig. 1

$$I_1 = \frac{2}{5}m_1(2r)^2 + m_1x^2 \quad \text{(Fig. 1)}$$

where m_1 is the mass of the solid sphere [applying the theorem of parallel axes] and is given by

$$m_1 = \frac{4}{3}\pi(2r)^3d. \quad \therefore\ I_1 = \frac{2}{5}\cdot\frac{4}{3}\pi(2r)^3d\cdot(2r)^2 + \frac{4}{3}\pi(2r)^3d\cdot x^2.$$

Again, the moment of inertia of the sphere of centre O' having the same size as the cavity about AB or $A'B'$ is given by

$$I_2 = \frac{2}{5}m_2r^2 + m_2(r \mp x)^2 \text{ [applying the theorem of parallel axes]}$$

where m_2 is the mass of the sphere of radius r and is given by

$$m_2 = \frac{4}{3}\pi r^3 d. \quad \therefore\ I_2 = \frac{2}{5}\cdot\frac{4}{3}\pi r^3 d\cdot r^2 + \frac{4}{3}\pi r^3 d(r \mp x)^2$$

Hence the required moment of inertia is given by

$$I = I_1 - I_2 = \frac{2}{5}\cdot\frac{4}{3}\pi(2r)^5 d + \frac{4}{3}\pi(2r)^3 d\cdot x^2 - \frac{2}{5}\cdot\frac{4}{3}\pi r^5 d - \frac{4}{3}\pi r^3 d(r^2 \mp 2rx + x^2)$$

$$\text{Or } 15I = 256\pi dr^5 + 160\pi dr^3x^2 - 8\pi dr^5 - 20\pi dr^5 \pm 40\pi dr^4x - 20\pi dr^3x^2$$

$$= 228\pi dr^5 + 140\pi dr^3x^2 \pm 40\pi dr^4x = 4\pi r^3 d[57r^2 + 35x^2 \pm 10rx]$$

4. A system consists of three identical spheres, each of mass 10 kg. and radius 10 cms. placed with their centres forming the vertices of an equilateral triangle of sides $30\sqrt{3}$ cm. Calculate the moment of inertia of the system about an axis passing through the C.G. of the system and perpendicular to the plane of the centres. **[C.U. (Hons.) 1978]**

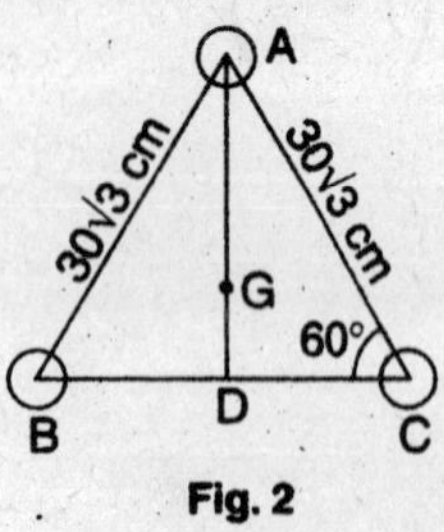

Fig. 2

From Fig. 2,

$$AD = AC\sin 60° = 30\sqrt{3}\cdot\frac{\sqrt{3}}{2} = 45 \text{ cms.}$$

$$\therefore\ AG = \frac{2}{3}AD = \frac{2}{3}\times 45 = 30 \text{ cms,}$$

where G denotes the centre of gravity of the system of particles through which the axis passes.

The moment of inertia of the system about an axis passing through G and perpendicular to the plane

= 3× [Moment of inertia of any sphere, say the sphere at A about the required axis].

= 3× [Moment of inertia of the sphere about its diameter parallel to the required axis + Mass of the sphere $\times(AG)^2$] (applying theorem of parallel axes)

$$= 3\left[\frac{2}{5}MR^2 + M(AG)^2\right],$$

where M = mass of the sphere, R = the radius of the sphere.

Hence the required moment of inertia

$$= 3\left[\frac{2}{5} \times 10 \times 10^3 \times 100 + 10 \times 10^3 \times (30)^2\right] = 2.82 \times 10^7 \text{ gm cm}^2.$$

X. Triangular Lamina

(i) *About one side of the triangle as axis :*

We consider a thin strip PQ of the triangular lamina ABC parallel to the side BC, about which the moment of inertia is to be determined (Fig. 4.7.19). The strip is of width dx and at a distance x from the axis BC. Let $BC = a$ and $AD = h$, the height of the triangle.

Moment of inertia of the strip about $BC = ydx\rho \cdot x^2$, where y is the length of the strip and ρ is the mass per unit area of the lamina. Now from the figure (4.7.19)

$$\frac{y}{(h-x)} = \frac{a}{h} \quad \text{or} \quad y = \frac{a}{h}(h-x)$$

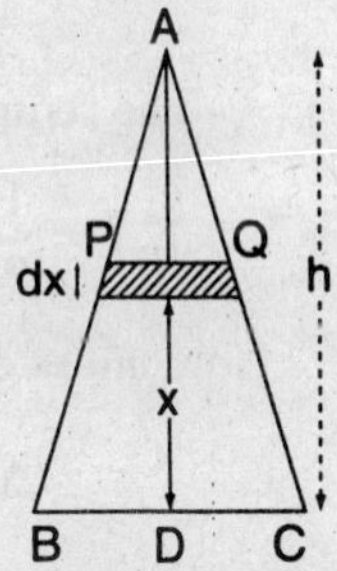

Fig. 4.7.19

$\therefore$ Moment of inertia of the strip about $BC = \frac{a}{h}(h-x)\rho x^2 dx$.

Hence the moment of inertia of the whole triangle is given by

$$I = \int \frac{a}{h}(h-x)\rho x^2 dx = \frac{a}{h}\rho\left[h\int_0^h x^2 dx - \int_0^h x^3 dx\right]$$

$$= \frac{a\rho}{h}\left[\frac{h^4}{3} - \frac{h^4}{4}\right] = \frac{a\rho h^3}{12} = \frac{1}{2}ah\rho \cdot \frac{h^2}{6} = \frac{Mh^2}{6}$$

where $M = \frac{1}{2}ah\rho$, the mass of the triangular lamina.

(ii) *About an axis through the centre of gravity and parallel to one side :*

In the triangular lamina ABC, G denotes the centre of gravity through which the axis XGX' parallel to the side BC passes (Fig. 4.7.20). It is required to calculate the moment of inertia of the triangular lamina about this axis. By the theorem of parallel axes, we have, moment of inertia of the triangular lamina about BC = moment of inertia of the triangular lamina about XGX' + mass of the triangular lamina $\times \left(\frac{h}{3}\right)^2$

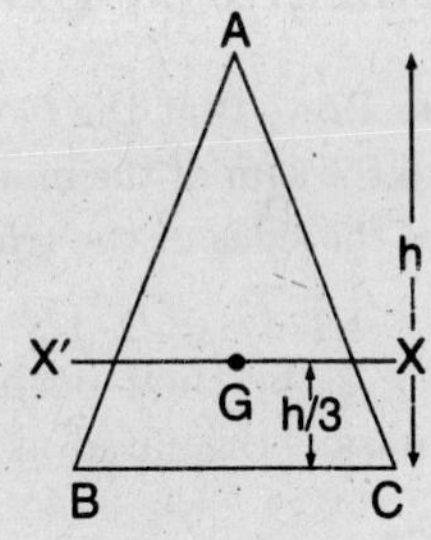

Fig. 4.7.20

$\therefore$ moment of inertia of the triangular lamina about XGX' = moment of inertia of the triangular lamina about BC − mass of the triangular lamina $\times \frac{h^2}{9} = \frac{Mh^2}{6} - \frac{Mh^2}{9} = \frac{Mh^2}{18}$.

(iii) *About an axis passing through the vertex A, and parallel to BC :*

Applying the theorem of parallel axes, the required moment of inertia of the lamina about PAP' parallel to BC = moment of inertia about $XGX' + M \times \left(\frac{2}{3}h\right)^2$

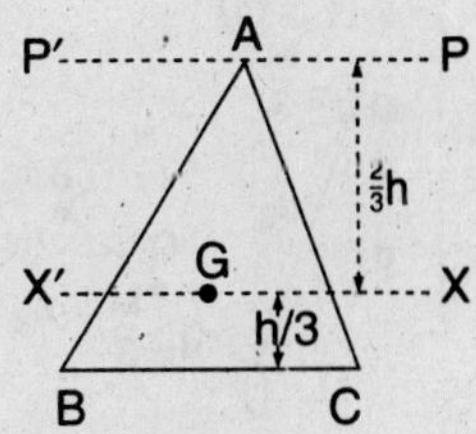

Fig. 4.7.21

$$= \frac{Mh^2}{18} + \frac{4Mh^2}{9} = \frac{Mh^2}{2} \text{ (Fig. 4.7.21).}$$

(iv) *About any axis in its plane through an angular point :*

Let BCD be the triangular lamina and DE be the axis through the angular point D in the plane of the triangle, about which the moment of inertia is to be considered. Let $DE = b$ and h_1, h_2 the perpendicular distances of the points C, B from the axis (Fig. 4.7.22).

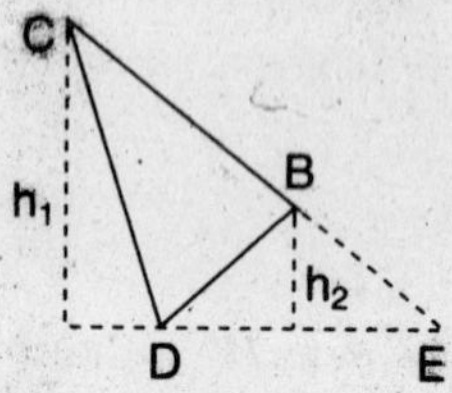

Fig. 4.7.22

Now, moment of inertia of the triangular lamina DEC about DE = moment of inertia of the lamina BED about DE + moment of inertia of the lamina BCD about DE.

Therefore, moment of inertia of ΔBCD about DE = moment of inertia of ΔDEC about DE − moment of inertia of ΔBED about $DE = \dfrac{M_1h_1^2}{6} - \dfrac{M_2h_2^2}{6}$.

Now, M_1 = mass of $\Delta DEC = \frac{1}{2}bh_1\sigma$, where σ = mass per unit area.

$$M_2 = \text{mass of } \Delta BED = \tfrac{1}{2}bh_2\sigma.$$

$\therefore$ The required moment of inertia,

$$I = \frac{1}{2}bh_1\sigma\cdot\frac{h_1^2}{6} - \frac{1}{2}bh_2\sigma\cdot\frac{h_2^2}{6} = \frac{b\sigma}{12}(h_1^3 - h_2^3) = \frac{b\sigma}{12}(h_1 - h_2)(h_1^2 + h_1h_2 + h_2^2)$$

Now, mass of

$$\Delta BCD = M_1 - M_2 = \frac{1}{2}bh_1\sigma - \frac{1}{2}bh_2\sigma = \frac{1}{2}b\sigma(h_1 - h_2) = M \text{ (say)}.$$

$$\therefore\ I = \frac{M}{6}(h_1^2 + h_1h_2 + h_2^2).$$

If the axis in question be parallel to BC, then $h_1 = h_2 = h$ (say), and we have $I = \dfrac{Mh^2}{2}$, which is the result already obtained in X (iii).

Illustrated Examples

1. Prove that the moment of inertia of a triangle about any axis in its plane or parallel to it, is equal to the sum of the moments of inertia, about the same axis, of three particles situated at the mid-points of the sides of the triangle—each particle having a mass equal to one-third the mass of the triangle.

Let O_1, O_2, O_3 be the mid-points of DB, BC and CD respectively of ΔBCD and let d_1, d_2, d_3 be their perpendicular distances from DE, the axis. We imagine three particles each of mass $\frac{M}{3}$ situated at O_1, O_2 and O_3 (Fig. 4.7.23). Then considering the sum of the moments of inertia, about DE, of these three particles, we have,

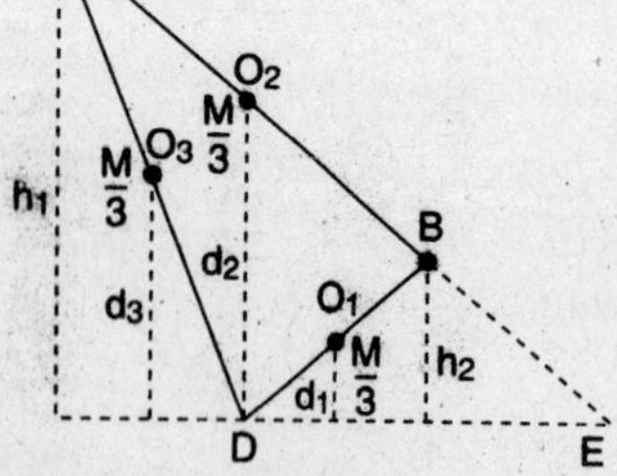

Fig. 4.7.23

$$\frac{Md_1^2}{3} + \frac{Md_2^2}{3} + \frac{Md_3^2}{3} = \frac{M}{3}(d_1^2 + d_2^2 + d_3^2)$$

$$= \frac{M}{3}\left[\left(\frac{h_2}{2}\right)^2 + \left(\frac{h_1 + h_2}{2}\right)^2 + \left(\frac{h_1}{2}\right)^2\right]$$

$$= \frac{M}{6}\,(h_1^2 + h_2^2 + h_1h_2)$$

which is the moment of inertia of the triangle BCD about DE—an axis in the plane of the triangle.

2. Prove that the moment of inertia of any triangle ABC about an axis perpendicular to the plane through A is equal to $\frac{M}{12}(3b^2+3c^2-a^2)$.

From the previous problem, we have, moment of inertia of the triangle ABC about any axis through A = sum of the moments of inertia of three particles $\left(\text{each of mass } \frac{M}{3}\right)$ situated at the mid-points O_1, O_2, O_3 of the three sides of the triangle, about the same axis

$$= \frac{M}{3}(O_1A)^2 + \frac{M}{3}(O_2A)^2 + \frac{M}{3}(O_3A)^2 \quad \text{(Fig. 4.7.24)}$$

[considering the axis perpendicular to the plane of the triangle and passing through A]

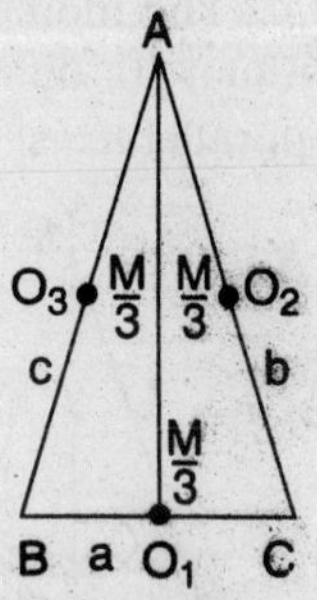

Fig. 4.7.24

$$= \frac{M}{3}\cdot\frac{2b^2+2c^2-a^2}{4} + \frac{M}{3}\cdot\frac{b^2}{4} + \frac{M}{3}\cdot\frac{c^2}{4}$$

$[\because AB^2 + AC^2 = 2BO_1^2 + 2O_1A^2$ or $c^2+b^2 = \frac{a^2}{2} + 2O_1A^2$

or $O_1A^2 = \frac{2b^2+2c^2-a^2}{4}]$

$$= \frac{M}{3}\left(\frac{3b^2+3c^2-a^2}{4}\right) = \frac{M}{12}(3b^2+3c^2-a^2)$$

XI. (a) Moment of Inertia of a Right Circular Solid Cone

(i) *About its axis of symmetry :*

The cone is supposed to be made up of a large number of co-axial circular discs parallel to the base, and we consider one such elementary circular disc of radius x, and width dy, at a distance y from the vertex of the solid cone, the disc being parallel to the base of the cone of radius r (Fig. 4.7.25). Now, the moment of inertia of the disc about the

$$\text{axis } YY' = \text{mass of the disc} \times \frac{x^2}{2} = \pi x^2 dy\rho\cdot\frac{x^2}{2} = \frac{\pi\rho}{2}x^4 dy$$

where ρ = mass per unit volume of the cone.

Now, from the figure (4.7.25), $\frac{x}{r} = \frac{y}{h}$ $\therefore x = \frac{r}{h}y$

$\therefore$ moment of inertia of the disc about

$$YY' = \frac{\pi\rho}{2}\cdot\frac{r^4}{h^4}y^4 dy.$$

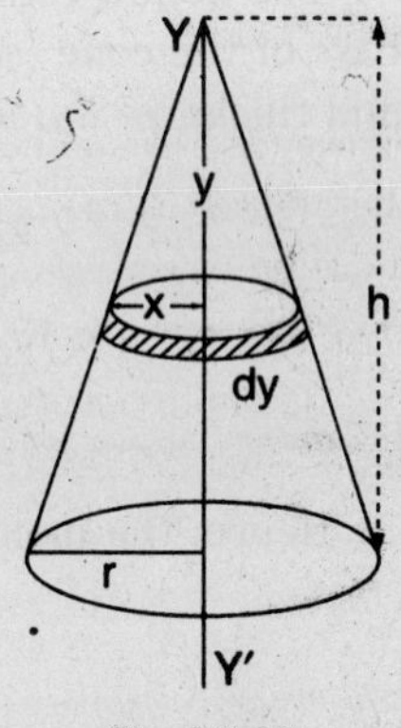

Fig. 4.7.25

$\therefore$ the moment of inertia of the whole cone about its axis of symmetry

$$= \frac{\pi\rho r^4}{2h^4}\int_0^h y^4 dy = \frac{\pi\rho r^4 h}{10} = \frac{1}{3}\pi r^2 h\rho\cdot\frac{3}{10}r^2 = \frac{3}{10}Mr^2$$

where $M = \frac{1}{3}\pi r^2 h\rho$ is the mass of the cone.

[Mass of the cone may be calculated in the following way :

We know, mass of the disc $= \pi x^2 dy\rho$

$\therefore$ mass of the cone, $M = \dfrac{\pi r^2 \rho}{h^2} \displaystyle\int_0^h y^2 dy = \frac{1}{3}\pi r^2 h\rho.$]

(ii) *About an axis passing through the vertex and perpendicular to the axis of symmetry.*

The moment of inertia of the elementary disc about XX' = moment of inertia of the disc about its diameter parallel to XX' + mass of the disc $\times y^2$ (Fig. 4.7.26). [by the theorem of parallel axes]

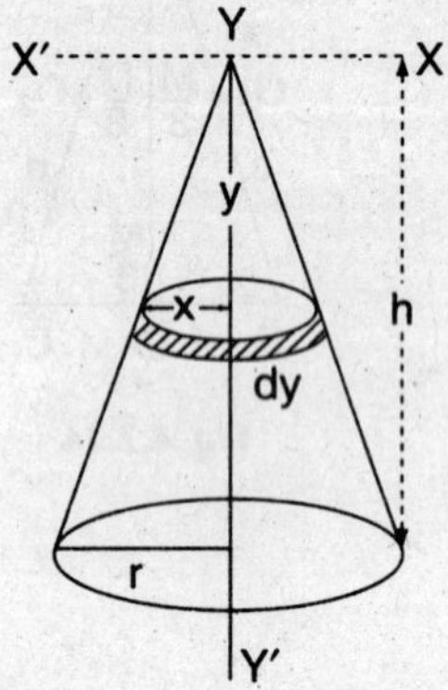

Fig. 4.7.26

$$= \pi\rho x^2 dy\cdot \frac{x^2}{4} + \pi\rho x^2 dy\cdot y^2 = \frac{\pi\rho}{4}x^4 dy + \pi\rho y^2\cdot x^2 dy$$

$$= \frac{\pi\rho r^4 y^4}{4h^4}dy + \pi\rho\frac{r^2}{h^2}\cdot y^4 dy \quad \left[\because x = \frac{r}{h}y\right]$$

$\therefore$ The moment of inertia of the whole cone about the required axis

$$= \frac{\pi\rho r^4}{4h^4}\int_0^h y^4 dy + \frac{\pi\rho r^2}{h^2}\int_0^h y^4 dy = \frac{\pi\rho r^4 h}{20} + \frac{\pi\rho r^2 h^3}{5}$$

$$= \frac{1}{3}\pi r^2 h\rho\cdot\frac{3}{20}r^2 + \frac{1}{3}\pi r^2 h\rho\cdot\frac{3}{5}h^2 = \frac{3}{20}Mr^2 + \frac{3}{5}Mh^2 = \frac{3}{5}M\left(\frac{r^2}{4} + h^2\right)$$

(b) Moment of inertia of a Hollow Cone or Conical Shell

(i) *About its axis of symmetry :*

The hollow cone is supposed to be divided up into a number of co-axial rings parallel to the base of the cone (same as in Fig. 4.7.25). The moment of inertia of one such ring of radius x, and thickness dy, at a distance y from the vertex about the axis in question

$$= \text{its mass} \times x^2 = 2\pi x dy\sigma\cdot x^2 \quad [\sigma \text{ is the mass per unit area}]$$

$$= 2\pi\sigma x^3 dy = 2\pi\sigma\cdot\frac{r^3}{h^3}y^3 dy$$

Hence, the moment of inertia of the entire hollow cone is given by

$$I = \frac{2\pi\sigma r^3}{h^3}\int_0^h y^3 dy = \frac{1}{2}\pi\sigma r^3 h = \pi r h\sigma\cdot\frac{r^2}{2} = \frac{Mr^2}{2}$$

where $M = \pi r h\sigma$ is the mass of the cone.

$$\left[\text{Mass of the hollow cone, } M = 2\pi\sigma\int_0^h x dy = 2\pi\sigma\frac{r}{h}\int_0^h y dy = \pi r h\sigma.\right]$$

(ii) *About an axis passing through the vertex and perpendicular to the axis of symmetry :*

The moment of inertia of the elementary ring about XX' (same as in Fig. 4.7.26)

$= $ its moment of inertia about its diameter parallel to XX'

$+$ mass of the ring $\times y^2$ [by the theorem of parallel axes]

$$= 2\pi x dy\sigma \cdot \frac{x^2}{2} + 2\pi x dy\sigma \cdot y^2 = \pi\sigma x^3 dy + 2\pi\sigma x y^2 dy$$

$$= \pi\sigma \frac{r^3}{h^3} y^3 dy + 2\pi\sigma \frac{r}{h} y^3 dy.$$

Therefore, the moment of inertia of the whole conical shell about XX' is given by

$$I = \pi\sigma \frac{r^3}{h^3} \int_0^h y^3 dy + 2\pi\sigma \frac{r}{h} \int_0^h y^3 dy = \pi r h\sigma \cdot \frac{r^2}{4} + \pi r h \sigma \cdot \frac{h^2}{2} = \frac{Mr^2}{4} + \frac{Mh^2}{2} = M\left(\frac{r^2}{4} + \frac{h^2}{2}\right)$$

Illustrative Examples

1. Prove that the moment of inertia of a truncated cone about its axis of symmetry is equal to

$$\frac{3}{10} M \frac{a^5 - b^5}{a^3 - b^3}$$

where M is the mass of the cone and a, b are the radii of its ends.

In Fig. 1 moment of inertia of the truncated cone $DBCE$ about its axis = moment of inertia of the cone ABC about the required axis – moment of inertia of the cone ADE about the same axis and is given by

$$I = \frac{3}{10} M_1 a^2 - \frac{3}{10} M_2 b^2 = \frac{3}{10} \cdot \frac{1}{3} \pi a^2 h_1 \rho \cdot a^2 - \frac{3}{10} \cdot \frac{1}{3} \pi b^2 h_2 \rho \cdot b^2$$

[where M_1 = mass of the cone $ABC = \frac{1}{3}\pi a^2 h_1 \rho$

and M_2 = mass of the cone $ADE = \frac{1}{3}\pi b^2 h_2 \rho$,

ρ being the density of the material]

$$= \frac{3}{10} \cdot \frac{1}{3} \pi a^5 k\rho - \frac{3}{10} \cdot \frac{1}{3} \pi b^5 k \rho = \frac{3}{10} \cdot \frac{1}{3} \pi \rho k (a^5 - b^5) \qquad (1)$$

$$\left[\because \frac{h_1}{h_2} = \frac{a}{b} \quad \therefore \frac{h_1}{a} = \frac{h_2}{b} = k \text{ (say)}\right]$$

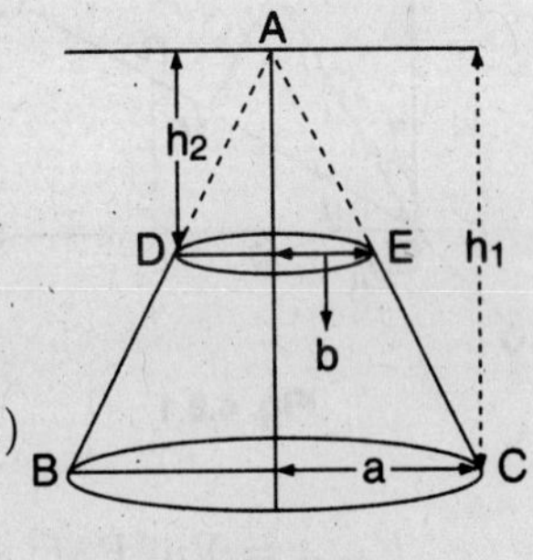

Fig. 1

Now, the mass of the truncated cone

$$M = M_1 - M_2 = \frac{1}{3}\pi\rho(a^2 h_1 - b^2 h_2) = \frac{1}{3}\pi\rho k (a^3 - b^3)$$

$$\therefore \frac{1}{3}\pi\rho k = \frac{M}{(a^3 - b^3)}$$

Substituting this value in the expression (1) for I, we have,

$$I = \frac{3}{10} M \frac{(a^5 - b^5)}{(a^3 - b^3)}.$$

2. Two equal right-circular conical shells of semi-vertical angle $\tan^{-1}\sqrt{2}$, are joined together apex to apex with their axes collinear. Prove that the moment of inertia of the combination about the common axis of symmetry is equal to the moment of inertia about any perpendicular axis passing through their common apex.

[C.U. (Hons.) 1949]

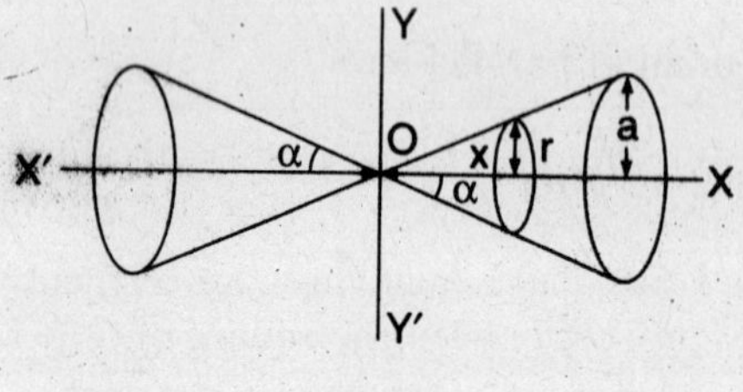

Fig. 2

We consider an elementary ring of mass m and radius r at a distance x from the apex O (Fig. 2). The moment of inertia of this ring about the common axis XOX' is mr^2. So the moment of inertia of the hollow cone about its axis $= \Sigma mr^2$.

Considering both the cones, the moment of inertia of the combination about the common axis $= 2\Sigma mr^2$.

Again the moment of inertia of the combination about the axis YOY'

$$= 2\left[\Sigma\frac{mr^2}{2} + \Sigma mx^2\right] \quad \text{[from the theorem of parallel axes]}$$

$$= 2\Sigma m\left(\frac{r^2}{2} + x^2\right) = 2\Sigma m\left[\frac{r^2}{2} + \frac{r^2}{2}\right] \quad \left[\because \tan\alpha = \frac{r}{x} = \sqrt{2} \text{ or, } x^2 = \frac{r^2}{2}\right]$$

$= 2\Sigma mr^2$, which solves the problem.

4.8 Momental Ellipsoid or Ellipsoid of Inertia

Let OX, OY and OZ be the three mutually perpendicular axes meeting at the point O (Fig. 4.8.1). It is required to find out the moment of inertia of a body about any line passing through the meeting point of the three mutually perpendicular concurrent axes.

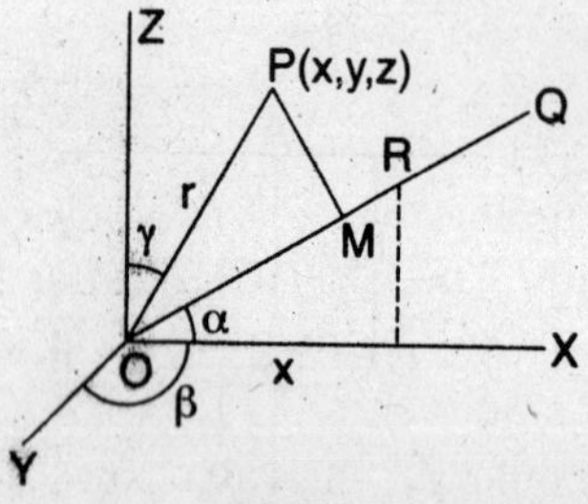

Fig. 4.8.1

Let OQ be an axis through O whose direction cosines i.e. $\cos\alpha, \cos\beta, \cos\gamma$ are respectively l, m and n. It is required to find out the moment of inertia of the body about OQ.

We consider an element of mass m' at $P(x, y, z)$ from which we draw perpendicular PM on OQ.

Then, the moment of inertia of the body about OQ is given by

$$I_{l,m,n} = \Sigma m' PM^2 = \Sigma m'(OP^2 - OM^2)$$

$$= \Sigma m'[r^2 - (lx + my + nz)^2]$$

$$[\because OM = \text{projection of } OP \text{ on } OQ = lx + my + nz]$$

$$= \Sigma m'[(x^2 + y^2 + z^2) - (lx + my + nz)^2] \quad [\because r^2 = x^2 + y^2 + z^2]$$

$$= \Sigma m'[x^2 + y^2 + z^2 - l^2x^2 - m^2y^2 - n^2z^2 - 2mnyz - 2nlzx - 2lmxy]$$

$$= \Sigma m'[(1 - l^2)x^2 + (1 - m^2)y^2 + (1 - n^2)z^2 - 2mnyz - 2nlzx - 2lmxy]$$

$$= \Sigma m'[(m^2 + n^2)x^2 + (n^2 + l^2)y^2 + (l^2 + m^2)z^2 - 2mnyz - 2nlzx - 2lmxy]$$

$$[\because l^2 + m^2 + n^2 = 1 \quad \therefore 1 - l^2 = m^2 + n^2, \text{ etc.}]$$

$$= \Sigma m'[l^2(y^2+z^2)+m^2(z^2+x^2)+n^2(x^2+y^2)-2mnyz-2nlzx-2lmxy]$$

$$= l^2\Sigma m'(y^2+z^2)+m^2\Sigma m'(z^2+x^2)+n^2\Sigma m'(x^2+y^2)$$
$$-2mn\Sigma m'yz-2nl\Sigma m'zx-2lm\Sigma m'xy$$

$$= l^2I_{xx}+m^2I_{yy}+n^2I_{zz}-2mnI_{yz}-2nlI_{zx}-2lmI_{xy}\cdots \tag{1}$$

where I_{xx}, I_{yy}, I_{zz} are the moments of inertia of the body about OX, OY, OZ respectively, and I_{yz}, I_{zx}, I_{xy} are the products of inertia referred to $Y-Z$, $Z-X$ and $X-Y$ axes respectively.

Now, along the line OQ we take a suitable distance OR (say) such that the moment of inertia of the body about OQ as axis, may be inversely proportional to the square of OR, i.e. $I \propto \dfrac{1}{OR^2}$ or we may write,

$$[l^2I_{xx}+m^2I_{yy}+n^2I_{zz}-2mnI_{yz}-2nlI_{zx}-2lmI_{xy}] \propto \frac{1}{OR^2}$$

$$\text{or} \quad [l^2I_{xx}+m^2I_{yy}+n^2I_{zz}-2mnI_{yz}-2nlI_{zx}-2lmI_{xy}] = \frac{Mk^4}{OR^2}$$

The constant Mk^4 is taken so as to make both sides of the equation dimensionally same, where k is some linear factor and M is the mass of the body.

$$\therefore \quad [l^2OR^2I_{xx}+m^2OR^2I_{yy}+n^2OR^2I_{zz}-2mnOR^2I_{yz}$$
$$-2nlOR^2I_{zx}-2lmOR^2I_{xy}] = Mk^4 = \text{constant}$$

$$\text{or,} \quad I_{xx}x^2+I_{yy}y^2+I_{zz}z^2-2I_{yz}yz-2I_{zx}zx-2I_{xy}xy = \text{constant}$$

[(x, y, z) being the co-ordinates of R referred to OX, OY, OZ, so that we have $x = lOR$, $y = mOR$ and $z = nOR$] (2)

(4.8.2) represents a surface of the second degree i.e. a quadric surface. Since OR is always finite and positive, the surface is a closed one and ellipsoid in nature. The locus of the point R is thus an ellipsoid, and this is called the **'momental ellipsoid'** or **'ellipsoid of inertia'** of the body at the point O.

Thus, at every point of the body, we may consider a similar momental ellipsoid. Now, we know from the properties of an ellipsoid that there can be found three perpendicular diameters such that if they are taken as the axes of co-ordinates, the resulting equation of the ellipsoid will have no terms involving yz, zx and xy, i.e., the terms containing products of inertia vanish. These new axes of co-ordinates are called the **principal axes of co-ordinates**. The momental ellipsoid referred to the principal axes may be written as

$$I'_{xx}x^2+I'_{yy}y^2+I'_{zz}z^2 = \text{constant} \tag{3}$$

where I'_{xx}, I'_{yy} and I'_{zz} are the moments of inertia of the body about the three principal axes of co-ordinates—also known as the **principal moments of inertia**

If we take $I = \frac{1}{OR^2}$ the equations (4.8.2) and (4.8.3) may be written in the form.

$$I_{xx}x^2+I_{yy}y^2+I_{zz}z^2-2I_{yz}yz-2I_{zx}zx-2I_{xy}xy = 1 \tag{4}$$

$$\text{and} \quad I'_{xx}x^2+I'_{yy}y^2+I'_{zz}z^2 = 1 \tag{5}$$

4.9 Plane Lamina

As a particular case of the preceding article we consider the case of a plane lamina (Fig. 4.9.1).

Let I_{xx}, I_{yy} be the moments of inertia of the lamina about OX, OY respectively and let I_{xy} be the product of inertia referred to the same two axes.

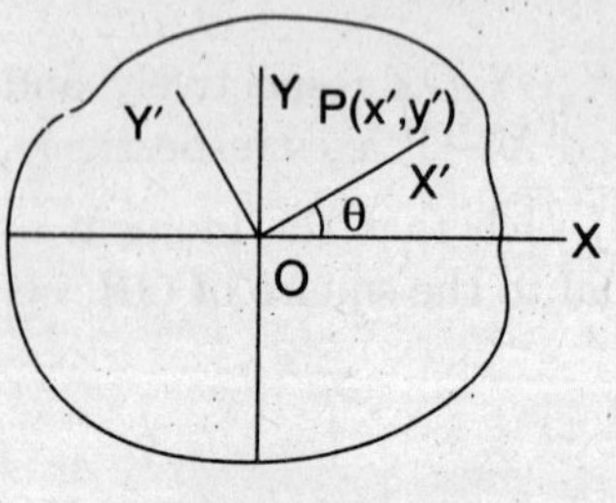

Fig. 4.9.1

[Since in the case of the lamina $I_{zz} = 0$, also $I_{yz} = I_{zx} = 0$.]

Here $I_{xx} = \Sigma my^2$, $I_{yy} = \Sigma mx^2$ and $I_{xy} = \Sigma mxy$.

We consider a point $P(x', y')$ referred to the new axes OX' and OY'. Then

$$x = x' \cos\theta - y' \sin\theta \tag{1}$$

$$y = x' \sin\theta + y' \cos\theta \tag{2}$$

where $\angle XOX' = \theta$.

Multiplying (4.9.1) by $\cos\theta$ and (4.9.2) by $\sin\theta$ throughout and adding, we have

$$x' = x\cos\theta + y\sin\theta. \tag{3}$$

Again, multiplying (4.9.1) by $\sin\theta$ and (4.9.2) by $\cos\theta$ throughout and substracting (4.9.1) from (4.9.2), we have

$$y' = y\cos\theta - x\sin\theta. \tag{4}$$

Hence, the moment of inertia of the lamina about $OX' = \Sigma my'^2$

$$= \Sigma m(y\cos\theta - x\sin\theta)^2 = \cos^2\theta\Sigma my^2 + \sin^2\theta\Sigma mx^2 - 2\sin\theta\cos\theta\Sigma mxy$$

$$= I_{xx}\cos^2\theta + I_{yy}\sin^2\theta - 2I_{xy}\sin\theta\cos\theta.$$

Similarly, moment of inertia about OY'

$$= \Sigma mx'^2 = I_{yy}\cos^2\theta + I_{xx}\sin^2\theta + 2I_{xy}\sin\theta\cos\theta.$$

The product of inertia referred to OX' and OY'

$$= \Sigma mx'y' = \Sigma m(x\cos\theta + y\sin\theta)(y\cos\theta - x\sin\theta)$$

$$= \Sigma m[y^2\sin\theta\cos\theta - x^2\sin\theta\cos\theta + xy(\cos^2\theta - \sin^2\theta)]$$

$$= (I_{xx} - I_{yy})\sin\theta\cos\theta + I_{xy}\cos 2\theta$$

$$= \frac{1}{2}(I_{xx} - I_{yy})\sin 2\theta + I_{xy}\cos 2\theta.$$

The moment of inertia of the plane lamina about a line passing through the intersection of OX and OY and perpendicular to the plane of the lamina

$$= \Sigma m(x^2 + y^2) = \Sigma mx^2 + \Sigma my^2$$

$$= I_{yy} + I_{xx} \quad \text{[From the theorem of perpendicular axes]}$$

In particular, if OX and OY refer to the principal axes, then I_{xx} and I_{yy} become the principal moments of inertia, and consequently the product of inertia $I_{xy} = 0$. In this case the moment of inertia about $OX' = I_{xx}\cos^2\theta + I_{yy}\sin^2\theta$ and moment of inertia about $OY' = I_{yy}\cos^2\theta + I_{xx}\sin^2\theta$.

4.9.1 Discussion on Principal Axes and Principal Moments of Inertia

We know that when a set of three mutually perpendicular axes, fixed in the body, coincides *with its symmetry axes,* all the products of inertia vanish. Such a property is not exclusive *for symmetric bodies only.* In fact for a body *with asymmetric shape,* it is always possible to have a set of three mutually perpendicular axes fixed at any point in the body such that the products of inertia referred to them vanish. These are called *the principal axes.* The inertia tensor† w.r.t these principal axes has evidently a diagonal form and *the diagonal elements of this inertia tensor* (the moments of inertia about these principal axes) are called the $PMI[I'_{xx}, I'_{yy}, I'_{zz}$ or I_1, I_2, I_3 for convenience].

We know that

$$I_1 = \Sigma m'(y^2 + z^2), \quad I_2 = \Sigma m'(z^2 + x^2), \quad I_3 = \Sigma m'(x^2 + y^2)$$

and hence

$$I_1 + I_2 \geq I_3, \quad I_2 + I_3 \geq I_1, \quad I_3 + I_1 \geq I_2.$$

We thus see that none of the three PMI can exceed the sum of the other two. A body, whose three PMI are all different i.e., $I_1 \neq I_2 \neq I_3$, is called *an asymmetrical top.* If any two are equal ($I_1 = I_2 \neq I_3$), then the body is called *a symmetrical top,* and if all the three are equal i.e., $I_1 = I_2 = I_3$, then the body is called *a spherical top.*

Determination of the PA for an asymmetrical top is very difficult. So we restrict ourselves to those bodies possessing some symmetry. Now the position of the CM and the directions of PA must display the same symmetry as the body. Suppose the body is such that $I_1 = I_2 \neq I_3$. Then the direction of one of the PA in the $X-Y$ plane may be chosen arbitrarily and the other one will be perpendicular to this. Further suppose the body has a plane of symmetry, then the CM must lie in this plane which also becomes the $X-Y$ plane. Obviously the third PA is normal to this plane of symmetry. In case the body is a spherical top, the three PA may be chosen as any three mutually perpendicular axes with CM as the origin. Here we state some useful results regarding symmetrical bodies.

(i) If a body is symmetrical w.r.t a plane, its CM lies in that plane.

(ii) If a body is symmetrical in two planes, its CM lies on their line of inter-section.

(iii) If a body has an axis of symmetry, its CM lies on that axis.

(iv) If a body is symmetrical in three planes with one common point, then that point is the CM.

(v) If a body has spherical symmetry about a point, then that point is its CM.

It is easy to explain why the PI will vanish for symmetrical bodies. Let $X-Y$ be one such plane of symmetry. Then for every (x, y, z) point, there will be another point $(x, y, -z)$—which will make I_{13} and I_{23} vanish. For another plane of symmetry say $Y-Z$ plane, similarly I_{12} and I_{13} will vanish. If l_1, m_1, n_1 be the direction cosines of the rotation axis relative to PAI, then (4.8.1) takes the form

$$\begin{aligned} I &= l_1^2 I_1 + m_1^2 I_2 + n_1^2 I_3 \qquad (5) \\ &= \left(1 - m_1^2 - n_1^2\right) I_1 + m_1^2 I_2 + n_1^2 I_3 \quad \because l_1^2 + m_1^2 + n_1^2 = 1 \\ &= I_1 - m_1^2(I_1 - I_2) - n_1^2(I_1 - I_3). \end{aligned}$$

†See Art. 4.12.

If $I_1 > I_2$ and $I_1 > I_3$, then $I < I_1$ and $I = I_1$ for rotation about X-axis (since $m_1 = n_1 = 0$).

If $I_1 < I_2$ and $I_1 < I_3$, then $I > I_1$ and $I = I_1$ for rotation about X-axis (since $m_1 = n_1 = 0$).

Therefore I cannot be larger than the largest PMI or smaller than the smallest one. So we can say that of the three PMI at a point one is the largest and another is the smallest of all the possible moments of inertia about the axes passing through the point. This property is sometimes utilised in determining the position of the PAI. The highest and the lowest values of I determine two axes and the third is perpendicular to both of them.

Illustrative Examples

1. Find the principal moments of inertia of a solid rectangular block with sides a, b, c about its principal axes through its centre of gravity—axes being parallel to the sides.

Here

$$I_{xx} = M\left[\frac{b^2 + c^2}{12}\right]; \quad I_{yy} = M\left[\frac{c^2 + a^2}{12}\right]; \quad I_{zz} = M\left[\frac{a^2 + b^2}{12}\right],$$

where M is the mass of the rectangular block (For calculation, see Art. 4.13 II).

2. Calculate the principal moments of inertia of a cylindrical rod of circular section in terms of its lengt · and diameter, and hence deduce the value of the moment of inertia round one of the generators of the cylinder. **[C.U. (Hons.) 1936]**

First part :

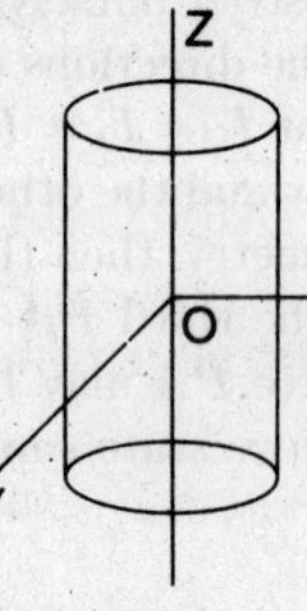

Fig. 1

$$\text{Here } I_{xx} = I_{yy} = M\left[\frac{l^2}{12} + \frac{r^2}{4}\right] = M\left[\frac{l^2}{12} + \frac{d^2}{16}\right]; \; I_{zz} = \frac{Mr^2}{2} = \frac{Md^2}{8}$$

where d is the diameter and l is the length of the cylinder.

Second part :

Here the moment of inertia about the generating line will be given by

$$\frac{Md^2}{8} + Mr^2 = \frac{Md^2}{8} + \frac{Md^2}{4} = \frac{3}{8}Md^2.$$

N.B. The equation to the ellipsoid of inertia at O in this case is

$$I_{xx}x^2 + I_{yy}y^2 + I_{zz}z^2 = 1 \quad \text{or} \quad M\left(\frac{l^2}{12} + \frac{r^2}{4}\right)(x^2 + y^2) + M\frac{r^2}{2}z^2 = 1.$$

which is the equation to a spheroid.

If $\frac{l^2}{12} + \frac{r^2}{4} = \frac{r^2}{2}$ i.e., $\frac{l^2}{12} = \frac{r^2}{4}$ i.e., $\frac{l}{\sqrt{3}} = r$, the above equation reduces to

$$\frac{Mr^2}{2}(x^2 + y^2 + z^2) = 1 \quad \text{or} \quad x^2 + y^2 + z^2 = a^2, \text{ where } a^2 = \frac{2}{Mr^2}$$

which represents a sphere.

3. Calculate the principal moments of inertia of a homogeneous cube at its centre of gravity. Hence find the moment of inertia about any axis passing through the centre of gravity. **[C.U. (Hons.) 1971]**

Here $a = b = c$, as the sides of the cube are equal.

$$\therefore\ I_{xx} = I_{yy} = I_{zz} = \frac{Ma^2}{6} \quad \text{(see Example 1).}$$

The moment of inertia about any axis passing through the centre of gravity is given by

$$I = I_{xx}(l^2 + m^2 + n^2) = I_{xx} = \frac{Ma^2}{6}.$$

4. Show that the momental ellipsoid at the centre of a homogeneous cube is a sphere. Calculate the radius of the sphere.

In case of a cube

$$I_{xx} = I_{yy} = I_{zz} = \frac{Ma^2}{6}$$

where M and a are the mass and side of the cube (see Illustrative Example-3). Hence from (4.8.5) we have the momental ellipsoid at centre O of the cube is

$$\frac{Ma^2}{6}(x^2 + y^2 + z^2) = 1 \quad \text{or} \quad (x^2 + y^2 + z^2) = \frac{6}{Ma^2} \quad \text{or} \quad x^2 + y^2 + z^2 = R^2$$

The above equation represents a sphere, of radius R given by

$$R^2 = \frac{6}{Ma^2}.$$

5. Calculate the moment of inertia of a homogeneous right circular solid cone about its generating line (i.e., a straight line on its sloping face). **[C.U. (Hons.) 1975 98 (Old Syll)]**

In this case, the length of the generating line $= \sqrt{r^2 + h^2}$. Here the axis of symmetry and other two axes mutually perpendicular to each other and perpendicular to the axis of symmetry at the vertex are the three principal axes of inertia

$$\therefore\ I_{yy} = \frac{3}{10}Mr^2; \quad I_{xx} = I_{zz} = \frac{3}{5}M\left(\frac{r^2}{4} + h^2\right)$$

So the moment of inertia about the generating line is given by

$$I = l^2 I_{xx} + m^2 I_{yy} + n^2 I_{zz} = l^2 I_{xx} + m^2 I_{yy} \quad (\because\ n = 0)$$

Now $l = \dfrac{r}{\sqrt{r^2 + h^2}}$ and $m = \dfrac{h}{\sqrt{r^2 + h^2}}$

$$\therefore\ I = \frac{3}{5}M\left(\frac{r^2}{4} + h^2\right)\frac{r^2}{(r^2 + h^2)} + \frac{3}{10}Mr^2\frac{h^2}{(r^2 + h^2)} = \frac{3}{10}Mr^2\frac{(r^2 + 4h^2)}{2(r^2 + h^2)} + \frac{3}{10}\frac{Mr^2h^2}{(r^2 + h^2)}$$

$$= \frac{3}{10}\frac{Mr^2}{(r^2 + h^2)}\left[\frac{1}{2}(r^2 + 4h^2) + h^2\right] = \frac{3}{20}\frac{Mr^2}{(r^2 + h^2)}(r^2 + 6h^2)$$

$$= \frac{3}{20}Mr^2\left[1 + \frac{5h^2}{(r^2 + h^2)}\right]$$

N.B. The equation to the ellipsoid of inertia in this case is

$$I_{xx}x^2 + I_{yy}y^2 + I_{zz}z^2 = 1$$

$$\text{or} \quad \frac{3}{5}M\left(\frac{r^2}{4}+h^2\right)x^2 + \frac{3}{10}Mr^2y^2 + \frac{3}{5}M\left(\frac{r^2}{4}+h^2\right)z^2 = 1$$

$$\text{or} \quad \frac{3}{5}M\left(\frac{r^2}{4}+h^2\right)(x^2+z^2) + \frac{3}{10}Mr^2y^2 = 1$$

which is the equation to a spheroid.

If $h = \frac{r}{2}$, the above expression reduces to

$$\frac{3}{5}M\frac{r^2}{2}(x^2+z^2) + \frac{3}{10}Mr^2y^2 = 1 \text{ or } \frac{3}{10}Mr^2(x^2+y^2+z^2) = 1$$

$$\text{or } x^2+y^2+z^2 = a^2 \text{ where } a^2 = \frac{10}{3Mr^2}.$$

This equation represents a sphere.

6. The three principal moments of inertia of a body at a point are given as 200, 300 and 450 gm-cm^2. Write down the equation of the ellipsoid of inertia at that point. What will be the kinetic energy and the angular momentum of the body when it is rotating about an axis passing through the point and having direction cosines $\frac{1}{\sqrt{2}}, \frac{1}{\sqrt{3}}$ and $\frac{1}{\sqrt{6}}$ respectively? The body is rotating at the rate of 100 revolutions per minute. If the body breaks up, which of the two—kinetic energy and angular momentum will remain constant? **[C.U. (Hons.) 1977]**

The equation to the ellipsoid of inertia at a point with reference to the principal axes is

$$I_{xx}x^2 + I_{yy}y^2 + I_{zz}z^2 = \text{constant}.$$

Here

$$I_{xx} = 200 \text{ gm-cm}^2; \quad I_{yy} = 300 \text{ gm-cm}^2; \quad I_{zz} = 450 \text{ gm-cm}^2$$

The above equation may then be written as

$$200x^2 + 300y^2 + 450z^2 = \text{constant}$$

$$\text{or} \quad 4x^2 + 6y^2 + 9z^2 = \text{constant}$$

This is the required equation.

From equation (4.10.3), we have,

$$\text{K.E., } T = \frac{1}{2}\left(I_{xx}w_x^2 + I_{yy}w_y^2 + I_{zz}w_z^2\right) = \frac{1}{2}\left[I_{xx}(\lambda w)^2 + I_{yy}(\mu w)^2 + I_{zz}(\nu w)^2\right]$$

[λ, μ, ν being the direction cosines of the axis]

$$= \frac{1}{2}w^2\left[I_{xx}\lambda^2 + I_{yy}\mu^2 + I_{zz}\nu^2\right] = \frac{1}{2}\left(\frac{10\pi}{3}\right)^2\left[200\cdot\frac{1}{2} + 300\cdot\frac{1}{3} + 450\cdot\frac{1}{6}\right]$$

$$\left[\because w = 2\pi n = 2\pi \times \frac{100}{60} = \frac{10\pi}{3} \text{ and } \lambda = \frac{1}{\sqrt{2}}, \mu = \frac{1}{\sqrt{3}}, \nu = \frac{1}{\sqrt{6}}\right]$$

$$= \frac{50\pi^2}{9} \times 275 = \frac{50 \times 9.87 \times 275}{9} = 15079.167 \text{ ergs}.$$

Angular momentum $\vec{L} = \hat{\imath}L_x + \hat{\jmath}L_y + \hat{k}L_z$ where $\hat{\imath}, \hat{\jmath}, \hat{k}$ are the unit vectors along the principal axes

$$\text{or } \vec{L} = \hat{\imath}I_{xx}w_x + \hat{\jmath}I_{yy}w_y + \hat{k}I_{zz}w_z = \hat{\imath}I_{xx}\lambda w + \hat{\jmath}I_{yy}\mu w + \hat{k}I_{zz}\nu w = w\left[\hat{\imath}I_{xx}\lambda + \hat{\jmath}I_{yy}\mu + \hat{k}I_{zz}\nu\right]$$

$$\therefore\ L = w\sqrt{I_{xx}^2\lambda^2 + I_{yy}^2\mu^2 + I_{zz}^2\nu^2} = \frac{10\pi}{3} \times \sqrt{(200)^2 \times \frac{1}{2} + (300)^2 \times \frac{1}{3} + (450)^2 \times \frac{1}{6}}$$

$$= 2881.37 \text{ gm-cm}^2/\text{sec}.$$

In case of rotational motion we know that angular momentum remains conserved. So even if the rotating body breaks up, angular momentum will be constant.

Vectorial Representation

4.10 Kinetic Energy of a Rigid Body in Terms of Angular Velocity and Principal Moments of Inertia

We consider a rigid body rotating about a fixed point O with angular velocity $\vec{\omega}$. The kinetic energy of a particle P of the body having velocity $\vec{v}$ and mass δm is $\frac{1}{2}\delta m v^2$. Hence the kinetic energy of the body is given by

$$T = \tfrac{1}{2}\Sigma\delta m v^2 \tag{1}$$

where the summation extends over all particles of which the body is composed. Let the position of the particle P be specified by the vector $\vec{r}$ relative to a fixed origin O. Then the position vector $\vec{r}$ is given by $\vec{r} = \hat{\imath}x + \hat{\imath}y + \hat{k}z$ where x, y, z are the co-ordinates of the point P and $\hat{\imath}$, $\hat{\jmath}$, $\hat{k}$ denote the rectangular unit vectors. Also resolving the vector $\vec{\omega}$ along the rectangular axes we have, $\vec{\omega} = \hat{\imath}\omega_x + \hat{\jmath}\omega_y + \hat{k}\omega_z$.

Hence,

$$\vec{v} = \vec{\omega} \times \vec{r} = \begin{vmatrix} \hat{\imath} & \hat{\jmath} & \hat{k} \\ \omega_x & \omega_y & \omega_z \\ x & y & z \end{vmatrix} = \hat{\imath}(\omega_y z - \omega_z y) + \hat{\jmath}(\omega_z x - \omega_x z) + \hat{k}(\omega_x y - \omega_y x). \tag{2}$$

$\therefore$ Substituting the magnitude of v^2 from (4.10.2) in (4.10.1), we have

$$\begin{aligned} T &= \tfrac{1}{2}\Sigma\delta m\left[(\omega_y z - \omega_z y)^2 + (\omega_z x - \omega_x z)^2 + (\omega_x y - \omega_y x)^2\right] \\ &= \tfrac{1}{2}\left[\omega_x^2\Sigma\delta m(y^2 + z^2) + \omega_y^2\Sigma\delta m(z^2 + x^2) + \omega_z^2\Sigma\delta m(x^2 + y^2)\right. \\ &\qquad \left. - 2\omega_y\omega_z\Sigma\delta m yz - 2\omega_z\omega_x\Sigma\delta m zx - 2\omega_x\omega_y\Sigma\delta m xy\right] \\ &= \tfrac{1}{2}\left[I_{xx}\omega_x^2 + I_{yy}\omega_y^2 + I_{zz}\omega_z^2 - 2\omega_y\omega_z I_{yz} - 2\omega_z\omega_x I_{zx} - 2\omega_x\omega_y I_{xy}\right] \end{aligned}$$

where I_{xx}, I_{yy}, I_{zz}, I_{yz}, I_{zx}, I_{xy} are the moments and products of inertia.

$$\begin{aligned} &= \tfrac{1}{2}\omega^2[I_{xx}l^2 + I_{yy}m^2 + I_{zz}n^2 - 2mnI_{yz} - 2nlI_{zx} - 2lmI_{xy}] \\ &= \tfrac{1}{2}\omega^2 I_{l,m,n}\ (\because \omega_x = l\omega, \omega_y = m\omega, \omega_z = n\omega) = \tfrac{1}{2}I_{l,m,n}\omega^2 \end{aligned}$$

If these axes are the principal axes of inertia, $I_{yz} = I_{zx} = I_{xy} = 0$

$$\therefore\ T = \tfrac{1}{2}\left[I_{xx}\omega_x^2 + I_{yy}\omega_y^2 + I_{zz}\omega_z^2\right]. \tag{3}$$

In the above equation I_{xx}, I_{yy}, I_{zz} are the principal moments of inertia and ω_x, ω_y, ω_z are the components of the angular velocity $\vec{\omega}$ in the directions of the principal axes of inertia.

4.11 Angular Momentum of a Rigid Body

The angular momentum $\vec{L}$ about any point O is the moment of $m\vec{v}$ about any point O.

Hence,

$$\vec{L} = \Sigma(\vec{r} \times \delta m\vec{v}) \tag{1}$$

where δm is the mass of a particle of the body, $\vec{r}$ its position vector and $\vec{v}$ its velocity. The summation extends over all particles in the body.

Now, $\vec{v} = \vec{\omega} \times \vec{r}$, where $\vec{\omega}$ is the angular velocity of the body.

Hence,

$$\vec{L} = \Sigma\delta m\,[\vec{r} \times (\vec{\omega} \times \vec{r})] = \Sigma\delta m\,[\vec{\omega}r^2 - (\vec{r}\cdot\vec{\omega})\vec{r}] \tag{2}$$

Now resolving the vectors $\vec{r}$ and $\vec{\omega}$ along $\hat{\imath}, \hat{\jmath}, \hat{k}$, we have

$$\vec{r} = \hat{\imath}x + \hat{\jmath}y + \hat{k}z$$

$$\vec{\omega} = \hat{\imath}\omega_x + \hat{\jmath}\omega_y + \hat{k}\omega_z$$

$$\therefore\ \vec{L} = \Sigma\delta m\left[(\hat{\imath}\omega_x + \hat{\jmath}\omega_y + \hat{k}\omega_z)(x^2 + y^2 + z^2) - (\hat{\imath}x + \hat{\jmath}y + \hat{k}z)\left\{(\hat{\imath}\omega_x + \hat{\jmath}\omega_y + \hat{k}\omega_z)\cdot(\hat{\imath}x + \hat{\jmath}y + \hat{k}z)\right\}\right]$$

Hence

$$L_x = I_{xx}\omega_x - I_{xy}\omega_y - I_{xz}\omega_z \tag{3}$$

$$L_y = -I_{yx}\omega_x + I_{yy}\omega_y - I_{yz}\omega_z \tag{4}$$

$$L_z = -I_{zx}\omega_x - I_{zy}\omega_y + I_{zz}\omega_z \tag{5}$$

If $\hat{\imath}$, $\hat{\jmath}$, $\hat{k}$ are the principal axes of inertia at O, then

$$I_{xy} = I_{yz} = I_{zx} = 0$$

$$\therefore\ L_x = I_{xx}\omega_x \quad L_y = I_{yy}\omega_y \quad L_z = I_{zz}\omega_z \tag{6}$$

Here I_{xx}, I_{yy}, I_{zz} are the principal moments of inertia at O.

If the body is constrained to rotate about a fixed axis along which the unit vector is $\hat{k}$, we have $\omega_x = \omega_y = 0$ and $\omega_z = \omega$ (say)

$$\therefore\ L_x = -I_{zx}\omega; \quad L_y = -I_{yz}\omega; \quad L_z = I_{zz}\omega.$$

Thus the angular momentum vector does not lie along the axis of rotation, unless the axis is a principal axis of inertia, although the component L_z is equal to the product of the moment of inertia about that axis and the angular velocity.

Illustrative Example

1. Show that $T = \frac{1}{2}\vec{L}\cdot\vec{\omega}$.

We know that

$$T = \frac{1}{2}\left[I_{xx}\omega_x^2 + I_{yy}\omega_y^2 + I_{zz}\omega_z^2\right] \quad \text{(from equation 4.10.3)}$$
$$= \frac{1}{2}\left[L_x\omega_x + L_y\omega_y + L_z\omega_z\right] \quad \text{(from equation 4.11.6)}$$
$$= \frac{1}{2}(\hat{\imath}L_x + \hat{\jmath}L_y + \hat{k}L_z)\cdot(\hat{\imath}\omega_x + \hat{\jmath}\omega_y + \hat{k}\omega_z) = \frac{1}{2}\vec{L}\cdot\vec{\omega}.$$

2. Show that $\frac{\partial T}{\partial \vec{\omega}} = \vec{L}$.

We have

$$T = \frac{1}{2}\vec{L}\cdot\vec{\omega} = \frac{1}{2}I\vec{\omega}\cdot\vec{\omega} = \frac{1}{2}I\vec{\omega}^2 \quad \therefore \quad \frac{\partial T}{\partial \vec{\omega}} = I\vec{\omega} = \vec{L}.$$

Alternative method

We know

$$T = \frac{1}{2}\left[I_{xx}\omega_x^2 + I_{yy}\omega_y^2 + I_{zz}\omega_z^2\right]$$

so that

$$\frac{\partial T}{\partial \omega_x} = I_{xx}\omega_x = L_x, \quad \frac{\partial T}{\partial \omega_y} = L_y \quad \text{and} \quad \frac{\partial T}{\partial \omega_z} = L_z$$

whence it follows that

$$\frac{\partial T}{\partial \omega_x}\hat{\imath} + \frac{\partial T}{\partial \omega_y}\hat{\jmath} + \frac{\partial T}{\partial \omega_z}\hat{k} = L_x\hat{\imath} + L_y\hat{\jmath} + L_z\hat{k} = \vec{L}.$$

Now

$$\frac{\partial T}{\partial \omega_x}\hat{\imath} = L_x\hat{\imath} \quad \text{or} \quad \frac{\hat{\imath}}{\hat{\imath}}\frac{\partial T}{\partial \omega_x}\hat{\imath} = L_x\hat{\imath} \quad \text{or} \quad \frac{\partial T}{\partial(\hat{\imath}\omega_x)} = L_x\hat{\imath} \quad \because \hat{\imath}\cdot\hat{\imath} = 1$$

similarly $\frac{\partial T}{\partial(\hat{\jmath}\omega_y)} = L_y\hat{\jmath}$ and $\frac{\partial T}{\partial(\hat{k}\omega_z)} = L_z\hat{k}$, so that in general we may write $\frac{\partial T}{\partial \vec{\omega}} = \vec{L}$.

4.12 Tensor of Inertia

We know that for a rigid body rotating about a fixed axis, the angular momontum is proportional to the angular velocity of the body and is given by

$$\vec{L} = I\vec{\omega}$$

where $\vec{L}$ is the angular momentum vector and $\vec{\omega}$ the angular velocity vector. The proportionality factor I is called the moment of inertia of the rotating body about the axis. Now the components of $\vec{L}$ along X, Y and Z axes are given by

$$\left.\begin{aligned} L_x &= I_{xx}\omega_x - I_{xy}\omega_y - I_{xz}\omega_z \\ L_y &= -I_{yx}\omega_x + I_{yy}\omega_y - I_{yz}\omega_z \\ L_z &= -I_{zx}\omega_x - I_{zy}\omega_y + I_{zz}\omega_z \end{aligned}\right\} \quad \text{[from (4.11.3), (4.11.4) and (4.11.5)]}$$

The nine coefficients I_{ij} can be written in an array often called the *tensor of inertia.* The nine elements of the tensor of inertia of a rigid body can be calculated if the shape of the rigid body be known. Now the K.E. of the rotating rigid body is

$$T = \frac{1}{2}[\omega_x^2\Sigma\delta m(y^2+z^2) + \omega_y^2\Sigma\delta m(z^2+x^2) + \omega_z^2\Sigma\delta m(x^2+y^2)$$
$$-2\omega_y\omega_z\Sigma\delta myz - 2\omega_z\omega_x\Sigma\delta mzx - 2\omega_x\omega_y\Sigma\delta mxy] \quad \text{(see Art. 4.10)}$$

Since $r^2 = x^2 + y^2 + z^2$, the K.E. may also be written in the form

$$T = \frac{1}{2}[\omega_x^2\Sigma\delta m(r^2-x^2) + \omega_y^2\Sigma\delta m(r^2-y^2) + \omega_z^2\Sigma\delta m(r^2-z^2)$$
$$-2\omega_y\omega_z\Sigma\delta myz - 2\omega_z\omega_x\Sigma\delta mzx - 2\omega_x\omega_y\Sigma\delta mxy]$$
$$= \frac{1}{2}\Sigma I_{ij}\omega_i\omega_j.$$

Here

$$\begin{aligned} I_{xx} &= \Sigma\delta m(r^2-x^2), \quad & I_{yz} &= +\Sigma\delta myz \\ I_{yy} &= \Sigma\delta m(r^2-y^2), \quad & I_{zx} &= +\Sigma\delta mzx \\ I_{zz} &= \Sigma\delta m(r^2-z^2), \quad & I_{xy} &= +\Sigma\delta mxy \end{aligned}$$

Hence the tensor of inertia can be written as

$$I_{ij} = \begin{bmatrix} \Sigma\delta m(r^2-x^2) & -\Sigma\delta mxy & -\Sigma\delta mxz \\ -\Sigma\delta myx & \Sigma\delta m(r^2-y^2) & -\Sigma\delta myz \\ -\Sigma\delta mzx & -\Sigma\delta mzy & \Sigma\delta m(r^2-z^2) \end{bmatrix}$$

In tensor notation, we may write

$$I_{ij} = \Sigma\delta m(r^2\delta_{ij} - r_ir_j),$$

where the Kronečker delta $\delta_{ij} = 1$ when $i = j$ and $\delta_{ij} = 0$, when $i \neq j$.

As $I_{ij} = I_{ji}$, it is called a symmetric tensor. Also I_{ij} is neither a scalar nor a vector. It is a tensor of the second rank.

The diagonal elements (I_{xx}, I_{yy}, I_{zz}) of the inertia tensor are known as *moments of inertia,* while the six off-diagonal elements are known as *products of inertia.*

Now writing

$$\left.\begin{aligned} I_{xy} &= -\Sigma\delta mxy = I_{yx} \\ I_{yz} &= -\Sigma\delta myz = I_{zy} \\ I_{zx} &= -\Sigma\delta mzx = I_{xz} \end{aligned}\right\} \text{ i.e. defining the products of inertia with an associated "minus" sign,}$$

the inertia tensor may be written as (avoiding minus signs)

$$I_{ij} = \begin{bmatrix} I_{xx} & I_{xy} & I_{xz} \\ I_{yx} & I_{yy} & I_{yz} \\ I_{zx} & I_{zy} & I_{zz} \end{bmatrix}$$

and (4.11.3), (4.11.4), (4.11.5) may also be written in the form

$$\begin{bmatrix} L_x \\ L_y \\ L_z \end{bmatrix} = \begin{bmatrix} I_{xx} & I_{xy} & I_{xz} \\ I_{yx} & I_{yy} & I_{yz} \\ I_{zx} & I_{zy} & I_{zz} \end{bmatrix} \begin{bmatrix} \omega_x \\ \omega_y \\ \omega_z \end{bmatrix} \text{ i.e., } \left.\begin{aligned} L_x &= I_{xx}\omega_x + I_{xy}\omega_y + I_{xz}\omega_z \\ L_y &= I_{yx}\omega_x + I_{yy}\omega_y + I_{yz}\omega_z \\ L_z &= I_{zx}\omega_x + I_{zy}\omega_y + I_{zz}\omega_z \end{aligned}\right\} \quad (1)$$

Equations for the principal moments of inertia and the directions of the principal axes :

A set of three mutually perpendicular axes having origin O, which are fixed in the body and rotating with it, and which are such that the products of inertia about them are zero, are called *the principal axes of inertia* of the body. An important property of a principal axis is that if a rigid body rotates about it, the direction of the angular momentum is the same as that of the angular velocity. Thus

$$\vec{L} = I\vec{\omega}$$

so that (4.12.1) may be written as

$$\begin{aligned} I_{xx}\omega_x + I_{xy}\omega_y + I_{xz}\omega_z &= I\omega_x \\ I_{yx}\omega_x + I_{yy}\omega_y + I_{yz}\omega_z &= I\omega_y \\ I_{zx}\omega_x + I_{zy}\omega_y + I_{zz}\omega_z &= I\omega_z \end{aligned} \quad \text{or} \quad \left.\begin{aligned} (I_{xx} - I)\omega_x + I_{xy}\omega_y + I_{xz}\omega_z &= 0 \\ I_{yx}\omega_x + (I_{yy} - I)\omega_y + I_{yz}\omega_z &= 0 \\ I_{zx}\omega_x + I_{zy}\omega_y + (I_{zz} - I)\omega_z &= 0 \end{aligned}\right\}. \quad (2)$$

The principal moments of inertia are found by setting the determinant of the coefficients of $\omega_x, \omega_y, \omega_z$ in (4.12.2) equal to zero, i.e,

$$\begin{bmatrix} (I_{xx} - I) & I_{xy} & I_{xz} \\ I_{yx} & (I_{yy} - I) & I_{yz} \\ I_{zx} & I_{zy} & (I_{zz} - I) \end{bmatrix} = 0 \quad (3)$$

This leads to a cubic equation in I having three real roots I_1, I_2, I_3, which are the *principal moments of inertia.* By putting $I = I_1$ in (4.12.2), we may obtain the ratio $\omega_x : \omega_y : \omega_z$, whence we may find the direction of ω, i.e., the direction of the principal axis corresponding to I_1. Similarly, by putting I_2 and I_3 in (4.12.2), we may determine the directions of the corresponding principal axes, as will be evident from the following illustrative example.

Illustrative Example

(a) Find the moments and products of inertia of a uniform square plate of side a about X, Y and Z axes, X and Y axes being taken as the adjacent sides of the plate and Z axis perpendicular to its plane.

(b) Find (i) the principal moments of inertia and (ii) the directions of the principal axes in this case.

[C.U. (Hons.) 1996, '99]

Considering an element $dxdy$ of the plate as shown in the Fig. 4.12.1, its moment of inertia about X-axis is $\rho y^2 dxdy$, ρ being the mass per unit area of the plate. So the moment of inertia of the entire plate about X-axis is given by

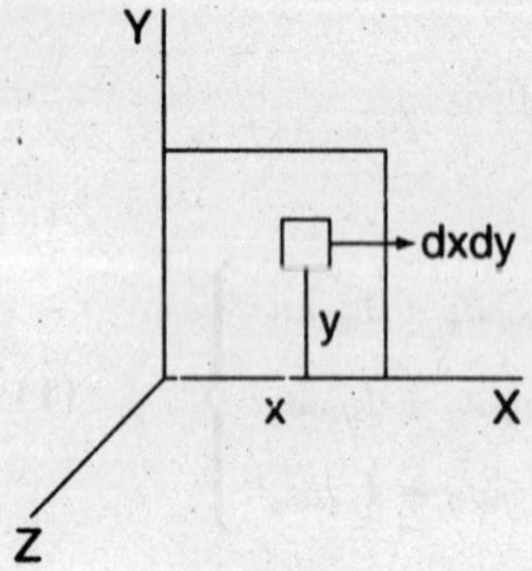

Fig. 4.12.1

$$I_{xx} = \int_{x=0}^{a}\int_{y=0}^{a} \rho y^2 dxdy = \frac{1}{3}\rho a^4 = \frac{1}{3}Ma^2 \tag{4}$$

($\because$ mass of the plate $M = \rho a^2$)

Similarly the moment of inertia of the plate about the Y-axis is given by

$$I_{yy} = \int_{x=0}^{a}\int_{y=0}^{a} \rho x^2 dxdy = \frac{1}{3}\rho a^4 = \frac{1}{3}Ma^2 \tag{5}$$

Now the moment of inertia of the plate about the Z-axis is

$$I_{zz} = I_{xx} + I_{yy} \quad \text{(from the theorem of perpendicular axes)}$$

$$= \frac{2}{3}Ma^2 \tag{6}$$

The product of inertia of the element $dxdy$ of the plate about X and Y axes is $\rho xydxdy$, and hence the product of inertia of the entire plate about X and Y axes is

$$I_{xy} = I_{yx} = -\int_{x=0}^{a}\int_{y=0}^{a} \rho xydxdy = -\frac{1}{4}\rho a^4 = -\frac{1}{4}Ma^2 \tag{7}$$

Again the product of inertia of the element $dxdy$ about the X and Z axes is the product of $\rho dxdy$ by the distances to the $X-Y$ and $Y-Z$ planes, which are 0 and x respectively, so that we may write

$$I_{xz} = I_{zx} = 0, \text{ and similarly } I_{yz} = I_{zy} = 0 \tag{8}$$

(b) (i) Substituting the above results in (4.12.3), we obtain

$$\begin{bmatrix} \left(\frac{1}{3}Ma^2 - I\right) & -\frac{1}{4}Ma^2 & 0 \\ -\frac{1}{4}Ma^2 & \left(\frac{1}{3}Ma^2 - I\right) & 0 \\ 0 & 0 & \left(\frac{2}{3}Ma^2 - I\right) \end{bmatrix} = 0 \tag{9}$$

$$\text{or} \quad \left[\left(\frac{1}{3}Ma^2 - I\right)\left(\frac{1}{3}Ma^2 - I\right) - \left(-\frac{1}{4}Ma^2\right)\left(-\frac{1}{4}Ma^2\right)\right]\left[\frac{2}{3}Ma^2 - I\right] = 0$$

$$\text{or} \quad \left[I^2 - \frac{2}{3}Ma^2 I + \frac{7}{144}M^2a^4\right]\left[\frac{2}{3}Ma^2 - I\right] = 0.$$

Solving we obtain the three roots of (4.12.9), given by

$$I_1 = \frac{1}{12}Ma^2, \quad I_2 = \frac{7}{12}Ma^2, \quad I_3 = \frac{2}{3}Ma^2$$

which are the principal moments of inertia.

(ii) Now to find the direction of the principal axis corresponding to I_1, we put $I = I_1 = \frac{1}{12}Ma^2$ in the following equations obtained from (4.12.2), written as

$$\left.\begin{aligned} \left(\frac{1}{3}Ma^2 - I\right)\omega_x - \frac{1}{4}Ma^2\omega_y + O\omega_z &= 0 \\ -\frac{1}{4}Ma^2\omega_x + \left(\frac{1}{3}Ma^2 - I\right)\omega_y + O\omega_z &= 0 \\ O\omega_x + O\omega_y + \left(\frac{2}{3}Ma^2 - I\right)\omega_z &= 0 \end{aligned}\right\} \tag{10}$$

The first two equations give $\omega_x = \omega_y$, while the third gives $\omega_z = 0$. Thus the direction of the principal axis is the same as the direction of the angular velocity vector,

$$\vec{\omega} = \omega_x\hat{\imath} + \omega_y\hat{\jmath} + \omega_z\hat{k} = \omega_x\hat{\imath} + \omega_y\hat{\jmath} = \omega_x(\hat{\imath} + \hat{\jmath}) \quad [\because \omega_x = \omega_y]$$

Thus the principal axis corresponding to I_1 is in the direction $(\hat{\imath} + \hat{\jmath})$. Similarly by putting $I = I_2 = \frac{7}{12}Ma^2$ in (4.12.10) we have $\omega_y = -\omega_x$, $\omega_z = 0$, so that the direction of the corresponding principal axis is given by $\vec{\omega} = \omega_x\hat{\imath} - \omega_x\hat{\jmath} = \omega_x(\hat{\imath} - \hat{\jmath})$ i.e., the principal axis in this case is in the direction $(\hat{\imath} - \hat{\jmath})$. Again by putting $I = I_3 = \frac{2}{3}Ma^2$ in (4.12.10), we have $\omega_x = 0, \omega_y = 0$, while ω_z is arbitrary. This gives $\vec{\omega} = \omega_z\hat{k}$, which shows that the third principal axis is in the direction $\hat{k}$.

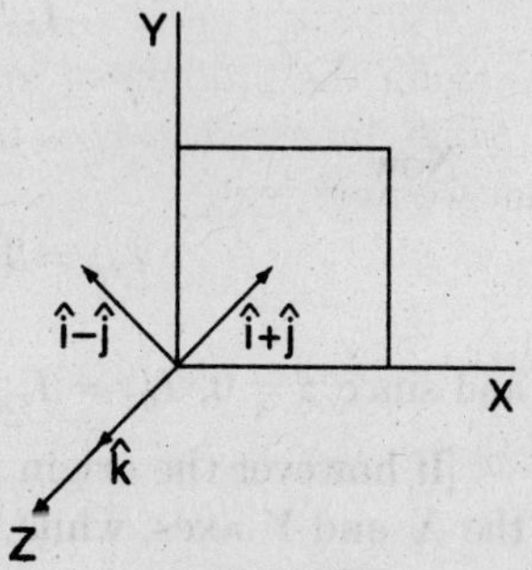

Fig. 4.12.2

The directions of the principal axes are shown by $(\hat{\imath}+\hat{\jmath})$, $(\hat{\imath}-\hat{\jmath})$ and $\hat{k}$ in Fig. 4.12.2. From the figure it is evident that these are mutually perpendicular and that $(\hat{\imath} + \hat{\jmath})$ and $(\hat{\imath} - \hat{\jmath})$ are in the directions of the diagonals of the square plate—which are the lines of symmetry.

The principal moments of inertia can also be determined from a knowledge of the lines of symmetry.

4.13 Moments and products of inertia and hence ellipsoids of inertia of some symmetrical bodies

I. *A thin rectangular lamina of length a and breadth b and mass per unit area ρ about axes as shown in* Fig. 4.13.1.

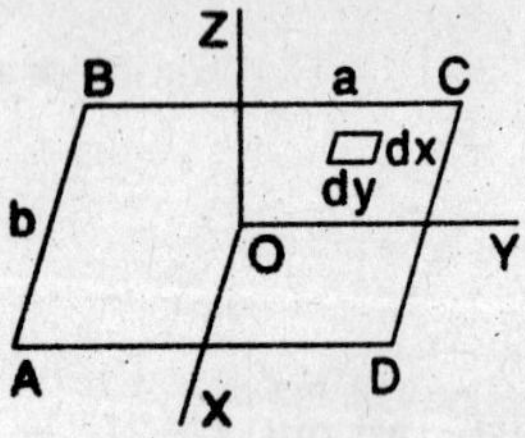

Fig. 4.13.1

Assuming uniform mass distribution, we may write

$$I_{xx} = \iint (y^2 + z^2)dxdy\rho = \iint y^2 dxdy\rho \quad (\because\ z = 0)$$

$$\text{or} \quad I_{xx} = \rho \int_{-a/2}^{a/2} y^2 dy \int_{-b/2}^{b/2} dx = \frac{Ma^2}{12}$$

where $M = ab\rho$ = mass of the lamina.

Similarly

$$I_{yy} = Mb^2/12$$

$$\text{and} \quad I_{zz} = \iint (x^2 + y^2)dxdy\rho = \rho \left[\int_{-b/2}^{b/2} x^2 dx \int_{-a/2}^{a/2} dy + \int_{-b/2}^{b/2} dx \int_{-a/2}^{a/2} y^2 dy \right]$$

$$= \frac{M}{12}(a^2 + b^2)$$

which may also be obtained from the theorem of perpendicular axes : $I_{zz} = I_{xx} + I_{yy}$. Again from the theorem of parallel axes, we may find

$$I_{AB} = I_{xx} + M\frac{a^2}{4} = \frac{Ma^2}{3}; \quad \text{and} \quad I_{BC} = I_{yy} + M\frac{b^2}{4} = \frac{Mb^2}{3}$$

and also the m.i. of the lamina about an axis passing through one of the corners and perpendicular to its plane is

$$I_{zz} + M(OA)^2 = I_{zz} + M\frac{a^2 + b^2}{4} = \frac{M(a^2 + b^2)}{3}.$$

Now

$$I_{xy} = I_{yx} = -\iint \rho xydxdy = -\rho \int_{-b/2}^{b/2} xdx \int_{-a/2}^{a/2} ydy = 0$$

and since $z = 0$, $I_{yz} = I_{zy} = 0$ and also $I_{zx} = I_{xz} = 0$.

[If however the origin is shifted to one corner B i.e., the two sides of the plate coincide with the X and Y axes, while Z-axis is perpendicular to the plane, we may write

$$I'_{xy} = I'_{yx} = -\rho \int_0^b xdx \int_0^a ydy = -\frac{1}{4}Mab$$

and $I'_{yz} = I'_{zy} = 0$ also $I'_{zx} = I'_{xz} = 0 \quad \because\ z = 0$.]

For the above rectangular plate with O at the centre, clearly I_{xx}, I_{yy}, I_{zz} represent the principal moments of inertia, and the ellipsoid of inertia at O is given by

$$I_{xx}x^2 + I_{yy}y^2 + I_{zz}z^2 = 1 \text{ or } \frac{Ma^2}{12}x^2 + \frac{Mb^2}{12}y^2 + \frac{M}{12}(a^2+b^2)z^2 = 1$$

$$\text{or } \frac{M}{12}\left[a^2x^2 + b^2y^2 + (a^2+b^2)z^2\right] = 1$$

which is the equation to a spheroid. If $a = b$ i.e., for a square plate the above equation reduces to

$$\frac{Ma^2}{12}\left[x^2 + y^2 + 2z^2\right] = 1$$

which also represents a spheroid.

II. ***A rectangular parallelopiped of sides a, b, c and mass per unit volume ρ about axes as shown in* Fig. 4.13.2 :**

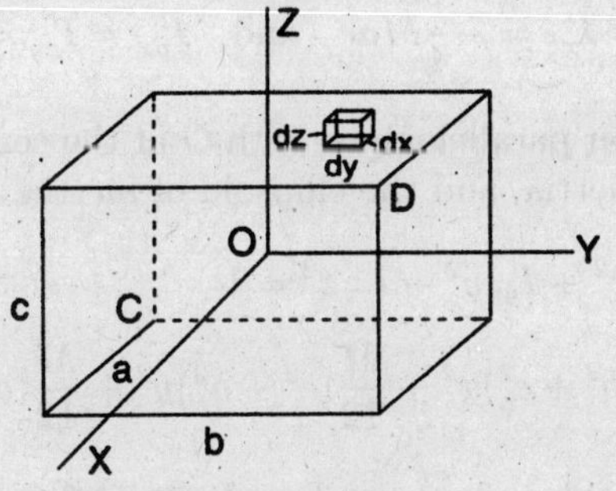

Fig. 4.13.2

Here, we have, assuming uniform mass distribution,

$$I_{xx} = \int (y^2+z^2)dxdydz\rho = \rho\left[\int_{-a/2}^{a/2} dx \int_{-b/2}^{b/2} y^2dy \int_{-c/2}^{c/2} dz + \int_{-a/2}^{a/2} dx \int_{-b/2}^{b/2} dy \int_{-c/2}^{c/2} z^2dz\right]$$

$$= \frac{M}{12}(b^2+c^2)$$

$M = \rho abc$ being the mass of the parallelopiped.

Similarly

$$I_{yy} = \frac{M}{12}(c^2+a^2) \quad \text{and} \quad I_{zz} = \frac{M}{12}(a^2+b^2).$$

Again

$$I_{xy} = I_{yx} = -\int xy\rho dxdydz = -\rho\int_{-a/2}^{a/2} xdx \int_{-b/2}^{b/2} ydy \int_{-c/2}^{c/2} dz = 0.$$

Similarly

$$I_{yz} = I_{zy} = 0 \quad \text{and} \quad I_{zx} = I_{xz} = 0.$$

[When the origin is shifted to one corner C, we have

$$I'_{xx} = \int (y^2 + z^2)dxdydz\rho$$

$$= \rho\left[\int_0^a dx \int_0^b y^2 dy \int_0^c dz + \int_0^a dx \int_0^b dy \int_0^c z^2 dz\right] = \frac{M}{3}(b^2 + c^2).$$

Similarly

$$I'_{yy} = \frac{M}{3}(c^2 + a^2) \quad \text{and} \quad I'_{zz} = \frac{M}{3}(a^2 + b^2).$$

Also

$$I'_{xy} = -\int xy\rho dxdydz = -\rho \int_0^a xdx \int_0^b ydy \int_0^c dz = -\frac{1}{4}Mab = I'_{yx}.$$

Similarly

$$I'_{zx} = I'_{xz} = -\frac{1}{4}Mac \quad \text{and} \quad I'_{yz} = I'_{zy} = -\frac{1}{4}Mbc.]$$

For the above rectangular parallelopiped with O at the centre, clearly I_{xx}, I_{yy}, I_{zz} represent the principal moments of inertia, and the ellipsoid of inertia at O is given by

$$I_{xx}x^2 + I_{yy}y^2 + I_{zz}z^2 = 1$$

$$\text{or} \quad \frac{M}{12}(b^2 + c^2)x^2 + \frac{M}{12}(c^2 + a^2)y^2 + \frac{M}{12}(a^2 + b^2)z^2 = 1$$

$$\text{or} \quad \frac{M}{12}\left[(b^2 + c^2)x^2 + (c^2 + a^2)y^2 + (a^2 + b^2)z^2\right] = 1$$

which is the equation to a general ellipsoid.

III. *A uniform circular disc of radius R and mass per unit area ρ about axes as shown in the Fig.* 4.13.3 :

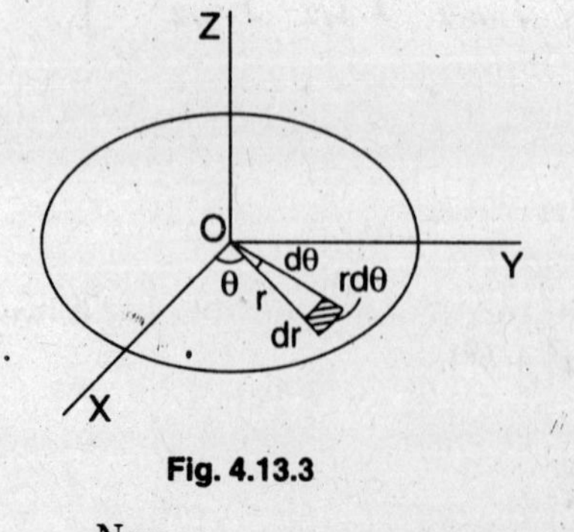

Fig. 4.13.3

Assuming uniform mass distribution and using cylindrical co-ordinates $x = r\cos\theta$, $y = r\sin\theta$ and $z = z$, and taking mass of the elementary shaded area as $rd\theta dr\rho$, we may write

$$I_{zz} = \int (x^2 + y^2)rd\theta dr\rho = \rho \int_0^R r^3 dr \int_0^{2\pi} d\theta$$

$$= \rho\frac{R^4}{4}2\pi = \frac{MR^2}{2},$$

$M = \pi R^2\rho$, being the mass of the disc.

Now

$$I_{xx} = \int (y^2 + z^2)rd\theta dr\rho = \int r^3 \sin^2\theta d\theta dr\rho \quad \text{(putting } z = 0 \text{ for the disc)}$$

$$= \rho\int_0^R r^3 dr \int_0^{2\pi} \sin^2\theta d\theta = \rho\frac{R^4}{4}\pi = \frac{MR^2}{4}.$$

Similarly it can be shown that $I_{yy} = \dfrac{MR^2}{4}$.

The result also follows from the theorem of perpendicular axes $I_{zz} = I_{xx} + I_{yy} = 2I_{xx}$ or $2I_{yy}$, I_{xx}, I_{yy} being the m.i. of the disc about a diameter. We can also show that $I_{xy} = I_{yz} = I_{zx} = 0$.

For the circular disc with O at the centre, clearly I_{xx}, I_{yy}, I_{zz} represent the principal moments of inertia, and the ellipsoid of inertia at O is written as

$$I_{xx}x^2 + I_{yy}y^2 + I_{zz}z^2 = 1 \quad \text{or} \quad \frac{MR^2}{4}(x^2+y^2) + \frac{MR^2}{2}z^2 = 0$$

which is the equation to an oblate spheroid, since $I_{xx} = I_{yy} < I_{zz}$.

IV. *A right circular cylinder of length l, radius R, mass per unit volume ρ about axes as shown in Fig. 4.13.4 :*

Assuming uniform mass distribution and using cylindrical co-ordinates $x = r\cos\theta$, $y = r\sin\theta$, $z = z$, we have

$$I_{zz} = \int (x^2+y^2) r d\theta dr dz \rho = \rho \int_0^R r^3 dr \int_0^{2\pi} d\theta \int_{-l/2}^{l/2} dz = \rho \frac{R^4}{4} 2\pi l = \frac{MR^2}{2},$$

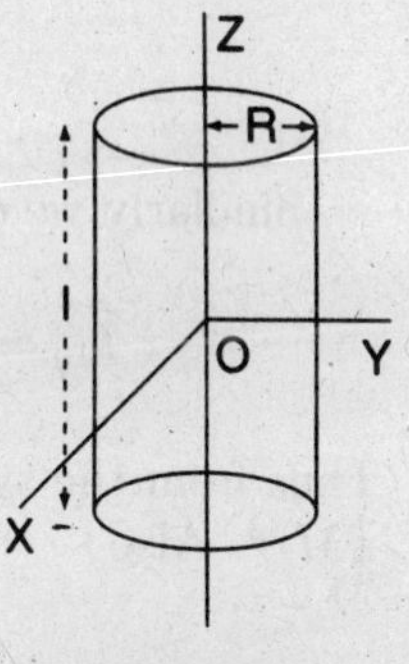

Fig. 4.13.4

$M = \pi R^2 l \rho$, being the mass of the cylinder. For a cylindrical tube of inner and outer radii R_1 and R_2,

$$I_{zz} = \rho\frac{(R_2^4 - R_1^4)}{4} 2\pi l = \pi\rho l(R_2^2 - R_1^2)\frac{(R_2^2+R_1^2)}{2} = M\frac{(R_2^2+R_1^2)}{2}.$$

Again $$I_{xx} = \int (y^2+z^2) r d\theta dr dz \rho = \int (r^2\sin^2\theta + z^2) r d\theta dr dz \rho$$

$$= \rho\left[\int_0^R r^3 dr \int_0^{2\pi} \sin^2\theta d\theta \int_{-l/2}^{l/2} dz + \int_0^R r dr \int_0^{2\pi} d\theta \int_{-l/2}^{l/2} z^2 dz\right]$$

$$= \rho\left[\frac{R^4}{4}\pi l + \frac{R^2}{2} 2\pi \frac{l^3}{12}\right] = M\left(\frac{R^2}{4} + \frac{l^2}{12}\right).$$

Similarly $I_{yy} = M\left(\dfrac{R^2}{4} + \dfrac{l^2}{12}\right)$—This also follows from symmetry.

For a tube of inner and outer radii R_1 and R_2,

$$I_{xx} = M\left[\frac{R_2^2+R_1^2}{4} + \frac{l^2}{12}\right] = I_{yy}.$$

[**N.B.** Putting $R = 0$, we may obtain the results of a thin rod, and also putting $l = 0$, we may obtain the results of a thin circular disc.]

Also

$$I_{xy} = I_{yx} = -\int xyr d\theta dr dz \rho = -\rho\int_0^R r^3 dr \int_0^{2\pi} \sin\theta\cos\theta d\theta \int_{-l/2}^{l/2} dz = 0.$$

Similarly $I_{yz} = I_{zy} = 0$; also $I_{zx} = I_{xz} = 0$.

Products of inertia being zero, for the right circular cylinder about axes at the centre O, clearly I_{xx}, I_{yy}, I_{zz} represent the principal moments of inertia and the ellipsoid of inertia at O can be calculated and shown to be a spheroid (see illustrative example 2 on page 265).

V. (a) ***A thin spherical shell of radius r, thickness dr, mass per unit volume ρ about axes as shown in Fig.*** **4.13.5**

Assuming uniform mass distribution, and using spherical polar co-ordinates,

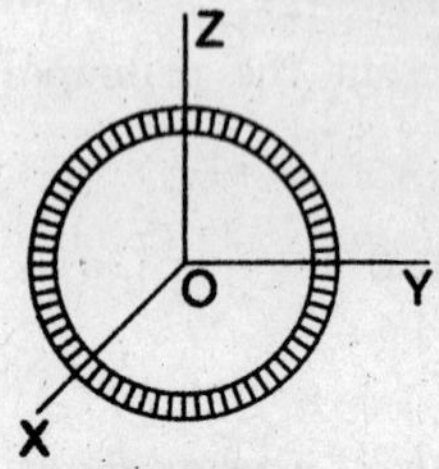

Fig. 4.13.5

$$x = r\sin\theta\cos\phi \qquad dV = r\sin\theta d\phi\cdot rd\theta\cdot dr$$
$$y = r\sin\theta\sin\phi \qquad = r^2\sin\theta d\theta d\phi dr$$
$$z = r\cos\theta$$

we have

$$I_{zz} = \int (x^2+y^2)\rho dV = \int \left[(r\sin\theta\cos\phi)^2 + (r\sin\theta\sin\phi)^2\right]\rho r^2\sin\theta d\theta d\phi dr$$
$$= \rho r^4 dr\int_0^{\pi}\sin^3\theta d\theta\int_0^{2\pi}d\phi = \rho r^4 dr\cdot\frac{4}{3}\cdot 2\pi = \frac{2}{3}Mr^2$$

$M = 4\pi r^2 dr\rho$, being the mass of the shell.

Similarly we can show that

$$I_{yy} = \int (z^2+x^2)\rho dV = \frac{2}{3}Mr^2 \quad\text{and}\quad I_{xx} = \int (y^2+z^2)\rho dV = \frac{2}{3}Mr^2.$$

Thus from the symmetry of the figure we have $I_{xx} = I_{yy} = I_{zz} =$ m.i. about any diameter = $\frac{2}{3}Mr^2$. Also

$$I_{xy} = -\int xy\rho dV = -\int r\sin\theta\cos\phi r\sin\theta\sin\phi\cdot\rho r^2\sin\theta d\theta d\phi dr$$
$$= -\rho r^4 dr\int_0^{\pi}\sin^3\theta d\theta\int_0^{2\pi}\sin\phi\cos\phi d\phi = 0 = I_{yx}.$$

Similarly $I_{yz} = I_{zy} = 0$ and $I_{zx} = I_{xz} = 0$—clearly for the shell at O, I_{xx}, I_{yy}, I_{zz} represent the principal moments of inertia, and the ellipsoid of inertia at O is

$$I_{xx}x^2 + I_{yy}y^2 + I_{zz}z^2 = 1 \text{ or } \frac{2}{3}Mr^2(x^2+y^2+z^2) = 1$$
$$\text{or } x^2+y^2+z^2 = a^2 \text{ where } a^2 = \frac{3}{2}Mr^2,$$

which represents a sphere.

(b) A uniform solid sphere of radius R, mass per unit volume ρ about axes as shown in Fig. 4.13.6.

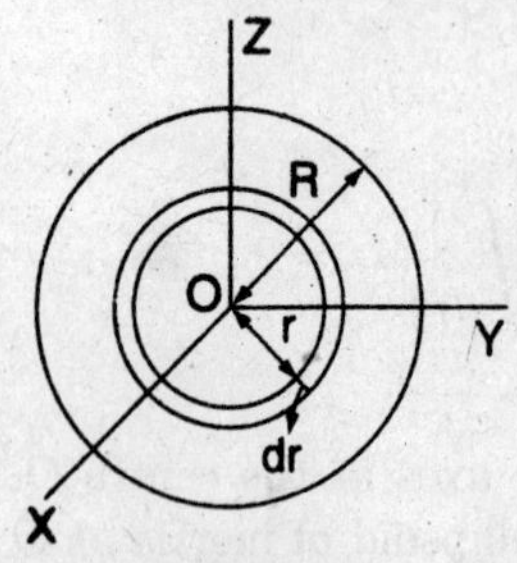

Fig. 4.13.6

Assuming the sphere to be divided into a large number of concentric shells of uniform mass distribution and considering one such elementary shell of radius r and width dr, we can proceed exactly as in (a), using spherical polar co-ordinates, and arrive at

$$I_{zz} = \rho\int_0^R r^4 dr\int_0^{\pi}\sin^3\theta d\theta\int_0^{2\pi}d\phi = \rho\frac{R^5}{5}\frac{4}{3}\cdot 2\pi = \frac{2}{5}MR^2,$$

$M = \frac{4}{3}\pi R^3\rho$ being the mass of the solid sphere.

From symmetry consideration $I_{xx} = I_{yy} = I_{zz}$ = m.i. about any diameter. We can otherwise also show that

$$I_{xx} = \int (y^2 + z^2)\rho dV = \frac{2}{5}MR^2 \quad \text{and} \quad I_{yy} = \int (z^2 + x^2)\rho dV = \frac{2}{5}MR^2.$$

Also as in (a) $I_{xy} = I_{yx} = 0$, $I_{yz} = I_{zy} = 0$ and $I_{zx} = I_{xz} = 0$. Clearly at O, the principal moments of inertia of the solid sphere are I_{xx}, I_{yy} and I_{zz} and the ellipsoid of inertia at O can also be calculated as in (a) and can be shown to be a sphere.

If however the above solid sphere rotates about a point O' on its surface as shown in Fig. 4.13.7 then the straight line joining O and O' being an axis of symmetry is a principal axis (X_3 say). Other two principal axes, being mutually perpendicular lines at O' lie in the tangent plane at O'. Therefore $I'_{xx} = I'_{yy} = \frac{2}{5}MR^2 + MR^2 = \frac{7}{5}MR^2$ from the theorem of parallel axes, and $I'_{zz} = \frac{2}{5}MR^2$. As $I'_{xx} = I'_{yy} > I'_{zz}$, the ellipsoid of inertia is a prolate spheroid.

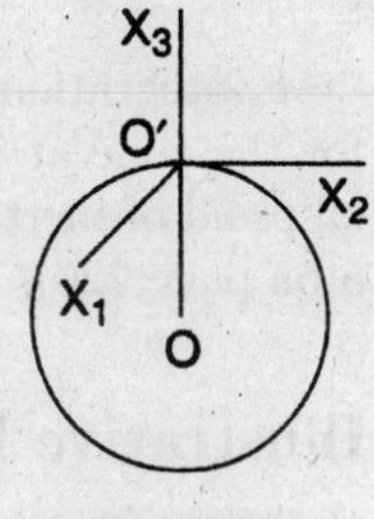

Fig. 4.13.7

VI. *A right circular solid cone of height h, radius of the base R, density of the material ρ, with origin at the vertex, about axes as shown in the Fig.* 4.13.8.

Assuming uniform mass distribution and using cylindrical co-ordinates, we have $x = r\cos\theta$, $y = r\sin\theta$, $z = z$, $dV = rd\theta drdz$. From the figure $\frac{r_0}{z} = \frac{R}{h}$ $\therefore\ r_0 = \frac{R}{h}z$.

$$\therefore\ I_{zz} = \int (x^2 + y^2)\rho dV = \int r^2 rd\theta drdz\rho = \rho \int_0^{r_0} r^3 dr \int_0^{2\pi} d\theta \int_0^h dz$$

$$= \rho\frac{r_0^4}{4}2\pi \int_0^h dz = \frac{\pi\rho}{2}\int_0^h \frac{R^4}{h^4}z^4 dz = \frac{\pi\rho}{2}\frac{R^4}{h^4}\frac{h^5}{5}$$

$$= \frac{3}{10}\cdot\frac{1}{3}\pi R^2 h\rho\cdot R^2 = \frac{3}{10}MR^2$$

where M = mass of the cone

$$= \int rd\theta drdz\rho = \rho\int_0^{r_0} rdr \int_0^{2\pi} d\theta \int_0^h dz = \pi\rho\frac{R^2}{h^2}\int_0^h z^2 dz = \frac{1}{3}\pi R^2 h\rho.$$

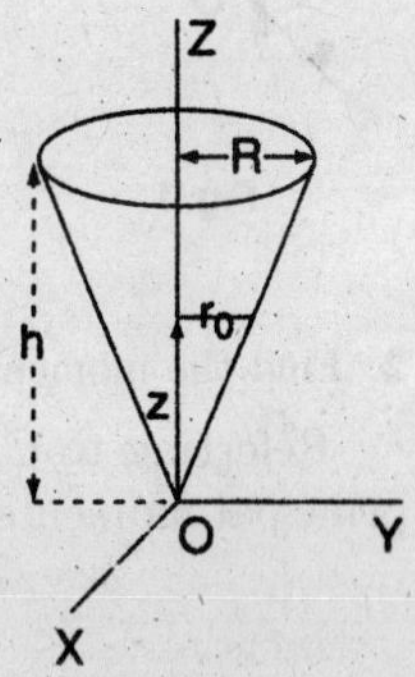

Fig. 4.13.8

Again

$$I_{xx} = \int (y^2 + z^2)\rho dV = \int (r^2\sin^2\theta + z^2) rd\theta drdz\rho$$

$$= \rho\left[\int_0^{r_0} r^3 dr \int_0^{2\pi}\sin^2\theta d\theta \int_0^h dz + \int_0^{r_0} rdr \int_0^{2\pi} d\theta \int_0^h z^2 dz\right]$$

$$= \rho\left[\frac{1}{4}\frac{R^4}{h^4}\int_0^h z^4 dz\cdot\pi + \frac{1}{2}\frac{R^2}{h^2}\int_0^h z^4 dz\cdot 2\pi\right] = \rho\left[\frac{\pi R^4}{4h^4}\frac{h^5}{5} + \frac{\pi R^2}{h^2}\frac{h^5}{5}\right]$$

$$= \frac{1}{3}\pi R^2 h\rho\cdot\frac{3}{20}R^2 + \frac{1}{3}\pi R^2 h\rho\frac{3}{5}h^2 = \frac{3}{5}M\left(\frac{R^2}{4} + h^2\right).$$

Similarly I_{yy} can be calculated and shown to be $\frac{3}{5}M\left(\frac{R^2}{4}+h^2\right)$ which also follows from symmetry.

Now

$$I_{xy} = -\int xy\rho dV = -\int r^2 \sin\theta\cos\theta r d\theta dr dz \rho$$

$$= -\rho\int_0^{r_0} r^3 dr \int_0^{2\pi} \sin\theta\cos\theta d\theta \int_0^h dz = 0 = I_{yx}$$

$\because$ θ integration vanishes. Similarly we can show $I_{yz} = I_{zy} = 0$ and $I_{zx} = I_{xz} = 0$. Clearly for the cone at its vertex O, I_{xx}, I_{yy}, I_{zz} represent the principal moments of inertia and the ellipsoid of inertia at O can be calculated and shown to be a spheroid (see illustrative example 5 on page 220).

Illustrative Examples

1. Find the moment of inertia of a thin uniform rod about an axis making an angle α with the rod and passing through its centre of mass. **[C.U. (Hons.) '70]**

For the axes as shown in the Fig. 1, x and z co-ordinates of every particle of the rod being zero, $I_{xy} = I_{yz} = I_{zx} = 0$. The axis OO' about which the moment of inertia of the thin rod is to be found, is chosen in the Y-Z plane, such that it makes an angle α with the positive direction of the Y-axis. The direction cosines of OO' are $l = 0$, $m = \cos\alpha$, $n = \sin\alpha$, and $I_{xx} = I_{zz} = \frac{Ml^2}{12}$ and $I_{yy} = 0$. Therefore m.i. of the thin rod about OO' is

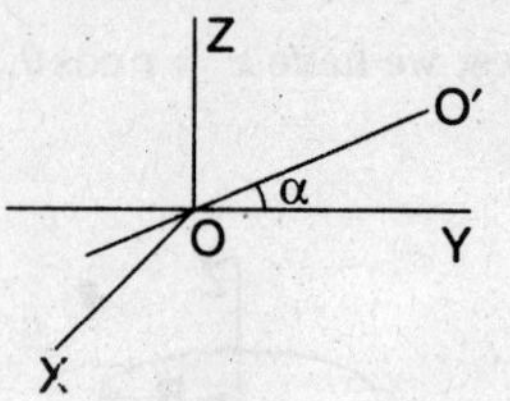

Fig. 1

$$I = l^2 I_{xx} + m^2 I_{yy} + n^2 I_{zz} = \frac{Ml^2}{12}\sin^2\alpha.$$

2. Find the moment of inertia of a thin rectangular plate about a diagonal. **[C.U. (Hons.) '90, 2002]**

Referring to Case I of Art. 4.13, we find that all the products of inertia are zero and the principal moments of inertia are

$$I_{xx} = \frac{Ma^2}{12}, \quad I_{yy} = \frac{Mb^2}{12}, \quad I_{zz} = \frac{M(a^2+b^2)}{12}.$$

The direction cosines of the diagonal BD are respectively given by $l = \frac{b}{\sqrt{a^2+b^2}}$, $m = \frac{a}{\sqrt{a^2+b^2}}$ and $n = 0$. So the m.i. of the rectangular plate about the diagonal BD is

$$I = l^2 I_{xx} + m^2 I_{yy} + n^2 I_{zz} = \frac{b^2}{a^2+b^2}\frac{Ma^2}{12} + \frac{a^2}{a^2+b^2}\frac{Mb^2}{12} = \frac{1}{6}\frac{Ma^2b^2}{(a^2+b^2)}.$$

N.B. For a thin square plate, $a = b$ and $I_{xx} = I_{yy} = \frac{Ma^2}{12}$. So the moment of inertia of a thin square plate about any axis passing through the centre of mass and arbitrarily oriented in the plane of the plate ($n = 0$) is given by

$$I = (l^2 + m^2)I_{xx} = I_{xx} = \frac{Ma^2}{12} \quad \because l^2 + m^2 = 1.$$

Thus the m.i. of a thin square plate about any axis in its plane and passing through the centre of mass is independent of the orientation of the axis.

3. Find the moment of inertia of a rectangular parallelopiped about a diagonal.

Referring to Case II of Art. 4.13, we find that all the products of inertia are zero, and the principal moments of inertia at O are

$$I_{xx} = \frac{M(b^2+c^2)}{12}, \quad I_{yy} = \frac{M}{12}(c^2+a^2), \quad I_{zz} = \frac{M}{12}(a^2+b^2).$$

The direction cosines of the diagonal CD are $l = \frac{a}{d}$, $m = \frac{b}{d}$ and $n = \frac{c}{d}$, where $d = \sqrt{a^2+b^2+c^2}$ is the length of the diagonal, so that the m.i. of the parallelopiped about the diagonal CD is

$$I = l^2 I_{xx} + m^2 I_{yy} + n^2 I_{zz}$$

$$= \frac{a^2}{(a^2+b^2+c^2)}\frac{M}{12}(b^2+c^2) + \frac{b^2}{(a^2+b^2+c^2)}\frac{M}{12}(c^2+a^2) + \frac{c^2}{(a^2+b^2+c^2)}\frac{M}{12}(a^2+b^2)$$

$$= \frac{M}{6}\frac{b^2c^2+c^2a^2+a^2b^2}{a^2+b^2+c^2}$$

N.B. For a cube $a = b = c$

$$\therefore\ I_{xx} = I_{yy} = I_{zz} = \frac{Ma^2}{6}.$$

Thus the m.i. of a cube about any axis passing through its centre of mass is $\frac{Ma^2}{6}$. This has already been shown in example 3 on page 219.

4.14 Routh's Rule

This rule is very helpful in finding the moment of inertia of symmetrical bodies. According to this rule, the moment of inertia of a body about any of the three mutually perpendicular axes of symmetry passing through the centre of mass of the body is given by

$$\frac{\text{mass of the body} \times \text{sum of the squares of the other two perpendicular semi axes}}{n}$$

where the denominator n is 3, 4 or 5 according as the body is a rectangle (or parallelopiped), a circle (or ellipse) or a sphere (or ellipsoid).

We shall deduce the following cases of moment of inertia by applying Routh's rule.

(i) *Moment of inertia of a uniform rectangular lamina about an axis passing through its centre and perpendicular to its plane :*

The moment of inertia of the rectangular lamina about the required axis is given by

$$I = \frac{M\left[\left(\frac{a}{2}\right)^2 + \left(\frac{b}{2}\right)^2\right]}{3} = \frac{M(a^2+b^2)}{12}$$

where M = mass of the lamina, and a, b are respectively the length and breadth of the lamina.

(ii) *Moment of inertia of a uniform circular disc about an axis passing through its centre and perpendicular to its plane :*

The moment of inertia of the circular disc about the axis under consideration is given by

$$I = \frac{M(r^2 + r^2)}{4} = \frac{Mr^2}{2}$$

where M = mass of the disc, and r its radius.

(iii) *Moment of inertia of a solid sphere about its diameter :*

The moment of inertia of the sphere about the required axis is given by

$$I = \frac{M(r^2 + r^2)}{5} = \frac{2}{5}Mr^2$$

where M = mass of the sphere and r is its radius.

4.15 Experimental Determination of Moment of Inertia

(i) *Torsional Pendulum :*

Torsional pendulum consists of a wire, one end of which is fixed at a torsion head T, and the other end carries a cradle C (Fig. 4.15.1). It is kept inside a box having glass walls and glass top. The cradle may be set to torsional oscillation by turning the torsion head. The time period of oscillation of the system can be determined accurately by means of lamp and scale arrangement. A spot of light is made to fall on a mirror M carried by the cradle and the reflected beam is received on a transluscent screen.

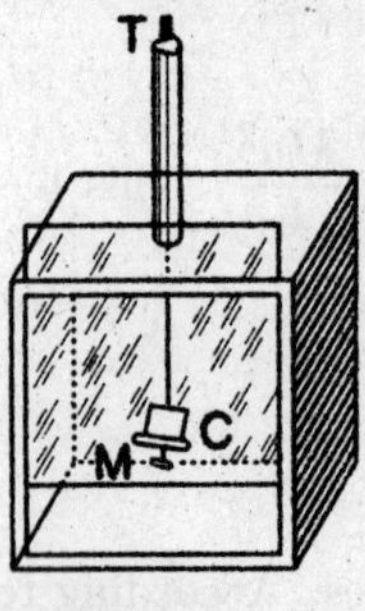

Fig. 4.15.1

If θ be a small angular displacement of the suspended system at any time, then the equation of motion of the suspended system may be written as

$$I_0 \frac{d^2\theta}{dt^2} = -\tau\theta$$

where τ is the restoring torque per unit angle of twist[‡] and I_0 is the moment of inertia of the suspended system about the wire

$$\therefore \quad \frac{d^2\theta}{dt^2} + \frac{\tau}{I_0}\theta = 0 \quad \text{or} \quad \frac{d^2\theta}{dt^2} + \omega^2\theta = 0 \quad \text{where } \omega^2 = \frac{\tau}{I_0}.$$

$\therefore$ The time period of oscillation[§] is given by

$$T_0 = \frac{2\pi}{\omega} = 2\pi\sqrt{\frac{I_0}{\tau}}. \qquad (1)$$

Now, the body of which the moment of inertia I_1 is to be determined (say, a rectangular bar) is placed on the cradle so that its centre of gravity lies on the prolongation of the suspension wire, and the time period of oscillation T_1 is found in the usual way.

[‡] See the Chapter on Elasticity.
[§] See the Chapter on SHM.

Next, another body of known moment of inertia I_2 is similarly placed on the cradle, and its time period T_2 is similarly determined.

Hence from the above two experiments,

$$T_1 = 2\pi\sqrt{\frac{I_0 + I_1}{\tau}} \tag{2}$$

$$\text{and} \quad T_2 = 2\pi\sqrt{\frac{I_0 + I_2}{\tau}}. \tag{3}$$

From equations (4.15.1), (4.15.2) and (4.15.3), we get

$$I_0 = \frac{\tau}{4\pi^2}T^2; \quad I_0 + I_1 = \frac{\tau}{4\pi^2}T_1^2; \quad I_0 + I_2 = \frac{\tau}{4\pi^2}T_2^2.$$

Hence, we get

$$I_1 = \frac{\tau}{4\pi^2}(T_1^2 - T^2) \tag{4}$$

$$\text{and} \quad I_2 = \frac{\tau}{4\pi^2}(T_2^2 - T^2) \tag{5}$$

$$\therefore \quad \frac{I_1}{I_2} = \frac{T_1^2 - T^2}{T_2^2 - T^2} \text{ or } I_1 = I_2\frac{T_1^2 - T^2}{T_2^2 - T^2}. \tag{6}$$

I_2 being known, I_1 can be calculated from equation (4.15.6).

(ii) *Inclined plane* :

We consider a body of mass m rolling down a plane (of inclination θ to the horizon) without slipping (Fig. 4.15.2). We suppose that the body starts from A and comes to the position A', traversing a distance x. The body acquires at the position A', a linear velocity v. The kinetic energy acquired by the body at A' is equal to the potential energy lost in falling from A to A'.

Now, K.E. at $A' = \frac{1}{2}mv^2 + \frac{1}{2}I\omega^2$, since the kinetic energy at A' is partly translational and partly rotational. Here I denotes the moment of inertia of the body about its axis of rotation and ω, the angular velocity.

$\because$ $I = mk^2, k$ being the radius of gyration of the body about the axis of rotation, and $\omega = \frac{v}{r}$, r being the radius of the body, we have, K.E. at A'

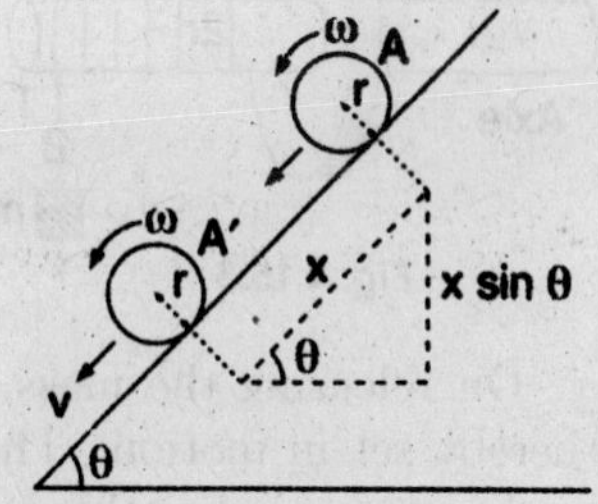

Fig. 4.15.2

$$= \frac{1}{2}mv^2 + \frac{1}{2}mk^2\frac{v^2}{r^2} = \frac{1}{2}mv^2\left(1 + \frac{k^2}{r^2}\right).$$

Again the loss of P.E. in falling from A to A'

$$= mgx\sin\theta \quad [\text{distance between the centres of } A \text{ and } A' = x]$$

$$\therefore \quad \frac{1}{2}mv^2\left(1 + \frac{k^2}{r^2}\right) = mgx\sin\theta \quad \text{or} \quad v^2 = \frac{2gx\sin\theta}{\left(1 + \frac{k^2}{r^2}\right)}.$$

Now, for a body starting from rest,

$$v^2 = 2fx, \quad \text{where } f \text{ is the linear acceleration at } A'.$$

$$\therefore\ 2fx = \frac{2gx\sin\theta}{1+\dfrac{k^2}{r^2}} \quad \text{or} \quad f = \frac{g\sin\theta}{1+\dfrac{k^2}{r^2}}. \tag{7}$$

From the above expression, it is evident that the linear acceleration is independent of the mass of the body. The following cases are of interest.

(a) *Solid sphere* $\quad k^2 = \frac{2}{5}r^2 \quad \therefore\ f = \frac{5}{7}g\sin\theta.$

(b) *Hollow sphere* $\quad k^2 = \frac{2}{3}r^2 \quad \therefore\ f = \frac{3}{5}g\sin\theta.$

(c) *Disc or Cylinder* $\quad k^2 = \frac{1}{2}r^2 \quad \therefore\ f = \frac{2}{3}g\sin\theta.$

(d) *Ring or Hollow cylinder* $\quad k^2 = r^2 \quad \therefore\ f = \frac{1}{2}g\sin\theta.$

It follows therefore that a solid sphere will roll down an inclined plane faster than a hollow sphere, disc, cylinder, etc. Measuring the value of x for time t, f can be determined from the relation $x = \frac{1}{2}ft^2$; f, being known, k^2 is determined from (4.15.7), whence the moment of inertia is calculated.

(iii) *Flywheel (first method)*

The flywheel, the moment of inertia of which is to be determined, is a heavy wheel capable of rotation about a horizontal axle, which is mounted on ball-bearings in order to minimise friction (Fig. 4.15.3). The arrangement is such that the axle is at a convenient height from the ground. A small peg is fixed at a suitable position on the axle and a small loop at the end of a cord is slipped over the peg—the other end of the cord carrying a suitable mass. The cord is wound over the axle.

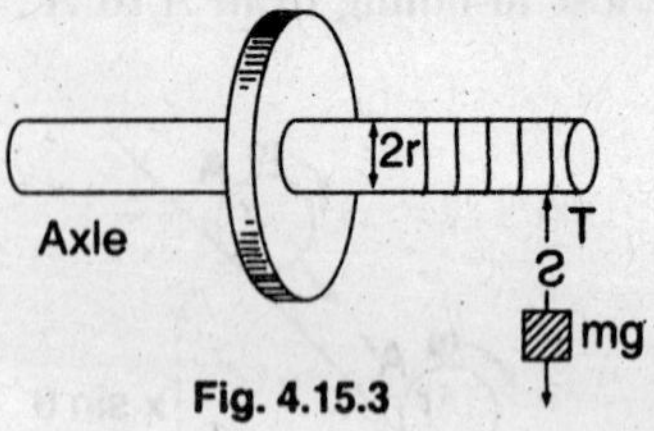

Fig. 4.15.3

On releasing the mass, it falls down under gravity unwinding the cord, and the wheel is thereby set in motion. The length of the cord is such that it is completely unwound as soon as the mass reaches the ground. Just before reaching the ground, the loss of potential energy of the mass is equal to mgh, where m is the mass of the body falling through a height h. The gain of kinetic energy at this position of the mass is $\frac{1}{2}mv^2$, where v is the linear velocity of the mass at the moment. Again, the gain of kinetic energy of the wheel is $\frac{1}{2}I\omega^2$, where I is the moment of inertia of the wheel about the axis of rotation, and ω is its angular velocity.

Therefore, from the principle of conservation of energy, we have

$$\tfrac{1}{2}mv^2 + \tfrac{1}{2}I\omega^2 = mgh$$

$$\text{or} \quad \tfrac{1}{2}mv^2 + \tfrac{1}{2}I\frac{v^2}{r^2} = mgh \quad [\because\ v = \omega r,\ r \text{ being the radius of the axle.}]$$

$$\therefore\ I = m\frac{(2gh - v^2)}{v^2}r^2$$

$$= m\frac{(2gh - 2fh)}{2fh}r^2 \quad [\because v^2 = 2fh, f \text{ being the linear acceleration of the body.}]$$

$$= \frac{m(g-f)}{f}r^2$$

$$= \frac{m(gt^2 - 2h)}{2h}r^2 \quad [\therefore\ h = \frac{1}{2}ft^2 \text{ or } f = \frac{2h}{t^2}, t \text{ being the time of fall.}]$$

Thus measuring the time of fall of the mass, I is determined.

To allow for the friction, if any, at the ball bearings, a suitbale mass m' is to be determined which will set the wheel just turning (i.e. the wheel will have rotation without acceleration). Hence in the above expression for I, m should be replaced by $(m - m')$ to allow for friction.

Second method

Considering the effect of friction in the previous experiment, the energy equation may be written as

$$\frac{1}{2}mv^2 + \frac{1}{2}I\omega^2 + \text{ work done against friction } = mgh.$$

In this method the mass m is detached as soon as it reaches the ground, and the subsequent turns n made by the wheel before coming to rest, is determined. Let C be the frictional couple; then the work done against friction for n turns $= 2\pi nC$.

This work done against friction is provided by the kinetic energy of the wheel

$$\therefore\ \frac{1}{2}I\omega^2 = 2\pi nC \qquad \therefore\ C = \frac{I\omega^2}{4\pi n}.$$

Again the total number of turns made by the wheel as the mass m falls through $h = \dfrac{h}{2\pi r}$.

Hence the angle turned through during the fall of the mass $= 2\pi\cdot\dfrac{h}{2\pi r} = \dfrac{h}{r}$

$\therefore$ work done against friction $= \dfrac{I\omega^2}{4\pi n}\cdot\dfrac{h}{r}$.

Hence the energy equation may be written as

$$\frac{1}{2}mv^2 + \frac{1}{2}I\omega^2 + \frac{I\omega^2 h}{4\pi nr} = mgh, \quad \text{or,} \quad \frac{1}{2}mv^2 + \frac{1}{2}I\frac{v^2}{r^2}\left(1 + \frac{h}{2\pi nr}\right) = mgh$$

$$I = \frac{2\left(mgh - \frac{1}{2}mv^2\right)r^2}{v^2\left(1 + \dfrac{h}{2\pi nr}\right)} = \frac{m(2gh - v^2)r^2}{v^2\left(1 + \dfrac{h}{2\pi nr}\right)} = \frac{m(g-f)r^2}{f\left(1 + \dfrac{h}{2\pi nr}\right)} = \frac{m(gt^2 - 2h)r^2}{2h\left(1 + \dfrac{h}{2\pi nr}\right)}$$

Everything on the right-hand side being known, I may be determined.

Illustrative Examples

1. A solid sphere rolls without slipping down a plane of inclination of 30° to the horizontal. Calculate its velocity after it has traversed a distance 5 ft. along the plane.

We know that $v^2 = \dfrac{2gx\sin\theta}{1+\dfrac{k^2}{r^2}}$.

In this case, $\theta = 30°$, $x = 5$ ft, $\dfrac{k^2}{r^2} = \dfrac{2}{5}$ and $g = 32$ ft/sec^2

$$\therefore\ v^2 = \frac{2 \times 32 \times 5 \times \sin 30°}{1+\frac{2}{5}} = \frac{320 \times \frac{1}{2}}{\frac{7}{5}} = 114.28, \quad \text{or, } v = 10.69 \text{ ft/sec.}$$

2. A body of mass m hangs from one end of a light rope which is wound on the surface of a horizontal cylinder of radius a, which is free to rotate about its axis. Prove that the body descends with an acceleration

$$f = g\Big/\left(1+\frac{I}{ma^2}\right)$$

where I is the moment of inertia of the cylinder about its axis. **[C.U. (Hons.) 1980]**

From Art. 4.15(iii), we have,

$$\frac{1}{2}mv^2 + \frac{1}{2}I\frac{v^2}{a^2} = mgh, \quad \text{or} \quad \frac{1}{2}I\frac{v^2}{a^2} = \frac{1}{2}m(2gh - v^2)$$

$$\text{or} \quad I = \frac{m(2gh - v^2)a^2}{v^2} = \frac{m(2gh - 2fh)a^2}{2fh} = \frac{m(g - f)a^2}{f}$$

[$\because v^2 = 2fh$, f being the linear acceleration of the body]

$$\text{or} \quad \frac{I.f}{ma^2} = g - f, \quad \text{or} \quad f + \frac{I.f}{ma^2} = g.$$

$$\therefore\ f = g\Big/\left(1+\frac{I}{ma^2}\right).$$

3. A cord is wound round the horizontal axle (whose radius is 1.2 cms) of a fly wheel, and a weight of 2 kgs is attached to the free-end of the cord. Starting from rest, the weight is released from the axle after falling through 100 cms. After the weight is released, the wheel is observed to make 14 turns in 5.8 sec. Neglecting friction, calculate (a) the kinetic energy of the weight at the moment of release, (b) moment of inertia of the fly wheel. **[C.H. Sc.]**

(a) Velocity of the weight at the moment of release $= \dfrac{2\pi rn}{t} = \dfrac{2\pi(1.2)14}{5.8}$ cms/sec.

Hence the kinetic energy $= \dfrac{1}{2}m(\text{velocity})^2 = \dfrac{1}{2} \times 2000\left[\dfrac{2\pi(1.2)14}{5.8}\right]^2 = 330902$ ergs.

(b) Neglecting friction, we have $\dfrac{1}{2}I\omega^2 + \dfrac{1}{2}mv^2 = mgh.$

Now,

$$mgh = 2000 \times 981 \times 100 = 1962 \times 10^5 \text{ ergs}$$

$$\frac{1}{2}mv^2 = 330902 \text{ ergs.} \quad \therefore\ \frac{1}{2}I\omega^2 = 1962 \times 10^5 - 3.309 \times 10^5 = 1958.691 \times 10^5$$

Now,

$$\omega = \frac{2\pi n}{t} = \frac{2\pi \times 14}{5.8}. \quad \therefore I = \frac{2 \times 1958.691 \times 10^5}{\left(\frac{2\pi \times 14}{5.8}\right)^2} = 17.04 \times 10^2 \text{ gm-cm.}^2$$

4. A catherine wheel when burning has its initial moment of inertia I reduced at a uniform rate K. If the couple G acting on it is constant, find the connection between the time t and (a) its angular velocity ω, (b) its angular acceleration $\frac{d\omega}{dt}$.

(a) From the problem we may write $\frac{dI_t}{dt} = -K$, where I_t is the moment of inertia of the catherine wheel at any time t

$$\therefore I_t = -Kt + A.$$

At time $t = 0$, $I_t = I$ (initial moment of inertia). Hence we have

$$I = A \quad \therefore I_t = -Kt + I.$$

Now

$$G = \frac{d}{dt}(I_t\omega) = \frac{d}{dt}[\omega(I - Kt)] \text{ or } \omega(I - Kt) = \int Gdt + B = Gt + B.$$

At $t = 0$, $\omega = 0$. $\therefore B = 0$.

So

$$Gt = \omega(I - Kt) \tag{1}$$

(b) From (1) we have $\omega = \dfrac{Gt}{(I - Kt)}$

$$\text{or } \frac{d\omega}{dt} = G\left[\frac{1}{I - Kt} + \frac{Kt}{(I - Kt)^2}\right] \quad [\because G \text{ is constant}]$$

$$= \frac{GI}{(I - Kt)^2} \text{ or } GI = (I - Kt)^2\frac{d\omega}{dt}$$

Exercises

1. (a) Explain what is meant by the moment of inertia and the radius of gyration of a body about an axis.

   Calculate the moment of inertia of a long thin rod about an axis through its midpoint and perpendicular to its length.

   (b) Find the moment of inertia of a thin uniform rod of mass m and length l about an axis passing through a point at a distance $\frac{l}{4}$ from one end of the rod and perpendicular to its length. **[Ans. $\frac{7}{48}ml^2$]**

   (c) Find the moment of inertia of a thin uniform rod about an axis passing through its centre and making an angle α with the rod. **[C.U. (Hons.) 1970] [Ans. $\frac{Ml^2}{12}\sin^2\alpha$]**

2. State and prove the theorem of parallel and perpendicular axes on moment of inertia.

   Calculate the moment of inertia of a thin circular disc about an axis tangential to the disc and perpendicular to its plane.

3. Calculate the moments of inertia in the following cases :

   (a) A circular lamina about an axis through its centre and perpendicular to its plane.

   (b) An elliptic disc about any of its principal axes.

   (c) A solid sphere about its diameter.

   (d) A triangular lamina about one of its sides.

4. Calculate the moment of inertia of a right circular solid cylinder about (a) its geometrical axis; (b) a straight line through its centre of gravity and perpendicular to its axis of symmetry.

   [C.U. (Hons.) 1989, 2007]

5. Calculate the moment of inertia of a sphere about its diameter.

   A sphere of mass m rolls without slipping with velocity v. Calculate the kinetic energy of the sphere.

6. Obtain an expression for the moment of inertia of a right circular solid cone about its generating line.

7. Find an expression for the moment of inertia of a rigid body about any axis.

   What do you understand by the ellipsoid of inertia at a point of the rigid body and its principal moments of inertia? What are the properties of the principal moments of inertia?

   [C.U. (Hons.) 1993]

8. Calculate the kinetic energy and the angular momentum of a rigid body spinning steadily about a fixed axis with angular velocity $\vec{\omega}$. Are the directions of $\vec{\omega}$ and $\vec{L}$ necessarily the same? Explain.

   [C.U. (Hons.) 1986, '88, '92]

9. Prove that the kinetic energy of rotation of a rigid body can be expressed in the form $T = \frac{1}{2}I_{ij}\omega_i\omega_j$ with the convention that repeated indices are summed over x, y, z and show that relative to any point in the rigid body it can be simplified to the form

$$T = \tfrac{1}{2}\left(I_1\omega_1^2 + I_2\omega_2^2 + I_3\omega_3^2\right).$$

   Interpret this result. Calculate the angular momentum $\vec{L}$ and show that $\frac{\partial T}{\partial \vec{\omega}} = \vec{L}$.

   [C.U. (Hons.) 1981, '84]

   (For answer to the last part see worked out examples)

10. Two equal homogeneous solid spheres each of radius r and mass M are attached by a rigid weightless rod so that the distance between their centres is $2a$. Calculate the moment of inertia of the system about (i) the connecting rod, and (ii) an axis perpendicular to the rod passing through its centre. **[C.U. (Hons.) 1980 (Old syll.)]** [**Ans.** (i) $\frac{4}{5}mr^2$ (ii) $\frac{4}{5}mr^2 + 2Ma^2$]

11. Consider two masses m_1 and m_2 separated by a distance r_0 from each other by a massless rigid rod. Show that the moment of inertia of the two masses about an axis through their centre of mass in a plane perpendicular to the rigid rod is given by

$$\frac{m_1 m_2}{m_1 + m_2} r_0^2$$

   [C.U. (Hons.) 1984]

   (See worked out examples)

12. Find the moment of inertia of a homogenous thin rectangular plate having sides $2a$ and a about a diagonal. **[C.U. (Hons.) 1990]** [**Ans.** $\frac{4}{15}Ma^2$]

   [**Hints.** Apply the formula for moment of inertia of a triangle about one side.]

13. Find the moment of inertia of a uniform vertical rod of mass m and length L about a horizontal axis passing through a point $\frac{1}{3}L$ from one end. **[C.U. (Hons.) 1994]** [**Ans.** $\frac{1}{9}mL^2$]

[**Hints.** Apply the theorem of parallel axes.]

14. Find the moment of inertia of a rectangular plate with sides a and b about a side.

[C.U. (Hons.) 1995] [**Ans.** $\frac{1}{3}Ma^2$ or $\frac{1}{3}Mb^2$]

[**Hints.** Apply the theorem of parallel axes or try otherwise.]

15. (a) Find the cartesian components of the angular momentum $\vec{L}$ of a rigid body rotating with angular velocity $\vec{\omega}$ about a fixed point. If the cartesian axes are oriented along the principal axes of inertia, how will the angular momentum be expressed? Show that the K.E of the body can then be expressed as $T = \frac{1}{2}\vec{\omega}\cdot\vec{L}$.

(b) Find the moment of inertia of a rigid body about an axis making angles α, β, γ with the principal axes. **[C.U. (Hons.) 1982, '93, 2006]**

16. Find the principal axes and the principal moments of inertia of a cylindrical tube of length L, mass M and inner and outer radii R_1 and R_2. **[C.U. (Hons.) 1983, cf. '84]**

(See Case IV of Art 4.13)

17. Show that the ellipsoid of inertia is a quadric surface with points at a distance from the origin corelated with radius of gyration about an axis in that direction. What are principal axes?

[C.U. (Hons.) 1985]

18. The three principal moments of inertia of a rigid body at a point are given as 20, 30 and 60 $\text{gm} - \text{cm}^2$. Calculate its K.E, when it is rotating at the rate of 100 revolutions per minute about an axis passing through the point, and having direction co-sines $\frac{1}{\sqrt{2}}, \frac{1}{\sqrt{3}}$ and $\frac{1}{\sqrt{6}}$. Derive the necessary expression for the K.E of rotation. **[C.U. (Hons.) 1987]**

[**Ans.** See worked out examples.]

19. Calculate the moment of inertia of a hollow sphere about a diameter and hence find the moment of inertia of a solid sphere also about a diameter.

A spherical cavity of radius $\frac{R}{2}$ lies inside a solid sphere of radius $2R$ and density ρ, such that their centres are separated by a distance $\frac{R}{2}$. Find the moment of inertia of this partly hollow sphere about an axis passing through its centre and being at right angles to the common diameter of the sphere and its cavity. **[C.U. (Hons.) 1988]**

20. Find a general expression for the moment of inertia of a rigid body about any axis passing through the origin in terms of moments and products of inertia of the body w.r.t rectangular cartesian co-ordinate axes. **[C.U. (Hons.) 1989]**

21. (a) If the moments of inertia of a rigid body about three perpendicular axes Ox, Oy, Oz and the products of inertia of the body about $x - y$, $y - z$, $z - x$ planes are known, find an expression for the ellipsoid of inertia of the body about O. What do you mean by principal axes and principal moments of inertia? Prove that the magnitude of the m.i. of a rigid body about an axis through O, cannot be greater than the greatest principal m.i. of the body through O, and cannot be lesser than the least principal m.i. of the body through O.

(b) Show that the m.i. of a solid right circular cone of uniform density, having mass M, vertical height H and radius of the base R, about its axis is $\frac{3}{10}MR^2$, and about an axis passing through its vertex and parallel to base is $\frac{3}{20}M(R^2 + 4H^2)$. What will be the relation between R and H, so that the ellipsoid of inertia at the vertex will be a sphere?

[C.U. (Hons.) 1990]

22. (a) Consider a three-dimensional uniform circular plate with the Z-axis perpendicular to its plane and the origin passing through the centre of the plate. Starting from the definition of the moments and products of inertia, show that the inertia tensor has the following form

$$\begin{bmatrix} I_{xx} & I_{xy} & 0 \\ I_{xy} & I_{yy} & 0 \\ 0 & 0 & I_{xx} + I_{yy} \end{bmatrix}$$

(b) Write down the K.E. expression in the principal axes system in terms of (i) angular momenta and (ii) moments of inertia.

A rigid body, with the principal moments of inertia at a point $I_x = 20$, $I_y = 30$ and $I_z = 40$ a.m.u A^2, is rotating at the rate of 50 revolutions per second about an axis passing through the point, and having direction cosines $\frac{1}{2}$, $\frac{1}{2}$ and $\frac{1}{\sqrt{2}}$. Calculate the K.E. of the body (1 a.m.u. $= 1.66 \times 10^{-24}$ gm, $1A = 10^{-8}$ cm). (See worked out examples)

(c) What is the relation between the principal moments of inertia components in the following cases? (i) a spherical top, (ii) a symmetrical top and (iii) an asymmetrical top.

Justify that a pyramidal NH_3 molecule, with the hydrogen atoms forming the base of the pyramid and the nitrogen at its peak, is a symmetric top. **[C.U. (Hons.) 1991]**

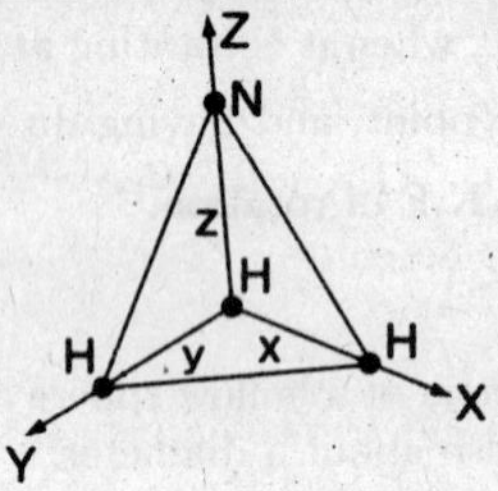

Answer to last part: Here $I_1 = m(y^2 + z^2)$, $I_2 = m(z^2 + x^2)$, $I_3 = m(x^2 + y^2)$, m being the mass of the NH_3 molecule; since the intermolecular distance between two H atoms is the same, $x = y$, so that $I_1 = m(x^2 + z^2)$, $I_2 = m(x^2 + z^2)$, $I_3 = m(x^2 + x^2) = 2mx^2$. $\therefore$ $I_1 = I_2 \neq I_3 \rightarrow$ hence NH_3 molecule is a symmetrical top.

23. (a) i. The products of inertia in the principal axes are zero—how is it realised in a symmetrical body?

ii. The magnitude of the m.i. of a body about an axis passing thorugh a point O cannot greater than the largest principal m.i., and cannot be less than the smallest principal m.i. of the body through O—Explain.

(b) A light flexible string is wrapped several times around a solid cylinder of mass 10 kg and diameter 0.06 m which can rotate without friction about a horizontal axis. The free-end of the string is pulled with a constant force of magnitude 5 Newton for a distance of 1 m. The cylinder is initially at rest.

i. What is the final K.E. of the cylinder?

ii. What is the final angular velocity?

iii. Calculate the torque and angular acceleration.

iv. What is the final velocity of the string? **[C.U. (Hons.) 1992]**
(See worked out examples)

24. A rigid body is rotating with an angular velocity ω about an axis lying in the Y-Z plane and making an angle of 45° with the Y-axis. Obtain the K.E. of the body in terms of the moments and products of inertia. Also obtain the components of angular momentum of the rigid body.

[C.U. (Hons.) 1993]

[**Hints.** Here

$\lambda = \cos 90^\circ = 0$	$\omega_x = \lambda\omega$
$\mu = \cos 45^\circ = \dfrac{1}{\sqrt{2}}$	$\omega_y = \mu\omega$
$\nu = \cos 45^\circ = \dfrac{1}{\sqrt{2}}$	$\omega_z = \nu\omega$

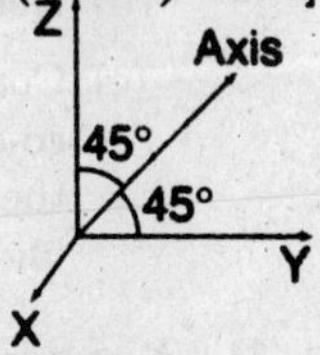

25. (a) A rigid body is having a rotational motion with one point of the body fixed. Obtain the relation between angular momentum $\vec{L}$ and angular velocity $\vec{\omega}$. Under what circumstances is $\vec{L}$ parallel to $\vec{\omega}$?
Write down the expression of K.E. in terms of the components of ω and moments and products of inertia.

(b) Suppose that in some units, the m.i. components are given as $I_{xx} = 2.0$, $I_{yy} = 2.0$, $I_{zz} = 5.0$, $I_{xy} = I_{yx} = 3.0$; other components are zero. Suppose that the rigid body is rotating with angular velocity $\vec{\omega}$ about an axis along $\hat{n} = \frac{1}{\sqrt{2}}\hat{\imath} + \frac{1}{\sqrt{2}}\hat{\jmath}$. Find the expression for the m.i. about the axis of rotation, and hence the K.E. Find also the direction and mgnitude of $\vec{L}$. **[C.U. (Hons.) 1995]**

(See worked out examples and also Q.39 of the Exercises)

26. A rectangular thin plate of dimensions a and $2a$ spins about an axis along the diagonal at a constant angular velocity ω.. Given $I_x = \frac{1}{12}ma^2$, $I_y = \frac{1}{3}ma^2$, (i) What is the direction and magnitude of the angular momentum? (ii) Calculate the torque on the axis of rotation.

[C.U. (Hons.) 1996 (Old Syll)]

27. (a) Find the relation between the components of the angular momentum and those of angular velocity for Eulerian motion of a rigid body. What are moments and products of inertia?

(b) What are principal moments of inertia? Show how you will find these quantities from a knowledge of the inertia tensor in an arbitrary frame of reference. **[C.U. (Hons.) 2006]**

(c) Find the moments and products of inertia of a uniform square plate of side 'a' about x, y, z axes, x and y being taken as the adjacent sides of the plate, and z perpendicular to its plane.

(d) Find the principal moments of inertia in this case. **[C.U. (Hons.) 1996, '99, 2007, 2008]**

[**Ans.** (a) See Art. 8.3; (b), (c), (d) ⇒ See worked out examples.]

28. (a) Motion of a rigid body is described in a co-ordinate system $Oxyz$. Show that the m.i. of the rigid about any axis through O is given by

$$I_{lmn} = I_{xx}l^2 + I_{yy}m^2 + I_{zz}n^2 + 2I_{xy}lm + 2I_{yz}mn + 2I_{zx}nl,$$

where (l, m, n) gives the direction co-sines of the axis; I_{xx}, I_{yy}, I_{zz} are the m.i. about OX, OY, OZ respectively, I_{xy}, I_{yz}, I_{zx} are the products of inertia about the pairs of planes.

(b) For a homogeneous sphere, show that (i) any three mutually perpendicular diameters form a set of principal axes, and (ii) the ellipsoid of inertia about the centre O is a sphere.

[C.U. (Hons.) 1997, '99]

29. (a) State and prove the parallel axes theorem of moment of inertia.

(b) Find the m.i. of a solid right circular cylinder of radius r and mass m about its axis. Hence find the distance travelled by its axis after it has rolled from rest without slipping for time t on a plane inclined at an angle θ with the horizontal. **[C.U. (Hons.) 1998]**

30. (a) Define ellipsoid of inertia and deduce on expression for it.

(b) Define principal axes and principal moments of inertia.

(c) Consider a right circular cone of mass M, height H and radius of the base R. Considering a co-ordinate system $Oxyz$ with origin at the vertex, and Z-axis on the axis of the cone, find the principal moments of inertia, principal axes, and m.i. of the cone about its generating line. [**Ans.** See worked out examples.] **[C.U. (Hons.) 1975, 1998 (old syll)]**

31. A sphere of radius a and mass m is rolling down a plane (inclined at an angle θ with the horizontal) without slipping. Find out a relation between the linear velocity and the distance travelled. **[C.U. (Hons.) 1999]**

32. (a) For rotational motion of rigid bodies, derive an expression for kinetic energy in terms of moment of inertia and angular velocity.

(b) Find the moment of inertia of a solid uniform right circular cone about an axis which is perpendicular to the axis of the cone and passes through the vertex.

(c) Consider a plate in the form of an isosceles right angled triangle. The length of the hopotenuse is 2 cm and surface density at a point is $\rho = ah$ where $a = 2$ gm/cm^3 and h is the distance of the point from the hypotenuse. Locate the centre of mass of the plate. **[C.U. (Hons.) 2000]**

[**Ans.** (c) See Exercises on Mechanics Chapter and also Appendix B–2.]

33. A thin rod of length l and mass M is suspended freely at its end. It is pulled aside and swung about a horizontal axis passing through its lowest position with angular speed ω. How high does its centre of mass rise above its lowest position? Neglect friction. [**Ans.** $h = (l\omega)^2/6$g.]

34. A metre stick is held vertically with one end on the floor and is then allowed to fall. Find the velocity of the other end when it hits the floor, assuming that the end on the floor does not slip. [**Ans.** 5.4 metres/sec.]

35. A cord is wrapped around the rim of a fly-wheel 0.5 m in radius and a steady downward pull of 50*nt* is exerted on the cord. The wheel is mounted in frictionless bearings on a horizontal shaft through its centre. The moment of inertia of the wheel is 4 kg-m^2.

(a) Compute the angular acceleration of the wheel. (b) Find the work done in unwinding 5 metre of the cord. (c) If a mass having a weight of 50*nt* hangs from the cord, compute the angular acceleration of the wheel. Why is this not the same as in part (a)?

[**Ans.** (a) 6.25 rad/sec^2, (b) 250 Joules, (c) 4.74 rad/sec^2; it is smaller than that in part (a), because the pull on the wheel is now smaller.]

36. (a) Find out the relation between the angular momentum and angular velocity of a rotating body and hence define the moment of inertia tensor. When are the angular velocity and angular momentum parallel?

(b) Find out the ratio of the moments of inertia of a thick spherical shell and a solid sphere of same mass and same outer radius, when the inner radius of the spherical shell is half of that of the outer.

Deduce the formula you use. (See worked out examples) [**Ans.** 31 : 28.]

[C.U. (Hons.) 2001]

37. (a) Set up the principal axes and find principal moments of inertia for a thin homogeneous square plate of mass m and sides of length a.

(b) For the above plate, calculate moment of inertia about a diagonal.

[C.U. (Hons.) 1990, 2002, 2007]

38. (a) Define moment of inertia and product of inertia for a rigid body.

(b) A rigid body rotates about an axis (with direction cosines l, m, n) with angular velocity $\omega(\omega_x, \omega_y, \omega_z)$. Show that the K.E. of rotation of the rigid body is given by $T_{\text{rot}} = \frac{1}{2}I_{l,m,n}\omega^2$, where $I_{l,m,n,}$ is the moment of inertia of the rigid body about the axis of rotation.

[C.U. (Hons.) 2002, 2007]

(c) A cylinder has a mass distribution that varies with distance from its axis. It rolls without slipping on a smooth horizontal plane. It is found that its translational K.E. = rotational K.E. Calculate moment of inertia of the cylinder. Show that the mass distribution is that of a thin spherical shell. **[C.U. (Hons.) 2002]**

[**Hints.** $M \propto a$ or $M = Ca$; also $\frac{1}{2}I\omega^2 = \frac{1}{2}Mv^2$ or $I\frac{v^2}{a^2} = Mv^2 = Cav^2$.

$\therefore I = Ca^3$; Now $\rho = \frac{M}{V}$ or $M = \pi a^2 l\rho = Ca$. $\therefore C = \pi\rho la$.

$\therefore I = \pi\rho la^4 = 4\pi a^2 l\rho\frac{a^2}{4}$ = mass of a thin spherical shell $\times \frac{a^2}{4}$.]

39. (a) Relative to a certain rectangular coordinate system (unit vectors $\hat{\imath}, \hat{\jmath}$ and $\hat{k}$) the moments and products of inertia of a rigid body are given by $I_{xx} = 3B$, $I_{yy} = 2B$, $I_{zz} = D$, $I_{yz} = -B = I_{zy}$ and the other components are zero.

What is the K.E. of rotation if the body rotates about the $\hat{\jmath}$ axis with constant angular velocity $\vec{\omega}$? Find also magnitude and direction of the angular momentum.

[**Ans.** (b) See Art. 4.13] **[C.U. (Hons.) 2003]**

[**Hints.** Here $\omega_x = 0, \omega_y = \omega$ and $\omega_z = 0$—put these values in the expression for K.E. Again putting these values along with the values of moments and products of inertia, find the magnitude of the angular momentum—angular momentum does not lie along the axis of rotation unless the axis is a principal axis.]

(b) Considering symmetry of (i) a rectangular plate, (ii) a solid cylinder and (iii) a solid sphere, find a set of principal axes and show them by sketches. Are your choices unique?

(c) State and explain the parallel and perpendicular axes theorems of moments of inertia.

[C.U. (Hons.) 2003]

(For answer to (b) see Art. 4.13)

40. Find the relation between the height and radius of a right circular cone when the moment of inertia of the cone becomes a minimum while rotating about an axis passing through its vertex and perpendicular to its axis.

[**Ans.** Here $I = \frac{3}{5}M\left(\frac{r^2}{4} + h^2\right)$; mass of the cone $= \frac{1}{3}\pi r^2 h\rho$ i.e. $r^2 h$ = constant = k (say).

$$\therefore \quad I = \frac{3}{5}\frac{M}{h}\left(\frac{r^2 h}{4} + h^3\right) = \frac{3}{5}M\left(\frac{k}{4h} + h^2\right)$$

$$\therefore \quad \frac{dI}{dh} = \frac{3}{5}M\left(k \cdot -\frac{1}{4h^2} + 2h\right) = 0$$

(from the condition of the problem) or $k = 8h^3$ or $r^2 h = 8h^3$ or $r^2 = 8h^2$.

$$\therefore \quad r = 2\sqrt{2}h \quad \text{or} \quad \frac{h}{r} = \frac{1}{2\sqrt{2}}]$$

41. (a) State the parallel axes theorem of moment of inertia.

(b) Figure shows a uniform wire of mass M and radius of each ring R. Calculate the moment of inertia of the system about a normal axis through O. (You may assume the formula for m.i. for a ring about a normal axis through its centre).

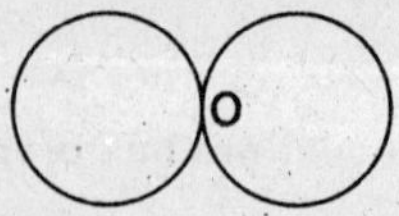

[**Ans.** M.I. of a ring about a normal axis through its centre is MR^2—Applying the theorem of parallel axes, the m.i. of one ring about a normal axis through O is $MR^2 + MR^2 = 2MR^2$; $\therefore$ for the whole system m.i. $= 2 \times 2MR^2 = 4MR^2$. **[C.U. (Hons.) 2004]**

(c) A homogeneous right circular cylinder has a height h and radius r. Set up a suitable principal axes system for the body. Calculate the principal moments of inertia. **[C.U. (Hons.) 2004]** (See Art. 4.13)

42. (a) For rotational motion of rigid bodies, derive an expression for K.E. in terms of m.i. and angular velocity and hence define the m.i. tensor. What does the expression for K.E. reduce to in the principal set of axes? In what respect are mass and m.i. comparable?

[C.U. (Hons.) 2007]

(b) Find the m.i. of a disc of radius a about an axis perpendicular to its surface and passing through a point at a distance $a/3$ from the centre. (let m be the mass).

[C.U. (Hons.) 2005]

[**Hints.** Apply the parallel axes theorem.]

43. (a) State the properties of principal m.i. of a rigid body. Show how you will find these quantities from a knowledge of the inertia tensor in arbitrary frame of reference.

(b) The m.i. of a rigid body are given by I_{xx}, I_{yy}, I_{zz} and the p.i. by I_{xy}, I_{yz}, I_{zx}. Find out the m.i. of the body about an axis whose direction cosines are given by (l, m, n).

[C.U. (Hons.) 2006]

44. (a) Explain the *Ellipsoid of inertia* and derive an expression for it in the case of a rigid body.

(b) Find the principal axes and p.m.i. of a cylindrical tube of length L, and M and inner and outer radii R_1 and R_2. (See Art. 4.13)

(c) State and prove the parallel axes theorem of m.i. **[C.U. (Hons.) 2007]**

Chapter 5

Simple Harmonic Motion

5.1 Periodic Motion

When a body is found to describe the same path repeatedly in some definite interval of time, its motion is said to be periodic, and the time taken by the body to complete its path once is called its *period.*

Instances of periodic motion are not far to seek. The motion of the earth round the sun is a periodic motion, and its period of motion is one year. The rotation of the earth about its own axis is a periodic motion, the period being 24 hours. The motion of the minute-hand of a clock is again a periodic motion, the period being one hour. Also the motion of a clock-pendulum is a periodic one.

5.2 Vibratory Motion

The periodic motion of a body is said to be oscillatory or vibratory when it is reversed in direction after a definite interval of time. The to-and-fro motion of a pendulum is an example of the motion of this type. It has to be noted that an oscillatory motion should always be periodic, but a periodic motion may not always be oscillatory in nature.

5.3 Simple Harmonic Motion

It is the simplest type of oscillatory motion of a body with the following characteristics :

(i) The motion is oscillatory and is repeated after equal intervals of time;

(ii) The restoring force acting on the body (or its acceleration) is always proportional to its displacement (measured along its path) from some fixed point on its path called its mean position;

(iii) Its acceleration is always directed towards its mean position.

The ideal case of a simple harmonic motion may be obtained by the motion on a diameter of the foot of the perpendicular dropped from the different positions of a point moving uniformly along the circumference of a circle.

We suppose a particle P moving along the circumference of a circle of radius a with uniform angular velocity ω. If we draw perpendicular PN from P on the diameter AOB (Fig. 5.3.1) for different positions of P in course of its motion, then the foot N of the perpendicular PN executes a to-and-fro motion along AOB with O as its mean position. This satisfies all the conditions of simple harmonic motion.

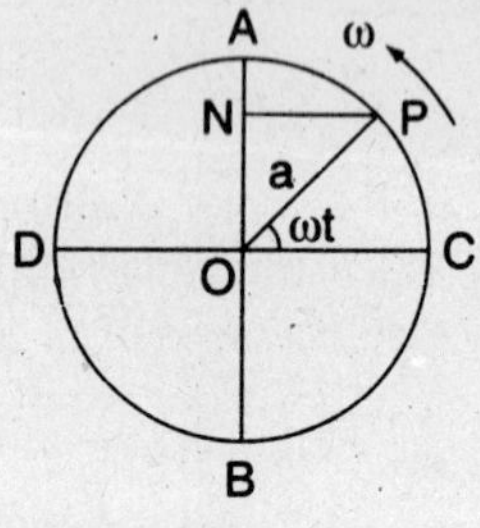

Fig. 5.3.1

Here the point P is called the *generating point*, and the circle along which it moves is known as the *circle of reference.*

When the foot N moves along the diameter AOB from O to A, from A to B and back to O, it is said to have completed its path once. In other words, it is said to have executed *one complete oscillation.*

5.4 Time-period

The time taken by N to complete one oscillation is called the *time-period* of oscillation. This is also the time taken by the particle P to move once round the circle of reference. If ω be the angular velocity (in radians per sec) with which the particle moves round the circle of reference, then the time-period T is given by the relation

$$\omega T = 2\pi, \quad \text{or,} \quad T = \frac{2\pi}{\omega}.$$

5.5 Frequency

The number of oscillations that N can execute in one second is called its *frequency* of oscillation. If n represents the frequency, T, the time-period, then

$$nT = 1, \quad \text{or,} \quad n = \frac{1}{T}.$$

5.6 Amplitude

The maximum displacement of N from its mean position along the path AOB given by OA or OB is called its *amplitude* of motion. Thus the amplitude of motion is the radius of the circle of reference.

5.7 Phase

The *phase* of a vibrating particle at any instant determines the state of displacement and motion of the particle at that instant. Thus any quantity which can completely specify both the displacement of a vibrating point and its direction of motion at any time will serve as a measure of phase at that time.

The *initial phase* of a vibrating particle is called epoch.

5.8 Calculation of Displacement of a Particle Executing S.H.M.

We consider any time t which elapses after the point N has passed through the mean position O, moving along OA (which is reckoned as the positive direction), during which the generating point on the circle of reference moves from C to P (Fig. 5.3.1). Then the displacement of N is given by

$$ON = y = OP \sin \omega t = a \sin \omega t = a \sin \frac{2\pi}{T} t,$$

where ωt is the phase at any instant t of the point N executing S.H.M.

The angle ωt completely determines the position and direction of motion of the vibrating point N.

If, instead, the generating point starts moving from C' (Fig. 5.8.1), the phase angle will be

$$\angle C'OP = \omega t, \quad \text{or,} \quad \angle COP = \omega t - \delta,$$

where $\delta = \angle C'OC$ is known as the initial phase angle or the *epoch* of the particle executing S.H.M.

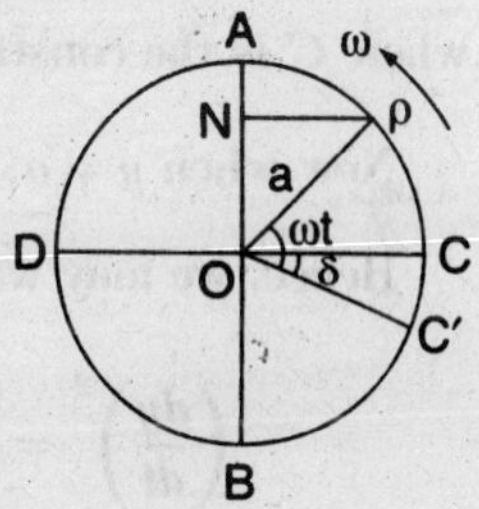

Fig. 5.8.1

Therefore, the displacement of N is given by

$$ON = y = a \sin(\omega t - \delta) = a \sin\left(\frac{2\pi}{T} t - \delta\right).$$

5.9 Calculation of Velocity

Differentiating the expression for displacement $y = a \sin \omega t$, with respect to t, we have

$$\text{Velocity } v = \text{rate of change of displacement} = \frac{dy}{dt} = a\omega \cos \omega t,$$

$$\text{or,} \quad v = a\omega\sqrt{1 - \sin^2 \omega t} = a\omega\sqrt{1 - \frac{y^2}{a^2}} = a\frac{2\pi}{T}\sqrt{1 - \frac{y^2}{a^2}} = \frac{2\pi}{T}\sqrt{a^2 - y^2}$$

when $y = 0$, i.e., at the mean position O, the velocity of the particle executing S.H.M. is maximum, and when $y = a$, i.e., at the extreme position A or B, the velocity is zero.

5.10 Calculation of Acceleration and Equation of Motion of S.H.M.

Acceleration f = rate of change of velocity $= \frac{dv}{dt} = -a\omega^2 \sin \omega t = -\omega^2 y$, which shows that acceleration is proportional to y, since ω is constant. Also the negative sign indicates that the acceleration is directed opposite to the displacement — which satisfies the conditions (ii) and (iii) for S.H.M.

Now,

$$\frac{dv}{dt} = -\omega^2 y, \quad \text{or,} \quad \frac{d^2y}{dt^2} + \omega^2 y = 0$$

which is the characteristic differential equation for S.H.M. The solution of the above equation may be done in the following way.

Multiplying both sides by $2\frac{dy}{dt}$, we have

$$2\frac{d^2y}{dt^2}\frac{dy}{dt} + 2\omega^2 y\frac{dy}{dt} = 0 \quad \text{or} \quad \frac{d}{dt}\left(\frac{dy}{dt}\right)^2 + \frac{d}{dt}(w^2y^2) = 0.$$

Integrating we have

$$\left(\frac{dy}{dt}\right)^2 + \omega^2y^2 = C,$$

where C is the constant of integration.

Now when $y = a$, $\frac{dy}{dt} = 0$. $\therefore\ \omega^2a^2 = C$.

Hence, we may write

$$\left(\frac{dy}{dt}\right)^2 = \omega^2(a^2 - y^2), \quad \text{or,} \quad \frac{dy}{dt} = \omega\sqrt{a^2 - y^2}, \quad \text{or,} \quad \frac{dy}{\sqrt{a^2 - y^2}} = \omega dt.$$

Integrating,

$$\sin^{-1}\frac{y}{a} = \omega t + \delta, \quad \text{or,} \quad y = a\sin(\omega t + \delta),$$

δ being the constant of integration.

The above equation may be graphically represented by plotting y against t, as shown in Fig. (5.10.1).

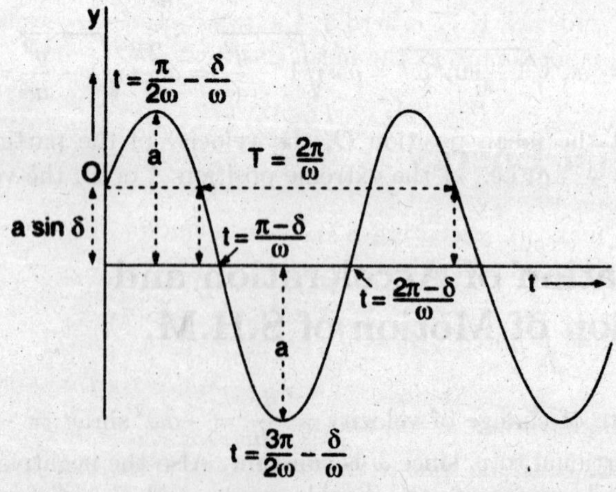

Fig. 5.10.1

At $t = 0$, $y = a\sin\delta$ which shows that the particle starts with an initial displacement $a\sin\delta$ as shown in the figure. Here the constant δ denotes the *initial phase* or *epoch*.

The displacement reaches its maximum value $+a$ in the positive direction at $t = \left(\frac{\pi}{2\omega} - \frac{\delta}{\omega}\right)$ and its maximum value $-a$ in the negative direction at $t = \left(\frac{3\pi}{2\omega} - \frac{\delta}{\omega}\right)$ respectively, its values being zero at $t = \left(\frac{\pi - \delta}{\omega}\right), \frac{2\pi - \delta}{\omega}$ and so on.

Thus the value of the displacement y oscillates between $+a$ are $-a$ about the mean position at which its value is zero. In other words, the particle executes a to-and-fro motion about the mean position and the motion is repeated after time T — which satisfies the first condition for S.H.M.

The *time-period* is given by $T = \frac{2\pi}{\omega}$.

Now, acceleration $f = \omega^2 y$ (considering only the magnitude)

$$\therefore \quad \omega = \sqrt{\frac{f}{y}}.$$

$$\therefore \quad T = \frac{2\pi}{\omega} = 2\pi\sqrt{\frac{y}{f}} = 2\pi\sqrt{\frac{\text{displacement}}{\text{acceleration}}}.$$

From the above expression, it follows that the time-period is independent of the *amplitude* of vibration. This is known as the *law of isochronism.*

5.11 Energy of a Particle Executing S.H.M.

Kinetic energy. Kinetic energy of a particle of mass m, executing S.H.M.

$$= \frac{1}{2}m \times (\text{velocity})^2 = \frac{1}{2}m \times \left(\omega\sqrt{a^2 - y^2}\right)^2 = \frac{1}{2}m\omega^2(a^2 - y^2).$$

At $y = a$, K.E. is zero, i.e., K.E. is zero at the two extreme positions. Again at $y = 0$, K.E. $= \frac{1}{2}m\omega^2 a^2$, i.e., K.E. is maximum at the mean position.

5.12 Potential Energy

P.E. of a particle of mass m, executing S.H.M. is calculated from the total amount of work done for the displacement y and is given by

$$\text{P.E.} = \int_0^y \text{force} \times dy = \int_0^y mf\,dy = \int_0^y m\omega^2 y\,dy = m\omega^2\frac{y^2}{2}.$$

Thus P.E. is maximum when $y = a$, and minimum (i.e., zero) when $y = 0$.

From the above, we conclude that K.E. is maximum when P.E. is minimum and *vice versa.*

And,

$$\text{K.E.} + \text{P.E.} = \frac{1}{2}m\omega^2(a^2 - y^2) + \frac{m\omega^2 y^2}{2} = \frac{1}{2}m\omega^2 a^2 = \text{constant}$$

which proves the *principle of conservation of energy.*

5.13 Graphical Representation of S.H.M.

We represent graphically the *phase-displacement* curve of a S.H.M. given by

$$y = a \sin \omega t = a \sin \frac{2\pi}{T} t.$$

We draw a circle of reference with radius equal to a, and take a reference point P — which is moving with uniform angular velocity ω over the circumference of the circle with period T same as that of the given S.H.M. As this point P moves over the circumference of the circle, it sweeps out angle ωt represented by $\angle COP$ in time t — which represents the phase of the S.H.M. The value of ωt changes from 0 to 2π when P completes one revolution. The distance of the foot of the perpendicular N drawn on AOB from different positions of P gives the corresponding displacement of the S.H.M.

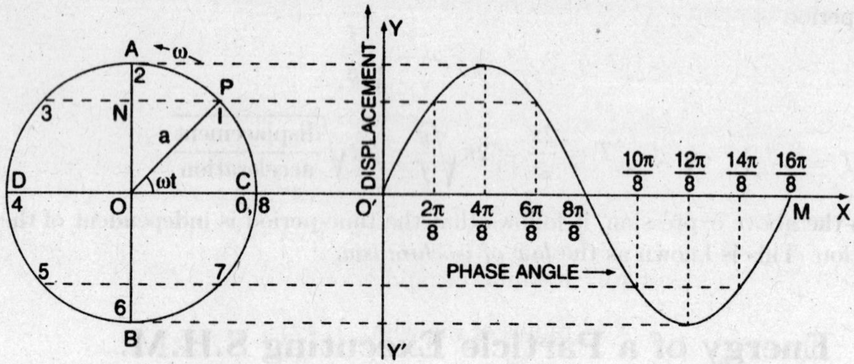

Fig. 5.13.1

For drawing the phase-displacement curve (Fig. 5.13.1), we divide the circle into 8 equal parts, and consider the point P to move over the circumference in the counter-clockwise direction starting from the mark 'zero'. Let us take two axes $O'X$, and $O'Y$ at right angles to each other. Now we plot displacements of N along $O'Y$ and the phase ωt or $\frac{t}{T}$ along $O'X$. We take a length $O'M$ along $O'X$, and divide it into 8 equal parts so that each division corresponds to $\frac{2\pi}{8}$ or 45°. When P is at 'zero' mark, the foot N coincides with the centre O of the circle of reference having zero displacement. When the particle is at 'mark' 2, the displacement of N is maximum and so on. When these displacements and phase-angles are plotted along the rectangular axes, a 'sine' graph is obtained as shown in the figure. It is to be noted that when the displacement of N is along OB, it is reckoned as negative, and plotted along $O'Y'$. The details of drawing the curve will be clear from the figure.

In the same way, we may represent graphically a S.H.M. given by $y = a \cos \omega t$, and the curve obtained is a characteristic cosine curve.

5.14 Examples of S.H.M.

Motion of a simple pendulum, and that of a vertical loaded spring are common examples of S.H.M. Also we shall come across various familiar examples of S.H.M., which will be discussed in the subsequent chapters. However a few interesting examples are given here.

I. Torsional pendulum

A cylindrical wire is fixed at one end, and carries a load in the form of a cylinder M at the bottom. Giving a slight rotation to M, the wire is twisted at the free end, and the system is made to oscillate about the wire as axis. If θ be the angle of twist imparted to the wire as shown in Fig. 5.14.1, then the deflecting couple $= I\dfrac{d^2\theta}{dt^2}$, I being the moment of inertia of the cylinder M about the wire. This active torque must be equal to restoring couple $C\theta$, where C is the restoring couple per unit angle of twist. So the equation of motion is

$$I\frac{d^2\theta}{dt^2} = -C\theta, \quad \text{or,} \quad \frac{d^2\theta}{dt^2} = -\frac{C}{I}\theta, \quad \text{or,} \quad \frac{d^2\theta}{dt^2} + \omega^2\theta = 0.$$

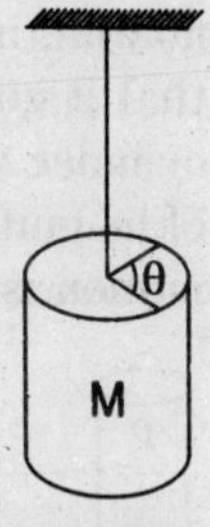

Fig. 5.14.1

where $\omega = \sqrt{\dfrac{C}{I}}$ which shows that the motion is simple harmonic in character and the time period is

$$T = \frac{2\pi}{\omega} = 2\pi\sqrt{\frac{I}{C}}.$$

Here $I = \dfrac{MR^2}{2}$, R being the radius of the cylinder of mass M and torsional rigidity $C = \dfrac{\pi n r^4}{2l}$, where

n = modulus of rigidity of the wire;

r = radius of the wire;

l = length of the wire;

II. Motion of a vertical loaded spring

We consider a weightless spring fixed at one end, and loaded at the other end by a mass m (Fig. 5.14.2). The spring extends a little, and the increased tension in it supports the load. Now the mass is pulled downwards through a small additional distance y from this position, and then released. The spring will exert a restoring pull on the mass, which for small extensions, is proportional to the extension, and may be written as Ky, K being the restoring force per unit extension of the spring. The equation of motion is given by

$$m\frac{d^2y}{dt^2} \text{ (deflecting force)} = -Ky, \quad \text{or,} \quad \frac{d^2y}{dt^2} + \omega^2 y = 0$$

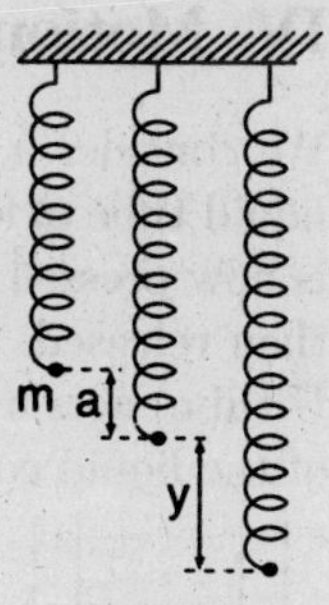

Fig. 5.14.2

where $\omega^2 = \dfrac{K}{m}$. This equation represents a simple harmonic motion with period

$$T = \frac{2\pi}{\omega} = 2\pi\sqrt{\frac{m}{K}}.$$

If the mass m extends the spring initially by a small distance a, then $mg = Ka$ or $K = \dfrac{mg}{a}$. Putting this value of K in the expression for T, we have

$$T = 2\pi\sqrt{\frac{a}{g}}.$$

If the spring has an effective mass m_0, then we may write

$$T = 2\pi\sqrt{\frac{m+m_0}{K}}.$$

III. Motion of a floating cylinder

We consider a cylinder of material of density ρ, length l and radius r. Let this cylinder be floating in a liquid of density σ, as shown in Fig. 5.14.3. The cylinder is now pressed such that it goes inside the liquid by a distance y below its position of rest and then released. The cylinder will displace a mass $\pi r^2 y\sigma$ of the liquid and hence will experience an upward thrust of magnitude $\pi r^2 y\sigma g$ — which is a restoring force proportional to displacement. The resulting motion is therefore simple harmonic and the equation of motion is

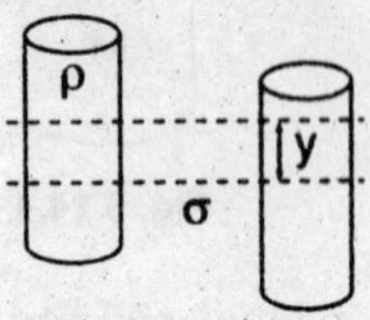

Fig. 5.14.3

$$M\frac{d^2y}{dt^2}\text{ (deflecting force)} = -\pi r^2 y\sigma g \quad (M = \text{mass of the cylinder}),$$

$$\text{or, } \pi r^2 l\rho\frac{d^2y}{dt^2} = -\pi r^2 y\sigma g, \quad \text{or, } \frac{d^2y}{dt^2} + \frac{\sigma g}{l\rho}y = 0,$$

$$\text{or, } \frac{d^2y}{dt^2} + \omega^2 y = 0 \text{ where } \omega^2 = \frac{\sigma g}{l\rho}$$

which is the differential equation of simple harmonic motioin, and the time period is

$$T = \frac{2\pi}{\omega} = 2\pi\sqrt{\frac{l\rho}{g\sigma}}.$$

If l_1 be the length of the cylinder inside the liquid while floating initially, then from the condition of floation, we have

$$\pi r^2 l_1 \sigma g = \pi r^2 l\rho g \quad \text{or} \quad l_1 = l\rho/\sigma. \quad \therefore\ T = 2\pi\sqrt{l_1/g}.$$

Due to viscosity of the liquid, the amplitude of motion falls off rapidly.

IV. Motion of a liquid in a U-tube

We consider a U-tube clamped vertically, and a liquid of density ρ is filled in it such that the liquid tube of length l has levels at A, A (Fig. 5.14.4). The liquid level in one limb of the U-tube is now pressed through a distance y, producing a difference $2y$ of heights in the two limbs, and then released. The weight of this "head" of liquid $\pi r^2 2y\rho g$ (r, being the uniform radius of the U-tube) acts as a restoring force, which is evidently proportional to displacement. The motion of the liquid column in the U-tube is, therefore simple harmonic, and the equation of motion is

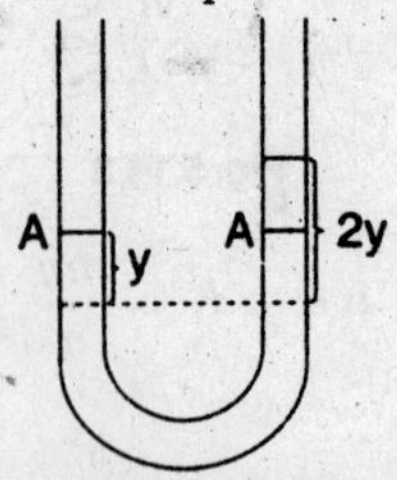

Fig. 5.14.4

$$M\frac{d^2y}{dt^2}\text{ (deflecting force)} = -\pi r^2\cdot 2y\cdot\rho g$$

$$(M = \text{mass of the column of liquid displaced} = \pi r^2 l\rho)$$

$$\text{or, } \pi r^2 l\rho\frac{d^2y}{dt^2} = -\pi r^2 2y\cdot\rho g \text{ or } \frac{d^2y}{dt^2} + \frac{2g}{l}y = 0$$

$$\text{or, } \frac{d^2y}{dt^2} + \omega^2 y = 0 \text{ where } \omega^2 = \frac{2g}{l}$$

which is the differential equation of s.h.m., and the time period is

$$T = \frac{2\pi}{\omega} = 2\pi\sqrt{\frac{l}{2g}}.$$

Here also the motion is quickly damped out as in the previous case.

V. Motion of a bar magnet, freely suspended, so as to move in a horizontal plane in the earth's magnetic field

Let the bar magnet AB swing round O (Fig. 5.14.5), about the mean position NS representing the magnetic meridian. The magnet is first made to turn through an angle θ to the position AB, and then released. The deflecting torque acting on the magnet at this positon is $I\frac{d^2\theta}{dt^2}$, I being the moment of inertia of the magnet about O. The forces mH, mH acting at the two ends of the magnet due to the horizontal component H of the earth's magnetic field at the place of observation, form a couple tending to restore the magnet to the meridian plane. The moment of this restoring couple $= mH \times AC = mH \times 2l\sin\theta$ (where $2l$ = length of the magnet) $= MH\sin\theta$ where $M = m \times 2l$ = magnetic moment of the magnet. The equation of motion of the magnet at the displaced position is

$$I\frac{d^2\theta}{dt^2} = -MH\sin\theta = -MH\theta, \text{ when } \theta = \text{very small}$$

$$\text{or,} \quad \frac{d^2\theta}{dt^2} + \frac{MH}{I}\theta = 0, \quad \text{or,} \quad \frac{d^2\theta}{dt^2} + \omega^2\theta = 0 \text{ where } \omega^2 = \frac{MH}{I}$$

which shows that the motion of the magnet is simple harmonic in nature, with a time period given by

$$T = \frac{2\pi}{\omega} = 2\pi\sqrt{\frac{I}{MH}}.$$

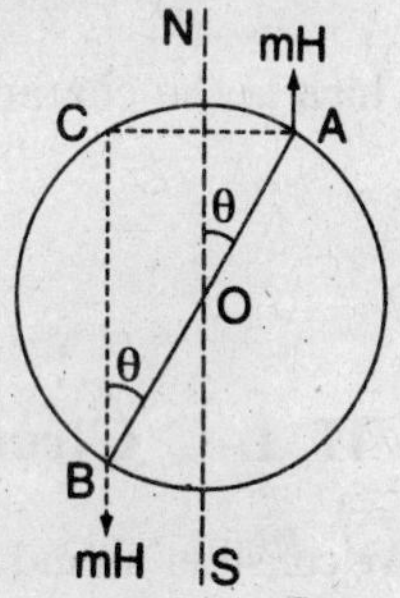

Fig. 5.14.5

N.B. Calculation of the angular displacement θ of the magnet at any instant. General solution of the above differential equation of s.h.m is

$$\theta = A\sin\omega t + B\cos\omega t.$$

To evaluate the constants A and B, we apply the initial condition, i.e, when $t = 0$, $\theta = 0$ — which gives $B = 0$.

$$\therefore \ \theta = A\sin\omega t.$$

Again we apply the condition — when $t = \frac{T}{4}$, $\theta = \theta_0$ = angular amplitude.

$$\therefore \quad \theta_0 = A\sin\left(\frac{2\pi}{T}\cdot\frac{T}{4}\right) = A,$$

so that the angular displacement of the magnet at any instant t is given by

$$\theta = \theta_0\sin\omega t = \theta_0\sin\sqrt{\frac{MH}{I}}t$$

VI. Motion of an electric dipole which is free to move in a uniform electric field

The dipole, when at rest, aligns itself in the direction of the field $\vec{E}$. It is made to turn through an angle θ to the position as shown in Fig. 5.14.6, and then released. The deflecting torque acting on the dipole at this position is $I\frac{d^2\theta}{dt^2}$, I being the moment of inertia of the dipole about O. The forces $q\vec{E}$, $q\vec{E}$, acting at the two ends of the dipole due to the electric field $\vec{E}$, form a couple, tending to restore the couple to its equilibrium position. The moment of this restoring couple is

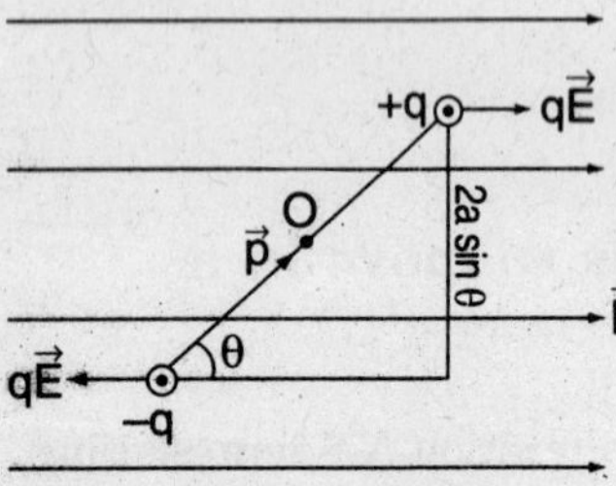

Fig. 5.14.6

$$qE(2a\sin\theta) = 2qaE\sin\theta = pE\sin\theta,$$

where $\vec{p}$ is the moment of the dipole of magnitude $2qa$, $2a$ being the separation between $+q$ and $-q$ forming the dipole. This restoring couple $\vec{p}\times\vec{E} = pE\sin\theta = pE\theta$ (when θ is very small) being proportional to θ, the dipole oscillates simple harmonically in the electric field, and the equation of motion at its displaced position is

$$I\frac{d^2\theta}{dt^2} = -pE\theta, \quad \text{or,} \quad \frac{d^2\theta}{dt^2} + \frac{pE}{I}\theta = 0 \quad \text{or,} \quad \frac{d^2\theta}{dt^2} + \omega^2\theta = 0, \quad \text{where } \omega^2 = \frac{pE}{I},$$

which is the characteristic differential equation of S.H.M. with time period given by

$$T = \frac{2\pi}{\omega} = 2\pi\sqrt{\frac{I}{pE}}.$$

VII. L-C Circuit — Electrical oscillation

We consider an inductor of self-inductance L, connected in series with a condenser of capacitance C, which has been charged to some potential $\frac{q}{C}$ by a source of e.m.f., the source being withdrawn after charging as shown in Fig. 5.14.7(a). The condenser starts to discharge, on closing the circuit through the inductor. After some time, the condenser gets completely discharged, and the p.d. between its plates decreases to zero. The current in the inductor, while growing, produces a magnetic field around it, by setting up lines of force, i.e., flux as shown in Fig. 5.14.7(b).

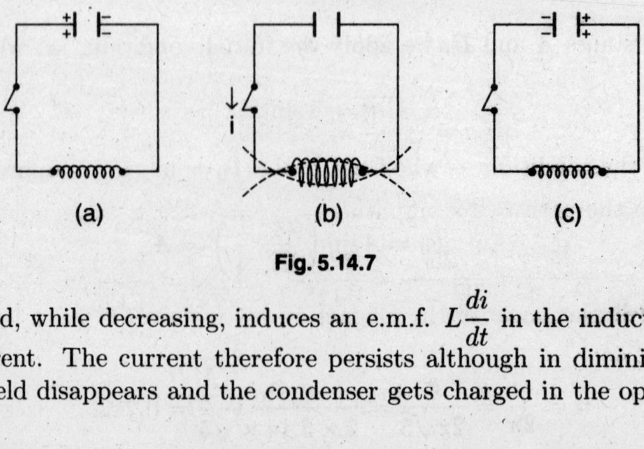

Fig. 5.14.7

The magnetic field, while decreasing, induces an e.m.f. $L\frac{di}{dt}$ in the inductance in the same direction as the current. The current therefore persists although in diminishing magnitude, until the magnetic field disappears and the condenser gets charged in the opposite sense as in Fig. 5.14.7(c).

The process now repeats itself in the reverse direction and continues itself in absence of any loss of energy. This process is called an "electrical oscillation". Now from energy point of view, the oscillation of an electrical circuit consists of a transfer of energy back and forth from the electric field of the condenser $\left(\frac{1}{2}CV^2 = \frac{1}{2}\frac{q^2}{C}\right)$ to the magnetic field of the inductor $\left(\frac{1}{2}Li_0^2\right)$. The total energy associated with the circuit remains constant.

Now for the above circuit, we may write the e.m.f. equation as

$$L\frac{di}{dt} + \frac{q}{C} = 0,$$

where $\frac{q}{C}$ is the e.m.f. across the condenser plates, q being the charge developed on the condenser; $L\frac{di}{dt}$ is the e.m.f. across the inductor. Taking $i = \frac{dq}{dt}$, we have

$$L\frac{d^2q}{dt^2} + \frac{q}{C} = 0, \quad \text{or,} \quad \frac{d^2q}{dt^2} + \frac{1}{LC}q = 0, \quad \text{or,} \quad \frac{d^2q}{dt^2} + \omega^2 q = 0, \text{ where } \omega^2 = \frac{1}{LC},$$

which is the characteristic differential equation of S.H.M. with time period given by

$$T = \frac{2\pi}{\omega} = 2\pi\sqrt{LC}.$$

Illustrative Examples

1. A particle is moving in a straight line with simple harmonic motion. Its velocity has the values 5 ft/sec and 4 ft/sec, when its distances from the centre-point of its motions are 2 ft and 3 ft respectively. Find the length of its path, the frequency of its oscillation, and the phase of its motion, when it is at a distance of 2 ft from the centre. **[London Higher School Certificate]**

We know that $v = \omega\sqrt{a^2 - y^2}$; when $v = 5$ ft/sec, $y = 2$ ft

$$\therefore\ 25 = \omega^2(a^2 - 4) \tag{1}$$

when $v = 4$ ft/sec, $y = 3$ ft

$$\therefore\ 16 = \omega^2(a^2 - 9) \tag{2}$$

From (1) and (2), we have

$$\frac{25}{16} = \frac{a^2 - 4}{a^2 - 9}, \text{ or, } 25a^2 - 225 = 16a^2 - 64, \quad \text{or,} \quad 9a^2 = 161,$$

$$\text{or, } a = \sqrt{\frac{161}{9}} = 4.216 \text{ ft}.$$

Now the length of the path = 2 × amplitude = $2a$ = 8.432 ft. From equation (1), we have

$$\omega^2 = \frac{25}{a^2 - 4} = \frac{25}{\frac{161}{9} - 4} = \frac{25 \times 9}{125} \quad \text{or} \quad \omega = \frac{5 \times 3}{\sqrt{125}} = \frac{3}{\sqrt{5}}.$$

∴ the frequency,

$$n = \frac{\omega}{2\pi} = \frac{3}{2\pi\sqrt{5}} = \frac{3}{2 \times 3.14 \times \sqrt{5}} = 0.2129.$$

We know $y = a \sin \omega t$, where ωt is the phase angle; when $y = 2$ ft.

$$\sin \omega t = \frac{2}{4.216} = 0.4767. \quad \therefore \omega t = 28°27'.$$

2. A particle of mass m executes S.H.M. in a straight line, with amplitude A and frequency n. If r_1 and r_2 be the displacements of the particle from the mean position, and the corresponding velocities be v_1 and v_2 respectively, show that

(a) frequency $n = \dfrac{1}{2\pi}\left[\dfrac{v_1^2 - v_2^2}{r_2^2 - r_1^2}\right]^{1/2}$ (b) amplitude $A = \left[\dfrac{v_1^2 r_2^2 - v_2^2 r_1^2}{v_1^2 - v_2^2}\right]^{1/2}$ and

(c) maximum K.E. $= \dfrac{1}{2}m\left[\dfrac{v_1^2 r_2^2 - v_2^2 r_1^2}{r_2^2 - r_1^2}\right]$

(a) Let the S.H.M. be represented by $r = A\sin(\omega t + \alpha)$, where A, ω and α are amplitude, angular frequency and phase constant respectively. At any instant, velocity of the particle is given by

$$v = \frac{dr}{dt} = A\omega\cos(\omega t + \alpha) = A\omega\sqrt{1 - \sin^2(\omega t + \alpha)} = \omega\sqrt{A^2 - r^2}$$

so that $v_1 = \omega\sqrt{A^2 - r_1^2}$ and $v_2 = \omega\sqrt{A^2 - r_2^2}$.

$$\therefore v_1^2 - v_2^2 = \omega^2(r_2^2 - r_1^2) \quad \text{or, } \omega^2 = \frac{v_1^2 - v_2^2}{r_2^2 - r_1^2}; \quad \therefore \text{ frequency } n = \frac{\omega}{2\pi} = \frac{1}{2\pi}\left[\frac{v_1^2 - v_2^2}{r_2^2 - r_1^2}\right]^{1/2}.$$

(b) Now from (a) we have,

$$v_1^2 = \omega^2(A^2 - r_1^2) = \left(\frac{v_1^2 - v_2^2}{r_2^2 - r_1^2}\right)(A^2 - r_1^2), \quad \text{or,} \quad A^2 - r_1^2 = v_1^2\frac{(r_2^2 - r_1^2)}{(v_1^2 - v_2^2)},$$

$$\text{or,} \quad A^2 = v_1^2\left(\frac{r_2^2 - r_1^2}{v_1^2 - v_2^2}\right) + r_1^2 = \frac{v_1^2 r_2^2 - v_2^2 r_1^2}{v_1^2 - v_2^2}. \quad \therefore \quad A = \left[\frac{v_1^2 r_2^2 - v_2^2 r_1^2}{v_1^2 - v_2^2}\right]^{1/2}.$$

$$\text{(c) Maximum K.E} = \frac{1}{2}mv_{\max}^2 = \frac{1}{2}m\omega^2 A^2 = \frac{1}{2}m\left(\frac{v_1^2 - v_2^2}{r_2^2 - r_1^2}\right)\left(\frac{v_1^2 r_2^2 - v_2^2 r_1^2}{v_1^2 - v_2^2}\right)$$

$$= \frac{1}{2}m\left(\frac{v_1^2 r_2^2 - v_2^2 r_1^2}{r_2^2 - r_1^2}\right).$$

3. A particle of mass m is attached to the mid-point of a string stretched vertically under constant tension T. The particle oscillates horizontally, but the ends are prevented from horizontal movements. If l be the length of the string, calculate the time period of small oscillation of the particle.

[C.U. (Hons.) 1981]

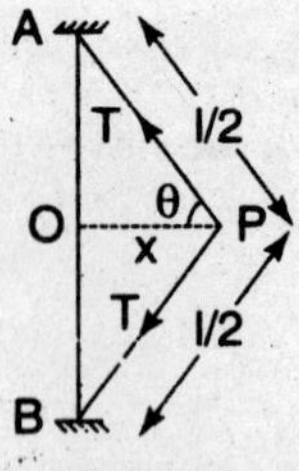

Fig. 1

Let the particle of mass m, attached to O, the mid-point of the vertical string AB, whose ends are fixed to rigid supports, be displaced horizontally to P by a small distance x (Fig. 1) and then released. The force acting on the body horizontally in the displaced position will be the horizontal components of the tension on the two halves of the string, given by

$$2T\cos\theta = 2T\frac{OP}{AP} = 2T\cdot\frac{x}{l/2} = \frac{4T}{l}x$$

acting along PO. This force, being proportional to x, is evidently the restoring force, so that the equation of motion of the particle is

$$m\frac{d^2x}{dt^2} \text{ (deflecting force)} = -\frac{4T}{l}x, \text{ or, } \frac{d^2x}{dt^2} + \frac{4T}{ml}x = 0,$$

$$\text{or, } \frac{d^2x}{dt^2} + \omega^2 x = 0 \quad \left(\text{where } \omega^2 = \frac{4T}{ml}\right)$$

which is the characteristic differential equation of S.H.M, with time period given by

$$T = \frac{2\pi}{\omega} = 2\pi\sqrt{\frac{ml}{4T}} = \pi\sqrt{\frac{ml}{T}}.$$

4. Two masses m and m' attached to the lower end of an elastic string, whose upper end is fixed, are hanging at rest. Now m' falls off. Show that the distance of m from the upper end of the string at time t is given by

$$a + b + c\cos\sqrt{\frac{g}{b}}t,$$

where a is the unstretched length of the string and b and c are the distances by which it would be stretched when supporting m and m' respectively.

When the mass m is attached at the lower end of the string, it extends through a distance b and reaches B. Now when the mass m' is also attached at the lower end, it further extends through a distance c and finally comes to rest at C. Now if m' falls off, the mass m begins to oscillate vertically about B. Equation of motion of the mass m with B as origin is written as

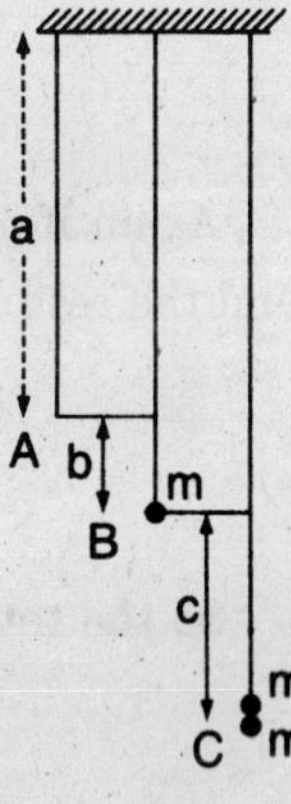

Fig. 2

$$m\frac{d^2y}{dt^2} = -Ky;$$

since the extension b is due to the mass m alone which is at rest at B, we may write

$$mg = Kb, \quad \text{or,} \quad K = \frac{mg}{b},$$

so that the above equation reduces to

$$\frac{d^2y}{dt^2} + \frac{g}{b}y = 0, \quad \text{or,} \quad \frac{d^2y}{dt^2} + \omega^2 y = 0$$

where $\omega^2 = g/b$.

The solution of this equation gives the displacement at time t

$$y = A\sin(\omega t + \alpha), A \text{ and } \alpha \text{ being constants,}$$

whence the velocity of mass m is given by

$$\frac{dy}{dt} = \omega\sqrt{A^2 - y^2}.$$

Now from initial conditions, at $t = 0, y = c$ and $\frac{dy}{dt} = 0$, we obtain $A = c$ and $\alpha = \pi/2$

$$\therefore\ y = c\cos\omega t = c\cos\sqrt{\frac{g}{b}}t.$$

Hence the total distance from the upper end of the string $= a + b + c\cos\sqrt{\frac{g}{b}}t.$

5. Show that the time period of oscillations of a body of mass M, suspended from a uniform spring of force constant K and mass m is given by

$$T = 2\pi\sqrt{\frac{M + \frac{m}{3}}{K}}.$$

Considering an element of length dx at a distance x from the upper (fixed) end of the spring, its K.E. is given by

$$\frac{1}{2}\frac{m}{l}dx\left(\frac{vx}{l}\right)^2,$$

where $\frac{m}{l}$ is the mass per unit length of the spring, velocity of the element $= \frac{vx}{l}$, v being the instantaneous velocity of the lower end of the spring—which is also the velocity of the mass M; here the displacement (hence the velocity) of any particle on the spring from its rest position is assumed to be directly proportional to its distance from the fixed end. So the K.E. of the whole spring is given by

$$\frac{1}{2}\frac{m}{l}\frac{v^2}{l^2}\int_0^l x^2dx = \frac{mv^2}{6};$$

also the K.E. of mass $M = \frac{1}{2}Mv^2$, so that the total K.E. of the system is

$$E_K = \frac{1}{2}Mv^2 + \frac{mv^2}{6} = \frac{1}{2}\left(M + \frac{m}{3}\right)v^2.$$

Again if y be the displacement of the mass M from its equilibrium position, then $v = \frac{dy}{dt}$, and the restoring force developed $= -Ky$, so that the P.E. of the system,

$$E_P = \int_0^y Kydy = \frac{1}{2}Ky^2.$$

So the total energy

$$E = E_K + E_p = \frac{1}{2}\left(M + \frac{m}{3}\right)\left(\frac{dy}{dt}\right)^2 + \frac{Ky^2}{2};$$

since the total energy of the system remains conserved, $\frac{dE}{dt} = 0$,

$$\text{or,}\quad \frac{d}{dt}\left[\frac{1}{2}\left(M + \frac{m}{3}\right)\left(\frac{dy}{dt}\right)^2 + \frac{Ky^2}{2}\right) = 0 \quad \text{or,}\quad \left(M + \frac{m}{3}\right)\frac{d^2y}{dt^2} + Ky = 0 \quad \left(\because \frac{dy}{dt} \neq 0\right)$$

$$\text{or,}\quad \frac{d^2y}{dt^2} + \frac{K}{\left(M + \frac{m}{3}\right)}y = 0, \quad \text{or,}\quad \frac{d^2y}{dt^2} + \omega^2 y = 0 \text{ where } \omega^2 = \frac{K}{M + \frac{m}{3}}$$

which is the characteristic differential equation of S.H.M. with a time period

$$T = \frac{2\pi}{\omega} = 2\pi\sqrt{\frac{M + \frac{m}{3}}{K}}.$$

6. A condenser of capacity $1\mu F$ is connected in series with an inductance 10 milli-henry. Calculate the natural frequency of the circuit.

Frequency of oscillation,

$$n = \frac{1}{2\pi\sqrt{LC}} = \frac{1}{2\pi\sqrt{10^{-2}\times 10^{-6}}} = \frac{10^4}{2\pi} = 1501 \text{ sec}^{-1}.$$

(Here $L = 10\times 10^{-3}$ Henry $= 10^{-2}$ Henry and $C = 10^{-6}$ Farad.)

7. A cylinder of length 'l' and cross-sectional area 'α' has an air-tight piston of mass 'm' and cross-section α inside it. If the piston be slightly pushed in, show that it will oscillate with simple harmonic motion and find its period.

If the piston be moved in through a distance x by the application of a force F, then

$$\text{compressional strain} = \frac{\text{Contraction in volume}}{\text{Original volume of gas in the cylinder}} = \frac{dx}{dl} = \frac{x}{l}$$

Now bulk modulus of elasticity

$$E = \frac{\text{Stress}}{\text{Strain}} = \frac{F/\alpha}{x/l} = \frac{Fl}{x\alpha}, \quad \text{or,} \quad F = \frac{E\alpha x}{l}.$$

This force, being proportional to displacement, is evidently restoring in nature, so that the equation of motion is

$$m\frac{d^2x}{dt^2} = -\frac{E\alpha}{l}x \quad \text{or,} \quad \frac{d^2x}{dt^2} + \frac{E\alpha}{ml}x = 0 \quad \text{or,} \quad \frac{d^2x}{dt^2} + \omega^2 x = 0 \quad \text{where } \omega^2 = \frac{E\alpha}{ml},$$

which is the characteristic differential equation of S.H.M., with period given by

$$T = \frac{2\pi}{\omega} = 2\pi\sqrt{\frac{ml}{E\alpha}}.$$

N.B. (i) For isothermal compression of air, $E = P$, $\therefore\ T = 2\pi\sqrt{\dfrac{ml}{P\alpha}}$, P being the pressure of air.

(ii) Again for adiabatic compression, $E = \gamma P$, where

$$\gamma = \frac{\text{specific heat of air at constant pressure}}{\text{specific heat of air at constant volume}}. \quad \therefore\ T = 2\pi\sqrt{\frac{ml}{\gamma P\alpha}}$$

8. A particle of mass m is attached to a horizental string of length l at a distance l_1 from one end. If F be the tension in the string, calculate the time period of transverse oscillations of the particle, assuming the amplitude of oscillation to be small. **[C.U. (Hons.) 1990]**

Assuming that the tension is uniform throughout the string, if the particle of mass m attached at a distance l_1 from one end of the string of length l be displaced as shown in the attached figure, through a small distance x vertically downwards, and then released, the restoring force that comes to play is given by $F\sin\theta_1 + F\sin\theta_2$.

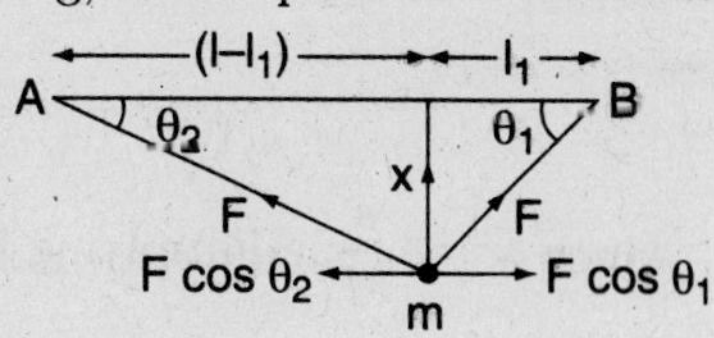

Fig. 3

If θ_1 and θ_2 are very small, we may write,

$$\sin\theta_1 = \tan\theta_1 = \frac{x}{l_1} \text{ and } \sin\theta_2 = \tan\theta_2 = \frac{x}{(l-l_1)},$$

so that the restoring force is given by

$$\left(\frac{F}{l_1}+\frac{F}{l-l_1}\right)x=\frac{Fl}{l_1(l-l_1)}x,$$

and the equation of motion is written as

$$m\frac{d^2x}{dt^2}=-\frac{Fl}{l_1(l-l_1)}x,\quad \text{or,}\quad \frac{d^2x}{dt^2}+\frac{Fl}{ml_1(l-l_1)}x=0$$

$$\text{or,}\quad \frac{d^2x}{dt^2}+\omega^2x=0 \quad \text{where}\quad \omega^2=\frac{Fl}{ml_1(l-l_1)}$$

which is the characteristic differential equation of S.H.M., with a period, given by

$$T=\frac{2\pi}{\omega}=2\pi\sqrt{\frac{ml_1(l-l_1)}{Fl}}.$$

9. A sphere of mass M, capable of producing an undamped S.H.M. along the X-axis is at rest. The stiffness constant is s. Another sphere of mass m moving with a velocity u_0, hits the first sphere elastically such that the line joining the centres is along the X-axis. Find the displacement of the first sphere at any subsequent instant of time in terms of M, m, u_0 and s. **[C.U. (Hons.) 1986]**

From the principle of conservation of linear momentum, we may write

$$mu_0=MV+mv, \tag{1}$$

where V and v are the respective velocities of the spheres of mass M and m after collision.

From (1), we have

$$(u_0-v)=\frac{MV}{m}. \tag{2}$$

Also from conservation of K.E. for the elastic collision between M and m, we may write

$$\frac{1}{2}mu_0^2=\frac{1}{2}MV^2+\frac{1}{2}mv^2,\quad \text{or,}\quad \frac{1}{2}m(u_0^2-v^2)=\frac{1}{2}MV^2,\quad \text{or,}\quad u_0^2-v^2=\frac{MV^2}{m}. \tag{3}$$

Dividing (3) by (2), we have

$$u_0+v=V. \tag{4}$$

From (2) and (4), we have

$$2u_0=V\left(1+\frac{M}{m}\right),\quad \text{or,}\quad V=\frac{2u_0}{\left(1+\frac{M}{m}\right)} \tag{5}$$

which is the velocity of M imparted by m at $t=0$, i.e., $x=0$.

Now the equation of motion of M is

$$M\frac{d^2x}{dt^2}=-sx,\quad \text{or,}\quad \frac{d^2x}{dt^2}=-\omega^2x,$$

where $\omega^2=\dfrac{s}{M}$. Multiplying both sides by $2\dfrac{dx}{dt}$ and integrating, we have

$$\left(\frac{dx}{dt}\right)^2=-\omega^2x^2+C,\quad C \text{ being the integration constant.}$$

Now at $x = 0$, i.e., $t = 0$, $\frac{dx}{dt} = V$, $\therefore\ C = V^2$

$$\therefore\ \left(\frac{dx}{dt}\right)^2 = -\omega^2x^2 + V^2, \quad \text{or,} \quad \frac{dx}{dt} = \pm\sqrt{V^2 - \omega^2x^2},$$

$$\text{or,}\ \pm\int\frac{dx}{\sqrt{V^2-\omega^2x^2}} = t + \text{constant}, \quad \text{or,} \quad \frac{1}{\omega}\sin^{-1}\frac{x}{V/\omega} = t + \text{constant}.$$

Now at $t = 0$, $x = 0$ which gives constant $= 0$

$$\therefore\ t = \frac{1}{\omega}\sin^{-1}\frac{x}{V/\omega}, \quad \text{or,} \quad x = \frac{V}{\omega}\sin\omega t = \frac{2u_0}{(1+\frac{M}{m})}\sqrt{\frac{M}{s}}\sin\sqrt{\frac{s}{M}}t \quad \text{using (5).}$$

5.15 Damped Harmonic Vibration

If we consider the fluid resistance to the motion of a particle oscillating in a fluid (air or liquid) to be proportional to its velocity, the equation of motion for small displacement is given by

$$\frac{d^2\vec{r}}{dt^2} = -k\frac{d\vec{r}}{dt} - \omega^2\vec{r}.$$

The term $k\frac{d\vec{r}}{dt}$ stands for the fluid resistance on a particle of unit mass. k may be defined as the fluid resistance or the damping effect to the motion of a particle of unit mass moving with unit velocity. If we restrict ourselves to rectilinear motion and call the co-ordinate y, then the equation of motion is

$$\frac{d^2y}{dt^2} + k\frac{dy}{dt} + \omega^2y = 0. \tag{5.15.1}$$

For solution of the above equation, we put

$$y = Ae^{pt}. \quad \therefore \quad \frac{dy}{dt} = Ape^{pt} \quad \text{and} \quad \frac{d^2y}{dt^2} = Ap^2e^{pt}.$$

Hence, from equation (5.15.1) we have,

$$Ae^{pt}(p^2+kp+\omega^2) = 0, \quad \text{or,} \quad p^2+kp+\omega^2 = 0, \quad \text{or,} \quad p = \frac{-k\pm\sqrt{k^2-4\omega^2}}{2}.$$

$$\therefore \quad y = Ae^{\frac{\left(-k+\sqrt{k^2-4\omega^2}\right)}{2}t} + Be^{\frac{\left(-k-\sqrt{k^2-4\omega^2}\right)}{2}t}. \tag{5.15.2}$$

The following three cases may occur in the solution.

(i) $k^2 > 4\omega^2$. In this case, only real quantities appear and the general solution is

$$y = e^{\frac{-kt}{2}}\left(Ae^{\frac{\sqrt{k^2-4\omega^2}}{2}t} + Be^{\frac{-\sqrt{k^2-4\omega^2}}{2}t}\right)$$

This is an aperiodic case. In this case for large damping the particle gradually comes to its equilibrium position, which it reaches at $t = \infty$. The displacement y decreases slowly as shown by the time displacement curve (Fig. 5.15.1).

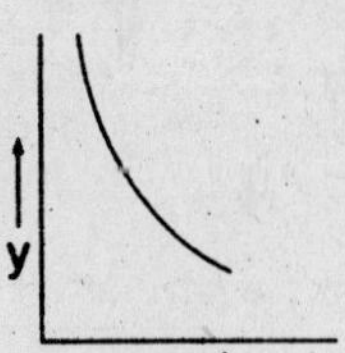

Fig. 5.15.1

(ii) $k^2 = 4\omega^2$. In this case the roots are equal. We may, however, suppose that the roots are not exactly equal and take the roots λ_1 and $\lambda_1 + h$, where $h \to 0$

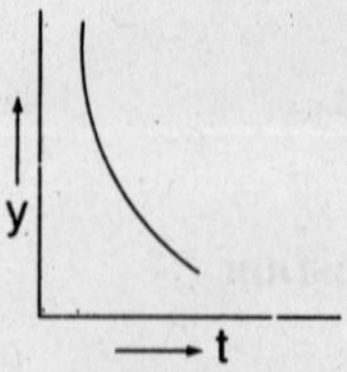

Fig. 5.15.2

$$\therefore\ y = Ae^{\lambda_1 t} + Be^{(\lambda_1+h)t} = e^{\lambda_1 t}\left[A + Be^{ht}\right] = e^{\lambda_1 t}[A + B(1+ht)]$$

[neglecting higher powers of h which are small]

$$= e^{\lambda_1 t}(C + Dt) = e^{-kt/2}(C + Dt),$$

where C and D are arbitrary constants. In this case the displacement falls from maximum value to zero very rapidly as shown in Fig. 5.15.2.

The motion is here dead-beat and it is critically damped.

(iii) $k^2 < 4\omega^2$. The general solution in this case, is given by

$$y = e^{\frac{-kt}{2}}\left[Ae^{j\frac{\sqrt{4\omega^2-k^2}}{2}t} + Be^{-j\frac{\sqrt{4\omega^2-k^2}}{2}t}\right] = e^{\frac{-kt}{2}}\left[Ae^{j\theta} + Be^{-j\theta}\right] \text{ where } \theta = \frac{\sqrt{4\omega^2 - k^2}}{2}t.$$

On separating the real from the imaginary and taking only the real part, we have

$$y = e^{\frac{-kt}{2}}[P\cos\theta + Q\sin\theta],$$

where P and Q are new constants.

Putting

$$P = R\cos\phi; \quad Q = R\sin\phi$$

$$y = Re^{\frac{-kt}{2}}\cos(\theta - \phi)$$

$$= Re^{\frac{-kt}{2}}\cos\left(\frac{\sqrt{4\omega^2 - k^2}}{2}t - \phi\right)\left[R = \sqrt{P^2 + Q^2} \text{ and } \tan\phi = \frac{Q}{P}\right]. \qquad (5.15.3)$$

Here y changes in sign alternately and we have periodic motion, but the amplitude diminishes continually due to the term $e^{\frac{-kt}{2}}$. The time period of oscillation is

$$T = \frac{2\pi}{\sqrt{\omega^2 - \frac{k^2}{4}}}$$

The curve showing the relation between y and t is shown in Fig. 5.15.3. It is clear that the oscillations die slowly or quickly according as k is small or large.

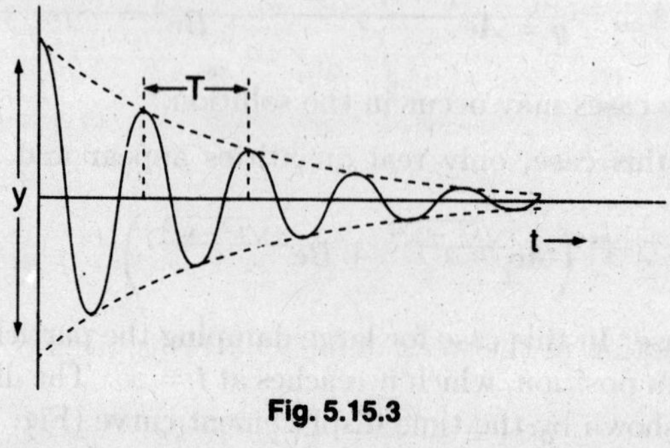

Fig. 5.15.3

Power dissipation when the damping is weak. In case the damping is weak, i.e., $k^2 \ll 4\omega^2$, the equation 5.15.3 may be written as

$$y = Re^{\frac{-kt}{2}} \cos\omega t,$$

considering the initial phase difference ϕ equal to zero.

$\therefore$ The kinetic energy of the oscillation of mass m is given by

$$T = \frac{1}{2}m\dot{y}^2 = \frac{1}{2}mR^2\left[-\omega e^{\frac{-kt}{2}}\sin\omega t - \frac{k}{2}e^{\frac{-kt}{2}}\cos\omega t\right]^2 = \frac{1}{2}mR^2e^{-kt}\left(\omega\sin\omega t + \frac{k}{2}\cos\omega t\right)^2$$

$$= \frac{1}{2}mR^2e^{-kt}\left(\omega^2\sin^2\omega t + \frac{k^2}{4}\cos^2\omega t + \omega k\sin\omega t\cos\omega t\right). \tag{5.15.4}$$

Hence taking time average over one vibration, the average kinetic energy of the oscillator is given by

$$T_{av} = \frac{\int_0^T \frac{1}{2}mR^2e^{-kt}\left(\omega^2\sin^2\omega t + \frac{k^2}{4}\cos^2\omega t + \frac{\omega k}{2}\sin 2\omega t\right)dt}{\int_0^T dt}$$

where T is the period of vibration

or, $T_{av} = \frac{1}{2}mR^2e^{-kt}\left(\frac{\omega^2}{2} + \frac{k^2}{8}\right)$

[e^{-kt} is taken to be constant since the amplitude of oscillation having the term e^{-kt} does not change in the cycle of motion]

or, $T_{av} = \frac{1}{2}mR^2e^{-kt}.\frac{\omega^2}{2} \quad \left[\because \omega^2 \gg \frac{k^2}{4}\right]$

$$= \frac{1}{4}mR^2e^{-kt}\omega^2.$$

Now the potential energy

$$V = \frac{1}{2}m\omega^2y^2 \text{ [see Art. 5.12]} = \frac{1}{2}m\omega^2R^2e^{-kt}\cos^2\omega t.$$

The average potential energy is obtained by taking average over one vibration and is

$$V_{av} = \frac{\frac{1}{2}m\omega^2R^2e^{-kt}\int_0^T \cos^2\omega t dt}{\int_0^T dt} = \frac{1}{4}m\omega^2R^2e^{-kt}.$$

Hence total energy

$$E = T + V = \frac{1}{4}m\omega^2R^2e^{-kt} + \frac{1}{4}m\omega^2R^2e^{-kt} = \frac{1}{2}m\omega^2R^2e^{-kt}.$$

The average dissipation in the weak damping limit is given by

$$p(t) = -\frac{dE}{dt} = \frac{1}{2}m\omega^2R^2ke^{-kt} = kE.$$

Quality factor Q

The quality factor is given by

$$Q = 2\pi\left(\frac{\text{energy stored}}{\text{time period} \times \text{energy loss in unit time}}\right) = \frac{2\pi E}{T \times p(t)} = \frac{2\pi E}{\frac{2\pi}{\omega} \times kE} = \frac{\omega}{k}$$

when k is small Q is high, i.e., for weak damping in an oscillation the quality factor should be high.

5.16 Forced Vibration

We consider an external periodic force $F_0 e^{i\omega t}$ acting on a particle of mass m subjected to a restoring force proportional to displacement and a resisting force proportional to its velocity. We may write the equation of motion as

$$\frac{d^2y}{dt^2} + k\frac{dy}{dt} + \omega_0^2 y = \frac{F_0}{m}e^{i\omega t}$$

ω_0 denotes the characteristic angular frequency of free and undamped vibration.

The above equation may be written in the form

$$\frac{d^2y}{dt^2} + k\frac{dy}{dt} + \omega_0^2 y = f_0 e^{i\omega t} \quad \left[\text{taking } \frac{F_0}{m} = f_0\right]. \tag{5.16.1}$$

Let us take the solution of the form

$$y = Ae^{i(\omega t - \phi)} \tag{5.16.2}$$

$$\therefore \quad \frac{dy}{dt} = Ai\omega e^{i(\omega t-\phi)} \quad \text{and} \quad \frac{d^2y}{dt^2} = -A\omega^2 e^{i(\omega t-\phi)}.$$

Then by equation (5.16.1), we have

$$-A\omega^2 e^{i(\omega t-\phi)} + kAi\omega e^{i(\omega t-\phi)} + A\omega_0^2 e^{i(\omega t-\phi)} = f_0 e^{i\omega t},$$

$$\text{or,} \quad A(-\omega^2 + ki\omega + \omega_0^2)e^{i(\omega t-\phi)} = f_0 e^{i\omega t}, \quad \text{or,} \quad A(\omega_0^2 - \omega^2 + ik\omega) = f_0 e^{i\phi},$$

$$\text{or,} \quad A(\omega_0^2 - \omega^2) + iAk\omega = \frac{F_0}{m}e^{i\phi}.$$

From the vector diagram (Fig. 5.16.1), we have

$$\sqrt{A^2(\omega_0^2 - \omega^2)^2 + A^2k^2\omega^2} = \frac{F_0}{m},$$

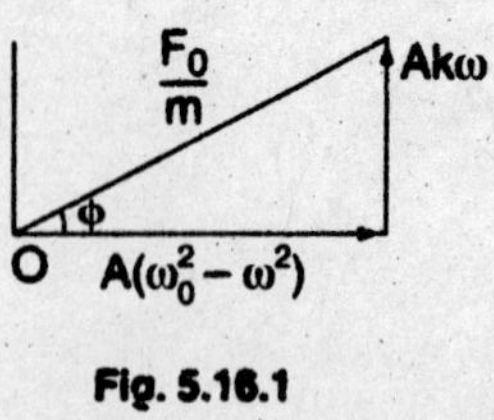

Fig. 5.16.1

$$\text{or,} \quad A = \frac{F_0}{m\sqrt{(\omega_0^2 - \omega^2)^2 + k^2\omega^2}} \quad \text{and} \quad \tan\phi = \frac{k\omega}{(\omega_0^2 - \omega^2)}.$$

$$\therefore \quad y = \frac{F_0}{m\sqrt{(\omega_0^2 - \omega^2)^2 + k^2\omega^2}}e^{i(\omega t-\phi)}, \tag{5.16.3}$$

where ϕ is given by the expression for $\tan\phi$. This solution is known as the 'particular integral' of the known non-homogeneous equation (5.16.1). The other part of the solution is known as complementary function which is obtained by the solution of equation.

$$\frac{d^2y}{dt^2} + k\frac{dy}{dt} + \omega_0^2 y = 0.$$

Hence, the general solution of equation (5.16.1) is

$$y = Re^{\frac{-kt}{2}} \cos\left(\frac{\sqrt{4\omega_0^2 - k^2}}{2}t - \phi\right) + \frac{F_0}{m\sqrt{(\omega_0^2 - \omega^2)^2 + k^2\omega^2}} e^{i(\omega t - \phi)} \tag{5.16.4}$$

The first part of the solution represents a damped vibration having the frequency of the characteristic free vibration of the damped system. This vibration dies out very soon due to the exponential term $e^{\frac{-kt}{2}}$.

However, if the damping effect is very small, the vibration continues for a longer time. The second part represents a permanent vibration with the same period as the period of vibration of the exciting force $F_0 e^{i\omega t}$. After some time when the first part of the solution becomes negligible, we may write the solution (5.16.4) as

$$y = \frac{F_0}{m\sqrt{(\omega_0^2 - \omega^2)^2 + k^2\omega^2}} e^{i(\omega t - \phi)} \tag{5.16.5}$$

Hennce, the kinetic energy of the particle of mass m at any instant is

$$\frac{1}{2}m\left(\frac{dy}{dt}\right)^2 = \frac{1}{2}m\frac{F_0^2\omega^2}{m^2(\omega_0^2 - \omega^2)^2 + m^2k^2\omega^2}\sin^2(\omega t - \phi)$$

[considering the real part of solution (5.16.5)].

Now, the maximum kinetic energy is equal to the total energy at any instant for steady harmonic motion.

So total energy

$$E = \frac{1}{2}m\frac{F_0^2\omega^2}{m^2(\omega_0^2 - \omega^2)^2 + m^2k^2\omega^2} = \frac{1}{2}m\omega^2 \cdot \frac{f_0^2}{(\omega_0^2 - \omega^2)^2 + k^2\omega^2} \qquad \left[\because \frac{F_0}{m} = f_0\right]$$

$$= \frac{\frac{1}{2}mf_0^2}{\left(\frac{\omega_0^2}{\omega} - \omega\right)^2 + k^2} = \frac{\frac{1}{2}mf_0^2}{\omega_0^2\left(\frac{\omega_0}{\omega} - \frac{\omega}{\omega_0}\right)^2 + k^2}. \tag{5.16.6}$$

If $\omega = \omega_0$, the energy becomes maximum for any given value of k. This is the case of resonance.

In absence of damping the energy becomes infinity at resonance.

Equation (5.16.6) may be written in the form

$$E = \frac{\frac{1}{2}mf_0^2}{\omega_0^2\Delta^2 + 4b^2}, \text{ where } \Delta = \frac{\omega_0}{\omega} - \frac{\omega}{\omega_0} \text{ and } k = 2b.$$

At resonance $\Delta = 0$ and the energy is maximum. Hence, we may write

$$E_{\max} = \frac{\frac{1}{2}mf_0^2}{4b^2}.$$

Plotting E against Δ we get the curves for different values of b as shown in Fig. 5.16.2. The energy is greater for smaller values of b for a particular value of the impressed force.

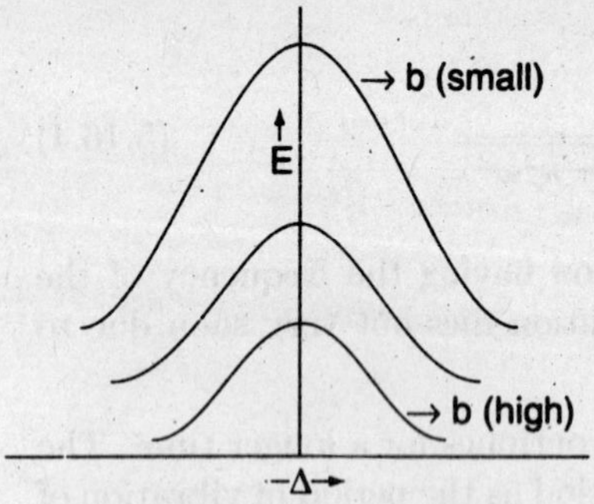

Fig. 5.16.2

Again,

$$\frac{1}{2}E_{\max} = \frac{\frac{1}{4}mf_0^2}{4b^2}.$$

If $E = \frac{1}{2}E_{\max}$, let $\Delta = \Delta_1$ then

$$\frac{\frac{1}{4}mf_0^2}{4b^2} = \frac{1}{2}\frac{mf_0^2}{\omega_0^2\Delta_1^2 + 4b^2}, \text{ or, } 8b^2 = (\omega_0^2\Delta_1^2 + 4b^2),$$

$$\text{or, } \omega_0^2\Delta_1^2 = 4b^2,$$

$$\text{or, } \Delta_1 = \pm\frac{2b}{\omega_0}, \quad \text{or, } \frac{1}{\Delta_1} = \pm\frac{\omega_0}{2b}$$

$\frac{1}{\Delta_1}$ is a measure of sharpness of resonance. Hence, it is defined as the reciprocal of Δ for which the energy is equal to half the energy at resonance. It should be noted that smaller the value of b, greater will be the sharpness of resonance.

Illustrative example

1. An elastic and mass-less spring exerts a restoring force $-Kx$ on a mass m attached to it and moving in the X-direction, where K is the spring constant and x is the displacement. If the vibrations suffer a small viscous damping (with damping constant b) proportional to velocity, show that for forced vibration with variable angular frequency ω, the amplitude is given by

$$a = \frac{a_0 b}{(b^2 + \delta^2)^{1/2}}$$

where a_0 is the amplitude of resonance $(\omega = \omega_0)$ and $\delta = (\omega - \omega_0)$. **[C.U. (Hons.) 1980]**

The equation of motion for forced vibration is

$$m\frac{d^2x}{dt^2} = -Kx - S\frac{dx}{dt} + Fe^{i\omega t}, \quad \text{or,} \quad \frac{d^2x}{dt^2} + \frac{K}{m}x + \frac{S}{m}\frac{dx}{dt} = \frac{F}{m}e^{i\omega t},$$

$$\text{or,} \quad \frac{d^2x}{dt^2} + 2b\frac{dx}{dt} + \omega_0^2 x = fe^{i\omega t},$$

where $2b = \frac{S}{m}$, b being the damping constant, $\omega_0^2 = \frac{K}{m}$ and $f = \frac{F}{m}$; ω_0 denotes the characteristic angular frequency of free and undamped vibration.

The general solution is given by

$$x = e^{-bt}\left(Ae^{\sqrt{b^2-\omega_0^2}\,t} + Be^{-\sqrt{b^2-\omega_0^2}\,t}\right) + \frac{fe^{j(\omega t-\alpha)}}{\sqrt{(\omega_0^2-\omega^2)^2 + 4b^2\omega^2}}$$

where $\tan\alpha = \frac{2b\omega}{(\omega_0^2 - \omega^2)}$ and A, B are constants.

As time advances, the natural vibration gradually diminishes and ultimately the system vibrates with the frequency of the external periodic force.

$$\therefore \quad x = \frac{fe^{j(\omega t-\alpha)}}{\sqrt{(\omega_0^2-\omega^2)^2+4b^2\omega^2}}.$$

Amplitude $a = \dfrac{f}{\sqrt{(\omega_0^2-\omega^2)+4b^2\omega^2}}$. At $\omega = \omega_0$ amplitude $a_0 = \dfrac{f}{2b\omega_0}$

$$\therefore \quad \frac{a}{a_0} = \frac{f}{\sqrt{(\omega_0^2-\omega^2)^2+4b^2\omega^2}} \times \frac{2b\omega_0}{f} = \frac{2\omega_0 b}{[(\omega_0+\omega)^2(\omega_0-\omega)^2+4b^2\omega^2)]^{1/2}}$$

$$= \frac{2\omega_0 b}{[(2\omega_0)^2\delta^2+b^2\cdot(2\omega_0)^2]^{1/2}} = \frac{b}{(\delta^2+b^2)^{1/2}} \quad [\text{putting } \omega = \omega_0 \text{ and } \omega-\omega_0=\delta]$$

$$\text{or,} \quad a = \frac{a_0 b}{(b^2+\delta^2)^{1/2}}$$

2. In a one dimensional motion of a mass 10 gm, it is acted on by a restoring force 10 dynes/cm, and a resisting force 2 dynes-sec/cm. (i) Find whether the motion is aperiodic or oscillatory; (ii) find the value of the resisting force which will make the motion critically damped; (iii) find the value of the mass for which the given forces will make the motion critically damped; (iv) suppose that the mass is given an initial impulse of 20 gm-cm.sec^{-1}, while at rest, obtain an expression for the displacement x at a time t. **[C.U. (Hons.) 1988,'90,'92,'94,'99]**

Writing the equation of motion as

$$m\frac{d^2x}{dt^2} = -K\frac{dx}{dt} - Sx, \quad \text{or,} \quad \frac{d^2x}{dt^2} + \frac{K}{m}\frac{dx}{dt} + \frac{S}{m}x = 0$$

$$\text{or,} \quad \frac{d^2x}{dt^2} + 2b\frac{dx}{dt} + \omega_0^2 x = 0. \qquad (1)$$

We have (i) $b = \dfrac{K}{2m} = \dfrac{2}{2\times 10} = 0.1$ dynes-sec/cm/gm ($\because$ $K = 2$ dyne-sec/cm, $m = 10$ gm)

and $\omega_0 = \sqrt{\dfrac{S}{m}} = \sqrt{\dfrac{10}{10}} = 1$ rad/sec ($\because$ $S = 10$ dynes/cm). So $b < \omega_0$ and the motion is oscillatory.

(ii) The motion will be critically damped when

$$b = \omega_0, \quad \text{or,} \quad b^2 = \omega_0^2, \quad \text{or,} \quad \frac{K^2}{4m^2} = \frac{S}{m}, \quad \text{or,} \quad K^2 = 4Sm,$$

$$\text{or,} \quad K = 2\sqrt{Sm} = 2\sqrt{10\times 10} = 20 \text{ dynes-sec/cm}.$$

(iii) $\dfrac{K^2}{4m^2} = \dfrac{S}{m}$ or $m = \dfrac{K^2}{4S} = \dfrac{(2)^2}{4\times 10} = 0.1$ gm.

(iv) For oscillatory motion. solution of equation (1) is

$$x = e^{-bt}\left(A\cos\sqrt{\omega_0^2-b^2}\,t + B\sin\sqrt{\omega_0^2-b^2}\,t\right)$$

From the condition of the problem, at $t = 0, x = 0$ which gives $A = 0$

$$\therefore \quad x = Be^{-bt} \sin \sqrt{\omega_0^2 - b^2}\, t$$

$$\text{or,} \quad \dot{x} = -Bbe^{-bt} \sin \sqrt{\omega_0^2 - b^2}\, t + B\sqrt{\omega_0^2 - b^2} e^{-bt} \cos \sqrt{\omega_0^2 - b^2}\, t.$$

Again from the condition of the problem, at $t = 0$, $\dfrac{dx}{dt} = \dot{x}_0 = v_0$, since the motion is excited by an impulsive force, we have

$$v_0 = B\sqrt{\omega_0^2 - b^2}, \quad \text{or,} \quad B = \frac{v_0}{\sqrt{\omega_0^2 - b^2}}.$$

Now impulse of the force = change in momentum of the mass

$$\text{or,} \quad 20 = mv - m \times 0 = mv_0 \quad (\because \text{ at } t = 0, v = v_0)$$

$$\therefore \quad v_0 = \frac{20}{m} = \frac{20}{10} = 2 \text{ cm/sec and } \sqrt{\omega_0^2 - b^2} = \sqrt{1 - 0.01} = \sqrt{0.99},$$

so that the displacement at time t is

$$x = Be^{-bt} \sin \sqrt{\omega_0^2 - b^2}\, t = \frac{2}{\sqrt{0.99}} e^{-0.1t} \sin \sqrt{0.99}\, t \text{ cm}.$$

3. Find in the above problem the time in which the amplitude falls to half value.

[Cf. C.U. (Hons.) 1997, 2007]

For oscillatory motion, the maximum displacements fall off with time. If x_1, x_2, x_3, x_4 etc. be the maximum successive displacements on both sides of the equilibrium position at times $t = \frac{T}{4}, \frac{T}{4} + \frac{T}{2}, \frac{T}{4} + T$ respectively (taking $t = 0$ when the impulse acts), then we can write

$$x_1 = x_0 e^{-b\frac{T}{4}}, \quad x_2 = x_0 e^{-b\left(\frac{T}{2} + \frac{T}{4}\right)}, \quad x_3 = x_0 e^{-b\left(T + \frac{T}{4}\right)} \text{ etc.}$$

Thus

$$\frac{x_1}{x_2} = \frac{x_2}{x_3} = \frac{x_3}{x_4} \cdots = \frac{x_{n-1}}{x_n} = e^{b\frac{T}{2}} = e^{\lambda}$$

where $\lambda = \dfrac{bT}{2}$ is known as the logarithmic decrement

$$\therefore \quad \frac{x_1}{x_n} = e^{bn\frac{T}{2}}.$$

According to the problem $x_n = \dfrac{x_1}{2}$ when $t = n\dfrac{T}{2}$.

$$\therefore 2 = e^{bt}, \quad \text{or,} \quad \log_e e^{bt} = \log_e 2, \quad \text{or,} \quad bt = 0.6931$$

$$\therefore t = \frac{0.6931}{0.1} = 6.931 \text{ secs.}$$

4. Show that for a forced vibration the total energy of the vibrating system is not constant, and also prove that in such a case

$$\frac{\text{average P.E}}{\text{average K.E}} = \frac{\omega_0^2}{\omega^2}$$

where $\omega_0 = \sqrt{S/m}$. **[C.U. (Hons.) 1989]**

The K.E of a particle of mass m in forced vibration is

$$\frac{1}{2}m\dot{x}^2 = \frac{1}{2}mA^2\omega^2 \sin^2(\omega t - \phi),$$

where the displacement $x = A\cos(\omega t - \phi)$ (considering the real part of the solution) and the P.E. is

$$\frac{1}{2}Sx^2 = \frac{1}{2}SA^2 \cos^2(\omega t - \phi),$$

so that the total energy is

$$E = \text{K.E} + \text{P.E.} = \frac{1}{2}mA^2\omega^2 \sin^2(\omega t - \phi) + \frac{1}{2}SA^2 \cos^2(\omega t - \phi)$$

which shows that E is a function of time, and is therefore not constant.

Now average K.E over a complete cycle

$$= \frac{1}{2}mA^2\omega^2 \cdot \frac{1}{T}\int_0^T \sin^2(\omega t - \phi)dt = \frac{1}{4}mA^2\omega^2 \quad \left[\because \int_0^T \sin^2(\omega t - \phi)dt = \frac{T}{2}\right]$$

Again average P.E. over a complete cycle

$$= \frac{1}{2}SA^2\frac{1}{T}\int_0^T \cos^2(\omega t - \phi)dt = \frac{1}{4}SA^2, \quad \left[\because \int_0^T \cos^2(\omega t - \phi)dt = \frac{T}{2}\right]$$

so that we may write

$$\frac{\text{average P.E.}}{\text{average K.E}} = \frac{S}{m\omega^2} = \frac{\omega_0^2}{\omega^2} \quad \left[\because \omega_0 = \text{natural angular frequency} = \sqrt{\frac{S}{m}}\right]$$

5. Show that in the steady state of forced vibration, the average rate of supply of energy by the driving force is equal to the average rate of work done against the frictional resistance.

[C.U. (Hons.) 1996, '98, 2006]

In order to maintain the steady state of forced vibration, the loss of energy owing to dissipation in each cycle due to frictional resistance must be made up by the energy supplied by the driving force. Let, at any instant, the force $F\sin\omega t$ move through a small distance δx in time δt. Work done is then given by $F\sin\omega t\delta x$, so that the time average rate of work done over a complete cycle is

$$\frac{1}{T}\int_0^T F\sin\omega t\frac{dx}{dt}dt = \frac{1}{T}\int_0^T F\sin\omega t \times \omega A\cos(\omega t - \alpha)dt$$

[Taking $x = A\sin(\omega t - \alpha)$ with the same notations used in example 1.]

$$= \frac{1}{T}\int_0^T F\omega A(\sin^2\omega t\sin\alpha + \sin\omega t\cos\omega t\cos\alpha)dt$$

$$= \frac{1}{T}\cdot F\omega A\frac{T}{2}\sin\alpha \quad \left[\because \int_0^T \sin^2\omega t dt = \frac{T}{2} \text{ and } \int_0^T \sin\omega t\cos\omega t dt = 0\right]$$

$$= \frac{F\omega A\sin\alpha}{2} = \frac{F\omega A}{2}\frac{2b\omega A}{f} = \frac{F\omega A}{2}\cdot\frac{K}{m}A\omega\cdot\frac{m}{F} = \frac{K}{2}A^2\omega^2.$$

Again the work done against frictional resistance for the displacement δx is $K\dfrac{dx}{dt}\delta x$, so that the time average rate of work done against frictional force is

$$\frac{1}{T}\int_0^T K\frac{dx}{dt}\frac{dx}{dt}dt = \frac{1}{T}\int_0^T KA^2\omega^2\cos^2(\omega t-\alpha)dt = \frac{1}{T}KA^2\omega^2\frac{T}{2} = \frac{K}{2}A^2\omega^2.$$

Thus the average rate of supply of energy by the driving force is equal to the average rate of work done against frictional resistance.

6. Calculate the mechanical impedance of a particle subjected to a restoring force proportional to displacement and a resisting force proportional to velocity. Find an expression for the velocity amplitude. When will it be maximum? **[C.U. (Hons.) 2005]**

Equation of motion of such a particle is

$$m\frac{d^2x}{dt^2}+K\frac{dx}{dt}+Sx = F\sin\omega t$$

and the steady state displacement is $x = A\sin(\omega t-\alpha)$, where

$$A = \frac{f}{\sqrt{(\omega_0^2-\omega^2)^2+4b^2\omega^2}},\quad \tan\alpha = \frac{2b\omega}{\omega_0^2-\omega^2},\quad \omega_0^2 = \frac{S}{m},\quad 2b = \frac{K}{m},\quad f = \frac{F}{m}$$

$$\therefore\ \frac{dx}{dt} = \frac{f\omega\cos(\omega t-\alpha)}{\sqrt{(\omega_0^2-\omega^2)^2+4b^2\omega^2}} = \frac{\frac{F}{m}\omega\cos(\omega t-\alpha)}{\sqrt{\left(\frac{S}{m}-\omega^2\right)^2+\frac{K^2\omega^2}{m^2}}} = \frac{\frac{F\omega}{m}\cos(\omega t-\alpha)}{\frac{\omega}{m}\sqrt{\frac{m^2}{\omega^2}\left(\frac{S}{m}-\omega^2\right)^2+K^2}}$$

$$= \frac{F\cos(\omega t-\alpha)}{\sqrt{\left(\frac{S}{\omega}-m\omega\right)^2+K^2}} = \frac{F\cos(\omega t-\alpha)}{Z} \tag{1}$$

where

$$Z = \sqrt{\left(\frac{S}{\omega}-m\omega\right)^2+K^2}$$

is the mechanical impedance. The velocity amplitude is then $\dfrac{F}{Z}$, F being the amplitude of the force. The displacement amplitude is evidently $\dfrac{F}{\omega Z}$. The mechanical impedance is then similar to the electrical impedance

$$\sqrt{\left(L\omega-\frac{1}{c\omega}\right)^2+R^2}$$

of an a.c. circuit consisting of an inductance L, capacitance C and a resistance R in series, when an a.c e.m.f of angular frequency ω is impressed on the circuit.

From the expression for steady state displacement, we may write

$$\frac{dx}{dt} = A\omega\cos(\omega t-\alpha) = A\omega\sin\left(\frac{\pi}{2}+\omega t-\alpha\right) = A\omega\sin(\omega t+\theta),$$

where

$$\theta = \frac{\pi}{2} - \alpha \quad \text{and} \quad \tan\theta = \cot\alpha = \frac{(\omega_0^2 - \omega^2)}{2b\omega}.$$

$$\therefore \quad \frac{dx}{dt} = \frac{F\sin(\omega t + \theta)}{Z} = \frac{F\sin(\omega t + \theta)}{\sqrt{\left(\frac{S}{\omega} - m\omega\right)^2 + K^2}}.$$

Now the velocity amplitude is maximum when Z is minimum. For given values of K and S, this will be so when

$$\frac{S}{\omega} = m\omega, \quad \text{or,} \quad \omega^2 = \frac{S}{m} = \omega_0^2.$$

Hence the velocity amplitude at resonance is F/K and $\tan\theta = 0$, i.e., $\theta = 0$. Thus the velocity and the periodic force are in phase at resonance, but at all other conditions, velocity is ahead of the force by a phase angle

$$\theta = \tan^{-1}\frac{(\omega_0^2 - \omega^2)}{2b\omega} = \tan^{-1}\left(\frac{\frac{S}{\omega} - m\omega}{K}\right).$$

7. Deduce an expression for power per cycle, and show that at velocity resonance, the power factor is unity.

We have shown in example 5, that

$$\text{Power} = \text{Time average rate of work done} = \bar{P} = \frac{F\omega A\sin\alpha}{2},$$

where F is the amplitude of the sinusoidal force, α the phase lag of displacement behind the force and A the displacement amplitude.

Writing

$$\sin\alpha = \sin(\frac{\pi}{2} - \theta) = \cos\theta,$$

θ being the phase of the velocity ahead of the force, we have

$$\bar{P} = \frac{F\omega A\cos\theta}{2}.$$

Now

$$x = A\sin(\omega t - \alpha). \quad \therefore \quad \frac{dx}{dt} = A\omega\cos(\omega t - \alpha),$$

so that velocity amplitude $= \omega A$ and $\bar{P} = \dfrac{F\omega A}{2}\cos\theta$.

The factor $\cos\theta$ is the power factor — a term usually applied in calculating power in a.c. circuits.

Now since

$$\tan\theta = \frac{\omega_0^2 - \omega^2}{2b\omega}, \quad \cos\theta = \frac{2b\omega}{\sqrt{(\omega_0^2 - \omega^2)^2 + 4b^2\omega^2}} = \sin\alpha,$$

so that

$$\bar{P} = \frac{F}{2}\omega A \frac{2b\omega}{\sqrt{(\omega_0^2-\omega^2)^2+4b^2\omega^2}}$$

$$= \frac{F}{2}\omega \frac{F/m}{\sqrt{(\omega_0^2-\omega^2)^2+4b^2\omega^2}} \cdot \frac{2b\omega}{\sqrt{(\omega_0^2-\omega^2)^2+4b^2\omega^2}}$$

$$\left(\text{putting } A = \frac{F/m}{\sqrt{(\omega_0^2-\omega^2)^2+4b^2\omega^2}}\right)$$

$$= \frac{F^2}{m}\frac{b\omega^2}{(\omega_0^2-\omega^2)^2+4b^2\omega^2} \tag{1}$$

$\bar{P}$ will be maximum, when for a given value of b, the denominator is minimum, i.e., when $\omega_0 = \omega$, and then we may write

$$P_{\text{res}} = P_{\text{max}} = \frac{F^2}{m}\cdot\frac{1}{4b} = \frac{F^2}{2K}.$$

Also when $\omega_0 = \omega$, the power factor

$$\cos\theta = \frac{2b\omega}{\sqrt{(\omega_0^2-\omega^2)^2+4b^2\omega^2}} = 1$$

so at velocity resonance power factor is unity.

8. If ω_1 and ω_2 be the half-power frequencies, and ω_0 is the resonant frequency of a forced system, show that $\omega_0^2 = \omega_1\omega_2$. **[C.U. (Hons.) 1987, 2006]**

We have already shown in problem (7) that the average power

$$\bar{P} = \frac{F^2}{m}\frac{b\omega^2}{(\omega_0^2-\omega^2)^2+4b^2\omega^2}$$

and for resonance $\omega_0 = \omega$, $\bar{P}$ becomes maximum and given by $P_{\text{max}} = \dfrac{F^2}{4bm}$. Let now ω be the angular frequency of the periodic force when

$$\bar{P} = \frac{P_{\text{max}}}{2}, \quad \text{i.e.,} \quad \frac{F^2}{m}\frac{b\omega^2}{(\omega_0^2-\omega^2)^2+4b^2\omega^2} = \frac{1}{2}\frac{F^2}{4bm}$$

which on simplification gives

$$(\omega_0^2-\omega^2)^2+4b^2\omega^2 = 8b^2\omega^2, \quad \text{or,} \quad \omega_0^4 - 2\omega_0^2\omega^2 + \omega^4 - 4b^2\omega^2 = 0,$$

$$\text{or,} \quad (\omega^2)^2 - (2\omega_0^2+4b^2)\omega^2 + (\omega_0^2)^2 = 0,$$

which is a quadratic equation having two roots ω_1^2 and ω_2^2 (say), so that we have

$$\omega_1^2\omega_2^2 = \omega_0^4, \quad \text{or,} \quad \omega_1\omega_2 = \omega_0^2.$$

N.B. (i) If the average power supplied to the driven system is plotted as a function of the frequency of the driving force of constant amplitude, a curve as shown in the attached fig. is obtained. The rapidity with which the average power diminishes as ω departs from the value at resonance is a *measure of sharpness of resonance.* It is evident from the figure that there are two values of ω — (i) one $\omega_1 < \omega_0$ and (ii) the other $\omega_2 > \omega_0$ for which $\bar{P} = \frac{1}{2}P_{\text{max}}$, i.e., $\frac{1}{2}P_{\text{res}}$. Calling these two as *half-power frequencies*, we may write $Q = \frac{\omega_0}{(\omega_2 - \omega_1)}$ as a measure of the sharpness of resonance. Q is known as the *quality factor;* $(\omega_2 - \omega_1)$ is called the *band width.* Also half-width of the resonance curve $= (\omega_0 - \omega_1) = (\omega_2 - \omega_0)$; so $(\omega_2 - \omega_1) = (\omega_2 - \omega_0) + (\omega_0 - \omega_1) = 2$ half-width, so that half-width $= \frac{\omega_0}{2Q}$. Again the log decrement $\lambda = \frac{\pi}{Q}$.

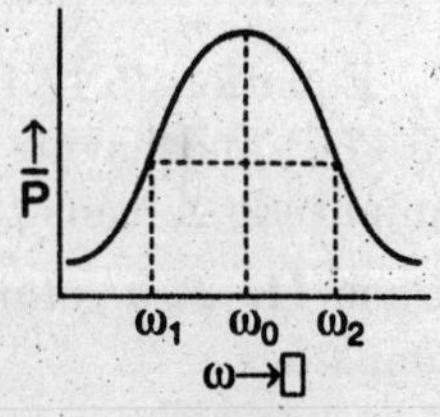

Fig. 4

$$\therefore\ Q = \frac{\pi}{\lambda};\ \text{so half-width} = \frac{\omega_0}{2Q} = \frac{\omega_0 \lambda}{2\pi} = \frac{\lambda}{2\pi}\ \text{if}\ \omega_0 = 1.$$

(ii) We know

$$A = \frac{f}{\sqrt{(\omega_0^2 - \omega^2)^2 + 4b^2\omega^2}},\quad \text{or,}\quad A^2 = \frac{f^2}{(\omega_0^2 - \omega^2)^2 + 4b^2\omega^2}.$$

Now maximum amplitude occurs at $\omega = \omega_0$

$$\therefore\quad A^2 = \frac{f^2}{4b^2\omega^2} = \frac{f^2}{4b^2\omega_0^2}\quad (\because\ \omega = \omega_0)$$

$$\therefore\quad A^2 = \frac{f^2 m}{4b^2 S}\quad \text{putting}\ \omega_0 = \sqrt{\frac{S}{m}}$$

i.e., the maximum of A^2 does neither depend on ω nor on ω_0.

5.17 Composition of Simple Harmonic Motions

In Light and Sound, we come across certain phenomena such as interference, beats, etc., where superposition of two simple harmonic motions is involved. We shall now discuss them both analytically and graphically.

Analytical treatment

Case I. *Superposition of two collinear simple harmonic motions of same frequency, but differing in amplitude and phase :*

Let the two collinear simple harmonic motions be represented by

$$x_1 = a \sin \omega t;\quad x_2 = b \sin(\omega t + \delta)$$

where a, b are the respective amplitudes, and δ is the phase difference between the two motions.

Hence, the resultant displacement is given from the principle of superposition, by

$$x = x_1 + x_2 = a \sin \omega t + b \sin(\omega t + \delta) = a \sin \omega t + b \sin \omega t \cos \delta + b \cos \omega t \sin \delta$$

$$= (a + b \cos \delta) \sin \omega t + b \sin \delta \cos \omega t = A \sin(\omega t + \phi) \tag{5.17.1}$$

where $a + b \cos \delta = A \cos \phi$ and $b \sin \delta = A \sin \phi$ so that by squaring and adding, we have

$$A^2 = a^2 + b^2 + 2ab \cos \delta. \quad \therefore \quad A = \sqrt{a^2 + b^2 + 2ab \cos \delta} \tag{5.17.2}$$

and by division, we have

$$\tan\phi = \frac{b\sin\delta}{a + b\cos\delta}. \tag{5.17.3}$$

Equation (5.17.1) represents a simple harmonic motion of amplitude A given by equation (5.17.2) and having the same frequency as that of the components. The resultant motion of amplitude A leads the first component by an angle ϕ given by equation (5.17.3).

Case II. *Two collinear simple harmonic motions of nearly equal frequencies but differing in amplitude :*

Let the two collinear simple harmonic motions be represented by

$$x_1 = a\sin\omega t = a\sin 2\pi n_1 t; \quad \text{and} \quad x_2 = b\sin 2\pi(n_1 + n_2)t$$

where a, b are the respective amplitudes, and n_1, $(n_1 + n_2)$ are the frequencies of the two motions, n_2 being very small.

The resultant motion is given by

$$\begin{aligned} x = x_1 + x_2 &= a\sin 2\pi n_1 t + b\sin 2\pi(n_1 + n_2)t \\ &= a\sin p_1 t + b\sin(p_1 + p_2)t \quad [\text{where } p_1 = 2\pi n_1 \text{ and } p_2 = 2\pi n_2] \\ &= a\sin p_1 t + b\sin p_1 t\cos p_2 t + b\cos p_1 t\sin p_2 t \\ &= (a + b\cos p_2 t)\sin p_1 t + b\sin p_2 t\cos p_1 t \\ &= A\cos\phi\sin p_1 t + A\sin\phi\cos p_1 t = A\sin(p_1 t + \phi) \end{aligned} \tag{5.17.4}$$

where $a + b\cos p_2 t = A\cos\phi$ and $b\sin p_2 t = A\sin\phi$ so that by squaring and adding, we get

$$A^2 = a^2 + b^2 + 2ab\cos p_2 t. \quad \therefore \quad A = \sqrt{a^2 + b^2 + 2ab\cos p_2 t}$$

and by division, we get

$$\tan\phi = \frac{b\sin p_2 t}{a + b\cos p_2 t}.$$

The resultant displacement given by expression (5.17.4) is not simple harmonic, since both A and ϕ change with time t :

The variation of the amplitude A with time t is of much importance in the study of the phenomenon of 'beats' in sound.

When $p_2 t = 0, 2\pi, 4\pi$, etc., A is maximum, and is equal to $(a + b)$.

Hence, the resultant amplitude reaches the maximum value after regular intervals of time $t = 2\pi/p_2$.

Again when $p_2 t = \pi$, 3π, 5π, etc., A is minimum, and is equal to $(a - b)$.

Thus the resultant amplitude attains its minimum value after regular intervals of time $t = 2\pi/p_2$.

Thus the resultant amplitude of motion passes alternately through its maximum and minimum values and the time between two successive maxima or minima of the amplitude A is

$$\frac{2\pi}{p_2} = \frac{1}{n_2} = \frac{1}{\text{difference of frequencies of the component motions}}.$$

Therefore, the number of two successive maxima or minima in one second, i.e., the number beats in one sec is n_2 which is the difference between the frequencies of the component motions.

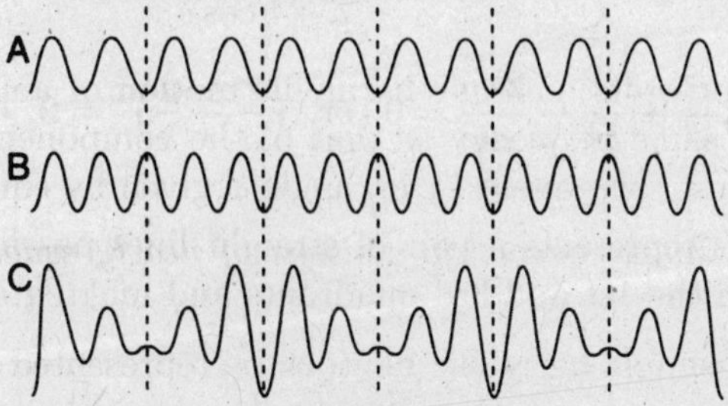

Fig. 5.17.1

The above result may be represented graphically, which explains the mechanism of formation of beats. In the figure (5.17.1), A and B represent two simple harmonic motions differing in frequencies in the ratio 4 : 5. C represents the resultant curve obtained by the superposition of the two component motions. Between two successive maxima or minima of C, the number of waves in A and B differ by 1. The two component waves meeting in the same phase give rise to maxima, and while meeting in opposite phases, give rise to minima — which, in case of sound waves, produces waxing and waning of sound known as 'beats'.

Case III. *Two rectangular simple harmonic motions of the same frequency, but differing in amplitude and phase (i.e., frequencies being in the ratio* 1 : 1).

Let the two simple harmonic motions acting at right angles to each other, be represented by

$$x = a\sin\omega t; \quad y = b\sin(\omega t + \delta).$$

The resultant motion is obtained as follows. From the above two equations, we have

$$\frac{x}{a} = \sin\omega t \quad \text{and} \quad \frac{y}{b} = \sin(\omega t + \delta) = \sin\omega t\cos\delta + \cos\omega t\sin\delta.$$

Since $\dfrac{x}{a} = \sin\omega t$ and $\cos\omega t = \sqrt{1 - \sin^2\omega t} = \sqrt{1 - \dfrac{x^2}{a^2}}$, we have,

$$\frac{y}{b} = \frac{x}{a}\cos\delta + \sqrt{1 - \frac{x^2}{a^2}}\sin\delta, \quad \text{or,} \quad \frac{y}{b} - \frac{x}{a}\cos\delta = \sqrt{1 - \frac{x^2}{a^2}}\sin\delta.$$

Now squaring both sides, we get

$$\frac{y^2}{b^2} + \frac{x^2}{a^2}\cos^2\delta - \frac{2xy}{ab}\cos\delta = \left(1 - \frac{x^2}{a^2}\right)\sin^2\delta \quad \text{or,} \quad \frac{x^2}{a^2} + \frac{y^2}{b^2} - \frac{2xy}{ab}\cos\delta = \sin^2\delta. \qquad (5.17.5)$$

This represents the equation to an ellipse inclined to the axes of the co-ordinates and contained within the rectangle of sides $2a$ and $2b$ as shown the figure (5.17.2). The major axis of the ellipse is inclined to the X-axis by an angle ϕ, given by

$$\tan 2\phi = \frac{2ab}{a^2 - b^2}\cos\delta$$

Hence, the resulting motion will be elliptic in nature.

By giving different values to δ, the following important cases may be discussed :

(a) When $\delta = 0$, i.e., when the two motions are in phase, we have

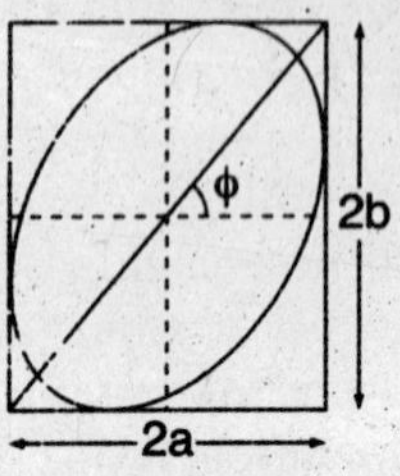

Fig. 5.17.2

$$\frac{x^2}{a^2} + \frac{y^2}{b^2} - \frac{2xy}{ab} = 0, \text{ or, } \left(\frac{x}{a} - \frac{y}{b}\right)^2 = 0, \text{ or, } \pm\left(\frac{x}{a} - \frac{y}{b}\right) = 0.$$

This represents a pair of straight lines passing through the origin, and lying in the 1st and 3rd quadrants and making an angle θ with the X-axis given by

$$\tan\theta = \frac{b}{a} \quad \text{(Fig. 5.17.3).}$$

(b) When $\delta = \pi$, i.e., when the two motions are in opposite phases, we have

$$\frac{x^2}{a^2} + \frac{y^2}{b^2} + \frac{2xy}{ab} = 0, \quad \text{or,} \quad \pm\left(\frac{x}{a} + \frac{y}{b}\right) = 0.$$

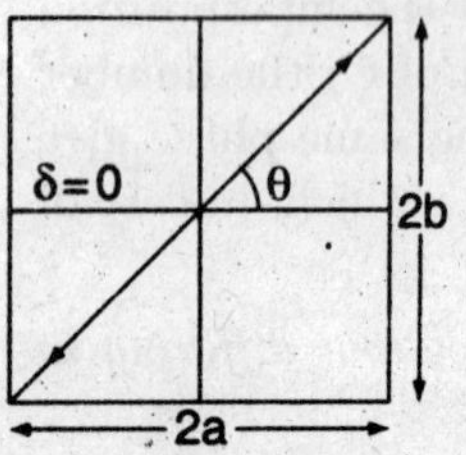

Fig. 5.16.3

This also represents a pair of straight lines, passing through the origin lying in the 2nd and 4th quadrants and inclined to the X-axis by an angle θ given by

$$\tan\theta = -\frac{b}{a}, \quad \text{as shown in Fig. 5.17.4.}$$

In the above two cases, if $a = b$, i.e., if the two motions have got the same amplitude, we have $\theta = +45°$ in the first case and $\theta = -45^0$ in the second case, i.e., in the second case, the straight line makes an angle of 135° with the positive direction of X-axis.

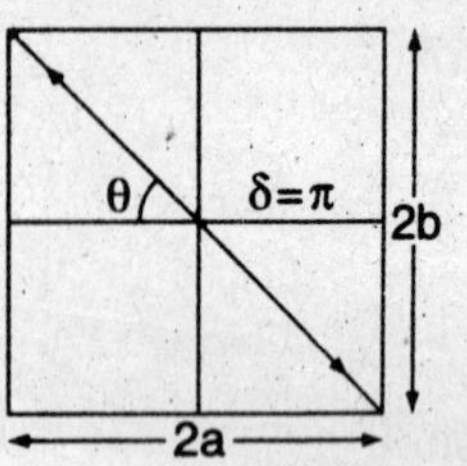

Fig. 5.16.4

(c) When $\delta = \frac{\pi}{2}$, i.e., when the phases of the two motions differ by $\frac{\pi}{2}$, we have

$$\frac{x^2}{a^2} + \frac{y^2}{b^2} = 1$$

which represents an ellipse with semi-axes a and b coinciding with the co-ordinate axes, i.e., the axes of the simple harmonic motions as shown in Fig. 5.17.5.

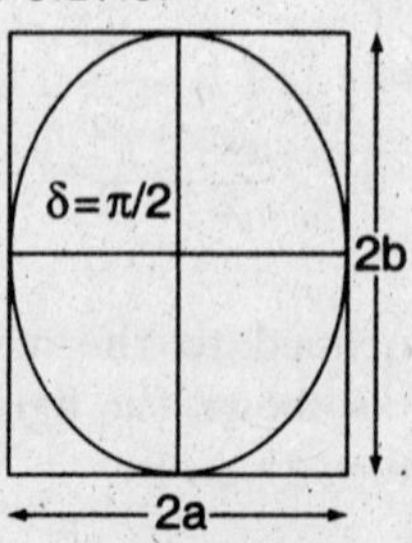

Fig. 5.17.5

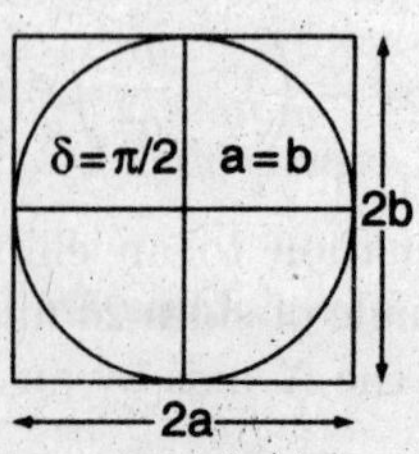

Fig. 5.17.6

If now, $a = b$, i.e., if the component motions have got the same amplitude, we have $x^2 + y^2 = a^2$ which shows that the resultant motion is a circuar one with radius a (Fig. 5.17.6).

5.18 Composition of Two Rectangular Simple Harmonic Motions differing in Amplitude, Phase and Frequency—Frequencies being in the Ratio 1 : 2

Let the two rectangular simple harmonic motions be represented by

$$x = a\sin\omega t;\quad y = b\sin(2\omega t + \delta).$$

The resultant motion is obtained as follows. From the above two equations we may write

$$\frac{x}{a} = \sin\omega t \quad \text{and} \quad \frac{y}{b} = \sin(2\omega t + \delta).$$

Now

$$\frac{y}{b} = \sin 2\omega t\cos\delta + \cos 2\omega t\sin\delta = 2\sin\omega t\cos\omega t\cos\delta + (1 - 2\sin^2\omega t)\sin\delta$$

$$= 2\frac{x}{a}\sqrt{1 - \frac{x^2}{a^2}}\cos\delta + \left(1 - 2\frac{x^2}{a^2}\right)\sin\delta$$

$$\left[\because \frac{x}{a} = \sin\omega t \text{ and } \cos\omega t = \sqrt{1 - \sin^2\omega t} = \sqrt{1 - \frac{x^2}{a^2}}\right]$$

or, $$\frac{y}{b} - \left(1 - \frac{2x^2}{a^2}\right)\sin\delta = 2\frac{x}{a}\sqrt{1 - \frac{x^2}{a^2}}\cos\delta.$$

Squaring both sides, we have

$$\frac{y^2}{b^2} + \left(1 - \frac{2x^2}{a^2}\right)^2\sin^2\delta - \frac{2y}{b}\left(1 - \frac{2x^2}{a^2}\right)\sin\delta = \frac{4x^2}{a^2}\left(1 - \frac{x^2}{a^2}\right)\cos^2\delta.$$

Adding $\left(1 - \frac{2x^2}{a^2}\right)^2\cos^2\delta$ to both sides, we have

$$\frac{y^2}{b^2} + \left(1 - \frac{2x^2}{a^2}\right)^2 - 2\frac{y}{b}\left(1 - \frac{2x^2}{a^2}\right)\sin\delta = 4\frac{x^2}{a^2}\left(1 - \frac{x^2}{a^2}\right)\cos^2\delta + \left(1 - \frac{2x^2}{a^2}\right)^2\cos^2\delta$$

$$= \cos^2\delta\left[4\frac{x^2}{a^2}\left(1 - \frac{x^2}{a^2}\right) + \left(1 - \frac{2x^2}{a^2}\right)^2\right] = \cos^2\delta.$$

The resultant motion is represented by the above equation. We consider the following particular cases.

(a) When $\delta = 0$, i.e., the two motions are in the same phase, we have

$$\frac{y^2}{b^2} + \left(1 - \frac{2x^2}{a^2}\right)^2 = 1,\quad \text{or,}\quad \frac{y^2}{b^2} - \frac{4x^2}{a^2} + \frac{4x^4}{a^4} = 0,\quad \text{or,}\quad \frac{y^2}{b^2} + \frac{4x^2}{a^2}\left(\frac{x^2}{a^2} - 1\right) = 0$$

which is a 4th degree equation and represents a curve having two loops as shown in Fig. 5.18.1. As the phase difference is gradually changed, the shape of the loop will also gradually change.

(b) When $\delta = \pi/2$, i.e., when the phases of the two motions differ by $\pi/2$, we have

$$\frac{y^2}{b^2} + \left(1 - \frac{2x^2}{a^2}\right)^2 - \frac{2y}{b}\left(1 - \frac{2x^2}{a^2}\right) = 0 \text{ or, } \frac{y^2}{b^2} + 1 - \frac{4x^2}{a^2} + \frac{4x^4}{a^4} - \frac{2y}{b} + \frac{4y}{b}\frac{x^2}{a^2} = 0$$

$$\text{or, } \left(\frac{y^2}{b^2} - \frac{2y}{b} + 1\right) + \frac{4x^4}{a^4} + \frac{4x^2}{a^2}\left(\frac{y}{b} - 1\right) = 0 \text{ or, } \left(\frac{y}{b} - 1\right)^2 + \left(\frac{2x^2}{a^2}\right)^2 + \frac{4x^2}{a^2}\left(\frac{y}{b} - 1\right) = 0$$

$$\text{or, } \left[\left(\frac{y}{b} - 1\right) + \frac{2x^2}{a^2}\right]^2 = 0.$$

This represents two coincident parabolas (as shown in Fig. 5.18.2) each of which is represented by

$$\left(\frac{y}{b} - 1\right) + \frac{2x^2}{a^2} = 0, \quad \text{or,} \quad \frac{2x^2}{a^2} = 1 - \frac{y}{b} \quad \text{or,} \quad x^2 = \frac{a^2}{2}\left(1 - \frac{y}{b}\right) = -\frac{a^2}{2b}(y - b).$$

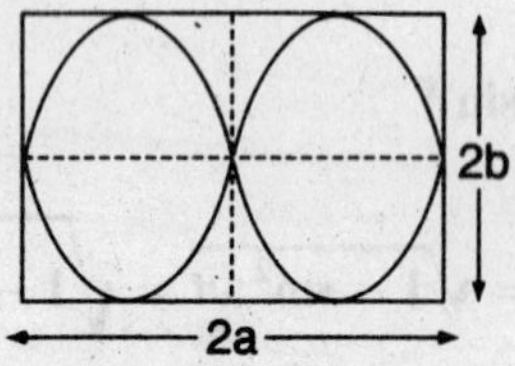

Fig. 5.18.1

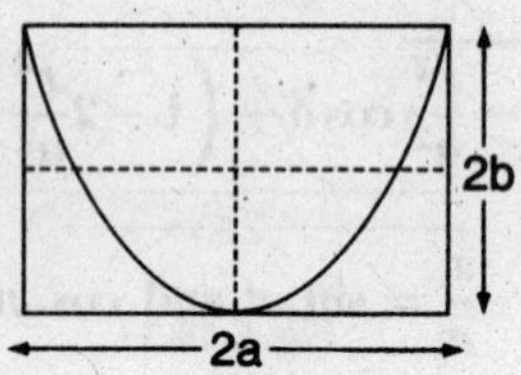

Fig. 5.18.2

5.19 Two Rectangular S.H. Motions differing in Amplitude, Phase and Frequency—Frequencies being in the Ratio 1 : 3

Let the two rectangular simple harmonic motions be represented by

$$x = a\sin\omega t, \quad y = b\sin(3\omega t + \delta).$$

From the above two equations, we may write

$$\frac{x}{a} = \sin\omega t, \quad \cos\omega t = \sqrt{1 - \frac{x^2}{a^2}}.$$

$$\therefore \quad \frac{y}{b} = \sin 3\omega t\cos\delta + \cos 3\omega t\sin\delta$$

$$= (3\sin\omega t - 4\sin^3\omega t)\cos\delta + (4\cos^3\omega t - 3\cos\omega t)\sin\delta$$

$$= \left(3\frac{x}{a} - 4\frac{x^3}{a^3}\right)\cos\delta + \left[4\left(1 - \frac{x^2}{a^2}\right)^{3/2} - 3\left(1 - \frac{x^2}{a^2}\right)^{1/2}\right]\sin\delta$$

$$= \left(\frac{3x}{a} - \frac{4x^3}{a^3}\right)\cos\delta + \left(1 - \frac{x^2}{a^2}\right)^{1/2}\left(1 - \frac{4x^2}{a^2}\right)\sin\delta.$$

Transposing and squaring

$$\left[\frac{y}{b}-\left(\frac{3x}{a}-\frac{4x^3}{a^3}\right)\cos\delta\right]^2=\left(1-\frac{x^2}{a^2}\right)\left(1-\frac{4x^2}{a^2}\right)^2\sin^2\delta.$$

The resultant motion is represented by the above equation. We now consider the following particular cases.

(a) When $\delta = 0$, i.e., when the two motions are in the same phase, the above equation reduces to

$$\left[\frac{y}{b}-\left(\frac{3x}{a}-\frac{4x^3}{a^3}\right)\right]^2=0,$$

which represents two coincident cubic curves (Fig. 5.19.1).

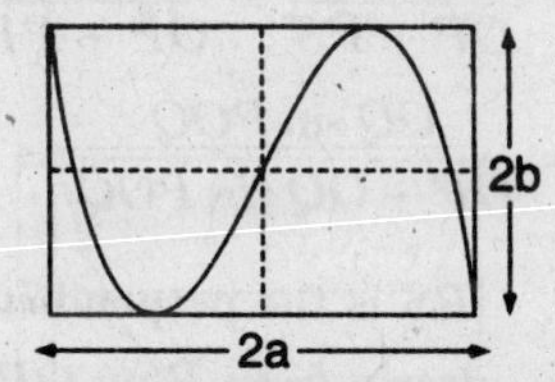

Fig. 5.19.1

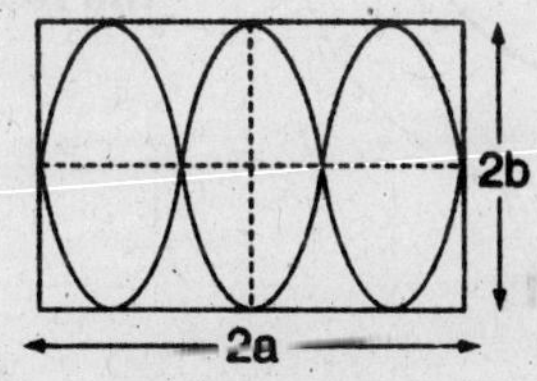

Fig. 5.19.2

(b) When $\delta = \pi/2$, i.e., when the phases of the two motions differ by $\pi/2$, we have

$$\left(1-\frac{x^2}{a^2}\right)\left(1-\frac{4x^2}{a^2}\right)^2-\frac{y^2}{b^2}=0$$

which is an equation of sixth degree, giving a curve of three loops (Fig. 5.19.2). As the phase difference is gradually changed, the shape of the loop will also gradually change.

In general, if the frequencies are in the ratio N, the curve representing the path of the particle will have N loops.

5.20 Graphical Treatment

5.20.1 Composition of two collinear simple harmonic motions of same period but differing in amplitude and phase

We consider two simple harmonic motions having the same period but different amplitudes and phases represented by two circles of reference of radius a and b (Fig. 5.20.1). In the figure P and Q represent the positions of two points on the circles rotating in an anti-clockwise direction with angular velocity ω. The displacements of the two S.H.Ms corresponding to two circles of reference are OB and OA, and their phase difference is $\angle QOX - \angle POX = \angle QOP$. Since the two component displacements are along the same straight line OX, resultant displacement of the two S.H.Ms by the principle of superposition $= OB + OA$. Now, we complete the parallelogram $OPRQO$. As PR is equal and parallel to OQ the projection of PR is equal to that of OQ.

$\therefore$ $BC = OA$, and so the resultant displacement $= OB + BC = OC =$ the projection of OR on XOX'. Thus we see that if R rotates in a circle of radius OR in anti-clockwise direction the projection of OR gives the resultant displacement of the two S.H.Ms. Hence, the resultant motion is a S.H.M denoted by the circle of reference of radius OR which gives the amplitude of the resultant motion. The amplitude is given by

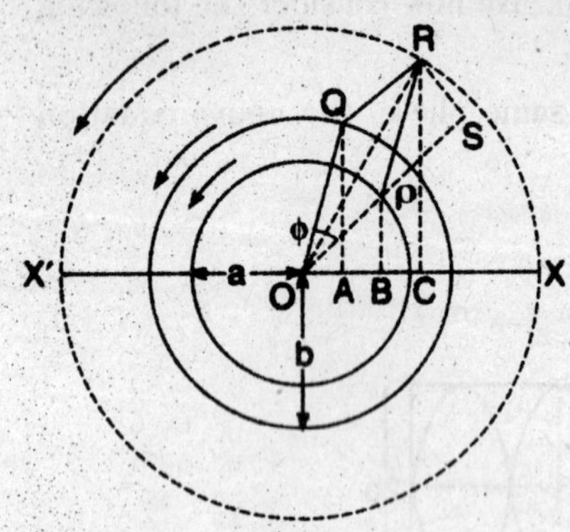

Fig. 5.18.1

$$OR = \sqrt{OP^2 + OQ^2 + 2 \cdot OP \cdot OQ \cos POQ}$$
$$= \sqrt{a^2 + b^2 + 2ab\cos\phi}.$$

The phase difference between the resultant S.H.M and the S.H.M of amplitude OP is given by

$$\tan ROP = \frac{RS}{OP + PS} = \frac{PR \sin RPS}{OP + PR \cos RPS}$$
$$= \frac{OQ \sin POQ}{OP + OQ \cos POQ} = \frac{b \sin\phi}{a + b\cos\phi} \qquad (5.20.1)$$

[RS is the perpendicular drawn from R on OP (produced).]

Two S.H.Ms of the same period may be added like vectors. We draw two straight lines OP and PR of magnitudes a and b making an external angle ϕ between them. RN is the perpendicular drawn from R on OP produced. Considering OP and PR as two vectors and applying the rule of vector addition it is clear that OR stands for the resultant of the two vectors which gives the resultant amplitude of the two S.H.Ms. The phase difference ϕ is the angle between the vectors (Fig. 5.20.2). The angle θ is the angle between the vectors OR and OP.

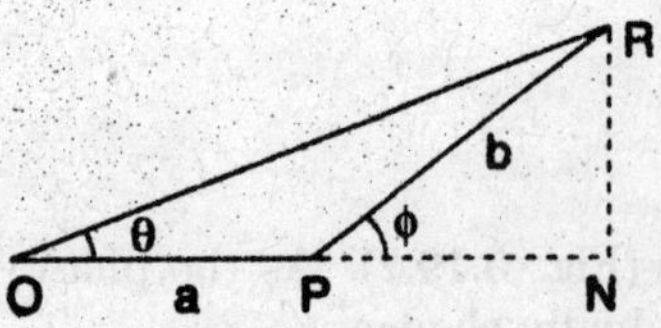

Fig. 5.20.2

5.20.2 Composition of two rectangular simple harmonic motions

Two simple harmonic motions acting at right angles to each other may be compounded graphically. We shall consider the following cases and find out the resultant motion graphically.

Case I. *Two simple harmonic motions having same time period and phase but different amplitudes :*

The two simple harmonic motions are represented by the two circles of motions as shown in Fig. 5.20.3. The radii of the two circles a and b are equal to the amplitudes of the two simple harmonic motions. We divide each circle into eight equal parts (since the ratio of the periods is 1 : 1). The starting points in the circles are marked zero. We then draw straight lines passing through the points of same number in the two circles parallel to two axes XOX' and YOY' along which the two simple harmonic motions take place. The points of intersection of these lines are marked and then joined. The intersecting points will be found to lie on a straight line which gives the path along which the resultant motion takes place, the arrow heads indicating the direction of motion about O. From the figure we see that the straight line AB is the diagonal of a rectangle having sides $2a$ and $2b$. The resultant motion is in a direction which makes an angle θ with the positive direction of X-axis such that $\tan\theta = \frac{a}{b}$. The amplitude of the resultant motion is OA or OB which is equal to $\sqrt{a^2 + b^2}$.

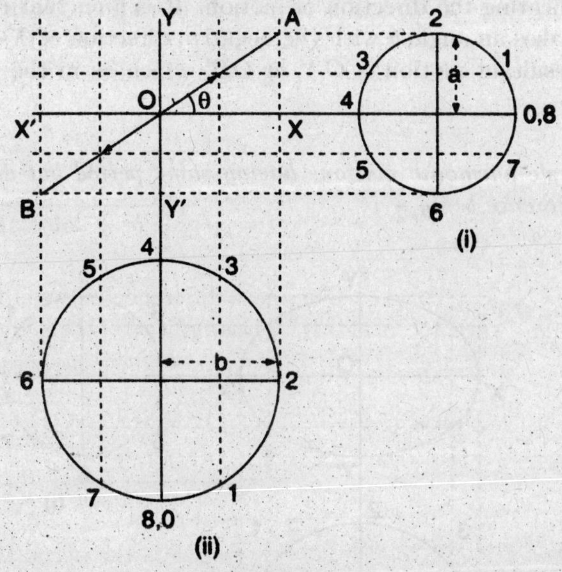

Fig. 5.20.3

Case II. ***Two simple harmonic motions having same period but differing in amplitude and phase, the phase difference being π :***

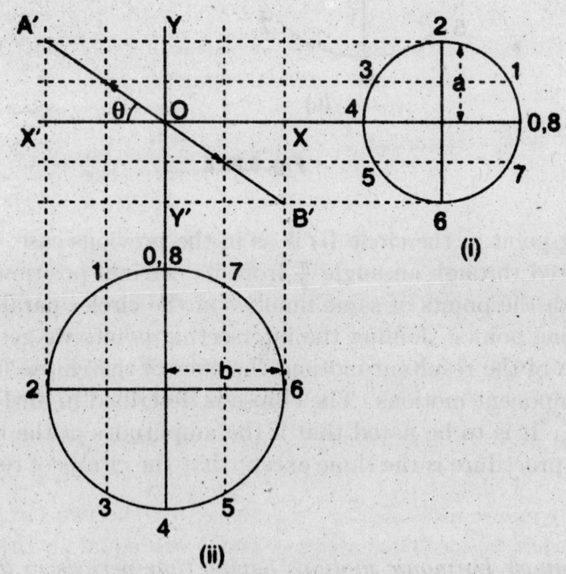

Fig. 5.20.4

Here the starting point in the circle (i) is as in the previous figure, but the starting point in the bigger circle (ii) is at the top of the circle instead of at the bottom as the starting point in this case has been shifted through an angle π. As before we divide the two circles into eight equal parts and draw straight lines through the points of same number in the circles parallel to two axes XOX' and YOY' (Fig. 5.20.4). Joining the points of intersection of these lines we get a straight line $A'B'$ which gives the path along which the resultant motion takes place, the arrow heads indicating the direction of motion. It is seen that the resultant motion is in a direction which makes an angle θ with the negative direction of X-axis such that $\tan\theta = \frac{a}{b}$. The amplitude of resultant motion is OA' or OB' which as in the previous case is equal to $\sqrt{a^2+b^2}$.

Case III. *Two simple harmonic motions having same period but differing in amplitude and phase, the phase difference being* $\frac{\pi}{2}$:

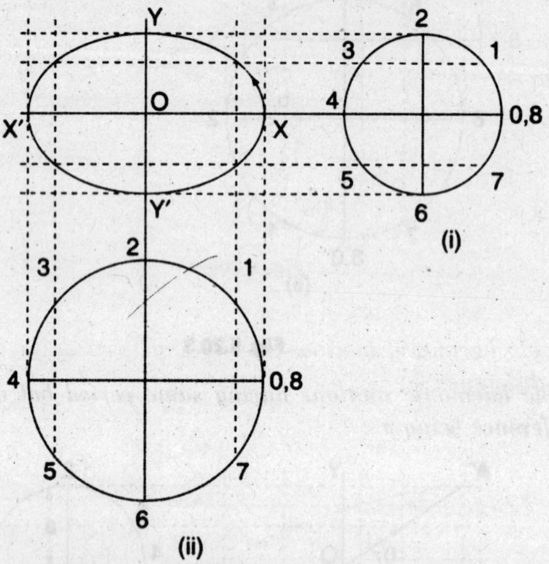

Fig. 5.20.5

Here the starting point in the circle (i) is as in the previous case, but the starting point in the circle (ii) is shifted through an angle $\frac{\pi}{2}$ from its original position. Drawing as before the straight lines through the points of same number in the circles parallel to XOX' and YOY', we get the intersecting points. Joining the intersecting points we get an ellipse (Fig. 5.20.5), which gives the path of the resultant motion, the axes of the ellipse being coincident with the directions of the component motions. The ellipse is described in anti-clockwise direction from the starting point X. It is to be noted that if the amplitudes of the two S.H.Ms are the same we get a circle. The procedure is the same except that the circles of reference of same diameter should be taken.

Case IV. *Two simple harmonic motions having time-periods in the ratio of* 2 : 1, *different amplitudes and no phase difference* :

In this case we divide the circles in the ratio 2 : 1. The circle (i) has been divided into eight equal parts and the circle (ii) into four equal parts. Proceeding as before we find that the resultant path is a pattern as shown in Fig. 5.20.6. The direction of the motion is indicated by the arrow heads.

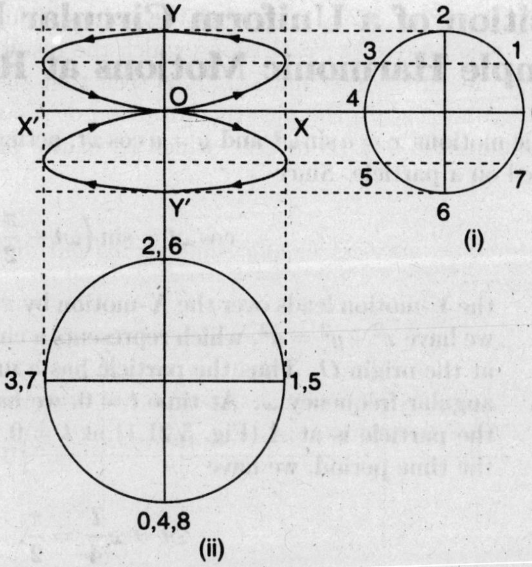

Fig. 5.20.6

Case V. *Two simple harmonic motions having time-periods in the ratio* 2 : 1, *different amplitudes and phase difference* $\frac{\pi}{2}$*:*

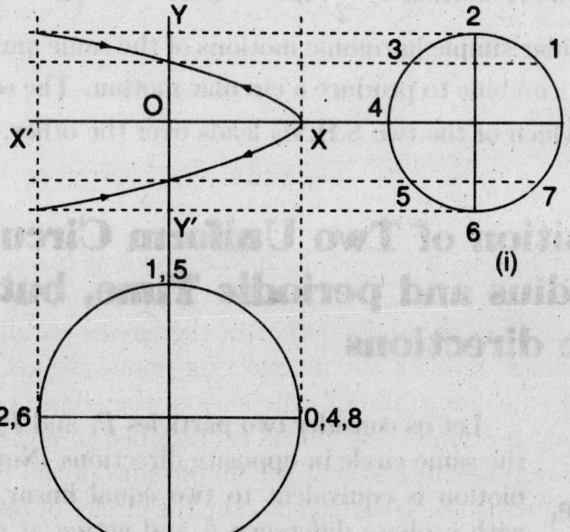

Fig. 5.20.7

Proceeding as in the previous case and shifting the starting point in the circle (ii) through angle $\frac{\pi}{2}$ we find that the resultant motion is a parabola as shown in Fig. 5.20.7, the directions of motion being indicated by arrow marks.

5.21 Composition of a Uniform Circular Motion from Two Simple Harmonic Motions at Right Angles

Let two simple harmonic motions $x = a\sin\omega t$ and $y = a\cos\omega t$, acting along X and Y axes be simultaneously impressed on a particle. Since

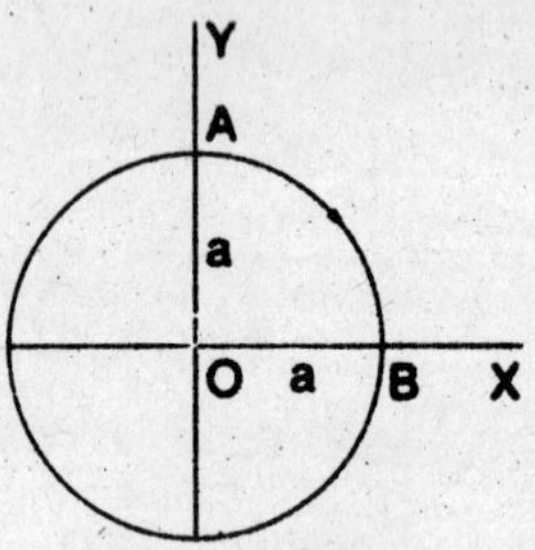

Fig. 5.21.1

$$\cos\omega t = \sin\left(\omega t + \frac{\pi}{2}\right),$$

the Y-motion leads over the X-motion by $\pi/2$. Squaring and adding, we have $x^2 + y^2 = a^2$, which represents a circle of radius a and centre at the origin O. Thus the particle has a uniform circular motion of angular frequency ω. At time $t = 0$, we have $x = 0$, and $y = a$. So the particle is at A (Fig. 5.21.1) at $t = 0$. At $t = T/4$, where T is the time period, we have

$$\omega t = \omega\frac{T}{4} = \frac{\pi}{2}.$$

Hence at $t = T/4$, $x = a$ and $y = 0$, so that the particle is at B. Thus the particle describes the circle in the *clock-wise* direction, i.e., the particle describes a *right handed* circular motion.

If the S.H.Ms impressed on the particle are $x = a\sin\omega t$, and $y = -a\cos\omega t$, the resulting motion of the particle will again be a uniform circular one, but the circle is described in the *counter clock-wise* direction, i.e., the particle describes a *left handed* circular motion. Here the Y-motion lags behind the X-motion by $\frac{\pi}{2}$, for $-\cos\omega t = \sin\left(\omega t - \frac{\pi}{2}\right)$.

Thus two perpendicular simple harmonic motions of the same amplitude and frequency, but differing in phase by $\frac{\pi}{2}$, combine to produce a circular motion. The sense of the circular motion described depends on which of the two S.H.Ms leads over the other.

5.22 Composition of Two Uniform Circular Motions of same radius and periodic Time, but described in opposite directions

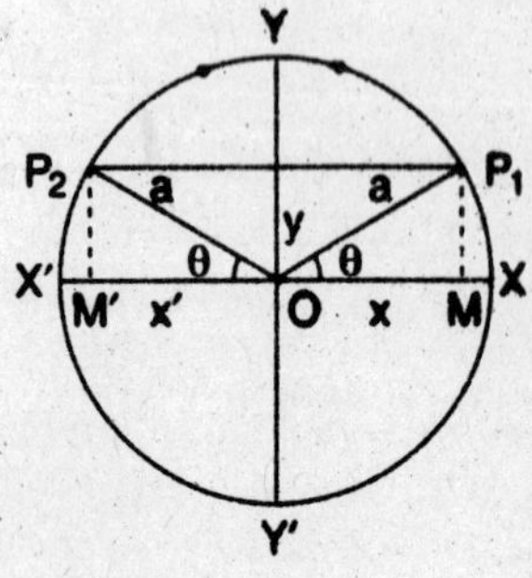

Fig. 5.22.1

Let us consider two particles P_1 and P_2 at any instant moving on the same circle in opposite directions. Now, we know that a circular motion is equivalent to two equal linear simple harmonic motions with a phase difference $\frac{\pi}{2}$ and acting at right angles to each other. Hence, the circular motion of the particle P_1 may be resolved along OX and OY. Similarly the circular motion of P_2 may be resolved along OX' and OY (Fig. 5.22.1). Now, the displacements of P_1 along OX and OY at any time are given by $x = a\cos\theta$ and $y = a\sin\theta$, where a is the radius of the circle and $\theta = \angle P_1OM = \angle P_2OM' = \omega t$ [ω is the angular velocity of P_1 and P_2]. The displacements of P_2 along OX' and OY are given by $x' = a\cos\theta$ and $y = a\sin\theta$. Since the displacements x and x' are in opposite directions, the resultant displacement is given by $2y = 2a\sin\theta = 2a\sin\omega t$, along OY. Thus the superposition of two equal circular motions in opposite directions gives rise to a linear simple harmonic motion along OY. The amplitude of the resultant S.H.M. is equal to the diameter of the circular motion and the periodic time remains the same.

5.23 Lissajous' Figures

Lissajous' figures are the curves formed by the superposition of two simple harmonic motions acting at right angles to each other. For two simple harmonic motions having different periods and with phase difference constantly changing the Lissajous' figure has the property of gradually covering the entire rectangle of sides, the ratio of which is equal to the ratio of the amplitudes a and b of the two S.H.M.s (Fig. 5.23.1).

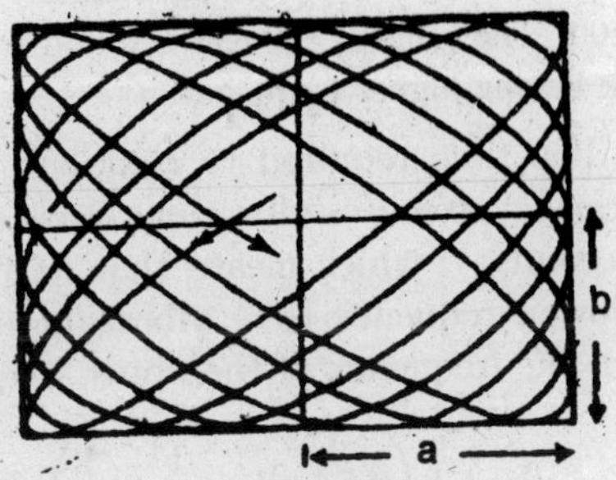

Fig. 5.23.1

Uses of Lissajous' Figures

1. Determination of the ratio of frequencies : The ratio of the frequencies can be easily determined from a curve. A horizontal and a vertical line are drawn to cut the curve, and the number of intersections, each makes with the curve are found. Then the ratio of the number of intersections on the horizontal line to that on the vertical line is as the ratio of the corresponding periods or as that of the vertical to the horizontal frequency.

2. Determination of the frequency of a tuning fork: In order to determine the unknown frequency of a tuning fork A, we take another tuning fork B of known frequency n. Now Lissajous' figures are formed. As the tuning forks differ slightly in frequency, the phase difference between the two changes. Let us suppose that a complete cycle of change takes place in t seconds. Then the difference in frequencies of the two tuning forks is $\frac{1}{t}$, so that the frequency of the first fork A is $\left(n \pm \frac{1}{t}\right)$.

Now a little wax is attached to the fork A. Let now the complete cycle of change take place in t_1 seconds. If now $t > t_1$, then the frequency of $A = n + \frac{1}{t}$. If however $t < t_1$, then the unknown frequency $= n - \frac{1}{t}$.

Illustrative Examples

1. Two tuning forks A and B of approximately the same frequency are used to obtain Lissajous' figures. It is observed that the figures go through a cycle of changes in 20 seconds. On slightly loading the fork B, the figures go through one cycle in 15 seconds. If A has a frequency 400 Hz, find the frequency of B before and after loading. **[C.U. (Hons.) '83]**

Before loading B, the period after which the Lissajous' figures are repeated is 20 seconds which means that the fork A gains or loses one vibration over B in 20 seconds. Frequency of $A = 400$ (given) Hz. Time for one complete cycle before loading B with wax $= 20$ seconds, so that the difference in frequencies of A and $B = \frac{1}{20} = 0.05$ Hz. Hence possible frequencies of B before loading are either $(400 + 0.05)$ or $(400 - 0.05)$. Now when B is loaded with wax, the difference in frequencies of the two forks is $\frac{1}{15} = 0.067$ Hz. Hence possible frequencies of B after loading, are $(400 + 0.067)$ or $(400 - 0.067)$.

Now loading with wax decreases the frequency of B, and as such the frequency difference between A and B increases when B is loaded. Evidently this may happen only if the frequency of B was originally less than that of A. So before loading, the frequency of $B = (400 - 0.05) = 399.95$ Hz, and that after loading $= (400 - 0.067) = 399.933$ Hz.

2. Two tuning forks are approximately an octave apart and the frequency of the fork of lower frequency is 100. The cycle of changes is found to be completed 5 times in 20 seconds. Find the frequency of the fork having higher frequency.

The time required to complete one cycle of changes is 4 seconds; so the fork of higher frequency executes vibrations more or less than double of that of the fork of lower frequency in 4 seconds — which means that its frequency is greater or less than double of that of the fork of lower frequency by $\frac{1}{4}$ vibration, so that, if the frequency of the lower frequency fork be n, then the frequency of the upper frequency fork is given by

$$2n \pm b \left(\text{where } b = \frac{1}{t} = \frac{1}{4}\right) = 2 \times 100 \pm \frac{1}{4} = 200 \pm 0.25 = 200.25, \quad \text{or,} \quad 199.75.$$

5.24 Experimental Methods for Producing Lissajous' Figures

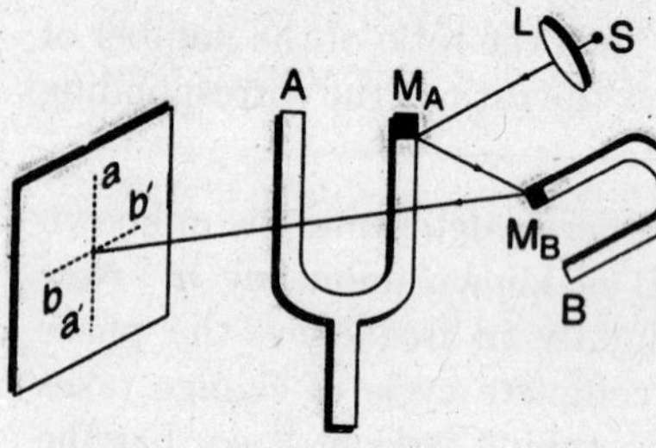

Fig. 5.24.1

(a) **Optical method.** A and B are two tuning forks, vibrating in perpendicular planes to each other. Light from a source S passes through the lens L and falls on the mirror M_A and then on M_B and finally on a screen. By adjusting any one fork, the phase difference δ can be adjusted, and different Lissajous' figures can be obtained on the screen as shown in Fig. 5.24.1.

(b) **Blackburn's pendulum.** It consists of a funnel F (filled with sand) having a fine nozzle (Fig. 5.24.2). $EACBDA$ is a string passing over the hooks at C and D. The length AE can be changed by moving the knot at A. When the pendulum is displaced about A along XX', it vibrates with a time period given by $T_1 = 2\pi\sqrt{\frac{l_1}{g}}$. When the pendulum is displaced about B along YY', it vibrates with a time period given by $T_2 = 2\pi\sqrt{\frac{l_2}{g}}$.

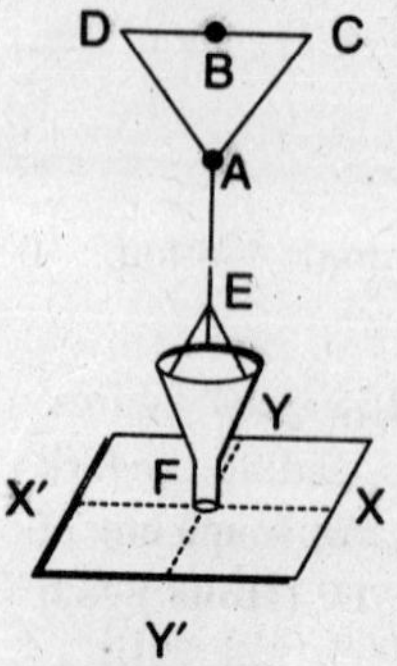

Fig. 5.24.2

When both the motions are given simultaneously, the nozzle of the funnel traces Lissajous' figures on a horizontal sheet of paper placed below the funnel. The shape and size of these figures will depend on the relative amplitudes and time periods of the simple harmonic motions.

(c) For demonstration of Lissajous' figures before an audience, the cathode ray oscilloscope (CRO) is the most suitable instrument.

The cathode ray tube consists of the electron gun, the horizontal and the vertical deflecting systems and the fluorescent screen (Fig. 5.24.3). The electron gun produces a narrow beam of high-speed electrons, and focusses the beam on the fluorescent screen, forming a visible spot S on the screen where the beam strikes. The electron beam passes, on its way from the electron gun to the fluorescent screen, through a set of parallel deflecting plates — the plates (V_1, V_2) deflecting the beam vertically and the set (H_1, H_2) horizontally. The voltage applied between the deflecting plates produces the necessary deflection—the voltage applied across V_1 and V_2 producing the vertical deflection of the spot on the screen, while that across H_1 and H_2 deflecting the spot horizontally.

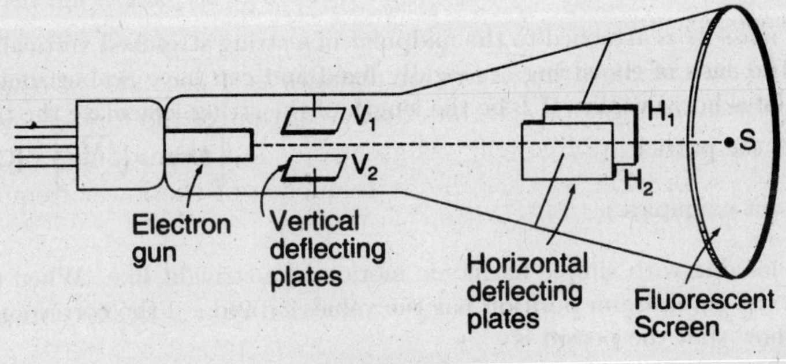

Fig. 5.24.3

For demonstrating Lissajous' figures on the CRO screen, two simple harmonic vibrations are first converted into sinusoidal voltages by means of two microphones. These voltages are amplified and then fed to the horizontal and the vertical deflecting plates of the cathode ray tube. The spot on screen moves under the simultaneous action of the two sinusoidal voltages at right angles and thus traces out the desired Lissajous' figure, visible on the fluorescent screen.

Exercises

1. Define a simple harmonic motion.

   Explain the following terms : (a) Periodic time; (b) Frequency; (c) Amplitude; (d) Phase.

2. Show that the acceleration of a particle executing simple harmonic motion is proportional to the displacement and directed towards the mean position.

3. Deduce the equation of motion of a particle executing simple harmonic motion.

4. Calculate the kinetic and potential energy of a particle executing a simple harmonic motion. Hence show that the total energy is constant.

5. Discuss the damped harmonic motion of a particle. Explain the condition of resonance.

6. Two collinear simple harmonic motions of nearly equal frequencies but differing in amplitude are superposed on each other. Calculate the resultant displacement. Hence explain the formation of beats. **[C.U. (Hons.) 2003, 2007 (old syll)]**

7. Find graphically or analytically the resultant of two simple harmonic motions of same frequency and of different amplitudes at right angles to each other and having a phase difference of $\frac{\pi}{2}$.

8. Find the resultant of two rectangular simple harmonic motions of same frequency, but differing in amplitude and phase. **[C.U. (Hons.) 2002, 2007]**

9. Show that a uniform motion in a circle is equivalent to two S.H.M.s at right angles to each other.

10. Show graphically the resultant of two rectangular S.H.M.s having time periods in the ratio 2 : 1, different amplitudes and same phase.

11. A cylindrical pole of mass M and area of cross-section A floats in a vertical position in a liquid of density ρ. If it is depressed slightly and then released, find the frequency of vertical oscillations. (See worked out examples.) $\left[\textbf{Ans. } \frac{1}{2\pi}\sqrt{\frac{A\rho g}{M}}\right]$ **[C.U. (Hons.) 1980, 2000, 2007]**

12. A particle of mass M is attached to the midpoint of a string stretched vertically under a constant tension T. The ends of the string are rigidly fixed and can have no horizontal movement. The particle oscillates horizontally. If L be the length of the string, calculate the time period of small oscillations of the particle. $\left[\textbf{Ans. } \pi\sqrt{\frac{ML}{T}}\right]$ **[C.U. (Hons.) 1981]**
(See worked out examples.)

13. A particle is moving with simple harmonic motion in a straight line. When the distance of the particle from the equilibrium position has the values x_1 and x_2, the corresponding velocities are u_1 and u_2. Show that the period is

$$2\pi\left\{(x_2^2 - x_1^2)/(u_1^2 - u_2^2)\right\}^{1/2}.$$

(See worked out examples.)

14. A particle of mass m is attached to a horizontal string of length l at a distance l_1 from one end. If F be the tension in the string, calculate the time period of transverse oscillations of the particle, assuming the amplitude of oscillation to be small. **[C.U. (Hons.) 1990]**
(See worked out examples.) $\left[\textbf{Ans. } T = 2\pi\sqrt{\frac{ml_1(l-l_1)}{Fl}}\right]$

15. A sphere of mass M, capable of producing an undamped S.H.M. along the X-axis is at rest. The stiffness constant is S. Another sphere of mass m, moving with a velocity u_0, hits the first sphere elastically, such that the line joining the centres is along the X-axis. Find the displacement of the first sphere at any subsequent instant of time in terms of M, m, u_0 and S. **[C.U. (Hons.) 1986]**
(See worked out examples.)

16. Show that the frequency of oscillations of a body of mass M suspended from a uniform spring of force constant K and mass m is given by

$$n = \frac{1}{2\pi}\sqrt{\frac{K}{M + \frac{m}{3}}}.$$

(See worked out examples.)

17. If the displacement of a moving point at any time is given by $x = a\cos\omega t + b\sin\omega t$, show that the motion is simple harmonic. If $a = 6, b = 8$ and $\omega = 3$, find the period, maximum velocity and maximum acceleration. [**Ans.** 2.09 secs, 30 cm/sec, 90 cm/sec^2]
(See worked out examples.)

18. Two masses m and m' attached to the lower end of an elastic string, whose upper end is fixed, are hanging at rest. Now m' falls off. Show that the distance of m from the upper end of the string at time t is given by $a+b+c\cos\sqrt{\frac{g}{b}}t$, where a is the unstretched length of the string and b and c are the distances by which it would be stretched, when supporting m and m' respectively.
(See worked out examples.)

19. A vertical U-tube of uniform cross-section contains water up to a height of 30 cms. Show that if the water on one side is depressed and then released, its motion up and down the two sides of the tube is simple harmonic and calculate its time period. **[Ans.** 1.098 secs]

(See worked out examples.)

20. Show that a bar magnet freely suspended so as to move in a horizontal plane in the earth's field, executes a simple harmonic motion. Hence calculate its time period and the angular displacement at any instant. **[Ans.** See Art. 5.14]

21. Show that the motion of an electric dipole, which is free to move in a uniform electric field is simple harmonic in character, and hence calculate its time period. **[Ans.** See Art. 5.14]

22. A vertical spring, fixed at the upper end, carries a mass m at the lower end. If the mass is pulled downwards through a small distance and then released, show that the spring will execute S.H.M and calculate its time period. **[Ans.** See Art. 5.14.]

23. Show that the equations governing the free undamped oscillations of the following systems are similar;

(a) a mass attached to a spring, (b) torsional oscillations, (c) electrical L-C circuit.

Find an expression for the natural frequency of any one of the above oscillatory systems.

[Ans. See Art. 5.14]

24. A cylinder of length l and cross-sectional area α has an air-tight piston of mass m and cross-section α inside it. If the piston be slightly pushed in, show that it will oscillate with S.H.M. and find its period. **[Ans.** See worked out examples.]

25. A conical buoy floats with its axis vertical and its apex points downwards in a big vessel containing water. If the weight of the buoy is 300 lbs and 1 cu.ft of water weights 64 lbs, find the time period of vertical oscillations, if the diameter of the base is 4 ft and the height of the cone is 5 ft.

[Ans. Let the buoy floating vertically have V of its volume inside the liquid of density ρ, and let r be the radius of the plane of floatation and h the depth of the buoy immersed as shown. Then from Archimedes' principle, weight of the body = weight of liquid displaced = $V\rho g$. Let the buoy be very slightly depressed through a very small vertical distance x. Assuming the additional liquid displaced to be cylindrical in cross-section (i.e., $A = \pi r^2$), the equation of motion is $m\frac{d^2x}{dt^2} = -Ax\rho g$, whence

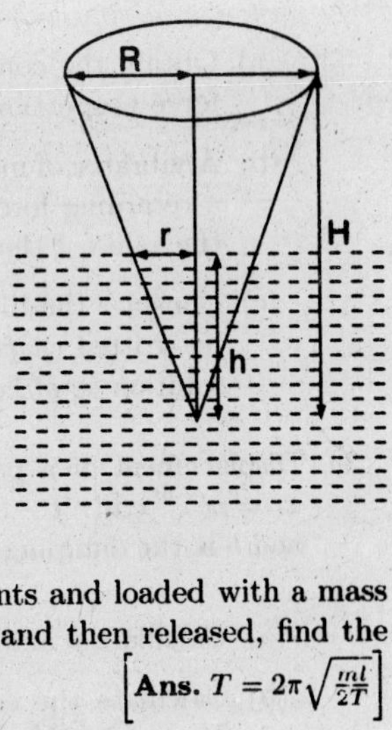

$$T = 2\pi\sqrt{\frac{m}{A\rho g}} = 2\pi\sqrt{\frac{V}{Ag}} = 2\pi\sqrt{\frac{\frac{1}{3}\pi r^2 h}{\pi r^2 g}} = 2\pi\sqrt{\frac{h}{3g}}.$$

Also mass of the buoy $= V\rho$, i.e., $300 = \frac{1}{3}\pi r^2 h \times 64$ or, $r^2 h = \frac{900}{64\pi}$; again from the fig., $\frac{r}{h} = \frac{R}{H} = \frac{2}{5}$ or $r = \frac{2}{5}h$ whence $h = 3.035$ ft.

$$\therefore\ T = 2\pi\sqrt{\frac{3.035}{3 \times 32}} = 1.117 \text{ sec.}]$$

26. A light string of length $2l$ is stretched horizontally between two points and loaded with a mass m at its mid-point. If it be pulled down through a small distance and then released, find the period of oscillation of the string. $\left[\textbf{Ans. } T = 2\pi\sqrt{\frac{ml}{2T}}\right]$

[Hints. Find K.E. and $P \cdot E$, then put $K \cdot E + P \cdot E$ = constant and then differentiate.]

27. In a one dimensional motion of a mass 10 gms, it is acted on by a restoring force 10 dynes/cm and a resisting force 2 dynes-sec/cm. (i) Find whether the motion is aperiodic or oscillatory, (ii) find the value of the resisting force which will make the motion critically damped, (iii) find the value of the mass for which the given forces will make the motion critically damped, (iv) suppose that the mass is given an initial impulse of 20 gm-cm $\sec^{-1}$, while at rest, obtain an expression for the displacement x at a time t.

[**Ans.** See worked out examples.] **[C.U. (Hons.) 1988, '90, '92, '94, '99]**

28. Find in the above problem the time in which the amplitude falls to half value.

[**Ans.** See worked out examples.] **[Cf. C.U. (Hons.) 1997, 2007 (old syll)]**

29. Show that in the steady state of forced vibration, the average rate of supply of energy by the driving force is equal to the average rate of work done against the frictional resistance.

[**Ans.** See worked out examples.] **[C.U. (Hons.) 1996, '98, 2006]**

30. Show that for a forced vibration, the total energy of the vibrating system is not constant, and also prove that in such a case

$$\frac{\text{average } P\cdot E}{\text{average } K\cdot E} = \frac{\omega_0^2}{\omega^2} \quad \text{where } \omega_0 = \sqrt{\frac{S}{m}}$$

[**Ans.** See worked out example.] **[C.U. (Hons.) 1989]**

31. (a) Calculate the mechanical impedance of a particle subjected to a restoring force proportional to displacement and a resisting force proportional to velocity. Find an expression for the velocity amplitude. When will it be maximum? **[C.U. (Hons.) 2000, 2005]**

(b) Deduce an expression for power per cycle, and show that at velocity resonance, the power factor is unity.

[**Ans.** See worked out examples.]

32. If ω_1 and ω_2 be the half-power frequencies, and ω_0 is the resonant frequency of a forced system, show that $\omega_0^2 = \omega_1\omega_2$. [**Ans.** See worked out examples.] **[C.U. (Hons.) 1987, 2006]**

33. (a) Obtain the condition of amplitude resonance in a forced vibration subject to a damping force proportional to the velocity.

(b) A vibrator of mass 1 gm is acted upon by a restoring force of 10^7 dynes/cm of displacement, a retarding force of 4×10^3 dynes/velocity and a driving force of $10^5 \cos \omega t$ dynes. Find the value of the maximum possible amplitude. (See Q. 35(c) of the Exercises)

(c) Consider the differential equation for plane waves and show that the general solution can be written as $f(ct \pm x)$, where the symbols have their usual meanings. Can one write the solution as $g(\pm ct \pm x)$ as well?

[C.U. (Hons.) 2001]

34. Displacement of a particle of mass m executing underdamped oscillatory motion is given by $x_t = \rho e^{-bt} \cos(n't - \alpha)$ where $n' = \sqrt{n^2 - b^2}$, n is the natural angular frequency of the particle and b is the damping factor:

(a) Calculate ρ and α when the motion starts with initial velocity V_0 only.

(b) Calculate the energy of vibration and show that for small damping it approximates to $E(t) = \frac{1}{2}mV_0^2 e^{-2bt}$.

(c) Draw $E(t) - t$ curve for the above small damping. **[C.U. (Hons.) 2002]**

Ans.

$$\text{Vel } V = \frac{dx_t}{dt} = \frac{d}{dt}\left[\rho e^{-bt}\cos(n't-\alpha)\right]$$

$$= \rho e^{-bt} \times -n'\sin(n't-\alpha) + \rho\cos(n't-\alpha)\times(-be^{-bt})$$

$$= -n'\rho e^{-bt}\sin(n't-\alpha) - b\rho e^{-bt}\cos(n't-\alpha)$$

(a) When the motion starts with initial vel V_0 i.e. at $t=0$, $V=V_0$; then $V_0 = -n'\rho \sin(-\alpha) - b\rho\cos(-\alpha) = -b\rho\cos\alpha + n'\rho\sin\alpha$.

Again initial displ $= 0$ at $t = 0$, which gives $\rho\cos(-\alpha) = 0$ or $\cos\alpha = 0 = \cos\frac{\pi}{2}$.

$\therefore\ \alpha = \pi/2 \because \rho \neq 0$.

With this value of α in the expression for V_0, we have

$$V_0 = -b\rho\times 0 + n'\rho\times 1 = n'\rho$$

$$\therefore\quad \rho = \frac{V_0}{n'} \quad \text{or} \quad \rho^2 = \frac{V_0^2}{n^2} \quad \text{or} \quad V_0^2 = \rho^2 n^2 \quad (\text{when } b = 0)$$

(b) Total energy $E = \text{K.E.} + \text{P.E.} = \frac{1}{2}mV^2 + \frac{1}{2}mn^2x^2$

Now,

$$T = \text{K.E.} = \tfrac{1}{2}mV^2 = \tfrac{1}{2}m\left[-n'\rho e^{-bt}\sin(n't-\alpha) - b\rho e^{-bt}\cos(n't-\alpha)\right]^2$$

$$= \tfrac{1}{2}m\Big[n'^2\rho^2e^{-2bt}\sin^2(n't-\alpha) + b^2\rho^2e^{-2bt}\cos^2(n't-\alpha)$$
$$+2n'\rho^2be^{-2bt}\sin(n't-\alpha)\cos(n't-\alpha)\Big]$$

$$T_{av} = \frac{\displaystyle\int_0^T \tfrac{1}{2}m\Big[n'^2\rho^2e^{-2bt}\sin^2(n't-\alpha) + b^2\rho^2e^{-2bt}\cos^2(n't-\alpha) + 2n'\rho^2be^{-2bt}\sin(n't-\alpha)\cos(n't-\alpha)\Big]dt}{\displaystyle\int_0^T dt}$$

$$= \tfrac{1}{2}m\left(\frac{n'^2\rho^2}{2} + \frac{b^2\rho^2}{2}\right)e^{-2bt} = \tfrac{1}{4}me^{-2bt}\left(n'^2\rho^2 + b^2\rho^2\right)$$

$$= \tfrac{1}{4}\rho^2me^{-2bt}n^2 \quad \left(\because n'^2 = n^2 - b^2 \text{ or } n^2 = n'^2 + b^2\right)$$

Again,

$$V = \text{P.E.} = \tfrac{1}{2}mn^2x^2 = \tfrac{1}{2}mn^2\left[\rho^2e^{-2bt}\cos^2(n't-\alpha)\right]$$

$$\therefore\quad V_{av} = \frac{\tfrac{1}{2}mn^2\rho^2e^{-2bt}\displaystyle\int_0^T\cos^2(n't-\alpha)dt}{\displaystyle\int_0^T dt} = \tfrac{1}{4}mn^2\rho^2e^{-2bt}$$

Hence the total energy

$$E = T_{av} + V_{av} = \tfrac{1}{4}\rho^2mn^2e^{-2bt} + \tfrac{1}{4}mn^2\rho^2e^{-2bt} = \tfrac{1}{2}\rho^2mn^2e^{-2bt} = \tfrac{1}{2}mV_0^2e^{-2bt}$$

$$\left[\because \frac{\int_0^T \cos^2(n't-\alpha)dt}{\int_0^T dt} = \frac{1}{2} \quad \text{and} \quad V_{0(av)}^2 = \rho^2 n^2\right]$$

(c) The graph will be represented by an exponential decay curve as shown in the figure.

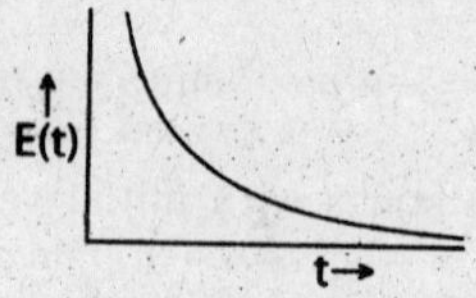

35. (a) Solve the equation of motion of a damped forced simple harmonic oscillator and derive the condition for velocity resonance. **[C.U. (Hons.) 2005]**

(b) Obtain the condition of amplitude resonance in forced vibration subject to a damping force proportional to velocity. **[C.U. (Hons.) 2005 (old syll)]**

(c) A vibrator of mass 1 gm is acted upon by a restoring force of 10^7 dyne/cm of displacement and a retarding force 4×10^3 dyne/velocity and a driving force of $10^5 \cos\omega t$ dynes. Find the value of the maximum possible amplitude. **[C.U. (Hons.) 2001, '05 (old syll)]**

Ans. Here mass = 1 gm, $k = 4 \times 10^3$ dynes, $s = 10^7$ dynes, $F = 10^5$ dynes (maximum value) so that the maximum displacement amplitude is given by

$$A_m = \frac{F}{k\left(\frac{s}{m} - \frac{k^2}{4m^2}\right)^{1/2}} = \frac{1}{4} \times 10^{-1} \times 6^{1/2} \text{ cm} \quad \text{(putting the given data)}$$

36. (a) What is the difference between forced vibration and coupled oscillation?

(b) Two mutually perpendicular oscillations are represented by

$$x(t) = A_1 \sin\omega_1 t, \quad y(t) = A_2 \sin(\omega_2 t + \phi)$$

If $\omega_2 = 2\omega_1$, sketch the Lissajous' figure for the motion when $\phi = \pi$.

(c) A horizontally mounted spring stretches 0.03 m when subjected to a force of 6.0 N. A 0.5 kg mass is attached to the free end and given an initial displacement of 0.015 m and an initial velocity of 0.4 m/s. Find the period, and amplitude qualitatively of the motion.

[C.U. (Hons.) 2008]

(**Ans.** See worked out examples)

Chapter 6

The Acceleration Due to Gravity

6.1 The Acceleration due to Gravity

The acceleration produced in a body due to the attractive force of the earth is known as the *acceleration due to gravity.* It is denoted by the letter g. Due to the rotation of the earth round its axis, different bodies placed at different latitudes are moving with different linear velocities, as a result of which, bodies near the equator moving with greater linear velocities will experience greater centrifugal force in comparison with that experienced by bodies near the poles and hence the acceleration due to gravity at different latitudes is affected. The acceleration due to gravity at a place is actually measured by the acceleration due to the earth's attractive force minus the centrifugal acceleration due to the earth's rotation. The value of g is also affected by other factors, viz. latitude, altitude and depth. Acceleration due to gravity may be accurately determined by pendulums.

6.2 Simple Pendulum

A simple pendulum consists of a heavy point mass suspended from a rigid support by means of an inextensible, weightless and perfectly flexible string.

In the figure 6.2.1, OA denotes a simple pendulum of length l in its undisturbed position, O being the point of suspension. On being displaced through an angle θ, the pendulum takes up the position OB. Now the force acting on the bob due to its weight is mg acting vertically downwards. This force may be resolved into two components—$mg\cos\theta$ and $mg\sin\theta$ as shown in the figure. The component $mg\cos\theta$ being balanced by the tension T in the string, the only effective restoring force is $mg\sin\theta$. Now the active force causing the displacement of the bob $= m\times$ linear acceleration of the bob, where m is the mass of the bob.

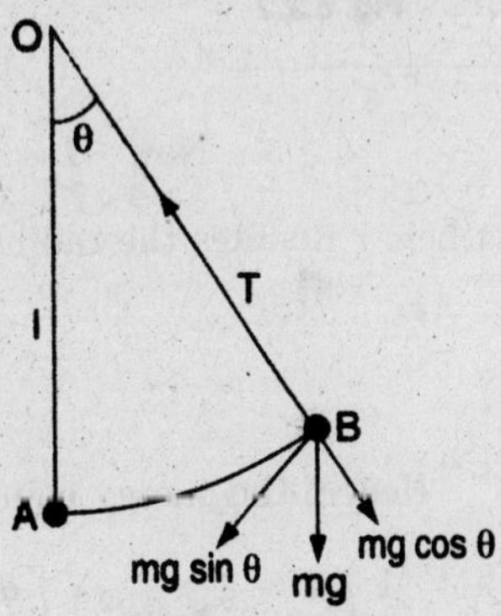

Fig. 6.2.1

Now, the linear acceleration $= l\dfrac{d^2\theta}{dt^2}$. $\therefore$ active force $= ml\dfrac{d^2\theta}{dt^2}$.

Thus the equation of motion of the pendulum can be written as

$$ml\frac{d^2\theta}{dt^2} = -mg\sin\theta.$$

When θ is small, we have

$$\frac{d^2\theta}{dt^2} + \omega^2\theta = 0, \quad \text{where } \omega = \sqrt{\frac{g}{l}}$$

which is the characteristic differential equation for simple harmonic motion*

$$\therefore \frac{2\pi}{T} = \sqrt{\frac{g}{l}} \quad \text{or} \quad T = 2\pi\sqrt{\frac{l}{g}},$$

where T is the time-period of oscillation of the pendulum.

Alternative Treatment

We may also find out the expression for time-period of a simple pendulum from energy consideration. Let the pendulum be displaced to OA making an angle α with OC, the undisturbed position. Here α denotes the angular amplitude. After time t, let it pass through OB, making an angle θ with OC. In coming from A to B, the bob of the pendulum falls through the vertical height $A'B'$. The loss of potential energy due to this vertical fall

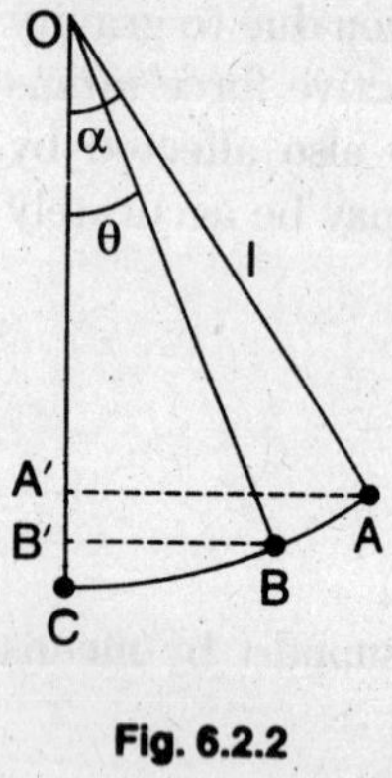

Fig. 6.2.2

$$= mgA'B', \; m \text{ being the mass of the bob}$$
$$= mg(OB' - OA')$$
$$= mgl(\cos\theta - \cos\alpha) \quad [l \text{ is the effective length of the pendulum.}]$$

Now this loss of potential energy is equal to gain in kinetic energy at the position OB.

Kinetic energy at OB is

$$\frac{1}{2}I\left(\frac{d\theta}{dt}\right)^2 \quad \left[\begin{array}{l} I \text{ is the moment of inertia of the} \\ \text{bob about an axis through } O. \end{array}\right]$$

$$\text{Now} \quad I = \frac{2}{5}mr^2 + ml^2 \quad [\text{applying the theorem of parallel axes}]$$

where r denotes the radius of the bob of the pendulum.

$$\therefore \text{K.E.} = \frac{1}{2}\left(\frac{2}{5}mr^2 + ml^2\right)\left(\frac{d\theta}{dt}\right)^2.$$

Hence, the energy equation may be written as

$$\frac{1}{2}\left(\frac{2}{5}mr^2 + ml^2\right)\left(\frac{d\theta}{dt}\right)^2 = mgl(\cos\theta - \cos\alpha) \quad \text{or} \quad \left(\frac{d\theta}{dt}\right)^2 = \frac{2mgl(\cos\theta - \cos\alpha)}{(\frac{2}{5}mr^2 + ml^2)}.$$

*See the Chapter on S.H.M.

Differentiating with respect to t and remembering that α is constant, we have

$$2\left(\frac{d\theta}{dt}\right)\frac{d^2\theta}{dt^2} = \frac{-2mgl}{\left(\frac{2}{5}mr^2 + ml^2\right)}\sin\theta\frac{d\theta}{dt}$$

$$\text{or} \quad \frac{d^2\theta}{dt^2} = -\frac{mgl}{\left(\frac{2}{5}mr^2 + ml^2\right)}\sin\theta = -\frac{mgl\theta}{\left(\frac{2}{5}mr^2 + ml^2\right)} \quad [\because \theta \text{ is small}]$$

whence

$$T = 2\pi\sqrt{\frac{\frac{2}{5}mr^2 + ml^2}{mgl}} = 2\pi\sqrt{\frac{\frac{2}{5}r^2 + l^2}{gl}}.$$

If the bob is considered to be a point mass,

$$T = 2\pi\sqrt{\frac{l}{g}}.$$

Now from the above expression for time-period we have,

$$g = 4\pi^2\frac{l}{T^2},$$

whence g may be determined.

When the value of $\frac{L}{T^2}$ at a place is 24.85 cm/sec^2 (say), the acceleration due to gravity at that place is given by

$$g = 981.1\ \text{cm/sec}^2.$$

An acceleration of 1 cm/sec^2 is called *gal* after the name of Galileo

$$\therefore\ g = 981.1\ \text{gal}.$$

6.3 Laws of Simple Pendulum

The motion of a simple pendulum is governed by the following laws known as the *laws of simple pendulum.*

Law 1 : The Law of Isochronism

The oscillations of a simple pendulum are isochronous. This means that the pendulum takes equal time to complete each oscillation whatever be the amplitude, provided the amplitude is small (within 4^0). In other words, the time-period is independent of the amplitude of oscillation.

Law 2 : The Law of Length

The period of oscillation of a pendulum at a given place is directly proportional to the square root of its length, i.e., $T \propto \sqrt{l}$, when g is constant.

Law 3 : The Law of Acceleration due to Gravity

The time-period of oscillation of a pendulum of a given length is inversely proportional to the square root of acceleration due to gravity, i.e., $T \propto \frac{1}{\sqrt{g}}$, when l is constant.

Law 4 : The Law of Mass

The time-period of oscillation of a pendulum is independent of the mass and material of the bob.

6.4 Seconds Pendulum

When a simple pendulum takes one second in swinging from one extreme position to the other, it is called a *seconds pendulum.* Hence, the period of oscillation of such a pendulum is 2 seconds. So, from the expression for time-period of oscillation, we may write

$$2 = 2\pi\sqrt{\frac{l}{g}} \quad \text{or} \quad l = \frac{g}{\pi^2}.$$

Thus, the length of the seconds pendulum changes from one place to another depending on the value of g.

Taking $g = 981$ cms/sec^2, we have $l = \frac{981}{\pi^2} = \frac{981}{9.87} = 99.39$ cms.

Taking $g = 32.2$ ft/sec^2, we have $l = \frac{32.2}{\pi^2} = \frac{32.2}{9.87} = 3.26$ ft.

6.5 Gain or Loss of Time by Pendulum Clock

The time-period T of a pendulum clock is affected by the change in length l or gravity g. The pendulum undergoes slow oscillation when the time-period increases and undergoes fast oscillation when the time-period decreases. Thus, due to change in l or g there is loss or gain of time.

We know

$$T = 2\pi\sqrt{\frac{l}{g}}. \quad \therefore \log T = \log 2\pi + \frac{1}{2}\log l - \frac{1}{2}\log g.$$

Differentiating, we have

$$\frac{dT}{T} = \frac{1}{2}\left(\frac{dl}{l} - \frac{dg}{g}\right) \tag{6.5.1}$$

If N be the number of oscillations executed by the pendulum per day, then $NT = 86400$ secs = constant.

$$\therefore \log N + \log T = \text{constant}.$$

Differentiating, we have

$$\frac{dN}{N} + \frac{dT}{T} = 0 \tag{6.5.2}$$

From (6.5.1) and (6.5.2), we get

$$\frac{dN}{N} = -\frac{1}{2}\left(\frac{dl}{l} - \frac{dg}{g}\right) = \frac{1}{2}\left(\frac{dg}{g} - \frac{dl}{l}\right) \tag{6.5.3}$$

Now, the number of oscillations executed by a seconds pendulum in a day is given by

$$N = \frac{1}{2} \times 86400 = 43200.$$

As in a seconds pendulum each complete oscillation is executed in two seconds, the number of seconds gained in a day is given by

$$\delta n = 2dN = N\left(\frac{dg}{g} - \frac{dl}{l}\right) = 43200\left(\frac{dg}{g} - \frac{dl}{l}\right). \tag{6.5.4}$$

(a) *Gain or Loss of Time due to Change in Length :*

In this case, if g is constant, $dg = 0$.

Hence, from (6.5.4) we get

$$\delta n = -43200\frac{dl}{l}.$$

Thus, as l increases, δn decreases, i.e., there is loss of time.

(b) *Gain or Loss of Time due to Change of Place :*

Here, if l is constant, $dl = 0$.

Hence, from (6.5.4) we get

$$\delta n = 43200\frac{dg}{g}.$$

Thus, as g increases, δn increases, i.e., there is gain of time.

Illustrative Examples

1. Supposing a pendulum to be constructed so as to beat second at a place where $g = 980$ cms/sec^2, how would you change its length so that it may beat seconds at a place where $g = 490$ cms/sec^2?

We know, $T = 2\pi\sqrt{\frac{l}{g}}$

Here, $T = 2$ secs, $g = 980$ cms/sec^2

$$\therefore\ 2 = 2\pi\sqrt{\frac{l}{980}}. \tag{1}$$

When $g = 490$ cms/sec^2, let the length of the pendulum be shortened to l' to beat seconds.

$$\therefore\ 2 = 2\pi\sqrt{\frac{l'}{490}}. \tag{2}$$

From (1) and (2), we have

$$\sqrt{\frac{l}{980}} = \sqrt{\frac{l'}{490}} \quad \text{or} \quad \frac{l'}{l} = \frac{490}{980} = \frac{1}{2} \quad \text{or} \quad l' = \frac{l}{2}.$$

Thus the length of the pendulum is shortened to half of its original value.

2. A pendulum of length l losses 5 secs in a day. By how much must it be shortened to keep correct time?

We know,

$$\delta n = -43200\frac{dl}{l}.$$

Here

$$\delta n = -5 \text{ secs} \quad \text{or} \quad 5 = 43200\frac{dl}{l} \quad \text{or} \quad dl = \frac{l}{8640}.$$

Hence, to keep correct time the length of the pendulum is to be shortened by $\frac{l}{8640}$.

3. A pendulum which beats second at a place where g = 980 cms/sec^2 is taken to place where g = 981 cms/sec^2. How many seconds will it lose or gain in a day?

We know,

$$\delta n = 43200\frac{dg}{g}$$

Here $g = 980$, $dg = 981 - 980 = 1$.

$$\therefore \ \delta n = 43200 \times \frac{1}{980} = \frac{4320}{98} = 44 \text{ secs (nearly)}.$$

Hence, the daily gain of the pendulum is nearly 44 secs.

6.6 Finite Arc Correction of a Pendulum : Anharmonic Oscillator

The calculation of the time-period considering θ to be very small is not applicable in case of a pendulum swinging with the formation of a finite arc, and the time-period in this case needs correction. This is called the *finite arc correction* of the period. We will now prove in the following way that the period of oscillation, when the pendulum swings with finite arc, is greater than the period that we have calculated considering θ to be small, and the system in such a case is known as an *anharmonic oscillator.*

We have the equation of motion of the pendulum given by

$$\frac{d^2\theta}{dt^2} = -\frac{g}{l}\sin\theta \text{ or } 2\frac{d\theta}{dt}\cdot\frac{d^2\theta}{dt^2} = -2\frac{g}{l}\sin\theta\frac{d\theta}{dt} \quad \left[\text{multiplying both sides by } 2\frac{d\theta}{dt}\right]$$

$$\text{or} \quad \frac{d}{dt}\left[\left(\frac{d\theta}{dt}\right)^2\right] = 2\frac{g}{l}\frac{d}{dt}(\cos\theta).$$

Integrating, we have

$$\left(\frac{d\theta}{dt}\right)^2 = 2\frac{g}{l}\cos\theta + C, \tag{6.6.1}$$

where C is the constant of integration.

Now, $\left(\frac{d\theta}{dt}\right) = 0$, when $\theta = \alpha$ (amplitude)

$$\therefore \quad 0 = \frac{2g}{l}\cos\alpha + C \quad \text{or} \quad C = -\frac{2g}{l}\cos\alpha$$

Substituting the value of C in equation (6.6.1), we have

$$\left(\frac{d\theta}{dt}\right)^2 = \frac{2g}{l}(\cos\theta - \cos\alpha) = \frac{4g}{l}\left(\sin^2\frac{\alpha}{2} - \sin^2\frac{\theta}{2}\right)$$

$$= \frac{4g}{l}\left(r^2 - \sin^2\frac{\theta}{2}\right) \quad \left[\text{putting } r = \sin\frac{\alpha}{2}\right]$$

$$= \frac{4g}{l}(r^2 - r^2\sin^2\phi) \quad \left[\text{putting } \sin\frac{\theta}{2} = r\sin\phi\right]$$

$$= \frac{4g}{l}r^2\cos^2\phi$$

$$\therefore \quad \frac{d\theta}{dt} = 2r\sqrt{\frac{g}{l}}\cos\phi \quad \text{or} \quad dt = \frac{1}{2r}\sqrt{\frac{l}{g}}\cdot\frac{d\theta}{\cos\phi}.$$

Since $\sin\frac{\theta}{2} = r\sin\phi.$

$$\therefore \quad \frac{1}{2}\cos\frac{\theta}{2}d\theta = r\cos\phi d\phi \quad \text{or} \quad d\theta = \frac{2r\cos\phi d\phi}{\cos\frac{\theta}{2}}$$

$$\therefore \quad dt = \frac{1}{2r}\sqrt{\frac{l}{g}}\cdot\frac{2r\cos\phi d\phi}{\cos\frac{\theta}{2}\cos\phi} = \sqrt{\frac{l}{g}}\cdot\frac{d\phi}{\cos\frac{\theta}{2}}$$

Again $\because \quad \sin\frac{\theta}{2} = r\sin\phi.$

$$\therefore \quad \cos\frac{\theta}{2} = \sqrt{1 - r^2\sin^2\phi}$$

$$\therefore \quad dt = \sqrt{\frac{l}{g}}(1 - r^2\sin^2\phi)^{-1/2}d\phi$$

$$= \sqrt{\frac{l}{g}}\left(1 + \frac{1}{2}r^2\sin^2\phi + \frac{1}{2}\cdot\frac{3}{4}r^4\sin^4\phi + \cdots\right)d\phi \quad \text{[expanding binomially]}.$$

Integrating this expression between the limits 0 to $\frac{T_0}{4}$ on the l.h.s. and 0 to $\frac{\pi}{2}$, i.e., 0 to α, on the r.h.s., we have

$$\int_0^{\frac{T_0}{4}} dt = \sqrt{\frac{l}{g}} \int_0^{\frac{\pi}{2}} \left(1 + \frac{1}{2} r^2 \sin^2 \phi + \frac{1}{2} \cdot \frac{3}{4} r^4 \sin^4 \phi + \cdots \right) d\phi$$

[where T_0 is the periodic time]

$$\text{or} \quad \frac{T_0}{4} = \sqrt{\frac{l}{g}} \left(\frac{\pi}{2} + \frac{1}{2} r^2 \cdot \frac{1}{2} \cdot \frac{\pi}{2} + \frac{1.3}{2.4} r^4 \cdot \frac{1.3}{2.4} \cdot \frac{\pi}{2} + \cdots \right)$$

$$\text{or} \quad T_0 = 4\sqrt{\frac{l}{g}} \cdot \frac{\pi}{2} \left[1 + \left(\frac{1}{2}\right)^2 r^2 + \left(\frac{1}{2}\right)^2 \left(\frac{3}{4}\right)^2 r^4 + \cdots \right]$$

$$= 2\pi \sqrt{\frac{l}{g}} \left(1 + \frac{1}{4} r^2\right) \quad \text{[considering only one correcting term]}$$

$$= T\left(1 + \frac{1}{4} \sin^2 \frac{\alpha}{2}\right) \tag{6.6.2}$$

where T is the period for an infinitely small amplitude, and T_0 is the observed period for an amplitude α.

If during an experiment, the amplitude falls from α_1 to α_2, where each is small, we may put

$$\sin^2 \frac{\alpha}{2} = \frac{\alpha^2}{4} = \frac{\alpha_1}{2} \cdot \frac{\alpha_2}{2} \quad \therefore \quad T_0 = T\left(1 + \frac{\alpha_1 \alpha_2}{16}\right) \tag{6.6.3}$$

or the corrected period

$$T = T_0 \left(1 - \frac{\alpha_1 \alpha_2}{16}\right).$$

Taking $\alpha = 8^0$

$$r = \sin \frac{\alpha}{2} = \sin 4^0 = 0.0698.$$

Substituting in equation (6.6.2), we get

$$T_0 = 2\pi \sqrt{\frac{l}{g}} \left[1 + \frac{1}{4}(0.0698)^2\right] = 2\pi \sqrt{\frac{l}{g}} (1.001218).$$

Again taking $\alpha = 50^0$,

$$r = \sin \frac{\alpha}{2} = \sin 25^0 = 0.4226$$

$$\therefore \ T_0 = 2\pi \sqrt{\frac{l}{g}} \left[1 + \frac{1}{4}(0.4226)^2\right] = 2\pi \sqrt{\frac{l}{g}} (1.04464).$$

Thus, in the first case, when the amplitude of oscillation is 8^0 the correction in taking only the first term of the expression for T_0 is nearly 0.12 per cent and in the second case, when the amplitude is as high as 50^0 the correction is nearly 4.4 per cent.

6.7 Simple Pendulum with a Movable Support

We consider a simple pendulum suspended from a movable support which can move along y-axis. Let the position of the point of support be at O at time $t = 0$ and is taken as the origin. After time $t = t$, let the position of the point of support be at O'. y_s denotes the positional co-ordinate of the point of support at this instant. Let M denote the mass of the support.

If θ be the angular displacement of the pendulum at time t, then the co-ordinates of the bob of the pendulum of mass m at that instant with respect to the origin O are

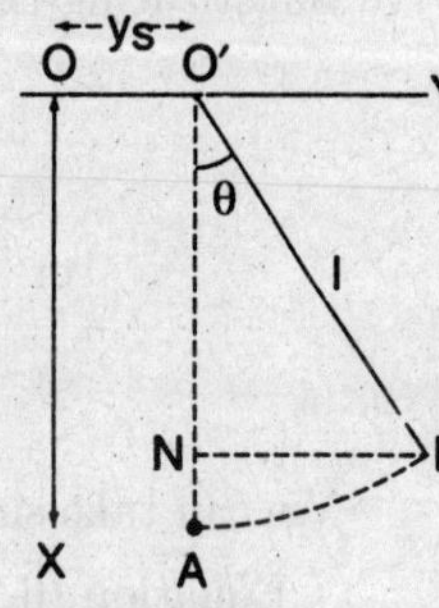

Fig. 6.7.1

$$x = O'N = l\cos\theta$$

$$y = OO' + BN = y_s + l\sin\theta \quad [\because OO' = y_s]$$

$$\therefore \dot{x} = -l\sin\theta\dot{\theta}$$

$$\dot{y} = \dot{y}_s + l\cos\theta\dot{\theta}.$$

Kinetic energy of the system is given by

$$T = \frac{1}{2}M\dot{y}_s^2 + \frac{1}{2}m(\dot{x}^2 + \dot{y}^2) = \frac{1}{2}M\dot{y}_s^2 + \frac{1}{2}ml^2\sin^2\theta\dot{\theta}^2 + \frac{1}{2}m(\dot{y}_s + l\cos\theta\dot{\theta})^2$$

$$= \frac{1}{2}M\dot{y}_s^2 + \frac{1}{2}ml^2\sin^2\theta\dot{\theta}^2 + \frac{1}{2}m\dot{y}_s^2 + \frac{1}{2}ml^2\cos^2\theta\dot{\theta}^2 + m\dot{y}_s l\cos\theta\dot{\theta}$$

$$= \frac{1}{2}M\dot{y}_s^2 + \frac{1}{2}ml^2\dot{\theta}^2 + \frac{1}{2}m\dot{y}_s^2 + m\dot{y}_s l\cos\theta\dot{\theta} = \frac{1}{2}(M+m)\dot{y}_s^2 + \frac{1}{2}ml^2\dot{\theta}^2 + ml\dot{y}_s\cos\theta\dot{\theta} \tag{6.7.1}$$

Potential energy of the system at the same instant is given by

$$V = mgAN = mg(O'A - O'N) = mg(l - l\cos\theta) = mgl(1-\cos\theta) \tag{6.7.2}$$

Lagrangian function is given by

$$L = T - V = \frac{1}{2}(M+m)\dot{y}_s^2 + \frac{1}{2}ml^2\dot{\theta}^2 + ml\dot{y}_s\cos\theta\dot{\theta} - mgl + mgl\cos\theta \tag{6.7.3}$$

[From 6.7.1 and 6.7.2]

$$\left.\begin{aligned} \therefore \quad \frac{\partial L}{\partial \dot{y}_s} &= (M+m)\dot{y}_s + ml\cos\theta\dot{\theta} \\ \frac{\partial L}{\partial y_s} &= 0 \\ \frac{\partial L}{\partial \dot{\theta}} &= ml^2\dot{\theta} + ml\dot{y}_s\cos\theta \\ \frac{\partial L}{\partial \theta} &= -ml\dot{y}_s\dot{\theta}\sin\theta - mgl\sin\theta \end{aligned}\right\} \tag{6.7.4}$$

From Lagrange's equation we have

$$\frac{d}{dt}\left(\frac{\partial L}{\partial \dot{y}_s}\right) - \frac{\partial L}{\partial y_s} = 0 \tag{6.7.5}$$

$$\frac{d}{dt}\left(\frac{\partial L}{\partial \dot{\theta}}\right) - \frac{\partial L}{\partial \theta} = 0. \tag{6.7.6}$$

Substituting the values from (6.7.4) in (6.7.5), we have

$$\frac{d}{dt}[(M+m)\dot{y}_s + ml\cos\theta\dot{\theta}] = 0$$

$$\text{or} \quad (M+m)\ddot{y}_s + ml\cos\theta\ddot{\theta} - ml\sin\theta\dot{\theta}^2 = 0. \tag{6.7.7}$$

Substituting the values in equation (6.7.6), we have

$$\frac{d}{dt}(ml^2\dot{\theta} + ml\dot{y}_s\cos\theta) + ml\dot{y}_s\sin\theta\dot{\theta} + mgl\sin\theta = 0$$

$$\text{or} \quad ml^2\ddot{\theta} + ml\ddot{y}_s\cos\theta - ml\dot{y}_s\sin\theta\dot{\theta} + ml\dot{y}_s\sin\theta\dot{\theta} + mgl\sin\theta = 0$$

$$\text{or} \quad l\ddot{\theta} + \ddot{y}_s\cos\theta + g\sin\theta = 0. \tag{6.7.8}$$

We can take $\sin\theta = \theta$ and $\cos\theta = 1$ when θ is very small.

$\therefore$ Equation (6.7.7) becomes

$$(M+m)\ddot{y}_s + ml\ddot{\theta} = 0 \quad [\text{neglecting } ml\theta\dot{\theta}^2 \text{ which is very small}]$$

$$\text{or} \quad \ddot{y}_s = -\frac{ml\ddot{\theta}}{M+m} \tag{6.7.9}$$

and equation (6.7.8) reduces to the form

$$l\ddot{\theta} + \ddot{y}_s + g\theta = 0 \text{ or } l\ddot{\theta} - \frac{ml\ddot{\theta}}{M+m} + g\theta = 0 \quad [\text{From 6.7.9}]$$

$$\text{or} \quad (M+m)l\ddot{\theta} - ml\ddot{\theta} + g(M+m)\theta = 0$$

$$\text{or} \quad Ml\ddot{\theta} + (M+m)g\theta = 0. \tag{6.7.10}$$

This gives the equation of motion of the pendulum.

The above equation may be written in the form

$$\ddot{\theta} + \frac{M+m}{M}\cdot\frac{g}{l}\theta = 0 \quad \text{or} \quad \ddot{\theta} + \omega^2\theta = 0, \text{ where } \omega^2 = \frac{M+m}{M}\frac{g}{l}.$$

$\therefore$ The time-period of oscillation is given by

$$T = 2\pi\sqrt{\frac{Ml}{(M+m)g}} \quad \text{or} \quad T = 2\pi\sqrt{\frac{l}{\left(1+\frac{m}{M}\right)g}}$$

If the support is considered to be rigid, its mass M should be taken to be infinite. So the time-period for a rigid support is

$$T = 2\pi\sqrt{\frac{l}{g}} \quad [\text{Putting } M = \infty]$$

This is same as the time-period of a simple pendulum with rigid support.

6.8 Conical Pendulum

When a simple pendulum is given motion in such a way that the bob of the pendulum describes a circular path uniformly on a horizontal plane and the revolving string traces out a conical surface, the pendulum is called a *conical pendulum.*

Let v be the uniform speed of the bob A and r be the radius of the circular path. At A, the forces maintaining equilibrium of the bob are :

(i) weight mg of the bob acting vertically downwards;

(ii) the centrifugal force $\frac{mv^2}{r}$ acting horizontally along OA;

(iii) the tension T of the string along AP making an angle θ with the vertical.

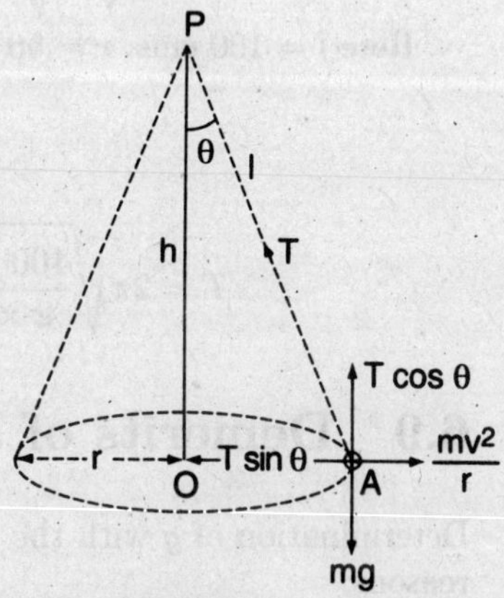

Fig. 6.8.1

The weight mg is balanced by the vertical component $T\cos\theta$ of the tension T of the string.

Hence,

$$mg = T\cos\theta. \qquad (6.8.1)$$

Again the horizontal component $T\sin\theta$ of the tension is balanced by the necessary centrifugal force $\frac{mv^2}{r}$, and we may write

$$\frac{mv^2}{r} = T\sin\theta \qquad (6.8.2)$$

From(6.8.1) and (6.8.2), we have

$$\tan\theta = \frac{v^2}{gr}. \qquad (6.8.3)$$

Now $v = r\omega$, where ω is the angular speed of the bob.

$$\therefore\ \tan\theta = \frac{r^2\omega^2}{gr} = \frac{\omega^2 r}{g}.$$

Now from the figure, $\tan\theta = \frac{r}{h}$.

$$\therefore\ \frac{r}{h} = \frac{\omega^2 r}{g} \quad \text{or} \quad \omega^2 = \frac{g}{h}.$$

$$\therefore\quad \omega = \sqrt{\frac{g}{h}} \quad \text{or} \quad \frac{2\pi}{T} = \sqrt{\frac{g}{h}} \quad \text{or} \quad T = 2\pi\sqrt{\frac{h}{g}} \quad \text{or} \quad T = 2\pi\sqrt{\frac{l\cos\theta}{g}}$$

[where $l = PA$, the effective length of the pendulum].

If θ is small, $T = 2\pi\sqrt{\frac{l}{g}}$.

That is, the time-period is independent of θ. In other words, the time-period for linear path of a pendulum is same as the time-period for circular path of the pendulum.

Illustrative Example

1. A sphere of mass 100 gms is fixed at one end of a string of length 1 metre. Calculate the periodic time for motion of the sphere along a horizontal circle of radius 50 cms ($g = 980$ cms/sec^2).

We know, $T = 2\pi\sqrt{\dfrac{l\cos\theta}{g}}$.

Here $l = 100$ cms, $r = 50$ cms.

$$\therefore \quad \sin\theta = \frac{50}{100} = \frac{1}{2} \quad \text{or} \quad \cos\theta = \frac{\sqrt{3}}{2}$$

$$\therefore \quad T = 2\pi\sqrt{\frac{100\times\sqrt{3}}{2\times 980}} = 2\pi\sqrt{\frac{1.73\times 100}{2\times 980}} = 6.28\sqrt{\frac{173}{2\times 980}} = 1.865 \text{ secs.}$$

6.9 Demerits of a Simple Pendulum

Determination of g with the help of a simple pendulum is not accurate owing to the following reasons:

(i) In actual simple pendulum we take a bob of finite dimension suspended by a thread, but in deducing the time-period $T = 2\pi\sqrt{\frac{l}{g}}$, we have ignored the dimension of the bob and also the moment of inertia of the thread which cannot be weightless.

(ii) The effect of resistance and buoyancy of air which appreciably affects the motion of the pendulum has not been taken into consideration.

(iii) In deducing the expression for time-period, we have considered θ to be very small, i.e., the motion takes place in a straight line, but actually the bob describes a finite arc. For finite arc the time-period increases with the increase of amplitude (vide Art. 6.6).

(iv) The bob has also a relative motion with respect to the thread due to the slackening of the thread when approaching the limit of swing.

(v) Errors may creep in due to the friction at the point of suspension which may damp the free movement of the pendulum.

6.10 Borda's Pendulum

This pendulum consists of a bob of large radius oscillating about the axis of suspension. The time-period, as deduced earlier (Alternative Treatment, Art. 6.2), is given by

$$T = 2\pi\sqrt{\frac{\frac{2}{5}r^2 + l^2}{lg}} = 2\pi\sqrt{\frac{l + \frac{2}{5}\frac{r^2}{l}}{g}}.$$

With this pendulum also, g cannot be determined accurately mainly due to the following reasons :

(i) The moment of inertia of the thread about the axis of suspension has been neglected;

(ii) The bob has relative motion with respect to the thread due to slackening effect of the thread when approaching the limit of swing.

6.11 Compound Pendulum

The defects of simple pendulum, as discussed earlier, may be removed by a compound pendulum which is a rigid body of arbitrary shape suspended from a horizontal axis.

In the figure 6.11.1, let O be a point through which the axis passes; G denotes the centre of gravity of the body and l denotes the distance of the centre of gravity of the body from the axis through O. In the rest position of the body, the line OG is vertical. On displacing the body through an angle θ the centre of gravity of the body comes to the position G' and there will be a couple of moment $mgl \sin\theta$ acting on the body due to its weight mg and the upward reaction at O, tending to restore the body to its original position. Hence, the equation of motion of the body, while it is free to oscillate about the axis, may be written as

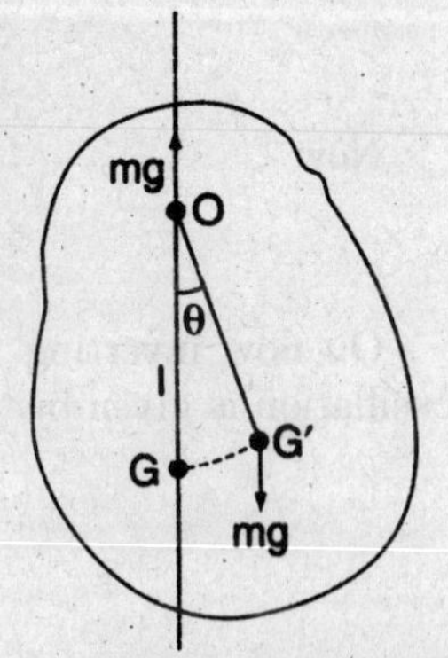

Fig. 6.11.1

$$I\frac{d^2\theta}{dt^2} = -mgl\sin\theta \quad \left[\begin{array}{l}\text{where } I \text{ is the moment of inertia of} \\ \text{the body about the axis in question}\end{array}\right]$$

$$= mgl\theta \quad \text{[when } \theta \text{ is small]}$$

$$\therefore \quad I\frac{d^2\theta}{dt^2} + mgl\theta = 0 \quad \text{or} \quad \frac{d^2\theta}{dt^2} + \frac{mgl}{I}\theta = 0$$

which shows that the motion is simple harmonic and the time-period is given by

$$T = 2\pi\sqrt{\frac{I}{mgl}}.$$

Now from the theorem of parallel axes

$$I = mk^2 + ml^2 = m(k^2 + l^2) \tag{6.11.1}$$

where mk^2 denotes the moment of inertia of the body about an axis through G and parallel to the axis of suspension.

Therefore,

$$T = 2\pi\sqrt{\frac{m(k^2+l^2)}{mgl}} = 2\pi\sqrt{\frac{k^2+l^2}{gl}}.$$

Now $\dfrac{k^2+l^2}{l} > l$, because k^2 is a positive quantity. We may take a length L which is numerically equal to $\dfrac{k^2+l^2}{l}$ as represented by OO' in the figure 6.11.2. L is called the length of an *equivalent simple pendulum* and the period of such a pendulum is

$$2\pi\sqrt{\frac{L}{g}}.$$

Fig. 6.11.2

From (6.11.1) we have

$$ml\left(l + \frac{k^2}{l}\right) = I \quad \text{or} \quad L = \frac{I}{ml}. \tag{6.11.2}$$

In this case, O is known as the *centre of suspension* and O' is called the *centre of oscillation*, and the pendulum in such case becomes a reversible one in the sense that the period of oscillation when O is the point of suspension is same as the period of oscillation when O' is the point of suspension.

The period of oscillation in the first case is given by

$$T_1 = 2\pi\sqrt{\frac{L}{g}} = 2\pi\sqrt{\frac{l+l_1}{g}}, \quad \text{where } l_1 = O'G.$$

Now

$$\frac{k^2+l^2}{l} = l + l_1 \quad \text{or} \quad k^2 = ll_1.$$

On now inverting the pendulum the point of suspension becomes O' and the period of oscillation is given by

$$T_2 = 2\pi\sqrt{\frac{k^2+l_1^2}{l_1 g}} = 2\pi\sqrt{\frac{ll_1+l_1^2}{l_1 g}} \quad [\because k^2 = ll_1]$$

$$= 2\pi\sqrt{\frac{l+l_1}{g}} = T_1.$$

The two periods are thus equal. This is known as the *reversibility* or *interchangeability of the centre of suspension* and *the centre of oscillation* of a compound pendulum.

Actually there are four points on the vertical line through G about which the pendulum oscillates with equal period. This may be proved in the following way.

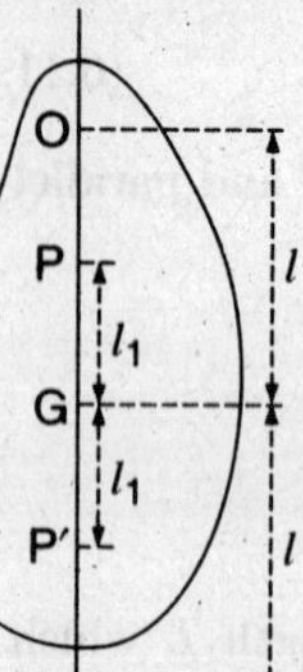

Fig. 6.11.3

We have,

$$T = 2\pi\sqrt{\frac{k^2+l^2}{lg}} \quad \text{or} \quad T^2 = 4\pi^2\frac{(k^2+l^2)}{gl} \quad \text{or} \quad l^2 - \frac{T^2 g}{4\pi^2}l + k^2 = 0$$

which is a quadratic equation in l with two positive roots. Thus for a given time-period T, there are wo values of l, i.e., there are two positions of O on the same side of G, for which the time-periods remain same. If we invert the pendulum, similarly it can be shown that l_1 also has two values, i.e., there are two positions of O' on the other side of G, for which the time-periods are the same. Hence, in all, there are four points on the vertical line through G—two on each side of G as shown in the figure 6.11.3. The four points on the vertical through G are O, P, P' and O'.

N.B. *Centre of suspension and centre of oscillation.* We consider the vertical plane which contains the centre of gravity of the compound pendulum, and is perpendicular to the axis of rotation. The point where the axis of rotation meets this plane (point O in Fig. 6.11.2) is called the *centre of suspension.* For a given centre of suspension O, there is a point O' at which the mass of the compound pendulum may be considered to the concentrated without causing any change in the periodic time. This point O' is called the *centre of oscillation* for the given centre of suspension O. The distance L of the centre of oscillation from the centre of suspension is equal to the length of the *equivalent simple pendulum.*

Condition for Minimum Time-Period

The time-period is given by

$$T = 2\pi\sqrt{\frac{k^2 + l^2}{lg}} \quad \text{or} \quad T^2 = 4\pi^2\frac{k^2 + l^2}{lg}.$$

Differentiating with respect to l, we have

$$2T\frac{dT}{dl} = \frac{4\pi^2}{g}\frac{d}{dl}\left(\frac{k^2 + l^2}{l}\right) = \frac{4\pi^2}{g}\frac{d}{dl}\left(l + \frac{k^2}{l}\right) = \frac{4\pi^2}{g}\left(1 - \frac{k^2}{l^2}\right)$$

$$\text{or} \quad T\frac{dT}{dl} = \frac{2\pi^2}{g}\left(1 - \frac{k^2}{l^2}\right).$$

Now for T to be minimum, we have $\frac{dT}{dl} = 0$. Therefore $1 - \frac{k^2}{l^2} = 0$ or $k^2 = l^2$.

$$\therefore \qquad T_{\min} = 2\pi\sqrt{\frac{2k^2}{gk}} = 2\pi\sqrt{\frac{2k}{g}}$$

If we record the time-periods of the pendulum corresponding to the different points above and below the centre of gravity as the centres of suspension, and plot in a graph of the time-periods against distances of these points from the centre of gravity, the curves as shown in the figure 6.11.4 are obtained.

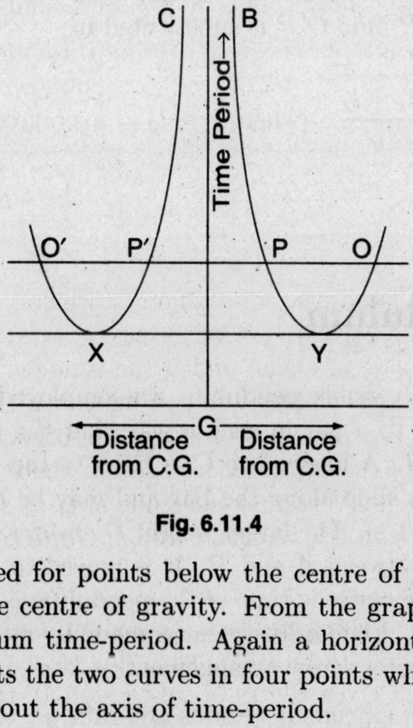

Fig. 6.11.4

The curve $CP'O'$ is obtained for points below the centre of gravity and the curve BPO is obtained for points above the centre of gravity. From the graph it follows that $XY = 2k$, whence we may get the minimum time-period. Again a horizontal line drawn above XY at right angles to the time axis cuts the two curves in four points which are placed by pairs, e.g., $P'P$ and $O'O$ symmetrically about the axis of time-period.

Then

$$T = 2\pi\sqrt{\frac{k^2 + l_1^2}{l_1 g}} = 2\pi\sqrt{\frac{k^2 + l_2^2}{l_2 g}}$$

where T = time-period given by the ordinate of any one of the points O', P', P or O;

l_1 = distance of O or O' from the axis of time-period and

l_2 = distance of P or P' from the axis of time-period.

$$\text{or} \quad \frac{T^2}{4\pi^2}gl_1 = k^2 + l_1^2 \quad \text{and} \quad \frac{T^2}{4\pi^2}gl_2 = k^2 + l_2^2$$

$$\therefore \quad \frac{T^2 g}{4\pi^2}(l_1 - l_2) = l_1^2 - l_2^2 \quad \text{or} \quad \frac{gT^2}{4\pi^2} = l_1 + l_2 \quad \text{or} \quad T = 2\pi\sqrt{\frac{l_1 + l_2}{g}}.$$

Thus the length of the equivalent simple pendulum in this case is $(l_1 + l_2)$ which is given by OP' or $O'P$. That is for the centre of suspension O, P' is the centre of oscillation and for the centre of suspension O', P is the centre of oscillation.

6.12 Bar Pendulum

The bar pendulum is a compound pendulum which consists of a heavy uniform rectangular bar along the length of which a series of holes at regular intervals above and below the centre of gravity of the bar are bored. The bar is supported by a horizontal knife-edge placed in turn in each of these holes. The time-periods of oscillation of the bar for these suspensions are measured. The graphical relation between the time-periods and the distances of the points of suspension from the centre of gravity is similar to that as shown in Fig. 6.11.4. From the curve, the average of the distances OP' and $O'P$ is substituted in

$$T = 2\pi\sqrt{\frac{l_1 + l_2}{g}} \quad [\text{where } l_1 + l_2 = \text{average distance}]$$

whence g may be obtained.

6.13 Kater's Pendulum

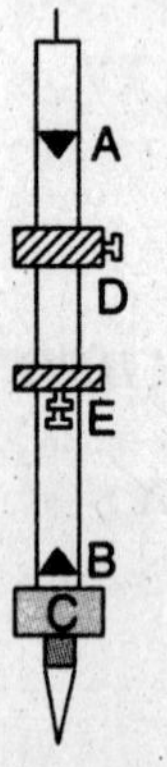

Fig. 6.13.1

The principle of reversible pendulum was employed by Kater. Two knife-edges A and B are fixed to a bar in such a way that the pendulum may be suspended from either A or B. A heavy bob C is fixed to the bar. The two weights D and E may be made to slide along the bar and may be clamped to it at any position necessary (Fig. 6.13.1). The larger weight D ensures that the centre of mass of the pendulum lies in between A and B. It is moved to various positions until one is found for which the periodic time of the pendulum is nearly same when suspended either from A or B. Final adjustment is made by moving the smaller weight E by means of a micrometer screw arrangement.

It is found to be extremely tedious and difficult to make the pendulum exactly reversible. It is usually adjusted so that the values of the two periods approach each other. We then adopt *Bessel's method* for the calculation of time-period which will be reckoned as the period of oscillation when the pendulum is made reversible.

6.14 Bessel's Computed Time-Period

It is not necessary to make the pendulum exactly reversible. Let T_1 and T_2 be the respective periods of oscillation while the pendulum oscillates about the knife-edges O_1 and O_2 (Fig. 6.14.1). Here T_1 and T_2 are not exactly equal, but slightly different. Let h_1 and h_2 be the distances of the centre of gravity G from O_1 and O_2 respectively. Then we may write

$$T_1 = 2\pi\sqrt{\frac{k^2+h_1^2}{gh_1}} \quad \text{or} \quad \frac{gT_1^2}{4\pi^2}h_1 = k^2+h_1^2. \tag{6.14.1}$$

Similarly,

$$T_2 = 2\pi\sqrt{\frac{k^2+h_2^2}{gh_2}} \quad \text{or} \quad \frac{gT_2^2}{4\pi^2}h_2 = k^2+h_2^2. \tag{6.14.2}$$

Fig. 6.14.1

Now, if $T_1 > T_2$, we have from (6.14.1) and (6.14.2)

$$\frac{g}{4\pi^2}\left(\frac{h_1T_1^2 - h_2T_2^2}{h_1-h_2}\right) = h_1+h_2$$

Let us put

$$T^2 = \frac{h_1T_1^2-h_2T_2^2}{h_1-h_2} \quad \therefore \quad \frac{gT^2}{4\pi^2} = h_1+h_2 \quad \text{or} \quad T = 2\pi\sqrt{\frac{h_1+h_2}{g}}. \tag{6.14.3}$$

T is named as *Bessel's computed time-period.* This hypothetical time-period is the time-period corresponding to a simple pendulum of length (h_1+h_2). Theoretically, calculating this time T, we get the value of g from the equation (6.14.3).

It would be better to express this computed time in a different form, because measurement of h_1 and h_2 separately involves greater error than the measurement of (h_1+h_2). So the error for separate measurement of h_1 and h_2 should be reduced to a minimum value and that may be done in the following way.

Let us put

$$r^2 = \frac{T_1^2+T_2^2}{2} \quad \text{and} \quad a^2 = \frac{T_1^2-T_2^2}{2}$$

so that

$$T_1^2 = r^2+a^2 \quad \text{and} \quad T_2^2 = r^2-a^2.$$

Substituting these values in the expression for T^2, we have

$$T^2 = \frac{T_1^2h_1 - T_2^2h_2}{h_1-h_2} = \frac{(r^2+a^2)h_1-(r^2-a^2)h_2}{h_1-h_2} = \frac{r^2(h_1-h_2)+a^2(h_1+h_2)}{h_1-h_2}$$

$$= r^2 + a^2\frac{(h_1+h_2)}{(h_1-h_2)} = \frac{T_1^2+T_2^2}{2} + \frac{T_1^2-T_2^2}{2}\cdot\frac{(h_1+h_2)}{(h_1-h_2} \tag{6.14.4}$$

The above expression involves comparatively small error in the calculation of computed time, because (h_1+h_2) is measurable with greater accuracy. The error that is involved in the measurement of h_1 and h_2 separately will not sensibly affect the computation of T, because (h_1-h_2) occurs in the coefficient of $(T_1^2-T_2^2)$ which is a very small fraction of $(T_1^2+T_2^2)$. Thus knowing (h_1+h_2) exactly, and h_1, h_2 approximately, we can compute the time period corresponding to the length (h_1+h_2). h_1 and h_2, and hence (h_1-h_2) can be obtained by locating the centre of gravity of the pendulum. This is done by balancing the pendulum horizontally.

Illustrative Examples

1. If the radius of the bob of a Borda's pendulum be 10 cms and the length of the pendulum be 1 metre, calculate the length of an equivalent simple pendulum.

In case of Borda's pendulum,

$$T = 2\pi\sqrt{\frac{l + \frac{\frac{2}{5}r^2}{l}}{g}}.$$

Hence the length of an equivalent simple pendulum is given by

$$L = l + \frac{\frac{2}{5}r^2}{l} = 100 + \frac{2}{5}\cdot\frac{(10)^2}{100} = 100.4 \text{ cms}.$$

2. Calculate the time-period of oscillation of a solid cylinder of length l and radius r about an axis perpendicular to its axis of symmetry and at a distance d from its centre of mass. Also calculate the minimum time-period.

Moment of inertia about an axis passing through its centre of mass and parallel to the axis in question is given by

$$I = M\left(\frac{l^2}{12} + \frac{r^2}{4}\right), \quad \text{where } M \text{ is the mass of the cylinder}$$

$$= Mk^2 \text{ (say)}.$$

Hence,

$$k = \sqrt{\frac{l^2}{12} + \frac{r^2}{4}} = \sqrt{\frac{l^2 + 3r^2}{12}}.$$

The length of the equivalent simple pendulum is, therefore,

$$\frac{d^2 + k^2}{d} = d + \frac{k^2}{d} = d + \frac{l^2 + 3r^2}{12d}$$

Hence, the time-period of oscillation is given by

$$T = 2\pi\sqrt{\frac{d + \frac{l^2+3r^2}{12d}}{g}}$$

The minimum time-period about the required axis is

$$T_{\min} = 2\pi\sqrt{\frac{2k}{g}} = 2\pi\left(\frac{l^2 + 3r^2}{3g^2}\right)^{1/4}$$

3. A uniform circular disc or a hoop is suspended by three equidistant threads each of length l attached to its rim. If the disc or the hoop is now given a small angular horizontal displacement and released, find the time-period of oscillation in each case.

The three points on the rim at which the strings are attached lie at the corners of an equilateral triangle. If T be the sum of the three tensions along the three threads, then the tension along each thread is equal to $\frac{T}{3}$.

The vertical displacement of the centre of mass is

$$z = l(1 - \cos\theta) = \frac{l\theta^2}{2} \text{ nearly} \quad [\because \theta \text{ is small}]$$

where θ denotes the angle which the string makes with the vertical.

Considering the vertical forces, we have

$$M\frac{d^2 z}{dt^2} = T\cos\theta - Mg,$$

where M is the mass of the suspended body.

When θ is small we may neglect the vertical motion of the body.

$$\therefore \quad 0 = T - Mg \quad \text{or} \quad T = Mg.$$

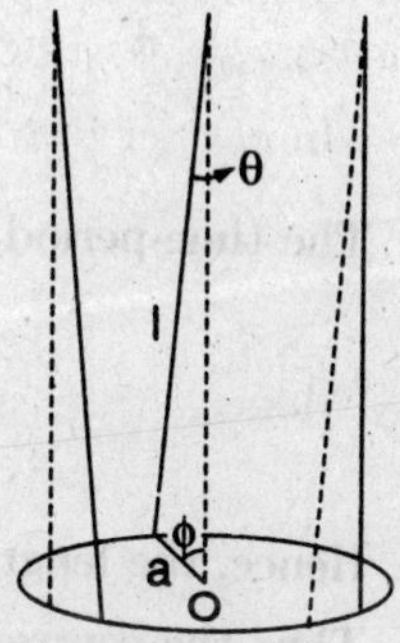

Fig. 1

We shall now consider the horizontal forces causing the horizontal motion. Let ϕ be the angle through which the circular disc or the hoop is displaced from its equilibrium position. The only horizontal forces are the horizontal components of the tensions. The horizontal component of the tension acting along each thread $= \frac{T}{3}\sin\theta = \frac{T}{3}\theta$, since θ is small. Its moment about the axis passing through the centre O of the hoop or disc $= \frac{T}{3}\theta \cdot a$, where a stands for the radius of the disc or the hoop. Hence, considering the effect of tensions along the three threads, the sum of the moments of the horizontal components of tensions is $3 \times \frac{T}{3}a\theta = Ta\theta$.

Hence, the equation of rotational motion of the disc or the hoop is given by

$$I\frac{d^2\phi}{dt^2} = -Ta\theta = -aMg\theta,$$

where I is the moment of inertia about the axis of rotation.

Now,

$$a\phi = l\theta \quad \text{or} \quad \theta = \frac{a\phi}{l}.$$

Hence, we have

$$I\frac{d^2\phi}{dt^2} = -\frac{Mga^2}{l}\phi \quad \text{or} \quad \frac{d^2\phi}{dt^2} + \frac{Mga^2}{Il}\phi = 0.$$

Hence, the time-period,

$$T = 2\pi\sqrt{\frac{Il}{gMa^2}}.$$

For a circular disc, $I = \frac{Ma^2}{2}$ $\quad \therefore T = 2\pi\sqrt{\frac{l}{2g}}$.

For a circular hoop $I = Ma^2$, $\quad \therefore T = 2\pi\sqrt{\frac{l}{g}}$.

4. A circular disc of radius 20 cms oscillates as a pendulum about a point on its circumference. Find the position or positions of the centre of oscillation. **[Punjab 1955]**

If M and r be the mass and radius of the disc respectively, then the moment of inertia of the disc about an axis passing through its centre and perpendicular to its plane is given by

$$Mk^2 = \frac{1}{2}Mr^2 \quad \therefore \quad k^2 = \frac{r^2}{2} = \frac{(20)^2}{2} = 200.$$

The time-period about the point on its circumference is given by

$$T = 2\pi\sqrt{\frac{k^2+r^2}{gr}} = 2\pi\sqrt{\frac{200+400}{20g}} = 2\pi\sqrt{\frac{30}{g}}.$$

Hence, the length of the equivalent simple pendulum is $L = 30$ cms.

Thus the centre of oscillation is 30 cms vertically below the point of suspension or 10 cms below the centre of the disc.

5. What is the distance between the centre of the suspension and the centre of oscillation on a uniform cylindrical metal bar used as a seconds pendulum?

(Diameter of the base = 1 cm, the density = 8 gms/c.c. and $g = 978$ cms/sec^2).

[Allahabad, 1949]

For a compound pendulum $T = 2\pi\sqrt{\dfrac{k^2+h^2}{gh}}$

Since it is used as a seconds pendulum,

$$T = 2 \text{ secs} \quad \therefore 2 = 2\pi\sqrt{\frac{k^2+h^2}{gh}} \quad \text{or} \quad \frac{k^2+h^2}{h} = \frac{g}{\pi^2} = \frac{978}{(3.14)^2} = 99.19 \text{ cms}.$$

$(k^2+h^2)/h$ denotes the distance between the centre of suspension and the centre of oscillation.

6. A cube of edge S and mass M is suspended vertically with one of its edges as the axis of suspension. Find the period for small vibrations. What is the length of the equivalent simple pendulum?

[C.U.(Hons.), 1995]

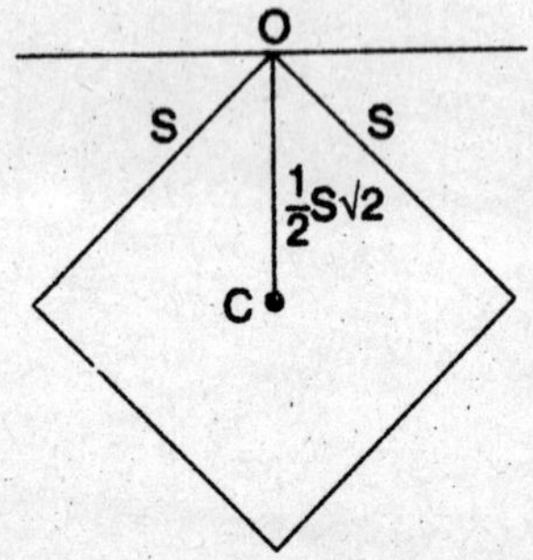

As the diagonal of a square of side S has length $= \sqrt{S^2+S^2} = S\sqrt{2}$, the distance OC of the centre of mass from the axis O as shown in the figure is $\frac{1}{2}S\sqrt{2}$. Now since the moment of inertia of a cube about an edge is the same as that of a square plate about a side, we have the moment of inertia of the system about an axis passing through O, and perpendicular to the plane of the paper given by

$$I = \frac{1}{3}M(S^2+S^2) = \frac{2}{3}MS^2$$

(applying the theorem of perpendicular axes).

Now the equation of motion, for small angular oscillation of the system is written as

$$I\frac{d^2\theta}{dt^2} = -Mg\left(\frac{1}{2}S\sqrt{2}\right)\theta \quad \text{or} \quad \frac{d^2\theta}{dt^2} + \frac{Mg\left(\frac{1}{2}S\sqrt{2}\right)}{I}\theta = 0 \quad \text{or} \quad \frac{d^2\theta}{dt^2} + \omega^2\theta = 0$$

where

$$\omega^2 = \frac{Mg\left(\frac{1}{2}S\sqrt{2}\right)}{I} = \frac{Mg\left(\frac{1}{2}S\sqrt{2}\right)}{\frac{2}{3}MS^2} = \frac{3\sqrt{2}g}{4S},$$

so that the time-period is

$$T = \frac{2\pi}{\omega} = 2\pi\sqrt{\frac{4S}{3\sqrt{2}g}} = 4\pi\frac{1}{\sqrt[4]{2}}\sqrt{\frac{S}{3g}}.$$

Again taking

$$T = 2\pi\sqrt{\frac{4S}{3\sqrt{2}g}}$$

and comparing with

$$T = 2\pi\sqrt{\frac{k^2+l^2}{lg}} = 2\pi\sqrt{\frac{L}{g}},$$

where $L = \dfrac{k^2+l^2}{l}$ = length of the equivalent simple pendulum, we have in this case

$$L = \frac{4S}{3\sqrt{2}} = \frac{2\sqrt{2}S}{3}.$$

7. A steel cylinder of length 20 cm and radius 3 cm is suspended by a brass wire 50 cm long and radius 2 mm. Calculate the length of the equivalent simple pendulum. The oscillations of the cylinder take place in a vertical plane and the cylinder and the wire are co-axial. (sp. gr. of steel and brass are 7.8 and 8.5 respectively). [C.U.(Hons.) 1989]

Let the centres of mass of the brass wire and the steel cylinder be G_1 and G_2 respectively as shown in the attached figure. Let the centre of mass of the whole system be at G which is at a distance x from G_2, i.e., $GG_2 = x$ cm and $GG_1 = (35 - x)$ cm, so that taking moments of the weights of the brass wire and steel cylinder about G_1 we may write

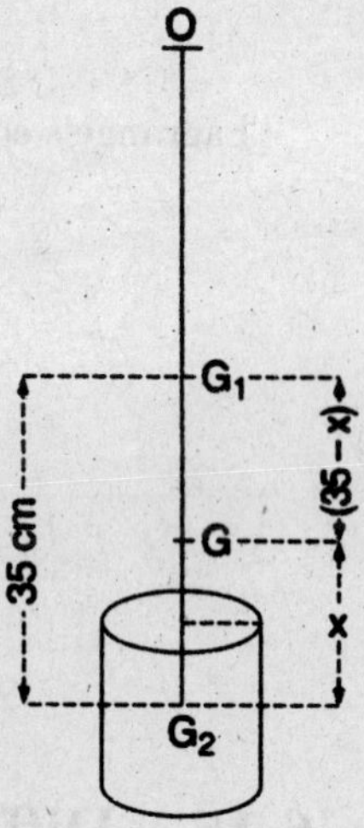

$$M_1 g(35 - x) = M_2 gx \qquad (1)$$

where M_1 = mass of the brass wire = $3.14 \times (0.2)^2 \times 50 \times 8.5$
= 53.38 gm

and M_2 = mass of the steel cylinder = $3.14 \times (3)^2 \times 20 \times 7.8$
= 4408.56

$\therefore$ from (1) $53.38(35 - x) = 4408.56 \times x$, whence $x = 0.418$ cm.

$\therefore$ $OG = 60 - 0.418 = 59.582$ cm $= l$.

Now length of the equivalent simple pendulum $L = \frac{I}{ml}$, where $m = (M_1 + M_2)$ and I = moment of inertia of the system about the axis through O

= M.I. of the wire about the axis through O + M.I. of the cylinder about the same axis

$$= \left[53.38\left\{\frac{(50)^2}{12} + \frac{(0.2)^2}{4}\right\} + 53.38 \times (25)^2\right] + \left[4408.56\left\{\frac{(20)^2}{12} + \frac{(3)^2}{4}\right\} + 4408.56 \times (60)^2\right]$$

(applying the theorem of parallel axes)

$$= [53.38(208.33 + 0.01) + 33362.50] + [4408.56(33.33 + 2.25) + 15870816.00]$$

$$= 11121.19 + 33362.50 + 156856.56 + 15870816.00 = 16072156.25 \text{ gm-cm}^2$$

$$\therefore \quad L = \frac{16072156.25}{(4408.56 + 53.38) \times 59.582} = 60.457 \text{ cm.}$$

8. A thin uniform rod of mass m and length R slides with its ends on a smooth vertical circle of radius R as shown in the figure. Find the period for small oscillations in terms of the given quantities.

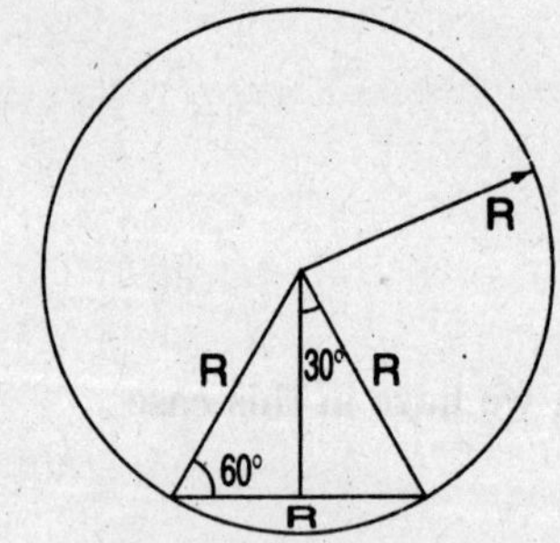

The ends of the rod form an equilateral triangle with the centre of the circle as shown in the attached figure. From the geometry of the figure, it follows that the centre of gravity of the rod will always be at a fixed distance $R\cos 30^0 = R\frac{\sqrt{3}}{2}$ from the centre of the vertical circle, and as such the length of the equivalent simple pendulum is $R\frac{\sqrt{3}}{2}$, and the time-period, $T = 2\pi\sqrt{\frac{R\sqrt{3}}{2g}}$ may be obtained in the following way.

K.E. $= T = \frac{1}{2}ml^2\dot{\theta}^2$ where $l = \frac{R\sqrt{3}}{2}$, and (assuming the zero level of P.E. to pass through the centre of the circle), P.E. $= V = -mgl\cos\theta$, so that the Lagrangian

$$L = T - V = \frac{1}{2}ml^2\dot{\theta}^2 + mgl\cos\theta,$$

whence we may write

$$\frac{\partial L}{\partial \dot{\theta}} = ml^2\dot{\theta}, \quad \frac{\partial L}{\partial \theta} = -mgl\sin\theta.$$

Lagrange's equation is

$$\frac{d}{dt}\left(\frac{\partial L}{\partial \dot{\theta}}\right) - \frac{\partial L}{\partial \theta} = 0 \quad \text{or} \quad ml^2\ddot{\theta} + mgl\sin\theta = 0$$

$$\text{or} \quad \ddot{\theta} + \frac{g}{l}\theta = 0 \quad (\text{putting } \sin\theta = \theta \text{ when } \theta \text{ is small})$$

$$\text{or} \quad \ddot{\theta} + \omega^2\theta = 0 \quad \text{where } \omega^2 = g/l$$

$$\therefore \quad T = \frac{2\pi}{\omega} = 2\pi\sqrt{\frac{l}{g}} = 2\pi\sqrt{\frac{R\sqrt{3}}{2g}}.$$

6.15 Different Corrections of Compound Pendulum

For very accurate measurement of g with the help of a compound pendulum, the following corrections are necessary.

6.15.1 Finite Arc Correction

When the pendulum oscillates with an angular amplitude greater than what is demanded for the oscillation to be exactly simple harmonic, a correction in the time-period is necessary. If

T_1 be the time-period corresponding to simple harmonic motion of the pendulum, and T_0 the observed time-period for finite arc of swing, it may be shown that

$$T_1 = T_0\left(1 - \frac{\alpha_1\alpha_2}{16}\right),$$

where α_1 and α_2 are the successive angular amplitudes. The loss of potential energy when the pendulum moves from A to $B = mgl(\cos\theta - \cos\alpha)$ [Fig. 6.2.2].

The gain in kinetic energy at $B = \frac{1}{2}I\omega^2$.

Hence, we have

$$\frac{1}{2}I\left(\frac{d\theta}{dt}\right)^2 = mgl(\cos\theta - \cos\alpha)$$

$$\text{or} \quad \frac{1}{2}m(k^2+l^2)\left(\frac{d\theta}{dt}\right)^2 = mgl(\cos\theta - \cos\alpha) \quad [\because\ I = m(k^2+l^2)].$$

Proceeding as in Art. 6.6, we have

$$T_0 = 2\pi\sqrt{\frac{k^2+l^2}{gl}}\left(1 + \frac{1}{4}\sin^2\frac{\alpha}{2}\right) = T_1\left(1 + \frac{1}{4}\sin^2\frac{\alpha}{2}\right)$$

$$\therefore \quad T_1 = T_0\left(1 + \frac{1}{4}\sin^2\frac{\alpha}{2}\right)^{-1} = T_0\left(1 - \frac{1}{4}\sin^2\frac{\alpha}{2}\right) \quad \text{[approximately]}.$$

If during an experiment, amplitude falls from α_1 to α_2, where each is small, we have

$$\sin^2\frac{\alpha}{2} = \frac{\alpha_1}{2}\cdot\frac{\alpha_2}{2} \quad \text{[approximately]}$$

$$\therefore \quad T_1 = T_0\left(1 - \frac{\alpha_1\alpha_2}{16}\right).$$

6.15.2 Air Correction

As the pendulum swings in air, it is necessary to investigate the corrections arising from the reaction of the medium. The effect of surrounding air on the motion of the pendulum can be separated into three parts.

(i) *The buoyancy.* This diminishes the weight of the pendulum by an amount equal to the weight of air it displaces.

(ii) *Consideration of air pendulum.* As some of the air moves with the pendulum, we may imagine a similar air pendulum, which increases the effective moment of inertia of the pendulum.

(iii) *The damping effect due to the viscous drag of air.* As a result of this effect the amplitude gets reduced and the time-period increases.

The corrections for error due to buoyancy and the inertia of air pendulum, can be effected in the following way :

Let m = mass of the pendulum,

μ = mass of displaced air,

h_1 = distance of the centre of gravity of the pendulum from the supporting knife-edge,

s_1 = distance of the centre of buoyancy of displaced air from the supporting knife-edge, (this is not equal to h_1 unless the pendulum is homogenous throughout).

Then the restoring couple $= mh_1 g \sin\theta - \mu g s_1 \sin\theta = g(mh_1 - \mu s_1)\sin\theta$

$$= g(mh_1 - \mu s_1)\theta \text{ [when } \theta \text{ is small]}$$

Let i_1 be the increase of moment of inertia due to the effect of air pendulum, so that the total moment of inertia about the knife-edge is given by

$$I = m(k^2 + h_1^2) + i_1.$$

Hence, the equation of motion of the pendulum may be written as

$$[m(k^2 + h_1^2) + i_1]\frac{d^2\theta}{dt^2} + g(mh_1 - \mu s_1)\theta = 0.$$

So, the time-period of the pendulum is,

$$T_1 = 2\pi\sqrt{\frac{m(k^2 + h_1^2) + i_1}{g(mh_1 - \mu s_1)}}$$

$$\therefore \frac{T_1^2 g}{4\pi^2} = \frac{m(k^2 + h_1^2) + i_1}{mh_1 - \mu s_1} = \frac{k^2 + h_1^2 + \frac{i_1}{m}}{h_1\left(1 - \frac{\mu \cdot s_1}{m \cdot h_1}\right)} = \frac{\left(k^2 + h_1^2 + \frac{i_1}{m}\right)}{h_1}\left(1 + \frac{\mu s_1}{mh_1}\right) \text{ [approximately]}$$

Neglecting the products of small quantities, we have

$$\frac{T_1^2 g}{4\pi^2} = \frac{k^2 + h_1^2}{h_1} + \frac{i_1}{mh_1} + \frac{k^2 + h_1^2}{h_1}\cdot\frac{\mu s_1}{mh_1}. \tag{6.15.1}$$

In the inverted position of the pendulum, we have similarly

$$\frac{T_2^2 g}{4\pi^2} = \frac{k^2 + h_2^2}{h_2} + \frac{i_2}{mh_2} + \frac{k^2 + h_2^2}{h_2}\cdot\frac{\mu s_2}{mh_2} \tag{6.15.2}$$

Multiplying (6.15.1) by h_1 and (6.15.2) by h_2 and then subtracting, we get

$$\frac{g}{4\pi^2}(T_1^2 h_1 - T_2^2 h_2) = (h_1^2 - h_2^2) + \frac{i_1 - i_2}{m} + \frac{k^2 + h_1^2}{h_1}\frac{\mu s_1}{m} - \frac{k^2 + h_2^2}{h_2}\cdot\frac{\mu s_2}{m}.$$

Now since T_1 and T_2 are nearly equal, we may put,

$$\frac{k^2+h_1^2}{h_1} = \frac{k^2+h_2^2}{h_2} = l = \text{the length of the equivalent simple pendulum.}$$

Hence, the above equation may be put in the form

$$\frac{g}{4\pi^2}(T_1^2h_1 - T_2^2h_2) = (h_1^2 - h_2^2) + \frac{i_1 - i_2}{m} + \frac{\mu l}{m}(s_1 - s_2)$$

or
$$\frac{g}{4\pi^2}\left(\frac{T_1^2h_1 - T_2^2h_2}{h_1 - h_2}\right) = (h_1 + h_2) + \frac{i_1 - i_2}{m(h_1 - h_2)} + \frac{\mu l}{m}\frac{s_1 - s_2}{h_1 - h_2} \quad (6.15.3)$$

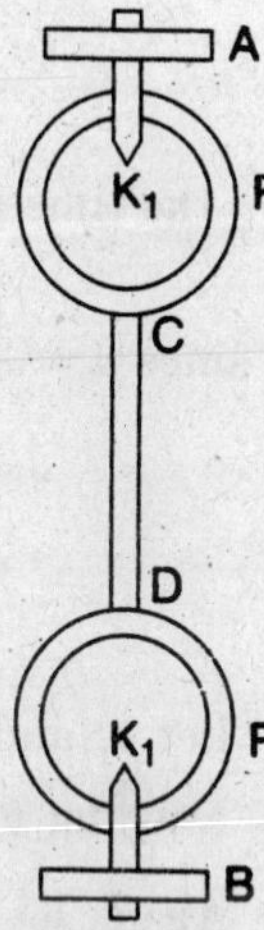

Fig. 6.15.1

$(s_1 - s_2)$ and $\frac{\mu}{m}$ can be determined with sufficient accuracy from the shape of the pendulum. Hence, the second correction term can be easily obtained. To determine the first correction term, another pendulum of identical shape, but of different mass can be constructed. From equation similar to (6.15.3) for the second pendulum, $(i_1 - i_2)$ may be obtained.

The correction terms on the right-hand side of the above expression 6.15.3 may be eliminated with the help of Repsold's. pendulum (Fig. 6.15.1), which is symmetrical in form about its centre of gravity, so that

$$s_1 = s_2 \quad \text{and} \quad i_1 = i_2.$$

Repsold's pendulum consists of a bar CD fixed into two rings R_1 and R_2 into which are screwed two short rods terminating in knife-edges K_1 and K_2, and carrying the two bobs A and B of which one is solid and the other is hollow. By moving the bobs upward or downward on the supporting stem, the time-periods of the pendulum about the knife edges may be made nearly equal.

6.15.3 Viscous Drag of Air

During the motion of the pendulum in air a viscous resistance comes into play between different layers of air moving at different rates, producing a damping effect on the motion of the pendulum—as a result of which, the amplitude gets reduced and the time-period increases. We may represent this effect by introducing a term proportional to the velocity in the equation of motion, which may thus be written as

$$\frac{d^2\theta}{dt^2} + \gamma\frac{d\theta}{dt} + \mu\theta = 0.$$

The solution is

$$\theta = Ae^{-\frac{\gamma t}{2}}\cos\left(\sqrt{\mu - \frac{\gamma^2}{4}}\,t - \alpha\right) \quad (6.15.4)$$

where A and α are constants.

The time-period is given by

$$T = \frac{2\pi}{\sqrt{\mu - \frac{\gamma^2}{4}}} = \frac{2\pi}{\sqrt{\mu}\left(1 - \frac{\gamma^2}{4\mu}\right)^{1/2}} = \frac{2\pi}{\sqrt{\mu}}\left(1 + \frac{\gamma^2}{8\mu}\right) \quad \text{(approximately)}$$

$$= T_0\left(1 + \frac{\gamma^2}{8\mu}\right),$$

where $T_0 = \dfrac{2\pi}{\sqrt{\mu}}$ represents the free period of the pendulum, i.e., when there is no air resistance

$$\therefore \quad \frac{T}{T_0} = \left(1 + \frac{\gamma^2}{8\mu}\right),$$

i.e., the time-period is increased by the viscosity of air in the ratio $\left(1 + \dfrac{\gamma^2}{8\mu}\right) : 1;$

or since $\mu = \dfrac{4\pi^2}{T^2}$ (nearly), the above ratio is $\left(1 + \dfrac{\gamma^2 T^2}{32\pi^2}\right) : 1.$

$$\therefore \quad T_0 = \frac{T}{\left(1 + \frac{\gamma^2 T^2}{32\pi^2}\right)} = T\left(1 - \frac{\gamma^2 T^2}{32\pi^2}\right) \quad \text{(approximately)}.$$

Let ρ_1 and ρ_2 be the successive amplitudes of the pendulum on the two sides of the centre. Now, let $\theta = \rho_1$, when $t = 0$, and hence $\rho_1 = A\cos\alpha$ [from (6.15.4)].
Again, let $\theta = \rho_2$, when $t = \dfrac{\pi}{\sqrt{\mu - \frac{\gamma^2}{4}}}$, and we have

$$\rho_2 = -A\exp\left[-\frac{\pi\gamma}{2\sqrt{\mu - \frac{\gamma^2}{4}}}\cos\alpha\right]$$

$$\therefore \quad \frac{\rho_1}{\rho_2} = \exp\left[\frac{\pi\gamma}{2\sqrt{\mu - \frac{\gamma^2}{4}}}\right] = \exp\left[\frac{\gamma}{4}\cdot\frac{2\pi}{\sqrt{\mu - \frac{\gamma^2}{4}}}\right] = e^{\frac{\gamma T}{4}}$$

The successive amplitudes being in opposite directions, the opposite signs of alternate amplitudes are omitted.

Thus, we have

$$\log_e\left(\frac{\rho_1}{\rho_2}\right) = \frac{\gamma T}{4}.$$

In one of Kater's experiments, the arc of a swing decreased from $1^0.41$ to $1^0.18$ in about 500 seconds, i.e., the swing decreased in the ratio 1.195 : 1.

$$\therefore \quad \left(\frac{\rho_1}{\rho_2}\right)^{500} = 1.195 \quad \text{or} \quad 500\log_e\left(\frac{\rho_1}{\rho_2}\right) = \log_e 1.195 = 0.178.$$

$$\therefore \quad 500\frac{\gamma T}{4} = 0.178 \quad \text{or} \quad \gamma T = \frac{4 \times 0.178}{500}$$

Since $T_0 = T\left(1 - \dfrac{\gamma^2 T^2}{32\pi^2}\right)$, the correction term is

$$\frac{\gamma^2 T^2}{32\pi^2} = \frac{4^2 \times (0.178)^2}{(500)^2 \times 32\pi^2} = 6 \times 10^{-9} \text{ (nearly)}.$$

Thus the correction term due to viscosity is very small, and so it is usually neglected.

6.15.4 Knife-edge Correction

Usually the knife-edges of a pendulum are not perfectly sharp, and the pendulum swings by a rolling contact of the knife-edge on its support. Approximately the edge of the knife may be regarded as cylindrical with O as the centre of curvature, and K the instantaneous point of contact of the knife-edge.

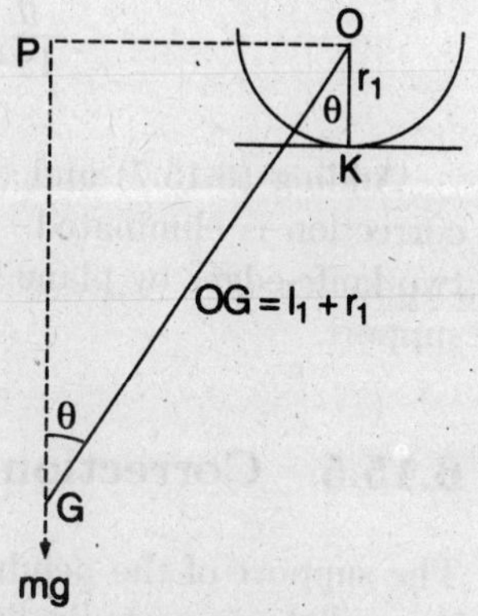

Fig. 6.15.2

From the figure it is clear that due to rolling of the knife-edge, the centre of suspension (point of contact) shifts to O and the distance of the centre of gravity becomes $(l_1 + r_1)$, where r_1 is the radius of curvature of the cylindrical surface of the knife-edge, and l_1, the distance of the centre of gravity from the knife-edge.

We have, therefore the restoring torque given by

$$mg \times OP = mg(l_1 + r_1)\sin\theta = mg(l_1 + r_1)\theta \quad \text{when } \theta \text{ is small.}$$

We have, therefore, the equation of motion of the pendulum given by

$$I\frac{d^2\theta}{dt^2} = -mg(l_1 + r_1)\theta \text{ or } m(k^2 + l_1^2)\frac{d^2\theta}{dt^2} = -mg(l_1 + r_1)\theta \text{ or } \frac{d^2\theta}{dt^2} + \frac{g(l_1 + r_1)}{k^2 + l_1^2}\theta = 0,$$

which is the characteristic equation for simple harmonic oscillation, and hence the time-period is given by

$$T_1 = 2\pi\sqrt{\frac{k^2 + l_1^2}{g(l_1 + r_1)}} = 2\pi\sqrt{\frac{k^2 + l_1^2}{gl_1\left(1 + \frac{r_1}{l_1}\right)}} = 2\pi\sqrt{\frac{k^2 + l_1^2}{gl_1}\left(1 - \frac{r_1}{l_1}\right)} \quad \text{(approximately)}$$

$\therefore$ Squaring and rearranging, we have

$$\frac{T_1^2 l_1}{4\pi^2}g = (k^2 + l_1^2)\left(1 - \frac{r_1}{l_1}\right). \tag{6.15.5}$$

Again, by using the second knife-edge, we have

$$\frac{T_2^2 l_2}{4\pi^2}g = (k^2 + l_2^2)\left(1 - \frac{r_2}{l_2}\right). \tag{6.15.6}$$

Subtracting (6.15.6) from (6.15.5), we have

$$\frac{g}{4\pi^2}(T_1^2 l_1 - T_2^2 l_2) = (l_1^2 - l_2^2) + r_2\left(\frac{k^2 + l_2^2}{l_2}\right) - r_1\left(\frac{k^2 + l_1^2}{l_1}\right) = (l_1^2 - l_2^2) + l(r_2 - r_1)$$

where $l = \dfrac{k^2 + l_1^2}{l_1} = \dfrac{k^2 + l_2^2}{l_2}$ = the length of the equivalent simple pendulum

$$\therefore \frac{g}{4\pi^2}\left(\frac{T_1^2 l_1 - T_2^2 l_2}{l_1 - l_2}\right) = (l_1 + l_2) + \frac{l}{l_1 - l_2}(r_2 - r_1). \tag{6.15.7}$$

If the knife-edges are interchanged, no other alteration being made, and if T_3, T_4 be the corresponding periods, we have

$$\frac{g}{4\pi^2}\left(\frac{T_3^2 l_1 - T_4^2 l_2}{l_1 - l_2}\right) = (l_1 + l_2) + \frac{l}{l_1 - l_2}(r_1 - r_2). \quad (6.15.8)$$

Adding (6.15.7) and (6.15.8), the terms involving r_1 and r_2 disappear so that knife-edge correction is eliminated. The effect of knife-edge curvature may be avoided by replacing the two knife-edges by plane bearings (flat plates) on the pendulum and a fixed knife-edge on the support.

6.15.5 Correction for Yielding of Support

The support of the pendulum must be sufficiently rigid, otherwise the support will be forced to oscillate co-periodically with the pendulum. The magnitude of error due to this cause was calculated by Pierce and others as follows.

Let O be the point of support of the knife-edge on the plane and G the centre of gravity of the pendulum. Let $OG = h_1$; let the point of support be displaced to O', when G is displaced to G'. Let GO and $G'O'$ be produced to meet at C. Evidently C is the effective centre of suspension. It means that due to yielding of support, the effective length of the pendulum is increased by an amount OC. Let $OC = \rho_1$. The acceleration of the centre of gravity is $h_1 \dfrac{d^2\theta}{dt^2}$ along the arc and $\dfrac{h_1^2 \left(\frac{d\theta}{dt}\right)^2}{h_1}$ or $h_1\left(\dfrac{d\theta}{dt}\right)^2$ towards the point of support.

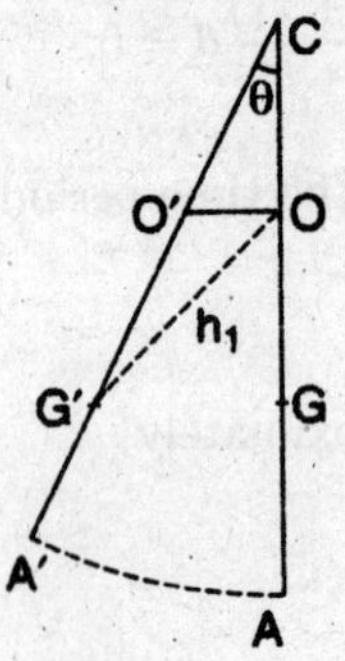

Fig. 6.15.3

Resolving these horizontally and vertically, we have horizontal acceleration

$$= \left[h_1 \frac{d^2\theta}{dt^2}\cos\theta - h_1\left(\frac{d\theta}{dt}\right)^2 \sin\theta\right] = \left[h_1\frac{d^2\theta}{dt^2} - h_1\left(\frac{d\theta}{dt}\right)^2\theta\right] \quad [\because \theta \text{ is small}]$$

and vertical acceleration

$$= \left[h_1\frac{d^2\theta}{dt^2}\sin\theta + h_1\left(\frac{d\theta}{dt}\right)^2\cos\theta\right] = \left[h_1\frac{d^2\theta}{dt^2}\theta + h_1\left(\frac{d\theta}{dt}\right)^2\right], \quad [\because \theta \text{ is small}]$$

Taking the horizontal acceleration into consideration; we have the force on the support in the horizontal direction given by

$$m\left[h_1\frac{d^2\theta}{dt^2} - h_1\left(\frac{d\theta}{dt}\right)^2\theta\right] = mh_1\frac{d^2\theta}{dt^2} \quad \left[\begin{array}{l}\text{neglecting the second term considering}\\ \theta \text{ to be extremely small}\end{array}\right]$$

where m is the mass of the pendulum.

Again the equation of motion of a compound pendulum is given by

$$\frac{d^2\theta}{dt^2} = -\frac{gh_1}{k^2 + h_1^2}\theta.$$

Hence, the force on the support in the horizontal direction is given by

$$p = \frac{mgh_1^2}{k^2 + h_1^2}\theta = \frac{mgh_1^2}{h_1h_2 + h_1^2}\theta \quad [\because k^2 = h_1h_2]$$

$$= \frac{mgh_1}{h_1 + h_2}\theta. \tag{6.15.9}$$

If α be the yield of the support per unit force in the horizontal direction, then the yield (Fig. 6.15.3).

$$OO' = \alpha p = \frac{mg\alpha h_1}{h_1 + h_2}\theta \quad \text{[from equation 6.15.9]}$$

Again, $OO' = \rho_1\theta \quad \therefore \rho_1 = \dfrac{mg\alpha h_1}{h_1 + h_2}$.

Hence,

$$\frac{\rho_1}{h_1} = \frac{mg\alpha}{h_1 + h_2} \quad \text{or} \quad \rho_1 h_1 = \frac{mg\alpha h_1^2}{h_1 + h_2}.$$

Similarly, in the inverted position,

$$\frac{\rho_2}{h_2} = \frac{mg\alpha}{h_1 + h_2} \quad \text{or} \quad \rho_2 h_2 = \frac{mg\alpha h_2^2}{h_1 + h_2}.$$

Hence,

$$\rho_1 h_1 - \rho_2 h_2 = \frac{mg\alpha}{h_1 + h_2}(h_1^2 - h_2^2) \quad \text{or} \quad \frac{\rho_1 h_1 - \rho_2 h_2}{h_1 - h_2} = mg\alpha. \tag{6.15.10}$$

Since in the erect position, the centre of suspension is raised by an amount ρ_1, the time-period T_1 in the erect position is given by

$$\frac{gT_1^2}{4\pi^2} = \frac{(h_1 + \rho_1)^2 + k^2}{h_1 + \rho_1} = (h_1 + \rho_1) + \frac{k^2}{h_1 + \rho_1}.$$

Similarly, in the inverted position, the time-period is given by

$$\frac{gT_2^2}{4\pi^2} = (h_2 + \rho_2) + \frac{k^2}{h_2 + \rho_2}$$

$$\therefore \quad \frac{g}{4\pi^2}(T_1^2 h_1 - T_2^2 h_2) = (h_1^2 - h_2^2) + (\rho_1 h_1 - \rho_2 h_2) + k^2\left(\frac{h_1}{h_1 + \rho_1} - \frac{h_2}{h_2 + \rho_2}\right)$$

$$= (h_1^2 - h_2^2) + (\rho_1 h_1 - \rho_2 h_2) + k^2\left(\frac{1}{1 + \frac{\rho_1}{h_1}} - \frac{1}{1 + \frac{\rho_2}{h_2}}\right)$$

Since ρ_1 and ρ_2 are small compared to h_1 and h_2, the last term in the above expression may be neglected.

$$\therefore \quad \frac{g}{4\pi^2}(T_1^2 h_1 - T_2^2 h_2) = (h_1^2 - h_2^2) + (\rho_1 h_1 - \rho_2 h_2)$$

$$\text{or} \quad \frac{g}{4\pi^2}\left(\frac{T_1^2 h_1 - T_2^2 h_2}{h_1 - h_2}\right) = (h_1 + h_2) + mg\alpha \quad \text{[from equation 6.15.10]}.$$

Again taking the vertical acceleration into consideration, we have the vertical force in the upward direction on the pendulum given by

$$m\left[h_1\frac{d^2\theta}{dt^2}\theta + h_1\left(\frac{d\theta}{dt}\right)^2\right] - mg$$

$$= m\left[-h_1\cdot\frac{gh_1}{k^2+h_1^2}\theta^2 + h_1\cdot 2gh_1\frac{(\cos\theta-\cos\alpha)}{k^2+h_1^2}\right] - mg$$

$$\left[\because \text{ from the energy equation we have } mgh_1(\cos\theta-\cos\alpha) = \frac{1}{2}m(k^2+h_1^2)\left(\frac{d\theta}{dt}\right)^2\right]$$

$$= m\left[-\frac{gh_1^2}{k^2+h_1^2}\theta^2 + gh_1^2\frac{(\alpha^2-\theta^2)}{k^2+h_1^2}\right] - mg \quad [\because \alpha \text{ and } \theta \text{ are small}]$$

$$= -\frac{2mgh_1^2}{k^2+h_1^2}\theta^2 + \frac{mgh_1^2}{k^2+h_1^2}\alpha^2 - mg$$

Considering only the variable part of the above expression, we have the vertical force given by

$$-\frac{2mgh_1^2}{k^2+h_1^2}\theta^2$$

which means a downward force of magnitude $\frac{2mgh_1^2}{k^2+h_1^2}\theta^2$ on the stand.

Since the force is of the second order in θ, the effect on the time-period is negligible in comparison with that due to the horizontal yield.

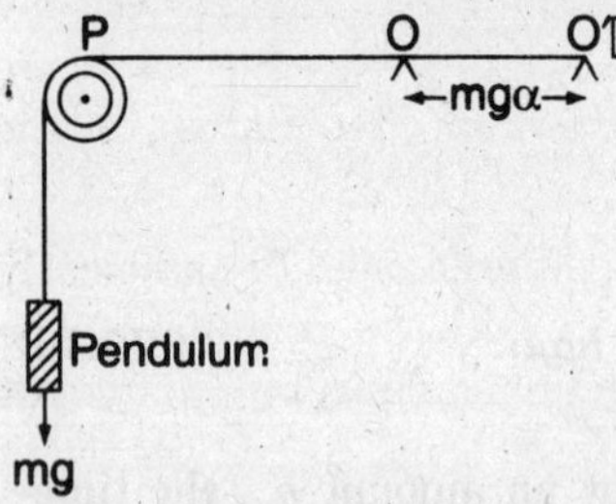

Fig. 6.15.4

Now the correction term $mg\alpha$ due to the horizontal yield may be found by hanging the pendulum with a string over a pulley, the end of the string being fixed to the support as shown in the figure 6.15.4. The displacement OO' which is $mg\alpha$ can be accurately determined by means of a microscope.

6.16 Summary of Corrections in the use of a Compound Pendulum

In order to determine accurately the value of g with the help of a compound pendulum, several corrections are to be made in the observed value T_0 of the periodic time. An account of the factors which make the corrections necessary, as also an outline of how the corrections can be made or avoided, is given below.

1. Finite arc correction. When the pendulum oscillates with an angular amplitude greater than what is demanded for the oscillation to be exactly simple harmonic, a correction in the time-period is necessary. If T be the time-period corresponding to simple harmonic motion of the pendulum, and T_0 the observed time-period for finite arc of swing, it may be shown that

$$T = T_0\left(1 - \frac{\alpha_1\alpha_2}{16}\right)$$

where α_1 and α_2 are the successive angular amplitudes in radians. This correction cannot be avoided, and should, therefore, be found out.

2. Air corrections. As the pendulum swings in air, three disturbing effects are introduced, all of which can be eliminated by working in vacuo.

(a) *The buoyancy effect.* This effectively diminishes the mass of the pendulum, and hence increases the period.

(b) *The air carried with the pendulum.* As some air moves with the pendulum, we may imagine a similar air pendulum, which increases the effective moment of inertia of the pendulum, and hence increases the period. For a pendulum symmetrical in form about the middle point, the corrections due to (a) and (b) can be shown to be zero. Repsold used such a pendulum (Fig. 6.15.1).

(c) *The damping effect due to the viscous drag on the pendulum.* As a result of this effect, the restoring couple is reduced, thereby increasing the amplitude and hence the time period increases. This correction is very small, and so it is usually neglected.

3. Knife-edge correction. Usually the knife-edges of a pendulum are not perfectly sharp, and the pendulum swings by a rolling contact of the knife-edge on its support. The edge of the knife may be regarded as cylindrical and the pendulum oscillates about the axis of the cylinder. The effect is to increase the length and hence the period. Correction due to this factor may be eliminated by making the radii of both knife-edges equal. The effect of knife-edge curvature may also be avoided by replacing the two knife-edges by plane bearings (flat plates) on the pendulum and a fixed knife-edge on the support.

4. Correction for yielding of support. The support of the pendulum must be sufficiently rigid, other-wise the support will be forced to oscillate co-periodically with the pendulum. The effect is a virtual increase in the value of the effective length of the pendulum, by an amount equal to the movement of the support produced by a horizontal force equal to the weight of the pendulum. This also increases the period. The correction term $mg\alpha$ due to horizontal yield (α being the yield of the support per unit force in the horizontal direction) may be found by hanging the pendulum with a string over a pulley, the end of the string being fixed to the support as shown in the figure 6.15.4. The displacement OO' which is $mg\alpha$ can be accurately determined by means of a microscope.

5. Temperature correction. Let T be the period corresponding to some standard temperature, say O^0C. Let it increase by dT when the temperature rises to t^0C. Then from (6.5.1), we may write $\frac{dT}{T} = \frac{dl}{2l}$ (as g is constant, $dg = 0$). If α be the co-efficient of linear expansion of the material of the pendulum rod, $dl = l\alpha t$, so that $\frac{dT}{T} = \frac{\alpha t}{2}$ or $dT = \frac{T\alpha t}{2}$.

6.17 Variation of *g*

In explaining the variation of g, we consider the following factors.

6.17.1 Effect of altitude

Let us suppose that the values of acceleration due to gravity on the surface of the earth and at a height h above the surface be g and g' respectively. Then, we have

$$\frac{g}{g'} = \frac{1/R^2}{1/(R+h)^2} = \frac{R^2 + 2Rh + h^2}{R^2} = \left(1 + \frac{2h}{R}\right), \quad \text{[approximately]}$$

where R is the radius of the earth.

$$\left[\because \text{ the acceleration due to gravity at any point is inversely proportional to the square of the distance of the point from the centre of the earth.}\right]$$

$$\therefore \quad \frac{g'}{g} = \left(1 + \frac{2h}{R}\right)^{-1} = \left(1 - \frac{2h}{R}\right) \quad \text{[approximately]}$$

$$\therefore \quad g' = g\left(1 - \frac{2h}{R}\right).$$

From the above expression, it is clear that greater the value of h, i.e., higher the altitude, the smaller will be the value of the acceleration due to gravity. Due to spheroidal shape of the earth, g' also depends on latitude. The expression for acceleration due to gravity at altitude h and latitude λ is given by $g' = g - \frac{2gh}{R}(1 + 0.00071 \cos 2\lambda)$. The relation between g' and g becomes more complicated due to the attractive action of elevated masses on the surface of the earth. Bouguer suggested the following correct expression,

$$g' = g\left(1 - \frac{2h}{R} + \frac{3hd}{2RD}\right),$$

where d is the density of the earth's surface constituents at the station under consideration and D is the mean density of the earth.

6.17.2 Effect of Depth

Let a body of mass m be taken at a depth h inside the earth (Fig. 6.17.1). Let us imagine a concentric sphere dividing the earth into two parts : (a) *internal sphere of radius* $(R - h)$, (b) *thick shell of thickness* h.

So far as the body at a depth h is concerned, it is just outside the internal sphere, and inside the thick shell. The attraction at any point inside a shell is zero (see Chapter VII). Therefore, the gravitational attraction on the body at a depth h is due only to the internal spherical mass of radius $(R - h)$. If F be the force of attraction experienced by the body, then from the law of gravitation we have

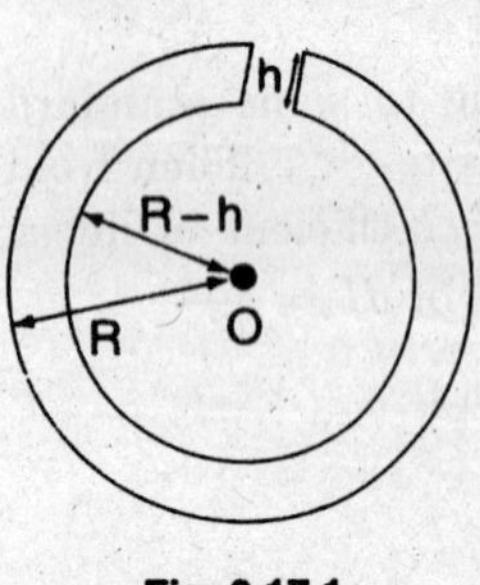

Fig. 6.17.1

$$F = G\frac{Mm}{(R - h)^2},$$

where M is the mass of the internal sphere and G is the gravitational constant.

Now, $M = \frac{4}{3}\pi(R - h)^3 D$, where D is the mean density of the earth

$$\therefore \; F = mg' = G\frac{\frac{4}{3}\pi(R - h)^3 Dm}{(R - h)^2},$$

where g' is the acceleration due to gravity at a depth h

$$\therefore \; g' = \frac{4}{3}\pi G(R - h)D.$$

Again, if the body is placed on the surface of the earth, the acceleration due to gravity is given by

$$g = G\frac{\frac{4}{3}\pi R^3 D}{R^2} = \frac{4}{3}\pi GRD$$

$$\therefore \quad \frac{g'}{g} = \frac{R-h}{R} = \left(1-\frac{h}{R}\right) \quad \text{or} \quad g' = g\left(1-\frac{h}{R}\right).$$

Therefore, as h increases, i.e., as the body is taken more and more inside the earth, g decreases and its value at the centre of the earth is zero, because at the centre $h = R$.

6.17.3 Effect of Latitude

The effect of latitude on the variation of g includes the following cases : (a) *the effect of rotation of the earth;* (b) *the effect due to shape of the earth.*

(a) Effect of Rotation of the Earth

In the figure 6.17.2, we consider the earth to be a perfect sphere rotating uniformly with angular velocity ω about the polar diameter NS. Let us consider a particle of mass m at M at latitude θ. Due to the rotation of the earth, the particle describes a circle of radius r. At M the forces acting are: (i) the centrifugal force $\frac{mv^2}{r}$ along MF; (ii) the force mg along MO. The resultant of these two forces is acting along MP and is given by

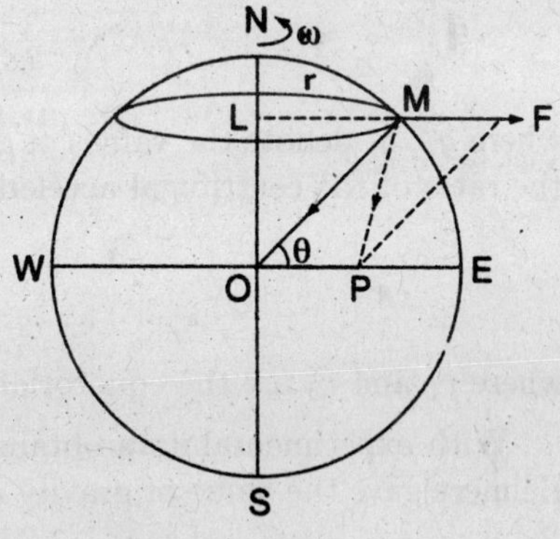

$$\overrightarrow{MP} = \sqrt{\overrightarrow{MO}^2 + \overrightarrow{OP}^2 - 2\overrightarrow{MO}\cdot\overrightarrow{OP}\cos MOP}$$

$$= \sqrt{m^2g^2 + \left(\frac{mv^2}{r}\right)^2 - 2mg\cdot\frac{mv^2}{r}\cos\theta}$$

$$= \sqrt{m^2g^2 + \frac{m^2R^4\cos^4\theta\omega^4}{R^2\cos^2\theta} - 2mg\frac{mR^2\cos^2\theta\omega^2\cdot\cos\theta}{R\cos\theta}}$$

$$= \sqrt{m^2g^2 + m^2R^2\cos^2\theta\omega^4 - 2m^2gR\cos^2\theta\omega^2}$$

[$\because\ v = r\omega = R\cos\theta\cdot\omega$, R being the radius of the earth.]

Fig. 6.17.2

The expression $m^2R^2\cos^2\theta\omega^4$ is negligible compared with other terms. Therefore,

$$\overrightarrow{MP} = \sqrt{m^2g^2 - 2m^2gR\omega^2\cos^2\theta} = mg\left(1 - \frac{2R\omega^2\cos^2\theta}{g}\right)^{1/2}$$

$$= mg\left(1 - \frac{R\omega^2\cos^2\theta}{g}\right) \quad \text{[approximately]}$$

This is the resultant force at M on the mass m.

Hence, the acceleration at M at latitude θ is given by

$$g_\theta = g\left(1 - \frac{R\omega^2 \cos^2\theta}{g}\right).$$

It should be noted that acceleration due to gravity is not directed along MO but along MP, the angle OMP being very small.

At the equator, $\theta = 0$, $\quad \therefore\ g_e = g\left(1 - \frac{R\omega^2}{g}\right)$

At the poles, $\theta = \frac{\pi}{2}$, $\quad \therefore\ g_p = g$.

Thus the apparent acceleration due to gravity is least at the equator and greatest at the poles.

Now $\dfrac{R\omega^2}{g} = \dfrac{1}{288}$ nearly, $\quad \therefore\ g_\theta = g\left(1 - \dfrac{\cos^2\theta}{288}\right)$.

This result is not in agreement with experimental results due to the spheroidal shape of the earth and non-homogeneity of the composition of the earth.

(b) Effect due to the Shape of the Earth

Clairaut assumed the earth's surface to be a spheroid of equilibrium such that a layer of water would spread uniformly over it. He assumed that the internal density was such that layers of uniform density are concentric and spheroids of similar nature. Clairaut's result may be written as

$$g_\lambda = g_e\left[1 + \left(\frac{5}{2}m - \epsilon\right)\sin^2\lambda\right]$$

where g_λ, g_e denote the values of gravity at the latitude λ and equator respectively; m denotes the ratio of the centrifugal acceleration and gravity at the equator; the ellipticity ϵ is given by

$$\epsilon = \frac{r_1 - r_2}{r_1}$$

where r_1 and r_2 are the equatorial and polar radii respectively.

With experimental data obtained in different experiments performed at different latitudes, Helmert gave the value of gravity as

$$g_\lambda = 978.00[1 + 0.005310\sin^2\lambda] \quad \text{and} \quad \varepsilon = \frac{1}{299}.$$

More recently, the International Geodetic Association adopted the following value of gravity

$$g_\lambda = 978.049[1 + 0.0052884\sin^2\lambda - 0.0000059\sin^2 2\lambda].$$

Illustrative Example

1. Show that if the earth were not rotating about its axis, the acceleration due to gravity at the equator would exceed its present value by 3.36 cms/sec^2; given that the radius of the earth = 6.37×10^8 cms and the angular velocity of the earth = 7.27×10^{-5} rad/sec. **[Agra 1959]**

We know, at the latitude θ the acceleration due to gravity is given by

$$g_\theta = g\left(1 - \frac{R\omega^2\cos^2\theta}{g}\right) \quad \text{or} \quad g - g_\theta = R\omega^2\cos\theta.$$

At the equator $\theta = 0$

$$\therefore \quad g - g_e = R\omega^2.$$

Here, g_e = the value of acceleration due to gravity at the equator considering the earth's rotation, and g = the value of gravity considering the earth to be stationary.

$\therefore$ the required increase of gravity is given by

$$g - g_e = 6.37 \times 10^8 \times (7.27)^2 \times 10^{-10} = 3.36 \text{ cms/sec}^2.$$

6.18 Different Methods for the Determination of g

The application of pendulum for the determination of g is a good example of the various methods used for this purpose. We shall consider a few other methods.

6.18.1 Atwood's Machine

This is a most direct method for the determination of g. In the ribbon-form of Atwood's machine (Fig. 6.18.1) a strip of paper or a ribbon S passes over a light pulley P running on ball bearings, and carries two equal weights, M at its two ends. The two weights carry identical ribbon below so that during the motion of the pulley with its ribbon and weights, no additional mass is transferred from one side of the pulley to the other. A steel strip L is rigidly clamped at one end, and carries at the other end a brush B impregnated with ink. The brush touches the paper ribbon stretched over the pulley. A small rider of mass m is placed above the mass M which rests on a platform A. Before the pulley is set in motion, the brush B is moved across the paper to mark a line which gives the starting point. Now the platform A is allowed to fall and the steel strip L is simultaneously set into vibration when the mass $(M+m)$ will move in the downward direction and M in the upward direction with acceleration a (say). Due to the motion of the ribbon, the vibrating brush will trace out wavy curve on the ribbon, the curve gradually lengthening out due to accelerated motion of the ribbon. Measuring the wave lengths from the curve we get the successive distances covered by the system in successive time-periods of the vibrating steel strip.

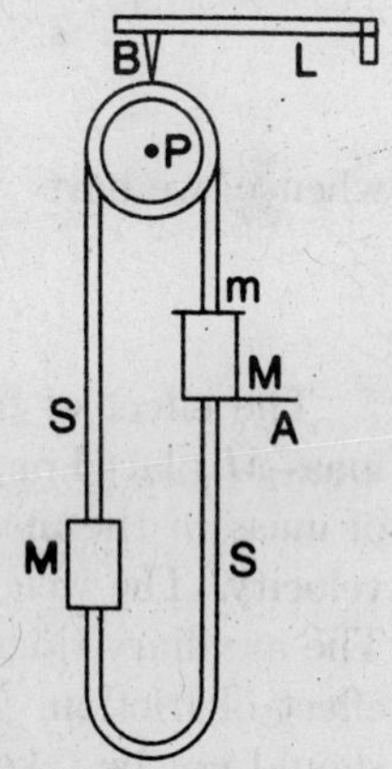

Fig. 6.18.1

Let x_1, x_2, x_3, $\cdots$ be the distances covered by the system in successive time periods of the steel strip (the distances being measured from the wavy curve); we have then

$$x_1 = \frac{1}{2}aT^2; \quad x_2 = aT\cdot T + \frac{aT^2}{2} = \frac{3}{2}aT^2; \quad x_3 = aT\cdot 2T + \frac{1}{2}aT^2 = \frac{5}{2}aT^2$$

where T = periodic time of the vibrating steel strip.

$$\therefore \ (x_2 - x_1) = (x_3 - x_2) = \cdots = aT^2.$$

Thus knowing T, a may be determined.

Again, if v be the velocity of the masses traversing a distance x, we have gain in kinetic energy of the system

$$= \frac{1}{2}I\omega^2 + \frac{1}{2}(2M+m)v^2$$

where I is the moment of inertia of the pulley about its axis of rotation and ω is the angular velocity at that instant.

Now the loss in potential energy of the system $= (M+m)gx - Mgx = mgx$

$$\therefore \quad \frac{1}{2}I\omega^2 + \frac{1}{2}(2M+m)v^2 = mgx \quad \text{or} \quad \frac{1}{2}I\frac{v^2}{r^2} + \frac{1}{2}(2M+m)v^2 = mgx$$

$$[\because v = r\omega, r \text{ being the radius of the pulley}]$$

$$\text{or} \quad \frac{v^2}{2}\left[\frac{I}{r^2} + (2M+m)\right] = mgx \quad \text{or} \quad v^2 = \frac{2mgx}{\left[\frac{I}{r^2} + (2M+m)\right]}.$$

Now $v^2 = 2ax$.

$$\therefore \quad 2ax = \frac{2mgx}{\left[\frac{I}{r^2} + (2M+m)\right]} \quad \text{or} \quad a = \frac{mg}{\left[\frac{I}{r^2} + (2M+m)\right]}$$

If the experiment be carried with two masses M' and M'' and the corresponding accelerations determined be a' and a'' respectively, then we have

$$a' = \frac{mg}{\left[\frac{I}{r^2} + (2M'+m)\right]}; \quad a'' = \frac{mg}{\left[\frac{I}{r^2} + (2M''+m)\right]}$$

whence, we have

$$g = \frac{2(M'-M'')}{m\left(\frac{1}{a'} - \frac{1}{a''}\right)}. \tag{6.18.1}$$

The effect of friction which retards the motion of the pulley is eliminated by loading the mass M placed on the platform by means of an auxiliary small rider. Removing the main rider of mass m the mass of auxiliary rider is adjusted until the system is given an initial uniform velocity. The weight of the auxiliary rider then neutralises the retarding force due to friction. The auxiliary rider is, therefore, kept on the mass throughout the experiment to eliminate the effect of friction. It is to be noted that the weight of this rider and the force due to friction should not be taken into consideration in calculating the value of g.

6.18.2 Inclined Plane

As we have seen in Chapter IV the acceleration of a body rolling down an inclined plane is given by

$$f = \frac{g\sin\theta}{1 + \frac{k^2}{r^2}}. \tag{6.18.2}$$

Knowing k from the shape of the body and f from the relation $x = \frac{1}{2}ft^2$, g can be easily determined from the expression (6.18.2).

6.18.3 Bifilar Suspension Method

When a heavy uniform rod is suspended horizontally by means of two equal vertical threads or cords at equal distances from the centre of gravity of the rod, the arrangement is called a ***bifilar suspension.*** The suspension may be made by means of two parallel or two non-parallel threads. We shall consider here the small oscillation of the rod in its own plane about the vertical axis through the centre of gravity of the rod.

We consider here the following cases of suspension : (a) suspension with parallel threads, (b) suspension with non-parallel threads.

(a) Parallel Threads

In the figure 6.18.2, AB denotes the rod displaced through a small angle ϕ, the rod being suspended by two equal threads of length l and separated by a distance $2a$. When the rod is in the equilibrium position CD, the threads are parallel to the vertical, but in the displaced position AB of the rod, the threads EF, GH get inclined to the vertical by an angle θ.

The tensions T acting upwards along the threads may be resolved into two components, one vertically and the other horizontally. The tensions resolved vertically upward along EF' and GH' are each of magnitude $T\cos\theta$ which balances the vertically downward force mg due to the weight of the rod in the equilibrium condition. Hence, for vertical equilibrium, we have

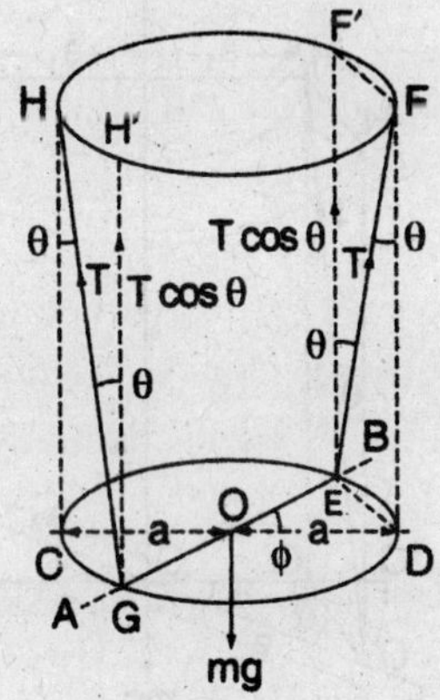

Fig. 6.18.2

$$2T\cos\theta = mg \quad \text{or} \quad T = \frac{mg}{2\cos\theta}.$$

Now from the figure $a\phi = l\theta$

$$\text{or } \cos\theta = \cos\left(\frac{a\phi}{l}\right) = 1 \; [\because \phi \text{ is small.}] \quad \therefore T = \frac{mg}{2}.$$

Now, the horizontal component of the tension is along ED and is given by

$$T\sin\theta = T\theta \quad [\because \theta \text{ is small}] = \frac{amg\phi}{2l} \quad \left[\because \theta = \frac{a\phi}{l}\right]$$

A similar force due to the horizontal component of the tension acts at the other end of the rod along GC. Since these two forces act at right angles to the axis of the rod, the moment of these forces about the axis of rotation is

$$= \frac{amg\phi}{2l}\cdot 2a = \frac{a^2mg\phi}{l}.$$

Since this couple tends to diminish the value of ϕ, it constitutes what is called the ***restoring couple.*** Hence, the equation of motion of the rod is given by

$$I\frac{d^2\phi}{dt^2} = -\frac{a^2mg}{l}\phi \quad \left[\text{where } I \text{ denotes the moment of inertia of the rod about its axis of rotation.}\right]$$

$$\text{or} \quad \frac{d^2\phi}{dt^2} + \frac{a^2mg}{Il}\phi = 0$$

from which $T = 2\pi\sqrt{\dfrac{Il}{a^2mg}}$, where T is the time-period of angular oscillation of the rod.

If we put $I = mk^2$ then,

$$T = 2\pi\sqrt{\frac{mk^2 l}{a^2 mg}} = 2\pi\frac{k}{a}\sqrt{\frac{l}{g}},$$

where k denotes the radius of gyration of the rod about the axis of rotation.

The above expression may be written as

$$g = \frac{4\pi^2 k^2 l}{a^2 T^2},$$

whence g may be determined.

(b) Non-Parallel Threads

In the figure 6.18.3, AB denotes the position of the rod displaced through a small angle ϕ, the rod being suspended by two threads EF and GH of equal lengths, but not parallel. Let the distance between the threads at the top and bottom be $2a_1$ and $2a_2$ respectively and let the length of each thread be l.

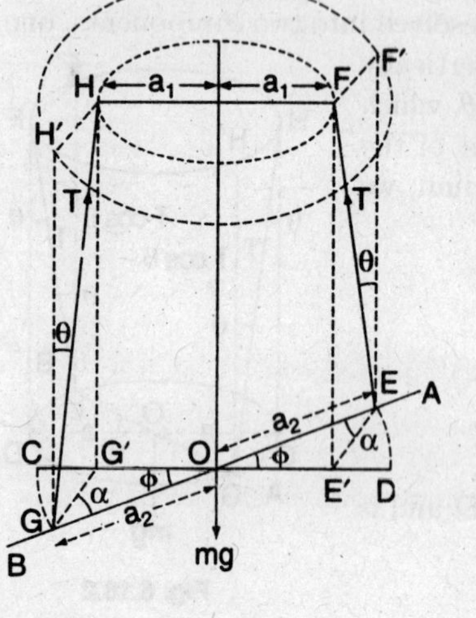

Fig. 6.18.3

Now, from the figure

$$\cos\theta = \frac{EF'}{EF} = \frac{EF'}{l}. \tag{6.18.3}$$

Now from the right-angled triangle EFF', we have

$$EF^2 = FF'^2 + EF'^2$$

$$\text{or} \quad EF'^2 = EF^2 - FF'^2 = l^2 - EE'^2$$

$$\text{or} \quad EF' = \sqrt{l^2 - EE'^2} \quad [\because EE' = FF']$$

Putting $EE' = x$, we have $EF' = \sqrt{l^2 - x^2}$.

Also, from the triangle EOE'

$$EE'^2 = (a_1^2 + a_2^2 - 2a_1 a_2 \cos\phi) = (a_2 - a_1)^2 \quad [\because \phi \text{ is small}]$$

$$\therefore \quad EE' = (a_2 - a_1) = x.$$

In addition

$$\cos\theta = \frac{\sqrt{l^2 - x^2}}{l} \quad \text{[from equation 6.18.3]}$$

For vertical equilibrium we may write

$$2T\cos\theta = mg \quad \text{or} \quad 2T\frac{\sqrt{l^2 - x^2}}{l} = mg \quad \text{or} \quad T = \frac{mgl}{2\sqrt{l^2 - x^2}}.$$

Now, the component of the tension along $EE' = T\sin\theta$.

Hence, the component perpendicular to OE

$$= T\sin\theta\sin\alpha = \frac{mgl}{2\sqrt{l^2-x^2}}\sin\theta\sin\alpha.$$

Therefore, the restoring torque tending to diminish the value of ϕ

$$= \frac{mgl}{2\sqrt{l^2-x^2}}\cdot 2a_2\sin\theta\sin\alpha = \frac{mgla_2}{\sqrt{l^2-x^2}}\cdot\sin\theta\sin\alpha$$

$$= \frac{mgla_2}{\sqrt{l^2-x^2}}\cdot\frac{x}{l}\sin\alpha \quad \left[\because \sin\theta = \frac{FF'}{l} = \frac{EE'}{l} = \frac{x}{l}\right] \tag{6.18.4}$$

But, from the triangle OEE',

$$\frac{a_1}{\sin\alpha} = \frac{EE'}{\sin\phi} = \frac{EE'}{\phi} = \frac{x}{\phi} \quad [\because \phi \text{ is small}]$$

$$\text{or} \quad \sin\alpha = \frac{a_1\phi}{x}. \tag{6.18.5}$$

Therefore, from (6.18.4) and (6.18.5) the restoring torque

$$= \frac{mgla_2}{\sqrt{l^2-x^2}}\cdot\frac{x}{l}\cdot\frac{a_1\phi}{x} = \frac{mga_1a_2\phi}{\sqrt{l^2-x^2}}.$$

Hence, the equation of motion of the rod is given by

$$I\frac{d^2\phi}{dt^2} = -\frac{mga_1a_2}{\sqrt{l^2-x^2}}\phi$$

where I is the moment of inertia of the rod about the vertical axis through the centre of gravity of the rod.

Hence, the time-period of oscillation of the rod is given by

$$t = 2\pi\sqrt{\frac{I\sqrt{l^2-x^2}}{mga_1a_2}} = 2\pi\sqrt{\frac{mk^2\sqrt{l^2-x^2}}{mga_1a_2}}$$

$$[\because I = mk^2, k \text{ being the radius of gyration}]$$

$$= \frac{2\pi k}{\sqrt{a_1a_2}}\sqrt{\frac{\sqrt{l^2-x^2}}{g}}.$$

Now, if y be the vertical distance between the ends of either suspension thread, then we have

$$y = EF' = \sqrt{l^2-x^2}. \quad \therefore \quad t = 2\pi\frac{k}{\sqrt{a_1a_2}}\sqrt{\frac{y}{g}} \tag{6.18.6}$$

In the case of bifilar suspension with parallel threads, we have

$$a_1 = a_2 = a \text{ (say)} \quad \text{and} \quad y = l. \quad \therefore \quad t = 2\pi\frac{k}{a}\sqrt{\frac{l}{g}}. \tag{6.18.7}$$

From the expression (6.18.6), we have

$$g = \frac{4\pi^2 k^2 y}{a_1 a_2 t^2}. \tag{6.18.8}$$

All the quantities in the right-hand side being known, g can be determined.

6.18.4 Body oscillating on a concave surface

When a sphere is displaced from its position of equilibrium on a concave surface and then released, it will roll backwards and forwards. It can be shown that, with sufficient small amplitude of oscillation, the sphere will undergo simple harmonic motion about the lowest point on the surface. We shall consider that friction between the sphere and the concave surface is sufficient to allow the sphere rolling without sliding.

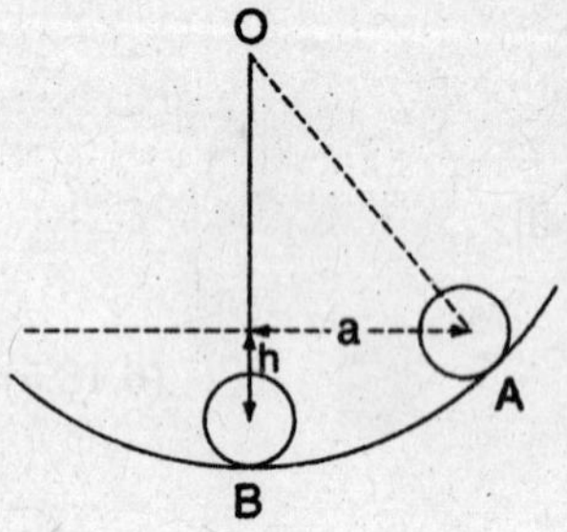

Fig. 6.18.4

Let us suppose that the sphere is released from rest at A (Fig. 6.18.4). When the body rolls from A to B, the loss of potential energy $= mgh$, where m is the mass of the body, and h is the vertical distance between A and B.

The gain in kinetic energy of the sphere at B

$$= \frac{1}{2}I\omega^2 + \frac{1}{2}mv^2 \qquad \left[\begin{array}{l} I \text{ is the moment of inertia of the sphere about its diameter,} \\ \omega \text{ is the angular velocity and } v \text{ is the linear velocity.} \end{array}\right]$$

$$= \frac{1}{2}\cdot\frac{2}{5}\frac{mr^2v^2}{r^2} + \frac{1}{2}mv^2 \quad (r \text{ is the radius of the sphere})$$

$$= \frac{7}{10}mv^2$$

$\therefore$ from the principle of conservation of energy,

$$\frac{7}{10}mv^2 = mgh \quad \text{or} \quad v^2 = \frac{10gh}{7}. \tag{6.18.9}$$

Now, if a be the amplitude of oscillation, then from the figure, we have

$$a^2 = h[2(R-r) - h],$$

where R denotes the radius of curvature of the concave surface

$$\therefore \quad h = \frac{a^2}{2(R-r)}, \quad \text{neglecting the term } h^2 \text{ which is very small.}$$

Substituting this value of h in (6.18.9), we have

$$v^2 = \frac{10g}{7}\cdot\frac{a^2}{2(R-r)} = \frac{5ga^2}{7(R-r)}, \quad \text{i.e., } v \text{ is proportional to } a.$$

This is the case of simple harmonic motion, where $\frac{v}{a} = \frac{2\pi}{T}$, T being the time-period of oscillation of the sphere;

$$\therefore \quad T = 2\pi \cdot \frac{a}{v} = 2\pi\sqrt{\frac{7(R-r)}{5g}}.$$

The above expression may be written in the form

$$g = \frac{28\pi^2(R-r)}{5T^2} \tag{6.18.10}$$

The radius of the concave surface is determined by a spherometer. Thus the value of g is determined.

N.B. It should be noted here that the oscillating sphere is equivalent to a compound pendulum having O as the point of suspension. The length of the equivalent simple pendulum is $7/5(R-r)$.

Illustrative Example

1. A thin metal rod is supported by parallel bifilar supports connected to its ends, each thread being equal to the length of the rod. Compare the periodic time of oscillation in a horizontal plane with that when the rod swings as a rigid pendulum about a horizontal axis through one end.

We know that the time period of a rod suspended by parallel bifilar suspension is given by

$$t = 2\pi\frac{k}{a}\sqrt{\frac{l}{g}}.$$

Here, the length of the thread is equal to the length of the rod. $\therefore\ l = 2a$.

$$\therefore \quad t = 2\pi \cdot \frac{2k}{2a}\sqrt{\frac{2a}{g}} = 2\pi\frac{2k}{\sqrt{lg}}.$$

Now, $k^2 = \frac{l^2}{12}$ $\quad \therefore\ k = \frac{l}{2\sqrt{3}}$

$$\therefore \quad t = \frac{2\pi}{\sqrt{3}}\sqrt{\frac{l}{g}} \tag{1}$$

Again, when the rod is used as a rigid pendulum oscillating about a horizontal axis, we have the time-period given by

$$t' = 2\pi\sqrt{\frac{k^2 + l'^2}{l'g}}.$$

Here l' is the distance between the point of suspension and the centre of gravity of the rod. Since, the centre of gravity of the rod is at its centre, we have $l' = l/2$.

$$\therefore \quad t' = 2\pi\frac{\sqrt{k^2 + \frac{l^2}{4}}}{\sqrt{\frac{l}{2}\cdot g}} = 2\pi\frac{\sqrt{\frac{l^2}{12} + \frac{l^2}{4}}}{\sqrt{\frac{l}{2}\cdot g}} \quad \left[\because\ k^2 = \frac{l^2}{12}\right] = \frac{2\sqrt{2}\pi}{\sqrt{3}}\sqrt{\frac{l}{g}}$$

$$\therefore \quad t : t' = 1 : \sqrt{2}.$$

6.19 Gravity Survey

In the preceding Article 6.17, we have seen that the value of gravity changes due to altitude, latitude, rotation and shape of the earth. Besides these factors, the value of gravity at a place changes also, although to a very small extent, by local causes such as mineral deposits, topography of the place, etc. Again as there is periodic deformation of the earth's surface, the value of gravity is also thereby affected.

The main object of 'gravity survey' is to determine the magnitude and direction of the gravity at various points on a sea-level surface and therefrom the possible distribution of matter in the earth. Hence, an idea of the internal structure of the earth may be obtained from a study of the 'gravity survey'.

As the variation of the value of gravity due to local and temporal changes is very small, the ordinary pendulum methods fail to determine this change very accurately. This is why more sensitive instruments are devised to measure this change. These instruments are discussed below.

6.19.1 Invariable Pendulum

Invariable Pendulums are used for place to place comparison of gravity where no marked local abnormalities exist. This pendulum is a rigid pendulum of invar-steel, suspended from a massive tri-pod in a partially evacuated vessel. The air pressure inside the vessel is standardised to make all air corrections of the pendulum constant. The variation of time-period of the pendulum due to change of temperature is very small owing to the use of invar-steel which has negligible co-efficient of expansion. The small change of period due to small change in temperature may be directly determined and thereby the pendulum is corrected for temperature.

The time-period of such a pendulum is first determined at a selected base station where the value of gravity is known. Let the time-period of the pendulum and the value of gravity at the station be denoted by T and g respectively. Then $T \propto \frac{1}{\sqrt{g}}$. Next, the time-period is determined at the field station where the value of gravity is to be determined. Let the time-period and the value of gravity at the field station be T' and g' respectively.

Then,

$$T' \propto \frac{1}{\sqrt{g'}}.$$

Therefore,

$$\frac{T^2}{T'^2} = \frac{g'}{g} \quad \text{or} \quad g' = g \cdot \frac{T^2}{T'^2}.$$

It should be noted that the results obtained will be of the same order of accuracy with which the time-period of the pendulum and other constants at the base station have been determined. Time-periods at the two stations may be measured very accurately by arranging precise time-signal broadcast at regular intervals.

Bullard accurately determined the value of g at a place in East Africa, by adopting the following technique :

Two pendulums were used — one at the base station Cambridge and other at the field station in East Africa. An agreed wireless signal (Morse signal) was recorded alongside the pendulum vibrations on a photographic plate at each station. This recording was repeated after an interval of one hour. The periods of the pendulum were compared with the time intervals of the Morse signals used.

6.19.2 Gravity Meter

"Gravity Meters or Balances" are used as prospecting instruments for the moderately accurate survey of abnormal conditions, e.g., high or low density surface constituents of the earth. We shall deal here with a few useful gravity meters.

(a) Quartz Gravity Meter

Threlfall and Pollock devised this gravity meter to determine the variation of gravity at different places. This instrument which is portable, consists of a fine quartz thread AB stretched horizontally—one end of the thread A is fixed, and the other end B can be twisted by means of a pointer P which moves over a circular scale (Fig. 6.19.1). A short metal rod M is fused across the thread near its centre. The metal rod is so loaded that its centre of gravity lies on one side of the thread.

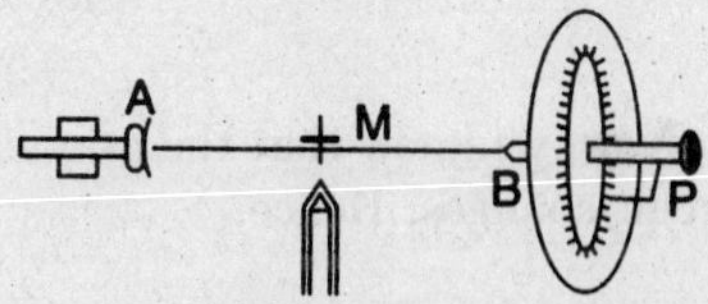

Fig. 6.19.1

The end B is twisted by means of the pointer to pull the metal rod into a horizontal position. This position is just stable, since slightest movement of the pointer will make the rod rotate. This position of the pointer P is noted on the circular scale.

If now, the instrument is taken at a place, where the value of g changes, it would be found that the rod is no longer in the horizontal position. By twisting the end B, the horizontal position of the rod is restored, and the position of the pointer is again noted. The difference between the two readings is a measure of the variation of the value of g. The instrument is calibrated by taking observations at two places where g is accurately known and hence it may be used for the measurement of g at other places. The instrument gives prompt and accurate readings when necessary corrections for the effect of temperature have been made.

(b) Boliden Gravity Meter

This is a more recent form of gravity meter in which two springs S, S, are fixed to a rod having a mass M and supporting the two plates C_2, C_3 below and above two other plates C_1, C_4. C_1C_2 and C_3C_4 form two parallel plate condensers (Fig. 6.19.2). The condensers are insulated from the frame of the instrument by means of two insulating slabs A and B. The condenser C_1C_2 is a part of an oscillatory circuit whose frequency is compared with that of a standard oscillator. Whenever there is a variation of g, there will be a proportional change in the gap between the two plates C_1, C_2 due to change of bending of the springs. This change of gap δt between the plates of the condenser (C_1, C_2) causes a proportional change δC of the capacity C of the condenser. Before change the capacity of the condenser was given by

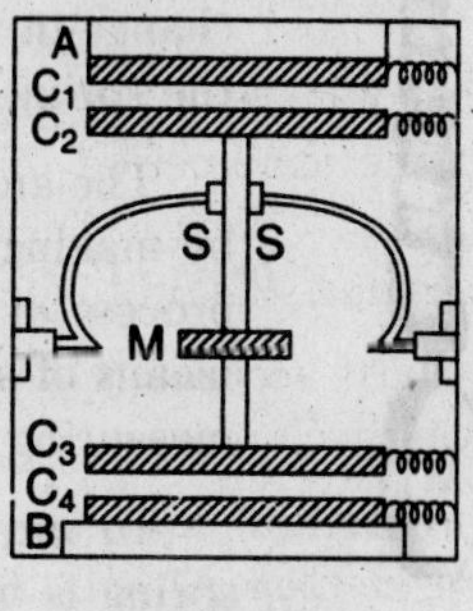

Fig. 6.19.2

$$C = \frac{KA}{4\pi t},$$

where K = the dielectric constant of the medium between C_1 and C_2.

Differentiating, we have

$$\delta C = -\frac{KA}{4\pi}\cdot\frac{\delta t}{t^2}.$$

The negative sign indicates that capacity decreases as the separation between the plates increases.

But considering only the magnitude, we have by actual division,

$$\frac{\delta C}{C} = \frac{\delta t}{t} \quad \text{or} \quad \delta C \propto \delta t.$$

Now the variation of gravity being proportional to the change of gap of the condenser, we may write

$$\delta g \propto \delta t. \quad \therefore \quad \delta g \propto \delta C.$$

Now, we know that the frequency of the oscillatory circuit changes when the capacity of the circuit changes. Hence,

$\delta N \propto \delta C$ [δN denotes the change of frequency]

$\therefore$ $\delta g \propto \delta N$ or $\delta g = \alpha \delta N$ [α denotes the constant of proportionality]

Thus knowing the change of frequency $\delta N, \delta g$ can be measured, provided α can be found out by calibration.

The instrument can be calibrated by applying known potential differences between the plates C_3, C_4. Now calculating the consequent forces of attraction between the plates and the corresponding changes of the frequencies of the oscillatory circuit with the condenser (C_3, C_4), a graph is plotted—from which the value of $\frac{\delta g}{\delta N} = \alpha$ is obtained.

Here δg corresponds to the force of attraction due to applied potential differences.

The sensitivity of the instrument has been found to be 1×10^{-3} cm/sec^2 which is 1 milligal. [1 milligal = 10^{-3} gal]

(c) Gulf Gravity Meter

Fig. 6.19.3

The gulf gravity meter is based on the principle of a spring balance. The lower end of the spring carries a fixed mass to which is attached a mirror M. The upper end is fixed to a torsion head. The change of gravity will cause a proportional change in the weight of the suspended system-resulting in the consequent pull of the spring and rotation of the mirror.

The angle of deviation of a beam of light reflected from the mirror is magnified by making the beam travel four times between the mirror and a reflector by the process of reflection. Finally, the image of the illuminated slit is observed by means of a microscope having an eye-piece fitted with a micrometer scale. Thus measuring the deviation, the variation of g can be detected.

To increase the sensitivity of this instrument, a flat metallic ribbon spiral spring is used. The sensitivity of the instrument is about 5×10^{-5} cm/sec^2 or 5×10^{-2} milligal.

6.19.3 Eötvös Balance

The accuracy of the invariable pendulum or the gravity meters is not sufficient to measure small variations in gravity caused by neighbouring masses, such as buildings or small geological deposits. For this purpose a more sensitive type of instrument, known as Eötvös balance, is used. Eötvös balance is used not only for sensitive measurement of small gravity variations, but also used for the measurement of the rate of change of g with distance in any horizontal direction, and other important quantities connected with the earth's gravitational field.

The essential parts of an Eötvös balance are shown in the figure 6.19.4. From a torsion-head T, hangs a wire S made of platinum-iridium alloy, about 30×10^{-3} to 40×10^{-3} mm in diameter and 25 to 60 cms in length. The suspension carries a horizontal rectangular beam of about 40 cms in length; one end of the beam carries a cylindrical weight A of platinum, gold or silver, of mass about 30 gms with its axis horizontal. From the other end of the beam is suspended another cylindrical weight B of mass about 25 gms by a platinum-iridium wire P of about 40 cms long. The suspension system carries a circular plane mirror M, a little above the level of the beam. A telescope and a scale are used in conjunction with the mirror to measure the deflection of the beam.

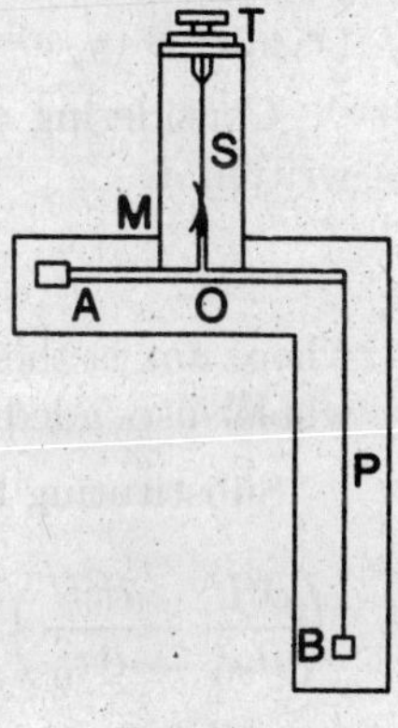

Fig. 6.19.4

If the value of gravity varies from point to point in the region surrounding the instrument, a torque will act on the system whose moment we can calculate in the following way.

Let the origin O of the co-ordinates be on the midpoint of the beam (Fig. 6.19.5). Let axes Ox, Oy and Oz be drawn so that Ox points towards the geographical north, Oy towards east and Oz downwards in the direction of the force of gravity.

Let g_0 be the force per unit mass placed at the origin along Oz. The force on a particle of mass m at the origin has components $0, 0, mg_0$ along the three axes. Let us assume that the earth's gravitational field has a potential U at any point (x, y, z). The component forces on unit mass at (x, y, z) are $\frac{\partial U}{\partial x}, \frac{\partial U}{\partial y}, \frac{\partial U}{\partial z}$ respectively. These forces can be written in the form

$$\frac{\partial U}{\partial x} = g_x; \quad \frac{\partial U}{\partial y} = g_y; \quad \frac{\partial U}{\partial z} = g_z.$$

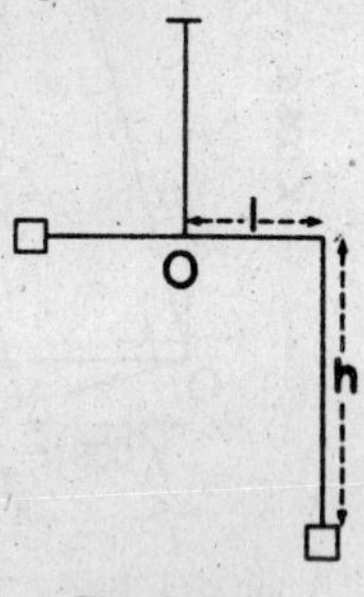

Fig. 6.19.5

The values of g_x, g_y, g_z at any point (x, y, z) very close to the origin may be calculated by Maclaurin's theorem, according to which, any function $f(x, y, z)$ can be expressed as

$$f(x, y, z) = f(0,0,0) + x\frac{\partial f}{\partial x_0} + y\frac{\partial f}{\partial y_0} + z\frac{\partial f}{\partial z_0}$$

$$+ \text{ smaller terms which are negligible.}$$

(Suffix 0 indicating the value of the differential co-efficient at the origin.)

Hence,

$$g_x = 0 + x\frac{\partial g_x}{\partial x_0} + y\frac{\partial g_x}{\partial y_0} + z\frac{\partial g_x}{\partial z_0} \quad \text{[approximately]}$$

$$= 0 + x\frac{\partial^2 U}{\partial x_0^2} + y\frac{\partial^2 U}{\partial x \partial y_0} + z\frac{\partial^2 U}{\partial x \partial z_0} \quad \text{[approximately]}$$

Similarly,

$$g_y = 0 + x\frac{\partial^2 U}{\partial y \partial x_0} + y\frac{\partial^2 U}{\partial y_0^2} + z\frac{\partial^2 U}{\partial y \partial z_0} \quad \text{[approximately]}$$

and $$g_z = g_0 + x\frac{\partial^2 U}{\partial z \partial x_0} + y\frac{\partial^2 U}{\partial z \partial y_0} + z\frac{\partial^2 U}{\partial z_0^2} \quad \text{[approximately]}$$

g_x, g_y, g_z are the forces per unit mass along Ox, Oy, Oz respectively at the point (x, y, z). Taking moments about Oz, the clockwise moment of the forces acting on the unit mass at (x, y, z) is $(g_y x - g_x y)$.

Considering a large body distributed over a certain space, the clockwise moment may be written as

$$\int (g_y x - g_x y) dm,$$

where dm is the mass concentrated at any point (x, y, z). The integration extends over the whole suspended system.

Substituting the values of g_x and g_y in the above expression we have the moment given by,

$$\left(\frac{\partial^2 U}{\partial y_0^2} - \frac{\partial^2 U}{\partial x_0^2}\right)\int xydm + \frac{\partial^2 U}{\partial x \partial y_0}\int (x^2 - y^2)dm + \frac{\partial^2 U}{\partial y \partial z_0}\int zxdm - \frac{\partial^2 U}{\partial x \partial z_0}\int yzdm. \quad (6.19.1)$$

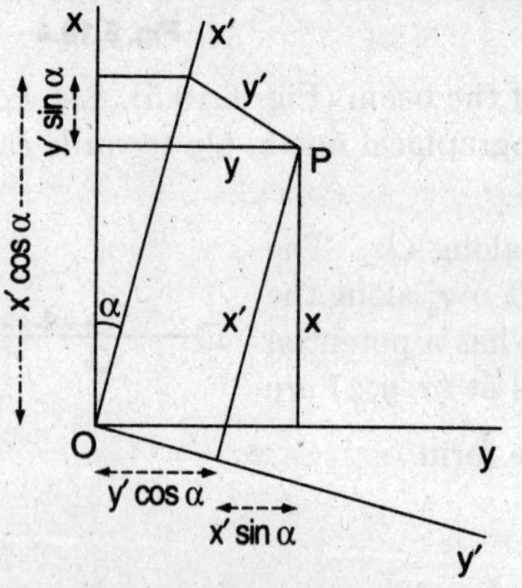

Fig. 6.19.6

In order to calculate the moments of the forces due to gravity acting on the suspended system of Eötvös balance, the different integrals in the above expression are to be evaluated. For this purpose, new axes of coordinates are selected, viz. Ox' along the geometrical axis of the beam, and Oy' horizontal and perpendicular to Ox'. The axes are so chosen that Ox' points towards the north, Oy' points to the east and Oz vertically downwards as before. In the equilibrium position of the beam, let the axis of the beam make an angle α with Ox and $(90° - \alpha)$ with Oy. Then, from Fig. 6.19.6.

$$x = x' \cos\alpha - y' \sin\alpha$$

$$y = x' \sin\alpha + y' \cos\alpha$$

Hence,

$$\left.\begin{aligned}
\int xydm &= \frac{1}{2}\sin 2\alpha \int (x'^2 - y'^2)dm + \cos 2\alpha \int x'y'dm \\
\int (x^2 - y^2)dm &= \cos 2\alpha \int (x'^2 - y'^2)dm - 2\sin 2\alpha \int x'y'dm \\
\int zxdm &= \cos\alpha \int zx'dm - \sin\alpha \int zy'dm \\
\int yzdm &= \sin\alpha \int zx'dm + \cos\alpha \int zy'dm.
\end{aligned}\right\} \quad (6.19.2)$$

The suspension system is so constructed that its mass is symmetrically distributed with respect to Ox' and also with respect to the vertical plane $x'Oz$ as shown in the figure 6.19.7. The symmetry with respect to Ox' makes $\int x'y'dm$ zero. This means that for every element of co-ordinates (x', y', z), there is another element of co-ordinates $(x', -y', z)$, so that the expression under the integral sign must have two parts $+x'y'dm$ and $-x'y'dm$—which make the integral zero.

Also, the beam and the weight on it are symmetrical with respect to the axis Ox' and hence for every element with co-ordinates (x', y', z) there is another element with co-ordinates $(x', y' - z)$ and as a result $\int x'zdm = 0$, because the expression inside the integral sign can be divided into pair of terms of the form $+x'zdm$ and $-x'zdm$. However, the lower weight B and the wire P contribute an amount mhl to the integral $\int x'zdm$, where m is their mass, h is the z-co-ordinate of the centre of gravity and l is its x' co-ordinate (Fig. 6.19.5). Symmetry about the plane $x'Oz$ also shows that $\int y'zdm = 0$, for it can be divided into pair of terms of the form $+y'zdm$ and $-y'zdm$.

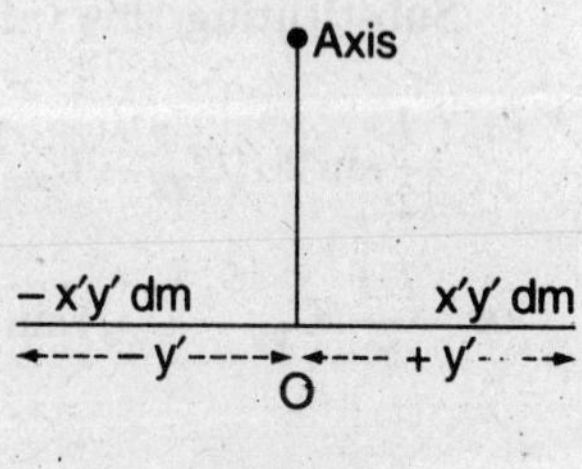

Fig. 6.19.7

$$\text{The remaining integral} = \int (x'^2 - y'^2)dm = \int (x'^2 + y'^2)dm \text{ (nearly)}$$

[$\because$ $y'^2 \ll x'^2$ for most element of mass]

$$= I,$$

where I is the moment of inertia of the suspension system about Oz. On substituting the values of the integrals thus obtained in the expression for the moments of forces due to gravity, we have the moment given by

$$\Gamma = \frac{I}{2}\sin 2\alpha\left(\frac{\partial^2 U}{\partial y_0^2} - \frac{\partial^2 U}{\partial x_0^2}\right) + I\cos 2\alpha\frac{\partial^2 U}{\partial x \partial y_0} + mhl\left(\cos\alpha\frac{\partial^2 U}{\partial y \partial z_0} - \sin\alpha\frac{\partial^2 U}{\partial x \partial z_0}\right). \quad (6.19.3)$$

If we put

$$U_{xx} = \frac{\partial^2 U}{\partial x_0^2}; \quad U_{yy} = \frac{\partial^2 U}{\partial y_0^2}; \quad U_{xy} = \frac{\partial^2 U}{\partial x \partial y_0}; \quad U_{yz} = \frac{\partial^2 U}{\partial y \partial z_0}; \quad U_{xz} = \frac{\partial^2 U}{\partial x \partial z_0},$$

the above expression becomes

$$\Gamma = \frac{I}{2}\sin 2\alpha(U_{yy} - U_{xx}) + I\cos 2\alpha U_{xy} + mhl(\cos\alpha U_{yz} - \sin\alpha U_{xz}) \quad (6.19.4)$$

U_{xz} and U_{yz} denote $\frac{\partial g}{\partial x_0}$ and $\frac{\partial g}{\partial y_0}$ respectively, i.e., the change of gravity in horizontal directions near the origin (change of gravity along horizontal directions means the change along the North and the East). The other quantities $(U_{yy} - U_{xx})$ and U_{xy} represent the quantitative properties of the gravitational field near the origin.

The moment Γ tends to displace the suspended system of the Eötvös balance as a result of which a restoring couple of moment $\tau\phi$ will be called into play in the suspension wire. τ denotes the restoring couple per unit twist and ϕ is the angle of twist. In the equilibrium position, we have, the moment Γ of the deflecting couple is equal to the moment $\tau\phi$ of the restoring couple, i.e., $\Gamma = \tau\phi$

$$\text{or} \quad \frac{I}{2}\sin 2\alpha(U_{yy} - U_{xx}) + I\cos 2\alpha U_{xy} + mhl(\cos\alpha U_{yz} - \sin\alpha U_{xz}) = \tau\phi. \quad (6.19.5)$$

Let d_1 be the scale reading corresponding to the deflection ϕ and d be the initial scale reading when there is no deflection. Then $d_1 - d = 2D\cdot\phi$, where D is the distance between the scale and the mirror.

$$\therefore \quad \phi = \frac{(d_1 - d)}{2D}. \tag{6.19.6}$$

Substituting this value of ϕ in the expression (6.19.5), we have

$$\frac{I}{2}\sin 2\alpha(U_{yy} - U_{xx}) + I\cos 2\alpha U_{xy} + mhl(\cos\alpha U_{yz} - \sin\alpha U_{xz}) = \frac{\tau(d_1 - d)}{2D}$$

or $$\frac{DI}{\tau}(U_{yy} - U_{xx})\sin 2\alpha + \frac{2DI}{\tau}U_{xy}\cos 2\alpha + \frac{2Dmhl}{\tau}U_{yz}\cos\alpha - \frac{2Dmhl}{\tau}U_{xz}\sin\alpha = d_1 - d. \tag{6.19.7}$$

The above expression may be written in the form

$$P\sin 2\alpha + Q\cos 2\alpha + R\sin\alpha + S\cos\alpha = d_1 - d, \tag{6.19.8}$$

where

$$P = \frac{DI}{\tau}(U_{yy} - U_{xx}); \quad Q = \frac{2DI}{\tau}U_{xy};$$

$$R = -\frac{2Dmhl}{\tau}U_{xz}; \quad S = \frac{2Dmhl}{\tau}U_{yz},$$

where d, P, Q, R and S are five unknown quantities to be evaluated.

For this, we require five equations of the form (6.19.8). We can write five equations of the form (6.19.8) for five different values of α corresponding to five separate values of d_1. In practice, the procedure is simplified by taking six values of α, such as

$$\alpha = 0°, \ 60°, \ 120°, \ 180°, \ 240°, \ 300°.$$

If d_1, d_2, d_3, d_4, d_5, d_6 are the corresponding values of d_1, then it can be shown that

$$\left.\begin{aligned} d_1 + d_3 + d_5 &= d_2 + d_4 + d_6 = 3d \\ 2\sqrt{3}P &= 2(d_2 - d_3) - (d_1 - d_4) \\ 2Q &= (d_1 + d_4 - 2d), \quad 2S = d_1 - d_4 \\ 2\sqrt{3}R &= 2(d_2 - d_5) - (d_1 - d_4) \end{aligned}\right\}. \tag{6.19.9}$$

Thus, the five unknown quantities, d, P, Q, R, S are determined. Knowing P, Q, R, S we may determine the following four quantities

$$U_{yy} - U_{xx} = \frac{P\tau}{DI}; \quad U_{xy} = \frac{Q\tau}{2DI}; \quad U_{xz} = -\frac{R\tau}{2Dmhl}; \quad U_{yz} = \frac{S\tau}{2Dmhl}.$$

The value of τ is determined by applying known deflecting couple to the system and measuring the angle produced. The moment of inertia I is determined by oscillation experiment. For estimation of the magnitudes of the above four quantities and hence to obtain an idea of necessary sensitiveness, we have applying Helmert's formula at latitude 50°30′ (London), $g = 981.1806$ cms/sec^2, and thus

$$\frac{\partial g}{\partial x_0} = U_{xz} = 7.9376 \times 10^{-9}, \quad U_{xy} = 0, \quad \frac{\partial g}{\partial y_0} = U_{yz} = 0, \quad U_{yy} - U_{xx} = 4.005 \times 10^{-9}.$$

In one experiment with Eötvös balance the constants $\frac{DI}{\tau}$ and $\frac{2Dmhl}{\tau}$ were 0.05162×10^9 and 0.14087×10^9 respectively which gives the difference in reading $d_1 - d = -1.3$, approximately for $\alpha = 90°$. In case of large local masses the difference in reading may be higher and can easily be noted.

The limit of accuracy with pendulum is about 1×10^{-4} C.G.S. units, while that with Eötvös balance is 1×10^{-9} C.G.S. units.

6.20 The Gravity Gradient and Horizontal Directive Tendency

In connection with gravity surveys it has been found necessary to introduce two other quantities connected with the earth's gravitational field. These are (a) the gravity gradient and (b) the horizontal directive tendency.

6.20.1 The Gravity Gradient

The term gravity gradient is an abbreviation for "maximum gradient of g in a horizontal direction near a point", where g is the vertical gravitational intensity at that point and is equal to $\frac{\partial U}{\partial z}$. The gravity gradient may, therefore, be written as $\frac{\partial g}{\partial s}$ which is the rate of variation of g per unit of legth measured along the direction of maximum rate of change of g. Since, it is a vector quantity, it can be resolved into two components $\frac{\partial g}{\partial s}\cos\phi$ and $\frac{\partial g}{\partial s}\sin\phi$ along Ox and Oy respectively, where ϕ denotes the angle which the vector $\frac{\partial g}{\partial s}$ makes with Ox.

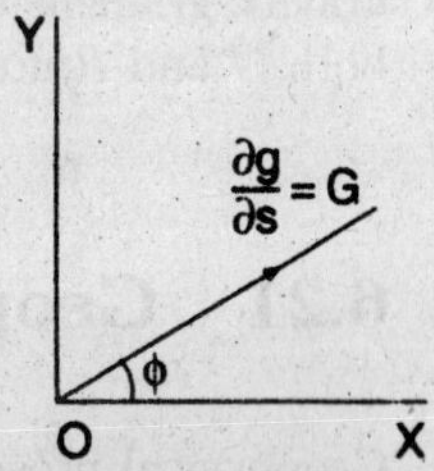

Fig. 6.20.1

Hence, we may write

$$\frac{\partial g}{\partial x} = \frac{\partial g}{\partial s}\cos\phi = G\cos\phi \quad \text{and} \quad \frac{\partial g}{\partial y} = \frac{\partial g}{\partial s}\sin\phi = G\sin\phi,$$

where G denotes the gravity gradient.

Now, $g = \frac{\partial U}{\partial z}$.

$$\therefore \quad \frac{\partial g}{\partial x} = \frac{\partial^2 U}{\partial x \partial z} = U_{xz} = G\cos\phi \quad \text{and} \quad \frac{\partial g}{\partial y} = \frac{\partial^2 U}{\partial y \partial z} = U_{yz} = G\sin\phi.$$

$$\therefore \quad G^2 = (U_{xz}^2 + U_{yz}^2) \quad \text{or} \quad G = (U_{xz}^2 + U_{yz}^2)^{1/2}. \tag{6.20.1}$$

Hence, knowing U_{xz} and U_{yz} from Eötvös balance experiment, G can be determined from the above expression.

6.20.2 Horizontal Directive Tendency

The horizontal directive tendency (abbreviated as H.D.T.) at any point denotes a curvature vector given by

$$R = g\left(\frac{1}{R_1} - \frac{1}{R_2}\right),$$

where R_1 and R_2 are the maximum and minimum radii of curvatures of gravitational equipotential surface or level surface at that point. It is to be noted that H.D.T. is not a true vector but a directed quantity. The direction of H.D.T. is conventionally taken as that horizontal direction in which the level surface has least vertical downward curvature; i.e., the radius of maximum curvature. Let the direction of R make an angle λ with OX; then it can be shown that

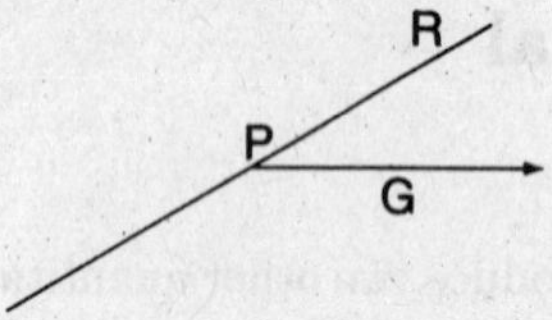

Fig. 16.20.2

$$\left.\begin{aligned} R\sin 2\lambda &= 2U_{xy} \\ R\cos 2\lambda &= U_{xx}^2 - U_{yy}^2. \end{aligned}\right\} \qquad (6.20.2)$$

In survey maps G is represented in magnitude and direction by an arrow-headed line drawn from the point P where we consider the value of G. R is represented in magnitude and direction by a line through P without any arrow (Fig. 6.20.2).

Also, joining the points where g has the same value we get curves kown as *isogams.* The gravity gradient G is always directed along the normal to these isogams. The dimensions of both G and R are $[T]^{-2}$ and are expressed in Eötvös units [1 Eötvös $= 10^{-9}/\text{sec}^2$].

6.21 Geophysical Prospecting

Geophysical prospecting is the location of geological deposits. This is done by means of the measurement of small variation of g. Electrical, magnetic and seismic quantities are associated with the deposits. Here, we are concerned with the measurement of variation of g, for which it is requried to measure the gravity gradient G and the horizontal directive tendency R at various points in the field under survey.

The measurements of *gravity gradient* and *horizontal directive tendency* are made with the help of Eötvös gravity balance. From equations (6.20.2) we may write

$$\tan 2\lambda = \frac{2U_{xy}}{U_{xx}^2 - U_{yy}^2}.$$

The two solutions λ_1 and λ_2 differ by $\pi/2$, and give the directions of the two principal axes of curvature of the level surface (i.e., the equipotential surface at the point). The values of G and R at various observing stations are plotted; the direction of λ_2 is such that *secant* $2\lambda_2$ and $U_{xx}^2 - U_{yy}^2$ are opposite in sign.

6.22 Change in the Direction of Force of Gravity with Time—The Horizontal Pendulum

It has been observed that as time passes, there is a change in the direction of the force due to gravity as a result of which a plumb line which sets itself in a direction normal to the gravitational equipotential surface, will change its direction. This change is due to deformation of the earth's surface. The change is so small that the deflection of the plumb line is at most 0.1 second of angle and hence the measurement of this change with plumb line is practically impossible.

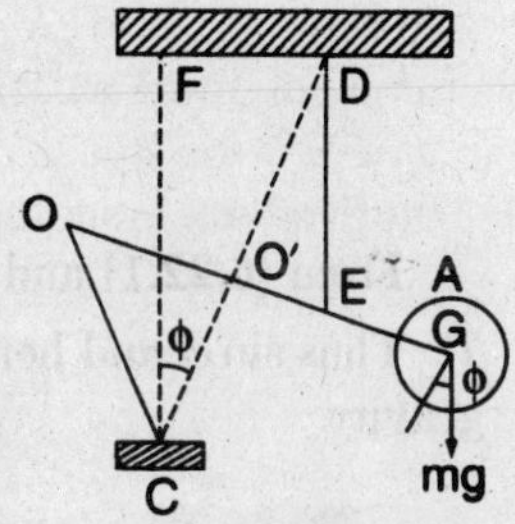

Fig. 6.22.1

Hengler devised what is called a *horizontal pendulum* to measure this small temporal change in the direction of the force of gravity.

A horizontal pendulum consists of a rod OA having a bob at its end A (Fig. 6.22.1). It is supported in an inclined position by two light strings DE and CO attached to rigid supports at C and D. The line joining C and D meets the pendulum at O' and makes an angle ϕ with the direction of the force of gravity. In the equilibrium condition the pendulum OA takes up a position in the plane parallel to the force gravity.

Now, let the bob of the pendulum be slightly displaced laterally towards the observer and then released, when the pendulum undergoes small oscillations slowly. The time-period of the pendulum may be calculated in the following way.

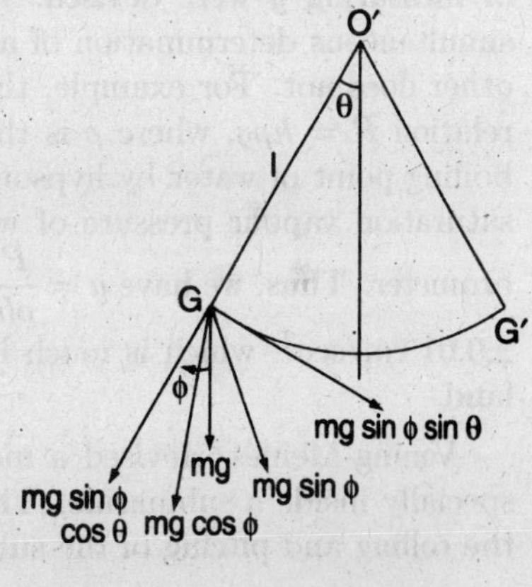

Fig. 6.22.2

The oscillation of the bob of centre of gravity G describes an arc GG' with O' as the centre of the arc (Fig. 6.22.2). The arc lies in a plane which makes an angle ϕ with a plane normal to the force of gravity. The force mg due to the weight of the bob acting in a vertical downward direction can be resolved into two components: (a) $mg\cos\phi$ in a direction perpendicular to the inclined plane and (b) $mg\sin\phi$ in the inclined plane of oscillation of the pendulum.

Further, the force $mg\sin\phi$ may be resolved into component $mg\sin\phi\sin\theta$ which is perpendicular to $O'G$. The moment of this force about an axis through O' is $mg\sin\phi\sin\theta\cdot l$, where $O'G = l$. This couple tends to diminish the value of θ, which constitutes what is called a *restoring couple*. Hence, the equation of motion of the pendulum is

$$I\frac{d^2\theta}{dt^2} = -mgl\sin\phi\sin\theta = -mgl\sin\phi\cdot\theta \quad [\because \theta \text{ is small, } \sin\theta = \theta],$$

where I is the moment of inertia of the pendulum about an axis through O'.

$$\therefore \quad \frac{d^2\theta}{dt^2} + \frac{mgl}{I}\sin\phi\cdot\theta = 0.$$

Putting $I = mk^2$, [k is the radius of gyration of the pendulum about an axis through O'] we have

$$\frac{d^2\theta}{dt^2} + \frac{gl}{k^2}\sin\phi\cdot\theta = 0.$$

Hence, the period of oscillation is given by

$$T = 2\pi\sqrt{\frac{k^2}{gl\sin\phi}} = \frac{2\pi k}{\sqrt{gl\sin\phi}}. \tag{6.22.1}$$

In practice, T is made large by making ϕ very small.

If $\phi = 90°$, the time-period is given by

$$T' = \frac{2\pi k}{\sqrt{gl}}. \tag{6.22.2}$$

From (6.22.1) and (6.22.2), we get $\frac{T'^2}{T^2} = \sin\phi$. From oscillation experiments we get T' and T. Thus $\sin\phi$ and hence ϕ is determined which gives the change in the direction of the force of gravity.

6.23 Measurement of g at Sea

As the vibrations of ships prohibited the use of pendulums on board, indirect methods of measuring g were devised. The method due to Hecker and Duffield depends upon the simultaneous determination of atmospheric pressure in two ways, of which one involves g and other does not. For example, the barometric height h gives the value of pressure P from the relation $P = h\rho g$, where ρ is the density of mercury. Again a simultaneous measurement of boiling point of water by hypsometer gives the value of P from the tables of temperature and saturation vapour pressure of water. The pressure P can also be measured by an anoeroid brometer. Thus, we have $g = \frac{P}{\rho h}$. In this method, the probable error in the value of g is about ± 0.01 cm/sec^2, which is much large in comparison with the value obtained by pendulum on land.

Vening-Meinesz devised a more accurate method for the measurement of g by pendulum specially inside a submarine. The pendulums are subject to following disturbances caused by the rolling and pitchig of the submarine:

(i) acceleration of the point of suspension along horizontal direction;

(ii) acceleration along vertical direction,

(iii) angular movement of the point of support which is known as *rocking;*

(iv) slipping of the knife-edges on the agate planes.

Of the four disturbances as stated above the first one is the most prominent. This can be completely eliminated by the oscillations of two similar half-second pendulums, i.e., pendulums having time-period of 1 sec. The oscillations of the pendulums are made in opposite phases in the same vertical plane from the same support. If θ_1, θ_2 be the angular displacements of the two pendulums, their equations of motion are

$$M(k^2 + l^2)\frac{d^2\theta_1}{dt^2} + Mgl\theta_1 + P = 0. \tag{6.23.1}$$

$$M(k^2 + l^2)\frac{d^2\theta_2}{dt^2} + Mgl\theta_2 + P = 0. \tag{6.23.2}$$

The term P gives the effect of horizontal acceleration of the support in the equations of motion of the pendulums.

From (6.23.1) and (6.23.2), we have

$$M(k^2+l^2)\left(\frac{d^2\theta_1}{dt^2}-\frac{d^2\theta_2}{dt^2}\right)+Mgl(\theta_1-\theta_2)=0. \tag{6.23.3}$$

The above equation is free from P. Thus $(\theta_1-\theta_2)$ gives the angular displacement of a pendulum undisturbed by the horizontal acceleration of the support.

As the vertical oscillation is caused by the alternate up-and-down motion of the vessel, the mean value of the accleration for a large number of observations may be considered negligible. The effect of disturbance due to angular movement of the support may be determined from the recorded values of the rocking angles, the correction due to this effect being very small.

In actual practice three-similar half-second pendulums are suspended from the same support and set oscillating together. By an optical device the difference between the angular deflections of the first and the second, and the difference between those of the second and the third pendulums, i.e., $(\theta_1-\theta_2)$ and $(\theta_2-\theta_3)$ are recorded on a sensitised paper along with time marks from two very accurate chronometers. Thus two sets of values of g are obtained from which the mean value of g is determined.

To eliminate the effect of magnetic disturbance, the pendulums are made up of brass instead of invar. To get rid of disturbance due to small angular movement of the vessel, the whole system is suspended in gymbals as found in a gyroscope. The effect of slight variation of temperature is avoided by keeping the apparatus containing the pendulum thermally isolated. The probable error in the values of g computed in a series of experiments is ± 0.0018 cm/sec^2.

Exercises

1. What is a simple pendulum? Deduce an expression for time-period of a simple pendulum. Explain how the time-period changes if the dimension of the bob is taken into consideration.

2. State the laws of simple pendulum. What is a seconds pendulum?

3. Explain the gain or loss of time by a pendulum clock due to change of place and length of the pendulum.

4. Show that the period of oscillation of a pendulum swinging with finite arc is greater than the period when the pendulum oscillates with small amplitude.

5. What are the demerits of a simple pendulum? Explain how the defects of a simple pedulum may be removed by a compound pendulum.

6. Describe a compound pendulum and show that the centres of oscillation and suspension for a given compound pendulum are interchangeable. What is the condition for minimum time-period of a compound pendulum?

7. Describe Kater's pendulum. Derive an expression for Bessel's computed time period.

8. State and explain the various corrections required to obtain correct time-period of a compound pendulum.

9. A simple pendulum with spherical bob of mass 1 kg, has a length of 100 cms, and hangs from the end of a metal rod which is fixed horizontally into a wall. If the rod is pulled sideways by 1 cm when a horizontal force of 1 kg acts parallel to the wall, estimate the percentage error in a measurement of g when this yielding of the support is neglected.

10. Give the theory of the compound pendulum. Define centre of oscillation and show that the centres of suspension and oscillation are reversible. Find the length of the equivalent simple pendulum. **[C.U. (Hons.) 2000, 06]**

In a reversible pendulum, the periods about the two knife-edges are t and $t+T$, where T is a small quantity. The knife-edges are distant l and l' from the centre of gravity of the pendulum. Prove that

$$l+l' = \frac{gt}{4\pi^2}\left(t+\frac{2l'}{l'-l}\cdot T\right)$$ **[Madras 1949]**

11. Explain how the value of g varies with altitude, depth and latitude.

12. Describe an Atwood's machine. Explain how you would determine the value of g with its help.

13. Explain how you would determine the vlue of g by the following methods: (a) Inclined plane, (b) Bifilar suspesion.

14. Write short notes: (a) Gravity survey, (b) Gravity meter, (c) Geophysical prospecting, (d) Value of g at sea.

15. Explain how Eötvös balance is used to measure accurately small variations in gravity. What are gravity gradient and horizontal directive tendency?

16. Show that the length of an equivalent simple pendulum for a uniform bar of length l suspended from one end is $2l/3$. **[C.U. 1981]**

17. Consider a compound pendulum in the form of a uniform rigid rod of length L, oscillating in a vertical plane. Obtain an expression for the time period of such a pendulum when the distance of the centre of supension from the centre of gravity is l.

Suppose that the length L above is 100 cm. Find out how far from the centre of rod, the centre of suspension should be, in order that the time period is minimum. **[C.U. (Hons.) 1985]**

18. A steel cylinder of length 20 cm and radius 3 cm is suspended by a brass wire 50 cm long and of radius 2 mm. Calculate the length of the equivalent simple pedulum. The oscillations of the cylinder take place in a vertical plane and the cylinder and the wire are co-axial. (Sp gr of steel and brass = 7.8 and 8.5 respectively.) [**Ans.** 60.457 cm.] **[C.U. (Hons.) 1989]**

(See worked out example.)

19. A particle of mass m is attached to a horizontal string of length l at a distance l_1 from one end. If F be the tension in the string, calculate the time period of transverse oscillations of the particle, assuming the amplitude of oscillation to be small. **[C.U. (Hons.) 1990]**

[**Hints.** See worked out example no. 8 at the end of Art. 5.14.] [**Ans.** $T = 2\pi\sqrt{\frac{m(l-l_1)l_1}{Fl}}$]

20. (a) Consider a homogeneous rigid body undergoing small oscillations about a horizontal axis. Find expressions for the time period (t) and the length (L) of the equivalent simple pendulum. When is t minimum? How are the expressions for t and L modified (if at all) when the body is inhomogeneous?

(b) In a simple pendulum, the string supporting the bob is replaced by a uniform rod of length l and mass M. Show that the time period of small oscillations is

$$2\pi\sqrt{\frac{2(M+3m)l}{3(M+2m)g}}$$ where m = mass of the bob. **[C.U. (Hons.) 1993, 1998, 2005]**

Answer to last part of (a) : For an inhomogeneous body, there will be a change of the effective CG of the body, with a corresponding change in l, so that L will change and as such t will change.

Answer to (b) : Considering the dimension of the bob to be negligible, we may write,

$$\text{active torque} = I\frac{d^2\theta}{dt^2} \quad \text{where } I = \left(ml^2 + \frac{M}{3}l^2\right)$$

and the moment of the restoring couple $= mgl\theta + Mg\frac{l}{2}\theta = l\left(mg + \frac{Mg}{2}\right)\theta$, so that the equation of motion is

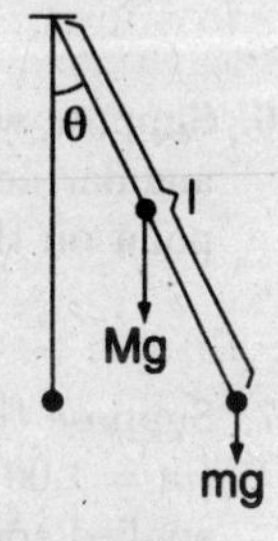

$$I\frac{d^2\theta}{dt^2} = -l\left(mg + \frac{Mg}{2}\right)\theta \text{ or } \frac{d^2\theta}{dt^2} + \frac{l\left(mg + \frac{Mg}{2}\right)}{I}\theta = 0.$$

Here

$$w^2 = \frac{l\left(mg + \frac{Mg}{2}\right)}{ml^2 + \frac{M}{3}l^2} = \frac{3(2m+M)g}{2(3m+M)l} \quad \text{whence } T = \frac{2\pi}{w} \text{ will give the result.}$$

Alternative Method: In falling from A to B, the loss of P.E $= mgA'B' + MgA''B''$

$$= mgl(\cos\theta - \cos\alpha) + Mg\frac{l}{2}(\cos\theta - \cos\alpha)$$

$$= gl(\cos\theta - \cos\alpha)\left(m + \frac{M}{2}\right)$$

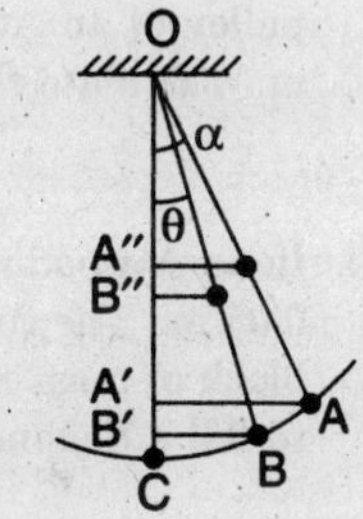

and gain in K.E at B is given by

$$\frac{1}{2}I\omega^2 = \frac{1}{2}l^2\left(m + \frac{M}{3}\right)\left(\frac{d\theta}{dt}\right)^2$$

where I = moment of inertia about the horizontal axis through $O = ml^2 + \frac{M}{3}l^2$ (from the theorem of parallel axis).

$\therefore$ the energy equation is

$$\frac{1}{2}l^2\left(m + \frac{M}{3}\right)\left(\frac{d\theta}{dt}\right)^2 = \left(m + \frac{M}{2}\right)gl(\cos\theta - \cos\alpha)$$

Differentiating w.r.t. t, remembering that $\cos\alpha$ = constant and θ = very small and then simplifying, we may write

$$\frac{d^2\theta}{dt^2} = -\frac{\left(m + \frac{M}{2}\right)g}{\left(m + \frac{M}{3}\right)l}\theta$$

whence the desired result follows.

21. Discuss the cases (a) when $M = 0$ and (b) when $m = 0$ in Problem 20(b).

22. A rectangular plate having edges of lengths a and b respectively hangs vertically from the edge of length a. Find (a) the period for small oscillations, and (b) the length of the equivalent simple pendulum. [**Ans.** (a) $2\pi\sqrt{2b/3g}$ (b) $\frac{2b}{3}$.]

23. A uniform solid sphere of radius a and mass M is suspended vertically downward from a point on its surface. (a) Find the period for small oscillations in a plane and (b) the length of the equivalent simple pendulum [**Ans.** (a) $2\pi\sqrt{7a/5g}$ (b) $\frac{7a}{5}$.]

24. A yo-yo consists of a cylinder of mass 80 gm around which a string of length 60 cm. is wound. If the end of the string is kept fixed, and the yo-yo is allowed to fall vertically starting from rest, find its speed when it reaches the end of the string. **[C.U. (Hons.) 1997]**

(For answer, see exercises of Chapter 3) [**Ans.** 280 cm/sec.]

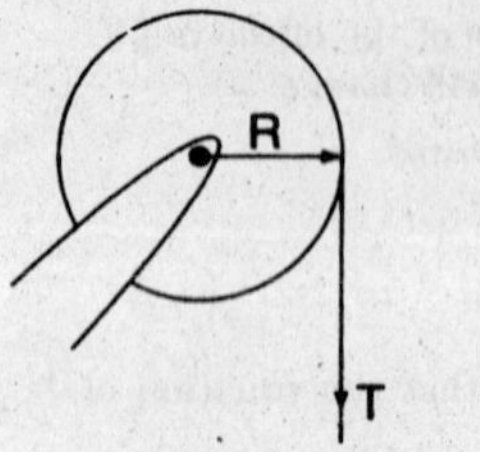

25. A uniform disc of radius $R = 0.5$ m and mass $M = 20$ kg can rotate without friction around a fixed horizontal shaft through its centre. A light cord is wound around the rim of the disc and a steady downward pull $T = 9.8$ nt is exerted on the cord. Find the angular acceleration of the disc, and the tangential acceleration of a point on the rim. Also find the angular velocity after 2 secs.

[**Ans.** 1.96 rad/sec^2, 0.98 metre/sec^2, 3.92 rad/sec.]

26. Suppose we hang a body of mass $m = 1.00$ kg from the cord in the previous problem. Find the angular acceleration of the disc ($M = 20$ kg, $R = 0.5$ m) and the tangential acceleration of a point on the rim of the disc. Also compute the tension in the cord.

[**Ans.** 1.78 rad/sec^2, 0.89 metre/sec^2, 8.91 nt.]

27. Suppose the disc ($M = 20$ kg, $R = 0.5$ m) in the last problem is initially at rest. As the body ($m = 1.00$ kg) is hung from the cord, the disc begins to rotate. Calculate the work done by the applied torque on the disc in 2.0 sec. Compute also the increase in the rotational K.E. of the disc.

[**Ans.** 15.8 joules, 15.8 Joules]

28. Two masses $m_1 = 12$ kg and $m_2 = 8$ kg are attached to the ends of a cord which passes over the pulley of an Atwood's machine. The mass of the pulley is $M = 10$ kg and its radius is $R = 0.1$ m. Calculate the tensions in the cord and the acceleration a of the system.

[**Ans.** $T_1 = 99$ nt, $T_2 = 91$ nt, $a = 1.57$ m/sec^2]

29. In an Atwood's machine, the pulley mounted in horizontal frictionless bearings has a radius $R = 0.05$ m. The cord passing over the pulley carries a block of mass $m_1 = 0.50$ kg at one end and a block of mass $m_2 = 0.46$ kg at the other. When set free from rest, the heavier block is observed to fall a distance of 0.75 m in 5 seconds. Compute the moment of inertia of the pulley.

[**Ans.** 0.0139 kg-m^2]

30. Show that a cylinder will slip on an inclined plane of inclination angle θ, if the coefficient of static friction between the plane and the cylinder is less than $\frac{1}{3}\tan\theta$.

31. The oxygen molecule has a mass of 5.30×10^{-26} kg and a moment of inertia of 1.94×10^{-46} kg-m^2 about an axis perpendicular to the internuclear axis and passing through the centre. The molecule in a gas has a mean speed of 500 m/sec., and its rotational K.E. is two-third of its translational K.E. Find its angular velocity.

[**Ans.** 6.75×10^{12} rad/sec]

Chapter 7

Gravitation

7.1 Newton's Law of Gravitation

Newton's law of gravitation is one of the greatest discoveries which scientific investigation ever yielded to mankind.

According to this law, *every particle of matter attracts every other particle with a force which is directly proportional to the product of the masses of the particles and inversely proportional to the square of the distance between them.*

Hence, if F be the force of attraction between the particles of masses m and m' separated by a distance r, them from this law, we have

$$F \propto \frac{mm'}{r^2}, \quad \text{or,} \quad F = G\frac{mm'}{r^2} \tag{7.1.1}$$

G is called the *gravitational constant* or *Newtonian constant*. If $m = m' = 1$, and $r = 1$, then $F = G$. G is thus numerically equal to the force of attraction between two unit masses separated by unit distance apart. In C.G.S. unit, unit mass should be taken 1 gm and unit distance 1 cm, force being expressed in dyne. The value of G in C.G.S. unit determined by recent experiments is 6.670×10^{-8}.

7.1.1 Dimension of G

The dimension of G is calculated as follows:

We know $F = G\dfrac{mm'}{r^2}$. $\therefore \quad G = \dfrac{Fr^2}{mm'}$.

Hence, the dimension of G is given by

$$[G] = \frac{\left[\dfrac{ML}{T^2}\right] \times [L^2]}{[M^2]} = [M^{-1}L^3T^{-2}]. \tag{7.1.2}$$

Hence, in C.G.S. unit $G = 6.670 \times 10^{-8}\,\text{cm}^3\,\text{gm}^{-1}\,\text{sec}^{-2}$.

If G' be the numerical value of G in F.P.S. unit, then we may write

$$G' \text{ ft}^3 \text{ lb}^{-1} \text{ sec}^{-2} = 6.670 \times 10^{-8} \text{ cm}^3 \text{ gm}^{-1} \text{ sec}^{-2}$$

or,
$$G' = 6.670 \times 10^{-8} \left(\frac{\text{cm}}{\text{ft}}\right)^3 \left(\frac{\text{gm}}{\text{lb}}\right)^{-1} = 6.670 \times 10^{-8} \left(\frac{\text{cm}}{\text{ft}}\right)^3 \left(\frac{\text{lb}}{\text{gm}}\right)$$

$$= 6.670 \times 10^{-8} \left(\frac{1}{30.48}\right)^3 (453.6) = 1.069 \times 10^{-9}. \tag{7.1.3}$$

Hence, in F.P.S. unit the value of G is expressed as $1.069 \times 10^{-9} \text{ ft}^3 \text{ lb}^{-1} \text{ sec}^{-2}$.

Similarly, in M.K.S. unit we have $G = 6.670 \times 10^{-11} m^3 \text{ kg}^{-1} \text{ sec}^{-2}$.

7.1.2 Astronomical Units

We can avoid continuous repetition of G by choosing a new unit of mass such that $G = 1$. Such unit is called *astronomical* or *theoretical unit*.

Now, from Newton's law, we have

$$F = G\frac{mm'}{r^2}, \quad \text{or,} \quad m'f = G\frac{mm'}{r^2}, \quad \text{or,} \quad f = \frac{Gm}{r^2},$$

where f is the acceleration produced in any mass m' by the attraction of a mass m at a distance r. If we write M for Gm, we get

$$f = \frac{M}{r^2}.$$

Let us now choose a unit of mass so that M represents the mass of the attracting body. This new unit is, therefore, the astronomical unit of mass. Thus we get

Gm astronomical units $= m$ gms,

or one astronomical unit $= \frac{1}{G}$ gm $= \frac{1}{6.68 \times 10^{-8}} = 1.55 \times 10^7$ gms.

Similarly, one astronomical unit $= \frac{1}{G}$ lbs $= \frac{1}{1.069 \times 10^{-9}} = 9.4 \times 10^8$ lbs.

7.2 Universality of the Law of Gravitation

The astronomical observations of the Danish astronomer Tycho Brahe on the motion of Mars led Kepler to formulate the laws of planetary motion. Newton gave us an explanation of planetary motion in terms of the force of gravitation which exists between the sun and the planet. Thus, Newton's law of gravitation holds even when r is large (except in the case, where relativistic corrections are appreciable). It should also be noted that Newton's law is valid even when r is very small, but ceases to hold when r is of the order of 10^{-7} cm. In this case, instead of attraction, there is repulsion, because if there is attraction between two atoms separated by a distance of the order of 10^{-7} cm, the atoms will superpose on each other which is practically impossible. Actually the force in this case is of electrical nature.

Thus we see that the law is applicable to distances ranging from smallest terrestrial to interplanetary one. It should also be noted that the law, unlike Inverse Square Law, as in the case of electrostatics and magnetism, is not influenced by the medium in which the force acts. It is also independent of the chemical composition of the attracting masses, and their temperatures. Thus we conclude that the law is universal.

7.3 Newton's Law of Gravitation and Einstein's Theory of Relativity

Einstein's general theory of relativity supplanted Newton's point of view according to which larger masses such as the sun produce in the surrounding space certain field of force which make the planets move along curved trajectories instead of straight lines. Einstein abandoned the very idea of force and explained the motion of the planets round the sun not in terms of any force, but in terms of the change of curvature of the surrounding space-time continuum due to the presence of the sun. Einstein's theory could account for the following phenomena while Newton's theory failed to do so.

(a) Einstein's theory predicted that the major axis of the planetary orbit should rotate slowly. Mercury is the only solar planet for which the predicted rotation can be observationally determined with certainty. The predicted rotation of the major axis of the orbit in case of Mercury is found to be 42.9 seconds of arc per century and the observed rotation is 43.5 seconds of arc per century. The agreement between the predicted and observed values of rotation is quite satisfactory.

(b) According to Einstein, the mass of a body moving with velocity v is given by

$$m = \frac{m_0}{\sqrt{1 - \frac{v^2}{c^2}}},$$

where m_0 stands for the rest mass and c for the velocity of light.

From the above expression it follows that as the velocity of the body becomes comparable with the velocity of light, the mass m is enormously increased. Thus the mass of a ray of light, according to Einstein, has a mass which, on account of its large velocity, can never be ignored. Hence, rays of light get deflected while passing near the gravitational field of the sun. According to Einstein's theory, this deflection is 1.75 seconds of arc. This prediction was tested by observations made at times of total eclipse on the apparent positions of stars whose light has passed close to the sun. The result obtained was in close agreement with the theory. It should be noted that the deflection of light calculated from Einstein's theory of relativity is twice as great as that calculated from Newton's theory considering corpuscular nature of light.

(c) The third crucial test of Einstein's general theory of relativity is the dependence of wave length of light on the gravitational potential of its source. Einstein showed that in case of light originating from the sun, the spectral lines have different wave lengths from those they would have, when emitted by some terrestrial source. The shift of spectral line due to the change of wave length is towards the red and hence the shift of spectral line is called *red-shift*. This red-shift predicted for the sun is about $\lambda \times 2.12 \times 10^{-6}$ for light of wave length λ which is exactly the value predicted by Einstein's theory.

In conclusion, we must say that Newton's law of gravitation is not discarded by Einstein's theory of relativity. It can be shown that Newton's theory of gravitation can be regarded as giving a first approximation to the results of Einstein's general theory of relativity. Actually Newton's theory holds good when the gravitational field is weak.

Illustrative Examples

1. A uniform sphere has a radius 2 cms. Find the percentage increase in its weight when a second sphere of radius 20 cms, and density 12 gms per c.c. is brought under it and nearly touching it. [$g = 981$ cms/sec^2 and $G = 6.67 \times 10^{-8}$ gm^{-1} cm^3 sec^{-2}]

The force of attraction between the two spheres is given by

$$F = G\frac{M.m}{(22)^2},$$

where M and m are the masses of large and small spheres respectively.

Now,

$$M = \frac{4}{3}\pi(20)^3 \times 12. \quad \therefore \quad F = \frac{6.67 \times 10^{-8} \times 4\pi(20)^3 \times 12 \times m}{3(22)^2}.$$

Writing $\dfrac{mg}{981}$ for m [$\because g = 981$], we have

$$F = \frac{6.67 \times 10^{-8} \times 4\pi(20)^3 \times 12 \times mg}{3(22)^2 \times 981}$$

$$\text{or,} \quad \frac{F}{mg} = \frac{6.67 \times 10^{-8} \times 4 \times 22 \times (20)^3 \times 12}{3(22)^2 \times 7 \times 981} = 5.65 \times 10^{-8}.$$

Here F denotes the force with which the smaller mass is attracted by the larger mass and mg denotes the weight of the smaller mass. Hence, the percentage increase in the weight of the smaller mass is

$$\frac{F}{mg} \times 100 = 5.65 \times 10^{-8} \times 100 = 5.65 \times 10^{-6}.$$

2. Find the gravitational force of attraction exerted between two spheres of mass 1 kg placed with their centres 20 cm apart. [g = 980 cms/sec^2, mass the earth = 4×10^{27} gms, radius of the earth = 6×10^8 cms.] [C.U. (Hons.) 1955]

Force of attraction between two spheres of masses m_1 and m_2, and separated by a distance r is given by

$$F = G\frac{m_1 m_2}{r^2}.$$

Here, $m_1 = m_2 = 1$ kg $= 1000$ gms; $r = 20$ cms.

$$\therefore \quad F = \frac{G.10^6}{(20)^2} \text{ dynes.}$$

Now, the weight of a body of mass m is the force with which the body is attracted towards the centre of the earth and is given by

$$mg = \frac{GMm}{R^2},$$

where M is the mass of the earth and R is its radius.

$$\therefore \quad g = \frac{GM}{R^2}, \quad \text{or,} \quad G = \frac{gR^2}{M}.$$

Here,

$$g = 980 \text{ cms/sec}^2.$$

$$M = 4 \times 10^{27} \text{ gms}$$

$$\text{and} \quad R = 6 \times 10^8 \text{ cms.}$$

$$\therefore \quad G = \frac{980 \times 36 \times 10^{16}}{4 \times 10^{27}} = 8.82 \times 10^{-8} \text{ C.G.S. unit.}$$

$$\therefore \quad F = \frac{8.82 \times 10^{-8} \times 10^6}{4 \times 10^2} = 2.205 \times 10^{-4} \text{ dyne.}$$

3. Two particles of masses m_1 and m_2, initially at rest at infinite distance from each other, move under the action of mutual gravitational pull. Show that at any instant their relative velocity of approach is $\sqrt{\frac{2G(m_1+m_2)}{R}}$, where R is their separation at that instant.

The gravitational force of attraction on m_1 due to m_2 at a distance r is $F_1 = \frac{Gm_1m_2}{r^2}$, so that the acceleration of m_1 is $a_1 = \frac{F_1}{m_1} = \frac{Gm_2}{r^2}$. Similarly the acceleration of m_2 due to m_1 is $a_2 = -\frac{Gm_1}{r^2}$ — the negative sign being given as a_2 is directed opposite to a_1. The relative acceleration of approach is

$$a = a_1 - a_2 = \frac{G(m_1+m_2)}{r^2}. \tag{1}$$

If v be the relative velocity, then $a = \frac{dv}{dt} = \frac{dv}{dr}\frac{dr}{dt}$, but $v = -\frac{dr}{dt}$ since r decreases with increasing t.

$$\therefore \quad a = -\frac{dv}{dr}v. \tag{2}$$

From (1) and (2)

$$vdv = -\frac{G(m_1+m_2)}{r^2}dr,$$

whence on integration, we get

$$\frac{v^2}{2} = \frac{G(m_1+m_2)}{r} + C.$$

Now according to the problem, $v = 0$ at $r = \infty$, which gives $C = 0$.

$$\therefore \quad v^2 = \frac{2G(m_1+m_2)}{r}.$$

Now let $v = v_R$ at an instant when $r = R$, then

$$v_R = \sqrt{\frac{2G(m_1+m_2)}{R}}.$$

4. Three identical bodies of mass M are located at the vertices of an equilateral triangle of side L. At what speed must they move if they all are to revolve under the influence of one another's gravity in a circular orbit circumscribing the triangle, while still preserving the equilateral triangle?

The three identical masses are situated at A, B, C, the vertices of the equilateral triangle ABC of side L and O is the centre of the circumscribing circle, whose radius is given by

$$R = \frac{L}{2}\sec 30^0 = \frac{L}{2}\times\frac{2}{\sqrt{3}} = \frac{L}{\sqrt{3}}.$$

Let v denote the speed of each mass M along the circle. Considering the motion of the mass at A, the force of gravitational attraction on it due to the masses at B and C are $\frac{GM^2}{L^2}$ along AB and $\frac{GM^2}{L^2}$ along AC. The resultant force along AD is given by

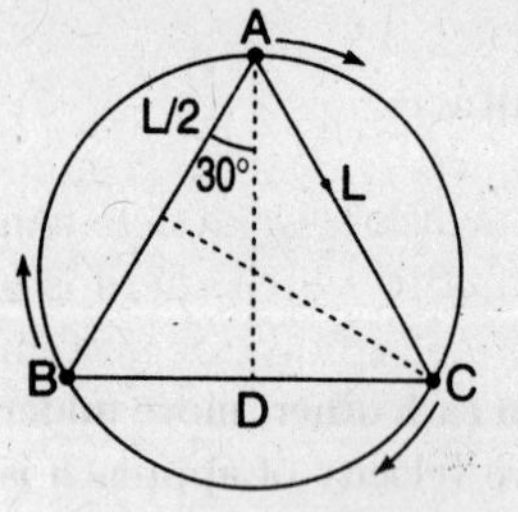

$$2\frac{GM^2}{L^2}\cos 30^0 = \frac{\sqrt{3}GM^2}{L^2}.$$

This force must be equal to the necessary centripetal force (required for maintaining the motion of the masses along the circumscribing circle while preserving the triangle), so that

$$\frac{\sqrt{3}GM^2}{L^2} = \frac{Mv^2}{R} = \frac{\sqrt{3}Mv^2}{L}\quad\left(\because R = \frac{L}{\sqrt{3}}\right),\quad\text{or,}\quad v = \sqrt{\frac{GM}{L}}.$$

5. The earth-moon distance is 3.8×10^5 km and the mass of the earth is 81 times that of the moon. Find out the distance from the earth to the point, where the gravitational fields due to the earth and the moon cancel out. Also calculate the separate fields there, given the radius of the earth = 6.4×10^3 km and $g = 9.8$ newton/m^2.

Let M_e = the mass of the earth, M_m = mass of the moon, R = the distance of the point from the earth's centre, and r = the distance of the point from the moon's centre. The gravitational field at the required point due to the earth is $\frac{GM_e}{R^2}$ (towards the earth) and that due to the moon is $\frac{GM_m}{r^2}$ (towards the moon). According to the problem, these two opposite fields should be equal, so that we have

$$\frac{M_e}{R^2} = \frac{M_m}{r^2},\quad\text{or,}\quad \frac{M_e}{M_m} = \frac{R^2}{r^2}.$$

But $\frac{M_e}{M_m} = 81$ (given). $\therefore\ \frac{R^2}{r^2} = 81$, or, $R = 9r$.

Now the distance between the earth and the moon is $R + r = 10r$, so that

$$r = \frac{1}{10}(R+r).\quad \therefore\ R = \frac{9}{10}(R+r) = \frac{9}{10}\times(3.8\times 10^5\text{ km}) = 3.42\times 10^5\text{ km}.$$

So the resultant field is zero at a distance of 3.42×10^5 km from the centre of the earth.

The gravitational field at this point due to earth alone is $f_e = \frac{GM_e}{R^2}$. If g be the acceleration due to gravity on the earth's surface, then $g = \frac{GM_e}{R_e^2}$, where R_e = radius of the earth

$$\text{or,}\quad f_e = \frac{gR_e^2}{R^2} = \frac{9.8\text{ newton/m}^2\times(6.4\times 10^3\text{ km})^2}{(3.42\times 10^5\text{ km})^2} = 3.43\times 10^{-3}\text{ newton/kg}\qquad(3)$$

which is also the separate field due to the moon at the neutral point.

7.4 Gravitational Field or Attraction or Intensity of the Field

By the attraction of a body or a system of particles at a point external to itself, we mean the force of attraction which the body or the system would exert on a particle of unit mass placed at that point. Hence, the attraction at any point due to a mass m at a distance r from it is given by

$$f = G\frac{m}{r^2}.$$

Thus there must be a definite value for the force f at every point at which the particle can be placed. We thus have the conception of the field of force at every point of which there is a force which is definite in magnitude and direction. Thus the field of force of a body or system of particles is a vector.

7.5 Attraction of a System of Particles

From the principle of superposition of fields of force, we have the force exerted at a point by a system of particles is the vector sum of the forces exerted by each of the particles separately.

Let the particles of masses m_1, m_2, m_3, etc. be situated at points P_1, P_2, P_3, etc. whose co-ordinates referred to rectangular axes OX, OY, OZ are (a_1, b_1, c_1), (a_2, b_2, c_2), (a_3, b_3, c_3), etc. (Fig. 7.5.1). It is required to find expressions for components f_x, f_y, f_z of the attraction of the system of particles at any point O' whose co-ordinates are (x, y, z). Let r_1, r_2, r_3, etc. denote the distances $O'P_1, O'P_2, O'P_3$, etc.

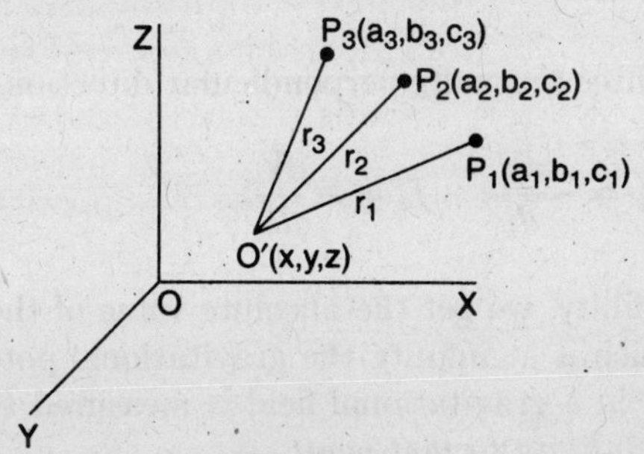

Fig. 7.5.1

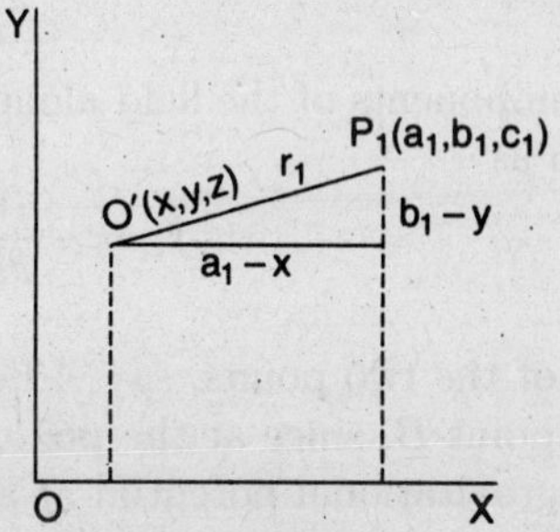

Fig. 7.5.2

Attraction at O' of mass m_1 at P_1 is given by $\dfrac{Gm_1}{r_1^2}$ directed along $O'P_1$ whose direction cosines from figure 7.5.2 are

$$\frac{a_1 - x}{r_1}, \quad \frac{b_1 - y}{r_1}, \quad \frac{c_1 - z}{r_1},$$

where $r_1^2 = (a_1 - x)^2 + (b_1 - y)^2 + (c_1 - z)^2$.

Therefore, the particle of mass m_1 exerts at O' a force whose components parallel to X, Y and Z axes are

$$\left[\frac{Gm_1}{r_1^2} \times \frac{(a_1 - x)}{r_1}\right], \left[\frac{Gm_1}{r_1^2} \times \frac{(b_1 - y)}{r_1}\right], \quad \text{and} \quad \left[\frac{Gm_1}{r_1^2} \times \frac{(c_1 - z)}{r_1}\right],$$

$$\text{i.e.,} \quad \frac{Gm_1(a_1 - x)}{r_1^3}, \frac{Gm_1(b_1 - y)}{r_1^3}, \quad \text{and} \quad \frac{Gm_1(c_1 - z)}{r_1^3}.$$

The other particles m_2, m_3, etc. make similar contributions, so that by the principle of superposition, we have

$$f_x = \sum \frac{Gm_s(a_s - x)}{r_s^3},$$

$$f_y = \sum \frac{Gm_s(b_s - y)}{r_s^3}$$

$$\text{and} \quad f_z = \sum \frac{Gm_s(c_s - z)}{r_s^3}, \quad \text{where } s = 1, 2, 3, \text{ etc.}$$

7.6 Gravitational Potential

In a gravitational field, to move a unit mass from one point A to another point B requires an expenditure of work. This amount of work done may be negative or positive according as the body is moved in the direction or against the direction of the force of attraction. This amount of work done is a measure of the potential difference between the two points A and B between which the unit mass is moved. Thus if at any point in the field, the attraction in a particular direction is f, then the work done in moving the unit mass through a very small distance ds, is $f.ds$ which is equal to the difference of potential between the points A and B at a distance ds apart (Fig. 7.6.1)

Fig. 7.6.1

$$\text{i.e.,} \quad f ds = V - (V + dV) = -dV, \quad \text{or,} \quad f = -\frac{dV}{ds}.$$

The components of the field along the three mutually perpendicular directions X, Y, Z may be written as

$$f_x = -\frac{dV}{dx}; \quad f_y = -\frac{dV}{dy}; \quad f_z = -\frac{dV}{dz}.$$

If one of the two points, say A be at infinity, we get the absolute value of the potential at the other point B, since at the point A which is at infinity the gravitational potential is zero. Thus the gravitational potential at a point in a gravitational field is measured *by the amount of work done in bringing a unit mass from infinity to that point.*

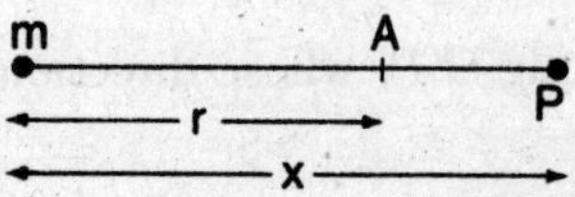

Fig. 7.6.2

Let the gravitational field be due to a particle of mass m. We are to calculate the gravitational potential at A which is a distance r from the particle. We consider a unit mass at P at a distance x from the particle (Fig. 7.6.2).

Now the force of attraction on the unit mass at P is $\frac{Gm}{x^2}$ and the work done in moving this unit mass through a distance dx in the direction of the force of attraction is $\frac{Gm}{x^2} \cdot dx$.

Hence, the total work done in moving the unit mass from infinity to the point A is

$$\int_{\infty}^{r} G\frac{m}{x^2} \cdot dx = Gm \int_{\infty}^{r} \frac{dx}{x^2} = -\frac{Gm}{r}.$$

Thus the potential at the point A is given by

$$V = -\frac{Gm}{r}.$$

The negative sign indicates that the work is done on the particle since the particle is moved in the direction of attraction.

7.7 Potential due to a System of Attracting Particles

Let us define a function by the relation

$$V = -\Sigma G\frac{m_s}{r_s}, \tag{7.7.1}$$

where

$$r_s^2 = (a_s - x)^2 + (b_s - y)^2 + (c_s - z)^2 \tag{7.7.2}$$

in the notation as used in Art. 7.5. Thus V is a function related to a system of attracting particles having a definite value at every point O' (Fig. 7.5.1) of the space external to the particles. It is a function of the co-ordinates (x, y, z) of the point O' and it is clearly a single valued function since it cannot have more than one value at each point O'. It also does not depend on the particular system of axes of reference.

By differentiation of the expression (7.7.1), we have

$$\frac{\partial V}{\partial x} = +\Sigma G\frac{m_s}{r_s^2}\cdot\frac{\partial r_s}{\partial x}.$$

Again from equation (7.7.2), by differentiation, we get

$$2r_s\frac{\partial r_s}{\partial x} = -2(a_s - x), \quad \text{or,} \quad \frac{\partial r_s}{\partial x} = -\frac{(a_s - x)}{r_s}. \quad \therefore \quad \frac{\partial V}{\partial x} = -\Sigma G\frac{m_s(a_s - x)}{r_s^3} = -f_x.$$

Similarly,

$$\frac{\partial V}{\partial y} = -\Sigma G\frac{m_s(b_s - y)}{r_s^3} = -f_y. \quad \text{and} \quad \frac{\partial V}{\partial z} = -\Sigma G\frac{m_s(c_s - z)}{r_s^3} = -f_z$$

— results already obtained in the previous article.

The function V is called the *potential of the system of attracting particles* or the *potential of the field of force.* We have proved that $-\frac{\partial V}{\partial x}, -\frac{\partial V}{\partial y}$ and $-\frac{\partial V}{\partial z}$ give the components of attraction in the directions of the axes and since directions of the axes can be chosen arbitrarily, it follows that the space derivative of the potential V in any direction gives the component of attraction in that direction. We may verify this in the following way:

Let $\frac{\partial V}{\partial s}$ denote the derivative of potential in the direction ds.

Therefore, we may write

$$\frac{dV}{ds} = \frac{\partial V}{\partial x}\frac{\partial x}{\partial s} + \frac{\partial V}{\partial y}\frac{\partial y}{\partial s} + \frac{\partial V}{\partial z}\frac{\partial z}{\partial s} = -(lf_x + mf_y + nf_z) = -f$$

$$= \text{field component in the direction } ds,$$

where

$$l = \frac{\partial x}{\partial s}, \quad m = \frac{\partial y}{\partial s}, \quad n = \frac{\partial z}{\partial s}$$

are the direction cosines of ds.

Thus we may define the potential at a point as a single-valued function of space, the derivative of which in a direction gives the intensity of the field in that direction.

In the vector notation, the above expression may be written as

$$\vec{f} = -\operatorname{grad} V = -\vec{\nabla} V.$$

7.8 Physical Interpretation of Potential

The total differential dV of the potential may be written in the form

$$dV = \frac{\partial V}{\partial x}dx + \frac{\partial V}{\partial y}dy + \frac{\partial V}{\partial z}dz = -(f_x dx + f_y dy + f_z dz).$$

Fig. 7.8.1

Hence, by integrating along the path from P to Q, we have

$$V_Q - V_P = -\int_P^Q \left(f_x \frac{\partial x}{\partial s} + f_y \frac{\partial y}{\partial s} + f_z \frac{\partial z}{\partial s} \right) ds.$$

On the right-hand side, the integral represents the work which the forces of attraction would perform upon a particle of unit mass as it moves along the path from P to Q, and the left-hand side which denotes the difference of potential between Q and P is therefore the work which the forces of attraction would perform upon a particle of unit mass as it moves along any path from P to Q. It is clear that the addition of a constant to the potential will not affect the values of the force components since the force components are obtained by differentiating the potential. Also by integrating the above expression, we get an expression for potential in terms of known force components together with a constant of integration. This constant may be so chosen as to make the potential vanish at infinite distance from the attracting matter. Based on this consideration, we conclude that the *potential at a given point due to a given attracting system is the work that would be done by the attractions of the system on a particle of unit mass as it moves along any path from an infinite distance up to the point considered.*

7.9 Equipotential Surface

We already known that the potential V of a given attracting system is a function of the co-ordinates (x, y, z). Then the equation $V(x, y, z) =$ constant represents a surface over which the potential remains constant. Such surfaces are called *equipotential surfaces.*

It follows from the definition of potential that only one such surface passes through any point of space so that no two equipotential surfaces can intersect. Also since V is constant over an equipotential surface, there is no difference of potential between any two points on this surface—as a result of which no work is done against the gravitational force in moving a unit mass along any path between the two points on this surface. Therefore, at every point on such a surface, the resultant attraction is normal to it through the point. That f (resultant attraction) is perpendicular to the equipotential surface may be shown in the following way.

Let $\vec{r} = x\hat{\imath} + y\hat{\jmath} + z\hat{k}$ be the position vector of any point $P(x, y, z)$ on the surface. Then $d\vec{r} = dx\hat{\imath} + dy\hat{\jmath} + dz\hat{k}$ lies in the tangent plane to the surface at P.

Now,

$$dV = \frac{\partial V}{\partial x}dx + \frac{\partial V}{\partial y}dy + \frac{\partial V}{\partial z}dz = 0 \quad [\because V(x, y, z) = \text{constant}.$$

$$\text{or,} \quad \left(\frac{\partial V}{\partial x}\hat{\imath} + \frac{\partial V}{\partial y}\hat{\jmath} + \frac{\partial V}{\partial z}\hat{k}\right)\cdot\left(dx\hat{\imath} + dy\hat{\jmath} + dz\hat{k}\right) = 0,$$

$$\text{or,} \quad \vec{\nabla}V\cdot d\vec{r} = 0, \quad \text{or,} \quad \vec{f}\cdot d\vec{r} = 0,$$

i.e., $\vec{f}$ is perpendicular to $d\vec{r}$ and therefore to the surface.

7.10 Lines of Force

A line of force in a gravitational field is a line such that the tangent at any point of it gives the direction of the resultant intensity at that point. Therefore, a line of force is at right angles to the equipotential surfaces at all their points of intersection, and conversely, a surface which intersects all lines of force at right angles is an equipotential surface.

7.11 Calculation of Potential—Spherical Bodies

7.11.1 Thin Spherical Shell

We consider a thin spherical shell of radius a and surface density ρ (Fig. 7.11.1). Let P be a point at a distance r from the centre O of the shell. We consider a slice $BB'CC'$ in the form of a ring with two planes close to each other and perpendicular to OP. So, the radius and the width of the ring are BD an BB' respectively. The area of such a slice is $(2\pi BD \times BB')$.

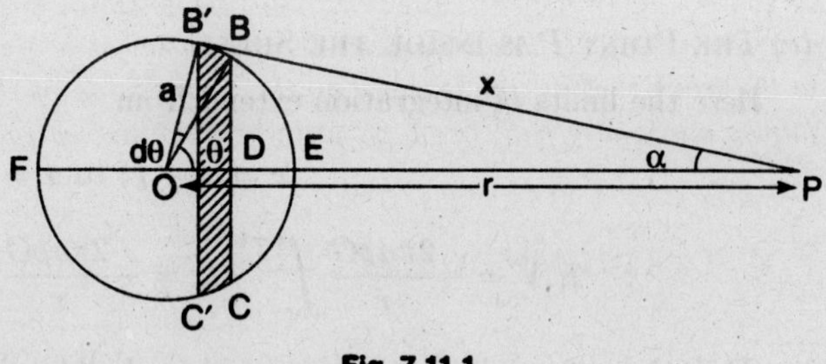

Fig. 7.11.1

Now,

$$BD = a\sin\theta \quad \text{and} \quad BB' = ad\theta.$$

Therefore, the mas of the slice $= 2\pi a\sin\theta\cdot ad\theta\cdot\rho = 2\pi a^2\rho\sin\theta d\theta$.

Hence, the potential at P due to this slice

$$dV = -G\frac{2\pi a^2\rho\sin\theta d\theta}{x}.$$

Now from the triangle BOP, we have

$$BP^2 = OP^2 + OB^2 - 2OP\cdot OB\cos\theta, \quad \text{or,} \quad x^2 = a^2 + r^2 - 2ar\cos\theta.$$

Differentiating we have

$$2xdx = 2ar\sin\theta d\theta. \quad \therefore \sin\theta d\theta = \frac{xdx}{ar}.$$

Therefore, substituting the value of $\sin\theta d\theta$, we have

$$dV = -G\frac{2\pi a\rho dx}{r}.$$

Hence, the potential for the whole shell is obtained by integrating the above expression and is given by

$$V = -\frac{2\pi a\rho G}{r}\int dx. \tag{7.11.1}$$

Now, we consider the following cases depending on the position of the point P.

(*a*) THE POINT P IS OUTSIDE THE SHELL

In this case, the limits of integration extend from $x = (r - a)$ to $x = (r + a)$. Hence,

$$V = -\frac{2\pi a\rho G}{r}\int_{r-a}^{r+a} dx = -\frac{2\pi a\rho G}{r}.2a = \frac{-4\pi a^2\rho G}{r} = -G\frac{M}{r},$$

where $M = 4\pi a^2\rho$, the mass of the shell. This is the value of the potential that a point mass M situated at O would have produced at P. That is, the shell behaves as if its whole mass were concentrated at its centre.

(*b*) THE POINT P IS ON THE SURFACE OF THE SHELL

Here the limits of integration extend from $x = 0$ to $x = 2a$.

$$\therefore V = -\frac{2\pi a\rho G}{r}\int_0^{2a} dx = -\frac{2\pi a\rho G}{r}.2a = -\frac{4\pi a^2\rho G}{r} = -\frac{GM}{a},$$

since $M = 4\pi a^2\rho$ and $r = a$, when P is on the shell.

(*c*) THE POINT P IS INSIDE THE SHELL

Here the limits of integration extend from

$$x = (a - r) \text{ to } x = (a + r)$$

$$\therefore V = -\frac{2\pi a\rho G}{r}\int_{a-r}^{a+r} dx = -\frac{2\pi a\rho G}{r}.2r = -\frac{4\pi a^2\rho G}{a} = -\frac{GM}{a}.$$

Thus the potential at any point inside a shell is constant and is equal to that on the surface.

Illustrative Example

1. The potentials within two homogeneous shells of same surface density are in the ratio 1 : 2. Calculate the ratio of their radii.

Let V_1 and V_2 be the potentials within two spherical shells of masses M_1 and M_2 and radii r_1 and r_2 respectively.

$$\therefore \quad V_1 = -\frac{GM_1}{r_1}, \quad V_2 = -\frac{GM_2}{r_2}.$$

Now, $M_1 = 4\pi r_1^2\sigma$ and $M_2 = 4\pi r_2^2\sigma$, where σ denotes the surface density of the shell.

$$\therefore\quad V_1 = -G\frac{4\pi r_1^2\sigma}{r_1} = -G4\pi r_1\sigma;\quad V_2 = -G\frac{4\pi r_2^2\sigma}{r_2} = -G4\pi r_2\sigma.\quad \therefore\quad \frac{V_1}{V_2} = \frac{r_1}{r_2} = \frac{1}{2}.$$

Thus the radii are in the ratio 1 : 2.

7.11.2 Uniform Solid Sphere

A solid sphere may be supposed to be made up of a number of thin uniform concentric spherical shells (Fig. 7.11.2), the masses of which may be supposed to be concentrated at the centre O.

(a) AT AN EXTERNAL POINT

Here, the potential due to all such shells at an external point P is given by

$$V = -\left[G\frac{m_1}{r} + G\frac{m_2}{r} + G\frac{m_3}{r} + \cdots\right],$$

where m_1, m_2, m_3, $\cdots$, etc. are the masses of the shells under consideration.

$$\therefore\quad V = -\frac{G}{r}(m_1 + m_2 + \cdots) = -\frac{GM}{r},$$

where M is the mass of the solid sphere.

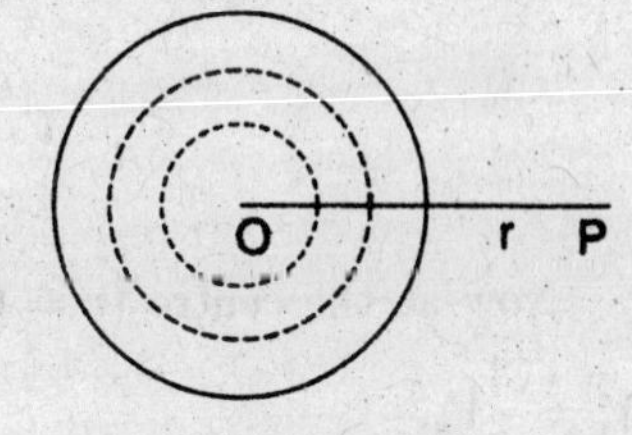

Fig. 7.11.2

(b) THE POINT IS ON THE SPHERE

The calculation may be done exactly in a similar way and it can be shown that the potential is given by $V = -\dfrac{GM}{a}$, where a is the radius of the solid sphere.

(c) AT AN INTERNAL POINT

At a point inside the sphere such as P, which is at a distance r from the centre O, it is required to find the potential. The point P, in this case, is external to a solid sphere of radius r and internal to a thick shell of radii a and r (Fig. 7.11.3). Let V_1 and V_2 denote the potentials at P due to these two parts.

$$\therefore\quad V_1 = -G\frac{\text{mass of the sphere of radius } r}{r}$$

$$= -G\frac{\frac{4}{3}\pi r^3\rho}{r} = -\frac{4}{3}\pi G r^2\rho,$$

where ρ is the density of the material of the sphere.

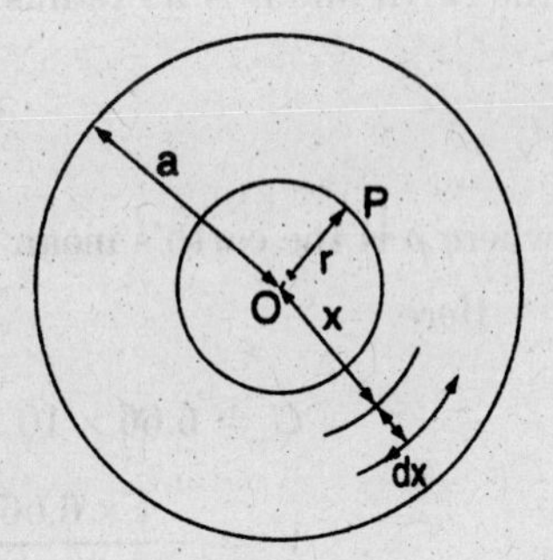

Fig. 7.11.3

To calculate V_2, we consider a thin concentric spherical shell of radius x and thickness dx, of which P is an internal point. Therefore, as we have seen earlier, that the potential at an internal point is equal to that on the surface; the potential at P due to the thin spherical shell under consideration is, given by

$$-G\frac{4\pi x^2 dx\rho}{x} = -G4\pi x dx\rho.$$

Hence, for the thick shell of radii a and r, the potential is given by

$$V_2 = -G4\pi\rho \int_r^a x\,dx = -G4\pi\rho \frac{(a^2 - r^2)}{2} = -2\pi G\rho(a^2 - r^2).$$

Hence, the potential at P due to the whole solid sphere is given by

$$V = V_1 + V_2 = -\frac{4}{3}\pi G\rho r^2 - 2\pi G\rho(a^2 - r^2)$$

$$= -\frac{2}{3}\pi G\rho[2r^2 + 3(a^2 - r^2)] = -\frac{2}{3}\pi G\rho(3a^2 - r^2).$$

Now, the mass of the solid sphere is given by

$$M = \frac{4}{3}\pi a^3 \rho. \quad \therefore \quad \rho = \frac{3M}{4\pi a^3}.$$

$$\therefore \quad V = -\frac{2}{3}\pi G \frac{3M}{4\pi a^3}(3a^2 - r^2) = -\frac{GM}{2a^3}(3a^2 - r^2).$$

Now at the centre ($r = 0$), $V_c = -\frac{3}{2}\frac{GM}{a}$ and at the surface $r = a$, $V_s = -\frac{GM}{a}$, so that $V_c = \frac{3}{2}V_s$.

Illustrative Example

1. The radius of the earth is 6.37×10^8 cm, its mean density 5.57 gms/cm^3 and the gravitational constant 6.66×10^{-8} C.G.S. unit. Calculate the earth's surface potential.

The potential on the surface of the earth is given by $V = -\frac{GM}{a}$, where M is the mass of the earth and a is its radius.

$$\therefore \quad V = -G\frac{\frac{4}{3}\pi a^3 \rho}{a} = -G\frac{4}{3}\pi a^2 \rho,$$

where ρ is the earth's mean density.

Here,

$$G = 6.66 \times 10^{-8} \text{ C.G.S. units}; \quad a = 6.37 \times 10^8 \text{ cm}; \quad \rho = 5.57 \text{ gms/cm}^3.$$

$$\therefore \quad V = -\frac{4 \times 6.66 \times 10^{-8} \times 22 \times (6.37)^2 \times 10^{16} \times 5.57}{3 \times 7}$$

$$= -\frac{4 \times 6.66 \times 22 \times (6.37)^2 \times 5.57 \times 10^8}{3 \times 7}$$

$$= -6.308 \times 10^{11} \text{ ergs/gm}.$$

This means that the work done in removing a unit mass from the surface of the earth to infinity is 6.308×10^{11} ergs.

7.11.3 Thick Spherical Shell

(*a*) AT AN EXTERNAL POINT

Here the treatment is same as in the case of solid sphere, and the potential is $V = -\dfrac{GM}{r}$, where M is the mass of the thick shell.

(*b*) AT A POINT ON THE SURFACE

The treatment is same as in the case of solid sphere, and the potential is given by

$$V = -\frac{GM}{a}.$$

(*c*) AT A POINT IN THE MATERIAL OF THE THICK SHELL

In this case, the point P is external to the spherical shell of internal radius a_1 and external radius r as shown in the figure 7.11.4.

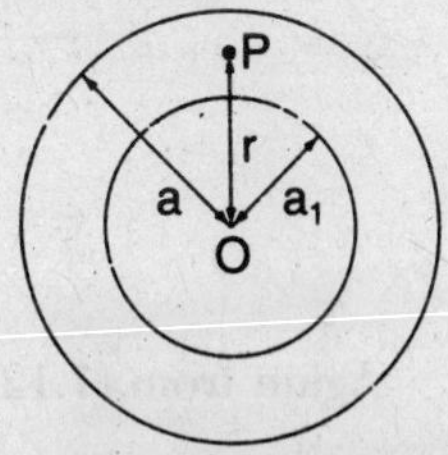

Fig. 7.11.4

Therefore,

$$V_1 = -G\frac{\frac{4}{3}\pi(r^3 - a_1^3)\rho}{r}.$$

Again, the point P is internal to a shell of radii r and a. Therefore, $V_2 = -2\pi G\rho(a^2 - r^2)$.

$\therefore$ The potential at P due to the thick shell is given by

$$V = V_1 + V_2 = -G\cdot\frac{4}{3}\pi\rho\frac{(r^3 - a_1^3)}{r} - 2\pi G\rho(a^2 - r^2) = -\frac{2}{3}\pi G\rho\left[\frac{2(r^3 - a_1^3)}{r} + 3(a^2 - r^2)\right]$$

$$= -\frac{2}{3}\frac{\pi G\rho}{r}[2(r^3 - a_1^3) + 3r(a^2 - r^2)] = -\frac{2}{3}\frac{\pi G\rho}{r}(3ra^2 - r^3 - 2a_1^3).$$

Now, M = mass of the thick shell $= \dfrac{4}{3}\pi(a^3 - a_1^3)\rho$. $\quad\therefore\quad \rho = \dfrac{3M}{4\pi(a^3 - a_1^3)}$.

$$\therefore\quad V = -\frac{2}{3}\cdot\frac{\pi G}{r}\cdot\frac{3M}{4\pi(a^3 - a_1^3)}\cdot(3ra^2 - r^3 - 2a_1^3) = -\frac{GM}{2r(a^3 - a_1^3)}\cdot(3ra^2 - r^3 - 2a_1^3).$$

(*d*) AT A POINT IN THE CAVITY OF THE THICK SHELL

We consider a constituent thin shell of radius x and thickness dx (Fig. 7.11.5). At a point in the cavity of the thick shell the potential due to this constituent shell is given by

$$dV = -G\frac{(\text{mass of this constituent shell})}{x} = -G\cdot\frac{4\pi x^2 dx\rho}{x} = -4\pi G\rho x dx.$$

$\therefore$ For the thick shell of radii a_1 and a, the potential is given by

$$V = -4\pi G\rho\int_{a_1}^{a} x dx = -4\pi G\rho\frac{(a^2 - a_1^2)}{2}$$

$$= -4\pi G\cdot\frac{3M}{4\pi(a^3 - a_1^3)}\cdot\frac{(a^2 - a_1^2)}{2}\quad\left[\because \rho = \frac{3M}{4\pi(a^3 - a_1^3)}\right]$$

$$= -\frac{3}{2}GM\cdot\frac{a + a_1}{a^2 + aa_1 + a_1^2}.$$

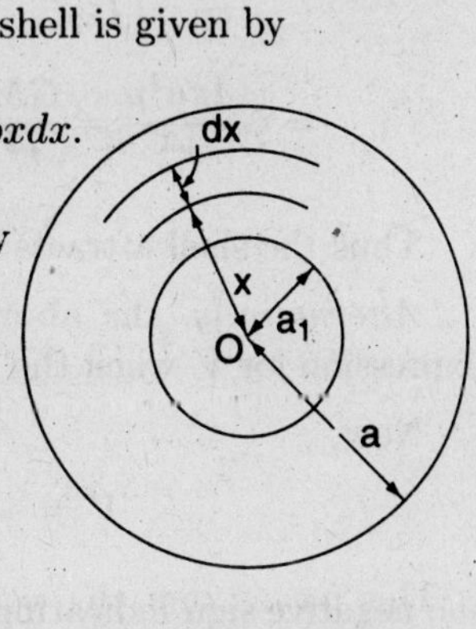

Fig. 7.11.5

7.12 Calculation of Attraction : Spherical Bodies

7.12.1 Thin Spherical Shell

With reference to the Fig. 7.11.1, the intensity or attraction at P due to the slice $BB'CC'$ is $G\dfrac{2\pi a^2 \rho \sin\theta d\theta}{x^2}$ directed along PB. The resultant intensity directed along PO is given by

$$df = G\frac{2\pi a^2 \rho \sin\theta d\theta}{x^2}\cos\alpha = G\frac{2\pi a^2 \rho \sin\theta d\theta}{x^2}\cdot\frac{r - a\cos\theta}{x}.$$

From figure 7.11.1 we have,

$$x^2 = a^2 + r^2 - 2ar\cos\theta, \tag{7.12.1}$$

$$\text{or,}\quad 2xdx = 2ar\sin\theta d\theta, \quad \text{or,} \quad \sin\theta d\theta = \frac{x\,dx}{ar}.$$

Again from (7.12.1),

$$x^2 - a^2 = r^2 - 2ar\cos\theta \text{ or, } x^2 - a^2 + r^2 = 2r(r - a\cos\theta) \text{ or, } (r - a\cos\theta) = \frac{x^2 - a^2 + r^2}{2r}.$$

$$\therefore\ df = G\frac{2\pi a^2\rho}{x^3}\cdot\frac{x\,dx}{ar}\cdot\frac{(x^2 - a^2 + r^2)}{2r} = G\frac{\pi a\rho}{r^2}\cdot\frac{(x^2 - a^2 + r^2)}{x^2}dx = G\frac{\pi a\rho}{r^2}\left[1 + \frac{(r^2 - a^2)}{x^2}\right]dx.$$

Hence, the attraction for the whole shell is obtained by integrating the above expression and is given by

$$f = G\frac{\pi a\rho}{r^2}\int\left[1 + \frac{(r^2 - a^2)}{x^2}\right]dx. \tag{7.12.2}$$

Now, we consider the following cases depending on the position of the point P.

(*a*) AT A POINT OUTSIDE THE SHELL

In this case, the limits of integration extend from $x = (r - a)$ to $x = (r + a)$.

$$f = \frac{G\pi a\rho}{r^2}\left[\int_{r-a}^{r+a} dx + \int_{r-a}^{r+a}\frac{(r^2 - a^2)}{x^2}dx\right] = \frac{G\pi a\rho}{r^2}\left[2a + (r^2 - a^2)\cdot\frac{2a}{(r^2 - a^2)}\right]$$

$$= G\frac{4\pi a^2\rho}{r^2} = \frac{GM}{r^2}, \quad \text{where } M = 4\pi a^2\rho, \text{ the mass of the shell.}$$

Thus the shell attracts an external particle as if its mass were concentrated at its centre.

Alternatively, the above expression for intensity can be obtained by differentiating the expression for V when the point P is outside the shell.

Now,

$$V = -\frac{GM}{r}. \quad \therefore\ f = -\frac{dV}{dr} = \frac{d}{dr}\left(\frac{GM}{r}\right) = -\frac{GM}{r^2}$$

the negative sign indicating that the attraction is towards O, i.e., along PO, but not along OP (the direction along which r increases).

(b) AT A POINT ON THE SHELL

Here the limits of integration extend from $x = 0$ to $x = 2a$.

Hence

$$f = \frac{G\pi a\rho}{r^2}\int_0^{2a}\left[1+\frac{(r^2-a^2)}{x^2}\right]dx.$$

But integration in this case is not possible, since the second term of the above integral is indeterminate, because on the surface $r = a$ and $x = 0$. Hence, to evaluate the above integral let us consider that the point P is situated not on the surface, but very near to the surface. Let $r = a + \delta$, where δ is very small. Hence the expression for intensity may be written as,

$$f = \frac{G\pi a\rho}{r^2}\left[\int_\delta^{2a+\delta} dx + \int_\delta^{2a+\delta}\frac{(a+\delta)^2-a^2}{x^2}\cdot dx\right] = \frac{G\pi a\rho}{r^2}\left[2a + \int_\delta^{2a+\delta}\frac{2a\delta}{x^2}dx\right]$$

$[\because\ \delta^2$ is small$]$

$$= \frac{G\pi a\rho}{r^2}\left[2a + 2a\delta\left(\frac{1}{\delta}-\frac{1}{2a+\delta}\right)\right] = \frac{G\pi a\rho}{r^2}\left[2a+2a\delta\cdot\frac{2a}{\delta(2a+\delta)}\right]$$

$$= \frac{G\pi a\rho}{r^2}\left[2a+\frac{4a^2}{2a+\delta}\right] = \frac{G4\pi a^2\rho}{a^2},\ \text{when } \delta = 0$$

$$= \frac{GM}{a^2}.$$

Alternatively, the above expression for attraction may be derived by simply differentiating the corresponding expression for potential. We know that the potential outside the surface of the shell is given by

$$V = -\frac{GM}{r}.\quad \therefore\ f = -\frac{dV}{dr} = \frac{d}{dr}\left(\frac{GM}{r}\right) = -\frac{GM}{r^2}.$$

This value of attraction continues up to the surface of the spherical shell and hence on the surface, the intensity is

$$f = -\frac{GM}{a^2}\quad [\because\ r = a \text{ on the surface}].$$

(c) AT A POINT INSIDE THE SHELL

Here the limits of integration extend from $x = (a-r)$ to $x = (a+r)$.

Hence,

$$f = G\frac{\pi a\rho}{r^2}\int_{a-r}^{a+r}\left[1+\frac{(r^2-a^2)}{x^2}\right]dx = \frac{G\pi a\rho}{r^2}\left[\int_{a-r}^{a+r}dx + (r^2-a^2)\int_{a-r}^{a+r}\frac{dx}{x^2}\right]$$

$$= \frac{G\pi a\rho}{r^2}\left[2r+(r^2-a^2)\cdot\frac{2r}{(a^2-r^2)}\right] = \frac{G\pi a\rho}{r^2}[2r-2r] = 0.$$

Thus there is no resultant attraction at any point inside the shell.

Alternatively, this value for attraction may be obtained as follows.

We know, potential at any internal point is given by

$$V = -G\frac{M}{a} = \text{constant.}\quad \therefore\quad f = -\frac{dV}{dr} = 0.$$

7.12.2 Uniform Solid Sphere

As explained before the intensity due to a solid sphere may be obtained by dividing the sphere into a number of thin uniform concentric shells, the masses of which may be supposed to be concentrated at the centre O (Fig. 7.11.2).

(*a*) AT AN EXTERNAL POINT

The resultant intensity due to all such shells is equal to the intensity due to the sphere directed along PO and is given by

$$f = \frac{Gm_1}{r^2} + \frac{Gm_2}{r^2} + \frac{Gm_3}{r^2} + \cdots,$$

where $m_1, m_2, m_3, \cdots$, etc. are the masses of the constituent shells.

$$\therefore\ f = \frac{G}{r^2}(m_1 + m_2 + m_3 + \cdots) = \frac{GM}{r^2},$$

where $M = m_1 + m_2 + m_3 + \cdots =$ mass of the solid sphere.

Alternatively, the above expression for intensity due to a solid sphere can be derived by differentiating the corresponding expression for potential with respect to r.

(*b*) AT A POINT ON THE SPHERE

The calculation may be done exactly in a similar way, and it may be shown that the intensity is given by

$$f = G\frac{M}{a^2}.$$

(*c*) AT AN INTERNAL POINT

At a point P inside the sphere which is at a distance r from the centre O, it is required to find the intensity (Fig. 7.12.1). In this case, the point P is external to a solid sphere of radius r, and internal to a thick shell of radii a and r.

The thick shell does not contribute any thing towards the intensity at its internal point P. Hence the resultant intensity at P is only due to the solid sphere of radius r and is given by

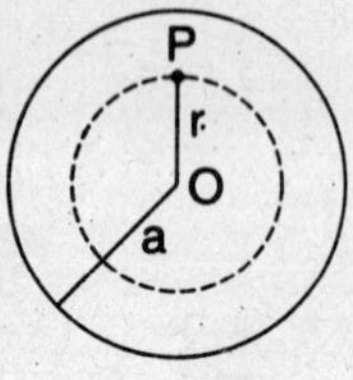

Fig. 7.12.1

$$f = G\frac{(\text{mass of the sphere of radius } r)}{r^2} \text{ along } PO$$

$$= G\cdot\frac{4}{3}\frac{\pi r^3\rho}{r^2} = \frac{4}{3}\pi Gr\rho = \frac{4}{3}\pi Gr\cdot\frac{3M}{4\pi a^3} \qquad \left[\because\ \rho = \frac{3M}{4\pi a^3}\right]$$

$$= \frac{GMr}{a^3}, \text{ where } M \text{ is the mass of the solid sphere.}$$

Thus the attraction is proportional to the distance r from the centre.

Alternatively, the expression for potential at any internal point due to a solid sphere is given by

$$V = -\frac{GM}{2a^3}(3a^2 - r^2). \quad \therefore \quad f = -\frac{dV}{dr} = \frac{GM}{2a^3}\cdot\frac{d}{dr}(3a^2 - r^2) = -\frac{GMr}{a^3}.$$

The negative sign only indicates that the attraction is not directed along increasing r, but towards the centre O.

7.12.3 Thick Spherical Shell

(*a*) AT AN EXTERNAL POINT

The treatment is same is in case of a solid sphere, and the intensity towards the centre of the shell at a distance r is given by

$$f = G\frac{M}{r^2}.$$

(*b*) AT A POINT ON THE SURFACE

In this case also as before the intensity towards the centre is given by

$$f = \frac{GM}{a^2} \quad [\because\ r = a, \text{ on the surface}]$$

(*c*) AT A POINT IN THE MATERIAL OF THE SHELL

In this case, the constituent shells which are external to the point P will contribute nothing towards the attraction at P. Hence the resultant intensity will be due to the shell of radii r and a_1. (Same as Fig. 7.11.4).

$\therefore$ the intensity

$$f = G\,\frac{(\text{mass of the shell of radii } r \text{ and } a_1)}{r^2} = G\cdot\frac{4}{3}\pi\frac{(r^3 - a_1^3)\rho}{r^2}$$

$$= G\frac{4}{3}\pi\frac{(r^3 - a_1^3)}{r^2}\cdot\frac{3M}{4\pi(a^3 - a_1^3)}$$

$$= \frac{GM}{r^2}\frac{(r^3 - a_1^3)}{(a^3 - a_1^3)} \quad \left[\because\ \rho = \frac{3M}{4\pi(a^3 - a_1^3)}, \text{ where } M \text{ is the mass of the shell}\right].$$

Alternatively, the expression for potential at a point in the material of a thick shell, is given by

$$V = -\frac{GM}{2r(a^3 - a_1^3)}\cdot(3ra^2 - r^3 - 2a_1^3)$$

We have by differentiating the above expression

$$f = -\frac{dV}{dr} = \frac{GM}{2(a^3 - a_1^3)}\cdot\frac{d}{dr}\left[\frac{3a^2r - r^3 - 2a_1^3}{r}\right] = \frac{GM}{2(a^3 - a_1^3)}\cdot\frac{d}{dr}\left(3a^2 - r^2 - 2a_1^3\cdot\frac{1}{r}\right)$$

$$= \frac{GM}{2(a^3 - a_1^3)}\cdot\left(-2r + \frac{2a_1^3}{r^2}\right) = -\frac{GM(r^3 - a_1^3)}{r^2(a^3 - a_1^3)}.$$

Hence, the intensity towards the centre O at a point in the material of the shell is

$$\frac{GM(r^3 - a_1^3)}{r^2(a^3 - a_1^3)}.$$

(*d*) AT A POINT IN THE CAVITY OF THE THICK SHELL

In this case the point P is within all the constituent thin shells and the resultant attraction at P is zero.

Alternatively, from the expression for potential, we have

$$V = -\frac{3}{2}GM\frac{(a + a_1)}{a^2 + aa_1 + a_1^2} = \text{constant.} \quad \therefore\ f = -\frac{dV}{dr} = 0.$$

7.13 Graphical representation of potential and attraction

The graphical variation of potential V and attraction f with the distance r from the centre O of a solid sphere and a thin shell is shown in the Figs. 7.13.1 and 7.13.2 respectively.

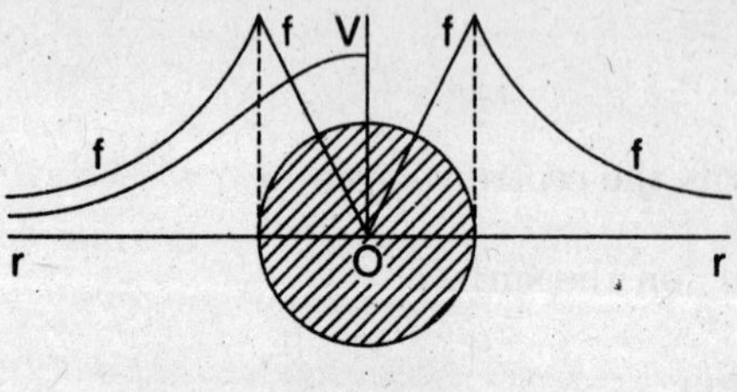

Fig. 7.13.1

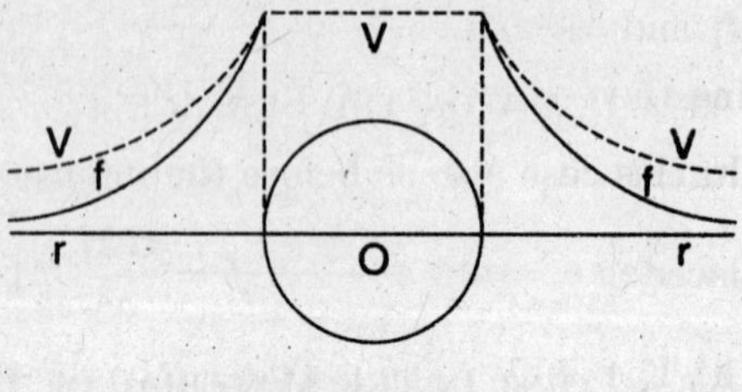

Fig. 7.13.2

Illustrative Examples

1. In a sphere the density varies inversely as the distance from the centre. Show that the attraction is the same at any two internal points.

We know that the intensity or attraction at an internal point of a solid sphere is given by $f = \frac{4}{3}\pi G r\rho$, where r is the distance of the internal point from the centre of the sphere and ρ is the density of the material of the sphere.

Here $\rho \propto \frac{1}{r}$. $\therefore\ \rho r =$ constant.

Hence $f =$ constant.

Thus intensity at any internal point is constant, i.e., intensity is the same at any two internal points.

2. A sphere is described on a radius of another sphere as diameter. If the latter sphere be a homogeneous sphere of mass M and radius a, show that the resultant attraction on the portion included in the smaller sphere is $\frac{1}{16}\frac{GM^2}{a^2}$.

Within the larger sphere and at the centre of the smaller sphere (as shown in Fig. 1), the field at B is of magnitude.

$$\frac{4}{3}G\pi\rho.BA.$$

Hence the portion included in the smaller sphere will be acted upon by a force given by

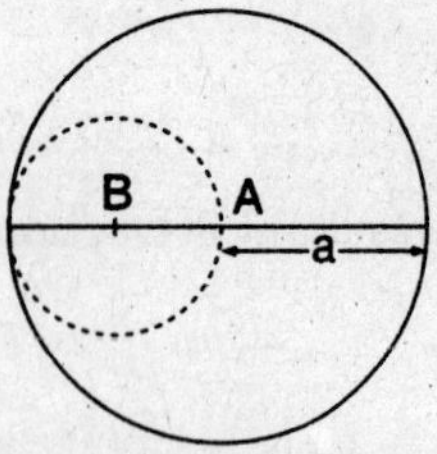

Fig. 1

$$F = \frac{4}{3}G\pi\rho.BA \times \text{mass of the smaller sphere}$$

$$= \frac{4}{3}G\pi\rho\left(\frac{a}{2}\right) \times \frac{4}{3}\pi\left(\frac{a}{2}\right)^3\rho = \left(\frac{4}{3}\right)^2 G\pi^2\rho^2\left(\frac{a}{2}\right)^4 = \frac{1}{9}G\pi^2\rho^2a^4$$

$$= \frac{1}{9}G\pi^2\frac{9M^2}{16\pi^2a^6}\cdot a^4 \qquad \left[\because M = \frac{4}{3}\pi a^3\rho. \quad \therefore\ \rho^2 = \frac{9M^2}{16\pi^2a^6}\right]$$

$$= \frac{1}{16}G\frac{M^2}{a^2}.$$

3. Two concentric spherical shells of uniform density, and having masses M_1 and M_2 are situated as shown in the figure 2. Find the force on a particle of mass m, when the particle is located at (i) $r = a$, (ii) $r = b$ and (iii) $r = c$. The distance r is measured from the centre of the shells.

(i) The point $r = a$ is external to both the shells of masses M_1 and M_2, which behave as if their masses M_1 and M_2 are concentrated at their common centre. The *magnitudes* of the gravitational fields at $r = a$ due to M_1 and M_2 are $\dfrac{GM_1}{a^2}$ and $\dfrac{GM_2}{a^2}$ and both are directed along the line joining the common centre and the point. Hence *the magnitude* of the resultant field is $\dfrac{G(M_1 + M_2)}{a^2}$ and hence the force on a particle of mass m placed at $r = a$ is $\dfrac{G(M_1 + M_2)m}{a^2}$.

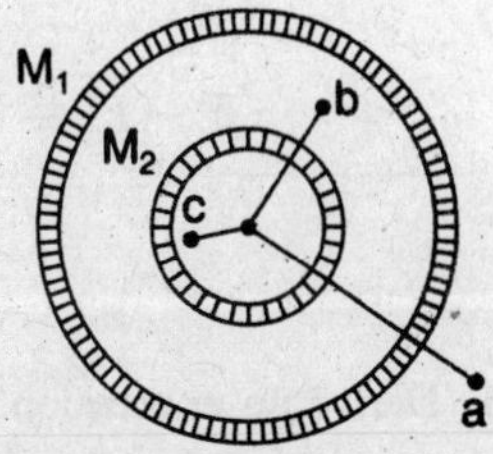

Fig. 2

(ii) The point $r = b$ is external to the shell of mass M_2 and internal to the shell of mass M_1. Now the field at an internal point due to a shell is zero. Hence the field at $r = b$ is due only to M_2 which behaves as if its entire mass is at its centre, and is given by $-\dfrac{GM_2}{b^2}$, and hence the force on a particle of mass m at $r = b$ is $-\dfrac{GM_2 m}{b^2}$.

(iii) The point $r = c$ is internal to both the shells of masses M_1 and M_2 respectively. Hence the field and also the force on a particle of mass m placed at $r = c$ will be zero.

4. A solid sphere of lead has mass M and radius R. A spherical hollow is drug out from it as shown in the figure, its boundary passing through the centre and also touching the boundary of the solid sphere. Find the gravitational force on a mass m placed at P which is distant r from O along the line of centres.

As shown in the figure 3, O is the centre of the solid sphere and O' the centre of the spherical hollow dug out. Since for the external point, a sphere behaves as if its entire mass is concentrated at its centre, the magnitude of the gravitational force on a mass m at P due to the solid sphere of mass M is $F = \dfrac{GMm}{r^2}$ along PO. Now the diameter of the smaller sphere, which had been dug out, is R, i.e., its radius $OO' = \frac{R}{2}$.

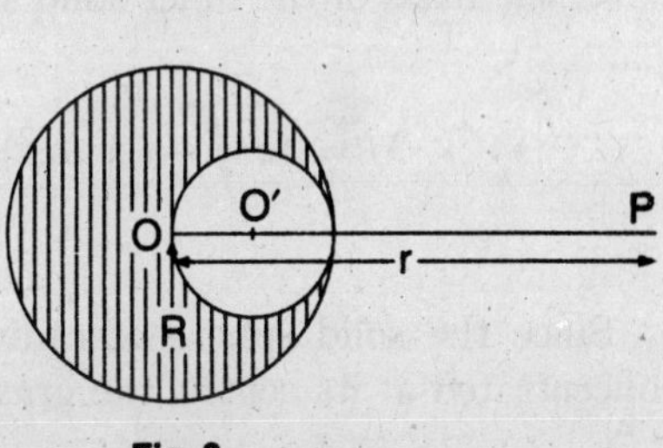

Fig. 3

The magnitude of the force on the particle of mass m at P due to this sphere of mass M', (had it been solid, i.e., before it is dug out) will be

$$F' = G\frac{M'm}{\left(r - \frac{R}{2}\right)^2} \text{ along } PO' \quad \left[\because \text{ distance } PO' = \left(r - \frac{R}{2}\right)\right].$$

Since the radius of this smaller solid sphere is half of that of the bigger solid sphere, we have $M' = M/8$, so that the above expression for F' reduces to

$$F' = \frac{GMm}{8\left(r - \frac{R}{2}\right)^2} \text{ along } PO'.$$

As both F and F' are in the same direction, the effective force on the particle of mass m due to hollowed sphere is

$$F - F' \equiv \left(\frac{GMm}{r^2} - \frac{GMm}{8r^2\left(1 - \frac{R}{2r}\right)^2} \right) \equiv \frac{GMm}{r^2}\left[1 - \frac{1}{8\left(1 - \frac{R}{2r}\right)^2}\right].$$

5. Derive an expression for the gravitational field inside a sphere of radius R, when the mass density at a point is $\rho = a + br^2$, where r is the distance of the point from the centre of the sphere.

In Fig. 4, O is the centre of the sphere of radius R and P is an internal point at which the gravitational field is to be found. Here $OP = r$. We now draw a concentric sphere through P. The point P is now external to the inner solid sphere of radius r, which alone is responsible for exerting attraction at P, since the outer spherical shell of internal radius r and external radius R (for which the point P is internal) exerts no attraction at P.

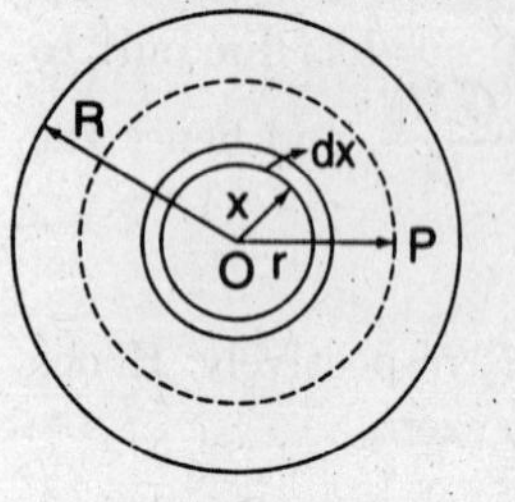

Fig. 4

To find out the mass of the inner solid sphere, we divide it into a large number of thin concentric shells and consider one such shell of radius x and infinitesimally small thickness dx, so that the mass of this elementary shell = its volume × density = $4\pi x^2 dx \times (a + bx)^2$ ($\because$ ρ at a distance x from the centre = $a + bx^2$).

$\therefore$ the mass of the inner solid sphere of radius r is

$$M = 4\pi \int_0^r (a + bx^2)x^2 dx = 4\pi \left[\frac{ax^3}{3} + \frac{bx^5}{5}\right]_0^r = 4\pi r^3 \left(\frac{a}{3} + \frac{br^2}{5}\right).$$

Since the solid sphere of radius r behaves at an external point P, as if its mass were concentrated at its centre, the gravitational field at P due to it is

$$F = -\frac{GM}{r^2} = -\frac{G}{r^2}\left[4\pi r^3 \left(\frac{a}{3} + \frac{br^2}{5}\right)\right] = -4\pi Gr\left(\frac{a}{3} + \frac{br^2}{5}\right).$$

This is the gravitational field at P due to the entire sphere of *variable* density.

7.14 Attraction between two portions of a solid sphere divided by a plane

To solve this problem, we are to make use of the theorem that the attraction between two parts of the same body is the same as the attraction of the whole on one part.

Let ρ be the density of the sphere and a its radius. Let AB be the intersection of the dividing plane with the plane of the paper and OD, the radius of the sphere drawn perpendicular to the dividing plane (Figs. 7.14.1). Let $OC = b$, $ON = x$ and $OP = r$. [P is any point in the sphere and PN is the perpendicular dropped from P on OD.] The attraction of the sphere at P is along PO and its magnitude is $\frac{4}{3}\pi G\rho r$. The component of this attraction along DO

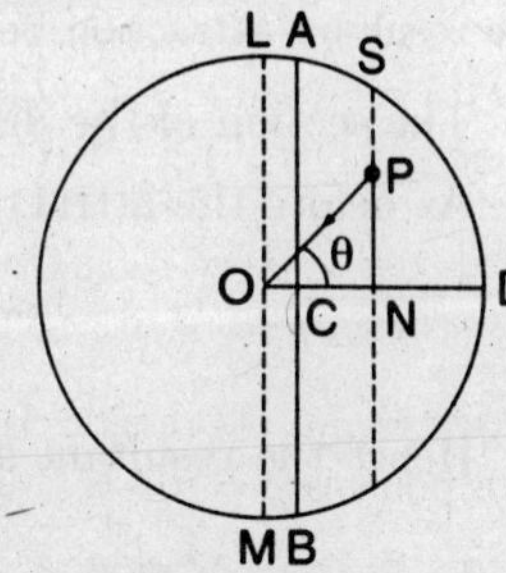

Fig. 7.14.1

$$= \frac{4}{3}\pi G\rho r\cos\theta = \frac{4}{3}\pi G\rho r\cdot\frac{x}{r} = \frac{4}{3}\pi G\rho x.$$

The other component along PN is ineffective in the sense that it contributes nothing towards attraction. Thus the component of attraction parallel to DO at every point in the plane through N at right angles to DO is the same, because x is the same in all cases from a diametrical plane parallel to AB, i.e., from the plane through LM.

Considering a disc between x and $x + dx$, we see that its radius $SN = a\sin\theta$ (Fig. 7.14.2).

$\therefore$ its area $= \pi a^2\sin^2\theta$ [a is the radius of the sphere]

$$= \pi a^2\left(1 - \frac{x^2}{a^2}\right)$$

$$= \pi(a^2 - x^2) \quad \left[\because \sin^2\theta = 1 - \cos^2\theta = 1 - \frac{x^2}{a^2}\right].$$

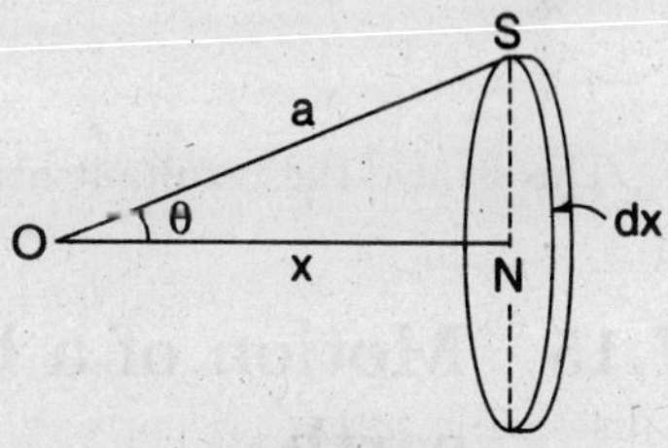

Fig. 7.14.2

Therefore the mass of the disc $= \pi(a^2 - x^2)dx\rho$.

Hence the attractive force of the sphere on the elementary disc in question is given by

$$\frac{4}{3}G\pi\rho x \times \pi\rho(a^2 - x^2)dx = \frac{4}{3}G\pi^2\rho^2 x(a^2 - x^2)dx.$$

Imagining the portion of the sphere ABD to be divided up into similar co-axial discs, the resultant attraction of the sphere on this portion or the attraction between two portions of the solid sphere is given by

$$\frac{4}{3}G\pi^2\rho^2\int_{x=b}^{x=a} x(a^2 - x^2)dx = \frac{4}{3}G\pi^2\rho^2\left[a^2\int_b^a x\,dx - \int_b^a x^3\,dx\right]$$

$$= \frac{4}{3}G\pi^2\rho^2\left[a^2\frac{(a^2 - b^2)}{2} - \frac{(a^4 - b^4)}{4}\right]$$

$$= \frac{4}{3}G\pi^2\rho^2\left(\frac{a^2 - b^2}{2}\right)^2 = \frac{1}{3}G\pi^2\rho^2(a^2 - b^2)^2$$

$$= \frac{1}{3}G\pi^2\cdot\frac{9M^2}{16\pi^2 a^6}(a^2 - b^2)^2$$

$$\left(\because \rho^2 = \frac{9M^2}{16\pi^2 a^6},\ M \text{ being the mass of the sphere}\right)$$

$$= \frac{3}{16}G\frac{M^2}{a^2}\left(1 - \frac{b^2}{a^2}\right)^2.$$

Illustrative Example

1. A uniform solid sphere of mass M and radius a is cut into two by a diametrical plane. Show that the resultant attraction between the two halves is given by $\dfrac{3}{16}G\dfrac{M^2}{a^2}$.

The section of the diametrical plane is represented by the dotted line LM in Fig. 7.14.1.

As before the attractive force of the sphere on the elementary disc is given by

$$\frac{4}{3}G\pi^2\rho^2 x(a^2 - x^2)dx.$$

Hence the resultant attraction of the sphere on either of the hemispheres is

$$\frac{4}{3}G\pi^2\rho^2 \int_0^a x(a^2 - x^2)dx = \frac{4}{3}G\pi^2\rho^2 \left[a^2 \int_0^a xdx - \int_0^a x^3 dx\right]$$

$$= \frac{4}{3}G\pi^2\rho^2 \left(\frac{a^4}{2} - \frac{a^4}{4}\right) = \frac{4}{3}G\pi^2\rho^2 \cdot \frac{a^4}{4}$$

$$= \frac{1}{3}G\pi^2 a^4 \cdot \frac{9M^2}{16\pi^2 a^6} = \frac{3}{16}G\frac{M^2}{a^2}.$$

This is also the resultant attraction between the two hemispheres.

7.15 Motion of a body in a tunnel bored through the earth

If a smooth tunnel be bored through the earth considered as a sphere, then the motion of a body when dropped in such a tunnel is a simple harmonic one and the time period of oscillation of the body may be calculated in the following way.

Let us suppose that the body is, in course of its motion along the tunnel, at P which is at a distance x from O, O being the foot of the perpendicular dropped from the centre C of the earth on the line of motion of the body.

The force on the body of mass m at P is

$$F = m \times \text{ intensity at } P = m \cdot \frac{4}{3}G\pi\rho r \text{ directed along } PC.$$

The component of this force along XOX', which is responsible for the motion of the body along the tunnel, is

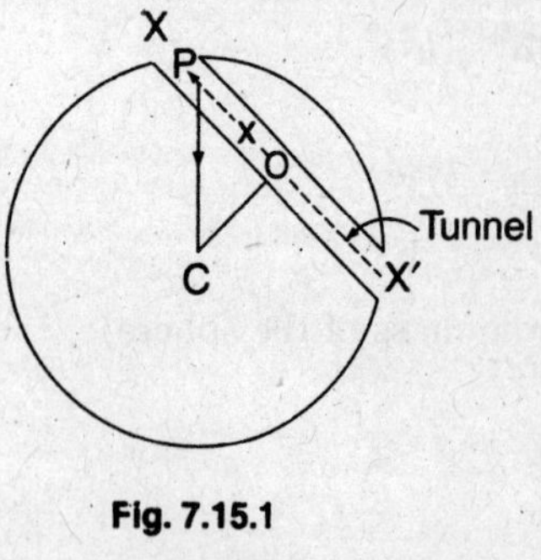

Fig. 7.15.1

$$F_x = m \cdot \frac{4}{3}G\pi\rho r \cdot \cos\theta \quad [\angle OPC = \theta] = m \cdot \frac{4}{3}G\pi\rho r \cdot \frac{x}{r}$$

$$= m \cdot \frac{4}{3}G\pi\rho x = kx, \quad \text{where } k = \tfrac{4}{3}mG\pi\rho = \text{constant}.$$

Therefore, while the body is moving along XOX', it is acted on by a force which is proportional to its distance from the mean position O, and directed towards the mean position as shown in the figure 7.15.1. Hence the motion of the body is a *simple harmonic* one.

The differential equation of such a motion may be written as

$$m\frac{d^2x}{dt^2} = -kx, \quad \text{or,} \quad \frac{d^2x}{dt^2} + \frac{k}{m}x = 0, \quad \text{or,} \quad \frac{d^2x}{dt^2} + \omega^2 x = 0.$$

Here

$$\omega^2 = \frac{k}{m} = \frac{4}{3}G\pi\rho. \quad \therefore\ \omega = \sqrt{\frac{4}{3}G\pi\rho}.$$

$\therefore$ The time period of such a motion is given by

$$T = \frac{2\pi}{\omega} = \frac{2\pi}{\sqrt{\frac{4}{3}G\pi\rho}} = \sqrt{\frac{3\pi}{G\rho}}.$$

Again, while the body is at the extreme position X, $r = a$, the radius of the earth. Hence the force on the body when it is at this extreme position, i.e., on the surface of the earth $= m \times$ Intensity on the surface $= m\cdot\frac{4}{3}\pi G\rho a.$

If g be the acceleration due to gravity on the surface, then this force may be equated to mg.

$$\therefore \quad \frac{4}{3}\pi G m\rho a = mg, \quad \text{or,} \quad G\rho = \frac{3g}{4\pi a}.$$

Substituting this value of $G\rho$ in the expression for time period, we have

$$T = \sqrt{\frac{3\pi}{3g/4\pi a}} = 2\pi\sqrt{\frac{a}{g}}.$$

This indicates that in whatever direction we bore the tunnel, the time period of oscillation will be the same.

It is to be noted that this periodic time is the same as that for a small satellite revolving round the earth close to its surface. The motion through the tunnel is the same as the projection of the circular motion of the satellite on the earth's diameter, coinciding with the tunnel.

If the numerical values of G and ρ or a and g are known, T can be calculated, as will be evident from the following illustrative example.

Illustrative Example

1. A body falls from the earth's surface through a frictionless tunnel passing through its centre. Assuming the earth to be a homogeneous sphere of radius 6400 kilometres and g = 980 cm/sec^2., calculate the time taken by the body to reach the earth's centre. **[C.U. (Hons.) 1954]**

As the time period is independent of the direction of the tunnel, in this case the time period of complete oscillation is

$$T = 2\pi\sqrt{\frac{a}{g}}.$$

Here

$$a = 6400 \text{ kilometres}$$
$$= 6400 \times 10^5 \text{ cm}$$
$$= 64 \times 10^7 \text{ cm}$$
$$g = 980 \text{ cm/sec}^2$$

$$\therefore \quad T = 2\pi\sqrt{\frac{64 \times 10^7}{980}} = 2 \times 3.14\sqrt{\frac{16 \times 4 \times 10^6}{2 \times 49}} = \frac{25.12}{7} \times 10^3\sqrt{2} \text{ secs.}$$

Hence the time taken to reach the earth's centre is

$$\frac{T}{4} = \frac{25.12 \times 10^3\sqrt{2}}{4 \times 7} = 1265 \text{ secs} = 21 \text{ minutes } 5 \text{ secs.}$$

2. A and B are the centres of two spherical bodies of masses m_1 and m_2 respectively. Find the point O at which the gravitational intensity vanishes. Show that if a particle at rest at O be given a very small displacement in the direction perpendicular to AB, it will execute simple harmonic motion. In case $m_1 = m_2 = m$ (say), show that the time period will be $T = \frac{\pi}{2}\sqrt{\frac{r^3}{Gm}}$, where $r = AB$.

[C.U. (Hons.) 1978]

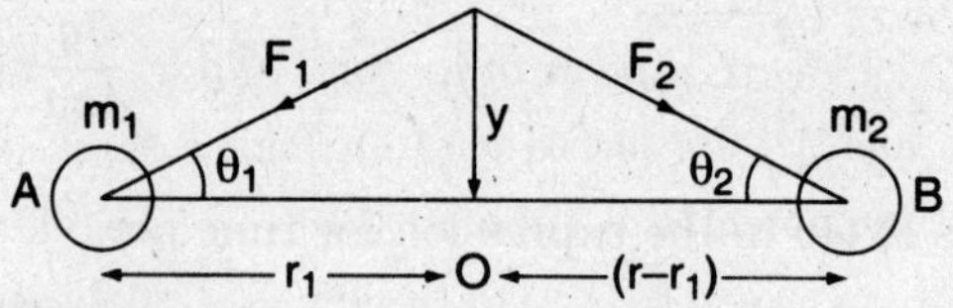

At O, let there be no gravitational intensity, i.e., Intensity at O due to m_1 = Intensity at O due to m_2.

$$\therefore \quad \frac{Gm_1}{r_1^2} = \frac{Gm_2}{(r-r_1)^2}, \quad \text{or,} \quad \frac{(r-r_1)^2}{r_1^2} = \frac{m_2}{m_1}, \quad \text{or,} \quad \frac{r-r_1}{r_1} = \sqrt{\frac{m_2}{m_1}},$$

$$\text{or,} \quad \frac{r}{r_1} = 1 + \sqrt{\frac{m_2}{m_1}}, \quad \text{or,} \quad r_1 = \frac{r}{1+\sqrt{\frac{m_2}{m_1}}}.$$

If $m_1 = m_2$, then $r_1 = \frac{r}{2}$.

The equation of motion of the particle of mass m' placed at O when, given a very small displacement y in the direction perpendicular to AB is written as

$$m'\frac{d^2y}{dt^2} = -(F_1 \sin\theta_1 + F_2 \sin\theta_2) \tag{1}$$

[The right hand side representing the restoring force in the vertical direction.]

Now since θ is very small, $\sin\theta \simeq \tan\theta \simeq \theta$.

Again,

$$F_1 = \frac{Gm_1m'}{r_1^2+y^2}, \quad \text{or,} \quad F_1\sin\theta_1 = F_1\tan\theta_1 = F_1\frac{y}{r_1} = \frac{Gm_1m'}{(r_1^2+y^2)}\cdot\frac{y}{r_1}.$$

Similarly,

$$F_2 \sin\theta_2 = \frac{Gm_2m'}{[(r-r_1)^2+y^2]}\cdot\frac{y}{(r-r_1)}.$$

Substituting these values in equation (1) we have the equation of motion

$$m'\frac{d^2y}{dt^2} = -m'\left[\frac{Gm_1}{(r_1^2+y^2)}\cdot\frac{y}{r_1} + \frac{Gm_2}{[(r-r_1)^2+y^2]}\cdot\frac{y}{r-r_1}\right].$$

Neglecting y^2, since y is very small we may write,

$$\frac{d^2y}{dt^2} = -\left[\frac{Gm_1}{r_1^3} + \frac{Gm_2}{(r-r_1)^3}\right]y,$$

which shows that the motion is simple harmonic in character.

If $m_1 = m_2 = m$ (say), then $r_1 = \dfrac{r}{2}$.

$$\therefore \quad \frac{d^2y}{dt^2} + \frac{16Gm}{r^3}y = 0, \quad \text{or,} \quad \frac{d^2y}{dt^2} + \omega^2 y = 0,$$

where $\omega^2 = \dfrac{16Gm}{r^3}$. $\therefore \quad \omega = \sqrt{\dfrac{16Gm}{r^3}}$, or, $T = \dfrac{2\pi}{4}\sqrt{\dfrac{r^3}{Gm}} = \dfrac{\pi}{2}\sqrt{\dfrac{r^3}{Gm}}$.

3. Two bodies of masses M_1 and M_2 are placed with distance d apart. Show that at a point, where the gravitational field is zero, the potential is given by $V = -\dfrac{G}{d}\left(M_1 + M_2 + 2\sqrt{M_1M_2}\right)$.

Let, at the point P at a distance r from M_1, the gravitational field be zero. Then the field at P due to M_1 is $-\dfrac{GM_1}{r^2}$ and the field at P due to M_2 is $-\dfrac{GM_2}{(d-r)^2}$, which is oppositely directed.

$$\therefore \quad -\frac{GM_1}{r^2} + \frac{GM_2}{(d-r)^2} = 0, \quad \text{or,} \quad \frac{r}{(d-r)} = \sqrt{\frac{M_1}{M_2}}, \quad \text{or,} \quad r\left(\sqrt{M_1}+\sqrt{M_2}\right) = d\sqrt{M_1}.$$

$$\therefore \quad r = \frac{d\sqrt{M_1}}{\sqrt{M_1}+\sqrt{M_2}} \quad \text{and} \quad (d-r) = \frac{d\sqrt{M_2}}{\sqrt{M_1}+\sqrt{M_2}}.$$

Now the potential at P due to M_1 is

$$-\frac{GM_1}{d\sqrt{M_1}}\left(\sqrt{M_1}+\sqrt{M_2}\right) = -\frac{G}{d}\left(M_1 + \sqrt{M_1M_2}\right)$$

and potential at P due to M_2 is

$$-\frac{GM_2}{d\sqrt{M_2}}\left(\sqrt{M_1}+\sqrt{M_2}\right) = -\frac{G}{d}\left(M_2 + \sqrt{M_1M_2}\right).$$

$\therefore$ The total potential at

$$P = -\frac{G}{d}\left[M_1 + M_2 + 2\sqrt{M_1M_2}\right].$$

7.16 Uniform shell bounded by non-intersecting eccentric spheres

7.16.1 Calculation of Potential

When the bounding spheres are not concentric, let O and O' be their centres and a, b be their radii; let r and r' be the distances of any point P from O and O', at which the potential is to be calculated.

(*a*) WHEN THE POINT P IS OUTSIDE BOTH THE SPHERES

In this case the potential at P due to the sphere of radius a is $\frac{-\frac{4}{3}G\pi a^3\rho}{r}$ and the potential due to the other sphere of radius b is $\frac{-\frac{4}{3}G\pi b^3\rho}{r'}$. Hence the potential at P due to the shell bounded by the two non-intersecting spheres is given by

$$V_1 = \left[\left(\frac{-\frac{4}{3}G\pi a^3\rho}{r}\right) - \left(\frac{-\frac{4}{3}G\pi b^3\rho}{r'}\right)\right] = -\frac{4}{3}\pi G\rho\left(\frac{a^3}{r} - \frac{b^3}{r'}\right).$$

(*b*) WHEN THE POINT P IS INSIDE THE MATERIAL OF THE SHELL

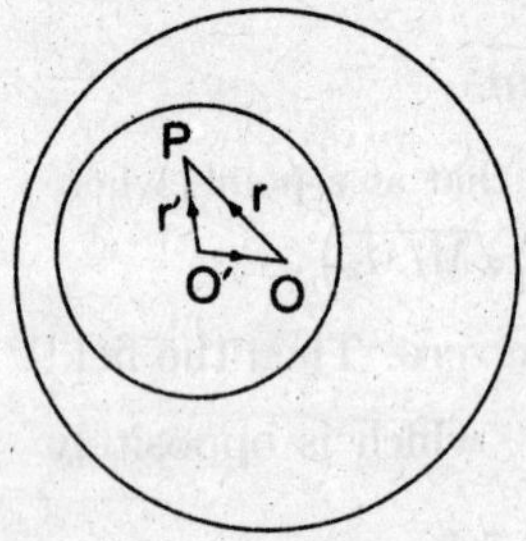

Fig. 7.16.1

In this case, the point P is inside the sphere of radius a, and outside the sphere of radius b. The potential at P due to the sphere of radius a is $-\frac{2}{3}G\pi\rho(3a^2 - r^2)$ and that due to the sphere of radius b is $\frac{-\frac{4}{3}\pi G b^3\rho}{r'}$. Hence the resultant potential which is the potential of the shell at P is given by

$$V_2 = -\frac{2}{3}\pi G\rho(3a^2 - r^2) + \frac{4}{3}\frac{\pi G b^3\rho}{r'}$$

$$= -\frac{2}{3}\pi G\rho\left(3a^2 - r^2 - \frac{2b^3}{r'}\right).$$

(*c*) WHEN THE POINT P IS INSIDE BOTH THE SPHERES, i.e., IN THE CAVITY (Fig. 7.16.1)

In this case, the point P being inside both the spheres, we have, the potential at P due to the sphere of radius a is $-\frac{2}{3}\pi G\rho(3a^2 - r^2)$ and that due to the sphere of radius b is $-\frac{2}{3}\pi G\rho(3b^2 - r'^2)$.

Hence the resultant potential at P, i.e., the potential due to the shell is given by

$$V_3 = -\frac{2}{3}\pi G\rho[(3a^2 - r^2) - (3b^2 - r'^2)] = -\frac{2}{3}\pi G\rho(3a^2 - r^2 - 3b^2 + r'^2).$$

7.16.2 Calculation of Intensity

Let us now take into consideration the attraction at P in the above three cases.

(*a*) WHEN THE POINT P IS OUTSIDE BOTH THE SPHERES

The attraction or intensity at P due to the sphere of radius a is $\frac{\frac{4}{3}\pi G a^3\rho}{r^2}$ along PO and the intensity at P due to the sphere of radius b is $\frac{\frac{4}{3}\pi G b^3\rho}{r'^2}$ along $O'P$.

(*b*) WHEN THE POINT P IS INSIDE THE MATERIAL OF THE SHELL

In this case the intensity at P due to the sphere of radius a is $\dfrac{\frac{4}{3}\pi G r^3 \rho}{r^2} = \dfrac{4}{3}\pi G r \rho$ along PO and the intensity at P due to the sphere of radius b is $\dfrac{\frac{4}{3}\pi G b^3 \rho}{r'^2}$ along $O'P$.

(*c*) THE POINT P IS INSIDE BOTH THE SPHERES, i.e., IN THE CAVITY

In this case, the point P is inside both the spheres. The component of resultant attraction along PO is $\frac{4}{3}\pi G\rho r$ and that along $O'P$ is $\frac{4}{3}\pi G\rho r'$. Hence the resultant intensity is $\frac{4}{3}\pi G\rho O'O$ in a direction parallel to $O'O$. Thus the intensity inside such a hollow shell is constant in magnitude and direction, i.e., the field of force inside such a shell is uniform.

Illustrative Example

1. If a spherical mass of radius a and density d has an eccentric spherical cavity of radius b, show that the attraction at any two points in the cavity is the same in magnitude and direction and find its value if the distance of the centre of the cavity from that of the sphere is c.

In this case attraction at any point P (Fig. 7.16.1) in the cavity is $\frac{4}{3}\pi Gd.\ O'O = \frac{4}{3}\pi Gdc$ ($\because O'O = c$) in a direction parallel to $O'O$.

Thus the attraction inside the cavity of such a shell is constant in magnitude and direction. Hence the attraction at any two points inside the cavity remains the same.

7.17 Calculation of potential and intensity in some special cases

7.17.1 Uniform Circular Disc or Plate

(*a*) POTENTIAL AT A POINT P ON THE AXIS

We consider a uniform circular plate of radius a and thickness t. P is a point on the axis of the plate at a distance r from the centre O of the disc. We are to calculate the potential at P.

Let us divide the disc into a number of concentric annuli or rings, and suppose one such annulus has radius x and width dx.

The mass of this annulus is $2\pi x dx\ \rho t$, where ρ is the density of the material of the disc. Therefore, the potential at P due to the annulus under consideration, is given by

$$dV = -G\frac{2\pi x dx \rho t}{\sqrt{r^2+x^2}}.$$

Hence the potential at P due to the whole disc is given by

$$V = -G.2\pi\rho t\int_0^a \frac{x\,dx}{\sqrt{r^2+x^2}} = -G\pi\rho t\int_0^a \frac{d(r^2+x^2)}{\sqrt{r^2+x^2}}$$

$$= -2G\pi\rho t\Big[(r^2+x^2)^{1/2}\Big]_0^a = -2G\pi\rho t\Big[(r^2+a^2)^{1/2} - r\Big]$$

$$= 2G\pi\rho t\Big[r - (r^2+a^2)^{1/2}\Big].$$

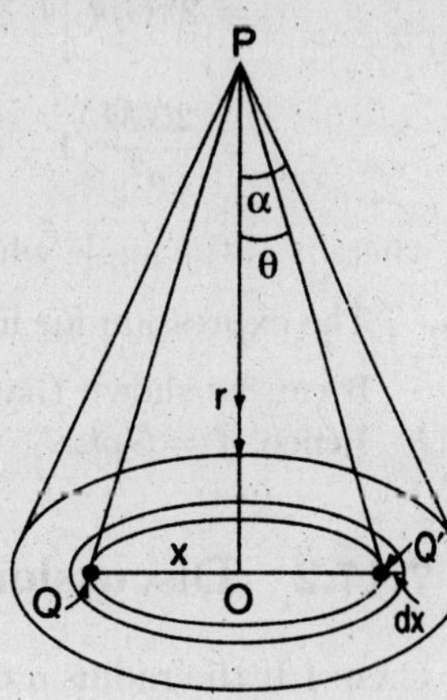

Fig. 7.17.1

Now M = mass of the disc $= \pi a^2 \rho t$. $\therefore \quad \rho t = \dfrac{M}{\pi a^2}$.

Hence, we may write

$$V = \frac{2GM}{a^2}\left[r - (r^2 + a^2)^{1/2}\right]$$

(b) ATTRACTION AT A POINT ON THE AXIS

In order to calculate the attraction at P, we consider two elements of masses dm at the two diametrically opposite positions Q and Q' as shown in the figure 7.17.1. Now the element dm at Q causes attraction on unit mass at P in the direction PQ and other mass at Q' causes attraction on the same unit mass of equal magnitude in the direction PQ'. These two attractive forces when resolved into two directions—one, along the axis PO and the other at right angles to it, we see from the figure that the axial components (i.e., components along PO) are additive and the perpendicular components cancel each other. Thus the attraction at P due to annulus in question is given by

$$df = G\Sigma\frac{dm\cos\theta}{PQ^2} = \frac{G\cos\theta}{PQ^2}.2\pi x\,dx\,\rho t \qquad [\Sigma dm = \text{mass of the ring of radius } x \text{ and thickness } dx = 2\pi x\,dx\,\rho t]$$

$$= 2\pi G\rho t\frac{r}{PQ^3}\cdot x\,dx \qquad \left[\because \cos\theta = \frac{r}{PQ}\right]$$

$$= 2\pi G\rho t\frac{r}{(r^2 + x^2)^{3/2}}\cdot x\,dx \qquad \left[\because PQ^3 = (r^2 + x^2)^{3/2}\right].$$

Therefore, the resultant attraction at P due to the whole disc directed along PO is given by

$$f = 2\pi G\rho tr\int_0^a \frac{x\,dx}{(r^2 + x^2)^{3/2}} = \pi G\rho tr\int_0^a \frac{d(r^2 + x^2)}{(r^2 + x^2)^{3/2}}$$

$$= -2\pi G\rho tr\left[(r^2 + x^2)^{-1/2}\right]_0^a = -2\pi G\rho tr\left[\frac{1}{(r^2 + a^2)^{1/2}} - \frac{1}{r}\right]$$

$$= 2\pi G\rho t\left[1 - \frac{r}{(r^2 + a^2)^{1/2}}\right] = \frac{2GM}{a^2}\left[1 - \frac{r}{(r^2 + a^2)^{1/2}}\right] \qquad \left[\because \rho = \frac{M}{\pi a^2 t}\right]$$

$$= \frac{2GM}{a^2}(1 - \cos\alpha),$$

where α is the angle which any radius of the disc subtends at P.

The expression for intensity may also be written in the form $f = 2\pi G\rho t(1 - \cos\alpha)$.

It can be shown that $2\pi(1 - \cos\alpha)$ is the solid angle ω subtended by the disc at the point P. Hence, $f = G\rho t\omega$.

7.17.2 Discussions

CASE I If the radius a of the disc becomes infinite, then $\alpha = \pi/2$; therefore,

$$f = \frac{2GM}{a^2} = \text{constant}.$$

Thus the attraction, in this case, is independent of the position of the point P from the disc. This result also holds good when the point P is very close to the disc unless the thickness of the plate is negligible.

CASE II If the point P is at a very large distance from the disc, then $\alpha \to 0$. $\therefore$ $f \to 0$.

CASE III A frustum of a right circular cone bounded by planes at right angles to its axis may be regarded as a pile of discs all subtending the same angle at the vertex. Hence the attraction due to such a frustum at the vertex of the cone of which the frustum is a part, is given by

$$f = 2\pi G\rho t(1 - \cos\alpha),$$

where t denotes the thickness of the frustum, and α the semivertical angle of the cone.

Alternatively, the attraction at P due to the circular disc may be obtained by differentiating the corresponding expression for potential. We have,

$$V = \frac{2GM}{a^2}\left[r - (r^2 + a^2)^{1/2}\right].$$

$$\therefore \quad f = -\frac{dV}{dr} = -\frac{2GM}{a^2}\cdot\frac{d}{dr}\left[r - (r^2 + a^2)^{1/2}\right] = -\frac{2GM}{a^2}\left[1 - \frac{r}{(r^2 + a^2)^{1/2}}\right].$$

Hence the intensity along PO is

$$\frac{2GM}{a^2}\left[1 - \frac{r}{(r^2 + a^2)^{1/2}}\right].$$

7.17.3 Uniform ring or Circular hoop—potential and attraction

We consider a uniform ring of radius a. P is a point on the axis of the ring at a distance r from the centre O of the ring. We are to calculate the potential and attraction at P.

Let us consider a small element of the ring of mass δm (Fig. 7.17.2) at Q.

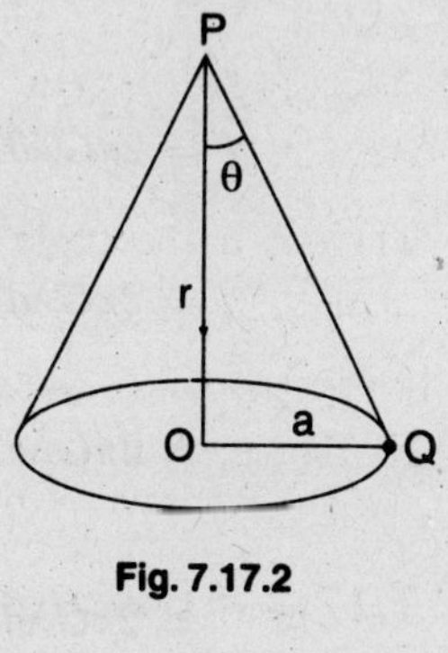

Fig. 7.17.2

Then the potential at P due to this element is given by

$$\delta V = -\frac{G\delta m}{PQ} = -\frac{G\delta m}{\sqrt{r^2 + a^2}}.$$

Hence the potential at P due to the whole ring is given by

$$V = -G\frac{\Sigma\delta m}{\sqrt{r^2 + a^2}} = -\frac{Gm}{\sqrt{r^2 + a^2}} \quad [\Sigma\delta m = m = \text{ mass of the ring}]$$

The attraction at P due to the ring is along PO and is given by

$$f = G\Sigma\frac{\delta m\cos\theta}{PQ^2} = \frac{Gmr}{PQ^3} = \frac{Gmr}{(r^2 + a^2)^{3/2}} \quad \left[\because \cos\theta = \frac{r}{PQ}\right].$$

Illustrative Examples

1. Prove that the potential of a circular disc of uniform surface density σ and radius a at a point on its circumference is $-4G\sigma a$.

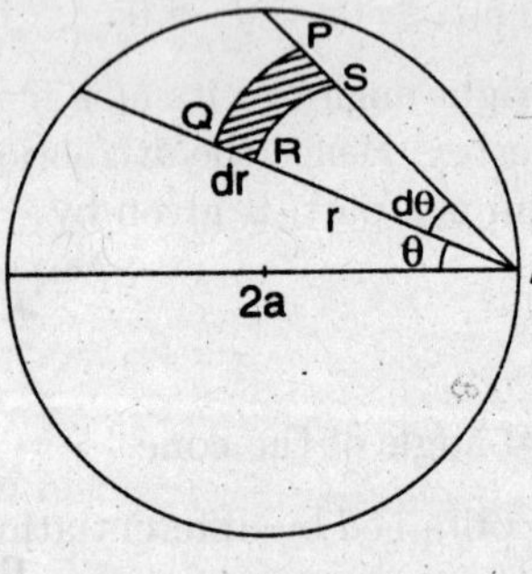

Fig. 1

The potential at any point A on the circumference of the disc due to the element $PQRS$ (Fig. 1) is given by

$$\delta V = -G\frac{(\text{mass of element } PQRS)}{r} = -G\frac{rd\theta dr\sigma}{r} = -G\sigma d\theta dr.$$

Hence the potential at A due to the circular disc is given by

$$V = -G\sigma\int_{-\pi/2}^{\pi/2}\int_0^{2a\cos\theta} d\theta dr = -G\sigma\int_{-\pi/2}^{\pi/2} 2a\cos\theta d\theta$$

$$= -2G\sigma a\Big[\sin\theta\Big]_{-\pi/2}^{\pi/2} = -4G\sigma a.$$

2. Prove that, if the surface density of circular disc at any point distant x from the centre is λx, then the attraction at a point on the axis at a distance d' is

$$\pi G\lambda d'\left\{\log\frac{\sqrt{(a^2+d'^2)}+a}{\sqrt{(a^2+d'^2)}-a} - \frac{2a}{\sqrt{(a^2+d'^2)}}\right\}.$$

We know that the resultant attraction at P due to a circular disc directed along PO (see Art. 7.17.1b) is given by

$$f = 2\pi G\rho tr\int_0^a \frac{x\,dx}{(r^2+x^2)^{3/2}}.$$

Here $\rho t = \lambda x =$ surface density

$r = d'$.

Hence the above expression may be written as

$$f = 2\pi G d'\lambda\int_0^a \frac{x^2dx}{(d'^2+x^2)^{3/2}}.$$

Let $x = d'\tan\theta$. $\therefore$ on differentiation we get

$$dx = d'\sec^2\theta d\theta.$$

$$\therefore\ f = 2\pi G\lambda d'\int\frac{d'^3\tan^2\theta.\sec^2\theta\,d\theta}{d'^3\sec^3\theta} = 2\pi G\lambda d'\int\frac{\tan^2\theta}{\sec\theta}d\theta = 2\pi G\lambda d'\int\frac{\sin^2\theta}{\cos\theta}d\theta$$

$$= 2\pi G\lambda d'\int\frac{(1-\cos^2\theta)}{\cos\theta}d\theta = 2\pi G\lambda d'\int(\sec\theta-\cos\theta)d\theta$$

$$= 2\pi G\lambda d'\left[\int\sec\theta d\theta - \int\cos\theta d\theta\right] = 2\pi G\lambda d'\int[\log(\sec\theta+\tan\theta)-\sin\theta]$$

$$= 2\pi G\lambda d'\left[\log\left(\frac{\sqrt{x^2+d'^2}}{d'}+\frac{x}{d'}\right) - \frac{x}{\sqrt{x^2+d'^2}}\right]_0^a$$

$$\left[\because\ \tan\theta = \frac{x}{d'},\quad \therefore\ \sec\theta = \frac{\sqrt{x^2+d'^2}}{d'} \text{ and } \sin\theta = \frac{\tan\theta}{\sec\theta} = \frac{x}{\sqrt{x^2+d'^2}}\right]$$

$$= 2\pi G\lambda d'\left[\log\left(\frac{\sqrt{a^2+d'^2}}{d'}+\frac{a}{d'}\right)-\frac{a}{\sqrt{a^2+d'^2}}\right]$$

$$= 2\pi G\lambda d'\left[\log\frac{\sqrt{a^2+d'^2}+a}{d'}-\frac{a}{\sqrt{a^2+d'^2}}\right]$$

$$= \pi G\lambda d'\left[2\log\frac{\sqrt{a^2+d'^2}+a}{d'}-\frac{2a}{\sqrt{a^2+d'^2}}\right]$$

$$= \pi G\lambda d'\left[\log\frac{(\sqrt{a^2+d'^2}+a)^2}{d'^2}-\frac{2a}{\sqrt{a^2+d'^2}}\right]$$

$$= \pi G\lambda d'\left[\log\frac{(\sqrt{a^2+d'^2}+a)^2}{(\sqrt{a^2+d'^2}+a)(\sqrt{a^2+d'^2}-a)}-\frac{2a}{\sqrt{a^2+d'^2}}\right]$$

$$= \pi G\lambda d'\left[\log\frac{\sqrt{a^2+d'^2}+a}{\sqrt{a^2+d'^2}-a}-\frac{2a}{\sqrt{a^2+d'^2}}\right].$$

7.17.4 Homogeneous Solid Cylinder

(a) POTENTIAL AT A POINT ON THE AXIS OF THE CYLINDER

Let ρ be the density of the cylinder and a be its radius (Fig. 7.17.3). Let us divide the cylinder into a number of co-axial discs. Let one such disc EF of thickness dr lie at a distance r from the point P at which the potential is being considered. Hence the potential at P due to the disc EF is given by

$$dV = 2\pi G\rho\left[r-\sqrt{r^2+a^2}\right]dr$$

[from the expression for potential due to a disc].

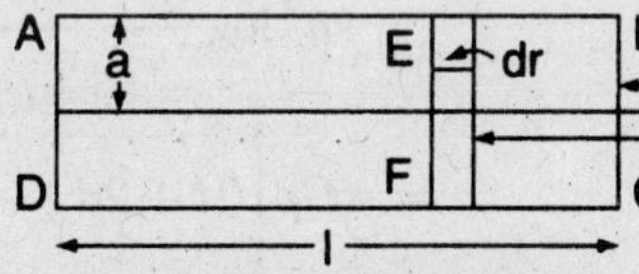

Fig. 7.17.3

Therefore, the potential at P due to the entire cylinder is given by

$$V = 2\pi G\rho\int_{r=d'}^{r=d'+l}\left(r-\sqrt{r^2+a^2}\right)dr = 2\pi G\rho\left[\left(\frac{r^2}{2}\right)_d^{d'+l}-\int_{d'}^{d'+l}\sqrt{r^2+a^2}dr\right].$$

Integration of the second term may be done in the following way.

Let

$$I = \int\sqrt{r^2+a^2}dr = r\sqrt{r^2+a^2}-\int\frac{r^2dr}{\sqrt{r^2+a^2}}. \qquad (7.17.1)$$

[integrating by parts]

Again

$$I = \int\frac{r^2+a^2}{\sqrt{r^2+a^2}}dr = \int\frac{r^2dr}{\sqrt{r^2+a^2}}+a^2\int\frac{dr}{\sqrt{r^2+a^2}}. \qquad (7.17.2)$$

From (7.17.1) and (7.17.2), we have

$$I = \frac{1}{2}\left[r\sqrt{r^2+a^2} + a^2\int \frac{dr}{\sqrt{r^2+a^2}}\right].$$

To evaluate the second integral of the above expression, let us put

$$r = a\tan\theta \quad \text{or,} \quad dr = a\sec^2\theta d\theta$$

$$\therefore \quad a^2\int \frac{dr}{\sqrt{r^2+a^2}} = a^2\int \sec\theta d\theta = a^2\log(\sec\theta + \tan\theta) = a^2\log\frac{r+\sqrt{r^2+a^2}}{a},$$

whence

$$I = \frac{1}{2}\left[r\sqrt{r^2+a^2} + a^2\log\frac{r+\sqrt{r^2+a^2}}{a}\right].$$

$$\therefore \quad V = 2\pi G\rho\left[\frac{r^2}{2} - \frac{1}{2}r\sqrt{r^2+a^2} - \frac{a^2}{2}\log\frac{r+\sqrt{r^2+a^2}}{a}\right]_{d'}^{d'+l}$$

$$= \pi G\rho\left[r^2 - r\sqrt{r^2+a^2} - a^2\log\frac{r+\sqrt{r^2+a^2}}{a}\right]_{d'}^{d'+l}$$

$$= \pi G\rho\Big[l(l+2d') - (d'+l)\sqrt{(d'+l)^2+a^2} + d'\sqrt{d'^2+a^2}$$

$$-a^2\log\frac{(d'+l)+\sqrt{(d'+l)^2+a^2}}{a} + a^2\log\frac{d'+\sqrt{d'^2+a^2}}{a}\Big]$$

$$= \pi G\rho\Big[l(l+2d') - (d'+l)\sqrt{(d'+l)^2+a^2}$$

$$+d'\sqrt{d'^2+a^2} - a^2\log\frac{(d'+l)+\sqrt{(d'+l)^2+a^2}}{d'+\sqrt{d'^2+a^2}}\Big]$$

$$= \pi G\rho\left[l(l+2d') - (d'+l)PA + d'.PB - a^2\log\frac{(d'+l)+PA}{d'+PB}\right]$$

$$[\because (d'+l)^2 + a^2 = PA^2 \text{ and } d'^2 + a^2 = PB^2]$$

$$= \frac{GM}{la^2}\left[l(l+2d') - (d'+l)PA + d'.PB - a^2\log\frac{(d'+l)+PA}{d'+PB}\right],$$

where $M = \pi a^2 l\rho$ = mass of the cylinder.

(b) ATTRACTION AT A POINT ON THE AXIS OF THE CYLINDER

Attraction at a point P due to the elementary disc EF (Fig. 7.17.3) is given by

$$df = 2\pi G\rho\left[1 - \frac{r}{\sqrt{r^2+a^2}}\right]dr \qquad \text{[from the expression for intensity due to a disc]}$$

Hence, the resultant attraction at P due to the entire cylinder is given by

$$f = 2\pi G\rho \int_{d'}^{d'+l} \left[1 - \frac{r}{\sqrt{r^2+a^2}}\right] dr = 2\pi G\rho \left[l - \left\{(a^2+r^2)^{1/2}\right\}_{d'}^{d'+l}\right]$$

$$= 2\pi G\rho \left[l - \left\{a^2 + (d'+l)^2\right\}^{1/2} + (a^2 + d'^2)^{1/2}\right] = 2\pi G\rho [l - PA + PB]$$

$$= \frac{2GM}{la^2}[l - PA + PB].$$

If l be infinite, $l - PA = l - (l + d') = -d'$.

Consequently the attraction of an infinite cylinder is given by

$$f = \frac{2GM}{la^2}[PB - d']$$

7.17.5 Thin uniform straight rod

(*a*) POTENTIAL DUE TO A THIN UNIFORM ROD AT ANY POINT

We consider any point P at which the potential due to the rod AB is to be considered (Fig. 7.17.4). Let m be the mass of the rod. Then mass per unit length of the rod is m/l, where l is the length of the rod.

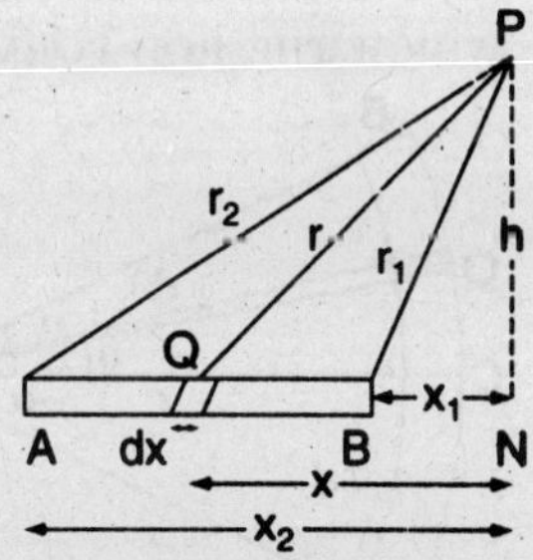

Fig. 7.17.4

We consider an element dx of the rod at Q the mass of which is $(m/l)dx$. Hence the potential at P due to this element is

$$dV = -\frac{G\frac{m}{l}dx}{r} = -\frac{Gmdx}{l\sqrt{h^2+x^2}}.$$

Therefore, the resultant potential due to the entire rod is given by

$$V = -\frac{Gm}{l}\int_{x=x_1}^{x=x_2} \frac{dx}{\sqrt{h^2+x^2}}$$

where $NQ = x$, $NB = x_1$, $NA = x_2$, $PN =$ perpendicular on AB produced $= h$

Let us put, $x = h\tan\theta$. $\therefore$ $dx = h\sec^2\theta d\theta$.

$$\therefore \quad V = -\frac{Gm}{l}\int \frac{h\sec^2\theta d\theta}{h\sec\theta} = -\frac{Gm}{l}\int \sec\theta d\theta = -\frac{Gm}{l}[\log(\sec\theta + \tan\theta)]$$

$$= -\frac{Gm}{l}\left[\log\frac{\sqrt{h^2+x^2}+x}{h}\right]_{x=x_1}^{x=x_2}$$

$$\left[\because \log(\sec\theta + \tan\theta) = \log\left(\sqrt{1+\frac{x^2}{h^2}} + \frac{x}{h}\right) = \log\frac{\sqrt{h^2+x^2}+x}{h}\right]$$

$$= -\frac{Gm}{l}\left[\log\frac{\sqrt{h^2+x_2^2}+x_2}{\sqrt{h^2+x_1^2}+x_1}\right] = -\frac{Gm}{l}\left[\log\frac{r_2+x_2}{r_1+x_1}\right],$$

where $r_1 = PB$ and $r_2 = PA$.

Now from the figure,

$$r_2^2 - x_2^2 = h^2 = r_1^2 - x_1^2 \quad \text{or,} \quad \frac{r_2 + x_2}{r_1 + x_1} = \frac{r_1 - x_1}{r_2 - x_2} = \frac{r_2 + x_2 + r_1 - x_1}{r_1 + x_1 + r_2 - x_2} = \frac{r_1 + r_2 + l}{r_1 + r_2 - l}.$$

$$\therefore \quad V = -\frac{Gm}{l} \log \frac{r_1 + r_2 + l}{r_1 + r_2 - l}.$$

If we draw an ellipse with A and B as foci then for all points on the ellipse the sum of the focal distances will be the same. Here for the point P on the surface of the ellipse the sum of the focal distances is $(r_1 + r_2)$. Thus for all points over the surface of this ellipse the potential is constant. The potential is also constant over the ellipsoid obtained by revolving this ellipse about the line joining the foci, i.e., about AB.

7.17.6 Uniform Rod bent into the form of a circular arc

(*a*) POTENTIAL AT THE POINT P WHICH IS THE CENTRE OF CURVATURE OF THE CIRCLE OF WHICH THE ROD FORMS A PART

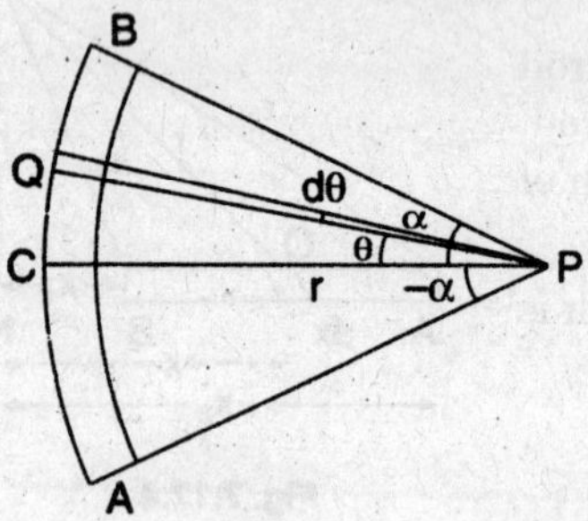

Fig. 7.17.5

Let ρ be the mass per unit length of the bent rod AB, C be the middle point of the rod (Fig. 7.17.5). Let $\angle CPB = \alpha$; we consider an element at Q between θ and $\theta + d\theta$. The potential at P due to this element is given by

$$dV = -G\frac{rd\theta\rho}{r} = -G\rho d\theta.$$

Hence the resultant potential at P due to the whole rod is given by

$$V = -G\rho \int_{-\alpha}^{+\alpha} d\theta = -G\rho.2\alpha.$$

Now, mass of the rod $M = 2\alpha r\rho$.

$$\therefore \quad 2\alpha\rho = \frac{M}{r}. \quad \therefore \quad V = -\frac{GM}{r}.$$

(*b*) ATTRACTION AT P

The attraction at P due to the element at Q is $\dfrac{Grd\theta\rho}{r^2} = \dfrac{G\rho d\theta}{r}$.

Now the component of this attraction along PC is given by

$$df = \frac{G\rho}{r} \cos\theta d\theta.$$

Hence the resultant attraction due to the entire rod is given by

$$f = \frac{G\rho}{r} \int_{-\alpha}^{+\alpha} \cos\theta d\theta = \frac{G\rho}{r}.2\sin\alpha.$$

Now

$$\rho = \frac{M}{2r\alpha}. \quad \therefore \quad f = \frac{GM}{r^2}\frac{\sin\alpha}{\alpha}.$$

7.17.7 Attraction at a point P due a thin uniform straight Rod

From the figure 7.17.6, it is clear that the attraction at P due to the element CD of a thin uniform straight rod AB is

$$\frac{Gm}{l}\frac{dx}{r^2},$$

where m/l denotes mass per unit length of the rod.

From the figure it follows that

$$x = h\tan\theta.$$

$$\therefore \quad dx = h\sec^2\theta d\theta = h\frac{r^2}{h^2}d\theta = \frac{r^2}{h}d\theta.$$

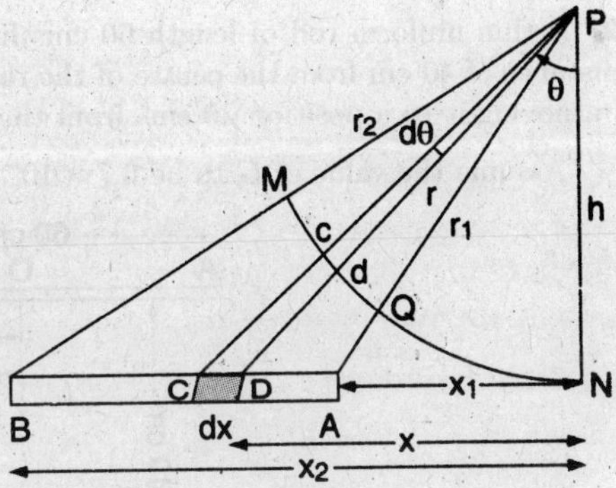

Fig. 7.17.6

$\therefore$ The attraction due to the element CD may be written as

$$\frac{Gm}{l}\cdot\frac{r^2}{h}d\theta\cdot\frac{1}{r^2} = \frac{Gmh}{h^2 l}d\theta.$$

Now $hd\theta = cd$ [$\because$ the arc MN is drawn with P as centre and h as radius]

Hence the expression for attraction becomes

$$\frac{Gm}{h^2 l}cd = \frac{Gm}{l}\cdot\frac{cd}{h^2}.$$

Thus we see that the attraction at P due to CD is the same as that at P due to *cd*, provided the mass per unit length in both the cases is the same.

Now *cd* is an element on the arc MQ drawn with centre P and radius h. It follows, therefore, that the attraction of the given rod at P is the same in all respects as that of a similar rod lying along the arc MQ. This attraction is in a direction which bisects the angle $\angle APB$ and is given by

$$f = \frac{2G\rho}{h}\sin\frac{APB}{2} \quad \text{[from the expression for intensity due to a bent rod].} \tag{7.17.3}$$

Now

$$\frac{m}{l} = \rho. \quad \therefore \quad f = \frac{2Gm}{hl}\sin\frac{APB}{2}. \tag{7.17.4}$$

Illustrative Example

1. Find the attraction produced by a long thin uniform rod of mass ρ per unit length at a distance r from the rod, if r is small compared with the length of the rod and the distance of each end of the rod is large compared with r.

In this case N (Fig. 7.17.6) lies between A and B at a great distance from each end. Then

$$\angle APB = \pi; \quad h = r \text{ and } \frac{m}{l} = \rho.$$

Hence from (7.17.3) the attraction which is nearly perpendicular to the rod is given by

$$f = \frac{2G\rho}{r}\sin\frac{\pi}{2} = \frac{2G\rho}{r}.$$

2. A thin uniform rod of length 60 cms has a mass of 60 gms. A point mass of 2 gms is placed at a distance of 40 cm from the centre of the rod along the direction of its length. Calculate the work done in moving it to a position 40 cms from the centre of the rod along a line perpendicular to its length.

Assume the value of G to be 6.7×10^{-8} c.g.s. units. **[C.U. (Hons.) 1976]**

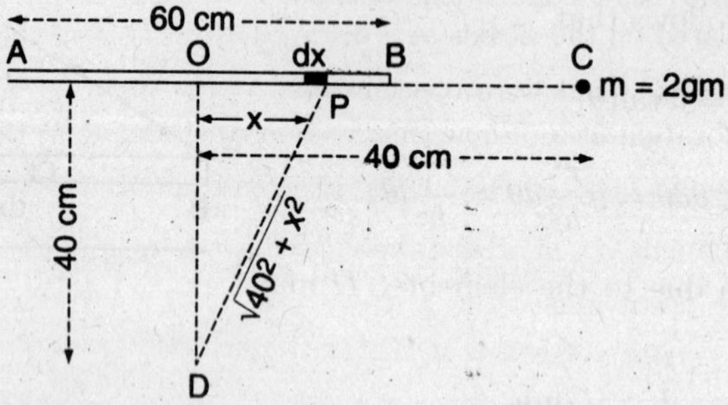

We consider an element dx of the rod at P at a distance x from the centre O of the rod. If ρ be the mass per unit length of the rod, then the potential at C due to the elementary mass at P is $-\dfrac{G\rho dx}{(40-x)}$ and also the potential at D due to the elementary mass at P is $-\dfrac{G\rho dx}{\sqrt{40^2+x^2}}$.

$\therefore$ Potential at C due to the whole rod AB

$$= -\int_{+30}^{-30} \frac{G\rho dx}{(40-x)}$$

and also the potential at D due to the whole rod AB

$$= -\int_{+30}^{-30} \frac{G\rho dx}{\sqrt{40^2+x^2}}.$$

$\therefore$ the potential difference between the points C and D

$$= -\int_{+30}^{-30} \frac{G\rho dx}{\sqrt{40^2+x^2}} + \int_{+30}^{-30} \frac{G\rho dx}{(40-x)}.$$

Hence the work done in moving a point mass of 2 gms from C to D

$$= 2\left[-\int_{+30}^{-30} \frac{G\rho dx}{\sqrt{40^2+x^2}} + \int_{+30}^{-30} \frac{G\rho dx}{(40-x)}\right]$$

$$= 2\left[G\rho \int_{-30}^{30} \frac{dx}{\sqrt{40^2+x^2}} + G\rho \int_{+30}^{-30} \frac{-d(40-x)}{(40-x)}\right]$$

$$= 2G\rho\left[\left(\log_e \left|x+\sqrt{40^2+x^2}\right|\right)_{-30}^{30} + \left(\log_e |40-x|\right)_{-30}^{+30}\right]$$

$$= 2G\rho[\log_e 80 - \log_e 20 + \log_e 10 - \log_e 70] = 2G\rho \log_e \frac{80 \times 10}{20 \times 70}$$

$$= 2G \times 1 \log_e\left(\frac{4}{7}\right) \quad \left[\because \rho = \frac{60}{60} = 1 \text{ gm/c.c.}\right]$$

$$= 2 \times 6.7 \times 10^{-8} \times (-0.56) \quad \left[\because \log_e \frac{4}{7} = -0.56\right]$$

$$= -7.5 \times 10^{-8} \text{ erg.}$$

$\therefore$ Work done by the field to move the mass m from C to $D = 7.5 \times 10^{-8}$ erg.

3. A uniform rod of mass M lies on the Z-axis between $z = 0$ and $z = l$. (i) Find the gravitational force on a point mass m placed on the Z-axis at a distance $a (a > l)$ from the origin.

(ii) If the whole mass of the rod were concentrated at the point z_0 such that the force on m were unchanged, find the location of z_0. Show that z_0 is located almost at the centre of the rod when $a \gg l$. **[C.U. (Hons.) 1991]**

(i) Considering an element dz at a distance z from the origin, its mass $= \frac{M}{l} dz$ and the gravitational force on the point mass m is

$$dF = \frac{G\frac{M}{l}dzm}{(a-z)^2},$$

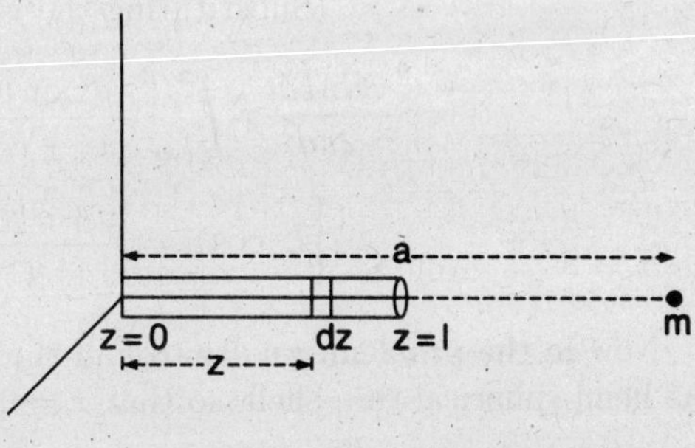

so that the total gravitational force is given by

$$F = \frac{GMm}{l}\int_0^l \frac{dz}{(a-z)^2} = -\frac{GMm}{l}\int_a^{(a-l)} \frac{dy}{y^2}$$

$$= \frac{GMm}{l}\left[\frac{1}{(a-l)} - \frac{1}{a}\right] = \frac{GMm}{a(a-l)}.$$

[Putting $y = a - z$, $dy = -dz$ and at $z = 0$, $y = a$ and at $z = l$, $y = (a - l)$]

(ii) If the whole mass of the rod be supposed to be concentrated at z_0, the force of attraction

$$= \frac{GMm}{(a-z_0)^2}.$$

Now according to the condition of the problem,

$$\frac{GMm}{(a-z_0)^2} = \frac{GMm}{a(a-l)}, \quad \text{or,} \quad (a-z_0)^2 = a(a-l),$$

$$\text{or,} \quad z_0^2 - 2az_0 + al = 0, \quad \text{or,} \quad z_0 = a \pm \sqrt{a^2 - al}.$$

Considering the negative sign

$$z_0 = a - \sqrt{a^2\left(1 - \frac{l}{a}\right)} = a - a\left(1 - \frac{l}{a}\right)^{1/2}$$

$$= a - a\left(1 - \frac{1}{2}\frac{l}{a}\right) \quad \left[\because a \gg l, \text{ higher powers of } l/a \text{ may be neglected}\right]$$

$$= \frac{l}{2},$$

so that z_0 is almost at the centre of the rod.

4. A hemi-sphere of mass M and radius a has a particle of mass m located at its centre. Find the force of attraction on it if the hemi-sphere is (i) a thin shell, (ii) a solid. **[C.U. (Hons.) 1989]**

(i) Let us consider a small element of surface ds at A [having co-ordinates (r, θ, ϕ)], on the thin shell, so that we may write $ds = a^2 \sin\theta d\theta d\phi$, and the total mass of the hemi-spherical thin shell is $M = 2\pi a^2 \sigma$, σ being the surface density of mass.

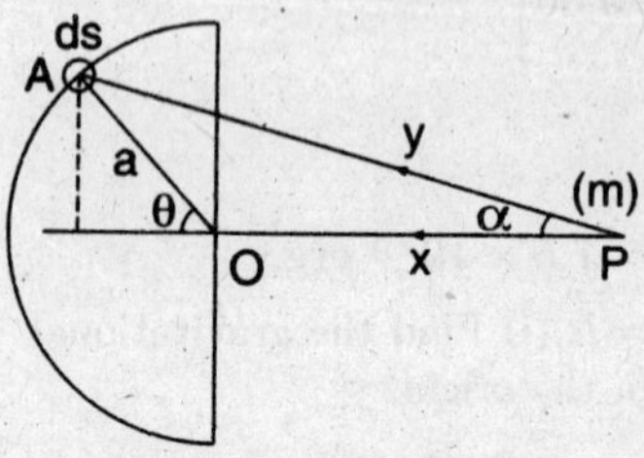

$\therefore \quad \sigma = \frac{M}{2\pi a^2}$. Let us first calculate the force of attraction due to a mass "m" placed at a point P at a distance x away from the centre of the hemi-spherical shell (for the sake of convenience).

The force of attraction along PO exerted by the mass m at P on the element ds at A is

$$\vec{F} = -\int_s \frac{Gm\sigma ds \cos\alpha}{y^2} = -\frac{GMm}{2\pi a^2} \int_{\theta=0}^{\theta=\pi/2} \int_{\phi=0}^{\phi=\pi} \frac{a^2 \sin\theta d\theta d\phi}{y^2} \cos\alpha$$

(Substituting the values of σ and ds)

$$= -\frac{GMm}{2\pi a^2}\pi \int_{\theta=0}^{\pi/2} \frac{a^2 \sin\theta (x + a\cos\theta) d\theta}{(x^2 + a^2 + 2ax\cos\theta)^{3/2}}$$

$$\left[\because \cos\alpha = \frac{x + a\cos\theta}{y} \text{ and } y^2 = x^2 + a^2 + 2ax\cos\theta\right]$$

Now in the problem, we are to find the attraction due to mass "m" placed at the centre of the hemi-spherical thin shell, so that $x = 0$ and we therefore obtain

$$\vec{F} = -\frac{GMm}{2a^2} \int_0^{\pi/2} \frac{a^2 \sin\theta \cdot a\cos\theta d\theta}{a^3} = -\frac{GMm}{2a^2} \int_{\theta=0}^{\pi/2} \frac{1}{2} \sin 2\theta d\theta$$

$$= -\frac{GMm}{2a^2}\frac{1}{2}\left[-\frac{\cos 2\theta}{2}\right]_0^{\pi/2} = -\frac{GMm}{4a^2}\left[\frac{\cos 2\theta}{2}\right]_{\pi/2}^{0} = -\frac{GMm}{4a^2}\left[\frac{1-(-1)}{2}\right] = -\frac{GMm}{4a^2}.$$

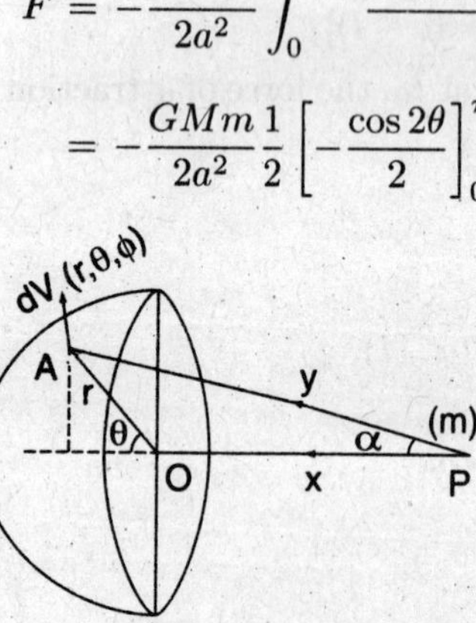

(ii) Let ρ be the volume density of the solid hemi-sphere,— then the mass M of the hemi-sphere is

$$M = \frac{2}{3}\pi a^3 \rho, \quad \text{or,} \quad \rho = \frac{3M}{2\pi a^3}.$$

Let us, for the sake of convenience, first calculate the force of attraction due to a mass "m" placed at P at a distance x from the centre of the hemi-sphere. Let us consider an element of volume dV at A having co-ordinates (r, θ, ϕ), so that $dV = r^2 \sin\theta d\theta d\phi dr$, and the force of attraction along PA is $-\int_V \frac{Gm\rho dV}{y^2}$ and its component along PO is $-\int_V \frac{Gm\rho dV}{y^2}\cos\alpha$

$$= -\frac{3}{2}\frac{GMm}{\pi a^3} \int_{\theta=0}^{\pi/2} \int_{\phi=0}^{\pi} \int_{r=0}^{a} \frac{r^2 \sin\theta d\theta d\phi dr}{y^2} \cos\alpha \quad \text{(substituting the values of } \rho \text{ and } dV)$$

$$= -\frac{3}{2}\frac{GMm}{\pi a^3} \int_{\theta=0}^{\pi/2} \int_{\phi=0}^{\pi} \int_{r=0}^{a} \frac{r^2 \sin\theta d\theta d\phi dr (x + r\cos\theta)}{(x^2 + r^2 + 2xr\cos\theta)^{3/2}} \quad \left[\because \cos\alpha = \frac{x + r\cos\theta}{y} \text{ and } y^2 = x^2 + r^2 + 2xr\cos\theta\right]$$

Now since the force is to be evaluated with the mass "m" at the centre of the hemi-sphere, x should be put equal to zero, so that the expression for attraction becomes

$$\vec{F} = -\frac{3}{2}\frac{GMm}{\pi a^3}.\pi \int_{\theta=0}^{\pi/2}\int_{r=0}^{a} \frac{r^3 \sin\theta\cos\theta d\theta dr}{r^3} = -\frac{3}{2}\frac{GMm}{a^3}\int_0^a dr \int_0^{\pi/2} \frac{\sin 2\theta}{2} d\theta$$

$$= -\frac{3}{2}\frac{GMm}{a^3} a \frac{1}{2}\left[\frac{\cos 2\theta}{2}\right]_{\pi/2}^{0} = -\frac{3}{4}\frac{GMm}{a^2}$$

7.18 Self-attracting System

Work done by the mutual attractive forces of the particles of a self-attracting system whilst the particles are brought from an infinite distance to the positions they occupy in the given system, is defined as the gravitational self-energy of the system.

Let the component particles of masses m_1, m_2, m_3, etc. be at A_1, A_2, A_3, etc., in the given system A (say) (Fig. 7.18.1). Let the distance between the masses m_1 and m_2 be r_{12}, and in general let the distance between m_s and m_t be r_{st}.

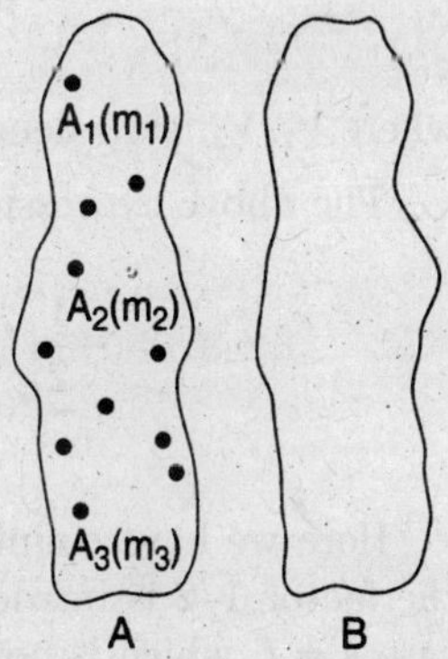

Fig. 7.18.1

We first bring m_1 from infinity to its assigned position A_1; the work done in this process is zero, for there are no particles of the system near enough to exert attraction on it. Next m_2 is brought from infinity to its position A_2; the work done on it = the potential of m_1 at $A_2 \times m_2$.

$$= -\frac{Gm_1}{r_{12}} \times m_2$$

—which is the gravitational P.E. of the pair of particles of masses m_1 and m_2—negative sign being used because the gravitational force is attractive.

Next, we bring m_3 to its position A_3, and the work done on it

$$= m_3 \times \text{ potential of } m_1 \text{ and } m_2 \text{ at } A_3$$

$$= -\left[\frac{Gm_1m_3}{r_{13}} + \frac{Gm_2m_3}{r_{23}}\right].$$

Next we bring m_4 to its position A_4, and hence the work done on it

$$= m_4 \times \text{potential of } m_1,\ m_2 \text{ and } m_3 \text{ at } A_4$$

$$= -\left[G\frac{m_1m_4}{r_{14}} + G\frac{m_2m_4}{r_{24}} + G\frac{m_3m_4}{r_{34}}\right]$$

and so on for the other particles of the system.

Hence the total work done in collecting all the particles from rest at infinite distances to their positions in the configuration A, i.e., the gravitational self energy,

$U =$ the sum of the above expressions for work or the work done in bringing m_1, m_2, m_3, m_4, etc., to their assigned positions A_1, A_2, A_3, etc.

$$= -\left\{G\frac{m_1 m_2}{r_{12}} + G\left[\frac{m_1}{r_{13}} + \frac{m_2}{r_{23}}\right] m_3 + G\left[\frac{m_1}{r_{14}} + \frac{m_2}{r_{24}} + \frac{m_3}{r_{34}}\right] m_4 + \cdots\right\}$$

$$= -\left[\frac{Gm_1 m_2}{r_{12}} + G\frac{m_1 m_3}{r_{13}} + \frac{Gm_2 m_3}{r_{23}} + G\frac{m_1 m_4}{r_{14}} + \cdots\right]$$

$$= -G \sum_{\substack{\text{allpairs}\\ s\neq t}} \frac{m_s m_t}{r_{st}}$$

—which is the P.E. of n particles due to their mutual gravitational attraction, being equal to the sum of the P.E. of all pairs of particles.

When the particles have all been brought to their positions in the configuration A, then

$$V_1 = -\left[\frac{Gm_2}{r_{12}} + \frac{Gm_3}{r_{13}} + \frac{Gm_4}{r_{14}} + \cdots\right]$$

$$V_2 = -\left[\frac{Gm_1}{r_{12}} + \frac{Gm_3}{r_{23}} + \frac{Gm_4}{r_{24}} + \cdots\right] \quad \text{and so on,}$$

where V_1, V_2, etc., denote the potentials of the system at A_1, A_2, A_3, etc.

The above expression for the gravitational self energy of n particles may be written as

$$U = -\frac{1}{2}G\sum_{s=1}^{s=n}\sum_{\substack{t=1\\ s\neq t}}^{t=n} \frac{m_s m_t}{r_{st}} = \frac{1}{2}(V_1 m_1 + V_2 m_2 + V_3 m_3 + \cdots) = \frac{1}{2}\Sigma V m.$$

Here we have counted each pair twice (once, for example, as 3, 4 and once as 4, 3); hence the factor 1/2 is introduced for correcting the over-counting. Further we have to exclude the case $s = t$, which is not a pair at all.

The above work done in bringing the particles from infinity to the configuration A may also be written as

$$\frac{1}{2}\Sigma V m = \frac{1}{2}\int V dm,$$

where V is the potential of the body A at any element dm of itself, and the integration is taken throughout the configuration A.

Conversely, the work done by the mutual attractive forces of the system as its particles are scattered to infinite distances from one another is $-\frac{1}{2}\int V dm.$

We can hence find the work done as the body changes from one configuration A to another configuration B.

This work = work done in the changing of its form from A to the state of diffusion at infinity + work done in collecting all the particles in a state of diffusion at infinity to the configuration B.

$$= -\frac{1}{2}\int V dm + \frac{1}{2}V' dm',$$

where the first integral is taken throughout the system A and the second integral is similarly taken throughout the second system B.

Let us now consider a system of n particles, each of same mass m, and separated from each other by the same average distance r. The gravitational self-energy is then

$$U = -\frac{1}{2}G\sum_{s=1}^{n}\sum_{\substack{t=1\\t\neq s}}^{n}\left(\frac{m^2}{r}\right)_{st} \qquad \text{(Putting } m_s = m_t = m \text{ and } r_{st} = r\text{)}$$

In fact on the right hand side, s comes n times, while t comes $(n-1)$ times. So we have

$$U = -\frac{1}{2}Gn(n-1)\frac{m^2}{r}.$$

Illustrative Examples

1. A self-attracting sphere of uniform density ρ and radius a changes to one of uniform density and radius b. Show that the work done by its mutual attractive forces is $\frac{3}{5}GM^2\left(\frac{1}{b}-\frac{1}{a}\right)$, where M is the mass of the sphere.

Here, the work done by the mutual attractive forces of the system as the particles which constitute the sphere of radius a are scattered to infinite distances from one another, is

$$W_1 = -\frac{1}{2}\int V\,dm.$$

Now, considering a point within the sphere at a distance x, the potential at this point within the sphere is

$$V = \frac{2}{3}G\pi\rho(3a^2 - x^2).$$

Let us now consider at this point a spherical shell of radius x and thickness dx. Then we have,

$$dm = 4\pi x^2 dx\rho. \quad \therefore \quad V\,dm = \frac{8}{3}G\pi^2\rho^2(3a^2 - x^2)x^2 dx.$$

$$\therefore \quad \int V\,dm = \frac{8}{3}G\pi^2\rho^2\int_0^a(3a^2 - x^2)x^2 dx = \frac{8}{3}G\pi^2\rho^2\left[3a^2\int_0^a x^2 dx - \int_0^a x^4 dx\right]$$

$$= \frac{8}{3}G\pi^2\rho^2\left(a^5 - \frac{a^5}{5}\right) = \frac{32}{15}G\pi^2\rho^2 a^5 = \frac{32}{15}G\pi^2\frac{9M^2}{16\pi^2 a^6}\cdot a^5$$

$$= \frac{6}{5}G\frac{M^2}{a} \quad \left[\because \rho^2 = \frac{9M^2}{16\pi^2 a^6}\right].$$

$$\therefore \quad W_1 = -\frac{1}{2}\int V\,dm = -\frac{3}{5}G\frac{M^2}{a}.$$

Similarly, if W_2 be the work done in bringing the particles from infinity to the second configuration to constitute a sphere of same mass but radius b, we have

$$W_2 = \frac{1}{2}\int V'dm' = \frac{3}{5}G\frac{M^2}{b}.$$

Therefore, the total work done is given by

$$W = W_1 + W_2 = \frac{3}{5}GM^2\left(\frac{1}{b} - \frac{1}{a}\right).$$

2. Show that if M be the mass of a sphere of radius a, the loss of gravitational potential energy in assembling the particles from a state of diffusion at an infinite distance is $\frac{3}{5}G\frac{M^2}{a}$, where G is the gravitational constant.

We suppose that the sphere is formed gradually by attracting from infinity one thin shell at a time. When the sphere has grown to a radius r, the potential at its surface is

$$-G\cdot\frac{\frac{4}{3}\pi r^3\rho}{r} = -\frac{4}{3}G\pi r^2\rho.$$

Now the work done in bringing an element of mass m in the form of a shell of radius r and thickness dr from infinity to the surface of this sphere will, therefore, be

$$-\frac{4}{3}G\pi\rho r^2 \times 4\pi r^2 dr\rho = -\frac{16}{3}G\pi^2\rho^2 r^4 dr.$$

Thus the total work done in forming the sphere by taking diffused masses from infinity, is given by

$$W = -\frac{16}{3}G\pi^2\rho^2\int_0^a r^4 dr = -\frac{16}{15}.G\pi^2\rho^2 a^5 = -\frac{16}{15}G\pi^2\cdot\frac{9M^2}{16\pi^2a^6}a^5 \quad \left[\because \rho^2 = \frac{9M^2}{16\pi^2a^6}\right]$$

$$= -\frac{3}{5}G\frac{M^2}{a}$$ (negative sign indicating loss of gravitational P.E.)

7.19 Laplace's Equation for Potential

Let V be the potential of a system of attracting particles at a point $P\ (x, y, z)$ *not in contact with the particles,* so that $V = -\Sigma\frac{Gm}{r}$, where m is the mass of the particle at (a, b, c) and r is the distance of P from the position of the mass m and is given by

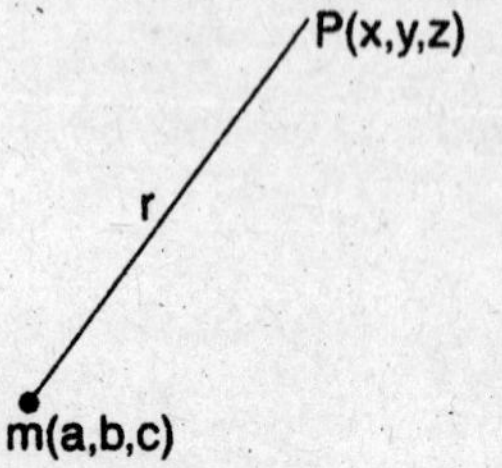

Fig. 7.19.1

$$r^2 = (x-a)^2 + (y-b)^2 + (z-c)^2.$$

Therefore,

$$2r\frac{\partial r}{\partial x} = 2(x-a), \quad \text{or,} \quad \frac{\partial r}{\partial x} = \frac{(x-a)}{r}.$$

Similarly,

$$\frac{\partial r}{\partial y} = \frac{(y-b)}{r} \quad \text{and} \quad \frac{\partial r}{\partial z} = \frac{(z-c)}{r}.$$

Now

$$V = -\Sigma\frac{Gm}{r}. \quad \therefore \quad \frac{\partial V}{\partial x} = \Sigma\frac{Gm}{r^2}\cdot\frac{\partial r}{\partial x} = \Sigma\frac{Gm}{r^3}(x-a) \quad \left[\text{putting the value of } \tfrac{\partial r}{\partial x}\right]$$

Similarly,

$$\frac{\partial V}{\partial y} = \Sigma \frac{Gm}{r^3}(y-b) \quad \text{and} \quad \frac{\partial V}{\partial z} = \Sigma \frac{Gm}{r^3}(z-c);$$

further

$$\frac{\partial^2 V}{\partial x^2} = \frac{\partial}{\partial x}\left[\Sigma \frac{Gm}{r^3}(x-a)\right] = \Sigma \frac{Gm}{r^3} - 3\Sigma \frac{Gm}{r^5}(x-a)^2.$$

Similarly,

$$\frac{\partial^2 V}{\partial y^2} = \Sigma \frac{Gm}{r^3} - 3\Sigma \frac{Gm}{r^5}(y-b)^2 \quad \text{and} \quad \frac{\partial^2 V}{dz^2} = \Sigma \frac{Gm}{r^3} - 3\Sigma \frac{Gm}{r^5}(z-c)^2.$$

Hence by addition, we get

$$\frac{\partial^2 V}{\partial x^2} + \frac{\partial^2 V}{\partial y^2} + \frac{\partial^2 V}{\partial z^2} = 3\Sigma \frac{Gm}{r^3} - 3\Sigma \frac{Gm}{r^5}.r^2 = 0$$

or, $\nabla^2 V = 0$; ∇ is called *Laplacian operator*,

or, $\vec{\nabla}.\vec{\nabla} V = 0,$

or, $\text{div grad } V = 0.$

This establishes Laplace's equation for potential.

Instead of considering isolated masses if we consider a continuous mass, we may arrive at the same result.

7.20 Poisson's Equation for Potential

Let the point P of co-ordinates (x, y, z) be inside the attracting matter. We describe a sphere of small radius R and centre (a, b, c), containing the point P. Since the sphere, we describe, is very small, we may regard the attracting mass within the sphere to be of uniform density ρ.

Now the matter which produces the potential V at P may be divided into two parts—the matter outside and the matter inside the small sphere.

Let V_1 denote the contribution towards potential at P by the matter outside the sphere, and V_2 the contribution towards potential at P by the matter inside the sphere. Since the point P is not in contact with the matter which produces the potential V_1, therefore, according to Laplace's equation $\nabla^2 V_1 = 0$ and V_2 being the potential at a point (x, y, z) inside a small sphere of radius R, we have,

$$V_2 = -\frac{2}{3}G\pi\rho(3R^2 - r^2),$$

where r is the distance between the centre of the sphere, and the point P.

We have therefore

$$\frac{\partial V_2}{\partial x} = \frac{\partial}{\partial x}\left[-\frac{2}{3}G\pi\rho(3R^2 - r^2)\right] = \frac{2}{3}\pi\rho G 2r\frac{\partial r}{\partial x} = \frac{4}{3}\pi G\rho(x-a) \quad \left[\because \frac{\partial r}{\partial x} = \frac{(x-a)}{r}\right].$$

$$\therefore \quad \frac{\partial^2 V_2}{\partial x^2} = \frac{4}{3}\pi G\rho.$$

Similarly, we have,

$$\frac{\partial^2 V_2}{\partial y^2} = \frac{4}{3}\pi G\rho; \frac{\partial^2 V_2}{dz^2} = \frac{4}{3}\pi G\rho$$

$$\therefore \quad \frac{\partial^2 V_2}{\partial x^2} + \frac{\partial^2 V_2}{\partial y^2} + \frac{\partial^2 V_2}{dz^2} = 4\pi G\rho, \quad \text{or,} \quad \nabla^2 V_2 = 4\pi G\rho.$$

Now V, at every point at which there is attracting matter of density ρ, is given by

$$V = V_1 + V_2.$$

$$\therefore \quad \nabla^2 V = \nabla^2 V_1 + \nabla^2 V_2 = 0 + 4\pi G\rho = 4\pi G\rho.$$

This result is known as *Poisson's equation for potential.*

7.21 The flux of gravitational intensity or attraction

Let P be a point in a gravitational field, where the gravitational intensity is f, acting along PA (Fig. 7.21.1). Through P, we draw any closed surface (the surface must be finite and have a unique outward drawn normal at every point), and consider a small element δs of this surface surrounding P. Through P, we draw PN normal to δs. Then f_n, the component of f along the outward drawn normal is $f\cos\theta$, where $\theta = \angle APN$.

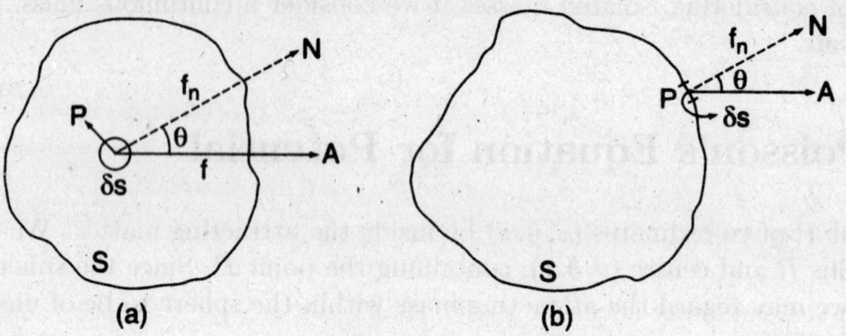

Fig. 7.21.1

The quantity $f_n\delta S$ or $(f\cos\theta)\delta S$ is termed the *flux of gravitational intensity or attraction* across the surface δs. Similarly,

$$\int f_n ds, \quad \text{or,} \quad \int f\cos\theta ds,$$

where the integration extends over the whole of the surface S, is the flux of gravitational intensity or attraction across that surface. In vector notation, the flux is $\int \vec{f}.\hat{n}ds$, where $\hat{n}$ is a unit vector along the outward drawn normal to δs.

7.21.1 Gauss's theorem

According to this theorem, the total flux of gravitational intensity or attraction over any closed surface, which has a unique outward drawn normal at every point, is equal to the product of $4\pi G$ and the mass enclosed by the surface, or zero if there be no mass inside the surface.

If ds denotes a small element of area on the closed surface and f_n the component of attraction along the normal to ds drawn outward, then according to Gauss's theorem.

$$\int f_n ds = 4\pi G \times (\text{mass enclosed by the surface}), \quad \text{or,} \quad \int \vec{f}.\hat{n} ds = 4\pi G \Sigma m.$$

Considering the field as due to a system of particles, we shall first consider a single particle. Let m be the mass of this particle situated at P outside the closed surface (Fig. 7.21.2).

Let a small cone be drawn with P as vertex. The cone cuts the closed surface in elements of area ds_1, ds_2, ds_3 and ds_4 at A, B, C and D respectively.

Let the outward normal components of attraction of m at A, B, C and D make angles $\theta_1, \theta_2, \theta_3$ and θ_4 with the line $PABCD$. Let the components of attraction along the outward normals be f_{n1}, f_{n2}, f_{n3} and f_{n4} respectively.

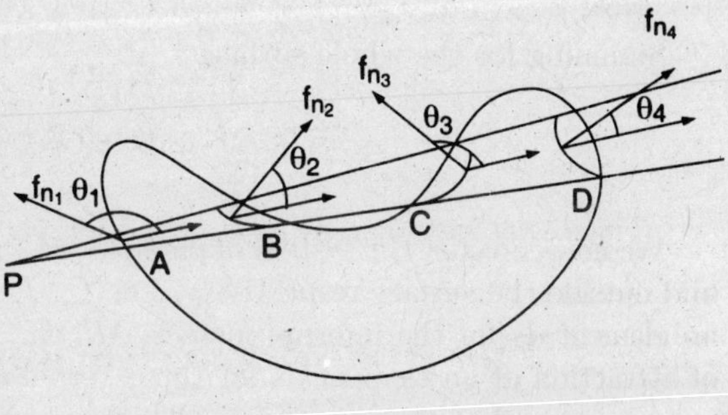

Fig. 7.21.2

Now the attraction at A due to the mass m at P is $-\dfrac{Gm}{PA^2}$ along AP and the component of this attraction along the outward normal

$$= \frac{Gm}{PA^2}\cos(\pi - \theta_1) = -\frac{Gm}{PA^2}\cos\theta_1 = f_{n_1}. \quad \therefore \quad f_{n1}ds_1 = -Gm\frac{ds_1\cos\theta_1}{PA^2} = -Gmdw,$$

where $dw = \dfrac{ds_1\cos\theta_1}{PA^2}$ = solid angle of the cone.

Similarly,

$$f_{n2} = \frac{Gm}{PB^2}\cos\theta_2. \quad \therefore \quad f_{n2}ds_2 = Gm\cdot\frac{ds_2\cos\theta_2}{PB^2} = Gmdw \quad \left[\because dw = \frac{ds_2\cos\theta_2}{PB^2}\right].$$

Also, in the same way it can be shown that

$$f_{n3}ds_3 = -Gmdw; \quad f_{n4}ds_4 = Gmdw.$$

$$\therefore f_{n1}ds_1 + f_{n2}ds_2 + f_{n3}ds_3 + f_{n4}ds_4 + \cdots = (Gmdw - Gmdw) + (Gmdw - Gmdw) + \cdots$$
$$= 0.$$

Thus the whole surface is divided into pairs of elements such as A and B, C and D, etc., and for each pair the contribution to the flux of gravitational intensity is zero. We may divide the whole surface into any number of pairs we like and since for each such pair the contribution to flux is zero, for the whole surface we may write

$$\int f_n ds = 0.$$

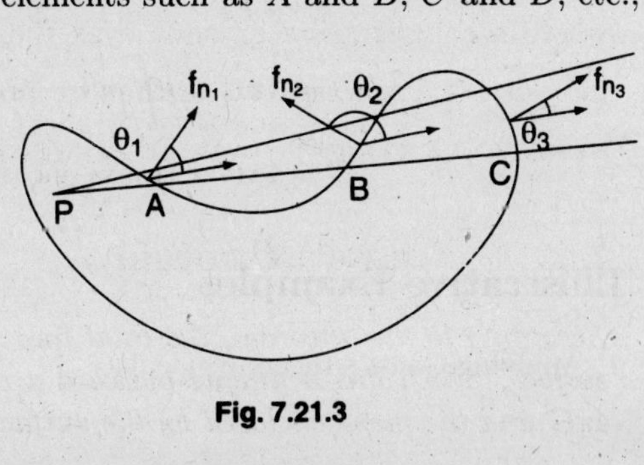

Fig. 7.21.3

We now consider a particle of mass M situated at P inside the closed surface, (Fig. 7.21.3).

In this case the contribution to the flux of force by the pair of elements at B and C is zero. Hence we shall consider only the flux at A.

The attraction at A due to the mass M at P is $-\dfrac{GM}{PA^2}$ along AP and the component of this attraction, along the outward normal, is given by

$$f_{n1} = \frac{GM}{PA^2}\cos\theta_1. \quad \therefore\ f_{n1}ds_1 = GM\frac{ds_1\cos\theta_1}{PA^2} = GM\,dw.$$

Summing for the whole surface

$$\int f_n ds = GM\int dw = 4\pi GM.$$

We now consider the system of particles M', M'', M''', ... and m', m'', m''', ... situated inside and outside the surface respectively. Let f'_n, f''_n, ... be the normal components of attraction at an element ds for the internal masses, M', M'', ... and F'_n, F''_n, ... be the normal components of attraction at an element ds for the external masses, m', m'', ...

$\therefore$ If f_n be the normal component of attraction of all these particles, then

$$f_n = (f'_n + f''_n + ...) + (F'_n + F''_n + ...)$$

$$\text{or,}\quad \int f_n ds = \int (f'_+ f''_n + ...)ds + \int (F'_n + F''_n + ...)ds$$

$$= 4\pi G \times (M' + M'' + ...) + 0$$

$$= 4\pi G \times (\text{total mass inside the surface}).$$

Alternative treatment

In vector notation, we have the resultant attraction $\vec{f} = \text{grad}\,V$.

Now the flux of the force of attraction over any closed surface S

$$= \iint \vec{f}_n ds = \iiint \text{div}\vec{f}dv \text{ (from divergence theorem)}$$

$$= \iiint \text{div grad}\,V\,dv = \iiint \vec{\nabla}.\vec{\nabla}V\,dv$$

$$= \iiint \nabla^2 V dv \quad \left[\because\ \vec{\nabla}.\vec{\nabla}V = \nabla^2 V\right]$$

$$= 4\pi G \iiint \rho dv \quad \text{[from Poisson's equation]}$$

$$= 4\pi G \times (\text{mass enclosed by the surface}).$$

Illustrative Examples

1. Applying Gauss's theorem calculate the attraction of an infinite rod at a distance r from the rod.

[C.U. (Hons.) 1984]

We consider a point P at a distance r from the rod. We are required to find out the intensity at the point P. We imagine a cylinder with the rod as axis and of radius r and length l (Fig. 1).

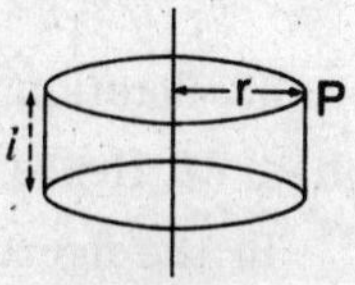

Fig. 1

The normal induction $\iint f_n ds$ over the plane faces of the cylinder is zero, since there is no component of attraction parallel to the rod. The normal induction over the curved surface of the cylinder $= 2\pi rlf$, where f is the intensity at P (since f is the normal force over the curved surface).

Then applying Gauss's theorem we may write $2\pi rlf = 4\pi G.\rho l$, where ρ is the mass per unit length of the rod. $\therefore f = \dfrac{2G\rho}{r}$.

2. Applying Gauss's theorem calculate the attraction inside and outside a spherical surface.

[C.U. (Hons.) 1988, '90]

We consider a spherical shell of mass M and radius a. We are required to find out the intensity or attraction at a point P inside the shell which is at a distance r from the centre O of the shell (Fig. 2). If f be the normal induction acting outward at any point on the sphere of radius r, then by Gauss's theorem we may write $4\pi r^2 f = 4\pi G\times$ (mass inside the surface of radius r) $= 4\pi G \times 0 = 0$. $\therefore f = 0$.

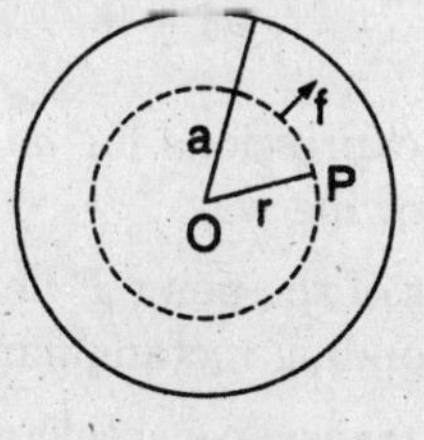

Fig. 2

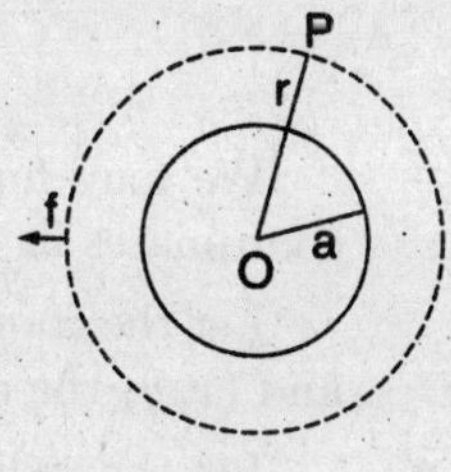

Fig. 3

Considering the point P outside the shell at a distance r from its centre O (Fig. 3), we may write in a similar way, by Gauss's theorem, $4\pi r^2 f = 4\pi G \times M$, or, $f = \dfrac{GM}{r^2}$, where M is the mass inside the sphere through P.

7.22 Kepler's laws

Kepler formulated the following three laws on the motion of planetary bodies.

(i) *The orbit of a planet is elliptic with the sun at one of the foci.*

(ii) *The line joining a planet to the sun traces out equal areas in equal times, i.e., the areal velocity of the radius vector is constant.*

(iii) *The square of the period of revolution of the planet round the sun is proportional to the cube of the major axis of the ellipse which the planet describes.*

7.22.1 Deduction of Kepler's laws from Newton's law of gravitation

Newton arrived at these laws from his well-known law of gravitation according to which two particles of masses M and m at a distance r apart attract one another with equal and opposite forces of magnitude $\frac{GMm}{r^2}$.

Considering the sun and a planet as bodies of masses M and m, the attraction of the sun on the planet produces an acceleration $\frac{GM}{r^2}$ and the attraction of the planet on the sun produces an acceleration $\frac{Gm}{r^2}$ which may be ignored without any serious departure from reality since $m \ll M$. Hence we may consider the sun to be at rest.

In the figure 7.22.1, O denotes the position of the sun. The point O is considered to be fixed and the line OX denotes an initial line.

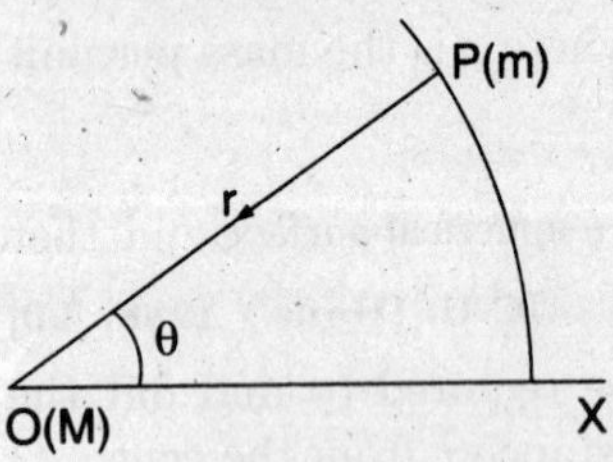

Fig. 7.22.1

Let us suppose that at any instant t the position P of the planet is given by the polar co-ordinates (r, θ). In this position the attraction of the sun produces an acceleration along PO on the planet given by

$$a_r = -\frac{GM}{r^2},$$

(the negative sign is due to the fact that the acceleration is towards the sun and r is directed away from the sun); a_r is called *radial acceleration.*

As there is no acceleration at right angles to OP, so the transverse acceleration

$$a_\theta = 0.$$

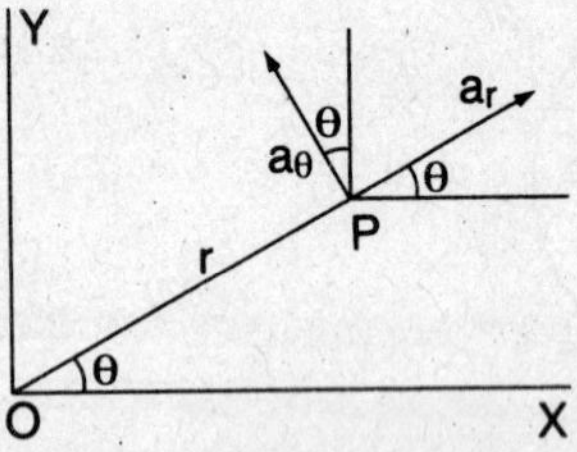

Fig. 7.22.2

We may find out expressions for a_r and a_θ referred to polar coordinates in the following way.

Let the coordinates of the point P be (r, θ) in polar coordinates and (x, y) the corresponding rectangular coordinates.

Let the radial and transverse accelerations, a_r and a_θ be acting along OP and perpendicular to OP respectively (Fig. 7.22.2).

Then the x-components of a_r and a_θ are given by

$$a_r \cos\theta + a_\theta \cos\left(\frac{\pi}{2} + \theta\right) = \frac{d^2x}{dt^2}, \quad \text{or,} \quad a_r \cos\theta - a_\theta \sin\theta = \frac{d^2x}{dt^2}.$$

Now $x = r\cos\theta$.

$$\therefore \quad \frac{dx}{dt} = \frac{dr}{dt}\cos\theta - r\sin\theta\frac{d\theta}{dt}.$$

$$\therefore \quad \frac{d^2x}{dt^2} = \cos\theta\left[\frac{d^2r}{dt^2} - r\left(\frac{d\theta}{dt}\right)^2\right] - \sin\theta\left[2\frac{dr}{dt}\frac{d\theta}{dt} + r\frac{d^2\theta}{dt^2}\right].$$

$$\therefore \quad a_r\cos\theta - a_\theta\sin\theta = \cos\theta\left[\frac{d^2r}{dt^2} - r\left(\frac{d\theta}{dt}\right)^2\right] - \sin\theta\left[2\frac{dr}{dt}\frac{d\theta}{dt} + r\frac{d^2\theta}{dt^2}\right] \qquad (7.22.1)$$

Again y-components of a_r and a_θ are given by

$$a_r\sin\theta + a_\theta\cos\theta = \frac{d^2y}{dt^2}.$$

Now $y = r\sin\theta$.

$$\therefore \quad \frac{dy}{dt} = \frac{dr}{dt}\sin\theta + r\cos\theta\frac{d\theta}{dt}.$$

$$\therefore \quad \frac{d^2y}{dt^2} = \cos\theta\left[2\frac{dr}{dt}\frac{d\theta}{dt} + r\frac{d^2\theta}{dt^2}\right] + \sin\theta\left[\frac{d^2r}{dt^2} - r\left(\frac{d\theta}{dt}\right)^2\right].$$

$$\therefore \quad a_r\sin\theta + a_\theta\cos\theta = \cos\theta\left[2\frac{dr}{dt}\frac{d\theta}{dt} + r\frac{d^2\theta}{dt^2}\right] + \sin\theta\left[\frac{d^2r}{dt^2} - r\left(\frac{d\theta}{dt}\right)^2\right]. \tag{7.22.2}$$

Now multiplying equation (7.22.1) by $\cos\theta$ and equation (7.22.2) by $\sin\theta$ and then adding, we have

$$a_r = \frac{d^2r}{dt^2} - r\left(\frac{d\theta}{dt}\right)^2.$$

Again on multiplying equation (7.22.1) by $\sin\theta$ and equation (7.22.2) by $\cos\theta$ and then subtracting, we have

$$a_\theta = r\frac{d^2\theta}{dt^2} + 2\frac{dr}{dt}\frac{d\theta}{dt}.$$

Hence we may write

$$\frac{d^2r}{dt^2} - r\left(\frac{d\theta}{dt}\right)^2 = -\frac{GM}{r^2}. \tag{7.22.3}$$

$$\text{or,} \quad m\frac{d^2r}{dt^2} - mr\left(\frac{d\theta}{dt}\right)^2 = ma_r = -\frac{GMm}{r^2} = f(r) = f\left(\frac{1}{u}\right). \tag{7.22.4}$$

and

$$r\frac{d^2\theta}{dt^2} + 2\frac{dr}{dt}\frac{d\theta}{dt} = 0. \tag{7.22.5}$$

Equation (7.22.5) may be written as

$$r^2\frac{d^2\theta}{dt^2} + 2r\frac{dr}{dt}\frac{d\theta}{dt} = 0, \text{ or, } \frac{d}{dt}\left(r^2\frac{d\theta}{dt}\right) = 0, \text{ or, } r^2\frac{d\theta}{dt} = \text{ constant } = h \text{ (say).} \tag{7.22.6}$$

$$\therefore \quad \frac{d\theta}{dt} = \frac{h}{r^2} = hu^2 \quad [\text{putting } \tfrac{1}{r} = u] \tag{7.22.7}$$

Now

$$\frac{dr}{dt} = \frac{d}{dt}\left(\frac{1}{u}\right) = -\frac{1}{u^2}\frac{du}{d\theta}\frac{d\theta}{dt} = -\frac{1}{u^2}\frac{du}{d\theta}hu^2 = -h\frac{du}{d\theta} \quad \text{(from 7.22.7)}$$

$$\therefore \quad \frac{d^2r}{dt^2} = -h\frac{d}{dt}\left(\frac{du}{d\theta}\right) = -h\frac{d^2u}{d\theta^2}\frac{d\theta}{dt} = -h^2u^2\frac{d^2u}{d\theta^2}. \tag{7.22.8}$$

Substituting the values of $\frac{d^2r}{dt^2}$ and $\frac{d\theta}{dt}$ in equation (7.22.3), we have

$$-h^2u^2\frac{d^2u}{d\theta^2} - \frac{1}{u}h^2u^4 = -\frac{GM}{r^2} = -GMu^2,$$

$$\left[-mh^2u^2\frac{d^2u}{d\theta^2} - m\frac{1}{u}(hu^2)^2 = -\frac{GMm}{r^2} = f(r) = f\left(\frac{1}{u}\right)\right.$$

$$\left.\text{or,} \quad \frac{d^2u}{d\theta^2} + u = -\frac{f\left(\frac{1}{u}\right)}{mh^2u^2} = -\frac{m}{L^2u^2}f\left(\frac{1}{u}\right) \quad \because L = mh\right]$$

$$\text{or,}\quad -h^2u^2\left(\frac{d^2u}{d\theta^2}+u\right) = -GMu^2,\quad \text{or,}\quad \frac{d^2u}{d\theta^2}+u=\frac{GM}{h^2}, \tag{7.22.9}$$

$$\text{or,}\quad \frac{d^2}{d\theta^2}\left(u-\frac{GM}{h^2}\right)+\left(u-\frac{GM}{h^2}\right)=0.$$

The solution of the above equation is given by

$$u-\frac{GM}{h^2}=A\cos(\theta+\beta)=A\cos\theta$$ [by suitably adjusting the initial conditions, β can be made zero]

$$\text{or}\quad u=\frac{GM}{h^2}+A\cos\theta=\frac{GM}{h^2}\left(1+\frac{Ah^2}{GM}\cos\theta\right)$$

$$\text{or}\quad \frac{1}{r}=\frac{GM}{h^2}\left(1+\frac{Ah^2}{GM}\cos\theta\right)\quad \text{or}\quad \frac{h^2/GM}{r}=1+\frac{Ah^2}{GM}\cos\theta.$$

This equation is of the form of the polar equation of a conic given by

$$\frac{l}{r}=1+\varepsilon\cos\theta,$$

where $l=\dfrac{h^2}{GM}$ = semi-latus rectum of the conic and $\varepsilon=\dfrac{h^2A}{GM}$ = eccentricity of the conic.

The form of the conic depends on A. Thus the orbit is an ellipse, a parabola or a hyperbola according as ε is less than, equal to, or greater than 1, i.e., the path will be ellipse, parabola or hyperbola according as

$$A \lesseqgtr \frac{GM}{h^2}.$$

Now it can be shown by actual calculation that $A<\dfrac{GM}{h^2}$. Also our common observation shows that every planet, during its course of revolution round the sun, comes to a definite point of its orbit after a regular interval of time. We therefore, may conclude that the orbit of the planet is a closed one. Of the three conics, ellipse is the only one which is closed. So the orbit of the planet is an ellipse.

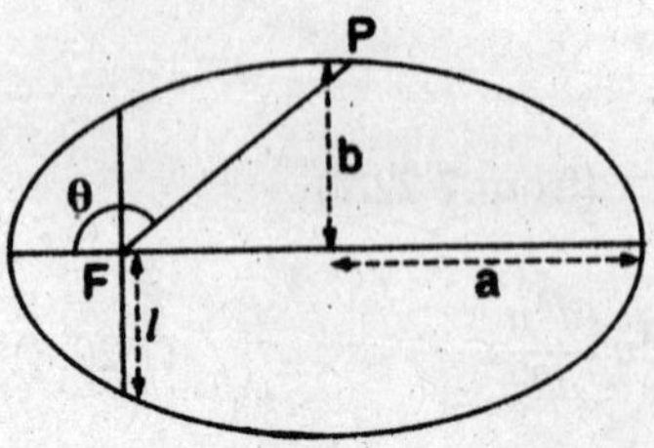

Fig. 7.22.3

For elliptic orbit, the origin of co-ordinates (r = 0) is situated at one focus F of the ellipse as shown in Fig. 7.22.3.

This establishes Kepler's first law—according to which the planet describes an elliptic orbit with the sun at one focus F—the eccentricity ε and the semi-latus rectum l of the ellipse being $\dfrac{h^2A}{GM}$ and $\dfrac{h^2}{GM}$ respectively.

N.B. From (7.22.6), multiplying both sides by m, we may write $mr^2\dfrac{d\theta}{dt}=mh=\text{constant}$ or $I\omega=\text{constant}$.

Thus for planetary motion, angular momentum remains conserved.

Alternative treatment

We consider the planet at the position P and its distance from the sun fixed at the point O is r at an instant.

Let v be the linear velocity of the planet at that instant. Hence its radial components $v_r = \frac{dr}{dt}$ and its transverse component is

$$v_\theta = r\frac{d\theta}{dt} \quad \text{(Fig. 7.22.4).}$$

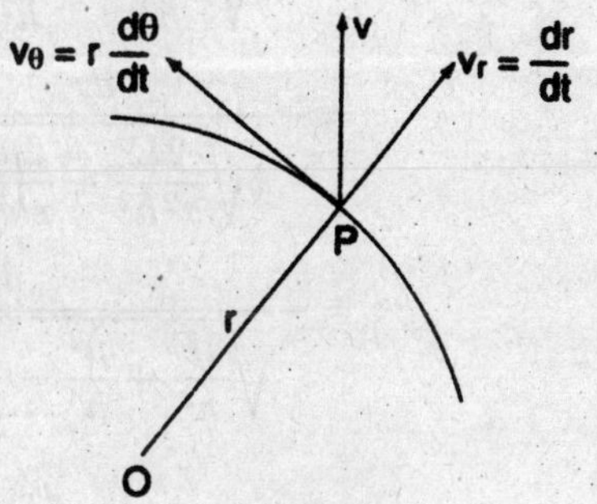

Fig. 7.22.4

$$\therefore\ v^2 = v_r^2 + v_\theta^2 = \left(\frac{dr}{dt}\right)^2 + \left(r\frac{d\theta}{dt}\right)^2 = \left(\frac{dr}{d\theta}\cdot\frac{d\theta}{dt}\right)^2 + \left(r\frac{d\theta}{dt}\right)^2$$

$$= \left(\frac{d\theta}{dt}\right)^2\left[\left(\frac{dr}{d\theta}\right)^2 + r^2\right] = \frac{h^2}{r^4}\left[\left(\frac{dr}{d\theta}\right)^2 + r^2\right]$$

$$\left[\because\ r^2\frac{d\theta}{dt} = h.\quad \therefore\ \left(\frac{d\theta}{dt}\right)^2 = \frac{h^2}{r^4}\right].$$

Now the kinetic energy per unit mass while the planet is at P

$$= \frac{1}{2}v^2 = \frac{1}{2}\cdot\frac{h^2}{r^4}\left[\left(\frac{dr}{d\theta}\right)^2 + r^2\right]$$

Again the potential energy per unit mass while the planet is at P

$$= \int_\alpha^r \frac{GM}{r^2}dr = -\frac{GM}{r} = -\frac{\mu}{r}$$

[$\mu = GM$, M being the mass of the sun].

$\therefore$ total energy per unit mass is given by

$$E = \text{K.E.} + \text{P.E.} = \frac{1}{2}\cdot\frac{h^2}{r^4}\left[\left(\frac{dr}{d\theta}\right)^2 + r^2\right] - \frac{\mu}{r},$$

$$\text{or,}\quad E + \frac{\mu}{r} = \frac{1}{2}\cdot\frac{h^2}{r^4}\left[\left(\frac{dr}{d\theta}\right)^2 + r^2\right],$$

$$\text{or,}\quad \left[\left(\frac{dr}{d\theta}\right)^2 + r^2\right] = \frac{2Er^4}{h^2} + \frac{2\mu r^4}{h^2 r} = \frac{2Er^4}{h^2} + \frac{2\mu r^3}{h^2},$$

$$\text{or,}\quad \left(\frac{dr}{d\theta}\right)^2 = \left(\frac{2Er^4}{h^2} + \frac{2\mu r^3}{h^2} - r^2\right)^* = \frac{2Er^4 + 2\mu r^3 - h^2r^2}{h^2},$$

$$\text{or,}\quad \left(\frac{d\theta}{dr}\right)^2 = \frac{h^2}{2Er^4 + 2\mu r^3 - h^2r^2}$$

$$\text{or,}\quad \frac{d\theta}{dr} = \frac{h}{\sqrt{2Er^4 + 2\mu r^3 - h^2r^2}},$$

---

* This expression has been obtained in Art. 3.14, Chapter 3.

$$\text{or,}\quad d\theta = \frac{h\,dr}{r\sqrt{2Er^2 + 2\mu r - h^2}} = \frac{dr}{r\sqrt{\frac{2Er^2}{h^2} + \frac{2\mu r}{h^2} - 1}}$$

$$= -\frac{r^2 dx}{r\sqrt{\frac{2Er^2}{h^2} + \frac{2\mu r}{h^2} - 1}} \quad \left[\text{putting } x = \tfrac{1}{r} \text{ we have, } dx = -\tfrac{1}{r^2}dr \text{ or } dr = -r^2 dx\right]$$

$$= -\frac{dx}{x\sqrt{\frac{2E}{x^2h^2} + \frac{2\mu}{xh^2} - 1}} = -\frac{dx}{\sqrt{\frac{2E}{h^2} + \frac{2\mu x}{h^2} - x^2}}$$

$$= -\frac{dx}{\sqrt{\frac{2E}{h^2} + \frac{\mu^2}{h^4} - \frac{\mu^2}{h^4} + \frac{2\mu x}{h^2} - x^2}} = -\frac{dx}{\sqrt{\frac{2E}{h^2} + \frac{\mu^2}{h^4} - \left(x^2 - \frac{2\mu x}{h^2} + \frac{\mu^2}{h^4}\right)}}$$

$$= -\frac{dx}{\sqrt{\frac{2E}{h^2} + \frac{\mu^2}{h^4} - \left(x - \frac{\mu}{h^2}\right)^2}}.$$

Putting

$$\frac{2E}{h^2} + \frac{\mu^2}{h^4} = a^2 \quad \text{and} \quad a\sin\phi = \left(x - \frac{\mu}{h^2}\right),$$

we have

$$a\cos\phi\, d\phi = dx$$

and

$$\sqrt{\frac{2E}{h^2} + \frac{\mu^2}{h^4} - \left(x - \frac{\mu}{h^2}\right)^2} = \sqrt{a^2 - a^2\sin^2\phi} = a\cos\phi.$$

$$\therefore \quad d\theta = -\frac{a\cos\phi\, d\phi}{a\cos\phi} = -d\phi.$$

$\therefore$ On integration, we have

$$\theta = -\phi + \beta, \text{ where } \beta \text{ is the constant of integration.}$$

$$= -\sin^{-1}\frac{\left(x - \frac{\mu}{h^2}\right)}{a} + \beta = -\sin^{-1}\frac{\left(x - \frac{\mu}{h^2}\right)}{\sqrt{\frac{2E}{h^2} + \frac{\mu^2}{h^4}}} + \beta,$$

$$\text{or,}\quad (\theta - \beta) = -\sin^{-1}\frac{\left(\frac{1}{r} - \frac{\mu}{h^2}\right)}{\sqrt{\frac{2E}{h^2} + \frac{\mu^2}{h^4}}}$$

$$\text{or,}\quad \sin(\theta - \beta) = -\frac{\left(\frac{1}{r} - \frac{\mu}{h^2}\right)}{\sqrt{\frac{2E}{h^2} + \frac{\mu^2}{h^4}}} = -\frac{\frac{1}{r} - \frac{\mu}{h^2}}{\frac{1}{h}\sqrt{2E + \frac{\mu^2}{h^2}}} = -\frac{\left(\frac{h}{r} - \frac{\mu}{h}\right)}{\frac{\mu}{h}\sqrt{\frac{2Eh^2}{\mu^2} + 1}}.$$

$$\therefore\ -\left(\frac{h}{r}-\frac{\mu}{h}\right)=\frac{\mu}{h}\sqrt{\frac{2Eh^2}{\mu^2}+1}\sin(\theta-\beta),$$

$$\text{or,}\quad \frac{h}{r}=\frac{\mu}{h}\left[1-\sqrt{\frac{2Eh^2}{\mu^2}+1}\sin(\theta-\beta)\right]$$

$$\text{or,}\quad \frac{h}{r}=\frac{\mu}{h}\left[1+\sqrt{\frac{2Eh^2}{\mu^2}+1}\cos\left(\theta-\beta+\frac{\pi}{2}\right)\right]$$

$$\text{or,}\quad \frac{1}{r}=\frac{\mu}{h^2}\left[1+\sqrt{\frac{2Eh^2}{\mu^2}+1}\cos(\theta-a)\right]\quad\left[\beta-\frac{\pi}{2}=a\right],$$

$$\text{or,}\quad r=\frac{\dfrac{h^2}{\mu}}{1+\sqrt{1+\dfrac{2Eh^2}{\mu^2}}\cos(\theta-a)}.$$

The above equation represents the polar equation of a conic, where

$$l=\frac{h^2}{\mu};\quad \varepsilon=\sqrt{1+\frac{2Eh^2}{\mu^2}}.$$

Hence

$$\frac{1}{r}=\frac{1+\varepsilon\cos(\theta-a)}{l}=\frac{1}{l}+\frac{\varepsilon}{l}\cos(\theta-a)\quad\text{and}\quad \varepsilon^2=1+\frac{2Eh^2}{\mu^2}.$$

For *repulsive force* μ is positive and l is negative. We shall now show that *in case of the repulsive force the total energy per unit mass is positive.*

From equation (7.22.9), we may write,

$$\frac{d^2u}{d\theta^2}+u=-\frac{GM}{h^2},$$

the negative sign in the right hand side of the above expression is due to repulsive force. Putting $GM=\mu$, the above equation reduces to the form

$$\frac{d^2u}{d\theta^2}+u=-\frac{\mu}{h^2},\ \text{or,}\ \frac{d^2}{d\theta^2}\left(u+\frac{\mu}{h^2}\right)+\left(u+\frac{\mu}{h^2}\right)=0,$$

$$\text{or,}\ u=\frac{1}{r}=-\frac{\mu}{h^2}+A\cos(\theta-a). \tag{7.22.10}$$

At $t=0$, $r=r_0$, $v=v_0$ and let ϕ be the angle between v_0 and r_0. Under this condition $\theta=0$.

$$\therefore\ (\dot{r})_{t=0}=v_0\cos\phi$$

$$(r\dot{\theta})_{t=0}=v_0\sin\phi$$

$$\text{and}\quad h=(r^2\dot{\theta})_{t=0}=r_0v_0\sin\phi.$$

At $t = 0$,

$$\frac{1}{r_0} = -\frac{\mu}{h^2} + A\cos\alpha. \tag{7.22.11}$$

From (7.22.10), we may write

$$\left(-\frac{1}{r^2}\cdot\frac{dr}{d\theta}\right)_{t=0} = A\sin\alpha, \quad \text{or,} \quad \left(-\frac{1}{r^2}\frac{dr}{dt}\Big/\frac{d\theta}{dt}\right)_{t=0} = A\sin\alpha.$$

$$\therefore \quad \left(\frac{dr}{dt}\right)_{t=0} = -\left(r^2\frac{d\theta}{dt}\right)_{t=0} A\sin\alpha, \quad \text{or,} \quad v_0\cos\phi = -hA\sin\alpha.$$

$$\therefore \quad A\sin\alpha = -\frac{v_0}{h}\cos\phi \quad \text{and} \quad A\cos\alpha = \frac{1}{r_0} + \frac{\mu}{h^2} \quad \text{[From 7.22.11]}.$$

$$\therefore \quad A^2 = \frac{\mu^2}{h^4} + \frac{2}{h^2}\left(\frac{\mu}{r_0} + \frac{v_0^2}{2}\right) \tag{7.22.12}$$

$$\left[\because \frac{1}{r_0} = \frac{v_0\sin\phi}{h}\right].$$

Total energy per unit mass at $t = 0$ is

$$E = E_{t=0} = \frac{1}{2}v_0^2 + \frac{\mu}{r_0} = \left(\frac{\mu}{r_0} + \frac{v_0^2}{2}\right) \tag{7.22.13}$$

$$\text{and } l = -\frac{h^2}{\mu} = -\frac{r_0^2 v_0^2 \sin^2\phi}{\mu}. \tag{7.22.14}$$

Thus the total energy E is positive as μ is positive. Hence $\varepsilon > 1$. Thus the orbit is the negative branch of hyperbola as in the case of the path of an alpha particle in Rutherford's experiment on α-particle scattering. In case of the *attractive force* μ is negative and l is positive. E may be negative or positive. The minimum value of ε^2 is zero.

Now in case of *attractive force* the total energy per unit mass is

$$E = \frac{1}{2}v^2 - \frac{\mu}{r}.$$

In order that the orbit is an ellipse $\varepsilon < 1$, i.e., E should be negative. E is negative when

$$\frac{1}{2}v^2 < \frac{\mu}{r}, \quad \text{or,} \quad v^2 < \frac{2\mu}{r}, \quad \text{i.e.,} \quad v^2 < \frac{2GM}{r}, \quad \text{i.e.,} \quad v^2 < v_e^2.$$

(v_e = velocity of escape).

"Total energy is negative" means that the potential energy is greater than the kinetic energy. When $\varepsilon = 0$, the orbit is circular. The *significance of the negative total energy* is that if we supply this amount of energy to the planet, then it will *leave the solar family for good.* We have for the orbit to be ellipse

$$\varepsilon = \sqrt{1 + \frac{2Eh^2}{\mu^2}} < 1,$$

i.c., E is negative and semi latus rectum

$$\frac{h^2}{\mu} = l = \frac{b^2}{a} = a(1-\varepsilon^2) = a\left[1 - 1 - \frac{2Eh^2}{\mu^2}\right] = -\frac{2aEh^2}{\mu^2},$$

where a = semi-major axis of the ellipse and b = semi-minor axis of the ellipse.

$$\therefore \quad a = -\frac{\mu}{2E}, \quad \text{or,} \quad E = -\frac{\mu}{2a}.$$

Again

$$E = \frac{v^2}{2} - \frac{\mu}{r}, \quad \text{or,} \quad v^2 = 2E + \frac{2\mu}{r} = -\frac{\mu}{a} + \frac{2\mu}{r} \quad \left(\text{putting } E = -\frac{\mu}{2a}\right) = \mu\left(\frac{2}{r} - \frac{1}{a}\right).$$

If the particle is projected at a distance r_0 with a velocity v_0, then we have

$$v_0^2 = \mu\left(\frac{2}{r_0} - \frac{1}{a}\right) < \frac{2\mu}{r_0}, \quad \text{i.e.,} \quad v_0 < \sqrt{\frac{2\mu}{r_0}}\text{—the escape velocity.}$$

Similarly for a hyperbolic orbit, we can show

$$v_0^2 = \mu\left(\frac{2}{r_0} + \frac{1}{a}\right), \quad \text{i.e.,} \quad v_0 > \sqrt{\frac{2\mu}{r_0}}\text{—the escape velocity,}$$

while for a parabolic orbit $a \to \infty$, so that

$$v_0^2 = \frac{2\mu}{r_0}, \quad \text{i.e.,} \quad v_0 = v_e\text{—the escape velocity.}$$

(In all the above expressions for v or v_0, we can put $\mu = \dfrac{K}{m}$, since $\mu = GM$ and $K = GMm$).

Thus we see that the orbit described by a planet is an ellipse when Newton's law of gravitation holds with the condition that the kinetic energy of a planet at any position in the orbit will be less than its potential energy. The above condition also implies that the velocity of projection is less than the escape velocity (since escape velocity $= \sqrt{\dfrac{2GM}{r}}$) at the point of projection.

The orbit will be a hyperbola if $\varepsilon > 1$, i.e., kinetic energy of the planet in the orbit is greater than the potential energy,

$$\text{or,} \quad v^2 > \frac{2GM}{r}.$$

The orbit will be a parabola if $\varepsilon = 1$, i.e., when kinetic energy = potential energy

$$\text{or,} \quad v^2 = \frac{2GM}{r}.$$

N.B. *Conservation of energy equation and differential equation of orbit from it :* We know that $v^2 = v_r^2 + v_\theta^2$ (Fig. 7.22.4).

$$\text{Now} \quad v_r = \frac{dr}{dt} = \frac{dr}{d\theta}\cdot\frac{d\theta}{dt} = \frac{dr}{d\theta}\frac{h}{r^2} \quad \left(\because r^2\frac{d\theta}{dt} = h\right),$$

$$\text{or,} \quad \frac{dr}{dt} = -h\frac{d}{d\theta}\left(\frac{1}{r}\right) = -h\frac{du}{d\theta} \quad \left(\because \frac{1}{r} = u\right).$$

$$\text{Again,} \quad v_\theta = r\frac{d\theta}{dt} = r\frac{h}{r^2} = hu.$$

Therefore, substituting these values of v_r and v_θ, we have

$$v^2 = h^2\left(\frac{du}{d\theta}\right)^2 + h^2u^2 = h^2\left\{\left(\frac{du}{d\theta}\right)^2 + u^2\right\}.$$

From the principle of conservation of energy we may write

$$T + V = E, \text{ or, } \frac{1}{2}mv^2 = E - V, \quad \text{or, } \quad v^2 = \frac{2}{m}(E - V),$$

$$\text{or, } \left(\frac{du}{d\theta}\right)^2 + u^2 = \frac{2}{mh^2}(E - V).$$

From the above we may write

$$E = \frac{1}{2}mh^2\left[\left(\frac{du}{d\theta}\right)^2 + u^2\right] + V(r).$$

Now, the central force field being conservative, the total mechanical energy in this field is constant. So differentiating both sides w.r.t. r, we have

$$\frac{dE}{dr} = 0 = \frac{mh^2}{2}\left[2\frac{du}{d\theta}\frac{d^2u}{d\theta^2}\frac{d\theta}{du}\frac{du}{dr} + 2u\frac{du}{dr}\right] + \frac{dV}{dr}$$

$$= \frac{mh^2}{2}\left[2\frac{d^2u}{d\theta^2}(-u^2) + 2u(-u^2)\right] + \frac{dV}{dr} \qquad \left(\because u = \frac{1}{r}, \quad \frac{du}{dr} = -\frac{1}{r^2} = -u^2\right)$$

$$= -\frac{mh^2}{2}2u^2\left[\frac{d^2u}{d\theta^2} + u\right] + \frac{dV}{dr} = -mh^2u^2\left(\frac{d^2u}{d\theta^2} + u\right) + \frac{dV}{dr}.$$

Now the central force acting on the particle is given by

$$f(r) = -\frac{dV}{dr} = -mh^2u^2\left(\frac{d^2u}{d\theta^2} + u\right),$$

$$\text{or, } \quad \frac{d^2u}{d\theta^2} + u = -\frac{f\left(\frac{1}{u}\right)}{mh^2u^2} = -\frac{m}{L^2u^2}f\left(\frac{1}{u}\right) \qquad (\because L = mh)$$

which is the differential equation of orbit of the particle moving in a central force field.

Second Law

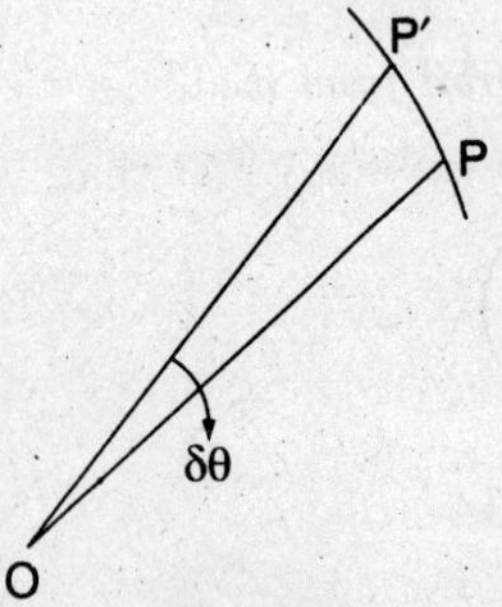

Fig. 7.22.5

In figure 7.22.5, let $\delta\theta$ be the angular displacement of the planet which moves from P to P' in a small interval of time δt. Therefore, area traced out by the line OP joining the planet and the sun in time $\delta t = \frac{1}{2}OP.\ PP' = \frac{1}{2}r.r\delta\theta$.

$$[\because OP = r \text{ and } PP' = r\delta\theta].$$

Hence the areal velocity

$$= \frac{1}{2}r^2 \lim_{\delta t \to 0}\frac{\delta\theta}{\delta t} = \frac{1}{2}r^2\frac{d\theta}{dt} = \frac{h}{2} = \text{ constant, (from the equation 7.22.6)}$$

which is Kepler's second law.

Third Law

If T be the time period of revolution of the planet round the sun, then

$$T = \frac{\text{Area of the elliptic orbit}}{\text{Areal velocity}} = \frac{\pi ab}{h/2} = \frac{2\pi ab}{h}$$

[a, b denote the semi-major and semi-minor axes of the elliptic orbit].

We know that in the case of an ellipse, the semi-latus rectum is given by $l = \dfrac{b^2}{a}$.

Also $l = \dfrac{h^2}{GM} = \dfrac{h^2}{\mu}$ $[GM = \mu \text{ (say)}]$.

$$\therefore \quad l = \frac{h^2}{\mu} = \frac{b^2}{a}, \quad \text{or,} \quad b^2 = \frac{a}{\mu}h^2, \quad \text{or,} \quad b = h\sqrt{\frac{a}{\mu}}.$$

$$\therefore \quad T = \frac{2\pi a h\sqrt{a/\mu}}{h} = 2\pi a\sqrt{\frac{a}{\mu}}, \quad \text{or,} \quad T^2 = 4\pi^2 a^2 \cdot \frac{a}{\mu} = \frac{4\pi^2 a^3}{\mu}, \quad \text{or,} \quad T^2 \propto a^3,$$

which establishes Kepler's third law.

7.23 Deduction of Newton's Inverse Square Law of Gravitation from Kepler's Laws

From Kepler's first law, it follows that a planet moves in an elliptic orbit, whose polar equation is given by

$$r = \frac{l}{1 + \varepsilon\cos\theta}, \quad \text{or,} \quad \frac{1}{u} = \frac{l}{1 + \varepsilon\cos\theta}, \quad \text{or,} \quad u = \frac{1}{l} + \frac{\varepsilon\cos\theta}{l}.$$

Now if a_r and a_θ represent the radial and transverse accelerations of the planet, then we know that

$$a_r = \frac{d^2r}{dt^2} - r\left(\frac{d\theta}{dt}\right)^2 \quad \text{and} \quad a_\theta = \frac{1}{r}\frac{d}{dt}\left(r^2\frac{d\theta}{dt}\right).$$

Now from Kepler's second law, areal velocity is constant,

$$\text{i.e.,} \quad \frac{1}{2}r^2\frac{d\theta}{dt} = \text{constant.} \quad \therefore \quad a_\theta = 0.$$

Thus there is only radial acceleration. Now we have already deduced in Art. 7.22.1 that

$$a_r = -h^2u^2\left(\frac{d^2u}{d\theta^2} + u\right).$$

Again from the equation of the ellipse, we have

$$\frac{du}{d\theta} = -\frac{\varepsilon\sin\theta}{l} \quad \text{and} \quad \frac{d^2u}{d\theta^2} = -\frac{\varepsilon\cos\theta}{l}.$$

Hence we may write,

$$a_r = -h^2u^2\left(-\frac{\varepsilon\cos\theta}{l} + \frac{1}{l} + \frac{\varepsilon\cos\theta}{l}\right) = -\frac{h^2u^2}{l} = -\frac{h^2}{l}\cdot\frac{1}{r^2} = -\frac{\mu}{r^2},$$

where $\mu = \frac{h^2}{l}$ = constant of proportionality. Thus we see that

$$a_r \propto \frac{1}{r^2}.$$

Thus the radial acceleration and hence the force of attraction between the sun and the planet is inversely proportional to the square of the distance between them. The negative sign in the expression for radial acceleration indicates that the force is directed towards the sun which is situated at one of the foci of the elliptic orbit.

Illustrative Example

1. Given that the distance of the planet Jupiter from the sun is 5.20 times the earth's distance, find the period of Jupiter's revolution round the sun in earth years. [C.U. (Hons.) 1980]

Let T_1 and T_2 be respectively the time periods of revolution of Jupiter and the earth round the sun; then from Kepler's third law we have,

$$\frac{T_1^2}{T_2^2} = \frac{a_1^3}{a_2^3},$$

where a_1 and a_2 are respectively the distances of Jupiter and the earth from the sun.

Here

$$T_2 = 1 \text{ year}; \quad \left(\frac{a_1}{a_2}\right) = 5.20.$$

$$\therefore \quad T_1^2 = (5.20)^3, \quad \text{or,} \quad T_1 = (5.20)^{3/2} = 11.86 \text{ earth years.}$$

2. A particle acted upon by a central force describes an orbit given by $r = a(1+\cos\theta)$, a being a constant. Show diagrammatically the nature of the orbit and find the nature of the force.

[C.U. (Hons.) 1980]

The equation $r = a(1+\cos\theta)$ of the orbit represents a Cardioide, the diagram of which is shown below (Fig. 1). In the figure at $A, \theta = 0$ and at O, $\theta = \pi$.

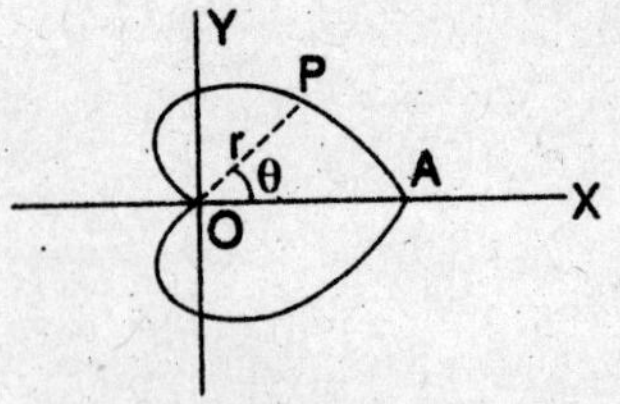

Fig. 1

$$\text{Now,} \quad r = a(1+\cos\theta) = 2a\cos^2\frac{\theta}{2},$$

$$\text{or,} \quad 2au = \sec^2\frac{\theta}{2} \quad [\text{Putting } 1/r = u]$$

$$\text{or,} \quad 2a\frac{du}{d\theta} = 2\sec\frac{\theta}{2}\cdot\frac{1}{2}\sec\frac{\theta}{2}\tan\frac{\theta}{2} = \sec^2\frac{\theta}{2}\tan\frac{\theta}{2}.$$

$$\therefore\quad 2a\frac{d^2u}{d\theta^2} = 2\sec\frac{\theta}{2}\cdot\frac{1}{2}\sec\frac{\theta}{2}\tan\frac{\theta}{2}\cdot\tan\frac{\theta}{2} + \frac{1}{2}\sec^2\frac{\theta}{2}.\sec^2\frac{\theta}{2} = \sec^2\frac{\theta}{2}\left(\tan^2\frac{\theta}{2} + \frac{1}{2}\sec^2\frac{\theta}{2}\right)$$

$$\text{or}\quad 2a\left(\frac{d^2u}{d\theta^2} + u\right) = \sec^2\frac{\theta}{2}\left(\tan^2\frac{\theta}{2} + \frac{1}{2}\sec^2\frac{\theta}{2}\right) + \sec^2\frac{\theta}{2}\qquad [\because\ 2au = \sec^2\theta/2]$$

$$\text{or,}\quad 2a\left(\frac{d^2u}{d\theta^2} + u\right) = \sec^2\frac{\theta}{2}\left[\tan^2\frac{\theta}{2} + \frac{1}{2}\sec^2\frac{\theta}{2} + 1\right] = \sec^2\frac{\theta}{2}\left[\frac{3}{2}\sec^2\frac{\theta}{2}\right]$$

$$= \frac{3}{2}\sec^4\frac{\theta}{2} = \frac{3}{2}.4a^2u^2 = 6a^2u^2.$$

$\therefore$ **if F be the central force,**

$$\frac{d^2u}{d\theta^2} + u = \frac{GM}{h^2} = \frac{GM}{r^2}\cdot\frac{r^2}{h^2} = \frac{F}{h^2u^2}\qquad \text{[See eqn. 7.22.9]},$$

$$\text{or,}\quad \frac{6a^2u^2}{2a} = \frac{F}{h^2u^2},\quad \text{or,}\quad F = 3ah^2u^4 = \frac{3ah^2}{r^4}.\quad \therefore\quad F \propto \frac{1}{r^4}.$$

3. Under the influence of a central force with centre at the point O, a particle moves in a circular orbit of radius R which passes through O. Show that the equation to the orbit is $r = 2R\cos\theta$ and hence find the law of force in terms of distance r from O. **[C.U. (Hons.) 1994, 2007]**

Let the particle P having polar co-ordinates (r, θ) describe a circle of diameter $2R$ under a central attractive force directed to any point O on its circumference. Let O be the pole and the diameter through O be taken as the initial line; then the polar equation of the circle passing through O is given by

$$r = 2R\cos\theta \qquad (1)$$

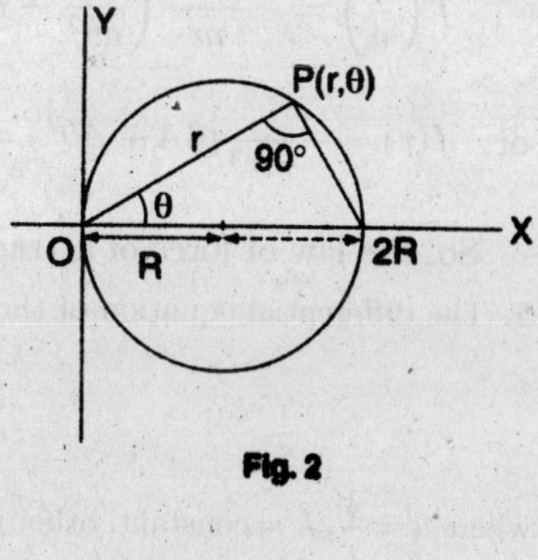

Fig. 2

($\because$ from the geometry of the fig. $\dfrac{r}{2R} = \cos\theta$.)

Putting $r = \dfrac{1}{u}$ in (1), we have

$$\frac{1}{u} = 2R\cos\theta,\quad \text{or,}\quad u = \frac{1}{2R}\sec\theta. \qquad (2)$$

Differentiating (2) w.r.t. θ, we have

$$\frac{du}{d\theta} = \frac{1}{2R}\sec\theta\tan\theta.$$

Differentiating again w.r.t. θ, we get

$$\frac{d^2u}{d\theta^2} = \frac{1}{2R}[\sec\theta\tan^2\theta + \sec^3\theta] = \frac{1}{2R}\sec\theta[\tan^2\theta + \sec^2\theta] = \frac{1}{2R}\sec\theta[\sec^2\theta - 1 + \sec^2\theta]$$

$$= \frac{1}{2R}\sec\theta[2\sec^2\theta - 1] = \frac{2}{2R}\sec^3\theta - \frac{1}{2R}\sec\theta = \frac{1}{R}\sec^3\theta - u$$

[from (2), putting $\frac{1}{2R}\sec\theta = u$]

$$\text{or,}\quad \frac{d^2u}{d\theta^2} + u = \frac{1}{R}\sec^3\theta. \qquad (3)$$

Now from the differential equation for the path of the particle in a central force field, we may write

$$f\left(\frac{1}{u}\right) = -mh^2u^2\left(\frac{d^2u}{d\theta^2}+u\right) = -mh^2u^2\frac{1}{R}\sec^3\theta \quad \text{[from (3)]}$$

$$= -8mh^2R^2u^5 \quad \text{[from (2)]}$$

$$\text{or,} \quad f(r) = -8mh^2R^2\frac{1}{r^5} = -\text{ constant} \times \frac{1}{r^5},$$

i.e., the law of force of attraction is an inverse fifth power of the distance from O.

4. The equation of the orbit of a particle moving under the influence of a central force is given as $r = A\theta^2$, where A is a constant. Obtain the force law. **[C.U. (Hons.) 1991]**

We have

$$r = A\theta^2, \quad \text{or,} \quad \frac{1}{r} = u = \frac{1}{A}\theta^{-2} = B\theta^{-2}. \quad \therefore \theta^{-2} = \frac{1}{Br} = \frac{A}{r}.$$

Differentiating, we get

$$\frac{du}{d\theta} = -2B\theta^{-3} \quad \text{and} \quad \frac{d^2u}{d\theta^2} = 6B\theta^{-4} = 6B\theta^{-2}\theta^{-2} = 6\frac{1}{r}\frac{A}{r} = \frac{6A}{r^2}.$$

Now from the differential equation for the path of the particle in a central force field, we may write,

$$f\left(\frac{1}{u}\right) = -\frac{L^2u^2}{m}\left(\frac{d^2u}{d\theta^2}+u\right) = -\frac{L^2u^2}{m}\left(\frac{6A}{r^2}+\frac{1}{r}\right) = -\frac{L^2}{mr^2}\frac{(6A+r)}{r^2} = -\frac{L^2}{mr^4}(6A+r),$$

$$\text{or,} \quad f(r) = -\frac{L^2}{mr^4}(6A + A\theta^2) = -\frac{L^2}{mr^4}A(6+\theta^2) = -\text{constant} \times \frac{1}{r^4}.$$

So, the law of force of attraction is an inverse fourth power of the radius vector.

5. The differential equation of the orbit of a particle of mass m under a central force is given by

$$\frac{d^2u}{d\theta^2} + u = -\frac{m}{L^2u^2}f\left(\frac{1}{u}\right),$$

where $u = \frac{1}{r}$, L = constant, other notations have usual significance. Use the above relation and consider the following: A particle moves in a central orbit described by $r = e^{-\alpha\theta}$, α is a positive constant, with force centre at O. Find the nature of the force as a function of r. **[C.U. (Hons.) 1997]**

Here we have

$$r = e^{-\alpha\theta}. \quad \therefore \quad u = e^{\alpha\theta} \quad \left(\because u = \frac{1}{r}\right).$$

$$\therefore \quad \frac{du}{d\theta} = \alpha e^{\alpha\theta}, \quad \frac{d^2u}{d\theta^2} = \alpha^2 e^{\alpha\theta} = \frac{\alpha^2}{r}.$$

Now from the equation of motion, we may write

$$f\left(\frac{1}{u}\right) = -\frac{L^2u^2}{m}\left(u + \frac{d^2u}{d\theta^2}\right) = -\frac{L^2}{mr^2}\left(\frac{1}{r}+\frac{\alpha^2}{r}\right) = -\frac{L^2}{mr^3}(1+\alpha^2).$$

$$\therefore \quad f(r) = -\frac{\text{constant}}{r^3}. \quad \because \quad \frac{L^2}{m}(1+\alpha^2) = \text{constant}.$$

$$\therefore f(r) \propto -\frac{1}{r^3}.$$

So the force is inversely proportional to the cube of radius vector and the nature of the force is attractive which is evident from the negative sign.

7.24 Velocity of Escape

We consider a body of mass m at a distance x from the centre of the earth (x being greater than R, the radius of the earth). The force with which the body is attracted by the earth towards its centre is given by

$$F = \frac{GMm}{x^2},$$

where M is the mass of the earth.

In order to displace the body by a very small distance dx against the force of attraction, the work done is

$$F dx = \frac{GMm}{x^2} \cdot dx.$$

Hence the total work that will be done in projecting the body from the surface of the earth to an infinite distance, i.e., away from gravitational influence is given by

$$W = \int_R^\infty \frac{GMm}{x^2} dx = -GMm\left[\frac{1}{x}\right]_R^\infty = \frac{GMm}{R}.$$

If the body in question be projected from the surface of the earth with kinetic energy which exceeds the above value of the potential energy, it will escape from the influence of the earth's gravitational field. Hence the minimum velocity of escape v_e is given by

$$\frac{1}{2}mv_e^2 = \frac{GMm}{R}, \quad \text{or,} \quad v_e = \sqrt{\frac{2GM}{R}} = \sqrt{2gR} \quad \left[\because g = \frac{GM}{R^2}\right],$$

where g is the acceleration due to gravity on the surface of the earth.

It should be noted that this is also the velocity which the body acquires if it is allowed to fall from infinity to the surface of the earth.

Using the values $R = 6.4 \times 10^8$ cm, $g = 980$ cm/sec^2, we get

$$v_e = \sqrt{2 \times 980 \times 6.4 \times 10^8} = 1.12 \times 10^6 \text{ cm per second.}$$

Thus the velocity of escape comes out to be 1.12×10^6 cm per second or about 7 miles per second or 25,200 miles per hour (nearly 25,000 miles/hr.).

Thus we see that with what a tremendous velocity the body is to be projected so that it will never return.

The escape velocity from the earth, i.e., about 7 miles per second, is more than thirty times the velocity of sound in air at 0°C, and about six times the root mean square velocity of hydrogen molecules at the same temperature. Even if the average velocity of the molecules of a gas is 25% or so of the escape velocity, an appreciable fraction of the molecules will have velocities above the escape velocity. Such fast molecules will soon be lost by escape from the earth's atmosphere. The earth was very hot in the past, and the gas molecules in its atmosphere had higher velocities. The molecules of the lightest gases, hydrogen and helium, had the highest velocities. They gradually escaped from the earth, while it was hot. This explains the rarity of hydrogen and helium in the earth's atmosphere.

The moon has a much smaller mass than the earth, and so also the planet mercury. They could not have retained any gaseous atmosphere about them because of the weakness of the pull exerted by them. There is evidence to support this view. The same thing applies to satellites of all planets.

7.25 Artificial Satellites

For hundreds of years the idea of cosmic space travel seemed to most people fantastic and impracticable. Although stories about the journey to other planets were written, still the idea of space travel remained only a dream. The whole world, therefore, was taken by surprise when on October 4, 1957, the world's first man-made satellite 'Sputnik' was launched in the Soviet Union. The 'Sputnik I' or the 'baby moon' as it was called, is spherical in form having diameter 58 cm and weight 83.6 kg. The 'Sputnik' was put into an elliptic orbit round the earth at a distance of 900 km from the earth. It made one full revolution round the earth in 96.2 minutes attaining a speed of 8 km/sec or nearly 5 miles per sec The satellite was taken to a planned height and put into the orbit at a predetermined point by means of multi-stage carrier rocket.

A month after the first artificial satellite was launched, the second artificial satellite 'Sputnik II' was put into orbit round the earth by the Soviet Union. 'Sputnik II' carried complicated scientific instruments for space research and also a dog passenger (Lika). The total weight of this artificial satellite was 508 kg. It was launched at a height of 1700 km and it revolved round the earth with a period of revolution 102 secs.

Soviet space-exploration was followed by the launching of the first American artificial satellite 'The Explorer' on January 31, 1958. This artificial satellite was of smaller size and weight.

Men all over the world watched with greater sensation when the Soviet Union put the first cosmonaut Yuri Gagarin into space in a space-ship and successfully retrieved the space-ship with the cosmonaut back.

In July, 1969, the American astronauts Niel Armstrong and Edwin Aldrin for the first time in the history of mankind landed on the moon. They began their return journey after their stay on the untrodden lunar surface, carrying valueless rocks and other mementoes for earthly beings to study for more knowledge about their closest neighbour moon.

It may be mentioned here that India put in orbit the first satellite (Aryabhatta) moving round the earth on April 19, 1975. India has launched a number of satellites and made much progress in the field of space research.

Artificial satellites are being made to carry an increasing number of instruments of various kinds. When a satellite is put into its orbit, these instruments transmit to the earth data on temperature, pressure, cosmic radiation, magnetic field etc. of the regions through which the satellite moves. Some transmit photographs of the clouds above the earth. These are expected to help in long range fore-casting of weather. Satellite links have brought the remote parts of the world closer to us, and have made it possible for us to view on the television screen in our room, whatever is happening anywhere in the world.

7.25.1 Condition to put a satellite into an orbit

We shall now investigate how the artificial satellite can continue to move round the earth like the moon. In order to avoid generation to heat by friction with air in the atmosphere which may destroy the satellite, the orbit of the satellite should lie outside the atmosphere. Again, at an altitude of 100 miles the density of air in the atmosphere becomes so small that at this height practically there is no friction between air and the satellite. That is why, the satellite should move at heights 100 miles or more above the surface of the earth if it is to orbit round the earth for a reasonable time. The nature of the orbit along which the satellite moves depends on the velocity with which the satellite is projected from the surface of the earth.

Let v be the velocity required to project the satellite of mass m to revolves close around the earth. Then we may write

$$G\frac{Mm}{R^2} = \frac{mv^2}{R} \tag{7.25.1}$$

$\frac{mv^2}{R}$ denotes the centrifugal force on the satellite which is balanced by the gravitational attractive force $\frac{GMm}{R^2}$ due to the earth. The radius of the earth R is taken to be the radius of the orbit of the satellite since the satellite revolves close around the earth.

$$\therefore \quad v^2 = \frac{GM}{R}, \quad \text{or,} \quad v = \sqrt{\frac{GM}{R}}; \tag{7.25.2}$$

v is also known as the *orbital speed of the satellite.*

If the satellite revolves well away from the earth, say at a height H above the earth's surface, then condition (7.25.1) or (7.25.2) still holds and we may write

$$v^2 = \frac{GM}{R+H}, \quad \text{or,} \quad v = \sqrt{\frac{GM}{R+H}},$$

which is the orbital speed in this case.

Again we know that the velocity of escape is given by

$$v_e = \sqrt{\frac{2GM}{R}}.$$

$$\therefore \quad \frac{v}{v_e} = \frac{1}{\sqrt{2}}, \quad \text{or,} \quad v = \frac{v_e}{\sqrt{2}} = \frac{25000}{\sqrt{2}} = 18000 \text{ miles/hour (nearly)}.$$

[$\because$ $v_e = 25000$ miles/hr. nearly].

Again we know, when $v < \sqrt{\frac{2GM}{R}}$, i.e., $v < 25000$ miles/hr., the orbit described by the satellite will be an ellipse. Hence for elliptic orbit of the satellite its velocity of projection should lie within the range, 18000 to 25000 miles per hour.

When $v = \sqrt{\dfrac{2GM}{R}} = 25000$ miles per hour, the satellite will move in a parabolic path and hence escapes from the earth.

Again when $v > \sqrt{\dfrac{2GM}{R}}$, i.e., $v > 25000$ miles/hr., the satellite moves along a hyperbolic path and escapes from the earth.

Thus to satisfy the condition that a satellite is put into an elliptic orbit with the centre of the earth at one of its foci, the velocity of projection of the satellite must lie between 18000 miles/hr. and 25000 miles/hr.

7.25.2 Period of a satellite revolving round the earth

If ω be the angular velocity of the satellite, which is supposed to revolve very close around the earth in an orbit of radius R (which is also taken as the radius of the earth in this case), we may write from (7.25.1) or (7.25.2), putting

$$v = \omega R, \quad \omega^2 = \frac{GM}{R^3}, \quad \text{or,} \quad \frac{4\pi^2}{T^2} = \frac{GM}{R^3} = \frac{g}{R},$$

since $\omega = \dfrac{2\pi}{T}$, T being the period of revolution of the satellite, and $g = \dfrac{GM}{R^2}$, so that we have

$$T^2 = 4\pi^2 \frac{R}{g}, \quad \text{or,} \quad T = 2\pi\sqrt{\frac{R}{g}}.$$

If, however, the satellite revolves well away from the earth, say, at a distance H above the earth's surface, condition (7.25.1) or (7.25.2) still holds. $(R+H)$ is the distance of the satellite from the centre of the earth. We may write from (7.25.1) or (7.25.2) putting $v =$ orbital speed $= \omega(R+H)$,

$$\omega^2 = \frac{GM}{(R+H)^3}, \quad \text{or,} \quad \frac{4\pi^2}{T^2} = \frac{gR^2}{(R+H)^3} \quad \left(\because \omega = \frac{2\pi}{T} \text{ and } \frac{GM}{R^2} = g\right)$$

$$\text{or,} \quad T^2 = \frac{4\pi^2}{g}\frac{(R+H)^3}{R^2}. \quad \therefore \quad T = 2\pi\frac{(R+H)}{R}\sqrt{\frac{R+H}{g}}.$$

The above expression for the time period may also be deduced by the following simple method.

A man in the satellite will be in a state of weightlessness, when the centrifugal force (equal and opposite to the centri-petal force, i.e., the force due to the centri-petal acceleration) acting on the man due to rotation of the satellite just balances the attraction exerted on him by the earth. Then if v be the orbital speed, we may write

$$\frac{mv^2}{(R+H)} = \frac{GMm}{(R+H)^2} = \frac{gR^2m}{(R+H)^2} \quad (\because GM = gR^2), \quad \text{or,} \quad v = \frac{R}{(R+H)}\sqrt{(R+H)g}. \tag{7.25.3}$$

If $H \ll R$, then the orbital speed $= \sqrt{Rg}$ approximately.

Again,

$$\text{Orbital speed} = \frac{\text{distance travelled in one revolution}}{\text{time for one revolution or period}}.$$

So

$$v = \frac{2\pi(R+H)}{T}, \quad \text{or,} \quad T = \frac{2\pi(R+H)}{v} = 2\pi\frac{(R+H)}{R}\sqrt{\frac{(R+H)}{g}} \quad \text{from (7.25.3).}$$

If $H \ll R$, the orbital period $= 2\pi\sqrt{\dfrac{R}{g}}$ approximately.

N.B. The orbital speed given by (7.25.3) can be written as $v = R\sqrt{\dfrac{g}{R+H}}$, which shows that greater is the height of the satellite above the earth's surface, smaller is the orbital speed. Again T can also be written as $T = \dfrac{2\pi(R+H)^{3/2}}{R\sqrt{g}}$, which shows that the period of the satellite is proportional to the $\frac{3}{2}$th power of its orbital radius $(R+H)$.

In the state of "weightlessness", the net force on all bodies inside the satellite is zero, and as such, they will appear to move without acceleration inside the satellite. In view of this, the value of "g" for a pendulum inside the satellite will be zero, and hence the simple pendulum experiment cannot be conducted there.

7.25.3 Geo-stationary Satellite

Communication satellites transmit signals across large distances by receiving these signals from one point on the earth's surface, and reflecting them down to another point. In order to provide a stationary target for the transmitted signals, these satellites must remain stationary at a point above the earth. Now a satellite would appear stationary over a point on the earth's equator, if its orbit is equatorial and circular at such a height over the earth that its period of revolution is 24 hours. Then it would be *synchronous* with earth's spin. Such a satellite is known as a ***geo-stationary*** satellite.

Illustrative Examples

1. A satellite is in a circular orbit around the earth. Show that the minimum additional energy needed by the satellite to escape from the gravitational pull of the earth is equal to its kinetic energy in the orbit. What will happen if its kinetic energy is suddenly doubled?

The K.E. of a satellite of mass m, revolving around the earth with velocity v is $\frac{1}{2}mv^2$. Again the energy needed by the satellite to escape from the earth's gravitational pull is $\frac{1}{2}mv_e^2$, so that the additional energy required

$$= \frac{1}{2}mv_e^2 - \frac{1}{2}mv^2 = \frac{1}{2}m(v_e^2 - v^2) = \frac{1}{2}m(2v^2 - v^2) \quad (\because\ v_e = \sqrt{2}v) = \frac{1}{2}mv^2,$$

which is the K.E. of the satellite in its orbit.

So if the K.E. in the orbit is suddenly doubled, the satellite would escape to infinity.

2. Find out an expression for the period T of an earth satellite orbiting in a circular path at height nR from the earth's surface, where R is the earth's radius. Compute the value of n which makes T equal to 1 day, given that $T = 1.41$ hours for $n = 0$.

The orbital period of an earth satellite at a height h above the earth's surface is

$$T = \frac{2\pi(R+h)}{R}\sqrt{\frac{R+h}{g}},$$

where R = earth's radius. (See Art. 7.25.2)

In the problem $h = nR$, so that

$$T = \frac{2\pi(R+nR)}{R}\sqrt{\frac{R+nR}{g}} = 2\pi\sqrt{\frac{R}{g}}(1+n)^{3/2}.$$

Now for $n = 0$, $T = 1.41$ hours,

$$\therefore \quad 2\pi\sqrt{\frac{R}{g}} = 1.41 \text{ hours.}$$

$$\therefore \quad T = 1.41(1+n)^{3/2} \text{ hours.}$$

Putting $T = 1$ day $= 24$ hours, we have

$$24 = 1.41(1+n)^{3/2}, \text{ or, } (1+n)^{3/2} = \frac{24}{1.41} = 17,$$

$$\text{or, } (1+n) = (17)^{2/3} = 6.61 \quad \text{(using Log Tables).}$$

$$\therefore \quad n = 5.61.$$

So a satellite orbiting the earth in a circular orbit at height $5.61R$ above the earth's surface will appear stationary over a point on the equator.

3. Two satellites of same mass are launched in the same orbit round the earth so as to rotate opposite to each other. They collide inelastically and stick together as wreckage. Obtain the total energy of the system before and after the collision. Discuss the subsequent motion of the wreckage.

The P.E. of a satellite in its orbit is $-\frac{GMm}{r}$ and the K.E. is $\frac{GMm}{2r}$. $\Big[$Since gravitational attraction $\frac{GMm}{r^2}$ = centri-petal force $\frac{mv^2}{r}$. $\therefore mv^2 = \frac{GMm}{r}$, or, $\frac{1}{2}mv^2 = \frac{GMm}{2r}\Big]$.

So the total energy is

$$E = \frac{GMm}{2r} - \frac{GMm}{r} = -\frac{GMm}{2r}.$$

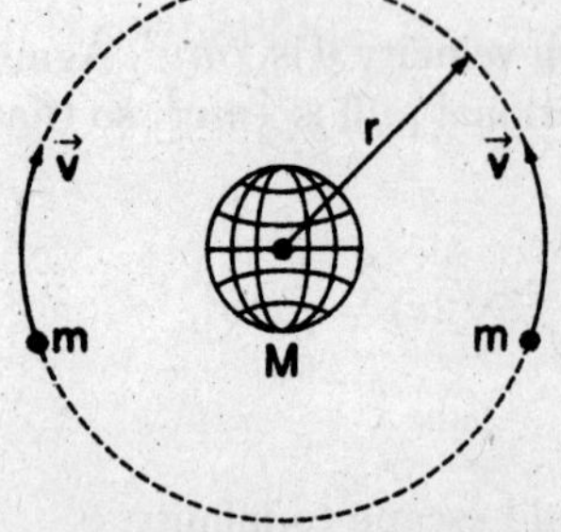

Here m = mass of the satellite, M = mass of the earth, and r = the orbital radius.

For the two satellites, each of mass m in the same orbit rotating opposite to each other as shown in the attached figure, the total energy

$$= -\frac{GMm}{2r} - \frac{GMm}{2r} = -\frac{GMm}{r}.$$

Let $\vec{v'}$ be the velocity of the wreckage after the collision. Then from conservation of momentum, we have, taking $\vec{v}$ as the orbital speed of the satellites,

$$m\vec{v} + m\vec{v} = (m+m)\vec{v'}, \quad \text{or,} \quad mv - mv = 2mv'. \quad \therefore \quad v' = 0.$$

Therefore the wreckage of mass $2m$ has no kinetic energy, but only potential energy. Hence the total energy after the collision will be given by $-\dfrac{GM(2m)}{r}$.

Since the velocity of the wreckage is zero, the centripetal force disappears and the wreckage falls down under gravity.

4. If the gravitational force between two bodies had been inversely proportional to the fourth power of the distance between them, then find out the escape velocity at the surface of the earth—given $g = 10\,\text{m/sec}^2$ and radius of the earth, $R = 6400 \times 10^3$ m. [C.U. (Hons.) 2001]

We consider a body of mass m at a distance r from the centre of the earth—here $r > R$, the radius of the earth.

According to the problem, the gravitational force between two bodies is inversely proportional to the fourth power of the distance between them—so that we have $F = \frac{GMm}{r^4}$ where M is the mass of the earth.

In order to displace the body by a very small distance dr against this force of attraction, the work done is $Fdr = \frac{GMm}{r^4}dr$, so that the total work that will be done in projecting the body from the surface of the earth to an infinite distance i.e. away from gravitational influence is given by

$$W = \int_R^\infty \frac{GMm}{r^4}dr = -\frac{1}{3}GMm\left[\frac{1}{r^3}\right]_R^\infty = \frac{1}{3}\frac{GMm}{R}$$

Therefore the escape velocity v_e is given by

$$\frac{1}{2}mv_e^2 = \frac{1}{3}\frac{GMm}{R^3} \quad \text{or} \quad v_e^2 = \frac{2}{3}\frac{GM}{R^3} = \frac{2}{3}gR \quad \left(\because mg = \frac{GMm}{R^4} \quad \text{or} \quad gR^4 = GM\right)$$

$$\therefore \quad v_e = \sqrt{\frac{2gR}{3}} = \sqrt{\frac{2 \times 10 \times 6400 \times 10^3}{3}} = 6.532 \times 10^3 \text{ m/sec}$$

(An alternate method has been given in the 'Exercises'.)

5. Consider a pair of stars of equal mass M rotating about their common C.M. The attraction between the stars is gravitational and the stars keep a separation l between them. Show that the time period of rotation of this double star system is given by $\pi l\sqrt{\frac{2l}{GM}}$ where G is the universal gravitational constant. [C.U. (Hons.) 2003]

We know that the motion of a two-particle system can be replaced by the motion of one single particle (w.r.t. each other—the other particle being kept fixed) having reduced mass $\mu = \frac{M \times M}{M+M} = \frac{M}{2}$. Since the gravitational attraction between the stars supplies the necessary centripetal force, we may write

$$\mu\omega^2 l = \frac{GMM}{l^2} \quad \text{or} \quad \omega^2 = \frac{GM^2}{\mu l^3} = \frac{2GM}{l^3} \quad \left(\because \mu = \frac{M}{2}\right) \quad \text{or} \quad \omega = \frac{1}{l}\sqrt{\frac{2GM}{l}}$$

$$\text{and} \quad T = \frac{2\pi}{\omega} = 2\pi l\sqrt{\frac{l}{2GM}} = \pi l\sqrt{\frac{2l}{GM}}$$

7.25.4 System of variable mass—motion of a rocket

We consider a system in which mass enters or leaves the system while we are observing. The rate of change of the total mass of the system, i.e., $\frac{dM}{dt}$ is positive when the mass enters and negative when it leaves the system. In the case of the motion of a rocket it burns fuel at a constant rate and ejects gas with a constant velocity. Hence there is decrease of mass of the rocket with the fuel, i.e., $\frac{dM}{dt}$ is negative.

Now we consider a system of mass whose centre of mass is moving with velocity $\vec{v}$ when seen from a particular reference frame. Let an external force F act on the system. After time δt let a mass δM be ejected from the system, its centre of mass moving with velocity $\vec{u}$ as seen by an observer. So the mass of the system is reduced to $M - \delta M$ and the velocity $\vec{v}$ of the centre of mass of the system is changed to $\vec{v} + \delta\vec{v}$. In case of rocket hot gas is ejected from its orifice at a fairly high speed, decreasing its mass and increasing its speed. The loss of mass is continuous during the burning process of the rocket. Here the external force $\vec{F}$ is the force of gravity and the resisting force of the atmosphere.

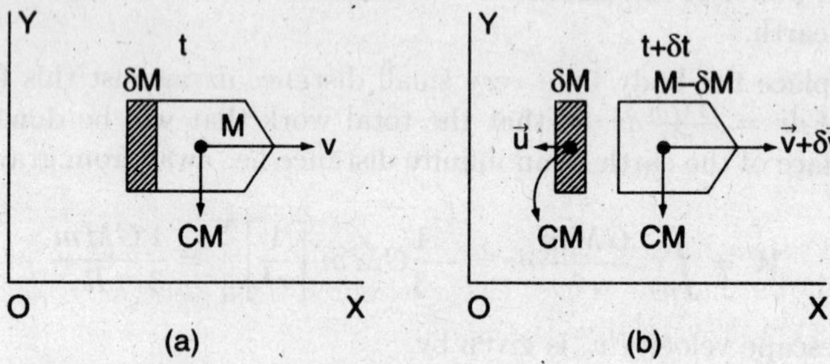

Fig. 7.25.1

Applying Newton's second law for the system in which the mass is not constant, we may write

$$\vec{F} = \frac{d\vec{P}}{dt},$$

where $\vec{P}$ denotes the momentum of the system. For finite time interval δt we can write,

$$\vec{F} = \frac{\delta\vec{P}}{\delta t} = \frac{\vec{P}_2 - \vec{P}_1}{\delta t},$$

where $\vec{P}_1$ and $\vec{P}_2$ denote respectively the initial and final momenta of the system.

Now

$$\vec{P}_1 = M\vec{v} \quad \text{and} \quad \vec{P}_2 = (M - \delta M)(\vec{v} + \delta\vec{v}) + \delta M\vec{u}.$$

$$\therefore \quad \vec{F} = \frac{[(M - \delta M)(\vec{v} + \delta\vec{v}) + \delta M\vec{u}] - [M\vec{v}]}{\delta t}$$

$$= \frac{M\delta\vec{v} - \delta M\vec{v} - \delta M\delta\vec{v} + \delta M\vec{u}}{\delta t}$$

$$= M\frac{\delta\vec{v}}{\delta t} + [\vec{u} - (\vec{v} + \delta\vec{v})]\frac{\delta M}{\delta t}. \qquad (7.25.1)$$

As δt approaches zero, $\dfrac{\delta\vec{v}}{\partial t}$ approaches $\dfrac{d\vec{v}}{dt}$. As $\dfrac{dM}{dt}$ is the change in mass with time, it is negative in this case. Hence as δt approaches zero, $\dfrac{\delta M}{\delta t}$ may be replaced by $-\dfrac{dM}{dt}$. So from equation (7.25.1) we obtain

$$\vec{F} = M\frac{d\vec{v}}{dt} + \vec{v}\frac{dM}{dt} - \vec{u}\frac{dM}{dt} \tag{7.25.2}$$

[$\because$ δv tends to zero as δt approaches zero]

$$\text{or,}\quad F = \frac{d}{dt}(M\vec{v}) - \vec{u}\frac{dM}{dt}. \tag{7.25.3}$$

In equation (7.25.1), $\vec{u} - (\vec{v} + \delta\vec{v})$ denotes the relative velocity of the ejected mass of the rocket with respect to the main body of the rocket. So equation (7.25.2) may be written as

$$M\frac{d\vec{v}}{dt} = \vec{F} + (\vec{u} - \vec{v})\frac{dM}{dt} \tag{7.25.4}$$

$$\text{or,}\quad M\frac{d\vec{v}}{dt} = \vec{F} + \vec{v}_{\text{rel}}\frac{dM}{dt}. \tag{7.25.5}$$

In the above equation $\vec{v}_{\text{rel}}\dfrac{dM}{dt}$ denotes the rate at which the momentum is transferred into or out of the system by the mass which the system has ejected or received. This may be considered as the force exerted on the system by the mass that is ejected or collected by the system. In case of rocket this is called the *thrust* and it is made as large as possible. The equation (7.25.4) shows that the rocket ejects as much mass per unit time as possible and so the speed of the ejected mass relative to the rocket should be as high as possible. The equation (7.25.4) may also be written in the form

$$\frac{M\,d\vec{v}}{dt} = \vec{F} + \vec{R}, \tag{7.25.6}$$

where $\vec{R} = \vec{v}_{\text{rel}}\dfrac{dM}{dt}$ which denotes the reaction force exerted on the system by the mass that leaves it.

If all external forces are neglected, i.e., $\vec{F} = 0$, we have from (7.25.5),

$$M\frac{d\vec{v}}{dt} = \vec{v}_{\text{rel}}\frac{dM}{dt}, \quad \text{or,}\quad d\vec{v} = \vec{v}_{\text{rel}}\frac{dM}{M}.$$

On integration from the time when the velocity is $\vec{v}_1$ and the mass of the rocket is M_1 to the time when the velocity is $\vec{v}_2$ and the mass of the rocket, is M_2, we have

$$\int_{\vec{v}_1}^{\vec{v}_2} d\vec{v} = \vec{v}_{\text{rel}}\int_{M_1}^{M_2}\frac{dM}{M},$$

$$\text{or,}\quad (\vec{v}_2 - \vec{v}_1) = -\vec{v}_{\text{rel}}\log_e\frac{M_1}{M_2} = -\vec{v}_{\text{rel}}\log_e\left(1 + \frac{M_1 - M_2}{M_2}\right).$$

This shows that the change in the speed of the rocket in any interval of time depends on the relative velocity of the ejected mass in the opposite direction and on the fraction of the mass ejected or exhausted during that interval of time.

If the rocket starts from rest, we may write $\vec{v}_1 = 0$.

$$\therefore \quad \vec{v}_2 = -\vec{v}_{\text{rel}} \log_e \left(\frac{M_1}{M_2}\right), \tag{7.25.7}$$

$$\text{or,} \quad \log_e \left(\frac{M_1}{M_2}\right) = -\frac{\vec{v}_2}{\vec{v}_{\text{rel}}}. \quad \therefore \quad \frac{M_1}{M_2} = e^{-\vec{v}_2/\vec{v}_{\text{rel}}}. \tag{7.25.8}$$

From equation (7.25.7) it follows that for attaining high speed $\log_e \dfrac{M_1}{M_2}$ should be as high as possible or the initial mass M_1 of the rocket should be as high possible when no external force acts on the rocket.

It should be noted that the above equation shows that the velocity of the rocket can increase to any value so that the final remaining mass M_2 is sufficiently small. Actually however, the ratio M_1/M_2 has a practical limit and it cannot be increased indefinitely. A single rocket, i.e., one-stage rocket will not, therefore, attain very high velocities. Multi-stage rockets are used for this purpose.

From relativistic mechanics we know that the rocket cannot be accelerated to a velocity equal to or greater than the velocity of light. When the rocket attains very high velocity compared to the velocity of light, we must take into consideration the variation of mass of the rocket with velocity and so the classical equations as deduced above will not be applicable.

Illustrative Examples on Rocket motion

1. A rocket weighing 30,000 lbs is fueled up on the launching pad and fired vertically and at burn-out weighs 15,000 lbs. Gases are exhausted at the rate of 10 slugs/sec with a velocity of 4,000 ft/sec relative to the rocket. Calculate the thrust on the rocket and its velocity at burn-out.

Thrust $= v_{\text{rel}} \dfrac{dM}{dt} =$ (4,000 ft/sec) (10 slugs/sec) = 40,000 lb-wt.

Just before burn-out the upward force = (40,000 − 15,000) = 25,000 lb-wt.

Initially the thrust = (40,000 − 30,000) lb−wt = 10,000 lb-wt.

Velocity $v = v_{\text{rel}} \log_e \dfrac{M_1}{M_2} = 4{,}000 \text{ ft/sec} \log_e \dfrac{30,000}{15,000} = (4{,}000 \text{ ft/sec}) \log_e 2 = 4392.21 \text{ ft/sec}$

2. (a) A rocket starts vertically upward with speed v_0. Show that its speed at a height h is given by $v_0^2 - v^2 = \dfrac{2gh}{1+\frac{h}{R}}$, where R is the radius of the earth and g the acceleration due to gravity at the earth's surface.

(b) A rocket has a velocity 50 kms/sec when at a distance of 1000 kms from the surface of the earth. What will be its velocity when it is 5000 kms away? Radius of the earth = 6.38×10^8 cms.

(c) Deduce an expression for the maximum height reached by a rocket fired with speed 0.9 times the escape velocity.

Ans. (a) K.E. at the earth's surface $= \dfrac{1}{2}mv_0^2$ (m = mass of the rocket) and gravitational P.E. at the earth's surface $= -\dfrac{GMm}{R}$ (M = mass of the earth). Again K.E. at a distance h from the earth's surface $= \dfrac{1}{2}mv^2$ (where v = velocity of the rocket at a distance h) and gravitational

P.E. at a distance of $h = -\dfrac{GMm}{(R+h)}$. Therefore from the principle of conservation of energy, we have

$$\frac{1}{2}mv_0^2 - \frac{GMm}{R} = \frac{1}{2}mv^2 - \frac{GMm}{(R+h)}.$$

$$\therefore \quad v_0^2 - v^2 = \frac{2GM}{R} - \frac{2GM}{(R+h)} = 2GM\left(\frac{1}{R} - \frac{1}{(R+h)}\right) = 2GM\frac{h}{R(R+h)}$$

$$= \frac{2gR^2}{R}\frac{h}{R+h} \quad (\because GM = gR^2)$$

$$= 2gh/\left(1+\frac{h}{R}\right). \tag{1}$$

(b) Apply the formula $v_1^2 - v_2^2 = 2gR^2\left(\dfrac{1}{h_1} - \dfrac{1}{h_2}\right)$ and get the result.

(c) Let $h_{\max}$ be the maximum height reached by the rocket, at which its velocity has been reduced to zero. So, substituting $v = 0$ and $h = h_{\max}$ in (1), we have

$$v_0^2 = \frac{2gh_{\max}}{1+\dfrac{h_{\max}}{R}}, \quad \text{or,} \quad v_0^2\left(1+\frac{h_{\max}}{R}\right) = 2gh_{\max},$$

$$\text{or,} \quad v_0^2 = h_{\max}\left(2g - \frac{v_0^2}{R}\right), \quad \text{or,} \quad h_{\max} = \frac{v_0^2}{2g - \dfrac{v_0^2}{R}}.$$

In the problem, $v_0 = 0.9\times$ escape velocity $= 0.9 \times \sqrt{2gR}$.

$$\therefore \quad h_{\max} = \frac{0.9\times 0.9\times 2gR}{2g - \dfrac{0.9\times 0.9\times 2gR}{R}} = \frac{1.62gR}{2g - 1.62g} = \frac{1.62gR}{0.38g} = 4.26R.$$

3. A rocket starts falling vertically from rest under constant gravity. From the instant it starts falling, it is also ejecting gas at a constant rate α in the vertically downward direction at a speed v_0 with respect to the rocket. Determine the speed of the rocket after time t. [C.U. (Hons.) 1992]

Applying equation 7.25.4 and writing $\vec{F} = M\vec{g}$, we have

$$M\frac{d\vec{v}}{dt} = M\vec{g} + \vec{v}_0\frac{dM}{dt}, \quad \text{or,} \quad d\vec{v} = \vec{g}dt + \vec{v}_0\frac{dM}{M} \tag{1}$$

$$\text{or,} \quad \int_{\vec{v}'}^{\vec{v}} d\vec{v} = \int_0^t \vec{g}dt + \vec{v}_0\int_{M_0}^{M}\frac{dM}{M}, \quad \text{or,} \quad \vec{v} - \vec{v}' = \vec{g}t + \vec{v}_0\log\frac{M}{M_0}, \tag{2}$$

where v' = velocity of the rocket when it is starting to eject gas, v = its velocity after time t, and M, M_0 = the masses of the system at $t = t$, $t = 0$.

If T be the life-time of the fuel, and if M_e and M_f be the masses of the empty vehicle, and fuel respectively, then $M_0 = M_e + M_f$, and at any time t, the spent up fuel $= \dfrac{M_f}{T}t$, so that the mass of the fuel remaining is given by

$$M_f - \frac{M_f}{T}t = M_f\left(1 - \frac{t}{T}\right).$$

$$\therefore \quad M = M_e + M_f\left(1 - \frac{t}{T}\right) = M_e + M_f - M_f\frac{t}{T} = M_0 - M_f\frac{t}{T} = M_0 - \alpha t,$$

where $\dfrac{M_f}{T} = \alpha =$ constant rate of ejection of the fuel.

$\therefore$ From (2), we get

$$\vec{v} - \vec{v}' = \vec{g}t + \vec{v}_0 \log_e\left(1 - \frac{\alpha t}{M_0}\right). \tag{3}$$

Since the term $\vec{v}_0$ is always opposite to $\vec{v}$, we can write in scalar notation,

$$v = v' + gt - v_0 \log_e\left(1 - \frac{\alpha t}{M_0}\right),$$

which is the speed of the rocket after time t.

4. Integrate the rocket equation to obtain v as a function of m, assuming a constant time rate of loss of mass.

Show, for a rocket starting initially from rest, with v' equal to 6800 ft/sec vertically downwards and a mass loss per second equal to $\frac{1}{60}$th of the initial mass, that in order to reach the escape velocity from the earth, the ratio of the mass of the fuel to the mass of the empty rocket must be almost 300. (Escape velocity from the earth = 11.2 km/sec). **[C.U. (Hons.) 1994, 2007]**

From the rocket equation

$$m\frac{dv}{dt} = -mg - v'\frac{dm}{dt},$$

we may write

$$\frac{dv}{dt} = -g - \frac{v'}{m}\frac{dm}{dt}. \tag{1}$$

Here m = mass of rocket, v = its forward velocity and v' = velocity of exhaust gas relative to rocket.

Now,

$$\frac{dv}{dt} = \frac{dv}{dm}\frac{dm}{dt} = \frac{dv}{dm}k,$$

where $k = \dfrac{dm}{dt}$ = constant time rate of loss of m, so that (1) reduces to

$$k\frac{dv}{dm} = -g - \frac{v'}{m}k,$$

which on integration within proper limits yields

$$\int_0^{v_r} dv = -\frac{g}{k}\int_{m_0}^{m_r} dm - v'\int_{m_0}^{m_r}\frac{dm}{m},$$

since at $t = 0$, $v = 0$, $m = m_0$ = the initial mass, and at $t = t_r$, the burn-out time (when the rocket is empty), $v = v_r$ = the final velocity of the rocket, $m = m_r$ the mass of the empty rocket. Therefore we have

$$v_r = \frac{g}{k}(m_0 - m_r) + v'\log_e\frac{m_0}{m_r} = \frac{g}{k}m_f + v'\log_e\left(1 + \frac{m_f}{m_r}\right),$$

where $m_f = (m_0 - m_r)$ = the initial mass of the fuel.

In our problem, $v = v_r = v_e = 11.2$ km/sec $= 6.95$ miles/sec; $v' = 6800$ ft/sec $= \frac{6800}{5280}$ miles/sec, $k = \frac{m_0}{60}$: substituting these values in the above expression, we have

$$6.95 = \frac{32.2 \times 60 m_f}{5280 \times m_0} + \frac{6800}{5280} \log_e \left(1 + \frac{m_f}{m_r}\right) = 0.36 \frac{m_f}{m_0} + 1.29 \ln \left(1 + \frac{m_f}{m_r}\right)$$

(writing $\ln = \log_e$).

Now $\frac{m_f}{m_0} < 1$ and $\frac{m_f}{m_r}$ is much greater than unity, so that we may write

$$6.95 \simeq 1.29 \ln \left(1 + \frac{m_f}{m_r}\right), \quad \text{or,} \quad \ln \left(1 + \frac{m_f}{m_r}\right) \simeq 5.7, \quad \text{or,} \quad \frac{m_f}{m_r} = e^{5.7} - 1 \simeq 298.$$

5. A rocket of mass 1000 kg is ready for a vertical take off. The exhaust velocity of its fuel is 4.5 km/sec. Deduce (a) the minimum rate of fuel ejection so that the rocket weight is just balanced, (b) the velocity acquired in $8S$ if the fuel ejection rate is 2.50 kg/sec. (You may neglect the effect of changing mass of the rocket in the given conditions). **[I.A.S. 1993]**

(a) Using notations of the previous problem, the rocket equation is

$$m \frac{dv}{dt} = -mg - v' \frac{dm}{dt}.$$

Here for $\left(\frac{dm}{dt}\right)_{\min}$ for just balancing the rocket weight, we should have

$$m \frac{dv}{dt} = 0, \quad \text{or,} \quad -\left(\frac{dm}{dt}\right)_{\min} = \frac{mg}{v'} = \frac{1000 \text{ kg} \times 9.8 \text{ m/sec}^2}{4.5 \times 10^3 \text{ m/sec}} = 2.18 \text{ kg/sec}.$$

(b) $m \frac{dv}{dt} = -mg - v' \frac{dm}{dt} = -1000 \times 9.8 - 4.5 \times 10^3 \times 2.5 = 21.05 \times 10^3$

or, $\left|\frac{dv}{dt}\right| = \frac{21.05 \times 10^3}{10^3} = 21.05$, so that we have

$$v = 21.05t \quad \text{(since at } t = 0,\ v = 0, \text{ the integration constant is zero)}$$
$$= 21.05 \times 8 = 168.4 \text{ m/sec.}$$

Illustrative Examples on Planetary motion, escape velocity, artificial satellites etc.

1. Calculate the velocity of escape in the case of the moon from the following data:

mean density of earth	= 5.52 gms/c.c.
mean density of moon	= 3.36 gms/c.c.
mean radius of earth	= 6380 km
mean radius of moon	= 1740 km
acceleration due to gravity at the earth's surface	= 980 cms/sec^2.

We know $g = \dfrac{GM}{R^2}$, where M and R denote the mass and the radius of the earth respectively.

$$\therefore \quad g = G\cdot\frac{\frac{4}{3}\pi R^3\rho}{R^2} = \frac{4}{3}\pi G R\rho,$$

where ρ is the mean density of the earth.

Again the velocity of escape from the moon is given by $v_e = \sqrt{\dfrac{2GM'}{R'}}$, where M' and R' denote respectively the mass and the radius of the moon.

$$\therefore \quad v_e = \sqrt{\frac{2G.4\pi R'^3\rho'}{3R'}} = \sqrt{\frac{8\pi G R'^2\rho'}{3}},$$

where ρ' denotes the mean density of the moon.

$$\therefore \quad v_e^2 = \frac{8\pi G R'^2\rho'}{3}.$$

$$\therefore \quad \frac{v_e^2}{g} = \frac{8\pi G R'^2\rho'}{3} \times \frac{3}{4\pi G R\rho} = \frac{2R'^2\rho'}{R\rho}, \quad \text{or,} \quad v_e = R'\sqrt{\frac{2g\rho'}{R\rho}}.$$

Here

$$R = 6380 \text{ km} = 6380 \times 10^5 \text{ cm} = 6.38 \times 10^8 \text{ cm}$$
$$R' = 1740 \text{ km} = 1740 \times 10^5 \text{ cm} = 1.74 \times 10^8 \text{ cm}$$
$$\rho = 5.52 \text{ gms/c.c.}$$
$$\rho' = 3.36 \text{ gms/c.c. and } g = 980 \text{ cms/sec}^2.$$

$$\therefore \quad v_e = 1.74 \times 10^8 \sqrt{\frac{2 \times 980 \times 3.36}{6.38 \times 10^8 \times 5.52}}$$

$$= 2.38 \times 10^5 \text{ cms/sec} = 2.38 \text{ km/sec.}$$

2. Estimate the mass of the sun, assuming the orbit of the earth round the sun to be a circle. The distance between the Sun and the Earth is 1.49×10^{13} cms, and $G = 6.66 \times 10^{-8}$ C.G.S. units.

[Bombay, 1962]

The force of attraction between the Sun and the Earth is given by

$$F = G\frac{Mm}{R^2},$$

where M = mass the Sun;

m = mass of the Earth;

R = radius of the Earth's orbit round the Sun.

Now as the Earth moves along a circular path of radius R round the Sun, it experiences a centrifugal force acting away from the centre of the orbit which is equal to $\dfrac{mv^2}{R}$, where v is the linear velocity of the Earth in its circular orbit.

$$\therefore \quad \frac{GMm}{R^2} = \frac{mv^2}{R}, \quad \text{or,} \quad M = \frac{v^2 R}{G}.$$

Now,

$$\frac{v}{R} = \omega = \frac{2\pi}{T},$$

where T is the period of revolution of the Earth round the Sun.

$$\therefore \quad v = \frac{2\pi}{T} R.$$

Hence

$$M = \left(\frac{2\pi}{T}\right)^2 \frac{R^3}{G} = \frac{4\pi^2}{T^2}\frac{R^3}{G}.$$

Here $R = 1.49 \times 10^{13}$ cms;

$T = 365$ days $= 365 \times 24 \times 3600$ secs.

$G = 6.66 \times 10^{-8}$ C.G.S. units.

$$\therefore \quad M = \frac{4 \times (3.14)^2 \times (1.49 \times 10^{13})^3}{(365 \times 24 \times 3600)^2 \times (6.66 \times 10^{-8})} = 19.72 \times 10^{34} \text{ gms.}$$

3. If a body falls freely in the earth's gravitational field from infinity, show that it attains the same velocity as that attained by a free fall from a height above the earth equal to the radius R under a constant acceleration of gravity 'g'. [C.U. (Hons.) 1963]

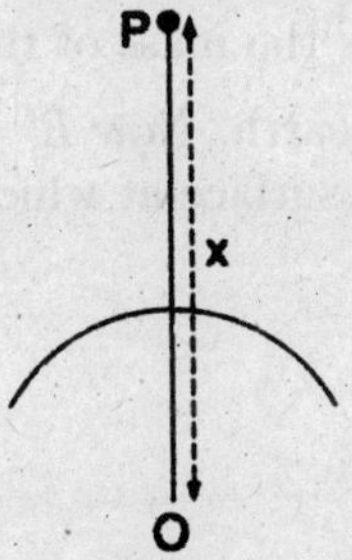

Fig. 1

Considering the body P at a distance x from the centre O of the earth, we may write the differential equation of motion of the body as

$$m\frac{d^2x}{dt^2} = -G\frac{Mm}{x^2}$$ where M and m denote respectively the masses of the earth and the body under consideration.

or, $$\frac{d^2x}{dt^2} = -\frac{GM}{x^2} = -\frac{\mu}{x^2} \quad [\text{putting } \mu = GM]$$

or, $$2\frac{dx}{dt}\cdot\frac{d^2x}{dt^2} = -2\frac{\mu}{x^2}\cdot\frac{dx}{dt} \quad \left(\text{multiplying both sides by } \frac{dx}{dt}\right).$$

Integrating both sides, we have

$$\left(\frac{dx}{dt}\right)^2 = \frac{2\mu}{x} + C,$$

where C is the constant of integration.

Now, at $x = \infty$, $\frac{dx}{dt} = 0$ [$\because$ the body starts from infinity].

This gives $C = 0$.

$$\therefore \quad \left(\frac{dx}{dt}\right)^2 = \frac{2\mu}{x}.$$

Again when $x = R$, the radius of the earth, $\frac{dx}{dt} = v$, the velocity at the surface of the earth.

$$\therefore \quad v^2 = \frac{2\mu}{R}, \quad \text{or,} \quad v = \sqrt{\frac{2\mu}{R}} = \sqrt{\frac{2GM}{R}}. \qquad (1)$$

Now if a body falls under a constant acceleration due to gravity g, we have the velocity attained by the body

$$v'^2 = u^2 + 2gh,$$

when $h = R, u = 0$.

$$\therefore \quad v'^2 = 2gR, \quad \text{or,} \quad v' = \sqrt{2gR}. \qquad (2)$$

Again,

$$g = \frac{GM}{R^2}, \quad \text{or,} \quad gR = \frac{GM}{R}.$$

Substituting this value of gR in (2) we have,

$$v' = \sqrt{\frac{2GM}{R}}. \qquad (3)$$

Thus $v = v'$.

4. A shell is projected horizontally from a rocket when it is at a height of 600 kms above the surface of the earth. Calculate the velocity of projection necessary for the shell to circle round the earth as a satellite at the above height.

[$G = 6.66 \times 10^{-8}$ C.G.S. units; Mass of earth = 6×10^{27} gms; Radius of earth = 6400 kms.]

[C.U. (Hons.) 1960]

In this case the velocity of projection is given by $v = \sqrt{\frac{GM}{R'}}$, where M is the mass of the earth and R' is the radius of the orbit along which the shell moves round the earth. Now $R' = R + h$, where R is the radius of the earth and h the height above the earth's surface at which the shell is projected.

$$\therefore \quad v = \sqrt{\frac{GM}{R+h}}.$$

Here

$$G = 6.66 \times 10^{-8} \text{ C.G.S. units.}$$

$$M = 6 \times 10^{27} \text{ gms.}$$

$$R = 6400\text{kms} = 6.4 \times 10^8 \text{ cms.}$$

$$h = 600 \text{ kms} = 0.6 \times 10^8 \text{ cms.}$$

$$\therefore \quad v = \sqrt{\frac{6.66 \times 10^{-8} \times 6 \times 10^{27}}{10^8 \times 7}}$$

$$= 7.55 \times 10^5 \text{ cms/sec.}$$

$$= 7.55 \text{ kms/sec.}$$

5. Assuming the moon to describe a circular orbit of radius 4×10^5 km round the earth in 27.3 days, calculate the periodic time of an artificial earth satellite near the earth's surface (Radius of the earth = 6400 km). **[C.U. (Hons.) 1972]**

If M be the mass of the earth, m, the mass of the moon which describes a circular orbit of radius r, then the force of attraction between the earth and moon is $G\frac{Mm}{r^2}$. This force supplies the necessary centripetal force (on the moon) $m\omega^2 r$, ω being the angular velocity of the moon in its orbit and given by $\frac{2\pi}{T}$, where T is the periodic time of the moon. For dynamic equilibrium,

$$G\frac{Mm}{r^2} = m\omega^2 r = m\left(\frac{2\pi}{T}\right)^2 r, \quad \text{or,} \quad T^2 = \frac{4\pi^2}{GM}r^3. \tag{1}$$

$$\therefore \quad T^2 \propto r^3.$$

Thus the square of the period of revolution of a satellite moving in a circular orbit round the earth is proportional to the cube of its distance from the centre of the earth.

Now in the case of the artificial satellite, we may write using eqn. (1),

$$T_1^2 = \frac{4\pi^2}{GM}r_1^3, \tag{2}$$

where T_1 = the periodic time of the satellite, and r_1 = radius of its orbit, which in this case is equal to the radius of the earth, as the satellite moves round very near the earth's surface.

From (1) and (2),

$$\frac{T_1^2}{T^2} = \frac{r_1^3}{r^3}. \quad \therefore T_1^2 = \frac{r_1^3}{r^3}T^2.$$

Here

$$r_1 = 6400 \text{ km} = 64 \times 10^7 \text{ cm}$$

$$r = 4 \times 10^5 \text{ km} = 4 \times 10^{10} \text{ cm.}$$

$$T = 27.3 \text{ days} = 27.3 \times 24 \times 60 \times 60 \text{ secs.}$$

Hence

$$T_1^2 = \left(\frac{64 \times 10^7}{4 \times 10^{10}}\right)^3 \cdot (27.3 \times 24 \times 60 \times 60)^2 = \left(\frac{16}{10^3}\right)^3 \times (27.3 \times 24 \times 60 \times 60)^2.$$

$$\therefore T_1 = 27.3 \times 24 \times 60 \times 60\sqrt{\left(\frac{16}{10^3}\right)^3} \text{ secs} = 1 \text{ hour } 19.58 \text{ min.}$$

6. State the conditions (without proof) that in a central inverse square field of force: (i) the orbit may be circular, (ii) the orbit may be parabolic. Two particles are moving in the same central inverse square field—one describes a circle of radius 6×10^3 kilometres and the other an ellipse of semi-axes 24×10^6 km and 3×10^6 km. Compare the time periods and areal velocities of the two particles. Find also the ratio of the maximum and minimum kinetic energies of the second particle.

(Standard formulae may be assumed). **[C.U. (Hons.) 1978]**

(i) *Condition for circular orbit*

Equation of orbit is $\frac{l}{r} = 1 + \varepsilon\cos\theta$. For the orbit to be circular, $\varepsilon = 0$, which gives $l = r$, i.e.,

$$\frac{h^2}{GM} = r, \text{ or, } h^2 = GMr, \quad \text{or,} \quad r^4\left(\frac{d\theta}{dt}\right)^2 = GMr \quad \left[\because r^2\frac{d\theta}{dt} = h\right]$$

$$\text{or, } r^2(rw)^2 = GMr, \quad \text{or,} \quad \frac{v^2}{r} = \frac{GM}{r^2}. \quad \therefore \frac{mv^2}{r} = \frac{GMm}{r^2},$$

i.e., the centrifugal force is balanced by the gravitational attractive force. Also from Art. 7.25.1 we know that for circular orbit

$$v = \frac{v_e}{\sqrt{2}}.$$

(ii) *For condition of parabolic orbit* see Arts. 7.25.1 and 7.22.1.

For the particle describing a circular orbit of radius r, we know that the square of the time period $T^2 \propto r^3$ (See worked example no. 5).

Also we know from Kepler's third law that for the particle describing an ellipse of semi-major axis a, the square of the time period $T'^2 \propto a^3$.

$$\therefore \quad \frac{T'^2}{T^2} = \frac{a^3}{r^3} = \left(\frac{24 \times 10^6}{6 \times 10^3}\right)^3, \text{ or, } \frac{T'^2}{T^2} = (4 \times 10^3)^3 \quad \left[\because a = 24 \times 10^6 \text{ km} \atop r = 6 \times 10^3 \text{ km}, \right.$$

$$\text{or, } \frac{T'}{T} = 8\sqrt{10} \times 10^4.$$

Areal velocity of the particle describing the circle

$$A = \frac{\pi r^2}{T}.$$

Also the areal velocity of the particle describing the ellipse

$$A' = \frac{\pi ab}{T'}.$$

$$\therefore \quad \frac{A}{A'} = \frac{r^2}{ab} \cdot \frac{T'}{T} = \frac{(6 \times 10^3)^2 \times 8\sqrt{10} \times 10^4}{(24 \times 10^6 \times 3 \times 10^6)} = \frac{4}{10^2}\sqrt{10} \quad [\because b = 3 \times 10^6 \text{ km}].$$

Again we know that

$$G\frac{Mm}{R^2} = \frac{mv^2}{R}, \quad \text{or, } \quad v^2 = \frac{GM}{R}, \quad \text{or, } \quad \frac{1}{2}mv^2 = \frac{1}{2}\frac{GMm}{R}, \quad \text{i.e., } \quad \text{K.E.} \propto \frac{1}{R}.$$

$$\therefore \text{K.E.}_{\text{max}} = \frac{1}{2}\frac{GMm}{b} \text{ and K.E.}_{\text{min}} = \frac{1}{2}\frac{GMm}{a}, \text{ or, } \frac{\text{K.E.}_{\text{max}}}{\text{K.E.}_{\text{min}}} = \frac{a}{b} = \frac{24 \times 10^6}{3 \times 10^6} = \frac{8}{1}.$$

7. Prove that the period of revolution of a small satellite around a planet in an orbit of radius just greater than the radius of the planet which is assumed spherical and of uniform density is independent of the radius of the planet. Find the period of revolution of the satellite if the density of the planet is 10 gms per c.c. Assume $G = 6.66 \times 10^{-8}$ C.G.S. units. **[C.U. (Hons.) 1980, '90]**

We know that

$$v^2 = \frac{GM}{R}, \quad \text{or,} \quad v^2 = \frac{4}{3}G\pi R^2 \rho,$$

where ρ is the density of the planet.

$$\therefore \quad v = R\left(\frac{4}{3}\pi G\rho\right)^{1/2}.$$

M = mass of the planet.

R = Radius of the orbit of the satellite = Radius of the planet.

If ω and T be the angular velocity and time period respectively, we have

$$\omega = \frac{2\pi}{T} = \frac{v}{R} = \left(\frac{4}{3}\pi G\rho\right)^{1/2} \quad ; \quad T = \frac{2\pi}{\left(\frac{4}{3}\pi G\rho\right)^{1/2}} = \left(\frac{3\pi}{G\rho}\right)^{1/2}.$$

This shows that T is independent of R. Again

$$\rho = 10 \text{ gms/c.c.}$$

$$G = 6.66 \times 10^{-8} \text{ C.G.S. units.}$$

$$\therefore \; T = \left(\frac{3 \times 3.14}{6.66 \times 10^{-8} \times 10}\right)^{1/2} = \left(\frac{3140}{222}\right)^{1/2} \times 10^3 = 3760 \text{ secs.}$$

8. A particle of mass m is moving in a plane under a repulsive inverse square law force $\frac{K}{r^2}$, where K is a positive real number and (r, θ) are the polar co-ordinates of the particle. Write down the equation of motion and solve it. What is the nature of the orbit? **[C.U. (Hons.) 1982]**

According to the problem, we have

$$\frac{d^2r}{dt^2} - r\left(\frac{d\theta}{dt}\right)^2 = \frac{K}{r^2}.$$

$$\therefore \; -h^2u^2\left(\frac{d^2u}{d\theta^2} + u\right) = u^2K,$$

$$\text{or,} \quad \frac{d^2u}{d\theta^2} + u = -\frac{K}{h^2} \quad \left[\because \; \frac{d^2r}{dt^2} = -h^2u^2\frac{d^2u}{d\theta^2} \text{ and } \frac{d\theta}{dt} = \frac{h}{r^2} = hu^2 \text{ (See Art. 7.22.1)}\right]$$

$$\text{or,} \quad \frac{d^2}{d\theta^2}\left(u + \frac{K}{h^2}\right) + \left(u + \frac{K}{h^2}\right) = 0,$$

which is the required equation of motion, and the solution is given by

$$u + \frac{K}{h^2} = A\cos(\theta - \theta_0), \quad \text{or,} \quad u = \frac{1}{r} = -\frac{K}{h^2} + A\cos(\theta - \theta_0). \tag{1}$$

[**N.B.** The solution may be obtained in the following way.

Writing u_1 for $\left(u + \frac{K}{h^2}\right)$, the equation reduces to

$$\frac{d^2u_1}{d\theta^2} + u_1 = 0.$$

Let $\frac{d}{d\theta} = D$, then we have $(D^2 + 1)u_1 = 0$, whence we get $D = \pm i$—which implies that the C.F. for the the solution of the above differential equation is

$$u_1(\text{C.F.}) = c_1e^{i\theta} + c_2e^{-i\theta} = (c_1 + c_2)\cos\theta + i(c_1 - c_2)\sin\theta.$$

Putting $(c_1 + c_2) = A\cos\theta_0$ and $i(c_1 - c_2) = A\sin\theta_0$, (where A and θ_0 are constants) we may write for the solution

$$u_1 = A\cos\theta\cos\theta_0 + A\sin\theta\sin\theta_0 = A\cos(\theta - \theta_0)$$

Here the P.I. = 0.]

The constants A and θ_0 are to be found from the initial conditions. At $t = 0$, $r = r_0$, $v = v_0$ and let α be the angle between v_0 and r_0. Under this condition, θ is taken as zero;

$$\therefore \quad \left(\frac{dr}{dt}\right)_{t=0} = v_0\cos\alpha; \quad \left[r\left(\frac{d\theta}{dt}\right)\right]_{t=0} = v_0\sin\alpha$$

and

$$h = \left[r^2\left(\frac{d\theta}{dt}\right)\right]_{t=0} = r_0 v_0\sin\alpha, \quad \text{or,} \quad \frac{1}{r_0} = \frac{v_0\sin\alpha}{h}.$$

From (1) at $t = 0$.

$$\frac{1}{r_0} = -\frac{K}{h^2} + A\cos\theta_0. \tag{2}$$

Also differentiating (1), we have

$$-\frac{1}{r^2}\cdot\frac{dr}{d\theta} = -A\sin(\theta - \theta_0).$$

$$\therefore \quad \left(-\frac{1}{r^2}\frac{dr}{d\theta}\right)_{t=0} = A\sin\theta_0, \quad \text{or,} \quad \left(-\frac{1}{r^2}\cdot\frac{dr/dt}{d\theta/dt}\right)_{t=0} = A\sin\theta_0.$$

$$\therefore \quad \left(\frac{dr}{dt}\right)_{t=0} = -\left(r^2\frac{d\theta}{dt}\right)_{t=0} A\sin\theta_0, \quad \text{or,} \quad v_0\cos\alpha = -hA\sin\theta_0.$$

$$\therefore \quad A\sin\theta_0 = -\frac{v_0}{h}\cos\alpha \quad \text{and} \quad A\cos\theta_0 = \frac{1}{r_0} + \frac{K}{h^2} \quad \text{(from 2).}$$

$$\begin{aligned}\therefore \quad A^2 &= \frac{v_0^2}{h^2}\cos^2\alpha + \frac{1}{r_0^2} + \frac{2K}{r_0 h^2} + \frac{K^2}{h^4} \\ &= \frac{v_0^2\cos^2\alpha}{h^2} + \frac{v_0^2\sin^2\alpha}{h^2} + \frac{2}{r_0}\frac{K}{h^2} + \frac{K^2}{h^4} \quad \left[\because \frac{1}{r_0} = \frac{v_0\sin\alpha}{h}\right] \\ &= \frac{v_0^2}{h^2} + \frac{2K}{r_0 h^2} + \frac{K^2}{h^4} = \frac{K^2}{h^4} + \frac{2}{h^2}\left(\frac{K}{r_0} + \frac{v_0^2}{2}\right).\end{aligned} \tag{3}$$

Now, total energy,

$$\begin{aligned}E &= \frac{1}{2}m\left[\left(\frac{dr}{dt}\right)^2 + \left(r\frac{d\theta}{dt}\right)^2\right] - m\int_\infty^r \frac{K}{r^2}dr \\ &= \frac{1}{2}mv^2 + \frac{Km}{r} \quad \left[\because v^2 = \left(\frac{dr}{dt}\right)^2 + \left(r\frac{d\theta}{dt}\right)^2\right] \\ &= \text{a constant quantity,}\end{aligned}$$

$$\therefore\ E = E_{t=0} = \frac{1}{2}mv_0^2 + \frac{Km}{r_0} = m\left(\frac{K}{r_0} + \frac{v_0^2}{2}\right), \quad \text{or,} \quad \frac{E}{m} = \frac{K}{r_0} + \frac{v_0^2}{2} = E'. \tag{4}$$

[E' = The total energy per unit mass]

$\therefore$ From (3) and (4), we have

$$A^2 = \frac{K^2}{h^4} + \frac{2E'}{h^2}.$$

Now from (1)

$$\frac{1}{r} = -\frac{K}{h^2} + A\cos(\theta - \theta_0), \quad \text{or,} \quad \frac{-h^2/K}{r} = 1 + \left(-\frac{Ah^2}{K}\right)\cos(\theta - \theta_0).$$

Comparing with

$$\frac{l}{r} = 1 + \varepsilon\cos(\theta - \theta_0)$$

we get

$$l = -\frac{h^2}{K} \quad \text{and} \quad \varepsilon = -\frac{Ah^2}{K} = Al.$$

$$\therefore\ \varepsilon^2 = l^2A^2 = \frac{h^4}{K^2}\left(\frac{K^2}{h^4} + \frac{2E'}{h^2}\right) = 1 + \frac{2E'h^2}{K^2}.$$

Since K is a positive quantity, E' is also positive.

$$\therefore\quad \varepsilon > 1.$$

Hence the orbit is the negative branch of a hyperbola, as in the case of the path of an α-particle in Rutherford's experiment on α-particle scattering.

9. A particle moves in a central field obeying the inverse square law. Show that for orbits of different radii, the angular momentum of the particle about the centre varies as the square root of the radius and the total energy varies inversely as the radius. **[C.U. (Hons.) 1977]**

For motion under inverse square law in a central force field, the equation of the orbit is that of a general conic given by [see Illustrative Example 8]

$$\frac{1}{r} = -\frac{K}{h^2} + A\cos(\theta - \theta_0), \tag{1}$$

where $\frac{K}{r^2}$ = force per unit mass; h = angular momentum per unit mass; A = an arbitrary constant; θ_0 = the angle which the radius vector makes at $t = 0$. Proceeding similarly as in example 8, we can write

$$\varepsilon^2 = 1 + \frac{2E'h^2}{K^2}.$$

Now, for a circular orbit $\varepsilon = 0$.

$$\therefore\quad E' = -\frac{K^2}{2h^2}.$$

Also from the previous example, we may write

$$A^2 = \frac{K^2}{h^4} + \frac{2E'}{h^2} = \frac{K^2}{h^4} + \frac{2}{h^2} \times \left(-\frac{K^2}{2h^2}\right) = 0.$$

$\therefore$ Equation (1) reduces to

$$\frac{1}{r} = -\frac{K}{h^2}, \quad \text{or,} \quad E' = -\frac{K^2}{2h^2} = -\frac{K}{h^2}\cdot\frac{K}{2} = \frac{K}{2r}, \quad \text{or,} \quad E' \propto \frac{1}{r}.$$

Thus the total energy varies inversely as the radius.

Now,

$$\frac{1}{r} = -\frac{K}{h^2}. \quad \therefore \quad h^2 = -Kr, \quad \text{or,} \quad h \propto \sqrt{r}.$$

So the angular momentum varies as the square root of the radius.

10. Two particles of masses M_1 and M_2 at a distance R apart rotate about their common centre of mass under the influence of gravity. Assuming the orbits to be circular, find the period of rotation of the particles. Also find the ratio of their kinetic energies. **[C.U. (Hons.) 1979, cf. 2003]**

Let r_1 and r_2 be the distances of M_1 and M_2 from the common centre of mass O; then we may write

$$M_1 r_1 = M_2 r_2. \tag{1}$$

If R be the distance between M_1 and M_2, then from figure (1),

$$r_1 = R - r_2 = R - \frac{M_1 r_1}{M_2}, \text{ or, } r_1\left(1 + \frac{M_1}{M_2}\right) = R, \text{ or, } r_1 = \frac{R}{1+\dfrac{M_1}{M_2}}. \tag{2}$$

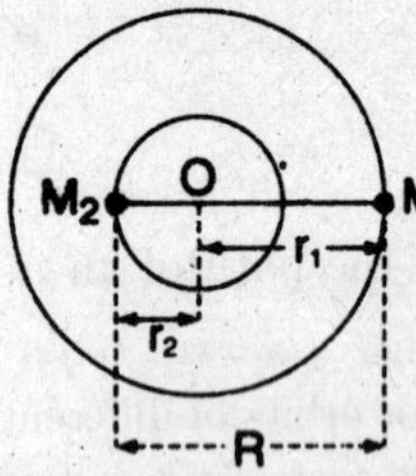

Fig. 1

Similarly,

$$r_2 = R - r_1 = R - \frac{M_2 r_2}{M_1}, \quad \text{or,} \quad r_2 = \frac{R}{1+\dfrac{M_2}{M_1}}. \tag{3}$$

Let the two particles rotate about their common centre of mass with angular velocity ω.

$$\therefore \; M_1 r_1 \omega^2 = \frac{GM_1M_2}{R^2}, \quad \text{or,} \quad \omega^2 r_1 = \frac{GM_2}{R^2},$$

$$\text{or,} \quad \omega^2 \frac{R}{1+\dfrac{M_1}{M_2}} = \frac{GM_2}{R^2} \quad \text{[from (2)]}$$

$$\text{or,} \quad \omega^2 = \frac{G(M_1+M_2)}{R^3}.$$

The time period for the particle of mass M_1

$$T_1 = \frac{2\pi}{\omega} = 2\pi\sqrt{\frac{R^3}{G(M_1+M_2)}} = 2\pi R\sqrt{\frac{R}{G(M_1+M_2)}}.$$

Both the particles will have the same time period. Hence the time period for the particle of mass M_2, $T_2 = T_1$. If the two particles are of equal mass M, then $T_1 = T_2 = \pi R\sqrt{\frac{2R}{GM}}$.

Now K.E. of mass M_1 is given by

$$\frac{1}{2}M_1w^2r_1^2 = \frac{1}{2}M_1w^2\frac{R^2}{\left(1+\frac{M_1}{M_2}\right)^2} = \frac{1}{2}\frac{w^2R^2M_1M_2^2}{(M_1+M_2)^2} = \frac{1}{2}\frac{\mu^2\omega^2R^2}{M_1}.$$

Similarly K.E. of mass M_2 is given by

$$\frac{1}{2}M_2w^2r_2^2 = \frac{1}{2}M_2w^2\frac{R^2}{\left(1+\frac{M_2}{M_1}\right)^2} = \frac{1}{2}\cdot\frac{w^2R^2M_2M_1^2}{(M_1+M_2)^2} = \frac{1}{2}\frac{\mu^2\omega^2R^2}{M_2}.$$

μ being the reduced mass of the two particle-system.

$$\therefore \quad \frac{\text{K.E. of mass } M_1}{\text{K.E. of mass } M_2} = \frac{M_2}{M_1}.$$

11. The co-ordinates x and y of a particle satisfy the differential equations

$$m\ddot{x} + He\dot{y} = eE$$

$$m\ddot{y} - He\dot{x} = 0,$$

where H, E, m and e are constants. Find the equation to the trajectory of the particle assuming $x = y = 0$ and $\dot{x} = \dot{y} = 0$ at $t = 0$. What is the nature of the trajectory? **[C.U. (Hons.) 1981]**

$$m\frac{d^2x}{dt^2} + He\frac{dy}{dt} = eE \tag{1}$$

$$m\frac{d^2y}{dt^2} - He\frac{dx}{dt} = 0. \tag{2}$$

From (2) we have

$$m\frac{d}{dt}\left(\frac{dy}{dt}\right) = He\frac{dx}{dt}, \text{ or, } m\int\frac{d}{dt}\left(\frac{dy}{dt}\right)dt = He\int\frac{dx}{dt}\cdot dt + C,$$

C being the constant of integration

$$\text{or, } m\int d\left(\frac{dy}{dt}\right) = He\int dx + C, \quad \text{or, } \quad m\frac{dy}{dt} = Hex + C.$$

At $t = 0, x = 0$ and $\frac{dy}{dt} = 0$. $\therefore \quad C = 0.$

Hence

$$m\frac{dy}{dt} = Hex. \quad \text{or, } \quad \frac{dy}{dt} = \frac{He}{m}x. \tag{3}$$

From (1) and (3) we may write

$$m\frac{d^2x}{dt^2} + He\cdot\frac{Hex}{m} = eE, \text{ or, } 2m\frac{dx}{dt}\cdot\frac{d^2x}{dt^2} + 2\frac{(He)^2}{m}\left(\frac{dx}{dt}\right)x = 2eE\frac{dx}{dt},$$

$$\text{or, } m\frac{d}{dt}\left[\left(\frac{dx}{dt}\right)^2\right] + \frac{(He)^2}{m}\frac{d}{dt}(x^2) = 2eE\frac{dx}{dt},$$

$$\text{or, } m\int d\left[\left(\frac{dx}{dt}\right)^2\right] + \frac{(He)^2}{m}\int d(x^2) = 2eE\int dx + C_1,$$

C_1 being the constant of integration.

$$\therefore \quad m\left(\frac{dx}{dt}\right)^2 + \frac{(He)^2}{m}x^2 = 2eEx + C_1.$$

Again from the given condition, at $t = 0$ $x = 0$, $\frac{dx}{dt} = 0$, we have $C_1 = 0$.

$$\therefore m\left(\frac{dx}{dt}\right)^2 + \frac{(He)^2}{m}x^2 = 2eEx, \text{ or, } \left(\frac{dx}{dt}\right)^2 = \frac{2eE}{m}x - \left(\frac{He}{m}\right)^2 x^2,$$

$$\text{or, } \left(\frac{dx}{dt}\right) = \pm\sqrt{\frac{2eE}{m}x - \left(\frac{He}{m}\right)^2 x^2},$$

$$\text{or, } \int\frac{dx}{\sqrt{\frac{2eE}{m}x - \left(\frac{He}{m}\right)^2 x^2}} = \int dt + C_2 = t + C_2 \quad (4)$$

C_2 being the constant of integration.

Let

$$\frac{He}{m}x - \frac{E}{H} = z, \quad \text{or, } \quad \frac{He}{m}dx = dz, \quad \text{or, } \quad dx = \frac{m}{He}dz.$$

$\therefore$ l.h.s. of equation (4)

$$= \frac{m}{He}\int\frac{dz}{\sqrt{\left(\frac{E}{H}\right)^2 - z^2}} = \frac{m}{He}\sin^{-1}\left(\frac{z}{E/H}\right)$$

$$= \frac{m}{He}\sin^{-1}\left(\frac{\frac{H^2e}{m}x - E}{E}\right) = \frac{m}{He}\sin^{-1}\left(\frac{H^2e}{mE}x - 1\right).$$

$\therefore$ Equation (4) may be written as

$$\frac{m}{He}\sin^{-1}\left(\frac{H^2e}{mE}x - 1\right) = t + C_2.$$

Again from the condition, $x = 0$ at $t = 0$, we have

$$C_2 = \frac{m}{He}\sin^{-1}(-1) = \frac{m}{He}\cdot\left(-\frac{\pi}{2}\right) = -\frac{m\pi}{2He}.$$

$$\therefore\quad \frac{m}{He}\sin^{-1}\left(\frac{H^2e}{mE}x - 1\right) = t - \frac{m\pi}{2He},$$

$$\text{or,}\quad t = \frac{m}{He}\left[\frac{\pi}{2} + \sin^{-1}\left(\frac{H^2e}{mE}x - 1\right)\right] = \frac{m}{He}\cos^{-1}\left(1 - \frac{H^2ex}{mE}\right),$$

$$\text{or,}\quad \cos\frac{Het}{m} = 1 - \frac{H^2ex}{mE},\quad \text{or,}\quad \frac{H^2ex}{mE} = 1 - \cos\frac{Het}{m},$$

$$\text{or,}\quad x = \frac{mE}{H^2e}\left(1 - \cos\frac{Het}{m}\right).$$

The above equation is of the form $x = a(1 - \cos\theta)$.

Again from (2)

$$\frac{d}{dt}\left(\frac{dy}{dt}\right) = \frac{He}{m}\frac{dx}{dt}.$$

Integrating we get

$$\frac{dy}{dt} = \frac{He}{m}x + C_3.$$

At $t = 0$, $x = 0$ and $\dot{y} = 0$, which gives $C_3 = 0$.

$$\therefore\quad \frac{dy}{dt} = \frac{He}{m}x = \frac{E}{H}\left(1 - \cos\frac{He}{m}t\right) \quad \text{(putting the value of } x\text{)}.$$

Integrating further we get

$$y = \frac{E}{H}\left(t - \frac{m}{He}\sin\frac{He}{m}t\right) + C_4.$$

At $t = 0$, $y = 0$ which gives $C_4 = 0$.

$$\therefore\quad y = \frac{E}{H}\left[t - \frac{m}{He}\sin\frac{He}{m}t\right] = \frac{mE}{H^2e}\left[\frac{He}{m}t - \sin\frac{He}{m}t\right].$$

The above equation is of the form

$$y = a(\theta - \sin\theta).$$

Hence we get a pair of solutions

$$x = \frac{mE}{H^2e}\left(1 - \cos\frac{He}{m}t\right)$$

$$y = \frac{mE}{H^2e}\left(\frac{He}{m}t - \sin\frac{He}{m}t\right)$$

which are the parametric equations of the cycloidal motion of a particle.

Hence the trajectory of the particle will be a cycloid.

7.26 Mass and Mean Density of the Earth

Let M be the mass of the earth and R be its radius. Let m be the mass of a body on the surface of the earth. Then the force with which the body of mass m is attracted towards the centre of the earth is given by

$$F = \frac{GMm}{R^2}.$$

Here, the earth being nearly spherical in shape, its mass may be supposed to be concentrated as its centre.

Now $F = mg$, where g is the acceleration generated if the body is allowed to fall freely.

$$\therefore \quad mg = \frac{GMm}{R^2}, \quad \text{or,} \quad g = \frac{GM}{R^2}.$$

Thus from the knowledge of R, g and G, M may be determined.

Again $M = \frac{4}{3}\pi R^3 \rho$, where ρ is the mean density of the earth. Thus from a knowledge of M and R, ρ may be determined.

The mass of the earth may be determined by the "mountain experiment" as will be described in connection with the determination of the constant of gravitation.

7.27 Airy's Mine.Experiment

In this experiment, the mean density ρ of the earth was compared with the density D of the surface constituents of the earth by determining the values of the acceleration due to gravity at the top and at the bottom of a mine.

Let g and g' be the values of the acceleration due to gravity at the top and at the bottom of a mine. Then

$$g = \frac{GM}{R^2} = G\frac{4}{3}\pi R\rho \quad \text{and} \quad g' = G\left[\frac{\frac{4}{3}\pi R^3\rho - 4\pi R^2 hD}{(R-h)^2}\right],$$

where h is the depth of the mine.

$$\therefore \quad \frac{g}{g'} = \frac{G.\frac{4}{3}\pi R\rho(R-h)^2}{G.4\pi R^2\left(\frac{R\rho}{3} - hD\right)} = \frac{(R-h)^2}{R} \cdot \frac{\rho}{(R\rho - 3hD)}$$

$$= \frac{(R-h)^2}{R} \cdot \frac{\rho}{R\rho}\left(1 - \frac{3hD}{R\rho}\right)^{-1} = \frac{(R-h)^2}{R^2}\left(1 + \frac{3hD}{R\rho}\right) \quad \text{(approximately)}$$

$$= \left(1 - \frac{2h}{R}\right)\left(1 + \frac{3hD}{R\rho}\right) \quad \text{(approximately)}$$

$$= 1 - \frac{2h}{R} + \frac{3hD}{R\rho} \quad \text{(approximately).}$$

From this expression, the mean density ρ may be determined if h are D are known. Airy was the first to perform the experiment at the Harton Coal pit in Sunderland in the year 1854.

The sample of the soil obtained in the process of boring the Coalmine-Shaft was examined and the value of D was found to be 2.5 gms/c.c.—from which the mean density of the earth was calculated; g and g' were determined by means of a pendulum. The mean density of the earth was calculated to be 6.5 gms/c.c. Airy's experiment gave unsatisfactory result due to the difficulty in determining the value of D accurately. Again taking $g = G\frac{4}{3}\pi R\rho$ and putting the experimentally determined values of g and ρ as shown above, we can find G.

Illustrative Examples

1. Given $G = 6.7 \times 10^{-8}$ C.G.S. units, the radius of the earth $= 6.4 \times 10^8$ cms and $g = 981$ cms/sec^2, calculate the mean density of the earth.

We know $g = \frac{GM}{R^2}$, where M is the mass of the earth and R is its radius.

Also $M = \frac{4}{3}\pi R^3 \rho$, where ρ is the mean density of the earth.

$$\therefore \quad g = \frac{4}{3}\pi G R \rho.$$

$$\therefore \quad \rho = \frac{3g}{4\pi G R}.$$

Hence

$$\rho = \frac{3 \times 981}{4 \times 3.14 \times 6.7 \times 10^{-8} \times 6.4 \times 10^8} = \frac{3 \times 981}{4 \times 3.14 \times 6.7 \times 6.4} = 5.46 \text{ gms/c.c.}$$

2. In a Mine Experiment the mean density calculated is 6.5 gms/c.c. and the average surface density in the vicinity of a mine shaft 800 metres deep is 2.5 gms/c.c. What is the acceleration due to gravity at the bottom of the shaft if the acceleration due to gravity on the surface be 981 cms/sec^2? Given radius of the earth = 6400 km.

We know,

$$\frac{g}{g'} = 1 - \frac{2h}{R} + \frac{3hD}{R\rho}.$$

Here

$$g = 981 \text{ cm/sec}^2; \quad h = 800 \text{ metres} = 8 \times 10^4 \text{ cms.}$$

$$R = 6.4 \times 10^8 \text{ cms.}; \quad D = 2.5 \text{ gms/c.c.}; \quad \rho = 6.5 \text{ gms/c.c.}$$

$$\therefore \quad \frac{981}{g'} = 1 - \frac{2 \times 8 \times 10^4}{6.4 \times 10^8} + \frac{3 \times 8 \times 10^4 \times 2.5}{6.4 \times 10^8 \times 6.5}$$

$$= 1 - 0.25 \times 10^{-3} + 0.144 \times 10^{-3}$$

$$= 1 - 0.000106 = 0.999894.$$

$$\therefore \quad g' = \frac{981}{0.999894} = 981.104 \text{ cm/sec}^2.$$

g' gives the value of acceleration due to gravity at the bottom of the shaft.

7.28 Determination of 'G'—Mountain Experiments

The early attempts to determine the gravitational constant were made by Bouger and Maskelyne. In their experiments, a large natural mass such as a mountain was selected. Its mass and shape were determined and its attraction on a plumb-line was compared with the attraction of the whole earth on the same plumb-line from which the gravitational constant was determined.

7.28.1 Bouger's Experiment

Bouger carried out the experiment in the Andes (Peru) in the year 1740. He measured the side-way attraction of Chimborazo, a mountain 20,000 feet high. The principle of the method may be understood from the figure (7.28.1).

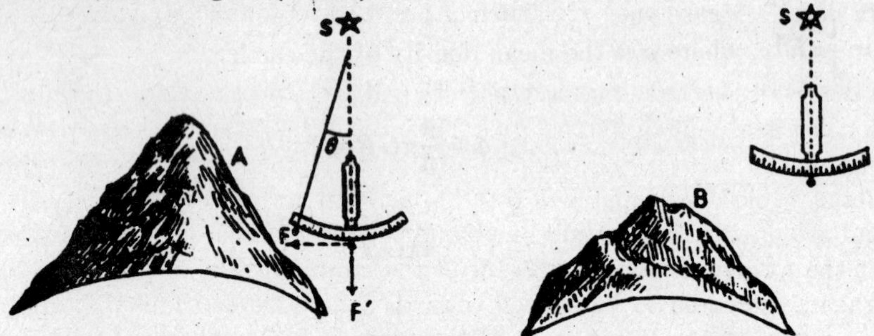

Fig. 7.28.1

A and *B* are two stations the former selected on the side of the mountain due south of its summit and latter in the same latitude, but some distance towards west away from its influence.

In the station *B*, let us suppose that a star *S* is directly overhead (just passing the meridian). Then a plumb-line hung vertically downward will be parallel to a telescope observing the star. In the station *A* the plumb-line is nearer the mountain and hence it will be attracted by the same. The deflection of the plumb-line was measured. Comparing the lateral pull of the mountain with the vertical pull of the earth, the gravitational constant was determined.

Let F and F' be the gravitational attractions on the plumb-line due to mountain and the earth respectively, and θ be the deflection of the plumb-line from the vertical.

Then

$$\tan\theta = \frac{F}{F'}; \text{ now } F = G\frac{mm'}{r^2},$$

where m is the mass of the plumb-line and m' that of the mountain; r denotes the distance of the centre of gravity of the mountain from the plumb-line.

Again $F' = mg$, g being the acceleration due to gravity at the place.

$$\therefore \quad \tan\theta = \frac{Gm'}{r^2 g} = \frac{GV\rho}{r^2 g},$$

where V is the volume and ρ is the density of the mountain

$$\therefore \quad G = \frac{gr^2\tan\theta}{V\rho}.$$

Thus to determine the value of G, a knowledge of the shape of the mountain and the position of its centre of gravity is required. Bouger made necessary observations in the most adverse conditions of snow and storm with great difficulty. He found that the plumb-line was drawn aside by about 8″. Had the mountain been of the density of the whole earth, Bouger calculated that the deflection of the plumb-line would have been twelve times as great proving thereby that the earth was twelve times as dense as the mountain which is far from the truth. Though his experiment failed to give a good result, yet Bouger deserves credit, because he was the first to show that the mountain really attracts and that the earth as a whole is denser than its outer surface. His experiment therefore proved that the earth was not a hollow shell or globe full of water as was conceived at that time.

7.28.2 Maskelyne's Experiment

Thirty years later, Maskelyne—the then English Astronomer Royal, repeated Bouger's experiment in a different manner.

Two stations were selected—one on the south and the other on the north slope of a mountain (Schiehallion) 3547 feet high in Perthshire (England) (Fig. 7.28.2). The experiment was carried out first at the south station, with the help of an instrument called *zenith sector*. A zenith sector consists of a telescope capable of rotation about a horizontal east-west axis at the object glass side, and provided with a plumb-line hanging from the axis over a graduated scale at the eye-piece. At the former station, in absence of the mountain, the star which would have been directly overhead, appeared to be shifted towards south due to the attraction of the plumb-line towards north. At the latter station, the same star was observed with the help of the zenith sector, and the star was found to be shifted slightly towards north due to the attraction of the plumb-line towards south. Thus the total shift of the star was double the deflection of the plumb-line at either station. The shift was carefully measured and it was about 55″, out of which a shift of 43″ was calculated to be due to the curvature of the earth's surface. Hence, the net shift due to gravitational pull of the mountain is (55″ − 43″) = 12″, i.e., the plumb-line at either station was deflected by 6″ (six seconds of arc) due to attraction of the mountain. This is the value of θ and hence G can be calculated by applying the formula as deduced in Bouger's experiment and was found to be 7.4×10^{-8} C.G.S. units. Had the mountain been of the same density as the earth, Maskelyne calculated that the deflection of the plumb-lines towards each other would have been 20″.9 which is 9/5 times the observed deflection (i.e., 12″). Hence, we conclude that the earth is 9/5 times as dense as the mountain. The density of the mountain was found to be 2.5 gms/c.c.; so the earth should have a density $(2.5 \times \frac{9}{5}) = 4.5$ gms/c.c. Based on a survey made thirty years later, the density of the mountain was determined more accurately which brought the density of the earth to 5 gms per c.c.

Fig. 7.28.2

7.29 Determination of G—Laboratory Methods

The early experiments as discussed above are of historical importance only, because the results obtained from these experiments are far from being accurate. The experiments conducted in the laboratory with the help of delicate instruments enable us to obtain the value of the gravitational constant with greater accuracy by measuring the attraction between two masses.

7.29.1 Cavendish's Experiment

In his experiment, Cavandish measured the mutual attraction between two lead spheres. The instrument used by him was originally devised by Mitchell who died before he could complete his work. Cavendish made a slight modification of the apparatus and carried out the experiment.

A long light rod AB (Fig. 7.29.1) carried a small lead sphere of mass m at each end and was fastened by two wires DA and DB attached to a stiff upright p. The system was suspended by a suspension wire CD from a torsion head. Two large lead spheres, each of mass M were suspended near the spheres on their opposite sides, so that the centres of the four spheres lay in the same horizontal plane. The attraction between the spheres produced a couple as a result of which the beam was deflected. Equilibrium condition was attained when the deflecting torque was balanced by the restoring torque called into play due to the twist of the suspension wire.

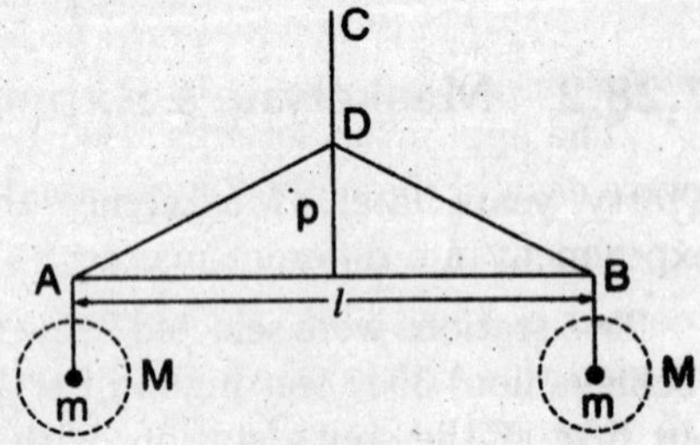

Fig. 7.29.1

Let θ be the deflection of the beam and d the distance between the centres of the inter-acting spheres, i.e., one large and one small sphere. Then the deflecting couple is $\frac{GMm}{d^2} \times l$, where l is the length of the beam. Now the restoring couple is $C\theta$, where C is the restoring torque per unit angle of twist. Therefore, in the equilibrium position

$$G\frac{Mm}{d^2}l = C\theta.$$

The value of C may be determined from the measurement of the time period of oscillation of the beam carrying the masses. The time period is given by

$$T = 2\pi\sqrt{\frac{I}{C}},$$

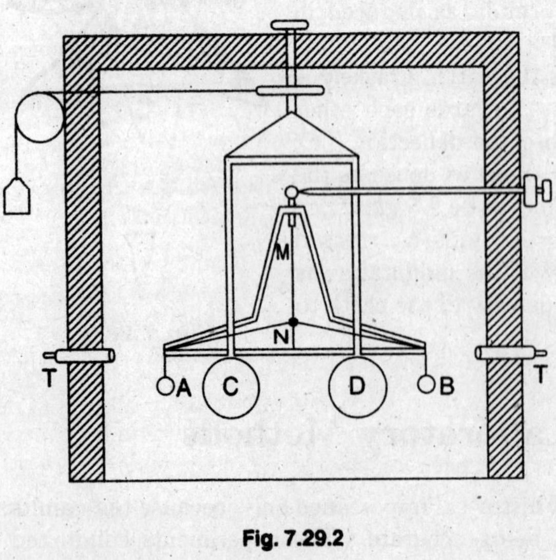

Fig. 7.29.2

where I is the moment of inertia of the suspended system about the axis of oscillation.

$$\therefore \quad C = \frac{4\pi^2 I}{T^2}.$$

Hence we may write,

$$\frac{GMml}{d^2} = \frac{4\pi^2 I}{T^2}.\theta. \quad \therefore \quad G = \frac{4\pi^2 I d^2}{MmlT^2}.\theta.$$

Also

$$g = G\cdot\frac{4}{3}\pi R\rho,$$

where R = radius of the earth and ρ = mean density of the earth.

$$\text{or,} \quad G = \frac{3g}{4\pi R\rho},$$

Hence,

$$\frac{3g}{4\pi R\rho} = \frac{4\pi^2 Id^2}{MmlT^2}\theta.$$

If g be measured with the help of a second's pendulum of length L, then we have,

$$2 = 2\pi\sqrt{\frac{L}{g}}, \quad \text{or,} \quad g = L\pi^2.$$

$$\therefore \quad \frac{3L\pi^2}{4\pi R\rho} = \frac{4\pi^2 Id^2}{MmlT^2}\theta \quad \text{or,} \quad \rho = \frac{3}{4}\cdot\frac{L}{\pi}\cdot\frac{1}{4R}\cdot\frac{lMm}{d^2}\cdot\frac{T^2}{I\theta}.$$

The apparatus actually used by Cavendish was large (Fig. 7.29.2). The small balls (A, B) were each of diameter 2 inches and the larger balls (C, D) were each of diameter 8 inches.

The beam carrying the small balls was suspended by the torsion wire MN made of silvered copper of 3 feet long; the beam was 6 feet long, the length of the wire being such that the time period was 420 seconds and the observations were made from outside to avoid disturbance due to temperature gradient, air currents, etc. The beam was provided at each end with a vernier moving over a fixed scale, and the readings were taken by telescopes T. By interchanging the positions of the large balls, the deflection was reversed and 2θ was obtained. The mean value of G and density as obtained by Cavendish by calculating from 29 readings are:

$$G = 6.754 \times 10^{-8} \text{ C.G.S. units}$$

$$\rho = 5.448 \text{ gms/c.c.}$$

Corrections were necessary for the following errors.

(a) Each large sphere also attracts the distant small sphere, producing an appreciable gravitational couple acting in the opposite direction.

(b) The spheres also attract the torsion beam tending to increase the deflecting couple.

(c) The rods carrying the large spheres are also attracted tending to increase the deflecting couple.

7.29.2 Boys' Modification of Cavendish's Experiment

This is the most important method of measuring the constant of gravitation. In Cavendish's experiment, the accuracy was limited by several factors which are as follows.

(a) The suspension wire was of larger diameter to support the heavy beam. Large diameter of the suspension wire results in large value of C [since $C = \frac{n\pi r^4}{2l}$, r and l being the radius and length of the wire and n the rigidity modulus of the wire] which causes comparatively small deflection.

(b) Measurement of θ by the vernier method was not accurate since the deflection is reduced by the frictional force called into play as the vernier scale slides over the main scale.

(c) Temperature gradients and disturbance due to the convection current could not be eliminated owing to the large dimension of the apparatus.

(d) The counter gravitational couple, i.e., the couple due to the influence of the large spheres on the distant small spheres also reduced the value of the deflection θ.

In Boys' experiment the apparatus was devised so as to eliminate the effects due to the above factors. In this experiment a small mirror strip S about 2.5 cms long which acted as the torsion bar was suspended by means of a fine quartz fibre from a torsion head T (Fig. 7.29.3). Inside a glass tube about 4 cms in diameter, were suspended two small gold balls (A and B) from the two ends of the mirror strip by similar quartz fibres. The two gold balls each of diameter 0.5 cm and mass 2.65 gms were suspended in such a way that one ball was about 15 cms above the level of the other. In an outer co-axial tube which could be rotated about the common axis, two large lead balls (C and D) each of diameter 11 cms and mass 7407 gms, were suspended so that the centres of one large ball and one small ball were at the same level, the distance between B and C being same as that between A and D.

The deflection was measured by the telescope and scale method. A half-millimetre scale was placed at a distance of about 7 metres from the mirror strip.

The experiment was performed by rotating the outer cylinder tube until the large lead balls lay on the opposite sides of the two gold balls so as to exert maximum deflection on the suspended system. The tube was then rotated so that the lead balls now lay on the other sides of the gold balls in a similar position again exerting maximum deflection in the opposite direction. The mean of these two deflections was taken. Let it be θ. The value of G was calculated as given below.

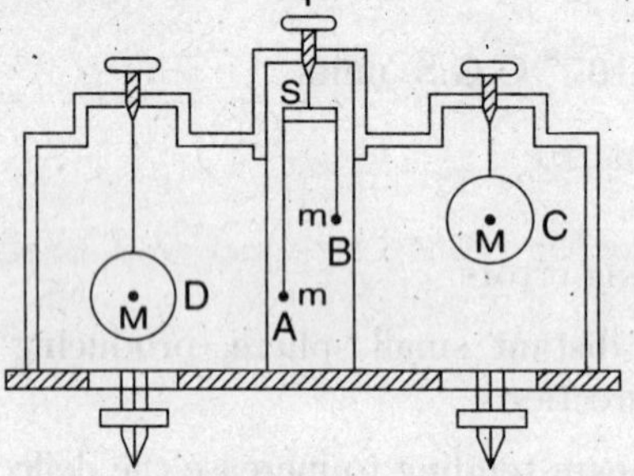

Fig. 7.29.3

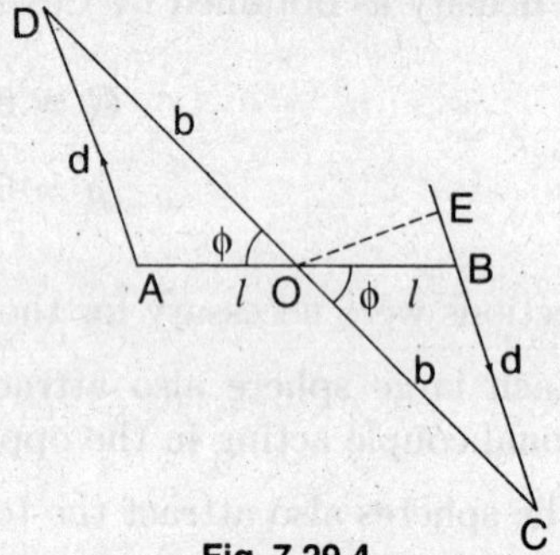

Fig. 7.29.4

The figure (7.29.4) represents a plan of the system after equilibrium position has been reached. A and B represent the centres of the gold balls, and C and D those of the lead balls. We shall neglect the attraction of the balls at different levels.

Let O be the mid-point of the line joining A and B, and l, the half length of the line AB, i.e., $OA = OB = l$. Let $OC = OD = b$. Also let $AD = BC = d$ and $\angle AOD = \angle BOC = \phi$. Let OE be the perpendicular dropped from O on CB produced.

In ΔOBC,

$$BC = \sqrt{OC^2 + OB^2 - 2OC.OB\cos\phi}, \quad \text{or,} \quad d = \sqrt{l^2 + b^2 - 2lb\cos\phi}.$$

Now the attraction between the two spheres along BC is given by,

$$F = \frac{GMm}{d^2},$$

where M and m are the masses of each large and small balls respectively. The moment of this attraction about O is

$$\frac{GMm}{d^2}.OE.$$

Considering an equal moment due to attraction along AD, the total moment of the attractive force becomes

$$2G\frac{Mm}{d^2}.OE.$$

Now from $\triangle OBC$, $\frac{\sin\phi}{\sin BCO} = \frac{BC}{OB} = \frac{d}{l}$. $\therefore$ $\sin BCO = \frac{l\sin\phi}{d}$.

Again in the right-angled triangle OEC, we have

$$OE = OC\sin ECO = OC\frac{l\sin\phi}{d} = \frac{bl\sin\phi}{d}.$$

$\therefore$ The total moment of the attractive force or the deflecting couple is

$$2G\frac{Mm}{d^2}\cdot\frac{bl\sin\phi}{d} = 2G\frac{Mmbl\sin\phi}{d^3} = 2G\frac{Mmbl\sin\phi}{(l^2+b^2-2lb\cos\phi)^{3/2}}.$$

The mirror strip comes to rest when the moment of the deflecting couple is just balanced by the moment of the restoring couple $C\theta$, where C is the torsional constant (or the torque per unit twist) of the quartz fibre and θ is the angular deflection of the suspension system from the equilibrium position.

$$\therefore\quad 2G\frac{Mmbl\sin\phi}{(l^2+b^2-2bl\cos\phi)^{3/2}} = C\theta, \quad \text{or,} \quad G = \frac{(l^2+b^2-2bl\cos\phi)^{3/2}}{2Mmbl\sin\phi}.C\theta.$$

The torsional constant C is obtained by measuring the time period T of the oscillation of the suspended system in absence of large masses, and we have,

$$T = 2\pi\sqrt{\frac{I}{C}},$$

where I is the moment of inertia of the suspended system which can be easily determined.

The value of G obtained by Boys was about 6.6576×10^{-8} C.G.S. units and the mean density $\rho = 5.5270$ gms per c.c.

7.29.3 Poynting's Method

G was measured by Poynting with the help of a common balance. The balance used was of the large bullion balance type (Fig. 7.29.5). It is strongly made of a gun-metal beam, and steel knife edges and planes.

From the ends of the beam are suspended two equal spherical balls (A and B) made of an alloy of lead and antimony—each of mass nearly 21.6 kgms. A spherical attracting body C of mass about 153 kgms, made of the same alloy is mounted on a turn-table so that it can be brought to a point vertically under A and then under B as shown in the figure. In each case, the downward attraction is given by

$$F = G\frac{Mm}{d^2},$$

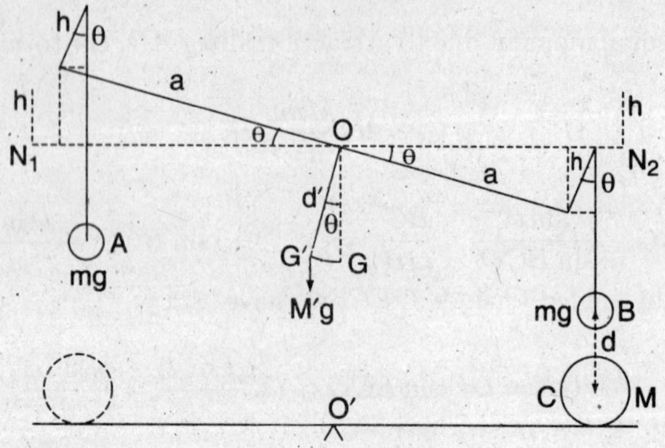

Fig. 7.29.5

where M = mass of the attracting body C, m = mass of A or B and d = distance between the centres of gravity of the large and small bodies; the downward force tilts the balance beam in the downward direction on the side, where the attraction is applied. The tilt in each case is measured by the suspension mirror of Kelvin as shown in figure (7.29.6). In this arrangement, m is a small mirror suspended by means of a bi-filiar suspension (a_1, a_2) from two horizontal brackets A and A' in level with each other with a small adjustable gap between them. The bracket A is movable being attached to the pointer P, and the bracket A' is fixed. Thus when the balance beam tilts, the wire a_1 turns with it making the mirror m turning about the stationary wire a_2. By adjusting the gap between A and A' suitably, the deflection of the beam is found to be magnified 150 times. The deflection is noted by lamp and scale arrangement. The scale is graduated in half millimetres, and is placed about 5 metres in front of the mirror. The image is observed from the room above by means of a vertical telescope fixed up in the ceiling of the room. The effect of air-draught is eliminated by using damping vanes V, suspended from the mirror, and kept immersed in oil. The change in the angular tilt due to moving C from under A to under B is a little more than one second. The same angle is produced by the addition of a weight of 0.0004 gm to one end of the balance beam.

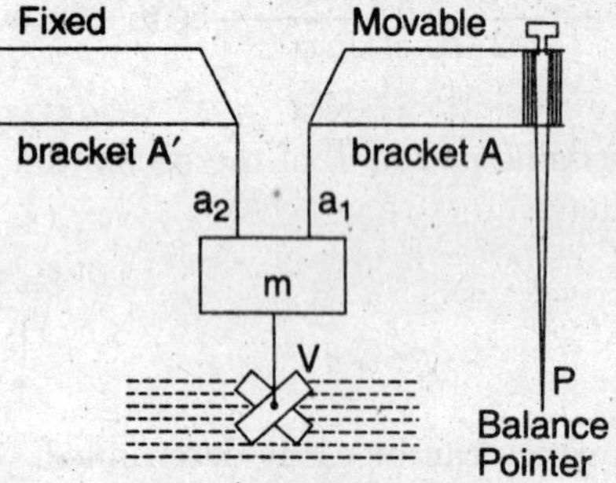

Fig. 7.29.6

Let h be the height of the end knife-edges N_1 and N_2 above the central knife-edge O, and let the beam be deflected through an angle θ by the attraction of the body C. Let a be the length of each arm of the balance, and let d' be the depth of the centre of gravity of the beam from the central knife-edge O, and M' be its mass. Then from the condition of equilibrium of the balance beam, we have,

$$(a\cos\theta - h\sin\theta)mg + M'gd'\sin\theta$$

$$= (a\cos\theta + h\sin\theta)mg + \frac{GMm}{d^2}(a\cos\theta + h\sin\theta),$$

or, $$(a\cos\theta - h\sin\theta)mg + M'gd'\sin\theta$$

$$= (a\cos\theta + h\sin\theta)\left(mg + \frac{GMm}{d^2}\right). \tag{7.29.1}$$

Again let the larger mass M be removed altogether, and a small mass m' be attached to B so that the angular twist is once more θ; hence we have,

$$(a\cos\theta - h\sin\theta)mg + M'gd'\sin\theta = (a\cos\theta + h\sin\theta)(mg + m'g) \tag{7.29.2}$$

from (7.29.1) and (7.29.2), we have,

$$m'g = \frac{GMm}{d^2}, \quad \text{or,} \quad G = \frac{m'd^2}{Mm}g.$$

The attraction of the body C on the balance beam is eliminated in the following way.

The masses A and B are raised by nearly 25 cms in a second experiment. Considering d_1 and d_2 the distances between the larger and the smaller masses in the two different experiments when m' and m'' are the corresponding masses attached to B, we have

$$\frac{GMm}{d_1^2} + F' = m'g \tag{7.29.3}$$

$$\text{and} \quad \frac{GMm}{d_2^2} + F' = m''g. \tag{7.29.4}$$

F' denotes the force exerted by the body C on the beam. From (7.29.3) and (7.29.4), we have

$$GMm\left(\frac{1}{d_2^2} - \frac{1}{d_1^2}\right) = (m'' - m')g, \quad \text{or,} \quad G = \frac{(m'' - m')g}{Mm\left(\frac{1}{d_2^2} - \frac{1}{d_1^2}\right)}.$$

The change of position of the body C is found to tilt the floor (i.e., the table) through an angle of about one-third of a second. This is eliminated by mounting a mass $\frac{M}{2}$ on the turn table at twice the distance of the body C from the turning point O' and diametrically opposite to C (not shown in Fig. 7.29.5).

Poynting's results were

$$G = 6.6984 \times 10^{-8} \text{ C.G.S. units.}$$

$$\rho = 5.4934 \text{ gms per c.c.}$$

7.29.4 Heyl's Method

The most accurate value of G was obtained by Heyl. He used the torsional balance method formerly used by Braun. The suspension system of a torsion balance consists of two equal spheres hanging with their centres at the same level. The system is allowed to describe small oscillations in the gravitational field of two large masses with their centres of gravity also at the same level. The whole system is enclosed in a large brass-container, resting on a plate-glass base. The air pressure inside is reduced to 2 mms of mercury to minimise the convection current. The attracting large masses are cylinders of steel (containing 0.90% carbon) of mass 66.3 kgms each, suspended from a system capable of rotation about a vertical axis mid-way between the two. In three different sets of experiments, the small masses are pairs of gold, platinum and optical glass spheres respectively each of mass about 50 gms. A torsion wire of tungsten (1 metre long and 0.025 mm diameter) is used to carry a light aluminium rod AB (of length 20.6 cm and mass 2.44 gms) from the ends of which two small spheres are hung (Fig. 7.29.7). Also there are two inclined copper wires and a tungsten wire so that almost the whole moment of inertia is in the spheres themselves.

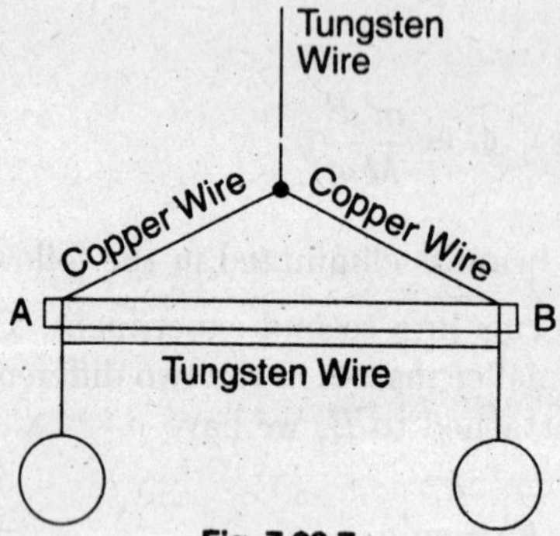

Fig. 7.29.7

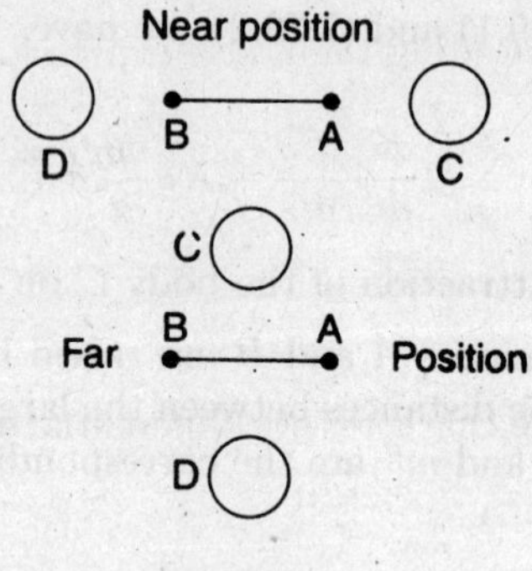

Fig. 7.29.8

The theory of the experiment can be understood by the attached diagram. The larger masses are placed in two different positions represented in figure (7.29.8).

Near Position

In this position the centres of the larger spheres are in one line with the centres of the smaller spheres.

Far Position

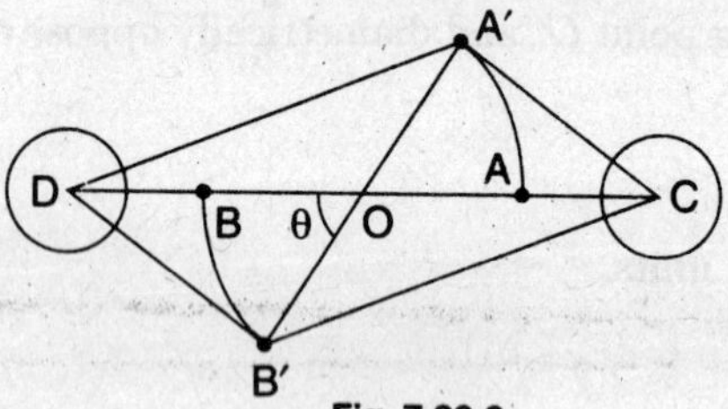

Fig. 7.29.9

In this position, the line joining the centres of larger spheres is at right angles to the line joining the centres of the smaller spheres.

Let us consider the total energy of the suspension system, when the larger masses are in the near position. Let the centres of smaller masses A and B be in the displaced positions A' and B' as shown in figure (7.29.9). The total energy of the system is due to —

(i) K.E. $= \frac{1}{2} I \left(\frac{d\theta}{dt}\right)^2$, where I is the moment of inertia of the suspension system about the axis of rotation and $\left(\frac{d\theta}{dt}\right)$ is the angular velocity;

(ii) P.E. due to the torsion of the fibre $= \frac{1}{2} K\theta^2$, where θ is the angular displacement and K is the torsional constant of the tungsten wire;

(iii) P.E. due to the reason that A is near to C and D, and B is near to D and C. We may write this as $\frac{GA_1\theta^2}{2}$, where A_1 is a constant involving the product of one small mass and one large mass, and a purely geometrical function of the distances OA and AC.

Hence if frictional losses are neglected, we may write,

$$\frac{1}{2}I\left(\frac{d\theta}{dt}\right)^2+\frac{1}{2}K\theta^2+\frac{GA_1\theta^2}{2}=\text{constant.}$$

Differentiating the above expression with respect to time, we have,

$$I\frac{d\theta}{dt}\cdot\frac{d^2\theta}{dt^2}+K\theta\frac{d\theta}{dt}+GA_1\theta\frac{d\theta}{dt}=0,\text{ or, }I\frac{d^2\theta}{dt^2}+(K+GA_1)\theta=0,$$

$$\text{or, }\frac{d^2\theta}{dt^2}+\frac{(K+GA_1)}{I}\theta=0,$$

which is the characteristic differential equation for simple harmonic motion, and the time period is given by

$$T_1=2\pi\sqrt{\frac{I}{(K+GA_1)}}. \tag{7.29.5}$$

When the larger masses are in the far position, we may deduce in the same way the expression for the time period which is given by

$$T_2=2\pi\sqrt{\frac{I}{(K+GA_2)}} \tag{7.29.6}$$

A_2 being a constant like A_1. From (7.29.5) we get,

$$\text{or, }\quad K+GA_1=\frac{4\pi^2}{T_1^2}.I,\quad\text{or, }\quad K=\frac{4\pi^2}{T_1^2}I-GA_1.$$

Similarly from (7.29.6), we get, $K=\dfrac{4\pi^2}{T_2^2}I-GA_2.$

Hence

$$\frac{4\pi^2}{T_1^2}I-GA_1=\frac{4\pi^2 I}{T_2^2}-GA_2,\quad\text{or,}\quad G(A_2-A_1)=4\pi^2 I\left(\frac{1}{T_2^2}-\frac{1}{T_1^2}\right),$$

$$\text{or,}\quad G=\frac{4\pi^2 I\left(\frac{1}{T_2^2}-\frac{1}{T_1^2}\right)}{(A_2-A_1)},\quad\text{whence } G \text{ may be determined.}$$

Values of G obtained are:

for platinum spheres, $G=6.664\times10^{-8}$ C.G.S. units;

for glass spheres, $G=6.674\times10^{-8}$ C.G.S. units.

$\therefore$ The mean value of $G=6.669\times10^{-8}$ C.G.S. units which is a little higher than the value obtained by Boys' and Braun's experiment.

The mean density $\rho=5.517$ gms per c.c.

We give below in a table the results of different experiments on G and ρ.

Year	Name of Experimentalist	Type of Experiment	G in C.G.S. units	ρ gm/c.c.
1775	Maskelyne and Hutton	Mountain	7.4×10^{-8}	5.0
1798	Cavendish	Torsion balance	6.754×10^{-8}	5.448
1854	Airy	Mine	5.7×10^{-8}	6.5
1891	Poynting	Sensitive bullion balance	6.698×10^{-8}	5.493
1895	Boys	Torsion balance	6.657×10^{-8}	5.527
1896	Eötvös	Torsion balance	6.66×10^{-8}	5.53
1930	Heyl	Torsion balance	6.670×10^{-8}	5.517

7.30 Variation in G

Experiments have been carried out by various investigators to see whether the value of G changes under different conditions and their results are summarised and given below.

7.30.1 Selectivity

Eötvös and others showed with the torsion balance that the value of G does not change with the change of the nature of attracting masses within the limit of experimental error. The same experiment proved also that G is independent of the state of chemical combination of the attracting masses. Even in case of masses of radioactive elements, the value of G was found to be unaffected.

7.30.2 Temperature

Using torsion balance of Boys' type Shaw showed that for temperature range from 0°C to 250°C the value of G remains unaffected within the limit of experimental error. Experiments of Poynting and Phillips established the same negative result.

7.30.3 Directivity

It was experimentally observed by various investigators that for anisotropic bodies such as crystals, the value of G obtained is independent of the orientation of the crystallographic axis within the limit of experimental error. That is to say, gravitational force is not directive even in the case of crystals.

7.30.4 Permeability

The effect of interposing different materials between the attracting masses was studied by Majorana, Austin, Thwing, Eötvös and his collaborators, but no change in the value of G within the limit of experimental error could be detected.

Illustrative Example

1. Two small spheres, each of radius 2 cms and mass 350 gms are suspended below the pans of a sensitive balance. A one milligram rider is moved 1/20th of a division along the right-hand beam (increasing the effective load in the right-hand pan by $\frac{1}{200}$ mg), and causes the pointer of the balance to be deflected by 10 divisions. A large sphere of mass 40 kgms and radius 10 cms is then placed under the left-hand sphere, nearly touching it. Calculate the new deflection of the pointer. Where must the large sphere be placed to reduce the deflection to zero? ($G = 6.67 \times 10^{-8}$ C.G.S. units, $g = 980$ cm sec^{-2}).

The force of attraction between the spheres is given by

$$F = \frac{6.67 \times 10^{-8} \times 350 \times 40 \times 10^3}{(10+2)^2} = 0.0065 \text{ dynes.}$$

Now the weight of $\frac{1}{200}$ mg, i.e., a force of $\frac{980}{200 \times 1000}$ or 0.0049 dynes causes a deflection of 10 divisions. Hence the force required for deflection through 1 division is 0.00049 dynes. The gravitational force overrides the effect of the weight of the rider by (0.0065 – 0.0049) = 0.0016 dynes. This force causes a deflection of $\frac{0.0016}{0.00049} = 3.3$ divisions on the other side.

The deflection will be zero when the gravitational force of the large sphere is balanced by the weight added to the rider. In this balancing condition let d be the distance between the centres of the small and large spheres.

$$\therefore \quad \frac{6.67 \times 10^{-8} \times 350 \times 40 \times 10^3}{d^2} = 0.0049,$$

$$\text{or,} \quad d = \sqrt{\frac{6.67 \times 10^{-8} \times 350 \times 40 \times 10^3}{0.0049}} = 13.8 \text{ cms.}$$

Hence the large sphere must be lowered by (13.8 – 12) = 1.8 cms.

Exercises

1. State Newton's Law of Gravitation. Is this law universally correct? Give reasons for your answer.

2. A spherical mass of 20 kg situated at the earth's surface is attracted by another spherical mass of 150 kg equal to the weight of 0.25 mg, when the centres of the masses are 30 cm apart. Calculate the mass of the earth assuming the radius of the earth to be 6×10^8 cm. [Ans. 4.8×10^{27} gm]

3. Explain what you mean by gravitational potential and intensity at a point. Find expressions for attraction and potential due to a system of particles.

4. What is the physical interpretation of potential? Explain what you mean by an equipotential surface.

5. Find an expression for the gravitational potential due to a thin spherical shell at a point outside the shell.

6. Calculate the gravitational field at any point inside a solid sphere.

7. Find an expression for intensity at any point on the axis of a circular disc. Explain how the intensity changes when the point is moved close to the disc.

8. Find expressions for potential and intensity due to a thin uniform straight rod.

9. Calculate the work done by the mutual attractive forces of the particles of a self-attracting system when the particles are brought from an infinite distance to the present configuration. Calculate the energy required for the formation of a sphere.

10. Calculate the amount of work done in scattering the constituent particles of a homogeneous sphere of mass 10 kg and radius 10 cm to infinity against the gravitational attraction.

$(G = 6.670 \times 10^{-8}$ C.G.S. units.) [**Ans.** 4 ergs.]

(See Art. 7.18 and worked out examples)

11. Deduce Poisson's equation and Laplace's equation for gravitational potential.

(See Art. 7.18 and worked out examples)

12. State and prove Gauss's theorem in a gravitational field. **[C.U. (Hons.) 1984, '88]**

13. State Kepler's laws on the motion of planetary bodies. Deduce Kepler's laws from Newton's law of gravitation. **[C.U. (Hons.) 1988, 2005]**

14. Assuming the planetary orbits in the solar system to be circular, find the period of rotation of Jupiter around the sun, the orbital radii of the earth and the Jupiter respectively being 1 and 5.2 astronomical units. [**Ans.** 11.86 years] **[C.U. (Hons.) 1980]**

(See worked out examples)

15. Explain how you would deduce Newton's Inverse Square Law of gravitation from Kepler's laws. **[C.U. (Hons.) 1988]**

16. Explain what you mean by velocity of escape. Calculate the velocity with which a body is to be projected from the surface of the earth so that it will never return.

17. Explain how the mean density of the earth was determined by Airy's Mine Experiment.

18. Mention the drawbacks in Cavendish's method for finding the value of G and discuss in detail Boys' method, bringing out clearly its advantages over Cavendish's method.

19. Describe Poynting's method for the determination of G. Explain the necessary corrections for finding G accurately.

20. Describe the most accurate method for the determination of the gravitation constant.

21. Prove that the total energy of a particle of mass m acted upon by a central force is given by

$$E = \frac{h^2}{2m}\left[u^2 + \left(\frac{du}{d\theta}\right)^2\right] + V(r),$$

where $V(r)$ is the potential energy and h is the angular momentum of the body, $u = 1/r$, r and θ being the polar co-ordinates of the particle. **[C.U. (Hons.) 1980, '85, '87, 2006]**

(For answer see worked out examples in Chapter 3)

22. A particle acted upon by a central force describes an orbit given by $r = a(1 + \cos\theta)$, a being a constant. Show diagrammatically the nature of the orbit and find the nature of the force.

(For answer see worked out examples) **[C.U. (Hons.) 1980]**

23. Prove that the period of revolution of a small satellite around a planet in an orbit of radius just greater than the radius of the planet, which is assumed spherical and of uniform density, is independent of the radius of the planet. Find the period of revolution of the satellite if the density of the planet is 10 g/c.c. Assume $G = 6.66 \times 10^{-8}$ C.G.S. (For Answer see worked out examples). **[C.U. (Hons.) 1980, '87, '90]**

24. Prove that Kepler's first and second laws lead to the conservation of angular momentum vector. **[C.U. (Hons.) 1980, '93, '95]**

(See Art. 7.22.1)

25. (a) Calculate the gravitational potential at a point within the cavity of a thick spherical shell of internal and external radii r_1 and r_2. **[C.U. (Hons.) 1981, '85]**

(b) A smooth straight narrow tunnel is imagined to be bored through the centre of the earth. A small body of mass m is dropped into the tunnel at one end on the surface of the earth. Assuming the earth to be a sphere of uniform density, prove that the body will execute s.h.m. between the opposite ends of the tunnel, and calculate its time period. How long will it take for the body to cross the tunnel? Assume $\rho = 5.5$ g/c.c. for the earth and $G = 6.67 \times 10^{-8}$ C.G.S. unit. (For answer see worked out examples). **[C.U. (Hons.) 1981]**

26. A particle of mass m is moving in a plane under a repulsive inverse square law force k/r^2, where k is a positive real number and (r, θ) are the polar co-ordinates of the particle. Write down the equation of motion and solve it. What is the nature of the orbit? **[C.U. (Hons.) 1982]**

(For answer see worked out examples)

27. (a) Write down the differential equation satisfied by the gravitational potential $V(x, y, z)$ at a point (x, y, z), where there is no distribution of matter.

(b) Calculate the gravitational potential due to a circular plate of uniform areal density ρ and radius a, at a distance r from the centre of the plate on the axis perpendicular to the plate at its centre. **[C.U. (Hons.) 1982]**

28. (a) Gauss's theorem for gravitation is

$$\oint_s \vec{g} \cdot d\vec{s} = -4\pi G \int_V \rho dV,$$

where ρ is the density of matter in the enclosed volume. What is g in the above equation? Use the above equation to calculate the magnitude and direction of the gravitational field at a point P inside a very long cylinder of uniform density. The point P is at a distance r from the axis.

(b) Describe briefly Heyl's method for the determination of G. **[C.U. (Hons.) 1984]**

29. (a) State Kepler's laws of planetary motion. Show from the second law that the force between the sun and a planet is a central force. **[C.U. (Hons.) 1984 (Spl), '86, 2005]**

(b) i. Show that the differential equation of motion of a particle under the influence of a central force $F(r)$ can be written as

$$\frac{d^2u}{d\theta^2} + u = -\frac{m}{L^2u^2}F\left(\frac{1}{u}\right).$$

where $u = 1/r$, L is the angular momentum and m is the mass of the particle. **[C.U. (Hons.) 1984 (Spl), '91, 2005, 07]**

[Ans. See Art. 7.22.1.**]**

ii. A particle moves along an orbit $r = A\cos\phi$ under the influence of a central field $F(r)$. Find the r dependence of the force. **[C.U. (Hons.) (Spl) 1984, 2007]**
(For answer see worked out examples)

30. An artificial satellite revolves round the earth near the earth's surface with a time period of 1 hour 20 minutes. Assuming the moon to describe a circular orbit of radius 4×10^5 km round the earth, calculate the time period of its revolution. Given the radius of the earth = 6.4×10^3 km. (For answer see worked out examples). **[C.U. (Hons.) 1985]**

31. (a) Show how you can conclude from Kepler's second law that the force acting on a planet is central.

(b) In an attractive inverse square central field, the equation of motion of a particle is

$$\dot{\vec{p}} = -\frac{k}{r^2}\frac{\vec{r}}{r}.$$

Taking $\dot{\vec{r}} = \hat{r}\dot{r} + \hat{\theta}r\dot{\theta}$, show that $\dot{\vec{p}} \times \vec{L} = \frac{d}{dt}mk\left(\frac{\vec{r}}{r}\right)$.

Hence, L being a constant of motion, show that $\vec{p} \times \vec{L} - mk\frac{\vec{r}}{r}$ is a constant of motion.

[C.U. (Hons.) 1986]

Answer: (a) From Kepler's second law, $\vec{r} \times \frac{d\vec{r}}{dt} = \text{constant}.$

$$\therefore \quad \frac{d}{dt}\left(\vec{r} \times \frac{d\vec{r}}{dt}\right) = 0, \quad \text{or,} \quad \vec{r} \times \frac{d^2\vec{r}}{dt^2} = 0 \quad \left(\because \frac{d\vec{r}}{dt} \times \frac{d\vec{r}}{dt} = 0\right).$$

Let $\vec{f}(r) = \frac{d^2\vec{r}}{dt^2}$ = force per unit mass of the planet; $\therefore\ \vec{r} \times \vec{f}(r) = 0$ — which implies that $\vec{f}(r)$ must be central, since moment of the force is zero.

(b) $$\frac{d}{dt}(\vec{p} \times \vec{L}) = \dot{\vec{p}} \times \vec{L} = \frac{mf(r)}{r}[\vec{r} \times (\vec{r} \times \dot{\vec{r}})] \qquad \because \vec{p} \times \dot{\vec{L}} = 0 \text{ and } \vec{L} = m(\vec{r} \times \dot{\vec{r}})$$

and writing $\dot{\vec{p}} = f(r)\frac{\vec{r}}{r}$

$$= \frac{mf(r)}{r}[(\vec{r}.\dot{\vec{r}})\vec{r} - r^2\dot{\vec{r}})] = \frac{mf(r)}{r}[r\dot{r}\vec{r} - r^2\dot{\vec{r}}] \quad \because \frac{1}{2}\frac{d}{dt}(\vec{r}.\vec{r}) = \vec{r}.\dot{\vec{r}} = r\dot{r}$$

$$= -mf(r)r^2\left[\frac{\dot{\vec{r}}}{r} - \frac{\dot{r}\vec{r}}{r^2}\right] = -mf(r)r^2\frac{d}{dt}\left(\frac{\vec{r}}{r}\right) = \frac{d}{dt}mk\left(\frac{\vec{r}}{r}\right),$$

$$\text{or,} \quad \frac{d}{dt}\left[\vec{p} \times \vec{L} - mk\left(\frac{\vec{r}}{r}\right)\right] = 0.$$

Thus for Kepler problem, there exists a conserved vector defined by $\vec{R} = \left(\vec{p} \times \vec{L} - \frac{mk\vec{r}}{r}\right)$ — which is a constant of motion. This is known as *Laplace-Runge-Lenz Vector.*

32. (a) Find the gravitational potential energy of a particle of mass m when it is placed (i) outside a homogeneous spherical shell of radius R and (ii) inside the spherical shell.

(b) Describe, with necessary theory, Hely's method of measurement of the Gravitational constant G. **[C.U. (Hons.) 1987, '90]**

33. State and prove Gauss's theorem in gravitation. Calculate the gravitational potential energy of a point mass m placed at a distance r from the centre of a solid sphere of mass M and radius R when (i) $r > R$, (ii) $r = R$ and (iii) $r < R$. **[C.U. (Hons.) 1988]**

34. What is a central force? Prove that the path of a particle under a central force lies in a plane. (Answer given in Mechanics Chapter). **[C.U. (Hons.) 1988, '91, '95, '99]**

35. Suppose that a rocket which starts from rest falls in a constant gravitational field. At the instant it starts to fall, it rejects gas at a constant rate α in the direction of the gravitational field and at speed v_0 w.r.t. the rocket. Find its speed after any time t. How far does the rocket travel in time t? **[C.U. (Hons.) 1989, '92]**

(See worked out examples)

36. (a) Find the potential energy or potential corresponding to the central force field defined by $\vec{F} = \sin \pi\gamma\hat{\gamma}_1$.

(b) A hemi-sphere of mass M and radius a has a particle of mass m located at its centre. Find the force of attraction on it if the hemi-sphere is (i) a thin shell, (ii) a solid. **[C.U. (Hons.) 1989]**

Answer: (a) Hints: Since $\vec{F}$ is a central force field $\vec{\nabla} \times \vec{F} = 0$ and as such $\vec{F}$ can be written as $\vec{F} = -\vec{\nabla}V$ (V = the potential)

$$\text{or,} \quad \left(\hat{\gamma}_1 \frac{\partial}{\partial\gamma_1} + \hat{\gamma}_2 \frac{\partial}{\partial\gamma_2} + \hat{\gamma}_3 \frac{\partial}{\partial\gamma_3}\right) V = -\vec{F} = -\sin \pi\gamma\hat{\gamma}_1 \quad \because \gamma = f(\gamma_1, \gamma_2, \gamma_3)$$

whence $V = \dfrac{\cos \pi\gamma}{\pi}$.

(b) For answer see worked out problems.

37. Starting from Gauss's theorem on gravitation, calculate the gravitational field outside and within the material of a spherical shell of uniform thickness. **[C.U. (Hons.) 1990]**

38. Show that in the gravitational field, the angular momentum and total energy of a planet are always conserved. Discuss the relevance of the principle of conservation of angular momentum with Kepler's second law. **[C.U. (Hons.) 1990, '99]**

39. The equation of the orbit of a particle moving under the influence of a central force $f(r)$ is $r = A\theta^2$, where A is a constant. Obtain the force law. **[C.U. (Hons.) 1991]**

(For answer see worked out examples)

40. A uniform rod of mass M lies on the Z-axis between $z = 0$ and $z = l$.

(a) Find the gravitational force on a point mass m placed on the Z-axis at a distance $a (a > l)$ from the origin.

(b) If the whole mass of the rod were concentrated at a point z_0 such that the force on m were unchanged, find the location of z_0. Show that z_0 is located almost at the centre of the rod when $a \gg l$. (For answer see worked out problems). **[C.U. (Hons.) 1991]**

41. (a) Derive Kepler's second law starting from the equation of motion. Mention any one approximation that is used in the derivation of Kepler's laws.

(b) A small satellite revolves round a spherical planet in a circular orbit of radius twice the radius of the planet. Find the period of revolution of the satellite, assuming that the average density of the material of the planet is 20 gm/c.c. ($G = 6.673 \times 10^{-8}$ c.c./gm-sec^2).

(For answer see worked out examples). **[C.U. (Hons.) 1992]**

42. (a) Find the expression for the velocity of escape of a particle from the surface of a spherical planet.

(b) Consider a homogeneous spherical shell of uniform thickness. Find the gravitational intensity (i) in the hollow region and (ii) on the outer surface. What will be these forces if this shell is placed within the hollow of another (larger) homogeneous spherical shell of uniform thickness? **[C.U. (Hons.) 1993]**

(For answer see worked out examples).

43. A meteorite falls towards the earth, starting from rest at a great distance h above the surface. Neglecting the air resistance, find its velocity on reaching the earth's surface, given a = the radius of the earth and g = the value of the acceleration due to gravity on the surface of the earth. **[C.U. (Hons.) 1994]**

Answer: P.E. at a height h above the earth's surface is $-\dfrac{GMm}{(a+h)}$ and that on earth's surface $= -\dfrac{GMm}{a}$, so that the loss in P.E. is given by

$$-\frac{GMm}{(a+h)} - \left(-\frac{GMm}{a}\right) = GMm\left(\frac{1}{a} - \frac{1}{a+h}\right) = \frac{GMmh}{a(a+h)}$$

Since this loss in P.E. must be equal to the gain in K.E. (as the meteorite starts from rest at the height h), we may write

$$\frac{1}{2}mv^2 = \frac{GMmh}{a(a+h)}, \quad \text{or,} \quad v^2 = \frac{2GMh}{a(a+h)} = \frac{2ga^2h}{a(a+h)} \qquad \because \frac{GM}{a^2} = g.$$

$$\therefore \quad v^2 = \frac{2gah}{(a+h)}, \quad \text{or,} \quad v = \sqrt{\frac{2gah}{(a+h)}}.$$

44. (a) Show that the equation of motion for a rocket projected vertically upward in a uniform gravitational field, neglecting atmospheric resistance is

$$m\frac{d\vec{v}}{dt} = -\vec{v}'\frac{dm}{dt} - mg\hat{k},$$

where m is the mass of the rocket, and v' is the velocity of escaping gases relative to the rocket and $\hat{k}$ is a unit vector in the vertically upward direction.

(b) Integrate the above equation to obtain v as a function of m, assuming a constant time rate of loss of mass. **[C.U. (Hons.) 1994, 2007]**

(c) Show, for a rocket starting initially from rest, with v' equal to 6800 ft/sec, vertically downwards and a mass loss per second equal to 1/60th of the initial mass, that in order to reach the escape velocity from the earth, the ratio of the mass of the fuel to the mass of the empty rocket must be almost 300. (Escape velocity from the earth = 11.2 km/sec).

(For answer see worked out examples) **[C.U. (Hons.) 1994]**

45. Give, with necessary theory, the outline of a precision method of determining G, the universal gravitational constant. **[C.U. (Hons.) 1994]**

46. (a) A particle of mass m moves under the action of a central field of force $F = f(r)$. Show that the differential equation of the path of the particle is

$$\frac{d^2u}{d\theta^2} + u = -\frac{f(1/u)}{mh^2u^2}, \quad \text{where } u = \frac{1}{r}.$$

(b) Show that if $f(r) = -\dfrac{k}{r^2}$, $k > 0$, the path is a conic section. **[C.U. (Hons.) 1994, 98]**

[**Ans.** See Art. 7.22.1.]

(c) Under the influence of a central force with centre at the point O, a particle moves in a circular orbit of radius R which passes through O. Show that the equation to the orbit is $r = 2R\cos\theta$, and hence find the law of force in terms of distance r from O.
(For answer see worked out examples) **[C.U. (Hons.) 1994]**

47. (a) If $\vec{R}(t)$ is the vector from the origin to the instantaneous position of a particle moving along a curve C, show that $\vec{R} \times d\vec{R}$ is equal to twice the area of the sector defined by the two vectors $\vec{R}(t), \vec{R}(t+dt) = \vec{R} + d\vec{R}$ and the arc of C, which they intercept.

(b) Hence show that (i) if $\vec{R} \times d\vec{R} = 0$, the vector $\vec{R}$ has a constant direction, (ii) if $\vec{R} \times \frac{d^2\vec{R}}{dt^2} = 0$, the particle moves so that its radius vector sweeps out equal areas in equal times. (For answer see Vector and Mechanics Chapters). **[C.U. (Hons.) 1995]**

48. Find the work required to be done to remove a particle from the surface of a spherical star of mass M and radius R, to infinity. Hence find an expression for the escape velocity v. Find the condition that the escape velocity will be equal to the velocity of light in empty space.

[C.U. (Hons.) 1995]

Ans: Hints: $v^2 = \frac{2GM}{R} = 2gR$ (G = gravitational constant of the star and g = acceleration due to gravity of the star); if m = mass of the partcle and E = its energy, then we have from relativistic mechanics, $E = mC^2$, or, $C^2 = E/m$. If $v = C$, then

$$\frac{E}{m} = \frac{2GM}{R} = 2gR, \quad \text{i.e.,} \quad E = \frac{2GMm}{R} = 2gmR.$$

49. (a) A mass m lies on the perpendicular through the centre of a uniform thin circular plate, of radius a, at a distance b from the centre. Find the force of attraction between the plate and the mass m. **[C.U. (Hons.) '95, 2000]**

(b) An artificial satellite revolves in a circular orbit about the earth at height H above the surface. Determine (i) the orbital speed and (ii) the orbital period, so that a man in the satellite is in a state of weightlessness. **[C.U. (Hons.) 1995]**

50. (a) Define central force. Prove that a central force field is conservative and find the corresponding potential energy of a particle in this field. **[C.U. (Hons.) 1996, 2008]**

[Ans. See Art. 3.14.]

(b) Writing $u = \frac{1}{r}$, show that the conservation of energy equation becomes

$$\left(\frac{du}{d\theta}\right)^2 + u^2 = \frac{2(E-V)}{mh^2},$$

where h is twice the areal velocity and V is the P.E.

(c) Differentiating the above relation, obtain

$$\frac{d^2u}{d\theta^2} + u = -\frac{f\left(\frac{1}{u}\right)}{mh^2u^2}.$$

(d) If a planet is to revolve around the sun in an elliptical orbit with the sun at a focus, show that the force of attraction must follow the inverse square law.

(e) Assuming an inverse square force to be $f(r) = -\frac{k}{r^2}$ $(k > 0)$, prove that the speed v of a particle moving in the above force field is given by

$$v = \frac{k}{m}\left(\frac{2}{r} \mp \frac{1}{a}\right), \quad (a \text{ is a constant}),$$

the upper and lower signs correspond to an elliptic and a hyperbolic path respectively.

(f) Hence deduce the value of the escape velocity. **[C.U. (Hons.) 1996, cf. '98]**

[Ans. See Art. 7.22.1.]

51. Consider the motion of a rocket under an external force $\vec{F}$. Let at any instant t, mass of the rocket be M and its velocity be $\vec{v}$ relative to some fixed co-ordinate system. Let the fuel be shot out with a constant velocity $\vec{u}$ relative to the rocket motion. Calculate the impulse $\vec{F}\Delta t$ and hence show that

$$M\frac{d\vec{v}}{dt} + \vec{u}\frac{dM}{dt} = \vec{F}.$$

[C.U. (Hons.) 1997, cf. '98, 2001]

52. (a) State and prove Gauss's theorem of gravitation. [C.U. (Hons.) 2000]

(b) Define gravitational self-energy. Imagine that a Galaxy consists of n stars each of mass m with an average separation r between every pair of them, i.e., $r_{ij} = r$ for all i and j. Calculate the gravitational self-energy of the Galaxy. (Ignore gravitational self-energy of individual stars). Given $n = 1.6 \times 10^{11}$, $m = 10^{33}$ gm, $r = 10^{23}$ cm, $G = 6.673 \times 10^{-8}$ $cm^3/gm - S^2$. [C.U. (Hons.) 1997, 2002, 2004, 2007]

(Apply expression for U of Art. 7.18.)

(c) Show that central orbit is a planer path. In which co-ordinate system would you prefer to describe such motion? Explain. [C.U. (Hons.) 1999]

(d) The differential equation of the orbit of a particle of mass m under a central force is given by

$$\frac{d^2u}{d\theta^2} + u = -\frac{m}{L^2u^2}f\left(\frac{1}{u}\right),$$

where $u = 1/r$, L = constant, other notations have usual significance.

Use the above relation and consider the following: a particle moves in a central orbit described by $r = e^{-\alpha\theta}$, α is a positive constant, with force centre at O. Find the nature of the force as a function of r. [C.U. (Hons.) 1997]

(For answer see worked out examples)

53. (a) Find out the gravitational intensity due to a solid uniform sphere inside and outside the sphere.

(b) An object is thrown vertically upward from the surface of the earth with a velocity v_0. What is its velocity at a high H above the surface? Find out the velocity with which it must be projected so that it never returns. The acceleration due to gravity on the surface of the earth is g. [C.U. (Hons.) 1999]

(For answer see worked out examples)

54. The P.E. function for the force between two atoms in a diatomic molecule can be expressed as $U(x) = \frac{a}{x^{12}} - \frac{b}{x^6}$, where a and b are positive constants and x is the distance between the two atoms. Derive an expression for the force between the two atoms and show that the two atoms repel each other for x greater than x_0. What is the value of x_0?

Ans. The force between the two atoms is the negative gradient of P.E., and is given by

$$F = -\frac{dU(x)}{dx} = -\frac{d}{dx}\left(\frac{a}{x^{12}} - \frac{b}{x^6}\right) = \frac{12a}{x^{13}} - \frac{6b}{x^7}$$

Let $F = 0$ when $x = x_0$, so that we have $\frac{12a}{x_0^{13}} - \frac{6b}{x_0^7} = 0$ or $x_0^6 = \frac{2a}{b}$ $\quad\therefore x_0 = \left(\frac{2a}{b}\right)^{1/6}$.

This is the equilibrium point — i.e., *a point of stable equilibrium.*

$$\therefore F = \frac{12a}{x^{13}} - \frac{6b}{x^7} = \frac{6b}{x^7}\left[\frac{2a/b}{x^6} - 1\right] = \frac{6b}{x^7}\left[\left(\frac{x_0}{x}\right)^6 - 1\right]$$

clearly, when $x < x_0, F$ is positive and the atoms repel each other. When $x > x_0, F$ is negative and the atoms attract each other.

55. The P.E. function for the force between two atoms in a diatomic molecule is expressed as $U(x) = \frac{a}{x^{10}} - \frac{b}{x^5}$ where a and b are positive constants and x is the distance between the atoms. (a) Calculate the distance x at which P.E. is a minimum. (b) Assume that one of the atoms remains at rest and that the other moves along x. Describe the possible motion. (c) The energy needed to break up the molecule into separated atoms ($x = \infty$) is called the ***dissociation energy***. What is the dissociation energy of the molecule?

Ans. (a) The value of x at which P.E. is minimum is obtained from $\frac{dU(x)}{dx} = 0$ whence $x = \left(\frac{2a}{b}\right)^{1/5}$.

(b) The force between the atoms is $F = -\frac{d}{dx}U(x) = \frac{10a}{x^{11}} - \frac{5b}{x^6}$ — which is zero when $x = \left(\frac{2a}{b}\right)^{1/5}$; when $x < \left(\frac{2a}{b}\right)^{1/5}$, the atoms repel each other; when $x > \left(\frac{2a}{b}\right)^{1/5}$, the atoms attract each other. If one of the atoms is fixed, then the other atom would oscillate about the equilibrium separation $(2a/b)^{1/5}$.

(c) The dissociation energy D_E is equal to the change in P.E. from its minimum value at equilibrium separation $x = (2a/b)^{1/5}$ to zero value at $x = \infty$

$$\therefore\ D_E = U(x=\infty) - U\left\{x = (2a/b)^{1/5}\right\} = 0 - \left\{\frac{a}{(2a/b)^2} - \frac{b}{2a/b}\right\} = b^2/4a.$$

If the K.E. at the equilibrium position is equal to or greater than this value, the molecule will dissociate.

56. (a) Consider the motion of a rocket under an external force $\vec{F}$. Let at any instant t the mass of the rocket be M, and its velocity $\vec{v}$, relative to some fixed co-ordinate system. Let the fuel be shot out with a constant velocity $\vec{u}$ relative to the rocket in motion. Calculate the impulse $\vec{F}\,\Delta t$ and hence show

$$M\frac{d\vec{v}}{dt} + \vec{u}\frac{dM}{dt} = \vec{F}.$$ **[C.U. (Hons.) 2001, 2003, 06]**

Integrate the above equation to obtain v as a function of M, assuming a constant time rate of loss of mass; hence find out the condition when the velocity attained by the rocket after a long time of flight will reach a maximum value. (See worked out examples 3 and 4 on page 472–73).

(b) Explain the terms gravitational potential and intensity. Find out the expressions of the intensity and potential due to a thin spherical shell at a point inside it.

[C.U. (Hons.) 2001]

57. If the gravitational force between two bodies had been inversely proportional to the fourth power of the distance between them, then find out the escape velocity at the surface of the earth given $g = 10\ \text{m/sec}^2$ and radius of the earth $= 6400 \times 10^3$ m. **[C.U. (Hons.) 2001]**

Ans. $F = \frac{GMm}{4^4} = -\frac{dU}{dr}$ [Here U = P.E., $r > R$ and $F \propto \frac{1}{r^4}$ given]

$$\therefore\quad U = -\int_R^{\infty} \frac{GMm}{r^4}dr = \frac{1}{3}GMm\left(\frac{1}{r^3}\right)_R^{\infty} = -\frac{GMm}{3R^3}$$

Now

$$|\text{P.E.}| = \text{K.E.}\quad \text{i.e.}\quad \frac{1}{3}\frac{GMm}{R^3} = \frac{1}{2}mv_e^2$$

$$\text{or } v_e^2 = \frac{2}{3}\frac{GM}{R^3} = \frac{2}{3}gR \quad \left(\because F = mg = \frac{GMm}{R^4} \quad \text{or} \quad gR^4 = GM\right)$$

$$\therefore\ v_e = \sqrt{\frac{2gR}{3}} = \sqrt{\frac{2 \times 10 \times 6400 \times 10^3}{3}} = 6.532\ \text{km/sec.}$$

58. (a) Consider a pair of starts of equal mass M rotating about their common C.M. The attraction between the stars is gravitational and the stars keep a separation l between them. Show

that the time period of rotation of this double star system is given by $\pi l\sqrt{\frac{2l}{GM}}$ where $G =$ the universal gravitational constant. **[C.U. (Hons.) 2003]**

Ans. The K.E. of the system = K.E.$_{\text{Transl.}}$ + K.E.$_{\text{Rot}}$ $= \frac{1}{2}\mu v^2 + \frac{1}{2}I\omega^2$

$$= \frac{M}{4}\omega^2 l^2 + \frac{M\omega^2 l^2}{4} = \frac{M\omega^2 l^2}{2}.$$

where $\mu =$ reduced mass $= \dfrac{M \times M}{M + M} = \dfrac{M}{2}$

and $I =$ m.i. about the axis passing through C.M. $= 2 \times M\left(\dfrac{l}{2}\right)^2 = M\dfrac{l^2}{2}$ and $v = \omega l$.

$$\left[\text{Position of C.M.} = \frac{M\frac{l}{2} + M\frac{l}{2}}{M + M} = \tfrac{l}{2} \text{ i.e. the point } O\right]$$

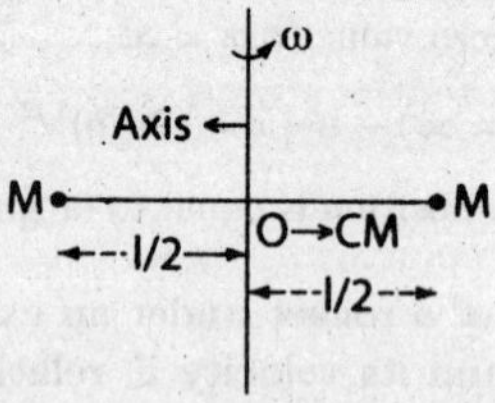

Again P.E. due to attraction $= -\frac{GMM}{l} = \frac{GM^2}{l}$ (neglecting the negative sign).
Since K.E. = P.E., we have

$$\frac{1}{2}M\omega^2 l^2 = \frac{GM^2}{l} \quad \text{or} \quad \omega^2 l^3 = 2GM \quad \text{or} \quad \omega = \sqrt{\frac{2GM}{l^3}} = \frac{2\pi}{T}$$

$$\text{or } T = 2\pi\sqrt{\frac{l^3}{2GM}} = \pi l\sqrt{\frac{4l}{2GM}} = \pi l\sqrt{\frac{2l}{GM}}$$

(b) Consider two concentric thin spherical shells of uniform surface densities of mass M_1 and M_2 and radii R_1 and R_2 respectively ($R_2 > R_1$). Find the gravitational force on a particle of mass m, when it is placed at a distance r from the centre and (i) $r > R_2$, (ii) $R_2 > r > R_1$, (iii) $r < R_1$. **[C.U. (Hons.) 2003, 2005]**

59. First part same as 56(a) above.

Consider now a case where the rocket with initial mass $(M_0 + m_0)$, where m_0 is the mass of the fuel, takes off vertically up in a uniform gravitational field downwards. What is the velocity of the rocket at an instant t_0 when all the fuel has been exhausted, in terms of M_0, m_0, t_0 and acceleration due to gravity g? **[C.U. (Hons.) 2003]**

Ans. Writing $(M_0 + m_0)$ for M and $\vec{F} =$ the gravitational force on the system $= -(M_0 + m_0)g$ in the equation of rocket motion, we have

$$(M_0 + m_0)\frac{d\vec{v}}{dt} + \vec{u}\frac{d}{dt}(M_0 + m_0) = -(M_0 + m_0)g$$

$$\text{or } \frac{d\vec{v}}{dt} + \vec{u}\frac{1}{(M_0 + m_0)}\frac{d}{dt}(M_0 + m_0) = -g$$

$$\text{or } d\vec{v} + \frac{\vec{u}}{(M_0 + m_0)}d(M_0 + m_0) = -gdt;$$

Integrating both sides within proper limits, remembering that at $t = t_i$, $\vec{v} = \vec{v}_i$ and at $t = t_0$, $m_0 = 0$, $\vec{v} = v_f$ we may write

$$\int_{\vec{v}_i}^{\vec{v}_f} d\vec{v} + \vec{u}\int_{m_0}^{0}\frac{d(M_0 + m_0)}{(M_0 + m_0)} = -g\int_{t_i}^{t_0} dt \quad \text{or} \quad (\vec{v}_f - \vec{v}_i) + \vec{u}\ln\frac{M_0}{(M_0 + m_0)} = -g(t_0 - t_i).$$

Now at $t_i = 0, \vec{v}_i = 0$ and we have $\vec{v}_f = -gt_0 - \vec{u}\ln\dfrac{M_0}{M_0 + \dot{m}}$.

60. (a) Show that for a particle moving in a central force field, the angular momentum about the force centre is conserved and it necessarily follows that (i) the orbit is a planer path, (ii) the areal velocity is constant. If the orbit be in XOY plane, what should be the direction of angular momentum vector? **[C.U. (Hons.) 2004, 2006]**

Ans. For a(i) see Art. 3.14.

(b) Show that the orbit equation of a particle moving under inverse square law of attractive force field is given by $\frac{1}{r} = \frac{1}{p}(1 + \epsilon\cos\theta)$ where the force field $\vec{F} = -\frac{k}{r^3}\vec{r}$. **[C.U. (Hons.) 2004]**

Ans. For a(ii) and (b) see Art. 7.22.1.

(c) Define gravitational potential. How is gravitational potential related to potential energy? Why is gravitational potential always negative? (Assume that potential at infinity is zero). **[C.U. (Hons.) 2004]**

61. A particle is moving in a circular orbit under inverse square force $f = -k/r^2$. If the value of k is suddenly reduced to half of its value, show that the path of the particle will be a parabola.

Ans. The problem can be easily solved by using Virial theorem which states that $\langle T\rangle = -\frac{1}{2}\langle V\rangle$ for inverse square force. Here the notation $\langle\ \ \rangle$ indicates the average value; T and V are K.E. and P.E. respectively. For a circular orbit $\langle T\rangle = T_{\text{instantaneous}}$. As k is *suddenly* decreased to half of its value, V is halved, but there will be no change in $T_{\text{inst.}}$. So the total energy becomes $E = T_{\text{inst.}} + \frac{V}{2}$. Now from Viral theorem, $T_{\text{inst.}} = -\frac{V}{2}$, so that $E = -\frac{V}{2} + \frac{V}{2} = 0$. Hence eccentricity $\epsilon = 1$ (using standard expression for ϵ in terms of E) and the path is a parabola.

62. An artificial earth satellite is revolving round the earth. It is in the form of a torrid (fig.) of radius 2.5 m. Calculate the angular velocity $\vec{\omega}$ of the satellite about its axis in order to generate artificial gravity ($g = 10\,\text{m/s}^2$) inside it. **[C.U. (Hons.) 2006]**

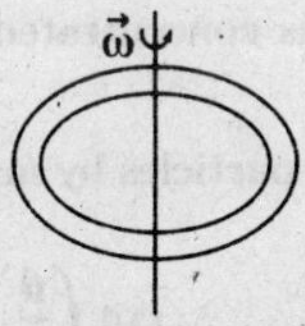

Ans. The artificial acceleration due to gravity g, which is required, will be generated by the centripetal acceleration produced by the rotation of the satellite (in the form of a toroid) with angular velocity $\vec{\omega}$ about its own axis. We know that the centripetal acceleration is $\vec{\omega}^2 r$, r being the radius of the toroid, so that we may write

$$\vec{\omega}^2 r = g \quad\text{or}\quad \vec{\omega}^2 = \frac{g}{r} = \frac{1000\ \text{cm/sec}^2}{250\ \text{cm}} = 4 \quad\text{or}\quad \vec{\omega} = 2$$

Now

$$\omega = 2\pi f (f = \text{r.p.s.}) \quad\text{or}\quad f = \frac{\omega}{2\pi} = \frac{2}{2\pi} = \frac{1}{\pi}\ \text{r.p.s.}$$

63. (a) State Gauss's theorem of gravitation.

(b) A uniform thin circular plate of surface density ρ subtends a solid angle ω at a point P on the axis. Calculate the gravitational intensity at P. **[C.U. (Hons.) 2006 (old syll.) 2007]**

(c) Show that the motion of a body under the action of a central force is confined to a plane.

(d) A body is thrown vertically upwards with an initial velocity V_0 from the surface of the moon. What is the maximum height attained by the body if the acceleration due to gravity on the surface of the moon is $g/6$? **[C.U. (Hons.) 2007]**

(See worked out examples)

64. (a) State Gauss's theorem of gravitation and derive Laplace's equation from it. Using this theorem, obtain the gravitational intensity due to a thick spherical shell at a point outside the shell. **[C.U. (Hons.) 2007]**

(b) The first and second stages of a two stage rocket have weight 100 kg and 10 kg and carry 800 kg and 90 kg of fuel supply. The velocity of the ejected gases relative to the rocket is 1.5 km/sec. Find the final velocity attained by the rocket. ($\log_e 10 = 2.3$). **[C.U. (Hons.) 2007]**

Hints. The final velocity attained by the first stage rocket will be the initial velocity of the second stage rocket.

65. (a) Why is the value of the escape velocity less at the moon's surface than that at the earth's surface?

(b) Two binary stars of masses 2×10^{21} kg and 3×10^{21} kg are 10^6 km apart, and are rotating about their C.M. Find the angular velocity ω—given $G = 6.67 \times 10^{-11}$ Nm^2 kg^{-2}. **[C.U. (Hons.) 2008]**

Ans. See worked out examples.

66. (a) Assuming that the earth is a uniform sphere of mass M and radius R, obtain an expression for the gravitational P.E. U of a body a mass m at a distance r from the centre of the earth. Make a plot of U versus r.

(b) Find the minimum muzzle speed need to shoot a projectile to escape from the earth completely. The radius of the earth is 6380 km and its mass is 5.97×10^{24} kg. Given $G = 6.67 \times 10^{-11}$ Nm^2/kg^2. (See worked out examples)

(c) How can one determine the mass of the earth if one knows the value of G and the radius of the earth? **[C.U. (Hons.) 2008]**

67. (a) What is Rutherford scattering? Explain how Rutherford was led to conclude that the positive charge of the atom is concentrated within a small region of dimensions 10^{-14} m. (See Appendix B.4.3) **[C.U. (Hons.) 2007]**

(b) In Rutherford scattering of α-particles by nuclei of atomic number Z, the angle of scattering θ is given by

$$\cot\left(\frac{\theta}{2}\right) = \frac{MV^2}{2Ze^2}b$$

where M is the mass of the α-particle, V its initial velocity and b the impact parameter. Calculate the number of particles scattered between the angles θ and $\theta + d\theta$ in terms of the incident flux and the above quantities. **[C.U. (Hons.) 2008]** (See Appendix B.4.3)

68. Deduce the equation of the orbit of a planet from energy consideration assuming the solution of the differential equation of motion of the planet.

Ans. From the figure 7.22.4 of Art. 7.22.1, we have, K.E. per unit mass is given by

$$T = \frac{1}{2}v^2 = \frac{1}{2}(v_r^2 + v_\theta^2) = \frac{1}{2}\left[\left(\frac{dr}{dt}\right)^2 + r^2\left(\frac{d\theta}{dt}\right)^2\right]$$

Now

$$\frac{dr}{dt} = \frac{dr}{d\theta}\frac{d\theta}{dt} = \frac{dr}{d\theta}\frac{h}{r^2} \qquad \left(\because r^2\frac{d\theta}{dt} = h\right)$$

$$\therefore \frac{dr}{dt} = -h\frac{d}{d\theta}\left(\frac{1}{r}\right) = -h\frac{du}{d\theta} \qquad \left(\because \frac{1}{r} = u\right)$$

so that

$$T = \frac{1}{2}h^2\left[\left(\frac{du}{d\theta}\right)^2 + u^2\right] \quad \left[\because \left(\frac{d\theta}{dt}\right)^2 = \frac{h^2}{r^4}\right]$$

Again P.E. per unit mass

$$V = \int_{\infty}^{r} \frac{GM}{r^2}dr = -\frac{GM}{r} = -\frac{\mu}{r} \quad \text{(putting GM} = \mu)$$

Now total energy

$$E = T + V \quad \text{or} \quad T = E - V \quad \text{or} \quad \frac{1}{2}h^2\left[\left(\frac{du}{d\theta}\right)^2 + u^2\right] = E + \mu u \tag{1}$$

From the solution of the differential equation of motion for the orbit, we may write

$$u = \frac{\mu}{h^2} + A\cos\theta = \frac{\mu}{h^2}\left(1 + \frac{h^2A}{\mu}\cos\theta\right) = \frac{\mu}{h^2}(1 + \epsilon\cos\theta)$$

$$\text{or} \quad \frac{h^2/\mu}{r} = 1 + \epsilon\cos\theta \quad \text{or} \quad \frac{l}{r} = 1 + \epsilon\cos\theta$$

which is the polar equation to a conic with $l = h^2/\mu$ and $\epsilon = \frac{h^2A}{\mu}$ in terms of h and μ.

Now differentiating

$$\left.\begin{aligned} u &= \frac{\mu}{h^2} + A\cos\theta \\ \text{we have} \quad \frac{du}{d\theta} &= -A\sin\theta \end{aligned}\right\} \tag{2}$$

From (1) we have

$$\left(\frac{du}{d\theta}\right)^2 + u^2 = \frac{2}{h^2}(E + \mu u)$$

$$\therefore A^2\sin^2\theta + \frac{\mu^2}{h^4} + A^2\cos^2\theta + \frac{2\mu}{h^2}A\cos\theta = \frac{2}{h^2}\left[E + \frac{\mu^2}{h^2} + A\mu\cos\theta\right] \quad \text{[using (2)]}$$

$$\text{or} \quad A^2 = \frac{2E}{h^2} + \frac{\mu^2}{h^4} = \frac{\mu^2}{h^4}\left(1 + \frac{2Eh^2}{\mu^2}\right) \quad \text{or} \quad A = \frac{\mu}{h^2}\left(1 + \frac{2Eh^2}{\mu^2}\right)^{1/2}$$

$$\therefore u = \frac{\mu}{h^2} + A\cos\theta = \frac{\mu}{h^2} + \frac{\mu}{h^2}\left(1 + \frac{2Eh^2}{\mu^2}\right)^{1/2}\cos\theta = \frac{\mu}{h^2}\left[1 + \left(1 + \frac{2Eh^2}{\mu^2}\right)^{1/2}\cos\theta\right]$$

$$= \frac{\mu}{h^2}[1 + \epsilon\cos\theta]$$

$$\text{or} \quad \frac{h^2/\mu}{r} = 1 + \epsilon\cos\theta \quad \text{or} \quad \frac{l}{r} = 1 + \epsilon\cos\theta$$

which is the polar equation to a conic

$$\left[\text{with } l = \frac{h^2}{\mu} \quad \text{and} \quad \epsilon = \left(1 + \frac{2Eh^2}{\mu^2}\right)^{1/2} \quad \text{in terms of total energy } E\right]$$

from energy consideration.

Chapter 8

Gyroscopic Motion

8.1 Precessional Torque

When a body moves in a circular path with uniform speed it has a linear velocity at any instant along the tangent to the circular path. The motion along the circular path is maintained due to an acceleration directed towards the centre of the circular path. It is this acceleration which constantly changes the direction of the velocity without altering its magnitude. Thus we have acceleration at constant speed. Similarly angular acceleration at constant angular speed may be obtained. In the latter case the plane of rotation changes direction at a certain rate, the rate of rotation about the axis of rotation remaining the same. This change in the plane of rotation is called *precession.* Precession is caused by a constant torque whose plane is always perpendicular to the instantaneous plane of rotation. The torque causing precession is called precessional torque.

In figure (8.1.1) PQ denotes the edge of a disc whose plane is revolving about an axis passing through O with angular velocity ω. Let the precessional rate of the plane be Ω. $P'Q'$ denotes the edge of the disc after a very short interval of time dt. When the edge of the disc is in the position PQ, the angular momentum $I\omega$ of the disc may be represented by the vector $\overrightarrow{OA}$ and when it is in the position $P'Q'$, the angular momentum is represented by the vector $\overrightarrow{OB}$. Now in the vector triangle OAB.

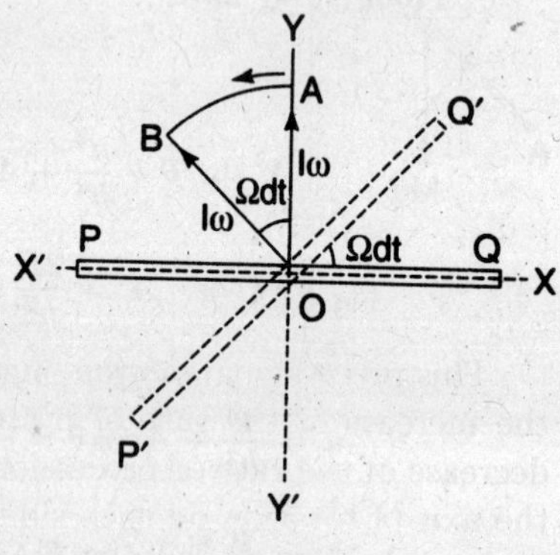

Fig. 8.1.1

$$\overrightarrow{OA} + \overrightarrow{AB} = \overrightarrow{OB}.$$

$$\therefore \ \overrightarrow{AB} = \overrightarrow{OB} - \overrightarrow{OA}$$

$\overrightarrow{AB}$ represents the vectorial change in angular momentum of the disc. The change in angular momentum is given by

$$AB = I\omega\Omega dt.$$

Hence the rate of change of angular momentum of the rotating disc is $I\omega\Omega$ which is equal to the torque applied to it. Thus the processional torque N is given by

$$N = I\omega\Omega, \quad \text{or,} \quad \Omega = \frac{N}{I\omega}. \tag{8.1.1}$$

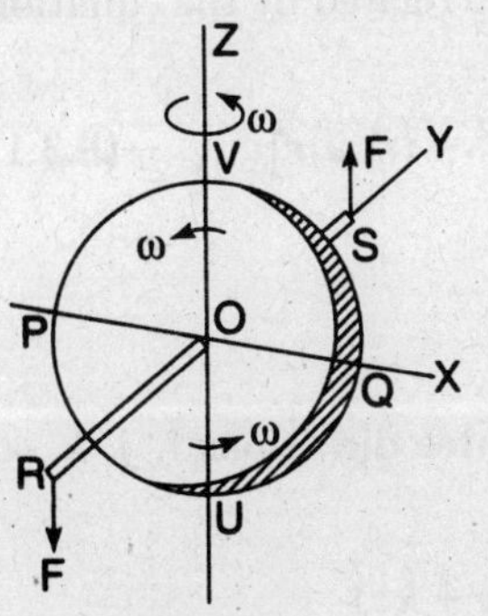

Fig. 8.1.2

It should be noted that the change in angular momentum $\overrightarrow{AB}$ is parallel to the plane of rotation and perpendicular to the axis of rotation.

If the axis of rotation of a body be along OY and the axis of torque along OX, then the body precesses about the third mutually perpendicular axis OZ. This fact illustrated in figure (8.1.2). The figure (8.1.2) shows a disc rotating about the axis OY with angular velocity ω. The two forces F, F form the couple about the axis OX and the precessional motion about the axis OZ will tend to bring P towards R and Q towards S. It is to be noted that if the disc were not rotating, the couple would tend to rotate the disc about the line PQ with V moving towards R and U towards S.

8.2 Gyroscope

Such a rotating symmetrical body as mentioned in Art. (8.1.1) with its axis of rotation supported at a point away from its C.G. and with the precessional rate of its axis of rotation maintained by a gravitational torque is called a *gyroscope* or *top*.

This is illustrated in figure (8.2.1) which represents a heavy disc rotating with angular velocity ω about its geometrical axis AOP which is supported at P by a vertical pivot BP.

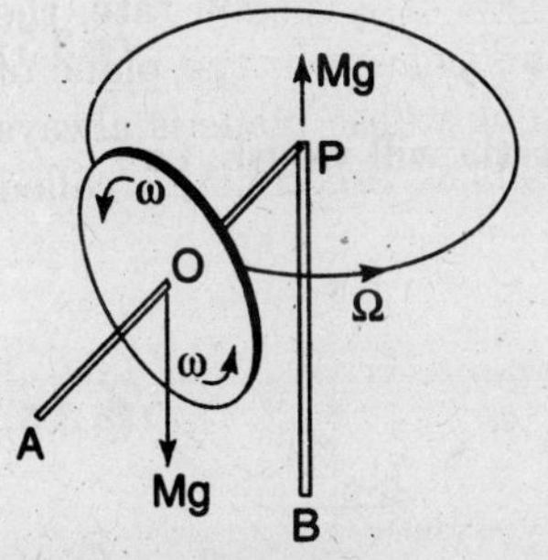

Fig. 8.2.1

The gravitational torque acting on the disc is $N' = Mgl$, where $l = OP$ and Mg is the weight of the disc. Hence the precessional rate which can be maintained by this torque is

$$\Omega = \frac{N'}{I\omega} = \frac{Mgl}{Mk^2\omega} = \frac{gl}{k^2\omega},$$

where k is the radius of gyration of the disc about AOP. The time period of its precessional motion is given by

$$T = \frac{2\pi}{\Omega} = 2\pi\frac{k^2\omega}{gl}. \qquad (8.2.1)$$

This precessional motion once started may be maintained by the gravitational torque. With the increase of the rate of precession the axis AOP rises while the same axis falls with the decrease of the rate of precession. This rise and fall of the axis of rotation or the oscillation of the axis of rotation on each side of steady precession is called *nutation*. This will be discussed in detail in the article on the general motion of a top.

8.3 Euler's Equation of Motion

We consider a rigid body constrained to rotate about a fixed point O. The body may be considered to be composed of a large number of particles. We consider one of such particles of mass δm having position vector $\vec{r}$ and velocity $\vec{v}$. Then the angular momentum of the particle about the fixed point is equal to $\vec{r} \times \delta m\vec{v}$. Hence the angular momentum of the whole body about O is given by $\vec{L} = \Sigma(\vec{r} \times \delta m\vec{v})$, the summation extending over all particles of which the body is composed.

Now the linear velocity vector $\vec{v}$ and the angular velocity vector $\vec{\omega}$ are related by the equation $\vec{v} = \vec{\omega} \times \vec{r}$.

$$\therefore \quad \vec{L} = \Sigma\delta m\left[\vec{r} \times (\vec{\omega} \times \vec{r})\right] = \Sigma\delta m\left[(\vec{r}\cdot\vec{r})\vec{\omega} - (\vec{r}\cdot\vec{w})\vec{r}\right] = \Sigma\delta m\left[\vec{\omega}r^2 - (\vec{r}\cdot\vec{\omega})\vec{r}\right]. \qquad (8.3.1)$$

Now resolving the vectors $\vec{r}$ ad $\vec{\omega}$ along $\hat{\imath}, \hat{\jmath}$ and $\hat{k}$ we have,

$$\vec{r} = \hat{\imath}x + \hat{\jmath}y + \hat{k}z; \quad \vec{\omega} = \hat{\imath}\omega_x + \hat{\jmath}\omega_y + \hat{k}\omega_z,$$

where ω_x, ω_y, ω_z are the components of the angular velocity in the three directions $\hat{\imath}$, $\hat{\jmath}$, $\hat{k}$, so that (8.3.1) may be written as

$$\vec{L} = \Sigma\delta m\left[(\hat{\imath}\omega_x + \hat{\jmath}\omega_y + \hat{k}\omega_z)(x^2 + y^2 + z^2) - (\hat{\imath}x + \hat{\jmath}y + \hat{k}z)\right.$$
$$\left.\{(\hat{\imath}\omega_x + \hat{\jmath}\omega_y + \hat{k}\omega_z)\cdot(\hat{\imath}x + \hat{\jmath}y + \hat{k}z)\}\right] \qquad (8.3.2)$$

$\therefore$ the component of $\vec{L}$ in the direction of $\hat{\imath}$ is given by

$$L_x = \Sigma\delta m[\omega_x(x^2 + y^2 + z^2) - x(\omega_x x + \omega_y y + \omega_z z)]$$
$$= \omega_x\Sigma\delta m(y^2 + z^2) - \omega_y\Sigma\delta mxy - \omega_z\Sigma\delta mxz = I_{xx}\omega_x - I_{xy}\omega_y - I_{xz}\omega_z. \qquad (8.3.3)$$

Similarly, we get the components of $\vec{L}$ along $\hat{\jmath}$ and $\hat{k}$ which are given by

$$L_y = -I_{yx}\omega_x + I_{yy}\omega_y - I_{yz}\omega_z \qquad (8.3.4)$$

$$L_z = -I_{zx}\omega_x - I_{zy}\omega_y + I_{zz}\omega_z. \qquad (8.3.5)$$

If $\hat{\imath}, \hat{\jmath}, \hat{k}$ are the principal axes of inertia at O, the products of inertia will vanish, i.e.,

$$I_{xy} = I_{yz} = I_{zx} = 0 = I_{yx} = I_{xz} = I_{zy}.$$

$$\therefore \quad \left.\begin{aligned} L_x &= I_{xx}\omega_x \\ L_y &= I_{yy}\omega_y \\ L_z &= I_{zz}\omega_z \end{aligned}\right\}. \qquad (8.3.6)$$

$$\therefore \quad \vec{L} = \hat{\imath}L_x + \hat{\jmath}L_y + \hat{k}L_z = \hat{\imath}I_{xx}\omega_x + \hat{\jmath}I_{yy}\omega_y + \hat{k}I_{zz}\omega_z.$$

We apply equation (3.40.1) and obtain

$$\dot{\vec{L}} = \frac{\delta\vec{L}}{\delta t} + \vec{\omega} \times \vec{L}$$
$$= (\hat{\imath}I_{xx}\dot{\omega}_x + \hat{\jmath}I_{yy}\dot{\omega}_y + \hat{k}I_{zz}\dot{\omega}_z) + (\hat{\imath}\omega_x + \hat{\jmath}\omega_y + \hat{k}\omega_z) \times (\hat{\imath}I_{xx}\omega_x + \hat{\jmath}I_{yy}\omega_y + \hat{k}I_{zz}\omega_z).$$

Now,

$$(\hat{\imath}\omega_x + \hat{\jmath}\omega_y + \hat{k}\omega_z) \times (\hat{\imath}I_{xx}\omega_x + \hat{\jmath}I_{yy}\omega_y + \hat{k}I_{zz}\omega_z)$$

$$= \begin{vmatrix} \hat{\imath} & \hat{\jmath} & \hat{k} \\ \omega_x & \omega_y & \omega_z \\ I_{xx}\omega_x & I_{yy}\omega_y & I_{zz}\omega_z \end{vmatrix}$$

$$= \hat{\imath}(I_{zz}\omega_y\omega_z - I_{yy}\omega_y\omega_z) + \hat{\jmath}(I_{xx}\omega_x\omega_z - I_{zz}\omega_x\omega_z) + \hat{k}(I_{yy}\omega_y\omega_x - I_{xx}\omega_y\omega_x).$$

$$\therefore \quad \dot{\vec{L}} = \hat{\imath}(I_{xx}\dot{\omega}_x - I_{yy}\omega_y\omega_z + I_{zz}\omega_y\omega_z) + \hat{\jmath}(I_{yy}\dot{\omega}_y - I_{zz}\omega_x\omega_z + I_{xx}\omega_x\omega_z) + \hat{k}(I_{zz}\dot{\omega}_z - I_{xx}\omega_y\omega_x + I_{yy}\omega_y\omega_x). \quad (8.3.7)$$

Now, $\dot{\vec{L}} = \vec{N}$, where $\vec{N}$ is the total moment of the external forces or the torque about the fixed point O. Hence the equation of motion of a rigid body with a fixed point is given by

$$\left.\begin{aligned} I_{xx}\dot{\omega}_x - (I_{yy} - I_{zz})\omega_y\omega_z &= N_x \\ I_{yy}\dot{\omega}_y - (I_{zz} - I_{xx})\omega_z\omega_x &= N_y \\ I_{zz}\dot{\omega}_z - (I_{xx} - I_{yy})\omega_x\omega_y &= N_z \end{aligned}\right\}. \quad (8.3.8)$$

The above equations are called *Euler's equations* of motion for a rigid body with a fixed point.

These equations are also applicable in case of the rotation of a free rigid body about its centre of mass. In this case I_{xx}, I_{yy}, I_{zz} are called the principal central moments of inertia of the body and $\omega_x, \omega_y, \omega_z$ denote the projections of the angular velocity of the body on its principal central axes of inertia. N_x, N_y, N_z are the moments of all external forces about the same axes.

If the moment of the external forces about the centre of mass of a rigid body is equal to zero, the rotation of the body about the centre of mass is said to be force-free rotation. In this case Euler's equations become

$$\left.\begin{aligned} I_{xx}\dot{\omega}_x - (I_{yy} - I_{zz})\omega_y\omega_z &= 0 \\ I_{yy}\dot{\omega}_y - (I_{zz} - I_{xx})\omega_z\omega_x &= 0 \\ I_{zz}\dot{\omega}_z - (I_{xx} - I_{yy})\omega_x\omega_y &= 0 \end{aligned}\right\}. \quad (8.3.9)$$

8.4 Kinetic Energy of Force-free Rotation of a Rigid Body

Multiplying the three equations (8.3.9) by ω_x, ω_y and ω_z respectively and adding them up, we have

$$I_{xx}\dot{\omega}_x\omega_x + I_{yy}\dot{\omega}_y\omega_y + I_{zz}\dot{\omega}_z\omega_z = 0, \quad \text{or,} \quad \frac{d}{dt}(I_{xx}\omega_x^2 + I_{yy}\omega_y^2 + I_{zz}\omega_z^2) = 0.$$

On integration we get,

$$I_{xx}\omega_x^2 + I_{yy}\omega_y^2 + I_{zz}\omega_z^2 = \text{constant}.$$

Now kinetic energy of the rotating body is given by

$$T = \frac{1}{2}\left(I_{xx}\omega_x^2 + I_{yy}\omega_y^2 + I_{zz}\omega_z^2\right). \quad \therefore \ 2T = \text{constant}.$$

This shows that the kinetic energy of a body under the action of no force is constant.

Again multiplying the first, second and third equations of (8.3.9) by $I_{xx}\omega_x$, $I_{yy}\omega_y$ and $I_{zz}\omega_z$ respectively and adding them up, we have

$$I_{xx}^2\omega_x\dot{\omega}_x + I_{yy}^2\omega_y\dot{\omega}_y + I_{zz}^2\omega_z\dot{\omega}_z = 0, \quad \text{or,} \quad \frac{d}{dt}\left(I_{xx}^2\omega_x^2 + I_{yy}^2\omega_y^2 + I_{zz}^2\omega_z^2\right) = 0.$$

Integrating the above expression, we have

$$I_{xx}^2\omega_x^2 + I_{yy}^2\omega_y^2 + I_{zz}^2\omega_z^2 = \text{constant}, \quad \text{or,} \quad L^2 = \text{constant}.$$

That $\vec{L}$ is constant has already been shown. (See Art. 3.34, Chapter 3). When the external torque $\vec{N} = 0$, $\vec{L}$ is constant—which means that both its magnitude and direction are constant.

8.5 Force-free Rotation of a Symmetrical Rigid Body

Let the axis of symmetry of the rigid body be taken as the Z—principal axis so that $I_{xx} = I_{yy} \neq I_{zz}$. Hence, Euler's equations, (8.3.9) reduce to the form

$$\left.\begin{aligned} I_{xx}\dot{\omega}_x &= (I_{xx} - I_{zz})\omega_y\omega_z \\ I_{yy}\dot{\omega}_y &= -(I_{xx} - I_{zz})\omega_z\omega_x \\ I_{zz}\dot{\omega}_z &= 0 \end{aligned}\right\}. \tag{8.5.1}$$

From the last of the above equations we have $\dot{\omega}_z = 0$, or $\omega_z =$ constant. Differentiating the first of the above equations with respect to time, we have

$$I_{xx}\ddot{\omega}_x = (I_{xx} - I_{zz})\omega_z\dot{\omega}_y, \quad \text{or,} \quad \ddot{\omega}_x = \frac{(I_{xx} - I_{zz})}{I_{xx}}\omega_z\dot{\omega}_y. \tag{8.5.2}$$

Substituting the value of $\dot{\omega}_y$ from the 2nd of the equations (8.5.1) in equation (8.5.2) we have

$$\ddot{\omega}_x = -\left[\frac{(I_{xx} - I_{zz})\omega_z}{I_{xx}}\right]^2 \omega_x. \tag{8.5.3}$$

This equation represents a simple harmonic motion having angular frequency given by

$$\Omega = \frac{I_{xx} - I_{zz}}{I_{xx}}\omega_z. \tag{8.5.4}$$

Hence the solution of equation (8.5.3) may be written as

$$\omega_x = A\sin\Omega t, \tag{8.5.5}$$

where A is some constant.

From equations (8.5.4) and (8.5.5) the first of the equations (8.5.1) becomes

$$A\Omega\cos\Omega t = \left(\frac{I_{xx} - I_{zz}}{I_{xx}}\right)\omega_z\omega_y = \Omega\omega_y, \quad \text{or,} \quad \omega_y = A\cos\Omega t. \tag{8.5.6}$$

From equations (8.5.5) and (8.5.6), we see that the vector $\hat{\imath}\omega_x + \hat{\jmath}\omega_y$ has a constant magnitude $\sqrt{(\omega_x^2 + \omega_y^2)} = A$, and rotates uniformly about the axis of symmetry with angular frequency Ω (Fig. 8.5.1). As ω_z is constant in magnitude the total angular velocity vector

$$\vec{\omega} = \hat{\imath}\omega_x + \hat{\jmath}\omega_y + \hat{k}\omega_z$$

is constant in magnitude and precesses about the axis of symmetry with same angular frequency Ω.

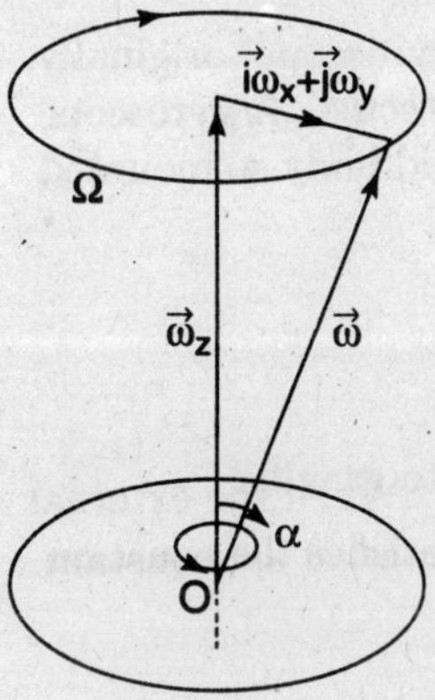

Fig. 8.5.1

From equation (8.5.4) it follows that smaller the difference between I_{xx} and I_{zz}, the slower will be the precession frequency Ω compared to the frequency ω of rotation. Let α be the angle between $\vec{\omega}_z$ and $\vec{\omega}$, i.e., between the axis of symmetry and axis of rotation.

$$\therefore \quad \tan\alpha = \frac{A}{\omega_z} = \text{constant}.$$

Hence, the axis of rotation describes a cone about the axis of symmetry, the semi-vertical angle of the cone being α. The cone thus described is called the *body cone.*

8.6 Precession of the Earth's Axis of Rotation

As the external torque acting on the earth is very weak, the diurnal rotation of the earth may be considered force-free. The polar axis is the axis of symmetry of the earth. The earth is slightly flattened at the poles and as a result the equatorial diameter is greater than the distance between the two poles. Hence I_{zz} is slightly greater than I_{xx} and their ratio is

$$\frac{I_{zz}}{I_{xx}} = 1.0030, \quad \text{or,} \quad \frac{I_{xx} - I_{zz}}{I_{xx}} = -0.0033.$$

The polar axis of the earth is not its axis of rotation. The axis of rotation precesses round the polar axis with angular frequency.

$$\Omega = \frac{I_{xx} - I_{zz}}{I_{xx}}\omega_z = -0.0033\omega_z.$$

$\therefore$ the magnitude of the precession angular frequency $\simeq \dfrac{\omega_z}{300}$.

Now ω_z is practically the same as the diurnal angular frequency.

Hence, the magnitude of the precession angular frequency $= \dfrac{1}{300} \times 1$ revolution per day, i.e., the period of precession is 300 days or 10 months. This means that an observer on the earth should observe that the axis of rotation traces out a circle round the North or the South pole once nearly every 300 days. Actually it is found that the amplitude of precession is very small. The axis of rotation never swings more than 15 feet from the North pole.

Moreover, the axis of rotation never describes a circular path. The path described is irregular and the period is 427 days instead of 300 days. The fluctuations are due to change in the mass distribution of the earth caused by atmospheric motion. The difference in precession period is due to the fact that the earth is not a completely rigid body.

8.7 Top Motion

The top motion is a familiar example of gyroscopic motion. The word 'gyroscope' originally meant an instrument by which the effect of the earth's rotation could be detected. A gyroscope or a gyrostat now denotes an instrument which is nothing but a rotating rigid body so mounted that it may change the direction of its angular velocity.

Theory of the Steady Precession of a Top

A top is a rigid body with an axis of symmetry acted on by the force due to gravity.

We know that the motion of a rigid body rotating with a fixed point O satisfies the equation

$$\dot{\vec{L}} = N, \quad \text{(see equation 3.34.2)},$$

where $\vec{L}$ is the angular momentum about O and $\vec{N}$ is the moment of the external force or torque about O. We shall here describe a steady precession of the top. In steady precession the axis of symmetry of the top describes with constant angular velocity a right circular cone with the vertical as axis and the top spins at the same time with constant angular velocity about its axis of symmetry.

In the adjoining vector diagram (8.7.1), let a = distance of the mass centre G from the fixed vertex O,

$\hat{n}$ = unit vector along the vertically upward direction,

$\hat{\imath}, \hat{\jmath}, \hat{k}$ = unit orthogonal vectors, $\hat{k}$ being along OG and $\hat{\imath}$ in the plane of $\hat{k}$ and $\hat{n}$,

θ = angle of inclination of OG to the vertical and m = mass of the top.

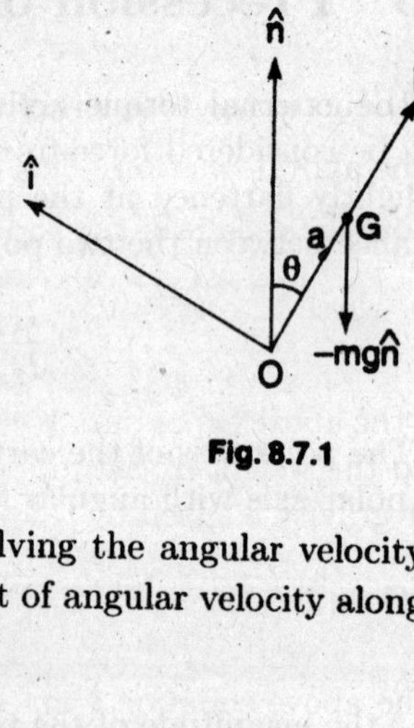

Fig. 8.7.1

From the diagram, we have

$$\hat{n} = \hat{\imath}\sin\theta + \hat{k}\cos\theta. \tag{8.7.1}$$

The angular velocity $\vec{\omega}$ of the top lies in the plane $(\hat{k}, \hat{n})$. Resolving the angular velocity along $\hat{\imath}$ and $\hat{k}$, we may write $\vec{\omega} = \hat{\imath}\omega_1 + \hat{k}s$, where ω_1 is the component of angular velocity along $\hat{\imath}$ and s denotes the spin of the top.

Now the velocity of the point G

$$= \vec{\omega} \times a\hat{k} = (\hat{\imath}\omega_1 + \hat{k}s) \times a\hat{k} = \hat{\imath} \times \hat{k}a\omega_1 = -\hat{\jmath}a\omega_1. \tag{8.7.2}$$

If we call Ω the angular velocity with which OG rotates about $\hat{n}$, then Ω denotes the precession. The velocity of the point G is then

$$\Omega\hat{n} \times a\hat{k} = \Omega a\hat{n} \times \hat{k} = \Omega a\sin\theta\,\hat{\imath} \times \hat{k} \quad \text{(from 8.7.1)}$$

$$= -\Omega a\sin\theta\hat{\jmath}. \tag{8.7.3}$$

Equating the expressions (8.7.2) and (8.7.3), we have

$$-\hat{\jmath}a\omega_1 = -\Omega a\sin\theta\hat{\jmath}, \quad \text{or,} \quad \omega_1 = \Omega\sin\theta. \tag{8.7.4}$$

When the precession is steady, θ, s and Ω are constants.

The angular momentum

$$\vec{L} = I_{xx}\hat{\imath}\omega_1 + I_{zz}\hat{k}s, \tag{8.7.5}$$

where I_{xx} = transverse moment of inertia at O and I_{zz} = axial moment of inertia at O. Substituting the expression for ω_1 from (8.7.4) in equation (8.7.5), we have

$$\vec{L} = I_{xx}\Omega\sin\theta\hat{\imath} + I_{zz}\hat{k}s. \tag{8.7.6}$$

The angular momentum vector $\vec{L}$ lies in the plane $(\hat{k}, \hat{n})$ and rotates rigidly with it. Hence $\dot{\vec{L}} = \Omega\hat{n} \times \vec{L}$ ($\because$ $\dot{\vec{L}}$ is the velocity of a point with position vector $\vec{L}$ in a rigid body which rotates with angular velocity $\Omega\hat{n}$)

$$\begin{aligned}
\text{or,} \quad \dot{\vec{L}} &= \Omega(\hat{\imath}\sin\theta + \hat{k}\cos\theta) \times (I_{xx}\Omega\sin\theta\hat{\imath} + I_{zz}\hat{k}s) \quad \text{[from (8.7.1) and (8.7.6)]} \\
&= \Omega(I_{zz}s\sin\theta\,\hat{\imath}\times\hat{k} + I_{xx}\Omega\sin\theta\cos\theta\,\hat{k}\times\hat{\imath}) \\
&= \Omega(-\hat{\jmath}I_{zz}s\sin\theta + \hat{\jmath}I_{xx}\Omega\sin\theta\cos\theta) \\
&= \Omega\sin\theta(I_{xx}\Omega\cos\theta - I_{zz}s)\hat{\jmath}.
\end{aligned}$$

$$\therefore \quad \vec{N} = \Omega\sin\theta(I_{xx}\Omega\cos\theta - I_{zz}s)\hat{\jmath}. \tag{8.7.7}$$

$\vec{N}$ denotes the moment of all the forces including the force due to gravity about O.

The above equation holds for steady precession with fixed values of θ, Ω and s.

Now the weight of the top at G has a moment about O which is equal to

$$a\hat{k} \times (-mg\hat{n}) = a\hat{k} \times (-mg\hat{\imath}\sin\theta - mg\hat{k}\cos\theta) = -mga\sin\theta\hat{k}\times\hat{\imath} = -mga\sin\theta\hat{\jmath}.$$

If the above expression is equal to $\vec{N}$ as in equation (8.7.7), then it follows that only the force due to the weight of the top maintains the motion. Hence the condition for steady precession under gravity is given by

$$\Omega\sin\theta(I_{xx}\Omega\cos\theta - I_{zz}s)\hat{\jmath} = -mga\sin\theta\hat{\jmath}, \quad \text{or,} \quad \Omega(I_{zz}s - I_{xx}\Omega\cos\theta) = mga. \tag{8.7.8}$$

The above equation connects the three constants θ, Ω and s. Hence there are two infinite sets of steady precession corresponding to arbitrary values of two out of the three constants θ, Ω and s. It is not possible to assign completely arbitrary values of two constants. These values should be such that the above equation gives a real value for the third constant.

From equation (8.7.8), we may write

$$s = \frac{I_{xx}\Omega\cos\theta}{I_{zz}} + \frac{mga}{I_{zz}\Omega}. \tag{8.7.9}$$

Thus we get an expression for the value of the spin. When Ω is small and s is large the above expression is approximately given by

$$s = \frac{mga}{I_{zz}\Omega}. \tag{8.7.10}$$

This expression may also be written in the form

$$s = \frac{mga}{mK^2\Omega} = \frac{ga}{K^2\Omega}, \tag{8.7.11}$$

where K is the radius of gyration about the axis of symmetry.

For a small gyroscope top the following are rough values for the quantities on which Ω depends—

$$a = 1.5 \text{ inches}, \quad K = 1 \text{ inch}.$$

If the top spins at 100 revolutions per second about its axis, then

$$s = 2\pi \times 100 = 200\pi.$$

$\therefore$ from equation (8.7.11), we have

$$\Omega = \frac{ga}{K^2 s} = \frac{32 \times 1.5 \times 12}{200\pi} = 0.916 \text{ radians/sec.}$$

Hence the axis makes one revolution about the vertical in $\dfrac{2\pi}{\Omega} = 6.85$ secs.

Illustrative Example

1. A uniform disc of mass 500 gms. and radius 5 cms. revolves about an axis through its centre and perpendicular to its plane at a rate of 50 revolutions per sec. Find the periodic time of the horizontal precessional motion of the axis if the flywheel is supported at a point on the axis 3 cms. from the centre of the disc and projected with axis horizontal without subsequent nutational motion. [$g = 981$ cms/sec^2.]

We know, $\Omega = \dfrac{ga}{K^2 s}$.

Here

$$g = 981 \text{ cms/sec}^2.$$

$$a = 3 \text{ cms}.$$

$$K^2 = \frac{5^2}{2} = 12.5 \text{ cms}^2.$$

$$s = 2\pi \times 50 = 100\pi.$$

$$\therefore \quad \Omega = \frac{981 \times 3}{12.5 \times 100\pi} = 0.749 \text{ radians/sec.}$$

Hence the periodic time of the horizontal precessional motion

$$= \frac{2\pi}{\Omega} = \frac{2\pi}{0.749} = 8.388 \text{ secs.}$$

8.8 General Motion of a Symmetrical Top

Let $\vec{i'}$, $\vec{j'}$, $\vec{k'}$ be fixed unit orthogonal vectors, $\hat{k}'$ being along the vertical upward direction. Let $\hat{\imath}, \hat{\jmath}, \hat{k}$ be other unit orthogonal vectors in which $\hat{k}$ points along OG, the axis of symmetry of the top. $\hat{\imath}, \hat{k}$ and $\hat{k}'$ are lying in the same plane. The vectors $\hat{\imath}, \hat{\jmath}, \hat{k}$ are not fixed but $\hat{k}$ is fixed in the top. In Fig. (8.8.1), θ denotes the angle between $\hat{k}$ and $\hat{k}'$ and ϕ denotes the angle between $\vec{i'}$ and the nodal line NN'. NN' is the line of intersection of the two planes $\hat{\imath}\hat{\jmath}$ and $\hat{\imath}'\hat{\jmath}'$.

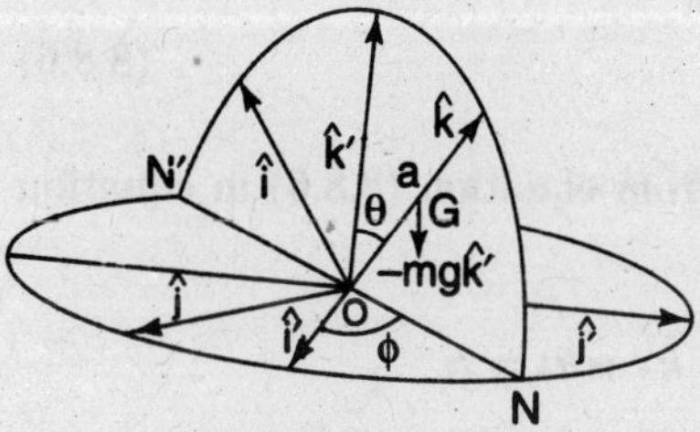

Fig. 8.8.1

Let the angular velocity of the top be given by

$$\vec{\omega} = \hat{\imath}\omega_1 + \hat{\jmath}\omega_2 + \hat{k}\omega_3, \tag{8.8.1}$$

where $\omega_1, \omega_2, \omega_3$ are the components of the angular velocity in the three directions $\hat{\imath}, \hat{\jmath}, \hat{k}$ and the angular velocity of the triad $\hat{\imath}, \hat{\jmath}, \hat{k}$ be given by

$$\vec{A} = \hat{\imath}A_1 + \hat{\jmath}A_2 + \hat{k}A_3. \tag{8.8.2}$$

Now the angular velocity about $\hat{k}'$ is $\dot{\phi}$ and its component along $\hat{\imath}$ is given by $A_1 = \dot{\phi}\sin\theta$.

Angular velocity about $\hat{\jmath}$ is $-\dot{\theta}$. $\therefore \quad A_2 = -\dot{\theta}$.

Angular velocity about $\hat{k}'$ having component along $\hat{k}$ is given by $A_3 = \dot{\phi}\cos\theta$.

Now the relative motion of the unit vectors $\hat{\imath}, \hat{\jmath}, \hat{k}$ and the top consists only of a rotation about $\hat{k}$. So, $A_1 = \omega_1 = \dot{\phi}\sin\theta$ and $A_2 = \omega_2 = -\dot{\theta}$.

The angular momentum

$$\vec{L} = I_{xx}\omega_1\hat{\imath} + I_{xx}\omega_2\hat{\jmath} + I_{zz}\omega_3\hat{k} \tag{8.8.3}$$

[$\because$ for a symmetrical top $I_{xx} = I_{yy}$].

Hence the rate of change of angular momentum is given by

$$\dot{\vec{L}} = I_{xx}\dot{\omega}_1\hat{\imath} + I_{xx}\dot{\omega}_2\hat{\jmath} + I_{zz}\dot{\omega}_3\hat{k} + \vec{A}\times\vec{L} \quad \text{[See equation (3.40.1)]}.$$

Now the moment of the weight of the top about O is given by

$$\vec{N} = a\hat{k}\times(-mg\hat{k}') = -mga\hat{k}\times\hat{k}' = -mga\sin\theta\hat{k}\times\hat{\imath} = -mga\sin\theta\hat{\jmath}. \tag{8.8.4}$$

Now, the motion of the top satisfies the equation $\dot{\vec{L}} = \vec{N}$.

$$\therefore \quad I_{xx}\dot{\omega}_1\hat{\imath} + I_{xx}\dot{\omega}_2\hat{\jmath} + I_{zz}\dot{\omega}_3\hat{k} + \vec{A}\times\vec{L} = -mga\sin\theta\hat{\jmath},$$

$$\text{or,} \quad I_{xx}\dot{\omega}_1\hat{\imath} + I_{xx}\dot{\omega}_2\hat{\jmath} + I_{zz}\dot{\omega}_3\hat{k} + (\hat{\imath}A_1 + \hat{\jmath}A_2 + \hat{k}A_3)\times(I_{xx}\omega_1\hat{\imath} + I_{xx}\omega_2\hat{\jmath} + I_{zz}\omega_3\hat{k})$$
$$= -mga\sin\theta\hat{\jmath}.$$

Equating the components in the direction of $\hat{k}$ we have

$$I_{zz}\dot{\omega}_3 + I_{xx}A_1\omega_2 - I_{xx}A_2\omega_1 = 0,$$

$$\text{or,} \quad I_{zz}\dot{\omega}_3 + I_{xx}\omega_1\omega_2 - I_{xx}\omega_2\omega_1 = 0,$$

$$\text{or,} \quad I_{zz}\dot{\omega}_3 = 0, \quad \text{or,} \quad \dot{\omega}_3 = 0.$$

$$\therefore \quad \omega_3 = \text{constant} = s \text{ (say)}.$$

Thus we see that the spin of the top ω_3 is constant.

From the figure it follows that the weight of the top has no moment about $\hat{k}'$. Hence the component of angular momentum is constant in this direction.

$$\therefore \quad \vec{L}\cdot\hat{k}' = \text{constant} = \alpha \text{ (say)}. \tag{8.8.5}$$

Now,

$$\hat{k}' = \hat{\imath}\sin\theta + \hat{k}\cos\theta. \tag{8.8.6}$$

Substituting the value of $\vec{L}$ from equation (8.8.3) and $\hat{k}'$ from equation (8.8.6) in equation (8.8.5) we have

$$(I_{xx}\omega_1\hat{\imath} + I_{xx}\omega_1\hat{\jmath} + I_{zz}\omega_3\hat{k})\cdot(\hat{\imath}\sin\theta + \hat{k}\cos\theta) = \alpha$$

$$\text{or,}\quad I_{xx}\omega_1\sin\theta + I_{zz}\omega_3\cos\theta = \alpha,$$

$$\text{or,}\quad I_{xx}\dot{\phi}\sin^2\theta + I_{zz}s\cos\theta = \alpha. \tag{8.8.7}$$

Now, the total energy of the top, E = kinetic energy T + potential energy V.

Here

$$T = \frac{1}{2}I_{xx}(\omega_1^2+\omega_2^2) + \frac{I_{zz}}{2}\omega_3^2 \quad\text{and}\quad V = mga\cos\theta.$$

$$\therefore\; E = \frac{1}{2}I_{xx}(\omega_1^2+\omega_2^2) + \frac{I_{zz}}{2}\omega_3^2 + mga\cos\theta = \frac{1}{2}I_{xx}(\dot{\phi}^2\sin^2\theta+\dot{\theta}^2) + \frac{I_{zz}}{2}s^2 + mga\cos\theta. \tag{8.8.8}$$

Equation (8.8.7) may be written in the form

$$I_{xx}\dot{\phi}\sin^2\theta = \alpha - I_{zz}s\cos\theta = \alpha - \beta\cos\theta, \quad\text{where } \beta = I_{zz}s. \tag{8.8.9}$$

From equation (8.8.8), we have

$$E = \frac{1}{2}I_{xx}\dot{\theta}^2 + \frac{1}{2}I_{xx}\dot{\phi}^2\sin^2\theta + \frac{I_{zz}}{2}s^2 + mga\cos\theta$$

$$= \frac{1}{2}I_{xx}\dot{\theta}^2 + \frac{1}{\sin^2\theta}\frac{(\alpha-\beta\cos\theta)^2}{2I_{xx}} + \frac{\beta^2}{2I_{zz}} + mga\cos\theta. \tag{8.8.10}$$

In this expression θ is the only variable. Solving the equation (8.8.10) we get the relation between θ and t. On substituting the value of θ in equation (8.8.9) we get $\dot{\phi}$. Without taking recourse to such a complicated solution, we can integrate the equation with the initial conditions that the top rotates about the axis $\hat{k}'$ with great speed at $t = 0$ and that the axis $\hat{k}'$ makes an angle θ_0 with the axis $\hat{k}$. Thus at $t = 0$, $\theta = \theta_0$, $\dot{\theta} = 0$ and $\dot{\phi} = 0$ and the values of the constants β, α and E are given by

$$\beta = I_{zz}s; \quad \alpha = \beta\cos\theta_0 \quad \text{[from equation (8.8.9)]}.$$

$$\therefore\quad E = \frac{\beta^2}{2I_{zz}} + mga\cos\theta_0 = \frac{1}{2}I_{xx}\dot{\theta}^2 + \frac{1}{\sin^2\theta}\frac{(\alpha-\beta\cos\theta)^2}{2I_{xx}} + \frac{\beta^2}{2I_{zz}} + mga\cos\theta. \tag{8.8.11}$$

Multiplying both sides of the above equation by $\dfrac{2}{I_{xx}}$, we have

$$\frac{2}{I_{xx}}mga\cos\theta_0 = \dot{\theta}^2 + \frac{1}{\sin^2\theta}\frac{(\alpha-\beta\cos\theta)^2}{I_{xx}^2} + \frac{2}{I_{xx}}mga\cos\theta,$$

$$\text{or,}\quad \dot{\theta}^2 = \frac{2mga}{I_{xx}}(\cos\theta_0-\cos\theta) - \frac{1}{\sin^2\theta}\frac{\beta^2(\cos\theta_0-\cos\theta)^2}{I_{xx}^2} \qquad [\because \alpha = \beta\cos\theta_0]$$

$$= \frac{2mga}{I_{xx}}(\cos\theta_0-\cos\theta) - \frac{1}{\sin^2\theta}\frac{I_{zz}^2s^2}{I_{xx}^2}(\cos\theta_0-\cos\theta)^2. \tag{8.8.12}$$

Putting

$$\frac{2mga}{I_{xx}} = p \quad\text{and}\quad \frac{I_{zz}s}{I_{xx}} = q,$$

the equation (8.8.12) may be written in the form

$$\dot{\theta}^2 = p(\cos\theta_0-\cos\theta) - \frac{q^2}{\sin^2\theta}(\cos\theta_0-\cos\theta)^2. \tag{8.8.13}$$

The above differential equation can be integrated by elliptic integral. We shall however try to obtain an approximate solution by assuming that the rotation about $\hat{k}'$ is very rapid. Under this condition a high torque will be required to change the angular momentum vector which lies along $\hat{k}'$ axis. Thus we consider a solution where angle θ differs very little from θ_0, i.e., $\theta = \theta_0 + \varepsilon(\varepsilon \ll \theta_0)$.

$$\therefore\quad \cos\theta = \cos\theta_0\cos\varepsilon - \sin\theta_0\sin\varepsilon = \cos\theta_0 - \varepsilon\sin\theta_0$$

$$\text{and}\quad \sin\theta = \sin\theta_0\cos\varepsilon + \cos\theta_0\sin\varepsilon = \sin\theta_0 + \varepsilon\cos\theta_0.$$

$$\therefore\quad \dot{\theta} = \dot{\varepsilon} \quad\text{and}\quad \cos\theta_0 - \cos\theta = \varepsilon\sin\theta_0.$$

Hence equation (8.8.13) may be written as

$$\dot{\varepsilon}^2 = p(\varepsilon\sin\theta_0) - \frac{q^2}{(\sin\theta_0+\varepsilon\cos\theta_0)^2}\varepsilon^2\sin^2\theta_0 = p\varepsilon\sin\theta_0 - q^2\varepsilon^2 \quad (\text{as } \varepsilon \ll \theta_0).$$

It should be noted that the term $q^2\varepsilon^2$ has been retained due to the factor q^2 which increases with s.

The above expression may be written in the form

$$\dot{\varepsilon}^2 = q^2\left(\frac{2p}{2q^2}\varepsilon\sin\theta_0 - \varepsilon^2\right) = q^2(2\varepsilon\varepsilon_1 - \varepsilon^2), \quad \text{where } \frac{p}{2q^2}\sin\theta_0 = \varepsilon_1,$$

$$\text{or,}\quad \frac{d\varepsilon}{dt} = q\sqrt{2\varepsilon\varepsilon_1-\varepsilon^2}, \quad\text{or,}\quad \int\frac{d\varepsilon}{\sqrt{2\varepsilon\varepsilon_1-\varepsilon^2}} = q\int dt,$$

$$\text{or,}\quad \cos^{-1}\frac{\varepsilon_1-\varepsilon}{\varepsilon_1} = qt + c, \ c \text{ being the constant of integration.}$$

At $t = 0, \theta = \theta_0$, i.e., $\varepsilon = 0$.

$$\therefore\quad \cos^{-1}1 = c, \quad\text{or,}\quad c = 0.$$

Hence

$$\cos^{-1}\frac{\varepsilon_1-\varepsilon}{\varepsilon_1}=qt,\quad \text{or,}\quad \varepsilon_1-\varepsilon=\varepsilon_1\cos qt,\quad \text{or,}\quad \varepsilon=\varepsilon_1(1-\cos qt). \tag{8.8.14}$$

$$\therefore\quad \theta=\theta_0+\varepsilon_1(1-\cos qt). \tag{8.8.15}$$

The above expression shows that θ undergoes small periodic change. The amplitude of the periodic change is smaller, greater the value of q, i.e., greater the value of the angular velocity of rotation. Periodic change of θ means the undulation of the axis of rotation of the top which is called *nutation*.

From equation (8.8.9), we may write

$$\dot{\phi}=\frac{\alpha-\beta\cos\theta}{I_{xx}\sin^2\theta}=\frac{\beta(\cos\theta_0-\cos\theta)}{I_{xx}\sin^2\theta}=\frac{I_{zz}s}{I_{xx}}\cdot\frac{\cos\theta_0-\cos\theta}{\sin^2\theta}$$

$$=q\frac{\cos\theta_0-\cos\theta}{\sin^2\theta}=\frac{q\varepsilon\sin\theta_0}{\sin^2\theta_0}=\frac{q\varepsilon}{\sin\theta_0}$$

$$=\frac{q\varepsilon_1(1-\cos qt)}{\sin\theta_0}\quad \text{[from equation (8.8.14) with the condition } \varepsilon\ll\theta_0] \tag{8.8.16}$$

Integrating the equation (8.8.16), we have,

$$\phi=\frac{\varepsilon_1 qt}{\sin\theta_0}-\frac{\varepsilon_1}{\sin\theta_0}\sin qt+\phi_0.$$

If we neglect for the moment the rapidly oscillating term containing $\sin qt$, we see that ϕ, i.e., the angle of rotation of the axis of the top round the vertical axis $\hat{k}'$ increases linearly with time. This motion is called *precession*.

The angular velocity of precession

$$=\frac{\varepsilon_1 q}{\sin\theta_0}=\frac{p\sin\theta_0}{2q^2}\cdot\frac{q}{\sin\theta_0}=\frac{p}{2q}=\frac{2mga}{I_{xx}}\Big/\frac{2I_{zz}s}{I_{xx}}=\frac{mga}{I_{zz}s}. \tag{8.8.17}$$

The above formula has already been obtained in the case of steady precession.

Let us now consider the expression

$$\varepsilon_1=\frac{p}{2q^2}\sin\theta_0,$$

to explain the meaning of the amplitude of nutation. It is clear from this expression that the amplitude of nutation ε_1 decreases with the increase of q, i.e., increase of spin s. As q^2 or s^2 occurs in the denominator of the above expression and s occurs as first order in equation (8.8.17), we conclude that for a rapidly spinning top the precession is much easier to observe than nutation. In figure (8.8.2) the precession and nutation of the axis of the top have been shown.

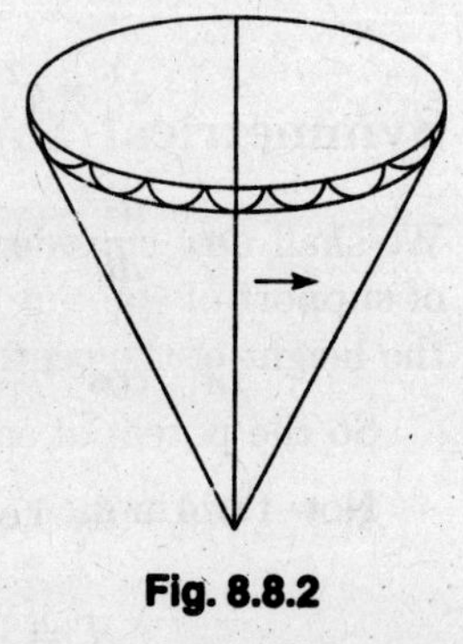

Fig. 8.8.2

8.9 Symmetrical Top (Alternative Treatment)—Eulerian Angles

The rotation of a rigid body can be described with the help of parameters which specify the position of the body. Such parameters are called Eulerian angles as shown in figure (8.9.1). The figure represents two co-ordinate systems $Oxyz$ and $Ox'y'z'$. The system $Oxyz$ is fixed and the system $Ox'y'z'$ is fixed in the rigid body. It is convenient to take the axes Ox', Oy' and Oz' along the principal axes of inertia through the point of support of the rigid body. In the figure the Eulerian angles are:

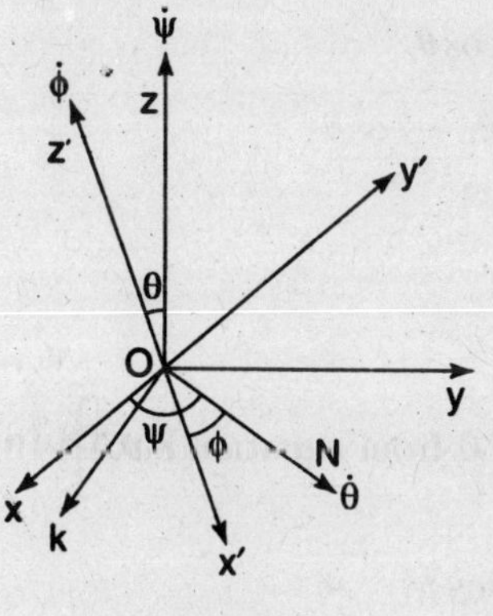

Fig. 8.9.1

$\theta =$ the angle between the axes Oz and Oz'. $\phi =$ the angle between the nodal line ON and the axis Ox'. The nodal line denotes the line of intersection between the planes xOy and $x'Oy'$;

$\psi =$ the angle between the nodal line and Ox.

The angular velocity vector $\dot{\psi}$ is directed along Oz, $\dot{\phi}$ along Oz' and $\dot{\theta}$ along the nodal line ON. Let $\omega_1; \omega_2$ and ω_3 be the angular velocity projections on the principal axes of inertia.

$$\therefore \quad \omega_3 = \dot{\phi} + \dot{\psi}\cos\theta. \tag{8.9.1}$$

We now draw a line Ok in the plane $x'Oy'$, the line Ok being perpendicular to ON.

Then from the figure it follows

$$\angle kO'x = \frac{\pi}{2} - \phi \quad \text{and} \quad \angle kOz = \frac{\pi}{2} + \theta.$$

Now, since all lines perpendicular to ON lie in the plane zOz', the line Ok must lie in the plane zOz'. The projection of $\dot{\psi}$ on Ok is $\dot{\psi}\cos(\pi/2+\theta) = -\dot{\psi}\sin\theta$ and the projection on Ox' is $-\dot{\psi}\sin\theta\cos(\pi/2-\phi) = -\dot{\psi}\sin\theta\sin\phi$.

Again the projection of $\dot{\psi}$ on Oy' is $\dot{\psi}\sin\theta\cos\phi$. The projection of $\dot{\theta}$ on Ox' and Oy' are $\dot{\theta}\cos\phi$ and $\dot{\theta}\sin\phi$. Hence we have,

$$\left.\begin{aligned} \omega_1 &= \dot{\theta}\cos\phi - \dot{\psi}\sin\theta\sin\phi \\ \omega_2 &= \dot{\theta}\sin\phi + \dot{\psi}\sin\theta\cos\phi \end{aligned}\right\}. \tag{8.9.2}$$

Symmetrical Top in a Gravitational Field

We shall now consider the motion of a symmetrical top in a gravitational field. Let the point of support of the top lie on its axis at a distance a below the centre of mass of the top. Then the height of the centre of mass above the point of support is given by $z = a\cos\theta$.

So the potential energy of the top is given by $V = mga\cos\theta$.

Now the kinetic energy of the top is

$$T = \frac{1}{2}I_{xx}\omega_1^2 + \frac{1}{2}I_{xx}\omega_2^2 + \frac{1}{2}I_{zz}\omega_3^2 = \frac{1}{2}I_{xx}(\omega_1^2+\omega_2^2) + \frac{1}{2}I_{zz}\omega_3^2.$$

The above expression when expressed in terms of Eulerian angles from equations (8.9.1) and (8.9.2) gives

$$T = \frac{1}{2}I_{xx}(\dot{\theta}^2 + \dot{\psi}^2 \sin^2\theta) + \frac{1}{2}I_{zz}(\dot{\phi} + \dot{\psi}\cos\theta)^2.$$

Now,

$$T - V = \frac{1}{2}I_{xx}(\dot{\theta}^2 + \dot{\psi}^2 \sin^2\theta) + \frac{1}{2}I_{zz}(\dot{\phi} + \dot{\psi}\cos\theta)^2 - mga\cos\theta = L. \tag{8.9.3}$$

$$\therefore \quad p_\phi = \frac{\partial L}{\partial \dot{\phi}} = I_{zz}(\dot{\phi} + \dot{\psi}\cos\theta)$$

and

$$p_\psi = \frac{\partial L}{\partial \dot{\psi}} = I_{xx}\dot{\psi}\sin^2\theta + I_{zz}(\dot{\phi} + \dot{\psi}\cos\theta)\cos\theta.$$

From Lagrange's equation

$$\frac{d}{dt}\left(\frac{\partial L}{\partial \dot{q}_k}\right) - \frac{\partial L}{\partial q_k} = 0,$$

we have,

$$\frac{d}{dt}\left(\frac{\partial L}{\partial \dot{\phi}}\right) = 0 \quad \text{and} \quad \frac{d}{dt}\left(\frac{\partial L}{\partial \dot{\psi}}\right) = 0 \quad \left[\because \frac{\partial L}{\partial \phi} = \frac{\partial L}{\partial \psi} = 0 \text{ from equation 8.9.3}\right]$$

$$\text{i.e.,} \quad \frac{\partial L}{\partial \dot{\phi}} = \text{constant} = \beta \text{ (say)} \quad \text{and} \quad \frac{\partial L}{\partial \dot{\psi}} = \text{constant} = \alpha \text{ (say)}.$$

$$\therefore \quad p_\phi = I_{zz}(\dot{\phi} + \dot{\psi}\cos\theta) = \text{constant} = \beta$$

$$\text{and} \quad p_\psi = I_{xx}\dot{\psi}\sin^2\theta + I_{zz}(\dot{\phi} + \dot{\psi}\cos\theta)\cos\theta = \text{constant} = \alpha$$

$$= I_{xx}\dot{\psi}\sin^2\theta + p_\phi\cos\theta,$$

$$\text{or,} \quad \dot{\psi} = \frac{p_\psi - p_\phi\cos\theta}{I_{xx}\sin^2\theta}. \tag{8.9.4}$$

Hence the energy equation gives

$$E = T + V = \frac{1}{2}I_{xx}(\dot{\theta}^2 + \dot{\psi}^2\sin^2\theta) + \frac{1}{2}I_{zz}(\dot{\phi} + \dot{\psi}\cos\theta)^2 + mga\cos\theta$$

$$= \frac{1}{2}I_{xx}\dot{\theta}^2 + \frac{(p_\psi - p_\phi\cos\theta)^2}{2I_{xx}\sin^2\theta} + \frac{1}{2}\frac{p_\phi^2}{I_{zz}} + mga\cos\theta$$

$$= \frac{1}{2}I_{xx}\dot{\theta}^2 + \frac{(\alpha - \beta\cos\theta)^2}{2I_{xx}\sin^2\theta} + \frac{\beta^2}{2I_{zz}} + mga\cos\theta.$$

This equation contains one variable θ. For solution of this equation we may proceed as in the previous article.

Now, the Lagrange's equation may be written in the form

$$\frac{d}{dt}\left(\frac{\partial L}{\partial \dot{\theta}}\right) - \frac{\partial L}{\partial \theta} = 0. \tag{8.9.5}$$

From equation (8.9.3), we have

$$\frac{\partial L}{\partial \dot\theta} = I_{xx}\dot\theta \quad \text{and} \quad \frac{\partial L}{\partial \theta} = I_{xx}\dot\psi^2 \sin\theta\cos\theta - I_{zz}(\dot\phi + \dot\psi\cos\theta)\dot\psi\sin\theta + mga\sin\theta.$$

Hence from equation (8.9.5), we have

$$\frac{d}{dt}(I_{xx}\dot\theta) - I_{xx}\dot\psi^2\sin\theta\cos\theta + I_{zz}(\dot\phi + \dot\psi\cos\theta)\dot\psi\sin\theta - mga\sin\theta = 0,$$

$$\text{or} \quad I_{xx}\ddot\theta = I_{xx}\dot\psi^2\sin\theta\cos\theta - I_{zz}(\dot\phi + \dot\psi\cos\theta)\dot\psi\sin\theta + mga\sin\theta. \tag{8.9.6}$$

The solution of equation (8.9.6) is very much complicated.

However, if we consider steady precession and ignore nutation, we may write $\dot\psi = \Omega =$ constant and $\theta =$ constant $= \theta_c$ (say).

$\therefore$ Equation (8.9.6) reduces to the form.

$$I_{xx}\Omega^2\sin\theta_c\cos\theta_c - I_{zz}w_3\Omega\sin\theta_c + mga\sin\theta_c = 0.$$

$$\text{or,} \quad \Omega^2 I_{xx}\cos\theta_c - \Omega I_{zz}w_3 + mga = 0,$$

$$\text{or,} \quad \Omega = \frac{I_{zz}w_3 \pm \sqrt{I_{zz}^2 w_3^2 - 4I_{xx}mga\cos\theta_c}}{2I_{xx}\cos\theta_c}. \tag{8.9.7}$$

Hence the time period of precession is given by

$$T = \frac{2\pi}{\Omega}.$$

From the expression (8.9.7) it follows that when $I_{zz}^2\omega_3^2 > 4I_{xx}mga\cos\theta_c$, for two values of Ω, there will be steady precession without any nutation.

The expression (8.9.7) may also be written in the form

$$\Omega = \frac{I_{zz}\omega_3 \pm I_{zz}\omega_3\sqrt{1 - \dfrac{4I_{xx}mga\cos\theta_c}{I_{zz}^2\omega_3^2}}}{2I_{xx}\cos\theta_c}$$

$$= \frac{I_{zz}\omega_3}{2I_{xx}\cos\theta_c}\left[1 \pm \sqrt{1 - \frac{4I_{xx}mga\cos\theta_c}{I_{zz}^2\omega_3^2}}\right]$$

$$= \frac{I_{zz}\omega_3}{2I_{xx}\cos\theta_c}\left[1 \pm \left(1 - \frac{2I_{xx}mga\cos\theta_c}{I_{zz}^2\omega_3^2} + \cdots\right)\right]$$

$$= \frac{I_{zz}\omega_3}{I_{xx}\cos\theta_c}, \quad \text{or,} \quad \frac{mga}{I_{zz}\omega_3}, \quad \text{very nearly.}$$

In the first of the above cases the precession Ω is very large and in the second case the precession is very small. The minimum value of the spin ω_3 which gives a real value for Ω is given by

$$I_{zz}^2\omega_3^2 = 4I_{xx}mga\cos\theta_c, \quad \text{or,} \quad \omega_3 = \frac{2}{I_{zz}}\sqrt{I_{xx}mga\cos\theta_c}.$$

8.10 Gyroscopic Applications

A rapidly revolving rigid body has got a tendency of preserving its axis of rotation, and this inherent property of maintaining the stability of direction by a rapidly rotating body has the following important applications.

8.10.1 The Gyroscopic Compass

It is essentially a gyrostat suspended in frictionless gimbals inside a supporting frame. When the gyrostat is made to rotate at a high speed, its axis, by virtue of the inherent directional stability of the gyrostat, always lies in the direction of the geographic meridian, i.e., its axis always lies along the earth's geographical north-south direction. Such a system when mounted on a ship or an aeroplane would remain undisturbed by the tossing, pitching, or any change in the direction of motion of the vessel. Due to its inherent directional stability, such a system is commonly used nowadays as a compass in ships and aeroplanes and particularly in submarines. A gyroscopic compass is preferable to an ordinary magnetic compass, as the former remains unaffected by magnetic disturbances.

8.10.2 Pendulum Gyrocompass

If a gyrostat, mounted as described above, is provided with an additional small weight, suitably suspended below the rotating fly-wheel, it is called the *Pendulum Gyrocompass.* If, due to some disturbance, its axis gets displaced, the small weight readily supplies the necessary restoring torque in order to bring its axis back to the original direction. A pendulum gyrocompass is shown in figure 8.10.1. D is a rotating fly-wheel mounted by its axle AB in a horizontal ring R. This horizontal ring carries a vertical stirrup, which is rigidly fixed to it, and loaded with a small weight W just below the centre O of the fly-wheel. The horizontal ring R is free to turn about the axis MN within a vertical ring R', which is also capable of turning freely about the vertical axis GH. For complete freedom of movement, the frame surrounding the vertical ring is carried in horizontal gimbals, of which E, F form one pair. It can be shown that this arrangement is stable only when the end A points truly north. Any disturbance of the axle AB pointing north-south direction, calls into play a directive force which restores the axle to its original position.

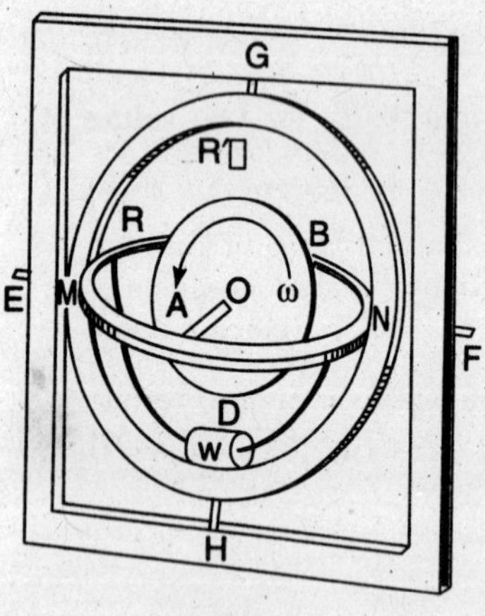

Fig. 8.10.1

8.10.3 Other Gyroscopic Applications

The barrel of a rifle or a gun is rifled by cutting spiral grooves inside it so as to impart to a bullet or a shot the necessary rapid spin about an axis in the direction of projection. This ensures an almost uniform linear motion of the bullet or the shot by improving its resistance to small deflective forces during its passage through air. The rolling of hoops and the riding of by-cycles are cases of statical instability. Both are possible only because of the gyroscopic effect. The gyroscopic action deflects their axes of rotation and thereby appropriately changes their planes of rotation, to counter-balance the disturbing effect due to gravity. In modern countries, the single track-train or the mono-rail is made possible by the stabilising action of gyrostats. Stabilising of ships is also done with the help of gyrostats. The modern aircraft appliances such as the automatic pilot, modern bomb sights, the artificial horizon, and the turn and bank indicators are all operated on gyrostat-controlled principles.

Exercises

1. Explain what you mean by precessional torque. Calculate the time period of precessional motion of a gyroscope.

2. Deduce Euler's dynamical equations of motion of a rigid body constrained to rotate about a fixed point. **[C.U. (Hons.) 2006]**

3. Discuss the force-free motion of a symmetrical rigid body.

4. What do you mean by Eulerian angles? How can the rotation of a rigid body be described with the help of Eulerian angles?

5. Discuss the motion of a symmetrical top and hence explain the precession and nutation of the axis of the top.

6. Write short notes on the following: (a) Gyrocompass, (b) Pendulum Gyrocompass.

7. Discuss some recent gyroscopic applications.

8. Find the relation between the components of the angular momentum and those of angular velocity for Eulerian motion of a rigid body. **[C.U. (Hons.) 1996]**

9. (a) Show that for force-free rotational motion of a rigid body, total angular momentum remains constant. (See Art. 8.4). **[C.U. (Hons.) 2006]**

   (b) Find the relation between components of the angular momentum and those of those of angular velocity for Eulerian motion of a rigid body. What are moments and products of inertia? (See Art. 8.3) **[C.U. (Hons.) 2006]**

10. Show from Euler's equation that the K.E. of a rigid body is conserved when the applied torque is zro. **[C.U. (Hons.) 2007]**

Chapter 9

Elasticity

9.1 Elastic Property

All bodies can be deformed by applying suitable forces. The deformation may be a change in size, shape or both. Elasticity is a property by virtue of which a body undergoing a change in size, shape or both due to application of external forces, tends to regain its original size and shape as soon as the external forces are withdrawn. Bodies which can regain completely their original conditions on withdrawal of the deforming forces, are said to be *perfectly elastic.* Bodies which do not show any tendency to regain their original condition are said to be *perfectly plastic.* There are no bodies which are perfectly elastic or perfectly plastic.

Here we consider the body to be within elastic limit, that is, whatever deformation the body will have due to the application of a system of forces in equilibrium, the body will regain its original form when the deforming forces are withdrawn. We shall consider only bodies which are homogeneous and isotropic, i.e., which have same properties at all points and in all directions.

9.2 Strain

The change in dimension of a body due to the application of a system of forces in equilibrium, is usually called *strain.* But the measure of strain is expressed in terms of change in dimension per unit dimension. To make the idea clear we take a simple illustration. Suppose a wire of length l is elongated by x with the application of a longitudinal force. In this case the linear strain of the wire is measured by x/l, i.e., the change in length per unit length.

9.3 Stress

When a system of forces, acting over a body is in equilibrium, the body undergoes deformation. Now this deformation is due to different particles of the body undergoing relative motion, and internal resisting forces come to play within the body to resist the relative motion of the particles. It is due to this internal resisting force, the body regains its original conditions when the external deforming forces are withdrawn. The internal resisting force that comes to play within the body during its deformation is generally called *stress,* but the measure of the stress at any point within a strained body is the internal force per unit area surrounding the point. If δF be the resisting force over an element of area δA surrounding any point O, we have stress P at O given by

$$P = \lim_{\delta A \to 0} \frac{\delta F}{\delta A} = \frac{dF}{dA}.$$

Since a force is a vector, so stress is also a vector. Hence stress at any point within a body may be resolved into three mutually perpendicular components of which one is normal to the surface surrounding O and the other two tangential to the surface.

The system of forces that will come into play within the body when it is strained may be shown in the following way by taking into consideration a unit cube within the strained body.

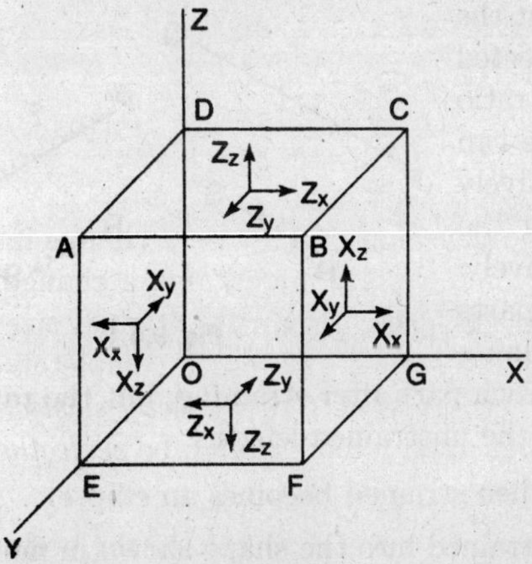

Fig. 9.3.1

In figure 9.3.1 $ABCD$, etc., is a unit cube within the strained body. OX, OY, OZ are the three rectangular axes coinciding with the three edges of the cube. We first consider two parallel faces $ADOE$ and $BCGF$ at right angles to the axis OX. Over these two surfaces the components of stresses are represented by X_x, X_y and X_z in which X_x represents the normal stress along X-axis. X_y and X_z represent the tangential stresses parallel to OY and OZ axes. We have altogether, over the six faces eighteen forces of which six are normal and twelve are tangential [In figure 9.3.1, only twelve stresses have been shown. Stresses Y_y, Y_x and Y_z on the parallel faces $ABFE$ and $DCGO$ have not been shown.] The tangential stresses tend to prevent the cube from rotation and the normal components of stresses tend to prevent the linear displacement of the cube. For equilibrium the oppositely directed normal components of the stresses must be equal. Hence the normal stress system should be reduced to three. Also to prevent rotation the tangential stress system should be reduced to three, because

$$X_z = Z_x, \; Z_y = Y_z, \; X_y = Y_x.$$

Thus we have got three normal stresses and three tangential stresses and hence we have got altogether six components in the stress system.

In a similar way the strain may be reduced to six components which will represent the effect of the stress system.

The normal components of strains give rise to linear expansion and are responsible for volume change. The tangential components give rise to rotation and cause relative displacements of different parallel layers in which the cube may be divided into three pairs of different planes. These tangential components of strains are responsible for the change in shape of the body. For equilibrium of the cube under consideration these strain components are reduced to the following six components, e_{xx}, e_{yy}, e_{zz}, $e_{xz} = e_{zx}$, $e_{xy} = e_{yx}$ and $e_{zy} = e_{yz}$.

The above consideration gives us the stress system in the case of general strain which includes both change in size and change in shape. If the strain in the body is such that there are only normal stress components existing and the tangential components are absent, there will remain volume strain only. Again when the normal components of the stress are absent, but the tangential components of the stress exist, there will be shearing strain, i.e., change in shape of the body.

9.4 Characteristics of Homogeneous Strain

It has already been mentioned that any change in the dimension of a body is called *qualitatively a strain.* The change may be in size, or in shape or in both. The strains are said to be homogeneous, when parts of a body originally equal, similar, and similarly situated, remain after the strain equal, similar and similarly situated. Although the body may change in volume, form and position, a straight line or a plane in a body in the unstrained condition will remain a straight line ad a plane in the strained condition. Equal and parallel straight lines will remain equal and parallel after straining, although they may change in position. A parallelogram remains a parallelogram after straining. An ellipse remains an ellipse. A circle may become an ellipse. Similarly a sphere may become an ellipsoid and a cube may become a parallelopiped.

From the definition of homogeneous strain we see that the ratio of the lengths of two parallel lines will remain unaffected by strain. We consider two parallel lines PQ and RS, the ratio of the lengths of the lines being $m : n$ (Fig. 9.4.1). We can divide PQ and RS into Nm and Nn equal parts respectively provided m and n are commensurable. After straining the parallel lines PQ and RS become $P'Q'$ and $R'S'$ respectively. $P'Q'$ and $R'S'$ will, therefore, consist of Nm and Nn parts respectively. It is evident from the figure that before straining the length of each part is not the same as the length of each part after straining, but the ratio of the strained lengths is $m : n$ which is same as that of the unstrained lengths.

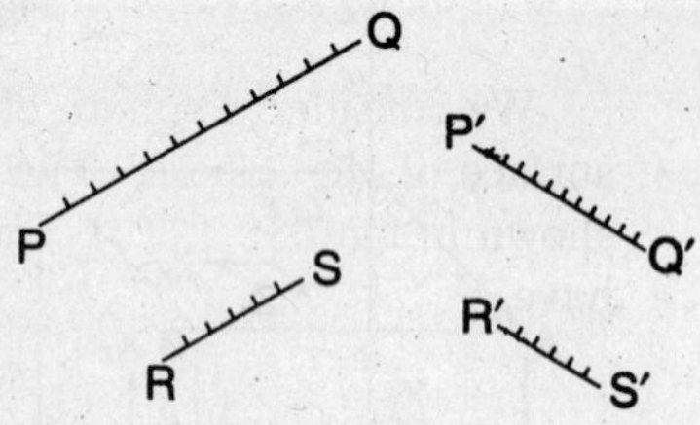

Fig. 9.4.1

From this result we shall at first show how a circle when strained becomes an ellipse.

We consider a circle having centre O [Fig. 9.4.2(a)] strained into the shape shown in figure 9.4.2(b).

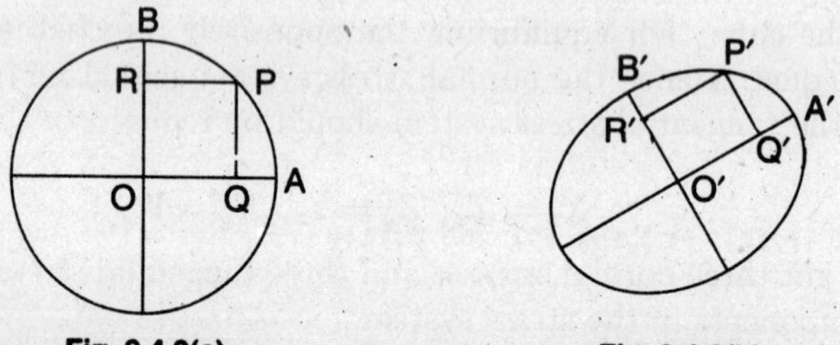

Fig. 9.4.2(a) Fig. 9.4.2(b)

We consider a point P on the circle and drop perpendiculars PQ and PR on OA and OB respectively. In the unstrained condition the ratio of the parallel lines PR to OA = the ratio of the parallel lines $P'R'$ to $O'A'$ in the strained condition, i.e.,

$$\frac{PR}{OA} = \frac{P'R'}{O'A'}.$$

Similarly $$\frac{PQ}{OB} = \frac{P'Q'}{O'B'} \quad \therefore \quad \frac{PR^2}{OA^2} + \frac{PQ^2}{OB^2} = \frac{P'R'^2}{O'A'^2} + \frac{P'Q'^2}{O'B'^2}$$

$$\text{or,} \quad \frac{PR^2 + PQ^2}{OA^2} = \frac{P'R'^2}{O'A'^2} + \frac{P'Q'^2}{O'B'^2} \quad [\because OA = OB]$$

$$\text{or,} \quad \frac{P'R'^2}{O'A'^2} + \frac{P'Q'^2}{O'B'^2} = 1 \quad [\because PR^2 + PQ^2 = OP^2 = OA^2]$$

Thus the point P' is situated on an ellipse having semimajor and semiminor axes $O'A'$ and $O'B'$ respectively. Thus figure 9.4.2(b) represents an ellipse.

We shall next show how a sphere is strained into an ellipsoid.

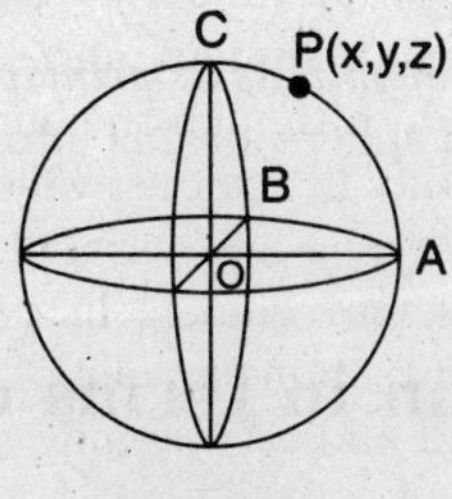

Fig. 9.4.3(a)

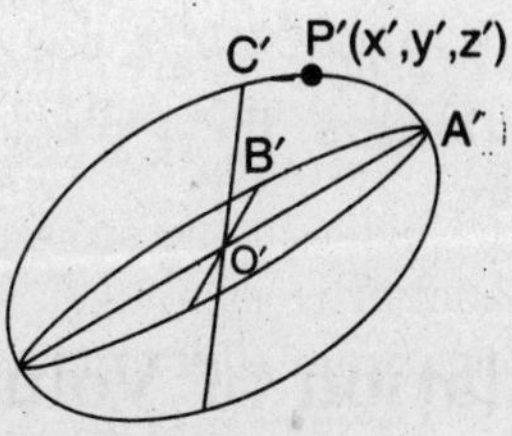

Fig. 9.4.3(b)

We consider a sphere of radius r and centre O. Let $P(x, y, z)$ be a point situated on the surface of the sphere. The sphere shown in figure 9.4.3(a) assumes the shape after strain as shown in figure 9.4.3(b). Since the ratio of the parallel lines remains unaltered after strain, we have

$$\frac{x}{r} = \frac{x'}{a'}, \quad \frac{y}{r} = \frac{y'}{b'}, \quad \frac{z}{r} = \frac{z'}{c'}.$$

In figure 9.4.3(b) $O'A'$, $O'B'$ and $O'C'$ have magnitudes a', b', c' and correspond to OA, OB and OC respectively and x', y', z' correspond to x, y, z in figure 9.4.3(a).

Now $x^2 + y^2 + z^2 = r^2$.

$$\therefore \quad x^2 + y^2 + z^2 = \frac{x'^2}{a'^2}r^2 + \frac{y'^2}{b'^2}r^2 + \frac{z'^2}{c'^2}r^2 = r^2, \quad \text{or,} \quad \frac{x'^2}{a'^2} + \frac{y'^2}{b'^2} + \frac{z'^2}{c'^2} = 1.$$

Thus the point $P'(x', y', z')$ is situated on an ellipsoid.

9.5 Principal Axes of Strain

At every point in a body, we may imagine three rectangular axes such that they remain mutually perpendicular to each other after strain and these three rectangular axes are called *principal axes of strain.* If these three axes have the same direction after straining as before, i.e., if OA, OB, OC be the positions of the three mutually perpendicular axes X, Y, Z before strain, then after strain the axes will be in the same position OA, OB, OC, we call this strain to be a *homogeneous pure strain.*

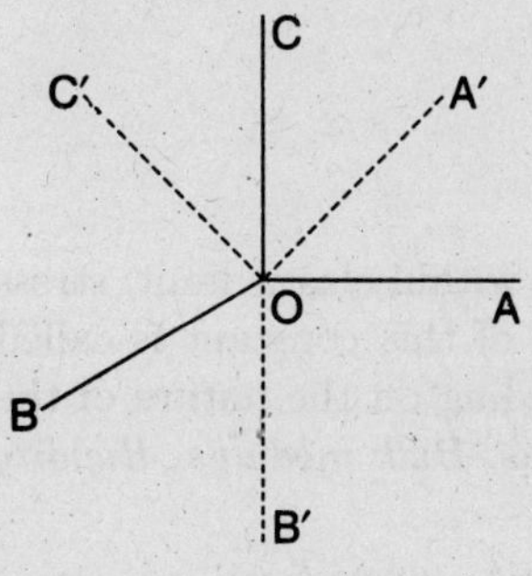

Fig. 9.5.1

If the axes X, Y, Z before strain occupy the positions OA, OB, OC, but after strain occupy the different positions OA', OB' and OC' (Fig. 9.5.1) and they remain at right angles to each other, a rotation is necessary to bring the axes to their original positions. In this case the strain is said to consist of a *pure strain* and a *rotation.*

The most general strain may be resolved into extensions along the three principal axes mutually perpendicular to each other (a compression may be expressed as a negative extension).

Let us take these directions as the axes of X, Y, Z respectively. If a line of unit length be drawn parallel to X-axis, after strain this will remain parallel, and suppose that after strain the length becomes $1 + e_{xx}$. Similarly unit lengths parallel to Y and Z axes become $1 + e_{yy}$ and $1 + e_{zz}$ after strain.

The quantities e_{xx}, e_{yy}, e_{zz} are called *principal elongations* or *principal strains* or *fractional elongations* along the axes. If $e_{xx} = e_{yy} = e_{zz}$, then a sphere strains into a sphere or any figure into a similar figure.

9.6 Calculation of Volume Strain in terms of the three Principal Strains

Let us consider a sphere of radius r in the unstrained condition. Now, in case of a homogeneous strain, the sphere is strained to an ellipsoid, the semi-axes of which are

in the direction of X-axis $r(1 + e_{xx})$,

in the direction of Y-axis $r(1 + e_{yy})$,

in the direction of Z-axis $r(1 + e_{zz})$,

and the volume of the ellipsoid is

$$\frac{4}{3}\pi\left[r^3(1 + e_{xx})(1 + e_{yy})(1 + e_{zz})\right] = \frac{4}{3}\pi r^3\left[1 + (e_{xx} + e_{yy} + e_{zz})\right] \text{ approximately}$$

$\therefore$ the change in volume after strain

$$\frac{4}{3}\pi r^3\left[1 + (e_{xx} + e_{yy} + e_{zz})\right] - \frac{4}{3}\pi r^3 = \frac{4}{3}\pi r^3(e_{xx} + e_{yy} + e_{zz}).$$

Therefore, the fractional change in volume or volume *dilatation* is given by

$$\delta = \frac{\frac{4}{3}\pi r^3(e_{xx} + e_{yy} + e_{zz})}{\frac{4}{3}\pi r^3} = e_{xx} + e_{yy} + e_{zz}.$$

When $e_{xx} = e_{yy} = e_{zz} = e$ (say), we have $\delta = 3e$, i.e., the cubical expansion is three times the linear expansion.

9.7 Hooke's Law

The fundamental law of elasticity is due to Hooke—which states that within elastic limit, stress is proportional to the strain. Thus $\frac{\text{stress}}{\text{strain}}$ is a constant. The value of this constant is called *modulus of elasticity.* There are different moduli of elasticity depending on the nature of the strains. We have the following moduli of elasticity: *Young's modulus, Bulk modulus, Rigidity modulus* and *Axial modulus.*

These moduli of elasticity are called *elastic constants of a substance.* Along with these four elastic constants we take into consideration one more elastic constant known as *Poisson's ratio.*

Young's modulus Y

$$= \frac{\text{applied load per unit cross-section}}{\text{increase in length per unit length}}$$

$$\text{i.e.,} \quad Y = \frac{\text{longitudinal stress}}{\text{longitudinal strain}} \text{ or longitudinal strain } (e) = \frac{\text{longitudinal stress}}{Y} = \frac{P}{Y}$$

If F be the applied tensile force which causes a body of length l and cross-section A to elongate by x in the direction of the force, the longitudinal stress is equal to F/A and the longitudinal strain is x/l. Therefore, we express Young's modulus as

$$Y = \frac{F/A}{x/l} = \frac{Fl}{Ax}$$

In absolute C.G.S. units, F/A is expressed in dynes/cm^2, while x/l is a pure number. Hence Young's modulus is expressed in dynes/cm^2. It is also expressed in lb wt per sq. inch or ton. wt per sq. inch.

The dimension of Y is

$$[Y] = \frac{MLT^{-2}.\frac{1}{L^2}}{L/L} = ML^{-1}T^{-2}$$

Bulk modulus or volume elasticity k

$$= \frac{\text{compressive or tensile force per unit area}}{\text{decrease or increase in volume per unit volume}}$$

$$\text{i.e., } k = \frac{\text{compressional or dilatational pressure}}{\text{volume strain}}$$

If P be the compressional or dilatational pressure, and $\frac{dV}{V}$, the volume strain, we have,

$$k = \frac{P}{\frac{dV}{V}},$$ [Here, since increase of P decreases V and vice versa, we should write $k = +P/-\frac{dV}{V}$]

where dV is the change in volume and V the original volume. In absolute C.G.S. units, P is expressed in dynes per sq. cm and $\frac{dV}{V}$ is a pure number. Hence k is expressed in dynes per sq. cm.

The dimension of k is

$$[k] = \frac{MLT^{-2}\frac{1}{L^2}}{\frac{L^3}{L^3}} = ML^{-1}T^{-2}$$

which is same as the dimension of Y.

Modulus of rigidity or shear modulus

$$n = \frac{\text{tangential force per unit area}}{\text{angular deformation}}, \quad \text{i.e.,} \quad n = \frac{\text{tangential stress}}{\text{angle of shear or shearing strain}}$$

A simple case of shearing strain which is responsible for change in shape of a cube when a tangential stress is acting over the top of the cube is illustrated here.

We consider one face $ABCD$ of a cube as shown in figure 9.7.1, the lower end of which is fixed. Let F be the tangential force applied to the upper surface of the cube which is responsible for the change in shape of the cube. The idea will be clear if we consider a pack of cards placed over a plane and apply a tangential force over the top of the pack. Each individual card will move relatively over the other in the direction of the tangential force applied, but the shape of the individual card will not change although the front of the pack instead of remaining rectangular will assume the shape of a parallelogram as shown by $A'B'CD$. This is a plain or a simple shear and $ABCD$ is the plane of shear. In this plane only change of shape occurs. Let θ be the angle of shear through which AD is displaced to the position $A'D$ due to relative motion of the plane surfaces parallel to the fixed plane DC. Expressing θ in circular measure, we have $\theta = \frac{AA'}{AD}$. If $AD = 1$, $\theta = AA'$. Thus the relative tangential displacement between two parallel planes separated by a unit distance is equal to the angle of shear. If f be the tangential stress, the modulus of rigidity is given by $n = \frac{f}{\theta}$. In absolute C.G.S. units f is expressed in dynes per sq. cm and θ is a pure number. Hence n is expressed in dynes per sq. cm. The dimension of n is $ML^{-1}T^{-2}$ which is the same as the dimensions of Young's modulus and bulk modulus of elasticity.

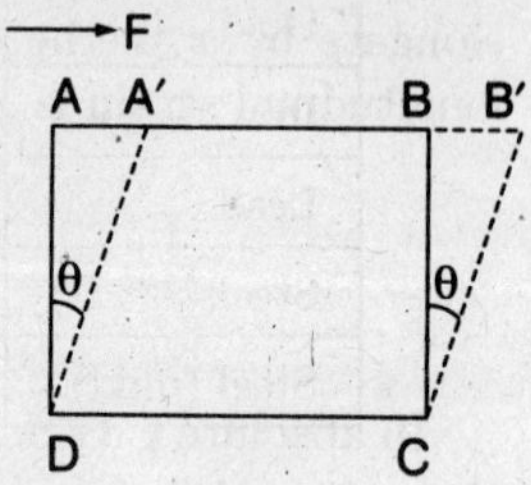

Fig. 9.7.1

$$\text{Axial modulus } \chi = \frac{\begin{array}{c}\text{applied force per unit area of cross-section}\\ \text{along a particular direction.}\end{array}}{\begin{array}{c}\text{increase in length per unit length in the}\\ \text{same direction while the strains in the other}\\ \text{two perpendicular directions are prevented.}\end{array}}$$

Axial modulus is also expressed in dynes per sq. cm. Its dimension is also $ML^{-1}T^{-2}$.

9.7.1 Poisson's Ratio

Careful measurement shows that when a body undergoes elongation in the direction of a tensile force, it experiences lateral contraction, i.e., contraction in a direction perpendicular to the direction of the tensile force. The ratio of lateral strain measured by decrease in width per unit width to the longitudinal strain is called *Poisson's ratio.*

Let us consider a wire of length L and diameter D which undergoes elongation l along the direction of the applied tensile force F. The wire will also undergo lateral contraction, i.e., the diameter of the wire will decrease. Let the decrease in diameter be d. Then the longitudinal strain $= \frac{l}{L}$ and lateral strain $= \frac{d}{D}$.

$$\therefore \text{ Poisson's ratio } \sigma = \frac{\text{lateral strain}}{\text{longitudinal strain}} = \frac{\text{lateral fractional contraction}}{\text{longitudinal fractional elongation}} = \frac{d/D}{l/L}.$$

$$\therefore \text{ Lateral fractional contraction} = \sigma \times \text{ longitudinal fractional elongation} = \sigma\frac{P}{Y}$$

It should be noted that if a compressive force is applied then there is lateral expansion. As σ is a ratio of two strains, it is a pure number.

Table for the Elastic Constants of Solids				
Material at 18°C	Young's modulus Y (in dynes/ sq.cm.$\times 10^{11}$)	Rigid modulus n (in dynes/ sq.cm.$\times 10^{11}$)	Bulk modulus k (in dynes/ sq.cm.$\times 10^{11}$)	Poisson's ratio σ
Aluminium	7.05	2.62	7.58	0.345
Copper	12.98	4.83	13.76	0.343
Gold	7.8	2.7	21.7	0.440
Iron	21.2	8.2	16.9	0.290
Lead	1.62	0.562	4.6	0.441
Steel (1% C)	21.0	8.10	16.88	0.293
Steel (mild)	21.0	8.12	16.78	0.291

Illustrative Examples

1. A flexible wire of length 2 metres and cross-section 1 sq.mm is stretched tight between two points A and B in the same horizontal plane. When a mass of 10 gms. is hung from mid-point of the wire, it is found that this point is depressed to a distance of 1 cm. below the line AB. Calculate Young's modulus for the material of the wire. **[cf. C.U.(Hons.) 1998]**

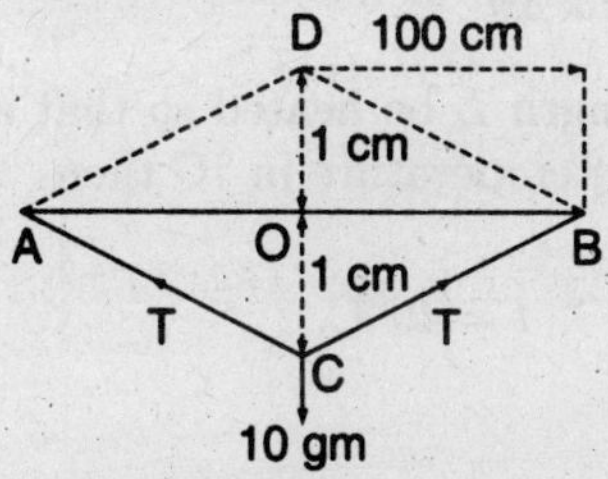

Fig. 1

Let AB be the wire stretched between two points A and B and let the mass of 10 gm. hung from O displace O to C. To find the tension T we complete the parallelogram $ACBDA$ (Fig. 1).

From the triangle of forces BCD giving the condition that C will be in equilibrium, we have

$$\frac{T}{10g} = \frac{BC}{CD} \quad \text{or,} \quad T = 10\frac{BC}{CD}g = \frac{10 \times 980 \times BC}{CD}.$$

Now

$$BC^2 = 100^2 + 1 = 100^2(1 + 0.0001)$$

$$\text{or,} \quad BC = 100\left(1 + \frac{1}{2} \times 0.0001\right) = 100 + 0.005 = 100.005 \text{ cm.}$$

$$\therefore \quad T = \frac{100.005 \times 9800}{2} \quad [\because CD = 2 \text{ cm.}]$$

$$= 5000.25 \times 98 \text{ dynes.}$$

Now

$$\text{stress} = \frac{T}{\text{area}} = \frac{5000.25 \times 98}{0.01} = 5000.25 \times 98 \times 10^2 - 5 \times 10^5 \times 98 \text{ (approx.)}$$

$$\text{strain} = \frac{100.005 - 100}{100} = 5 \times 10^{-5}$$

$$\therefore \quad Y = \frac{\text{stress}}{\text{strain}} = \frac{5 \times 10^5 \times 98}{5 \times 10^{-5}} = 9.8 \times 10^{11} \text{ dynes/sq.cm.}$$

2. You are given 200 c.c. of air at a pressure due to 760 mm of mercury. On increasing the pressure by that due to 1 mm of mercury without change of temperature the volume is observed to decrease by 0.263 c.c. Find the co-efficient of volume elasticity. **[Calcutta, 1952]**

Co-efficient of volume elasticity or bulk modulus of elasticity $= \frac{P}{dV/V}$. Here P is the compressional pressure and is given by $P = 0.1 \times 981 \times 13.6$ dynes/cm^2 and the volume strain

$$\frac{dV}{V} = \frac{0.263}{200} = 1.315 \times 10^{-3}.$$

$$\therefore \text{ co-efficient of volume elasticity} = \frac{0.1 \times 981 \times 13.6}{1.315 \times 10^{-3}} = 1.014 \times 10^6 \text{ dynes/cm}^2.$$

3. The tension of a steel wire of radius 0.5 mm stretched between two rigid supports 1 metre apart is found to be 3.77×10^6 dynes when the temperature is 20°C. At what temperature will the tension on the wire just vanish? Assume Young's modulus for steel $= 2 \times 10^{12}$ dynes per sq.cm. and the co-efficient of linear expansion of steel $= 1.2 \times 10^{-5}$ per °C. **[C.U.(Hons.) 1981]**

Here, radius of the wire, $r = 0.5$ mm $= 5 \times 10^{-2}$ cm, tension T at 20°C $= 3.77 \times 10^6$ dynes.

Young's modulus $Y = 2 \times 10^{12}$ dynes/cm^2, the co-efficient of linear expansion of steel $=$ 1.2×10^{-5} per °C.

$$\text{Now } Y = \frac{T/\pi r^2}{\text{linear strain}}.$$

$$\therefore \text{ linear strain} = \frac{T}{\pi r^2} \cdot \frac{1}{Y} = \frac{3.7 \times 10^6}{3.14 \times 25 \times 10^{-4} \times 2 \times 10^{12}} = 2.4 \times 10^{-4}.$$

Now the tension on the wire will just vanish, if the wire of length L be heated so that an elongation of length l occurs in its length L. If 't' be the required temperature in °C then

$$\alpha = \frac{l}{L(t-20)}, \quad \text{or,} \quad t - 20 = \frac{2.4 \times 10^{-4}}{1.2 \times 10^{-5}} = 20 \quad \therefore t = 40°\text{C}.$$

9.8 Elastic Properties of Matter

Let us consider a vertical metal wire the upper end of which is fixed and the lower end carries a scale-pan. If we measure the increase or elongation of the wire when different loads are placed in the scale-pan and plot a graph between the load per unit cross-section of the unstretched wire and the extension per unit length of the unstretched wire, we obtain a curve as shown in figure 9.8.1.

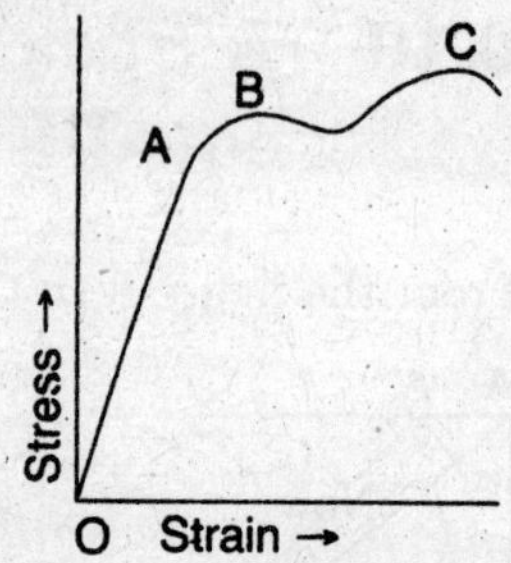

Fig. 9.8.1

The first part of the curve OA shows that extension per unit length or strain is proportional to the load per unit area or stress. Since the curve passes through the origin O, it follows that the strain is zero when the stress is zero. Up to the point A Hooke's law holds and also up to the point A we see that when we unload the wire, the stretched wire shortens until its length is the same as it was before the weight was given on the scale-pan. This means that the wire is perfectly elastic up to A. When we go beyond this range we exceed what is known as *elastic limit.* When the stress is greater than the value represented by OA, the curve takes a bend and on removing the weights at any point here, the wire does not regain its original length. Under this condition the wire is said to have acquired a *permanent set.* At B a stage is reached where a great elongation begins faster than between A and B which is shown by the bending of the curve BC. The point B is called the *yield point.* Hence yield point is that point at which the great extension starts. Beyond the yield point the elongation increases very rapidly with load and there is rapid contraction of the area of cross-section of the wire until the breaking point C is reached. The maximum stress before snapping corresponds to a point on the curve called the *breaking stress.* The corresponding strain is called *breaking strain.*

Even within elastic limit different bodies show difference in their behaviour. Some solids regain their original condition immediately after the removal of deforming force, while the others take considerable time to regain their original condition. This delay in regaining the original condition of the body is known as *elastic after effect.* Glass shows this effect to a considerable extent while in quartz fibre, phosphor bronze, silver and gold this effect is practically absent.

9.9 Viscosity of Metals and Elastic Fatigue

We consider two vertical wires one of steel and the other of zinc having same length and diameter carrying bars of same weight and dimension suspended vertically from a rigid support. The bars are set into oscillation and it is found that oscillations of different wires take place at different rates. The steel wire oscillates for a long time, whereas the zinc wire oscillates for a short time. This decay in oscillation is not wholly or mainly due to resistance of the air for the effect of air resistance will be the same in case of both the wires. It is due to the dissipation of energy when the different parts of the wire are in relative motion. This property, from analogy with the case of liquids and gases are called *viscosity* of the metal.

A remarkable property of the viscosity of a metal has been discovered by Lord Kelvin. It is found that the oscillation decays at a faster rate for a wire oscillating continuously than that for a fresh wire. If the wire oscillating continuously be given a rest for some time and then allowed to oscillate again, the oscillations will not die away so rapidly as before. This shows that the continuously oscillating wire behaves as if it has got tired or fatigued and therefore can continue oscillation with great difficulty. This condition is expressed by the term *elastic fatigue.*

9.10 A Shear is Equivalent to a Compression and an equal Extension at Right Angles to each other

From the figure 9.10.1, it clear that in the case of a plane shear all the lines parallel to the diagonal DB of the cube get elongated and all the lines parallel to CA get shortened. From the diagram, $DB = DE$ and $DC = DB\cos 45° = DB\cdot\frac{1}{\sqrt{2}}$ or $DB = \sqrt{2}DC = \sqrt{2}l$, where l is the length of each side of the cube.

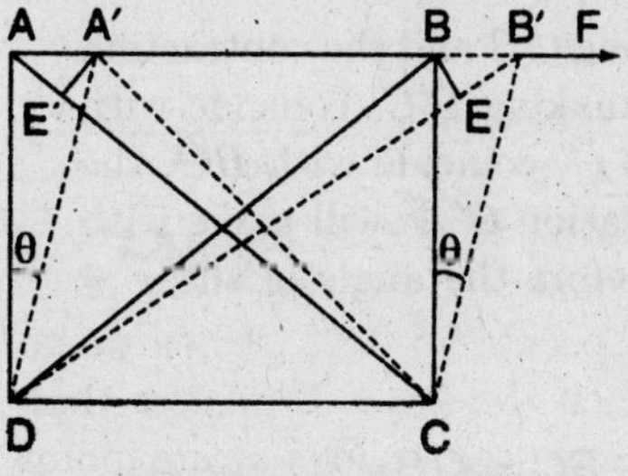

Fig. 9.10.1

Now

$$EB' = BB'\cos\angle EB'B = BB'\cos 45° \quad [\angle EB'B = 45°]$$

$$= \frac{BB'}{\sqrt{2}}.$$

But

$$BB' = BC \cdot \theta = l\theta. \quad \therefore EB' = \frac{l\theta}{\sqrt{2}}.$$

Now, linear elongation of DB is EB'. Let e be the linear extensional strain

$$\therefore e = \frac{EB'}{DE} = \frac{EB'}{DB} = \frac{l\theta}{\sqrt{2}} \Big/ \sqrt{2}l = \frac{\theta}{2}$$

or $\theta = 2e$. Thus θ is twice the linear extensional strain or fractional elongation. Similarly, the diagonal AC contracts by AE', and the linear contractional strain or fractional contraction is found to be equal to $\frac{\theta}{2}$. Thus a simple shear is produced as a result of equal amount of fractional linear elongation and fractional linear contraction at right angles to each other, either (extension or contraction) being inclined at an angle of 45° to the direction of the shearing displacement.

Conversely, *simultaneous equal extension and contraction at right angles to each other are equivalent to a shear.*

This may be deduced in the following way.

In figure 9.10.2 AB and CD denote two axes perpendicular to each other along which extension and contraction take place. Let

$$OA = OB = OC = OD = 1.$$

Hence before strain $ACBDA$ is a square and after strain it becomes a parallelogram $A'C'B'D'A'$.

$$\therefore OB' = OB + BB' = 1 + e$$

$$OC' = OC - CC' = 1 - e$$

where BB' and CC' denote respectively the extension along OB and contraction along OC. Both elongation and contraction have been taken equal to e.

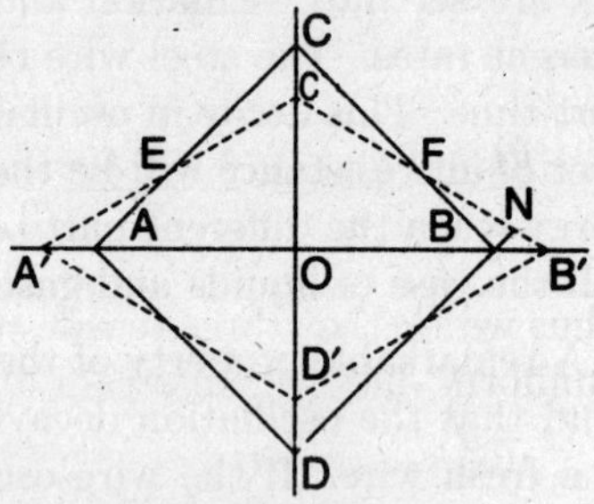

Fig. 9.10.2

Now

$$B'C'^2 = OB'^2 + OC'^2 = (1+e)^2 + (1-e)^2 = 2 + 2e^2$$

$$= 2 \text{ (considering } e \text{ to be very small, its square is neglected).}$$

Also, $BC^2 = OB^2 + OC^2 = 2. \quad \therefore B'C' = BC.$

Hence we can rotate $B'C'A'D'$ so that $B'C'$ coincides with BC.

The relation between the angle of shear θ and the elongation e along OB and the contraction e along OC can be found in the following way. Before the rotation making $B'C'$ coincide with BC, the angle made by CA with $C'A'$ was CEC'. Now to make $B'C'$ coincide with BC, the system has to be rotated through an angle CFC', so that after rotation $C'A'$ will make with CA the angle $CEC' + CFC'$. Now since $\angle CEC' = \angle CFC'$, therefore the angle of shear $= 2\angle CFC' = 2\angle B'FB$. The angle of shear may also be written as

$$\theta = 2\frac{BN}{BF} = 2\frac{BB' \sin 45°}{BF} = \frac{2BB'}{\frac{1}{2}BC\sqrt{2}} = \frac{2BB'}{OB} = 2e \quad (\because BC = OB\sqrt{2}).$$

Now, the fractional elongation along OB is $\frac{BB'}{OB} = e = \frac{\theta}{2}$.

Similarly it can be shown that the contraction along OC is also equal to half the angle of shear.

Thus, we see that simultaneous elongation and contraction at right angles to each other are equivalent to a shear.

Illustrative Examples

1. A substance is subjected to tensile strains equal to a, b, c in the directions of the rectangular axes x, y, z respectively. Prove that the resultant strain is equivalent to that due to superposition of a uniform dilatational strain equal to $a+b+c$, together with two shears, one equal to $\left[\frac{2}{3}(a+b+c) - 2c\right]$ in the plane $(x - z)$, and another shear $\left[\frac{2}{3}(a+b+c) - 2b\right]$ in the plane $(x - y)$.

Let δ be the uniform dilatational strain, and θ_1 be the angle of shear in the $(x - y)$ plane and θ_2, the angle of shear in the $(x - z)$ plane. We know that, a uniform dilatational strain is equal to three times the linear strain

$$\therefore \quad a = \frac{\delta}{3} + \frac{\theta_1}{2} + \frac{\theta_2}{2}, \quad b = \frac{\delta}{3} - \frac{\theta_1}{2}, \quad c = \frac{\delta}{3} - \frac{\theta_2}{2} \quad \text{so that } a+b+c = \delta.$$

$$\therefore \quad \frac{\theta_1}{2} = \frac{\delta}{3} - b = \frac{a+b+c}{3} - b, \text{ or, } \theta_1 = \frac{2}{3}(a+b+c) - 2b \text{ and } \theta_2 = \frac{2}{3}(a+b+c) - 2c.$$

Thus we see that three tensile strains a, b, c along the axes x, y, z respectively are equivalent to a uniform dilatational strain and two shears, one in the plane $(x - y)$ and another in the plane $(z - x)$.

2. A substance is subjected to tensile strains equal to 3×10^{-4}, 2×10^{-4} and 5×10^{-4} in the directions of the axes x, y, z respectively. Resolve the resultant strain into a uniform dilatational strain together with two shears, one in the plane $(x - y)$ and the other in the plane $(x - z)$.

If δ be the uniform dilatational strain and, θ_1 and θ_2 be the shears in the $(x - y)$ and $(x - z)$ plane respectively, we have, from the results of example 1 above,

$$3 \times 10^{-4} = \frac{\delta}{3} + \frac{\theta_1}{2} + \frac{\theta_2}{2}; \quad 2 \times 10^{-4} = \frac{\delta}{3} - \frac{\theta_1}{2}; \quad 5 \times 10^{-4} = \frac{\delta}{3} - \frac{\theta_2}{2}$$

so that $\delta = 10 \times 10^{-4}$.

$$\therefore \quad \frac{\theta_1}{2} = \frac{\delta}{3} - 2 \times 10^{-4} = \frac{10}{3} \times 10^{-4} - 2 \times 10^{-4}$$

$$\text{or,} \quad \theta_1 = 10^{-4}\left(\frac{20}{3} - 4\right) = 2.66 \times 10^{-4} \text{ radian}$$

$$\text{and} \quad \frac{\theta_2}{2} = -\left(5 \times 10^{-4} - \frac{\delta}{3}\right) = -\left(5 - \frac{10}{3}\right) \times 10^{-4}$$

$$\text{or,} \quad \theta_2 = -\left(10 - \frac{20}{3}\right) \times 10^{-4} = --3.33 \times 10^{-4} \text{ radian.}$$

9.11 Relations connecting Y, k, n and σ and to show that σ lies between $+0.5$ and -1

We consider a cube in a strained medium. The edges of the cube are parallel to the three rectangular axes OX, OY and OZ. Suppose the cube is subjected to a uniform tensile stress P over each of the faces (Fig. 9.11.1). We have the linear strain along X-axis due to tensile stress along X-axis given by $\frac{P}{Y}$. Linear strain along X-axis due to tensile stress along Y-axis is given by $-\frac{\sigma P}{Y}$. Also the linear strain along X-axis due to tensile stress along Z-axis is given by $-\frac{\sigma P}{Y}$. Similarly, by considering the two other pairs of tensile stresses along Y and Z axes, we have the resultant linear strain along X or Y or Z axis given by

$$e = \frac{P}{Y} - \frac{2\sigma P}{Y} = \frac{P}{Y}(1 - 2\sigma).$$

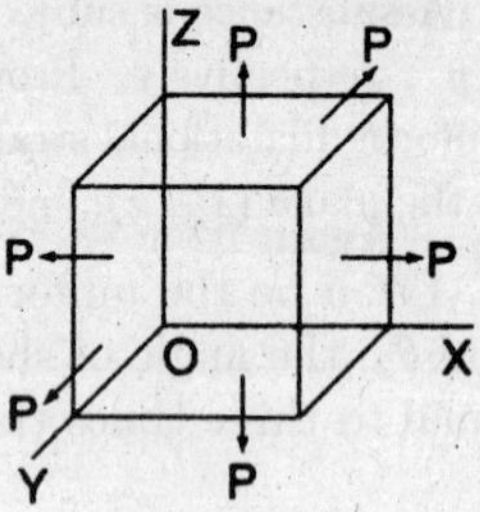

Fig. 9.11.1

If δ be the uniform volume strain, we have,

$$\delta = 3e = \frac{3P}{Y}(1 - 2\sigma).$$

Now,

$$\delta = \frac{P}{k} = \frac{3P}{Y}(1 - 2\sigma), \quad \text{or,} \quad \frac{Y}{k} = 3(1 - 2\sigma). \tag{9.11.1}$$

which is the relation connecting Y, k and σ.

Next we suppose that the cube is subjected to a tensile stress along X-axis and an equal compressional stress along Y-axis (Fig. 9.11.2). We therefore have linear strain along X-axis due to tensile stress along X-axis given by $\frac{P}{Y}$ and the linear strain along X-axis due to compressional stress along Y-axis is given by $+\frac{\sigma P}{Y}$. Therefore, the resultant linear strain along X-axis is

$$\frac{P}{Y} + \frac{\sigma P}{Y} = \frac{P}{Y}(1 + \sigma).$$

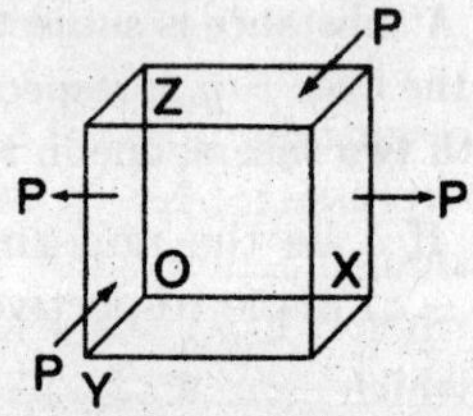

Fig. 9.11.2

Similarly, the resultant linear strains along Y and Z axes are

$$-\frac{P}{Y} - \frac{\sigma P}{Y} = -\frac{P}{Y}(1 + \sigma); \quad -\frac{\sigma P}{Y} + \frac{\sigma P}{Y} = 0 \text{ respectively.}$$

If θ be the angle of shear in the $X - Y$ plane, we have $\frac{\theta}{2} = \frac{P}{Y}(1 + \sigma)$.

But $\theta = \frac{P}{n}$.

$$\therefore \frac{P}{n} = \frac{2P}{Y}(1 + \sigma), \quad \text{or,} \quad \frac{Y}{n} = 2(1 + \sigma), \tag{9.11.2}$$

which is the relation connecting Y, n and σ.

From (9.11.2), we have,

$$\frac{Y - 2n}{2n} = \sigma \tag{9.11.3}$$

which shows that σ is expressed in terms of Young's modulus and rigidity modulus.

From (9.11.1) and (9.11.2), we have

$$\frac{Y}{k} = 3 - 6\sigma \quad \text{and} \quad \frac{Y}{n} = 2 + 2\sigma.$$

Adding we have,

$$Y\left(\frac{1}{k} + \frac{1}{n}\right) = 5 - 4\sigma = 5 - 4\left(\frac{Y - 2n}{2n}\right) \text{ or } Y\left(\frac{1}{k} + \frac{3}{n}\right) = 9 \text{ or } Y = \frac{9nk}{n + 3k}, \qquad (9.11.4)$$

which is the relation connecting Y, k and n. Equation (9.11.4) can be written in different forms like

$$\frac{1}{k} + \frac{3}{n} = \frac{9}{Y} \quad \text{and} \quad \frac{1}{3k} + \frac{1}{n} = \frac{3}{Y}$$

Again from (9.11.1) and (9.11.2), we have

$$\frac{Y}{3k} = 1 - 2\sigma; \quad \frac{Y}{2n} = 1 + \sigma.$$

Subtracting and re-arranging we have

$$\sigma = Y\left(\frac{3k - 2n}{18kn}\right).$$

But

$$Y = \frac{9nk}{n + 3k}. \quad \therefore \sigma = \frac{9nk}{(n + 3k)} \times \frac{(3k - 2n)}{18kn} = \frac{3k - 2n}{6k + 2n}. \qquad (9.11.5)$$

Here σ is expressed terms of bulk modulus and modulus of rigidity.

Again, from (9.11.1) and (9.11.2), we may write $3k(1 - 2\sigma) = 2n(1 + \sigma)$. Now n and k are essentially positive quantities and so in the left-hand expression $(1 - 2\sigma)$ and in the right-hand expression $(1 + \sigma)$ of the above relation will be either negative or positive. When σ is greater than 0.5, left-hand side becomes negative, but the right-hand side remains positive. Again when σ is less than -1, left-hand side becomes positive, but right-hand side remains negative, which cannot be the case. Therefore, σ cannot be greater than 0.5 or less than -1. We do not know any metallic material whose Poisson's ratio is negative. The Poisson's ratio for most of the substances lies between 0.2 and 0.4.

9.12 Relation between Axial Modulus and other Elastic Constants

We consider a cube in a strained medium. Suppose, P is the tensile stress along X-axis. As a result of this stress, there will be linear extensional strain along X-axis and contractional strains along Y-axis and Z-axis. Now, we imagine a suitable tensile stress P_1, say, along both Y and Z-axes such that they will counteract the contractional strains. When this is the case there will be no linear strain along Y and Z axes. Let a, b, c be the extensional strains along X, Y, Z axes respectively. In this case $b = c = 0$.

Now linear extensional strain along X-axis due to P is $\frac{P}{Y}$. Linear contractional strain along X-axis due to P_1 along Y-axis is $\frac{\sigma P_1}{Y}$ and linear contractional strain along X-axis due to P_1 along Z-axis is $\frac{\sigma P_1}{Y}$.

Therefore, the resultant linear extensional strain along X-axis is

$$a = \frac{P}{Y} - \frac{2\sigma P_1}{Y} = \frac{1}{Y}(P - 2\sigma P_1). \tag{9.12.1}$$

Next, linear extensional strain along Y-axis due to P_1 is $\frac{P_1}{Y}$. Linear contractional strain along Y-axis due to P along X-axis is $\frac{\sigma P}{Y}$ and linear contractional strain along Y-axis due to P_1 along Z-axis is $\frac{\sigma P_1}{Y}$.

Therefore, the resultant linear extensional strain along Y-axis is

$$b = \frac{P_1}{Y} - \frac{\sigma P}{Y} - \frac{\sigma P_1}{Y} \tag{9.12.2}$$

but, since $b = 0$, we have $P_1 = \frac{P\sigma}{(1-\sigma)}$.

$$\therefore \quad a = \frac{1}{Y}\left[P - 2\sigma \cdot \frac{P\sigma}{(1-\sigma)}\right] \quad \text{[from eqn. (9.12.1)]}$$

$$= \frac{P}{Y}\left[\frac{1-\sigma-2\sigma^2}{(1-\sigma)}\right] = \frac{P}{Y}\frac{(1+\sigma)(1-2\sigma)}{(1-\sigma)} \tag{9.12.3}$$

$\therefore$ axial modulus

$$\chi = \frac{P}{a} = \frac{P}{\frac{P}{Y}\frac{(1+\sigma)(1-2\sigma)}{(1-\sigma)}} = \frac{Y(1-\sigma)}{(1+\sigma)(1-2\sigma)}. \tag{9.12.4}$$

Putting $Y = 2n(1+\sigma)$, we may write

$$\chi = \frac{2n(1+\sigma)(1-\sigma)}{(1+\sigma)(1-2\sigma)} = \frac{2n(1-\sigma)}{(1-2\sigma)} = \frac{2n\left(1 - \frac{3k-2n}{6k+2n}\right)}{\left(1 - 2\cdot\frac{3k-2n}{6k+2n}\right)} \quad \text{[from eqn. (9.11.5)]}$$

$$= 2n \cdot \frac{3k+4n}{6n} = \frac{3k+4n}{3} = k + \frac{4}{3}n.$$

It follows from the relation (9.12.3) that the elongation of a bar when prevented from contracting laterally is less than the elongation of the same bar if the sides of the bar were free in the ratio $\left(1 - \frac{2\sigma^2}{1-\sigma}\right) : 1$, in both the cases, of course, the bar being under the action of the same tensile stress along its length. In case of a steel bar for which $\sigma = 0.267$, the elongation of a bar with fixed sides is about $\frac{4}{5}$ times the elongation of the same bar with free sides.

9.12.1 Relations connecting Y, k, n and σ (Alternative Treatment)

Relations connecting the elastic constants can be established on the basis of the "Principle of Superposition", according to which the effects produced by the different stresses acting at a point are quite independent of one another, or more correctly, it is assumed that each component stress produces the same effect if acting alone, as it contributes to the final result when acting in conjunction with others. The applicability of this principle is confirmed by experiment.

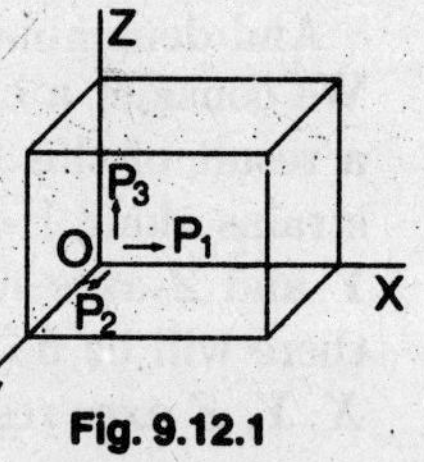

Fig. 9.12.1

Since for isotropic substances, the directions of the axes of strain will be those of the normal stresses, the most general homogeneous stress at a point will be P_1, P_2, P_3 along OX, OY, OZ respectively (Fig. 9.12.1).

Hence $X_x = P_1$, $Y_y = P_2$ and $Z_z = P_3$ while $X_y = Y_z = Z_x = 0$.

The strains corresponding to these stresses are given by

$$\left.\begin{aligned} e_{xx} &= \frac{P_1}{Y} - \frac{\sigma P_2}{Y} - \frac{\sigma P_3}{Y} = \frac{1}{Y}[P_1 - \sigma(P_2 + P_3)] \\ e_{yy} &= \frac{P_2}{Y} - \frac{\sigma P_3}{Y} - \frac{\sigma P_1}{Y} = \frac{1}{Y}[P_2 - \sigma(P_3 + P_1)] \\ e_{zz} &= \frac{P_3}{Y} - \frac{\sigma P_1}{Y} - \frac{\sigma P_2}{Y} = \frac{1}{Y}[P_3 - \sigma(P_1 + P_2)] \end{aligned}\right\} \tag{9.12.5}$$

and $e_{xy} = e_{yz} = e_{zx} = 0$.

The above equations may be written in the form.

$$\begin{aligned} P_1 - \sigma P_2 - \sigma P_3 &= Y e_{xx} \\ -\sigma P_1 + P_2 - \sigma P_3 &= Y e_{yy} \\ -\sigma P_1 - \sigma P_2 + P_3 &= Y e_{zz} \end{aligned}$$

whence,

$$P_1 = \frac{\begin{vmatrix} Ye_{xx} & -\sigma & -\sigma \\ Ye_{yy} & 1 & -\sigma \\ Ye_{zz} & -\sigma & 1 \end{vmatrix}}{\begin{vmatrix} 1 & -\sigma & -\sigma \\ -\sigma & 1 & -\sigma \\ -\sigma & -\sigma & 1 \end{vmatrix}}$$

Numerator of the above expression on the right hand side

$$\begin{aligned} &= Ye_{xx}\begin{vmatrix} 1 & -\sigma \\ -\sigma & 1 \end{vmatrix} + \sigma\begin{vmatrix} Ye_{yy} & -\sigma \\ Ye_{zz} & 1 \end{vmatrix} - \sigma\begin{vmatrix} Ye_{yy} & 1 \\ Ye_{zz} & -\sigma \end{vmatrix} \\ &= Ye_{xx}(1-\sigma^2) + \sigma(Ye_{yy} + \sigma Ye_{zz}) - \sigma(-\sigma Ye_{yy} - Ye_{zz}) \\ &= Ye_{xx}(1-\sigma^2) + \sigma[Ye_{yy}(1+\sigma) + Ye_{zz}(1+\sigma)] \\ &= Y(1+\sigma)[e_{xx}(1-\sigma) + \sigma(e_{yy} + e_{zz})] \\ &= Y(1+\sigma)(1-\sigma)\left[e_{xx} + \frac{\sigma}{(1-\sigma)}(e_{yy} + e_{zz})\right]. \end{aligned}$$

And denominator

$$\begin{aligned} &= 1\begin{vmatrix} 1 & -\sigma \\ -\sigma & 1 \end{vmatrix} + \sigma\begin{vmatrix} -\sigma & -\sigma \\ -\sigma & 1 \end{vmatrix} - \sigma\begin{vmatrix} -\sigma & 1 \\ -\sigma & -\sigma \end{vmatrix} \\ &= (1-\sigma^2) + \sigma(-\sigma - \sigma^2) - \sigma(\sigma^2 + \sigma) = (1+\sigma)(1-\sigma-2\sigma^2) \\ &= (1+\sigma)^2(1-2\sigma) \end{aligned}$$

$$\therefore\ P_1 = \frac{(1-\sigma)Y}{(1+\sigma)(1-2\sigma)}\left[e_{xx} + \frac{\sigma}{(1-\sigma)}(e_{yy}+e_{zz})\right] \tag{9.12.6}$$

$$\begin{aligned}
\text{or } P_1 &= \frac{Y}{(1+\sigma)(1-2\sigma)}e_{xx} - \frac{\sigma Y}{(1+\sigma)(1-2\sigma)}e_{xx} + \frac{\sigma Y}{(1+\sigma)(1-2\sigma)}e_{yy} + \frac{\sigma Y}{(1+\sigma)(1-2\sigma)}e_{zz} \\
&= \left[\frac{Y}{(1+\sigma)(1-2\sigma)}e_{xx} - \frac{2\sigma Y}{(1+\sigma)(1-2\sigma)}e_{xx}\right] \\
&\quad + \left[\frac{\sigma Y}{(1+\sigma)(1-2\sigma)}e_{xx} + \frac{\sigma Y}{(1+\sigma)(1-2\sigma)}e_{yy} + \frac{\sigma Y}{(1+\sigma)(1-2\sigma)}e_{zz}\right] \\
&= \frac{Y}{(1+\sigma)}e_{xx} + \frac{\sigma Y}{(1+\sigma)(1-2\sigma)}(e_{xx}+e_{yy}+e_{zz}) \\
&= 2n'e_{xx} + \lambda\delta
\end{aligned} \tag{9.12.7}$$

where $2n' = \dfrac{Y}{(1+\sigma)}$; $\delta = (e_{xx}+e_{yy}+e_{zz}$; $\lambda = \dfrac{\sigma Y}{(1+\sigma)(1-2\sigma)}$.

Similarly it can be shown that

$$\left.\begin{aligned}
P_2 &= \frac{(1-\sigma)Y}{(1+\sigma)(1-2\sigma)}\left[e_{yy} + \frac{\sigma}{(1-\sigma)}(e_{zz}+e_{xx})\right] \\
\text{and}\quad P_3 &= \frac{(1-\sigma)Y}{(1+\sigma)(1-2\sigma)}\left[e_{zz} + \frac{\sigma}{(1-\sigma)}(e_{xx}+e_{yy})\right]
\end{aligned}\right\} \tag{9.12.8}$$

Thus P_1, P_2 and P_3 may written in the form

$$\left.\begin{aligned}
P_1 &= 2n'e_{xx} + \lambda\delta \\
P_2 &= 2n'e_{yy} + \lambda\delta \\
P_3 &= 2n'e_{zz} + \lambda\delta
\end{aligned}\right\}. \tag{9.12.9}$$

Adding equations (9.12.9), we get

$$P_1 + P_2 + P_3 = \delta[2n' + 3\lambda].$$

If $P_1 = P_2 = P_3 = P$, $\delta = \frac{3P}{2n'+3\lambda}\cdot \delta$ measures, to a first order, the fractional change in volume, i.e., the volume strain, since it is the sum of the principal elongations under a uniform pressure (compression or dilatation) in all directions.

Now, the bulk modulus $k = \frac{P}{\delta}$.

Hence

$$k = \frac{P}{\frac{3P}{2n'+3\lambda}} = \lambda + \frac{2}{3}n' = \frac{\sigma Y}{(1+\sigma)(1-2\sigma)} + \frac{Y}{3(1+\sigma)} = \frac{Y(1+\sigma)}{3(1+\sigma)(1-2\sigma)} = \frac{Y}{3(1-2\sigma)}$$

$$\text{or}\quad Y = 3k(1-2\sigma) \tag{9.12.10}$$

This gives us a relation between bulk modulus, Young's modulus and Poisson's ratio. It now remains to identity n' with n, the rigidity modulus.

Considering a cube (as shown in figure 9.12.2) which is subjected to a simple stress of magnitude P on four of its sides, the stress across the diagonal plane $ACHF$ will be compressive and of magnitude P. Across the diagonal plane $BDIG$, it will be tensile or extensive and of magnitude P. Thus taking the axes OX, OY, OZ as parallel to O_1B, O_1A, O_1O_2, the stress system is given by $X_x = P$, $Y_y = -P$, $Z_z = 0$; $X_y = Y_z = Z_x = 0$ and the corresponding strains

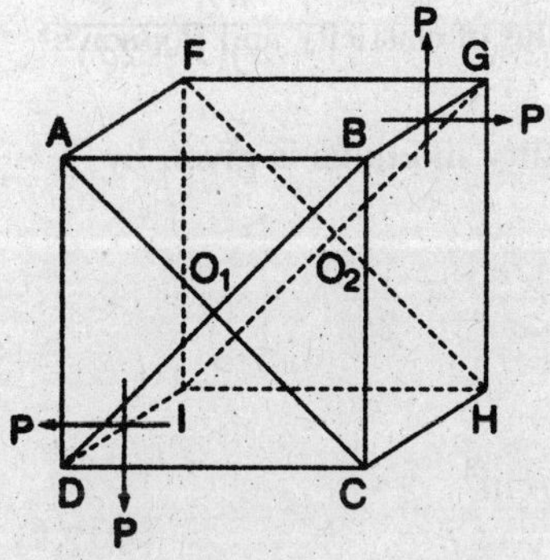

Fig. 9.12.2

$$e_{xx} = -e_{yy} = \frac{(1+\sigma)P}{Y}, \quad e_{zz} = 0.$$

After the strain has taken place

$$\tan DA'O_1' = \frac{DO_1'}{A'O_1'} = \frac{(1+e_{xx})DO_1}{(1+e_{yy})AO_1} = \frac{1+e_{xx}}{1+e_{yy}} \tag{9.12.11}$$

Now shearing strain (which is the strain produced in the right angle DAB) is given by

$$\theta = \angle DA'B' - \angle DAB = 2\angle DA'O_1' - \frac{\pi}{2}$$

where O_1' denotes the point of intersection of the lines $A'C$ and DB' in the strained cube (Fig. 9.10.1)

$$\therefore \quad \tan\frac{\theta}{2} = \tan\left(\angle DA'O_1' - \frac{\pi}{4}\right) = \frac{\tan DA'O_1' - 1}{1+\tan DA'O_1'} = \frac{\frac{1+e_{xx}}{1+e_{yy}} - 1}{1+\frac{1+e_{xx}}{1+e_{yy}}}$$

$$= \frac{e_{xx} - e_{yy}}{2 + e_{xx} + e_{yy}} = \frac{1}{2}(e_{xx} - e_{yy})$$

[neglecting e_{xx}, e_{yy} in the denominator which are very small in comparison with 2].

Therefore, for small angle of shear,

$$\theta = e_{xx} - e_{yy} = \frac{P}{Y}(1+\sigma) + \frac{P}{Y}(1+\sigma) = \frac{2P}{Y}(1+\sigma). \tag{9.12.12}$$

Now, modulus of rigidity n is defined by

$$n = \frac{P}{\theta} = \frac{Y}{2(1+\sigma)} \tag{9.12.13}$$

and by comparison with n' in equation (9.12.7) we see that $n' = n$.

Eliminating σ from equations (9.12.10) and (9.12.13), we obtain

$$Y = \frac{9nk}{3k+n} \quad \text{or} \quad \frac{1}{Y} = \frac{3k+n}{9nk} = \frac{1}{3n} + \frac{1}{9k}.$$

Also eliminating Y from equations (9.12.10) and (9.12.13), we get

$$\sigma = \frac{3k-2n}{6k+2n}.$$

Illustrative Examples

1. Young's modulus of a substance is equal to 7 × 10^{11} dynes/cm^2, and the rigidity modulus for the same substance is equal to 3 × 10^{11} dynes/cm^2. Calculate the bulk modulus of elasticity and Poisson's ratio for the substance.

The relation involving Young's modulus, bulk modulus and rigidity modulus is given by

$$Y = \frac{9nk}{3k + n}.$$

Here

$$Y = 7 \times 10^{11} \text{ dynes/cm}^2; \quad n = 3 \times 10^{11} \text{ dynes/cm}^2$$

$$\therefore \quad 7 \times 10^{11} = \frac{9 \times 3 \times 10^{11} \times k}{3k + 3 \times 10^{11}};$$

$$(21k + 21 \times 10^{11}) = 27k \quad \text{or} \quad k = 3.5 \times 10^{11} \text{ dynes/cm}^2.$$

Again, $Y = 2n(1 + \sigma)$.

$$\therefore \quad 7 = 2 \times 3(1 + \sigma) \quad \text{or} \quad \sigma = \frac{1}{6} = 0.166.$$

2. A body is under stress along three perpendicular axes OX, OY, OZ such that there is no linear strain along OY or OZ. It is then stressed along the same axes so that the linear strains along OY and OZ are equal and there is no strain along OX. Compare the ratio of the stress to strain along OX in the first case with that along OY or OZ in the second.

When there is no linear strain along OY or OZ axis, the linear strain along OX is given by

$$a = \frac{P}{Y}\left[\frac{1 - \sigma - 2\sigma^2}{1 - \sigma}\right]; \quad \text{[see Art.(9.12): eqn. (9.12.3)]}$$

$\therefore$ The ratio of stress to strain along OX is given by

$$\frac{P}{a} = \frac{Y(1 - \sigma)}{1 - \sigma - 2\sigma^2}.$$

Again the linear strain along OY or OZ is given by

$$b = \frac{P_1}{Y} - \frac{\sigma P}{Y} - \frac{\sigma P_1}{Y}, \quad \text{[see Art. (9.12): eqn. (9.12.2)]}$$

when there is no linear strain along OX we have

$$\frac{1}{Y}(P - 2\sigma P_1) = 0 \quad \text{or} \quad P = 2\sigma P_1$$

$$\therefore \quad b = \frac{P_1}{Y} - \frac{\sigma P_1}{Y} - \frac{2\sigma^2 P_1}{Y} = \frac{P_1}{Y}(1 - \sigma - 2\sigma^2)$$

$\therefore$ The ratio of stress to strain is $\frac{P_1}{b} = Y/(1 - \sigma - 2\sigma^2)$.

Thus, the ratio of stress to strain in the first case : the ratio of stress to strain in the second case : : $(1-\sigma) : 1$.

3. A body is under stress along three mutually perpendicular directions OX, OY and OZ. (i) Deduce the condition under which the linear strains along OX and OY will be equal while that along OZ will be zero. (ii) Also find the condition under which the linear strains along OX and OY will be zero.

[C.U. (Hons.) 1980]

From (9.12.5) we have,

$$e_{xx} = \frac{1}{Y}[P_1 - \sigma(P_2 + P_3)]; \; e_{yy} = \frac{1}{Y}[P_2 - \sigma(P_3 + P_1)]; \; e_{zz} = \frac{1}{Y}[P_3 - \sigma(P_1 + P_2)].$$

(i) According to the problem

$$e_{zz} = \frac{1}{Y}[(P_3 - \sigma(P_1 + P_2)] = 0 \text{ or } P_3 = \sigma(P_1 + P_2) \text{ or } \sigma = \frac{P_3}{(P_1 + P_2)}. \quad (1)$$

Also

$$e_{xx} = e_{yy} \text{ or } \frac{1}{Y}[P_1 - \sigma(P_2 + P_3) = \frac{1}{Y}[P_2 - \sigma(P_3 + P_1)]$$

$$\text{or } P_1 - P_2 = \sigma(P_2 + P_3) - \sigma(P_3 + P_1) = \sigma(P_2 - P_1)$$

$$\text{or } \sigma = -1. \quad (2)$$

From (1) and (2) we have

$$\frac{P_3}{(P_1 + P_2)} = -1 \quad \text{or} \quad P_1 + P_2 = -P_3 \quad \text{or} \quad P_1 + P_2 + P_3 = 0$$

which is the required condition.

(ii) Here $e_{xx} = 0$ and $e_{yy} = 0$

$$\therefore \quad \frac{1}{Y}[P_1 - \sigma(P_2 + P_3)] = 0 \quad \text{or} \quad \sigma = \frac{P_1}{P_2 + P_3}.$$

Also

$$\frac{1}{Y}[P_2 - \sigma(P_3 + P_1)] = 0 \quad \text{or} \quad \sigma = \frac{P_2}{P_1 + P_3}.$$

$$\therefore \quad \frac{P_1}{P_2 + P_3} = \frac{P_2}{P_1 + P_3} \quad \text{or} \quad P_1^2 + P_1P_3 = P_2^2 + P_2P_3$$

$$\text{or} \quad P_1^2 - P_2^2 = P_2P_3 - P_1P_3 \quad \text{or} \quad (P_1 - P_2)(P_1 + P_2) = -P_3(P_1 - P_2)$$

$$\text{or} \quad (P_1 - P_2)(P_1 + P_2 + P_3) = 0.$$

$\therefore$ either $P_1 = P_2$ or $P_1 + P_2 + P_3 = 0$ is the required condition.

9.13 Torsion of a Cylinder or Wire

We consider a cylinder of length l and radius a (Fig. 9.13.1). Let the lower end of the cylinder be clamped and at the upper end a torque is applied. As a result of the torque the cylinder is twisted through a certain angle.

If we imagine the cylinder in question to be divided up into a number of co-axial discs, then as a result of twist these discs will turn relative to one another about the axis of the cylinder and the amount of turning of a disc depends on its position from the fixed end. The disc which is remote from the fixed end will turn more about the axis than the one nearer to the fixed end. We consider now an imaginary co-axial cylinder of radius r. A straight line AB is considered over the surface of this imaginary cylinder parallel to its axis OO'. After twisting the line gets inclined (as $A'B$) to the original line. The angle between the two lines measures the angle of shear. AA' represents the amount of turn of an element originally at A on the surface of the cylinder; A' represents the displaced position of the element. We have, therefore, $l\theta = AA'$. As r is the distance of the element from the axis of the cylinder and ϕ the angle of twist, we have, $r\phi = AA'$.

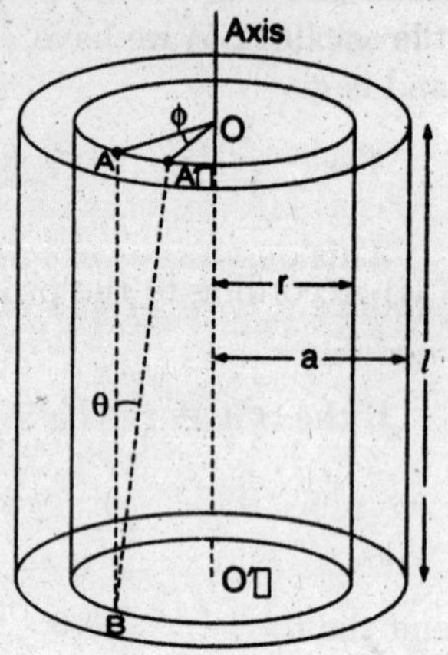

Fig. 9.13.1

$$\therefore \quad r\phi = l\theta \quad \text{or} \quad \theta = \frac{r\phi}{l}.$$

Now, due to elastic property of the material the relative turning of the co-axial disc gives rise to a restoring torque which tends to prevent the twisting of the cylinder.

To calculate this restoring torque we proceed in the following way:

We imagine again the cylinder to be divided up into a number of co-axial cylindrical shells. We take into consideration one such shell of radius r and thickness dr. The section of this shell looks like an annulus of radius r and thickness dr. On the face of this shell at the position A we imagine an element of length dx and breadth dr (Fig. 9.13.2).

Over this element let F be the tangential force that comes to play due to relative turning

$$\therefore \text{ Tangential stress} = \frac{F}{drdx}$$

$$\therefore \; n = \frac{F/drdx}{\theta} = \frac{F/drdx}{r\phi/l} = \frac{Fl}{\phi rdrdx}$$

where n is the rigidity modulus of the material of the cylinder

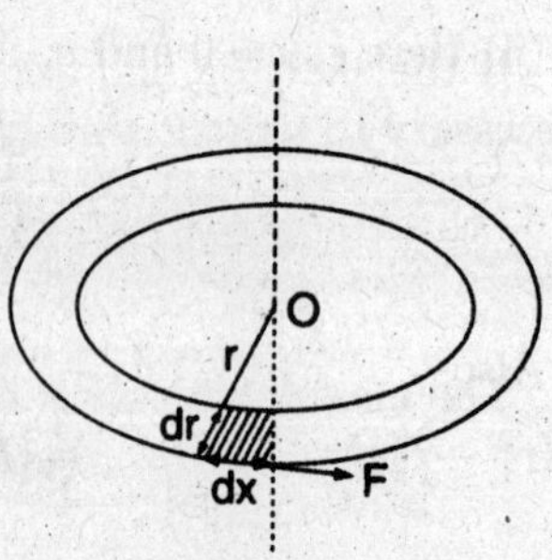

Fig. 9.13.2

$$\therefore \quad F = \frac{n\phi}{l}rdrdx$$

The moment of this force about the axis of the cylinder is given by

$$F\cdot r = \frac{n\phi}{l}r^2drdx.$$

Therefore, the restoring torque over the entire surface of the annulus is given by

$$\delta\Gamma = \frac{n\phi}{l}r^2dr\Sigma dx = \frac{2\pi n\phi}{l}r^3dr \quad [\because \Sigma dx = 2\pi r]$$

Therefore, the total restoring torque that comes to play during the twisting of the entire cylinder is

$$\Gamma = \frac{2\pi n\phi}{l}\int_0^a r^3 dr = \frac{2\pi n\phi}{l}\cdot\frac{a^4}{4} = \frac{\pi n a^4}{2l}\phi \tag{9.13.1}$$

The above expression for torque remains unchanged if we consider that the upper end is fixed and at the lower end a torque is applied. The restoring torque Γ is equal in magnitude to the applied torque. $\frac{\Gamma}{\phi}$ is called the *torsional constant* or the *torsional rigidity* τ of the cylinder and is given by

$$\tau = \frac{n\pi a^4}{2l} \tag{9.13.2}$$

This quantity is of great importance in torsional oscillations. Again the torque required to produce a twist of one radian in a cylinder of unit length, i.e., $\frac{1}{2}\pi n a^4$ is called the *torsional modulus.*

If the cylinder is hollow of outer and inner radii r_1 and r_2 respectively, then

$$\Gamma = \frac{2\pi n\phi}{l}\int_{r_2}^{r_1} r^3 dr = \frac{n\pi\phi}{2l}(r_1^4 - r_2^4)$$

and the torsional rigidity in this case is

$$\frac{n\pi(r_1^4 - r_2^4)}{2l}. \tag{9.13.3}$$

9.13.1 A Hollow shaft is stronger than a Solid one of the same mass

We consider a solid cylindrical shaft of length l, radius r, and of a material of rigidity modulus n. The torque Γ necessary to twist it through an angle of ϕ radian is

$$\Gamma = \frac{\pi n r^4\phi}{2l} = \frac{\pi n r^2(r^2\phi)}{2l}. \tag{9.13.4}$$

If the shaft be hollow with internal and external radii r_2 and r_1 respectively, the torque necessary to twist it through the same angle of ϕ radian will be

$$\Gamma' = \frac{\pi n(r_1^4 - r_2^4)\phi}{2l} = \frac{\pi n(r_1^2 - r_2^2)(r_1^2 + r_2^2)\phi}{2l}. \tag{9.13.5}$$

The masses (volume × density) and lengths of the hollow and solid shafts are equal, so that $\pi(r_1^2 - r_2^2)l\rho = \pi r^2 l\rho$, where ρ is the density of the material of the shaft.

$$\therefore \quad r_1^2 - r_2^2 = r^2 \tag{9.13.6}$$

Substituting r^2 for $(r_1^2 - r_2^2)$ in (9.13.5) and comparing with (9.13.4) we get

$$\frac{\Gamma'}{\Gamma} = \frac{r_1^2 + r_2^2}{r^2} \tag{9.13.7}$$

But we have already obtained $r_1^2 - r_2^2 = r^2$, so that $r_1^2 + r_2^2 > r^2$. Therefore from (9.13.7), we have

$$\frac{\Gamma'}{\Gamma} > 1 \quad \text{or} \quad \Gamma' > \Gamma$$

i.e., the torque necessary to twist a hollow cylinder is greater than the torque necessary to twist a solid cylinder of the same mass, length and material through the same angle. Hence a greater amount of energy will be stored in the hollow cylindrical shaft when twisted. For the same reason, the shaft in the form of a hollow cylinder will be stronger than a solid shaft.

9.14 Torsional Rigidities of Bars of Non-circular Cross-sections—St. Venant's Approximate Formula

The quantity $\dfrac{\Gamma l}{n\phi}$ in the expression (9.13.1) is equal to the geometrical moment of inertia of the section of the cylinder of radius a about its axis of symmetry. Here the geometrical moment of inertia $AK^2 = \pi a^2 . \dfrac{a^2}{2} = \dfrac{\pi a^4}{2}$, but in all other cases, where the section is other than circular, the quantity $\dfrac{\Gamma l}{n\phi}$ is less than AK^2. This has been studied by St. Venant.

When the rod is of elliptic cross-section with semi axes a and b

$$\frac{\Gamma l}{n\phi} = \frac{\pi a^3 b^3}{a^2 + b^2} \text{ and } AK^2 = \frac{\pi ab(a^2 + b^2)}{4}$$

When the cross-section of the rod is a square of each side equal to $2a$

$$\frac{\Gamma l}{n\phi} = 2.2492a^4 \text{ and } AK^2 = \frac{8a^4}{3} = 2.667a^4.$$

For a rectangular cross-section of sides $2a$ and $2b$

$$\frac{\Gamma l}{n\phi} = \frac{16ab^3}{3} - \frac{4^5 b^4}{\pi^5}\left\{\tan h\frac{\pi a}{2b} + \frac{1}{3^5}\tan h\frac{3\pi a}{2b} + \frac{1}{5^5}\tan h\frac{5\pi a}{2b} + \cdots\right\}$$

If $a > 3b$, the series of hyperbolic tangents can differ only in the fourth significant figure from $1 + \frac{1}{3^5} + \frac{1}{5^5} + \cdots$, the sum of which is approximately 1.0045.

Thus for a flat strip (where $a > 3b$), we have

$$\frac{\Gamma l}{n\phi} = \frac{16ab^3}{3} - \frac{4^5 b^4}{\pi^5} \times 1.0045 = \frac{16}{3}ab^3\left(1 - \frac{192}{\pi^5}\frac{b}{a} \times 1.0045\right) = \frac{16}{3}ab^3\left(1 - 0.630\frac{b}{a}\right).$$

For a rod having cross-section in the form of an equilateral triangle

$$\frac{\Gamma l}{n\phi} = \frac{9\sqrt{3}}{5}a^4,$$

where $3a$ is the height of the triangle and $AK^2 = \sqrt{3}\, a^4$.

Illustrative Examples

1. Two cylindrical shafts have the same length and mass and are made of the same material. One is solid, while the other, which is hollow, has an external radius twice the internal radius. Compare their torsional rigidities. **[C.U. (Hons.) '92]**

The mass of the solid cylinder of radius r, length l and density $\rho = \pi r^2 l\rho$. The mass of the hollow cylinder of same length and density having external radius $2r_1$ and internal radius r_1

$$= \pi[(2r_1)^2 - r_1^2]l\rho = 3\pi r_1^2 l\rho$$

$\because$ the masses of the two cylindrical shafts are same, we may write

$$\pi r^2 l\rho = 3\pi r_1^2 l\rho \quad \text{or} \quad r^2 = 3r_1^2 \tag{1}$$

Now the torsional rigidity of the solid cylindrical shaft is given by

$$\tau_1 = \frac{n\pi r^4}{2l}$$

For the hollow cylindrical shaft the torsional rigidity is given by

$$\tau_2 = \frac{n\pi[(2r_1)^4 - r_1^4]}{2l} = \frac{15n\pi r_1^4}{2l}. \quad \therefore \frac{\tau_1}{\tau_2} = \frac{r^4}{15r_1^4} = \frac{9}{15} = \frac{3}{5} \quad \left[\because \frac{r^4}{r_1^4} = 9 \text{ from (1)}\right]$$

2. (a) Prove that if a number of rods of torsional rigidities $\tau_1, \tau_2, \tau_3, \cdots$, etc., are joined end to end, the torsional rigidity of the combination is given by **[C.U. (Hons.) 2007]**

$$\frac{1}{\tau} = \frac{1}{\tau_1} + \frac{1}{\tau_2} + \frac{1}{\tau_3} + \cdots.$$

We know that the applied torque is given by

$$\Gamma = \frac{n\pi a^4\phi}{2l}.$$

Hence for rods having lengths, radii, rigidity moduli and angles of twist given respectively by l_1, a_1, n_1, ϕ_1; l_2, a_2, n_2, ϕ_2; $\cdots$; etc., we have

$$\Gamma = \frac{\pi n_1 a_1^4 \phi_1}{2l_1} = \frac{\pi n_2 a_2^4 \phi_2}{2l_2} = \frac{\pi n_3 a_3^4 \phi_3}{2l_3} = \cdots$$

$$\therefore \quad \phi_1 = \frac{2\Gamma l_1}{\pi n_1 a_1^4} = \frac{\Gamma}{\tau_1} \quad \left[\because \tau_1 = \frac{n_1 \pi a_1^4}{2l_1}\right]$$

$$\phi_2 = \frac{2\Gamma l_2}{\pi n_2 a_2^4} = \frac{\Gamma}{\tau_2} \quad \left[\because \tau_2 = \frac{n_2 \pi a_2^4}{2l_2}\right]$$

$$\phi_3 = \frac{2\Gamma l_3}{\pi n_3 a_3^4} = \frac{\Gamma}{\tau_3} \quad \left[\because \tau_3 = \frac{n_3 \pi a_3^4}{2l_3}\right]$$

$$\therefore \quad \phi_1 + \phi_2 + \phi_3 + \cdots = \Gamma\left(\frac{1}{\tau_1} + \frac{1}{\tau_2} + \frac{1}{\tau_3} + \cdots\right) \quad \text{or} \quad \frac{\Gamma}{\tau} = \Gamma\left(\frac{1}{\tau_1} + \frac{1}{\tau_2} + \frac{1}{\tau_3} + \cdots\right),$$

where τ is the torsional rigidity of the combination.

$$\therefore \quad \frac{1}{\tau} = \frac{1}{\tau_1} + \frac{1}{\tau_2} + \cdots$$

(b) Applying the result of 2(a) or otherwise, find the torsional rigidity of a uniformly tapering wire (end radii a and b, $a > b$) of length l in terms of the rigidity modulus of the material of the wire.

The radius of the uniformly tapering wire at a distance x from the fixed end B of radius b is

$$r = b + \frac{(a-b)}{l}x.$$

If the torsional rigidity of the elementary portion of the rod of thickness dx at a distance x from the fixed end B be τ_x and $d\phi$ be the angle of twist between the ends of this element, the applied torque Γ is given by $\Gamma = \tau_x d\phi = \dfrac{\pi n r^4}{2dx} d\phi$

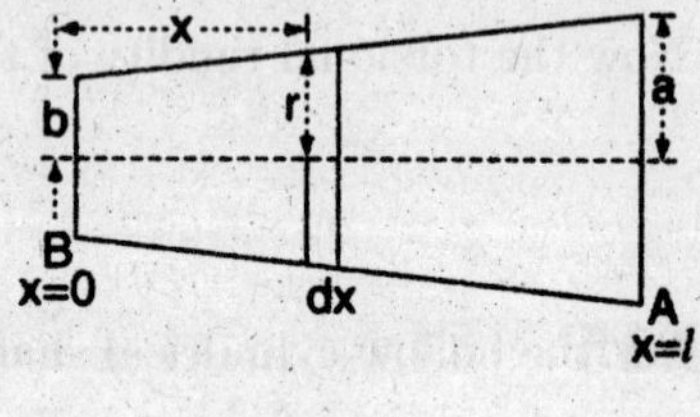

$$\therefore \quad \frac{1}{\Gamma}\int_0^{\phi} d\phi = \int_0^l \frac{2dx}{\pi n r^4} = \frac{2}{\pi n}\int_0^l \frac{dx}{\left[b + \frac{(a-b)}{l}x\right]^4}. \qquad (1)$$

To integrate, let us put

$$z = b + \frac{(a-b)}{l}x. \quad \therefore \quad dz = \frac{(a-b)}{l}dx$$

and when $x = 0$, $z = b$, also when $x = l$, $z = a$

$$\therefore \quad \frac{\phi}{\Gamma} = \frac{2}{\pi n}\frac{l}{(a-b)}\int_b^a z^{-4}dz = -\frac{2}{3\pi n}\frac{l}{(a-b)}\left(\frac{1}{z^3}\right)_b^a = -\frac{2}{3\pi n}\frac{l}{(a-b)}\left(\frac{1}{a^3} - \frac{1}{b^3}\right)$$

$$= \frac{2}{3\pi n}\frac{l}{(a-b)}\left(\frac{1}{b^3} - \frac{1}{a^3}\right) = \frac{2}{3\pi n} l \frac{a^2 + ab + b^2}{a^3 b^3}.$$

Therefore the torsional rigidity of the tapering wire is

$$\tau = \frac{\Gamma}{\phi} = \frac{3}{2}\frac{\pi n a^3 b^3}{l}\frac{1}{a^2 + ab + b^2}. \qquad (2)$$

We can also obtain the same expression by applying the result of 2(a) directly.

$$\frac{1}{\tau} = \sum \frac{1}{\tau_x} = \int_0^l \frac{2dx}{\pi n r^4} = \frac{2}{\pi n}\int_0^l \frac{dx}{\left[b + \frac{(a-b)}{l}x\right]^4}$$

whence we may get the expression (2)

3. A solid cylindrical rod has half the length as a hollow cylindrical rod (of the same material and same mass) which has an external radius $\sqrt{2}$ times the internal radius. Show that the torsional rigidities of the two rods are in the ratio of 8 : 3. **[C.U. (Hons.) 1979, cf. 1992, 2008]**

The mass of the solid cylinder of radius r, length l and density $\rho = \pi r^2 l \rho$. The mass of the hollow cylinder of same density and length $2l$ and having external radius $\sqrt{2}\, r_1$ and internal radius r_1

$$= \pi\left[\left(\sqrt{2}\, r_1\right)^2 - r_1^2\right] \times 2l\rho = 2\pi r_1^2 l\rho$$

$\because$ the masses of the two cylinders are same, we may write

$$\pi r^2 l\rho = 2\pi r_1^2 l\rho \quad \text{or} \quad r^2 = 2r_1^2.$$

The torsional rigidity of the solid cylinder is given by $\tau = \frac{n\pi r^4}{2l}$.

The torsional rigidity of the hollow cylinder is given by

$$\tau' = \frac{n\pi}{4l}\left[\left(\sqrt{2}\,r_1\right)^4 - r_1^4\right] = \frac{n\pi}{4l}\cdot(3r_1^4)$$

$$\therefore \quad \frac{\tau}{\tau'} = \frac{2r^4}{3r_1^4} = \frac{2}{3} \times 4 \quad \left[\because \frac{r^4}{r_1^4} = 4\right]$$

$$= \frac{8}{3}$$

$$\therefore \tau : \tau' :: 8 : 3.$$

9.15 Elasticities of Liquids and Gases

A solid can resist both shearing and volume strain. A liquid, however, cannot resist a shearing strain, but it can resist volume strain like solids. A gaseous substance also cannot resist shearing strain like liquids, but can resist volume strain to a comparatively small extent. Thus a solid possesses modulus of rigidity, Young's modulus and bulk modulus and a fluid (liquid and gas) possesses only bulk modulus of elasticity. In a fluid though there is no tangential stress, yet in practice, relative motion of the fluid is accompanied by tangential forces tending to resist the motion which persists as long as the motion continues. The tangential force called into play when the fluid is in motion is due to a property known as *viscosity of the fluid.* Viscosity may be regarded as a transient type of rigidity.

9.16 Isothermal and Adiabatic Elasticities

For a perfect gas the relation between the pressure P, volume V and absolute temperature T is given by $PV = RT$.

When the temperature is constant, we may write the above equation as PV = constant.

Differentiating the above equation, we have

$$PdV + VdP = 0 \quad \text{or} \quad P = -V\left(\frac{dP}{dV}\right) = k_T,$$

where k_T denotes the bulk modulus of elasticity of the gas under isothermal condition. When however the gas is adiabatically compressed, the heat generated due to compression remains in the gas and the relation between P and V for a perfect gas is then given by

$$PV^\gamma = \text{constant},$$

where γ is the ratio of the sp. ht. of the gas at constant pressure to its sp. ht. at constant volume.

Differentiation of the above equation gives

$$P\gamma V^{\gamma-1} + V^\gamma \frac{dP}{dV} = 0 \quad \text{or} \quad \gamma P = -V\left(\frac{dP}{dV}\right) = k_S,$$

where k_s denotes the bulk modulus of elasticity of the gas under adiabatic condition. The ratio of the adiabatic elasticity to the isothermal elasticity is thus given by $\frac{k_S}{k_T} = \gamma$; γ depends on the nature of the gas and it is 1.66 for a monatomic gas and 1.41 for air and other diatomic gases.

9.17 Bending of Beams

When a beam is bent by an applied torque, tensile forces act on some layers of the beam, and compressional forces act on some other layers as a result of which filaments of the beam nearest the outside curve of the bent beam are extended and the filaments nearest the inside curve get compressed. Internal forces come into play to counteract the effect of bending.

9.17.1 Internal bending moment

Let us consider a portion of the bent beam and imagine the beam to be divided up into straight parallel longitudinal filaments. Before bending all the filaments are of same lengths and the sections of this portion at the ends are parallel. But after bending the two end faces are inclined to each other at an angle ϕ as shown in Fig. 9.17.1

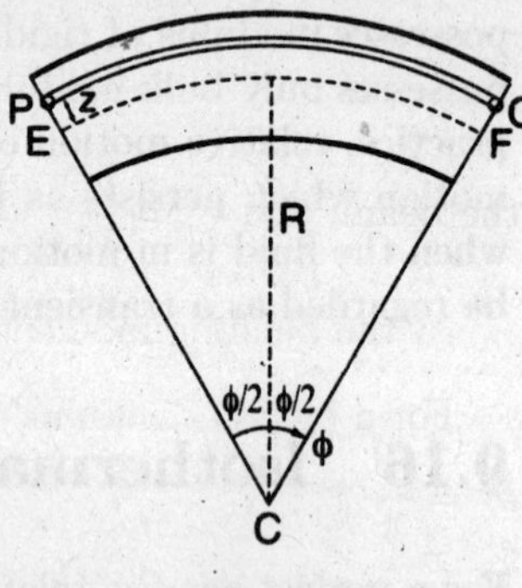

Fig. 9.17.1

From the bending of the beam it appears that some of the filaments get shortened and some elongated and between them there is a surface, the filaments over which maintain the original length. This surface is called *neutral surface* and the section of this surface in the plane of the figure is called *neutral axis* as shown by the dotted line and a straight line through C perpendicular to the plane of the figure is called the *axis of bending*. Centres of curvature of different bent filaments constituting the beam lie on this line.

Let us consider a filament at a distance z from the neutral axis beyond the centre of curvature. In this case, this filament gets elongated. Let R be the radius of curvature of the neutral axis. Then the original length of the filament in question is $R\phi$ and the present elongated length is $(R + z)\phi$. Therefore, the linear extension of the filament $= (R + z)\phi - R\phi = z\phi$.

The linear extensional strain $= \dfrac{z\phi}{R\phi} = \dfrac{z}{R}$.

Let α be the sectional area of the filament in question and f be the longitudinal force that comes to play across the section of the filament to resist elongation. We have then $f = Y\alpha\frac{z}{R}$, where Y is the Young's modulus of the material of the beam.

Thus the filament in question is acted on by a force f to resist its elongation that arises out of the elastic properties of the material. Similarly for a filament distant z below the neutral axis towards the centre, a resisting force will come into play to resist shortening of the filament. These two forces are oppositely directed. Since the two filaments are at equal distances from the neutral axis and on opposite sides of it, they constitute a couple acting across the section. Taking different pairs of equidistant filaments, we see that across the section a system of resisting couples come into play and the resultant of these system of couples constitutes what is called *internal bending moment.* This is equal in magnitude to the external bending moment.

Let us now calculate the moment of the force f (as already mentioned) at a distance z from the neutral axis. The moment of this force is $f \cdot z$ about the neutral axis, and $\Sigma f \cdot z$ is the internal resisting moment that opposes the external bending moment. Now, we know that

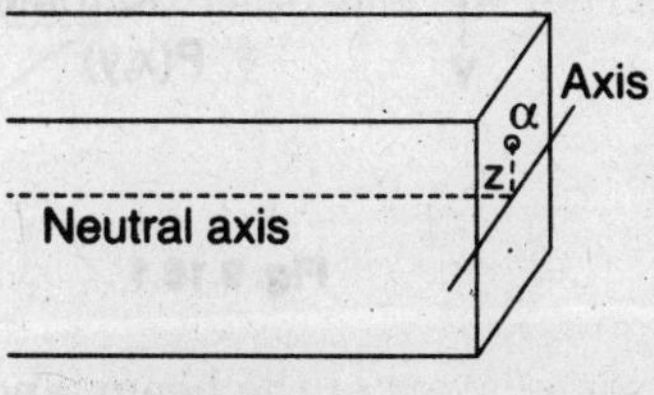

Fig. 9.17.2

$$f = \frac{Y\alpha z}{R}. \quad \therefore \Sigma f z = \frac{Y}{R}\Sigma \alpha z^2,$$

but $\Sigma \alpha z^2$ signifies the geometrical moment of inertia of the section of the beam about the axis passing through the centre of the section, coinciding with the plane of the section, and parallel to the breadth of the beam as shown in Fig. 9.17.2. $\Sigma \alpha z^2$ is equal to Ak^2, where A is the sectional area of the beam and k the radius of gyration.

Therefore, the moment of the internal resisting couple $= \frac{YAk^2}{R}$ = the *internal bending moment.*

The quantity YAk^2 is called the ***flexural rigidity*** of the beam and is defined as the external bending moment required to produce unit radius of curvature. It measures the *resistance of the beam* to bending.

The general equation of bending is therefore written as $\dfrac{YAk^2}{R}$ = applied bending moment or external bending moment. (9.17.1)

For a beam of rectangular cross-section, $A = ab$, where a is the breadth, b the thickness of the beam, and $k^2 = \dfrac{b^2}{12}$.

$\therefore$ the bending moment for rectangular cross-section $= \dfrac{Yab^3}{12R}$.

For a beam of circular cross-section $A = \pi r^2$, where r is the radius of the cross-section and $k^2 = \dfrac{r^2}{4}$.

$\therefore$ the bending moment for circular cross-section $= \dfrac{Y\pi r^4}{4R}$.

N.B. Beam bent by a constant couple: From our above discussions, it is clear that the beam will be bent into an arc of a circle when the torque is uniform throughout the beam. We consider a beam AB (Fig. 9.17.3) bent by applying equal couples (M, M) at its two ends A and B. For equilibrium of the portion AC, the internal torque applied on AC by BC must be equal and opposite to the torque M applied at A. If the radius of curvature at C be R, then

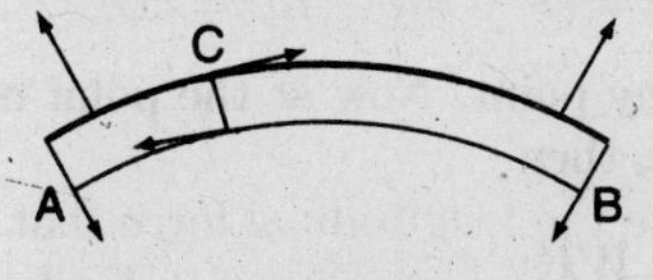

Fig. 9.17.3

$$\frac{YAk^2}{R} = M \quad \text{or} \quad R = \frac{YAk^2}{M} \tag{9.17.2}$$

C being any arbitrary section of the beam.

The radius of curvature R of the beam will be same throughout the beam, i.e., the beam will be bent into an arc of a circle of radius R given by (9.17.2).

9.18 Cantilever

A cantilever is a uniform beam fixed horizontally at one end and loaded at the other end. We consider a light cantilever fixed at O and loaded at the free end by a concentrated weight W.

Let $P(x, y)$ be any point on the beam and l be the length of the beam. The applied bending moment due to the load W about the point P which is at a distance x from the fixed end $= W(l - x)$.

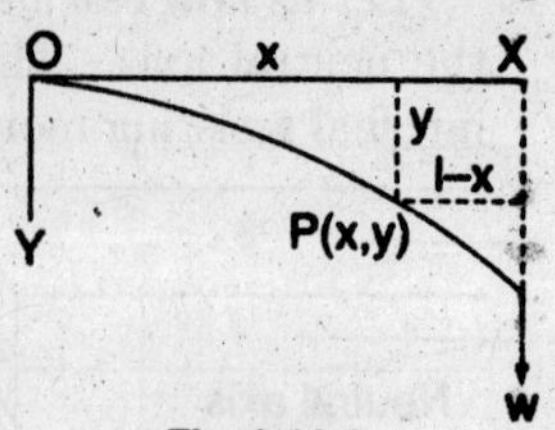

Fig. 9.18.1

Hence the bending moment equation is given by

$$\frac{YAk^2}{R} = W(l - x) \quad \left[\because \text{ the cantilever, being light is supposed to be weight-less}\right]$$

The left-hand side represents the internal bending moment.

$$\text{or} \quad YAk^2\frac{d^2y}{dx^2} = W(l - x) \quad \left[\text{when the bending is small the curvature } \tfrac{1}{R} \text{ at } P \text{ may be taken to be equal to } \tfrac{d^2y}{dx^2}.\right]$$

On integration, we have,

$$YAk^2\frac{dy}{dx} = W\left(lx - \frac{x^2}{2}\right) + C_1, \; C_1 \text{ being the integration constant.}$$

Now, $\frac{dy}{dx} = 0$, when $x = 0$, because at the fixed end the bar remains horizontal.

$$\therefore \quad C_1 = 0.$$

We have then,

$$YAk^2\frac{dy}{dx} = W\left(lx - \frac{x^2}{2}\right).$$

On further integration, we have

$$YAk^2y = W\left(\frac{lx^2}{2} - \frac{x^3}{6}\right) + C_2, \; C_2 \text{ being another integration constant.}$$

Again, $y = 0$, when $x = 0$, and therefore, $C_2 = 0$

$$\therefore YAk^2y = W\left(\frac{lx^2}{2} - \frac{x^3}{6}\right).$$

This gives the expression for the depression of the beam at any point. Now at the point of loading, $x = l$, and let δ be the depression of the beam. We have, then

$$YAk^2\delta = W\left(\frac{l^3}{2} - \frac{l^3}{6}\right) = \frac{Wl^3}{3} \quad \text{or} \quad \delta = \frac{Wl^3}{3YAk^2}. \tag{9.18.1}$$

If the beam be of rectangular cross-section of breadth a and thickness b, we have

$$Ak^2 = \frac{ab^3}{12}. \quad \therefore \quad \delta = \frac{Wl^3}{3Y \cdot \frac{ab^3}{12}} = \frac{4Wl^3}{Yab^3} \tag{9.18.2}$$

Now, due to the load W each element of the beam experiences a shearing stress of magnitude $\frac{W}{ab}$ as shown in Fig. 9.18.2.

As a result of this stress there will be depression at each point of the beam. If l be the length of the beam and δ_1 be the depression at the end due to shear, we have,

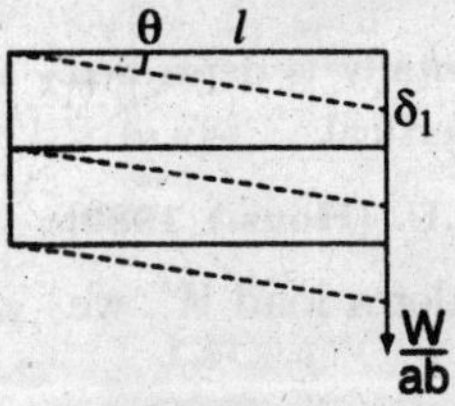

Fig. 9.18.2

$$\frac{\delta_1}{l} = \theta \quad \therefore\ n = \frac{\frac{W}{ab}}{\frac{\delta_1}{l}} = \frac{Wl}{\delta_1 ab},$$

n being the modulus of rigidity and θ the shearing strain

$$\therefore\ \delta_1 = \frac{Wl}{nab}. \quad \therefore\ \frac{\delta_1}{\delta} = \frac{Wl}{nab}\bigg/\frac{4Wl^3}{Yab^3} = \frac{Y}{4n}\left(\frac{b}{l}\right)^2 = \frac{Y}{n}\left(\frac{b}{2l}\right)^2.$$

Now, b is negligibly small in comparison with l. Hence δ_1 is insignificant in comparison with δ. This is why, we usually do not take into consideration the depression of the cantilever due to shear.

In the above deduction we have ignored the weight of the cantilever. Let us now suppose that the beam has weight w per unit length and carries a concentrated load W at its free end (Fig. 9.18.3).

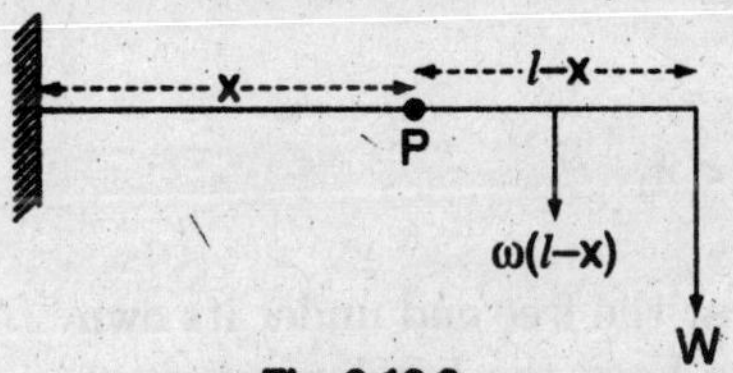

Fig. 9.18.3

Then the bending moment equation at P is given by

$$YAk^2\frac{d^2y}{dx^2} = W(l-x) + \frac{\omega(l-x)(l-x)}{2}$$

$$= W(l-x) + \frac{\omega}{2}(l-x)^2$$

$$\text{or}\quad YAk^2\frac{d^2y}{dx^2} = W(l-x) + \frac{\omega}{2}(l^2 + x^2 - 2lx).$$

Integrating, we have,

$$YAk^2\frac{dy}{dx} = W\left(lx - \frac{x^2}{2}\right) + \frac{\omega}{2}\left(l^2x + \frac{x^3}{3} - lx^2\right) + C_1.$$

On further integration, we have

$$YAk^2y = W\left(\frac{lx^2}{2} - \frac{x^3}{6}\right) + \frac{\omega}{2}\left(\frac{l^2x^2}{2} + \frac{x^4}{12} - \frac{lx^3}{3}\right) + C_1x + C_2.$$

Now, at $x = 0$, $\frac{dy}{dx} = 0$ and $y = 0$

$$\therefore\ C_1 = C_2 = 0$$

$$\therefore\ YAk^2y = W\left(\frac{lx^2}{2} - \frac{x^3}{6}\right) + \frac{\omega}{2}\left(\frac{l^2x^2}{2} + \frac{x^4}{12} - \frac{lx^3}{3}\right)$$

If δ be the depression of the beam at the free end where the load is applied, we have

$$YAk^2\delta = W\left(\frac{l^3}{2} - \frac{l^3}{6}\right) + \frac{\omega}{2}\left(\frac{l^4}{2} + \frac{l^4}{12} - \frac{l^4}{3}\right) = \frac{Wl^3}{3} + \frac{\omega l}{2}\cdot\frac{l^3}{4} = \frac{Wl^3}{3} + \frac{W_1l^3}{8}$$

where W_1 is the weight of the beam

$$\therefore\ \delta = \frac{l^3}{3YAk^2}\left(W + \frac{3}{8}W_1\right) \tag{9.18.3}$$

Illustrative Examples

1. The free end of a cantilever beam of rectangular cross-section mounted horizontally is depressed less when the longer side of the rectangle is vertical than when the shorter side is vertical.

[C.U. (Hons.) 1980]

Case I: Considering a light cantilever being depressed at the free end under a load W, we have the depression at the free end given by

$$\delta_1 = \frac{4Wl^3}{Yab^3},$$

where l is the length, a is the breadth, and b is the thickness of the cantilever beam—the smaller side b (considering $a > b$) being kept vertical.

Again when the longer side of the rectangle is vertical, the depression at the free end is given by

$$\delta_2 = \frac{4Wl^3}{Yba^3},$$

so that we have

$$\frac{\delta_1}{\delta_2} = \frac{a^2}{b^2}, \quad \text{i.e.,} \quad \frac{\delta_1}{\delta_2} > 1 \quad \because a > b \text{ or } \delta_2 < \delta_1.$$

Case II: Considering a very heavy cantilever being depressed at the free end under its own weight ωl (ω being the weight per unit length of the beam), we have the bending moment equation about the point P at a distance x from the fixed end, given by

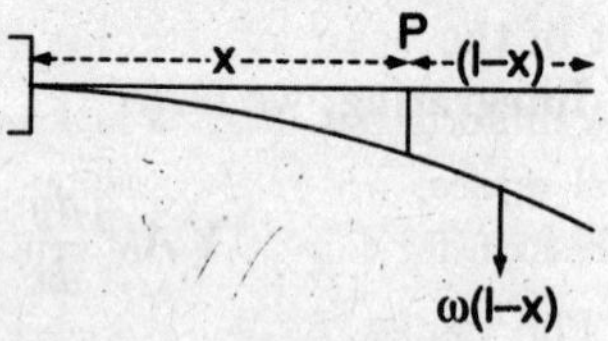

$$YAk^2 \frac{d^2y}{dx^2} = \frac{\omega}{2}(l-x)^2,$$

all other notations having usual significance.

Integrating the above equation twice, and putting the usual boundary conditions, we get the depression at the free end given by

$$YAk^2\delta = \frac{\omega l^4}{8} \quad \text{or} \quad \delta = \frac{\omega l^4}{8YAk^2}.$$

When the breadth is a and the depth is b, i.e., b is vertical, then $Ak^2 = \frac{ab^3}{12}$, so that

$$\delta_1 = \frac{\omega l^4 \times 12}{8Yab^3}. \tag{1}$$

Again when the breadth is b and depth is a, i.e., a is vertical then $Ak^2 = \frac{ba^3}{12}$, so that

$$\delta_2 = \frac{\omega l^4 \times 12}{8Yba^3} \tag{2}$$

From (1) and (2), we have $\frac{\delta_1}{\delta_2} = \frac{a^2}{b^2} > 1 \quad \because a > b$, i.e., $\delta_1 > \delta_2$ or $\delta_2 < \delta_1$.

2. Show that, for the same cross-sectional areas, the beam of a square cross-section is stiffer than one of circular cross-section, of the same length and material. Show also that for a given load, the depressions in the two cases are in the ratio $3 : \pi$.

The depression δ at the loaded end of a light canti-lever beam, whose other end is clamped horizontally, is given by $\delta = \frac{4Wl^3}{Yab^3}$, where W is the load; l, the length of the beam, and a and b are respectively the breadth and depth of its section. If the beam has a square section, $a = b$, and the corresponding depression at the loaded end is

$$\delta_1 = \frac{4Wl^3}{Yb^4}. \tag{1}$$

Again if the beam has a circular cross-section of radius r, then $Ak^2 = \frac{\pi r^4}{4}$, and the corresponding depression at the free end is

$$\delta_2 = \frac{Wl^3}{3YAk^2} = \frac{Wl^3}{3Y\frac{\pi r^4}{4}} = \frac{4Wl^3}{3Y\pi r^4} \tag{2}$$

From (1) and (2), we have

$$\frac{\delta_1}{\delta_2} = \frac{3\pi r^4}{b^4}.$$

But from the condition of the problem, the cross-sectional areas of the two beams are equal, so that we may write $b^2 = \pi r^2$ and hence

$$\frac{\delta_1}{\delta_2} = \frac{3\pi r^4}{(\pi r^2)^2} = \frac{3}{\pi} < 1 \quad \therefore \delta_1 < \delta_2.$$

Thus the depression at the loaded end of the beam of square cross-section is smaller than that of the beam of circular cross-section. Hence the former is stiffer than the latter.

3. A uniform bar $ABCD$ is supported horizontally at B and C, where $AB = \frac{1}{4}BC = CD = l$. If two equal masses m, m are suspended at A and D, the bar rises by height h in the middle. Deduce an expression for h in terms of relevant quantities.

The bar $ABCD$ is placed horizontally upon two knife-edges at B and C. The bar bends on suspending weights mg, mg at A and D, and the bent position of the bar is shown in the figure. The reaction at each knife-edge is also mg. The couple acting at each end is

$$\Gamma = mgl \quad \text{where} \quad l = AB = CD.$$

The bar bends into a circular arc of radius R. Let us complete the circle. EF is the elevation (h). Now, from the property of the circle, we have

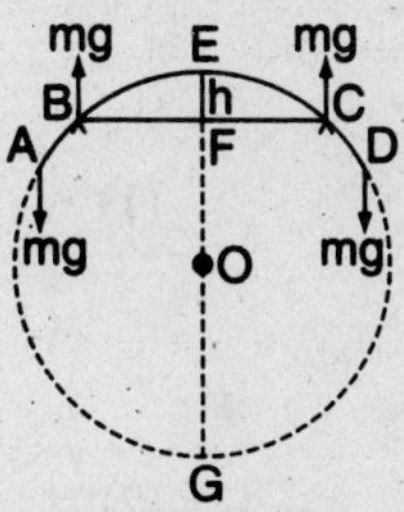

$$EF \times FG = BF^2 \quad \text{or} \quad h(2R - h) = (2l)^2 \quad \left(\because\ BF = \frac{1}{2}BC = 2l\right)$$

$$\text{or} \quad h \times 2R = 4l^2 \quad (\text{neglecting } h^2)$$

$$\text{or} \quad h = \frac{2l^2}{R} \tag{1}$$

Now when the bar is bent, a restoring couple $\frac{YAk^2}{R}$ is developed, where Y is the Young's modulus and Ak^2 is the geometrical moment of inertia of the section of the bar. This restoring couple is equal and opposite to the external couple Γ, so that we may write

$$\frac{YAk^2}{R} = \Gamma = mgl \quad \text{or} \quad R = \frac{YAk^2}{mgl}$$

Putting this value of R in (1), we get

$$h = \frac{2mgl^3}{YAk^2}$$

which is the required expression

4. A steel bar of length 2 metres and having a square cross-section of side 1 cm is supported at one end and at the middle. Find the deflection of the free end, when a load of 3 kg is placed half way between the supports (Young's modulus for steel = 2×10^{12} dynes cm^{-2}). [C.U. (Hons.) '87]

At any point P at a distance x from the supported end of the bar, the bending moment equation is given by

$$YAk^2 \frac{d^2y}{dx^2} = W\left(\frac{l}{2} - x\right) - \frac{W}{2}(l - x) \qquad \left[\begin{array}{l}\text{Here } l = \text{half the length} \\ \text{of the steel bar} = 1 \text{ metre}\end{array}\right]$$

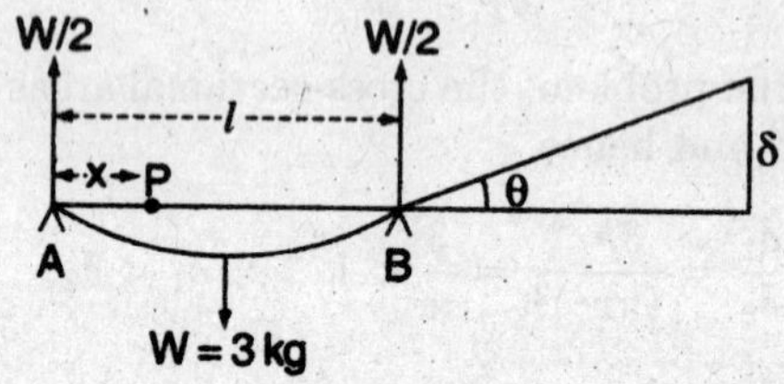

Integrating, we have

$$YAk^2 \frac{dy}{dx} = W\left(\frac{lx}{2} - \frac{x^2}{2}\right) - \frac{W}{2}\left(lx - \frac{x^2}{2}\right) + C_1.$$

Now at $x = \frac{l}{2}, \frac{dy}{dx} = 0$ which gives,

$$0 = W\left(\frac{l^2}{4} - \frac{l^2}{8}\right) - \frac{W}{2}\left(\frac{l^2}{2} - \frac{l^2}{8}\right) + C_1$$

whence we have $C_1 = \frac{Wl^2}{16}$

$$\therefore\ YAk^2 \frac{dy}{dx} = W\left(\frac{lx}{2} - \frac{x^2}{2}\right) - \frac{W}{2}\left(lx - \frac{x^2}{2}\right) + \frac{Wl^2}{16},$$

so that we may write

$$YAk^2 \left(\frac{dy}{dx}\right)_{\text{at } x = l} = W\left(\frac{l^2}{2} - \frac{l^2}{2}\right) - \frac{W}{2}\left(l^2 - \frac{l^2}{2}\right) + \frac{Wl^2}{16}$$

$$= \frac{Wl^2}{16} - \frac{Wl^2}{4} = -\frac{3}{16}Wl^2$$

Now

$$\left(\frac{dy}{dx}\right)_{\text{at } x = l} = \tan\theta = \frac{\delta}{l}$$

$$\therefore \frac{\delta}{l} = \frac{3}{16}\frac{Wl^2}{YAk^2} \quad \text{(considering } \delta \text{ to be negative)}$$

$$\therefore \delta = \frac{3Wl^3}{16YAk^2}.$$

Here $W = 3$ kg $= 3 \times 9.8N = 3 \times 9.8 \times 10^5$ dynes, $l = 1$ m $= 10^2$cm, $Y = 2 \times 10^{12}$ dynes/cm^2, $A = 1$ cm^2, $k^2 = \frac{a^2}{12} = \frac{1}{12}$cm^2

$$\therefore \delta = \frac{3 \times 3 \times 9.8 \times 10^5 \times 10^6 \times 12}{16 \times 2 \times 10^{12}} = \frac{27 \times 9.8}{8} \times 10^{-1} = 33.075 \times 10^{-1} = 3.3075 \text{ cm.}$$

9.19 Reciprocal Theorem

Reciprocal theorem in the case of a cantilever of negligible weight means that the depression at any point P due to a load applied at the free end point Q is the same as the depression at Q produced by a similar load at P.

From Figs. 9.19.1(a) and (b) it will be proved that if by applying a load W at Q, the depression at the point P be δ, then by applying the same load at P the depression at Q will also be δ.

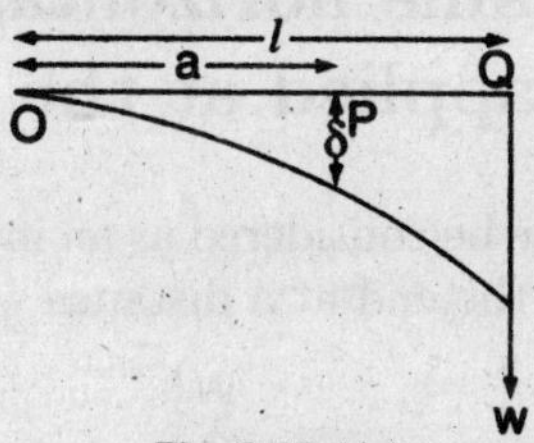

Fig. 9.19.1(a)

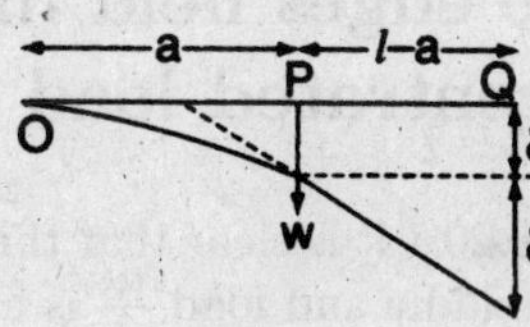

Fig. 9.19.1(b)

The general expression for depression in the case of a cantilever is given by

$$YAk^2y = W\left(\frac{lx^2}{2} - \frac{x^3}{6}\right)$$

Suppose $y = \delta$, when $x = a$

$$\therefore \quad YAk^2\delta = Wa^2\left(\frac{l}{2} - \frac{a}{6}\right) = \frac{Wa^2}{6}(3l - a) \quad \text{or} \quad \delta = \frac{Wa^2}{6YAk^2}(3l - a) \tag{9.19.1}$$

Let us now calculate the depression at Q when the load W is applied at P [Fig. 9.19.1(b)].

When the load is applied at P, the portion OP is bent, but the portion PQ remains straight and is tangential to the curve at P as shown by the dotted line. We have, therefore, the depression at Q in this case given by $\delta_1 + \delta_2$ as shown in the figure, where δ_1 is the depression at P when load is applied at P and δ_2 is the product of the length PQ and the slope of the curve at P. We shall first find out the depression δ_1 at P when load is applied at P. The depression δ_1 is given by

$$\delta_1 = \frac{Wa^3}{3YAk^2}$$

Again, we know that in case of a cantilever $YAk^2\dfrac{dy}{dx} = W\left(lx - \dfrac{x^2}{2}\right)$.

Here $x = a, l = a$

$$\therefore \quad YAk^2\frac{dy}{dx} = \frac{Wa^2}{2} \quad \text{or} \quad \frac{dy}{dx} = \frac{Wa^2}{2YAk^2}.$$

$$\therefore \quad \delta_2 = (l-a)\frac{dy}{dx} = \frac{Wa^2}{2YAk^2}(l-a) \quad \because \tan\theta = \frac{dy}{dx} = \frac{\delta_2}{l-a}$$

$$\therefore \text{ The depression at } Q = \delta_1 + \delta_2 = \frac{Wa^3}{3YAk^2} + \frac{Wa^2(l-a)}{2YAk^2} = \frac{Wa^2}{YAk^2}\left(\frac{a}{3} + \frac{l}{2} - \frac{a}{2}\right)$$

$$= \frac{Wa^2}{6YAk^2}(3l - a) = \delta.$$

It should be noted that we have considered one of the points (the point Q) at the end of the cantilever. The theorem, however, is a general one and holds for any position of the points P and Q.

9.20 A weightless beam supported near the ends on two knife edges held in the same horizontal plane and a concentrated load *W* is applied at the mid-point

From the Figure 9.20.1 it is clear that this case may be considered as an inverted cantilever with fixed end at the middle and load $\frac{W}{2}$ is applied at the end at a distance $\frac{l}{2}$ from the mid-point.

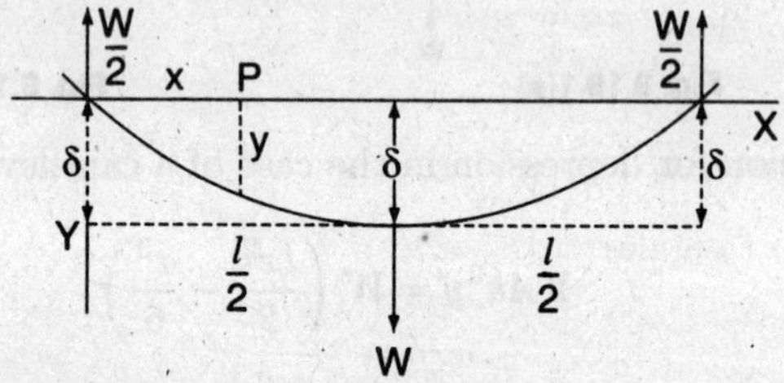

Fig. 9.20.1

Now, the expression for the depression at the loaded end of a light cantilever is, $\delta = \frac{Wl^3}{3YAk^2}$, and when the beam is of rectangular cross-section

$$\delta = \frac{4Wl^3}{Yab^3}$$

In the problem this expression for δ should be modified by putting $\frac{W}{2}$ for W and $\frac{l}{2}$ for l. Hence, we have

$$\delta = \frac{4\frac{W}{2}\left(\frac{l}{2}\right)^3}{Yab^3} = \frac{Wl^3}{4Yab^3} \qquad (9.20.1)$$

Expression (9.20.1) provides us with an easy and accurate laboratory method for determining the young's modulus of a material available in the form of a rectangular beam.

We may find out a similar expression for δ independently by taking into consideration the bending moment equation for any point P, which is

$$YAk^2\frac{d^2y}{dx^2} = W\left(\frac{l}{2} - x\right) - \frac{W}{2}(l - x)$$

On integration we have

$$YAk^2\frac{dy}{dx} = W\left(\frac{lx}{2} - \frac{x^2}{2}\right) - \frac{W}{2}\left(lx - \frac{x^2}{2}\right) + C_1 \tag{9.20.2}$$

On further integration we have

$$YAk^2y = W\left(\frac{lx^2}{4} - \frac{x^3}{6}\right) - \frac{W}{2}\left(\frac{lx^2}{2} - \frac{x^3}{6}\right) + C_1x + C_2 \tag{9.20.3}$$

We have the boundary conditions, $y = 0$, when $x = 0$

and $\dfrac{dy}{dx} = 0$, when $x = \dfrac{l}{2}$.

The first condition gives $C_2 = 0$ and, therefore, from the second condition, we have

$$0 = W\left(\frac{l^2}{4} - \frac{l^2}{8}\right) - \frac{W}{2}\left(\frac{l^2}{2} - \frac{l^2}{8}\right) + C_1 \text{ whence } C_1 = \frac{Wl^2}{16}.$$

$\therefore$ from equation (9.20.3), we have

$$YAk^2y = W\left(\frac{lx^2}{4} - \frac{x^3}{6}\right) - \frac{W}{2}\left(\frac{lx^2}{2} - \frac{x^3}{6}\right) + \frac{Wl^2}{16}x.$$

At the mid-point we have, $y = \delta$ and $x = \frac{l}{2}$

$$\therefore \quad YAk^2\delta = W\left(\frac{l^3}{16} - \frac{l^3}{48}\right) - \frac{W}{2}\left(\frac{l^3}{8} - \frac{l^3}{48}\right) + \frac{Wl^3}{32} = \frac{Wl^3}{48} \quad \text{or} \quad \delta = \frac{Wl^3}{48YAk^2} \tag{9.20.4}$$

when the beam is of rectangular cross-section

$$\delta = \frac{Wl^3}{4Yab^3} \tag{9.20.5}$$

From equation (9.20.2), we have

$$YAk^2\frac{dy}{dx} = W\left(\frac{lx}{2} - \frac{x^2}{2}\right) - \frac{W}{2}\left(lx - \frac{x^2}{2}\right) + \frac{Wl^2}{16} = \frac{Wl^2}{16} - \frac{Wx^2}{4}.$$

At the left end of the beam $x = 0$.

$$\therefore \quad \frac{dy}{dx} = \frac{Wl^2}{16YAk^2}.$$

At the right end of the beam, $x = l$;

$$\therefore \quad \frac{dy}{dx} = -\frac{3Wl^2}{16YAk^2} \tag{9.20.6}$$

Equation (9.20.6) has been employed to solve the illustrative example no. 4 on page 563.

9.20.1 Beam supported near the ends on two knife edges held in the same horizontal plane, and a concentrated load W is applied at the mid-point, weight of the beam being taken into consideration

Let us suppose that the beam has weight ω per unit length, and carries a concentrated load W at its mid-point. Considering the equilibrium of the beam under the different forces distributed as shown in the Fig. 9.20.2, we may write the bending moment equation at P, given by

$$YAk^2\frac{d^2y}{dx^2} = (W+\omega l)\left(\frac{l}{2}-x\right) - \frac{W+\omega l}{2}(l-x) + \omega x\cdot\frac{x}{2}$$

$$= (W+\omega l)\left(\frac{l}{2}-x-\frac{l}{2}+\frac{x}{2}\right) + \frac{\omega x^2}{2} = -(W+\omega l)\frac{x}{2} + \frac{\omega x^2}{2} \tag{9.20.7}$$

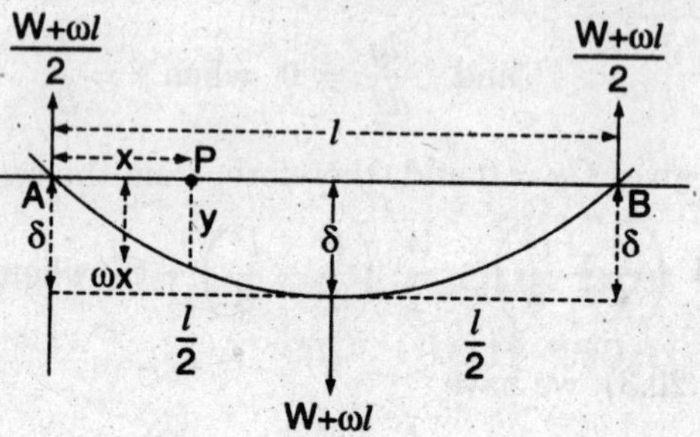

Fig. 9.20.2

Integrating, we have

$$YAk^2\frac{dy}{dx} = -(W+\omega l)\frac{x^2}{4} + \frac{\omega x^3}{6} + C_1 \tag{9.20.8}$$

on further integration, we have

$$YAk^2 y = -(W+\omega l)\frac{x^3}{12} + \frac{\omega x^4}{24} + C_1 x + C_2. \tag{9.20.9}$$

Now we are to apply the following two boundary conditions for the problem, (i) $y = 0$, when $x = 0$ (ii) $\frac{dy}{dx} = 0$, when $x = \frac{l}{2}$.

The first condition gives $C_2 = 0$ and therefore applying the second condition, we have from (9.20.8)

$$0 = -(W+\omega l)\frac{l^2}{16} + \frac{\omega l^3}{48} + C_1$$

$$\therefore\ C_1 = (W+\omega l)\frac{l^2}{16} - \frac{\omega l^3}{48} = \frac{Wl^2}{16} + \frac{\omega l^3}{24}$$

Putting this value of C_1 in (9.20.9), we have

$$YAk^2 y = -(W+\omega l)\frac{x^3}{12} + \frac{\omega x^4}{24} + \left(\frac{Wl^2}{16} + \frac{\omega l^3}{24}\right)x.$$

Putting $x = \frac{l}{2}$ and $y = \delta$ in the above expression, we obtain the depression δ of the beam at its mid-point given by

$$\therefore\ YAk^2\delta = -(W+\omega l)\frac{l^3}{96} + \frac{\omega l^4}{384} + \left(\frac{Wl^2}{16} + \frac{\omega l^3}{24}\right)\frac{l}{2}$$

$$= Wl^3\left(-\frac{1}{96} + \frac{1}{32}\right) + \omega l^4\left(-\frac{1}{96} + \frac{1}{384} + \frac{1}{48}\right)$$

$$= \frac{Wl^3}{48} + \omega l^4\frac{5}{384} = \frac{l^3}{48}\left(W + \frac{5}{8}\omega l\right)$$

$$= \frac{l^3}{48}\left(W + \frac{5}{8}W_1\right) \quad \text{(where } W_1 = \omega l = \text{weight of the beam)}$$

$$\therefore\ \delta = \frac{l^3}{48YAk^2}\left(W + \frac{5}{8}W_1\right) \tag{9.20.10}$$

9.21 Uniform bending of a beam equally loaded at two free ends, and supported on two knife edges at equal distances from the two ends

It can be shown in the following way that the beam has uniform bending moment at every point in this case, and hence it will bend in the form of an arc of a circle as shown in fig. 9.21.1.

We consider a beam AB resting on two horizontal knife edges K_1 and K_2. Let equal forces F act vertically downwards at points near the ends of the beam, and equidistant from the knife edges. Dotted curve in the figure indicates the neutral axis. We consider now a section at P at a distance d from the knife edge K_2. Now, the moment of the force F at B about P is given by $F(D+d)$ in the clockwise sense and the moment of the force F at K_2 about P is Fd, in the anti-clockwise sense. Therefore the resultant bending moment at $P = F(D+d) - Fd = FD = MgD$, where Mg is the weight of each load attached to the free ends.

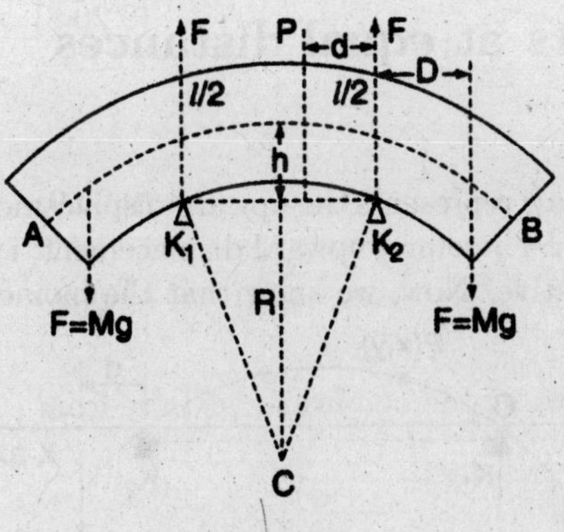

Fig. 9.21.1

Thus it is clear that at every point the bending moment is constant and so the beam bends in the form of an arc of a circle.

For equilibrium of the portion PB, internal bending moment that comes into play due to rest of the beam PA, must be equal to this applied bending moment. Now, the internal bending moment at P is given by $\frac{YAk^2}{R}$. Hence for the equilibrium of the portion PB, we have

$$\frac{YAk^2}{R} = MgD. \tag{9.21.1}$$

If h be the upward deflection of the beam at the mid-point, we have

$$h(2R-h)=\left(\frac{l}{2}\right)^2 \text{ (approximately)},$$

where R is the radius of curvature of the neutral surface and l, the length of the beam between the knife edges.

Since the radius of curvature is considerably large in comparison with h, the above expression may be reduced to the form

$$2Rh=\frac{l^2}{4} \quad \text{or} \quad \frac{1}{R}=\frac{8h}{l^2}$$

$\therefore$ equation (9.21.1) then becomes

$$\frac{YAk^2\cdot 8h}{l^2}=MgD \quad \text{or} \quad Y=\frac{l^2gD}{8Ak^2}\cdot\frac{M}{h} \qquad (9.21.2)$$

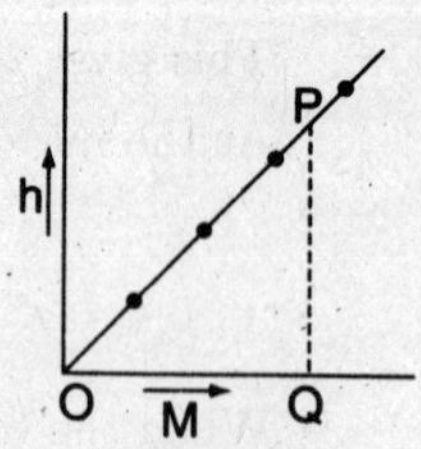

Fig. 9.21.2

which gives an expression for the value of Young's modulus for the material of the beam. From this expression it follows that M varies as h; so a number of readings should be taken for the deflection h corresponding to different masses suspended from A and B. The curve of h against M will be a straight line passing through the origin.

Take a convenient point P on the curve as shown in figure 9.21.2. PQ gives h and OQ gives the corresponding value of M. Substituting the values of h and M in equation (9.21.2), we get the value Y.

For accurate measurement of Y, see Art. 9.30.1(c).

9.22 Upward displacement at any point of a beam bent in the form of a circular arc, convex outward by suspending two equal weights at equal distances from the two knife edges

We consider a point P whose co-ordinates are (x, y). Here y represents the upward displacement of the point P. In the figure Y-axis is drawn downwards. Therefore, upward displacement, i.e., the value of y at any point in this case is reckoned negative. Now, we know that the moment equation is given by

$$\frac{YAk^2}{R}=Mgd, \text{ but } \frac{1}{R}=-\frac{d^2y}{dx^2} \text{ (approx.)};$$

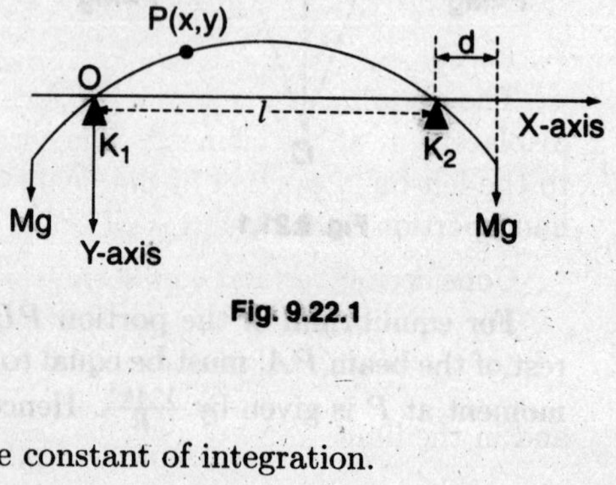

Fig. 9.22.1

the curvature is considered negative, since Y-axis is drawn in the downward direction; therefore,

$$-YAk^2\frac{d^2y}{dx^2}=Mgd$$

On integration, we have

$$-YAk^2\frac{dy}{dx}=Mgd\cdot x+C_1, C_1 \text{being the constant of integration.}$$

On further integration, we have

$$-YAk^2y = \frac{1}{2}Mgd\cdot x^2 + C_1x + C_2, C_2 \text{ being another constant of integration.}$$

Now, when $x = \frac{l}{2}$, i.e., at the mid-point $\frac{dy}{dx} = 0$.
We have, therefore, $C_1 = -\frac{Mgdl}{2}$.
Again when $x = 0, y = 0$, which gives $C_2 = 0$

$$\therefore \quad -YAk^2y = \frac{1}{2}Mgd\cdot x^2 - \frac{Mgdlx}{2} \quad \text{or} \quad YAk^2y = \frac{1}{2}Mgd(lx - x^2) \tag{9.22.1}$$

This gives the value of the upward displacement y at any point P of the beam.
At the mid-point $x = \frac{l}{2}, y = h$

$$\therefore \quad YAk^2h = \frac{1}{2}Mgd\frac{l^2}{4} = \frac{Mgdl^2}{8} \quad \text{or} \quad h = \frac{Mgdl^2}{8YAk^2} \tag{9.22.2}$$

When the beam is of rectangular cross-section $Ak^2 = \frac{ab^3}{12}$. $\therefore$ $h = \frac{3Mgdl^2}{2Yab^3}$.

9.23 General method of determining deflection due to bending of a beam—application to calculate deflections in different cases

Let ω be the weight per unit length of the beam. Let us consider an element δx, the left edge of which is at a distance x from the origin and the right edge at a distance $x + \delta x$ as shown in Fig. 9.23.1.

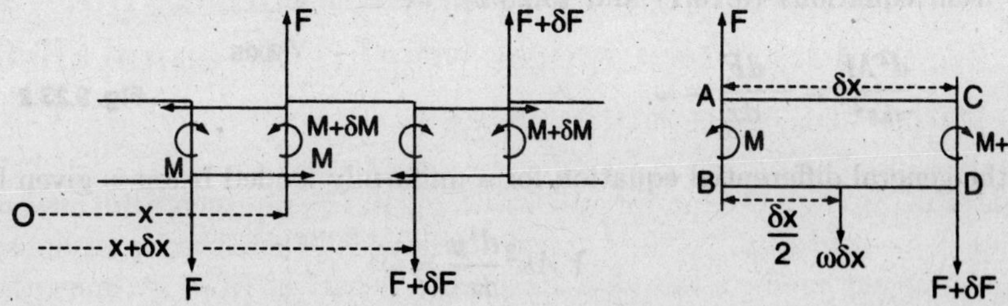

Fig. 9.23.1

The shearing stress at the left face is F and that at the right face is $F + \delta F$ during upward displacement of the element. The corresponding internal resisting moment at the left face due to the left-hand portion of the beam is M in the counter clockwise direction and at the right-hand portion of the beam is $M + \delta M$ in the clockwise direction.

Considering vertical equilibrium of the element δx, the force equation becomes

$$F - \omega\delta x - (F + \delta F) = 0 \quad \therefore \quad \frac{\delta F}{\delta x} = -\omega$$

and in the limit

$$\frac{dF}{dx} = -\omega \quad \text{or} \quad \omega = -\frac{dF}{dx} \tag{9.23.1}$$

Considering moment equation (taking moments about A since there is no rotation) we may write

$$M - \omega\delta x\frac{\delta x}{2} - (M+\delta M) - (F+\delta F)\delta x = 0$$

or $\delta M = -F\delta x$ $\left[\text{neglecting the product of the infinitesimally small quantities such as } (\delta x)^2 \text{ and } \delta F\delta x\right]$

or $\dfrac{\delta M}{\delta x} = -F$ or $\dfrac{dM}{dx} = -F$ in the limit or $F = -\dfrac{dM}{dx}$ (9.23.2)

Combining (9.23.1) and (9.23.2) we get

$$\frac{d^2M}{dx^2} = \omega \quad (9.23.3)$$

We shall use the above relations in the following problems relating to bending.

N.B. The signs in the above equations are consequences of the sign convention we have adopted. A different convention may drop one of the negative signs.

(i) *A uniformly loaded beam supported horizontally on two knife edges*

We now take the case of a uniformly loaded beam of weight ω per unit length supported horizontally on two knife edges K_1 and K_2 as shown in Fig. 9.23.2.

We know that the bending moment at any point $P(x, y)$ of the beam is $YAk^2\dfrac{d^2y}{dx^2}$.

$\therefore \quad M = YAk^2\dfrac{d^2y}{dx^2}$ or $\dfrac{d^2M}{dx^2} = YAk^2\dfrac{d^4y}{dx^4}$.

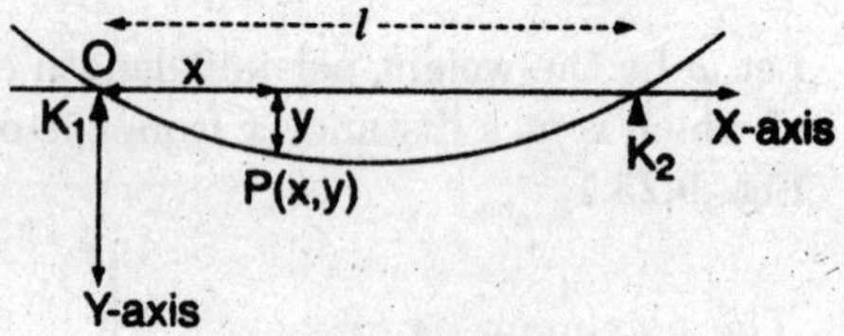

Fig. 9.23.2

Again, from equations (9.23.1) and (9.23.2), we have

$$\frac{d^2M}{dx^2} = -\frac{dF}{dx} = \omega.$$

Hence the general differential equation for a uniformly loaded beam is given by

$$YAk^2\frac{d^4y}{dx^4} = \omega \quad (9.23.4)$$

Integrating four times in succession, we get

$$YAk^2\frac{d^3y}{dx^3} = \omega x + C_1 \quad (9.23.5)$$

$$YAk^2\frac{d^2y}{dx^2} = \frac{\omega x^2}{2} + C_1x + C_2 \quad (9.23.6)$$

$$YAk^2\frac{dy}{dx} = \frac{\omega x^3}{6} + \frac{C_1x^2}{2} + C_2x + C_3 \quad (9.23.7)$$

$$YAk^2y = \frac{\omega x^4}{24} + \frac{C_1x^3}{6} + \frac{C_2x^2}{2} + C_3x + C_4 \quad (9.23.8)$$

C_1, C_2, C_3 and C_4 are the four constants of integration.

Evaluation of the four constants : $y = 0$ at $x = 0$, gives $C_4 = 0$ [from eqn. (9.23.8)]
Again

$$\frac{dy}{dx} = \text{maximum} \quad \text{or} \quad \frac{d^2y}{dx^2} = 0 \quad \text{at } x = 0, l$$

For $x = 0$, we have $C_2 = 0$ [from eqn. (9.23.6)]

For $x = l, 0 = \frac{\omega l^2}{2} + C_1 l$ [from eqn. (9.23.6)] or $C_1 = -\frac{\omega l}{2}$.

Again, $\frac{dy}{dx} = 0$, at $x = \frac{l}{2}$ gives

$$0 = \frac{1}{6}\cdot\frac{\omega l^3}{8} + \frac{1}{2}C_1\cdot\frac{l^2}{4} + C_3 \quad \text{[from eqn. (9.23.7)]}$$

$$\text{or} \quad 0 = \frac{\omega l^3}{48} - \frac{\omega l^3}{16} + C_3 \quad \text{or} \quad C_3 = \frac{\omega l^3}{24}$$

Thus we get the following values of the four constants

$$C_1 = -\frac{\omega l}{2}, \quad C_2 = 0, \quad C_3 = \frac{\omega l^3}{24}, \quad C_4 = 0$$

We have, therefore,

$$YAk^2 y = \frac{\omega x^4}{24} - \frac{\omega l x^3}{12} + \frac{\omega l^3 x}{24} \quad \text{[from eqn. (9.23.8)]}$$

$$= \frac{\omega x}{24}(x^3 - 2x^2 l + l^3) = \frac{\omega x}{24}(l - x)(l^2 + lx - x^2)$$

$$\text{or} \quad y = \frac{\omega x(l - x)(l^2 + lx - x^2)}{24YAk^2}$$

The maximum depression δ is clearly at the mid-point of the beam where $x = \frac{l}{2}$, and is given by

$$\delta = \frac{\omega\frac{l}{2}\left(l - \frac{l}{2}\right)}{24YAk^2}\cdot\left(l^2 + \frac{l^2}{2} - \frac{l^2}{4}\right) = \frac{5}{384}\cdot\frac{\omega l^4}{YAk^2} = \frac{5W_1 l^3}{384YAk^2}$$

where W_1 is the total weight of the beam.

(ii) *A uniformly loaded beam clamped horizontally at both ends*

We consider a uniformly loaded beam of length l clamped at both ends. Let ω be the mass per unit-length of the beam. Then the differential equation for the depression of the beam is given by

$$YAk^2\frac{d^4y}{dx^4} = \omega$$

Integrating four times in succession, we get the four equations from (9.23.5) to (9.23.8).

To evaluate the four constants we use the conditions

$$y = 0 \quad \text{and} \quad \frac{dy}{dx} = 0 \quad \text{at } x = 0, l$$

$\therefore$ $C_3 = 0$ [from eqn. (9.23.7)] and $C_4 = 0$ [from eqn. (9.23.8)]

At $x = l$, by equations (9.23.7) and (9.23.8), we have

$$\frac{1}{6}\omega l^3 + \frac{1}{2}C_1 l^2 + C_2 l = 0$$

$$\text{and} \quad \frac{1}{24}\omega l^4 + \frac{1}{6}C_1 l^3 + \frac{1}{2}C_2 l^2 = 0 \quad \text{or} \quad \frac{1}{6}\omega l^2 + \frac{1}{2}C_1 l + C_2 = 0 \tag{9.23.9}$$

$$\text{and} \quad \frac{1}{12}\omega l^2 + \frac{1}{3}C_1 l + C_2 = 0 \tag{9.23.10}$$

Subtracting (9.23.10) from (9.23.9), we have

$$\frac{1}{6}C_1 l = -\frac{\omega l^2}{12} \quad \text{or} \quad C_1 = -\frac{\omega l}{2}.$$

Putting the value of C_1 in (9.23.7) and putting $x = l$, we have

$$\frac{1}{6}\omega l^3 - \frac{1}{4}\omega l^3 + C_2 l = 0 \quad \text{or} \quad C_2 = \frac{\omega l^2}{12}.$$

Substituting the values of C_1 and C_2 in (9.23.8), we have

$$YAk^2 y = \frac{\omega x^4}{24} - \frac{\omega l x^3}{12} + \frac{\omega l^2 x^2}{24} = \frac{\omega x^2}{24}(x^2 - 2lx + l^2) = \frac{\omega x^2}{24}(l - x)^2.$$

Therefore, the depression δ at the mid-point is given by

$$YAk^2\delta = \frac{\omega\left(\frac{l}{2}\right)^2}{24}\left(l - \frac{l}{2}\right)^2 = \frac{\omega l^4}{384YAk^2} = \frac{W_1 l^3}{384YAk^2}$$

where W_1 is the weight of the beam.

Thus the depression at the centre of the beam is only one-fifth of what it would be if the same beam be freely supported horizontally near the ends.

(iii) *Uniform beam supported horizontally near the ends on two knife edges and a concentrated load W is applied at any point of the beam, the weight of the beam being neglected:*

Here, since $\omega = 0$, we may write $YAk^2\dfrac{d^4y}{dx^4} = 0$.

Integrating four times in succession, we have

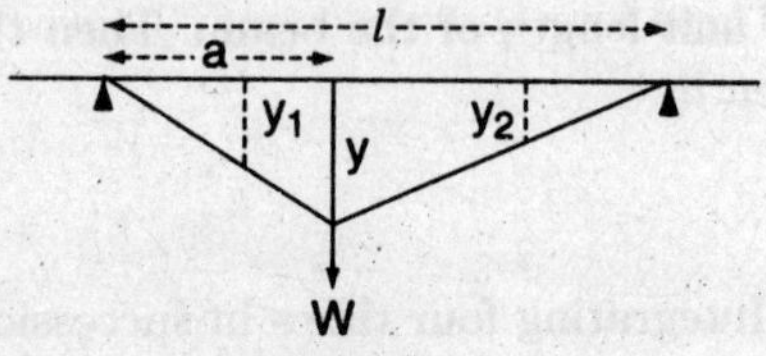

Fig. 9.23.4

$$YAk^2\frac{d^3y}{dx^3} = A; \qquad YAk^2\frac{d^2y}{dx^2} = Ax + B;$$

$$YAk^2\frac{dy}{dx} = \frac{Ax^2}{2} + Bx + C; \quad YAk^2 y = \frac{Ax^3}{6} + \frac{Bx^2}{2} + Cx + D.$$

A, B, C, D being the four constants of integration.

We have now two sets of equations, one set from $x = 0$ to $x = a$ (the point of loading) and another set from $x = a$ to $x = l$ (Fig. 9.23.4).

Thus between $x = 0$ and $x = a$, we have

$$YAk^2 \frac{d^3y_1}{dx^3} = A_1 \tag{9.23.11}$$

$$YAk^2 \frac{d^2y_1}{dx^2} = A_1x + B_1 \tag{9.23.12}$$

$$YAk^2 \frac{dy_1}{dx} = \frac{A_1x^2}{2} + B_1x + C_1 \tag{9.23.13}$$

$$YAk^2 y_1 = \frac{A_1x^3}{6} + \frac{B_1x^2}{2} + C_1x + D_1 \tag{9.23.14}$$

A_1, B_1, C_1, D_1 being the four constants for the portion of the beam between $x = 0$ and $x = a$.

Between $x = a$ and $x = l$, we have

$$YAk^2 \frac{d^3y_2}{dx^3} = A_2 \tag{9.23.15}$$

$$YAk^2 \frac{d^2y_2}{dx^2} = A_2x + B_2 \tag{9.23.16}$$

$$YAk^2 \frac{dy_2}{dx} = \frac{A_2x^2}{2} + B_2x + C_2 \tag{9.23.17}$$

$$YAk^2 y_2 = \frac{A_2x^3}{6} + \frac{B_2x^2}{2} + C_2x + D_2 \tag{9.23.18}$$

A_2, B_2, C_2, D_2 being the four constants for the portion of the beam between $x = a$ and $x = l$.

We are now to evaluate the eight constants in these two sets of equations.

The conditions at the point of loading are:

(1) the difference between the shearing forces on the two sides of the point of loading is equal to the load W;

(2) the bending moment remains unchanged on passing the point of loading;

(3) the value of $\frac{dy}{dx}$ remains unchanged on passing the point of loading;

(4) the depression also remains unaffected on passing the point of loading.

These four conditions may be written as

$$\left.\begin{aligned} &F_2 - F_1 = W; \; M_1 = M_2; \\ &\frac{dy_1}{dx} = \frac{dy_2}{dx}; \quad y_1 = y_2 \end{aligned}\right\} \text{ at the point of loading.}$$

From these conditions we have

$$A_1 - A_2 = -W \tag{9.23.19}$$

$$B_1 - B_2 = Wa \tag{9.23.20}$$

$$C_1 - C_2 = -\frac{Wa^2}{2} \tag{9.23.21}$$

$$D_1 - D_2 = \frac{Wa^3}{6} \tag{9.23.22}$$

Since

$$F_1 = \frac{dM_1}{dx} = YAk^2\frac{d^3y_1}{dx^3}; \qquad F_2 = \frac{dM_2}{dx} = YAk^2\frac{d^3y_2}{dx^3};$$

$$M_1 = YAk^2\frac{d^2y_1}{dx^2}; \qquad M_2 = YAk^2\frac{d^2y_2}{dx^2}.$$

Considering end conditions, we have

$$y_1 = 0; \quad \frac{d^2y_1}{dx^2} = 0 \text{ at } x = 0$$

$$\text{and } \; y_2 = 0; \quad \frac{d^2y_2}{dx^2} = 0 \text{ at } x = l.$$

At $x = 0$, $y_1 = 0$—which gives $D_1 = 0$ [from (9.23.14)]

$$\therefore D_2 = -\frac{Wa^3}{6} \quad \text{[from (9.23.22)]} \tag{9.23.23}$$

At $x = 0$, $\dfrac{d^2y_1}{dx^2} = 0$—which gives $B_1 = 0$ [from (9.23.12)]

$$\therefore B_2 = -Wa \quad \text{[from (9.23.20)]} \tag{9.23.24}$$

At $x = l$, $\dfrac{d^2y_2}{dx^2} = 0$—which gives $0 = A_2l + B_2$ [from (9.23.16)]

$$\therefore A_2 = -\frac{B_2}{l} = \frac{Wa}{l} \quad \text{[from (9.23.24)]} \tag{9.23.25}$$

$$\therefore A_1 = \frac{Wa}{l} - W = -\frac{W}{l}(l-a) \quad \text{[from (9.23.19)]} \tag{9.23.26}$$

At $x = l$, $y_2 = 0$—which gives

$$0 = \frac{1}{6}A_2l^3 + \frac{1}{2}B_2l^2 + C_2l + D_2 \quad \text{[from (9.23.18)]}$$

Substituting the values of A_2, B_2 and D_2 in the above equation we have

$$0 = \frac{Wal^2}{6} - \frac{1}{2}Wal^2 + C_2l - \frac{Wa^3}{6} \quad \text{whence, } C_2 = \frac{Wa}{6l}(2l^2 + a^2)$$

$$\therefore C_1 = \frac{Wa}{6l}(2l^2 + a^2) - \frac{Wa^2}{2} \quad \text{[from (9.23.21)]}$$

$$= \frac{Wa}{6l}(l-a)(2l-a) \tag{9.23.27}$$

Thus we get the following values of the eight constants

$$A_1 = -\frac{W}{l}(l-a); \quad B_1 = 0; \quad C_1 = \frac{Wa}{6l}(l-a)(2l-a); \quad D_1 = 0;$$

$$A_2 = \frac{Wa}{l}; \quad B_2 = -Wa; \quad C_2 = \frac{Wa}{6l}(2l^2+a^2); \quad D_2 = -\frac{Wa^3}{6}.$$

$$\therefore\ YAk^2y_1 = -\frac{W}{6l}(l-a)x^3 + \frac{Wa}{6l}(l-a)(2l-a)x$$

$$= \frac{W}{6l}[a(2l-a)(l-a)x - (l-a)x^3] \tag{9.23.28}$$

This gives the depression y_1 at any point in the first segment of the beam, i.e., from $x = 0$ to $x = a$.

When the load is applied at the mid-point $a = l/2$, the deflection δ at the mid-point is obtained by putting $x = l/2$.

$$\therefore\ YAk^2\delta = \frac{W}{6l}\left[\frac{l}{2}\cdot\frac{3l}{2}\cdot\frac{l}{2}\cdot\frac{l}{2} - \frac{l}{2}\cdot\frac{l^3}{8}\right] = \frac{Wl^3}{48} \quad \text{or} \quad \delta = \frac{Wl^3}{48YAk^2}$$

the same result which we obtained independently and also from consideration of inverted cantilever (Art. 9.20). Similarly the depression y_2 in the second segment (from $x = a$ to $x = l$) is

$$YAk^2y_2 = \frac{Wa}{6l}x^3 - \frac{Wa}{2}x^2 + \frac{Wa}{6l}(2l^2+a^2)x - \frac{Wa^3}{6} \tag{9.23.29}$$

whence $\delta = \dfrac{Wl^3}{48YAk^2}$, putting $x = l/2$ and $a = l/2$.

(iv) *Uniform beam supported horizontally near the ends on two knife edges and a concentrated load is applied at any point of the beam, the weight of the beam being taken into consideration.*

Let ω be the weight per unit length of the beam. Hence

$$YAK^2\frac{d^4y}{dx^4} = \omega.$$

As before we get two sets of equations, one set from $x = 0$ to $x = a$ and another set from $x = a$ to $x = l$. Thus between $x = 0$ and $x = a$ we have

$$YAK^2\frac{d^3y_1}{dx^3} = \omega x + A_1 \tag{9.23.30}$$

$$YAK^2\frac{d^2y_1}{dx^2} = \frac{\omega x^2}{2} + A_1x + B_1 \tag{9.23.31}$$

$$YAK^2\frac{dy_1}{dx} = \frac{\omega x^3}{6} + \frac{A_1x^2}{2} + B_1x + C_1 \tag{9.23.32}$$

$$YAK^2y_1 = \frac{\omega x^4}{24} + \frac{A_1x^3}{6} + \frac{B_1x^2}{2} + C_1x + D_1 \tag{9.23.33}$$

Between $x = a$ and $x = l$, we have,

$$YAK^2\frac{d^3y_2}{dx^3} = \omega x + A_2 \tag{9.23.34}$$

$$YAK^2\frac{d^2y_2}{dx^2} = \frac{\omega x^2}{2} + A_2x + B_2 \tag{9.23.35}$$

$$YAK^2\frac{dy_2}{dx} = \frac{\omega x^3}{6} + \frac{A_2x^2}{2} + B_2x + C_2 \tag{9.23.36}$$

$$YAK^2y_2 = \frac{\omega x^4}{24} + \frac{A_2x^3}{6} + \frac{B_2x^2}{2} + C_2x + D_2 \tag{9.23.37}$$

Now we are to calculate the values of the eight constants. We apply the four conditions at the point of loading as in the previous problem. From these conditions we have

$$A_1 - A_2 = -W \tag{9.23.38}$$

$$B_1 - B_2 = Wa \tag{9.23.39}$$

$$C_1 - C_2 = -\frac{Wa^2}{2} \tag{9.23.40}$$

$$D_1 - D_2 = \frac{Wa^3}{6} \tag{9.23.41}$$

At $x = 0$, $y_1 = 0$, $\dfrac{d^2y_1}{dx^2} = 0$ and at $x = l$, $y_2 = 0$ and $\dfrac{d^2y_2}{dx^2} = 0$.

From (9.23.33) applying the condition at $x = 0$, $y_1 = 0$, we have $D_1 = 0$ and so

$$D_2 = -\frac{Wa^3}{6}. \tag{9.23.42}$$

At $x = 0$ since $\dfrac{d^2y_1}{dx^2} = 0$ we have from (9.23.31), $B_1 = 0$

$$\therefore\ B_2 = -Wa.$$

The condition at $x = l$, $\dfrac{d^2y_2}{dx^2} = 0$ gives

$$0 = \frac{\omega l^2}{2} + A_2l + B_2 \quad \text{or} \quad A_2 = -\frac{B_2}{l} - \frac{\omega l}{2} = +\frac{Wa}{l} - \frac{\omega l}{2}. \tag{9.23.43}$$

$$\therefore\quad A_1 = -W + A_2 = -W + \frac{Wa}{l} - \frac{\omega l}{2} = -\frac{W}{l}(l-a) - \frac{\omega l}{2}. \tag{9.23.44}$$

At $x = l$, $y_2 = 0$

$$\therefore\quad 0 = \frac{\omega l^4}{24} + \frac{A_2l^3}{6} + \frac{B_2l^2}{2} + C_2l + D_2.$$

Substituting the values of A_2, B_2 and D_2 in the above equation, we have

$$0 = \frac{\omega l^4}{24} + \left(\frac{Wa}{l} - \frac{\omega l}{2}\right)\frac{l^3}{6} - \frac{Wal^2}{2} + C_2l - \frac{Wa^3}{6},$$

whence we get,

$$C_2 = \frac{Wa}{6l}(a^2+2l^2)+\frac{\omega l^3}{24}.$$

$$\therefore \quad C_1 = C_2 - \frac{Wa^2}{2} = \frac{Wa}{6l}(a^2+2l^2)+\frac{\omega l^3}{24}-\frac{Wa^2}{2} = \frac{Wa}{6l}(l-a)(2l-a)+\frac{\omega l^3}{24}.$$

Thus we get

$$A_1 = -\frac{W}{l}(l-a)-\frac{\omega l}{2}; \qquad A_2 = \frac{Wa}{l}-\frac{\omega l}{2};$$

$$B_1 = 0; \qquad B_2 = -Wa;$$

$$C_1 = \frac{Wa}{6l}(l-a)(2l-a)+\frac{\omega l^3}{24}; \qquad C_2 = \frac{Wa}{6l}(a^2+2l^2)+\frac{\omega l^3}{24};$$

$$D_1 = 0; \qquad D_2 = -\frac{Wa^3}{6}.$$

Hence we may write

$$YAK^2y_1 = \frac{\omega x^4}{24}-\frac{x^3}{6}\left\{\frac{W}{l}(l-a)+\frac{\omega l}{2}\right\}+x\left\{\frac{Wa}{6l}(l-a)(2l-a)+\frac{\omega l^3}{24}\right\} \tag{9.23.45}$$

This gives the depression at any point in the first segment of the beam from $x = 0$ to $x = a$.

$\therefore$ when the load is applied at the midpoint, the deflection δ at the midpoint is given by

$$YAK^2\delta = \frac{\omega l^4}{16\times 24}-\frac{l^3}{48}\left(\frac{W}{l}\cdot\frac{l}{2}+\frac{\omega l}{2}\right)+\frac{l}{2}\left\{\frac{W}{6l}\cdot\frac{l}{2}\cdot\frac{l}{2}\cdot\frac{3}{2}l+\frac{\omega l^3}{24}\right)=\frac{Wl^3}{48}+\frac{5\omega l^4}{384}$$

$$= \frac{Wl^3}{48}+\frac{5W_1l^3}{384} \quad [\omega l = W_1 = \text{weight of the beam}]$$

$$= \frac{l^3}{48}\left(W+\frac{5W_1}{8}\right)$$

$$\text{or} \quad \delta = \frac{l^3}{48YAK^2}\left(W+\frac{5W_1}{8}\right)$$

Similarly we get

$$YAK^2y_2 = \frac{\omega x^4}{24}+\frac{x^3}{6}\left(\frac{Wa}{l}-\frac{\omega l}{2}\right)-\frac{x^2}{2}Wa+x\cdot\left\{\frac{Wa}{6l}(a^2+2l^2)+\frac{\omega l^3}{24}\right\}-\frac{Wa^3}{6}. \tag{9.23.46}$$

This gives the depression at any point in the second segment of the beam from $x = a$ to $x = l$, whence

$$\delta = \frac{l^3}{48YAK^2}\left(W+\frac{5W_1}{8}\right),$$

putting $x = l/2$ and $a = l/2$.

(v) *Uniform beam clamped horizontally at both ends and a concentrated load W is applied at any point, the weight of the beam being neglected:*

In this case we have two sets of equations similar to what we considered in the case of a bar supported near the ends with concentrated load at any point.

The values of the eight constants in the two sets of equations (9.23.11) to (9.23.18) [Art. 9.23 (iii)] are determined in the following way.

The conditions at the point of loading are given by

$$F_2 - F_1 = W; \quad M_1 = M_2; \quad \frac{dy_1}{dx} = \frac{dy_2}{dx}; \quad y_1 = y_2.$$

From these conditions, we have

$$A_1 - A_2 = -W \tag{9.23.47}$$

$$B_1 - B_2 = Wa \tag{9.23.48}$$

$$C_1 - C_2 = -\frac{Wa^2}{2} \tag{9.23.49}$$

$$D_1 - D_2 = \frac{Wa^3}{6} \tag{9.23.50}$$

Considering the clamped end conditions, we have

$$y_1 = 0; \quad \frac{dy_1}{dx} = 0 \text{ at } x = 0$$

$$\text{and} \quad y_2 = 0; \quad \frac{dy_2}{dx} = 0 \text{ at } x = l.$$

Since $y_1 = 0$ at $x = 0$, we have $D_1 = 0$ from (9.23.14).

At $x = 0$, $\frac{dy_1}{dx} = 0$—which gives $C_1 = 0$ [from (9.23.13)]

$$\therefore \quad C_2 = \frac{Wa^2}{2} \text{ and } D_2 = -\frac{Wa^3}{6} \quad \text{[from (9.23.21) and (9.23.22)]}$$

From (9.23.17) and (9.23.18) we may write

$$\frac{1}{2}A_2l^2 + B_2l + \frac{1}{2}Wa^2 = 0$$

$$\text{and} \quad \frac{1}{6}A_2l^3 + \frac{1}{2}B_2l^2 + \frac{1}{2}Wa^2l - \frac{Wa^3}{6} = 0 \quad \left[\because \text{ at } x = l, \frac{dy_2}{dx} = y_2 = 0\right]$$

$$\text{or} \quad A_2l^3 + 2B_2l^2 + Wa^2l = 0 \tag{9.23.51}$$

$$\text{and} \quad A_2l^3 + 3B_2l^2 + 3Wa^2l - Wa^3 = 0 \tag{9.23.52}$$

Subtracting (9.23.51) from (9.23.52), we have

$$B_2l^2 = Wa^3 - 2Wa^2l \quad \text{or} \quad B_2 = \frac{Wa^2}{l^2}(a - 2l).$$

$$\therefore \quad B_1 = Wa + B_2 = \frac{Wa}{l^2}(l^2 + a^2 - 2al) \quad \text{[from (9.23.48)]}$$

$$= \frac{Wa}{l^2}(l-a)^2.$$

Substituting the values of B_2, C_2 and D_2 in (9.23.18), we have

$$\frac{A_2 l^3}{6} + \frac{Wa^2}{l^2}(a-2l)\frac{l^2}{2} + \frac{1}{2}Wa^2 l - \frac{Wa^3}{6} = 0 \quad [\because \text{ at } x = l,\ y_2 = 0]$$

$$\text{or } A_2 = \frac{Wa^2}{l^3}(3l - 2a).$$

$$\therefore \quad A_1 = -W + A_2 = -W + \frac{Wa^2}{l^3}(3l - 2a) \quad \text{[from (9.23.47)]}$$

$$= -\frac{W}{l^3}(l^3 + 2a^3 - 3a^2 l) = -\frac{W(l-a)}{l^3}(l^2 + al - 2a^2)$$

$$= -\frac{W(l-a)^2}{l^3}(l + 2a).$$

Thus we have the following values of the constants

$$A_1 = -\frac{W(l-a)^2}{l^3}(l+2a); \qquad A_2 = \frac{Wa^2}{l^3}(3l-2a);$$

$$B_1 = \frac{Wa}{l^2}(l-a)^2; \qquad B_2 = \frac{Wa^2}{l^2}(a-2l);$$

$$C_1 = 0; \qquad C_2 = \frac{Wa^2}{2};$$

$$D_1 = 0; \qquad D_2 = -\frac{Wa^3}{6}.$$

Thus we have, the depression in the first segment (from $x = 0$ to $x = a$) given by,

$$YAk^2 y_1 = -\frac{W(l-a)^2}{6l^3}(l+2a)x^3 + \frac{Wa}{2l^2}(l-a)^2 x^2. \tag{9.23.53}$$

If δ be the depression at the mid-point when the load W is at the middle, we have $a = l/2$, $x = l/2$.

$$\therefore \quad YAk^2\delta = -\frac{W}{6l^3}\frac{l^2}{4}\cdot 2l \cdot \frac{l^3}{8} + \frac{Wl}{4l^2}\cdot\frac{l^2}{4}\cdot\frac{l^2}{4} = \frac{Wl^3}{192} \quad \text{or} \quad \delta = \frac{Wl^3}{192YAk^2}.$$

Similarly

$$YAk^2 y_2 = \frac{Wa^2}{6l^3}(3l-2a)x^3 - \frac{Wa^2}{2l^2}(2l-a)x^2 + \frac{Wa^2 x}{2} - \frac{Wa^3}{6} \tag{9.23.54}$$

gives the depression in the second segment (from $x = a$ to $x = l$).

The depression at the mid-point when the load is at the mid-point is

$$YAk^2\delta = \frac{Wl^2}{24l^3}\cdot 2l\cdot\frac{l^3}{8} - \frac{Wl^2}{8l^2}\cdot\frac{3l}{2}\frac{l^2}{4} + \frac{W}{2}\cdot\frac{l^2}{4}\cdot\frac{l}{2} - \frac{W}{6}\frac{l^3}{8} = \frac{Wl^3}{192} \quad \text{or} \quad \delta = \frac{Wl^3}{192YAk^2}.$$

Considering the weight of the beam W_1 it can be shown that

$$\delta_1 = \frac{l^3}{192YAK^2}\left(W + \frac{W_1}{2}\right).$$

9.24 Transverse Vibration of a Loaded Cantilever

If the free end of a cantilever is loaded by a weight W, the depression at this end is given by

$$\delta = \frac{(W + \frac{3}{8}W_1)\,l^3}{3YAk^2} \tag{9.24.1}$$

Let us now suppose that the canti-lever is depressed further through a small distance y by applying a force F at the free end. We have then

$$y + \delta = \frac{\{(W + \frac{3}{8}W_1) + F\}\,l^3}{3YAk^2}. \tag{9.24.2}$$

From (9.24.1) and (9.24.2), we have

$$F = \frac{3YAk^2}{l^3}y$$

When the depressing force F is removed, the elastic force $-F$ comes into play. This force causes motion of the weight in the upward direction along Y-axis and the equation of motion of the load of mass M is given by

$$(M + m)\frac{d^2y}{dt^2} = -F,$$

where m is the correction term assumed constant due to the mass of the cantilever

$$\therefore \quad (M + m)\frac{d^2y}{dt^2} + F = 0 \quad \text{or} \quad \frac{d^2y}{dt^2} + \frac{3YAk^2}{l^3(M + m)}y = 0.$$

This equation represents a S.H.M. of period given by

$$T = 2\pi\sqrt{\frac{(M + m)l^3}{3YAk^2}}.$$

It is to be noted that if the mass of the cantilever is negligible, then $m = 0$

$$\therefore \quad T = 2\pi\sqrt{\frac{Ml^3}{3YAk^2}}. \tag{9.24.3}$$

If for a number of values of M, the corresponding values of T are obtained and M plotted against T^2, we get a straight line passing through the origin if the above theory is correct. The slope of the curve is $\dfrac{4\pi^2 l^3}{3YAk^2}$, from which Y may be determined.

9.25 Bending of Beams with Considerable Curvature

If a beam is bent with considerable curvature, the curvature cannot be taken to be equal to $\frac{d^2y}{dx^2}$.

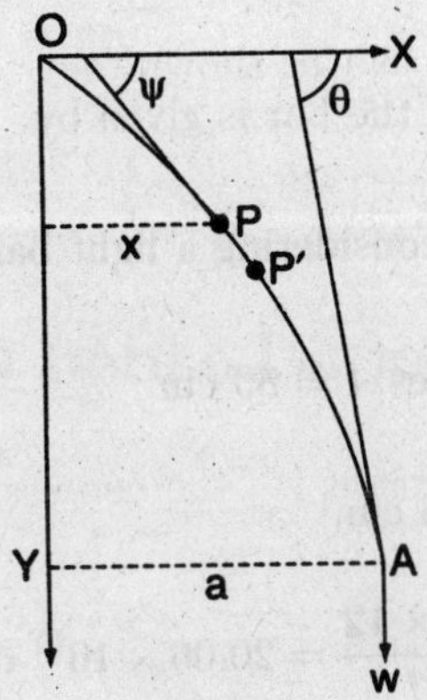

Fig. 9.25.1

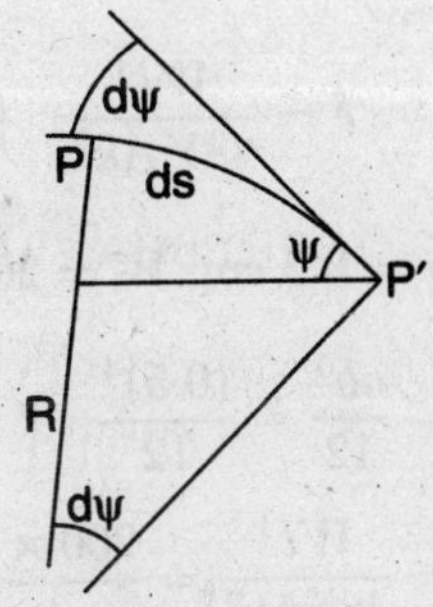

Fig. 9.25.2

We consider a flexible cantilever OA at the free end of which a load W is attached which produces considerable bending of the beam. Consider any point P on the cantilever with co-ordinates (x, y) with reference to the origin O. The moment equation at a point P of the cantilever is given by

$$\frac{YAk^2}{R} = W(a - x) \tag{9.25.1}$$

$1/R$ denotes the curvature at P and 'a' is the horizontal distance between the fixed and the loaded ends of the cantilever. If the angle between the tangents at P and P' be $d\psi$ and ds be the elementary length PP', then

$$\frac{1}{R} = \frac{d\psi}{ds} = \frac{d\psi}{dx}\cdot\frac{dx}{ds} = \frac{d\psi}{dx}\cos\psi$$

Substituting the value of $\frac{1}{R}$ in (9.25.1), we have

$$W(a - x) = YAk^2\frac{d\psi}{dx}\cos\psi \quad \text{or} \quad W(a - x)dx = YAk^2\cos\psi d\psi.$$

Integrating between the points O and A, we have

$$\int_0^a W(a - x)dx = YAk^2\int_0^\theta \cos\psi d\psi,$$

where θ is the angle which the tangent at A makes with the horizontal.

$$\text{or} \quad \frac{Wa^2}{2} = YAk^2\sin\theta \tag{9.25.2}$$

If the load be increased such that the tangent at A becomes vertical, then $\theta = \pi/2$ and hence

$$\frac{W}{2}a_0^2 = YAk^2 \quad \text{or} \quad a_0 = \sqrt{\frac{2YAk^2}{W}},$$

where a_0 denotes the horizontal distance between the fixed and the loaded ends when the tangent at A is vertical.

Illustrative Examples

1. A bar 80 cm long and 0.5 cm² in cross-section is supported horizontally at its ends. The depression in the middle produced by a load of 200 gms is 2 mm. Calculate Young's modulus of the material.

[Bombay, 1953]

In this case the depression at the mid-point of the bar is given by

$$\delta = \frac{Wl^3}{48YAk^2} \quad \text{[Art. 9.20—considering a light bar]}$$

Here $\delta = 2\text{ mm} = 0.2\text{ cm}$, $W = 200 \times 981$ dynes, $l = 80$ cm

$$Ak^2 = \frac{ab^3}{12} = \frac{(0.5)^4}{12} \quad [\because a = b = 0.5\text{ cm.}]$$

$$\therefore Y = \frac{Wl^3}{48\delta Ak^2} = \frac{200 \times 981 \times (80)^3 \times 12}{48 \times 0.2 \times (0.5)^4} = 20.06 \times 10^{11}\text{ dynes/cm}^2.$$

2. A light horizontal beam is supported at its ends on two knife edges and a concentrated load of weight W is applied at the mid-point of the beam. Draw the shearing force diagram and the bending moment diagram appropriate to this problem.

The beam is supported on the knife edges K_1 and K_2 at its ends and is loaded by the weight W at its centre [Fig. 1(i)]. We consider the shearing force F and the bending moment M at a section P which is at a distance x from the knife edge K_1 [Fig. 1(ii)]. Considering the equilibrium of the portion K_1P of the beam, we have

$$\frac{W}{2} + F = 0 \text{ and } M - Fx = 0.$$

$$\therefore F = -\frac{W}{2} \text{ and } M = Fx = -\frac{W}{2}x. \quad (1)$$

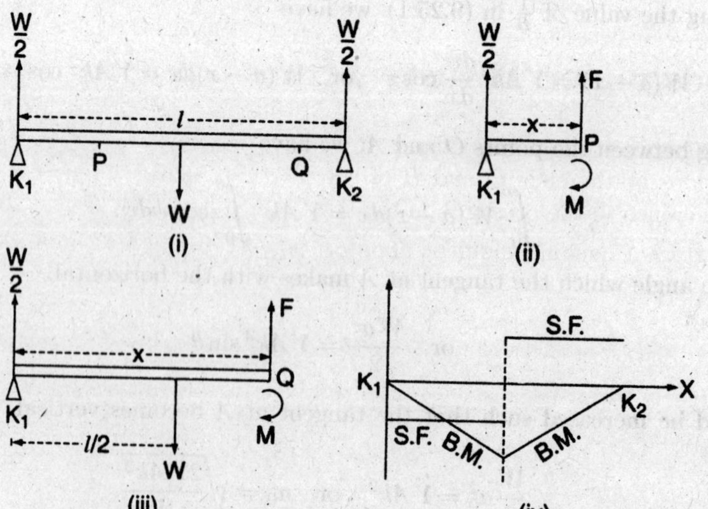

Fig. 1

The negative sign indicates that the shearing force at the point P on the portion K_1P is acting downwards, while the bending moment is acting in the anti-clockwise direction. Also considering shearing stress and the bending moment at a section Q which is at a distance x from K_1, we have for equilibrium of the portion K_1Q [Fig. 1(iii)] of the beam,

$$\frac{W}{2} + F = W \text{ and } M - Fx + \frac{Wl}{2} = 0.$$

$$\therefore F = \frac{W}{2} \text{ and } M = -\frac{Wl}{2} + \frac{Wx}{2}. \quad (2)$$

Plotting F against x and M against x from (1) and (2), we get the shearing force diagram and the bending moment diagram as shown in Fig. 1(iv).

3. **A light cantilever beam 1 metre long, of circular cross-section (r = 1 cm) is clamped horizontally at one end and carries a load of 2 kg at the free-end which produces a very small bending of the beam. The free end is depressed and then released so that it begins to vibrate. Calculate the time period of the vibration, neglecting damping. Young's modulus for the beam = 2×10^{12} dynes per cm^2.**

[C.U. (Hons.) 1980 (Old Syll.)]

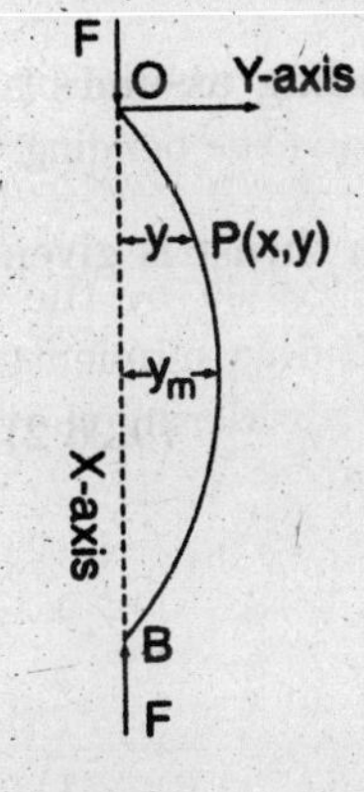

Fig. 9.26.1

From the equation 9.24.3 we have the time period of vibration of a light cantilever given by

$$T = 2\pi\sqrt{\frac{Ml^3}{3YAK^2}}$$

Here $M = 2$ kg $= 2 \times 10^3$ gm, $l = 1$ metre $= 10^2$ cm

$$AK^2 = \frac{\pi r^4}{4} = \frac{3.14 \times (1)^4}{4} = \frac{3.14}{4} \quad (\because r = 1 \text{ cm})$$

$$Y = 2 \times 10^{12} \text{ dynes/cm}^2$$

$$\therefore \quad T = 2 \times 3.14\sqrt{\frac{2 \times 10^3 \times 10^6 \times 4}{3 \times 2 \times 10^{12} \times 3.14}} = 0.1294 \text{ sec.}$$

9.26 Rods under Thrust—Struts

A strut is a long thin straight rod subject to a thrust. Let two equal forces F be applied in opposite directions at the ends O and B of a strut (Fig. 9.26.1). Let us consider any point $P(x, y)$ with reference to the origin O on the strut which is slightly bent.

Now, the external bending moment about P is — Fy, and for equilibrium condition, we have

$$YAk^2\frac{d^2y}{dx^2} = -Fy \quad (9.26.1)$$

the negative sign is taken because $\frac{d^2y}{dx^2}$ is negative, since it follows from the figure that $\frac{dy}{dx}$ decreases as x increases.

$$\therefore \quad YAk^2\frac{d^2y}{dx^2} + Fy = 0 \text{ or } \frac{d^2y}{dx^2} + \frac{F}{YAk^2}y = 0$$

$$\text{or } \frac{d^2y}{dx^2} + m^2y = 0, \text{ where } m^2 = \frac{F}{YAk^2}.$$

The solution of this equation is

$$y = A\cos mx + A'\sin mx, \quad A \text{ and } A' \text{ being constants.}$$

Let l be the length of the rod. Then the boundary conditions are $y = 0$ when $x = 0$, and $y = 0$ when $x = l$.

The first condition gives $A = 0$ and the second condition gives $A'\sin ml = 0$.

Hence either $A' = 0$ or $\sin ml = 0$.

$A' = 0$, means that there is no deflection of the strut and the strut is therefore a straight one.

Considering $\sin ml = 0$, we have

$$ml = s\pi[s = 1, 2, 3, \cdots, \text{etc.}] \quad \text{or} \quad m = \frac{s\pi}{l}.$$

Thus $y = A'\sin\frac{s\pi x}{l}$, which shows that the shape of the rod after bending assumes a harmonic or a sine curve.

When $s = 1$, the minimum length l to assume the above shape for a given thrust is given by $l = \frac{\pi}{m}$

$$\therefore y = A'\sin\frac{\pi x}{l} = y_{\text{max}}\sin\frac{\pi x}{l} \quad \left[\because \text{ at } x = \frac{l}{2}, y_{\text{max}} = A'\right] \tag{9.26.2}$$

Again, we know

$$m^2 = \frac{F}{YAk^2} = \frac{\pi^2}{l^2} \quad [\text{for } s = 1]$$

$$\text{or} \quad Fl^2 = \pi^2 YAk^2 = \text{constant} \tag{9.26.3}$$

$$\text{or} \quad l \propto \frac{1}{\sqrt{F}}$$

Thus for a rod of given cross-section and given material we note that the length of the rod to assume a harmonic shape is inversely proportional to the square root of the given thrust. Hence if the length of the rod be longer, comparatively small thrust should be applied at the ends of the rod to keep it in equilibrium after assuming the harmonic shape.

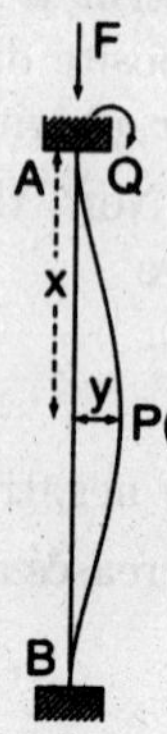

Fig. 9.26.2

When

$$m = \frac{2\pi}{l}, \frac{3\pi}{l}, \cdots, \text{etc.},$$

equilibrium of the rod is again obtained and we obtain various harmonic shapes corresponding to multiples of half a sine curve.

Struts clamped at both ends: We consider a strut with two ends fixed (Fig. 9.26.2). At the fixed ends the bending moment is not zero. Let Q be the moment at the end A. Then the moment of the forces on AP at $P(x, y)$ is $-Fy + Q$.

Equating the above expression with the internal bending moment $YAk^2\frac{d^2y}{dx^2}$ at P, we have

$$YAk^2\frac{d^2y}{dx^2} = -Fy + Q = -F\left(y - \frac{Q}{F}\right) = -Fs \quad \left[\text{putting } y - \tfrac{Q}{F} = s\right]$$

$$\text{or}\quad YAk^2\frac{d^2s}{dx^2} = -Fs \quad \left[\because\ \frac{d^2y}{dx^2} = \frac{d^2s}{dx^2}\right].$$

The solution of this equation is

$$s = C\cos mx + D\sin mx$$

$$\text{or}\quad y = \frac{Q}{F} + C\cos mx + D\sin mx \tag{9.26.4}$$

$$\text{or}\quad \frac{dy}{dx} = -Cm\sin mx + Dm\cos mx \tag{9.26.5}$$

Now the end conditions are

$$y = 0 \text{ at } x = 0 \text{ and } x = l;$$

$$\text{and}\quad \frac{dy}{dx} = 0 \text{ at } x = 0 \text{ and } x = l.$$

These conditions give

$$\frac{Q}{F} + C = 0 \text{ and } \frac{Q}{F} + C\cos ml + D\sin ml = 0 \tag{9.26.6}$$

$$\text{and}\quad Dm = 0 \text{ and } -Cm\sin ml + Dm\cos ml = 0 \tag{9.26.7}$$

From the first three of the above equations we have

$$C(1 - \cos ml) = 0 \tag{9.26.8}$$

and from the last two equations $C\sin ml = 0$.

If the rod is bent at all, C cannot be zero. Hence we have

$$\left.\begin{aligned}\sin ml &= 0\\ \cos ml &= 1\end{aligned}\right\} \tag{9.26.9}$$

Since the above two equations are to be simultaneously satisfied, we have

$$ml = 2\pi, 4\pi, 6\pi, \text{ etc.}$$

$$\therefore\ \sqrt{\frac{F}{YAk^2}}\,l = 2\pi, 4\pi, 6\pi, \text{ etc. or } \frac{F}{YAk^2} = \frac{4\pi^2}{l^2} \text{ or } \frac{16\pi^2}{l^2} \text{ or } \frac{36\pi^2}{l^2} \text{ etc.} \tag{9.26.10}$$

We know that $C = -\frac{Q}{F}$—hence the final value of y is given by

$$y = \frac{Q}{F} - \frac{Q}{F}\cos mx \quad [\text{from (9.26.4)}]$$

$$= \frac{Q}{F}(1 - \cos mx) = \frac{Q}{F}\left(1 - \cos\frac{2\pi}{l}x\right) \text{ or } \frac{Q}{F}\left(1 - \cos\frac{4\pi}{l}x\right) \text{ etc.}$$

From equation (9.26.10) it follows that the smallest force required to bend a rod with clamped ends is four times as great as the smallest force required to bend the same rod when the ends are not clamped.

9.27 Strain Energy

When a body is strained, internal forces due to elastic property of the body are called into play and hence work is done upon the body against the internal forces. This work is stored up within body as potential energy which is called *strain energy.* We consider here the following cases of strain energy.

(a) Stretching of a wire (i.e., for tensile or longitudinal strain)

We consider a wire of length L stretched by an amount l by a stretching force F. Now, for a small increase in length dl of the wire, the work done will be $F \cdot dl$. Hence the work done for stretching the wire by a length l is given by

$$W = \int_0^l F \cdot dl = \int_0^l \frac{Y\alpha}{L} l\,dl \quad \left[\because Y = \frac{F}{\alpha}\frac{L}{l}, \text{ where } \alpha \text{ is the cross-section of the wire}\right]$$

$$\text{or} \quad W = \frac{Y\alpha}{L}\int_0^l l\,dl = \frac{Y\alpha l^2}{2L} = \frac{1}{2}Fl = \frac{1}{2} \times \text{stretching force} \times \text{amount stretched}$$

$\therefore$ Work done per unit volume, i.e., the energy density due to longitudinal strain

$$= \frac{1}{2}\frac{Fl}{L\alpha} = \frac{1}{2}\left(\frac{F}{\alpha}\right)\left(\frac{l}{L}\right) = \frac{1}{2} \times \text{stress} \times \text{strain}$$

$$= \frac{1}{2} \times \text{Young's modulus} \times (\text{strain})^2. \tag{9.27.1}$$

(b) In the case of *volume strain,* proceeding exactly as above, we can show, that if the volume V of a substance be subjected to an external pressure p and the corresponding volume change be v, the work done for this volume change is

$$W = \frac{1}{2}\frac{k}{V}v^2 = \frac{1}{2}\frac{kv}{V}v = \frac{1}{2}p \times v \quad \left[\because p = \frac{kv}{V}, \ k \text{ being the bulk modulus}\right]$$

$$= \frac{1}{2} \times \text{increase in pressure} \times \text{decrease in volume}$$

Again we may write for the work done,

$$W = \frac{1}{2}\left(\frac{kv}{V}\right)\left(\frac{v}{V}\right)V = \frac{1}{2} \times \text{volume stress} \times \text{volume strain} \times \text{volume}$$

so that the energy density due to volume strain

$$= \frac{1}{2} \times \text{volume stress} \times \text{volume strain} = \frac{1}{2} \times \text{bulk modulus} \times (\text{volume strain})^2$$

which is similar to (9.27.1).

(c) Again in the case of *shearing strain*, the rigidity modulus $n = \frac{F/\alpha}{\theta}$, where F = the tangential force applied, α = the cross-sectional area (i.e., $\frac{F}{\alpha}$ = the shearing stress developed) and θ = the angle of shear

$$\therefore\ F = n\alpha\theta.$$

Considering this shear θ to occur by steps of $d\theta$, the total work done in the process is

$$W = \int_0^{\theta} F\cdot a d\theta,$$

where $ad\theta$ = linear displacement corresponding to an elementary shear $d\theta$, 'a' being the linear dimension of the body perpendicular to the applied force.

Hence

$$W = n\alpha a\int_0^{\theta}\theta d\theta = \frac{1}{2}n\alpha a\theta^2 = \frac{1}{2}n\alpha\theta\times a\theta = \frac{1}{2}\times \text{ applied torque}\times\text{ shear strain}$$

Again we may write for the work done,

$$W = \frac{1}{2}n\alpha a\theta^2 = \frac{1}{2}\times n\theta\times\theta\times\alpha a = \frac{1}{2}\times\text{ shearing stress}\times\text{ shearing strain}\times\text{ volume},$$

so that the energy density due to shearing strain,

$$= \frac{1}{2}\times\text{ shearing stress}\times\text{ shearing strain} = \frac{1}{2}\times\text{ shear modulus}\times(\text{shearing strain})^2$$

which is similar to (9.27.1).

9.27.1 Bending of a beam: Strain Energy

We suppose that due to bending a filament of length dl at a distance z from the neutral axis extends by e. The work done in stretching the filament by e is equal to the potential energy stored up in the filament. Now work done in stretching the filament by de

$$= \text{force along the filament}\times de$$

$$= \text{stress}\times\text{area of cross-section of the filament}\times de$$

$$= \text{stress}\times\alpha\times de\ (\alpha \text{ is the area of cross-section})$$

$$= Y\cdot\frac{e}{dl}\alpha\cdot de \quad \left[\because\ Y = \frac{\text{stress}}{\text{strain}} = \frac{\text{stress}}{e/dl}\right]$$

The work done in stretching the filament by e

$$= \int_0^{e}\frac{Y\alpha}{dl}\cdot ede = \frac{Y\alpha}{dl}\frac{e^2}{2}$$

Now the linear strain of the filament in question is given by $\frac{e}{dl} = \frac{z}{R}$.

$\therefore$ the work done in stretching the filament by e

$$= \frac{Y\alpha}{dl}\cdot\frac{1}{2}\frac{z^2}{R^2}(dl)^2 = \frac{Y\,dl}{2R^2}\cdot\alpha z^2$$

Hence the potential energy stored up in a beam of length dl due to bending is given by

$$dV = \frac{Y\,dl}{2R^2}\Sigma\alpha z^2 = \frac{Y\,dl}{2R^2}Ak^2 \tag{9.27.2}$$

If we consider G as the bending moment, then

$$G = \frac{YAk^2}{R} \quad \text{or} \quad \frac{1}{R^2} = \frac{G^2}{(YAk^2)^2} \tag{9.27.3}$$

Substituting (9.27.3) in (9.27.2) we have,

$$dV = \frac{dl}{2}\frac{G^2}{(YAk^2)^2}\cdot YAk^2 = \frac{G^2dl}{2YAk^2}$$

Therefore, the potential energy stored up in the whole beam of length l is

$$V = \frac{G^2}{2YAk^2}\int_0^l dl = \frac{G^2 l}{2YAk^2} \tag{9.27.4}$$

For a light cantilever, the potential energy stored up in the whole cantilever of length l is

$$V = \frac{G^2 l}{2YAK^2} = \left(\frac{YAK^2}{R}\right)^2 \times \frac{l}{2YAK^2} \quad \left(\because G = \frac{YAK^2}{R}\right)$$

$$= \frac{1}{2}\frac{YAK^2 l}{R^2} = \frac{1}{2}\frac{Wl^2}{R} \quad \left(\because \frac{YAK^2}{R} = Wl\right)$$

Again

$$V = \frac{G^2 l}{2YAK^2} = \frac{W^2l^2\cdot l}{2YAK^2} \quad \left(\because G = \frac{YAK^2}{R} = Wl\right)$$

$$= \frac{W^2l^3}{2YAK^2} = \frac{W}{2}\frac{Wl^3}{YAK^2}$$

$$= \frac{3W\delta}{2} \quad \left[\because YAK^2\delta = \frac{Wl^3}{3} \text{ or } \frac{Wl^3}{YAK^2} = 3\delta,\ \delta \text{ being the depression at the loaded end of the cantilever.}\right]$$

Illustrative Example

Calculate the work that has to be done to make a hoop out of a steel band of length $l = 2$ m, width $b =$ 6 cm and thickness $h = 2$ mm, assuming the process to take place within elastic range of the material.

The potential energy of bending of the band of length dl is given from (9.27.2) by

$$dV = \frac{1}{2}\frac{YAK^2}{R^2}dl.$$

When the steel band is bent into a circle of radius R, the potential energy of the entire beam is

$$V = \frac{YAK^2}{2R^2}\int_0^l dl = \frac{YAK^2 l}{2R^2},$$

since R is a constant throughout the bent band. Using $2\pi R = l$ or $R = \frac{l}{2\pi}$ and $AK^2 = \frac{bh^3}{12}$ in the expression for potential energy, we have

$$V = \frac{Y\frac{bh^3}{12}l}{2\cdot\frac{l^2}{4\pi^2}} = \frac{\pi^2}{6}\frac{bh^3Y}{l} = \frac{9.86\times 6\times(0.2)^3\times 2\times 10^{12}}{6\times 200} = 0.08\times 10^{10}\ \text{ergs} = 0.08\ \text{KJ}$$

where $Y = 2\times 10^{12}$ dynes/cm^2.

9.27.2 Twisting of a rod of circular cross-section: Strain Energy

We know from the theory of the torsion of a cylinder that the torsional couple is given by

$$C = \frac{n\pi a^4}{2l}\phi$$

$\therefore$ the work done and hence the energy stored up in twisting the cylinder by $d\phi$ is

$$dV = Cd\phi = \frac{n\pi a^4}{2l}\phi d\phi$$

$\therefore$ the total work done or the energy stored up in twisting the cylinder by ϕ_0 is

$$V = \frac{n\pi a^4}{2l}\int_0^{\phi_0}\phi d\phi = \frac{n\pi a^4}{4l}\phi_0^2$$

If C_0 be the torque when ϕ_0 is the angle of twist, we have

$$C_0 = \frac{n\pi a^4}{2l}\phi_0. \quad \therefore \quad V = \frac{1}{2}C_0\phi_0 = \frac{C_0^2 l}{n\pi a^4} \tag{9.27.5}$$

9.28 Flat Spiral Springs

Fig. 9.28.1

By a flat spiral spring we mean a uniform wire wound over a cylinder in such a way that the plane of each turn of the wire is at right angles to the axis of the cylinder. To consider the stretching and motion of such a spring, we take one end of the spring bent twice at right angles and fixed and the other end of the spring is also bent twice at right angles and carries the necessary load for its stretching. The vertical portion of both the ends of the wire coincides with the axis of the cylinder as shown in Fig. 9.28.1.

In the case of a flat spiral spring, only torsional couple comes to play.

Let R be the radius of the cylinder which is the radius of each turn of the spring and 'a' the radius of the wire of which the spiral spring is made. If W be the load applied to the lower end of the spring, the torque applied is given by

$$C = WR.$$

Now in the case of a flat spring, only torsional energy we are to consider because no bending moment comes to play. Therefore the potential energy V stored up in the spiral is given by

$$V = \frac{C^2 l}{n\pi a^4} = \frac{W^2 R^2 l}{n\pi a^4},$$

where l is the length of the spiral portion of the wire.

Now if x be the vertical displacement when load is W, then the work done by the load is

$$\int_0^x W\,dx = V = \frac{W^2 R^2 l}{n\pi a^4}$$

Differentiating we have,

$$W = \frac{2WR^2 l}{n\pi a^4} \cdot \frac{dW}{dx} \text{ or } \frac{dx}{dW} = \frac{2R^2 l}{n\pi a^4}$$

Integtrating we have,

$$x = \frac{2R^2 l}{n\pi a^4} W + K, \quad \text{where } K \text{ is the constant of integration.}$$

When $W = 0$, $x = 0$ $\therefore$ $K = 0$. Thus

$$x = \frac{2WR^2 l}{n\pi a^4} \tag{9.28.1}$$

The above expression gives a relation between x and W. From the above equation (9.28.1) we see that by measuring the displacement x, we can measure the modulus of rigidity which is given by

$$n = \frac{2WR^2 l}{\pi a^4 x} \tag{9.28.2}$$

Again we may express the value of the potential energy arising out of torsion in terms of x. We know that

$$V = \frac{W^2 R^2 l}{n\pi a^4} \text{ and } W = \frac{n\pi a^4 x}{2R^2 l} \quad \text{[from equation (9.28.1)]}$$

$$\therefore \quad V = \frac{(n\pi a^4)^2 x^2}{(2R^2 l)^2} \cdot \frac{R^2 l}{n\pi a^4} = \frac{n\pi a^4}{4R^2 l} x^2. \tag{9.28.3}$$

In this case the weight W is lowered by x and the C.G. of the spring by $x/2$. This corresponds to a decrease of potential energy equal to $\left(Wx + \frac{W_1 x}{2}\right)$, where W_1 is the weight of the spring and hence the potential energy is

$$\frac{n\pi a^4}{4R^2 l} x^2 - Wx - \frac{W_1 x}{2}$$

For consideration of the vertical oscillation of the flat spiral spring we are to calculate both the kinetic and potential energies in the following way:

The oscillation of the spiral spring is made possible by slightly pulling down the load W attached to the lower end of the spring and subsequently releasing it.

Kinetic energy due to the moving weight W is $\frac{1}{2}\frac{W}{g}\left(\frac{dx}{dt}\right)^2$.

Now as the load moves, the end of the spring attached to the load also moves with the same velocity as that of the load. But the fixed end of the spring is always at rest and hence at any point on the spring the velocity is proportional to its distance below the fixed point of the spring. At any point on the spring which is at a distance s below the fixed end, the velocity is $\frac{s}{l}\cdot\frac{dx}{dt}$. Let us consider an element of length ds at any point whose mass is ρds, where ρ is the linear density of the wire. Thus the kinetic energy of the element is

$$\frac{1}{2}\rho ds\left(\frac{s}{l}\frac{dx}{dt}\right)^2 = \frac{1}{2}\rho\left(\frac{dx}{dt}\right)^2\frac{s^2}{l^2}ds$$

Therefore, the kinetic energy associated with the spring is

$$\frac{1}{2}\rho\left(\frac{dx}{dt}\right)^2\frac{1}{l^2}\int_0^l s^2 ds = \frac{1}{2}\rho\left(\frac{dx}{dt}\right)^2\cdot\frac{1}{l^2}\cdot\frac{l^3}{3} = \frac{1}{6}\rho l\left(\frac{dx}{dt}\right)^2 = \frac{1}{6}\frac{W_1}{g}\left(\frac{dx}{dt}\right)^2$$

where W_1 is the weight of the spring.

$\therefore$ total kinetic energy is

$$\frac{1}{2}\cdot\frac{W}{g}\left(\frac{dx}{dt}\right)^2 + \frac{1}{6}\cdot\frac{W_1}{g}\left(\frac{dx}{dt}\right)^2 = \frac{1}{2g}\left(W + \frac{W_1}{3}\right)\left(\frac{dx}{dt}\right)^2$$

$\because$ the sum of the potential and kinetic energies is constant, we have

$$\frac{1}{2g}\left(W + \frac{W_1}{3}\right)\left(\frac{dx}{dt}\right)^2 + \frac{n\pi a^4}{4R^2 l}x^2 - Wx - \frac{W_1 x}{2} = \text{constant.}$$

Differentiating with respect to t, we have

$$\frac{1}{g}\left(W + \frac{W_1}{3}\right)\frac{dx}{dt}\cdot\frac{d^2x}{dt^2} + \frac{n\pi a^4}{2R^2 l}\cdot x\frac{dx}{dt} - W\frac{dx}{dt} - \frac{W_1}{2}\frac{dx}{dt} = 0$$

$$\text{or}\quad \frac{d^2x}{dt^2} + g\frac{n\pi a^4}{2R^2 l\left(W + \frac{W_1}{3}\right)}x - \frac{\left(W + \frac{W_1}{2}\right)g}{\left(W + \frac{W_1}{3}\right)} = 0$$

which is the differential equation of periodic motion.

The time period of vertical oscillation of the loaded spring is given by

$$T = 2\pi\sqrt{\frac{2R^2 l\left(W + \frac{W_1}{3}\right)}{gn\pi a^4}}$$

Thus knowing the period of oscillation, n may be determined.

9.28.1 Angular oscillation

If the tangent to the lower end of the spring is twisted through an angle ϕ, the bend per unit length is ϕ/l. In this case bending moment $G = \frac{YAk^2}{R}$ comes to play and the equation of motion of the spring undergoing angular oscillations is given by

$$(I_1 + I_2)\frac{d^2\phi}{dt^2} = -\frac{YAk^2}{R} = -\frac{\phi}{l}YAk^2 \quad \left[\because \frac{l}{R} = \phi\right],$$

where I_1 and I_2 are the moments of inertia of the load and the spring respectively

$$\therefore \quad \frac{d^2\phi}{dt^2} + \frac{YAk^2}{l(I_1 + I_2)}\phi = 0.$$

Hence the time period of angular oscillation is given by

$$T = 2\pi\sqrt{\frac{l(I_1 + I_2)}{YAk^2}} = 2\pi\sqrt{\frac{4l\left(I_1 + \frac{W_1R^2}{3}\right)}{Y\pi a^4}} \quad \left[\because I_2 = \frac{W_1}{3}R^2 \text{ and } Ak^2 = \frac{\pi a^4}{4}\right].$$

Illustrative example

A spiral spring whose diameter is 3 cms consists of 200 turns of wire of radius 1.0 mm, the length of the spring when hanging vertically being 50 cm. The rigidity modulus of the wire is 8×10^{11} dynes per square cm. Calculate the depression of the free end due to torsion, when a weight of 50 gms is hung on the lower end, the upper end being fixed.

The depression due to torsion is given by $x = \frac{2WR^2l}{n\pi a^4}$ [from equation (9.28.1)].

Here

$$W = 50 \times 981 \text{ dynes}; \quad R = \frac{3}{2} \text{ cms} = 1.5 \text{ cms}$$

$$l = 200 \times 2\pi R = 600\pi \text{ cms}; \quad n = 8 \times 10^{11} \text{ dynes/cm}^2$$

$$a = 1 \text{ mm} = 0.1 \text{ cm}$$

$$\therefore \quad x = \frac{2 \times 50 \times 981 \times (1.5)^2 \times 600\pi}{8 \times 10^{11} \times \pi \times (0.1)^4} = 1.655 \text{ cms}.$$

9.29 Non-flat Spiral Spring

By a non-flat spring we mean a uniform wire wound over a cylinder in such a way that the plane of each turn of the coil makes a constant angle α with the horizontal plane. Fig. 9.29.1 shows the vertical section of the spiral spring, R being the radius of the cylinder over which the wire is wound. The figure shows how the couples are acting when one end of the spring is fixed and the other end is loaded with W. We consider any point P on the spring. Here the applied couple is WR about the horizontal tangent PT to the cylinder at P. This is equivalent to the bending moment $G = WR\sin\alpha$ about PN at right angles to the turn of the wire at P. Now the twisting couple $C = WR\cos\alpha$ about the portion of the wire PQ. The sense of the couples is as shown in the figure.

When the load undergoes a vertical displacement through x, the work done by it accounts for the potential energy in the wire due to bending and twisting and hence we may write out the energy-equation in the following way:

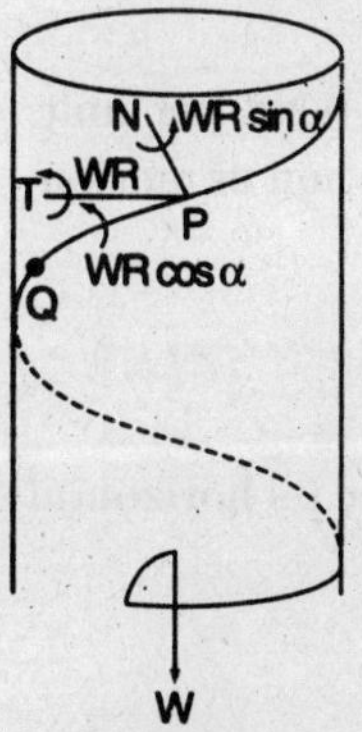

Fig. 9.29.1

Let V_1 be the potential energy due to bending and V_2 the potential energy due to torsion. We know $V_1 = \frac{G^2 l}{2YAk^2}$, where l is the length of the spiral portion of the wire and Ak^2 is the geometrical moment of inertia of the section of the wire. $Ak^2 = \frac{\pi a^4}{4}$ for a wire of circular section.

$$\therefore \quad V_1 = \frac{W^2R^2\sin^2\alpha l}{2Y\cdot\frac{\pi a^4}{4}} = \frac{2W^2R^2\sin^2\alpha l}{Y\pi a^4}. \tag{9.29.1}$$

Now,

$$V_2 = \frac{C^2 l}{n\pi a^4} = \frac{W^2R^2\cos^2\alpha l}{n\pi a^4}.$$

If x be the vertical displacement of the spring when load is W, the work done by W is $\int_0^x W dx$, which must be equal to the sum of the potential energies due to bending and torsion.

$$\therefore \int_0^x W dx = V_1 + V_2 = \frac{2W^2R^2 l\sin^2\alpha}{Y\pi a^4} + \frac{W^2R^2 l\cos^2\alpha}{n\pi a^4}$$

Differentiating both sides with respect to x, we have

$$W = \frac{2R^2 l\sin^2\alpha}{Y\pi a^4}\cdot 2W\frac{dW}{dx} + \frac{R^2 l\cos^2\alpha}{n\pi a^4}\cdot 2W\frac{dW}{dx} \quad \text{or} \quad \frac{dx}{dW} = \frac{4R^2 l\sin^2\alpha}{Y\pi a^4} + \frac{2R^2 l\cos^2\alpha}{n\pi a^4}$$

Integrating we have,

$$x = \frac{4WR^2 l\sin^2\alpha}{Y\pi a^4} + \frac{2WR^2 l\cos^2\alpha}{n\pi a^4} + K$$

K being the constant of integration.

Now when $W = 0, x = 0$; $\therefore K = 0$

$$\therefore \quad x = \frac{2WR^2 l}{\pi a^4}\left(\frac{2\sin^2\alpha}{Y} + \frac{\cos^2\alpha}{n}\right)$$

In addition to the vertical displacement there is an angular displacement in the horizontal plane.

In this case the torsional couple is $WR\cos\alpha$

$$\therefore \quad WR\cos\alpha = \frac{n\pi a^4}{2l}\phi \quad \text{or} \quad \phi = \frac{2WRl\cos\alpha}{n\pi a^4}$$

where ϕ is the angle of twist at the end of the wire. Now the angular displacement takes place in a plane at right angles to PQ and hence this plane makes an angle $\frac{\pi}{2} - \alpha$ with the horizontal. Therefore, the component of this angular displacement in the horizontal plane is given by

$$\beta = \phi\cos\left(\frac{\pi}{2} - \alpha\right) = \phi\sin\alpha = \frac{2WRl\sin\alpha\cos\alpha}{n\pi a^4} = \frac{WRl\sin 2\alpha}{n\pi a^4}$$

which coils up the spring as it acts inwards. Again, if θ be the angular deflection for the whole length of the spring due to the bending couple $WR\sin\alpha$, we have

$$\frac{YAk^2\theta}{l} = WR\sin\alpha \quad \left[\because \frac{\theta}{l} = \frac{1}{R} = \text{curvature of bending}\right]$$

$$\text{or} \quad \theta = \frac{WRl\sin\alpha}{YAk^2} = \frac{4WRl\sin\alpha}{Y\pi a^4}$$

This angular displacement θ due to bending is in the plane of the coil and hence its horizontal component γ is given by

$$\gamma = \theta\cos\alpha = \frac{4WRl\sin\alpha\cos\alpha}{Y\pi a^4} = \frac{2WRl\sin 2\alpha}{Y\pi a^4}.$$

This causes the spring to uncoil as it acts outwards

$$\therefore \beta - \gamma = \frac{WRl\sin 2\alpha}{\pi a^4}\left(\frac{1}{n} - \frac{2}{Y}\right)$$

when $\frac{1}{n} > \frac{2}{Y}$ or $Y > 2n$, we see that the spring coils up. When $\frac{1}{n} < \frac{2}{Y}$, the spring uncoils. In reality, in case of most metals $Y > 2n$ and so generally non-flat springs coil up when loaded.

9.30 Experimental Determination of Elastic Constants

9.30.1 Young's modulus

(*a*) *Vernier method:* This is a simple method of measuring the Young's modulus of a wire. The apparatus used in this method consists of two identical long vertical wires A and B of the same material and cross-section suspended from the same support EF. The wire A carries a millimetre scale S and is stretched by a fixed weight W at its free end. The experimental wire B carries a vernier V which slides over the millimetre scale S with minimum friction when the experimental wire is stretched. The lower end of the wire B carries a hanger H on which a fixed load is always kept which is called the *dead load.*

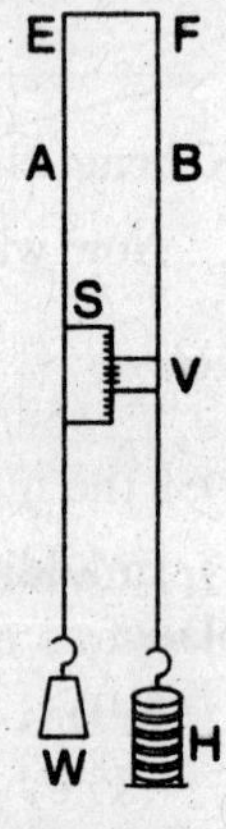

Fig. 9.30.1

To start with the experiment weights are successively placed on the hanger in regular increasing order and the extension of the wire at each step is noted from the position of the vernier against the millimetre scale. It should be noted that the maximum weight put on the hanger should not exceed the elastic limit of the wire. Extensions (negative) are also noted when the weights are diminished in the order in which they were increased. For each load, the mean of the two extensions, one for load increasing and another for load diminishing, is determined. Thus the extensions of the experimental wire for different loads are found out. A graph is next plotted with load against extension. The graph is a straight line passing through the origin. The radius r of the experimental wire is measured by means of a micrometer screw gauge. The length l of experimental wire is measured by means of a micrometer screw gauge. The length l of experimental wire from its fixed support to the point of attachment to the vernier is also measured. If m be the mass of a load on the hanger for which the extension is x, then the stretching force $F = mg$ and strain $= \frac{x}{l}$.

The cross-section of the wire is $A = \pi r^2$.

Hence the Young's modulus for the material of the experimental wire is

$$Y = \frac{Fl}{Ax} = \frac{mgl}{\pi r^2 x}$$

$\left(\frac{m}{x}\right)$ may be determined from the load-extension graph.

(*b*) *Ewing's extensometer:* When the specimen, the Young's modulus of which is to be determined, is a bar, Ewing's extensometer may be used. Ewing's extensometer consists of two metal arms QP and SR which are pivoted at C and D by screws on the specimen bar AB (Fig. 9.30.2). The arm QP is bent round to form a vertical arm which carries a micrometer screw at its lower end. The tip T of the micrometer screw rests in a V-shaped groove cut at the end S of the arm SR. A microscope M provided with a micrometer scale is fitted at the lower end of an arm which is pivoted at its upper end P. The microscope is focussed on a horizontal scratch at the end R of the arm SR.

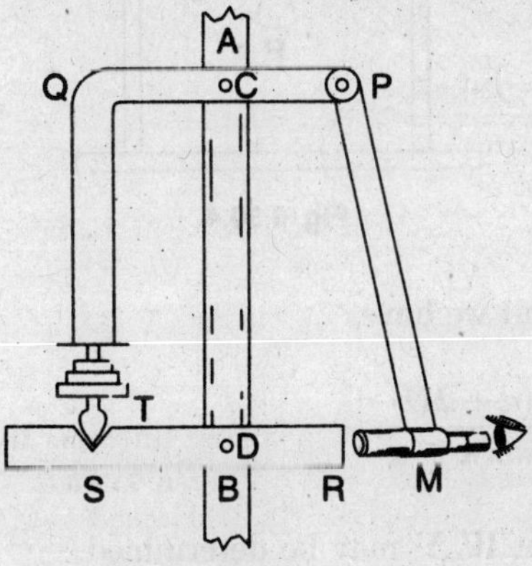

Fig. 9.30.2

To start with the experiment the two arms QP and SR are made horizontal by means of the micrometer screw. The experimental bar is now stretched downwards by means of a testing machine as a result of which SR will be tilted a little about T as fulcrum and the end R will be displaced downwards. The downward displacement of R can be observed with the help of the microscope and is measured by its micrometer scale.

In fig. 9.30.3 RR' denotes the downward shift of the end R. Now the increase in length of the experimental bar due to stretching is DD'. From the similar triangles TRR' and TDD' we have,

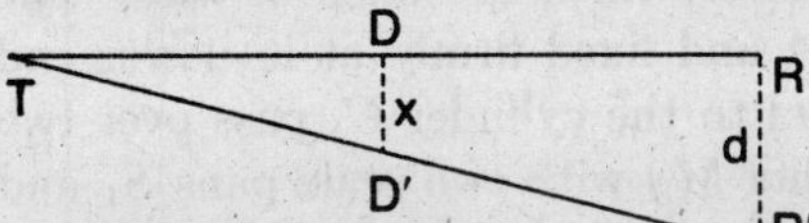

Fig. 9.30.3

$$\frac{DD'}{RR'} = \frac{TD}{TR} \quad \text{or} \quad DD' = RR'\frac{TD}{TR} \quad \text{or} \quad x = d\frac{TD}{TR}$$

where x is the increase in length of the experimental bar and d is the downward shift of the end R. Thus knowing d, TD and TR, x can be determined. Knowing the elongation x of the bar, the Young's modulus for the material of the bar may be determined in the usual way as already described.

(*c*) *König's method:* The method, due to König, consists of measuring the angle through which the free ends of a beam supported at both ends are turned when the beam is loaded at its centre.

The beam AB is placed horizontally on two steel knife edges K_1, K_2 (Fig. 9.30.4). A pair of pillars carrying two small mirrors M_1, M_2 are fixed normally near the ends of the beam. At the middle of the beam there is a frame R which carries a knife edge K, the sharp end of which rests on the beam. The lower part of the frame carries a hook H which in turn carries a hanger P. Light from a scale S suffers reflection first from the mirror M_2 and then from the mirror M_1. The scale is viewed by means of a telescope T. The angle θ through which each free end of the beam is turned is determined by measuring the difference in scale readings in the telescope before and after putting the load on the hanger. If d be the horizontal distance between the two mirrors, s the difference between the scale readings and D is the horizontal distance of the scale from the mirror M_2, then

$$s = (2d + 4D)\theta \quad \text{or} \quad \theta = \frac{s}{2d + 4D}. \qquad (9.30.1)$$

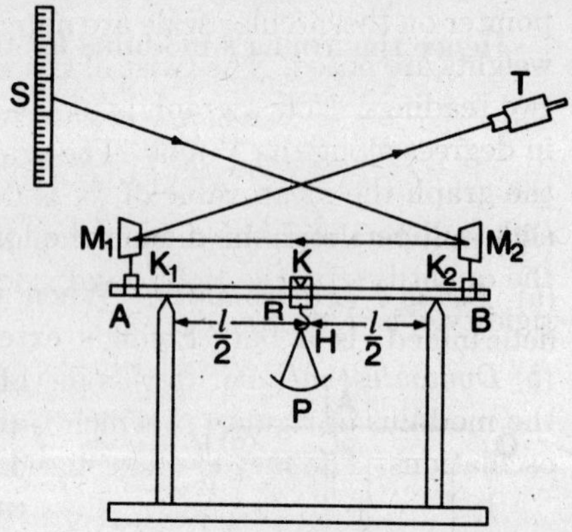

Fig. 9.30.4

Now

$$\tan\theta = \frac{dy}{dx} = \frac{Wl^2}{16YAK^2} \qquad (9.30.2)$$

where l is the distance between the knife edges K_1, K_2. [see Art. 9.20]

We may equate (9.30.1) and (9.30.2), θ being small;

$$\therefore \quad \frac{s}{2d + 4D} = \frac{Wl^2}{16YAK^2}$$

If the beam is of rectangular cross-section, $AK^2 = \frac{ab^3}{12}$, and we have

$$\frac{3Wl^2}{4Yab^3} = \frac{s}{2(d + 2D)} \quad \text{or} \quad Y = \frac{3Wl^2(d + 2D)}{2ab^3 s}.$$

Thus, knowing the change in scale reading s for the weight W, Y may be determined.

In this method greater accuracy is obtained as the change in scale reading s may be made larger than the depression produced by the load.

9.30.2 Modulus of rigidity

(a) *Statical method:* This method is suitable for comparatively thick specimen of wire. The specimen AB is a thick wire clamped at its upper end A and fixed firmly at its lower end to a cylinder C (Fig. 9.30.5). Two flexible cords, attached to the cylinder C, pass over two freely running pulleys P_1 and P_2. The pulleys carry weights Mg with two scale pans S_1 and S_2 at the ends of the cords and the weights act tangentially to the cylinder C. Hence the couple applied to the end B of the specimen wire is $2MgR$, where R is the radius of the cylinder C. The angle of twist ϕ at the end B is directly read on the horizontal circular scale S by the movement over the scale of a pointer attached to the wire at B. The angle of twist may also be measured by noting the deflection from a mirror attached to the wire by the usual lamp and scale arrangement. We have then $2MgR = \frac{n\pi a^4\phi}{2l}$, where a is the radius of the wire, l its length and n the modulus of rigidity for its material.

$$\therefore n = \frac{2MgDl}{\pi a^4 \phi},$$

where D is the diameter of the cylinder.

If the twist ϕ be measured in degrees; then ϕ (in radians) $= \frac{\pi\phi^0}{180}$.

Then

$$n = \frac{360gDl}{\pi^2 a^4}\left(\frac{M}{\phi^0}\right) \qquad (9.30.3)$$

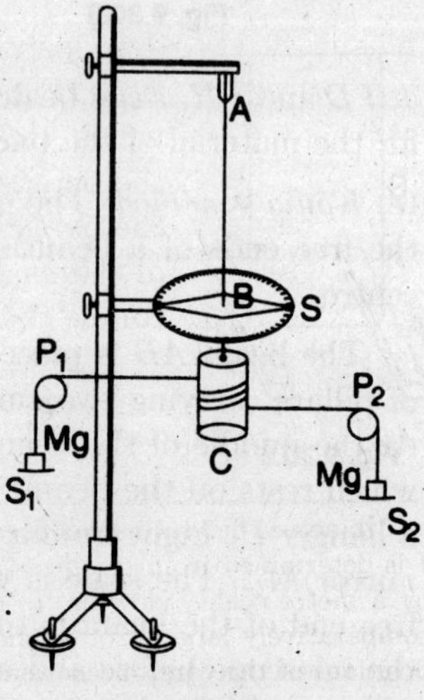

Fig. 9.30.5

Various equal weights are placed on the two scale pans S_1 and S_2 and the readings of the pointer on the circular scale are noted at each time. Readings for both increasing and decreasing weights are noted. The twist of the wire for each load is determined from the mean of the above two readings. Now a graph is plotted with loads along the X-axis and angles of twist measured in degrees along the Y-axis. The graph is a straight line which passes through the origin. From the graph the mean value of $\frac{M}{\phi^0}$ is determined. The diameter of the cylinder is measured by a slide callipers and the diameter of the specimen wire is measured by a screw gauge. Thus all the quantities on the right hand side of equation (9.30.3) are known and hence the modulus of rigidity n is determined.

(*b*) *Dynamical method:* In this method a cylinder or a disc is suspended from one end of a wire, the modulus of rigidity of which is to be determined. The system is made to execute torsional oscillations. The period of oscillation is given by

$$T = 2\pi\sqrt{\frac{I}{\tau}} \tag{9.30.4}$$

where I is the moment of inertia of the suspended body about the wire and τ is the couple per unit twist of the wire. (See Torsional Pendulum of Art. 5.14).

From equation (9.30.4) we may write

$$\tau = \frac{4\pi^2 I}{T^2} \quad \text{or} \quad \frac{n\pi a^4}{2l} = \frac{4\pi^2 I}{T^2}$$

where l and a are respectively the length and radius of the wire

$$\text{or} \quad n = \frac{8\pi I l}{T^2 a^4}$$

Thus knowing T, a, I and l, the modulus of rigidity n for the wire can be easily calculated.*

9.30.3 Searle's method for the determination of Young's modulus and modulus of rigidity

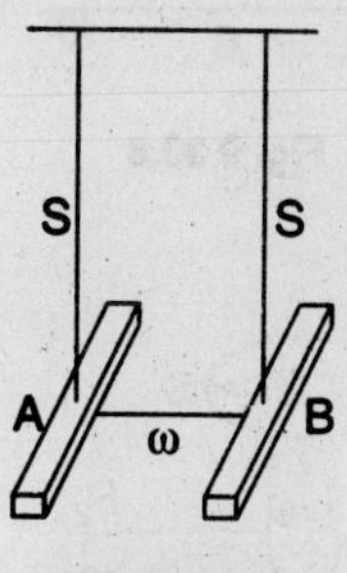

Fig. 9.30.6

Two identical bars A and B are suspended horizontally by means of two strings S, S, (Fig. 9.30.6). A uniform wire w, the values of Y and n for the material of which are to be determined, is fixed between the centres of the two bars. For the determination of Y for the material of the wire, we bring the two ends of the bars closer to each other, so as to bend the wire to form an arc of a circle of radius R (say) (Fig. 9.30.7). Due to bending of the wire, internal resisting couple that comes to play is given by

$$\frac{YAk^2}{R} = \frac{YAk^2}{l}2\theta \quad \left[\because \frac{l}{2}/R = \theta\right].$$

*In actual determination of n by this method, T is determined with the help of a stop watch in the usual way. I is determined from the mass and dimensions of the cylinder ($I = MR^2/2$, R being its radius) l is measured by a metre scale. Now as 'a' occurs in the expression as a^4, a slight error in its measurement will cause a comparatively large error in the final result. So 'a' is very carefully determined by a screw gauge at different positions of the wire, measurements being taken at each place twice at right angled directions, and mean of the readings gives 'a'.

When the ends of the bar are released, this resisting couple comes to play as the restoring torque and the bars will undergo simple harmonic oscillation. We may, therefore, write the equation of motion of each bar as

$$I\frac{d^2\theta}{dt^2} = -\frac{2YAk^2}{l}\theta \quad \left[\begin{array}{l} I \text{ is the moment of inertia of} \\ \text{each bar about the string.}\end{array}\right]$$

$$\text{or} \quad \frac{d^2\theta}{dt^2} + \frac{2YAk^2}{Il}\theta = 0$$

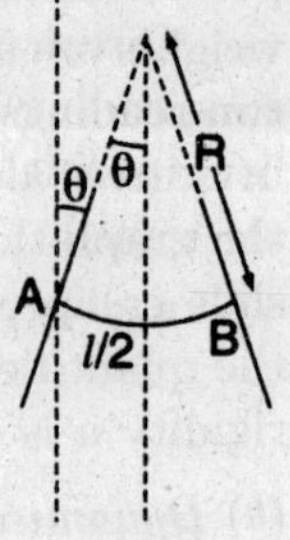

Fig. 9.30.7

The time-period T_1 of oscillation is given by

$$\left(\frac{2\pi}{T_1}\right)^2 = \frac{2YAk^2}{Il} \quad \text{or} \quad T_1 = 2\pi\sqrt{\frac{Il}{2YAk^2}} = 2\pi\sqrt{\frac{Il}{2Y\frac{\pi r^4}{4}}} = 2\pi\sqrt{\frac{2Il}{Y\pi r^4}}$$

where r is the radius of the wire.

$$\therefore\ T_1^2 = \frac{4\pi^2 \cdot 2Il}{Y\pi r^4} = \frac{8\pi Il}{Yr^4} \quad \text{or} \quad Y = \frac{8\pi Il}{T_1^2 r^4} \tag{9.30.5}$$

All the quantities on the right hand side of the above equation being known, Y is determined.

For the determination of the modulus of rigidity of the wire, one of the bars A is fixed horizontally and a torque is applied at the other bar B (say) to twist the wire (Fig. 9.30.8). Now the torsional couple that comes to play here, serves as the restoring torque. We have then the equation of motion of the bar B about its axis of suspension given by

$$I\frac{d^2\theta}{dt^2} = -\frac{n\pi r^4}{2l}\theta \quad \text{or} \quad \frac{d^2\theta}{dt^2} + \frac{n\pi r^4}{2Il}\theta = 0.$$

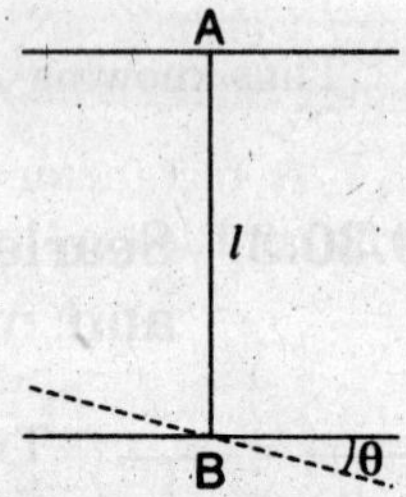

Fig. 9.30.8

So the bar B executes simple harmonic motion about the axis of suspension AB. The time period T_2 of oscillation is given by

$$\left(\frac{2\pi}{T_2}\right)^2 = \frac{n\pi r^4}{2Il} \quad \text{or} \quad T_2^2 = \frac{8\pi Il}{nr^4} \quad \text{or} \quad n = \frac{8\pi Il}{T_2^2 r^4}. \tag{9.30.6}$$

Thus n is determined.

Again we have from equations (9.30.5) and (9.30.6)

$$\frac{Y}{n} = \frac{8\pi Il}{T_1^2 r^4} \times \frac{T_2^2 r^4}{8\pi Il} = \frac{T_2^2}{T_1^2}$$

Now, we know

$$1 + \sigma = \frac{Y}{2n} = \frac{T_2^2}{2T_1^2} \quad \text{or} \quad \sigma = \frac{T_2^2 - 2T_1^2}{2T_1^2}$$

Thus σ is determined.

9.30.4 Determination of Poisson's ratio

(*a*) *Direct method:* In this method Poisson's ratio is determined by measuring lateral and longitudinal strains directly. The lateral strain is measured by a micrometer screw-gauze and longitudinal strain may be determined by Vernier method or by an extensometer as already described.

(*b*) *Bar method:* When a flat bar of rectangular cross-section is bent lengthwise, it will also be bent breadthwise forming anticlastic surfaces as shown in Fig. 9.30.9. It has already been shown that the longitudinal extensional strain of a filament PQ at a distance z above the neutral axis EF is z/R_1, where R_1 is the radius of curvature of the neutral axis. Hence the transverse filament QS at a distance z below the neutral axis FG (curvature here being in opposite direction) will suffer contractional strain z/R_2, where R_2 is the radius of anticlastic curvature. In the figure O and O' denote the centres of curvature of EF and FG respectively.

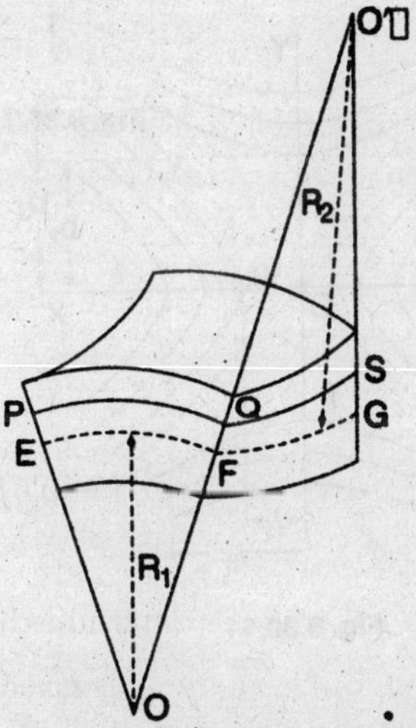

Fig. 9.30.9

$$\therefore \sigma = \frac{\text{lateral strain}}{\text{longitudinal strain}} = \frac{z/R_2}{z/R_1} = \frac{R_1}{R_2}$$

Thus measuring the radii of curvatures R_1 and R_2, Poisson's ratio σ may be determined. R_1 and R_2 may be directly determined by clamping pointers to the bar and observing the distances and angles traversed when a given couple is applied to the bar.

(*c*) *Cornu's method:* Poisson's ratio for good reflectors such as glass or highly polished metal may be determined by this method.

A rectangular bar of the specimen is placed symmetrically on two knife edges K_1 and K_2 and loaded by suspending two equal weights at equal distances from the two knife edges (Fig. 9.30.10). A plane optical test plate P is placed in contact with the bar.

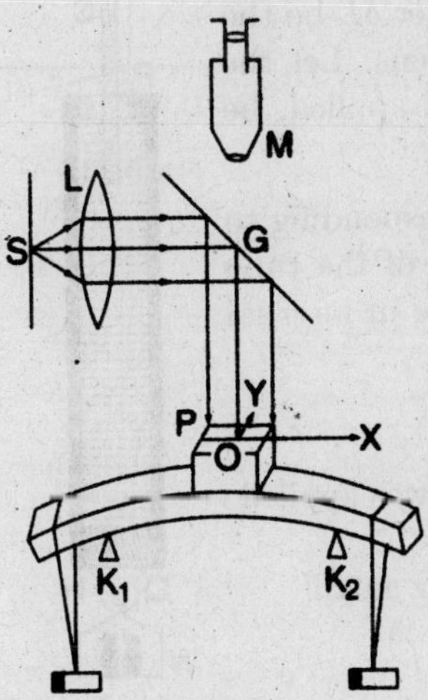

Fig. 9.30.10

Monochromatic light from an extended source S is rendered parallel by means of a lens L and is incident on a glass plate G inclined at an angle of 45° to the vertical. A portion of the light reflected from the glass plate G is incident perpendicularly on the test plate P. Fringes which are rectangular hyperbolas formed due to interference between the rays reflected from the upper surface of the test plate and those reflected from the upper surface of the specimen bar may be viewed with the help of a travelling microscope M. If l_1 and l_n be the distances of the first and the nth fringe from the origin O measured along OX, i.e., along the length of the beam (Fig. 9.30.11), we have

$$\frac{l_n^2}{R_1} = n\lambda \text{ and } \frac{l_1^2}{R_1} = \lambda. \quad \therefore \frac{l_n^2 - l_1^2}{R_1} = (n-1)\lambda, \tag{9.30.7}$$

where R_1 is the radius of curvature along the length of the beam.

Similarly, if b_1 and b_n be the distances of the first and the nth fringe from the origin O measured along OY, i.e., along the breadth of the beam (Fig. 9.30.11), we have,

$$\frac{b_n^2 - b_1^2}{R_2} = (n-1)\lambda, \tag{9.30.8}$$

where R_2 is the radius of curvature along the breadth of the beam.

From equations (9.30.7) and (9.30.8) we may write

$$\frac{l_n^2 - l_1^2}{R_1} = \frac{b_n^2 - b_1^2}{R_2} \quad \text{or} \quad \frac{R_1}{R_2} = \sigma = \frac{l_n^2 - l_1^2}{b_n^2 - b_1^2}. \tag{9.30.9}$$

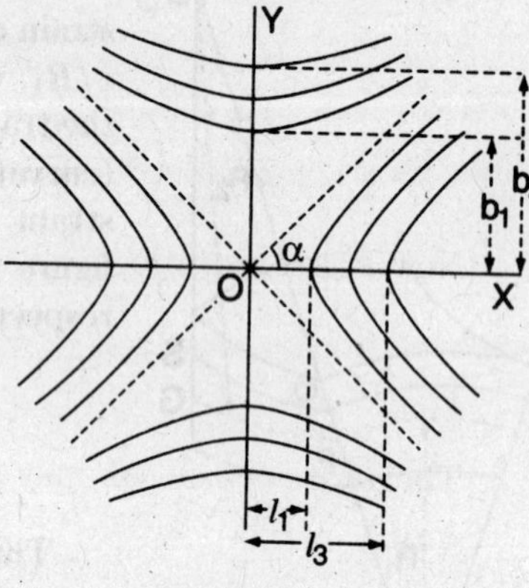

Fig. 9.30.11

Considering the 1st and the 3rd fringe (say), the distances l_1, l_3, b_1 and b_3 (Fig. 9.30.11) are measured by the travelling microscope and hence σ is determined. If the asymptotes of rectangular hyperbolas make an angle α with X-axis, then it can be shown

$$\sigma = \frac{R_1}{R_2} = \cot^2\alpha.$$

The angle α may be measured directly by a goniometer eyepiece.

(d) *Poisson's ratio for India rubber:* In Fig. 9.30.12, PQ denotes a thin walled India rubber tube of length nearly one metre and diameter 2 cms. The upper end of the tube is closed by a rubber cork through which passes a capillary tube C. The lower end of the tube is closed and carries a hook H. From the hook any desired load may be suspended. The rubber tube and part of the capillary tube are filled with water. The water should be made free from air-bubbles. The apparatus is supported by a rigid stand. When a suitable load W is suspended from the hook H, the length and hence the internal volume of the tube increases. This causes fall in the level of the meniscus in the capillary tube C. Let δL be the increase in length of the tube and δh be the fall in level of the meniscus. Let the length, diameter and volume of the rubber tube before the load was applied, be L, D and V respectively.

Let δV be the increase in internal volume of the rubber tube corresponding to an increase in length by an amount δL. Now the increase in length of the tube is coupled with decrease in cross-section of the tube. Let the decrease in internal cross-section be δs.

$$\therefore V + \delta V = (s - \delta s)(L + \delta L),$$

where s stands for the internal cross-section of the tube before the load was applied.

$$\text{or} \quad V + \delta V = sL + s\delta L - \delta sL \qquad [\text{neglecting } \delta s \cdot \delta L \text{ which is very small}]$$

$$= V + s\delta L - \delta sL$$

$$\text{or} \quad \delta V = s\delta L - \delta sL \tag{9.30.10}$$

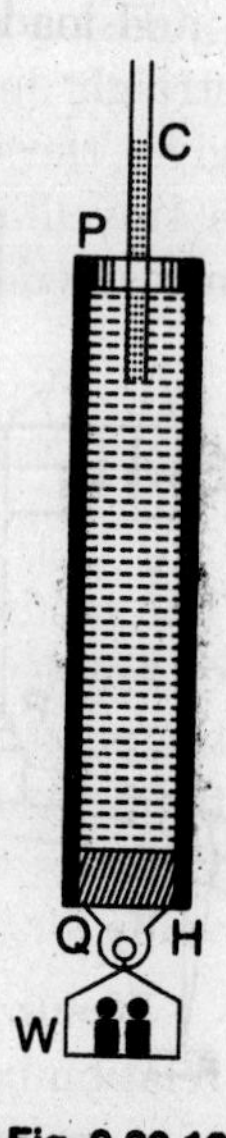

Fig. 9.30.12

Now

$$s = \frac{\pi D^2}{4} \tag{9.30.11}$$

or $$\delta s = \frac{2\pi D \delta D}{4} = \frac{\pi D}{2}\delta D = 2s\frac{\delta D}{D} \quad \text{[from equation (9.30.11)]} \tag{9.30.12}$$

Substituting this value of δs in (9.30.10), we have

$$\delta V = s\delta L - 2sL\frac{\delta D}{D} \quad \text{or} \quad \frac{\delta V}{\delta L} = s - \frac{2sL}{D}\cdot\frac{\delta D}{\delta L}$$

$$\therefore \quad \frac{\delta D}{\delta L} = \left(s - \frac{\delta V}{\delta L}\right) \Big/ \frac{2sL}{D} = \frac{D}{2L} - \frac{\delta V}{\delta L}\frac{D}{2sL} = \frac{D}{2L}\left(1 - \frac{1}{s}\frac{\delta V}{\delta L}\right). \tag{9.30.13}$$

Now Poisson's ratio

$$\sigma = \frac{\delta D/D}{\delta L/L} = \frac{1}{2}\left(1 - \frac{1}{s}\frac{\delta V}{\delta L}\right) \quad \text{[from equation (9.30.13)]}$$

The above expression may also be written in the form

$$\sigma = \frac{1}{2}\left(1 - \frac{1}{\pi r^2}\frac{\delta V}{\delta L}\right) \tag{9.30.14}$$

where r is the internal radius of the tube.

Again $\delta V = \pi a^2 \delta h$, where a is the internal radius of the capillary tube. Hence equation (9.30.14) may be expressed as

$$\sigma = \frac{1}{2}\left(1 - \frac{1}{\pi r^2}\frac{\pi a^2 \delta h)}{\delta L}\right) = \frac{1}{2}\left(1 - \frac{a^2}{r^2}\cdot\frac{\delta h}{\delta L}\right). \tag{9.30.15}$$

Using different loads the corresponding values of δh and δL are plotted. The slope of the straight line so obtained gives the value of $\frac{\delta h}{\delta L}$, which when substituted in equation (9.30.15) gives the value of σ. The values of a and r are determined at different points by means of a travelling microscope and their mean values are taken. The changes δL and δh are also measured with the help of a travelling microscope.

9.31 Variation of Elasticity with Temperature

The moduli of elasticity decrease, in general, with the rise of temperature. For small ranges of temperature the relation is nearly linear. Lee and Shave, and Andrews found that at temperatures within 150°C of the melting point of a material, Young's modulus Y and temperature T are related exponentially and the relation is $Y = Y_1 e^{-b_1 T}$, where b_1 has one value for temperatures up to half the absolute temperature of the melting point and the other value for higher temperatures.

Kohlrausch and Loomis observed for temperatures between 15°C and 100°C that the relation between modulus of rigidity and temperature is given by

$$n_T = n_0(1 - aT - bT^2)$$

Harton determined the modulii of rigidity in cases of a number of metallic wires and quartz fibres between temperatures $16°C$ and $100°C$. In case of pure copper and steel, a linear relation was obtained. The relation was, however, approximately linear in case of silver. In case of invar alloy the elasticity remains unaffected by change in temperature.

Exercises

1. Define Yonung's modulus, Bulk modulus, Modulus of rigidity, axial modulus and Poisson's ratio.

2. Show that a shear is equivalent to an elongational strain and an equal compressional strain at right angles to each other **[C.U. (Hons.) '81, '96, '98, 2003, 2005, 2007]**

3. Show that for a homogeneous isotropic medium $Y = 2n(1+\sigma)$. **[C.U. (Hons.) '98, 2003, 2005]**

4. (a) If Y, k and σ represent Young's modulus, Bulk modulus and Poisson's ratio respectively, prove that $Y = 3k(1-2\sigma)$. State the limiting values of σ.

[C.U. (Hons.) '91, '93, 2001, 2005 (Old Syll.)]

(b) If stresses in three mutually perpendicular directions in an isotropic body be $2 \times 10^8, 4 \times 10^8$ and 3×10^8 dynes cm^{-2}, what will be the volume strain?

(Young's modulus for the material $= 20 \times 10^{11}$ dynes cm^{-2}, Poisson's ratio for the material $= 0.3$). **[Ans.** 1.8×10^{-4}**] [C.U. (Hons.) '77, '95]**

[Hints. Use equations (9.12.5)].

5. Define Young's modulus, Bulk modulus and modulus of rigidity. If E, k and n represent these moduli respectively, prove the relation.

$$E = \frac{9nk}{3k+n}.$$ **[Allahabad '43]**

6. Deduce the relation between axial modulus, bulk modulus and modulus of rigidity.

7. (a) Define modulus of rigidity. Explain clearly the dynamical method for the determination of modulus of rigidity of a wire.

(b) A cylinder of mass 1 kg. and radius 10 cm. is suspended along its axis by a 100 cm. long wire of radius 1 mm. The cylinder is given a twist and oscillates with a time period 2.12 sec. Find the modulus of rigidity of the material of the suspension wire.

[C.U. (Hons.) '88]

[Hints. Use the formula for n in the dynamical method.]

8. Derive an expression for the couple required to bend a uniform straight metallic strip into an arc of a circle of small curvature. **[A.U. '50]**

9. A uniform beam is supported horizontally near the ends on two knife edges and a concentrated load W is applied at any point of the beam. Neglecting the weight of the beam, calculate the depression at any point of the beam.

10. (a) A uniform beam is clamped horizontally at both ends and a concentrated load is applied at the midpoint of the beam. Neglecting the weight of the beam calculate the sag at the midpoint of the beam.

(b) Calculate the depression at the mid-point of a heavy beam clamped at both ends when a concentrated load is applied at the mid-point of the beam. **[Ans.** $\delta = \frac{l^3}{192YAK^2}\left(W + \frac{W_1}{2}\right)$]

11. Deduce an expression for the vertical oscillation of a loaded flat spring.

12. Describe Searle's method for the determination of Young's modulus and modulus of rigidity.

13. Describe a method for the determination of Poisson's ratio of glass.

14. Describe König's method for the determination of Young's modulus of a beam.

15. Describe cornu's method for the determination of Poisson's ratio of a polished metal.

16. Describe a method for the determination of Poisson's ratio for India rubber.

17. Show that the smallest force required to bend a strut clamped at the ends is four times as the smallest force required to bend the same strut when the ends are not clamped.

18. A light beam of length l and cross-sectional area α is rigidly fixed at one end and carries a concentrated load W at the other end. Deduce an expression for the strain energy of the beam.

[C.U. (Hons.) '91]

19. If the bulk modulus and the simple rigidity of a substance are k and n, show that its simple elongational elasticity is given by $k + \frac{4}{3}n$. **[C.U. (Hons.) '89]**

20. A steel bar of length 2 metres and having a square cross-section of side 1 cm is supported at one end and at the middle. Find the deflection of the free end when a load of 3 kg is placed half-way between the supports. (Young's modulus for steel $= 2 \times 10^{12}$ dynes cm^{-2})

[**Ans.** 3.3075 cm] **[C.U. (Hons.) '87]**

(See worked out examples)

21. A steel wire of length 100 cm and of square cross-section (cross-sectional area 0.25 cm^2) is bent in the form of a circle and the two ends welded together. The wire circle is heated until its temperature rises by 180°C and the circle is then slipped over a wheel which fits exactly. Estimate the pressure exerted on the wheel when the circle cools down to its original temperature (Y for steel $= 1.95 \times 10^{12}$ dynes cm^{-2} and coefficient of linear expansion of steel $\alpha = 1.10 \times 10^{-5}$ per°C).

[C.U. (Hons.) '86]

[**Ans.** The wire circle, on being heated until its temperature rises by 180°C, fits exactly over the wheel, so that the fractional increase in length of the wire $\left(\frac{\delta L}{L}\right)$ on heating is exactly equal to the tensile strain developed in it on cooling down to its original temperature. So on cooling from θ_2 to θ_1 the tensile stress developed in it is given by $p = Y\frac{\delta L}{L}$ where δL is the increase in length (of the wire circle of initial length L) due to heating.

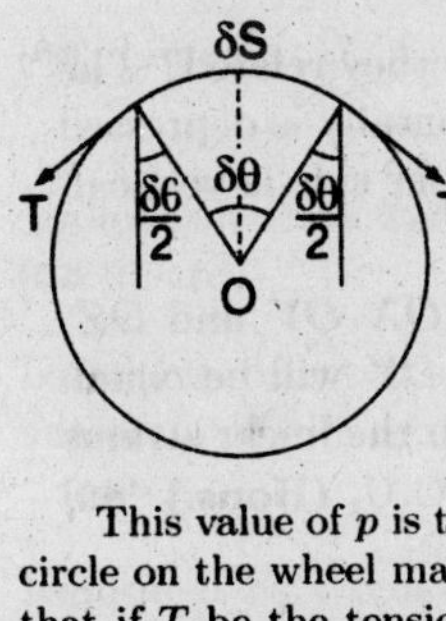

Again, the coefficient of linear expansion of steel is, from definition, given by

$$\alpha = \frac{\delta L}{L(\theta_2 - \theta_1)} \quad \text{or} \quad \delta L = \alpha L(\theta_2 - \theta_1).$$

Substituting this value of δL in the expression for p we have

$$p = Y\alpha(\theta_2 - \theta_1) = 1.95 \times 10^{12} \times 11 \times 10^{-6} \times 180 = 38.61 \times 10^8 \text{ dynes cm}^{-2}.$$

This value of p is the circumferential stress in the wire circle, and the inward pressure P exerted by the wire circle on the wheel may be calculated in terms of p, in the following way; cosidering the attached figure, we see that if T be the tension developed in the ring when it is cooled, the total inward (acting towards O) force on the element δS is

$$2T \sin\frac{\delta\theta}{2} = T\delta\theta, \quad (\text{where } \delta\theta \text{ is very small i.e., } \sin\tfrac{\delta\theta}{2} = \tfrac{\delta\theta}{2})$$

so that the inward pressure is given by

$$P = \frac{T\delta\theta}{r\delta\theta \cdot a} = \frac{T}{ar} = \frac{pa\delta r}{ar} = \frac{p\delta r}{r}$$

where a, r and δr are respectively the breadth, radius and thickness of the ring and p is the circumferential stress in the wire circle

$$\therefore \quad P = \frac{p\delta r}{r} = \frac{2\pi p \delta r}{L} \quad \left(\because 2\pi r = L \text{ or } r = \frac{L}{2\pi}\right)$$

$$= \frac{2 \times 3.14 \times 38.61 \times 10^8}{100} \times 0.5 \quad (\because L = 100 \text{ cm and } \delta r = \sqrt{0.25} = 0.5 \text{ cm})$$

$$= 1.212 \times 10^8 \text{ dynes cm}^{-2}.]$$

22. A horizontal wire of length $2l$ and cross-section α is tightly fixed to rigid supports at both ends. A weight W is suspended from the mid-point of the wire. Show that the depression d of the mid-point is given by $d^3 = \frac{Wl^3}{Y\alpha}$ where Y = Young's modulus of the material of the wire.

[C.U. (Hons.) '85, cf. 98]

[**Ans.** From the figure $2T\sin\theta = W$ or $T = \frac{W}{2\sin\theta}$. Now stress $= \frac{T}{\alpha}$ and strain $= \frac{\sqrt{l^2+d^2}-l}{l}$

$$= \frac{\sqrt{l^2+d^2}}{l} - 1 = \left(1 + \frac{d^2}{l^2}\right)^{1/2} - 1 = \frac{d^2}{2l^2} \quad \text{(neglecting smaller terms)}$$

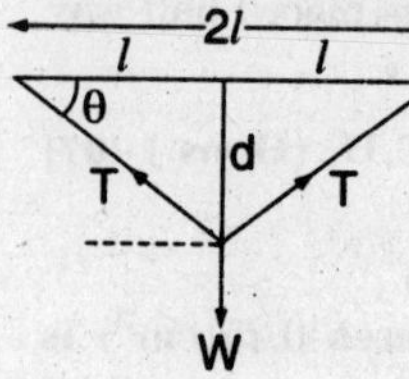

$$\therefore \quad Y = \frac{T/\alpha}{d^2/2l^2} = \frac{T2l^2}{\alpha d^2} = \frac{W}{2\sin\theta}\,\frac{2l^2}{\alpha d^2}$$

$$= \frac{W}{2}\frac{l}{d}, \frac{2l^2}{\alpha d^2} \quad \left(\because \sin\theta \simeq \tan\theta = \frac{d}{l} \text{ when } \theta \text{ is small}\right)$$

$$= \frac{Wl^3}{\alpha d^3} \text{ whence the result follows.}]$$

23. Consider a rod of circular cross-section, of length l and radius a. The volume of the rod is not changed when the rod is stretched. Show that the Poisson's ratio has the value $\frac{1}{2}$ in this case.

[C.U. (Hons.) '84]

[**Ans.** Volume of the rod $= \pi a^2 l$. If α and β be the longitudinal and lateral strains, then the increased length $= l(1+\alpha)$ and the decreased radius $= a(1-\beta)$, so that the new volume $= \pi a^2(1-\beta)^2 l(1+\alpha) = \pi a^2 l(1-\beta)^2(1+\alpha)$.

From the condition of the problem, we have $\pi a^2 l = \pi a^2 l(1-\beta)^2(1+\alpha)$ or $1+\alpha = (1-\beta)^{-2} = 1+2\beta$ since β is very small. $\therefore$ Poisson's ratio $\sigma = \frac{\beta}{\alpha} = \frac{1}{2}$.]

24. (a) Define flexural rigidity and internal bending moment of a beam. How are they related? The free-end of a cantilever beam of rectangular cross-section mounted horizontally is depressed less when the longer side of the rectangle is vertical than when the shorter side is vertical. Explain.

(b) A body is under stress along three mutually perpendicular directions OX, OY and OZ. (i) Deduce the condition under which the linear strains along OX and OY will be equal, while that along OZ will be zero. (ii) Also find the condition under which the linear strains along OX and OY will be zero. **[C.U. (Hons.) '80]**
(See worked out examples).

25. The tension of a steel wire of radius 0.5 mm stretched between two rigid supports 1 m apart is found to be 3.77×10^6 dynes when the temperature is 20°C. At what temperature will the tension on the wire just vanish? Assume Young's modulus for steel $= 2 \times 10^{12}$ dynes/cm^2 and the coefficient of linear expansion of steel $= 1.2 \times 10^{-5}$ per°C.

[Ans: 40°C—see worked out examples] [C.U. (Hons.) '81]

26. (a) Calculate the torsional modulus of a wire of radius R and length l. if the material of the wire has a shear modulus G and the potential energy stored. **[C.U. (Hons.) 99, 2007]**

(b) A body of moment of inertia I is suspended from the same wire. Find its time period of oscillation. **[C.U. (Hons.) '82]**
(See torsional pendulum of Art. 5.14).

27. Define Young's modulus, Poisson's ratio, bulk modulus and shear modulus of an isotropic elastic body. Derive two independent relations connecting any three of these quantities. How many of these four quantities are independent? Explain

One litre of glycerine gets reduced in volume by 0.21 cm^3 under a pressure of 10 kg/cm^2; calculate its bulk modulus. **[C.U. (Hons.) '83, '85, '87, 2001, cf. 2004]**

[**Hints.** $K = \frac{P}{\frac{\delta V}{V}}$ Here $P = 10 \times 9.8 \times 10^5$ dynes/cm^2 $\delta V = 0.21$ and $V = 1000$ cm^3.]

28. (a) A uniform beam with a square cross-section (width and depth equal to l) and length L, has one end anchored in a vertical wall. Assuming small bending, find the deflection of its free end due to its own weight in terms of relevant parameters.
(See worked out examples).

(b) Show that the isothermal bulk modulus of a perfect gas in equilibrium is equal to its pressure. **[C.U. (Hons.) sp. '83]**

29. Show that the depression of a cantilever at any point P due to a load at Q is the same as that at Q when the load is at P. Q represents the free end of the cantilever.
(See Reciprocal Theorem). **[C.U. (Hons.) '84, '94, '96, 2002, 03, 2007, 2006 (old syll.)]**

30. (a) Explain clearly how an internal torque develops due to bending of a beam. Show that for a small bending of a homogeneous beam of uniform cross-section, the torque N can be expressed as $N = YAK^2\frac{d^2y}{dx^2}$ where Y, A and K are the Young's modulus, cross-section and radius of gyration respectively.

(b) A uniform beam of length l and of rectangular cross-section (area of cross-section is A) is clamped horizontally at one end. A weight W is suspended from the free end. W is so large that the weight of the beam can be neglected. Find the depression at the free end due to shear alone. (Modulus of rigidity $= n$). **[C.U. (Hons.) Sp. '84, '94, '96, 2002]**

31. A metal rod of length L and cross-section s suffers a small longitudinal strain. Obtain an expression for the potential energy due to this strain if the Young's modulus of the material is Y.
[C.U. (Hons.) '86, 2000]

32. Find the depression of a cantilever beam of uniform cross-section and weight W, when loaded at the free end by a weight W_0. What is the strain energy in this case?
[C.U. (Hons.) '87, '99, 2004]

33. A cylindrical wire of length l and radius r is clamped at one end and twisted at the other end by a torque about its axis. Find the modulus of rigidity of the material of the wire. Show that the strain energy of the twisted wire is $\frac{1}{2}C_m\theta_m$, where C_m is the torque for the maximum twist θ_m. What is torsional rigidity? **[C.U. (Hons.) '80, '88, '92, '97]**

34. Calculate the amount of work done in stretching by 1 mm a piece of steel wire 1 sq. mm. in cross-section and 1 metre in length; Young's modulus of steel being 2×10^{12} c.g.s units.
[**Ans.** Apply the formula $W = \frac{Y\alpha l^2}{2L}$] **[C.U. (Hons.) '89]**

35. (a) What do you mean by internal bending moment of a beam? Find an expression for it.
The two ends of a light beam rest on two knife-edges and a weight is placed at its mid-point. Find the depression of the mid-point of the beam.

(b) What is meant by torsional oscillation? Discuss how the modulus of rigidity of the material of a long wire can be determined, using torsional oscillation. **[C.U. (Hons.) '90]**

36. Two uniform cylinders, one solid and the other hollow, have the same length and mass and are made of the same material. The hollow cylinder has external radius twice the internal radius. Find the ratio of torsional rigidities of the two cylinders. Deduce any formula you use.

[**Ans.** See worked out examples] **[C.U. (Hons.) '92, '99, 2007]**

37. (a) A beam of weight W_0 is supported horizontally over knife-edges placed at its ends. A load W is placed at its mid-point. Calculate the depression of the beam at an arbitrary point.

(b) What are the values of Young's modulus and Poisson's ratio for a liquid?

[C.U. (Hons.) '93, 2004]

[**Ans.** For incompressible liquids, there will be no longitudinal strain for any finite longitudinal stress in that direction. Hence

$$Y = \frac{\text{Longitudinal stress}}{\text{Longitudinal strain}} = \infty(\text{infinity})$$

For a liquid filled inside a very light deformable cylinder of radius r and length l, volume $= \pi r^2 l =$ constant or $\pi r^2 dl + 2\pi r l dr = 0$ or $\pi r^2 dl = -2\pi r l dr$ or $-\frac{dr}{r} = \frac{1}{2}\frac{dl}{l}$. Hence $\sigma = -\frac{\frac{dr}{r}}{\frac{dl}{l}} = \frac{1}{2}$]

38. Assuming that the stress-strain relations in an elastic solid are linear, deduce, by the application of the principle of superposition, the relations connecting the Young's modulus, Poisson's ratio and the moduli of bulk and rigidity of an isotropic body. Also find the axial modulus for pure elongational strain in a single direction, when the lateral strains are zero. **[C.U. (Hons.) '95]**

39. (a) A cantilever is loaded at its free end. The free end is now further displaced and set free. Calculate the period of its oscillation. Also describe how you would utilise this method of oscillation to find out Young's modulus of the material of the cantilever.

(b) A circular rod, supported on horizontal knife-edges, 40 cm apart, has a length of 60 cm and carries a load of 10 kg suspended from each end. The radius of the rod is 4 mm and when the loads are applied, the centre rises by 4.8 mm. Find the Young's modulus for the rod.

[C.U. (Hons.) '96 (Old. Syll.)]

[**Ans.** Apply (9.21.2) or (9.22.2) i.e., $Y = \frac{l^2 d}{8AK^2}\frac{Mg}{h}$ Here $AK^2 = \frac{\pi r^4}{4}$, $r = 0.4$ cm, $l = 60$ cm, $d = 10$ cm, $h = 0.48$ cm $Mg = 10 \times 9.8 \times 10^5$ dynes/cm^2 $= 9.8 \times 10^6$ dynes/cm^2]

40. (a) Consider a homogeneous beam of uniform cross-section; ω is its weight per unit length. Over any section at x (Fig.), the shearing force is F and internal bending moment is N. Establish (i) $\omega = -\frac{dF}{dx}$, (ii) $F = -\frac{dN}{dx}$.

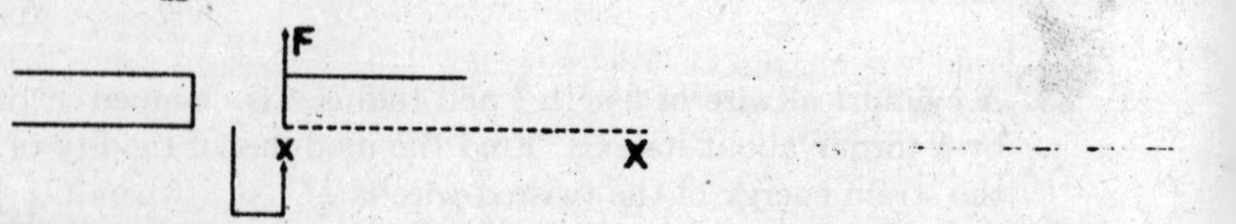

(b) Bulk modulus of water is 2.3×10^5 dynes/cm^2. How much pressure is needed to compress a sample of water by 0.1%? (See worked out examples) **[C.U. (Hons.) '97, 2006]**

41. A steel wire of length 1 m and diameter 1 mm is fixed horizontally at the two ends. A weight of 10 kg is placed at the middle. Calculate the depression of the weight due to elongation of the wire. Also calculate the resulting strain energy. (Young's modulus $= 2 \times 10^{12}$ dynes/cm^2).

[**Ans.** See worked out problems] **[C.U. (Hons.) '98]**

42. A circular disc of 10 cms radius and mass 1 kg is suspended in a horizontal plane by a vertical wire attached to its centre. If the diameter of the wire is 1 mm, its length 1.5 m and the period of torsional vibration of the disc is 5 sec, find the rigidity modulus of the material of the wire.

[**Hints:** Use the formula for 'n' in the dynamical method] [**Ans.** 1.206×10^{12} dynes/cm^2]

43. A thin uniform wire of radius r is clamped horizontally between rigid supports A and B, so that it is just taut, and the length of the wire between A ad B is $2l$. A mass m is hung from the mid-point C and displaces C through a vertical distance y, small compared with l. Find expressions for the stress and strain in the wire, and for young's modulus, in terms of m, l, r and y, neglecting in each case squares and higher powers of y/l in comparison with unity.

(See worked out examples) [**Ans.** $\frac{mgl}{2\pi r^2 y}, \frac{y^2}{2l^2}, \frac{mgl^3}{\pi r^2 y^3}$.] [**Oxford and Cambridge**]

44. Calculate the depression at the mid-point of a heavy beam rigidly fixed at both ends, when a concentrated load W is applied at the mid-point of the beam.

Such a beam is known as a *built-in* or *encastre beam,* as shown in the attached figure with a load W at the middle. At both ends of the beam, $y = \frac{dy}{dx} = 0$ and also at $x = l/2$, i.e., the mid-point, $\frac{dy}{dx} = 0$. The torque M acting at each end is shown is the figure. Considering any point P at a distance x from one fixed end, we may write for equilibrium of this portion of the beam $(0 < x < l/2)$.

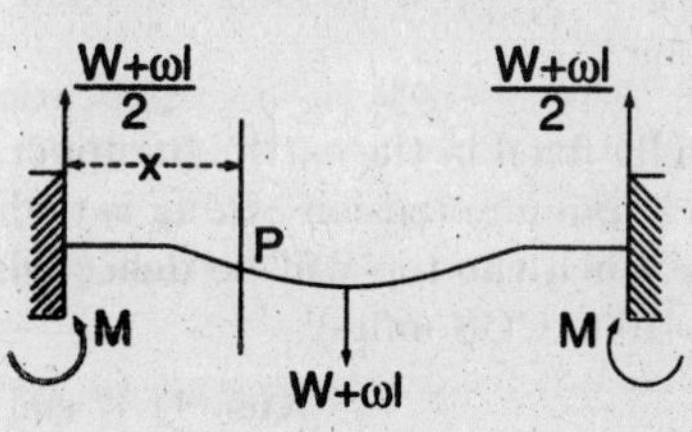

$$F - \frac{W + wl}{2} + wx = 0$$

(where F is the shearing force in the plane through P)

$$\text{or} \quad F = \frac{W}{2} + w\left(\frac{l}{2} - x\right)$$

$$\text{or} \quad \frac{dM}{dx} = -F = -\frac{W}{2} - w\left(\frac{l}{2} - x\right), \quad \text{whence we have}$$

$$YAk^2\frac{d^3y}{dx^3} = -\frac{W}{2} - w\left(\frac{l}{2} - x\right) \quad \left(\because M = YAk^2\frac{d^2y}{dx^2}\right). \tag{1}$$

Integrating three times in succession, we have

$$YAk^2\frac{d^2y}{dx^2} = -\frac{Wx}{2} - w\left(\frac{lx}{2} - \frac{x^2}{2}\right) + C_1 \tag{2}$$

$$YAk^2\frac{dy}{dx} = -\frac{Wx^2}{4} - w\left(\frac{lx^2}{4} - \frac{x^3}{6}\right) + C_1x + C_2 \tag{3}$$

$$YAk^2y = -\frac{Wx^3}{12} - w\left(\frac{lx^3}{12} - \frac{x^4}{24}\right) + C_1\frac{x^2}{2} + C_2x + C_3. \tag{4}$$

From (3), applying the condition at $x = 0$, $\frac{dy}{dx} = 0$, we get $C_2 = 0$.

Again from (3), applying the condition at $x = \frac{l}{2}$, $\frac{dy}{dx} = 0$, we have

$$-\frac{Wl^2}{16} - w\left(\frac{l^3}{16} - \frac{l^3}{48}\right) + C_1\frac{l}{2} = 0 \quad \text{or} \quad C_1 = \frac{Wl}{8} + \frac{wl^2}{12}.$$

Also from (4), applying the condition $y = 0$ at $x = 0$, we have $C_3 = 0$, ad so the equation (4) reduces to

$$YAk^2y = -\frac{Wx^3}{12} - w\left(\frac{lx^3}{12} - \frac{x^4}{24}\right) + \frac{1}{2}\left(\frac{Wl}{8} + \frac{wl^2}{12}\right)x^2$$

$$= W\left(\frac{lx^2}{16} - \frac{x^3}{12}\right) + w\left(\frac{x^4}{24} - \frac{lx^3}{12} + \frac{l^2x^2}{24}\right).$$

The depression δ at the mid-point is obtained by putting $y = \delta$ and $x = \frac{l}{2}$.

$$\therefore \quad YAk^2\delta = W\left(\frac{l^3}{64} - \frac{l^3}{96}\right) + wl^4\left(\frac{1}{16 \times 24} - \frac{1}{96} + \frac{1}{96}\right)$$

$$= \frac{Wl^3}{192} + \frac{W_1 l^3}{384} = \frac{l^3}{192}\left(W + \frac{1}{2}W_1\right) \quad (\because W_1 = wl).$$

$$\therefore \quad \delta = \frac{l^3}{192Ak^2}\left(W + \frac{1}{2}W_1\right). \tag{5}$$

45. Prove that if a number of rods are joined end to end, the torsional rigidity of the combination is given by the same formula as that for a number of resistances in parallel. Hence or otherwise find the torsional rigidity of a tapering wire of length l, rigidity modulus n and radii a and b of the ends. **[C.U. (Hons.) 2007]**

(See worked out examples).

46. A man weighing 70 kg stands on the middle of a wooden plank 4 metres long, 30 cms broad, and 4 cms deep, supported at its ends. What will be his time of swing? Young's modulus of wood = 10^{11} dynes cm^{-2}. [**Ans.** 0.48 sec]

[**Hints:** For a bar supported at its ends and loaded at the middle, $\delta = \frac{Wl^3}{4Yab^3}$. Depressing the plank further by x, find the net downward force, and derive the eqn. for s.h.m.]

47. A vertical steel rod of circular cross-section of radius 1 cm is rigidly fixed in the earth. Its upper end is 3 metres from the earth. A thick string which can stand a maximum tension of 2 kg is tied at the upper end of the rod and pulled horizontally. Find how much the top will be deflected before the string snaps. (Y for steel = 2×10^{12} CGS units, g = 1000 CGS units).

[**Ans.** 11.47 cm]

[**Hints:** The rod is like a vertical cantilever since its lower end is fixed and a force of 2 kg is applied at its upper end. The deflection of the upper end is $\delta = \frac{Wl^3}{3YAk^2}$].

48. A steel wire 2 mm in diameter is just stretched between two points at a temperature of 25°C. Determine the tension when the temperature falls to 15°C. (Coefficient of linear expansion of steel = 11×10^{-6} per °C, Young's modulus for steel = 21×10^{11} dynes cm^{-2}.)

[**Ans.** 7.26×10^6 dynes]

49. Calculate the strain potential energy of a horizontal rod 1 m long and 0.4 sq cm in cross-section fixed at both ends at a temperature of 25°C when it has been cooled down to 10°C. (Coefficient of linear expansion for the material of the rod = 11×10^{-6} per °C and young's modulus for it = 2.1×10^{12} dynes cm^{-2}) [**Ans.** 1.14×10^6 ergs]

50. Calculate the force necessary to stretch by 1 mm a rod of copper 100 cm long and 2 mm radius.

If the energy stored in the stretched rod were converted into heat and imparted to the rod, what rise of temperature would result?

(Using CGS units, assume $J = 4.2 \times 10^7$ and for copper Young's modulus = 12×10^{11}, density = 9, specific heat 0.1) **(Leverpool Univ)** [**Ans.** 154 kg-wt, $\frac{1}{126}$°C]

51. Two cylinders A and B of radii r and $2r$ are soldered co-axially. The free end of A is clamped and the free end of B is twisted by an agle θ. Calculate the twist at the junction, taking the materials of the two cylinders to be the same and lengths equal.

[**Ans.** If a couple Γ be applied at the free end of B to produce a twist θ, then let ϕ be the twist produced at the junction of the two cylinder A and B. The action of this couple is thus to produce (i) a twist ϕ between the clamped end and the free end of the cylinder A and (ii) a relative twist of $(\theta - \phi)$ between the jucntion-end and the free end of B. Hence for the cylinder A, we have $\Gamma = \frac{\pi n r^4}{2l}\phi$, where n = rigidity modulus of A, r its radius and l its length.

Similarly for the cylinder B, $\Gamma = \frac{\pi n (2r)^4}{2l}(\theta - \phi)$, where $2r$ = radius of B and l its legth, so that we may write

$$\frac{16\pi n r^4}{2l}(\theta - \phi) = \frac{\pi n r^4}{2l}\phi \quad \text{or} \quad 16(\theta - \phi) = \phi \quad \text{whence } \phi = \frac{16}{17}\theta.$$

52. Find the amount of work done in twisting a steel wire of radius 1 mm and length 25 cms through an angle of 45°, modulus of rigidity of steel being 8×10^{11} CGS units. [**Ans.** 1.551×10^6 ergs]

53. A body suspended symmetrically from the lower end of a wire 100 cm long and 1.22 mm in diameter oscillates about the wire as axis with a period of 1.25 sec. If the modulus of rigidity of the material of the wire be 8×10^{11} dynes cm^{-2}, calculate the moment of inertia of the body about the axis of rotation. [**Ans.** 6.885×10^3 gm-cm^2] [**I.A.S.**]

54. A gold wire of 0.32 mm in diameter elongates by 1 mm when stretched by a force of 330 gm-wt and twists through 1 radian when equal and opposite torques of 145 dynes are applied at its ends. Find the value of Poisson's ratio for gold (g = 981 cms per sec^2). [**Ans.** 0.430]

55. A light elastic string is suspended vertically from a point and carries a heavy mass at its lower end, which stretches it through a distance l cm. Show that the vertical oscillations of the system are simple harmonic in nature and of time period equal to that of a simple pendulum of length l cm.

[**Ans.** Stress = $\frac{F}{A}$, where $F = mg$, A = area of cross-section, m = mass hanging. Let L = original length of the string.

If the string is pulled down further through a small distance x, resultant downward force = $mg - T$, where T = tension of the string at this moment. Hence

$$m\frac{d^2x}{dt^2} = mg - T. \tag{1}$$

If μ = modulus of elasticity, we have

$$\mu = \frac{T/A}{(l+x)/L} \quad \text{or} \quad T = \frac{\mu(l+x)}{L}A. \tag{2}$$

Also originally F = stretching force, l = stretch, and L = original length, so that

$$F = \mu\frac{l}{L}A \quad \text{or} \quad \mu = \frac{FL}{lA}. \tag{3}$$

Substituting this vlaue of μ in (2), we have

$$T = \frac{FL(l+x)A}{lLA} = \frac{mg(l+x)}{l}.$$

Putting this value of T in (1), we have

$$m\frac{d^2x}{dt^2} = mg - \frac{mg(l+x)}{l} = mg\left(1 - \frac{l+x}{l}\right) = -\frac{mg}{l}x$$

$$\text{or} \quad \frac{d^2x}{dt^2} = -\frac{g}{l}x \quad \text{or} \quad \frac{d^2x}{dt^2} + \frac{g}{l}x = 0$$

which is the characteristic differential equation for simple harmonic motion. Hence time period, $T = 2\pi\sqrt{l/g}$, which is same as that of a simple pendulum of length l.]

56. (a) Find out the expression for the internal bending moment of a beam.

[C.U. (Hons.) 2001, 2002, 2005]

(b) Let a horizontal beam of length 1 m and mass per unit length 2.0 gm per cm be fixed at one end, while the free end is loaded with a weight of 1 kg. Find the torque at a section at a distance of 40 cm from the fixed end. **[C.U. (Hons.) 2001, '05]**

[**Ans.** Here

$$\frac{d^2M}{dx^2} = \omega \quad \text{or} \quad d\left(\frac{dM}{dx}\right) = \omega dx$$

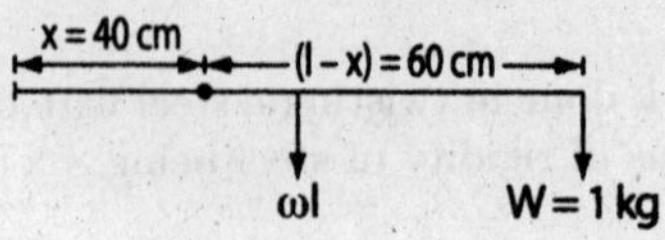

Integrating we have

$$\frac{dM}{dx} = \omega x + C_1;$$

now at $x = 0$,

$$\frac{dM}{dx} = 0 \quad \therefore C_1 = 0 \quad \text{or} \quad \frac{dM}{dx} = \omega x;$$

Integrating again we have

$$M = \frac{\omega x^2}{2} + C_2;$$

again at $x = 0$, $M = 0$.

$$\therefore \quad C_2 = 0 \quad \therefore \quad M = \frac{\omega x^2}{2},$$

so that the net torque at a section at a distance of 40 cm from the fixed end as shown in the figure is given by

$$M = \frac{\omega x^2}{2} + W(l - x)$$

Here $\omega = 2$ gm/cm, $x = 40$ cm, $(l - x) = 60$ cm, $W = 1000$ gm.
$\therefore M = 40 \times 40 + 1000 \times 60 = 1600 + 60{,}000 = 61{,}600$ dyne-cm.]

57. (a) **A light beam** OA **is fixed rigidly at** O **and loaded at the free-end A (fig.) Assume bending is small**

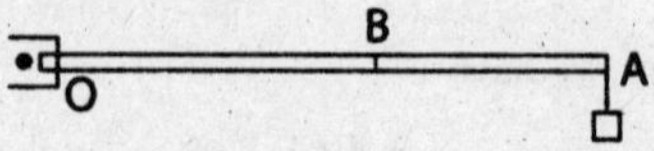

i. **Draw a section of the beam in bent position. Define neutral layer and discuss the deformation of the different layers in relation to the neutral layer.**
[C.U. (Hons.) 2002]

Section of the beam in bent position is shown in the attached figure. $C \rightarrow$ **centre of curvature of the neutral layer** NN'; $R \rightarrow$ **radius of curvature of the neutral layer;** $CC' \rightarrow$ **Axis of bending. Layers above** NN' **are elongated, while layers below** NN' **are shortened.**

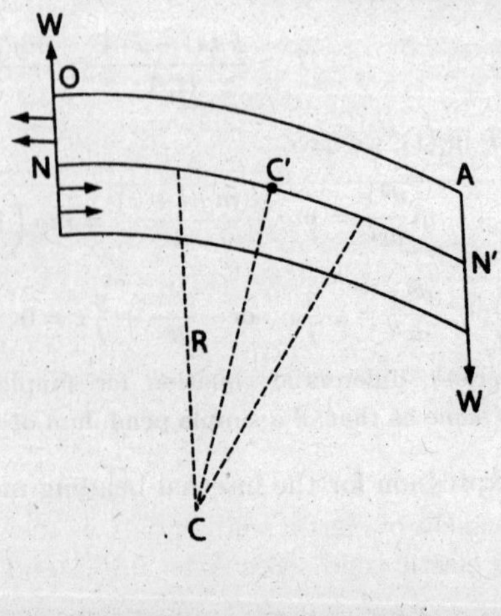

For definition of neutral layer, see Art. 9.17.1.

ii. Find an expression for the internal bending moment over a section B about the neutral layer. (See Art. 9.17.1)

iii. Show that the depression at B due to load at A is equal to the depression at A due to load at B. (See Art. 9.19) **[C.U. (Hons.) 2002]**

(b) A bar of cross-section A is subjected to two equal and opposite tensile forces T, T at its ends (fig.). Consider a plane AB making an angle θ with the normal section:

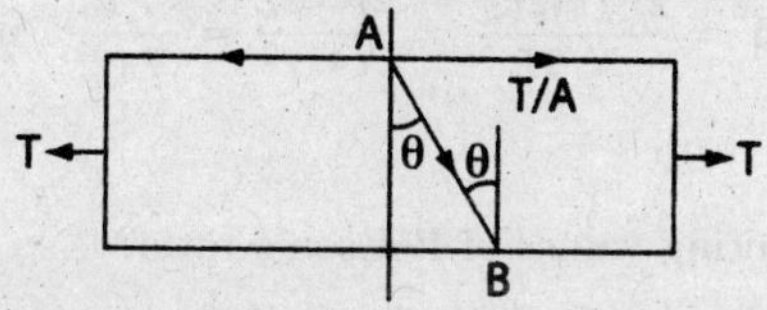

i. Calculate the shearing stress along AB.

ii. For what value of θ will the shearing stress be maximum?

iii. Given that the beam cracks due to shear, state along which section it will crack.

(i) The shearing stress T/A will act along the upper surface of the bar (as shown by arrow heads) so that the shearing stress along AB making an angle θ with the normal section $= \frac{T}{A}\operatorname{cosec}\theta$.

(ii) $\frac{T}{A}\operatorname{cosec}\theta$ will be maximum when $\theta = 90°$ i.e. along the surface.

(iii) If the beam cracks due to shear, it should be along the normal section, because on opposite sides of the normal section shearing stresses of maximum value act.

58. (a) Define the terms shear stress, shear strain and modulus of rigidity. What do you understand by the term *elastic limit*? **[C.U. (Hons.) 2000, 2006 (old syll.), 2007]**

(b) A force $\vec{F}$ stretches a wire by an amount Δl. Find the amount of work done.

[C.U. (Hons.) 2000, 2006 (old syll.)]

(See Art. 9.27)

(c) A wire of diameter 1 mm is kept taut by suspending some weight at its end. When the temperature of the wire is decreased by 20 °C, find the amount of additional weight to be put on to keep the length of the wire unchanged. ($Y = 2 \times 10^{12}$ cgs unit, $\alpha = 1 \times 10^{-5}/°\text{C}$)

[C.U. (Hons.) 2000]

Answer to part (c): Here

$$\alpha = \frac{l}{L \times 20} \quad \text{or} \quad \frac{l}{L} = 20 \times 10^{-5}$$

$$Y = \frac{Mg/\pi r^2}{l/L} = \frac{Mg}{\pi r^2} \times \frac{1}{20 \times 10^{-5}} \quad \text{or} \quad 2 \times 10^{12} = \frac{M \times 980}{\pi \times (0.05)^2 \times 20 \times 10^{-5}}$$

$$\text{or} \quad M = \frac{2 \times 10^{12} \times 3.14 \times 25 \times 10^{-4} \times 20 \times 10^{-5}}{980} = \frac{50 \times 10^3 \times 3.14 \times 20}{980} \text{ gm}$$

$$= \frac{314 \times 10^3}{98 \times 10^3} \text{ kg} = 3.204 \text{ kg}$$

59. Bulk modulus of a liquid is 3×10^5 dynes/cm^2. How much pressure is needed to compress the sample of liquid by 0.2%? **[C.U. (Hons.) 2004]**

[Ans. See Q. 27 above**]**

60. Define geometrical moment of inertia and bending moment. Derive from first principle, the strain energy function of an elastic solid. (See Arts. 9.17.1 and 9.27) **[C.U. (Hons.) 2007]**

61. A wire of length L and radius r is fixed at one end. A force F applied at the free-end produces an extension x; what is the extension produced, in another wire of the same material but of length $2L$, radius $2r$ and subjected to a force $2F$? **[C.U. (Hons.) 2007]**

[**Ans.** For the first rod,

$$Y = \frac{F/\pi r^2}{x/L} = \frac{FL}{x\pi r^2} \quad \text{or} \quad x = \frac{FL}{\pi r^2 Y}$$

For the second rod,

$$Y = \frac{2F/\pi(2r)^2}{X/2L} = \frac{2F \cdot 2L}{X 4\pi r^2} = \frac{FL}{X\pi r^2} \quad \text{or} \quad X == \frac{FL}{\pi r^2 Y}$$

$\therefore x = X$.

62. (a) What are the limiting values of Poisson's ratio?

(b) A solid cylindrical rod of radius r has half the length of a hollow cylindrical rod of the same material and the same mass. The external radius of the hollow rod is $\sqrt{2}$ times its internal radius. Compare the torsional rigidities of the two rods. **[Ans.** 8 : 3]

(For answer see worked examples) **[C.U. (Hons.) 1979, cf. 1992, 2008]**

Chapter 10

Surface Tension

10.1 Introduction

Careful observation shows that some insects walk over the surface of a liquid with small depressions at the regions of contact. Also a needle when carefully dropped on the surface of water floats with a small depression formed below and around the needle. It is also our common experience that a liquid drop assumes a spherical shape. These phenomena as mentioned above lead us to believe the skin-like behaviour of the surface of a liquid, i.e., as if a thin stretched membrane is covering the liquid surface. The real existence of such a stretched memberane with corresponding tension over the surface requires that the tension should increase with the stretching of the surface, but in the case of surface tension of a liquid, we find that it remains constant with stretching—which is contrary to the behaviour of an actual stretched membrane.

The contractile character of the surface layer of a liquid is not due to any real existence of surface tension, but it is due to the inter-molecular actions resulting in the existence of molecular surface energy, and the contractile character of the surface as evidenced is not due to any tension coming to play but due to the fact that the surface molecules of a system in equilibrium tend to attain minimum potential energy. The spherical shape of a mercury globule is due to the above fact since the area of the sphere is minimum for a given volume of the substance.

Such a contractile property of a liquid surface leads to the development of the forces of surface tension at the boundary, and because of the contractile charcter of a liquid surface, the mass of the liquid behaves as if it is surrounded by a stretched membrane. So we may consider the surface tension force to lie in the plane of the surface, acting perpendicular to any line drawn on the surface, and like stress in solids, this force is exerted by one portion of the surface on the other and is a pull.

Althrough the various surface tension phenomena may be explained clearly from the standpoint of molecular theory, yet the assumption of the existence of a fictitious surface tension as if existing over the surface enables us to solve many problems more conveniently. Hence we express surface tension as if it exerts over the surface of a liquid in terms of the *force per unit length exerted in the tangential plane to the surface and in a direction at right angles to an element of a line drawn over the surface.*

If we imagine a line AB (Fig. 10.1.1) drawn in any direction in the liquid surface, the surface on either side of this line exerts a pulling force on the surface on the other side of the line. This force lies in the plane of the surface and is at right angles to the line AB. The magnitude of this force per unit length of AB is taken as a measure of surface tension of the liquid. Thus if F be the total force on either side of the line AB of length l, then the surface tension

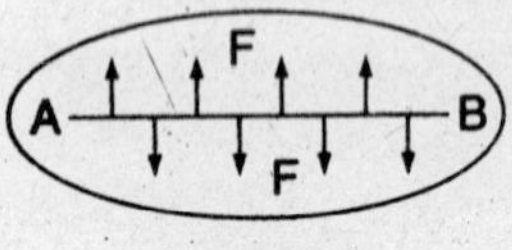

Fig. 10.1.1

$$S = \frac{F}{l}, \quad \text{or,} \quad S = F \text{ if } l = 1.$$

In S.I. system its unit is Newton/metre, and the C.G.S. system, the unit is dyne/cm.

The dimension of surface tension is

$$[S] = \frac{[M][L]}{[T^2]} \cdot \frac{1}{[L]} = MT^{-2}.$$

The most important thing that we should bear in mind in this connection is that the cohesive force in between the molecules of a liquid gives rise to the property known as *surface tension* and in manifestation of this property the free surface of a liquid always behaves like a stretched membrane and tends to contract to the smallest area possible as if it is in a state of tension. The ascent of water or the descent of mercury in a capillary tube is due to the difference in the magnitudes of the cohesive force of the liquid molecules and the adhesive force between the liquid and the material of the tube. The glass wall (of the capillary glass tube) immediately above the edge of water attracts water molecules towards it by adhesion. These molecules in turn attract nearby water molecules, pulling them up by cohension and as a result, water continues to rise in the tube, until the upward pull is equalised by the weight of the liquid pulled up. It is in the same process that liquids are soaked up by wicks and blotting paper, and water is drawn from the wet sub-soil into the top layers through the fine pores of the soil.

To explain the contractile tendency of a liquid surface, we see that a molecule of the liquid, well below the surface, is attracted by cohesive forces exerted by all other molecules surrounding it. In the bulk of the liquid, these attractive forces balance one another and there will be no resultant force on the molecule under consideration. Now for a molecule on or near the surface, the downward pull is greater than the upward pull, and as such the molecules on or near the surface are pulled inwards, and as a result the surface area tends to decrease. The surface of a liquid thus acquires a contractile property like a stretched membrane.

10.1.1 Alternative definition of Surface Tension—Surface Energy and Surface Tension

The contractile character of a liquid surface shows that there is free energy associated with it and so work must be done to stretch the surface. Thus it follows that every surface molecule must possess potential energy, i.e., energy by virtue of its position. This energy per unit area of a surface is called its *intrinsic surface energy*. The fact that a liquid surface is the seat of potential energy can be clearly understood when a soap film is ruptured, for the liquid is projected in all directions with a considerable velocity (here, the potential energy is converted into kinetic energy).

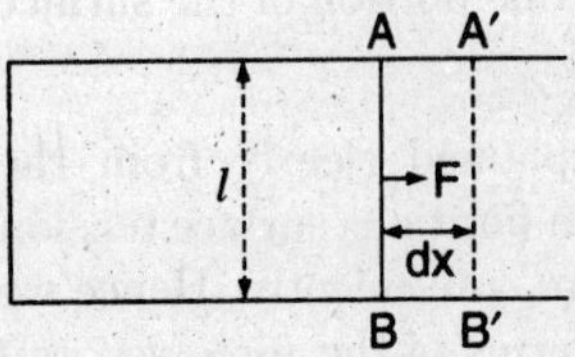

Fig. 10.1.2

We consider a film in a frame in which AB is a movable cross-piece which can move over the two parallel sides as shown in figure 10.1.2. It is due to the surface tension that the cross-piece

AB is attracted inwards with a force which is $2Sl$, since there are two free surfaces on both sides of the film. So, to hold the film in position, we are to apply a force in the opposite direction given by $F = 2Sl$.

If now the film is stretched by a very small length dx, the work done due to this stretching is given by

$$W = Fdx = 2Sldx = S.2ldx = S.\text{ increase in area.}$$

$$\therefore \quad S = \frac{\text{work done}}{\text{increase in area}}.$$

The surface tension S is thus numerically equal to the mechanical work done in creating unit area of the film surface. Therefore, S is numerically equal to the (mechanical part) surface energy per unit area, since the work done accounts for the increase in surface energy. From this we cannot jump to the conclusion that surface tension is equal to the surface energy per unit area. In fact, surface energy per unit area is greater than the work done per unit area. This is due to the fact that when a surface film is stretched, it tends to have lowering in temperture, and the external source (say, surrounding air) readily supplies the requisite heat energy to keep the temperature of the surface unchanged. Thus, surface energy per unit area = energy supplied to unit area by external source to keep the temperature of the surface constant + the mechanical work done in creating unit area by stretching the surface

$$\text{i.e.,} \quad E = h + S, \tag{10.1.1}$$

where E is the surface energy per unit area and h the energy supplied to unit area by external source.

For example, we know that surface tension of water at 0°C is 76 dynes per cm. but the surface energy per unit area at 0°C is about 117 ergs. If $E = S$, then we should expect that surface energy at 0°C is 76 ergs/cm^2, but this is not the actual case as is found from the above data. Hence $E > S$ as given in the relation (10.1.1).

From the above discusssions, we may now define intrinsic surface energy and surface tensi on as given below.

Intrinsic Surface Energy

When the surface of a liquid is extended, molecules from the interior rise to the surface against the inward cohesive forces of attraction. The necessary mechanical work done on the molecules is stored as potential energy in the new surface formed. Further, when the surface is extended, its temperature tends to fall. So, heat flows into it from the surroundings to keep its temperature constant, and is added to its energy. Thus the surface possesses potential energy due to mechanical work, and also heat energy is absorbed from the surroundings. This total energy per unit surface area is called the *intrinsic surface energy.*

Thus the intrinsic surfce energy E consists of a *mechanical* part S and a *thermal* part h. The mechanical part, which is numerically equal to the surface tension S, is called the *free surface energy.*

Hence the surface tension of a liquid may also be defined as *the mechanical work done in increasing the surface area by unit amount under constant temperature conditions*—which is the free surface energy. So it may also be expressed in erg/cm^2 or in Joule/(metre)^2.

We may now calculate h from the following thermodynamical consideration by treating a surface film to a cyclic change of operations as shown in figure 10.1.3.

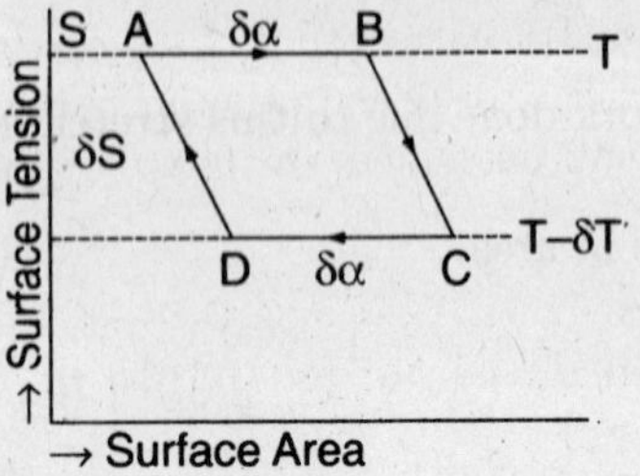

Fig. 10.1.3

Let us suppose that the initial condition of the film is represented by A, i.e., the initial condition is expressed in terms of surface tension S at absolute temperature T and area α. We consider that the film is stretched isothermally from a condition represented by A to that represented by B. Then, we treat the film adiabatically till it comes to the position C and its absolute temperature is $(T - \delta T)$. We now allow the film to contract isothermally at that temperature till it arrives at D. Finally, the film is allowed to expand adiabatically until it comes back to the original condition A.

The cycle is a reversible one, and hence the external work done is represented by the area $ABCD$. Now this area is the difference between the work done by the film when it contracts isothermally and the work done on the film during isothermal stretching. When the film contracts isothermally at $(T - \delta T)$, the work done by the film is

$$\left(S - \frac{dS}{dT}\delta T\right)\delta\alpha.$$

Again, the work done on the film during isothermal stretching at T is $S\delta\alpha$. Therefore, the net external work done by the film during the whole cycle of operations is

$$\left(S - \frac{dS}{dT}\delta T\right)\delta\alpha - S\delta\alpha,$$

because the work done during adiabatic contraction is equal and opposite to the work done during adiabatic expansion. If h be the heat energy supplied to unit area from outside source at temperature T, the external energy supplied during isothermal expansion is $h\delta\alpha$. Hence, from the laws of thermodynamics, we may write,

$$\frac{\text{net external work done during a complete cycle}}{\text{heat energy absorbed at higher temperature}} = \frac{\delta T}{T}$$

or, $$\frac{\left(S - \frac{dS}{dT}\delta T\right)\delta\alpha - S\delta\alpha}{h\delta\alpha} = \frac{\delta T}{T},$$

whence $h = -T\dfrac{dS}{dT}$.

$$\therefore \quad E = S - T\frac{dS}{dT}.$$

The quantity $\frac{dS}{dT}$ is called the temperature coefficient of surface tension. From the above result we see that $E = S$, when (i) $T = 0$, i.e., at the absolute zero and (ii) $\frac{dS}{dT} = 0$, i.e., when the surface tension does not change with a change in temperature.

It should be remembered that in the case of almost all liquids, surface tension decreases with the rise of temperature, and thus $\frac{dS}{dT}$ negative. Hence E is greater than S.

10.1.2 Case 1. Spraying of a Liquid into Droplets

When a liquid is sprayed, the surface area of the droplets formed is larger than that of the liquid sprayed. This requires mechanical work to be done—which is stored as potential energy in the new surface formed. Hence energy is spent in spraying as will be evident in the following discussion.

Expression for the work necessary to break up a liquid drop of radius R into n identical smaller drops

When a liquid drop is broken into a number of identical smaller drops, the surface area *increases*. Hence work is to be done against the surface tension force. Let us first calculate the increase in surface area.

Let r be the radius of each of the n smaller drops. Since the total volume of all the n smaller drops is the same as the volume of the original drop of radius R, we may write

$$n \times \frac{4}{3}\pi r^3 = \frac{4}{3}\pi R^3. \quad \therefore\ r = \frac{R}{(n)^{1/3}}.$$

Thus, increase in surface area $= n \times 4\pi r^2 - 4\pi R^2 = 4\pi(nr^2 - R^2)$

$$= 4\pi\left[n\frac{R^2}{n^{2/3}} - R^2\right] = 4\pi R^2[n^{1/3} - 1].$$

The work necessary is therefore, $W =$ Surface tension $\times$ increase in surface area

$$= S \times 4\pi R^2[n^{1/3} - 1] = 4\pi R^2 S(n^{1/3} - 1). \qquad (10.1.2)$$

10.1.3 Case 2. Inverse Phenomenon of Case 1

Again, when a number of droplets of a liquid coalesce to form a single drop, the surface area decreases and hence the surface energy also decreases. The decrease in surface energy is equal to the energy of the new decreased surface formed. We suppose that n identical droplets of water, each of radius r, coalesce to form a single drop of radius R. Then, the surface area of n droplets $= n \times 4\pi r^2$, and the surface area of the combined single drop $= 4\pi R^2$.

$\therefore$ decrease in surface area $= n \times 4\pi r^2 - 4\pi R^2 = 4\pi(nr^2 - R^2)$.

If W be the energy released, we have

$$W = \text{decrease in surface area} \times \text{Surface tension}$$
$$= 4\pi(nr^2 - R^2) \times S. \qquad (10.1.3)$$

This energy is spent in increasing the temperature of the water drop. Let θ be the rise in temperature. Then, if both the specific heat and density of water are equal to unity, the quantity of heat Q absorbed by the drop is given by

$$Q = \text{mass} \times \text{specific heat} \times \theta$$
$$= \frac{4}{3}\pi R^3 \times \text{density} \times \text{specific heat} \times \theta = \frac{4}{3}\pi R^3\theta. \qquad (10.1.4)$$

From the first law of thermodynamics, we have $W = JQ$, where J = mechanical equivalent of heat.

So from (10.1.3) and (10.1.4), we may write

$$4\pi(nr^2 - R^2)S = J \times \frac{4}{3}\pi R^3\theta$$

$$\text{or,}\quad \theta = \frac{3S}{J}\left(\frac{nr^2 - R^2}{R^3}\right) = \frac{3S}{J}\left(\frac{nr^2}{R^3} - \frac{1}{R}\right). \tag{10.1.5}$$

Again, since the total volume of n droplets is equal to that of the combined drop, we have

$$n \times \frac{4}{3}\pi r^3 = \frac{4}{3}\pi R^3,\quad \text{or,}\quad nr^3 = R^3,\quad \text{or,}\quad \frac{nr^2}{R^3} = \frac{1}{r}.$$

Putting this value of $\frac{nr^2}{R^3} = \frac{1}{r}$ in (10.1.5), we get

$$\theta = \frac{3S}{J}\left(\frac{1}{r} - \frac{1}{R}\right).$$

Illustrative Examples

1. Calculate the energy per square centimetre of a mercury surface at 0°C, if the surface tension of mercury at this temperature is equal to 441 dynes per cm and its value decreases at the rate of 0.379 dynes per cm per degree centrigrade rise in temperature. Calculate also the energy per square centimetre of water at 0°C; given surface tension of water at 0°C = 76 dynes per cm and temperature coefficient of surface tension of water = 0.15 dynes per cm per degree centrigrade.

In both the case, $\frac{dS}{dT}$ is negative.

Therefore, for mercury

$$E = S + T\frac{dS}{dT} = 441 + 273 \times 0.379 = 544.467 \text{ ergs per sq cm,}$$

for water

$$E = S + T\frac{dS}{dT} = 76 + 273 \times 0.15 = 116.95 \text{ ergs per sq cm.}$$

2. A soap film 0.001 mm thick is at the temperature of melting of ice. Calculate the fall in temperature of the film due to stretching it adiabatically until its area is doubled. Assume that the specific heat of soap solution and its density are each numerically equal to unity, and that the surface tension of the solution decreases at the rate of 0.15 dynes per cm per degree centrigrade rise of temperature.

Let A square centimetre be the area of the film. Then $2A$ sq cm will be the total area of the film, since there are two free surfaces. As the area is doubled the increase in area is $2A$ sq cm. Therefore, the energy absorbed

$$= \frac{2hA}{J} \text{ cals,}\quad [h \to \text{heat energy absorbed per unit area}]$$

where J is the mechanical equivalent of heat. This is supplied at the expense of the internal energy of the film and hence its temperature will fall. Let θ°C be the fall of temperature, so that the heat supplied

$$= ms\theta \text{ (from calorimetric consideration),}$$

where m = mass of the film

$$= 0.0001 \times A \times \rho \quad [\rho = \text{density of the film}]$$

$$= 0.0001 \times A \times 1 \quad [\because \rho = 1]$$

and s = its specific heat = 1

$$\therefore \; ms\theta = \frac{2hA}{J} \quad \text{or} \quad 0.0001 \times A \times \theta = \frac{2hA}{J}$$

$$\text{or} \quad \theta = \frac{2h}{0.0001 \times J} = \frac{2T\frac{dS}{dT}}{0.0001 \times J} \quad \left[\because h = T\frac{dS}{dT}\right]$$

$$= \frac{2 \times 273 \times 0.15}{0.0001 \times 4.2 \times 10^7} \quad [\because J = 4.2 \times 10^7 \text{ ergs/cal.}]$$

$$= 0.0195°\text{C} = 19.5 \times 10^{-3}°\text{C}.$$

3. If the rate of change of surface energy of a liquid with temperature is directly proportional to the absolute temperature T, show that the surface tension $S = AT^2 + BT + C$, where A, B and C are constants.

From the condition of the problem, we have

$$\frac{dE}{dT} \propto T. \quad \therefore \frac{dE}{dT} = KT, \text{ where } K \text{ is a contant of proportionality.}$$

We know

$$E = S - T\frac{dS}{dT} \quad \text{or} \quad \frac{dE}{dT} = \frac{dS}{dT} - \frac{dS}{dT} - T\frac{d^2S}{dT^2} \quad [\text{differentiating both sides with respect to } T]$$

$$\text{or} \; -T\frac{d^2S}{dT^2} = \frac{dE}{dT} = KT \quad \text{or} \quad \frac{d^2S}{dT^2} = -K.$$

On integration, we have

$$\frac{dS}{dT} = -KT + B \quad [B \text{ being the constant of integration}].$$

On further integration, we have

$$S = -\frac{K}{2}T^2 + BT + C.$$

Writing A for $-\frac{K}{2}$, the above expression may be written as

$$S = AT^2 + BT + C.$$

4. A glass tube of circular cross-section is closed at one end. This end is weighted and the tube floats vertically in water with the heavy end downwards. How far below the water surface is the end of the tube? Given: outer radius of the tube 0.14 cm, mass of weighted tube = 0.2 gm, surface tension of water = 73 dynes/cm and g = 980 cm/sec^2. **[C.U. (Hons.) Old Syll. '80]**

Let l be the length of the tube inside water. The forces acting on the tube are:

(i) Upward thrust of water $= \pi r^2 l \times 1 \times 980 = \frac{22}{7} \times (0.14)^2 l \times 980 = 60.368l$ dynes;

(ii) weight of the tube acting downwards = mg = $0.2 \times 980 = 196$ dynes;

(iii) force due to surface tension acting downwards $= 2\pi r S = 2 \times \frac{22}{7} \times 0.14 \times 73$
$= 64.24$ dynes.

Since the tube is in equilibrium, upward force = sum of the downward forces or $60.368l = 196 + 64.24 = 260.24$.

$$\therefore \quad l = \frac{260.24}{60.368} = 4.31 \text{ cm.}$$

5. A hydrometer of mass 50 gm is floating in water with 3 cm of its stem lying above the water level. The diameter of the stem is 0.4 cm. What length of the stem would be outside the water level, had there been no surface tension of water? Assume that the surface tension of water is 72 dynes/cm and the angle of contact is zero. **[C.U. (Hons.) 1990]**

The force due to surface tension acting downwards $= 2\pi r S = 2 \times 3.14 \times 0.2 \times 72$ dynes
$= 90.432$ dynes.

If l be the length of the stem depressed under water due to the above force, then the upward thrust of water is $\pi r^2 l \times 1 \times 980$ dynes. Since the hydrometer is in equilibrium, the upward thrust due to the displaced water is balanced by the downward force due to surface tension, so that we may write

$$\pi r^2 l \times 980 = 90.432 \quad \text{or} \quad l = \frac{90.432}{3.14 \times (0.2)^2 \times 980} = 0.734 \text{ cm.}$$

Thus the hydrometer dips more by 0.734 cm due to surface tension. So in absence of surface tension, the length of the stem outside water $= 3 + 0.734 = 3.734$ cm.

6. Calculate the work done in spraying a spherical drop of mercury of radius 1 mm into a million identical droplets. The surface tension of mercury is 550 dynes/cm.

The work done in spraying a liquid drop of radius R into n droplets is, from (10.1.2), given by $W = 4\pi R^2 S(n^{1/3} - 1)$.

Here $R = 1$ mm $= 0.1$ cm, $S = 550$ dynes/cm and $n = 10^6$.

$$\therefore \quad W = 4 \times 3.14 \times (0.1)^2 \times 550 \times \left[(10^6)^{1/3} - 1\right]$$
$$= 4 \times 3.14 \times 0.01 \times 550 \times 99 = 6839 \text{ ergs.}$$

7. Calculate the amount of energy needed to break a drop of water 2 mm in diameter into 10^9 droplets of equal size. Surface tension of water is 73×10^{-3} nt/m.

The work done in spraying a liquid drop of radius R into n identical droplets is given by $W = 4\pi R^2 S(n^{1/3} - 1)$.

Here $R = 1$ mm $= 0.1$ cm, $n = 10^9$, $S = 73 \times 10^{-3}$ nt/m $= 73 \times 10^{-3} \times \frac{10^5}{10^2}$ dynes/cm $= 73$ dynes/cm.

$$\therefore \quad W = 4 \times 3.14 \times (0.1)^2 \times 73 \left[(10^9)^{1/3} - 1\right] = 4 \times 3.14 \times 0.01 \times 73 \times 999$$
$$= 9.160 \times 10^3 \text{ ergs} = 9.160 \times 10^{-4} \text{ Joule.}$$

8. Calculate the loss of energy if 1000 drops of water each of diameter 2 mm coalesce to form one large drop. The surface tension of water is 72 dynes/cm.

Let R cm be the radius of the large drop formed when 1000 droplets, each of radius 1 mm coalesce. Since the volume remains unchanged, we have

$$\frac{4}{3}\pi R^3 = 1000 \times \frac{4}{3}\pi(0.1)^3. \quad \therefore \quad R = (1000)^{1/3} \times (0.1) \text{ cm} = 1 \text{ cm}.$$

Now the surface area of 1000 droplets $= 1000 \times 4\pi(0.1)^2 = 40\pi$ cm^2 and that of the large drop $= 4\pi$ (1 cm)^2 $= 4\pi$ cm^2, so that the decrease in surface area $= 40\pi - 4\pi = 36\pi$ cm^2.

$\therefore$ energy liberated $=$ decrease in surface area $\times$ surface tension $=$ (36 $\times$ 3.14 cm^2) $\times$ 72 dynes/cm $=$ 8139 ergs.

9. Two drops of a liquid, each of radius r, coalesce to form a larger drop. Derive an expression for the rise in temperature.

When two or more small drops of a liquid coalesce to form a single large drop, the surface area decreases and hence the surface energy also decreases. The energy lost appears as heat, resulting in the rise in temperature of the drop.

Let R be the radius of the larger drop formed. Since in the process volume remains unchanged, we have

$$\frac{4}{3}\pi R^3 = 2 \times \frac{4}{3}\pi r^3 \quad \text{or} \quad R = 2^{1/3}r.$$

Now decrease in surface area $= 2 \times 4\pi r^2 - 4\pi R^2 = 4\pi[2r^2 - R^2] = 4\pi\left[2r^2 - 2^{2/3}r^2\right]$

$$= 8\pi r^2\left[1 - 2^{-1/3}\right].$$

$\therefore$ the energy released

$$W = S \times 8\pi r^2\left(1 - 2^{-1/3}\right). \tag{1}$$

This energy is spent in increasing the temperature of the liquid. Let θ be the rise in temperature. If ρ be the density and s be the specific heat of the liquid, the quantity of heat Q absorbed by the drop is given by

$$Q = \text{mass} \times \text{sp. ht} \times \text{rise in temp.}$$

$$= \frac{4}{3}\pi R^3\rho \times s \times \theta = \frac{4}{3}\pi(2r^3)\rho s\theta \quad (\because R = 2^{1/3}r). \tag{2}$$

Now from the first law of thermodynamics, we have $W = JQ$, J being the mechanical equivalent of heat. Putting the values of W and Q from (1) and (2), we may write

$$S \times 8\pi r^2\left[1 - 2^{-1/3}\right] = J \times \frac{4}{3}\pi(2r^3)\rho s\theta. \quad \therefore \quad \theta = \frac{3S}{rJ\rho s}\left[1 - 2^{-1/3}\right].$$

10.2 Angle of Contact

If a glass plate is dipped vertically inside a liquid which wets the plate (as in the case of water, alcohol, etc.), the level of the liquid rises up a little along the side of the plate [Fig. 10.2.1]. Again when a plate is dipped vertically inside a liquid which does not wet the wall of the plate (as in the case of mercury). the level of the liquid is depressed a little along the region of contact as shown in fig. 10.2.2.

So when a solid is dipped into a liquid, the liquid surface near the wall of contact is not horizontal—it is either a little bit elevated or depressed, depending on the nature of the liquid. In the case of water which wets a glass plate—here attraction between liquid, i.e., water molecules called *cohesion* is less than their attraction to the sides i.e. solid glass molecules called *adhesion* and *hence the wetting*—the liquid is drawn a little up along the wall (fig. 10.2.1), while for a liquid like mercury—here *reverse is true* and *hence no wetting*—the liquid surface is a bit depressed near the surface of the solid (fig. 10.2.2). If we take a section of the liquid surface near the solid-face, we get a continuous curve—known as the *capillary curve.*

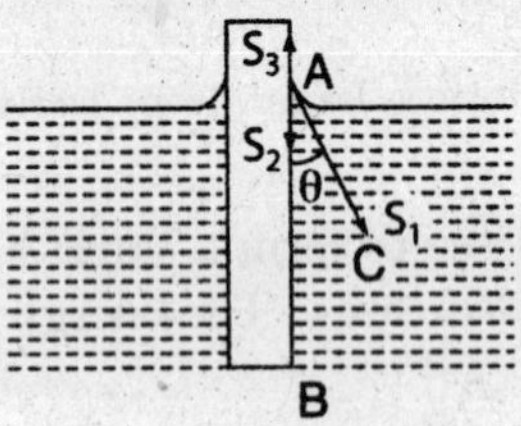

Fig. 10.2.1

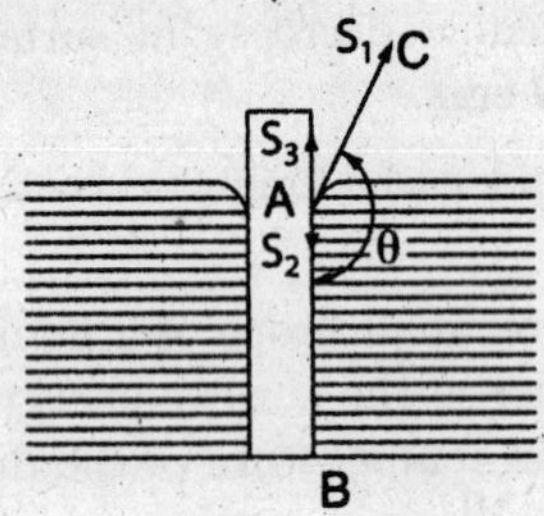

Fig. 10.2.2

We consider the point A which is the point of contact of the capillary curve i.e. the liquid surface with the solid surface. The angle BAC which the tangent AC to the liquid surface at A makes with AB, the solid surface inside the liquid, is called *the angle of contact* between the solid and the liquid.

In the case of water, alcohol, etc., this angle is actute, whereas in case of meercury, etc., this angle is obtuse.

It should be noted that for pure water and clean glass, the angle of contact is zero. It is also nearly zero in case of benzene or alcohol in contact with glass. For ordinary water and glass the angle of contact is about 18° and for mercury and glass it is about 140°. For most liquids and glass the angle of contact is less than 90°. The angle of contact depends upon the nature of the liquid and the solid, the medium above the free surface of the liquid and is independent of the angle of inclination of the solid.

10.3 General consideration of interfacial Surface Tension

Since surface tension is a molecular phenomenon, i.e., surface tension arises out of molecular attractions, we have no reason to think that surface tension is present only in the case of a liquid. It also exists in the interfaces of liquids and gases, liquids and solids and solids and gases.

Let us consider a drop of a liquid on the surface of a solid.

Let S_1 = interfacial surface tension between the liquid and air,

S_2 = interfacial surface tension between the liquid and the solid

and S_3 = interfacial surface tension between the solid and air.

Hence for equilibrium of the liquid drop, we may write

$$S_1 \cos\theta + S_2 = S_3 \quad \text{or} \quad \cos\theta = \frac{S_3 - S_2}{S_1}, \quad \text{(figures 10.3.1 and 10.3.2)}$$

where θ is the angle of contact of the liquid with the solid. So for equilibrium of the liquid drop resting on the surface of the solid, the condition is that $\frac{S_3-S_2}{S_1}$ must not be greater than unity since $\cos\theta$ cannot be greater than unity.

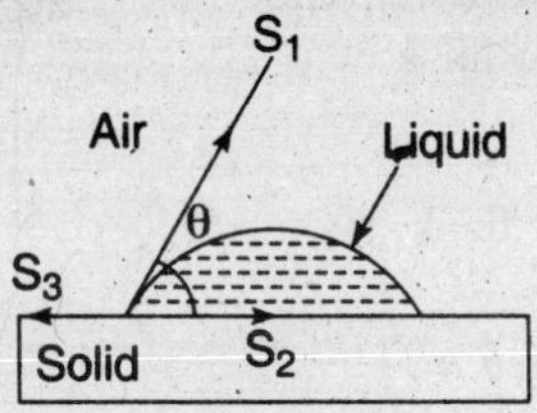

Fig. 10.3.1

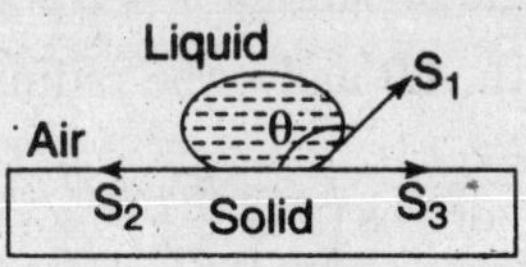

Fig. 10.3.2

When θ is acute [Fig. 10.3.1], $\cos\theta$ is positive, and therefore, $S_3 > S_2$; on the otherhand if θ is obtuse [Fig. 10.3.2], $\cos\theta$ is negative and therefore, $S_3 < S_2$.

If $(S_3 - S_2) > S_1$, $\cos\theta$ becomes greater than unity. Equilibrium is not then possible and the liquid drop at the region of contact must move continually resulting in complete wetting of the wall of the solid.

10.4 Neumann's Triangle

If the three tensions S_1, S_2 and S_3 are such that a triangle can be formed with the sides proportional to the magnitudes of these tensions, then equilibrium will be possible. Such a triangle is called *Neumann's triangle*. But when $(S_3 - S_2) > S_1$ or $S_3 > (S_1 + S_2)$, the formation of such a triangle is not possible, and the equilibrium is lost resulting in complete wetting of the solid surface.

The above consideration is applicable when a liquid drop is placed on the surface of another liquid (the two liquids being immiscible), as shown in figure 10.4.1. Here the three interfacial surface tensions are given by

S_1 = surface tension between air and liquid II,

S_2 = surface tension between liquid I and liquid II,

and S_3 = surface tension between air and liquid I.

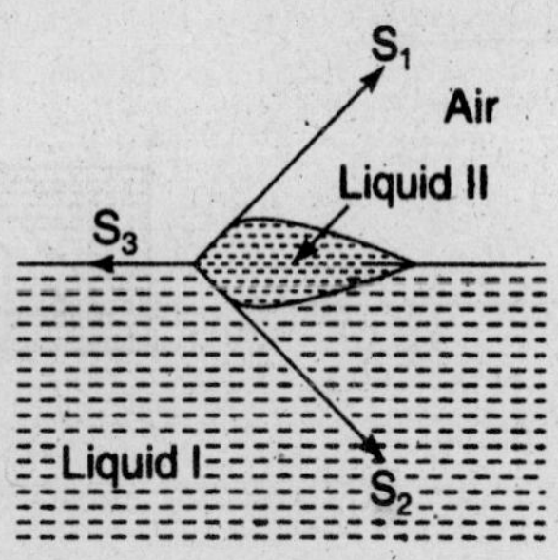

Fig. 10.4.1

In this case, if $S_3 > (S_1 + S_2)$, the equilibrium is lost and the molecules at the edge of the drop move continually in the direction of S_3 resulting in the spreading of the liquid drop II on the surface of the liquid I. For example, in case of a drop of olive oil on water surface, since S_3 = 72 dynes/cm, S_1 = 32 dynes/cm, S_2 = 30 dynes/cm, the drop spreads all over the surface of water.

10.5 Measurement of Angle of Contact

(i) The angle of contact between mercury and glass may be determined in the following way.

An inverted spherical flask is nearly filled with clean mercury and the level of mercury in the flask is adjusted by raising or lowering the reservoir R [Fig. 10.5.1] unitl there is no curved portion where the mercury meets glass, i.e., the mercury surface is horizontal up to the wall of the flask.

The required angle of contact $\theta = \angle BAC$.

If $2l$ be the length AB and r the radius of the flask, then

$$\cos \angle OAN = \frac{l}{r} \text{ or } \cos\left(\theta - \frac{\pi}{2}\right) = \frac{l}{r} \quad \text{or} \quad \sin\theta = \frac{l}{r}$$

$$\text{or } \theta = \sin^{-1}\left(\frac{l}{r}\right).$$

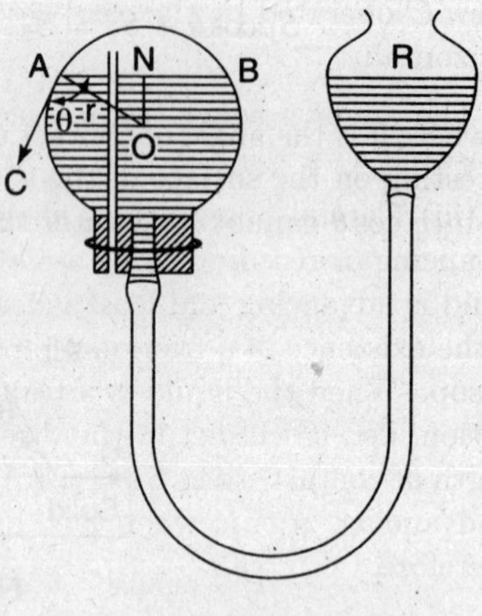

Fig. 10.5.1

It should be noted that θ in this case must lie between $\pi/2$ and π. Thus by measuring l and r, the angle of contact is determined.

(ii) Another simple method for the determination of the angle of contact is to immerse a solid plate in an inclined manner inside mercury contained in a trough as shown in figure 10.5.2. The inclination of the glass plate is adjusted until the liquid surface meets the glass plate at O without any curvature. From the figure, we have

$$\tan\phi = \frac{AC}{OC}, \quad \text{i.e.,} \quad \phi = \tan^{-1}\frac{AC}{OC}.$$

The angle of contact is given by $(\pi - \phi)$.

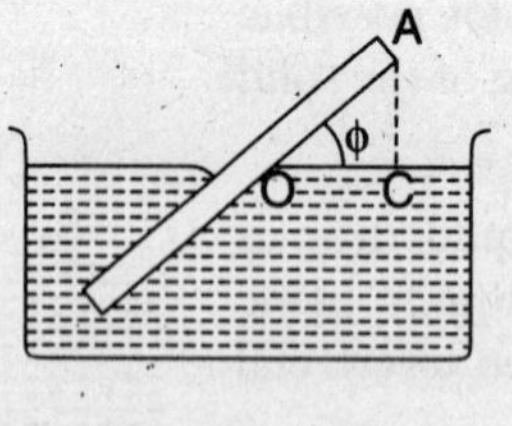

Fig. 10.5.2

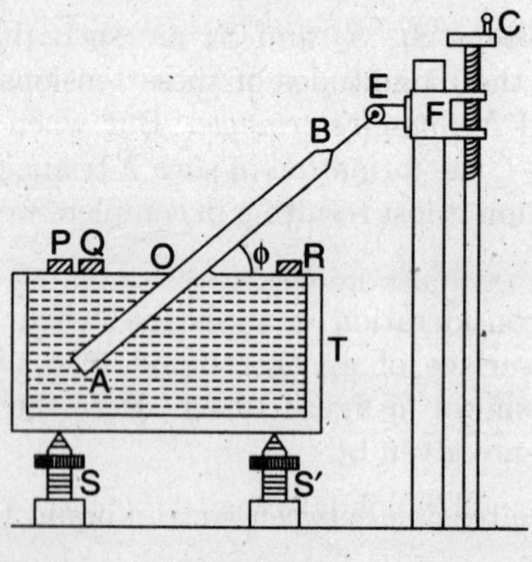

Fig. 10.5.3

A similar arrangement was employed by N.K. Adam for measuring the angle of contact between water and glass coated with wax [Fig. 10.5.3].

A glass trough T coated inside with wax is filled with water above the level of its sides whch have been made flat on the top. The top surface is made perfectly horizontal by adjusting the screws S and S'. P, Q and R are rectangular pieces of glass coated with wax. By moving the glass pieces P and Q across the surface of the liquid from the right side to the position as shown in the figure, the surface is made free from contamination. AB is a glass plate coated with wax and held in a clamp which may be rotated about a horizontal axis through E. The plate is turned at a suitable angle of inclination and its position is adjusted by means of the screw C operated by the carriage F until the water surface meeting the glass plate is perfectly horizontal.

The angle ϕ which is the angle between the plate and the horizontal surface of water is measured with the help of a protractor whence the angle of contact is determined.

(iii) Careful observations show that the angle of contact varies according as the liquid is advancing or receding along a plate. The angle of contact is greatest at those points where the liquid is advancing and least where it is receding. This variation of the angle of contact is due to the existence of a frictional force between the solid and the liquid as assumed by Adam and Jessop. When the liquid is advancing the frictional force called into play tends to oppose the motion, i.e., it will act in the direction of S_2 [Fig. 10.3.1]. If μ be the frictional force per unit length of contact, then $S_3 = \mu + S_2 + S_1 \cos\theta_A$, when θ_A is the angle of contact when the liquid is advancing. Again, when the liquid is receding the frictional force acts in the direction of S_3. Therefore,

$$S_3 + \mu = S_2 + S_1 \cos\theta_R,$$

where θ_R is the angle of contact when the liquid is receding.

From these two equations eliminating μ, we get

$$S_3 - S_2 - S_1 \cos\theta_A = S_2 - S_3 + S_1 \cos\theta_R$$

$$\text{or} \quad S_3 = S_2 + S_1 \left(\frac{\cos\theta_A + \cos\theta_R}{2}\right). \tag{10.5.1}$$

Again we know

$$S_3 = S_2 + S_1 \cos\theta, \tag{10.5.2}$$

where θ is the angle of contact in absence of any friction.

$\therefore$ from (10.5.1) and (10.5.2), we have

$$\cos\theta = \frac{\cos\theta_A + \cos\theta_R}{2}.$$

This expression may also be written in the form

$$2\cos\theta = \cos\theta_A + \cos\theta_R = 2\cos\frac{(\theta_A + \theta_R)}{2}\cos\frac{(\theta_A - \theta_R)}{2}.$$

If $(\theta_A - \theta_R)$ is small which is actually the case, then $\cos\dfrac{\theta_A - \theta_R}{2} \to 1$.

$$\therefore \quad \cos\theta = \cos\left(\frac{\theta_A + \theta_R}{2}\right) \quad \text{or} \quad \theta = \frac{\theta_A + \theta_R}{2}.$$

Thus, if the angles of contact θ_A and θ_R are determined, then the mean value of these angles gives the true angle of contact θ.

(iv) This method was devised by Ablett for the determination of the angle of contact between a liquid and a solid, and to show the variation of angles of contact when the liquid is advancing or receding over the solid.

The essential part of the apparatus consists of a solid cylinder of about 3 inches in diameter and 3 inches long. The cylinder is carefully coated with paraffin wax and then turned smooth when it is cooled. It is then mounted in such a way that it can move about a horizontal axis with various speeds. The cylinder is partly immersed in a liquid contained in a rectangular glass vessel (Fig. 10.5.4). One side of the vessel is provided with horizontal slits SS'.S' consists of a fine slot in a black paper covering the wall of this side.

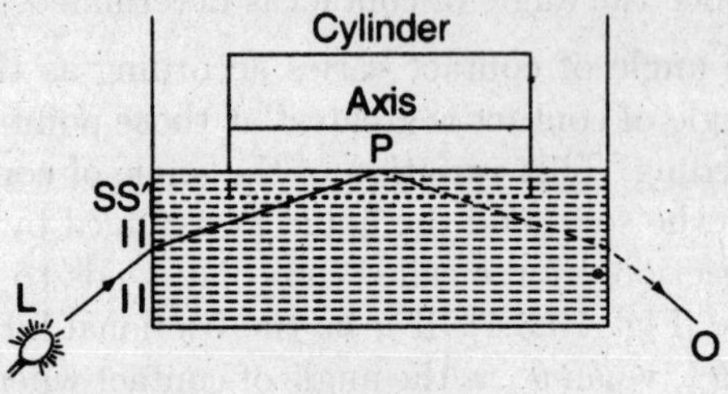

Fig. 10.5.4

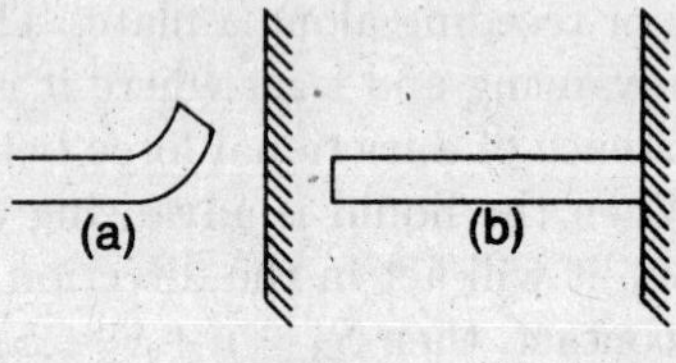

Fig. 10.5.5

Parallel rays of light from a lamp L pass obliquely upward at first through the adjustable slit S and then through the slit S', and fall finally on the point P (say) of the liquid surface near its contact with the cylinder. The rays reflected from this point in the downward direction are observed through the other side of the vessel at O. If the liquid surface meeting the cylinder be curved the image of the slit appears curved to the observer as shown in figure 10.5.5(a). The depth of the liquid in the vessel is adjusted until the liquid surface meets the cylinder horizontally. When this is achieved, the image of the slit will appear straight as shown in figure 10.5.5(b).

In figure 10.5.6, O denotes the centre of a cross-section of the cylinder and θ, the angle of contact. It can be shown from simple geometry that $\angle AOB = (\pi - \theta)$.

$$\therefore \quad \cos(\pi - \theta) = \frac{OB}{OA} \quad \text{or} \quad \cos\theta = \frac{h-r}{r},$$

where r denotes the radius of the cylinder and h the height of the liquid level from the lowest generator of the cylinder.

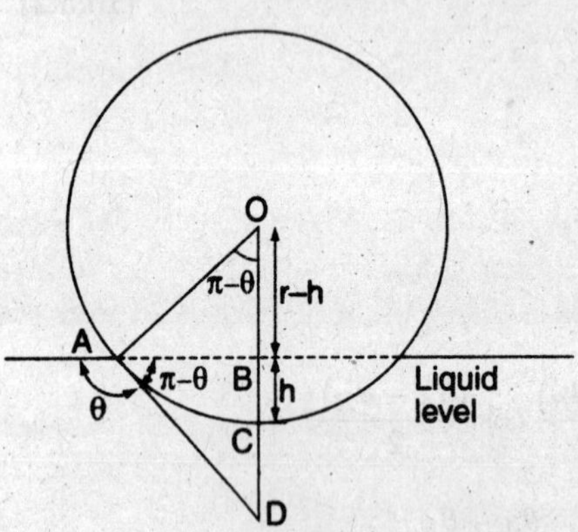

Fig. 10.5.6

The angles of contact θ_A and θ_R are measured, in a similar experiment, by rotating the cylinder in the anticlockwise and clockwise directions respectively. The mean value of θ_A corresponding to the case of liquid advancing over the solid was 113°9′. The mean value of θ_R corresponding to the case of liquid receding over the solid was 96°20′.

$$\therefore \quad \theta = \frac{\theta_A + \theta_R}{2} = \frac{113°9' + 96°20'}{2} = 104°34'.$$

For peripheral speeds below 0.44 mm/sec, θ_A and θ_R depend upon the speed.

10.6 Shape of Liquid Meniscus in a Capillary Tube

The shape of liquid meniscus in a capillary tube may be concave or convex according as the liquid wets or does not wet the wall of the tube. We shall investigate the factors determining the shape of the liquid meniscus in a capillary tube of glass in the following way.

Let a capillary tube of glass be dipped vertically in a liquid meeting its surface at P (Figs 10.6.1, 10.6.2 and 10.6.3). Then a liquid molecule at P in contact with the tube there, will be attracted outward by the solid molecules of the tube close to it due to the *force of adhesion,* and attracted inwards by the near molecules of the liquid, due to the *force of cohesion;* the net force actually acting on it will, therefore, be the resultant of these two forces of adhesion and cohesion.

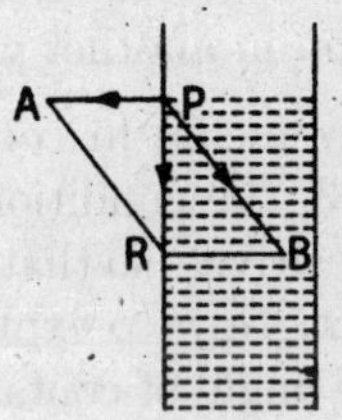

Fig 10.6.1

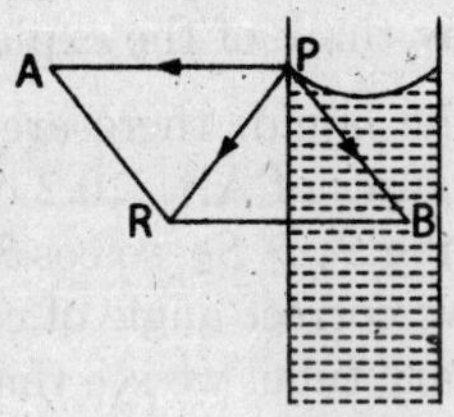

Fig 10.6.2

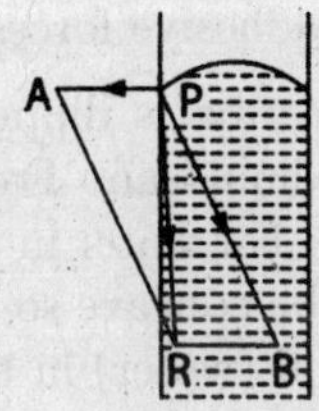

Fig 10.6.3

Now, the resultant force of adhesion acts at right angles to the tube at the point P. Let it be represented by the straight line PA and, the resultant force of cohesion acts at an angle of 45° to the vertical, and is represented by the straight line PB, so that, the two forces acting on the molecule are inclined to each other at an angle of 135° (Fig. 10.6.1). Their resultant, obtained by the ordinary law of parallelogram of forces, is represented by the diagonal PR of the parallelogram $PARB$, with PA and PB as adjacent sides. Its direction will naturally depend upon the relative magnitudes of the two forces PA and PB, and the following different cases arise:

(i) If PA/PB be equal to $1/\sqrt{2}$, i.e., if $PB = \sqrt{2} \cdot PA$, the resultant, (PR) will lie along the vertical, as shown in Fig. 10.6.1.

(ii) If PB be smaller than $\sqrt{2}PA$, the resultant (PR) will lie outside the liquid, as shown in Fig. 10.6.2.

(iii) If PB be greater than $\sqrt{2} \cdot PA$, then PR, will lie inside the liquid, as shown in Fig. 10.6.3.

What happens to the molecule P, happens to all other molecules in contact with the glass of the tube. And since a liquid cannot permanently withstand a shearing stress, its surface at every point will be at right angles to the resultant force there, when the liquid attains the position of equilibrium.

Thus, in the *first case,* when the resultant force PR acts along the vertical, i.e., when $PB = \sqrt{2} \cdot PA$, or, the *cohesive force is $\sqrt{2}$ times the adhesive force,* the molecules of the liquid are neither raised nor lowered and the ***liquid surface remains flat or plane.***

In the second case, when PR lies outside the liquid, i.e., when $PB < \sqrt{2} \cdot PA$, or the cohesive force is less than $\sqrt{2}$ times the adhesive force, the molecules of the liquid near the walls of the tube are raised up against the tube, those in the middle remaining practically unaffected, thus making the liquid surface *concave upwards,* as in the case of water and other liquids which wet the walls of the tube.

And, when PR lies inside the liquid, i.e., when $PB > \sqrt{2} \cdot PA$, or, the cohesive force is greater than $\sqrt{2}$ times the adhesive force, the liquid molecules near the walls of the tube are depressed, making the surface convex upwards, as in the case of mercury and other liquids, which do not wet the walls of the tube.

10.7 Capillarity—Elevation or Depression of a liquid in a Capillary Tube

The phenomenon of ascent or descent of a liquid in a capillary tube is known as *capillarity.* In case of ascent, the liquid meniscus is concave; while in the case of descent, it is convex. The physical explanation of such a phenomenon has been given in Art. 10.6 from consideration of cohesive and adhesive forces. We may think of the explanation in another way.

When the tube is dipped into the liquid, there are three media in contact—liquid-solid, solid-air, and liquid-air. From the figures of Art. 10.2 or 10.3, the condition of equilibrium is $\cos\theta = \frac{S_3 - S_2}{S_1}$. For water in a glass-tube $S_3 > S_2$; so $\cos\theta =$ positive—so that $0 < \theta < \frac{\pi}{2}$. So the surface must be concave so as to have correct angle of contact. Now to explain why there will be rise of liquid (water) in the capillary tube, we see that the angle of contact necessitates the surface to be curved with concavity upwards. Once it becomes concave, the external pressure and the internal pressure develop a difference, so that the water surface inside cannot be in level with that outside. Since an exposed surface tends to have minimum area, (i.e. to have minimum energy due to the property of surface tension) the concave meniscus within the tube therefore, tends to flatten and as a result the equilibrium of the vertical forces is lost, i.e., the vertical component of the force in the upward direction becomes greater than the force due to the weight of the liquid elevated, causing the liquid to rise until the two forces balance each other.

Thus the effect of surface tension is accounted for by taking a force equal to S acting upwards tangential to the liquid surface (fig. 10.7.1) in the tube. This tangential force resolved vertically upwards supports a certain column of liquid to reach the condition of equilibrium. We may look into the problem of ascent from a different angle also. So long as S_3 is not balanced by S_2 and the appropriate component of S_1, equilibrium is not possible. S_3 is usually large compared to S_2 and S_1—which means that there is an effective force upwards which is balanced only after a certain length of water column is dragged inside the tube.

Similarly when a liquid has a convex meniscus, the flattening tendency of the meniscus causes depression of the liquid. The depression continues until the downward force causing depression of the liquid balances the hydrostatic pressure due to the depth below the free surface.

10.7.1 Rise of a liquid in a Capillary Tube

We consider a uniform capillary tube dipped vertically inside a liquid which wets the walls of the tube. The liquid rises and stands at a height h say, above the surface of the liquid in the trough. Let θ be the angle of contact (which is acute in this case) and r the radius of the tube. Due to surface tension a force acts inward along the tangent to the liquid surface at each point of contact of the liquid with the wall of the tube (as shown by the thick line inside the liquid in fig. 10.7.1). According to Newton's third law, a reactionary force is called into play acting in the opposite direction. The vertical component of this reaction acting vertically upward is $S\cos\theta$ at each point of contact, S being the surface tension of the liquid.

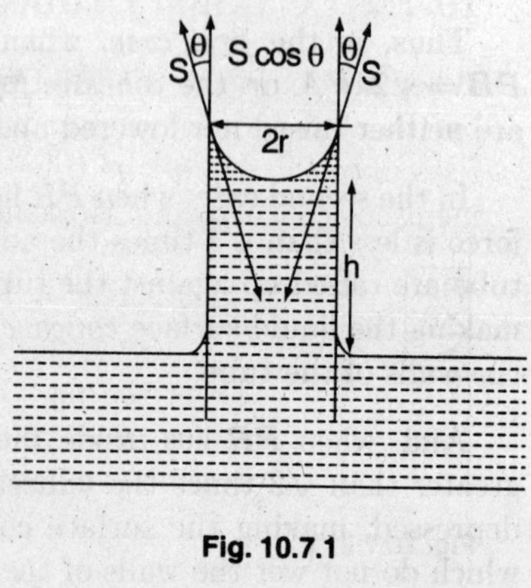

Fig. 10.7.1

Hence the total vertical upward force $= 2\pi r S\cos\theta$, since the liquid meets the tube in a circle of length $2\pi r$. The liquid rises in the tube due to this force, until it is balanced by the weight of the liquid cylinder of height h and the weight of a small volume of liquid in the meniscus as shown by the shaded region.

Now weight of the liquid cylinder of height $h = \pi r^2 h\rho g$, where ρ is the density of the liquid.

Again weight of the liquid in the shaded region of the meniscus = weight of a liquid cylinder of radius r and height r — weight of a liquid hemisphere of radius r.

$$= \pi r^3 \rho g - \frac{2}{3}\pi r^3 \rho g = \rho g\left(\pi r^3 - \frac{2}{3}\pi r^3\right) \quad \text{[considering the meniscus to be hemispherical in shape]}$$

$$= \frac{1}{3}\pi r^3 \rho g.$$

Hence, for equilibrium, we may write

$$2\pi r S\cos\theta = \pi r^2 h\rho g + \frac{1}{3}\pi r^3 \rho g = \pi r^2 \rho g\left(h + \frac{r}{3}\right) \quad \text{or} \quad S = \frac{r\rho g\left(h + \frac{r}{3}\right)}{2\cos\theta}. \tag{10.7.1}$$

Case I. If $\theta = 0$ as in the case of water and alcohol, we have

$$S = \frac{r\rho g\left(h + \frac{r}{3}\right)}{2}. \tag{10.7.2}$$

Case II. If $\theta > 90°$ as in the case of mercury, $\cos\theta$ becomes negative; hence h becomes negative, i.e., mercury cannot rise in a capillary tube, but is depressed in the capillary tube below the surface of mercury outside the cpillary tube. The same argument holds in case of water in a capillary tube inside of which is coated with paraffin wax, since θ is greater than 90° in this case.

Since in a capillary tube $h \gg r$, equation (10.7.1) may be written as

$$S = \frac{rh\rho g}{2\cos\theta}$$

or, for a given liquid, rh = constant, i.e., the elevation of a liquid in a capillary tube is inversely proportional to the radius of the tube. This is known as *Jurin's law*. The graph of r against h is a rectangular hyperbola.

10.7.2 Capillary tube of insufficient length

When the length of the tube is greater than h, the liquid rises in the tube to a height h given by the above equation. But when the length of the tube is smaller than h, say h', the liquid rises upto the top of the tube, spreads there, until the angle of contact increases to make the meniscus concave, and as result the radius of curvature increases to a new value R', such that

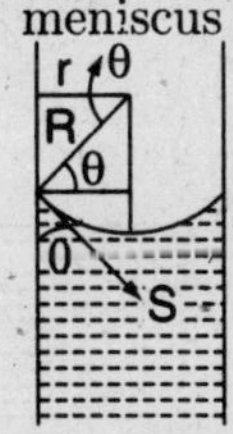

Fig. 10.7.2

$$R'h' = Rh = \frac{2S}{\rho g} = \text{constant. This may be proved as follows.}$$

From the fig. 10.7.2, we have $r = R\cos\theta$ (R = radius of curvature of liquid meniscus, θ = angle of contact). Putting this value of r in the capillary rise expression, we have

$$Rh = \frac{2S}{\rho g} = \text{constant.}$$

10.7.3 Energy changes associated with capillary rise

When a liquid rises against gravity in a capillary tube, its potential energy increases. To provide this energy, the surface energy of the system decreases. Let us explain how this happens.

There are three media—glass, liquid and air. Hence there are three surface boundaries: liquid-air boundary, glass-air boundary and glass-liquid boundary. Each boundary has its own surface tension, which is different from the other and is equal to the free energy per unit area of that surface. Let S_1, S_2, S_3 be the surface tensions of liquid-air, glass-air and liquid-glass boundaries respectively.

As the liquid rises in the capillary tube, the liquid-glass boundary increases in area by α (say), while glass-air boundary decreases by the same amount. Therefore increase in free energy of liquid-glass boundary $= S_3\alpha$ and decrease in free energy of glass-air boundary $= S_2\alpha$. Since $S_3 < S_2$, there is a net decrease of $(S_2 - S_3)\alpha$ in the free-energy of the system. A part of this lost energy is used up in overcoming the viscous forces which arise during the rise of the liquid in the capillary tube, and the remaining part appears as the gravitational potential energy of the raised column.

10.7.4 Expression for the capillary rise

The adjoining figure (10.7.4) shows respectively the stages—(a) when the liquid has not risen at all, (b) when the liquid has risen to a height x and (c) when the liquid has risen to a final height h—which is the equilibrium stage.

If r be the radius of the tube, then due to rise of liquid, the increase in area of liquid-glass boundary or the decrease in area of glass-air boundary is $2\pi rh$. Therefore, the net loss of free energy of the system is

$$(S_2 - S_3)2\pi rh. \tag{10.7.3}$$

Now considering the equilibrium of all the three media: glass, liquid and air [Fig. 10.7.3(c)], we have

$$S_2 = S_1 \cos\theta + S_3 \quad \text{or} \quad S_2 - S_3 = S_1 \cos\theta,$$

where θ is the angle of contact. Putting this value of $(S_2 - S_3)$ in (10.7.3), the net loss in the free energy is

$$S_1 \cos\theta(2\pi rh). \tag{10.7.4}$$

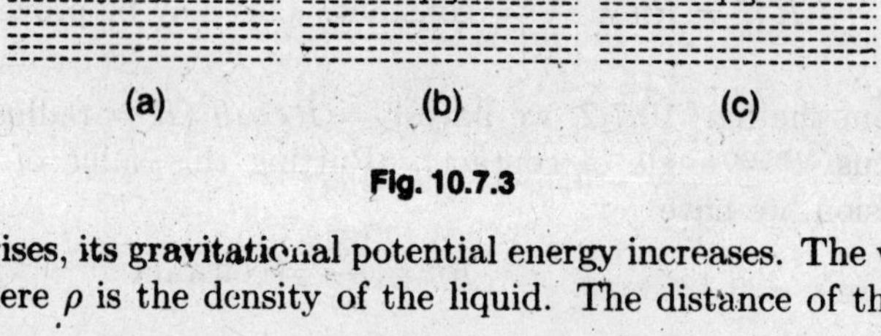

Fig. 10.7.3

Now as the liquid rises, its gravitational potential energy increases. The weight of the raised column is $\pi r^2 h\rho g$, where ρ is the density of the liquid. The distance of the centre of gravity of the raised column from the general level of the liquid outside the tube is $h/2$. Hence its gravitational potential energy is

$$\pi r^2 h\rho g\frac{h}{2} = \frac{1}{2}\pi r^2 \rho g h^2. \tag{10.7.5}$$

Again, as the liquid rises, viscous forces arise. The amount of energy spent in doing work against these forces is calculated by assuming a downward force F to act upon the liquid column as it moves from the position (a) to the position (c), F being just sufficient at all stages to prevent acceleration. At the intermediate stage (b), we shall have $F = \pi r^2 \rho g(h - x)$, since $\pi r^2 \rho g(h - x)$ is the weight of the liquid which on being added to the liquid in (b) gives the stage (c), when no further rise takes place. The work done or the energy spent against F in making the liquid rise through a height dx is Fdx. Hence the total energy spent when the liquid rises to the final height h is

$$\int_0^h F dx = \int_0^h \pi r^2 \rho g(h - x)dx = \pi r^2 \rho g\left(hx - \frac{x^2}{2}\right]_0^h$$

$$= \frac{1}{2}\pi r^2 \rho g h^2 = \pi r^2 h\rho g \cdot \frac{h}{2} = mg\frac{h}{2}, \tag{10.7.6}$$

which shows that *half of the free surface energy is dissipated by viscous forces in capillary rise*—which warms the liquid.

Equating (10.7.4) to (10.7.5) plus (10.7.6), we have

$$S_1 \cos\theta(2\pi r h) = \frac{1}{2}\pi r^2 \rho g h^2 + \frac{1}{2}\pi r^2 \rho g h^2$$

$$\text{or} \quad S_1 \cos\theta(2\pi r h) = \pi r^2 \rho g h^2 \quad \text{or} \quad h = \frac{2S_1 \cos\theta}{r\rho g},$$

which is the expression for the capillary rise.

Illustrative Example

The tube of a mercury barometer is 3.0 mm in diameter. What error is introduced into the reading because of surface tension (S)? For mercury S = 465 dynes per cm, $\theta = 128°$, density = 13.6 gm/cm^3; $g = 980$ cms/sec^2). **[C.U. (Hons.) 1965]**

The depression h of the mercury column as a result of surface tension will give rise to the error in the reading.

Now,

$$h + \frac{r}{3} = \frac{2S\cos\theta}{r\rho g}$$

$$\text{or} \quad h = \frac{2S\cos\theta}{r\rho g} - \frac{r}{3} = \frac{2 \times 465 \times \cos 128°}{0.15 \times 13.6 \times 980} - \frac{0.15}{3} \quad \left[\because r = \tfrac{3}{2}\text{ mm} = 0.15\text{ cm}\right]$$

$$= \frac{-2 \times 465 \times \cos 52°}{0.15 \times 13.6 \times 980} - 0.05 = -0.3363\text{ cm}.$$

Hence the depression $h = 0.3363$ cm.

10.8 Pressure on a Curved Membrane of Uniform Tension

We consider an element $PQRT$ of a curved membrance (the surface of the membrance in this case being synclastic) of principal radii of curvature r_1 and r_2 respectively (Fig. 10.8.1). It is evident that an excess pressure acts on the concave side of the membrane to counterbalance the effect due to surface tension.

If p be the magnitude of this excess pressure, the force over the element $= pxy$, where xy is the surface area of the element. Let this element be now pushed normally outwards through δz, so that the element reaches the position shown by $P'Q'R'T'$. Then the work done by the excess pressure $= pxy\delta z$.

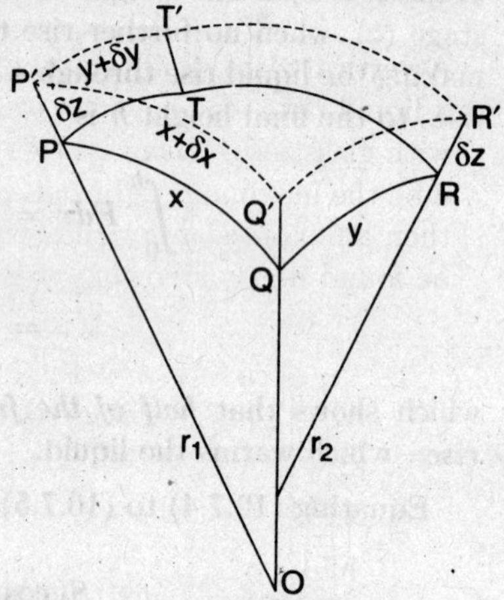

Fig. 10.8.1

Now, if S be the surface tension, then the work done for creating the new surface $= S\delta(xy)$, where $\delta(xy)$ denotes the increase in surface area. If the free surface be on both sides of the element, the total work done $= 2S\delta(xy)$.

Thus, we have,

$$pxy\delta z = 2S\delta(xy) = 2S(x\delta y + y\delta x).$$

Now from the figure,

$$\frac{x+\delta x}{x} = \frac{r_1+\delta z}{r_1} \quad \text{or} \quad \frac{\delta x}{x} = \frac{\delta z}{r_1}. \quad \therefore \quad \delta x = \frac{x}{r_1}\delta z.$$

Similarly,

$$\delta y = \frac{y}{r_2}\delta z.$$

$$\therefore \quad pxy\delta z = 2S\left(\frac{xy}{r_2}\delta z + \frac{xy}{r_1}\delta z\right) = 2Sxy\delta z\left(\frac{1}{r_1}+\frac{1}{r_2}\right).$$

$$\therefore \quad p = 2S\left(\frac{1}{r_1}+\frac{1}{r_2}\right). \tag{10.8.1}$$

Case I. If the surface be a spherical one, as in the case of a spherical bubble (soap bubble).

$$r_1 = r_2 = r(\text{say }). \quad \therefore \quad p = \frac{4S}{r}. \tag{10.8.2}$$

Case II. If the bubble be a cylindrical one, one of the radii of curvature r_2 (say) is infinity, and if

$$r_1 = r, \quad p = \frac{2S}{r}. \tag{10.8.3}$$

Case III. If instead of a synclastic surface, we consider an anti-clastic one as shown in Fig. 10.8.2, one of the radii of curvature r_2 say, is negative, and we have,

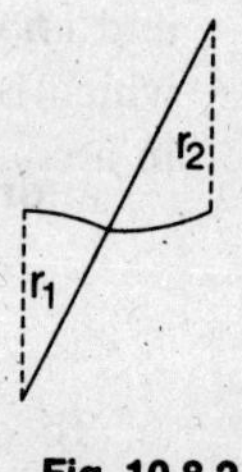

Fig. 10.8.2

$$p = 2S\left(\frac{1}{r_1}-\frac{1}{r_2}\right).$$

Case IV. If there is only one surface, as in the case of a liquid drop or an air bubble in a liquid, we have,

$$p = S\left(\frac{1}{r_1} + \frac{1}{r_2}\right) \tag{10.8.4}$$

provided the surface be a synclastic one. For a spherical liquid drop or an air bubble in a liquid,

$$r_1 = r_2 = r \text{ say, and } p = \frac{2S}{r}. \tag{10.8.5}$$

10.8.1 Rise of a Liquid in a Conical Capillary tube

A vertical glass capillary tube with a slightly conical bore just dips with its apex upwards into a liquid (like water, which wets its walls) of density ρ, surface tension S and angle of contact with glass zero. There is a very small hole at the apex, so that air can escape from the tube. If a be the maximum radius of cross-section of the tube, l its length and g the intensity of gravity, then an expression for the height h to which the liquid rises in the conical capillary tube may be found in the following way.

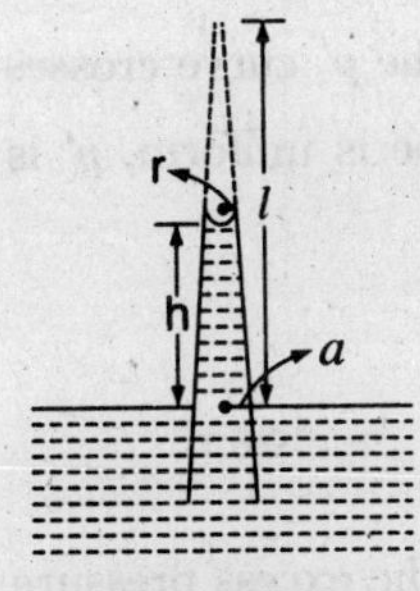

Fig. 10.8.3

We suppose that the liquid takes up the position as shown in fig. 10.8.3, and that r is the radius of curvature of the meniscus. Then from the geometry of the figure, we have

$$\frac{r}{(l-h)} = \frac{a}{l} \quad \text{or} \quad r = \frac{a}{l}(l-h)$$

and at the meniscus

$$\frac{2S}{r} = g\rho h \text{ or } \frac{2S}{g\rho} = hr \quad \text{or} \quad h\frac{a}{l}(l-h) = \frac{2S}{g\rho}$$

$$\text{or } ah - \frac{a}{l}h^2 = \frac{2S}{g\rho} \quad \text{or} \quad h^2 - lh + \frac{2Sl}{g\rho a} = 0$$

whence we have

$$h = \frac{l}{2} \pm \sqrt{\frac{l^2}{4} - \frac{2Sl}{g\rho a}}. \quad \therefore (h_2, h_1)_{h_2 > h_1} = \frac{l}{2} \pm \frac{l}{2}\sqrt{1 - \frac{8S}{g\rho a l}}$$

when l is very large

$$\sqrt{1 - \frac{8S}{g\rho a l}} \simeq 1 - \frac{4S}{g\rho a l}. \quad \therefore \left[h\right]_{l\to\infty} = \frac{l}{2} - \frac{l}{2}\left(1 - \frac{4S}{g\rho a l}\right) = \frac{2S}{g\rho a} = h_1 \tag{10.8.6}$$

which is the usual capillary rise expression (with $\theta = 0$) for a uniform capillary of radius a, and is obtained by considering the negative sign. This suggests that it is the lower height h_1 which is the stable one, and as such the positive sign is rejected. Therefore the stable height in a conical capillary is

$$h = \frac{l}{2}\left[1 - \sqrt{1 - \frac{8S}{g\rho a l}}\right]. \tag{10.8.7}$$

To establish the above fact conclusively, we are to discuss the equilibrium of the above two possible positions, and for that purpose we consider the liquid when its meniscus is at a height h, r being the radius of the corresponding cross-section of the tube—the liquid column is not necessarily in equilibrium. Let

$$g\rho h = p, \quad \frac{2S}{r} = \frac{2Sl}{a(l-h)} = p'.$$

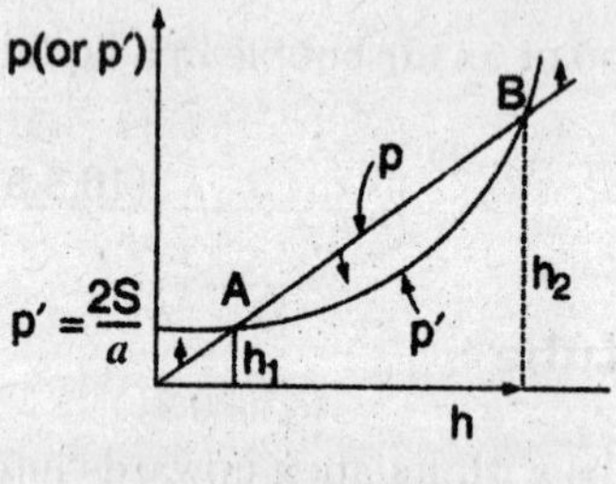

Fig. 10.8.4

For equilibrium, $p = p'$. If $p > p'$, the meniscus will fall; if $p < p'$, the meniscus will rise.

Now plotting p and p' both against h, we see that there are two points of intersection viz. A and B, corresponding to the roots h_1 and h_2 of the equation $p = p'$. Between A and B for any given h, $p > p'$, so that a meniscus placed at this level would fall—it means that surface tension effects are not large enough to support the liquid column.

Below A, $p < p'$ and above B also, $p < p'$, so that in each case the meniscus would rise.

Now A corresponds to a stable (equilibrium) position of the liquid in the tube, because a small displacement on either side is followed by a return of the meniscus to its undisturbed position. On the other hand B corresponds to an unstable position, as a small displacement on either side is followed by a greater displacement.

The existence of two equilibrium positions depends on whether or not the p' curve crosses the straight line for p. Imaginary roots occur if $\frac{2S}{g\rho al} > \frac{1}{4}$, while if the tube is uniform, p' is parallel to the h-axis and there is only one position of equilibrium.

10.9 Shape of Films

(i) When the pressure inside a film is same as the pressure outside it, then the excess pressure $p = 0$.

$$\therefore \quad \frac{1}{r_1} + \frac{1}{r_2} = 0, \quad \text{or} \quad r_1 = -r_2.$$

Thus the principal radii of curvature at any point are equal in magnitude but opposite in directions.

This type of surface may be obtained by dipping the broad end of one funnel into soap solution and bringing another funnel in contact with it and then drawing them apart.

A section of such a film surface is shown in figure 10.9.1. In this sectional diagram, P is any point on the curve and O is the centre of curvature in the plane of the diagram. OP is produced to meet the axis of symmetry $P'Q'$ at O'. Then PO' is the other radius of curvature in a plane at right angles to the plane of the diagram.

Here $PO = -PO'$.

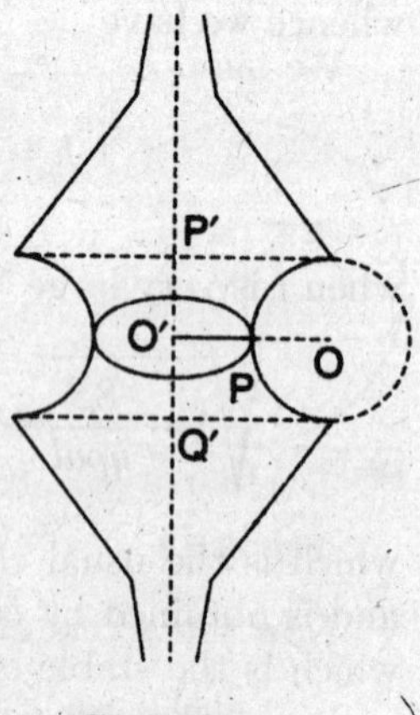

Fig. 10.9.1

(ii) If a film formed between two funnels, be given an additional pressure inside, the film may not assume exactly a cylindrical shape, but a shape whose section in the plane of the paper is shown in fig. 10.9.2. In figure 10.9.2, E and F are the points of inflexion. We consider a point P on the surface EDF. Let us draw a normal at P which cuts the axis of symmetry $P'Q'$ at O. Then OP becomes the radius of curvature of the surface at P in a plane at right angles to the plane of the figure. $O'P$ represents the radius of curvature at P of the surface in the plane of the figure.

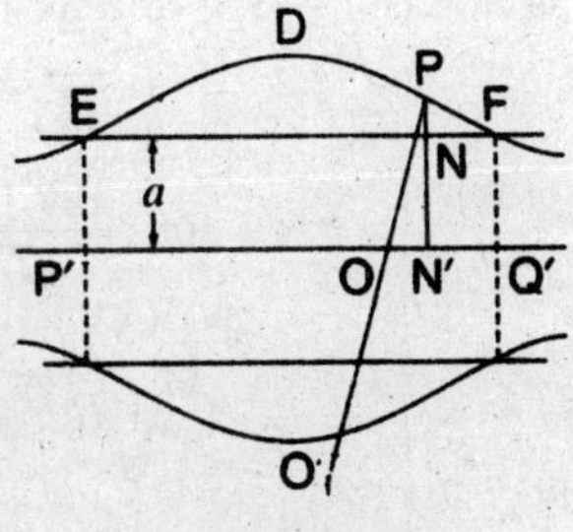

Fig. 10.9.2

Let $OP = r$ and $O'P = R$. If p be the excess pressure inside the film, we have

$$p = 2S\left(\frac{1}{R} + \frac{1}{r}\right).$$

Now at the point of inflexion F (say), $R = \infty$ and $r = a$; then $p = \dfrac{2S}{a}$.

Therefore, we may write

$$\frac{2S}{a} = 2S\left(\frac{1}{R} + \frac{1}{r}\right). \quad \therefore \quad \frac{1}{a} = \frac{1}{R} + \frac{1}{r}.$$

Now from the figure, $r = OP = PN'$ (approx), since EPF does not differ much from a straight line.

$$\therefore \quad r = PN + NN' = \delta + a, \quad \text{where } \delta = PN.$$

$$\therefore \quad \frac{1}{a} = \frac{1}{R} + \frac{1}{\delta + a} \quad \text{or} \quad \frac{1}{R} = \frac{1}{a} - \frac{1}{\delta + a} = \frac{\delta}{a(\delta + a)} = \frac{\delta}{a^2} \text{ (approx.)} \qquad (10.9.1)$$

$\therefore \quad \dfrac{1}{R} \propto \delta$, since a is constant.

Thus the reciprocal of the radius of curvature at a point P on the curve is proportional to the distance of the point from the horizontal line EF. The above equation represents the locus of a point situated near the centre of a circle of radius a, when the the circle rolls on a straight line. This can be proved in the following way.

We suppose a circle rolling (without slipping) with uniform angular velocity w along a straight line. Let Y be the centre of the circle and X any position of the moving point near the centre. C denotes the point of contct of the circle with the line $P'Q'$ along which it moves. XN is the perpendicular dropped on the line CY produced. If v be the velocity of the moving point X, and R the radius of curvature of the path it describes, then $\dfrac{v^2}{R}$ = acceleration of X along the normal to the path. Since the system is turning about C, the angular velocity at X is given by $\dfrac{v}{XC} = w$. Hence XC is the normal to the path traced by the point X.

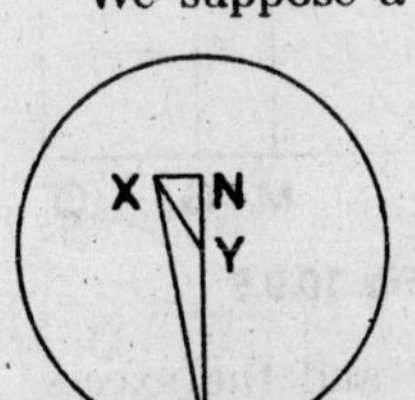

Fig. 10.9.3

Now the acceleration of X

= the acceleration of Y + the acceleration of X with respect to Y

= the acceleration of X with respect to Y

[$\because$ the acceleration of Y is zero as the centre Y moves uniformly along a straight line.]

Now, since X describes a circle round Y, the acceleration of X with respect to Y is w^2XY.

$\therefore$ the acceleration of X with respect to Y is along XY.

Hence the acceleration along XC, i.e., along the normal to the path $= w^2 XY \cos YXC$.

$$\therefore \quad \frac{v^2}{R} = w^2.XY \cos YXC$$

$$\text{or} \quad \frac{XC^2.w^2}{R} = w^2.XY \cos YXC \quad [\because v = XC.w]$$

$$\text{or} \quad \frac{1}{R} = \frac{XY}{XC^2} \cos YXC = \frac{XY \cos XYN}{XC^2}$$

$$[\because \angle YXC \text{ is very small } \therefore \angle YXC = \angle XYN]$$

$$= \frac{YN}{a^2} \text{ (approximately)},$$

since XC is nearly equal to a, the radius of the rolling circle.

Thus $\frac{1}{R}$ is proportional to the distance of the moving point X from the straight line described by the centre Y. Thus we see that the path described by a point fixed near the centre of a circle rolling on a straight line possesses this property. The shape of the curve is shown in fig. 10.9.2. The distance between the points of inflexion E and F is πa, i.e., half the circumference of the rolling circle (Fig. 10.9.4).

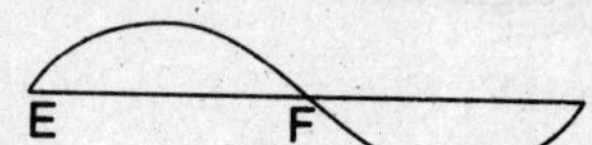

Fig. 10.9.4

The following typical cases may occur depending on the distance between the ends of the two funnels between which the film is collected.

Case I *Film of length less than half the circumference of the rolling circle, and is convex outwards.* In Fig. 10.9.5 C and D represent the position of the funnel, DG, the radius of the funnel and FQ' the radius of the rolling circle. If p_1 be the excess pressure inside the film over the outside pressure, then the excess pressure at any point L on the curve is given by

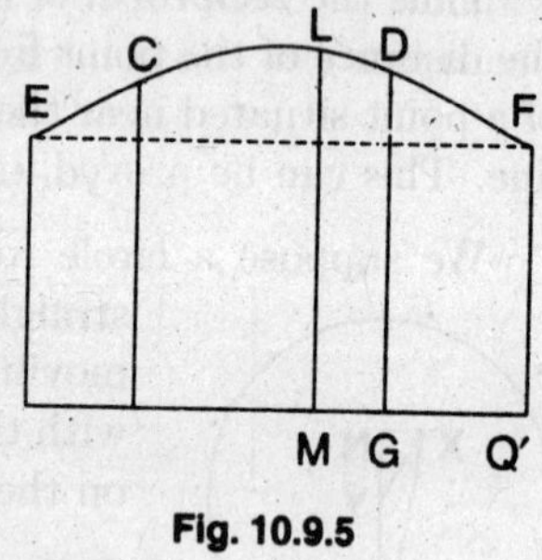

Fig. 10.9.5

$$p_1 = 2S\left(\frac{1}{R} + \frac{1}{LM}\right)$$

where R is the radius of curvature of the curve at L. If we consider the point at F instead of the point at L, then we have,

$$\frac{1}{R} = 0 \text{ and hence } p_1 = \frac{2S}{FQ'}$$

Now if the film were exactly cylindrical, CD would have been straight and the excess pressure p, in the case, would be

$$p = \frac{2S}{DG}$$

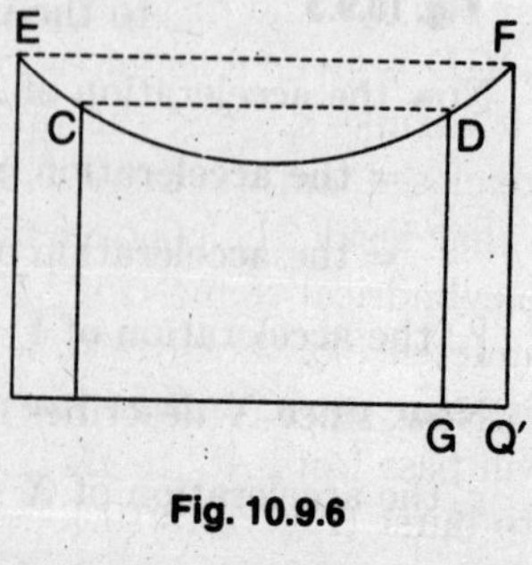

Fig. 10.9.6

Now $DG > FQ'$ $\therefore$ $p_1 > p$. Hence the pressure in the film which bulges out is greater than the pressure in the straight film.

Case II. *Film of length less than half the circumference of the rolling circle, but concave outwards.* In this case, excess pressure p_2 inside the film at F is

$$\frac{2S}{FQ'} \quad \text{(Fig. 10.9.6)}$$

Had the film been exactly cylindrical, the excess pressure would have been given by

$$p = \frac{2S}{DG}.$$

But $DG < FQ'$. $\therefore\ p > p_2$. Thus the pressure in this case is less than the pressure in the straight film.

So, when we take two films — one concave outwards and the other convex outwards, both being of length less than a semi-circle of the rolling or generating circle, and when these two films are communicated to each other, the system will be stable forming ultimately both the films cylindrical. Since $p_1 > p_2$, the convex film compresses to a cylindrical form, and the concave one bulges out to a cylindrical form due to flow of air from the first to the second film.

Case III. *A cylindrical film of length greater than half the circumference of the rolling circle and convex outwards at the middle.*

Here excess pressure p_3 inside the film is $p_3 = \dfrac{2S}{FQ'}$ and $p = \dfrac{2S}{DG}$ (Fig. 10.9.7)

$$\because\ FQ' > DG. \quad \therefore\ p_3 < p.$$

Case IV. *Film of length greater than half the circumference of the rolling circle with concavity outward at the middle.* In this case, the excess pressure $p_4 = \dfrac{2S}{FQ'}$ and $p = \dfrac{2S}{DG}$ (Fig. 10.9.8). But $DG > FQ'$

$$\therefore\ p_4 > p.$$

When two such films as mentioned in case III and case IV are communicated to each other, air will flow from the concave film to the convex one since $p_4 > p_3$. Thus ultimately, the concave film collapses and the convex film bulges out to burst, and the system cannot be stable.

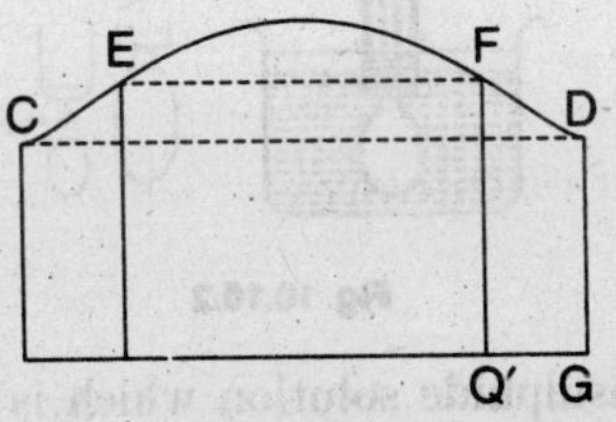

Fig. 10.9.7

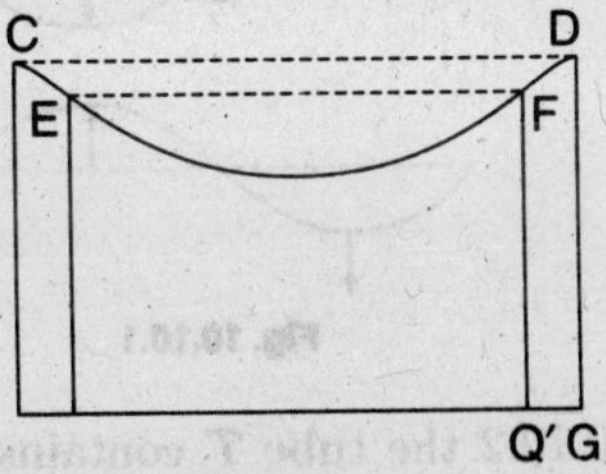

Fig. 10.9.8

These results have been confirmed by experimental arrangement due to Rucker and Reinold as shown in fig. 10.9.9 and fig. 10.9.10. Cylindrical soap films are formed between glass tubes AC and BD. The distance AC or BD is made less than half the circumference of the glass tube. BD is given a slight extra pressure so that it bulges out and AC given a less pressure so that it shrinks keeping the taps T_1 and T_2 open and the tap T closed. Next the taps T_1 and T_2 are closed and the tap T is kept open. As the pressure inside the convex film BD is greater than that inside the concave film AC, air will pass from BD to AC until the films assume the same cylindrical shape (Fig. 10.9.9). If the distance AC or BD is made greater than half the circumference of the glass tube, a similar experiment will yield different result. In this case the air pressure in AC will be greater than the air pressure in BD and when the tap T is opened, air will pass from AC to BD. The film AC will, therefore collapse and the film BD will bulge out to burst (Fig. 10.9.10.).

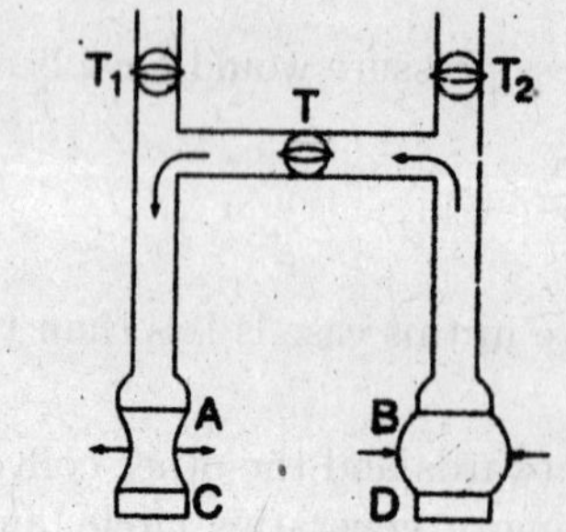

Fig. 10.9.9

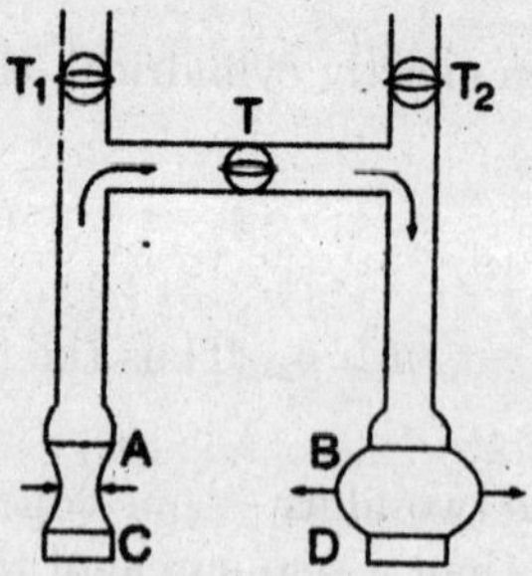

Fig. 10.9.10

10.10 Instability of Cylindrical Films

We now imagine the bubbles AB and CD joined together to form a long cylindrical bubble as shown in figure 10.10.1. The length of the bubble AD is greater than the circumference of the cylinder. In this case the bubble is unstable, because any slight disturbance which causes shrinkage of the portion CD will result in bulging out of the portion AB due to the flow of air from CD to AB. Ultimately the portion CD collapses with the formation of a large bubble at A and a small bubble at D.

The instability of a cylindrical film may be shown by the following experiment.

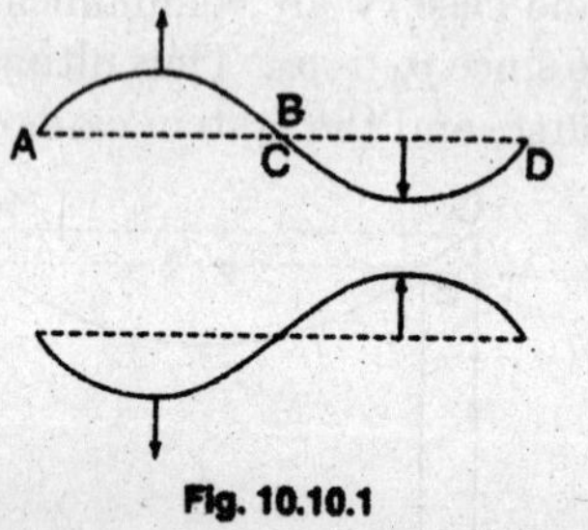

Fig. 10.10.1

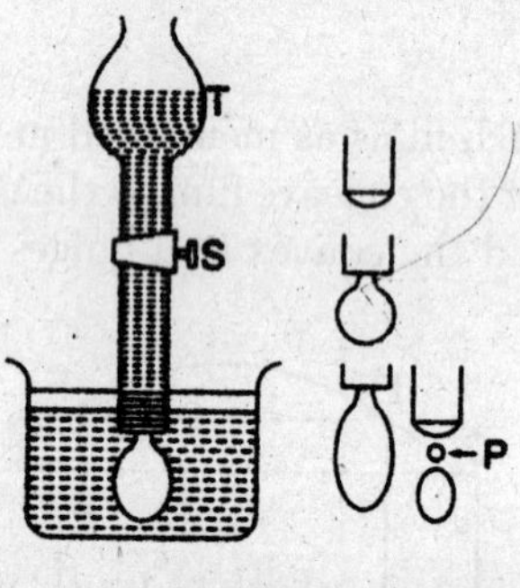

Fig. 10.10.2

In figure 10.10.2 the tube T contains carbon disulphide solution which is allowed to flow very slowly in a vessel of water by adjusting the stopcock S. The drops formed at different stages are shown in the figure. The drop finally breaks up into two drops. The smaller drop is called *Plateau's spherule.*

Illustrative Examples

1. Calculate the difference between the pressures inside and outside a spherical soap bubble of diameter 1 inch blown with a soap solution of surface tension 25 dynes per cm. [C.U. 1957]

Excess pressure p in a spherical soap bubble of radius r is given by

$$p = \frac{4S}{r}$$

In this case, $r = \frac{1}{2}$ inch $= 1.27$ cm. and $S = 25$ dynes per cm.

$$\therefore\ p = \frac{4 \times 25}{1.27} = 78.74 \text{ dynes/cm}^2.$$

2. The interiors of a cylindrical and a spherical soap buble are put into communication. Determine the ratio of the radii of the sphere and the cylinder in order that the bubbles may be in equilibrium.

The two bubbles in communication should be in equilibrium when pressure of air inside both the bubbles is same, i.e., the excess pressure inside the cylindrical bubble over the pressure outside is same as the excess of pressure inside the spherical bubble over that outside. Let r_1 and r_2 be the radii of the sphere and the cylinder respectively. Then, for equilibrium of the two, we have,

$$p = \frac{4S}{r_1} = \frac{2S}{r_2} \quad \therefore\ \frac{r_1}{r_2} = \frac{2}{1}$$

Thus the radius of the sphere should be twice the radius of the cylinder.

3. There is a minute circular hole at the bottom of a small hollow vessel. The vessel has to be immersed in water to a depth of 40 cms., before any water penetrates inside. Find the radius of the hole, if the surface tension and density of water be 73 dynes per cm and 1 gm per c.c. respectively.

[C.U. (Hons.) 1963, '88, cf '98, 2003]

Water cannot penetrate till the hydrostatic pressure is greater than the excess of pressure that comes into play due to surface tension. At the limiting condition when water just penetrates, these two pressures should be equal. That is

$$h\rho g = \frac{2S}{r} \quad (r \text{ being the radius of the hole})$$

$$\text{or} \quad r = \frac{2S}{h\rho g}.$$

In this case $S = 73$ dynes per cm, $h = 40$ cms and $\rho = 1$ gm/cc

$$\therefore\ r = \frac{2 \times 73}{40 \times 1 \times 981} = 0.0037 \text{ cm} = 3.7 \times 10^{-3} \text{ cm}.$$

4. Two spherical soap bubbles of diameter 10 cms and 6 cms respectively are formed, one at each end of a narrow horizontal glass tube. What is the pressure difference between the ends of the tube, if the surface tension of the soap solution be 30 dynes per cm? What happens when the bubbles are put in pressure communication with each other through the tube? **[C.U. (Hons.) 1961]**

If the excess of pressure inside the bubbles of radii r_1 and r_2 respectively be p_1 and p_2, then

$$p_1 = \frac{4S}{r_1} \quad \text{and} \quad p_2 = \frac{4S}{r_2}$$

Here $r_1 = 5$ cms, $r_2 = 3$ cms

$$\therefore\ r_2 < r_1 \text{ and so } p_2 > p_1.$$

Hence the required pressure difference is given by

$$p_2 - p_1 = 4S\left(\frac{1}{r_2} - \frac{1}{r_1}\right) = 4 \times 30\left(\frac{1}{3} - \frac{1}{5}\right) = 16 \text{ dynes/cm}^2.$$

In this case, since $p_2 > p_1$, the air from inside the smaller bubble will rush through the horizontal narrow tube towards the larger bubble. So the smaller bubble gradually shrinks and the larger one gradually bulges out. This will continue and ultimately the smaller one collapses and the larger one bulges out to burst. The system cannot thus be a stable one.

5. Two soap bubbles of radii a and b coalesce to form a single bubble of radius c. If the external pressure is B, show that the surface tension is given by

$$\frac{B(c^3 - a^3 - b^2)}{4(a^2 + b^2 - c^2)}.$$

[C.U. (Hons.) 1965, '84, '91, 2000]

Inside pressure of the bubble of radius a is

$$p_1 = B + \frac{4S}{a}, \quad S \text{ being the surface tension of the soap solution.}$$

Similarly inside pressure of the bubble of radius b is $p_2 = B + \dfrac{4S}{b}$ and inside pressure of the bubble of radius c is $p_3 = B + \dfrac{4S}{c}$.

Let m_1, m_2 and m_3 be the masses of air inside the three bubbles respectively. Now since the total mass of air in the first two bubbles is equal to that in the third one, we may write

$$m_3 = m_1 + m_2 \tag{1}$$

Now

$$m_1 = \frac{4}{3}\pi a^3 \rho_1 \quad [\rho_1 \text{ being the density of air in the first bubble}]$$

$$= \frac{4}{3}\pi a^3 k p_1 = \frac{4}{3}\pi a^3 k\left(B + \frac{4S}{a}\right) \quad \left[\begin{array}{l}\because \text{ the density is proportional to} \\ \text{pressure according to Boyle's law,} \\ \therefore\ \rho_1 = kp_1,\ k \text{ being a constant.}\end{array}\right]$$

Similarly,

$$m_2 = \frac{4}{3}\pi b^3 k p_2 = \frac{4}{3}\pi b^3 k\left(B + \frac{4S}{b}\right)$$

$$m_3 = \frac{4}{3}\pi c^3 k p_3 = \frac{4}{3}\pi c^3 k\left(B + \frac{4S}{c}\right)$$

Substituting the values of m_1, m_2 and m_3 in eqn. (1), we have

$$\frac{4}{3}\pi k\left[c^3\left(B + \frac{4S}{c}\right)\right] = \frac{4}{3}\pi k\left[a^3\left(B + \frac{4S}{a}\right) + b^3\left(B + \frac{4S}{b}\right)\right]$$

$$\text{or } c^3B + 4Sc^2 = a^3B + 4Sa^2 + b^3B + 4Sb^2$$

$$\text{or } 4S(a^2 + b^2 - c^2) = B(c^3 - a^3 - b^3)$$

$$\text{or } S = \frac{B(c^3 - a^3 - b^3)}{4(a^2 + b^2 - c^2)}.$$

6. A soap bubble containing air, of radius r_1 is raised from $t_1^\circ C$ to $t_2^\circ C$ when the radius changes to r_2. Set up the equation to determine r_2 in terms of r_1, t_1, t_2, the surface tensions S_1 and S_2 of the soap solution at the two temperatures and P_0 the constant external pressure (The equation for r_2 need not be solved). **[C.U. (Hons.) 1977]**

When a soap bubble of radius r_1 at temperature $t_1^\circ C$ containing air is raised to temperature $t_2^\circ C$, the air inside expands and the bubble is enlarged against the constant external pressure P_0. Applying the combination of Boyle's law and Charle's law we may write in this case

$$\frac{P_1V_1}{T_1} = \frac{P_2V_2}{T_2} \tag{1}$$

Now inside pressure of the bubble of radius r_1 is $P_1 = P_0 + \dfrac{4S_1}{r_1}$, where S_1 is the surface tension of soap solution at temperature $t_1^\circ C$; and

V_1 = volume of the bubble at temperature $t_1^\circ C = \frac{4}{3}\pi r_1^3$; again we have

P_2 = inside pressure of bubble of radius $r_2 = P_0 + \dfrac{4S_2}{r_2}$,

S_2 being the surface tension at temperature $t_2^\circ C$ and V_2 = volume of the bubble at $t_2^\circ C = \frac{4}{3}\pi r_2^3$.

$\therefore$ from (1) we may write

$$\frac{\left(P_0 + \frac{4S_1}{r_1}\right)\frac{4}{3}\pi r_1^3}{t_1+273} = \frac{\left(P_0 + \frac{4S_2}{r_2}\right)\frac{4}{3}\pi r_2^3}{t_2+273}$$

$$\text{or}\quad (P_0r_1 + 4S_1)r_1^2 \times (t_2+273) = (P_0r_2 + 4S_2)r_2^2 \times (t_1+273)$$

which is the required equation. From this equation r_2 may be determined in terms of r_1, t_1, t_2, S_1, S_2 and P_0.

7. Calculate the amount of work done if a soap bubble is slowly enlarged from a radius of 10 cms to a radius of 20 cms. [S = 30 dynes/cm.] Explain why the amount of work necessary for this enlargement is less than that if the increase were made at a faster rate. **[C.U. (Hons.) '84, cf. 2008]**

Let r be the instantaneous radius of the bubble in the process of enlargement and let it increase to $r + dr$ at some other time, so that, the increase in the surface area

$$= 4\pi(r+dr)^2 - 4\pi r^2 = 8\pi r dr \quad [\text{neglecting } 4\pi(dr)^2 \text{ which is small}]$$

Now work done for this increase is given by $S \times 8\pi r dr$, S being the surface tension of soap solution.

Since there are two free surfaces, the work done is $\delta W = 16\pi S r dr$.

Therefore the total work necessary for the enlargement of the bubble from a radius of 10 cms to a radius of 20 cms is given by

$$W = 16\pi S\int_{10}^{20} r dr = \frac{16\pi S(20^2-10^2)}{2} = 8 \times 3.14 \times 30 \times 300 = 2.26 \times 10^5 \text{ ergs.}$$

If the enlargement is made slowly, the process becomes isothermal. When the bubble increases from a radius of 10 cms to a radius of 20 cms slowly, surface area increases. So some work is to be done against surface tension—which becomes the surface energy. However if the process be made faster, there is fall of temperature as a result of enlargement. So some energy goes to compensate the fall of temperature. Hence mechanical work is less than that in the previous case.

8. If the latent heat of water is 600 cals per gm, its density 1 gm per c.c. and its surface tension 73 dynes per cm, calculate the minimum radius a drop can have, if it is free to evaporate but prevented from receiving heat.

Since the drop is not allowed to receive heat from outside, the surface energy per unit area of the drop should, therefore, be numerically equal to its surface tension. The limiting condition for a drop to exist is that the amount of heat necessary for the evaporation of a certain quantity of liquid from the drop should be equal to the loss of surface energy due to evaporation.

If the drop has a radius r and if it is reduced by δr, volume of liquid evaporated

$$= \frac{4}{3}\pi r^3 - \frac{4}{3}\pi(r-\delta r)^3 = \frac{4}{3}\pi[r^3 - (r-\delta r)^3]$$

$$= \frac{4}{3}\pi \cdot 3r^2\delta r = 4\pi r^2 \delta r \quad \text{[neglecting the square and cube of } \delta r\text{]}$$

$\therefore$ mass of the liquid evaporated $= 4\pi r^2\delta r$ [$\because$ density $= 1$ gm/cc].

Hence heat required to evaporate the above quantity of liquid $= 600 \times 4\pi r^2 \delta r$ cals.

Now loss of energy

$$= [4\pi r^2 - 4\pi(r-\delta r)^2]S, \; S \text{ being the surface tension of water}$$

$$= 4\pi \cdot S \cdot 2r\delta r \quad \text{[neglecting the term containing square of } \delta r\text{]}$$

$$= 8\pi S r \delta r$$

$$\therefore \quad \frac{8\pi S r\delta r}{J} = 600 \times 4\pi r^2 \delta r, \; J \text{ being the mechanical equivalent of heat}$$

$$\text{or} \quad r = \frac{2S}{J \times 600} = \frac{2 \times 73}{4.2 \times 10^7 \times 600} = 5.79 \times 10^{-9} \text{ cm}$$

9. A round drop of water has an excess pressure of 1 atmosphere due to its surface tension. If the surface tension of water is 72 dynes/cm, calculate the radius of the drop. **[C.U. (Hons.) 1983]**

Excess pressure in a round drop is given by $p = \frac{2S}{r}$.

Here $p = 1$ atmosphere $= 1.013 \times 10^6$ dynes/cm^2, $S = 72$ dynes/cm

$$\therefore \; r = \frac{2S}{p} = \frac{2 \times 72}{1.013 \times 10^6} = 1.421 \times 10^{-4} \text{ cm.}$$

10. A small drop of water has a radius equal to r cm. Calculate the decrease in the surface energy of the drop when its radius is diminished by a small quantity δr cm due to evaporation, and find the condition that the surface energy lost may suffice to provide the latent heat necessary for the evaporation.

Let E be the surface energy per unit area, i.e., per sq. cm.

$\therefore$ loss of surface energy $= [4\pi r^2 - 4\pi(r-\delta r)^2]E = 8\pi E r\delta r$ ergs (approx.).

Now mass of water evaporated due to diminution of radius by δr cm $= 4\pi r^2 \delta r$ gm (density of water $= 1$ gm/cc).

$\therefore$ the corresponding heat energy necessary for the evaporation of the above mass of water $= L \times 4\pi r^2 \delta r$ cals, where L is the latent heat of evaporation.

Thus evaporation may continue when loss in surface energy is greater than the latter quantity of heat, i.e.,

$$\frac{8\pi E r \delta r}{J} > L \times 4\pi r^2 \delta r \text{ or } \frac{2E}{J} > Lr \text{ or } r < \frac{2E}{LJ}.$$

11. **Calculate the radius of the largest drop of water that can evaporate at 0°C without heat being communicated to it. Surface energy of water at 0°C = 117 ergs per sq. cm. Latent heat of evaporation of water at 0°C = 606 cals per gm.** **[C.U. (Hons.) 1949]**

The condition for evaporation from the surface of a drop is that the radius of the drop is less than $\frac{2E}{LJ}$.

Hence for the maximum radius of the drop, r should be equal to $\frac{2E}{LJ}$

$$\therefore r = \frac{2E}{LJ}$$

Here $E = 117$ ergs per sq cm, $L = 606$ cals per gm, $J = 4.2 \times 10^7$ ergs per calorie

$$\therefore r = \frac{2 \times 117}{606 \times 4.2 \times 10^7} = 0.09194 \times 10^{-7} = 9.194 \times 10^{-9} \text{ cm}.$$

12. **Two spherical soap bubbles, A and B, of radii 3 cm and 5 cm coalesce so as to have a portion of their surface in common. Calculate the radius of curvature of this common surface.**

Let r be the radius of curvature of the common surface of the soap bubbles A and B. The excess pressure in A above atmospheric is $\frac{4S}{3}$, while that in B is $\frac{4S}{5}$. Thus there is an excess pressure on the A-side of the common surface above that on the B-side by an amount

$$\left(\frac{4S}{3} - \frac{4S}{5}\right).$$

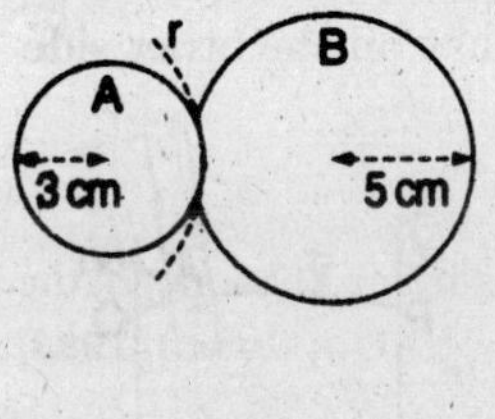

But this excess pressure must be equal to $\frac{4S}{r}$. Hence we have

$$\frac{4S}{r} = \frac{4S}{3} - \frac{4S}{5} \quad \text{or} \quad \frac{1}{r} = \frac{1}{3} - \frac{1}{5} \quad \text{or} \quad r = 7.5 \text{ cm}$$

and the common surface is concave towards the bubble of smaller radius.

13. **Two spherical soap bubbles coalesce to form a bigger bubble without any leakage of air. If V be the change in volume of the contained air and S the change in total surface area, show that $3PV + 4ST = 0$ where T is the surface tension of the soap solution and P is the atmospheric pressure.**

Let r_1 and r_2 be the radii of the two soap bubbles before they coalesce and R the radius of the combined bubble. In each case, the pressure outside the bubble is atmospheric pressure P. If p_1, p_2 be the pressures inside the bubbles before they coalesce and p the pressure inside the combined bubble, we have

$$p_1 = P + \frac{4T}{r_1}; p_2 = P + \frac{4T}{r_2} \quad \text{and} \quad p = P + \frac{4T}{R}.$$

If V_1, V_2 and V' be the corresponding volumes, we have

$$V_1 = \frac{4}{3}\pi r_1^3; \quad V_2 = \frac{4}{3}\pi r_2^3 \quad \text{and} \quad V' = \frac{4}{3}\pi R^3.$$

If the temperature remains constant, we have from Boyle's law $p_1V_1 + p_2V_2 = pV'$

$$\text{or} \quad \left(P + \frac{4T}{r_1}\right)\frac{4}{3}\pi r_1^3 + \left(P + \frac{4T}{r_2}\right)\frac{4}{3}\pi r_2^3 = \left(P + \frac{4T}{R}\right)\frac{4}{3}\pi R^3$$

$$\text{or} \quad P\left(\frac{4}{3}\pi r_1^3 + \frac{4}{3}\pi r_2^3 - \frac{4}{3}\pi R^3\right) + \frac{4T}{3}\left(4\pi r_1^2 + 4\pi r_2^2 - 4\pi R^2\right) = 0 \tag{1}$$

But

$$\frac{4}{3}\pi r_1^3 + \frac{4}{3}\pi r_2^3 - \frac{4}{3}\pi R^3 = V \text{ (the change in volume)}$$

$$\text{and} \quad 4\pi r_1^2 + 4\pi r_2^2 - 4\pi R^2 = S \text{ (change in total surface area).}$$

Therefore (1) becomes

$$PV + \frac{4T}{3}S = 0 \quad \text{or} \quad 3PV + 4ST = 0.$$

10.11 Shape of an interfacial surface

We consider first two incompressible liquids in contact; the lighter one is of density σ_1 and the denser one of density σ_2. The attached figure 10.11.1 gives the principal section of such a system. In the figure the curve PQ is a section of the interfacial surface, and we shall find an equation to the curve in the following way.

We consider two co-ordinate axes OX, OY with O as origin. We now take a point A on PQ, whose ordinate is y, say, and we imagine an elementary area $\delta\alpha$ surrounding the point A. Let p be the difference in pressures between the two points, one inside PQ, and the other outside PQ. Therefore, the excess pressure at A on the concave side of PQ over that on the other side is given by

$$p = S\left(\frac{1}{r_1} + \frac{1}{r_2}\right)$$

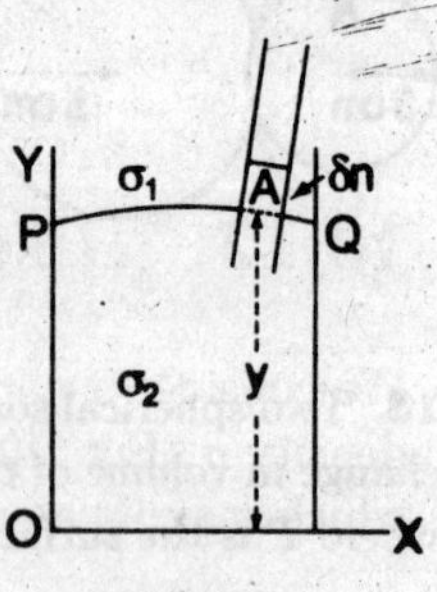

Fig. 10.11.1

where r_1 and r_2 are the principal radii of curvature of the surface at A. Let us now imagine the element to suffer a virtual displacement normally through δn; then the work done

$$= p\delta\alpha\delta n = S\left(\frac{1}{r_1} + \frac{1}{r_2}\right)\delta\alpha\delta n = S\left(\frac{1}{r_1} + \frac{1}{r_2}\right)dV$$

where $\delta V = \delta\alpha\delta n =$ an element of volume.

Therefore, the total work done in making the displacement of the entire interfacial surface through δn is given by

$$\iiint S\left(\frac{1}{r_1} + \frac{1}{r_2}\right)dV$$

Let us now consider the change in the gravitational potential energy due to the above displacement.

When there was no displacement, the potential energy at A of the elementary mass contained in volume δV should be written as $\sigma_1\delta V gy$.

When the elementary area $\delta\alpha$ moves through δn, the volume δV is now filled with the liquid of density σ_2. Therefore, the potential energy in this case is given by $\sigma_2 \delta V gy$.

$\therefore$ the change in gravitational potential energy is $(\sigma_2 - \sigma_1)gy\delta V$.

Hence the total change in gravitational potential energy is $\iiint (\sigma_2 - \sigma_1)gydV$.

Since the system is in equilibrium, we may write

$$\iiint S\left(\frac{1}{r_1} + \frac{1}{r_2}\right) dV = \iiint (\sigma_2 - \sigma_1)gydV$$

$$\text{or} \quad \iiint \left[S\left(\frac{1}{r_1} + \frac{1}{r_2}\right) - (\sigma_2 - \sigma_1)gy\right] dV = 0.$$

Since the liquid is incompressible

$$\iiint dV = 0 \quad \text{[i.e., the total change in volume is equal to zero]}$$

Hence,

$$S\left(\frac{1}{r_1} + \frac{1}{r_2}\right) - (\sigma_2 - \sigma_1)gy = \text{constant.}$$

This equation may be regarded as the differential equation of the surface having section PQ.

The above reasoning holds in the case when one of the media is an incompressible liquid, and the other is a gas (air, say), for

$$\iiint dV = 0$$

also in this condition. Putting $\sigma = \sigma_2 - \sigma_1$, we obtain the equation for liquid-air interface given by

$$S\left(\frac{1}{r_1} + \frac{1}{r_2}\right) - \sigma gy = \text{constant.}$$

10.12 Rise of a liquid along an inclined solid surface

We consider a perfectly clean glass plate BC dipping in an inclined manner inside a liquid of density σ (Fig. 10.12.1). AEB represents the principal section of the free surface of the liquid which may be assumed horizontal at a great distance from the plate.

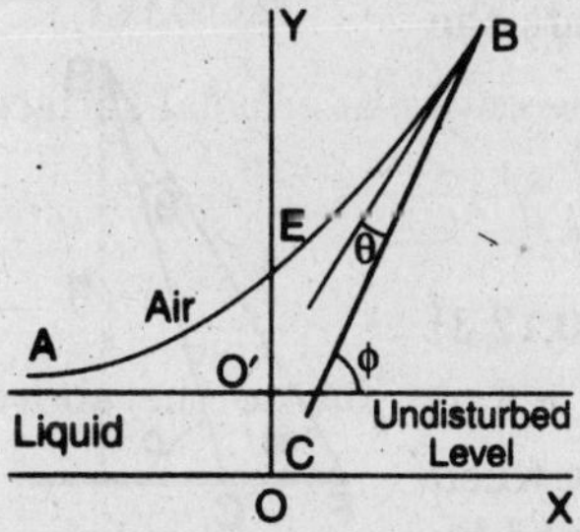

Fig. 10.12.1

The general equation to the curve is given by

$$S\left(\frac{1}{r_1} + \frac{1}{r_2}\right) - \sigma gy = \text{constant.} \qquad (10.12.1)$$

In this case the interfacial surface is cylindrical and one of the radii of curvature is infinite. We have, therefore, the modified equation to the curve in the present case given by

$$\frac{S}{r} - \sigma gy = \text{constant,}$$

where r is the radius of curvature in the plane of the figure.

Now if we shift our origin to the undisturbed level at a great distance from the plate, we have at the origin, $y = 0$ and $r = \infty$. Therefore, from the above expression we have, constant = 0, and hence

$$\frac{S}{r} = \sigma g y.$$

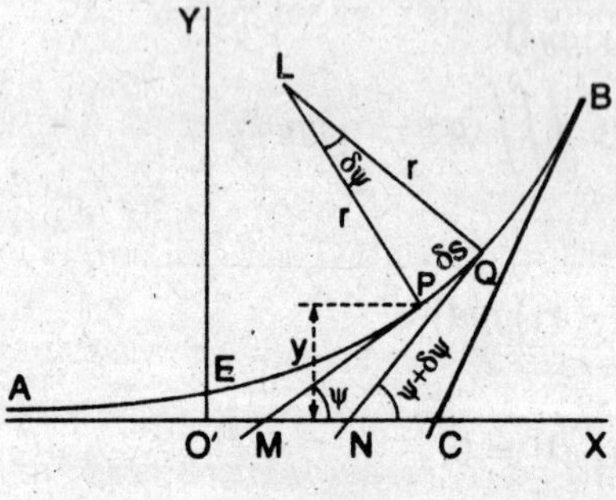

Fig. 10.12.2

We now consider the tangents PM, QN and normals PL, QL drawn at P and Q (Fig. 10.12.2). The two points P and Q on the curve are at an elementary distance δs apart. Let the tangents PM and QN make angles ψ and $\psi + \delta\psi$ respectively with the axis $O'X$. Let y be the ordinate of the point P. The normals meet at the centre of curvature L of the curve PQ. Now $\angle PLQ = \delta\psi$, $PL = QL = r$ and $PQ = \delta s$.

From the figure, we have

$$r\delta\psi = \delta s \quad \text{or} \quad \frac{1}{r} = \frac{\delta\psi}{\delta s} \quad \therefore\ S\frac{\delta\psi}{\delta s} = \sigma g y;$$

again $\sin\psi = \frac{\delta y}{\delta s}$

$$\therefore\ \frac{1}{\delta s} = \frac{\sin\psi}{\delta y} \quad \therefore\ S\sin\psi\delta\psi = \sigma g y \delta y.$$

Integrating, we have,

$$-S\cos\psi = \sigma g\frac{y^2}{2} + \text{constant} \quad \text{or} \quad 2S\cos\psi + \sigma g y^2 = \text{constant}$$

Now, when $y = 0, \psi = 0 \quad \therefore$ constant $= 2S$.

Hence we may write

$$2S\cos\psi + \sigma g y^2 = 2S \quad \text{or} \quad g\sigma y^2 = 2S(1 - \cos\psi) \tag{10.12.2}$$

This is the equation of the curve AEB.

Discussions

Case I. If the plate makes an angle ϕ with the horizontal, a tangent BF (Fig. 10.12.3) to the curve at B where the liquid meets the plate, makes an angle ψ with the horizontal, given by $\phi = \psi + \theta$ where θ is the angle of contact or $\psi = \phi - \theta$.

Therefore, we have, $g\sigma h^2 = 2S[1 - \cos(\phi - \theta)]$, where h represents the ordinate of B.

When the plate is dipped vertically $\phi = 90°$, and we have,

$$g\sigma h^2 = 2S(1 - \sin\theta) \quad \text{or} \quad h = \sqrt{\frac{2S}{g\sigma}(1 - \sin\theta)} \tag{10.12.3}$$

which determines the height to which the liquid rises against the plate when dipped vertically. If θ be known, the surface tension may be determined by measuring the height h.

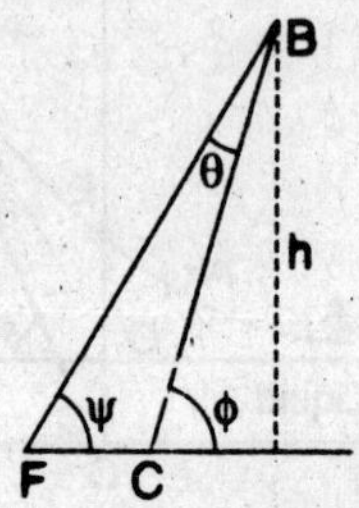

Fig. 10.12.3

For water, the angle of contact $\theta = 0$

$$\therefore\ h = \sqrt{\frac{2S}{g\sigma}}$$

Taking $S = 70$ dynes/cm, $g = 981$ cm/sec^2 and $\sigma = 1$, for water,

$$h = \sqrt{\frac{140}{981}} = 3.77 \text{ mm}.$$

Case II. If the plate is turned horizontal and the liquid touches the lower surface of the plate (Fig. 10.12.4) at the point of contact B then, $\phi = 180°$

$$\therefore\ g\sigma h^2 = 2S[1 - \cos(180° - \theta)] = 2S(1 + \cos\theta).$$

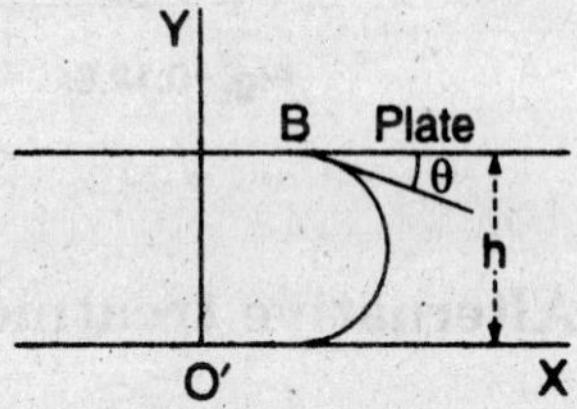

Fig. 10.12.4

In addition, if $\theta = 0$, we have

$$g\sigma h^2 = 4S \quad \text{or} \quad \frac{S}{g\sigma} = \frac{h^2}{4}.$$

The expression $\frac{S}{g\sigma}$ is often called *Laplace's constant.*

Case III. When a sessile drop or bubble resting on a horizintal plate or pendent drop or bubble projecting from a horizontal plane is large, a central portion of its surface P (Fig. 10.12.5) does not sensibly deviate from the cylindrical shape and a tangent drawn from a point on this portion is vertical. We have in this case,

$$g\sigma h'^2 = 2S(1 + \cos 90°) = 2S \quad [\text{here, when } \theta = 90°,\ y = h'] \tag{10.12.4}$$

Again

$$g\sigma h^2 = 2S(1 + \cos\theta) = 4S\cos^2\frac{\theta}{2}.$$

$$\text{or} \quad \cos^2\frac{\theta}{2} = \frac{g\sigma h^2}{4S} = \frac{g\sigma h^2}{2g\sigma h'^2} = \frac{h^2}{2h'^2} \quad \text{or} \quad \cos\frac{\theta}{2} = \frac{h}{\sqrt{2}\,h'} \tag{10.12.5}$$

Here h denotes the thickness of the bubble or drop.

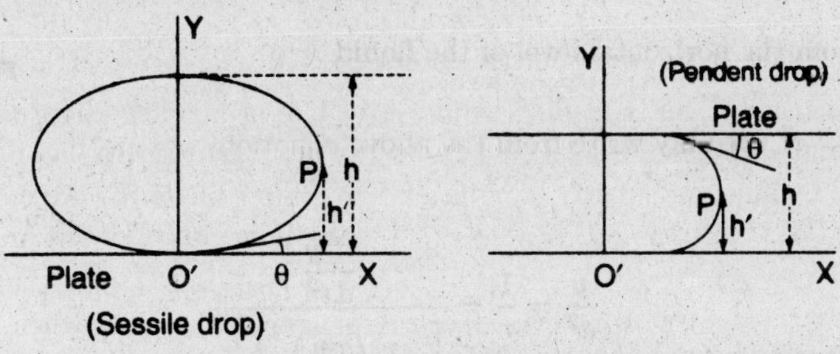

Fig. 10.12.5

It should be noted that in case of a sessile drop or bubble the angle of contact = $180° - \theta$.

Equations (10.12.3) and (10.12.4) lead us to the practical determination of surface tension of a liquid in the form of a large drop as well as the angle of contact.

The surface tension of mercury in the form of a large drop on a glass plate may be determined by a method devised by Quincke. The point P (Fig. 10.12.5) where the tangent is vertical is located by an optical method and then h' is determined. Thus knowing h', surface tension of mercury is determined from the expression (10.12.3). For experimental details see Art. 10.22.

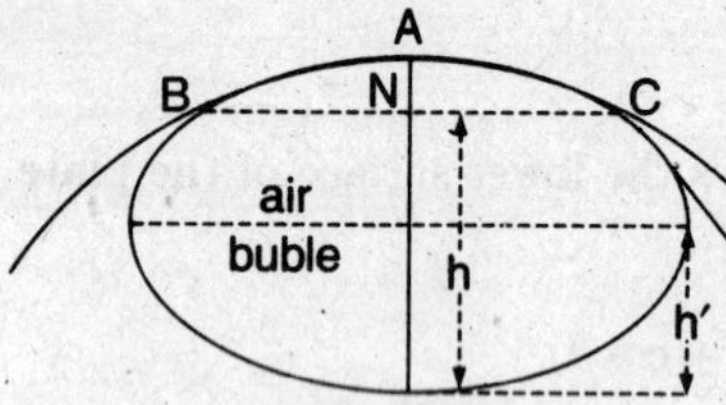

Fig. 10.12.6

In the case of bubbles, it is more convenient to use the concave surface of a large lens instead of a glass plate. In this case we use the two equations (10.12.3) and (10.12.4). We must remember that h is the depth of the bubble below the horizontal plane through the circle of contact of the liquid with the glass as shown by the dotted line BNC and h' should be measured from the bottom of the bubble (Fig. 10.12.6).

Alternative treatment

10.12.1 Rise of a liquid along the side of a plane plate dipped vertically

We consider a perfectly clean glass plate dipped vertically in a liquid of density σ, the angle of contact of the liquid with the plate being acute as shown in the figure. The surface of the liquid close to the glass assumes a definite shape as shown by the curve DA (Fig. 10.12.7). A section of the surface obtained by a vertical plane perpendicular to the glass plate gives the shape assumed by the liquid surface at the region of contact with the glass plate—which is known as the *capillary curve*.

We consider on the curve DA a point $P(x, y)$, where one of the radii of curvature is infinity and the other is r (say) which is the radius of curvature of APD at P. If p be the difference of pressures between the two sides of the surface at P, we have

$$p = \frac{S}{r} = g\sigma y;$$

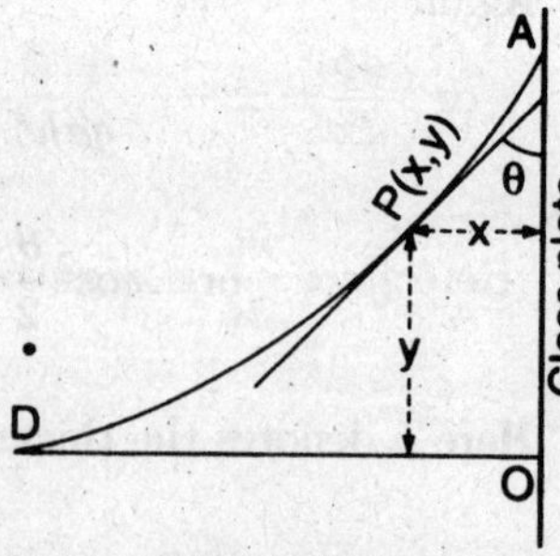

Fig. 10.12.7

y is measured from the horizontal level of the liquid.

Putting $\dfrac{S}{g\sigma} = a^2$, we may write from the above equation,

$$\frac{y}{a^2} = \frac{1}{r} = \frac{\dfrac{d^2y}{dx^2}}{\left[1 + \left(\dfrac{dy}{dx}\right)^2\right]^{3/2}}.$$

Let $\dfrac{dy}{dx} = z$

$$\therefore \quad \frac{y}{a^2} = \frac{\dfrac{dz}{dx}}{(1+z^2)^{3/2}} = \frac{\dfrac{dz}{dy}\cdot\dfrac{dy}{dx}}{(1+z^2)^{3/2}} = \frac{z\dfrac{dz}{dy}}{(1+z^2)^{3/2}} \quad \text{or} \quad \frac{zdz}{(1+z^2)^{3/2}} = \frac{ydy}{a^2}.$$

Integrating we obtain

$$-\frac{1}{(1+z^2)^{1/2}} = \frac{y^2}{2a^2} + C,$$

where C is the constant of integration.

Now at D, $y = 0$ and $z = \dfrac{dy}{dx} = 0$. $\quad \therefore C = -1$.

Hence

$$-\frac{1}{(1+z^2)^{1/2}} = \frac{y^2}{2a^2} - 1 = \frac{y^2 - 2a^2}{2a^2} \tag{10.12.6}$$

Squaring both sides,

$$\frac{1}{1+z^2} = \left(\frac{y^2 - 2a^2}{2a^2}\right)^2 \quad \text{or} \quad z^2 = \left(\frac{2a^2}{y^2 - 2a^2}\right)^2 - 1 = \frac{(2a^2)^2 - (y^2 - 2a^2)^2}{(y^2 - 2a^2)^2} = \frac{4a^2y^2 - y^4}{(2a^2 - y^2)^2}$$

$$\text{or} \quad z = \frac{dy}{dx} = \pm\frac{\sqrt{4a^2y^2 - y^4}}{(2a^2 - y^2)} = \pm\frac{y}{2a^2 - y^2}\sqrt{(4a^2 - y^2)}.$$

Let $y = 2a \sin\phi$

$$\therefore \quad \frac{dy}{dx} = 2a\cos\phi\frac{d\phi}{dx}$$

$$\therefore \quad 2a\cos\phi\frac{d\phi}{dx} = \pm\frac{2a\sin\phi\sqrt{4a^2 - 4a^2\sin^2\phi}}{2a^2 - 4a^2\sin^2\phi} \quad \text{or} \quad 2a\cos\phi\frac{d\phi}{dx} = \pm\frac{2a\sin\phi\cdot 2a\cos\phi}{2a^2(1 - 2\sin^2\phi)}$$

$$\text{or} \quad \frac{d\phi}{dx} = \pm\frac{\sin\phi}{a(1 - 2\sin^2\phi)} \quad \text{or} \quad \frac{(1 - 2\sin^2\phi)}{\sin\phi}d\phi = \pm\frac{dx}{a}.$$

Here since $\dfrac{dy}{dx}$ is negative, we consider only the negative sign, and we may write

$$\left(\frac{1}{\sin\phi} - 2\sin\phi\right)d\phi = -\frac{dx}{a}$$

Integrating the above expression on both sides we have,

$$\int \operatorname{cosec}\phi d\phi - 2\int \sin\phi d\phi = \frac{x}{a} + C_1 \quad \text{or} \quad \log\tan\frac{\phi}{2} + 2\cos\phi = -\frac{x}{a} + C_1$$

where C_1 is the constant of integration.

Now at A where $x = 0$, y has its maximum value given by $\phi = 90°$, and putting this condition in this equation, we have $C_1 = 0$

$$\therefore 2\cos\phi + \log\tan\frac{\phi}{2} = -\frac{x}{a}.$$

This gives the equation to the capillary curve. If $\phi = 0$ or $y = 0$, $x = \infty$, so that the capillary curve becomes asymptotic to the liquid surface.

If θ be the angle of contact, then $\left(\frac{dy}{dx}\right)$ at $A = -\cot\theta$ and hence from equation (10.12.6), we have,

$$-\frac{1}{\sqrt{1+\cot^2\theta}} = \frac{(OA)^2}{2a^2} - 1 \quad \text{or} \quad OA^2 = 2a^2\left(1 - \frac{1}{\operatorname{cosec}\theta}\right) = 2a^2(1-\sin\theta).$$

Substituting the value of a^2 the above expression may be written as

$$OA^2 = \frac{2S}{\sigma g}(1-\sin\theta) \quad \text{or} \quad OA = \sqrt{\frac{2S(1-\sin\theta)}{\sigma g}}$$

If $\theta = 0$, $\quad OA = \sqrt{\frac{2S}{\sigma g}}$.

From the above expression surface tension of the liquid may be found.

Illustrative Example

A vertical glass plate is dipped vertically in turpentine. Calculate the thickness of the meniscus of the turpentine surface in contact with the vertical glass plate. The angle of contact of turpentine with glass = 17°; surface tension of turpentine = 27 dynes/cm; density of turpentine = 0.87 gm/cc.

We know that

$$y = \sqrt{\frac{2S(1-\sin\theta)}{\sigma g}}$$

Here y denotes the thickness of the meniscus in contact with the vertical glass plate.

$$\theta = 17°, \ \sigma = 0.87 \text{ gm/cc}, \ S = 27 \text{ dynes/cm}$$

$$y = \sqrt{\frac{2 \times 27(1-\sin 17°)}{0.87 \times 981}} = 0.211 \text{ cm} = 2.11 \text{ mm}.$$

10.12.2 Sessile drop—Theory of Quincke's method

A sessile drop is a sufficiently large drop having a shape intermediate between a sphere and a horizontal plane. A section of such a drop (mercury) resting on a horizontal plate is shown in figure 10.12.8. The drop is under the combined action of gravity and surface tension. As the drop is sufficiently large, its upper surface is horizontal and its thickness is small compared with its diameter. At any point $A(x, y)$ on the surface of the drop a tangent is drawn which makes an angle ψ with the X-axis. The pressure at A is given by $p = P + g\sigma y$ where P is the external pressure and y is the vertical distance of A below the X-axis.

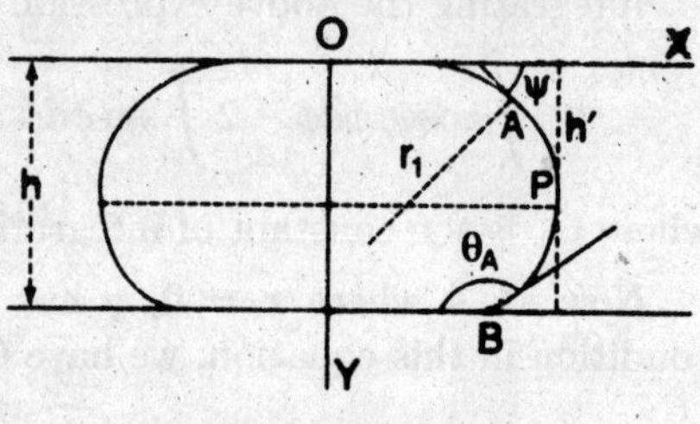

Fig. 10.12.8

Now if r_1 and r_2 be the principal radii of curvature at A in and perpendicular to the plane of the figure, then

$$p = S\left(\frac{1}{r_1}+\frac{1}{r_2}\right)+P$$

$$\because r_2 \gg r_1, \quad \therefore \quad p = \frac{S}{r_1}+P = g\sigma y + P \quad \text{or} \quad \frac{S}{r_1} = g\sigma y \tag{10.12.7}$$

Now $\dfrac{1}{r_1} = \dfrac{d\psi}{ds}$, where ds is an element of the curvature at A

$$\text{or} \quad \frac{1}{r_1} = \frac{d\psi}{dy}\cdot\frac{dy}{ds} = \frac{d\psi}{dy}\cdot\sin\psi \quad [\because dy/ds = \sin\psi]$$

Substituting this value of $\frac{1}{r_1}$ in equation (10.12.7), we have

$$S\frac{d\psi}{dy}\cdot\sin\psi = g\sigma y \;\text{ or }\; S\int_0^{\psi}\sin\psi d\psi = g\sigma\int_0^{y} y dy \;\text{ or }\; S(1-\cos\psi) = \frac{g\sigma y^2}{2}$$

$$\therefore \quad g\sigma y^2 = 2S(1-\cos\psi) \tag{10.12.8}$$

At P, $\psi = 90$ and $y = h'$

$$\therefore \quad g\sigma h'^2 = 2S \tag{10.12.9}$$

At B, $\psi = \theta_A$ and $y = h$

$$\therefore \quad g\sigma h^2 = 2S(1-\cos\theta_A) \tag{10.12.10}$$

The same treatment is applicable in case of a large gas bubble entrapped under a horizontal plate within a liquid as shown in figure (10.12.9). In this case the angle of contact $\theta_A < 90°$.

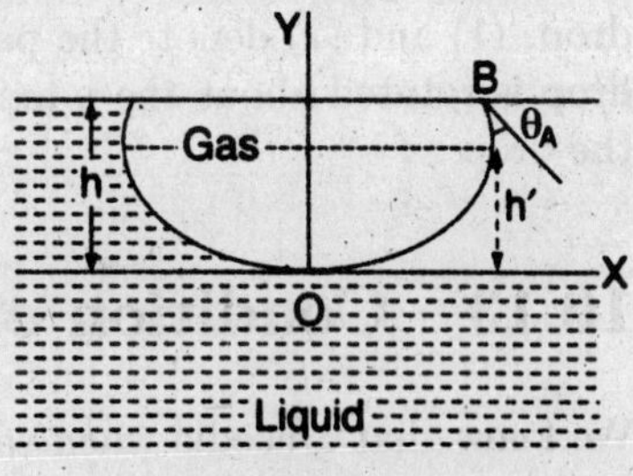

Fig. 10.12.9

Illustrative Example

1. A large drop of mercury is formed on a horizontal plane. Derive an equation for the shape of the boundary curve in a vertical section through the centre of the drop. **[C.U. (Hons.) 1974]**

In case of a sessile drop, i.e., a large drop resting on a horizontal plane we have from equation (10.12.8)

$$\frac{g\sigma y^2}{2} = S(1-\cos\psi) = 2S\sin^2\frac{\psi}{2}$$

$$\text{or} \quad y^2 = \frac{4S}{\sigma g}\sin^2\frac{\psi}{2} = 2a^2\sin^2\frac{\psi}{2} \text{ where } a = \sqrt{\frac{2S}{\sigma g}}$$

$$\therefore \quad y = \sqrt{2}a\sin\frac{\psi}{2} \tag{1}$$

Now

$$\frac{dy}{dx} = \tan\psi \quad \text{or} \quad dx = dy\cot\psi.$$

From (1)

$$dy = \sqrt{2}a \cdot \frac{1}{2} \cos\frac{\psi}{2} \cdot d\psi = \sqrt{2}a \cdot \cos\frac{\psi}{2}\frac{d\psi}{2}$$

$$\therefore \quad dx = \sqrt{2}a \cos\frac{\psi}{2}\frac{d\psi}{2} \cot\psi$$

$$\text{or} \quad dx = \sqrt{2}a \cos\frac{\psi}{2} \cdot \frac{\cos\psi}{\sin\psi} \cdot \frac{d\psi}{2} = \sqrt{2}a \cos\frac{\psi}{2} \cdot \frac{d\psi}{2}\left(\frac{1 - 2\sin^2\frac{\psi}{2}}{2\sin\frac{\psi}{2}\cos\frac{\psi}{2}}\right)$$

$$= \frac{a}{\sqrt{2}}\left[\operatorname{cosec}\frac{\psi}{2} - 2\sin\frac{\psi}{2}\right]\frac{d\psi}{2}.$$

Integrating we have

$$x = \frac{a}{\sqrt{2}}\left[\log_e \tan\frac{\psi}{4} + 2\cos\frac{\psi}{2} + A\right] \tag{2}$$

where A is the constant of integration.

Also

$$y = \sqrt{2}a \sin\frac{\psi}{2} \quad \text{from (1)}$$

Here x, y are the co-ordinates of any point on the boundary of the vertical section of the drop. (1) and (2) denote the parametric equations for the shape of the boundary curve. If the drop is rotated about the y-axis we get the surface of revolution which will give the shape of the drop.

10.13 Condition of equilibrium of a floating needle

We know that a needle when carefully dropped on the surface of water floats.

In figure 10.13.1, a section of the needle (perpendicular to its axis) floating on the surface of water is shown. The different forces acting on the needle are

(a) the vertical downward force due to the weight of the needle;

(b) the surface tensions at F and H;

(c) the thrust of water on the immersed portion FGH of the needle which is equal to the weight of the volume of water having area of cross-section $DEHGFD$;

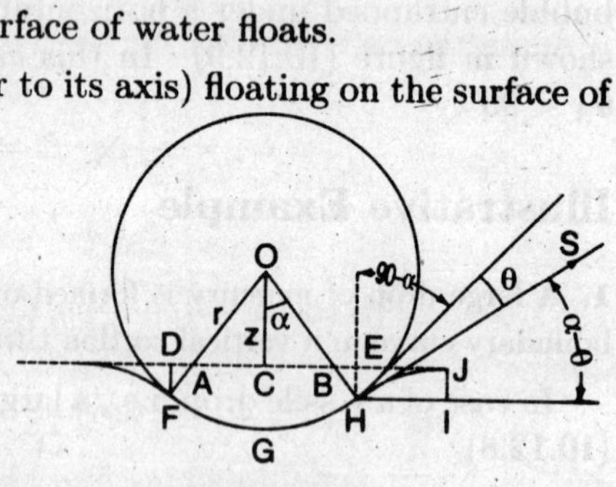

Fig. 10.13.1

The vertical upward force due to surface tensions at F and $H = 2S\sin(\alpha - \theta)$.

Now the area

$$DEHGFD = \Delta OFH - \Delta OAB + \Delta FAD + \Delta HEB$$

$$= \Delta OFH - \Delta OAB + 2\Delta FAD \quad [\because \Delta FAD = \Delta HEB]$$

$$= \left[\frac{1}{2}r \cdot r \cdot 2\alpha - z \cdot z\tan\alpha + 2 \cdot \frac{1}{2}(r\cos\alpha - z)(r\cos\alpha - z)\tan\alpha\right]$$

$$= [r^2\alpha - z^2\tan\alpha + (r\cos\alpha - z)^2\tan\alpha]$$

Here θ denotes the angle of contact between the liquid and the tangent to the needle at H.

r and z denote respectively the radius of the needle and the height OC of its axis from the free level, and $\alpha = \frac{1}{2} \times$ angle FOH.

Considering unit length of the needle, the length being considered perpendicular to the plane of the diagram, the volume of the liquid of cross-section $DEHGFD$

$$= [r^2\alpha - z^2 \tan\alpha + (r\cos\alpha - z)^2 \tan\alpha]$$

Hence the weight of this volume of water $= \rho g[r^2\alpha - z^2 \tan\alpha + (r\cos\alpha - z)^2 \tan\alpha]$, where ρ is the density of the liquid.

In the equilibrium condition considering vertical forces, we have, the total vertical upward force = vertical downward force.

Now the total vertical upward force

$$= 2S\sin(\alpha - \theta) + \rho g[r^2\alpha - z^2 \tan\alpha + (r\cos\alpha - z)^2 \tan\alpha]$$

The vertical downward force is due to the weight of the needle and is equal to mg, where m is the mass per unit length of the needle.

Thus, we have,

$$2S\sin(\alpha - \theta) + \rho g[r^2\alpha - z^2 \tan\alpha + (r\cos\alpha - z)^2 \tan\alpha] = mg$$

$$\text{or} \quad 2S\sin(\alpha - \theta) + \rho g[r^2\alpha + r^2 \sin\alpha\cos\alpha - 2zr\sin\alpha] = mg \qquad (10.13.1)$$

Considering the horizontal forces, we have in the equilibrium condition, the horizontal component of surface tension at H + horizontal thrust on cross-section of area $JI \times 1$ or JI = surface tension S at J.

Now the horizontal component of the surface tension along $HI = S\cos(\alpha - \theta)$.

Again the horizontal thrust on the area JI

$$= \frac{JI}{2}\rho g \cdot JI = \frac{1}{2}\rho g(JI)^2 = \frac{1}{2}\rho g(r\cos\alpha - z)^2$$

$$\therefore \quad S\cos(\alpha - \theta) + \frac{1}{2}\rho g(r\cos\alpha - z)^2 = S \quad \text{(for horizontal equilibrium)}$$

$$\text{or} \quad \rho g(r\cos\alpha - z)^2 = 2S[1 - \cos(\alpha - \theta)] = 4S\sin^2\frac{(\alpha - \theta)}{2} \qquad (10.13.2)$$

Equations (10.13.1) and (10.13.2) determine the condition of equilibrium of the floating needle.

In case of water, we may put $\rho = 1$ and $\theta = 0$ whence these two equations become

$$2S\sin\alpha + g(r^2\alpha + r^2 \sin\alpha\cos\alpha - 2zr\sin\alpha) = mg \qquad (10.13.3)$$

$$g(r\cos\alpha - z)^2 = 4S\sin^2\frac{\alpha}{2} \qquad (10.13.4)$$

Eliminating z from the above equations we get the condition of equilibrium of the floating needle in terms of r and α.

10.14 Rise of a liquid between two vertical plane plates making a small angle with each other

The attached figure shows the two plates $OEFGO$ and $OEF'G'O$ making an angle α radian with each other—with the liquid between them.

Let the angle of contact of the liquid with the plates be zero, i.e., we imagine that the two plates are of glass and the liquid is water.

In figure 10.14.1, KMN and KPQ represent the curves of contact of the upper surface of the liquid with the plates. It can be proved in the following way that each of the curves of contact is approximately a rectangular hyperbola.

We consider an element of liquid of thickness δx bounded by the vertical planes $MPAD$ and $NQBC$ at a distance x from the origin O. Equation to the curve of contact will be obtained from the consideration of the equilibrium of this element of liquid.

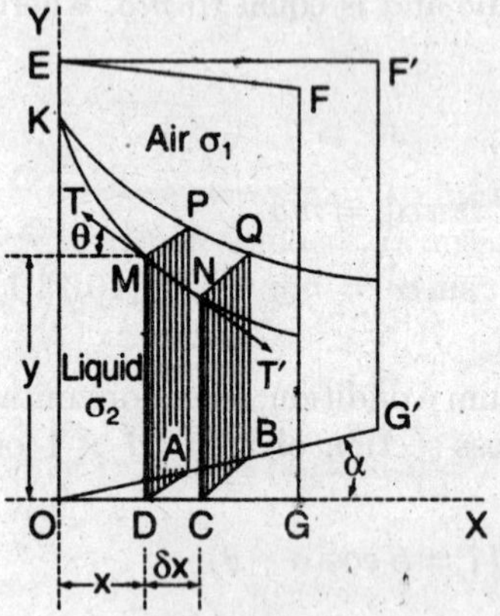

Fig. 10.14.1

Now the volume of this element $= y\cdot x\alpha\cdot \delta x$, where $x\alpha = DA$ (approximately), $y = MD = PA$ and $DC = \delta x$.

If σ be the effective density of the liquid, i.e., $\sigma = \sigma_2 - \sigma_1$, weight of this element of liquid $= y\alpha x\delta x\sigma g$. σ_1 and σ_2 are respectively the absolute densities of air and the liquid.

This is acting downwards and this must be balanced by the upward force that comes to play due to surface tension.

The liquid on the left hand side of PM will act tangentially with a force $S\cdot PM = S\cdot DA = Sx\alpha$ along MT which is the tangent at M. We resolve this force into horizontal and vertical components, of which the vertical component is $Sx\alpha\sin\theta$. This is the effective component acting in the upward direction. This vertical component is evidently a function of x. By drawing a tangent at N we get similarly the direction of attractive force on NQ by the liquid on the right hand side of NQ. From the figure it is clear that at this point, the vertical component of this attractive force is acting downwards. Since the vertical component is a function of x, we have the resultant vertical component $= S\alpha\dfrac{d}{dx}(x\sin\theta)\delta x$.

This quantity is the resultant of the forces of surface tension on MP and NQ when resolved vertically, and it is acting vertically downwards.

Again the forces due to surface tension acting on MN and PQ are equal in magnitude and directed normal to MN and PQ, which make angles θ with the vertical, and hence their resultant resolved vertically upwards is

$$2S\cdot MN\cos\theta = 2S\delta x \quad [\because\ MN\cos\theta = \delta x]$$

Hence the resultant upward force due to surface tension on the element of liquid is given by

$$2S\delta x - S\alpha\frac{d}{dx}(x\sin\theta)\delta x$$

Hence for equilibrium of the element of liquid under consideration we have

$$y\alpha\sigma g x\delta x = \left[2S - S\alpha\frac{d}{dx}(x\sin\theta)\right]\delta x \quad \text{or} \quad xy = \frac{2S}{g\sigma\alpha} - \frac{S}{g\sigma}\cdot\frac{d}{dx}(x\sin\theta) \qquad (10.14.1)$$

This may be considered as the equation to the curve KMN or KPQ.

If we neglect the second term of the above equation as a first approximation, we have

$$xy = \frac{2S}{g\sigma\alpha} = \text{constant.}$$

This shows that the curve KMN or KPQ is approximately a rectangular hyperbola. The second term on the right hand side of equation (10.14.1) may be neglected on the assumption that the forces due to surface tension acting on MP and NQ are equal and opposite.

10.14.1 Rise of a liquid between two vertical parallel plates

In figure 10.14.2, a vertical section of a liquid rising between the two vertical plates AB and CD has been shown. If we consider sections parallel to each other and perpendicular to the plates, the profile curve EFG remains unchanged.

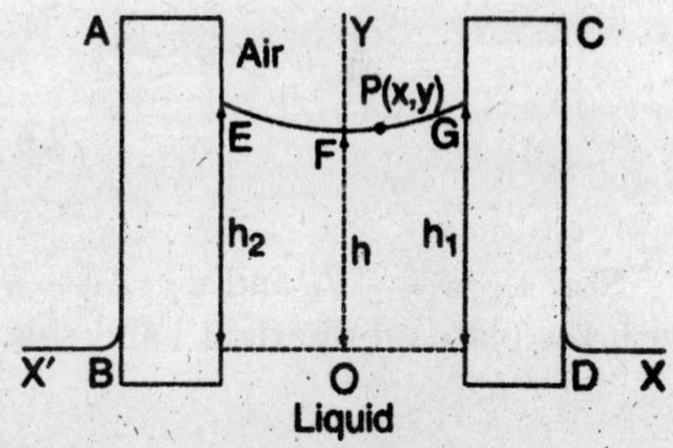

Fig 10.14.2

In the general level $X'X$ we consider the origin O midway between the plates corresponding to the rectangular axes OX or OY. At any point $P(x, y)$ on the profile curve, we have

$$S\left(\frac{1}{r_1} + \frac{1}{r_2}\right) - g\sigma y = \text{constant}$$

where σ is the effective density of the liquid. As one of the radii is infinite, we have

$$\frac{S}{r} - g\sigma y = \text{constant,}$$

where r is the radius of curvature at P in the plane of the diagram. If the origin lies in the general level as shown in the figure, the constant becomes equal to zero;

$$\therefore \ \frac{S}{r} = g\sigma y.$$

Now $r\delta\psi = \delta s$ and $\delta y = \delta s \sin\psi$ [see Art. 10.12, Fig. 10.12.2]

$$\therefore \ \frac{S}{r}\delta y = S\cdot\frac{\delta\psi}{\delta s}\cdot\delta s\sin\psi = S\sin\psi\delta\psi = g\sigma y\delta y$$

Integrating, we have

$$-S\cos\psi = \frac{g\sigma y^2}{2} + K,$$

where K is the constant of integration.

At $F, y = h$ and $\psi = 0$ [$\because$ the tangent drawn at F is parallel to the axis $X'X$, $\therefore \ \psi = 0$]

$$\therefore \ -S = \frac{g\sigma h^2}{2} + K \quad \text{or} \quad K = -S - \frac{g\sigma h^2}{2}$$

Hence,

$$-S\cos\psi = \frac{g\sigma y^2}{2} - S - \frac{g\sigma h^2}{2} \text{ or } S(1-\cos\psi) = \frac{g\sigma}{2}(y^2 - h^2)$$

$$\text{or } y^2 = \frac{2S}{g\sigma}(1-\cos\psi) + h^2 \qquad (10.14.2)$$

This equation is the differential equation to the profile curve.

Let the tangents drawn at the points of contact of the liquid with the plate on the right hand side and the left hand side make angles ψ_1 and ψ_2 with the axis $X'X$ respectively and let the ordinates of the corresponding points of contact be h_1 and h_2 respectively.

Then by equation (10.14.2), we have,

$$h_1^2 = \frac{2S}{g\sigma}(1-\cos\psi_1) + h^2$$

$$h_2^2 = \frac{2S}{g\sigma}(1-\cos\psi_2) + h^2$$

Now $\psi_1 = \frac{\pi}{2} - \theta_1$ and $\psi_2 = \frac{\pi}{2} - \theta_2$, where θ_1 and θ_2 are the angles of contact of the liquid with the plate on the right hand side and the left hand side respectively.

$$\therefore\ h_1^2 = \frac{2S}{g\sigma}(1-\sin\theta_1) + h^2 \qquad (10.14.3)$$

$$h_2^2 = \frac{2S}{g\sigma}(1-\sin\theta_2) + h^2. \qquad (10.14.4)$$

Subtracting (10.14.4) from (10.14.3), we have,

$$(h_1^2 - h_2^2) = \frac{2S}{g\sigma}(\sin\theta_2 - \sin\theta_1)$$

This gives the relation between h_1 and h_2 i.e., the ordinates of the points of contact.

10.14.2 Rise of water between two vertical parallel plates close to each other

As the distance of separation between the plates is very small, we may consider the length of the profile curve ACB to be semi-circular.

Considering the equilibrium of the elevated portion of water of thickness l, the thickness being measured perpendicular to the plane of the paper, we may write,

the weight of the water column of height h and thickness $l+$ weight of the liquid in the meniscus denoted by the shaded portion of thickness $l =$ the sum of the two upward forces due to surface tension.

Now the weight of the water column of height h and thickness $l = hdl\sigma g = hdlg$ where d is the distance of separation between the plates. [$\because\ \sigma = 1$ for water]

Again weight of the water in the meniscus denoted by the shaded region of thickness l = weight of water in the rectangular slice $AMNBA$ of thickness l — weight of the semi-circular slice ACB of thickness l. [MN is a tangent drawn through the lowest point C of the profile curve.]

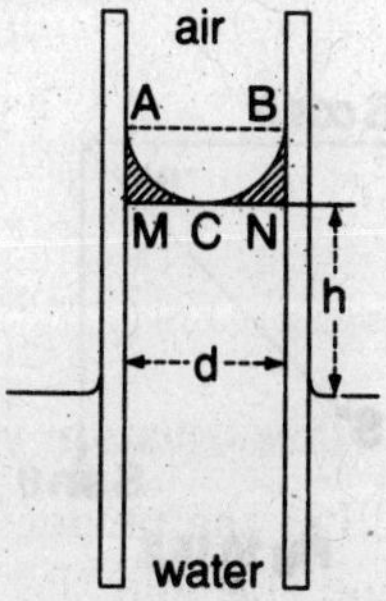

Fig 10.14.3

$$= \frac{d}{2}\cdot d\cdot lg - \frac{1}{2}\frac{\pi d^2}{4}lg = \frac{d^2}{2}lg - \frac{\pi d^2 l}{8}g = gd^2 l\left(\frac{1}{2} - \frac{\pi}{8}\right)$$

Also the upward forces due to surface tension

$$= Sl + Sl \quad [\because \text{ the angle of contact is taken to be zero}]$$

$$= 2Sl$$

For equilibrium of the elevated portion of water, we may therefore write,

$$hdlg + gd^2 l\left(\frac{1}{2} - \frac{\pi}{8}\right) = 2Sl$$

$$\text{or} \quad hdg + gd^2\left(\frac{1}{2} - \frac{\pi}{8}\right) = 2S \quad \text{or} \quad h + d\left(\frac{1}{2} - \frac{\pi}{8}\right) = \frac{2S}{dg}$$

$$\text{or} \quad h = \frac{2S}{dg} - d\left(\frac{1}{2} - \frac{\pi}{8}\right) = \frac{2S}{dg} - d\left(\frac{7\pi}{44} - \frac{\pi}{8}\right) = \frac{2S}{dg} - \frac{3d\pi}{88} = \frac{2S}{dg} - 0.107d.$$

10.15 Force between two parallel plates separated by a thin layer of liquid, due to surface tension

In figure 10.15.1, AE and BF denote two parallel plates separated by a thin film of liquid (water, say), which wets them. We, therefore, consider the angle of contact to be zero. Let d be the distance between the two plates and D the diameter of the area of the plate wet by the liquid.

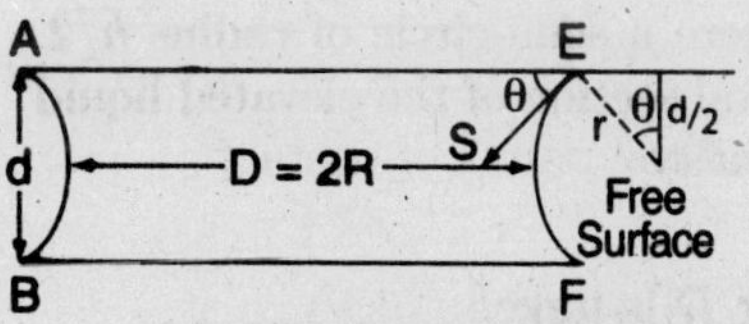

Fig 10.15.1

The radii of curvature of the free surface of the liquid are $d/2$ and $D/2$ approximately (in and perpendicular to the plane of the diagram). Hence the pressure inside the film is less than the pressure outside by a quantity $= S\left(\frac{1}{r} - \frac{1}{R}\right)$; ($\cos\theta = \frac{d/2}{r}$ or $r = \frac{d}{2}$ when θ is small).

$$= S\left(\frac{1}{d/2} - \frac{1}{D/2}\right) = 2S\left(\frac{1}{d} - \frac{1}{D}\right);$$

usually D is large in comparison with d;

$$\therefore \text{ the above difference of pressures} = \frac{2S}{d}(\text{approx.})$$

Hence the plate AE of diameter D, say, which is wetted will be urged towards the plate BF with a force given by

$$\frac{2S}{d}\cdot\frac{\pi D^2}{4} = \frac{\pi S D^2}{2d}. \qquad (10.15.1)$$

If the angle of contact is not zero, but equal to θ, the expression for force should be modified in the following way:

If α be the area of the film between the plates, and p its circumference, the plates will be pressed together with a force given by

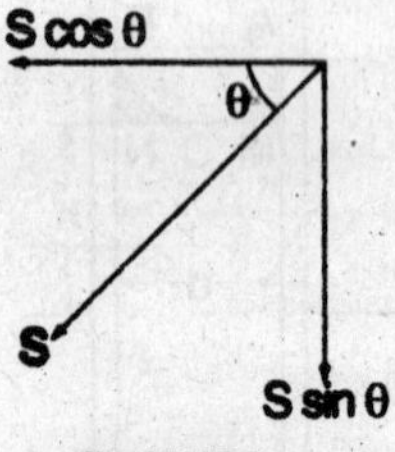

Fig 10.15.2

$$\frac{2S\alpha\cos\theta}{d} + pS\sin\theta. \qquad (10.15.2)$$

Here $S\sin\theta$ denotes the vertical component of surface tension which urges the plates further towards each other (Fig 10.15.2)

and $\left[\frac{2S\alpha\cos\theta}{d}\right]$

is the force pushing the plates together due to the excess of atmospheric pressure outside the plates over the liquid pressure between them.

10.16 Force to pull a horizontal plate upwards from a liquid surface

If a flat plate is placed on the surface of a liquid, and then raised through a small distance, a layer of liquid will adhere to the plate and will thus be elevated as shown in the figure 10.16.1. To separate the plate from the liquid purely by a normal pull will necessitate a considerable force. If F represents the normal pull, we have F = force due to the difference of pressures just above and below the plate $AB = P\alpha - \alpha(P - g\sigma h) = \alpha g\sigma h$ where P = atmospheric pressure; α = area of the plate; σ = density of the liquid.

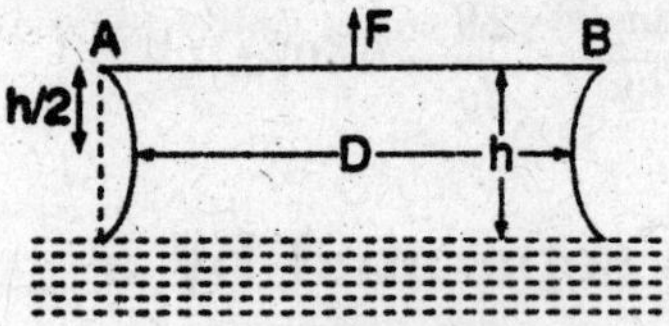

Vertical section

Fig. 10.16.1

The term h denotes the height up to which the liquid will be elevated before the plate is separated (i.e., before the liquid film is ruptured), and it depends on the surface tension of the liquid as is considered below.

Let us assume that the diameter of the plate is large, so that if the liquid wets the plate; the trace of the liquid meniscus in a vertical plane is everywhere a semi-circle of radius $h/2$. Hence the pressure inside the film at a point where the horizontal section of the elevated liquid is least, is less than the outside atmospheric pressure by a quantity

$$= S\left(\frac{1}{h/2} - \frac{1}{D/2}\right) = \frac{2S}{h} \quad [\because \text{ the diameter } D \text{ is large.}]$$

This pressure difference must be $\frac{1}{2}g\sigma h$ if the shape of the meniscus is assumed semi-circular. Thus, for horizontal equilibrium

$$\therefore \frac{1}{2}g\sigma h = \frac{2S}{h} \quad \text{or} \quad h^2 = \frac{4S}{g\sigma} \quad \text{or} \quad h = 2\sqrt{\frac{S}{g\sigma}}$$

$$\therefore \text{ the normal pull is } F = \alpha g\sigma h = \alpha g\sigma\cdot 2\sqrt{\frac{S}{g\sigma}} = 2\alpha\sqrt{Sg\sigma} \qquad (10.16.1)$$

[An elegant method to find h is by considering the horizontal equilibrium of a slice of the elevated liquid, 1 cm wide (Fig. 10.16.2). Two forces, each equal to S, act tangentially to the meniscus, while the hydro-static pressure on the vertical rectangular end of the slice acts in the same direction and is given by $h\left(P - \frac{1}{2}g\sigma h\right)$. The external pressure P acting in the opposite direction, produces a force Ph. So for horizontal equilibrium,

$$Ph = h\left(P - \frac{1}{2}g\rho h\right) + 2S \text{ or } \frac{1}{2}g\sigma h^2 = 2S$$

$$\text{or } h = 2\sqrt{\frac{S}{g\sigma}}]$$

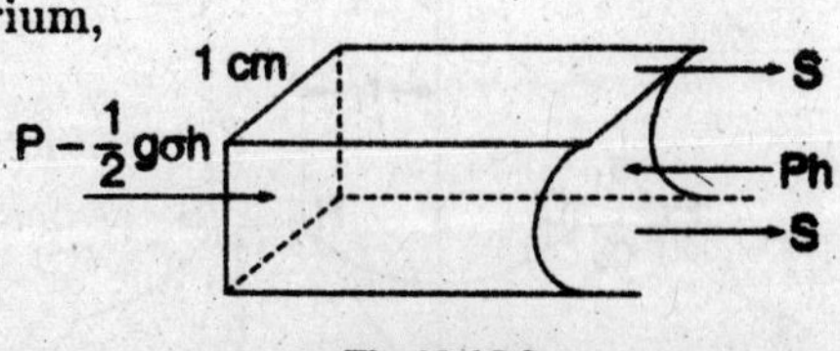

Fig 10.16.2

F is measured by means of a suitable balance. Thus knowing F, we may determine S.

This formula is, however, approximately true and the correct expression for the formula was given by Ferguson in the form

$$F = 2\alpha g\sigma\left[B - \frac{B^2}{3r} - \frac{B^3\sqrt{2}}{3r^2}\right]$$

where $B = \sqrt{\frac{S}{g\sigma}}$ and r is the radius of the plate.

It should be noted that we have assumed the plate to be completely wetted by the liquid, and hence the above expression will have to be modified in case the liquid has a finite angle of contact.

Illustrative Examples

1. Two plates each of area 10 sq cm are separated by a thin layer of water. If the distance of separation between the plates is 0.01 mm. calculate the force urging the plates together ($S = 75$ dynes/cm).

We know from (10.15.1) that

$$F = \frac{2S}{d}\cdot\frac{\pi D^2}{4}$$

Here $S = 75$ dynes/cm, $d = 0.01$ mm $= 0.001$ cm

$$\frac{\pi D^2}{4} = \text{area of the plate} = 10 \text{ sq cm.}$$

$$\therefore F = \frac{2 \times 75 \times 10}{0.001} = 1.5 \times 10^6 \text{ dynes.}$$

2. Calculate the vertical pull in gm-wt required to detach a horizontal circular plate of area 50 sq cm from the surface of a liquid of density 1 gm/cc and angle of contact zero. ($S = 30$ dynes/cm, $g = 980$ cm/sec^2) **[cf C.U. (Hons.) '81, '94]**

The required detaching force is from (10.16.1)

$$F = 2\alpha\sqrt{Sg\sigma}.$$

Here $\alpha = 50$ sq cm, $S = 30$ dynes/cm, $g = 980$ cm/sec^2, $\sigma = 1$ gm/cc.

$$\therefore F = 100\sqrt{30 \times 980 \times 1} \text{ dynes} = \frac{100}{980}\sqrt{30 \times 980} \text{ gm-wt} = 17.5 \text{ gms-wt.}$$

10.17 Liquid pulled upwards by a horizontal flat ring

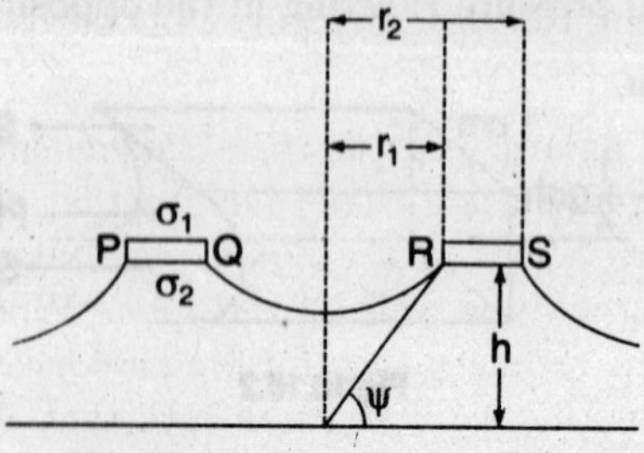

Fig 10.17.1

The figure 10.17.1 shows the central vertical section of a flat ring pulling a liquid upwards. The two cross-sections of the material of the ring PQ and RS are rectangular. Let the tangents to the profile curve at P, Q, R, S make an angle ψ with the horizontal.

The downward pull only due to surface tension acting round the circumference of each circle is given by

$$2\pi r_1 S \sin\psi + 2\pi r_2 S \sin\psi = 2\pi(r_1 + r_2) S \sin\psi$$

Now the area of the surface of the annulus in contact with the liquid $= \pi r_2^2 - \pi r_1^2 = \pi(r_2^2 - r_1^2)$ and the pressure at a point on the lower face of the ring $= (P - g\sigma h)$, where P is the atmospheric pressure.

Hence, the upward force due to liquid pressure on the lower face of the ring is given by

$$(P - g\sigma h)\pi(r_2^2 - r_1^2)$$

and the downward force due to atmospheric pressure over the upper surface of the ring is given by

$$P\pi(r_2^2 - r_1^2)$$

Hence, the resultant vertical downward force other than that due to surface tension and weight of the ring is given by

$$\pi P(r_2^2 - r_1^2) - \pi(r_2^2 - r_1^2)(P - g\sigma h) = \pi(r_2^2 - r_1^2) g\sigma h.$$

Now neglecting the weight of the ring, if F be the vertical upward pull just sufficient to separate the ring from the surface of the liquid, we have

$$F = 2\pi S(r_1 + r_2)\sin\psi + \pi(r_2^2 - r_1^2) g\sigma h;$$

when $\psi = 90°$, and r_1 and r_2 are nearly equal and each equal to r, say, we have then $F = 4\pi r S$.

This approximate expression for F is often used to measure surface tension of a liquid for which $\psi = 90°$. F means the maximum extra force required to detach a ring of rectangular ribbon or circular wire. Ring method has been applied by Harkins, Young, Cheng and other investigators for the determination of the surface tension of a liquid. For experimental details see Art. (10.22).

10.18 Surface Tension and Vapour Pressure

Lord Kelvin showed for the first time that due to surface tension vapour pressure in equilibrium with a curved surface is not the same as the pressure of vapour in equilibrium with a flat one. That means if we take a liquid in a closed vessel, saturation pressure over the flat surface of the liquid will not be the same as the saturation pressure corresponding to the surface being concave outwards or convex outwards.

It may be proved in the following way that the saturation pressure of the vapour corresponding to the surface concave outwards is less than the saturation pressure corresponding to the flat surface. Again, saturation pressure corresponding to a surface convex outwards is greater than that corresponding to a flat surface.

Let us take two arrangements as shown in the attached figures. In both the arrangements, we take same kind of liquid, and the vessel containing the liquid is a closed one, so that the space above the surface of the liquid contains nothing but vapour of the liquid. We take a capillary tube such that the liquid under consideration can wet the walls of the tube. When such a tube is dipped in the liquid, the liquid will rise in the capillary tube forming a concave meniscus at the top as shown in fig. 10.18.1(a). We take another capillary tube such that the liquid in question does not wet the walls of the tube. When such a tube is dipped into the liquid, instead of rise, the liquid column will be pressed down in the capillary tube and the meniscus will be convex as shown in fig 10.18.1(b). Both the arrangements may be possible in case of water by taking two glass capillary tubes, one of which has a thin paraffin coating inside.

Let ρ be the density of the liquid under consideration; σ, the density of vapour above the liquid and r, the radius of the bore of the tube. Considering fig. 10.18.1(a), we see that at the plane surface of the liquid, pressure at a point just outside the surface and that just inside it are equal; but this is not the case near the meniscus. Let P be the saturation pressure of the vapour at the flat surface. Then considering fig. 10.18.1(a), the vapour pressures at A, just outside the concave surface, at a height h from the surface of the liquid is given by

$$P_1 = P - hg\sigma \tag{10.18.1}$$

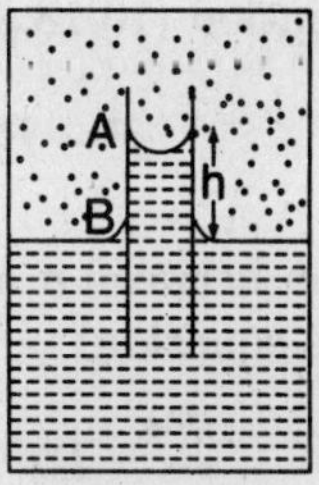

Fig 10.18.1(a)

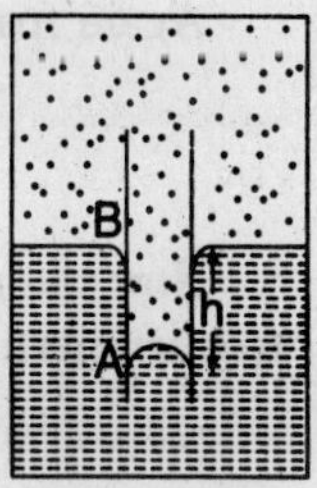

Fig 10.18.1(b)

Again the pressure at A within the liquid just below the concave surface, is given by

$$P_2 = P - hg\rho \tag{10.18.2}$$

Therefore, the difference of pressures at points just outside the meniscus and just inside it, is given by

$$P_1 - P_2 = hg(\rho - \sigma)$$

But this is equal to $\frac{2S}{r}$, where r is the radius of curvature of the concave meniscus, which is also the radius of the bore of the tube.

$$\therefore\ hg(\rho - \sigma) = \frac{2S}{r} \quad \text{or} \quad hg = \frac{2S}{r} \cdot \frac{1}{\rho - \sigma}$$

Substituting this value of hg in (10.18.1), we get,

$$P_1 = P - \frac{2S}{r} \cdot \frac{\sigma}{\rho - \sigma} \tag{10.18.3}$$

Thus we see that due to surface tension in this case, the saturation pressure of vapour over a curved surface is less than that over a flat surface by

$$\frac{2S}{r} \cdot \frac{\sigma}{\rho - \sigma}$$

Again considering fig. 10.18.1(b), we have,

$$P_1 = P + hg\sigma \tag{10.18.4}$$

$$\text{and} \quad P_2 = P + hg\rho. \tag{10.18.5}$$

Now for surface convex outwards,

$$P_2 - P_1 = \frac{2S}{r}. \quad \therefore\ hg(\rho - \sigma) = \frac{2S}{r} \quad \text{or} \quad hg = \frac{2S}{r}\cdot\frac{1}{\rho - \sigma}.$$

Hence, substituting this value of hg in (10.18.4), we get

$$P_1 = P + \frac{2S}{r}\cdot\frac{\sigma}{\rho - \sigma} \tag{10.18.6}$$

Thus it is clear from the above expression that when the surface is convex outwards, due to surface tension, saturation pressure of vapour over it is greater than that over the flat surface by

$$\frac{2S}{r}\cdot\frac{\sigma}{\rho - \sigma}$$

Again from equation (10.18.3), we may write

$$P - P_1 = \frac{2S}{r}\cdot\frac{\sigma}{\rho - \sigma} = \frac{2S}{r}\cdot\frac{\sigma}{\rho} \quad [\because\ \rho \gg \sigma] \tag{10.18.7}$$

Now the relation between σ and the pressure P is given by (assuming the saturated vapour to obey perfect gas equation)

$$\frac{P}{\sigma} = RT \quad \text{or} \quad \sigma = \frac{P}{RT}. \therefore\ P - P_1 = \frac{2S}{r}\cdot\frac{P}{RT\rho}.$$

$$\therefore\ \frac{P - P_1}{P} = \frac{2S}{r}\cdot\frac{1}{RT\rho} \tag{10.18.8}$$

Left hand side of equation (10.18.8) signifies the relative change in the vapour pressure for a given surface tension and given radius of curvature of the meniscus.

10.18.1 Surface Tension and Evaporation

From the expression (10.18.6) it follows that the vapour pressure of a spherical drop of liquid exceeds that of a plane surface by an amount $\frac{2S}{r}\cdot\frac{\sigma}{\rho}$. If r is very small, then this amount will be very large. This means that if, for some reasons a very small condensation droplet forms, the vapour pressure of the liquid in this droplet will become so high that there will be immediate evaporation and as a result the droplet will disappear rapidly. Thus, even when the vapour is supersaturated, condensation may not take place unless there are some nuclei on which the drops may form initially with comparatively large radius. As the droplets are formed on the particles of dust, the droplets start with comparatively large radius which reduces the vapour pressure to a reasonable value so that the droplets have little tendency to evaporate. A droplet can grow in size only when the vapour pressure in the air is greater than that corresponding to the saturated vapour pressure for its surface. When the air becomes supersaturated, the drops may grow in size. This explains the fact that due to fall in temperature when the atmosphere becomes supersaturated, there is deposition of water around the particles of dust in the atmosphere–leading to the formation of fog or cloud.

Illustrative Example

Calculate the minimum aqueous vapour pressure required to prevent the evaporation of a drop of water of radius 10^{-4} mm at 0°C, given density of saturated aqueous vapour at 0°C = 4.8×10^{-6} gm/cc; surface tension of water at 0°C = 76 dynes per cm; saturated aqueous vapour pressure at 0°C over a plane surface = 6×10^3 dynes/cm^2. **[C.U. (Hons.) 1954]**

From equation (10.18.6) we have the minimum aqueous vapour pressure given by

$$P_1 = P + \frac{2S}{r} \cdot \frac{\sigma}{\rho - \sigma}.$$

Here $r = 10^{-4}$ mm $= 10^{-5}$ cm; $S = 76$ dynes/cm, $\rho = 1$ gm/cc; $\sigma = 4.8 \times 10^{-6}$ gm/cc and $P = 6 \times 10^3$ dynes/cm^2.

$$\therefore\ P_1 = 6 \times 10^3 + \frac{2 \times 76}{10^{-5}} \cdot \frac{4.8 \times 10^{-6}}{1 - 4.8 \times 10^{-6}} = 6 \times 10^3 + 152 \times 4.8 \times 10^{-1}(1 - 4.8 \times 10^{-6})^{-1}$$

$$= 6 \times 10^3 + 152 \times 0.48(1 + 4.8 \times 10^{-6}) \text{ [approx.]}$$

$$= 6.073 \times 10^3 \text{ dynes/sq cm.}$$

10.18.2 Alternative Treatment

Considering fig. 10.18.1(b), we may write

$$P_1 = P + \int g\sigma dh \text{ and } P_2 = P + \int g\rho dh. \quad \therefore \quad P_2 - P_1 = g \int (\rho - \sigma) dh$$

Again

$$P_2 - P_1 = \frac{2S}{r}. \quad \therefore g \int (\rho - \sigma) dh = \frac{2S}{r} \tag{10.18.9}$$

Now

$$dP = g\sigma dh \quad \text{or} \quad dh = \frac{dP}{g\sigma} \tag{10.18.10}$$

Again we know that,

$$\frac{P}{\sigma} = RT \quad \text{or} \quad \sigma = \frac{P}{RT} \quad \left[\begin{array}{l}\text{assuming that the saturated vapour} \\ \text{obeys perfect gas equation}\end{array}\right]$$

$$\therefore\ g \int (\rho - \sigma) \frac{dP}{g\sigma} = \frac{2S}{r} \quad \text{[from 10.18.9 and 10.18.10]}$$

$$\text{or} \quad \int (\rho - \sigma) \frac{dP}{P/RT} = \frac{2S}{r} \quad \text{or} \quad \int (\rho - \sigma) \frac{dP}{P} \cdot RT = \frac{2S}{r}$$

$$\text{or} \quad RT\rho \int \frac{dP}{P} = \frac{2S}{r} \quad [\because \rho \gg \sigma]$$

$$\text{or} \quad \int_P^{P_1} \frac{dP}{P} = \frac{2S}{r\rho} \cdot \frac{1}{RT} \quad \text{or} \quad \log_e \frac{P_1}{P} = \frac{2S}{r\rho RT} \tag{10.18.11}$$

This expression gives the ratio of the vapour pressures P_1 and P for equilibrium with the curved surface of radius r, and a flat surface respectively. It should be noted that the left hand sides of the expressions (10.18.8) and (10.18.11) are approximately equal.

Wilson showed that dust-free moist air also will suffer condensation with the formation of fine mist when expanded adiabatically to a volume 1.38 times its original volume. From vapour pressure table it follows that this is equivalent to creating eight times supersaturation of water vapour.

Wilson also showed that gaseous ions can serve as nuclei for condensation, if the supersaturation exceeds a certain value. If a droplet is given an electrical charge e, then electrical potential energy of such a droplet is $\frac{e^2}{2r}$, where r is the radius of the droplet. The tendency of this energy to attain a minimum value, helps the droplet to grow. Now the potential energy of the droplet due to surface tension is $4\pi r^2 S$. As this energy attains minimum value, the radius of the droplet decreases. Thus the effect of the charge is to counteract the force due to surface tension, i.e., to reduce the surface tension of the liquid constituting the droplet. As the excess vapour pressure is proportional to surface tension, the effect of the charge is to reduce the vapour pressure and thus a charged droplet does not evaporate so rapidly as an uncharged one.

Illustrative Example

If dust-free moist air is suddenly expanded so that the saturated vapour pressure over a drop is eight times that over a plane surface, calculate the radius of the drop, given room temperature = 18°C, $S = 73$ dynes/cm, $R = 4.61 \times 10^6$ ergs/°C.

We know

$$\log_e \frac{P_1}{P} = \frac{2S}{r\rho RT}$$

Here $\frac{P_1}{P} = 8$, $\rho = 1$ gm/cc, $T = 291$°K, $S = 73$ dynes/cm, $R = 4.61 \times 10^6$ ergs/°C.

$$\therefore\ 2.303 \times \log_{10} 8 = \frac{2 \times 73}{r \times 4.61 \times 10^6 \times 291}$$

$$\text{or}\quad r = \frac{2 \times 73}{2.303 \times 4.61 \times 10^6 \times 291 \times \log_{10} 8} = 5.22 \times 10^{-8}\ \text{cm.}$$

N.B. The above value of r gives the magnitude of molecular dimension and lends support to the view that a vapour contains a certain proportion of co-aggregated molecules which act as nuclei for condensation.

10.19 Surface Tension and Boiling

It is an expermental fact that a pure liquid may be raised to a temperature which exceeds much its normal boiling point, if no dissolved gas acts as nuclei for the bubbles of vapour to form. When however the bubble formation occurs, the growth of the bubble is very rapid, since the surface tension is increasingly insufficient to counteract the excess pressure inside the bubble and hence boiling with bumping occurs.

In the normal process of boiling, the bubble is in equilibrium when

$$P_1 + P_2 = P + \frac{2S}{r} \tag{10.19.1}$$

where P_1 = pressure of vapour inside the bubble; P_2 = pressure of gas inside the bubble P = external pressure.

The left hand side of the above expression represents the total internal pressure and the right hand side represents the total outside pressure.

Now at a given temperature,

$$P_2 \propto \frac{1}{\text{volume of the bubble}} \propto \frac{1}{\frac{4}{3}\pi r^3} \quad \text{or} \quad P_2 \propto \frac{1}{r^3} \quad \text{or} \quad P_2 = \frac{K}{r^3}, \quad \text{where } K \text{ is a constant.}$$

$\therefore$ from (10.19.1), we have

$$P_1 + \frac{K}{r^3} = P + \frac{2S}{r} \quad \text{or} \quad P_1 - P = \frac{2S}{r} - \frac{K}{r^3} \tag{10.19.2}$$

Now at the equilibrium condition,

$$\frac{d}{dr}(P_1 - P) = 0 \text{ or } \frac{d}{dr}\left(\frac{2S}{r} - \frac{K}{r^3}\right) = 0$$

$$\text{or} \quad -\frac{2S}{r^2} + \frac{3K}{r^4} = 0 \text{ or } 3K = 2Sr^2 \text{ or } K = \frac{2}{3}Sr^2$$

$\therefore$ from (10.19.2), we have,

$$P_1 - P = \frac{2S}{r} - \frac{2}{3}Sr^2 \cdot \frac{1}{r^3} = \frac{2S}{r} - \frac{2S}{3r} = \frac{4S}{3r} = \frac{2}{3} \times \frac{2S}{r}$$

Hence, boiling begins with a bubble of radius r, when the vapour pressure in it exceeds the outside pressure by the amount $\frac{4S}{3r}$, i.e., boiling will start as soon as the excess pressure amounts to two-third of the capillary pressure.

Illustrative Example

At 100°C, the saturation vapour pressure of water increases by 27 mm of mercury per degree centigrade. If water under an external pressure of 76 cms of mercury has to be superheated to 100.6^{C} before boiling occurs, calculate the radius of the largest air-bubble available as a vaporisation locus. The surface tension of water at 100°C is 58.4 dynes per cm.

We know, $P_1 - P = \frac{4}{3} \cdot \frac{S}{r}$.

Here $P_1 - P = 27 \times 0.6 = 16.2$ mm of mercury $= 1.62$ cm $= (1.62 \times 13.6 \times 981)$ dynes/sq.cm; $S = 58.4$ dynes/cm.

$$\therefore \quad 1.62 \times 13.6 \times 981 = \frac{4}{3} \cdot \frac{58.4}{r} \quad \text{or} \quad r = \frac{4 \times 58.4}{3 \times 1.62 \times 13.6 \times 981} = 3.6 \times 10^{-3} \text{ cm.}$$

10.20 Explanation of Surface Tension: Laplace's theory of capillary action

In the interior of a liquid, the molecular attractions act equally in all directions. Near the surface of the liquid, there is no symmetry of the attracting force. Let us suppose that the attraction between two molecules becomes negligible when the distance between their centres is C. With some particular molecule as centre, if a sphere of radius C is drawn, the molecule 1 will not be affected by the molecules outside the sphere, but those molecules inside the sphere will attract molecule 1 in question equally in all directions so that the resultant force on the molecule 1 is zero. In case of the molecule 2 or 3 a part of the *'sphere of influence'* lies outside the liquid, and hence this part of the sphere will contain molecules of vapour or gas. Thus there is a resultant downward force on the molecule. This force is due to the shaded part of the sphere of radius C and centre of 2 or 3. The remainder of the liquid in the sphere exerts no resultant force on the molecule 2 or 3. In case of molecule 4, it is to be noted that the downward force due to the shaded portion of the liquid is maximum (fig. 10.20.1). So the attraction is confined to molecules lying in a layer of thickness C below the surface and increases as the surface of the liquid is approached. As a result of this downward pull on the surface molecules, the surface layer will be under a state of tension which is the cause of the phenomenon of surface tension. Thus we conclude that the surface tension is a molecular phenomenon.

Fig 10.20.1

Surface tension was accounted for by Laplace from the molecular phenomena. The law of force between the molecules is not known with any certainty. This law does not follow the inverse square law of gravitation; rather it is likely that instead of inverse square, it varies inversely as the eighth power of the distance between the molecules. Also, the chemical property of the liquid contributes to the nature of the force.

According to Laplace, if we consider two point masses m and m' separated by a distance r apart, the force of attraction between them is given by $F = mm'\psi(r)$.

Here the nature of the function $\psi(r)$ is not known; it is assumed that $\psi(r)$ and hence the force falls to zero value when r exceeds a certain value C. C is called the *'range of molecular attraction'*. The value of C is very small. Further with the increase of r, the force decreases much more rapidly than what is demanded according to Newton's law of gravitation. Thus this force is different from the gravitational force.

From the above consideration of this attractive force, let us find out an expression for the force on a mass m at a distance x from the surface of the liquid.

We consider a thin layer of liquid of thickness dr at a distance r from the point mass m (Fig. 10.20.2). Let m' be the mass per unit area of the layer. We have, then, according to Laplace's law, the force of attraction given by $\delta F = mm'\psi(r)$.

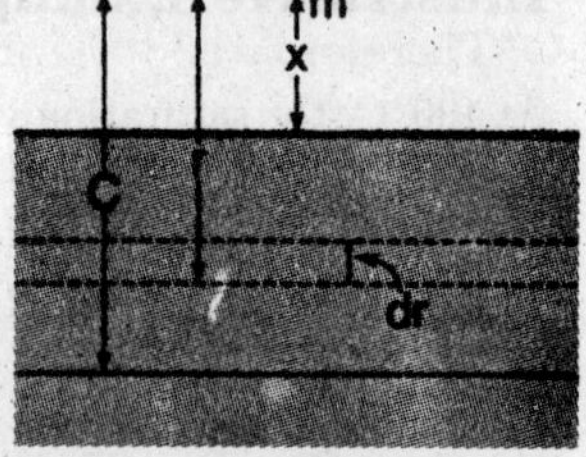

Fig 10.20.2

We have taken unit area because the range of molecular action is so small that the liquid beyond this area cannot reasonably have any contribution towards this attractive force on m.

Now

$$m' = 1 \cdot dr \cdot \rho = \rho dr. \quad \therefore \; \delta F = m\rho\psi(r)dr,$$

where ρ is the density of the liquid. From similar consideration, contribution towards attraction will be the summation of the above expression between the limits $r = C$ and $r = x$. Therefore, the force of attraction on m is given by

$$F = \int_x^C m\rho\psi(r)dr = m\rho \int_x^C \psi(r)dr \tag{10.20.1}$$

Integration of the above expression is not possible because the nature of the function is not known. But it is a fact that the value of this expression is constant when $r = C$. Hence

$$\int_x^C \psi(r)dr$$

is some function of x. Let us call it $\phi(x)$. Thus the force of attraction of the whole liquid on m may be written as $F = m\rho\phi(x)$.

If a liquid is divided along a plane and the two parts are pulled apart perpendicular to this plane, two new surfaces are created.

As the separation produces two new surfaces, the total work done in the operation is equal to $2S$, where S is the surface tension of the liquid. Based on this consideration, we may find out, in the following way, an expression for surface tension.

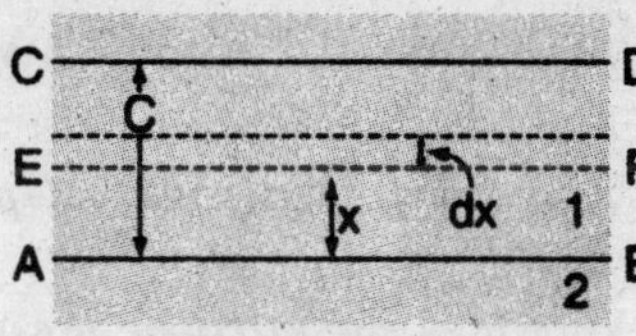

Fig 10.20.3

Let AB be a plane imagined inside the liquid to divide it into two portions (1 and 2). We consider a thin layer EF at a distance x from AB (Fig. 10.20.3). Let m be the mass per unit area of this layer. Now, the force on m by the portion of the liquid below AB is given by $m\rho\phi(x)$.

For this layer to move through a distance dx away from AB, the work done is $m\rho\phi(x)dx$. Therefore, the work done to remove the layer from x to C is given by

$$\delta W = m\rho \int_x^C \phi(x)dx.$$

Let us put p for $\int_x^C \phi(x)dx$.

$$\therefore\ \delta W = m\rho p = \rho^2 p dx \quad [\because\ m = 1 \cdot dx \cdot \rho]$$

Therefore, the total work done in moving all the layers between AB and CD is given by

$$W = \rho^2 \int_0^C p dx$$

which is numerically equal to $2S$ since two new surfaces are built.

$$\therefore\ 2S = \rho^2 \int_0^C p dx = \rho^2 \Big[px\Big]_0^C - \rho^2 \int_0^C x\frac{dp}{dx} \cdot dx$$

Now

$$p = \int_x^C \phi(x)dx. \quad \therefore\ \frac{dp}{dx} = -\phi(x)$$

and $\Big[px\Big]_0^C = 0$, because $px = 0$, when $x = 0$ and $p = \int_x^C \phi(x)dx = 0$, when $x = C$

$$\therefore\ 2S = \rho^2 \int_0^C x\phi(x)dx \qquad (10.20.2)$$

which is the required expression for surface tension.

10.20.1 Estimation of the range of molecular action (C)

Now considering the quantity

$$\int_0^C x\phi(x)dx$$

we see that both the force and the work become zero when x exceeds C which is the greatest value of x entering into the expression for the work. Hence, although we cannot integrate this expression, yet we can say that the maximum value of this expression is

$$\int_0^C C\phi(x)dx$$

Therefore, the maximum value of the expression for surface tension is

$$\rho^2 \int_0^C C\phi(x)dx = C\rho^2 \int_0^C \phi(x)dx = CK, \text{ where } K = \rho^2 \int_0^C \phi(x)dx \qquad (10.20.3)$$

is known as the *intrinsic pressure* of a liquid

$$\therefore CK > 2S \qquad \left[\because C\rho^2 \int_0^C \phi(x)dx > \rho^2 \int_0^C x\phi(x)dx\right]$$

$$\text{or} \quad C > \frac{2S}{K}.$$

This ascertains the *range of molecular action* for a given liquid. For example, in case of water,

$$S = 74 \text{ dynes per cm.}, K = 1.16 \times 10^{10} \text{ dynes per sq cm.*}$$

∴ For water

$$C > \frac{2 \times 74}{1.16 \times 10^{10}} \quad \text{or} \quad C > 1.27 \times 10^{-8} \text{ cm.}$$

Thus the minimum value of C is 1.27×10^{-8} cm. This value is in fair agreement with molecular distances as obtained from other experiments.

10.20.2 Intrinsic pressure (K) and its estimation

By intrinsic pressure of a liquid we mean the force of attraction that comes to play normally to the surface per unit area when one portion of the liquid is drawn apart from the rest.

We consider an imaginary plane AB dividing the liquid into two portions as shown in fig. 10.20.3 and also a layer of liquid EF of thickness dx at a distance x from AB, and let m be the mass per unit area of the layer. Now the force of attraction on unit area of the layer by the liquid below AB is given by $\delta F = m\rho\phi(x)$.

Similar attractive forces will be on layers existing between AB and CD. Hence the total force of attraction per unit area acting normally is given by

$$F = \rho^2 \int_0^C \phi(x)dx = K \qquad (10.20.4)$$

*see the numerical example on page 671.

which gives the measure of the intrinsic pressure.

It may be shown in the following way that this intrinsic pressure K is numerically equal to half the amount of work or energy that is to be spent for removing unit volume of liquid from the interior of the liquid to the free space above the surface of the liquid.

Again, the amount of work done for removing unit volume of liquid from the interior of the liquid to the free space above the surface of the liquid

= the amount of work done for removing unit volume of the liquid from depth C to the surface + the amount of work done for removing unit volume of the liquid from the surface to a distance C above the surface.

Let us first calculate the amount of work done for carrying unit volume of the liquid from its interior at a depth C to its surface.

We consider a mass m at P (Fig. 10.20.4). Then the force of attraction on m due to the liquid below it is given by the general expression

$$m\rho \int_0^C \psi(r)dr$$

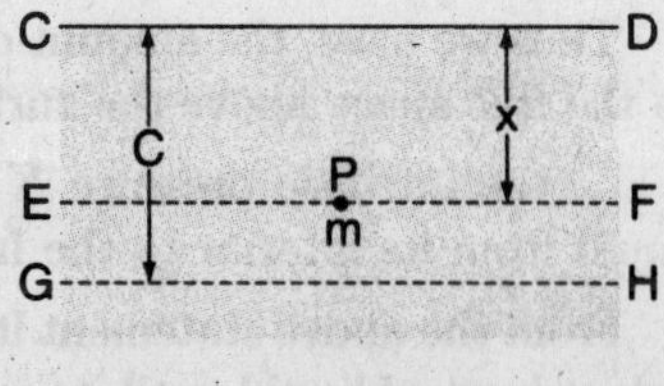

Fig 10.20.4

[from Laplace's law, force of attraction $= mm'\psi(r)$, where m' = mass per unit area of the layer of thickness dr at a distance r below P. Now $m' = \rho dr$.

$\therefore$ force of attraction of the layer on $m = m\rho\psi(r)dr$.

$\therefore$ total force of all the layers extending between $r = 0$ and $r = C$ is given by,

$$m\rho \int_0^C \psi(r)dr] \quad \text{acting in the downward direction.}$$

Again the liquid above P will attract m in the upward direction, and the force of attraction of the liquid above P is given by

$$m\rho \int_0^x \psi(r)dr$$

Therefore, the resultant force of attraction on m is given by

$$F = m\rho \int_0^C \psi(r)dr - m\rho \int_0^x \psi(r)dr = m\rho \int_x^C \psi(r)dr = m\rho\phi(x) \text{ acting downwards.}$$

Hence the work done in moving the mass through a distance dx is given by

$$\delta W = F \cdot dx = m\rho\phi(x)dx$$

Therefore, total work done in moving the mass from the interior at a depth C to the surface [i.e., from GH to CD, fig. 10.20.4] is given by

$$W = m\rho \int_0^C \phi(x)dx = v\rho^2 \int_0^C \phi(x)dx \quad [m = v\rho, \text{ where } v \text{ is the volume of the mass}];$$

when v = unity,

$$W = \rho^2 \int_0^C \phi(x)dx = K. \tag{10.20.5}$$

Thus we see that the work done in removing unit volume of the liquid from interior at a depth C to the surface is numerically equal to the intrinsic pressure of the liquid.

Again we consider a mass m in the process of its removal from the surface of the liquid to a distance x from the surface. Now at this position, the force on it is given by $m\rho\phi(x)$.

Then the work done in moving this mass through a distance $dx = m\rho\phi(x)dx$.

$\therefore$ the total work done in removing this mass from the surface to a distance C above it

$$= \int_0^C m\rho\phi(x)dx = v\rho^2 \int_0^C \phi(x)dx.$$

When $v =$ unity, we have the total work done for removing unit volume of the liquid from its surface to a distance C above it

$$= \rho^2 \int_0^C \phi(x)dx = K \quad \text{which is same as (10.20.5)}$$

Thus we have, the amount of work done for removing unit volume of liquid from its interior to the free space above the surface of the liquid $= 2K$

$\therefore$ the intrinsic pressure $K = \frac{1}{2}\times$ (the amount of work done for removing unit volume of liquid from its interior to the free space above the surface of the liquid).

From the above statement it follows that K is equal to half the work required to disintegrate unit volume of liquid until its molecules are so far apart that they exert no longer any attraction upon one another. In other words K is equal to half the internal work required to vaporise unit volume of the liquid.

The above conclusion may be put to test by taking into consideration the following concrete numerical example.

In case of water at 15°C, its density is taken approximately unity; the latent heat of evaporation at 15°C is 586 cals per gm. If water vapour at 15°C obeys gas laws, its volume per gm. calculated from a vapour pressure of 1.278 cm of mercury is 78030 cc. The evaporation of this volume pushes back the vapour already present whose pressure is

$$\frac{1.278}{76} \times 1.013 \times 10^6 \text{ dynes per sq cm}$$

[taking the standard atmospheric pressure = 1.013×10^6 dynes per sq cm]

Therefore, the work done

$$= \frac{1.278}{76} \times 1.013 \times 10^6 \times 78030 \text{ ergs} = \frac{1.278 \times 1.013 \times 10^6 \times 78030}{76 \times 4.182 \times 10^7} = 31.76 \text{ cals.}$$

$\therefore$ the resultant internal latent heat $= 586 - 31.76 = 554.24$ cals/gm.

Hence the internal work required to evaporate unit volume of water (here unit mass = unit volume, since density of water is unity).

$$= 2K = 554.24 \times 4.182 \times 10^7 \text{ ergs/cc} = 2.32 \times 10^{10} \text{ dynes per sq cm}$$

$$\text{or} \quad K = 1.16 \times 10^{10} \text{ dynes per sq cm.}$$

The quantity $\frac{a}{v^2}$ in Van der Waals' equation also affords a method of estimating intrinsic pressure. The value of intrinsic pressure obtained by considering Van der Waals' equation is of the same order of magnitude as found in the above numerical example.

10.21 Dependence of Surface Tension on Temperature

The surface tension of all liquids decreases linearly with rise of temperature when the temperature range is small. Surface tension S at temperature t is given by $S = S_0(1 - \alpha t)$, where S_0 is the value of surface tension at 0°C and α is the temperature coefficient of surface tension. The above expression may also be written in the form

$$\frac{dS}{dt} = -S_0\alpha = -k \quad [k = S_0\alpha \text{ say}]$$

At critical temperature the value of surface tension is zero. The interface between the liquid and the vapour disappears at this temperature and so the absolute value of surface tension depends on how far the temperature is from its critical value. According to Van der Waals the surface tension at absolute temperature T is given by

$$S = S_0\left(1 - \frac{T}{T_C}\right)^{3/2},$$

where T_C is the critical temperature and S_0 is a constant for the liquid. Differentiating this expression with respect to T, we have,

$$\frac{dS}{dT} = -\frac{3S_0}{2T_C}\left(1 - \frac{T}{T_C}\right)^{1/2} \quad \text{or} \quad \frac{1}{S_0}\cdot\frac{dS}{dT} = -\frac{3}{2T_C}\left(1 - \frac{T}{T_C}\right)^{1/2}. \quad \therefore \quad \frac{dS}{dT} = 0, \text{ when } T = T_C.$$

Thus both S and $\frac{dS}{dT}$ are zero at critical temperature.

A modified formula, according to Ferguson, is given by

$$S = S_0\left(1 - \frac{T}{T_C}\right)^{n},$$

where n is a constant for a single liquid, but varies slightly from liquid to liquid. The mean value of n is about 1.21.

Eötvös' law

Let M be the molecular weight of a liquid and ρ its density; the surface area occupied by a gm-molecule, assuming the molecules are symmetrical in shape, is proportional to $\left(\frac{M}{\rho}\right)^{2/3}$. This surface is called molar surface. Surface energy in the molar surface is, therefore, proportional to $S\left(\frac{M}{\rho}\right)^{2/3}$ and is known as *molar energy.* According to Eötvös' law

$$S\left(\frac{M}{\rho}\right)^{2/3} = K(T_C - T),$$

where K is a universal constant known as Eötvös' constant. The approximate value of K is 2.2. The above relation was corrected by Ramsey and Shields giving

$$S\left(\frac{M}{\rho}\right)^{2/3} = K(T_C - T - \delta).$$

The constant δ has a value between 6 and 8 for most liquids. S denotes the surface tension of the liquid at the absolute temperature T and T_C, the critical temperature. From this relation it follows that surface tension is zero when $T = T_C - \delta$. Differentiating the above expression, we have,

$$\frac{d}{dT}\left[S\left(\frac{M}{\rho}\right)^{2/3}\right] = -K,$$

where K is a constant which, as already stated, is approximately equal to 2.2 for many liquids.

Illustrative Examples

1. Molecular wt. of ether is 74 and its density at 0°C 0.737 gm/cc; if the surface tension of ether at 0°C is 19.2 dynes/cm and the Eötvös' constant is 2.16, calculate approximately the critical temperature of ether.

We know,

$$S\left(\frac{M}{\rho}\right)^{2/3} = K(T_C - T)$$

Hence at temperature 0°C, we may write

$$S_0\left(\frac{M}{\rho}\right)^{2/3} = K(T_C - T_0) \qquad (1)$$

where S_0 is the surface tension at 0°C, i.e., at the absolute temperature T_0.

Here $M = 74$, $\rho = 0.737$ gm/cc, $K = 2.16$ and $S_0 = 19.2$ dynes/cm.

$$\therefore \log\left(\frac{M}{\rho}\right)^{2/3} = \frac{2}{3}\log\left(\frac{M}{\rho}\right) = \frac{2}{3}\log\left(\frac{74}{0.737}\right) = \frac{2}{3}\log 100 \text{ (approx.)} = \frac{4}{3} = 1.333$$

$$\therefore \text{ antilog } 1.333 = 21.54 = \left(\frac{M}{\rho}\right)^{2/3}$$

Thus we have from (1)

$$19.2 \times 21.54 = 2.16(T_C - 273) \quad \text{or} \quad T_C - 273 = \frac{19.2 \times 21.54}{2.16}$$

$$\text{or} \quad T_C = \frac{19.2 \times 21.54}{2.16} + 273 \quad \text{or} \quad T_C = 192°\text{C (approximately)}.$$

2. If a graph of surface tension S versus the Kelvin temperature T is drawn with the origin at $T = 0$, $S = 0$, prove that the intercept of the tangent to the curve with the S-axis at any point is equal to the surface energy per unit area at the corresponding temperature. **[C.U. (Hons.) 1980]**

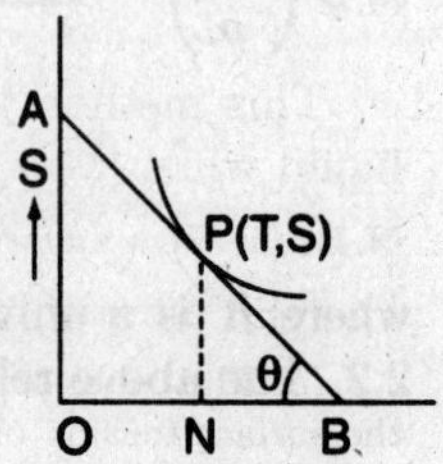

Fig 1

From Fig. 1, we may write

$$\tan\theta = \frac{PN}{NB} = \frac{OA}{OB} = -\frac{dS}{dT}. \quad \therefore \quad OA = -OB\frac{dS}{dT} \qquad (1)$$

$$\text{and} \quad NB = -\frac{dT}{dS}\cdot PN = -\frac{dT}{dS}S \qquad (2)$$

From (1), we have

$$OA = -OB\frac{dS}{dT} = -(ON + NB)\frac{dS}{dT} = -(T + NB)\frac{dS}{dT} = -\left(T - \frac{dT}{dS}S\right)\frac{dS}{dT} \quad \text{[from (2)]}$$

$$= S - T\frac{dS}{dT} = \text{the surface energy per unit area at temperature } T.$$

10.22 Experimental Determination of Surface Tension

10.22.1 Capillary rise method (Laboratory method)

A few capillary tubes of different bores are drawn, and are fixed parallel to each other on a strip of glass-plate by means of wax; a needle is also fixed to the glass-plate parallel to the tubes as shown in figure 10.22.1. The glass-plate is adjusted with the help of a clamp in such a way that the tubes remain vertical with their lower ends immersed in the experimental liquid contained in a trough—the tip of the needle just touching the surface of the liquid.

Liquid rises to different heights depending on the bores of the capillary tubes—greater the bore, less will be the height and *vice versa.* With the help of a travelling microscope, the readings corresponding to the upper end of the needle, and those of the horizontal parts of the concave meniscus of the liquid in different tubes are taken. The length of the needle is also measured by means of the travelling microscope. The difference beween the reading of the upper tip of the needle and a meniscus reading for a tube subtracted from the length of the needle (or added to the length of the needle in case the tip of the needle lies below the meniscus) gives the height of the liquid in the tube. Thus the heights h for different tubes are measured. The positions of the meniscus of the liquid in the tubes are marked by ink, and the capillary tubes are cut at these marked positions (since the bores of the tubes may not be uniform throughout) and their diameters are measured by means of the travelling microscope in mutually perpendicular directions, whence the radii r of the tubes are obtained.

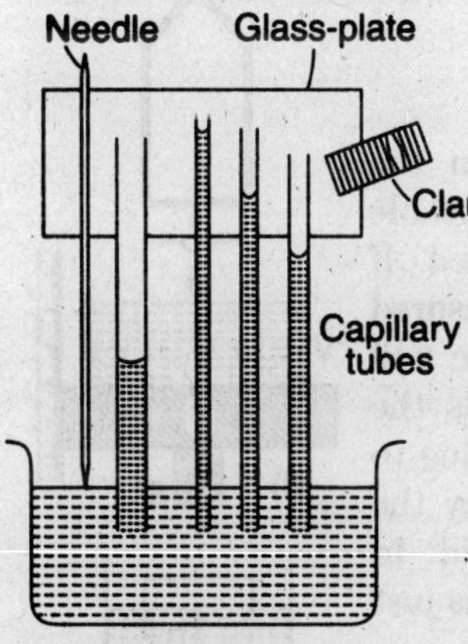

Fig 10.22.1

A curve is drawn with h as ordinate and r as abscissa. It is a rectangular hyperbola which verifies Jurin's law. A suitable point on the curve is chosen, for which r and h are found out. Hence surface tension may be determined from the relation

$$S = \frac{r\left(h + \frac{r}{3}\right)\rho g}{2}$$

This method is usually employed in the laboratory to determine the surface tension of a liquid which wets glass walls.

N.B. (a) Here angle of contact has been considered zero, which is the case with many liquids.

(b) It essential to make the tubes perfectly vertical, which may be done with the help of a plumb-line.

(c) The tubes as also the liquid should be made free from all contaminations which may appreciably change the surface tension of the liquid.

(d) Since surface tension depends on temperature, the temperature of the experimental liquid should be noted.

(e) For slightly wider tubes the meniscus may be treated as having a semi-elliptical section, and the corrected expression for S is given by

$$S = \frac{r\rho gh}{2}\left[1 + \frac{r}{3h} - 0.111\frac{r^2}{h^2} + 0.0741\frac{r^3}{h^3}\right]$$

10.22.2 Balance method

This method is suitable for finding the surface tension of a liquid which wets the glass.

A glass plate G is supported from below one pan of a balance and then the balance is equilibriated. The experimental liquid is taken in a vessel V which is placed on an adjustable table T [Fig. 10.22.2(i)]. The adjustable table is raised up by means of a screw S' till the glass plate touches the surface of the liquid. When this is done the equilibrium condition of the balance is disturbed as the glass plate is pulled down due to surface tension of the liquid. Now, additional masses are placed on the other pan of the balance until the glass plate is withdrawn from the liquid. The downward pull due to surface tension acting on the plate $= 2Sl + 2St = 2S(l+t)$, where l and t are respectively the length and thickness of the plate at its lower end. This is equal to mg, where m is the mass added to the pan.

Hence

$$2S(l+t) = mg \quad \text{or} \quad S = \frac{mg}{2(l+t)}.$$

Thus S may be determined.

Alternatively, without adding masses to the other pan when the equilibrium condition of the balance is disturbed, the adjustable table is raised up until the equilibrium condition of the balance is restored. If h be the height through which the table is raised [the height is measured by a suitable scale shown in [Fig. 10.22.2(i)] then the glass plate will be immersed inside the liquid to a depth h [Fig. 10.22.2(ii)]. As the glass plate is immersed inside the liquid there is an upward force due to buoyancy effect which is equal to the wt. of the liquid displaced by the glass plate and is given by $lth\rho g$, where ρ is the density of the liquid. In the equilibrium condition the downward pull due to surface tension is just balanced by the upward force due to buoyancy.

$$\therefore\ 2S(l+t) = lth\rho g \quad \text{or} \quad S = \frac{lth\rho g}{2(l+t)}$$

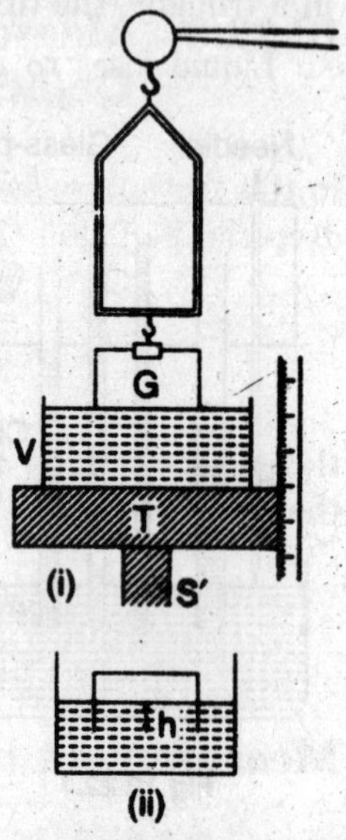

Fig. 10.22.2

Instead of using a glass plate, a glass or wire frame may be supported from the lower surface of the balance. The arrangement is so made that when the balance is equilibriated, the horizontal portion of the frame is at a distance of about 5 mm above the general surface of the liquid. The frame is then completely immersed inside the liquid and then masses are placed on the other pan until the horizontal portion of the frame is again at a distance of 5 mm above the general level. Then $2Sl = mg$, where l is the length of the horizontal portion and m is the additional mass. Hence

$$S = \frac{mg}{2l}.$$

This method is suitable for determining the surface tension of soap solution. It may also be used to determine surface tension of water or a solution. For accurate result the vessel should be completely filled with the liquid so that the surface of the liquid may be made free from contamination with the help of waxed pieces of glass as explained in Art. [10.5(ii)].

Illustrative Example

A glass plate of length 4 cm, breadth 2 cm. and thickness 2 mm is supported from below the pan of a balance with its longer side horizontal and touching the surface of a liquid of surface tension 74 dynes/cm. What is the weight to be added on the other pan to detach the plate from the liquid surface?

We use the equation,

$$2S(l+t) = mg$$

Here $l = 4$ cms, $t = 2$ mm $= 0.2$ cm, $S = 74$ dynes/cm

$$\therefore\ 2 \times 74(4+0.2) = m \times 981$$

$$\text{or}\quad m = \frac{74 \times 8.4}{981} = 0.6336 \text{ gm} = 634 \text{ mgms. (nearly).}$$

10.22.3 Measurement of surface tension and angle of contact by Quincke's method

In this method surface tension and angle of contact of mercury obtained in the form of a sessile drop can be found. From the relation

$$S = \frac{g\sigma h'^2}{2} \quad \text{[see eqn. (10.12.9)]}$$

the surface tension of mercury can be found if h' is known. Thus knowing S we can find out the angle of contact θ_A from the relation

$$g\sigma h^2 = 2S(1-\cos\theta_A) \quad \text{[see eqn. (10.12.10)] if } h \text{ is known.}$$

Measurement of h'

To measure h' the arrangement as shown in figure 10.22.3 is made. Light from an incandescent lamp B is focussed, by a lens L and a plane glass plate G (the glass plate makes an angle of 45° to the horizontal) acting as a mirror, on the edge of the mercury drop and is reflected from there. A short-focus microscope M which is mounted horizontally near the drop is raised or lowered until a bright thin horizontal line is in coincidence with the horizontal cross-wire of the eye piece of the microscope. This image is evidently formed by direct reflection from the vertical surface of the drop. The reading of the microscope is then noted. The microscope is then raised up to focus the upper surface of the drop on which some lycopodium powder is sprinkled. The reading of the microscope is again noted. The difference between these two readings gives the value of h'.

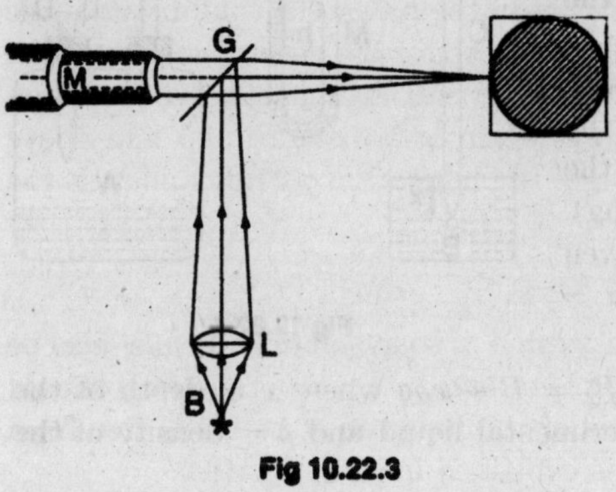

Fig 10.22.3

Measurement of *h*

The whole height h of the drop is measured either by a spherometer or by a microscope. The microscope is first focussed on the top and then on the bottom of the drop. The difference between the readings of the microscope at these two positions gives the value of h.

In deducing the expressions in equations (10.12.9) and (10.12.10), we have assumed that $r_2 \gg r_1$ or $\frac{1}{r_2} \ll \frac{1}{r_1}$ and hence we have neglected $\frac{1}{r_2}$. This is applicable in case of a very large drop, but in actual case the drops which are used for the measurement of surface tension and angle of contact are not so large and as such we cannot ignore the curvature $\frac{1}{r_2}$. For this reason some correction is necessary. Ferguson gave a more accurate formula given by

$$h'^2 = a^2 + 0.606\frac{a^3}{r} \tag{10.22.1}$$

where r = maximum horizontal radius of the drop and $a = \sqrt{\dfrac{2S}{\sigma g}}$.

As a first approximation Ferguson put $a^2 = h'^2$ in the second term of the right hand side of the above equation and calculated a better value of a^2. The value of a^2 thus obtained was again substituted in the small term and a new value of a^2 was obtained. The process was repeated until a constant value of a^2 was found whence S is calculated.

In one case of a bubble of air in tap water Ferguson obtained the value of surface tension of tap water from (10.22.1) by the method of successive approximation as already mentioned. The surface tension was found to be 73.65 dynes/cm at 8°C.

10.22.4 Jaeger's method

The excess pressure inside a spherical bubble of air within a liquid has been utilised in this method.

The apparatus consists of a capillary tube C placed vertically downwards in a vessel B containing the experimental liquid (Fig. 10.22.4). The capillary tube is connected to a manometer M containing a liquid of known density and also to a Woulfe's bottle W fitted with a dropping funnel F. Water is poured into the funnel, and allowed to run slowly into the Woulfe's bottle by means of a stop-cock, as a result of which a difference of pressure is created as shown by the manometer. When the pressure difference as indicated by the manometer attains an appropriate value, a bubble breaks off from the tip of the capillary tube with a corresponding change in the manometer reading. Next, another bubble begins to form and the maximum pressure difference recorded by the manometer is h, say (when the radius of curvature of the bubble is maximum, i.e., just at the point of breaking). Hence the maximum pressure, inside the bubble is given by $P_1 = B + g\sigma h$, where B = atmospheric pressure; σ = density of liquid in the manometer.

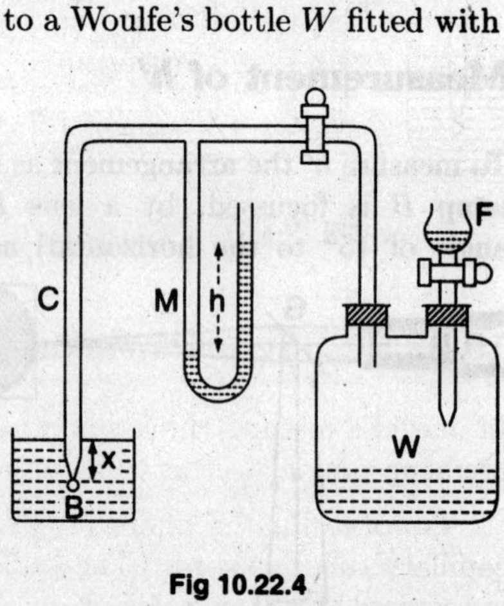

Fig 10.22.4

The pressure just outside the bubble is given by $P_2 = B + x\rho g$ where x = depth of the tip of the capillary tube below the surface of the experimental liquid and ρ = density of the experimental liquid.

$\therefore$ the excess pressure inside the bubble is given by

$$P = P_1 - P_2 = g(\sigma h - \rho x) = \frac{2S}{r},$$

where r is the radius of the bubble

$$\therefore \quad S = \frac{gr(\sigma h - \rho x)}{2} \tag{10.22.2}$$

It is to be noted that at the breaking condition of the bubble, its radius is not exactly equal to the radius of the orifice of the capillary tube, but is a function of the radius of the orifice for a particular tube and steady rate of bubbling. Writing $f(r)$ for r, we have

$$S = \frac{f(r)g(\sigma h - \rho x)}{2}$$

Since $f(r)$ is unknown, absolute value of the surface tension cannot be determined from the above expression, but a comparison of surface tensions at different temperatures may be made since $f(r)$ remains same.

The attached figure 10.22.5 shows the modified experimental arrangement of Jaeger's method.

Here the liquid is contained in a comparatively long narrow bulb so that it contains only small quantity of liquid. A platinum or a glass tube is fitted vertically and sealed at the top of this bulb so that its one end is dipped in the liquid and the other end is connected to a reservoir of pure nitrogen. Manometric connection is made as shown in the figure. Excess pressure in this case is directly measured from the manometric reading. According to Shroedinger the expression for surface tension is given by

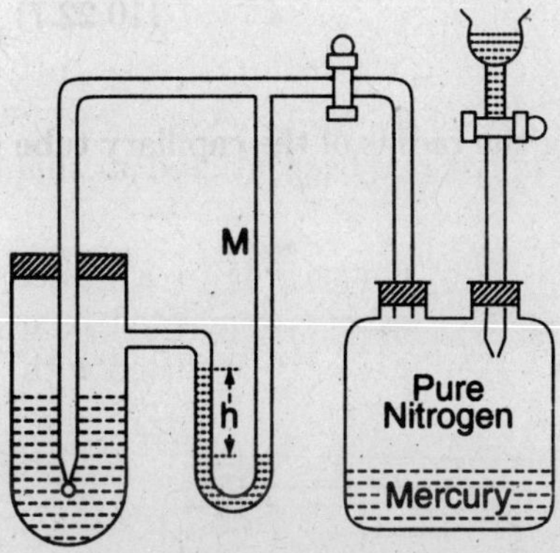

Fig 10.22.5

$$S = \frac{1}{2}rp\left[1 - \frac{2}{3}\cdot\frac{r}{h} - \frac{1}{6}\left(\frac{r}{h}\right)^2\right]$$

where p is the excess pressure and $h = \frac{p}{(\rho-\rho')g}$, where ρ and ρ' give the density of the liquid in the manometer and that of the issuing gas respectively.

10.22.5 Sugden's theory and experiment

In Jaeger's method the assumption that instability and detachment occur when the bubbles are hemispherical seems to be unjustified. Sugden solved this problem.

Theory. In the figure (10.22.6) the central section of a bubble formed at the tip of a capillary tube is shown. The capillary tube dips inside a liquid the surface tension of which is required. At any point A on the surface of a bubble we may write the general equation (10.12.1) as

$$S\left(\frac{1}{r_1} + \frac{1}{r_2}\right) + \sigma g y = \text{const.} \tag{10.22.3}$$

The minus sign before σgy has been made plus because the point A lies below the general level. At the lowest point B of the bubble if $r_1 = r_2 = b$, we have

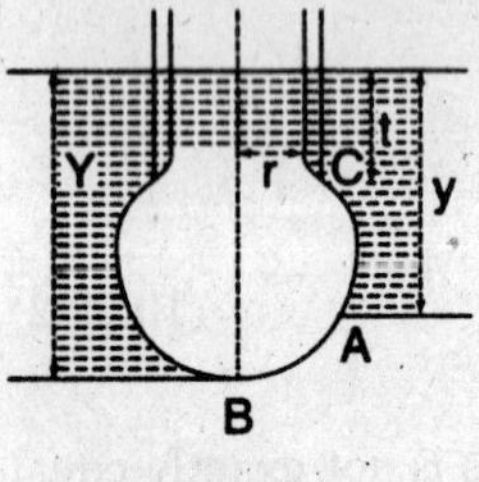

Fig 10.22.6

$$\frac{2S}{b} + \sigma gY = \text{const.} \tag{10.22.4}$$

From (10.22.3) and (10.22.4) we may write

$$S\left(\frac{1}{r_1}+\frac{1}{r_2}\right)+\sigma gy = \frac{2S}{b}+\sigma gY \quad \text{or} \quad S\left(\frac{1}{r_1}+\frac{1}{r_2}\right) = \frac{2S}{b}+\sigma g(Y-y) \tag{10.22.5}$$

At the end of the capillary tube, i.e., at the point C, $Y - y = Y - t$. Hence we have

$$S\left(\frac{1}{r_1}+\frac{1}{r_2}\right) = \frac{2S}{b}+\sigma g(Y-t)$$

$$\text{or} \quad S\left(\frac{1}{r_1}+\frac{1}{r_2}\right) = \frac{2S}{b}+\sigma gz \quad [\text{putting } (Y-t)=z] \tag{10.22.6}$$

If we write the excess pressure $S\left(\frac{1}{r_1}+\frac{1}{r_2}\right) = g\sigma h$, where h stands for positive depth, the equation (10.22.6) reduces to the form

$$g\sigma h = \frac{2S}{b} + g\sigma z \tag{10.22.7}$$

Multiplying both sides of the above equation by $\frac{r}{a^2}$, r being the radius of the capillary tube we have,

$$\frac{g\sigma hr}{a^2} = \frac{2S}{b}\cdot\frac{r}{a^2} + g\sigma\frac{rz}{a^2}.$$

Dividing both sides by $g\sigma$ we have

$$\frac{hr}{a^2} = \frac{2S}{g\sigma}\cdot\frac{r}{ba^2} + \frac{rz}{a^2}$$

Putting $\frac{2S}{g\sigma} = a^2$, we have,

$$\frac{hr}{a^2} = \frac{r}{b} + \frac{rz}{a^2} \quad \text{or} \quad \frac{hr}{a^2} = \frac{r}{b} + \frac{r}{a}\cdot\frac{z}{b}\cdot\frac{b}{a}$$

$$\text{or} \quad \frac{r}{X} = \frac{r}{b} + \frac{r}{a}\cdot\frac{z}{b}\cdot\frac{b}{a} \quad \left[\text{putting } X = \frac{a^2}{h}\right]$$

$$\text{or} \quad \frac{r}{X} = \frac{r}{b} + \frac{r}{a}\cdot\frac{z}{b}\cdot\sqrt{\frac{\beta}{2}} \quad \left[\text{putting } \frac{2b^2}{a^2} = \beta\right] \tag{10.22.8}$$

Now

$$X = \frac{a^2}{h} = \frac{2S}{g\sigma h} \quad \text{or} \quad \frac{2S}{X} = g\sigma h.$$

$\therefore$ X denotes the radius of the bubble. So if the radius of the bubble is equal to the radius of the capillary tube, we have $X = r$. For a given liquid and orifice of the capillary tube the values of r, S and σ are fixed. Hence r/a must have a fixed value. If the excess pressure in the bubble changes, both the shape and size of the bubble will change. For a certain pressure, $h, X, b, r/X$ and so β will remain constant. Bashforth and Adam's table gives for a fixed value of β the corresponding values of r/b and z/b. From this table for any value of β, r/X may be calculated from the equation (10.22.8).

Now

$$\frac{r}{X} = \frac{rh}{a^2} = \frac{rhg\sigma}{2S}. \quad \therefore \quad S = \frac{1}{2}g\sigma hX \tag{10.22.9}$$

Thus surface tension of the liquid may be calculated from the above equation. Calculation shows that for a fixed value of r/a if the size of the bubble steadily increases, r/X passes through maximum and then decreases. When r/X attains the maximum value, the bubble is detached. At this stage the excess pressure inside the bubble is maximum. Sugden gives four-figure tables of maximum values of r/X corresponding to the values of r/a. From Sugden's table it is found that the condition under which the bubble breaks away depends on the radius of the tube. If from the point C on the surface of the bubble a normal is drawn inwards and if it makes an angle φ_c with the vertical axis of the bubble then it is found that in case of tube of very small radius when $\varphi_c = 90°$, the bubble breaks away. In this case the radius of the bubble is equal to the radius of the capillary tube and the shape of the bubble is nearly hemispherical as we observe in the case of Jaeger's experiment. For a tube of radius 0.35 cm the bubble breaks away when $\varphi_c = 160°$.

Experiment

Sugden used the apparatus shown in Fig. 10.22.7 to measure the maximum pressures to detach bubbles from the lower ends of two capillary tubes of different radii immersed to the same depth in a liquid. The experimental liquid is kept in the vessel A. The vessel A is surrounded by water bath which is screened from the other parts of the apparatus. The temperature of the experimental liquid can be adjusted with the help of the water bath. B and C are two capillary tubes of different radii kept vertically with their ends at the same level inside the experimental liquid. The open end of the vessel A is closed by a rubber stopper and the two tubes B and C pass through the stopper. The upper part of A is connected to a water manometer M and a mercury suction bulb S with a capillary end E.

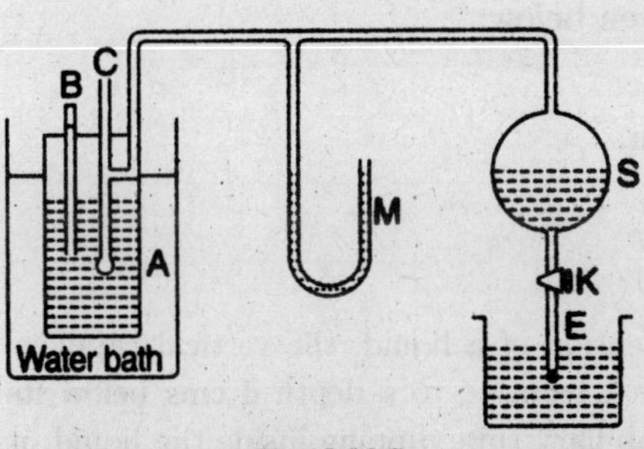

Fig 10.22.7

At first the upper end of one of the tubes say B is closed by a stopper and the tap K is opened. When the tap is opened, mercury in the suction bulb S will flow down through the tube having the capillary end, as a result of which the pressure in the tube A will be reduced. Hence air will rush into the open tube C and a series of bubbles will form and break away at its lower end. The maximum pressure during formation of bubbles is read from the manometer.

For the two tubes of radii r_1 and r_2 the maximum excess pressures are given by

$$p_1 = g\sigma H_1 \text{ and } p_2 = g\sigma H_2.$$

H_1 and H_2 are the values of H for the two tubes obtained from the manometer. Here σ is the density of the experimental liquid minus the density of air.

Now $g\sigma H_1$ = pressure due to the depth t of the lower end of the open tube below the surface of the experimental liquid + excess pressure inside the lower end of the tube.

$$\therefore \quad g\sigma H_1 = g\sigma t_1 + g\sigma h_1 \quad \left[\because \text{ excess pressure } S\left(\frac{1}{r_1}+\frac{1}{r_2}\right) = g\sigma h\right]$$

$$\text{or} \quad H_1 = t + h_1 = t + \frac{a^2}{X_1}$$

Similarly

$$H_2 = t + \frac{a^2}{X_2}.$$

$$\therefore \quad H_1 - H_2 = a^2\left(\frac{1}{X_1} - \frac{1}{X_2}\right) \quad \text{or} \quad a^2 = \frac{H_1 - H_2}{1/X_1 - 1/X_2} \qquad (10.22.10)$$

If we put $X_1 = r_1$ and $X_2 = r_2$ in the above equation we get the approximate value of a^2, i.e., $S\left(\because\ a^2 = \frac{2S}{g\sigma}\right)$. From the value a^2 the values of r_1/a and r_2/a may be calculated. From Sugden's table knowing r_1/a and r_2/a the maximum values of r_1/X_1 and r_2/X_2 are determined. From this we get the values of X_1 and X_2. Substituting the values of X_1 and X_2 in (10.22.10) we get better value of a^2. With this value of a again the values of r_1/a and r_2/a are calculated and in a similar way we get from Sugden's table the values of r_1/X_1 and r_2/X_2 and hence X_1 and X_2.

Substituting the values of X_1 and X_2 in equation (10.22.10) we get more accurate value of a^2. This process is repeated several times until we get the constant value of a^2.

Since

$$a^2 = \frac{2S}{g\sigma}, \quad \therefore \quad S = \frac{a^2 \cdot g\sigma}{2}$$

Thus knowing a^2 and σ, the surface tension of the liquid may be determined.

Sugden's results at two different temperatures are given below:

Surface tension of water at 20°C = 72.91 dynes/cm.

Surface tension of Benzene at 20°C = 28.86 dynes/cm.

Illustrative Example

1. In a Jaeger type of experiment to measure the surface tension of a liquid, the vertical capillary tube of radius 0.5 mm was dipped inside a liquid of density 1.1 gm/cc to a depth 4 cms below its surface. When an air bubble was formed at the end of the capillary tube dipping inside the liquid, it was observed from manometric reading that the pressure inside the bubble exceeded the atmospheric pressure by 0.547 cm of mercury. Calculate the surface tension of the liquid.

The surface tension of the liquid is given by

$$S = \frac{gr(\sigma h - \rho x)}{2}$$

Here r = 0.5 mm = 0.05 cm, h = 0.547 cm, σ = 13.6 gm/cc, ρ = 1.1 gm/cc, x = 4 cms,

$$S = \frac{981 \times 0.05}{2}(13.6 \times 0.547 - 1.1 \times 4) = 74.53 \text{ dynes/cm}.$$

10.22.6 *U*-tube arrangement for measuring surface tension of a liquid at different temperatures

This method is used to compare the surface tensions of a liquid at different temperatures. The apparatus consists of a *U*-tube with limbs of different cross-sections. The experimental liquid is taken in the *U*-tube and the same is mounted inside a suitable temperature bath.

In fig. 10.22.8 the pressure just below the meniscus in the limb AB, inside the liquid is

$$P - \frac{2S}{r_1}; \quad [P \text{ is the atmospheric pressure}]$$

pressure just below the meniscus in the limb CD, inside the liquid is

$$P - \frac{2S}{r_2} \quad \left[\begin{array}{l} r_1 \text{ and } r_2 \text{ are the radii of} \\ \text{the bores of the two limbs} \end{array}\right]$$

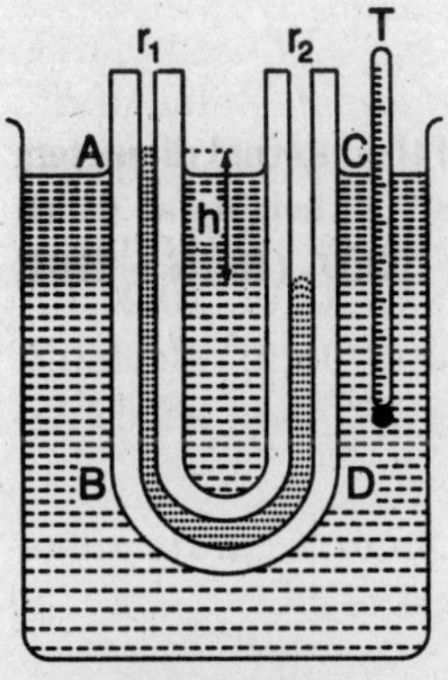

Fig 10.22.8

Therefore the difference of pressures in the two limbs of the *U*-tube is

$$p = 2S\left(\frac{1}{r_1} - \frac{1}{r_2}\right).$$

But this is equal to $h\rho g$, where h is the difference in the levels of liquid in the two limbs of tube and ρ the density of the liquid in the tube.

Therefore,

$$2S\left(\frac{1}{r_1} - \frac{1}{r_2}\right) = h\rho g \text{ or } S = \frac{h\rho g}{2\left(\frac{1}{r_1} - \frac{1}{r_2}\right)}.$$

The values of S are found for different temperatures t and a graph is then plotted between the temperatures t and the corresponding values of S. From the nature of the graph shown in fig. 10.22.9, we see that surface tension of the liquid decreases with rise of temperature and becomes zero at a particular temperature t_c known as the critical temperature which depends on the nature of the liquid. Taking two points $A(t_1, S_1)$ and $B(t_2, S_2)$ on the graph, we see that the gradient of the graph gives us the temperature coefficient of surface tension i.e.,

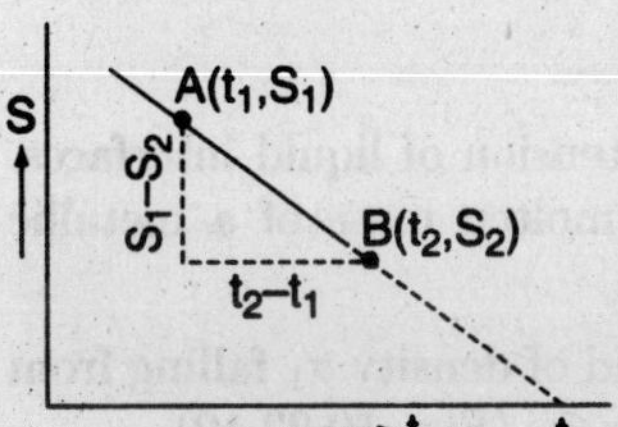

Fig 10.22.9

$$\frac{dS}{dt} = \frac{S_1 - S_2}{t_2 - t_1}$$

We assume here that the angle of contact of the liquid is zero. The method may also be used to determine the surface tension of mercury. In this case there will be depression in the tube of smaller bore since the mercury meniscus is convex upward.

$$\therefore \quad \frac{2S\cos\theta}{r_1} - \frac{2S\cos\theta}{r_2} = h\rho g,$$

where θ is the angle the surface of the liquid makes with the glass or the supplement of the angle of contact

$$\text{or} \quad S = \frac{h\rho g}{2} \Big/ \left(\frac{1}{r_1} - \frac{1}{r_2}\right)\cos\theta.$$

Illustrative Examples

1. A *U*-tube whose ends are open and whose limbs are vertical, contains an oil of specific gravity 0.85 and surface tension 28 dynes per cm. If one limb has a diameter of 2.2 mm and the other a diameter of 0.8 mm, what is the difference in levels of the oil in the two limbs? (Assume the angle of contact between the oil and the glass is zero.)

We know that

$$S = \frac{h\rho g}{2\left(\frac{1}{r_1} - \frac{1}{r_2}\right)} \quad \therefore\ h = \frac{2S\left(\frac{1}{r_1} - \frac{1}{r_2}\right)}{\rho g}$$

Here $S = 28$ dynes/cm, $r_2 = \frac{2.2}{2}$ mm $= 0.11$ cm, $r_1 = \frac{0.8}{2}$ mm $= 0.04$ cm, $\rho = 0.85$ gm/cc

$$\therefore\ h = \frac{2 \times 28\left(\frac{1}{0.04} - \frac{1}{0.11}\right)}{0.85 \times 981} = \frac{891}{0.85 \times 981} = 1.07 \text{ cm.}$$

2. A *U*-tube is supported with its limbs vertical and is partly filled with water. If the internal diameters of the limbs are respectively equal to 1 cm and 1 mm, what will be the difference in heights at which the water stands in the two limbs? Surface tension of water = 70 dynes/cm. [C.U. (Hons.) 1982]

The difference in heights is given by

$$h = \frac{2S}{\rho g}\left(\frac{1}{r_1} - \frac{1}{r_2}\right)$$

Here $S = 70$ dynes/cm, $r_1 = 0.5$ cm, $r_2 = 0.05$ cm, $\rho = 1$ gm/cc

$$\therefore\ h = \frac{2 \times 70}{981}\left(\frac{1}{0.05} - \frac{1}{0.5}\right) = 2.56 \text{ cms.}$$

10.22.7 Drop-weight method

This method is probably the best means of measuring the surface tension of liquid interfaces. Drop-weight method is suitable for measuring surface tension of molten mass of a metallic substance or alloy.

In this experiment, the weight mg is found for a drop of one liquid of density σ_1 falling from a tube of external radius r, dipping inside a second liquid of density σ_2 (Fig. 10.22.10).

We shall first consider that the detachment of the drop is approximately a case of statical equilibrium. This means that just before the detachment of the pendulous drop from the tube, the shape of the drop attached to the tube is cylindical, (not shown in the figure) and the equation of equilibrium of that portion of the drop which is hanging is given by

$$mg\left(\frac{\sigma_1 - \sigma_2}{\sigma_1}\right) + P\pi r^2 = 2\pi r S_{12} \qquad (10.22.11)$$

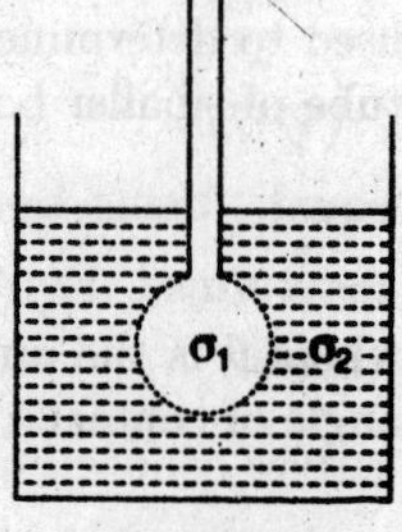

Fig 10.22.10

where $mg\left(\frac{\sigma_1-\sigma_2}{\sigma_1}\right)$ = the apparent weight of the drop in the liquid of density σ_2 acting downwards;

S_{12} = interfacial surface tension;

$P\pi r^2$ = the excess force arising out of excess of pressure inside the drop due to surface tension acting downwards; and $2\pi r S_{12}$ = the upward force due to surface tension.

Now

$$P = \frac{S_{12}}{r} \text{ (the drop being cylindrical in shape)}$$

$\therefore$ the equation (10.22.11) may be written as

$$mg\left(\frac{\sigma_1 - \sigma_2}{\sigma_1}\right) + \frac{S_{12}}{r}\cdot \pi r^2 = 2\pi r S_{12} \quad \text{or} \quad mg\left(\frac{\sigma_1 - \sigma_2}{\sigma_1}\right) = \pi r S_{12}$$

$$\therefore \quad m = \frac{\pi r S_{12}}{g(\sigma_1 - \sigma_2)}\cdot \sigma_1.$$

But the detachment is essentially a dynamical problem. The dimensional analysis gives,

$$m = \frac{S_{12} r \sigma_1}{g(\sigma_1 - \sigma_2)} \times \text{ some non-dimensional variable function.}$$

The same problem was attacked by Harkins and Brown. They considered a drop of liquid detached from the lower end of a rod or tube and falling under gravity. Let the lower end or the tip of the tube be of external radius r. The weight mg of the drop may be regarded as a function of the volume V of the drop, the radius r of the tip, the surface tension S and the density ρ of the liquid. Hence we may write

$$mg = f(V, r, S, \rho)$$

This expression may also be written in the form,

$$F = KV^x r^y S^z \rho^q,$$

where K is a constant having no dimension. The dimensions on the two sides of the above equation being the same, we may write

$$[MLT^{-2}] = K[L^3]^x [L]^y [MT^{-2}]^z [ML^{-3}]^q = KL^{3x+y-3q} M^{z+q} T^{-2z}$$

Equating the powers of L, M and T on both sides, we have, $3x + y - 3q = 1$, $z + q = 1$ and $z = 1$

$$\therefore q = 0 \quad \text{and} \quad x = \frac{1-y}{3}.$$

$$\therefore \quad mg = KV^{(1-y)/3} r^y S = KrSV^{\frac{1-y}{3}} r^{y-1} = KrS\left(\frac{r}{V^{\frac{1}{3}}}\right)^{y-1} = rS\phi\left(\frac{r}{V^{\frac{1}{3}}}\right) \qquad (10.22.12)$$

$\phi\left(\frac{r}{V^{1/3}}\right)$ is a non-dimensional function of the variables as the dimensions of rS and mg are equal. Using liquids of known surface tensions, the values of $\frac{mg}{rS}$ together with $\frac{r}{V^{1/3}}$ were determined. Repetition of the experiment with the same liquid but other tips gave sets of such values. A curve connecting $\frac{r}{V^{1/3}}$ and $\frac{mg}{rS}$, i.e., $\phi\left(\frac{r}{V^{1/3}}\right)$ was plotted. Precisely the same curves connecting $\frac{r}{V^{1/3}}$ and $\phi\left(\frac{r}{V^{1/3}}\right)$ were obtained for four liquids such as, water, benzene, carbontetrachloride and ethylene dibromide. These curves which were superposed on each other could be used to find the surface tension of an unknown liquid by determining the mass m of the liquid and the radius r of the tube. Thus $\frac{r}{V^{1/3}}$ was calculated and the value of $\phi\left(\frac{r}{V^{1/3}}\right)$ was determined from the graph. Thus knowing $\phi\left(\frac{r}{V^{1/3}}\right)$, S may be determined from the relation

$$S = \frac{mg}{r\phi\left(\frac{r}{V^{\frac{1}{3}}}\right)}$$

Lord Rayleigh using water drops in air showed that the value of the non-dimensional variable function was nearly constant for a large variation in r. The value of this constant was found to be 3.8.

Thus

$$m = \frac{3.8 S_{12} r \sigma_1}{g(\sigma_1 - \sigma_2)} = \frac{3.8 S_{12} r}{g} \quad \left[\begin{array}{l}\text{Since } \sigma_1 = 1 \text{ for water and } \sigma_2, \\ \text{the density of air may be neglected}\end{array}\right]$$

$$\text{or} \quad S_{12} = \frac{mg}{3.8r} \tag{10.22.13}$$

Here S_{12} represents the surface tension of water-air interface. The weight of one drop in an actual experiment is determined by collecting a known number of drops, and then taking the weight of it. The radius r of the orifice of the tube through which the drops fall, is determined carefully by means of a travelling microscope. Hence surface tension S_{12} may be determined from the above equation and from the knowledge of m and r.

Illustrative Example

In a drop weight method for the determination of surface tension between water and air, a glass tube of external diameter 2 mm is used, and 100 drops of water are collected. The mass of these drops is 2.8 gms. Find the surface tension of water in air.

We know,

$$S_{12} = \frac{mg}{3.8r}$$

Here $r = 1$ mm $= 0.1$ cm, $m =$ weight of one drop $= \frac{2.8}{100} = 0.028$ gms, $g = 980$ cm/sec^2

$$\therefore\ S_{12} = \frac{0.028 \times 980}{3.8 \times 0.1} = 72.21 \text{ dynes/cm.}$$

10.22.8 Rayleigh's jet method

The breaking up of a liquid jet from a horizontal orifice into spherical drops was utilised by Rayleigh in determining the surface tension of a liquid. On careful observation, it is found that the drops in their formative stage oscillate about their mean spherical form. In fig. 10.22.11 the two points B and C lie on one such cyclic path. The time period of vibration of a drop about its mean spherical form is given by

$$t = \pi\sqrt{\frac{r^3\sigma}{2S}}$$

where r = radius of the drop, i.e., the radius of the orifice of the horizontal tube; σ = density of the drop of liquid and S = surface tension of the liquid to be determined.

This may be deduced dimensionally in the following way.

Considering the time period to be a function of r, S, σ, we may write $t = Kr^a\sigma^bS^c$ where K is a dimension-less constant.

$$\text{or} \quad [T] = K[L]^a[ML^{-3}]^b\left[\frac{MLT^{-2}}{L}\right]^c = KL^{(a-3b)}M^{(b+c)}T^{-2c}$$

which gives $(a-3b)=0$, $(b+c)=0$ and $-2c=1$, whence $c=-\frac{1}{2}$, $b=\frac{1}{2}$ and $a=\frac{3}{2}$

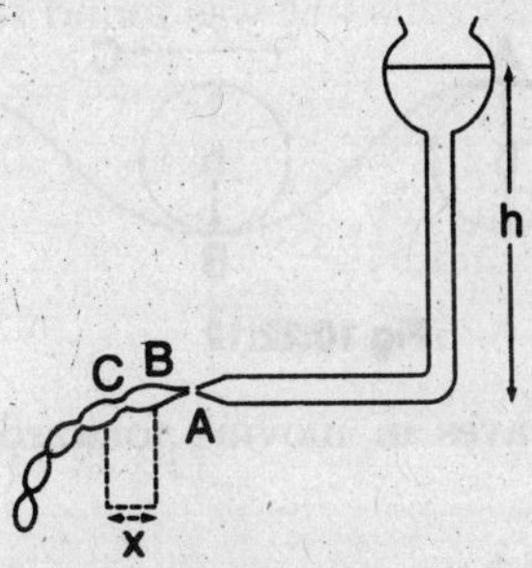

Fig 10.22.11

$$\therefore\ t = Kr^{\frac{3}{2}}\sigma^{\frac{1}{2}}S^{-\frac{1}{2}} = K\sqrt{\frac{r^3\sigma}{S}}$$

The value of the constant K was found experimentally to be $\frac{\pi}{\sqrt{2}}$

$$\therefore\ t = \pi\sqrt{\frac{r^3\sigma}{2S}}$$

Now, the jet follows a parabolic path issuing from the horizontal tube at A, and the horizontal velocity at A is given by $v = \sqrt{2gh}$, where h is the height of the liquid head above the horizontal.

Now, if x be the distance between B and C we have $x = vt$. $\therefore\ t = \frac{x}{v} = \frac{x}{\sqrt{2gh}}$.

Hence

$$\pi\sqrt{\frac{r^3\sigma}{2S}} = \frac{x}{\sqrt{2gh}}.$$

Squaring,

$$\frac{\pi^2r^3\sigma}{2S} = \frac{x^2}{2gh} \qquad \therefore\ S = \frac{\pi^2ghr^3\sigma}{x^2}$$

The distance x can be measured with the help of a travelling microscope. Measuring h and r, S can be determined. Lenard applied the oscillation of a drop to determine the surface tension of a liquid by taking instantaneous photographs of the drops.

Illustrative Example

A large drop of aniline of radius 3 cms and specific gravity 0.976 is suspended in water at 75°C and floats in equilibrium. When slightly deformed, the drop oscillates about its mean spherical form with a periodic time of 5.25 secs. Calculate the interfacial surface tension.

We know that the time period of vibration of a drop about its mean spherical form is given by

$$t = \pi\sqrt{\frac{r^3\sigma}{2S}} \quad \text{or} \quad t^2 = \pi^2.\frac{r^3\sigma}{2S} \quad \text{or} \quad S = \frac{\pi^2r^3\sigma}{2t^2}$$

Here $r = 3$ cms, $\sigma = 0.976$ gm/cm^3, $t = 5.25$ secs.

$$\therefore\ S = \frac{9.87\times(3)^3\times0.976}{2\times(5.25)^2} = 4.71\ \text{dynes/cm.}$$

10.22.9 Rayleigh's ripple method

Theory. The velocity with which waves travel over the surface of a liquid depends on both gravity and surface tension of the liquid. The relation between velocity, surface tension and gravity may be determined in the following way.

Velocity of gravity waves on a liquid

For the sake of simplicity, we consider the gravity waves formed by the drops of liquid in or near the surface describing circular path in a vertical plane, and we consider the drops to execute motion in the counter clockwise direction, and that the waves are propagated in the direction from left to right with the formation of crests and troughs as shown in the fig. 10.22.12.

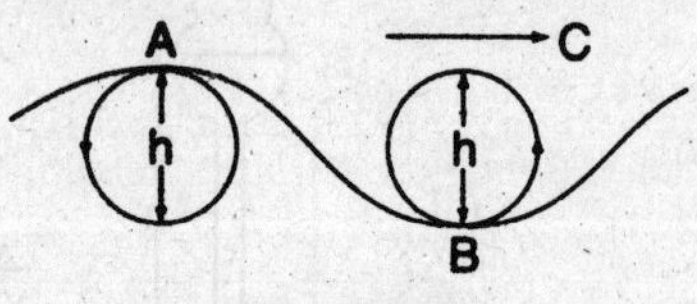

Fig 10.22.12

Let C be the velocity of the waves in the horizontal direction, and let r be the radius of the circle the drops describe in the counter clockwise direction. Let T be the period of motion of the drop; this is also the time taken by the waves in moving forward through a distance equal to the wave length λ.

Now on the crest A, let v_1 be the velocity of the particle. Then we have,

$$v_1 = C - \frac{2\pi}{T}r, \quad \text{or} \quad v_1^2 = C^2 + \frac{4\pi^2 r^2}{T^2} - \frac{4\pi Cr}{T}.$$

At the trough B, similarly, the velocity of the particle is given by

$$v_2 = C + \frac{2\pi}{T}r, \quad \text{or} \quad v_2^2 = C^2 + \frac{4\pi^2 r^2}{T^2} + \frac{4\pi Cr}{T}. \quad \therefore \quad v_2^2 - v_1^2 = \frac{8\pi Cr}{T}$$

Again from Bernoulli's theorem

$$v_2^2 - v_1^2 = 2gh = 4gr \quad [\because h = 2r]$$

$$\therefore \quad \frac{8\pi Cr}{T} = 4gr \quad \text{or} \quad \frac{2\pi C}{\lambda/C} = g \quad [\because T = \lambda/C]$$

$$\therefore \quad \frac{2\pi C^2}{\lambda} = g. \quad \therefore C^2 = \frac{g\lambda}{2\pi} \quad \text{or} \quad C = \sqrt{\frac{g\lambda}{2\pi}}.$$

Effect of surface tension on the velocity of gravity waves

We consider now a section of the liquid at a point P on the crest whose height is y and breadth is dx and thickness at right angles to the plane of the figure is unity (Fig. 10.22.13). We have the weight of this column given by $y\rho dxg$, where ρ is the density of the liquid. Had it been the case that the gravity force is the only force acting here, this column would have been acted on by the above force, but the existence of surface tension brings in additional force as discussed below.

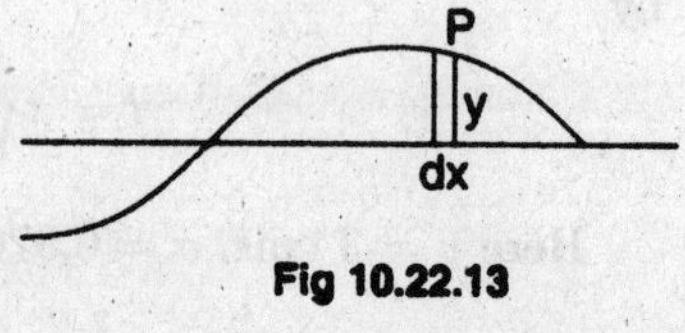

Fig 10.22.13

We know that when the surface of a liquid is a curved one, a difference of pressure comes to play in the direction from concave side to convex one. In figure 10.22.14 the magnitude of this pressure in the direction NP is $\frac{S}{R}$, since the wave surface is a cylindrical one. Hence the magnitude of the force arising out of surface tension in the direction NP is given by

$$\frac{S}{R}\cdot ds \times 1 = \frac{S}{R}\cdot ds$$

and in the direction PN this force is $-\frac{S}{R}ds$. The vertical downward component of this force $= -\frac{S}{R}\cdot ds\cos\theta$

$$= -\frac{S}{R}\cdot ds\cdot\frac{dx}{ds} = -\frac{S}{R}dx \qquad \left[\because \cos\theta = \frac{dx}{ds}\right]$$

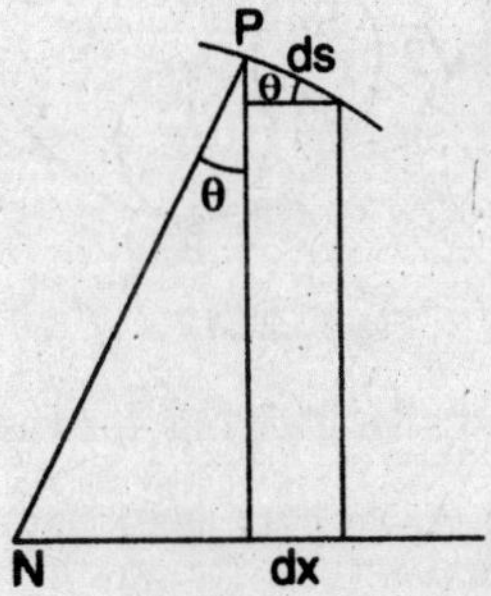

Fig 10.22.14

Hence the resultant force on this column due to gravity and surface tension is given by

$$yg\rho dx - \frac{S}{R}dx;$$

again we know, $\frac{1}{R} = \frac{d^2y}{dx^2}$, when the curvature is small as in the present case. Now, the equation to a simple harmonic wave (i.e., capillary curve) travelling over a liquid is given by

$$y = a\sin\left(\frac{2\pi x}{\lambda} + b\right)$$

where y = the ordinate of any point above the undisturbed level, λ = the wave length, a = the amplitude, x = the abscissa of the point measured from some arbitrary origin, b = some constant,

whence

$$\frac{d^2y}{dx^2} = -\frac{4\pi^2}{\lambda^2}y.$$

Thus the resultant vertical pull on the column is given by

$$yg\rho dx - \frac{S}{R}dx = yg\rho dx + \frac{4\pi^2}{\lambda^2}\cdot Sydx = y\rho dx\left(g + \frac{4\pi^2 S}{\lambda^2\rho}\right)$$

From the above expression, we see that the effect of surface tension is, as if, to change g to $\left(g + \frac{4\pi^2 S}{\lambda^2\rho}\right)$. Hence the velocity of waves, controlled by both gravity and surface tension is obtained by changing g to $\left(g + \frac{4\pi^2 S}{\lambda^2\rho}\right)$.

Hence

$$C = \sqrt{\frac{\lambda}{2\pi}\left(g + \frac{4\pi^2 S}{\lambda^2\rho}\right)} \tag{10.22.14}$$

Now $C = n\lambda$, where n = the frequency of the waves

$$\therefore\ n\lambda = \sqrt{\frac{\lambda}{2\pi}\left(g + \frac{4\pi^2 S}{\lambda^2\rho}\right)} \quad \text{or} \quad n^2\lambda^2 = \frac{\lambda}{2\pi}\left(g + \frac{4\pi^2 S}{\lambda^2\rho}\right)$$

$$\text{or} \quad \frac{2\pi S}{\lambda\rho} = n^2\lambda^2 - \frac{\lambda g}{2\pi} \quad \text{or} \quad S = \frac{n^2\lambda^3\rho}{2\pi} - \frac{\lambda^2 g\rho}{4\pi^2}. \tag{10.22.15}$$

Discussions : From the expression (10.22.14), we see that the velocity of propagation of waves is infinite when λ is zero or infinite. Again C is proportional to the square root of an expression consisting of the sum of the two terms g and $\frac{4\pi^2 S}{\lambda^2 \rho}$ whose product is constant. Therefore it follows that the expression will be minimum when the two terms are equal. Thus for minimum value of C,

$$g = \frac{4\pi^2 S}{\lambda^2 \rho} \quad \text{or} \quad \lambda^2 = \frac{4\pi^2 S}{g\rho} \quad \text{or} \quad \lambda = 2\pi\sqrt{\frac{S}{g\rho}}$$

$$\therefore C_{\min} = \sqrt{\frac{\lambda}{2\pi}\cdot 2g} = \left[\left(\frac{S}{g\rho}\right)^{1/2}\right]^{1/2}(2g)^{1/2} = \frac{S^{1/4}}{(g\rho)^{1/4}}\cdot(2g)^{1/2} = \sqrt{2}\left(\frac{Sg}{\rho}\right)^{1/4}$$

In the case of water $S = 75$ dynes/cm and $\lambda = 1.7$ cms, and we have

$$C_{\min} = \sqrt{2}\left(\frac{Sg}{\rho}\right)^{\frac{1}{4}} = \sqrt{2}(Sg)^{\frac{1}{4}} \quad [\because \rho = 1]$$

$$= \sqrt{2}\left(S\cdot\frac{4\pi^2 S}{\lambda^2}\right)^{\frac{1}{4}} = \sqrt{2}\left(\frac{2\pi S}{\lambda}\right)^{\frac{1}{2}} = \left(\frac{4\pi S}{\lambda}\right)^{\frac{1}{2}}$$

$$= \left(\frac{4\times 3.14\times 75}{1.7}\right)^{1/2} = 23 \text{ cms/sec.}$$

Hence no waves can travel over the surface of water with a velocity smaller than 23 cms per sec. For any velocity greater than this, it is possible to find a wave length λ such that waves of this length will travel with the given velocity. From the expression 10.22.14 it follows that for smaller wavelength, the surface tension term predominates and the velocity increases with the decrease of wavelength. For large wavelength the gravitational term predominates and the velocity increases with the increase of wavelength. For wavelengths below that coresponding to the minimum velocity, the disturbances are called *ripples*, and for wavelengths above that corresponding to the minimum velocity the disturbances are called waves. A wave is, therefore, controlled mainly by gravity, whereas a ripple is controlled mainly by surface tension.

Experiment

(a) Rayleigh's method : Lord Rayleigh first applied the measurement of ripples to the determination of surface tension of a liquid. Ripples are generated by the motion of a glass plate attached to the lower prong of an electrically driven tuning fork dipping into the liquid under test. To make the ripples visible, light reflected from the surface of the liquid, when ripples are formed, is brought to a focus, so that the ripples may be observed. In order to see the ripples distinctly, intermittent illumination whose period should be the same as that of the ripples, is employed so that the ripples will appear to the eye stationary in position. The illumination can be made intermittent by keeping in front of the source of light a small tin-plate which is attached rigidly to the prong of a second tuning fork having the same frequency as that of the exciting fork, and the second fork is also driven electromagnetically.

The ripples are thus made distinct. The number of ripples between two points at a known distance apart is counted, and the wavelength λ is determined therefrom. If T be the time period of vibration of the exciting tuning fork, we have

$$CT = \lambda \quad \text{or} \quad C = \frac{\lambda}{T}.$$

Thus the velocity C is also determined.

Hence applying the formula

$$S = \frac{n^2\lambda^3\rho}{2\pi} - \frac{\lambda^2 g\rho}{4\pi^2} = \frac{\lambda^3\rho}{2\pi T^2} - \frac{\lambda^2 g\rho}{4\pi^2} \tag{10.22.16}$$

surface tension S can be determined. In the present experiment, the second term is small compared with the first term, and may be neglected for approximate calculation of surface tension.

(b) Modern methods : (i) An improvement of the method introduced by Lord Rayleigh has been made by Ghosh, Banerjee and Dutta. This method having several advantages is given below.

The liquid under test is taken in a shallow rectangular porcelain trough of dimensions $10'' \times 6'' \times 1\frac{1}{2}''$. An electrically maintained tuning fork of frequency 100 vibrations per second is mounted above the liquid with its prongs horizontal — one above the other — so that they vibrate in a vertical plane. To the lower prong of the tuning fork is attached a dipper which is a blade, 3″ long of polished silver or aluminium (Fig. 10.22.15). The plane of the dipper is vertical, but perpendicular to the plane of vibration of the prongs. The dipper by its up and down vibrations, in and out of the liquid, due to vibration of the prong, exites the liquid surface giving rise to the formation of the ripples. The two trains of ripples thus formed being reflected from the walls of the trough produce stationary ripples.

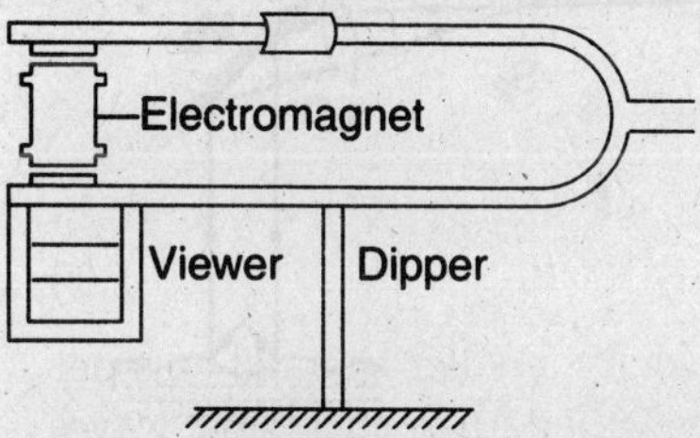

Fig 10.22.15

Also attached to the same prong as the dipper but nearer the end, is a viewer, which is a rectangular frame work of metal having two horizontal steel wires along its length — the plane of the viewer being vertical and parallel to the plane of vibration of the prongs. The viewer is meant for the ripples to be seen and photographed, and the wavelength to be measured therefrom.

For the measurement of λ the following arrangement is made. A parallel beam of light is allowed to proceed in a plane perpendicular to the plane of vibration of the prongs, and to be reflected by the ripple-curved surface of the liquid. The reflected rays pass obliquely upward through the wires of the viewer. An observer receiving this light will see the shadwows of the two wires wavy in nature instead of straight. To enable the number of ripples formed within a measured length to be counted, a metal plate with two fine notches at a measured distance apart is fixed in the plane of the wires of the viewer. The two fine notches along with the shadow ripples is photographed from which the exact wavelength can be calculated with the help of a micrometer scale. We assume here that the wavelength of the shadow ripples is equal to the wavelength of the real ripples.

The frequency n of the tuning fork is accurately determined by using a chronographic recorder which is tuned in unison with the fork — the recorder receiving signal from a standard clock at intervals of one second. Thus knowing λ and n, the surface tension of the liquid of density ρ is calculated from

$$S = \frac{\lambda^3 n^2\rho}{2\pi} - \frac{\lambda^2 g\rho}{4\pi^2}$$

(ii) We shall here consider an outline of the experimental arrangement devised by R.C. Brown.

In the figure 10.22.16 L_1 and L_2 denote two loudspeakers, and R_1 and R_2 two reeds tuned to the supply frequency and actuated by the loudspeakers. S_1 and S_2 are two slits. D is a dipper one end of which is fixed with R_2 and the other touches the liquid surface. Due to A.C. supply through the loudspeaker, the reed R_2 and hence the dipper D vibrates up and down producing small waves or ripples on the surface of the liquid under test. The other reed R_1 vibrates to and fro with the concave mirror M_1 attached to it. Light from the source A is first focussed by the lens L on the slit S_1 and then on the slit S_2 by the concave mirror M_1, S_2 being in the focal plane of the lens L'. Thus a beam of parallel light falls on the plane mirror M_2 from which the reflected parallel beam of light is incident on the surface of the liquid. As a result of the motion of R_1, light from the source A will pass through the slits intermittently. Thus the intermittent reflected light from the surface of the liquid under test will be viewed through the telescope T.

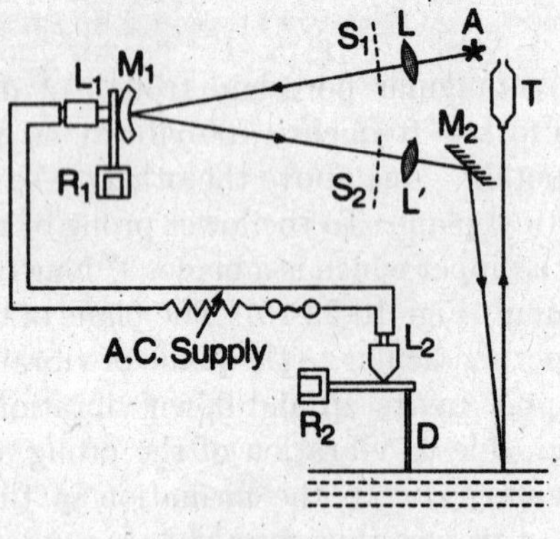

Fig 10.22.16

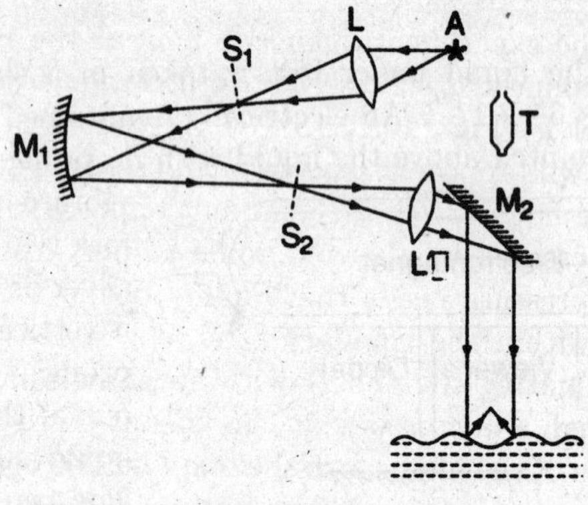

Fig 10.22.17

From the figure 10.22.17 it is clear that the reflected light from the experimental liquid converges and hence through the telescope we shall observe equispaced bright lines. The distance d between two consecutive lines is equal to $\lambda/2$ as found from the following simple consideration.

For the formation of bright lines in the undisturbed position, a trough behind will move and occupy the place of the trough ahead of it in the time interval between two intermittent flashes. But the time between two intermittent flashes is half the time period of the vibrating reed attached to the mirror M_1. Hence if C be the velocity of the wave propagation, we have,

$$\frac{CT}{2} = d.$$

But $CT = \lambda$. $\quad \therefore\ \lambda/2 = d$.

Thus measuring λ we can find S from equation 10.22.15.

Illustrative Example

A light stylus attached to the prong of a tuning fork vibrating 100 times/sec. dips slightly in a trough full of mercury and generates ripples of wavelength $\lambda = 0.278$ cm. Assuming $g = 980$ cms/sec^2, and density of mercury = 13.6 gms/cm^3, calculate the surface tension of mercury and also the wavelength of the slowest moving ripple on mercury surface. [C.U. (Hons.) 1950]

We know the expression for surface tension is

$$S = \frac{\lambda^3 n^2 \rho}{2\pi} - \frac{\lambda^2 g \rho}{4\pi^2}$$

Here $\lambda = 0.278$ cms, $\rho = 13.6$ gm/cm^3, $g = 980$ cm/sec^2, $n = 100$ vibrations/sec

$$\therefore\ S = \frac{(0.278)^3 \times (100)^2 \times 13.6}{2 \times 3.14} - \frac{(0.278)^2 \times 980 \times 13.6}{4 \times 9.86} = 438.89 \text{ dynes/cm.}$$

Again, the wavelength of the slowest moving ripple is given by

$$\lambda = 2\pi\sqrt{\frac{S}{g\rho}} = 2 \times 3.14\sqrt{\frac{438.89}{980 \times 13.6}} = 1.14 \text{ cms.}$$

(c) Ferguson's method : This method is suitable specially in the case when the experimental liquid is available only in a very small quantity.

The experimental liquid is kept at the end A of a capillary tube AB mounted horizontally and connected successively to a manometer M' and a Woulfe's bottle fitted with a dropping funnel F (Fig. 10.22.18). If the pressures at the two ends of the liquid drop are equal, the drop will appear as shown in the separate figure. Here the radii of curvature at the two ends of the liquid drop are equal but oppositely directed. The pressure in the Woulfe's bottle may be increased by allowing liquid to pass into the Woulfe's bottle through the dropping funnel. The pressure inside the Woulf's bottle is so adjusted that the meniscus of the liquid under test within the capillary tube at the end A becomes perfectly plane. This is examined by illuminating the end A with an electric lamp L and observing the light reflected therefrom through a microscope M. So long as there is curvature of the liquid meniscus at the end A, an image of the filament of the lamp will be observed through the microscope, but when curvature vanishes, i.e., the meniscus becomes plane, the image also vanishes giving uniform illumination.

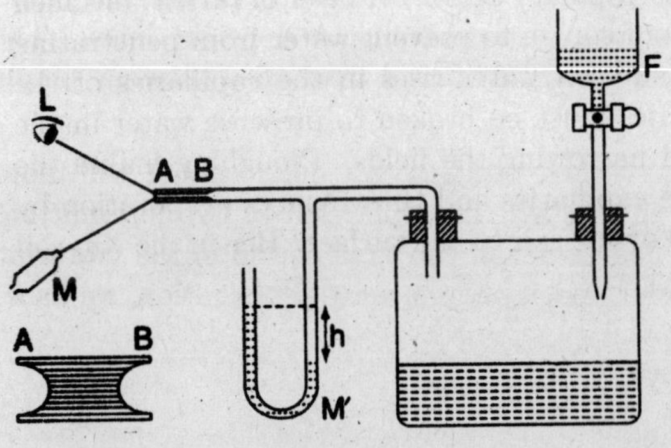

Fig 10.22.18

Under the circumstances there would be no pressure difference on the two sides of A due to surface tension. The excess pressure at the other end B of the capillary tube is given by the manometer reading and is equal to $h\rho g$ where h is the difference of levels of the liquid columns in the manometer and ρ is the density of the manometric liquid. At B the pressure at a point just outside the meniscus exceeds that at a point just inside the meniscus by $\frac{2S}{r}$, where r is the radius of curvature of the meniscus at B and this is equal to $gh\rho$

$$\therefore\ gh\rho = \frac{2S}{r} \quad \text{or} \quad S = \frac{rgh\rho}{2}$$

Here r is taken as the radius of the capillary tube.

10.23 Some practical aspects of surface tension

(a) Sharp edges of glass rods and tubes are rounded merely by heating them until they begin to melt, as surface tension then pulls the edges into a circular shape. A liquid of lower surface tension spreads more easily, and as such soap is added to spraying mixtures. Soap solution has a surface tension smaller than that of pure water. The spreading quality of paint or solder depends on its surface tension. It is for the above reason that a small quantity of oil poured over the surface of pure water, spreads over the entire surface.

Large waves are created when violent wind blows over the sea. To calm such waves, sailors in ancient times, used to pour oil over the sea. The wind draws the oil in the forward direction, leaving comparatively pure water (having greater surface tension than oily water) behind. The force due to surface tension in the backward direction becomes greater than that in the forward direction, and prevents the waves from rising very high.

(b) When camphor particles are sprinkled over a clean surface of water, they are found to dance on the surface vigorously. The reason is that the camphor particles dissolve in water and the surface tension of the solution is less than that of pure water. Due to the irregularity in the shape of the particles, the camphor dissolves more at one side than at the other. The surface tension on the former side is therefore much reduced than that on the other side. So the camphor particles are pulled towards the region of higher surface tension. Whenever the particles go over to the other side, the same thing is repeated again. Thus they appear to dance on the surface of water.

If however a small quantity of oil be poured upon the surface of water in the vicinity of the camphor particles, their movements immediately cease, since the oil film formed on the water surface reduces the surface tension below that of the camphor solution.

(c) The surface tension of water prevents its passing through the pores in umbrella or tent fabric. Touching the inside of an umbrella or a tent having spherical water drops rolling outside over the cloth, will break the surface of water, and the umbrella or the tent begins to leak. Campers are therefore advised not to touch the inside of their tent canvas during rain.

(d) Capillary phenomena are widely found in nature. Water is drawn from the wet sub-soil into the top layers through the fine pores of the soil. Sandy soil is dry because sub-soil water cannot rise far through the large inter-spaces between sand particles. The hygroscopic nature of cotton-wool, textiles, soils and concrete is due to capillary action. A layer of tarfelt, bitumen or other water-proof material is used in damp-proof course to prevent water from penetrating along the walls into residential buildings. The sub-soil water rises in the capillaries of the soil and evaporates at the surface. These capillaries must be broken to preserve water in the sub-soil. This is actually done by ploughing and harrowing the fields. Ploughing makes the soil loose, thereby increasing the diameters of the capillaries and thus reduces evaporation by preventing the rise of water from the lower layers of the soil to the surface. Hence the sub-soil moisture is retained much longer.

Exercises

1. Define surface tension, surface energy and angle of contact. Deduce a relation between surface tension and surface energy.

   [C.U. (Hons.) 1980, '82, '84, '86, '88, '96, 98, 2000, 04, 05, 07, 06 (old syll.)]

2. Describe a method of measuring angle of contact between mercury and glass.

3. Describe Ablett's method of measuring the angle of contact beteen a liquid and a solid. How would you show the variation of angles of contact when the liquid is advancing or receding over the solid?

4. (a) What is capillary action? Explain why water rises in a narrow glass tube. Deduce an expression for capillary rise h of a liquid of density ρ (θ = angle of contact). Describe the capillary tube method of measuring the surface tension of a liquid.

[C.U. (Hons.) 1982, '83, '88]

(b) If the surface tension of mercury is 520 dynes per cm, its angle of contact with glass 140°, its density 13.6 gms per cc, find the maximum length of the mercury column which can be supported in a capillary tube of radius 0.01 cm kept vertically. (Negative sign in the answer shows that there will be depression; $\cos 140° = -0.7660$) [**Ans.** −1.82 cm.]

5. (a) Show that the excess pressure acting on the curved surface of a curved membrane is given by $p = 2T\left(\frac{1}{r_1} + \frac{1}{r_2}\right)$, where r_1 and r_2 are the radii of curvature and T the surface tension of the membrane. **[C.U. (Hons.) 1981, '83, '84 (sp), '85, '87, 98, 2000]**

(b) The pressure inside a soap bubble of radius 1 cm balances 1.4 mm column of oil of specific gravity 0.80. Calculate the surface tension of the soap solution. [**Ans.** 27.44 dynes/cm.]

6. Discuss the shape of films.

7. Calculate the rise of a liquid along an inclined solid surface. How would you determine the surface tension and the angle of contact in case of a liquid in the form of a large drop?

[C.U. (Hons.) 1985]

8. Calculate the rise of a liquid along the side of a plane plate dipped vertically.

9. Discuss the condition of equilibrium of a floating needle. **[C.U. (Hons.) 1989]**

10. Discuss the rise of a liquid between two vertical parallel plates.

11. Calculate the force to pull a horizontal plate upwards from a liquid surface.

[C.U. (Hons.) 1994]

12. Obtain the relation between the vapour pressure over a curved surface and that over a flat surface. **[C.U. (Hons.) 1986, '95]**

13. Discuss the formation of clouds. **[C.U. (Hons.) 1999]**

14. Discuss Laplace's theory of capillary action.

15. Describe Jaeger's method for the determination of surface tensions at different temperatures.

16. Describe a method of determining the surface tension of molten mass of a metallic substance.

17. Describe a modern method for the determination of surface tension of a liquid.

18. Describe an outline of R.C. Brown's method for the determination of surface tension of a liquid.

19. How would you determine the surface tension of a liquid when available only in a small quantity?

20. Describe Sugden's experiment for the determination of surface tension of a liquid.

21. The diameters of the two arms of a U-tube partially filled with water are 10 mm and 1 mm. If surface tension of water is 72 dynes/cm, calculate the difference in heights of the liquid levels in the two arms. [**Ans.** 2.64 cm.] **[C.U. (Hons.) 1982]**

(See worked out examples)

22. Describe Rayleigh's jet method of determining the surface tension of a liquid.

23. Explain the effect of surface tension on the velocity of gravity waves. Describe Rayleigh's ripple-method for the determination of surface tension of a liquid.

24. One gram of mercury is placed between two plane sheets of glass which are pressed together until the mercury forms a circular disc of uniform thickness and 5 cm. radius. Calculate the force exerted on the upper plate by the mercury if its surface tension be 440 dynes/cm, its angle of contact 140° and its density 13.6 gms/cc. **[C.U. (Hons.) 1989, '96]**
(**Hints.** Apply the expression (10.15.2) of Art. 10.15 using angle of contact). [**Ans.** 66016 dynes force of repulson.]

25. A hydrometer of mass 50 gm is floating in water with 3 cm of its stem lying above the water level. The diameter of the stem is 0.4 cm. What length of the stem would be outside the water level, had there been no surface tension of water? Assume that the surface tension of water is 72 dynes/cm and the angle of contact is zero. **[C.U. (Hons.) 1990]** [**Ans.** 3.734 cm.]
(See worked out examples)

26. A cylindrical vessel of mass 500 gms and outer radius 10 cms is floating in water. It has a circular hole of radius 0.2 mm at the bottom surface. Calculate the maximum weight that can be placed in the vessel without submerging it. Surface tension of water is 73 dynes/cm.
[C.U. (Hons.) 1993, 2007 (old syll)]

Hints. At the critical condition when water just begins to trickle inside the vessel of radius R through the hole of radius r, we may write, referring to the attached fig.

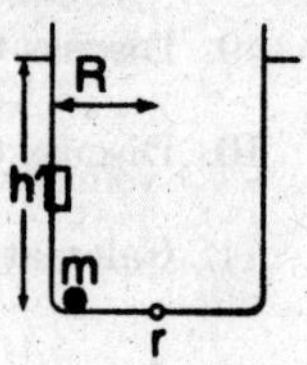

$$\pi R^2 h' \rho g + 2\pi R S = (500 + m)g \tag{1}$$

$$\text{and} \quad h' \rho g = \frac{2S}{r} \tag{2}$$

From (1) and (2) we can calculate m, the maximum weight that can be placed in the vessel without submerging it.

27. A soap bubble hanging in air from the end of an open glass tube shrinks by the action of its surface tension. Show that it is possible to prevent the bubble from collapsing by giving it a strong electric charge. Calculate the limiting diameter, if air can stand a maximum electric field of 20,000 volts/cm (surface tension = 48.5 dynes/cm). **[C.U. (Hons.) 1995]**

Answer. If the pressure due to surface tension $\frac{2S}{r}$ acting inwards is balanced by the mechanical pressure due to electrification $\frac{2\pi\sigma^2}{\varepsilon_0}$ acting normally outwards, collapsing can be prevented. Here r = limiting radius, σ = surface density of charge, and $\varepsilon_0 = 1$ $\therefore$ $2\pi\sigma^2 = \frac{2S}{r}$.

If Q be the total charge on the sphere, then the potential of the sphere is $\frac{Q}{r}$ and $\sigma = \frac{Q}{4\pi r^2}$
$\therefore\ 2\pi\left(\frac{Q}{4\pi r^2}\right)^2 = \frac{2S}{r}$ or $Q^2 = 16\pi r^3 S$

$$\therefore \text{ potential } V = \frac{Q}{r} = 4\sqrt{\pi r S} \tag{1}$$

Also we know that the potential V and field E at the surface of a charged sphere are the same as for a point charge of the same magnitude placed at the centre. Therefore, we may write,

$$V = \frac{Q}{r} \text{ and } E = \frac{Q}{r^2} \text{ whence we have } V = Er \tag{2}$$

$\therefore$ from (1) and (2)

$$Er = 4\sqrt{\pi r S} \quad \text{or} \quad E^2 r^2 = 16\pi r S \quad \text{or} \quad r = \frac{16\pi S}{E^2}$$

Here $E = 20,000$ volts/cm $= \frac{20,000}{300}$ esu $= \frac{200}{3}$ esu, $S = 48.5$ dynes/cm.

Put and get the result—$r = 0.55$ cm.

Alternative Method: We know that force per unit area due to electrification

$$F = \frac{1}{2}\epsilon_0 E^2 N/m^2 \quad [E \text{ is the electrostatic force}]$$

Now putting $\epsilon_0 = \frac{1}{4\pi}$, we have

$$\therefore \quad F = \frac{1}{8\pi} E^2 \text{ dynes/cm}^2 \text{ in the outward direction.}$$

Due to surface tension, the excess pressure is $\frac{2S}{r}$.

In the balanced condition $\frac{E^2}{8\pi} = \frac{2S}{r}$.

$$\therefore \quad r = \frac{16\pi S}{E^2}$$

Here $E = 20{,}000$ V/cm $= \frac{200}{3}$ esu unit.

$$\therefore \quad r = \frac{16 \times 3.14 \times 48.5 \times 9}{40000} = 0.548 \simeq 0.55 \text{ cm.}$$

28. Calculate the vertical force necessary to detach a horizontal flat circular plate of radius 4 cm from the surface of a liquid of surface tension 75 dynes/cm, density 1 gm/cc. Take the angle of contact between the plate and the liquid to be zero. **[C.U. (Hons.) 1994]**

[**Ans.** See worked out examples of Art. 10.16.]

29. A soap bubble of radius 1 cm and surface tension 20 dynes/cm is charged to a potential of 6000 volts. Find the increase in the radius. **[C.U. (Hons.) 1987]**

[**Answer.** If the radius of the soap bubble changes from r_1 to r_2 slowly due to charging, the process may be assumed to be isothermal. We also assume that the surface tension of the soap film remains unaffected by charging. Now the pressure inside the soap bubble in the uncharged state is $\left(P + \frac{4S}{r_1}\right)$, and let the volume of the bubble in the uncharged state be V_1 (i.e., $\frac{4}{3}\pi r_1^3$). Again in the charged state, the pressure inside the bubble is $\left(P + \frac{4S}{r_2} - \frac{V^2}{8\pi r_2^2}\right)$ and its volume V_2 (i.e., $\frac{4}{3}\pi r_2^3$), where the outward electro-static pressure due to charging is $\frac{2\pi\sigma^2}{\epsilon_0} = \frac{2\pi}{\epsilon_0}\left(\frac{Q}{4\pi r_2^2}\right)^2 = \frac{\epsilon_0 V^2}{8\pi r_2^2}$ where $V = \frac{Q}{\epsilon_0 r_2}$ is the electro-static potential due to a charge Q on a sphere of radius r_2 and σ is the charge per unit area of the sphere.

For air $\epsilon_0 = 1$

From the condition of the problem, we may therefore write

$$\left(P + \frac{4S}{r_1}\right)\frac{4}{3}\pi r_1^3 = \left(P + \frac{4S}{r_2} - \frac{V^2}{8\pi r_2^2}\right)\frac{4}{3}\pi r_2^3$$

$$\text{or} \quad Pr_1^3 + 4Sr_1^2 = Pr_2^3 + 4Sr_2^2 - \frac{V^2 r_2}{8\pi}$$

$$\text{or} \quad P(r_2^3 - r_1^3) + 4S(r_2^2 - r_1^2) - \frac{V^2 r_2}{8\pi} = 0. \tag{1}$$

When the change in radius is very small, $r_2 = r_1 + r$, where $r \ll r_1$ and we may write from (1)

$$P\left[r_1^3 + 3r_1^2 r + 3r_1 r^2 + r^3 - r_1^3\right] + 4S\left[r_1^2 + 2r_1 r + r^2 - r_1^2\right] - \frac{V^2 r_1}{8\pi} - \frac{V^2 r}{8\pi} = 0.$$

Neglecting terms containing r^3 and r^2, we have

$$(P3r_1^2 r + 8Sr_1 r) - \frac{V^2 r_1}{8\pi} - \frac{V^2 r}{8\pi} = 0$$

$$\text{or} \quad r\left[3Pr_1^2 + 8Sr_1 - \frac{V^2}{8\pi}\right] = \frac{V^2 r_1}{8\pi}$$

$$\text{or} \quad r = \frac{V^2 r_1}{8\pi\left[3Pr_1^2 + 8Sr_1 - \frac{V^2}{8\pi}\right]} \tag{2}$$

In the given problem, $P = 76 \times 13.6 \times 980 = 10^6$ dynes/cm^2, $V = 6000$ volts $= \frac{6000}{300} = 20$ esu of potential, $r_1 = 1$ cm, $S = 20$ dynes/cm. Putting these values in (2), we have

$$r = \frac{400}{8\pi\left[3\times10^6 + 160 - \frac{400}{8\pi}\right]}$$

$$= \frac{16}{3\times10^6 + 160 - 16} = \frac{16}{3\times10^6 + 0.000144\times10^6}$$

$$= \frac{16}{3.000144} \times 10^{-6} = 5.2\times10^{-6}\ \text{cm.}]$$

30. Calculate the amount of energy needed to break a drop of water 1 mm in radius into 10^6 droplets of equal size, taking surface tension of water = 72 dynes/cm. [**Ans.** 896 ergs.]
(See worked out examples)

31. Calculate the loss of energy if 1000 drops of water, each of diameter 2 mm coalesce to form one large drop. The surface tension of water is 72 dynes/cm. [**Ans.** 8139 ergs.]
(See worked out examples)

32. The vertical force necessary to detach a horizontal flat circular plate of radius 5 cm from the surface of a liquid of density 1 gm/cc is found to be 42.55 gm-wt. Assuming the angle of contact to be zero, calculate the surface tension of the liquid.
Deduce the necessary formula. [**Ans.** See worked out examples.] **[C.U. (Hons.) 1981]**

33. A round drop of water has an excess pressure of 1 atmosphere due to its surface tension. If the surface tension of water is 72 dynes/cm, calculate the radius of the drop. **[C.U. (Hons.) 1983]**
[**Ans.** See worked out examples.]

34. (a) Explain briefly the phenomenon of surface tension in terms of molecular interaction. **[C.U. (Hons.) 1983, '86, 90, 92, 94, 98, 03, 05]**
(b) A vessel has a small circular hole (radius 0.1 mm) at the bottom. To what height in the vessel can water be filled without any leakage through the bottom? Surface tension of water is 72 dynes/cm. **[C.U. (Hons.) 1983, cf.'84 (spl),'88, cf. 98]**
[**Ans.** See worked out examples.]

35. (a) Explain why very minute liquid drops evaporate as soon as they are formed.
(b) The radius of a soap bubble is slowly increased from 2 cm to 4 cm. Calculate the mechanical work done (surface tension of soap solution = 30 dynes per cm). Suppose that the process (of increasing the radius) is done quickly. Will the mechanical work required be the same? Explain. **[C.U. (Hons.) 1984]**
[**Ans. Ist part :** (See worked out examples)
Second part : (See worked out examples)

36. Two soap bubbles of radii r_1 and r_2 coalesce to form a bubble of radius r_3. Show that the surface tension of soap solution is given by

$$S = \frac{P\left(r_3^3 - r_1^3 - r_2^3\right)}{4\left(r_1^2 + r_2^2 - r_3^2\right)}$$

where P is the external pressure. **[C.U. (Hons.) 1984 (sp), '91, 2000]**

37. In between two parallel plates separated by a small distance d, certain amount of liquid is spread over a circular area of radius $R (R \gg d)$. Show that, if the angle of contact of the liquid is less than 90°, there will be attraction between the two plates. (The expression for pressure difference across a curved surface can be assumed). **[C.U. (Hons.) 1985, '96]**
(See Art. 10.15)

38. Obtain an expression for the height of rise of a liquid in a capillary tube. Show that one half of the potential energy of the liquid column is lost when the liquid rises in the capillary tube and account for this loss of energy. **[C.U. (Hons.) 1987]**

(See Arts. 10.7.3 and 10.7.4)

39. (a) Explain surface tension on the basis of molecular theory. Find the relation between the surface tension and the total surface energy.

[C.U. (Hons.) 1990, '92, '94, '98, 03, 05 (old syll)]

(b) Discuss with mathematical theory, how the surface tension of a liquid can be measured by the capillary rise method. **[C.U. (Hons.) 1990]**

40. (a) Obtain an expression for the pressure difference existing on the two sides of a curved liquid surface. **[C.U. (Hons.) 1991, '94, '98, 03, 06, 07]**

(b) A clean glass plate is introduced vertically into a liquid for which the angle of contact is acute. Find the equation of the capillary curve formed by the liquid.

[C.U. (Hons.) 1991, 2007 (old syll)]

(See Art. 10.12.1)

41. A liquid of density ρ rises to a height h in a uniform capillary tube of radius r. Assuming that the angle of contact θ is acute, find the formula for surface tension. If the radius of the capillary tube varies slightly with height, how will the above formula be modified?

(For answer to the second part, see Art. 10.8.1) **[C.U. (Hons.) 1992, '99]**

42. Discuss the influence of the curvature of a liquid surface on the saturation vapour pressure and explain the function of small nuclei in the formation of clouds. **[C.U. (Hons.) 1995, '99]**

43. (a) Consider a sessile drop (fairly large drop of liquid) resting on a plane glass plate. Discuss qualitatively the effect of surface tension and gravity on the shape of the drop.

(b) The capillary ascent (h) of a liquid in a narrow tube of length l, ($l > h$), is approximately given by the relation $T = \frac{1}{2}\frac{rh\rho g}{\cos\theta}$, where we ignore volume of liquid in the meniscus. Notations have usual significance. Imagine now that the length l of the tube be gradually decreased from $l > h$ to $l \to 0$. Discuss how the shape of the meniscus will change. Find the value of θ when $l \to 0$. **[C.U. (Hons.) 1997]**

[**Answer to (b)** — See Art 10.7.2 and add the following :

For a tube of length $l > h$, we have seen $r = R\cos\theta$ and $T = \frac{1}{2}\frac{rh\rho g}{\cos\theta}$ can be written as $Rh = \frac{2T}{\rho g}$ = const. i.e., the liquid just spreads over the walls of tube at the top and its meniscus acquires a new radius of curvature such that Rh = constant. So if the length of the tube l is gradually decreased from $l > h$ to $l \to 0$ by several steps, then at each step we will see that both θ and R go on increasing with the above condition Rh = const. and ultimately when $l \to 0$, h also tends to zero, i.e., $R \to \infty$, i.e., the meniscus becomes plane and not concave. The top of the tube is in the same line as the liquid level; then θ assumes the value given by $\cos\theta \to 0$ or $\theta \to 90°$.]

44. (a) How does surface tension of a liquid normally vary with temperature?

[C.U. (Hons.) 2006 (old syll) 2007]

(b) An empty vessel with a small hole (diameter 0.1 mm) at the bottom is gradually immersed in a liquid of surface tension 42 dynes/cm. Calculate the depth at which the liquid will start penetrating into the vessel. (Density of the liquid = 0.975 gm/cc).

[C.U. (Hons.) 1998, 2003, 05 (old syll), 06 (old syll), cf. 07]

[**Ans.** See worked out problems.]

45. Find out the minimum vapour pressure inside a bubble of radius r inside a liquid for boiling to occur. The external pressure is P and the surface tension of the liquid is S.

(For answer, see Art. 10.19) **[C.U. (Hons.) 1999]**

46. A cylindrical film with closed ends is formed between two wire rings of radius 2 cms and distance apart 5 cms. Find the volume of air enclosed by the film.

[**Ans.** The volume of air inside the cylindrical film is $V = V_1 + 2V_2$, where V_1 is the volume of air in the cylinder with flat ends and V_2 is the volume of air enclosed between one spherical end and one flat end as shown in the attached figure 1.

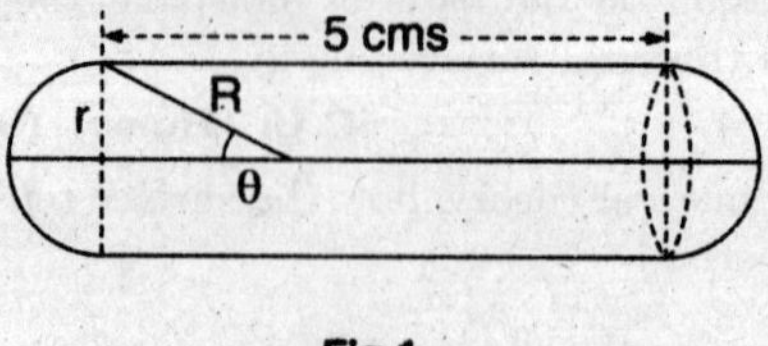

Fig 1

Now $V_1 = \pi r^2 l = 3.14 \times 2^2 \times 5 = 62.80$ cc.

For V_2, we proceed as follows:

Since the pressure of air is same everywhere within the film, we have $\frac{4S}{R} = \frac{2S}{r}$, R being the radius of curvature of the spherical surface at the end

$$\therefore\ R = 2r = 4 \text{ cms}$$

and from the figure $\sin\theta = \frac{r}{R} = \frac{1}{2}$ $\quad\therefore\ \theta = 30°$.

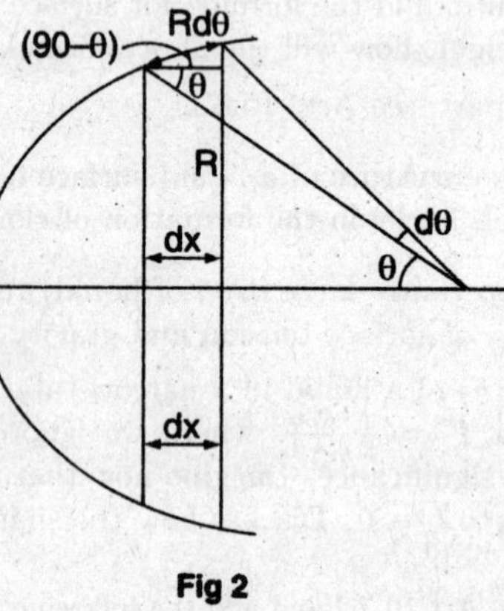

Fig 2

To calculate V_2, we divide the spherical portion (of any one end of the cylinder) into a number of co-axial circular discs, and calculate the volume of one such disc. Then the sum of the volumes of all such discs will give V_2. Taking one such disc at a position θ as shown in the figure 2, its volume is given by

$$dV_2 = \pi R^2 \sin^2\theta dx = \pi R^2 \sin^2\theta \cdot R\sin\theta d\theta$$

$$\therefore\ dV_2 = \pi R^3 \sin^3\theta d\theta \qquad \left[\begin{array}{rl} \because \cos(90-\theta) = & \frac{dx}{Rd\theta} \\ \therefore\quad dx = & R\sin\theta d\theta \end{array}\right]$$

$$= \pi R^3 \frac{1}{4}(3\sin\theta - \sin 3\theta)d\theta$$

$$\therefore\ V_2 = \frac{\pi R^3}{4}\left[3\int_0^\theta \sin\theta d\theta - \int_0^\theta \sin 3\theta d\theta\right]$$

$$= \frac{\pi R^3}{4}\left[3\left(-\cos\theta\right)_0^{30°} - \frac{1}{3}\left(-\cos 3\theta\right)_0^{30°}\right]$$

$$= \frac{\pi R^3}{4}\left[3\left(\cos\theta\right)_{30°}^{0°} - \frac{1}{3}\left(\cos 3\theta\right)_{30°}^{0°}\right] = \frac{\pi R^3}{4} \times 0.07$$

$$= 3.14 \times 16 \times 0.07 = 3.52 \text{ cc.}$$

$$\therefore\ 2V_2 = 2 \times 3.52 = 7.04 \text{ cc}$$

Hence the total volume of air enclosed by the film is $V = V_1 + 2V_2 = 62.80 + 7.04 = 69.84$ cc.]

47. A loop of cotton thread is placed on a soap film and the film is broken inside the loop. If the thread forms a circle of radius $r = 2$ cms, find the tension in the thread (surface tension S of soap film is 25 dynes/cm.)

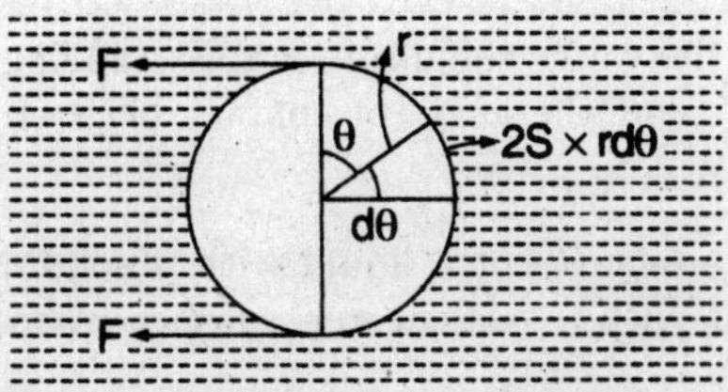

[Hints for Answer. Considering the equilibrium of the right hand semi-circular portion of the loop, we see that the forces acting towards left are the tensions F at the two ends of the vertical diameter, and are balanced by the component of the force acting towards right on the segment $rd\theta$ of the semi-circle (perpendicular to $rd\theta$) given by $2S \times rd\theta$ ($\because$ the film has two free surfaces). So for equilibrium of the right hand semi-circle, we should have, with reference to the attached figure

$$2F = \int_0^{\pi} 2Srd\theta \sin\theta = 2rS\int_0^{\pi} \sin\theta d\theta = 2rS[-\cos\theta]_0^{\pi} = 4rS$$

$\therefore$ the tension in the loop is $F = 2rS = 2 \times 2 \times 25 = 100$ dynes.]

48. The sten of a common hydrometer is a circular cylinder of diameter 2 mm. It floats, with its stem wetted in alcohol whose specific gravity is 0.796 and surface tension 25.5 dynes/cm. Calculate how much deeper it floats than if alcohol had zero surface tension.

(Cambridge Higher School Certificate)

(See worked out examples) **[Ans.** 0.653 cm.]

49. A large soap bubble of radius R contains inside it a smaller soap bubble of radius r. If the smaller bubble now bursts isothermally with no leakage of air from the system as a whole, so that a new bubble of radius a is formed, show that the radii of the three bubbles are connected by the relation $P(a^3 - R^3) + 4T(a^2 - R^2 - r^2) = 0$, where P = atmospheric pressure and T = surface tension of soap solution.

(See worked out examples).

50. Ripples are produced on the surface of a liquid with the help of a tuning fork of frequency 60 and the wave length observed is 0.5 cm. Calculate the surface tension of the liquid. (Density of the liquid is 1.0 gm/cm^3 and $g = 980$ cm/sec^2) **[Ans.** 65.8 dynes/cm]

[Apply the expression (10.22.15)]

51. A liquid of specific gravity 1.5 is observed to rise 3.0 cm in a capillary tube of diameter 0.50 mm, and the liquid wets the surface of the tube calculate the excess pressure inside a spherical bubble of 1.0 cm diameter blown from the same liquid. **[Ans.** 440 dynes/cm^2]

52. (a) Explain the terms angle of contact and wetting. How does the angle of contact depend on adhesive and cohesive forces when a liquid is in contact with a solid? See Art. 10.1 and 10.6. **[C.U. (Hons.) 01, 05 (old syll), 07 (old syll)]**

(b) Find out the expression for the angle of contact when a liquid, solid and a gas meet at a point. See Art. 10.3 **[C.U. (Hons.) 2001, 05 (old syll)]**

(c) If the angle of contact of glass and a liquid of specific gravity 0.5 is 60° and the liquid rises to a height of 2 cm in a glass capillary tube of radius 0.05 cm, find out the surface tension of the liquid. (See worked out examples) **[C.U. (Hons.) 2001]**

(d) If a vessel has a hole of radius r at its bottom, show that the liquid inside will come out of the vessel if its depth exceeds $h = \frac{2\gamma}{r\rho g}$, where γ is the surface tension of the liquid, ρ its density and g the acceleration due to gravity. (See worked out examples)

[C.U. (Hons.) 2001, 06 (old syll), 05 (old syll) 2007]

53. Find out the height of the water column in a glass capillary tube of radius 1 mm. The surface tension is 73 dynes/cm and the angle of contact between water and glass is 0°. Deduce the formula you use. **[C.U. (Hons.) 1999]**

54. (a) Define angle of contact (θ) of a liquid with a solid surface. Draw sections of the liquid surfaces in contact with a vertical solid surface when (i) $\theta < \frac{\pi}{2}$ and (ii) $\theta > \frac{\pi}{2}$ (See Art. 10.2) **[C.U. (Hons.) 2002]**

(b) A sessile drop $\theta > \frac{\pi}{2}$ rests on a horizontal surface, which the liquid doe not wet. Draw a vertical section of the drop. In this plane set up the coordinate axes OX and OY with O as the highest point on the drop and OY vertically downward.

With the above geometry, set up the differential equation giving the shape of the drop. (Ignore density of the gas surrounding the drop.) For answer see Art. 10.12.2 and the worked out problem on the Art.

(c) Calculate the surface energy of a spherical liquid drop of radius a and surface tension T. What happens to the surface energy when the drop evaporates? **[C.U. (Hons.) 2002]**

Ans. The surface area of the liquid drop of radius a is $4\pi a^2$. and the work done in blowing i.e. the surface energy (mechanical part of the work done) is $2T \times 4\pi a^2 = 8\pi T a^2$ where T = surface tension of the liquid and the factor 2 comes because there are two free surfaces of the bubble. When the drop evaporates, its surface energy decreases i.e. evaporation takes place at the cost of decrease of surface energy i.e. diminution in size of the drop takes place (necessary latent heat coming from the surrounding atmosphere)

(d) An isolated tiny spherical drop evaporates at a constant temperature. Calculate the radius of the drop from the following data: density of the liquid is 0.5 gm/cc, latent heat of evaporation is 50 cal/gm and surface tension is 50 dynes/cm.

[**Ans.** See worked out examples on pages 642 and 643. **[C.U. (Hons.) 2002]**

55. (a) Discuss briefly the molecular theory of surface tension. Establish also that surface tension can be defined as the potential energy per unit area of surface under the condition of constant temperature. **[C.U. (Hons.) 2003, 05 (old syll)]**

(b) Calculate the work done when a spherical water drop breaks down into 10^6 spherical droplets of equal size. Given that the surface tension of water is 72 dynes/cm. (See worked out examples) **[C.U. (Hons.) 2003]**

(c) An empty vessel with a small hole at the bottom (diameter 0.1 mm) is gradually immersed in water. Calculate the depth at which the vessel has to be immersed, so that water starts penetrating the vessel. (Surface tension of water = 72 dynes/cm) **[C.U. (Hons.) 2003]**

[**Ans.** See worked out examples.]

56. (a) Why is very clean air, free from any dust particle, not favourable to the formation of fog? See Art. 10.18.1.

(b) Two glass discs each of radius R and mass M. are wetted with water and put together, so that the thickness of the water layer between them it t. The system is placed horizontally on a table and assume the watting to be complete. Calculate the normal force required to be applied on the upper disc to separate them out. What is the lower limit of the mass M to separate them out in such a way? (Assume the acceleration due to gravity be g and mass of water layer to be negligible and surface tension of water be S).

[C.U. (Hons.) 2007]

[**Ans.** Applying (10.15.1) we have the upper plate (or radius R) which is wetted will be urged towards the lower one with a normal force given by,

$$F = \frac{2S}{t}\pi R^2 = \frac{2\pi R^2 S}{t}$$

The lower limit of mass M is such that $mg \leq F$

$$\text{i.e.} \quad Mg \leq \frac{2\pi R^2 S}{t} \quad \text{or} \quad M \leq \frac{2\pi R^2 S}{tg}$$

57. (a) Why is it preferable to use hot water to cold water for washing clothes?

[**Hints.** Surface tension of soap solution prepared with hot water is much less than that of the solution prepared with the cold water—which helps remove the dust and other particles from the pores of clothes.]

(b) What do you mean by saturated vapour pressure at a particular temperature?

(c) A spherical soap bubble (surface tension = 26 dynes/cm) is slowly enlarged from a radius of 1 cm to a radius of 10 cm. Calculate the work done (See worked out problems).

[C.U. (Hons.) 2008]

Answer to part (b): When a liquid evaporates, its molecules leave the free surface of the liquid and form a vapour which exerts a pressure upon the surface. This pressure is called *vapour pressure* and depends on the number of molecules per unit volume in the space above the surface. Along with evaporation, some vapour molecules return back to the liquid surface. This phenomenon is called *condensation.* The vapour pressure attains a *maximum* steady value when the number of molecules returning to the surface per second is equal to the number leaving it per second. The vapour pressure is then called the *saturation vapour pressure* at that temperature.

N.B. The saturation vapour pressure is connected with curvature of the liquid surface. A molecule leaves the surface when it possesses sufficient KE to overcome the attraction of its neighbouring molecules. Let us draw the sphere of molecular attraction round a molecule in a *convex* surface and also round a molecule in a plane surface. We see that the number of molecules hindering evaporation is less with the convex surface than with the plane surface. Consequently the number of molecules leaving a convex surface per second is greater than that if the surface were plane. Hence saturation vapour pressure over a convex surface must be greater than that over a plane surface. By similar reasoning, the saturation vapour pressure over a concave surface should be less than that over a plane surface.

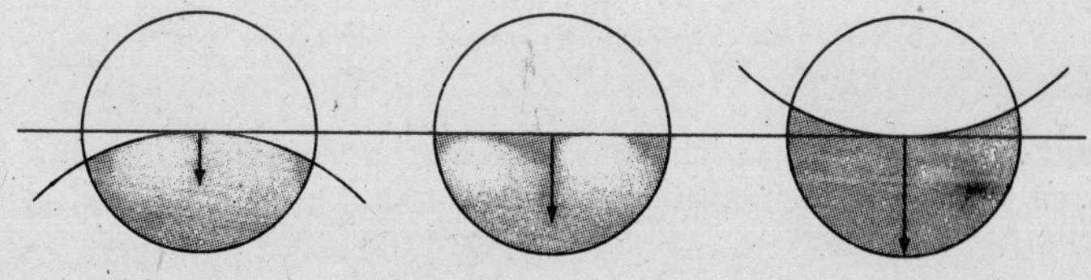

58. Calculate the pressure inside a small air bubble of radius 0.01 mm situated at a depth of 20 cm below the free surface of a liquid of density 0.8 gm/cm^3 and surface tension 75 dynes/cm, taking the atmospheric pressure to be 76 cm of mercury.

Ans. Pressure outside the bubble is atmospheric plus that due to 20 cm long column of liquid of density 0.8 gm/cm^3 i.e. $(76 \times 13.6 \times 980) + 20 \times 0.8 \times 980 = 1.03 \times 10^6$ dynes/cm^2. Hence the pressure inside the bubble $= 1.03 \times 10^6 + \frac{2S}{r}$ (excess pressure due to surface tension S of a liquid in which the bubble is formed) $= 1.03 \times 10^6 + \frac{2 \times 75}{0.001} = 1.03 \times 10^6 + 0.15 \times 10^6 = 1.18 \times 10^6$ dynes/cm^2.

59. For a certain large sessile drop of mercury on glass, it was found that the depth of the longest diameter below the highest point in the drop is $\sqrt{3}/2$ times the maximum height of the drop. Show that the angle of contact for mercury on glass is $\pi - \cos^{-1}(1/3)$.

Ans. Using (10.12.8) and (10.12.9) we have

$$\frac{g\sigma h^2}{2} = S(1 - \cos\theta_A) \quad \text{and} \quad \frac{g\sigma h'^2}{2} = S$$

Dividing the two expressions and rearranging, we have

$$\cos\theta_A = 1 - \frac{h^2}{h'^2}$$

Here

$$h' = \frac{\sqrt{3}}{2}h \quad \text{or} \quad \frac{h^2}{h'^2} = \frac{4}{3} \quad \therefore \quad \cos\theta_A = 1 - \frac{h^2}{h'^2} = 1 - \frac{4}{3} = -\frac{1}{3}$$

$$\text{or} \quad \cos(\pi - \theta_A) = 1/3 \quad \text{or} \quad \pi - \theta_A = \cos^{-1}(1/3) \quad \text{or} \quad \theta_A = \pi - \cos^{-1}(1/3)$$

Chapter 11

Viscosity

11.1 Introduction

When a liquid flows slowly and steadily over a fixed horizontal surface, i.e. when its flow is stream line, its layer in contact with the fixed surface is stationary and the velocity of each layer increases with its distance from the fixed surface as shown in Fig. 11.1.1.

We consider any particular layer AB of the liquid. The layer immediately below it moves slower than it, whereas the layer just above it moves faster than it. The difference in velocities between the adjacent layers will give rise to internal friction as a result of which the faster layer tends to accelerate the slower one and vice-versa. This internal friction by virtue of which a liquid opposes the relative motion between its different layers is known as viscosity of the liquid.

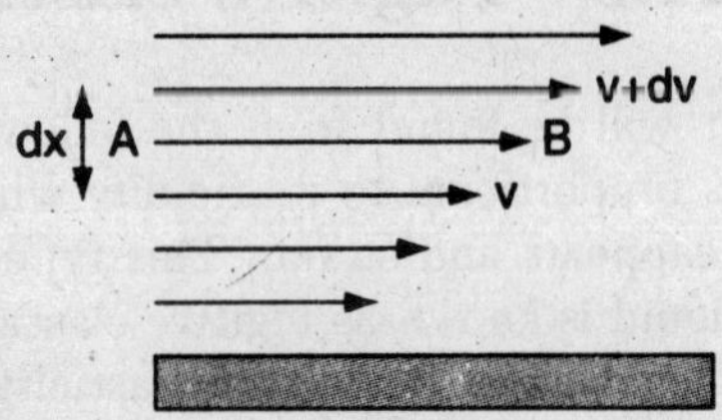

Fig. 11.1.1

When a liquid is at rest, we do not observe any shape elasticity or rigidity in it, but when the liquid is in orderly motion (as for example, stream-line motion), it has been found that there comes into play a tangential stress between any two layers of the liquid which are moving relative to each other. The appearance of this tangential stress, which retards the fastly moving layer and accelerates the slowly moving layer is analogous to tangential stress, that comes into play during shearing strain.

11.2 Coefficient of viscosity

If f represents the tangential stress between two layers of a liquid which are moving relative to each other, then according to Newton's law of viscous resistance,

$$f \propto \text{the velocity gradient} \frac{dv}{dx},$$

where dv is the change of velocity between two layers separated by a distance dx.

$$\therefore \quad f = \eta \frac{dv}{dx}$$

η is the constant of proportionality known as the coefficient of viscosity of the liquid. If $\frac{dv}{dx} = 1$, $f = \eta$. Thus the coefficient of viscosity of a liquid is the tangential stress per unit velocity gradient.

11.2.1 Dimension of η

$$\eta = \frac{f}{\frac{dv}{dx}} = \frac{F/A}{\frac{dv}{dx}} \quad \text{[}f = \text{force per unit area } F/A\text{, where } F \text{ is the tangential force and } A \text{ the area over which it acts.]}$$

$$\therefore \ [\eta] = \frac{[MLT^{-2}]/[L^2]}{[LT^{-1}]/[L]} = [ML^{-1}T^{-1}].$$

The C.G.S. unit of the coefficient of viscosity is known as *Poise* after the name of Poiseuille. A coefficient of viscosity of 1 Poise means that a force of 1 dyne is required to maintain a velocity difference of 1 cm./sec. between two layers each of area 1 sq. cm. and 1 cm. apart.

11.3 Fugitive elasticity

It will be found from the following consideration that the coefficient of viscosity of a liquid is proportional to its rigidity which has its existence for short duration, and then breaks and reappears and so on. This type of short-duration intermittent elasticity that is present in a liquid is known as fugitive elasticity. According to Maxwell, the viscosity in a liquid is due to the existence of fugitive elasticity in it.

From consideration of shape elasticity, tangential stress, $f = n\theta$, where n is the modulus of rigidity and θ, the angle of shear.

$$\text{or } f = n \tan\theta = n\frac{dy}{dx} \quad (\because \theta \text{ is small})$$

Again,

$$f = \eta\frac{dv}{dx} = \eta\frac{d}{dx}\left(\frac{dy}{dt}\right) = \eta\frac{d}{dt}\left(\frac{dy}{dx}\right) = \eta\frac{d\theta}{dt} \quad \left(\because \theta = \tan\theta = \frac{dy}{dx}\right)$$

$\frac{d\theta}{dt}$ represents the rate at which the shear breaks, and let it be proportional to the angle of shear. We may, therefore, write $\frac{d\theta}{dt} = \lambda\theta$, where λ is a constant.

We have therefore

$$f = \eta\lambda\theta = n\theta \text{ or } \eta = \frac{n}{\lambda} \quad \therefore \ \eta \propto n.$$

Thus we see that the coefficient of viscosity is proportional to the rigidity. The quantity $\frac{1}{\lambda}$ is called the time of relaxation of the medium which measures the time taken by the shear to disappear, if its formation is discontinued.

The above consideration leads us to conclude that the appearance of viscous force during orderly motion of a liquid is due to the existence of the so-called intermittent shear elasticity called "fugitive elasticity", as is evidenced during a "drakes and ducks" play.

11.4 Streamline and Turbulent motion—Critical velocity

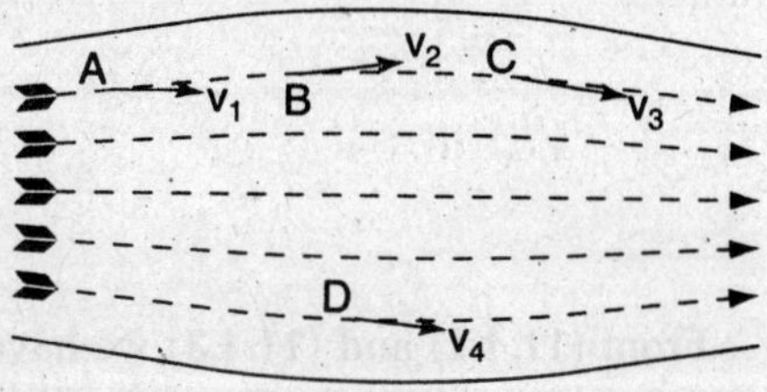

Fig. 11.4.1

A liquid which flows such that its velocity at any point is always the same in magnitude and direction is said to have a *steady* or *streamline flow*. We consider the motion of a liquid in a pipe as shown in Fig. 11.4.1. Particles at the points A, B, C etc. have velocities v_1, v_2, v_3 etc. respectively at a given time. If, however, with the passage of time, the velocity of whatever particle happens to be at A is still v_1, that at B is still v_2 and so on, the liquid will be said to have a *streamline flow*. The path followed by particles in such a flow is called a *streamline*. No two streamlines can cross each other. Streamlines have the further property that the tangent to a line at any point gives the direction of flow of the fluid at that point. So there is similarity between streamlines and the lines of force in a magnetic or electric field. When the streamlines crowd together, the velocity of flow increases. Cases of streamline flow are rather limited; they occur for small velocities and in narrow vessels. The flow of a liquid is more often *unsteady* or *turbulent*. In turbulent motion, velocity of the particles at a given point varies with time both in magnitude and direction in an irregular manner. The paths of the particles continually change and portions of the liquid acquire a rotatory motion forming eddies and vortices—which move down the stream.

In the case of flow of a liquid in a tube, a difference of pressure is to be maintained between its ends. The layer of liquid in contact with the wall of the tube is at rest, and the velocity of layers increases towards the axis of the tube. The stream line flow is maintained when this velocity is below a certain limit known as the *critical velocity*. When this velocity is reached, different parts of the liquid no longer travel in straight lines along the tube but follow zig-zag paths. When the velocity further increases, the orderly motion is completely lost, and becomes *turbulent*.

Osborne Reynolds showed that the critical velocity v is related to the density of the liquid ρ, its coefficient of viscosity η and the radius of the tube r, by the equation

$$v = \frac{K\eta}{\rho r}$$

where K is called Reynolds number, which is a non-dimensional variable (as shown in Art. 1.5) and its value is approximately 1000 for narrow tubes. In general, for a liquid of high viscosity, the stream-line motion is continued even when the velocity is high. On the other hand high density of the liquid as well as the wide bore of the tube helps the motion to be turbulent. The expression for critical velocity may be deduced by the method of dimensions in the following way.

Since the critical velocity v depends on η, ρ, and r, we have

$$v = K\eta^x \rho^y r^z, \text{ where } K \text{ is a dimensionless number.}$$

Putting the dimensions of the quantities involved in the expression, we have

$$[LT^{-1}] = [ML^{-1}T^{-1}]^x [ML^{-3}]^y [L]^z = [M^{x+y}] [L^{-x-3y+z}] [T^{-x}].$$

Equating the powers of $[M]$, $[L]$ and $[T]$ on both sides of the above equation we have,

$$x + y = 0 \tag{11.4.1}$$

$$-x - 3y + z = 1 \tag{11.4.2}$$

$$\text{and} \quad -x = -1$$

$$\text{or } x = 1. \tag{11.4.3}$$

From (11.4.1) and (11.4.3) we have $y = -1$. Hence from (11.4.2) we have $z = -1$

$$\therefore \quad v = K\eta\rho^{-1}r^{-1} = \frac{K\eta}{\rho r}.$$

11.5 Experimental demonstration of stream line and turbulent motion

Osborne Reynolds demonstrated the stream line and turbulent flow of a liquid with the help of a simple experiment. The experimental arrangement is shown in Fig. 11.5.1.

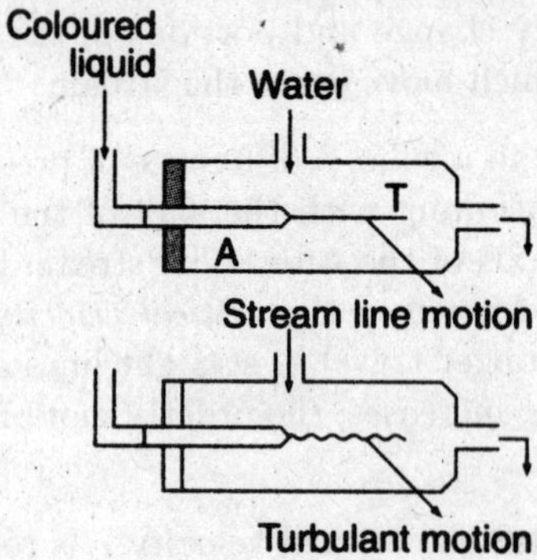

Fig. 11.5.1

In the figure T is a wide glass tube through which water or any uncoloured liquid flows. A is a narrow tube fitted along the axis of the wide tube through which coloured liquid flows. When the velocity of water through the wide tube is low, a streak of coloured liquid is seen along the axis of the tube. This represents a stream line motion. When the velocity is increased beyond its critical velocity, it is found that the coloured streak takes a wavy appearance and then spreads out forming vortex. This represents a turbulent flow.

11.6 Poiseuille's formula for flow of an incompressible liquid through a horizontal narrow tube

We consider the orderly or streamline motion of a liquid through a narrow tube. Here the stream lines are parallel to the axis of the tube and hence there is no radial flow of the liquid in the tube. The pressure will be constant over any given cross-section, but it varies from point to point along the length of the tube. It is assumed that the liquid in contact with the walls of the tube is at rest, i.e. there is no slipping between the liquid and the walls. Thus the velocity of the liquid gradually decreases radially from the axis towards the walls of the tube, wherc the velocity is considered zero.

When the steady flow of the liquid in the tube is attained, let v be the velocity at a distance r from the axis of the tube (Fig. 11.6.1). Then the velocity gradient at this point is $\frac{dv}{dr}$. Hence the viscous drag per unit area, i.e. the tangential resisting force per unit area is $\eta\frac{dv}{dr}$. This force acts over the unit area of the surface of the linear cylinder of liquid of radius r in a direction opposite to the pressure gradient. Hence the total resisting force over the surface of the liquid cylinder is $\eta\frac{dv}{dr}\cdot 2\pi rl$.

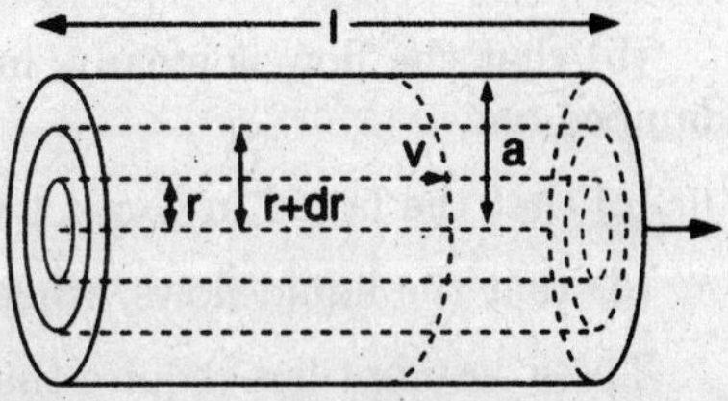

Fig. 11.6.1

Now, let us suppose that P is the difference of pressure between the ends of the tube. Hence the active force that arises out of this difference of pressure is $P\pi r^2$. This force tends to accelerate the liquid cylinder. When the steady condition is attained, this accelerating force is balanced by the retarding force arising out of viscosity.

Hence we have,

$$P\pi r^2 = -\eta\frac{dv}{dr}\cdot 2\pi rl \quad [\because \tfrac{dv}{dr} \text{ is -ve as } v \text{ decreases with increasing } r]$$

$$\text{or } r\,dr = -\frac{2\eta l}{P}\,dv.$$

Integrating the above expression between the limits $r = r$ and $r = a$, we have

$$\int_r^a r\,dr = -\frac{2\eta l}{P}\int_v^0 dv \qquad [\because v = 0 \text{ at } r = a]$$

$$\text{or } \frac{a^2 - r^2}{2} = \frac{2\eta l}{P}v, \text{ or } v = \frac{P}{4\eta l}(a^2 - r^2) \tag{11.6.1}$$

This gives the velocity at any distance r from the axis of the tube. The velocity profile or the distribution curve of the advancing liquid is thus a parabola as shown by the dotted curve in the figure.

Now let dV be the volume of liquid which flows through the cylindrical shell per second between the radii r and $r + dr$ (Fig. 11.6.1).

Then we have,

$$dV = 2\pi r\,dr\,v = \frac{\pi P}{2\eta l}(a^2 - r^2)r\,dr.$$

Therefore, the total volume that passes through the tube in unit time is

$$V = \frac{\pi P}{2\eta l}\int_0^a (a^2 - r^2)r\,dr = \frac{\pi P}{2\eta l}\left[a^2\int_0^a r\,dr - \int_0^a r^3\,dr\right] = \frac{\pi P}{2\eta l}\left[\frac{a^4}{2} - \frac{a^4}{4}\right] = \frac{\pi P a^4}{8\eta l} \tag{11.6.2}$$

Now, $P = P_1 - P_2$, where P_1 and P_2 are the pressures at the two ends of the tube.

$$\therefore\ V = \frac{\pi(P_1 - P_2)a^4}{8\eta l} \tag{11.6.3}$$

This is known as Poiseuille's equation.

11.6.1 Alternative treatment

We consider a right cylindrical horizontal tube of radius a and length l, through which an incompressible liquid is driven, without turbulence, by the application of a steady pressure difference P between the ends of the tube. It will be assumed:

(a) that the flow is everywhere parallel to the axis of the tube,

(b) that the flow is steady, initial disturbances due to acceleration from rest having been damped out,

(c) that the liquid in contact with the wall of the tube is at rest,

(d) that the liquid flows, when it is subjected to a very small shearing force.

These assumptions require that the pressure gradient shall be constant along the tube. Let the axis of the tube be taken as the X-axis, so that the ends of the tube are defined by $x = 0$ and $x = l$, the pressures there being p_1 and p_2 where $p_1 > p_2$. Then $(p_1 - p_2) = P$. Since the pressure gradient is constant, we have $\frac{dp}{dx} = \alpha$ (say) which on integration gives $p = \alpha x + \beta, \beta$ being a constant of integration. Now at $x = 0$, $p = p_1$, so that $\beta = p_1$ and at $x = l, p = p_2$, so that $p_2 = \alpha l + p_1$.

Hence we may write

$$\frac{dp}{dx} = -\frac{(p_1 - p_2)}{l} = -\frac{P}{l} \tag{11.6.4}$$

The negative sign indicates that the pressure decreases in the direction of x increasing.

We consider the portion of the liquid in the tube bounded by two co-axial cylindrical surfaces of radii r and $r + \delta r$ and by two planes at right angles to the axis of the tube at a distance δx (Fig. 11.6.2).

The tangential force due to viscosity exerted by the liquid nearer to the axis, on the cylindrical element of radius r and length δx is $2\pi r \delta x \eta \frac{dv}{dr}$, acting in the direction to diminish v, i.e. acting in the negative direction of X axis. Similarly the tangential force exerted by the liquid outside the annular cylinder, on the cylindrical element of same length and radius $r + \delta r$ is

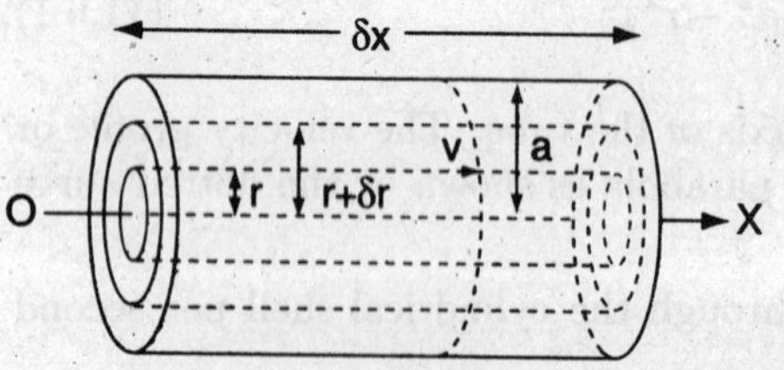

Fig. 11.6.2

$$2\pi r \delta x \eta \frac{dv}{dr} + \frac{d}{dr}\left(2\pi r \delta x \eta \frac{dv}{dr}\right) \delta r$$

acting in the direction to increase v, i.e. acting in the positive direction of X-axis.

Hence the resultant tangential force due to viscosity tending to increase v

$$= \frac{d}{dr}\left(2\pi r \delta x \eta \frac{dv}{dr}\right) \delta r.$$

Besides the tangential force due to viscosity we are to consider the pressures acting over the ends of the cylindrical element between radii r and $r + \delta r$. If $\frac{dp}{dx}$ denotes the pressure gradient, then the ends of the cylinder in question will be acted on by forces due to pressures which are given by $2\pi r\, \delta r\, p$ in the positive direction of X-axis and $\left[2\pi r\, \delta r \left(p + \frac{dp}{dx}\delta x\right)\right]$, in the negative direction of X-axis. Since the motion is steady, the liquid moves without any acceleration. Hence the force tending to increase v must be equal to that diminishing it

$$\therefore \frac{d}{dr}\left(2\pi r\delta x\eta\frac{dv}{dr}\right)\delta r = 2\pi r\,\delta r\,\frac{dp}{dx}\delta x$$

$$\text{or}\quad \eta\frac{d}{dr}\left(r\frac{dv}{dr}\right) = r\frac{dp}{dx} \quad \text{or} \quad \frac{dp}{dx} = \frac{\eta}{r}\frac{d}{dr}\left(r\frac{dv}{dr}\right) \quad \text{or} \quad -\frac{Pr}{l\eta} = \frac{d}{dr}\left(r\frac{dv}{dr}\right) \quad \text{using (11.6.4).}$$

Integrating we have

$$r\frac{dv}{dr} = -\frac{Pr^2}{2\eta l} + A, \quad \text{where } A \text{ is the constant of integration}$$

$$\text{or}\quad \frac{dv}{dr} = -\frac{Pr}{2\eta l} + \frac{A}{r}.$$

Integrating again, we have

$$v = -\frac{Pr^2}{4\eta l} + A\log r + B,$$

where B is another constant of integration. Since the velocity is not infinite at $r = 0$, i.e. along the axis of the tube, A must vanish. Hence the above expression reduces to

$$v = -\frac{Pr^2}{4\eta l} + B.$$

Now, at $r = a, v = 0$.

$$\therefore\quad B = \frac{Pa^2}{4\eta l}.$$

Hence

$$v = \frac{P}{4\eta l}(a^2 - r^2).$$

Next we proceed in a similar way as in the previous article to deduce Poiseuille's equation.

11.7 Flow of an incompressible liquid through the annular space between two concentric cylinders

We consider a thin wire of radius b placed co-axially in a narrow tube of length l and radius a.

We know, the velocity at a distance r from the axis is

$$v = -\frac{Pr^2}{4\eta l} + A\log r + B \tag{11.7.1}$$

Now at $r = b, v = 0$, which gives

$$0 = -\frac{Pb^2}{4\eta l} + A\log b + B \tag{11.7.2}$$

Again at $r = a, v = 0$, which gives

$$0 = -\frac{Pa^2}{4\eta l} + A\log a + B \tag{11.7.3}$$

From (11.7.2) and (11.7.3), we have

$$\frac{P(a^2-b^2)}{4\eta l} = A(\log a - \log b) \quad \text{or} \quad A = \frac{P(a^2-b^2)}{4\eta l(\log a - \log b)} = \frac{P(a^2-b^2)}{4\eta l \log a/b}.$$

Substituting the value of A in (11.7.3), we get

$$B = \frac{Pa^2}{4\eta l} - \frac{P(a^2-b^2)\log a}{4\eta l \log\frac{a}{b}}.$$

Substituting the values of A and B in (11.7.1), we have

$$v = -\frac{Pr^2}{4\eta l} + \frac{P(a^2-b^2)\log r}{4\eta l \log\frac{a}{b}} + \frac{Pa^2}{4\eta l} - \frac{P(a^2-b^2)\log a}{4\eta l\log\frac{a}{b}} = \frac{P}{4\eta l}\left[(a^2-r^2) + \frac{(a^2-b^2)\log\frac{r}{a}}{\log\frac{a}{b}}\right].$$

$\therefore$ the volume of the liquid which flows per second through the annular space is given by

$$V = \int_b^a 2\pi r dr v = \frac{P\pi}{2\eta l}\int_b^a \left[(a^2-r^2) + \frac{(a^2-b^2)\log\frac{r}{a}}{\log\frac{a}{b}}\right] r\,dr$$

$$= \frac{P\pi}{2\eta l}\left[a^2\int_b^a r\,dr - \int_b^a r^3\,dr + \frac{(a^2-b^2)}{\log\frac{a}{b}}\int_b^a \log\frac{r}{a} r\,dr\right]$$

$$= \frac{P\pi}{2\eta l}\left[a^2\left(\frac{a^2}{2}-\frac{b^2}{2}\right) - \left(\frac{a^4}{4}-\frac{b^4}{4}\right) + \frac{(a^2-b^2)}{\log\frac{a}{b}}\left\{\int_b^a \log r\cdot r\,dr - \log a\int_b^a r\,dr\right\}\right]$$

$$= \frac{P\pi}{2\eta l}\left[\frac{a^4}{4} - \frac{a^2b^2}{2} + \frac{b^4}{4} + \frac{(a^2-b^2)}{\log\frac{a}{b}}\left\{\frac{b^2}{2}\log\frac{a}{b} - \frac{(a^2-b^2)}{4}\right\}\right]$$

$$= \frac{P\pi}{2\eta l}\left[\frac{a^4}{4} - \frac{a^2b^2}{2} + \frac{b^4}{4} + \frac{b^2(a^2-b^2)}{2} - \frac{(a^2-b^2)^2}{4\log\frac{a}{b}}\right]$$

$$= \frac{P\pi}{2\eta l}\left[\frac{a^4}{4} - \frac{b^4}{4} - \frac{(a^2-b^2)^2}{4\log\frac{a}{b}}\right] = \frac{P\pi}{8\eta l}(a^2-b^2)\left[(a^2+b^2) - \frac{(a^2-b^2)}{\log\frac{a}{b}}\right].$$

11.8 Corrections to Poiseuille's formula

Poiseuille's equation

$$V = \frac{\pi P a^4}{8\eta l}$$

is approximately true since two important factors have been neglected in the deduction of the above expression. In the first place the pressure difference P is utilised partly in communicating kinetic energy to the liquid and in the second place it is not proper to assume that there is no acceleration of the liquid along the axis of the tube. As a matter of fact there is some acceleration near the inlet end of the tube and it does not become zero until an appreciable length of the tube has been traversed. To correct for this a factor $1.64a$ should be added to the length l of the tube.

To find out kinetic energy correction let us suppose that p is the effective pressure difference which overcomes viscosity. Now the work done against viscous force in one second is pV, and kinetic energy given to the liquid of density ρ per second

$$= \int_0^a \frac{1}{2}\rho 2\pi r\, dr v \cdot v^2 = \int_0^a \rho\pi v^3 r\, dr = \int_0^a \rho\pi \left(\frac{p}{4\eta l}\right)^3 (a^2 - r^2)^3 r\, dr$$

$$= \rho\pi \left(\frac{p}{4\eta l}\right)^3 \int_0^a (a^2 - r^2)^3 r\, dr = \rho\pi \left(\frac{p}{4\eta l}\right)^3 \cdot \frac{a^8}{8}.$$

The total loss of energy per second is therefore

$$pV + \rho\pi \left(\frac{p}{4\eta l}\right)^3 \cdot \frac{a^8}{8}$$

and this is equal to PV

$$\therefore \quad pV + \rho\pi \left(\frac{p}{4\eta l}\right)^3 \frac{a^8}{8} = PV$$

$$\text{or} \quad pV + \frac{\rho}{\pi^2 a^4}\left(\frac{\pi p a^4}{8\eta l}\right)^3 = PV$$

$$\text{or} \quad pV + \frac{\rho V^3}{\pi^2 a^4} = PV \qquad \left[\because V = \frac{\pi p a^4}{8\eta l}\right]$$

$$\text{or} \quad p = P - \frac{\rho V^2}{\pi^2 a^4}.$$

This correction has been tested experimentally by Hagenbach, Couette and Wilberforce, but it has been found that it is only approximately true. The correction should be

$$p = P - \frac{k\rho V^2}{\pi^2 a^4},$$

where k is a constant whose value depends upon the form of the apparatus. The value of k is nearly equal to unity. For most accurate work its value may be obtained by calibration. When we apply these two corrections, Poiseuille's equation becomes.

$$V = \frac{\pi a^4}{8\eta(l + 1.64a)}\left(P - \frac{k\rho V^2}{\pi^2 a^4}\right)$$

11.9 Downward flow of a liquid through a vertical narrow tube

We consider a vertical narrow tube of axis AB (Fig. 11.9.1) through which the liquid is flowing steadily. Within the tube we consider a cylindrical shell of liquid having radii r and $r + \delta r$ and length δz. Then the viscous force acting on the inner wall of the shell in the vertical upward direction is $2\pi r\delta z\,\eta\frac{dv}{dr}$. The viscous force acting on the outer wall in the vertical downward direction is

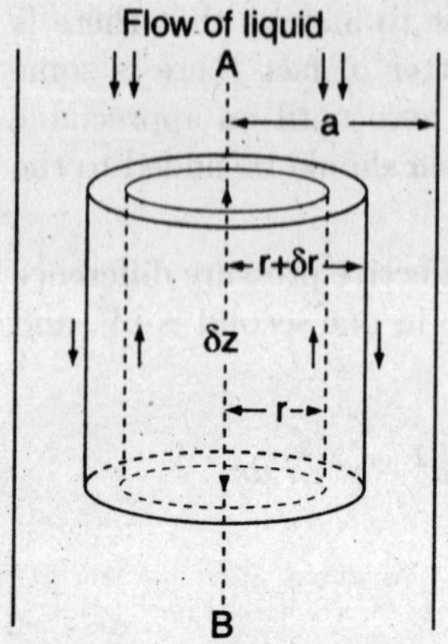

Fig. 11.9.1

$$2\pi r\delta z\,\eta\frac{dv}{dr} + \frac{d}{dr}\left(2\pi r\delta z\,\eta\frac{dv}{dr}\right)\delta r.$$

Now the force due to pressure p on the upper annular flat surface of the element of the liquid cylinder acting in the vertical downward direction is $2\pi r\,\delta r\,p$.

Similarly, the force due to pressure on the lower annular flat surface of the liquid cylinder acting in the upward direction is

$$2\pi r\,\delta r\left(p + \frac{dp}{dz}\delta z\right).$$

In the steady state the liquid moves without any acceleration and hence the resultant downward force is zero.

Thus

$$-2\pi r\delta z\,\eta\frac{dv}{dr} + 2\pi r\delta z\,\eta\frac{dv}{dr} + \frac{d}{dr}\left(2\pi r\delta z\,\eta\frac{dv}{dr}\right)\delta r$$

$$+2\pi r\,\delta r\,\delta z\rho g + 2\pi r\,\delta r\,p - 2\pi r\,\delta r\left(p + \frac{dp}{dz}\delta z\right) = 0.$$

[$2\pi r\,\delta r\,\delta z\rho g$ denotes the vertical downward force due to the weight of the cylindrical shell of the liquid; ρ denotes the density of the liquid.]

$$\text{or}\quad 2\pi\delta z\,\eta\frac{d}{dr}\left(r\frac{dv}{dr}\right)\delta r + 2\pi r\,\delta r\,\delta z\rho g - 2\pi r\,\delta r\frac{dp}{dz}\delta z = 0$$

$$\text{or}\quad \eta\frac{d}{dr}\left(r\frac{dv}{dr}\right) + r\rho g = r\frac{dp}{dz} \quad\text{or}\quad \frac{dp}{dz} = \eta\frac{1}{r}\frac{d}{dr}\left(r\frac{dv}{dr}\right) + \rho g.$$

If P_1 and P_2 be the pressures at the inlet and outlet ends of the tube AB of length l, then considering the pressure gradient $\frac{dp}{dz}$ to be constant we have

$$\frac{dp}{dz} = -\frac{(P_1 - P_2)}{l} \quad\therefore\quad -\frac{(P_1 - P_2)}{l} = \eta\frac{1}{r}\frac{d}{dr}\left(r\frac{dv}{dr}\right) + \rho g$$

$$\text{or}\quad \left[\rho g + \frac{(P_1 - P_2)}{l}\right] r = -\eta\frac{d}{dr}\left(r\frac{dv}{dr}\right).$$

On integration, we have

$$\left[\rho g + \frac{(P_1 - P_2)}{l}\right]\frac{r^2}{2} + K = -\eta r\frac{dv}{dr}$$

where K is the constant of integration. Integrating again, we have

$$\left[\rho g + \frac{(P_1 - P_2)}{l}\right]\frac{r^2}{4} + K\log r + K' = -\eta v.$$

where K' is another constant of integration. Now, $K=0$, since v cannot be infinite when $r=0$.

$$\therefore \quad \left[\rho g + \frac{(P_1 - P_2)}{l}\right]\frac{r^2}{4} + K' = -\eta v.$$

Again, when $r = a, v = 0$

$$\therefore \quad K' = -\left[\rho g + \frac{(P_1 - P_2)}{l}\right]\frac{a^2}{4}.$$

Hence

$$\eta v = \left[\rho g + \frac{(P_1 - P_2)}{l}\right]\frac{(a^2 - r^2)}{4} \quad \text{or} \quad v = \left[\rho g + \frac{(P_1 - P_2)}{l}\right]\frac{(a^2 - r^2)}{4\eta}.$$

Hence the volume of the liquid flowing per sec. through the vertical tube AB is given by

$$V = \int_0^a 2\pi r\, dr\, v = \frac{\pi}{2\eta}\left[\rho g + \frac{(P_1 - P_2)}{l}\right]\int_0^a (a^2 - r^2) r\, dr = \frac{\pi a^4}{8\eta}\left[\rho g + \frac{(P_1 - P_2)}{l}\right]$$

In the Fig. 11.9.2 a vertical viscometer is shown. The vertical narrow tube AB is connected to a liquid container. If h be the height of the liquid surface in the container above A, then the pressure at the inlet A is given by $P_1 = \pi + \rho g h$, where π is the barometric pressure. The pressure as the outlet is $P_2 = \pi$.

Hence $P_1 - P_2 = \rho g h$

$$\therefore \quad V = \frac{\pi a^4}{8\eta}\left[\rho g + \frac{\rho g h}{l}\right] = \frac{\pi a^4}{8\eta l}\cdot \rho g(l + h) = \frac{\pi a^4}{8\eta l}\cdot \rho g h' \qquad (11.9.1)$$

where h' is the height of the surface of the liquid in the container above the outlet B.

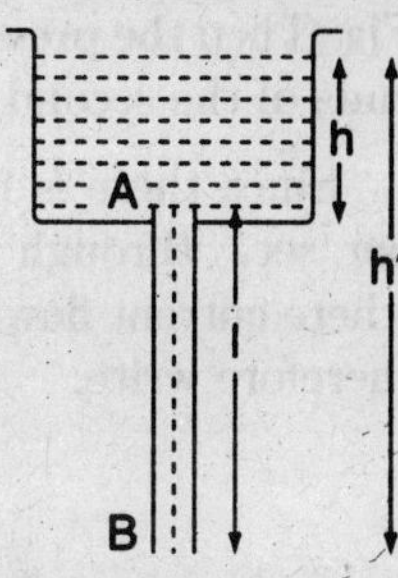

Fig. 11.9.2

11.10 Flow of a liquid through an inclined narrow tube

If the tube is inclined to the vertical at an angle θ (Fig. 11.10.1) then considering the component of gravity g along the axis of the tube, we have

$$V = \frac{\pi a^4}{8\eta}\left[\rho g \cos\theta + \frac{(P_1 - P_2)}{l}\right] = \frac{\pi a^4}{8\eta}\left[\rho g \cos\theta + \frac{\rho g h}{l}\right]$$

$$= \frac{\pi a^4}{8\eta}\left[\frac{\rho g(l\cos\theta + h)}{l}\right] = \frac{\pi a^4}{8\eta l}\rho g h' \quad (\because h' = l\cos\theta + h).$$

Thus in the case of the flow of liquid through a vertical narrow tube or an inclined narrow tube, we require the height of the liquid h' for the determination of the viscosity of the liquid.

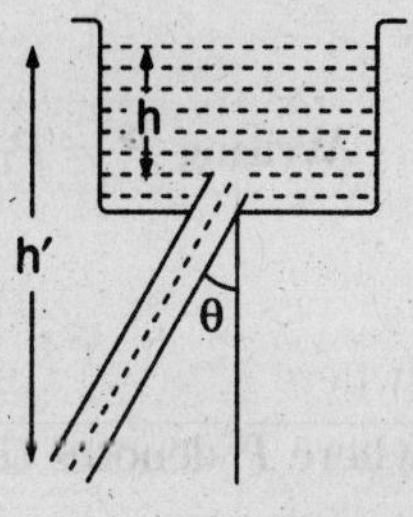

Fig. 11.10.1

11.11 Flow of a liquid through capillaries in series and in parallel

Poiseuille's formula given by (11.6.2) may be written as

$$V = \frac{P}{8\eta l/\pi a^4} \tag{11.11.1}$$

which may be compared with Ohm's law for flow of electric current ($i = E/R$). Here the rate of flow V of the liquid corresponds to current i, the pressure difference P to the potential difference E and $8\eta l/\pi a^4$ to the resistance R, so that the problem of flow of a liquid through capillaries in series may be treated as flow of current through resistances connected in series.

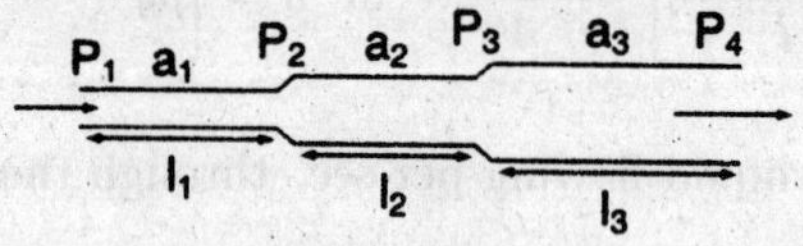

Fig. 11.11.1

We consider an arrangement consisting of a number of capillaries of radii a_1, a_2 and a_3, having lengths l_1, l_2 and l_3 respectively joined in series (Fig. 11.11.1). Let P_1 and P_4 be the pressures at the extreme ends and P_2, P_3 the pressures at the junctions indicated in the Fig. Then the pressure difference across the ends of the first capillary is $(P_1 - P_2)$, that across the ends of the second capillary $(P_2 - P_3)$ and that across the ends of the third capillary $(P_3 - P_4)$.

Since there is no accumulation of the liquid at the junctions, the volume of liquid flowing per sec. through all the capillaries must be the same. This is exactly similar to the case where current flowing through a number of resistances connected in series is the same. We may therefore write.

$$V = \frac{\pi(P_1 - P_2)a_1^4}{8\eta l_1} = \frac{\pi(P_2 - P_3)a_2^4}{8\eta l_2} = \frac{\pi(P_3 - P_4)a_3^4}{8\eta l_3}.$$

Whence we have

$$P_1 - P_2 = \frac{8\eta V l_1}{\pi a_1^4};\ P_2 - P_3 = \frac{8\eta V l_2}{\pi a_2^4};\ P_3 - P_4 = \frac{8\eta V l_3}{\pi a_3^4}.$$

Adding the above equations, we have

$$P_1 - P_4 = \frac{8\eta V}{\pi}\left[\frac{l_1}{a_1^4} + \frac{l_2}{a_2^4} + \frac{l_3}{a_3^4}\right].$$

Writing $P = P_1 - P_4$, the above equation reduces to

$$V = \frac{\pi P}{8\eta} \Big/ \left[\frac{l_1}{a_1^4} + \frac{l_2}{a_2^4} + \frac{l_3}{a_3^4}\right] \tag{11.11.2}$$

where P denotes the pressure difference across the composite tube.

(11.11.2) gives us the expression for the volume of liquid flowing per second through the capillary tubes connected in series.

If, now, the same pressure difference P is maintained across the ends of *three capillaries connected* in parallel, then the volume of liquid flowing per sec. through them is given by

$$V = V_1 + V_2 + V_2 = \frac{\pi P a_1^4}{8\eta l_1} + \frac{\pi P a_2^4}{8\eta l_2} + \frac{\pi P a_3^4}{8\eta l_3} = \frac{\pi P}{8\eta}\left(\frac{a_1^4}{l_1} + \frac{a_2^4}{l_2} + \frac{a_3^4}{l_3}\right) \quad (11.11.3)$$

Putting $R = 8\eta l/\pi a^4$; equation (11.11.1) is written as

$$V = \frac{P}{R}.$$

Hence *for series arrangement,* we have from equation (11.11.2)

$$V = \frac{P}{R_1 + R_2 + R_3} \quad (11.11.4)$$

where

$$R_1 = \frac{8\eta l_1}{\pi a_1^4}, \quad R_2 = \frac{8\eta l_2}{\pi a_2^4}$$

etc. are the resistances offered by the individual tubes to the flow of the liquid.

Similarly *for parallel arrangement,* from (11.11.3), we have

$$V = P\left(\frac{1}{R_1} + \frac{1}{R_2} + \frac{1}{R_3}\right) \quad (11.11.5)$$

Thus the effective resistance R of three capillaries in series to viscous flow is

$$R = R_1 + R_2 + R_3,$$

while the same in the case of three capillaries in parallel is

$$\frac{1}{R} = \frac{1}{R_1} + \frac{1}{R_2} + \frac{1}{R_3}.$$

These expressions are analogous to those for electrical resistances connected in series and parallel respectively.

Illustrative examples

1. Water flows through a horizontal tube of length 20 cm. and internal radius 0.081 cm. under a constant head of the liquid 20 cm. high. In 12 minutes 864 c.c. of liquid issues from the tube. Calculate the viscosity coefficient for water. The density of water is 1 gm., per c.c. and $g = 981$ cm. per sec per sec. Also verify that the conditions for streamline flow exist. **[C.U. (Hons.) 1965, '69, '01]**

We know,

$$\eta = \frac{\pi P a^4}{8Vl}.$$

Here,

$$P = 20 \times 981 \times 1 \text{ dynes per sq. cm.}$$

$$a = 0.081 \text{ cm.}$$

$$V = \frac{864}{12 \times 60} \text{ c.c.}$$

$$l = 20 \text{ cm.}$$

$$\therefore \ \eta = \frac{3.14 \times 20 \times 981 \times (0.081)^4 \times 720}{8 \times 864 \times 20} = 0.0114 \text{ Poise.}$$

The maximum velocity of flow which is along the axis of the tube is

$$v_{\max} = \frac{Pa^2}{4\eta l} \quad (\because r = 0)$$

$$= \frac{20 \times 981 \times (0.081)^2}{4 \times 0.0114 \times 20} = 117 \text{ cm./sec. (approx.)}$$

Again critical velocity

$$v_c = \frac{K\eta}{\rho a} = \frac{1000\eta}{\rho a} = \frac{1000 \times 0.0114}{1 \times 0.081} = 170 \text{ cm./sec. (approx.)}$$

Thus $v_{\max} < v_c$ and the condition for stream line motion exists.

2. A cylindrical vessel has three horizontal capillary tubes of the same bore; each 50 cm. long, attached to its sides at heights 2.5 cm., 7.5 cm. and 15 cm. respectively from the base. The level of the liquid in the vessel is maintained at 25 cm. from the bottom by supply from outside. What should be the length of a single overflow tube having double the cross-section of the bore, which when attached horizontally at the base of the cylinder, will require supply of liquid at the same rate to maintain the same liquid height in the cylinder? **[C.U. (Hons.) 1980]**

Let r_1 = radius of the single out-flow tube; r_2 = radius of each of the three capillary tubes. From the given conditon

$$\pi r_1^2 = 2\pi r_2^2 \text{ or } r_1 = \sqrt{2} r_2 \text{ or } \left(\frac{r_1}{r_2}\right)^4 = 4.$$

Now, the volume of liquid passing per second through the single out-flow tube of length l, is

$$V = \frac{\pi P r_1^4}{8\eta l} = \frac{\pi \cdot 25 \cdot \rho g \cdot r_1^4}{8\eta l},$$

where ρ is the density of the liquid.

Again volumes of liquid passing per second through the capillary tubes each of length 50 cm. are

$$V_1 = \frac{\pi(25-2.5)\rho g r_2^4}{8\eta \cdot 50} = \frac{\pi \cdot 22.5 \rho g \cdot r_2^4}{8\eta \cdot 50}$$

$$V_2 = \frac{\pi(25-7.5)\rho g r_2^4}{8\eta \cdot 50} = \frac{\pi \cdot 17.5 \rho g \cdot r_2^4}{8\eta \cdot 50}$$

$$V_3 = \frac{\pi(25-15)\rho g r_2^4}{8\eta \cdot 50} = \frac{\pi \cdot 10 \cdot \rho g \cdot r_2^4}{8\eta \cdot 50}.$$

According to the problem $V = V_1 + V_2 + V_3$

$$\therefore \quad \frac{\pi \cdot 25 \cdot \rho g r_1^4}{8\eta l} = \frac{\pi \rho g r_2^4}{8\eta \cdot 50}(22.5 + 17.5 + 10) \text{ or } l = \left(\frac{r_1}{r_2}\right)^4 \times 25 = 4 \times 25 = 100 \text{ cms.}$$

3. **Liquid flows in stream line motion through a horizontal tube of length l and radius a under an effective pressure difference P. If a cylindrical coaxial rod of radius $\frac{a}{\sqrt{2}}$ and length l is inserted into the tube, calculate the percentage of reduction of the rate of flow for the same pressure difference between the ends.** **[cf. C.U. (Hons.) 1996]**

In this case the volume of the liquid flowing per second through the tube is given by

$$V = \frac{\pi P a^4}{8\eta l};$$

when the co-axial rod is inserted into the tube, the volume of the liquid flowing per second through the annular space is given by

$$V' = \frac{P\pi}{8\eta l}(a^2 - b^2)\left[(a^2 + b^2) - \frac{(a^2 - b^2)}{\log \frac{a}{b}}\right].$$

Here, $b = \frac{a}{\sqrt{2}}$.

$$\therefore \quad V' = \frac{P\pi}{8\eta l}\left(a^2 - \frac{a^2}{2}\right)\left[\left(a^2 + \frac{a^2}{2}\right) - \frac{\left(a^2 - \frac{a^2}{2}\right)}{\log\sqrt{2}}\right] = \frac{P\pi a^4}{8\eta l} \cdot \frac{1}{2}\left[\frac{3}{2} - \frac{1}{2 \times 2.303 \log_{10}\sqrt{2}}\right]$$

$$= \frac{P\pi a^4}{8\eta l} \cdot \frac{1}{2}[1.5 - 1.442] = 0.029 \cdot \frac{P\pi a^4}{8\eta l}.$$

The rate of flow is reduced by

$$V - V' = \frac{P\pi a^4}{8\eta l} - 0.029\frac{P\pi a^4}{8\eta l} = 0.971\frac{P\pi a^4}{8\eta l}.$$

$\therefore$ the percentage of reduction of the rate of flow

$$= \frac{V \quad V'}{V} \times 100 = 0.971 \times 100 = 97.1\%.$$

4. **Two vessels of equal cross-section, α are joined near their bases by a horizontal narrow tube of length l and internal radius r. Initially the liquid surfaces are at heights $3h$ and h respectively, above**

the capillary tube. Calculate the time taken for the difference in levels to become h, if the coefficient of viscosity is η and density d. The flow is assumed to be slow. **[C.U. (Hons.) 1963, 98]**

Let x be the depression of the level in time t (say) in one vessel; x will then be the corresponding rise of the level in the other vessel, so that $x\alpha$ c.c. volume of liquid passes through the capillary tube in time t. Hence the average volume of liquid that passes per second through the narrow tube is

$$\frac{d}{dt}(x\alpha) = \alpha\frac{dx}{dt}.$$

Now difference of pressure at the instant

$$= (3h - x)\,dg - (h + x)\,dg = 2dg(h - x).$$

Therefore, applying Poiseuille's formula, we have

$$\alpha\frac{dx}{dt} = \frac{\pi 2dg(h-x)r^4}{8\eta l}, \text{ or } \frac{dx}{(h-x)} = \frac{\pi dg\,r^4}{4\eta l\alpha}\cdot dt.$$

Now for the difference in levels to become h, the limits of integration should be $x = 0$ to $x = \frac{h}{2}$.

$$\therefore \quad \int_0^{h/2}\frac{dx}{(h-x)} = \frac{\pi dg\,r^4}{4\eta l\alpha}\int_0^t dt \text{ or } \Big[-\log(h-x)\Big]_0^{h/2} = \frac{\pi dg\,r^4}{4\eta l\alpha}t$$

$$\text{or} \quad \log_e 2 = \frac{\pi dg\,r^4 t}{4\eta l\alpha} \text{ or } t = \frac{4\eta l\alpha\log_e 2}{\pi dg\,r^4}.$$

5. Two vertical cylindrical vessels, each of diameter 10 cm. are connected at the base by a narrow horizontal tube of length 50 cm. and internal diameter 2 mms. The vessel contains water such that its levels stand 25 cms. and 5 cms. respectively above the connecting tube. After what time will the water level in the first vessel stand 17.5 cm. above the connecting tube? Deduce the necessary formula. Assume $\eta = 0.01$ poise for water. **[C.U. (Hons.) 1979]**

From illustrative Example 4, we have

$$\frac{d}{dt}(x\alpha) = \alpha\frac{dx}{dt} = \pi(5)^2\cdot\frac{dx}{dt}.$$

Now the difference of pressures at the instant

$$= (25 - x)\times 1\times g - (5 + x)\times 1\times g$$

$$= 2g(10 - x).$$

Hence applying Poiseuille's formula, we have

$$25\pi\frac{dx}{dt} = \frac{\pi\times 2g(10-x)r^4}{8\eta l}.$$

Here

$$l = 50 \text{ cm.}$$
$$r = 0.1 \text{ cm.}$$
$$\eta = 0.01$$

$$\therefore \quad 25\pi \frac{dx}{dt} = \frac{\pi \times g(10-x)(0.1)^4}{4 \times 0.01 \times 50}$$

$$\int dt = \frac{5 \times 10^5}{g} \int \frac{dx}{(10-x)}.$$

For the difference of levels to become 17.5 cm., the limits of integration should be $x = 0$ to $x = \frac{17.5}{2}$ cm. = 8.75 cm.

$$\therefore \quad \int_0^t dt = \frac{5 \times 10^5}{g} \int_0^{8.75} \frac{dx}{(10-x)} = \frac{5 \times 10^5}{g} \Big[\log_e (10-x) \Big]_{8.75}^{0} = \frac{5 \times 10^5}{g} \log_e \frac{10}{1.25}$$

$$= \frac{5 \times 10^5}{g} \times 2.302 \times \log_{10} 8 = 1060 \text{ secs.} = 17.6 \text{ minutes.}$$

6. For the determination of viscosity of an oil, a capillary tube of 2 mm. bore and 30 cm. length is kept in a vertical position and a large funnel is fitted on the top of it. The oil is poured into the funnel and its level is maintained at 35 cm. from the bottom of the capillary. If 2 c.c. of this oil flow out in 6 minutes and 10 c.c. of it weigh 9 gm. calculate the viscosity of the oil.

From Article 11.9, we have

$$V = \frac{\pi a^4}{8\eta l} \rho g h'.$$

Here

$$V = \frac{2}{6 \times 60} \text{ c.c./sec.}$$

$$a = 0.1 \text{ cm.} \qquad l = 30 \text{ cm.}$$

$$h' = 35 \text{ cm.} \qquad \rho = \frac{9}{10} \text{ gm./c.c.}$$

$$\therefore \quad \eta = \frac{\pi a^4 \rho g h'}{8Vl} = \frac{3.14 \times (0.1)^4 \times 9 \times 6 \times 60 \times 980 \times 35}{8 \times 10 \times 2 \times 30} = 7.27 \text{ poise.}$$

7. Three capillaries of same length but internal radii $3r, 4r, 5r$ are connected in series and a liquid flows through them in stream line condition. If the pressure difference across the third capillary is 8.1 mm., find the pressure difference across the first capillary.

Since the tubes are connected in series

$$V_1 = V_2 = V_3$$

(where V_1, V_2 and V_3 are the rates of flow through individual capillaries)

$$\text{or} \quad \frac{\pi p_1 (3r)^4}{8\eta l} = \frac{\pi p_2 (4r)^4}{8\eta l} = \frac{\pi p_3 (5r)^4}{8\eta l} \quad \text{(applying Poiseuille's formula)}$$

p_1, p_2, p_3 being the pressure differences across the first, second and third capillaries respectively.

$$\text{or} \quad 81p_1 = 256p_2 = 625p_3 \quad \text{or} \quad p_1 = \frac{625}{81}p_3.$$

According to the problem, $p_3 = 0.81$ cm.

$$\therefore \quad p_1 = \frac{625}{81} \times 0.81 = 62.5 \text{ cm.}$$

11.12 Torque acting on a cylinder immersed in a rotating fluid

We consider a cylinder suspended inside a co-axial cylinder, the space between the two cylinders being filled with a viscous liquid. Now, the outer cylinder is made to rotate about the common axis with constant angular velocity. We are required to calculate the couple or torque exerted on the inner cylinder due to rotation of the outer cylinder. Let a and b be the radii of the inner and outer cylinder respectively and let the fluid cover a length l of the cylinders. The innermost layer of the fluid will be at rest, since the inner cylinder is at rest. The outer most layer is rotating with angular velocity Ω which is the speed of rotation of the outer cylinder.

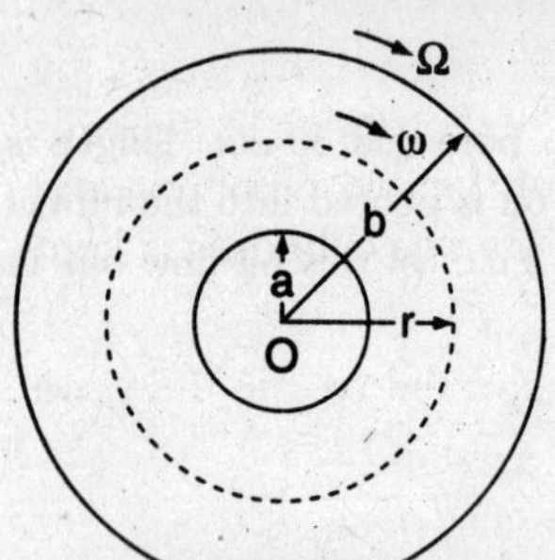

Fig. 11.12.1

We consider a fluid cylinder of radius r, rotating with angular velocity w (say). Then its linear velocity is rw and the velocity gradient at a distance r from the centre O (Fig.11.12.1) is $\frac{d}{dr}(rw) = r\frac{dw}{dr} + w$.

We consider only the first term which is effective in producing vicosity effects.

Hence the viscous force on the side of the fluid cylinder is given by

$$F = \eta \cdot 2\pi r l \cdot r\frac{dw}{dr} = 2\pi\eta l r^2 \frac{dw}{dr}.$$

Hence the viscous torque acting on the fluid cylinder

$$\Gamma = F \cdot r = 2\pi\eta l r^3 \frac{dw}{dr}.$$

Since the fluid in the space between the inner and outer cylinder is in a steady state, the torque on the inner cylinder is also equal to Γ. Now, the expression for Γ may be put in the form

$$\Gamma \frac{dr}{r^3} = 2\pi\eta l dw.$$

Integrating the above expression between the limits defined by $r = a$ and $r = b$, we have

$$\Gamma \int_a^b \frac{dr}{r^3} = 2\pi\eta l \int_0^\Omega dw \quad \text{or} \quad \frac{\Gamma}{2}\left(\frac{1}{a^2} - \frac{1}{b^2}\right) = 2\pi\eta l\Omega$$

$$\Gamma = \frac{4\pi\eta l\Omega a^2 b^2}{(b^2 - a^2)}. \tag{11.12.1}$$

This torque Γ acts on the inner cylinder in a clockwise direction.

When the inner cylinder rotates with angular velocity Ω, let Γ be the clockwise torque causing the rotation of the inner cylinder; then in the steady state the torque on the fluid cylinder is given by

$$\Gamma = -2\pi\eta l r^3 \frac{dw}{dr}$$

[Here $\frac{dw}{dr}$ is negative as w decreases with increasing r]

$$\therefore \quad -\Gamma \int_a^b \frac{dr}{r^3} = 2\pi\eta l \int_\Omega^0 dw \quad \text{or} \quad \frac{\Gamma}{2}\left[\frac{1}{b^2} - \frac{1}{a^2}\right] = -2\pi\eta l\Omega \quad \text{or} \quad \Gamma = \frac{4\pi\eta l\Omega a^2 b^2}{(b^2 - a^2)} \tag{11.12.2}$$

In this treatment the torque on the base of the cylinder has not been taken into consideration. Actually there is a torque Γ' acting on the base of the cylinder which may be eliminated by taking in turn two different lengths of the cylinder acted on by the fluid. The torque Γ' depends on the values of a, b and the distance of separation between the bases of the two cylinders, and the torque will remain unchanged if the distance between the bases of the two cylinders is kept constant and may be expressed by $\Gamma' = f(a, b)$.

Hence considering two different lengths l_1, l_2 of the inner cylinder covered by the fluid, we have the torques in the two cases given by

$$\Gamma_1 = \frac{4\pi\eta\Omega a^2 b^2}{b^2 - a^2} \cdot l_1 + f(a, b); \quad \Gamma_2 = \frac{4\pi\eta\Omega a^2 b^2}{b^2 - a^2} \cdot l_2 + f(a, b)$$

Hence,

$$\Gamma_1 - \Gamma_2 = \frac{4\pi\eta\Omega a^2 b^2}{b^2 - a^2}(l_1 - l_2) \tag{11.12.3}$$

To eliminate the end correction, two fixed cylinders having the same diameter as the inner cylinder may also be placed coaxially one above and the other below the inner cylinder, the gap between the fixed and the inner cylinder being very small. This method of eliminating the end-effects applies to gases and liquids in which the inner cylinder is completely immersed.

11.13 Viscosity of high viscous liquids

According to Stokes' law the resisting force due to viscosity acting on a small sphere falling through a liquid of infinite extent is given by $F = 6\pi a\eta v$, where v is the terminal velocity of the sphere and a its radius.

In the case of resisted fall the body attains a terminal velocity which remains constant when the retarding force due to viscosity is equal to the effective gravitational force.

Hence

$$\frac{4}{3}\pi a^3(\rho-\sigma)g = 6\pi a\eta v \text{ or } \eta = \frac{2}{9}\frac{a^2(\rho-\sigma)g}{v} \tag{11.13.1}$$

where ρ and σ are the densities of the sphere and liquid respectively.

Now, the viscous resisting force or the viscous drag $6\pi a\eta v$ has been deduced for a sphere falling in a liquid of infinite extent. In a practical case, when the liquid is confined in a vessel, say, a cylindrical vessel, correction should be applied due to boundaries of the walls and bottom of the cylinder and the corrected formula for η when the liquid is confined in a cylinder is given by

$$\eta = \frac{2}{9}\cdot\frac{a^2(\rho-\sigma)g}{v\left(1+2.4\frac{a}{R}\right)\left(1+3.3\frac{a}{h}\right)} \tag{11.13.2}$$

where R is the radius of the cylinder and h the height of the liquid.

Here, $v\left(1+2.4\frac{a}{R}\right)$ is the velocity of the sphere in the fluid of infinite width. This correction was made by Ladenburg.

$v\left(1+3.3\frac{a}{h}\right)$ denotes the velocity in the fluid of infinite depth.

The expression for viscous drag $F = 6\pi a\eta v$ may be deduced by the method of dimensions in the following way.

Since F depends on a, η and v, we may write $F = Ka^x\eta^y v^z$, where K is a dimensionless number. Putting the dimensions of different quantities involved in the above expression, we have

$$[MLT^{-2}] = L^x\left[ML^{-1}T^{-1}\right]^y\left[LT^{-1}\right]^z = M^y\cdot L^{x-y+z}\cdot T^{-y-z}.$$

Equating the indices of M, L and T, we have

$$\begin{aligned} y &= 1 \\ x-y+z &= 1 \\ -(y+z) &= -2 \end{aligned}$$

whence, $z = 1$ and $x = 1$ $\therefore$ $F = Ka\eta v$.

From mathematical analysis it was shown by Stokes that

$$K = 6\pi \quad \therefore \quad F = 6\pi a\eta v.$$

11.13.1 Discussions

From equation (11.13.1), we have $v = \dfrac{2}{9}\dfrac{ga^2(\rho-\sigma)}{\eta}$.

This shows that the terminal velocity is proportional to the square of the radius of the sphere and the velocity distribution curve of the advancing spherical body i.e., the curve of critical velocity against square of the radius is a straight line passing through the origin as shown in Fig (11.13.1). Again the curve of v against a is a parabola as shown in the figure (11.13.2).

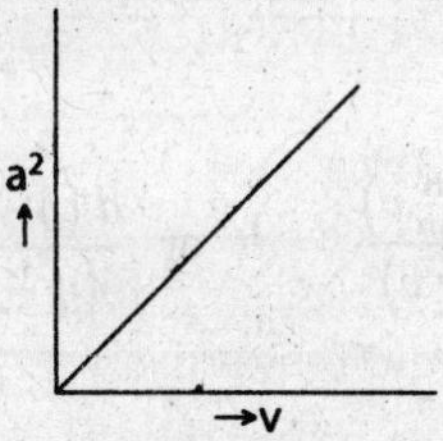

Fig. 11.13.1

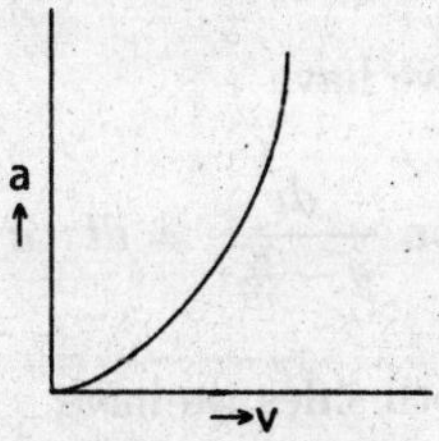

Fig. 11.13.2

In the case of a rain drop having radius 0.001 cm. falling through air for which $\eta = 1.8 \times 10^{-4}$ C.G.S. unit, the terminal velocity is given by

$$v = \frac{2 \times 981 \times (0.001)^2}{9 \times 1.8 \times 10^{-4}}$$ [Taking $\rho = 1$, and neglecting the density of air σ.]

or $v = 1.2$ cm./sec.

This explains the slow rate at which tiny drops of water which form clouds fall through the atmosphere. It should be noted that the terminal velocity is independent of the pressure of the gas through which the drops fall since η is independent of pressure.

On the other hand, bigger rain drops fall rapidly through the air. Stokes' formula does not hold if the radius of the drop is more than about 0.01 cm., because in that case the velocity of the falling drop exceeds a certain critical value as a result of which the motion of the surrounding fluid becomes turbulent. At this stage the viscous resistance is not proportional to the velocity, but proportional to the square of the velocity.*

From equation (11.13.1) it follows that if $\sigma > \rho$, i.e. when the density of the fluid through which the body falls is greater than the density of the body, the terminal velocity v becomes negative. This means that the body moves through the fluid in the upward direction. This explains why air bubbles in water or in any other liquid rise up.

11.14 Motion of a body falling through a viscous medium : Terminal Velocity

If a body be subjected to a viscous force proportional to its instantaneous velocity the equation of motion of the body will be given by

$$m\frac{d^2x}{dt^2} = mg - F \text{ or } m\frac{d^2x}{dt^2} = mg - K\frac{dx}{dt} \quad [\because \text{ the viscous force } F \propto \tfrac{dx}{dt}]$$

$$\text{or} \quad \frac{d^2x}{dt^2} = g - \frac{K}{m}\frac{dx}{dt} \quad \text{or} \quad \frac{d}{dt}\left(\frac{dx}{dt}\right) = g - \frac{K}{m}\frac{dx}{dt}.$$

*The resisting force F depends on the density ρ of the fluid, the velocity v of the body w.r.t. the fluid and the maximum area of cross-section A perpendicular to the direction of flow. So we may write $F = K\rho^x v^y A^z$, K being a dimensionless factor. Putting the dimensions on both sides, we have

$$[MLT^{-2}] = [ML^{-3}]^x[LT^{-1}]^y[L^2]^z = M^x L^{-3x+y+2z}T^{-y}$$

whence $x = 1, y = 2$ and $-3x + y + 2z = 1$ or $z = 1$, so that $F = KA\rho v^2$—K depends on the shape of the body.

Putting $\frac{dx}{dt} = v$, we have

$$\frac{dv}{dt} = g - \frac{K}{m}v \text{ or } \frac{dv}{g - \frac{K}{m}v} = dt \text{ or } -\frac{m}{K}\frac{d\left(g - \frac{K}{m}v\right)}{\left(g - \frac{K}{m}v\right)} = dt \text{ or } \frac{d\left(g - \frac{K}{m}v\right)}{\left(g - \frac{K}{m}v\right)} = -\frac{K}{m}dt.$$

Integrating on both sides we have

$$\log_e\left(g - \frac{K}{m}v\right) = -\frac{K}{m}t + C.$$

At $t = 0$, the velocity $v = 0$; hence $\log_e g = C$

$$\therefore \log_e \frac{\left(g - \frac{K}{m}v\right)}{g} = -\frac{K}{m}t \text{ or } \left(1 - \frac{K}{mg}v\right) = e^{-\frac{K}{m}t} \text{ or } v = \frac{mg}{K}\left(1 - e^{-\frac{K}{m}t}\right) \quad (11.14.1)$$

If $\frac{Kt}{m}$ is sufficiently large, the above equation may be written as $v = \frac{mg}{K}$, which shows that v acquires a constant value $\frac{mg}{K}$.

This is the maximum velocity of the falling body and is known as the *terminal velocity.*

The graph of t against v as shown in the Fig. (11.14.1) shows that v grows exponentially and approaches asymptotically to $v_T = mg/k$.

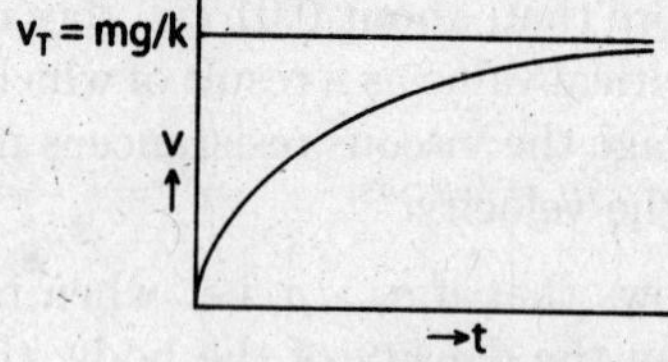

Fig. 11.14.1

We may obtain the relation between the displacement x and the time t. Equation (11.14.1) may be written in the form

$$dx = \frac{mg}{K}\left(1 - e^{-\frac{K}{m}t}\right) dt.$$

Integrating we have

$$x = \frac{mg}{K}\left(t + \frac{m}{K}e^{-\frac{K}{m}t}\right) + C'.$$

At $t = 0$. $x = 0$, so that we have

$$\therefore 0 = \left(\frac{mg}{K}\right) \cdot \frac{m}{K} + C' \text{ or } C' = -\frac{m^2 g}{K^2}$$

$$\therefore x = \frac{mg}{K}\left(t + \frac{m}{K}e^{-\frac{K}{m}t}\right) - \frac{m^2 g}{K^2} = \frac{m^2 g}{K^2}\left(\frac{K}{m}t + e^{-\frac{K}{m}t} - 1\right).$$

Illustrative examples

1. An air bubble, radius 1 cm., is allowed to rise through a long cylindrical column of syrup of radius 5 cm., and travels at a steady rate of 0.21 cm. per second. If the density of the syrup is 1.47 gm. per c.c. find its viscosity at the temperature of the experiment.

In case of a highly viscous liquid, applying Stokes' law, we have,

$$\eta = \frac{2}{9}\frac{(\rho - \sigma)ga^2}{v\left(1 + 2.4\frac{a}{R}\right)\left(1 + 3.3\frac{a}{h}\right)}.$$

Here, the cylinder being long, the correcting factor for the bottom of the cylinder, i.e. $\left(1 + 3.3\frac{a}{h}\right)$ may be neglected. Again since the bubble is of negligible density and moving in the upward direction, the effective force in the upward direction is $\frac{4}{3}\pi a^3 \rho g$ (ρ being the density of the syrup) and the viscous force here acts in the downward direction. Therefore, the above formula may be written as

$$\eta = \frac{2}{9} \cdot \frac{\rho g a^2}{v\left(1 + 2.4 \cdot \frac{a}{R}\right)}.$$

Here $\rho = 1.47$ gm./c.c., $a = 1$ cm., $g = 981$ cm./sec.^{2}, $v = 0.21$ cm. per sec.; and $R = 5$ cm.

$$\therefore\ \eta = \frac{2}{9}\frac{1.47 \times 981 \times 1}{0.21\left(1 + 2.4 \times \frac{1}{5}\right)} = 1.03 \times 10^3 \text{ Poise.}$$

2. Emery powder particles are stirred up in a beaker of water 10 cm. deep. Assuming the particles to be spherical, find the radius of the largest particle remaining in suspension after 24 hours. Take the density of emery as 4 gm. per c.c. and the viscosity of water as 0.010 Poise.

[C.U. (Hons.) 93, cf. 95, 2000, 2006 (old syll)]

We already know that

$$\frac{4}{3}\pi a^3(\rho - \sigma)g = 6\pi\eta a v \quad \therefore\ a^2 = \frac{9}{2}\frac{\eta v}{(\rho - \sigma)g}.$$

Here ρ = density of emery particles = 4 gm./c.c.

σ = density of water = 1 gm./c.c.

η = coefficient of viscosity of water = 0.010 Poise

v = terminal velocity of emery particles = $\frac{10}{24 \times 3600}$ cm./sec.

$g = 981$ cm./sec.^{2}

and a = radius of the largest particle, which is to be found out.

$$\therefore\ a^2 = \frac{9}{2}\frac{0.010 \times 10}{24 \times 3600 \times (4 - 1) \times 981} = \frac{9 \times 0.1 \times 10^{-2}}{6 \times 24 \times 36 \times 981} = \frac{0.1 \times 10^{-2}}{24 \times 24 \times 981}$$

$$a = \frac{10^{-1}}{24}\sqrt{\frac{0.1}{981}} = 4.21 \times 10^{-5} \text{ cm.}$$

3. Two circular horizontal discs each of radius 5 cm. and distance apart one millimetre are separated by a layer of oil of viscosity 1.01 C.G.S. unit. The upper disc is fixed while the lower one is revolved in vertical frictionless bearings by the tengential pull of a flexible string, the tension in which is 6000 dynes, which unwraps from a cylinder, radius 2 cm. coaxial with the disc and fastened rigidly to it. Find the time taken by the rotating system to make one complete revolution. Deduce the formula you would use.

Let the lower disc be made to rotate with angular velocity w. We consider a layer of oil between the two discs, the velocity gradient of whcih at a distance r from the common axis of the discs will be on an average $\frac{rw}{x}$, where x is the distance of separation of the two discs. So the torque acting on an annulas of radius r and width dr, of the upper disc is equal to $\eta \cdot 2\pi r\, dr \frac{rw}{x} r$. Hence the total torque acting on the upper disc is given by

$$\tau = \int_0^a \frac{2\pi\eta w r^3 dr}{x} = \frac{\pi\eta w a^4}{2x}$$

where a is the radius of the upper disc.

Here

$$\eta = 1.01 \text{ C.G.S. } x = 1 \text{ mm.} = 0.1 \text{ cm.}$$

$$a = 5 \text{ cm.} \qquad \tau = 6000 \times 2 = 12000 \text{ dyne-cm.}$$

Now,

$$w = \frac{2x\tau}{\pi\eta a^4} \text{ or } T = \frac{2\pi^2\eta a^4}{2x\tau} = \frac{\pi^2\eta a^4}{x\tau} \quad [T = \text{time for one complete revolution}]$$

$$= \frac{(3.14)^2 \times 1.01 \times 625}{0.1 \times 12000} = 5.188 \text{ secs.}$$

11.15 Flow of a gas through a narrow tube

Though the definition of the coefficient of viscosity of a liquid applies equally to a gas, yet difficulty arises when we use Poiseuille's equation in the case of a gas. As a liquid is supposed to be incompressible, its density is indpendent of pressure but a gas being compressible has its density proportional to pressure. Thus in case of a liquid the volume flowing through any section of the tube in a given time is constant, but in the case of a gas the mass, instead of the volume, flowing through any section in a given time is constant.

Considering the mass of the gas which crosses any section of the tube in one second to be constant, we may write.

ρV = constant, where ρ is the density of the gas and V the volume of the gas flowing in unit time through the section under consideration. Now, since $\rho \propto P$, where P is the uniform pressure over the section, we have PV = constant.

We take an element of length dx of the tube and dP the difference of pressure between the ends of the element. The element considered is so small that the change of density of the gas within the element is practically negligible and we may apply for the element Poiseuille's equation for the flow of liquid. Thus, we have the volume of gas leaving the element in one second given by

$$V = -\frac{\pi a^4}{8\eta}\frac{dP}{dx}.$$

The negative sign indicates that the pressure decreases with the increase of x.

Let P_1 be the pressure of the gas at the inlet end, and V_1 the volume entering the tube per second. We have, then

$$P_1V_1 = PV = -\frac{\pi a^4}{8\eta} \cdot P \cdot \frac{dP}{dx} \therefore \; P_1V_1 dx = -\frac{\pi a^4}{8\eta} \cdot P\, dP.$$

Integrating between the limits 0 to l, we have

$$P_1V_1l = -\frac{\pi a^4}{8\eta}\int_{P_1}^{P_2} P\,dP = -\frac{\pi a^4}{8\eta}\frac{(P_2^2 - P_1^2)}{2} \quad \text{or} \quad P_1V_1 = \frac{\pi a^4(P_1^2 - P_2^2)}{16\eta l} \tag{11.15.1}$$

[P_2 = pressure at the outlet end of the tube.]

In deducing the above formula it has been assumed that there is no relative motion between the wall of the tube and the adjacent layer of gas. Actually, there is appreciable relative motion or 'slipping' when the pressure is low. This fact reduces the value of the coefficient of viscosity. Considering the corrections for kinetic energy and slipping effect, it has been shown by Erk that the above formula becomes

$$\eta = \frac{\pi a^4}{16lP_1V_1}(P_1^2 - P_2^2)\left(1 + \frac{4\lambda}{a}\right) - \frac{\rho V_1}{8\pi l}\left(k + \log_e \frac{P_1}{P_2}\right),$$

where λ is called the *slipping coefficient* which is constant for a particular gas and k is a constant, the value of which depends upon the form of the apparatus.

Illustrative example

A soap bubble of radius 4 cm. and surface tension 30 dynes per cm. is blown at the end of a tube of length 10 cm. and internal diameter 0.200 cm. If the viscosity of air is 1.85 × 10^{-4} C.G.S. units, find the time taken by the bubble to be reduced to a radius of 2 cm. [cf. C.U. (Hons.) 1976, '92]

We know that for the flow of a gas through a narrow tube of length l and internal radius a,

$$P_1V_1 = \frac{\pi(P_1^2 - P_2^2)a^4}{16\eta l}.$$

Here P_1 = pressure at the inlet end of the tube = $\frac{1}{2}$ (pressure when the bubble is of radius r_1+ pressure when the bubble is of radius r_2).

$$= \frac{1}{2}\left(B + \frac{4S}{r_1} + B + \frac{4S}{r_2}\right)$$

[B = Atmospheric pressure and S = Surface tension of soap solution.]

$$= B + 2S\left(\frac{r_1 + r_2}{r_1r_2}\right) = B + 2 \times 30\left(\frac{4+2}{4 \times 2}\right) = (B + 45) \text{ dynes per sq. cm.}$$

Again

$$P_2 = B = \text{atmospheric pressure}$$

$$\eta = 1.85 \times 10^{-4}\text{C.G.S. units}$$

$$a = 0.1 \text{ cm. and } l = 10 \text{ cm.}$$

$$\because \quad P_1V_1 = \frac{\pi(P_1^2 - P_2^2)a^4}{16\eta l}$$

$$\therefore \quad (B+45)V_1 = \frac{3.14\left[(B+45)^2 - B^2\right](0.1)^4}{16 \times 1.85 \times 10^{-4} \times 10}$$

$$\text{or } V_1 = \frac{(2B+45)45}{(B+45)} \times \frac{3.14 \times 10^{-4}}{160 \times 1.85 \times 10^{-4}} = 45\left[1 + \frac{1}{(1+\frac{45}{B})}\right] \times \frac{3.14}{160 \times 1.85}$$

$$= 45\left(2 - \frac{45}{B}\right) \times \frac{3.14}{160 \times 1.85} \quad \text{[Neglecting higher powers of } \tfrac{45}{B} \text{ which are small.]}$$

$$= 90 \times \frac{3.14}{160 \times 1.85} \quad (\text{neglecting } \tfrac{45}{B}).$$

Now the change in the volume of the bubble due to flow of air

$$= \frac{4}{3}\pi(4^3 - 2^3) = \frac{4}{3} \times 3.14 \times 56.$$

Hence the required time for the bubble to be reduced to a radius of 2 cm.

$$= \frac{\frac{4}{3} \times 3.14 \times 56}{90 \times 3.14} \times 160 \times 1.85 \text{ secs.} = 4.094 \text{ minutes} = 4 \text{ min. } 56 \text{ sec.}$$

An alternative method has been given in the Exercises of this chapter.

11.16 Dependence of viscosity on temperature

11.16.1 Liquids

Coefficient of viscosity of liquids diminishes very rapidly with the rise of temperature. Although many empirical formulae have been suggested to represent the variation of viscosity with temperature, yet no such formula can demand a great degree of accuracy. Thorpe and Rodger determined the coefficients of viscosity of a large number of liquids at different temperatures and found that the formula given by Slotte agreed best with the experimental results.

The empirical formula of Slotte is

$$\eta = \frac{A}{(1+Bt)^n},$$

where η is the coefficient of viscosity of the liquid at temperature t. The constants A, B, n depend on the nature of the liquid. This formula is applicable to pure liquids and does not apply to the case of mixtures of chemical compounds such as oils.

Andrade suggested a viscosity-temperature relation for liquids given by

$$\eta\rho^{-\frac{1}{3}} = Ae^{\frac{c\rho}{T}}$$

where ρ is the density of the liquid and T the absolute temperature, and A and C are constants. This agrees fairly well with experimental results except in the cases of water and some alcohol.

11.16.2 Gases

Coefficient of viscosity of gases increases with the rise of temperature. From kinetic theory we have the relation between η and the absolute temperature T given by $\eta = KT^{1/2}$, where K is a constant.

It has been found experimentally that the variation of η for gases with temperature agrees well with Sutherland's theoretical formula, given by

$$\eta = \eta_0 \frac{\alpha T^{1/2}}{1 + \frac{S}{T}}$$

where η_0 is the coefficient at 0°C; α and S are constants. S is called Sutherland's constant. This formula agrees fairly well for many gases at moderate temperature but at high and low temperatures the formula fails. Koch observed that the viscosity of mercury vapour varies more rapidly with the temperature. According to him the viscosity—temperature relation for mercury vapour is given by $\eta = \alpha T^{1.6}$, where α is a constant.

11.17 Effect of pressure on the viscosity

11.17.1 Liquids

Viscosity of liquids increases with pressure at a rapidly increasing rate. The rate of increase of viscosity with pressure varies from liquid to liquid. The behaviour of water is quite different in character from that of other liquids. The viscosity of water decreases with the increase of pressure up to a few hundred atmospheres.

11.17.2 Gases

It can be shown from the kinetic theory of gases that the coefficient of viscosity of gases is independent of the pressure when the pressure is moderate. When the pressure is low, it can be shown from the kinetic theory that the coefficient of viscosity is proportional to pressure. These predictions of the kinetic theory agree fairly well with experimental results.

Bridgman has shown that at high pressures, the coefficient of viscosity of gases increases with pressure.

11.18 Experimental determination of the viscosity of liquids

11.18.1 Viscosity of water

Poiseuille's equation

$$V = \frac{P\pi a^4}{8\eta l}$$

is utilised for the determination of the viscosity of a liquid such as water which can be obtained in large quantity. The apparatus used is shown in the attached Fig. 11.18.1.

It consists of a moderately large vessel A in which a constant water level is maintained by an inflow tap T and the outlet tube $B.CD$ is a capillary tube fitted horizontally to the vessel at a depth h below the free surface of water in the vessel. The volume V of water flowing through the capillary tube per second is measured by collecting water in a beaker in a known time. The pressure difference between the two ends of the capillary tube is given by $P = \rho gh$, where ρ is the density of the liquid.

If water is used, $\rho = 1$ and $P = gh$. Hence $V = \frac{\pi gha^4}{8\eta l}$, where l and a denote the length and internal radius of the capillary tube respectively.

$$\therefore \quad \eta = \frac{\pi gha^4}{8Vl}.$$

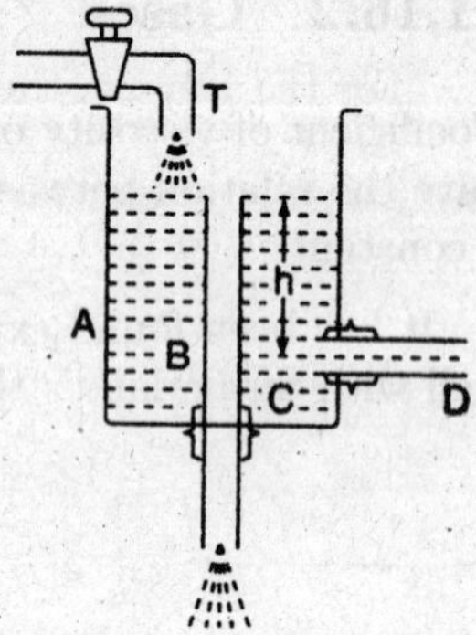

Fig. 11.18.1

Thus all the quantities on the right-hand side of the above expression being known, η is determined.

The essential condition for the success of this experiment is that the motion of the liquid must be stream-line. To ensure this, the pressure difference should not be very high. In actual practice, the pressure difference is varied by increasing or decreasing the value of h at regular steps and the volume of water is collected for a measured time corresponding to each pressure difference. A graph is obtained plotting h against V. The graph is a straight line passing through the origin in case of stream-line flow.

It should be noted that the pressure difference P is partly utilised in communicating kinetic energy to the liquid. The corrected value of the pressure to be inserted in Poiseuilles' eqauation is

$$P - \frac{\rho V^2}{\pi^2 a^4} = \rho gh - \frac{\rho V^2}{\pi^2 a^4} = g\left(h - \frac{V^2}{g\pi^2 a^4}\right), \quad \text{when } \rho = 1.$$

To correct for the effect of acceleration at the inlet end of the tube, a factor 1.64a should be added to the length of the tube.

Hence

$$\eta = \frac{\pi g\left(h - \frac{V^2}{g\pi^2 a^4}\right) a^4}{8V(l + 1.64a)}.$$

11.18.2 Viscosity of a liquid by Couette's method

In this method corrections to Poiseuille's equation are eliminated. The experimental arrangement is as shown in Fig. 11.18.2. Liquid through the tube T from a constant head supply flows through two capillary tubes of different lengths l_1 and l_2, but of same diameter. The pressure difference between the ends of the tube $A = h_1\rho g$, where ρ is the density of the liquid. The pressure difference between the ends of the other tube $B = h_2\rho g$.

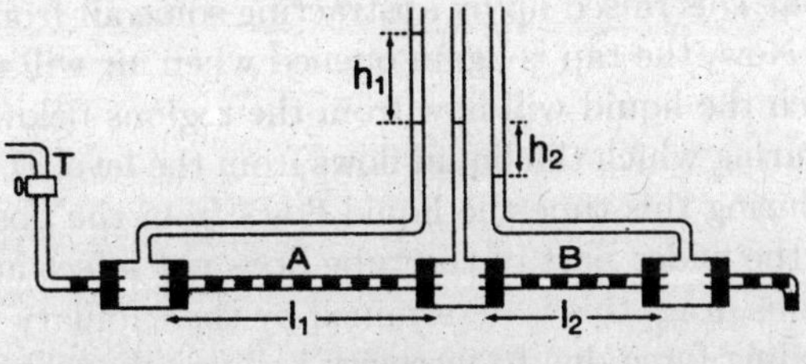

Fig. 11.18.2

For the tube A, the volume of the liquid flowing per second is given by

$$V = \frac{\pi a^4(h_1\rho g - \alpha)}{8\eta(l_1 + K)} \left[\alpha = \frac{\rho V^2}{\pi^2 a^4},\quad K = 1.64a\right] \tag{11.18.1}$$

Similarly for the tube B, we have

$$V = \frac{\pi a^4(h_2\rho g - \alpha)}{8\eta(l_2 + K)}. \tag{11.18.2}$$

From (11.18.1) and (11.18.2), we have $8\eta V(l_1 - l_2) = \pi a^4 \rho g(h_1 - h_2)$.

Thus α and K are eliminated.

Hence

$$\eta = \frac{\pi a^4 \rho g(h_1 - h_2)}{8V(l_1 - l_2)}.$$

11.18.3 Measurement of variation of viscosity with temperature

The apparatus used is shown in Fig. 11.18.3. The capillary tube CD is immersed in a vessel A containing the experimental liquid. The wider tube DEF is joined to the capillary tube and siphons the liquid from the vessel A to the beaker. The volume of the liquid flowing per second throgh the capillary tube is measured by collecting it in the beaker. A bent pin R is attached to the tube so that by lowering the tube $FEDC$ during the flow of liquid, the level of the liquid may be maintained at the tip of the pin. The excess pressure causing the flow is given by $P = \rho gh$, where h is the vertical distance between D and F. Raising the liquid in the vessel A to different temperatures, the viscosity of the liquid at various temperatures may be determined. It is found that the viscosity decreases with the rise of temperature.

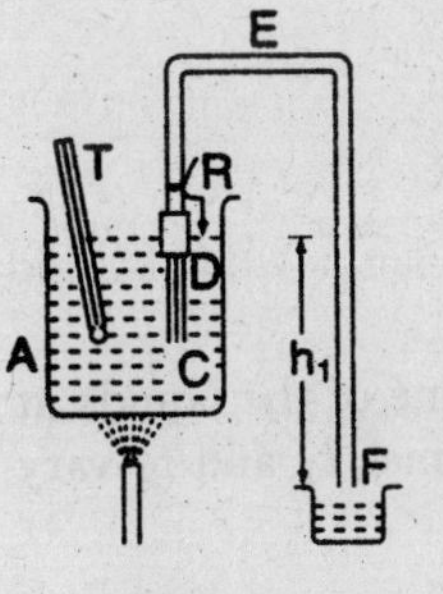

Fig. 11.18.3

11.18.4 Viscometer for comparison of viscosities

This viscometer is due to Ostwald and is used for comparing viscosities of two liquids or of the same liquid at different temperatures.

The apparatus consists of a bent tube having a capillary portion AB and two bulbs P and Q. It is kept immersed in a suitable bath. One of the experimental liquids is at first introduced into the tube. The amount of liquid introduced is adjusted to occupy the part DCB between the marks D and B with the tap T closed as shown in Fig. 11.18.4. On opening the tap the liquid level at D is raised up by abstracting some air from the tube and then the tap is closed. Now, the tap is again opened when air will rush inside the tube as a result of which the liquid will flow from the regions below D and above B. Let t_1 be the time during which the liquid flows from the level D to C and from the level B to E, i.e. during this time the liquid flows from the position DAB to the position CAE. As the wider part of the tube does not affect appreciably the rate of flow due to viscous drag, this is determined by the capillary portion AB of the tube. [As the retarding force due to viscosity is inversely proportional to the fourth power of the radius of the tube; the effect of viscosity on the rate of flow in the wider part of the tube is practically nil.]

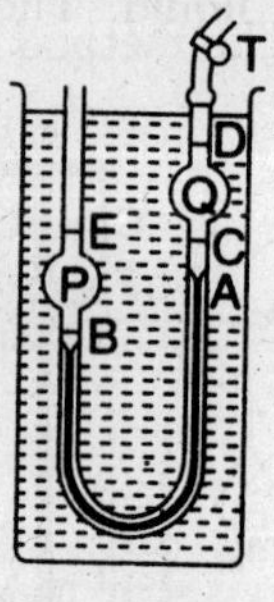

Fig. 11.18.4

The volume of the liquid flowing per second is given by

$$V_1 = \frac{\pi P_1 a^4}{8\eta_1 l}$$

where l is the length and a the radius of the capillary tube. P_1 denotes the pressure difference between the ends of the capillary tube and η_1 is the coefficient of viscosity of the liquid. The pressure difference P_1 is changing during the flow of the liquid and for every position of the liquid, the pressure difference is proportional to the density of the liquid, i.e. $P_1 \propto \rho_1$, where ρ_1 is the density of the liquid.

$\therefore\ P_1 = K\rho_1$, where K is a constant.

$$\therefore\ V_1 = \frac{K\pi\rho_1 a^4}{8\eta_1 l}. \tag{11.18.3}$$

If the experiment is repeated with a second liquid of density ρ_2 and coefficient of viscosity η_2, then the rate of flow of this liquid is given by

$$V_2 = \frac{K\pi\rho_2 a^4}{8\eta_2 l} \tag{11.18.4}$$

$\therefore$ from (11.18.3) and (11.18.4), we have

$$\frac{V_1}{V_2} = \frac{\rho_1}{\rho_2}\cdot\frac{\eta_2}{\eta_1}.$$

V_1 and V_2 also vary during the flow, but the times for the flow of unit volume of the liquids at corresponding positions vary inversely as V_1 and V_2 and hence the total times t_1 and t_2 vary inversely as V_1 and V_2,

$$\text{i.e. } t_1 \propto \frac{1}{V_1} \text{ and } t_1 \propto \frac{1}{V_2}. \quad \therefore\ \frac{t_2}{t_1} = \frac{V_1}{V_2} = \frac{\rho_1}{\rho_2}\cdot\frac{\eta_2}{\eta_1} \text{ or } \frac{\eta_1}{\eta_2} = \frac{t_1}{t_2}\cdot\frac{\rho_1}{\rho_2}.$$

Thus, knowing the densities ρ_1, ρ_2 of the two liquids and, t_1 and t_2, the times for their corresponding flows in the tube, the viscosities of the two liquids can be compared.

11.18.5 Measurement of viscosity of a liquid by Searle's viscometer (Co-axial cylinder method)

The method is suitable for the measurement of viscosity coefficients of highly viscous liquids.

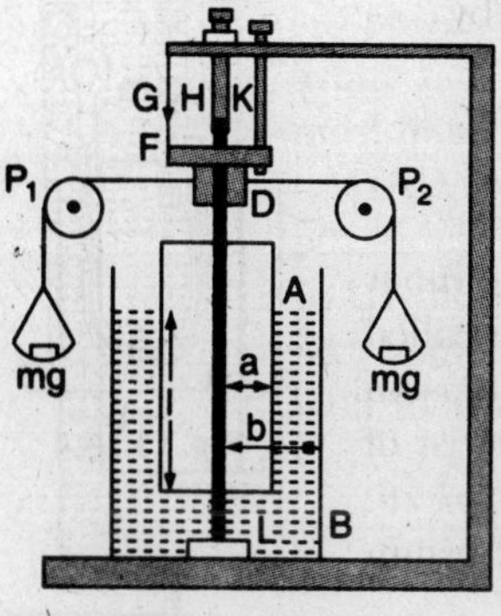

Fig. 11.18.5

A sectional diagram of Searle's viscometer is shown in the attached Fig. 11.18.5. In the figure the inner cylinder A of radius a is a solid cylinder fixed to an axle HL which is freely pivoted at its ends H and L. The axle carries a drum D and a disc F so that the cylinder may be made to rotate under the actions of two weights mg attached to the ends of a string wound round the drum and passing over two ball-bearing pulleys P_1 and P_2. The plate F is used in conjunction with an index G. This is used to measure the time period of rotation of the inner cylinder. The locking key K is used to fix the position of the inner cylinder. The outer cylinder B of radius b contains the experimental liquid and a length l of the inner cylinder A is acted upon by the viscous drag coming into play due to rotation of the inner cylinder.

If Γ be the torque causing the rotation of the inner cylinder due to the action of the two weights mg, then we have from eqn. (11.12.2)

$$\Gamma = \frac{4\pi\eta l\Omega a^2 b^2}{b^2 - a^2}$$

where Ω is the angular velocity of the rotating cylinder and η the coefficient of viscosity of the liquid. Now, $\Gamma = mgd$, where d is the diameter of the drum.

$$\therefore\ mgd = \frac{4\pi\eta l\Omega a^2 b^2}{(b^2 - a^2)} \quad \text{or } \eta = \frac{gd(b^2 - a^2)}{4\pi\Omega a^2 b^2} \cdot \frac{m}{l}.$$

Again $\frac{2\pi}{\Omega} = T$, where T is the time period of rotation of the inner cylinder.

$$\therefore\ \eta = \frac{gd(b^2 - a^2)}{8\pi^2 a^2 b^2} \cdot \frac{mT}{l}. \qquad (11.18.5)$$

From this expression it follows that for a given liquid, $\frac{mT}{l}$ = constant, i.e. if we plot mT against l, by taking different values of mT and l in an experiment, we should get a straight line passing through the origin, but actually we get a straight line which makes an intercept α on the X axis in the negative side of the origin (Fig. 11.18.6). The intercept α gives the end correction and so in the expression for η, l should be corrected to $l + \alpha$.

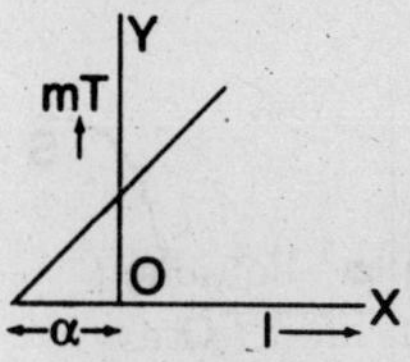

Fig. 11.18.6

$$\therefore\ \eta = \frac{gd(b^2 - a^2)}{8\pi^2 a^2 b^2} \cdot \frac{mT}{(l + \alpha)}.$$

11.18.6 Alternative arrangement for co-axial cylinder method

The apparatus used is shown diagrammatically in Fig. 11.18.7. The outer cylinder B which contains the liquid is rotated with constant angular velocity by means of a small electric motor. The inner cylinder A is suspended co-axially inside the cylinder B by a long and fine torsion wire. The suspension wire carries a mirror to measure the angle of rotation of the inner cylinder (caused by the rotation of the outer cylinder) by the usual lamp and scale arrangement. Let l be the length of the inner cylinder immersed in the liquid and let Ω be the angular velocity of the outer cylinder. As the outer cylinder rotates, a torque Γ is transmitted through the liquid to the inner cylinder due to the viscosity of the liquid which is given by

$$\Gamma = \frac{4\pi\eta l\Omega a^2 b^2}{(b^2 - a^2)}.$$

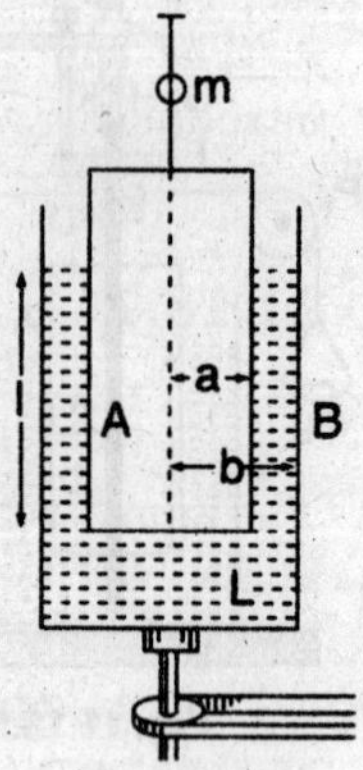

Fig. 11.18.7

This constitutes the deflecting torque which causes rotation of the inner cylinder. When this deflecting torque is equal to the restoring torque called into play due to the twist of the suspension fibre, the inner cylinder attains an equilibrium condition. If τ be the torsional constant per unit angle of twist of the suspension wire, then the restoring torque due to torsional couple $= \tau\theta$, where θ is the angle of twist of the suspension wire. Hence in the equilibrium condition of the inner cylinder, we have,

$$\Gamma = \tau\theta \text{ or } \frac{4\pi\eta\Omega a^2 b^2 l}{(b^2 - a^2)} = \tau\theta.$$

Considering the torque on the base of the cylinder the above equation takes the form

$$\frac{4\pi\eta\Omega a^2 b^2}{(b^2 - a^2)} l + f(a, b) = \tau\theta \quad \text{[See Art. 11.12]}$$

The experiment is repeated with the same liquid but with different lengths and the corresponding values of θ are measured by a lamp and scale arrangement. Let l_1, l_2 be the lengths of the liquid covering the inner cylinder and θ_1, θ_2 be the corresponding values of the angle of twist, then

$$\frac{4\pi\eta\Omega a^2 b^2}{(b^2 - a^2)} l_1 + f(a, b) = \tau\theta_1; \quad \frac{4\pi\eta\Omega a^2 b^2}{(b^2 - a^2)} l_2 + f(a, b) = \tau\theta_2.$$

Hence

$$\frac{4\pi\eta\Omega a^2 b^2}{(b^2 - a^2)}(l_1 - l_2) = \tau(\theta_1 - \theta_2) \text{ or } \eta = \frac{\tau(b^2 - a^2)(\theta_1 - \theta_2)}{4\pi\Omega a^2 b^2 (l_1 - l_2)} \tag{11.18.6}$$

τ is determined from the torsional oscillations of the inner cylinder in absence of the liquid. The time period of torsional oscillation is given by

$$T = 2\pi\sqrt{\frac{I}{\tau}},$$

where I is the moment of inertia of the inner cylinder about its axis of suspension.

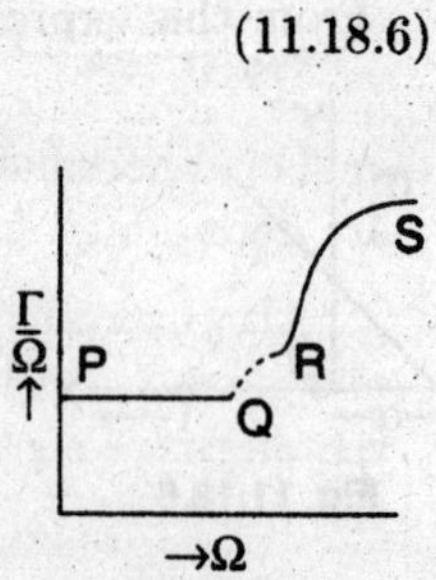

Fig. 11.18.8

In deducing the equation 11.18.6 we have assumed the stream-line flow of the liquid. It is observed experimentally that if the angular velocity Ω exceeds a certain limit Ω_c (called the critical angular velocity) the stream-line motion is lost and there is formation of vortex. So the angular velocity must be kept small, so that Γ/Ω is constant. This may be clearly seen from the attached curve between Ω and Γ/Ω (Fig. 11.18.8). Γ/Ω remains constant up to Q, beyond which it varies as shown by the dotted curve, in a some what uncertain and irregular manner, and at higher values of Ω, the relation takes the form $(\alpha\Omega + \beta\Omega^2)$, where α and β are constants. Experiment is therefore to be performed in the range where Γ/Ω is constant. Ω_c is related to η, the separation distance x between the cylinders and the density ρ of the liquid as shown on page 9.

The end corrections due to torque on the base of the cylinder A may be avoided by introducing two thick discs near the ends of the cylinder which act as guard ring.

11.18.7 Determination of the viscosity of a transparent viscous liquid by Stokes' method

The apparatus used consists of a tall cylinder containing the given liquid. The cylinder is closed at its ends by rubber corks R and S. The tube T is inserted inside the cylinder through the cork R in such a way that the axis of the tube coincides with the axis of the cylinder. The cylinder is kept immersed in another vessel V containing water as shown in Fig. 11.18.9. By this arrangement the variation of temperature of the liquid in the cylinder due to transference of heat may be avoided. The temperature may also be kept uniform by stirring the water confined in the vessel by a stirrer S'. Two strips of paper L_1 and L_2 are pasted round the cylinder so that the liquid column in the cylinder is divided into threc equal parts.

Some steel ball-bearings[†] are taken and their diameters are measured with the help of a screw gauge. The balls are thoroughly wetted with the experimental liquid and then introduced inside the tube T by means of a funnel so that the balls fall centrally through the liquid in the cylinder. When a ball reaches the mark L_1, it attains a terminal velocity. The time t taken by the ball to traverse a distance L_1L_2 is measured. Then the terminal velocity of fall of the ball $v = \frac{L_1L_2}{t}$. Thus, the terminal velocities of different balls are measured. Next the densities ρ and σ of the ball and the liquid respectively are determined by a specific gravity bottle. The internal radius R of the cylinder and the depth h of the liquid in the cylinder are then determined. Then for any ball with terminal velocity v and radius a the coefficient of viscosity of the liquid is determined from the equation (11.13.2), which is given by

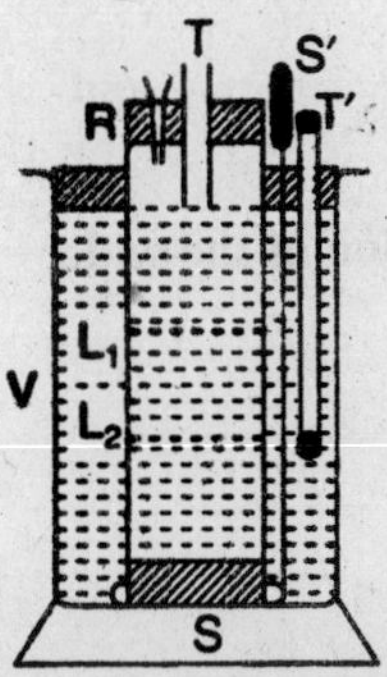

Fig. 11.18.9

$$\eta = \frac{2}{9}\frac{a^2(\rho - \sigma)g}{v(1 + 2.4a/R)(1 + 3.3a/h)}. \tag{11.18.7}$$

For different balls with different values of v and a, the values of the coefficient of viscosity are determined from which we get the mean value of η.

Alternatively, a graph may be drawn with a^2 along the X-axis and corrected v [corrected $v = v(1+2.4a/R)(1+3.3a/h)$] along the Y-axis—which will be a straight line passing through the origin. This proves the validity of Stokes' law. Taking a point on the straight line we get the value of

$$\frac{a^2}{v(1 + 2.4a/R)(1 + 3.3a/h)}$$

and substituting this value in equation (11.18.7), we get the value of η.

†For a less viscous liquid spheres of much smaller size are required and these may be obtained by blowing melted Wood's metal through the fine aperture of a glass tube into cold water.

11.19 Experimental determination of viscosity of gases

(a) The apparatus used for the measurement of viscosity of a gas is a U-tube as shown in Fig. 11.19.1.

The tube CD is of uniform capillary bore of length l and radius a. The excess of pressure at the end C of the capillary tube is produced by means of a pellet P of mercury of weight mg in the wider tube AEC. Let r be the radius of the wider tube and Q the volume between the marks A and E. Let t be the time in seconds taken for the lower surface of the pellet to pass from A to E respectively. Therefore, the volume of the gas that enters per second through the inlet end C of the capillary tube is given by $V_1 = \frac{Q}{t}$ and the pressure at the inlet end of the capillary tube is given by

$$P_1 = B + b - \delta,$$

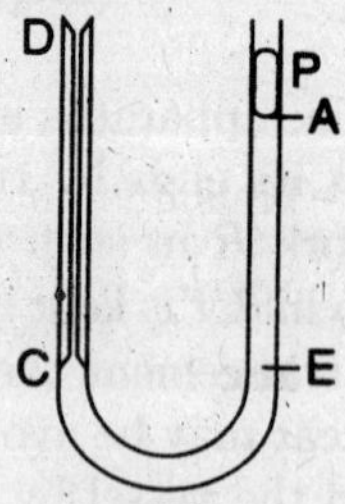

Fig. 11.19.1

where B = atmospheric pressure;

$b = \frac{mg}{\pi r^2}$ = pressure due to the pellet;

δ = sticking coefficient or virtual reduction of excess pressure due to friction between the pellet and the wider tube.

The pressure at the outlet end D of the capillary tube is given by

$$P_2 = B.$$

Hence using equation (11.15.1), we have

$$(B + b - \delta)\frac{Q}{t} = \frac{\pi a^4}{16\eta l}\left[(B + b - \delta)^2 - B^2\right].$$

If L be the length between the marks A and E and α the internal cross-section of the wider tube, then

$$Q = L\alpha = \pi r^2 L.$$

Hence we have,

$$\frac{(B + b - \delta)}{t}\pi r^2 L = \frac{\pi a^4}{16\eta l}\left[(B + b - \delta)^2 - B^2\right]$$

$$\text{or } r^2 L = \frac{a^4 t}{16\eta l}\left[(B + b - \delta) - \frac{B^2}{(B + b - \delta)}\right]$$

$$\text{or } r^2 L = \frac{a^4 t}{16\eta l}\left[(B + b - \delta) - \frac{B}{\left(1 + \frac{b - \delta}{B}\right)}\right]$$

$$= \frac{a^4 t}{16\eta l}\left(B + b - \delta - B + b - \delta - \frac{b^2}{B}\right)$$

$$= \frac{a^4 t}{16\eta l}\left(2b - \frac{b^2}{B} - 2\delta\right) \quad \text{[neglecting the terms containing } \tfrac{\delta}{B}\text{]}$$

$$\text{or } \eta = \frac{a^4 t}{16 r^2 L l}\left(2b - \frac{b^2}{B} - 2\delta\right) = \frac{a^4 t}{8 r^2 L l}\left(b - b^2 2B - \delta\right) \qquad (11.19.1)$$

Thus η is determined if δ is known.

The above expression may also be written in the form

$$\left(b - \frac{b^2}{2B} - \delta\right) t = \eta \times \frac{8r^2 Ll}{a^4} = \text{constant} \tag{11.19.2}$$

which enables us to determine the sticking coefficient experimentally by taking separately two pellets of mercury of different weights and measuring time in each case for the pellet to fall from A to E. If b_1 and b_2 be the pressures due to the two different pellets and t_1, t_2 be the corresponding times of fall, then

$$\left(b_1 - \frac{b_1^2}{2B} - \delta\right) t_1 = \left(b_2 - \frac{b_2^2}{2B} - \delta\right) t_2. \tag{11.19.3}$$

Thus δ is determined. Substituting this value of δ in equation (11.19.1) η is determined.

(b) *Searle's method for the determination of the viscosity of a gas*

The following experiment for measuring the coefficient of viscosity of air was conducted in the year 1904 in Poynting laboratories at Birmingham.

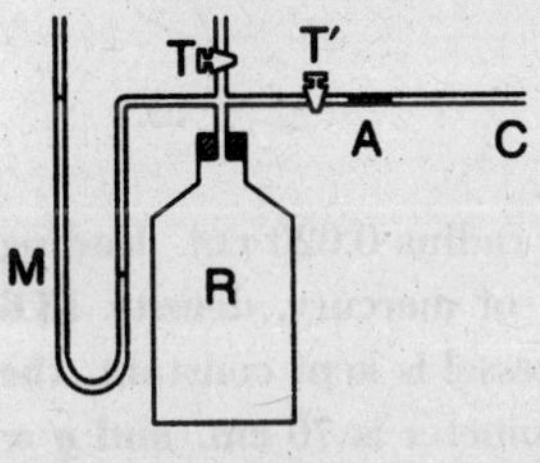

Fig. 11.19.2

The attached Fig. 11.19.2 shows a large vessel R of capacity about 45 litres. Air is filled in the vessel by means of bicycle pump connected to the tap T till the pressure inside the vessel is raised to 30 cms. of water above the atmospheric pressure as recorded by the manometer M. A capillary tube AC of length l and radius a is connected to the vessel through the tap T'. The tap T' is kept open for a measured time and then it is closed. The fall of pressure is noted in the manometer M.

Let dV be the volume of air flowing through the capillary tube in time dt. Then, from equation (11.15.1) we have,

$$P\frac{dV}{dt} = \frac{P^2 - B^2}{16\eta l} \cdot \pi a^4,$$

where B is the atmospheric pressure and P is the pressure in the vessel.

But,

$$(P - dP)(V + dV) = PV,$$

where dP is the fall in pressure due to the flow of air of volume dV and V is the volume of the vessel

$$\therefore \quad PV + P\,dV - V\,dP = PV$$

$$\text{or} \quad P\,dV = V\,dP \quad [\text{neglecting the product of } dP \text{ and } dV]$$

$$\therefore \quad V\,dP = \frac{P^2 - B^2}{16\eta l}\, dt \cdot \pi a^4$$

$$\text{or} \quad V \cdot \frac{dP}{P^2 - B^2} = \frac{\pi a^4}{16\eta l}\, dt.$$

If the pressure in the vessel falls from P_1 to P_2 in time t, we have,

$$V\int_{P_1}^{P_2}\frac{dP}{P^2-B^2}=\frac{\pi a^4}{16\eta l}\int_0^t dt$$

$$\text{or}\quad \frac{V}{2B}\int_{P_1}^{P_2}\left(\frac{dP}{P+B}-\frac{dP}{P-B}\right)=\frac{\pi a^4}{16\eta l}\cdot t$$

$$\text{or}\quad \frac{V}{2B}\left(\log\frac{P_2+B}{P_1+B}-\log\frac{P_2-B}{P_1-B}\right)=\frac{\pi a^4}{16\eta l}t$$

$$\text{or}\quad \frac{V}{2B}\log\frac{(P_2+B)(P_1-B)}{(P_1+B)(P_2-B)}=\frac{\pi a^4}{16\eta l}t$$

$$\text{or}\quad \eta=\frac{B\pi a^4 t}{8Vl\log\frac{(P_2+B)(P_1-B)}{(P_1+B)(P_2-B)}} \tag{11.19.4}$$

Here the small change in the volume V due to the movement of the manometric liquid has been ignored.

Illustrative Example

A glass bulb of volume 500 c.c. has a capillary tube of length 40 cm., and radius 0.020 cm. leading from it. The bulb is filled with hydrogen at an initial pressure of 86 cm. of mercury, density 13.6 gm. per c.c., and it is found that if the volume of the gas remaining in the vessel is kept constant, the pressure falls to 80 cm. of mercury in 25.4 seconds. If the height of the barometer is 76 cm. and g = 981 cm. per sec. per sec., find the viscosity of hydrogen.

Let the initial pressure be due to a height h_1 cm. of mercury of density ρ above the normal barometric height H cm. of mercury, and the final pressure be due to a height h_2 cm. of mercury above the normal barometric height.

Substituting these values in 11.19.4, we have

$$\because\ P_1=(H+h_1)\rho g,\ \ P_2=(H+h_2)\rho g,\ \ B=H\rho g$$

$$\eta=\frac{H\rho g\pi a^4 t}{8lV\log\frac{(H+h_2+H)\rho g\cdot(H+h_1-H)\rho g}{(H+h_1+H)\rho g\cdot(H+h_2-H)\rho g}}$$

$$=\frac{H\rho g\pi a^4 t}{8lV\log\frac{(2H+h_2)h_1}{(2H+h_1)h_2}}.$$

Here

H = 76 cm., V = 500 c.c.,

ρ = 13.6 gms./c.c., $h_1 = 86 - 76 = 10$ cm.,

a = 0.020 cm., $h_2 = 80 - 76 - 4$ cm.,

g = 981 cm./sec^2, l = 40 cm.

t = 25.4 sec.,

Substituting these values in the above expression we have

$$\therefore\ \eta = \frac{(76 \times 13.6 \times 981) \times 3.14 \times (0.02)^4 \times 25.4}{8 \times 500 \times 40 \log_{10}\left(\frac{10}{4} \cdot \frac{156}{162}\right) \times 2.303} = \frac{1.013 \times 10^6 \times 3.14 \times 2^4 \times 10^{-8} \times 25.4}{8 \times 2 \times 10^4 \times \log_{10} \frac{65}{27} \times 2.303}$$

$$= \frac{1.103 \times 3.14 \times 10^{-2} \times 25.4}{10^4 \times 0.3815 \times 2.303} = \frac{1.013 \times 3.14 \times 25.4}{3.815 \times 2.303} \times 10^{-5}$$

$$= 9.21 \times 10^{-5} \text{ C.G.S. units.}$$

(c) *Rankine's experiment for the measurement of viscosity of a gas*

This experiment is suitable for studying the effect of pressure on the viscosity of a gas. The apparatus is shown in the Fig. 11.19.3, where $CGFAED$ is a glass tube about 3 mm. diameter having a capillary CD. A small mercury pellet B is introduced into the part FA of the tube. When the tube is held vertically, the mercury pellet produces an excess pressure and it starts falling under its own weight forcing some of the gas into capillary tube CD. The rate of flow of the gas into the capillary tube is determined by noting the time taken by the pellet to fall through a measured length. I_1, I_2 are two fixed marks such that the volume V_1 of CGI_1 is equal to that of DEI_2. Let t be the time taken by the pellet to fall from I_1 to I_2. Let V_2 be the volume $CGFAED$. To start with, if the apparatus be placed horizontally, the pressure of the gas will be uniform. Let this pressure be P. Now, if ρ be the density of the gas at unit pressure, then the density of the gas at pressure P is $P\rho$. Hence the mass of the gas enclosed in $CGFAED = P\rho V_2$ which remains constant. At the beginning of the interval t, let P_1 be the pressure of the gas above I_1; then the pressure of the gas below I_1 is given by $P_2 = P_1 + \frac{mg}{\alpha}$, where m is the mass of the pellet and α is the internal cross-section of the tube FA. Now, as the mass of the gas remains constant, we may write

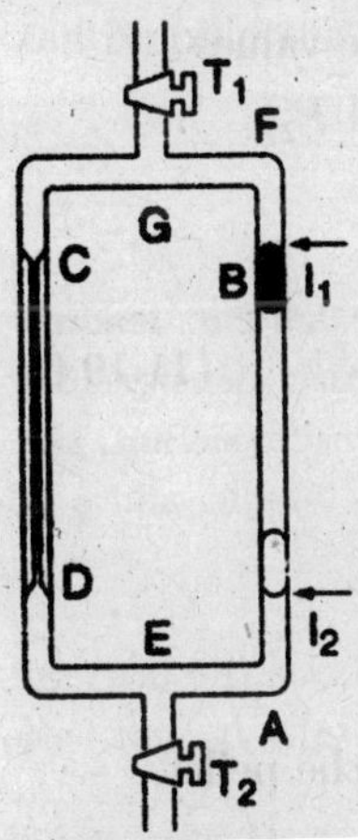

Fig. 11.19.3

$$P\rho V_2 = P_1\rho V_1 + \rho(V_2 - V_1)\left(P_1 + \frac{mg}{\alpha}\right) \text{ or } P_1 = P - \frac{mg}{\alpha} + \frac{mg}{\alpha} \cdot \frac{V_1}{V_2}$$

$$\text{or } P_1 + \frac{mg}{\alpha} = P_2 = P + \frac{mg}{\alpha} \cdot \frac{V_1}{V_2}.$$

At the end of the time t in which the pellet falls down to the position I_2, the volume of the gas above the mercury pellet now becomes $(V_2 - V_1)$. Then the pressure of the gas above the mercury pellet is given by

$$P_1' = P - \frac{mg}{\alpha} + \frac{mg}{\alpha} \cdot \frac{V_2 - V_1}{V_2}$$

and the pressure of the gas below the mercury pellet is given by

$$P_2' = P + \frac{mg}{\alpha} \cdot \frac{V_2 - V_1}{V_2}.$$

Hence the mass of the gas which has passed through the capillary tube in time t

$$= \rho P_2(V_2 - V_1) - \rho P_2' V_1 = \rho\left(P + \frac{mg}{\alpha} \cdot \frac{V_1}{V_2}\right)(V_2 - V_1) - \rho\left(P + \frac{mg}{\alpha} \cdot \frac{V_2 - V_1}{V_2}\right)V_1$$

$$= P\rho(V_2 - 2V_1).$$

Thus the average rate of flow per second is

$$\frac{P\rho}{t}(V_2 - 2V_1). \tag{11.19.5}$$

But, we know

$$P_2V = \frac{(P_2^2 - P_1^2)\pi a^4}{16\eta l},$$

where V is the volume of the gas flowing per unit time inside the capillary tube when the pellet is at I_1; l and a denote respectively the length and radius of the capillary tube CD.

Also

$$P_2'V' = \frac{(P_2'^2 - P_1'^2)\pi a^4}{16\eta l},$$

where V' is the volume of the gas flowing per unit time when the pellet is at I_2.

Hence the average rate of flow

$$= \frac{\pi a^4}{16\eta l}\left[\frac{(P_2^2 - P_1^2) + (P_2'^2 - P_1'^2)}{2}\right] = \frac{\pi a^4}{16\eta l}\cdot\frac{mg}{\alpha}\cdot 2P$$ [Substituting the values of P_1, P_2, P_1' and P_2'].

The mass of the gas flowing through the capillary tube in unit time

$$= \frac{\pi a^4}{16\eta l}\cdot\frac{mg}{\alpha}2P\rho. \tag{11.19.6}$$

From (11.19.5) and (11.19.6) we have

$$\frac{V_2 - 2V_1}{t} = \frac{\pi a^4}{8\eta l}\cdot\frac{mg}{\alpha}$$

$V_2 - 2V_1$ is the volume between the two marks I_1 and I_2 minus the volume of the pellet.

The above expression may be written in the form

$$\eta = \frac{mg\pi a^4 t}{8\alpha l(V_2 - 2V_1)}, \tag{11.19.7}$$

whence the value of η may be determined.

In this experiment also, a correction should be made due to frictional resistance experienced by the mercury pellet as it glides along the wall of the tube. Due to frictional resistance the pressure due to the weight of the mercury pellet becomes less than $\frac{mg}{\alpha}$. This change in pressure may be due to changes in the angle of contact between mercury and glass. The effect is eliminated by making observations both with the pellet intact and broken into two and three equal segments along the length. It is found experimentally that the effect is doubled when the pellet is divided into two segments and trebled when it is divided into three segments.

Let β be the frictional resistance which is a certain fraction of $\frac{mg}{\alpha}$. If t_1, t_2 and t_3 be the observed times of fall of the pellet when it is in one segment and when broken into two and three segments respectively, it drives the same volume of gas through the capillary tube. Then, we have,

$$(1-\beta)t_1 = (1-2\beta)t_2 = (1-3\beta)t_3 \text{ or } \beta = \frac{t_2 - t_1}{2t_2 - t_1} = \frac{t_3 - t_1}{3t_3 - t_1}. \tag{11.19.8}$$

Thus β is determined. Hence we should write $\frac{mg}{\alpha}(1-\beta)$ in place of $\frac{mg}{\alpha}$ in equation (11.19.7) to correct for the effect of friction owing to sticking of the mercury pellet.

The following table gives the values of η for some liquids and gases at different temperatures.

Table for coefficients of viscosity of liquids and gases

Liquid	Temp. (°C)	η (Poise)	Gas	Temp. (°C)	$\eta \times 10^6$ (Poise)
	0	0.01786			
Water	20	0.01002	Air	0	171
	30	0.00798		20	181
	10	1.38			
Olive oil	20	0.84	Carbon	0	138
	30	0.52	dioxide	20	146
	0	0.01681			
Mercury	20	0.01552	Hydrogen	0	84
	30	0.01499		20	88
	10	0.0653			
Aniline	20	0.0440	Oxygen	0	192
	30	0.0318		20	200
	10	0.0075			
Benzene	20	0.00647	Helium	0	186
	30	0.00562		20	194
	10	24.2			
Castor oil	20	9.86	Argon	0	212
	30	4.52		20	222

The table shows that in case of liquids the coefficient of viscosity decreases with increase of temperature, whereas it increases in case of gases with increase of temperature. The physical explanation of both the phenomena has been given in the Exercises (Question 23(b)) of Chapter 14.

11.20 Non-Newtonian liquids

We know that according to Newton's law, for the steady flow of a viscous fluid, the shearing stress is proportional to the velocity gradient, i.e. $f = \eta\frac{dv}{dz}$. Here the coefficient of viscosity is constant at a given temperature. Liquids obeying this law are called Newtonian liquids. It is found that in case of some liquids such as colloids, paints, blood, etc., the shearing stress is not proportional to the velocity gradient. These liquids which do not obey Newton's law are called non-Newtonian liquids. In case of non-Newtonian liquids the ratio of shearing stress to velocity gradient is called apparent viscosity which will have a fixed value for a given value of the velocity gradient. Apparent viscosity decreases with the increase of velocity gradient. For a Newtonian liquid flowing through a capillary tube the velocity is maximum along the axis of the tube and decreases towards the wall of the tube and becomes zero at the wall. For a non-Newtonian liquid on the other hand, the velocity is practically constant throughout the tube and vanishes in a steep gradient at the wall.

There is another important difference between a Newtonian and a non-Newtonian liquid. A Newtonian liquid begins to move under the action of smallest possible shearing stress whereas a non-Newtonian liquid can withstand small shearing stress without moving. A non-Newtonian liquid such as a gelatine solution called *sol* becomes *gel* and behaves to some extent like a solid possessing yield point and elasticity when left for some time. The *gel* can be transformed into a *sol* by heating or by shaking. A substance possessing the properties of a *gel* which may be transformed into a *sol* when disturbed is called *thixotrophic.* A good paint possesses this property. When a paint is kept undisturbed, it remains as *gel,* but when shaken it is transformed into a *sol* and so it can be easily brushed over a surface and the paint brushed on the surface is transformed into a *gel,* and thereby the flow of the paint over the surface is stopped.

Exercises

1. Define coefficient of visosity of a Newtonian fluid. What is its dimension?
[C.U. (Hons.) '82, '85, '88, 2001, 03, 05]

2. Distinguish between stream line and turbulent motion.

3. Define coefficient of viscosity and describe a method of determining the coefficient of viscosity of a liquid. **[C.U. '58]**

4. (a) Deduce Poiseuille's formula for the flow of a viscous fluid through a narrow horizontal tube mentioning the assumptions made. What are the corrections to Poiseuille's formula?
[C.U. (Hons.) '82, '84, '85, '87, '89, '92, '94, '96, '99, 07]

   (b) In an experiment with Poiseuille's apparatus the following data were obtained.
   Volume of water issuing per minute = 7.06 c.c.
   Head of water = 34.1 cms.
   Length of tube = 56.45 cm.
   Radius of tube = 0.0514 cm.
   Find the coefficient of viscosity of water. [Ans. 0.01377 Poise]

5. Discuss the dependence of viscosity of liquids and gases on temperature and pressure.

6. Calculate the volume of liquid flowing per second through a vertical tube. Explain how you would determine the viscosity of a liquid with the help of a vertical viscometer.

7. Describe the rotating cylinder method of determing the viscosity of a liquid. Deduce the necessary formula. Explain how end corrections can be made. **[C.U. (Hons.) '88, '91, '95]**

8. (a) Deduce an expression for the viscous drag (Stokes' law) by the method of dimensions in the case of a small sphere falling through a viscous liquid. State the assumptions clearly. Explain how this expression is utilised to determine the viscosity of a highly viscous liquid.
[C.U. (Hons.) '80, '83, '91, '96, '97, '99, 04]

   (b) Deduce the equation of motion of a body falling through a viscous medium and subject to a viscous force proportional to its instantaneous velocity. **[C.U. (Hons.) '83, '94, 01, 04]**

   (c) A soap bubble of radius 2 cm. is blown at the end of a capillary tube of length 72 cm. and of radius 1 mm. The radius of the bubble is found to be reduced to 1 cm. in one minute. Calculate the coefficient of viscosity of air assuming the surface tension of the film to be 72 dynes/cm.

   [Ans. 0.00024 C.G.S. unit] (See worked out examples) **[C.U. (Hons.) 1976]**

9. Deduce Poiseuille's equation in the case of a gas.

10. Describe how you would compare the viscosities of two liquids or of the same liquid at different temperatures.

11. Describe Rankine's experiment for the measurement of the viscosity of a gas.

12. A plate of area 200 cm^2 rests on a layer of castor oil 1 mm thick. The coefficient of viscosity of castor oil is 15.5 poise. Calculate the force required to move the plate horizontally with a speed of 4 cm./sec. **[C.U. (Hons.) '86, '91, cf. 03]**

[Ans. 1.24 Newton] (Apply the formula $F = \eta A \frac{dv}{dx}$)

13. The Reynolds number (R) for the flow of a fluid through a right circular cylindrical tube of radius a is defined by $R = 2\rho av/\eta$, where ρ = density of the fluid, v = average velocity of the fluid over a cross-section and η = coefficient of viscosity. (In the question of '87, v = velocity along the axis of the tube).

Given that turbulence sets in when $R \geq 2200$, estimate the pressure difference between the ends of the tube which will lead to the onset of turbulent flow of water in a pipe of radius 1 cm. and length 2.5 m. Given η for water = 0.01 poise. (l = 5 m in the question of '87)·

[C.U. (Hons.) '87, '90]

[**Hints.** (i) For the question of 1987, v = velocity of the fluid along the axis of the tube $= v_{max} = \frac{Pa^2}{4\eta l}$ and from the given definition of R, $v = \frac{\eta R}{2a\rho}$.

$\therefore$ Equating the above two expressions, we have

$$\frac{Pa^2}{4\eta l} = \frac{\eta R}{2a\rho} \quad \text{or} \quad P = \frac{2\eta^2 lR}{a^3\rho}$$

The data of all the quantities on the r.h.s. being known, P can be calculated.]

(ii) For the question of 1990, v = average velocity of the fluid over a cross-section $= v_{av}$.

$$= \frac{1}{\pi a^2}\int_0^a v_r 2\pi r dr = \frac{1}{\pi a^2}\int_0^a \frac{(a^2 - r^2)P}{4\eta l} 2\pi r dr$$

$$= \frac{P}{2a^2\eta l}\left[\int_0^a a^2 r dr - \int_0^a r^3 dr\right] = \frac{P}{2a^2\eta l}\left(\frac{a^4}{2} - \frac{a^4}{4}\right)$$

$$= \frac{Pa^2}{8\eta l}$$

This expression can be obtained in a very simple way. Velocity is maximum at the axis given (by $\frac{Pa^2}{4\eta l}$ and is zero at the fixed end at the tube.

$$\therefore \quad v_{av} = \frac{1}{2}\left(\frac{Pa^2}{4\eta l} + 0\right) = \frac{Pa^2}{8\eta l}$$

Now from the given definition of R,

$$v_{av} = v = \frac{\eta R}{2a\rho}$$

Equating the above two expressions, we may write

$$\frac{Pa^2}{8\eta l} = \frac{\eta R}{2a\rho} \quad \text{or} \quad P = \frac{4\eta^2 lR}{a^3\rho}$$

The data of all the quantities on the right hand side being known, P can be calculated.]

14. Write short notes on: (a) fugitive elasticity, (b) critical velocity and (c) non-Newtonian liquid.

15. A cylindrical vessel has three horizontal capillary tubes of the same bore, each 50 cm long, attached to its sides at the heights 2.5 cm, 7.5 cm and 15 cm respectively from the base. The level of the liquid in the vessel is maintained at 25 cm from the bottom by supply from outside. What should be length of a single out-flow tube having double the cross-section of the bore, which when attached horizontally at the base of the cylinder, will require supply of liquid at the same rate to maintain the same liquid height in the cylinder? **[C.U. (Hons.) '80]**

(See worked out examples)

16. A horizontal tube of diameter 2 mm and length 50 cm is connected to the bottom of a cubical tank of side 100 cm, containing water of viscosity 0.01 poise. The tank is initially full. Water is then allowed to flow out through the tube. Find the time after which the tank will be one quarter full ($g = 980$ cm/sec^2). **[C.U. (Hons.) '84]**

Answer: Let h be the height of the liquid in the tank above the horizontal tube at any instant t. The pressure difference across the ends of the tube at that instant is $p = h\rho g$, ρ being the density of the liquid. If the height of the liquid falls through dh in a very small time-interval dt, then the volume of the liquid flowing through the tube in time dt is Adh, where A is the area of cross-section of the tank. Therefore the rate of flow of the liquid at the instant t is $-\frac{d}{dt}(Adh)$. The negative sign indicates that h decreases with increase in t. Applying Poiseuille's formula, we have

$$-\frac{d(Adh)}{dt} = \frac{\pi h\rho g.a^4}{8\eta l} \quad \text{or} \quad dt = -\frac{8\eta l A}{\pi\rho g a^4}\frac{dh}{h}.$$

If t be the time required by initial level at H in the tank to fall to $H/4$, we have,

$$\int_0^t dt = -\frac{8\eta l A}{\pi\rho g a^4}\int_H^{H/4}\frac{dh}{h} = -\frac{8\eta l A}{\pi\rho g a^4}\Big[\log_e h\Big]_H^{H/4}$$

$$\text{or} \quad t = -\frac{8\eta l A}{\pi\rho g a^4}\left(\log_e\frac{H}{4} - \log_e H\right) = \frac{8\eta l A}{\pi\rho g a^4}\log_e 4.$$

In the problem, $\eta = 0.01$, $l = 50$ cm, $A = (100 \times 100) = 10^4$ sq cm.

$\pi = 3.14$, $\rho = 1$, $g = 980$ cm/sec^2, $a = 0.1$, $\log_e 4 = \log_{10} 4 \times 2.303$; so we can easily calculate t.

17. (a) What is meant by a Newtonian fluid? Give an example of a non-Newtonian fluid.

(b) What is Reynold's number? Purely from a dimensional analysis, relate Reynold's number to critical velocity and other relevant pameters.

(c) Consider a 50 cm long tube, through which water ($\eta = 0.01$ poise) is flowing. The first 25 cm of the tube is of radius 1 mm and the remaining 25 cm is of radius 2mm. If the pressure difference maintained at the two ends is 20 cm of water, calculate the amount of water flowing out per second through the tube. **[C.U. (Hons.) '84 (Spl) '04]**
(Apply the formula for capillaries in series).

18. Three narrow tubes of equal lengths but of different radii are connected in series. Use Poiseuille's formula (without any correction) to obtain an expression for the amount of liquid flowing through the tubes per second when the pressure difference between the two ends of the series is P. How will this expression be modified if the tubes are connected in parallel? **[C.U. (Hons.) '85]**

19. If water is flowing down a horizontal tube of circular cross-section of radius 1 mm under pressure gradient of 10 dynes/cm^3, find the velocity of flow at a point 0.5 mm from the centre of the tube, given that the coefficient of viscosity of water is 0.01 gm cm^{-1} sec^{-1}.

(See worked out examples) **[C.U. (Hons.) '89, '96 (old syl.)]**

20. Distinguish between a Newtonian and a non-Newtonian fluid, giving examples.

[C.U. (Hons.) 84, '91, 03, 05, 07 (old)]

21. At the end of a tube of length 10 cm and internal diameter 2 mm, a soap bubble of radius 4 cm and surface tension 30 dynes/cm is blown. The other end of the tube is now opened. Neglecting compressibility of air, find the time taken by the bubble to be reduced to a radius of 2 cm. (Coefficient of viscosity of air = 1.85×10^{-4} poise). [C.U. (Hons.) '92]

(See worked out example on page 664)

Alternative method: Pressure difference between the ends of the tube

$$P = \frac{4S}{r} \tag{1}$$

Again from Poiseuille's equation

$$V = \frac{\pi P a^4}{8\eta l} \tag{2}$$

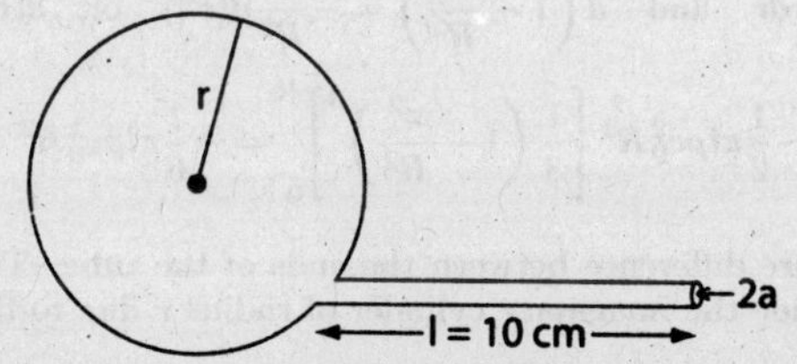

But

$$V = -\frac{d}{dt}\left(\frac{4}{3}\pi r^3\right)$$

so that we may write

$$\frac{d}{dt}\left(\frac{4}{3}\pi r^3\right) = -\frac{\pi a^4}{8\eta l}\frac{4S}{r} \quad \text{[using (1) and (2)]} \tag{3}$$

$$\text{or } \frac{4}{3}\pi \cdot 3r^2 \cdot \frac{dr}{dt} = -\frac{\pi a^4}{8\eta l}\frac{4S}{r}$$

$$\text{or } \frac{dr}{dt} = -\left(\frac{a^4 S}{8\eta l}\right)\frac{1}{r^3} \quad \text{or} \quad \int_0^T dt = -\left(\frac{8\eta l}{Sa^4}\right)\int_{r_1}^{r_2} r^3 dr$$

$$\text{or } T = -\frac{8\eta l}{Sa^4}\cdot\frac{1}{4}\left(r_2^4 - r_1^4\right) = \frac{2\eta l}{Sa^4}\left(r_1^4 - r_2^4\right)$$

$$= \frac{2 \times 1.85 \times 10^{-4} \times 10}{30 \times 10^{-4}}\left[(4)^4 - (2)^4\right]$$

$$= \frac{2 \times 1.85 \times 10 \times 240}{30} = \frac{2 \times 185 \times 24}{30}$$

$$= 296 \text{ secs} = 4 \text{ min } 56 \text{ secs.}$$

22. (a) For a rotating cylinder viscometer, derive the expression for the coefficient of viscosity. Estimate the end correction at the lower end. Discuss what happens if the speed of the rotating cylinder is increased progressively.

(b) In a breaker of water 10 cm deep, powder particles are dropped and stirred well. Assuming the particles to be spherical, calculate the radius of the largest particle that will remain in suspension after 24 hours. Given that density of the material of powder is 4 gm/cc and the coefficient of viscosity of water is 0.010 poise. [C.U. (Hons.) '93, cf. '95, 2000, 06 (old)]

(See worked out examples)

23. Liquid of density ρ and coefficient of viscosity η flows in stream-line through a capillary tube of radius R and length l. The velocity of the liquid at a distance r from the axis of the tube is given by $v(r) = v_0\left(1 - \frac{r^2}{R^2}\right)$.

Find (i) the K.E. of the liquid within the volume of the tube, (ii) the pressure difference between the ends of the tube. [C.U. (Hons.) '94]

Ans. (i) K.E. of the liquid in the annular space between two imaginary cylinders of radii r and $r + dr$ is given by

$$dT = \frac{1}{2}(2\pi r dr l\rho)v_0^2\left(1 - \frac{r^2}{R^2}\right)^2$$

$\therefore$ K.E. of the liquid within the volume of the tube is written as

$$T = \frac{1}{2}\pi l\rho v_0^2\int_{r=0}^{R} d(r^2)\left(1 - \frac{r^2}{R^2}\right)^2 = -\frac{1}{2}\pi l\rho v_0^2 R^2\int_{r=0}^{R}\left(1 - \frac{r^2}{R^2}\right)^2 d\left(1 - \frac{r^2}{R^2}\right)$$

$$\because d(r^2) = 2rdr \quad \text{and} \quad d\left(1 - \frac{r^2}{R^2}\right) = -\frac{1}{R^2}d(r^2) \quad \text{or} \quad d(r^2) = -R^2 d\left(1 - \frac{r^2}{R^2}\right)$$

$$\therefore \quad T = -\frac{1}{2}\pi l\rho v_0^2 R^2\left[\frac{1}{3}\left(1 - \frac{r^2}{R^2}\right)^3\right]_0^R = -\frac{1}{6}\pi l\rho v_0^2 R^2 \times (-1) = \frac{1}{6}\pi l\rho v_0^2 R^2$$

(ii) Let P be the pressure difference between the ends of the tube. Then the horizontal force acting on the liquid contained inside the imaginary cylinder of radius r due to the pressure difference P is $\pi r^2 P$.

Also the opposing force on the above cylinder of liquid due to the viscous force is $\eta 2\pi r l\frac{dv}{dr}$.

Again since there is no acceleration of the liquid in the cylinder, the net horizontal force must be zero;

$$\text{i.e. } \pi r^2 P + \eta 2\pi r l\frac{dv}{dr} = 0 \text{ or } r^2 P + \eta 2 r l v_0\left(-\frac{2r}{R^2}\right) = 0 \because v = v_0\left(1 - \frac{r^2}{R^2}\right) \therefore \frac{dv}{dr} = v_0\left(-\frac{2r}{R^2}\right)$$

$$\text{or } r^2 P = \frac{4\eta l v_0 r^2}{R^2} \quad \therefore \quad P = \frac{4\eta l v_0}{R^2}$$

24. Indicate the advantages which the rotating cylinder method possesses over Poiseuille's method for finding the viscosity of gases. [C.U. (Hons.) '95]

25. A thin wire of radius b is placed co-axially in a narrow tube of length l and radius a. Find the rate of volume of liquid flowing through the annular space between the wire and the tube, when a pressure difference P is maintained between the ends of the tube, if η is the coefficient of viscosity of the liquid and the steady flow condition exists. [C.U. (Hons.) '96, 07 (old syll)]

26. Set up the equation of motion of a spherical body as it descends through a fluid. Calculate its terminal velocity v_T. Draw a curve of V_T vs r, r = radius of the body.

[C.U. (Hons.) '97, 2001, 04, 05 (old syll)]

27. (a) Explain briefly the idea of critical velocity and Reynold's number.

[C.U. (Hons.) '98, 2000, 04]

(b) Two identical vertical cylinderical vessels of cross-section α are joined at their bases by a narrow horizontal tube of length l and internal radius r. A liquid of density ρ and viscosity coefficient η is flowing slowly from one vessel to the other through the tube. Initially the liquid surfaces in the two vessels are at heights $3h$ and h respectively above the capillary tube. Calculate the time taken for the difference of levels to become h. After what time will the levels be equal? [C.U. (Hons.) '98]

(See worked out examples)

28. Calculate the maximum velocity of an oil drop of radius 10^{-4} cm falling in air. Given density of oil = 0.9 gm/cc, viscosity of air = 1.8×10^{-4} gm per cm per sec, g = 980 cm/sec². Neglect air bouyancy.

Answer: The drop (radius a, density ρ) falling in air (density σ) acquires maximum (terminal) velocity v, when the resultant driving (gravity) force equals the retarding viscous force i.e., when

$$\frac{4}{3}\pi a^3(\rho-\sigma)g = 6\pi\eta a v,$$

where η is the viscosity of air. If the air boyancy is ignored i.e. $\sigma = 0$, we have

$$\frac{4}{3}\pi a^3 \rho g = 6\pi\eta a v \quad \text{or} \quad v = \frac{2}{9}\frac{a^2\rho g}{\eta} = \frac{2\times(10^{-4})^2\times 0.9\times 980}{9\times 1.8\times 10^{-4}} = 1.1\times 10^{-2} \text{ cm/sec.}$$

29. Two equal drops of water are falling through air with a steady velocity of 5 cm/sec. If the drops coalesce, what will be the new velocity?

Answer Let a be the radius of each drop. The steady terminal velocity v of a drop of radius a and density ρ, falling through air is given by $v = \dfrac{2}{9}\dfrac{a^2(\rho-\sigma)g}{\eta}$ where σ is the density and η the viscosity of air. Thus

$$v \propto a^2 \quad \text{or} \quad v = ka^2.$$

Now volume of each drop $= \frac{4}{3}\pi a^3$, so that the volume of the coalesced drop $= 2\times\frac{4}{3}\pi a^3 = \frac{4}{3}\pi\left[2^{1/3}a\right]^3$.

$\therefore$ radius of the coalesced drop $= 2^{1/3}a$.

Hence the new terminal velocity of the coalesced drop is given by

$$v' = k\left[2^{1/3}a\right]^2 = 2^{2/3}ka^2 = 2^{2/3}v = 2^{2/3}\times 5 = 7.94 \text{ cm/sec.}$$

30. A cylindrical vessel of height h is kept maintained full of water. Three capillary tubes of same radius a and same length l are fitted horizontally in the wall of the cylinder at distances $\frac{h}{2}$, $\frac{3h}{4}$ ad h measured from the top. Water can flow through these tubes. Show that the length of a single out-flow tube of the same radius which can replace the three tubes, when fitted at the bottom is $\frac{4l}{9}$.

31. A gas bubble of diameter 2 cm rises steadily through a solution of density 1.75 gm/cc at the rate of 0.35 cm/sec. Calculate the coefficient of viscosity of the solution. Neglect the density of the gas. **[Ans: 1.09×10^3 poise]**

(See worked out examples)

32. A body of mass m is falling through a viscous medium, having resistance proportional to its instantaneous velocity. If the body also experiences an upward force due to buoyancy, find out its terminal velocity. Obtain a relation connecting its displacement after a time t. If the resistance of the medium is directly proportional to the square of the velocity of the body, find its terminal velocity and also its displacement after time t. Assume that the body is not rotating during its fall through the viscous medium. **[C.U. (Hons.) 1983, 94, 01, 04]**

Answer: (a) In the treatment given in Art. 11.14 we have ignored the buoyancy-effect of the medium, and also assumed that the body is not rotating during fall, and that for small velocities, the motion is stream-line. Considering the upward force due to buoyancy, the equation of motion of the body may be written as

$$m\frac{dv}{dt} = mg - m'g - Kv, \; m' \text{ being the mass of the fluid displaced by the body}$$

$$\text{or} \quad \frac{dv}{dt} = g_e - kv \text{ where } g_e = \left(1 - \frac{m'}{m}\right) g$$

$$= \text{the effective acceleration due to gravity on the body and } k = \frac{K}{m}$$

$$\text{or} \quad \frac{dv}{g_e - kv} = dt, \text{ which on integration gives } -\tfrac{1}{k}\log_e(g_e - kv) = t + C$$

Now at $t = 0$, $v = 0$ — which gives $C = -\frac{1}{k}\log_e g_e$, so that we have

$$\frac{1}{k}\log_e g_e - \frac{1}{k}\log_e(g_e - kv) = t \quad \text{or} \quad \log_e \frac{g_e - kv}{g_e} = -kt$$

$$\text{or} \quad \frac{g_e - kv}{g_e} = e^{-kt} \quad \text{or} \quad v = \frac{dx}{dt} = \frac{g_e}{k}(1 - e^{-kt}),$$

whence the terminal or limiting speed is g_e/k, since as t increases, e^{-kt} becomes negligible compared with unity, and v will then have a constant value g_e/k.

From the above expression for velocity, we can easily find the relation between x and t (by simple integration and making the integration constant zero with suitable initial conditions), the relation being

$$x = \frac{g_e}{k^2}\left(kt + e^{-kt} - 1\right)$$

K depends on the shape of the body and k is related to viscosity through K.

(b) If however the resisting force of the medium is directly proportional to the square of the velocity of the body, the equation of motion of the body may be written as

$$m\frac{dv}{dt} = mg - m'g - \alpha v^2$$

$$\text{or} \quad m\frac{dv}{dt} = mg_e - \alpha v^2 = \alpha(V^2 - v^2)$$

where $g_e = \left(1 - \frac{m'}{m}\right) g$ and considering that $mg_e = \alpha V^2$ from the condition of the problem, so that we have

$$\int \frac{dv}{V^2 - v^2} = \int \frac{\alpha}{m} dt \quad \text{or} \quad \frac{1}{2V}\log_e\left(\frac{V + v}{V - v}\right) = \frac{\alpha}{m} t + C_1.$$

Now at $t = 0$, $v = 0$ — which gives the integration constant $C_1 = 0$

$$\therefore \; t = \frac{m}{2\alpha V}\log_e\left(\frac{V + v}{V - v}\right) = \frac{V}{2g_e}\log_e\left(\frac{V + v}{V - v}\right)$$

whence on simplification, we get

$$v = V\tanh\frac{g_e t}{V}$$

If the initial velocity of the body is less than V, its velocity will go on increasing with time until it reaches a value V — known as the **terminal velocity** of the body.

Writing $v = \frac{dx}{dt}$ in the above equation we can show that the relation between x and t is given by

$$x = \frac{V^2}{g_e}\log_e\cosh\frac{g_e t}{V}$$

33. Water flows in a stream-line through a horizontal pipe of length l and of cross-sectional radius a from a vertical tank of cross-section A. Find the time when the height of the level in the tank becomes half, if η be the coefficient of viscosity. (**Ans.** $t = \frac{8l\eta A}{\pi \rho g a^4} \log_e 2$)

(See worked out examples)

34. A cylindrical vessel of radius 0.05 m has at its bottom a horizontal capillary tube of length 0.2 m and internal radius 0.4 mm. The vessel is completely filled with water, and then the water is allowed to flow out of the capillary. How long will it take for the water level to fall to half its initial height in the vessel? The viscosity of water is 0.001 MKS unit ($g = 9.8$ m/sec^2, density of water $= 1 \times 10^3$ Kg/m^3, $\log_e 2 = 0.6930$). [**Ans.** 1.1×10^4 **secs** $\approx$ **3 hours**)

(See worked out examples)

35. Prove that if the level of water in a reservoir falls from mark A to B in 120 secs, when it flows through a capillary, and in case of a liquid of specific gravity 0.9, the level falls from A to B in 100 secs, then the ratio of the viscosity of the liquid to that of water is 3 : 4.

(See worked out examples)

36. A horizontal capillary of diameter 2 mm and length 20 cm is connected to a tank of water and 0.2 cc of water flows out per sec. What will be the rate of flow of water if another capillary of length 10 cm and diameter 1 mm is joined in series with the first capillary? (**Ans.** 0.022 cc/sec)

(See worked out examples)

37. A tube of radius R and length L is connected in series with another of radius $R/2$ and length $L/4$. If the pressure across the two tubes taken together is p, calculate the pressures across the tubes separately. (**Ans.** $p/5, 4p/5$)

(See worked out examples)

38. Two capillaries of lengths 70 cm and 30 cm and radii 0.2 mm and 0.1 mm respectively are connected in series. If a liquid enters the first capillary at a pressure of 80 cm of mercury and leaves the second capillary at 70 cm of mercury, find the pressure at the junction of the two capillaries. (**Ans.** 78.73 cm of Hg)

(See worked out examples)

39. Three capillaries of lengths 8L, 0.2L and 2L and radii, $r, 0.2r$ and $0.5r$ respectively are connected in series. If the total pressure across the system in an experiment is p, calculate the pressure across the shortest (middle) capillary. (**Ans.** $\frac{25}{33}p$)

(See worked out examples)

40. An air bubble of 1 cm radius rises through a long column of liquid of density 1.47×10^3 kg/m^3 with a steady velocity of 0.21 cm/sec. Find the viscosity of the liquid ($g = 9.8$ m/sec^2). Neglect the density of air. (**Ans.** 1.52×10^2 Kg/(m–sec)

(See worked out examples)

41. With what terminal velocity will an air bubble 1.0 mm in diameter rise in a liquid of viscosity 150 centipoise and density 0.90 gm/cm^3? (**Ans.** 0.33 cm/sec)

(See worked out examples)

42. Determine the radius of a drop of water falling through air, if the terminal velocity of the drop is 1.2 cm/sec. Assume the coefficient of viscosity of air $= 1.8 \times 10^{-4}$ cgs units and the density of air $= 1.21 \times 10^{-3}$ gm/cc. (**Ans.** 0.001 cm/sec)

(See worked out examples)

43. (a) What is meant by a non-Newtonian liquid?

(b) State clearly the assumptions required to derive Poiseuille's formula for the flow of a liquid through a narrow horizontal tube. [C.U. (Hons.) 01, 05, 08]

(c) 864 ml of water flows in 12 mins. through a horizontal capillary tube of length 20 cm and radius 0.081 cm under 20 cm water pressure. Find out the coefficient of viscosity of water. What is the Reynolds number corresponding to the above flow of water?
(See worked out examples) [C.U. (Hons.) 01, 05]

44. (a) Define stream line motion and stream line for a fluid flow. Justify the statement—"Two stream-lines cannot intersect." [C.U. (Hons.) 2002, 05, 06 (old syll)]

(b) A Newtonian liquid is maintained between two large horizontal plates and is made to flow in a stream line motion along a horizontal path. A uniform velocity gradient 30/sec. is maintained in the vertical direction. Calculate the tangential stress necessary to maintain the uniform flow. [C.U. (Hons.) 2002]

[**Hints.** Apply the formula $f = \eta \frac{dv}{dx}$]

45. A square lamina of area 100 cm^2 is placed on a liquid layer of 2 mm. thickness and the force required to pull of lamina horizontally with a velocity 4 cm/sec. is 0.2 N. Find the coefficient of viscosity of the liquid. [C.U. (Hons.) 2003]

[**Hints.** Apply the formula $F = \eta A \frac{dv}{dx}$]

46. Explain why large rain drops come down very fast. [C.U. (Hons.) 04]
(See Art. 11.13.1)

47. Water flows steadily through a composite system consisting of two narrow tubes A and B joined end to end. A has a length 25 cm and radius 0.03 cm, while the tube B has a length 10 cm and radius 0.04 cm. If the total pressure difference between the extreme ends be 16 cm of water, calculate the pressure difference between the ends of the tube A. [C.U. (Hons.) 2006]
(See worked out problems)

48. (a) How does the coefficient of viscosity of a liquid normally vary with temperature?

(b) Explain the terms 'flow line' and 'stream line'.

(c) Define Reynold's number. Discuss how the estimation of Reynold's number helps to determine whether a motion is streamline or turbulent.

(d) Coefficient of viscosity of a liquid A is double that of another liquid B. Which liquid will flow more readily and why? [C.U. (Hons.) 2007]

[**Hints.** Liquid B will flow more readily because coefficient of viscosity of B is less than that of A. For reasons see Art. 11.1 and 11.3.]

49. (a) "Coefficient of viscosity of glycerine is 8.4 poise"—Explain.

(b) A liquid of coefficient of viscosity η flows steadily through a cylindrical tube of radius r and length l under a pressure P. Show that its velocity at a point inside the tube situated at a distance x from the axis is given by $v = \frac{P}{4\eta l}(r^2 - x^2)$. [C.U. (Hons.) 2008]
(See Art. 11.6)

Chapter 12

Fluid Mechanics

12.1 Introduction

Fluid mechanics deals with the laws of equilibrium and motion of a fluid (liquid or gas) and also the interaction of the fluid with the solids along which it flows. In fluid mechanics we do not take into consideration, the molecular structure of the fluid, but consider the fluid as a continuous medium.

Fluid statics is one branch of fluid mechanics which deals with the conditions and laws of equilibrium of a fluid under the action of external forces. *Fluid dynamics* is another branch of fluid mechanics which deals with the laws of motion of the fluid and its interaction with solids. The fluid under consideration may be incompressible or compressible. Usually a liquid is treated as an *incompressible* fluid since the effect of pressure on the density of the liquid can be neglected. A gas is called a *compressible* fluid as the pressure in this case, can affect the density of the gas. A fluid is called an *ideal fluid* when it has no viscosity which causes friction. In practice, we call a fluid to be an ideal one when we ignore the viscosity of the fluid for simplicity of calculation. So we can regard a fluid medium as a *perfect fluid,* incapable of exerting shearing stress, and whether at rest or in motion, such that the pressure it exerts on any surface in contact with it is always normal to the surface, and consequently, the pressure at any point in such a fluid is the same in every direction.

12.2 Fluid statics

Here we shall consider the following types of external forces acting on an element of volume of the fluid.

(i) Mass force or body force. (ii) Surface force.

Mass force or body force: This force acting on the elementary volume of the fluid does not depend on the existence of the other portions of the fluid, but depends on the mass of the element. In fact, the force is proportional to the mass of the element.

Let us consider that the body force acting on unit mass of the fluid is $\vec{F}$. Then the force acting on the elementary volume dV of the fluid $= \vec{F}\rho dV$, where ρ is the density of the fluid.

The vector $\vec{F}$ is called the *intensity of the body force field.* In case of gravity we may write $\vec{g}$ for $\vec{F}$. When the body force is conservative, we may write

$$\vec{F} = -\operatorname{grad}\phi, \text{ where}$$

$$\operatorname{grad}\phi = \hat{\imath}\frac{\partial\phi}{\partial x} + \hat{\jmath}\frac{\partial\phi}{\partial y} + \hat{k}\frac{\partial\phi}{\partial z},$$

ϕ being a scalar function. $\phi(x, y, z, t)$ is called *body force potential.*

Surface force: An element of the fluid is acted on by this force due to the adjacent particles of the rest of the fluid. The surface force on unit area over which it acts is called *average pressure intensity.* In equilibrium condition the tangential components of the surface force in a fluid are zero and only the normal components of the surface force exist. The pressure intensity at any point in a fluid is the same in all directions.

12.3 Condition of equilibrium of a fluid

Let us consider a container of liquid at rest. We choose an infinitesimal rectangular parallelopiped fluid in shape within the container. Let us now consider a point $Q(x, y, z)$ within the rectangular parallelopiped.

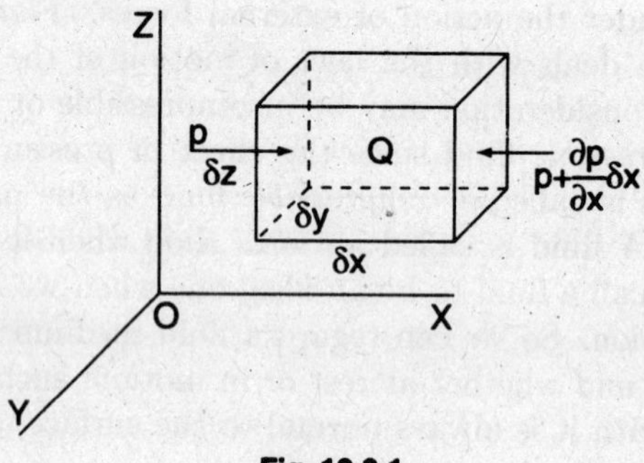

Fig. 12.3.1

Let the components of the body force $\vec{F}$ at Q be F_x, F_y and F_z. Now, the force due to pressure p on the area $\delta y\delta z$ of the parallelopiped along the X-axis $= p\delta y\delta z$.

The same force on the opposite face of the parallelopiped

$$= \left(p + \frac{\partial p}{\partial x}\delta x\right)\delta y\delta z$$

acting along the negative direction of X-axis. Hence the resultant force due to pressure along X-axis

$$= p\delta y\delta z - \left(p + \frac{\partial p}{\partial x}\delta x\right)\delta y\delta z = -\frac{\partial p}{\partial x}\delta x\delta y\delta z.$$

For equilibrium under the action of the body force we may write

$$F_x\rho\delta x\delta y\delta z - \frac{\partial p}{\partial x}\delta x\delta y\delta z = 0 \text{ or } F_x - \frac{1}{\rho}\frac{\partial p}{\partial x} = 0. \tag{12.3.1}$$

Similarly we have,

$$F_y - \frac{1}{\rho}\frac{\partial p}{\partial y} = 0 \tag{12.3.2}$$

$$\text{and} \quad F_z - \frac{1}{\rho}\frac{\partial p}{\partial z} = 0 \tag{12.3.3}$$

Multiplying the above equations by $\hat{\imath}, \hat{\jmath}, \hat{k}$ respectively and then adding, we have

$$\hat{\imath}F_x + \hat{\jmath}F_y + \hat{k}F_z - \frac{1}{\rho}\left(\hat{\imath}\frac{\partial p}{\partial x} + \hat{\jmath}\frac{\partial p}{\partial y} + \hat{k}\frac{\partial p}{\partial z}\right) = 0$$

$$\text{or} \quad \vec{F} - \frac{1}{\rho}\,\text{grad}\, p = 0. \tag{12.3.4}$$

We know that if the force $\vec{F}$ be the gradient of a scalar, then the curl of the force vanishes (Art. 3.13). So, in this case curl $\vec{F} = 0$. Hence for equilibrium under the action of the body force, the force must be conservative.

If the force be due to gravity, $F_x = F_y = 0$ and $F_z = -g$ [Here we take the Z-axis vertically upward and the X- and Y-axes in the horizontal plane].

$\therefore$ from equation (12.3.3) we have

$$-g - \frac{1}{\rho}\frac{dp}{dz} = 0 \quad \text{or} \quad \frac{dp}{\rho} = -g dz. \tag{12.3.5}$$

For an incompressible fluid at constant temperature, the density ρ is independent of pressure and (12.3.5) gives on integration,

$$p = -\rho g z + k,$$

where k is the constant of integration. At $z = 0, p = p_0$ (say); hence we have

$$p = -\rho g z + p_0. \tag{12.3.6}$$

This equation is the basic equation of hydrostatics for incompressible fluids. Here p_0 denotes the external pressure on the surface of the fluid. From the equation (12.3.6) it follows that $p - p_0$ does not depend on p_0. This means that pressure exerted by external forces on the fluid is transmitted equally in all directions. This is what is known as *Pascal's principle*.

From the equation $p = -\rho g z + k$, we can also establish *Archimedes' principle* for a body immersed in a fluid as shown in the Fig. 12.3.2. Considering the pressures p_1 and p_2 on the upper and lower surfaces of the body respectively, we may write

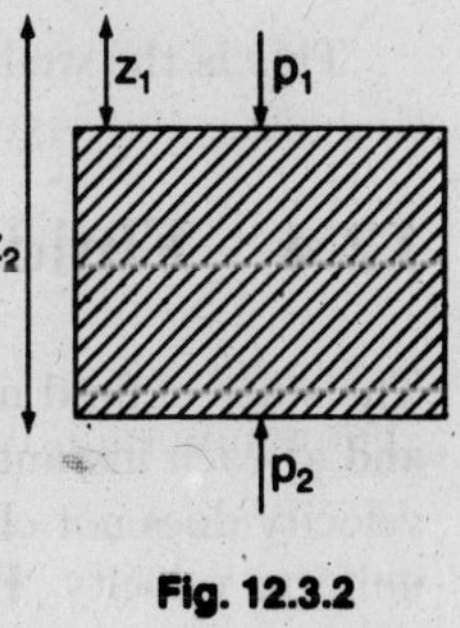

Fig. 12.3.2

$$p_1 = -\rho g z_1 + k$$

$$p_2 = -\rho g z_2 + k$$

whence we get

$$(p_1 - p_2) = \rho g (z_2 - z_1)$$

where z_1 and z_2 are the depths of the upper and lower surfaces of the body from the surface of the fluid.

Now multiplying the above equation by the cross-sectional area α of the body, we have

$$\alpha(p_1 - p_2) = \rho g \alpha(z_2 - z_1).$$

Or thrust = weight of the fluid displaced, acting upward, ($\because p_2 > p_1$).

This is what is known as *Archimedes' principle.*

It may be mentioned here that equation (12.3.5) can be directly obtained from Bernoulli's equation (12.8.19) by putting $v = 0, h = z$ and then differentiating and rearranging the terms.

For gases at constant temperature, the density is proportional to the pressure (by Boyle's law), i.e.

$$\rho \propto p \text{ or } \frac{\rho}{\rho_0} = \frac{p}{p_0}.$$

Hence equation (12.3.5) gives

$$\frac{dp}{p} = -\frac{\rho_0}{p_0} g dz.$$

Integrating between $z = 0$ (on the surface of the earth) and $z = z$, we have,

$$\log \frac{p}{p_0} = -\frac{\rho_0}{p_0} g z,$$

p_0 being the value of p at $z = 0$

$$\text{or } p = p_0 e^{-\left(\frac{\rho_0}{p_0}\right) g z}. \tag{12.3.7}$$

The equation (12.3.7) is due to Halley. This expression gives the observed pressures at different altitudes. It is to be noted that this formula holds provided the temperature remains the same at all altitudes.

Taking $\rho = \frac{Mp}{RT}$, for a perfect gas, [M = molecular weight of the gas] we have from equation (12.3.5)

$$\frac{dp}{dz} = -\rho g = -\frac{Mp}{RT} g \quad \text{or} \quad \frac{dp}{p} = -\frac{Mg}{RT} dz$$

$$\text{or} \quad \log p = -\frac{Mg}{RT} z + \log p_0 \tag{12.3.8}$$

[taking the constant of integration $k = \log p_0$]

$$\text{or } p = p_0 e^{-\frac{Mg}{RT} z}. \tag{12.3.9}$$

This is the well-known *law of atmospheres.*

12.4 Fluid dynamics

The state of fluid motion is completely defined only if the flow velocity is known at every point and at each instant. In practice we consider the problem of stationary states in which the flow velocity does not change with time. For example, let us consider the flight of an aeroplane with uniform velocity. Here we may consider the aeroplane stationary and the air flowing past the plane with uniform velocity, i.e. the flow of air around the plane is steady. In the unsteady state in which case the velocity changes with time, the paths described by single particles of the fluid in course of time are called *lines of motion*. *A stream line* is an imaginary line where the tangent at each point at a given instant coincides with the velocity vector of the fluid at this point. For steady motion of the fluid, the stream lines coincide with the paths of the fluid particles. When a fluid flows in stream lines in a tube, these stream lines are parallel to the axis of the tube. Aggregation of these stream lines constitutes what is known as a *Stream tube*.

12.5 General equations of fluid motion—Euler's equation of motion

We can make transformation from fluid statics to fluid dynamics by introducing the inertial or effective force per unit volume $-\rho\vec{f}$ by d' Alembert's principle (Art. 3.18). By this principle we can use statical rules to problems in dynamics. Using the equation (12.3.4) in fluid statics and considering equilibrium between inertial force, body force and pressure force we may write

$$\rho\vec{F} - \rho\vec{f} - \text{grad}\, p = 0 \text{ or } \rho\vec{F} - \rho\frac{d^2\vec{r}}{dt^2} - \text{grad}\, p = 0 \text{ or } \frac{d^2\vec{r}}{dt^2} = \vec{F} - \frac{1}{\rho}\text{grad}\, p. \qquad (12.5.1)$$

This equation is called *Euler's equation of motion*. In determining the velocity vector $\vec{v}$ and hence the acceleration vector $\vec{f}$ or $\frac{d^2\vec{r}}{dt^2}$ we shall use Eulerian method, according to which the fluid motion is determined by specifying the velocity in space at each instant of time, i.e.

$$\vec{v} = \phi(\vec{r}, t),$$

where

$$\vec{v} = \hat{i}v_x + \hat{j}v_y + \hat{k}v_z \text{ and}$$

$$\vec{r} = \hat{i}x + \hat{j}y + \hat{k}z,$$

x, y and z are called *Eulerian variables*.

The components of the acceleration vector $\vec{f}$ of the particles of the fluid are

$$f_x = \frac{dv_x}{dt} = \frac{\partial v_x}{\partial t} + \frac{\partial v_x}{\partial x}v_x + \frac{\partial v_x}{\partial y}v_y + \frac{\partial v_x}{\partial z}v_z$$

$$f_y = \frac{dv_y}{dt} = \frac{\partial v_y}{\partial t} + \frac{\partial v_y}{\partial x}v_x + \frac{\partial v_y}{\partial y}v_y + \frac{\partial v_y}{\partial z}v_z$$

$$f_z = \frac{dv_z}{dt} = \frac{\partial v_z}{\partial t} + \frac{\partial v_z}{\partial x}v_x + \frac{\partial v_z}{\partial y}v_y + \frac{\partial v_z}{\partial z}v_z$$

Thus from equation (12.5.1) using the above relations, the component forms of the general equation of fluid motion may be written as

$$\left.\begin{aligned}\frac{\partial v_x}{\partial t}+\frac{\partial v_x}{\partial x}v_x+\frac{\partial v_x}{\partial y}v_y+\frac{\partial v_x}{\partial z}v_z&=F_x-\frac{1}{\rho}\frac{\partial p}{\partial x}\\ \frac{\partial v_y}{\partial t}+\frac{\partial v_y}{\partial x}v_x+\frac{\partial v_y}{\partial y}v_y+\frac{\partial v_y}{\partial z}v_z&=F_y-\frac{1}{\rho}\frac{\partial p}{\partial y}\\ \frac{\partial v_z}{\partial t}+\frac{\partial v_z}{\partial x}v_x+\frac{\partial v_z}{\partial y}v_y+\frac{\partial v_z}{\partial z}v_z&=F_z-\frac{1}{\rho}\frac{\partial p}{\partial z}\end{aligned}\right\}\qquad(12.5.2)$$

Multiplying the three equations (12.5.2) respectively by unit vectors, $\hat{\imath}, \hat{\jmath}$ and $\hat{k}$ and then adding we have,

$$\hat{\imath}\frac{\partial v_x}{\partial t}+\hat{\jmath}\frac{\partial v_y}{\partial t}+\hat{k}\frac{\partial v_z}{\partial t}+\left[\left(\hat{\imath}v_x+\hat{\jmath}v_y+\hat{k}v_y\right)\cdot\left(\hat{\imath}\frac{\partial}{\partial x}+\hat{\jmath}\frac{\partial}{\partial y}+\hat{k}\frac{\partial}{\partial z}\right)\right]$$

$$\left(\hat{\imath}v_x+\hat{\jmath}v_y+\hat{k}v_z\right)=\left(\hat{\imath}F_x+\hat{\jmath}F_y+\hat{k}F_2\right)-\frac{1}{\rho}\left(\hat{\imath}\frac{\partial p}{\partial x}+\hat{\jmath}\frac{\partial p}{\partial y}+\hat{k}\frac{\partial p}{\partial z}\right)$$

or $$\frac{\partial\vec{v}}{\partial t}+\left(\vec{v}\cdot\vec{\nabla}\right)\vec{v}=\vec{F}-\frac{1}{\rho}\vec{\nabla}p.\qquad(12.5.3)$$

Now, from vector formula we have,

$$\vec{\nabla}\left(\vec{A}\cdot\vec{B}\right)=\left(\vec{B}\cdot\vec{\nabla}\right)\vec{A}+\vec{B}\times\left(\vec{\nabla}\times\vec{A}\right)+\left(\vec{A}\cdot\vec{\nabla}\right)\vec{B}+\vec{A}\times\left(\vec{\nabla}\times\vec{B}\right)$$

[See Art. 2.24]

For $\vec{A}=\vec{B}=\vec{v}$, we have

$$\vec{\nabla}\left(\vec{v}\cdot\vec{v}\right)=2\left(\vec{v}\cdot\vec{\nabla}\right)\vec{v}+2\vec{v}\times\left(\vec{\nabla}\times\vec{v}\right)$$

or $$\frac{1}{2}\vec{\nabla}\left(v^2\right)=\left(\vec{v}\cdot\vec{\nabla}\right)\vec{v}+\vec{v}\times\left(\vec{\nabla}\times\vec{v}\right)$$

or $$\left(\vec{v}\cdot\vec{\nabla}\right)\vec{v}=\frac{1}{2}\vec{\nabla}\left(v^2\right)-\vec{v}\times\left(\vec{\nabla}\times\vec{v}\right)$$

Using this relation, equation (12.5.3) becomes

$$\frac{\partial\vec{v}}{\partial t}+\frac{1}{2}\vec{\nabla}\left(v^2\right)-\vec{v}\times\left(\vec{\nabla}\times\vec{v}\right)=\vec{F}-\frac{1}{\rho}\vec{\nabla}p\qquad(12.5.4)$$

Equation (12.5.4) may also be written in the form

$$\frac{\partial\vec{v}}{\partial t}+\frac{1}{2}\,\mathrm{grad}\,v^2-\left[\vec{v}\,\mathrm{curl}\,\vec{v}\right]=\vec{F}-\frac{1}{\rho}\,\mathrm{grad}\,p\qquad(12.5.5)$$

{[$\vec{v}\,\mathrm{curl}\,\vec{v}$] indicates cross product of $\vec{v}$ and $\mathrm{curl}\,\vec{v}$}.

For stationary flow $\frac{\partial \vec{v}}{\partial t} = 0$ and so equation (12.5.5) may be written as

$$\frac{1}{2}\,\text{grad}\,v^2 - [\vec{v}\,\text{curl}\,\vec{v}] = \vec{F} - \frac{1}{\rho}\,\text{grad}\,p. \tag{12.5.6}$$

Equation (12.5.5) or (12.5.6) which is known as *Euler's equation,* is the fundamental equation of fluid dynamics.

It is to be noted that equation (12.5.6) has been deduced neglecting the effect of viscosity of the fluid.

12.6 Equation of continuity

Besides the equation (12.5.5) or (12.5.6) which is the fundamental equation of fluid dynamics, we have another equation called the *equation of continuity.* The equation of continuity is an expression of the law of conservation of mass in fluid mechanics.

We consider an element of volume of the fluid rectangular parallelopiped in shape of edges $\delta x, \delta y$ and δz (Fig. 12.3.1). Let the components of velocity of the fluid at Q be v_x, v_y and v_z. Now the mass of the fluid flowing in through the area $\delta y \delta z$ of the element per second parallel to the X-axis

$= \rho v_x \delta y \delta z$ and the mass flowing out through the opposite face

$$= \left\{\rho v_x + \frac{\partial}{\partial x}(\rho v_x)\delta x\right\}\delta y \delta z$$

Hence the net mass flowing out per second parallel to the X-axis

$$= \frac{\partial}{\partial x}(\rho v_x)\delta x \delta y \delta z.$$

Similarly the net mass flowing out per second parallel to the Y-axis

$$= \frac{\partial}{\partial y}(\rho v_y)\delta x \delta y \delta z$$

and that flowing out per second parallel to the Z-axis

$$= \frac{\partial}{\partial z}(\rho v_z)\delta x \delta y \delta z.$$

Hence the total mass flowing out of the parallelopiped per second

$$= \left[\frac{\partial}{\partial x}(\rho v_x) + \frac{\partial}{\partial y}(\rho v_y) + \frac{\partial}{\partial z}(\rho v_z)\right]\delta x \delta y \delta z$$

$$= \left(\vec{\nabla} \cdot \rho \vec{v}\right)\delta x \delta y \delta z.$$

Now according to the law of conservation of mass, the mass flowing out per second must be equal to the decrease in the amount of fluid per second within the element of volume. Hence

$$-\frac{\partial \rho}{\partial t}\delta x \delta y \delta z = \left(\vec{\nabla} \cdot \rho \vec{v}\right)\delta x \delta y \delta z \quad \text{or} \quad \frac{\partial \rho}{\partial t} + \vec{\nabla} \cdot \rho \vec{v} = 0. \tag{12.6.1}$$

This equation may also be written as

$$\frac{\partial \rho}{\partial t} + \operatorname{div}(\rho \vec{v}) = 0. \tag{12.6.2}$$

This is the *equation of continuity*.

For incompressible fluids the equation (12.6.2) is simplified to

$$\operatorname{div} \vec{v} = 0 \quad \left(\because \frac{\partial \rho}{\partial t} = 0\right).$$

We know that when $\operatorname{curl} \vec{F} = 0$, the force $\vec{F}$ may be expressed as the gradient of a scalar. Similarly if $\operatorname{curl} \vec{v} = 0, \vec{v}$ may be expressed as the gradient of a scalar. Now, for an irrotational flow $\operatorname{curl} \vec{v} = 0$ and so $\vec{v} = -\operatorname{grad} \phi$, where ϕ is called *velocity potential.* The equation of continuity for incompressible fluids then becomes div $\vec{v}$ = div grad $\phi = 0$ or $\nabla^2 \phi = 0$ and the Euler's hydrodynamical equation (12.5.6) becomes

$$\frac{1}{2} \operatorname{grad} v^2 = \vec{F} - \frac{1}{\rho} \operatorname{grad} p. \tag{12.6.3}$$

The velocity potential ϕ which satisfies the differential Laplace's equation $\nabla^2 \phi = 0$ represents the velocity potential for an irrotational fluid motion. When ϕ is determined by solving the above equation, the velocity field $\vec{v}$ may be obtained by taking the gradient of ϕ.

Thus the pressure distribution may be calculated from equation (12.6.3).

12.7 Bernoulli's equation and Conservation of energy

If the body force $\vec{F}$ is derived from a potential U, we have

$$\vec{F} = -\operatorname{grad} U = -\vec{\nabla} U.$$

Hence equation (12.5.4) for steady flow turns out to be

$$\frac{1}{2}\vec{\nabla}\left(v^2\right) - \vec{v} \times \left(\vec{\nabla} \times \vec{v}\right) = -\vec{\nabla} U - \frac{1}{\rho}\vec{\nabla} p. \tag{12.7.1}$$

For irrotational flow of an incompressible fluid we have,

$$\vec{\nabla} \times \vec{v} = 0 \text{ and } \frac{1}{\rho}\vec{\nabla} p = \vec{\nabla}\frac{p}{\rho}.$$

Hence equation (12.7.1) may be written as

$$\frac{1}{2}\vec{\nabla}\left(v^2\right) + \vec{\nabla}\frac{p}{\rho} + \vec{\nabla} U = 0 \quad \text{or} \quad \vec{\nabla}\left(\frac{1}{2}v^2 + \frac{p}{\rho} + U\right) = 0$$

$$\text{or} \quad \frac{1}{2}v^2 + \frac{p}{\rho} + U = \text{constant}. \tag{12.7.2}$$

This is Bernoulli's equation, which expresses the principle of conservation of mechanical energy, since $\frac{1}{2}v^2$ is the K.E. per unit mass, $U = gh$ is the gravitational P.E. per unit mass at a height h above the datum level, and $\left(U + \frac{p}{\rho}\right)$ is the P.E. function per unit mass.

12.8 Bernoulli's theorem

Bernoulli's theorem states that when an incompressible non-viscous fluid flows in stream line motion, the sum of the pressure head, gravitational head, and kinetic energy head remains constant throughout its motion. When stated mathematically, we may write,

$$\frac{p}{\rho g} + \frac{v^2}{2g} + h = H(\text{const.}) \tag{12.8.1}$$

[Putting $U = gh$ in (12.7.2) we get this expression.]

where $\frac{p}{\rho g}$ = pressure head, $\frac{v^2}{2g}$ = kinetic energy head, h = gravitational head and H = total head.

Deduction of Bernoulli's equation (Alternative treatment)

In order to deduce the equation (12.8.1), we shall make the following assumptions:

(i) the liquid is incompressible, i.e. the density ρ is independent of pressure p;

(ii) the liquid is non-viscous, i.e. there is no loss of energy in overcoming friction;

(iii) the velocity of the fluid is below the critical value, so that the motion is stream line, and there is no loss of energy due to formation of vortex, eddies, etc.;

(iv) there are no sudden contraction, expansion and no sharp bending—which may cause loss of energy.

In order to deduce Bernoulli's theorem, let us take a point P inside the moving fluid, and its co-ordinates are expressed by (x, y, z, t) in the cartesian system, the fourth variable being the time variable 't'. Let q be the velocity of the liquid at the point P, and let u, v, w be the components of velocity along X, Y, Z axes respectively. So these components of velocity are functions of (x, y, z, t). As time 't' varies, there is change in velocity at the point P, and hence acceleration is produced. If we want to find out the acceleration at P, it is not sufficient to find out the rate of change of velocity with time only, but we must also consider whether a small volume of liquid, as it is moving onwards at the point P, does so with a velocity it had on arriving at P or it had some additional velocity.

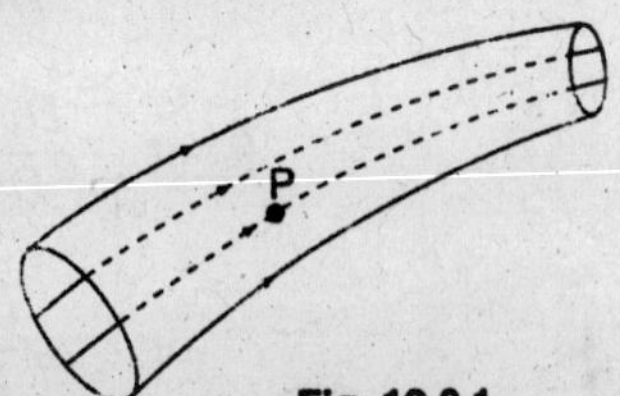

Fig. 12.8.1

Now the velocity component $v_x = f(x, y, z, t)$. Let the velocity be $v_x + dv_x$ after an interval of time dt, and then this velocity is a function of $(x + v_x\,dt, y + v_y\,dt, z + v_z\,dt, t + dt)$;

$$\text{or } v_x + dv_x = f[x + v_x\,dt, y + v_y\,dt, z + v_z\,dt, t + dt]$$

$$= f(x, y, z, t) + v_x\frac{\partial f}{\partial x}dt + v_y\frac{\partial f}{\partial y}dt + v_z\frac{\partial f}{\partial z}dt + \frac{\partial f}{\partial t}dt$$

$$= v_x + v_x\frac{\partial v_x}{\partial x}dt + v_y\frac{\partial v_x}{\partial y}dt + v_z\frac{\partial v_x}{\partial z}dt + \frac{\partial v_x}{\partial t}dt$$

$$\therefore \quad \frac{dv_x}{dt} = v_x\frac{\partial v_x}{\partial x} + v_y\frac{\partial v_x}{\partial y} + v_z\frac{\partial v_x}{\partial z} + \frac{\partial v_x}{\partial t} \tag{12.8.2}$$

Hence the x-component of acceleration is

$$\left(\frac{\partial v_x}{\partial t}+v_x\frac{\partial v_x}{\partial x}+v_y\frac{\partial v_x}{\partial y}+v_z\frac{\partial v_x}{\partial z}\right).$$

In the above expression, the first term denotes the change in velocity with respect to time at the point $P(x,y,z,t)$ inside the moving fluid. The last three terms denote the change in velocity as the point P is moving onwards and thereby (x,y,z) is changed.

Let us put,

$$\left(\frac{\partial}{\partial t}+v_x\frac{\partial}{\partial x}+v_y\frac{\partial}{\partial y}+v_z\frac{\partial}{\partial z}\right)=\frac{D}{Dt}. \tag{12.8.3}$$

Thus writing the above left-hand expression as an operator $\frac{D}{Dt}$, the (x,y,z) components of acceleration are

$$\frac{Dv_x}{Dt},\ \frac{Dv_y}{Dt},\ \frac{Dv_z}{Dt}.$$

Let F_x, F_y, F_z be the components of the impressed force per unit mass at (x,y,z) and p the pressure at this point. Then the resulting force acting on an element of volume $dx\,dy\,dz$ in the X-direction

$$=-\frac{\partial p}{\partial x}\,dx\,dy\,dz+F_x\rho\,dx\,dy\,dz.$$

Again the resulting force

$$=\frac{Dv_x}{Dt}\rho\,dx\,dy\,dz.$$

Hence

$$\frac{Dv_x}{Dt}\rho\,dx\,dy\,dz=-\frac{\partial p}{\partial x}\,dx\,dy\,dz+F_x\rho\,dx\,dy\,dz$$

or
$$\frac{Dv_x}{Dt}+\frac{1}{\rho}\frac{\partial p}{\partial x}-F_x=0$$

or
$$v_x\frac{\partial v_x}{\partial x}+v_y\frac{\partial v_x}{\partial y}+v_z\frac{\partial v_x}{\partial z}+\frac{\partial v_x}{\partial t}+\frac{1}{\rho}\frac{\partial p}{\partial x}-F_x=0 \tag{12.8.4}$$

Similarly

$$v_x\frac{\partial v_y}{\partial x}+v_y\frac{\partial v_y}{\partial y}+v_z\frac{\partial v_y}{\partial z}+\frac{\partial v_y}{\partial t}+\frac{1}{\rho}\frac{\partial p}{\partial y}-F_y=0 \tag{12.8.5}$$

and
$$v_x\frac{\partial v_z}{\partial x}+v_y\frac{\partial v_z}{\partial y}+v_z\frac{\partial v_z}{\partial z}+\frac{\partial v_z}{\partial t}+\frac{1}{\rho}\frac{\partial p}{\partial z}-F_z=0 \tag{12.8.6}$$

These are Euler's dynamical equations.

Let us now assume the steady state condition so that

$$\frac{\partial v_x}{\partial t}=0,\ \frac{\partial v_y}{\partial t}=0,\ \frac{\partial v_z}{\partial t}=0.$$

$\therefore$ the above equations reduce to

$$v_x \frac{\partial v_x}{\partial x} + v_y \frac{\partial v_x}{\partial y} + v_z \frac{\partial v_x}{\partial z} + \frac{1}{\rho}\frac{\partial p}{\partial x} - F_x = 0 \qquad (12.8.7)$$

$$v_x \frac{\partial v_y}{\partial x} + v_y \frac{\partial v_y}{\partial y} + v_z \frac{\partial v_y}{\partial z} + \frac{1}{\rho}\frac{\partial p}{\partial y} - F_y = 0 \qquad (12.8.8)$$

$$v_x \frac{\partial v_z}{\partial x} + v_y \frac{\partial v_z}{\partial y} + v_z \frac{\partial v_z}{\partial z} + \frac{1}{\rho}\frac{\partial p}{\partial z} - F_z = 0 \qquad (12.8.9)$$

Case I

Let us first suppose that the motion is *irrotational,* so that we may, from the definition of irrotational flow, express the velocity components in terms of a single potential function ϕ, such that

$$v_x = -\frac{\partial \phi}{\partial x}, \quad v_y = -\frac{\partial \phi}{\partial y}, \quad v_z = -\frac{\partial \phi}{\partial z}$$

$$\therefore \quad \frac{\partial v_x}{\partial y} = -\frac{\partial}{\partial y}\left(\frac{\partial \phi}{\partial x}\right) = -\frac{\partial^2 \phi}{\partial x \partial y} \qquad [\because d\phi \text{ is a perfect differential}]$$

$$= \frac{\partial}{\partial x}\left(-\frac{\partial \phi}{\partial y}\right) = \frac{\partial v_y}{\partial x}$$

$$\text{Similarly,} \quad \frac{\partial v_y}{\partial z} = \frac{\partial v_z}{\partial y}, \quad \frac{\partial v_z}{\partial x} = \frac{\partial v_x}{\partial z}. \qquad (12.8.10)$$

It follows, therefore, that in case of irrotational motion

$$\frac{\partial v_y}{\partial x} - \frac{\partial v_x}{\partial y} = 0, \quad \frac{\partial v_z}{\partial y} - \frac{\partial v_y}{\partial z} = 0, \quad \text{and} \quad \frac{\partial v_x}{\partial z} - \frac{\partial v_z}{\partial x} = 0.$$

Now multiplying both sides of eqn. (12.8.7) by dx and using relations (12.8.10) we have

$$v_x \frac{\partial v_x}{\partial x}\, dx + v_y \frac{\partial v_y}{\partial x}\, dx + v_z \frac{\partial v_z}{\partial x}\, dx + \frac{1}{\rho}\frac{\partial p}{\partial x}\, dx - F_x\, dx = 0. \qquad (12.8.11)$$

Similarly from eqns. (12.8.8) and (12.8.9) we have

$$v_x \frac{\partial v_x}{\partial y}\, dy + v_y \frac{\partial v_y}{\partial y}\, dy + v_z \frac{\partial v_z}{\partial y}\, dy + \frac{1}{\rho}\frac{\partial p}{\partial y}\, dy - F_y\, dy = 0 \qquad (12.8.12)$$

$$v_x \frac{\partial v_x}{\partial z}\, dz + v_y \frac{\partial v_y}{\partial z}\, dz + v_z \frac{\partial v_z}{\partial z}\, dz + \frac{1}{\rho}\frac{\partial p}{\partial z}\, dz - F_z\, dz = 0 \qquad (12.8.13)$$

Adding up (12.8.11), (12.8.12) and (12.8.13), and remembering the relations,

$$\frac{\partial v_x}{\partial x}dx + \frac{\partial v_x}{\partial y}dy + \frac{\partial v_x}{\partial z}dz = dv_x$$

$$\frac{\partial v_y}{\partial x}dx + \frac{\partial v_y}{\partial y}dy + \frac{\partial v_y}{\partial z}dz = dv_y$$

$$\frac{\partial v_z}{\partial x}dx + \frac{\partial v_z}{\partial y}dy + \frac{\partial v_z}{\partial z}dz = dv_z$$

we have,

$$v_x\,dv_x + v_y\,dv_y + v_z\,dv_z + \frac{dp}{\rho} - [F_x\,dx + F_y\,dy + F_z\,dz] = 0$$

$$\text{or}\quad d\left(\frac{1}{2}v^2\right) + \frac{dp}{\rho} + \left[\frac{\partial U}{\partial x}dx + \frac{\partial U}{\partial y}dy + \frac{\partial U}{\partial z}dz\right] = 0$$

where $v^2 = v_x^2 + v_y^2 + v_z^2$, and

$$F_x = -\frac{\partial U}{\partial x},\quad F_y = -\frac{\partial U}{\partial y},\quad F_z = -\frac{\partial U}{\partial z},$$

U being the potential energy per unit mass at (x, y, z) with respect to the forces acting at the point.

Since

$$dU = \frac{\partial U}{\partial x}dx + \frac{\partial U}{\partial y}dy + \frac{\partial U}{\partial z}dz,$$

the above equation may be written as

$$d\left(\frac{1}{2}v^2\right) + \frac{dp}{\rho} + dU = 0. \tag{12.8.14}$$

Integrating, we get

$$\frac{1}{2}v^2 + \frac{p}{\rho} + U = F(t),$$

$F(t)$ being an arbitrary function of time. [$\because$ we integrate with respect to space only]

The term $F(t)$ is without influence on the resultant pressure, and it is frequently omitted. If the fluid is incompressible, the above equation becomes $\frac{1}{2}v^2 + \frac{p}{\rho} + U =$ constant throughout the entire tube of flow, which is Bernoulli's theorem for irrotational flow in the case of the steady stream-lined flow of a non-viscous incompressible fluid. We know that in irrotational flow, the velocity components can be derived from a potential function ϕ. That is why such flow is frequently called *potential* flow.

Case II

Let the motion be *not irrotational*, i.e. when the expressions

$$\frac{\partial v_y}{\partial x} - \frac{\partial v_x}{\partial y},\quad \frac{\partial v_z}{\partial y} - \frac{\partial v_y}{\partial z}\quad \text{and}\quad \frac{\partial v_x}{\partial z} - \frac{\partial v_z}{\partial x}\quad \text{do not vanish.}$$

Multiplying both sides of equation (12.8.7) by v_x, we have

$$v_x v_x \frac{\partial v_x}{\partial x} + v_x v_y \frac{\partial v_x}{\partial y} + v_x v_z \frac{\partial v_x}{\partial z} + v_x \frac{1}{\rho}\frac{\partial p}{\partial x} - F_x v_x = 0$$

or $$v_x \frac{\partial}{\partial x}\left[\frac{1}{2}v_x^2\right] + v_y \frac{\partial}{\partial y}\left[\frac{1}{2}v_x^2\right] + v_z \frac{\partial}{\partial z}\left[\frac{1}{2}v_x^2\right] + v_x \frac{\partial}{\partial x}\int\frac{\partial p}{\rho} + v_x \frac{\partial U}{\partial x} = 0$$

or $$\left[v_x \frac{\partial}{\partial x} + v_y \frac{\partial}{\partial y} + v_z \frac{\partial}{\partial z}\right]\frac{1}{2}\left(v_x^2\right) + v_x \frac{\partial}{\partial x}\left[\int\frac{\partial p}{\rho} + U\right] = 0. \qquad (12.8.15)$$

Also from (12.8.8) and (12.8.9), proceeding similarly, we get,

$$\left[v_x \frac{\partial}{\partial x} + v_y \frac{\partial}{\partial y} + v_z \frac{\partial}{\partial z}\right]\frac{1}{2}\left(v_y^2\right) + v_y \frac{\partial}{\partial y}\left[\int\frac{\partial p}{\rho} + U\right] = 0 \qquad (12.8.16)$$

and $$\left[v_x \frac{\partial}{\partial x} + v_y \frac{\partial}{\partial y} + v_z \frac{\partial}{\partial z}\right]\frac{1}{2}\left(v_z^2\right) + v_z \frac{\partial}{\partial z}\left[\int\frac{\partial p}{\rho} + U\right] = 0. \qquad (12.8.17)$$

Adding up (12.8.15), (12.8.16) and (12.8.17), we have

$$\left[v_x \frac{\partial}{\partial x} + v_y \frac{\partial}{\partial y} + v_z \frac{\partial}{\partial z}\right]\frac{1}{2}v^2 + \left[v_x \frac{\partial}{\partial x} + v_y \frac{\partial}{\partial y} + v_z \frac{\partial}{\partial z}\right]\left[\int\frac{\partial p}{\rho} + U\right] = 0$$

or $$\left[v_x \frac{\partial}{\partial x} + v_y \frac{\partial}{\partial y} + v_z \frac{\partial}{\partial z}\right]\left[\frac{1}{2}v^2 + \int\frac{\partial p}{\rho} + U\right] = 0$$

or $$v\frac{\partial}{\partial S}\left[\frac{1}{2}v^2 + \frac{p}{\rho} + U\right] = 0 \qquad (12.8.18)$$

where $\frac{\partial}{\partial S}$ denotes the differentiation along a stream line, and it follows from general mathematical relation that

$$v\frac{\partial}{\partial S} = v_x \frac{\partial}{\partial x} + v_y \frac{\partial}{\partial y} + v_z \frac{\partial}{\partial z}.$$

Integrating the expression (12.8.18), we have $\frac{1}{2}v^2 + \frac{p}{\rho} + U =$ constant, for a particular stream line—the value of the constant depending upon the particular stream-line chosen. The above is the expression for Bernoulli's theorem for the rotational flow.

Discussion: In the above expression,

(i) $\frac{v^2}{2}$ denotes K.E. for unit mass,

(ii) U denotes P.E. for unit mass, and

(iii) $\frac{p}{\rho}$ denotes pressure energy, i.e. energy required to move the liquid against the pressure without imparting any velocity to the liquid.

When the external force is gravitational, i.e. for a gravitational field, $U = gh$, where h is the vertical height above the sea-level, and the above equation takes the form

$$\frac{v^2}{2} + \frac{p}{\rho} + gh = C_1. \qquad (12.8.19)$$

Expression (12.8.19) may be modified to

$$\frac{v^2}{2g} + \frac{p}{\rho g} + h = \frac{C_1}{g} = H. \qquad (12.8.20)$$

Here h = the gravitational head; $\frac{p}{\rho g}$ = the pressure head; and $\frac{v^2}{2g}$ = K. E. head, i.e. the velocity head, and the total head H must be a height of vertical distance. Equation (12.8.20) is therefore dimensionally constant on both sides at all points in a stream-line flow.

12.9 Experimental arrangement for the illustration of Bernoulli's theorem

As shown in Fig. 12.9.1, A is a vessel in which a constant water level is maintained. B, C, D are horizontal tubes of different cross-sections fitted horizontally to the vessel through which water is flowing. When the end of the tube D is kept closed it is found that the water levels in the vertical tubes, E, F, G connected respectively to B, C, D remain same. On the other hand, on opening the tube D, water will be found to rise to different heights in the tubes E, F, G. Let P_B, P_C, P_D and V_B, V_C, V_D be the pressures and velocities at B, C and D respectively. Since the datum level coincides with the axis of the tube, i.e. $h = 0$, we have, from Bernoulli's theorem.

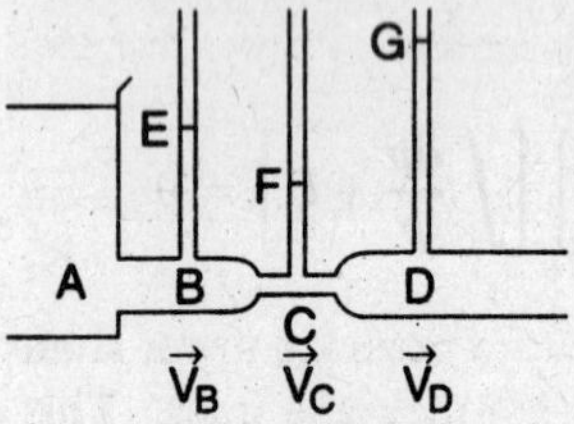

Fig. 12.9.1

$$P_B + \frac{1}{2}V_B^2 = P_C + \frac{1}{2}V_C^2 = P_D + \frac{1}{2}V_D^2 \quad [\because \rho = 1, \text{ for water}] \qquad (12.9.1)$$

As there is no accumulation of water in any tube, the rate of flow of water in all tubes remains unaltered.

Hence $V_B\alpha_B = V_C\alpha_C = V_D\alpha_D$, where α_B, α_C, α_D are the cross-sections of the tubes B, C, D respectively. In the figure $\alpha_D > \alpha_B > \alpha_C$

$$\therefore \; V_D < V_B < V_C.$$

Hence from equation (12.9.1) it follows that $P_D > P_B > P_C$.

If h_1, h_2, h_3 be the heights up to which water rises in the vertical tubes E, F, G, then $h_3 > h_1 > h_2$. From the above discussions, Bernoulli's theorem may also be, qualitatively, expressed by saying that *where the speed of a fluid is high, the pressure is low and where the speed is low, the pressure is high.*

We shall now discuss some other illustrations of Bernoulli's theorem.

(i) Sometimes the roof of a house is found to the blown off in a gale. In a gale when wind blows with tremendous velocity over the roof of a house, the pressure above the roof is reduced to a great extent as a result of which the pressure inside the house becomes very large in comparison with the pressure over the roof. This high difference of pressure causes the roof to be blown off.

(ii) It is found that a 'ping-pong' ball when placed over a stream of water jet remains in the stream without falling. The velocity of water jet causes the pressure within it reduced and so the high atmospheric pressure keeps the ball in the jet at a fixed position. In the fixed position the ball turns round due to the motion of the water jet.

(iii) When air is blown over the upper surface of a sheet of paper, the pressure in the air-stream above the paper will be reduced and so the excess pressure below the sheet will force the paper upward.

(iv) When two ships move side by side and parallel to each other, they tend to come closer and closer to each other. The water entrapped between the ships moves faster because of the narrower space. The velocity of water between the ships is thus greater than that away from the ships. As a result, the pressure of water between the ships is smaller than that on the far sides of the ships. The excess pressure on the far sides tends to bring the ships closer.

(v) A sprayer which is used for spraying disinfectants, germicides or perfumes is another familiar illustration. The rapidly moving stream of air reduces the pressure on the liquid in the tube below atmospheric pressure, so that the latter pushes the liquid up the tube, to be blown out in the form of a spray.

(vi) In bombed cities during the wat, it is found that broken glass pieces from windows litter the streets, but not the interior of houses. The reason is that the bomb blast sets up powerful air current down the street. The pressure in air current is reduced to one below atmospheric from Bernoulli's principle. The air in houses, which was at atmospheric pressure, blows out the windows.

(vii) It is dangerous for a person to stand near a fast moving train, because he may be drawn towards it.

(viii) In a bunsen burner, the gas jet from the central nozzle produces a pressure lower than atmospheric about the nozzle. So the atmospheric pressure forces air through the inlet. This mixes with the gas and burns it.

12.10 Applications of Bernoulli's theorem

12.10.1 Venturimeter

Venturimeter is used to determine the rate of flow of water i.e., the volume of water flowing per unit time through a pipe. It is a practical application of Bernoulli's theorem. The attached figure (12.10.1) shows diagrammatically the principle of a venturimeter. It consists, in fact, of two truncated conical pipes connected by means of a cylindrical pipe called the "*throat*". The rise of the liquid up to different heights in the tubes attached respectively to the conical and the cylindrical pipes as shown in the figure indicates the pressures at different points. From the following consideration, we may calculate the rate of flow of water in the tube.

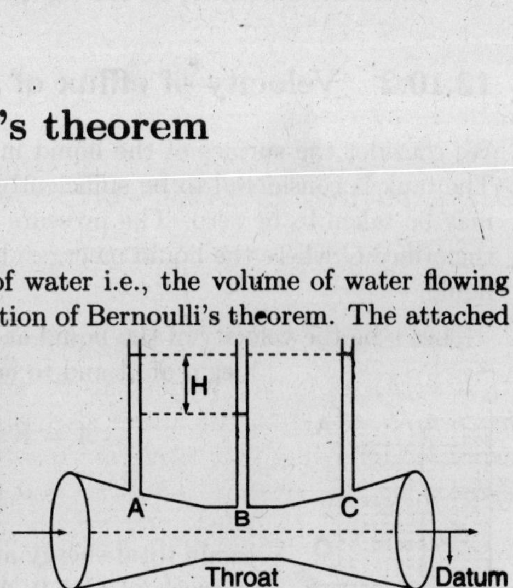

Fig. 12.10.1

Let P_1, u_1 be the pressure, velocity at A; and let P, u be the pressure, velocity at B. Since the datum level coincides with the axis of the tube, we have from Bernoulli's theorem $h = 0$, and therefore, we may write

$$\frac{1}{2}u_1^2 + \frac{P_1}{\rho} = \frac{1}{2}u^2 + \frac{P}{\rho}$$

putting $h = 0$, $p = P$ and $v = u$ in (12.8.19)

$$\frac{u^2}{2} + \frac{P}{\rho} = \text{constant.}$$

$\therefore$ for points A and B

$$\frac{u_1^2}{2} + \frac{P_1}{\rho} = \frac{u^2}{2} + \frac{P}{\rho}$$

Now from the figure,

$$P_1 - P = H\rho g$$

So we may write,

$$Hg = \frac{P_1 - P}{\rho} = \frac{u^2 - u_1^2}{2}$$

Again let s_1 and s represent the areas of sections at A and B respectively. The volume of liquid that passes across, in unit time, will then be given by

$$Q = s_1 u_1 \text{ or } u_1 = \frac{Q}{s_1} \text{ and } Q = su \text{ or } u = \frac{Q}{s}.$$

On substituting these values of u_1 and u, we have,

$$Hg = \frac{1}{2}(u^2 - u_1^2) = \frac{1}{2}\left[\frac{Q^2}{s^2} - \frac{Q^2}{s_1^2}\right] = \frac{Q^2}{2}\left[\frac{s_1^2 - s^2}{s^2 s_1^2}\right].$$

$$\therefore Q^2 = 2Hg\frac{s^2 s_1^2}{s_1^2 - s^2} \text{ or, } Q = ss_1\sqrt{\frac{2Hg}{s_1^2 - s^2}} \tag{12.10.1}$$

Now all the quantities on the right hand side being known, Q can be measured.

12.10.2 Velocity of efflux of a liquid—Torricelli's Theorem

We consider the surface of the liquid in a tank at a height h above the level of the orifice O. The tank is considered to be sufficiently wide, so that the velocity at the surface of the liquid may be taken to be zero. The pressure at the surface is clearly atmospheric. The pressure at the orifice O where the liquid emerges, being also atmospheric, plays no part in the flow of the liquid.

Let v be the velocity of the liquid at the level of the orifice. Considering the tube of flow to begin at A and to end at O, we may write, total energy at

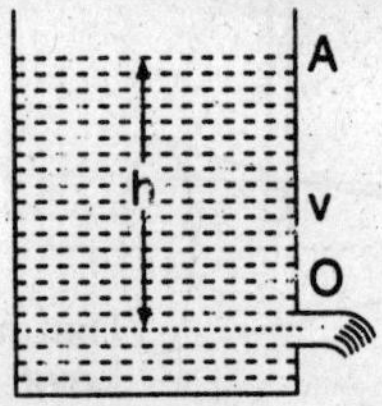

Fig. 12.10.2

$$A = \text{K.E.} + \text{P.E.} + \text{Pressure energy}$$

$$= 0 + gh + 0 = gh \text{ [K.E.} = 0\text{, since } v = 0 \text{ at } A]$$

again total energy at $O = \frac{1}{2}v^2 + 0 + 0$, since the pressure and the potential energies at $O = 0$. As the total energy remains unchanged, we may write

$$gh = \frac{1}{2}v^2 \text{ or, } v = \sqrt{2gh}. \tag{12.10.2}$$

The above expression gives the velocity of efflux of the liquid at the orifice. Thus, we see that the velocity of efflux of a liquid through an orifice is equal to the velocity attained by a body in falling freely from the surface of the liquid to the orifice.

This is known as *Torricelli's theorem.*

From the above, it follows that if a nozzle were attached to the orifice so as to send a jet upwards, the jet would reach the same level as the surface inside the vessel. Owing to viscosity of the liquid, this never happens. The velocity is always less than $\sqrt{2gh}$. If α be the area of the opening, the calculated volume of the liquid flowing out per unit time is

$$\alpha v = \alpha\sqrt{2gh} \tag{12.10.3}$$

Since the stream-lines converge as they approach the orifice, the cross-section of the stream continues to diminish for a short distance outside the tank. It is the area of the smallest cross-section, known as the *vena contracta,* which should be used in equation (12.10.3). For a sharp-edged circular opening, the area of the vena contracta is about 0.624 times (i.e., 62.4%) the area of the orifice. This fraction is called the *coefficient of contraction.*

12.10.3 Pitot tube

Pitot tube is used to determine the rate of flow of water as is done by a venturimeter. It consists of two vertical tubes having small apertures at their lower ends. The aperture of the tube AB is parallel to the direction of flow of water and that of CD is facing the flow of water [Fig. 12.10.3]. The rise of water column in the tube AB measures the static pressure at B. As the flow of water is stopped in the plane of the aperture at D within the tube CD, we may write

$$p_B + \frac{1}{2}v^2 = p_D + 0 \text{ or, } p_D - p_B = \frac{1}{2}v^2,$$

where p_D and p_B denote the pressures at D and B. Now $p_D - p_B = hg$, where h is the difference of levels of water in the two tubes. Pitot tube is also used to measure the flow of gas. It is widely used in aeroplanes to determine air speed which gives the speed of aeroplane relative to air.

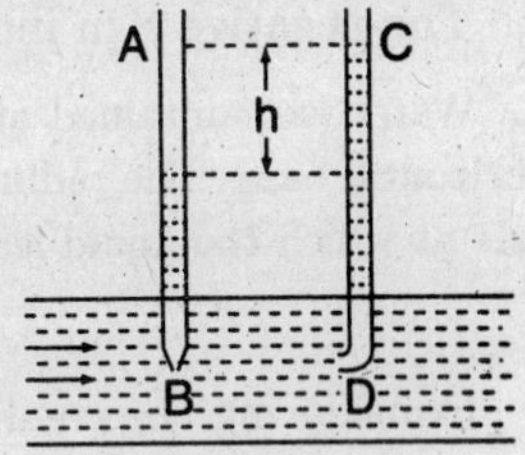

Fig. 12.10.3

Illustrative Examples

1. Assuming the atmosphere to be perfectly dry, calculate the fall in temperature of the atmosphere per kilometer change in height of the atmosphere. ($\gamma = 1.4$; mean molecular weight of air = 29, $R = 8.3 \times 10^7$ ergs/gm. mol/degree and $g = 980$ cm./sec.^{2}) **[Rajasthan 1957]**

From equation (12.3.8) we have

$$\frac{dp}{p} = -\frac{M}{RT}g\,dz. \tag{1}$$

Now the adiabatic relation between T and P is given by

$$Tp^{\frac{1-\gamma}{\gamma}} = \text{const. or, } \log T + \frac{1-\gamma}{\gamma}\log p = \text{constant.}$$

Differentiating the above expression we have

$$\frac{dT}{T} = \frac{(\gamma-1)}{\gamma}\frac{dp}{p} \text{ or, } dT = \frac{(\gamma-1)}{\gamma}\frac{dp}{p}T = -\frac{(\gamma-1)}{\gamma}\frac{Mg}{R}dz.$$

Here

$$\gamma = 1.4 \qquad dz = 10^3\text{m} = 10^5\text{cm.}$$

$$M = 29 \qquad R = 8.3 \times 10^7 \text{ ergs/gm. mol/degree.}$$

$$g = 980 \text{ cm./sec.}^2$$

Hence

$$dT = -\frac{(1.4-1)}{1.4} \times \frac{29 \times 980 \times 10^5}{8.3 \times 10^7} = -\frac{2}{7} \times \frac{29 \times 980 \times 10^5}{8.3 \times 10^7} = -9.7°.$$

Hence the fall in temperature is 9.7°.

2. **Water flows along a horizontal tube of which the cross-section is not constant. Calculate the change in pressure when the velocity of flow changes from 10 cm./sec. to 20 cm./sec.**

For horizontal flow Bernoulli's equation takes the form

$$\frac{1}{2}v_1^2 + \frac{p_1}{\rho} = \frac{1}{2}v_2^2 + \frac{p_2}{\rho}.$$

Here $\rho = 1$ gm./c.c. $\quad \therefore p_2 - p_1 = \frac{1}{2}(v_1^2 - v_2^2) = -150$ dynes/cm.^{2}.

The negative sign indicates that due to increase of velocity the pressure decreases.

3. **Water is maintained at a height of 15 cm. is a vessel which has a small circular hole in its thin horizontal base. The radius of the hole is 1 mm and the area of the free surface is 30 sq. cm. Find the rate at which the liquid leaves the vessel if the coefficient of contraction is 0.624.**

[C.U. (Hons.) 1981]

Considering the coefficient of contraction, equation (12.10.1) may be written as

$$Q = 0.624 s s_1 \sqrt{\frac{2Hg}{s_1^2 - s^2}}.$$

Here

$$s = \pi(0.1)^2 = 0.0314 \text{ sq. cm.}$$

$$s_1 = 30 \text{ sq. cm.}; \quad H = 15 \text{ cms.}$$

$$\therefore Q = 0.624 \times 30 \times 0.0314 \sqrt{\frac{2 \times 15 \times 980}{900 - 0.000986}} = 3.36 \text{ c.c./sec.}$$

4. **Calculate the velocity of efflux of kerosine oil from a tank in which the pressure is 50 lbs. wt. per sq. inch above the atmospheric pressure. The density of kerosine is 48 lbs. per cubic ft.**

The velocity of efflux of kerosine oil is given by $v = \sqrt{2gh}$.

Now the pressure due to the height h of the level of kerosine is $p = \rho gh$.

If p is expressed in pound weight then $p = \rho h$.

Here

$$p = 50 \text{ lbs. wt/sq. inch} = 50 \times 144 \text{ lbs. wt/ft.}^2; \quad \text{and } \rho = 48 \text{ lbs./ft.}^3$$

$$h = \frac{p}{\rho} = \frac{50 \times 144}{48} \text{ ft.}; \quad v = \sqrt{\frac{2 \times 32 \times 50 \times 144}{48}} = 97.97 \text{ ft./sec.}$$

5. Water stands at a depth h in a large open tank whose side-walls are vertical. A hole is made in one of the walls at a depth x below the water surface. Calculate the horizontal distance d from the wall at which the emerging stream of water strikes the ground. What should be the value of x which makes d maximum? **[C.U. (Hons.) 1979]**

The velocity of the emerging water at the orifice is given by $v = \sqrt{2gx}$ which is the horizontal velocity only. If t be the time taken to strike the ground, then

$$(h - x) = \frac{1}{2}gt^2.$$

$\therefore$ the required horizontal distance

$$d = v.t. = \sqrt{2gx} \cdot \sqrt{\frac{2(h-x)}{g}} = 2\sqrt{x(h-x)}.$$

For 'd' to be either maximum or minimum, we should have

$$\frac{d}{dx}\left[2\sqrt{x(h-x)}\right] = 0 \text{ or, } 2\left[\frac{1}{2}\sqrt{h-x} \cdot \frac{1}{\sqrt{x}} - \frac{1}{2}\frac{\sqrt{x}}{\sqrt{h-x}}\right] = 0$$

$$\text{or, } \frac{\sqrt{h-x}}{\sqrt{x}} = \frac{\sqrt{x}}{\sqrt{h-x}}$$

$$\text{or, } h - x = x \text{ or, } x = \frac{h}{2}.$$

Hence d will be maximum or minimum when $x = \frac{h}{2}$

Again

$$\frac{d^2}{dx^2}\left[2\sqrt{x(h-x)}\right] = \frac{d}{dx}\left[\frac{\sqrt{h-x}}{\sqrt{x}} - \frac{\sqrt{x}}{\sqrt{h-x}}\right] = \frac{d}{dx}\left[\frac{h-2x}{\sqrt{x(h-x)}}\right] = \frac{d}{dx}\left[\frac{2(h-2x)}{d}\right]$$

$$= \frac{-4d - 2(h-2x)\frac{d}{dx}(d)}{d^2} = \text{negative, when } x = \frac{h}{2}$$

$\therefore$ The horizontal distance d is maximum when $x = h/2$.

12.11 Bernoulli's equation for gases

Bernoulli's equation is also applicable to ideal gases. As the density of a gas depends on pressure, Bernoulli's equation for gases is

$$\frac{1}{2}v^2 + \int \frac{dp}{\rho} + U = \text{constant}.$$

For gravitational field the above equation is written as

$$\frac{1}{2}v^2 + \int \frac{dp}{\rho} + gz = \text{ constant.} \tag{12.11.1}$$

Let us now calculate $\int \frac{dp}{\rho}$ for an ideal gas for both isothermal and adiabatic processes.

Isothermal process: In this case $\frac{p}{\rho}$ = constant = K (say)

$$\therefore \quad \int \frac{dp}{\rho} = K \int \frac{dp}{p} = K \log p + K_1,$$

where K_1 is the constant of integration. Hence Bernoulli's equation in isothermal processes is

$$\frac{1}{2}v^2 + K \log p + gz = \text{ constant.} \tag{12.11.2}$$

Adiabatic process: In this process

$$pv^\gamma = \text{ constant, where } \gamma = \frac{C_p}{C_v}$$

$$= \frac{\text{Specific heat of gas at constant pressure}}{\text{Specific heat of gas at constant volume}}.$$

Now $v \propto \frac{1}{\rho}$

$$\therefore \quad p\left(\frac{1}{\rho}\right)^\gamma = \text{ constant } = k \text{ (say) or, } \frac{1}{\rho} = k^{1/\gamma}\left(\frac{1}{p}\right)^{1/\gamma}$$

$$\therefore \quad \int \frac{dp}{\rho} = k^{1/\gamma} \int p^{-1/\gamma} dp = \frac{\gamma k^{1/\gamma}}{\gamma - 1} p^{\frac{\gamma-1}{\gamma}} + \text{ constant.}$$

Now $p^{1/\gamma}\rho^{-1} = k^{1/\gamma}$

$$\therefore \int \frac{dp}{\rho} = \frac{\gamma}{\gamma - 1}\frac{p}{\rho} + \text{ constant.}$$

Hence Bernoulli's equation takes the form

$$\frac{1}{2}v^2 + \frac{\gamma}{\gamma - 1}\frac{p}{\rho} + gz = \text{ constant.} \tag{12.11.3}$$

in adiabatic processes.

12.12 Efflux of gases

Let p_1 and ρ_1 be the pressure and density of a gas in a vessel and suppose that the gas flows through an orifice into a gaseous atmosphere where p_2, ρ_2 are the respective pressure and density. If there are no external forces, we have from Bernoulli's equation,

$$\frac{1}{2}v_1^2 + \frac{\gamma}{\gamma-1}\frac{p_1}{\rho_1} = \frac{1}{2}v_2^2 + \frac{\gamma}{\gamma-1}\frac{p_2}{\rho_2}$$

or, $$\frac{1}{2}(v_2^2 - v_1^2) = \frac{\gamma}{\gamma-1}\left(\frac{p_1}{\rho_1} - \frac{p_2}{\rho_2}\right) = \frac{\gamma p_1}{(\gamma-1)\rho_1}\left(1 - \frac{p_2}{p_1}\frac{\rho_1}{\rho_2}\right).$$

Now,

$$\frac{p_1}{\rho_1^\gamma} = \frac{p_2}{\rho_2^\gamma} \quad \therefore \quad \frac{\rho_1}{\rho_2} = \left(\frac{p_1}{p_2}\right)^{1/\gamma}$$

$$\therefore \quad \frac{1}{2}(v_2^2 - v_1^2) = \frac{\gamma p_1}{(\gamma-1)\rho_1}\left[1 - \frac{p_2}{p_1} \times \left(\frac{p_1}{p_2}\right)^{1/\gamma}\right] = \frac{\gamma p_1}{(\gamma-1)\rho_1}\left[1 - \left(\frac{p_2}{p_1}\right)^{\frac{\gamma-1}{\gamma}}\right]$$

From the above expression it follows that an increase in velocity of the gas flowing out is accompanied by a decrease in pressure. This principle is utilised in ordinary vacuum cleaner.

12.13 Derivation (alternative), from the principle of conservation of energy, of Bernoulli's equation for variation of pressure along a stream line

Method 1. Considering a tube of flow (Fig. 12.13.1) in a frictionless i.e., non-viscous, incompressible fluid in stream-line motion, let the velocity at any cross-section of the tube be uniform throughout the section.

Again we consider any two sections A and B (Fig. 12.13.1) having different cross-sectional areas, and let the velocities at A and B be v_1 and v_2, the heights of the sections above a convenient datum (reference) plane be h_1 and h_2, and the pressures of the liquid be respectively p_1 and p_2.

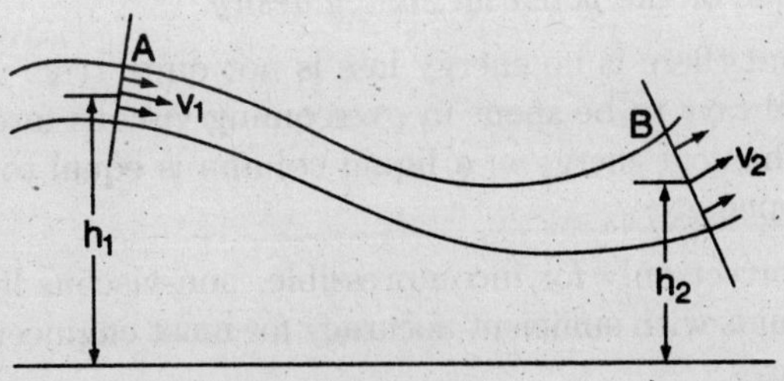

Fig. 12.13.1

Since there is no accumulation of the liquid any where in the tube, the same mass m will pass through any section of the tube in a given time. The work done in bringing this mass to the conditions existing at A consists of three parts: (i) work required to raise it to the height h_1 above the datum plane $= mgh_1$ — which is the *potential energy;* (ii) work done in imparting K.E. to it $= \frac{1}{2}mv_1^2$; (iii) work done in forcing the volume $\frac{m}{\rho}$ (ρ being the density of the liquid) into a region where the pressure is $p_1 = p_1\frac{m}{\rho}$ — which is the *pressure energy.*

Now the liquid is non-viscous — so it does no work and as such the energy that enters section A in a given time is also equal to the energy which leaves section B in the same time, so that we may write

$$mgh_1 + \frac{1}{2}mv_1^2 + \frac{p_1 m}{\rho} = mgh_2 + \frac{1}{2}mv_2^2 + p_2\frac{m}{\rho}.$$

Dividing throughout by mg, we have

$$h_1 + \frac{v_1^2}{2g} + \frac{p_1}{\rho g} = h_2 + \frac{v_2^2}{2g} + \frac{p_2}{\rho g}, \tag{12.13.1}$$

which gives the corresponding values per unit weight of the fluid, and means that along a stream line,

$$h + \frac{v^2}{2g} + \frac{p}{\rho g} = \text{constant} \tag{12.13.2}$$

or *Elevation* (or gravity) head + *velocity* (or dynamic) head + *pressure* (or static) heat = *Total head.*

(12.13.1) or (12.13.2) is the mathematical formulation of Bernoulli's theorem in the case of an incompressible fluid. Physically we may say that if no work is done on (or by) an incompressible liquid as it flows, the total head along a stream-line remains constant (or that the sum of the potential, kinetic and pressure energies remains constant along a stream line). The term *head,* used in this context, represents potential due to a height of liquid.

Equation (12.13.1) or (12.13.2) may also be written in the form

$$p + hg\rho + \frac{1}{2}\rho v^2 = \text{constant}. \tag{12.13.3}$$

Here p may be looked upon as the *pressure energy,* being the work, which unit volume of the fluid could perform if it were allowed to move to a place where the pressure is zero. $hg\rho$ represents the potential energy of unit volume and is the work it could do if it fell to the level from which h is measured. $\frac{1}{2}\rho v^2$ is the kinetic energy of unit volume and is the work it could do if it were brought to rest. Writing $\phi = (p + hg\rho)$, we may say that ϕ is the potential energy function per unit volume or the *potential energy desity.*

The assumption that there is no energy loss is not quite true. All liquids possess viscosity and some energy has always to be spent in overcoming viscous forces. This energy appears as heat. The gain of mechanical energy of a liquid column is equal to the work done on it minus the heat energy developed.

While rigorously correct only for incompressible, non-viscous liquids, the theorem may be applied to ordinary liquids with sufficient accuracy for most engineering purposes. The theorem is also applicable to gases.

Method 2. For equilibrium under the action of a body force acting on a fluid element per unit volume of a non-viscous fluid, we may write from (12.3.4)

$$\vec{F} - \vec{\nabla}p = 0, \tag{12.13.4}$$

where $\vec{F}$ is the external body force per unit volume acting on the fluid. In the gravitational force field $\vec{F} = \rho\vec{g}$, ρ being the density of the fluid and $\vec{g}$ the acceleration due to gravity.

Now gravitational force field being conservative, $\vec{F}$ can be expressed as $\vec{F} = -\rho\vec{\nabla}V$ where $V = gh$ is the gravitational potential energy per unit mass. p in (12.13.4) is the pressure of the surrounding fluid. The force acting on unit mass of the fluid is therefore, written as

$$-\vec{\nabla}V - \frac{\vec{\nabla}p}{\rho} = -\vec{\nabla}\left(V + \int \frac{dp}{\rho}\right)$$

ρ being a function of pressure p only. Now writing

$$\phi = V + \int \frac{dp}{\rho},$$

we may say that ϕ is the potential energy function for unit mass. Again kinetic energy for unit mass of a fluid moving with velocity v is $\frac{1}{2}v^2$. So applying the principle of conservation of mechanical energy for the flow of a non-viscous fluid, we may write

$$\frac{1}{2}v^2 + V + \int \frac{dp}{\rho} = \text{constant} \qquad (12.13.5)$$

$V = gh$ being the gravitational potential energy per unit mass at a height h above the reference level. Equation (12.3.5) is the Bernoulli's theorem.

Exercises

1. Deduce Euler's equation of motion for a non-viscous liquid.

2. Deduce Bernoulli's equation of motion for rotational flow as well as for irrotational flow of a liquid. What is Venturimeter? Explain its use.

3. Derive an expression for the equation of continuity of an ideal fluid of density ρ. What is the form of this equation, when the fluid is incompressible? **[C.U. (Hons.) '82, '84, 2000, 02, 05]**

4. Deduce an expression for the velovity of efflux of a liquid.

5. Derive Bernoulli's equation in case of gases.

6. Obtain an expression for the velocity of efflux of a gas.

7. How would you experimentally demonstrate the truth of Bernoulli's theorem? Mention some important applications of Bernoulli's theorem.

8. Describe a Pitot tube. How would you determine the rate of flow of water with its help?

9. Calculate the velocity of efflux of water from an orifice in a tank which is at a depth 10 ft. from the surface of water. **[25.3 ft/sec.]**

10. A vessel with a leak is filled with a gas of density ρ_1 and pressure p_1. Find an expression for the velocity of the escaping gas if its pressure and density be p_2 and ρ_2 respectively, when the gas issues out through the orifice. The pressure and volume changes may be assumed to occur adiabatically. (ratio of the specific heats = γ) **[C.U. Hons. 76, 78]**

    (See worked out examples)

11. A perfect gas of molecular weight M at a temperature T is flowing steadily and isothermally through a narrow pipe of varying cross-section and varying height (h) above the ground. Show that the pressure p, velocity v of the gas and cross-section s of the tube at any two arbitrary points (referred to as 1 and 2) in the tube are related by $p_1v_1s_1 = p_2v_2s_2$ and $v_1^2 - v_2^2 = 2g(h_2 - h_1) + \frac{2RT}{M}\log_e\left(\frac{v_1s_1}{v_2s_2}\right)$. **[C.U. Hons. 84(Spl.)]**

    (See worked out examples)

12. Starting from Bernoulli's theorem, prove Torricelli's theorem. What is vena contracta? Explain the working principle of a venturimeter. **[C.U. (Hons.) '80]**

13. (a) Estabslish Bernoulli's theorem. How is it connected with the principle of conservation of energy? Discuss qualitatively its application to explain the lift on an air-plane wings.

    (b) Water is maintained at a height of 20 cm in a vessel which has a small circular hole in its thin horizontal base. The radius of the hole is 2 mm and the area of the free surface is 30 sq cm. Find the rate at which the liquid leaves the vessel if the coefficient of contraction is 0.625. **[C.U. (Hons.) '81]**

    (See worked out examples)

14. (a) Show that Bernoulli's equation for steady, non-viscous and incompressible flow of a liquid, is consistent with the law of pressure change with height in a fluid at rest.

    (b) A cylindrical container of large cross-section is filled with water. Show that the liquid will flow out of an orifice in the wall of the container with speed $v = \sqrt{2gh}$, where h is the height of the water level above the orifice. **[C.U. (Hons.) '83 (Spl)]**

    (See worked out examples)

15. (a) Deduce Bernoulli's equation for an ideal irrotational stream-line fluid flow from the principle of conservation of energy. Writing down the Bernoulli's equatin for the case of an incompressible fluid, explain that the pressure can be considered to be a kind of "potential energy density". **[C.U. (Hons.) '84, '90, '93]**

    (b) Consider a vertical tube through which water is flowing down. The radius of the tube at a point P is 1 cm and the radius at Q which is situated 40 cm below P is 0.5 cm. Find the rate of flow of water through the tube if the pressure differs by 39.3 cm of water between P and Q. **[C.U. (Hons.) '84]**

    (See worked out examples)

16. State Bernoulii's theorem for steady flow of an ideal compressible fluid. Write down the form relevant to an incompressible fluid. **[C.U. (Hons.) 2002]**

    Discuss how this theorem is utilised in a venturimeter for measuring the rate of flow of a liquid through a tube. **[C.U. (Hons.) '84 (Spl)]**

17. (a) Establish Bernoulli's equation for an incompressible fluid, when the flow is stream-line and irrotational, and the extenal force is the gravitational force.

    (b) To what form does Bernoulli's equation reduce when the fluid is at rest? Show how Pascal's law of transmission of fluid pressure and Archimedes' principle can be obtained from this. **[C.U. (Hons.) '86]**

18. For variation of pressure along a stream-line, derive the Bernoulli's theorem from the conservation of energy principle. **[C.U. (Hons.) '88, '90, '99]**

19. (a) What is an ideal fluid?

    (b) A fluid is at rest under an external force $\vec{F}$ and a pressure gradient force. If the pressure be p at a point (x, y, z), show that at equilibrium $\vec{F} = \vec{\nabla}p$. Utilize this to establish Archimedes' principle. **[C.U. (Hons.) 2004]**

(c) Set up Euler's equation of motion for continuous flow of an ideal fluid. Hence deduce Bernoulli's theorem for steady stream-line motion of a fluid.

(d) Use the result of (b) to show that under a constant external force field (time independent), pressure difference between two given points in the fluid at rest is also constant.

[C.U. (Hons.) '97, 2003, 06 (old syll)]

20. Steam is rushing from a boiler through a conical pipe, the diameters of the ends of which are α and β respectively. If V and v be the corresponding velocities of the steam, and if the motion be supposed to be that of divergence from the vertex of the cone, prove that

$$\frac{v}{V} = \frac{\alpha^2}{\beta^2} e^{\left(\frac{v^2 - V^2}{2k}\right)}$$

where k is pressure divided by density which is constant.

21. A mass of liquid rotates with angular velocity ω about a vertical axis. Find the equation of the free surface, neglecting restraining action of the supporting surfaces. (**Ans.** $\frac{p}{\rho} + \frac{1}{2}\omega^2(x^2 + y^2) + gz$ = const.)

22. If the flow of a liquid of sp gr 1.00 is taking place through a horizontal tapering tube of circular cross-section, the diameters of the entrance and exit ends being 10.0 cm and 7.5 cm respectively, calculate the velocities of the flow at the two ends, between which a differential pressure gauge shows a pressure difference of 35 cm of Hg. (Sp gr of Hg = 13.6) (**Ans.** 561 cm/sec, 998 cm/sec)

23. Figure shows the vertical section of a Venturimeter tube. Calculate the velocity of the liquid (flowing through the verturimeter) at the narrow end from the following data: $A_1 = 0.8$ cm^2, $A_2 = 0.2$ cm^2, $h = 8$ cm, $g = 980$ cm/sec^2. [C.U. (Hons.) 2002]

Ans. Apply equation (12.10.1)

$$Q = A_1 A_2 \sqrt{\frac{2hg}{A_1^2 - A_2^2}}$$

to find Q and then $v_2 = \frac{Q}{A_2}$.

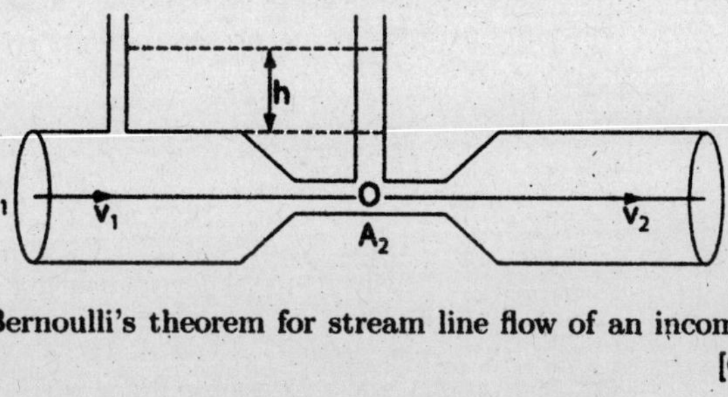

24. (a) State and prove Bernoulli's theorem for stream line flow of an incompressible non-viscous fluid. [C.U. (Hons.) 04, 05]

(b) For an ideal fluid consider two neighbouring points along a stream line. For these two points different terms have a relationship between them given by

$$g\delta h + v\delta v + \frac{\delta p}{\rho} = 0 \tag{1}$$

where the notations have usual significance. In case of a gas we can neglect the effect of h. Now consider an adiabatic process $\left[\frac{p}{\rho^\gamma} = c\right]$ and prove that

$$\frac{v^2}{2g} + Ap^n = \text{constant} \tag{2}$$

where A is some constant and $n = 1 - \frac{1}{\gamma}$.

(c) In case of an isothermal process, what will be the form of equation (2)?

[C.U. (Hons.) 2004]

(For answers to (b) and (c) see Art. 12.11)

25. (a) Define stream-line motion of a fluid. State Bernoulli's equation for such motion.

(b) Deduce Pascal's law of transmission of pressure from the condition of equilibrium of an incompressible fluid under a body force.

(c) Explain the terms flow line and stream line.

(d) Derive Bernoulli's equation for an ideal fluid.

(e) A gasoline storage tank of cross-sectional area A_1 is partially filled to a height h. The space above the gasoline contains air at pressure p_0. The gasoline flows out, through a short pipe of cross-sectional area A_2, at the bottom of the tank. Derive expressions for the flow speed (speed at which gasoline flows out of the tank) and the volume flow rate assuming $A_2 \ll A_1$. **[C.U. (Hons.) 2007]**

(For answer see Art. 12.10.1)

26. (a) Why is it dangerous to stand near a railway line when a fast train is passing by?

(b) A horizontal tube has radii 0.5 cm and 0.3 cm at two places. For flowing water, the pressure difference at the two places is 1 cm of water. What is the rate of flow?

[C.U. (Hons.) 2008]

Hints. Apply equation 12.10.1 $Q = SS_1\sqrt{\frac{2\,Hg}{S_1^2 - S^2}}$. Here $S_1 = \pi(0.5)^2$ cm^2, $S^2 = \pi(0.3)^2$ cm^2, $H = 1$ cm, $g = 980$ cm/sec^2—put these values and calculate Q.

Chapter 13

Osmosis and Diffusion

13.1 Osmosis

Various experiments show that when two different liquids or two solutions of different concentrations are separated by a membrane, one liquid will pass through the membrane with greater rapidity than the other. If the property of the membrane be such that it allows certain liquids or solutions to pass through partially or wholly and prevents the passage of other liquids or solutions, it would then be called a *semi-permeable membrane*. The process of spontaneous flow of one liquid or a solution of lower concentration into the other of higher concentration through a semi-permeable membrane is called *Osmosis*.

If an animal bladder, full of alcohol, is immersed in pure water, it swells, but if it is first filled with pure water and then immersed in alcohol, it shrinks. This shows that the animal bladder allows water to pass through but not alcohol. In the first case, water enters the bladder and thus it swells, whereas in the second case water leaves the bladder and hence it shrinks.

13.2 Abbe Nollet's Experiment—Demonstration of osmotic pressure

The phenomenon of osmosis, i.e. the passage of a liquid through membranes was first observed by Abbe Nollet, who closed the lower wide mouth of a thistle funnel by a semi-permeable membrane (say Pig's bladder) and filled it partly with a solution of cane sugar and introduced it in a vessel containing pure water as shown in the Fig. 13.2.1, so that the level of solution inside it is the same as that of water in the vessel.

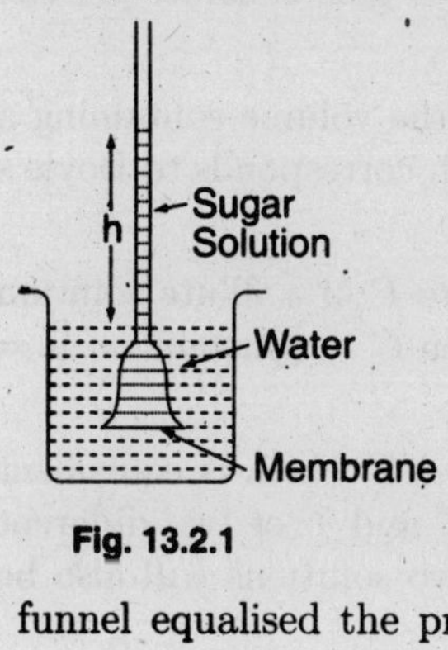

Fig. 13.2.1

As there is no hydrostatic pressure on the semi-permeable membrane, it allows water to pass through it freely into the funnel to make the solution dilute, but prevents the sugar molecules from coming out of the membrane into the water of the vessel. The level of the solution in the stem of the funnel was found to rise up until the hydrostatic pressure exerted by the raised column of solution in the funnel equalised the pressure, driving the water into the funnel and thereby further flow of water from the vessel into the funnel was prevented. The equilibrium is, however, dynamic, i.e. in a given time the number of water molecules entering the funnel is equal to the number of molecules leaving it.

The pressure which must be applied to the solution to prevent the flow of the solvent into it through a semi-permeable membrane is called the *osmotic pressure* of the solution. In the above experiment, if h be the height of the raised column of the solution in the funnel above the level of water in the vessel, and ρ be the density of solution in the stem, then the hydro-static pressure of the raised column of solution is $h\rho g$. Since this hydro-static pressure balanced the osmotic pressure of the solution, we get

Osmotic pressure of the solution $= h\rho g$.

Measurement of osmotic pressure with such simple apparatus is not reliable for the membrane used is not truly *semi-permeable.*

Semi-permeable membranes and the phenomenon of osmotic pressure play a great role in connection with the properties of cells in plants and animals.

13.3 Van't Hoff's Equation and the laws of osmotic pressure

From the experimental data collected by Pfeffer and others on osmotic pressure, Van't Hoff deduced the following relation (for non-electrolytic solutions) which embodies the laws of osmotic pressure of dilute solutions.

If m gms. of a solute of molecular weight M be present in V c.c. of its solution at temperature T^0A, then the osmotic pressure P (in dynes per sq. cm. of the solution) is given by

$$PV = \frac{m}{M}RT = n\,RT \tag{13.3.1}$$

where $n = \frac{m}{M}$ = number of gm.-molecules of solute present in V c.c. of the solution and R is a constant.

From the relation (13.3.1) we have the following laws:

Law I—Temperature T remaining constant, the osmotic pressure P of a dilute solution of a given substance is directly proportional to its concentration ($C = \frac{m}{V}$ = gms of solute per c.c. of the solution), i.e. $P \propto C$ when T = constant.

But as the concentration of a solution is inversely proportional to the volume containing a given mass of the solute, we may write $P \propto \frac{1}{V}$ or PV = constant which corresponds to Boyle's law for gases.

Law II—Concentration C remaining constant, the osmotic pressure P of a dilute solution is directly proportional to its absolute temperature T, i.e. $P \propto T$ when C = constant or $\frac{P}{T}$ = constant which corresponds to Charle's law for gases.

Combining the above two laws, we have $\frac{PV}{T}$ = constant or $PV = KT$ which is equivalent to the relation (13.3.1). From this relation, we learn that when P, V and T of two different solutions are same, the number of gm-molecules (n) present in the two solutions will also be the same.

Thus we can write:

Law III—Equal volumes of all non-electrolytic dilute solutions having the same osmotic pressure P and same temperature T contain equal number n of gm-molecules of solute.

This corresponds to Avogadro's law for gases.

13.3.1 Conclusion—Solution behaves like a gas

The osmotic pressure P of a solution of known volume V at a known temperature T, containing m gms. of a solute of molecular weight M can be accurately determined from experiment given in the next article. If the values of P, V, T, m and M be substituted in the relation (13.3.1), the value of R comes out as 8.7×10^7 ergs—which agrees fairly well with the value of molar gas constant of a perfect gas. The relation (13.3.1) is, therefore, identical with the iedal gas equation $PV = RT$. So, we may conclude that the ideal gas equation $PV = RT$ is also valid for substances in dilute solution, provided P denotes the osmotic pressure. Moreover, dilute solutions, as explained above, obey various laws corresponding to Boyle's law, Charle's law and Avogadro's law for gases. Hence a dilute solution behaves like a gas.

The above conclusion may also be stated as—the osmotic pressure of a non-electrolytic dilute solution is equal to the pressure which the solute molecules would exert if they were transformed into gas molecules at the same temperature and kept in a space of volume equal to that of the solution.

13.4 Determination of osmotic pressure

The semi-permeable membrane generally used for experimental work is cupric ferro-cyanide supported in the walls of a porous pot. The porous pot is filled with a solution of copper sulphate and placed in a vessel containing a solution of potassium ferro-cyanide. The two solutions meet in the walls of the pot and a precipitate of cupric-ferro-cyanide is formed there. The porous pot is then thoroughly washed.

13.4.1 Pfeffer's method

The porous pot P thus prepared, is filled with the desired solution and a manometer M is fitted into it as shown in Fig. 13.4.1. A part of the manometer is also filled with the solution to drive out any air. To start with, the level of mercury in the two limbs of the U-tube of the manometer is the same. The pot P is then placed in a vessel V containing water.

Osmosis takes place and water diffuses into the pot, thereby raising the pressure of the solution. The process continues until the excess of pressure as shown by the manometer becomes equal to the osmotic pressure of the solution.

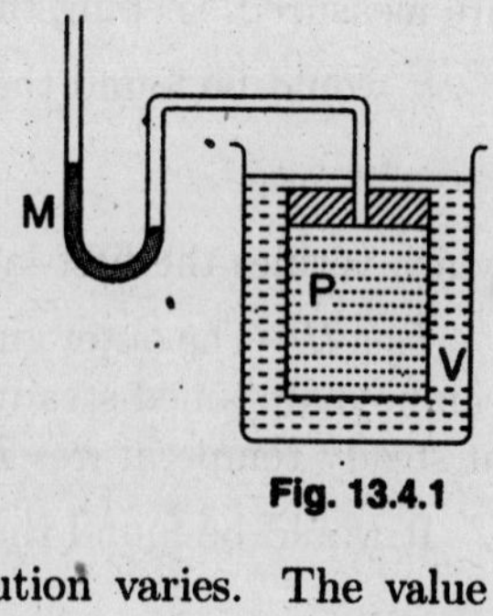

Fig. 13.4.1

The method suffers from the following drawbacks:

(i) As water enters the porous pot, the concentration of the solution varies. The value of osmotic pressure obtained is for the final concentration and not for the concentration with which we have started the experiment.

(ii) The final concentration of the solution is not uniform. The value of osmotic pressure obtained is for the concentration in the immediate neighbourhood of the membrane. It is difficult to know the concentration there.

13.4.2 Berkeley and Hartley's apparatus

An accurate method of measuring the osmotic pressure of a solution is that due to Berkeley and Hartley.

The apparatus consists of a porous tube CD having a semi-permeable membrane of cupric-ferrocyanide formed in it as explained above. Two side tubes A and B of brass are fitted into it. The tube A has a stop-cock T_1 in it and a glass capillary tube E is fitted vertically in B. The stopcock T_1 is opened and the tube is filled with the solvent so that the liquid stands in the capillary tube as shown in Fig. 13.4.2. The tube CD is placed inside a gun-metal casting$G$, having a stop-cock T_2 and a side-tube F. The solution under examination is filled into the casting and pressure is applied through F be means of a piston, so that the level of liquid in the capillary E just remains stationary. The applied pressure is equal to the osmotic pressure since the small hydro-static pressure on the solvent in CD is almost negligible.

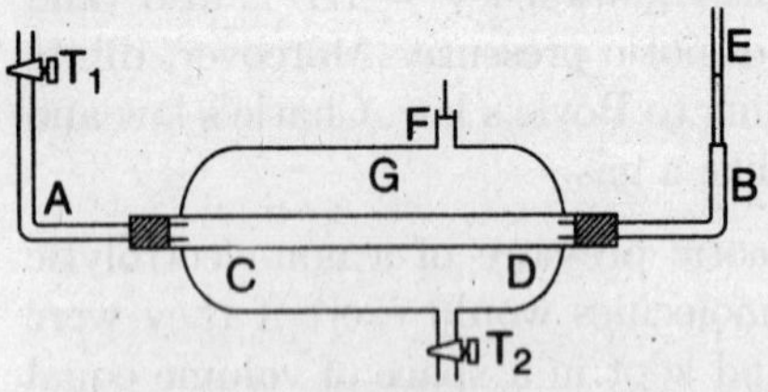

Fig. 13.4.2

13.4.3 Measurement of osmotic pressure

The solvent within the porous tube CD will tend to flow into the solution through the semi-permeable membrane of copper ferrocyanide, due to which the solvent meniscus in E tends to fall. Now a force is applied on the piston in the side tube F until the meniscus in E remains stationary. This will happen when the applied pressure is equal to the sum of the osmotic pressure of the solution and the small hydro-static pressure due to a certain height of the solution above the level of solvent.

13.4.4 Verification of the laws of osmotic pressure

Law I—The gun-metal tube is filled one after another with two solutions of the same substance but of different concentrations C_1 and C_2. The osmotic pressures P_1 and P_2 for the two solutions are measured, keeping the temperature of the two solutions same.

It would be found that

$$\frac{P_1}{P_2} = \frac{C_1}{C_2}$$

which verifies the first law.

Law II—The outer gun-metal tube is surrounded by a jacket in which different temperatures can be maintained steady. The osmotic pressures P_1 and P_2 of the same solution are measured at steady temperatures $T_1^0 A$ **and** $T_2^0 A$ respectively.

It would be found that

$$\frac{P_1}{P_2} = \frac{T_1}{T_2}$$

which verifies the second law.

Law III—The gun-metal tube is filled one after another with the same volume of two solutions of two different substances containing the same number of gm-molecules of the two substances. The osmotic pressures of the two solutions will be found to be the same at a constant temperature—which verifies the third law.

N.B. In the derivation of the laws of osmotic pressure and their verifications, the dilute solutions are considered to be non-electrolytic in character. It has been found from observations that solutions of electrolytes exert greater osmotic pressures than those of non-electrolytes. This is due to the splitting up of a molecule into two or more ions. In such a case, $PV = iRT, i$ being a constant depending upon the degree of dissociation and known as Van't Hoff factor.

13.5 Theories of osmotic pressure and semi-permeability

13.5.1 Theory of osmotic pressure

From the stand point of kinetic theory, osmotic pressure, like gas pressure, is due to the bombardments of molecules on the wall of the semi-permeable membrane. The number of collisions of solvent molecules on the wall of the membrane facing the solution, will be less than the number of collisions occurring on the wall of the membrane facing the solvent. Hence, the solvent molecules from the side of the solvent flow through the membrane in an attempt to make the number of collisions of solvent molecules on the two sides of the membrane equal.

According to the hydro-static theory, the passage of solvent molecules into the solution is supposed to be due to the existence of a force of attraction between the molecules of the solvent and the solute.

13.5.2 Theory of semi-permeability

There is, probably, no satisfactory theory to explain the semi-permeability of the membrane, although several theories have been put forward. The most generally applicable is the theory of preferential solubility. According to this theory, one of the substances (solvent in the case of osmosis), dissolves in the membrane, passes through it and finally comes out on the other side, i.e. on the side of solution. This theory is supported by experiment, and gives a reasonable picture of the phenomenon occurring in semi-permeable membranes.

13.6 Relations between the osmotic pressure and the other properties of a solution

13.6.1 Relation between osmotic pressure and vapour pressure

The vapour pressure of a solution is less than that of a pure solvent. This lowering of vapour pressure in the case of a non-electrolytic solution, is governed by Raoult's law, according to which the vapour pressure of a solution is proportional to the mol-fraction of the solvent present in the solution.

If N and n be respectively the numbers of gm-molecules of the solvent and the solute present in the solution, then for the solution, the mol-fraction of the solvent will be $\frac{N}{N+n}$, while for the solvent alone, the mol-fraction of the solvent will be $\frac{N}{N+0} = 1$. Hence if p_0 and p_s be respectively the vapour pressures of the pure solvent and solution at temperature T^0A, then

$$\frac{p_0}{p_s} = 1 \div \frac{N}{N+n} = \frac{N+n}{N} \quad \text{or,} \quad 1 - \frac{p_s}{p_0} = 1 - \frac{N}{N+n}$$

$$\text{or,} \quad \frac{p_0 - p_s}{p_0} = \frac{n}{N+n}. \tag{13.6.1}$$

If P be the osmotic pressure of the solution at temperature T^0A, then it can be shown that

$$\frac{p_0 - p_s}{p_0} = \frac{MP}{\rho RT} \tag{13.6.2}$$

Here M is the molecular weight of the solvent, while ρ is the density of dilute solution which may, approximately, be taken to be equal to the density of solvent.

13.6.2 Relation between osmotic pressure and boiling point

The boiling point of a solution is greater than that of the pure solvent. The relation between the osmotic pressure P and the boiling point T_B^0A of the solvent is given by

$$P = \frac{L\rho dT}{T_B} \tag{13.6.3}$$

Here L and ρ are respectively the latent heat of vaporisation (in ergs per gm) and density of the solvent and dT is the elevation of the boiling point of the solution. For a non-electrolytic dilute solution, we have

$$P = \frac{mRT_B}{VM} \tag{13.6.4}$$

from eqn. (13.3.1).

From (13.6.3) and (13.6.4) we have

$$\frac{mRT_B}{VM} = \frac{L\rho dT}{T_B} \quad \text{or,} \quad dT = \frac{mRT_B^2}{VML\rho} = \frac{mRT_B^2}{MLw} \quad (\text{where } w = V\rho). \tag{13.6.5}$$

Here V represents the volume of the solution. If there be no expansion or contraction of solution, then V will also represent the volume of the solvent and $V\rho = w =$ mass of solvent in which m gms of solute (of molecular weight M) are dissolved.

Now molecular elevation (θ^0) of the boiling point is the increase of boiling point of a solvent by dissolving 1 gm-molecule of solute in 100 gms of solvent at temperature T_B^0A. Hence when $m = M$ and $w = 100$ gms, $dT = \theta^0$.

Therefore,

$$\theta = \frac{MRT_B^2}{ML \times 100} = \frac{RT_B^2}{100L} \quad \text{or,} \quad \frac{RT_B^2}{L} = \theta \times 100.$$

Hence eqn. (13.6.5) may be written as

$$dT = \frac{100m\theta}{M \times w} \tag{13.6.6}$$

13.6.3 Relation between osmotic pressure and freezing point

The freezing point of a solution is less than that of its solvent. The depression dT of the freezing point of a solution by dissolving m gms of a substance in w gms of solvent is also given by equations (13.6.5) and (13.6.6).

In this case, L will represent the latent heat of solidification of solvent and θ the molecular depression of the freezing point.

13.6.4 Relation between osmotic pressure and association or dissociation of molecules

If there be association or dissociation of molecules in the solution, then the molecular concentration of the substance in the solution will decrease or increase. For these solutions, we have already seen that the relation between the osmotic pressure P, volume V of solution containing one gm-molecule of solute and temperature $T^0 A$ is given by

$$PV = iRT$$

i being a factor known as Van't Hoff factor.

For abnormal solutions (i.e. solutions containing associated or dissociated molecules), i is the ratio of the experimental value of the molecular depression of the freezing point to its normal value (for unassociated or undissociated substance). Thus when one gm-molecule of sodium chloride is dissolved in 100 gms of water, the depression of freezing point is 36 while the value for normal substance (unassociated or undissociated substance) is 18.5. Hence $i = \frac{36}{18.5} = 1.94$. By measuring osmotic pressure, the degree of association or dissociation of molecules can be calculated.

13.6.5 Illustrative examples

1. 10 gms. of sugar of molecular weight 360 is dissolved in 1 litre of water at temperature 20°C. If $R = 8.31 \times 10^7$ dynes-cms/°C, calculate the osmotic pressure of the solution.

From equation(13.3.1) we have

$$P = \frac{m}{MV}RT.$$

Here, $m = 10$ gms; $V = 1000$ c.c.; $M = 360$; $T = 293^0$ K.

$$\therefore\ P = \frac{10 \times 8.31 \times 10^7 \times 293}{360 \times 1000} = 6.76 \times 10^5 \text{ dynes per sq. cm.}$$

2. Calculate the elevation of boiling point of water when one gram-molecule of a salt is dissolved in 100 c.c. of water.

The osmotic pressure of a solution is given by

$$P = \frac{m}{MV}RT.$$

Here $m = M$ gms; $V = 100$ c.c.; $T = 373^0$ K.

$$\therefore\ P = \frac{8.31 \times 10^7 \times 373}{100} = 3.10 \times 10^8 \text{ dynes per sq. cm.}$$

Now from equation (13.6.3) the elevation of boiling point is given by

$$dT = \frac{PT_B}{L\rho}.$$

Taking $L = 537 \times 4.2 \times 10^7$ ergs/gm, $T_B = 373^0 K$ and $\rho = 0.96$, we have,

$$dT = \frac{3.10 \times 10^8 \times 373}{0.96 \times 537 \times 4.2 \times 10^7} = 5.34^0\text{C}.$$

For a concentration of 1 gm-molecule per 100 c.c., the elevation is called *molecular elevation of boiling point.*

13.7 Diffusion

If we take a solution of copper sulphate in a beaker and water is poured in it in such a way that the line of demarcation between the liquid and the solvent is clearly seen and the solution is not disturbed, we shall observe that although the layer separating the liquid is very distinct initially, but in course of days the blue colour of the copper sulphate solution gradually spreads and the line of demarcation disappears. This process in which the molecules of the solute (copper sulphate) move upward from regions of greater concentration to regions of lower concentration is called *diffusion.*

Diffusion also takes place in gases. To demonstrate the diffusion of gases we take a tall jar having a moveable glass plate near the middle of the jar. Now denser gas like CO_2 is taken in the lower half of the jar, the upper half being filled with a comparatively light gas like hydrogen. On removing the glass plate it will be found that heavier gas CO_2 moves slowly upwards and the molecules of the two gases intermix with each other and finally we get a mixture of two gases of uniform density. This process is called *inter-diffusion.*

Diffusion is also found among solids. For example, when sheets of lead and gold are kept in intimate contact with each other for a pretty long time, diffusion of one into the other takes place. Mercury diffuses into many metals such as lead, copper, etc. and make them brittle. More recently it has been detected that radioactive lead can diffuse into ordinary lead when they are kept in contact with each other for a long time. This process is known as self-diffusion.

Again when a vessel is occupied by a single gas in non-uniform concentration, a process of self-diffusion goes on according to Fick's law (See Art. 13.8), but the measurement of the corresponding *coefficient of self-diffusion* is impeded by the fact that an absolute identity of particle characteristics would prevent the movement of any group of them being observed. This difficultly has been avoided without serious error by a number of means. Boardman and Wild chose pairs of gases e.g. nitrogen with carbon monoxide, or nitrous oxide with carbondioxide, which have equal values of molecular diameter and fields of attraction so that their inter diffusion was equivalent to the self-diffusion of either. Other experimenters have used stable isotopes e.g. uranium hexafluoride of which one part was enriched with U_{238} and the other with U_{235}. The current progress of diffusion was sampled at intervals by admitting a small amount of gas in a particular part of apparatus to mass spectrographic analysis.

From the above we conclude that Diffusion is the phenomenon arising out of transference of mass due to difference in concentration and results finally in the equilibrium distribution in concentration. Diffusion may be of two different types: *self-diffusion*—when it occurs in a single gaseous system and *inter-diffusion* when two different gases communicate with each other. We shall concern ourselves in Kinetic theory with *self-diffusion* only.

13.8 Fick's Law

Fick formulated the law of diffusion which states that the rate of diffusion in any direction is proportional to the concentration gradient of the solute in that direction. Concentration gradient means the rate of change of concentration with distance.

There is a close analogy between the process of diffusion and the conduction of heat through a solid. We know that the amount of heat Q passing in unit time through a solid of area A at a place where the temperature gradient is $\frac{d\theta}{dx}$ [temperature gradient is the rate of change of temperature θ with distance x] is given by

$$Q = -KA\frac{d\theta}{dx},$$

where K is the thermal conductivity of the material of the solid. The negative sign shows that θ decreases with the increase of x.

Fick showed that the diffusion takes place from a region of greater concentration to a region of lower concentration in the same way as the flow of heat takes place from a region of higher temperature to a region of lower temperature. According to Fick's law, the mass M of a dissolved substance diffusing in unit time is proportional to the area A through which diffusion occurs and to the rate at which the concentration C, of the dissolved substance diminishes in a direction at right angles to the area A. Hence we may write

$$M = -DA\frac{dC}{dx} \tag{13.8.1}$$

D is called the coefficient of diffusion and $\frac{dC}{dx}$ is the concentration gradient. D depends on the nature of the solute in a given solvent and also on its concentration. *Coefficient of diffusion* D is defined as the *number of molecules* that cross unit area (i.e. $A = 1$) in unit time for unit concentration gradient (i.e. $\frac{dC}{dx} = 1$).

13.9 Differential equation for diffusion

We consider a tall jar of cross-sectional area A containing a solution. Let the concentration of the solution in a plane at a height x be C. Hence, the mass of the solute crossing the plane in unit time is $-DA\frac{dC}{dx}$. Let the concentration of the solution in a plane at a height $x + \delta x$ be $C + \delta C$. Then the mass of the solute leaving the plane at a height $x + \delta x$ is

$$-DA\frac{d}{dx}\left(C + \frac{dC}{dx}.\delta x\right) = -DA\frac{dC}{dx} - DA\frac{d^2C}{dx^2}\delta x.$$

Hence the mass of the solute added in unit time to the volume $A\delta x$ of the solution

$$= -DA\frac{dC}{dx} - \left(-DA\frac{dC}{dx} - DA\frac{d^2C}{dx^2}\delta x\right) = DA\frac{d^2C}{dx^2}\delta x.$$

$\therefore$ the rate of change of concentration

$$= DA\frac{d^2C}{dx^2}\delta x / A\delta x = D\frac{d^2C}{dx^2}.$$

Denoting the rate of change of concentration by $\frac{dC}{dt}$, we have,

$$\frac{dC}{dt} = D\frac{d^2C}{dx^2}. \tag{13.9.1}$$

Thus Fick's law is expressed as a differential equation.

To verify the expression (13.9.1), Fick cemented a tube vertically into a vessel containing crystals of sodium chloride. Water was poured into the tube. The tube with the vessel at its lower end was kept inside a water bath until the rate of diffusion becomes steady. For steady diffussion, $\frac{dC}{dt} = 0$.

Hence

$$D\frac{d^2C}{dx^2} = 0 \text{ or } \frac{d^2C}{dx} = 0 \text{ or } C = ax + b \tag{13.9.2}$$

where a and b are constants. A small glass bulb was dipped at various depths inside the solution contained in the tube and the densities at different depths of the solution were determined by measuring the weights of the bulb at these depths. From the relation between density and concentration we can determine the concentrations at different depths whence the expression (13.9.2) was verified.

13.10 Determination of the coefficient of diffusion

We consider a tall cylinder filled with water. The solution, the coefficient of diffusion of which is to be determined, is introduced below the water. The plane $x = -l_1$ shows the plane of separation between the water and the solution. $x = 0$ and $x = -l$ denote the upper and lower ends of the cylinder.

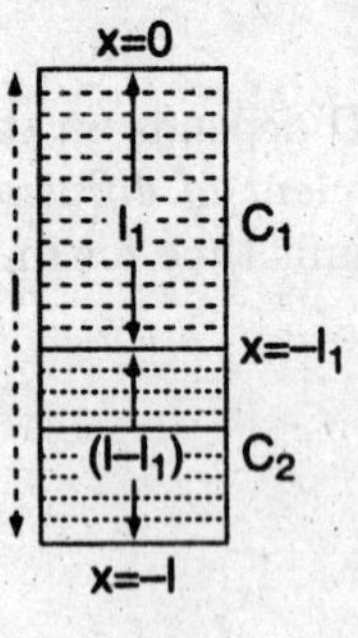

Fig. 13.10.1

Now, the solution of the differential equation (13.9.1) is

$$C = A_0 + [A\cos mx + B\sin mx]e^{-m^2Dt}. \tag{13.10.1}$$

Let us assume that C_1 and C_2 are the initial concentrations of the solute at regions from $x = 0$ to $x = -l_1$ and $x = -l_1$ to $x = -l$ respectively.

$$\text{At } x = 0, \quad \frac{dC}{dx} = 0 \quad [\because \text{ no solute passes across the upper end}] \tag{13.10.2}$$

Now, from equation (13.10.1) we have,

$$\frac{dC}{dx} = [-Am\sin mx + Bm\cos mx]e^{-m^2Dt}$$

$$\text{or } 0 = Bme^{-m^2Dt} \text{ [applying the condition (13.10.2)]}$$

$$\text{i.e. } B = 0.$$

Hence equation (13.10.1) becomes,

$$C = A_0 + A\cos mxe^{-m^2Dt}. \tag{13.10.3}$$

Again

$$\frac{dC}{dx} = 0 \text{ at } x = -l. \tag{13.10.4}$$

Hence from (13.10.3) and (13.10.4) we have

$$0 = -Am \sin ml \bar{e}^{m^2 Dt} \quad \text{or} \quad \sin ml = 0 \quad \text{or} \quad ml = s\pi,$$

where $s = 1, 2, 3$ etc.

$\therefore$ we may write equation (13.10.1) as

$$C = A_0 + \sum_{s=1}^{\infty} A_s \cos \frac{s\pi}{l} x e^{-\left(\frac{s\pi}{l}\right)^2 Dt} \tag{13.10.5}$$

Now at time $t = 0$, let $C = C_0$

$$\therefore \quad C_0 = A_0 + \sum_{s=1}^{\infty} A_s \cos \frac{s\pi x}{l} = C_1 + C_2 \tag{13.10.6}$$

$$\text{or} \quad \int_{-l}^{-l_1} C_2 dx + \int_{-l_1}^{0} C_1 dx = \int_{-l}^{0} A_0 dx + \int_{-l}^{0} \Sigma A_s \cos \frac{s\pi x}{l} dx$$

$$\text{or} \quad C_2(l - l_1) + C_1 l_1 = A_0 l \quad \left[\because \int_{-l}^{0} \Sigma \cos \frac{s\pi x}{l} dx = 0\right]$$

$$\text{or} \quad A_0 = \frac{C_1 l_1}{l} + \frac{C_2 l_2}{l} = \frac{C_1 l_1 + C_2 l_2}{l} \quad [(l - l_1) = l_2] \tag{13.10.7}$$

To find A_s we multiply both sides of equation (13.10.6) by $\cos \frac{s\pi x}{l} dx$ and then integrating we have

$$\int_{-l}^{-l_1} C_2 \cos \frac{s\pi x}{l} dx + \int_{-l_1}^{0} C_1 \cos \frac{s\pi x}{l} dx = A_0 \int_{-l}^{0} \cos \frac{s\pi x}{l} dx + \int_{-l}^{0} \Sigma A_s \cos^2 \frac{s\pi x}{l} dx$$

$$\text{or} \quad \frac{C_2 l}{s\pi}\left[\sin \frac{s\pi x}{l}\right]_{-l}^{-l_1} + \frac{C_1 l}{s\pi}\left[\sin \frac{s\pi x}{l}\right]_{-l_1}^{0} = \frac{A_s l}{2} \left[\because \int_{-l}^{0} \cos \frac{s\pi x}{l} dx = 0\right]$$

$$\text{or} \quad \frac{C_1 l}{s\pi} \cdot \sin \frac{s\pi l_1}{l} - \frac{C_2 l}{s\pi} \sin \frac{s\pi l_1}{l} = \frac{A_s l}{2}$$

$$\text{or} \quad \frac{l}{s\pi}(C_1 - C_2) \sin \frac{s\pi l_1}{l} = \frac{A_s l}{2}$$

$$\text{or} \quad A_s = \frac{2(C_1 - C_2)}{s\pi} \sin \frac{s\pi l_1}{l}. \tag{13.10.8}$$

Hence from (13.10.7) and (13.10.8) equation (13.10.5) becomes

$$C = \frac{C_1 l_1 + C_2 l_2}{l} + \sum_{s=1}^{s=\infty} \frac{2(C_1 - C_2)}{s\pi} \sin \frac{s\pi l_1}{l} \cos \frac{s\pi x}{l} e^{-\left(\frac{s\pi}{l}\right)^2 Dt}.$$

We may take $C_1 = 0$ and $C_2 = c$ (say) and $l_1 = l_2 = l/2$

$$\therefore\ C = \frac{c}{2} - \sum_{s=1}^{s=\infty} \frac{2c}{s\pi} \sin\frac{s\pi}{2} \cos\frac{s\pi x}{l} e^{-\left(\frac{s\pi}{l}\right)^2 Dt} = \frac{c}{2}\left[1 - \frac{4}{\pi}\sum_{s=1}^{\infty}\frac{1}{s}\sin\frac{s\pi}{2}\cos\frac{s\pi x}{l}e^{-\left(\frac{s\pi}{l}\right)^2 Dt}\right]$$

$$= \frac{c}{2}\left[1 - \frac{4}{\pi}\left\{\sin\frac{\pi}{2}\cos\frac{\pi x}{l}e^{-\left(\frac{\pi}{l}\right)^2 Dt} + 0 + \frac{1}{3}\sin\frac{3\pi}{2}\cos\frac{3\pi x}{l}e^{-\left(\frac{3\pi}{l}\right)^2 Dt} + 0 + \cdots\right\}\right].$$

Considering t to be large, the above equation may be written in the form

$$C = \frac{c}{2}\left[1 - \frac{4}{\pi}\cos\frac{\pi x}{l}e^{-\frac{\pi^2 Dt}{l^2}}\right]. \tag{13.10.9}$$

To find the value of D a fixed plane in the solution is chosen and the values of concentration c in this plane at different times are determined. Now, from equation (13.10.9) we have,

$$\frac{\pi}{2c}\left(\frac{c}{2} - C\right) = \cos\frac{\pi x}{l}e^{-\frac{\pi^2 Dt}{l^2}}.$$

If we take a point at a distance x in the upper half of the cylinder so that $\frac{c}{2} > C$ at all times in practice, then

$$\left(\frac{c}{2} - C\right) = \frac{2c}{\pi}\cos\frac{\pi x}{l}e^{-\frac{\pi^2 Dt}{l^2}} = be^{-\frac{\pi^2 Dt}{l^2}},\ \left[b = \frac{2c}{\pi}\cos\frac{\pi x}{l}\right]$$

where b is positive.

The above expression may also be written in the form

$$\log_e\left(\frac{c}{2} - C\right)\Big/ b = -\frac{\pi^2 Dt}{l^2}.$$

If we plot $\log_e\left(\frac{c}{2} - C\right)/b$ against t at which the concentration c is measured, we get a straight line, the slope of which is $-\frac{\pi^2 D}{l^2}$. Measuring the slope the value of D is obtained.

Again, if we take a point at a distance x in the lower half of the cylinder, then $\frac{c}{2} < C$ at all times in practice; so we may write

$$\left(C - \frac{c}{2}\right) = -\frac{2c}{\pi}\cos\frac{\pi x}{l}e^{-\frac{\pi^2 Dt}{l^2}} = +\frac{2c}{\pi}\cos\pi\frac{(l-x)}{l}e^{-\frac{\pi^2 Dt}{l^2}} = b'e^{-\frac{\pi^2 Dt}{l^2}}\ \left[b' = \frac{2c}{\pi}\cos\pi\frac{(l-x)}{l}\right]$$

where b' is positive

$$\therefore\ \log_e\left(C - \frac{c}{2}\right)\Big/ b' = e^{-\frac{\pi^2 Dt}{l^2}}.$$

Thus plotting $\log_e\left(C - \frac{c}{2}\right)/b'$ against t, we can determine D in a similar way. It is to be noted that the coefficient of diffusion D depends on the concentration and so the value of D deduced above is an approximate one.

Kelvin measured the densities of different layers of the solutions by introducing into it a series of beads of different known densities and noting their equilibrium positions and thus the distribution of concentration was measured. The method suffers from the following disadvantages:

(a) Some salt from the solution crystallises on the beads.

(b) Air bubbles are formed on their surfaces. This factor tends to change the buoyancy of beads and thus the results obtained will be liable to errors.

The other methods used to measure concentration are based on:

(a) measurement of refractive index of the solution;

(b) measurement of the rotation of the plane of polarisation for optically active solutions such as sugar solutions etc.;

(c) measurement of contact potential difference.

13.10.1 Inter-diffusion of Solids

It is found that two solids kept in contact diffuse into one another despite the fact that solids have a definite crystalline structure. Fick's law also applies to solids. When sheets of gold and lead are placed in contact for a longtime particles of one are found to diffuse into the other.

Exercises

1. Define osmosis and osmotic pressure. Describe an experiment for the determination of osmotic pressure of a solution.
2. State the laws of osmotic pressure. How would you verify the laws experimentally?
3. Discuss the relations between the osmotic pressure, the boiling point and the freezing point of a solution.
4. Give some account of the phenomenon of diffusion.
5. State and explain Fick's law. Derive a differential expression for diffusion.
6. Define coefficient of diffusion. Describe a method for the determination of the coefficient of diffusion.
7. One gram of cane-sugar (molecular weight 342 gm.) is dissolved in 500 c.c. of water. Calculate the value of osmotic pressure of the solution at 0^0C. [**Ans.** 1.32×10^5 dynes per sq. cm.]
8. What is meant by self-diffusion? Explain with an example. Define the coefficient of self-diffusion. **[C.U. (Hons.) 2004]**

Chapter 14

Kinetic Theory of Matter

14.1 Introduction

It is supposed that the molecules constituting a body are in a perpetual state of agitation. Many physical properties of matter can be explained by the motion of the molecules. According to the kinetic theory the molecules in a solid state are closely packed together in a geometric pattern and each molecule vibrates about a fixed position in its pattern. A solid may be heated to a temperature such that the movement of the molecules may be greater than the vibration of the molecules about their mean position. Thus due to heating the kinetic energy possessed by the molecules, constituting the body increases causing violent vibration of the molecules and on further increasing the temperature, the agitation or the vibration of the molecules becomes so high that there is disruption of the positions of the molecules in the pattern. At this stage the molecules temporarily form a pattern coming close to a few other molecules and then break away again. The matter is then said to be in a liquid state. In a liquid the molecules near the surface are bound strongly together due to a phenomenon known as surface tension of the liquid. If the liquid is now heated, the motion of the molecules becomes sufficiently high to destroy the binding force of the molecules in the surface and the molecules in large number are thrown asunder in space. The matter is now said to be in a gaseous state. The distances between the molecules are so large at this stage that the molecules practically exert no influence on each other. Under this condition there is no vibratory motion of the molecules, and the molecules move in a straight line for a short time and encounter with other molecules and then again move in a straight line in a different direction for a short interval of time.

14.2 Kinetic Theory of Gases—Pressure exerted by a Perfect Gas

The assumptions on which the mechanical picture of a perfect gas is based are as follows:

(i) A perfect gas consists of a large number of molecules which are considered to be rigid and spheres of the same mass.

(ii) The size of a gas molecule is assumed to be negligible compared to the distance traversed by a molecule between any two successive collisions—called the free path.

(iii) The molecules are considered to be perfectly elastic spheres and no forces of attraction or repulsion exist between the molecules and the walls of the container. In other words, the whole of the molecular energy is kinetic.

(iv) In any steady state, the molecules incessantly collide with each other, but assemble uniformly throughout the containing vessel and so the density of the gas remains unaffected.

(v) The molecules move in a straight line with uniform velocity between two sucessive collisions.

(vi) The time during which a collision takes place is negligible in comparison with the time taken to traverse the free path.

(vii) Lastly, the molecules are assumed to be geometrical mass points so that the volume occupied by them is negligible compared to the volume of the vessel.

14.3 Expression for Pressure of a Perfect Gas

We consider a perfect gas enclosed in a vessel. The molecules of the gas are moving with all possible velocities. At any moment some molecules move slowly and some move very rapidly so that the speed of the molecules ranges from zero to infinity. Moreover, the velocities of the molecules can be represented by suitable vectors.

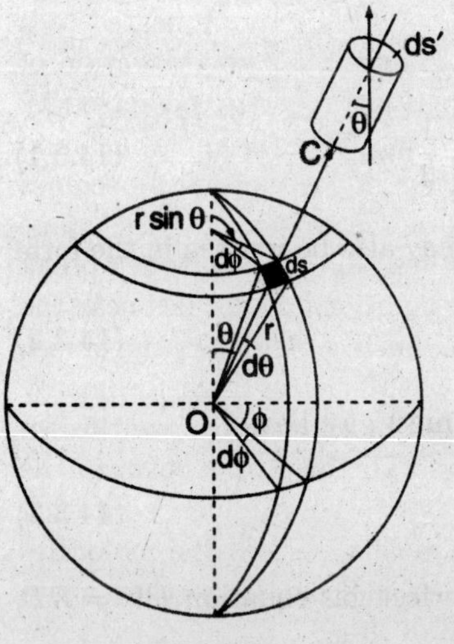

Fig. 14.3.1

Let us now consider some molecules of the gas having velocity vector C directed from the point O (Fig. 14.3.1). Our task will be to find out the number of molecules having velocity vectors C and $C+dC$ and directions within the solid angle $d\Omega$ about the vector C formed by lines drawn from O and touching the edge of the surface ds, the area ds on the surface of the sphere of radius r being formed by two circles of latitude differing by $d\theta$, and two circles of longitude differing by $d\phi$. This number is given by $dN' = \frac{d\Omega}{4\pi} \cdot dN$, where dN is the number of molecules with velocities C and $C + dC$ and $d\Omega = \frac{ds}{r^2} = \frac{r \sin\theta\, d\phi \cdot r\, d\theta}{r^2}$ [Fig. 14.3.1] $= \sin\theta\, d\theta\, d\phi$.

$$\therefore dN' = \frac{dN \sin\theta\, d\theta\, d\phi}{4\pi}.$$

We shall now calculate the number of molecules striking the area ds' of the wall of the container in a small interval of time dt during which the molecules do not suffer collisions. We consider a cylinder of length Cdt and base ds' as shown in the figure. If V be the volume of the container and dV the volume of the cylinder, the fraction $\frac{dV}{V}$ of the molecules will be contained within the cylinder and hence the number of molecules having velocities between C and $C+dC$ and directions within $d\Omega$, i.e. directions between $\theta, \theta+d\theta$, and $\phi, \phi+d\phi$ striking ds' in time dt

$$= dN' \frac{dV}{V} = \frac{dN' C\, dt\, ds' \cos\theta}{V} \quad [\because\ dV = C\, dt\, ds' \cos\theta]$$

Now, the normal component of incident momentum of a molecule of mass m moving with velocity C towards the area ds' making an angle θ to the normal is $-mC\cos\theta$ and the normal component of the reflected momentum is $mC\cos\theta$. Hence the change in momentum suffered by each molecule is $-2mC\cos\theta$ and the total change in momentum suffered by molecules having velocities between C and $C+dC$ and directions between $\theta, \theta+d\theta$ and $\phi, \phi+d\phi$

$$= \frac{dN}{4\pi}\sin\theta\, d\theta\, d\phi \frac{C\, dt}{V}\cos\theta\, ds'(-2mC\cos\theta) = -\frac{dN}{2\pi V}mC^2\, dt\, ds' \cos^2\theta \sin\theta\, d\theta\, d\phi.$$

$\therefore$ the change in momentum per unit area and per unit time suffered by all the molecules striking the wall coming from all directions

$$= -\frac{m}{2\pi V}\int_0^\infty C^2\, dN \int_0^{\pi/2} \cos^2\theta \sin\theta\, d\theta \int_0^{2\pi} d\phi$$

$$= \frac{m}{2\pi V}\int_0^\infty C^2\, dN \int_0^{\pi/2} \cos^2\theta d(\cos\theta) \int_0^{2\pi} d\phi = -\frac{m}{3V}\int_0^\infty C^2\, dN.$$

The above expression is equal to the pressure exerted by the molecules on the wall. As the pressure acts in a direction opposite to the direction in which the change in momentum takes place we may write

$$P = \frac{m}{3V}\int_0^\infty C^2\, dN.$$

Now the mean square velocity

$$\overline{c^2} = \frac{1}{N}\int_0^\infty C^2\, dN \quad \therefore \quad P = \frac{1}{3}\cdot\frac{N}{V}m\overline{c^2} = \frac{1}{3}nm\overline{c^2} \tag{14.3.1}$$

where n is the number molecules per c.c. The equation (14.3.1) may also be written in the form

$$PV = \frac{2}{3}\cdot\frac{1}{2}mN\overline{c^2}. \tag{14.3.2}$$

If we consider a gram molecule of the gas which occupies a volume V, we have

$$PV = \frac{2}{3}\cdot\frac{1}{2}M\overline{c^2} \tag{14.3.3}$$

where M is the molecular weight of the gas. Considering the perfect gas equation $PV = RT$, we may write

$$RT = \frac{2}{3}\cdot\frac{1}{2}M\overline{c^2} = \frac{2}{3}\times \begin{array}{l}\text{kinetic energy of one gram}\\ \text{molecule of the gas}\end{array} \tag{14.3.4}$$

The above expression may also be written in the form

$$\overline{c^2} = \frac{3RT}{M}. \tag{14.3.5}$$

Thus we see that $\overline{c^2}$ is proportional to the absolute temperature. Again the kinetic energy is proportional to $\overline{c^2}$ and so proportional to the absolute temperature.

The equation (14.3.5) shows that at absolute zero $\overline{c^2}$ becomes zero, i.e. at absolute zero all motions cease which is not the actual case. This is because that the perfect gas equation does not hold good up to absolute zero. According to quantum theory a substance possesses some energy known as zero-point energy at absolute zero.

The equation (14.3.5) may be written in the form

$$\frac{1}{2}M\overline{c^2} = \frac{3}{2}RT \text{ or } \frac{1}{2}mN\overline{c^2} = \frac{3}{2}RT \text{ or } \frac{1}{2}m\overline{c^2} = \frac{3}{2}\frac{R}{N}T = \frac{3}{2}KT$$

where $K = \frac{R}{N}$ = Boltzmann's constant.

Thus the mean kinetic energy per molecule = $\frac{3}{2}KT$.

Again

$$P = \frac{1}{3}nm\overline{c^2} = nKT \text{ from (14.3.6).} \tag{14.3.6}$$

The above equation is known as Clapeyron's equation.

Illustrative Example

1. Assuming the density of hydrogen to be 0.00008987 gm. per c.c., find the mean square velocity of hydrogen molecules at N.T.P.

We know that,

$$\overline{c^2} = \frac{3RT}{M} \text{ or } \overline{c^2} = \frac{3PV}{M} = \frac{3P}{\rho},$$

where ρ is the density of hydrogen gas,

$$\text{or } \sqrt{\overline{c^2}} = RMS \text{ velocity } = \sqrt{3P/\rho}.$$

Here,

$$P = 76 \times 13.6 \times 981 \text{ dynes/sq. cm.}$$

$$\rho = 0.00008987 \text{ gm. per c.c.}$$

$$\therefore \sqrt{\overline{c^2}} = \sqrt{\frac{3 \times 76 \times 13.6 \times 981}{0.00008987}} = 1.839 \times 10^5 \text{ cm./sec.}$$

14.4 Deductions from the Pressure Expression

The following laws valid for ideal gases can be deduced from the kinetic theory of gases.

14.4.1 Boyle's law

From the equation (14.3.3) we have

$$PV = \frac{1}{3}M\overline{c^2}.$$

When temperature remains constant $\overline{c^2}$ remains constant. Hence PV = constant. Thus, at constant temperature the product of the pressure and the volume of a gas remains constant, which is Boyle's law.

14.4.2 Charle's law

We know that $PV = \frac{1}{3}M\overline{c^2}$.

Now $\overline{c^2}$ is proportional to the absolute temperature, hence $PV \propto T$.

$\therefore$ $V \propto T$ when P is constant, i.e. at constant pressure the volume of a given mass of a gas is proportional to its absolute temperature, which is Charle's law.

14.4.3 Avogadro's law

This law states that equal volumes of different ideal gases under identical conditions of temperature and pressure contain the same number of molecules. If there are two gases of same volume and at the same pressure, we may write

$$P = \frac{1}{3}\frac{N_1}{V}m_1\overline{c_1^2} = \frac{1}{3}\frac{N_2}{V}m_2\overline{c_2^2} \tag{14.4.1}$$

where the subscripts 1 and 2 are meant for the first and the second gases respectively. Considering constant temperature of the two gases we may write

$$\frac{1}{2}m_1\overline{c_1^2} = \frac{1}{2}m_2\overline{c_2^2}$$

$\therefore$ from equation (14.4.1) we have $N_1 = N_2$.

Thus at the same temperature and pressure the two gases contain the same number of molecules which is Avogadro's law. The actual number of molecules per c.c. at N.T.P. is 2.687×10^{19}. This number is known as *Loschmidt's number*, but the number of atoms in a gm. atom or the number of molecules in a gm-molecule is called the *Avogadro's number* which is 6.02×10^{23}.

14.4.4 Dalton's law

This law states that the pressure in a mixture of gases is equal to the sum of the pressures exerted separately by the constitutent gases occupying the same volume at a particular temperature. If $m_1, m_2, m_3, \cdots$ and $n_1, n_2, n_3, \cdots$ be respectively the masses and numbers of molecules per c.c. of different gases, then the resultant pressure of the mixture of gases is given by

$$p = \frac{1}{3}m_1n_1\overline{c_1^2} + \frac{1}{3}m_2n_2\overline{c_2^2} + \frac{1}{3}m_3n_3\overline{c_3^2} + \cdots$$

But the partial pressures are

$$p_1 = \frac{1}{3}m_1n_1\overline{c_1^2}, \ p_2 = \frac{1}{3}m_2n_2\overline{c_2^2} \cdots \text{ etc.}$$

Hence $p = p_1 + p_2 + \cdots$, which is Dalton's law.

14.5 Maxwell's Law of Distribution of Molecular Velocities

If the molecules in the given volume of a gas possess the same velocity at any instant, the velocities may be disturbed at the next instant due to collisions with other molecules as a result of which there will be a non-uniform distribution of velocities of the molecules. As the number of molecules in a gas per c.c. is very large and the size of the molecules is too small, it is needless to consider the behaviour of individual molecules. Actually the problem is a statistical one and so Maxwell applied the laws of probability in deducing the law of distribution of molecular velocities.

We, now, consider a gas in a state of thermal equilibrium and assume that the gas molecules obey the following conditions:

(a) The number of molecules in any volume of the gas remains the same and that the velocities in any direction are equal. This assumption demands that there will be no aggregation of the molecules anywhere and that the convection in the gas is absent.

(b) The velocities along any three co-ordinate axes are independent of each other.

(c) The distribution of velocities is free from any other influences. The probability of velocity of a molecule having velocities between the limits u and $u + du$ is $f(u)\,du$.

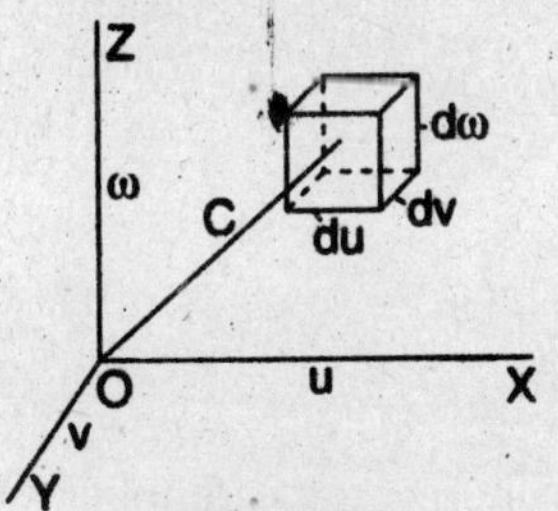

Fig. 14.5.1

With these assumptions Maxwell has deduced, as given below, the distribution law which is not changed due to collisions.

We represent the velocity components u, v, w along OX, OY, OZ axes (Fig. 14.5.1) of a molecule moving with velocity C. The relation between C and the component velocities is given by

$$C^2 = u^2 + v^2 + w^2. \tag{14.5.1}$$

Now, the probability that a molecule has a component velocity lying between u and $u + du$ along X-axis is $f(u)\,du$. Similarly, the probabilities that the component velocities of the molecule lie between v and $v + dv$ and w and $w + dw$ along Y and Z axes respectively are $f(v)\,dv$ and $f(w)\,dw$. As these three components are independent, the probability that they occur simultaneously is $f(u)\,f(v)\,f(w)\,du\,dv\,dw$.

If C is the resultant velocity, the probability may be expressed as $F(C)\,du\,dv\,dw$.

For convenience we write the above expression as $\phi(C^2)\,du\,dv\,dw$, where ϕ denotes an appropriate form of the function. Hence

$$f(u)f(v)f(w)\,du\,dv\,dw = \phi(C^2)\,du\,dv\,dw$$

$$\therefore\ f(u)f(v)f(w) = \phi(C^2) = \phi(u^2 + v^2 + w^2). \tag{14.5.2}$$

For a particular value of $C, \phi(C^2) =$ constant.

$$\therefore\ d\left[\phi(C^2)\right] = 0$$

$$\text{or}\ \ d\left[f(u)f(v)f(w)\right] = 0$$

$$\text{or}\ \ f(u)d\left[f(v)f(w)\right] + f(v)f(w)f'(u)\,du = 0$$

$$\text{or}\ \ f(w)f(u)f'(v)\,dv + f(u)f(v)f'(w)\,dw + f(v)f(w)f'(u)\,du = 0.$$

$f'(u), f'(v)$ and $f'(w)$ denote the derivatives of $f(u), f(v)$ and $f(w)$ respectively.

Dividing the preceding expression by $f(u)f(v)f(u)$ we have,

$$\frac{f'(u)}{f(u)}\,du + \frac{f'(v)}{f(v)}\,dv + \frac{f'(w)}{f(w)}\,dw = 0. \tag{14.5.3}$$

Again differentiating the equation (14.5.1), we have

$$u\,du + v\,dv + w\,dw = 0. \tag{14.5.4}$$

Multiplying the equation (14.5.4) by λ and then adding it to equation (14.5.3), we have

$$\left[\frac{f'(u)}{f(u)} + \lambda u\right] du + \left[\frac{f'(v)}{f(v)} + \lambda v\right] dv + \left[\frac{f'(w)}{f(w)} + \lambda w\right] dw = 0.$$

Since the velocity components are independent of each other, the above equation is zero if only each of the terms of the equation is separately equal to zero.

$$\therefore\; \frac{f'(u)}{f(u)}\,du + \lambda u\,du = 0 \;\text{ or }\; \frac{f'(u)}{f(u)}\,du = -\lambda u\,du.$$

On integration we have,

$$\log_e f(u) = -\frac{\lambda u^2}{2} + \log A$$

where $\log A$ is the constant of integration

$$\therefore\; \frac{f(u)}{A} = e^{-\frac{\lambda u^2}{2}} \;\text{ or }\; f(u) = Ae^{-\frac{\lambda u^2}{2}}.$$

Similarly,

$$f(v) = Ae^{-\frac{\lambda v^2}{2}} \;\text{ and }\; f(w) = Ae^{-\frac{\lambda w^2}{2}}.$$

From the definition of probability we may write

$$\int_{-\infty}^{+\infty} f(u)\,du = 1, \;\text{ or }\; A\int_{-\infty}^{+\infty} e^{-\frac{\lambda u^2}{2}}\,du = 1$$

$$\text{or}\;\; A\sqrt{\frac{2\pi}{\lambda}} = 1 \;\left[\because \int_{-\infty}^{+\infty} e^{-bu^2}\,du = \sqrt{\frac{\pi}{b}}\right] \;\text{ or }\; A = \sqrt{\frac{\lambda}{2\pi}}$$

$$\therefore\; f(u)f(v)f(w)\,du\,dv\,dw = \sqrt{\frac{\lambda^3}{8\pi^3}}\cdot e^{-\frac{\lambda}{2}(u^2+v^2+w^2)}\,du\,dv\,dw.$$

Now, $C^2 = u^2 + v^2 + w^2$ and $du\,dv\,dw$ represents a volume element in the velocity diagram (Fig. 14.5.1). In spherical polar co-ordinates (r, θ, ϕ), $du\,dv\,dw$ can be replaced by

$$dC\cdot Cd\theta\cdot C\sin\theta d\phi \quad \text{(Fig. 14.5.2).}$$

Hence

$$f(u)f(v)f(w)\,du\,dv\,dw = \sqrt{\frac{\lambda^3}{8\pi^3}}\cdot e^{-\frac{\lambda}{2}C^2}\,C^2\,dC \times \sin\theta\,d\theta\,d\phi. \tag{14.5.5}$$

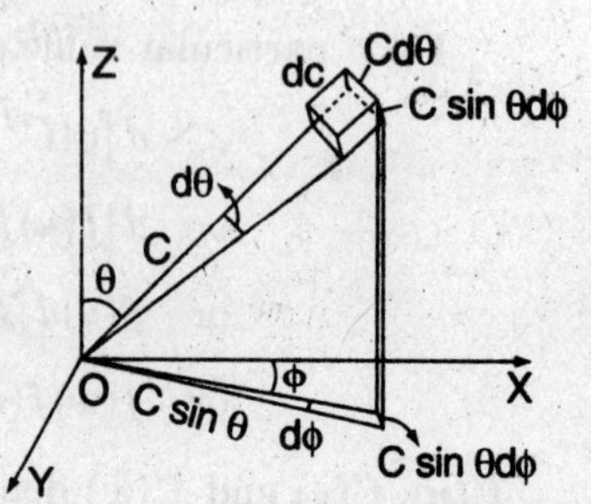

Fig. 14.5.2

This expression gives the probability of the velocity C ending in the volume element $du\,dv\,dw$ or lying between C and $C + dC, \theta$ and $\theta + d\theta$, and ϕ and $\phi + d\phi$. To find out the probability of a velocity lying between C and $C + dC$ the expression must be integrated for all values of θ and ϕ between 0 to π and 0 to 2π respectively. Since there are n molecules per c.c., the number having velocities lying between C and $C + dC$ is given by

$$dn_C = n \cdot \sqrt{\frac{\lambda^3}{8\pi^3}} e^{-\frac{\lambda}{2}C^2} C^2\, dC \int_0^{\pi} \sin\theta\, d\theta \int_0^{2\pi} d\phi = 4\pi n \sqrt{\frac{\lambda^3}{8\pi^3}} e^{-\frac{\lambda}{2}C^2}\, dC, \tag{14.5.6}$$

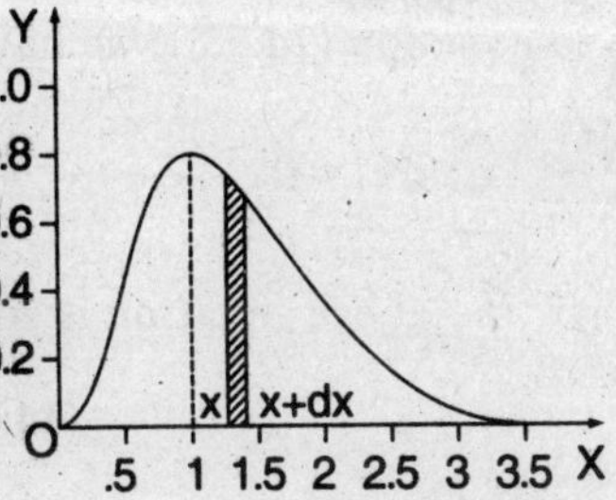

Fig. 14.5.3

Putting $\frac{\lambda}{2}C^2 = x^2$, we have

$$dn_C = \frac{4n}{\sqrt{\pi}} x^2 e^{-x^2} dx.$$

It is now possible to represent Maxwell's law graphically. If we plot $y = \frac{4}{\sqrt{\pi}} e^{-x^2}\, x^2$ against x, the curve obtained is shown in Fig. 14.5.3.

Then the number of molecules dn having velocities lying between x and $x + dx$ is proportional to the shaded area. The ordinate (Fig. 14.5.3) y gives the fraction of number of the molecules with velocity corresponding to x. It is seen from the curve that the probability of the number of molecules is greatest when $x = 1$, but the probability is comparatively low for $x = 2$ or $x = \frac{1}{2}$. Thus the whole gas can be treated as having the most probable velocity.

Maxwell's law of distribution of velocities is unsatisfactory because of its assumption regarding independence of the three velocity components. The proof was later modified by Maxwell himself and by Boltzmann and other investigators. The method of collisions developed by Maxwell and Boltzmann is not superior to Maxwell's original deduction. The method of statistical mechanics developed by Jeans and others is free from objections.

14.6 Calculation of Average Velocity, Mean Square Velocity and the Most Probable Velocity on the basis of Maxwell's Velocity Distribution Law

14.6.1 Average velocity

The mean or the average velocity of all the molecules is given by

$$\bar{c} = \frac{1}{n}\int_0^{\infty} C\, dn_C = 2\sqrt{\frac{\lambda^3}{2\pi}} \int_0^{\infty} e^{-\frac{\lambda}{2}C^2} C^3\, dC.$$

Now,

$$\int_0^{\infty} e^{-\frac{\lambda}{2}C^2} C^3\, dC = \frac{1}{2(\lambda/2)^2} = \frac{2}{\lambda^2} \quad \left[\because \int_0^{\infty} e^{-bC^2} C^3\, dC = \frac{1}{2b^2}\right]$$

$$\therefore\ \bar{c} = \frac{4}{\lambda^2} \cdot \sqrt{\frac{\lambda^3}{2\pi}} = \frac{4}{\sqrt{2\pi\lambda}} = 2\sqrt{\frac{2}{\pi\lambda}}. \tag{14.6.1}$$

14.6.2 Mean square velocity

It is defined as

$$\overline{c^2} = \frac{1}{n}\int_0^\infty C^2\, dn_C = 2\sqrt{\frac{\lambda^3}{2\pi}}\int_0^\infty e^{-\frac{\lambda}{2}C^2} C^4\, dC.$$

Now,

$$\int_0^\infty e^{-\frac{\lambda}{2}C^2} C^4\, dC \quad \left[\because \int_0^\infty e^{-bx^2} x^4 dx = \frac{3}{8}\sqrt{\frac{\pi}{b^5}}\right] = \frac{3\sqrt{\pi}}{8(\lambda/2)^{5/2}}$$

$$\therefore\ \overline{c^2} = 2\sqrt{\frac{\lambda^3}{2\pi}}\cdot\frac{3\sqrt{\pi}}{8(\lambda/2)^{5/2}} = \frac{3}{\lambda}.$$

∴ the root mean square velocity is given by

$$C_r = \sqrt{\overline{c^2}} = \sqrt{\frac{3}{\lambda}}. \tag{14.6.2}$$

Now

$$\frac{\bar{c}}{C_r} = 2\cdot\sqrt{\frac{2}{3\pi}} \quad \text{or} \quad \bar{c} = 0.921 C_r. \tag{14.6.3}$$

14.6.3 The value of C, for which dn_C is maximum, is the most probable velocity C_p

and it may be obtained by differentiating equation (14.5.6) with respect to **C** and then equating the result to zero. Hence we may write

$$\frac{dn_C}{dC} = A\left[2e^{-\frac{\lambda}{2}C_p^2}C_p - \lambda C_p^3 e^{-\frac{\lambda}{2}C_p^2}\right] = 0 \left[A = 4\pi n\sqrt{\frac{\lambda^3}{8\pi^3}}\right]$$

$$\text{or}\quad 2e^{-\frac{\lambda}{2}C_p^2}C_p = \lambda C_p^3 e^{-\frac{\lambda}{2}C_p^2} \quad \text{or} \quad C_p^2 = 2/\lambda \quad \text{or} \quad C_p = \sqrt{\frac{2}{\lambda}} \tag{14.6.4}$$

$$\therefore\ \frac{C_p}{C_r} = \sqrt{2/3}.$$

N.B. $\frac{dn_C}{dC} = 0$ gives the condition for dn_C to be either maximum or minimum. We may however show that $\frac{d^2n_C}{dC^2}$ is negative, and so the value of C_p as given above corresponds to maximum of dn_C.

14.7 Evaluation of the Constant λ and Discussion of the Different Results

We know that $P = \frac{1}{3}nm\overline{c^2}$ substituting for $\overline{c^2}, 3/\lambda$ we have $P = \frac{1}{3}nm\cdot\frac{3}{\lambda} = \frac{nm}{\lambda}$ but $P = nKT$.

Hence we have

$$\frac{nm}{\lambda} = nKT \quad \text{or} \quad \lambda = \frac{m}{KT}.$$

$\therefore$ equation (14.5.6) may be written as

$$dn_C = 4\pi n\left(\frac{m}{2\pi KT}\right)^{3/2} e^{-\frac{mC^2}{2KT}} C^2 dC. \tag{14.7.1}$$

Equation (14.7.1) shows that there will be molecules with all velocities between 0 and ∞. A simple consequence of this is that molecules with a velocity greater than the velocity of escape (given by $C_e = 2gr$, where g is the acceleration due to gravity and r the radius of the planet) from the planet will permanently leave the planet's atmosphere. For earth $C_e = 11.2$ km./sec. and it is 2.4 km./sec. for moon. This accounts for the loss of or rarified atmospheres in many planets and the moon. Also since $\lambda = \frac{m}{KT}$, the loss is more at higher temperatures and for lighter gases.

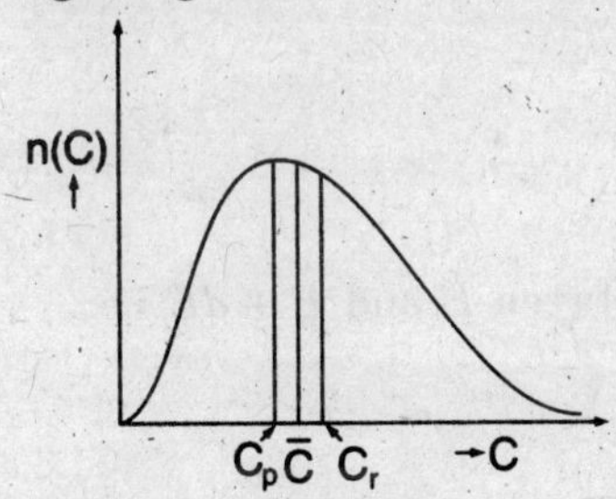

Fig. 14.7.1

Again putting $\lambda = m/KT$, we have from (14.6.1), (14.6.2) and (14.6.4)

$$\bar{c} = \sqrt{\frac{8KT}{\pi m}}, \quad C_r = \sqrt{\frac{3KT}{m}} \quad \text{and} \quad C_p = \sqrt{\frac{2KT}{m}},$$

so that the ratio of the average, r.m.s. and the most probable velocities at a given temperature is

$$\bar{c} : C_r : C_p = \sqrt{\frac{8KT}{\pi m}} : \sqrt{\frac{3KT}{m}} : \sqrt{\frac{2KT}{m}} = \sqrt{\frac{8}{\pi}} : \sqrt{3} : \sqrt{2}.$$

Therefore,

$$\frac{\bar{c}}{C_r} = \sqrt{\frac{8}{3\pi}} = 0.921, \quad \frac{C_p}{C_r} = \sqrt{\frac{2}{3}} = 0.817 \quad \text{or} \quad C_p : \bar{c} : C_r = 1 : 1.128 : 1.224$$

Now the velocity of a molecule with average energy of $\frac{3}{2}KT$ is

$$C_r = \sqrt{\overline{c^2}} = \sqrt{\frac{3KT}{m}}$$

since $\overline{\frac{1}{2}mc^2} = \frac{3}{2}KT$. This velocity is denoted as C_r, because it is the square root of the average of the squared molecular velocities—the r.m.s. velocity and is not the same as the simple arithmetic average velocity $\bar{c}$. To illustrate the difference between the two kinds of average, we consider an assembly of two molecules—one with a velocity of 1m/sec., and the other with a velocity of 3m/sec. The average velocity of the two molecules is

$$\bar{c} = \frac{c_1 + c_2}{2} = \frac{1+3}{2} = 2\text{m/sec.}$$

whereas the r.m.s. velocity

$$C_r = \sqrt{\frac{c_1^2 + c_2^2}{2}} = \sqrt{\frac{1^2 + 3^2}{2}} = 2.24\text{m/sec.}$$

In fact we have, from Maxwell's distribution law, the relationship between $\bar{c}$ and C_r which is

$$C_r = \sqrt{\frac{3\pi}{8}}\bar{c} = 1.09\bar{c}$$

which shows that r.m.s. velocity is about 9 per cent greater than the arithmetic average velocity.

Again the most probable velocity C_p is the velocity with which maximum number of molecules move in a gas at equilibrium, obeying Maxwell's distribution law.

14.8 Energy Distribution of Molecules

We know that $E = \frac{1}{2}mC^2$;

$$\left.\begin{aligned} \therefore\ dE &= mC\,dC \\ \therefore\ dC &= \frac{dE}{mC} = \frac{dE}{m\sqrt{\frac{2E}{m}}} = \frac{dE}{\sqrt{2mE}} \end{aligned}\right\} \qquad (14.8.1)$$

Now, from Maxwell's velocity distribution law, the number of molecules having velocity lying between C and $C + dC$ is from (14.7.1)

$$dn_C = 4\pi n\left(\frac{m}{2\pi KT}\right)^{3/2} e^{-mC^2/2KT}C^2 dC$$

using (14.8.1), the number of molecules dn_E having energy lying between E and $E + dE$ i.e. the energy distribution among the molecules is given by

$$dn_E = 4\pi n\left(\frac{m}{2\pi KT}\right)^{3/2} e^{-E/KT}\frac{2E}{m}\frac{dE}{\sqrt{2mE}}$$

$$\text{or}\quad \frac{dn_E}{n} = 2(E/\pi)^{1/2}(KT)^{-3/2}e^{-E/KT}dE \qquad (14.8.2)$$

which is $M - B$ energy distribution expression.

The $M - B$ energy distribution curves for two different temperatures are shown in Fig. 14.8.1 where $\frac{dn_E}{n}$ is plotted as a function of E/KT. The curves first rise according to $E^{1/2}$, reach their maxima and then fall according to $e^{-E/KT}$.

For the most probable energy the left hand side of (14.8.2) should be maximum, for which $\frac{d}{dE}\left(\frac{dn_E}{n}\right)$ should be equal to zero i.e.

$$\frac{2}{\sqrt{\pi}}(KT)^{-3/2}\left[\frac{1}{2}E^{-1/2}e^{-E/KT} + E^{1/2}\left(-\frac{1}{KT}\right)e^{-E/KT}\right] = 0$$

which gives $E_p = \frac{1}{2}KT$ (putting $E = E_p$ for the most probable energy).

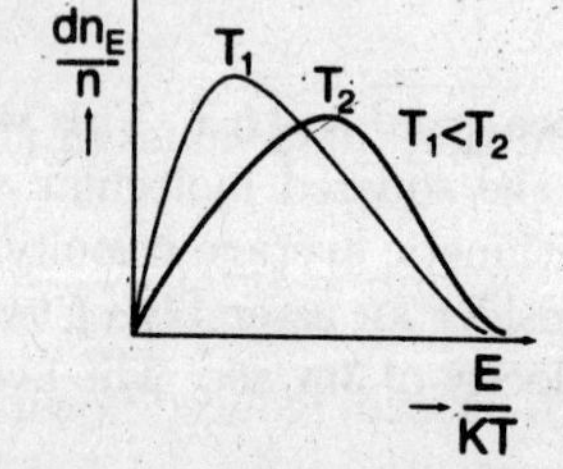

Fig. 14.8.1

Now $\frac{d}{dE}\left(\frac{dn_E}{n}\right) = 0$ gives the condition for energy to be either maximum or minimum. We may however show that

$$\frac{d^2}{dE^2}\left(\frac{dn_E}{n}\right)$$

is negative for the above value of E_p and therefore the value of E_p as found above corresponds to a maximum of $\frac{dn_E}{n}$. It is to be noted here that

$$C_p^2 = \frac{2KT}{m} \quad\text{or}\quad \frac{1}{2}mC_p^2 = KT$$

which is the energy of a molecule moving with the most probable velocity.

14.9 Temperature Dependence and Reduced form of Velocity Distribution

Temperature dependence:

From (14.7.1) we have

$$\frac{dn_C}{n} = 4\pi(m/2\pi KT)^{3/2}e^{-mC^2/2KT}C^2dC.$$

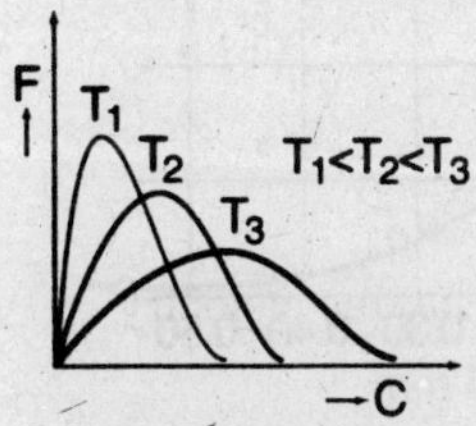

Fig. 14.9.1

Integrating the above expression w.r.t. c for all possible values, we should get the result equal to 1.

i.e. $$\int_0^\infty 4\pi(m/2\pi KT)^{3/2}e^{-mC^2/2KT}C^2dC = 1 \text{ or } \int_0^\infty FdC = 1 \tag{14.9.1}$$

Obviously the area under the distribution curve of molecular velocity—plot of F v.s. C—is unity. But from Fig. 14.9.1 we know that the maximum of the curve occurs at $C_p = \sqrt{2KT/m}$. So with increase in temperature, the peak of the curve should get shifted to a higher value of C. But since from (14.9.1) the area is fixed being equal to unity, the curve would get broadened with increase in temperature. Fig. 14.9.1 shows three curves giving the broadening of the distribution curve with rise in temperature. We may thus regard temperature as a measure of the broadening of the distribution curve.

14.9.1 Reduced Form of Maxwell's law

Writing $C_0 = \frac{C}{C_p}$, C_0 being the so-called *reduced velocity,* we may show in the following way that Maxwell velocity distribution law can be written in a form which is independent of temperature

$$C_0 = \frac{C}{C_p} = \frac{C}{\sqrt{2KT/m}} \text{ or } C = \sqrt{\frac{2KT}{m}}C_0 \text{ and } dC = \sqrt{\frac{2KT}{m}}dC_0.$$

Putting these values in (14.7.1), we have

$$dn_C = 4\pi n\left(\frac{m}{2\pi KT}\right)^{3/2} e^{-C_0^2}\left(\frac{2KT}{m}\right)^{3/2} C_0^2dC_0 = \frac{4}{\sqrt{\pi}}ne^{-C_0^2}C_0^2\,dC_0 \tag{14.9.2}$$

which is the reduced form of Maxwell's velocity distribution law and is independent of temperature.

14.10 Experimental Verification of Maxwell's Law of Distribution of Velocities

Maxwell's distribution law has been experimentally verified both directly and indirectly. The earliest type of verification of the distribution law was made from the broadening effect of spectral lines emitted by light atoms. The most important direct method for the verification of the law was due to Estermann, Simpson and Stern. The experiment was carried out in the year 1947. The schematic diagram of the apparatus used is shown in Fig. 14.10.1.

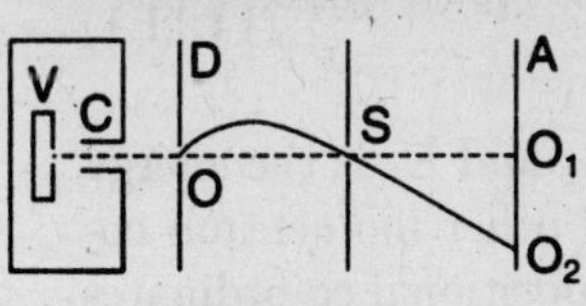

Fig. 14.10.1

A beam of cesium atoms emerges from a fine slit C in the oven V enclosed in a highly evacuated chamber. Most of the cesium atoms are stopped by the diaphragm D, but some pass through the slit O constituting almost a narrow horizontal beam which is finally collimated by the slit S and impinge on a hot tungsten wire. The cesium atoms leave the wire as positive ions and are collected by a negatively charged plate (not shown in the figure). Thus the plate current gives directly the number of cesium atoms striking the wire.

As the beam is deflected by gravity, the path of the cesium atoms passing through O and S is parabolic as shown in the figure. Plotting the plate current against the deflection $\delta(\delta = O_1O_2)$ of the ion we get a curve as shown in the Fig. 14.10.2.

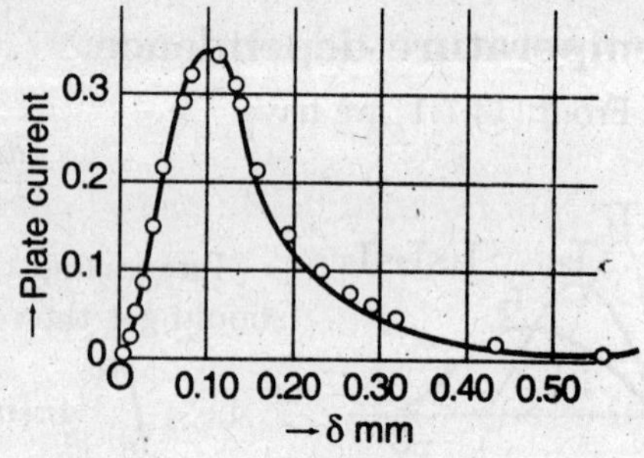

Fig. 14.10.2

In the figure the dots are experimental points and the smooth curve is obtained from Maxwell's distribution formula. It is to be noted here that smaller the velocity of an ion, greater will be the deflection and *vice-versa.*

14.11 Law of Equipartition of Energy

According to this law, in a dynamical system in thermal equilibrium the total energy of the system is distributed equally amongst its various degrees of freedom and the mean energy for each degree of freedom is equal to $\frac{1}{2}KT$, where K is Boltzmann's constant and T is the temperature of the system. This law was derived by Maxwell and later Boltzmann showed that this law holds for rotational and vibrational motions also.

The molecules in a monatomic gas have only translatory motion and hence each molecule has three velocities u, v, w which are perpendicular to and independent of each other. Thus we get three independent squared terms in the expression for kinetic energy for each molecule and so each molecule of a monatomic gas has three degrees of freedom.

In case of a diatomic gas, however, the motion of the molecules is not translatory only but the two atoms constituting the molecule separated by a fixed distance can rotate about an axis passing through the centre of gravity of the molecule and at right angles to the line joining the atoms. In this case the expression for kinetic energy contains three squared terms for translational motion and two squared terms for rotational motion. Hence in all the molecule possesses five degrees of freedom. In addition to the rotational motion, the atoms constituting the molecule can have vibrational or oscillatory motion along the line joining the atoms which gives additional squared terms in the expression for energy.

We shall now derive the law of equipartition of energy.

In the general case for a particle the energy is given by

$$E = ap_1^2 + E_1. \tag{14.11.1}$$

Here ap_1^2 is the energy corresponding to the momentum co-ordinate p_1 and E_1 is the energy corresponding to the other co-ordinates, i.e. positional co-ordinate and other momentum co-ordinate; a is independent of p_1, but in general may be a function of the positional co-ordinates q's. Boltzmann showed that the number of particles with their motions defined by the positional co-ordinate range q_1, $q_1 + dq_1$; q_2, $q_2 + dq_2$; $\cdots$, $q_n + dq_n$ and momentum co-ordinate range p_1, $p_1 + dp_1$; p_2, $p_2 + dp_2$; $\cdots$, $p_n + dp_n$ is given by

$$dn = Ane^{-\frac{E}{KT}} dq_1 dq_2 \cdots dp_1 dp_2 \cdots$$

Hence from (14.11.1), we have

$$dn = Ane^{-(ap_1^2+E_1)/KT} dq_1 dq_2 \cdots dp_1 dp_2 \cdots$$

Now from the definition of probability the constant A is given by

$$\int \cdots \int Ae^{-(ap_1^2+E_1)/KT} dq_1 dq_2 \cdots dp_1 dp_2 \cdots = 1$$

$$\text{or} \quad A = \frac{1}{\int \cdots \int e^{-(ap_1^2+E_1)/KT} dq_1 dq_2 \cdots dp_1 dp_2 \cdots} \tag{14.11.2}$$

Now the mean energy corresponding to p_1 is given by

$$\bar{E} = \frac{1}{n}\int ap_1^2\, dn = \frac{1}{n}\int \cdots \int Anap_1^2 \bar{e}^{(ap_1^2+E_1)/KT} dq_1 dq_2 \cdots dp_1 dp_2 \cdots$$

$$= \frac{\int \cdots \int ap_1^2 e^{-(ap_1^2+E_1)/KT} dq_1 dq_2 \cdots dp_1 dp_2 \cdots}{\int \cdots \int e^{-(ap_1^2+E_1)/KT} dq_1 dq_2 \cdots dp_1 dp_2 \cdots}$$

[Substituting the value of A from eqn. (14.11.2)]

$$= \frac{\int ap_1^2 e^{-(ap_1^2)/KT} dp_1 \int \cdots \int e^{-E_1/KT} dq_1 dq_2 \cdots dp_2 \cdots}{\int e^{-(ap_1^2)/KT} dp_1 \int \cdots \int e^{-E_1/KT} dq_1 dq_2 \cdots dp_2 \cdots}$$

$$= \frac{\int_{-\infty}^{+\infty} ap_1^2 e^{-(ap_1^2)/KT} dp_1}{\int_{-\infty}^{+\infty} e^{-(ap_1^2)/KT} dp_1},$$

[$\because$ the variables $p_1, p_2 \cdots q_1, q_2 \cdots$ are all independent variables, the integral may be considered as the product of two integrals.]

The limits of integration extend from $-\infty$ to $+\infty$.

Let $\frac{ap_1^2}{KT} = u$.

$$\therefore \quad \frac{2ap_1 dp_1}{KT} = du \text{ or } dp_1 = \frac{1}{2}\left(\frac{KT}{a}\right)^{\frac{1}{2}} u^{-\frac{1}{2}} du \quad \left[\because p_1 = \left(\frac{KT}{a}u\right)^{\frac{1}{2}}\right].$$

Hence

$$\bar{E} = \frac{\int_0^{+\infty} KTu^{\frac{1}{2}} e^{-u}\, du}{\int_0^{+\infty} u^{-\frac{1}{2}} e^{-u}\, du} = KT \cdot \frac{\frac{1}{2}\Gamma\left(\frac{1}{2}\right)}{\Gamma\left(\frac{1}{2}\right)}$$

$$\left[\because \int_0^\infty e^{-x}x^{n-1} = \text{Gamma function } \Gamma(n)\right.$$

$$\therefore \int_0^\infty u^{\frac{1}{2}}e^{-u}\,du = \Gamma(3/2) \text{ and } \int_0^\infty u^{-\frac{1}{2}}e^{-u}\,du = \Gamma\left(\frac{1}{2}\right) \text{ and } \Gamma(3/2)$$

$$\left. = \Gamma\left(\frac{1}{2}+1\right) = \frac{1}{2}\Gamma\left(\frac{1}{2}\right)\right]$$

$$\text{or } \bar{E} = \frac{1}{2}KT.$$

Thus the mean kinetic energy for each degree of freedom corresponding to the momentum p_1 is $\frac{1}{2}KT$. Similar values may be obtained for other momenta also. Thus the mean energy for each momentum co-ordinate per degree of freedom is $\frac{1}{2}KT$. It is to be noted that the momentum p_1 may be due to translational, rotational and vibrational motions.

14.12 Degrees of Freedom and Ratio of Specific Heats

If we consider a gram-molecule of a gas having n degrees of freedom, then the energy of the system is given by

$$E = \frac{n}{2}NKT. \tag{14.12.1}$$

N being the number of molecules in a gram-molecule of the gas.

If dQ be the quantity of heat given to one gram-molecule of a gas, then from the first law of thermodynamics we have,

$$dQ = dE + P\,dV \tag{14.12.2}$$

where dE is the increase in internal energy of the gas and $P\,dV$ is the external work, i.e. the work done by the gas due to expansion by a volume dV against the pressure P. If C_v be the molar specific heat of the gas at constant volume, then

$$C_v = \left(\frac{dQ}{dT}\right)_v = \left(\frac{dE}{dT}\right)_v = \frac{d}{dT}\left(\frac{n}{2}NKT\right) = \frac{nNK}{2} = \frac{nR}{2}. \tag{14.12.3}$$

Again equation (14.12.2) may be written in the form

$$dQ = C_v\,dT + P\,dV. \tag{14.12.4}$$

Since

$$PV = RT \quad \therefore \quad P\,dV = R\,dT - V\,dP.$$

Hence from equation (14.12.2) we have,

$$dQ = C_v\,dT + R\,dT - V\,dP \quad \text{or} \quad \left(\frac{dQ}{dT}\right)_p = C_p = C_v + R$$

$$\therefore \quad C_p - C_v = R. \tag{14.12.5}$$

Now

$$\frac{C_p}{C_v} = \gamma = \frac{C_v + R}{C_v} = 1 + \frac{R}{C_v} = 1 + \frac{2}{n} \tag{14.12.6}$$

[from equation (14.12.3)].

For monatomic gases, $n = 3$.

$$\therefore \quad \gamma = 1 + \frac{2}{3} = \frac{5}{3} = 1.66.$$

This agrees with experimental results for monatomic gases like argon, helium, etc.

For diatomic gases, $n = 5$

$$\therefore \quad \gamma = 1 + \frac{2}{5} = 1.4.$$

This value agrees approximately with the experimental results for hydrogen, nitrogen, oxygen, etc.

For triatomic gases, $n = 6$

$$\therefore \quad \gamma = 1 + \frac{2}{6} = 1.33.$$

This value agrees approximately with the experimental results for gases like carbon dioxide, hydrogen sulphide, etc.

14.13 Mean Free Path

The mean free path of a molecule is the average distance traversed by a molecule between successive collisions. In defining the mean free path of the molecules we assume that the molecules are like rigid elastic spheres having definite dimensions and that there is no force of attraction or repulsion between the molecules. The molecules of the gas collide with each other as a result of which they deviate from their straight-line paths. The distance between the centres of the molecules up to which the molecules will pass without changing the direction of path may be taken as the effective diameter of the molecule. Let this diameter be σ. When the distance between the centres of the molecules is less than σ, collisions occur.

Let us now assume that all the molecules of the gas are at rest except one molecule which is moving with velocity C. Hence there will be a volume $\pi\sigma^2 C$ which contains all the molecules with which the moving molecule will collide per second. If n be the number of molecules per unit volume, then the moving molecule will collide with the number of molecules $\pi\sigma^2 Cn$ per second, i.e. the number of collisions per second is $\pi\sigma^2 Cn$ when the total distance traversed is C. Hence the average distance traversed by a molecule between successive collisions or the mean free path is given by

$$\lambda = \frac{C}{\pi\sigma^2 Cn} = \frac{1}{\pi\sigma^2 n}. \tag{14.13.1}$$

Actually the molecules possess all possible velocities, the distribution of velocities being governed by Maxwell's law. So, a molecule will collide with $\pi\sigma^2 \bar{r} n$ molecules per second instead of $\pi\sigma^2 Cn$, $\bar{r}$ being the mean relative velocity of the molecule with respect to the other molecules. Hence the mean free path is given by

$$\lambda = \frac{1}{\pi\sigma^2 n} \cdot \frac{C}{\bar{r}}. \tag{14.13.2}$$

As a first approximation, it was Clausius who assumed that all the molecules move with same velocity. Let the relative velocity of the molecule moving with the velocity C_1 with respect to another molecule moving with the velocity C_2 be r. Hence we may write

$$r = (C_1^2 + C_2^2 - 2C_1C_2\cos\theta)^{\frac{1}{2}},$$

where θ is the angle between the directions of C_1 and C_2. According to Clausius' assumption

$$C_1 = C_2 = C \quad \therefore \quad r = C\sqrt{2(1-\cos\theta)}.$$

Now, the number of molecules travelling within the solid angle lying between θ and $\theta + d\theta$ [Fig. 14.13.1] is

$$\frac{n \cdot 2\pi r\sin\theta \cdot r\,d\theta}{4\pi r^2} = \frac{n}{2}\sin\theta\,d\theta.$$

Hence the mean relative velocity of all the molecules is given by

$$\bar{r} = \frac{1}{n}\int_0^\pi \frac{nC}{2}\sqrt{2(1-\cos\theta)}\sin\theta\,d\theta = 2C\int_0^\pi \sin^2\frac{\theta}{2}\cos\frac{\theta}{2}\,d\theta$$

$$= 4C\int_0^\pi \sin^2\frac{\theta}{2}d\left(\sin\frac{\theta}{2}\right) = \frac{4C}{3} \text{ or } \frac{C}{\bar{r}} = \frac{3}{4}$$

$$\therefore \lambda = \frac{3}{4\pi\sigma^2 n}. \tag{14.13.3}$$

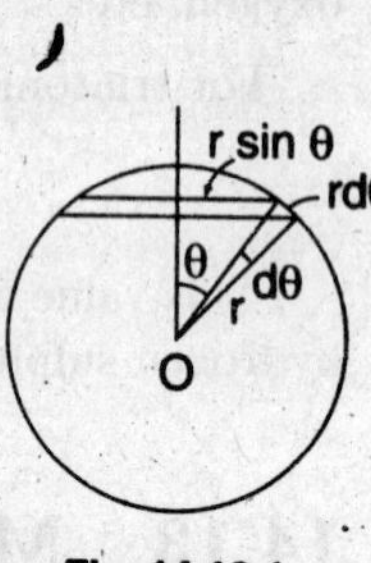

Fig. 14.13.1

Applying Maxwell's law for the distribution of velocities the above equation may further be corrected when the mean free path becomes

$$\lambda = \frac{1}{\sqrt{2}\pi\sigma^2 n}. \tag{14.13.4}$$

Now, since the pressure of the gas is proportional to n, the mean free path λ is inversely proportional to the pressure of the gas. At normal temperature and pressure the value of λ for oxygen is 0.633×10^{-5} cm. and for nitrogen 0.588×10^{-5} cm., but at a pressure of 10^{-4} mm. of mercury, the mean free path of most of the gases lies between 5 to 10 cms. Jeans took into consideration the persistence velocity on collision and deduced the formula

$$\lambda = \frac{1}{\sqrt{2}} \cdot \frac{1.319}{\pi\sigma^2 n}.$$

Considering intermolecular forces Sutherland deduced

$$\lambda = \frac{1.402}{\sqrt{2}\pi\sigma^2 n(1 + A/T)},$$

where A is a constant for a particular gas and T is the absolute temperature. For an electron, the dimension of which is very small in comparison with the molecular dimension, the expression for mean free path should be modified. We can treat the electron to be a point and so the molecular sphere of exclusion in this case, should be of radius $\sigma/2$ instead of σ as in the case of a molecule. Also as the velocity of the electron is very large in comparison with that of the molecules, the correction for relative velocity need not be taken into consideration. Hence the mean free path of an electron

$$\lambda_e = \frac{1}{\pi(\sigma/2)^2 n}. \tag{14.13.5}$$

From (14.13.4) and (14.13.5), we have

$$\lambda_e = 4\sqrt{2}\lambda. \tag{14.13.6}$$

14.14 Probability of no collision and the Law of Distribution of Free paths—Expression for mean free path from it

The probability P_x that a molecule moving with velocity C does not suffer a collision in describing a distance x is a function of x and can be written as $f(x)$. The probability P_{x+dx} that the molecule will describe a path $x + dx$ is therefore $f(x + dx)$.

Now $f(x + dx)$ may be expanded in the form.

$$f(x+dx) = f(x) + dx f'(x) + \frac{(dx)^2}{\lfloor 2} f''(x) + \cdots\cdots \quad \text{[from Taylor's theorem]}$$

$$= f(x) + f'(x)dx \quad \text{[neglecting the terms containing higher powers of } dx]$$

$$\therefore \quad P_{x+dx} = P_x + \frac{dP_x}{dx} dx.$$

Now according to the theorem of probability $P_{x+dx} = P_x \cdot P_{dx}$, i.e. the probability that a molecule will traverse the distance $x + dx$ without suffering a collision is equal to the product of the respective probabilities when the molecule will traverse the distances x and dx without collision.

$$\therefore \quad P_x \cdot P_{dx} = P_x + \frac{dP_x}{dx} dx \tag{14.14.1}$$

When a molecule has described a path x, the probability that there is no collision in traversing a further distance dx is given by

$$P_{dx} = \left(1 - \frac{dx}{\lambda}\right)$$

$\therefore$ from (14.14.1) we have,

$$f(x)\left(1 - \frac{dx}{\lambda}\right) = f(x) + \frac{df(x)}{dx} dx \text{ or, } -f(x)\frac{dx}{\lambda} = f'(x)dx \text{ or, } \frac{f'(x)}{f(x)} = -\frac{1}{\lambda}$$

or, $\log f(x) = -\dfrac{x}{\lambda} + \log A$ [on integration] where $\log A$ is the constant of integration

or, $\log \dfrac{f(x)}{A} = -x/\lambda$ or, $f(x) = Ae^{-x/\lambda}$. $\therefore$ $P_x = Ae^{-x/\lambda}$

Now when $x = 0, P_x = 1$. $\therefore$ $A = 1$.

Hence $P_x = e^{-x/\lambda}$—which gives the law of distribution of free paths. If the total number of molecules be N_0, and N the number having a free path x, then the probability of this free path is

$$\frac{N}{N_0} = P_x = f(x) = e^{-x/\lambda} \quad \text{or} \quad N = N_0 e^{-x/\lambda} \qquad (14.14.2)$$

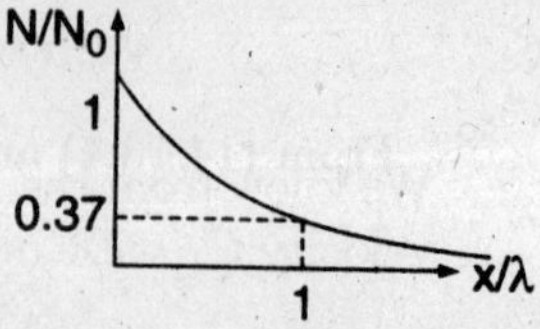

Fig 14.14.1

(14.14.2) is the well-known *survival* equation, which shows that the number of molecules falls off exponentially with the distance x. The curve obtained by plotting $\frac{N}{N_0}$ against $\frac{x}{\lambda}$ is shown in the attached fig. (14.14.1).

From the above law of distribution of free paths, we can find an expression for mean free path, with the simplifying assumption that all molecules except the one under consideration are at rest. Let this particular molecule be projected into the gas, and let us suppose that the gaseous mass is divided into parallel layers of infinitesimal thickness d, the thickness being so small that it does not contain more than one molecule (Fig. 14.14.2). Along its path, the moving molecule will collide with such other molecules in the layer, whose centres lie within a distance σ from the path traced out by the centre of this molecule, σ being the diameter of the molecule or in other words, it will collide with those whose centres are contained in a circle of radius σ and area $\pi\sigma^2$ described about the centre of the moving molecule. $\pi\sigma^2$ is called the *mutual collision cross-section* of the incident and the stationary molecules.

Fig 14.14.2

So if there are N molecules in a layer, the total collision cross-section is $\pi\sigma^2 N$,

$$\therefore \quad \text{probability of collision} = \frac{\text{Collision cross-section}}{\text{Area of the layer}} = \frac{\pi\sigma^2 N}{\text{Area of the layer}}$$

$$\therefore \text{ probability of no collision} = 1 - \frac{\pi\sigma^2 N}{\text{Area of the layer}}$$

$$= 1 - \frac{\pi\sigma^2 N d}{\text{Volume of the layer}} \quad \left[\begin{array}{c}\because \text{area of a layer} = \text{its} \\ \text{volume/its thickness}\end{array}\right]$$

$$= 1 - \pi\sigma^2 n d \quad \left[\begin{array}{l}\text{where } n = \text{number of molecules} \\ \text{per unit volume of the layer}\end{array}\right]$$

As the molecule crosses a thickness x, $\frac{x}{d}$ layers of the gas are crossed over, and the probability that it does not suffer collision during this passage is $(1 - \pi\sigma^2 nd)^{x/d}$. When $d \to 0$, the probability of no collision within a distance x is given by

$$\lim_{d\to 0}(1 - \pi\sigma^2 nd)^{x/d} = \lim_{d\to 0}\left[(1 - \pi\sigma^2 nd)^{1/\pi\sigma^2 nd}\right]^{\pi\sigma^2 nx}$$

$$= \left(\frac{1}{e}\right)^{\pi\sigma^2 nx} = e^{-\pi\sigma^2 nx} \quad \left(\because \lim_{x\to 0}(1-x)^{1/x} = \frac{1}{e}\right)$$

But we know that the above probability is $e^{-x/\lambda}$ where λ is the mean free path

$$\therefore \; e^{-x/\lambda} = e^{-\pi\sigma^2 nx} \quad \text{or} \quad \lambda = \frac{1}{\pi\sigma^2 n}$$

which is the same as (14.13.1).

14.15 Deduction of the Expression for the mean free path from Maxwell's law of the Distribution of velocities

We know from the article 14.13 that the average relative velocity of the molecules moving with velocity C_1 with respect to the molecules moving with velocity C_2 is given by

$$\bar{r} = \frac{1}{n}\int_0^{\pi} \frac{n}{2}(C_1^2 + C_2^2 - 2C_1C_2\cos\theta)^{\frac{1}{2}} \sin\theta \, d\theta$$

$$= \frac{1}{2}\int_0^{\pi} (C_1^2 + C_2^2 - 2C_1C_2\cos\theta)^{\frac{1}{2}} \sin\theta \, d\theta.$$

Let

$$(C_1^2 + C_2^2 - 2C_1C_2\cos\theta)^{\frac{1}{2}} = z.$$

$\therefore$ differentiating both sides we have,

$$\sin\theta \, d\theta = \frac{z dz}{C_1C_2}.$$

Hence

$$\bar{r} = \frac{1}{2C_1C_2}\int z^2 \, dz = \left[\frac{1}{6C_1C_2}(C_1^2 + C_2^2 - 2C_1C_2\cos\theta)^{3/2}\right]_0^{\pi}.$$

If $C_1 > C_2$, we have

$$\bar{r} = \frac{1}{6C_1C_2}\left[(C_1 + C_2)^3 - (C_1 - C_2)^3\right] = \frac{2C_2(3C_1^2 + C_2^2)}{6C_1C_2} = \frac{3C_1^2 + C_2^2}{3C_1} \qquad (14.15.1)$$

but if $C_1 < C_2$ we have

$$\bar{r} = \frac{1}{6C_1C_2}\left[(C_1 + C_2)^3 - (C_2 - C_1)^3\right] = \frac{2C_1(3C_2^2 + C_1^2)}{6C_1C_2} = \frac{3C_2^2 + C_1^2}{3C_2} \qquad (14.15.2)$$

Now, the probability that the molecules of velocity C_2 exist is $4\pi na^3e^{-bC_2^2}C_2^2\,dC_2$, where

$$a = \sqrt{\frac{m}{2\pi KT}} \quad \text{and} \quad b = \frac{m}{2KT}$$

[from Maxwell's distribution formula].

Hence, the average relative velocity of the molecule of velocity C_1 with respect to all other molecules moving with velocity C_2 is given by

$$\bar{R}' = \frac{\int_0^{\infty} \bar{r}\cdot 4\pi na^3e^{-bC_2^2}C_2^2\,dC_2}{n} \quad \text{or,} \quad \bar{R}' = \int_0^{\infty} \bar{r}\cdot 4\pi a^3e^{-bC_2^2}C_2^2\,dC_2.$$

In the above expression $\bar{r}$ has two different values according as $C_1 > C_2$ or $C_1 < C_2$. Hence we may write

$$\bar{R}' = \int_0^{C_1} \frac{3C_1^2 + C_2^2}{3C_1}\cdot 4\pi a^3e^{-bC_2^2}C_2^2\,dC_2 + \int_{C_1}^{\infty} \frac{3C_2^2 + C_1^2}{3C_2}\cdot 4\pi a^3e^{-bC_2^2}C_2^2\,dC_2$$

[from (14.15.1) and (14.15.2)].

The limit of integration 0 to C_1 is for the region $C_1 > C_2$ and C_1 to ∞ for the region $C_1 < C_2$. Now the average relative velocity of any molecule with respect to all other molecules moving in all possible directions is given by

$$\bar{R} = \frac{\int_0^\infty \bar{R}' \cdot 4\pi n a^3 e^{-bC_1^2} C_1^2 \, dC_1}{n}$$

$$\text{or, } \bar{R} = \int_0^\infty 4\pi a^3 e^{-bC_1^2} C_1^2 \, dC_1 \int_0^{C_1} \frac{3C_1^2 + C_2^2}{3C_1} 4\pi a^3 e^{-bC_2^2} C_2^2 \, dC_2$$

$$+ \int_0^\infty 4\pi a^3 e^{-bC_1^2} C_1^2 \, dC_1 \int_{C_1}^\infty \frac{3C_2^2 + C_1^2}{3C_2} 4\pi a^3 e^{-bC_2^2} C_2^2 \, dC_2$$

$$= 16\pi^2 a^6 \left(\int_0^\infty e^{-bC_1^2} C_1^2 \, dC_1 \int_0^{C_1} \frac{3C_1^2 + C_2^2}{3C_1} e^{-bC_2^2} C_2^2 \, dC_2 \right.$$

$$\left. + \int_0^\infty e^{-bC_1^2} C_1^2 \, dC_1 \int_{C_1}^\infty \frac{3C_2^2 + C_1^2}{3C_2} e^{-bC_2^2} C_2^2 \, dC_2 \right)$$

$$= 16\pi^2 a^6 \cdot \frac{1}{4\sqrt{2}} \left(\frac{\pi}{b^7} \right)^{\frac{1}{2}}$$

$$\left[\because \text{ the value of the integrals within the bracket } = \frac{1}{4\sqrt{2}} \left(\frac{\pi}{b^7} \right)^{\frac{1}{2}}. \right]$$

$$= \frac{4}{\sqrt{2\pi b}} \quad \left[\because \ a = \sqrt{\frac{b}{\pi}} \right] \tag{14.15.3}$$

$$\text{or, } \bar{R} = \frac{4}{\sqrt{2\pi \cdot \frac{m}{2KT}}} = \frac{4\sqrt{KT}}{\sqrt{\pi m}} = \sqrt{2}\sqrt{\frac{8KT}{\pi m}}$$

$$= \sqrt{2}\bar{c}, \text{ where } \bar{c} \text{ is the average velocity.} \tag{14.15.4}$$

Hence Maxwell's mean free path is given by

$$\lambda = \frac{\bar{c}}{\pi n \sigma^2 \cdot \sqrt{2}\bar{c}} = \frac{1}{\sqrt{2}\pi n \sigma^2} \quad [\because \text{ in this case } C = \bar{c}]$$

which has already been discussed in the article 14.13.

14.16 Experimental determination of the mean free path

Born's method. The first direct experimental method for determining the mean free path is due to Born, who employed the arrangement as shown in fig. 14.16.1. It consists essentially of a quartz tube Q in which silver is heated electrically at S. The silver evaporates and escapes through the narrow slit R. It then passes through a circular hole in each of the four brass discs P_1, P_2, P_3, P_4, placed one above the other, the distance between the successive discs being 1 cm. Each of these discs carries a glass quadrant G, such that the apex of the quadrant lies at the centre of the hole (shown separately at P) and each quadrant is displaced through a right angle relative to the next to ensure that each would receive one-fourth of the incident beam. The whole apparatus can be evacuated to any desired pressure by means of a vacuum pump and the resulting pressure can be determined with the help of a McLeod gauge or a Knudsen manometer.

The discs are cooled by a freezing mixture bath K and consequently the silver atoms get deposited on the quadrants. The relative amounts of silver deposited are obtained by a photo-metric comparison of the densities of the deposits.

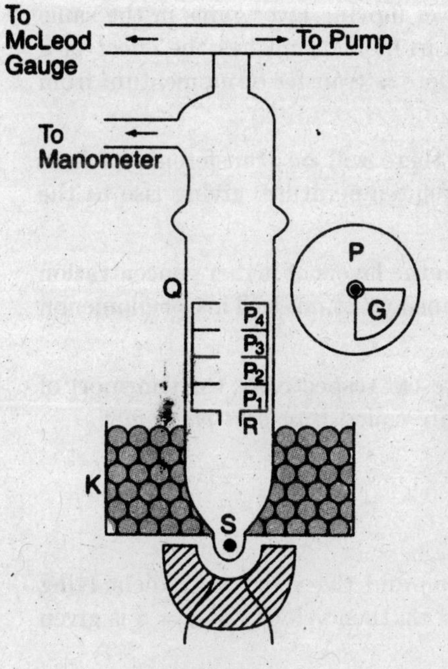

Fig 14.16.1

Now if d, d_0 represent the densities of the deposits on any of the upper and the next lower quadrant respectively, x the distance between them and λ the mean free path, it follows from the law of distribution of free paths that

$$d = d_0 e^{-x/\lambda} \tag{14.16.1}$$

In practice, however, the beam spreads on the quadrants and the best procedure is to find the density d_{10} on a quadrant when all the gas in the vessel is pumped out. The free path is then equal to the length of the chamber. Next the gas is introduced, producing the free path λ and the density d_1 on the same quadrant is found out. Then

$$d_1 = d_{10} e^{-x_1/\lambda} \tag{14.16.2}$$

x_1 being the distance of the quadrant from the slit R.

Similarly for another quadrant,

$$d_2 = d_{20} e^{-x_2/\lambda} \tag{14.16.3}$$

Taking logarithm of both sides of (14.16.2), we have

$$\ln d_1 = \ln d_{10} - \frac{x_1}{\lambda} \quad \text{or} \quad \frac{x_1}{\lambda} = \ln d_{10} - \ln d_1 = \ln \frac{d_{10}}{d_1} \tag{14.16.4}$$

Similarly from (14.16.3), we have

$$\frac{x_2}{\lambda} = \ln \frac{d_{20}}{d_2} \tag{14.16.5}$$

From (14.16.4) and (14.16.5) we have

$$\lambda = (x_2 - x_1) \ln \frac{d_2 d_{10}}{d_1 d_{20}} = \ln \frac{d_2 d_{10}}{d_1 d_{20}} \quad [\because\ (x_2 - x_1) = 1 \text{ cm}] \tag{14.16.6}$$

whence λ, the mean free path can be experimentally determined. Born found that for $p = 5.8 \times 10^{-3}$ mm, $\lambda = 1.7$ cm and for $p = 4.5 \times 10^{-3}$ mm, $\lambda = 2.4$ cm. Hence the product $p\lambda$ is constant as predicted by the theory. Thus the mean free path can be experimentally determined and the law of distribution of free paths verified. Later Bielz improved upon Born's arrangement and obtained similar results.

14.17 Transport phenomena in gases

Before the steady condition of a gas is attained, the following phenomena either singly or jointly occur.

(i) When there is a relative motion between the layers of the gas, a force comes into play which retards the motion of the faster moving layer and accelerates the slower moving layer. This gives rise to the phenomenon known as the *viscosity* of gases. The faster layer loses molecules and this loss equivalent to a loss of momentum has the effect of a force tending to retard the motion of the layer. In a similar manner the slower moving layer gains in the same time molecules and this gain which is equivalent to a gain in momentum has the effect of a force tending to accelerate the motion of the layer. Thus, there is transfer of momentum from layer to layer.

(ii) As the temperature may vary from layer to layer, there will be transference of heat energy from the layer of higher temperature to that of lower temperature, giving rise to the phenomenon of thermal *conduction*.

(iii) If the molecular density varies, molecules diffuse from the layers of higher concentration to the layers of lower concentration. In this case there is transport of mass. This phenomenon is known as diffusion.

It is obvious that viscosity, conduction and diffusion represent respectively, the transport of momentum, heat energy and mass ad for this reason, they are called *transport phenomena.*

14.17.1 Viscosity from Kinetic theory

We consider a gas moving along the X-axis with velocity u_0 and the velocity gradient lying along the Z-axis. [Fig. 14.17.1]. If the velocity u is u_0 at $z = z_0$, then velocity at $z = z$ is given by $u = u_0 + \frac{du}{dz}z$.

Before the steady state of the gas has been attained, we consider the molecules crossing the plane $X - Y$. Those crossing the plane $X - Y$ in a downward direction will possess on an average greater momentum than that of the molecules at $z = z_0$. In a similar way, the molecules crossing the plane $X-Y$ in an upward direction will on an average have less momentum than that of the molecules at $z = z_0$. Thus, there is a net gain of momentum in the downward direction.

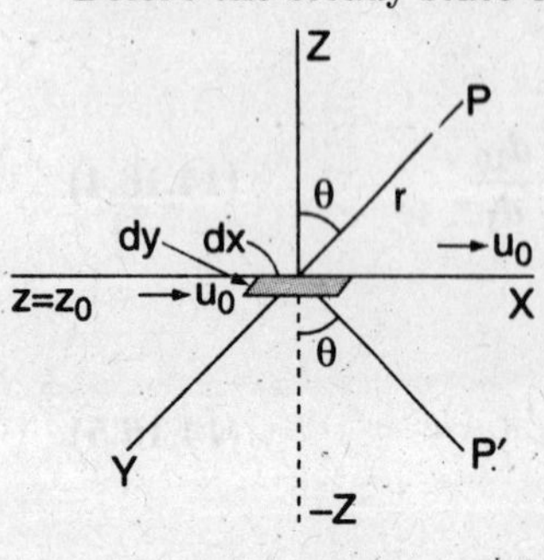

Fig. 14.17.1

We shall now calculate the net transfer of momentum per unit time through unit area which will give the force per unit area due to the velocity gradient across the plane $X - Y$.

We consider a small element of volume dv at P, at a distance r from the area $dx\,dy$ and we shall find the rate of transfer of momentum in a downward direction through the small area $dx\,dy$ due to this volume dv at P. Similarly, the upward transfer of momentum through $dx\,dy$ due to a small element of volume dv' at P' should be taken into consideration. Let n be the number of molecules per c.c. Also we assume that on the average a single molecule experiences $\frac{\bar{c}}{\lambda}$ number of collisions in unit time, where $\bar{c}$ is the average velocity and λ is the mean free path. In time dt, $n\,dv$ molecules in the volume dv experience $n\frac{\bar{c}}{\lambda}dv\,dt$ collisions, i.e. in time dt the number of new paths started by the molecules leaving dv is $\frac{n\bar{c}}{\lambda}\,dv\,dt$. Of this number of molecules moving towards $dx\,dy$ in time dt, those moving in a direction making an angle θ with the Z-axis, are given by

$$\frac{n\bar{c}}{\lambda}\,dv\,dt\frac{dx\,dy\cos\theta}{4\pi r^2}.$$

Of this number, those moving towards $dx\,dy$ in time dt without suffering collision are

$$\frac{n\bar{c}}{\lambda}\,dv\,dt\frac{dx\,dy\cos\theta}{4\pi r^2}e^{-\frac{r}{\lambda}}.$$

Thus, the momentum transferred through $dx\,dy$ in time dt in the downward direction is

$$= \frac{n\bar{c}}{\lambda}\,dv\,dt\frac{dx\,dy\cos\theta}{4\pi r^2}e^{-r/\lambda}m\left(u_0 + r\cos\theta\frac{du}{dz}\right),$$

where m is the mass of each molecule and $z = r\cos\theta$.

Now, the elementary volume expressed in spherical polar co-ordinates is given by

$$dv = dr(r\sin\theta\,d\phi)(r\,d\theta) = r^2\sin\theta\,dr\,d\theta\,d\phi.$$

$\therefore$ the momentum transferred through $dx\,dy$ in time dt from the space above the plane $X-Y$ in the downward direction.

$$= \frac{nm\bar{c}}{\lambda}\frac{dx\,dy\,dt}{4\pi}\left[u_0\int_0^\infty e^{-r/\lambda}dr\int_0^{\pi/2}\cos\theta\sin\theta\,d\theta\int_0^{2\pi}d\phi\right.$$
$$\left. + \frac{du}{dz}\int_0^\infty re^{-r/\lambda}dr\int_0^{\pi/2}\cos^2\theta\sin\theta\,d\theta\int_0^{2\pi}d\phi\right]. \tag{14.17.1}$$

Similarly, momentum transferred through $dx\,dy$ in time dt from the space below the plane $X-Y$ in the upward direction.

$$= \frac{nm\bar{c}}{\lambda}\cdot\frac{dx\,dy\,dt}{4\pi}\left[u_0\int_0^\infty e^{-r/\lambda}dr\int_0^{\pi/2}\cos\theta\sin\theta\,d\theta\int_0^{2\pi}d\phi\right.$$
$$\left. - \frac{du}{dz}\int_0^\infty r\,e^{-r/\lambda}dr\int_0^{\pi/2}\cos^2\theta\sin\theta\,d\theta\int_0^{2\pi}d\phi\right] \tag{14.17.2}$$

[Since $u = u_0 - \frac{du}{dz}r\cos\theta$ at $z = -z$]

Hence the net transfer of momentum through $dx\,dy$ in time dt in the downward direction is obtained by subtracting (14.17.2) from (14.17.1) and is given by

$$M = \frac{nm\bar{c}}{\lambda}\cdot\frac{dx\,dy\,dt}{2\pi}\left[\frac{du}{dz}\int_0^\infty re^{-r/\lambda}dr\int_0^{\pi/2}\cos^2\theta\sin\theta\,d\theta\int_0^{2\pi}d\phi\right].$$

Now

$$\int_0^\infty r\bar{e}^{r/\lambda}dr = \lambda^2;\quad \int_0^{\pi/2}\cos^2\theta\sin\theta\,d\theta = \frac{1}{3};\quad \int_0^{2\pi}d\phi = 2\pi$$

$$\therefore\quad M = \frac{nm\bar{c}}{\lambda}\cdot\frac{dx\,dy\,dt}{2\pi}\left[\frac{du}{dz}\cdot\lambda^2\cdot\frac{1}{3}\cdot 2\pi\right]\quad\text{or;}\quad \frac{M}{dx\,dy\,dt} = \frac{nm\bar{c}\lambda}{3}\cdot\frac{du}{dz},$$

i.e. tangential stress or the viscous drag per unit area

$$f = \frac{1}{3} nm\bar{c}\lambda \frac{du}{dz}$$

or, the coefficient of viscosity of the gas,

$$\eta = \frac{f}{\frac{du}{dz}} = \frac{1}{3} nm\bar{c}\lambda = \frac{1}{3}\rho\bar{c}\lambda = \frac{1}{3\sqrt{2}} \cdot \frac{m\bar{c}}{\pi\sigma^2} \tag{14.17.3}$$

where ρ is the density of the gas.

Chapman deduced the following formula following Maxwell's method,

$$\eta = 0.49 m\bar{c}/\sqrt{2}\pi\sigma^2.$$

It is to be noted from the expression (14.17.3) that η does not depend on pressure, but depends on $\bar{c}$, i.e. on temperature of the gas.

14.17.2 Thermal conductivity from kinetic theory

In this case as the temperature varies from layer to layer of a gas thermal energy is transferred from layers of higher temperature to the layers of lower temperature.

Let T_0 be the temperature at the plane $z = z_0$ (Fig. 14.17.1) and so above the plane $X - Y$ at a distance z, the temperature is $T_0 + z\frac{dT}{dz}$ and below the plane $X - Y$ at a distance z, the temperature is $T_0 - z\frac{dT}{dz}$.

We shall now calculate the net transfer of thermal energy per unit time through unit area.

As before we consider a small element of volume dv at P and shall find the rate of flow of energy in the downward direction through a small area $dx\,dy$ due to the volume dv at P. Proceeding in a similar way as in the case of viscosity, the amount of energy transferred through the area $dx\,dy$ in time dt from the volume dv in the downward direction is

$$\frac{n\bar{c}}{\lambda}\, dv\, dt \frac{dx\, dy}{4\pi r^2} \cos\theta e^{-r/\lambda} mC_v \left(T_0 + r\cos\theta \frac{dT}{dz}\right),$$

since each molecule leaving dv will carry an amount of energy $mC_v\ (T_0 + r\cos\theta\frac{dT}{dz})$, where C_v is the specific heat of the gas and m is the mass of each molecule. Hence the rate of flow of energy through $dx\,dy$ from the space above the plane $X - Y$ in the downward direction is

$$\frac{nm\bar{c}}{\lambda} C_v \cdot \frac{dx\,dy}{4\pi} \left[T_0 \int_0^{\infty} e^{-r/\lambda} dr \int_0^{\pi/2} \cos\theta \sin\theta\, d\theta \int_0^{2\pi} d\phi \right.$$

$$\left. + \frac{dT}{dz} \int_0^{\infty} r e^{-r/\lambda} dr \int_0^{\pi/2} \cos^2\theta \sin\theta\, d\theta \int_0^{2\pi} d\phi \right] \tag{14.17.4}$$

$$[\because\ dv = r^2 \sin\theta dr\, d\theta\, d\phi].$$

Similarly, the rate of flow of energy through $dx\,dy$ from the space below $X-Y$ in the upward direction

$$= \frac{nm\bar{c}C_v}{\lambda}\frac{dx\,dy}{4\pi}\left[T_0\int_0^\infty e^{-r/\lambda}dr\int_0^{\pi/2}\cos\theta\sin\theta\,d\theta\int_0^{2\pi}d\phi\right.$$
$$\left.-\frac{dT}{dz}\int_0^\infty re^{-r/\lambda}dr\int_0^{\pi/2}\cos^2\theta\sin\theta\,d\theta\int_0^{2\pi}d\phi\right]. \tag{14.17.5}$$

Hence the net rate of flow of energy through $dx\,dy$ is obtained by subtracting (14.17.5) from (14.17.4) and is given by

$$\frac{dE}{dt} = \frac{nm\bar{c}C_v}{\lambda}\cdot\frac{dx\,dy}{2\pi}\left[\frac{dT}{dz}\int_0^\infty re^{-r/\lambda}dr\int_0^{\pi/2}\cos^2\theta\sin\theta\,d\theta\int_0^{2\pi}d\phi\right]$$
$$= \frac{nm\bar{c}C_v}{\lambda}\cdot\frac{dx\,dy}{2\pi}\left[\frac{dT}{dz}\cdot\lambda^2\cdot\frac{1}{3}\cdot 2\pi\right] = \frac{nm\bar{c}C_v\lambda}{3}\cdot\frac{dT}{dz}\cdot dx\,dy.$$

Now, the thermal conductivity K of the gas is defined as

$$K = \frac{\frac{dE}{dt}}{dx\,dy\frac{dT}{dz}} = \frac{nm\bar{c}C_v\lambda}{3} = \frac{1}{3}\rho\bar{c}\lambda C_v$$

$$\text{or, } K = \eta C_v.\ \text{[from (14.17.3)]} \quad \text{or} \quad \frac{K}{\eta C_v} = 1 \tag{14.17.6}$$

A rigorous deduction introduces a numerical factor in the above relation which becomes $K = \epsilon\eta C_v$, where ϵ varies from 1 to 2.57 depending on the mathematical analysis. Chapman obtained for molecules with translational energy only and spherical symmetry, i.e. for monatomic gases $K = 2.5\eta C_v$.

For diatomic gases, $\epsilon = 1.9$.

14.17.3 Diffusion in gases

Gases diffuse more rapidly than liquids due to their high molecular speed. A law of the same form as Fick's law is also applicable to gases. Let us consider that the concentration of molecules in a gas is n which remains constant along the horizontal $X-Y$ plane but varies along the vertical Z-axis. Experiments show that the number of molecules crossing unit area per second in the direction of the Z-axis is given by $N = -D\frac{dn}{dz}$, where D is a constant known as the coefficient of diffusion of the gas which depends on the nature of the gas.

We now consider two different gases A and B placed one above the other having mean free paths and mean molecular velocities $\lambda_1, \bar{c}_1$ and $\lambda_2, \bar{c}_2$ respectively. When a steady state is attained, molecular concentrations of the two gases in a horizontal plane will be constant. Let n_1, n_2 be the molecular concentrations of the two gases at any plane $z = z_0$. As the pressure is constant throughout the gas, then from Avogadro's law

$$n_1 + n_2 = n = \text{ constant.}$$

$$\therefore \quad \frac{dn_1}{dz} = -\frac{dn_2}{dz} = \alpha \text{ (say) or, } n_1 = \alpha z + \beta$$

and $n_2 = -\alpha z + \beta$, where β is the constant of integration.

As in the cases of viscosity and thermal conduction we consider an elementary volume dv (Fig. 14.17.1) at P at a distance r from the area $dx\,dy$. Considering n_1 as the number of molecules of type A per c.c. in the plane $z = z_0$, the number of molecules moving from the volume dv towards $dx\,dy$ in the downward direction in unit time without suffering collision is

$$\frac{\bar{c}_1}{\lambda_1}\, dv \frac{dx\,dy}{4\pi r^2} \cos\theta e^{-r/\lambda_1}\left(n_1 + r\cos\theta \frac{dn_1}{dz}\right) \quad \text{[See Art. 14.17.1]}$$

Similarly, the number of molecules moving from the elementary volume dv at P' towards $dx\,dy$ in the upward direction without suffering collision is

$$\frac{\bar{c}_1}{\lambda_1}\, dv \frac{dx\,dy}{4\pi r^2} \cos\theta\; e^{-r/\lambda_1}\left(n_1 - r\cos\theta \frac{dn_1}{dz}\right).$$

Hence the number of molecules crossing the area $dx\,dy$ in unit time in the downward direction

$$= \iiint \frac{\bar{c}_1}{\lambda_1}\, dv \frac{dx\,dy}{4\pi r^2} \cos\theta e^{-r/\lambda_1} \cdot 2r\cos\theta \frac{dn_1}{dz}$$

$$= \iiint \frac{\bar{c}_1}{\lambda_1} \cdot r^2 \sin\theta\, dr\, d\theta\, d\phi \cdot \frac{dx\,dy}{4\pi r^2} \cos\theta e^{-r/\lambda_1} \cdot 2r\cos\theta \frac{dn_1}{dz}$$

$$[\because \; dv = dr(r\sin\theta\, d\phi)(r\, d\theta)]$$

$$= \frac{dx\,dy\,\bar{c}_1}{2\pi\lambda_1} \cdot \frac{dn_1}{dz}\left[\int_0^\infty r e^{-r/\lambda_1} dr \int_0^{\pi/2} \cos^2\theta \sin\theta\, d\theta \int_0^{2\pi} d\phi\right]$$

$$= \frac{dx\,dy}{2\pi\lambda_1} \cdot \frac{dn_1}{dz} \cdot \bar{c}_1 \lambda_1^2 \cdot \frac{1}{3} \cdot 2\pi$$

$$= dx\,dy \cdot \frac{1}{3}\bar{c}_1\lambda_1 \frac{dn_1}{dz} = dx\,dy \cdot \frac{1}{3}\alpha\bar{c}_1\lambda_1 \qquad \left[\because \; \frac{dn_1}{dz} = \alpha\right]$$

$\therefore$ the number of molecules of the gas of type A crossing unit area in unit time in the downward direction is given by

$$\nu_1 = \frac{1}{3}\alpha\bar{c}_1\lambda_1.$$

Similarly, the number of molecules of the gas of type B crossing unit area in unit time in the downward direction is given by

$$\nu_2 = -\frac{1}{3}\alpha\bar{c}_2\lambda_2 \quad \left[\because \; \frac{dn_2}{dz} = -\alpha\right].$$

Thus, there will be a resultant transfer of molecules $\nu_1 + \nu_2$ in the downward direction. This will cause a change in the molecular density which is not possible and hence the gas as a whole moves with a mass velocity w in the direction of z increasing to neutralise the transfer of molecules.

Let T_1, T_2 be the actual number of molecules for the gases A and B respectively moving in the downward direction. Hence we have,

$$T_1 = \nu_1 - wn_1; \quad T_2 = \nu_2 - wn_2.$$

Now, as the resultant transfer is zero

$$T_1 + T_2 = 0 \text{ or, } (\nu_1 + \nu_2) - w(n_1 + n_2) = 0 \text{ or, } w = \frac{\nu_1 + \nu_2}{n_1 + n_2}$$

$$\therefore T_1 = \nu_1 - \frac{\nu_1 + \nu_2}{n_1 + n_2} n_1 = \frac{\nu_1 n_2 - \nu_2 n_1}{n_1 + n_2} = \frac{1}{3} \frac{\alpha(n_2 \lambda_1 \bar{c}_1 + n_1 \lambda_2 \bar{c}_2)}{n_1 + n_2} = \frac{1}{3} \frac{dn_1}{dz} \frac{(n_2 \lambda_1 \bar{c}_1 + n_1 \lambda_2 \bar{c}_2)}{n_1 + n_2}.$$

Now, by definitión $T_1 = D\frac{dn_1}{dz}$. [As T_1 is the number of molecules moving in the negative direction, in the expresion for T_1, negative sign is omitted.]

Hence,

$$D = \frac{1}{3} \frac{(n_2 \lambda_1 \bar{c}_1 + n_1 \lambda_2 \bar{c}_2)}{n_1 + n_2}. \tag{14.17.7}$$

The equation (14.17.7) is known as Meyer's formula for the coefficient of interdiffusion of two gases.

To find the coefficient of self diffusion of a gas, we put $\lambda_1 = \lambda_2 = \lambda$ (say), $\bar{c}_1 = \bar{c}_2 = \bar{c}$ (say). Hence, $D = \frac{1}{3}\lambda\bar{c} = \frac{\eta}{\rho}$. [from (14.17.3)] (14.17.8)

Thus the coefficient of self-diffusion of a gas is the ratio of the coefficient of viscosity to the density of the gas.

14.18 General treatment for the transport phenomena

Let us picture a cloud of molecules moving in all directions and suppose that any magnitude G of the physical entity (momentum, heat energy or mass) be associated to each molecule — the value of G varying progressively from one region of the gas to another. We take a plane XY called the observational plane and assume that the value of G is the same at all points on this plane. We now choose suitable co-ordinate axes such that the gradient $\left(\frac{dG}{dz}\right)$ of the magnitude G lies along the Z-axis and we choose a small area $dxdy$ in the observational plane. Consequently molecules, lying in the positive direction of Z, have higher values of the magnitude G than those of the observational plane and will therefore tend to equalise this difference by transferring G. Similar reasoning is applied for the molecules lying in the negative direction of Z.

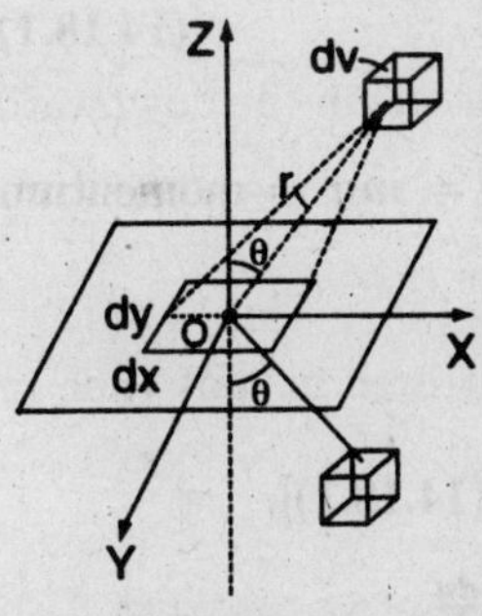

Fig 14.18.1

We assume that N = number of molecules per unit volume; dv = a small volume at a distance r from the origin O — 'r' making an angle θ with the Z-axis; $\bar{z} = \frac{\bar{c}}{\lambda}$ = the collision frequency, where λ is the average distance between two successive collisions, i.e., the mean free path and $\bar{c}$, the average velocity.

In time 'dt', the number of collisions in the volume dv is $N\bar{z}dvdt$ = the number of molecules starting new paths. Of these, the fraction $dxdy\cos\theta/4\pi r^2$ move in the desired direction; again of this group of molecules, only the fraction $e^{-r/\lambda}$ can reach the surface '$dxdy$' without suffering collisions.

$\therefore$ the number of molecules reaching the small area '$dxdy$' without suffering collisions is

$$N dv \bar{z} dt \frac{dxdy \cos\theta}{4\pi r^2} e^{-r/\lambda}.$$

Expressing dv in spherical polar co-ordinates, the above expression may be written as

$$Ne^{-r/\lambda}\bar{z}dt\frac{dxdy\cos\theta}{4\pi r^2}r^2\sin\theta d\theta d\phi dr = \frac{N\bar{z}dxdydt}{4\pi}\sin\theta\cos\theta d\theta d\phi e^{-r/\lambda}dr$$

Assuming the value of the physical entity 'G' to retain the same magnitude correspnding to the last collision, the transport of the physical property G from all the space above the observational plane is given by

$$G_{\downarrow} = \frac{N\bar{c}dxdydt}{4\pi\lambda}\int_0^{\pi/2}\int_0^{2\pi}\int_0^{\infty}\left(G + \frac{dG}{dz}r\cos\theta\right)\sin\theta\cos\theta d\theta d\phi e^{-r/\lambda}dr$$

Similarly transport of the physical property 'G' from all the space below the observational plane is given by

$$G_{\uparrow} = \frac{N\bar{c}dxdydt}{4\pi\lambda}\int_0^{\pi/2}\int_0^{2\pi}\int_0^{\infty}\left(G - \frac{dG}{dz}r\cos\theta\right)\sin\theta\cos\theta d\theta d\phi e^{-r/\lambda}dr$$

$\therefore$ the net transport of the physical property 'G' to the observational plane from the space below and above it is written as

$$G_{\downarrow} - G_{\uparrow} = \frac{2N\bar{c}dxdydt}{4\pi\lambda}\left(\frac{dG}{dz}\right)\int_0^{\pi/2}\sin\theta\cos^2\theta d\theta\int_0^{2\pi}d\phi\int_0^{\infty}re^{-r/\lambda}dr$$

Now

$$\int_0^{\pi/2}\sin\theta\cos^2\theta d\theta = \frac{1}{3};\ \int_0^{2\pi}d\phi = 2\pi \text{ and } \int_0^{\infty}re^{-r/\lambda}dr = \lambda^2$$

$$\therefore G_{\downarrow} - G_{\uparrow} = \frac{1}{3}\lambda N\bar{c}dxdydt\left(\frac{dG}{dz}\right) \quad (14.18.1)$$

Viscosity. Viscosity is due to transport of momentum, so that $G = mu =$ momentum parallel to X-axis

$$\therefore \quad \frac{dG}{dz} = m\frac{du}{dz} = m \times \text{velocity gradient}.$$

Now, viscous drag = net force in time $dt = \frac{1}{3}\lambda\bar{c}Nmdxdydt\frac{du}{dz}$ [from (14.18.1)].

$\therefore$ tangential stress = net force per unit area in unit time $= \frac{1}{3}\lambda\bar{c}Nm\frac{du}{dz}$.

Hence, the coefficient of viscosity,

$$\eta = \frac{\text{tangential stress}}{\text{velocity gradient}} = \frac{1}{3}\lambda\bar{c}Nm = \frac{1}{3}\rho\bar{c}\lambda$$

where $Nm = \rho =$ the density of the gas.

Conduction. Conductivity is due to transport of heat energy, and as such

$$G = mC_vT \text{ and } \frac{dG}{dz} = mC_v\frac{dT}{dz} = mC_v \times \text{ temperature gradient},$$

so that the net heat energy flowing across unit area in unit time is, from (14.18.1) given by

$$\frac{1}{3}\lambda\bar{c}NmC_v\frac{dT}{dz}.$$

$\therefore$ the coefficient of thermal conductivity,

$$K = \frac{\text{the net flow of heat energy across unit area in unit time}}{\text{the temperature gradient}} = \frac{1}{3}\lambda\bar{c}\rho C_v = \eta C_v.$$

Diffusion. Diffusion is due to transport of mass. Let us suppose that the column consists of a single gas, so that there will be self-diffusion— the molecules in the upper half are originally coloured red and those in the lower half are coloured blue. If

n = molecular density of both kinds of molecules;

n_1 = molecular density of the red molecules;

n_2 = molecular density of the blue molecules;

then $n = n_1 + n_2$, so that $G = \dfrac{n_1}{n_1 + n_2} = \dfrac{n_1}{n}$.

$$\therefore \quad \frac{dG}{dz} = \frac{1}{n}\frac{dn_1}{dz} = \frac{1}{N}\frac{dN_1}{dz} \quad [\text{putting } n = N, n_1 = N_1]$$

From (14.18.1), the coefficient of self-diffusion,

$$D = \frac{G}{dxdydt\frac{dN_1}{dz}} = \frac{1}{3}\bar{c}\lambda = \frac{\eta}{\rho}.$$

Discussion. $\frac{\eta}{\rho}$ — known as the *kinematic viscosity*, has the same dimension as the *diffusion coefficient* and is equal to it numerically. *Kinematic viscosity* is also called the *velocity diffusion coefficient*, since η is an impulse flux related to a unit velocity gradient, and the product of ρ and the velocity is the impulse of a unit volume, and so the ratio of impulse flux η to the density ρ is the *velocity flux*.

From $K = \eta C_v$ and $D = \frac{\eta}{\rho}$, we have $D = \frac{K}{\rho C_v}$

$\frac{K}{\rho C_v}$ depends on the properties of a gas and gives the rate of levelling out of temperature and is called the *coefficient of temperature conductivity*. Now since K determines the *flux* of the quantity of heat transported by a gas and ρC_v is the heat capacity of a unit volume of the gas, $\frac{K}{\rho C_v}$ is obviously the flux of the temperature at unit temperature gradient, i.e., the *coefficient of temperature diffusion* ($\because \ \Delta T = \Delta Q/C$, the ratio of the flux of the quantity of heat to the heat capacity). The quantitative relations between the transport coefficients D, K and η also support the view that transport phenomena truly represent the processes of diffusion of a substance, the temperature and the velocity respectively. Also any of the transport coefficients on being measured experimentally gives us the mean free path of a molecule and therefore the size of a molecule.

Illustrative examples

1. Determine the mean free path and collision frequency for air molecules at standard temperature and pressure, given that the viscosity $\eta = 1.7 \times 10^{-4}$. C.G.S. units, mean velocity $= 4.5 \times 10^4$ cm/sec and density 1.29 gm/litre.

From the relation (14.17.3) we have,

$$\lambda = \frac{3\eta}{\rho \bar{c}} = \frac{3 \times 1.7 \times 10^{-4}}{1.29 \times 10^{-3} \times 4.5 \times 10^4} = 8.78 \times 10^{-6} \text{ cm.}$$

Collision frequency

$$= \frac{\bar{c}}{\lambda} = \frac{4.5 \times 10^4}{8.78 \times 10^{-6}} = 5.12 \times 10^9 \text{ per second.}$$

2. A shower of 10^4 molecules, each molecule originally moving with the same velocity, traverses a gas. Find the number of molecules which will remain undeflected even after traversing a length twice the mean free path. **[C.U. (Hons.) 1976]**

From Article 14.14 we know that

$$P_x = e^{-x/\lambda} \text{ or, } \frac{N}{N_0} = e^{-x/\lambda},$$

where N_0 is the total number of molecules and N the number having a freee path x.

In this problem $N_0 = 10^4, x = 2\lambda$

$$\therefore \; N = 10^4 e^{-2\lambda/\lambda} = 10^4 e^{-2}$$

$$= 10^4 \times 0.1353 = 1353.$$

14.19 Brownian movement

In 1827 Robert Brown, an English botanist, observed through a high power microscope that small particles such as pollen grains in suspension in a liquid show irregular motion in all directions. The observation was extended to a large number of colloidal solutions and the same phenomenon was observed. The suspended particles under ultra-microscope look like stars of light moving irregularly. This irregular motion is called Brownian movement of the particles. Investigations preceding the work of Perrin established the following facts.

(i) The motions of the suspended particles are irregular and random.

(ii) The motions are not affected by jarking or shaking of the bottle.

(iii) The greater the viscosity of the liquid, slower the motion of the suspended particles.

(iv) The smaller the size of the particles, the greater the motion.

(v) The motions are continuous and eternal.

In case of some variety of colloidal quartz particles in liquid suspension which are thousands of years old, Brownian motion is still exhibited.

Since the motion is eternal, the only source of energy of this motion is due to the thermal energy of the molecules. The variation of the motion with viscosity of the liquid, size of the particles and temperature shows that it is due to unbalanced forces by the impact of the liquid molecules. When the suspended particles are sufficiently large, on the average equal number of molecules strike them on all sides at each instant. Hence the particles are bombarded more or less uniformly. In this case, the motion of a particle will be governed by both gravity and the viscosity of the liquid. In case of smaller particles, according to laws of probability, the number of molecules striking on all sides may not be equal. The inequality in the number of molecular impacts on different sides gives rise to a force causing the suspended particles to move and so the motion will be governed by the viscosity of the liquid.

14.20 Brownian Movements and the Law of Atmospheres

We have seen in fluid statics (Art. 12.3) that $p = p_0 e^{-\frac{Mgz}{RT}}$, which is known as *the law of atmospheres.*

Let n and n_0 be the number of molecules or particles per unit volume at heights z and z_0 respectively. Since n and n_0 are directly proportional to the pressures p and p_0, we have,

$$n = n_0 e^{-\frac{Mg}{RT}(z-z_0)} \tag{14.20.1}$$

$$= n_0 c^{-\frac{mNg}{RT}(z-z_0)} \tag{14.20.2}$$

where N is the Avogadro number and m is the mass of a single molecule. This is known as Laplace's law of atmospheres and it shows how the number of molecules per unit volume in a gaseous atmosphere varies with the distance from the earth's surface. The density of the molecules decreases as the height increases. The rate of decrease is greater, greater the mass of the molecules and smaller the temperature. From equation (14.20.2) knowing g, z, R and T for given values of n/n_0 and z, the Avogadro number may be determined. The expression (14.20.2) also shows that the height required to produce a given rarefaction varies with the molecular weight of the gas, i.e. the nature of the gas. In case of colloidal solution the weight of each colloidal particle is reduced due to buoyancy of the liquid and so mg in equation (14.20.2) should be replaced by $mg\left(1 - \frac{d}{D}\right)$, where d is the density of the liquid and D that of the particle

$$\therefore \frac{n}{n_0} = e^{-\frac{Nm}{RT}\left(1-\frac{d}{D}\right)g(z-z_0)} = e^{-\frac{NVD}{RT}\left(1-\frac{d}{D}\right)g(z-z_0)} = e^{-\frac{NV}{RT}(D-d)g(z-z_0)},$$

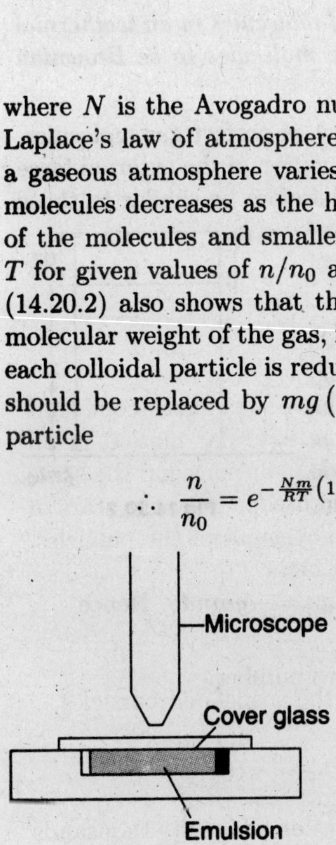

Fig 14.20.1

where V is the volume of the particle.

$$\text{or,} \quad \log\frac{n_0}{n} = \frac{NV}{RT}(D-d)g(z-z_0). \tag{14.20.3}$$

To test the above equation Perrin used colloidal solution of gum mastic and gum gamboge. Gamboge was treated with alcohol and then the clear solution was poured in water. In this way an emulsion containing spherical globules was formed.

Perrin obtained spherical globules of gamboge of same size from the emulsion by fractional centrifuging. An emulsion containing the separated gamboge particles of identical shape was taken in a small cell which was kept at constant temperature by means of water bath. The cell was then placed on the stage of a microscope (Fig. 14.20.1). The field of view of the microscope was restricted by using a small diaphragm. By raising or lowering the microscope the values of n at different levels were determined. The density D of the particles was then determined by three different methods. For the same suspension the three methods gave for one case the density of the particles 1.1942 gm/cm^3, 1.194 gm/cm^3 and 1.195 gm/cm^3. To find the volume V of a particle the radius of the particle was determined also by three methods. In one method the radius of the particle was determined by counting the number of particles in a given area. Another method for the determination of mass instead of volume V was by direct weighing. The most convenient method of measuring the radius of the particle is by using Stokes' law. If r be the radius of the particle, then the volume is given by $V = \frac{4}{3}\pi r^3$.

Hence equation (14.20.3) may be written in the form

$$N = \frac{3RT}{4\pi r^3(D-d)g(z-z_0)} \log \frac{n_0}{n}. \qquad (14.20.4)$$

Thus measuring r, D and d, N may be computed knowing the values of $\frac{n}{n_0}$ and z with the help of the microscope. Also from these experimental data it was possible to establish the law of atmospheres. The Avogadro number N was found to have a value between 6.5×10^{23} and 7.2×10^{23}.

Alternative method for the derivation of the law of distribution of molecules in an isothermal vertical column of a gas under the action of gravity, assuming the molecules to be Brownian particles.

As the molecules of a solute dispersed in a weak solution behave as perfect gas molecules, the Brownian particles form a gas which is in equilibrium under the action of the external force field due to earth's gravitation. The distribution of pressure will therefore be similar to those prevailing in the case of gases of the earth's atmosphere. The law of distribution in an isothermal vertical column of the gas composed of Brownian particles can be obtained in the following way.

We consider a layer of the gas bounded by the surfaces at heights z and $z+dz$ on which the pressures are p and $p+dp$ respectively (Fig. 14.20.2) and let us suppose that z increases upwards. We consider the equilibrium of unit area of the layer. If the density of the gas at z is ρ, the force due to gravity acting on the layer is $g\rho dz$. The net force on the layer is $(p+dp-p+g\rho dz)$ which should be zero for equilibrium.

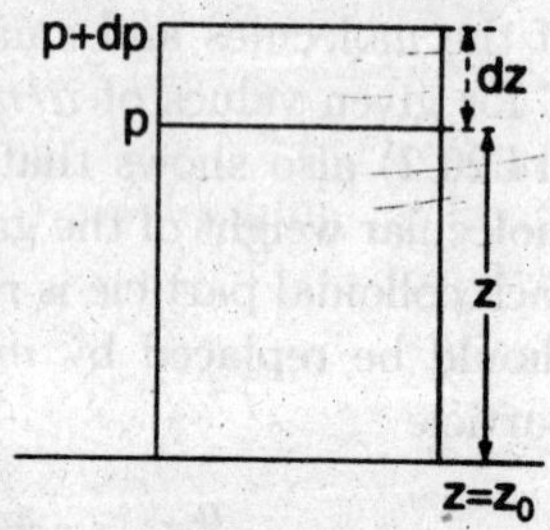

Fig 14.20.2

Hence $dp = -g\rho dz$

Now for a perfect gas, $p = nKT$, and $\rho = mn$, so that $dp = KTdn = -gmndz$. Hence

$$\frac{dn}{n} = -\frac{mg}{KT}dz = -\frac{Nm}{RT}gdz, \quad N \text{ being the Avogadro number.}$$

Integrating we obtain

$$n = n_0 e^{-\frac{Nmg}{RT}(z-z_0)}$$

where n and n_0 denote the number of particles per unit volume at the heights z and z_0 respectively. Now the effective mass of a suspended particle, due to buoyancy of the surrounding fluid, is $\frac{4}{3}\pi r^3(D-d)$, where r = radius of the particle, assuming it to be spherical, D = density of the particles, and d = density of the inter-granular fluid. So from the above expression, we may write

$$N = \frac{3RT}{4\pi r^3(D-d)g(z-z_0)} \log_e \frac{n_0}{n}$$

which is same as (14.20.4) and provides us with a convenient method of determining the Avogadro number.

14.21 The displacement of the particles undergoing Brownian movements and verification of Einstein's equation

It has been already mentioned that a particle of smaller size in suspension in a liquid experiences unbalanced forces due to molecular impacts, as a result of which the particle moves in a zigzag way through the liquid and is opposed by the forces due to the viscosity of the liquid. The theory of this motion was first deduced by Einstein and Smoluchowsky independently in 1905. We shall first consider here a treatment due to Langevin.

Let X be the unbalanced force acting on a particle along x-axis due to molecular impact. The equation of motion of the particle is then given by

$$m\frac{d^2x}{dt^2} = -k\frac{dx}{dt} + X \tag{14.21.1}$$

where $k\frac{dx}{dt}$ denotes the retarding force due to the viscosity of the liquid and 'm' the mass of the particle.

According to Stokes' law, $k = 6\pi a\eta$, a being the radius of the particle which has been considered spherical and η, the coefficient of viscosity of the liquid. We shall now change the equation (14.21.1) so that it contains the terms x^2 and $\left(\frac{dx}{dt}\right)^2$ as we are interested with the magnitude of displacement without its sign. Multiplying both sides of equation (14.21.1) by x, we have

$$mx\frac{d^2x}{dt^2} = -kx\frac{dx}{dt} + Xx \tag{14.21.2}$$

Now

$$\frac{d^2}{dt^2}(x^2) = 2\left[x\frac{d^2x}{dt^2} + \left(\frac{dx}{dt}\right)^2\right]. \quad \therefore \quad x\frac{d^2x}{dt^2} = \frac{1}{2}\frac{d^2}{dt^2}(x^2) - \left(\frac{dx}{dt}\right)^2.$$

Also

$$x\frac{dx}{dt} = \frac{1}{2}\frac{d}{dt}(x^2).$$

Hence we have from equation (14.21.2)

$$\frac{1}{2}m\frac{d^2}{dt^2}(x^2) - m\left(\frac{dx}{dt}\right)^2 = -\frac{k}{2}\frac{d}{dt}(x^2) + Xx. \tag{14.21.3}$$

If we average over a large number of particles, Xx will have equal probability of positive and negative signs. Hence $\overline{Xx} = 0$.

Now the mean kinetic energy of the particle along x-axis only is $\frac{1}{2}m\left(\frac{dx}{dt}\right)^2 = \frac{1}{2}\frac{R}{N}T$, where the symbols have their usual meanings. Hence equation (14.21.3) may be written as

$$\frac{m}{2}\frac{d^2}{dt^2}\overline{(x^2)} - \frac{RT}{N} = -\frac{k}{2}\frac{d}{dt}\overline{(x^2)}.$$

Putting $\frac{d}{dt}\overline{(x^2)} = z$, the above equation becomes,

$$\frac{m}{2}\cdot\frac{dz}{dt} - \frac{RT}{N} = -\frac{kz}{2} \text{ or, } \frac{dz}{dt} - \frac{2RT}{mN} = -\frac{k}{m}z$$

$$\text{or, } \frac{dz}{dt} = \frac{2RT}{mN} - \frac{k}{m}z \quad \text{or,} \quad \frac{d\left(\frac{2RT}{mN} - \frac{k}{m}z\right)}{\left(\frac{2RT}{mN} - \frac{k}{m}z\right)} = -\frac{k}{m}dt.$$

Integrating the above expression between the limits from $z = 0$ to $z = z$ and $t = 0$ to $t = t$, we have,

$$\log\left(\frac{\frac{2RT}{mN}}{\frac{2RT}{mN} - \frac{k}{m}z}\right) = \frac{k}{m}t \text{ or, } \frac{2RT}{mN} - \frac{k}{m}z = \frac{2RT}{mN}e^{-\frac{k}{m}t}$$

$$\text{or, } z = \frac{2RT}{Nk}\left(1 - e^{-\frac{k}{m}t}\right)$$

$$\text{or, } \frac{d}{dt}\overline{(x^2)} = \frac{2RT}{Nk}\left(1 - e^{-\frac{k}{m}t}\right) = \frac{RT}{3\pi a\eta N}\left(1 - e^{-\frac{k}{m}t}\right). \quad (14.21.4)$$

Now for a finite interval of time $e^{-\frac{k}{m}t}$ is negligible since $\frac{k}{m}$ is of the order of 10^5. Hence the equation (14.21.4) becomes

$$\frac{d}{dt}\overline{(x^2)} = \frac{RT}{3\pi a\eta N}$$

Integrating we have,

$$\overline{(\Delta x)^2} = \frac{RT}{N}\frac{\tau}{3\pi a\eta} \quad (14.21.5)$$

where $\overline{(\Delta x)^2}$ denotes the mean value of the squared displacements in time τ. (14.21.5) is Einstein's equation which shows that the mean squared displacement of a Brownian particle during time τ along any axis is proportional to this time. The most surprising result of Einstein's formula is that $\overline{(\Delta x)^2}$ is independent of mass of the particle, and Perrin showed that this prediction of the theory is satisfied to a high degree of precision.

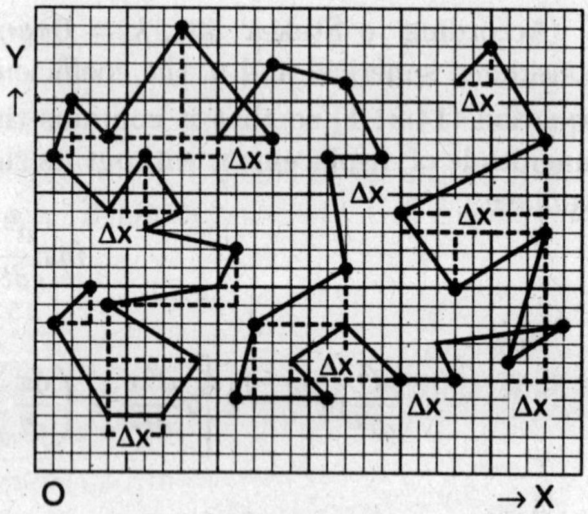

Fig 14.21.1

Perrin measured the value of $\overline{(\Delta x)^2}$ using camera lucida, the field of view of which had a graph paper where 16 divisions represented 5.0×10^{-3} cm. and the time interval τ chosen was 30 secs, From successive positions of the particle in an interval of time 30 secs the value of $\overline{\Delta x}$ can be computed [Figure (14.21.1)]. Perrin verified the dependence of $\overline{(\Delta x)^2}$ on τ, T, η and a as in equation (14.21.5) and obtained $N = 6.88 \times 10^{23}$ for gamboge of very uniform quality. We have thus a convincing proof of molecular collisions and hence the existence of molecules—in glaring support of the molecular kinetic theory.

14.22 Einstein's theory of translational Brownian movement

We shall consider here a simplified treatment due to Einstein. Einstein calculated the mean squared displacement of a Brownian particle for an interval of time τ from the consideration of the diffusion of the particles in the medium due to their irregular motions and also due to the difference in osmotic pressures between different parts having gradient in concentration.

Let us now calculate the coefficient of diffusion D from irregular motion. For the sake of simplicity let us assume that every particle will undergo displacement Δ in time τ along x-axis. Let us now imagine a cylinder with axis parallel to the x-axis and of length Δ. Let the molecular concentrations at the two end faces be n_1 and n_2 respectively and the area of each face be S which gives the cross-sectional area of the cylinder. The number of particles entering the cylinder crossing the left end surface in time τ is $\frac{1}{2}n_1\Delta S$, as half of the number of particles contained in the cylinder is coming from the left of the end face crossing it in time τ.

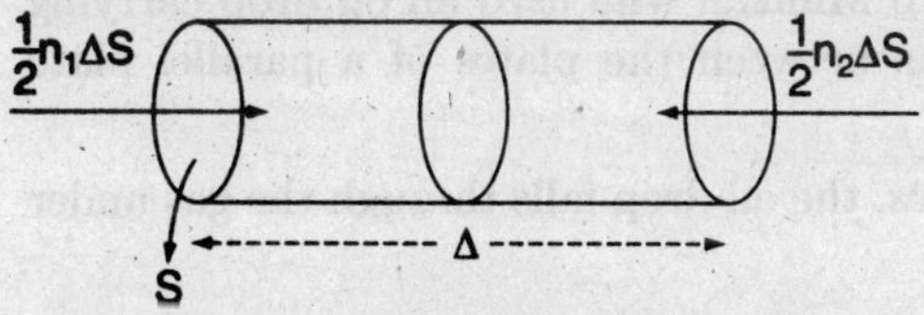

Fig 14.22.1

In a similar way the number of particles entering the cylinder crossing the right end face is $\frac{1}{2}n_2\Delta S$. Hence the number crossing the middle face of the cylinder to the right in time τ is $\frac{1}{2}(n_1 - n_2)\Delta S$. This number is equal to $-D\frac{dn}{dx}S\tau$.

$$\therefore\quad \frac{1}{2}(n_1 - n_2)\Delta S = -D\frac{dn}{dx}S\tau.$$

But,

$$n_1 - n_2 = -\frac{dn}{dx}\Delta \quad\therefore\quad -\frac{1}{2}\frac{dn}{dx}\Delta^2 S = -D\frac{dn}{dx}S\tau. \quad\therefore\quad D = \frac{\Delta^2}{2\tau}. \tag{14.22.1}$$

Let us now calculate the coefficient of diffusion due to osmotic pressure. The osmotic pressures at the left and right end faces are $p_1 = n_1KT$ and $p_2 = n_2KT$ respectively. Hence the cylinder experiences a force $(p_1 - p_2)S = KT(n_1 - n_2)S$ to the right.

If n be the mean concentration, then the number of particles contained in the cylinder is $n\Delta S$. Hence each particle contained in the cylinder will experience a force given by

$$F = \frac{KT(n_1 - n_2)S}{n\Delta S} = \frac{KT}{\Delta}\cdot\frac{(n_1 - n_2)}{n}.$$

Now, since the particle moves in a viscous medium we have, from Stokes' law,

$$F = 6\pi a\eta v = \frac{KT}{\Delta}\cdot\frac{(n_1 - n_2)}{n} \tag{14.22.2}$$

Now,

$$n_2 = n_1 + \frac{dn}{dx}\cdot\Delta \quad\therefore\quad \frac{n_1 - n_2}{n} = -\frac{1}{n}\frac{dn}{dx}\Delta.$$

Hence equation (14.22.2) becomes

$$6\pi a\eta v = -\frac{KT}{n\Delta}\cdot\frac{dn}{dx}\Delta = -\frac{KT}{n}\frac{dn}{dx} \quad\text{or,}\quad nv = -\frac{KT}{6\pi a\eta}\frac{dn}{dx}. \tag{14.22.3}$$

The above expression gives the number of particles moving per second per unit area which is equal to $-D\frac{dn}{dx}$.

Hence

$$D\frac{dn}{dx} = \frac{KT}{6\pi a\eta}\frac{dn}{dx} \quad \text{or,} \quad D = \frac{RT}{N}\cdot\frac{1}{6\pi a\eta}. \tag{14.22.4}$$

∴ substituting D from (14.22.1), we have

$$\Delta^2 = \frac{RT}{N}\cdot\frac{\tau}{3\pi a\eta}. \tag{14.22.5}$$

To verify the relation (14.22.4) Brillouin used a suspension of gamboge in glycerine and measured the coefficient of diffusion D and calculated the value of N using the same relation.

14.23 Translational Brownian movement in gases

The validity of Einstein's equation for particles suspended in gases was first established by Ehrenhaft. The most important experiment is due to Millikan who used an oil drop carrying electric charge and allowed it to fall through a gas between the plates of a parallel plate condenser.

When no electric field is applied between the plates, the oil drop falls through the gas under gravity. We have, then, from Stokes' law.

$$mg = kv_1 \quad \text{where } k = 6\pi a\eta \tag{14.23.1}$$

and v_1 is the terminal velocity of fall of the oil drop under gravity. When the electric field F is applied between the plates in such a way that it opposes the motion of the drop, we have,

$$mg - Fq = kv_2 \tag{14.23.2}$$

where v_2 denotes the observed velocity of fall and q is the charge carried by the drop.

From (14.23.1) and (14.23.2), we have

$$\frac{mg - Fq}{mg} = \frac{v_2}{v_1}. \quad \therefore \quad \frac{Fq}{mg} = \frac{v_1 - v_2}{v_1} = \frac{k(v_1 - v_2)}{mg} \quad \text{[from (14.23.1)]}$$

$$\text{or,} \quad k = \frac{Fq}{(v_1 - v_2)}. \tag{14.23.3}$$

As the charge carried by each drop is some multiple of the electronic charge e, we may write $q = ne$.

Hence equation (14.23.3) becomes

$$k = \frac{Fne}{(v_1 - v_2)}. \tag{14.23.4}$$

Millikan made observations with a large number of oil drops carrying different charges and obtained the least common divisor for $\dfrac{k(v_1 - v_2)}{F}$ which gives the electronic charge e.

Let this be $\frac{k}{F}(v_1 - v_2)_0$. Hence

$$\frac{k}{F}(v_1 - v_2)_0 = e \quad \text{or,} \quad k = 6\pi a\eta = \frac{Fe}{(v_1 - v_2)_0}. \tag{14.23.5}$$

Now, from equation (14.21.5), we have,

$$\overline{(\Delta x)^2} = \frac{2RT}{N} \cdot \frac{\tau}{6\pi a\eta}.$$

Hence from (14.23.5) we have,

$$\overline{(\Delta x)^2} = \frac{2RT}{N} \cdot \frac{\tau(v_1 - v_2)_0}{Fe}.$$

This relation may also be used to get the value of Faraday constant given by

$$Ne = \frac{2RT}{\overline{(\Delta x)^2}} \cdot \frac{\tau(v_1 - v_2)_0}{F}. \tag{14.23.6}$$

With 1,735 observations of displacements on a drop with unit charge, Millikan and Fletcher got $Ne = 2.88 \times 10^{14}$ e.s.u.

The value of Faraday constant thus obtained agrees fairly with the value obtained by electrolysis for a univalent ion. Thus, the theory of Brownian movement in the case of gases is established. Substituting the value of e in the expression (14.23.6) we have, $N = 6.06 \times 10^{23}$. The advantage of this method over that with liquid emulsion is that a single drop may be observed for a pretty long time. Also by using gases of low viscosity at low pressure, the values of $\overline{(\Delta x)^2}$ were found to be 50 times as large as those in the case of liquids for the same drop.

14.24 Rotational Brownian movement

Brownian particles also undergo rotational motion. By a similar analysis, it can be shown that the mean squared angular displacement of the particles due to rotational Brownian movement is given by

$$\overline{\alpha^2} = \frac{RT}{N} \cdot \frac{\tau}{4\pi\eta a^3} \tag{14.24.1}$$

The above relation was verified by Perrin who obtained the value of N by measuring $\overline{\alpha^2}$. The value of the Avogadro number obtained from rotational Brownian movement is given by $N = 6.5 \times 10^{23}$.

Illustrative Example

Observations on the Brownian movement in water showed that the horizontal displacements of a given particle in eleven successive intervals of 30 sec. were 0, 5.6, −4.7, −10.8, 6.6, −9.8, −11.2, −4.0, 15.0, 19.1, 16.0 × 10^{-4} cm. The temperature was 20°C at which the viscosity of water is 0.0100 C.G.S. and the radius of the particle was 1.05 × 10^{-6} cm. If $R = 8.32 \times 10^7$, obtain the value of the Avogadro number.

$$\overline{(\Delta x)^2} = \frac{10^{-8}[(0)^2 + (5.6)^2 + (4.7)^2 + (10.8)^2 + (6.6)^2 + (9.8)^2 + (11.2)^2 + (4.0)^2 + (15.0)^2 + (19.1)^2 + (16)^2]}{11}$$

$$= 117.89 \times 10^{-8} \text{ cm.}^2.$$

Now,

$$N = \frac{RT}{(\Delta x)^2} \cdot \frac{\tau}{3\pi a\eta} \quad \text{[from equation (14.21.5)]}$$

Here

$$R = 8.32 \times 10^7 \text{ C.G.S. units}; \quad T = 293° \text{ K}; \quad \tau = 30 \text{ secs.}$$

$$a = 1.05 \times 10^{-6} \text{ cm.}; \quad \eta = 0.0100 \text{ C.G.S. unit}$$

$$\therefore N = \frac{8.32 \times 10^7 \times 293 \times 30}{117.89 \times 10^{-8} \times 3 \times 3.14 \times 1.05 \times 10^{-6} \times 0.0100} = 5.72 \times 10^{23}.$$

14.25 Van der Waals' equation of state

In the deduction of the equation of state $pV = RT$ of perfect gases, dimensions of the molecules were neglected. Also it was assumed that there is no force of attraction or repulsion between the molecules. For real gases however these assumptions are not valid. It was Clausius who pointed out that the dimensions of the molecules should not be ignored though the molecules are of extremely small size. When the size of the molecules is taken into consideration, Clausius showed that the volume of the gas V should be replaced by $V - b$, where b is the volume correction term which depends upon the total volume occupied by the molecules.

Calculation of 'b'

We can imagine each molecule of the gas to be surrounded by the sphere of influence of radius σ such that the centre of no other molecule can penetrate it. Let V be the total volume of the gas which contains n molecules. Then $(n-1)\frac{4}{3}\pi\sigma^3$ is excluded for centres of other molecules. Hence, the actual volume free from the spheres of influence is $V - (n-1)\frac{4}{3}\pi\sigma^3 = V - \frac{4}{3}\pi\sigma^3 n$, since n is large. So, the real molecular density is given by

$$n' = \frac{n}{V - \frac{4}{3}\pi\sigma^3 n}. \tag{14.25.1}$$

Now, the centres of the molecules of a gas enclosed in a vessel can approach within a distance $\sigma/2$ from the walls of the container, but the sphere of influence of a molecule extends to the wall if the molecule be at a distance σ from the wall. Hence the centre of no other molecule can lie within the hemisphere of volume $\frac{2}{3}\pi\sigma^3$. Let dV be a volume taken near the wall, then $n' \cdot \frac{2}{3}\pi\sigma^3$ is the fraction of the volume per c.c. not available for the centres of other molecules. Let dV' be the actual amount of dV existing as free space. Then,

$$dV - dV' = dV \cdot n' \cdot \frac{2}{3}\pi\sigma^3 \quad \text{or,} \quad dV' = dV\left(1 - \frac{2}{3}\pi\sigma^3 n'\right) = dV\left(1 - \frac{\frac{2}{3}\pi\sigma^3 n}{V - \frac{4}{3}\pi\sigma^3 n}\right).$$

The actual number of molecules in the volume dV is

$$n'\,dV' = \frac{n}{V - \frac{4}{3}\pi\sigma^3 n} \cdot dV\left(1 - \frac{\frac{2}{3}\pi\sigma^3 n}{V - \frac{4}{3}\pi\sigma^3 n}\right) = \frac{n\,dV}{V - 2b}\left(1 - \frac{b}{V - 2b}\right),$$

where $b = \frac{2}{3}\pi\sigma^3 n$. Now multiplying the numerator and denominator by $\left(1+\frac{b}{V-2b}\right)$, we may write

$$n'\,dV' = \frac{n\,dV}{(V-2b)} \frac{\left(1-\frac{b^2}{(V-2b)^2}\right)}{\left(1+\frac{b}{V-2b}\right)} = \frac{n\,dV}{V-b} \quad \left[\because \ \frac{b^2}{(V-2b)^2} \text{ is very small}\right].$$

Thus, the effect of the size of the molecules is to reduce the volume by b. Now $b = 4 \times \frac{4}{3}\pi n\left(\frac{\sigma}{2}\right)^3$. Hence the volume is reduced by an amount which is equal to four times the total volume of the molecules present in the volume V of the gas.

Hirn first introduced the idea of the forces of attraction between the molecules and with this idea it can be shown that the measured pressure is less than the actual pressure. A molecule in the interior of an enclosure experiences forces equally in all directions due to the surrounding gas molecules and hence the resultant force acting on the molecule is zero. But this is not the case when we consider the molecule near the wall of the container. In this case the number of molecules on one side of the molecule in question will be greater than that on the other side and the molecule will experience a dragging force acting normally away from the wall of the container. Thus, the measured pressure is less than the actual pressure exerted by a perfect gas.

The measured pressure p, therefore, should be increased by p', where p' is called the *intrinsic pressure* arising out of the dragging force as mentioned above. p' may be calculated in the following way.

The dragging force acting on a molecule is proportional to the number of molecules per c.c. of the gas. Again the force acting on unit area of the surface will be proportional to the product of the dragging force acting on a molecule and the number of molecules striking on unit area of the surface per second. Hence the dragging force acting on unit area (p') is proportional to n^2. Now, $n \propto \frac{1}{V}$,

$\therefore \ n^2 \propto \frac{1}{V^2}$. Hence $p' \propto \frac{1}{V^2}$ or $p' = \frac{a}{V^2}$, where a is a constant. Thus in the case of real gases the perfect gas equation is changed into

$$(p+p')(V-b) = RT \quad \text{or} \quad \left(p+\frac{a}{V^2}\right)(V-b) = RT. \tag{14.25.2}$$

The above equation was first deduced by Van der Waals and is known as Van der Waals' equation of state for a real gas—valid for 1 gm-molecule of the gas*.

14.26 Deduction of Van der Waals' equation from Virial Theorem

We shall now deduce Van der Waals' equation from the Virial Theorem of Clausius. Considering the gas to be composed of particles, the equation of dynamics may be applied to their motion.

---

*If n gm-moles of a gas be present in volume v, the volume of one gm-mole would be v/n, so that we have

$$\left(P+\frac{a}{(v/n)^2}\right][v/n-b] = RT \quad \text{or} \quad \left(P+\frac{n^2 a}{v^2}\right)(v-nb) = nRT$$

which is Vander Waals' equation for n gm-moles of a gas. 'a' is expressed in atm-litre^2/mole^2 and 'b' is expressed in litre/mole units.

We can write the equations of motion of the molecules of the gas as

$$m\frac{d^2x}{dt^2} = X;\ m\frac{d^2y}{dt^2} = Y;\ m\frac{d^2z}{dt^2} = Z$$

where x, y, z are the co-ordinates of the centre of any molecule measured with respect to a certain system of axes and X, Y, Z are the components of external forces acting on the molecule.

Multiplying the above equations by x, y, z respectively and then adding, we have

$$m\left(x\frac{d^2x}{dt^2} + y\frac{d^2y}{dt^2} + z\frac{d^2z}{dt^2}\right) = Xx + Yy + Zz. \qquad (14.26.1)$$

Now

$$\frac{1}{2}\frac{d^2}{dt^2}(x^2) = \left[x\frac{d^2x}{dt^2} + \left(\frac{dx}{dt}\right)^2\right] \quad \text{or,}\ x\frac{d^2x}{dt^2} = \frac{1}{2}\frac{d^2(x^2)}{dt^2} - \left(\frac{dx}{dt}\right)^2.$$

Similarly

$$y\frac{d^2y}{dt^2} = \frac{1}{2}\frac{d^2(y^2)}{dt^2} - \left(\frac{(dy)}{dt}\right)^2;\ \text{and}\ z\frac{d^2z}{dt^2} = \frac{1}{2}\frac{d^2(z^2)}{dt^2} - \left(\frac{dz}{dt}\right)^2$$

Hence equation (14.26.1) becomes

$$\frac{m}{2}\frac{d^2}{dt^2}(x^2 + y^2 + z^2) - m\left[\left(\frac{dx}{dt}\right)^2 + \left(\frac{dy}{dt}\right)^2 + \left(\frac{dz}{dt}\right)^2\right] = (Xx + Yy + Zz)$$

$$\text{or,}\quad \frac{m}{4}\frac{d^2}{dt^2}(x^2 + y^2 + z^2) - \frac{m}{2}\left[\left(\frac{dx}{dt}\right)^2 + \left(\frac{dy}{dt}\right)^2 + \left(\frac{dz}{dt}\right)^2\right] = \frac{1}{2}(Xx + Yy + Zz).$$

For all the molecules of the gas, the terms of the above equation must be summed up and so we have,

$$\frac{m}{4}\sum\frac{d^2}{dt^2}(x^2 + y^2 + z^2) - \frac{1}{2}\sum mc^2 = \frac{1}{2}\sum(Xx + Yy + Zz).$$

Multiplying both sides of the above equation by dt and averaging over a comparatively long interval of time τ, we have

$$\frac{m}{4\tau}\sum\int_0^\tau \frac{d^2}{dt^2}(x^2 + y^2 + z^2)\,dt - \frac{1}{\tau}\sum\int_0^\tau \frac{m}{2}c^2\,dt = \frac{1}{2\tau}\sum\int_0^\tau (Xx + Yy + Zz)\,dt$$

$$\text{or,}\quad \frac{m}{4\tau}\sum\left[\frac{d}{dt}(x^2 + y^2 + z^2)\right]_0^\tau - \frac{1}{\tau}\sum\int_0^\tau \frac{m}{2}c^2\,dt = \frac{1}{2\tau}\sum\int_0^\tau (Xx + Yy + Zz)\,dt. \qquad (14.26.2)$$

Now, the first term in the above equation is vanishingly small as the values of

$$\left[\frac{d}{dt}(x^2 + y^2 + z^2)\right]_0^\tau$$

are of the same order at $t = 0$ and $t = \tau$, τ being large. The second and the third terms in the above equation remain finite for these terms contain τ both in the numerator and denominator. Now,

$$\frac{1}{\tau}\sum\int_0^{\tau}\frac{m}{2}c^2\,dt$$

denotes the total kinetic energy of the molecules averaged over a finite time. Also the term

$$\frac{1}{2\tau}\sum\int_0^{\tau}(Xx+Yy+Zz)\,dt$$

denotes the sum

$$\frac{1}{2}\sum(Xx+Yy++Zz)$$

averaged over the time τ. Hence equation (14.26.2) may be written as

$$\frac{1}{2}\sum\overline{mc^2} = -\frac{1}{2}\sum\overline{(Xx+Yy+Zz)} \tag{14.26.3}$$

The bar denotes the time average from 0 to τ. The expression on the right hand side of the equation (14.26.3) is known as the 'Virial' of the mechanical system.

We can deduce the equation of a perfect gas from the above theorem. We consider the gas enclosed in a parallelopiped of sides a, b, c parallel to the co-ordinate axes. For a perfect gas there is no force of attraction between the molecules. Let us consider two walls of the parallelopiped parallel to Y-Z plane and for one wall $x = 0$ and for the other wall $x = a$, i.e. the two walls are separated by a distance a. Let p be the pressure exerted by the gas. Now, for the wall at $x = 0, p$ is in the positive direction of x-axis and for the second wall at $x = a, p$ is in the negative direction of x-axis. Then for the 1st wall the value of $\Sigma\overline{Xx}$ at $x = 0$ is $p \times bc \times 0 = 0$ and for the second wall the value of $\Sigma\overline{Xx}$ at $x = a$ is $-p \times bc \times a = -pV$, where V is the volume of the parallelopiped. Similar values are obtained for the other two pairs of walls.

Hence

$$-\frac{1}{2}\Sigma\overline{(Xx+Yy+Zz)} = \frac{3pV}{2}.$$

Thus from Virial Theorem, we have

$$\frac{1}{2}\sum\overline{mc^2} = \frac{3pV}{2}. \tag{14.26.4}$$

Now, $\frac{1}{2}\Sigma\overline{mc^2}$ is the average value of the total kinetic energy in a given volume. Hence we may write

$$\frac{1}{2}\Sigma\overline{mc^2} = \frac{1}{2}mn\overline{c^2},$$

where n is the number of molecules in the given volume

$$\therefore\ \frac{1}{2}mn\overline{c^2} = \frac{3pV}{2} \quad \text{or,}\quad pV = \frac{1}{3}mn\overline{c^2}.$$

This is the equation for a perfect gas.

We shall now consider the case of a real gas. In this case intermolecular forces acting between the molecules should be taken into consideration in calculating the Virial of the system. Let

$\phi(r)$ be the force between two molecules, the centres of which are located respectively at x, y, z and x', y', z'. The distance between the molecules is then given by

$$r^2 = (x - x')^2 + (y - y')^2 + (z - z')^2.$$

The components of forces acting on the molecules are given by

$$X = \phi(r)\frac{x - x'}{r}; \quad Y = \phi(r)\frac{y - y'}{r}; \quad Z = \phi(r)\frac{z - z'}{r}$$

$$\text{and} \quad X' = \phi(r)\frac{x' - x}{r}; \quad Y' = \phi(r)\frac{y' - y}{r}; \quad Z' = \phi(r)\frac{z' - z}{r}$$

$\therefore$ the total contribution to virial by the intermolecular forces

$$= -\frac{1}{2}\sum\left[\left\{\phi(r)\frac{x - x'}{r}x - \phi(r)\frac{x - x'}{r}x'\right\} + \left\{\phi(r)\frac{y - y'}{r}y - \phi(r)\frac{y - y'}{r}y'\right\}\right.$$

$$\left. + \left\{\phi(r)\frac{z - z'}{r}z - \phi(r)\frac{z - z'}{r}z'\right\}\right]$$

$$= -\frac{1}{2}\sum\frac{\phi(r)}{r}\left[(x - x')^2 + (y - y')^2 + (z - z')^2\right] = -\frac{1}{2}\sum r\phi(r),$$

where the summation extends over all pairs of molecules. Hence, considering intermoleculer forces we have, from Virial Theorem,

$$\frac{1}{2}m n\overline{c^2} = \frac{3pV}{2} - \frac{1}{2}\sum r\phi(r). \tag{14.26.5}$$

We shall have to find out now the value of $\Sigma r\phi(r)$ for all pairs of molecules. Now if N be the total number of molecules, then the number of pairs of molecules is $\frac{N}{2}(N - 1) = \frac{N^2}{2}$ (approximately). The probability that one pair of molecules will have the distance of separation lying between r and $r + dr$ is $\frac{4\pi r^2 dr}{V}$, where V is the total available space, i.e. the volume of the gas. Hence, the number of pairs having the separation lying between r and $r + dr$ is $\frac{4\pi r^2 dr}{V} \cdot \frac{N^2}{2}$. Now as the potential energy is different for different separation, the number of pairs having the separation r

$$= \frac{4\pi r^2 dr}{V} \cdot \frac{N^2}{2} e^{-E_r/KT} = \frac{2\pi r^2 N^2}{V} e^{-E_r/KT}\, dr,$$

where E_r is the potential energy of a molecule at a distance r from another molecule, K and T being respectively the Boltzmann constant and the absolute temperature. Hence we may write

$$\frac{1}{2}\sum r\phi(r) = \frac{\pi N^2}{V}\int r^3\phi(r)e^{-E_r/KT}\, dr.$$

Considering repulsive virial for the gas, the above expression may be written in the form

$$\frac{1}{2}\sum r\phi(r) = \frac{\pi N^2}{V}\int_{\sigma-\delta}^{\sigma} r^3\phi(r)e^{-E_r/KT}\, dr.$$

The limits of integration extend from $\sigma - \delta$ to σ, where σ is the molecular diameter and δ is a very small quantity, the repulsive force being infinite at $\sigma - \delta$ and zero at σ.

As σ and $\sigma - \delta$ differ by a very small quantity, r under the integral sign not occurring in the functional term may be replaced by σ, a constant. Hence, the above expression is written as

$$\frac{1}{2}\sum r\phi(r) = \frac{\pi\sigma^3 N^2}{V}\int_{\sigma-\delta}^{\sigma} \phi(r)e^{-E_r/KT}\,dr.$$

Also, we know

$$\phi(r) = -\frac{dE_r}{dr}$$

$$\therefore\ \frac{1}{2}\sum r\phi(r) = -\frac{\pi\sigma^3 N^2}{V}\int_{\infty}^{0} e^{-E_r/KT}\,dE_r = \frac{\pi\sigma^3 N^2}{V}\cdot KT\left[e^{-E_r/KT}\right]_{\infty}^{0} = \frac{\pi\sigma^3 N^2 KT}{V}.$$

If $\phi'(r)$ be the attractive force, then the virial due to the attractive force is

$$\frac{1}{2}\sum r\phi'(r) = \frac{\pi N^2}{V}\int_{\sigma}^{\infty} r^3\phi'(r)e^{-E_r/KT}\,dr.$$

In this case the limits of integration range from $r = \sigma$ to $r = \infty$, the attractive force decreasing rapidly with distance and acting beyond σ.

The above expression may also be written as

$$\frac{1}{2}\sum r\phi'(r) = \frac{\pi N^2}{V}\int_{\sigma}^{\infty} r^3\phi'(r)e^{-\frac{1}{KT}\int_r^\infty \phi'(r)dr}\,dr \quad \because\ E_r = \int_r^\infty \phi'(r)\,dr.$$

As $\phi'(r)$ in the above expression is not known, it is not possible to calculate the above expression correctly. However, the integral will have a term containing σ and also an exponential term containing KT. Hence the integral will be a function of temperature T. Let this function be $f(T)$ and the expression for the virial due to the attractive force is given by

$$\frac{1}{2}\sum r\phi'(r) = \frac{\pi N^2}{V}f(T) \quad \left[\text{Here } f(T) = \int_{\sigma}^{\infty} r^3\phi'(r)e^{-\frac{1}{KT}\int_r^\infty \phi'(r)dr}\,dr\right]$$

Considering both the repulsive and attractive forces, we have from equation (14.26.5),

$$\frac{1}{2}mn\overline{c^2} = \frac{3pV}{2} - \frac{\pi\sigma^3 N^2 KT}{V} - \frac{\pi N^2}{V}f(T)$$

$$\text{or,}\quad \frac{3}{2}RT = \frac{3}{2}pV - \frac{\pi\sigma^3 N^2 KT}{V} - \frac{\pi N^2}{V}f(T) \quad \left[\because\ \frac{1}{2}mn\overline{c^2} = 3/2RT\right]$$

$$\text{or,}\quad RT = pV - \frac{2}{3}\frac{\pi\sigma^3 N^2 KT}{V} - \frac{2}{3}\frac{\pi N^2}{V}f(T) = pV - \frac{2}{3}\frac{\pi\sigma^3 N^2 VKT}{V^2} - \frac{2}{3}\frac{\pi N^2 V}{V^2}f(T)$$

$$\therefore\quad \frac{RT}{V} = p - \frac{2}{3}\frac{\pi\sigma^3 N^2 KT}{V^2} - \frac{2}{3}\frac{\pi N^2}{V^2}f(T)$$

$$\text{or,}\quad \left(p - \frac{\frac{2}{3}\pi N^2 f(T)}{V^2}\right) = \frac{RT}{V} + \frac{\frac{2}{3}\pi\sigma^3 N^2 KT}{V^2}$$

putting $b = \frac{2}{3}\pi\sigma^3 N$, we have

$$\left(p - \frac{\frac{2}{3}\pi N^2 f(T)}{V^2}\right) = \frac{RT}{V} + \frac{bNKT}{V^2} = \frac{RT}{V} + \frac{bRT}{V^2} = \frac{RT}{V}(1 + b/V).$$

It we put $-\frac{2}{3}\pi N^2 f(T) = a$, the above expression becomes

$$\left(p+\frac{a}{V^2}\right) = \frac{RT}{V}(1+b/V) \text{ or, } \left(p+\frac{a}{V^2}\right)\frac{V}{(1+b/V)} = RT \text{ or, } \left(p+\frac{a}{V^2}\right)V(1-b/V) = RT,$$

(neglecting higher powers of b/V)

$$\therefore \quad \left(p+\frac{a}{V^2}\right)(V-b) = RT.$$

The above relation represents the behaviour of CO_2 but the results for other gases do not agree with the above relation. Also the isothermals obtained from Van der Waals' equation do not tally with experimental results due to Andrews.

The constants a and b of Van der Waals' equation are found to vary in case of different substances and to vary for a particular substance with temperature.

Diameter of a molecule

The diameter σ of a molecule may be determined from the constant b in Van der Waals' equation.

From Van der Waals' equation

$$\left(p+\frac{a}{V^2}\right)(V-b) = RT,$$

we have by differentiation, taking V to be constant,

$$(V-b)\left(\frac{dp}{dT}\right)_v = R \text{ or, } \left(\frac{dp}{dT}\right)_v = \frac{R}{(V-b)} = \frac{1}{T}\left(p+\frac{a}{V^2}\right)$$

$$\text{or, } \frac{1}{p}\left(\frac{dp}{dT}\right)_v = \frac{1}{T}\left(1+\frac{a}{pV^2}\right). \tag{14.26.6}$$

From the above expression it follows that if the pressure coefficient of the gas at constant volume be experimentally determined, then 'a' may be determined provided T, p and V are known. Again differentiating the expression for Van der Waals' equation, considering p to be constant we have,

$$\left(p+\frac{a}{V^2}\right)\left(\frac{dV}{dT}\right)_p - (V-b)\frac{2a}{V^3}\left(\frac{dV}{dT}\right)_p = R$$

$$\text{or, } \left(\frac{dV}{dT}\right)_p\left[p+\frac{a}{V^2}-(V-b)\frac{2a}{V^3}\right] = R$$

$$\text{or, } \left(\frac{dV}{dT}\right)_p\left[p+\frac{a}{V^2}-(V-b)\frac{2a}{V^3}\right] = \frac{\left(p+\frac{a}{V^2}\right)(V-b)}{T}$$

$$\text{or, } \left(\frac{dV}{dT}\right)_p\left(p-\frac{a}{V^2}\right) = \frac{\left(p+\frac{a}{V^2}\right)(V-b)}{T} \quad \left(\text{neglecting } \frac{2ab}{V^3} \text{ which is very small}\right)$$

$$\text{or, } \left(\frac{dV}{dT}\right)_p\left(1-\frac{a}{pV^2}\right) = \frac{V\left(1+\frac{a}{pV^2}\right)(1-b/V)}{T}$$

$$\text{or, } \frac{1}{V}\left(\frac{dV}{dT}\right)_p = \frac{1}{T}\frac{\left(1+\frac{a}{pV^2}\right)(1-b/V)}{(1-a/pV^2)} = \frac{1}{T}\left(1+\frac{a}{pV^2}\right)\left(1-\frac{a}{pV^2}\right)^{-1}(1-b/V)$$

$$= \frac{1}{T}\left(1+\frac{2a}{pV^2}\right)(1-b/V) \quad \text{[neglecting small terms]}$$

$$= \frac{1}{T}\left(1+\frac{2a}{pV^2}-b/V\right) \quad \text{[neglecting small terms]} \qquad (14.26.7)$$

Thus knowing the volume coefficient of the gas at constant pressure, b may be determined from (14.26.7) as 'a' can be determined from equation (14.26.6). Hence the molecular diameter σ may be determined from the relation $b = \frac{2}{3}\pi N\sigma^3$. Also, since the coefficient of viscosity η of a gas can be experimentally determined, the molecular diameter σ may also be determined from the relation $\eta = \frac{0.49m\bar{c}}{\sqrt{2}\pi\sigma^2}$ (see Art. 14.17.1).

Exercises

1. Deduce an expression for the pressure of a gas from the kinetic theory of gases.

2. Deduce Boyle's law, Charles' law, Avogadro's law and Dalton's law from the kinetic theory of gases.

3. Obtain Maxwell's distribution law of velocities for an ideal gaseous system. Justify the law on experimental basis. **[C.U. (Hons.) 1955, '57]**

4. Find expressions for average velocity, mean square velocity and the most probable velocity, assuming Maxwell's law of distribution of velocities.

5. State the law of equipartition of energy. Prove that for a monatomic gas, the value of γ is 5/3 and for a diatomic gas it is 7/3. Deduce the formula you would use.

6. What do you mean by the mean free path of the molecules of gas? Show that it is equal to $\frac{1}{\sqrt{2}\pi\sigma^2 n}$.

7. Deduce an expression for the mean free path of the molecules of a gas applying Maxwell's law of distribution of velocities. **[C.U. (Hons.) '94]**

8. Explain what you mean by the viscosity of a gas. Deduce an expression for the viscosity of a gas in terms of the mean free path. Discuss the variation of viscosity with temperature.

9. Obtain the relation $K = \eta C_v$, the symbols bearing usual meanings. How far does this relation tally with experimental results? **[C.U. (Hons.) 2007]**

10. What is diffusion? Deduce an expression for the coefficient of interdiffusion of two gases.

11. What is Brownian motion? Deduce Einstein's formula for the translational Brownian motion of particles suspended in a liquid. How is the Avogadro number determined from it? **[C.U. (Hons.) 2004, 06]**

12. What is Brownian motion? Describe a suitable arrangement for the demonstration of Brownian motion of particles and adduce reasons for the conclusion that the motion is the result of irregular impacts on the particles by the molecules of the liquid in heat motion. Give a brief survey of Perrin's work on Brownian motion and his determination of N on the basis of his results. **[C.U. (Hons.) 1951, '55]**

13. Give an account of the translational Brownian movement in gases.

14. Deduce Van der Waals' equation for real gases. Mention some of the defects of this equation.

15. Deduce Maxwell's velocity distribution law. Hence obtain an expression for the energy distribution of the gas molecules and calculate the most probable energy for a Maxwellian distribution. Compare the graphical natures of the distribution functions in the two cases. What does, the total area under the above curves signify? **[C.U. (Hons.) 1980, 96, 07]**

16. (a) Deduce an expression for the most probable velocity of the molecules of a gas. Also deduce the mean and the r.m.s. velocities and express them in terms of the most probable velocity. Indicate the three velocities in a diagram showing the velocity distribution of the gas molecules. **[C.U. (Hons.) 1980]**

(b) The four atoms in the NH_3 (ammonia) molecule are located at the four corners of a tetrahedron. What is the total number of translational, rotational, and vibrational degrees of freedom of this molecule? On the basis of equipartition law, calculate C_v and γ for NH_3 gas. **[C.U. (Hons.) 1980]**

[**Ans.** NH_3 molecule being a non-linear arrangement of atoms, number of translational degrees of freedom $(t) = 3$, number of rotational degrees (r) of freedom $= 3$.

Number of vibrational degrees (v) of freedom for (i) elastic NH_3 molecule $= 3n - 6 = 6$, n being the total number of atoms in the NH_3 molecule, (ii) for rigid NH_3 molecule $v = 0$. Again molar specific heat at constant volume $C_v = \left[\frac{3}{2} + \frac{r}{2} + v\right] R$—for rigid NH_3 molecule $v = 0$, $r = 3$; $\therefore$ $C_v = 3R$ and $C_p = 4R$ and $\gamma = \frac{4}{3} = 1.33$ and for elastic NH_3 molecule, $r = 3$, $v = 6$. $\therefore C_v = 9R$ and $C_p = 10R$ and $\gamma = \frac{10}{9} = 1.11$.]

17. The coefficients of viscosity at room temperature and atmospheric pressure of two different monatomic gases of atomic weights μ_1 and μ_2 are η_1 and η_2 respectively. Show that the ratio of the thermal conductivities of these gases is given by $\frac{K_1}{K_2} = \frac{\eta_1}{\eta_2}\frac{\mu_2}{\mu_1}$. **[C.U. (Hons.) 1984]**

[**Ans.** $K = \eta c_v (c_v =$ specific heat at constant volume for 1 gm of the gas)

$= \frac{\eta c_v'}{m} (c_v' = mc_v, c_v'$ being the specific heat at constant volume for 1 molecule of the gas and $m =$ mass of a molecule)

$= \frac{\eta C_v}{\mu} (C_v = Nc_v', C_v$ being the molar specific heat at constant volume, N the Avagadro's number and $\mu = Nm =$ the molecular weight of the gas).

For the two monatomic gases $K_1 = \eta_1 \frac{C_v}{\mu_1}$ and $K_2 = \eta_2 \frac{C_v}{\mu_2}$ whence the required result follows.]

18. Using the law of equipartition, obtain the average energies per molecule in the following cases. (a) a gas of mon-atomic molecules, (b) a gas of diatomic molecules where the two atoms can be thought to be connected by a rigid weightless rod, (c) a gas of diatomic molecules, (d) a gas of triatomic linear molecules, (e) a gas of non-linear triatomic molecules. **[C.U. (Hons.) 1984, 07]**

[**Ans.** Average energy per molecule of a polyatomic gas, $E_v = (\frac{3}{2} + \frac{r}{2} + v)KT$, symbols having usual meanings.

(a) Here $r = 0$, $v = 0$, $\therefore E_v = \frac{3}{2}KT$; (b) for diatomic rigid molecules $r = 2$, $v = 0$ $\therefore E_{av} = \frac{5}{2}KT$; (c) for diatomic elastic molecules, $r = 2$, $v = 3n - 5 = 1$ $\therefore E_{av} = \frac{7}{2}KT$ (d) (i) for triatomic linear rigid molecules, $r = 2$, $v = 0$, $\therefore E_{av} = 5/2KT$; (ii) for triatomic linear elastic molecules, $r = 2, v = 3n - 5 = 4; \therefore E_{av} = \frac{13}{2}KT$ (e) (i) for triatomic non-linear rigid molecules, $r = 3$, $v = 0$. $\therefore E_{av} = 3KT$; (ii) for triatomic non-linear elastic molecules, $r = 3$, $v = 3n - 6 = 3$ $\therefore E_{av} = 6KT$.]

19. The coefficient of viscosity of a gas at temperatures $-20^0, 0^0, 100^0, 180^0$, and 300^0 are respectively 12.8×10^{-6}, 14.0×10^{-6}, 18.6×10^{-6}, 22.1×10^{-6}, and 26.7×10^{-6} MKS units. Check how far these data are in agreement with the expected variation of η with T as deduced from the kinetic theory and account for the discrepancies, if any. **[C.U. (Hons.) 1980]**

[**Ans.** $\eta = \frac{1}{3}mn\bar{c}\lambda = \frac{1}{3}mn\left(\frac{8KT}{\pi m}\right)^{\frac{1}{2}}\frac{1}{\sqrt{2}\pi\sigma^2 n} = \frac{2}{3}\frac{1}{\pi^{3/2}\sigma^2}\sqrt{mKT}$, i.e., $\eta \propto \sqrt{T}$ or $\frac{\eta_T}{\eta_0} = \left(\frac{T}{273}\right)^{\frac{1}{2}}$.

Theoretical values: For temperatures	Observed Values
−20°C and 0°C	
$\eta_{-20}/\eta_0 = \left(\frac{253}{273}\right)^{1/2} = 0.9625$	$\eta_{-20}/\eta_0 = \frac{12.8 \times 10^{-6}}{14.0 \times 10^{-6}} = 0.914$
100°C and 0°C	
$\eta_{100}/\eta_0 = \left(\frac{373}{273}\right)^{1/2} = 1.169$	$\eta_{100}/\eta_0 = \frac{18.6 \times 10^{-6}}{14.0 \times 10^{-6}} = 1.33$
180°C and 0°C	
$\eta_{180}/\eta_0 = \left(\frac{453}{273}\right)^{1/2} = 1.288$	$\eta_{180}/\eta_0 = \frac{22.1 \times 10^{-6}}{14.0 \times 10^{-6}} = 1.579$
Lastly 300°C and 0°C	
$\eta_{300}/\eta_0 = \left(\frac{573}{273}\right)^{1/2} = 1.450$	$\eta_{300}/\eta_0 = \frac{26.7 \times 10^{-6}}{14.0 \times 10^{-6}} = 1.912$

It is evident from above that the experimentally observed values show that η increases with increase in temperature more rapidly than the theoretically calculated values. The increase in η does not follow $T^{\frac{1}{2}}$ law. This is because we generally consider the molecules as rigid elastic spheres without any mutual interaction. Infact, however, they are surrounded by a force-field repulsive in character. When T increases, so does $\bar{c}$ and the molecules penetrate into each other's force field to some extent. The molecules thus have their effective size decreased at higher temperatures resulting in an increase in the value of η.]

20. In an experiment on Brownian motion, the root mean squared displacement in the horizontal direction of a minute spherical particle suspended in water was found to be 1.1×10^{-3} cm. over a 30 sec interval at 27°C at which the coefficient of viscosity of water is 10^{-2} c.g.s. unit. If the gas constant $R = 8.32 \times 10^7$ ergs/°K mole and Avogadro's number $N = 6.02 \times 10^{23}$, calculate the radius of the particle.

[**Ans.** 1.091×10^{-5} cm.] **[C.U. (Hons.) 1981]**

21. (a) Show that the pressure per unit volume p of a gas is equal to $\frac{2}{3}E$, where E is the mean energy of the gas molecules per unit volume. **[C.U. (Hons.) 1982]**

(b) Show that the root mean square value of a Cartesian component of molecular velocity is equal to $\sqrt{\frac{P}{\rho}}$, where P is the pressure and ρ is the density of the gas.

[C.U. (Hons.) 1984 (Spl.)]

22. What is the most probable kinetic energy E_m of a molecule? Is it equal to $\frac{1}{2}mv_m^2$ where v_m is the most probable speed of the molecule? Give reasons for your answer. **[C.U. (Hons.) 1985]**
(For answer, see Art. 14.8)

23. (a) Show how the relation $\eta = \frac{1}{3}mn\lambda\bar{c}$ implies that η is independent of n and hence of pressure at a given temperature. Under what assumptions is this independence valid? Show also that as $n \to 0$, η becomes proportional to n.

(b) Explain physically why the viscosity of a liquid decreases with rise in temperature, while for a gas, η increases with T. **[C.U. (Hons.) 1986]**

[**Ans.** (a) Combining $\eta = \frac{1}{3}mn\bar{c}\lambda = \frac{1}{3}\rho\bar{c}\lambda$ and $\lambda = \frac{1}{\sqrt{2}\pi\sigma^2 n}$ we have $\eta = \frac{m\bar{c}}{3\sqrt{2}\pi\sigma^2}$.

Since none of $m, \bar{c}, \sigma$ is a function of pressure, η appears to be independent of pressure or density of the gas. This conclusion has been verified experimentally from a low pressure of few mm. of mercury to several atmospheres. At extremely low pressures, λ becomes comparable with the dimensions of the vessel, and thereafter it cannot increase any further and becomes constant. Any further decrease of pressure reduces ρ and hence η, because λ becomes constant. Under these conditions, molecular collisions become rare and the very assumption made in the deduction of the expression for η no longer holds. Now the effects of individual molecules must be simply additive and therefore η will be proportional to density. Thus at very low pressures η should rapidly decrease with decrease of pressure.

The departure from the conclusion—pressure independence of η—is also observed at extremely high pressures when λ becomes comparable with molecular diameters and the transport is not through λ but through a distance greater than λ. The effect of this is to cause η/ρ to vary.

(b) The viscosity of liquids generally decreases with rise in temperature i.e. a liquid becomes more free-flowing at higher temperature. This is in sharp contrast with the behaviour of gases whose viscosity increases with rise of temperature. This indicates that the mechanism in the two cases is quite different. In gases it is a momentum transport phenomenon from layer to layer caused by the kinetic motion of the gas molecules which has been likened to the jumping of a group of people from one train to another to and fro, between two trains in parallel unequal motion. It is evident that such groups would retard the motion of the faster train relative to the slower one. In the above analogy, raising of temperature may be simulated by more frequent to and fro jumps by this group of people; this would cause more retardation i.e. higher viscosity.

Liquid viscosity on the other hand, is due to internal cohesion or intermolecular attraction and so decreases with rise in temperature. This is easy to understand from the "hole theory" of liquids. There are more "holes" i.e. more free volume at higher temperature and so the molecules can flow more freely just as a loose crowd can pass through a door more easily than a dense crowd.]

24. The thermal conductivity of helium gas at NTP is $1.4 \times 10^{-3} J\ cm^{-1} S^{-1} deg^{-1}$. At what pressure at 0°C is the mean free path 1 cm. long? **[C.U. (Hons.) 1987]**

[**Ans.** Hints Apply $K = \eta C_v, \eta = \frac{1}{3}\ mn\ \bar{c}\lambda$

$$C_v = \frac{xR}{2} (x = \text{degrees of freedom} = 3 \text{ for } He);$$

$$\bar{c} = \sqrt{\frac{8KT}{\pi m}} = \sqrt{\frac{8RT}{\pi Nm}};$$

$$R/N = 1.38 \times 10^{-16};$$

$$P = n\frac{R}{N}T \text{ or } n = \frac{PN}{RT} \text{ and } m = \frac{M}{N} = \frac{4}{N}.$$

[**Ans.** 15.9 dynes/cm.^{2}.]

25. Show that the Maxwell's law of distribution of molecular speed is independent of temperature, if the most probable velocity is taken as the unit of measuring speed. **[C.U. (Hons.) 1990]**
(For answer, see Art. 14.9.1)

26. A gas molecule of diameter 10^{-8} cm. and mass 3×10^{-24} gm. escapes from a furnace having temperature 4000°K with the r.m.s. speed into a chamber containing cold argon atoms each of diameter 3×10^{-8} cm. at a density of 4×10^{19} atoms per cc:

(a) What is the speed of the escaping molecule?

(b) What is the initial number of collisions per unit time experienced by the molecule? Deduce any formula you use for your calculation. **[C.U. (Hons.) 1991]**

[**Hints for Ans.** $C_r = \sqrt{\frac{3KT}{m}}$. Taking $K = 1.38 \times 10^{-16}$ and the given values of m and T, $C_r = 7.43 \times 10^5$ cm./sec. Again from $\lambda = \frac{1}{\sqrt{2}\pi n\sigma^2}$ putting the values of n and σ, we get λ. ∴ number of collisions per unit time experienced by the molecule $= \frac{C_r}{\lambda}$.]

27. Particles of mass 6.2×10^{-14} gms are suspended in a liquid at 27^0C and are observed to have a root-mean-square speed of 1.4 cm./sec. Calculate the Avogadro number from the principle of equipartition of energy and these data. **[C.U. (Hons.) 1991]**

28. Calculate the molecular K.E of 1 gm. of hydrogen gas at 50^0C, given the molecular weight of hydrogen is 2 and the gas constant R is 8.3×10^7 ergs per gm. molecule. **[C.U. (Hons.) 1989]**

29. Calculate the molecular translational and rotational energy of 1 gm. of hydrogen gas at 50^0C, assuming the molecules to be dumb-bell shaped. Also estimate therefrom the specific heat at constant volume. R (ideal gas constant) = 8.314 Joules per mole per ^{0}K. **[C.U. (Hons.) 1993]**

30. The density of hydrogen at N.T.P. is 8.96×10^{-5} gm. per cc. Calculate the root mean square velocity of an oxygen molecule under the same conditions. **[C.U. (Hons.) 1994]**

Hints. From $p = \frac{1}{3}\rho C_r^2$, using the values of ρ and normal pressure, we can find C_r — then using $C_r = \sqrt{\frac{3RT}{M}}$ (where R = molar gas constant, and M = gm-molecular mass of hydrogen), we can find out R – then again from $C_r = \sqrt{\frac{3RT}{M}}$, we can find out C_r of an oxygen molecule.

31. (a) Explain clearly the meaning of the term mean free path. Obtain an expression for the mean free path of the moelcules of a Maxwellian gas.

(b) Describe an experiment for the measurement of the mean free path of the molecules of a gas. **[C.U. (Hons.) '94, '99]**

32. Show that the mean square displacement per unit time of colloidal particles suspended in a liquid is directly proportional to the absolute temperature of the liquid. **[C.U. (Hons.) '94]**

33. (a) Viscosity, conductivity and diffusion are called transport phenomena. Explain what is transported in each case. Obtain, from kinetic theory of gases, an expression for the viscosity of a gas. How can the dependence of this viscosity on pressure and temperature be explained? **[C.U. (Hons.) '95, '99, 2005, 08]**

(b) Deduce the law of distribution of molecules, in an isothermal vertical column of a gas, assuming the molecules to be Brownian particles. Indicate briefly how Perrin used this result to determine the Avogadro number. **[C.U. (Hons.) '95, '98, 06 (old syll)]**

34. (a) Deduce Maxwell's velocity distribution law. Hence find the energy distribution law. Find an expression for the most probable speed of the molecules.

[C.U. (Hons.) '96, 2000, 02, 06 (old syll)]

(b) State the law of equipartition of energy. Calculate the total molecular translational and rotational energy of 1 gm. of hydrogen gas at 50°C and 100°C, assuming the molecules to be dumb-bell shaped. Will the specific heat at constant volume be the same at these two temperatures? **[C.U. (Hons.) 96, 03]**

(c) The coefficient of viscosity of a gas is 16.6×10^{-6}NS^{-1}m^{-1}. Calculate the mean free path, frequency of collisions and the diameter of the gas molecules. Given $\bar{v}$ = 450 mS^{-1}, $\rho = 1.25$ kg m^{-3} and number density is 2.7×10^{25} molecules m^{-3}.

[C.U. (Hons.) '96] [**Ans.** See worked out examples]

35. Colloidal particles are suspended in a liquid. Using Einstein's theory find the temperature dependence of the mean square displacement per unit time. What can you say about the mean displacement per unit time? **[C.U. (Hons.) 96, 2000, 02]**

An experiment with such particles suspended in a liquid of temperature 80°C, shows the mean squared displacement per unit time to be 1.5 cm^2/sec. Find the radius of a suspended particle. ($N = 6.023 \times 10^{23}$/mole, $\eta = 0.01$ poise)

[C.U. (Hons.) '96, 2000] [**Ans.** See worked out examples]

36. (a) Explain the conductivity of a gas on the basis of the kinetic theory. How is it related to the viscosity? Show that it is independent of the pressure of the gas.

[C.U. (Hons.) 2000, 02, 07]

(b) If 'u' be a cartesian component of the molecular velocity (say, the x component), show that the ratio of the square of the velocity of sound in an ideal gas to the mean square value of 'u' equals the ratio of the specific heat at constant pressure to the specific heat at constant volume. **[C.U. (Hons.) '97, cf. '08]**

[**Answer to part (b).** From $P = \frac{1}{3}\rho C_r^2$ (where P = pressure of the gas, ρ = density of the gas, C_r = r.m.s speed of the gas molecules), we may write

$$C_r = \sqrt{\frac{3P}{\rho}} = \sqrt{\frac{3}{\gamma}\cdot\frac{\gamma P}{\rho}} = \sqrt{\frac{3}{\gamma}}C \qquad \left[\begin{array}{l} C = \text{velocity of sound,} \\ \gamma = \dfrac{\text{specific heat at constant pressure}}{\text{specific heat at constant volume}} \end{array}\right]$$

i.e., $C_r^2 = \frac{3}{\gamma}C^2$ $\quad\cdots\quad$ (1) Now $C_r^2 = u^2 + v^2 + w^2 = 3u^2$ if $u = v = w$, u, v, w being the three cartesian components of the molecular velocity, so that from (1), we have $C^2/u^2 = \gamma$.

37. A shower of molecules, each originally moving with the same velocity, traverses a gas. Calculate the fraction of molecules which will be travelling undeflected even after travelling (i) 0.2 and (ii) 2 times the mean free path. **[C.U. (Hons.) '97]** [**Ans.** See worked out examples.]

38. (a) Give the kinetic interpretation of pressure and calculate the pressure exerted by a perfect gas. How is the pressure expected to vary with height in an isothermal vertical column of a gas? (Answer to last part: See Art. 12.3 and 14.20). **[C.U. (Hons.) 98, 03, 05]**

(b) Define degrees of freedom of a system. State the classical equipartition theorem.

(c) Apply the theorem and calculate the following:

i. total internal energy of one *gm*-mole of a perfect gas;

ii. molar specific heat of the gas at constant volume;

iii. a linear symmetric tri-atomic molecule as shown in the attached figure is in thermal equilibrium with similar molecules. Apart from translation and rotation, the end-atoms can undergo longitudinal vibrations. Calculate the average energy of such a molecule at temperature T. How does the average energy change when the vibrational modes are absent? **[C.U. (Hons.) '98]** [**Ans.** See worked out examples.]

39. (a) What is meant by speed distribution law? Show that the probability that a molecule may have speed lying between u and $u+du$ in a definite direction is given by $f(u)du = ae^{-bu^2}du$, a and b are constants for the system at a temperature T. $\quad\ldots\quad$ (1)

Plot $f(u)$ vs u. Find the mean square speed $\overline{u^2}$.

(b) A vessel contains n molecules of a perfect gas at temperature T, mass of each molecule is m. Given that $a = \sqrt{m/2\pi KT}$ and $b = m/2KT$, use equation (1) and show that the number of molecules having speed lying between c and $c + dc$ is given by

$$dn_c = n_c dc = 4\pi n a^3 e^{-bc^2} c^2 dc = 4\pi n \left(\frac{m}{2\pi KT}\right)^{3/2} e^{-\frac{mc^2}{2KT}} c^2 dc$$

[C.U. (Hons.) 98, 04, 07]

(c) Plot n_c vs c for two different temperatures T and $4T$ on the same graph and calculate the most probable speed c_m and the number of molecules nc_m having speed c_m. Indicate c_m on the graph.

(d) What is the significance of the total area under the curve and the c-axis? On the graph shade the area for speed limits of the molecules between $0.5c_m$ and $2c_m$.

(e) Estimate the percentage of the total molecules whose speeds lie between $c_m \pm 0.01c_m$.

[C.U. (Hons.) 98, 2004, cf. 07]

40. (a) One *gm*-mole of a perfect gas at N.T.P. occupies 22.5 litres. Use the above data and make a rough estimate of average separation between the molecules at N.T.P.

(b) Given that the diameter of a molecule $\simeq 3 \times 10^{-9}$cm, and assuming that the interaction between the molecules extends over 4 molecular lengths (assumption), explain from the result of (a), whether molecules at N.T.P. on the average, may be treated as non-interacting or not.

(c) Consider an isothermal vertical column of a gas composed of Brownian particles. Show that the number density of the particles varies with height (h) according to the equation $n = n_0 e^{-\frac{Mg}{RT}h}$ where n_0 is the number density at $h = 0$, M is the molecular weight. Other notations have usual significance. **[C.U. (Hons.) '98]**

41. Explain Boyle's law, Avogadro's hypothesis and Dalton's law of particle pressures, using kinetic interpretation of pressure. **[C.U. (Hons.) 2000]**

42. Prove that the number of molecules striking unit area of the container wall per unit time is equal to $\dfrac{n\bar{c}}{4}$, where n is the number of molecules per unit volume and by definition $\bar{c} = \dfrac{1}{n}\displaystyle\int_0^\infty c dn_c$.

[C.U. (Hons.) 2008]

[**Ans.** To compute the number of molecules of a gas that strike (or cross) a surface per unit area per unit time, we proceed as follows.

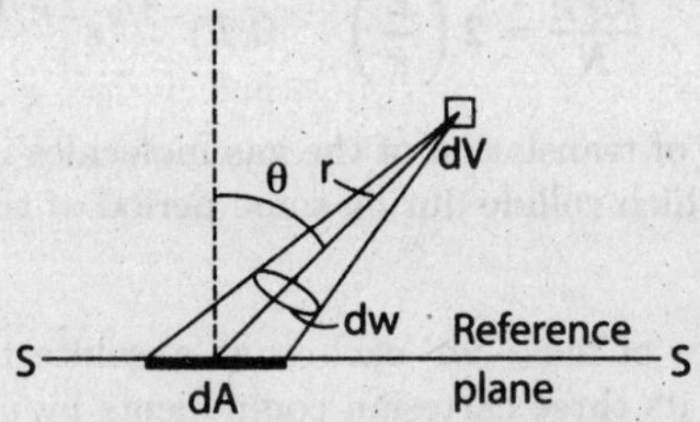

Let SS be the surface of one of the walls of the container, dA being an elemental area of it. We consider dV, an elemental volume at (r, θ, ϕ). The number of molecules in $dV = dn_c dV$, where dn_c is the number density of molecules with velocity c and $c + dc$. Now, how many of these molecules will move towards dA is determined by the solid angle $d\omega$ subtended by dA at dV. This number is

$$\frac{dn_c dV}{4\pi} d\omega = \frac{dn_c dV}{4\pi}\frac{dA\cos\theta}{r^2} = \frac{dn_c}{4\pi}(r^2 \sin\theta d\theta d\phi dr)\frac{dA\cos\theta}{r^2}$$

$$= \frac{dn_c dA \sin\theta\cos\theta d\theta d\phi dr}{4\pi}$$

$\because d\omega = dA\cos\theta/r^2$ and $dV = r^2 \sin\theta d\theta d\phi dr$ = volume element in spherical polar coordinates.

Again to find out how many of these will strike dA in time dt, coming from all possible directions and all possible distances, we have to integrate the above expression for different variables using proper limits, and the result is

$$dN = \frac{dA}{4\pi}\int_0^{\pi/2} \sin\theta\cos\theta d\theta \int_0^{2\pi} d\phi \int_0^{cdt} dr \int_0^\infty dn_c$$

(molecules which would have to cover a distance $> cdt$ cannot strike dA in time dt)

$$= \frac{dAdt}{4\pi} \times \frac{1}{2} \times 2\pi \times \int_0^\infty cdn_c = \frac{1}{4}dAdt\int_0^\infty cdn_c = \frac{n\bar{c}}{4}dAdt$$

$$\left[\because \bar{c} = \text{average velocity} = \frac{n_1c_1 + n_2c_2 + \cdots}{n_1 + n_2 + \cdots} = \frac{\Sigma n_i c_i}{\Sigma n_i} = \frac{\int_0 cdn_c}{n}\right.$$

where per unit volume, there are n_1 molecules moving with velocity c_1, n_2 with c_2 and etc.

$$\therefore \int_0^\infty c dn_c = n\bar{c}$$

So the number of molecules striking unit area of a surface per unit time is $\frac{1}{4}n\bar{c}$, where $\bar{c}$ is the average velocity.

43. Show that the number of molecules striking unit area of a wall per unit time is equal to $\dfrac{P}{\sqrt{2\pi mkT}}$, P being the pressure and m is the mass of a molecule, k is the Boltzmann constant and T the absolute temperature.

[**Ans.** The number molecules striking unit area per second is $dN = \frac{1}{4}n\bar{c}$ where $\bar{c} =$ the mean velocity.

$$\therefore \quad dN = \frac{1}{4}n\sqrt{\frac{8kT}{\pi m}} \qquad \because \bar{c} = \sqrt{\frac{8kT}{\pi m}}$$

$$= \frac{1}{4}\left(\frac{p}{kT}\right)\sqrt{\frac{8kT}{\pi m}} \qquad \because p = nkT$$

$$= p/\sqrt{2\pi mkT}$$

44. Show that the fraction of the number of molecules whose kinetic energy lies in the range E and $E + dE$ is

$$\frac{dN_E}{N} = 2\left(\frac{E}{\pi}\right)^{1/2}(kT)^{-3/2}e^{-E/kT}dE.$$

45. Show that the mean K.E. of translation of the gas molecules in a given volume is three-quarters of that of the molecules which collide during some period of time with the wall of the containing vessel.

46. A perfect gas of molecules of mass 'm' each is at absolute temperature T. The velocity of a molecule is denoted by $\vec{v}$, its three cartesian components by v_x, v_y and v_z and its speed by v.

(a) Find an expression for the average number of molecules per unit volume with x component of velocity lying between v_x and $v_x + dv_x$ irrespective of the values of the other two components. State the assumptions you have made.

(b) Write down the expression for the average number of molecules per unit volume when x component of velocity lies between v_x and $v_x + dv_x$ and y component of velocity lies between v_y and $v_y + dv_y$ irrespective of the value of v_z. **[C.U. (Hons.) 2001, 05]**

(c) Show by a rough sketch the variation of average number of molecules with v_x for the case (a) above. What is the most probable value of v_x in case (a)?

(d) What are the following mean values in case (a)? $\bar{v}_x, \overline{v_x^2}$. **[C.U. (Hons.) 2001]**

47. (a) What do you mean by the law of equipartition of energy?

Consider an ideal gas of diatomic molecules where each molecule is like two mass points connected by a massless spring. The spring can only have longitudinal (along the line joining the two atoms) vibration. The gas is at absolute temperature T. Utilize the law of equipartition of energy to obtain an expression for the total energy per mole.

[C.U. (Hons.) 2001]

(b) i. Derive an expression connecting the pressure of an ideal gas with its density and the mean square speed of its molecules. Show that this expression is consistent with the equation of state of an ideal gas of monatomic molecules. **[C.U. (Hons.) 01, 05]**

ii. Write down an expression for the total kinetic energy (K.E.) per unit volume in an ideal monatomic gas at absolute temperature T. How does this K.E. vary with temperature if pressure is maintained constant? **[C.U. (Hons.) 2001]**

48. Obtain the distribution law of molecular speed for a two dimensional gas. For answer see Art. 2.24 of Gupta-Roy's "Thermal Physics". **[C.U. (Hons.) 2002]**

49. (a) State and explain how thermal conductivity of an ideal gas depends on the density of the gas at a fixed temperature.

(b) Ignoring the difference between r.m.s. speed and average speed, show that the ratio of the pressure to the viscosity coefficient gives approximately the number of collisions per unit time for a molecule of a gas (Given r.m.s. speed $= \sqrt{\frac{3kT}{m}}$).

Hints. $P = \frac{1}{3}\rho\overline{c^2}$ and $\eta = \frac{1}{3}\rho\bar{c}\lambda$.

$$\therefore \quad \frac{P}{\eta} = \frac{\overline{c^2}}{\bar{c}\lambda} = \frac{\bar{c}}{\lambda} = n$$

the collision frequency taking $\sqrt{\overline{c^2}} = \bar{c}$ according to the question.

(c) What is transported in diffusion process? Define the diffusion coefficient.

[C.U. (Hons.) 2002]

50. A shower of molecules, each originally moving with the same velocity, traverses a gas.

(a) Calculate the fraction of molecules which will be travelling undeflected after traversing 0.693 times the mean free path.

(b) Calculate the distance after which only one percent of the incident molecules remain undeflected. **[C.U. (Hons.) 2003]**

[**Ans.** See worked out problems]

51. (a) Explain what meant by the *mean free path* (λ) of molecules in a gas. Show that the probability of a gas molecule traversing a distance x without collision is $e^{-x/\lambda}$.

[C.U. (Hons.) 04, 06 (old syll), 07]

(b) The pressure within a cathode ray tube is 10^{-8} atmospheric pressure.

The mean free path of air molecules at atmospheric pressure and at room temperature (27°C) is 10^{-5} cm. Estimate the mean free path of air molecules within the cathode ray tube at room temperature. **[C.U. (Hons.) 04]**

52. What do you mean by self-diffusion? Define the coefficient of self-diffusion D.

[C.U. (Hons.) 04, 06]

In an ideal gas, there exists a concentration gradient only along the z-direction (i.e. $\frac{dn}{dz} \neq 0$, $\frac{dn}{dx} = 0$, $\frac{dn}{dy} = 0$; $n \rightarrow$ number of molecules per unit volume). Using kinetic theory of gas, find out an expression for the coefficient of self-diffusion D. Show the nature of dependence of D on temperature. **[C.U. (Hons.) 2004]**

53. State the law of equipartition of energy. Apply this law to find an expression for the total energy of one gm-mole of a monoatomic ideal gas at absolute temperature T. Find also the expression for the molar specific heat at constant volume. **[C.U. (Hons.) 2005]**

54. What do you mean by the mean free path (λ) of molecules in a gas? The mean free path in a gas is 5.0 cm. Among 10,000 free-paths of those molecules, how many are between 4.9 cm and 5.1 cm? **[C.U. (Hons.) 2006, cf. 2007]**

Hints. Use the formula $\frac{N}{N_0} = e^{-x/\lambda}$.

55. (a) Define degrees of freedom (f) of a gas molecule. Show that for an ideal gas

$$\gamma = \frac{C_p}{C_v} = 1 + \frac{2}{f}$$

where symbols have usual meanings. **[C.U. (Hons.) 06, 08]**

(b) 2 moles of an ideal gas, being initially at a temperature 273°K were isothermally expanded to 4 times its initial volume. Then the gas was heated at constant volume to initial pressure. The total amount of heat absorbed by the gas during the entire process is 40.3 kJ. Find the degrees of freedom of the gas molecule. ($R = 8.3$ J.mol^{-1}.K^{-1}). **[C.U. (Hons.) 2006]**

Hints. Total heat absorbed $= C_v dt + pdV = C_v dT + RT \ln \frac{V_2}{V_1}$.

Assume the 2 moles of the ideal gas at STP. Then apply the above relation to calculate C_V—then find C_p and hence f.

56. (a) Write down the expression for pressure (P) exerted by a gas according to kinetic theory and show that $\overline{u^2} = \frac{P}{\rho}$, where u is the x-component of the molecular speed and ρ is the density of the gas. **[C.U. (Hons.) 2007]**

(b) Show that the law of equipartition of energy demands that the motion of relatively large Brownian particles are practically unnoticeable. **[C.U. (Hons.) 2007]**

[**Ans.** To understand why the mean thermal speed of a relatively large Brownian particle is not an observable quantity we see that the mean translational K.E. of the particle, considered as a giant molecule in equilibrium with fluid molecules, is $\frac{3}{2}$ kT. For a typical particle of dimensions $\sim 10^5$ cm, the computed speed is ~ 1 cm/sec. The observed r.m.s. displacement in a 30 sec interval is $\sim 10^{-3}$ cm in a liquid and $\sim 10^{-2}$ cm in a gas and are thus several orders small compared with the thermal speeds.]

57. A thermally insulated box is separated into two compartments (of volumes V_1 and V_2 respectively) by a membrane. The first compartment contains an ideal gas at temperature T, while the second one is empty (i.e. in vacuum). The membrane is suddenly removed and the gas fills both the compartments and reaches equilibrium. What is the final temperature of the gas?

[C.U. (Hons.) 2007]

(Try yourself)

58. (a) $f(u)du$ is the probability that a gas molecule may have speed lying between u and $u + du$ along x-direction. Plot the nature of variation of $f(u)$ with u.

(b) State Dulong and Petit's law. Prove it from the law of equipartition of energy.

[C.U. (Hons.) 2008]

59. (a) Calculate the number of molecules striking unit area of a container-wall in unit time. (Molecule-molecule collision may be ignored)

[**Ans.** See answer to Q.42 above.]

(b) Establish Avagadro's hypothesis starting from the kinetic molecular theory for gases.

(c) The density of H_2 gas in a container under 76 cm of Hg pressure is 9×10^{-5} gm/ml. Calculate the r.m.s. velocity of the H_2 molecules within the container. (Density of Hg = 13.6 gm/ml and $g = 9.8$ m/s^2). **[C.U. (Hons.) 2008]**

(Use the relevant formula—see worked out examples)

Chapter 15

Exhaust Pumps and Gauges

15.1 Production and measurement of low pressures

High vacuum technique plays a very important role in modern development of science. Its use in the manufacture of electronic valves, X-ray equipments and in certain branches of chemical industry is well known. In high vacuum pumping arrangement, pressure is first reduced to 1 mm. of mercury by means of a backing pump such as rotary oil pumps and then the pressure may further be reduced to a very low value from 10^{-4} mm. to 10^{-7} mm. by means of molecular or diffusion-condensation pumps.

The millimetre of mercury is generally used as the unit of pressure in high vacuum technique. A pressure of 1 mm. of mercury is called 1 torr after the name of Torricelli. It has been possible to reduce the pressure lower than 10^{-13} torr by modern technique.

15.2 The Speed of a pump

The speed of a pump at a given pressure is defined as the volume of the gas exhausted per second. Hence $S = -\frac{dV}{dt}$, where S is the speed of the pump and $\frac{dV}{dt}$ is the rate of change in volume of the gas in an enclosure at any instant. Assuming constant temperature we have from Boyle's law.

$$(P - P_0)V = (V + dV)[(P - P_0) - dP]$$

where P_0 = the limiting pressure,

P = pressure of the gas at any instant in the vessel to be evacuated,

V = volume of the vessel.

The above equation may be written in the form

$$(P - P_0)dV = VdP \quad \text{[neglecting the product } dPdV \text{ which is very small]}$$

$$\text{or,} \quad dP = (P - P_0)\frac{dV}{V} \quad \text{or,} \quad \frac{dP}{dt} = \frac{(P - P_0)}{V}\frac{dV}{dt} = -\frac{S(P - P_0)}{V}$$

$$\text{or,} \quad \frac{S}{V}\int_{t_1}^{t_2} dt = -\int_{P_1}^{P_2} \frac{dP}{P - P_0},$$

where P_1 and P_2 are the respective pressures at the instants t_1 and t_2.

$$\therefore \quad \frac{S}{V}(t_2 - t_1) = \log_e \frac{P_1 - P_0}{P_2 - P_0} = 2.303 \log_{10} \frac{P_1 - P_0}{P_2 - P_0}$$

$$\text{or,} \quad S = 2.303 \frac{V}{(t_2 - t_1)} \log_{10} \frac{P_1 - P_0}{P_2 - P_0}. \tag{15.2.1}$$

If the limiting pressure P_0 is very low, we have from the above equation

$$S = 2.303 \frac{V}{(t_2 - t_1)} \log_{10} \frac{P_1}{P_2} \tag{15.2.2}$$

This is known as Gaede's equation. From the above expression, we have

$$\log_e \frac{P_1}{P_2} = \frac{S(t_2 - t_1)}{V} = \frac{S}{V}t, \text{ where } t_2 - t_1 = t. \quad \therefore \quad \frac{P_2}{P_1} = e^{-\frac{S}{V}t}. \tag{15.2.3}$$

If $t = \frac{V}{S}$, then

$$\frac{P_2}{P_1} = \frac{1}{e} = \frac{1}{2.718} \simeq 0.37, \quad \text{or,} \quad P_2 = 0.37 P_1,$$

i.e. during the time $t = V/S$, the pressure reduces to 0.37 times the initial pressure.

15.3 Conductance of a tube

If Q be the volumetric flow rate of a gas through a tube due to the pressure difference δP, then the conductance of the tube is given by

$$C = \frac{Q}{\delta P} \quad \left[Q = -p\frac{dV}{dt}\right].$$

The above expression may also be written in the form

$$Q = \frac{\delta P}{1/C}.$$

If we compare the pressure difference with potential difference and volumetric flow rate with current in an electric circuit, then $\frac{1}{C}$ is equivalent to electric resistance. As in an electric circuit, here also if the tubes are arranged in series and in parallel we may write

$$\frac{1}{C} = \frac{1}{C_1} + \frac{1}{C_2} + \frac{1}{C_3} + \cdots \quad \text{and} \quad C = C_1 + C_2 + C_3 + \cdots$$

The conductance of a tube depends on its diameter d. Actually $C \propto d^3$.

As the dimensions of the speed S of a pump and C are same, we can write the same series and parallel equations in both the cases.

So if a pump having speed S be connected to a vessel through a tube having conductance C, then the effective speed of the pump is given by $\frac{1}{S'} = \frac{1}{S} + \frac{1}{C}$.

From the above equation it follows that the effective speed of the pump or the effective rate of pumping is large when the conductance of the connecting tube is large. Hence, the connecting tube should be of large diameter in order that the pumping rate is increased.

15.4 Water-jet pumps

These pumps are used in many laboratories for removing air from a vessel. Originally water-jet pump was made of glass and so got easily fractured and hence the modern water-jet pumps are made of metal.

In this pump water from a constant pressure supply enters the pump at A. A thick wire B is inserted along the path of the water stream to make the flow of water turbulent. The side tube T is connected to the vessel to be exhausted. The exit jet E and the 'choke tube' C are connected to the vessel to be evacuated through the side tube T (Fig. 15.4.1).

Small air bubbles are trapped by the water-jet as it passes from the exit jet to the choke tube and are finally carried away by the water stream. The pumping speed is about 20 c.c. per second and the lowest pressure of the vessel exhausted is about 1 cm. of mercury.

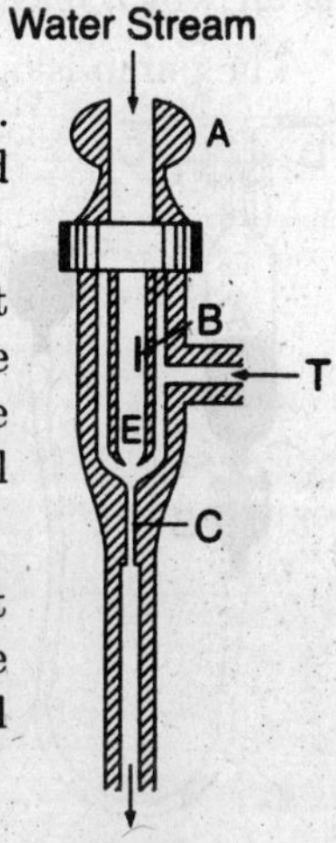

Fig 15.4.1

15.5 Sprengel pump

This pump is used in conjunction with water pump. The pump is shown in fig. (15.5.1). AB and CD are capillary tubes of diameters 0.15 cm. and 0.5 cm. respectively. The two capillary tubes widen out into bulbs nearly 2 cms. in diameters near B and C. The vessel to be evacuated is connected to the side tube T. Pellets of mercury fall from the jet J and entrap bubbles of gas which are carried away by the pellets through the tube CD into the trough G. The supply of mercury in J is maintained from the reservoir R. The reservoir is in communication with a water pump. The capillary tube AB passes through the reservoir R and dips into the trough containing mercury. The lower end of the capillary tube AB is connected to a side tube K. When the water pump operates, air is drawn through the side tube K and carries pellets of mercury with it. When air with mercury pellets reaches the upper end of the capillary tube AB, mercury is transferred to the reservoir and air passes to the water pump. Thus the level of mercury in the reservoir is maintained. The rate of flow of air through the side tube K is controlled by a clip attached to K.

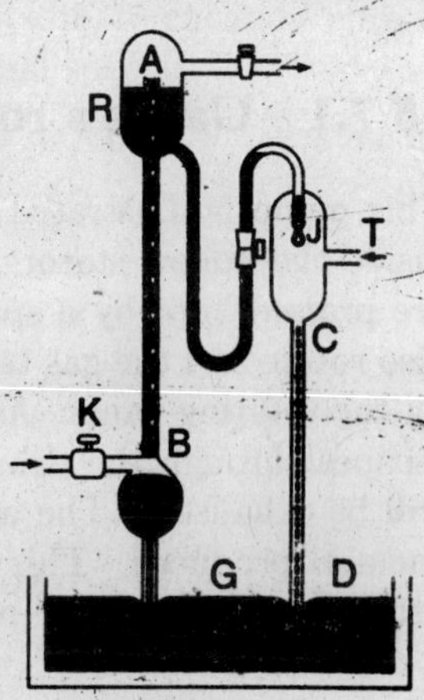

Fig 15.5.1

15.6 Geissler pump

Geissler pump is used when a lower pressure is required. A Geissler pump which is often called a Toepler pump is shown in fig. (15.6.1). 'A' denotes a cylindrical barrel of capacity 200 c.c. having a two-way capillary tap T at its upper end. The barrel may be put into communication with the vessel to be exhausted through the tube C by adjusting the tap T. The tap may also be adjusted to connect the barrel to the tube D which is open to atmosphere. B is a smaller barrel at the lower end of the barrel A having a side tap K. The barrel B is connected to the mercury reservoir R by means of a rubber tube. The function of the side tube K is to remove the air which may enter the apparatus and get entrapped in B.

The pump is operated by raising the reservoir with the tap T so adjusted that the barrel A is connected to the atmosphere through the tube D.

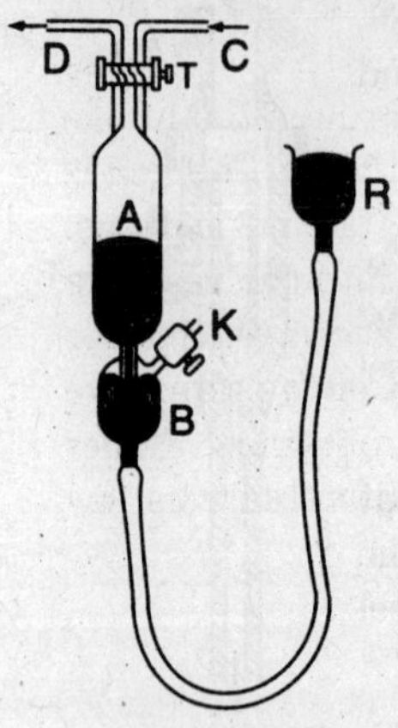

Fig 15.6.1

When mercury will fill the barrel completely, the tube T is closed. Again the position of the reservoir is adjusted and the tap is turned so that the barrel A is put in communication with the vessel to be exhausted through the tube C. The pressure of the gas in the vessel together with the gas in the tube C forces down the mercury in A. The reservoir R is lowered until the gas from the vessel fills the barrel A. The reservoir is next raised with the tap T closed until the pressure of the gas in A exceeds the atmospheric pressure. At this stage the tap is adjusted until A is connected to the atmosphere. Thus the gas may be removed from the barrel A. By repeating this process several times, the vessel may be exhausted. It is to be noted that when the space to be exhausted is greater than 200 c.c., Geissler pump should be used.

15.7 Rotary oil pumps

There are two types of rotary oil pumps, viz., (a) Gaede's rotary oil pump, (b) Cenco rotary oil pump.

15.7.1 Gaede's rotary oil pump

This pump is illustrated in fig. (15.7.1). The cylindrical solid rotor C revolves eccentrically inside the hollow stator S. A slot is cut across the rotor diametrically and two vanes V, V' are pressed hard by a spring on the inner wall of the stator. As the rotor rotates, the vanes also rotate and the gas taken inside the stator through the intake I is trapped and compressed by the rotating vanes. As the pressure of the compressed gas is sufficiently increased, the gas escapes through the exhaust valve E. Thus the vessel to which the inlet tube is connected will be exhausted. The whole pump is immersed in oil to prevent leakage of gas into the high vacuum produced. This pump is used to exhaust a vessel of large capacity and produces a pressure as low as 10^{-3} mm. of mercury.

Fig 15.7.1

Fig 15.7.2

15.7.2 Cenco rotary oil pump

A 'Cenco' pump is shown in fig. (15.7.2). In this pump a steel rotor R is mounted eccentrical to a shaft and rotates about the shaft inside the steel stator S. A vane V operated by a spring arm S' presses against the rotor and slides upon it when the rotor rotates. With the rotation of the rotor the gas from the vesel to be exhausted enters the stator through the intake I. The gas inside the stator is compressed by the rotation of the rotor and finally the gas is expelled through the exhaust valve E. The whole pump is immersed in oil to prevent leakage. A special type of valve is fitted to the intake tube to prevent oil from getting into the vessel to be exhausted when the pump is stopped. The maximum speed of the pump is about 6 litres per minute and the vacuum obtained is about 10^{-3} mm. of mercury. When two Cenco pumps are mounted in series on a common shaft driven by a motor, a lower pressure is obtained. Other types of this pump commercially known as Megavac' and 'Hypervac' produce pressure as low as 10^{-4} mm. of mercury.

15.8 Molecular pump

A molecular pump works on the principle that a dragging force is exerted on the molecules of a gas by a rotating surface close to a fixed one. In fig. 15.8.1 a Gaede molecular pump is shown.

R denotes a drum rotating inside an outer cylinder S. The wall of the rotating surface is close to that of the fixed surface, the space or the clearance between the two surfaces being 0.03 mm. The interspace in the region between the inlet tube I and the outlet tube O is, however, comparatively large. As the pressure is low which is attained by a backing pump, the mean free path of the molecules is large in comparison with the interspace between the surfaces and so the gas molecules make more collisions with the wall of the rotating cylinder than among themselves. The direction of rotation of the rotating cylinder is such that the gas molecules coming in contact with the wall of the rotor are dragged from the vessel to be exhausted with the speed of the rotor and transferred into the backing pump. The difference of pressures at the inlet and outlet tubes is proportional to the angular velocity of the rotor. The final pressure depends on fore-vacuum pressure and the angular velocity of the rotor. With fore-vacuum of 1 mm. and speed of rotation 12,000 revolutions per minute, the final pressure obtained is about 10^{-6} mm.

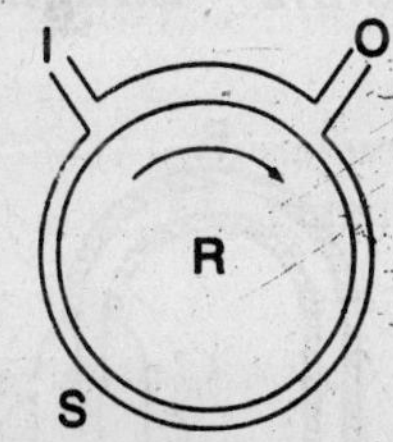

Fig 15.8.1

15.9 Diffusion-Condensation pumps

The diffusion pump is used for producing high vacuum. As diffusion and condensation play an important role in operating the pump, it is also called the condensation pump.

15.9.1 Gaede's diffusion pump

To understand the principle of action of this pump we consider the fig. (15.9.1).

Through the glass tube AB a stream of mercury, a readily condensable vapour, is passing. The vessel V connected to AB is called the "vapour trap". The porous plug P divides the space between AB and the vessel V. The vessel V is connected to the apparatus to be evacuated which is already at a low pressure.

The mercury vapour stream carries along with it any gas present in the tube AB and thereby the gas pressure in the tube is reduced much. Then the gas from the apparatus to be evacuated diffuses into the mercury vapour through the porous plug. The diffusion continues so long as the pressure of air at C is greater than the partial pressure in AB. Mercury vapour diffusing towards C through the porous plug is condensed in the vapour trap V by liquid air. This causes reduction of gas pressure in the vessel to be evacuated.

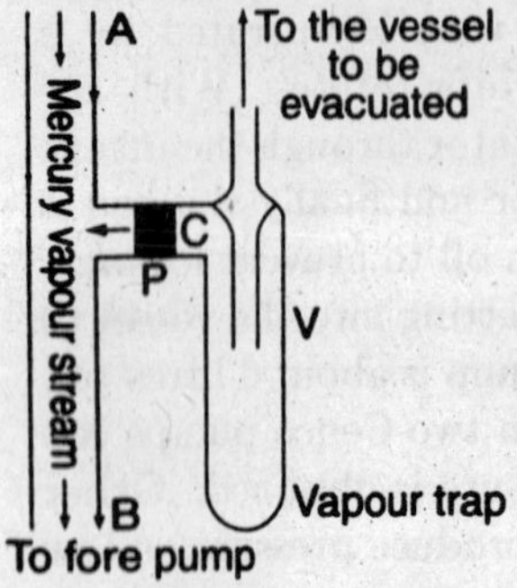

Fig 15.9.1

In an actual experiment it is unnecessary to have a porous plug in lieu of which a very narrow cylindrical slit is used. Gaede showed that for a pump of this type to attain maximum efficiency, the width of the slit must be of the same order of magnitude as the mean free path of the molecules of the gas. Vacua of less than 10^{-6} mm. may be obtained with the help of this pump. Although there is no limit to the vacuum attained by a diffusion pump, in practice the following disadvantages arise: (i) slow exhaust speed; (ii) regulation of the temperature of mercury vapour is required.

The pump works after the fore-vacuum is obtained with the help of a rotary pump.

15.9.2 Langmuir's pump

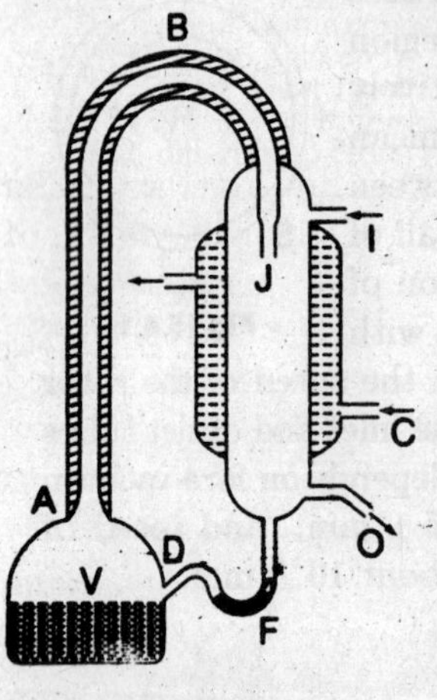

Fig 15.9.2

Langmuir eliminated the above disadvantages. In Langmuir pump there is arrangement for cooling the mercury vapour thoroughly at the stream so that back diffusion of mercury vapour towards C is completely eliminated. Moreover in this apparatus there is no restriction over the size of the slit used for condensation purpose.

A Langmuir's pump is shown in fig. 15.9.2. The vessel V contains some mercury. When it is boiled its vapour rises along the bent tube AB and passes to a wide tube C through a jet J. The tube AB is surrounded by asbestos so that the mercurry may not condense before reaching the jet. The tube C is also surrounded by a jacket through which cold water is circulated so that mercury vapour in the tube C may condense. The condensed mercury vapour is collected in the bend F and returns finally to the vessel V through the tube D. As the mercury vapour collects in F, it cannot pass directly from the vessel V to the tube C. The vessel to be evacuated is connected to the inlet tube I. The outlet tube O is connected to an auxiliary rotary pump. When the mercury vapour stream passes out through the jet, the gas coming from the vessel to be evacuated through the inlet tube diffuses through the mercury vapour. So long as the partial pressure of gas near the tube I is greater than that near the jet, the process of diffusion continues. The diffused gas in the tube C is driven out by condensed vapour and by the operation of the auxiliary pump the gas is removed through the outlet tube O. The diffusion of mercury towards the tube I is prevented by making the aperture of the tube I very small.

The pump is very efficient. It reduces a pressure of 1 mm to 10^{-5} mm in 80 sec. with a speed of nearly 4 litres/sec.

Methods other than pumping are used to bring down the vacuum below 10^{-6} mm. The vessel to be evacuated is connected to a tube surrounded by liquid air and containing recently heated cocoanut charcoal. The gas is adsorbed by charcoal and a pressure of the order of 10^{-7} mm. may be obtained. This process is known as *Sorption process.*

Gases such as nitrogen, carbondioxide, ammonia and hydrogen may be removed by this process.

Hydrogen may also be eliminated by palladium or platinum black. Oxygen is also adsorbed by tungsten. The adsorption of nitrogen by charcoal is called physical adsorption and the adsorption of oxygen by tungsten is known as chemical adsorption. The substances which can adsorb gases by chemical adsorption or by any chemical process are called Getter.

15.10 Pressure Gauges

A simple manometer cannot measure low pressures accurately. In all high vacuum-work, pressure gauges are used for the measurement of low pressures. The following gauges are used for the measurement of pressures below 10^{-1} mm.

15.10.1 McLeod gauge

McLeod gauge gives the absolute value of the pressure of a gas. The principle of the gauge is to isolate a known volume of the gas, the pressure of which is to be determined and then it is compressed to a small known volume at which the pressure is measured. Applying Boyle's law the initial pressure of the gas isolated is determined. The apparatus is shown in the fig. 15.10.1.

It consists of a closed graduated capillary tube A sealed to a bulb D at its lower end. The volume of the bulb depends on the pressure to be measured. B is another capillary tube of same diameter to eliminate the error due to surface tension. C, a wider tube joined to B and connected below the bulb D communicates to the high vacuum chamber. The mercury reservoir R is connected to the lower glass tube E of the gauge by a rubber tubing. A scale is fixed behind the tubes A and B. Lowering the reservoir when the mercury level falls below E, the pressure of the gas in the bulb D is made equal to the pressure in the chamber which is to be measured. Let this pressure be P and the volume of the gas in the bulb D and the capillary tube be V. Now the reservoir is raised until the trapped gas is compressed to a very small volume in the capillary tube A. Let this volume be v and the corresponding pressue be p. Then by Boyle's law, we have,

$$PV = pv \quad \text{or} \quad P = \frac{pv}{V}.$$

Now, $p = P + h$, where P is the unknown pressure expressed in mms. of mercury and h, the difference in mercury levels in the tubes A and B expressed in mms. Also $v = l\alpha$, where l is the length of the volume of the compressed gas trapped in the capillary tube A and α is its cross-section. Hence

$$P = \frac{(P+h)l\alpha}{V} \quad \text{or} \quad P = \frac{hl\alpha}{V} \quad [\because P \ll h]. \qquad (15.10.1)$$

Thus the unknown low pressure P is determined.

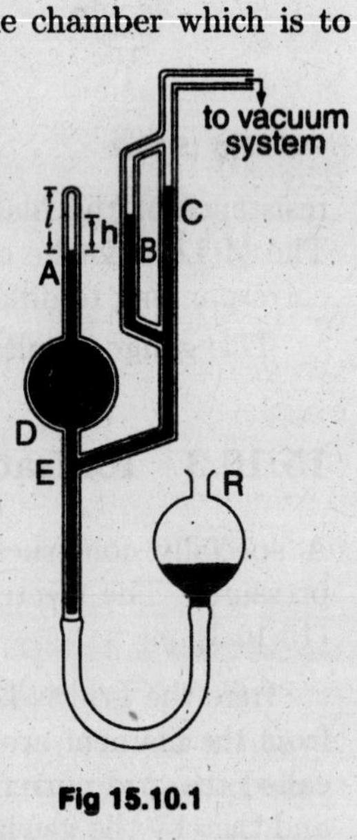

Fig 15.10.1

An alternative method of using McLeod gauge is that the mercury in the tube B is brought to such a position that it is in level with the top of the tube A. Let H be the length of the volume of the compressed gas expressed in mms. in the capillary tube under this condition. We have, then, from Boyle's law

$$PV = (p+H)H\alpha = H^2\alpha \quad [\because\ p \ll H] \quad \text{or} \quad P = \frac{\alpha}{V}H^2. \qquad (15.10.2)$$

From the expression (15.10.1) it follows that we have a non-uniform scale of pressure. The scale is placed behind the tubes A and B. In this case a quadratic scale with zero mark at the upper end is placed behind the tubes. The great advantage of this method is that the scale opens out for lower pressures and gets crowded for comparatively high pressures.

This gauge is used to measure a pressure of the order of 10^{-5} mm.

15.10.2 Pirani gauge

It is a confirmed experimental fact that, at low pressures below 0.1 mm. of mercury, when the mean free path of molecules becomes comparable with the dimension of the containing vessel, the rate of loss of heat by conduction through a gas is proportional to the pressure, i.e. the rate $= \gamma P$, where γ is the constant of proportionality.

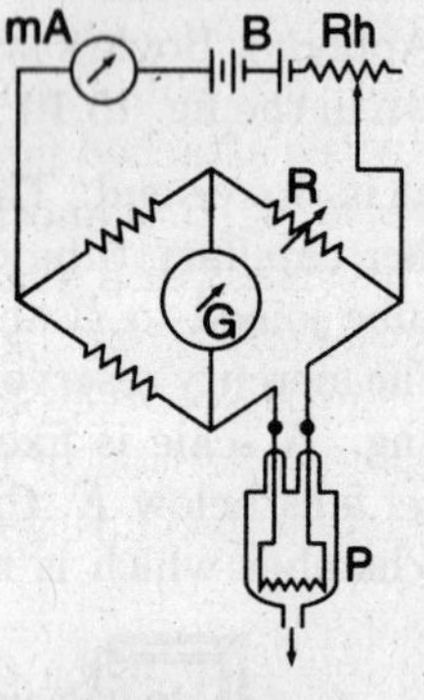

Fig 15.10.2

In Pirani gauge a heated metallic filament is fitted in a small vessel connected to a pump system and a McLeod gauge. The metallic filament in the gauge is so chosen that it must have a high temperature coefficient of resistance. The gauge is P placed in one arm of a Wheatstone's Bridge's (fig. 15.10.2). The resistances in the three arms of the bridge are made up of manganin of low temperature coefficient of resistance. R is a variable resistance, the value of which is so adjusted that there is no deflection of the galvanometer G. Current is supplied by a battery of cells and is adjusted by means of a rheostat. A milliammeter in series with the rheostat records the heating current. When the pump works, the pressure of the gas within the vessel containing the filament decreases as a result of which the heat conduction from the filament of the gauge decreases. Hence the temperature and so the resitance increases. The value of the resistance of the filament at a certain pressure is determined by the Wheatstone's Bridge. The McLeod gauge connected to the Pirani gauge is then calibrated by measuring pressures corresponding to different values of the resitances.

This gauge is suitable for measuring pressures in the range 10^{-2} mm. – 10^{-4} mm.

15.10.3 Ionisation gauge

A specially constructed triode is used as an ionisation gauge for the measurement of low pressure. The electrical circuit required for the operation of the guage is shown in figure (15.10.3).

Here the grid is kept at a positive potential with respect to the filament and so electrons from the filament are attracted towards the grid. The accelerated electrons constitute what is called the grid current. Some of the attracted electrons pass through the meshes of the grid and thereby the gas in the space between the plate and the grid is ionised. The plate P is kept at a negative potential and so the positive ions will be attracted to the plate, but the electrons in the space between G and P will be repelled by the plate. The current due to positive ions is recorded by a sensitive micro-ammeter. Below a pressure of 10^{-4} mm. the ionisation current is proportional to the pressure of the gas in the exhaust system. This gauge is not an absolute gauge. It is calibrated for series of pressures. It is to be noted that if the tube connected to the exhaust system be narrow, then the guage should be kept close to the exhaust system.

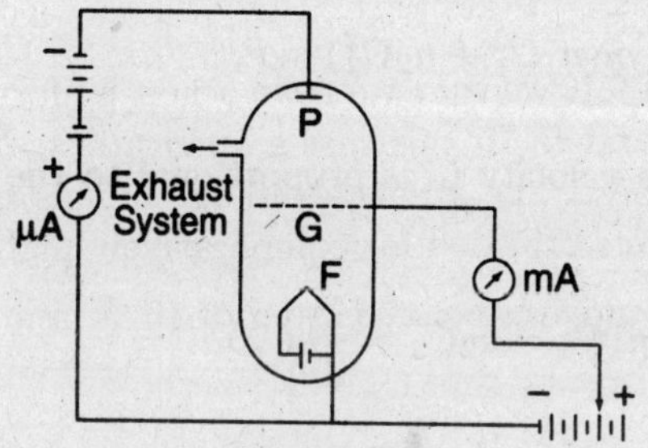

Fig 15.10.3

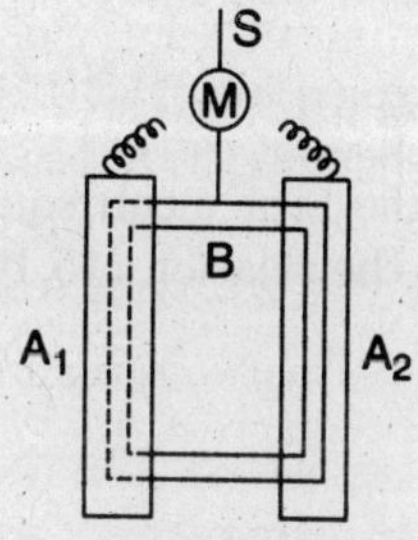

Fig 15.10.4

15.10.4 Knudsen gauge

This guage can be used to measure pressures ranging from 10^{-3} mm. to 10^{-7} mm. and it is called an absolute manometer as it gives pressure directly in dynes. Thus, no calibration of this gauge is required. The principle of Knudsen guage can be explained by the attached figure (15.10.4). In the figure A_1, A_2 are two fixed plates which are electrically heated. B is another plate kept at a temperature lower than that of A_1, A_2 and is suspended by means of a quartz fibre S which carries a mirror M. When it is suspended in an enclosure containing rarefied gas, the impingement of gas molecules from the sides of the fixed plates A_1 and A_2 on the plate B causes the plate B to be repelled away from the plates A_1 and A_2. The deflection of the plate B is measured by the usual lamp and scale arrangement. The deflection depends upon the pressure of the gas in the enclosure.

We shall now consider the theory of the instrument which is applicable only when the distance between the plates A_1, A_2 and B is small in comparison with the mean-free path of the gas molecules. Let T_1, T_2 be the temperatures of the fixed plates A_1, A_2 and the movable plate$B$ respectively, and let the mean velocities and the root mean square velocities of the gas molecules corresponding to temperatures T_1 and T_2 be $\bar{c}_1, \bar{c}_2$ and C_1, C_2 respectively. Let n_1 be the number of molecules per c.c. moving towards B and n_2 be the number of molecules per c.c. moving towards A_1, A_2. As there is no accumulation of the molecules between the plates, the number of molecules per c.c. moving in one direction must be equal to that moving in the opposite direction. Hence we may write

$$\frac{1}{2}n_1\bar{c}_1 = \frac{1}{2}n_2\bar{c}_2 = \frac{1}{4}n\bar{c} \tag{15.10.3}$$

where n is the molecular density of the gas molecules outside the plates and $\bar{c}$ is their average velocity.

Now, the pressure exerted by the molecules from A_1, A_2 on B is $\frac{1}{3}n_1mC_1^2$, where m is the mass of each molecule. Similarly, the pressure exerted by molecules from B on A_1, A_2 is $\frac{1}{3}n_2mC_2^2$. Hence the total pressure due to the gas molecules in the space between the plates causing the plate B to be repelled is

$$\frac{1}{3}m(n_1C_1^2 + n_2C_2^2).$$

Also, the movable plate B is forced towards A_1, A_2 by the pressure P due to the molecules outside the plates. Hence the resultant pressure repelling the plate B is

$$\frac{1}{3}m(n_1C_1^2 + n_2C_2^2) - P. \tag{15.10.4}$$

Now, since the root mean square velocity C is proportional to the average velocity $\bar{c}$, we may write from the relation (15.10.3)

$$n_1C_1 = n_2C_2 = \frac{1}{2}nC.$$

Hence from (15.10.4)the resultant pressure on B is given by

$$F = \frac{1}{3}m \cdot \frac{1}{2}nC(C_1 + C_2) - P = \frac{1}{2} \cdot \frac{1}{3}mnC^2\left(\frac{C_1 + C_2}{C}\right) - P = \frac{P}{2}\left(\frac{C_1 + C_2}{C}\right) - P.$$

If the temperature of the gas is kept at that of the moving plate B, then $C = C_2$.

$$\therefore \quad F = \frac{P}{2}\left(1 + \frac{C_1}{C_2}\right) - P = \frac{P}{2}\left(1 + \sqrt{\frac{T_1}{T_2}}\right) - P$$

[$\because$ C^2 is proportional to the absolute temperature T]

$$= \frac{P}{2}\left(1 + \sqrt{\frac{T_1}{T_2}} - 2\right) = \frac{P}{2}\left(\frac{\sqrt{T_1} - \sqrt{T_2}}{\sqrt{T_2}}\right)$$

$$\text{or,} \quad P = 2F \cdot \frac{\sqrt{T_2}}{\sqrt{T_1} - \sqrt{T_2}}$$

$$\text{or,} \quad P = \frac{2F \cdot \sqrt{T_2}(\sqrt{T_1} + \sqrt{T_2})}{T_1 - T_2} = \frac{2F(\sqrt{T_1T_2} + T_2)}{T_1 - T_2}.$$

If the difference between T_1 and T_2 is not large, the above expression may be written in the form

$$P = 4F \cdot \frac{T_2}{T_1 - T_2}. \tag{15.10.5}$$

From the deflection of the moving plate, the torsional constant of the suspension fibre and the other constants of the apparatus, F can be determined and thus the pressure within the enclosure is measured.

Illustrative Example

1. The fixed metallic strips in a Knudsen gauge are maintained at 200°C. If the room temperature rises from 15°C to 20°C, what is the ratio of the deflections at the different temperatures?

From the relation (15.10.5), we have

$$F = \frac{P(T_1 - T_2)}{4T_2}.$$

As $P \propto C_2^2 \propto T_2$ and $F \propto \theta$, the deflection, we have from the above expression

$$\frac{F}{F'} = \frac{\theta}{\theta'} = \frac{T_1 - T_2}{T_1 - T_2'},$$

where θ, θ' are the deflections corresponding to temperatures T_1, T_2 and T_1, T_2' of the fixed plates and the room respectively.

Here

$$T_1 = 273 + 200 = 473°K$$

$$T_2 = 273 + 15 = 288°K$$

$$T_2' = 273 + 20 = 293°K$$

$\therefore$ the ratio of the deflections is given by

$$\frac{\theta}{\theta'} = \frac{473 - 288}{473 - 293} = \frac{185}{180} = \frac{37}{36}.$$

Exercises

1. Define the speed S of a pump. Show that $S = \frac{V}{t_2 - t_1} \log_e \frac{P_1 - P_0}{P_2 - P_0}$

2. Describe a Water-jet pump with a neat sketch and explain its action. What is its pumping speed?

3. Describe with a diagram Cenco rotary oil pump and explain its action. What is the order of vacuum attained by such a pump?

4. Explain how Gaede's diffusion pump works. What are its disadvantages? Indicate how Langmuir removed the disadvantages.

5. Draw a diagram of a McLeod gauge and explain its action.

6. Explain the principle of action of a Pirani gauge. What is its range of measurement?

7. Explain with a diagram the principle of working of an Ionisation gauge. Indicate the pressure ranges covered by the guage.

8. Describe the principle of action of a Kundsen gauge. Why is it called an absolute manometer?

Chapter 16

Fourier Analysis

16.1 Fourier's Theorem

We know from the principle of superposition that a number of collinear simple harmonic vibrations of commensurate periods, when combined together, will produce a periodic motion. The French mathematician, Fourier, enunciated a converse theorem of this, which can be briefly stated as—

"*any single-valued periodic function whatever, can be expressed as a summation of simple harmonic terms having frequencies which are multiples of that of the given function.*"

The theorem has two limitations regarding, the nature of the complex vibration, viz.,

(i) that the displacement must be single-valued and continuous; and

(ii) that the displacement must always have a finite value.

It is obvious that these limitations do not apply in the case of sound waves, because a particle cannot have two displacements at the same instant of time nor can its displacement be infinite.

The theorem deals with the synthesis of a complex periodic vibration from simple harmonic terms, and also indicates a method by which a complex vibration can be analysed into its component simple harmonic vibrations.

Fourier's theorem has ample applications in different branches of heat and acoustics. In acoustics, it is applied for the analysis of musical notes. A complex musical *note* may be graphically represented as a periodic function. Fourier's theorem states that it is made up of a number of simple harmonic functions which represent pure *tones*.

16.2 Fourier Series

We consider a purely arbitrary process which is repeated every T seconds. Let this process be represented by $x = f(t)$.

Hence

$$f(t+T) = f(t). \qquad (16.2.1)$$

Let us assume that (i) $f(t)$ is single valued, finite and bounded in the interval $(-\pi, \pi)$; (ii) $f(t)$ is piece-wise continuous but may have a finite number of discontinuities, (iii) $f(t)$ has atmost a finite number of maxima and minima—in the interval of one oscillation and (iv) $f(t+2\pi) = f(t)$ for values of t outside the interval $(-\pi, \pi)$. A function which satisfies these conditions in a given interval is said to be *piece-wise regular* in that interval. Under the above set of conditions, which are known as *Dirichlet's conditions,* the function $f(t)$ may be represented over a complete period, and hence from $t = -\infty$ to $t = +\infty$, except at the discontinuities, by*a series of simple harmonic functions, the frequencies of which are integral multiples of the fundamental frequency. Such a series is called the *Fourier's Series* and may be expressed by the following complex representation:

$$\begin{aligned} f(t) &= a_0 + a_1 e^{jwt} + a_2 e^{2jwt} + \cdots + a_r e^{rjwt} + \cdots \\ &\quad a_{-1} e^{-jwt} + a_{-2} e^{-2jwt} + \cdots + a_{-r} e^{-rjwt} + \cdots \\ &= \sum_{r=-\infty}^{r=+\infty} a_r e^{rjwt} \end{aligned} \tag{16.2.2}$$

where $w = \frac{2\pi}{T}$.

Since the left-hand side of the above equation is real, the co-efficients of the series on the right-hand side must not involve imaginary terms.

16.3 Evaluation of the Constants a_0 and a_r

To obtain a_0, we are to integrate both sides of the expression (16.2.2) over one complete period, i.e., from 0 to T, whence we may write

$$\begin{aligned} \int_0^T f(t)\,dt &= \int_0^T \left(\sum_{r=-\infty}^{r=+\infty} a_r e^{rjwt} \right) dt = \sum_{r=-\infty}^{r=+\infty} a_r \int_0^T e^{rjwt}\,dt = \sum_{r=-\infty}^{r=+\infty} a_r \left[\frac{e^{rjwt}}{jrw} \right]_0^T \\ &= \sum_{r=-\infty}^{r=+\infty} a_r \frac{1}{jrw} \left[e^{j2\pi r} - 1 \right] \quad (\because wT = 2\pi) \\ &= \sum_{r=-\infty}^{r=+\infty} a_r \frac{1}{jrw} [\cos 2\pi r + j \sin 2\pi r - 1] \\ &= 0, \text{ if } r \neq 0. \end{aligned}$$

If $r = 0$, we have from

$$\int_0^T f(t)\,dt = \sum_{r=-\infty}^{r=+\infty} a_r \int_0^T e^{rjwt}\,dt,$$

$$\int_0^T f(t)\,dt = a_0 \int_0^T dt = a_0 T \text{ or, } a_0 = \frac{1}{T}\int_0^T f(t)\,dt = \overline{f(t)} \tag{16.3.1}$$

where $\overline{f(t)}$ denotes the mean value of $f(t)$.

*The proof of expressing $f(t)$ in a Fourier series under this very general condition is beyond the scope of this book.

To obtain a_r we multiply both sides of the equation (16.2.2) by e^{-jrwt} and integrate from 0 to T, whence we get

$$\int_0^T f(t)e^{-jrwt}\,dt = a_r\int_0^T dt = a_rT \quad \text{or,} \quad a_r = \frac{1}{T}\int_0^T f(t)e^{-jrwt}\,dt. \tag{16.3.2}$$

We may also write from the above expression

$$a_{-r} = \frac{1}{T}\int_0^T f(t)e^{jrwt}\,dt. \tag{16.3.3}$$

Now, the equation (16.2.2) may be written in the form

$$f(t) = \sum_{r=-\infty}^{r=-1} a_r e^{jrwt} + a_0 + \sum_{r=1}^{r=\infty} a_r e^{jrwt} = \sum_{r=\infty}^{r=1} a_{-r}e^{-jrwt} + a_0 + \sum_{r=1}^{r=\infty} a_r e^{jrwt}$$

[on replacing r by $-r$]

$$= a_0 + \sum_{r=1}^{r=\infty}(a_r e^{jrwt} + a_{-r}e^{-jrwt}). \tag{16.3.4}$$

The above expression may also be written in the form

$$f(t) = a_0 + \sum_{r=1}^{r=\infty}(a_r + a_{-r})\cos rwt + j(a_r - a_{-r})\sin rwt. \tag{16.3.5}$$

Let $a_r + a_{-r} = A_r$, $j(a_r - a_{-r}) = B_r$ and $a_0 = \frac{A_0}{2}$, so that the equation (16.3.5) takes the form

$$\begin{aligned} f(t) &= \frac{A_0}{2} + \sum_{r=1}^{r=\infty} A_r \cos rwt + \sum_{r=1}^{r=\infty} B_r \sin rwt \\ &= \frac{A_0}{2} + A_1\cos wt + A_2\cos 2wt + \cdots + A_r\cos rwt + \cdots \\ &\quad + B_1\sin wt + B_2\sin 2wt + \cdots + B_r \sin rwt + \cdots \end{aligned} \tag{16.3.6}$$

The above expression gives the usual real form of the Fourier series.

16.4 Evaluation of the constants A_r and B_r

$$\begin{aligned} A_r = a_r + a_{-r} &= \frac{1}{T}\int_0^T f(t)e^{-jrwt}\,dt + \frac{1}{T}\int_0^T f(t)e^{jrwt}\,dt \\ &= \frac{1}{T}\int_0^T f(t)(e^{-jrwt} + e^{jrwt})\,dt = \frac{2}{T}\int_0^T f(t)\cos rwt\,dt \end{aligned} \tag{16.4.1}$$

$$\begin{aligned} B_r = j(a_r - a_{-r}) &= \frac{1}{T}\int_0^T j\{f(t)e^{-jrwt}\,dt - f(t)e^{jrwt}\,dt\} \\ &= -\frac{1}{T}\int_0^T f(t)j(e^{+jrwt} - e^{-jrwt})\,dt = \frac{2}{T}\int_0^T f(t)\sin rwt\,dt. \end{aligned} \tag{16.4.2}$$

16.5 Fourier Series involving phase angles

Again putting $A_r = C_r \cos\delta_r$ and $B_r = C_r \sin\delta_r$, the equation (16.3.6) may also be written in the form

$$f(t) = \frac{A_0}{2} + \sum_{r=1}^{r=\infty} C_r \cos(rwt - \delta_r) \tag{16.5.1}$$

where $C_r = \sqrt{A_r^2 + B_r^2}$ and $\tan\delta_r = \frac{B_r}{A_r}$.

The equation (16.5.1) may also be written in the form

$$f(t) = \frac{A_0}{2} + \sum_{r=1}^{r=\infty} C_r \sin(rwt - \delta_r + \pi/2). \tag{16.5.2}$$

The equation (16.5.1) or (16.5.2) gives the form of the Fourier series involving phase angles. With the help of the equations (16.2.2), (16.3.6), (16.5.1) and (16.5.2) we can analyse or synthesise any complex form of vibration. The constant amplitude a_0 or $A_0/2$ is a measure of the displacement of the axis of the vibration curve from the axis of the co-ordinates. In case the ordinate of the curve coincides with the axis of the curve, the constant vanishes.

Illustrative Example

1. A quantity $x = f(t)$ has the value a from $t = 0$ to $t = T/2$ and $-a$ from $t = T/2$ to $t = T$. Express x as a Fourier expression in t.

Here

$$a_0 = \frac{1}{T}\int_0^T f(t)\,dt = \frac{1}{T}\int_0^{T/2} f(t)\,dt + \frac{1}{T}\int_{T/2}^{T} f(t)\,dt = \frac{a}{T}\int_0^{T/2} dt - \frac{a}{T}\int_{T/2}^{T} dt$$

$$= \frac{a}{T}\cdot\frac{T}{2} - \frac{a}{T}\left(T - \frac{T}{2}\right) = 0.$$

$$\therefore\ A_0 = 2a_0 = 0.$$

Again,

$$A_r = \frac{2}{T}\int_0^T f(t)\cos rwt\,dt = \frac{2}{T}\int_0^{T/2} a\cos rwt\,dt - \frac{2}{T}\int_{T/2}^{T} a\cos rwt\,dt$$

$$= \frac{2a}{T}\left[\frac{\sin rwt}{rw}\right]_0^{T/2} - \frac{2a}{T}\left[\frac{\sin rwt}{rw}\right]_{T/2}^{T} = 0.$$

Also

$$B_r = \frac{2}{T}\int_0^T f(t)\sin rwt\,dt = \frac{2a}{T}\int_0^{T/2} \sin rwt\,dt - \frac{2a}{T}\int_{T/2}^{T} \sin rwt\,dt$$

$$= \frac{2a}{T}\left[\frac{-\cos rwt}{rw}\right]_0^{T/2} - \frac{2a}{T}\left[\frac{-\cos rwt}{rw}\right]_{T/2}^{T} = \frac{2a}{rwT}[\cos rwt]_{T/2}^{0} + \frac{2a}{rwT}[\cos rwt]_{T/2}^{T}$$

$$= \frac{a}{r\pi}[1 - \cos r\pi] + \frac{a}{r\pi}[\cos 2r\pi - \cos r\pi] = \frac{a}{r\pi} + \frac{a}{r\pi}\cos 2r\pi - \frac{2a}{r\pi}\cos r\pi.$$

$$\therefore \ x = f(t) = \sum_{r=1}^{r=\infty} B_r \sin rwt = \sum_{r=1}^{r=\infty} \frac{a}{r\pi}(1 + \cos 2r\pi - 2\cos r\pi)\sin rwt$$

$$= \frac{4a}{\pi}\sin wt + \frac{a}{2\pi} \times 0 \sin 2wt + \frac{4a}{3\pi}\sin 3wt + \frac{a}{4\pi} \times 0 \sin 4wt + \frac{4a}{5\pi}\sin 5wt + \cdots$$

$$= \frac{4a}{\pi}\left(\sin wt + \frac{\sin 3wt}{3} + \frac{\sin 5wt}{5} + \cdots\right)$$

16.6 Fourier Representation of Even and Odd Functions

If $f(x)$ defined in the interval $-\pi < x < \pi$ is *even*, the Fourier series will have cosine terms only, and as such, is called *cosine series*, while for odd functions Fourier series will have sine terms only, and is known *as sine series.* This may be proved as follows.

An *even function* is defined as $f(x) = f(-x)$.

Let

$$f(x) = a_0 + \sum_1^\infty a_r \cos rx + \sum_1^\infty b_r \sin rx.$$

Here

$$a_0 = \frac{1}{2\pi}\int_{-\pi}^{\pi} f(x)dx = \frac{1}{2\pi}\left[\int_{-\pi}^{0} f(x)dx + \int_0^{\pi} f(x)dx\right]$$

$$= \frac{1}{2\pi}\left[-\int_0^{-\pi} f(x)dx + \int_0^{\pi} f(x)dx\right]$$

$$= \frac{1}{2\pi}\left[\int_0^{\pi} f(-x)dx + \int_0^{\pi} f(x)dx\right] \quad \left[\begin{array}{l}\text{putting } -x \text{ for } x\\ \text{in the first integral}\end{array}\right]$$

$$= \frac{1}{\pi}\int_0^{\pi} f(x)dx \quad [\text{since } f(x) = f(-x)]$$

Now

$$a_r = \frac{1}{\pi}\int_{-\pi}^{\pi} f(x)\cos rx dx = \frac{1}{\pi}\left[-\int_0^{-\pi} f(x)\cos rx dx + \int_0^{\pi} f(x)\cos rx dx\right]$$

$$= \frac{1}{\pi}\left[\int_0^{\pi} f(-x)\cos(-rx)dx + \int_0^{\pi} f(x)\cos rx dx\right] \quad \left[\begin{array}{l}\text{putting } -x \text{ for } x\\ \text{in the first term}\end{array}\right]$$

$$= \frac{2}{\pi}\int_0^{\pi} f(x)\cos rx dx \quad [\text{since } f(-x) = f(x) \text{ and } \cos(-rx) = \cos rx]$$

Again

$$b_r = \frac{1}{\pi}\int_{-\pi}^{\pi} f(x)\sin rx dx = \frac{1}{\pi}\left[-\int_0^{-\pi} f(x)\sin rx dx + \int_0^{\pi} f(x)\sin rx dx\right]$$

$$= \frac{1}{\pi}\left[\int_0^{\pi} f(-x)\sin(-rx)dx + \int_0^{\pi} f(x)\sin rx dx\right] \quad \left[\begin{array}{l}\text{putting } -x \text{ for } x\\ \text{in the first term}\end{array}\right]$$

$$= 0 \quad [\text{since } f(-x) = f(x) \text{ but } \sin(-rx) = -\sin rx]$$

So

$$f(x) = \frac{1}{\pi}\int_0^{\pi} f(x)dx + \sum_{r=1}^{\infty}\left[\frac{2}{\pi}\int_0^{\pi} f(x)\cos rx dx\right]\cos rx$$

Thus the Fourier series expansion of an even periodic function consists of only the cosine terms and a constant term, all the sine terms being absent.

Again an *odd* function is defined as $f(x) = -f(-x)$ and proceeding exactly as in the above case, we can prove that

$$a_0 = 0, a_r = 0 \quad \text{and} \quad b_r = \frac{2}{\pi}\int_0^{\pi} f(x)\sin rx dx.$$

So, the Fourier series expansion of an odd periodic function consists of only the sine terms, the constant term and all the cosine terms being zero.

In case of even functions, $f(x) = f(-x)$, i.e., $f(\pi) = f(-\pi)$. So in the interval $-\pi < x < \pi$, it is represented by the attached graph (Fig. 16.6.1)

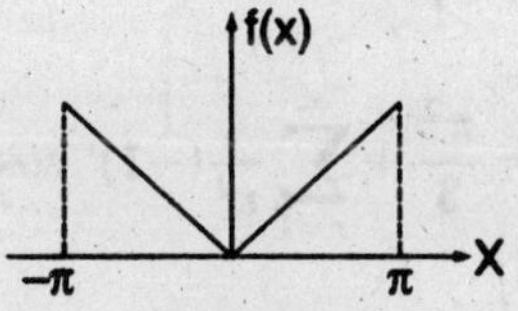

Fig 16.6.1

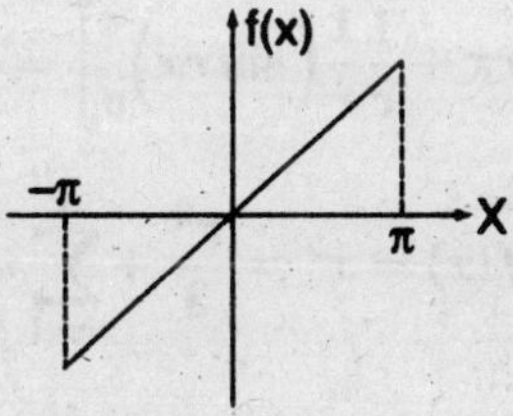

Fig 16.6.2

In case the function is odd, $f(x) = -f(-x)$, i.e., $f(\pi) = -f(-\pi)$. So in the interval $-\pi < x < \pi$, it is represented by the attached graph (Fig. 16.6.2)

16.6.1 Evaluation of Riemann Zeta Function by the method of Fourier Expansion

The Riemann Zeta functions denoted by $\xi(2)$ and $\xi(4)$ and in general denoted by $\xi(k)$ are defined as

$$\xi(2) = \sum_1^{\infty}\frac{1}{r^2} = \frac{1}{1^2} + \frac{1}{2^2} + \frac{1}{3^2} + \cdots\infty$$

$$\xi(4) = \sum_1^{\infty}\frac{1}{r^4} = \frac{1}{1^4} + \frac{1}{2^4} + \frac{1}{3^4} + \cdots\infty$$

and in general, as

$$\xi(k) = \sum_1^{\infty}\frac{1}{r^k} = \frac{1}{1^k} + \frac{1}{2^k} + \frac{1}{3^k} + \cdots\infty$$

Sums of the above series can be evaluated by the method of Fourier expansion as explained in the following way.

We start with the function $f(x)$ defined as, $f(x) = x^2$, $(-\pi \leq 0 \leq \pi)$; now $f(x)$ being an even function, $[\because f(-x) = (-x)^2 = x^2 = f(x)]$ we have $b_r = 0$.

$$\therefore x^2 = a_0 + \sum a_r \cos rx \quad \text{(see Art. 16.6)}$$

Again we have already proved in Art. 16.6 that

$$a_0 = \frac{1}{\pi}\int_0^{\pi} f(x)dx = \frac{1}{\pi}\int_0^{\pi} x^2 dx = \frac{\pi^2}{3}$$

Also we have proved in Art. 16.6 that

$$a_r = \frac{2}{\pi}\int_0^{\pi} f(x)\cos rx dx = \frac{2}{\pi}\int_0^{\pi} x^2 \cos rx dx = \frac{2}{\pi}\left[\frac{x^2}{r}\left(\sin rx\right)_0^{\pi} - 2\int_0^{\pi} x \cdot \frac{1}{r}\sin rx dx\right]$$

$$= \frac{2}{\pi}\left[0 - \frac{2}{r}\int_0^{\pi} x\sin rx dx\right] = -\frac{4}{r\pi}\int_0^{\pi} x\sin rx dx = -\frac{4}{r\pi}\left[-\frac{x}{r}\left(\cos rx\right)_0^{\pi} + \int_0^{\pi}\frac{1}{r}\cos rx dx\right]$$

$$= -\frac{4}{r\pi}\left[-\frac{\pi}{r}\cos r\pi + \frac{1}{r}\frac{1}{r}\left(\sin rx\right)_0^{\pi}\right] = \frac{4}{r^2}\cos r\pi = \frac{4}{r^2}(-1)^r.$$

$$\therefore f(x) = x^2 = \frac{\pi^2}{3} + \sum_{r=1}^{\infty} a_r \cos rx = \frac{\pi^2}{3} + \sum_{r=1}^{\infty}\frac{4}{r^2}(-1)^r \cos rx$$

Putting $x = \pi$, we get

$$\pi^2 = \frac{\pi^2}{3} + \sum_1^{\infty}\frac{4}{r^2}(-1)^r\cos r\pi = \frac{\pi^2}{3} + \sum_1^{\infty}\frac{4}{r^2}(-1)^r\cdot(-1)^r = \frac{\pi^2}{3} + 4\sum_1^{\infty}\frac{1}{r^2} = \frac{\pi^2}{3} + 4\xi(2)$$

$$\text{or } \xi(2) = \sum_1^{\infty}\frac{1}{r^2} = \frac{1}{4}\left(\pi^2 - \frac{\pi^2}{3}\right) = \frac{1}{4}\cdot\frac{2}{3}\pi^2 = \frac{\pi^2}{6}$$

To find $\xi(4)$, let us take $f(x) = x^4$

$$\therefore a_0 = \frac{1}{\pi}\int_0^{\pi} f(x)dx = \frac{1}{\pi}\int_0^{\pi} x^4 dx = \frac{\pi^4}{5}$$

Since $f(x)$ is an even function, $b_r = 0$

$$\therefore f(x) = x^4 = a_0 + \sum_1^{\infty} a_r \cos rx \quad \text{where } a_0 = \frac{\pi^4}{5}$$

Now we have

$$a_r = \frac{2}{\pi}\int_0^{\pi} x^4 \cos rx dx = \frac{2}{\pi}(-1)^r\left[\frac{4\pi^3}{r^2} - \frac{24\pi}{r^4}\right] = (-1)^r\left[\frac{8\pi^2}{r^2} - \frac{48}{r^4}\right]$$

$$\therefore f(x) = x^4 = \frac{\pi^2}{5} + \sum_1^{\infty}(-1)^r\left(\frac{8\pi^2}{r^2} - \frac{48}{r^4}\right)\cos rx$$

Putting $x = \pi$, we get

$$\pi^4 = \frac{\pi^4}{5} + \sum_1^\infty (-1)^r \left(\frac{8\pi^2}{r^2} - \frac{48}{r^4}\right) \cos r\pi = \frac{\pi^4}{5} + \sum_1^\infty (-1)^r \left(\frac{8\pi^2}{r^2} - \frac{48}{r^4}\right) \cdot (-1)^r$$

$$= \frac{\pi^4}{5} + \sum_1^\infty \left(\frac{8\pi^2}{r^2} - \frac{48}{r^4}\right) = \frac{\pi^4}{5} + 8\pi^2 \sum_1^\infty \frac{1}{r^2} - 48 \sum_1^\infty \frac{1}{r^4}$$

or $$\frac{4\pi^4}{5} = 8\pi^2 \xi(2) - 48\xi(4) = 8\pi^2 \cdot \frac{\pi^2}{6} - 48\xi(4)$$

or $$48\xi(4) = \left(\frac{4}{3} - \frac{4}{5}\right)\pi^4 = \frac{8}{15}\pi^4 \quad \text{or} \quad \xi(4) = \frac{\pi^4}{90}.$$

16.7 Examples illustrating 'Analysis and Synthesis' by Fourier's Theorem

Example 1: Square Wave

A periodic square wave has the displacement curve as shown in the fig. 16.7.1. This type of wave is utilised in signals (e.g. television signals).

Here the displacement is

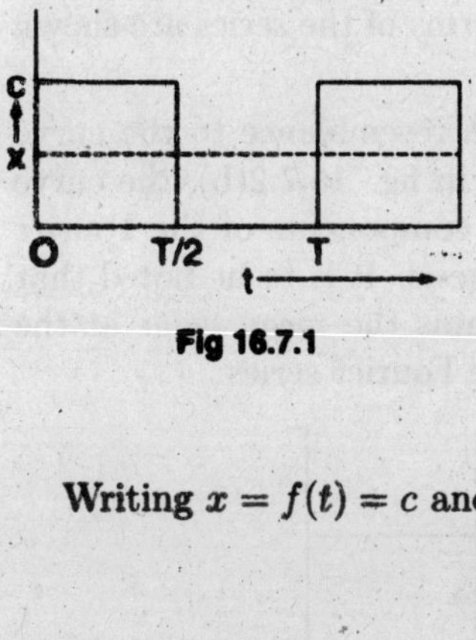

Fig 16.7.1

$$x = f(t) = c \text{ (constant) from } t = 0 \text{ to } t = T/2$$

$$\text{and } x = f(t) = 0 \text{ from } t = \frac{T}{2} \text{ to } t = T,$$

when the time-axis of co-ordinates is taken through the lowest point of the displacement curve. From (16.3.1), A_0 may be obtained as follows:

$$a_0 = \frac{A_0}{2} = \frac{1}{T}\int_0^T f(t)\, dt.$$

Writing $x = f(t) = c$ and changing the limits, we may write

$$\frac{A_0}{2} = \frac{1}{T}\int_0^{T/2} c\, dt = \frac{c}{2} \quad \text{or} \quad A_0 = c$$

which means that the axis of the displacement curve is the line $x = \frac{c}{2}$.

Again from (16.4.1) we have,

$$A_r = \frac{2}{T}\int_0^T f(t) \cos r\omega t\, dt$$

which, on writing $x = c$ and changing the limits, reduces to

$$A_r = \frac{2}{T}\int_0^{T/2} c \cos r\omega t\, dt = \frac{2c}{T}\int_0^{T/2} \cos r \cdot \frac{2\pi}{T} t\, dt = \frac{2c}{T} \frac{\left[\sin \frac{2\pi r t}{T}\right]_0^{T/2}}{\frac{2\pi r}{T}} - \frac{c}{\pi r}\left[\sin \frac{2\pi r t}{T}\right]_0^{T/2} = 0 \tag{16.7.1}$$

which means that all the cosine terms of the Fourier series have zero amplitude and consequently disappear altogether.

Also from (16.4.2) we have,

$$B_r = \frac{2}{T}\int_0^T f(t)\sin rwt\,dt$$

which, on writing $x = f(t) = c$ and changing the limits, becomes

$$B_r = \frac{2}{T}\int_0^{T/2} c\sin r\cdot\frac{2\pi}{T}t\,dt = \frac{2c}{T}\frac{\left[-\cos r\frac{2\pi}{T}t\right]_0^{T/2}}{r\cdot\frac{2\pi}{T}} = \frac{c}{\pi r}[-\cos r\pi + 1] \qquad (16.7.2)$$

when r is even, $\cos r\pi = +1$ and $B_r = 0$;

when r is odd, $\cos r\pi = -1$ and $B_r = \frac{2c}{\pi r}$.

Consequently, all the even terms of the sine series disappear, and we are left with the odd terms having amplitude coefficients respectively given by

$$\frac{2c}{\pi}\left(1, \frac{1}{3}, \frac{1}{5}, \frac{1}{7}, \frac{1}{9}, \text{ etc.}\right).$$

The complete series therefore, becomes

$$x = f(t) = \frac{c}{2} + \frac{2c}{\pi}\left[\sin wt + \frac{1}{3}\sin 3wt + \frac{1}{5}\sin 5wt + \cdots + \frac{1}{r}\sin rwt\right] \qquad (16.7.3)$$

Figures 16.7.2(a) and 16.7.2(b) show how the addition of these terms can be indicated graphically. They also indicate how with the addition of successive terms, the resultant curve takes the shape of a square wave. In fig. 16.7.2(a), the first three terms of the series are shown independently, and as a resultant curve (dotted curve).

It is evident that even in this case, the resultant bears a rough resemblance to the curve to be analysed. By the addition of 15 terms of the series as shown in fig. 16.7.2(b), the curve is almost perfectly produced. The superposition of the sinusoidal components of the Fourier series to reproduce a periodic wave form is known as *Fourier Synthesis.* It is to be noted that at the discontinuities ($t = 0$ and $t = T/2$), the Fourier series attains the *mean value* at the discontinuity. In this case it is $c/2$. This is a *general property* of the Fourier series.

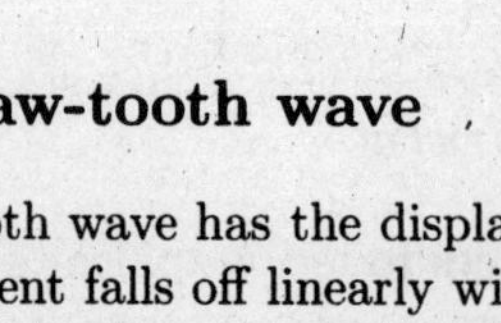

Fig 16.7.2(a) Addition of 3 terms

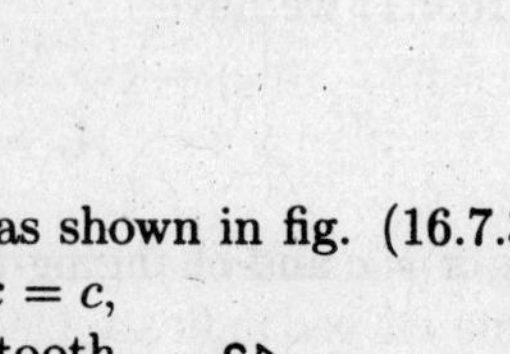

Fig 16.7.2(b) Addition of 15 terms

Example 2: Saw-tooth wave

A periodic saw-tooth wave has the displacement curve as shown in fig. (16.7.3), which shows that the displacement falls off linearly with time from $x = c$, when $t = 0$ to $x = 0$ when $t = T$, i.e., for such a saw-tooth wave, we may write,

$$x = f(t) = c\left(1 - \frac{t}{T}\right),$$

when the time-axis passes through the lowest points of the periodic curve.

Fig 16.7.3

From (16.3.1), putting the value of x, we have

$$A_0/2 = \frac{c}{T}\int_0^T \left(1-\frac{t}{T}\right) dt = \frac{c}{T}\left[t-\frac{t^2}{2T}\right]_0^T = \frac{c}{T}\cdot\frac{T}{2} = \frac{c}{2} \tag{16.7.4}$$

which gives the ordinate of the displacement curve. Also from (16.4.1), putting the value of x, we may write

$$\begin{aligned} A_r &= \frac{2c}{T}\int_0^T \left(1-\frac{t}{T}\right)\cos r\frac{2\pi}{T}t\,dt \\ &= \frac{2c}{T}\left[\left(1-\frac{t}{T}\right)\frac{T}{2\pi r}\sin\frac{2\pi r}{T}t\right]_0^T - \frac{2c}{T}\int_0^T\left(-\frac{1}{T}\right)\frac{T}{2\pi r}\sin\frac{2\pi r}{T}t\,dt \\ &= 0 + \frac{2c}{T}\left[\frac{1}{2\pi r}\left(-\frac{T}{2\pi r}\cos\frac{2\pi r}{T}t\right)\right]_0^T = 0. \end{aligned} \tag{16.7.5}$$

Thus all the cosine terms of the Fourier series have zero amplitude, and consequently disappear altogether.

Again from (16.4.2), putting the value of x, we may write

$$\begin{aligned} B_r &= \frac{2c}{T}\int_0^T\left(1-\frac{t}{T}\right)\sin\frac{2\pi rt}{T}\,dt \\ &= \frac{2c}{T}\left[\left(1-\frac{t}{T}\right)\left(-\frac{T}{2\pi r}\right)\cos\frac{2\pi r}{T}t\right]_0^T - \frac{2c}{T}\int_0^T\left(-\frac{1}{T}\right)\left(-\frac{T}{2\pi r}\right)\cos\frac{2\pi r}{T}t\,dt \\ &= \frac{2c}{T}\frac{T}{2\pi r} - \frac{2c}{T}\frac{1}{2\pi r}\left[\frac{T}{2\pi r}\sin\frac{2\pi r}{T}t\right]_0^T = \frac{c}{\pi r} - 0 = \frac{c}{\pi r}. \end{aligned}$$

Thus the amplitude coefficients of the sine series are

$$\frac{c}{\pi}\left[1, \frac{1}{2}, \frac{1}{3}, \frac{1}{4}, \cdots, \frac{1}{r}\right].$$

The Fourier's series in this case is therefore

$$f(t) = \frac{c}{2} + \frac{c}{\pi}\left[\sin wt + \frac{1}{2}\sin 2wt + \frac{1}{3}\sin 3wt + \cdots + \frac{1}{r}\sin rwt\right]. \tag{16.7.6}$$

The addition of successive terms of this series is indicated graphically in figs. [16.7.4(a), (b), (c), (d)]. The curves indicate clearly the convergence of the series, i.e. the curves show how with the addition of successive terms, the resultant curve gradually approaches the shape of the saw-tooth wave. In other words, the resemblance between the resultant curve and the curve under analysis becomes closer, as more and more terms are employed as will be evident from the figures. It is to be noted that at the discontinuities (i.e. at $t = 0$ or $t = T$), the series attains the mean of the values on the two sides of the discontinuity. In this case, this value is $c/2$.

Such type of wave-form is found in the ripple potential of a rectifier and also in the ideal time-base arrangement.

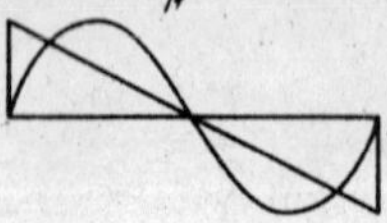
Fig 16.7.4(a) 1 term

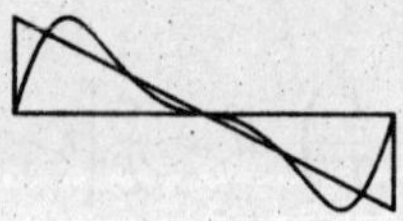
Fig 16.7.4(b) 2 terms

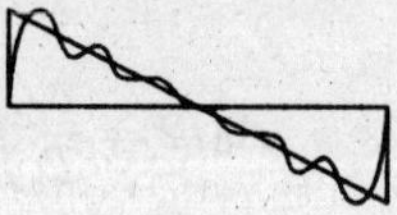
Fig 16.7.4(c) 3 terms

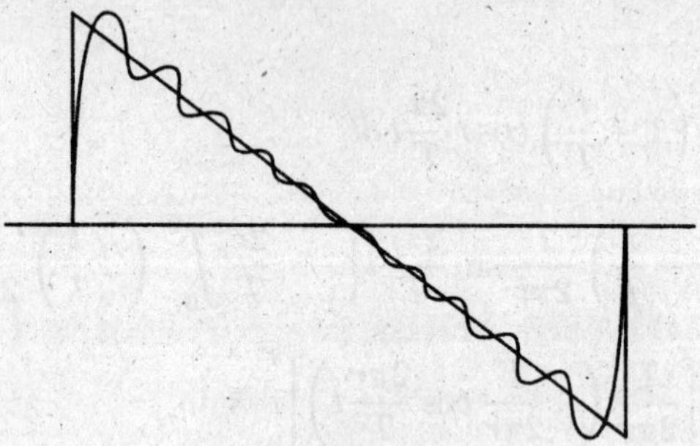
Fig 16.7.4(d) 10 terms

16.8 Expansion of a periodic function $f(x)$

Unlike in Art. (16.2) we consider that the function of x is not periodic with time T. A Fourier series may be employed to express a periodic function $f(x)$ which is any continuous function of the independent variable x, defined in an interval $(-\pi, \pi)$ as a linear relation between the sine terms and cosine terms of the same period and is given by

$$f(x) = \frac{A_0}{2} + A_1 \cos x + A_2 \cos 2x + \cdots + B_1 \sin x + B_2 \sin 2x + \cdots$$

$$= \frac{A_0}{2} + \sum_{r=1}^{r=\infty} A_r \cos rx + \sum_{r=1}^{r=\infty} B_r \sin rx. \tag{16.8.1}$$

To obtain the values of the constants A_0, A_r and B_r we first integrate both sides of the equation (16.8.1) between the limits $-\pi$ and π, whence we get

$$\int_{-\pi}^{\pi} f(x)dx = \frac{A_0}{2}\int_{-\pi}^{\pi} dx \quad \left[\because \int_{-\pi}^{\pi} \cos rx\, dx = 0 \text{ and } \int_{-\pi}^{\pi} \sin rx\, dx = 0\right]$$

$$\text{or } A_0 = \frac{1}{\pi}\int_{-\pi}^{\pi} f(x)\, dx. \tag{16.8.2}$$

Again multiplying both sides of the relation (16.8.1) by $\cos rx$ and integrating between the limits $-\pi$ and π, we have,

$$\int_{-\pi}^{\pi} f(x) \cos rx\, dx = A_r \int_{-\pi}^{\pi} \cos^2 rx\, dx \quad [\because \text{ the other integrals vanish}]$$

$$= \frac{A_r}{2} \cdot 2\pi = \pi A_r \quad \text{or } A_r = \frac{1}{\pi}\int_{-\pi}^{\pi} f(x) \cos rx\, dx. \tag{16.8.3}$$

Similarly multiplying both sides of the relation (16.8.1) by $\sin rx$ and integrating between the limits $-\pi$ and π, we have,

$$B_r = \frac{1}{\pi}\int_{-\pi}^{\pi} f(x)\sin rx\,dx. \tag{16.8.4}$$

The coefficients remain unchanged when the limits of integration are 0 and 2π. Hence

$$f(x) = \frac{1}{2\pi}\int_{-\pi}^{+\pi} f(x)dx + \sum_{r=1}^{r=\infty}\frac{\cos rx}{\pi}\int_{-\pi}^{\pi} f(x)\cos rx\,dx + \sum_{r=1}^{r=\infty}\frac{\sin rx}{\pi}\int_{-\pi}^{\pi} f(x)\sin rx\,dx. \tag{16.8.5}$$

N.B. If $f(x)$ is not periodic, Fourier expansion holds only within the given interval ($-\pi$ to $+\pi$). Outside this interval, the Fourier series is periodic having the same period i.e. 2π, but it does not agree with $f(x)$.

16.9 Extension of the Range

Let us suppose that the range over which the function $f(x)$ is to be expanded extends from $x = -l$ to $x = +l$.

Let

$$\frac{xl}{\pi} = z \quad \therefore \quad dx = \frac{\pi}{l}dz.$$

So when $x = -\pi, z = -l$ and when $x = +\pi, z = +l$.

Also

$$f(x) = \phi(z).$$

Hence from (16.8.5), we have

$$\begin{aligned}\phi(z) &= \frac{1}{2\pi}\int_{-l}^{l}\phi(z)\frac{\pi}{l}dz + \sum_{r=1}^{r=\infty}\frac{1}{\pi}\cos\frac{r\pi z}{l}\int_{-l}^{+l}\phi(z)\frac{\pi}{l}\cos\frac{r\pi z}{l}dz \\ &\quad + \sum_{r=1}^{r=\infty}\frac{1}{\pi}\sin\frac{r\pi z}{l}\int_{-l}^{+l}\phi(z)\frac{\pi}{l}\sin\frac{r\pi z}{l}dz \\ &= \frac{1}{2l}\int_{-l}^{+l}\phi(z)dz + \sum_{r=1}^{r=\infty}\frac{1}{l}\cos\frac{r\pi z}{l}\int_{-l}^{+l}\phi(z)\cos\frac{r\pi z}{l}dz \\ &\quad + \sum_{r=1}^{r=\infty}\frac{1}{l}\sin\frac{r\pi z}{l}\int_{-l}^{+l}\phi(z)\sin\frac{r\pi z}{l}dz.\end{aligned}$$

If the independent variable x is identified with z and $f(x)$ is identified with $\phi(z)$, then over the range between $x = -l$ to $x = +l$, the above expression may be written in the form,

$$\begin{aligned}f(x) &= \frac{1}{2l}\int_{-l}^{+l} f(x)dx + \sum_{r=1}^{r=\infty}\frac{1}{l}\cos\frac{r\pi x}{l}\int_{-l}^{+l} f(x)\cos\frac{r\pi x}{l}dx \\ &\quad + \sum_{r=1}^{r=\infty}\frac{1}{l}\sin\frac{r\pi x}{l}\int_{-l}^{+l} f(x)\sin\frac{r\pi x}{l}dx.\end{aligned} \tag{16.9.1}$$

16.10 Half-Range Series

Sometimes a function $f(x)$ in the range $(0, l)$, having half the period of $2l$, is required to be expanded in a Fourier series. Since the nature of the function outside the range $0 < x < l$ is immaterial, we may extend the function in the range $-l < x < 0$, so that the function becomes odd or even. The Fourier series of $f(x)$ will thus contain either the sine terms only or the cosine terms only.

If $f(x)$ is to be expanded in a *half-range sine series* in $0 < x < l$, we are to consider $f(x) = -f(x)$ in the range $-l < x < 0$. The function is then *odd* in the entire period ($-l$ to l). The Fourier expansion of $f(x)$ is then

$$\left.\begin{aligned} f(x) &= \sum_{r=1}^{\infty} b_r \sin\frac{r\pi x}{l} \\ \text{where} \quad b_r &= \frac{2}{l}\int_0^l f(x)\sin\frac{r\pi x}{l}dx \end{aligned}\right\} \tag{16.10.1}$$

On the other hand, if $f(x)$ is to be expanded as a half-range cosine series in $0 < x < l$, we are to choose $f(x) = f(-x)$ in the range $-l < x < 0$, making it even over the entire period. Then

$$\left.\begin{aligned} f(x) &= a_0 + \sum_{r=1}^{\infty} a_r \cos\frac{r\pi x}{l} \\ \text{where} \quad a_0 &= \frac{1}{l}\int_0^l f(x)dx \\ \text{and} \quad a_r &= \frac{2}{l}\int_0^l f(x)\cos\frac{r\pi x}{l}dx \end{aligned}\right\} \tag{16.10.2}$$

16.11 Advantages of Fourier Series

One of the advantages of Fourier representation over some other representations, such as Taylor's series expansion, is that it may be used to represent discontinuous functions, where all orders of derivatives need not exist whereas Taylors expansion is possible in case all orders of derivatives do exist. Secondly in Fourier expansion, the even and odd functions are conveniently expressed as cosine and sine series respectively.

Fourier expansions are valid over the given basic interval. Unless the function $f(x)$ is known to be periodic with a period equal to the basic interval, there is no assurance whatever that the Fourier representation will have any meaning outside the basic interval.

In addition to the advantages of representing discontinuous and periodic functions, there is a third advantage in using Fourier series. This is realised in the solution of equation of motion of an oscillating particle subject to a periodic driving force. The Fourier expansion of the driving force then gives us the fundamental term and a series of harmonics.

16.12 Importance of Fourier's Theorem

This theorem is of great importance for its application in the study of analysis and synthesis of vibrations. Different harmonic components in a given vibration can be found out with the help of Fourier's theorem. Again for choosing different harmonic components to produce vibration of a given period, we must take help of this theorem. The harmonic component having the lowest frequency is called the "fundamental", the components having frequencies different from that of the fundamental are called "overtones", and the overtones whose frequencies are multiplies of the fundamental frequency are known as "harmonics".

The quality of a musical note depends on the number of overtones present and their relative positions. So the quality of a musical note can be determined with the help of Fourier's theorem after proper analysis of the note. Fourier's theorem also finds its application in the analysis of the output of a rectifier circuit and hence also in the designing of suitable filter circuits.

Illustrative Examples

1. An alternating e.m.f. of sine form is applied to a full-wave rectifier. Obtain an expression for the rectified current in harmonic series.

Here, since the function I is defined between $wt = 0$ and $wt = \pi$, we have

$$f(t) = I = I_0 \sin wt = \frac{A_0}{2} + \sum_{r=1}^{r=\infty} A_r \cos 2rwt + \sum_{r=1}^{r=\infty} B_r \sin 2rwt. \quad (1)$$

From the relation (16.3.1) we have,

$$\frac{A_0}{2} = \frac{I_0}{T}\int_0^{\pi} \sin wt\, dt = \frac{I_0}{wT}[\cos wt]_{\pi}^{0} = \frac{2I_0}{\pi} \quad [\because wT = \pi].$$

Also multiplying both sides of equation (1) by $\cos 2rwt$ and integrating between the limits 0 and π, we have

$$\int_0^{\pi} I_0 \sin wt \cos 2rwt\, dt = \int_0^{\pi}\left[\frac{A_0}{2} + \sum_{r=1}^{r=\infty} A_r \cos 2rwt + \sum_{r=1}^{r=\infty} B_r \sin 2rwt\right]\cos 2rwt\, dt. \quad (2)$$

Now the l.h.s. of the equation (2)

$$= \frac{I_0}{2}\int_0^{\pi}\{\sin(2r+1)wt + \sin(1-2r)wt\}\, dt = -\frac{I_0}{2}\left\{-\frac{2}{2r+1} - \frac{2}{1-2r}\right\}$$

$$= I_0\left\{\frac{1}{2r+1} - \frac{1}{2r-1}\right\} = -\frac{2I_0}{4r^2-1},$$

where $r = 1, 2, 3, \cdots$

and the r.h.s of the equation(2) $= \frac{A_r}{2}\pi$

$$\therefore\; A_r = -\frac{4I_0}{\pi}\cdot\frac{1}{4r^2-1}.$$

It can be shown that the coefficient B_r vanishes.

$$\therefore \; I_0 \sin wt = \frac{2I_0}{\pi}\left(1 - \frac{2}{3}\cos 2wt - \frac{2}{15}\cos 4wt - \frac{2}{35}\cos 6wt \cdots\right)$$

$$= \frac{4I_0}{\pi}\left(\frac{1}{2} - \frac{\cos 2wt}{3} - \frac{\cos 4wt}{15} - \frac{\cos 6wt}{35} \cdots\right)$$

which is the required expression for the rectified current in harmonic series.

2. A triangular wave is shown in the attached figure for which we have at

$$t = 0, \quad y = 0,$$

$$t = \frac{T}{2}, \quad y = A,$$

$$t = T, \quad y = 0.$$

Apply Fourier's analysis to determine its simple harmonic components.

To find out the equation of the line OR, we consider a point P of co-ordinates (t, y) on the line, so that from congruent triangles OPQ and ORN, we have,

$$\frac{PQ}{OQ} = \frac{RN}{ON} \quad \text{or} \quad \frac{y}{t} = \frac{A}{T/2} \quad \text{or} \quad y = \frac{2At}{T}$$

is the equation of the line OR for $t = 0$ to $t = T/2$.

Similarly to find out the equation of the line RT, we consider on it a point S having co-ordinates (t, y), so that considering congruent triangles, as before, we may show that $y = 2A\left(1 - \frac{t}{T}\right)$ is the equation of the line RT for $t = T/2$ to $t = T$.

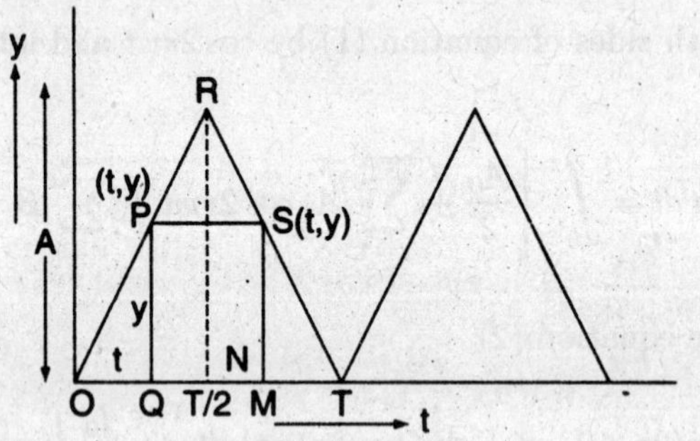

Now

$$a_0 = \frac{1}{T}\int_0^T y\,dt = \frac{1}{T}\int_0^{T/2} \frac{2At}{T}dt + \frac{1}{T}\int_{T/2}^{T} 2A\left(1 - \frac{t}{T}\right)dt$$

$$= \frac{2A}{T^2}\left[\frac{t^2}{2}\right]_0^{T/2} + \frac{2A}{T}\left\{(t)_{T/2}^{T} - \frac{1}{T}\left(\frac{t^2}{2}\right)_{T/2}^{T}\right\} = \frac{A}{4} + \frac{2A}{T}\left(\frac{T}{2} - \frac{1}{2T}\left(\frac{3}{4}T^2\right)\right)$$

$$= \frac{A}{4} + \frac{A}{4} = \frac{A}{2}$$

Again

$$a_r = \frac{2}{T}\int_0^T y\cos r\omega t dt = \frac{2}{T}\left[\int_0^{T/2} \frac{2At}{T}\cos r\omega t dt + \int_{T/2}^T 2A\left(1-\frac{t}{T}\right)\cos r\omega t dt\right]$$

$$= \frac{4A}{T^2}\left[\int_0^{T/2} t\cos r\omega t dt + \int_{T/2}^T (T-t)\cos r\omega t dt\right]$$

$$= \frac{4A}{T^2}\left[T\int_{T/2}^T \cos r\omega t dt + \int_0^{T/2} t\cos r\omega t dt - \int_{T/2}^T t\cos r\omega t dt\right] \qquad (1)$$

Now

$$\int_{T/2}^T \cos r\omega t dt = \left[\frac{\sin r\omega t}{r\omega}\right]_{T/2}^T = \frac{1}{r\omega}[\sin 2r\pi - \sin r\pi] = 0$$

$$\int_0^{T/2} t\cos r\omega t dt = \left[t\frac{\sin r\omega t}{r\omega}\right]_0^{T/2} - \left[\int_0^{T/2} 1\cdot\frac{\sin r\omega t}{r\omega}dt\right]$$

$$= \frac{1}{r\omega}\left[\left\{\frac{T}{2}\sin r\pi - 0\right\} - \left\{\int_0^{T/2} \sin r\omega t dt\right\}\right]$$

$$= \frac{1}{(r\omega)^2}\left[\cos r\omega t\right]_0^{T/2} = \frac{1}{(r\omega)^2}[\cos r\pi - \cos 0]$$

$$\text{and} \quad \int_{T/2}^T t\cos r\omega t dt = \frac{1}{(r\omega)^2}[\cos 2r\pi - \cos r\pi].$$

So from (1) we have

$$a_r = \frac{4A}{T^2}\left[T.0 + \frac{1}{(r\omega)^2}(\cos r\pi - \cos 0) - \frac{1}{(r\omega)^2}(\cos 2r\pi - \cos r\pi)\right]$$

$$= \frac{4A}{r^2\omega^2T^2}[2\cos r\pi - 2\cos 0] = \frac{2A}{r^2\pi^2}(\cos r\pi - \cos 0)$$

$$\therefore \quad a_r = -\frac{4A}{r^2\pi^2} \text{ when } r = 1, 3, 5, \cdots.$$

$$\text{and} \quad a_r = 0 \text{ when } r = 2, 4, 6, \cdots.$$

Again

$$b_r = \frac{2}{T}\int_0^T y\sin r\omega t dt = \frac{2}{T}\left[\int_0^{T/2} \frac{2At}{T}\sin r\omega t dt + \int_{T/2}^T 2A\left(1-\frac{t}{T}\right)\sin r\omega t dt\right]$$

$$= \frac{4A}{T^2}\left[T\int_{T/2}^T \sin r\omega t dt + \int_0^{T/2} t\sin r\omega t dt - \int_{T/2}^T t\sin r\omega t dt\right]$$

Solving each integral separately, we have

$$T\int_{T/2}^{T} \sin r\omega t dt = \frac{T}{r\omega}(\cos r\pi - \cos 2r\pi)$$

$$\int_{0}^{T/2} t\sin r\omega t dt = -\frac{T}{2r\omega}\cos r\pi \text{ and } \int_{T/2}^{T} t\sin r\omega t dt = \frac{T}{r\omega}\left[\frac{1}{2}\cos r\pi - \cos 2r\pi\right]$$

Substituting these values, we have, $b_r = 0$

$$\therefore\ y = a_0 + \sum_{r=1,3,5}^{\infty} a_r \cos r\omega t$$

$$= \frac{A}{2} - \frac{4A}{\pi^2}\left[\frac{1}{1^2}\cos\omega t + \frac{1}{3^2}\cos 3\omega t + \frac{1}{5^2}\cos 5\omega t + \cdot\right] \quad (\text{putting } r = 1, 3, 5, \cdots)$$

3. Expand the function

$$f(x) = \begin{cases} 0, & -\pi \le x \le 0 \\ x, & 0 \le x \le \pi \end{cases}$$

$$f(x + 2\pi) = f(x)$$

in a Fourier series. From this expansion, show that

$$1 + \frac{1}{3^2} + \frac{1}{5^2} + \cdots = \frac{\pi^2}{8}$$

[C.U. (Hons) 2000]

Expanding $f(x)$ here in a half-range cosine series in the range $0 < x < \pi$, we have the Fourier expansion given by [$\because$ $f(x) = 0$ in the range $-\pi < x < 0$]

$$f(x) = x = a_0 + \sum_{r=1}^{\infty} a_r \cos rx$$

Here
$$a_0 = \frac{1}{\pi}\int_0^{\pi} x dx = \frac{\pi}{2},$$

$$a_r = \frac{2}{\pi}\int_0^{\pi} x\cos rx dx = \frac{2}{\pi}\left[\left(x\frac{\sin rx}{r}\right)_0^{\pi} - \int_0^{\pi} 1\cdot\frac{\sin rx}{r}dx\right]$$

$$= \frac{2}{\pi}\left[0 + \frac{1}{r^2}\left(\cos rx\right)_0^{\pi}\right] = \frac{2}{\pi r^2}(\cos r\pi - \cos 0)$$

$$\therefore\ a_r = -\frac{4}{\pi r^2} \text{ for odd values of } r, \text{ i.e., } r = 1, 3, 5, \cdots.$$

and $a_r = 0$ for even values of r, i.e., $r = 2, 4, 6, \cdots$.

Hence the coefficients or cosines of even multiplies are zero.

$$\therefore\ x = \frac{\pi}{2} - \frac{4}{\pi}\left(\cos x + \frac{1}{3^2}\cos 3x + \frac{1}{5^2}\cos 5x + \cdots\right)$$

Putting $x = 0$, we have

$$0 = \frac{\pi}{2} - \frac{4}{\pi}\left(1 + \frac{1}{3^2} + \frac{1}{5^2} + \cdots\right) \quad \text{or} \quad \frac{\pi^2}{8} = 1 + \frac{1}{3^2} + \frac{1}{5^2} + \cdots.$$

4. Obtain the Fourier series for a function of $f(x)$ where

$$f(x) = \begin{cases} \cos x \text{ for } & 0 \le x \le \pi \\ -\cos x \text{ for } & -\pi \le x \le 0 \end{cases}$$

Let

$$f(x) = a_0 + \sum_{r=1}^{\infty}(a_r \cos rx + b_r \sin rx)$$

$$a_0 = \frac{1}{2\pi}\int_{-\pi}^{\pi} f(x)dx = \frac{1}{2\pi}\int_{-\pi}^{0} -\cos x dx + \frac{1}{2\pi}\int_{0}^{\pi} \cos x dx$$

$$= \frac{1}{2\pi}\Big(-\sin x\Big)_{-\pi}^{0} + \frac{1}{2\pi}\Big(\sin x\Big)_{0}^{\pi} = 0$$

$$a_r = \frac{1}{\pi}\int_{-\pi}^{\pi} f(x)\cos rx dx = 0$$

$$b_r = \frac{1}{\pi}\int_{-\pi}^{\pi} f(x)\sin rx dx = \frac{4r}{\pi(r^2-1)}[1 + (-1)^r]$$

$= 0$ for odd values of r

$= \dfrac{8r}{\pi(r^2-1)}$ for even values of r

$$\therefore f(x) = \frac{8}{\pi}\left[\frac{2}{1.3}\sin 2x + \frac{4}{3.5}\sin 4x + \cdots\right]$$

5. Find the Fourier series for the function $f(x)$ defined as

$$f(x) = \begin{cases} -k, & -\pi < x < 0 \\ k, & 0 < x < \pi \end{cases}$$

Hence show that

$$\frac{\pi}{4} = 1 - \frac{1}{3} + \frac{1}{5} - \frac{1}{7} + \cdots$$

[C.U. (Hons.) 2008]

Let $f(x) = a_0 + \sum a_r \cos rx + \sum b_r \sin rx$.

Here

$$a_0 = \frac{1}{2\pi}\int_{-\pi}^{\pi} f(x)dx = \frac{1}{2\pi}\int_{-\pi}^{0} -kdx + \frac{1}{2\pi}\int_{0}^{\pi} kdx = 0$$

$$a_r = \frac{1}{\pi}\int_{-\pi}^{\pi} f(x)\cos rx dx = \frac{1}{\pi}\int_{-\pi}^{0} -k\cos rx dx + \frac{1}{\pi}\int_{0}^{\pi} k\cos rx dx = 0$$

$$b_r = \frac{1}{\pi}\left[\int_{-\pi}^{0} -k\sin rx dx + \int_{0}^{\pi} k\sin rx dx = \frac{2k}{r\pi}(1-\cos r\pi)\right]$$

$$\therefore\ b_r = 0 \quad \text{for even values of } r$$

$$= \frac{4k}{r\pi} \quad \text{for odd values of } r$$

$$\therefore\ f(x) = \sum_{1,3,5}^{\infty} \frac{4k}{r\pi}\sin rx = \frac{4k}{\pi}\left(\sin x + \frac{\sin 3x}{3} + \frac{\sin 5x}{5} + \cdots\right)$$

Putting $x = \frac{\pi}{2}$, we have

$$f\left(\frac{\pi}{2}\right) = k = \frac{4k}{\pi}\left(1 - \frac{1}{3} + \frac{1}{5} - \frac{1}{7} + \cdots\right) \quad \text{or} \quad \frac{\pi}{4} = 1 - \frac{1}{3} + \frac{1}{5} - \frac{1}{7} + \cdots$$

Exercises

1. State Fourier's theorem. What are the "limitations" under which it holds?

2. Write down the Fourier series in a general form, and calculate the coefficients of the series.

3. Expand a square wave into a Fourier series.

4. Expand a saw-tooth wave into a Fourier series.

5. If $y = x^2$ over the range $x = 0$ to $x = l$, find a Fourier expansion for y for points within the above-mentioned limits.

[**Ans.** $\left[\frac{l^3}{3} + \left(\frac{l^2}{\pi^2}\right)\left\{\cos\frac{\pi x}{l} + \left(\cos\frac{2\pi x}{l}\right)/4 + \cdots\right\} - \left(\frac{l^2}{\pi}\right)\left\{\sin\frac{\pi x}{l} + \left(\sin\frac{2\pi x}{l}\right)/2 + \cdots\right\}\right]$

6. The current in a circuit with a half-wave rectifier is given by $i = I_0 \sin wt$ between the limits $wt = 0$ to $wt = \pi$ and $i = 0$ between the limits $wt = \pi$ to $wt = 2\pi$. Find an expression for the rectified current in harmonic series.

[**Ans.** $\left[i = \left(\frac{I_0}{2\pi}\right)\{2 + \pi\sin wt - 4(\cos 2wt)/3 \cdots + (\cos 4wt)/15 + \cdots\}\right]$

7. Express $f(x) = x$ as a Fourier series in the interval $-\pi < x < \pi$. **[C.U. (Hons.) 1980, '95, 98]**

[**Ans.** Writing the series as

$$f(x) = a_0 + \sum_{r=1}^{\infty} a_r \cos rx + \sum_{r=1}^{\infty} b_r \sin rx$$

we have

$$a_0 = \frac{1}{2\pi}\int_{-\pi}^{\pi} x\,dx = \frac{1}{2\pi}\left(\frac{x^2}{2}\right)_{-\pi}^{\pi} = 0$$

$$a_r = \frac{1}{\pi}\int_{-\pi}^{\pi} x\cos rx\,dx = \frac{1}{\pi}\left[x\frac{\sin rx}{r}\Bigg|_{-\pi}^{\pi} - \frac{1}{r}\int_{-\pi}^{\pi}\sin rx\,dx\right] = 0$$

but

$$b_r = \frac{1}{\pi}\int_{-\pi}^{\pi} x \sin rx\,dx = \frac{1}{\pi}\left[-x\frac{\cos rx}{r}\Big|_{-\pi}^{\pi} + \frac{1}{r}\int_{-\pi}^{\pi}\cos rx\,dx\right]$$

$$= (-1)^{r+1}\frac{2}{r} = \frac{2}{r} \text{ (when } r \text{ is odd)}$$

$$= -\frac{2}{r} \text{ (when } r \text{ is even)}$$

$$\therefore\ f(x) = 2\left(\sin x - \frac{\sin 2x}{2} + \frac{\sin 3x}{3} - \frac{\sin 4x}{4} + \cdots\right).$$

The series consists of only the sine terms. This is also evident since $f(x) = x$ is an odd function.

8. Expand $f(x) = x^2, 0 < x < 2\pi$ in a Fourier series when the period is 2π, to show that

$$f(x) = \frac{4\pi^2}{3} + \sum_{n=1}^{\infty}\left(\frac{4}{n^2}\cos nx - \frac{4\pi}{n}\sin nx\right)$$ **[C.U. (Hons.) 1986, '90]**

Let

$$f(x) = a_0 + \sum_{n=1}^{\infty}(a_n \cos nx + b_n \sin nx),$$

so that we have

$$a_0 = \frac{1}{2\pi}\int_0^{2\pi} f(x)\,dx = \frac{1}{2\pi}\int_0^{2\pi} x^2\,dx = \frac{1}{2\pi}\left(\frac{x^3}{3}\right)_0^{2\pi} = \frac{4\pi^2}{3}$$

$$a_n = \frac{1}{\pi}\int_0^{2\pi} x^2 \cos nx\,dx = \frac{1}{\pi}\left[\frac{x^2}{n}\sin nx\Big|_0^{2\pi} - \int_0^{2\pi}\frac{2x}{n}\sin nx\,dx\right]$$

$$= -\frac{2}{n\pi}\int_0^{2\pi} x \sin nx = \frac{4}{n^2} \quad (n = \text{ any integer})$$

and

$$b_n = \frac{1}{\pi}\int_0^{2\pi} x^2 \sin nx\,dx = \frac{1}{\pi}\left[-\frac{x^2}{n}\cos nx\Big|_0^{2\pi} + \frac{2}{n}\int_0^{2\pi} x\cos nx\,dx\right]$$

$$= -\frac{4\pi}{n} \quad (n = \text{ any integer}).$$

The function $f(x) = x^2$ being even, the series should have only the cosine terms and the constant term, but since it is periodic, it has all the terms present.

Hence the result.

9. Expand $f(x) = A$, where A is a constant in the range $0 < x < \pi$ in a Fourier series. Hence show that

$$\frac{\pi}{4} = 1 - \frac{1}{3} + \frac{1}{5} - \frac{1}{7} + \cdots$$ **[cf. C.U. (Hons.) 1987, '96]**

Expanding $f(x)$ here in a half-range sine series in the range $0 < x < \pi$, we have the Fourier series given by

$$f(x) = \sum_{r=1}^{\infty} b_r \sin rx \text{ where } b_r = \frac{2}{\pi}\int_0^{\pi} f(x)\sin rx\,dx.$$

Now

$$b_r = \frac{2}{\pi}\int_0^{\pi} f(x)\sin rx dx = \frac{2A}{\pi}\left(-\frac{\cos rx}{r}\right)_0^{\pi} = \frac{2A}{\pi r}(1-\cos r\pi)$$

$$= \frac{4A}{\pi r} \quad (\text{if } r \text{ is odd, } \cos r\pi = -1)$$

$$= 0 \quad (\text{if } r \text{ is even, } \cos r\pi = 1)$$

Therefore we have

$$f(x) = A = \frac{4A}{\pi}\sum_{r=1,3,5,\cdots}^{\infty}\frac{1}{r}\sin rx = \frac{4A}{\pi}\left[\sin x + \frac{1}{3}\sin 3x + \frac{1}{5}\sin 5x + \cdots\right]$$

Putting $x = \frac{\pi}{2}$, we get

$$\frac{\pi}{4} = 1 - \frac{1}{3} + \frac{1}{5} - \frac{1}{7} + \cdots$$

10. Show that the function

$$f(x) = \begin{cases} 0 \text{ for } -\pi \le x < 0 \\ x \text{ for } 0 \le x < \pi \end{cases}$$

can be expanded in a Fourier series as

$$f(x) = \frac{\pi}{4} - \frac{2}{\pi}\sum_{n=1}^{\infty}\frac{\cos(2n-1)x}{(2n-1)^2} - \sum_{n=1}^{\infty}(-1)^n\frac{\sin nx}{n}.$$

[C.U. (Hons.) 1984 (Spl.) '91]

Let

$$f(x) = a_0 + \sum_{n=1}^{\infty} a_n \cos nx + \sum_{n=1}^{\infty} b_n \sin nx,$$

so that we have

$$a_0 = \frac{1}{2\pi}\int_{-\pi}^{\pi} f(x)dx = \frac{1}{2\pi}\int_{-\pi}^{0} 0.dx + \frac{1}{2\pi}\int_0^{\pi} x\,dx = \frac{\pi}{4}$$

$$a_n = \frac{2}{2\pi}\int_{-\pi}^{0} 0.\cos nx\,dx + \frac{2}{2\pi}\int_0^{\pi} x\cos nx\,dx$$

$$= \frac{1}{\pi}\left[x\frac{\sin nx}{n} - 1\left(-\frac{1}{n^2}\cos nx\right)\right]_0^{\pi}$$ [applying Kronečker's formula as given and discussed in problem no. 12 of the Exercises]

$$= \frac{1}{\pi n^2}(\cos n\pi - 1)$$

$$b_n = \frac{1}{\pi}\int_{-\pi}^{0} 0\sin nx\,dx + \frac{1}{\pi}\int_0^{\pi} x\sin nx$$

$$= \frac{1}{\pi}\left[x\left(-\frac{\cos nx}{n}\right) - 1\left(-\frac{1}{n^2}\sin nx\right)\right]_0^{\pi} = -\frac{1}{n}\cos n\pi$$

$$= +\frac{1}{n} \quad (\text{for } n = \text{ odd integers})$$

$$= -\frac{1}{n} \quad (\text{for } n = \text{ even intergers})$$

$$\therefore\ b_n = (-1)^n\left(-\frac{1}{n}\right)$$

$$\therefore f(x) = \frac{\pi}{4} + \frac{1}{\pi}\sum_{n=1}^{\infty}\frac{\cos(n\pi - 1)}{n^2}\cos nx - \sum_{n=1}^{\infty}(-1)^n\frac{\sin nx}{n}$$

$$= \frac{\pi}{4} + \frac{1}{\pi}\left[-\frac{2\cos x}{1^2} - \frac{2\cos 3x}{3^2} - \frac{2\cos 5x}{5^2}\cdots\right] - \sum_{n=1}^{\infty}(-1)^n\frac{\sin nx}{n} \quad (1)$$

$$= \frac{\pi}{4} - \frac{2}{\pi}\sum_{n=1}^{\infty}\frac{\cos(2n-1)x}{(2n-1)^2} - \sum_{n=1}^{\infty}(-1)^n\frac{\sin nx}{n}.$$

N.B. At $x = 0, f(x) = 0$, so that from (1) we have

$$\therefore 0 = \frac{\pi}{4} - \frac{2}{\pi}\left[\frac{1}{1^2} + \frac{1}{3^2} + \frac{1}{5^2} + \cdots\right] - 0 \text{ or } \frac{2}{\pi}\left[\frac{1}{1^2} + \frac{1}{3^2} + \frac{1}{5^2} + \cdots\right] = \frac{\pi}{4} \text{ or } 1 + \frac{1}{3^2} + \frac{1}{5^2}\cdots = \frac{\pi^2}{8}.$$

11. Analyse, with the help of Fourier's theorem, a square periodic wave given by the following form

$$y = A \text{ (constant) for } 0 \le t \le \tfrac{T}{2}$$

$$y = 0 \qquad \text{for } \tfrac{T}{2} \le t \le T.$$

What harmonics are present in the expansion? **[C.U. (Hons.) 1988]**

(For answer see Art. 16.7 Example 1)

12. Obtain Fourier Series expansion of the following function with period 2π.

$$f(x) = \begin{cases} x(\pi - x) & \text{for } 0 \le x \le \pi \\ x(\pi + x) & \text{for } -\pi \le x \le 0 \end{cases}$$

Hence prove that

$$1 - \frac{1}{3^3} + \frac{1}{5^3} - \frac{1}{7^3} + \cdots = \frac{\pi^3}{32}.$$ **[C.U. (Hons.) 1992]**

Let

$$f(x) = a_0 + \sum_{n=1}^{\infty}a_n\cos nx + \sum_{n=1}^{\infty}b_n\sin nx,$$

so that we have

$$a_0 = \frac{1}{2\pi}\int_{-\pi}^{\pi}f(x)\,dx = \frac{1}{2\pi}\int_{-\pi}^{0}x(\pi + x)\,dx + \frac{1}{2\pi}\int_0^{\pi}x(\pi - x)\,dx$$

$$= \frac{1}{2\pi}\left[\pi\frac{x^2}{2} + \frac{x^3}{3}\right]_{-\pi}^{0} + \frac{1}{2\pi}\left[\pi\frac{x^2}{2} - \frac{x^3}{3}\right]_0^{\pi} = 0$$

$$a_n = \frac{1}{\pi}\int_{-\pi}^{0}x(\pi + x)\cos nx\,dx + \frac{1}{\pi}\int_0^{\pi}x(\pi - x)\cos nx\,dx$$

$$= \frac{\pi}{\pi}\int_{-\pi}^{0}x\cos nx\,dx + \frac{1}{\pi}\int_{-\pi}^{0}x^2\cos nx\,dx + \frac{\pi}{\pi}\int_0^{\pi}x\cos nx\,dx - \frac{1}{\pi}\int_0^{\pi}x^2\cos nx\,dx$$

$$= \left[\frac{x\sin nx}{n} - (1)\frac{(-\cos nx)}{n^2}\right]_{-\pi}^{0} + \frac{1}{\pi}\left[x^2\frac{\sin nx}{n} - (2x)\frac{(-\cos nx)}{n^2} + (2)\frac{(-\sin nx)}{n^3}\right]_{-\pi}^{0}$$

$$+ \left[x\frac{\sin nx}{n} - (1)\frac{(-\cos nx)}{n^2}\right]_0^{\pi} - \frac{1}{\pi}\left[x^2\frac{\sin nx}{n} - (2x)\frac{(-\cos nx)}{n^2} + (2)\frac{(-\sin nx)}{n^3}\right]_0^{\pi}$$

[using Kronecker's formula $\int_0^a f(x)\phi(x)\,dx$ (where $f(x)$ is a polynomial of x and $\phi(x)$ is $\cos nx$ or $\sin nx$) $= [f(x)\phi_1(x) - f'(x)\phi_2(x) + f''(x)\phi_3(x)\cdots$ so on$]_0^a$ where $f'(x), f''(x)$ etc. are the first, second etc. derivatives of $f(x)$ and $\phi_1(x), \phi_2(x)$ etc are the first, second etc. integrals of $\phi(x)$]

$$\therefore\ a_n = \frac{1}{n^2} - \frac{\pi\sin n\pi}{n} - \frac{\cos n\pi}{n^2} + \frac{1}{\pi}\left[\pi^2\frac{\sin n\pi}{n} + \frac{2\pi\cos n\pi}{n^2} - \frac{2}{n^3}\sin n\pi\right]$$

$$+\frac{\pi\sin n\pi}{n} + \frac{\cos n\pi}{n^2} - \frac{1}{n^2} - \frac{1}{\pi}\left[\frac{\pi^2\sin n\pi}{n} + \frac{2\pi\cos n\pi}{n^2} - \frac{2}{n^3}\sin n\pi\right]$$

$$= 0.$$

Again,

$$b_n = \frac{1}{\pi}\int_{-\pi}^0 x(\pi + x)\sin nx\,dx + \frac{1}{\pi}\int_0^\pi x(\pi - x)\sin nx\,dx$$

$$= \frac{\pi}{\pi}\int_{-\pi}^0 x\sin nx\,dx + \frac{1}{\pi}\int_{-\pi}^0 x^2\sin nx\,dx + \frac{\pi}{\pi}\int_0^\pi x\sin nx\,dx - \frac{1}{\pi}\int_0^\pi x^2\sin nx\,dx$$

$$= \frac{1}{\pi}\int_{-\pi}^0 x^2\sin nx\,dx - \frac{1}{\pi}\int_0^\pi x^2\sin nx\,dx\ \left[\because \int_{-\pi}^0 x\sin nx\,dx = -\int_0^\pi x\sin nx\,dx\right]$$

On integration, using Kronecker's formula, and putting the limits, we have

$$b_n = \frac{4}{\pi n^3}(1 - \cos n\pi) = 0\ \text{(for } n = \text{ even integer or zero)}$$

$$= \frac{8}{n^3\pi}\ \text{(for } n = \text{ odd integer)}$$

$$\therefore\ f(x) = \sum_{n=1,3,5\text{ etc.}}^{\infty}\frac{8}{n^3\pi}\sin nx\ \text{(for } n = \text{odd integers only)}$$

$$= \frac{8}{\pi}\left[\frac{\sin x}{1^3} + \frac{\sin 3x}{3^3} + \frac{\sin 5x}{5^3} + \cdots\right].$$

Putting $x = \frac{\pi}{2}$, we have

$$f\left(\frac{\pi}{2}\right) = \frac{8}{\pi}\left[\sin\frac{\pi}{2} + \frac{\sin\frac{3\pi}{2}}{3^3} + \frac{\sin 5\frac{\pi}{2}}{5^3} + \cdots\right].$$

Also

$$f\left(\frac{\pi}{2}\right) = \frac{\pi}{2}\left(\pi - \frac{\pi}{2}\right) = \frac{\pi^2}{4}\quad \left[\text{putting } \tfrac{\pi}{2} = x \text{ in } f(x) = x(\pi - x)\right]$$

Equating the above two expressions, the desired result follows.

13. A sinusoidal voltage $E\sin wt$ is passed through a half-wave rectifier which clips the negative portion of the work. The resulting periodic function is given by

$$f(t) = \begin{cases} 0 & \text{for } -\pi/2 < t \le 0 \\ E\sin wt & \text{for } 0 < t \le \frac{\pi}{2} \end{cases}$$

where $T = \frac{2\pi}{w}$. Develop the function in a Fourier series. [C.U. (Hons.) '94]

Let $f(t) = a_0 + \sum_{n=1}^{\infty}(a_n \cos nwt + b_n \sin nwt)$, so that we have

$$a_0 = \frac{1}{T}\int_{-T/2}^{T/2} f(t)\,dt = \frac{1}{T}\left[\int_{-T/2}^{0} 0.dt + \int_{0}^{T/2} E\sin wt\,dt\right] = \frac{E}{\pi}$$

$$a_n = \frac{2}{T}\int_{-T/2}^{0} 0.\cos nwt\,dt + \frac{2}{T}\int_{0}^{T/2} E\sin wt\cos nwt\,dt$$

$$= \frac{2E}{2T}\int_{0}^{T/2}\left[\sin(wt+nwt)+\sin(wt-nwt)\,dt\right]$$

$$= \frac{E}{T}\left\{-\frac{\cos(n+1)wt}{w(n+1)}+\frac{\cos(n-1)wt}{w(n-1)}\right\}_0^{T/2} = \frac{E}{wT}\left[\frac{1-\cos(n+1)\pi}{n+1}-\frac{1-\cos(n-1)\pi}{n-1}\right]$$

$$= -\frac{E}{\pi}\,\frac{1+\cos n\pi}{n^2-1}.$$

Obviously a_1 remains undefined for $n = 1$, and may be evaluated as follows

$$a_1 = \frac{2}{T}\int_{-T/2}^{0} 0\cdot\cos wt\,dt + \frac{2}{T}\int_{0}^{T/2} E\sin wt\cos wt\,dt = \frac{2E}{2T}\int_{0}^{T/2}\sin 2wt\,dt = 0.$$

Now,

$$b_n = \frac{2}{T}\int_{-T/2}^{0} 0\cdot\sin nwt\,dt + \frac{2}{T}\int_{0}^{T/2} E\sin wt\sin nwt\,dt$$

$$= \frac{2E}{2T}\int_{0}^{T/2}\left[\cos(n-1)wt-\cos(n+1)wt\right]dt$$

$$= \frac{E}{wT}\left[\frac{\sin(n-1)wt}{(n-1)}-\frac{\sin(n+1)wt}{n+1}\right]_0^{T/2}$$

$$= \frac{E}{2\pi}\left[\frac{\sin(n-1)\pi}{(n-1)}-\frac{\sin(n+1)\pi}{(n+1)}\right] = 0.$$

For $n = 1$, b_1 remains undefined, which is evaluated as follows

$$b_1 = \frac{2}{T}\int_{0}^{T/2} E\sin wt\sin wt\,dt = \frac{2E}{2T}\int_{0}^{T/2}(1-\cos 2wt)dt = \frac{E}{2}.$$

Substituting the values of a_0, a_n and b_n in the series, we get the result.

14. Determine the Fourier series expansion of the following function

$$f(x) = \begin{cases} a \text{ for } -\pi \le x \le 0 \\ b \text{ for } 0 \le x \le \pi \end{cases}$$

(See worked out examples) [C.U. (Hons.) '83]

15. Derive the Fourier series expansion of the function $f(x) = A\cos(kx)$. [C.U. (Hons.) '93]
(See worked out examples).

16. The function $f(x) = x^2$ is defined within the interval $-\pi \le x \le \pi$ and outside it is periodic.

(a) Show whether the function is even or odd within the interval $-\pi \leq x \leq \pi$.

(b) Expand $f(x)$ in a Fourier series to show

$$f(x) = \frac{\pi^2}{3} + \sum_{n=1}^{\infty} \frac{4}{n^2}(-1)^n \cos nx$$

(c) Use the above result to show that, $\displaystyle\sum_{n=1}^{\infty} \frac{1}{n^2} = \frac{\pi^2}{6}$

(d) Can you expand $f(x) = \tan x$ in a Fourier series? Explain. **[C.U. (Hons.) '97, 07]**

(For answers to parts from (a) to (c) see Art 16.6.1)

Answer to (d): $f(x) = \tan x$ cannot be expanded in a Fourier series since the function does not satisfy Dirichlet's condition of having finite number of discontinuities in the interval ($-\pi$ to π) i.e. there are infinite number of discontinuities of $f(x) = \tan x$ in the interval ($-\pi$ to π) as shown in the attached Fig 1.

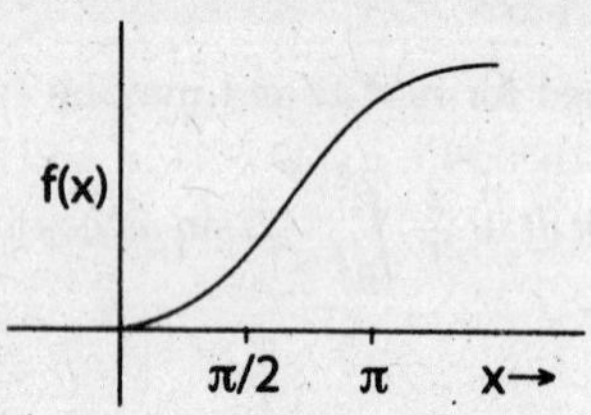

Fig. 1

17. State Fourier's theorem, mentioning the conditions of validity for discontinuous functions. Expand the function

$$f(x) = x \text{ for } -\pi < x < \pi$$
$$f(x + 2\pi) = f(x)$$

in a Fourier series. Is this series differentiable every-where? **[C.U. (Hons.) '98, 07]**

(For answer see worked out examples and Dirichlet's conditons)

18. (a) Show that for an odd function, the Fourier series reduces to Fourier sine series. (See Art. 16.6)

(b) Expand the function

$$f(x) = |x| \text{ for } -\pi \leq x \leq \pi$$
$$f(x + 2\pi) = f(x)$$

in a Fourier series. Is this series differentiable everywhere?

[C.U. (Hons.) '99, 07, cf Burd '95]

Ans. Let

$$f(x) = a_0 + \sum_{r=1}^{\infty} a_r \cos rx + \sum_{r=1}^{\infty} b_r \sin rx$$

since $f(x) = |x|$ is an even function, all the sine terms disappear, i.e., $b_n = 0$. Also $f(x) = |x|$ means that

$$f(x) = \begin{cases} -x, & -\pi \leq x \leq 0 \\ x, & 0 \leq x \leq \pi \end{cases}$$

Now

$$a_0 = \frac{1}{2\pi}\int_{-\pi}^{\pi} x dx = \frac{2}{2\pi}\int_0^{\pi} x dx = \frac{1}{\pi}\cdot\frac{\pi^2}{2} = \frac{\pi}{2}$$

Also

$$a_r = \frac{1}{\pi}\int_{-\pi}^{\pi} x\cos rx = \frac{2}{\pi}\int_0^{\pi} x\cos rx = \frac{2}{\pi}\left[\left(\frac{x\sin rx}{r}\right)_0^{\pi} - \int_0^{\pi} 1\cdot\frac{\sin rx}{r}dx\right]$$

$$= \frac{2}{\pi}\left[0 + \left(\frac{\cos rx}{r^2}\right)_0^{\pi}\right] = \frac{2}{\pi r^2}(\cos r\pi - \cos 0) = -\frac{4}{\pi r^2} \text{ (for odd values of } r)$$

$$= 0 \quad \text{(for even values of } r)$$

$$\therefore\ f(x) = |x| = \frac{\pi}{2} - \frac{4}{\pi}\left[\frac{1}{1^2}\cos x + \frac{1}{3^2}\cos 3x + \cdots\right]$$

Putting $x = 0$, we have

$$\frac{\pi^2}{8} = \frac{1}{1^2} + \frac{1}{3^2} + \frac{1}{5^2} + \cdots$$

19. State Fourier's theorem. Is this theorem, applicable at the point of discontinuity of a function? Expand the function

$$f(x) = \begin{cases} 0, & -\pi \le x \le 0 \\ x, & 0 \le x \le \pi \end{cases}$$

$$f(x + 2\pi) = f(x)$$

in a Fourier series. From this expansion, show that

$$1 + \frac{1}{3^2} + \frac{1}{5^2} + \cdots = \frac{\pi^2}{8}$$

[C.U. (Hons.) 2000, 05, 06 (old syll), 07]

(For answer see Dirichlet's conditions and solved problems)

20. Find the value of $\sum_{n=1}^{\infty}\frac{1}{n^2}$ using Fourier's series. **[C.U. (Hons.) 1997, 07]**

(See Art. 16.6.1) **[Answer.** $\frac{\pi^2}{6}$]

21. Prove that $\sum_{n=1}^{\infty}\frac{1}{(2n-1)^2} = \frac{\pi^2}{8}$, using Fourier's series.

22. Using Fourier's series, prove that $\sum_{n=1}^{\infty}\frac{1}{n^4} = \frac{\pi^4}{90}$

(See Art. 16.6.1)

23. (a) If the Fourier series expansion of a function $f(x)$ contains only sine terms, what kind of function is $f(x)$? Is the converse true?

[**Ans.** Odd function—see Art. 16.6]

(b) Let $F(x)$ have a Fourier series expansion

$$F(x) = \sum_{n=1}^{\infty} a_n \cos nx + \sum_{n=1}^{\infty} b_n \sin nx;$$

then prove that

$$< F^2(x) > \equiv \frac{1}{2\pi}\int_{-\pi}^{+\pi} F^2(x)dx = \sum_{n=1}^{\infty} \frac{a_n^2 + b_n^2}{2}.$$

[C.U. (Hons.) 2001, 05 (old syll)]

[**Hints.** From the given expansion, we may write

$$a_n = \frac{1}{\pi}\int_{-\pi}^{\pi} F(x)\cos nx dx \quad \text{and} \quad b_n = \frac{1}{\pi}\int_{-\pi}^{\pi} F(x)\sin nx dx$$

$$\therefore \quad \sum_{n=1}^{\infty} \frac{a_n^2 + b_n^2}{2} = \frac{\frac{1}{\pi^2}\int_{-\pi}^{\pi} F^2(x)\cos^2 nx dx + \frac{1}{\pi^2}\int_{-\pi}^{\pi} F^2(x)\sin^2 nx dx}{2} = \frac{1}{2\pi^2}\int_{-\pi}^{\pi} F^2(x)dx$$

whence the desired result follows.]

24. The function $f(x)$ is expanded in Fourier series

$$f(x) = \frac{a_0}{2} + \sum_{n=1}^{\infty}\left(a_n \cos\frac{n\pi x}{l} + b_n \sin\frac{n\pi x}{l}\right)$$

(a) State the Dirichhet's conditions imposed upon the function $f(x)$

(b) Evaluate b_n's when $f(x)$ is an even function. **[C.U. (Hons.) 2002]**

25. Determine the Fourier series expansion of the function $f(x) = \pi^2 - x^2$ for $-\pi \le x \le \pi$.

[C.U. (Hons.) 2003]

Hints. Let

$$f(x) = a_0 + \sum_{n=1}^{\infty} a_n \cos nx + \sum_{n=1}^{\infty} b_n \sin nx$$

so that we may write

$$a_0 = \frac{1}{2\pi}\int_{-\pi}^{\pi}(\pi^2 - x^2)dx; \quad a_n = \frac{1}{\pi}\int_{-\pi}^{\pi}(\pi^2 - x^2)\cos nx dx \quad \text{and} \quad b_n = \frac{1}{\pi}\int_{-\pi}^{\pi}(\pi^2 - x^2)\sin nx dx$$

Calculate and obtain the result.

26. (a) State the Fourier theorem for expansion of a periodic function.

(b) State the Dirichlet's conditions imposed on the function for Fourier expansion to be possible.

(c) Show that if $y = f(x)$, $-L \le x \le L$ be an even function, then its Fourier series is a Fourier cosine series.

(d) Can $y = \tan x$ be expanded in a Fourier series? Explain. **[C.U. (Hons.) 2004]**
(See worked out examples)

27. (a) State the Dirichlet's conditions for Fourier series expansion of a function $f(x)$.

(b) Let $f(x)$ have a convergent Fourier series expansion

$$f(x) = \frac{1}{2}a_0 + \sum_{n=1}^{\infty}(a_n \cos nx + b_n \sin nx);$$

prove that

$$\frac{1}{\pi}\int_{-\pi}^{\pi}[f(x)]^2 dx = \frac{1}{2}a_0^2 + \sum_{n=1}^{\infty}(a_n^2 + b_n^2)$$

Hints. Proceed as in question 23 above. **[C.U. (Hons.) 2007]**

28. Find the Fourier series expansion for the function $f(x)$ defined by

$$
\begin{aligned}
f(x) &= x \text{ for } 0 < x < \pi \\
&= -x \text{ for } -\pi < x < 0, \\
f(x + 2\pi) &= f(x)
\end{aligned}
$$

[C.U. (Hons.) 2008]

(For answer see worked out examples)

Chapter 17

Differential Equations

17.1 Classification of differential equations

An equation involving the differential coefficients or differentials of an unknown function is a differential equation. The solution of a differential equation is to find out the unknown function under certain conditions. When the unknown function is a function of only one variable, the differential equation is called an *ordinary differential equation* such as

$$\frac{dy}{dx} = \frac{1}{x}.$$

If, however, the unknown function be a function of two or more independent variables, the differential equation is called a *partial differential equation,* such as

$$\frac{\partial z}{\partial x} + \frac{\partial z}{\partial y} = x^2.$$

The *order* of a differential equation is the order of the highest derivative or differential which occurs in the equation. The *degree* of a differential equation which can be written as a polynomial in the derivatives is the degree of the highest order which occurs in the equation. Considering the equation

$$\left(\frac{d^2y}{dx^2}\right)^2 + \left(\frac{dy}{dx}\right)^3 + \alpha y = x^2$$

we see that it is of the second order and also of the second degree. An equation of the form $M(x,y)\,dx + N(x,y)\,dy = 0$ will be *homogeneous* if and only if $M(x,y)$ and $N(x,y)$ are homogeneous functions of the same degree. For example, let us consider the equation $(x^2+y^2)\,dx - 2xy\,dy = 0$ which may be written as

$$\frac{dy}{dx} = \frac{x^2+y^2}{2xy}.$$

This shows that the ratio of two homogeneous functions of the same degree is a homogeneous function of degree zero.

Again the equation $(x^2+y^2)\,dx - (xy^2 - y^3)\,dy = 0$ is *inhomogeneous* because the functions M and N here are not of the same degree.

Illustrative Examples

1. Solve $(x^2+y^2)\,dx - 2xy\,dy = 0$.

This equation can be written as $\dfrac{dy}{dx} = \dfrac{x^2+y^2}{2xy}$.

Putting $y = vx$, we have $\dfrac{dy}{dx} = v + x\dfrac{dv}{dx}$.

$$\therefore \quad v + x\frac{dv}{dx} = \frac{x^2+y^2}{2xy} = \frac{x^2+v^2x^2}{2vx^2} = \frac{1+v^2}{2v} \quad \text{or,} \quad x\frac{dv}{dx} = \frac{1-v^2}{2v}$$

$$\therefore \quad \frac{dx}{x} - \frac{2v}{1-v^2}dv = 0.$$

We thus see that by the above substitution, the equation has been reduced to a form in which the variables have become separable.

Now, on integration, we have

$$\ln x + \ln(1-v^2) = \ln A, \quad (\ln A \text{ being the constant of integration})$$

$$\text{or,} \quad x(1-v^2) = A.$$

Putting $v = \frac{y}{x}$ and simplifying, the solution is $x^2 - y^2 = Ax$.

2. Solve $(2x - 5y + 3)\,dx - (2x + 4y - 6)\,dy = 0$.

The above equation may be written as

$$\frac{dy}{dx} = \frac{2x-5y+3}{2x+4y-6}.$$

This equation appears to be a homogeneous one, but it is not so, and can be readily reduced to the homogeneous form by the following substitution.

Let us put $x = x' + h$ and $y = y' + k$ where h and k are constants, so that $dx = dx'$ and $dy = dy'$, and we may write the above equation as

$$\frac{dy'}{dx'} = \frac{2x'-5y'+2h-5k+3}{2x'+4y'+2h+4k-6}.$$

We may now choose the values of h and k in such a way that $2h-5k+3=0$ and $2h+4k-6=0$ whence we get $h=1$ and $k=1$.

Thus, the equation reduces to

$$\frac{dy'}{dx'} = \frac{2x'-5y'}{2x'+4y'}$$

which is now homogeneous, and can be solved by putting $y' = vx'$, so that

$$\frac{dy'}{dx'} = v + x'\frac{dv}{dx'},$$

and we may therefore write

$$v + x'\frac{dv}{dx'} = \frac{2-5v}{2+4v} \quad \text{or,} \quad x'\frac{dv}{dx'} = \frac{2-5v}{2+4v} - v = \frac{2-7v-4v^2}{2+4v}$$

which on simplification reduces to

$$\frac{dx'}{x'} + \frac{4}{3}\frac{dv}{4v-1} + \frac{2}{3}\frac{dv}{v+2} = 0.$$

Integrating we have,

$$\ln x' + \frac{1}{3}\ln(4v-1) + \frac{2}{3}\ln(v+2) = \ln C_1 \quad \text{or,} \quad x'^3(4v-1)(v+2)^2 = C.$$

Replacing v by $\frac{y'}{x'}$, we get $(4y' - x')(y' + 2x')^2 = C$.

Also replacing x' by $(x-1)$ and y' by $(y-1)$, we obtain the solution as

$$(4y - x - 3)(y + 2x - 3)^2 = C.$$

3. Solve $\dfrac{dy}{dx} = \dfrac{6x - 2y - 7}{3x - y + 4}$.

The above equation cannot be solved by the substitution of $x = x' + h$ and $y = y' + k$, since in this case, the set of equations which give the values of h and k, cannot be solved, because the determinant

$$\begin{vmatrix} 6 & -2 \\ 3 & -1 \end{vmatrix}$$

of the coefficients of the variables is equal to zero.

The above equation may be reduced to one with variables separable by putting $3x - y = v$.

Then $3 - \dfrac{dy}{dx} = \dfrac{dv}{dx}$ and the given equation reduces to

$$\frac{dv}{dx} = 3 - \frac{2v-7}{v+4} = \frac{v+19}{v+4} \quad \text{or,} \quad dx = \frac{v+4}{v+19}dv = \left(1 - \frac{15}{v+19}\right)dv$$

Integrating we have, $x + C = v - 15\ln(v + 19)$.

Restoring the value of v, we get the solution in the form $2x - y - 15\ln(3x - y + 19) = C$.

17.2 Linear Differential Equations of the First order

(a) The equation

$$\frac{dy}{dx} + Py = Q. \tag{1}$$

where P and Q are functions of x alone or constants, is called a *linear differential equation of the first order.*

If $Q \equiv 0$, then the equation (17.2.1) is called a *linear homogeneous equation* and when $Q \neq 0$, the equation is called a *linear inhomogeneous equation.*

The solution of the equation (17.2.1) may be found by multiplying both sides of the equation by $e^{\int P\,dx}$.

$$\therefore \quad \frac{dy}{dx}e^{\int P\,dx} + Pye^{\int P\,dx} = Qe^{\int P\,dx} \quad \text{or,} \quad \frac{d}{dx}\left(ye^{\int P\,dx}\right) = Qe^{\int P\,dx}.$$

Integrating we have

$$ye^{\int P\,dx} = \int Qe^{\int P\,dx}\,dx + C.$$

Thus, the solution is

$$y = e^{-\int P\,dx}\left[\int Qe^{\int P\,dx}\,dx + C\right].$$

Here the factor $e^{\int P\,dx}$ is called an *integrating factor,* because when both sides of the equation (17.2.1) are multiplied by this factor, we get an equation which can be solved by simple integration.

Illustrative Examples

1. Solve $\cos x\dfrac{dy}{dx} + y\sin x = 1.$

Dividing both sides by $\cos x$, the equation reduces to

$$\frac{dy}{dx} + y\tan x = \sec x. \tag{1}$$

Here $P = \tan x$ and $Q = \sec x$

Hence the integrating factor is $e^{\int P\,dx} = e^{\int \tan x\,dx} = e^{\log\sec x} = \sec x.$

Multiplying both sides of the equation (1) by $\sec x$, we have

$$\sec x\frac{dy}{dx} + y\tan x\sec x = \sec^2 x \quad \text{or,} \quad \frac{d}{dx}(y\sec x) = \sec^2 x$$

which on integration gives, $y\sec x = \tan x + C.$

2. Solve $\dfrac{dy}{dx} + y\cos x = \dfrac{1}{2}\sin 2x.$

Here $P = \cos x$ and the integrating factor is $e^{\int P\,dx} = e^{\int \cos x\,dx} = e^{\sin x}.$

Therefore, multiplying both sides of the equation by $e^{\sin x}$, we have

$$e^{\sin x}\frac{dy}{dx} + ye^{\sin x}\cos x = \frac{1}{2}\sin 2xe^{\sin x} \quad \text{or,} \quad \frac{d}{dx}\left(ye^{\sin x}\right) = \frac{1}{2}\sin 2xe^{\sin x}$$

which on integration gives

$$ye^{\sin x} = \frac{1}{2}\int e^{\sin x}\sin 2x\,dx + C = \frac{1}{2}\int e^{\sin x}2\sin x\cos x\,dx + C = \int e^{\sin x}\sin x\,d(\sin x) + C.$$

Now to evaluate the integral on the right-hand side, let us put $\sin x = z$, so that $\cos x\,dx = dz$ and the integral may be written as

$$\int ze^z\,dz = z\int e^z\,dz - \int\left[\int e^z\,dz\right]dz = ze^z - \int e^z\,dz$$

$$= ze^z - e^z = e^z(z-1) = e^{\sin x}(\sin x - 1).$$

Thus we have the solution $ye^{\sin x} = e^{\sin x}(\sin x - 1) + C.$

3. One of the basic equations in electric circuits is $L\frac{di}{dt} + Ri = E(t)$ where $E(t)$ denotes the e.m.f. which is a function of time. Solve the above equation when $E(t) = E_0$ and the initial current is i_0.

We may then write, $L\frac{di}{dt} + Ri = E_0$ or, $\frac{di}{dt} + \frac{R}{L}i = \frac{E_0}{L}$.

Here $P = \frac{R}{L}$ and the integrating factor is $e^{\int P dt} = e^{\int \frac{R}{L} dt} = e^{\frac{R}{L}t}$.

Multiplying both sides of the equation by $e^{\frac{R}{L}t}$, we have

$$e^{\frac{R}{L}t}\frac{di}{dt} + \frac{R}{L}ie^{\frac{R}{L}t} = \frac{E_0}{L}e^{\frac{R}{L}t} \quad \text{or} \quad \frac{d}{dt}\left(i\,e^{\frac{R}{L}t}\right) = \frac{E_0}{L}e^{\frac{R}{L}t}.$$

Integrating we have

$$i\,e^{\frac{R}{L}t} = \frac{E_0}{L}\int e^{\frac{R}{L}t}dt + C = \frac{E_0}{R}e^{\frac{R}{L}t} + C. \quad \therefore \quad i = \frac{E_0}{R} + Ce^{-\frac{R}{L}t}.$$

Now when $t = 0, i = i_0$, then $C = i_0 - \frac{E_0}{R}$, so that the solution is

$$i = \frac{E_0}{R}\left(1 - e^{-\frac{R}{L}t}\right) + i_0 e^{-\frac{R}{L}t}.$$

It is to be noted that as $t \to \infty, i = \frac{E_0}{R}$, a constant.

(b) The equation

$$\frac{dy}{dx} + Py = Qy^n \tag{17.2.2}$$

in which P and Q are functions of x alone and n is any number, is not directly linear, but may be transformed into the linear form on suitable substitution as shown below.

Dividing both sides by y^n, the above equation reduces to

$$y^{-n}\frac{dy}{dx} + Py^{-n+1} = Q.$$

Let us now put $y^{-n+1} = v$, so that $(-n+1)y^{-n}\frac{dy}{dx} = \frac{dv}{dx}$ and we may therefore write the equation as

$$\frac{dv}{dx} + (1-n)Pv = (1-n)Q$$

which is linear in v, and can therefore be solved by the usual method. An equation of the above form is called *Bernoulli's equation.*

Illustrative examples

1. Solve $\frac{dy}{dx} - y = xy^5$.

Dividing both sides by y^5, we get $y^{-5}\frac{dy}{dx} - y^{-4} = x$.

Let us now put $y^{-4} = v$, so that $y^{-5}\frac{dy}{dx} = -\frac{1}{4}\frac{dv}{dx}$, and the above equation reduces to

$$-\frac{1}{4}\frac{dv}{dx} - v = x \quad \text{or,} \quad \frac{dv}{dx} + 4v = -4x$$

which is linear in v. Here the integrating factor is $e^{4\int dx} = e^{4x}$.

Multiplying both sides by e^{4x} and integrating, we have

$$ve^{4x} = -4\int xe^{4x}\,dx = -xe^{4x} + \frac{1}{4}e^{4x} + C \quad \text{or,} \quad y^{-4}e^{4x} = -xe^{4x} + \frac{1}{4}e^{4x} + C$$

or, $\dfrac{1}{y^4} = -x + \dfrac{1}{4} + Ce^{-4x}$ is the required solution.

2. Solve $\dfrac{dy}{dx} + x\sin 2y = x^3\cos^2 y$.

Dividing both sides by $\cos^2 y$, we have $\sec^2 y\dfrac{dy}{dx} + 2x\tan y = x^3$.

Putting $\tan y = z$, we have $\sec^2 y\dfrac{dy}{dx} = \dfrac{dz}{dx}$ and the above equation may be written as $\dfrac{dz}{dx} + 2xz = x^3$ which is linear in z. Here the integrating factor is $e^{\int P\,dx} = e^{\int 2x\,dx} = e^{x^2}$.

Multiplying both sides by e^{x^2} and integrating we have

$$ze^{x^2} = \int x^3 e^{x^2}\,dx + C = \frac{1}{2}e^{x^2}(x^2-1) + C \quad \text{(integrating by parts).}$$

Thus the solution is $e^{x^2}\tan y = \frac{1}{2}e^{x^2}(x^2-1) + C$.

(c) Let us take an equation of the form $\dfrac{d^2y}{dx^2} + P\left(\dfrac{dy}{dx}\right)^2 = Q$ where $P$and Q are functions of x.

We can reduce this equation to the linear form in the following way.

Putting $z = \left(\dfrac{dy}{dx}\right)^2$, we have $2\dfrac{dy}{dx}\dfrac{d^2y}{dx^2} = \dfrac{dz}{dx}$ or, $\dfrac{d^2y}{dx^2} = \dfrac{1}{2}\dfrac{dz}{dy}$.

Hence the given equation becomes $\dfrac{dz}{dy} + 2Pz = 2Q$ which is linear in z and can be solved by the usual method.

17.3 Exact and Inexact differential equations: Use of suitable integrating factors for reducing inexact equations to exact ones

The differential equation

$$M dx + N dy = 0 \tag{17.3.1}$$

where M and N are functions of x, y, is said to be *exact* if the l.h.s. of (17.3.1) is an exact differential and if Cauchy's condition

$$\frac{\partial M}{\partial y} = \frac{\partial N}{\partial x} \tag{17.3.2}$$

is satisfied. But most often equations of the form (17.3.1) are not exact satisfying the condition (17.3.2), but they can be made exact by multiplying them with suitable factors known as *integrating factors.* The equations can then be solved by simple integration as will be evident from the following illustrative examples.

Illustrative Examples

1. Solve the following equation by the use of suitable integrating factor $xdy - ydx + x^2dx = 0$.

[C.U. (Hons.) 1981]

The given equation may be written as $xdy + (x^2 - y)dx = 0$ whence

$$M = x^2 - y \quad \text{or} \quad \frac{\partial M}{\partial y} = -1 \quad \text{and} \quad N = x \quad \text{or} \quad \frac{\partial N}{\partial x} = +1.$$

So the equation is not exact, but can be made exact as shown below. The equation may be written as

$$xdy - ydx = -x^2dx \quad \text{or} \quad \frac{xdy - ydx}{x^2} = -dx \quad \left(\text{multiplying both sides by } \tfrac{1}{x^2}\right)$$

$$\text{or} \quad d\left(\frac{y}{x}\right) = -dx \quad \text{or} \quad \frac{y}{x} = -x + c \quad \text{or} \quad \frac{y}{x} + x = c$$

is the required solution, and $\frac{1}{x^2}$ is the integrating factor, obtained by inspection of the presence of the group of terms $(xdy - ydx)$, as being part of an exact differential.

2. Solve the differential equation $(y + xy^2)dx - xdy = 0$. **[C.U. (Hons.) 1999]**

Here $M = y + xy^2$ or $\dfrac{\partial M}{\partial y} = 1 + 2xy$ and $N = -x$ or $\dfrac{\partial N}{\partial x} = -1$.

So the equation is not exact, but can be made exact as shown below. The equation may be written as

$$ydx - xdy = -xy^2dx \quad \text{or} \quad \frac{ydx - xdy}{y^2} = -xdx \quad \left(\text{multiplying both sides by } \tfrac{1}{y^2}\right)$$

$$\text{or} \quad d\left(\frac{x}{y}\right) = -xdx \quad \text{or} \quad \frac{x}{y} = -\frac{x^2}{2} + C \quad \text{or} \quad \frac{x}{y} + \frac{x^2}{2} = C$$

is the required solution and $\frac{1}{y^2}$ is the integrating factor, obtained by inspection of the presence of the group of terms $(ydx - xdy)$, as being part of an exact differential.

3. Solve the differential equation $xdx + ydy = (x^2 + y^2)dx$.

Here the integrating factor is $\left(\frac{1}{x^2+y^2}\right)$ and the equation is written as

$$\frac{xdx + ydy}{x^2 + y^2} = dx \quad \text{or} \quad \frac{1}{2}\frac{d(x^2 + y^2)}{(x^2 + y^2)} = dx \quad \text{or} \quad \frac{d(x^2 + y^2)}{(x^2 + y^2)} = 2dx.$$

Integrating both sides we have

$$\ln(x^2 + y^2) = 2x + \ln C \quad (\ln C \text{ being the integration constant})$$

$$= \ln e^{2x} + \ln C \quad \text{or} \quad \ln(x^2 + y^2) = \ln Ce^{2x}$$

$\therefore\ x^2 + y^2 = Ce^{2x}$ is the required solution.

Here also the integrating factor is obtained by inspection of the presence of a certain group of terms, as being part of an exact differential.

N.B. In general, however, an integrating factor can be found from the following two simple rules.

(i) If $\frac{\frac{\partial M}{\partial y}-\frac{\partial N}{\partial x}}{N}=f(x)$, a function of x alone, then $e^{\int f(x)dx}$ is an integrating factor of (17.3.1).

(ii) Again if $\frac{\frac{\partial M}{\partial y}-\frac{\partial N}{\partial x}}{M}=-g(y)$, a function of y alone, then $e^{\int g(y)dy}$ is an integrating factor of (17.3.1)

Applying rule (i) to illustrative example 1 above, we see that

$$\frac{\frac{\partial M}{\partial y}-\frac{\partial N}{\partial x}}{N}=\frac{-2}{x}=f(x),$$

so that the integrating factor is

$$e^{\int -\frac{2}{x}dx}=e^{-2\int\frac{dx}{x}}=e^{-2\log x}=e^{\log\frac{1}{x^2}}=\frac{1}{x^2}.$$

Similarly we can verify the same for the other two illustrative examples.

17.4 Solution of second order differential equations

In Physics, the second order differential equation occurs frequently, and we should know the different techniques of solving such an equation. We shall consider here different methods other than the advanced technique of solving second order differential equations by integral transform method, which is beyond the scope of this book. The methods to be followed for solving a second order differential equation are (i) trial solution method, (ii) differential operator method, (iii) series solution method.

17.4.1 Second Order Linear Differential Equations with Constant Coefficients

This type of differential equation is represented by

$$\frac{d^2y}{dx^2}+a\frac{dy}{dx}+by=f(x) \tag{17.4.1}$$

where a and b are constant coefficients and $f(x)$ is a function of x only.

(17.4.1) is a non-homogeneous second order differential equation while with the right hand side equal to zero, the above equation reduces to a homogeneous one, which will be solved by the trial solution method as discussed in Art. 17.4.2.

Before discussing the solution of the above equation, let us consider the various properties of the operator D, which stands for $\frac{d}{dx}$, so that

$$Dy=\frac{dy}{dx};\quad \text{also}\quad \frac{d^2}{dx^2}=D^2\quad \text{and so}\quad D^2y=\frac{d^2y}{dx^2}.$$

If u and v are the differential functions of x, we have the following fundamental laws of algebra.

(i) *Distributive Law:* $D(u+v) = Du + Dv$.

(ii) *Index Law:* $D^m D^n u = D^{m+n} u = D^n D^m u$, if m and n are positive integers.

(iii) $D(Ku) = KDu$, provided k is constant.

(iv) It is to be noted that D does not commute with u, i.e. $D(uv) \neq uDv$.

(v) *Inverse operator* D^{-1}.

If $u = Dv$, then $v = D^{-1}u$.

The symbol D^{-1} is therefore equivalent to an integration.

Again $u = Dv = DD^{-1}u$.

Let us now consider the operator $(D-\alpha)^{-1}$ where α is a constant.

We may write $(D-\alpha)\left[(D-\alpha)^{-1}u\right] = u$.

Let $(D-\alpha)v = u \quad \therefore v = (D-\alpha)^{-1}u$.

Now

$$(D-\alpha)^{-1} = \frac{1}{-\alpha\left(1-\frac{D}{\alpha}\right)} = -\frac{1}{\alpha}\left[1+\frac{D}{\alpha}+\frac{D^2}{\alpha^2}+\frac{D^3}{\alpha^3}+\cdots\right].$$

Illustrative Examples

1. Find the values of (a) $D^{-1}(3x^2)$, (b) $D^{-1}(e^x)$

(a) $D^{-1}(3x^2) = 3D^{-1}(x^2) = x^3$.

(b) $D^{-1}(e^x) = e^x$.

2. Find the value of $(D-\alpha)^{-1}x^5$

$$\begin{aligned}(D-\alpha)^{-1}x^5 &= -\frac{1}{\alpha}\left[1+\frac{D}{\alpha}+\frac{D^2}{\alpha^2}+\frac{D^3}{\alpha^3}+\frac{D^4}{\alpha^4}+\frac{D^5}{\alpha^5}+\cdots\right]x^5 \\ &= -\frac{1}{\alpha}\left[x^5+\frac{5}{\alpha}x^4+\frac{20}{\alpha^2}x^3+\frac{60}{\alpha^3}x^2+\frac{120}{\alpha^4}x+\frac{120}{\alpha^5}\right].\end{aligned}$$

(vi) *To prove that* $f(D)e^{\lambda x} = f(\lambda)e^{\lambda x}$, where $f(p)$ is a rational function of p and is given by

$$f(p) = a_0 + a_1 p + a_2 p^2 + a_3 p^3 + \cdots + b_1 p^{-1} + b_2 p^{-2} + b_3 p^{-3} + \cdots$$

where a_1, a_2, a_3, etc. and b_1, b_2, b_3, etc. are independent of each other.

Since $De^{\lambda x} = \lambda e^{\lambda x}$, $D^{-1}e^{\lambda x} = \lambda^{-1}e^{\lambda x}$ and more generally $D^n e^{\lambda x} = \lambda^n e^{\lambda x}$ where n is a positive or negative integer, we may, therefore, write,

$$\begin{aligned}f(D)e^{\lambda x} &= [a_0 + a_1 D + a_2 D^2 + \cdots + b_1 D^{-1} + b_2 D^{-2} + \cdots]e^{\lambda x} \\ &= [a_0 + a_1\lambda + a_2\lambda^2 + \cdots + b_1\lambda^{-1} + b_2\lambda^{-2} + \cdots]e^{\lambda x} = f(\lambda)e^{\lambda x}.\end{aligned}$$

(vii) *To show that* $f(D)e^{\lambda x}V = e^{\lambda x}f(D+\lambda)V$, where V is any rational function of x.

We know that

$$D(e^{\lambda x}V) = \lambda e^{\lambda x}V + e^{\lambda x}DV = e^{\lambda x}(D+\lambda)V \tag{17.4.2}$$

which may also be written as

$$e^{-\lambda x}D(e^{\lambda x}V) = (D+\lambda)V \tag{17.4.3}$$

so that the effect of operating on $e^{\lambda x}V$ with $[e^{-\lambda x}D]$ is to give $(D+\lambda)V$.

Again

$$\begin{aligned} D^2(e^{\lambda x}V) &= D[\lambda e^{\lambda x}V + e^{\lambda x}DV] = \lambda^2 e^{\lambda x}V + \lambda e^{\lambda x}DV + \lambda e^{\lambda x}DV + e^{\lambda x}D^2V \\ &= e^{\lambda x}(D+\lambda)^2 V. \end{aligned}$$

Hence we may write $D^n(e^{\lambda x}V) = e^{\lambda x}(D+\lambda)^n V$ where n is a positive integer.

Also we may operate again and write

$$\begin{aligned} D^{n+1}(e^{\lambda x}V) &= D\left[e^{\lambda x}(D+\lambda)^n V\right] = e^{\lambda x}(D+\lambda)\left[(D+\lambda)^n V\right] \text{ using equation (17.4.2)} \\ &= e^{\lambda x}(D+\lambda)^{n+1}V. \end{aligned} \tag{17.4.4}$$

Thus it is found to be true for $(n+1)$. Hence it is true for all positive integers. Thus, the result follows.

In the above problem, if V is a constant K (say), then we should have

$$f(D)e^{\lambda x}K = e^{\lambda x}f(D+\lambda)K = Ke^{\lambda x}f(D+\lambda)1.$$

We now proceed to discuss the solution of the equation (17.4.1).

Let y_1 be a particular value of y which satisfies the equation (17.4.1), and let $y = y_1 + y_2$, so that the equation (17.4.1) may be written as

$$\left(\frac{d^2y_1}{dx^2} + a\frac{dy_1}{dx} + by_1\right) + \left(\frac{d^2y_2}{dx^2} + a\frac{dy_2}{dx} + by_2\right) = f(x).$$

As y_1 satisfies the equation (17.4.1), we may write

$$\frac{d^2y_1}{dx^2} + a\frac{dy_1}{dx} + by_1 = f(x) \tag{17.4.5}$$

and hence we have,

$$\frac{d^2y_2}{dx^2} + a\frac{dy_2}{dx} + by_2 = 0. \tag{17.4.6}$$

The quantity y_1 is called the *Particular Integral* which is any solution of the equation (17.4.5). The quantity y_2 is called the *Complementary Function.* This is the solution of the equation (17.4.6) or in other words, the solution of the equation (17.4.1) when its right-hand side is made zero. The sum of the Complementary Function (C.F.) and the Particular Integral (P.I.) gives the complete solution of the equation (17.4.1).

17.4.2 Determination of the Complementary Function: Application of differential operator cum trial solution method

Let us now find out the complementary function. Making the right-hand side of the equation (17.4.1) zero, we have

$$\frac{d^2y}{dx^2} + a\frac{dy}{dx} + by = 0, \tag{17.4.7}$$

which is a homogeneous second order linear differential equation with constant coefficients.

Using operator D which stands for $\frac{d}{dx}$, we have

$$D^2 = \frac{d^2}{dx^2} \text{ so that } Dy = \frac{dy}{dx} \quad \text{and} \quad D^2y = \frac{d^2y}{dx^2}.$$

Equation (17.4.7) may therefore be written as

$$D^2y + a\,Dy + by = 0 \quad \text{or,} \quad (D^2 + aD + b)y = 0. \tag{17.4.8}$$

To solve the equation (17.4.8), let $y \simeq e^{\lambda x}$ be a trial solution.

$$\therefore \quad Dy \simeq \lambda e^{\lambda x} \quad \text{and} \quad D^2y \simeq \lambda^2 e^{\lambda x},$$

so that from the Eqn. (17.4.8) we have

$$(\lambda^2 + a\lambda + b)e^{\lambda x} = 0 \quad \text{or,} \quad \lambda^2 + a\lambda + b = 0 \quad \because e^{\lambda x} \neq 0$$

$$\therefore \quad \lambda = \frac{-a \pm \sqrt{a^2 - 4b}}{2}.$$

Hence the roots of the above equation are given by

$$\lambda_1 = \frac{-a + \sqrt{a^2 - 4b}}{2} \quad \text{and} \quad \lambda_2 = \frac{-a - \sqrt{a^2 - 4b}}{2}$$

$$\therefore \quad \lambda_1 + \lambda_2 = -a \text{ and } \lambda_1\lambda_2 = b.$$

Equation (17.4.8) may thus be written as

$$[D^2 - (\lambda_1 + \lambda_2)D + \lambda_1\lambda_2]\, y = 0$$

$$\text{or,} \quad [D(D - \lambda_1) - \lambda_2(D - \lambda_1)]\, y = 0$$

$$\text{or} \quad (D - \lambda_1)(D - \lambda_2)y = 0. \tag{17.4.9}$$

Putting $u = (D - \lambda_2)y$, the equation (17.4.9) becomes

$$(D - \lambda_1)u = 0 \quad \text{or,} \quad \frac{du}{dx} - \lambda_1 u = 0 \quad \text{or,} \quad \frac{du}{u} = \lambda_1\, dx.$$

Integrating we have,

$$\ln u = \lambda_1 x + \ln A \quad \text{or,} \quad \ln \frac{u}{A} = \lambda_1 x \quad \text{or,} \quad u = Ae^{\lambda_1 x},$$

where A is an arbitrary constant.

$$\therefore \quad (D - \lambda_2)y = Ae^{\lambda_1 x}, \quad \text{or,} \quad \frac{dy}{dx} - \lambda_2 y = A_e^{\lambda_1 x}.$$

Multiplying both sides by the integrating factor $e^{\int -\lambda_2\, dx}$, we have,

$$e^{-\lambda_2 x}\frac{dy}{dx} - \lambda_2 e^{-\lambda_2 x} y = Ae^{(\lambda_1 - \lambda_2)x} \quad \text{or,} \quad \frac{d}{dx}\left[ye^{-\lambda_2 x}\right] = Ae^{(\lambda_1 - \lambda_2)x}$$

$$\text{or,} \quad ye^{-\lambda_2 x} = A\int e^{(\lambda_1 - \lambda_2)x}\, dx$$

$$\therefore \quad ye^{-\lambda_2 x} = \frac{A}{(\lambda_1 - \lambda_2)} e^{(\lambda_1-\lambda_2)x} + B \quad \text{or,} \quad y = Ce^{\lambda_1 x} + Be^{\lambda_2 x} \tag{17.4.10}$$

where $C = \dfrac{A}{\lambda_1 - \lambda_2}$.

Case 1. If λ_1 and λ_2 are real and distinct, then the relation (17.4.10) is the complementary function, since it satisfies the eqn. (17.4.7) and contains two independent arbitrary constants equal in number to the order of the equation.

Case 2. If $\lambda_1 = \lambda_2 = \alpha$ (say), then the relation (17.4.10) reduces to $y = (C + B)e^{\alpha x}$, which is not the general solution, since it involves only one independent constant (as $B + C = C_1$ say) but the equation is of the second order. To find the general solution in this case, we proceed as follows.

Since $\lambda_1 = \lambda_2 = \alpha$, the differential equation (17.4.7) now assumes the form

$$\frac{d^2y}{dx^2} - 2\alpha\frac{dy}{dx} + \alpha^2 y = 0.$$

Let $y = e^{\alpha x}v$, where v is a function of x, be a trial solution of this equation. Substituting this value of y in the left-hand side of the above equation, we have $e^{\alpha x}\dfrac{d^2v}{dx^2} = 0$ or, $\dfrac{d^2v}{dx^2} = 0$ since $e^{\alpha x} \neq 0$.

Integrating twice, we get $v = C_1 + C_2 x$.

Hence the solution of equation (17.4.7) in this case is $y = (C_1 + C_2 x)e^{\alpha x}$ which is the general solution, since it contains two independent arbitrary constants.

The general solution $y = (C_1 + C_2 x)e^{\alpha x}$ when $\lambda_1 = \lambda_2$ may also be derived in the following way.

From (17.4.9), we may write, since $\lambda_1 = \lambda_2$, $(D - \lambda_1)^2 y = 0$.

Hence $(D - \lambda_1)z = 0$ where $(D - \lambda_1)y = z$ or, $z = C_2 e^{\lambda_1 x}$.

Substituting this value of z in $(D - \lambda_1)y = z$, we get

$$\frac{dy}{dx} - \lambda_1 y = C_2 e^{\lambda_1 x}.$$

Here the integrating factor is $\bar{e}^{\lambda_1 x}$, so that

$$ye^{-\lambda_1 x} = \int C_2\, dx = C_1 + C_2 x \quad \therefore \quad y = (C_1 + C_2 x)e^{\lambda_1 x}.$$

Writing $\lambda_1 = \lambda_2 = \alpha$ (say), the general solution is $y = (C_1 + C_2 x)e^{\alpha x}$.

Case 3. If $\lambda_1 = \alpha + i\beta$ and $\lambda_2 = \alpha - i\beta$, i.e. if both λ_1 and λ_2 are complex, then the solution (17.4.10) is given by $y = Ce^{(\alpha+i\beta)x} + Be^{(\alpha-i\beta)x}$.

By adjusting the arbitrary constants, the above solution can be expressed in a more convenient form not involving imaginary expressions and that can be done in the following way:

$$y = e^{\alpha x}[Ce^{i\beta x} + Be^{-i\beta x}] = e^{\alpha x}[C(\cos\beta x + i\sin\beta x) + B(\cos\beta x - i\sin\beta x)]$$

$$= e^{\alpha x}[(C + B)\cos\beta x + i(C - B)\sin\beta x] = e^{\alpha x}[P\cos\beta x + Q\sin\beta x]$$

where $P = C + B$ and $Q = i(C - B)$ are the arbitrary constants which may have any real values.

Again by putting $A\cos\delta$ for P and $A\sin\delta$ for Q, the solution may be written as $y = Ae^{\alpha x}\cos(\beta x - \delta)$.

Also by putting $A\sin\delta$ for P and $A\cos\delta$ for Q, we may have $y = Ae^{\alpha x}\sin(\beta x + \delta)$ where A and δ are the two arbitrary constants.

17.5 Determination of the Particular Integral

To find the particular integral, we can use either *the method of undetermined coefficients* (for the case where a and b are constants) or the more general *method of variation of parameters* or *the operator method.* As the general method of variation of parameters is laborious and as such will not be very convenient and effective for the students, we shall confine our attention to the other two methods.

17.5.1 Method of undetermined coefficients

The method of undetermined coefficients for obtaining the particular integral is particularly well adapted when the driving function $f(x)$ of the inhomogeneous equation (17.4.1 or 17.4.5) is a sum of terms such as sines, cosines, exponentials, powers of x and their products whose derivatives are combinations of a finite number of functions. In this case, we assume for y_p a linear combination of all terms entering with undetermined coefficients and then substitute it into the equation and equate coefficients of like terms.

This may be illustrated by the following examples

Illustrative examples

1. Consider the differential equation $\dfrac{d^2y}{dx^2} + 3\dfrac{dy}{dx} + 2y = x^3 + x$. To obtain the particular integral, let us assume a general polynomial of the third degree of the form $y_p = ax^3 + bx^2 + cx + d$.

Substituting this into the above equation, we obtain

$$(6ax + 2b) + 3(3ax^2 + 2bx + c) + 2(ax^3 + bx^2 + cx + d) = x^3 + x.$$

Equating coefficients of like powers of x on both sides, we have

$$2a = 1, 9a + 2b = 0, 6a + 6b + 2c = 1, 2b + 3c + 2d = 0$$

Solving these equations, we obtain $a = \dfrac{1}{2}, b = -\dfrac{9}{4}, c = \dfrac{23}{4}, d = -\dfrac{51}{8}$, so that the particular integral is

$$y_p = \frac{1}{2}x^3 - \frac{9}{4}x^2 + \frac{23}{4}x - \frac{51}{8} = \frac{1}{8}(4x^3 - 18x^2 + 46x - 51).$$

2. Consider the differential equation $\dfrac{d^3y}{dx^3} + y = e^{2x}\cos 3x$. To obtain the particular integral, we assume $y_p = ae^{2x}\cos 3x + be^{2x}\sin 3x$; substituting this into the above equation, we obtain

$$(9b - 45a)e^{2x}\cos 3x - (9a + 45b)e^{2x}\sin 3x = e^{2x}\cos 3x.$$

Equating the coefficients of like terms on both sides, we have $(9b - 45a) = 1$ and $(9a + 45b) = 0$.

Solving for a and b, we have $a = -\frac{5}{234}$, $b = \frac{1}{234}$, so that the particular integral is

$$y_p = \frac{e^{2x}}{234}(\sin 3x - 5\cos 2x)$$

N.B. If however the driving function $f(x)$ contains terms that appear in the homogeneous solution (i.e., the complementary function), the above method should be altered in the following manner. The term or family of terms that are included in the assumed particular integral, which are also contained in the homogeneous solution, should be multiplied by x^m, where m is of large enough order, so that all of these assumed particular integral terms are of the form xy_h.

3. As an illustrative example, we consider the following differential equation

$$\frac{d^2y}{dx^2} + \frac{dy}{dx} = 2x + 3\cos x + e^{-x}. \tag{17.5.1}$$

If the usual procedure were followed, the particular integral should be assumed to have the form

$$y_p = Ax + B + C\cos x + D\sin x + Ee^{-x} \tag{17.5.2}$$

However since the complementary function is $y_h = C_1 + C_2e^{-x}$, i.e. the homogeneous solution contains a term having e^{-x} in it, the assumed particular integral should be modified, according to the rule stated above, to

$$y_p = Ax^2 + Bx + C\cos x + D\sin x + Exe^{-x} \tag{17.5.3}$$

The reason for this modification is to make the assumed particular integral linearly independent of the complementary function.

Substituting (17.5.3) into (17.5.1) and equating like coefficients on both sides, we obtain $A = 1, B = -2, C = -\frac{3}{2}, D = \frac{3}{2}, E = -1$.

The complete solution of (17.5.1) may now be written as

$$y = y_h + y_p = C_1 + C_2e^{-x} + x^2 - 2x - \frac{3}{2}\cos x + \frac{3}{2}\sin x - xe^{-x}$$

If (17.5.2) had been used instead of (17.5.3), it would have given us that $A = 0$ and E and B would be undetermined. This would mean that the terms x^2 and xe^{-x} would not have been found as part of y_p.

4. Solve

$$(D^2 - 4D + 4)y = x^3e^{2x} + xe^{2x} \tag{17.5.4}$$

Since for the homogeneous equation $(D^2 - 4D + 4)y = 0$ or $(D - 2)^2y = 0$, the roots are equal, we have the complementary function given by $y_h = (C_1 + C_2x)e^{2x}$. Now e^{2x} is a part of $f(x)$ and also occurs in the $C.F$ corresponding to a root of multiplicity two, so that as an assumed particular integral, we take

$$y_p = Ax^5e^{2x} + Bx^4e^{2x} + Cx^3e^{2x} + Ex^2e^{2x} \tag{17.5.5}$$

It is to be noted that terms involving xe^{2x} and e^{2x} are not included, since they appear in the $C.F$ with arbitrary coefficients. Then from (17.5.5)

$$Dy_p = 2Ax^5e^{2x} + (5A + 2B)x^4e^{2x} + (4B + 2C)x^3e^{2x}$$
$$+(3C + 2E)x^2e^{2x} + 2Exe^{2x}$$
$$D^2y_p = 4Ax^5e^{2x} + (20A + 4B)x^4e^{2x} + (20A + 16B + 4C)x^3e^{2x}$$
$$+(12B + 12C + 4E)x^2e^{2x} + (6C + 8E)xe^{2x} + 2Ee^{2x}$$
$$\therefore (D^2 - 4D + 4)y_p = 20Ax^3e^{2x} + 12Bx^2e^{2x} + 6Cxe^{2x} + 2Ee^{2x} = x^3e^{2x} + xe^{2x}$$

Equating coefficients of like terms on both sides, we obtain $20A = 1$, $12B = 0$, $6C = 1$, $2E = 0$, whence we have $A = \frac{1}{20}$, $B = 0$, $C = \frac{1}{6}$ and $E = 0$

$$\therefore y_p = \frac{1}{20}x^5e^{2x} + \frac{1}{6}x^3e^{2x}$$

and the complete solution is written as

$$y = y_h + y_p = C_1e^{2x} + C_2xe^{2x} + \frac{1}{20}x^5e^{2x} + \frac{1}{6}x^3e^{2x}.$$

17.5.2 Operator method

For finding the particular integral by the operator method, we consider the equation

$$(D^2 + aD + b)y = f(x).$$

Hence

$$y_p = \frac{f(x)}{D^2 + aD + b}.$$

Evaluation of the right-hand side involves a knowledge of the properties of the operator D and their applications at appropriate places—as will be evident from the following illustrative examples which illustrate the methods of finding both the complementary function and the particular integral.

Illustrative Examples

1. Solve $\frac{d^2y}{dx^2} + \frac{dy}{dx} - 6y = 0.$

The equation may be written as $(D^2 + D - 6)y_h = 0$ or, $(D - 2)(D + 3)y_h = 0$.

Here the roots are $\lambda_1 = 2$ and $\lambda_2 = -3$, which are real and distinct, and the solution is $y_h = c_1e^{2x} + c_2e^{-3x}$.

2. Solve $\frac{d^2y}{dx^2} - 2a\frac{dy}{dx} + a^2y = 0.$

The equation may be written as $(D^2 - 2aD + a^2)y_h = 0$ or, $(D - a)^2y_h = 0$.

Here the roots λ_1 and λ_2 are equal, that is the equation has repeated roots and $\lambda_1 = \lambda_2 = a$ (say).

So the solution is $y_h = (c_1 + c_2x)e^{ax}$.

3. Solve $\frac{d^2y}{dx^2} + 2\frac{dy}{dx} + 5y = 0.$

The equation may be written as $(D^2 + 2D + 5)y_h = 0.$

Here the roots are complex and are given by $\lambda_1 = -1 + 2i$ and $\lambda_2 = -1 - 2i.$

So the solution, i.e. the complementary function is $y_h = C_1e^{(-1+2i)x} + C_2e^{(-1-2i)x}$ which can be put in the form $y_h = e^{-x}(A\cos 2x + B\sin 2x)$ by adjusting the arbitrary constants.

4. Solve $\frac{d^2y}{dt^2} + w^2y = 0.$

Writing $\frac{d}{dt} = D$, the equation may be written as $(D^2 + w^2)y_h = 0.$

Here the equation for λ is $\lambda^2 + w^2 = 0$, which gives $\lambda_1 = iw$ and $\lambda_2 = -iw.$

The solution is

$$\begin{aligned} y_h &= C_1e^{iwt} + C_2e^{-iwt} = (C_1 + C_2)\cos wt + i(C_1 - C_2)\sin wt \\ &= P\cos wt + Q\sin wt \qquad (1) \end{aligned}$$

where $P = (C_1 + C_2)$ and $Q = i(C_1 - C_2)$ are the two arbitrary constants.

Again by putting $A\cos\delta$ for P and $A\sin\delta$ for Q, the solution (1) may be written as

$$y_h = A\cos(wt - \delta). \qquad (2)$$

Also by putting $A\sin\delta$ for P and $A\cos\delta$ for Q, we may have

$$y_h = A\sin(wt + \delta) \qquad (3)$$

where A and δ are any two arbitrary constants.

Any one of the above three relations may be taken as the solution of $\frac{d^2y}{dt^2} + w^2y = 0$, which is the standard differential equation for a particle executing simple harmonic motion. Linear equation and its solution thus find a very important application in the above problem.

5. Solve $\frac{d^2y}{dt^2} + k\frac{dy}{dt} + w^2y = 0.$

The equation may be written as $(D^2 + kD + w^2)y_h = 0.$

Here the equation for λ is $\lambda^2 + k\lambda + w^2 = 0$ or $\lambda = \frac{-k \pm \sqrt{k^2 - 4w^2}}{2}.$

If $k^2 > 4w^2$, the solution is

$$y_h = C_1e^{\frac{-k+\sqrt{k^2-4w^2}}{2}t} + C_2e^{\frac{-k-\sqrt{k^2-4w^2}}{2}t} = e^{-\frac{k}{2}t}\left[C_1e^{\sqrt{\frac{k^2}{4}-w^2}\,t} + C_2e^{-\sqrt{\frac{k^2}{4}-w^2}\,t}\right].$$

Again if $k^2 < 4w^2$, then $\sqrt{k^2 - 4w^2} = i\sqrt{4w^2 - k^2}$, and the solution is

$$\begin{aligned} y_h &= C_1e^{\frac{-k+i\sqrt{4w^2-k^2}}{2}t} + C_2e^{\frac{-k-i\sqrt{4w^2-k^2}}{2}t} = e^{-\frac{k}{2}t}\left[C_1e^{\frac{i\sqrt{4w^2-k^2}}{2}t} + C_2e^{\frac{-i\sqrt{4w^2-k^2}}{2}t}\right] \\ &= e^{-\frac{k}{2}t}\left[(C_1 + C_2)\cos\sqrt{w^2 - \frac{k^2}{4}}\,t + i(C_1 - C_2)\sin\sqrt{w^2 - \frac{k^2}{4}}\,t\right] \\ &= e^{-\frac{k}{2}t}\left[P\cos\sqrt{w^2 - \frac{k^2}{4}}\,t + Q\sin\sqrt{w^2 - \frac{k^2}{4}}\,t\right] \end{aligned}$$

where P and Q are any two arbitrary constants. Again by putting $A\cos\phi$ for P and $A\sin\phi$ for Q, the solution reduces to

$$y_h = Ae^{-\frac{k}{2}t}\cos\left(\sqrt{w^2 - \frac{k^2}{4}}\,t - \phi\right)$$

where A and ϕ are constants.

The solution given above is a very important one, since it represents the solution of the standard differential equation for damped harmonic motion.

6. Solve $(D^2+4)y_h = x^2$.

To obtain the complementary function, let us put $(D^2+4)y_h = 0$.

Here the roots are $\lambda_1 = +2i$ and $\lambda_2 = -2i$, and hence the complementary function $y_h = A\cos 2x + B\sin 2x$.

Again the particular integral

$$y_p = \frac{1}{D^2+4}x^2 = \frac{1}{4(1+\frac{1}{4}D^2)}x^2 = \frac{1}{4}\left(1+\frac{1}{4}D^2\right)^{-1}x^2$$

$$= \frac{1}{4}\left[1 - \frac{1}{4}D^2 + \frac{1}{16}D^4 - \cdots\right]x^2 = \frac{1}{4}\left(x^2 - \frac{1}{2}\right).$$

$\therefore$ the total solution is $y = y_h + y_p = A\cos 2x + B\sin 2x + \frac{1}{4}\left(x^2 - \frac{1}{2}\right)$.

7. Solve $\frac{d^2y}{dx^2} - 6\frac{dy}{dx} + 9y = x^2$.

The equation may be written as $(D^2 - 6D + 9)y = x^2$

For the complementary function, let us put $(D^2 - 6D + 9)y_h = 0$ or, $(D-3)^2 y_h = 0$.

Here $\lambda_1 = \lambda_2 = 3$; so the complementary function is $y_h = (C_1 + C_2x)e^{3x}$.

Again for the particular integral, we may write

$$y_p = \frac{x^2}{(3-D)^2} = \frac{1}{9}\left[1 - \frac{D}{3}\right]^{-2}x^2 = \frac{1}{9}\left[1 + \frac{2}{3}D + \frac{1}{3}D^2 + \cdots\right]x^2 = \frac{1}{9}\left[x^2 + \frac{4}{3}x + \frac{2}{3}\right]$$

$\therefore$ the total solution is $y = y_h + y_p = (C_1 + C_2x)e^{3x} + \frac{1}{9}\left(x^2 + \frac{4}{3}x + \frac{2}{3}\right)$.

8. Solve $(D^2 - 3D + 2)y = e^x$.

To obtain the complementary function, let us put

$$(D^2 - 3D + 2)y_h = 0 \quad \text{or,} \quad (D-1)(D-2)y_h = 0.$$

Here the roots of the equation are $\lambda_1 = 1$ and $\lambda_2 = 2$, and therefore the complementary function is $y_h = C_1e^x + C_2e^{2x}$. For the particular integral, let us write

$$y_p = \frac{1}{(D-1)}\cdot\frac{1}{(D-2)}e^x.$$

Now let
$$u = \frac{1}{(D-2)}e^x, \text{ so that } (D-2)u = e^x \quad \text{or,} \quad \frac{du}{dx} - 2u = e^x.$$

Multiplying both sides by the integrating factor $e^{-2\int dx}$, i.e. e^{-2x} and then integrating, we may write
$$ue^{-2x} = \int e^x e^{-2x}\,dx = \int e^{-x}\,dx = -e^{-x} \quad \therefore \quad u = -e^x.$$

Now
$$y_p = \frac{1}{(D-1)}u, \quad \text{or,} \quad (D-1)y_p = u \quad \text{or,} \quad \frac{dy}{dx} - y_p = -e^x.$$

Also multiplying both sides by the integrating factor $e^{-\int dx}$, i.e. e^{-x} and then integrating, we have
$$y_p e^{-x} = \int -e^x e^{-x}\,dx \quad \text{or,} \quad y_p = e^x \int -e^x e^{-x}\,dx = -xe^x.$$

Thus, the particular integral is $-xe^x$, so that the total solution is $y = C_1e^x + C_2e^{2x} - xe^x$.

9. Find a particular integral of $(D-a)(D-b)y = Q$.

Here the particular integral is given by
$$y_p = \frac{1}{(D-a)} \cdot \frac{1}{(D-b)}Q.$$

Let $\dfrac{1}{(D-b)}Q = u$ so that $(D-b)u = Q$ or, $\dfrac{du}{dx} - bu = Q$.

Multiplying both sides by the integrating factor e^{-bx} and then integrating, we have
$$ue^{-bx} = \int Qe^{-bx}\,dx \quad \text{or,} \quad u = e^{bx}\int Qe^{-bx}\,dx.$$

Again,
$$y_p = \frac{1}{(D-a)}u, \text{ so that } (D-a)y_p = u \quad \text{or,} \quad \frac{dy_p}{dx} - ay_p = u = e^{bx}\int Qe^{-bx}\,dx.$$

Multiplying both sides by the integrating factor e^{-ax} and then integrating, we may write
$$y_p e^{-ax} = \int e^{bx}e^{-ax}\int Qe^{-bx}\,dx\,dx. \quad \therefore \quad y_p = e^{ax}\int e^{(b-a)x}\int Qe^{-bx}(dx)^2.$$

10. Solve $(D^2 - 3D + 2)y = e^{5x}$.

To obtain the complementary function, let us put
$$(D^2 - 3D + 2)y_h = 0 \text{ or, } (D-1)(D-2)y_h = 0.$$

Here the roots of the equation are 1 and 2, so that the complementary function
$$y_h = C_1e^x + C_2e^{2x}.$$

Again the particular integral is written as
$$y_p = \frac{1}{(D-1)} \cdot \frac{1}{(D-2)}e^{5x}.$$

To obtain the value of the particular integral, we may follow any of the methods given below.

First method

Applying the final formula obtained in example 9 above, we may write

$$y_p = \frac{1}{(D-1)} \cdot \frac{1}{(D-2)} e^{5x} = e^x \int e^{(2-1)x} \int e^{5x} e^{-2x} (dx)^2$$

(putting $a = 1, b = 2$ and $Q = e^{5x}$)

$$= e^x \int e^x \int e^{3x} (dx)^2 = e^x \int e^x \frac{1}{3} e^{3x}\, dx = \frac{1}{3} e^x \int e^{4x}\, dx = \frac{1}{12} e^{5x}.$$

Second method

$$y_p = \frac{1}{(D-1)(D-2)} e^{5x} = \left[-\frac{1}{(D-1)} + \frac{1}{(D-2)} \right] e^{5x} = -\frac{1}{(D-1)} e^{5x} + \frac{1}{(D-2)} e^{5x}$$

Now let $u = \frac{1}{(D-1)} e^{5x}$, so that $(D-1)u = e^{5x}$ or, $\frac{du}{dx} - u = e^{5x}$.

Multiplying both sides by the integrating factor e^{-x} and then integrating, we have

$$ue^{-x} = \int e^{5x} \cdot e^{-x}\, dx \quad \therefore \quad u = e^x \int e^{5x} \cdot e^{-x}\, dx.$$

Similarly putting $v = \frac{1}{(D-2)} e^{5x}$ and proceeding in a similar way, we have,

$$v = e^{2x} \int e^{5x} \cdot e^{-2x}\, dx$$

$$\therefore \quad y_p = -e^x \int e^{5x} e^{-x}\, dx + e^{2x} \int e^{5x} \cdot e^{-2x}\, dx = -\frac{1}{4} e^x e^{4x} + \frac{1}{3} e^{2x} e^{3x} = \frac{1}{12} e^{5x}.$$

Thus, the total solution is $y = y_h + y_p = C_1 e^x + C_2 e^{2x} + \frac{1}{12} e^{5x}$.

11. Solve $(D-3)^2 y = 2e^{4x}$.

For the complementary function, let us put $(D-3)^2 y_h = 0$.

Here $\lambda_1 = \lambda_2 = 3$, and so the complementary function is $y_h = (C_1 + C_2 x) e^{3x}$.

For the particular integral, we have

$$y_p = \frac{1}{(D-3)^2} 2e^{4x} = \frac{2e^{4x}}{(4-3)^2} \text{ (by putting } D = 4)$$

$$= 2e^{4x}.$$

$\therefore$ the total solution is $y = y_h + y_p = (C_1 + C_2 x) e^{3x} + 2e^{4x}$.

12. Solve $(D-2)^2 y = 6e^{2x}$.

For the complementary function, let us put $(D-2)^2 y_h = 0$.

Here $\lambda_1 = \lambda_2 = 2$, and so the complementary function is $y_h = (C_1 + C_2 x) e^{2x}$.

For the particular integral, we have

$$y_p = \frac{1}{(D-2)^2}6e^{2x} = 6e^{2x}\frac{1}{(D+2-2)^2} \quad \text{(putting } D+2 \text{ for } D)$$

$$= 6e^{2x}\frac{1}{D^2}1 = 6e^{2x}\frac{x^2}{2} = 3x^2e^{2x}.$$

$\therefore$ the total solution is $y = y_h + y_p = (C_1 + C_2x)e^{2x} + 3x^2e^{2x}$.

13. Solve $(D^2 - 5D + 6)y = xe^x$.

Here the complementary function is $y_h = C_1e^{2x} + C_2e^{3x}$.

For the particular integral, we have

$$y_p = \frac{xe^x}{D^2 - 5D + 6} = e^x\frac{1}{(D+1)^2 - 5(D+1) + 6}x$$

$\Big(\because$ in the general case $(D^2 + aD + b)y = e^{kx}X$ we have

$$y_p = \frac{1}{D^2 + aD + b}e^{kx}X = e^{kx}\frac{1}{(D+k)^2 + a(D+k) + b}X\Big) = e^x\left[\frac{1}{2}\left(1 - \frac{3}{2}D + \frac{1}{2}D^2\right)^{-1}\right]x$$

$$= \frac{1}{2}e^x\left(1 + \frac{3}{2}D + \cdots\right)x = \frac{1}{2}e^x\left(x + \frac{3}{2}\right).$$

Hence the total solution is $y = y_h + y_p = C_1e^{2x} + C_2e^{3x} + \frac{1}{2}e^x\left(x + \frac{3}{2}\right)$.

14. Solve $(D^2 - 4D + 4)y = x^3e^{2x}$ or, $(D-2)^2y = x^3e^{2x}$.

Here the complementary function is $y_h = (C_1 + C_2x)e^{2x}$.

For the particular integral, we have

$$y_p = \frac{1}{(D-2)^2}x^3e^{2x} = e^{2x}\frac{1}{D^2}x^3 \quad [\text{putting } (D+2) \text{ for } D]$$

$$= e^{2x}\frac{x^5}{20}.$$

$\therefore$ the total solution is $y = y_h + y_p = (C_1 + C_2x)e^{2x} + \frac{1}{20}e^{2x}x^5$.

15. Solve $(D^2 + 4)y = \sin 3x$.

Here the complementary function is $y_h = C_1\cos 2x + C_2\sin 2x$, since $\lambda = \pm 2i$ and the particular integral is written as

$$y_p = \frac{1}{(D^2+4)}\sin 3x = \frac{1}{-(3)^2 + 4}\sin 3x = -\frac{1}{5}\sin 3x$$

(replacing D^2 by $-(3)^2$, since $D(\sin 3x) = 3\cos 3x$, $D^2(\sin 3x) = -3^2\sin 3x$)

$\therefore$ the total solution is $y = y_h + y_p = C_1\cos 2x + C_2\sin 2x - \frac{1}{5}\sin 3x$.

16. Solve $(D^2+1)y = \cos 2x$.

Here the complementary function is $y_h = C_1 \cos x + C_2 \sin x$, since $\lambda = \pm i$, and for the particular integral, we have

$$y_p = \frac{1}{D^2+1}\cos 2x = \frac{1}{-(2)^2+1}\cos 2x = -\frac{1}{3}\cos 2x.$$

$\therefore$ the general solution is $y = y_h + y_p = C_1 \cos x + C_2 \sin x - \frac{1}{3}\cos 2x$

17. Solve $(D^2+1)y = \cos x$.

As in Ex. 16, we have the complementary function given by $y_h = C_1 \cos x + C_2 \sin x$, but the method of finding the particular integral, as employed in Ex. 16 fails here. We may however, substitute, the exponential value of $\cos x$ and proceed. Alternatively, we may proceed as follows.

Let

$$y = \frac{1}{D^2+1}\cos x \quad \text{and} \quad z = \frac{1}{D^2+1}\sin x.$$

$$\therefore \quad y + iz = \frac{1}{D^2+1}(\cos x + i\sin x) = \frac{1}{D^2+1}e^{ix} = e^{ix}\frac{1}{(D+i)^2+1}\cdot 1 = e^{ix}\frac{1}{2iD+D^2}\cdot 1$$

$$= e^{ix}\frac{1}{2iD}\left(1+\frac{D}{2i}\right)^{-1}\cdot 1 = e^{ix}\frac{1}{2i}\frac{1}{D}\cdot 1 = e^{ix}\frac{x}{2i} = \frac{x}{2i}(\cos x + i\sin x).$$

$\therefore$ equating the real part, $y_p = \frac{x}{2}\sin x$

$\therefore$ the total solution is $y = y_h + y_p = C_1 \cos x + C_2 \sin x + \frac{x}{2}\sin x$.

18. Solve $(D^2+3D-4)y = \sin 2x$.

Here the complementary function is $y_h = C_1 e^{x} + C_2 e^{-4x}$, and for the particular integral, we have

$$y_p = \frac{1}{D^2+3D-4}\sin 2x = \frac{1}{(D-1)(D+4)}\sin 2x.$$

The operator here is not of the form $\frac{1}{F(D)^2}$, and so the short method does not apply. However, we may use either of the following procedures to shortern the work.

(a) $$y_p = \frac{1}{(D-1)(D+4)}\sin 2x = \frac{(D+1)(D-4)}{(D^2-1)(D^2-16)}\sin 2x = \frac{1}{100}(D^2-3D-4)\sin 2x$$

[replacing D^2 by $-(2)^2$]

$$= \frac{1}{100}(-4\sin 2x - 6\cos 2x - 4\sin 2x) = -\frac{1}{50}(4\sin 2x + 3\cos 2x).$$

(b) $$y_p = \frac{1}{D^2+3D-4}\sin 2x = \frac{1}{(-4+3D-4)}\sin 2x = \frac{1}{3D-8}\sin 2x = \frac{3D+8}{9D^2-64}\sin 2x$$

$$= -\frac{1}{100}(3D+8)\sin 2x = -\frac{1}{100}(6\cos 2x + 8\sin 2x) = -\frac{1}{50}(4\sin 2x + 3\cos 2x).$$

$\therefore$ the total solution is $y = y_h + y_p = C_1 e^{x} + C_2 e^{-4x} - \frac{1}{50}(4\sin 2x + 3\cos 2x)$.

19. Solve $(D^2 - 2D + 5)y = 10 \sin x$.

For the complementary function, we have here $\lambda = 1 \pm 2i$.

$\therefore$ the complementary function is $y_h = e^x(C_1 \cos 2x + C_2 \sin 2x)$.

For the particular integral, we have

$$y_p = \frac{1}{D^2 - 2D + 5} 10 \sin x = \frac{(D^2+5)+2D}{(D^2+5)^2 - 4D^2} 10 \sin x = \frac{D^2 + 2D + 5}{(-1^2+5)^2+4} 10 \sin x$$

$$= \frac{1}{2}(D^2 + 2D + 5) \sin x = \frac{1}{2}(-\sin x + 2\cos x + 5 \sin x) = 2 \sin x + \cos x.$$

$\therefore$ the total solution is $y = y_h + y_p = e^x(C_1 \cos 2x + C_2 \sin 2x) + 2 \sin x + \cos x$.

17.6 Applications of Linear Equations to problems in Physics

1. Find the time required for a sum of money to double itself at 5% per annum compounded continuously.

If x be the amount after t years, then according to the problem, we may write

$$\frac{dx}{dt} = 0.05x \text{ or } \frac{dx}{x} = 0.05dt.$$

Integrating we have

$$\ln x = 0.05t + \ln k \text{ or } x = ke^{0.05t}.$$

Now at $t = 0$, $x = k$, so that at $t = t, x = 2k$

$$\therefore 2k = ke^{0.05t} \text{ or } \ln 2 = 0.05t \text{ or } t = \frac{2.303 \times \log_{10} 2}{0.05} = 13.86 \text{ years}.$$

2. According to Newton's law of cooling, the rate at which a substance cools in moving air is proportional to the difference between the temperature of the substance and that of the air. If the temperature of the air is 30° and the substance cools from 100° to 70° in 15 minutes, find when the temperature will be 40°.

Let T be the temperature of the substance at time t minutes. we have then

$$\frac{dT}{dt} = -k(T - 30) \text{ or } \frac{dT}{T - 30} = -kdt$$

(We have chosen the constant of proportionality as $-k$. It will be found that k is positive, but if $+k$ is used, it will be found that k is negative)

Integrating between the limits $t = 0$, $T = 100$ and $t = 15$, $T = 70$, we have

$$\int_{100}^{70} \frac{dT}{T - 30} = -k \int_0^{15} dt \text{ or } \ln 40 - \ln 70 = -15k \text{ or } 15k = \ln \frac{7}{4} = 0.56.$$

Integrating again between the limits $t = 0, T = 100$ and $t = t, T = 40$, we have

$$\int_{100}^{40} \frac{dT}{T-30} = -k \int_0^t dt \text{ or } \ln 10 - \ln 70 = -kt \text{ or } kt = \ln 7$$

$$\text{or } \quad 15kt = 15 \ln 7 \text{ or } t = \frac{15 \ln 7}{0.56} = 52 \text{ min.}$$

3. In a certain culture of bacteria, the rate of increase is proportional to the number present. (a) If it is found that the number doubles in 4 hours, how many may be expected at the end of 12 hours? (b) If there are 10^4 at the end of 3 hours and 4×10^4 at the end of 5 hours, how many were there in the beginning?

Let x denote the number of bacteria at time t hours. We have then

$$\frac{dx}{dt} = kx \quad \text{or} \quad \frac{dx}{x} = kdt \tag{1}$$

(a) Integrating (1) between the limits $t = 0, x = x_0$ and $t = 4, x = 2x_0$, we may write

$$\int_{x_0}^{2x_0} \frac{dx}{x} = k \int_0^4 dt \quad \text{or} \quad \ln 2x_0 - \ln x_0 = 4k \quad \text{or} \quad 4k = \ln 2$$

Again integrating (1) between the limits $t = 0, x = x_0$ and $t = 12, x = x$, we have

$$\int_{x_0}^{x} \frac{dx}{x} = k \int_0^{12} dt \quad \text{or} \quad \ln \frac{x}{x_0} = 12k = 3(4k) = 3 \ln 2 = \ln 8$$

whence $x = 8x_0$, that is, there are bacteria which are 8 times the original number.

(b) Integrating (1) between the limits $t = 3, x = 10^4$ and $t = 5, x = 4 \times 10^4$, we have

$$\int_{10^4}^{4\times 10^4} \frac{dx}{x} = k \int_3^5 dt \quad \text{or} \quad \ln 4 = 2k \quad \text{or} \quad k = \ln 2$$

Again integrating (1) between the limits $t = 0, x = x_0$, and $t = 3, x = 10^4$, we have

$$\int_{x_0}^{10^4} \frac{dx}{x} = k \int_0^3 dt \quad \text{or} \quad \ln \frac{10^4}{x_0} = 3k = 3 \ln 2 = \ln 8$$

$\therefore \ \frac{10^4}{x_0} = 8$ or $x_0 = \frac{10^4}{8}$ which gives the original number of bacteria.

4. A certain chemical dissolves in water at a rate proportional to the product of the amount undissolved and the difference between the concentration in a saturated solution and the concentration in the actual solution. In 100 gms of a saturated solution, it is known that 50 gms of the substance are dissolved. If when 30 gms. of the chemical are agitated with 100 gms of water, 10 gms are dissolved in 2 hours, how much will be dissolved in 5 hours?

Let x gms of the chemical remain undissolved after t hours. At this time the concentration of the actual solution is $\frac{30-x}{100}$ and that of a saturated solution is $\frac{50}{100}$. Then according to the problem,

$$\frac{dx}{dt} = kx\left(\frac{50}{100} - \frac{30-x}{100}\right) = kx\frac{x+20}{100}$$

$$\text{or} \quad \frac{dx}{x(x+20)} = \frac{k}{100}dt \quad \text{or} \quad \frac{dx}{x} - \frac{dx}{x+20} = \frac{k}{5}dt.$$

Integrating between $t = 0, x = 30$ and $t = 2, x = 30 - 10 = 20$, we have

$$\int_{30}^{20} \frac{dx}{x} - \int_{30}^{20} \frac{dx}{x+20} = \frac{k}{5}\int_0^2 dt \quad \text{or} \quad \ln\frac{2}{3} - \ln\frac{4}{5} = \frac{2}{5}k \quad \text{or} \quad k = \frac{5}{2}\ln\frac{5}{6} \quad \text{or} \quad k = -0.46.$$

Again integrating between $t = 0, x = 30$ and $t = 5, x = x$, we have

$$\int_{30}^{x} \frac{dx}{x} - \int_{30}^{x} \frac{dx}{x+20} = \frac{k}{5}\int_0^5 dt \text{ or } \ln\frac{5x}{3(x+20)} = k = -0.46$$

$$\text{or} \quad \frac{x}{x+20} = \frac{3}{5}e^{-0.46} = 0.38$$

whence $x = 12$. Thus the amount dissolved after 5 hours is $30 - 12 = 18$ gms.

5. Find the time required for a cylindrical tank of radius 8 ft and height 10 ft to empty through a round hole of radius 1 inch in the bottom of the tank, given that water will issue from such a hole with velocity approximately $v = 4.8\sqrt{h}$ ft/sec, h being the depth of the water in the tank.

The volume of water which runs out per second may be thought of as that of a cylinder 1 inch in radius and of height v. Hence the volume which runs out in time dt sec. is

$$\pi\left(\frac{1}{12}\right)^2\left(4.8\sqrt{h}\right)dt = \frac{\pi}{144}\left(4.8\sqrt{h}\right)dt.$$

If dh be the corresponding drop in the water level in the tank, the volume of water which runs out is also given by $\pi(8)^2 dh = 64\pi dh$.

Hence

$$\frac{\pi}{144}\left(4.8\sqrt{h}\right)dt = -64\pi dh \quad \text{or} \quad dt = -\frac{64 \times 144}{4.8}\frac{dh}{\sqrt{h}} = -1920\frac{dh}{\sqrt{h}}$$

Integrating between $t = 0, h = 10$ and $t = t, h = 0$, we have

$$\int_0^t dt = -1920\int_{10}^{0}\frac{dh}{\sqrt{h}} \quad \text{or} \quad t = -3840\sqrt{h}\Big]_{10}^{0} = 3840\sqrt{10} \text{ sec.} = 3 \text{ hr. } 22 \text{ min.}$$

6. A mass m, free to move along the X-axis is attracted towards the origin with a force proprotional to its distance from the origin. Find the motion, (a) if it starts from rest at $x = x_0$, and (b) if it starts at $x = x_0$ with initial velocity v_0 moving away from the origin.

Let x denote the distance from the origin to the mass at time t; then we may write

$$m\frac{d^2x}{dt^2} = -sx \quad \text{or} \quad \frac{d^2x}{dt^2} + w^2x = 0 \quad \text{where} \quad w^2 = \frac{s}{m}.$$

As already done in Ex. 4 in Art. 17.5.2, the solution of the above differential equation is

$$x = C_1 \sin wt + C_2 \cos wt. \tag{1}$$

Differentiating (1) with respect to t, we may write

$$v = -wC_2 \sin wt + wC_1 \cos wt. \tag{2}$$

(a) When $t=0$, $x=x_0$ and $v=0$. Then $C_1=0$ from (2), $C_2=x_0$ from (1) and $x=x_0 \cos wt$.

(b) When $t=0, x=x_0$ and $v=v_0$. Then $C_2=x_0, C_1=\dfrac{v_0}{w}$ and $x=\dfrac{v_0}{w}\sin wt + x_0 \cos wt$.

In (a), the motion is simple harmonic of amplitude x_0 and period $\dfrac{2\pi}{w}$.

In (b) the motion is also simple harmonic of amplitude $\dfrac{\sqrt{v_0^2+w^2x_0^2}}{w}$ and period $\dfrac{2\pi}{w}$.

7. If in problem 6 the medium offers a resistance which is proportional to the instantaneous velocity of the body, find the motion.

The equation of motion of the body may then be written as

$$m\frac{d^2x}{dt^2}=-b\frac{dx}{dt}-sx \quad \text{or,} \quad \frac{d^2x}{dt^2}+k\frac{dx}{dt}+w^2x=0 \quad \text{where } \frac{b}{m}=k \text{ and } w^2=\frac{s}{m}.$$

This differential equation represents a damped harmonic motion, and the solution has already been done in Ex. 5 in Art 17.5.2.

8. If an external periodic force $F\sin pt$ or $F\cos pt$ or more generally Fe^{ipt} acts on the body of mass m, which when displaced is subjected to a restoring force proportional to displacement and a frictional force proportional to velocity, find the motion of the body.

The equation of motion of the body may be written as

$$m\frac{d^2x}{dt^2}=-b\frac{dx}{dt}-sx+Fe^{ipt} \quad \text{or,} \quad \frac{d^2x}{dt^2}+k\frac{dx}{dt}+w^2x=fe^{ipt}$$

where $k=\frac{b}{m}$, $w^2=\frac{s}{m}$, $f=\frac{F}{m}$ and $p=$ the angular frequency of the applied periodic force.

Let $x=x_1$ (the complementary function) which satisfies the equation

$$\frac{d^2x_1}{dt^2}+k\frac{dx_1}{dt}+w^2x_1=0 \tag{1}$$

and also let $x=x_2$ (the particular integral) which satisfies the equation

$$\frac{d^2x_2}{dt^2}+k\frac{dx_2}{dt}+w^2x_2=fe^{ipt} \tag{2}$$

so that

$$\frac{d^2}{dt^2}(x_1+x_2)+k\frac{d}{dt}(x_1+x_2)+w^2(x_1+x_2)=fe^{ipt}.$$

The solution x_1 of equation (1) is

$$Ae^{-\frac{k}{2}t}\cos\left(\sqrt{w^2-\frac{k^2}{4}}\,t-\phi\right)$$

where A and ϕ are constants, and is same as in the case of resisted oscillation, and has already been done in Ex. 5 in Art. 17.5.2.

For x_2, the particular integral, we may proceed as follows:
Eqn. (2) may be written as $(D^2 + kD + w^2)x_2 = fe^{ipt}$.

$$\therefore \quad x_2 = \frac{fe^{ipt}}{D^2 + kD + w^2} = \frac{fe^{ipt}}{i^2p^2 + kip + w^2} \text{ (replacing } D \text{ by } ip \text{ and } D^2 \text{ by } i^2p^2)$$

$$= \frac{fe^{ipt}[(w^2 - p^2) - kip]}{[(w^2 - p^2) + kip][(w^2 - p^2) - kip]} = \frac{fe^{ipt}[(w^2 - p^2) - kip]}{(w^2 - p^2)^2 + k^2p^2}$$

$$= \frac{fe^{ipt}}{\sqrt{(w^2 - p^2)^2 + k^2p^2}} \cdot \frac{(w^2 - p^2) - ikp}{\sqrt{(w^2 - p^2)^2 + k^2p^2}}$$

$$= \frac{fe^{ipt}}{\sqrt{(w^2 - p^2)^2 + k^2p^2}}(\cos\alpha - i\sin\alpha), \text{ where}$$

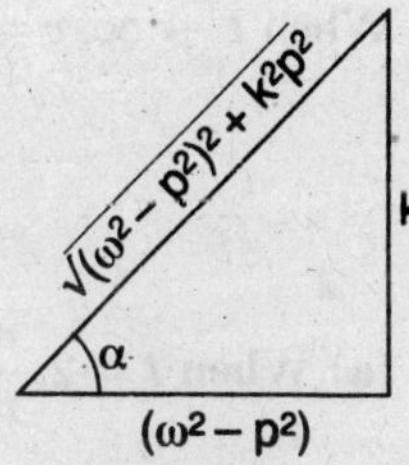

Fig 17.6.1

$$\cos\alpha = \frac{(w^2 - p^2)}{\sqrt{(w^2 - p^2)^2 + k^2p^2}} \text{ and}$$

$$\sin\alpha = \frac{kp}{\sqrt{(w^2 - p^2)^2 + k^2p^2}} \text{ and hence } x_2 = \frac{fe^{i(pt-\alpha)}}{\sqrt{(w^2 - p^2)^2 + k^2p^2}}.$$

Now, if the force is $F \sin pt$, then

$$x_2 = \frac{f}{\sqrt{(w^2 - p^2)^2 + k^2p^2}} \sin(pt - \alpha)$$

and if the force is $F \cos pt$, then

$$x_2 = \frac{f}{\sqrt{(w^2 - p^2)^2 + k^2p^2}} \cos(pt - \alpha).$$

Thus the complete solution is given by

$$x = x_1 + x_2 = Ae^{-\frac{k}{2}t} \cos\left[\sqrt{w^2 - \frac{k^2}{4}}\, t - \phi\right] + \frac{f}{\sqrt{(w^2 - p^2)^2 + k^2p^2}} e^{i(pt-\alpha)}.$$

9. A body of mass m falls from rest in a medium for which the resistance (lb) is proportional to the square of the velocity (ft/sec). If the terminal velocity is 150 ft/sec, find (a) the velocity at the end of 2 seconds, and (b) the time required for the velocity to become 100 ft/sec.

Let v denote the velocity of the body at time t seconds. Net force on the body = weight of the body$-$resistance, and the equation of motion is

$$m\frac{dv}{dt} = mg - Kv^2 \tag{1}$$

Taking $g = 32$ ft/sec^2, it is seen that some simplification is possible by choosing $K = 2mk^2$. Then equation (1) reduces to

$$\frac{dv}{dt} = 2(16 - k^2v^2) \quad \text{or} \quad \frac{dv}{k^2v^2 - 16} = -2dt$$

On integration, we have

$$\ln \frac{kv-4}{kv+4} = -16kt + \ln C \quad \text{or} \quad \frac{kv-4}{kv+4} = Ce^{-16kt}$$

When $t = 0, v = 0$; then $C = -1$

$$\therefore \frac{kv-4}{kv+4} = -e^{-16kt} \tag{2}$$

When $t \to \infty, v = 150$, then $e^{-16kt} = 0$, $\therefore k = \frac{2}{75}$ and equation (2) becomes

$$\frac{v-150}{v+150} = -e^{-0.43t}$$

(a) When $t = 2, \dfrac{v-150}{v+150} = -e^{-0.86} = -0.423$ or $v = 61$ ft/sec.

(b) When $v = 100$, $e^{-0.43t} = 0.2 = e^{-1.6}$ or $t = 3.7$ sec.

10. A man and a parachute weighing ω lbs falls from rest under the force of gravity. If the resistance of air is proportional to the speed v, find the speed as a function of time. Find also the limiting speed v_0.

[C.U. (Hons.) 1989]

Equation of motion here is

$$\omega\frac{d^2y}{dt^2} = \omega g - k\frac{dy}{dt} \quad \text{or} \quad \frac{d^2y}{dt^2} = g - k'\frac{dy}{dt} \text{ where } k' = \frac{k}{\omega}.$$

Integrating we have $\dfrac{dy}{dt} = gt - k'y + C_1$. Now at $t = 0$, $\dfrac{dy}{dt} = 0$ and $y = 0$, whence we have $C_1 = 0$

$$\therefore \frac{dy}{dt} + k'y = gt \tag{1}$$

Multiplying both sides of (1) by the integrating factor $e^{k't}$ $\left(e^{\int k'dt} = e^{k't}\right)$, we have

$$\frac{d}{dt}\left(ye^{k't}\right) = gte^{k't}$$

Integrating we have

$$ye^{k't} = g\left(\frac{t}{k'}e^{k't} - \int \frac{1}{k'}e^{k't}dt\right) + C_2 = g\left(\frac{t}{k'}e^{k't} - \frac{1}{k'^2}e^{k't}\right) + C_2.$$

$$\therefore y = \frac{g}{k'}\left(t - \frac{1}{k'}\right) + C_2e^{-k't}$$

Now $y = 0$ at $t = 0$; $\therefore C_2 = \frac{g}{k'^2}$

$$\therefore y = \frac{gt}{k'} + \frac{g}{k'^2}\left(e^{-k't} - 1\right) \text{ is the required solution.}$$

From the above expression we may obtain

$$\frac{dy}{dt} = \frac{g}{k'} - \frac{g}{k'}e^{-k't} = \frac{g}{k'}(1 - e^{-k't}).$$

Again at $t = \infty, e^{-k't} = 0$; $\therefore \left(\frac{dy}{dt}\right)_{\max} = \frac{g}{k'} = \frac{g\omega}{k}$ = limiting or terminal velocity.

11. A parachutist is falling with speed 176 ft/sec. when his parachute opens. If the air resistance is $\omega v^2/256$ lb, where ω is the total weight of the man and parachute, find his speed as a function of time t after the parachute opened.

Here the net force on the system = weight of the system—air resistance; so the equation of motion is

$$\frac{\omega}{g}\frac{dv}{dt} = \omega - \frac{\omega v^2}{256} \quad \text{or} \quad \frac{dv}{v^2 - 256} = -\frac{dt}{8} \text{ (taking } g = 32 \text{ ft/sec}^2)$$

Integrating between the limits $t = 0, v = 176$ and $t = t, v = v$, we have

$$\int_{176}^{v} \frac{dv}{v^2 - 256} = -\frac{1}{8}\int_0^t dt \text{ or } \frac{1}{32}\ln\frac{v - 16}{v + 16}\bigg]_{176}^{v} = -\frac{t}{8}\bigg]_0^t \quad \text{or} \quad \ln\frac{v - 16}{v + 16} - \ln\frac{5}{6} = -4t$$

$$\text{or} \quad \frac{v - 16}{v + 16} = \frac{5}{6}e^{-4t} \quad \text{or} \quad v = 16\frac{6 + 5e^{-4t}}{6 - 5e^{-4t}}.$$

Now at $t = \infty, e^{-4t} = 0$ $\therefore v_{\max} = 16$ ft/sec which is the approximate constant speed (attained quickly by the parachutist)—known as the limiting or terminal speed.

12. A mass m is projected vertifically upward from O with initial velocity v_0. Find the maximum height reached, assuming that the resistance of air is proportional to the velocity.

[C.U. (Hons), 1979]

Let x denote the distance of the mass m from O at time t. The mass is acted upon by two forces—the gravitational force of magnitude mg and the resistance of magnitude $sv = s\frac{dx}{dt}$ (Fig. 17.6.2), each directed downwards. Hence using, mass × acceleration = net force, we may write

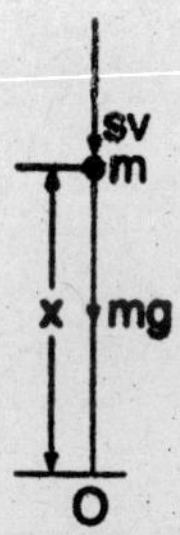

$$m\frac{d^2x}{dt^2} = -mg - s\frac{dx}{dt} \quad \text{or} \quad \frac{d^2x}{dt^2} + k\frac{dx}{dt} = -g \quad \text{where } k = \frac{s}{m}.$$

The solution is

$$x = C_1 + C_2e^{-kt} - \frac{g}{k}t \tag{1}$$

Fig 17.6.2

and is obtained as follows.

Writing $\frac{dx}{dt} = v$, we have $\frac{dv}{dt} + kv = -g$

$$\text{or} \quad \frac{dv}{dt}e^{kt} + kve^{kt} = -ge^{kt} \text{ (Multiplying both sides by } I.F = e^{\int k\,dt} = e^{kt})$$

$$\text{or} \quad \frac{d}{dt}(ve^{kt}) = -ge^{kt} \quad \text{or} \quad \int d(ve^{kt}) = \int -ge^{kt}dt + A$$

$$\text{or} \quad ve^{kt} = -\frac{g}{k}e^{kt} + A \quad \therefore v = \frac{dx}{dt} = -\frac{g}{k} + Ae^{-kt}$$

$$\text{or} \quad \int dx = \int -\frac{g}{k}dt + \int Ae^{-kt}dt + C_1$$

$$\text{or} \quad x = -\frac{g}{k}t - \frac{A}{k}e^{-kt} + C_1 = C_1 + C_2e^{-kt} - \frac{g}{k}t \quad (\text{where } C_2 = -\tfrac{A}{k})$$

(Alternatively the same solution can be obtained by the operator method, as shown below. Let $x = e^{\lambda t}$ be a trial solution of the auxiliary equation $(D^2 + kD)x = 0$

$\therefore\ (\lambda^2 + k\lambda)e^{\lambda t} = 0$ or $\lambda^2 + k\lambda = 0\ \because\ e^{\lambda t} \neq 0$ or $\lambda = 0$ or $-k\quad \therefore\ CF = A + Be^{-kt}$

Again

$$P.I = \frac{1}{D^2 + kD}(-g) = \frac{1}{kD}\left(1 + \frac{D}{k}\right)^{-1}(-g) = \frac{1}{kD}\left[1 - \frac{D}{k} + \frac{D^2}{k^2} - \cdots\right](-g)$$

$$= \frac{1}{k}\left[\frac{1}{D} - \frac{1}{k}\right](-g) = \frac{1}{k}\left[\frac{1}{D}(-g)\right] + \frac{g}{k^2}$$

So the total solution

$$x = CF + PI = A + Be^{-kt} + \frac{g}{k^2} - \frac{g}{k}t$$

$$= \left(A + \frac{g}{k^2}\right) + Be^{-kt} - \frac{g}{k}t = C_1 + C_2e^{-kt} - \frac{g}{k}t$$

writing $A + \frac{g}{k^2} = C_1$ and $B = C_2$).

Differentiating (1) with respect to t, we have

$$v = \frac{dx}{dt} = -kC_2e^{-kt} - \frac{g}{k} \tag{2}$$

when $t = 0$, $x = 0$ and $v = v_0$; then $C_1 + C_2 = 0$, and

$$v_0 = -kC_2 - \frac{g}{k} \quad \text{whence} \quad C_1 = -C_2 = \frac{v_0}{k} + \frac{g}{k^2}$$

Substituting these values in (1), we have

$$x = \frac{1}{k^2}(g + kv_0)\left(1 - e^{-kt}\right) - \frac{g}{k}t.$$

The maximum height is reached when $v = 0$; then from (2) we have

$$e^{-kt} = \frac{-g}{k^2C_2} = \frac{g}{g + kv_0}, \quad \text{so that } t = \frac{1}{k}\ln\frac{g + kv_0}{g}.$$

Then the maximum height is

$$x = \frac{1}{k^2}(g + kv_0)\left(1 - \frac{g}{g + kv_0}\right) - \frac{g}{k}\left(\frac{1}{k}\ln\frac{g + kv_0}{g}\right) = \frac{1}{k}\left[v_0 - \frac{g}{k}\ln\frac{g + kv_0}{g}\right].$$

13. In an electric circuit containing an inductance L and a resistance R in series, a source of alternating e.m.f. $E_0 \sin wt$ is impressed. Calculate the current i at any instant.

The differential equation of such a circuit is written as $L\dfrac{di}{dt} + Ri = E_0 \sin wt$.

Writing $D \equiv \frac{d}{dt}$, we have $(LD + R)i = E_0 \sin wt$.

$$\therefore \quad i = \frac{E_0 \sin wt}{LD + R} = \frac{E_0(R - LD) \sin wt}{R^2 - L^2D^2} = \frac{E_0(R \sin wt - Lw \cos wt)}{R^2 + L^2w^2} \quad (\text{writing } D^2 = -w^2)$$

$$= \frac{E_0(A \sin wt \cos\phi - A \cos wt \sin\phi)}{R^2 + L^2w^2}$$

$$= \frac{E_0 A \sin(wt - \phi)}{R^2 + L^2w^2}$$

$$= \frac{E_0}{\sqrt{R^2 + L^2w^2}} \sin(wt - \phi) = i_0 \sin(wt - \phi). \quad (1)$$

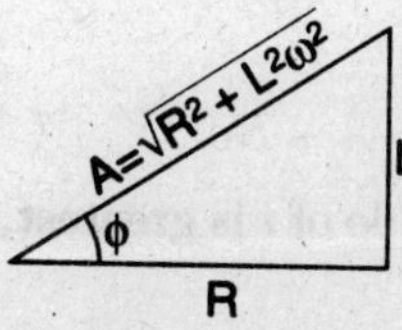

Fig 17.6.3

[where $R = A\cos\phi$ and $Lw = A\sin\phi$ so that $A = \sqrt{R^2 + L^2w^2}$ and $\tan\phi = \frac{Lw}{R}$, as shown in the attached vector diagram]

Here $Lw = X_L$ is the inductive reactance and $Z = \sqrt{R^2 + L^2w^2}$ is the impedance of the circuit. ϕ denotes the phase difference between the e.m.f. and the current.

The solution given by (1) is only a part of the complete solution, and is known as the *particular integral.* The other part known as the *complementary function* is $i = Ce^{-\frac{R}{L}t}$, C being a constant. Obviously with increase of time t, the term $e^{-\frac{R}{L}t}$ gradually and quickly becomes vanishingly small. Hence this part is not finally effective.

14. For an electric circuit containing a resistance R, an inductance L and a capacitance C in series, the alternating e.m.f. impressed is $E_0 \sin wt$. Derive the expression for the steady-state current.

If at any instant 't', q be the charge on the condenser, and i be the strength of the current, we may write

$$L\frac{di}{dt} + Ri + \frac{q}{C} = E_0 \sin wt.$$

Differentiating,

$$L\frac{d^2i}{dt^2} + R\frac{di}{dt} + \frac{i}{C} = E_0 w \cos wt \left(\because \frac{dq}{dt} = i\right) \text{ or, } \left(LD^2 + RD + \frac{1}{C}\right) i = E_0 w \cos wt. \quad (1)$$

The required steady-state current is the particular integral of (1).

$$\therefore \quad i = \frac{E_0 w \cos wt}{LD^2 + RD + \frac{1}{C}} = \frac{E_0 w \cos wt}{-Lw^2 + RD + \frac{1}{C}} = \frac{E_0 w \cos wt}{RD - \left(Lw - \frac{1}{Cw}\right) w}$$

$$= \frac{wE_0(RD + Xw)}{R^2D^2 - X^2w^2} \cos wt \left[\text{writing } X \text{ for } \left(Lw - \frac{1}{Cw}\right)\right]$$

$$= \frac{wE_0}{-R^2w^2 - X^2w^2}(-Rw \sin wt + Xw \cos wt) = \frac{E_0}{R^2 + X^2}(R \sin wt - X \cos wt)$$

$$= \frac{E_0}{R^2 + X^2}(A \sin wt \cos\phi - A \cos wt \sin\phi) = \frac{E_0}{\sqrt{R^2 + X^2}} \sin(wt - \phi)$$

$$= i_0 \sin(wt - \phi) \left(\text{writing, } \frac{E_0}{\sqrt{R^2 + X^2}} = i_0\right). \quad (2)$$

[Putting $R = A\cos\phi$ and $X = A\sin\phi$ so that $A = \sqrt{R^2+X^2}$ and $\tan\phi = \frac{X}{R}$, as shown in the attached vector diagram]

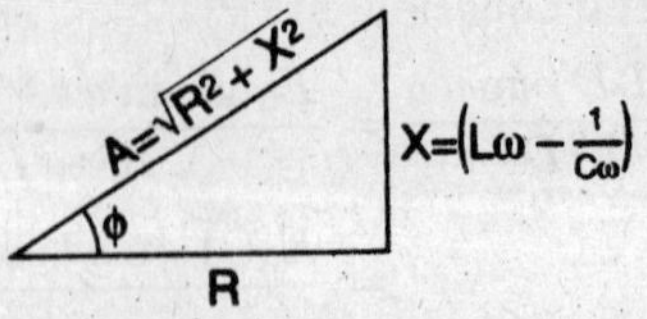

Fig 17.6.4

X is called the *effective reactance of the circuit.* When $X = 0$, the amplitude of i is greatest, and the circuit is in resonance, and then $w = \dfrac{1}{\sqrt{CL}}$.

$Z = \sqrt{R^2+X^2}$ is the impedance of the circuit. ϕ is the difference of phase between the e.m.f. and the current.

The solution given by (2) is only a part of the complete solution, and is known as the *particular integral.* The other part known as the *complementary function* is $i = C_1e^{\lambda_1 t} + C_2e^{\lambda_2 t}$, where C_1 and C_2 are the two constants and λ_1, λ_2 are the roots of $\left(LD^2 + RD + \frac{1}{C}\right)i = 0$. λ_1 and λ_2 are essentially negative. Hence the complementary function dies off quickly with the increase of time 't', and may therefore be neglected.

15. The circuit consisting of an inductance L, a condenser of capacitance C, and an e.m.f. E in series is known as an *harmonic oscillator.* Find q and i when $E = E_0\cos wt$ and the initial conditions are $q = q_0$, $i = i_0$ when $t = 0$.

The differential equation in this case may be written as

$$L\frac{d^2q}{dt^2} + \frac{q}{C} = E_0\cos wt \quad \text{or,} \quad \frac{d^2q}{dt^2} + \frac{q}{CL} = \frac{E_0}{L}\cos wt \quad \text{or,} \quad \left(D^2 + \frac{1}{CL}\right)q = \frac{E_0}{L}\cos wt. \quad (1)$$

We shall consider the following two cases

(a) when $w \neq \dfrac{1}{\sqrt{CL}}$ and (b) when $w = \dfrac{1}{\sqrt{CL}}$

(a) The complete solution of (1) is

$$q = A\cos\frac{1}{\sqrt{CL}}t + B\sin\frac{1}{\sqrt{CL}}t + \frac{E_0}{L}\frac{1}{D^2 + \frac{1}{CL}}\cos wt$$

$$= A\cos\frac{1}{\sqrt{CL}}t + B\sin\frac{1}{\sqrt{CL}}t + \frac{E_0C}{1 - w^2CL}\cos wt$$

$$\text{and} \quad i = \frac{1}{\sqrt{CL}}\left(-A\sin\frac{1}{\sqrt{CL}}t + B\cos\frac{1}{\sqrt{CL}}t\right) - \frac{E_0Cw}{1 - w^2CL}\sin wt.$$

Using initial conditions,

$$A = q_0 - \frac{E_0C}{1 - w^2CL} \quad \text{and} \quad B = \sqrt{CL}i_0.$$

Then

$$q = \left(q_0 - \frac{E_0C}{1 - w^2CL}\right)\cos\frac{1}{\sqrt{CL}}t + \sqrt{CL}i_0\sin\frac{1}{\sqrt{CL}}t + \frac{E_0C}{1 - w^2CL}\cos wt$$

and $i = i_0 \cos \frac{1}{\sqrt{CL}}t - \frac{1}{\sqrt{CL}}\left(q_0 - \frac{E_0C}{1-w^2CL}\right)\sin \frac{1}{\sqrt{CL}}t - \frac{E_0Cw}{1-w^2CL}\sin wt.$

(b) In this case, the differential equation is

$$\frac{d^2q}{dt^2} + w^2q = \frac{E_0}{L}\cos wt.$$

Then the complete solution is

$$q = A\cos wt + B\sin wt + \frac{E_0}{2Lw}t\sin wt$$

$$\text{and} \quad i = w(-A\sin wt + B\cos wt) + \frac{E_0}{2L}\left(\frac{1}{w}\sin wt + t\cos wt\right).$$

From initial conditions we have, $A = q_0$ and $B = \frac{i_0}{w}$.

Substituting these values of Aand B, we have,

$$q = q_0\cos wt + \frac{i_0}{w}\sin wt + \frac{E_0}{2Lw}t\sin wt$$

$$\text{and} \quad i = i_0\cos wt - q_0 w\sin wt + \frac{E_0}{2L}\left(\frac{1}{w}\sin wt + t\cos wt\right).$$

In case (b) the frequency of the e.m.f. is the natural frequency of the oscillator, that is, the frequency when there is no e.m.f. The circuit is in resonance since the reactance $X = Lw - \frac{1}{Cw} = 0$ when $w = \frac{1}{\sqrt{CL}}$. The term $\frac{E_0t}{2L}\cos wt$ shows that the amplitude decreases with increase of t, and that subsequently such a circuit will be destroyed.

16. **A spring of negligible weight hangs vertically. A mass m is attached to the other end. If the mass is moving with velocity v_0 when the spring is unstretched, find the velocity v as a function of the stretch x.**

According to Hooke's law, the force opposing the spring to stretch, i.e., the spring force is proportional to the stretch, so that we may write—net force on the body = weight of the body − spring force.

∴ the equation of motion is

$$m\frac{dv}{dt} = mg - kx \quad \text{or} \quad m\frac{dv}{dx}\frac{dx}{dt} = mg - kx \quad \text{or} \quad mv\frac{dv}{dx} = mg - kx,$$

which gives on integration $mv^2 = 2mgx - kx^2 + C$. Now at $x = 0, v = v_0$, $\therefore C = mv_0^2$.

$$\therefore\ mv^2 = 2mgx - kx^2 + mv_0^2 \text{ is the required solution.}$$

17. **A spring for which $k = 48$ lb/ft. hangs in a vertical position with its upper end fixed. A mass, weighing 16 lb is attached to the lower end. After coming to rest, the mass is pulled down 2 inches and released. Discuss the resulting motion of the mass, neglecting air resistance.**

Taking the origin at the centre of gravity of the mass, after coming to rest, let x be the change in position of the mass at time t. The net force at time t is the restoring force $-kx$

(spring force) corresponding to the change x in the position of the mass, so that the equation of motion is

$$\frac{16}{g}\frac{d^2x}{dt^2} = -48x \quad \text{or} \quad \frac{d^2x}{dt^2} + 96x = 0 \text{ (taking } g = 32 \text{ ft/sec}^2\text{)}$$

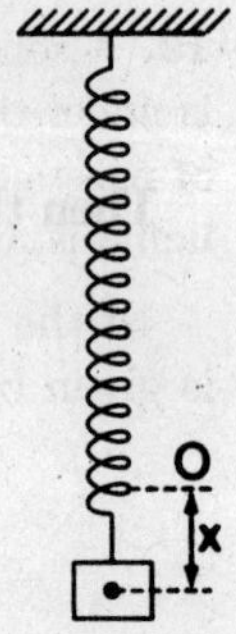

Fig 17.6.5

Solving we obtain $x = C_1 \sin\sqrt{96}\,t + C_2 \cos\sqrt{96}\,t$, which gives on differentiation,

$$v = \frac{dx}{dt} = \sqrt{96}\left(C_1 \cos\sqrt{96}\,t - C_2 \sin\sqrt{96}\,t\right).$$

Now at $t = 0, x = \frac{1}{6}$ and $v = 0$, whence we have $C_2 = \frac{1}{6}$ and $C_1 = 0$

$$\therefore\ x = \frac{1}{6}\cos\sqrt{96}\,t,$$

which represents a simple harmonic motion with period $\frac{2\pi}{\sqrt{96}} = 0.641$ sec, frequency $= \frac{\sqrt{96}}{2\pi} = 1.56$ cycles/sec and amplitude $= \frac{1}{6}$ ft.

18. Two equal masses are suspended at the end of a weightless elastic spring of stifness k. One mass falls off. Solve the differential equation of the motion of the remaining mass. **[C.U. (Hons.) 1980]**

k is the restoring force per unit length. $\therefore\quad kl = 2\text{mg}$ or, $k = 2\text{mg}/l$.

Now the upward force due to tension when one mass is dropped is $k(l - x)$, and this force is balanced by an upward force mf and a downward force mg, so that the equation of motion is

$$mf = k(l - x) - mg = kl - kx - mg, \text{ or, } mf + kx = kl - mg = 2mg - mg = mg.$$

$$\therefore\ m\frac{d^2x}{dt^2} + kx = mg, \quad \text{or, } \frac{d^2x}{dt^2} + \frac{k}{m}x = g \text{ or, } \frac{d^2x}{dt^2} + w_0^2 x = g \text{ (putting } w_0^2 = \tfrac{k}{m}\text{)}$$

which on solution gives $x = A\cos w_0 t + B \sin w_0 t + \dfrac{g}{w_0^2}$.

At $t = 0, x = l$ which gives $l = A + \dfrac{g}{w_0^2}$.

Now $\dfrac{dx}{dt} = -Aw_0 \sin w_0 t + Bw_0 \cos w_0 t$.

At $t = 0, \dfrac{dx}{dt} = 0$ which gives $B = 0$. $\quad\therefore\ x = A\cos w_0 t + \dfrac{g}{w_0^2}$.

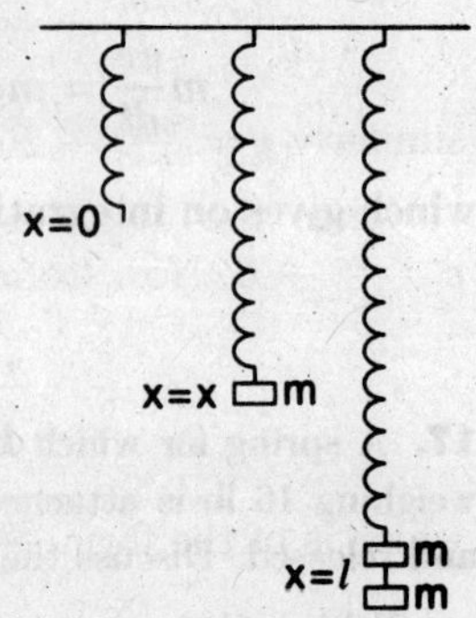

Fig 17.6.6

Putting $k = \dfrac{2mg}{l}$, i.e., $\dfrac{k}{m} = \dfrac{2g}{l} = w_0^2$ we can write the solution in a more general form as

$$x = A\cos\left(\sqrt{\frac{2g}{l}}t + \delta\right) + \frac{g}{2g}\cdot l = A\cos\left(\sqrt{\frac{2g}{l}}t + \delta\right) + \frac{l}{2} \qquad (1)$$

Now at $t = 0, x = l$ which gives $l = A\cos\delta + \dfrac{l}{2}$. $\quad\therefore\ A\cos\delta = \dfrac{l}{2}$.

Also at $t = 0, \dfrac{dx}{dt} = 0$. Hence from (1) we get $\delta = 0$. $\quad\therefore\ A = \dfrac{l}{2}$.

So the solution may be written as

$$x = \frac{l}{2}\cos\sqrt{\frac{2g}{l}}t + \frac{l}{2} = \frac{l}{2}\left(\cos\sqrt{\frac{2g}{l}}t + 1\right).$$

19. A solid sphere is given some initial velocity in any direction in a liquid. As it moves through it, it is under the vertical downward force due to gravity and an upward force due to buoyancy and a force of resistance due to viscosity of the liquid proportional to its velocity. Show that if the depth of the liquid is unlimited, the motion will be ultimately vertically downwards. **[cf. C.U. (Hons.) 1994, '99]**

If the position vector of the body at any instant of time t be $\vec{r}$, then its equation of motion is given by

$$m\frac{d\vec{V}}{dt} = \frac{4}{3}\pi a^3(\rho - \sigma)g\hat{k} - b\frac{d\vec{r}}{dt} \qquad \left[\begin{array}{l} a \to \text{radius of the solid sphere,} \\ \rho \to \text{density of the body,} \\ \sigma \to \text{density of the liquid;} \end{array}\right]$$

The component forms of the above equation are

$$m\frac{du}{dt} = -b\frac{dx}{dt} \tag{1}$$

$$m\frac{dv}{dt} = -b\frac{dy}{dt} \tag{2}$$

$$m\frac{d\omega}{dt} = \frac{4}{3}\pi a^3(\rho - \sigma)g - b\frac{dz}{dt} \tag{3}$$

The initial velocity is given at the origin and its components along x, y, z axes are respectively u_0, v_0, ω_0.

Integrating (1) we have

$$m\int_{u_0}^{u} du = -b\int_0^x dx \quad \text{or} \quad (u - u_0) = -\frac{b}{m}x \quad \text{or} \frac{dx}{dt} + \frac{b}{m}x = u_0$$

Multiplying both sides by the integrating factor $e^{\int \frac{b}{m}dt} = e^{\frac{b}{m}t}$, we have

$$\frac{d}{dt}(xe^{bt/m}) = u_0e^{bt/m} \quad \text{or} \quad xe^{bt/m} = \frac{m}{b}u_0e^{bt/m} + A \quad \text{or} \quad x = \frac{m}{b}u_0 + Ae^{-bt/m};$$

similarly $y = \frac{m}{b}v_0 + Be^{-bt/m}$.

Thus the displacements x and y remain unchanged in magnitude after a time when $e^{-bt/m} \to 0$. Now from (3) we have

$$\frac{d\omega}{dt} = \frac{(\rho - \sigma)}{\rho}g - \frac{b}{m}\frac{dz}{dt} \quad \text{or} \quad \frac{d\omega}{dt} + \frac{b}{m}\omega = \frac{(\rho - \sigma)}{\rho}g$$

which is of the form $\frac{dy}{dx} + Py = Q$. Therefore the solution is

$$\omega = \frac{m}{b}\frac{(\rho - \sigma)}{\rho}g + Ce^{-bt/m}.$$

So after a time when $e^{-bt/m} \to 0$, the velocity of the body is $\omega = \frac{m}{b}\frac{(\rho-\sigma)}{\rho}g$ which is a constant and is obviously directed downwards.

20. A vessel of volume V contains a gas that is kept at a constant temperature. The gas slowly leaks out of a small hole of area A. The outside pressure is low so that no molecules leak back. Prove that the pressure at any time t is given by $p = P_0 e^{-kt}$ where P_0 is the initial pressure, and calculate k in terms of V, A and $\bar{c}$, the average velocity.

From kinetic theory of matter, we know that the number of molecules striking unit area of the wall (of a container) per second is $\frac{n\bar{c}}{4}$, so that the velocity of flow of the gas is $\frac{\bar{c}}{4}$.

Since the gas is kept at a constant temperature, we may write

$$d(pV) = 0 \quad \text{or} \quad pdV + Vdp = 0 \quad \text{or} \quad dV = -\frac{V}{p}dp \quad \text{or} \quad \frac{dp}{p} = -\frac{dV}{V}.$$

Again dV = velocity of flow × time × area = $\frac{\bar{c}A}{4}dt$.

$$\therefore \quad \frac{dp}{p} = -\frac{\bar{c}A}{4V}dt,$$

whence we obtain on integration $\ln p = -\frac{\bar{c}A}{4V}t + C_1$, C_1 being the integration constant.

Now at $t = 0, p = P_0$, $\therefore \; C_1 = \ln P_0$.

Putting this value of C_1 and on simplification we obtain

$$\ln\frac{p}{P_0} = \ln e^{-\frac{\bar{c}A}{4V}t} \text{ or } p = P_0 e^{-\frac{\bar{c}A}{4V}t} = P_0 e^{-kt} \text{ where } k = \frac{\bar{c}A}{4V}.$$

17.7 Power Series Solutions of Differential Equations

We shall now consider differential equations belonging to such a form which cannot be solved by the methods already discussed. For this purpose we shall discuss here the series solutions of differential equations.

Let us try to find the solution of a second order differential equation

$$y'' = F(x, y, y') \tag{17.7.1}$$

which satisfies the initial conditions

$$(y)_{x=x_0} = y_0 \quad \text{and} \quad (y')_{x=x_0} = y_0' \tag{17.7.2}$$

Let us suppose that the solution $y = f(x)$ exists and may be given in the form of a Taylor's series,

$$y = f(x) = f(x_0) + (x - x_0)f'(x_0) + \frac{(x - x_0)^2}{2!}f''(x_0) + \cdots. \tag{17.7.3}$$

When $x_0 = 0$, we have Maclaurin's series

$$f(x) = f(0) + xf'(0) + \frac{x^2}{2!}f''(0) + \cdots$$

We shall have to find $f(x_0), f'(x_0), f''(x_0) \cdots$

From conditions (17.7.2), we have $f(x_0) = y_0$ and $f'(x_0) = y'_0$.

Also from equation (17.7.1), we have $f''(x_0) = (y'')_{x=x_0} = F(x, y_0, y'_0)$.

Differentiating equation (17.7.1) on both sides with respect to x

$$y''' = F'_x(x, y, y') + F'_y(x, y, y')y' + F'_{y'}(x, y, y')y''; \tag{17.7.4}$$

substituting the value $x = x_0$ into the right-hand side, we have $f'''(x_0) = (y''')_{x=x_0}$.

Similarly we have, $f^{IV}(x_0) = \left(y^{IV}\right)_{x=x_0}$ and so on.

In equation (17.7.3) we put the values of these derivatives and obtain the solution of the equation.

Illustrative Examples

1. Find the solution of the equation $y'' = -yx^2$ which satisfies the initial conditions $(y)_{x=0} = 1$, $(y')_{x=0} = 0$.

Here, we have $f(0) = y_0 = 1$ and $f'(0) = y'_0 = 0$

Hence from the given equation, we have $(y'')_{x=0} = f''(0) = 0$.

Also $y''' = -y'x^2 - 2xy$ [differentiating the given equation $y'' = -yx^2$ with respect to x.]

$\therefore \quad (y''')_{x=0} = f'''(0) = 0.$

Further, $y^{IV} = -y''x^2 - 2xy' - 2xy' - 2y = -y''x^2 - 4xy' - 2y$.

$\therefore \quad (y^{IV})_{x=0} = f^{IV}(0) = -2.$

So differentiating the given equation $y'' = -yx^2$. k times on both sides, we have $y^{(k+2)} = -y^kx^2 - 2kxy^{(k-1)} - k(k-1)y^{(k-2)}$; when $x = 0$, we have $y_0^{(k+2)} = -k(k-1)y_0^{(k-2)}$.

If we put $k + 2 = n$, the above expression becomes $y_0^n = -(n-3)(n-2)y_0^{(n-4)}$.

Hence, we have

$$\begin{aligned}
f^{IV}(0) &= y_0^{IV} = -1.2\\
f^{V}(0) &= y_0^{V} = -2.3.y'_0 = 0\\
f^{VI}(0) &= y_0^{VI} = -3.4.y''_0 = 0\\
f^{VII}(0) &= y_0^{VII} = -4.5.y'''_0 = 0\\
f^{VIII}(0) &= y_0^{VIII} = -5.6.y_0^{IV} = -5.6.(-2) = (-1)^2(1.2)(5.6)\\
f^{IX}(0) &= y_0^{IX} = -6.7.y_0^{V} = 0\\
f^{X}(0) &= y_0^{X} = -7.8.y_0^{VI} = 0\\
f^{XI}(0) &= y_0^{XI} = -8.9.y_0^{VII} = 0\\
f^{XII}(0) &= y_0^{XII} = -9.10.y_0^{VIII} = (-1)^3(1.2)(5.6)(9.10)\\
&\cdots\cdots\cdots\cdots\\
f^{4k}(0) &= y_0^{4k} = (-1)^k(1.2)(5.6)(9.10)\cdots[(4k-3)(4k-2)].
\end{aligned}$$

Putting the values of derivatives in Maclaurin's series we have,

$$y = 1 - 1.2\frac{x^4}{4!} + (1.2)(5.6)\frac{x^8}{8!} - (1.2)(5.6)(9.10)\frac{x^{12}}{12!} + \cdots$$

$$+(-1)^k\frac{x^{4k}}{4k!}(1.2)(5.6)\cdots[(4k-3)4k-2)].$$

When the equation is linear, it is convenient to use the series

$$y = a_0 + a_1x + a_2x^2 + \cdots + a_nx^n$$

in the differential equation as will be evident from the following illustrative example.

2. Solve the equation $y'' - 2xy' - 4y = 0$ which satisfies the initial conditions $(y)_{x=0} = 0$ and $(y')_{x=0} = 1$.

For solution of the above equation, we put $y = a_0 + a_1x + a_2x^2 + \cdots + a_nx^n + \cdots$, where $x = 0$, $(y)_{x=0} = a_0 = 0$ and $(y')_{x=0} = a_1 = 1$.

Hence,

$$y = x + a_2x^2 + a_3x^3 + a_4x^4 + \cdots + a_nx^n + \cdots$$
$$y' = 1 + 2a_2x + 3a_3x^2 + 4a_4x^3 + \cdots + na_nx^{n-1} + \cdots$$
$$y'' = 2a_2 + 6a_3x + 12a_4x^2 + \cdots + n(n-1)a_nx^{n-2} + \cdots$$

Substituting these values in the given differential equation, we have

$$2a_2 + 6a_3x + 12a_4x^2 + \cdots + n(n-1)a_nx^{n-2} + \cdots$$
$$- 2x - 4a_2x^2 - 6a_3x^3 - \cdots - 2na_nx^n - \cdots$$
$$- 4x - 4a_2x^2 - 4a_3x^3 - \cdots - 4a_nx^n - \cdots = 0.$$

In order that this series vanishes for all values of x in some region surrounding $x = 0$, it is necessary and sufficient that the coefficient of each power of x should vanish.

Thus, $2a_2 = 0$, hence $a_2 = 0$; $(6a_3 - 2 - 4) = 0$, hence $a_3 = 1$; $(12a_4 - 4a_2 - 4a_2) = 0$, hence $a_4 = 0$; $n(n-1)a_n = (n-2)2a_{n-2} + 4a_{n-2}$, hence $a_n = \frac{2a_{n-2}}{n-1}$.

So, $a_5 = \frac{2a_3}{4} = \frac{2.1}{4} = \frac{1}{2!}$; $a_6 = 0$; $a_7 = \frac{2a_5}{6} = \frac{1}{3}\cdot\frac{1}{2!} = \frac{1}{3!}$; $a_8 = 0$; $a_9 = \frac{2a_7}{8} = \frac{1}{4!}$.

$$\therefore \quad a_{2k} = 0 \quad\text{and}\quad a_{2k+1} = \frac{1}{k!}; \quad \therefore \quad y = x + \frac{x^3}{1!} + \frac{x^5}{2!} + \frac{x^7}{3!} + \cdots + \frac{x^{2k+1}}{k!}.$$

The above series converges for all values of x.

3. Solve the equation $y' - x^2 - e^y = 0$ which satisfies the condition $(y)_{x=0} = 0$.

For solution of the above equation, we put

$$y = a_0 + a_1x + a_2x^2 + a_3x^3 + \cdots + a_nx^n. \tag{1}$$

Applying the initial condition we have $a_0 = 0$.

Hence the equation (1) reduces to the from

$$y = a_1 x + a_2 x^2 + a_3 x^3 + \cdots + a_n x^n \tag{2}$$

$$\therefore \quad y' = a_1 + 2a_2 x + 3a_3 x^2 + \cdots + n a_n x^{n-1} \tag{3}$$

$$e^y = 1 + y + \frac{1}{2!}y^2 + \frac{1}{3!}y^3 + \frac{1}{4!}y^4 + \cdots \frac{y^n}{n!} + \cdots \tag{4}$$

Substituting the value of y from (2) in (4) we have

$$e^y = 1 + a_1 x + \left(a_2 + \frac{a_1^2}{2}\right)x^2 + \left(a_3 + a_1 a_2 + \frac{a_1^3}{6}\right)x^3$$
$$+ \left(a_4 + \frac{a_2^2}{2} + a_1 a_3 + \frac{a_1^2 a_2}{2} + \frac{a_1^4}{24}\right)x^4 + \cdots$$

Substituting the values of y' and e^y in the differential equation, we have,

$$(a_1 - 1) + (2a_2 - a_1)x + \left(3a_3 - 1 - a_2 - \frac{1}{2}a_1^2\right)x^2 + \left(4a_4 - a_3 - a_1 a_2 - \frac{a_1^3}{6}\right)x^3$$
$$+ \left(5a_5 - a_4 - \frac{1}{2}a_2^2 - a_1 a_3 - \frac{1}{2}a_1^2 a_2 - \frac{1}{24}a_1^4\right)x^4 + \cdots = 0.$$

Equating the coefficient of each power of x to zero we have

$$a_1 - 1 = 0 \quad \text{or} \quad a_1 = 1; \quad a_2 = \frac{a_1}{2} = \frac{1}{2}$$

$$a_3 = \frac{1}{3}\left(1 + a_2 + \frac{1}{2}a_1^2\right) = \frac{2}{3}; \quad a_4 = \frac{1}{4}\left(a_3 + a_1 a_2 + \frac{a_1^3}{6}\right) = \frac{1}{3}$$

$$a_5 = \frac{1}{5}\left(a_4 + \frac{1}{2}a_2^2 + a_1 a_3 + \frac{a_1^2 a_2}{2} + \frac{a_1^4}{24}\right) = \frac{17}{60}.$$

Hence the solution is given by

$$y = x + \frac{x^2}{2} + \frac{2}{3}x^3 + \frac{1}{3}x^4 + \frac{17}{60}x^5 + \cdots$$

17.8 Gamma function

Gamma function is defined as

$$\Gamma(n) = \int_0^\infty e^{-x} x^{n-1}\, dx \tag{17.8.1}$$

Replacing n by $n+1$, we have

$$\Gamma(n+1) = \int_0^\infty e^{-x} x^n\, dx.$$

Integrating (by parts) the right hand side, we obtain

$$\Gamma(n+1) = \left|x^n e^{-x}\right|_{\infty}^{0} + n\int_0^{\infty} x^{n-1}e^{-x}\,dx = n\Gamma(n). \tag{17.8.2}$$

Proceeding as above repeatedly, we get

$$\begin{aligned}\Gamma(n+1) &= n\Gamma(n) = n(n-1)\Gamma(n-1) = n(n-1)(n-2)\Gamma(n-2)\cdots \\ &= n(n-1)(n-2)\cdots 3.2.\Gamma(1).\end{aligned} \tag{17.8.3}$$

But

$$\Gamma(1) = \int_0^{\infty} e^{-x}\,dx = \left|e^{-x}\right|_{\infty}^{0} = 1.$$

$$\therefore\quad \Gamma(n+1) = n(n-1)(n-2)(n-3)\cdots 3.2.1 = n!.$$

Illustrative example

Evaluate $\int_{-\infty}^{\infty} e^{-x^2}x^{2n}\,dx$

$$\int_{-\infty}^{\infty} e^{-x^2}x^{2n}\,dx = 2\int_0^{\infty} e^{-x^2}x^{2n}\,dx.$$

Let $x^2 = z$. $\quad\therefore\ 2x\,dx = dz$.

Hence we get,

$$\int_{-\infty}^{\infty} e^{-x^2}x^{2n}\,dx = 2\cdot\frac{1}{2}\int_0^{\infty} e^{-z}z^{n-\frac{1}{2}}\,dz = \Gamma\left(n+\frac{1}{2}\right)$$

[From the definition of Γ-function].

Proceeding repeatedly as in equation (17.8.2) and (17.8.3) we obtain

$$\Gamma\left(n+\frac{1}{2}\right) = \left(n-\frac{1}{2}\right)\left(n-\frac{3}{2}\right)\cdots\frac{1}{2}\Gamma\left(\frac{1}{2}\right) = \left(n-\frac{1}{2}\right)\left(n-\frac{3}{2}\right)\cdots\frac{1}{2}\sqrt{\pi}.$$

[It can be shown that $\Gamma\left(\frac{1}{2}\right) = \sqrt{\pi}$]

17.9 Frobenius' Method for Power Series Solutions of Differential Equations

Very often we come across, in different branches of Physics, some second order linear homogeneous differential equations viz. Bessel's differential equation, Legendre's differential equation, Hermite's differential equation etc. Frobenius method—applicable for solving such differential equations—consists in expressing one variable as a series of ascending or descending powers of another variable. But before we proceed to solve any equation by the power series method, we must first be sure about whether any such series solution exists for the differential equation. The criterion for testing the existence may be known with reference to an important theorem called *Frobenius-Fuchs' theorem.*

A second order linear homogeneous differential equation is written as

$$a_0(x)\frac{d^2y}{dx^2} + a_1(x)\frac{dy}{dx} + a_2(x)y = 0 \qquad (17.9.1)$$

$$\text{or} \quad \frac{d^2y}{dx^2} + \frac{a_1(x)}{a_0(x)}\frac{dy}{dx} + \frac{a_2(x)}{a_0(x)}y = 0$$

$$\text{or} \quad \frac{d^2y}{dx^2} + X_1\frac{dy}{dx} + X_2 y = 0 \qquad (17.9.2)$$

(17.9.2) is the rationalised form of equation (17.9.1).

If (17.9.2) possesses singularity at $x = x_0$, the solution can be developed in a power-series, about the point $x = x_0$, having a finite number of terms, when and only when $(x - x_0)X_1(x_0)$ and $(x - x_0)^2 X_2(x_0)$ remain finite—this is known as *Frobenius-Fuch's theorem.*

If the functions $(x-x_0)X_1$ and $(x-x_0)^2X_2$ are both analytic at x_0, then x_0 is called a *regular singular point* or we may say that the singularity is *non-essential.* If they are not analytic at x_0, then the singularity is *essential* or *irremovable.* Fuch's theorem applies to non-essential singularity. Thus series solution method is valid for its application to a differential equation having no singularity or if there be any singularity at all, it will still be valid for its application, if the singularity is non-essential. Now to clarify the meaning of analytic function, we may say that a function $f(x)$ is analytic at x_0 if its Taylor's expansion about x_0, i.e.,

$$\sum_{n=0}^{\infty} \frac{f^n(x_0)}{n!}(x - x_0)^n$$

exists and converges to $f(x)$ in some open interval $a < x_0 < b$.

Let us consider a differential equation *with no singularity* given by

$$\frac{d^2y}{dx^2} + x\frac{dy}{dx} + (x^2 + 2)y = 0.$$

Here $X_1 = x, X_2 = (x^2 + 2)$, and the power series solution in powers of x, which implies $x_0 = 0$, is possible, because $x_0 = 0$ is an ordinary point. So the solution can be written as $y = \sum a_r x^{k+r}$ with $a_0 \neq 0$.

Again we consider a differential equation *with singularity* given by

$$\frac{d^2y}{dx^2} + \frac{y}{x^3} = 0$$

There is singularity at $x = 0$, but if this singularity is a non-essential or removable one, the series solution method may still be applied.

Here $X_2 = \frac{1}{x^3}$

$$\therefore\ (x - 0)^2 X_2(x)\Big]_{x=0} = x^2 \cdot \frac{1}{x^3}\Big]_{x=0} = \infty;$$

so the singularity is essential and cannot be removed, and as such series solution method cannot be applied.

But for the differential equation given by

$$\frac{d^2y}{dx^2}+\frac{1}{x}\frac{dy}{dx}-\frac{1}{x^2}y=0,$$

we obtain $X_1=\frac{1}{x}, X_2=-\frac{1}{x^2}$

$$\therefore\ \left[(x-0)\frac{1}{x}\right]\to 1$$

as $x\to 0$; also $\left[(x-0)^2\cdot -\frac{1}{x^2}\right]\to -1$ as $x\to 0$ showing that the singularity is removable and hence series solution method can be applied in this case.

17.10 Bessel's differential equation and its solution

Bessel's equation is an important differential equation which often occurs in the analysis of the vibration of a circular membrane, diffraction by a circular aperture etc. Separation of three dimensional wave equation as well as Laplace's equation in circular cylindrical co-ordinates and separation of Helmholtz's equation in spherical co-ordinates (into independent variable separable form) lead to Bessel's differential equation—solutions of which are called *Bessel's functions.* Bessel's functions appear in a wide variety of physical problems.

Bessel's differential equation is of the form

$$x^2y''+xy'+(x^2-n^2)y=0. \tag{17.10.1}$$

This equation has a singularity at $x=0$, but this singularity can easily be shown to be a non-essential one, and therefore its solution can be obtained as a power series developed about this point.

In order to solve it in a series of ascending powers of x, let us assume that the series solution is

$$y=\sum_{r=0}^{\infty}a_rx^{k+r}.\quad\therefore\quad y'=\sum_{r=0}^{\infty}a_r(k+r)x^{k+r-1};\quad y''=\sum_{r=0}^{\infty}a_r(k+r)(k+r-1)x^{k+r-2}.$$

Substituting these values in (17.10.1), we have

$$x^2\sum_{r=0}^{\infty}a_r(k+r)(k+r-1)x^{k+r-2}+x\sum_{r=0}^{\infty}a_r(k+r)x^{k+r-1}+(x^2-n^2)\sum_{r=0}^{\infty}a_rx^{k+r}=0$$

$$\text{or,}\quad\left[\{(k+r)(k+r-1)+(k+r)-n^2\}x^{k+r}+x^{k+r+2}\right]a_r=0. \tag{17.10.2}$$

The expression (17.10.2) being an identity, we shall equate the coefficients of various powers of x to zero. Equating the coefficient of the lowest power of x to zero (putting $r=0$), we have the *indicial equation*

$$a_0(k^2-n^2)=0;\quad\because\quad a_0\neq 0,\quad\text{we have}\quad k=\pm n. \tag{17.10.3}$$

Equating the coefficient of x^{k+1} (putting $r=1$) to zero, we have $a_1\left\{(k+1)^2-n^2\right\}=0$ but, from (17.10.3) $(k+1)^2-n^2\neq 0$; hence $a_1=0$.

Equating the coefficient of the general term x^{k+r+2} to zero, we have

$$\{(k+r+2)^2 - n^2\}a_{r+2} + a_r = 0 \quad \text{or,} \quad a_{r+2} = -\frac{a_r}{(k+r+2+n)(k+r+2-n)}.$$

which is the *recurrence relation* between the coefficients a_r.

Case I: If $k = +n$,

$$a_2 = -\frac{a_0}{(n+2+n)(n+2-n)} = -\frac{a_0}{2(2n+2)}; \quad [\text{putting } r=0]$$

$$a_4 = -\frac{a_2}{4(2n+4)} = \frac{a_0}{2.4(2n+2)(2n+4)} \quad [\text{putting } r=2];$$

$$a_6 = -\frac{a_4}{6(2n+6)} = -\frac{a_0}{2.4.6(2n+2)(2n+4)(2n+6)}; \quad [\text{putting } r=4]$$

$$a_{2r} = \frac{(-1)^r a_0}{2.4.6\cdots 2r \cdots (2n+2)(2n+4)\cdots(2n+2r)}.$$

Also $a_1 = a_3 = a_5 = \cdots = 0$.

Hence the series solution is

$$y = a_0 x^n \left[1 - \frac{x^2}{2(2n+2)} + \frac{x^4}{2.4(2n+2)(2n+4)} - \cdots \right.$$
$$\left. + \frac{(-1)^r x^{2r}}{2.4.6\cdots 2r \cdots (2n+2)(2n+4)\cdots(2n+2r)}\right]$$

$$\text{or,} \quad y = a_0 x^n \sum_{r=0}^{\infty} \frac{(-1)^r x^{2r}}{2^r (r)!.2^r (n+1)\cdots(n+r)}.$$

If $a_0 = \frac{1}{2^n \Gamma(n+1)}$, the solution is called $J_n(x)$, which is the Bessel's function of the 1st kind of order n and is given by

$$J_n(x) = \frac{x^n}{2^n \Gamma(n+1)} \sum_{r=0}^{\infty} \frac{(-1)^r x^{2r}}{2^{2r}(r)!(n+1)(n+2)\cdots(n+r)}$$

$$= \sum_{r=0}^{\infty} (-1)^r \left(\frac{x}{2}\right)^{n+2r} \frac{1}{r!\Gamma(n+r+1)}. \tag{17.10.4}$$

Case II: If $k = -n$, the solution is

$$J_{-n}(x) = \sum_{r=0}^{\infty} (-1)^r \left(\frac{x}{2}\right)^{-n+2r} \frac{1}{r!\Gamma(-n+r+1)},$$

for n not equal to an integer. The general solution of Bessel's differential equation is given by

$$\boldsymbol{y = A\, J_n(x) + B\, J_{-n}(x).} \tag{17.10.5}$$

If we put $s = r - n$, where s is a positive integer, then

$$J_{-n}(x) = \sum_{s=0}^{\infty}(-1)^{s+n}\left(\frac{x}{2}\right)^{2s+n} \cdot \frac{1}{(n+s)!\Gamma(s+1)} = (-1)^n \sum_{s=0}^{\infty}(-1)^s\left(\frac{x}{2}\right)^{2s+n} \frac{1}{(n+s)!\Gamma(s+1)}$$

$$= (-1)^n \sum_{s=0}^{\infty}\left(\frac{x}{2}\right)^{2s+n} \frac{1}{s!\Gamma(n+s+1)} = (-1)^n J_n(x). \tag{17.10.6}$$

Thus, when n is not an integer,

$$y = A\,J_n(x) + B\,J_{-n}(x) = A\,J_n(x) + B(-1)^n J_n(x)$$

$$= [A + B(-1)^n]J_n(x) = C\,J_n(x),$$

where C is an arbitrary constant. This is not the general solution of Bessel's differential equation which must contain two linearly independent functions multiplied by an arbitrary constant.

Now, we consider the Bessel's function $Y_n(x)$ of order n of the second kind given by

$$Y_n(x) = \frac{1}{\sin n\pi}\left[\cos n\pi\, J_n(x) - J_{-n}(x)\right]. \tag{17.10.7}$$

If n is not an integer, the function $Y_n(x)$ is dependent on $J_n(x)$ and since it is a linear combination of $J_n(x)$ and $J_{-n}(x)$, it is the solution of Bessel's differential equation of order n. If n is an integer, we have from (17.10.6) and (17.10.7)

$$Y_n(x) = \frac{\cos n\pi\, J_n(x) - (-1)^n J_n(x)}{\sin n\pi} = \frac{0}{0}.$$

Thus, when n is an integer, $Y_n(x)$ is defined as

$$Y_n(x) = \lim_{r \to n} \frac{J_s(x)\cos r\pi - J_{-s}(x)}{\sin r\pi}.$$

Weber gave the solution of the differential equation (17.10.1), when $n = 0$ as

$$Y_0(x) = \frac{2}{\pi}\left[J_0(x)\left(\log\frac{x}{2} + \gamma\right) + \left(\frac{x}{2}\right)^2 - \frac{\left(1+\frac{1}{2}\right)\left(\frac{x}{2}\right)^4}{(2!)^2} + \left(1 + \frac{1}{2} + \frac{1}{3}\right)\frac{\left(\frac{x}{2}\right)^6}{(3!)^2}\cdots\right]$$

where γ is called *Euler's constant* defined by

$$\gamma = \lim_{n\to\infty}\left(1 + \frac{1}{2} + \frac{1}{3} + \cdots + \frac{1}{n} - \log n\right) = 0.5772157.$$

Again, when n is any positive integer,

$$Y_n(x) = \frac{2}{\pi}\left(\gamma + \log\frac{x}{2}\right)J_n(x) - \frac{1}{\pi}\sum_{r=0}^{\infty}(-1)^r \frac{\left(\frac{x}{2}\right)^{n+2r}}{r!(n+r)^{-1}} \cdot \left(\sum_{m=1}^{n+r} m^{-1} + \sum_{m=1}^{r} m^{-1}\right)$$

$$-\frac{1}{\pi}\sum_{r=0}^{n-1}\left(\frac{x}{2}\right)^{-n+2r}\frac{(n-r-1)!}{r!}.$$

For $r = 0$, we may write

$$\sum_{m=1}^{n} m^{-1} \text{ instead of } \sum_{m=1}^{n+r} m^{-1} + \sum_{m=1}^{r} m^{-1}.$$

The presence of the logarithmic term in the expression for function $Y_n(x)$ shows that the functions are infinite at $x = 0$. The general solution for Bessel's differential equation may therefore be written as $y = C_1 J_n(x) + C_2 Y_n(x)$, where C_1 and C_2 are arbitrary constants.

17.10.1 Generating Function for $J_n(x)$

$J_n(x)$, the Bessel function of first kind of order n, is the coefficient of t^n in the expansion of the function $e^{\frac{x}{2}(t-\frac{1}{t})}$. That is

$$e^{\frac{x}{2}(t-\frac{1}{t})} = \sum_{n=0}^{\infty} J_n(x) t^n \tag{17.10.8}$$

Proof. Expanding the exponentials in Maclaurin's series, we have

$$e^{\frac{xt}{2}} = \sum_{r=0}^{\infty} \frac{x^r t^r}{2^r r!} \quad \text{and} \quad e^{-\frac{x}{2t}} = \sum_{s=0}^{\infty} (-1)^s \frac{x^s}{2^s} \cdot \frac{t^{-s}}{s!}$$

Multiplying these series, we obtain

$$e^{\frac{x}{2}(t-\frac{1}{t})} = \sum_{r=0}^{\infty} \frac{x^r t^r}{2^r r!} \cdot \sum_{s=0}^{\infty} \frac{(-1)^s x^s}{2^s t^s s!}$$

Replacing r by $n+s$, we get the coefficient of t^n as

$$\left.\begin{aligned} \sum_{s=0}^{\infty} \frac{x^{n+s}}{2^{n+s}(n+s)!} \cdot \frac{(-1)^s x^s}{2^s s!} &= \sum_{s=0}^{\infty} (-1)^s \frac{1}{s!(n+s)!} \left(\frac{x}{2}\right)^{n+2s} \quad (n \geq 0) \\ &= J_n(x) \text{ (using 17.10.4)} \end{aligned}\right\}. \tag{17.10.9}$$

The coefficient of t^{-n} is obtained by putting $s = n + r$. We have then the coefficient of t^{-n} as

$$\left.\begin{aligned} \sum_{r=0}^{\infty} \frac{x^r}{2^r r!} \cdot \frac{(1)^{n+r} x^{n+r}}{2^{n+r}(n+r)!} &= (-1)^n \sum_{r=0}^{\infty} (-1)^r \left(\frac{x}{2}\right)^{n+2r} \frac{1}{r!(n+r)!} \\ &= (-1)^n J_n(x) = J_{-n}(x) \text{ (using 17.10.6)} \end{aligned}\right\}. \tag{17.10.10}$$

Combining (17.10.9) and (17.10.10) we get the coefficient of t^n in $e^{\frac{x}{2}(t-\frac{1}{t})}$ as $J_n(x)$ for $-\infty \leq n \leq \infty$. Thus

$$\sum_{n=-\infty}^{\infty} J_n(x) t^n = e^{\frac{x}{2}(t-\frac{1}{t})}.$$

Thus (17.10.8) is proved.

Using this equation, it is possible to calculate the Bessel's functions of different orders (for n integer). Thus the function $e^{\frac{x}{2}\left(t-\frac{1}{t}\right)}$ is known as *the generating function* for Bessel's functions. The generating function equation (17.10.8) also enables us to express trigonometric functions as expansions containing Bessel's functions.

17.10.2 Recurrence formula for $J_n(x)$

We have from (17.10.4)

$$J_n(x) = \sum_{r=0}^{\infty} (-1)^r (x/2)^{n+2r} \frac{1}{r!\Gamma(n+r+1)}$$

$$\therefore \quad J_n'(x) = \sum_{r=0}^{\infty} (-1)^r \cdot \frac{1}{r!\Gamma(n+r+1)} \frac{(n+2r)}{2} (x/2)^{n+2r-1}$$

$$\text{or,} \quad xJ_n'(x) = \sum_{r=0}^{\infty} (-1)^r \frac{(n+2r)}{r!\Gamma(n+r+1)} (x/2)^{n+2r}$$

$$\text{or,} \quad xJ_n'(x) = nJ_n(x) + \sum_{r=0}^{\infty} (-1)^r \frac{(n+2r)}{r!\Gamma(n+r+1)} \cdot (x/2)^{n+2r} - nJ_n(x)$$

$$= nJ_n(x) + \sum_{r=0}^{\infty} (-1)^r \frac{(n+2r)}{r!\Gamma(n+r+1)} (x/2)^{n+2r}$$

$$- \sum_{r=0}^{\infty} (-1)^r \frac{n}{r!\Gamma(n+r+1)} \cdot (x/2)^{n+2r}$$

$$= nJ_n(x) + \sum_{r=0}^{\infty} (-1)^r \frac{2r}{r!\Gamma(n+r+1)} (x/2)^{n+2r}$$

$$= nJ_n(x) + x \sum_{r=0}^{\infty} (-1)^r \frac{r}{r!\Gamma(n+r+1)} \left(\frac{x}{2}\right)^{n+2r-1}$$

$$= nJ_n(x) + x \sum_{r=0}^{\infty} \frac{(-1)^r}{(r-1)!\Gamma(n+r+1)} (x/2)^{n+2r-1}.$$

If we put $r = s+1$, we have

$$xJ_n'(x) = nJ_n(x) + x \sum_{r=0}^{\infty} \frac{(-1)^{s+1}}{s!\Gamma(n+s+2)} (x/2)^{n+2s+1}$$

$$= nJ_n(x) - x \sum_{r=1}^{\infty} \frac{(-1)^s}{s!\Gamma(n+s+2)} (x/2)^{n+2s+1} = nJ_n(x) - xJ_{n+1}(x). \tag{17.10.11}$$

Similarly, it can be proved that

$$xJ_n'(x) + nJ_n(x) = xJ_{n-1}(x). \tag{17.10.12}$$

Adding (17.10.11) and (17.10.12), we have

$$2xJ_n'(x) = x\{J_{n-1}(x) - J_{n+1}(x)\} \quad \text{or} \quad 2J_n'(x) = J_{n-1}(x) - J_{n+1}(x). \tag{17.10.13}$$

If $n = 0$, we have from (17.10.13),

$$2J_0'(x) = J_{-1}(x) - J_1(x) = (-1)J_1(x) - J_1(x) = -2J_1(x) \quad \text{or,} \quad J_0'(x) = -J_1(x).$$

Multiplying (17.10.11) by x^{-n-1}, we have

$$x^{-n}J_n'(x) = n\,x^{-n-1}J_n(x) - x^{-n}J_{n+1}(x)$$

$$\text{or,} \quad x^{-n}J_n'(x) - nx^{-n-1}J_n(x) = -x^{-n}J_{n+1}(x)$$

$$\text{or,} \quad \frac{d}{dx}[x^{-n}J_n(x)] = -x^{-n}J_{n+1}(x). \tag{17.10.14}$$

Similarly, it can be shown that

$$\frac{d}{dx}[x^{n}J_n(x)] = +x^{n}J_{n-1}(x). \tag{17.10.15}$$

Again subtracting (17.10.12) from 17.10.11), we have

$$xJ_n'(x) - xJ_n'(x) - nJ_n(x) = nJ_n(x) - xJ_{n+1}(x) - xJ_{n-1}(x)$$

$$\text{or,} \quad \frac{2n}{x}J_n(x) = J_{n+1}(x) + J_{n-1}(x).$$

N.B. Instead of (17.10.1), Bessel's equation of nth order also occurs very often in another standard form given by

$$\frac{d^2y}{dx^2} + \frac{1}{x}\frac{dy}{dx} - \left(1 - \frac{n^2}{x^2}\right)y = 0 \tag{17.10.16}$$

Putting $x = iz$, we have

$$\frac{dz}{dx} = \frac{1}{i} \quad \text{and} \quad \frac{dy}{dx} = \frac{dy}{dz}\cdot\frac{dz}{dx} = \frac{1}{i}\frac{dy}{dz}$$

$$\text{and} \quad \frac{d^2y}{dx^2} = \frac{d}{dx}\left(\frac{1}{i}\frac{dy}{dz}\right) = \frac{1}{i}\frac{d^2y}{dz^2}\frac{dz}{dx} = \frac{1}{i^2}\frac{d^2y}{dz^2} = -\frac{d^2y}{dz^2}$$

With these substitutions, (17.10.16) becomes

$$-\frac{d^2y}{dz^2} + \frac{1}{iz}\frac{1}{i}\frac{dy}{dz} - \left\{1 - \frac{n^2}{(iz)^2}\right\}y = 0 \quad \text{or} \quad \frac{d^2y}{dz^2} + \frac{1}{z}\frac{dy}{dz} + \left(1 + \frac{n^2}{z^2}\right)y = 0. \tag{17.10.17}$$

Since the complete solution of (17.10.16) is $y = C_1J_n(x) + C_2Y_n(x)$, we have the solution of (17.10.17) (obtained by putting $x = iz$) given by $y = C_1J_n(iz) + C_2Y_n(iz)$, which after much simplification is written as $y = C_1I_n(x) + C_2K_n(x)$ where $I_n(x)$ = modified Bessel's function of the first kind of order n and K_n = modified Bessel's function of the second kind of order n.

17.11 Legendre's differential equation and its solution

It is a very important differential equation and finds application in the solution of Laplace's equation in spherical polar co-ordinates. These solutions are very important for solving potential and field problems in electro-statics.

This equation is of the form

$$(1-x^2)y'' - 2xy' + n(n+1)y = 0. \tag{17.11.1}$$

To test the existence of power series solution, equation (17.11.1) is written in the equivalent rationalised form as

$$y'' - \frac{2x}{(1-x^2)}y' + \frac{n(n+1)}{(1-x^2)}y = 0.$$

Here $X_1 = -\dfrac{2x}{1-x^2}$, $X_2 = \dfrac{n(n+1)}{1-x^2}$. So X_1 and X_2 are both finite for $x = 0$; hence y can be written as a series of ascending or descending powers of x, but $x = \pm 1 = x_0$ is a point of singularity, and

$$(x-x_0)X_1 \to -(x-x_0)\cdot\frac{2x}{(1-x^2)}\Bigg]_{x=x_0=1} = \frac{2x}{x+1} \to 1, \text{ finite}$$

$$(x-x_0)^2X_2 \to (x-1)^2\cdot\frac{n(n+1)}{(1+x)(1-x)}\Bigg]_{x=x_0=1} = -\frac{x-1}{x+1}n(n+1) \to 0, \text{ finite,}$$

so that power series solution is valid for $x = 0$, as well as for $x = 1$.

As in the case of Bessel's differential equation, here also we assume an infinite series solution of the form (considering y as a series of descending powers of x)

$$y = \sum_{r=0}^{\infty} a_r x^{k-r}. \quad \therefore \quad y' = \sum_{r=0}^{\infty} a_r(k-r)x^{k-r-1} \text{ and } y'' = \sum_{r=0}^{\infty} a_r(k-r)(k-r-1)x^{k-r-2}.$$

Substituting these values in (17.11.1), we have

$$\sum_{r=0}^{\infty}\left[(1-x^2)a_r(k-r)(k-r-1)x^{k-r-2} - 2xa_r(k-r)x^{k-r-1} + n(n+1)a_r x^{k-r}\right] = 0$$

$$\text{or,} \quad \sum_{r=0}^{\infty}\left[(k-r)(k-r-1)x^{k-r-2} + \{n(n+1) - 2(k-r) - (k-r)(k-r-1)\}x^{k-r}\right]a_r = 0$$

$$\text{or,} \quad \sum_{r=0}^{\infty}\left[(k-r)(k-r-1)x^{k-r-2} + \{n(n+1) - (k-r)(k-r+1)\}x^{k-r}\right]a_r = 0. \tag{17.11.2}$$

In the above relation the coefficient of every power of x must vanish. When $r = 0$, we have equating the coefficient of x^k to zero, $[n(n+1) - k(k+1)]a_0 = 0$, (where a_0 is the coefficient of the first term of the series which cannot be equal to zero)—known as the *indicial equation.*

$\therefore n(n+1) - k(k+1) = 0$ or, $n^2 - k^2 + n - k = 0$ or, $(n-k)(n+k+1) = 0$.

Hence

$$k = n \quad \text{or} \quad k = -n-1. \tag{17.11.3}$$

Putting $r = 1$, we equate the coefficient of x^{k-1} to zero and hence get

$$\{n(n+1) - (k-1)k\}a_1 = 0 \quad \text{from (17.11.2)}$$

$$\because \ \{n(n+1) - (k-1)k\} \neq 0 \quad \text{from (17.11.3)}. \quad \therefore \quad a_1 = 0.$$

Equating the coefficient of x^{k-r}, the general term in (17.11.2) to zero, we have

$$(k-r+2)(k-r+1)a_{r-2} + \{n(n+1) - (k-r)(k-r+1)\}a_r = 0$$

$$\therefore \ a_r = -\frac{(k-r+2)(k-r+1)}{n(n+1) - (k-r)(k-r+1)}a_{r-2}$$

which is the *recurrence relation.*

when $k = n$, the recurrence formula is

$$a_r = -\frac{(n-r+2)(n-r+1)}{n^2+n-(n-r)(n-r+1)}a_{r-2} = -\frac{(n-r+2)(n-r+1)}{r(2n-r+1)}a_{r-2}.$$

Again when $k = -n-1$

$$a_r = -\frac{(-n-r+1)(-n-r)}{n^2+n-(n+r+1)(n+r)}a_{r-2} = \frac{(n+r-1)(n+r)}{r(2n+r+1)}a_{r-2}.$$

Thus when $k = n$

$$a_2 = -\frac{n(n-1)}{2(2n-1)}a_0 \ \text{(putting } r = 2\text{)};$$

$$a_3 = -\frac{(n-1)(n-2)}{3(2n-2)}a_1 = 0, \ \text{since } a_1 = 0;$$

$$a_4 = -\frac{(n-2)(n-3)}{4(2n-3)}a_2 = +\frac{n(n-1)(n-2)(n-3)}{2.4(2n-1)(2n-3)}a_0;$$

Similarly, it can be shown that $a_5 = a_7 = a_9 = \cdots = 0$.

In general

$$a_{2r} = (-1)^r \frac{n(n-1)(n-2)(n-3)\cdots(n-2r+1)}{2.4\cdots 2r(2n-1)(2n-3)\cdots(2n-2r+1)}a_0. \tag{17.11.4}$$

Hence for $k = n$

$$y = \left[a_0x^n + a_1x^{n-1} + a_2x^{n-2} + \cdots\right]$$

$$= \left[a_0x^n - \frac{n(n-1)}{2(2n-1)}a_0x^{n-2} + \frac{n(n-1)(n-2)(n-3)a_0x^{n-4}}{2.4(2n-1)(2n-3)} - \cdots\right]$$

$$= a_0\left[x^n - \frac{n(n-1)}{2(2n-1)}x^{n-2} + \frac{n(n-1)(n-2)(n-3)}{2.4(2n-1)(2n-3)}x^{n-4} - \cdots\right] \tag{17.11.5}$$

Here a_0 is an arbitrary constant. If the infinite series giving the solution of a given differential equation reduces into a finite series, then the solution is called the *Polynomial.* Legendre Polynomials of the first kind are denoted by $P_n(x)$ and those of second kind are denoted by $Q_n(x), n$ being a positive integer. When the arbitrary constant

$$a_0 = \frac{1.3.5\cdots(2n-1)}{n!},$$

the above solution is known as the *Polynomial* given by

$$P_n(x) = \frac{1.3.5\cdots(2n-1)}{n!}\left[x^n - \frac{n(n-1)}{2(2n-1)}x^{n-2} + \frac{n(n-1)(n-2)(n-3)}{2.4(2n-1)(2n-3)}x^{n-4}\cdots\right]$$

$$= \frac{(2n)!}{(2.4.6\cdots 2n)n!}\left[x^n - \frac{n(n-1)}{2(2n-1)}x^{n-2} + \frac{n(n-1)(n-2)(n-3)}{2.4(2n-1)(2n-3)}x^{n-4} - \cdots\right]$$

$$= \frac{(2n)!}{2^n \cdot n!n!}\left[x^n - \frac{n(n-1)}{2(2n-1)}x^{n-2} + \frac{n(n-1)(n-2)(n-3)}{2.4(2n-1)(2n-3)}x^{n-4} - \cdots\right]. \quad (17.11.6)$$

When $n = 0, 1, 2, 3, 4, 5, 6$ etc. we have,

$$\left.\begin{aligned}
P_0(x) &= 1 \\
P_1(x) &= \frac{2!}{2.1!1!}[x] = x \\
P_2(x) &= \frac{4!}{2^2.2!2!}\left[x^2 - \frac{2}{2.3}\right] = \frac{3}{2}\left(x^2 - \frac{1}{3}\right) = \frac{3}{2}x^2 - \frac{1}{2} \\
P_3(x) &= \frac{6!}{2^3.3!3!}\left[x^3 - \frac{3.2}{2.5}x\right] = \frac{5}{2}\left(x^3 - \frac{3}{5}x\right) = \frac{5}{2}x^3 - \frac{3}{2}x. \\
&\text{Similarly it can be shown that} \\
P_4(x) &= \frac{35}{8}x^4 - \frac{15}{4}x^2 + \frac{3}{8} \\
P_5(x) &= \frac{63}{8}x^5 - \frac{35}{4}x^3 + \frac{15}{8}x \text{ and so on.}
\end{aligned}\right\} \quad (17.11.7)$$

When $k = -n-1$, we have the recurrence formula given by

$$a_r = \frac{(n+r-1)(n+r)}{r(2n+r+1)}a_{r-2}.$$

Hence,

$$a_2 = \frac{(n+1)(n+2)}{2(2n+3)}a_0;$$

a_3 will contain a_1 and so is equal to zero. In this way it can be shown that

$$a_5 = a_7 = a_9 \cdots = 0; \quad a_4 = \frac{(n+3)(n+4)}{4(2n+5)}a_2 = \frac{(n+1)(n+2)(n+3)(n+4)}{2.4(2n+3)(2n+5)}a_0.$$

In general

$$a_{2r} = \frac{(n+1)(n+2)\cdots(n+2r)}{2.4\cdots 2r(2n+3)(2n+5)\cdots(2n+2r+1)}a_0.$$

Hence for $k = -n-1$, the series solution is

$$y = a_0\left[x^{-n-1} + \frac{(n+1)(n+2)}{2(2n+3)}x^{-n-3} + \frac{(n+1)(n+2)(n+3)(n+4)}{2.4(2n+3)(2n+5)}x^{-n-5} + \cdots\right].$$

When $a_0 = \frac{n!}{1.3.5\cdots(2n+1)}$, the solution is denoted by the Polynomial $Q_n(x)$ and is given by

$$Q_n(x) = \frac{n!}{1.3.5\cdots(2n+1)}$$
$$\left[x^{-n-1} + \frac{(n+1)(n+2)}{2(2n+3)}x^{-n-3} + \frac{(n+1)(n+2)(n+3)(n+4)}{2.4(2n+3)(2n+5)}x^{-n-5} + \cdots\right].$$

The general solution of Legendre's equation is $y = A\,P_n(x) + B\,Q_n(x)$, where A, B are arbitrary constants.

Legendre polynomials defined by equation (17.11.7) occur very frequently in a large number of topics in physical, engineering and applied mathematics. They are used to find the solutions of Laplace's and Helmholtz's equations in spherical polar co-ordinates and hence have their direct applications in the study of potential and thermal distribution in spherically symmetrical bodies. They are also used to express the distribution function of angular correlation of two γ-rays emitted in the successive transitions of a nuclei, and also in the study of scattering problems in the special case of radial incidence.

17.11.1 Generating Function for Legendre's Polynomial $P_n(x)$

Legendre's polynomial $P_n(x)$ is the coefficient of y^n in the expansion of the function

$$z = f(x,y) = (1-2xy+y^2)^{-1/2} = [1-(2xy-y^2)]^{-1/2}$$

that is

$$(1-2xy+y^2)^{-1/2} = \sum_{n=0}^{\infty} y^n P_n(x) \tag{17.11.8}$$

Proof. Using binomial theorem, the function $[1-(2xy-y^2)]^{-1/2}$ can be expanded as

$$[1-(2xy-y^2)]^{-1/2} = [1-y(2x-y)]^{-1/2}$$
$$= 1 + \frac{y}{2}(2x-y) + \frac{1.3}{2.4}y^2(2x-y)^2 + \frac{1.3.5}{2.4.6}y^3(2x-y)^3 + \cdots + \frac{1.3.5\cdots(2n-1)}{2.4.6\cdots 2n}y^n(2x-y)^n.$$

Collecting the coefficients of y^n in the above expansion, we obtain

$$\frac{1.3.5\cdots(2n-1)}{2.4.6\cdots 2n}(2x)^n - \frac{1.3.5\cdots(2n-3)}{2.4.6\cdots(2n-2)}(2x)^{n-2}\,{}^{n-1}C_1 + \frac{1.3.5\cdots(2n-5)}{2.4.6\cdots(2n-4)}(2x)^{n-4}\,{}^{n-2}C_2 - \cdots$$
$$= \frac{1.3.5\cdots(2n-1)}{n!}\left[x^n - \frac{n(n-1)}{2(2n-1)}x^{n-2} + \frac{n(n-1)(n-2)(n-3)}{2.4.(2n-1)(2n-3)}x^{n-4} - \cdots\right]$$
$$= P_n(x) \quad \text{[using (17.11.6)]}$$

$$\therefore\ z = (1-2xy+y^2)^{-1/2} = \sum_{n=0}^{\infty} y^n P_n(x) = P_0(x) + yP_1(x) + y^2P_2(x) + \cdots$$

which proves the equation (17.11.8).

Therefore by expanding the function $(1-2xy+y^2)^{-1/2}$ binomially in ascending powers of y, we can obtain the Legendre's polynomials of different order as the coefficients of the corresponding powers of y. Thus the function $(1-2xy+y^2)^{-1/2}$ is said to be the *generating function* for the Legendre's polynomial $P_n(x)$.

We can now derive the following useful and direct conclusions from (17.11.8).

(a)

$$P_n(1) = 1 \tag{17.11.9}$$

Proof. From (17.11.8) we can write (by putting $x = 1$) that $P_n(1)$ is the coefficient of y^n in the expansion of $(1-2y+y^2)^{-1/2}$, i.e., $\left[(1-y)^2\right]^{-1/2}$, i.e., $(1-y)^{-1}$. Thus $P_n(1)$ = coefficient of y^n in $(1+y+y^2+\cdots+y^n+\cdots) = 1$, which is however a distinguishing property of Legendre's polynomials.

(b)

$$P_n(-1) = (-1)^n \tag{17.11.10}$$

Proof. This result follows directly from (17.11.8) on putting $x = -1$, so that we may write

$$(1+2y+y^2)^{-1/2} = \sum_{n=0}^{\infty} P_n(-1)y^n$$

$$\text{or} \quad \sum_{n=0}^{\infty} P_n(-1)y^n = \left\{(1+y)^2\right\}^{-1/2} = (1+y)^{-1} = \sum_{n=0}^{\infty}(-1)^n y^n.$$

So equating the coefficients of y^n on both sides, we get the result (17.11.10). Again from (17.11.9) and (17.11.10), it is obvious that $P_n(-1) = (-1)^n P_n(1)$.

(c)

$$P_n(-x) = (-1)^n P_n(x). \tag{17.11.11}$$

In the calculation of the generating function for $P_n(x)$, we have already shown that

$$P_n(x) = \frac{1.3.5\cdots(2n-1)}{n!}\left[x^n - \frac{n(n-1)}{2(2n-1)}x^{n-2} + \cdots\right]$$

$$= \frac{(2n)!}{(2.4.6\cdots 2n)n!}\left[x^n - \frac{n(n-1)}{2(2n-1)}x^{n-2} + \cdots\right] = \frac{(2n)!}{2^n\cdot n!n!}\left[x^n - \frac{n(n-1)}{2(2n-1)}x^{n-2} + \cdots\right]$$

Replacing x by $-x$, we may write

$$P_n(-x) = \frac{(2n)!}{2^n\cdot n!n!}\left[(-1)^n x^n - \frac{n(n-1)}{2(2n-1)}(-1)^{n-2}x^{n-2} + \cdots\right]$$

$$= \frac{(2n)!}{2^n\cdot n!n!}\left[(-1)^n x^n - \frac{n(n-1)(-1)^n}{2(2n-1)(-1)^2}x^{n-2} + \cdots\right]$$

$$= (-1)^n\frac{(2n)!}{2^n\cdot n!n!}\left[x^n - \frac{n(n-1)}{2(2n-1)}x^{n-2} + \cdots\right]$$

$$= (-1)^n P_n(x), \ n \text{ being an integer.}$$

(17.11.11) may alternatively be derived more conveniently from (17.11.8) with the knowledge that the generating function $(1-2xy+y^2)^{-1/2} = f(x,y)$ remains unchanged when x is replaced by $-x$ and y is replaced by $-y$. That is $f(x,y) = f(-x,-y)$, so that we may write

$$(1-2xy+y^2)^{-1/2} = \left[1-2(-x)(-y)+(-y)^2\right]^{-1/2}$$

$$= \sum_{n=0}^{\infty} P_n(-x)(-y)^n = \sum_{n=0}^{\infty}(-1)^n P_n(-x)y^n$$

Combining this equation with (17.11.8), we have

$$\sum_{n=0}^{\infty} P_n(x)y^n = \sum_{n=0}^{\infty}(-1)^n P_n(-x)y^n,$$

whence on equating the coefficients, we obtain

$$P_n(x) = (-1)^n P_n(-x) \quad \text{or} \quad P_n(-x) = (-1)^n P_n(x)$$

n being an integer.

N.B. If $n = 2m$, we have $P_{2m}(-x) = P_{2m}(x)$. If however $n = (2m+1)$, we have $P_{2m+1}(-x) = -P_{2m+1}(x)$.

17.11.2 Rodrigues' formula for the Legendre's Polynomials

We can deduce an important expression for the polynomials $P_n(x)$ from Legendre's differential equation.

Let $u = (x^2-1)^n$.

$$\therefore \quad \frac{du}{dx} = 2nx(x^2-1)^{n-1}.$$

Again

$$(1-x^2)\frac{du}{dx} = -2nx(x^2-1)^n = -2nx\,u \quad \text{or,} \quad (1-x^2)\frac{du}{dx} + 2nx\,u = 0. \tag{17.11.12}$$

Differentiating the above equation with respect to x, we have

$$(1-x^2)\frac{d^2u}{dx^2} - 2x\frac{du}{dx} + 2nx\frac{du}{dx} + 2nu = 0 \quad \text{or,} \quad (1-x^2)\frac{d^2u}{dx^2} + 2(n-1)x\frac{du}{dx} + 2nu = 0. \tag{17.11.13}$$

Differentiating the equation (17.11.13) r times in succession, we have

$$(1-x^2)\frac{d^2u_r}{dx^2} + 2(n-r-1)x\frac{du_r}{dx} + (2n-r)(r+1)u_r = 0.$$

Here,

$$u_r = \frac{d^r u}{dx^r}. \tag{17.11.14}$$

When we consider the particular case $r = n$, equation (17.11.14) reduces to

$$(1-x^2)\frac{d^2u_n}{dx^2} - 2x\frac{du_n}{dx} + n(n+1)u_n = 0 \tag{17.11.15}$$

where

$$u_n = \frac{d^n u}{dx^n} = \frac{d^n}{dx^n}(x^2-1)^n.$$

The equation (17.11.15) is Legendre's equation. u_n is a polynomial of degree n. The solution of Legendre's equation is given by

$$P_n(x) = Cu_n = C\frac{d^n}{dx^n}(x^2-1)^n \tag{17.11.16}$$

where C is a constant to be determined.

[$\because$ Legendre's equation has one and only one distinct solution of the form $P_n(x), P_n(x)$ is a constant multiple of u_n.]

Considering the highest power of x on each side of the equation (17.11.16), we have

$$\frac{(2n)!}{2^n(n!)^2}\cdot x^n = C\frac{d^n}{dx^n}x^{2n} = C\frac{(2n)!}{n!}x^n \quad \text{or,} \quad C = \frac{1}{2^n n!}.$$

Substituting this value of the constant C in (17.11.16), we have,

$$P_n(x) = \frac{1}{2^n n!}\frac{d^n}{dx^n}(x^2-1)^n.$$

This is Rodrigues' formula for the Legendre's polynomials.

17.11.3 Properties of Legendre's polynomial

(1) **Orthogonality of $P_n(x)$:** The Legendre's polynomials are orthogonal. Considering two polynomials $P_n(x)$ and $P_m(x)$, we can show that if $m \neq n$,

$$\int_{-1}^{1} P_n(x)P_m(x)dx = 0$$

which is the condition of orthogonality.

(2) From the orthogonal property, we can also deduce another important relation

$$\int_{-1}^{1} [P_n(x)]^2 dx = \frac{2}{(2n+1)} \quad \text{for } n = 0, 1, 2 \text{ etc.}$$

(3) **Expansion of a function $f(x)$ in a series of Legendre's polynomials:**

A function $f(x)$ which is sectionally continuous in the interval $(-1, 1)$ with its derivatives—also sectionally continuous in every interval interior to $(-1, 1)$ may be expanded in a series of the form

$$f(x) = \sum_{0}^{\infty} a_n P_n(x).$$

To determine the coefficient a_n, we multiply throughout by $P_n(x)$ and integrate between -1 and $+1$, so that we obtain

$$\int_{-1}^{1} f(x)P_n(x)dx = a_n \int_{-1}^{1} [P_n(x)]^2 dx = \frac{2a_n}{(2n+1)}.$$

The general coefficient a_n can therefore be written as

$$a_n = \frac{(2n+1)}{2}\int_{-1}^{1} f(x)P_n(x)dx$$

17.11.4 Associated Legendre's equation

A more general form of a differential equation different from the usual Legendre's equation is the *associated Legendre's equation*—frequently occurring in mathematical physics for the solution of some potential problems, and is written as

$$(1-x)^2y'' - 2xy' + \left[n(n+1) - \frac{m^2}{(1-x^2)}\right]y = 0 \quad (17.11.17)$$

This equation differs from (17.11.1) only by the presence of the term containing m. The particular solution which is of physical significance is given by

$$y = (1-x^2)^{m/2}\frac{d^m}{dx^m}[P_n(x)] \quad (17.11.18)$$

Solution (17.11.18) can be obtained by series solution method, but we can show in a simple way that (17.11.18) is a solution of (17.11.17).

We start with Legendre's equation

$$(1-x^2)y'' - 2xy' + n(n+1)y = 0$$

Differentiating this equation m times w.r.t x, we get

$$(1-x^2)\frac{d^2u}{dx^2} - 2x(m+1)\frac{du}{dx} + (n-m)(n+m+1)u = 0 \quad \text{writing } u = \frac{d^my}{dx^m}. \quad (17.11.19)$$

Since $P_n(x)$ is a solution of Legendre's equation, we should have

$$u = \frac{d^m}{dx^m}P_n(x)$$

as the solution of (17.11.19).

If we put $y = u(1-x^2)^{m/2}$ in equation (17.11.19), we obtain

$$(1-x^2)\frac{d^2y}{dx^2} - 2x\frac{dy}{dx} + \left[n(n+1) - \frac{m^2}{(1-x^2)}\right]y = 0$$

Thus $y = (1-x^2)^{m/2}\frac{d^m}{dx^m}[P_n(x)]$ is a solution of the associated Legendre's equation and is termed as *associated Legendre's polynomial* $P_n^m(x)$

$$\therefore\ P_n^m(x) = (1-x^2)^{m/2}\frac{d^m}{dx^m}[P_n(x)].$$

17.11.5 Recurrence formula for $P_n(x)$

Let

$$z = (1 - 2xy + y^2)^{-\frac{1}{2}} = \sum y^n P_n(x)$$

$$\text{or} \quad z^2(1 - 2xy + y^2) = 1$$

Differentiating w.r.t y, we obtain

$$2z\frac{dz}{dy}(1 - 2xy + y^2) + z^2(2y - 2x) = 0$$

$$\text{or} \quad \frac{dz}{dy}(1 - 2xy + y^2) + z(y - x) = 0$$

$$\text{or} \quad (1 - 2xy + y^2)\sum ny^{n-1}P_n(x) + (y - x)\sum y^n P_n(x) = 0$$

Equating the coefficients of y^{n-1} on both sides, we obtain

$$nP_n(x) - 2x(n-1)P_{n-1}(x) + (n-2)P_{n-2}(x) + P_{n-2}(x) - xP_{n-1}(x) = 0$$

$$\text{or} \quad nP_n(x) = (2n-1)xP_{n-1}(x) - (n-1)P_{n-2}(x) \tag{17.11.20}$$

17.12 Hermite's Differential Equation and its solution

The differential equation written as

$$y'' - 2xy' + 2\lambda y = 0 \tag{17.12.1}$$

where λ is a non-negative constant is called *Hermite's differential equation.* This differential equation finds application in statistics and wave mechanics. The power series solution is applicable in this case, because $x = x_0$ is a regular singular point of the equation i.e., the equation has a non-essential singularity. So we assume

$$y = \sum_{r=0}^{\infty} a_r x^{k+r}$$

as the series solution of the above equation, so that

$$y' = \sum_{r=0}^{\infty}(k+r)a_r x^{k+r-1} \quad \text{and} \quad y'' = \sum_{r=0}^{\infty}(k+r)(k+r-1)a_r x^{k+r-2}.$$

Substituting these values in (17.12.1), we have

$$\sum_{r=0}^{\infty}\left[(k+r)(k+r-1)x^{k+r-2} - 2(k+r)x^{k+r} + 2\lambda x^{k+r}\right]\, a_r = 0$$

$$\text{or} \quad \sum_{r=0}^{\infty}\left[(k+r)(k+r-1)x^{k+r-2} - 2(k+r-\lambda)x^{k+r}\right]\, a_r = 0. \tag{17.12.2}$$

The expression (17.12.2) being an identity, we shall equate the coefficients of various powers of x to zero. Equating the coefficient of the lowest power of x to zero (putting $r = 0$), we obtain

$$k(k-1)a_0 = 0 \quad (\text{coefficient of } x^{k-2}).$$

This equation is called the *indicial equation* and determines k. The two values of k which are permissible are

$$k = 0 \quad \text{or} \quad k = 1. \tag{17.12.3}$$

($\because a_0 \neq 0$, as it is the coefficient of the first term).

Also we have

$$(k+1)ka_1 = 0 \quad (\text{coefficient of } x^{k-1}). \tag{17.12.4}$$

But equation (17.12.4) with equation (17.12.3) is valid only if $a_1 = 0$ or $k = 0$ or both are zero.

Now equating to zero the coefficient of the general term i.e. x^{k+r} in (17.12.2), we obtain

$$a_{r+2}(k+r+2)(k+r+1) - 2a_r(k+r-\lambda) = 0$$

$$\text{i.e.,} \quad a_{r+2} = \frac{2(k+r-\lambda)}{(k+r+2)(k+r+1)}a_r = \frac{2(k+r)-2\lambda}{(k+r+2)(k+r+1)}a_r \tag{17.12.5}$$

which is the *recursion* formula or the *recurrence* relation between the coefficients, from which, we may, starting with a_0, find out the coefficients. Now we shall discuss the two choices for solution since $k = 0$ or $k = 1$.

Choice 1. When $k = 0$, equation (17.12.5) gives

$$a_{r+2} = \frac{2r-2\lambda}{(r+2)(r+1)}a_r \tag{17.12.6}$$

whence we obtain

$$a_2 = -\frac{2\lambda}{2!}a_0, \quad a_4 = \frac{4-2\lambda}{4.3}a_2 = -\frac{(4-2\lambda)\cdot 2\lambda}{4.3.2!}a_0 = \frac{2^2(-2+\lambda)\lambda}{4!}a_0.$$

Thus

$$a_{2m} = \frac{(-2)^m\lambda(\lambda-2)\cdots(\lambda-2m+2)}{2m!}a_0.$$

Similarly odd coefficients are given by

$$a_3 = \frac{2-2\lambda}{3.2}a_1 = -\frac{2(\lambda-1)}{3!}a_1$$

$$a_5 = \frac{6-2\lambda}{5.4}a_3 = -\frac{2(\lambda-1)(6-2\lambda)}{5.4.3!}a_1 = (-2)^2\frac{(\lambda-1)(\lambda-3)}{5!}a_1 \quad \text{and so on.}$$

Thus the general term is given by

$$a_{2m+1} = \frac{(-2)^m(\lambda-1)(\lambda-3)\cdots(\lambda-2m+1)}{(2m+1)!}a_1.$$

$$\therefore\ y = a_0 + a_1x + a_2x^2 + a_3x^3 + \cdots \quad \left(\text{Putting } k = 0 \text{ in } y = \sum_{r=0}^{\infty} a_r x^{k+r}\right)$$

$$= a_0\left[1 - \frac{2\lambda}{2!}x^2 + \frac{2^2\lambda(\lambda-2)}{4!}x^4 + \cdots + \frac{(-2)^m\lambda(\lambda-2)\cdots(\lambda-2m+2)}{2m!}x^{2m}\right]$$

$$+a_1\left[x - \frac{2(\lambda-1)}{3!}x^3 + \frac{2^2(\lambda-1)(\lambda-3)}{5!}x^5 + \cdots\right.$$

$$\left. + \frac{(-2)^m(\lambda-1)(\lambda-3)\cdots(\lambda-2m+1)}{(2m+1)!}x^{2m+1}\right] \text{ for } a_1 \neq 0 \tag{17.12.7}$$

$$\text{and } y = a_0\left[1 - \frac{2\lambda}{2!}x^2 + \frac{2^2\lambda(\lambda-2)}{4!}x^4 + \cdots + \frac{(-2)^m\lambda(\lambda-2)\cdots(\lambda-2m+2)}{2m!}x^{2m}\right] \text{ for } a_1 = 0$$

(17.12.8)

$$= u \text{ (say)}$$

Choice 2 When $k = 1$, we have from (17.12.5)

$$a_{r+2} = \frac{2(r+1) - 2\lambda}{(r+3)(r+2)}a_r.$$

Putting $r = 1, 3, \cdots$, etc. in the above relation, we obtain

$$a_3 = a_5 = \cdots = 0 \quad [\because a_1 = 0 \text{ from (17.12.4)}]$$

Again putting $r = 0, 2, 4$ etc. in the above relation, we have

$$a_2 = \frac{2-2\lambda}{3.2}a_0 = -\frac{2(\lambda-1)}{3!}a_0$$

$$a_4 = \frac{6-2\lambda}{5.4}a_2 = \frac{2^2(\lambda-1)(\lambda-3)}{5!}a_0 \text{ and so on.}$$

Thus the general term is given by

$$a_{2m} = \frac{(-2)^m(\lambda-1)(\lambda-3)\cdots(\lambda-2m+1)}{(2m+1)!}a_0$$

$$\therefore\ y = \sum_{r=0}^{\infty} a_r x^{r+1} = \sum_{r=0}^{\infty} a_r x x^r = a_0x + a_2x^3 + a_4x^5 + \cdots + a_{2m}x^{2m+1}$$

$$= a_0\left[x - \frac{2(\lambda-1)}{3!}x^3 + \frac{2^2(\lambda-1)(\lambda-3)}{5!}x^5 + \cdots\right.$$

$$\left. + \frac{(-2)^m(\lambda-1)(\lambda-3)\cdots(\lambda-2m+1)}{(2m+1)!}x^{2m+1}\right]$$

$$= v \text{ (say)} \tag{17.12.9}$$

Inspection of (17.12.7) and (17.12.9) shows that (17.12.9) is the part of solution as given by (17.12.7). But both of them are the solutions of the same differential equation so that (17.12.9) cannot be a part of solution (17.12.7).

Hence $a_1 = 0$ and the solution in the case $k = 0$ must be given by (17.12.8). In view of this, the general solution of Hermite's differential equation is, by the superposition of (17.12.8) and (17.12.9), $y = Au + Bv$ where A and B are arbitrary constants and u, v are given by (17.12.8) and (17.12.9).

17.13 Hermite's Polynomials and their generating function

Evidently when λ is an even integer, equation (17.12.8) gives an even polynomial of degree n. Putting $\lambda = n$, where n is an even integer and

$$a_0 = (-1)^{n/2}\frac{n!}{(n/2)!},$$

we get the coefficient of x^n in (17.12.8)

$$= (-1)^{n/2}\frac{n!}{(n/2)!}\frac{(-2)^{n/2}n(n-2)\cdots(n-n+2)}{n!} = 2^n\frac{\frac{n}{2}\left(\frac{n}{2}-1\right)\cdots 1}{(n/2)!} = 2^n.$$

Similarly the coefficient of x^{n-2}

$$= (-1)^{n/2}\frac{n!}{(n/2)!}(-2)^{(n-2)/2}n(n-2)\cdots(n-n+2+2) = -\frac{n(n-1)}{1!}2^{n-2} \text{ etc.}$$

$$\therefore\ (y)_{\lambda=n} = y_n = (2x)^n - \frac{n(n-1)}{1!}(2x)^{n-2} + \frac{n(n-1)(n-2)(n-3)}{2!}(2x)^{n-4}\cdots + (-1)^{n/2}\frac{n!}{(n/2)!}.$$

This polynomial is known as *the Hermite's polynomial* of degree n and is denoted by

$$H_n(x) = (2x)^n - \frac{n(n-1)}{1!}(2x)^{n-2} + \frac{n(n-1)(n-2)(n-3)}{2!}(2x)^{n-4} + \cdots (-1)^{n/2}\frac{n!}{(n/2)!}$$

$$\text{or}\quad H_n(x) = \sum_{r=0}^{(n/2)}(-1)^r\frac{n!}{r!(n-2r)!}(2x)^{n-2r}$$

where

$$\left(\frac{n}{2}\right) = \begin{cases} n/2 & \text{if } n \text{ is even} \\ \frac{1}{2}(n-1) & \text{if } n \text{ is odd} \end{cases} \tag{17.13.1}$$

Values of Hermite's polynomials of different order (even and odd) can be obtained from (17.13.1).

The *generating function* for a polynomial is a function $W(t, x)$ which has a series expansion in the variable t of the form

$$W(t, x) = \sum_{n=0}^{\infty}\phi_n(x)t^n$$

where $\phi_n(x)$ is the required polynomial. The *generating function* for the Hermite's polynomials is

$$e^{2tx-t^2} = \sum_{n=0}^{\infty} \frac{t^n}{n!} H_n(x) \tag{17.13.2}$$

[*Proof.* We have

$$e^{2tx-t^2} = e^{2tx} e^{-t^2} = \sum_{r=0}^{\infty} \frac{(2tx)^r}{r!} \cdot \sum_{s=0}^{\infty} \frac{(-t^2)^s}{s!} = \sum_{r,s=0}^{\infty} (-1)^s \frac{(2x)^r}{r!s!} t^{r+2s}$$

$$\therefore \text{ coefficient of } t^n \text{ (for fixed value of } s) = (-1)^s \frac{(2x)^{n-2s}}{(n-2s)!s!} \text{ (putting } r + 2s = n).$$

But the total coefficient of t^n is obtained by summing over all allowed values of s, (for $r = n - 2s \geq 0$); $\therefore\ n - 2s \geq 0$, i.e., $s \leq \frac{n}{2}$.

So we can say, that if n is even, s goes from 0 to $\frac{n}{2}$ and if n is odd, s goes from 0 to $(n-1)/2$. Hence the required coefficient of t^n

$$= \sum_{s=0}^{(n/2)} (-1)^s \frac{(2x)^{n-2s}}{(n-2s)!s!} = \frac{H_n(x)}{n!}$$

$$\therefore\ e^{2tx-t^2} = \sum_{n=0}^{\infty} \frac{t^n}{n!} H_n(x) \text{ i.e., } e^{x^2-(t-x)^2} = \sum_{n=0}^{\infty} \frac{t^n}{n!} H_n(x)]$$

which may be written as

$$e^{x^2-(t-x)^2} = e^{2tx-t^2} = \frac{H_0(x)}{0!} t^0 + \frac{H_1(x)}{1!} t^1 + \frac{H_2(x)}{2!} t^2 + \cdots + \frac{H_n(x)}{n!} t^n + \frac{H_{n+1}(x)}{(n+1)!} t^{n+1}.$$

Now differentiating, both sides partially w.r.t. 't', n times and then putting $t = 0$, we have

$$\frac{H_n(x)}{n!} n! = \left[\frac{\partial^n}{\partial t^n} e^{-(t-x)^2}\right]_{t=0} \cdot e^{x^2}.$$

Putting $(t - x) = u$ i.e. at $t = 0, x = -u$ and hence $\frac{\partial}{\partial t} \equiv \frac{\partial}{\partial u}$, we may write

$$\left[\frac{\partial^n}{\partial t^n} e^{-(t-x)^2}\right]_{t=0} = \frac{\partial^n}{\partial u^n}\left(e^{-u^2}\right) = (-1)^n \frac{\partial^n}{\partial x^n}\left(e^{-x^2}\right) = (-1)^n \frac{d^n}{dx^n}\left(e^{-x^2}\right).$$

$$\therefore\quad H_n(x) = (-1)^n e^{x^2} \frac{d^n}{dx^n}\left(e^{-x^2}\right) \tag{17.13.3}$$

which is the differential form of Hermite's polynomials and is known as *Rodrigues' formula*.

Putting $n = 0, 1, 2, 3$ in (17.13.3) we have

$$H_0(x) = e^{x^2} \cdot e^{-x^2} = 1,$$

$$H_1(x) = (-1)e^{x^2}\frac{d}{dx}\left(e^{-x^2}\right) = 2x,$$

$$H_2(x) = (-1)^2 e^{x^2}\frac{d^2}{dx^2}\left(e^{-x^2}\right) = e^{x^2}\frac{d}{dx}\left(-2xe^{-x^2}\right) = 4x^2 - 2,$$

$$H_3(x) = 8x^3 - 12x,$$

$$H_4(x) = 16x^4 - 48x^2 + 12,$$

$$H_5(x) = 32x^5 - 160x^3 + 120x, \text{ and so on.}$$

The solution of the problem of linear harmonic oscillator in wave mechanics involves the application of the solution of Hermite's differential equation and hence of Hermite's polynomials.

17.13.1 Recurrence Formulae for $H_n(x)$ and to show that $H_n(x)$ is a solution of Hermite's equation

Hermite's equation is $y'' - 2xy' + 2ny = 0$ for integral values taking $\lambda = n$. Also we have

$$e^{2tx-t^2} = \sum_{n=0}^{\infty} \frac{H_n(x)t^n}{n!}. \tag{17.13.4}$$

(a) Differentiating partially w.r.t x, we have

$$2te^{2tx-t^2} = \sum_{n=0}^{\infty} H_n'(x)\frac{t^n}{n!} \quad \text{or} \quad 2t\sum_{n=0}^{\infty}\frac{H_n(x)t^n}{n!} = \sum_{n=0}^{\infty} H_n'(x)\frac{t^n}{n!}$$

whence, on equating the coefficients of t^n on both sides, we have

$$2\frac{n!}{(n-1)!}H_{n-1}(x) = H_n'(x) \quad \text{or} \quad 2nH_{n-1}(x) = H_n'(x) \tag{17.13.5}$$

(b) Differentiating, partially w.r.t. t, both sides of (17.13.4), we obtain

$$2(x-t)e^{2tx-t^2} = \sum_{n=1}^{\infty} H_n(x)\frac{t^{n-1}}{(n-1)!} \quad (\because n = 0 \text{ corresponds to making the r.h.s. equal to zero})$$

$$\text{or} \quad 2x\sum_{n=0}^{\infty} H_n(x)\frac{t^n}{n!} - 2\sum_{n=0}^{\infty} H_n(x)\frac{t^{n+1}}{n!} = \sum_{n=1}^{\infty} H_n(x)\frac{t^{n-1}}{(n-1)!}$$

Equating the coefficients of t^n on both sides, we obtain

$$2x\frac{H_n(x)}{n!} - 2\frac{H_{n-1}(x)}{(n-1)!} = \frac{H_{n+1}(x)}{n!}, \text{ i.e., } 2xH_n(x) = 2nH_{n-1}(x) + H_{n+1}(x) \tag{17.13.6}$$

(c) Eliminating $H_{n-1}(x)$ from (17.13.5) and (17.13.6), we have

$$2xH_n(x) = H'_n(x) + H_{n+1}(x) \quad \text{or} \quad H'_n(x) = 2xH_n(x) - H_{n+1}(x) \tag{17.13.7}$$

(d) Differentiating (17.13.7) w.r.t. x, we obtain

$$H''_n(x) = 2xH'_n(x) + 2H_n(x) - H'_{n+1}(x)$$

Putting $H'_{n+1}(x) = 2(n+1)H_n(x)$ obtained from (17.13.5) on replacing n by $(n+1)$, we have

$$H''_n(x) = 2xH'_n(x) + 2H_n(x) - 2(n+1)H_n(x)$$

$$\text{or} \quad H''_n(x) - 2xH'_n(x) + 2nH_n(x) = 0 \tag{17.13.8}$$

which clearly shows that $y = H_n(x)$ is a solution of Hermite's differential equation.

(e) As in the case of Legendre's polynomials $P_n(x)$, we can also prove here that

$$H_n(-x) = (-1)^n H_n(x).$$

We know that $e^{2tx-t^2} = e^{2tx} \cdot e^{-t^2}$

$$\text{or} \quad e^{2tx-t^2} = \left[\sum_{n=0}^{\infty} \frac{(2x)^n}{n!} t^n\right]\left[\sum_{n=0}^{\infty} (-1)^n \frac{t^{2n}}{n!}\right] = \sum_{n=0}^{\infty}\sum_{k=0}^{\infty}\left\{\frac{(-1)^k (2x)^{n-2k}}{k!(n-2k)!} \cdot t^n\right\}$$

$$\therefore\ H_n(x) = \sum_{k=0}^{n/2} \frac{(-1)^k n!(2x)^{n-2k}}{k!(n-2k)!} \tag{17.13.9}$$

$$\because\ e^{2tx-t^2} = \sum_{n=0}^{\infty} \frac{H_n(x)t^n}{n!}$$

Replacing x by $-x$, we obtain

$$H_n(-x) = \sum_{k=0}^{n/2} \frac{(-1)^k n!(-2x)^{n-2k}}{k!(n-2k)!} = \sum_{k=0}^{n/2} \frac{(-1)^k (-1)^{n-2k} n!(2x)^{n-2k}}{k!(n-2k)!}$$

$$= (-1)^n \sum_{k=0}^{n/2} \frac{(-1)^k n!(2x)^{n-2k}}{k!(n-2k)!} = (-1)^n H_n(x) \quad \text{[using (17.13.9)]}$$

17.13.2 Orthogonal property of $H_n(x)$

Hermite's polynomials are orthogonal. Considering two polynomials $H_n(x)$ and $H_m(x)$, we can show that

$$\int_{-1}^{1} H_n(x)H_m(x)dx = 0 \quad (\text{if } m \neq n)$$

which is the condition of orthogonality

Exercises

1. Explain, with illustrations, the following terms:

(a) ordinary and partial differential equations;

(b) order and degree of a differential equation;

(c) homogeneous and inhomogeneous differential equations;

(d) integrating factor.
(See Art. 17.1)

2. (a) Illustrate, with reference to a physical problem, the method of solving a linear differential equation of the first order.

(b) Give, with an example, the method of solving the Bernoulli's equation.

(c) Solve

i. $(x^2 + y^2)\,dx = xy\,dy$.

ii. $\dfrac{dy}{dx} = \dfrac{6x - 2y - 7}{2x + 3y - 6}$.

iii. $(2x - 2y + 5)\,dy - (x - y + 3)\,dx = 0$.

iv. $x\,dx + y\,dy + \dfrac{x\,dy - y\,dx}{x^2 + y^2} = 0$.

v. $x\cos(y/x)(y\,dx + x\,dy) = y\sin(y/x)(x\,dy - y\,dx)$.

vi. $\cos^2 x\dfrac{dy}{dx} + y = \tan x$.

vii. $\dfrac{dy}{dx} + \dfrac{1-2x}{x^2}y = 1$.

viii. $\dfrac{dy}{dx} + \dfrac{x}{1-x^2}y = x\sqrt{y}$

ix. $\dfrac{dy}{dx} + y = y^2(\cos x - \sin x)$.

(See worked out examples)

3. If an electric circuit contains a resistance R (ohms) and a condenser of capacitance C (farads) in series and an e.m.f. E (volts), the charge q (coulombs) on the condenser is given by $R\dfrac{dq}{dt} + \dfrac{q}{C} = E$.

If $R = 10$ ohms, $C = 10^{-3}$ farad and $E(t) = 100\sin 120\pi t$ volts,

(a) find q, assuming that $q = 0$ when $t = 0$;

(b) use $i = \dfrac{dq}{dt}$ to find i, assuming that $i = 5$ amperes when $t = 0$.
(See worked out examples)

4. Write down the ordinary second order linear differential equation with constant coefficients. Illustrate in connection with the solution of such an equation the meaning of (i) complementary function, (ii) particular integral.

(See Art. 17.4.1)

5. Explain why the general solution of a second order differential equation should contain two arbitrary constants. Indicate how these constants are determined. Illustrate your answer with an example.

6. Solve

(i) $\dfrac{d^2y}{dx^2} - 7\dfrac{dy}{dx} + 12 = 0.$ (ii) $(D^2 - 4D + 4)y = 0.$ (iii) $(D^2 + D)y = 0.$

(iv) $\dfrac{d^2y}{dx^2} - 4\dfrac{dy}{dx} + 3y = 2e^{3x}.$ (v) $\dfrac{d^2y}{dx^2} + y = \sin x.$ (vi) $(D^2 - 2kD + k^2)y = e^x.$

(vii) $(D^2 - 2D + 1)y = e^x.$

7. Show that the solution of $\dfrac{d^2x}{dt^2} + k\dfrac{dx}{dt} + \mu x = 0$ is

$$x = e^{-\frac{k}{2}t}(A\cos nt + B\sin nt) \text{ if } k^2 < 4\mu \text{ and } n^2 = \mu - \frac{k^2}{4}.$$

(See worked out examples)

8. A particle of mass m is acted on by a restoring force proportional to its displacement and a resisting force proportional to its instantaneous velocity. Write down its equation of motion.

Taking the restoring force per unit displacement as s, and the resisting force per unit velocity as r, discuss the nature of the motion when (i) $r^2 > 4sm$, (ii) $r^2 = 4sm$ and (iii) $r^2 < 4sm$.

(See worked out examples and the chapter on SHM)

9. If, in Q. 8, the particle is acted on by a force $F\sin wt$, what will be the solution?

Discuss the contributions of the C.F. and P.I. to the solution of the motion.

10. An alternating e.m.f. $E_0 \sin wt$ is impressed in a circuit containing an inductance L and a resistance R. Write down the differential equation for the instantaneous current i, and solve it for the particular integral by the operator method. Also obtain the general solution by adding the complementary function.

(See worked out examples)

11. A galvanometer coil oscillates in a closed circuit of resistance R. Its equation of motion is given by $I\dfrac{d^2\theta}{dt^2} + \dfrac{k}{R}\dfrac{d\theta}{dt} + c\theta = 0$ where I is the moment of inertia of the coil, C is the restoring torque per unit twist of the suspension and k is a constant and θ the angular displacement.

Discuss how the nature of motion of the coil will change as R is gradually reduced from a large to a small value.

(See worked out examples)

12. Solve $y'' - 2x^2y' + 4xy = x^2 + 2x + 2$ in powers of x.

[*Ans.* $y = a_0\left(1 - \frac{2}{3}x^3 - \frac{2}{45}x^6 \cdots\right) + a_1\left(x - \frac{x^4}{6} - \frac{1}{63}x^7 + \cdots\right) + \frac{x^2}{2} + \frac{x^3}{3} + \frac{x^4}{12} + \frac{x^6}{45} + \cdots$]

13. Explain how an inexact differential may be transformed into an exact differential by the use of integrating factors.

Solve the following equation by the use of suitable integrating factors

$$x\,dy - y\,dx + x^2\,dx = 0.$$ **[C.U. (Hons.) 1981]**

(For answer see Art. 17.3).

14. Solve (a) $\dfrac{d^2y}{dx^2} - 3\dfrac{dy}{dx} + 2y = x^3$; (b) $\dfrac{d^2y}{dx^2} + y = e^x \cos x$. **[C.U. (Hons.) 1983]**

[*Ans.* (a) For PI,

$$y = \frac{1}{D^2 - 3D + 2}x^3 = \frac{1}{(D-2)(D-1)}x^3 = \left[\frac{1}{(1-D)} - \frac{1}{(2-D)}\right]x^3$$

$$= \left[(1-D)^{-1} - \frac{1}{2}\left(1 - \frac{D}{2}\right)^{-1}\right]x^3 = \left[(1 + D + D^2 + D^3) - \frac{1}{2}\left(1 + \frac{D}{2} + \frac{D^2}{4} + \frac{D^3}{8}\right)\right]x^3$$

$$= \left[\frac{1}{2} + \frac{3}{4}D + \frac{7}{8}D^2 + \frac{15}{16}D^3\right]x^3 = \frac{x^3}{2} + \frac{9x^2}{4} + \frac{21x}{4} + \frac{45}{8}.$$

(b) For PI,

$$y = \frac{1}{D^2+1}e^x \cos x = e^x \frac{1}{[(D+1)^2+1]}\cos x = e^x \frac{1}{(D^2+2D+2)}\cos x = e^x \frac{[(D^2+2) - 2D]}{[(D^2+2)^2 - 4D^2]}\cos x$$

$$= e^x \frac{D^2 - 2D + 2}{D^4 + 4}\cos x = \frac{e^x}{5}(D^2 - 2D + 2)\cos x = \frac{e^x}{5}(2\sin x + \cos x).$$

15. Solve $y\dfrac{d^2y}{dx^2} = \left(\dfrac{dy}{dx}\right)^2$. **[C.U. (Hons.) 1986, '89, '93]**

Dividing both sides by y^2 and rearranging,

$$\frac{1}{y}\frac{d^2y}{dx^2} - \frac{1}{y^2}\left(\frac{dy}{dx}\right)^2 = 0 \quad \text{or,} \quad \frac{d}{dx}\left(\frac{1}{y}\frac{dy}{dx}\right) = 0.$$

Integrating,

$$\frac{1}{y}\frac{dy}{dx} = a \quad \text{or} \quad \frac{dy}{y} = a\,dx.$$

Again integrating

$$\ln y = ax + \ln c \quad \text{or} \quad \ln\frac{y}{c} = \ln e^{ax}. \quad \therefore \quad y = ce^{ax} \text{ is the solution.}$$

16. Solve $\dfrac{\partial^2 Z(x,y)}{\partial x \partial y} = x^2 y$, subject to the condition $z(x,0) = x^2$ and $z(1,y) = \cos y$.

[C.U. (Hons.) 1984]

[*Ans:* The given equation is written as $\dfrac{\partial}{\partial x}\left(\dfrac{\partial Z}{\partial y}\right) = x^2 y$ which on integration gives

$$\frac{\partial Z}{\partial y} = \frac{yx^3}{3} + C \quad \text{or,} \quad \partial Z = \frac{yx^3}{3}\partial y + C\partial y.$$

Integrating again $Z = \dfrac{x^3y^2}{6} + Cy + D$. Now when $y = 0, Z = x^2 \quad \therefore \quad D = x^2$

Again when $x = 1, Z = \cos y$

$$\therefore \quad \cos y = \frac{y^2}{6} + Cy + 1 \text{ (putting } D = x^2) \quad \text{or,} \quad C = \frac{6\cos y - y^2 - 6}{6y}$$

$$\therefore \quad Z = \frac{x^3y^2}{6} + \frac{6\cos y - y^2 - 6}{6y}y + x^2$$

$$\text{or,} \quad 6Z = x^3y^2 + 6\cos y - y^2 - 6 + 6x^2 = (x^3 - 1)y^2 + 6x^2 + 6\cos y - 6$$

is the solution.

17. (a) Solve the boundary value problem $\frac{\partial u}{\partial x} = 4\frac{\partial u}{\partial y}$ given $u(0, y) = 8e^{-3y}$ by the method of separation of variables. **[C.U. (Hons.) 1986]**

(b) Find the general solution of $y\frac{d^2y}{dx^2} = \left(\frac{dy}{dx}\right)^2$. **[C.U. (Hons.) 1986]**

[*Ans.* (a) Let $u = X(x)Y(y)$, so that $\frac{\partial u}{\partial x} = Y\frac{dX}{dx}$ and $\frac{\partial u}{\partial y} = X\frac{dY}{dy}$ and the given equation may be written as

$$Y\frac{dX}{dx} = 4X\frac{dY}{dy} \quad \text{or,} \quad \frac{1}{X}\frac{dX}{dx} = \frac{4}{Y}\frac{dY}{dy} = m \text{ (say)} \quad \therefore \frac{1}{X}\frac{dX}{dx} = m \quad \text{or,} \quad \frac{dX}{X} = m\,dx$$

$$\text{(Integrating)} \quad \ln X = mx + \ln C \quad \therefore X = Ce^{mx}.$$

Again

$$\frac{4}{Y}\frac{dY}{dy} = m \quad \text{or,} \quad \frac{dY}{Y} = \frac{m}{4}dy = n\,dy\left(n = \frac{m}{4}\right).$$

Integrating

$$\ln Y = ny + \ln D \quad \therefore Y = De^{ny}, \quad \text{so that we may write}$$

$$u = CDe^{mx}e^{ny} = Le^{mx}e^{ny} \quad [L = CD].$$

From the condition of the problem, we have

$$8e^{-3y} = Le^{ny} \quad \text{or,} \quad L = 8e^{-3y-ny} = 8e^{-y(3+n)}$$

$$\therefore u = Le^{mx}e^{ny} = 8e^{-y(3+n)}e^{mx}e^{ny} = 8e^{mx-3y}$$

is the solution.

Alternatively we can proceed as follows

$$u(o, y) = 8e^{-3y} \quad \therefore \frac{\partial u}{\partial y} = -24e^{-3y} \therefore \frac{\partial u}{\partial x} = 4\frac{\partial u}{\partial y} = -96e^{-3y}$$

$\therefore u = \int du = -96e^{-3y}x + A$. Now at $x = 0$, $u = 8e^{-3y}$

$$\therefore A = 8e^{-3y} \quad \therefore u = 8e^{-3y}(1 - 12x)$$

(b) Let $\frac{dy}{dx} = p$

$$\therefore \frac{d^2y}{dx^2} = \frac{dp}{dx} = \frac{dp}{dy}\frac{dy}{dx} = p\frac{dp}{dy}$$

so the given equation may be written as

$$yp\frac{dp}{dy} = p^2 \quad \text{or} \quad \frac{dp}{p} = \frac{dy}{y}$$

Integrating we have,

$$\ln p = \ln y + \ln C \quad \text{or} \quad p = Cy \therefore \frac{dy}{dx} = Cy \quad \text{or} \quad \frac{dy}{y} = Cdx.$$

Integrating again, we have $\ln y = Cx + C_1 = \ln e^{Cx+C_1}$

$$\therefore y = e^{Cx+C_1} = Ae^{Cx} \text{ where } e^{C_1} = A.$$

18. (a) Solve the following Bernoulli's equation $\frac{dy}{dx} + \frac{1}{x} = \frac{e^y}{x^2}$. **[C.U. (Hons.) 1988]**

(b) If $du = (x^2 + y^2 + 4)xdx + (x^2 - y^2 + 9)ydy$, show that du is an exact differential and find u. **[C.U. (Hons.) 1988]**

(c) Solve $\frac{d^2y}{dx^2} - 5\frac{dy}{dx} + 6y = 2e^x$ when $x = 0, y = 1$ and $\frac{dy}{dx} = 1$. **[C.U. (Hons.) 1988]**

[*Ans.* (a) Dividing both sides by e^y, we may write

$$e^{-y}\frac{dy}{dx} + \frac{1}{x}e^{-y} = \frac{1}{x^2} \quad \text{or,} \quad -\frac{dz}{dx} + \frac{z}{x} = \frac{1}{x^2} \quad \text{or,} \quad \frac{dz}{dx} + \frac{-z}{x} = -\frac{1}{x^2},$$

putting

$$e^{-y} = z \quad \text{or,} \quad \frac{dz}{dx} = -e^{-y}\frac{dy}{dx}.$$

$\therefore$ the I.F. $= e^{-\int \frac{1}{x}dx} = e^{-\log x} = e^{\log x^{-1}} = \frac{1}{x}$.

Now proceed and get the result.

(b) Let $du = Mdx + Ndy$ where $M = (x^2 + y^2 + 4)x$ and $N = (x^2 - y^2 + 9)y$ are functions of x and y.

Now

$$\frac{\partial M}{\partial y} = 2xy \quad \text{and} \quad \frac{\partial N}{\partial x} = 2xy$$

$\therefore$ du is an exact differential.

Now

$$\int Mdx = \int (x^2 + y^2 + 4)xdx = \frac{x^4}{4} + \frac{x^2y^2}{2} + 2x^2 \quad \text{(keeping } y = \text{constant)}$$

Again

$$\int Ndy = \int (x^2 - y^2 + 9)ydy = \int (-y^2 + 9)ydy \quad \left[\begin{array}{l}\text{retaining only those terms}\\ \text{which do not contain } x\end{array}\right]$$

$$= -\frac{y^4}{4} + 9\frac{y^2}{2}$$

$\therefore$ the solution is

$$u = \frac{x^4}{4} - \frac{y^4}{4} + \frac{x^2y^2}{2} + 2x^2 + \frac{9y^2}{2} + C.$$

(c) For CF, we put the r.h.s $= 0$, so that the equation is $(D^2 - 5D + 6)y = 0$ for which the auxiliary equation is $(\lambda^2 - 5\lambda + 6) = 0$ or $\lambda = 2, 3$

$$\therefore\ C.F. = Ae^{2x} + Be^{3x}; \quad P.I. = \frac{2e^x}{D^2 - 5D + 6} = \frac{2e^x}{1 - 5 + 6} = e^x \quad \text{(putting } D = 1)$$

$\therefore$ The general solution is $y = Ae^{2x} + Be^{3x} + e^x$.

Now at $x = 0$, $y = 1$ $\quad \therefore\ 1 = A + B + 1$ or $A + B = 0$.

Again at $x = 0$, $\frac{dy}{dx} = 1$ $\quad \therefore\ 1 = 2A + 3B + 1$ or $2A + 3B = 0$

Solving we get $A = 0, B = 0$. $\quad \therefore\ y = e^x$ is the required solution.

19. Obtain the general solution of the equation

$$\frac{\partial^2 y}{\partial t^2} = C^2\frac{\partial^2 y}{\partial x^2} - 2k\frac{\partial y}{\partial t}$$

by the method of separation of variables. **[C.U. (Hons.) 1988, '90, '92]**

[*Ans.* Let $y = X(x)T(t)$ be the general solution, so that

$$\frac{\partial^2 y}{\partial x^2} = T\frac{d^2X}{dx^2}, \quad \frac{\partial^2 y}{\partial t^2} = X\frac{d^2T}{dt^2} \text{ and } \frac{\partial y}{\partial t} = X\frac{dT}{dt}.$$

So the given equation can be written as

$$X\frac{d^2T}{dt^2} = C^2T\frac{d^2X}{dx^2} - 2kX\frac{dT}{dt} \quad \text{or,} \quad \left(\frac{d^2T}{dt^2} + 2k\frac{dT}{dt}\right)X = C^2T\frac{d^2X}{dx^2}$$

$$\text{or,} \quad \frac{1}{T}\left(\frac{d^2T}{dt^2} + 2k\frac{dT}{dt}\right) = \frac{C^2}{X}\frac{d^2X}{dx^2} = -\lambda^2 \text{(say)}$$

where λ = a constant so that taking

$$\frac{C^2}{X}\frac{d^2X}{dx^2} = -\lambda^2 \quad \text{or} \quad \frac{d^2X}{dx^2} + \frac{\lambda^2}{C^2}X = 0, \quad \text{whose solution is}$$

$$X = A\cos\frac{\lambda}{C}x + B\sin\frac{\lambda}{C}x \quad (A, B \text{ are constants}). \tag{1}$$

Again taking

$$\frac{1}{T}\left(\frac{d^2T}{dt^2} + 2k\frac{dT}{dt}\right) = -\lambda^2 \quad \text{or,} \quad \frac{d^2T}{dt^2} + 2k\frac{dT}{dt} + \lambda^2 T = 0. \tag{2}$$

Let $T \simeq e^{mt}$ be a trial solution, so that $e^{mt}(m^2 + 2km + \lambda^2) = 0$.

But $e^{mt} \neq 0$. $\therefore \quad m^2 + 2km + \lambda^2 = 0$

$$\text{or,} \quad m = -\frac{2k \pm \sqrt{4k^2 - 4\lambda^2}}{2} = -k \pm \sqrt{k^2 - \lambda^2}$$

$$\therefore \quad T = Ce^{(-k+\sqrt{k^2-\lambda^2}t} + De^{(-k-\sqrt{k^2-\lambda^2})t} \tag{3}$$

(C, D are constants).

Hence the general solution of the given equation is

$$Y = X(x)T(t) = \left(A\cos\frac{\lambda}{C}x + B\sin\frac{\lambda}{C}x\right)\left(Ce^{(-k+\sqrt{k^2-\lambda^2})t} + De^{-k-\sqrt{k^2-\lambda^2})t}\right)$$

where A, B, C, D are arbitrary constants.

N.B. Now solution of (2) gives rise to three cases (a) $k > \lambda$ (large damping); (b) $k \simeq \lambda$ (critical damping) and (c) $k < \lambda$ (small damping).

(Case (c) is the only relevant one for the displacement of a viscously damped string whose equation is given in the problem for solution, and the solution is obtained from (3) as

$$T = e^{-kt}\left[Ce^{i\sqrt{\lambda^2-k^2}t} + De^{-i\sqrt{\lambda^2-k^2}t}\right]$$

which may be simplified as in case 3 of Art. 17.4.2 so that we may write $T = ae^{-kt}\cos\left(\sqrt{\lambda^2 - k^2}t - \delta\right)$ with proper substitutions involving δ.

$\therefore$ the general solution is reduced to

$$y = \left(A\cos\frac{\lambda}{C}x + B\sin\frac{\lambda}{C}x\right)\left\{ae^{-kt}\cos\left(\sqrt{\lambda^2 - k^2}t - \delta\right)\right\}$$

20. Solve the differential equation $\frac{dy}{dx} + \ln x^y = 0$ subject to the condition $y \to 1$ when $x \to 0$.

[C.U. (Hons.) 1982]

The given equation is written as $\frac{dy}{dx} + y\ln x = 0$, so that the

$$I.F = e^{\int \ln x dx} = e^{x\ln x - x} = e^{\ln x^x - x} = x^x e^{-x}.$$

Multiplying both sides by the I.F. and integrating we have $yx^x e^{-x} = C$ or $y = Ce^x x^{-x}$; now $x \to 0$, $y \to 1$ gives $C = 1$ $\therefore$ $y = e^x x^{-x}$ is the solution.

21. Obtain a general solution of the differential equation $\frac{d^2y}{dx^2} + k\frac{dy}{dx} = x$ where k is a constant.

[C.U. (Hons.) 1982]

For C.F., we put the r.h.s = 0, so that the equation is $(D^2 + kD)y = 0$, for which the auxiliary equation is $(\lambda^2 + k\lambda) = 0$, i.e., $\lambda = 0$ or $-k$

$$\therefore \; C.F = A + Be^{-kx}$$

Again

$$P.I. = \frac{x}{D^2 + kD} = \frac{x}{kD\left(1 + \frac{D}{k}\right)} = \frac{1}{kD}\left(1 - \frac{D}{k} + \frac{D^2}{k^2} - \frac{D^3}{k^3} + \cdots\right)x$$

$$= \left(\frac{1}{kD} - \frac{1}{k^2} + \frac{D}{k^3} - \cdots\right)x = \frac{x^2}{2k} - \frac{x}{k^2} + \frac{1}{k^3}$$

$\therefore$ the general solution is

$$y = A + Be^{-kx} + \frac{x^2}{2k} - \frac{x}{k^2} + \frac{1}{k^3}.$$

22. Obtain the general solution of the following differential equation $\frac{d^2y}{dx^2} + 3\frac{dy}{dx} + 2y = 3e^{2x}$.

[C.U. (Hons.) 1985]

For C.F., we put the r.h.s = 0, so that the equation is $(D^2 + 3D + 2)y = 0$, for which the auxiliary equation is $\lambda^2 + 3\lambda + 2 = 0$ or $\lambda = -2, -1$

$$\therefore\ C.F. = Ae^{-2x} + Be^{-x}$$

Again

$$P.I. = \frac{1}{D^2 + 3D + 2}3e^{2x}.$$

Simply putting $D = 2$, we have

$$P.I. = 3e^{2x}\frac{1}{4 + 6 + 2}\cdot 1 = \frac{1}{4}e^{2x}$$

$\therefore$ the general solution is $y = Ae^{-2x} + Be^{-x} + \frac{1}{4}e^{2x}$.

23. Find the general solution of $\frac{d^2y}{dx^2} - 5\frac{dy}{dx} + 6y = 0$. [C.U. (Hons.) 1989]

(See worked out examples). [*Ans.* $y = C_1e^{3x} + C_2e^{2x}$.]

24. A simple harmonic oscillator of mass m, oscillating along X-axis is subjected to a restoring force of magnitude kx and a retarding force of magnitude $r\frac{dx}{dt}$. Set up its differential equation of motion, and find the solution. (See worked out examples)

Discuss the nature of the motion for different relative values of k and r.

[C.U. (Hons.) 1985, '90]

[*Hints.* **Answer to the last part:** (a) For $k > r$ (large damping) the motion will be overdamped; (b) for $k \simeq r$ (critical damping) the motion will be critically damped; (c) for $k < r$ (small damping) the motion will be oscillatory—For details see chapter 5—the article on "Damped harmonic vibration".]

25. A man and a parachute weighing ω lbs. fall from rest under the force of gravity. If the resistance of air is proportional to the speed v, find the speed as a function of the time t. Find also the "limiting velocity" v_0.

(See worked out examples). [C.U. (Hons.) 1989]

26. Obtain the general solution of the following differential equation $\frac{d^2y}{dx^2} + 5\frac{dy}{dx} + 6y = e^x$.

(See worked out examples). [C.U. (Hons.) 1991]

27. Consider the equation of motion of a damped harmonic oscillator $m\ddot{x} + 2b\dot{x} + \omega_0^2x = 0$, where m, b, ω_0 are constants. Solve this equation and discuss the nature of motion for $b > \omega_0$, $b < \omega_0$ and $b = \omega_0$. Also find the rate of dissipation of energy. [C.U. (Hons.) 1992]

(For answer to second and third parts, see Chapter 5—the Article on "Damped harmonic vibration".)

28. (a) The equation of motion of a forced harmonic oscillator with spring constant k and a small damping constant β is $m\ddot{x} + \beta\dot{x} + kx = F\cos\omega t$.
Find the amplitude and the phase angle by which the displacement x leads the applied force F when the natural frequency $\omega_0 = \left(\frac{k}{m}\right)^{1/2}$ is: (i) much greater than the driving frequency ω; (ii) equal to ω; (iii) much less than ω.
Hence show that the half-width of the resonance curve is approximately equal to the log-decrement divided by 2π.

(b) Prove that the maximum of the square of the amplitude does not occur either at the natural frequency ω_0 of damped motion or at ω;

(c) Comment on the fact that the oscillator has maximum energy at $\omega = \omega_0$.

[C.U. (Hons.) 1987]

(For answer to part (a) see worked out examples; for answers to (b) and (c) see Chapter 5—the Article on "Forced vibration".

29. (a) Find the most general solution of the equation $y\dfrac{d^2y}{dx^2} = \left(\dfrac{dy}{dx}\right)^2$.

(See worked out examples).

(b) Given that $du = xdy - ydx - 2\sqrt{x^2 - y^2}dx$. Solve the equation $du = 0$ to obtain y in terms of x. Is du a perfect differential?

(c) Solve the equation $(x^2 - 2xy)\dfrac{dy}{dx} + (x^2 - 3xy + 2y^2) = 0$. **[C.U. (Hons.) 1993]**

[*Ans.* (b) Putting $du = 0$ and rearranging the terms, we may write

$$xdy = \left(y + 2\sqrt{x^2 - y^2}\right) dx \quad \text{or} \quad x\frac{dy}{dx} = y + 2\sqrt{x^2 - y^2}$$

Let $y = vx$. $\therefore \dfrac{dy}{dx} = v + x\dfrac{dv}{dx}$, so that the above equation is written as

$$\left(v + x\frac{dv}{dx}\right) x = vx + 2x\sqrt{1 - v^2} = x\left(v + 2\sqrt{1 - v^2}\right) \quad \text{or} \quad x\frac{dv}{dx} = 2\sqrt{1 - v^2} \quad \text{or} \quad \frac{dv}{\sqrt{1 - v^2}} = 2\frac{dx}{x}.$$

Integrating we have

$$\sin^{-1}\frac{y}{x} = 2\ln x + \ln c = \ln cx^2. \quad \therefore\ y = x\sin(\ln cx^2).$$

To show whether du is a perfect differential, find u by integrating the given equation, and then find the values of $\dfrac{\partial^2 u}{\partial x \partial y}$ and $\dfrac{\partial^2 u}{\partial y \partial x}$. If these two values are equal, then du is a perfect differential, but if these two values are not equal, then du is not a perfect differential.

(c) Let $y = vx$. $\therefore \dfrac{dy}{dx} = v + x\dfrac{dv}{dx}$, so that the given equation may be written as

$$(x^2 - 2vx^2)\left(v + x\frac{dv}{dx}\right) + (x^2 - 3vx^2 + 2v^2x^2) = 0$$

$$\text{or} \quad x^2(1 - 2v)\left(v + x\frac{dv}{dx}\right) + x^2(1 - 3v + 2v^2) = 0$$

$$\text{or} \quad v + x\frac{dv}{dx} = -\frac{(1 - 3v + 2v^2)}{(1 - 2v)} = -\frac{(1 - 2v)(1 - v)}{(1 - 2v)} = v - 1$$

$$\therefore\ x\frac{dv}{dx} = -1 \quad \text{or} \quad dv = -\frac{dx}{x} \quad \text{or} \quad v = -\ln x + \ln c$$

$$\therefore\ \frac{y}{x} = \ln\frac{c}{x} \quad \text{or} \quad y = x\ln\frac{c}{x} = \ln(cx^{-1})^x.$$

30. A particle falls in a vertical line under gravity (supposed constant) and the force of air resistance to its motion is proportional to its velocity. Find the velocity of the particle at any instant of its motion in terms of the limiting velocity.

(See worked out examples). **[C.U. (Hons.) 1994, cf. 1983]**

31. Find the solution of the following differential equation which satisfies the given conditions:

$$\frac{d^2x}{dt^2} + 3\frac{dx}{dt} + 2x = 0$$

where $x = 0$ and $\frac{dx}{dt} = 1$ at $t = 0$. (See worked out examples) **[C.U. (Hons.) 1996 (old syll)]**

32. Illustrate the power series method of solving a second order linear differential equation by finding the series solution of the Legendre's differential equation and find the general expression for the Legendre polynomial $P_n(x)$. Show also that $P_n(1) = 1$ and $P_n(-1) = (-1)^n$.

For answer see Article 17.11 and 17.11.1. **[C.U. (Hons.) 1996]**

33. (a) Show that $f_1(x) = Ae^{mx}$ and $f_2(x) = Be^{nx}$, where $m \neq n$, are linearly independent.

(b) Find the general solution of any one of the following differential equations.

i. $\frac{d^2y}{dx^2} + 6\frac{dy}{dx} + 9y = 0$ where at $x = 0$, $y = 0$ and $\frac{dy}{dx} = 0$.

ii. $\frac{d^2y}{dt^2} + 25y = 0$ where at $t = 0, y = 0$ and $\frac{dy}{dt} = 10$ cm/s. **[C.U. (Hons.) 1997]**

[*Ans.* (a)

$$f_1(x)f_2(x) = ABe^{(m+n)x} = Ce^{(m+n)x}. \tag{1}$$

Also

$$\frac{f_1(x)}{f_2(x)} = \frac{A}{B}e^{(m-n)x} = De^{(m-n)x} \tag{2}$$

$$\text{and} \quad f_1'(x) = Ame^{mx}, \quad f_2'(x) = Bne^{nx}$$

$$\therefore \quad \frac{f_1'(x)}{f_2'(x)} = \frac{mf_1(x)}{nf_2(x)} \tag{3}$$

so that from (1), (2) and (3) we may conclude that since $m \neq n$, $f_1(x)$ and $f_2(x)$ are linearly independent.

(b) (i) The equation is $(D^2 + 6D + 9)y = 0$, so that the auxiliary equation is $\lambda^2 + 6\lambda + 9 = 0$ or $(\lambda + 3)^2 = 0$, i.e., the roots are equal, i.e., $\lambda_1 = \lambda_2 = -3$

$\therefore$ solution is $y = (A + Bx)e^{-3x}$, so that $y = 0$ at $x = 0$ gives $A = 0$.

Again $\frac{dy}{dx} = -3(A + Bx)e^{-3x} + Be^{-3x}$, so that $\frac{dy}{dx} = 0$ at $x = 0$ gives $-3A + B = 0$ $\therefore B = 0$.

$\therefore$ the solution is $y = 0 \cdot e^{-3x} = 0$, i.e., we have no solution for y.

(ii) The equation is $(D^2 + 25)y = 0$, so that the auxiliary equation is $\lambda^2 + 25 = 0$, whence the complex roots are $\lambda_1 = 5i, \lambda_2 = -5i$, and the solution is $y = A\cos 5t + B\sin 5t$, A and B are constants. Now the condition at $t = 0, y = 0$ gives $A = 0$.

Again $\frac{dy}{dt} = -5A\sin 5t + 5B\cos 5t$, so that applying the condition at $t = 0$, $\frac{dy}{dt} = 10$ cm/s, we have, $10 = 5B$ or $B = 2$. So the complete solution is $y = 2\sin 5t$.

34. (a) Given that $du = xdy - (2y + x^3e^x)dx$, solve the equation $du = 0$. Is du a perfect differential?

(b) Solve the equation $y'' + 9y = 2\sin 3x$, subject to the initial condition $y = 1$, $y' = 0$ at $x = 0$. **[C.U. (Hons.) 1998]**

(a) (i) Putting $du = 0$, we may write the equations as

$$\frac{dy}{dx} = \frac{2y}{x} + x^2e^x \quad \text{or} \quad \frac{1}{x^2}dy - \frac{2}{x^3}ydx = e^x dx \text{ (dividing both sides by } x^2 \text{ and re-arranging)}$$

$$\text{or} \quad \int d\left(\frac{y}{x^2}\right) = \int e^x dx + C \quad \text{or} \quad \frac{y}{x^2} = e^x + C, \text{ so that } y = x^2e^x + Cx^2 = x^2(e^x + C).$$

(ii) To show whether du is a perfect differential, let us first find u by integrating the given equation as shown below:

$$\int du = x\int dy - 2y\int dx - \int x^3e^x dx + C$$

$$\text{or} \quad u = xy - 2xy - \left[x^3\int e^x dx - \int\left\{3x^2\int e^x dx\right\}dx\right] + C$$

$$= -xy - \left[x^3e^x - \int 3x^2e^x dx\right] + C$$

$$= -xy - x^3e^x + 3\left[x^2\int e^x dx - \int\left\{2x\int e^x dx\right\}dx\right] + C$$

$$= -xy - x^3e^x + 3x^2e^x - 6\int xe^x dx + C$$

$$= -xy - x^3e^x + 3x^2e^x - 6\left[x\int e^x dx - \int\left\{1\int e^x dx\right\}dx\right] + C$$

$$= -xy - x^3e^x + 3x^2e^x - 6xe^x + 6e^x + C$$

whence

$$\frac{\partial u}{\partial x} = -y - 3x^2e^x - x^3e^x + 6xe^x + 3x^2e^x - 6e^x - 6xe^x + 6e^x$$

$$= -y - x^3e^x \quad \therefore \frac{\partial^2 u}{\partial y \partial x} = -1.$$

Again $\dfrac{\partial u}{\partial y} = -x$. $\quad\therefore \dfrac{\partial^2 u}{\partial x \partial y} = -1$. $\quad\therefore$ du is a perfect differential.

(b) For CF we may write $(D^2 + 9)y = 0$ so that $\lambda = \pm 3i$.

$\therefore CF = A\cos 3x + B\sin 3x$.

For PI, let $y = \dfrac{2}{D^2 + 9}\cos 3x$ and $z = \dfrac{2}{D^2 + 9}\sin 3x$ so that

$$y + iz = \frac{2}{D^2 + 9}e^{i3x} = 2e^{i3x}\frac{1}{(D + 3i)^2 + 9}1 \text{ (putting } D = D + 3i)$$

$$= 2e^{i3x}\frac{1}{6iD + D^2} = 2e^{i3x}\frac{1}{6iD}\left(1 + \frac{D}{6i}\right)^{-1} \cdot 1$$

$$= 2e^{i3x} \cdot \frac{1}{6i}\frac{1}{D} \cdot 1 = 2e^{i3x} \cdot \frac{x}{6i} = \frac{x}{3i}(\cos 3x + i\sin 3x)$$

Equating the imaginary part we have

$$PI = \frac{x}{3}\cos 3x$$

$\therefore$ the complete solution is

$$y = A\cos 3x + B\sin 3x + \frac{x}{3}\cos 3x$$

Now for the initial condition at $x = 0, y = 1$, we have $1 = A$ or $A = 1$.

Again for the initial condition at $x = 0, y' = 0$, we have

$$y' = -3A\sin 3x + 3B\cos 3x - x\sin 3x + \frac{\cos 3x}{3}$$

$\therefore$ the condition $x = 0, y' = 0$ gives $0 = 3B + \frac{1}{3}$ or $B = -\frac{1}{9}$

$$\therefore y = \cos 3x - \frac{1}{9}\sin 3x + \frac{x}{3}\cos 3x$$

is the required final solution.

35. Starting from Hermite's equation $y'' - 2xy' + 2ny = 0$ (n = constant integer), derive by Frobenius' method the standard form of Hermite's polynomial $H_n(x)$. Hence prove that $H_n'(x) = 2nH_{n-1}(x)$. **[C.U. (Hons.) 1998, 2005, 07]**

(See Art. 17.12, 17.13 and 17.13.1)

36. (a) Use the power series solution method to solve the Legendre's differential equation

$$(1 - x^2)\frac{d^2y}{dx^2} + 2x\frac{dy}{dx} + \lambda y = 0.$$

(See Art. 17.11)

(b) Solve the differential equation $(y + xy^2)dx - xdy = 0$.

(See worked out examples). **[C.U. (Hons.) 1999, 2006]**

37. Solve the equation $y'' - 2y' + y = 4e^{3x} + x$, subject to the boundary condition $y = 0, y' = 0$ at $x = 0$. **[C.U. (Hons.) 2000]**

For C.F, putting r.h.s = 0, we have $(D^2 - 2D + 1)y = 0$, so that the auxiliary equation is $\lambda^2 - 2\lambda + 1 = 0$ or $(\lambda - 1)^2 = 0$ or $\lambda_1 = \lambda_2 = +1$, i.e., the roots are equal; $\therefore$ C.F. $y = (A + Bx)e^x$.

Now

$$P.I = \frac{1}{(D^2 - 2D + 1)}(4e^{3x} + x) = \frac{4e^{3x}}{D^2 - 2D + 1} + \frac{x}{D^2 - 2D + 1}$$

$$= \frac{4e^{3x}}{9 - 6 + 1} + \frac{x}{(D-1)^2} \text{ (putting } D = 3 \text{ in the first term)}$$

$$= e^{3x} + (1 - D)^{-2}x = e^{3x} + (1 + 2D + \cdots)x = e^{3x} + x + 2.$$

So, the complete solution is

$$y = C.F. + P.I. = (A + Bx)e^x + e^{3x} + x + 2 \quad (1)$$

Now

$$y' = \frac{dy}{dx} = (A + Bx)e^x + Be^x + 3e^{3x} + 1 \quad (2)$$

Applying boundary conditions, we have from (1), $0 = A + 1 + 2 = A + 3$ or $A = -3$ and from (2), $0 = A + B + 3 + 1$ or $B = -1$ putting $A = -3$.

$\therefore$ Complete solution $y = (-3 - x)e^x + e^{3x} + x + 2$.

38. A body moves in a straight line so that its velocity exceeds by 2 its distance from a fixed point of the line. If $v = 5$ when $t = 0$, find the equation of motion. [*Ans.* $x = 5e^t - 2$.]

[*Hints.* $\frac{dx}{dt} = x + 2$.]

39. Find the time required for a sum of money to double itself at 5% per annum compounded continuously. [*Ans.* 13.9 years.]

[*Hints.* $\frac{dx}{dt} = 0.05x$, where x is the amount after t years.]

40. Radium decomposes at a rate proportional to the amount present. If half the original amount disappears in 1600 years, find the percentage lost in 100 years. [*Ans.* 9.5%.]

[*Hints.* $-\dfrac{dN}{dt} = \lambda N$ or $\dfrac{dN}{N} = -\lambda dt$. Integrating we get $\ln N = -\lambda t + \ln A$. At $t = 0, N = N_0$ (say); so $\ln N_0 = \ln A$ $\quad \therefore \ln N = -\lambda t + \ln N_0$ or $N = N_0 e^{-\lambda t}$—then proceed in the usual way.]

41. If, when the temperature of the air is 20°C, a certain substance cools from 100°C to 60°C in 10 minutes, find the temperature after 40 minutes. [*Ans.* 25°C.]

(See worked out examples)

42. Find the time required for a square tank of side 6 ft and depth 9 ft to empty through a one inch circular hole in the bottom. (Assume as in worked out problem no. 5, $v = 4.8\sqrt{h}$ ft/sec, h being, the depth of the water in the tank). [*Ans.* 137 min.]

43. A brick wall ($k = 0.0012$) is 30 cm thick. If the inner surface is at a temperature of 20°C and the outer is at 0°C, find the temperature in the wall as a function of the distance from the outer surface and the heat loss per day through a square metre. [*Ans.* $T = \dfrac{2x}{3}$, 691,000 cals.]

[*Hints.* Under certain conditions, the constant quantity Q cals per sec. of heat flowing through a wall is $Q = -kA\dfrac{dT}{dx}$ where k is the thermal conductivity of the material, $A(cm^2)$ is the area of a face of the wall perpendicular to the direction of flow and T is the temperature $x(cm)$ from that face such that T decreases as x increases.]

44. Water is flowing into a vertical cylindrical tank of radius 6 ft and height 9 ft at the rate of 6π ft^3/min and is escaping through a hole 1 inch in diameter in the bottom. Find the time required to fill the tank. [*Ans.* 65 min.]

[*Hints.* $\left(\dfrac{\pi}{10} - \dfrac{\pi}{(24)^2}4.8\sqrt{h}\right) dt = 36\pi dh$.]

45. A funnel 10 inches in diameter at the top and 1 inch in diameter at the bottom is 24 inches deep. If initially full of water, find the time required to empty. [*Ans.* 13.7 sec.]

46. The differential equation of a circuit containing a resistance R, capacitance C and e.m.f. $e = E_0 \sin \omega t$ is $R\dfrac{di}{dt} + \dfrac{i}{C} = \dfrac{de}{dt}$. Asssuming R, C, E_0, ω to be constants, find the current i at time t. [*Ans.* $i = \dfrac{E_0 C\omega}{1 + R^2C^2\omega^2}(\cos \omega t + RC\omega \sin \omega t) + C_1 e^{-t/RC}$.]

(See worked out examples)

47. If the population of a country doubles in 50 years, in how many years will it treble under the assumption that the rate of increase is proportional to the number of inhabitants?

[*Hints.* $\dfrac{dy}{dt} = ky$ where y denotes the population at time t years.) [*Ans.* 79 years]

48. Solve the equation $y'' + 4y' + y = x^2 + 2x$. **[C.U. (Hons.) 2001, 05 (old syll)]**

For CF, putting r.h.s. $= 0$, we have auxiliary equation given by $\lambda^2 + 4\lambda + 1 = 0$ whence $\lambda = \dfrac{-4 \pm \sqrt{16 - 4}}{2} = -2 \pm \sqrt{3}$

$$\therefore CF = e^{-2x}\left(Ae^{\sqrt{3}x} + Be^{-\sqrt{3}x}\right)$$

Again

$$PI = \frac{1}{(D^2+4D+1)}(x^2+2x) = \frac{1}{(D^2+4D+1)}x^2 + \frac{1}{(D^2+4D+1)}\cdot 2x$$

$$= \left[1+(D^2+4D)\right]^{-1}\cdot x^2 + \left[1+(D^2+4D)\right]^{-1}\cdot 2x$$

$$= \{1-(D^2+4D)+\cdots\}\cdot(x^2) + \{1-(D^2+4D)+\cdots\}\cdot(2x)$$

$$= (x^2-2-8x)+(2x-8) = x^2-6x-10$$

$\therefore$ the complete solution is given by

$$y = CF + PI = e^{-2x}\left(Ae^{\sqrt{3}x} + Be^{-\sqrt{3}x}\right) + x^2 - 6x - 10$$

49. If the solution $y(x)$ of Hermite's differential equation is written as $y(x) = \sum_{r=0}^{\infty} a_r x^{k+r}$, then show that the allowed values of k are zero and one only. **[C.U. (Hons.) 2001]**

(For answer see Art. 17.12)

50. (a) i. Set up the one dimensional equation of motion of a particle of mass m subjected to a force such that its potential energy is $\frac{1}{2}kx^2$. If a damping force $F = bv$ is also present, where v is the velocity and b a constant, find out the resultant motion. Discuss the case when $\frac{b}{2m} = \sqrt{\frac{k}{m}}$.

ii. Compute the total (kinetic + potential) energy. Is it a constant? Justify your answer. **[C.U. (Hons.) 2001]**

(For answer see worked out problems on Mechanics, i.e. chapter 3)

(b) A particle is subjected to a frictional force so that its velocity decreases with time according to $v = c(t-t_1)^2$, where t_1 is the time when it stops and c a constant. Find the force $F(v)$ and the distance travelled before it stops. **[C.U. (Hons.) 2001]**

[*Hints.* $F = \frac{d}{dt}(mv) = m\frac{d}{dt}[C(t-t_1)^2]$; also $m\frac{dv}{dt} = -kv$—then proceed in the usual way with the boundary condition at $t = t_1$, $v = 0$.]

(See worked out problems on Mechanics, i.e. Chapter 3)

51. Solve any one: (i) $\frac{d^2y}{dx^2} - 2\frac{dy}{dx} + y = 2\cos x$, (ii) $(x + y\cos x)dx + \sin x\,dy = 0$.

[C.U. (Hons.) 2002]

Ans. (i) For C.F., putting r.h.s. $= 0$, we have $D^2 - 2D + 1 = 0$, so that the auxiliary equation is $\lambda^2 - 2\lambda + 1 = 0$ or $(\lambda-1)^2 = 0$ or $\lambda_1 = \lambda_2 = +1$ i.e. the roots are equal.

$\therefore$ C.F. $= (A+Bx)e^x$. Again for P.I. we may write

$$P.I. = \frac{2\cos x}{D^2-2D+1} = \frac{2\cos x}{-(1)^2-2D+1} = -\frac{\cos x}{D} = -\sin x [\because D^2 = -(1)^2]$$

so the complete solution is $y = CF + PI = (A+Bx)e^x - \sin x$.

(ii) The given equation may be written as

$$xdx + y\cos x dx + \sin x dy = 0 \quad \text{or} \quad xdx + d(y\sin x) = 0$$

$$\text{or} \quad \int xdx + \int d(y\sin x) = c_1 \quad \text{or} \quad \frac{x^2}{2} + y\sin x - c_1 = 0$$

$$\text{or } x^2 + 2y\sin x + c = 0 \quad (\text{putting } c = -2c_1)$$

is the solution.

52. Solve the differential equation $(y - xy^2)dx - (x + x^2y)dy = 0$. **[C.U. (Hons.) 2003]**

Ans. The given equation is written as

$$ydx - xy^2dx - xdy - x^2ydy = 0 \quad \text{or} \quad ydx - xdy - xy(ydx + xdy) = 0$$

$$\text{or} \quad \frac{ydx - xdy}{y^2} - \frac{xy}{y^2}(ydx + xdy) = 0 \quad \text{(Dividing both sides by } y^2\text{)}$$

$$\text{or} \quad d\left(\frac{x}{y}\right) - \frac{x}{y}d(xy) = 0$$

$$\text{or} \quad \frac{d(x/y)}{x/y} - d(xy) = 0 \quad \text{(again dividing both sides by } x/y\text{)}$$

$$\text{or} \quad \int \frac{d(x/y)}{x/y} - \int d(xy) = c_1$$

$$\text{or} \quad \ln\left(\frac{x}{y}\right) - xy + c = 0 \quad \text{is the required solution (putting } c = -c_1\text{)}$$

53. Solve the differential equation

$$\frac{2x}{y^3}dx + \frac{y^2 - 3x^2}{y^4}dy = 0$$

[C.U. (Hons.) 2004]

Ans. The given equation may be written as

$$\frac{2x}{y^3}dx + \frac{1}{y^2}dy - 3\frac{x^2}{y^4}dy = 0 \quad \text{or} \quad d\left(\frac{x^2}{y^3}\right) = -\frac{1}{y^2}dy$$

$$\text{or} \quad \int d\left(\frac{x^2}{y^3}\right) = \int -\frac{1}{y^2}dy + c \quad \text{or} \quad \frac{x^2}{y^3} = \frac{1}{y} + c$$

is the required solution.

54. Solve $\dfrac{d^2y}{dx^2} + x\dfrac{dy}{dx} + y = 0$ by the method of Frobenius. (1)

Ans. Let us assume

$$y = \sum_{r=0}^{\infty} a_r x^{k+r}$$

as the power series solution with $a_0 \neq 0$, r = positive and k = a constant.

Putting the values of $\dfrac{d^2y}{dx^2}$ and $\dfrac{dy}{dx}$ in (1), we may write

$$\sum_{r=0}^{\infty} a_r(k+r)(k+r-1)x^{k+r-2} + \sum_{r=0}^{\infty} a_r(k+r+1)x^{k+r} = 0 \qquad (2)$$

This identity holds good for all values of r and this is possible only when the coefficient of each and every power of x on the l.h.s. is zero individually. (This follows from the analysis of *uniqueness of power series method*). Now r is never negative—so the lowest power of x is $(k-2)$ and its coefficient (putting $r = 0$) $a_0k(k-1) = 0$.

This is known as the *indicial* equation; since $a_0 \neq 0$. $\therefore k = 0$ or 1.

Equating the coefficient of x^{k-1} to zero—which is obtained by putting $r = 1$ in the first term of equation (2)—we have $a_1(k+1)k = 0$ which gives $a_1 = 0$ for $k = -1$, but for $k = 0$, a_1 may or may not be zero.

Equating the coefficient of x^{k+r} to zero (which is obtained by replacing r by $(r+2)$ in the first term, the second term remaining the same) we obtain

$$a_{r+2}(k+r+2)(k+r+1)+a_r(k+r+1)=0 \quad \text{or} \quad a_{r+2}=-\frac{(k+r+1)}{(k+r+2)(k+r+1)}a_r=-\frac{a_r}{(k+r+2)}$$

which is the *recurrence relation* between the coefficients a_r.

Case 1. For $k=0$, $a_{r+2}=-\dfrac{a_r}{r+2}$. Hence for $r=0,1,2,3,\ldots,$

$$a_2=-\frac{a_0}{2},\quad a_4=-\frac{a_2}{4}=\frac{a_0}{2.4},\quad a_6=-\frac{a_4}{6}=-\frac{a_0}{2.4.6}$$

$$\therefore\quad a_{2r}=(-1)^r\frac{a_0}{2.4.6\cdots 2r}$$

and

$$a_3=-\frac{a_1}{1.3},\quad a_5=-\frac{a_3}{5}=\frac{a_1}{1.3.5},\quad a_7=-\frac{a_5}{7}=-\frac{a_1}{1.3.5.7}$$

$$\therefore\quad a_{2r+1}=(-1)^r\frac{a_1}{1.3.5.7\cdots(2r+1)}$$

So

$$y=a_0\left[1-\frac{x^2}{2}+\frac{x^4}{2.4}-\cdots+(-1)^r\frac{x^{2r}}{2.4.6\cdots 2r}\right]$$

$$+a_1x\left[1-\frac{x^2}{1.3}+\frac{x^4}{1.3.5}-\cdots+(-1)^r\frac{x^{2r}}{1.3.5\cdots(2r+1)}\right] \quad (3)$$

is the solution of equation (1) for $k=0$.

Case 2. For $k=1$, $a_{r+2}=-\dfrac{a_r}{r+3}$; so for $r=0,1,2,3,\ldots,$ we have

$$a_2=-\frac{a_0}{1.3},\quad a_4=-\frac{a_2}{5}=\frac{a_0}{1.3.5}\cdots$$

$$\therefore a_{2r}=(-1)^r\frac{a_0}{1.3.5\cdots(2r+1)}$$

and $a_1=0$ and also $a_3=a_5=\cdots=0$

So

$$y=a_0x\left[1-\frac{x^2}{1.3}+\frac{x^4}{1.3.5}-\cdots+(-1)^r\frac{x^{2r}}{1.3.5\cdots(2r+1)}\right] \quad (4)$$

Clearly the solution (4) is included in the second part of (3) except that a_0 is replaced by a_1 and hence in order that equation (1) may have two independent solutions, a_1 must be zero even if $k=0$ and then equation (3) reduces to

$$y=a_0\left[1-\frac{x^2}{2}+\frac{x^4}{2.4}-\cdots+(-1)^r\frac{x^{2r}}{2.4.6\cdots 2r}\right]$$

The complete solution is therefore written as

$$y=A\left[1-\frac{x^2}{2}+\frac{x^4}{2.4}-\cdots+(-1)^r\frac{x^{2r}}{2.4.6\cdots 2r}\right]$$

$$+Bx\left[1-\frac{x^2}{1.3}+\frac{x^4}{1.3.5}-\cdots+(-1)^r\frac{x^{2r}}{1.3.5\cdots(2r+1)}\right]$$

where A and B are arbitrary constants.

55. Solve

$$\frac{d^2y}{dx^2} + 2xy' + (x^2+2)y = 0 \tag{1}$$

by the *method of Frobenius*.

Ans. Let us assume

$$y = \sum_{r=0}^{\infty} a_r x^{k+r} \tag{2}$$

as the power series solution with $a_0 \neq 0$, $r =$ positive and $k =$ a constant.

Putting y'', y' and y from (2) in (1) we may write

$$\sum_{r=0}^{\infty}(k+r)(k+r-1)a_r x^{k+r-2} + 2\sum_{r=0}^{\infty}(k+r)a_r x^{k+r} + \sum_{r=0}^{\infty} a_r x^{k+r+2} + 2\sum_{r=0}^{\infty} a_r x^{k+r} = 0$$

$$\text{or } \sum_{r=0}^{\infty}(k+r)(k+r-1)a_r x^{k+r-2} + 2\sum_{r=0}^{\infty}(k+r+1)a_r x^{k+r} + \sum_{r=0}^{\infty} a_r x^{k+r+2} = 0 \tag{3}$$

This identity holds good for all values of r and this is possible only when the coefficient of any power of x is zero. Equating the coefficient of the lowest power of x to zero (lowest power of x is $(k-2)$ for $r=0$), we have the *indicial* equation $a_0 k(k-1) = 0$ or $k=0$ and $k=1$.

Equating the coefficient of x^{k-1} to zero (putting $r=1$ in the first term) we have $a_1(k+1)k = 0$. Thus for $k=1$, $a_1 = 0$ and for $k=0$, a_1 may or may not be zero. Equating the coefficient of x^{k+r} to zero, (which is obtained by replacing r by $(r+2)$ in the first term and r by $(r-2)$ in the last term) we get

$$(k+r+2)(k+r+1)a_{r+2} + 2(k+r+1)a_r + a_{r-2} = 0 \tag{4}$$

which is *the recurrence* relation between the coefficients a_r.

Case 1. For $k=0$, we have from (4) $(r+2)(r+1)a_{r+2} + 2(r+1)a_r + a_{r-2} = 0$

$$\text{or} \quad a_{r+2} = -\frac{2(r+1)a_r + a_{r-2}}{(r+2)(r+1)}$$

For $r=0$,

$$a_2 = -\frac{2a_0}{2} = -a_0$$

For $r=1$,

$$a_3 = -\frac{2.2a_1}{3.2} = -\frac{2}{3}a_1$$

For $r=2$,

$$a_4 = -\frac{2.3a_2 + a_0}{4.3} = -\frac{-6a_0 + a_0}{12} = \frac{5}{12}a_0$$

For $r=3$,

$$a_5 = -\frac{2.4a_3 + a_1}{5.4} = -\frac{-8.\frac{2}{3}a_1 + a_1}{20} = \frac{13}{60}a_1 \quad \text{and so on.}$$

$$\therefore \quad y = \sum_{r=0}^{\infty} a_r x^r = a_0 + a_1 x + a_2 x^2 + a_3 x^3 + \cdots$$

$$= a_0\left(1 - x^2 + \frac{5}{12}x^4 - \cdots\right) + a_1\left(x - \frac{2}{3}x^3 + \frac{13}{60}x^5 - \cdots\right) \tag{5}$$

Case 2. For $k = 1$, the recurrence relation (4) reduces to

$$(r+3)(r+2)a_{r+2} + 2(r+2)a_r + a_{r-2} = 0 \quad \text{or} \quad a_{r+2} = -\frac{2(r+2)a_r + a_{r-2}}{(r+3)(r+2)}$$

For $r = 0$,

$$a_2 = -\frac{4a_0}{3.2} = -\frac{2}{3}a_0$$

For $r = 2$,

$$a_4 = -\frac{2.4a_2 + a_0}{5.4} = -\frac{-8.\frac{2}{3}a_0 + a_0}{20} = \frac{13}{60}a_0$$

and so on. Again $a_1 = a_3 = a_5 = \cdots = 0$ for $k = 1$.

So

$$y = a_0 x\left(1 - \frac{2}{3}x^2 + \frac{13}{60}x^4 - \cdots\right) \tag{6}$$

The solution (6) is included in the second part of (5) except that a_0 is replaced by a_1 and hence in order that equation (1) may have two independent solutions, a_1 must be zero even if $k = 0$ and then solution (5) reduces to

$$y = a_0\left(1 - x^2 + \frac{5}{12}x^4 - \cdots\right).$$

The complete solution is therefore written as

$$y = A\left(1 - x^2 + \frac{5}{12}x^4 - \cdots\right) + Bx\left(1 - \frac{2}{3}x^2 + \frac{13}{60}x^4 - \cdots\right)$$

where A and B are arbitrary constants.

56. Solve

$$\frac{d^2y}{dt^2} + \omega^2 y = 0 \tag{1}$$

by the method of Frobenius.

Ans. Let us assume

$$y(t) = \sum_{r=0}^{\infty} a_r t^{k+r} \tag{2}$$

as the power series solution with $a_0 \neq 0$, r = positive and k = a constant; then

$$\frac{d^2y}{dt^2} = \sum_{r=0}^{\infty} a_r(k+r)(k+r-1)t^{k+r-2} \tag{3}$$

Using (2) and (3) in (1), we may write

$$\sum_{r=0}^{\infty} a_r(k+r)(k+r-1)t^{k+r-2} + \omega^2 \sum_{r=0}^{\infty} a_r t^{k+r} = 0 \tag{4}$$

$$\text{or} \quad a_0 k(k-1)t^{k-2} + a_1 k(k+1)t^{k-1} + \sum_{r=0}^{\infty}\left[a_{r+2}(k+r+2)(k+r+1) + \omega^2 a_r\right] t^{k+r} = 0 \tag{5}$$

[1st term is obtained by putting $r = 0$, 2nd term by putting $r = 1$ and the 3rd term or the general term by putting $r = r+2$ in (4).]

Since equation (5) must be valid for all values of t, we require that

$$a_0 k(k-1) = 0 \quad \text{(indicial equation)} \tag{6}$$

$a_1 k(k+1) = 0$ and $a_{r+2}(k+r+2)(k+r+1) + \omega^2 a_r = 0$

$$\text{or} \quad a_{r+2} = -\frac{\omega^2}{(k+r+2)(k+r+1)} a_r \tag{7}$$

Equation (7) is the *recursion* formula, i.e. the *recurrence relation* between the coefficients. The solutions of the indicial equation (6) are $k = 0$ and $k = 1$ ($\because a_0 \neq 0$). For $k = 0$, a_1 is arbitrary and we obtain two independent series solutions from $k = 0$ since $a_0 \neq 0$ by hypothesis.

In this case the recursion formula reduces to

$$a_{r+2} = -\frac{\omega^2}{(r+2)(r+1)} a_r$$

Taking even values of r, we have

for $r = 0$,

$$a_2 = -\frac{\omega^2 a_0}{2.1} = -\frac{\omega^2 a_0}{2!},$$

for $r = 2$,

$$a_4 = -\frac{\omega^2 a_2}{4.3} = -\frac{\omega^2}{4.3}\left(-\frac{\omega^2 a_0}{2!}\right) = \frac{\omega^4}{4!} a_0$$

For $r = 4$,

$$a_6 = -\frac{\omega^2}{6.5} a_4 = -\frac{\omega^2}{6.5}\left(\frac{\omega^4}{4!} a_0\right) = -\frac{\omega^6 a_0}{6!}$$

Again taking odd values of r, we have

for $r = 1$,

$$a_3 = -\frac{\omega^2}{3.2} a_1 = -\frac{\omega^2}{3!} a_1,$$

for $r = 3$,

$$a_5 = -\frac{\omega^2}{5.4} a_3 = -\frac{\omega^2}{5.4}\left(-\frac{\omega^2 a_1}{3!}\right) = \frac{\omega^4}{5!} a_1$$

for $r = 5$,

$$a_7 = -\frac{\omega^2}{7.6} a_5 = -\frac{\omega^2}{7.6}\left(\frac{\omega^4 a_1}{5!}\right) = -\frac{\omega^6 a_1}{7!}$$

The general solution now becomes

$$y(t) = a_0\left[1 - \frac{(\omega t)^2}{2!} + \frac{(\omega t)^4}{4!} - \frac{(\omega t)^6}{6!} + \cdots\right] + \frac{a_1}{\omega}\left[\omega t - \frac{(\omega t)^3}{3!} + \frac{(\omega t)^5}{5!} - \frac{(\omega t)^7}{7!} + \cdots\right]$$

$$\text{or} \quad y(t) = a_0 \cos \omega t + \frac{a_1}{\omega} \sin \omega t$$

is the required solution.

57. Solve

$$x^2 \frac{d^2 y}{dx^2} - 6y = 0$$

by the method of Frobenius. **[C.U. (Hons.) 2006]**

[**Hints.** The given equation may be written as

$$\frac{d^2 y}{dx^2} - \frac{6}{x^2} y = 0 \quad \text{or} \quad \frac{d^2 y}{dx^2} + \omega^2 y = 0$$

putting $\omega^2 = -6/x^2$. Then proceed as in example 56.]

58. Find one solution of the following differential equation by the method of Frobenius

$$(1-x^2)\frac{d^2y}{dx^2} - 2x\frac{dy}{dx} + l(l+1)y = 0$$

where l is a positive integer. **[C.U. (Hons.) 2003]**

(See Art. 17.11)

59. Show that

$$\int_{-1}^{1} P_l(x)P_m(x)dx = 0 \quad \text{for } l \neq m;$$

$P_l(x)$ is the Legendre polynomial of order l. (See Art. 17.11.3). **[C.U. (Hons.) 2006]**

60. (a) Use Frobenius method to solve the following differential equation for Hermite Polynmial around $x = 0$

$$\frac{d^2y}{dx^2} - 2x\frac{dy}{dx} + 2ny = 0$$

where n is a non-negative integer. (See Art. 17.11).

(b) Legendre polynomials may be expressed as

$$(1-2tx+t^2)^{-1/2} = \sum_{n=0}^{\infty} P_n(x)t^n$$

Use the above relation to show that

(i) $P_n(-x) = (-1)^n P_n(x)$

(ii) $(l+1)P_{l+1}(x) = (2l+1)xP_l(x) - lP_{l-1}(x)$. **[C.U. (Hons.) 2007]**

(See Art. 17.11.1 and 17.11.5)

61. Solve the equation $y'' + 2y' - y = 4e^{3x} + x$ subject to the boundary condition $y = 0$, $y' = 0$ at $x = 0$. **[C.U. (Hons.) 2007]**

(See worked out examples)

62. Solve the equation

$$(1-x^2)\frac{d^2y}{dx^2} - 2x\frac{dy}{dx} + 6y = 0 \text{ around } x = 0$$

[C.U. (Hons.) 2008]

[**Hints.** This is Legendre's differential equation]

63. Find whether $d\phi$ is an exact differential where $d\phi = (x^3 - 4xy)dx + (y^2 - 2x)dy$.

[C.U. (Hons.) 2007]

(See worked out problems)

Chapter 18

Some Important Types of Partial Differential Equations occurring in Physics and their Solutions

18.1 Introduction

It has already been mentioned that the equations involving partial derivatives are called *partial differential equations.* The following second order partial differential equations are the basic equations of mathematical physics.

18.1.1 Three Dimensional Laplace's equation in Cartesian Co-ordinates

Let $u(x, y, z)$ be a harmonic function* of three variables x, y and z. Considering u to be the gravitational potential at a point not occupied by attracting matter (as in Art. 7.19), we have already deduced

$$\nabla^2 u = \frac{\partial^2 u}{\partial x^2} + \frac{\partial^2 u}{\partial y^2} + \frac{\partial^2 u}{\partial z^2} = 0. \tag{18.1.1}$$

This is Laplace's equation in three dimensional cartesian form. This is the most important partial differential equation in applied mathematics. The equation is also satisfied by the following functions:

(i) the magnetic potential in free space;

(ii) the electric potential in the steady flow of electric current in solid conductors;

(iii) the temperature in the thermal equilibrium of solids;

(iv) the velocity potential in homogeneous liquids in case of irrotational flow in hydrodynamics;

*By harmonic function we mean a function which possesses continuous partial derivations of the first and second orders and satisfies Laplace's equation.

(v) the electro-static potential in uniform dielectric;

We give below the derivation of Laplace's equation in the cases of two important functions occurring in the steady flow of electric current in solid conductors and the velocity potential in homogeneous liquids in hydrodynamics.

18.1.2 The electric potential in the steady flow of electric current

We consider a volume V in a homogeneous medium. Let an electric current pass through it. The density of the electric current at any point is given by

$$\vec{J}(x,y,z) = J_x\hat{\imath} + J_y\hat{\jmath} + J_z\hat{k}.$$

We assume that the electric current density is independent of time and there is no source of current in the volume in question. Thus the flux of $\vec{J}$ through any closed surface S inside the volume V is equal to zero. Hence we may write

$$\iint_S \vec{J}\cdot d\vec{s} = \iint_S \vec{J}\cdot\hat{n}ds = 0,$$

where $\hat{n}$ is the unit vector in the outward normal direction.

Hence we may write from Gauss's theorem

$$\iint_S \vec{J}\cdot\vec{n}ds = \iiint_V (\text{div}\vec{J})\,dV = 0 \quad \text{or} \quad \text{div}\vec{J} = 0. \tag{18.1.2}$$

Now the electric force $\vec{E}$ in the medium of conductivity k is given by

$$\vec{J} = k\vec{E}, \quad \text{from Ohm's law.} \tag{18.1.3}$$

From Maxwell's electromagnetic theory, we have

$$\text{curl}\,\vec{E} = -\frac{\partial\vec{B}}{\partial t},$$

where $\vec{B}$ is the magnetic induction. Now if the process is stationary, the vector $\vec{B}$ will be independent of time and hence $\frac{\partial\vec{B}}{\partial t} = 0$.

$$\therefore \quad \text{curl}\,\vec{E} = 0$$

or $\vec{E} = \text{grad}\,\phi$, where ϕ is a scalar function called the *electric potential energy.*

Hence from (18.1.3), we may write

$$\vec{J} = k\,\text{grad}\,\phi. \tag{18.1.4}$$

$\therefore$ from (18.1.2), we have

$$\text{div}(k\,\text{grad}\,\phi) = 0 \text{ or } k\text{div}(\text{grad}\,\phi) = 0;$$

$$\text{or} \quad \left(\hat{\imath}\frac{\partial}{\partial x} + \hat{\jmath}\frac{\partial}{\partial y} + \hat{k}\frac{\partial}{\partial z}\right)\cdot\left(\hat{\imath}\frac{\partial\phi}{\partial x} + \hat{\jmath}\frac{\partial\phi}{\partial y} + \hat{k}\frac{\partial\phi}{\partial z}\right) = 0$$

$$\text{or} \quad \frac{\partial^2\phi}{\partial x^2} + \frac{\partial^2\phi}{\partial y^2} + \frac{\partial^2\phi}{\partial z^2} = 0. \tag{18.1.5}$$

We thus arrive at the Laplace's equation.

18.1.3 The potential flow of a fluid

We know that the equation of continuity of a compressible liquid is

$$\frac{\partial \rho}{\partial t} + \text{div}(\rho \vec{v}) = 0. \qquad (18.1.6)$$

If the liquid under consideration is incompressible, $\rho = \text{const.}$, i.e.

$$\frac{\partial \rho}{\partial t} = 0 \quad \therefore \quad \text{div}(\vec{v}) = 0. \qquad (18.1.7)$$

If the velocity $\vec{v}$ is expressed in terms of the velocity potential function ϕ, we may write $\vec{v} = \text{grad}\,\phi$.

Hence the equation (18.1.7) may be written as

$$\text{div}(\text{grad}\,\phi) = 0 \quad \text{or} \quad \vec{\nabla}\cdot(\vec{\nabla}\phi) = 0 \quad \text{or} \quad \nabla^2\phi = 0, \quad \text{i.e.} \quad \frac{\partial^2\phi}{\partial x^2} + \frac{\partial^2\phi}{\partial y^2} + \frac{\partial^2\phi}{\partial z^2} = 0.$$

Thus the velocity potential function satisfies Laplace's equation.

Illustrative Example

State what is meant by an irrotational vector field and explain how it can be related to a scalar field. Derive the equation satisfied by such a scalar field when the parent vector field is also solenoidal. What is the name of this partial differential equation? Give two examples from physics where this equation is satisfied.

[C.U. (Hons.) 1983]

The vector field, the curl of which is equal to zero is called an *irrotational vector field.* If $\vec{F}$ be the irrotational vector field then curl $\vec{F} = 0$.

Now since the curl of a gradient always vanishes, $\vec{F}$ must be gradient of a scalar ϕ.

$$\therefore \quad \vec{F} = -\,\text{grad}\,\phi = -\vec{\nabla}\phi$$

ϕ is called the *potential energy.*

Now if the vector field $\vec{F}$ is solenoidal, we have

$$\vec{\nabla}\cdot\vec{F} = 0 \quad \text{or} \quad \vec{\nabla}\cdot\vec{\nabla}\phi = 0 \quad \text{or} \quad \nabla^2\phi = 0.$$

This equation is called *Laplace's equation.* For two examples of Laplace's equation see Articles 18.1.2 and 18.1.3.

18.2 Method of separation of variables

It is assumed that $u(x, y, z) = X(x)Y(y)Z(z)$ which means that the function of many variables is a product of the functions, each of which is a function of a single variable. It is to be noted that this method is applicable only in the case of a homogeneous equation. The method is illustrated in the solution of Laplace's three-dimensional equation.

18.3 Solution of three-dimensional Laplace's Equation in Cartesian Co-ordinates

The equation is

$$\frac{\partial^2 u}{\partial x^2}+\frac{\partial^2 u}{\partial y^2}+\frac{\partial^2 u}{\partial z^2}=0. \tag{18.3.1}$$

Considering the variables to be separable, we assume

$$u=X(x)Y(y)Z(z), \tag{18.3.2}$$

where $X(x)$ is a function of x only, $Y(y)$ is a function of y only, and $Z(z)$ is a function of z only.

From (18.3.2) we have,

$$\frac{\partial^2 u}{\partial x^2}=YZ\frac{d^2X}{dx^2},\quad \frac{\partial^2 u}{\partial y^2}=ZX\frac{d^2Y}{dy^2}\quad\text{and}\quad \frac{\partial^2 u}{\partial z^2}=XY\frac{d^2Z}{dz^2}.$$

Substituting these results in (18.3.1) and then re-arranging, we have,

$$\frac{1}{X}\frac{d^2X}{dx^2}+\frac{1}{Y}\frac{d^2Y}{dy^2}+\frac{1}{Z}\frac{d^2Z}{dz^2}=0. \tag{18.3.3}$$

The variables are now separated. Now the relation (18.3.3) being of the form

$$F_1(X)+F_2(Y)+F_3(Z)=0,$$

will be true only if F_1, F_2, F_3 are constant functions since each of F_1, F_2, F_3 is a function of only one independent variable. Let us call these constants $-n^2$, $-m^2$, $+p^2$, such that $p^2=m^2+n^2$.

We may, therefore, write

$$\frac{1}{X}\frac{d^2X}{dx^2}=-n^2\quad\text{or}\quad\frac{d^2X}{dx^2}+n^2X=0.$$

This equation is a linear differential equation with constant coefficient and its solution is

$$X=A_1\cos nx+A_2\sin nx. \tag{18.3.4}$$

Also

$$\frac{1}{Y}\frac{d^2Y}{dy^2}=-m^2\quad\text{or}\quad\frac{d^2Y}{dy^2}+m^2Y=0$$

which gives

$$Y=A_3\cos my+A_4\sin my. \tag{18.3.5}$$

Again

$$\frac{1}{Z}\frac{d^2Z}{dz^2}=p^2\quad\text{or}\quad\frac{d^2Z}{dz^2}-p^2Z=0$$

which gives

$$Z=Ee^{pz}+Fe^{-pz}. \tag{18.3.6}$$

Thus the general solution of (18.3.1) is

$$u = (A_1 \cos nx + A_2 \sin nx)(A_3 \cos my + A_4 \sin my) \quad (Ee^{pz} + FE^{-pz}) \tag{18.3.7}$$

where $p^2 = m^2 + n^2$.

It is to be noted that for verification we can easily show that Laplace's equation $\nabla^2 u = 0$ is satisfied by

$$u = \frac{l}{\left[(x-a)^2 + (y-b)^2 + (z-c)^2\right]^{1/2}} \tag{18.3.8}$$

where l is a constant and (a, b, c), the co-ordinates of a fixed point.

18.3.1 Three-dimensional Laplace's Equation in Cylindrical Co-ordinates

We know that the cartesian form of the three-dimensional Laplace's equation is

$$\nabla^2 u = \frac{\partial^2 u}{\partial x^2} + \frac{\partial^2 u}{\partial y^2} + \frac{\partial^2 u}{\partial z^2} = 0. \tag{18.3.9}$$

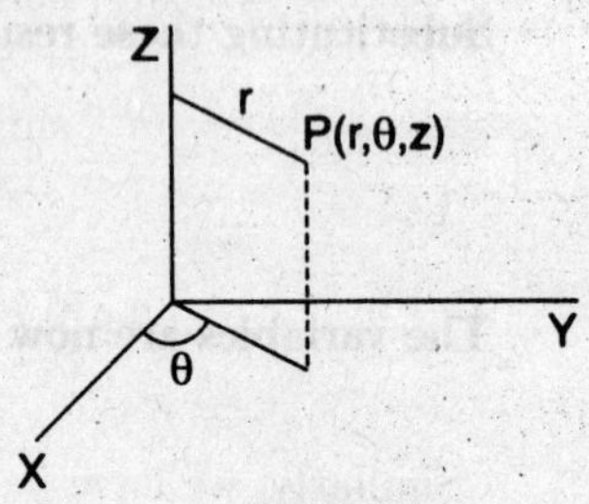

Fig 18.3.1

Expression (18.3.9) can be transformed to the cylindrical form by using cylindrical co-ordinates (r, θ, z), as explained in fig. (18.3.1), transformation equations being

$$\left.\begin{aligned} &x = r\cos\theta,\ y = r\sin\theta,\ z = z \\ &\text{so that } r = \sqrt{x^2+y^2}, \\ &\theta = \tan^{-1}\tfrac{y}{x} \text{ and } z = z \end{aligned}\right\}. \tag{18.3.10}$$

From (18.3.10), we have

$$\frac{\partial x}{\partial r} = \cos\theta; \quad \frac{\partial r}{\partial x} = \frac{x}{\sqrt{x^2+y^2}} = \frac{r\cos\theta}{r} = \cos\theta$$

$$\frac{\partial y}{\partial r} = \sin\theta; \quad \frac{\partial r}{\partial y} = \frac{y}{\sqrt{x^2+y^2}} = \frac{r\sin\theta}{r} = \sin\theta;$$

$$\frac{\partial \theta}{\partial x} = \frac{1}{1+\frac{y^2}{x^2}}\left(-\frac{y}{x^2}\right) = -\frac{x^2}{x^2+y^2}\cdot\frac{y}{x^2} = -\frac{r\sin\theta}{r^2} = -\frac{\sin\theta}{r}.$$

Similarly $\dfrac{\partial \theta}{\partial y} = \dfrac{\cos\theta}{r}$.

Now

$$\frac{\partial u}{\partial x} = \frac{\partial u}{\partial r}\frac{\partial r}{\partial x} + \frac{\partial u}{\partial \theta}\frac{\partial \theta}{\partial x} + \frac{\partial u}{\partial z}\frac{\partial z}{\partial x} = \frac{\partial u}{\partial r}\cos\theta + \frac{\partial u}{\partial \theta}\left(-\frac{\sin\theta}{r}\right).$$

Also

$$\frac{\partial u}{\partial y} = \frac{\partial u}{\partial r}\frac{\partial r}{\partial y} + \frac{\partial u}{\partial \theta}\frac{\partial \theta}{\partial y} + \frac{\partial u}{\partial z}\frac{\partial z}{\partial y} = \frac{\partial u}{\partial r}\sin\theta + \frac{\partial u}{\partial \theta}\frac{\cos\theta}{r}.$$

Thus

$$\frac{\partial u}{\partial x} = \cos\theta\frac{\partial u}{\partial r} - \frac{\sin\theta}{r}\frac{\partial u}{\partial \theta}$$ [Here z has the same value as that before the transformation.]

$$\frac{\partial u}{\partial y} = \sin\theta\frac{\partial u}{\partial r} + \frac{\cos\theta}{r}\frac{\partial u}{\partial \theta}.$$

So we have the following equivalence of cartesian and polar operators:

$$\frac{\partial}{\partial x} \equiv \left(\cos\theta\frac{\partial}{\partial r} - \frac{\sin\theta}{r}\frac{\partial}{\partial \theta}\right); \quad \frac{\partial}{\partial y} \equiv \left(\sin\theta\frac{\partial}{\partial r} + \frac{\cos\theta}{r}\frac{\partial}{\partial \theta}\right).$$

Again

$$\frac{\partial^2 u}{\partial x^2} = \frac{\partial}{\partial x}\left(\frac{\partial u}{\partial x}\right) = \left(\cos\theta\frac{\partial}{\partial r} - \frac{\sin\theta}{r}\frac{\partial}{\partial \theta}\right)\left(\cos\theta\frac{\partial u}{\partial r} - \frac{\sin\theta}{r}\frac{\partial u}{\partial \theta}\right)$$

$$= \cos\theta\left(\cos\theta\frac{\partial^2 u}{\partial r^2} - \frac{\sin\theta}{r}\frac{\partial^2 u}{\partial r\partial\theta} + \frac{\partial u}{\partial\theta}\frac{\sin\theta}{r^2}\right)$$

$$-\frac{\sin\theta}{r}\left(\cos\theta\frac{\partial^2 u}{\partial\theta\partial r} - \sin\theta\frac{\partial u}{\partial r} - \frac{1}{r}\frac{\partial u}{\partial\theta}\cos\theta - \frac{1}{r}\sin\theta\frac{\partial^2 u}{\partial\theta^2}\right)$$

$$= \cos^2\theta\frac{\partial^2 u}{\partial r^2} - \frac{2\sin\theta\cos\theta}{r}\frac{\partial^2 u}{\partial r\partial\theta} + \frac{\sin^2\theta}{r^2}\frac{\partial^2 u}{\partial\theta^2} + \frac{\sin^2\theta}{r}\frac{\partial u}{dr} + \frac{2\sin\theta\cos\theta}{r^2}\frac{\partial u}{\partial\theta}.$$

Similarly, we have

$$\frac{\partial^2 u}{\partial y^2} = \sin^2\theta\frac{\partial^2 u}{\partial r^2} + \frac{2\sin\theta\cos\theta}{r}\frac{\partial^2 u}{\partial r\partial\theta} + \frac{\cos^2\theta}{r^2}\frac{\partial^2 u}{\partial\theta^2} + \frac{\cos^2\theta}{r}\frac{\partial u}{\partial r} - \frac{2\sin\theta\cos\theta}{r^2}\frac{\partial u}{\partial\theta}$$

so that

$$\nabla^2 u = \frac{\partial^2 u}{\partial x^2} + \frac{\partial^2 u}{\partial y^2} + \frac{\partial^2 u}{\partial z^2} = \frac{\partial^2 u}{\partial r^2} + \frac{1}{r}\frac{\partial u}{\partial r} + \frac{1}{r^2}\frac{\partial^2 u}{\partial\theta^2} + \frac{\partial^2 u}{\partial z^2} = 0 \tag{18.3.11}$$

is the cylindrical form of equation (18.3.9).

18.3.2 Solution of three-dimensional Laplace's Equation in Cylindrical Co-ordinates

In order to solve equation (18.3.11) by the method of separation of variables, let us assume that

$$u(r,\theta,z) = R(r)\Theta(\theta)Z(z) \tag{18.3.12}$$

R, Θ, Z being functions of r, θ, z alone.

From (18.3.12) we have

$$\frac{\partial u}{\partial r} = \Theta Z\frac{dR}{dr}, \quad \frac{\partial^2 u}{\partial r^2} = \Theta Z\frac{d^2R}{dr^2}, \quad \frac{\partial^2 u}{\partial\theta^2} = RZ\frac{d^2\Theta}{d\theta^2} \quad \text{and} \quad \frac{\partial^2 u}{\partial z^2} = R\Theta\frac{d^2 Z}{dz^2}.$$

Substituting these values in equation (18.3.11), we may write

$$\Theta Z\frac{d^2R}{dr^2}+\frac{1}{r}\frac{dR}{dr}\Theta Z+\frac{1}{r^2}RZ\frac{d^2\Theta}{d\theta^2}+R\Theta\frac{d^2Z}{dz^2}=0.$$

Dividing throughout by $R\Theta Z$, we get

$$\frac{1}{R}\frac{d^2R}{dr^2}+\frac{1}{Rr}\frac{dR}{dr}+\frac{1}{r^2\Theta}\frac{d^2\Theta}{d\theta^2}+\frac{1}{Z}\frac{d^2Z}{dz^2}=0 \tag{18.3.13}$$

$$\text{or,}\quad \frac{1}{R}\left(\frac{d^2R}{dr^2}+\frac{1}{r}\frac{dR}{dr}\right)+\frac{1}{r^2}\frac{1}{\Theta}\frac{d^2\Theta}{d\theta^2}+\frac{1}{Z}\frac{d^2Z}{dz^2}=0. \tag{18.3.14}$$

The first three terms of (18.3.14) being independent of z, the last term should also be independent of z, i.e., the last term must be a constant;

$$\therefore\quad \frac{d^2Z}{dz^2}=CZ,\ C \text{ being a constant.} \tag{18.3.15}$$

Again since the first two and the last terms of (18.3.13) are independent of θ, the third term should also be independent of θ, and as such we may put

$$\frac{d^2\Theta}{d\theta^2}=D\Theta,\ D \text{ being a constant.} \tag{18.3.16}$$

Equation (18.3.14) then becomes

$$r^2\frac{d^2R}{dr^2}+r\frac{dR}{dr}+(D+Cr^2)R=0. \tag{18.3.17}$$

Putting $C=k^2$. (18.3.15) becomes $\dfrac{d^2Z}{dz^2}=k^2Z$ whose solution is $Z=A_1e^{kz}+B_1e^{-kz}$.

Also putting $D=-n^2$, (18.3.16) becomes $\dfrac{d^2\Theta}{d\theta^2}=-n^2\Theta$ whose solution is

$$\Theta=A_2\cos n\theta+B_2\sin n\theta.$$

Equation (18.3.17) then reduces to

$$r^2\frac{d^2R}{dr^2}+r\frac{dR}{dr}+(k^2r^2-n^2)R=0.$$

Writing $\nu=kr$, this may be written as

$$\nu^2\frac{d^2R}{d\nu^2}+\nu\frac{dR}{d\nu}+(\nu^2-n^2)R=0 \tag{18.3.18}$$

which is *Bessel's differential equation of order* n and its solutions are known as ***cylindrical functions*** or *Bessel's functions of order* n.

Hence if $J_n(\nu)$ be a Bessel function of order n, then the solution of Laplace's equation is

$$u(r,\theta,z)=R(r)\Theta(\theta)Z(z)=J_n(kr)e^{\pm in\theta}e^{\pm kz} \tag{18.3.19}$$

which is known as a *cylindrical harmonic,* and this harmonic is symmetrical about Z-axis if $n = 0$.

We can also find Bessel's differential equation in another standard form instead of (18.3.18) and that may be done in the following way.

Starting from (18.3.13), since the variables are separated, we may write

$$\frac{1}{Z}\frac{d^2Z}{dz^2} = \lambda^2 \quad \text{or,} \quad \frac{d^2Z}{dz^2} - \lambda^2 Z = 0. \tag{18.3.20}$$

Also

$$\frac{1}{\Theta}\frac{d^2\Theta}{d\theta^2} = -n^2 \quad \text{or,} \quad \frac{d^2\Theta}{d\theta^2} + n^2\Theta = 0. \tag{18.3.21}$$

$$\left.\begin{aligned} &(18.3.20) \text{ yields } Z = e^{\pm\lambda z} \\ &(18.3.21) \text{ yields } \Theta = e^{\pm in\theta} \end{aligned}\right\} \tag{18.3.22}$$

or in other words, the solutions of (18.3.20) and (18.3.21) are

$$\left.\begin{aligned} Z &= A_1e^{\lambda z} + B_1e^{-\lambda z} \\ \Theta &= A_2\cos n\theta + B_2\sin n\theta \end{aligned}\right\}. \tag{18.3.23}$$

Equation (18.3.14) then reduces to

$$\frac{d^2R}{dr^2} + \frac{1}{r}\frac{dR}{dr} + \left(\lambda^2 - \frac{n^2}{r^2}\right)R = 0 \tag{18.3.24}$$

which is the standard form of Bessel's equation of order n, and assumes the form

$$\frac{d^2R}{dx^2} + \frac{1}{x}\frac{dR}{dx} + \left(1 - \frac{n^2}{x^2}\right)R = 0 \quad [\text{on putting } \lambda r = x] \tag{18.3.25}$$

[$\lambda r = x$ or $r = \frac{1}{\lambda}x$ or $dr = \frac{1}{\lambda}dx$, so that we may write $\frac{dR}{dr} = \lambda\frac{dR}{dx}$ and (18.3.24) is written as

$$\frac{d^2R}{dr^2} + \frac{1}{r}\frac{dR}{dr} + \lambda^2\left(1 - \frac{n^2}{\lambda^2r^2}\right)R = 0$$

$$\text{or} \quad \lambda^2\frac{d^2R}{dx^2} + \frac{\lambda^2}{x}\frac{dR}{dx} + \lambda^2\left(1 - \frac{n^2}{\lambda^2r^2}\right)R = 0$$

$$\left\{\because \frac{d}{dr}\left(\frac{dR}{dr}\right) = \lambda\frac{d}{dx}\left(\frac{\lambda dR}{dx}\right) \text{ and } \frac{1}{r}\frac{dR}{dr} = \frac{\lambda}{x}\frac{\lambda dR}{dx}\right\}$$

$$\text{or} \quad \frac{d^2R}{dx^2} + \frac{1}{x}\frac{dR}{dx} + \left(1 - \frac{n^2}{x^2}\right)R = 0]$$

The general solutions of (18.3.25) are

$$R = A_3J_n(\lambda r) + B_3J_{-n}(\lambda r) \quad [\text{for fractional values of } n] \tag{18.3.26}$$

$$\text{and} \quad R = A_3J_n(\lambda r) + B_3Y_n(\lambda r) \quad [\text{for integral values of } n] \tag{18.3.27}$$

Thus the solutions of (18.3.11) with the help of (18.3.23), (18.3.26) and (18.3.27) are

$$u(r,\theta,z) = (A_1 e^{\lambda z} + B_1 e^{-\lambda z})(A_2 \cos n\theta + B_2 \sin n\theta)[A_3 J_n(\lambda r) + B_3 J_{-n}(\lambda r)] \quad (18.3.28)$$

$$\text{and} \quad u(r,\theta,z) = (A_1 e^{\lambda z} + B_1 e^{-\lambda z})(A_2 \cos n\theta + B_2 \sin n\theta)[A_3 J_n(\lambda r) + B_3 Y_n(\lambda r)]. \quad (18.3.29)$$

Corollary 1. The general solution of (18.3.24) may be written as

$$R = A_{\lambda n} J_n(\lambda r) + B_{\lambda n} Y_n(\lambda r) \quad (18.3.30)$$

where $A_{\lambda n}$ and $B_{\lambda n}$ are constants.

$A_{\lambda n}$ and $B_{\lambda n}$ are determined from the physical conditions of the problem. As for example, if the solution is to be finite for the region including the pole ($r = 0$), $B_{\lambda n}$ must be equal to zero, because $Y_n(\lambda r) \to \infty$ as $r \to 0$, which can be shown from the series of $Y_n(x)$ and hence the solution of (18.3.11) is given by

$$u(r,\theta,z) = \sum_{\lambda}\sum_{n} A_{\lambda n} J_n(\lambda r) e^{\pm\lambda z \pm in\theta} = \sum_{\lambda}\sum_{n} A_{\lambda n} J_n(\lambda r) e^{\pm(\lambda z + in\theta)}. \quad (18.3.31)$$

Corollary 2. If $\lambda = n = 0$, we have

$$\frac{d^2R}{dr^2} + \frac{1}{r}\frac{dR}{dr} = 0, \quad \frac{d^2\Theta}{d\theta^2} = 0 \quad \text{and} \quad \frac{d^2Z}{dz^2} = 0.$$

Integrating the first equation we find $R = A_1 \log_e r + B_1$, where A_1 and B_1 are constants which may be determined from the boundary conditions of the problem. Integrating the second and the third equations, we have, $\Theta = A_2\theta + B_2$, $Z = A_3 z + B_3$.

Hence the complete solution is

$$u = (A_1 \log_e r + B_1)(A_2\theta + B_2)(A_3 z + B_3). \quad (18.3.32)$$

18.4 Three-Dimensional Laplace's Equation in Spherical Polar Co-ordinates

We know that the cartesian form of the three-dimensional Laplace's equation is

$$\nabla^2 u = \frac{\partial^2 u}{\partial x^2} + \frac{\partial^2 u}{\partial y^2} + \frac{\partial^2 u}{\partial z^2} = 0. \quad (18.4.1)$$

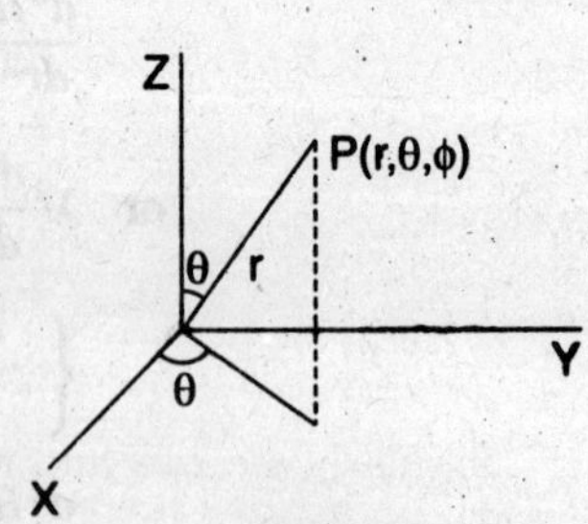

Fig 18.4.1

Expression (18.4.1) can be transformed to the spherical polar form by using spherical polar co-ordinates (r,θ,ϕ) as explained in figure (18.4.1).

The transformation equations here are

$$\left.\begin{aligned} &x = r\sin\theta\cos\phi,\ y = r\sin\theta\sin\phi,\ z = r\cos\theta \\ &\text{so that } r^2 = x^2 + y^2 + z^2,\ \phi = \tan^{-1}\frac{y}{x} \\ &\text{and } \tan\theta = \frac{\sqrt{x^2+y^2}}{z} \end{aligned}\right\}. \quad (18.4.2)$$

From (18.4.2) we may write

$$\frac{\partial r}{\partial x}=\frac{x}{r}=\sin\theta\cos\phi;\quad \frac{\partial r}{\partial y}=\frac{y}{r}=\sin\theta\sin\phi;\quad \frac{\partial r}{\partial z}=\frac{z}{r}=\cos\theta;$$

$$\frac{\partial \theta}{\partial x}=\frac{\cos\theta\cos\phi}{r};\quad \frac{\partial \theta}{\partial y}=\frac{\cos\theta\sin\phi}{r};\quad \frac{\partial \theta}{\partial z}=-\frac{\sin\theta}{r};$$

$$\frac{\partial \phi}{\partial x}=-\frac{\sin\phi}{r\sin\theta};\quad \frac{\partial \phi}{\partial y}=\frac{\cos\phi}{r\sin\theta};\quad \frac{\partial \phi}{\partial z}=0.$$

With the above values, we may have

$$\frac{\partial u}{\partial x}=\frac{\partial u}{\partial r}\frac{\partial r}{\partial x}+\frac{\partial u}{\partial \theta}\frac{\partial \theta}{\partial x}+\frac{\partial u}{\partial \phi}\frac{\partial \phi}{\partial x}=\sin\theta\cos\phi\frac{\partial u}{\partial r}+\frac{\cos\theta\cos\phi}{r}\frac{\partial u}{\partial \theta}-\frac{\sin\phi}{r\sin\theta}\frac{\partial u}{\partial \phi}.$$

Thus the equivalence between cartesian and spherical polar operators is given by

$$\frac{\partial}{\partial x}\equiv\sin\theta\cos\phi\frac{\partial}{\partial r}+\frac{\cos\theta\cos\phi}{r}\frac{\partial}{\partial \theta}-\frac{\sin\phi}{r\sin\theta}\frac{\partial}{\partial \phi},$$

whence we may write

$$\begin{aligned}\frac{\partial^2 u}{\partial x^2}&=\frac{\partial}{\partial x}\left(\frac{\partial u}{\partial x}\right)=\left(\sin\theta\cos\phi\frac{\partial}{\partial r}+\frac{\cos\theta\cos\phi}{r}\frac{\partial}{\partial \theta}-\frac{\sin\phi}{r\sin\theta}\frac{\partial}{\partial \phi}\right)\\&\quad\left(\sin\theta\cos\phi\frac{\partial u}{\partial r}+\frac{\cos\theta\cos\phi}{r}\frac{\partial u}{\partial \theta}-\frac{\sin\phi}{r\sin\theta}\frac{\partial u}{\partial \phi}\right)\\&=\sin^2\theta\cos^2\phi\frac{\partial^2 u}{\partial r^2}+\frac{2\sin\theta\cos\theta\cos^2\phi}{r}\frac{\partial^2 u}{\partial r\partial\theta}\\&\quad-\frac{2\sin\theta\cos\theta\cos^2\phi}{r^2}\frac{\partial u}{\partial\theta}-\frac{2\sin\phi\cos\phi}{r}\frac{\partial^2 u}{\partial r\partial\phi}\\&\quad+\frac{\sin\phi\cos\phi}{r^2}\frac{\partial u}{\partial\phi}+\frac{\cos^2\theta\cos^2\phi}{r}\frac{\partial u}{\partial r}\\&\quad+\frac{\cos^2\theta\sin^2\phi}{r^2}\frac{\partial^2 u}{\partial\theta^2}-\frac{2\cos\theta\sin\phi\cos\phi}{r^2\sin\theta}\frac{\partial^2 u}{\partial\theta\partial\phi}\\&\quad+\frac{\cos^2\theta\sin\phi\cos\phi}{r^2\sin^2\theta}\frac{\partial u}{\partial\phi}+\frac{\sin^2\phi}{r}\frac{\partial u}{\partial r}+\frac{\cos\theta\sin^2\phi}{r^2\sin\theta}\frac{\partial u}{\partial\theta}\\&\quad+\frac{\sin^2\phi}{r^2\sin^2\theta}\frac{\partial^2 u}{\partial\phi^2}+\frac{\sin\theta\cos\phi}{r^2\sin\phi}\frac{\partial u}{\partial\phi}.\end{aligned}$$

Similarly we can find out a similar expression for $\frac{\partial^2 u}{\partial y^2}$ by replacing ϕ by $\left(\frac{\pi}{2}+\phi\right)$ in the above expression for $\frac{\partial^2 u}{\partial x^2}$; also we have

$$\begin{aligned}\frac{\partial^2 u}{\partial z^2}&=\frac{\partial}{\partial z}\left(\frac{\partial u}{\partial z}\right)=\left(\cos\theta\frac{\partial}{\partial r}-\frac{\sin\theta}{r}\frac{\partial}{\partial\theta}\right)\left(\cos\theta\frac{\partial u}{\partial r}-\frac{\sin\theta}{r}\frac{\partial u}{\partial\theta}\right)\\&=\cos^2\theta\frac{\partial^2 u}{\partial r^2}-\frac{2\sin\theta\cos\theta}{r}\frac{\partial^2 u}{\partial r\partial\theta}+\frac{2\sin\theta\cos\theta}{r}\frac{\partial u}{\partial\theta}+\frac{\sin^2\theta}{r}\frac{\partial u}{\partial r}+\frac{\sin^2\theta}{r^2}\frac{\partial^2 u}{\partial r^2}.\end{aligned}$$

Substituting these values of

$$\frac{\partial^2 u}{\partial x^2}, \quad \frac{\partial^2 u}{\partial y^2} \quad \text{and} \quad \frac{\partial^2 u}{\partial z^2}$$

in equation (18.4.1), we have on simplification,

$$\frac{\partial^2 u}{\partial r^2} + \frac{2}{r}\frac{\partial u}{\partial r} + \frac{1}{r^2}\frac{\partial^2 u}{\partial \theta^2} + \frac{\cot\theta}{r^2}\frac{\partial u}{\partial \theta} + \frac{1}{r^2\sin^2\theta}\frac{\partial^2 u}{\partial \phi^2} = 0. \tag{18.4.3}$$

which is the spherical polar form of (18.4.1). Expression (18.4.3) may also be written as

$$\frac{1}{r^2}\left[\frac{\partial}{\partial r}\left(r^2\frac{\partial u}{\partial r}\right) + \frac{1}{\sin\theta}\frac{\partial}{\partial \theta}\left(\sin\theta\frac{\partial u}{\partial \theta}\right) + \frac{1}{\sin^2\theta}\frac{\partial^2 u}{\partial \phi^2}\right] = 0. \tag{18.4.4}$$

18.4.1 Solution of three-dimensional Laplace's Equation in Spherical Polar Co-ordinates

Equation (18.4.3) may also be written in an equivalent form as

$$r^2\frac{\partial^2 u}{\partial r^2} + 2r\frac{\partial u}{\partial r} + \frac{1}{\sin\theta}\frac{\partial}{\partial \theta}\left(\sin\theta\frac{\partial u}{\partial \theta}\right) + \frac{1}{\sin^2\theta}\frac{\partial^2 u}{\partial \phi^2} = 0. \tag{18.4.5}$$

In order to solve the equation (18.4.3) or (18.4.5) by the method of separation of variables, let us assume that

$$u(r, \theta, \phi) = R(r)\Theta(\theta)\Phi(\phi) = R(r)S(\theta, \phi). \tag{18.4.6}$$

R, Θ, Φ being functions of r, θ, ϕ alone. The function $S(\theta, \phi)$ is called a *surface harmonic.* The function Θ, when ϕ is constant is called a *zonal* surface harmonic.

From (18.4.6) we have

$$\frac{\partial u}{\partial r} = \Theta\Phi\frac{dR}{dr}; \quad \frac{\partial^2 u}{\partial r^2} = \Theta\Phi\frac{d^2R}{dr^2}; \quad \frac{\partial u}{\partial \theta} = R\Phi\frac{d\Theta}{d\theta};$$

$$\frac{\partial^2 u}{\partial \theta^2} = R\Phi\frac{d^2\Theta}{d\theta^2} \quad \text{and} \quad \frac{\partial^2 u}{\partial \phi^2} = R\Theta\frac{d^2\Phi}{d\phi^2}.$$

Substitution of these values in (18.4.3) and (18.4.5) and then division by $R\Theta\Phi$ on both sides give us

$$\frac{1}{R}\frac{d^2R}{dr^2} + \frac{2}{Rr}\frac{dR}{dr} + \frac{1}{r^2\Theta}\frac{d^2\Theta}{d\theta^2} + \frac{\cot\theta}{r^2\Theta}\frac{d\Theta}{d\theta} + \frac{1}{r^2\sin^2\theta}\frac{1}{\Phi}\frac{d^2\Phi}{d\phi^2} = 0 \tag{18.4.7}$$

and

$$\left[\frac{r^2}{R}\frac{d^2R}{dr^2} + \frac{2r}{R}\frac{dR}{dr} + \frac{1}{\Theta\sin\theta}\frac{d}{d\theta}\left(\sin\theta\frac{d\Theta}{d\theta}\right)\right]\sin^2\theta = -\frac{1}{\Phi}\frac{d^2\Phi}{d\phi^2} = \lambda^2 \text{ (say)}. \tag{18.4.8}$$

From(18.4.8) we have,

$$-\frac{1}{\Phi}\frac{d^2\Phi}{d\phi^2} = \lambda^2 \quad \text{or} \quad \frac{d^2\Phi}{d\phi^2} + \lambda^2\Phi = 0$$

which gives

$$\Phi = Ce^{\pm i\lambda\phi} \tag{18.4.9}$$

and $\displaystyle \frac{1}{R}\frac{d}{dr}\left(r^2\frac{dR}{dr}\right) = -\frac{1}{\Theta\sin\theta}\frac{d}{d\theta}\left(\sin\theta\frac{d\Theta}{d\theta}\right) + \frac{\lambda^2}{\sin^2\theta} = n(n+1)$ (say),

from which we may write

$$\frac{d}{dr}\left(r^2\frac{dR}{dr}\right) - n(n+1)R = 0 \tag{18.4.10}$$

and $$\frac{1}{\sin\theta}\frac{d}{d\theta}\left(\sin\theta\frac{d\Theta}{d\theta}\right) + \left[n(n+1) - \frac{\lambda^2}{\sin^2\theta}\right]\Theta = 0. \tag{18.4.11}$$

Considering (18.4.7), if we write

$$\frac{1}{\Phi}\frac{d^2\Phi}{d\phi^2} = -\lambda^2 \quad \text{and} \quad \frac{r^2}{R}\frac{d^2R}{dr^2} + \frac{2r}{R}\frac{dR}{dr} = n(n+1),$$

then we get $\Phi = Ce^{\pm i\lambda\phi}$ which is same as (18.4.9) and

$$r^2\frac{d^2R}{dr^2} + 2r\frac{dR}{dr} - n(n+1)R = 0$$

which is same as (18.4.10).

Also we get

$$\frac{d^2\Theta}{d\theta^2} + \cot\theta\frac{d\Theta}{d\theta} + \left[n(n+1) - \frac{\lambda^2}{\sin^2\theta}\right]\Theta = 0 \tag{18.4.12}$$

which is same as (18.4.11).

Now, as the equation (18.4.10) is homogeneous, we may put $r = e^s$. Hence the equation becomes

$$\frac{d^2R}{ds^2} + \frac{dR}{ds} - n(n+1)R = 0$$

$\Big[r = e^s$ or $\frac{dr}{ds} = e^s$

$$\therefore \frac{dr}{dR} = \frac{dr}{ds}\cdot\frac{ds}{dR} = e^s\frac{ds}{dR} \quad \text{or} \quad \frac{dR}{dr} = \frac{1}{e^s}\frac{dR}{ds} = \frac{1}{r}\frac{dR}{ds}$$

$$\therefore \frac{d^2R}{dr^2} = \frac{1}{r}\frac{d^2R}{ds^2}\cdot\frac{ds}{dr} + \frac{dR}{ds}\times\left(-\frac{1}{r^2}\right) = \frac{1}{r^2}\frac{d^2R}{ds^2} - \frac{1}{r^2}\frac{dR}{ds}$$

$$\therefore r^2\frac{d^2R}{dr^2} + 2r\frac{dR}{dr} - n(n+1)R = 0, \quad \text{or,} \quad r^2\left\{\frac{1}{r^2}\frac{d^2R}{ds^2} - \frac{1}{r^2}\frac{dR}{ds}\right\} + 2r\frac{1}{r}\frac{dR}{ds} - n(n+1)R = 0$$

or, $\displaystyle \frac{d^2R}{ds^2} + \frac{dR}{ds} - n(n+1)R = 0\Big]$, or, $\{D^2 + D - n(n+1)\}R = 0 \quad \left[D \equiv \frac{d}{ds}\right]$

or, $(D-n)(D+n+1)R = 0$.

This gives

$$R = Ae^{ns} + Be^{-(n+1)s} = Ar^n + \frac{B}{r^{n+1}}. \tag{18.4.13}$$

Again if we put $\cos\theta = \mu$, in equation (18.4.11), we have, since

$$\frac{d\Theta}{d\theta} = \frac{d\Theta}{d\mu}\cdot\frac{d\mu}{d\theta} = -\sin\theta\frac{d\Theta}{d\mu} \quad \text{or,} \quad \frac{1}{\sin\theta}\frac{d}{d\theta} \equiv -\frac{d}{d\mu}.$$

$$\therefore \quad \frac{d}{d\mu}\left(\sin^2\theta\frac{d\Theta}{d\mu}\right) + \left[n(n+1) - \frac{\lambda^2}{\sin^2\theta}\right]\Theta = 0 \quad \text{[from (18.4.11)]}$$

$$\text{or,} \quad \frac{d}{d\mu}\left[(1-\mu^2)\frac{d\Theta}{d\mu}\right] + \left[n(n+1) - \frac{\lambda^2}{1-\mu^2}\right]\Theta = 0. \tag{18.4.14}$$

The equation (18.4.14) is *Legendre's associated equation* and has its solution given by

$$\Theta = C_n P_n^\lambda(\mu) + D_n Q_n^\lambda(\mu) \tag{18.4.15}$$

$$= C_n P_n^\lambda(\cos\theta) + D_n Q_n^\lambda(\cos\theta). \tag{18.4.16}$$

C_n and D_n are determined from the physical condition of the problem. $P_n^\lambda(\mu)$ and $Q_n^\lambda(\mu)$ are called associated *Legendre polynomials* of the λth order and nth degree. The associated Legendre polynomials of the first and second kinds are respectively given by

$$P_n^\lambda(\mu) = (1-\mu^2)^{\lambda/2}\ \frac{d^\lambda P_n(\mu)}{dx^\lambda}$$

$$Q_n^\lambda(\mu) = (1-\mu^2)^{\lambda/2}\ \frac{d^\lambda Q_n(\mu)}{dx^\lambda}.$$

Summing up for all n, we have the general solution of (18.4.3) given by

$$u(r,\theta,\phi) = \sum_{n=0}^{n=\infty}\left(A_n r^n + \frac{B_n}{r^{n+1}}\right)\Theta e^{\pm i\lambda\phi}. \tag{18.4.17}$$

Corollary 1. If $\lambda = 0$, the equation (18.4.14) becomes

$$(1-\mu^2)\frac{d^2\Theta}{d\mu^2} - 2\mu\frac{d\Theta}{d\mu} + n(n+1)\Theta = 0 \tag{18.4.18}$$

which is Legendre's equation. The solution is given by $\Theta = C_n P_n(\mu) + D_n Q_n(\mu)$.

Thus,

$$u(r,\theta,\phi) = \sum_{n=0}^{\infty}\left(A_n r^n + \frac{B_n}{r^{n+1}}\right)\{C_n P_n(\cos\theta) + D_n Q_n(\cos\theta)\}\,\Phi_0. \tag{18.4.19}$$

When $D_n = 0$, the expresion (18.4.19) becomes

$$u(r,\theta,\phi) = \sum_{n=0}^{\infty} C_n\left(A_n r^n + \frac{B_n}{r^{n+1}}\right)P_n(\cos\theta)\Phi_0 = \sum_{n=0}^{\infty}\left(A_n' r^n + \frac{B_n'}{r^{n+1}}\right)P_n(\cos\theta)\Phi_0 \tag{18.4.20}$$

$$\begin{bmatrix} A_n C_n = A_n' \\ B_n C_n = B_n' \end{bmatrix}.$$

Here Φ_0 denotes the value of Φ when $\lambda = 0$ and is given by $\Phi_0 = c_1\phi + c_2$ which follows from (18.4.8).

It is to be noted that for integral values of n

$$P_n(\mu) = \frac{1}{2^n n!}\frac{d^n}{d\mu^n}(\mu^2 - 1)^n.$$

Corollary 2. If there is axial symmetry about the Z-axis, then u depends only on r and θ, and so (18.4.3) reduces to

$$\frac{\partial^2 u}{\partial r^2} + \frac{2}{r}\frac{\partial u}{\partial r} + \frac{1}{r^2}\frac{\partial^2 u}{\partial \theta^2} + \frac{\cot\theta}{r^2}\frac{\partial u}{\partial \theta} = 0. \tag{18.4.21}$$

The solution of (18.4.21) is

$$u(r,\theta) = \sum_{n=0}^{\infty}\left(A_n r^n + \frac{B_n}{r^{n+1}}\right)\{C_n P_n(\cos\theta) + D_n Q_n(\cos\theta)\}. \tag{18.4.22}$$

In case $D_n = 0$ for specific boundary conditions, then

$$u(r,\theta) = \sum_{n=0}^{\infty}\left(A_n r^n + \frac{B_n}{r^{n+1}}\right) C_n P_n(\cos\theta) \tag{18.4.23}$$

$$= \sum_{n=0}^{\infty}\left(A'_n r^n + \frac{B'_n}{r^{n+1}}\right) P_n(\cos\theta). \tag{18.4.24}$$

18.5 Solution of two-dimensional Laplace's Equation in Cartesian Co-ordinates—Problem of Steady-State flow of heat

The equation is

$$\frac{\partial^2 u}{\partial x^2} + \frac{\partial^2 u}{\partial y^2} = 0. \tag{18.5.1}$$

This equation is used in the study of problems dealing with magnetic and electric fields, stationary thermal state, problems in hydrodynamics etc.

Replacing u by θ, the temperature which is independent of time, equation (18.5.1) gives

$$\frac{\partial^2 \theta}{\partial x^2} + \frac{\partial^2 \theta}{\partial y^2} = 0. \tag{18.5.2}$$

This equation can be solved by the method of separation of variables as will be evident from the following problems.

Problem 1. Find the steady-state temperature distribution in a thin plate bounded by the lines $x = 0, x = l, y = 0$ and $y = \infty$, assuming that heat cannot escape from either surface of the plate, the edges $x = 0$ and $x = l$ being kept at a temperature zero and also the lower edge $y = 0$ is kept at temperature $F(x)$ and the edge $y = \infty$ at temperature zero:

In order to determine the steady-state temperature in the present problem, we are to solve equation (18.5.2) with the conditions given by (i) $\theta(0, y) = 0$; (ii) $\theta(l, y) = 0$; (iii) $\theta(x, 0) = F(x)$; and (iv) $\theta(x, \infty) = 0$. (Fig. 1)

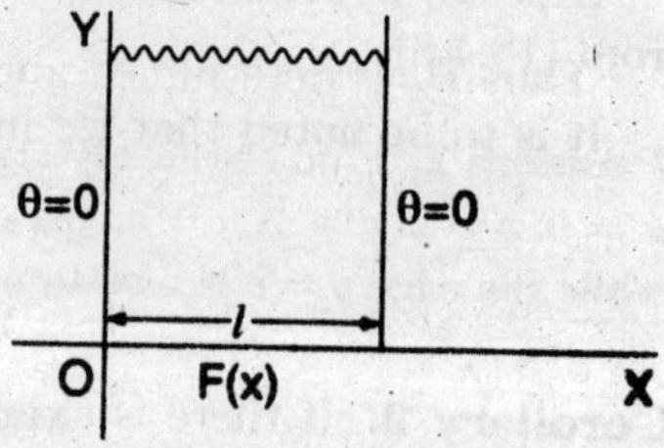

Fig 1

In order to solve (18.5.2) by the method of separation of variables, let us assume that $\theta(x, y) = X(x)Y(y)$,

X, Y being functions of x, y alone, so that

$$\frac{\partial^2\theta}{\partial x^2} = Y\frac{d^2X}{dx^2} \quad \text{and} \quad \frac{\partial^2\theta}{\partial y^2} = X\frac{d^2Y}{dy^2}.$$

Substituting these values in (18.5.2), we get on simplification,

$$\frac{1}{X}\frac{d^2X}{dx^2} = -\frac{1}{Y}\frac{d^2Y}{dy^2} = -\lambda^2 \text{ (say), (since the variables are separated)}$$

whence we get $\dfrac{1}{X}\dfrac{d^2X}{dx^2} = -\lambda^2$ or, $\dfrac{d^2X}{dx^2} + \lambda^2 X = 0$ which gives $X = A_1\cos\lambda x + A_2\sin\lambda x$

$$\text{and} \quad \frac{1}{Y}\frac{d^2Y}{dy^2} = \lambda^2 \quad \text{or,} \quad \frac{d^2Y}{dy^2} - \lambda^2 Y = 0 \tag{1}$$

which gives

$$Y = A_3e^{\lambda y} + A_4e^{-\lambda y}. \tag{2}$$

$$\begin{aligned}\therefore\quad \theta(x, y) &= (A_1\cos\lambda x + A_2\sin\lambda x)(A_3e^{\lambda y} + A_4e^{-\lambda y})\\ &= e^{-\lambda y}(A\cos\lambda x + B\sin\lambda x) + e^{\lambda y}(C\cos\lambda x + D\sin\lambda x)\end{aligned} \tag{3}$$

where A, B, C, D are arbitrary constants. We shall now apply the boundary conditions.

From the condition (iv), $\theta = 0$ when $y = \infty$. Hence we have $C = D = 0$. Again from (i) we have $\theta = 0$ when $x = 0$, and so we have from (3) $A = 0$. Also from (ii) we get at $x = l, \theta = 0$, for all positive values of y.

Hence

$$B\sin\lambda l = 0 \text{ or } \sin\lambda l = 0 \quad [\because B \neq 0 \text{ for a non-trivial solution}]$$

$$\text{or} \quad \lambda l = s\pi, \text{ i.e., } \lambda = \frac{s\pi}{l}, \quad \text{where } s = 0, 1, 2, 3, \cdots.$$

Hence the solution for each value of s is

$$\theta_s = B_s e^{-\frac{s\pi y}{l}}\sin\frac{s\pi x}{l} \quad \text{where } B_s \text{ is an arbitrary constant.} \tag{4}$$

Considering the sum of all possible solutions, we may write

$$\theta = \sum_{s=1}^{\infty} B_s e^{-\frac{s\pi y}{l}}\sin\frac{s\pi x}{l}. \tag{5}$$

From the condition (iii), we have $\theta = F(x)$ at $y = 0$; hence from (5) we have

$$F(x) = \sum_{s=1}^{\infty} B_s\sin\frac{s\pi x}{l}$$

$$\therefore \int_0^l F(x)\sin\frac{s\pi x}{l}\,dx = \int_0^l B_s \sin^2\frac{s\pi x}{l}\,dx = \frac{B_s l}{2} \quad \text{or,} \quad B_s = \frac{2}{l}\int_0^l F(x)\sin\frac{s\pi x}{l}\,dx. \tag{6}$$

Thus B_s is determined and hence we get the solution of the problem from equation (5).

Problem 2. Find the steady-state temperature distribution of a finite thin plate bounded by the lines $x = 0, x = a, y = 0, y = b$, the edges $x = 0, x = a$ and $y = 0$ being maintained at zero temperature, while the edge $y = b$ is maintained at temperature $F(x)$.

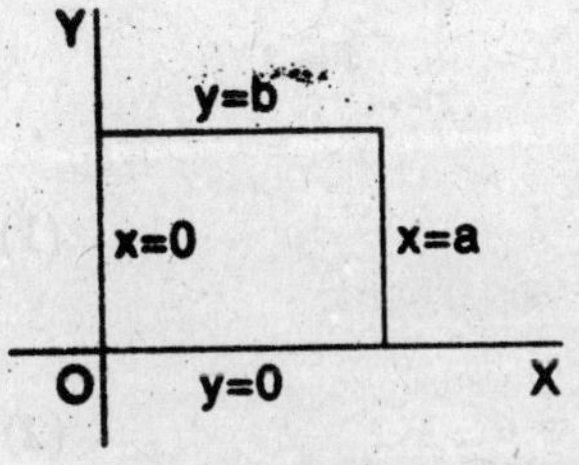

Fig 2

In order to solve this problem we are to solve equation (18.5.2) with the conditions given by (i) $\theta(0,y) = 0$; (ii) $\theta(a,y) = 0$; (iii) $\theta(x,0) = 0$; (iv) $\theta(x,b) = F(x)$; (Fig. 2)

We start with the solution (3) of Problem 1 which is given by

$$\theta(x,y) = e^{-\lambda y}(A\cos\lambda x + B\sin\lambda x) + e^{\lambda y}(C\cos\lambda x + D\sin\lambda x).$$

In the above equation applying the condition (i) we have

$$A = C = 0. \quad \therefore\ \theta(x,y) = Be^{-\lambda y}\sin\lambda x + De^{\lambda y}\sin\lambda x. \tag{1}$$

Also the condition (ii) gives the possible values of λ as in the previous problem, given by $\lambda = \frac{s\pi}{a}$.

We have then

$$\theta_s(x,y) = e^{-\frac{s\pi y}{a}} B_s \sin\frac{s\pi x}{a} + e^{\frac{s\pi y}{a}} D_s \sin\frac{s\pi x}{a} = \left(e^{-\frac{s\pi y}{a}} B_s + e^{\frac{s\pi y}{a}} D_s\right)\sin\frac{s\pi x}{a}.$$

Now at $y = 0, \theta = 0$ (condition iii)

Hence

$$(B_s + D_s)\sin\frac{s\pi x}{a} = 0. \quad \therefore\ B_s + D_s = 0 \quad \text{or} \quad B_s = -D_s.$$

Hence we can write

$$\theta_s(x,y) = B_s\left(e^{-\frac{s\pi y}{a}} - e^{\frac{s\pi y}{a}}\right)\sin\frac{s\pi x}{a} = 2B_s\sinh\frac{s\pi y}{a}\sin\frac{s\pi x}{a} = C_s\sinh\frac{s\pi y}{a}\sin\frac{s\pi x}{a}$$

where $C_s = 2B_s$.

Summing up for all possible values of s, we have

$$\theta(x,y) = \sum_{s=1}^{s=\infty} C_s\sinh\frac{s\pi y}{a}\sin\frac{s\pi x}{a}. \tag{2}$$

From the condition (iv), we have at $y = b, \theta = F(x)$. Applying this condition to the above equation, we have

$$F(x) = \sum_{s=1}^{s=\infty} C_s\sinh\frac{s\pi b}{a}\sin\frac{s\pi x}{a}.$$

Hence

$$C_s\sinh\frac{s\pi b}{a} = \frac{2}{a}\int_0^a F(x)\sin\frac{s\pi x}{a}\,dx \quad \text{or,} \quad C_s = \frac{2}{a\sinh\left(\frac{s\pi b}{a}\right)}\cdot\int_0^a F(x)\sin\frac{s\pi x}{a}\,dx.$$

For use of the general solution (18.4.22) we consider the following Illustrative Examples.

Illustrative Examples

1. If a spherical surface S of radius a is kept at a fixed distribution of electric potential of the form $u = F(\theta)$ and if the space inside and outside the spherical surface be assumed to be free of charges, find the potential outside and inside the spherical surface.

Here we have, $u = F(\theta)$ when $r = a$.

Potential outside the Spherical Surface: Since $\lim_{r\to\infty} u = 0$, i.e. the potential vanishes at infinity, the boundary condition demands that we cannot have any positive power of r in the general solution (18.4.22).

Hence $A'_n = 0$.

$$\therefore \; u = \sum_{n=0}^{\infty} \frac{B'_n}{r^{n+1}} P_n(\cos\theta), \text{ for } r > a \tag{1}$$

$$\text{or,} \quad F(\theta) = f(\cos\theta) = \sum_{n=0}^{\infty} \frac{B'_n}{a^{n+1}} P_n(\cos\theta), \text{ for } r = a. \tag{2}$$

Let us put $\cos\theta = u$; hence, we may write

$$f(u) = \sum_{n=0}^{\infty} \frac{B'_n}{a^{n+1}} P_n(u) \quad \text{or,} \quad a^{n+1} f(u) = \sum_{n=0}^{\infty} B'_n P_n(u)$$

$$\text{or,} \quad a^{n+1} \int_{-1}^{1} f(u) P_n(u)\, du = \int_{-1}^{1} \sum_{n=0}^{\infty} B'_n P_n(u) \cdot P_n(u)\, du = B'_n \int_{-1}^{1} P_n^2(u)\, du = \frac{2B'_n}{2n+1}$$

whence

$$B'_n = \frac{2n+1}{2} a^{n+1} \int_0^{\pi} F(\theta) P_n(\cos\theta) \sin\theta \, d\theta \quad \because \int_{-1}^{1} P_n^2(u)\, du = \frac{2}{2n+1}.$$

Thus, knowing B'_n, we get the potential outside the spherical surface from (1).

Potential inside the spherical surface: As the potential inside the spherical surface cannot be infinite, the general solution must not have any term containing negative power of r.

$\therefore \; B'_n = 0$, and so we have

$$u = \sum_{n=0}^{\infty} A'_n r^n P_n(\cos\theta), \text{ for } r < a \tag{3}$$

$$\therefore \quad F(\theta) = \sum_{n=0}^{\infty} A'_n a^n P_n(\cos\theta) \quad [\because \; u = F(\theta) \text{ at } r = a]$$

As in the previous case we have,

$$A'_n = \frac{2n+1}{2a^n} \int_0^{\pi} F(\theta) P_n(\cos\theta) \sin\theta \, d\theta.$$

Thus, knowing A'_n we get the potential inside the spherical surface from (3).

2. Find the potential outside a charged conducting sphere of radius a.

Here, as $r \to \infty$, $u \to 0$ and at $r = a$, $u = v$ (say).

The solution of equation (18.4.3) is

$$u(r, \theta, \phi) = \sum_{n=0}^{\infty} (A_n r^n + B_n / r^{n+1}) \left[C_n P_n^{\lambda}(\cos\theta) + D_n Q_n^{\lambda}(\cos\theta)\right] e^{\pm i\lambda\phi}.$$

Here u is independent of θ and ϕ, i.e., $\lambda = n = 0$ whence the solution becomes $u = \left(A + \frac{B}{r}\right)$, where A and B are constants to be determined from the boundary conditions of the problem. From the condition $u \to 0$ as $r \to \infty$, we have $A = 0$ and from the condition $u = v$, at $r = a$, we have $u = \frac{B}{a}$ or $B = av$. $\therefore\ u = \frac{av}{r}$.

3. Find the the potential at any point when a conducting sphere of radius a is placed in a uniform electric field of intensity E along Z-direction.

The boundary conditions of the given problem are

(i) $u = 0$ at $r = a$ (taking the origin at the centre of the sphere);

(ii) $u = -Ez = -Er\cos\theta$ when $r \to \infty$.

From the boundary conditions, it follows that u is independent of ϕ, but contains the term $\cos\theta$, i.e., $\lambda = 0$.

The solution is

$$u = \sum \left(A_n r^n + \frac{B_n}{r^{n+1}}\right) C_n P_n(\cos\theta).$$

From condition (i)

$$0 = \sum \left(A'_n a^n + \frac{B'_n}{a^{n+1}}\right) P_n(\cos\theta). \tag{1}$$

From condition (ii)

$$-Er\cos\theta = \Sigma A'_n r^n P_n(\cos\theta)$$

$$\text{or,}\quad -ErP_1(\cos\theta) = \Sigma A'_n r^n P_n(\cos\theta). \tag{2}$$

From eqn. (2), equating the coefficients of $P_1(\cos\theta)$ on both sides, we get, $A'_1 r = -Er$. $\therefore$ $A'_1 = -E$.

Equating the coefficients of $P_n(\cos\theta)$ on both sides, we get $A'_n = 0$ for $n = 0$ and for $n > 1$.

Now we know

$$\int_{-1}^{+1} P_n(\cos\theta) P_m(\cos\theta) d(\cos\theta) = \frac{2}{2n+1}\delta_n^m,$$

where

$$\delta_n^m = 0 \text{ for } m \neq n. \qquad [\delta_n^m = \text{Kronečker's delta}]$$
$$= 1 \text{ for } m = n.$$

Multiplying eqn. (1) by $P_n(\cos\theta)$ and integrating between the limits -1 and $+1$, we have

$$A'_n a^n \int_{-1}^{+1} P_n^2(\cos\theta) d(\cos\theta) + \frac{B'_n}{a^{n+1}} \int_{-1}^{+1} P_n^2(\cos\theta) d(\cos\theta) = 0$$

$$\text{or,}\quad A'_n a^n \frac{2}{2n+1} + \frac{B'_n}{a^{n+1}} \cdot \frac{2}{2n+1} = 0. \quad \therefore\quad B'_n = -A'_n a^{2n+1}.$$

Now we know that $A'_n = 0$ (except at $n = 1$); $\therefore \quad B'_n = 0$ (except at $n = 1$).

$$\therefore \quad B'_1 = -A'_1 a^3 = +Ea^3.$$

$\therefore$ potential at any point

$$u(r,\theta) = \left(A'_1 r + \frac{B'_1}{r^2}\right) P_1(\cos\theta) = \left(-Er + \frac{Ea^3}{r^2}\right)\cos\theta = -Er\cos\theta\left[1 - \frac{a^3}{r^3}\right].$$

18.6 Application of Partial Derivatives to Thermodynamical problems

We shall here discuss the state of a system and the changes which occur in passing from one state to another. For this purpose we introduce certain variable quantities in order to specify a particular state. We shall here proceed to determine certain relations between the variables, for which we must clearly understand the meaning of what is called *a perfect or a complete differential*. A quantity is said to be a perfect differential if its actual change in value during transition from one state to another is independent of the path traversed and depends only upon the initial and the final states. Hence it is a single-valued function of its state. In other words it has only one value for each definite state irrespective of how that state is attained. Such a quantity can, therefore, be integrated when the limits of initial and final states are known in order to find out any change produced in it.

We suppose that a certain quantity z depends on two other quantities x and y, i.e., $z = f(x,y)$. Let us now choose infinitesimal changes dx and dy in the variables x and y. We draw the elementary rectangle $ABCD$ (fig. 18.6.1), the co-ordinates of A, B, C, D being respectively (x,y), $(x,y+dy)$, $(x+dx,y+dy)$, $(x+dx,y)$. We consider the path ABC from A to C. If the value of the quantity is z at A, its value at B will be

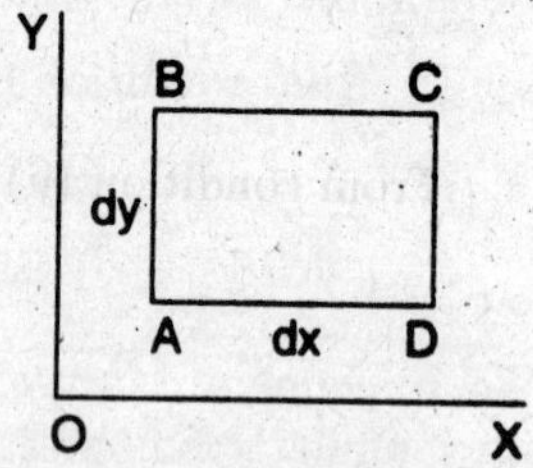

Fig 18.6.1

$$\left(z + \frac{\partial z}{\partial y}\,dy\right).$$

The value of z at C is

$$z + \frac{\partial z}{\partial y}dy + \frac{\partial}{\partial x}\left(z + \frac{\partial z}{\partial y}dy\right)dx = z + \frac{\partial z}{\partial y}dy + \frac{\partial z}{\partial x}dx + \frac{\partial^2 z}{\partial x \partial y}dx\,dy. \qquad (18.6.1)$$

Next we consider the path ADC from A to C. The value of z at D is

$$\left(z + \frac{\partial z}{\partial x}dx\right)$$

and the value of z at C is

$$z + \frac{\partial z}{\partial x}dx + \frac{\partial}{\partial y}\left(z + \frac{\partial z}{\partial x}dx\right)dy = z + \frac{\partial z}{\partial x}dx + \frac{\partial z}{\partial y}dy + \frac{\partial^2 z}{\partial y \partial x}dx\,dy. \qquad (18.6.2)$$

If dz be a perfect differential, the relations (18.6.1) and (18.6.2) must be equal.

Hence we have

$$\frac{\partial^2 z}{\partial x \partial y} = \frac{\partial^2 z}{\partial y \partial x}. \tag{18.6.3}$$

It is to be noted that this is the condition that must be satisfied for dz to be a perfect differential.

From the 1st Law of Thermodynamics we have

$$dQ = dU + p\,dV \tag{18.6.4}$$

and from the 2nd Law of Thermodynamics, we have

$$dQ = T dS \tag{18.6.5}$$

where U represents the internal energy, S the entropy, T the absolute temperature, V, the volume, p, the pressure and dQ an infinitesimal quantity of heat, supplied to a thermodynamic system.

From (18.6.4) and (18.6.5) we have $dU = T\,dS - p\,dV$, S and V being the variables. Taking partial derivatives, we have

$$\left(\frac{\partial U}{\partial S}\right)_V = T \quad \text{and} \quad \left(\frac{\partial U}{\partial V}\right)_S = -p.$$

Now we know that the change in internal energy dU is a perfect differential as it is completely determined for any state when we know the values of the co-ordinates which define that state.

$\therefore$ from condition (18.6.3) we may write,

$$\frac{\partial^2 U}{\partial S \partial V} = \frac{\partial^2 U}{\partial V \partial S} \quad \text{or,} \quad \frac{\partial}{\partial V}\left(\frac{\partial U}{\partial S}\right) = \frac{\partial}{\partial S}\left(\frac{\partial U}{\partial V}\right) \quad \text{or,} \quad \left(\frac{\partial T}{\partial V}\right)_S = -\left(\frac{\partial p}{\partial S}\right)_V$$

where is one of Maxwell's four important thermodynamic relations. The other three relations may similarly be obtained.

18.6.1 Wave equation

(a) Three-dimensional wave equation is represented by

$$\frac{\partial^2 u}{\partial t^2} = c^2 \left(\frac{\partial^2 u}{\partial x^2} + \frac{\partial^2 u}{\partial y^2} + \frac{\partial^2 u}{\partial z^2}\right) \tag{18.6.6}$$

where c is a constant having dimension of velocity, t is the time, u is the entity under consideration and x, y, z are the co-ordinates w.r.t. a certain frame of reference. The entity under consideration may be mechanical displacement or the potential for the flow of electric current in solid conductors or the field components of an electromagnetic wave. The above equation is involved in the study of the propagation of plane waves of large amplitude.

(b) Two-dimensional wave equation is represented by

$$\frac{\partial^2 u}{\partial t^2} = c^2 \left(\frac{\partial^2 u}{\partial x^2} + \frac{\partial^2 u}{\partial y^2}\right). \tag{18.6.7}$$

This equation is involved in the study of the vibration of rectangular and circular membrances.

(c) One-dimensional wave equation is represented by

$$\frac{\partial^2 u}{\partial t^2} = c^2 \frac{\partial^2 u}{\partial x^2}. \tag{18.6.8}$$

This equation is involved in the study of the transverse vibrations of a stretched string, the longitudinal vibrations of rods, electrical oscillations in wires etc. For derivation of this equation, the student is advised to consult any standard book on sound.

18.6.2 Fourier Equation for Heat Conduction

(a) Three-dimensional equation for heat conduction is represented by

$$\frac{\partial u}{\partial t} = \alpha^2 \left(\frac{\partial^2 u}{\partial x^2} + \frac{\partial^2 u}{\partial y^2} + \frac{\partial^2 u}{\partial z^2} \right) \tag{18.6.9}$$

where $\alpha^2 = h$, is known as *diffusivity*.

This equation is used in the study of the propagation of heat.

(b) One-dimensional equation for heat conduction is represented by

$$\frac{\partial u}{\partial t} = \alpha^2 \frac{\partial^2 u}{\partial x^2}. \tag{18.6.10}$$

18.7 Solution of one-dimensional wave equation by D'Alembert's method

We may obtain a general solution of (18.6.8) by a symbolic method in the following way: Let us write

$$D_x \equiv \frac{\partial}{\partial x} \quad \text{and} \quad D_x^2 \equiv \frac{\partial^2}{\partial x^2}.$$

Hence equation (18.6.8) may be written in the symbolic form

$$\frac{\partial^2 \xi}{\partial t^2} = c^2 \frac{\partial^2 \xi}{\partial x^2} = (cD_x)^2 \xi \tag{18.7.1}$$

where ξ is the displacement and x is the position at which the displacement is ξ.

Let

$$(cD_x) = a. \quad \therefore \frac{\partial^2 \xi}{\partial t^2} = a^2 \xi. \tag{18.7.2}$$

Here we shall treat the differential equation as an ordinary differential equation with constant coefficient.

The solution of (18.7.2) is given by $\xi = Ae^{at} + Be^{-at}$, where A, B are arbitrary constants. Thus the equation (18.7.2) is formally satisfied by $\xi = f_1(x)e^{cD_x t} + f_2(x)e^{-cD_x t}$; A and B have been taken to be the arbitrary functions of x as the integration has been done with respect to t.

Again,

$$e^{cD_x t} = e^{hD_x} = 1 + \frac{hD_x}{1!} + \frac{h^2 D_x^2}{2!} + \cdots \quad \text{where } ct = h.$$

Also,

$$f(x+h) = f(x) + \frac{h}{1!} D_x f(x) + \frac{h^2}{2!} D_x^2 f(x) + \cdots \quad \text{(from Taylor's series)}.$$

$$\therefore \quad f(x+h) = f(x) \cdot e^{hD_x}.$$

Thus we have, $f_1(x) e^{cD_x t} = f_1(x+ct)$.

Putting $h = -ct$, we have similarly $f_2(x) e^{-cD_x t} = f_2(x-ct)$.

Thus the general solution of one-dimensional wave equation is $\xi = f_1(x+ct) + f_2(x-ct)$.

The solution $\xi = f_2(x-ct)$ represents a wave travelling with velocity c in the positive direction along the X-axis. To prove this, let ξ' be the displacement at a point x' at time t'. At time $t''(t'' > t')$, let the displacement at x'' be ξ' so that the form of the waves remains unchanged. Then

$$f_2(x' - ct') = \xi' = f_2(x'' - ct'') \quad \text{or} \quad x' - ct' = x'' - ct'' \quad \text{or} \quad c = \frac{x'' - x'}{t'' - t'}.$$

Therefore, since $t'' > t'$, $x'' > x'$, i.e. the wave advances in the positive direction of the X-axis.

Similarly, the solution $\xi = f_1(x+ct)$ represents a wave travelling with velocity c in the negative direction along the X-axis.

18.7.1 Solution by the method of separation of variables or the Fourier method

Let us assume

$$\xi(x,t) = X(x)T(t) \tag{18.7.3}$$

where X is a function of x alone and T that of t alone. We have then

$$\frac{\partial^2 \xi}{\partial x^2} = T \frac{d^2 X}{dx^2} \quad \text{and} \quad \frac{\partial^2 \xi}{\partial t^2} = X \frac{d^2 T}{dt^2}.$$

Substituting these results in (18.7.1), we have

$$X \frac{d^2 T}{dt^2} = c^2 T \frac{d^2 X}{dx^2} \quad \text{or} \quad \frac{1}{X}\frac{d^2 X}{dx^2} = \frac{1}{c^2 T}\frac{d^2 T}{dt^2} \quad \text{(dividing throughout by } XTc^2).$$

Now that the variables are separated, we may write

$$\frac{1}{X}\frac{d^2 X}{dx^2} = \frac{1}{c^2 T}\frac{d^2 T}{dt^2} = \lambda \text{ (say)},$$

whence we get

$$\left.\begin{aligned} \frac{d^2 X}{dx^2} &= \lambda X \\ \text{and} \qquad \frac{d^2 T}{dt^2} &= c^2 \lambda T \end{aligned}\right\} \tag{18.7.4}$$

where λ is the constant of separation.

We now consider the following three cases:

(i) $\lambda = 0$, so that from (18.7.4) we have

$$\frac{d^2X}{dx^2} = 0, \quad \frac{d^2T}{dt^2} = 0$$

which give

$$X = Ax + B; \quad T = Ct + D. \tag{18.7.5}$$

(ii) $\lambda = \mu^2$, so that from (18.7.4), we have

$$\frac{d^2X}{dx^2} - \mu^2 X = 0; \quad \frac{d^2T}{dt^2} - \mu^2 c^2 T = 0$$

which give

$$\left.\begin{aligned} X &= Ae^{\mu x} + Be^{-\mu x} \\ T &= Ce^{\mu ct} + De^{-\mu ct} \end{aligned}\right\} \tag{18.7.6}$$

where A, B, C and D are arbitrary constants.

(iii) $\lambda = -\mu^2$, so that from (18.7.4), we have

$$\frac{d^2X}{dx^2} + \mu^2 X = 0, \quad \frac{d^2T}{dt^2} + \mu^2 c^2 T = 0$$

which give

$$\left.\begin{aligned} X &= A\cos\mu x + B\sin\mu x \\ T &= C\cos\mu ct + D\sin\mu ct \end{aligned}\right\}. \tag{18.7.7}$$

Let us now apply the boundary conditions

$$\xi(0, t) = 0 \tag{18.7.8}$$

$$\xi(l, t) = 0 \tag{18.7.9}$$

$$\xi(x, 0) = F(x) \tag{18.7.10}$$

$$\left(\frac{\partial \xi}{\partial t}\right)_{t=0} = \phi(x) \tag{18.7.11}$$

From (18.7.8) and (18.7.9), it follows that

$$\xi(0, t) = X(0)T(t) = 0 \quad [\because \xi(x, t) = X(x)T(t)]$$

$$\text{and} \quad \xi(l, t) = X(l)T(t) = 0.$$

Hence either $T(t) = 0$ or $X(0) = 0$ and $X(l) = 0$.

Thus from (18.7.5) when $x = 0$, we have, $B = 0$ and $X(l) = 0 = Al + B$. $\therefore A = 0$.

Also from (18.7.6) when $x = 0$, we have $X(0) = 0 = A + B$ and $X(l) = 0 = Ae^{\mu l} + Be^{-\mu l}$ which give $A = B = 0$.

In either case $A = B = 0$ gives $X(x) = 0$.

Hence $\xi(x,t) = X(x)T(t) = 0$.

Thus the solutions (18.7.5) and (18.7.6) fail and so the equations (18.7.4) give the solution (18.7.7). Substituting the expressions for $X(x)$ and $T(t)$ from (18.7.7) into (18.7.3) we have,

$$\xi(x,t) = (A\cos\mu x + B\sin\mu x)(C\cos\mu ct + D\sin\mu ct).$$

We now apply the conditions (18.7.8) and (18.7.9) to determine the constants A and B. We have then $\xi(0,t) = 0 = A(C\cos\mu ct + D\sin\mu ct)$. This gives $A = 0$.

Again $\xi(l,t) = 0 = B\sin\mu l(C\cos\mu ct + D\sin\mu ct)$. Hence $B\sin\mu l = 0$.

$$\therefore \quad \sin\mu l = 0 \quad [\because B \neq 0];$$

$$\text{or,} \quad \mu l = n\pi \quad \text{or,} \quad \mu = \frac{n\pi}{l}.$$

Thus the solution of (18.7.3) satisfying the boundary conditions (18.7.8) and (18.7.9), is given by

$$\xi_n(x,t) = \sin\frac{n\pi x}{l}\left(C_n\cos\frac{n\pi ct}{l} + D_n\sin\frac{n\pi ct}{l}\right). \tag{18.7.12}$$

The constant B is included in C_n and D_n.

Since the equation (18.7.1) is linear and homogeneous, the sum of the solutions is also a solution and hence the function represented by the series

$$\xi(x,t) = \sum_{n=1}^{\infty}\xi_n(x,t) = \sum_{n=1}^{\infty}\left(C_n\cos\frac{n\pi ct}{l} + D_n\sin\frac{n\pi ct}{l}\right)\sin\frac{n\pi x}{l}. \tag{18.7.13}$$

The above solution must satisfy the initial condition (18.7.10). Hence applying this initial condition in (18.7.13) we have,

$$\xi(x,0) = F(x) = \sum_{n=1}^{\infty} C_n\sin\frac{n\pi x}{l}. \tag{18.7.14}$$

The RHS of (18.7.14) being a Fourier expansion, we have

$$C_n = \frac{2}{l}\int_0^l F(x)\sin\frac{n\pi x}{l}\,dx. \tag{18.7.15}$$

Again from the initial condition (18.7.11), we have, since

$$\left(\frac{\partial\xi}{\partial t}\right) = \sum_{n=1}^{\infty}\sin\frac{n\pi x}{l}\left[-\frac{n\pi c}{l}C_n\sin\frac{n\pi ct}{l} + \frac{n\pi c}{l}D_n\cos\frac{n\pi ct}{l}\right],$$

$$\left(\frac{\partial\xi}{\partial t}\right)_{t=0} = \sum_{n=1}^{\infty} D_n\frac{n\pi c}{l}\sin\frac{n\pi x}{l} = \phi(x).$$

The Fourier coefficients of this series are given by

$$D_n \frac{n\pi c}{l} = \frac{2}{l} \int_0^l \phi(x) \sin \frac{n\pi x}{l}\, dx \quad \text{or,} \quad D_n = \frac{2}{n\pi c} \int_0^l \phi(x) \sin \frac{n\pi x}{l}\, dx. \tag{18.7.16}$$

Thus in the series (18.7.13), the coefficients C_n and D_n are determined, and we get the complete solution of equation (18.7.1). If we put $\phi(x) = 0$ (i.e., the initial velocity = 0), then $D_n = 0$ and hence equation (18.7.13) reduces to

$$\xi(x,t) = \sum_{n=1}^{\infty} C_n \cos \frac{n\pi ct}{l} \sin \frac{n\pi x}{l} = \sum_{n=1}^{\infty} \frac{C_n}{2} \sin \frac{n\pi}{l}(x - ct) + \sum_{n=1}^{n=\infty} \frac{C_n}{2} \sin \frac{n\pi}{l}(x + ct).$$

The above equation is of the form

$$\xi(x,t) = f(x - ct) + f(x + ct) \tag{18.7.17}$$

which is the solution of the one-dimensional wave equatiion (18.7.1).

18.8 Solution of two-dimensional wave equation in rectangular and polar co-ordinates

18.8.1 Transverse vibrations of a rectangular membrane (rectangular co-ordinates)

We consider the transverse vibrations of a uniform rectangular membrane. Here we shall consider a theoretical membrane which is a perfectly flexible and infinitely thin lamina of solid matter and of uniform thickness stretched in all directions by a tension so great that it remains sensibly unaltered during the vibrations. Hence we consider the tension τ and density ρ as constants. The free vibration of such a membrane is given by the differential equation

$$\frac{\partial^2 \xi}{\partial t^2} = c^2 \left(\frac{\partial^2 \xi}{\partial x^2} + \frac{\partial^2 \xi}{\partial y^2} \right) \quad \text{where } c^2 = \frac{\tau}{\rho}. \tag{18.8.1}$$

We shall now suppose that the boundary of the membrane is the rectangle formed by the co-ordinate axes and the lines $x = a$, $y = b$. For every point on the boundary $\xi = 0$, i.e., $\xi = 0$ for $x = 0$, $x = a$, $y = 0$, $y = b$.

Hence,

$$\xi(0,y,t) = 0, \quad \xi(a,y,t) = 0, \quad \xi(x,0,t) = 0, \quad \xi(x,b,t) = 0. \tag{18.8.2}$$

Also the initial displacement and velocity are given by

$$\xi(x,y,0) = F(x,y) \tag{18.8.3}$$

$$\text{and} \quad \dot{\xi}_t(x,y,0) = \left[\frac{\partial}{\partial t} \xi(x,y,t) \right]_{t=0} = \phi(x,y). \tag{18.8.4}$$

Let us assume that

$$\xi(x,y,t) = X(x)Y(y)T(t) \tag{18.8.5}$$

where X is the function dependent on x only, Y on y only and T on t only.

Equation (18.8.1) then gives

$$\frac{1}{X}\frac{d^2X}{dx^2} + \frac{1}{Y}\frac{d^2Y}{dy^2} = \frac{1}{c^2T}\frac{d^2T}{dt^2}. \tag{18.8.6}$$

Here the variables have been separated, and hence we may write choosing proper constants of separation:

(i) $\frac{1}{X}\frac{d^2X}{dx^2} = 0; \frac{1}{Y}\frac{d^2Y}{dy^2} = 0; \frac{1}{c^2T}\frac{d^2T}{dt^2} = 0$, which on solving give

$$\left.\begin{aligned} X &= A_1x + B_1 \\ Y &= A_2y + B_2 \\ T &= A_3t + B_3 \end{aligned}\right\}. \tag{18.8.7}$$

(ii) $\frac{1}{X}\frac{d^2X}{dx^2} = \lambda_1^2; \frac{1}{Y}\frac{d^2Y}{dy^2} = \lambda_2^2; \frac{1}{c^2T}\frac{d^2T}{dt^2} = \lambda^2$

[so that $\lambda^2 = (\lambda_1^2 + \lambda_2^2)c^2$], which on solving give

$$\left.\begin{aligned} X &= A_1e^{\lambda_1 x} + B_1e^{-\lambda_1 x} \\ Y &= A_2e^{\lambda_2 y} + B_2e^{-\lambda_2 y} \\ T &= A_3e^{c\lambda t} + B_3e^{-c\lambda t} \end{aligned}\right\}. \tag{18.8.8}$$

(iii) $\frac{1}{X}\frac{d^2X}{dx^2} = -\lambda_1^2; \frac{1}{y}\frac{d^2Y}{dy^2} = -\lambda_2^2; \frac{1}{c^2T}\frac{d^2T}{dt^2} = -\lambda^2$

[so that $\lambda^2 = (\lambda_1^2 + \lambda_2^2)c^2$], which on solving give

$$\left.\begin{aligned} X &= A_1\cos\lambda_1 x + B_1\sin\lambda_1 x, \\ Y &= A_2\cos\lambda_2 y + B_2\sin\lambda_2 y, \\ T &= A_3\cos\lambda ct + B_3\sin\lambda ct \end{aligned}\right\}. \tag{18.8.9}$$

Now from (18.8.2), it is evident that $X(0) = X(a) = 0; Y(0) = 0 = Y(b)$, so that the solutions (18.8.7) and (18.8.8) cannot give the solution of (18.8.1). However, in (18.8.9), the function $X(x)$ is a linear combination of $\sin\lambda_1 x$ and $\cos\lambda_1 x$ in which the co-sine term is to be rejected in view of the first condition $\xi(0, y, t) = 0$. We are thus left with $X(x) = B_1\sin\lambda_1 x = 0$. As $\xi(a, y, t) = 0$ gives $X(a) = 0$, so that $\sin\lambda_1 a = 0 = \sin m\pi$, m being a positive integer. Thus $\lambda_1 = \frac{m\pi}{a}$, and similarly $Y = B_2\sin\lambda_2 y = 0$ gives $\lambda_2 = \frac{n\pi}{b}$ for integral values of n.

Thus, the solution of (18.8.1) is of the form

$$(A\cos\lambda_{mn}t + B\sin\lambda_{mn}t)\sin\frac{m\pi x}{a}\sin\frac{n\pi y}{b} \tag{18.8.10}$$

where A and B are constants. The constants B_1, B_2, A_3, B_3 are included in A and B; λ_{mn} is given by

$$\lambda_{mn}^2 = \left(\frac{m\pi c}{a}\right)^2 + \left(\frac{n\pi c}{b}\right)^2. \tag{18.8.11}$$

Such quantities are termed as *eigen values* or *characteristic values* of the vibrating membranes.

Considering different constants A and B for each choice of m and n and their superposition we have the general solution of (18.8.1) given by

$$\xi(x, y, t) = \sum_{m=1}^{\infty}\sum_{n=1}^{\infty}(A_{mn}\cos\lambda_{mn}t + B_{mn}\sin\lambda_{mn}t)\sin\frac{m\pi x}{a}\sin\frac{n\pi y}{b}. \tag{18.8.12}$$

Now applying the condition (18.8.3), i.e., for $t = 0$, $\xi(x, y, 0) = F(x, y)$, we have

$$F(x, y) = \sum_{m=1}^{\infty}\sum_{n=1}^{\infty} A_{mn}\sin\frac{m\pi x}{a}\sin\frac{n\pi y}{b}. \tag{18.8.13}$$

This is called *Fourier double sine series.*

Values of A_{mn} and B_{mn}: Multiplying the above double series on both sides by $\sin\frac{m\pi x}{a}\sin\frac{n\pi y}{b}$ and then integrating over the rectangle from $x = 0$ to $x = a$ and from $y = 0$ to $y = b$, we have

$$A_{mn} = \frac{4}{ab}\int_0^a\int_0^b F(x, y)\sin\frac{m\pi x}{a}\sin\frac{n\pi y}{b}\,dx\,dy. \tag{18.8.14}$$

Again differentiating the equation (18.8.12) with respect to t and applying the condition (18.8.4), we have

$$\dot{\xi}(x, y, 0) = \phi(x, y) = \sum_{m=1}^{\infty}\sum_{n=1}^{\infty}\lambda_{mn}B_{mn}\sin\frac{m\pi x}{a}\sin\frac{n\pi y}{b}.$$

Multiplying the above expression on both sides by $\sin\frac{m\pi x}{a}\sin\frac{n\pi y}{b}$ and then integrating over the rectangle from $x = 0$ to $x = a$ and from $y = 0$ to $y = b$, we have

$$B_{mn} = \frac{4}{ab\,\lambda_{mn}}\int_0^a\int_0^b \phi(x, y)\sin\frac{m\pi x}{a}\sin\frac{n\pi y}{b}\,dx\,dy. \tag{18.8.15}$$

The frequencies of the natural vibrations or eigen values are found by ascribing different integral values to m and n in the expression

$$\frac{\lambda_{mn}}{2\pi} = \frac{c}{2}\sqrt{\frac{m^2}{a^2} + \frac{n^2}{b^2}} \quad \text{[obtained from equation (18.8.11)].} \tag{18.8.16}$$

In case of gravest mode or fundamental mode, i.e., the mode of lowest frequency, $m = n = 1$. Hence

$$\frac{\lambda_{11}}{2\pi} = \frac{c}{2ab}\sqrt{a^2 + b^2}.$$

If $a = b$, i.e., for a square membrane,

$$\frac{\lambda_{11}}{2\pi} = \frac{c}{a}\frac{1}{\sqrt{2}}. \tag{18.8.17}$$

18.8.2 Vibrations of a circular membrane (Polar co-ordinates)

In equation

$$\frac{\partial^2 \xi}{\partial t^2} = c^2 \left(\frac{\partial^2 \xi}{\partial x^2} + \frac{\partial^2 \xi}{\partial y^2} \right),$$

if we put $x = r\cos\theta$ and $y = r\sin\theta$, we get the polar form of the above equation, which is

$$\frac{\partial^2 \xi}{\partial t^2} = c^2 \left(\frac{\partial^2 \xi}{\partial r^2} + \frac{1}{r}\frac{\partial \xi}{\partial r} + \frac{1}{r^2}\frac{\partial^2 \xi}{\partial \theta^2} \right). \tag{18.8.18}$$

To obtain the solution of the above equation by the method of separation of variables, we assume that

$$\xi(r, \theta, t) = R(r)\Theta(\theta)T(t) \tag{18.8.19}$$

where R, Θ, T are respectively the functions of r, θ and t alone. From (18.8.19) we have,

$$\frac{\partial \xi}{dr} = \Theta T \frac{dR}{dr}; \quad \frac{\partial^2 \xi}{\partial r^2} = \Theta T \frac{d^2R}{dr^2};$$

$$\frac{\partial^2 \xi}{\partial \theta^2} = RT \frac{d^2\Theta}{d\theta^2}; \quad \frac{\partial^2 \xi}{\partial t^2} = R\Theta \frac{d^2T}{dt^2}.$$

Then equation (18.8.18) with these substitutions gives

$$\frac{1}{c^2T}\frac{d^2T}{d\theta^2} = \frac{1}{R}\frac{d^2R}{dr^2} + \frac{1}{rR}\frac{dR}{dr} + \frac{1}{r^2\Theta}\frac{d^2\Theta}{d\theta^2} = -\lambda^2 \text{ (say)}, \tag{18.8.20}$$

since the variables have been separated. To make the above equation independent of Θ, we put

$$\frac{1}{\Theta} \cdot \frac{d^2\Theta}{d\theta^2} = -n^2. \tag{18.8.21}$$

Hence the differential equation (18.8.20) becomes

$$\frac{1}{R}\frac{d^2R}{dr^2} + \frac{1}{rR} \cdot \frac{dR}{dr} - \frac{n^2}{r^2} = -\lambda^2 \quad \text{or,} \quad \frac{d^2R}{dr^2} + \frac{1}{r}\frac{dR}{dr} + \left(\lambda^2 - \frac{n^2}{r^2}\right) R = 0. \tag{18.8.22}$$

This gives the ordinary differential equation for R and the ordinary differential equations for Θ and T are given by

$$\frac{d^2\Theta}{d\theta^2} + n^2\Theta = 0 \tag{18.8.23}$$

$$\text{and} \quad \frac{d^2T}{dt^2} + c^2\lambda^2 T = 0. \tag{18.8.24}$$

The solution of (18.8.23) is $\Theta = Ke^{\pm in\theta}$, where $n = 0, 1, 2 \cdots$ and K is an arbitrary constant. Equation (18.8.22) is Bessel's equation and its general solution is

$$R(r) = A\,J_n(\lambda r) + BY_n(\lambda r) \quad A \text{ and } B \text{ being arbitrary constants.} \tag{18.8.25}$$

Thus we see that the solution of (18.8.22) involves two distinct functions $J_n(\lambda r)$ and $Y_n(\lambda r)$, each multiplied by respective arbitrary constants A and B. The Bessel's function $Y_n(\lambda r) \to \infty$ when $r \to 0$ and hence we take $B = 0$. So, the solution becomes $R(r) = A\,J_n(\lambda r)$.

Now the boundary condition requires that $\xi(a, \theta, t) = 0$ when $r = a$, 'a' being the radius of the circular membrane.

$$\therefore \quad R(a) = A\,J_n(\lambda a) = 0 \quad \text{or} \quad J_n(\lambda a) = 0.$$

This equation gives the admissible values of λ.

Now, the solution of (18.8.24) is $T = P\cos\lambda ct + Q\sin\lambda ct$.

Hence the general solution of equation (18.8.18) is

$$\xi(r, \theta, t) = \sum_{m=1}^{\infty}\sum_{n=1}^{\infty}(P_{mn}\cos\lambda_{mn}ct + Q_{mn}\sin\lambda_{mn}ct)e^{\pm in\theta}J_n(\lambda_{mn}r). \tag{18.8.26}$$

When $n = 0$, the equation giving the admissible values of λ is given by $J_0(\lambda a) = 0$. From a table of Bessel's functions we see that this equation is satisfied when λa has values 2.404, 5.520, 8.654, 11.792, 14.931, 18.071, 21.212, 24.353, 27.494 etc. Hence the fundamental angular frequency is given by $\omega_1 = \frac{2.404}{a}c$. The other possible frequencies are given by the above numbers multiplied by c/a.

18.9 Solution of one-dimensional Fourier equation for heat conduction by the method of separation of variables

A parabolic equation of the type

$$\frac{\partial^2 u}{\partial x^2} = \frac{1}{k}\frac{\partial u}{\partial t} \tag{18.9.1}$$

is said to be a one-dimensional *diffusion equation.* Here k is the diffusivity and $u(x, t)$ is the temperature at any point of a solid at time t. Since $u = u(x, t)$ we may suppose that

$$u = X(x)T(t), \tag{18.9.2}$$

where $X(x)$ is a function of x alone and $T(t)$ is a function of t alone.

From (18.9.2) we have

$$\frac{\partial^2 u}{\partial x^2} = T\frac{d^2X}{dx^2} \quad \text{and} \quad \frac{\partial u}{\partial t} = X\frac{dT}{dt}$$

Putting these values in (18.9.1), we have

$$\frac{1}{X}\frac{d^2X}{dx^2} = \frac{1}{kT}\frac{dT}{dt} \tag{18.9.3}$$

The variables having now been separated, we assume that the L.H.S of (18.9.3) is independent of t and the R.H.S. independent of x, so that each side of (18.9.3) can be equated to some constant known as *constant of separation.*

According as this constant is negative or positive, we arrive at the following two cases:

Case 1. Let this separation constant be $-n^2$. Then from (18.9.3) we have

$$\frac{1}{X}\frac{d^2X}{dx^2} = -n^2 \quad \text{or} \quad \frac{d^2X}{dx^2} + n^2X = 0 \tag{18.9.4}$$

$$\text{and} \quad \frac{1}{kT}\frac{dT}{dt} = -n^2 \quad \text{or} \quad \frac{dT}{dt} + n^2kT = 0 \tag{18.9.5}$$

(18.9.4) gives on integration $X = a\cos(nx + \delta)$, a and δ being arbitrary constants and (18.9.5) gives on integration $T = be^{-kn^2t}$, b being an arbitrary constant.

These two solutions when combined in accordance with (18.9.2), i.e., $u = X(x)T(t)$ yield

$$u = A\cos(nx + \delta)e^{-kn^2t} \tag{18.9.6}$$

where $A(= ab)$, n and δ are constants and may be determined from the boundary conditions.

Case 2. Let the separation constant be n^2, then from (18.9.3) we have

$$\frac{d^2X}{dx^2} - n^2X = 0 \quad \text{and} \quad \frac{dT}{dt} - n^2kT = 0,$$

which give on integration $X = a\cos h(nx + \delta)$ and $T = be^{kn^2t}$, so that the combined solution of (18.9.1) is

$$u = A\cos h(nx + \delta)e^{kn^2t}. \tag{18.9.7}$$

The solutions (18.9.6) and (18.9.7) can be expressed as

$$u = A_n\cos(nx + \delta_n)e^{-kn^2t} \tag{18.9.8}$$

$$\text{and} \quad u = A_n\cos h(nx + \delta_n)e^{kn^2t} \tag{18.9.9}$$

where the constants A_n and δ_n correspond to a particular value of n.

Since these are the solutions of (18.9.1) for all values of n, we may therefore take the summation over all values of n, and write the solution of (18.9.1) as

$$u(x,t) = \sum_{n=0}^{\infty} A_n\cos(nx + \delta_n)e^{-kn^2t} \tag{18.9.10}$$

$$\text{and} \quad u(x,t) = \sum_{n=0}^{\infty} A_n\cos h(nx + \delta_n)e^{kn^2t} \tag{18.9.11}$$

according as the separation constant is negative or positive.

N.B. Now applying the boundary conditions (i) $u = 0$ when $x = 0$ and (ii) $u = 0$ when $x = l$ for all values of t, to the solution given by (18.9.6), we have

$$u(o,t) = A\cos\delta e^{-kn^2t} = 0$$

$$\text{and} \quad u(l,t) = A\cos(nl + \delta)e^{-kn^2t} = 0,$$

so that $A\cos\delta = 0$ and $A\cos nl\cos\delta - A\sin nl\sin\delta = 0$, which yield $\delta = \frac{\pi}{2}$ and $n = \frac{\pi}{l}, \frac{2\pi}{l}, \frac{3\pi}{l}, \cdots$.

Here $n = \frac{\pi}{l}, \frac{2\pi}{l}, \frac{3\pi}{l}, \cdots$ are called *eigen values.* With $n = \frac{\pi}{l}$, the solution is

$$u(x,t) = A\sin\frac{\pi x}{l}e^{-k\frac{\pi^2}{l^2}t}.$$

18.10 Solution of three-dimensional Wave Equation

18.10.1 Cartesian Co-ordinates

The equation is

$$\nabla^2\xi = \frac{\partial^2\xi}{\partial x^2} + \frac{\partial^2\xi}{\partial y^2} + \frac{\partial^2\xi}{\partial z^2} = \frac{1}{c^2}\frac{\partial^2\xi}{\partial t^2}. \tag{18.10.1}$$

Let us consider, for solution of (18.10.1), the boundary conditions given by

(i) $\frac{\partial\xi}{\partial x} = 0$ when $x = 0, x = a$; (ii) $\frac{\partial\xi}{\partial y} = 0$ when $y = 0, y = a$;

(iii) $\frac{\partial\xi}{\partial z} = 0$ when $z = 0, z = a$; (iv) $\xi \neq 0$ at $t = 0$.

In order to solve (18.10.1) by the method of separation of variables, let us assume that

$$\xi(x, y, z, t) = X(x)Y(y)Z(z)T(t) \tag{18.10.2}$$

where X is a function of x alone, Y that of y alone, Z that of z alone and T that of t alone.

From (18.10.2) we have

$$\frac{\partial^2\xi}{\partial x^2} = YZT\frac{d^2X}{dx^2}; \quad \frac{\partial^2\xi}{\partial y^2} = XZT\frac{d^2Y}{dy^2};$$

$$\frac{\partial^2\xi}{\partial z^2} = XYT\frac{d^2Z}{dz^2}; \quad \text{and} \quad \frac{\partial^2\xi}{\partial t^2} = XYZ\frac{d^2T}{dt^2}.$$

Substituting these values in equation (18.10.1) and dividing both sides by $XYZT$, we get

$$\frac{1}{X}\frac{d^2X}{dx^2} + \frac{1}{Y}\frac{d^2Y}{dy^2} + \frac{1}{Z}\frac{d^2Z}{dz^2} = \frac{1}{c^2T}\frac{d^2T}{dt^2}. \tag{18.10.3}$$

The variables having been separated, each of the four terms in (18.10.3) should be constant, and as such we may write

$$\frac{1}{X}\frac{d^2X}{dx^2} = -\lambda_1^2 \quad \text{or,} \quad \frac{d^2X}{dx^2} + \lambda_1^2X = 0$$

which gives $X = A_1\cos\lambda_1x + B_1\sin\lambda_1x$ as solution; again

$$\frac{1}{Y}\frac{d^2Y}{dy^2} = -\lambda_2^2 \quad \text{or,} \quad \frac{d^2Y}{dy^2} + \lambda_2^2Y = 0$$

which gives $Y = A_2\cos\lambda_2y + B_2\sin\lambda_2y$ as solution; also

$$\frac{1}{Z}\frac{d^2Z}{dz^2} = -\lambda_3^2 \quad \text{or,} \quad \frac{d^2Z}{dz^2} + \lambda_3^2Z = 0,$$

which gives $Z = A_3\cos\lambda_3z + B_3\sin\lambda_3z$ as solution; and

$$\frac{1}{c^2T}\cdot\frac{d^2T}{dt^2} = -\lambda^2 \quad \text{or} \quad \frac{d^2T}{dt^2} + \lambda^2c^2T = 0$$

which gives $T = A_4\cos\lambda ct + B_4\sin\lambda ct$ as solution, where $\lambda^2 = \lambda_1^2 + \lambda_2^2 + \lambda_3^2$ and A_1, A_2, A_3, A_4, B_1, B_2, B_3, B_4 are arbitrary constants.

Hence the solution of (18.10.1) is

$$\xi(x,y,z,t) = (A_1\cos\lambda_1 x + B_1\sin\lambda_1 x)(A_2\cos\lambda_2 y + B_2\sin\lambda_2 y)$$
$$(A_3\cos\lambda_3 z + B_3\sin\lambda_3 z)(A_4\cos\lambda ct + B_4\sin\lambda ct). \quad (18.10.4)$$

This equation may also be written in the form

$$\xi(x,y,z,t) = Ae^{\pm i(\lambda_1 x+\lambda_2 y+\lambda_3 z+\lambda ct)}.$$

For real values of λ_1, λ_2, λ_3, the solution represents a plane wave moving in the direction of the vector $\vec{\lambda}$ $(\lambda_1, \lambda_2, \lambda_3)$.

Now, putting $\frac{\partial\xi}{\partial x} = 0$ when $x = 0$ we obtain $B_1 = 0$; again putting $\frac{\partial\xi}{\partial y} = 0$ when $y = 0$ we obtain $B_2 = 0$; also putting $\frac{\partial\xi}{\partial z} = 0$ when $z = 0$ we obtain $B_3 = 0$; also $\xi \neq 0$ at $t = 0$, but $\dot{\xi} = 0$ at $t = 0$—which gives $B_4 = 0$; hence, we have the solution given by

$$\begin{aligned}\xi(x,y,z,t) &= A_1\cos\lambda_1 x . A_2\cos\lambda_2 y . A_3\cos\lambda_3 z . A_4\cos\lambda ct \\ &= A\cos\lambda_1 x\cos\lambda_2 y\cos\lambda_3 z\cos\lambda ct. \quad (18.10.5)\end{aligned}$$

Now from (18.10.5), we have

$$\frac{\partial\xi}{\partial x} = -A\lambda_1\sin\lambda_1 x\cos\lambda_2 y\cos\lambda_3 z\cos\lambda ct.$$

The boundary condition (i) gives $\frac{\partial\xi}{\partial x} = 0$, when $x = a$; hence we have

$$\sin\lambda_1 a\cos\lambda_2 y\cos\lambda_3 z\cos\lambda ct = 0$$

i.e., $\sin\lambda_1 a = 0$ or $\lambda_1 a = n_1\pi$ or, $\lambda_1 = \frac{n_1\pi}{a}$, n_1 being an integer.

Similarly with the boundary conditions (ii) and (iii)

$\frac{\partial\xi}{\partial y} = 0$, when $y = a$ and $\frac{\partial\xi}{\partial z} = 0$, when $z = a$, we have

$$\lambda_2 = \frac{n_2\pi}{a} \quad\text{and}\quad \lambda_3 = \frac{n_3\pi}{a} \quad\text{and}\quad \lambda = \sqrt{\lambda_1^2+\lambda_2^2+\lambda_3^2} = \frac{\pi}{a}\left(n_1^2+n_2^2+n_3^2\right)^{1/2}.$$

Hence the expression (18.10.5) reduces to

$$\xi(x,y,z,t) = A\cos\frac{n_1\pi}{a}x\cos\frac{n_2\pi}{a}y\cos\frac{n_3\pi}{a}z\cos\left[\left(n_1^2+n_2^2+n_3^2\right)^{1/2}\frac{\pi ct}{a}\right].$$

So, the general solution is

$$\xi(x,y,z,t) = \sum_{n_1=1}^{\infty}\sum_{n_2=1}^{\infty}\sum_{n_3=1}^{\infty} A_{n_1n_2n_3}\cos\frac{n_1\pi}{a}x\cos\frac{n_2\pi}{a}y\cos\frac{n_3\pi}{a}z$$
$$\cos\left[\left(n_1^2+n_2^2+n_3^2\right)^{1/2}\frac{\pi ct}{a}\right]. \quad (18.10.6)$$

18.10.2 Spherical Polar Co-ordinates

The three-dimensional wave equation in cartesian co-ordinates given by (18.10.1) can be transformed to one in spherical polar co-ordinates (r, θ, ϕ) by using the transformation equations: $x = r\sin\theta\cos\phi$, $y = r\sin\theta\sin\phi$ and $z = r\cos\theta$—as has been done in transforming the cartesian form of Laplace's equation into one in spherical polar co-ordinates. The wave equation in spherical polar co-ordinates is of the form,

$$\frac{\partial^2\xi}{\partial r^2} + \frac{2}{r}\frac{\partial\xi}{\partial r} + \frac{1}{r^2}\frac{\partial^2\xi}{\partial\theta^2} + \frac{\cot\theta}{r^2}\frac{\partial\xi}{\partial\theta} + \frac{1}{r^2\sin^2\theta}\frac{\partial^2\xi}{\partial\phi^2} = \frac{1}{c^2}\cdot\frac{\partial^2\xi}{\partial t^2}. \tag{18.10.7}$$

As usual we assume

$$\xi(r,\theta,\phi,t) = R(r)\Theta(\theta)\Phi(\phi)T(t). \tag{18.10.8}$$

Hençe equation (18.10.7) takes the form

$$\frac{1}{R}\frac{d^2R}{dr^2} + \frac{2}{Rr}\frac{dR}{dr} + \frac{1}{r^2\Theta}\frac{d^2\Theta}{d\theta^2} + \frac{\cot\theta}{r^2\Theta}\frac{d\Theta}{d\theta} + \frac{1}{r^2\sin^2\theta}\frac{1}{\Phi}\frac{d^2\Phi}{d\phi^2} = \frac{1}{c^2T}\frac{d^2T}{dt^2}. \tag{18.10.9}$$

We may write

$$\left.\begin{aligned} \frac{1}{c^2T}\cdot\frac{d^2T}{dt^2} &= -\lambda^2 \\ \frac{1}{\Phi}\frac{d^2\Phi}{d\phi^2} &= -k^2 \end{aligned}\right\} \tag{18.10.10}$$

and

$$\frac{1}{\Theta}\frac{d^2\Theta}{d\theta^2} + \frac{\cot\theta}{\Theta}\frac{d\Theta}{d\theta} = \frac{k^2}{\sin^2\theta} - n(n+1) \tag{18.10.11}$$

since the variables are separated. Hence equation (18.10.9) reduces to

$$\frac{1}{R}\cdot\frac{d^2R}{dr^2} + \frac{2}{Rr}\frac{dR}{dr} + \frac{k^2}{r^2\sin^2\theta} - \frac{n(n+1)}{r^2} - \frac{k^2}{r^2\sin^2\theta} = -\lambda^2$$

$$\text{or,}\quad \frac{1}{R}\frac{d^2R}{dr^2} + \frac{2}{Rr}\frac{dR}{dr} = \frac{n(n+1)}{r^2} - \lambda^2. \tag{18.10.12}$$

The solutions of (18.10.10) are $T = e^{\pm i\lambda ct}$; $\Phi = e^{\pm ik\phi}$.

Also equation (18.10.11) may be written as

$$\frac{d^2\Theta}{d\theta^2} + \cot\theta\frac{d\Theta}{d\theta} + \left[n(n+1) - \frac{k^2}{\sin^2\theta}\right]\Theta = 0$$

$$\text{or,}\quad \frac{1}{\sin\theta}\left[\sin\theta\frac{d^2\Theta}{d\theta^2} + \cos\theta\frac{d\Theta}{d\theta}\right] + \left[n(n+1) - \frac{k^2}{\sin^2\theta}\right]\Theta = 0$$

$$\text{or,}\quad \frac{1}{\sin\theta}\cdot\frac{d}{d\theta}\left(\sin^2\theta\frac{d\Theta}{d\theta}\right) + \left[n(n+1) - \frac{k^2}{\sin^2\theta}\right]\Theta = 0. \tag{18.10.13}$$

Now, by putting $\cos\theta = \mu$, we have shown (in Art. 18.4.1) that

$$\frac{1}{\sin\theta}\cdot\frac{d}{d\theta} \equiv -\frac{d}{d\mu}.$$

Hence the equation (18.10.13) reduces to

$$\frac{d}{d\mu}\left[(1-\mu^2)\frac{d\Theta}{d\mu}\right]+\left[n(n+1)-\frac{k^2}{\sin^2\theta}\right]\Theta=0.$$

This is Legendre's *associated equation* and its solution is

$$\Theta = A_n P_n^k(\mu) + B_n Q_n^k(\mu) = A_n P_n^k(\cos\theta) + B_n Q_n^k(\cos\theta).$$

If the solution is to be finite when $\mu \to 0, B_n$ must be zero and hence the above solution can be written as $\Theta = A_n P_n^k(\cos\theta)$.

Now, the equation (18.10.12) may be written in the form

$$\frac{1}{R}\frac{d^2R}{dr^2}+\frac{2}{Rr}\frac{dR}{dr}+\lambda^2-\frac{n(n+1)}{r^2}=0 \quad \text{or,} \quad \frac{d^2R}{dr^2}+\frac{2}{r}\frac{dR}{dr}+\left\{\lambda^2-\frac{n(n+1)}{r^2}\right\}R=0.$$

If we put $R = r^{-\frac{1}{2}}S(r)$, this equation is reduced to Bessel's equation of the form

$$\frac{d^2S}{dr^2}+\frac{1}{r}\frac{dS}{dr}+\left\{\lambda^2-\frac{(n+\frac{1}{2})^2}{r^2}\right\}S=0. \tag{18.10.14}$$

Hence S, i.e., R is determined.

Thus R, Θ, Φ and T are determined. Substituting these values in equation (18.10.8) we get the solution of the wave equation in spherical polar co-ordinates.

18.10.3 Cylindrical Polar Co-ordinates

The cartesian form of the three-dimensional wave equation given by (18.10.1) can be transformed to one in cylindrical polar co-ordinates (r, θ, z) by using the transformation equations $x = r\cos\theta, y = r\sin\theta$ and $z = z$, as has been done in the case of the transformation of the cartesian form of Laplace's equation into one in cylindrical polar co-ordinates.

Cylindrical polar form of the three-dimensional wave equation is

$$\frac{\partial^2\xi}{\partial r^2}+\frac{1}{r}\frac{\partial\xi}{\partial r}+\frac{1}{r^2}\frac{\partial^2\xi}{\partial\theta^2}+\frac{\partial^2\xi}{\partial z^2}=\frac{1}{c^2}\frac{\partial^2\xi}{\partial t^2}. \tag{18.10.15}$$

In order to solve the above equation by the method of separation of variables, let us assume that

$$\xi(r,\theta,z,t) = R(r)\Theta(\theta)Z(z)T(t). \tag{18.10.16}$$

From (18.10.16) we have

$$\frac{\partial^2\xi}{\partial r^2}=\Theta ZT\frac{d^2R}{dr^2}; \quad \frac{\partial^2\xi}{\partial\theta^2}=RZT\frac{d^2\Theta}{d\theta^2}; \quad \frac{\partial^2\xi}{\partial z^2}=R\Theta T\frac{d^2Z}{dz^2};$$

$$\frac{\partial^2\xi}{\partial t^2}=R\Theta Z\frac{d^2T}{dt^2}; \quad \frac{\partial\xi}{\partial r}=\Theta ZT\frac{dR}{dr}.$$

Substituting these values in (18.10.15) and then dividing both sides by $R\Theta ZT$, we get

$$\frac{1}{R}\left(\frac{d^2R}{dr^2}+\frac{1}{r}\frac{dR}{dr}\right)+\frac{1}{r^2\Theta}\frac{d^2\Theta}{d\theta^2}+\frac{1}{Z}\frac{d^2Z}{dz^2}=\frac{1}{c^2T}\frac{d^2T}{dt^2} \tag{18.10.17}$$

in which the variables have been separated.

Let us put

$$\left.\begin{aligned}\frac{1}{\Theta}\frac{d^2\Theta}{d\theta^2}&=-k^2\\ \frac{1}{Z}\frac{d^2Z}{dz^2}&=-k_1^2\\ \frac{1}{T}\frac{d^2T}{dt^2}&=-c^2\lambda^2\end{aligned}\right\}. \tag{18.10.18}$$

$\therefore$ equation (18.10.17) becomes

$$\frac{1}{R}\left(\frac{d^2R}{dr^2}+\frac{1}{r}\frac{dR}{dr}\right)-\frac{k^2}{r^2}=k_1^2-\lambda^2=-\alpha^2 \text{ (say)}$$

$$\text{or}\quad \frac{d^2R}{dr^2}+\frac{1}{r}\frac{dR}{dr}+\left(\alpha^2-\frac{k^2}{r^2}\right)R=0 \tag{18.10.19}$$

which is Bessel's equation of order k and argument αr.

On putting $\alpha r = x$, the above equation takes the form

$$\frac{d^2R}{dx^2}+\frac{1}{x}\frac{dR}{dx}+\left(1-\frac{k^2}{x^2}\right)R=0. \tag{18.10.20}$$

Solutions of (18.10.18) and (18.10.19) are

$$\Theta = A_1\cos k\theta + B_1\sin k\theta;\quad Z = A_2\cos k_1z + B_2\sin k_1z;$$

$$T = A_3\cos\lambda ct + B_3\sin\lambda ct;\quad R = A_4J_k(\alpha r)+B_4Y_k(\alpha r) = A_4J_k(\alpha r)$$

$$[\because\ Y_k(\alpha r)\to\infty \text{ when } r\to 0,\quad \therefore\quad B_4=0]$$

Hence the complete solution is

$$\begin{aligned}\xi(r,\theta,z,t) &= (A_1\cos k\theta + B_1\sin k\theta)(A_2\cos k_1z + B_2\sin k_1z)\\ &\quad (A_3\cos\lambda ct + B_3\sin\lambda ct)A_4J_k(\alpha r).\end{aligned} \tag{18.10.21}$$

Illustrative Examples

1. Show that the wave equation

$$\frac{\partial^2u}{\partial t^2}=c^2\left(\frac{\partial^2u}{\partial x^2}+\frac{\partial^2u}{\partial y^2}+\frac{\partial^2u}{\partial z^2}\right)$$

may be written in the form

$$\frac{\partial^2}{\partial t^2}(ru)=c^2\frac{\partial^2}{\partial r^2}(ru).$$

Hence show that the general solution is of the form

$$u = \frac{1}{r}\{f(r-ct)+g(r+ct)\}.$$ **[C.U. (Hons.) 2002]**

We know that

$$\text{div grad } u = \nabla^2 u = \frac{\partial^2 u}{\partial x^2}+\frac{\partial^2 u}{\partial y^2}+\frac{\partial^2 u}{\partial z^2}.$$

Again

$$\text{grad } u(r) = \frac{\vec{r}}{r}\frac{\partial u}{\partial r},$$

the function u depending on the distance r only from the origin.

$$\therefore \quad \text{div grad } u(r) = \text{div grad } \frac{\vec{r}}{r}\frac{\partial u}{\partial r} = \text{div grad } (\phi\vec{r}), \quad \text{where } \phi = \frac{1}{r}\frac{\partial u}{\partial r}$$

$$\text{or, div grad } u(r) = (\vec{\nabla}\phi)\cdot\vec{r}+\phi(\vec{\nabla}\cdot\vec{r}) = \vec{\nabla}\left(\frac{1}{r}\frac{\partial u}{\partial r}\right)\cdot\vec{r}+\left(\frac{1}{r}\frac{\partial u}{\partial r}\right)(\vec{\nabla}\cdot\vec{r})$$

$$= \frac{\vec{r}\cdot\vec{r}}{r}\left(\frac{1}{r}\frac{\partial^2 u}{\partial r^2}-\frac{1}{r^2}\frac{\partial u}{\partial r}\right)+\frac{3}{r}\frac{\partial u}{\partial r} \quad \left[\because \quad \vec{\nabla}\cdot\vec{r}=3\right]$$

$$= \frac{r^2}{r}\left(\frac{1}{r}\frac{\partial^2 u}{\partial r^2}-\frac{1}{r^2}\frac{\partial u}{\partial r}\right)+\frac{3}{r}\frac{\partial u}{\partial r} = \frac{\partial^2 u}{\partial r^2}+\frac{2}{r}\frac{\partial u}{\partial r}.$$

Hence the wave equation becomes

$$c^2\left(\frac{\partial^2 u}{\partial r^2}+\frac{2}{r}\frac{\partial u}{\partial r}\right) = \frac{\partial^2 u}{\partial t^2}.$$

This equation may be written in the form

$$c^2\frac{\partial^2}{\partial r^2}(ru) = \frac{\partial^2}{\partial t^2}(ru) \quad \text{or} \quad \frac{\partial^2}{\partial t^2}(ru) = (c\,D_r)^2 ru,$$

where

$$D_r \equiv \frac{\partial}{\partial r} \quad \text{and} \quad D_r^2 \equiv \frac{\partial^2}{\partial r^2}.$$

The above equation is of the form

$$\frac{\partial^2}{\partial t^2}(ru) = a^2(ru) \quad [\text{putting } a = (cD_r)].$$

Proceeding as in Art. 18.7.1, the solution of the above equation is

$$ru = f(r-ct)+g(r+ct) \quad \text{or,} \quad u = \frac{1}{r}\{f(r-ct)+g(r+ct)\}.$$

This gives the general solution of the spherical wave equation. The solution gives two spherical waves, one moving away from the origin and the other towards the origin. The term $\frac{1}{r}$ in the solution shows that the amplitude diminishes linearly with the distance of the outward moving spherical wave.

2. Write down the three-dimensional equation of heat flow through a medium of thermal conductivity k. Show that under steady state condition, this reduces to the Laplace's equation $\nabla^2 u = 0$ where $u(r)$ denotes the temperature. Using spherical polar co-ordinates, verify that this has the solution of the form $u(r) = Q/4\pi kr + B$ where Q is the rate at which heat flows out isothermally from a point source at $r = 0$ and B is a constant. If the temperatures on the two concentric spheres of radii r_1 and r_2 with the point source at the centre be u_1 and u_2 respectively, show that $Q = \dfrac{4\pi k(u_1 - u_2)r_1 r_2}{r_2 - r_1}$.

[C.U. (Hons.) 1980]

The three-dimensional equation of heat flow through a medium is represented by

$$\frac{\partial u}{\partial t} = \alpha^2\left(\frac{\partial^2 u}{\partial x^2} + \frac{\partial^2 u}{\partial y^2} + \frac{\partial^2 u}{\partial z^2}\right) \quad \text{where } \alpha^2 = h = \frac{k}{\rho s},$$

k, ρ, s being respectively the thermal conductivity, density, specific heat and diffusivity of the medium.

In the steady state condition,

$$\frac{\partial u}{\partial t} = 0; \quad \therefore \quad \frac{\partial^2 u}{\partial x^2} + \frac{\partial^2 u}{\partial y^2} + \frac{\partial^2 u}{\partial z^2} = \nabla^2 u = 0$$

which is Laplace's equation; and may be written, using the spherical polar co-ordinates (r, θ, ϕ), as

$$r^2\frac{\partial^2 u}{\partial r^2} + 2r\frac{\partial u}{\partial r} + \frac{1}{\sin\theta}\frac{\partial}{\partial\theta}\left(\sin\theta\frac{\partial u}{\partial\theta}\right) + \frac{1}{\sin^2\theta}\frac{\partial^2 u}{\partial\phi^2} = 0. \tag{1}$$

In the present problem, heat flows radially from the source at the centre, and equation (1) will be independent of θ and ϕ.

So equation (1) reduces to

$$r^2\frac{d^2 u}{dr^2} + 2r\frac{du}{dr} = 0 \quad \text{or} \quad \frac{d}{dr}\left(r^2\frac{du}{dr}\right) = 0.$$

Integrating we get, $r^2\dfrac{du}{dr} = A$, A being the constant of integration or $du = A\dfrac{dr}{r^2}$.

On integration, we have

$$u = -\frac{A}{r} + B, \tag{2}$$

B being another constant of integration.

Now,

$$Q = -k \cdot 4\pi r^2\frac{du}{dr} = -k \cdot 4\pi r^2 \cdot \frac{A}{r^2} = -4\pi kA \quad \text{or} \quad A = -\frac{Q}{4\pi k}. \tag{3}$$

From (2),

$$u = \frac{Q}{4\pi kr} + B.$$

Let the values of u at $r = r_1$ and $r = r_2$ be u_1 and u_2 respectively.

$\therefore$ from (2),

$$u_1 = -\frac{A}{r_1} + B \quad \text{and} \quad u_2 = -\frac{A}{r_2} + B \quad \text{or} \quad u_1 - u_2 = A\left(\frac{1}{r_2} - \frac{1}{r_1}\right) = A\frac{r_1 - r_2}{r_1 r_2}$$

$$\text{or} \quad Q = \frac{4\pi k(u_1 - u_2)r_1 r_2}{r_2 - r_1} \quad \text{[putting the value of } A \text{ from (3)].}$$

3. Using Laplace's equation in cylindrical co-ordinates, solve the problem of cylindrical heat flow, and find an expression for the temperature at any distance r from the axis.

The three dimensional equation of heat flow through a medium is given by

$$\frac{\partial u}{\partial t} = \alpha^2\left(\frac{\partial^2 u}{\partial x^2} + \frac{\partial^2 u}{\partial y^2} + \frac{\partial^2 u}{\partial z^2}\right)$$

where $\alpha^2 = h = \frac{k}{\rho s}$, k, ρ, s and h being respectively the thermal conductivity, density, specific heat and diffusivity of the medium.

In the steady state,

$$\frac{\partial u}{\partial t} = 0 \quad \therefore \quad \frac{\partial^2 u}{\partial x^2} + \frac{\partial^2 u}{\partial y^2} + \frac{\partial^2 u}{\partial z^2} = 0$$

which is Laplace's equation in Cartesian co-ordinates. This may be written, using the cylindrical co-ordinates (r, θ, z), as

$$\frac{\partial^2 u}{\partial r^2} + \frac{1}{r}\frac{\partial u}{\partial r} + \frac{1}{r^2}\frac{\partial^2 u}{\partial \theta^2} + \frac{\partial^2 u}{\partial z^2} = 0. \tag{1}$$

In the present problem, heat flows radially from the source placed along the axis, and therefore equation (1) will be independent of θ and z. So equation (1) reduces to

$$\frac{d^2 u}{dr^2} + \frac{1}{r}\frac{du}{dr} = 0 \quad \text{or} \quad \frac{1}{r}\frac{d}{dr}\left(r\frac{du}{dr}\right) = 0.$$

Integrating we get $r\frac{du}{dr} = A$, A being the constant of integration

$$\therefore \quad \frac{du}{dr} = \frac{A}{r}. \tag{2}$$

Now the amount of heat Q flowing per second across an isothermal surface is (from the definition of thermal conductivity) given by

$$Q = -2\pi r l k\frac{du}{dr} = -2\pi l k A \quad \text{[using (2)].} \tag{3}$$

To evaluate A, we are to integrate (2).

$$\therefore \quad \int du = A\int \frac{dr}{r} + B \quad \text{or} \quad u = A\log_e r + B \tag{4}$$

B being another integration constant.

When the steady state is attained, let $u = u_1$ at $r = r_1$ and $u = u_2$ at $r = r_2$.

$\therefore$ from (4)

$$u_1 = A\log_e r_1 + B \tag{5}$$

$$\text{and} \quad u_2 = A\log_e r_2 + B \tag{6}$$

so that from (5) and (5), $u_2 - u_1 = A\log_e \frac{r_2}{r_1}$

$$\therefore \quad A = -\frac{(u_1 - u_2)}{\log_e(r_2/r_1)}. \tag{7}$$

Multiplying (5) by $\log_e r_2$ and (6) by $\log_e r_1$, subtracting one from the other, and then on simplification, we have

$$B = \frac{u_1 \log_e r_2 - u_2 \log_e r_1}{\log_e(r_2/r_1)}. \tag{8}$$

From (3) and (7) we have

$$Q = 2\pi lk \cdot \frac{(u_1 - u_2)}{\log_e(r_2/r_1)}.$$

Again substituting the values of A from (7) and B from (8) in (4), the temperature distribution along the radial direction is given by

$$u = -\frac{(u_1 - u_2)}{\log_e(r_2/r_1)}\log_e r + \frac{u_1 \log_e r_2 - u_2 \log_e r_1}{\log_e(r_2/r_1)}$$

$$= \frac{1}{\log_e(r_2/r_1)}\left[u_1 \log_e r_2 - u_2 \log_e r_1 - (u_1 - u_2)\log_e r\right].$$

4. Write down the Laplace's equation on a plane in terms of the polar co-ordinates (r, θ). Solve it by the method of separation of variables and write the general expression of the solution which is finite at $r = 0$ and single valued in θ. [C.U. (Hons.) 1983, '85, '91, '93]

Two dimensional Cartesian form of Laplace's equation is

$$\frac{\partial^2 u}{\partial x^2} + \frac{\partial^2 u}{\partial y^2} = 0$$

which can be transformed to the polar form, [see Art. 18.3.1] as

$$\frac{\partial^2 u}{\partial r^2} + \frac{1}{r}\frac{\partial u}{\partial r} + \frac{1}{r^2}\frac{\partial^2 u}{\partial \theta^2} = 0. \tag{1}$$

Equation (1) may also be written as

$$\frac{1}{r}\frac{\partial}{\partial r}\left(r\frac{\partial u}{\partial r}\right) + \frac{1}{r^2}\frac{\partial^2 u}{\partial \theta^2} = 0. \tag{2}$$

In order to solve equation (2) by the method of separation of variables, let us assume that

$$u(r, \theta) = R(r)\Theta(\theta) \tag{3}$$

R and Θ being functions of r and θ alone.

From (3), we have

$$\frac{\partial u}{\partial r}=\Theta\frac{dR}{dr},\quad \frac{\partial^2 u}{\partial r^2}=\Theta\frac{d^2R}{dr^2}\quad\text{and}\quad \frac{\partial^2 u}{\partial \theta^2}=R\frac{d^2\Theta}{d\theta^2}.$$

Substituting these values in (2), we may write

$$\frac{1}{r}\frac{\partial}{\partial r}\left(r\Theta\frac{dR}{dr}\right)+\frac{1}{r^2}R\frac{d^2\Theta}{d\theta^2}=0\quad\text{or}\quad \frac{\Theta}{r}\frac{dR}{dr}+\Theta\frac{d^2R}{dr^2}+\frac{R}{r^2}\frac{d^2\Theta}{d\theta^2}=0$$

$$\text{or}\quad \frac{1}{Rr}\frac{dR}{dr}+\frac{1}{R}\frac{d^2R}{dr^2}+\frac{1}{r^2\Theta}\frac{d^2\Theta}{d\theta^2}=0\quad\text{(dividing both sides by } R\Theta)$$

$$\text{or}\quad \frac{1}{R}\left(r^2\frac{d^2R}{dr^2}+r\frac{dR}{dr}\right)=-\frac{1}{\Theta}\frac{d^2\Theta}{d\theta^2}\quad\text{(multiplying both sides by } r^2 \text{ and on simplification)} \tag{4}$$

Since the left hand side of equation (4) is a function of r alone and the right hand side is a funciton of θ alone, we conclude that each should be equal to a quantity independent of (r,θ) and let this constant be n^2, so that we have

$$\frac{1}{R}\left(r^2\frac{d^2R}{dr^2}+r\frac{dR}{dr}\right)=n^2 \tag{5}$$

$$\text{and}\quad \frac{1}{\Theta}\frac{d^2\Theta}{d\theta^2}=-n^2. \tag{6}$$

The variables are now separated. Considering (5) we may write

$$\frac{d^2R}{dr^2}+\frac{1}{r}\frac{dR}{dr}-\frac{n^2}{r^2}R=0$$

whose solution is

$$R=Ar^n+Br^{-n}\ \text{if}\ n\neq 0. \tag{7}$$

This is the radial part. A and B are arbitrary constants.

Again taking (6) we may write

$$\frac{d^2\Theta}{d\theta^2}+n^2\Theta=0$$

which is the well-known *equation of simple harmonic motion.* Its solution is

$$\Theta=C\cos n\theta+D\sin n\theta\ \text{if}\ n\neq 0. \tag{8}$$

This is the θ-part. C and D are arbitrary constants.

If $n=0$, then the corresponding solutions are given by $R=A_0\log_e r+B_0$ and $\Theta=C_0\theta+D_0$ where A_0, B_0, C_0, D_0 are arbitrary constants.

n is called the *degree of harmonic,* and these solutions of Laplace's equation are called *circular harmonics.* Circular harmonics are of great importance in the solution of two dimensional problems having cylindrical symmetry. The circular harmonics are then

$$\left.\begin{aligned} u_0 &= (A_0\log_e r+B_0)(C_0\theta+D_0)\quad \text{for degree zero}\\ u_n &= (A_nr^n+B_nr^{-n})(C_n\cos n\theta+D_n\sin n\theta)\quad \text{for degree } n\end{aligned}\right\}. \tag{9}$$

In general

$$u = \sum R_n \Theta_n$$

since the sum of the above solutions for different n with different constants is also a solution of Laplace's equation.

5. An infinite uncharged conducting cylinder of radius 'a' is set with its axis (in Fig. along Z-axis) perpendicular to a uniform electric field (along X-axis) of strength $\vec{E}$. Find the potential in the region out-side the cylinder. **[C.U. (Hons.) 1989]**

This problem can be solved by the use of circular harmonics as shown below.

Due to introduction of the cylinder, there will be local distortion of the field. The boundary conditions of the problem are:

(i) $V = 0$ at $r = a$ ($\because$ the cylinder is perfectly conducting, it is an equipotential surface, and we may take its potential as zero).

(ii) $V = -\vec{E} \cdot \vec{r} = -Er\cos\theta$ at $r \to \infty$ ($\because$ the potential will remain finite at infinity and equal to that due to the original field).

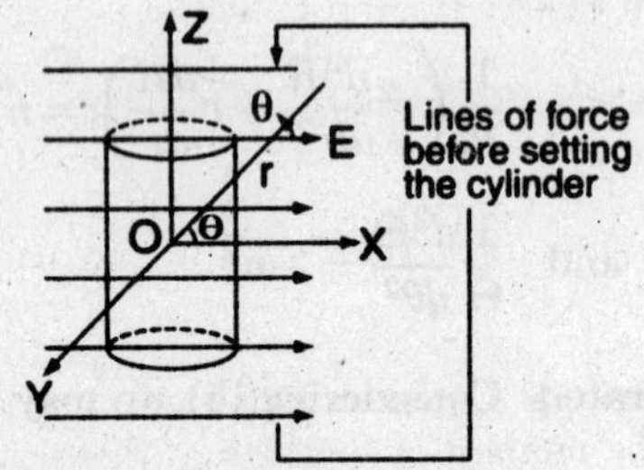

Here V is not a function of z but a funciton of (r, θ), and must satisfy Laplace's equation, which is of the form

$$\frac{1}{r}\frac{\partial}{\partial r}\left(r\frac{\partial V}{\partial r}\right) + \frac{1}{r^2}\frac{\partial^2 V}{\partial \theta^2} = 0. \quad (1)$$

The general solution of (1) is

$$V = \sum (A_n r^n + B_n r^{-n}))C_n \cos n\theta + D_n \sin n\theta). \quad (2)$$

From conditions (i) and (ii), we may write

$$0 = \sum (A_n a^n + B_n a^{-n})(C_n \cos n\theta + D_n \sin n\theta) \quad (3)$$

$$\text{and} \quad -Er\cos\theta = \sum A_n r^n (C_n \cos n\theta + D_n \sin n\theta). \quad (4)$$

From (4) it follows that A_n should be zero except for $n = 1$ and $D_n = 0$ and for that we should have

$$A_1 C_1 = -E. \quad (5)$$

Again from (3)

$$A_n C_n a^n + B_n C_n a^{-n} = 0 \quad \text{or} \quad B_n C_n = -A_n C_n a^{2n}. \quad (6)$$

We have seen that A_n should be zero except for $n = 1$; similarly it can be shown that $B_n = 0$ except for $n = 1$, and for that we should have from (5) and (6)

$$B_1C_1 = -A_1C_1a^2 = Ea^2. \tag{7}$$

Hence from (2) we may write

$$V = (A_1r + B_1r^{-1})(C_1\cos\theta) = (A_1C_1r + B_1C_1r^{-1})\cos\theta$$

$$= \left(-Er + \frac{Ea^2}{r}\right)\cos\theta \quad \text{[using (5), (6) and (7)]}$$

$$= -Er\left(1 - \frac{a^2}{r^2}\right)\cos\theta.$$

Exercises

1. Deduce Laplace's equation in case of electric potential in the steady flow of electric current.

2. How would you transform Cartesian Laplace's equation to: (a) the spherical polar form and (b) the cylindrical form?

3. Apply the method of separation of variables to the solution of three-dimensional Laplace's equation in cartesian co-ordinates.

4. Write down the three-dimensional Laplace's equation in cylindrical co-ordinates. How would you solve the equation by the method of separation of variables? What will be the solution if $\lambda = n = 0$?

   Calculate the electrostatic potential at a point due to a uniformly charged cylinder of infinite length.

5. Solve Laplace's equation expressed in spherical polar co-ordinates. What will be the general solution if there is axial symmetry about Z-axis?

6. Write down Laplace's equation in two dimensional rectangular co-ordinates. What is the general solution of the equation?

7. Explain the meaning of a perfect or a complete differential. Deduce one of Maxwell's four important relations.

8. Write down the wave equation in one dimension in rectangular co-ordinates. Solve the equation by D' Alembert's method or by the method of separation of variables. Interpret the solution.

9. Write down the two-dimensional wave equation in rectangular co-ordinates. Hence explain the vibration of a rectangular membrane and calculate the eigen values of the vibrating membrane.

10. Write down the two-dimensional wave equation in polar co-ordinates, and solve it to find the eigen values and eigen functions in case of a circular membrane.

11. Write down the three-dimensional wave equation in cartesian co-ordinates. Show that the solution represents a plane wave.

12. Write down the three-dimensional wave equations in spherical co-ordinates and cylindrical co-ordinates and find out their solutions.

13. A function $V(x, y, z)$ of the Cartesian Coordinates (x, y, z) of a point satisfies the equation $\nabla^2 V = 0$. Obtain possible solutions of this equation by the method of separation of variables. Write down a solution satisfying the condition $V(x, y, z = 0) = 0$ and $V \neq 0$ at points for which $z \neq 0$. **[C.U. (Hons.) 1982]**

[*Hints.* The solution is obtained from (18.3.7) and is written as

$$V = (A_1 \cos nx + A_2 \sin nx)(A_3 \cos my + A_4 \sin my)(Ee^{pz} + Fe^{-pz})$$

Now $V_{(x,y,z=0)} = 0 \quad \therefore\ E + F = 0$ or $E = -F$

$$\therefore\ V = E(e^{pz} - e^{-pz})(A_3 \cos my + A_4 \sin my)(A_1 \cos nx + A_2 \sin nx)$$

showing that $V \neq 0$ at points for which $z \neq 0$.]

14. A function $\phi(x, t)$ satisfies the wave equation where x is the Cartesion co-ordinate of a field point and t is the time. Write down the differential equation. Obtain possible solutions of this equation by the method of separation of variables. Consider only those solutions which vary simple harmonically with time. **[C.U. (Hons.) 1982]**

[*Hints.* The solution is obtained from (18.7.7) and is written as

$$\phi(x, t) = (A \cos \mu x + B \sin \mu x)(C \cos \mu ct + D \sin \mu ct)$$

which with appropriate substitutions, may be written as

$$\phi = a \sin \mu c \left(t \pm \frac{x}{c}\right) \quad \text{or} \quad a \cos \mu c \left(t \pm \frac{x}{c}\right)$$

these obviously represent solutions which vary simple harmonically with time.]

15. A rectangular stretched membrane of sides a and b, having its edges parallel to the X and Y axes and bounded rigidly at the edges is given a slight deformation in the Z-direction perpendicular to its plane. The differential equation for z is $\dfrac{\partial^2 z}{\partial t^2} = c^2 \left(\dfrac{\partial^2 z}{\partial x^2} + \dfrac{\partial^2 z}{\partial y^2}\right)$ where c is a constant. Solve the equation by the method of separation of variables, assuming the initial conditons $z = f(x, y)$ and $\dfrac{dz}{dt} = 0$ at $t = 0$. **[C.U. (Hons.) 1981]**

(See Art. 18.8.1)

16. Suppose that the following differential equation refers to a problem of two dimensional steady flow of heat $\dfrac{\partial^2 T}{\partial x^2} + \dfrac{\partial^2 T}{\partial y^2} = 0$.

Solve for T with the boundary conditons

$$T(o, y) = 0, T(a, y) = 0, T(x, \infty) = 0, T(x, 0) = \sin \frac{\pi x}{a}.$$ **[C.U. (Hons.) 1984]**

(See Art. 18.5)

17. (a) Solve the equation $\dfrac{\partial^2 u}{\partial t^2} = c^2 \dfrac{\partial^2 u}{\partial x^2}$ where c^2 is a constant, by the method of separation of variables under the boundary conditions $u(0, t) = 0$, and and $u(l, t) = 0$ for all t.

(b) If the initial conditions at $t = 0$ are $u(x, 0) = f(x)$ and $\left.\dfrac{\partial u}{\partial t}\right]_{t=0} = g(x)$, determine the magnitudes of the arbitrary coefficients in the solution obtained in (a).

[C.U. (Hons.) 1987]

[*Hints.* (a) The solution is obtained from (18.7.12) and is written as

$$u(x, t) = \sin \frac{n\pi x}{t} \left(C_n \cos \frac{n\pi ct}{l} + D_n \sin \frac{n\pi ct}{l}\right).$$

(b) $u(x,0) = f(x)$, i.e., $f(x) = C_n \sin \frac{n\pi x}{l}$ or $C_n = \frac{f(x)}{\sin \frac{n\pi x}{l}}$

$$\frac{\partial u}{\partial t} = \sin \frac{n\pi x}{l}\left(-C_n \frac{n\pi c}{l} \sin \frac{n\pi ct}{l} + D_n \frac{n\pi c}{l} \cos \frac{n\pi ct}{l}\right)$$

$$\left.\frac{\partial u}{\partial t}\right]_{t=0} = \sin \frac{n\pi x}{l}\left(D_n \frac{n\pi c}{l}\right) = g(x) \quad \therefore\ D_n = \frac{g(x)}{\frac{n\pi c}{l} \sin \frac{n\pi x}{l}}$$

18. The displacement y of a viscously damped string is given by the equation $\frac{\partial^2 y}{\partial t^2} = c^2 \frac{\partial^2 y}{\partial x^2} - 2k \frac{\partial y}{\partial t}$. Find a general solution, of the above equation by the method of separation of variables.

(See worked out examples of Chapter 17) **[C.U. (Hons.) 1988, '90, '92]**

19. (a) Write down Laplace's equation in two dimensions in rectangular cartesian co-ordinates (x, y). Transform this to plane polar co-ordinates (r, θ) and hence show how this equation may be solved by the method of separation of variables in terms of circular harmonics.

[C.U. (Hons.) 1989, '93]

(See worked out examples)

(b) An infinite uncharged conducting cylinder of radius 'a' is set with its axis perpendicular to a uniform electric field of strength E. Find the potential in the region out-side the cylinder.

[C.U. (Hons.) 1989]

(See worked out examples)

20. (a) Write down Laplaces equation $\nabla^2 V = 0$ in spherical polar co-ordinates. Find its general solution by the method of separation of variables. **[C.U. (Hons.) 1995]**

(b) Apply the above solution to the case of a charged conducting sphere of radius R, for which a boundary condition is $V = V_0$ for $r = R$. Show that at all interior points $r < R, V = V_0$ and at all exterior points, $r > R, V = \frac{V_0 R}{r}$.

Ans. From the worked out example 2 on page 982, we have

$$u = A + \frac{B}{r}$$

Here for $r = R$, $u = V(r) = V_0$

$\because V$ must be finite (given) we have

$$V_{in}(r) = V(r < R) = C \text{ (say)}$$

$$\text{and} V_{out}(r) = V(r > R) = A + \frac{B}{r}$$

Applying the boundary condition, we get

$$V_{in}(r)\Big|_{r=R} = V_{out}(r)\Big|_{r=R}$$

$\therefore C = A + \frac{B}{r}$; Now $C = V_0 \because V_{in}(r) = V_0$

$$\therefore A + \frac{B}{r} = V_0.$$

Again from the condition $u = V(r) \to 0$ as $r \to \infty$, we have $A = 0$, so that we have $\frac{B}{r} = V_0$.

Again the condition $u = V_0$ at $r = R$ gives

$$V_0 = \frac{B}{R} \quad \text{or} \quad B = V_0 R$$

$$\therefore u = V(r) = \frac{B}{r} = \frac{V_0 R}{r}$$

21. Write down the Laplace's equation on a plane in terms of the polar co-ordinates (r, θ). Solve it by the method of separation of variables and write down the general expression of the solution, which is finite at $r = 0$ and single-valued in θ. **[C.U. (Hons.) 1983, '85, '91, '93]**

(See worked out examples)

22. (a) Write down Laplace's equation $\nabla^2 V = 0$ in cylindrical polar co-ordinates. Find its general solution by the method of separation of variables. **[C.U. (Hons.) 1996 (old syll.)]**

(b) Write down the general wave equation in three dimensions. Express it in terms of cylindrical polar co-ordinates and find the general solution by the method of separation of variables. **[C.U. (Hons.) 1996]**

23. (a) A scalar field is defined by $\phi = \phi(x, y)$. Show that $\dfrac{\partial^2 \phi}{\partial y \partial x} = \dfrac{\partial^2 \phi}{\partial x \partial y}$. **[C.U. (Hons.) 97, 04, 06 (old syll)]**

(b) Laplace's equation in two dimensions $\dfrac{\partial^2 \phi}{\partial x^2} + \dfrac{\partial^2 \phi}{\partial y^2} = 0$ has the solution

$$\phi(x, y) = (A_1 \cos kx + A_2 \sin kx)\left(B_1 e^{ky} + B_2 e^{-ky}\right) \tag{1}$$

where A_1, A_2, B_1, B_2 and k are constants.

Now consider the following boundary conditions: (i) $\phi(0, y) = 0$; (ii) $\phi(a, y) = 0$ and $\phi(x, y \to \infty) = 0$.

Show that for the above boundary conditions, equation (1) gives the general solution of the form

$$\phi(x, y) = \sum_{n=1}^{\infty} A_n \sin \frac{n\pi x}{a} e^{-\frac{n\pi y}{a}}.$$

(c) Explain why summation occurs in the final form. **[C.U. (Hons.) 1997, cf. 04]**

(For answer see worked out examples of Art. 18.5)

24. Find out the general solution of the following partial differential equation by the method of separation of variables

$$\frac{\partial^2 u}{\partial x^2} = \alpha^2 \frac{\partial u}{\partial t} = \frac{1}{k} \frac{\partial u}{\partial t} \text{ (putting } \alpha^2 = \frac{1}{k}) \qquad \textbf{[C.U. (Hons.) 1999, 2005]}$$

[See Art. (18.9)]

25. Draw a right handed Cartesian co-ordinate system $OXYZ$ and take a point $P(x, y, z)$. Show in the diagram the cylindrical polar and spherical polar co-ordinates of P. Express x, y, z in terms of these two co-ordinate systems. **[C.U. (Hons.) 2000]**

(See Arts 3.7, 3.8, 18.3.1 and 18.4)

26. By the method of separation of variables, reduce Laplace's equation in rectangular Cartesian co-ordinate system to ordinary differential equations. **[C.U. (Hons.) 2000, 05, 07 (old syll)]**

(See Art. 18.3.)

27. Determine the condition under which the following differential equation can be solved using the method of separation of variables $C_1\dfrac{\partial\phi(x,t)}{\partial t}+C_2\nabla^2\phi(x,t)+V(x,t)\phi(x,t)=0$, where C_1 and C_2 are constants. **[C.U. (Hons.) 2001]**

[*Hins.* Due to the presence of the third term, the method cannot be applied.

In absence of the third term, the equation resembles that of the three dimensional equation for heat conduction. Again if the third term consists of a first order partial derivative w.r.t. to t, while the first term consists of a second order derivative w.r.t. t and the second term a second order derivative w.r.t. x, then the method will work as in the case of viscously damped string.]

28. Consider the differential $\dfrac{\partial u}{\partial t}=a^2\dfrac{\partial^2 u}{\partial x^2}$ where a is a constant. Find the solution $u(x,t)$ for $t>0$ subject to the condition that $u(x,t=0)=\phi(x)$ and $u(x,t\to\infty)$ is finite. **[C.U. (Hons.) 2001]**

(See worked out examples)

29. (a) Given $\psi=\psi(x,y,z)$, write $\nabla^2\psi=0$ in spherical polar coordinate system.

(b) Assume separation of variables i.e. $\psi=R(r)\Theta(\theta)\Phi(\phi)$. Hence show that Laplace's equation can be decoupled into three total differential equations.

(c) Find solution for the ϕ equation, subject to the condition $\phi(\phi+2\pi)=\phi(\phi)$.

[C.U. (Hons.) 2002]

(d) Solve $\nabla^2\xi=\dfrac{1}{C^2}\dfrac{\partial^2\xi}{\partial t^2}$ where $\xi=\xi(\vec{r},t)$ for harmonic wave. Show that the solution represents a travelling wave whose amplitude fall off inversely as distance r.

[C.U. (Hons.) 2002, 06 (old syll), 07]

(See worked out examples)

30. (a) For the scalar field $\phi=\phi(x,y)$ prove that $\dfrac{\partial^2\phi}{\partial x\partial y}=\dfrac{\partial^2\phi}{\partial y\partial x}$.

[C.U. (Hons.) 2004, 06 (old syll)]

(See worked out examples)

(b) Laplace's equation in two dimensions $\dfrac{\partial^2\phi}{\partial x^2}+\dfrac{\partial^2\phi}{\partial y^2}=0$ has solution

$$\phi(x,y)=\left(A_1e^{kx}+A_2e^{-kx}\right)(B_1\cos ky+B_2\sin ky) \tag{1}$$

where A_1, B_1, A_2, B_2 and k are constants.

Now consider the following boundary conditions

(i) $\phi(x,0)=0$, (ii) $\phi(x,a)=0$, (iii) $\phi(x\to\infty,y)=0$, (iv) $\phi(0,y)=F(x,y)$.

Show that for the above conditions equation (1) gives the general solution of the form

$$\phi(x,y)=\sum_{n=1}^{\infty}B_n\sin\frac{n\pi y}{a}e^{-\frac{n\pi x}{a}}$$

$$=\sum_{n=1}^{\infty}\left[\frac{2}{a}\int_0^a F(x,y)\sin\frac{n\pi y}{a}dy\right]\sin\frac{n\pi y}{a}e^{-\frac{n\pi x}{a}}$$

[C.U. (Hons.) 2004, cf. 1997]

(See Art. 18.5 and worked out examples).

31. By the method of separation of variables, reduce Laplace's equation in rectangular Cartesian coordinate system to ordinary differential equations. **[C.U. (Hons.) 2005, 07 (old syll)]**

(See Art. 18.3)

32. Solve by the method of separation of variables

$$3\frac{\partial u}{\partial x} + 2\frac{\partial u}{\partial y} = 0, \quad u(x, 0) = 4e^{-x}$$

[C.U. (Hons.) 2006]

(See worked out examples of Chapter 17).

33. Solve the following equation by the method of separation of variables

$$\frac{\partial^2 V}{\partial x^2} + \frac{\partial^2 V}{\partial y^2} = 0$$

with (i) $V = 0$ when $y = 0$, (ii) $V = 0$ when $y = \pi$ (iii) $V = V_0$ when $x = 1$ and (iv) $V(-x, y) = V(x, y)$. **[C.U. (Hons.) 2007]**

(See Art. 18.5)

34. (a) Express the divergence operator in spherical polar coordinates starting from the expression of the same in Cartesian coordinates. (See Art. 18.4).

(b) Solve the partial differential equation $\frac{\partial u}{\partial t} = \frac{\partial^2 u}{\partial x^2}$ by the method of separation of variables in the region $0 < x < 4$ for $t > 0$. Given $u(0, t) = 0$, $u(4, t) = 0$, and $u(x, 0) = 2x$. **[C.U. (Hons.) 2008]**

(For answer see worked out examples of the Exercises of Chapter 17)

Chapter 19

Matrices

19.1 Introduction

A matrix is a set of mathematical objects written in a compact form and defined in a convenient way so that it can obey the ordinary laws of Algebra. The matrix notation found its origin in the consideration of linear equations and transformations.

19.1.1 Linear equations and matrix notation

We consider the following linear equations

$$\left.\begin{aligned} a_1x + b_1y &= k_1 \\ \text{and} \quad a_2x + b_2y &= k_2 \end{aligned}\right\} \text{ where } \begin{vmatrix} a_1 & b_1 \\ a_2 & b_2 \end{vmatrix} \neq 0.$$

The above equations may be put in the following compact form

$$\begin{bmatrix} a_1 & b_1 \\ a_2 & b_2 \end{bmatrix} \begin{bmatrix} x \\ y \end{bmatrix} = \begin{bmatrix} k_1 \\ k_2 \end{bmatrix}$$

The array of elements $\begin{bmatrix} a_1 & b_1 \\ a_2 & b_2 \end{bmatrix}$ is called a **matrix.** It consists of 2 rows and 2 columns, so that it is called a *square matrix* of order 2×2.

In general, *a matrix is a rectangular array of elements arranged in rows and columns. If the numbers of rows and columns be m and n, the matrix is said to be of order $m \times n$ (which is read as m by n)*

$\therefore$ the matrix $[a_1b_1c_1]$ is said to be of order 1×3. Since it contains only one row, it is also termed as a *row matrix.*

Similarly the matrix $\begin{bmatrix} a_1 \\ a_2 \end{bmatrix}$ is a matrix of order 2×1. It is also known as *a column matrix.*

From the above consideration, the set of linear equations

$$\begin{aligned} a_1x + b_1y + c_1z &= k_1 \\ a_2x + b_2y + c_2z &= k_2 \\ a_3x + b_3y + c_3z &= k_3 \end{aligned}$$

may be put in the following matrix notation

$$\begin{bmatrix} a_1 & b_1 & c_1 \\ a_2 & b_2 & c_2 \\ a_3 & b_3 & c_3 \end{bmatrix} \begin{bmatrix} x \\ y \\ z \end{bmatrix} = \begin{bmatrix} k_1 \\ k_2 \\ k_3 \end{bmatrix}.$$

The matrix

$$\begin{bmatrix} a_{11} & a_{12} & a_{13} & \cdots & a_{1n} \\ a_{21} & a_{22} & a_{23} & \cdots & a_{2n} \\ a_{31} & a_{32} & a_{33} & \cdots & a_{3n} \\ \cdots & \cdots & \cdots & \cdots & \cdots \\ \cdots & \cdots & \cdots & \cdots & \cdots \\ a_{m1} & a_{m2} & a_{m3} & \cdots & a_{mn} \end{bmatrix}$$

is put in the form $[a_{ij}]$ where the suffixes ij indicate that the element occurs in the ith row and jth column.

19.2 Definitions

(a) Rectangular matrix: If $m \neq n$, the matrix is called a *rectangular matrix.*

(b) Square matrix: If $m = n$, the corresponding matrix is said to be a *square matrix.*

(c) Row matrix: If the elements are all arranged in only one row, the matrix is said to be a *row matrix.* If the number of elements is n, it is $1 \times n$ matrix.

$\therefore \begin{bmatrix} a_1 & a_2 & a_3 & \cdots & a_n \end{bmatrix}$ is a row matrix.

(d) Column matrix: If the elements are all arranged in only one column, the matrix is said to be a *column matrix.* If the number of elements is m, it is $m \times 1$ matrix.

$\therefore \begin{bmatrix} a_1 \\ a_2 \\ \vdots \\ a_n \end{bmatrix}$ is a column matrix.

(e) Singular and non-singular matrices: A square matrix A is said to be singular or non-singular according as the determinant $|A| = 0$ or $|A| \neq 0$.

(f) Sub-matrix: A matrix formed by deleting some rows or some columns or some of both rows and columns of a matrix A, is called a *sub-matrix* or *a block-matrix* of A.

(g) Equality of matrices: Two matrices $[a_{ij}]$ and $[b_{ij}]$ are equal, if they have the same number of rows and columns where $a_{ij} = b_{ij}$ for all i, j.

19.3 Matrix addition

Let us consider the equations

$$\left.\begin{aligned} y_1 &= a_{11}x_1 + a_{12}x_2 \\ y_2 &= a_{21}x_1 + a_{22}x_2 \end{aligned}\right\} \tag{1}$$

$$\left.\begin{aligned} z_1 &= b_{11}x_1 + b_{12}x_2 \\ z_2 &= b_{21}x_1 + b_{22}x_2 \end{aligned}\right\} \tag{2}$$

$$\left.\begin{aligned} u_1 &= y_1 + z_1 \\ u_2 &= y_2 + z_2 \end{aligned}\right\} \tag{3}$$

Then

$$\left.\begin{aligned} u_1 &= (a_{11}+b_{11})x_1 + (a_{12}+b_{12})x_2 \\ \text{and} \quad u_2 &= (a_{21}+b_{21})x_1 + (a_{22}+b_{22})x_2 \end{aligned}\right\}. \tag{4}$$

Thus the equations (19.3.4) are obtained by simple addition of (19.3.1) and (19.3.2). Now, let the equations (19.3.1) be put in the form $y = A \cdot x$ where $A = \begin{bmatrix} a_{11} & a_{12} \\ a_{21} & a_{22} \end{bmatrix} = [a_{ij}]$ and the equations (19.3.2) be put as $z = B \cdot x$ where $B = \begin{bmatrix} b_{11} & b_{12} \\ b_{21} & b_{22} \end{bmatrix} = [b_{ij}]$.

Then

$$u = [a_{ij} + b_{ij}]\, x \quad \text{or} \quad [a_{ij}]\, x + [b_{ij}]\, x = [a_{ij} + b_{ij}]\, x$$

$$\text{or} \quad [a_{ij}] + [b_{ij}] = [a_{ij} + b_{ij}].$$

This law of addition may be extended to the matrices of any order. Thus the sum of two matrices each of order $m \times n$, is a matrix of the same order $m \times n$, each of whose elements is the sum of the corresponding elements of the given matrices.

DEFINITION: *Two matrices are said to be conformable for addition, when both have the same number of rows and the same number of columns. Otherwise the sum does not exist.*

19.4 Scalar multiplication

The product of an $m \times n$ matrix A and scalar quantity c is also an $m \times n$ matrix, of which every element is the product of the corresponding element of the given matrix A and the given scalar quantity. It is denoted by Ac or cA. Thus

$$\text{if } A = \begin{bmatrix} a_{11} & a_{12} & a_{13} \\ a_{21} & a_{22} & a_{23} \\ a_{31} & a_{32} & a_{33} \end{bmatrix}, \quad \text{then } Ac = \begin{bmatrix} ca_{11} & ca_{12} & ca_{13} \\ ca_{21} & ca_{22} & ca_{23} \\ ca_{31} & ca_{32} & ca_{33} \end{bmatrix}$$

19.5 Subtraction of matrices

Two matrices A and B are conformable for subtraction if they have the same number of rows and the same number of columns. The subtraction of B from A is the sme as the addition of A and $-B$.

i.e., $A - B = A + (-B)$

$$\text{i.e.,}\quad \begin{bmatrix} a_{11} & a_{12} \\ a_{21} & a_{22} \end{bmatrix} - \begin{bmatrix} b_{11} & b_{12} \\ b_{21} & b_{22} \end{bmatrix} = \begin{bmatrix} a_{11} & a_{12} \\ a_{21} & a_{22} \end{bmatrix} + \begin{bmatrix} -b_{11} & -b_{12} \\ -b_{21} & -b_{22} \end{bmatrix}$$

$$= \begin{bmatrix} a_{11} - b_{11} & a_{12} - b_{12} \\ a_{21} - b_{21} & a_{22} - b_{22} \end{bmatrix}.$$

19.6 Null Matrix

A matrix each of whose elements is zero is said to be a null matrix.

$\therefore$ $A - A$ is a null matrix, and it is denoted by [0]. It is clear that $A = O + A = A + O$.

19.7 Laws of Matrix Algebra

19.7.1 Commutative law of addition

If the matrices A and B are conformable for adition, then $A + B = B + A$.

Proof. Let $A = [a_{ij}]$ and $B = [b_{ij}]$ be the two matrices of the same order. Then

$$\begin{aligned} A + B &= [a_{ij}] + [b_{ij}] = [a_{ij} + b_{ij}] = [b_{ij} + a_{ij}] \\ &= [b_{ij}] + [a_{ij}] = B + A. \end{aligned}$$

19.7.2 Associative law of addition

If the matrices A, B, C are conformable for addition, then $(A + B) + C = A + (B + C)$.

Proof. Let $A = [a_{ij}], B = [b_{ij}], C = [c_{ij}]$ be the three matrices of the same order. Then we may write,

$$\begin{aligned} (A + B) + C &= ([a_{ij}] + [b_{ij}]) + [c_{ij}] = [a_{ij} + b_{ij}] + [c_{ij}] = [(a_{ij} + b_{ij}) + c_{ij}] \\ &= [a_{ij} + b_{ij} + c_{ij}] = [a_{ij} + (b_{ij} + c_{ij})] = [a_{ij}] + [b_{ij} + c_{ij}] \\ &= [a_{ij}] + ([b_{ij}] + [c_{ij}]) = A + (B + C). \end{aligned}$$

19.7.3 Commutative law of scalar multiplication

To show that $Ac = cA$ where A is a matrix and c is a scalar quantity.

Proof. If A be a matrix and c be a scalar, then from definition $Ac = cA$. Hence the commutative law of scalar multiplication holds.

19.7.4 Associative law of scalar multiplication

If A be a matrix and l, m be two scalars, then $l(Am) = (lA)m$.

Proof. Let $A = [a_{ij}]$, then $l(Am) = l([a_{ij}]\,m) = l([a_{ij}m]) = [la_{ij}m] = [lma_{ij}]$.

Again

$$\begin{aligned}(lA)m &= (l[a_{ij}])\,m = ([la_{ij}])\,m = [la_{ij}m] = [lma_{ij}]\\ \therefore\ l(Am) &= (lA)m.\end{aligned}$$

19.7.5 Distributive law for scalar multiplication

If A, B be two matrices and k be a scalar quantity, then

$$k(A+B) = kA + kB \text{ and } (A+B)k = Ak + Bk.$$

Proof. Let $A = [a_{ij}], B = [b_{ij}]$, then we may write

$$\begin{aligned}k(A+B) &= k([a_{ij}] + [b_{ij}]) = k\,[a_{ij} + b_{ij}] = [k(a_{ij} + b_{ij})] = [ka_{ij} + kb_{ij}]\\ &= [ka_{ij}] + [kb_{ij}] = k\,[a_{ij}] + k\,[b_{ij}] = kA + kB.\end{aligned}$$

Again

$$\begin{aligned}(A+B)k &= ([a_{ij}] + [b_{ij}])\,k = [a_{ij} + b_{ij}]\,k = [(a_{ij} + b_{ij})k] = [a_{ij}k + b_{ij}k]\\ &= [a_{ij}k] + [b_{ij}k] = [a_{ij}]\,k + [b_{ij}]\,k = Ak + Bk.\end{aligned}$$

19.8 Matrix multiplication

We consider two linear transformations as follows:

$$\left.\begin{aligned}y_1 &= a_{11}x_1 + a_{12}x_2\\ y_2 &= a_{21}x_1 + a_{22}x_2\end{aligned}\right\} \qquad (1)$$

where $A = \begin{bmatrix} a_{11} & a_{12}\\ a_{21} & a_{22}\end{bmatrix}$

and

$$\left.\begin{aligned}z_1 &= b_{11}y_1 + b_{12}y_2\\ z_2 &= b_{21}y_1 + b_{22}y_2\end{aligned}\right\} \qquad (2)$$

where $B = \begin{bmatrix} b_{11} & b_{12}\\ b_{21} & b_{22}\end{bmatrix}$.

Then

$$\left.\begin{aligned}z_1 &= b_{11}(a_{11}x_1 + a_{12}x_2) + b_{12}(a_{21}x_1 + a_{22}x_2)\\ &= (b_{11}a_{11} + b_{12}a_{21})x_1 + (b_{11}a_{12} + b_{12}a_{22})x_2\\ z_2 &= b_{21}(a_{11}x_1 + a_{12}x_2) + b_{22}(a_{21}x_1 + a_{22}x_2)\\ &= (b_{21}a_{11} + b_{22}a_{21})x_1 + (b_{21}a_{12} + b_{22}a_{22})x_2\end{aligned}\right\}. \qquad (3)$$

Now if we denote (19.8.1) by $y = Ax$, (19.8.2) by $z = By$, and (19.8.3) by $z = Cx$, then C is the matrix given by

$$C = \begin{bmatrix} b_{11}a_{11} + b_{12}a_{21} & b_{11}a_{12} + b_{12}a_{22} \\ b_{21}a_{11} + b_{22}a_{21} & b_{21}a_{12} + b_{22}a_{22} \end{bmatrix}.$$

But

$$z = By = B(Ax) = BAx = Cx \quad \therefore \ BA = C$$

or

$$\begin{bmatrix} b_{11} & b_{12} \\ b_{21} & b_{22} \end{bmatrix} \begin{bmatrix} a_{11} & a_{12} \\ a_{21} & a_{22} \end{bmatrix} = \begin{bmatrix} b_{11}a_{11} + b_{12}a_{21} & b_{11}a_{12} + b_{12}a_{22} \\ b_{21}a_{11} + b_{22}a_{21} & b_{21}a_{12} + b_{22}a_{22} \end{bmatrix}.$$

Thus the multiplication of matrices is defined. It is obvious that *the matrices B and A are conformable for the product BA, if the number of columns of B is equal to the number of rows of A.*

DEFINITION. *If B be an $m \times n$ matrix and A be an $n \times p$ matrix, then the product BA is an $m \times p$ matrix. The (ij)th element of the product BA is obtained by multiplying the elements of ith row of B and the elements of jth column of A and considering the sum of the products so formed.*

19.9 Laws of Matrix multiplication

19.9.1 Commutative law

Matrix multiplication is not, in general, commutative.

If A and B are two matrices, then $AB \neq BA$, in general.

Proof. Let us consider the (i, j)th elements of AB and BA. The (i, j)th element of AB is the inner product of the ith row of A and the jth column of B. But the (i, j)th element of BA is the inner product of the ith row of B and the jth column of A. In general, they are different from each other. Hence the result.

19.9.2 Associative law

If the matrices A, B, and C be conformable for the products AB and BC, then $(AB)C = A(BC)$.

Proof. We consider the following example.

Let

$$A = \begin{bmatrix} 1 & 2 \\ 3 & 4 \end{bmatrix}, \quad B = \begin{bmatrix} 2 & 1 \\ 3 & 5 \end{bmatrix}, \quad C = \begin{bmatrix} 2 & 3 \\ 1 & 1 \end{bmatrix}.$$

Then

$$AB = \begin{bmatrix} 1.2 + 2.3 & 1.1 + 2.5 \\ 3.2 + 4.3 & 3.1 + 4.5 \end{bmatrix} = \begin{bmatrix} 8 & 11 \\ 18 & 23 \end{bmatrix}.$$

$$\therefore \ (AB)C = \begin{bmatrix} 8 & 11 \\ 18 & 23 \end{bmatrix} \begin{bmatrix} 2 & 3 \\ 1 & 1 \end{bmatrix} = \begin{bmatrix} 8.2 + 11.1 & 8.3 + 11.1 \\ 18.2 + 23.1 & 18.3 + 23.1 \end{bmatrix} = \begin{bmatrix} 27 & 35 \\ 59 & 77 \end{bmatrix}.$$

Again

$$BC = \begin{bmatrix} 2.2+1.1 & 2.3+1.1 \\ 3.2+5.1 & 3.3+5.1 \end{bmatrix} = \begin{bmatrix} 5 & 7 \\ 11 & 14 \end{bmatrix}$$

$$\therefore\ A(BC) = \begin{bmatrix} 1 & 2 \\ 3 & 4 \end{bmatrix}\begin{bmatrix} 5 & 7 \\ 11 & 14 \end{bmatrix} = \begin{bmatrix} 1.5+2.11 & 1.7+2.14 \\ 3.5+4.11 & 3.7+4.14 \end{bmatrix} = \begin{bmatrix} 27 & 35 \\ 59 & 77 \end{bmatrix}$$

$$\therefore\ (AB)C = A(BC).$$

COROLLARY. *If A is a square matrix, then the product AA is denoted by A^2, the product AAA is denoted by A^3 etc.*

19.9.3 Distributive law

For matrices A, B, C, show that $(A+B)C = AC+BC$, where the matrices are conformable for above multipication and addition.

Proof. Let

$$A = \begin{bmatrix} 1 & 2 \\ 3 & 4 \end{bmatrix},\ B = \begin{bmatrix} 2 & 1 \\ 3 & 5 \end{bmatrix},\ C = \begin{bmatrix} 2 & 3 \\ 1 & 1 \end{bmatrix}.$$

Then

$$A+B = \begin{bmatrix} 1+2 & 2+1 \\ 3+3 & 4+5 \end{bmatrix} = \begin{bmatrix} 3 & 3 \\ 6 & 9 \end{bmatrix}$$

$$\therefore\ (A+B)C = \begin{bmatrix} 3 & 3 \\ 6 & 9 \end{bmatrix}\begin{bmatrix} 2 & 3 \\ 1 & 1 \end{bmatrix} = \begin{bmatrix} 3.2+3.1 & 3.3+3.1 \\ 6.2+9.1 & 6.3+9.1 \end{bmatrix} = \begin{bmatrix} 9 & 12 \\ 21 & 27 \end{bmatrix}.$$

Again

$$AC+BC = \begin{bmatrix} 1.2+2.1 & 1.3+2.1 \\ 3.2+4.1 & 3.3+4.1 \end{bmatrix} + \begin{bmatrix} 2.2+1.1 & 2.3+1.1 \\ 3.2+5.1 & 3.3+5.1 \end{bmatrix}$$

$$= \begin{bmatrix} 4 & 5 \\ 10 & 13 \end{bmatrix} + \begin{bmatrix} 5 & 7 \\ 11 & 14 \end{bmatrix} = \begin{bmatrix} 4+5 & 5+7 \\ 10+11 & 13+14 \end{bmatrix} = \begin{bmatrix} 9 & 12 \\ 21 & 27 \end{bmatrix}.$$

$$\therefore\ (A+B)C = AC+BC.$$

Similarly it can be shown that $A(B+C) = AB+AC$.

19.10 Transpose of a matrix

The matrix $\tilde{A}$ or A^T is said to be the transpose of the matrix A, if the rows of $\tilde{A}$ or A^T are the same as the columns of A.

$\therefore$ If A be an $m \times n$ matrix, $\tilde{A}$ or A^T will be an $n \times m$ matrix. The (i,j)th element of A will be equal to the (j,i)th element of $\tilde{A}$ or A^T. Thus if $A = [a_{ij}]$, then $\tilde{A} = A^T = [a_{ji}]$.

19.11 Properties

(i) If A^T be the *transpose of the matrix* A, then $(A^T)^T = A$.

Let $A = [a_{ij}]$; then $A^T = [a_{ji}]$, $\therefore$ $(A^T)^T = [a_{ij}] = A$.

(ii) $(cA)^T = cA^T$ where *c is a scalar quantity.*

Let

$$A = [a_{ij}] \quad \therefore \quad cA = c[a_{ij}] = [ca_{ij}],$$

so that we may write

$$(cA)^T = [ca_{ji}] = c[a_{ji}] = cA^T.$$

(iii) $(A+B)^T = A^T + B^T$ *where the matrices* A, B *are of the same order:*

Let

$$A = [a_{ij}], \quad B = [b_{ij}],$$

then

$$\begin{aligned} A^T &= [a_{ji}], \quad B^T = [b_{ji}]. \\ \therefore \quad A+B &= [a_{ij}] + [b_{ij}] = [a_{ij} + b_{ij}] \\ \therefore \quad (A+B)^T &= [a_{ji} + b_{ji}] = [a_{ji}] + [b_{ji}] = A^T + B^T. \end{aligned}$$

(iv) $(AB)^T = B^T A^T$, *where* A, B *are conformable for the product* AB.

Proof. Let

$$A = \begin{bmatrix} a_{11} & a_{12} \\ a_{21} & a_{22} \end{bmatrix}, \quad B = \begin{bmatrix} b_{11} & b_{12} \\ b_{21} & b_{22} \end{bmatrix};$$

then we have

$$\begin{aligned} A^T &= \begin{bmatrix} a_{11} & a_{21} \\ a_{12} & a_{22} \end{bmatrix}, \quad B^T = \begin{bmatrix} b_{11} & b_{21} \\ b_{12} & b_{22} \end{bmatrix} \\ \therefore \quad AB &= \begin{bmatrix} a_{11}b_{11} + a_{12}b_{21} & a_{11}b_{12} + a_{12}b_{22} \\ a_{21}b_{11} + a_{22}b_{21} & a_{21}b_{12} + a_{22}b_{22} \end{bmatrix} \\ \therefore \quad (AB)^T &= \begin{bmatrix} a_{11}b_{11} + a_{12}b_{21} & a_{21}b_{11} + a_{22}b_{21} \\ a_{11}b_{12} + a_{12}b_{22} & a_{21}b_{12} + a_{22}b_{22} \end{bmatrix}. \end{aligned}$$

Again

$$B^T A^T = \begin{bmatrix} b_{11} & b_{21} \\ b_{12} & b_{22} \end{bmatrix} \begin{bmatrix} a_{11} & a_{21} \\ a_{12} & a_{22} \end{bmatrix} = \begin{bmatrix} b_{11}a_{11} + b_{21}a_{12} & b_{11}a_{21} + b_{21}a_{22} \\ b_{12}a_{11} + b_{22}a_{12} & b_{12}a_{21} + b_{22}a_{22} \end{bmatrix}$$

$$\therefore \quad (AB)^T = B^T A^T.$$

(v) *If A is a square matrix, then det* $A = \det A^T$

Since the determinant remains the same for an interchange of rows and columns, the result is obvious.

19.12 Diagonal matrix

A square matrix all of whose elements except the diagonal ones are zero, is called a *diagonal* matrix.

For example $\begin{bmatrix} d_1 & 0 & 0 \\ 0 & d_2 & 0 \\ 0 & 0 & d_3 \end{bmatrix}$ is a diagonal matrix.

A given matrix by suitable transformation can be made a diagonal one, the process being known as *diagonalisation.*

19.12.1 Scalar matrix

If in a diagonal matrix, all the diagonal elements are equal, it is called a *scalar matrix.* Thus $\begin{bmatrix} d & 0 & 0 \\ 0 & d & 0 \\ 0 & 0 & d \end{bmatrix}$ is a scalar matrix.

19.12.2 Unit matrix or Identity matrix

A scalar matrix of order n is said to be a unit matrix, if each of the diagonal elements is equal to unity. It is also known as *identity matrix.*

Thus $I = \begin{bmatrix} 1 & 0 & 0 \\ 0 & 1 & 0 \\ 0 & 0 & 1 \end{bmatrix}$ is a *unit matrix* of order 3.

PROPERTIES. (i) *If A is a square matrix and I is the unit matrix of the same order, then $AI = IA = A$.*

Let

$$A = \begin{bmatrix} a_{11} & a_{12} & a_{13} \\ a_{21} & a_{22} & a_{23} \\ a_{31} & a_{32} & a_{33} \end{bmatrix} \text{ and } I = \begin{bmatrix} 1 & 0 & 0 \\ 0 & 1 & 0 \\ 0 & 0 & 1 \end{bmatrix}.$$

Then

$$AI = \begin{bmatrix} a_{11}.1 + a_{12}.0 + a_{13}.0 & a_{11}.0 + a_{12}.1 + a_{13}.0 & a_{11}.0 + a_{12}.0 + a_{13}.1 \\ a_{21}.1 + a_{22}.0 + a_{23}.0 & a_{21}.0 + a_{22}.1 + a_{23}.0 & a_{21}.0 + a_{22}.0 + a_{23}.1 \\ a_{31}.1 + a_{32}.0 + a_{33}.0 & a_{31}.0 + a_{32}.1 + a_{33}.0 & a_{31}.0 + a_{32}.0 + a_{33}.1 \end{bmatrix}$$

$$= \begin{bmatrix} a_{11} & a_{12} & a_{13} \\ a_{21} & a_{22} & a_{23} \\ a_{31} & a_{32} & a_{33} \end{bmatrix} = A.$$

Similarly $IA = A$. $\therefore$ $AI = IA = A$.

(ii) *If $[d]$ be a scalar matrix and I be the unit matrix of the same order, then $[d] = dI$.*

Let

$$[d] = \begin{bmatrix} d & 0 & 0 \\ 0 & d & 0 \\ 0 & 0 & d \end{bmatrix} \text{ and } I = \begin{bmatrix} 1 & 0 & 0 \\ 0 & 1 & 0 \\ 0 & 0 & 1 \end{bmatrix}$$

$$\therefore\ dI = d\begin{bmatrix} 1 & 0 & 0 \\ 0 & 1 & 0 \\ 0 & 0 & 1 \end{bmatrix} = \begin{bmatrix} d\times 1 & d\times 0 & d\times 0 \\ d\times 0 & d\times 1 & d\times 0 \\ d\times 0 & d\times 0 & d\times 1 \end{bmatrix} = \begin{bmatrix} d & 0 & 0 \\ 0 & d & 0 \\ 0 & 0 & d \end{bmatrix} = [d].$$

19.13 Adjoint matrix

If $A = [a_{ij}]$ be a square matrix and if $B = [A_{ij}]$ be a matrix whose elements are the co-factors of the corresponding elements of A in det A (or $|A|$), then the transpose of B is called the adjoint or adjugate of A and it may be denoted by $[A_{ji}]$ or adj A.

Let

$$A = \begin{bmatrix} a_{11} & a_{12} & a_{13} \\ a_{21} & a_{22} & a_{23} \\ a_{31} & a_{32} & a_{33} \end{bmatrix}$$

and let $A_{11}, A_{12}, A_{21}, A_{22} \cdots$ be the co-factors of $a_{11}, a_{12}, a_{21}, a_{22}, \cdots$ in

$$|A| = \begin{vmatrix} a_{11} & a_{12} & a_{13} \\ a_{21} & a_{22} & a_{23} \\ a_{31} & a_{32} & a_{33} \end{vmatrix}, \text{ then adj. } A = \begin{bmatrix} A_{11} & A_{21} & A_{31} \\ A_{12} & A_{22} & A_{32} \\ A_{13} & A_{23} & A_{33} \end{bmatrix}.$$

19.13.1 Reciprocal matrix

If A be a non-singular matrix, then the matrix $\frac{\text{adj } A}{|A|}$ is called the reciprocal of A.

$$\therefore \text{ Reciprocal of } A = \begin{bmatrix} \frac{A_{11}}{|A|} & \frac{A_{21}}{|A|} & \frac{A_{31}}{|A|} \\ \frac{A_{12}}{|A|} & \frac{A_{22}}{|A|} & \frac{A_{32}}{|A|} \\ \frac{A_{13}}{|A|} & \frac{A_{23}}{|A|} & \frac{A_{33}}{|A|} \end{bmatrix}.$$

PROPERTY. The product of a matrix A and its adjoint is commutative and it is a scalar matrix whose diagonal element is $|A|$ i.e., $A.(\text{adj } A) = |A|.I = (\text{adj } A).A$.

Let

$$A = \begin{bmatrix} a_{11} & a_{12} & a_{13} & \cdots \\ a_{21} & a_{22} & a_{23} & \cdots \\ a_{31} & a_{32} & a_{33} & \cdots \\ \cdots & \cdots & \cdots & \cdots \end{bmatrix}, \text{ then adj } A = \begin{bmatrix} A_{11} & A_{21} & A_{31} & \cdots \\ A_{12} & A_{22} & A_{32} & \cdots \\ A_{13} & A_{23} & A_{33} & \cdots \\ \cdots & \cdots & \cdots & \cdots \end{bmatrix}$$

where $A_{11}, A_{12}\cdots$ are the co-factors of $a_{11}, a_{12}\cdots$ i.e, $|A|$. The (i,j)th element of $A(\text{adj}\,A)$ = inner product of the ith row of A and the jth column of adj A.

$$= a_{i1}A_{j1} + a_{i2}A_{j2} + \cdots + a_{in}A_{jn} = 0 \text{ or } |A|$$

according as $i \neq j$ or $i = j$

$$\therefore \quad A(\text{adj}\,A) = [|A|] = |A|.I.$$

In a similar way, $(\text{adj}\,A)\cdot A = [|A|] = |A|\cdot I$. Hence the result.

COROLLARY 1. If $|A| \neq 0, A\cdot\left(\frac{\text{adj}\,A}{|A|}\right) = I = \left(\frac{\text{adj}\,A}{|A|}\right)\cdot A.$

COROLLARY 2. $|\,\text{adj}\,A| = |A|^{n-1}$ where A is a square matrix of order n.

Proof. $\det(A.\,\text{adj}\,A) = \det[|A|]$ or $|A|\cdot|\,\text{adj}\,A| = |A|^n$ or $|\,\text{adj}\,A| = |A|^{n-1}$.

19.14 Symmetric matrix

If a square matrix A is equal to its transpose, then A is said to be symmetric i.e. if $A = A^T$, then A is symmetric.

Let

$$A = \begin{bmatrix} 0 & 1 \\ 1 & 0 \end{bmatrix}, \text{ then } A^T = \begin{bmatrix} 0 & 1 \\ 1 & 0 \end{bmatrix} = A$$

i.e. $a_{ij} = a_{ji}$ or the elements of A are identical with those of A^T.

19.14.1 Skew (or anti-symmetric) matrix

If a square matrix A is equal to the negative of its transpose i.e., if $A = -A^T$, then A is said to be skew or anti-symmetric matrix.

Let

$$A = \begin{bmatrix} 0 & -1 \\ 1 & 0 \end{bmatrix}, \text{ then } A^T = \begin{bmatrix} 0 & 1 \\ -1 & 0 \end{bmatrix},$$

so that A is a skew matrix i.e., $a_{ij} = -a_{ji}$ or the elements of A are equal to the negative of the elements in its transpose.

19.14.2 Complex conjugate matrix

If A be a square matrix, then its complex conjugate A^* is such that the elements of A^* are the complex conjugate of the elements of A.

Thus if

$$A = \begin{bmatrix} 1+i & 2-3i \\ 2 & 3i \end{bmatrix}, \text{ then } A^* = \begin{bmatrix} 1-i & 2+3i \\ 2 & -3i \end{bmatrix}.$$

If $A = A^*$, then A is real.

19.14.3 Hermitian conjugate

If A be a square matrix, then its Hermitian conjugate $A^\dagger$ is obtained as a transpose of the complex conjugate of A. Thus if $A^\dagger = [A^*]^T$, then $A^\dagger$ is the Hermitian conjugate of the matrix A. Thus if

$$A = \begin{bmatrix} 1+i & 2-3i \\ 2 & 3i \end{bmatrix}, \text{ then } A^* = \begin{bmatrix} 1-i & 2+3i \\ 2 & -3i \end{bmatrix}.$$

$$\therefore \; A^\dagger = [A^*]^T = \begin{bmatrix} 1-i & 2 \\ 2+3i & -3i \end{bmatrix}.$$

It also follows that $A^\dagger = [A^*]^T = [A^T]^*$, since the two operations of complex conjugation and transposition are carried out one after another on the matrix A, the order of the two operations is immaterial.

19.14.4 Hermitian matrix

The matrix A will be called *Hermitian matrix,* if $A^\dagger$, called its *Hermitian conjugate is equal to A i.e., $A = A^\dagger$.*

Thus if

$$A = \begin{bmatrix} 0 & -i \\ i & 0 \end{bmatrix}, \text{ then } A^* = \begin{bmatrix} 0 & i \\ -i & 0 \end{bmatrix}. \quad \therefore [A^*]^T = \begin{bmatrix} 0 & -i \\ i & 0 \end{bmatrix} = A;$$

hence A is a Hermitian matrix.

19.14.5 Unitary matrix

If a square matrix A and its Hermitian conjugate $A^\dagger$ be such that $AA^\dagger = I$ (unit matrix), then A is said to be a unitary matrix.

Thus if

$$A = \begin{bmatrix} 0 & 1 \\ 1 & 0 \end{bmatrix}, \text{ then } A^* = \begin{bmatrix} 0 & 1 \\ 1 & 0 \end{bmatrix}. \quad \therefore [A^*]^T = \begin{bmatrix} 0 & 1 \\ 1 & 0 \end{bmatrix} = A.$$

That is A is a Hermitian matrix.

Now

$$A^\dagger = [A^*]^T = \begin{bmatrix} 0 & 1 \\ 1 & 0 \end{bmatrix}. \quad \therefore AA^\dagger = \begin{bmatrix} 0 & 1 \\ 1 & 0 \end{bmatrix}\begin{bmatrix} 0 & 1 \\ 1 & 0 \end{bmatrix} = \begin{bmatrix} 1 & 0 \\ 0 & 1 \end{bmatrix} = I.$$

Hence A is a unitary matrix.

19.14.6 Co-factor matrix

A co-factor matrix is defined as $A = [a]^c = [a^{ij}]$.

Let

$$[a] = \begin{bmatrix} a_{11} & a_{12} & a_{13} \\ a_{21} & a_{22} & a_{23} \\ a_{31} & a_{32} & a_{33} \end{bmatrix},$$

then we may write

$$[a]^c = \begin{bmatrix} a^{11} & a^{12} & a^{13} \\ a^{21} & a^{22} & a^{23} \\ a^{31} & a^{32} & a^{33} \end{bmatrix}.$$

where

$$a^{11} = (-1)^{1+1} \begin{vmatrix} a_{22} & a_{23} \\ a_{32} & a_{33} \end{vmatrix}, \quad a^{21} = (-1)^{2+1} \begin{vmatrix} a_{12} & a_{13} \\ a_{32} & a_{33} \end{vmatrix} \quad \text{and so on.}$$

19.15 Orthogonal matrix

A square matrix A is said to be *orthogonal* if $AA^T = I$.

Thus let

$$A = \begin{bmatrix} 0 & 1 \\ 1 & 0 \end{bmatrix}, \text{ then } A^T = \begin{bmatrix} 0 & 1 \\ 1 & 0 \end{bmatrix},$$

so that we have

$$AA^T = \begin{bmatrix} 0 & 1 \\ 1 & 0 \end{bmatrix} \begin{bmatrix} 0 & 1 \\ 1 & 0 \end{bmatrix} = \begin{bmatrix} 1 & 0 \\ 0 & 1 \end{bmatrix} = I.$$

19.15.1 Properties of an Orthogonal matrix

(i) ***Every orthogonal matrix is non-singular.***

Let A be an orthogonal matrix. Then $A^T A = I$.

$$\therefore \ |A^T A| = |I| \text{ or } |A^T||A| = 1.$$

$$\therefore \ |A| \neq 0 \text{ i.e. } A \text{ is non-singular.}$$

(ii) ***Unit matrix is an orthogonal matrix.***

Let $I = \begin{bmatrix} 1 & 0 & 0 \\ 0 & 1 & 0 \\ 0 & 0 & 1 \end{bmatrix}$ be the unit matrix; then

$$I^T = \begin{bmatrix} 1 & 0 & 0 \\ 0 & 1 & 0 \\ 0 & 0 & 1 \end{bmatrix}. \quad \therefore \ I^T I = \begin{bmatrix} 1 & 0 & 0 \\ 0 & 1 & 0 \\ 0 & 0 & 1 \end{bmatrix} \begin{bmatrix} 1 & 0 & 0 \\ 0 & 1 & 0 \\ 0 & 0 & 1 \end{bmatrix} = \begin{bmatrix} 1 & 0 & 0 \\ 0 & 1 & 0 \\ 0 & 0 & 1 \end{bmatrix} = I.$$

$\therefore$ I is an orthogonal matrix.

(iii) *If A be an orthogonal matrix, then $|A| = \pm 1$.*

We have

$$A^T A = I \text{ or } |A^T A| = |I| \text{ or } |A^T||A| = 1$$

$$\text{or } |A|^2 = 1 \text{ or } |A| = \pm 1.$$

(iv) *The product of two orthogonal matrices is orthogonal.*

Let A and B be two orthogonal matrices.

$\therefore$ $A^T A = I$ and $B^T B = I$.

$$\begin{aligned} \therefore (AB)^T (AB) &= (B^T A^T)(AB) = B^T(A^T A)B \\ &= B^T I B = B^T(IB) = B^T B = I. \end{aligned}$$

$\therefore$ AB is also orthogonal.

(v) *The transpose of an orthogonal matrix is orthogonal.*

Let the matrix A be orthogonal. Then $A^T A = I$.

Let the matrix B be such that $AB = I$.

Then

$$A^T = A^T I = A^T(AB) = (A^T A) B = IB = B$$

$$\therefore \quad AB = AA^T = I$$

$$\therefore \quad (AA^T)^T = I^T = I \text{ or } (A^T)^T A^T = I$$

$$\therefore \quad A^T \text{ is orthogonal.}$$

(vi) *The inverse of an orthogonal matrix is orthogonal.*

Let A be an orthogonal matrix. $\therefore$ $A^T A = I$.

Let B be the inverse of A. $\therefore$ $AB = BA = I$.

$$\begin{aligned} \therefore (AB)^T(AB) &= (B^T A^T)(AB) = B^T (A^T A) B \\ &= B^T I B = B^T(IB) = B^T B. \end{aligned}$$

But $(AB)^T(AB) = I^T I = I$ $\therefore$ $B^T B = I$, so that B is orthogonal.

19.16 Matrix division—Inverse of a matrix

If the matrices A and B are such that $AB = BA = I$, then B is said to be the inverse of A and is written as A^{-1}.

NOTE. Since A and B are conformable for the products AB and BA, where $AB = BA$, we conclude that A and B must be square matrices of the same order. Thus a matrix having an inverse must be a square matrix.

PROPERTIES. (i) *The inverse of a matrix is unique:*

Proof. If possible, let each of B and C be the inverse of the matrix A. Then

$$AB = BA = I \tag{1}$$

$$\text{and } AC = CA = I. \tag{2}$$

From associative law of multiplication of matrices,

$$C(AB) = (CA)B. \tag{3}$$

But from (19.16.1), $C(AB) = CI = C$ and from (19.16.2) $(CA)B = IB = B$.

$\therefore$ from (19.16.3) $C = B$. Hence the result.

(ii) *The necessary and sufficient condition that the inverse of a square matrix exists is that the matrix is non-singular.*

(a) Let B be the inverse of a square matrix A.

Then

$$AB = BA = I \text{ or } |AB| = |I| \text{ or } |A| \cdot |B| = 1 \text{ or } |A| \neq 0$$

i.e. A is non-singular. Thus the condition is necessary.

(b) Next let the matrix be non-singular i.e. $|A| \neq 0$.

Then

$$A \cdot \left(\frac{\text{adj } A}{|A|}\right) = \left(\frac{\text{adj } A}{|A|}\right) \cdot A = I.$$

$\therefore\ B = \frac{\text{adj } A}{|A|}$ exists i.e. the inverse of A exists.

NOTE. It is clear that $B = \frac{\text{adj } A}{|A|}$ is the reciprocal of A.

$\therefore$ *the reciprocal of a matrix is the same as its inverse.*

(iii) *If the matrices A, B are conformable for the product AB, then $(AB)^{-1} = B^{-1}A^{-1}$.*

We have

$$(AB)\left(B^{-1}A^{-1}\right) = A\left(BB^{-1}\right)A^{-1} = AIA^{-1} = (AI)A^{-1} = AA^{-1} = I.$$

Again

$$\left(B^{-1}A^{-1}\right)(AB) = B^{-1}\left(A^{-1}A\right)B = B^{-1}IB = B^{-1}(IB) = B^{-1}B = I$$

$$\text{i.e.} \quad (AB)\left(B^{-1}A^{-1}\right) = I = \left(B^{-1}A^{-1}\right)(AB).$$

$\therefore\ B^{-1}A^{-1}$ is the inverse of AB or in other words $(AB)^{-1} = B^{-1}A^{-1}$.

(iv) From the property discussed in (ii) (a) and (b) above, and the definition of inverse matrix, we know that if the determinant $|A|$ of the square matrix $A = [a_{ij}]$ is not zero, the matrix A is said to be non-singular and its inverse B is defined as $AB = I$. The defining equation for B is

$$B = \frac{[a^{ji}]}{|A|}$$

where $[a^{ji}]$ is the transpose of the co-factor matrix. Thus if

$$A = \begin{bmatrix} a_{11} & a_{12} & a_{13} \\ a_{21} & a_{22} & a_{23} \\ a_{31} & a_{32} & a_{33} \end{bmatrix},$$

so that

$$|A| = \begin{vmatrix} a_{11} & a_{12} & a_{13} \\ a_{21} & a_{22} & a_{23} \\ a_{31} & a_{32} & a_{33} \end{vmatrix}$$

and

$$[a^{ji}] = \begin{bmatrix} a^{11} & a^{12} & a^{13} \\ a^{21} & a^{22} & a^{23} \\ a^{31} & a^{32} & a^{33} \end{bmatrix}^T = \begin{bmatrix} a^{11} & a^{21} & a^{31} \\ a^{12} & a^{22} & a^{32} \\ a^{13} & a^{23} & a^{33} \end{bmatrix}$$

where

$$a^{11} = (-1)^{1+1} \begin{vmatrix} a_{22} & a_{23} \\ a_{32} & a_{33} \end{vmatrix} = (a_{22}a_{33} - a_{23}a_{32}).$$

$$a^{21} = (-1)^{2+1} \begin{vmatrix} a_{12} & a_{13} \\ a_{32} & a_{33} \end{vmatrix} = -(a_{12}a_{33} - a_{13}a_{32}) = (a_{32}a_{13} - a_{12}a_{33}).$$

From what we have discussed above, we may write

$$A^{-1} = \frac{\text{adj}\, A}{|A|} = \frac{[a^{ji}]}{|A|}, \quad [a^{ji}]$$

being the transpose of the co-factor matrix, and the co-factors having been defined and evaluated as shown in our above discussion.

19.17 Trace of a matrix

The sum of the diagonal elements of a matrix A is called the trace of A. Thus if, $A = \begin{bmatrix} 3 & 5 \\ 4 & 6 \end{bmatrix}$, then the trace of A is $3 + 6 = 9$.

The trace of the product of two matrices A and B is independent of the order of multiplication $T_r(AB) = T_r(BA)$.

19.17.1 The derivative of a matrix

If A be a matrix, then $\frac{dA}{dx} = \left[\frac{d}{dx}(a_{ij})\right]$ i.e., the derivative of a matrix w.r.t. a variable x is equal to the matrix of the derivatives of each element w.r.t. x separately. Thus

$$\frac{d}{dx}\begin{bmatrix} 2x & 3x^2 & 4x^3 \\ e^{-x} & 0 & 5x^2 \end{bmatrix} = \begin{bmatrix} 2 & 6x & 12x^2 \\ -e^{-x} & 0 & 10x \end{bmatrix}$$

19.17.2 The integral of a matrix

If A be a matrix, then $\int A\,dx = \left[\int a_{ij}dx\right]$ i.e., the integral of a matrix w.r.t. a variable x is equal to the matrix of the integrals of each element w.r.t. x separately.

Thus,

$$\int \begin{bmatrix} 2x & 3x^2 & 4x^3 \\ e^{-x} & 0 & 5x^2 \end{bmatrix} dx = \begin{bmatrix} x^2+C_1 & x^3+C_2 & x^4+C_3 \\ -e^{-x}+C_4 & C & \frac{5}{3}x^3+C_5 \end{bmatrix}$$

$$= \begin{bmatrix} x^2 & x^3 & x^4 \\ -e^{-x} & C & \frac{5}{3}x^3 \end{bmatrix} + \begin{bmatrix} C_1 & C_2 & C_3 \\ C_4 & 0 & C_5 \end{bmatrix}$$

19.17.3 Minor and Rank

If $m-r$ rows and $n-r$ columns of an $m \times n$ matrix be deleted, a square matrix of order r, remains. The determinant of this square matrix with a positive or negative sign prefixed to it is called a minor of order r of the given matrix.

If there exists at least one minor of order r which does not vanish, where each minor of order $(r+1)$ vanishes, then the number r is said to be the rank of the given matrix.

19.18 Applications of the matrix method

19.18.1 Solution of problems involving matrix algebra

Illustrative Examples

1. Given

$$A = \begin{bmatrix} 2 & 3 & 4 \\ 5 & 6 & 7 \\ 8 & 5 & 11 \end{bmatrix} \text{ and } B = \begin{bmatrix} 3 & -2 & 3 \\ 5 & 4 & 3 \\ 1 & 3 & 2 \end{bmatrix}$$

find $A+B$, $A-B$ and $2A+3B$.

Ans. (i) $A+B = \begin{bmatrix} 2+3 & 3-2 & 4+3 \\ 5+5 & 6+4 & 7+3 \\ 8+1 & 5+3 & 11+2 \end{bmatrix} = \begin{bmatrix} 5 & 1 & 7 \\ 10 & 10 & 10 \\ 9 & 8 & 13 \end{bmatrix}$.

(ii) $A-B = A+(-B) = \begin{bmatrix} 2 & 3 & 4 \\ 5 & 6 & 7 \\ 8 & 5 & 11 \end{bmatrix} + \begin{bmatrix} -3 & 2 & -3 \\ -5 & -4 & -3 \\ -1 & -3 & -2 \end{bmatrix}$

$$= \begin{bmatrix} 2-3 & 3+2 & 4-3 \\ 5-5 & 6-4 & 7-3 \\ 8-1 & 5-3 & 11-2 \end{bmatrix} = \begin{bmatrix} 1 & 5 & 1 \\ 0 & 2 & 4 \\ 7 & 2 & 9 \end{bmatrix}.$$

(iii) $2A+3B = 2\begin{bmatrix} 2 & 3 & 4 \\ 5 & 6 & 7 \\ 8 & 5 & 11 \end{bmatrix} + 3\begin{bmatrix} 3 & -2 & 3 \\ 5 & 4 & 3 \\ 1 & 3 & 2 \end{bmatrix} = \begin{bmatrix} 4 & 6 & 8 \\ 10 & 12 & 14 \\ 16 & 10 & 22 \end{bmatrix} + \begin{bmatrix} 9 & -6 & 9 \\ 15 & 12 & 9 \\ 3 & 9 & 6 \end{bmatrix}$

$$= \begin{bmatrix} 4+9 & 6-6 & 8+9 \\ 10+15 & 12+12 & 14+9 \\ 16+3 & 10+9 & 22+6 \end{bmatrix} = \begin{bmatrix} 13 & 0 & 17 \\ 25 & 24 & 23 \\ 19 & 19 & 28 \end{bmatrix}$$

2. Find AB where $A = \begin{bmatrix} a & b & c \end{bmatrix}$ and $B = \begin{bmatrix} x \\ y \\ z \end{bmatrix}$.

Ans. Here A is an 1×3 matrix and B is a 3×1 matrix. So they are conformable for the product

$$\therefore \quad AB = \begin{bmatrix} a & b & c \end{bmatrix} \begin{bmatrix} x \\ y \\ z \end{bmatrix} = [ax+by+cz]$$

3. If $A = \begin{bmatrix} 1 & 2 & 3 \\ 2 & 4 & 3 \\ 3 & 4 & 6 \end{bmatrix}$, $B = \begin{bmatrix} 0 & -1 & 3 \\ -2 & 1 & 4 \\ 3 & -2 & -1 \end{bmatrix}$, show that $AB \neq BA$.

Ans.

$$AB = \begin{bmatrix} 1\times 0+2(-2)+3\times 3 & 1(-1)+2\times 1+3(-2) & 1\times 3+2\times 4+3(-1) \\ 2\times 0+4(-2)+3\times 3 & 2(-1)+4\times 1+3(-2) & 2\times 3+4\times 4+3(-1) \\ 3\times 0+4(-2)+6\times 3 & 3(-1)+4\times 1+6(-2) & 3\times 3+4\times 4+6(-1) \end{bmatrix}$$

$$= \begin{bmatrix} 5 & -5 & 8 \\ 1 & -4 & 19 \\ 10 & -11 & 19 \end{bmatrix}$$

Again,

$$BA = \begin{bmatrix} 0 & -1 & 3 \\ -2 & 1 & 4 \\ 3 & -2 & -1 \end{bmatrix} \begin{bmatrix} 1 & 2 & 3 \\ 2 & 4 & 3 \\ 3 & 4 & 6 \end{bmatrix}$$

$$= \begin{bmatrix} 0\times 1+(-1)\times 2+3\times 3 & 0\times 2+(-1)\times 4+3\times 4 & 0\times 3+(-1)\times 3+3\times 6 \\ (-2)\times 1+1\times 2+4\times 3 & (-2)\times 2+1\times 4+4\times 4 & (-2)\times 3+1\times 3+4\times 6 \\ 3\times 1+(-2)\times 2+(-1)\times 3 & 3\times 2+(-2)\times 4+(-1)\times 4 & 3\times 3+(-2)\times 3+(-1)\times 6 \end{bmatrix}$$

$$= \begin{bmatrix} 7 & 8 & 15 \\ 12 & 16 & 21 \\ -4 & -6 & -3 \end{bmatrix}$$

$\therefore \ AB \neq BA.$

The difference between the two matrices AB and BA is known as the *commutator* of A and B and is denoted by $[A, B] = AB - BA$. It should be clear that $[B, A] = -[A, B]$. If in particular AB is equal to BA, the two matrices A and B are said to *commute* with each other.

4. Find adj A and A^{-1} where $A = \begin{bmatrix} 1 & -1 & 2 \\ 2 & -2 & 1 \\ 3 & 3 & -1 \end{bmatrix}$.

Ans. Let B be the matrix each of whose element is the co-factor of the corresponding element of A; i.e., the co-factor matrix of $A = B$.

$$\text{or, } A^c = B = \begin{bmatrix} \begin{vmatrix} -2 & 1 \\ 3 & -1 \end{vmatrix} & -\begin{vmatrix} 2 & 1 \\ 3 & -1 \end{vmatrix} & \begin{vmatrix} 2 & -2 \\ 3 & 3 \end{vmatrix} \\ -\begin{vmatrix} -1 & 2 \\ 3 & -1 \end{vmatrix} & \begin{vmatrix} 1 & 2 \\ 3 & -1 \end{vmatrix} & -\begin{vmatrix} 1 & -1 \\ 3 & 3 \end{vmatrix} \\ \begin{vmatrix} -1 & 2 \\ -2 & 1 \end{vmatrix} & -\begin{vmatrix} 1 & 2 \\ 2 & 1 \end{vmatrix} & \begin{vmatrix} 1 & -1 \\ 2 & -2 \end{vmatrix} \end{bmatrix}$$

$$= \begin{bmatrix} 2-3 & 3+2 & 6+6 \\ 6-1 & -1-6 & -3-3 \\ -1+4 & 4-1 & -2+2 \end{bmatrix} = \begin{bmatrix} -1 & 5 & 12 \\ 5 & -7 & -6 \\ 3 & 3 & 0 \end{bmatrix};$$

interchanging rows and columns of B, we have the transpose of the co-factor matrix i.e., adj A, given by

$$\text{adj } A = (A^c)^T = \begin{bmatrix} -1 & 5 & 3 \\ 5 & -7 & 3 \\ 12 & -6 & 0 \end{bmatrix}.$$

Also $|A| = 1(2-3) - 1(3+2) + 2(6+6) = -1 - 5 + 24 = 18$.

$$\therefore\ A^{-1} = \frac{\text{adj } A}{|A|} = \frac{1}{18}\begin{bmatrix} -1 & 5 & 3 \\ 5 & -7 & 3 \\ 12 & -6 & 0 \end{bmatrix} = \begin{bmatrix} -\frac{1}{18} & \frac{5}{18} & \frac{1}{6} \\ \frac{5}{18} & -\frac{7}{18} & \frac{1}{6} \\ \frac{2}{3} & -\frac{1}{3} & 0 \end{bmatrix}.$$

5. If I be the unit matrix of the third order, prove that $I^2 = I$.

Ans. We have $I = \begin{bmatrix} 1 & 0 & 0 \\ 0 & 1 & 0 \\ 0 & 0 & 1 \end{bmatrix}$.

$$\therefore\ I^2 = I.I. = \begin{bmatrix} 1 & 0 & 0 \\ 0 & 1 & 0 \\ 0 & 0 & 1 \end{bmatrix}\begin{bmatrix} 1 & 0 & 0 \\ 0 & 1 & 0 \\ 0 & 0 & 1 \end{bmatrix}$$

$$= \begin{bmatrix} 1\times1+0\times0+0\times0 & 1\times0+0\times1+0\times0 & 1\times0+0\times0+0\times1 \\ 0\times1+1\times0+0\times0 & 0\times0+1\times1+0\times0 & 0\times0+1\times0+0\times1 \\ 0\times1+0\times0+1\times0 & 0\times0+0\times1+1\times0 & 0\times0+0\times0+1\times1 \end{bmatrix}$$

$$= \begin{bmatrix} 1 & 0 & 0 \\ 0 & 1 & 0 \\ 0 & 0 & 1 \end{bmatrix} = I.$$

6. If $A = \begin{bmatrix} 1 & 1 \\ \alpha & \beta \\ \alpha^2 & \beta^2 \end{bmatrix}$ find AA^T.

Ans. We have $A^T = \begin{bmatrix} 1 & \alpha & \alpha^2 \\ 1 & \beta & \beta^2 \end{bmatrix}$.

$$\therefore AA^T = \begin{bmatrix} 1 & 1 \\ \alpha & \beta \\ \alpha^2 & \beta^2 \end{bmatrix} \begin{bmatrix} 1 & \alpha & \alpha^2 \\ 1 & \beta & \beta^2 \end{bmatrix}$$

$$= \begin{bmatrix} 1\times1+1\times1 & 1\times\alpha+1\times\beta & 1\times\alpha^2+1\times\beta^2 \\ \alpha\times1+\beta\times1 & \alpha\times\alpha+\beta\times\beta & \alpha\times\alpha^2+\beta\times\beta^2 \\ \alpha^2\times1+\beta^2\times1 & \alpha^2\times\alpha+\beta^2\times\beta & \alpha^2\times\alpha^2+\beta^2\times\beta^2 \end{bmatrix}$$

$$= \begin{bmatrix} 2 & (\alpha+\beta) & \alpha^2+\beta^2 \\ \alpha+\beta & \alpha^2+\beta^2 & \alpha^3+\beta^3 \\ \alpha^2+\beta^2 & \alpha^3+\beta^3 & \alpha^4+\beta^4 \end{bmatrix}.$$

7. Find the values of x, y, z if the matrix $\begin{bmatrix} 0 & 2y & z \\ x & y & -z \\ x & -y & z \end{bmatrix}$ is orthogonal.

Ans. Let

$$A = \begin{bmatrix} 0 & 2y & z \\ x & y & -z \\ x & -y & z \end{bmatrix} \quad \therefore A^T = \begin{bmatrix} 0 & x & x \\ 2y & y & -y \\ z & -z & z \end{bmatrix}$$

$$\therefore A^T A = \begin{bmatrix} 0 & x & x \\ 2y & y & -y \\ z & -z & z \end{bmatrix} \begin{bmatrix} 0 & 2y & z \\ x & y & -z \\ x & -y & z \end{bmatrix} = \begin{bmatrix} 2x^2 & 0 & 0 \\ 0 & 6y^2 & 0 \\ 0 & 0 & 3z^2 \end{bmatrix}.$$

But A being an orthogonal matrix, $A^T A = I$.

$$\therefore \begin{bmatrix} 2x^2 & 0 & 0 \\ 0 & 6y^2 & 0 \\ 0 & 0 & 3z^2 \end{bmatrix} = \begin{bmatrix} 1 & 0 & 0 \\ 0 & 1 & 0 \\ 0 & 0 & 1 \end{bmatrix} \quad \text{or, } 2x^2 = 1,\ 6y^2 = 1,\ 3z^2 = 1$$

$$\therefore \quad x = \pm\frac{1}{\sqrt{2}},\ y = \pm\frac{1}{\sqrt{6}},\ z = \pm\frac{1}{\sqrt{3}}.$$

8. If $A = \begin{bmatrix} 1 & 2 & 1 \\ -1 & -1 & 0 \\ 1 & 0 & 0 \end{bmatrix}$, find A^3.

Ans.

$$A^2 = A.A = \begin{bmatrix} 1 & 2 & 1 \\ -1 & -1 & 0 \\ 1 & 0 & 0 \end{bmatrix}\begin{bmatrix} 1 & 2 & 1 \\ -1 & -1 & 0 \\ 1 & 0 & 0 \end{bmatrix}$$

$$= \begin{bmatrix} 1-2+1 & 2-2+0 & 1+0+0 \\ -1+1+0 & -2+1+0 & -1+0+0 \\ 1+0+0 & 2+0+0 & 1+0+0 \end{bmatrix} = \begin{bmatrix} 0 & 0 & 1 \\ 0 & -1 & -1 \\ 1 & 2 & 1 \end{bmatrix}$$

$$\therefore A^3 = A^2.A = \begin{bmatrix} 0 & 0 & 1 \\ 0 & -1 & -1 \\ 1 & 2 & 1 \end{bmatrix}\begin{bmatrix} 1 & 2 & 1 \\ -1 & -1 & 0 \\ 1 & 0 & 0 \end{bmatrix}$$

$$= \begin{bmatrix} 0+0+1 & 0+0+0 & 0+0+0 \\ 0+1-1 & 0+1+0 & 0+0+0 \\ 1-2+1 & 2-2+0 & 1+0+0 \end{bmatrix} = \begin{bmatrix} 1 & 0 & 0 \\ 0 & 1 & 0 \\ 0 & 0 & 1 \end{bmatrix} = I, \text{ i.e., } A^3 = I.$$

19.18.2 Solution of a set of linear equations by matrix method

A set of linear equations can be put as $AX = B$ in matrix notations. Then multiplying both sides by A^{-1}, we get

$$A^{-1}AX = A^{-1}B \text{ or, } X = A^{-1}B$$

which gives the required solution. Linear equations can be solved by the matrix method as shown in the following illustrative example.

Illustrative Example

Solve the following equations by the matrix method

$$\begin{aligned} x - 2y + 3z &= 5 \\ 4x + 3y + 4z &= 7 \\ x + y - z &= -4. \end{aligned}$$

Ans. Let

$$A = \begin{bmatrix} 1 & -2 & 3 \\ 4 & 3 & 4 \\ 1 & 1 & -1 \end{bmatrix}, \; X = \begin{bmatrix} x \\ y \\ z \end{bmatrix}, \text{ and } B = \begin{bmatrix} 5 \\ 7 \\ -4 \end{bmatrix}.$$

Then the above equations may be put, in matrix notation, as

$$AX = B \tag{1}$$

Now, $|A| = 1 \times (-3-4) - 2(4+4) + 3(4-3) = -20$.

Let C be the co-factor matrix obtained by replacing each element of A by its corresponding co-factor, then

$$C = \begin{bmatrix} -3-4 & 4+4 & 4-3 \\ 3-2 & -1-3 & -2-1 \\ -8-9 & 12-4 & 3+8 \end{bmatrix} = \begin{bmatrix} -7 & 8 & 1 \\ 1 & -4 & -3 \\ -17 & 8 & 11 \end{bmatrix}.$$

Interchanging rows and columns of C, we have the transpose of the co-factor matrix i.e., adj A given by

$$\text{adj}\,A = C^T = \begin{bmatrix} -7 & 1 & -17 \\ 8 & -4 & 8 \\ 1 & -3 & 11 \end{bmatrix}$$

$$\therefore\ A^{-1} = \frac{\text{adj}\,A}{|A|} = -\frac{1}{20}\begin{bmatrix} -7 & 1 & -17 \\ 8 & -4 & 8 \\ 1 & -3 & 11 \end{bmatrix} = \begin{bmatrix} \frac{7}{20} & -\frac{1}{20} & \frac{17}{20} \\ -\frac{8}{20} & \frac{4}{20} & -\frac{8}{20} \\ -\frac{1}{20} & \frac{3}{20} & -\frac{11}{20} \end{bmatrix}.$$

Now from (19.18.1) $AX = B$

$$\text{or, } A^{-1}AX = A^{-1}B \text{ or, } X = A^{-1}B$$

$$\text{or, } \begin{bmatrix} x \\ y \\ z \end{bmatrix} = \begin{bmatrix} \frac{7}{20} & -\frac{1}{20} & \frac{17}{20} \\ -\frac{8}{20} & \frac{4}{20} & -\frac{8}{20} \\ -\frac{1}{20} & \frac{3}{20} & -\frac{11}{20} \end{bmatrix}\begin{bmatrix} 5 \\ 7 \\ -4 \end{bmatrix} = \begin{bmatrix} \frac{35}{20} & -\frac{7}{20} & -\frac{68}{20} \\ -\frac{40}{20} & +\frac{28}{20} & +\frac{32}{20} \\ -\frac{5}{20} & +\frac{21}{20} & +\frac{44}{20} \end{bmatrix} = \begin{bmatrix} -2 \\ 1 \\ 3 \end{bmatrix}$$

$$\therefore\ x = -2,\ y = 1 \text{ and } z = 3.$$

19.18.3 Solution of the problem of linear inversion

Let us consider a set of linear equations

$$\left.\begin{aligned} u &= a_1x + b_1y + c_1z \\ v &= a_2x + b_2y + c_2z \\ w &= a_3x + b_3y + c_3z \end{aligned}\right\}. \tag{19.18.1}$$

Our problem is to express x, y, z in terms of u, v and w and that can be done in the following way.

The above set of equations (19.18.1) can be written, in the equivalent matrix notation, as

$$B = \begin{bmatrix} u \\ v \\ w \end{bmatrix} = \begin{bmatrix} a_1 & b_1 & c_1 \\ a_2 & b_2 & c_2 \\ a_3 & b_3 & c_3 \end{bmatrix}\begin{bmatrix} x \\ y \\ z \end{bmatrix} = AX. \tag{19.18.2}$$

The determinant of the matrix A is called the *transformation matrix,* and if its value is not equal to zero, then its inverse i.e., A^{-1} exists and this will produce the required linear inversion, as will be evident from the following.

Now

$$|A| = \begin{vmatrix} a_1 & b_1 & c_1 \\ a_2 & b_2 & c_2 \\ a_3 & b_3 & c_3 \end{vmatrix}$$

whose value is not zero, so that A^{-1} exists and may be calculated as follows.

Let $[A^{ij}]$ be the co-factor matrix obtained by replacing each element of A by its corresponding co-factor; then

$$[A^{ij}] = \begin{bmatrix} a^{11} & a^{12} & a^{13} \\ a^{21} & a^{22} & a^{23} \\ a^{31} & a^{32} & a^{33} \end{bmatrix}$$

where $a^{11} = (-1)^{1+1} \begin{vmatrix} b_2 & c_2 \\ b_3 & c_3 \end{vmatrix}$, $a^{12} = (-1)^{1+2} \begin{vmatrix} a_2 & c_2 \\ a_3 & c_3 \end{vmatrix}$ etc.

Interchanging rows and columns, we have the transpose of the co-factor matrix i.e., adj A given by $[A^{ji}] = \text{adj}\, A$, so that the inverse of the matrix A is given by

$$A^{-1} = \frac{\text{adj}\, A}{|A|}.$$

Multiplying both sides of equation (19.18.2) by A^{-1}, we get

$$A^{-1}B = A^{-1}AX \text{ or, } X = A^{-1}B \tag{19.18.3}$$

which will give the required solution.

Illustrative example

Let

$$\begin{aligned} u &= -x - y + 4z \\ v &= 3x + 4y - 8z \\ w &= -2x + z. \end{aligned}$$

In order to express x, y, z in terms of u, v and w i.e., to invert the above set of linear equations, we may write the above equations in the matrix equivalent form as

$$\begin{bmatrix} u \\ v \\ w \end{bmatrix} = \begin{bmatrix} -1 & -1 & 4 \\ 3 & 4 & -8 \\ -2 & 0 & 1 \end{bmatrix} \begin{bmatrix} x \\ y \\ z \end{bmatrix}.$$

Here

$$A = \begin{bmatrix} -1 & -1 & 4 \\ 3 & 4 & -8 \\ -2 & 0 & 1 \end{bmatrix};$$

then the corresponding co-factor matrix obtained by replacing each element of A by its corresponding co-factor, is given by

$$A^c = [A^{ij}] = \begin{bmatrix} 4 & 13 & 8 \\ 1 & 7 & 2 \\ -8 & 4 & -1 \end{bmatrix},$$

so that interchanging rows and columns, we have the transpose of the co-factor matrix given by

$$\text{adj}\, A = (A^c)^T = \left[A^{ji}\right] = \begin{bmatrix} 4 & 1 & -8 \\ 13 & 7 & 4 \\ 8 & 2 & -1 \end{bmatrix}$$

and

$$|A| = (-1) \times (4-0) - 1(16-3) + 4(0+8) = -4 - 13 + 32 = 15.$$

$$\therefore\ A^{-1} = \frac{\text{adj}\, A}{|A|} = \frac{1}{15}\begin{bmatrix} 4 & 1 & -8 \\ 13 & 7 & 4 \\ 8 & 2 & -1 \end{bmatrix} = \begin{bmatrix} \frac{4}{15} & \frac{1}{15} & -\frac{8}{15} \\ \frac{13}{15} & \frac{7}{15} & \frac{4}{15} \\ \frac{8}{15} & \frac{2}{15} & -\frac{1}{15} \end{bmatrix}.$$

Therefore the inverse transformation is given by

$$\begin{bmatrix} \frac{4}{15} & \frac{1}{15} & -\frac{8}{15} \\ \frac{13}{15} & \frac{7}{15} & \frac{4}{15} \\ \frac{8}{15} & \frac{2}{15} & -\frac{1}{15} \end{bmatrix}\begin{bmatrix} u \\ v \\ w \end{bmatrix} = \begin{bmatrix} x \\ y \\ z \end{bmatrix}$$

$$\begin{aligned} x &= \frac{4}{15}u + \frac{1}{15}v - \frac{8}{15}w \\ \text{or,}\quad y &= \frac{13}{15}u + \frac{7}{15}v + \frac{4}{15}w \\ z &= \frac{8}{15}u + \frac{2}{15}v - \frac{1}{15}w \end{aligned} \qquad \begin{aligned} 15x &= 4u + v - 8w \\ \text{or,}\quad 15y &= 13u + 7v + 4w \\ 15z &= 8u + 2v - w \end{aligned}$$

which are the linearly inverted relations.

19.18.4 Diagonalisation of a matrix

In physics, we very often come across problems in which a given matrix requires to be diagonalised where only the diagonal elements survive. The problem of finding the moment of inertia of a rigid body about a specified axis is an example.

The angular momentum of a rigid body is given by

$$[L] = [I_n][w] \qquad (19.18.4)$$

where $[L]$ is a column vector, $[w]$ is also a column vector but $[I_n]$ is a square matrix of order n called the *inertia matrix* defined by*

$$[I_n] = \begin{bmatrix} A & -F & -E \\ -F & B & -D \\ -E & -D & C \end{bmatrix},$$

where A, B, C are the moments of intertia about X, Y and Z axes respectively and D, E, F denote the products of inertia referred to $Y-Z, Z-X$ and $X-Y$ axes respectively. *If now the axes be chosen in such a way that the terms D, E, F involving the products of interia vanish, then the non-diagonal terms of the inertia matrix become zero and only the diagonal terms survive. The three axes so chosen are called the Principal axes of inertia, and the process of finding the Principal axes of inertia, so that the terms involving the products of inertia vanish, is known as the process of diagonalisation of the inertia matrix.* After such diagonalisation, the diagonal terms are known as the *principal moments of inertia* and are denoted as I_1, I_2, and I_3. The matrix equation (19.18.4) can now be written as

$$\begin{bmatrix} L_1 \\ L_2 \\ L_3 \end{bmatrix} = \begin{bmatrix} I_1 & 0 & 0 \\ 0 & I_2 & 0 \\ 0 & 0 & I_3 \end{bmatrix} \begin{bmatrix} w_1 \\ w_2 \\ w_3 \end{bmatrix}. \tag{19.18.5}$$

Therefore, w.r.t. the principal axes, the relations connecting the angular momenta components and angular velocity components are written as

$$L_1 = I_1 w_1, \quad L_2 = I_2 w, \quad L_3 = I_3 w_3. \tag{19.18.6}$$

Hence, it is essential to know the technique of how to diagonalise a matrix, such as the inertia matrix as discussed above.

19.18.5 Diagonalisation Technique

Let us consider a square matrix $A \equiv [a_{ij}]$ of order n, and let X be a column matrix (or vector) of the same order, such that

$$AX = \lambda X \tag{19.18.7}$$

where λ is a scalar. Thus the transformation matrix A operates on the column matrix to produce the effect of scalar multiplication of the column vector by λ. The *column vector X* is then called an *eigen-vector* (characteristic or proper) of the transformation matrix A; λ is called an *eigen-value* of A corresponding to the column vector X. The problem of finding the eigen-values and the corresponding eigen-vectors of a matrix is known as the *eigen-value problem*.

Equation (19.18.7) may be explicitly written as

$$\begin{bmatrix} a_{11} & a_{12} & \cdots & a_{1n} \\ a_{21} & a_{22} & \cdots & a_{2n} \\ \cdots & \cdots & \cdots & \cdots \\ a_{n1} & a_{n2} & \cdots & a_{nn} \end{bmatrix} \begin{bmatrix} x_1 \\ x_2 \\ \vdots \\ x_n \end{bmatrix} = \lambda \begin{bmatrix} x_1 \\ x_2 \\ \vdots \\ x_n \end{bmatrix}$$

*See Chapter 4 for details.

$$\left.\begin{aligned} a_{11}x_1 + a_{12}x_2 + \cdots + a_{1n}x_n &= \lambda x_1 \\ a_{21}x_1 + a_{22}x_2 + \cdots + a_{2n}x_n &= \lambda x_2 \\ \cdots\cdots\cdots\cdots\cdots\cdots\cdots\cdots \\ a_{n1}x_1 + a_{n2}x_2 + \cdots + a_{nn}x_n &= \lambda x_n \end{aligned}\right\}. \quad \text{or,} \tag{19.18.8}$$

Re-arranging the terms, the above set of equations reduces to

$$\begin{aligned} (a_{11} - \lambda)x_1 + a_{12}x_2 + \cdots + a_{1n}x_n &= 0 \\ a_{21}x_1 + (a_{22} - \lambda)x_2 + \cdots + a_{2n}x_n &= 0 \\ \cdots\cdots\cdots\cdots\cdots\cdots\cdots\cdots \\ a_{n1}x_1 + a_{n2}x_2 + \cdots + (a_{nn} - \lambda)x_n &= 0 \end{aligned}$$

which can be written, in the matrix form, as

$$\begin{bmatrix} (a_{11} - \lambda) & a_{12} & \cdots & a_{1n} \\ a_{21} & (a_{22} - \lambda) & \cdots & a_{2n} \\ \cdots & \cdots & \cdots & \cdots \\ a_{n1} & a_{n2} & \cdots & (a_{nn} - \lambda) \end{bmatrix} \begin{bmatrix} x_1 \\ x_2 \\ \vdots \\ x_n \end{bmatrix} = 0 = \begin{bmatrix} 0 \\ 0 \\ \vdots \\ 0 \end{bmatrix}$$

$$\text{or,} \quad [A - \lambda I][X] = [0] = 0, \tag{19.18.9}$$

where I is a unit matrix of order n.

The equation (19.18.9) being a system of n simultaneous linear equations in the n paramaters x_i has always a trivial solution $x_i = 0$, but will also have non-trivial solutions, if and only if $\det(A - \lambda I) = |D_\lambda| = 0$

$$\text{i.e.,} \quad |D_\lambda| = \begin{vmatrix} (a_{11} - \lambda) & a_{12} & \cdots & a_{1n} \\ a_{21} & (a_{22} - \lambda) & \cdots & a_{2n} \\ \cdots & \cdots & \cdots & \cdots \\ a_{n1} & a_{n2} & \cdots & (a_{nn} - \lambda) \end{vmatrix} = 0. \tag{19.18.10}$$

From (19.18.10) we can definitely say that $|D_\lambda|$ is a polynomial (of degree n) in λ, and it is called the *characteristic polynomial* of the matrix A. This determinantal equation (19.18.10) is called *the characteristic equation* of A. Now the above determinant can be expanded as

$$|D_\lambda| = C_0\lambda^n + C_1\lambda^{n-1} + C_2\lambda^{n-2} + \cdots + C_n = 0 \tag{19.18.11}$$

where the coefficients C_0, C_1 etc. are functions of the elements of A. Since we obtain the polynomial equation (19.18.11) from the determinantal equation (19.18.10), this polynomial equivalent equation is also called *the characteristic equation.*

Again, since the polynomial is of degree n, it must have n roots λ_1, λ_2, λ_3 etc. which make the polynomials vanish, so that we can write

$$C_0\lambda^n + C_1\lambda^{n-1} + \cdots + C_n = (\lambda_1 - \lambda)(\lambda_2 - \lambda)\cdots(\lambda_n - \lambda).$$

Considering a square matrix of order 3, we may write

$$C_0\lambda^3 + C_1\lambda^2 + C_2\lambda + C_3 = (\lambda_1 - \lambda)(\lambda_2 - \lambda)(\lambda_3 - \lambda)$$

whence we have

$$\begin{aligned} C_0 &= -1 = (-1)^3, \\ C_1 &= (\lambda_1 + \lambda_2 + \lambda_3) = (-1)^{3-1}(\lambda_1 + \lambda_2 + \lambda_3), \\ C_2 &= -(\lambda_1\lambda_2 + \lambda_2\lambda_3 + \lambda_3\lambda_1) = (-1)^{3-2}(\lambda_1\lambda_2 + \lambda_2\lambda_3 + \lambda_3\lambda_1), \\ C_3 &= \lambda_1\lambda_2\lambda_3. \end{aligned}$$

So for the general case of order n, we can write

$$\begin{aligned} C_0 &= (-1)^n, \\ C_1 &= (-1)^{n-1}(\lambda_1 + \lambda_2 + \cdots + \lambda_n), \\ &\cdots\cdots\cdots\cdots \\ C_n &= \lambda_1\lambda_2\cdots\lambda_n. \end{aligned} \tag{19.18.12}$$

Thus, we see that $\lambda_1, \lambda_2, \lambda_3 \cdots \lambda_n$ *are the values of* λ *for which* $\det(A - \lambda I)$ *vanishes, and hence equation* (19.18.9) *will possess non-trivial solutions. In other words, we have the result that every square matrix has exactly as many eigen-values as its order and these eigen-values are the roots of its characteristic polynomial.*

Again, it can be verified by an inspection of (19.18.10) that

$$\left.\begin{aligned} C_0 &= (-1)^n, \\ C_1 &= (-1)^{n-1}(a_{11} + a_{22} + \cdots + a_{nn}), \\ &\cdots\cdots\cdots\cdots \\ C_n &= D(0) = \det[A] \end{aligned}\right\}. \tag{19.18.13}$$

Comparing (19.18.12) and (19.18.13), we have the following two important results regarding the eigen-values of a matrix:

(a) $$\lambda_1 + \lambda_2 + \cdots + \lambda_n = a_{11} + a_{22} + \cdots + a_{nn} \equiv Tr[A] \tag{19.18.14}$$

which shows that *the sum of the roots of the characteristic equation or the sum of the eigen-values is equal to the sum of the diagonal elements of* $[A]$ i.e., equal to the trace of $[A]$ or $Tr[A]$.

(b) $$\lambda_1\lambda_2\cdots\lambda_n = \det[A] \tag{19.18.15}$$

which shows that the product of the eigen-values is equal to the determinant of the matrix A.

Illustrative Examples

1. Find the eigen-values of the matrix $A = \begin{bmatrix} 2 & -2 & 3 \\ 2 & -1 & 0 \\ -1 & 1 & 1 \end{bmatrix}$.

The characteristic polynomial is

$$D(\lambda) = \det(A - \lambda I) = \begin{vmatrix} 2-\lambda & -2 & 3 \\ 2 & -1-\lambda & 0 \\ -1 & 1 & 1-\lambda \end{vmatrix}$$

$$= -\lambda^3 + 2\lambda^2 - 6\lambda + 5 = (1-\lambda)(\lambda^2 - \lambda + 5).$$

The eigen-values of A, which are the solutions of the equation $D(\lambda) = 0$, are therefore, $1, \left(1 + i\sqrt{19}\right)/2$ and $\left(1 - i\sqrt{19}\right)/2$. We can also verify equations (19.18.14) and (19.18.15) in this case.

2. If A is non-singular, show that the eigen-values of A^{-1} are the reciprocals of those of A and every eigen-vector of A is also an eigen-vector of A^{-1}.

Let λ be an eigen-value of A corresponding to the eigen-vector X, so that

$$AX = \lambda X. \tag{1}$$

Since A^{-1} exists, we multiply equation (1) by A^{-1} to obtain

$$A^{-1}AX = A^{-1}\lambda X \quad \text{or,} \quad X = \lambda A^{-1}X. \tag{2}$$

Since A is non-singular, $\lambda \neq 0$.

Hence, dividing equation (2) by λ, we have

$$A^{-1}X = \left(\frac{1}{\lambda}\right)X. \tag{3}$$

Since this must be true for every eigen-value of A, the result follows.

3. Find the eigen-values and the eigen-vectors of the matrix

$$A = \begin{bmatrix} q & p & p \\ p & q & p \\ p & p & q \end{bmatrix} \tag{1}$$

where p and q are scalars, $(p \neq 0)$.

The characteristic polynomial of the given matrix is

$$\det(A - \lambda I) = \begin{vmatrix} q-\lambda & p & p \\ p & q-\lambda & p \\ p & p & q-\lambda \end{vmatrix}$$

$$= (q - p - \lambda)\left[\lambda^2 - (2q+p)\lambda + (q^2 + qp - 2p^2)\right]$$

$$= (q - p - \lambda)^2(q + 2p - \lambda). \tag{2}$$

The eigen-values of A are therefore

$$\lambda_1 = q - p, \ \lambda_2 = q - p, \ \lambda_3 = q + 2p. \tag{3}$$

Here the eigen-value $(q - p)$ has multiplicity 2, while the eigen-value $(q + 2p)$ has multiplicity 1.

To obtain the eigen-vectors, we first take $\lambda = \lambda_3 = q + 2p$ and solve the equation $(A - \lambda_3 I)X = 0$. We have

$$\begin{bmatrix} -2p & p & p \\ p & -2p & p \\ p & p & -2p \end{bmatrix} \begin{bmatrix} x_1 \\ x_2 \\ x_3 \end{bmatrix} = \begin{bmatrix} 0 \\ 0 \\ 0 \end{bmatrix} \tag{4}$$

$$\left.\begin{aligned} -2px_1 + px_2 + px_3 &= 0 \\ \text{or, } px_1 - 2px_2 + px_3 &= 0 \\ px_1 + px_2 - 2px_3 &= 0 \end{aligned}\right\}. \tag{5}$$

The rank of the matrix in equation (4) is 2, so that only two of the equations (5) will be independent and the third equation does not give any additional constraint. Solving any two of equations (5), we get

$$x_3 = x_2 = x_1 \quad (x_1 \text{ being arbitrary}). \tag{6}$$

Taking $x_1 = a$, the eigen-vector corresponding to the eigen-value $\lambda_3 = q + 2p$ can be written as

$$X_3 = \left\{ a \quad a \quad a \right\} \tag{7}$$

or, choosing $a = 1$,

$$X_3 = \left\{ 1 \quad 1 \quad 1 \right\}. \tag{8}$$

Again taking $\lambda = q - p$, and solving the equation $[A - (q - p)I]X = 0$, we have

$$\begin{bmatrix} p & p & p \\ p & p & p \\ p & p & p \end{bmatrix} \begin{bmatrix} x_1 \\ x_2 \\ x_3 \end{bmatrix} = \begin{bmatrix} 0 \\ 0 \\ 0 \end{bmatrix}. \tag{9}$$

Now the matrix of equation (9) is of rank 1, which is consistent with the fact that the eigen-value $q - p$ has multiplicity 2. There will be only one independent equation and it is of no use to write all the three equations. The equation to be solved is

$$px_1 + px_2 + px_3 = 0 \tag{10}$$

whose solution, since $p \neq 0$, is

$$x_3 = -x_1 - x_2 \quad (x_1 \text{ and } x_2 \text{ being arbitrary}) \tag{11}$$

which involves a two-fold arbitrariness. x_1 and x_2 may be given any values whatsoever according to our choice. Let us first choose $x_1 = 1, x_2 = 0$ to obtain the first eigen-vector

$$X_1 = \left\{ 1 \quad 0 \quad -1 \right\} \tag{12}$$

To obtain the second eigen-vector, we may choose $x_1 = 0, x_2 = 1$, so that $x_3 = -1$, which gives the eigen-vector

$$X_2 = \left\{ 0 \quad 1 \quad -1 \right\}. \tag{13}$$

It is to be noted that the most general eigen-vector having the eigen-value $q-p$ is $\left\{ a \;\; b \;\; -a-b \right\}$ with any arbitrary values of a and b. However this will depend linearly on X_1 and X_2. In fact it can be shown that

$$\left\{ a \;\; b \;\; -a-b \right\} = aX_1 + bX_2 \tag{14}$$

The solution is, finally, therefore:

(i) for eigen-value $q+2p$, eigen-vector is $\left\{ 1 \;\; 1 \;\; 1 \right\}$

(ii) for eigen-value $q-p$ with multiplicity 2, a set of two linearly independent eigen-vectors is $\left\{ 1 \;\; 0 \;\; -1 \right\}$, $\left\{ 0 \;\; 1 \;\; -1 \right\}$.

If *normalised* eigen-vectors are required, the above eigen-vectors can be normalised to give

$$X_1 = \frac{1}{\sqrt{2}}\{1 \quad 0 \quad -1\}, \quad X_2 = \frac{1}{\sqrt{2}}\{0 \quad 1 \quad -1\}, \quad X_3 = \frac{1}{\sqrt{3}}\{1 \quad 1 \quad 1\}.$$

It can be verified that the operation of the matrix A of equation (1) on any of its eigen-vectors gives a constant multiple of the same eigen-vector. Thus,

$$\begin{bmatrix} q & p & p \\ p & q & p \\ p & p & q \end{bmatrix} \begin{bmatrix} a \\ a \\ a \end{bmatrix} = \begin{bmatrix} (q+2p)a \\ (q+2p)a \\ (q+2p)a \end{bmatrix} = (q+2p) \begin{bmatrix} a \\ a \\ a \end{bmatrix} \tag{15}$$

and

$$\begin{bmatrix} q & p & p \\ p & q & p \\ p & p & q \end{bmatrix} \begin{bmatrix} a \\ b \\ -a-b \end{bmatrix} = \begin{bmatrix} (q-p)a \\ (q-p)b \\ (p-q)(a+b) \end{bmatrix} = (q-p) \begin{bmatrix} a \\ b \\ -a-b \end{bmatrix}. \tag{16}$$

N.B. X_1, X_2 and X_3 are column vectors, as mentioned earlier.

19.18.6 The Process of Diagonalisation of a matrix

With some knowledge about the eigen-values and eigen-vectors as illustrated in the above examples, we now consider the process of diagonalisation.

Let A be a square matrix of order n with elements a_{ij}, which has n linearly independent eigen-vectors which may be denoted by X_i. Let λ_i be the corresponding eigen-values so that $AX_i = \lambda_i X_i$.

If we denote the eigen-vectors X_i by column vectors in the form $X_i = \left\{ x_{1i} \;\; x_{2i} \;\; \cdots \;\; x_{ni} \right\}$, then the eigen-value equation can be written in the matrix form

$$\begin{bmatrix} a_{11} & a_{12} & \cdots & a_{1n} \\ a_{21} & a_{22} & \cdots & a_{2n} \\ \vdots & \vdots & \cdots & \vdots \\ a_{n1} & a_{n2} & \cdots & a_{nn} \end{bmatrix} \begin{bmatrix} x_{1i} \\ x_{2i} \\ \vdots \\ x_{ni} \end{bmatrix} = \lambda_i \begin{bmatrix} x_{1i} \\ x_{2i} \\ \vdots \\ x_{ni} \end{bmatrix}. \tag{19.18.16}$$

Writing explicitly the jth component of the equation from the above, we have

$$\sum_{k=1}^{n} a_{jk} x_{ki} = \lambda_i x_{ji}. \tag{19.18.17}$$

Let us now construct a matrix P of order $(n \times n)$ whose columns are the eigen-vectors x_i i.e.,

$$P = \begin{bmatrix} x_{11} & \cdots & x_{1i} & \cdots & x_{1n} \\ x_{21} & \cdots & x_{2i} & \cdots & x_{2n} \\ \vdots & & & & \\ x_{n1} & \cdots & x_{ni} & \cdots & x_{nn} \end{bmatrix}, \quad (P)_{ji} = x_{ji}. \tag{19.18.18}$$

Since $\det P \neq 0$, the matrix P will have its inverse i.e., P^{-1} exists.

Let us now state that the *matrix* $P^{-1}AP$ *is diagonal whose diagonal elements are the eigen-values of* A. To prove this, let us define a diagonal matrix A_1 whose diagonal elements are λ_i i.e.,

$$A_1 = \begin{bmatrix} \lambda_1 & \cdots & \cdots & 0 \\ 0 & \lambda_2 & \cdots & 0 \\ \cdots & \cdots & \cdots & \cdots \\ 0 & \cdots & \cdots & \lambda_n \end{bmatrix}. \tag{19.18.19}$$

We are to prove that

$$P^{-1}AP = A_1. \tag{19.18.20}$$

Multiplying equation (19.18.20) from the left by P, we obtain

$$AP = PA_1. \tag{19.18.21}$$

Taking the jith element of the left hand side of the above equation, we get

$$(AP)_{ji} = \sum_{k=1}^{n} (A)_{jk}(P)_{ki} = \sum_{k=1}^{n} a_{jk}x_{ki} = \lambda_i x_{ji}. \tag{19.18.22}$$

[using equation (19.18.17)]

Now the jith element of the right hand side of equation (19.18.21) is given by

$$(PA_1)_{ji} = \sum_{k=1}^{n} (P)_{jk}(A_1)_{ki} = \sum_{k=1}^{n} x_{jk}\lambda_i\delta_{ki} = \lambda_i x_{ji}. \tag{19.18.23}$$

Thus the equality in equation (19.18.21) is established. That is equations (19.18.22) and (19.18.23) clearly prove the validity of equations (19.18.20) and (19.18.21). This proves our statement. This process is called *the diagonalisation of a matrix.* Thus P is the *diagonalising* matrix which diagonalises the given matrix A, the process of obtaining the diagonal matrix being represented as $P^{-1}AP = A_1$, where P is the matrix formed out of the eigen-vectors and P^{-1} is its inverse.

Illustrative Examples

1. Diagonalise the matrix of equation (1), given in illustrative example 3 under Art. 19.18.5.

We construct a matrix whose columns are the eigen-vectors of the given matrix A of equation (1) of illustrative example 3 under Art. 19.18.5. From equations (8), (12) and (13) obtained in the same example, we have

$$P = \begin{bmatrix} 1 & 0 & 1 \\ 0 & 1 & 1 \\ -1 & -1 & 1 \end{bmatrix}. \tag{1}$$

the inverse of which is found to be

$$P^{-1} = \frac{1}{3}\begin{bmatrix} 2 & -1 & -1 \\ -1 & 2 & -1 \\ 1 & 1 & 1 \end{bmatrix}. \tag{2}$$

We may then write

$$P^{-1}AP = \frac{1}{3}\begin{bmatrix} 2 & -1 & -1 \\ -1 & 2 & -1 \\ 1 & 1 & 1 \end{bmatrix}\begin{bmatrix} q & p & p \\ p & q & p \\ p & p & q \end{bmatrix} \times \begin{bmatrix} 1 & 0 & 1 \\ 0 & 1 & 1 \\ -1 & -1 & 1 \end{bmatrix} = \begin{bmatrix} q-p & 0 & 0 \\ 0 & q-p & 0 \\ 0 & 0 & q+2p \end{bmatrix}. \tag{3}$$

It is to be noted that $P^{-1}AP$ is a diagonal matrix whose diagonal elements are just the eigen-values of A.

Thus we can state the following theorem:

If a matrix of order n has n linearly independent eigen-vectors, then it is related through a similarity transformation to a diagonal matrix whose diagonal elements are the eigen-values of the matrix.

It is important to mention here that the matrix P which is capable of diagonalising a given matrix is not unique. This is because we can arrange the eigen-vectors $X_1, X_2 \cdots X_n$ *in any order* in the construction of P. In example 1 above, for instance, instead of choosing $P = \begin{bmatrix} X_1 & X_2 & X_3 \end{bmatrix}$, we could have choosen

$$Q = \begin{bmatrix} X_3 & X_1 & X_2 \end{bmatrix} = \begin{bmatrix} 1 & 1 & 0 \\ 1 & 0 & 1 \\ 1 & -1 & -1 \end{bmatrix} \tag{4}$$

whence we would have obtained

$$Q^{-1}AQ = \begin{bmatrix} q+2p & 0 & 0 \\ 0 & q-p & 0 \\ 0 & 0 & q-p \end{bmatrix} \tag{5}$$

which is again a diagonal matrix whose diagonal elements are the eigen-values of A. The order of the eigen-values in expression (5) is different from that in expression (3), and in each case, the order of the eigen-values corresponds to the order of the eigen-vectors in the construction of the diagonalising matrix. We can therefore state the general rule that *in the process of diagonalisation $P^{-1}AP = A_1$, the order of the eigen-values in A_1 corresponds to the order of the eigen-vectors of A in constructing P.*

2. Diagonalise the matrix $A = \begin{bmatrix} 3 & 4 & 2 \\ 4 & 3 & -3 \\ 0 & 0 & 1 \end{bmatrix}$.

The characteristic polynomial is

$$D(\lambda) = \det(A - \lambda I) = \begin{vmatrix} 3-\lambda & 4 & 2 \\ 4 & 3-\lambda & -3 \\ 0 & 0 & 1-\lambda \end{vmatrix} = (1-\lambda)[(3-\lambda)^2 - 16]$$

which is a polynomial of degree 3 in λ and hence admits of three roots λ_1, λ_2 and λ_3.

The eigen-values of A, which are the solutions of $D(\lambda) = (1-\lambda)\left[(3-\lambda)^2 - 16\right] = 0$ are therefore $\lambda_1 = 1$, $\lambda_2 = -1$, and $\lambda_3 = 7$.

So the diagonalising matrix P will be such that

$$P^{-1}AP = \begin{bmatrix} 1 & 0 & 0 \\ 0 & -1 & 0 \\ 0 & 0 & 7 \end{bmatrix}.$$

So to find the diagonalising matrix, the following steps should be followed one after another.

Step I: *The eigen-values of the matrix to be diagonalised are to be found.*

Step II: *The corresponding eigen-vectors which are linearly independent are to be found.*

Step III: *With the eigen-vectors as the columns, the diagonalising matrix is to be constructed.*

Step IV: *The diagonal form of the given matrix A is to be found from $P^{-1}AP = A_1$.*

The following illustrative example will make the method clear.

3. Diagonalise the matrix given by $A = \begin{bmatrix} -9 & 4 & 4 \\ -8 & 3 & 4 \\ -16 & 8 & 7 \end{bmatrix}$.

Step I: *To find the eigen-values of A,* we have the characteristic equation given by

$$\det(A - \lambda I) = \begin{vmatrix} -9-\lambda & 4 & 4 \\ -8 & 3-\lambda & 4 \\ -16 & 8 & 7-\lambda \end{vmatrix} = 0$$

$$\text{or,} \quad \begin{vmatrix} -1-\lambda & 4 & 4 \\ -1-\lambda & 3-\lambda & 4 \\ -1-\lambda & 8 & 7-\lambda \end{vmatrix} = 0 \quad (\text{using } C_1' = C_1 + C_2 + C_3)$$

$$\text{or,} \quad -(1+\lambda)\begin{vmatrix} 1 & 4 & 4 \\ 1 & 3-\lambda & 4 \\ 1 & 8 & 7-\lambda \end{vmatrix} = 0$$

$$\text{or,} \quad (1+\lambda)\begin{vmatrix} 0 & (1+\lambda) & 0 \\ 1 & (3-\lambda) & 4 \\ 1 & 8 & (7-\lambda) \end{vmatrix} = 0 \quad (\text{using } R_1' = R_1 - R_2)$$

$$\text{or,} \quad (1+\lambda)\begin{vmatrix} 0 & (1+\lambda) & 0 \\ 0 & -(\lambda+5) & -(3-\lambda) \\ 1 & 8 & (7-\lambda) \end{vmatrix} = 0 \quad (\text{using } R_2' = R_2 - R_3)$$

$$\text{or,} \quad (1+\lambda)\,[-(1+\lambda)(3-\lambda)] = 0$$

$$\text{or,} \quad (\lambda+1)(\lambda+1)(\lambda-3) = 0$$

giving the three roots $\lambda_1 = -1$, $\lambda_2 = -1$ and $\lambda_3 = 3$.

We see here that the eigen-value -1 has multiplicity 2, while the eigen-value 3 has multiplicity 1.

Step II: *To obtain the eigen-vectors,* we first take $\lambda = \lambda_3 = 3$ and solve the equation $(A - \lambda_3 I)\,X = 0$. We have then

$$\begin{bmatrix} -12 & 4 & 4 \\ -8 & 0 & 4 \\ -16 & 8 & 4 \end{bmatrix} \begin{bmatrix} x_1 \\ x_2 \\ x_3 \end{bmatrix} = \begin{bmatrix} 0 \\ 0 \\ 0 \end{bmatrix}$$

$$\text{or,} \quad \begin{bmatrix} -12 & 4 & 4 \\ 4 & -4 & 0 \\ -16 & 8 & 4 \end{bmatrix} \begin{bmatrix} x_1 \\ x_2 \\ x_3 \end{bmatrix} = \begin{bmatrix} 0 \\ 0 \\ 0 \end{bmatrix} \quad (\text{using } R_2' = R_2 - R_1)$$

$$\text{or,} \quad \begin{bmatrix} -12 & 4 & 4 \\ 4 & -4 & 0 \\ -4 & 4 & 0 \end{bmatrix} \begin{bmatrix} x_1 \\ x_2 \\ x_3 \end{bmatrix} = \begin{bmatrix} 0 \\ 0 \\ 0 \end{bmatrix} \quad (\text{using } R_3' = R_3 - R_1)$$

$$\text{or,} \quad \begin{bmatrix} -12 & 4 & 4 \\ 4 & -4 & 0 \\ 0 & 0 & 0 \end{bmatrix} \begin{bmatrix} x_1 \\ x_2 \\ x_3 \end{bmatrix} = \begin{bmatrix} 0 \\ 0 \\ 0 \end{bmatrix}. \qquad (1)$$

The rank of the matrix in (1) is 2 and hence only two of the equations obtained from (1) will be independent and will give one independent solution. From (1) we may write

$$\left.\begin{aligned} -12x_1 + 4x_2 + 4x_3 &= 0 \\ 4x_1 - 4x_2 &= 0 \end{aligned}\right\} \qquad (2)$$

whence we have $x_1 = x_2$. Since x_1 and x_2 are arbitrary, we may write $x_1 = 1$ and $x_2 = 1$, which on substitution in (2) gives $x_3 = 2$

$$\therefore \; X_3 = \begin{bmatrix} 1 \\ 1 \\ 2 \end{bmatrix}. \qquad (3)$$

To obtain the other two eigen-vectors, let us take $\lambda = \lambda_1 = -1$ or, $\lambda = \lambda_2 = -1$.

With $\lambda = \lambda_1 = -1$, we are to solve the equation $(A - \lambda_1 I)X = 0$. We have then

$$\begin{bmatrix} -8 & 4 & 4 \\ -8 & 4 & 4 \\ -16 & 8 & 8 \end{bmatrix} \begin{bmatrix} x_1 \\ x_2 \\ x_3 \end{bmatrix} = \begin{bmatrix} 0 \\ 0 \\ 0 \end{bmatrix}$$

$$\text{or,} \quad \begin{bmatrix} -8 & 4 & 4 \\ 0 & 0 & 0 \\ -16 & 8 & 8 \end{bmatrix} \begin{bmatrix} x_1 \\ x_2 \\ x_3 \end{bmatrix} = \begin{bmatrix} 0 \\ 0 \\ 0 \end{bmatrix} \quad \text{(using } R_2' = R_2 - R_1)$$

$$\text{or,} \quad \begin{bmatrix} -8 & 4 & 4 \\ 0 & 0 & 0 \\ 0 & 0 & 0 \end{bmatrix} \begin{bmatrix} x_1 \\ x_2 \\ x_3 \end{bmatrix} = \begin{bmatrix} 0 \\ 0 \\ 0 \end{bmatrix} \tag{4}$$

(using $R_3' = R_3 - 2R_1$).

The rank of the matrix in (4) is 1, which is consistent with the fact that the eigen-value -1 has multiplicity 2. There will be only one independent equation which may be written as $-8x_1 + 4x_2 + 4x_3 = 0$, whence $2x_1 = x_2 + x_3$ (x_2 and x_3 being arbitrary), which involves a two-fold arbitrariness. x_2 and x_3 may be given any values whatsoever according to our choice. Choosing $x_2 = 1$, $x_3 = 1$, we have $x_1 = 1$; again choosing $x_2 = 1$, $x_3 = -1$, we have $x_1 = 0$.

$\therefore$ the eigen-vectors are

$$X_1 = \begin{bmatrix} 1 \\ 1 \\ 1 \end{bmatrix}, \quad X_2 = \begin{bmatrix} 0 \\ 1 \\ -1 \end{bmatrix}.$$

Step III and **Step IV:** Hence the diagonalising matrix P is constructed with the eigen-vectors as columns, so that

$$P = \begin{bmatrix} 1 & 0 & 1 \\ 1 & 1 & 1 \\ 1 & -1 & 2 \end{bmatrix}$$

and the corresponding co-factor matrix is

$$P^c = \begin{bmatrix} 3 & -1 & -2 \\ -1 & 1 & 1 \\ -1 & 0 & 1 \end{bmatrix}$$

and therefore the transpose of the co-factor matrix, which is also its adjoint is

$$\text{adj}\, P = (P^c)^T = \begin{bmatrix} 3 & -1 & -1 \\ -1 & 1 & 0 \\ -2 & 1 & 1 \end{bmatrix}$$

so that we may write

$$P^{-1} = \frac{(P^c)^T}{|P|} = \frac{\begin{bmatrix} 3 & -1 & -1 \\ -1 & 1 & 0 \\ -2 & 1 & 1 \end{bmatrix}}{1} = \begin{bmatrix} 3 & -1 & -1 \\ -1 & 1 & 0 \\ -2 & 1 & 1 \end{bmatrix}$$

$$\therefore\ A_1 = P^{-1}AP = P^{-1}\begin{bmatrix} -9 & 4 & 4 \\ -8 & 3 & 4 \\ -16 & 8 & 7 \end{bmatrix}\begin{bmatrix} 1 & 0 & 1 \\ 1 & 1 & 1 \\ 1 & -1 & 2 \end{bmatrix} = P^{-1}\begin{bmatrix} -1 & 0 & 3 \\ -1 & -1 & 3 \\ -1 & 1 & 6 \end{bmatrix}$$

$$= \begin{bmatrix} 3 & -1 & -1 \\ -1 & 1 & 0 \\ -2 & 1 & 1 \end{bmatrix}\begin{bmatrix} -1 & 0 & 3 \\ -1 & -1 & 3 \\ -1 & 1 & 6 \end{bmatrix} = \begin{bmatrix} -1 & 0 & 0 \\ 0 & -1 & 0 \\ 0 & 0 & 3 \end{bmatrix}.$$

Thus the diagonalised matrix consists of diagonal elements which are the eigen-values of the given matrix.

19.18.7 Cayley-Hamiltons' Theorem

The theorem states that every square matrix satisfies its own characteristic equation. Let us verify the theorem for the matrix $A = \begin{bmatrix} 3 & 1 \\ -1 & 2 \end{bmatrix}$.

The characteristic equation is

$$\det(A - \lambda I) = \begin{vmatrix} 3-\lambda & 1 \\ -1 & 2-\lambda \end{vmatrix} = 0 \quad \text{or,} \quad \lambda^2 - 5\lambda + 7 = 0.$$

According to Cayley-Hamiltons' Theorem $[A]^2 - 5[A] + 7[I] = 0$.

Now

$$[A]^2 = \begin{bmatrix} 3 & 1 \\ -1 & 2 \end{bmatrix} \times \begin{bmatrix} 3 & 1 \\ -1 & 2 \end{bmatrix} = \begin{bmatrix} 8 & 5 \\ -5 & 3 \end{bmatrix} \quad \text{and} \quad 5[A] = \begin{bmatrix} 15 & 5 \\ -5 & 10 \end{bmatrix}$$

$$\therefore\ [A]^2 - 5[A] + 7[I] = \begin{bmatrix} 8 & 5 \\ -5 & 3 \end{bmatrix} - \begin{bmatrix} 15 & 5 \\ -5 & 10 \end{bmatrix} + 7\begin{bmatrix} 1 & 0 \\ 0 & 1 \end{bmatrix} = \begin{bmatrix} 0 & 0 \\ 0 & 0 \end{bmatrix} = 0.$$

Exercises

1. Determine the matrices A and B where $A+2B = \begin{bmatrix} 1 & 2 & 0 \\ 6 & -3 & 3 \\ -5 & 3 & 1 \end{bmatrix}$ and $2A - B = \begin{bmatrix} 2 & 1 & 5 \\ 2 & -1 & 6 \\ 0 & 1 & 2 \end{bmatrix}$.

2. Solve for x, y, z, t where $\begin{bmatrix} y+z & x+z \\ 7-t & 6-z \end{bmatrix} = \begin{bmatrix} 9-t & 8-t \\ x+y & x+y \end{bmatrix}$.

[*Ans.* Since the given matrices are equal, they are identical, term by term.

$$\therefore \begin{aligned} y+z &= 9-t, \\ x+z &= 8-t, \\ 7-t &= x+y, \\ 6-z &= x+y. \end{aligned} \quad \text{or,} \quad \left.\begin{aligned} y+z+t &= 9 \\ x+z+t &= 8 \\ x+y+t &= 7 \\ x+y+z &= 6. \end{aligned}\right\} \quad \begin{matrix} (1) \\ (2) \\ (3) \\ (4) \end{matrix}$$

Adding,

$$3(x+y+z+t) = 30 \quad \text{or,} \quad x+y+z+t = 10. \qquad (5)$$

Subtracting (1), (2), (3), (4) separately from (5), we get the values $x = 1$, $y = 2$, $z = 3$ and $t = 4$.

3. Show by an example that matrix multiplication is not commutative.

4. If A be any square matrix whatever, of third order, and $E = \begin{bmatrix} 1 & 0 & 0 \\ 0 & 1 & 0 \\ 0 & 0 & 1 \end{bmatrix}$, show that $AE = EA = A$.

5. If $A = \begin{bmatrix} 0 & 1 & 2 \\ 1 & 2 & 3 \\ 3 & 1 & 1 \end{bmatrix}$. and $I = \begin{bmatrix} 1 & 0 & 0 \\ 0 & 1 & 0 \\ 0 & 0 & 1 \end{bmatrix}$, find the matrix such that $AB = I$.

6. What do you mean by a transposed matrix? If a matrix is equal to its transpose, what is the nature of the matrix?

7. If A is a square matrix, then show that $A + A^T$ is a symmetric matrix.

8. What do you mean by a Hermitian matrix? What are its basic properties?

9. Explain, with examples, what you mean by complex conjugate matrix, transposed matrix and Hermitian conjugate matrix.

10. What do you mean by a *Singular* matrix and a *non-singular* matrix?

11. What do you mean by adjoint of a matrix?

12. What is the necessary and sufficient condition that a matrix A has its inverse?

13. If a matrix A is invertible, how can you find the inverse?

[*Ans.* Steps which are to be followed to find the inverse of a matrix A, are:

Step I: The determinant of the matrix is to be found. If its value is not equal to zero, it will have its inverse.

Step II: The co-factor matrix is to be found.

Step III: The transpose of the co-factor matrix i.e., adj A is to be found next.

Step IV: The inverse matrix is then given by $A^{-1} = \frac{\text{adj } A}{|A|}$.

14. Compute the inverse of the following matrix $\begin{bmatrix} 0 & 1 & 2 \\ 1 & 2 & 3 \\ 3 & 1 & 1 \end{bmatrix}$.

15. If $A = \begin{bmatrix} 1 & -2 & 1 \\ 2 & -1 & 0 \\ 1 & 0 & 0 \end{bmatrix}$, verify that $A^3 = A^2.A = A.A^2$ and hence find A^{-1}.

16. If the matrix $A = \begin{bmatrix} 3 & -3 & 4 \\ 2 & -3 & 4 \\ 0 & -1 & 1 \end{bmatrix}$, prove that $A^3 = A^{-1}$.

17. If $A = \begin{bmatrix} 1 & 2 & 5 \\ -1 & 3 & 4 \end{bmatrix}$ and $B = \begin{bmatrix} 3 & -2 & 1 \\ 0 & -1 & 4 \\ 5 & 2 & -1 \end{bmatrix}$, find whether $A + B$, AB and BA exist. State reasons where the operations are not possible.

18. For the matrices $A = \begin{bmatrix} 1 & 1 & 1 \\ 1 & 2 & 3 \\ 1 & 4 & 9 \end{bmatrix}$ and $B = \begin{bmatrix} 2 & 5 & 3 \\ 3 & 1 & 2 \\ 1 & 2 & 1 \end{bmatrix}$, show that $(AB)^{-1} = B^{-1}A^{-1}$.

19. Find the adjoint and inverse of $\begin{bmatrix} \cos\theta & \sin\theta \\ -\sin\theta & \cos\theta \end{bmatrix}$.

[*Ans.* Let $A = \begin{bmatrix} \cos\theta & \sin\theta \\ -\sin\theta & \cos\theta \end{bmatrix}$.

Let B be the matrix whose elements are the co-factors of the corresponding elements of A.

$$\therefore \quad B = A^c = \begin{bmatrix} \cos\theta & \sin\theta \\ -\sin\theta & \cos\theta \end{bmatrix}$$

Interchanging rows and columns, we get

$$(A^c)^T = \text{adj } A = \begin{bmatrix} \cos\theta & -\sin\theta \\ \sin\theta & \cos\theta \end{bmatrix}$$

Again $|A| = \cos^2\theta + \sin^2\theta = 1$.

$$\therefore \quad A^{-1} = \frac{\text{adj } A}{|A|} = \begin{bmatrix} \cos\theta & -\sin\theta \\ \sin\theta & \cos\theta \end{bmatrix}.$$

20. Find the values of x, y, z, t for which the matrices $\begin{bmatrix} x+y & y-z \\ 5-t & 7+x \end{bmatrix}$ and $\begin{bmatrix} t-x & z-t \\ z-y & x+z+t \end{bmatrix}$ may be equal.

21. What is meant by an orthogonal matrix? Verify that the matrix defined by $A = \frac{1}{3}\begin{bmatrix} 1 & 2 & 2 \\ 2 & 1 & -2 \\ -2 & 2 & -1 \end{bmatrix}$ is orthogonal.

[*Ans.* A will be called an *orthogonal matrix,* if its premultiplication by its transpose gives a unitary matrix i.e. $A^TA = I$.

Now

$$A^T = \frac{1}{3}\begin{bmatrix} 1 & 2 & -2 \\ 2 & 1 & 2 \\ 2 & -2 & -1 \end{bmatrix},$$

so that

$$A^T A = \frac{1}{9}\begin{bmatrix} 1 & 2 & -2 \\ 2 & 1 & 2 \\ 2 & -2 & -1 \end{bmatrix}\begin{bmatrix} 1 & 2 & 2 \\ 2 & 1 & -2 \\ -2 & 2 & -1 \end{bmatrix} = \frac{1}{9}\begin{bmatrix} 9 & 0 & 0 \\ 0 & 9 & 0 \\ 0 & 0 & 9 \end{bmatrix} = \begin{bmatrix} 1 & 0 & 0 \\ 0 & 1 & 0 \\ 0 & 0 & 1 \end{bmatrix} = I.$$

22. Show that the adj of a diagonal matrix of order 3 is also a diagonal matrix.

[*Ans.* We consider a diagonal matrix A defined as $A = \begin{bmatrix} a & 0 & 0 \\ 0 & b & 0 \\ 0 & 0 & c \end{bmatrix}$, so that we have

$$\det A = |A| = \begin{vmatrix} a & 0 & 0 \\ 0 & b & 0 \\ 0 & 0 & c \end{vmatrix} = abc \neq 0;$$

hence the matrix A is non-singular and has its invrse. To find the co-factor matrix, we see that the co-factors of $|A|$ are

$$a^{11} = (-1)^2\begin{vmatrix} b & 0 \\ 0 & c \end{vmatrix} = bc, \quad a^{21} = (-1)^3\begin{vmatrix} 0 & 0 \\ 0 & c \end{vmatrix} = 0,$$

$$a^{12} = (-1)^3\begin{vmatrix} 0 & 0 \\ 0 & c \end{vmatrix} = 0, \quad a^{22} = (-1)^4\begin{vmatrix} a & 0 \\ 0 & c \end{vmatrix} = ca,$$

$$a^{13} = (-1)^4\begin{vmatrix} 0 & b \\ 0 & 0 \end{vmatrix} = 0, \quad a^{23} = (-1)^5\begin{vmatrix} a & 0 \\ 0 & 0 \end{vmatrix} = 0,$$

$$a^{31} = (-1)^4\begin{vmatrix} 0 & 0 \\ b & 0 \end{vmatrix} = 0, \quad a^{32} = (-1)^5\begin{vmatrix} a & 0 \\ 0 & 0 \end{vmatrix} = 0,$$

$$a^{33} = (-1)^6\begin{vmatrix} a & 0 \\ 0 & b \end{vmatrix} = ab,$$

$\therefore$ the co-factor matrix $A^c = \begin{bmatrix} bc & 0 & 0 \\ 0 & ca & 0 \\ 0 & 0 & ab \end{bmatrix}$.

$\therefore$ transpose of the co-factor matrix i.e. $(A^c)^T = \text{adj } A = \begin{bmatrix} bc & 0 & 0 \\ 0 & ca & 0 \\ 0 & 0 & ab \end{bmatrix}$ which is a diagonal matrix.

23. From the matrix equation $AX = B$, find X where $A = \begin{bmatrix} 1 & -1 & 0 \\ 0 & 1 & -1 \\ 1 & 1 & 1 \end{bmatrix}$ and $B = \begin{bmatrix} 2 \\ 1 \\ 7 \end{bmatrix}$.

State the order of the matrix X.

[*Ans.* Since A is a 3×3 matrix, X must be a matrix of order 3×1, such that the product is a 3×1 matrix.

Let $X = \begin{bmatrix} x \\ y \\ z \end{bmatrix}$; then from $AX = B$, we may write $\begin{bmatrix} 1 & -1 & 0 \\ 0 & 1 & -1 \\ 1 & 1 & 1 \end{bmatrix}\begin{bmatrix} x \\ y \\ z \end{bmatrix} = \begin{bmatrix} 2 \\ 1 \\ 7 \end{bmatrix}$ whence we get

$$x - y = 2 \quad (1)$$

$$y - z = 1 \quad (2)$$

$$x + y + z = 7. \quad (3)$$

Subtracting (2) from (1)

$$x - 2y + z = 1. \tag{4}$$

Sibtracting (4) from (3) $3y = 6$ or $y = 2$. $\therefore$ from (1) $x - y = 2$ or $x - 2 = 2$ or $x = 4$.

Again from (2) $y - z = 1$ or $2 - z = 1$ or $z = 1$. $\therefore$ the required matrix $X = \begin{bmatrix} 4 \\ 2 \\ 1 \end{bmatrix}$.

24. Solve, by matrix method, the following equations

$$\begin{aligned} -x_1 + 3x_3 + 1 &= 0 \\ 2x_1 - x_2 - 4x_3 - 2 &= 0 \\ x_2 + 2x_3 - 4 &= 0. \end{aligned}$$

[*Ans.* The equations may be written as

$$\left.\begin{aligned} x_1 + 0x_2 - 3x_3 &= 1 \\ 2x_1 - x_2 - 4x_3 &= 2 \\ 0x_1 + x_2 + 2x_3 &= 4 \end{aligned}\right\}. \tag{1}$$

Let

$$A = \begin{bmatrix} 1 & 0 & -3 \\ 2 & -1 & -4 \\ 0 & 1 & 2 \end{bmatrix}, \quad X = \begin{bmatrix} x_1 \\ x_2 \\ x_3 \end{bmatrix}, \quad B = \begin{bmatrix} 1 \\ 2 \\ 4 \end{bmatrix}.$$

Then from (1)

$$AX = B \text{ or, } A^{-1}AX = A^{-1}B \text{ or, } X = A^{-1}B. \tag{2}$$

Now the co-factor matrix of A is given by

$$A^c = \begin{bmatrix} -2+4 & 0-4 & 2+0 \\ -3-0 & 2+0 & 0-1 \\ 0-3 & -6+4 & -1+0 \end{bmatrix} = \begin{bmatrix} 2 & -4 & 2 \\ -3 & 2 & -1 \\ -3 & -2 & -1 \end{bmatrix}.$$

Interchanging rows and columns of A^c, we have the transpose of A^c i.e.,

$$\text{adj } A = (A^c)^T = \begin{bmatrix} 2 & -3 & -3 \\ -4 & 2 & -2 \\ 2 & -1 & -1 \end{bmatrix}.$$

Also we have

$$|A| = 1(-2+4) + 0(0-4) - 3(2+0) = 2 + 0 - 6 = -4$$

$$\therefore A^{-1} = \frac{\text{adj } A}{|A|} = -\frac{1}{4}\begin{bmatrix} 2 & -3 & -3 \\ -4 & 2 & -2 \\ 2 & -1 & -1 \end{bmatrix} = \begin{bmatrix} -\frac{1}{2} & \frac{3}{4} & \frac{3}{4} \\ 1 & -\frac{1}{2} & \frac{1}{2} \\ -\frac{1}{2} & \frac{1}{4} & \frac{1}{4} \end{bmatrix}.$$

$\therefore$ from (2)

$$\begin{bmatrix} x_1 \\ x_2 \\ x_3 \end{bmatrix} = \begin{bmatrix} -\frac{1}{2} & \frac{3}{4} & \frac{3}{4} \\ 1 & -\frac{1}{2} & \frac{1}{2} \\ -\frac{1}{2} & \frac{1}{4} & \frac{1}{4} \end{bmatrix} \begin{bmatrix} 1 \\ 2 \\ 4 \end{bmatrix}.$$

$$\therefore x_1 = -\frac{1}{2} + \frac{3}{2} + 3 = 4, \quad x_2 = 1 - 1 + 2 = 2 \quad \text{and} \quad x_3 = -\frac{1}{2} + \frac{1}{2} + 1 = 1.$$

25. Solve by matrix method the following equations

$$x + y + z = 6$$

$$x + 2y - 3z = -4$$

$$5x - y + 2z = 9.$$

26. Diagonalise the inertia matrix $\begin{bmatrix} A & -F & -E \\ -F & B & -D \\ -E & -D & C \end{bmatrix}$.

[*Ans.* The characteristic equation is

$$\begin{vmatrix} (A-\lambda) & -F & -E \\ -F & (B-\lambda) & -D \\ -E & -D & (C-\lambda) \end{vmatrix} = 0$$

or, $(A-\lambda)[(B-\lambda)(C-\lambda) - D^2] + F[-F(C-\lambda) - DE] - E[FD + E(B-\lambda)] = 0$

or, $(A-\lambda)(B-\lambda)(C-\lambda) - D^2A + \lambda D^2 - F^2C + F^2\lambda - FDE - FDE - E^2B + E^2\lambda = 0$

or, $ABC - AB\lambda - AC\lambda + A\lambda^2 - BC\lambda + B\lambda^2 + C\lambda^2 - \lambda^3 - D^2A + \lambda D^2 - F^2C + F^2\lambda$
$-2FDE - E^2B + E^2\lambda = 0$

or, $\lambda^3 - (A+B+C)\lambda^2 - (D^2 + E^2 + F^2 - BC - CA - AB)\lambda$
$-(ABC - 2FDE - AD^2 - BE^2 - CF^2) = 0.$

If $\lambda_1, \lambda_2, \lambda_3$ be the roots of the above equation, then we may write

$$\lambda^3 - (A+B+C)\lambda^2 - (D^2 + E^2 + F^2 - BC - CA - AB)\lambda$$
$$-(ABC - 2FDE - AD^2 - BE^2 - CF^2)$$
$$= (\lambda - \lambda_1)(\lambda - \lambda_2)(\lambda - \lambda_3).$$

Equating the coefficients of $\lambda^3, \lambda^2, \lambda$ and the constant terms, we would obtain the values of $\lambda_1, \lambda_2, \lambda_3$ in terms of A, B, C, D, E, F which would be the eigen-values. Then the diagonalised form of the inertia matrix will be

$$\begin{bmatrix} \lambda_1 & 0 & 0 \\ 0 & \lambda_2 & 0 \\ 0 & 0 & \lambda_3 \end{bmatrix} = \begin{bmatrix} I_1 & 0 & 0 \\ 0 & I_2 & 0 \\ 0 & 0 & I_3 \end{bmatrix},$$

so that $I_1 = \lambda_1$, $I_2 = \lambda_2$, and $I_3 = \lambda_3$, where I_1, I_2, I_3 are the principal moments of inertia.

27. Show that the matrix $A = \begin{bmatrix} 1 & -6 & -4 \\ 0 & 4 & 2 \\ 0 & -6 & -3 \end{bmatrix}$ is a diagonalisable matrix. Find the diagonalising matrix and the diagonal matrix.

28. Verify Cayley-Hamiltons' theorem for the matrix $A = \begin{bmatrix} -11 & -10 & 5 \\ 5 & 4 & -5 \\ -20 & -20 & 4 \end{bmatrix}$.

29. If A is a Hermitian matrix and P is an orthogonal matrix, then show that $P^{-1}AP$ is symmetric.

30. Let $A = \begin{bmatrix} -2 & 1 & 3 \\ 0 & -1 & 1 \\ 1 & 2 & 0 \end{bmatrix}$; show that A is non-singular. Determine adj A and A^{-1}.

31. If (l_1, m_1, n_1), (l_2, m_2, n_2) and (l_3, m_3, n_3) be the direction consines of three mutually perpendicular lines referred to orthogonal Cartesian system, then show that the matrix A having those direction cosines as elements and defined as $A = \begin{bmatrix} l_1 & m_1 & n_1 \\ l_2 & m_2 & n_2 \\ l_3 & m_3 & n_3 \end{bmatrix}$ is an orthogonal matrix.

32. Find the eigen-values and eigen-vectors of the matrix $A = \begin{bmatrix} 8 & -6 & 2 \\ -6 & 7 & -4 \\ 2 & -4 & 3 \end{bmatrix}$.

33. Obtain the characteristic equation of the matrix $A = \begin{bmatrix} 1 & 0 & 2 \\ 0 & 2 & 1 \\ 2 & 0 & 3 \end{bmatrix}$ and verify that it is satisfied by A and hence find its inverse.

34. Verify Cayley-Hamilton's theorem for the matrix $A = \begin{bmatrix} 1 & \sqrt{2} & 0 \\ \sqrt{2} & -1 & 0 \\ 0 & 0 & 1 \end{bmatrix}$ and use the result to find A^{-1}.

35. Transform the matrix $A = \begin{bmatrix} 8 & -12 & 5 \\ 15 & -25 & 11 \\ 24 & -42 & 19 \end{bmatrix}$ into a doagonal form.

36. Solve, by the matrix method, the following equations

(a)
$$\begin{aligned} x + y + z &= 0 \\ 2x - y - 3z &= 2 \\ 3x - 5y + 4z &= 3; \end{aligned}$$

(b)
$$\begin{aligned} x_1 + x_3 &= -1 \\ -2x_1 + x_2 &= 1 \\ -x_2 + x_3 &= -5. \end{aligned}$$

37. (a) Consider the following transformation in three dimensions:

$$\begin{aligned} x' &= x\cos\theta + y\sin\theta \\ y' &= -x\sin\theta + y\cos\theta \\ z' &= z. \end{aligned}$$

i. Write down the transformation matrix.
ii. Denoting that matrix by $A(\theta)$, show that $A(\theta_1)A(\theta_2) = A(\theta_1 + \theta_2)$.
iii. Is $A(\theta)$ unitary? Justify your answer.
iv. What are the eigen-values of $A(\theta)$?

(b) Show that for a hermitian matrix

i. the eigen-values are real;
ii. it remains hermitian under a unitary transformation. **[C.U. (Hons.) 1996]**

[*Ans.* (a) (i) The transformation matrix is written as

$$A(\theta) = \begin{bmatrix} \cos\theta & \sin\theta & 0 \\ -\sin\theta & \cos\theta & 0 \\ 0 & 0 & 1 \end{bmatrix} \quad (1)$$

(ii) $A(\theta_1)A(\theta_2) = \begin{bmatrix} \cos\theta_1 & \sin\theta_1 & 0 \\ -\sin\theta_1 & \cos\theta_1 & 0 \\ 0 & 0 & 1 \end{bmatrix} \times \begin{bmatrix} \cos\theta_2 & \sin\theta_2 & 0 \\ -\sin\theta_2 & \cos\theta_2 & 0 \\ 0 & 0 & 1 \end{bmatrix}$

$$= \begin{bmatrix} \cos\theta_1\cos\theta_2 - \sin\theta_1\sin\theta_2 & \cos\theta_1\sin\theta_2 + \sin\theta_1\cos\theta_2 & 0 \\ -\sin\theta_1\cos\theta_2 - \cos\theta_1\sin\theta_2 & -\sin\theta_1\sin\theta_2 + \cos\theta_1\cos\theta_2 & 0 \\ 0 & 0 & 1 \end{bmatrix}$$

[applying the rule of multiplication of matrices]

$$= \begin{bmatrix} \cos(\theta_1+\theta_2) & \sin(\theta_1+\theta_2) & 0 \\ -\sin(\theta_1+\theta_2) & \cos(\theta_1+\theta_2) & 0 \\ 0 & 0 & 1 \end{bmatrix} = A(\theta_1+\theta_2) \text{ from (1)}$$

(iii) From (1) we may write $A^*(\theta) = \begin{bmatrix} \cos\theta & \sin\theta & 0 \\ -\sin\theta & \cos\theta & 0 \\ 0 & 0 & 1 \end{bmatrix}$ where $A^*(\theta)$ denotes the complex conjugate of $A(\theta)$. If $A^\dagger(\theta)$ be the hermitian conjugate of $A(\theta)$, then

$$A^\dagger(\theta) = [A^*(\theta)]^T = \begin{bmatrix} \cos\theta & \sin\theta & 0 \\ -\sin\theta & \cos\theta & 0 \\ 0 & 0 & 1 \end{bmatrix}^T = \begin{bmatrix} \cos\theta & -\sin\theta & 0 \\ \sin\theta & \cos\theta & 0 \\ 0 & 0 & 1 \end{bmatrix}$$

$$\therefore\ A^\dagger(\theta)A(\theta) = \begin{bmatrix} \cos\theta & -\sin\theta & 0 \\ \sin\theta & \cos\theta & 0 \\ 0 & 0 & 1 \end{bmatrix} \times \begin{bmatrix} \cos\theta & \sin\theta & 0 \\ -\sin\theta & \cos\theta & 0 \\ 0 & 0 & 1 \end{bmatrix}$$

$$= \begin{bmatrix} \cos^2\theta + \sin^2\theta & \sin\theta\cos\theta - \sin\theta\cos\theta & 0 \\ \sin\theta\cos\theta - \sin\theta\cos\theta & \sin^2\theta + \cos^2\theta & 0 \\ 0 & 0 & 1 \end{bmatrix}$$

[applying the rule of multiplication of matrices]

$$= \begin{bmatrix} 1 & 0 & 0 \\ 0 & 1 & 0 \\ 0 & 0 & 1 \end{bmatrix} = I_3,\ I_3 \text{ being a unit matrix of order 3.}$$

Thus from definition $A(\theta)$ is a unitary matrix.

(iv) For the matrix $A(\theta)$, the characteristic polynomial is given by

$$D(\lambda) = \det\{A(\theta) - \lambda I\},\ \lambda \text{ being the eigen-value of } A(\theta).$$

$$= \begin{vmatrix} \cos\theta - \lambda & \sin\theta & 0 \\ -\sin\theta & \cos\theta - \lambda & 0 \\ 0 & 0 & 1-\lambda \end{vmatrix} = (1-\lambda)\left\{(\cos\theta - \lambda)^2 + \sin^2\theta\right\},$$

so that the characteristic equation, whose solutions give the eigen-values λ, is

$$D(\lambda) = (1-\lambda)\left\{(\cos\theta-\lambda)^2 + \sin^2\theta\right\} = 0$$

$\therefore$ either $(1-\lambda) = 0$ or $\lambda = 1$ or $(\cos\theta - \lambda)^2 + \sin^2\theta = 0$, i.e., $(\lambda - \cos\theta)^2 = -\sin^2\theta$

$$\therefore\quad \lambda - \cos\theta = \pm i\sin\theta \quad (i = \sqrt{-1})$$

$$\text{or}\quad \lambda = \cos\theta \pm i\sin\theta = e^{\pm i\theta}$$

$\therefore$ the eigen-values are 1, $e^{i\theta}$ and $e^{-i\theta}$.

(b) (i) Let A be a hermitian matrix

$$\therefore \quad A^\dagger = A$$

Let $\bar{X}$ be the eigen-vector of A, belonging to the eigen-value λ

$$\therefore \ A\bar{X} = \lambda\bar{X} \tag{1}$$

Multiplying (1) from left by $\bar{X}^\dagger$, we obtain

$$\bar{X}^\dagger A\bar{X} = \bar{X}^\dagger \lambda \bar{X} \tag{2}$$

where $\bar{X}^\dagger = (\bar{X}^*)^T$

$$\text{or} \quad \left(\bar{X}^\dagger A\bar{X}\right)^\dagger = \left(\bar{X}^\dagger \lambda \bar{X}\right)^\dagger$$

$$\text{or} \quad \bar{X}^\dagger A^\dagger (\bar{X}^\dagger)^\dagger = \bar{X}^\dagger \lambda^\dagger (\bar{X}^\dagger)^\dagger \quad \left[\text{using } (ABC)^\dagger = C^\dagger B^\dagger A^\dagger\right]$$

$$\text{or} \quad \bar{X}^\dagger A^\dagger \bar{X} = \bar{X}^\dagger \lambda^\dagger \bar{X}; \tag{3}$$

since A is hermitian, $A^\dagger = A$, so that from (3) we obtain

$$\bar{X}^\dagger A\bar{X} = \bar{X}^\dagger \lambda^* \bar{X} \quad (\because \lambda^\dagger = \lambda^*) \tag{4}$$

From (2) and (4) we may write

$$(\lambda - \lambda^*)\bar{X}^\dagger \bar{X} = 0. \tag{5}$$

But $\bar{X}$ being the eigen-vector of A is obviously a non-null vector, so that $\bar{X}^\dagger \bar{X} \neq 0$. Physically $\bar{X}^\dagger \bar{X}$ represents the (length)^2 of the vector $\bar{X}$. Hence from (5) we have $\lambda - \lambda^* = 0$ or $\lambda = \lambda^*$ i.e., the eigen-values of the hermitian matrix are real.

(ii) Let $H = -\begin{bmatrix} \cos\theta & i\sin\theta \\ i\sin\theta & \cos\theta \end{bmatrix}$ be a hermitian matrix. We are to prove that H also remains hermitian under a unitary transformation, i.e., H^{-1} is also hermitian.

Here

$$H^{-1} = \frac{\text{adj}H}{|H|} = \begin{bmatrix} \cos\theta & i\sin\theta \\ i\sin\theta & \cos\theta \end{bmatrix} \quad \because |H| = -1$$

Let $H^{-1} = P$ (say), so that we have

$$P = \begin{bmatrix} \cos\theta & i\sin\theta \\ i\sin\theta & \cos\theta \end{bmatrix}$$

$$\therefore P^\dagger = (P^*)^T = \begin{bmatrix} \cos\theta & i\sin\theta \\ i\sin\theta & \cos\theta \end{bmatrix}^T = \begin{bmatrix} \cos\theta & i\sin\theta \\ i\sin\theta & \cos\theta \end{bmatrix} = P$$

Thus $P^\dagger = P$, i.e., P is also a hermitian matrix. Therefore a hermitian matrix also remains hermitian under a unitary transformation.

N.B. We consider a linear transformation in which the transformation matrix is a *square* one so that in the equation $y = Ax$, x and y are column vectors of order $n \times 1$ and A is of order $n \times n$. If the matrix A of transformation is unitary (satisfying $AA^\dagger = I = A^\dagger A$), it is said to be a *unitary transformation*.

Using $AA^\dagger = I = A^\dagger A$, $y = Ax$ can be written as

$$y^\dagger y = x^\dagger A^\dagger A x = x^\dagger x \quad \text{or} \quad \sum_{i=1}^{n} y_i^* y_i = \sum_{i=1}^{n} x_i^* x_i$$

which shows that *the norm of a vector remains invariant under a unitary transformation.*

Again in our discussion on "diagonalisation of a matrix", we have shown that $B = P^{-1}AP$ is a diagonal matrix whose diagonal elements are just the eigen-values of A. The matrices A and B of the above relation are said to be related by a *similarity transformation. They perform the same job in the same vector space on vectors referred to different co-ordinate systems.* The inverse transformation of the above relation, i.e., $A = PBP^{-1}$ is also a similarity transformation. If in particular, the matrix P with which a similarity transformation is performed happens to be *unitary*, it is called a *unitary transformation.* If further P is a real orthogonal matrix it is called an *orthogonal transformation.* We have thus $B = P^{-1}AP$, $P^{-1} = P^\dagger$ *for unitary transformation and* $B = P^{-1}AP$, $P^{-1} = P^T$ *for orthogonal transformation.*]

38. Find the eigen-values and eigen-vectors of the matrix $A = \begin{bmatrix} 1 & 1 \\ 0 & 1 \end{bmatrix}$. **[C.U.(Hons.)'97]**

[*Ans.* Let $\bar{X}$ be an eigen-vector of the given matrix A, belonging to the eigen-value λ; the eigen-vector equation is then $A\bar{X} = \lambda\bar{X}$ or $A\bar{X} = \lambda I\bar{X}$, I being a unit matrix of order 2 or $(A - \lambda I)\bar{X} = \bar{0}$, for which non-null solution exists only if $\det(A - \lambda I) = 0$. Now

$$\det(A - \lambda I) = \begin{vmatrix} 1-\lambda & 1 \\ 0 & 1-\lambda \end{vmatrix} = (1-\lambda)^2$$

$\therefore$ the characteristic or secular equation is

$$D(\lambda) = \det(A - \lambda I) = 0 \quad \text{or} \quad (1-\lambda)^2 = 0 \text{ i.e.,} \lambda = 1, 1$$

i.e., the eigen-values of the given matrix are $\lambda_1 = 1$ and $\lambda_2 = 1$.

Let $\bar{X}_1 = \begin{pmatrix} x_{11} \\ x_{21} \end{pmatrix}$ be the eigen-vectors of A belonging to the eigen-value $\lambda_1 = 1$, so that we may write

$$(A - \lambda_1 I)\bar{X}_1 = \bar{0} \quad \text{or} \quad \begin{bmatrix} 1-1 & 1 \\ 0 & 1-1 \end{bmatrix}\begin{bmatrix} x_{11} \\ x_{21} \end{bmatrix} = \begin{bmatrix} 0 \\ 0 \end{bmatrix}$$

$$\text{or} \quad \begin{bmatrix} 0 & 1 \\ 0 & 0 \end{bmatrix}\begin{bmatrix} x_{11} \\ x_{21} \end{bmatrix} = \begin{bmatrix} 0 \\ 0 \end{bmatrix} \quad \therefore\ 0.x_{11} + 1.x_{21} = 0 \text{ or } x_{21} = 0.$$

Here since x_{11} is indeterminate, it can be chosen arbitrarily. So let $x_{11} = a$ (say)

$$\therefore\ \bar{X}_1 = \begin{bmatrix} a \\ 0 \end{bmatrix} = a\begin{bmatrix} 1 \\ 0 \end{bmatrix} \quad \left(\begin{array}{l}\text{where } a \text{ is considered} \\ \text{to be any constant.}\end{array}\right)$$

So an infinite number of solutions will be possible. One typical solution is $\bar{X}_1 = \begin{pmatrix} 1 \\ 0 \end{pmatrix}$. Similarly the other eigen-vector belonging to the eigen-value $\lambda_2 = 1$ is given by $\bar{X}_2 = \begin{pmatrix} 1 \\ 0 \end{pmatrix}$. Therefore $\lambda_1 = 1 = \lambda_2$ and $\bar{X}_1 = \begin{pmatrix} 1 \\ 0 \end{pmatrix} = \bar{X}_2$.]

39. Find the eigen-values and eigen-vectors of the matrix $\begin{pmatrix} 1 & 0 & 0 \\ 0 & 0 & 1 \\ 0 & 1 & 0 \end{pmatrix}$. Is this matrix unitary?

[C.U. (Hons.) 1998]

[*Ans.* (a) Let $\bar{X}$ be the eigen-vector of the given matrix A, belonging to the eigen-value λ; then the matrix satisfies the eigen-vector equation given by $A\bar{X} = \lambda\bar{X} = \lambda I\bar{X}$, I being a unit matrix of order 3 or $(A - \lambda I)\bar{X} = \bar{0}$, for which non-null solution exists if and only if $\det(A - \lambda I) = 0$. Now

$$\det(A - \lambda I) = \begin{vmatrix} 1-\lambda & 0 & 0 \\ 0 & -\lambda & 1 \\ 0 & 1 & -\lambda \end{vmatrix} = (1-\lambda)(\lambda^2 - 1)$$

$\therefore$ the characteristic or secular equation of the given matrix A is

$$\det(A - \lambda I) = (1-\lambda)(\lambda^2 - 1) = 0$$

$$\text{or} \quad (1-\lambda)(\lambda - 1)(\lambda + 1) = 0 \tag{1}$$

$\therefore$ the eigen-values of the given matrix are $\lambda_1 = 1$, $\lambda_2 = 1$ and $\lambda_3 = -1$.

Let $\bar{X}_1 = \begin{pmatrix} x_{11} \\ x_{21} \\ x_{31} \end{pmatrix}$ be an eigen-vector of A belonging to the eigen-value $\lambda_1 = 1$.

$$\therefore \quad (A - \lambda_1 I)\bar{X}_1 = \bar{0} \quad \text{or} \quad \begin{pmatrix} 1-1 & 0 & 0 \\ 0 & -1 & 1 \\ 0 & 1 & -1 \end{pmatrix} \begin{pmatrix} x_{11} \\ x_{21} \\ x_{31} \end{pmatrix} = \begin{pmatrix} 0 \\ 0 \\ 0 \end{pmatrix}$$

$$\text{or} \quad \begin{pmatrix} 0 & 0 & 0 \\ 0 & -1 & 1 \\ 0 & 1 & -1 \end{pmatrix} \begin{pmatrix} x_{11} \\ x_{21} \\ x_{31} \end{pmatrix} = \begin{pmatrix} 0 \\ 0 \\ 0 \end{pmatrix}$$

whence we have

$$-x_{21} + x_{31} = 0 \tag{i}$$

$$\text{and} \quad x_{21} - x_{31} = 0 \tag{ii}$$

The two equations (i) and (ii) are identical and we may therefore write

$$x_{21} = x_{31} = a \text{ (say)}.$$

Here x_{11} is indeterminate—so it may have any value. Let $x_{11} = b$ (say),

$$\therefore \quad \bar{X}_1 = \begin{pmatrix} b \\ a \\ a \end{pmatrix} \quad \text{where } a \text{ and } b \text{ are constants.}$$

So an infinite number of solutions will be possible.

Let $a = 1$ and $b = 0$, so that one typical solutions is $\bar{X}_1 = \begin{pmatrix} 0 \\ 1 \\ 1 \end{pmatrix}$. Again since $\lambda_2 = \lambda_1 = 1$, another eigen-vector $\bar{X}_2 = \begin{pmatrix} 0 \\ 1 \\ 1 \end{pmatrix}$.

Let $\bar{X}_3 = \begin{pmatrix} x_{13} \\ x_{23} \\ x_{33} \end{pmatrix}$ be another eigen-vector of A, corresponding to the eigen-value $\lambda_3 = -1$, so that we may write

$$(A - \lambda_3 I)\bar{X}_3 = \bar{0} \quad \text{or} \quad \begin{pmatrix} 2 & 0 & 0 \\ 0 & 1 & 1 \\ 0 & 1 & 1 \end{pmatrix} \begin{pmatrix} x_{13} \\ x_{23} \\ x_{33} \end{pmatrix} = \begin{pmatrix} 0 \\ 0 \\ 0 \end{pmatrix}$$

whence we have $2x_{13} = 0$ or $x_{13} = 0$

$$x_{23} + x_{33} = 0 \tag{iii}$$

$$\text{and} \quad x_{23} + x_{33} = 0 \tag{iv}$$

Equations (iii) and (iv) being identical, we have $x_{23} = -x_{33}$; let $x_{33} = a_1$, then $x_{23} = -a_1$ where a_1 is any constant.

$$\therefore \quad \bar{X}_3 = \begin{pmatrix} 0 \\ -a_1 \\ a_1 \end{pmatrix} = a_1 \begin{pmatrix} 0 \\ -1 \\ 1 \end{pmatrix}$$

So an infinite number of solutions will be possible, of which one typical solution is $\bar{X}_3 = \begin{pmatrix} 0 \\ -1 \\ 1 \end{pmatrix}$ (choosing $a_1 = 1$.)

(b) A matrix A is said to be unitary if $AA^T = I$.

Here

$$AA^T = \begin{pmatrix} 1 & 0 & 0 \\ 0 & 0 & 1 \\ 0 & 1 & 0 \end{pmatrix}\begin{pmatrix} 1 & 0 & 0 \\ 0 & 0 & 1 \\ 0 & 1 & 0 \end{pmatrix}^T = \begin{pmatrix} 1 & 0 & 0 \\ 0 & 0 & 1 \\ 0 & 1 & 0 \end{pmatrix}\begin{pmatrix} 1 & 0 & 0 \\ 0 & 0 & 1 \\ 0 & 1 & 0 \end{pmatrix} = \begin{pmatrix} 1 & 0 & 0 \\ 0 & 1 & 0 \\ 0 & 0 & 1 \end{pmatrix} = I_3$$

I_3 being a unit matrix of order 3.

$\therefore$ the matrix A is unitary.]

40. Find the eigen-values and a set of orthonormal eigen-vectors of the following matrix

$$A = \begin{pmatrix} 4 & -1 & -1 \\ -1 & 4 & -1 \\ -1 & -1 & 4 \end{pmatrix}.$$

[C.U. (Hons.) 2000]

[*Ans.* Let $\bar{X}$ be an eigen-vector of the given matrix, belonging to the eigen-value λ; then the eigen-vector equation is written as

$$A\bar{X} = \lambda\bar{X} \quad \text{or} \quad A\bar{X} = \lambda I\bar{X},$$

I being a unit or identity matrix of order 3

$$\text{or} \quad (A - \lambda I)\bar{X} = \bar{0},$$

for which non-zero solution exists only if $\det(A - \lambda I) = 0$, which is the characteristic or secular equation, whose solutions give the eigen-values of A.

Now

$$\det(A - \lambda I) = 0 \quad \text{or} \quad \left|\begin{pmatrix} 4 & -1 & -1 \\ -1 & 4 & -1 \\ -1 & -1 & 4 \end{pmatrix} - \lambda\begin{pmatrix} 1 & 0 & 0 \\ 0 & 1 & 0 \\ 0 & 0 & 1 \end{pmatrix}\right| = 0 \quad \text{or} \quad \begin{vmatrix} 4-\lambda & -1 & -1 \\ -1 & 4-\lambda & -1 \\ -1 & -1 & 4-\lambda \end{vmatrix} = 0$$

$$\text{or} \quad (4-\lambda)\{(4-\lambda)^2 - 1\} + 1\{-(4-\lambda) - 1\} - 1(1 + 4 - \lambda) = 0$$

$$\text{or} \quad (4-\lambda)(15 - 8\lambda + \lambda^2) + (\lambda - 5) - (5 - \lambda) = 0$$

$$\text{or} \quad 60 - 32\lambda + 4\lambda^2 - 15\lambda + 8\lambda^2 - \lambda^3 + 2\lambda - 10 = 0$$

$$\text{or} \quad -\lambda^3 + 12\lambda^2 - 45\lambda + 50 = 0 \quad \text{or} \quad \lambda^3 - 12\lambda^2 + 45\lambda - 50 = 0$$

$$\text{or} \quad (\lambda - 2)(\lambda^2 - 10\lambda + 25) = 0 \quad \text{or} \quad (\lambda - 2)(\lambda - 5)(\lambda - 5) = 0 \tag{1}$$

so that the eigen-values of the given matrix A are $\lambda_1 = 2$; $\lambda_2 = 5$ and $\lambda_3 = 5$.

Let $\bar{X}_1 = \begin{pmatrix} x_{11} \\ x_{21} \\ x_{31} \end{pmatrix}$ be an eigen-vector of A corresponsing to the eigen-value $\lambda_1 = 2$; then the eigen-vector $\bar{X}_1$ will be given by the non-zero solutions of the equation

$$(A - \lambda I)\bar{X}_1 = \bar{0} \quad \text{or} \quad (A - 2I)\bar{X}_1 = \bar{0}$$

$$\text{or} \quad \left[\begin{pmatrix} 4 & -1 & -1 \\ -1 & 4 & -1 \\ -1 & -1 & 4 \end{pmatrix} - \begin{pmatrix} 2 & 0 & 0 \\ 0 & 2 & 0 \\ 0 & 0 & 2 \end{pmatrix}\right]\begin{bmatrix} x_{11} \\ x_{21} \\ x_{31} \end{bmatrix} = \begin{bmatrix} 0 \\ 0 \\ 0 \end{bmatrix}$$

$$\text{or} \quad \begin{bmatrix} 4-2 & -1 & -1 \\ -1 & 4-2 & -1 \\ -1 & -1 & 4-2 \end{bmatrix}\begin{bmatrix} x_{11} \\ x_{21} \\ x_{31} \end{bmatrix} = \begin{bmatrix} 0 \\ 0 \\ 0 \end{bmatrix} \quad \left(\begin{array}{l}\text{using the rule for} \\ \text{subtraction of matrices}\end{array}\right)$$

$$\text{or} \quad \begin{pmatrix} 2 & -1 & -1 \\ -1 & 2 & -1 \\ -1 & -1 & 2 \end{pmatrix}\begin{pmatrix} x_{11} \\ x_{21} \\ x_{31} \end{pmatrix} = \begin{bmatrix} 0 \\ 0 \\ 0 \end{bmatrix}$$

$$\text{or} \quad \begin{pmatrix} 2 & -1 & -1 \\ -1 & 2 & -1 \\ 0 & -3 & 3 \end{pmatrix}\begin{pmatrix} x_{11} \\ x_{21} \\ x_{31} \end{pmatrix} = \begin{bmatrix} 0 \\ 0 \\ 0 \end{bmatrix} \quad [\text{using } R_3' = R_3 - R_2]$$

whence we obtain the equivalent linear equations given by

$$2x_{11} - x_{21} - x_{31} = 0 \quad (2)$$

$$-x_{11} + 2x_{21} - x_{31} = 0 \quad (3)$$

$$\text{and} \quad 0x_{11} - 3x_{21} + 3x_{31} = 0 \quad (4)$$

From (4) we have $x_{21} = x_{31}$, which when substituted in (2) gives $x_{11} = x_{21}$

$$\therefore \quad x_{11} = x_{21} = x_{31} = a \text{ (say)};$$

so for the eigen-value $\lambda_1 = 2$, the eigen-vector $\bar{X}_1 = a\begin{pmatrix}1\\1\\1\end{pmatrix}$ where a is a scalar capable of having any integral value including unity. When $a = 1$, $\bar{X}_1 = \begin{pmatrix}1\\1\\1\end{pmatrix}$ which is a set of orthonormal eigen-vectors.

[**N.B.** If the scalar product of two vectors is zero, the vectors are said to be *orthogonal.* A set of vectors each of which is orthogonal to all the remaining vectors of the set is called an *orthogonal set.* If each of the vectors is further normalised to unity, it is called an *orthonormal set.* A vector can be normalised by dividing it by its *norm*—which is the length of the vector in familiar language. If the norm of a vector is unity, it is called a *unit vector* or *normalised vector.* If $\vec{u}$ be any vector, then $\frac{\vec{u}}{|\vec{u}|}$ is the *normalised vector* parallel to $\vec{u}$.]

Again corresponding to the eigen-values $\lambda_2 = 5 = \lambda_3$, the eigen-vectors $\bar{X}_2$ or $\bar{X}_3$ of A will be obtained from $(A - 5I)\bar{X}_2 = \bar{0}$ or $(A - 5I)\bar{X}_3 = \bar{0}$. Considering any one, we have

$$(A - 5I)\bar{X}_2 = \bar{0} \quad \text{or} \quad \left[\begin{pmatrix}4 & -1 & -1\\ -1 & 4 & -1\\ -1 & -1 & 4\end{pmatrix} - 5\begin{pmatrix}1 & 0 & 0\\ 0 & 1 & 0\\ 0 & 0 & 1\end{pmatrix}\right]\begin{bmatrix}x_{12}\\ x_{22}\\ x_{32}\end{bmatrix} = \begin{bmatrix}0\\0\\0\end{bmatrix}$$

$$\text{or} \quad \begin{pmatrix}-1 & -1 & -1\\ -1 & -1 & -1\\ -1 & -1 & -1\end{pmatrix}\begin{pmatrix}x_{12}\\ x_{22}\\ x_{32}\end{pmatrix} = \begin{bmatrix}0\\0\\0\end{bmatrix}$$

$$\text{or} \quad \begin{pmatrix}-1 & -1 & -1\\ 0 & 0 & 0\\ 0 & 0 & 0\end{pmatrix}\begin{pmatrix}x_{12}\\ x_{22}\\ x_{32}\end{pmatrix} = \begin{bmatrix}0\\0\\0\end{bmatrix} \quad \left(\begin{array}{l}\text{using } R_3' = R_3 - R_2\\ \text{and } R_2' = R_2 - R_1\end{array}\right)$$

which is equivalent to $-x_{12} - x_{22} - x_{32} = 0$ or $x_{12} = -(x_{22} + x_{32})$.

Since x_{22} and x_{32} are independent, an infinite number of solutions will be possible; choosing $x_{22} = 1$, $x_{32} = 1$, we have $x_{12} = -2$, so that one typical solution is $\bar{X}_2 = \begin{pmatrix}-2\\1\\1\end{pmatrix}$. Again choosing $x_{22} = 1$, $x_{32} = -1$, we have $x_{12} = 0$, so that another typical solution is $\bar{X}_2 = \begin{pmatrix}0\\1\\-1\end{pmatrix}$ which is a set of orthonormal eigen-vectors. Similarly we may proceed to obtain the solutions of $\bar{X}_3$—which is the eigen-vector belonging to the eigen-value λ_3 of A.]

41. (a) Find out the eigen-values and eigen-vectors of the matrix $A = \begin{pmatrix}2 & -1 & -1\\ -1 & 2 & -1\\ -1 & -1 & 2\end{pmatrix}$ (proceed exactly as in question 40).

[*Ans.* Eigen-values 0, 3, 3.]

(b) Show that all the eigen-values of a Hermitian matrix are real.

[For answer see question no. 37.(b)(i)] **[C.U. (Hons.) 1999]**

42. (a) Find the trace and determinant of the matrix $A = \begin{bmatrix} 0 & -i \\ i & 0 \end{bmatrix}$. Find the eigen-values using these values.

(b) Show that A does not commute with $B = \begin{bmatrix} 0 & 1 \\ 1 & 0 \end{bmatrix}$.

(c) If a matrix A is hermitian and $A^2 = 1$, show that A is also unitary.

[C.U. (Hons.) 2001, 2005 (old syll)]

Ans. (a)

$$Tr \cdot (A) = 0 + 0 = 0 \tag{1}$$

$$\text{and} \quad \det(A) = \begin{vmatrix} 0 & -i \\ i & 0 \end{vmatrix} = 0 + i^2 = -1 \tag{2}$$

If λ_1 and λ_2 be the eigen values, then $\lambda_1 + \lambda_2 = \text{Tr}.A = 0$ from (1) and $\lambda_1\lambda_2 = \det(A) = -1$ from (2). Solving we get either $\lambda_1 = 1, \lambda_2 = -1$ or $\lambda_1 = -1, \lambda_2 = 1$.

(b) Here $AB = \begin{bmatrix} 0 & -i \\ i & 0 \end{bmatrix} \begin{bmatrix} 0 & 1 \\ 1 & 0 \end{bmatrix} = \begin{bmatrix} -i & 0 \\ 0 & i \end{bmatrix}$.

Again

$$BA = \begin{bmatrix} 0 & 1 \\ 1 & 0 \end{bmatrix} \begin{bmatrix} 0 & -i \\ i & 0 \end{bmatrix} = \begin{bmatrix} i & 0 \\ 0 & -i \end{bmatrix}$$

so that $AB \neq BA$. $\therefore A$ does not commute with B.

(c) Here $A^2 = 1$ or $A.A = 1$. Now we know that since A is hermitian, $A = A^\dagger$, so that $AA^\dagger = 1$ i.e. the matrix A is unitrary.

43. (a) If $AB = BA$ prove that A and B are square matrices of same dimensions.

[C.U. (Hons.) 2002, 07]

Ans. Answer is obvious since A and B will be conformable for the product AB or BA if the m and n of A and B are equal i.e. they are both square matrices of the same dimensions—otherwise the product does not exist.

(b) Find the eigenvalues and the eigenvectors of the square matrix $\begin{pmatrix} 5 & 2 \\ 2 & 2 \end{pmatrix}$.

[C.U. (Hons.) 2002, 2007]

(See worked out examples)

44. Find the eigenvalues and a set of orthogonal eigenvectors of the following matrix $\begin{pmatrix} 10 & 0 & 0 \\ 0 & 6 & -4 \\ 0 & -4 & 6 \end{pmatrix}$.

[C.U. (Hons.) 2003]

(See worked out examples)

45. Find the eigenvalues and the eigenvectors for the matrix $A = \begin{bmatrix} 1 & 1 \\ 4 & 1 \end{bmatrix}$. **[C.U. (Hons.) 2004]**

(See worked out examples)

46. Express $A = \begin{pmatrix} 0 & -i \\ i & 0 \end{pmatrix}$ as the sum of a symmetric and an anti-symmetric matrix.

[C.U. (Hons.) 2005 (old syll)]

(See Art. 19.14 and 19.14.1—construct a symmetric and an anti-symmetric matrix and then take their sum).

47. Find the eigenvalues and eigenvectors of the matrix $\begin{bmatrix} 1 & 0 & 0 \\ 0 & 0 & 2 \\ 0 & 2 & 0 \end{bmatrix}$. Is the matrix unitary?

[C.U. (Hons.) 2005]

(See worked out examples).

48. If a matrix B commutes with a diagonal matrix A, no two elements of which are equal, show that B is a diagonal matrix. **[C.U. (Hons.) 2006]**

Ans. Let A be a diagonal matrix of order n whose elements are

$$(A)_{ij} = a_i\delta_{ij} \tag{1}$$

where a_i are scalars such that $a_i \neq a_j$, if $i \neq j$ and $1 \leq i,\ j \leq n$. Let the ijth element of the matrix B be b_{ij}. We are given that

$$AB = BA \tag{2}$$

Taking the ijth element of both sides of (2), we have

$$\sum_{p=1}^{n}(A)_{ip}(B)_{pj} = \sum_{p=1}^{n}(B)_{ip}(A)_{pj} \tag{3}$$

On using (1) this becomes

$$\sum_{p=1}^{n} a_i\delta_{ip}b_{pj} = \sum_{p=1}^{n} b_{ip}a_j\delta_{pj}$$

$$\text{or}\quad a_ib_{ij} = b_{ij}a_j \quad \text{or} \quad (a_i - a_j)b_{ij} = 0 \tag{4}$$

This shows that if $i \neq j$, then $b_{ij} = 0$. The only elements of B which are likely to be different from zero are the diagonal elements b_{ii} and b_{jj} for $1 \leq i \leq n$, proving that B is a diagonal matrix.

49. Show that every square matrix can be uniquely written as the sum of a hermitian and an anti-hermitian matrix.

Ans. Let $A \equiv [a_{ij}]$ be a square matrix of order n. Let A be expressible as

$$A = B + C \tag{1}$$

where B is a hermitian matrix and C an anti-hermitian—both of order n. Taking the hermitian conjugate of both sides of (1), we get

$$A^\dagger = B - C \tag{2}$$

Adding and subtracting (1) and (2), we find

$$B = (A + A^\dagger)/2, \quad C = (A - A^\dagger)/2 \tag{3}$$

Taking hermitian conjugate of (3), we have

$$B^\dagger = (A^\dagger + A)/2 = B, \quad C^\dagger = (A^\dagger - A)/2 = -C \tag{4}$$

So it is clear that B is hermitian and C is anti-hermitian, since we know that a matrix is hermitian if $A^\dagger = A$ or $a_{ij}^* = a_{ji}$, $1 \leq i,\ j \leq n$ and again a matrix is anti-hermitian if $A^\dagger = -A$ or $a_{ij}^* = -a_{ji}$, $1 \leq i, j \leq n$.

50. (a) Given $A = \begin{bmatrix} 1 & -2 \\ -2 & -2 \end{bmatrix}$, determine the eigenvalues and normalized eigenvectors of A.

[C.U. (Hons.) 2006]

(See worked out examples)

(b) Is A is real symmetric matrix? (See Art. 19.14 and worked out examples)

(c) Find a similarity transformation that diagonalises the matrix. (See worked out examples)

[C.U. (Hons.) 2006 (old syll)]

51. (a) Show that the product of two symmetric matrices is symmetric if and only if they commute.

[C.U. (Hons.) 2007]

Ans. If A and B are two symmetric matrices, then taking the product of these two matrices, we get

$$[AB]^T = A^T B^T = AB = BA \quad \text{(commutative rule applied)}$$

$$\because A = A^T, \quad B = B^T \quad \text{(proved)}$$

(b) Show that the trace of a matrix remains invariant under a similarity transformation.

[C.U. (Hons.) 2007]

(See worked out examples)

52. (a) Find the eigenvalues and eigenvectors of the matrix $\begin{bmatrix} 0 & 1 \\ 1 & 0 \end{bmatrix}$.

(b) Show that a Hermitian matrix remains Hermitian under a unitary transformation.

(c) Prove that the eigenvalues of a Hermitian matrix are all real.

(d) Show that the length of a vector in real space remains invariant under an orthogonal transformation. **[C.U. (Hons.) 2007]**

(See worked out examples).

53. (a) Find the eigenvalues and normalised eigenvectors of the matrix $\begin{bmatrix} 1 & 1 \\ 1 & 1 \end{bmatrix}$.

(b) Find the given values and a set of orthonormal eigenvectors of the matrix $\begin{bmatrix} 4 & -1 & -1 \\ -1 & 4 & -1 \\ -1 & -1 & 4 \end{bmatrix}$.

[C.U. (Hons.) 2007]

(See worked out examples)

54. (a) Given a unitary matrix U, show that $U^{-1}AU$ is hermitian if A is hermitian.

Ans. Taking hermitian conjugate of $U^{-1}AU$, we have $\left(U^{-1}AU\right)^{\dagger} = \left(U^{-1}\right)^{\dagger} A^{\dagger}U^{\dagger}$; since $UU^{\dagger} = I$ and $A = A^{\dagger}$ from the conditions of the problem, the r.h.s. $= UAU^{-1}$.

(b) Find the eigenvalues and eigenvectors of the matrix $\begin{bmatrix} 1 & 3 \\ 3 & 1 \end{bmatrix}$.

(See worked out examples)

(c) If a matrix is both hermitian and unitary, show that all its eigen values are ± 1.

[C.U. (Hons.) 2008]

Ans. Let U be a unitary matrix and let x be an eigenvector of U with the eigenvalue λ, so that

$$Ux = \lambda x \tag{1}$$

Taking the hermitian conjugate of both sides, we have

$$x^\dagger U^\dagger = \lambda^* x^\dagger \tag{2}$$

Multiplying (1) from the left by (2), we get

$$x^\dagger U^\dagger U x = \lambda\lambda^* x^\dagger x \tag{3}$$

Since U is unitary, $U^\dagger U = I$, so that equation (3) reduces to

$$x^\dagger x \left[|\lambda|^2 - 1\right] = 0$$

Now $x^\dagger x$ is the square of the norm of x and hence cannot vanish unless x is a null vector. Since we are interested in non-zero vectors x which satisfy (1), we must have $|\lambda|^2 = 1$ or $\lambda = \pm 1$, proving the desired result.

55. Show that the eigen-vectors corresponding to distinct eigen-values of a hermitian matrix are orthogonal.

Ans. Let x_1 and x_2 be two eigen-vectors of the hermitian matrix H corresponding to the distinct eigen-values λ_1 and λ_2 respectively, so that

$$Hx_1 = \lambda_1 x_1 \tag{1}$$

$$Hx_2 = \lambda_2 x_2 \tag{2}$$

Taking the hermitian conjugate of (2), we have

$$x_2^\dagger H = \lambda_2 x_2^\dagger \tag{3}$$

where we have used the fact that λ_2 is real. Multiplying (1) from the left by $x_2^\dagger$, (3) from the right by x_1 and subtracting, we have

$$(\lambda_1 - \lambda_2) x_2^\dagger x_1 = 0 \tag{4}$$

Since $\lambda_1 \neq \lambda_2$, we have $x_2^\dagger x_1 = 0$, showing that the vectors x_1 and x_2 are orthogonal.

56. If S and A are unitrary matrices, show that $S^{-1}AS$ is also a unitary matrix.

[C.U. (Hons.) 2007]

Ans. Proceed exactly as shown in problem 54(a) given above.

Chapter 20

Infinite Sequences and Infinite Series

20.1 Introduction

We shall deal, in the following sections, with the exposition of some of the properties of infinite sequences and series, and particular attention will be given to power series. The subject of infinite series is of extreme importance in applied mathematics, because it makes possible the numerical solutions of many important physical problems. The solutions of certain differential equations which occur frequently in the mathematical solutions of many physical problems, are expressed in terms of infinite series, and a study of the properties of these solutions requires a knowledge of the manner in which infinite series may be manipulated. Hence it is essential for the students of applied science to acquire an intelligent understanding of the subject.

Sequence : Let us consider the following arrays of numbers or terms :

(i) $1, 2, 3, 4, \cdots, n, \cdots$ (ii) $\frac{1}{2}, \frac{1}{3}, \frac{1}{4}, \frac{1}{5}, \cdots, \frac{1}{(n+1)}, \cdots$

(iii) $1, 3, 5, 7, \cdots, (2n-1), \cdots$ (iv) $1, 4, 9, 16, 25, \cdots, n^2, \cdots$

(v) $x, x^2, \frac{x^3}{1\times2}, \frac{x^4}{1\times2\times3}, \cdots, \frac{nx^n}{n!}, \cdots$

[$n!$ being the factorial number given by, $n! = 1 \times 2 \times 3 \times \cdots (n-1) \times n$]

In each of the above collections of numbers or terms, there is a definite rule or law which connects a particular term with its particular position. If $u_1, u_2, u_3, \cdots$ denote the 1st, 2nd, 3rd, $\cdots$ terms of the set, the suffixes $1, 2, 3, \cdots$ denoting their positions, the nth term u_n can also be found. The nth terms of the above sets are respectively n, $\frac{1}{(n+1)}$, $(2n-1)$, n^2, $\frac{nx^n}{n!}$, and the successive terms of any set can be obtained from its nth term by assigning to n, the values $1, 2, 3, \cdots$. These types of collections of arranged quantities are called *sequences.*

Definition. *A sequence is a succession of numbers or terms* $u_1, u_2, u_3, \cdots, u_n, \cdots$, *where to every value of* n, *there corresponds a particular number or term of the collection, so that* u_n *is a single-valued function of* n. A sequence is said to be *finite or infinite* according as the number of terms in it is finite or infinite. In our discussions, we shall consider only infinite sequences. A sequence containing the terms $u_1, u_2, u_3, \cdots$ is denoted by any of the symbols (u_n), $\{u_n\}$, and seq. u_n.

20.2 Different Types of Sequences

In general there are three types of sequences, *viz.*, (i) convergent sequence, (ii) divergent sequence, (iii) oscillatory sequence.

(i) Convergent sequence : Let us consider the sequence $(u_n) = 2, \frac{3}{2}, \frac{4}{3}, \cdots, \frac{(n+1)}{n}, \cdots,$ where we see that the terms gradually approach a particular number 1. As we assign to n greater and greater values, the difference between the corresponding term and the number 1 becomes less and less.

In particular, let $n = 10$; then $u_n = u_{10} = \frac{11}{10}$; also for $n = 100$, $u_n = u_{100} = \frac{101}{100}$, so that we have $u_{10} - 1 = \frac{1}{10}$ and $u_{100} - 1 = \frac{1}{100}$.

Thus it is possible to move as near to 1 as we please by choosing the term u_n for sufficiently large values of n. Here we shall say that the sequence (u_n) converges to the limit 1.

Definition. *Given any pre-assigned positive number ε, however small, if we find a number m (depending on ε) such that $|u_n - l| < \varepsilon$ for all $n \geq m$, then the sequence (u_n) is said to converge to the limit l, where l is a finite quantity.*

And we write $\lim\limits_{n\to\infty} u_n = l$ or $u_n \to l$ as $n \to \infty$.

The finite number l is called the *limit of the sequence.* In the example given above, 1 is the limit of the sequence.

(ii) Divergent sequence : A sequence (u_n) which is not convergent, according to the above definition, is called a *divergent sequence.*

Definition. *Given any preassigned positive quantity N, however large, if we find a positive number m, such that $u_n > N$ for all $n \geq m$, then (u_n) is said to diverge to $+\infty$.*

And we write $\lim\limits_{n\to\infty} u_n = +\infty$ or $u_n \to +\infty$ as $n \to \infty$.

Similarly, *given any preassigned positive quantity N, however large, if we find a positive number m, such that $u_n < -N$ or $-u_n > N$ for all $n \geq m$, then (u_n) is said to diverge to $-\infty$.*

And we write $\lim\limits_{n\to\infty} u_n = -\infty$ or $u_n \to -\infty$ as $n \to \infty$.

(iii) Oscillatory sequence : *If a sequence is neither convergent nor divergent, but has all its terms lying between two numbers, finite or infinite, the sequence is said to be oscillatory.* When the terms of a sequence oscillate between two finite numbers, the sequence is said to oscillate finitely. Otherwise, it is said to oscillate infinitely.

Thus an oscillatory sequence is said to oscillate finitely if there exists a finite number M, such that $|u_n| \leq M$, for all values of n. In the case of an infinitely oscillatory sequence, no such finite number M can be found.

20.3 Monotone Sequence

If $u_{n+1} \geq u_n$ for all values of n, the sequence (u_n) is said to be *monotone ascending or increasing.*

Here $u_1 \leq u_2 \leq u_3 \leq \cdots$, so that the successive terms of the sequence gradually increase.

If $u_{n+1} > u_n$ for all values of n, the sequence (u_n) is said to be *strictly monotone ascending or increasing.*

Here $u_1 < u_2 < u_3 < \cdots$

Similarly if $u_{n+1} \leq u_n$ for all values of n, the sequence (u_n) is said to be monotone descending or decreasing.

Again, if $u_{n+1} < u_n$ for all values of n, the sequence (u_n) is said to be *strictly monotone descending or decreasing.*

Illustrative Example

1. Show that the sequence (u_n) is convergent where $u_n = \frac{2n+1}{2n}$.

Here the sequence is $\frac{3}{2}, \frac{5}{4}, \frac{7}{6}, \cdots$.

Now $|u_n - 1| = \left|\frac{2n+1}{2n} - 1\right| = \left|\frac{1}{2n}\right| = \frac{1}{2n}$ which is less than ε for $n > \frac{1}{2\varepsilon}$.

Choosing $m = \frac{1}{2\varepsilon}$, we have $|u_n - 1| < \varepsilon$ for $n > m$.

$\therefore \quad \lim_{n\to\infty} u_n = 1$, i.e., (u_n) is convergent and converges to the limit 1.

We give below some important properties of sequences : (a) *A monotone sequence is convergent if it is bounded;* (b) *If a monotone sequence is unbounded, it diverges to* $+\infty$ *or* $-\infty$; (c) *A convergent sequence has a unique limit;* (d) *A convergent sequence is bounded;* (e) *Cauchy's general principle of convergence : The necessary and sufficient condition that the sequence* (u_n) *is convergent is that corresponding to a given preassigned positive number* ε, *however small, we find a positive integer* m *such that* $|u_{n+p} - u_n| < \varepsilon$ *for* $n \geq m$, *where* p *is a positive integer.*

20.4 Series—Infinite Series

If $u_1, u_2, u_3, \cdots, u_n, \cdots$ be the terms of a sequence (u_n), then the expression $u_1+u_2+\cdots+u_n = \sum u_n$ (which is the indicated sum of the terms of the sequence) is called a *series.* If the number of terms is limited, the series is said to be *finite.* If however the number of terms is unlimited, the series to said to be *infinite.* The general term or nth term u_n is the expression that indicates the law of formation of the terms of the series.

The expression $\sum_{n=1}^{\infty} u_n$ denotes an infinite series, where the nth term u_n is a single-valued function of n. The sum of the first n terms of the series is called the *nth partial sum of the series* and is denoted by S_n, so that

$$S_1 = u_1,\ S_2 = u_1 + u_2,\ S_3 = u_1 + u_2 + u_3,\ \cdots,\ S_n = u_1 + u_2 + u_3 + \cdots + u_n.$$

20.5 Different Types of Infinite Series

Infinite series are divided into three types : (i) convergent series, (ii) divergent series, (iii) oscillatory series.

Definition

(i) Convergent series. *If* S_n *denotes the nth partial sum of the infinite series* $\sum_{n=1}^{\infty} u_n$, *then the series is said to be convergent when* $S_n \to l$ *as* $n \to \infty$, *where* l *is a finite number. This number* l *is called the sum of the infinite series.*

(ii) Divergent series. *An infinite series* $\sum_{n=1}^{\infty} u_n$ *is said to be divergent when its nth partial sum* S_n *is such that* $S_n \to \pm\infty$ *as* $n \to \infty$.

(iii) Oscillatory series. *If an infinite series be such that it does not tend to a definite limit, but by grouping its terms in different ways, the corresponding sums tend to two different numbers, finite or infinite, then the series is said to be an oscillatory one.* When the series tends to two finite numbers, it is said to oscillate finitely. If however, the two numbers are not finite, it is said to oscillate infinitely.

20.6 Geometric Series

The power series of n terms

$$S_n = a + ar + ar^2 + \cdots + ar^{n-1}. \tag{1}$$

is called the *geometric series,* of which r is the common ratio. We shall show in the following way that *the above geometric series is convergent or divergent according as* $|r| < 1$ *or* $|r| \geq 1$ *and it oscillates finitely or infinitely according as* $r = -1$ *or* $r < -1$.

Multiplying (20.6.1) on both sides by r, we obtain

$$rS_n = ar + ar^2 + ar^3 + \cdots + ar^n. \tag{2}$$

Subtracting (20.6.1) from (20.6.2) we have $rS_n - S_n = ar^n - a$

$$\text{or} \quad S_n = \frac{a(1-r^n)}{(1-r)} = \frac{a}{1-r} - \frac{ar^n}{1-r} \tag{3}$$

Now if $|r| < 1$, i.e., $-1 < r < 1$, then $r^n \to 0$ as $n \to \infty$, so that $S_n \to \frac{a}{1-r}$ as $n \to \infty$. Hence if $|r| < 1$, the sum S_n of a geometric series approaches a limit as the number of terms is increased indefinitely. In this case the infinite series represented by (20.6.1) is said to be *convergent,* and its sum is $\lim_{n\to\infty} S_n = \frac{a}{1-r} = S$.

If $|r| > 1$, then $r^n \to \infty$ as $n \to \infty$, so that from (20.6.3) $|S_n| \to \infty$ as $n \to \infty$, i.e., $|S_n|$ will increase without limit. In this case, the series is said to be *divergent.*

Again if $|r| = 1$, $S_n = a + a + a + \cdots$ to n terms, so that $|S_n| \to \infty$ as $n \to \infty$ and the series is *divergent.*

If $|r| = -1$, we come across an interesting situation. In this case the geometric series becomes

$$S_n = a - a + a - a + \cdots \text{ to } n \text{ terms.}$$

In this case, if n is even $S_n = 0$ and if n is odd, $S_n = a$. As n increases indefinitely, the absolute value of S_n does not increase indefinitely but still S_n does not approach a limit but oscillates finitely between 0 and a. A series of this sort is called an *oscillating series.* It is also a divergent series.

Again if $|r| < -1$, then for odd values of n, $(1 - r^n) \to \infty$, as $n \to \infty$ and for even values of n, $(1 - r^n) \to -\infty$, as $n \to \infty$.

$\therefore$ S_n from (20.6.3) oscillates infinitely between $-\infty$ and ∞.

Illustrative Examples

1. Test the convergence of the series $1+\frac{1}{2}+\frac{1}{2^2}+\frac{1}{2^3}+\cdots$

Here the nth partial sum

$$S_n = 1+\frac{1}{2}+\frac{1}{2^2}+\cdots+\frac{1}{2^{n-1}} = \frac{1-\frac{1}{2^n}}{1-\frac{1}{2}} = 2\left(1-\frac{1}{2^n}\right)$$

Now if $n\to\infty$, $\frac{1}{2^n}\to 0$, so that we may write

$$\lim_{n\to\infty} S_n = \lim_{n\to\infty} 2\left(1-\frac{1}{2^n}\right) = 2.$$

Therefore the given series is convergent and its sum is 2.

2. Test the series $1+2+3+4+\cdots$

Here $S_n = 1+2+3+4+\cdots+n = \frac{n(n+1)}{2}$, which tends to ∞ as $n\to\infty$. So the series is divergent.

3. Test the series $1-1+1-1+1-1+\cdots$

Here $S_n = 1-1+1-1+\cdots$ to n terms.

If n is even, $S_n = 0$ and if n is odd, $S_n = 1$.

$\therefore$ $S_n \to 0$ or 1 as $n\to\infty$, so that the series oscillates finitely between 0 and 1.

20.7 Nature of Infinite Series

(i) *The nature of an infinite series remains unchanged if a finite number of terms is added to it or removed from it.*

Proof. Let $\sum u_n = u_1+u_2+u_3+\cdots$ be convergent and let l be its sum. If S_n denotes its nth partial sum, then

$$\lim_{n\to\infty} S_n = \lim_{n\to\infty}(u_1+u_2+\cdots+u_n) = l$$

Let k be the sum of a finite number of terms. If k be added to $\sum u_n$, then

$$\sum u_n + k = \lim_{n\to\infty} S_n + k = l+k, \text{ which is finite.}$$

Thus the resulting series is convergent.

If now k be subtracted from $\sum u_n$, then $\sum u_n - k = \lim\limits_{n\to\infty} S_n - k = l-k$, which is also finite.

Thus the resulting series is convergent. Similar result will follow in the case when the series is divergent or oscillatory.

(ii) *The nature of an infinite series remains unchanged, if all the terms of the series be multiplied by the same constant.*

This property can be best understood from the following illustrative examples :

(a) *If* $\sum\limits_1^\infty u_n = s$, *show that* $\sum\limits_1^\infty a u_n = as$.

Let S_n be the nth partial sum of the series $\sum u_n$, i.e., $S_n = \sum_{r=1}^{r=n} u_r$. Since $\sum u_n$ converges to s, we have $S_n \to s$ as $n \to \infty$.

$\therefore\ aS_n \to as$, as $n \to \infty$, where a is a constant, i.e., $\sum au_n$ converges to as.

(b) *If the series* $\sum_1^\infty u_n$ *and* $\sum_1^\infty v_n$ *converge and their sums are* s *and* t *respectively, then show that the series* $\sum_1^\infty (u_n + v_n)$ *converges and its sum is* $s + t$.

Let S_n and T_n be the nth partial sums of the series $\sum u_n$ and $\sum v_n$ respectively. Then we may write

$$S_n = \sum_{r=1}^{n} u_r, T_n = \sum_{r=1}^{n} v_r,$$

so that $S_n \to s$ as $n \to \infty$ and $T_n \to t$ as $n \to \infty$.

But $\lim_{n\to\infty} [S_n + T_n] = \lim_{n\to\infty} S_n + \lim_{n\to\infty} T_n = s + t$

$$\therefore \quad \sum (u_n + v_n) \text{ converges to } (s + t).$$

20.8 General Principle of Convergence for Series

In applied mathematics, convergent series are of utmost importance. It is thus necessary to have a means of testing a series for convergence or divergence.

The necessary and sufficient condition that the series $\sum u_n$ *converges is that corresponding to a preassigned positive number* ε, *however small, there exists a positive number* m *such that*

$$|S_{n+p} - S_n| < \varepsilon, \quad \textit{i.e.,} \quad |u_{n+1} + u_{n+2} + u_{n+3} + \cdots + u_{n+p}| < \varepsilon$$

for all $n \geq m$, p *being any positive integer, where* $S_n = \sum_{r=1}^{n} u_r$.

From the above, it is obvious that

$$|S_n - S_{n-1}| < \varepsilon \quad \text{or} \quad |u_n| < \varepsilon \quad \text{or} \quad \lim_{n\to\infty} u_n = 0 \tag{1}$$

which is a necessary condition for convergence. But this condition is not sufficient. However, the condition

$$(u_{n+1} + u_{n+2} + \cdots + u_{n+p}) \to 0$$

for all p when $n \to \infty$ is sufficient.

That is, in a convergent series, the terms must approach zero as a limit. If however the nth term of a series does not approach zero as n becomes infinite, we know at once that the series is divergent.

We have stated that (20.8.1) is a necessary condition for convergence but it is not *sufficient*. That is, even if the nth term does approach zero, we cannot state that the series is convergent. As for example, we consider the series for which

$$S_n = 1 + \frac{1}{2} + \frac{1}{3} + \cdots + \frac{1}{n}$$

Here we have $\lim_{n\to\infty} u_n = \lim_{n\to\infty} \frac{1}{n} = 0.$

Therefore (20.8.1) is fulfilled. However, we shall prove later that this series is divergent.

20.9 The Comparison Test

Although the use of the preceding general principle in determining the convergence or divergence of a series is fundamental, we shall now turn to the development of special tests that are, as a rule, easier to apply than the general principle. In many cases, the question of convergence or divergence of a given series may be answered by comparing the given series with one whose character is known.

Test for convergence : Let us test the series

$$U = u_1 + u_2 + u_3 + \cdots + u_n + \cdots, \tag{1}$$

where all the terms of (20.9.1) are positive. Now if a convergent series of positive terms

$$V = v_1 + v_2 + v_3 + \cdots + v_n + \cdots \tag{2}$$

can be found whose terms are never less than the corresponding terms in the series (20.9.1), then (20.9.1) is a convergent series and its sum does not exceed that of (20.9.2).

To prove this statement, let

$$U_n = u_1 + u_2 + \cdots + u_n \quad \text{and} \quad V_n = v_1 + v_2 + \cdots + v_n.$$

Since by hypothesis, (20.9.2) is convergent, we have $\lim_{n\to\infty} V_n = V$. Now since $V_n < V$ and $U_n < V_n$, it follows that $U_n < V$. Therefore, by the general principle U_n approaches a limit and the series (20.9.1) is convergent.

As an example of this test, we consider the series

$$U = 1 + \frac{1}{2^2} + \frac{1}{3^3} + \frac{1}{4^4} + \cdots \tag{3}$$

This series can be compared with the series

$$V = 1 + \frac{1}{2} + \frac{1}{2^2} + \frac{1}{2^3} + \frac{1}{2^4} + \cdots \tag{4}$$

The latter series is a geometric series and is known to be convergent. Now the terms of (20.9.4) are never less than the corresponding terms of (20.9.3). Hence it follows that the series (20.9.3) is also convergent.

Test for divergence : By the use of the comparison principle, it is also possible to test a series for divergence.

Let it be required to test the series

$$U = u_1 + u_2 + u_3 + \cdots + u_n + \cdots \tag{5}$$

where all the terms of (20.9.5) are positive, and are never less than the corresponding terms of a series of positive terms

$$V = v_1 + v_2 + v_3 + \cdots + v_n + \cdots \tag{6}$$

which is known to be divergent. Then (20.9.5) is a divergent series.

By the use of this principle, we may prove that the harmonic series

$$U = 1 + \frac{1}{2} + \frac{1}{3} + \frac{1}{4} + \cdots + \frac{1}{n} + \cdots \tag{7}$$

is divergent. To show this, we re-write (20.9.7) in the form

$$U = 1 + \frac{1}{2} + \left(\frac{1}{3} + \frac{1}{4}\right) + \left(\frac{1}{5} + \frac{1}{6} + \frac{1}{7} + \frac{1}{8}\right) + \left(\frac{1}{9} + \frac{1}{10} + \cdots + \frac{1}{16}\right) + \cdots \tag{8}$$

We now compare this series with the series

$$V = \frac{1}{2} + \frac{1}{2} + \left(\frac{1}{4} + \frac{1}{4}\right) + \left(\frac{1}{8} + \frac{1}{8} + \frac{1}{8} + \frac{1}{8}\right) + \left(\frac{1}{16} + \cdots + \frac{1}{16}\right) + \cdots \tag{9}$$

Now the terms of (20.9.8) are never less than the terms of (20.9.9). But the series (20.9.9) is divergent, since the sum of the terms in each pair of parentheses is $\frac{1}{2}$. So the sum of these terms increases without limit as the number of terms becomes infinite. Hence the series (20.9.8) is also divergent.

Thus the two principles discussed above in determining the convergence or divergence of a series may be generalised as follows :

(i) *If* $\sum u_n$ *and* $\sum v_n$ *be two series of positive terms and if* $u_n \leq v_n$ *for all n, then* $\sum u_n$ *is convergent if* $\sum v_n$ *is convergent.*

(ii) *If* $\sum u_n$ *and* $\sum v_n$ *be two series of positive terms and if* $u_n \geq v_n$ *for all n, then* $\sum u_n$ *is divergent if* $\sum v_n$ *is divergent.*

N.B. *An infinite series having its terms all positive or all negative is divergent if each term of the series is greater than a given finite quantity :* Let $\sum u_n$ be the given infinite series and let S_n be its nth partial sum. Then we may write

$$S_n = u_1 + u_2 + u_3 + \cdots + u_n.$$

If each term of the series $\sum u_n$ be greater than a given finite quantity k, then

$$S_n > k + k + k + \cdots \text{ to } n \text{ terms, i.e., } S_n > nk.$$

But as $n \to \infty$, $nk \to \infty$, so that $S_n \to \infty$ as $n \to \infty$.

Hence the result.

20.10 Cauchy's Integral Test

For the comparison test we require that at least a few types of convergent series should be known. For the establishment of such types and for the test of many series of positive terms, Cauchy's integral test is useful, and may be stated as follows :

Let

$$u_1 + u_2 + u_3 + \cdots = \sum_{n=1}^{\infty} u_n \tag{1}$$

be a series of positive terms, such that $u_{n+1} < u_n$. Now, if there exists a positive function $f(n)$, decreasing for $n > 1$ and such that $f(n) = u_n$, then the given series converges if the integral $I = \int_1^\infty f(n)dn$ exists, and the series diverges if the integral does not exist.

As an example of the application of this test, we consider the series

$$\frac{1}{1^p} + \frac{1}{2^p} + \frac{1}{3^p} + \cdots + \frac{1}{n^p} = \sum_{n=1}^{\infty} \frac{1}{n^p} \tag{2}$$

For applying Cauchy's integral test, let $f(n) = \frac{1}{n^p}$.

Now we have

$$I = \int_1^\infty \frac{dn}{n^p} = \begin{cases} \frac{1}{(1-p)} n^{(1-p)}\Big|_1^\infty & \text{if } p \neq 1 \\ \log_e n\Big|_1^\infty & \text{if } p = 1 \end{cases}$$

We thus see that I exists if $p > 1$ and does not exist if $p \leq 1$. Hence the series converges if $p > 1$ and diverges if $p \leq 1$. The series (20.10.2) is a very useful one for use in comparison with others.

20.11 D'Alembert's Ratio Test

If the two series $\sum u_n$ and $\sum v_n$ be such that the ratio $\frac{u_n}{v_n} \leq k$ for all values of n, where k is a finite quantity, then $\sum u_n$ will be convergent if $\sum v_n$ is convergent.

Proof. Since $\sum v_n$ is convergent, the sum of the series $\sum v_n$ tends to a finite limit l as $n \to \infty$.

Now from the given condition, $u_n \leq kv_n$

$$\therefore \quad u_1 \leq kv_1, u_2 \leq kv_2, u_3 \leq kv_3, \cdots$$

$$\therefore \quad \sum u_n = u_1 + u_2 + u_3 + \cdots$$

$$\leq k(v_1 + v_2 + v_3 + \cdots)$$

$$\leq kl,$$

whence we may say that $\sum u_n$ is convergent, since all the terms of the series $\sum u_n$ being positive, it cannot be an oscillatory one. So it is convergent.

N.B. Similarly it can be shown that if $\frac{u_n}{v_n} \geq m$ for all n, where m is a finite quantity, the series $\sum u_n$ will be divergent, if $\sum v_n$ is divergent.

20.12 Cauchy's Ratio Test

In the infinite geometric series,

$$a + ar + ar^2 + \cdots + ar^n + \cdots$$

the ratio of the consecutive general terms ar^n and ar^{n+1} is the common ratio r. We have seen that this series is convergent when $|r| < 1$ and divergent for other values. We now consider a ratio test that may be applied to any series as discussed below.

Let

$$u_1 + u_2 + u_3 + \cdots + u_n + u_{n+1} + \cdots \tag{1}$$

be an infinite series of positive terms, of which the consecutive general terms u_n and u_{n+1} form the test ratio given by $\left|\frac{u_{n+1}}{u_n}\right| =$ test ratio. Now the limit of this test ratio when n becomes infinite is given by $\rho = \lim_{n\to\infty} \left|\frac{u_{n+1}}{u_n}\right|$, provided the limit exists. We have then the following conditions :

(i) when $\rho < 1$, the series is convergent;

(ii) when $\rho > 1$, the series is divergent;

(iii) when $\rho = 1$, the ratio test gives no information on convergence or divergence.

Proof. (i) When $\rho < 1$, we can choose n so large, say, $n \geq m$, that from the definition of a limit, the ratio $\frac{u_{n+1}}{u_n}$ will differ from ρ by as little as we please, and will therefore be less than a proper fraction $r < 1$. Hence $u_{m+1} < u_m r, u_{m+2} < u_m r^2, u_{m+3} < u_m r^3$ and so on.

We thus see that after the term u_m, each term of the series (20.12.1) is less than the corresponding term of the geometric series

$$u_m(1 + r + r^2 + \cdots) \tag{2}$$

which may written as $u_m \cdot \frac{1}{1-r}$ $(\because\ r < 1)$—which is finite.

Thus the series (20.12.2) and therefore also the series (20.12.1) are convergent, so that we may generalise that *an infinite series is convergent if the ratio of each term to its preceding term is less than or equal to a fixed number which is less than unity. That is if* $\sum u_n$ *be the given series and* r *the given fixed number less than 1, the condition of convergence of* $\sum u_n$ is $\frac{u_{n+1}}{u_n} \leq r < 1$ for all values of n.

(ii) If $\rho > 1$, we can proceed with the same line of reasoning as in (i) and show that the series (20.12.1) is divergent. Alternatively, we can proceed as follows. Let S_n be the nth partial sum of $\sum u_n$. Then

$$\begin{aligned} S_n &= u_1 + u_2 + u_3 + \cdots + u_n \\ &= u_1\left[1 + \frac{u_2}{u_1} + \frac{u_3}{u_1} + \frac{u_4}{u_1} + \cdots \text{ to } n \text{ terms}\right] \\ &= u_1\left[1 + \frac{u_2}{u_1} + \frac{u_3}{u_2}\cdot\frac{u_2}{u_1} + \frac{u_4}{u_3}\cdot\frac{u_3}{u_2}\cdot\frac{u_2}{u_1} + \cdots \text{ to } n \text{ terms}\right] \\ &\geq u_1\left[1 + r + r^2 + r^3 + \cdots \text{ to } n \text{ terms}\right] \end{aligned}$$

i.e., $S_n \geq \frac{u_1(r^n-1)}{(r-1)}$ which tends to ∞ as $n \to \infty$.

$$\therefore\quad \sum u_n \text{ is divergent.}$$

Thus *an infinite series* $\sum u_n$ *is divergent if* $\frac{u_{n+1}}{u_n} \geq r > 1$ *for all values of* n.

(iii) For $\rho = 1$, the test fails, as will be evident from the consideration of the series

$$\frac{1}{1^p} + \frac{1}{2^p} + \frac{1}{3^p} + \cdots + \frac{1}{n^p} = \sum_{n=1}^{\infty} \frac{1}{n^p}. \tag{3}$$

Here, we have the test ratio

$$= \left|\frac{u_{n+1}}{u_n}\right| = \left(\frac{n}{n+1}\right)^p = \left[1 - \frac{1}{(n+1)}\right]^p$$

whence

$$\rho = \lim_{n\to\infty}\left[1 - \frac{1}{(n+1)}\right]^p = (1)^p = 1.$$

Thus whatever value p may have, ρ is equal to unity. But in Art. 20.10, it was shown that for $p > 1$, the series converges and for $p \le 1$, the series diverges. Thus it is evident that $\rho = 1$ for both convergent and divergent series.

It is to be noted that for convergence it is not enough that the test ratio should be less than unity for all values of n. The test definitely requires that the limit of the test ratio, i.e.,

$$\rho = \lim_{n\to\infty}\left|\frac{u_{n+1}}{u_n}\right|$$

should be less than unity.

For example, in the series

$$1 + \frac{1}{2} + \frac{1}{3} + \frac{1}{4} + \cdots + \frac{1}{n} = \sum_{n=1}^{\infty}\frac{1}{n}$$

the test ratio is always less than unity, but the limit, i.e., ρ is however equal to unity.

Illustrative Example

Show that the infinite series $\sum \frac{1}{n^p}$ is convergent or divergent according as $p > 1$ or ≤ 1.

Case I. Let $p > 1$. The given series is $\sum \frac{1}{n^p} = \frac{1}{1^p} + \frac{1}{2^p} + \frac{1}{3^p} + \cdots$

Since $p > 1, 3^p > 2^p$ or $\frac{1}{3^p} < \frac{1}{2^p}$ $\therefore$ $\frac{1}{2^p} + \frac{1}{3^p} < \frac{1}{2^p} + \frac{1}{2^p} = \frac{2}{2^p} = \frac{1}{2^{p-1}}$.

Similarly, $\frac{1}{4^p} + \frac{1}{5^p} + \frac{1}{6^p} + \frac{1}{7^p} < \frac{1}{4^p} + \frac{1}{4^p} + \frac{1}{4^p} + \frac{1}{4^p} = \frac{4}{4^p} = \frac{1}{4^{p-1}}$

$$\begin{aligned}\therefore \quad \sum \frac{1}{n^p} &= \frac{1}{1^p} + \left(\frac{1}{2^p} + \frac{1}{3^p}\right) + \left(\frac{1}{4^p} + \frac{1}{5^p} + \frac{1}{6^p} + \frac{1}{7^p}\right) + \cdots \\ &< 1 + \frac{1}{2^{p-1}} + \frac{1}{4^{p-1}} + \frac{1}{8^{p-1}} + \cdots \\ &= 1 + r + r^2 + r^3 + \cdots,\end{aligned}$$

where $r = \frac{1}{2^{p-1}}$, which is less than 1 as $p > 1$, so that we may write $\sum \frac{1}{n^p} < \frac{1}{1-r} (\because \ r < 1)$ which is finite.

Thus $\sum \frac{1}{n^p}$ is convergent for $p > 1$.

Case II. Let $p = 1$. The given series is then written as

$$\sum \frac{1}{n^p} = 1 + \frac{1}{2} + \frac{1}{3} + \frac{1}{4} + \cdots$$

$$= 1+\frac{1}{2}+\left(\frac{1}{3}+\frac{1}{4}\right)+\left(\frac{1}{5}+\frac{1}{6}+\frac{1}{7}+\frac{1}{8}\right)+\cdots$$

$$> 1+\frac{1}{2}+\left(\frac{1}{4}+\frac{1}{4}\right)+\left(\frac{1}{8}+\frac{1}{8}+\frac{1}{8}+\frac{1}{8}\right)+\cdots$$

i.e., $> 1+\frac{1}{2}+\frac{1}{2}+\frac{1}{2}+\cdots$ which tends to ∞ as $n \to \infty$, so that the series $\sum \frac{1}{n^p}$ is divergent for $p = 1$.

Case III. Let $p < 1$; then $2^p < 2$ or $\frac{1}{2^p} > \frac{1}{2}$. Similarly, $\frac{1}{3^p} > \frac{1}{3}, \frac{1}{4^p} > \frac{1}{4}$ and so on, so that we may write

$$\sum \frac{1}{n^p} = \frac{1}{1^p}+\frac{1}{2^p}+\frac{1}{3^p}+\frac{1}{4^p}+\cdots$$

$$> 1+\frac{1}{2}+\frac{1}{3}+\frac{1}{4}+\cdots$$

which is divergent from the same reasoning as in Case II.

$\therefore$ the series $\sum \frac{1}{n^p}$ is divergent for $p < 1$.

Special Cases A. If two series of positive terms $\sum u_n$ and $\sum v_n$ be such that $\lim\limits_{n\to\infty} \frac{u_n}{v_n} = l$, where l is a finite quantity, the series are both convergent or both divergent.

Proof. I. First, let $\sum v_n$ be convergent. Since $\lim\limits_{n\to\infty} \frac{u_n}{v_n} = l$, we may choose a positive number ε, however small, such that $\frac{u_n}{v_n} < (l+\varepsilon) = m$ (say); i.e., $\frac{u_n}{v_n} < m$ for all n or $u_n < mv_n$ for all n.

$$\therefore \quad u_1 + u_2 + u_3 + \cdots < m(v_1 + v_2 + v_3 + \cdots)$$

$$\therefore \quad \sum u_n < m \sum v_n \text{ for all } n.$$

Now since $\sum v_n$ is convergent, it follows that $\sum u_n$ is also convergent.

II. Next, let $\sum v_n$ be divergent, then since $\lim\limits_{n\to\infty} \frac{u_n}{v_n} = l$, we may choose a positive number ε, however small, such that $\frac{u_n}{v_n} > l - \varepsilon = k$ (say) or $u_n > kv_n$ for all n.

$$\therefore \quad u_1 + u_2 + u_3 + \cdots > k(v_1 + v_2 + v_3 + \cdots) \quad \text{or} \quad \sum u_n > k \sum v_n.$$

Hence $\sum u_n$ is also divergent.

B. If $\sum u_n$ and $\sum v_n$ be two series of positive terms such that $\frac{u_{n+1}}{u_n} \leq \frac{v_{n+1}}{v_n}$ for all n, then $\sum u_n$ will be convergent if $\sum v_n$ is convergent. Again if $\frac{u_{n+1}}{u_n} \geq \frac{v_{n+1}}{v_n}$ for all n, then $\sum u_n$ will be divergent if $\sum v_n$ is divergent.

Proof : Case I. Let $\sum v_n$ be convergent having l for its limit. From the given condition, we have

$$\frac{u_{n+1}}{u_n} \leq \frac{v_{n+1}}{v_n} \quad \text{or} \quad \frac{u_{n+1}}{v_{n+1}} \leq \frac{u_n}{v_n} \text{ for all } n;$$

$$\text{i.e.,} \quad \frac{u_{n+1}}{v_{n+1}} \leq \frac{u_n}{v_n} \leq \frac{u_{n-1}}{v_{n-1}} \leq \cdots \frac{u_3}{v_3} \leq \frac{u_2}{v_2} \leq \frac{u_1}{v_1} = m \text{ (say)}$$

$$\therefore \quad u_1 + u_2 + u_3 + \cdots \leq m(v_1 + v_2 + v_3 + \cdots)$$

i.e., $\sum u_n \le m \sum v_n = ml$, so that we may conclude that $\sum u_n$ is convergent.

Case II. Let $\sum v_n$ be divergent. From the given condition, we have

$$\frac{u_{n+1}}{u_n} \ge \frac{v_{n+1}}{v_n} \quad \text{or} \quad \frac{u_{n+1}}{v_{n+1}} \ge \frac{u_n}{v_n} \text{ for all } n;$$

$$\text{i.e.,} \quad \frac{u_{n+1}}{v_{n+1}} \ge \frac{u_n}{v_n} \ge \frac{u_{n-1}}{v_{n-1}} \ge \cdots \ge \frac{u_3}{v_3} \ge \frac{u_2}{v_2} \ge \frac{u_1}{v_1} = k \text{ (say)}$$

$$\therefore \quad u_1 + u_2 + u_3 + \cdots \ge k(v_1 + v_2 + v_3 + \cdots)$$

i.e., $\sum u_n \ge k \sum v_n$, so that $\sum u_n$ is divergent since $\sum v_n$ is divergent.

20.13 Cauchy's Root Test

The series $\sum u_n$ is convergent if $\lim_{n\to\infty} \sqrt[n]{u_n}$ exists and is $\le r < 1$ and is divergent if $\lim_{n\to\infty} \sqrt[n]{u_n}$ is $\ge r > 1$.

Proof. (i) We have here $\sqrt[n]{u_n} \le r < 1$, so that $u_n \le r^n < 1$ for all n.

$$\begin{aligned} \therefore \quad \sum u_n &= u_1 + u_2 + u_3 + \cdots \\ &\le r + r^2 + r^3 + \cdots \end{aligned}$$

i.e., $\sum u_n \le \frac{r}{1-r}$ ($\because\ r < 1$), which is finite.

$\therefore$ the series $\sum u_n$ is convergent.

(ii) In this case, we have $\sqrt[n]{u_n} \ge r > 1$, so that $u_n \ge r^n > 1$ for all n.

$$\begin{aligned} \therefore \quad \sum u_n &= u_1 + u_2 + u_3 + \cdots \\ &\ge r + r^2 + r^3 + \cdots \end{aligned}$$

But $r + r^2 + r^3 + \cdots + r^n = \frac{r^{n+1}-1}{r-1} - 1 = \frac{r^{n+1}-r}{r-1}$ which $\to \infty$ as $n \to \infty$.

$\therefore \sum u_n$ is divergent.

20.14 Raabe's Test

The series $\sum u_n$ of positive terms is convergent or divergent according as

$$\lim_{n\to\infty} n\left[\frac{u_n}{u_{n+1}} - 1\right] > \text{ or } < 1.$$

Nothing can be ascertained when the given limit equals unity.

Proof. Let us consider the auxiliary series $\sum v_n = \sum \frac{1}{n^p}$. If $p > 1$, we have already proved that $\sum v_n$ is convergent, so that from Art. 20.12, $\sum u_n$ is convergent if

$$\frac{u_n}{u_{n+1}} \ge \frac{v_n}{v_{n+1}} \quad \text{i.e., if} \quad \frac{u_n}{u_{n+1}} \ge \frac{(n+1)^p}{n^p} = \left(1 + \frac{1}{n}\right)^p$$

$$\text{i.e., if} \quad \frac{u_n}{u_{n+1}} \geq 1 + \frac{p}{n} + \frac{p(p-1)}{2!n^2} + \frac{p(p-1)(p-2)}{3!n^3} + \cdots$$

$$\text{i.e., if} \quad n\left(\frac{u_n}{u_{n+1}} - 1\right) \geq p + \frac{p(p-1)}{2!n} + \frac{p(p-1)(p-2)}{3!n^2} + \cdots$$

$\therefore \sum u_n$ is convergent if

$$\lim_{n\to\infty}\left\{n\left(\frac{u_n}{u_{n+1}} - 1\right)\right\} \geq \lim_{n\to\infty}\left\{p + \frac{p(p-1)}{2!n} + \frac{p(p-1)(p-2)}{3!n^2} + \cdots\right\}$$

$$\text{i.e., if} \quad \lim_{n\to\infty}\left\{n\left(\frac{u_n}{u_{n+1}} - 1\right)\right\} \geq p. \quad \text{But from supposition, } p > 1.$$

$\therefore \sum u_n$ is convergent if $\lim\limits_{n\to\infty}\left\{n\left(\frac{u_n}{u_{n+1}} - 1\right)\right\} > 1.$

Similarly it can be shown that $\sum u_n$ is divergent when the limit is < 1. Again by equating the given limit to unity, an absurd result follows, so that nothing can be ascertained.

20.15 Cauchy's Logarithmic Test

The infinite series $\sum u_n$ is convergent or divergent according as

$$\lim_{n\to\infty}\left\{n\log\frac{u_n}{u_{n+1}}\right\} > 1 \quad \text{or} \quad < 1.$$

Proof. Let us consider the auxiliary series $\sum v_n = \sum \frac{1}{n^p}$. If $p > 1$, we have already proved that $\sum v_n$ is convergent, so that from Art. 20.12, the given series $\sum u_n$ is convergent if $\frac{u_n}{u_{n+1}} \geq \frac{v_n}{v_{n+1}}$, i.e., if $\frac{u_n}{u_{n+1}} \geq \left(\frac{n+1}{n}\right)^p = \left(1 + \frac{1}{n}\right)^p$.

Taking logarithm of both sides and multiplying by n throughout, we obtain that the series $\sum u_n$ will be convergent if

$$n\log\frac{u_n}{u_{n+1}} \geq np\log\left(1 + \frac{1}{n}\right).$$

But $np\log\left(1 + \frac{1}{n}\right) = np\left[\frac{1}{n} - \frac{1}{2n^2} + \frac{1}{3n^3} - \cdots\right] = p\left[1 - \frac{1}{2n} + \frac{1}{3n^2} - \cdots\right]$

If $n \to \infty$, then $\sum u_n$ is convergent if $\lim\limits_{n\to\infty}\left\{n\log\left(\frac{u_n}{u_{n+1}}\right)\right\} \geq p$. But by hypothesis, $p > 1$, so that the series $\sum u_n$ is convergent if $\lim\limits_{n\to\infty}\left\{n\log\frac{u_n}{u_{n+1}}\right\} > 1$. Similarly $\sum u_n$ is divergent if the limit < 1.

20.16 Cauchy's Condensation Test

If $f(n)$ is a positive and monotone decreasing function of n and a is an integer > 1, then the two series $\sum f(n)$ and $\sum a^n f(a^n)$ are both convergent or both divergent.

Proof. Let

$$\sum f(n) = \{f(1)+f(2)+\cdots+f(a)\}+\{f(a+1)+f(a+2)+\cdots+f(a^2)\}$$
$$+\{f(a^2+1)+f(a^2+2)+\cdots+f(a^3)\}+\cdots$$
$$+\{f(a^{n-1}+1)+f(a^{n-1}+2)+\cdots+f(a^n)\}+\cdots$$

If u_n denotes the nth group, then $u_n = f(a^{n-1}+1)+f(a^{n-1}+2)+\cdots+f(a^n)$, of which the number of terms is $(a^n - a^{n-1})$. Obviously $\sum u_n$ denotes the above series $\sum f(n)$. Since $f(n)$ is a positive and monotone decreasing function of n, we may write,

$$(a^n - a^{n-1})f(a^n) \le u_n \le (a^n - a^{n-1})f(a^{n-1})$$

or $$\left(1-\frac{1}{a}\right)\{a^n f(a^n)\} \le u_n \le (a-1)\{a^{n-1}f(a^{n-1})\}$$

or $$\left(1-\frac{1}{a}\right)\sum a^n f(a^n) \le \sum u_n \le (a-1)\sum a^{n-1}f(a^{n-1}) \tag{1}$$

But $\sum u_n$ denotes the given series. Therefore from (20.16.1) $\sum u_n$, i.e., the given series $\sum f(n)$ is convergent or divergent according as $\sum a^n f(a^n)$ is convergent or divergent.

20.17 Alternating Series

A series of which the constituent terms are alternately positive and negative is called an *alternating series.*

We consider the alternating series

$$S = u_1 - u_2 + u_3 - u_4 + \cdots \tag{1}$$

We shall prove that the above series is convergent if each term of the series is numerically less than its preceding term and if the terms decrease indefinitely such that $\lim_{n\to\infty} u_n = 0$.

Proof. In the above series, $u_1 > u_2 > u_3 > u_4 > \cdots$ and $u_1, u_2, u_3, \cdots$ are all positive. Now when n is even, we may write S_n as

$$S_n = (u_1 - u_2) + (u_3 - u_4) + \cdots + (u_{n-1} - u_n) \tag{2}$$

which is obviously positive since each bracketted expression is positive and

$$S_n = u_1 - (u_2 - u_3) - (u_4 - u_5) - \cdots - (u_{n-2} - u_{n-1}) - u_n \tag{3}$$

which is less than the finite quantity u_1, since each bracketted expression is positive.

Therefore when n increases through even values, (20.17.2) shows that S_n increases and (20.17.3) shows that S_n is always less than u_1. Hence S_n approaches a limit L. But S_{n+1} also approaches this limit L, since $S_{n+1} = S_n + u_{n+1}$ and $\lim_{n\to\infty} u_{n+1} = 0$.

Thus when n increases through all integral values, $S_n \to L$ and the series is convergent.

The above proof leads to a very important consequence which is stated as :

The error made by terminating a convergent alternating series at any term does not exceed numerically the value of the first of the terms discarded.

For example, from the Maclaurin series expansion of the function $\ln(1+x)$, we have

$$\ln(1+x) = x - \frac{x^2}{2} + \frac{x^3}{3} - \frac{x^4}{4} + \cdots$$

Putting $x = 1$, the sum of the series is written as

$$S = 1 - \frac{1}{2} + \frac{1}{3} - \frac{1}{4} + \frac{1}{5} - \cdots = \ln 2 = 0.693 \tag{4}$$

to three decimal places. Now the sum of the first 10 terms of the seies (20.17.4) is 0.646 and the value of the series differs from this by less than one-eleventh.

20.18 Absolute and Conditional Convergence

If the positive and negative terms of an infinite series are taken all positively, and if the sum of this new series formed tends to a finite limit as $n \to \infty$, i.e., the new series is convergent, then the given series is said to be *absolutely* or unconditionally convergent. Mathematically if $\sum u_n$ be an infinite series of positive and negative terms, then $\sum u_n$ is said to be ***absolutely convergent*** if $\sum |u_n|$ is convergent.

Again if $\sum u_n$ is convergent but $\sum |u_n|$ is not convergent, then $\sum u_n$ is said to be *conditionally* convergent.

For example, the series

$$1 - \frac{1}{2^2} + \frac{1}{3^3} - \frac{1}{4^4} + \frac{1}{5^5} - \cdots + (-1)^{n-1}\frac{1}{n^n} \tag{1}$$

is absolutely convergent, since the series

$$1 + \frac{1}{2^2} + \frac{1}{3^3} + \frac{1}{4^4} + \cdots + \frac{1}{n^n}$$

is convergent. Again the alternating series

$$1 - \frac{1}{2} + \frac{1}{3} - \frac{1}{4} + \frac{1}{5} - \cdots = \ln 2 \tag{2}$$

is *conditionally* convergent, since the harmonic series is divergent.

In a conditionally convergent series, it is not always permissible to change the order of the terms or to group the terms together in brackets in an arbitrary manner. These operations may change the sum of such a series or may change a convergent series into a divergent one or vice versa. As an example, let us again consider the convergent series

$$1 - \frac{1}{2} + \frac{1}{3} - \frac{1}{4} + \cdots + \frac{1}{2n+1} - \frac{1}{2n+2} + \cdots \tag{3}$$

whose sum is equal to the limit of the expression $\displaystyle\sum_{n=0}^{n=\infty}\left(\frac{1}{2n+1} - \frac{1}{2n+2}\right)$.

We can now arrange the terms of this series in a new manner, (by putting two negative terms after each positive term) as given below :

$$1-\frac{1}{2}-\frac{1}{4}+\frac{1}{3}-\frac{1}{6}-\frac{1}{8}+\cdots+\frac{1}{2n+1}-\frac{1}{4n+2}-\frac{1}{4n+4}+\cdots \tag{4}$$

The sum of this convergent series is the limit of the expression given by

$$\sum_{n=0}^{n=\infty}\left(\frac{1}{2n+1}-\frac{1}{4n+2}-\frac{1}{4n+4}\right).$$

Now from the identity

$$\frac{1}{2n+1}-\frac{1}{4n+2}-\frac{1}{4n+4}=\frac{1}{2}\left(\frac{1}{2n+1}-\frac{1}{2n+2}\right),$$

it is evident that the sum of the series (20.18.3) is half the sum of the series (20.18.4).

We can therefore generalise that for a given series which is convergent but not absolutely convergent, it is possible to arrange the terms in such a way that the new series converges to any preassigned number whatsoever.

20.19 Power Series

A series of the form given by

$$S=a_0+a_1x+a_2x^2+a_3x^3+\cdots+a_nx^n+\cdots, \tag{1}$$

where the coefficients a_0, a_1, a_2, $\cdots$ are independent of x, is called a power series in x. A power series in x may converge for all values of x or for no value of x except $x=0$; or it may converge for some values of x different from 0 and may be divergent for other values.

Interval of Convergence : Taking the ratio of the $(n+1)$th and nth terms of the above power series (20.19.1), we obtain

$$\frac{a_{n+1}x^{n+1}}{a_nx^n}=\frac{a_{n+1}}{a_n}x$$

Now if the coefficients of the series be such that $\lim\limits_{n\to\infty}\left|\frac{a_{n+1}}{a_n}\right|=L$, where L is a definite number, then Cauchy's test ratio of Art. 20.12 is given by $\rho=xL$. From Cauchy's ratio test, we consider two cases :

(1) If $L=0$, then the series (20.19.1) converges for all values of x, since in this case $\rho=0$.

(2) If $L\neq 0$, the series will converge if xL is numerically <1, i.e., when x lies in the interval $-\frac{1}{|L|}<x<\frac{1}{|L|}$ and will diverge for values of x outside this interval. This interval is called the *interval of convergence*. The end points of this interval must be examined separately, which will be evident from the following example.

We consider the power series given by

$$x-\frac{x^2}{2^2}+\frac{x^3}{3^2}-\frac{x^4}{4^2}+\cdots+(-1)^{n-1}\frac{x^n}{n^2}+\cdots. \tag{2}$$

Here we have

$$L = \lim_{n\to\infty}\left|\frac{a_{n+1}}{a_n}\right| = \lim_{n\to\infty}\frac{n^2}{(n+1)^2} = +1,$$

so that this series converges when x is numerically < 1 and diverges when x is numerically > 1. Now to examine the end points, if we put $x = 1$ in (20.19.1), we obtain

$$1 - \frac{1}{2^2} + \frac{1}{3^2} - \frac{1}{4^2} + \cdots,$$

which is an alternating series that converges. If we now put $x = -1$ in (20.19.1), we have

$$-1 - \frac{1}{2^2} - \frac{1}{3^2} - \frac{1}{4^2} - \cdots,$$

which is a convergent series, as can be seen by comparison with the p series given by

$$\frac{1}{1^p} + \frac{1}{2^p} + \frac{1}{3^p} + \cdots + \frac{1}{n^p} \cdots = \sum_{n=1}^{\infty}\frac{1}{n^p}, \text{ when } p > 1.$$

Thus we see that the series (20.19.1) has the interval of convergence $-1 \leq x \leq +1$.

20.20 Theorems regarding Power Series

In the following, we shall state some fundamental theorems which are of importance in the use of power series in applied mathematics. We shall discuss and explain them one by one, with reference to examples, without going into their rigorous proofs.

Theorem 1. Sum of two convergent power series : Let

$$S_1(x) = a_0 + a_1x + a_2x^2 + \cdots + a_nx^n + \cdots \tag{1}$$

be a power series convergent for $|x| < r_1$ and

$$S_2(x) = b_0 + b_1x + b_2x^2 + \cdots + b_nx^n + \cdots \tag{2}$$

be another power series convergent for $|x| < r_2$. If now we consider r to be a number smaller than both r_1 and r_2, such that both of the above two power series converge for $|x| < r$, then the sum of $S_1(x)$ and $S_2(x)$ is represented by the power series

$$S_1(x) + S_2(x) = (a_0 + b_0) + (a_1 + b_1)x + (a_2 + b_2)x^2 + \cdots + (a_n + b_n)x^n + \cdots \tag{3}$$

which is convergent at least for $|x| < r$. That is, two power series may be added term by term and the resulting series converges to $S_1(x) + S_2(x)$ for those values of x for which both of the given series converge. For example, let

$$S_1(x) \;=\; e^x = 1 + x + \frac{x^2}{2!} + \cdots + \frac{x^n}{n!} + \cdots \tag{4}$$

$$\text{and}\quad S_2(x) \;=\; e^{-x} = 1 - x + \frac{n^2}{2!} - \cdots + (-1)^n\frac{x^n}{n!} + \cdots \tag{5}$$

We have then

$$\begin{aligned} S_1(x) + S_2(x) &= e^x + e^{-x} \\ &= 2\left[1 + \frac{x^2}{2!} + \cdots + \frac{1+(-1)^n}{2} \cdot \frac{x^n}{n!} + \cdots\right] \\ &= 2\cosh x \end{aligned} \tag{6}$$

In this case, both the series $S_1(x)$ and $S_2(x)$ converge for all values of x and hence their sum converges for all values of x.

Theorem 2. The Product of two convergent power series : If $S_1(x) = \sum_{n=0}^{\infty} a_n x^n$ and $S_2(x) = \sum_{n=0}^{\infty} b_n x^n$ both converge for $|x| < r$, then the product $S_1(x)S_2(x)$ is given by the power series

$$S_1(x)S_2(x) = a_0b_0 + (a_0b_1 + a_1b_0)x + (a_0b_2 + a_1b_1 + a_2b_0)x^2 + \cdots$$

which also converges at least for $|x| < r$.

That is, one power series may be multiplied by another to obtain the expansion of the product of two functions. The coefficients are formed just as in the multiplication of two polynomials. The product series is convergent, whenever the two given series are convergent.

As an illustration of this theorem, we consider the two series

$$S_1(x) = \sinh x = x + \frac{x^3}{3!} + \frac{x^5}{5!} + \cdots + \frac{x^{2n-1}}{(2n-1)!} + \cdots$$

$$\text{and} \quad S_2(x) = \cosh x = 1 + \frac{x^2}{2!} + \frac{x^4}{4!} + \cdots + \frac{x^{2n-2}}{(2n-2)!} + \cdots$$

We may then write

$$\begin{aligned} S_1(x)S_2(x) &= \sinh x \cosh x = x + \left(\frac{1}{2!} + \frac{1}{3!}\right)x^3 + \left(\frac{1}{4!} + \frac{1}{3!2!} + \frac{1}{5!}\right)x^5 + \cdots \\ &= x + \frac{2}{3}x^3 + \frac{2}{15}x^5 + \cdots \\ &= \frac{1}{2}\left[2x + \frac{(2x)^3}{3!} + \frac{(2x)^5}{5!} + \cdots\right] \\ &= \frac{1}{2}\sinh 2x \end{aligned}$$

Since the two given series converge for $|x| < 1$, the product series is also convergent for $|x| < 1$.

Theorem 3. The quotient of two convergent series : If $S_1(x) = \sum_{n=0}^{\infty} a_n x^n$ and $S_2(x) = \sum_{n=0}^{\infty} b_n x^n$ both converge for $|x| < r$ and if $b_0 \neq 0$, then the quotient, obtained by dividing the series for $S_1(x)$ by that for $S_2(x)$, is represented by the series

$$\frac{S_1(x)}{S_2(x)} = \frac{a_0}{b_0} + \frac{a_1b_0 - a_0b_1}{b_0^2}x + \frac{a_2b_0^2 - a_1b_0b_1 + a_0b_1^2 - a_0b_0b_2}{b_0^3}x^2 + \cdots$$

In this case, we can draw no conclusion regarding the region of convergence of the quotient series from a knowledge of the regions of convergence of the series $S_1(x)$ and $S_2(x)$.

This may be illustrated by considering the following two series :

$$S_1(x) = \sin x = x - \frac{x^3}{3!} + \frac{x^5}{5!} - \cdots + \frac{(-1)^{n-1}x^{2n-1}}{(2n-1)!} + \cdots$$

$$\text{and} \quad S_2(x) = \cos x = 1 - \frac{x^2}{2!} + \frac{x^4}{4!} - \cdots + \frac{(-1)^{n-1}x^{2n-2}}{(2n-2)!} + \cdots$$

Dividing $S_1(x)$ by $S_2(x)$, we have

$$\frac{S_1(x)}{S_2(x)} = \frac{\sin x}{\cos x} = x + \frac{x^3}{3} + \frac{2x^5}{15} + \cdots = \tan x.$$

One important point to be noted here is that the series for $\sin x$ and $\cos x$ converge for all values of x, but that for $\tan x$ is convergent only for $|x| < \pi/2$.

Theorem 4. Substitution of one series into another : Let the series

$$z = a_0 + a_1y + a_2y^2 + \cdots + a_ny^n + \cdots \tag{7}$$

converge for $|y| < r_1$ and let the series

$$y = b_0 + b_1x + b_2x^2 + \cdots + b_nx^n + \cdots \tag{8}$$

converge for $|x| < r_2$.

Now if $|b_0| < r_1$, then we may substitute for y in the series (20.20.7) its value in terms of x from the series (20.20.8) and thus obtain z as a power series in x. This will converge if x is sufficiently small.

In the special case that the given series for z converges for all values of y, the series for z in terms of x may then always be found, and this series will converge for all values of $|x| < r_2$.

As an example we consider the expansion of $e^{\cos x}$ as a power series in x. In this case, we may write

$$z = e^y = 1 + \frac{y}{1!} + \frac{y^2}{2!} + \cdots + \frac{y^n}{n!} + \cdots \tag{9}$$

$$\text{and} \quad y = \cos x = 1 - \frac{x^2}{2!} + \frac{x^4}{4!} - \frac{x^6}{6!} + \cdots \tag{10}$$

In this case, the series (20.20.9) and (20.20.10) converge for all values of x. We now form the various powers of y, which may be written as

$$y^2 = 1 - x^2 + \frac{1}{3}x^4 - \cdots$$

$$y^3 = 1 - \frac{3}{2}x^2 + \frac{7}{8}x^4 - \cdots$$

$$y^4 = 1 - 2x^2 + \frac{5}{3}x^4 - \cdots$$

Substituting these into (20.20.9), we have

$$\begin{aligned} e^y &= 1+\left(1-\frac{x^2}{2}+\frac{x^4}{24}-\cdots\right)+\frac{1}{2}\left(1-x^2+\frac{x^4}{3}-\cdots\right) \\ &\quad +\frac{1}{6}\left(1-\frac{3}{2}x^2+\frac{7}{8}x^4-\cdots\right)+\frac{1}{24}\left(1-2x^2+\frac{5}{3}x^4-\cdots\right) \\ &= \left(1+1+\frac{1}{2}+\frac{1}{6}+\frac{1}{24}+\cdots\right)-\left(\frac{1}{2}+\frac{1}{2}+\frac{1}{4}+\frac{1}{12}+\cdots\right)x^2 \\ &\quad +\left(\frac{1}{24}+\frac{1}{6}+\frac{7}{48}+\frac{5}{72}+\cdots\right)x^4+\cdots \end{aligned}$$

Hence

$$e^y = e^{\cos x} = 2\frac{17}{24}-1\frac{1}{3}x^2+\frac{61}{144}x^4 \tag{11}$$

It should be noted that the coefficients in this series really correspond to those of an infinite one, and that the final values here given are only the approximate values found by taking into consideration, the first few terms of each series. This situation arises whenever $b_0 \neq 0$ in (20.20.10). However, it is sometimes possible to make an elementary change that simplifies the final result, and as such in the above case, we may write

$$e^{\cos x} = e^{\cos x-1+1} = e^{\cos x-1}e = e^u e = e\left(1+u+\frac{u^2}{2!}+\frac{u^3}{3!}+\cdots\right), \tag{12}$$

$$\text{where}\quad u = \cos x - 1 = -\frac{x^2}{2!}+\frac{x^4}{4!}-\frac{x^6}{6!}+\cdots$$

Raising u to the various required powers, we can find out u^2, u^4, etc. and then substitute these results into (20.20.12), whence we have

$$e^{\cos x} = e\left(1-\frac{x^2}{2}+\frac{x^4}{6}-\frac{31}{720}x^6+\cdots\right). \tag{13}$$

The coefficients are now exact, and the computation of the successive terms is much simpler than what we have done in the previous method.

Theorem 5. Differentiation of a power series : If the series given by

$$S_1(x) = a_0+a_1x+a_2x^2+\cdots+a_nx^n+\cdots \tag{14}$$

converges for $|x| < r$, then the derivatives of $S_1(x)$ may be obtained by term by term differentiation of the series (20.20.14) in the form

$$\frac{d}{dx}S_1(x) = a_1+2a_2x+3a_3x^2+\cdots+na_nx^{n-1}+\cdots \tag{15}$$

which also converges for $|x| < r$.

As an example, we consider the following series

$$\sin x = x-\frac{x^3}{3!}+\frac{x^5}{5!}-\cdots+(-1)^{n-1}\frac{x^{2n-1}}{(2n-1)!}+\cdots \tag{16}$$

which on differentiation gives

$$\begin{aligned}\frac{d}{dx}(\sin x) &= 1-\frac{x^2}{2!}+\frac{x^4}{4!}-\cdots+(-1)^{n-1}\frac{x^{2n-2}}{(2n-2)!}+\cdots \\ &= \cos x.\end{aligned} \tag{17}$$

In this case both the series (20.20.16) and (20.20.17) converge for all values of x.

Theorem 6. Integration of a power series : If the series $S_1(x)=\sum_{n=0}^{\infty} a_n x^n$ converges for $|x|<r$, then the integral of $S_1(x)$ may be found by integrating the above series term by term in the form

$$\int_0^x S_1(x)dx = \sum_{n=0}^{\infty}\frac{a_n}{n+1}x^{n+1};$$

This new series also converges for $|x|<r$.

For example, we consider the series given by

$$\frac{1}{1+x^2} = 1-x^2+x^4-\cdots+(-1)^{n-1}x^{2n-2}+\cdots \tag{18}$$

which on integration yields

$$\begin{aligned}\int_0^x \frac{dx}{1+x^2} &= x-\frac{x^3}{3}+\frac{x^5}{5}-\cdots+(-1)^{n-1}\frac{x^{2n-1}}{2n-1}+\cdots \\ &= \tan^{-1} x.\end{aligned} \tag{19}$$

Since the series (20.20.18) is convergent for $|x|<1$, the series for $\tan^{-1} x$, i.e., (20.20.19) is also convergent in this interval.

Theorem 7. Equality of power series : If we have

$$\sum_{n=0}^{n=\infty} a_n x^n = \sum_{n=0}^{n=\infty} b_n x^n,$$

both converging for $|x|<r$, then the coefficients of like powers of the two series must be equal. That is, we must have $a_s=b_s$, $s=0,1,2,3,\cdots$

It then follows that if a function is expanded in a certain interval by different methods, the series obtained must be identical.

Theorem 8. Reversion of a power series : In certain applications of mathematical analysis to physical problems, it is necessary to *reverse* a power series. Let the given series be

$$y = ax+bx^2+cx^3+dx^4+ex^5+fx^6+gx^7+\cdots, \tag{20}$$

where it is stipulated that the coefficient a is not equal to zero.

It is required to find the coefficients of the series

$$x = Ay+By^2+Cy^3+Dy^4+Ey^5+Fy^6+Gy^7+\cdots \tag{21}$$

In order to determine the coefficients A, B, C, etc., the series (20.20.21) is substituted into the series (20.20.20) and the coefficients of like powers of y are equated to each other. This procedure yields the following equations :

$$A = \frac{1}{a}, \qquad B = -\frac{b}{a^3}, \qquad C = \frac{2b^2 - ac}{a^5}$$

$$D = \frac{5abc - a^2d - 5b^3}{a^7}, \qquad E = \frac{6a^2bd + 3a^2c^2 + 14b^4 - a^3e - 21ab^2c}{a^9}$$

$$F = \frac{7a^3be + 7a^3cd + 84ab^3c - a^4f - 28a^2b^2d - 28a^2bc^2 - 42b^5}{a^{11}}$$

$$G = \frac{\begin{array}{c} 8a^4bf + 8a^4ce + 4a^4d^2 + 120a^2b^3d + 180a^2b^2c^2 + 132b^6 - a^5g \\ -36a^3b^2e - 72a^3bcd - 12a^3c^3 - 330ab^4c \end{array}}{a^{13}}$$

20.21 Use of Series for the Computation of Functions

In many cases, the series expansion of a function gives a direct method for the computation of the numerical value of a function for a certain given value of the argument. Tables of functions may, in this way, be computed.

For example, to compute the value of sin 10°, we make use of the Maclaurin series expansion of the sine function. We may then write

$$\sin x = x - \frac{x^3}{3!} + \frac{x^5}{5!} - \frac{x^7}{7!} + \cdots$$

In this case $x = 10° = \frac{\pi}{18}$ radians. Hence we have

$$\sin\left(\frac{\pi}{18}\right) = \frac{\pi}{18} - \left(\frac{\pi}{18}\right)^3 \frac{1}{3!} + \left(\frac{\pi}{18}\right)^5 \frac{1}{5!} - \left(\frac{\pi}{18}\right)^7 \frac{1}{7!} + \cdots = 0.1736.$$

Since this is an alternating series, we know that the error introduced by stopping at this term is less than $\left(\frac{\pi}{18}\right)^9 \frac{1}{9!}$.

As another example, we consider a series useful in the computation of logarithms. The Maclaurin series expansion of the function $\ln(1+x)$ may be written as

$$\ln(1+x) = x - \frac{x^2}{2} + \frac{x^3}{3} - \frac{x^4}{4} + \cdots \tag{1}$$

Putting $x = -x$ in (20.21.1), we have

$$\ln(1-x) = -x - \frac{x^2}{2} - \frac{x^3}{3} - \frac{x^4}{4} - \cdots \tag{2}$$

From (20.21.1) and (20.21.2) we may write

$$\ln\frac{1+x}{1-x} = \ln(1+x) - \ln(1-x) = 2\left(x + \frac{1}{3}x^3 + \frac{1}{5}x^5 + \frac{1}{7}x^7 + \cdots\right). \tag{3}$$

This series converges when $|x| < 1$.

Now putting $x = \frac{1}{2n+1}$, (where $n > 0$), we have $\frac{1+x}{1-x} = \frac{n+1}{n}$, so that $|x| < 1$ for all values of $n > 0$. Substituting this value into (20.21.3) we have

$$\ln(n+1) = \ln n + 2\left[\frac{1}{2n+1} + \frac{1}{3}\frac{1}{(2n+1)^3} + \frac{1}{5}\frac{1}{(2n+1)^5} + \cdots\right] \quad (4)$$

This series (20.21.4) converges for all positive values of n and is well adapted for computation. For example, putting $n = 1$, we have, from (20.21.4)

$$\ln 2 = 2\left[\frac{1}{3} + \frac{1}{3} \times \frac{1}{3^3} + \frac{1}{5} \times \frac{1}{3^5} + \cdots\right] = 0.69315.$$

Again if $n = 2$, we have

$$\ln 3 = \ln 2 + 2\left(\frac{1}{5} + \frac{1}{3} \times \frac{1}{5^3} + \frac{1}{5} \times \frac{1}{5^5} + \cdots\right) = 1.09861.$$

In this way, we may compute the natural logarithms of any number. If we require common logarithms of numbers, denoted by $\log n$, we may write

$$\log n = \frac{\ln n}{\ln 10} = \frac{\ln n}{2.30259}$$

For example, putting $n = 2$, we have

$$\log 2 = \frac{\ln 2}{\ln 10} = \frac{0.69315}{2.3025} = 0.3010.$$

Illustrative Examples

1. Consider the convergence of the following sequences :

(i) $\{u_n\} = \frac{1}{2}, \frac{1}{3}, \frac{1}{4}, \cdots$. (ii) $\{u_n\} = (1 + \frac{1}{n})$.

Ans. (i) Here $u_n = \frac{1}{n+1}$

$\therefore\ |u_n - 0| = \left|\frac{1}{n+1} - 0\right| = \left|\frac{1}{n+1}\right| = \frac{1}{n+1}$, which is less than ε for $n > \frac{1}{\varepsilon} - 1$.

$\therefore$ for any preassigned positive value of ε, however small, we can find a sufficiently large value of n such that

$$|u_n - 0| < \varepsilon. \therefore \lim_{n\to\infty} u_n = 0 \text{ i.e., the sequence } \{u_n\} \text{ is convergent to the limit } 0.$$

(ii) Here we have $u_n = 1 + \frac{1}{n}$;

$\therefore\ |u_n - 1| = |1 + \frac{1}{n} - 1| = |\frac{1}{n}| = \frac{1}{n}$, which is less than ε for $n > \frac{1}{\varepsilon}$.

$\therefore$ for any preassigned positive value of ε, however small, we can find a sufficiently large value of n such that

$$|u_n - 1| < \varepsilon \text{ or } \lim_{n\to\infty} u_n = 1, \text{ i.e., the sequence } \{u_n\} \text{ converges to the limit } 1.$$

2. Show that the following sequence is convergent $\frac{1}{2}, \frac{5}{4}, \frac{7}{8}, \frac{17}{16}, \cdots$

Ans. Here $u_n = 1 + (-1)^n \cdot \frac{1}{2^n}$, so that $u_n \to 1$ as $n \to \infty$;

$\therefore$ the sequence $\{u_n\}$ converges to the limit 1.

3. Test the convergence of the sequence, $1, 0, -\frac{1}{3}, 0, \frac{1}{5}, 0, -\frac{1}{7}, \cdots$

Ans. Here the sequence may be put in the form

$$\sin\frac{\pi}{2}, \frac{1}{2}\sin\pi, \frac{1}{3}\sin\frac{3\pi}{2}, \frac{1}{4}\sin 2\pi, \frac{1}{5}\sin\frac{5\pi}{2}, \cdots$$

so that $u_n = \frac{1}{n}\sin\frac{n\pi}{2}$.

The sequence is obviously bounded whose bounds are 1 and $-\frac{1}{3}$.

Again $|u_n - 0| = \left|\frac{1}{n}\sin\frac{n\pi}{2}\right| = \frac{1}{n}\left|\sin\frac{n\pi}{2}\right| \le \frac{1}{n}$ $\quad$ [since $\left|\sin\frac{n\pi}{2}\right| \le 1$]

i.e., $|u_n - 0| < \varepsilon$ for $n > \frac{1}{\varepsilon}$ or $u_n \to 0$ as $n \to \infty$.

$\therefore$ the given sequence converges to the limit 0.

4. Show that the sequence $u_n = (-1)^n\left(1 + \frac{1}{n}\right)$ is oscillatory.

Ans. The sequence is $-2, \frac{3}{2}, -\frac{4}{3}, \frac{5}{4}, \cdots$, where the odd terms are negative and even terms are positive.

Let n be odd. Then $|u_n + 1| = \left|-\left(1 + \frac{1}{n}\right) + 1\right| = \left|-\frac{1}{n}\right| = \frac{1}{n}$

$\therefore \quad |u_n + 1| < \varepsilon$ for $n > \frac{1}{\varepsilon}$. Thus $u_n \to -1$ as $n \to \infty$.

Next, let n be even; then $|u_n - 1| = \left|1 + \frac{1}{n} - 1\right| = \left|\frac{1}{n}\right| = \frac{1}{n}$

$\therefore \quad |u_n - 1| < \varepsilon$ for $n > \frac{1}{\varepsilon}$. Thus $u_n \to 1$ as $n \to \infty$.

Again for all n, $|u_n| \le 2$.

$\therefore$ the sequence $\{u_n\}$ oscillates finitely between -1 and $+1$.

5. Find the limit of the sequence $\sqrt{2}, \sqrt{2 + \sqrt{2}}, \sqrt{2 + \sqrt{2 + \sqrt{2}}}, \cdots$

Ans. Let u_n and u_{n-1} denote respectively the nth and $(n-1)$th terms of the given sequence. Then

$$u_n = \sqrt{2 + u_{n-1}} \quad \text{or} \quad \lim_{n\to\infty} u_n = \sqrt{2 + \lim_{n\to\infty} u_{n-1}} \qquad (1)$$

If l be the limit of the given sequence, then $\lim_{n\to\infty} u_n = l = \lim_{n\to\infty} u_{n-1}$, so that from (1) we have $l = \sqrt{2 + l}$ or $l^2 - l - 2 = 0$ or $(l - 2)(l + 1) = 0$ or $l = 2$ or -1.

But it is evident that the sequence cannot be negative

$\therefore$ the limit of $\{u_n\}$ is 2.

6. Discuss the convergence or divergence of the sequence (u_n) of which the nth term is given by $u_n = x^n$.

Ans. Case I. Let $|x| < 1$, i.e., $-1 < x < 1$.

Here $|u_n| = |x^n| = |x|^n \to 0$ as $n \to \infty$. Therefore (u_n) is convergent and its limit is 0.

Case II. Let $x = 1$. Then $u_n = x^n = 1$, so that every term of the sequence is 1. Therefore (u_n) is convergent and its limit is 1.

Case III. Let $x = -1$. Then $u_n = x^n = (-1)^n$, so that $u_n = -1$ for odd values of n and $u_n = +1$ for even values of n.

$\therefore \quad (u_n)$ is oscillating finitely between -1 and $+1$.

Case IV. Let $x > 1$. Here n may be chosen sufficiently large, so that $u_n = x^n$ may be made to exceed any positive number, however large.

$\therefore \quad u_n \to \infty$ as $n \to \infty$, so that (u_n) is divergent.

Case V. Let $x < -1$. Here if n is even, u_n may be made to exceed any positive number, however large, by properly choosing a sufficiently large even value of n. Also if n is odd, $-u_n$ may be made to exceed any positive number, however large, by choosing properly a sufficiently large odd value of n.

$$\therefore \quad u_n \to +\infty \text{ as } n \to \infty \text{ for even values of } n,$$

$$\text{and} \quad u_n \to -\infty \text{ as } n \to \infty \text{ for odd values of } n.$$

$\therefore \quad (u_n)$ is oscillating infinitely between $-\infty$ and $+\infty$.

7. If $u_n = 1 + \frac{1}{2} + \frac{1}{3} + \cdots + \frac{1}{n} - \log n$, and $v_n = 1 + \frac{1}{2} + \frac{1}{3} + \cdots + \frac{1}{n-1} - \log n$, then (u_n) is monotone descending and (v_n) is monotone ascending and each of them converges to the same limit l, where $0.3 < l < 1$.

Ans. In the first case $u_{n+1} - u_n = \frac{1}{n+1} - \log(n+1) + \log n = \frac{1}{n+1} - \log\left(1 + \frac{1}{n}\right)$.

Similarly in the second case, $v_{n+1} - v_n = \frac{1}{n} - \log(n+1) + \log n = \frac{1}{n} - \log\left(1 + \frac{1}{n}\right)$.

But $\frac{1}{n+1} < \log\left(1 + \frac{1}{n}\right) < \frac{1}{n}$. $\therefore$ $u_{n+1} - u_n < 0$ and $v_{n+1} - v_n > 0$.

$\therefore \quad (u_n)$ is monotone descending and (v_n) is monotone ascending.

Again $\frac{1}{n} > \log\left(1 + \frac{1}{n}\right)$ or $\frac{1}{n} > \log(n+1) - \log n$.

Putting $n = 1, 2, 3, \cdots, n$ and then adding, we get

$$1 + \frac{1}{2} + \frac{1}{3} + \cdots + \frac{1}{n} > (\log 2 - \log 1) + (\log 3 - \log 2)$$
$$+ (\log 4 - \log 3) + \cdots + [\log(n+1) - \log n]$$

$$\text{or} \quad 1 + \frac{1}{2} + \frac{1}{3} + \cdots + \frac{1}{n} > \log(n+1)$$

Now since n is positive, $\log(n+1) > \log n$.

$$\therefore \quad 1 + \frac{1}{2} + \frac{1}{3} + \cdots + \frac{1}{n} > \log n$$

$$\therefore \quad u_n = 1 + \frac{1}{2} + \frac{1}{3} + \cdots + \frac{1}{n} - \log n = \text{positive}$$

Thus it is seen that (u_n) is positive and monotone descending.

$\therefore \quad (u_n)$ is convergent. Let l be its limit; then $l \geq 0$.

But $u_1 = 1$ and (u_n) is monotone descending. $\therefore \quad l < 1$.

Also $u_n - v_n = \frac{1}{n}$. $\therefore$ $\lim\limits_{n \to \infty} (u_n - v_n) = 0$, so that (v_n) also converges to the same limit l.

Again $v_2 = 1 - \log 2 > 0.3$. $\therefore$ $l > 0.3$, so that $0.3 < l < 1$.

8. If $u_n = \left(1 + \frac{1}{n}\right)^n$, show that the sequence (u_n) is monotone ascending and bounded converging to the limit e, where $e = 1 + \frac{1}{1!} + \frac{1}{2!} + \frac{1}{3!} + \cdots$ to ∞.

Ans. Binomially expanding, we may write

$$u_n = \left(1 + \frac{1}{n}\right)^n = 1 + n \cdot \frac{1}{n} + \frac{n(n-1)}{2!} \cdot \frac{1}{n^2} + \frac{n(n-1)(n-2)}{3!} \cdot \frac{1}{n^3} + \cdots$$

$$= 1 + 1 + \frac{1}{2!}\left(1 - \frac{1}{n}\right) + \frac{1}{3!}\left(1 - \frac{1}{n}\right)\left(1 - \frac{2}{n}\right) + \cdots \quad (1)$$

Now if n increases, the number of terms of the r.h.s. of (1) increases and each of its terms (except the 1st and the 2nd) also increases.

$\therefore$ (u_n) is monotone ascending.

Again from (1)

$$u_n < 1+1+\frac{1}{2!}+\frac{1}{3!}+\cdots+\frac{1}{n!} \qquad (2)$$

$$< 1+1+\frac{1}{2}+\frac{1}{2^2}+\cdots+\frac{1}{2^{n-1}}$$

$$\text{or} \quad u_n < 1+\frac{1-\frac{1}{2^n}}{1-\frac{1}{2}} = 1+2\left(1-\frac{1}{2^n}\right) = 3-\frac{1}{2^{n-1}}$$

$\therefore$ $u_n < 3$ but $u_n > 2$. $\therefore$ $2 < u_n < 3$, i.e., (u_n) is bounded. Thus (u_n) is monotone ascending and bounded. Therefore it must be convergent. Let k be the limit of (u_n).

From (2)

$$\lim_{n\to\infty} u_n \le \lim_{n\to\infty}\left(1+1+\frac{1}{2!}+\frac{1}{3!}+\cdots\right) = e \quad \therefore \quad k \le e \qquad (3)$$

Also for $m > n$, we obtain from (1) that for any particular n,

$$u_m = \left(1+\frac{1}{m}\right)^m > 1+1+\frac{1}{2!}\left(1-\frac{1}{m}\right)+\frac{1}{3!}\left(1-\frac{1}{m}\right)\left(1-\frac{2}{m}\right)+\cdots$$

$$+\frac{1}{n!}\left(1-\frac{1}{m}\right)\left(1-\frac{2}{m}\right)\cdots\left(1-\frac{n-1}{m}\right) \qquad (4)$$

But as $m \to \infty$, $u_m \to k$, so that from (4) we have

$$k \ge 1+1+\frac{1}{2!}+\frac{1}{3!}+\cdots+\frac{1}{n!}.$$

Now taking the limit when $n \to \infty$, we may write

$$k \ge \lim_{n\to\infty}\left(1+1+\frac{1}{2!}+\frac{1}{3!}+\cdots\right) = e, \text{ i.e., } k \ge e \qquad (5)$$

$\therefore$ from (3) and (5), we have $k = e$, i.e., $\lim_{n\to\infty}\left(1+\frac{1}{n}\right)^n = e$.

9. If $u_n = \sqrt[n]{n}$, prove that the sequence (u_n) is convergent and its limit is equal to 1.

Ans. Let $\sqrt[n]{n} = 1+x$ or $n = (1+x)^n$.

But $(1+x)^n = 1+nx+\frac{n(n-1)}{2!}x^2+\cdots > \frac{n(n-1)}{2}x^2$, i.e., $n > \frac{n(n-1)}{2}x^2$ or $x^2 < \frac{2}{n-1}$, i.e., $x < \left(\frac{2}{n-1}\right)^{1/2}$.

But $x = \sqrt[n]{n}-1$. $\therefore$ $|\sqrt[n]{n}-1| < \left(\frac{2}{n-1}\right)^{1/2} < \varepsilon$ (say), if $\left(\frac{n-1}{2}\right)^{1/2} > \frac{1}{\varepsilon}$ or $\frac{n-1}{2} > \frac{1}{\varepsilon^2}$ or if $n > 1+\frac{2}{\varepsilon^2}$, i.e., $|u_n - 1| < \varepsilon$ for $n > 1+\frac{2}{\varepsilon^2}$, so that we may conclude that (u_n) is convergent and its limit is equal to 1.

10. Test the following series for convergence or divergence :

(i) $1+\frac{1}{2}+\frac{1}{3}+\frac{1}{4}+\cdots$ (ii) $1+\frac{1}{1.2}+\frac{1}{2.2^2}+\frac{1}{3.2^3}+\cdots$ **[C.U. (Hons.) 2005]**

Ans. (i) Here

$$\begin{aligned}\sum u_n &= 1+\frac{1}{2}+\frac{1}{3}+\frac{1}{4}+\frac{1}{5}+\cdots+\frac{1}{n}+\cdots\\ &= 1+\frac{1}{2}+\left(\frac{1}{3}+\frac{1}{4}\right)+\left(\frac{1}{5}+\frac{1}{6}+\frac{1}{7}+\frac{1}{8}\right)+\cdots\\ &> 1+\frac{1}{2}+\left(\frac{1}{4}+\frac{1}{4}\right)+\left(\frac{1}{8}+\frac{1}{8}+\frac{1}{8}+\frac{1}{8}\right)+\cdots\end{aligned}$$

i.e., $\sum u_n > 1+\frac{1}{2}+\frac{1}{2}+\frac{1}{2}+\cdots$, which tends to ∞ as $n\to\infty$, so that $\sum u_n$ is divergent.

(ii) Here $\sum u_n = 1+\frac{1}{1.2}+\frac{1}{2.2^2}+\frac{1}{3.2^3}+\cdots$

Let us compare the series $\sum u_n$ with another series $\sum v_n$, where

$$\sum v_n = 1+\frac{1}{2}+\frac{1}{2^2}+\frac{1}{2^3}+\cdots$$

Obviously $u_n < v_n$ for all $n > 2$. But $\sum v_n$ is a geometric series with its common ratio $\frac{1}{2}<1$.

$\therefore$ $\sum v_n$ is convergent, so that $\sum u_n$ is also convergent.

11. Find whether the series $\sum u_n$ defined by $u_n = \sqrt{n^2+1}-n$ is convergent or divergent.

Ans. Here

$$\begin{aligned}u_n &= n\left\{\left(1+\frac{1}{n^2}\right)^{1/2}-1\right\}\\ &= n\left\{1+\frac{1}{2}\cdot\frac{1}{n^2}+\frac{1}{2!}\cdot\frac{1}{2}\cdot\left(\frac{1}{2}-1\right)\cdot\frac{1}{n^4}+\cdots-1\right\}\\ &= \frac{1}{2n}-\frac{1}{8n^3}+\cdots\end{aligned}$$

Let $\sum v_n$ be another series, where $v_n = \frac{1}{n}$.

$$\therefore\quad \frac{u_n}{v_n} = \left(\frac{1}{2n}-\frac{1}{8n^3}+\cdots\right)\Big/\frac{1}{n} = \frac{1}{2}-\frac{1}{8n^2}+\cdots$$

$\therefore$ $\lim_{n\to\infty}\frac{u_n}{v_n}=\frac{1}{2}$, so that $\sum u_n$ and $\sum v_n$ are both convergent or both divergent. But from Ex.10(1), the series $\sum v_n$ is divergent. Therefore $\sum u_n$ is also divergent.

12. Test the convergence of the series $1-\frac{1}{2}+\frac{1}{3}-\frac{1}{4}+\cdots$ **[C.U. (Hons.) 2005]**

Ans. Here

$$\begin{aligned}S &= \left(1-\frac{1}{2}\right)+\left(\frac{1}{3}-\frac{1}{4}\right)+\left(\frac{1}{5}-\frac{1}{6}\right)+\cdots\\ &= \frac{1}{2}+\frac{1}{12}+\frac{1}{30}+\cdots\\ &= \text{sum of positive terms} = \text{positive.}\end{aligned}$$

Again

$$S = 1-\left(\frac{1}{2}-\frac{1}{3}\right)-\left(\frac{1}{4}-\frac{1}{5}\right)-\left(\frac{1}{6}-\frac{1}{7}\right)-\cdots$$
$$= 1-\left[\frac{1}{6}+\frac{1}{20}+\frac{1}{42}+\cdots\right]<1$$

$\therefore$ S is positive and < 1. $\therefore$ the series is convergent.

13. Test the convergence or divergence of the series $x+\frac{x^2}{2}+\frac{x^3}{3}+\frac{x^4}{4}+\cdots$ **[C.U. (Hons.) 2005, 07]**

Ans. Here $u_n = \frac{x^n}{n}$ and $u_{n+1} = \frac{x^{n+1}}{n+1}$, so that we may write

$$\frac{u_{n+1}}{u_n} = \frac{n}{n+1}\cdot x = \left(1-\frac{1}{n+1}\right)\cdot x$$

$$\therefore \quad \lim_{n\to\infty}\frac{u_{n+1}}{u_n} = \lim_{n\to\infty}\left(1-\frac{1}{n+1}\right)\cdot x = x$$

$\therefore$ the series is convergent if $x<1$ and divergent if $x>1$.

Also, if $x=1$, $S = 1+\frac{1}{2}+\frac{1}{3}+\frac{1}{4}+\cdots$ is divergent from Ex.10(i).

14. Test for convergence the following series, $\frac{1}{2}+\frac{2}{3^2}+\frac{2^2}{4^3}+\frac{2^3}{5^4}+\cdots$

Ans. Here $u_n = \frac{2^{n-1}}{(n+1)^n} = \frac{1}{2}\frac{2^n}{(n+1)^n}$, so that we may write

$$\lim_{n\to\infty}\sqrt[n]{u_n} = \lim_{n\to\infty}\left[\frac{1}{\sqrt[n]{2}}\cdot\frac{2}{(n+1)}\right] = 0 \text{ which is } < 1.$$

$\therefore$ the series is convergent.

15. Test the following series for convergence or divergence, $\frac{2}{1}+\frac{3}{4}+\frac{4}{9}+\frac{5}{16}+\cdots$

Ans. Here $u_n = \frac{n+1}{n^2}$. Let us now consider another series $\sum v_n = 1+\frac{1}{2}+\frac{1}{3}+\frac{1}{4}+\cdots$, where $v_n = \frac{1}{n}$.

$$\therefore \quad \frac{u_n}{v_n} = \frac{(n+1)}{n^2}\Big/\frac{1}{n} = \frac{n+1}{n},$$

so that we may write $\lim\limits_{n\to\infty}\frac{u_n}{v_n} = \lim\limits_{n\to\infty}\dfrac{n+1}{n} = 1$. Therefore the series $\sum u_n$ and $\sum v_n$ are both convergent or both divergent. But the series $\sum v_n$ is divergent from Ex.10(i).

$\therefore$ the given series $\sum u_n$ is also divergent.

16. Test for convergence or divergence the following series.

$$x+\frac{1}{2}\cdot\frac{x^3}{3}+\frac{1.3}{2.4}\cdot\frac{x^5}{5}+\frac{1.3.5}{2.4.6}\cdot\frac{x^7}{7}+\cdots$$

Ans. Here

$$u_n = \frac{1.3.5\cdots(2n-3)}{2.4.6\cdots(2n-2)}\cdot\frac{x^{2n-1}}{2n-1}$$

$$\text{and} \quad u_{n+1} = \frac{1.3.5\cdots(2n-3)\cdot(2n-1)}{2.4.6\cdots(2n-2)\cdot 2n}\cdot\frac{x^{2n+1}}{2n+1},$$

so that we may write

$$\frac{u_{n+1}}{u_n} = \frac{(2n-1)^2}{2n}\cdot\frac{x^2}{2n+1} = \frac{4n^2\left(1-\frac{1}{2n}\right)^2}{4n^2\left(1+\frac{1}{2n}\right)}\cdot x^2 = \frac{\left(1-\frac{1}{2n}\right)^2}{\left(1+\frac{1}{2n}\right)}\cdot x^2$$

which tends to x^2 as $n \to \infty$.

$\therefore$ by Cauchy's ratio test, the series is convergent or divergent according as $x^2 < 1$ or > 1, i.e., according as $|x| < 1$ or > 1.

When $x = 1$, the ratio test fails. Then we may write

$$n\left[\frac{u_n}{u_{n+1}} - 1\right] = n\left[\frac{2n(2n+1)}{(2n-1)^2} - 1\right] = \frac{n(4n^2+2n-4n^2+4n-1)}{(2n-1)^2} = \frac{6n^2-n}{(2n-1)^2}$$

$$= \frac{6n^2\left(1-\frac{1}{6n}\right)}{4n^2\left(1-\frac{1}{2n}\right)^2} = \frac{3}{2}\frac{1-\frac{1}{6n}}{\left(1-\frac{1}{2n}\right)^2}$$

which tends to $\frac{3}{2}$ as $n \to \infty$.

$$\therefore \quad \lim_{n\to\infty} n\left[\frac{u_n}{u_{n+1}} - 1\right] = \frac{3}{2}, \text{ i.e., } > 1. \quad \therefore \text{ the series is convergent.}$$

When $x = -1$, all the terms of the series are reversed in sign as compared to the above case when $x = 1$. $\therefore$ the series is convergent.

Thus combining the above results, we conclude that the given series is convergent or divergent according as $|x| \le 1$ or $|x| > 1$.

17. Test for convergence or divergence the series $\sum \frac{1}{n(\log n)^p}$.

Ans. Here let $f(n) = \frac{1}{n(\log n)^p}$, so that, we may write

$$a^n f(a^n) = a^n \frac{1}{a^n(\log a^n)^p} = \frac{1}{(n\log a)^p} = \frac{1}{n^p(\log a)^p}$$

$\therefore$ $\sum f(n)$ is convergent or divergent according as $\sum \frac{1}{n^p(\log a)^p}$ is convergent or divergent.

But for a fixed value of p, we have $(\log a)^p = $ constant.

$\therefore$ $\sum f(n)$ is convergent or divergent according as $\sum \frac{1}{n^p}$ is convergent or divergent, i.e., according as $p > 1$ or ≤ 1.

18. Show that the series

$$\frac{1+a}{2+a} + \frac{(1+a)(2+a)}{(2+a)(4+a)} + \frac{(1+a)(2+a)(3+a)}{(2+a)(4+a)(6+a)} + \cdots$$

is convergent if $a \ge 0$.

Ans. (i) Let $a > 0$. Now for any integer $k \ne 1$, we have $\frac{k+a}{2k+a} < \frac{1+a}{2+a}$.

$$\therefore \quad \frac{2+a}{4+a} < \frac{1+a}{2+a}, \frac{3+a}{6+a} < \frac{1+a}{2+a} \cdots$$

$$\therefore \quad \frac{1+a}{2+a} + \frac{(1+a)(2+a)}{(2+a)(4+a)} + \frac{(1+a)(2+a)(3+a)}{(2+a)(4+a)(6+a)} + \cdots$$

$$< \frac{1+a}{2+a} + \left(\frac{1+a}{2+a}\right)^2 + \left(\frac{1+a}{2+a}\right)^3 + \cdots$$

$$\text{i.e.,} \quad < r + r^2 + r^3 + \cdots \quad \left[\text{putting } r = \frac{1+a}{2+a}\right]$$

But $r = \frac{1+a}{2+a} < 1$ for all positive values of a.

$$\therefore \quad r + r^2 + r^3 + \cdots \to \frac{1}{1-r} - 1 = \frac{r}{1-r}$$

$\therefore$ the given series is $< \frac{r}{1-r}$ which is finite.

$\therefore$ the given series is convergent for $a > 0$.

(ii) Let $a = 0$, so that the series is $\sum u_n = \frac{1}{2} + \frac{1.2}{2.4} + \frac{1.2.3}{2.4.6} + \cdots$

Here $u_n = \frac{1.2.3\cdots n}{2.4.6\cdots 2n} = \frac{1}{2^n}$. Similarly, $u_{n+1} = \frac{1}{2^{n+1}}$.

$\therefore \quad \frac{u_{n+1}}{u_n} = \frac{2^n}{2^{n+1}} = \frac{1}{2}$ which is < 1. Therefore for $a = 0$, the given series is convergent.

Exercises

1. Show that the sequence whose nth term is $2 + \frac{(-1)^n}{2^n}$ is convergent and its limit is 2.

[**Hints.** Show $|u_n - 2| \to 0$ as $n \to \infty$]

2. Discuss the behaviour of the sequence $\frac{1}{n} + 2(-1)^n$ as $n \to \infty$.

Ans. Here $u_n = \frac{1}{n} + 2(-1)^n$ so that the sequence is

$$-1, \frac{5}{2}, -\frac{5}{3}, \frac{9}{4}, -\frac{9}{5}, \frac{13}{6}, \cdots$$

If n be odd, then

$$|u_n + 2| = \left|\frac{1}{n} - 2 + 2\right| = \left|\frac{1}{n}\right| = \frac{1}{n} < \varepsilon \text{ for } n > \frac{1}{\varepsilon}$$

so that $u_n \to -2$ as $n \to \infty$. Again, if n be even, then

$$|u_n - 2| = \left|\frac{1}{n} + 2 - 2\right| = \left|\frac{1}{n}\right| = \frac{1}{n} < \varepsilon \text{ for } n > \frac{1}{\varepsilon},$$

so that $u_n \to 2$ as $n \to \infty$.

$\therefore \quad (u_n)$ oscillates finitely between -2 and $+2$.

3. Discuss the behaviour of the sequence $\{u_n\}$ as $n \to \infty$, where $u_n = \frac{1}{n} \sin \frac{n\pi}{2}$. [**Ans.** Convergent.]

4. Prove that the sequence $\{u_n\}$ is convergent if $u_n = \left(1 + \frac{1}{n+1}\right)^{n+1}$. [**Ans.** See Ex.8 of worked out problems.]

5. Prove that

(a) $\lim_{n\to\infty} \frac{x^n}{\lfloor n} = 0$ for all values of x and

(b) $\lim_{n\to\infty} \frac{x^n}{n} = 0$ for $|x| < 1$.

Ans. (a) Here $u_n = \frac{x^n}{\lfloor n}$; then $u_{n+1} = \frac{x^{n+1}}{\lfloor n+1}$.

$$\therefore \quad \frac{u_{n+1}}{u_n} = \frac{x^{n+1}}{\lfloor n+1} \cdot \frac{\lfloor n}{x^n} = \frac{x}{n+1}$$

$$\therefore \quad \lim_{n\to\infty} \left|\frac{u_{n+1}}{u_n}\right| = \lim_{n\to\infty} \left|\frac{x}{n+1}\right| = 0,$$

so that $\lim_{n\to\infty} u_n = \lim_{n\to\infty} \frac{x^n}{\lfloor n} = 0$ for all values of x.

(b) Proceed exactly as in (a).

6. If $S_n = \frac{1}{1.2}+\frac{1}{2.3}+\frac{1}{3.4}+\cdots+\frac{1}{n(n+1)}$, show that $\{S_n\}$ is a bounded monotone increasing sequence.

Ans. Here

$$S_n = \frac{1}{1.2}+\frac{1}{2.3}+\frac{1}{3.4}+\cdots+\frac{1}{n(n+1)}$$

$$\therefore \quad S_{n+1} = \frac{1}{1.2}+\frac{1}{2.3}+\frac{1}{3.4}+\cdots+\frac{1}{n(n+1)}+\frac{1}{(n+1)(n+2)}$$

$$\therefore \quad S_{n+1}-S_n = \frac{1}{(n+1)(n+2)} > 0.$$

$\therefore$ $S_{n+1} > S_n$, i.e., $\{S_n\}$ is monotone increasing.

Again

$$S_n = \frac{1}{1.2}+\frac{1}{2.3}+\frac{1}{3.4}+\cdots+\frac{1}{n(n+1)}$$

$$= \left(1-\frac{1}{2}\right)+\left(\frac{1}{2}-\frac{1}{3}\right)+\left(\frac{1}{3}-\frac{1}{4}\right)+\cdots+\left(\frac{1}{n}-\frac{1}{n+1}\right)=1-\frac{1}{n+1}$$

Putting $n = 1, 2, 3, 4, \cdots$, we get the sequence as $\frac{1}{2}, \frac{2}{3}, \frac{3}{4}, \frac{4}{5}, \cdots$

Thus every term of the sequence lies between $\frac{1}{2}$ and 1. Thus $\{S_n\}$ is bounded. Hence the result.

7. Test the convergence of the sequence, where $u_n = \{2+(-1)^n \cdot \frac{1}{n}\}$.

Ans. $\{u_n\}$ converges to the limit 2.

[**Hints.** Find $|u_n - 2|$ and show that $|u_n - 2| \to 0$ as $n \to \infty$.]

8. Test the following series for convergence or divergence :

(a) $1+2+3+4+\cdots$

(b) $1+\frac{1}{1.2}+\frac{1}{1.2.3}+\frac{1}{1.2.3.4}+\cdots$

(c) $\frac{1}{\underline{|2}}+\frac{2}{\underline{|3}}+\frac{3}{\underline{|4}}+\frac{4}{\underline{|5}}+\cdots$

Ans. (a) Here $\sum u_n = \frac{n(n+1)}{2} \to \infty$ as $n \to \infty$. $\therefore$ $\sum u_n$ is divergent.

(b) Here $u_n = \frac{1}{\underline{|n}}$, $u_{n+1} = \frac{1}{\underline{|n+1}}$

$\therefore$ $\frac{u_{n+1}}{u_n} = \frac{1}{n+1} \to 0$ as $n \to \infty$. $\therefore$ $\sum u_n$ is convergent.

(c) Proceed as in (b).

9. Test the following series for convergence or divergence :

(a) $\frac{1}{1.2.3}+\frac{1}{2.3.4}+\frac{1}{3.4.5}+\cdots$

(b) $\frac{14}{1^3}+\frac{24}{2^3}+\frac{34}{3^3}+\cdots$

Ans. (a) Here

$$u_n = \frac{1}{n(n+1)(n+2)}$$

so that

$$u_{n+1} = \frac{1}{(n+1)(n+2)(n+3)}.$$

Applying Raabe's test,

$$\lim_{n\to\infty} n\left[\frac{u_n}{u_{n+1}}-1\right] = \lim_{n\to\infty} n\left[\frac{(n+1)(n+2)(n+3)}{n(n+1)(n+2)}-1\right]$$

$$= \lim_{n\to\infty} n\left[\frac{n+3}{n}-1\right] = 3 \text{ which is } > 1.$$

∴ the given series is convergent.

(b) Here

$$u_n = \frac{10n+4}{n^3} \quad \text{and} \quad u_{n+1} = \frac{10(n+1)+4}{(n+1)^3} = \frac{10n+14}{(n+1)^3}.$$

Now, applying Raabe's test, it can be shown that

$$\lim_{n\to\infty} n\left(\frac{u_n}{u_{n+1}} - 1\right] = 2 > 1. \quad \therefore \sum u_n \text{ is convergent.}$$

10. Examine the convergence of the following infinite series :

(a) $\sum \frac{n-1}{n^3+1}$.

(b) $\sum \frac{1}{\sqrt{n^2+1+n}}$

Ans. (a) Here $u_n = \frac{n-1}{n^3+1}$; let us consider another series $\sum v_n$, where $v_n = \frac{1}{n^2}$, so that

$$v_n - u_n = \frac{1}{n^2} - \frac{n-1}{n^3+1} = \frac{n^2+1}{n^2(n^3+1)}$$

which is positive for all values of n.

∴ $u_n < v_n$ for all n. Again, since $\sum v_n$ is convergent, from Art. 20.11, we conclude that $\sum u_n$ is also convergent.

(b) Here

$$u_n = \frac{1}{\sqrt{n^2+1}+n} = \frac{\sqrt{n^2+1}-n}{\left\{\sqrt{n^2+1}+n\right\}\left\{\sqrt{n^2+1}-n\right\}} = \sqrt{n^2+1}-n.$$

Then proceed exactly in the same way as in Ex.11 of worked out problems.

11. Test for convergence or divergence the following series whose nth term is

(a) $(n^3+1)^{1/3} - n$

(b) $\frac{n}{(n+1)^3}$

(c) $\frac{(n+1)^p}{\underline{|n+1}}$.

Ans. (a) Here

$$\begin{aligned} u_n &= n\left\{\left(1+\frac{1}{n^3}\right)^{1/3} - 1\right\} \\ &= n\left\{1 + \frac{1}{3}\cdot\frac{1}{n^3} + \frac{1}{\underline{|2}}\cdot\frac{1}{3}\cdot\left(\frac{1}{3}-1\right)\cdot\frac{1}{n^6} + \cdots - 1\right\} \\ &= n\left(\frac{1}{3n^3} - \frac{1}{9n^6} + \cdots\right\} = \frac{1}{3n^2} - \frac{1}{9n^5} + \cdots \end{aligned}$$

Considering another series $\sum v_n = \frac{1}{n^2}$, we may write

$$\frac{u_n}{v_n} = \frac{1}{3} - \frac{1}{9n^3} + \cdots \quad \therefore \quad \lim_{n\to\infty} \frac{u_n}{v_n} = \frac{1}{3}.$$

∴ $\sum u_n$ and $\sum v_n$ are both convergent or both divergent. But since $\sum v_n$ is convergent, from Art. 20.11, we conclude that $\sum u_n$ is also convergent.

(b) Proceed exactly as in (a)

(c) Here

$$u_n = \frac{(n+1)^p}{\underline{|n+1}}; \text{ also } u_{n+1} = \frac{(n+2)^p}{\underline{|n+2}}.$$

Then show that $\frac{u_{n+1}}{u_n} \to 0$ as $n \to \infty$, so that $\sum u_n$ is convergent.

12. Test the convergence of the series $\sum u_n$, where $u_n = \frac{n^{n^2}}{(1+n)^{n^2}}$.

Ans. By Cauchy's root test, we have

$$\sqrt[n]{u_n} = \left(\frac{n}{1+n}\right)^{\frac{n^2}{n}} = \left(\frac{n}{1+n}\right)^n = \frac{1}{(1+\frac{1}{n})^n}.$$

$$\therefore \quad \lim_{n\to\infty} \sqrt[n]{u_n} = \frac{1}{\lim\limits_{n\to\infty}\left(1+\dfrac{1}{n}\right)^n} = \frac{1}{e}$$

since $e > 2$, $\frac{1}{e} < \frac{1}{2}$, i.e., < 1. $\quad\therefore\ \sum u_n$ is convergent.

13. Test the series $\frac{1}{2} + \frac{1}{2+1} + \frac{1}{2^2+1} + \frac{1}{2^3+1} + \cdots$ ad. inf. for convergence or divergence.

Ans. Here $u_n = \frac{1}{2^{n-1}+1}$ and $u_{n+1} = \frac{1}{2^n+1}$. Now find $\frac{u_{n+1}}{u_n}$ and show that

$$\lim_{n\to\infty} \frac{u_{n+1}}{u_n} = \frac{1}{2}, \text{ i.e., } < 1. \quad \therefore \sum u_n \text{ is convergent.}$$

14. Examine the convergence or divergence of

$$1 + \frac{1}{2^2} + \frac{2^2}{3^3} + \frac{3^3}{4^4} + \cdots$$

Ans. Since the removal of the first term does not affect the nature of the series, we get by neglecting the first term

$$u_n = \frac{n^n}{(n+1)^{n+1}} = \frac{1}{(n+1)}\left(\frac{n}{n+1}\right)^n = \frac{1}{n+1}\cdot\frac{1}{(1+\frac{1}{n})^n}.$$

$$\therefore \quad \lim_{n\to\infty} u_n = \lim_{n\to\infty} \frac{1}{(n+1)}\frac{1}{e}$$

Now let us consider another series of which the nth term is $v_n = \frac{1}{n}$.

$$\therefore \quad \lim_{n\to\infty} \frac{u_n}{v_n} = \lim_{n\to\infty} \frac{n}{(n+1)e} = \lim_{n\to\infty} \frac{1}{(1+\frac{1}{n})\,e} = \frac{1}{e}$$

which is finite. Hence the series $\sum u_n$ and $\sum v_n$ are both convergent or both divergent. But the series $\sum v_n = \sum \frac{1}{n}$ is divergent.

$$\therefore \sum u_n \text{ is also divergent.}$$

15. Test the following series whose nth term is

(a) $\sqrt{n^2+1} - n$

(b) $n^n/(n+1)^{n+1}$

Ans. (a) See Ex.11 of worked out problems. (b) See sum no. 14 of the Exercises.

16. Examine the convergence or divergence of the series

$$x + \frac{x^2}{2} + \frac{x^3}{3} + \frac{x^4}{4} + \cdots$$

Ans. See Ex.13 of worked out problems.

17. Show that the series

$$1 + \frac{2^p}{2!} + \frac{3^p}{3!} + \frac{4^p}{4!} + \cdots$$

is convergent for all values of p.

Ans. Here

$$u_n = \frac{n^p}{n!} \quad \text{and} \quad u_{n+1} = \frac{(n+1)^p}{(n+1)!}.$$

Now find $\frac{u_n}{u_{n+1}}$ and show that

$$\lim_{n\to\infty} \frac{u_n}{u_{n+1}} > 1, \text{ so that } \sum u_n \text{ is convergent.}$$

18. Test for convergence the series

$$1 + \frac{1+\alpha}{1+\beta} + \frac{(1+\alpha)(2+\alpha)}{(1+\beta)(2+\beta)} + \frac{(1+\alpha)(2+\alpha)(3+\alpha)}{(1+\beta)(2+\beta)(3+\beta)} + \cdots$$

where $\beta - \alpha \neq 1$.

Ans. Here

$$u_n = \frac{(1+\alpha)(2+\alpha)\cdots(n-1+\alpha)}{(1+\beta)(2+\beta)\cdots(n-1+\beta)}$$

$$\text{and} \quad u_{n+1} = \frac{(1+\alpha)(2+\alpha)\cdots(n-1+\alpha)(n+\alpha)}{(1+\beta)(2+\beta)\cdots(n-1+\beta)(n+\beta)}.$$

$$\therefore \quad n\left(\frac{u_n}{u_{n+1}} - 1\right) = n\left[\frac{n+\beta}{n+\alpha} - 1\right] = \frac{n(\beta-\alpha)}{n+\alpha} = \frac{\beta-\alpha}{1+\frac{\alpha}{n}},$$

so that applying Raabe's test,

$$\lim_{n\to\infty} n\left(\frac{u_n}{u_{n+1}} - 1\right) = (\beta - \alpha);$$

$\therefore$ the series $\sum u_n$ is convergent if $\beta - \alpha > 1$, i.e., $\beta > 1 + \alpha$.

19. Examine whether the following series is convergent or divergent :

$$\frac{x^2}{2} + \frac{2}{3}\cdot\frac{x^4}{4} + \frac{2.4}{3.5}\cdot\frac{x^6}{6} + \frac{2.4.6}{3.5.7}\cdot\frac{x^8}{8} + \cdots \text{ where } x > 0.$$

Ans. Here

$$u_n = \frac{2.4.6\cdots(2n-2)}{3.5.7\cdots(2n-1)}\cdot\frac{x^{2n}}{2n}$$

$$\text{and} \quad u_{n+1} = \frac{2.4.6\cdots(2n-2)\cdot 2n}{3.5.7\cdots(2n-1)(2n+1)}\cdot\frac{x^{2n+2}}{(2n+2)}$$

so that

$$\frac{u_{n+1}}{u_n} = \frac{(2n)^2}{(2n+1)(2n+2)}\cdot x^2 = \frac{4n^2}{4n^2\left(1+\frac{1}{2n}\right)\left(1+\frac{2}{2n}\right)}\cdot x^2.$$

$$\text{and} \quad \lim_{n\to\infty}\frac{u_{n+1}}{u_n} = x^2.$$

$\therefore$ the series $\sum u_n$ is convergent or divergent according as $x^2 < 1$ or > 1, i.e., $|x| < 1$ or > 1.

20. Find whether the following series is convergent or divergent :

$$x^2 + \frac{2^2}{3.4}\cdot x^4 + \frac{2^2\cdot 4^2}{3.4.5.6}\cdot x^6 + \frac{2^2.4^2.6^2}{3.4.5.6.7.8}\cdot x^8 + \cdots \text{ where } x > 0.$$

Ans. Hints. Find u_n and u_{n+1}. Then show that

$$\lim \frac{u_{n+1}}{u_n} \to x^2 \text{ as } n \to \infty$$

exactly as in sum no.19 above. Also, if $x = 1$, then show that

$$n\left(\frac{u_n}{u_{n+1}} - 1\right) \to \frac{3}{2} \text{ as } n \to \infty.$$

$\therefore$ $\sum u_n$ is convergent.

21. Find whether the following series is convergent or divergent :

$$1 + \frac{2x}{\lfloor 2} + \frac{3^2x^2}{\lfloor 3} + \frac{4^3x^3}{\lfloor 4} + \cdots$$

Ans. Here

$$u_n = \frac{n^{n-1}x^{n-1}}{\lfloor n} \quad \text{and} \quad u_{n+1} = \frac{(n+1)^n x^n}{\lfloor n+1},$$

so that we may write

$$\frac{u_{n+1}}{u_n} = \frac{(n+1)^n x}{n^{n-1}.(n+1)} = \left(\frac{n+1}{n}\right)^{n-1} \cdot x = \left(1 + \frac{1}{n}\right)^{n-1} .x \to ex \text{ as } n \to \infty.$$

$\therefore$ $\sum u_n$ is convergent or divergent according as $ex < 1$ or > 1, i.e., $x < \frac{1}{e}$ or $> \frac{1}{e}$.

22. Test the convergence of the series

$$1 + \frac{3}{7} \cdot x + \frac{3.6}{7.10}x^2 + \frac{3.6.9}{7.10.13}x^3 + \cdots$$

Ans. Here

$$u_n = \frac{3.6.9 \cdots (3n-3)}{7.10.13 \cdots (3n+1)} x^{n-1} \quad \text{and} \quad u_{n+1} = \frac{3.6.9 \cdots (3n-3).3n}{7.10.13 \cdots (3n+1).(3n+4)} x^n$$

$$\therefore \quad \frac{u_{n+1}}{u_n} = \frac{3n}{3n+4} \cdot x = \frac{1}{1 + \frac{4}{3n}} \cdot x \to x \text{ as } n \to \infty.$$

$\therefore$ $\sum u_n$ is convergent or divergent according as $x < 1$ or > 1.

For $x = 1$,

$$n\left(\frac{u_n}{u_{n+1}} - 1\right] = n\left(\frac{3n+4}{3n} - 1\right) = n \cdot \frac{4}{3n} = \frac{4}{3} > 1.$$

$\therefore$ $\sum u_n$ is convergent.

23. Examine the convergence of the series

$$1 + \frac{x}{2} + \frac{\lfloor 2}{3^2}x^2 + \frac{\lfloor 3}{4^3}x^3 + \cdots$$

Ans. Here

$$u_n = \frac{\lfloor n-1}{n^{n-1}} x^{n-1} \quad \text{and} \quad u_{n+1} = \frac{\lfloor n}{(n+1)^n} x^n.$$

Then show that

$$\frac{u_{n+1}}{u_n} \to \frac{x}{e} \text{ as } n \to \infty.$$

$\therefore$ $\sum u_n$ is convergent or divergent according as $\frac{x}{e} < 1$ or > 1, i.e., $x < e$ or $> e$.

24. Find whether the following series are convergent or divergent :

(a) $\dfrac{x}{1}+\dfrac{1}{2}\cdot\dfrac{x^3}{3}+\dfrac{1.3}{2.4}\cdot\dfrac{x^5}{5}+\dfrac{1.3.5}{2.4.6}\cdot\dfrac{x^7}{7}+\cdots$

(b) $\left(\dfrac{2^2}{1^2}-\dfrac{2}{1}\right)^{-1}+\left(\dfrac{3^3}{2^3}-\dfrac{3}{2}\right)^{-2}+\left(\dfrac{4^4}{3^4}-\dfrac{4}{3}\right)^{-3}+\cdots$

Ans. (a) See Ex.16 of worked out problems.

(b) Here

$$u_n=\left[\left(\frac{n+1}{n}\right)^{n+1}-\frac{n+1}{n}\right]^{-n}.$$

$$\therefore\quad \sqrt[n]{u_n}=\left(\frac{n+1}{n}\right)^{-1}\left[\left(\frac{n+1}{n}\right)^{n}-1\right]^{-1}=\frac{n}{n+1}\left[\left(1+\frac{1}{n}\right)^{n}-1\right]^{-1}$$

$$=\frac{1}{1+\frac{1}{n}}\cdot\frac{1}{(1+\frac{1}{n})^n-1}\to\frac{1}{e-1}\text{ as }n\to\infty.\quad\text{But }\frac{1}{e-1}<1.$$

$\therefore\ \sum u_n$ is convergent.

25. Examine whether the following series is convergent or divergent

$$\frac{5}{1.2.4}+\frac{7}{2.3.5}+\frac{9}{3.4.6}+\frac{11}{4.5.7}+\cdots$$

Hints. Here

$$u_n=\frac{2(n+1)+1}{n.(n+1)\cdot(n+3)}$$

and then proceed in the usual way.

26. (a) Are the following series convergent or divergent? (Proof not required)

i. $1+\frac{1}{2}+\frac{1}{3}+\frac{1}{4}+\cdots$

ii. $1-\frac{1}{2}+\frac{1}{3}-\frac{1}{4}+\cdots$ **[C.U. (Hons.) 2005]**

(See worked out examples)

(b) Prove that the series

$$x+\frac{x^2}{2}+\frac{x^3}{3}+\cdots$$

is convergent for $x<1$ and divergent for $x>1$. **[C.U. (Hons.) 2005, 07]**

(See worked out examples)

27. (a) Show that the expansion of $\log(1+x)$ in ascending powers of x is convergent when $|x|<1$.

(See worked out examples)

(b) Test the series

$$1^2+2^2x+3^2x^2+4^2x^3+\cdots$$

for convergence for $x<1, x>1$ and $x=1$. **[C.U. (Hons.) 2007]**

(See worked out examples)

(c) Examine the convergence of the series

$$\sum_{n=1}^{\infty}\frac{x^n}{n!}$$

[C.U. (Hons.) 2008]

(See worked out examples)

Chapter 21

Complex-Valued Functions—Complex Variables

21.1 Introduction

Since the subject of complex variables is fundamental to the study of the subject of applied mathematics, a solid background in this area is essential for the application of many useful methods of analysis to obtain solutions to a wide range of practical problems.

The following chapter deals with a review and brief summary of some of the useful results of complex function theory. It is assumed that the students for whom this book is written, have some familiarity with the algebra of complex numbers. For a thorough coverage of this topic, the students are advised to consult books on Algebra for advanced students and mathematical physics.

Complex numbers : Any ordered pair of real numbers x, y—written as (x, y) is called a *complex number.* Thus the complex number $z = x + jy$, where $j = (-1)^{1/2}$, is defined as the ordered pair (x, y) of real numbers x and y, and is written as $z = (x, y)$.

The real numbers x and y are called *the real and imaginary parts* of the complex number z. It is customary to write $R(z) = x$, and $I(z) = y$, where $R(z)$ stands for the real part of z and $I(z)$ stands for the imaginary part of z.

The quantity $\sqrt{(x^2 + y^2)}$ taken with the positive sign is defined as the *absolute value* or *modulus* of z—written as $|z|$. Thus $|z| = \sqrt{x^2 + y^2}$, so that $|z| = 0$ if and only if $x = 0$ and $y = 0$.

The term *imaginary number* does not mean that such a number does not exist or exists in the imagination of people. The letter j is merely a symbol which denotes *imaginary unit* just as 1 denotes the unit of real numbers. Thus the imaginary number jy means y units of an imaginary number just as x means x units of a real number.

It is not appropriate to call the expression $x + jy$ an imaginary number. It is called a *complex number,* which is an ordered pair consisting of x real units and y imaginary units. In the expression $x + jy$, the sign + does not indicate addition as understood by the casual reader. What the expression $x + jy$ means is that x real units are superimposed in order by y imaginary units to get a complex number $z = x + jy$.

21.2 Geometrical interpretation of the imaginary number j

We draw a circle of unit radius with centre O as the origin. This circle intersects the X-axis at A, A' and the Y-axis at B, B' respectively as shown in Fig. 21.2.1. Real numbers will be represented by points on the X-axis—called the *real axis*. Since $OA = OA' = 1$, the points A and A' denote the numbers 1 and -1 respectively.

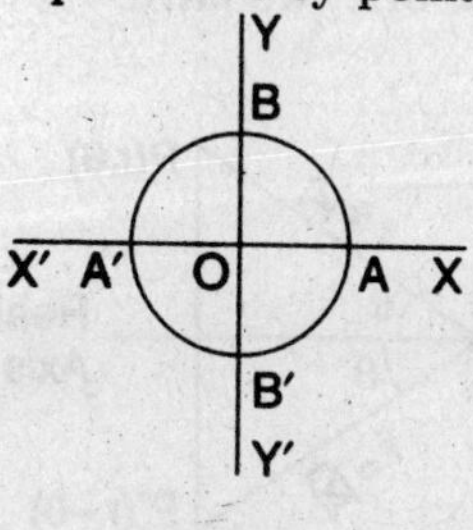

Fig 21.2.1

Now $1 \times (-1) = -1$. Also $\angle AOA' = 2$ right angles. Hence the multiplication by -1 may be regarded as an operation by which OA turns through 2 right angles in the anti-clockwise sense. Again from definition $j \times j = \sqrt{-1} \times \sqrt{-1} = -1$. $\therefore$ multiplication by j may be looked upon as an operation of turning OA through one right angle in the anti-clockwise sense. But by this operation, OA comes to the position OB, so that the point A on the X-axis is brought to the point B on the Y-axis.

Also $1 \times \sqrt{-1} = \sqrt{-1} = j$. $\therefore$ the point B denotes the imaginary number j. Since the point B is on the Y-axis, the Y-axis is considered as the imaginary axis. All purely imaginary numbers may thus be represented by points on the Y-axis.

Note. Multiplication of a real number by j is thus looked upon as an operation of turning the magnitude of the number through one right angle in the anti-clockwise sense, but multiplication of a complex number by j represents a vector obtained by rotating the complex number through one right angle in the positive direction. This may be proved as follows.

If z represents a complex number with its modulus r and amplitude or argument θ, then $z = r(\cos\theta + j\sin\theta)$. Again $j = \cos\frac{\pi}{2} + j\sin\frac{\pi}{2}$.

$$\therefore \quad jz = \left(\cos\frac{\pi}{2} + j\sin\frac{\pi}{2}\right) r(\cos\theta + j\sin\theta) = r\left[\cos\left(\theta + \frac{\pi}{2}\right) + j\sin\left(\theta + \frac{\pi}{2}\right)\right]$$

which shows that jz represents a vector obtained by rotating the vector z through a right angle in the positive direction.

21.3 Graphical representation of complex numbers—Argand Diagram

It is sometimes convenient to employ a graphical representation of a complex variable (or complex number). By plotting x, the real part of z, as the abscissa, and y, the imaginary part of z, as the ordinate, we obtain the *complex plane* or *Argand diagram* as shown in the adjoining figure. Any point $P(x, y)$ in this plane can be labelled as

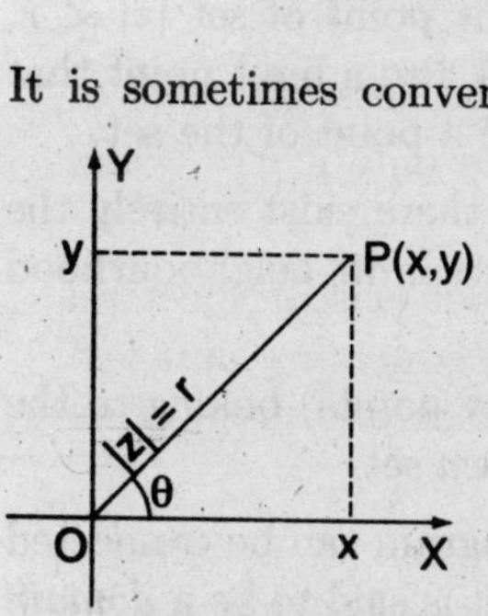

Fig 21.3.1

$$z = x + jy, \tag{1}$$

where z is called the *complex co-ordinate* of the point P. The X-axis of this plane is called the *real axis* and Y-axis is called the *imaginary* axis. It is obvious that the point (x, y) does not coincide with the point (y, x) except for the special case when $x = y$.

In the chapter on "vector analysis", we have seen that the points in the X-Y plane can be identified with the two-dimensional position vector $\vec{r} = x\hat{\imath} + y\hat{\jmath}$. As a result, two-dimensional vector analogue can be developed for many topics of our complex analysis.

From Fig. 21.3.1, we can write that length $OP = (x^2+y^2)^{1/2} = |z| =$ modulus of $z = x+jy$. If we introduce polar co-ordinates (r, θ) for the point P, we have then $x = r\cos\theta$, $y = r\sin\theta$ and also from (21.3.1) we may write $z = x + jy = r(\cos\theta + j\sin\theta) = re^{j\theta}$, where

$$r = (x^2 + y^2)^{1/2} = |z| \tag{2}$$

is taken positive and defined as modulus of z. The angle $\theta = \tan^{-1}\frac{y}{x}$ is called the *angle* or *argument* of z. If P' be the point, whose complex co-ordinate is conjugate of z, we have

$$\bar{z} = x - jy = re^{-j\theta} = r\{\cos(-\theta) + j\sin(-\theta)\}. \tag{3}$$

Thus in Argand diagram, complex numbers come in conjugate pairs. Such a pair of points forms what are known as *mirror images* of each other, with X-axis as the mirror. Therefore, the conjugate of z, i.e., $\bar{z}$ is called the *reflected image* of z in real axis as shown in the adjoining figure. It is obvious from (21.3.3) that x-co-ordinates of P and P' are the same, while their y-coordinates are negatives of each other. In polar co-ordinates z and $\bar{z}$ have the same r value, but their θ values are negatives of each other.

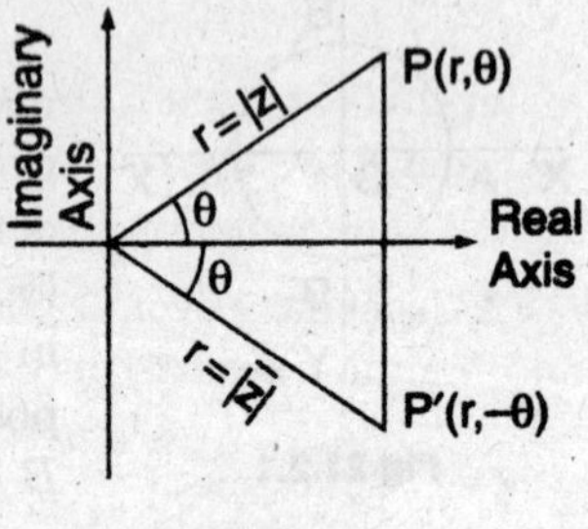

Fig 21.3.2

21.4 Functions of a complex variable

All of the elementary functions of real variables may be extended to the complex plane—replacing the real variable x by the complex variable z. Before giving a formal definition of the function of a complex variable, let us define some useful terms.

Set of points. A set of points in the Argand diagram is a collection of points finite or infinite in number.

Neighbourhood of a point. Let z_0 be a point in the Argand diagram. A set of all the points z such that $|z - z_0| < \varepsilon$, where ϵ is an arbitrarily chosen small positive number, is defined as the neighbourhood of the point z_0.

Limit point of a set. A point z_0, every neighbourhood of which contains a point of set S other than z_0, is defined as a limit point of the set S of points in the Argand plane. For example, each point on the circumference of circle $|z| = r$, is the limit point of set $|z| < r$. These points do not belong to the set. But each point inside the circle is also a limit point that belongs to the set. Thus the limit point of a set may not necessarily be a point of the set.

If z_0 be a limit of the set S, such that in the neighbourhood of z_0, there exist entirely the points of the set S, it is defined as interior or inner point. If all the points in its neighbourhood do not belong to the set S, it is said to be the boundary limit point.

A set is said to be *closed*, if all of its limit points (inner or boundary points) belong to the set. If a set consists of entirely the interior points, it is known as an *open* set.

Domain. If every pair of points of a set of points in the Argand diagram can be connected by a polygonal arc, every point of which is a point of the set, then the set is said to be a *domain* or *region. Open domain* is open connected set of points. When the boundary points of the set are also added to an open domain, it becomes a *closed domain.*

We may now give a *formal definition of a function of complex variables.* Let x and y be a pair of real variables such that $z = x + jy$, and u and v be a pair of real functions such that

$w = u + jv$, then w is *said to be the function of complex variable* z, *and written as* $w = f(z)$, *if to every value of* z *in a certain domain* D, *there correspond one or more definite values of* w. In case w has only one value for each value of z in the given domain D, w is said to be *uniform* or *single-valued function* of z, and if it takes more than one value for some or all values of z in D, then w is known as a *many-valued or multiple-valued function of* z. Thus the function $w = u + jv$ of the complex variable $z = x + jy$—also known as the *complex function*—consists of an ordered pair of real functions of real variables, i.e.

$$w = f(z) = u(x, y) + jv(x, y). \tag{1}$$

In this expression the functions u and v are real functions of the real variables x and y. Just as x and y are the real and imaginary parts of z, $u(x, y)$ and $v(x, y)$ are the real and imaginary parts of the complex function $w(z)$, i.e., $u(x, y) = Rw(z)$, $v(x, y) = Iw(z)$.

As specific examples, we consider the complex functions $\sin z$ and e^z—for which the real and imaginary parts may be found as follows :

$$\begin{aligned} \sin z = \sin(x + jy) &= \sin x \cos jy + \cos x \sin jy = \sin x \cosh y + j \cos x \sinh y \qquad (2) \\ &\quad (\text{using } \cos jy = \cosh y \text{ and } \sin jy = j \sinh y) \end{aligned}$$

From (21.4.1) and (21.4.2) we may write, $u(x, y) = \sin x \cosh y$ and $v(x, y) = \cos x \sinh y$.

Again, for e^z, we have

$$e^z = e^{x+jy} = e^x e^{jy} = e^x(\cos y + j \sin y)$$

whence we find that $u(x, y) = e^x \cos y$ and $v(x, y) = e^x \sin y$.

Path of a complex variable. In the case of a function $f(z)$ of a real variable, the independent variable z takes values corresponding to the points lying on the X-axis. But in the case of the function $f(z)$ of a complex variable, the independent variable z takes values lying in a plane (called the z-*plane*), so that path of z is a polygonal arc joining the initial and final points of z. This path may be a straight line or a curve. So *path of a complex variable* z *is either a straight line or a curve.*

Now in order that w in (21.4.1) may be a function of z, some conditions must be satisfied, and from the theory of functions of complex variables, it follows that $w = f(z)$ *must be continuous and shall have a unique derivative.*

Continuity. In what we have discussed above, we have seen that $w = f(z)$ is defined as a function of a complex variable z, if to every value of z in a certain domain (or region) correspond one or more values of w. This definition is on the same model as the usual definition of a function of a real variable. Therefore definition of continuity for the function of a complex variable is exactly the same as that for the function of a real variable. Thus *the function* $f(z)$ *of a complex variable* z *is said to be continuous at the point* z_0 *if, for a given positive number* ε, *we can find a number* δ *such that* $|f(z) - f(z_0)| < \varepsilon$ *for all points in the domain* D *satisfying* $|z - z_0| < \delta$, where in general δ depends upon ε and z_0. If however δ is independent of z_0, then $f(z)$ is called *uniformly continuous* in the domain D. Thus for *uniform continuity* of the function $f(z)$, it is possible to find a number $\eta(\varepsilon)$ independent of z_0 such that $|f(z) - f(z_0)| < \varepsilon$ for every pair of points z, z_0 of the domain D for which $|z - z_0| < \eta(\varepsilon)$.

Note. In the case of the continuity of the function of a complex variable, z approaches z_0 along any path whatsoever, while for the function of a real variable, z approaches z_0 along X-axis only.

Differentiability. Since the mode of definition of continuity is the same both in the cases of the functions of real and complex variables, definition for the derivative of a complex function $f(z)$ is identical with the definition for the derivative of a real function. Hence following the definition suggested in books of differential calculus, we may write

$$f'(z_0) = \lim_{z \to z_0} \frac{f(z) - f(z_0)}{z - z_0} \tag{3}$$

and we may say that $f(z)$ is differentiable if the limit on the right exists. *This limit, if it exists, is called the derivative or the differential coefficient of $f(z)$ at z_0 and is denoted by $f'(z_0)$.* [Value of the limit is independent of the amplitude Δz as $z \to z_0$. In case this limit depends on the amplitude Δz, the function $f(z)$ is not differentiable. Again if the function $f(z)$ is differentiable at $z = z_0$, it is necessarily continuous at this point, for

$$|f(z_0 + \Delta z) - f(z_0)| \to 0 \text{ as } \Delta z \to 0, \text{ if } \lim_{\Delta z \to 0} \frac{f(z_0 + \Delta z) - f(z_0)}{\Delta z} \text{ exists]}$$

More precisely the above definition is restated as follows :

For any given positive number ε, we can find a number δ such that

$$\left| \frac{f(z) - f(z_0)}{z - z_0} - f'(z_0) \right| < \varepsilon$$

provided that $0 < |z - z_0| < \delta$, along whatever path z approaches z_0 in the domain D of Argand diagram.

Remark. In order that a single-valued function $f(z)$ defined in a domain D of Argand diagram may be differentiable at any point z_0 of D, it is not enough that

$$\lim_{z \to z_0} \frac{fz) - f(z_0)}{z - z_0}$$

exists; what is required further is that this limit must be *unique and finite*, that is, it must have the same value along whatever path z approaches z_0, as may be illustrated by the following example :

Example. *If* $f(z) = \frac{x^3 y(y - jx)}{x^6 + y^2}$, $z = 0$, $f(0) = 0$, *prove that* $\frac{f(z) - f(0)}{z} \to 0$ *as* $z \to 0$ *along any radius vector but not as* $z \to 0$ *in any manner.*

Ans. If $z \to 0$ along the radius vector $y = mx$, then

$$\lim_{z \to 0} \frac{f(z) - f(0)}{z} = \lim_{z \to 0} \left[\frac{x^3 y(y - jx)}{x^6 + y^2} \cdot \frac{1}{z} \right]$$

$$= \lim_{z \to 0} \left[\frac{-jx^3 y(x + jy)}{x^6 + y^2} \cdot \frac{1}{x + jy} \right] \quad (\because z = x + jy)$$

$$= \lim_{x \to 0} \frac{-jx^3 y}{x^6 + y^2} = \lim_{x \to 0} \frac{-jmx^4}{x^6 + m^2 x^2} \quad \text{(when } y = mx\text{)}$$

$$= \lim_{x \to 0} \frac{-jmx^2}{x^4 + m^2} = 0.$$

Again, if $z \to 0$ along the curve $y = x^3$, then

$$\begin{aligned}\lim_{z\to 0}\frac{f(z)-f(0)}{z} &= \lim_{z\to 0}\left[\frac{x^3y(y-jx)}{x^6+y^2}\cdot\frac{1}{z}\right]\\ &= \lim_{x\to 0}\frac{-jx^6}{x^6+x^6} \quad (\text{when } y = x^3)\\ &= \lim_{x\to 0}\left(-\frac{j}{2}\right) = -\frac{j}{2}.\end{aligned}$$

Hence the result.

21.5 Analytic Function

Definition. A single-valued function $f(z)$ is *analytic* (or *regular* or *holomorphic* or *monogenic*) in a domain D, if it has a *unique derivative* at every point of the domain. Thus an analytic function in a domain is differentiable at all points of the domain. If we say that the function $f(z)$ is analytic at a point $z = z_0$, this means $f(z)$ has a derivative at every point inside a small circle about $z = z_0$. So a *regular point* of $f(z)$ is a point at which $f(z)$ is analytic.

A function $f(z)$ may be differentiable (analytic) in a domain except possibly for a finite number of points. These points are called *singular points* or *singularities* of the function $f(z)$ in that domain. Thus a *singular point* or *singularity* of $f(z)$ is a point at which $f(z)$ is not analytic. It is called an *isolated singular point* if $f(z)$ is analytic everywhere else inside some small circle about the singular point.

Explanation with examples. Meaning of $f(z)$ having a derivative at a point z_0 is that $f'(z)$ defined by (21.4.3) has the same value no matter how we approach z_0. Although there may be an infinite number of ways by which we can approach a point z_0, $f'(z)$ is always the same. For example we consider a function $f(z) = z^2$; then

$$f'(z) = \frac{d}{dz}(z^2) = \lim_{\Delta z\to 0}\frac{(z+\Delta z)^2 - z^2}{\Delta z} = \lim_{\Delta z\to 0}(2z+\Delta z) = 2z$$

which is independent of how Δz tends to zero. Thus z^2 is an analytic function. Again we consider the function $f(z) = |z|^2$ as another example. Then

$$\lim_{\Delta z\to 0}\frac{\Delta f}{\Delta z} = \lim_{\Delta z\to 0}\frac{|z+\Delta z|^2 - |z|^2}{\Delta z}.$$

Now numerator of the above expression is always real and the denominator $\Delta z = \Delta x + j\Delta y$, has different values depending on our mode of approach. For example if change in z is due to that in real variable only, then $\Delta y = 0$ and $\Delta z = \Delta x$; again if change in z is due to that in imaginary part only, i.e., along a vertical line in Argand diagram, then $\Delta x = 0$, and $\Delta z = j\Delta y$, and along other similar directions, Δz is some complex number, and therefore, $\lim\limits_{\Delta z\to 0}\frac{\Delta f}{\Delta z}$ has different values for different directions of approach to z_0, that is, $|z|^2$ is not analytic.

From the above examples of analytic and non-analytic functions, we cannot, definitely, tell whether a function has a derivative except taking recourse to equation (21.4.3). The following discussions will answer this question.

21.6 Necessary and sufficient conditions for $f(z)$ to be analytic

(a) Necessary conditions for $f(z)$ to be analytic. If $w = f(z) = u(x,y) + jv(x,y)$ is analytic in a domain, then in that domain

$$\frac{\partial u}{\partial x} = \frac{\partial v}{\partial y} \quad \text{and} \quad \frac{\partial u}{\partial y} = -\frac{\partial v}{\partial x} \tag{1}$$

or in short $u_x = v_y$ and $u_y = -v_x$, which are the required *necessary conditions* for the function $f(z)$ to be analytic in the domain. These relations or equations are the famous Cauchy-Riemann conditions—which were discovered by Cauchy, and used extensively by Riemann in the theory of analytic functions. These conditions are necessary for the existence of a derivative of $f(z)$.

Proof. We have a complex function $f(z)$ of $z = x + jy$, such that $w = f(z) = u(x,y) + jv(x,y)$, where $u(x,y)$ and $v(x,y)$ are real functions of real variables x and y. Since $z = x + jy$, we have $\Delta z = \Delta x + j\Delta y$.

Let $w = f(z) = u(x,y) + jv(x,y)$ be analytic inside a domain. So differential coefficient of w at any point z of this domain must exist; i.e.,

$$\lim_{\Delta z \to 0} \frac{\Delta w}{\Delta z} = \lim_{\Delta z \to 0} \frac{f(z + \Delta z) - f(z)}{\Delta z}$$

exists and is unique as $\Delta z \to 0$ along any path we choose.

Now, let us choose that $\Delta z \to 0$ along the real axis, so that $\Delta y = 0$ and $\Delta z = \Delta x$. Hence in this case,

$$\lim_{\Delta z \to 0} \frac{\Delta w}{\Delta z} = \lim_{\Delta x \to 0} \frac{\Delta w}{\Delta x}.$$

Again, let us choose that $\Delta z \to 0$ along the imaginary axis, so that $\Delta x = 0$ and $\Delta z = j\Delta y$. Hence in this case

$$\lim_{\Delta z \to 0} \frac{\Delta w}{\Delta z} = \lim_{\Delta y \to 0} \frac{\Delta w}{j\Delta y}.$$

Since $\lim\limits_{\Delta z \to 0} \frac{\Delta w}{\Delta z}$ exists and is unique (because w is analytic), it follows that

$$\lim_{\Delta x \to 0} \frac{\Delta w}{\Delta x} \quad \text{and} \quad \lim_{\Delta y \to 0} \frac{\Delta w}{j\Delta y}$$

exist and are the same.

Hence

$$\frac{dw}{dz} = \frac{\partial w}{\partial x} = \frac{1}{j}\frac{\partial w}{\partial y};$$

now

$$\frac{\partial w}{\partial x} = \frac{\partial u}{\partial x} + j\frac{\partial v}{\partial x}, \quad \because \quad w = u + jv$$

$$\text{and} \quad \frac{\partial w}{\partial y} = \frac{\partial u}{\partial y} + j\frac{\partial v}{\partial y}, \quad \because \quad w = u + jv.$$

We have just shown above that

$$\frac{\partial w}{\partial y} \text{ and } \frac{\partial w}{\partial x} \text{ exist,}$$

so that

$$\frac{\partial u}{\partial x}, \ \frac{\partial v}{\partial x}, \ \frac{\partial u}{\partial y}, \ \frac{\partial v}{\partial y}$$

must also exist. Also we have

$$\frac{\partial w}{\partial x} = \frac{1}{j}\frac{\partial w}{\partial y}$$

$$\therefore \quad \frac{\partial u}{\partial x} + j\frac{\partial v}{\partial x} = \frac{1}{j}\left(\frac{\partial u}{\partial y} + j\frac{\partial v}{\partial y}\right) = -j\left(\frac{\partial u}{\partial y} + j\frac{\partial v}{\partial y}\right) = \frac{\partial v}{\partial y} - j\frac{\partial u}{\partial y} \tag{2}$$

Now equating real and imaginary parts, we have

$$\left.\begin{aligned} \frac{\partial u}{\partial x} &= \frac{\partial v}{\partial y} \quad \text{and} \quad \frac{\partial u}{\partial y} = -\frac{\partial v}{\partial x} \\ \text{or, in short } u_x &= v_y \quad \text{and} \quad u_y = -v_x \end{aligned}\right\} \tag{3}$$

which are Cauchy-Riemann conditions necessary for the function $f(z)$ to be analytic in the given domain.

(b) **Sufficient conditions for $f(z)$ to be analytic.** We have seen above that the truth of the Cauchy-Riemann equations (21.6.3) is a necessary condition for the function $f(z)$ to be analytic in a domain, but is not a sufficient condition. What is further required is that the *four partial derivatives u_x, u_y, v_x, v_y must exist and must be continuous at all points of the domain.* Thus for the existence of derivative of a function, the sufficient requirement is continuity of these partial derivatives, satisfying Cauchy-Riemann conditions (21.6.3).

Proof. It is obvious that

$$\begin{aligned} \Delta w &= \Delta f(z) = \Delta u + j\Delta v \\ &= \frac{\Delta u}{\Delta x}\Delta x + \frac{\Delta u}{\Delta y}\Delta y + j\frac{\Delta v}{\Delta x}\Delta x + j\frac{\Delta v}{\Delta y}\Delta y \\ &= \left(\frac{\Delta u}{\Delta x} + j\frac{\Delta v}{\Delta x}\right)\Delta x + \left(\frac{\Delta u}{\Delta y} + j\frac{\Delta v}{\Delta y}\right)\Delta y. \end{aligned}$$

Again the derivative of $f(z)$ with respect to z is given by

$$\begin{aligned} \frac{df(z)}{dz} = f'(z) &= \lim_{\Delta z \to 0}\frac{\Delta w}{\Delta z} = \lim_{\substack{\Delta x \to 0 \\ \Delta y \to 0}} \frac{\left(\frac{\Delta u}{\Delta x} + j\frac{\Delta v}{\Delta x}\right)\Delta x + \left(\frac{\Delta u}{\Delta y} + j\frac{\Delta v}{\Delta y}\right)\Delta y}{\Delta x + j\Delta y} \\ &= \lim_{\substack{\Delta x \to 0 \\ \Delta y \to 0}} \frac{\left(\frac{\Delta u}{\Delta x} + j\frac{\Delta v}{\Delta x}\right)}{\left(1 + j\frac{\Delta y}{\Delta x}\right)}\left[1 + j\frac{\Delta y}{\Delta x}\frac{\left(\frac{\Delta v}{\Delta y} - j\frac{\Delta u}{\Delta y}\right)}{\left(\frac{\Delta u}{\Delta x} + j\frac{\Delta v}{\Delta x}\right)}\right] \\ &= \lim_{\substack{\partial x \to 0 \\ \partial y \to 0}} \frac{\frac{\partial u}{\partial x} + j\frac{\partial v}{\partial x}}{1 + j\frac{\partial y}{\partial x}}\left[1 + j\frac{\partial y}{\partial x}\frac{\frac{\partial v}{\partial y} - j\frac{\partial u}{\partial y}}{\frac{\partial u}{\partial x} + j\frac{\partial v}{\partial x}}\right] \end{aligned} \tag{4}$$

Now, from (21.6.2) above, we have

$$\frac{\frac{\partial v}{\partial y} - j\frac{\partial u}{\partial y}}{\frac{\partial u}{\partial x} + j\frac{\partial v}{\partial x}} = 1$$

and hence equation (21.6.4) above reduces to

$$\frac{df(z)}{dz} = \frac{\partial u}{\partial x} + j\frac{\partial v}{\partial x} = \frac{\partial f(z)}{\partial x} \tag{5}$$

which shows that the derivative of $f(z)$ with respect to z is independent of the direction of approach in the complex plane as long as the derivatives are continuous; if the derivatives are not continuous, $\frac{\partial u}{\partial x}$, etc. would have different values depending upon how the derivatives were evaluated. This would again result in a non-unique first derivative and the function would not be analytic.

Equation (21.6.5) implies that if these partial derivatives are continuous, $f'(z)$ exists and possesses a finite unique value, i.e., the function is necessarily analytic.

To illustrate the Cauchy-Reimann conditions, we consider the following two simple examples:

(i) Let $f(z) = z^2$; then the real part is $u(x,y) = x^2 - y^2$ and the imaginary part is $v(x,y) = 2xy$; thus

$$\frac{\partial u}{\partial x} = \frac{\partial v}{\partial y} = 2x$$

$$\text{and} \quad \frac{\partial u}{\partial y} = -\frac{\partial v}{\partial x} = -2y$$

and hence the partial derivatives satisfy Cauchy-Riemann differential equations. It is also evident that these partial derivatives are continuous at all points in a domain of the complex plane. We therefore conclude that the function $f(z) = z^2$ is analytic.

(ii) Let $f(z) = \bar{z} = x - jy$—so that $u = x$, $v = -y$. Applying Cauchy-Riemann conditions, we obtain

$$\frac{\partial u}{\partial x} = 1 \neq \frac{\partial v}{\partial y}$$

so that Cauchy-Riemann equations are not satisfied and $f(z) = \bar{z}$ is not an analytic function of z. It is interesting to note that $f(z) = \bar{z}$ is continuous everywhere, but nowhere analytic.

We shall now prove an interesting result that an analytic function $f(z)$ is independent of $\bar{z}$. For this purpose, we consider a complex variable $z = x + jy$—then $\bar{z} = x - jy$, so that $x = \frac{1}{2}(z + \bar{z})$, $y = \frac{1}{2j}(z - \bar{z})$.

We have then

$$\frac{\partial x}{\partial z} = \frac{1}{2}, \quad \frac{\partial y}{\partial z} = \frac{1}{2j}, \quad \frac{\partial x}{\partial \bar{z}} = \frac{1}{2}, \quad \frac{\partial y}{\partial \bar{z}} = -\frac{1}{2j}.$$

Again we may write

$$\begin{aligned}\frac{df}{d\bar{z}} &= \frac{du}{d\bar{z}} + j\frac{dv}{d\bar{z}} = \frac{\partial u}{\partial x}\frac{\partial x}{\partial \bar{z}} + \frac{\partial u}{\partial y}\frac{\partial y}{\partial \bar{z}} + j\frac{\partial v}{\partial x}\frac{\partial x}{\partial \bar{z}} + j\frac{\partial v}{\partial y}\frac{\partial y}{\partial \bar{z}}\\ &= \frac{1}{2}\frac{\partial u}{\partial x} - \frac{1}{2j}\frac{\partial u}{\partial y} + j\left[\frac{1}{2}\frac{\partial v}{\partial x} - \frac{1}{2j}\frac{\partial v}{\partial y}\right]\\ &= \frac{1}{2}\left(\frac{\partial u}{\partial x} - \frac{\partial v}{\partial y}\right) + \frac{j}{2}\left(\frac{\partial v}{\partial x} + \frac{\partial u}{\partial y}\right) = 0 \quad \text{using (21.6.3)}\end{aligned}$$

which shows that $f(z)$ is independent of $\bar{z}$.

21.7 Polar form of Cauchy-Riemann conditions

We know that $z = x + jy = re^{j\theta} = r(\cos\theta + j\sin\theta)$ so that the co-ordinates x, y are given in terms of polar co-ordinates r, θ as $x = r\cos\theta$, $y = r\sin\theta$.

$$\begin{aligned} \therefore \quad \frac{\partial r}{\partial x} &= \frac{x}{r} = \cos\theta, \frac{\partial r}{\partial y} = \frac{y}{r} = \sin\theta, \\ \frac{\partial \theta}{\partial x} &= -\frac{\sin\theta}{r} \quad \text{and} \quad \frac{\partial \theta}{\partial y} = \frac{\cos\theta}{r} \end{aligned}$$

Thus

$$\frac{\partial u}{\partial x} = \frac{\partial u}{\partial r}\frac{\partial r}{\partial x} + \frac{\partial u}{\partial \theta}\frac{\partial \theta}{\partial x} = \frac{\partial u}{\partial r}\cos\theta - \frac{\partial u}{\partial \theta}\frac{\sin\theta}{r}$$

Similarly

$$\begin{aligned} \frac{\partial u}{\partial y} &= \frac{\partial u}{\partial r}\sin\theta + \frac{\partial u}{\partial \theta}\frac{\cos\theta}{r} \\ \frac{\partial v}{\partial x} &= \frac{\partial v}{\partial r}\cos\theta - \frac{\partial v}{\partial \theta}\frac{\sin\theta}{r} \\ \frac{\partial v}{\partial y} &= \frac{\partial v}{\partial r}\sin\theta + \frac{\partial v}{\partial \theta}\frac{\cos\theta}{r} \end{aligned}$$

Substituting these values in Cauchy-Riemann differential equations, we obtain

$$\begin{aligned} \frac{\partial u}{\partial r}\cos\theta - \frac{\partial u}{\partial \theta}\frac{\sin\theta}{r} &= \frac{\partial v}{\partial r}\sin\theta + \frac{\partial v}{\partial \theta}\frac{\cos\theta}{r} \\ \text{and} \quad \frac{\partial u}{\partial r}\sin\theta + \frac{\partial u}{\partial \theta}\frac{\cos\theta}{r} &= -\frac{\partial v}{\partial r}\cos\theta + \frac{\partial v}{\partial \theta}\frac{\sin\theta}{r} \end{aligned}$$

which when simplified give

$$\left.\begin{aligned} \frac{\partial u}{\partial r} &= \frac{1}{r}\frac{\partial v}{\partial \theta}, \text{ i.e., } u_r = \frac{1}{r}v_\theta \\ \text{and} \quad \frac{\partial u}{\partial \theta} &= -r\frac{\partial v}{\partial r}, \text{ i.e., } u_\theta = -rv_r \end{aligned}\right\} \tag{1}$$

These are Cauchy-Riemann conditions in polar form. The derivatives of the function $w =$

$f(z)$ in the polar form may be derived as follows :

$$\begin{aligned}
\frac{dw}{dz} = \frac{\partial w}{\partial x} &= \frac{\partial w}{\partial r}\frac{\partial r}{\partial x} + \frac{\partial w}{\partial \theta}\frac{\partial \theta}{\partial x} = \frac{\partial w}{\partial r}\cos\theta - \frac{\sin\theta}{r}\frac{\partial w}{\partial \theta} \\
&= \frac{\partial w}{\partial r}\cos\theta - \frac{\sin\theta}{r}\left(\frac{\partial u}{\partial \theta} + j\frac{\partial v}{\partial \theta}\right) \quad (\because\ w = u + jv) \\
&= \frac{\partial w}{\partial r}\cos\theta - \frac{\sin\theta}{r}\left(-r\frac{\partial v}{\partial r} + jr\frac{\partial u}{\partial r}\right) \left[\begin{array}{l}\text{using Cauchy-Riemann} \\ \text{equations (21.7.1) in polar form}\end{array}\right] \\
&= \frac{\partial w}{\partial r}\cos\theta - \sin\theta\left(-\frac{\partial v}{\partial r} + j\frac{\partial u}{\partial r}\right) \\
&= (\cos\theta - j\sin\theta)\frac{\partial w}{\partial r} \quad \left(\text{writing}\frac{\partial w}{\partial r} = \frac{\partial u}{\partial r} + j\frac{\partial v}{\partial r}\right) \qquad (2) \\
&= -\frac{j}{r}(\cos\theta - j\sin\theta)\frac{\partial w}{\partial \theta}
\end{aligned}$$

21.8 Laplace's equations

Cauchy-Riemann equations are

$$\frac{\partial u}{\partial x} = \frac{\partial v}{\partial y} \quad \text{and} \quad \frac{\partial u}{\partial y} = -\frac{\partial v}{\partial x}$$

$$\therefore \quad \frac{\partial^2 u}{\partial x^2} = \frac{\partial^2 v}{\partial x \partial y} \quad \text{and} \quad \frac{\partial^2 u}{\partial y^2} = -\frac{\partial^2 v}{\partial y \partial x}$$

so that by adding we may write

$$\frac{\partial^2 u}{\partial x^2} + \frac{\partial^2 u}{\partial y^2} = 0,$$

since

$$\frac{\partial^2 v}{\partial x \partial y} = \frac{\partial^2 v}{\partial y \partial x} \text{ because } \frac{\partial u}{\partial x}, \text{ etc., are continuous.}$$

$$\therefore \quad \nabla^2 u = \frac{\partial^2 u}{\partial x^2} + \frac{\partial^2 u}{\partial y^2} = 0, \text{ similarly } \nabla^2 v = \frac{\partial^2 v}{\partial x^2} + \frac{\partial^2 v}{\partial y^2} = 0$$

which are Laplace's equations in a plane—solutions of which are called ***harmonic solutions***. Thus for an analytic function $w = f(z) = u + jv$, u and v—both satisfying Laplace's equation—are harmonic functions.

21.9 Zeros of an analytic complex function

A zero of an analytic function $f(z)$ is defined to be a value of $z = z_0$ such that $f(z_0) = 0$.

Now the function $f(z)$ is analytic everywhere inside a circle whose centre is the point $z = z_0$, if its Taylor's expansion about z_0, i.e.,

$$f(z) = \sum_{n=0}^{\infty} \frac{f^n(z_0)}{n!}(z - z_0)^n$$

exists and converges to $f(z)$ in some open interval $a < z_0 < b$.

Now the above Taylor's expansion gives

$$f(z) = f(z_0) + (z - z_0)f'(z_0) + \frac{(z - z_0)^2}{2!}f''(z_0) + \cdots + \frac{(z - z_0)^n}{n!}f^n(z_0).$$

Here, if $f(z_0) = 0$ and $f'(z_0) \neq 0$, then $f(z)$ is said to have a *simple zero* at $z = z_0$.

If however $f(z_0) = f'(z_0) = f''(z_0) = \cdots = f^{n-1}(z_0) = 0$ and $f^n(z_0) \neq 0$, then the point $z = z_0$ is known as *zero of order n* of the function $f(z)$, since then

$$\begin{aligned} f(z) &= \frac{(z - z_0)^n}{n!}f^n(z_0) + \frac{(z - z_0)^{n+1}}{(n+1)!}f^{n+1}(z_0) + \cdots \\ &= (z - z_0)^n \left[\frac{f^n(z_0)}{n!} + \frac{(z - z_0)}{(n+1)!}f^{n+1}(z_0) + \cdots\right] \\ &= (z - z_0)^n \phi(z), \qquad (1) \\ \text{where} \quad \phi(z) &= \frac{f^n(z_0)}{n!} + \frac{(z - z_0)}{(n+1)!}f^{n+1}(z_0) + \cdots \\ &= \text{a non-zero and analytic function at } z = z_0 \end{aligned}$$

It follows from (21.9.1) that the point $z = z_0$ may be defined as a *zero of order n* of the function $f(z)$ if $f(z) = (z - z_0)^n \phi(z)$ for $f(z)$ being analytic and non-zero function at $z = z_0$.

If we draw a small circle C about z_0 as centre, then within it $\phi(z) \neq 0$. Also for $z \neq z_0$ within C, $(z - z_0)^n \neq 0$ and hence z_0 is the only zero of the function $f(z)$ within C. As such the zero point $z = z_0$ is *isolated.* But z_0 being arbitrary may be any zero point of $f(z)$ and therefore all the zeros of an analytic function $f(z)$ are isolated.

21.10 Singularities of an analytic function

A point, where the function $f(z)$ ceases to be analytic (regular) is called the *singular point* of $f(z)$. We can also say that $f(z)$ has a *singularity* at the point.

The function $f(z)$ is said to be *singular at infinity* if $f\left(\frac{1}{z}\right)$ is singular at $z = 0$.

Isolated singularity. If $f(z)$ has a singularity at $z = z_0$ and if there is no other singularity within a small circle surrounding the point $z = z_0$, then $z = z_0$ is said to be an *isolated singularity* of the function $f(z)$; otherwise it is called *non-isolated.*

Kinds of singularity. We consider a function $f(z)$ which is analytic within a domain D, except at the point $z = z_0$, which is an isolated singularity. We now draw a circle C_1 with its centre at $z = z_0$ and radius as small as we please, and another large concentric circle C_2 of any radius R, lying wholly within the domain D. Now in the annulus between these two circles, $f(z)$ is analytic—hence if z is any point of this annulus, then we can expand $f(z)$ by Laurent's theorem as

$$f(z) = \sum_{n=0}^{\infty} a_n(z - z_0)^n + \sum_{n=1}^{\infty} b_n(z - z_0)^{-n} \quad [0 < |(z - z_0)| < R], \qquad (1)$$

where

$$a_n = \frac{1}{2\pi j}\int_{C_1}\frac{f(z)}{(z-z_0)^{n+1}}dz \quad \text{and} \quad b_n = \frac{1}{2\pi j}\int_{C_2}(z-z_0)^{n-1}f(z)dz.$$

The term

$$\left\{\sum_{n=1}^{\infty} b_n(z-z_0)^{-n}\right\}$$

on the right hand side of (21.10.1) is called the *principal part* of the expansion of $f(z)$ at the isolated singularity $z = z_0$.

In relation to the above principal part, there may be three possibilities :

(i) All the coefficients b_n may be zero, i.e., there will be no term in the principal part, so that the function is equal to the analytic function except at $z = z_0$. Such a singularity is called *non-essential* or *removable singularity* of $f(z)$.

(ii) There may be infinite number of terms in the principal part. Then the point $z = z_0$ is said to be an *essential* or *non-removable* singularity of $f(z)$.

(iii) The principal part may contain a finite number of terms—then the function $f(z)$ is said to have a *pole* or *singularity of order m*,

We thus see that the above three cases give rise to three kinds of singularity.

(i) *Non-essential or removable singularity : When all the coefficients b_n are zero,* we then have

$$f(z) = \sum_{n=0}^{\infty} a_n(z-z_0)^n,$$

which is analytic for $|z - z_0| < R$, except at the point $z = z_0$. Let $\phi(z)$ be the sum function of the power series

$$\sum_{n=0}^{\infty} a_n(z-z_0)^n.$$

Now $\phi(z)$ differs from $f(z)$ only for $z = z_0$, where there is singularity. To avoid this singularity $z = z_0$, we can suitably define $f(z)$ at $z = z_0$ so that we have

$$\begin{aligned}\phi(z) &= f(z) \text{ for } 0 < |z-z_0| < R\\ &= a_0 \text{ for } z = z_0.\end{aligned}$$

This type of *singularity, which can be made to disappear by defining the function suitably,* is called *non-essential* or *removable singularity.* This is a rather artificial sort of singularity and is of little importance in the theory of functions.

For example the function $\frac{\sin(z-z_0)}{(z-z_0)}$ has removable singularity at $z = z_0$ because

$$\begin{aligned}\frac{\sin(z-z_0)}{(z-z_0)} &= \frac{1}{(z-z_0)}\left[(z-z_0) - \frac{(z-z_0)^3}{3!} + \frac{(z-z_0)^5}{5!} - \frac{(z-z_0)^7}{7!} + \cdots\right]\\ &= 1 - \frac{(z-z_0)^2}{3!} + \frac{(z-z_0)^4}{5!} - \frac{(z-z_0)^6}{7!} + \cdots\end{aligned}$$

has no term containing negative power of $(z - z_0)$.

However this singularity can be removed and the function can be made analytic by defining $\frac{\sin(z-z_0)}{(z-z_0)} = 1$ at $z = z_0$.

(ii) *Essential or non-removable singularity (or isolated essential singularity)* : If there be infinite number of terms in the principal part of (21.10.1), that is, if the series $b_1(z-z_0)^{-1} + b_2(z-z_0)^{-2} + \cdots$ does not terminate, then $(z-z_0)$ is called an *isolated essential* singularity of the function $f(z)$.

For example the function $\sin\left(\frac{1}{z-z_0}\right)$ has essential singularity at $z = z_0$, because

$$\sin\left(\frac{1}{z-z_0}\right) = \frac{1}{z-z_0} - \frac{1}{3!(z-z_0)^3} + \frac{1}{5!(z-z_0)^5} - \cdots$$

has infinite number of terms with negative powers of $(z-z_0)$.

(iii) **Pole:** If there be finite number of terms in the principal part of (21.10.1), that is, if the series $b_1(z-z_0)^{-1} + b_2(z-z_0)^{-2} + \cdots + b_m(z-z_0)^{-m}$ consists of finite number of terms, then $z = z_0$ is said to be a pole of order m of the function $f(z)$.

When $m = 1$, the pole is said to be a *simple pole.*

For example the function $\frac{\sin(z-z_0)}{(z-z_0)^4}$ has a pole at $z = z_0$, because

$$\begin{aligned}\frac{\sin(z-z_0)}{(z-z_0)^4} &= \frac{1}{(z-z_0)^4}\left[(z-z_0) - \frac{1}{3!}(z-z_0)^3 + \frac{1}{5!}(z-z_0)^5 - \frac{1}{7!}(z-z_0)^7 + \cdots\right] \\ &= \frac{1}{(z-z_0)^3} - \frac{1}{3!(z-z_0)} + \frac{1}{5!}(z-z_0) - \frac{1}{7!}(z-z_0)^3 + \cdots\end{aligned}$$

has finite number of terms (here the first two terms only) with negative powers of $(z-z_0)$.

We have seen above that in the neighbourhood of the pole $z = z_0$ of order m, the function

$$\begin{aligned}f(z) &= \sum_{n=0}^{\infty} a_n(z-z_0)^n + \frac{b_1}{(z-z_0)} + \frac{b_2}{(z-z_0)^2} + \cdots + \frac{b_m}{(z-z_0)^m} \\ &= \frac{1}{(z-z_0)^m}\left[\sum_{n=0}^{\infty} a_n(z-z_0)^{n+m} + \{b_m + b_{m-1}(z-z_0) + \cdots \right. \\ &\qquad \left. + b_2(z-z_0)^{m-2} + b_1(z-z_0)^{m-1}\}\right] \\ &= \frac{1}{(z-z_0)^m}\phi(z),\end{aligned}$$

where

$$\phi(z) = \sum_{n=0}^{\infty} a_n(z-z_0)^{n+m} + b_m + b_{m-1}(z-z_0) + \cdots + b_1(z-z_0)^{m-1}.$$

Obviously, $\phi(z) \to b_m$ as $z \to z_0$; hence $\phi(z)$ is analytic in the neighbourhood of the pole $z = z_0$; when $m = 1$, the pole is known as a *simple pole.*

For a *pole of order m* at $z = z_0$, the limiting value of the function $f(z)$ is obtained, as follows, by considering the principal part in its Laurent's expansion:

$$\lim_{z\to z_0} f(z) = \lim_{z\to z_0}\sum_{n=1}^{m} b_n(z-z_0)^{-n} \quad \text{or,} \quad \lim_{z\to z_0}|f(z)| = \lim_{z\to z_0}\left|\sum_{n=1}^{m} b_n(z-z_0)^{-n}\right|$$

$$= \lim_{z\to z_0}\left|\frac{b_m}{(z-z_0)^m}+\frac{b_{m-1}}{(z-z_0)^{m-1}}+\cdots+\frac{b_1}{(z-z_0)}\right|$$

$$= \lim_{z\to z_0}\frac{1}{|(z-z_0)^m|}\left|\{b_m+b_{m-1}(z-z_0)+\cdots+b_n(z-z_0)^{m-n}+\cdots+b_1(z-z_0)^{m-1}\}\right.$$

$$\geq \lim_{z\to z_0}\frac{1}{|(z-z_0)^m|}\{|b_m|+|b_{m-1}(z-z_0)|+\cdots+|b_1(z-z_0)^{m-1}|\}$$

$$\geq \lim_{z\to z_0}\frac{1}{|(z-z_0)^m|}|b_m| \geq \infty, \text{ i.e., } |f(z)|\to\infty \text{ as } z\to z_0.$$

Conclusively, if $f(z)$ is analytic in a given domain except at a point $z = z_0$, and $\lim_{z\to z_0} f(z) \to \infty$, then $f(z)$ is said to have a *pole* or *singularity* at $z = z_0$. For the pole of order m, we have $\lim_{z\to z_0}(z-z_0)^m f(z) = b_m = \text{constant}(\neq 0)$ and the value of the function is given by $\frac{\phi(z)}{(z-z_0)^m}$.

In case there does not exist a value of m such that $\lim_{z\to z_0}(z-z_0)^m f(z)$ is finite, then the point $z = z_0$ is an *essential* or *non-removable singularity of* $f(z)$.

If an analytic function $f(z)$ has a pole of order m at the point $z = z_0$, then $\frac{1}{f(z)}$ has a zero of order m at this point.

$z = z_0$ will be a pole of order m, if $(z-z_0)^m f(z)$ is analytic and non-zero at $z = z_0$, so that $\phi(z) = \frac{1}{(z-z_0)^m f(z)}$ is analytic and non-zero.

$\therefore \quad \frac{1}{f(z)} = (z-z_0)^m\phi(z)$—whence it follows that $\frac{1}{f(z)}$ has a zero of order m at $z = z_0$. A similar procedure will show that if $\frac{1}{f(z)}$ has a pole of order m, then $f(z)$ will have a pole of order m. It also follows that since zeros are isolated, the poles must be isolated.

21.11 The point at infinity

In treating the complex variables, it is convenient to regard infinity as a point. The point infinity is considered by putting $z = \frac{1}{z_0}$ in $f(z)$, so that the behaviour of $f(z)$ at infinity is examined by the behaviour of $f\left(\frac{1}{z_0}\right)$ at $z_0 = 0$.

Consequently, $f(z)$ is analytic or has a zero or has a simple pole or has an essential singularity, etc., at infinity according as $f\left(\frac{1}{z_0}\right)$ is analytic or has a zero or has a simple pole or has an essential singularity at $z_0 = 0$, e.g., if $f\left(\frac{1}{z_0}\right)$ has a zero of order m at $z_0 = 0$, then writing $\phi(z_0)$ for $f\left(\frac{1}{z_0}\right)$, its expansion at $z_0 = 0$ by Taylor's theorem is

$$f\left(\frac{1}{z_0}\right) = \phi(z_0) = a_m z_0^m + a_{m+1}z_0^{m+1} + \cdots$$

so that the corresponding expansion of $f(z)$ at $z = \infty$ is

$$f(z) = \frac{a_m}{z^m} + \frac{a_{m+1}}{z^{m+1}} + \cdots,$$

which does not contain constant terms and terms containing positive powers of z and hence $f(z)$ has a zero of order m at infinity.

Similarly if $f\left(\frac{1}{z_0}\right)$ has a pole of order m at $z_0 = 0$, then by Laurent's expansion, we have

$$f\left(\frac{1}{z_0}\right) = \phi(z_0) = \sum_{n=0}^{\infty} a_n z_0^n + \sum_{n=1}^{m} b_n z_0^{-n},$$

so that the corresponding expansion of $f(z)$ at $z = \infty$ is

$$f(z) = \sum_{n=1}^{m} b_n z^n + \sum_{n=0}^{\infty} a_n z^{-n}$$

which has $\sum_{n=1}^{m} b_n z^n$ as its principal part at $z = \infty$ and this principal part being a finite series of positive powers of z at $z = z_0$, shows that $f(z)$ has a pole of order m at infinity.

21.12 Evaluation of residues

If $f(z)$ is analytic in the neighbourhood of a point $z = z_0$, then by Cauchy's theorem, we have $\int f(z)dz = 0$ for any contour C in that neighbourhood. But if $f(z)$ possesses a pole or an essential isolated singularity at $z = z_0$ and z_0 lies inside the contour C, then the value of the above integral will in general be different from zero. In such cases, we can represent the function $f(z)$ by Laurent's series expansion, i.e.,

$$f(z) = \sum_{n=0}^{\infty} a_n (z - z_0)^n + \sum_{n=1}^{\infty} \frac{b_n}{(z - z_0)^n} \quad [\text{for } 0 < |z - z_0| < R],$$

where R is the distance from z_0 to the nearest singular point of $f(z)$ and

$$b_n = \frac{1}{2\pi j} \int_C (z - z_0)^{n-1} f(z)dz$$

$$\therefore \quad b_1 = \frac{1}{2\pi j} \int_C f(z)dz \quad \text{or,} \quad \int_C f(z)dz = 2\pi j b_1 \tag{1}$$

the integration being taken round C in *the anti-clockwise direction* in the domain $0 < |z - z_0| < R$ and the point $z = z_0$ is contained in its interior. The coefficient b_1 is called the *residue* of $f(z)$ at $z = z_0$ and is denoted as

$$b_1 = \operatorname*{Res}_{z=z_0} f(z).$$

In case $f(z)$ has a *simple pole at a point* $z = z_0$, then the Laurent's series expansion gives

$$f(z) = \sum_{n=0}^{\infty} a_n(z - z_0)^n + \frac{b_1}{(z - z_0)}$$

or $$(z - z_0)f(z) = b_1 + \sum_{n=0}^{\infty} a_n(z - z_0)^{n+1} \quad \therefore \quad \lim_{z \to z_0} (z - z_0)f(z) = b_1$$

i.e., $$\operatorname*{Res}_{z=z_0} f(z) = \lim_{z \to z_0} (z - z_0)f(z)$$

Thus we may say that *in the case of a simple pole at* $z = z_0$, $\lim_{z\to z_0} (z - z_0)f(z)$ *is the residue of* $f(z)$ *at the pole.*

Residue at infinity. If $f(z)$ has an isolated singularity at infinity and if C is a large circle enclosing within it all other singularities of $f(z)$. Then we define the residue of $f(z)$ at infinity to be the value of

$$\frac{1}{2\pi j} \int_C f(z)dz,$$

where the integral is taken along C in a *clockwise direction*, or we may define the residue of $f(z)$ at $z = \infty$ as

$$-\frac{1}{2\pi j} \int_C f(z)dz,$$

the integration being taken round C *in an anti-clockwise direction.*

Let $f(z)$ have a pole of order m at infinity. Then $f(1/z)$ has a pole of order m at $z = 0$. As such $f(1/z)$ can be expanded in a Laurent's series for values of z in the annulus $0 < |z| < \rho$, where ρ is small

$$\text{i.e.,} \quad f\left(\frac{1}{z}\right) = \sum_{n=0}^{\infty} a_n z^n + \sum_{n=1}^{m} b_n \left(\frac{1}{z}\right)^n. \tag{2}$$

Replacement of $\frac{1}{z}$ by z yields the expansion of $f(z)$ in the neighbourhood of $z = \infty$,

$$f(z) = \sum_{n=1}^{m} b_n z^n + \sum_{n=0}^{\infty} a_n z^{-n} \tag{3}$$

so that the residue of $f(z)$ at infinity

$$= -\frac{1}{2\pi j} \int_C f(z)dz \quad \text{(by definition of residue)}$$

$$= -\frac{1}{2\pi j} \int_C \left[\sum_{n=1}^{m} b_n z^n + \sum_{n=0}^{\infty} a_n z^{-n}\right] dz \quad \text{from (21.12.3)}$$

$$= -\frac{1}{2\pi j} \int_C \frac{a_1}{z} dz \quad \left(\text{other integrals vanish, being of the form } \int_C \frac{dz}{z^p}, p \neq 1\right)$$

$$= -\frac{1}{2\pi j} a_1 \int_C \frac{dz}{z} = -\frac{1}{2\pi j} a_1 2\pi j \quad \because \int_C \frac{dz}{z} = 2\pi j$$

$$= -a_1$$

= *the negative of the coefficient of* $\frac{1}{z}$ *in the expansion of* $f(z)$ *for values of* z *in the neighbourhood of* $z = \infty$.

21.13 Methods of finding the residue at a finite pole

In the following we shall discuss the methods for finding the residue at a simple pole (i.e., of order 1) and also that at a pole of order m.

(a) *Residue at a simple pole.* (i) Let $f(z)$ have a simple pole at a point $z = z_0$; then the corresponding Laurent's series expansion is of the form

$$f(z) = \frac{b_1}{(z - z_0)} + a_0 + a_1(z - z_0) + a_2(z - z_0)^2 + \cdots (b_1 \neq 0)$$

i.e., $$(z - z_0)f(z) = b_1 + a_0(z - z_0) + a_1(z - z_0)^2 + a_2(z - z_0)^3 + \cdots$$

Taking limit when $z \to z_0$, we have

$$b_1 = \lim_{z \to z_0} (z - z_0)f(z),$$

where b_1 is the residue at $z = z_0$. Thus the residue of $f(z)$ at $z = z_0$ is given by

$$\lim_{z \to z_0} (z - z_0)f(z).$$

(ii) If $f(z)$ has a simple pole at $z = z_0$, then we may put $f(z) = \{\phi(z)/\psi(z)\}$, where $\phi(z)$ and $\psi(z)$ are both analytic at $z = z_0$ and $\phi(z_0) \neq 0$, and $\psi(z)$ has a simple zero at $z = z_0$. So $\psi(z)$ can be expanded by Taylor's series in the form

$$\begin{aligned}\psi(z) &= \psi[(z - z_0) + z_0] = \psi(z_0) + (z - z_0)\psi'(z_0) + \frac{(z - z_0)^2}{2!}\psi''(z_0) + \cdots \\ &= (z - z_0)\psi'(z_0) + \frac{(z - z_0)^2}{2!}\psi''(z_0) + \cdots \quad [\because \psi(z_0) = 0]\end{aligned}$$

We may then write

$$\begin{aligned}\lim_{z \to z_0} (z - z_0)f(z) &= \lim_{z \to z_0} \frac{(z - z_0)\phi(z)}{(z - z_0)\psi'(z_0) + \frac{1}{2}(z - z_0)^2\psi''(z_0) + \cdots} \\ &= \lim_{z \to z_0} \frac{\phi(z)}{\psi'(z_0) + \frac{1}{2}(z - z_0)\psi''(z_0) + \cdots} = \{\phi(z_0)/\psi'(z_0)\}\end{aligned}$$

Thus at a simple pole $z = z_0$, we also have residue $= \{\phi(z_0)/\psi'(z_0)\}$.

(b) *Residue at a pole of order m.* If $f(z)$ has a pole of order m, where $m > 1$ at a point $z = z_0$, then the expansion of $f(z)$ about $z = z_0$ in a Laurent's series is given by

$$\left.\begin{aligned} f(z) &= \frac{b_m}{(z-z_0)^m} + \frac{b_{m-1}}{(z-z_0)^{m-1}} + \cdots + \frac{b_2}{(z-z_0)^2} + \frac{b_1}{(z-z_0)} \\ &\quad + a_0 + a_1(z-z_0) + \cdots, \quad (\text{where } b_m \neq 0) \end{aligned}\right\} \tag{1}$$

Multiplying both sides by $(z - z_0)^m$, we obtain

$$\begin{aligned} (z-z_0)^m f(z) &= b_m + b_{m-1}(z-z_0) + \cdots + b_2(z-z_0)^{m-2} + b_1(z-z_0)^{m-1} \\ &\quad + a_0(z-z_0)^m + a_1(z-z_0)^{m+1} + \cdots \end{aligned}$$

We observe here that the residue b_1 is the coefficient of $(z-z_0)^{m-1}$ in the Taylor's expansion of the function $\phi(z) = (z - z_0)^m f(z)$ with centre $z = z_0$. Again the coefficient of $(z - z_0)^{m-1}$ in the expansion of $\phi(z)$ is $\phi^{m-1}(z_0)/(m-1)!$

$$\therefore \quad \text{Residue} = b_1 = \frac{\phi^{m-1}(z_0)}{(m-1)!} = \frac{1}{(m-1)!} \lim_{z \to z_0} \left[\frac{d^{m-1}}{dz^{m-1}}\{(z-z_0)^m f(z)\}\right].$$

Thus, when $z = z_0$ *is a pole of order m*, then the residue of $f(z)$ at $z = z_0$ is given by

$$\{\phi^{m-1}(z_0)/(m-1)!\}.$$

If we put $z = z_0 + t$ in (21.13.1), we have

$$f(z_0 + t) = \frac{b_m}{t^m} + \frac{b_{m-1}}{t^{m-1}} + \cdots + \frac{b_2}{t^2} + \frac{b_1}{t} + a_0 + a_1 t + \cdots, \tag{2}$$

where b_1 is the residue of $f(z)$ at $z = z_0$, whatever be the order of the pole at $z = z_0$. Here b_1 is the coefficient of $\frac{1}{t}$ in the expansion (21.13.2).

Thus the formal rule to find the residue of $f(z)$ at a point $z = z_0$ which may be a pole of any order is

(i) putting $z = z_0 + t$ in $f(z)$, we are to expand it in powers of t, where t is a sufficiently small quantity;

(ii) coefficient of $\frac{1}{t}$ is the residue of $f(z)$ at $z = z_0$.

Illustrative Examples

1. Show that the function $f(z) = \sin x \cosh y + j \cos x \sinh y$ is continuous as well as analytic everywhere.

Ans. Since u and v are both rational functions of x and y whose denominators are non-zero for all values of x and y, therefore u and v are both continuous everywhere, so that $f(z)$ is also continuous everywhere. Again we have

$$\frac{\partial u}{\partial x} = \cos x \cosh y, \quad \frac{\partial u}{\partial y} = \sin x \sinh y$$

$$\text{and} \quad \frac{\partial v}{\partial x} = -\sin x \sinh y, \quad \frac{\partial v}{\partial y} = \cos x \cosh y$$

The four partial derivatives are also rational functions of x and y with non-zero denominators for all values of x and y; therefore, they are continuous everywhere.

Again

$$\frac{\partial u}{\partial x} = \frac{\partial v}{\partial y} \quad \text{and} \quad \frac{\partial u}{\partial y} = -\frac{\partial v}{\partial x}$$

i.e., Cauchy-Riemann conditions are satisfied. Thus the four partial derivatives being continuous everywhere and Cauchy-Riemann conditions being satisfied, the function $f(z)$ is analytic everywhere.

2. Show that the function $f(z) = xy + jy$ is everywhere continuous but is not analytic.

Ans. Let $f(z) = u + jv$, where $u(x, y) = xy$ and $v(x, y) = y$.

Thus we see that both u and v being polynomials in x and y are continuous everywhere, so that $f(z)$ is continuous everywhere.

Again

$$\frac{\partial u}{\partial x} = y, \quad \frac{\partial u}{\partial y} = x, \quad \frac{\partial v}{\partial x} = 0, \quad \frac{\partial v}{\partial y} = 1,$$

$$\text{so that} \quad \frac{\partial u}{\partial x} \neq \frac{\partial v}{\partial y} \quad \text{and} \quad \frac{\partial u}{\partial y} \neq -\frac{\partial v}{\partial x},$$

i.e., the Cauchy-Riemann conditions are not satisfied anywhere. Hence $f(z)$ is not analytic at any point.

3. Show that the function $f(z) = e^{-z^{-4}} (z \neq 0)$ and $f(0) = 0$ is not analytic at $z = 0$, although Cauchy-Riemann conditions are satisfied. How would you explain this?

Ans. Here $u + jv = e^{-(x+jy)^{-4}}$

$$\text{or,} \quad u + jv = e^{-\frac{(x-jy)^4}{(x^2+y^2)^4}} = e^{-\frac{1}{(x^2+y^2)^4}\{(x^4+y^4-6x^2y^2)-j4xy(x^2-y^2)\}}$$

$$\therefore \quad u = e^{-\frac{x^4+y^4-6x^2y^2}{(x^2+y^2)^4}} \cos\frac{4xy(x^2-y^2)}{(x^2+y^2)^4}, \; v = -e^{-\frac{x^4+y^4-6x^2y^2}{(x^2+y^2)^4}} \sin\frac{4xy(x^2-y^2)}{(x^2+y^2)^4}$$

At $z = 0$,

$$\frac{\partial u}{\partial x} = \lim_{x\to 0} \frac{u(x,0)-u(0,0)}{x} = \lim_{x\to 0} \frac{e^{-x^{-4}}}{x} = 0$$

$$\frac{\partial u}{\partial y} = \lim_{y\to 0} \frac{u(0,y)-u(0,0)}{y} = \lim_{y\to 0} \frac{-e^{-y^{-4}}}{y} = 0$$

$$\frac{\partial v}{\partial x} = \lim_{x\to 0} \frac{v(x,0)-v(0,0)}{x} = \lim_{x\to 0} \frac{0}{x} = 0$$

$$\frac{\partial v}{\partial y} = \lim_{y\to 0} \frac{v(0,y)-v(0,0)}{y} = \lim_{y\to 0} \frac{0}{y} = 0$$

Hence Cauchy-Riemann conditions are satisfied at $z = 0$; but

$$f'(0) = \lim_{z\to 0} \frac{f(z) - f(0)}{z} = \lim_{z\to 0} \frac{e^{-z^{-4}}}{z}$$

$$= \lim_{z\to 0} \frac{e^{-r^{-4}}}{re^{j(\pi/4)}} \text{ if } z \to 0 \text{ along } z = re^{j\frac{\pi}{4}}$$

$$= \infty$$

which shows that $f'(z)$ does not exist at $z = 0$.

$\therefore$ $f(z)$ is not analytic at $z = 0$.

Note. $f(z)$ may not be analytic at $z = z_0$ if merely the Cauchy-Riemann conditions are satisfied there. In addition to this $f'(z)$ must exist and should be unique at z_0 and at every point in the neighbourhood of z_0.

4. Examine the nature of the function

$$f(z) = \frac{x^2y^5(x+jy)}{x^4+y^{10}} (z \neq 0) \quad \text{and} \quad f(0) = 0$$

in the region including the origin.

Ans. Here

$$u + jv = \frac{x^2y^5(x+jy)}{x^4+y^{10}} \quad \therefore \quad u = \frac{x^3y^5}{x^4+y^{10}}, \quad v = \frac{x^2y^6}{x^4+y^{10}}.$$

At the origin

$$\frac{\partial u}{\partial x} = \lim_{x\to 0} \frac{u(x,0) - u(0,0)}{x} = \lim_{x\to 0} \frac{0-0}{x} = 0$$

$$\frac{\partial u}{\partial y} = \lim_{y\to 0} \frac{u(0,y) - u(0,0)}{y} = \lim_{y\to 0} \frac{0-0}{y} = 0.$$

Similarly

$$\frac{\partial v}{\partial x} = 0 = \frac{\partial v}{\partial y}$$

Hence Cauchy-Riemann conditions are satisfied at the origin; but

$$\lim_{z\to 0} \frac{f(z) - f(0)}{z} = \lim_{\substack{x\to 0\\ y\to 0}} \left[\frac{x^2y^5(x+jy)}{x^4+y^{10}} - 0\right] \frac{1}{x+jy} \quad \text{(since } f(0) = 0 \text{ and } z = x + jy)$$

$$= \lim_{\substack{x\to 0\\ y\to 0}} \frac{x^2y^5}{x^4+y^{10}} = \lim_{x\to 0} \frac{x^2m^5x^5}{x^4+m^{10}x^{10}} \quad \text{if } z \to 0 \text{ along the radius vector } y = mx$$

$$= \lim_{x\to 0} \frac{m^5x^3}{1+m^{10}x^6} = 0.$$

Again,

$$\lim_{z\to 0} \frac{f(z) - f(0)}{z} = \lim_{\substack{x\to 0\\ y\to 0}} \frac{x^2y^5}{x^4+y^{10}} = \lim_{x\to 0} \frac{x^2x^2}{x^4+x^4} \quad \text{if } z \to 0 \text{ the curve } y^5 = x^2$$

$$= \frac{1}{2},$$

which shows that $f'(0)$ does not exist. Hence $f(z)$ is not analytic at the origin, although Cauchy-Riemann conditions are satisfied there.

5. Show that the function $f(z) = \sqrt{(|xy|)}$ is not regular at the origin, although Cauchy-Riemann equations are satisfied at the point.

Ans. Let the function be $f(z) = u(x,y) + jv(x,y)$, so that we may here write, equating real and imaginary parts, $u(x,y) = \sqrt{(|xy|)}$ and $v(x,y) = 0$.

Now at the origin,

$$\frac{\partial u}{\partial x} = \lim_{x\to 0} \frac{u(x,0) - u(0,0)}{x} = \lim_{x\to 0} \frac{0-0}{x} = 0$$

$$\frac{\partial u}{\partial y} = \lim_{y\to 0} \frac{u(0,y) - u(0,0)}{y} = \lim_{y\to 0} \frac{0-0}{y} = 0.$$

Similarly,

$$\frac{\partial v}{\partial x} = 0 = \frac{\partial v}{\partial y} \quad \text{at } x = 0, y = 0.$$

Hence Cauchy-Riemann conditions are satisfied at the origin.

Again

$$f'(0) = \lim_{z\to 0} \frac{f(z) - f(0)}{z} = \lim_{\substack{x\to 0\\ y\to 0}} \frac{\sqrt{(|xy|)}}{x + jy} \quad \text{(putting } z = x + jy\text{)}$$

$$= \lim_{x\to 0} \frac{\sqrt{(|mx^2|)}}{x + jmx} \quad \text{if } z \to 0 \text{ along } y = mx$$

$$= \frac{\sqrt{|m|}}{1 + jm}$$

i.e., the value of $f'(0)$ depends on m, so that $f'(0)$ is not unique. Hence $f(z)$ is not analytic at the origin, although Cauchy-Riemann equations are satisfied there. So it is not regular there.

6. Prove that the function $f(z) = u + jv$, where

$$f(z) = \frac{x^3(1+j) - y^3(1-j)}{x^2 + y^2} (z \neq 0), \quad f(0) = 0$$

is continuous and that Cauchy-Riemann equations are satisfied at the origin, yet $f'(z)$ does not exist there.

Ans. Equating real and imaginary parts we have here

$$u = \frac{x^3 - y^3}{x^2 + y^2}, \quad v = \frac{x^3 + y^3}{x^2 + y^2} \quad \text{(when } z \neq 0\text{)}.$$

Herc we see that both u and v are rational and finite for all values of $z \neq 0$; so u and v are continuous at all those points for which $z \neq 0$. Hence $f(z)$ is continuous, where $z \neq 0$.

At the origin $u = 0$, $v = 0$ (since $f(0) = 0$ given). Hence u and v are both continuous at the origin; consequently, $f(z)$ is continuous at the origin, so that $f(z)$ is continuous everywhere.

Now at the origin

$$\frac{\partial u}{\partial x} = \lim_{x\to 0}\frac{u(x,0)-u(0,0)}{x} = \lim_{x\to 0}\left(\frac{x}{x}\right) = 1$$

$$\frac{\partial u}{\partial y} = \lim_{y\to 0}\frac{u(0,y)-u(0,0)}{y} = \lim_{y\to 0}\left(\frac{-y}{y}\right) = -1$$

Similarly

$$\frac{\partial v}{\partial x} = 1 \text{ and } \frac{\partial v}{\partial y} = 1$$

From the above we may write $\frac{\partial u}{\partial x} = \frac{\partial v}{\partial y}$ and $\frac{\partial u}{\partial y} = -\frac{\partial v}{\partial x}$, i.e., Cauchy-Riemann conditions are satisfied at $z = 0$.

Again

$$f'(0) = \lim_{z\to 0}\frac{f(z)-f(0)}{z}$$

$$= \lim_{\substack{x\to 0\\ y\to 0}}\left[\frac{x^3-y^3+j(x^3+y^3)}{x^2+y^2}\cdot\frac{1}{x+jy}\right] \quad \left[\begin{array}{l}\text{putting } z = x+jy\\ \text{and } f(0)=0\end{array}\right]$$

Now if $z \to 0$ along $y = x$, then

$$f'(0) = \lim_{x\to 0}\frac{x^3-x^3+j(x^3+x^3)}{x^2+x^2}\cdot\frac{1}{x+jx} = \lim_{x\to 0}\frac{jx}{x+jx} = \frac{j}{1+j} = \frac{j(1-j)}{2} = \frac{1+j}{2}$$

Again if $z \to 0$ along $y = 0$, we have then

$$f'(0) = \lim_{x\to 0}\frac{x^3(1+j)}{x^3} = 1+j$$

Thus we see that $f'(0)$ is not unique, i.e., its values are not the same as $z \to 0$ along different curves. Hence $f'(z)$ does not exist at the origin.

7. If $u(x,y) = x^2 - y^2$ is the real part of an analytic function $f(z) = u + jv$, find v.

Ans. From the value of u, we have

$$\nabla^2 u = \frac{\partial^2 u}{\partial x^2} + \frac{\partial^2 u}{\partial y^2} = 2 - 2 = 0,$$

i.e., u satisfies Laplace's equation, so that u is a harmonic function.

To find v, let us start with the Cauchy-Riemann condition $\frac{\partial v}{\partial y} = \frac{\partial u}{\partial x}$. Integrating partially w.r.t. y, we obtain

$$v(x,y) = 2xy + g(x), \tag{1}$$

where the constant of integration is independent of y.

To find $g(x)$, we differentiate (1) partially w.r.t. x, whence we have

$$\frac{\partial v}{\partial x} = 2y + \frac{d}{dx}g(x) = -\frac{\partial u}{\partial y} \quad \text{(Cauchy-Riemann condition)}$$

$$= 2y \quad \text{(from the given value of } u\text{)}.$$

Thus

$$v = 2xy + c \quad [\text{putting } g(x) = \text{const.} = c \text{ (say)}]$$

$$\text{or} \quad f(z) = x^2 - y^2 + 2jxy + jc = (x + jy)^2 + c' = z^2 \text{ for } c' = 0 (c' = jc).$$

8. Prove that the function $u = x^3 - 3xy^2 + 3x^2 - 3y^2 + 1$ satisfies Laplace's equation and determine the corresponding regular function $u + jv$.

Ans. Here

$$\frac{\partial u}{\partial x} = 3x^2 - 3y^2 + 6x. \quad \therefore \quad \frac{\partial^2 u}{\partial x^2} = 6x + 6 \tag{1}$$

Again

$$\frac{\partial u}{\partial y} = -6xy - 6y \quad \therefore \quad \frac{\partial^2 u}{\partial y^2} = -6x - 6 \tag{2}$$

Adding (1) and (2), we have $\frac{\partial^2 u}{\partial x^2} + \frac{\partial^2 u}{\partial y^2} = 0$ which is Laplace's equation. Thus u satisfies Laplace's equation, and is therefore a harmonic function. Now Cauchy-Riemann conditions are

$$\frac{\partial u}{\partial x} = \frac{\partial v}{\partial y} \quad \text{and} \quad \frac{\partial u}{\partial y} = -\frac{\partial v}{\partial x}$$

$$\begin{aligned}
\therefore \quad dv &= \frac{\partial v}{\partial x}dx + \frac{\partial v}{\partial y}dy \\
&= -\frac{\partial u}{\partial y}dx + \frac{\partial u}{\partial x}dy \quad \text{(by Cauchy-Riemann equations)} \\
&= (6xy + 6y)dx + (3x^2 - 3y^2 + 6x)dy \\
&= (6xy\ dx + 3x^2 dy) - 3y^2 dy + (6y\ dx + 6x\ dy)
\end{aligned}$$

Integrating, we have $v = 3x^2y - y^3 + 6xy + c$, where c is an absolute or complex constant. Thus

$$\begin{aligned}
u + jv &= (x^3 - 3xy^2 + 3x^2 - 3y^2 + 1) + j(3x^2y - y^3 + 6xy + c) \\
&= (x + jy)^3 + 3(x + jy)^2 + 1 + jc = z^3 + 3z^2 + 1 + jc \\
&= z^3 + 3z^2 + c', \quad \text{where } c' = (1 + c)
\end{aligned}$$

9. Prove that the curves u = constant and v = constant intersect at right angles.

Ans. We know that the curves intersect at right angles if the tangents to them at their points of intersection are at right angles. Differentiating partially the given equations of the curves, we have

$$du = \frac{\partial u}{\partial x}dx + \frac{\partial u}{\partial y}dy = 0 \quad \text{and} \quad dv = \frac{\partial v}{\partial x}dx + \frac{\partial v}{\partial y}dy = 0.$$

Therefore tangents of the angles formed by the tangents drawn to the curves at their points of intersection with the real axes are respectively given by

$$\left(\frac{dy}{dx}\right)_1 = -\frac{\partial u}{\partial x}\Big/\frac{\partial u}{\partial y} \quad \text{and} \quad \left(\frac{dy}{dx}\right)_2 = -\frac{\partial v}{\partial x}\Big/\frac{\partial v}{\partial y}.$$

Now

$$\left(\frac{dy}{dx}\right)_1 \left(\frac{dy}{dx}\right)_2 = \frac{\frac{\partial u}{\partial x}\frac{\partial v}{\partial x}}{\frac{\partial u}{\partial y}\frac{\partial v}{\partial y}} = -1 \quad \text{(by Cauchy-Riemann conditions)}$$

which shows that the two tangents are at right angles if the function is regular, so that the two curves intersect at right angles.

10. Show that $\nabla^2|u|^p = p(p-1)|u|^{p-2}|f'(z)|^2$.

Ans. Here if u be positive $|u| = u$ and if u be negative $|-u| = u$. Considering first u to be positive, we may write

$$\frac{\partial}{\partial x}u^p = pu^{p-1}\frac{\partial u}{\partial x}. \quad \therefore \quad \frac{\partial^2 u^p}{\partial x^2} = p(p-1)u^{p-2}\left(\frac{\partial u}{\partial x}\right)^2 + pu^{p-1}\frac{\partial^2 u}{\partial x^2}$$

Similarly, we have

$$\frac{\partial^2 u^p}{\partial y^2} = p(p-1)u^{p-2}\left(\frac{\partial u}{\partial y}\right)^2 + pu^{p-1}\frac{\partial^2 u}{\partial y^2}$$

Adding the above two results and noting that

$$\nabla^2 \equiv \frac{\partial^2}{\partial x^2} + \frac{\partial^2}{\partial y^2},$$

we have

$$\begin{aligned}\nabla^2 u^p &= pu^{p-1}\nabla^2 u + p(p-1)u^{p-2}\left[\left(\frac{\partial u}{\partial x}\right)^2 + \left(\frac{\partial u}{\partial y}\right)^2\right] \\ &= p(p-1)u^{p-2}\left[\left(\frac{\partial u}{\partial x}\right)^2 + \left(\frac{\partial u}{\partial y}\right)^2\right] \qquad (1)\end{aligned}$$

(since from Laplace's equation $\nabla^2 u = 0$)

Now, if $w = f(z) = u + jv$, then

$$f'(z) = \frac{\partial u}{\partial x} + j\frac{\partial v}{\partial x} \quad \therefore \quad |f'(z)|^2 = \left(\frac{\partial u}{\partial x}\right)^2 + \left(\frac{\partial v}{\partial x}\right)^2 \qquad (2)$$

so that from (1) and (2) it follows that

$$\nabla^2 u^p = p(p-1)u^{p-2}|f'(z)|^2 \quad \text{or,} \quad \nabla^2|u|^p = p(p-1)|u|^{p-2}|f'(z)|^2$$

Again if u be negative say u_1, i.e., $|-u| = u_1$, then as before we may write

$$\nabla^2 u_1^p = p(p-1)u_1^{p-2}|f'(z)|^2 \quad \text{or,} \quad \nabla^2|u_1|^p = p(p-1)|u_1|^{p-2}|f'(z)|^2$$

11. Find the residue of $\frac{1}{(z^2+z_0^2)^2}$ at $z = jz_0$.

Ans. First method. We have

$$f(z) = \frac{1}{(z^2+z_0^2)^2} = \frac{1}{[(z+jz_0)(z-jz_0)]^2}.$$

Here $z = jz_0$ is a pole of order two and the residue at $z = jz_0$ is

$$\lim_{z\to jz_0} \frac{d}{dz}\left[(z-jz_0)^2\frac{1}{(z-jz_0)^2(z+jz_0)^2}\right]$$

$$= \lim_{z\to jz_0} \frac{d}{dz}\left[\frac{1}{(z+jz_0)^2}\right] = \lim_{z\to jz_0}\left[\frac{-2}{(z+jz_0)^3}\right]$$

$$= \frac{-2}{(2jz_0)^3} = \frac{-2}{8(-j)z_0^3} = \frac{1}{4jz_0^3}$$

Second method. Putting $z = jz_0 + t$ in the given function $f(z)$, we have

$$f(z) = \frac{1}{(z^2+z_0^2)^2} = \frac{1}{(z+jz_0)^2(z-jz_0)^2} = \frac{1}{t^2(t+2jz_0)^2}$$

$$= -\frac{1}{4z_0^2t^2}\left(1+\frac{t}{2jz_0}\right)^{-2} = -\frac{1}{4z_0^2t^2}\left[1-\frac{t}{jz_0}+\dots\right]$$

So the coefficient of $\frac{1}{t}$ in the above expression is $\frac{1}{4jz_0^3}$. Hence the residue at $z = jz_0$ is $\frac{1}{4jz_0^3}$.

12. Find the residue of $\frac{2z}{(z+4)(z-1)^2}$ at $z = 1$.

Ans. We have here $f(z) = \frac{2z}{(z+4)(z-1)^2}$.

Here also $z = 1$ is a pole of order two, so that the residue at $z = 1$ is given by

$$\lim_{z\to 1}\left[\frac{d}{dz}\{(z-z_0)^2f(z)\}\right] = \lim_{z\to 1}\frac{d}{dz}\left(\frac{2z}{z+4}\right) = \lim_{z\to 1}\frac{d}{dz}\left[2-\frac{8}{z+4}\right]$$

$$= \lim_{z\to 1}\frac{8}{(z+4)^2} = \frac{8}{25}$$

13. Find the residue of $\frac{z^3}{(z-1)^4(z-2)(z-3)}$ at $z = 1$.

Ans. $f(z) = \frac{z^3}{(z-1)^4(z-2)(z-3)}$ has a pole of order four at $z = 1$. So the residue at $z = 1$ is

$$\lim_{z\to 1}\frac{1}{3!}\frac{d^3}{dz^3}[(z-1)^4f(z)]$$

$$= \lim_{z\to 1}\frac{1}{3!}\frac{d^3}{dz^3}\left[\frac{z^3}{(z-2)(z-3)}\right] = \lim_{z\to 1}\frac{1}{3!}\frac{d^3}{dz^3}\left[z+5-\frac{8}{(z-2)}+\frac{27}{(z-3)}\right]$$

$$= \lim_{z\to 1}\frac{1}{3!}\frac{d^2}{dz^2}\left[1+\frac{8}{(z-2)^2}-\frac{27}{(z-3)^2}\right] = \lim_{z\to 1}\frac{1}{3!}\frac{d}{dz}\left[-\frac{16}{(z-2)^3}+\frac{54}{(z-3)^3}\right]$$

$$= \lim_{z\to 1}\frac{1}{3!}\left[\frac{48}{(z-2)^4}-\frac{162}{(z-3)^4}\right] = \frac{1}{6}\left[48-\frac{162}{16}\right] = 8-\frac{27}{16} = \frac{101}{16}$$

14. Find the residues at the poles of the function $\frac{z^4}{(c^2+z^2)^4}$.

Ans. Here

$$f(z) = \frac{z^4}{(c^2+z^2)^4} = \frac{z^4}{(z-jc)^4(z+jc)^4}.$$

Evidently $z = jc$ and $z = -jc$ are the two poles each of order 4.

To find the residue at the pole $z = jc$, let us put $z = jc + t$, then we may write

$$f(z) = \frac{(t+jc)^4}{t^4(2jc+t)^4} = \frac{(t+jc)^4\left(1+\frac{t}{2jc}\right)^{-4}}{t^4(2jc)^4}$$

Now

$$\begin{aligned}\frac{1}{t^4}(t+jc)^4 &= \frac{1}{t^4}(t^4+4jct^3-6c^2t^2-4jc^3t+c^4)\\ &= 1+\frac{4jc}{t}-\frac{6c^2}{t^2}-\frac{4jc^3}{t^3}+\frac{c^4}{t^4}\end{aligned}$$

Again

$$\frac{1}{(2jc)^4}\left(1+\frac{t}{2jc}\right)^{-4} = \frac{1}{16c^4}\left[1-\frac{4t}{2jc}+\frac{10}{(2jc)^2}t^2-\frac{20}{(2jc)^3}t^3+\cdots\right]$$

Therefore the coefficient of $\frac{1}{t}$ in $f(z)$ is

$$\frac{1}{16c^4}\left(4jc-12jc+10jc-\frac{5}{2}jc\right) = -\frac{j}{32c^3}$$

which is the residue at the pole $z = jc$.

Again writing $-c$ for c in the above result, we have the residue at the other pole $z = -jc$, given by $\frac{j}{32c^3}$.

15. Show that $f(z) = \frac{1}{z^4+2z^2+1}$ has two double poles.

Ans. $f(z)$ is analytic everywhere except at the points, where

$z^4+2z^2+1 = 0$, i.e., $(z^2+1)^2 = 0$, i.e., $(z+j)^2(z-j)^2 = 0$, i.e., $z = j, j, -j, -j$.

Now

$$\begin{aligned}\lim_{z\to j}(z-j)^2\frac{1}{z^4+2z^2+1} &= \lim_{z\to j}(z-j)^2\frac{1}{(z-j)^2(z+j)^2}\\ &= \frac{1}{(2j)^2} = -\frac{1}{4} \quad \text{(non-zero and finite)}\end{aligned}$$

$\therefore$ $z = j$ is a double pole. Similarly

$$\begin{aligned}\lim_{z\to -j}(z+j)^2\frac{1}{z^4+2z^2+1} &= \lim_{z\to -j}(z+j)^2\frac{1}{(z+j)^2(z-j)^2}\\ &= \frac{1}{(-2j)^2} = -\frac{1}{4} \quad \text{(non-zero and finite)}\end{aligned}$$

So $z = -j$ is also a double pole.

Exercises

1. Explain, with an example, what you mean by a complex function.

2. Explain, with an example, what you mean by an analytic function. Show that the function $f(z) = |z|^2$ has a unique derivative at the origin, but nowhere else.

3. Derive the Cauchy-Riemann conditions for a function of a complex variable to be analytic.

4. Show that the function $f(z) = xy + jy$ is everywhere continuous but is not analytic.

[**Ans.** See worked out examples.]

5. If $u(x, y) = x^2 - y^2$ is the real part of an analytic function $f(z) = u + jv$, find v.

[**Ans.** See worked out examples.]

6. Show that the functions $u = \frac{1}{2}\log(x^2 + y^2)$ and $v = \cosh x \cosh y$ are harmonic functions.

[**Ans.** Show that they satisfy Laplace's equation.]

7. Prove that the function $f(z) = u + jv$, where $f(z) = \frac{x^3(1+j)-y^3(1-j)}{x^2+y^2} (z \neq 0)$, $f(0) = 0$ is continuous and that Cauchy-Riemann conditions are satisfied at the origin, yet $f'(z)$ does not exist there.

[**Ans.** See worked out examples.]

8. Show that the function $f(z) = e^{-z^{-4}} (z \neq 0), f(0) = 0$ is not analytic at $z = 0$, although Cauchy-Riemann conditions are satisfied at that point. How would you explain this? [**Ans.** See worked out examples.]

9. Examine the nature of the function $f(z) = \frac{x^2y^5(x+jy)}{x^4+y^{10}} (z \neq 0), f(0) = 0$ in the region including the origin.

[**Ans.** See worked out examples.]

10. Show that the function $f(z) = \sqrt{|xy|}$ is not regular at the origin, although Cauchy-Riemann equations are satisfied at the point. [**Ans.** See worked out examples.]

11. Prove that the function $u = x^3 - 3xy^2 + 3x^2 - 3y^2 + 1$ satisfies Laplace's equation and determine the corresponding regular function $u + jv$. [**Ans.** See worked out examples.]

12. Prove that the curves $u =$ constant and $v =$ constant intersect at right angles.

[**Ans.** See worked out examples.]

13. Show that, if $f(z) = u + jv$ is an analytic function of $z = x + jy$,

$$\nabla^2|u|^p = p(p-1)|u|^{p-2}|f'(z)|^2.$$ [**Ans.** See worked out examples.]

14. What is meant by singularity of a function $f(z)$ at a point? Explain essential singularity and non-essential singularity in this context. Explain what you mean by pole of a function $f(z)$.

15. Explain what you mean by a simple pole and a pole of order m.

16. Explain what you mean by residue of $f(z)$ at $z = z_0$.

17. Discuss the methods of finding (a) the residue at a simple pole, (b) the residue at a pole of order m.

18. Show that the residue of a function $f(z)$ at infinity is the negative of the coefficient of $\frac{1}{z}$ in the expansion of $f(z)$ for values of z in the neighbourhood of $z = \infty$. [**Ans.** See Art. 21.12.]

19. (a) Find the residue of $\frac{1}{(z^2+z_0^2)^2}$ at $z = jz_0$. [**Ans.** See worked out examples.]

(b) Find the residue of $\frac{z^2}{z^2+z_0^2}$ at $z = jz_0$. [Proceed as in (a), **Ans.** $\frac{z_0 j}{2}$.]

20. Find the residue of $\frac{2z}{(z+4)(z-1)^2}$ at $z = 1$. **[Ans.** See worked out examples.]

21. (a) Find the residue of $\frac{z^3}{(z-1)^4(z-2)(z-3)}$ at $z = 1$. **[Ans.** See worked out examples.]

(b) Find the residue of $\frac{z^4}{(z-1)^4(z-2)(z-3)}$ at $z = 1$. [Proceed as in (a), **Ans.** 175/16.]

22. Find the residues at the poles of the function $\frac{z^4}{(c^2+z^2)^4}$. **[Ans.** See worked out examples.]

23. Show that $f(z) = \frac{1}{z^4+2z^2+1}$ has two double poles. **[Ans.** See worked out examples.]

24. Find the residue of $\frac{z}{(z-a)(z-b)}$ at infinity.

[Ans. Residue of a function $f(z)$ at infinity is given by $\lim_{z\to\infty}[-zf(z)]$

$$\therefore \quad \lim_{z\to\infty}\left[-\frac{z^2}{(z-a)(z-b)}\right] = \lim_{z\to\infty}\left[\frac{-1}{\left(1-\frac{a}{z}\right)\left(1-\frac{b}{z}\right)}\right] = -1.]$$

25. Find the residue of $\frac{z^3-z^2+1}{z^3}$ at infinity.

[Ans. In this case $\lim_{z\to\infty}[-zf(z)] = \lim_{z\to\infty}\left[-\frac{z^3-z^2+1}{z^2}\right) = \infty$ and hence the required limit does not exist. Now, let us take $f(z)$ as $f(z) = \frac{1}{z^3} - \frac{1}{z} + 1$ in which the coefficient of $(+\frac{1}{z})$ is -1. But the negative of the coefficient of $\frac{1}{z}$ in the expansion of $f(z)$ for values of z in the neighbourhood of $z = \infty$ is the residue at infinity (see Art. 21.12). Therefore, the required residue is 1.]

26. (a) Given a function $f(x)$, assume that the Taylor series for $f(x)$,

$$f(x) = a_0 + a_1(x-a) + a_2(x-a)^2 + a_3(x-a)^3 + \cdots + a_n(x-a)^n + \cdots$$

Calculate $f(a), f'(a), f''(a) \cdots f^n(a)$ where $f^n(x)$ is the nth derivative of $f(x)$ at a.

(b) Write the Taylor series of $f(x)$ about $x = a$.

(c) Hence define an analytic function in terms of expansion in Taylor series.

[C.U. (Hons.) 2006]

Ans. (i)

$$f(x) = a_0 + a_1(x-a) + a_2(x-a)^2 + a_3(x-a)^3 + \cdots + a_n(x-a)^n + \cdots \quad (1)$$

Putting $x = a$ in (1) we have $f(a) = a_0$.

Differentiating (1) we have

$$f'(x) = a_1 + 2a_2(x-a) + 3a_3(x-a)^2 + \cdots \quad (2)$$

Putting $x = a$, we have $f'(a) = a_1$

Differentiating (2) we have

$$f''(x) = 2a_2 + 6a_3(x-a) + 12a_4(x-a)^2 + \cdots \quad (3)$$

Putting $x = a$, we get $f''(a) = 2a_2$

Differentiating (3), we have

$$f'''(x) = 6a_3 + 24a_4(x-a) + \cdots \quad (4)$$

Putting $x = a$, we get, $f'''(a) = 6a_3$ and so on

For answers to (b) and (c), see Art. 21.9.

27. (a) Prove that $|\sin(z)| \geq |\sin(x)|$, where $z = x + iy$.

(b) Find the real part, modulus, and argument of

$$f(z) = \frac{1+z}{1-z}, \quad \text{where } z = x + iy$$

(c) Expand $\frac{1}{x-2}$ in a Taylor series about the point $x = 1$. What is the radius of convergence of this series? **[C.U. (Hons.) 2008]**

For answers to (a) and (b) see Arts. 21.1, 21.2 and 21.3.

For answer to (c) see Art. 21.9.

28. Test whether the function z^2 is anatilytic. **[C.U. (Hons.) 2007]**

(See worked out examples)

Chapter 22

Functions of Several Real Variables

22.1 Real-valued Function of a Real Variable

We suppose that x is a real variable over the domain S (a set of real numbers), that is, x can assume as its value any member of S. Now for each value of x belonging to S, let there exist a *unique real number* y. Then we say that y is a *real-valued function* of the real variable x whose domain is S. The domain of y is then called the *range of the function*. Here x is often called the *independent* variable and y, the *dependent* variable and we say that the *dependent variable* y *is a function* f *of the independent variable* x—written as $y = f(x)$ over $S. f(x)$ is also interpreted as the functional value at x. For example, electrical resistance r of a copper rod depends on its temperature (θ)—it is expressed as $r = f(\theta)$. The function f here can be expressed in algebraic or exponential or trigonometric or hyperbolic or parametric form or in any other suitable form. Students are supposed to be familiar with the methods (given in any standard book of *Differential Calculus*) for finding the existence of limit, conditions of continuity and differentiability and actual first and higher order differential coefficients of such functions.

22.2 Functions in Higher Dimensional Space, i.e., Functions of more than One Variable

From the above, we are now acquainted with the meaning of the function of a single real variable. We now extend this idea to define a function of several real variables or what may be considered as, a function of a point in Euclidean vector space R^n. However, our main concern will be with *two and three dimensional spaces*.

[**N.B. Euclidean Vector Space R^n:** (a) A point in two-dimensional space is an *ordered pair* of real numbers (x, y), represented by a point P in the XY-plane. The set of all such two-dimensional points forms what we call *a two-dimensional Euclidean space* R^2.

(b) A point in three-dimensional space is an *ordered triple* of three real numbers (x, y, z)—represented by a point P in our usual XYZ-space. The set of all such three-dimensional points forms what we call *the three-dimensional Euclidean space* R^3.

(c) We can generalise the above notion and consider an *ordered n-tuple* of real numbers $(x_1, x_2, x_3, \cdots, x_n)$ and call it a point P in n-dimensional space. The set of all such n-dimensional points is called *n-dimensional Euclidean space* R^n.

We often write $\vec{x} = (x_1, x_2, \cdots, x_n)$ and call it *a point* $\vec{x} \in R^n$ with n co-ordinates $x_1, x_2, \cdots, x_n$ or *a vector* $\vec{x}$ with n-components $x_1, x_2, \cdots, x_n$ (x_k is the kth component of the vector $\vec{x}$ or kth co-ordinate of the point $\vec{x}$).

Higher dimensional spaces arise in the study of Relativity, Statistical and Quantum Mechanics. Even infinite-dimensional spaces occur in Quantum Mechanics.]

Definition. Let S_1 be a set of real numbers and S_2 a set of points in an Euclidean Vector Space R^n. We suppose that for each point $P \in S_2$, there corresponds a unique value w in S_1. The set of all pairs (P, w), where $P \in S_2$ and w is the corresponding value in S_1, is called a *function*, denoted by f, defined on R^n.

The functional values are denoted by $w = f(P)$ or equivalently $w = f(x, y)$, $w = f(x, y, z)$ $\cdots$ $w = f(x_1, x_2, x_3, \cdots, x_n)$ according to the dimensions of the space. The set of P's for which f is defined is called the *domain of the function*, denoted by D. The set of functional values (i.e., the set of w's which correspond to some P in D) is called the *range* of f—denoted by $R(f)$.

Examples of functions of two or three variables

(a) $z = f(x, y) = x + y$ is defined over the entire XY-plane.

(b) $z = f(x, y) = \frac{xy}{x^2+y^2}$ is defined over the entire XY-plane, excluding the only point $(0, 0)$.

(c) $z = f(x, y) = \sqrt{x + y - 1}$ is defined over the region $x + y \geq 1$, which includes the points of the line $x + y = 1$.

(d) $z = f(x, y) = \frac{1}{\sqrt{x+y-1}}$ is defined over the region $x + y > 1$, which does not include the points of the line $x + y = 1$ (open region).

(e) $z = \sqrt{1 - x^2 - y^2}$ is defined over the *closed* circular region $x^2 + y^2 \leq 1$.

(f) $z = \log(1 - x^2 - y^2)$ is defined over the region $x^2 + y^2 < 1$ ($\log 0$ is undefined; so points on the circle $x^2 + y^2 = 1$ are to be excluded from the circular domain) (open region).

(g) *Linear function.* $u = f(x, y) = ax + by + c$ is defined for all points (x, y) in the XY-plane.

General polynomial in x and y contains a sum of a finite number of terms like $a_{mn}x^m y^n$ (m, n are positive integers or zero). All such polynomials are defined over the entire XY-plane.

(h) *Rotational functions.* We consider the quotient of two polynomials $f(x, y)/g(x, y)$. They are defined where $g(x, y) \neq 0$, as for example

$$u = \frac{ax + by + c}{\alpha x + \beta y + \gamma}$$

is defined for all points (x, y) in the XY-plane, excepting points which lie on the line $\alpha x + \beta y + \gamma = 0$.

(i) *Algebraic functions.* $z_1 = \sqrt{(x-2)(y+3)}$; $z_2 = \sqrt{x^2 + 2y^2 - 4}$.

Here the function z_1 is defined at all ponts (x, y) except where the expression under the radical sign is negative. Thus z_1 is defined in the closed region formed by the first and third quadrants into which the lines $x = 2, y = -3$ divide the plane as shown by the shaded region in the attached figure (Fig. 22.2.1).

The function z_2 is defined where $x^2 + 2y^2 \geq 4$, i.e., for points on or outside the ellipse $\frac{x^2}{4} + \frac{y^2}{2} = 1$.

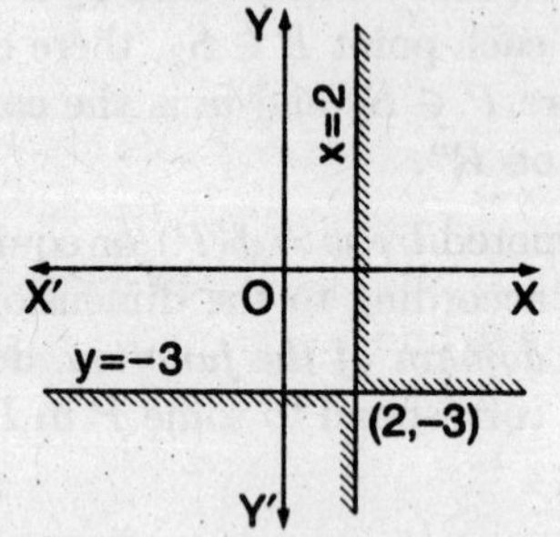

Fig 22.2.1

(j) $f(x, y, z) = x^2y + y^2z + z^2x$ is a function of three independent variables x, y, z. It is defined for all points $(x, y, z) \in R^3$.

In the study of *Thermodynamics* and *Behaviour of Real Gases*, we very often come across functions of real variables. For example, the thermodynamic state of a gaseous system can be specified by its pressure p, volume V and temperature T. But out of these three real variables p, V, T, only two are dependent variables and the third may be considered as their function. For a homogeneous body in the solid, liquid or gaseous state, the general equation of state may be expressed as

$$p = f(V, T), \quad \text{or,} \quad f(p, V, T) = 0. \tag{1}$$

The form of the above equation is precisely known only in the case of a perfect gas, viz., $pV = RT$. Though the nature of the relationship between the above-mentioned real variables is not precisely known for other eases, still we can proceed a good deal towards the study of the behaviour a body or a system, through methods of differential calculus (as will be discussed in the following articles), if only the meaning of the above mathematical function is properly understood.

22.3 Limit of a Function of Two Variables: Simultaneous Limit or Double Limit

A function f of two independent variables x and y, is said to tend to a limit A as the point (x, y) tends to a given point (a, b), if, to each $\varepsilon > 0$, there can be found one $\delta > 0$, such that $|f(x, y) - A| < \varepsilon$, for all points (x, y) of the domain which belongs to some δ-neighbourhood N

of (a, b). The inequality may not be satisfied when $x = a$, $y = b$.

$$\left[\begin{array}{l} N \text{ may be a } \textit{square} \text{ neighbourhood like } 0 < |x-a| < \delta, 0 < |y-b| < \delta \\ \text{or, } N \text{ may be a } \textit{circular} \text{ neighbourhood like } 0 < (x-a)^2 + (y-b)^2 < \delta^2 \\ \text{or, } N \text{ may be any other neighbourhood} \end{array}\right]$$

We may then write $\lim\limits_{(x,y)\to(a,b)} f(x, y) = A$.

Illustrative Examples

1. Let $f(x, y) = \frac{x^2y^2}{x^2+y^2}$. Using ε-δ definition, establish $\lim\limits_{(x,y)\to(0,0)} \dfrac{x^2y^2}{x^2+y^2} = 0$.

Ans. Let $\varepsilon > 0$ be given. We are to find a value of $\delta > 0$, such that in some δ-neighbourhood of $(0,0)$ for all points (x, y),

$$\left|\frac{x^2y^2}{x^2+y^2} - 0\right| < \varepsilon \quad \text{or,} \quad \frac{x^2y^2}{x^2+y^2} < \varepsilon$$

Now, clearly $x^2 < x^2 + y^2, y^2 < x^2 + y^2$, so that

$$\frac{x^2y^2}{x^2+y^2} < \frac{(x^2+y^2)^2}{x^2+y^2} = x^2 + y^2 < \varepsilon, \text{ if } 0 < x^2 + y^2 < \delta^2 \text{ where } \delta = \sqrt{\varepsilon}.$$

Thus we have found out $\delta = \sqrt{\varepsilon}$ to satisfy the requirements of the definition. Therefore

$$\lim_{(x,y)\to(0,0)} \frac{x^2y^2}{x^2+y^2} = 0.$$

2. Establish that $\lim\limits_{(x,y)\to(0,0)} xy\dfrac{x^2-y^2}{x^2+y^2} = 0$.

Ans. Let $\varepsilon > 0$ be given. To find a δ-neighbourhood of $(0,0)(\delta > 0)$ such that in that neighbourhood N, we have

$$\left|xy\frac{x^2-y^2}{x^2+y^2} - 0\right| < \varepsilon, \text{ i.e., } \left|xy\frac{x^2-y^2}{x^2+y^2}\right| < \varepsilon$$

But

$$\left|xy\frac{x^2-y^2}{x^2+y^2}\right| = |x||y|\frac{|x^2-y^2|}{|x^2+y^2|}$$

and clearly $|x| < \sqrt{x^2+y^2}$, $|y| < \sqrt{x^2+y^2}$ and $\frac{|x^2-y^2|}{|x^2+y^2|} < 1$.

$$\therefore \quad \left|xy\frac{x^2-y^2}{x^2+y^2} - 0\right| = |x||y|\frac{|x^2-y^2|}{|x^2+y^2|} < \sqrt{x^2+y^2}.\sqrt{x^2+y^2} = x^2 + y^2 < \varepsilon$$

$$\text{i.e.,} \quad \left|xy\frac{x^2-y^2}{x^2+y^2} - 0\right| < \varepsilon, \text{ if } 0 < x^2 + y^2 < \delta^2 \text{ where } \delta = \sqrt{\varepsilon}.$$

$\therefore$ the requirements of the definition of limit are met and hence the limit exists and is equal to zero.

Non-existence of Limit: Let

$$f(x, y) = \frac{xy}{x^2 + y^2}.$$

The domain of f is the whole XY-plane punctured at the origin $(0, 0)$. For existence of limit, we are to examine the values of f near $(0, 0)$. Introducing polar co-ordinates $x = \rho\cos\theta$, $y = \rho\sin\theta$, f takes the form

$$f(x, y) = \frac{xy}{x^2 + y^2} = \frac{\rho\cos\theta\cdot\rho\sin\theta}{\rho^2} = \frac{1}{2}\sin 2\theta.$$

Here, of course, $\rho = \sqrt{x^2 + y^2}$, and as $(x, y) \to (0, 0), \rho \to 0$.

From the above, it is clear that f is constant on each straight line through the origin $(y = x\tan\theta)$ and that, in general, the constant is different on different lines. Clearly then, the values of f cannot be made close to any constant A by simply restricting ρ. Thus

$$\lim_{(x,y)\to(0,0)} \frac{xy}{x^2 + y^2}$$

does not exist in this case.

N.B. It is easy to see that if $\lim\limits_{(x,y)\to(a,b)} f(x, y) = A$, and if $y = \phi(x)$ be any function of x whatsoever, such that $\phi(x) \to b$ as $x \to a$, then $f\{x, \phi(x)\} \to A$ as $x \to a$.

So, if we can determine two functions $\phi_1(x)$, $\phi_2(x)$ such that the limits of $f\{x, \phi_1(x)\}$ and $f\{x, \phi_2(x)\}$ are different as $x \to a$, then we can conclude that the simultaneous limit $\lim\limits_{(x,y)\to(a,b)} f(x, y) = A$ *does not exist.*

Illustrative Examples

1. Let

$$f(x, y) = \begin{cases} 1 & \text{when } xy \neq 0 \\ 0 & \text{when } xy = 0 \end{cases}$$

To show that $\lim\limits_{(x,y)\to(0,0)} f(x, y)$ does not exist, we choose $y = 0 = \phi_1(x)$ and $y = x = \phi_2(x)$, so that $f\{x, \phi_1(x)\} = f(x, 0) = 0$ and $f\{x, \phi_2(x_1)\} = f(x, x) = 1$ since $f(x, 0) \to 0$ as $x \to 0$ and $f(x, x) \to 1$ as $x \to 0$.

$\therefore \quad \lim\limits_{(x,y)\to(0,0)} f(x, y)$ does not exist.

2. Let $f(x, y) = (x + y)\frac{y+(x+y)^2}{y-(x+y)^2}$.

To show that $\lim\limits_{(x,y)\to(0,0)} f(x, y)$ does not exist, we choose $y = 0 = \phi_1(x)$ and $y = x^2 = \phi_2(x)$, so that

$$f\{x, \phi_1(x)\} = f(x, 0) = -x$$

$$\text{and} \quad f\{x, \phi_2(x)\} = f(x, x^2) = \frac{(1 + x)(2 + 2x + x^2)}{-(2 + x)}.$$

Again, $\lim\limits_{x\to 0} f(x, 0) = 0$ and $\lim\limits_{x\to 0} f(x, x^2) = -1$, so that the required limit does not exist.

22.4 Repeated or Iterated Limits

Let $f(x, y)$ be defined in a certain neighbourhood of (a, b). Then $\lim\limits_{x \to a} f(x, y), y$ remaining constant, when exists, will be different for different values of y, and in fact, this limit will be a function of y, say $\phi(y)$.

If, then, $\lim\limits_{y \to b} \phi(y)$ exists and is equal to l_1, we may write

$$\lim_{y \to b} \left\{ \lim_{x \to a} f(x, y) \right\} = l_1.$$

l_1 is called *repeated* or *iterated* limit of $f(x, y)$ as $x \to a$ and then $y \to b$.

A change in the order of obtaining limits may produce a change in the final result, say l_2 instead of l_1.

Thus we may obtain another repeated limit l_2 (say) written as

$$\lim_{x \to a} \left\{ \lim_{y \to b} f(x, y) \right\} = l_2.$$

There is no reason why l_1 should be always equal to l_2. What is worth noticing is that even when both repeated limits exist and have the same value, the simultaneous or double limit may not exist.

Illustrative Examples

1. Verify that the double limit $\lim\limits_{(x,y) \to (0,0)} \dfrac{x + y}{x - y}$ does not exist.

Ans. Along the line $y = mx$, $f(x, y) = \frac{x+y}{x-y} = \frac{1+m}{1-m}$.

Thus, by setting $(x, y) \to (0, 0)$ along a suitable line, $f(x, y)$ will approach any value k specified conveniently. Hence $f(x, y)$ does not tend to a unique limit as $(x, y) \to (0, 0)$.

But we observe that both repeated limits exist, although they have different values.

$$\lim_{y \to 0} \left\{ \lim_{x \to 0} \left(\frac{x + y}{x - y} \right) \right\} = \lim_{y \to 0} \left(\frac{y}{-y} \right) = -1$$

Again

$$\lim_{x \to 0} \left\{ \lim_{y \to 0} \left(\frac{x + y}{x - y} \right) \right\} = \lim_{x \to 0} \left(\frac{x}{x} \right) = 1.$$

Even when both repeated limits exist and have the same value, the double limit may not exist. The following example will illustrate this fact.

2. Verify that for $f(x, y) = \frac{xy}{xy + x - y}$, repeated limits exist and are equal, but still the double limit does not exist.

Ans. Here

$$\lim_{x \to 0} \left\{ \lim_{y \to 0} f(x, y) \right\} = \lim_{x \to 0} \left(\frac{0}{0 + x} \right) = 0$$

Again

$$\lim_{y \to 0} \left\{ \lim_{x \to 0} f(x, y) \right\} = \lim_{y \to 0} \left(\frac{0}{0 - y} \right) = 0$$

Thus the repeated limits exist and are equal

But to show that $\lim_{(x,y)\to(0,0)} f(x,y)$ does not exist, we choose $\phi_1(x)$ as $y = 0$ and $\phi_2(x)$ as $y = x$.

$$\therefore \quad f\{x, \phi_1(x)\} = 0 \text{ as } x \to 0 \text{ and } f\{x, \phi_2(x)\} = \frac{x^2}{x^2+0} = 1 \text{ as } x \to 0$$

$$\therefore \quad \lim_{x\to 0} f\{x, \phi_1(x)\} = 0 \text{ and } \lim_{x\to 0} f\{x, \phi_2(x)\} = 1$$

So the double limit does not exist.

N.B. From the above what we have observed is that *in case the double limit exists, the two repeated limits, if exist, must necessarily be equal but the converse is not true.* However, *if the repeated limits are not equal, then the double limit cannot exist.*

22.5 Continuity in R^2 for a function f of two variables x and y at a point (a, b) in the domain D of definition of the function f

We assume that (a, b) is a limit point of D, so that in any neighbourhood of (a, b), there exist an infinite number of points of D where f is defined. Now f is said to be continuous at (a, b), if to each $\varepsilon > 0$, there exists a positive number δ, such that at all points (x, y) in some δ-neighbourhood N of (a, b)—say in $(x-a)^2 + (y-b)^2 < \delta^2$ or in $|x-a| < \delta$ and $|y-b| < \delta$, we have

$$|f(x,y) - f(a,b)| < \varepsilon \tag{1}$$

N.B. The above condition (22.5.1) is satisfied if $(a, b) \in D$ but not a limit point of D, i.e., if (a, b) is an isolated point of D. If (a, b) is a limit point of D, then f is continuous at (a, b) if $\lim_{(x,y)\to(a,b)} f(x,y) = f(a,b)$ or for $(a+h, b+k) \in D$, $\lim_{(h,k)\to(0,0)} f(a+h, b+k) = f(a,b)$.

A function f is said to be continuous in a certain region if it is continuous at every point of the region in the sense described above.

Salient Points to remember about Continuity

A. If $\lim_{(x,y)\to(a,b)} f(x,y)$ does not exist, the question of continuity at (a, b) does not arise at all.

B. If $f(x, y)$ is continuous at (a, b) and $f(a, b) \neq 0$, then there exists a certain neighbourhood of (a, b) such that $f(x, y)$ has the same sign as that $f(a, b)$ for every point of that neighbourhood. As an immediate consequence, we get that if $f(x, y)$ is zero in every neighbourhood of (a, b) where $f(x, y)$ is continuous, then $f(a, b)$ is equal to zero.

C. If a function f of two variables x, y is continuous at (a, b), then the function $f(x, b)$, which is a function of a single variable x (y remaining constant at b) is continuous at $x = a$. Similarly, $f(a, y)$, a function of a single variable y is continuous at $y = b$.

D. If f and g are continuous functions in a region R, then $f \pm g$ and fg are also continuous, and f/g is continuous at all points where $g \neq 0$.

Polynomials are everywhere continuous: Rational functions f/g are continuous except where $g = 0$. Continuous functions of continuous functions are continuous—thus $\sin xy$ and $e^{-(x^2+y^2)}$ are continuous.

E. A function f of two variables x and y is said to be uniformly continuous in a region R, open or closed, when for any $\varepsilon > 0$, we can find a value $\delta > 0$, so that if (x_1, y_1) and (x_2, y_2) are any two points of R for which $|x_1 - x_2| < \delta$ and $|y_1 - y_2| < \delta$, then

$$|f(x_1, y_1) - f(x_2, y_2)| < \varepsilon$$

A function $f(x, y)$ which is continuous in a closed region R, is bounded there and takes its least and greatest values at least once in the region R.

For proofs of all the above points, students are advised to consult any standard book on *Differential Calculus.*

Illustrative Examples on Continuity

1. Show that the function

$$f(x,y) = \begin{cases} xy\dfrac{x^2 - y^2}{x^2 + y^2} & \text{where } x^2 + y^2 \neq 0 \\ 0 & \text{where } x = 0, y = 0 \end{cases}$$

is continuous at $(0, 0)$.

Ans.

$$|f(x,y) - f(0,0)| = |xy|\frac{|x^2 - y^2|}{|x^2 + y^2|} < |x||y| < \sqrt{x^2 + y^2}\sqrt{x^2 + y^2} < \varepsilon$$

when $x^2 + y^2 < \delta^2, \delta = \sqrt{\varepsilon}$ [see Ex.2 of art. 22.3].

$\therefore \quad \lim\limits_{(x,y)\to(0,0)} f(x,y) = f(0,0)$, i.e., f is continuous at $(0, 0)$.

2. Prove that the function

$$f(x,y) = \begin{cases} \dfrac{xy}{\sqrt{x^2 + y^2}}, & (x,y) \neq (0,0) \\ 0, & (x,y) = (0,0) \end{cases}$$

is continuous at $(0, 0)$.

Ans. Let $x = r\cos\theta, y = r\sin\theta$.

Then

$$\left|\frac{xy}{\sqrt{x^2 + y^2}} - 0\right| = r|\cos\theta\sin\theta| \leq r = \sqrt{x^2 + y^2} < \varepsilon,$$

if $x^2 < \frac{\varepsilon^2}{2}, y^2 < \frac{\varepsilon^2}{2}$, i.e., if $|x| < \frac{\varepsilon}{\sqrt{2}}, |y| < \frac{\varepsilon}{\sqrt{2}}$.

$$\therefore \quad \lim_{(x,y)\to(0,0)} \frac{xy}{\sqrt{x^2 + y^2}} = 0 = f(0,0)$$

so that f is continuous at $(0, 0)$.

3. Prove that the function

$$f(x,y) = \begin{cases} \dfrac{x^3 + y^3}{x - y}, & \text{when } x \neq y \\ 0, & \text{when } x = y \end{cases}$$

is not continuous at $(0, 0)$.

Ans. We shall prove that $\lim\limits_{(x,y)\to(0,0)} f(x,y)$ does not exist and as such $f(x,y)$ cannot be continuous at $(0, 0)$.

$$f(x,y) = \frac{x^3+y^3}{x-y} = \frac{x^3+(x-mx^3)^3}{mx^3} = \frac{1+(1-mx^2)^3}{m}$$

(putting $y = x - mx^3$, so that as $x \to 0$, $y \to 0$)

$$\therefore \quad \lim_{(x,y)\to(0,0)} = \lim_{x\to 0} \frac{1+(1-mx^2)^3}{m} = \frac{2}{m}$$

so that the limit is different for different choices of m, i.e., the limit can become any arbitrary value k specified conveniently and is, therefore, not unique.

$\therefore$ the limit does not exist and f is not continuous at $(0, 0)$.

22.6 Directional Derivative : Partial Derivative—Differentiability and Differential

Let $Q(x+h, y+k)$ be a point in the neighbourhood N of $P(x, y)$ and α be the angle that the line joining (x, y) and $(x+h, y+k)$ makes with the positive direction of X-axis. If $\rho = \sqrt{h^2+k^2}$, then $h = \rho\cos\alpha$, $k = \rho\sin\alpha$ and as $Q \to P$, $\rho \to 0$.

Then

$$\lim_{\rho\to 0} \frac{f(x+\rho\cos\alpha, y+\rho\sin\alpha) - f(x,y)}{\rho}$$

if exists, is called the *derivative* of $f(x, y)$ at (x, y) in the direction α and is denoted by $D_\alpha f(x, y)$.

If $\alpha = 0$, the derivative is denoted by $\frac{\partial f}{\partial x}$ and when $\alpha = \frac{\pi}{2}$ the derivative is $\frac{\partial f}{\partial y}$. These are the partial derivatives of f at (x, y) w.r.t. x and y respectively.

Thus

$$\lim_{h\to 0} \frac{f(x+h,y) - f(x,y)}{h} \quad \text{if exists, is } \frac{\partial f}{\partial x}$$

$$\text{and} \quad \lim_{k\to 0} \frac{f(x,y+k) - f(x,y)}{k} \quad \text{if exists, is } \frac{\partial f}{\partial y}.$$

A function $f(x, y)$ is said to be differentiable at (x, y) if it is well defined in the neighbourhood of (x, y) and the increment $\Delta f = f(x+h, y+k) - f(x, y)[(x+h, g+k)$ being a point in the neighbourhood of (x, y), where $f(x, y)$ is defined] can be expressed in the form

$$\Delta f = Ah + Bk + \varepsilon\rho, \tag{1}$$

where A and B are independent of h, k and $\varepsilon \to 0$ as $\rho \to 0$ $\left(\rho = \sqrt{h^2+k^2}\right)$. If the function is differentiable at (x, y), $\frac{\partial f}{\partial x}$ and $\frac{\partial f}{\partial y}$ exist at (x, y) and $A = \frac{\partial f}{\partial x}$, $B = \frac{\partial f}{\partial y}$. Moreover, the function is continuous at (x, y). Thus, differentiability implies continuity. Thus if a function is not continuous at a point, it cannot be differentiable there.

Illustrative Examples

1. The function

$$f(x,y) = \begin{cases} \dfrac{xy}{x^2+y^2}, & \text{when } x^2+y^2 \neq 0 \\ 0, & \text{when } x^2+y^2 = 0 \end{cases}$$

is not differentiable at $(0,0)$, because the function is not continuous at $(0,0)$.

Let $(x,y) \to (0,0)$ along $y = mx$; then we have

$$\lim_{(x,y)\to(0,0)} f(x,y) = \lim_{x\to 0} \frac{x^2 m}{(1+m^2)x^2} = \frac{m}{1+m^2}$$

which being a function of m has different values for different m, so that $\lim\limits_{(x,y)\to(0,0)} f(x,y)$ does not exist and hence the question of continuity does not arise. The function is, therefore, not differentiable at $(0,0)$ for differentiability would imply continuity.

2. $f(x,y) = |x| + |y|$ is continuous but not differentiable at $(0,0)$.

Here $f(0,0) = 0$ and $|f(x,y) - f(0,0)| = |x| + |y| < \varepsilon$ if $|x| < \delta = \frac{\varepsilon}{2}$ and $|y| < \delta = \frac{\varepsilon}{2}$.

$\therefore \quad f(x,y)$ is continuous at $(0,0)$.

But $\lim\limits_{x\to 0} \dfrac{f(x,0) - f(0,0)}{x} = \lim\limits_{x\to 0} \dfrac{|x|}{x}$ does not exist.

$\therefore$ partial derivatives at $(0,0)$ do not exist.

If the function were differentiable, the partial derivatives would exist. From (22.6.1) we can write

$$\Delta f = \frac{\partial f}{\partial x} h + \frac{\partial f}{\partial y} k + \varepsilon\rho$$

The expression $\frac{\partial f}{\partial x} h + \frac{\partial f}{\partial y} k$ is called the *differential* of f and is denoted by df or the total differential of f.

Thus differentiability would imply $\lim\limits_{\rho\to 0} \frac{\Delta f - df}{\rho}$ tends to 0. That is, $(\Delta f - df)$ is $0(\rho)$.

Note. A function will have differential if it is differentiable.

3. $f(x,y) = \begin{cases} \dfrac{x^3 - y^3}{x^2+y^2}, & (x,y) \neq (0,0) \\ 0, & (x,y) = (0,0) \end{cases}$ is not differentiable at $(0,0)$.

Here

$$\lim_{h\to 0} \frac{f(0+h,0) - f(0,0)}{h} = \lim_{h\to 0} \frac{h^3}{h^3} = 1. \quad \therefore \quad f_x(0,0) = 1.$$

Also

$$f_y(0,0) = -1, \Delta f = f(h,k) - f(0,0) = \frac{h^3 - k^3}{h^2+k^2},$$

so that we have

$$\frac{\Delta f - (hf_x(0,0) + kf_y(0,0))}{\sqrt{h^2+k^2}} = \frac{\frac{h^3-k^3}{h^2+k^2} - (h-k)}{\sqrt{h^2+k^2}} = \frac{h^2k - k^2h}{(h^2+k^2)^{3/2}}$$

$$\therefore \quad \lim_{\rho\to 0} \frac{\Delta f - (hf_x(0,0) + kf_y(0,0))}{\rho} = \lim_{(h,k)\to(0,0)} \frac{h^2k - k^2h}{(h^2+k^2)^{3/2}}$$

does not exist, because, if $(h, k) \to (0,0)$ along $k = mh$, the limiting value will be $\frac{m-m^2}{(1+m^2)^{3/2}}$ which is not unique.

But the function is continuous at the origin, for

$$|f(x,y) - f(0,0)| = \left|\frac{x^3 - y^3}{x^2 + y^2}\right| \leq \frac{|x^3| + |y^3|}{x^2 + y^2} \leq 2\sqrt{x^2 + y^2} < \varepsilon$$

if $x^2 + y^2 < \delta^2$ where $\delta = \frac{\varepsilon}{2}$.

It is important to note that the directional derivative of a function $f(x, y, z)$ of three independent variables in the direction η having direction cosines (l, m, n) is the scalar product of vectors $\left(\frac{\partial f}{\partial x}, \frac{\partial f}{\partial y}, \frac{\partial f}{\partial z}\right)$ and (l, m, n), provided $\frac{\partial f}{\partial x}, \frac{\partial f}{\partial y}$ and $\frac{\partial f}{\partial z}$ exist.

A Sufficient Condition for Differentiability: If D be the domain where $f : R^2 \to R$ is defined and $(a, b) \in D$, such that

(i) both the partial derivatives f_x and f_y exist at (a, b);

(ii) one of the partial derivatives is continuous at (a, b), then f is differentiable at (a, b). The proof may be found in any standard book on *Differential Calculus.* The following example will illustrate the above condition.

Show that

$$f(x,y) = \begin{cases} xy\dfrac{x^2 - y^2}{x^2 + y^2}, & \text{if } (x,y) \neq (0,0) \\ 0, & \text{if } (x,y) = (0,0) \end{cases}$$

is differentiable at $(0, 0)$.

We are to show that f satisfies the sufficient condition of differentiability.

Here

$$\lim_{h\to 0} \frac{f(0+h,0) - f(0,0)}{h} = \lim_{h\to 0} \frac{0-0}{h} = 0.$$

Also,

$$\lim_{k\to 0} \frac{f(0,0+k) - f(0,0)}{k} = \lim_{k\to 0} \frac{0-0}{k} = 0.$$

$$\therefore \quad f_x(0,0) = 0 = f_y(0,0)$$

When $(x, y) \neq (0, 0)$

$$f_x(x,y) = \frac{x^4 y + 4x^2 y^3 - y^5}{(x^2 + y^2)^2}.$$

$$\therefore \quad |f_x(x,y) - f_x(0,0)| \leq \frac{6\left(x^2 + y^2\right)^{5/2}}{\left(x^2 + y^2\right)^2} = 6\left(x^2 + y^2\right)^{1/2} < \varepsilon$$

if $(x^2 + y^2) < \delta^2 = \frac{\varepsilon^2}{36}$

$\therefore$ $f_x(x, y)$ is continuous at $(0, 0)$. So is $f_y(x, y)$,

$\because$ $f(x, y)$ is differentiable at $(0, 0)$.

The Conditions of Differentiability are not Necessary: We consider the following example to show that differentiability of f at (a,b) may not imply continuity of f_x or f_y at (a,b).

$$f(x,y)=\begin{cases} x^2\sin\dfrac{1}{x}+y^2\sin\dfrac{1}{y}, & \text{when } xy\neq 0\\[2ex] x^2\sin\dfrac{1}{x}, & \text{when } x\neq 0, y=0\\[2ex] y^2\sin\dfrac{1}{y}, & \text{when } x=0, y\neq 0\\[2ex] 0, & \text{when } x=y=0\end{cases}$$

We are to show that in this case neither f_x nor f_y is continuous at $(0,0)$, but the function is differentiable at $(0,0)$.

Here

$$\lim_{h\to 0}\frac{f(h,0)-f(0,0)}{h}=\lim_{h\to 0}h\sin\frac{1}{h}=0 \quad \therefore \quad f_x(0,0)=0.$$

Similarly, $f_y(0,0)=0$.

When $x\neq 0$, $f_x(x,y)=2x\sin\frac{1}{x}-\cos\frac{1}{x}$.

When $y\neq 0$, $f_y(x,y)=2y\sin\frac{1}{y}-\cos\frac{1}{y}$.

Then

$$f_x(x,y)=\begin{cases} 2x\sin\frac{1}{x}-\cos\frac{1}{x}, & \text{when } x\neq 0\\ 0, & \text{when } x=0\end{cases}$$

But by Cauchy's condition for existence of limit, $\lim\limits_{x\to 0}\cos\frac{1}{x}$ does not exist.

Therefore, $f_x(x,y)$ is not continuous at $(0,0)$.

By the same argument $f_y(x,y)$ is not continuous at $(0,0)$, but we see that

$$\Delta f=h^2\sin\frac{1}{h}+k^2\sin\frac{1}{k} \quad \text{and} \quad hf_x(0,0)+kf_y(0,0)=0.$$

$$\therefore \quad \lim_{(h,k)\to(0,0)}\frac{\Delta f-(hf_x(0,0)+kf_y(0,0))}{\sqrt{h^2+k^2}}=0,$$

because

$$\left|\frac{h^2\sin\frac{1}{h}+k^2\sin\frac{1}{k}}{\sqrt{h^2+k^2}}\right|\leq\frac{h^2+k^2}{\sqrt{h^2+k^2}}=\sqrt{h^2+k^2},$$

which can be made always less than ε in any arbitrary neighbourhood of $(0,0)$.

Thus f is differentiable at $(0,0)$, although the sufficiency of the condition is not satisfied.

22.7 Higher Order Partial Derivatives

We have defined f_x and f_y at (a,b).

Now, if (i) $\lim\limits_{h\to 0} \dfrac{f_x(a+h,b)-f_x(a,b)}{h}$,

(ii) $\lim\limits_{h\to 0} \dfrac{f_y(a+h,b)-f_y(a,b)}{h}$,

(iii) $\lim\limits_{k\to 0} \dfrac{f_x(a,b+k)-f_x(a,b)}{k}$ and

(iv) $\lim\limits_{k\to 0} \dfrac{f_y(a,b+k)-f_y(a,b)}{k}$ all exist, the limiting values are f_{xx}, f_{xy}, f_{yx} and f_{yy} at (a,b) respectively.

Now (ii) can be written as

$$\begin{aligned} f_{xy}(a,b) &= \lim_{h\to 0}\frac{1}{h}\left[\lim_{k\to 0}\frac{f(a+h,b+k)-f(a+h,b)}{k} - \lim_{k\to 0}\frac{f(a,b+k)-f(a,b)}{k}\right] \\ &= \lim_{h\to 0}\lim_{k\to 0}\frac{F(h,k)}{hk} \end{aligned}$$

where

$$F(h,k) = f(a+h,b+k) - f(a+h,b) - f(a,b+k) + f(a,b) \tag{1}$$

In the same way, from (iii), we have

$$f_{yx}(a,b) = \lim_{k\to 0}\lim_{h\to 0}\frac{F(h,k)}{hk}$$

Thus $f_{yx}(a,b)$ and $f_{xy}(a,b)$ are, in fact, the repeated limits of the same expression taken in different order.

Since repeated limits, if exist, may be different, $f_{yx}(a,b)$ and $f_{xy}(a,b)$ may be different.

Illustrative Examples

1. For the function

$$f(x,y) = \begin{cases} x^2\tan^{-1}\left(\frac{y}{x}\right) - y^2\tan^{-1}\left(\frac{x}{y}\right), & \text{when } x\neq 0, y\neq 0 \\ 0 & \text{elsewhere,} \end{cases}$$

show that $f_{xy}(0,0) \neq f_{yx}(0,0)$.

Here $F(h,k)$ as defined earlier in (22.7.1) is given by

$$\begin{aligned} F(h,k) &= f(h,k) - f(h,0) - f(0,k) + f(0,0) \\ &= h^2\tan^{-1}\left(\frac{k}{h}\right) - k^2\tan^{-1}\left(\frac{h}{k}\right), \quad \text{so that we have} \end{aligned}$$

$$\frac{F(h,k)}{hk} = \frac{1}{h}\left\{\frac{h^2\tan^{-1}\frac{k}{h} - k^2\tan^{-1}\left(\frac{h}{k}\right)}{k}\right\}.$$

$$\therefore \quad \lim_{h\to 0}\lim_{k\to 0}\frac{F(h,k)}{hk} = \lim_{h\to 0}\frac{1}{h}\left[\lim_{k\to 0}\frac{h^2\tan^{-1}\frac{k}{h} - k^2\tan^{-1}\frac{h}{k}}{k}\right]. \tag{1}$$

Now,

$$\lim_{k\to 0} \frac{h^2 \tan^{-1} \frac{k}{h}}{k} \qquad \left[\frac{0}{0} \text{ form}\right]$$

$$= \lim_{k\to 0} h^2 \times \frac{1}{1+\frac{k^2}{h^2}} \times \frac{1}{h} = \lim_{k\to 0} \frac{h^3}{h^2+k^2} = h.$$

Also,

$$\lim_{k\to 0} \frac{k^2 \tan^{-1} \frac{h}{k}}{k} = \lim_{k\to 0} k \tan^{-1}\left(\frac{h}{k}\right) \qquad [0\times\infty \text{ form}]$$

$$= \lim_{k\to 0} \frac{\tan^{-1} \frac{h}{k}}{\frac{1}{k}} \qquad \left[\frac{\infty}{\infty} \text{ form}\right]$$

$$= \lim_{k\to 0} \frac{\frac{1}{1+\frac{h^2}{k^2}}\left(-\frac{h}{k^2}\right)}{-\frac{1}{k^2}} = \lim_{k\to 0} \frac{k^2 h}{h^2+k^2} = 0$$

$\therefore$ both the limits

$$\lim_{k\to 0} \frac{h^2 \tan^{-1}\left(\frac{k}{h}\right)}{k} \quad \text{and} \quad \lim_{k\to 0} \frac{k^2 \tan^{-1}\left(\frac{h}{k}\right)}{k} \text{ exist.}$$

$\therefore$ the value of the expression within bracket in (1) is h.

$$\therefore \quad \lim_{h\to 0} \lim_{k\to 0} \frac{F(h,k)}{hk} = \lim_{h\to 0} \frac{h}{h} = 1.$$

Again,

$$\lim_{k\to 0} \lim_{h\to 0} \frac{F(h,k)}{hk},$$

proceeding as above, can be shown to be equal to

$$\lim_{k\to 0} \frac{1}{k} \left[\lim_{h\to 0} \frac{h^2 \tan^{-1}\left(\frac{k}{h}\right) - k^2 \tan^{-1}\left(\frac{h}{k}\right)}{h}\right] = \lim_{k\to 0} \frac{-k}{k} = -1.$$

$$\therefore \quad f_{xy}(0,0) \neq f_{yx}(0,0).$$

2. Let

$$f(x,y) = \begin{cases} (x^2+y^2)\tan^{-1}\frac{y}{x}, & \text{when } x \neq 0 \\ \frac{\pi}{2} y^2, & \text{when } x = 0. \end{cases}$$

Show that $f_{yx}(0,0) \neq f_{xy}(0,0)$.

Here

$$\lim_{h\to 0} \frac{f(h,0)-f(0,0)}{h} = \lim_{h\to 0} \frac{0-0}{h} = 0.$$

$$\therefore \quad f_x(0,0) = 0.$$

Again

$$\lim_{h\to 0}\frac{f(h,k)-f(0,k)}{h}.$$

$$= \lim_{h\to 0}\frac{(h^2+k^2)\tan^{-1}\frac{k}{h}-\frac{\pi}{2}k^2}{h} \qquad \left[\frac{0}{0}\text{ form}\right]$$

$$= \lim_{h\to 0}\left\{2h\tan^{-1}\frac{k}{h}+(h^2+k^2)\frac{1}{1+\frac{k^2}{h^2}}\left(-\frac{k}{h^2}\right)\right\}$$

$$= \lim_{h\to 0}\left\{2h\tan^{-1}\frac{k}{h}-k\right\} = \lim_{h\to 0} 2h\tan^{-1}\frac{k}{h}-k \qquad (1)$$

Now,

$$\lim_{h\to 0} h\tan^{-1}\frac{k}{h} \qquad [0\times\infty\text{ form}]$$

$$= \lim_{h\to 0}\frac{\tan^{-1}\frac{k}{h}}{\frac{1}{h}} \qquad \left[\frac{\infty}{\infty}\text{ form}\right]$$

$$= \lim_{h\to 0}\frac{\frac{1}{1+\frac{k^2}{h^2}}\times\left(-\frac{k}{h^2}\right)}{\left(-\frac{1}{h^2}\right)} = \lim_{h\to 0}\frac{h^2k}{h^2+k^2} = 0$$

$\therefore$ from (1) $f_x(0,k)=-k$.

$$\therefore \quad \lim_{k\to 0}\frac{f_x(0,k)-f_x(0,0)}{k} = \lim_{k\to 0}-\frac{k}{k} = -1.$$

$$\therefore \quad f_{yx}(0,0)=-1.$$

Proceeding in the same way it can be shown that $f_y(h,0)=h$, and $f_y(0,0)=0$.

$$\therefore \quad \lim_{h\to 0}\frac{f_y(h,0)-f_y(0,0)}{h} = \lim_{h\to 0}\frac{h}{h} = 1.$$

$$\therefore \quad f_{xy}(0,0)=1.$$

Thus,

$$f_{yx}(0,0)\neq f_{xy}(0,0).$$

22.8 On the Equality of f_{yx} and f_{xy} at $(a,b)\in D$

We have the following two theorems that will provide sufficient conditions for continuity, differentiability and existence of partial differentials.

22.8.1 Theorem of Schwarz

Let D be the domain of definition of $f(x,y)$ and $(a,b) \in D$. Let further

(i) f_y exist in some neighbourhood of (a,b); and

(ii) f_{yx} be continuous at (a,b),

then f_{xy} exists at (a,b) and $f_{yx}(a,b) = f_{xy}(a,b)$.

22.8.2 Theorem of Young

If f_x and f_y exist in some neighbourhood of (a,b) and f_x, f_y are both differentiable at (a,b), then $f_{xy}(a,b) = f_{yx}(a,b)$.

Proofs of the above two theorems can be found in any standard book on *Differential Calculus*.

It should be noted that the conditions of the above two theorems are by no means necessary for $f_{xy}(a,b) = f_{yx}(a,b)$.

Illustrative Examples

1. Let

$$f(x,y) = \begin{cases} \dfrac{x^2y^2}{x^2+y^2}, & \text{when } x^2+y^2 \neq 0 \\ 0 & \text{when } x^2+y^2 = 0 \end{cases}$$

Here,

$$\lim_{h\to 0} \frac{f(h,0)-f(0,0)}{h} = \lim_{h\to 0} \frac{0-0}{h} = 0$$

$$\text{and} \quad \lim_{k\to 0} \frac{f(0,k)-f(0,0)}{k} = \lim_{k\to 0} \frac{0-0}{k} = 0.$$

$$\therefore \quad f_x(0,0) = 0 = f_y(0,0).$$

Now

$$\lim_{h\to 0} \frac{f(h,k)-f(h,0)}{h} = \lim_{h\to 0} \frac{h^2k^2}{h(h^2+k^2)} = \lim_{h\to 0} \frac{hk^2}{h^2+k^2} = 0,$$

since

$$\left|\frac{hk^2}{h^2+k^2}\right| = \frac{|h||k^2|}{h^2+k^2} \leq |h| \leq \sqrt{h^2+k^2} < \varepsilon \text{ if } h^2+k^2 < \delta^2 = (\varepsilon^2).$$

$$\therefore \quad f_x(0,k) = 0.$$

Similarly, $f_y(h,0) = 0$.

$$\therefore \quad \lim_{k\to 0} \frac{f_x(0,k)-f_x(0,0)}{k} = \lim_{k\to 0} \frac{0-0}{k} = 0 \quad \text{and} \quad \lim_{h\to 0} \frac{f_y(h,0)-f_y(0,0)}{h} = 0.$$

$$\therefore \quad f_{xy}(0,0) = f_{yx}(0,0) = 0,$$

but we see that conditions of neither Schwarz's theorem nor Young's theorem are satisfied.

2. Let

$$f_{xy} = \begin{cases} \dfrac{8x^3y^3}{(x^2+y^2)^3}, & \text{when } (x,y) \neq (0,0) \\ 0, & \text{when } (x,y) = (0,0). \end{cases}$$

Let $(x,y) \to (0,0)$ along $y = mx$, so that we may write

$$\lim_{(x,y)\to(0,0)} f_{xy}(x,y) = \frac{8m^3}{(1+m^2)^3}, \quad \text{which shows that the limit is not unique.}$$

$$\therefore \quad \lim_{(x,y)\to(0,0)} f_{xy} \text{ does not exist.}$$

$\therefore$ f_{xy} is not continuous at $(0,0)$.

Similarly, f_{yx} is also not continuous at $(0,0)$. Thus Schwarz's theorem does not hold.

It remains to be shown that conditions of Young's theorem are also violated.

Here,

$$\lim_{h\to 0} \frac{f_x(h,0) - f_x(0,0)}{h} = 0. \quad \therefore \quad f_{xx}(0,0) = 0.$$

If f_x were differentiable at $(0,0)$ we could establish that

$$\frac{\{f_x(h,k) - f_x(0,0)\} - \{hf_{xx}(0,0) + kf_{yx}(0,0)\}}{\sqrt{h^2+k^2}} \to 0 \quad \text{as } (h,k) \to (0,0).$$

If $x^2 + y^2 \neq 0$, $f_x(x,y) = \dfrac{2xy^4}{(x^2+y^2)^2}$.

$$\therefore \quad \frac{f_x(h,k) - f_x(0,0) - hf_{xx}(0,0) - kf_{yx}(0,0)}{\sqrt{h^2+k^2}} = \frac{2hk^4}{(h^2+k^2)^{5/2}}$$

Now,

$$\frac{2hk^4}{(h^2+k^2)^{5/2}} \to 0 \text{ as } (h,k) \to (0,0)$$

is not true as because

$$\lim_{(h,k)\to(0,0)} \frac{2hk^4}{(h^2+k^2)^{5/2}}$$

does not exist if $(h,k) \to (0,0)$ along $k = mh$.

$\therefore$ f_x is not differentiable at $(0,0)$. Similarly, f_y is also not differentiable at $(0,0)$.

3. For the function

$$f(x,y) = \begin{cases} (x^2+y^2)\log(x^2+y^2), & (x,y) \neq (0,0) \\ 0, & (x,y) = (0,0) \end{cases}$$

$$\text{show that} \quad f_{yx}(0,0) = f_{xy}(0,0)$$

but conditions of Schwarz's theorem are not satisfied.

Here,

$$\begin{aligned}\lim_{h\to 0}\frac{f(h,0)-f(0,0)}{h} &= \lim_{h\to 0}\frac{h^2\log h^2}{h} = \lim_{h\to 0} h\log h^2\\ &= \lim_{h\to 0}\frac{\log h^2}{\frac{1}{h}} \quad \left[\frac{\infty}{\infty}\text{ form}\right]\\ &= \lim_{h\to 0}(-2h) = 0.\end{aligned}$$

$$\therefore\quad f_x(0,0) = 0.$$

Similarly, $f_y(0,0) = 0$.

Again

$$\begin{aligned}\lim_{h\to 0}\frac{f(h,k)-f(0,k)}{h} &= \lim_{h\to 0}\frac{(h^2+k^2)\log(h^2+k^2)-k^2\log k^2}{h} \quad [\div\text{ form}]\\ &= \lim_{h\to 0}(2h\log(h^2+k^2)+2h) = 0.\end{aligned}$$

$$\therefore\quad f_x(0,k) = 0.$$

$$\therefore\quad \lim_{k\to 0}\frac{f_x(0,k)-f_x(0,0)}{k} = 0.$$

$$\therefore\quad f_{yx}(0,0) = 0.$$

Proceeding in the same way we get

$$f_y(h,0) = 0 \quad\text{and}\quad \lim_{h\to 0}\frac{f_y(h,0)-f_y(0,0)}{h} = 0 \quad\therefore\quad f_{xy}(0,0) = 0$$

When $(x,y)\neq(0,0)$

$$f_{yx} = \frac{4xy}{x^2+y^2} = f_{xy} \quad\therefore\quad f_{xy} = \begin{cases}\dfrac{4xy}{x^2+y^2}, & \text{when } (x,y)\neq(0,0)\\ 0, & \text{when } (x,y)=(0,0)\end{cases}$$

is not continuous at $(0,0)$ because of non-existence of limit of f_{yx} as $(x,y)\to(0,0)$.

22.9 Function of Three Variables

We consider the function $f(u,v,w)$ which is a differentiable function of u,v,w where u,v and w are again differentiable functions of independent variables x,y and z.

The partial derivatives $\frac{\partial f}{\partial x}$, $\frac{\partial f}{\partial y}$ and $\frac{\partial f}{\partial z}$ are given by (applying chain rule),

$$\begin{aligned}\frac{\partial f}{\partial x} &= \frac{\partial f}{\partial u}\frac{\partial u}{\partial x}+\frac{\partial f}{\partial v}\frac{\partial v}{\partial x}+\frac{\partial f}{\partial w}\frac{\partial w}{\partial x};\\ \frac{\partial f}{\partial y} &= \frac{\partial f}{\partial u}\frac{\partial u}{\partial y}+\frac{\partial f}{\partial v}\frac{\partial v}{\partial y}+\frac{\partial f}{\partial w}\frac{\partial w}{\partial y};\\ \text{and}\quad \frac{\partial f}{\partial z} &= \frac{\partial f}{\partial u}\frac{\partial u}{\partial z}+\frac{\partial f}{\partial v}\frac{\partial v}{\partial z}+\frac{\partial f}{\partial w}\frac{\partial w}{\partial z}.\end{aligned}$$

Multiplying these equations by dx, dy, dz respectively and then adding, we obtain

$$\begin{aligned} df &= f_x dx + f_y dy + f_z dz \\ &= \frac{\partial f}{\partial u}\left(\frac{\partial u}{\partial x}dx + \frac{\partial u}{\partial y}dy + \frac{\partial u}{\partial z}dz\right) + \frac{\partial f}{\partial v}\left(\frac{\partial v}{\partial x}dx + \frac{\partial v}{\partial y}dy + \frac{\partial v}{\partial z}dz\right) \\ &\quad + \frac{\partial f}{\partial w}\left(\frac{\partial w}{\partial x}dx + \frac{\partial w}{\partial y}dy + \frac{\partial w}{\partial z}dz\right) \\ &= \frac{\partial f}{du}du + \frac{\partial f}{dv}dv + \frac{\partial f}{dw}dw = f_u du + f_v dv + f_w dw, \end{aligned}$$

which shows that df has the same form as if u, v, w were independent variables.

Illustrative Example

1. If $Pdx + Qdy + Rdz$ can be made a perfect differential of some function of x, y, z by multiplying each term by a common factor, then

$$P\left(\frac{\partial Q}{\partial x} - \frac{\partial R}{\partial y}\right) + Q\left(\frac{\partial R}{\partial x} - \frac{\partial P}{\partial z}\right) + R\left(\frac{\partial P}{\partial y} - \frac{\partial Q}{\partial z}\right) = 0$$

Here we have

$$du = \frac{\partial u}{\partial x}dx + \frac{\partial u}{\partial y}dy + \frac{\partial u}{\partial z}dz, \text{ then } P, Q, R \text{ are proportional to } \frac{\partial u}{\partial x}, \frac{\partial u}{\partial y}, \frac{\partial u}{\partial z}.$$

That is

$$\mu P = \frac{\partial u}{\partial x} \qquad (1)$$

$$\mu Q = \frac{\partial u}{\partial y} \qquad (2)$$

$$\mu R = \frac{\partial u}{\partial z} \qquad (3)$$

On differentiating (1) w.r.t. y and z, (2) w.r.t. z and x, and (3) w.r.t. x and y we get

$$P\frac{\partial \mu}{\partial y} + \mu\frac{\partial P}{\partial y} = \frac{\partial^2 u}{\partial y \partial x} = \mu\frac{\partial Q}{\partial x} + Q\frac{\partial \mu}{\partial x} \qquad (4)$$

$$Q\frac{\partial \mu}{\partial z} + \mu\frac{\partial Q}{\partial z} = \frac{\partial^2 u}{\partial y \partial z} = R\frac{\partial \mu}{\partial y} + \mu\frac{\partial R}{\partial y} \qquad (5)$$

$$P\frac{\partial \mu}{\partial z} + \mu\frac{\partial P}{\partial z} = \frac{\partial^2 u}{\partial z \partial x} = R\frac{\partial \mu}{\partial x} + \mu\frac{\partial R}{\partial x} \qquad (6)$$

On rearranging the terms, we have

$$\begin{aligned} \mu\left(\frac{\partial P}{\partial y} - \frac{\partial Q}{\partial x}\right) &= Q\frac{\partial \mu}{\partial x} - P\frac{\partial \mu}{\partial y} \\ \mu\left(\frac{\partial Q}{\partial z} - \frac{\partial R}{\partial y}\right) &= R\frac{\partial \mu}{\partial y} - Q\frac{\partial \mu}{\partial z} \\ \mu\left(\frac{\partial R}{\partial x} - \frac{\partial P}{\partial z}\right) &= P\frac{\partial \mu}{\partial z} - R\frac{\partial \mu}{\partial x}. \end{aligned}$$

Multiplying the above equations by R, P and Q respectively and then adding we get the result,

$$P\left(\frac{\partial Q}{\partial x}-\frac{\partial R}{\partial y}\right)+Q\left(\frac{\partial R}{\partial x}-\frac{\partial P}{\partial z}\right)+R\left(\frac{\partial P}{\partial y}-\frac{\partial Q}{\partial z}\right)=0$$

22.10 Homogeneous Function

If a function $f(x,y,z)$ of three independent variables x, y, z be such that $f(tx,ty,tz) = t^n f(x,y,z)$ for every $t(>0)$, then $f(x,y,z)$ is said to be a *homogeneous function of degree* n or equivalently, $f(x,y,z)$ is said to be a homogeneous function of degree n if we can write

$$f(x,y,z)=x^n\Phi\left(\frac{y}{x},\frac{z}{x}\right),$$

x being any one of the variables appearing in the function.

22.10.1 Euler's Theorem on Homogeneous Function

If $u = f(x,y,z)$ be a homogeneous function in x, y, z of degree n having continuous partial derivatives, then

$$x\frac{\partial u}{\partial x}+y\frac{\partial u}{\partial y}+z\frac{\partial u}{\partial z}=nu.$$

Proof. As $f(x,y,z)$ is a homogeneous function of degree n we can write $f(tx,ty,tz) = t^n f(x,y,z)$.

Differentiating both sides with respect to t

$$\frac{\partial f}{\partial(tx)}x+\frac{\partial f}{\partial(ty)}y+\frac{\partial f}{\partial(tz)}z=nt^{n-1}f(x,y,z)$$

Now putting $t = 1$, we get the result.

As an immediate consequence differentiating again w.r.t. t we get

$$\left(x\frac{\partial}{\partial(tx)}+y\frac{\partial}{\partial(ty)}+z\frac{\partial}{\partial(tz)}\right)\left(x\frac{\partial f}{\partial(tx)}+y\frac{\partial f}{\partial(ty)}+z\frac{\partial f}{\partial(tz)}\right)=n(n-1)t^{n-2}f(x,y,z).$$

Putting $t = 1$, we get

$$\left(x\frac{\partial}{\partial x}+y\frac{\partial}{\partial y}+z\frac{\partial}{\partial z}\right)^2 f(x,y,z)=n(n-1)f(x,y,z)$$

22.10.2 Converse of Euler's Theorem

If a differentiable function f of three variables x, y, z satisfies the relation

$$x\frac{\partial f}{\partial x}+y\frac{\partial f}{\partial y}+z\frac{\partial f}{\partial z}=nf(x,y,z)$$

then f will be a homogeneous function of x, y, z of degree n.

Proof. Let us put $u = \frac{x}{z}, v = \frac{y}{z}$ and $w = z$, then $f(x, y, z) = F(u, v, w)$. We have

$$x\frac{\partial f}{\partial x} = x\left\{\frac{\partial F}{\partial u}\frac{\partial u}{\partial x} + \frac{\partial F}{\partial v}\frac{\partial v}{\partial x} + \frac{\partial F}{\partial w}\frac{\partial w}{\partial x}\right\} = u\frac{\partial F}{\partial u}$$

$$x\frac{\partial f}{\partial y} = y\left\{\frac{\partial F}{\partial u}\frac{\partial u}{\partial y} + \frac{\partial F}{\partial v}\frac{\partial v}{\partial y} + \frac{\partial F}{\partial w}\frac{\partial w}{\partial y}\right\} = v\frac{\partial F}{\partial v}$$

$$z\frac{\partial f}{\partial z} = z\left\{\frac{\partial F}{\partial u}\frac{\partial u}{\partial z} + \frac{\partial F}{\partial v}\frac{\partial v}{\partial z} + \frac{\partial F}{\partial w}\frac{\partial w}{\partial z}\right\} = -u\frac{\partial F}{\partial u} - v\frac{\partial F}{\partial v} + w\frac{\partial F}{\partial w}$$

$$\therefore \quad x\frac{\partial f}{\partial x} + y\frac{\partial f}{\partial y} + z\frac{\partial f}{\partial z} = nu, \quad \text{transforms to } w\frac{\partial F}{\partial w} = nF.$$

On integration, $F = w^n \cdot A$, where A is a constant free from w but may involve u and v. Let us put $A = \Phi(u, v)$.

We get, therefore,

$$F = w^n\Phi\left(\frac{x}{z}, \frac{y}{z}\right) = z^n\Phi\left(\frac{x}{z}, \frac{y}{z}\right).$$

$\therefore$ F, i.e., f is a homogeneous function of degree n.

Illustrative Examples

1. If f is a homogeneous function of nth degree in x, y, z and $F = \Phi(u, v, w)$ where $u = \frac{\partial f}{\partial x}$, $v = \frac{\partial f}{\partial y}$, $w = \frac{\partial f}{\partial z}$ are differentiable, then

$$u\frac{\partial F}{\partial u} + v\frac{\partial F}{\partial v} + w\frac{\partial F}{\partial w} = \frac{n}{n-1}f.$$

Here, we shall first prove that u, v, w are homogeneous functions of degree $n - 1$.

From the given condition, we may write

$$x\frac{\partial f}{\partial x} + y\frac{\partial f}{\partial y} + z\frac{\partial f}{\partial z} = nf, \quad \text{or,} \quad xu + yv + zw = nf. \tag{1}$$

Differentiating (1) w.r.t. x

$$u + x\frac{\partial u}{\partial x} + y\frac{\partial v}{\partial x} + z\frac{\partial w}{\partial x} = n\frac{\partial f}{\partial x} = nu$$

$$\text{or,} \quad x\frac{\partial u}{\partial x} + y\frac{\partial}{\partial x}\left(\frac{\partial f}{\partial y}\right) + z\frac{\partial}{\partial x}\left(\frac{\partial f}{\partial z}\right) = (n-1)u$$

$$\text{or,} \quad x\frac{\partial u}{\partial x} + y\frac{\partial}{\partial y}\left(\frac{\partial f}{\partial x}\right) + z\frac{\partial}{\partial z}\left(\frac{\partial f}{\partial x}\right) = (n-1)u \quad \left[\because \frac{\partial^2 f}{\partial x\partial y} = \frac{\partial^2 f}{\partial y\partial x}\right]$$

$$\text{or,} \quad x\frac{\partial u}{\partial x} + y\frac{\partial u}{\partial y} + z\frac{\partial u}{\partial z} = (n-1)u$$

$\therefore$ u is also a homogeneous function of degree $n - 1$.

Similarly, each of v and w is a homogeneous function of degree $n-1$. Since each of x, y and z is a function of u, v, w we get

$$u = \frac{\partial f}{\partial x} = \frac{\partial F}{\partial u}\cdot\frac{\partial u}{\partial x} + \frac{\partial F}{\partial v}\cdot\frac{\partial v}{\partial x} + \frac{\partial F}{\partial w}\cdot\frac{\partial w}{\partial x}$$

$$v = \frac{\partial f}{\partial y} = \frac{\partial F}{\partial u}\cdot\frac{\partial u}{\partial y} + \frac{\partial F}{\partial v}\cdot\frac{\partial v}{\partial y} + \frac{\partial F}{\partial w}\cdot\frac{\partial w}{\partial y}$$

$$w = \frac{\partial f}{\partial z} = \frac{\partial F}{\partial u}\cdot\frac{\partial u}{\partial z} + \frac{\partial F}{\partial v}\cdot\frac{\partial v}{\partial z} + \frac{\partial F}{\partial w}\cdot\frac{\partial w}{\partial z}.$$

$\therefore$ from (1)

$$\begin{aligned} nf = xu + yv + zw &= \frac{\partial F}{\partial u}\left(x\frac{\partial u}{\partial x} + y\frac{\partial u}{\partial y} + z\frac{\partial u}{\partial z}\right) + \frac{\partial F}{\partial v}\left(x\frac{\partial v}{\partial x} + y\frac{\partial v}{\partial y} + z\frac{\partial v}{\partial z}\right) \\ &\quad + \frac{\partial F}{\partial w}\left(x\frac{\partial w}{\partial x} + y\frac{\partial w}{\partial y} + z\frac{\partial w}{\partial z}\right) \\ &= (n-1)\left\{u\frac{\partial F}{\partial u} + v\frac{\partial F}{\partial v} + w\frac{\partial F}{\partial w}\right\}. \end{aligned}$$

$$\therefore \quad \frac{n}{n-1}f = u\frac{\partial F}{\partial u} + v\frac{\partial F}{\partial v} + w\frac{\partial F}{\partial w}$$

2. If $x^2 = vw, y^2 = uw, z^2 = uv$ and $\Phi(u,v,w) = f(x,y,z)$, then $xf_x + yf_y + zf_z = u\Phi_u + v\Phi_v + w\Phi_w$.

From the given relation, we have

$$y^2z^2 = u^2vw = u^2x^2.$$

$$\therefore \quad u = \pm\frac{yz}{x}, \quad \text{similarly,} \quad v = \pm\frac{zx}{y}, \quad \text{and} \quad w = \pm\frac{xy}{z}$$

$\therefore$ each of u, v, w is a homogeneous function of x, y, z of degree 1, so that we may write

$$x\frac{\partial u}{\partial x} + y\frac{\partial u}{\partial y} + z\frac{\partial u}{\partial z} = u$$

$$x\frac{\partial v}{\partial x} + y\frac{\partial v}{\partial y} + z\frac{\partial v}{\partial z} = v$$

$$\text{and} \quad x\frac{\partial w}{\partial x} + y\frac{\partial w}{\partial y} + z\frac{\partial w}{\partial z} = w.$$

Now

$$f_x = \frac{\partial \Phi}{\partial u}\cdot\frac{\partial u}{\partial x} + \frac{\partial \Phi}{\partial v}\cdot\frac{\partial v}{\partial x} + \frac{\partial \Phi}{\partial w}\cdot\frac{\partial w}{\partial x}.$$

Similar expressions may be obtained for f_y and f_z

$$\begin{aligned} \therefore \quad xf_x + yf_y + zf_z &= \frac{\partial \Phi}{\partial u}\left(x\frac{\partial u}{\partial x} + y\frac{\partial u}{\partial y} + z\frac{\partial u}{\partial z}\right) + \frac{\partial \Phi}{\partial v}\left(x\frac{\partial v}{\partial x} + y\frac{\partial v}{\partial y} + z\frac{\partial v}{\partial z}\right) \\ &\quad + \frac{\partial \Phi}{\partial w}\left(x\frac{\partial w}{\partial x} + y\frac{\partial w}{\partial y} + z\frac{\partial w}{\partial z}\right) \\ &= u\Phi_u + v\Phi_v + w\Phi_w \end{aligned}$$

22.11 Jacobians

If $u_1, u_2, \cdots, u_n$ are functions of n variables $x_1, x_2, \cdots, x_n$ having partial derivatives of first order at every point of their common domain of definition, then the determinant

$$\begin{vmatrix} \dfrac{\partial u_1}{\partial x_1} & \dfrac{\partial u_1}{\partial x_2} & \cdots & \dfrac{\partial u_1}{dx_n} \\ \dfrac{\partial u_2}{\partial x_1} & \dfrac{\partial u_2}{\partial x_2} & \cdots & \dfrac{\partial u_2}{dx_n} \\ \dfrac{\partial u_n}{\partial x_1} & \dfrac{\partial u_n}{\partial x_2} & \cdots & \dfrac{\partial u_n}{dx_n} \end{vmatrix} \quad \text{or in short} \quad \begin{vmatrix} \dfrac{\partial u_1}{\partial x_1} & \dfrac{\partial u_2}{\partial x_2} & \cdots & \dfrac{\partial u_n}{dx_n} \end{vmatrix}$$

is called the *Jacobian* of $u_1, u_2, \cdots, u_n$ w.r.t. $x_1, x_2, \cdots, x_n$ and is denoted by

$$\frac{\partial(u_1, \cdots, u_n)}{\partial(x_1, \cdots, x_n)} \quad \text{or,} \quad J_{1,2,\cdots,n}.$$

22.11.1 Chain Rule for Jacobians

If u_1, u_2, u_3 are functions of y_1, y_2, y_3 and y_1, y_2, y_3 are functions of x_1, x_2, x_3, then

$$\frac{\partial(u_1, u_2, u_3)}{\partial(x_1, x_2, x_3)} = \frac{\partial(u_1, u_2, u_3)}{\partial(y_1, y_2, y_3)} \frac{\partial(y_1, y_2, y_3)}{\partial(x_1, x_2, x_3)}$$

Proof. By chain rules, $\dfrac{\partial u_i}{\partial x_j} = \displaystyle\sum_{r=1}^{n} \dfrac{\partial u_i}{\partial y_r}\dfrac{\partial y_r}{\partial x_j} \quad i, j = 1, 2, 3.$

$$\text{Now, R.H.S.} = \begin{vmatrix} \dfrac{\partial u_1}{\partial y_1} & \dfrac{\partial u_1}{\partial y_2} & \dfrac{\partial u_1}{\partial y_3} \\ \dfrac{\partial u_2}{\partial y_1} & \dfrac{\partial u_2}{\partial y_2} & \dfrac{\partial u_2}{\partial y_3} \\ \dfrac{\partial u_3}{\partial y_1} & \dfrac{\partial u_3}{\partial y_2} & \dfrac{\partial u_3}{\partial y_3} \end{vmatrix} \begin{vmatrix} \dfrac{\partial y_1}{\partial x_1} & \dfrac{\partial y_1}{\partial x_2} & \dfrac{\partial y_1}{\partial x_3} \\ \dfrac{\partial y_2}{\partial x_1} & \dfrac{\partial y_2}{\partial x_2} & \dfrac{\partial y_2}{\partial x_3} \\ \dfrac{\partial y_3}{\partial x_1} & \dfrac{\partial y_3}{\partial x_2} & \dfrac{\partial y_3}{\partial x_3} \end{vmatrix}$$

$$= \begin{vmatrix} \displaystyle\sum_{r=1}^{3} \dfrac{\partial u_1}{\partial y_r}\dfrac{\partial y_r}{\partial x_1} & \displaystyle\sum \dfrac{\partial u_1}{\partial y_r}\dfrac{\partial y_r}{\partial x_2} & \displaystyle\sum \dfrac{\partial u_1}{\partial y_r}\dfrac{\partial y_r}{\partial x_3} \\ \displaystyle\sum \dfrac{\partial u_2}{\partial y_r}\dfrac{\partial y_r}{\partial x_1} & \displaystyle\sum \dfrac{\partial u_2}{\partial y_r}\dfrac{\partial y_r}{\partial x_2} & \displaystyle\sum \dfrac{\partial u_2}{\partial y_r}\dfrac{\partial y_r}{\partial x_3} \\ \displaystyle\sum \dfrac{\partial u_3}{\partial y_r}\dfrac{\partial y_r}{\partial x_1} & \displaystyle\sum \dfrac{\partial u_3}{\partial y_r}\dfrac{\partial y_r}{\partial x_2} & \displaystyle\sum \dfrac{\partial u_3}{\partial y_r}\dfrac{\partial y_r}{\partial x_3} \end{vmatrix}$$

$$= \begin{vmatrix} \dfrac{\partial u_1}{\partial x_1} & \dfrac{\partial u_1}{\partial x_2} & \dfrac{\partial u_1}{\partial x_3} \\ \dfrac{\partial u_2}{\partial x_1} & \dfrac{\partial u_2}{\partial x_2} & \dfrac{\partial u_2}{\partial x_3} \\ \dfrac{\partial u_3}{\partial x_1} & \dfrac{\partial u_3}{\partial x_2} & \dfrac{\partial u_3}{\partial x_3} \end{vmatrix} = \frac{\partial(u_1, u_2, u_3)}{\partial(x_1, x_2, x_3)}$$

As an important corollary we get,

$$\frac{\partial(x_1,x_2,x_3)\partial(y_1,y_2,y_3)}{\partial(y_1,y_2,y_3)\partial(x_1,x_2,x_3)} = 1.$$

This result follows from the chain rule by putting $u_1 = x_1, u_2 = x_2$ and $u_3 = x_3$.

$$\text{L.H.S.} = \frac{\partial(u_1,u_2,u_3)\partial(y_1,y_2,y_3)}{\partial(y_1,y_2,y_3)\partial(x_1,x_2,x_3)} = \begin{vmatrix} \dfrac{\partial u_1}{\partial x_1} & \dfrac{\partial u_1}{\partial x_2} & \dfrac{\partial u_1}{\partial x_3} \\ \dfrac{\partial u_2}{\partial x_1} & \dfrac{\partial u_2}{\partial x_2} & \dfrac{\partial u_2}{\partial x_3} \\ \dfrac{\partial u_3}{\partial x_1} & \dfrac{\partial u_3}{\partial x_2} & \dfrac{\partial u_3}{\partial x_3} \end{vmatrix} = 1$$

22.11.2 Jacobian of Implicit Functions

If $f(u,v,x,y) = 0$ and $g(u,v,x,y) = 0$, then

$$\frac{\partial(u,v)}{\partial(x,y)} = \frac{\partial(f,g)}{\partial(x,y)} \Big/ \frac{\partial(f,g)}{\partial(u,v)} \quad \text{when} \quad \frac{\partial(f,g)}{\partial(u,v)} \neq 0$$

Differentiating $f(u,v,x,y) = 0$ and $g(u,v,x,y) = 0$ w.r.t. x and y we get

$$\frac{\partial f}{\partial x} + \frac{\partial f}{\partial u}\cdot\frac{\partial u}{\partial x} + \frac{\partial f}{\partial v}\cdot\frac{\partial v}{\partial x} = 0$$

$$\frac{\partial f}{\partial y} + \frac{\partial f}{\partial u}\cdot\frac{\partial u}{\partial y} + \frac{\partial f}{\partial v}\cdot\frac{\partial v}{\partial y} = 0$$

$$\frac{\partial g}{\partial x} + \frac{\partial g}{\partial u}\cdot\frac{\partial u}{\partial x} + \frac{\partial g}{\partial v}\cdot\frac{\partial v}{\partial x} = 0$$

$$\frac{\partial g}{\partial y} + \frac{\partial g}{\partial u}\cdot\frac{\partial u}{\partial y} + \frac{\partial g}{\partial v}\cdot\frac{\partial v}{\partial y} = 0$$

Now,

$$\frac{\partial(f,g)}{\partial(u,v)}\cdot\frac{\partial(u,v)}{\partial(x,y)} = \begin{vmatrix} \dfrac{\partial f}{\partial u} & \dfrac{\partial f}{\partial v} \\ \dfrac{\partial g}{\partial u} & \dfrac{\partial g}{\partial v} \end{vmatrix} \begin{vmatrix} \dfrac{\partial u}{\partial x} & \dfrac{\partial u}{\partial y} \\ \dfrac{\partial v}{\partial x} & \dfrac{\partial v}{\partial y} \end{vmatrix}$$

$$= \begin{vmatrix} \dfrac{\partial f}{\partial u}\cdot\dfrac{\partial u}{\partial x} + \dfrac{\partial f}{\partial v}\cdot\dfrac{\partial v}{\partial x} & \dfrac{\partial f}{\partial u}\cdot\dfrac{\partial u}{\partial y} + \dfrac{\partial f}{\partial v}\cdot\dfrac{\partial v}{\partial y} \\ \dfrac{\partial g}{\partial u}\cdot\dfrac{\partial u}{\partial x} + \dfrac{\partial g}{\partial v}\cdot\dfrac{\partial v}{\partial x} & \dfrac{\partial g}{\partial u}\cdot\dfrac{\partial u}{\partial y} + \dfrac{\partial g}{\partial v}\cdot\dfrac{\partial v}{\partial y} \end{vmatrix}$$

$$= \begin{vmatrix} -\dfrac{\partial f}{\partial x} & -\dfrac{\partial f}{\partial y} \\ -\dfrac{\partial g}{\partial x} & -\dfrac{\partial g}{\partial y} \end{vmatrix} = \frac{\partial(f,g)}{\partial(x,y)}$$

Hence the result follows.

In the general case if $F_i(u_1, u_2, \cdots, u_n, x_1, x_2, \cdots, x_n) = 0, i = 1, 2, \cdots, n$, then

$$\frac{\partial(u_1, \cdots, u_n)}{\partial(x_1, \cdots, x_n)} = (-1)^n \frac{\partial(F_1, \cdots, F_n)}{\partial(x_1, \cdots, x_n)} \Big/ \frac{\partial(F_1, \cdots, F_n)}{\partial(u_1, \cdots, u_n)}. \tag{22.11.1}$$

Illustrative Examples

1. If the roots of the equation in λ

$$(\lambda - x)^3 + (\lambda - y)^3 + (\lambda - z)^3 = 0$$

are u, v, w, then show that

$$\frac{\partial(u, v, w)}{\partial(x, y, z)} = -2\frac{(y-z)(z-x)(x-y)}{(v-w)(w-u)(u-v)}$$

The equation is

$$\lambda^3 - (x+y+z)\lambda^2 + (x^2+y^2+z^2)\lambda - \frac{1}{3}(x^3+y^3+z^3) = 0 \text{ if}$$

$$\lambda^3 - p\lambda^2 + q\lambda - r = 0, \tag{1}$$

where

$$p = x+y+z, \quad q = x^2+y^2+z^2, \quad r = \frac{1}{3}(x^3+y^3+z^3) \tag{2}$$

Since u, v, w are the roots of (1), we have

$$\left.\begin{aligned} u+v+w &= p \\ uv+vw+uw &= q \\ uvw &= r \end{aligned}\right\} \tag{3}$$

From (2), we may write

$$\frac{\partial(p, q, r)}{\partial(x, y, z)} = \begin{vmatrix} 1 & 1 & 1 \\ 2x & 2y & 2z \\ x^2 & y^2 & z^2 \end{vmatrix} = 2(y-z)(z-x)(x-y)$$

From (3), we may also write

$$\frac{\partial(p, q, r)}{\partial(u, v, w)} = \begin{vmatrix} 1 & 1 & 1 \\ v+w & w+u & u+v \\ vw & uw & uv \end{vmatrix} = -(u-v)(v-w)(w-u)$$

$$\frac{\partial(u, v, w)}{\partial(x, y, z)} = \frac{\partial(u, v, w)}{\partial(p, q, r)} \cdot \frac{\partial(p, q, r)}{\partial(x, y, z)} = -2\frac{(y-z)(z-x)(x-y)}{(u-v)(v-w)(w-u)}$$

2. The above example can also be solved by using Jacobian for implicit functions.

Here, $u+v+w=x+y+z$, $uv+vw+wu=x^2+y^2+z^2$, $uvw=\frac{1}{3}(x^3+y^3+z^3)$.
Let

$$\begin{aligned}\Phi_1 &\equiv (u+v+w-x-y-z)=0\\ \Phi_2 &\equiv uv+vw+wu-x^2-y^2-z^2=0\\ \Phi_3 &\equiv uvw-\frac{1}{3}(x^3+y^3+z^3)=0.\end{aligned}$$

$\therefore$ from (22.11.1)

$$\frac{\partial(u,v,w)}{\partial(x,y,z)}=(-1)^3\frac{\partial(\Phi_1,\Phi_2,\Phi_3)}{\partial(x,y,z)}\Bigg/\frac{\partial(\Phi_1,\Phi_2,\Phi_3))}{\partial(u,v,w)}$$

Calculating separately the Jacobians, the quotient gives the result.

3. Show that the relations

$$\begin{aligned}F_1(u_1,u_2,u_3,x_1,x_2,x_3) &= 0\\ F_2(u_1,u_2,u_3,x_1,x_2,x_3) &= 0\\ F_3(u_1,u_2,u_3,x_1,x_2,x_3) &= 0\end{aligned}$$

can be transformed as

$$\begin{aligned}\Phi_1(x_1,x_2,x_3,u_1) &= 0\\ \Phi_2(x_2,x_3,u_1,u_2) &= 0\\ \Phi_3(x_3,u_1,u_2,u_3) &= 0\end{aligned}$$

From the property of Jacobian of implicit functions, we have

$$\begin{aligned}\frac{\partial(u_1,u_2,u_3)}{\partial(x_1,x_2,x_3)} &= (-1)^3\frac{\partial(\Phi_1,\Phi_2,\Phi_3)}{\partial(x_1,x_2,x_3)}\Bigg/\frac{\partial(\Phi_1,\Phi_2,\Phi_3)}{\partial(u_1,u_2,u_3)}\\ &= (-1)^3\frac{\dfrac{\partial\Phi_1}{\partial x_1}\dfrac{\partial\Phi_2}{\partial x_2}\dfrac{\partial\Phi_3}{\partial x_3}}{\dfrac{\partial\Phi_1}{\partial u_1}\dfrac{\partial\Phi_2}{\partial u_2}\dfrac{\partial\Phi_3}{\partial u_3}}\end{aligned}$$

$$\left[\frac{\partial(\Phi_1,\Phi_2,\Phi_3)}{\partial(x_1,x_2,x_3)}=\begin{vmatrix}\dfrac{\partial\Phi_1}{\partial x_1} & \dfrac{\partial\Phi_1}{\partial x_2} & \dfrac{\partial\Phi_1}{\partial x_3}\\ 0 & \dfrac{\partial\Phi_2}{\partial x_2} & \dfrac{\partial\Phi_2}{\partial x_3}\\ 0 & 0 & \dfrac{\partial\Phi_3}{\partial x_3}\end{vmatrix}=\frac{\partial\Phi_1}{\partial x_1}\frac{\partial\Phi_2}{\partial x_2}\frac{\partial\Phi_3}{\partial x_3}\right.$$

Similarly,

$$\frac{\partial(\Phi_1,\Phi_2,\Phi_3)}{\partial(u_1,u_2,u_3)}=\begin{vmatrix}\frac{\partial\Phi_1}{\partial u_1} & 0 & 0\\ \frac{\partial\Phi_2}{\partial u_1} & \frac{\partial\Phi_2}{\partial u_2} & 0\\ \frac{\partial\Phi_3}{\partial u_1} & \frac{\partial\Phi_3}{\partial u_2} & \frac{\partial\Phi_3}{\partial u_3}\end{vmatrix}=\frac{\partial\Phi_1}{\partial u_1}\frac{\partial\Phi_2}{\partial u_2}\frac{\partial\Phi_3}{\partial u_3}$$

4. If $x = r\cos\theta$, $y = r\sin\theta$, find $\frac{\partial(x,y)}{\partial(r,\theta)}, \frac{\partial(r,\theta)}{\partial(x,y)}$. Show that

$$\frac{\partial r}{\partial x} = \frac{\partial x}{\partial r},\quad \frac{\partial\theta}{\partial x}=\frac{1}{r^2}\cdot\frac{\partial x}{\partial\theta}$$

$$\frac{\partial r}{\partial y} = \frac{\partial y}{\partial r},\quad \frac{\partial\theta}{\partial y}=\frac{1}{r^2}\cdot\frac{\partial y}{\partial\theta}$$

Here we have

$$x = r\cos\theta, y = r\sin\theta \tag{1}$$

$$\therefore\quad \frac{\partial x}{\partial r}=\cos\theta,\quad \frac{\partial x}{\partial\theta}=-r\sin\theta,\quad \frac{\partial y}{\partial r}=\sin\theta,\quad \frac{\partial y}{\partial\theta}=r\cos\theta.$$

From (1),

$$r=\sqrt{x^2+y^2},\quad \theta=\tan^{-1}\left(\frac{y}{x}\right) \tag{2}$$

$$\therefore\quad \frac{\partial r}{\partial x}=\frac{2x}{2\sqrt{x^2+y^2}}=\frac{x}{\sqrt{x^2+y^2}},\ \frac{\partial r}{\partial y}=\frac{y}{\sqrt{x^2+y^2}}$$

$$\frac{\partial\theta}{\partial x}=\frac{1}{1+\frac{y^2}{x^2}}\left(-\frac{y}{x^2}\right)=-\frac{y}{x^2+y^2}$$

$$\frac{\partial\theta}{\partial y}=\frac{1}{1+\frac{y^2}{x^2}}\left(\frac{1}{x}\right)=\frac{x}{x^2+y^2}.$$

$$\therefore\quad \frac{\partial(x,y)}{\partial(r,\theta)}=\begin{vmatrix}\frac{\partial x}{\partial r} & \frac{\partial x}{\partial\theta}\\ \frac{\partial y}{\partial r} & \frac{\partial y}{\partial\theta}\end{vmatrix}=\begin{vmatrix}\cos\theta & -r\sin\theta\\ \sin\theta & r\cos\theta\end{vmatrix}=r(\cos^2\theta+\sin^2\theta)=r$$

and

$$\frac{\partial(r,\theta)}{\partial(x,y)}=\begin{vmatrix}\frac{\partial r}{\partial x} & \frac{\partial r}{\partial y}\\ \frac{\partial\theta}{\partial x} & \frac{\partial\theta}{\partial y}\end{vmatrix}=\begin{vmatrix}\frac{x}{\sqrt{x^2+y^2}} & \frac{y}{\sqrt{x^2+y^2}}\\ -\frac{y}{x^2+y^2} & \frac{x}{x^2+y^2}\end{vmatrix}$$

$$=\frac{x^2+y^2}{(x^2+y^2)^{3/2}}=\frac{1}{\sqrt{x^2+y^2}}=\frac{1}{r}.$$

Again

$$\begin{aligned}
\frac{\partial r}{\partial x} &= \cos\theta = \frac{\partial x}{\partial r} \\
\frac{\partial \theta}{\partial x} &= -\frac{y}{x^2+y^2} = \frac{-r\sin\theta}{r^2} = \frac{1}{r^2}\frac{\partial x}{\partial \theta} \\
\frac{\partial r}{\partial y} &= \frac{y}{\sqrt{x^2+y^2}} = \frac{r\sin\theta}{r} = \sin\theta = \frac{\partial y}{\partial r} \\
\frac{\partial \theta}{\partial y} &= \frac{x}{x^2+y^2} = \frac{r\cos\theta}{r^2} = \frac{1}{r^2}\frac{\partial y}{\partial \theta}
\end{aligned}$$

22.11.3 Some Important Cases

Two important formulae are stated below:

1.

$$\iint_{R_{xy}} f(x,y)dxdy = \iint_{R_{uv}} f[x(u,v), y(u,v)] \left|\frac{\partial(x,y)}{\partial(u,v)}\right| dudv. \tag{22.11.2}$$

2.

$$\begin{aligned}
\iiint_{R_{xyz}} f(x,y,z)dxdydz &= \iiint_{R_{uvw}} f[x(u,v,w), y(u,v,w), z(u,v,w)] \\
&\quad \times \left|\frac{\partial(x,y,z)}{\partial(u,v,w)}\right| dudvdw
\end{aligned} \tag{22.11.3}$$

22.11.4 Deductions of Equations (22.11.2) and (22.11.3) in Some Special Cases

A. Cartesian (x, y) to polar (r, θ), $x = r\cos\theta$, $y = r\sin\theta$.

$$\therefore \quad J = \frac{\partial(x,y)}{\partial(r,\theta)} = r.$$

$$\therefore \quad \iint_{R_{xy}} f(x,y)dxdy = \iint_{R_{r\theta}} f(r\cos\theta, r\sin\theta) r dr d\theta \tag{22.11.4}$$

B. Cartesian (x, y, z) to cylindrical (r, θ, z), $x = r\cos\theta$, $y = r\sin\theta$, $z = z$.

$$\therefore \quad J = \frac{\partial(x,y,z)}{\partial(r,\theta,z)} = r.$$

$$\therefore \quad \iiint_{R_{xyz}} f(x,y,z)dxdydz = \iiint_{R_{r\theta z}} f(r\cos\theta, r\sin\theta, z) r dr d\theta dz \tag{22.11.5}$$

C. Cartesian (x, y, z) to spherical polar (r, θ, ϕ), $x = r\sin\theta\cos\phi$, $y = r\sin\theta\sin\phi$, $z = r\cos\theta$.

$$\therefore \quad J = \frac{\partial(x,y,z)}{\partial(r,\theta,\phi)} = r^2\sin\theta.$$

$$\therefore \quad \iiint_{R_{xyz}} f(x,y,z)dxdydz = \iiint_{R_{r\theta\phi}} f(r\sin\theta\cos\phi, r\sin\theta\sin\phi, r\cos\theta)$$
$$\times r^2 \sin\theta dr d\theta d\phi \qquad (22.11.6)$$

Illustrative Examples

1. Find the Jacobian of the transformation from rectangular Cartesian co-ordinates (x, y, z) to spherical polar co-ordinates (r, θ, ϕ). Also evaluate the integral

$$\iiint_V (x^2 + y^2 + z^2) dxdydz,$$

where V is the volume of a sphere with centre at the origin and radius R. **[B.U. (Hons) 1992]**

Here Cartesian co-ordinates (x, y, z) are transformed to spherical polar co-ordinates (r, θ, ϕ).

$$\therefore \quad x = r\sin\theta\cos\phi, \quad y = r\sin\theta\sin\phi, \quad z = r\cos\theta.$$

$$\begin{aligned}
\therefore \quad \frac{\partial x}{\partial r} &= \sin\theta\cos\phi, \frac{\partial y}{\partial r} = \sin\theta\sin\phi, \frac{\partial z}{\partial r} = \cos\theta \\
\frac{\partial x}{\partial \theta} &= r\cos\theta\cos\phi, \frac{\partial y}{\partial \theta} = r\cos\theta\sin\phi, \frac{\partial z}{\partial \theta} = -r\sin\theta \\
\frac{\partial x}{\partial \phi} &= -r\sin\theta\sin\phi, \frac{\partial y}{\partial \phi} = r\sin\theta\cos\phi, \frac{\partial z}{\partial \phi} = 0.
\end{aligned}$$

$$\begin{aligned}
\therefore \quad J &= \frac{\partial(x,y,z)}{\partial(r,\theta,\phi)} = \begin{vmatrix} \sin\theta\cos\phi & r\cos\theta\cos\phi & -r\sin\theta\sin\phi \\ \sin\theta\sin\phi & r\cos\theta\sin\phi & r\sin\theta\cos\phi \\ \cos\theta & -r\sin\theta & 0 \end{vmatrix} \\
&= \cos\theta \begin{vmatrix} r\cos\theta\cos\phi & -r\sin\theta\sin\phi \\ r\cos\theta\sin\phi & r\sin\theta\cos\phi \end{vmatrix} + r\sin\theta \begin{vmatrix} \sin\theta\cos\phi & -r\sin\theta\sin\phi \\ \sin\theta\sin\phi & r\sin\theta\cos\phi \end{vmatrix} \\
&= r^2\sin\theta\cos^2\theta \begin{vmatrix} \cos\phi & -\sin\phi \\ \sin\phi & \cos\phi \end{vmatrix} + r^2\sin^3\theta \begin{vmatrix} \cos\phi & -\sin\phi \\ \sin\phi & \cos\phi \end{vmatrix} \\
&= r^2\sin\theta\cos^2\theta + r^2\sin^3\theta = r^2\sin\theta(\cos^2\theta + \sin^2\theta) \\
&= r^2\sin\theta.
\end{aligned}$$

2nd Part

$$\iiint_V (x^2+y^2+z^2)dxdydz = \int_{r=0}^{R}\int_{\theta=0}^{\pi}\int_{\phi=0}^{2\pi} r^2\cdot r^2\cdot \sin\theta dr d\theta d\phi$$

$$= \int_{\phi=0}^{2\pi}\left[\int_{\theta=0}^{\pi}\left[\int_{r=0}^{R} r^4 dr\right]\sin\theta d\theta\right]d\phi$$

$$= \frac{R^5}{5}\int_{\phi=0}^{2\pi}\int_{\theta=0}^{\pi}\sin\theta d\theta d\phi = -\frac{R^5}{5}\int_{\phi=0}^{2\pi}\left[\cos\theta\right]_0^{\pi}d\phi$$

$$= \frac{2R^5}{5}\int_{\phi=0}^{2\pi}d\phi = \frac{4\pi R^5}{5}$$

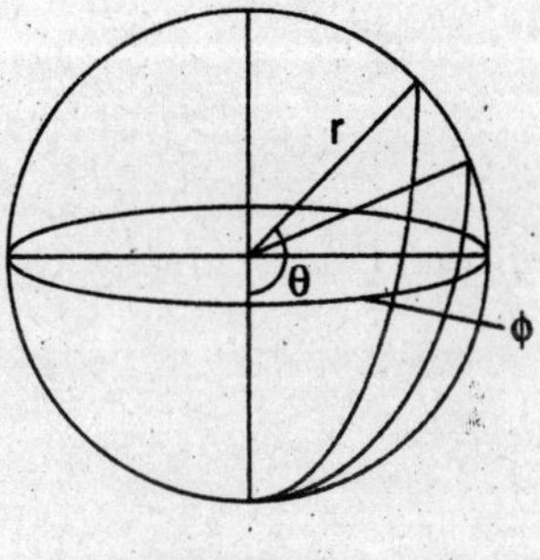

Fig 22.11.1

2. Using the transformation $u = x + y$ and $v = x - 2y$, evaluate $\int_R (x+y)^2 dxdy$, where R is the parallelogram in the XY-plane with vertices $(1,0), (3,1), (2,2), (0,1)$. **[BPH 1997]**

Using transformation

$$u = x + y, \quad v = x - 2y \tag{1}$$

the point $A(1,0)$ is transformed to $A(1+0, 1-2.0)$, i.e., $A(1,1)$; $B(3,1)$ is transformed to $B(3+1, 3-2.1)$, i.e., $B(4,1)$.

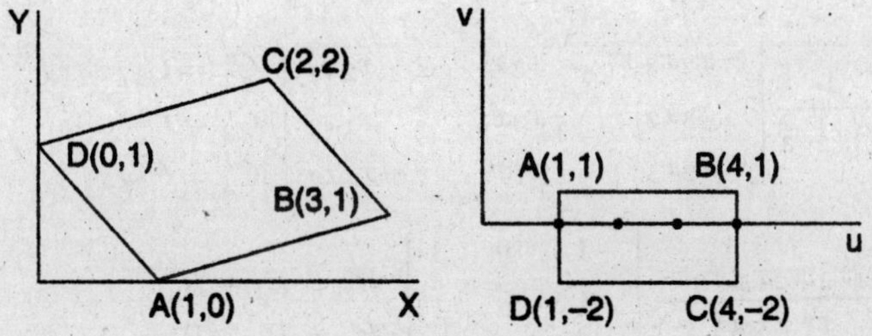

Fig 22.11.2

Similarly, $C(2,2)$ is transformed to $C(4,-2)$ and $D(0,1)$ to $D(1,-2)$.

Again, from (1), solving for x and y we get,

$$x = \frac{1}{3}(2u+v), y = \frac{1}{3}(u-v).$$

$$\therefore \quad J = \frac{\partial(x,y)}{\partial(u,v)} = \begin{vmatrix} \frac{\partial x}{\partial u} & \frac{\partial x}{\partial v} \\ \frac{\partial y}{\partial u} & \frac{\partial y}{\partial v} \end{vmatrix} = \begin{vmatrix} \frac{2}{3} & \frac{1}{3} \\ \frac{1}{3} & \frac{-1}{3} \end{vmatrix} = -\frac{1}{3}.$$

$$\therefore \int_R (x+y)^2 dxdy = \int_{u=1}^{4}\int_{v=-2}^{1} u^2|J|dudv = \frac{1}{3}\int_{u=1}^{4}\int_{v=-2}^{1} u^2 dudv$$

$$= \frac{1}{3}\int_{v=-2}^{1}\left[\int_{u=1}^{4} u^2 du\right] dv = \frac{1}{3}\int_{-2}^{1}\left[\frac{u^3}{3}\right]_1^4 dv$$

$$= 7\int_{-2}^{1} dv = 7\big[v\big]_{-2}^{1} = 21.$$

3. A transformation from (x_1, x_2, x_3) to (y_1, y_2, y_3) is defined by

$$y_1 = \frac{x_2 x_3}{x_1}, \quad y_2 = \frac{x_3 x_1}{x_2}, \quad y_3 = \frac{x_1 x_2}{x_3}.$$

Find the Jacobian $\dfrac{\partial(y_1, y_2, y_3)}{\partial(x_1, x_2, x_3)}$ of the transformation.

$$\frac{\partial(y_1,y_2,y_3)}{\partial(x_1,x_2,x_3)} = \begin{vmatrix} \frac{\partial y_1}{\partial x_1} & \frac{\partial y_1}{\partial x_2} & \frac{\partial y_1}{\partial x_3} \\ \frac{\partial y_2}{\partial x_1} & \frac{\partial y_2}{\partial x_2} & \frac{\partial y_2}{\partial x_3} \\ \frac{\partial y_3}{\partial x_1} & \frac{\partial y_3}{\partial x_2} & \frac{\partial y_3}{\partial x_3} \end{vmatrix} = \begin{vmatrix} \frac{-x_2x_3}{x_1^2} & \frac{x_3}{x_1} & \frac{x_2}{x_1} \\ \frac{x_3}{x_2} & \frac{-x_3x_1}{x_2^2} & \frac{x_1}{x_2} \\ \frac{x_2}{x_3} & \frac{x_1}{x_3} & \frac{-x_1x_2}{x_3^2} \end{vmatrix} \quad \begin{matrix} R_1' = R_1 \cdot x_1^2 \\ R_2' = R_2 \cdot x_2^2 \\ R_3' = R_3 \cdot x_3^2 \end{matrix}$$

$$= \frac{1}{x_1^2 x_2^2 x_3^2}\begin{vmatrix} -x_2x_3 & x_3x_1 & x_2x_1 \\ x_3x_2 & -x_3x_1 & x_1x_2 \\ x_2x_3 & x_1x_3 & -x_1x_2 \end{vmatrix} \quad \begin{matrix} C_1' = C_1/x_2x_3 \\ C_2' = C_2/x_3x_1 \\ C_3' = C_3/x_1x_2 \end{matrix}$$

$$= \frac{x_1x_2.x_2x_3.x_3x_1}{x_1^2.x_2^2.x_3^2}\begin{vmatrix} -1 & 1 & 1 \\ 1 & -1 & 1 \\ 1 & 1 & -1 \end{vmatrix} \quad \begin{matrix} R_1' = R_1 + R_3 \\ R_2' = R_2 + R_3 \end{matrix}$$

$$= \begin{vmatrix} 0 & 2 & 0 \\ 2 & 0 & 0 \\ 1 & 1 & -1 \end{vmatrix} = 2\begin{vmatrix} 0 & 2 \\ -1 & 1 \end{vmatrix} = 4.$$

22.12 Functional Relationship

Let u, v, w be three functions of 3 independent variables x, y, z. *The necessary and sufficient condition* that there may exist a relation between u, v and w is that $\dfrac{\partial(u,v,w)}{\partial(x,y,z)} = 0$.

22.12.1 The Condition is Necessary

Let the relation be of the form $F(u, v, w) = 0$. Differentiating w.r.t. x, y, z we get,

$$\frac{\partial F}{\partial u}\frac{\partial u}{\partial x} + \frac{\partial F}{\partial v}\frac{\partial v}{\partial x} + \frac{\partial F}{\partial w}\frac{\partial w}{\partial x} = 0$$

$$\frac{\partial F}{\partial u}\frac{\partial u}{\partial y} + \frac{\partial F}{\partial v}\frac{\partial v}{\partial y} + \frac{\partial F}{\partial w}\frac{\partial w}{\partial y} = 0$$

$$\frac{\partial F}{\partial u}\frac{\partial u}{\partial z} + \frac{\partial F}{\partial v}\frac{\partial v}{\partial z} + \frac{\partial F}{\partial w}\frac{\partial w}{\partial z} = 0$$

The result follows by eliminating $\frac{\partial F}{\partial u}, \frac{\partial F}{\partial v}, \frac{\partial F}{\partial w}$ from the above three relations.

22.12.2 The Sufficiency of the Condition

We assume that $\dfrac{\partial(u, v, w)}{\partial(x, y, z)} = 0$; it remains to be shown that there exists a relation between u, v, w.

The equations which connect u, v, w and x, y, z can be transformed to the form (by process of elimination) given by

$$\Phi_1(x, y, z, u) = 0$$

$$\Phi_2(y, z, u, v) = 0$$

$$\Phi_3(z, u, v, w) = 0$$

From the property of Jacobian of implicit functions we have,

$$\frac{\partial(u, v, w)}{\partial(x, y, z)} = (-1)^3 \frac{\partial(\Phi_1, \Phi_2, \Phi_3)}{\partial(x, y, z)} \Bigg/ \frac{\partial(\Phi_1, \Phi_2, \Phi_3)}{\partial(u, v, w)}$$

$$= -\frac{\dfrac{\partial \Phi_1}{\partial x}\dfrac{\partial \Phi_2}{\partial y}\dfrac{\partial \Phi_3}{\partial z}}{\dfrac{\partial \Phi_1}{\partial u}\dfrac{\partial \Phi_2}{\partial v}\dfrac{\partial \Phi_3}{\partial w}} \quad \text{[by Ex.3 at the end of Art. 22.11.2]}$$

Since,

$$\frac{\partial(u, v, w)}{\partial(x, y, z)} = 0,$$

we obtain

$$\frac{\partial \Phi_1}{\partial x}\frac{\partial \Phi_2}{\partial y}\frac{\partial \Phi_3}{\partial z} = 0,$$

so that we may say that if $\dfrac{\partial \Phi_1}{\partial x} = 0$, then Φ_1 is independent of x, that is, Φ_1 is a function of y, z and u. Then by elimination of y and z one by one from $\Phi_1 = 0$, $\Phi_2 = 0$, $\Phi_3 = 0$, we get the relation between u, v and w.

If $\dfrac{\partial \Phi_2}{\partial y} = 0$, Φ_2 is independent of y and then by elimination of z from $\Phi_2 = 0$ and $\Phi_3 = 0$ we get the required relation.

If $\dfrac{\partial \Phi_3}{\partial z} = 0$, then the relation is obtained from $\Phi_3 = 0$. Thus the proposition is proved.

Illustrative Examples

1. **If** $u = \dfrac{x+y}{z}$, $v = \dfrac{y+z}{x}$, $w = \dfrac{y(x+y+z)}{xz}$, **show that** u, v, w **are functionally related and find the relation between them**

$$\frac{\partial(u,v,w)}{\partial(x,y,z)} = \begin{vmatrix} \frac{1}{z} & \frac{1}{z} & -\frac{x+y}{z^2} \\ -\frac{y+z}{x^2} & \frac{1}{x} & \frac{1}{x} \\ -\frac{y^2+yz}{x^2 z} & \frac{x+2y+z}{xz} & -\frac{(x+y)y}{xz^2} \end{vmatrix}$$

$$= \frac{1}{x^4 z^4}\begin{vmatrix} z & z & -(x+y) \\ -(y+z) & x & x \\ -yz(y+z) & xz(x+2y+z) & -xy(x+y) \end{vmatrix}$$

$$= \frac{1}{x^4 z^4}\begin{vmatrix} z & 0 & -(x+y+z) \\ -(y+z) & x+y+z & x+y+z \\ -yz(y+z) & z(x+y)^2+z^2(x+y) & (x+y+z)(z-x)y \end{vmatrix}$$

$$= \frac{(x+y+z)^2}{x^4 z^4}\begin{vmatrix} z & 0 & -1 \\ -(y+z) & 1 & 1 \\ -yz(y+z) & z(x+y) & yz-xy \end{vmatrix}$$

$$= \frac{(x+y+z)^2}{x^4 z^4}\begin{vmatrix} 0 & 0 & -1 \\ -y & 1 & 1 \\ -yz(y+z) & z(x+y) & yz-xy \end{vmatrix} = 0$$

Therefore, u, v, w are functionally related.

Here, $uv - 1 = \frac{y(x+y+z)}{xz} = w$, so that $uv = 1 + w$ is the relation.

2. Show that $ax^2 + 2hxy + by^2$ and $Ax^2 + 2Hxy + By^2$ are independent unless

$$\frac{a}{A} = \frac{h}{H} = \frac{b}{B}.$$

Let $u = ax^2 + 2hxy + by^2$ and $v = Ax^2 + 2Hxy + By^2$.

$$\therefore \quad \frac{\partial(u,v)}{\partial(x,y)} = 4\begin{vmatrix} ax+hy & hx+by \\ Ax+Hy & Hx+By \end{vmatrix} = 4\{(aH-Ah)x^2 + (aB-Ab)xy + (Bh-bH)y^2\}.$$

If u and v are not independent, then $\frac{\partial(u,v)}{\partial(x,y)} = 0$.

$$\therefore \quad (aH-Ah)x^2 + (aB-Ab)xy + (Bh-bH)y^2 = 0$$

The coefficients of x^2 and y^2 must be zero, since the variables x and y are independent.

$\therefore \quad aH - Ah = 0$ or, $\frac{a}{A} = \frac{h}{H}$

Also, $Bh - bH = 0$ or, $\frac{h}{H} = \frac{b}{B}$

$\therefore$ unless $\frac{a}{A} = \frac{h}{H} = \frac{b}{B}$, u and v are independent.

3. If u, v are two polynomials in x and y which are homogeneous and of nth degree, then

$$n(udv - vdu) = \frac{\partial(u, v)}{\partial(x, y)}(xdy - ydx).$$

Given that u and v are homogeneous in x and y of degree n, u and v possess continuous partial derivatives. Therefore, by Euler's theorem

$$x\frac{\partial u}{\partial x} + y\frac{\partial u}{\partial y} = nu, \quad x\frac{\partial v}{\partial x} + y\frac{\partial v}{\partial y} = nv.$$

Again,

$$du = \frac{\partial u}{\partial x}dx + \frac{\partial u}{\partial y}dy \quad \text{and} \quad dv = \frac{\partial v}{\partial x}dx + \frac{\partial v}{\partial y}dy.$$

$$\begin{aligned}\therefore \quad nudv - nvdu &= \left(x\frac{\partial u}{\partial x} + y\frac{\partial u}{\partial y}\right)\left(\frac{\partial v}{\partial x}dx + \frac{\partial v}{\partial y}dy\right) - \left(x\frac{\partial v}{\partial x} + y\frac{\partial v}{\partial y}\right)\left(\frac{\partial u}{\partial x}dx + \frac{\partial u}{\partial y}dy\right) \\ &= \left(\frac{\partial u}{\partial x}\frac{\partial v}{\partial y} - \frac{\partial v}{\partial x}\frac{\partial u}{\partial y}\right)(xdy - ydx) = \frac{\partial(u, v)}{\partial(x, y)}(xdy - ydx).\end{aligned}$$

22.13 Implicit Function : Existence Theorem

Let $f(x, y)$ be a function of two variables. The functional equation $f(x, y) = 0$ may or may not determine y as a function of x. Even if the equation determines y as a function of x, such a function is not necessarily unique.

Let $y = \Phi(x)$ be a root of the equation $f(x, y) = 0$ such that the values of x for which $\Phi(x)$ is defined, are given by $f[x, \Phi(x)] = 0$ which holds identically. Then $y = \Phi(x)$ is an *implicit function* defined by $f(x, y) = 0$.

We state, without proof, the theorem regarding existence of implicit function.

22.13.1 Existence Theorem (for Two Variables)

Let $f(x, y)$ be a function of two variables x and y and $(a, b) \in D$, the domain of definition of $f(x, y)$ such that

(i) $f(a, b) = 0$,

(ii) f_x and f_y exist and are continuous in a certain neighbourhood of (a, b) and

(iii) $f_y(a, b) \neq 0$;

then there exists a rectangle $(a - h, a + h; b - k, b + k)$ about (a, b) such that for every value of x in $[a - h, a + h]$, the equation $f(x, y) = 0$ determines y uniquely as $y = \Phi(x)$ lying in the interval $[b - k, b + k]$ with the properties:

(i) $b = \Phi(a)$,

(ii) $f(x, \Phi(x)) = 0$ for every x in $[a - h, a + h]$,

(iii) $\Phi(x)$ is derivable and both $\Phi(x)$ and $\Phi'(x)$ are continuous in $[a - h, a + h]$ and $\Phi'(x) = -\frac{f_x}{f_y}$ at (a, b).

Illustrative Examples

Examine the following equations for the existence of unique implicit function near indicated points; also find the first derivative of the solution whenever these exist:

1. $y^2 + xy^3 + x^2 = 0$ near $(0, 0)$.

Here, $f(x, y) = y^2 + xy^3 + x^2$

$f(0, 0) = 0$ holds. $f_x = y^3 + 2x$, $f_y = 2y + 3xy^2$.

Here, f_x and f_y are both continuous near $(0, 0)$, but both f_x and f_y vanish at $(0, 0)$.

$\therefore$ y cannot be determined as a function of x, nor x can be determined as a function of y.

2. $y^3 + x^3 - 3xy + y = 0$ near $(0, 0)$.

$f(x, y) = y^3 + x^3 - 3xy + y$, $f(0, 0) = 0$

$f_x = 3x^2 - 3y$, $f_y = 3y^2 - 3x + 1$

$\therefore$ f_x and f_y are continuous in any neighbourhood of $(0, 0)$.

Again $f_y(0, 0) = 1 \neq 0$

$\therefore$ y can be determined uniquely as a function of x.

$$\left(\frac{dy}{dx}\right)_{(0,0)} = -\left(\frac{f_x}{f_y}\right)_{(0,0)} = 0.$$

3. $y^2 - yx^2 - 2x^5 = 0$ near $(0, 0)$ and $(1, -1)$.

$f(x, y) = y^2 - yx^2 - 2x^5$, $f(0, 0) = 0$.

Also, $f(1, -1) = 0$, $f_x = -2xy - 10x^4$, $f_y = 2y - x^2$.

$\therefore$ f_x and f_y are continuous in any neighbourhood of $(0, 0)$ and $(1, -1)$.

$f_y(0, 0) = 0$, but $f_y(1, -1) = -3 \neq 0$.

$\therefore$ y can be determined uniquely as x near $(1, -1)$ but not near $(0, 0)$.

$$y = \frac{x^2 \pm \sqrt{x^4 + 8x^5}}{2}$$

At $x = 1$,

$$y = \frac{x^2 - \sqrt{x^4 + 8x^5}}{2} \quad \text{gives } y = -1.$$

Hence the unique value of y is

$$y = \frac{x^2 - \sqrt{x^4 + 8x^5}}{2},$$

$$\text{and} \quad \left(\frac{dy}{dx}\right)_{(1,-1)} = -\left(\frac{f_x}{f_y}\right)_{(1,-1)} = -\left(\frac{-8}{-3}\right) = -\frac{8}{3}.$$

Exercises

1. Use ε-δ definition to justify the following statements on limits:

(a) $\lim\limits_{(x,y)\to(0,0)} \dfrac{xy}{\sqrt{x^2+y^2}} = 0$

(b) $\lim\limits_{(x,y)\to(0,0)} \dfrac{x^3-y^3}{x^2+y^2} = 0$

(c) $\lim\limits_{(x,y)\to(0,0)} \dfrac{x^2y^2}{x^2+y^2} = 0$

(d) $\lim\limits_{(x,y)\to(0,0)} y^2 \sin\dfrac{1}{x} = 0$

(e) $\lim\limits_{(x,y)\to(0,0)} \dfrac{x^2-y^2}{x^2+y^2} = 0$

(f) $\lim\limits_{(x,y)\to(0,0)} e^{-(x^2+y^2)} = 0$

(g) $\lim\limits_{(x,y)\to(0,0)} e^{-1/x^2+y^2} = 1.$

2. (a) Let $f(x,y) = \begin{cases} 1, & \text{if } xy \neq 0 \\ 0, & \text{if } xy = 0 \end{cases}$

Show that the two repeated limits exist at $(0,0)$ and are equal, but the simultaneous limit $\lim\limits_{(x,y)\to(0,0)} f(x,y)$ does not exist.

(b) Let $f(x,y) = \begin{cases} xy\dfrac{x^2-y^2}{x^2+y^2}, & \text{where } x^2+y^2 \neq 0 \\ 0, & \text{where } x^2+y^2 = 0 \end{cases}$.

Show that here double limit and both repeated limits exist at $(0,0)$.

(c) Let $f(x,y) = \begin{cases} x\sin\dfrac{1}{y} + y\sin\dfrac{1}{x}, & xy \neq 0 \\ 0, & xy = 0 \end{cases}$

Show that at $(0,0)$ the double limit exists, but the repeated limits do not exist.

3. Show that the following limits do not exist:

(a) $\lim\limits_{(x,y)\to(0,0)} \dfrac{x^2y}{x^4+y^2}$ (put $y = mx^2$),

(b) $\lim\limits_{(x,y)\to(0,0)} \dfrac{x^3+y^3}{x-y}$ (put $y = x - mx^3$),

(c) $\lim\limits_{(x,y)\to(0,0)} \dfrac{xy^3}{x^2+y^6}$ (put $x = my^3$),

(d) $\lim\limits_{(x,y)\to(0,0)} (x+y)\dfrac{y+(x+y)^2}{y-(x+y)^2}$,

(e) $\lim\limits_{(x,y)\to(0,0)} \dfrac{x^2y^4}{(x^2+y^4)^2}$ (put $x = my^2$).

4. (a) Show that the function $f(x,y) = \begin{cases} xy\dfrac{x^2-y^2}{x^2+y^2}, & \text{if } (x,y) \neq (0,0) \\ 0, & \text{if } (x,y) = (0,0) \end{cases}$

is continuous at $(0,0)$.

(b) Let $f(x,y) = \begin{cases} \dfrac{2xy}{x^2+y^2}, & \text{where } x^2+y^2 \neq 0 \\ 0, & \text{where } x^2+y^2 = 0 \end{cases}$

Prove that f is a continuous function of either variable when the other variable is given a fixed value. Is the function continuous at $(0,0)$. **[C.U. (Math Hons) 1990, '96]**

(c) Show that $f(x,y) = \begin{cases} \dfrac{x^3-y^3}{x^2+y^2}, & \text{where } (x,y) \neq (0,0) \\ 0, & \text{if } (x,y) = (0,0) \end{cases}$

is continuous at $(0,0)$, possesses partial derivatives at $(0,0)$, but is not differentiable at $(0,0)$. Explain the reason for failure of differentiability at $(0,0)$.

[C.U. (Math Hons) 1993, '98]

5. Show that the function f defined in the neighbourhood of $(0,0)$ by $f(x,y) = |x|+|y|$ is continuous, but not differentiable at $(0,0)$. **[C.U. (Math Hons) 1990]**

6. (a) A function f of two variables admits of a differential at a point (a,b) in its domain. Show that f is continuous at (a,b) and both $\dfrac{\partial f}{\partial x}$ and $\dfrac{\partial f}{\partial y}$ exist at that point.

(b) Show that the function f defined by $f(x,y) = \begin{cases} \dfrac{xy^2}{x^2+y^4}, & \text{if } x \neq 0 \\ 0, & \text{if } x = 0 \end{cases}$

possesses first order partial derivatives at $(0,0)$, yet it is not differentiable at $(0,0)$.

[C.U. (Math Hons) 1995]

7. (a) If $f(x,y) = \begin{cases} \dfrac{x^3+y^3}{x-y}, & x \neq y \\ 0, & x = y \end{cases}$, show that $f(x,y)$ is not continuous at $(0,0)$, but f_x and f_y exist at $(0,0)$. [See Ex. 3(b) above.] **[C.U. (Math Hons) 1997]**

(b) Let $f(x,y) = \begin{cases} x^2 \sin\dfrac{1}{x} + y^2 \sin\dfrac{1}{y}, & \text{when neither } x = 0 \text{ nor } y = 0 \\ x^2 \sin\dfrac{1}{x}, & \text{when } x \neq 0, y = 0 \\ y^2 \sin\dfrac{1}{y}, & \text{when } y \neq 0, x = 0 \\ 0, & \text{when } x = y = 0 \end{cases}$

Show that neither f_x nor f_y is continuous at $(0,0)$, but the function is differentiable at $(0,0)$. In what way is continuity of f_x and f_y related to differentiability?

[C.U. (Math Hons) 1997, 2000]

8. Let $f(x,y) = \begin{cases} (x^2+y^2)\tan^{-1}\dfrac{y}{x}, & \text{when } x \neq 0 \\ \dfrac{\pi}{2}y^2, & \text{when } x = 0 \end{cases}$

Show that $f_{xy}(0,0) \neq f_{yx}(0,0)$.

9. A function f is defined as follows:

$$f(x,y) = \begin{cases} \dfrac{x^2y^2}{x^2+y^2}, & \text{for all } (x,y) \neq (0,0) \\ 0, & \text{when } (x,y) = (0,0) \end{cases}$$

Show that (a) f does not satisfy the conditions of Schwarz's theorem or Young's theorem, but (b) $f_{xy} = f_{yx}$ at $(0,0)$. **[C.U. (Math Hons) 1995, 2002]**

10. If $f(x,y) = \begin{cases} xy, & \text{when } |x| \geq |y| \\ -xy, & \text{when } |x| < |y| \end{cases}$, show that $f_{xy}(0,0) \neq f_{yx}(0,0)$. Which conditions of Schwarz's theorem are not satisfied by f? **[C.U. (Math Hons) 1996]**

11. The rectangular co-ordinates (x,y,z) of a point in space are given in terms of its spherical polar co-ordinates (r,θ,ϕ) by the equations $x = r\sin\theta\cos\phi$, $y = r\sin\theta\sin\phi$ and $z = r\cos\theta$. Compute the nine partial derivatives of x, y, z w.r.t. r, θ, ϕ and vice versa.

Show that $\dfrac{\partial x}{\partial r}\dfrac{\partial r}{\partial x} + \dfrac{\partial x}{\partial \theta}\dfrac{\partial \theta}{\partial x} + \dfrac{\partial x}{\partial \phi}\dfrac{\partial \phi}{\partial x} = 1$.

12. If $V = \cos^{-1}\dfrac{x+y}{\sqrt{x}+\sqrt{y}}$, then verify that $\cos V = \dfrac{x+y}{\sqrt{x}+\sqrt{y}}$ is a homogeneous function of x, y of degree $\dfrac{1}{2}$. Hence prove that $x\dfrac{\partial V}{\partial x} + y\dfrac{\partial V}{\partial y} + \dfrac{1}{2}\cot V = 0$.

Ans. $\cos V = \dfrac{x+y}{\sqrt{x}+\sqrt{y}} = f(x,y)$ say, so that we may write

$$f(tx,ty) = \frac{t(x+y)}{\sqrt{t}(\sqrt{x}+\sqrt{y})} = t^{1/2}f(x,y).$$

$\therefore$ $\cos V$ is a homogeneous function in x, y of degree $\dfrac{1}{2}$.

Again, using Euler's theorem, we obtain

$$x\frac{\partial}{\partial x}(\cos V) + y\frac{\partial}{\partial y}(\cos V) = \frac{1}{2}\cos V$$

$$\text{or,} \quad -x\sin V\frac{\partial V}{\partial x} - y\sin V\frac{\partial V}{\partial y} = \frac{1}{2}\cos V$$

$$\text{or,} \quad x\frac{\partial V}{\partial x} + y\frac{\partial V}{\partial y} + \frac{1}{2}\cot V = 0$$

We may similarly obtain

(a) if $u = \tan^{-1}\dfrac{x^3+y^3}{x-y}$, then $x\dfrac{\partial u}{\partial x} + y\dfrac{\partial u}{\partial y} = \sin 2u$,

(b) if $u = \sin^{-1}\sqrt{\dfrac{x^{1/3}+y^{1/3}}{y^{1/2}+y^{1/2}}}$, then $x\dfrac{\partial u}{\partial x} + y\dfrac{\partial u}{\partial y} + \dfrac{1}{12}\tan u = 0$.

13. Consider the function f defined by

$$f(x,y) = \begin{cases} xy\dfrac{x^2-y^2}{x^2+y^2}, & \text{where } x^2+y^2 \neq 0 \\ 0, & \text{where } x^2+y^2 = 0 \text{ (i.e., } x=0, y=0) \end{cases}$$

Show that $f_{xy}(0,0) = 1$, $f_{yx}(0,0) = -1$, so that $f_{xy} \neq f_{yx}$ at $(0,0)$.

[C.U. (Math Hons) 1991, '93, '99]

14. Let z be a differentiable function of x and y, and let $x = r\cos\theta$, $y = r\sin\theta$. Prove that

(a) $\left(\frac{\partial z}{\partial r}\right)^2 + \frac{1}{r^2}\left(\frac{\partial z}{\partial \theta}\right)^2 = \left(\frac{\partial z}{\partial x}\right)^2 + \left(\frac{\partial z}{\partial y}\right)^2$.

(b) $\frac{\partial^2 z}{\partial r^2} + \frac{1}{r}\frac{\partial z}{\partial r} + \frac{1}{r^2}\frac{\partial^2 z}{\partial \theta^2} = \frac{\partial^2 z}{\partial x^2} + \frac{\partial^2 z}{\partial y^2}$. **[C.U. (Math Hons) 1988, '99]**

15. A function $f(x, y)$ having continuous second order partial derivatives, when expressed in terms of the new variables u and v defined by $x = \frac{1}{2}(u+v)$ and $y^2 = uv$, becomes $g(u, v)$; prove that

$$\frac{\partial^2 g}{\partial u \partial v} = \frac{1}{4}\left(\frac{\partial^2 f}{\partial x^2} + 2\frac{x}{y}\frac{\partial^2 f}{\partial x \partial y} + \frac{\partial^2 f}{\partial y^2} + \frac{1}{y}\frac{\partial f}{\partial y}\right).$$ **[C.U. (Math Hons) 1999]**

16. Let $y = F(x, t)$, where F is a differentiable function of two independent variables x and t which are related to two new variables u, v by the relations

$$u = x + ct, v = x - ct;$$

prove that the partial differential equation (*wave equation in Mathematical Physics*)

$$\frac{\partial^2 y}{\partial x^2} - \frac{1}{c^2}\frac{\partial^2 y}{dt^2} = 0 \text{ can be transformed into } \frac{\partial^2 y}{\partial u \partial v} = 0.$$

Hence obtain the general solution $y = g(x - ct) + h(x + ct)$ (g, h are arbitrary functions.

(See worked out examples of chapter 18)

17. The roots of the equation in $\lambda, (\lambda - x)^3 + (\lambda - y)^3 + (\lambda - z)^3 = 0$ are u, v, w. Prove that the Jacobian

$$\frac{\partial(u, v, w)}{\partial(x, y, z)} = -2\frac{(y-z)(z-x)(x-y)}{(v-w)(w-u)(u-v)}$$ **[C.U. (Math Hons) 1995]**

(See worked out examples)

18. If λ, μ, ν be the roots (assumed to be all real) of the following equation in k

$$\frac{x}{a+k} + \frac{y}{b+k} + \frac{z}{c+k} = 1; \text{ prove that } \frac{\partial(x, y, z)}{\partial(\lambda, \mu, \nu)} = -\frac{(\mu - \nu)(\nu - \lambda)(\lambda - \mu)}{(b-c)(c-a)(a-b)}.$$

19. Show that the functions $u = x + y - z, v = x - y + z, w = x^2 + y^2 + z^2 - 2yz$ are not independent. Find the relation among them.

20. Show that the functions $u = x + y + z, v = xy + yz + zx, w = x^3 + y^3 + z^3 - 3xyz$ are not independent, but they are related by $u^3 = 3uv + w$. **[C.U. (Math Hons) 1992, '98]**

21. Show that the expression $ax^2 + by^2 + cz^2 + 2fyz + 2gzx + 2hxy$ can be resolved into linear factors if

$$\begin{vmatrix} a & h & g \\ h & b & f \\ g & f & c \end{vmatrix} = 0.$$

22. Show that the transformation $x = r\cos\theta, y = r\sin\theta$ reduces the equation

$$xy\left(\frac{\partial^2 u}{\partial x^2} - \frac{\partial^2 u}{\partial y^2}\right) - (x^2 - y^2)\frac{\partial^2 u}{\partial x \partial y} = 0 \text{ to } r\frac{\partial^2 u}{\partial r \partial \theta} - \frac{\partial u}{\partial \theta} = 0.$$ **[C.U. (Math Hons) 1995]**

23. If $H(x, y)$ be a homogeneous function of x and y of degree n, having continuous partial derivatives and $u(x, y) = (x^2 + y^2)^{-\frac{n}{2}}$, show that

$$\frac{\partial}{\partial x}\left(H\frac{\partial u}{\partial x}\right) + \frac{\partial}{\partial y}\left(H\frac{\partial u}{\partial y}\right) = 0.$$ **[C.U. (Math Hons) 1996]**

24. (a) Let $u(x, y)$ and $v(x, y)$ be two functions having continuous first order partial derivatives at any point (x, y) of their domain and $\dfrac{\partial(u, v)}{\partial(x, y)} = 0$. Show that $u(x, y)$ and $v(x, y)$ are functionally dependent.

(b) If $u(x, y) = \phi(xy) + \sqrt{xy}\psi\left(\dfrac{y}{x}\right), x \neq 0, y \neq 0$, where ϕ and ψ are twice differentiable, prove that $x^2\dfrac{\partial^2 u}{\partial x^2} - y^2\dfrac{\partial^2 u}{\partial y^2} = 0$. Mention the necessity of the condition $x \neq 0, y \neq 0$.

[C.U. (Math Hons) 1996]

25. Transform the equation $x^2\dfrac{\partial z}{\partial x} + y\dfrac{\partial z}{\partial y} = z^2$, by introducing new independent variables $u = x$ and $v = \dfrac{1}{y} - \dfrac{1}{x}$ and the new function $w = \dfrac{1}{z} + \dfrac{1}{x}$. **[C.U. (Math Hons) 1996]**

26. If $u = \log(x^3 + y^3 + z^3 - 3xyz)$, show that $u_{xx} + u_{yy} + u_{zz} = -\dfrac{3}{(x+y+z)^2}$.

27. Let $\phi(x, y) = \begin{cases} (x^2 + y^2)\log(x^2 + y^2), & \text{when } x^2 + y^2 \neq 0 \\ 0, & \text{when } x^2 + y^2 = 0. \end{cases}$

Show that $\phi_{xy}(0, 0) = \phi_{yx}(0, 0)$, although neither ϕ_{xy} nor ϕ_{yx} is continuous at $(0, 0)$.

[C.U. (Math Hons) 1997]

[**N.B.** This example shows that Schwarz's conditions are not necessary for equality of mixed order derivatives of a function $f : R^2 \to R$ at a point (a, b) in the domain of definition.]

28. If z be a function of two variables x, y and if $x = c\cosh u \cos v, y = c\sinh u \sin v$, ($c$ is a real number), show that

$$\frac{\partial^2 z}{\partial u^2} + \frac{\partial^2 z}{\partial v^2} = \frac{1}{2}c^2(\cosh 2u - \cos 2v)\left(\frac{\partial^2 z}{\partial x^2} + \frac{\partial^2 z}{\partial y^2}\right)$$

(Assume that the second order partial derivatives of z are continuous.)

[C.U. (Math Hons) 1990, '97]

29. Let $f(x, y) = \begin{cases} x\sin\dfrac{1}{y} + y\sin\dfrac{1}{x}, & xy \neq 0 \\ 0, & xy = 0. \end{cases}$

Show that the simultaneous limit exists at the origin, but the repeated limit does not exist.

[C.U. (Math Hons) 1998]

30. If $f(0) = 0$ and $f'(x) = \dfrac{1}{1+x^2}$, prove without using the method of integration, that

$$f(x) + f(y) = f\left(\frac{x+y}{1-xy}\right), xy \neq 0.$$

31. If f be a differentiable homogeneous function of degree r, and if g and h be differentiable homogeneous functions both of degree s, then prove that the composite function $f(g, h)$ is homogeneous of degree rs. **[C.U. (Math Hons) 1998]**

32. If $u = \dfrac{x+y}{1-xy}$ and $v = \tan^{-1} x + \tan^{-1} y$, find $\dfrac{\partial(u, v)}{\partial(x, y)}$. Are u and v functionally related? If so, find the relationship. **[C.U. (Math Hons) 1999]**

33. When is a function $f(x, y)$ said to be differentiable at the point (x, y)? Prove that the function $f(x, y) = \sqrt{|xy|}$ is not differentiable at the point $(0, 0)$, but that $\frac{\partial f}{\partial x}$ and $\frac{\partial f}{\partial y}$ both exist at the origin and have the value 0. **[C.U. (Math Hons) 1999]**

34. Show that $2xy - \log_e xy = 2$ determines y uniquely as a function of x near the point $(1, 1)$ and find $\frac{dy}{dx}$ at $(1, 1)$. **[C.U. (Math Hons) 1999]**

35. If $f(x, y) = \begin{cases} x \sin \frac{1}{y} + \frac{x^2 - y^2}{x^2 + y^2}, & \text{if } y \neq 0 \\ 0, & \text{if } y = 0, \end{cases}$

show that $\lim\limits_{y \to 0} \lim\limits_{x \to 0} f(x, y)$ exists, but neither $\lim\limits_{(x,y) \to (0,0)} f(x, y)$ nor $\lim\limits_{x \to 0} \lim\limits_{y \to 0} f(x, y)$ exists.

[C.U. (Math Hons) 2000]

36. (a) State Euler's theorem on homogeneous function of two variables. If $u = \tan^{-1} \frac{x^3 + y^3}{x - y}$, prove that $x^2 \frac{\partial^2 u}{\partial x^2} + 2xy \frac{\partial^2 u}{\partial x \partial y} + y^2 \frac{\partial^2 u}{\partial y^2} = (1 - 4 \sin^2 u) \sin 2u$.

(b) If w be a function of u and v, where $u = x^2 - y^2 - 2xy, v = y$, then prove that $(x + y) \frac{\partial w}{\partial x} + (x - y) \frac{\partial w}{\partial y} = 0$ is equivalent to $\frac{\partial w}{\partial v} = 0$. **[C.U. (Math Hons) 2000]**

37. Prove that the following three functions u, v, w are functionally related and find the relation connecting them, when $u = \frac{x}{y - z}, v = \frac{y}{z - x}, w = \frac{z}{x - y}$. **[C.U. (Math Hons) 2000]**

[**Hints.** Show that $\frac{\partial(u, v, w)}{\partial(x, y, z)} = 0$ and then show that $uv + vw + wu = -1$.]

38. (a) Show that $y^2 - x^2 y - 2x^5 = 0$ determines y uniquely as a function of x near the point $(1, -1)$ and find $\frac{dy}{dx}$ at $(1, -1)$. **[C.U. (Math Hons) 2000, '01]**

(b) Prove using the method of Jacobian that $\tan^{-1} x + \tan^{-1} y = \tan^{-1} \frac{x + y}{1 - xy}$.

[C.U. (Math Hons) 2000]

39. If $u = (1 - 2xy + y^2)^{-1/2}$, prove that $\frac{\partial}{\partial x} \left\{ (1 - x^2) \frac{\partial u}{\partial x} \right\} + \frac{\partial}{\partial y} \left\{ y^2 \frac{\partial u}{\partial y} \right\} = 0$.

[C.U. (Math Hons) 2001]

40. (a) Let $f(x, y) = \begin{cases} \frac{x^4 + y^4}{x - y}, & \text{when } x \neq y \\ 0, & \text{when } x = y, \end{cases}$

Show that $\frac{\partial f}{\partial x}$ and $\frac{\partial f}{\partial y}$ exist at $(0, 0)$, but f is not continuous at $(0, 0)$. State a set of sufficient conditions for the continuity of a function $f(x, y)$ at an interior point of its domain of definition.

(b) Let $f : D \to R (D \in R^2)$ be a function and (a, b) be an interior point of D. If $\frac{\partial f}{\partial x}$ exists at (a, b) and $\frac{\partial f}{\partial y}$ is continuous at (a, b), then prove that f is differentiable at (a, b).

[C.U. (Math Hons) 2001]

41. (a) If $f(x, y)$ be a function of two variables x and y, where $x = u^2v$ and $y = v^2u$, then show that

$$2x^2\frac{\partial^2 f}{\partial x^2} + 2y^2\frac{\partial^2 f}{\partial y^2} + 5xy\frac{\partial^2 f}{\partial x \partial y} = uv\frac{\partial^2 f}{\partial u \partial v} - \frac{2}{3}\left(u\frac{\partial f}{\partial u} + v\frac{\partial f}{\partial v}\right).$$

(The relevant partial derivatives are assumed to be continuous.)

(b) Let (a, b) be an interior point of the domain of definition of $f(x, y)$. Let f_x and f_y be both differentiable at (a, b). Prove that $f_{xy}(a, b) = f_{yx}(a, b)$.

(c) Let $f(x, y, z)$ be a homogeneous function of degree n having continuous second order partial derivatives; prove that each of f_x, f_y, f_z is a homogeneous function of degree $(n - 1)$.

(d) Let $f : D \to R(D \in R^2)$ be a function and (a, b) be an interior point of D. State the conditions under which $f(x, y) = 0$ will determine y uniquely as a function of x near the point (a, b).

[C.U. (Math Hons) 2001]

42. Let u, v be functions of ξ, η, ζ having continuous first order partial derivatives and ξ, η, ζ be functions of x and y having continuous first order partial derivatives; prove that

$$\frac{\partial(u,v)}{\partial(x,y)} = \frac{\partial(u,v)}{\partial(\xi,\eta)}\frac{\partial(\xi,\eta)}{\partial(x,y)} + \frac{\partial(u,v)}{\partial(\eta,\zeta)}\frac{\partial(\eta,\zeta)}{\partial(x,y)} + \frac{\partial(u,v)}{\partial(\zeta,\xi)}\frac{\partial(\zeta,\xi)}{\partial(x,y)}.$$

[C.U. (Math Hons) 2001]

43. (a) Let f be a real-valued function of two variables and (a, b) be an interior point of the domain of definition of f. Assume that both the first order partial derivatives of f exist at (a, b) and one of them is bounded in some neighbourhood of (a, b). Prove that f is continuous at (a, b).

(b) If $f(x, y) = \begin{cases} \dfrac{xy}{\sqrt{x^2 + y^2}}, & \text{for all } (x, y) \neq (0, 0) \\ 0, & \text{when } (x, y) = (0, 0), \end{cases}$

show that f is not differentiable at (0, 0). **[C.U. (Math Hons) 2002]**

44. (a) You are given a differentiable function $f(x, y)$. Prove that if the variables x and y are homogeneous linear functions of X and Y, then the obtained function $F(X, Y)$ is related with the given function as follows:

$$x\frac{\partial f}{\partial x} + y\frac{\partial f}{\partial y} = X\frac{\partial F}{\partial x} + Y\frac{\partial F}{\partial y}.$$

(b) You are given a differentiable function ϕ defined by $\phi(cx - az, cy - bz) = 0$; express z as a differentiable function of x and y and prove that

$$a\frac{\partial z}{\partial x} + b\frac{\partial z}{\partial y} = c.$$

[C.U. (Math Hons) 2002]

45. Let u_1, u_2, u_3 be three real-valued functions of three real variables x_1, x_2, x_3. Assume that all the first order partial derivatives are continuous. If now there exists a functional relation between u_1, u_2, u_3, then prove that

$$\frac{\partial(u_1, u_2, u_3)}{\partial(x_1, x_2, x_3)} = 0.$$

[C.U. (Math Hons) 2002]

46. Can you apply the Implicit Function Theorem to the equation $b^2x^2 + a^2y^2 = a^2b^2$ to solve for y near $(\pm a, 0)$? Justify your answer. Solve the equation for y near the point (α, β) on it for which $\beta > 0$ and find $\dfrac{dy}{dx}$ at that point. **[C.U. (Math Hons) 2002]**

Chapter 23

Partial Differentiation

23.1 Introduction

Many of the fundamental laws of physics are expressed most simply by differential equations. For example the motion of a particle of mass m, when acted upon by a force whose components along the three axes of a Cartesian system of reference frame are F_x, F_y and F_z, is given by a system of *ordinary differential* equations

$$m\frac{d^2x}{dt^2} = F_x, \quad m\frac{d^2y}{dt^2} = F_y, \quad m\frac{d^2z}{dt^2} = F_z.$$

The physical laws governing the distribution of temperature in solids, the propagation of electricity in cables, the distribution of velocities in moving fluids, etc., are expressed in terms of partial differential equations. It is therefore necessary that the students of applied mathematics should have a fair knowledge of the fundamental definitions and operations involving partial differentiation.

23.2 Partial Derivatives

In physics, very often we are to deal with several variables. As a specific example, in the case of a gas where pressure, volume and temperature are to be studied, we find that any one of them can be expressed in terms of the other two. In general a quantity $F(x, y, z)$ is said to be a function of the three variables x, y, z if the value of F is determined by the values of x, y and z. If, for example, x, y, z are the Cartesian co-ordinates of a certain point in space, then $F(x, y, z)$ may be the temperature at that point, and as x, y and z take on other values, $F(x, y, z)$ will give the temperature in the region under consideration.

23.2.1 Continuity

The function $F(x, y, z)$ is continuous at a point (a, b, c), for which it is defined if

$$\lim_{\substack{x\to a\\ y\to b\\ z\to c}} F(x, y, z) = F(a, b, c)$$

independently of the manner in which x approaches a, y approaches b, and z approaches c.

Now for a given $F(x, y, z)$, it is possible to keep y and z constant and allow x to vary; thus F is reduced to a function of x only, and may have a derivative defined and computed in the usual way. This derivative is called the *partial derivative* of F with respect to x. Therefore, by definition, we have

$$\frac{\partial F}{\partial x} = \lim_{h \to 0} \frac{F(x+h, y, z) - F(x, y, z)}{h} = F_x,$$

the symbol $\frac{\partial F}{\partial x}$ denoting the partial derivative.

Again, if we keep x and z constant, F is reduced to a function of y alone—whose derive is the partial derivative of F w.r.t. y and is written as

$$\frac{\partial F}{\partial y} = F_y = \lim_{k \to 0} \frac{F(x, y+k, z) - F(x, y, z)}{k}.$$

In the same manner we define the partial derivative of F w.r.t. z, and write it as

$$\frac{\partial F}{\partial z} = F_z = \lim_{q \to 0} \frac{F(x, y, z+q) - F(x, y, z)}{q}.$$

If $F(x, y, z)$ has partial derivatives at each point of a domain, then those derivatives are themselves functions of x, y and z, and may have further partial derivatives which are called the *second partial derivatives of the function F*. For example

$$\frac{\partial}{\partial x}\left(\frac{\partial F}{\partial x}\right) = \frac{\partial^2 F}{\partial x^2} = F_{xx},$$

$$\frac{\partial}{\partial y}\left(\frac{\partial F}{\partial x}\right) = \frac{\partial^2 F}{\partial y \partial x} = F_{yx},$$

$$\frac{\partial}{\partial y}\left(\frac{\partial F}{\partial z}\right) = \frac{\partial^2 F}{\partial y \partial z} = F_{yx}, \text{ etc.}$$

23.2.2 Order of differentiation

$\frac{\partial^2 F}{\partial x \partial y}$ denotes the derivative of $\frac{\partial F}{\partial y}$ w.r.t. x, while $\frac{\partial^2 F}{\partial y \partial x}$ denotes the derivative of $\frac{\partial F}{\partial x}$ w.r.t. y. Now it may be shown that if $F(x, y)$ and its derivatives $\frac{\partial F}{\partial x}$ and $\frac{\partial F}{\partial y}$ are *continuous*, then the order of differentiation is immaterial, and we have

$$\frac{\partial^2 F}{\partial x \partial y} = \frac{\partial^2 F}{\partial y \partial x}, \text{ etc.}$$

In general $\partial^{p+q} F/(\partial x^p \partial y^q)$ signifies the result of differentiating $F(x, y, z)$ p times w.r.t. x and q times w.r.t. y, the order of differentiation being immaterial. Similar result may be extended to any number of variables.

Illustrative Examples

1. Find F_x, F_y, F_{xx}, F_{xy}, F_{yx} and F_{yy} where $F(x, y) = ax^2 + 2hxy + by^2$.

Ans. Here $F_x = 2ax + 2hy, F_y = 2hx + 2by, F_{xx} = 2a, F_{xy} = 2h, F_{yx} = 2h, F_{yy} = 2b$.

Here we find $F_{xy} = F_{yx}$. In most cases we will observe this equality. But it is in general not true.

2. If $F(x, y) = xy(x^2-y^2)/(x^2+y^2)$, for $(x, y) \neq (0,0)$ and $F(0,0) = 0$, show that $F_{xy}(0,0) \neq F_{yx}(0,0)$.

Ans. We know

$$F_{xy} = \lim_{h\to 0} \frac{F_y(x+h, y) - F_y(x, y)}{h}.$$

$$\therefore \quad F_{xy}(0,0) = \lim_{h\to 0} \frac{F_y(h,0) - F_y(0,0)}{h}. \tag{1}$$

Now

$$F_y(0,0) = \lim_{k\to 0} \frac{F(0,k) - F(0,0)}{k} = \lim_{k\to 0} \frac{0}{k} = 0$$

$$\text{and} \quad F_y(h,0) = \lim_{k\to 0} \frac{F(h,k) - F(h,0)}{k} = \lim_{k\to 0} \frac{hk(h^2-k^2)}{k(h^2+k^2)} = h.$$

$\therefore$ from (1) we have

$$F_{xy}(0,0) = \lim_{h\to 0} \frac{h-0}{h} = 1.$$

Again, we know

$$F_{yx} = \lim_{k\to 0} \frac{F_x(x, y+k) - F_x(x, y)}{k}.$$

$$\therefore \quad F_{yx}(0,0) = \lim_{k\to 0} \frac{F_x(0,k) - F_x(0,0)}{k}. \tag{2}$$

Now

$$F_x(0,0) = \lim_{h\to 0} \frac{F(h,0) - F(0,0)}{h} = \lim_{h\to 0} \frac{0}{h} = 0$$

$$\text{and} \quad F_x(0,k) = \lim_{h\to 0} \frac{F(h,k) - F(0,k)}{h} = \lim_{h\to 0} \frac{hk(h^2-k^2)}{h(h^2+k^2)} = -k.$$

$\therefore$ from (2) we have

$$F_{yx}(0,0) = \lim_{k\to 0} \frac{-k-0}{k} = -1.$$

Therefore, we see that $F_{xy}(0,0) \neq F_{yx}(0,0)$.

We now state the conditions for which the equality holds.

23.2.3 Conditions for equality of F_{xy} and F_{yx}

(1) If (a, b) be a point in the domain of definitions of $F(x, y)$, so that $F_x(x, y)$ and $F_y(x, y)$ are differentiable at (a, b), then $F_{xy}(a, b) = F_{yx}(a, b)$.

(2) If (a, b) be a point in the domain of $F(x, y)$ so that $F_x(x, y)$ exists in a certain neighbourhood of (a, b) and $F_{xy}(x, y)$ is continuous at (a, b), then $F_{xy}(a, b) = F_{yx}(a, b)$.

3. Show that $\nabla^2 F = 0$, when $F(x, y) = \log(x^2 + y^2) + \tan^{-1}\left(\frac{y}{x}\right)$.

Ans. Here $F(x,y) = \log(x^2+y^2) + \tan^{-1}\left(\frac{y}{x}\right)$.

$$\therefore \quad \frac{\partial F}{\partial x} = \frac{2x}{x^2+y^2} + \frac{1}{1+\frac{y^2}{x^2}}\left(\frac{-y}{x^2}\right) = \frac{2x}{x^2+y^2} - \frac{y}{x^2+y^2}.$$

So

$$\frac{\partial^2 F}{\partial x^2} = \frac{(x^2+y^2)2 - 2x(2x)}{(x^2+y^2)^2} + \frac{2xy}{(x^2+y^2)^2} = \frac{2y^2-2x^2+2xy}{(x^2+y^2)^2}.$$

Again

$$\frac{\partial F}{\partial y} = \frac{2y}{x^2+y^2} + \frac{1}{1+\frac{y^2}{x^2}} \cdot \frac{1}{x} = \frac{2y+x}{x^2+y^2}.$$

$$\therefore \quad \frac{\partial^2 F}{\partial y^2} = \frac{(x^2+y^2)2-(2y+x)2y}{(x^2+y^2)^2} = \frac{2x^2-2y^2-2xy}{(x^2+y^2)^2} = -\frac{\partial^2 F}{\partial x^2}$$

so that

$$\frac{\partial^2 F}{\partial x^2} + \frac{\partial^2 F}{\partial y^2} = 0, \quad \text{i.e.,} \quad \nabla^2 F = 0.$$

The operator ∇^2 is known as Laplacian operator.

23.3 Total Differential

Let us consider a function $z = F(x,y)$. If both x and y are allowed slight changes Δx and Δy respectively, then let the corresponding change in z be Δz. Then we may write

$$\begin{aligned} \Delta z &= F(x+\Delta x, y+\Delta y) - F(x,y) \\ &= F(x+\Delta x, y+\Delta y) - F(x, y+\Delta y) + F(x, y+\Delta y) - F(x,y) \end{aligned} \tag{23.3.1}$$

By Lagrange's Mean Value theorem, we know

$$F(x+\Delta x) - F(x) = \Delta x F'(x+\theta\Delta x), 0<\theta<1 \tag{23.3.2}$$

Now, from (23.3.1), in the expression $F(x+\Delta x, y+\Delta y) - F(x, y+\Delta y)$, the variable y remains unchanged. So we may apply M.V. theorem and the derivative will be w.r.t. x partially.

So

$$F(x+\Delta x, y+\Delta y) - F(x, y+\Delta y) = \Delta x F_x(x+\theta_1\Delta x, y+\Delta y), 0<\theta_1<1 \tag{23.3.3}$$

$$\text{and} \quad F(x, y+\Delta y) - F(x,y) = \Delta y F_y(x, y+\theta_2\Delta y), 0<\theta_2<1 \tag{23.3.4}$$

Now, since $F_x(x,y)$ and $F_y(x,y)$ are considered to be continuous, we have

$$F_x(x+\theta_1\Delta x, y+\Delta y) = F_x(x,y) + c_1 \tag{23.3.5}$$

$$\text{and} \quad F_y(x, y+\theta_2\Delta y) = F_y(x,y) + c_2 \tag{23.3.6}$$

where c_1 and c_2 both tend to zero as Δx and Δy tend to zero. So equations (23.3.3) and (23.3.4) are reduced to the form

$$\begin{aligned} F(x+\Delta x, y+\Delta y) - F(x, y+\Delta y) &= [F_x(x,y)+c_1]\Delta x \\ \text{and} \quad F(x, y+\Delta y) - F(x,y) &= [F_y(x,y)+c_2]\Delta y \end{aligned}$$

Hence from (23.3.1), we obtain

$$\Delta z = F_x(x,y)\Delta x + F_y(x,y)\Delta y + c_1\Delta x + c_2\Delta y \tag{23.3.7}$$

Again the differential of z is defined as

$$dz = \frac{\partial F}{\partial x}\Delta x + \frac{\partial F}{\partial y}\Delta y \tag{23.3.8}$$

So equation (23.3.7) reduces to

$$\Delta z = dz + c_1\Delta x + c_2\Delta y \tag{23.3.9}$$

Now, as $\Delta x \to 0$, $c_1 \to 0$ and as $\Delta y \to 0$, $c_2 \to 0$, then

$$\Delta z = dz \tag{23.3.10}$$

Taking Δx and Δy as dx and dy respectively, we obtain from (23.3.8)

$$dz = \frac{\partial F}{\partial x}dx + \frac{\partial F}{\partial y}dy. \tag{23.3.11}$$

The term dz is known as the total differential of the function $z = F(x, y)$.

Illustrative Examples

1. Calculate the error in z where $z = x^2y^3$, and the errors in x and y are Δx and Δy respectively.

Ans. Since the errors in x and y are given as Δx and Δy respectively, the correct values of x and y are written as $x + \Delta x$ and $y + \Delta y$ respectively. If Δz be the corresponding error in z, then

$$\begin{aligned} z+\Delta z &= (x+\Delta x)^2(y+\Delta y)^3 \\ &= (x^2+2x\Delta x)(y^3+3y^2\Delta y) \end{aligned} \quad \left[\begin{array}{l}\text{neglecting the terms containing} \\ \text{higher orders of } \Delta x \text{ and } \Delta y\end{array}\right]$$

$$\text{or,} \quad z+\Delta z = x^2y^3 + 2xy^3\Delta x + 3x^2y^2\Delta y \quad \left[\begin{array}{l}\text{neglecting the term containing} \\ \text{the product of } \Delta x \text{ and } \Delta y\end{array}\right]$$

$\therefore$ the error in $z = \Delta z = 2xy^3\Delta x + 3x^2y^2\Delta y$.

2. The rate of flow (V) of a liquid through a capillary tube of radius r and length l at a pressure difference P between its ends is given by $V = \frac{\pi P r^4}{8\eta l}$, where η is the coefficient of viscosity of the liquid. In an experiment, the errors in the measurement of P, r, l and V are 1%, 1.5%, 0.5% and 2% respectively. Evaluate the error in the measurement of η. **[B.U. (Hons.) '95]**

Ans. Here $V = \dfrac{\pi P r^4}{8\eta l}$ or, $\eta = \dfrac{\pi P r^4}{8lV}$.

Now, let $\Delta P, \Delta r, \Delta l$ and ΔV be the errors in P, r, l and V respectively. So according to the given data, $\Delta P = \frac{P}{100}, \Delta r = \frac{3r}{200}, \Delta l = \frac{l}{200}$ and $\Delta V = \frac{V}{50}$. If $\Delta\eta$ be the corresponding error in η, we may write

$$\begin{aligned}
\eta + \Delta\eta &= \frac{\pi}{8}(P+\Delta P)(r+\Delta r)^4(l+\Delta l)^{-1}(V+\Delta V)^{-1} \\
&= \frac{\pi P r^4}{8lV}\left(1+\frac{\Delta P}{P}\right)\left(1+\frac{\Delta r}{r}\right)^4\left(1+\frac{\Delta l}{l}\right)^{-1}\left(1+\frac{\Delta V}{V}\right)^{-1} \\
&= \frac{\pi P r^4}{8lV}\left(1+\frac{\Delta P}{P}\right)\left(1+4\frac{\Delta r}{r}\right)\left(1-\frac{\Delta l}{l}\right)\left(1-\frac{\Delta V}{V}\right) \\
&\qquad \text{[neglecting higher order terms in } \Delta \text{ as small]} \\
&= \eta\left(1+\frac{\Delta P}{P}+4\frac{\Delta r}{r}\right)\left(1-\frac{\Delta l}{l}-\frac{\Delta V}{V}\right) \quad \text{[neglecting smaller terms]} \\
&= \eta\left[1+\frac{\Delta P}{P}+4\frac{\Delta r}{r}-\frac{\Delta l}{l}-\frac{\Delta V}{V}\right] \quad \text{[again neglecting smaller terms]}
\end{aligned}$$

$$\begin{aligned}
\therefore \quad \Delta\eta &= \eta\left[\frac{\Delta P}{P}+4\frac{\Delta r}{r}-\frac{\Delta l}{l}-\frac{\Delta V}{V}\right] \\
&= \eta\left[\frac{1}{100}+\frac{3}{50}-\frac{1}{200}-\frac{1}{50}\right] = \frac{9\eta}{200}.
\end{aligned}$$

$\therefore$ Error in $\eta = \frac{9}{200} \times 100 = 4.5\%$.

3. The density of the material of a uniform circular cylinder is determined by measuring its mass m, length l and diameter d. Calculate the density (in kg-m^{-3}) and its error from the following data: $m = 47.36 \pm 0.01$ g, $l = 15.28 \pm 0.05$ mm, $d = 21.37 \pm 0.04$ mm. **[B.U. (Hons.) 2001]**

Ans. Let V and D be the volume and density of the cylinder respectively.

$$\therefore \quad V = \frac{\pi}{4}d^2 l = \frac{\pi}{4}\left(\frac{21.37}{10^3}\right)^2\left(\frac{15.28}{10^3}\right) = 5477.75 \times 10^{-9}\text{m}^3.$$

Again, let the errors in m, l, d, V and D be Δm, Δl, Δd, ΔV and ΔD respectively.

Now

$$D = \frac{m}{V} = \frac{47.36 \times 10^{-3}}{5477.75 \times 10^{-9}}\text{ kg/m}^3 = 8.646 \times 10^3 \text{ kg/m}^3.$$

Also, we have

$$\begin{aligned}
V + \Delta V &= \frac{\pi}{4}(d+\Delta d)^2(l+\Delta l) = \frac{\pi}{4}d^2 l\left(1+\frac{\Delta d}{d}\right)^2\left(1+\frac{\Delta l}{l}\right) \\
&= V\left(1+2\frac{\Delta d}{d}\right)\left(1+\frac{\Delta l}{l}\right). \quad \text{[neglecting higher orders of } \Delta\text{]}.
\end{aligned}$$

$$\therefore \quad \Delta V = V\left(2\frac{\Delta d}{d}+\frac{\Delta l}{l}\right) \quad \text{or,} \quad \frac{\Delta V}{V} = 2\frac{\Delta d}{d}+\frac{\Delta l}{l}$$

$$= \pm 2\times\frac{0.04}{21.37}\pm\frac{0.05}{15.28} = \pm 0.007$$

Finally, we have

$$D+\Delta D = \frac{m+\Delta m}{V+\Delta V} = \frac{m}{V}\left(1+\frac{\Delta m}{m}\right)\left(1+\frac{\Delta V}{V}\right)^{-1}$$

$$= D\left(1+\frac{\Delta m}{m}\right)\left(1-\frac{\Delta V}{V}\right) \quad \text{[neglecting higher order terms].}$$

$$\therefore \quad \Delta D = D\left(\frac{\Delta m}{m}-\frac{\Delta V}{V}\right) \quad \text{[neglecting smaller terms]}$$

Putting the values, we obtain

$$\Delta D = 8.646\times 10^3\left(\pm\frac{0.01}{47.36}\mp 0.007\right)$$

$$= \mp 8.646(0.007-0.00021)\times 10^3 = \mp 58.705.$$

$\therefore$ error in density $= \pm 58.705\,\text{kg/m}^3$.

23.4 Taylor's Theorem for Functions of Two Variables

If a function $F(x, y)$ possesses continuous partial derivatives of nth order in the neighbourhood of a point (a, b), then for any point $(a+h, b+k)$ in the neighbourhood of (a, b),

$$F(a+h,b+k) = F(a,b)+\left(h\frac{\partial}{\partial x}+k\frac{\partial}{\partial y}\right)F(a,b)+\frac{1}{2!}\left(h\frac{\partial}{\partial x}+k\frac{\partial}{\partial y}\right)^2 F(a,b)+\cdots$$

$$+\frac{1}{(n-1)!}\left(h\frac{\partial}{\partial x}+k\frac{\partial}{\partial y}\right)^{n-1}F(a,b)+\frac{1}{n!}\left(h\frac{\partial}{\partial x}+k\frac{\partial}{\partial y}\right)^n$$

$$F(a+\theta h, b+\theta k) \quad [0<\theta<1] \qquad (23.4.1)$$

Proof. Let $x = a+th$ and $y = b+tk, 0 \le t \le 1$, so that we may write $F(x, y) = F(a+th, b+tk) = \phi(t)$ (say).

$$\therefore \quad \phi^{(1)}(t) = \frac{dF}{dt} = \frac{\partial F}{\partial x}\frac{\partial x}{\partial t}+\frac{\partial F}{\partial y}\frac{\partial y}{\partial t} \quad \left[\begin{array}{l}\text{this result will be proved}\\ \text{in the next article}\end{array}\right]$$

$$= h\frac{\partial F}{\partial x}+k\frac{\partial F}{\partial y} = \left(h\frac{\partial}{\partial x}+k\frac{\partial}{\partial y}\right)F;$$

$$\phi^{(2)}(t) = \frac{d^2F}{dt^2} = \frac{d}{dt}\left(\frac{dF}{dt}\right) = h\frac{\partial}{\partial x}\left(h\frac{\partial}{\partial x}+k\frac{\partial}{\partial y}\right)F+k\frac{\partial}{\partial y}\left(h\frac{\partial}{\partial x}+k\frac{\partial}{\partial y}\right)F$$

$$= \left(h^2\frac{\partial^2}{\partial x^2}+2hk\frac{\partial^2}{\partial x\partial y}+k^2\frac{\partial^2}{\partial y^2}\right)F = \left(h\frac{\partial}{\partial x}+k\frac{\partial}{\partial y}\right)^2 F.$$

Similarly, we can show that $\phi^{(n)}(t) = \left(h\frac{\partial}{\partial x} + k\frac{\partial}{\partial y}\right)^n F$, where $F = F(a+th, b+tk)$.

Now, by Maclaurin's theorem, we have

$$\begin{aligned}\phi(t) &= \phi(0) + \frac{t}{1!}\phi^{(1)}(0) + \frac{t^2}{2!}\phi^{(2)}(0) + \cdots \\ &\quad + \frac{t^{n-1}}{(n-1)!}\phi^{(n-1)}(0) + \frac{t^n}{n!}\phi^{(n)}(\theta t), 0 < \theta < 1\end{aligned}$$

For $t = 1$

$$\phi(1) = \phi(0) + \phi^{(1)}(0) + \frac{1}{2!}\phi^{(2)}(0) + \cdots + \frac{1}{(n-1)!}\phi^{(n-1)}(0) + \frac{1}{n!}\phi^{(n)}(\theta), 0 < \theta < 1 \quad (23.4.2)$$

Now, $\phi(1) = F(a+h, b+k)$ and $\phi(0) = F(a, b)$.

$$\begin{aligned}\therefore \quad \phi^{(1)}(0) &= \left(h\frac{\partial}{\partial x} + k\frac{\partial}{\partial y}\right)F(a,b), \\ \phi^{(2)}(0) &= \left(h\frac{\partial}{\partial x} + k\frac{\partial}{\partial y}\right)^2 F(a,b), \cdots \\ \text{and} \quad \phi^{(n)}(\theta) &= \left(h\frac{\partial}{\partial x} + k\frac{\partial}{\partial y}\right)^n F(a+\theta h, b+\theta k)\end{aligned}$$

Putting these values in (23.4.2), we obtain

$$\begin{aligned}F(x+h, y+k) &= F(x,y) + \left(h\frac{\partial}{\partial x} + k\frac{\partial}{\partial y}\right)F(x,y) + \frac{1}{2!}\left(h\frac{\partial}{\partial x} + k\frac{\partial}{\partial y}\right)^2 F(x,y) + \cdots \\ &\quad + \frac{1}{n!}\left(h\frac{\partial}{\partial x} + k\frac{\partial}{\partial y}\right)^n F(x+\theta h, y+\theta k), \quad 0 < \theta < 1 \qquad (23.4.3)\end{aligned}$$

which is Taylor's series or expansion for a function of two variables.

From (23.4.3) Taylor's series or expansion for a function of a single variable may be written, with the Lagrangian form of the remainder, as

$$\begin{aligned}F(x+h) &= F(x) + \frac{h}{1!}F'(x) + \frac{h^2}{2!}F''(x) + \cdots \\ &\quad + \frac{h^n}{n!}F^{(n)}(x) + \frac{h^{n+1}}{(n+1)!}F^{(n+1)}(x+\theta h), \quad 0 < \theta < 1 \qquad (23.4.4)\end{aligned}$$

N.B. In the above derivation of Taylor's formula, it is assumed that $F(x)$ possesses a continuous nth derivative. The term

$$R_{n+1} = \frac{h^{n+1}}{(n+1)!}F^{(n+1)}(x+\theta h)$$

is called the *remainder* after $(n+1)$ terms. It may happen that $F(x)$ possesses derivatives of all orders and that $\lim_{n\to\infty} R_{n+1} = 0$. In that case we have the convergent infinite series

$$F(x+h) = F(x) + \frac{hF'(x)}{1!} + \frac{h^2F''(x)}{2!} + \cdots + \frac{h^nF^{(n)}(x)}{n!}. \qquad (23.4.5)$$

If we put $x = 0$ and $h = x$ in (23.4.5) we obtain

$$F(x) = F(0) + \frac{xF'(0)}{1!} + \frac{x^2F''(0)}{2!} + \cdots + \frac{x^nF^{(n)}(0)}{n!}. \tag{23.4.6}$$

which is known as Maclaurin's series.

As an example, let us obtain the Maclaurin series expansion of the function $F(x) = e^x$. In this case, we have $F(0) = 1, F'(0) = 1 \cdots F^n(0) = 1$.

Hence

$$e^x = 1 + \frac{x}{1!} + \frac{x^2}{2!} + \frac{x^3}{3!} + \cdots + \frac{x^n}{n!} \tag{23.4.7}$$

This series is seen to converge for all values of x.

As another example, let it be required to expand the function $F(x) = \cos x$ into a Maclaurin series. In this case, we have

$$F(x) = \cos x, F'(x) = -\sin x, F''(x) = -\cos x, F^{(n)}(x) = \cos\left(x + \frac{n\pi}{2}\right).$$

Substituting these into (6), we obtain

$$\cos x = 1 - \frac{x^2}{2!} + \frac{x^4}{4!} - \frac{x^6}{6!} + \cdots + (-1)^{n-1}\frac{x^{(2n-2)}}{(2n-2)!} \tag{23.4.8}$$

This series is seen to converge for all values of x.

23.4.1 Symbolic Form of Taylor's Series Expansion of Functions of Two or More Variables

A very useful and convenient form of Taylor's expansion may be obtained by the use of the symbolic operator D defined by

$$D_x = \frac{d}{dx} \quad D_x^2 = \frac{d^2}{dx^2} \quad \cdots \quad D_x^n = \frac{d^n}{dx^n} \tag{23.4.9}$$

By the use of the derivative operator D_x, defined by (23.4.9), we may write Taylor's expansion in the form

$$F(x+h) = F(x) + \frac{h}{1!}D_xF(x) + \frac{h^2}{2!}D_x^2F(x) + \cdots + \frac{h^n}{n!}D_x^nF(x) + \cdots \tag{23.4.10}$$

where $D_x^rF(x_0) = \frac{d^r}{dx^r}F(x)$ at $x = x_0$.

We may write (23.4.10) in the form

$$F(x+h) = \left(1 + \frac{hD_x}{1!} + \frac{h^2D_x^2}{2!} + \cdots + \frac{h^nD_x^n}{n!} + \cdots\right)F(x). \tag{23.4.11}$$

However, if we put $x = hD_x$ in (23.4.7), we obtain

$$e^{hD_x} = 1 + \frac{hD_x}{1!} + \frac{h^2D_x^2}{2!} + \cdots + \frac{h^nD_x^n}{n!} + \cdots \tag{23.4.12}$$

Hence we may write (23.4.11) in the symbolic form

$$F(x+h) = e^{hD_x}F(x). \tag{23.4.13}$$

This form of Taylor's expansion is compact and easy to remember. By the use of (23.4.13) we may deduce the form of Taylor's expansion for a function of two or more variables.

Let $F(x,y)$ be a function of the two independent variables x and y, and let it have continuous partial derivatives of all orders. Now, if we hold y constant, we have by (23.4.13)

$$F(x+h,y) = e^{hD_x}F(x,y) \tag{23.4.14}$$

where in this case D_x^r has the significance

$$D_x^r = \frac{\partial^r}{\partial x^r}$$

since we are holding y constant.

In the same way, if we hold x constant, we have

$$F(x,y+k) = e^{kD_y}F(x,y) \tag{23.4.15}$$

where

$$D_y^r = \frac{\partial^r}{\partial y^r}.$$

If we operate on (23.4.14) with e^{kD_y}, we obtain

$$\begin{aligned} e^{kD_y}F(x+h,y) &= F(x+h,y+k) \\ &= e^{kD_y}e^{hD_x}F(x,y) \\ &= e^{hD_x+kD_y}F(x,y) \end{aligned} \tag{23.4.16}$$

We thus have the important result that

$$\begin{aligned} F(x+h,y+k) &= e^{hD_x+kD_y}F(x,y) \\ &= \left[1+\frac{hD_x+kD_y}{1!}+\frac{(hD_x+kD_y)^2}{2!}+\cdots+\frac{(hD_x+kD_y)^n}{n!}+\cdots\right]F(x,y) \\ &= F(x,y)+h\frac{\partial F}{\partial x}+k\frac{\partial F}{\partial y}+\frac{1}{2!}\times\left(h^2\frac{\partial^2 F}{\partial x^2}+2hk\frac{\partial^2 F}{\partial x\partial y}+k^2\frac{\partial^2 F}{\partial y^2}\right)+\cdots \end{aligned} \tag{23.4.17}$$

Equation (23.4.17) is Taylor's expansion of a function of two variables. The symbolic derivation given above is based on the fact that the operators D_x and D_y commute with constants and satisfy many of the basic laws of algebra and hence may be treated in many ways as if they were algebraic quantities.

By the same reasoning, if we have a function $F(x_1,x_2,x_3,\cdots,x_n)$ of the n variables $(x_1,x_2,\cdots,x_n)$ that has continuous partial derivatives of all orders, we have

$$F(x_1+h_1,x_2+h_2,\cdots,x_n+h_n) = e^{h_1D_1+h_2D_2+\cdots+h_nD_n}F(x_1,x_2,\cdots,x_n) \tag{23.4.18}$$

where

$$D_m^r = \frac{\partial^r}{\partial x_m^r},\ D_m^r D_n^s = \frac{\partial^{r+s}}{\partial x_m^r \partial x_n^s}, \quad \text{etc.} \tag{23.4.19}$$

Series (23.4.18) is Taylor's expansion, in the symbolic form, of a function of n variables. This form of Taylor's expansion is of great usefulness in applied mathematics.

Illustrative Example

1. Expand $\sin(xy)$ in powers of $(x-1)$ and $\left(y-\frac{\pi}{2}\right)$ up to second degree terms.

Ans.

$$\begin{aligned}
F(x,y) &= \sin(xy) & \therefore\ F\left(1,\tfrac{\pi}{2}\right) &= 1\\
F_x(x,y) &= y\cos(xy) & \therefore\ F_x\left(1,\tfrac{\pi}{2}\right) &= 0\\
F_y(x,y) &= x\cos(xy) & \therefore\ F_y\left(1,\tfrac{\pi}{2}\right) &= 0\\
F_{xx}(x,y) &= -y^2\sin(xy) & \therefore\ F_{xx}\left(1,\tfrac{\pi}{2}\right) &= -\tfrac{\pi^2}{4}\\
F_{xy}(x,y) &= \cos(xy) - xy\sin(xy) & \therefore\ F_{xy}\left(1,\tfrac{\pi}{2}\right) &= -\tfrac{\pi}{2}\\
\text{and}\quad F_{yy}(x,y) &= -x^2\sin(xy) & \therefore\ F_{yy}\left(1,\tfrac{\pi}{2}\right) &= -1
\end{aligned}$$

$\therefore\ \sin(xy) = F(x,y) = 1 - \frac{\pi^2}{8}(x-1)^2 - \frac{\pi}{2}(x-1)\left(y-\frac{\pi}{2}\right) - \frac{1}{2}\left(y-\frac{\pi}{2}\right)^2$ (up to 2nd degree terms).

23.5 Differentiation of Composite Functions: Chain Rule

As a simple case of composite functions, let us consider $F = F(x,y)$ where x and y are both functions of the independent variable t,that is,

$$x = x(t), \quad y = y(t) \tag{23.5.1}$$

Now, if t is given an increment Δt, then x and y receive increments Δx, Δy, and F receives an increment ΔF given by

$$\Delta F = F(x+\Delta x, y+\Delta y) - F(x,y). \tag{23.5.2}$$

Now, by Taylor's expansion, we have

$$\begin{aligned}
F(x+\Delta x, y+\Delta y) &= e^{\Delta x D_x + \Delta y D_y} F(x,y)\\
&= F(x,y) + \frac{\partial F}{\partial x}\Delta x + \frac{\partial F}{\partial y}\Delta y + \delta_1 \Delta x + \delta_2 \Delta y
\end{aligned} \tag{23.5.3}$$

where $\lim\limits_{\substack{\Delta x\to 0\\ \Delta y\to 0}} \delta_1 = 0 \quad \lim\limits_{\substack{\Delta x\to 0\\ \Delta y\to 0}} \delta_2 = 0.$

Hence

$$\Delta F = \frac{\partial F}{\partial x}\Delta x + \frac{\partial F}{\partial y}\Delta y + \delta_1 \Delta x + \delta_2 \Delta y. \tag{23.5.4}$$

Dividing this by Δt and taking the limit as $\Delta t \to 0$, we have

$$\lim_{\Delta t\to 0}\frac{\Delta F}{\Delta t} = \lim_{\Delta t\to 0}\left(\frac{\partial F}{\partial x}\frac{\Delta x}{\Delta t} + \frac{\partial F}{\partial y}\frac{\Delta y}{\Delta t} + \delta_1\frac{\Delta x}{\Delta t} + \delta_2\frac{\Delta y}{\partial t}\right) \tag{23.5.5}$$

Now, as $\Delta t \to 0$, $\Delta x \to 0$ and $\Delta y \to 0$, and if t is the only independent variable, we have

$$\lim_{\Delta t\to 0}\frac{\Delta F}{\Delta t} = \frac{dF}{dt},\ \lim_{\Delta t\to 0}\frac{\Delta x}{\Delta t} = \frac{dx}{dt},\ \text{etc.} \tag{23.5.6}$$

Therefore, provided the functions $x(t)$ and $y(t)$ are differentiable, (23.5.5) becomes

$$\frac{dF}{dt} = \frac{\partial F}{\partial x}\frac{dx}{dt} + \frac{\partial F}{\partial y}\frac{dy}{dt}. \tag{23.5.7}$$

If there are other independent variables besides t, then we must use the notation

$$\lim_{\Delta t\to 0}\frac{\Delta F}{\Delta t} = \frac{\partial F}{\partial t},\ \lim_{\Delta t\to 0}\frac{\Delta x}{\Delta t} = \frac{\partial x}{\partial t},\ \text{etc.} \tag{23.5.8}$$

and we have

$$\frac{\partial F}{\partial t} = \frac{\partial F}{\partial x}\frac{\partial x}{\partial t} + \frac{\partial F}{\partial y}\frac{\partial y}{\partial t}. \tag{23.5.9}$$

This formula may be extended to the case where F is a function of any number of variables $x, y, z, \cdots$ and $x, y, z, \cdots$, etc., are functions of the variables $t, r, s, p, \cdots$, etc.

The results may be stated in the following form: If F is a function of the n variables x_1, $x_2, \cdots, x_n$ so that

$$F = F(x_1, x_2, x_3, \cdots, x_n)$$

and each variable x is a function of the single variable t so that $x_r = x_r(t)$, $r = 1, 2, 3, \cdots, n$ then

$$\frac{dF}{dt} = \frac{\partial F}{\partial x_1}\frac{dx_1}{dt} + \frac{\partial F}{\partial x_2}\frac{dx_2}{dt} + \cdots + \frac{\partial F}{\partial x_n}\frac{dx_n}{dt}. \tag{23.5.10}$$

If, however, each variable x is a function of the p variables $t_1, t_2, \cdots, t_p$ so that

$$x_r = x_r(t_1, t_2, \cdots, t_p), \quad r = 1, 2, 3, \cdots, n \tag{23.5.11}$$

then

$$\left.\begin{aligned}\frac{\partial F}{\partial t_1} &= \frac{\partial F}{\partial x_1}\frac{\partial x_1}{\partial t_1} + \frac{\partial F}{\partial x_2}\frac{\partial x_2}{\partial t_1} + \cdots + \frac{\partial F}{\partial x_n}\frac{\partial x_n}{\partial t_1}\\ \frac{\partial F}{\partial t_s} &= \frac{\partial F}{\partial x_1}\frac{\partial x_1}{\partial t_s} + \frac{\partial F}{\partial x_2}\frac{\partial x_2}{\partial t_s} + \cdots + \frac{\partial F}{\partial x_n}\frac{\partial x_n}{\partial t_s},\quad \text{etc.}\end{aligned}\right\} \tag{23.5.12}$$

23.5.1 Second and Higher Derivatives

As an illustration of the manner in which higher derivatives may be computed from these fundamental formulae, let us differentiate (23.5.7) on the assumption that x and y are functions of the single variable t. We, therefore, have

$$\frac{d^2F}{dt^2} = \frac{d}{dt}\left(\frac{\partial F}{\partial x}\right)\frac{dx}{dt} + \frac{\partial F}{\partial x}\frac{d^2x}{dt^2} + \frac{d}{dt}\left(\frac{\partial F}{\partial y}\right)\frac{dy}{dt} + \frac{\partial F}{\partial y}\frac{d^2y}{dt^2} \tag{23.5.13}$$

Now, since $\partial F/\partial x$ and $\partial F/\partial y$ are functions of x and y, we apply (23.5.7) to $\partial F/\partial x$ and $\partial F/\partial y$ instead of F and obtain

$$\frac{d}{dt}\left(\frac{\partial F}{\partial x}\right) = \frac{\partial^2 F}{\partial x^2}\frac{dx}{dt} + \frac{\partial^2 F}{\partial x \partial y}\frac{dy}{dt} \tag{23.5.14}$$

and
$$\frac{d}{dt}\left(\frac{\partial F}{\partial y}\right) = \frac{\partial^2 F}{\partial x \partial y}\frac{dx}{dt} + \frac{\partial^2 F}{\partial y^2}\frac{dy}{dt} \tag{23.5.15}$$

Substituting these in (23.5.13), we have

$$\frac{d^2F}{dt^2} = \frac{\partial^2 F}{\partial x^2}\left(\frac{dx}{dt}\right)^2 + 2\frac{\partial^2 F}{\partial x \partial y}\frac{dx}{dt}\frac{dy}{dt} + \frac{\partial^2 F}{\partial y^2}\left(\frac{dy}{dt}\right)^2 + \frac{\partial F}{\partial x}\frac{d^2x}{dt^2} + \frac{\partial F}{\partial y}\frac{d^2y}{dt^2} \tag{23.5.16}$$

Expressions for the third and higher order derivatives may be found in a similar manner.

23.5.2 Change of Variables

An important application of Eq. (23.5.9) is its use in changing variables; for example, let

$$F = F(x, y) \tag{23.5.17}$$

and it is desired to replace x and y by the polar coordinates r and θ given by

$$x = r\cos\theta, \quad y = r\sin\theta \tag{23.5.18}$$

Then F becomes a function of r and θ, and we have by (23.5.9)

$$\left.\begin{aligned}\frac{\partial F}{\partial r} &= \frac{\partial F}{\partial x}\frac{\partial x}{\partial r} + \frac{\partial F}{\partial y}\frac{\partial y}{\partial r} = \frac{\partial F}{\partial x}\cos\theta + \frac{\partial F}{\partial y}\sin\theta \\ \frac{\partial F}{\partial \theta} &= \frac{\partial F}{\partial x}\frac{\partial x}{\partial \theta} + \frac{\partial F}{\partial y}\frac{\partial y}{\partial \theta} = \frac{\partial F}{\partial x}(-r\sin\theta) + \frac{\partial F}{\partial y}r\cos\theta\end{aligned}\right\} \tag{23.5.19}$$

Solving these equations for $\partial F/\partial x$ and $\partial F/\partial y$, we have

$$\left.\begin{aligned}\frac{\partial F}{\partial x} &= \frac{\partial F}{\partial r}\cos\theta - \frac{\partial F}{\partial \theta}\frac{\sin\theta}{r} \\ \frac{\partial F}{\partial y} &= \frac{\partial F}{\partial r}\sin\theta + \frac{\partial F}{\partial \theta}\frac{\cos\theta}{r}\end{aligned}\right\} \tag{23.5.20}$$

The second order derivatives may be computed by Eq. (23.5.16).

23.6 Alternative Treatment of Total Differential

Let us consider $F = F(x, y)$ a function of two variables x and y. Now let us give x an increment Δx and y an increment Δy. Then, as was stated in Eq. (23.5.4), F takes an increment ΔF, where

$$\Delta F = \frac{\partial F}{\partial x}\Delta x + \frac{\partial F}{\partial y}\Delta y + \delta_1\Delta x + \delta_2\Delta y \tag{23.6.1}$$

provided $\partial F/\partial x$ and $\partial F/\partial y$ are continuous.

In general the third term is an infinitesimal of higher order than the first term, and the fourth term is in general a higher-order infinitesimal than the second. We take the first two terms of (23.6.1) and call them the *differentials* of F and write

$$dF = \frac{\partial F}{\partial x}\Delta x + \frac{\partial F}{\partial y}\Delta y. \tag{23.6.2}$$

The definition is completed by saying that if x and y are *independent* variables

$$dx = \Delta x, \quad dy = \Delta y, \tag{23.6.3}$$

then (23.6.2) takes the form

$$dF = \frac{\partial F}{\partial x}dx + \frac{\partial F}{\partial y}dy \tag{23.6.4}$$

This expression is called the *total differential* of $F(x, y)$. This definition may be extended to the case where F is a function of the n *independent variables* $x_1, x_2, \cdots, x_n$ to obtain

$$dF = \frac{\partial F}{\partial x_1}dx_1 + \frac{\partial F}{\partial x_2}dx_2 + \cdots + \frac{\partial F}{\partial x_n}dx_n. \tag{23.6.5}$$

This definition (23.6.5) has been based on the assumption that x and y are independent variables. Let us now examine the case where this is not true. Let us suppose that x and y are functions of the three independent variables u, v, w, so that

$$x = x(u, v, w); \quad y = y(u, v, w) \tag{23.6.6}$$

Now, since u, v and w are independent, we have

$$\left.\begin{aligned} dx &= \frac{\partial x}{\partial u}du + \frac{\partial x}{\partial v}dv + \frac{\partial x}{\partial w}dw \\ dy &= \frac{\partial y}{\partial u}du + \frac{\partial y}{\partial v}dv + \frac{\partial y}{\partial w}dw \end{aligned}\right\} \tag{23.6.7}$$

And since F is a function of u, v and w, we have

$$dF = \frac{\partial F}{\partial u}du + \frac{\partial F}{\partial v}dv + \frac{\partial F}{\partial w}dw. \tag{23.6.8}$$

But, by Eqs. (23.5.12), we have

$$\left.\begin{aligned} \frac{\partial F}{\partial u} &= \frac{\partial F}{\partial x}\frac{\partial x}{\partial u} + \frac{\partial F}{\partial y}\frac{\partial y}{\partial u} \\ \frac{\partial F}{\partial v} &= \frac{\partial F}{\partial x}\frac{\partial x}{\partial v} + \frac{\partial F}{\partial y}\frac{\partial y}{\partial v} \\ \frac{\partial F}{\partial w} &= \frac{\partial F}{\partial x}\frac{\partial x}{\partial w} + \frac{\partial F}{\partial y}\frac{\partial y}{\partial w} \end{aligned}\right\} \tag{23.6.9}$$

Substituting these equations into (23.6.8), we have

$$\begin{aligned} dF &= \frac{\partial F}{\partial x}\left(\frac{\partial x}{\partial u}du + \frac{\partial x}{\partial v}dv + \frac{\partial x}{\partial w}dw\right) + \frac{\partial F}{\partial y}\left(\frac{\partial y}{\partial u}du + \frac{\partial y}{\partial v}dv + \frac{\partial y}{\partial w}dw\right) \\ &= \frac{\partial F}{\partial x}dx + \frac{\partial F}{\partial y}dy \end{aligned} \tag{23.6.10}$$

This is the same as (23.6.4). It thus follows that the differential of a function F of the variables $x_1, x_2, \cdots, x_n$ has the form (23.6.5) *whether the variables $x_1, x_2, \cdots, x_n$ are independent or not.*

Let us now consider the case where

$$F(x_1, x_2, \cdots, x_n) = C \tag{23.6.11}$$

where C is a constant. This relation cannot exist when $x_1, x_2, \cdots, x_n$ are independent variables unless $F = C$. Let us suppose that $x_1, x_2, \cdots, x_n$ are functions of *independent* variables $u_1, u_2, \cdots, u_m$. Hence F may be regarded as a function of the variables $u_1, u_2, \cdots, u_m$ and we write

$$F(u_1, u_2, \cdots, u_m) = C. \tag{23.6.12}$$

Hence

$$dF = \frac{\partial F}{\partial u_1}du_1 + \frac{\partial F}{\partial u_2}du_2 + \cdots + \frac{\partial F}{\partial u_m}du_m. \tag{23.6.13}$$

Now, since $u_1, u_2, \cdots, u_m$ are *independent* variables, u_1 may be changed without changing the value of the other variables or the value of F since $F = C$. Therefore

$$F(u_1 + \Delta u_1, u_2, u_3, \cdots, u_m) = C. \tag{23.6.14}$$

Hence

$$\frac{\partial F}{\partial u_1} = \lim_{\Delta u_1 \to 0} \frac{F(u_1 + \Delta u_1, u_2, \cdots, u_m) - F(u_1, u_2, \cdots, u_m)}{\Delta u_1} = 0. \tag{23.6.15}$$

In the same manner we may prove that

$$\frac{\partial F}{\partial u_r} = 0, \quad r = 2, 3, \cdots, m \tag{23.6.16}$$

Hence, as a consequence of (23.6.13), we have

$$dF = 0 \tag{23.6.17}$$

and by (23.6.5) we have

$$dF = \frac{\partial F}{\partial x_1}dx_1 + \frac{\partial F}{\partial x_2}dx_2 + \cdots + \frac{\partial F}{\partial x_n}dx_n = 0. \tag{23.6.18}$$

23.7 Differential of Implicit Functions

If we have the relation

$$F(x,y) = 0, \tag{23.7.1}$$

we are accustomed to say that this equation defines y as an *implicit* function of x and is equivalent to the equation

$$y = \phi(x) \tag{23.7.2}$$

If the functional relation (23.7.1) is simple, then we can actually solve (23.7.1) to obtain y in the form (23.7.2). For example, we consider

$$x^2 + y^2 - a^2 = 0 \tag{23.7.3}$$

This equation may be solved for y to give

$$y = \pm\sqrt{a^2 - x^2}. \tag{23.7.4}$$

However, if Eq. (23.7.1) is complicated, it is in general not possible to solve it for y. It may be shown that y in (23.7.1) satisfies the definition of a function of x in the sense that when x is given, (23.7.1) determines a value of y. It is convenient and enables us to differentiate (23.7.1) with respect to either x or y without solving the equation explicitly for x or y.

To differentiate (23.7.1), let us take its first differential. As a special case of (23.6.18) we have

$$dF = \frac{\partial F}{\partial x}dx + \frac{\partial F}{\partial y}dy = 0 \tag{23.7.5}$$

and hence

$$\frac{dy}{dx} = -\frac{\partial F/\partial x}{\partial F/\partial y} = y' \tag{23.7.6}$$

This may be written in the form

$$\frac{\partial F}{\partial x} + \frac{\partial F}{\partial y}y' = 0 \tag{23.7.7}$$

To obtain the second derivative, let

$$\phi = \left(\frac{\partial F}{\partial x} + \frac{\partial F}{\partial y}y'\right) \tag{23.7.8}$$

Applying (23.7.7) to ϕ, we have

$$\frac{\partial \phi}{\partial x} + \frac{\partial \phi}{\partial y}y' = 0 \tag{23.7.9}$$

Now

$$\frac{\partial \phi}{\partial x} = \frac{\partial^2 F}{\partial x^2} + \frac{\partial^2 F}{\partial x \partial y}y' + \frac{\partial F}{\partial y}\frac{\partial y'}{\partial x} \tag{23.7.10}$$

and

$$\frac{\partial \phi}{\partial y}y' = \frac{\partial^2 F}{\partial x \partial y}y' + \frac{\partial^2 F}{\partial y^2}(y')^2 + y'\frac{\partial F}{\partial y}\frac{\partial y'}{\partial y} \tag{23.7.11}$$

Substituting in (23.7.9), we obtain

$$\frac{\partial^2 F}{\partial x^2} + 2\frac{\partial^2 F}{\partial x \partial y} y' + \frac{\partial^2 F}{\partial y^2}(y')^2 + \frac{\partial F}{\partial y} y'' = 0 \tag{23.7.12}$$

since

$$\frac{\partial y'}{\partial x} + y'\frac{\partial y'}{\partial y} = \frac{dy'}{dx} = y''.$$

Repeating this process, we may find the derivatives y''', y'''', etc., provided the partial derivatives of $F(x, y)$ exist and provided that $\partial F/\partial y \neq 0$.

23.7.1 One equation, more than two variables

The equation

$$F(x, y, z) = 0 \tag{23.7.13}$$

defines any one of the variables, for example, x, in terms of the other two. If we take the differential of (23.7.13), we have

$$dF = \frac{\partial F}{\partial x} dx + \frac{\partial F}{\partial y} dy + \frac{\partial F}{\partial z} dz = 0 \tag{23.7.14}$$

If we put $y =$ const., then $dy = 0$, and we have

$$\left(\frac{dz}{dx}\right)_y = -\frac{\partial F/\partial x}{\partial F/\partial z} \tag{23.7.15}$$

where the subscript denotes that y is held constant. This is less ambiguous than the notation $\partial z/\partial x$. If $x =$ const., we have

$$\left(\frac{dy}{dz}\right)_x = -\frac{\partial F/\partial z}{\partial F/\partial y} \tag{23.7.16}$$

If $z =$ const, we obtain

$$\left(\frac{dx}{dy}\right)_z = -\frac{\partial F/\partial y}{\partial F/\partial x} \tag{23.7.17}$$

Multiplying Eqs. (23.7.15), (23.7.16) and (23.7.17) together, we have

$$\left(\frac{dx}{dy}\right)_z \left(\frac{dy}{dz}\right)_x \left(\frac{dz}{dx}\right)_y = -1 \tag{23.7.18}$$

This is sometimes written in the form

$$\left(\frac{\partial x}{\partial y}\right)\left(\frac{\partial y}{\partial z}\right)\left(\frac{\partial z}{\partial x}\right) = -1 \tag{23.7.19}$$

23.7.2 Chain Rule—Alternative Treatment

Let $z = F(x, y)$ where x and y are independent variables and $x = x(t)$, $y = y(t)$ where t is an independent variable. Let for an increment Δt of t, x and y also have increments Δx and Δy respectively, so that the ultimate increment in z is Δz.

Now if $z = F(x, y)$ and $\frac{\partial z}{\partial x}$, $\frac{\partial z}{\partial y}$ are continuous, we may write

$$\Delta z = \frac{\partial z}{\partial x}\Delta x + \frac{\partial z}{\partial y}\Delta y + c_1\Delta x + c_2\Delta y \quad \text{(from equation 23.3.7)}$$

Again, from equations (23.3.8), (23.3.9) and (23.3.10), we have

$$\Delta z = dz = \frac{\partial z}{\partial x}\Delta x + \frac{\partial z}{\partial y}\Delta y$$

$$\therefore \quad \frac{\Delta z}{\Delta t} = \frac{\partial z}{\partial x}\frac{\Delta x}{\Delta t} + \frac{\partial z}{\partial y}\frac{\Delta y}{\Delta t} \tag{23.7.20}$$

Now, if x and y are differentiable w.r.t. t, i.e., by taking limits as $\Delta t \to 0$, we may write

$$\frac{dz}{dt} = \frac{\partial z}{\partial x}\frac{dx}{dt} + \frac{\partial z}{\partial y}\frac{dy}{dt} \tag{23.7.21}$$

The above relation can be extended to three or more variables. Then for a function $u(x, y, z)$ where $x = x(t)$, $y = y(t)$ and $z = z(t)$, we may write

$$\frac{du}{dt} = \frac{\partial u}{\partial x}\frac{dx}{dt} + \frac{\partial u}{\partial y}\frac{dy}{dt} + \frac{\partial u}{\partial z}\frac{dz}{dt} \tag{23.7.22}$$

Again, if we put $t = x$ in (23.7.21), we have

$$\frac{dz}{dx} = \frac{\partial z}{\partial x} + \frac{\partial z}{\partial y}\frac{dy}{dx} \tag{23.7.23}$$

Now, if $z = F(x, y) = c$ (constant), we have $\frac{dz}{dx} = 0$, so that (23.7.23) reduces to

$$\frac{\partial z}{\partial x} + \frac{\partial z}{\partial y}\frac{dy}{dx} = 0 \quad \text{or,} \quad F_x + F_y\frac{dy}{dx} = 0 \quad \text{or,} \quad \frac{dy}{dx} = -\frac{F_x}{F_y} \tag{23.7.24}$$

Illustrative Example

1. If $u = F\left(\frac{x}{y}, \frac{y}{z}, \frac{z}{x}\right)$, prove that $x\frac{\partial u}{\partial x} + y\frac{\partial u}{\partial y} + z\frac{\partial u}{\partial z} = 0$.

Ans. Let $\frac{x}{y} = p$, $\frac{y}{z} = q$ and $\frac{z}{x} = r$; then $u = F(p, q, r)$, so that we may write

$$\frac{\partial u}{\partial x} = \frac{\partial u}{\partial p}\frac{\partial p}{\partial x} + \frac{\partial u}{\partial q}\frac{\partial q}{\partial x} + \frac{\partial u}{\partial r}\frac{\partial r}{\partial x} = \frac{\partial u}{\partial p}\cdot\left(\frac{1}{y}\right) + \frac{\partial u}{\partial q}\cdot(0) + \frac{\partial u}{\partial r}\left(\frac{-z}{x^2}\right).$$

$$\therefore \quad x\frac{\partial u}{\partial x} = \frac{\partial u}{\partial p}\frac{x}{y} - \frac{\partial u}{\partial r}\frac{z}{x} = p\frac{\partial u}{\partial p} - r\frac{\partial u}{\partial r}. \tag{1}$$

Again

$$\frac{\partial u}{\partial y} = \frac{\partial u}{\partial p}\frac{\partial p}{\partial y} + \frac{\partial u}{\partial q}\frac{\partial q}{\partial y} + \frac{\partial u}{\partial r}\frac{\partial r}{\partial y} = \frac{\partial u}{\partial p}\left(\frac{-x}{y^2}\right) + \frac{\partial u}{\partial q}\left(\frac{1}{z}\right) + \frac{\partial u}{\partial r}(0).$$

$$\therefore \quad y\frac{\partial u}{\partial y} = -\frac{\partial u}{\partial p}\left(\frac{x}{y}\right) + \frac{\partial u}{\partial q}\left(\frac{y}{z}\right) = q\frac{\partial u}{\partial q} - p\frac{\partial u}{\partial p}. \tag{2}$$

Similarly

$$z\frac{\partial u}{\partial z} = r\frac{\partial u}{\partial r} - q\frac{\partial u}{\partial q}. \tag{3}$$

Adding (1), (2) and (3), we obtain

$$x\frac{\partial u}{\partial x} + y\frac{\partial u}{\partial y} + z\frac{\partial u}{\partial z} = 0.$$

23.8 Maxima and Minima

Quite frequently in the application of mathematics to science it is necessary to determine the maximum or minimum values of a function of one or more variables.

Let us consider a function F of the single variable x, so that $F = F(x)$.

A *maximum* of $F(x)$ is a value of $F(x)$ which is *greater* than those immediately preceding or immediately following, while a *minimum* of $F(x)$ is a value of $F(x)$ which is *less* than those immediately preceding or following. In defining and discussing maxima and minima of $F(x)$, it is assumed that $F(x)$ and all necessary derivatives are continuous and single-valued functions of x.

In order to determine whether the function $F(x)$ has a maximum or a minimum at a point $x = a$, we may use Taylor's expansion of a function of one variable, in the form

$$F(x+h) = F(x) + \frac{h}{1!}F'(x) + \frac{h^2}{2!}F''(x) + \frac{h^3}{3!}F'''(x) + \cdots \tag{23.8.1}$$

Let $x = a$ be the critical point under consideration, and then we may write

$$\Delta(h) = F(a+h) - F(a) = \frac{h}{1!}F'(a) + \frac{h^2}{2!}F''(a) + \frac{h^3}{3!}F'''(a) + \cdots \tag{23.8.2}$$

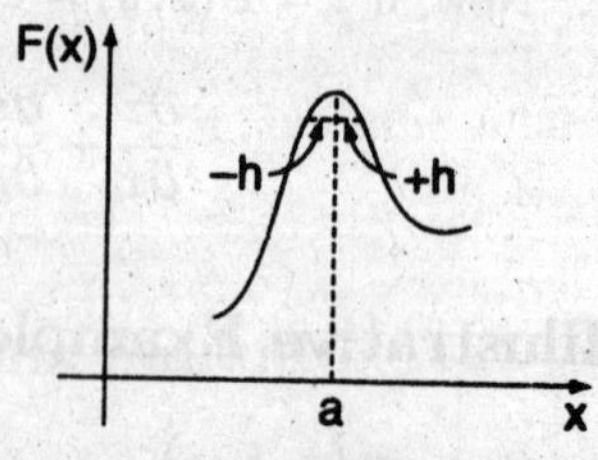

Fig 23.8.1

$\Delta(h)$ is thus the change in the value of the function when the argument of the function is changed by h. This is illustrated graphically, in the case where $F(a)$ is a maximum, by Fig. 23.8.1.

Now, if $x = a$ is a point at which $F(x)$ has either a maximum or a minimum, we shall have for h sufficiently small

$$\text{Sign of } \Delta(h) = \text{Sign of } \Delta(-h) \tag{23.8.3}$$

since for a maximum, if we move to the left of the critical point or to the right of the critical point, the function will decrease and for a minimum point the function will increase. Now if $F'(x) \neq 0$, then for h sufficiently small $\Delta(h) \simeq hF'(a)$ since the higher-order terms in (23.8.2) may be neglected. In the same way $\Delta(-h) \simeq -hF'(a)$. Hence, in order for (23.8.3) to be satisfied, we must have

$$F'(a) = 0 \tag{23.8.4}$$

at either a *maximum* or a *minimum*. Now at a maximum $\Delta(h)$ must be negative and at a minimum $\Delta(h)$ must be positive for either positive or negative values of h.

Hence, if (23.8.4) is satisfied, for h sufficiently small,

$$\Delta(h) \simeq \frac{h^2}{2!}F''(a). \tag{23.8.5}$$

Since h^2 is always positive, then it is evident that at a *maximum* we must have

$$F''(a) < 0 \tag{23.8.6}$$

and at a *minimum*

$$F''(a) > 0 \tag{23.8.7}$$

Let us suppose that

$$F''(a) = 0 \quad \text{and} \quad F'''(a) \neq 0 \tag{23.8.8}$$

Then for h sufficiently small

$$\Delta(h) \simeq \frac{h^3}{3!}F'''(a) \tag{23.8.9}$$

Since, if $F'''(a) \neq 0$, the expression (23.8.9) changes sign with h, we cannot have a maximum or a minimum. If, however,

$$F'(a) = 0, \quad F''(a) = 0, \quad F'''(a) = 0 \tag{23.8.10}$$

then for h sufficiently small

$$\Delta(h) = \frac{h^4}{4!}F^{IV}(a) \tag{23.8.11}$$

Hence $F(a)$ will be a *maximum* if

$$F^{IV}(a) < 0 \tag{23.8.12}$$

and a *minimum* if

$$F^{IV}(a) > 0, \text{ etc.} \tag{23.8.13}$$

23.8.1 Maxima and Minima of Functions of Two Variables

We define the maximum and minimum values of a function $F(x, y)$ of two variables x, y in the following manner:

$F(a, b)$ is a *maximum* of $F(x, y)$ when, for all sufficiently small positive or negative values of h and k,

$$\Delta(h, k) = F(a + h, b + k) - F(a, b) < 0 \tag{23.8.14}$$

$F(a, b)$ is a *minimum* of $F(x, y)$ when, for all small positive or negative values of h and k,

$$\Delta(h, k) = F(a + h, b + k) - F(a, b) > 0 \tag{23.8.15}$$

By Taylor's expansion of a function of two variables, we have, if the necessary derivatives exist and are continuous,

$$F(x+h, y+k) = F(x, y) + h\frac{\partial F}{\partial x} + k\frac{\partial F}{\partial y} + \frac{1}{2!}\left(h^2\frac{\partial^2 F}{\partial x^2} + 2hk\frac{\partial^2 F}{\partial x \partial y} + k^2\frac{\partial^2 F}{\partial y^2}\right) + \cdots \quad (23.8.16)$$

Putting $F_x = \partial F/\partial x, F_y = \partial F/\partial y$,

$$A = \frac{\partial^2 F}{\partial x^2}, \quad B = \frac{\partial^2 F}{\partial x \partial y}, \quad C = \frac{\partial^2 F}{\partial y^2} \quad (23.8.17)$$

at the point $x = a$, $y = b$, we may write from (23.8.16)

$$\Delta(h, k) = hF_x(a, b) + kF_y(a, b) + \frac{1}{2!}(h^2 A + 2hkB + k^2 C) + \cdots \quad (23.8.18)$$

It is thus evident that for small values of h and k, in order for $\Delta(h, k)$ to have the same sign independently of the signs of h and k, it is necessary for the coefficients of h and k in (23.8.18) to vanish. This gives

$$\frac{\partial F}{\partial x} = 0, \frac{\partial F}{\partial y} = 0 \quad (23.8.19)$$

evaluated at $x = a, y = b$ as a required condition for *either* a maximum or a minimum. Such a point is known as a *stationary* point. If the conditions (23.8.19) are satisfied, then $\Delta(h, k)$ reduces to

$$\Delta(h, k) = \frac{1}{2!}(h^2 A + 2hkB + k^2 C) + \cdots \quad (23.8.20)$$

To facilitate the discussion, we make use of the identity

$$Ah^2 + 2Bhk + Ck^2 = \frac{(Ah + Bk)^2 + (AC - B^2)k^2}{A}. \quad (23.8.21)$$

We may then write (23.8.20) in the form

$$\Delta(h, k) = \frac{(Ah + Bk)^2 + (AC - B^2)k^2}{2!A} \quad (23.8.22)$$

The sign of $\Delta(h, k)$ given by (23.8.22) is *independent* of the signs of h and k provided that

$$AC - B^2 > 0 \quad (23.8.23)$$

$$\text{or,} \quad AC - B^2 = 0 \quad (23.8.24)$$

This may be seen since $(Ah + Bk)^2$ is always positive or zero; therefore, if $AC - B^2$ is negative, the numerator of (23.8.22) will be positive when $k = 0$ and negative when $Ah + Bk = 0$.

Therefore, a *second* condition for a maximum or a minimum of $F(x, y)$ is that at the point in question

$$AC > B^2 \quad \text{or,} \quad \frac{\partial^2 F}{\partial x^2}\frac{\partial^2 F}{\partial y^2} > \left(\frac{\partial^2 F}{\partial x \partial y}\right)^2 \quad (23.8.25)$$

An investigation of the exceptional cases when (23.8.24) is satisfied or when

$$A = B = C = 0 \quad (23.8.26)$$

is beyond the scope of this discussion. When condition (23.8.25) is satisfied at $x = a, y = b$, we see that $F(a, b)$ will have a *maximum* when

$$\frac{\partial^2 F}{\partial x^2} < 0, \quad \frac{\partial^2 F}{\partial y^2} < 0 \tag{23.8.27}$$

evaluated at $x = a, y = b$. $F(a, b)$ will have a *minimum* when

$$\frac{\partial^2 F}{\partial x^2} > 0, \quad \frac{\partial^2 F}{\partial y^2} > 0 \tag{23.8.28}$$

evaluated at $x = a, y = b$.

If

$$AC - B^2 < 0, \tag{23.8.29}$$

then $F(x, y)$ has *neither* a maximum nor a minimum. The points where $F_x = 0$, $F_y = 0$, but $F(x, y)$ does not possess maxima or minima are called ***saddle points***. Here $AC - B^2 < 0$. By a similar course of reasoning we may obtain the conditions for maxima and minima of functions of three or more variables.

Illustrative Examples

1. Examine $F(x, y) = x^2y + xy^2 - axy$ for maximum and minimum values.

Ans. Here we have

$$\frac{\partial F}{\partial x} = (2x + y - a)y, \quad \frac{\partial F}{\partial y} = (2y + x - a)x \tag{1}$$

$$\frac{\partial^2 F}{\partial x^2} = 2y, \quad \frac{\partial^2 F}{\partial y^2} = 2x, \quad \frac{\partial^2 F}{\partial x \partial y} = 2x + 2y - a \tag{2}$$

The conditions (23.8.19) are

$$(2x + y - a)y = 0, \quad (2y + x - a)x = 0 \tag{3}$$

Condition (23.8.25) is

$$4xy > (2x + 2y - a)^2 \tag{4}$$

The system of Eqs. (3) has the four solutions

$$\left.\begin{array}{llll} x = 0 & x = a & x = 0 & x = \dfrac{a}{3} \\ y = 0 & y = 0 & y = a & y = \dfrac{a}{3} \end{array}\right\} \tag{5}$$

Only the last values satisfy (4), and a maximum or a minimum of $F(x, y)$ is located at

$$x = \frac{a}{3}, \quad y = \frac{a}{3} \tag{6}$$

If a is positive, $\partial^2 F/\partial x^2$ is positive when $y = a/3$; therefore

$$F\left(\frac{a}{3}, \frac{a}{3}\right) = -\frac{a^3}{27} \tag{7}$$

is a minimum. If a is negative, $\partial^2 F/\partial x^2$ is negative when $y = a/3$; hence $-a^3/27$ is a maximum.

2. Examine, for extreme values of $F(x, y) = x^3 + y^2 - 3xy$.

Ans. Here, $F(x, y) = x^3 + y^2 - 3xy$, so that we have $F_x = 3x^2 - 3y$ and $F_y = 2y - 3x$. Putting $F_x = 0$ and $F_y = 0$, we may obtain $x^2 = y$ and $y = \frac{3}{2}x$. Solving we get the stationary points as $(0, 0)$, $(\frac{3}{2}, \frac{9}{4})$.

Again, $A = F_{xx} = 6x$, $B = F_{xy} = -3$, $C = F_{yy} = 2$.

At $(\frac{3}{2}, \frac{9}{4})$, $A = 9$, $B = -3$, $C = 2$. $\therefore$ $AC - B^2 = 9 > 0$. So $F(x, y)$ is minimum at $(\frac{3}{2}, \frac{9}{4})$, as $A = 9 > 0$.

Again, at $(0, 0)$, $A = 0$, $B = -3$, $C = 2$, $\therefore$ $AC - B^2 = -9 < 0$. So, $(0, 0)$ is a saddle point.

3. Find the maxima and minima of the function $F(x, y) = x^2y^2 - 5x^2 - 8xy - 5y^2$.

Ans. Here, $F(x, y) = x^2y^2 - 5x^2 - 8xy - 5y^2$.

$$\therefore \quad F_x = 2xy^2 - 10x - 8y \text{ and } F_y = 2x^2y - 8x - 10y$$

Putting $F_x = 0$ and $F_y = 0$, we have

$$xy^2 - 5x - 4y = 0 \tag{1}$$

$$\text{and} \quad x^2y - 4x - 5y = 0 \tag{2}$$

Multiplying (1) by x and (2) by y, and then subtracting, we get, $x^2 = y^2$, or, $x = \pm y$. Using this in (1) and (2) we obtain the stationary points as $(0, 0)$, $(1, -1)$, $(-1, 1)$, $(3, 3)$, $(-3, -3)$.

Again, $A = F_{xx} = 2y^2 - 10$, $B = F_{xy} = 4xy - 8$, $C = F_{yy} = 2x^2 - 10$.

$\therefore$ at $(0, 0)$

$$A = -10,\ B = -8,\ C = -10 \text{ and } AC - B^2 = 100 - 64 = 36 > 0,\ A < 0$$

$\therefore$ $F(x, y)$ is maximum at $(0, 0)$.

At $(1, -1)$

$$A = -8,\ B = -12,\ C = -8 \text{ and } AC - B^2 = 64 - 144 = -80 < 0$$

$\therefore$ $(1, -1)$ is a saddle point.

Similarly, we can show that $(-1, 1)$, $(3, 3)$ and $(-3, -3)$ are all saddle points.

23.9 Lagrange's Method of Undetermined Multipliers

In a great many practical problems it is desired to find the maximum or minimum value of a function of certain variables when the variables are not independent but are connected by some given relation.

For example, let it be required to find the maximum or minimum value of the function

$$u = u(x, y, z) \tag{23.9.1}$$

and let the variables x, y and z be connected by the relation

$$\phi(x, y, z) = 0 \text{ — known as the constraint equation} \tag{23.9.2}$$

In principle it should be possible to solve (23.9.2) for one variable in terms of the other two. This variable may then be eliminated by substitution into (23.9.1). Then the maximum or minimum values of u may be investigated by the methods discussed above since, by the elimination, u has been reduced to a function of two variables. In many cases the solution of (23.9.2) for any of the variables may be extremely difficult or impossible. To deal with such cases, Lagrange used an ingenious device which is known in the literature as the *method of undetermined multipliers.* This method will now be discussed.

If u is to have a maximum or minimum, it is necessary that

$$\frac{\partial u}{\partial x} = 0, \quad \frac{\partial u}{\partial y} = 0, \quad \frac{\partial u}{\partial z} = 0. \tag{23.9.3}$$

Hence

$$\frac{\partial u}{\partial x}dx + \frac{\partial u}{\partial y}dy + \frac{\partial u}{\partial z}dz = 0 \tag{23.9.4}$$

Differentiating the functional relation (23.9.2), we obtain

$$\frac{\partial \phi}{\partial x}dx + \frac{\partial \phi}{\partial y}dy + \frac{\partial \phi}{\partial z}dz = 0 \tag{23.9.5}$$

Let us now multiply (23.9.5) by the parameter λ and add the resulting equation to (23.9.4). We thus obtain

$$\left(\frac{\partial u}{\partial x} + \lambda\frac{\partial \phi}{\partial x}\right)dx + \left(\frac{\partial u}{\partial y} + \lambda\frac{\partial \phi}{\partial y}\right)dy + \left(\frac{\partial u}{\partial z} + \lambda\frac{\partial \phi}{\partial z}\right)dz = 0 \tag{23.9.6}$$

This equation will be satisfied if

$$\frac{\partial u}{\partial x} + \lambda\frac{\partial \phi}{\partial x} = 0, \quad \frac{\partial u}{\partial y} + \lambda\frac{\partial \phi}{\partial y} = 0, \quad \frac{\partial u}{\partial z} + \lambda\frac{\partial \phi}{\partial z} = 0. \tag{23.9.7}$$

The three Eqs. (23.9.7) and Eq. (23.9.2) furnish four equations to determine the proper values of the variables x, y, z and λ to assure the maximum or minimum value of the function u.

Illustrative Examples

1. Find the volume of the largest parallelopiped that can be inscribed in the ellipsoid

$$\frac{x^2}{a^2} + \frac{y^2}{b^2} + \frac{z^2}{c^2} = 1. \tag{1}$$

Ans. In this case the function to be maximized is $u = 8xyz$ and Eqs. (23.9.7) are

$$\left.\begin{aligned} 8yz + \frac{2\lambda}{a^2}x &= 0 \\ 8xz + \frac{2\lambda}{b^2}y &= 0 \\ 8xy + \frac{2\lambda}{c^2}z &= 0 \end{aligned}\right\} \tag{2}$$

Multiplying the first equation by x, the second one by y and the last one by z, and adding them, we obtain

$$3u + 2\lambda\left(\frac{x^2}{a^2} + \frac{y^2}{b^2} + \frac{z^2}{c^2}\right) = 0. \tag{3}$$

In view of (1) this becomes

$$3u + 2\lambda = 0 \tag{4}$$

Hence

$$2\lambda = -3u \tag{5}$$

From the first equation of (2) multiplied by x, we have

$$u\left(1 - \frac{3}{a^2}x^2\right) = 0 \quad \text{or,} \quad x = \frac{a}{\sqrt{3}}$$

Similarly,

$$y = \frac{b}{\sqrt{3}}, \quad z = \frac{c}{\sqrt{3}}.$$

Hence the required maximum volume is $u = \dfrac{8abc}{3\sqrt{3}}$.

2. Using the method of Lagrange multiplier, find the maximum of $F = 4xyz$, subject to the constraint $x^2 + y^2 + z^2 - a^2 = 0$. **[B.U. (Hons.) 1997]**

Ans. Constraint equation is

$$x^2 + y^2 + z^2 - a^2 = 0 \tag{1}$$

Let us now consider

$$F(x, y, z) = 4xyz + \lambda(x^2 + y^2 + z^2 - a^2). \tag{2}$$

$$\therefore \quad \frac{\partial F}{\partial x} = 4yz + 2\lambda x = 0 \tag{3}$$

$$\frac{\partial F}{\partial y} = 4zx + 2\lambda y = 0 \tag{4}$$

$$\text{and} \quad \frac{\partial F}{\partial z} = 4xy + 2\lambda z = 0 \tag{5}$$

Comparing the values of λ in (3) and (4) we get $\lambda = -\frac{2yz}{x} = -\frac{2zx}{y}$ which implies $y^2 = x^2$, i.e., $y = x$.

Similarly, from (4) and (5), we have $x = z$.

$\therefore$ we have

$$x = y = z \tag{6}$$

Now, from (1), we get

$$x = y = z = \pm\frac{a}{\sqrt{3}} \tag{7}$$

Again the value of $F = 4xyz$ will lie between $-\frac{4a^3}{3\sqrt{3}}$ and $\frac{4a^3}{3\sqrt{3}}$.

So the maximum value of F will be $4a^3/3\sqrt{3}$.

3. Using the method of Lagrange multiplier, find the area of the largest rectangle which can be inscribed in a semi-circle of radius R with one of the larger sides of the rectangle coinciding with the diameter.

[B.U. (Hons.) 1996]

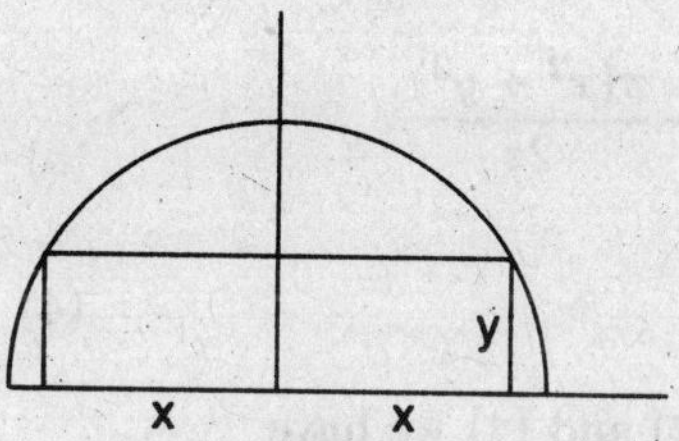

Fig 23.9.1

Ans. If the bounding diameter be along the X-axis and the inscribed area be a rectangle, it must be symmetric about the Y-axis, and as such let the length of the rectangle be $2x$ and breadth be y and area be S. So we may write

$$S = 2xy \tag{1}$$

and the equation of constraint is

$$x^2 + y^2 = R^2 \tag{2}$$

Now, let $F(x, y) = 2xy + \lambda(x^2 + y^2 - R^2)$, so that

$$\frac{\partial F}{\partial x} = 2y + 2\lambda x \quad \text{and} \quad \frac{\partial F}{\partial y} = 2x + 2\lambda y.$$

Writing $\frac{\partial F}{\partial x} = \frac{\partial F}{\partial y} = 0$, we obtain

$$y + \lambda x = 0 \quad \text{and} \quad x + \lambda y = 0 \tag{3}$$

whence on equating λ, we get $-\frac{y}{x} = -\frac{x}{y}$ or, $x^2 = y^2$ or, $x = y$.

We have then from (2)

$$x = \frac{R}{\sqrt{2}} = y \tag{4}$$

$\therefore$ from (1), $S = 2\frac{R}{\sqrt{2}}\frac{R}{\sqrt{2}} = R^2$.

Here we observe that when $y = 0$, the area $S = 0$; again as y increases, area also increases and when y increases to its maximum value, $x = 0$ and the area S is also zero. Between these two stages at a certain level given by (4), area is maximum. So the maximum area $= R^2$.

4. Find the height of a right circular cylinder with greatest volume that can be inscribed in a sphere of radius R.

[B.U. (Hons.) 1993]

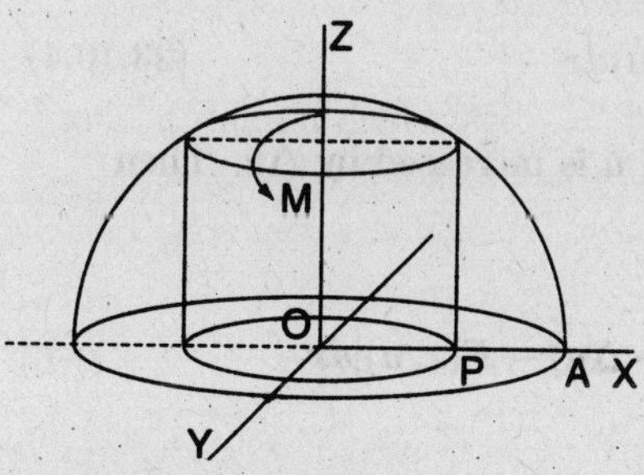

Fig 23.9.2

Ans. Here, as shown by the attached figure, we have $OA = R$ (given). Let $OP = a =$ radius of the base of the cylinder, $OM = z =$ height of the cylinder; then the volume of the cylinder

$$V = \pi a^2 z = \pi(x^2 + y^2)z \tag{1}$$

($\because$ the base of the cylinder is circular, $x^2 + y^2 = a^2$)

Equation to the sphere is

$$x^2 + y^2 + z^2 - R^2 = 0 \tag{2}$$

which is the constraint equation.

Now, let

$$F(x,y,z) = \pi(x^2+y^2)z + \lambda(x^2+y^2+z^2-R^2) \tag{3}$$

$$\therefore \quad \frac{\partial F}{\partial x} = 2\pi xz + 2x\lambda = 0 \quad \text{or,} \quad \lambda = -\pi z$$

$$\frac{\partial F}{\partial y} = 2\pi yz + 2y\lambda = 0 \quad \text{or,} \quad \lambda = -\pi z$$

$$\text{and} \quad \frac{\partial F}{\partial z} = \pi(x^2+y^2) + 2z\lambda = 0 \quad \text{or,} \quad \lambda = -\frac{\pi(x^2+y^2)}{2z}$$

whence we obtain

$$\frac{x^2+y^2}{2z} = z \quad \text{or,} \quad x^2+y^2 = 2z^2 \tag{4}$$

For stationary points, condition (4) is to be satisfied. From (2) and (4) we have

$$z = \frac{R}{\sqrt{3}}. \quad \therefore \; V_{\max} = \pi \cdot \frac{2R^2}{3} \cdot \frac{R}{\sqrt{3}} = \frac{2\pi R^3}{3\sqrt{3}}$$

and the height of the cylinder with maximum volume $= z = \frac{R}{\sqrt{3}}$.

23.10 Differentiation of a Definite Integral

It is frequently required to differentiate a definite integral with respect to its limits or with respect to some parameter. Let $F(x,u)$ be a continuous function of x, u and consider

$$\phi(u) = \int_{a(u)}^{b(u)} F(x,u)dx \tag{23.10.1}$$

where u is a parameter appearing in the integrand and we assume that the limits a and b of the definite integral are continuous functions of the parameter u, so that

$$a = a(u), \quad b = b(u) \tag{23.10.2}$$

The integral, therefore, defines a function $\phi(u)$ of the parameter u. We shall now show that differentiation of the function $\phi(u)$ yields the important equation

$$\frac{d\phi}{du} = \int_a^b \frac{\partial F}{\partial u}dx + F(b,u)\frac{db}{du} - F(a,u)\frac{da}{du} \tag{23.10.3}$$

To establish this equation, let u be given an increment Δu in (23.10.1). Hence

$$\phi(u+\Delta u) = \int_{a+\Delta a}^{b+\Delta b} F(x,u+\Delta u)dx \tag{23.10.4}$$

where Δa and Δb are the increments that a and b take when u is increased by Δu. Then

$$\begin{aligned}\Delta\phi &= \phi(u+\Delta u) - \phi(u) \\ &= \int_{a+\Delta a}^{a} F(x,u+\Delta u)dx + \int_a^b [F(x,u+\Delta u) - F(x,u)]dx \\ &\quad + \int_b^{b+\Delta b} F(x,u+\Delta u)dx\end{aligned} \tag{23.10.5}$$

Now by the concept of the definite integral of a continuous function, between the limits x_1 and x_2, we have

$$\int_{x_1}^{x_2} F(x)dx = F(x_0)(x_2 - x_1) \tag{23.10.6}$$

where x_0 is some intermediate point between x_2 and x_1, given by $x_1 < x_0 < x_2$.

This may be seen intuitively by the concept of the integral (23.10.6) giving the area under the curve $F(x)$ between the points x_1 and x_2. In this case $F(x_0)$ is a mean ordinate such that when it is multiplied by the length $x_2 - x_1$, it gives the same area as that given by the integral. We may apply Eq. (23.10.6) to the first integral of (23.10.5) and obtain

$$\int_{a+\Delta a}^{a} F(x, u + \Delta u)dx = F(t_1, u + \Delta u)[a - (a + \Delta a)] = -F(t_1, u + \Delta u)\Delta a \tag{23.10.7}$$

where $a + \Delta a > t_1 > a$.

In the same way the last integral of (23.10.5) may be expressed in the form

$$\int_{b}^{b+\Delta b} F(x, u + \Delta u)dx = \Delta b = \Delta b F(t_2, u + \Delta u) \tag{23.10.8}$$

where $b < t_2 < b + \Delta b$.

Now

$$\lim_{\Delta u \to 0} \int_a^b \frac{[F(x, u + \Delta u) - F(x, u)]dx}{\Delta u} = \int_a^b \frac{\partial F}{\partial u} dx. \tag{23.10.9}$$

Hence, if we divide (23.10.5) by Δu, then using the results (23.10.7), (23.10.8) and (23.10.9), and realising that

$$\lim_{\Delta u \to 0} t_1 = a, \quad \lim_{\Delta u \to 0} t_2 = b, \tag{23.10.10}$$

we may write

$$\frac{d\phi}{du} = \int_a^b \frac{\partial F}{\partial u} dx + F(b, u)\frac{db}{du} - F(a, u)\frac{da}{du}. \tag{23.10.11}$$

This is the required result.

In the special case where a and b are fixed, we have

$$\frac{d\phi}{du} = \int_a^b \frac{\partial F}{\partial u} dx \tag{23.10.12}$$

If $F(x)$ does not contain the parameter u, and $b = u$, $a =$ const. we have

$$\frac{\partial}{\partial b} \int_a^b F(x)dx = F(b). \tag{23.10.13}$$

In the same way, differentiation with respect to the lower limit gives

$$\frac{\partial}{\partial a} \int_a^b F(x)dx = -F(a) \tag{23.10.14}$$

These equations are useful in evaluating certain definite integrals.

23.11 Integration under the Integral Sign—Multiple Integrations

The possibility of differentiating under the integral sign leads to the converse possibility of integration. Let

$$\phi(u) = \int_a^b F(x,u)dx \tag{23.11.1}$$

where a and b are constants. Multiplying by du and integrating with respect to u between u_0 and u, we have

$$\int_{u_0}^{u} \phi(u')du' = \int_{u_0}^{u} du' \int_a^b F(x,u')dx = Q(u). \tag{23.11.2}$$

The integrations are to be carried out first with respect to x and then with respect to u. Now let us consider

$$P(u) = \int_a^b dx \int_{u_0}^{u} F(x,u')du' \tag{23.11.3}$$

We are required to show that

$$P(u) = Q(u). \tag{23.11.4}$$

Let us differentiate (23.11.3) with respect to u. By the results of the last article, when we carry out the differentiation on the right under the integral sign, we obtain

$$\frac{\partial}{\partial u}\int_{u_0}^{u} F(x,u')du' = F(x,u) \tag{23.11.5}$$

where the differentiation has been carried out with respect to the upper limit. Hence

$$\frac{dP}{du} = \int_a^b F(x,u)dx = \phi(u) \tag{23.11.6}$$

Therefore

$$\int_{u_0}^{u} \frac{dP}{du'}du' = \int_{u_0}^{u} \phi(u')du' = P(u) - P(u_0). \tag{23.11.7}$$

But, from (23.11.3), we have

$$P(u_0) = 0 \tag{23.11.8}$$

Hence

$$P(u) = \int_{u_0}^{u} \phi(u')du' = Q(u) \tag{23.11.9}$$

or,

$$\int_a^b dx \int_{u_0}^{u} F(x,u')du' = \int_{u_0}^{u} du' \int_a^b F(x,u')dx \tag{23.11.10}$$

which was to be proved. This shows the possibility of *interchanging the order of multiple integrations.*

As an example of the use of the concept of integrating under the integral sign, we consider $F(x,u) = x^u$ where $u > -1$.

Now

$$\int_0^1 F(x,u)dx = \int_0^1 x^u dx = \frac{1}{u+1}.$$

Now multiplying by du, and integrating between a and b, we have

$$\int_a^b du \int_0^1 x^u dx = \int_a^b \frac{du}{u+1} = \ln\frac{b+1}{a+1}.$$

But

$$\int_a^b du \int_0^1 x^u dx = \int_0^1 dx \int_a^b x^u du = \ln\frac{b+1}{a+1}.$$

Now

$$\int_a^b x^u du = \frac{x^b - x^a}{\ln x}.$$

Hence

$$\int_0^1 \frac{x^b - x^a}{\ln x} dx = \ln\frac{b+1}{a+1}.$$

23.12 Evaluation of Certain Definite Integrals

The evaluation of the following definite integrals will illustrate some of the general principles discussed in Art. 23.11. Let us consider the integral

$$I = \int_0^\infty \frac{\sin bx}{x} dx.$$

If we change the sign of b, then the sign of the integral is changed. Substitution of $b = 0$ causes the integral to vanish. However, if we put $bx = y$, $b > 0$, then we obtain

$$I = \int_0^\infty \frac{\sin y}{y} dy.$$

This shows that the integral does not depend on b but is a constant. Considered as a function of b, it has a discontinuity at $b = 0$.

Let us consider the integral

$$\int_0^\infty e^{-kx} dx = -\frac{e^{-kx}}{k}\bigg|_0^\infty = \frac{1}{k} \quad \text{if } k > 0$$

If we consider k to be the complex number $k = a + jb$, then we have

$$\int_0^\infty e^{-(a+jb)x} dx = \frac{1}{a+jb} = \frac{a-jb}{a^2+b^2}, \quad a > 0.$$

Separating the real and imaginary parts, we obtain the two integrals

$$\int_0^\infty e^{-ax} \cos bx dx = \frac{a}{a^2+b^2} \qquad (23.12.1)$$

and

$$\int_0^\infty e^{-ax} \sin bx dx = \frac{b}{a^2+b^2} \qquad (23.12.2)$$

Let us integrate (23.12.1) with respect to b and obtain

$$\int_0^b db \int_0^\infty e^{-ax}\cos bx dx = \int_0^\infty e^{-ax}dx \int_0^b \cos bx db$$

$$= \int_0^\infty e^{-ax}\frac{\sin bx}{x}dx = a\int_0^b \frac{db}{a^2+b^2} = \tan^{-1}\frac{b}{a} \quad (23.12.3)$$

Putting $a = 0$ in (23.12.3), we obtain the result

$$\int_0^\infty \frac{\sin bx}{x}dx = \frac{\pi}{2}, \quad b > 0 \quad (23.12.4)$$

Again the integral

$$I_1 = \int_0^\infty e^{-x^2}dx$$

occurs very frequently in many branches of applied mathematics, particularly in the theory of probability. This integral represents the area of the so-called *probability curve*

$$u = \bar{e}^{x^2}. \quad (23.12.5)$$

Since the indefinite integral cannot be found except by a development in series, we are led to employ a certain device to evaluate the definite integral. Since the variable of integration in a definite integral is of no importance, we have

$$I_1 = \int_0^\infty e^{-x^2}dx, \quad I_1 = \int_0^\infty e^{-y^2}dy. \quad (23.12.6)$$

Multiplying these integrals together, we obtain

$$I_1^2 = \int_0^\infty e^{-x^2}dx \int_0^\infty e^{-y^2}dy = \int_0^\infty\int_0^\infty e^{-x^2-y^2}dxdy. \quad (23.12.7)$$

It is permissible to introduce e^{-x^2} under the sign of integration since x and y are to be considered as independent variables. If we consider x and y as the coordinates of a Cartesian reference frame and let z be a vertical coordinate, then the double integral (23.12.7) will represent the volume of a solid of revolution bounded by the surface $z = e^{-x^2-y^2}$. We may find this volume by introducing polar coordinates. Then the element of area in the XY plane is $ds = rdrd\theta$ where $x^2+y^2 = r^2$.

There may be some questions concerning the validity of this process since the double integral (23.12.7) is the limit given by

$$I_1^2 = \lim_{R\to\infty}\int_0^R\int_0^R e^{-x^2-y^2}dxdy \quad (23.12.8)$$

which represents the volume over a square in the XY plane of sides each equal to R (see Fig. 23.12.1).

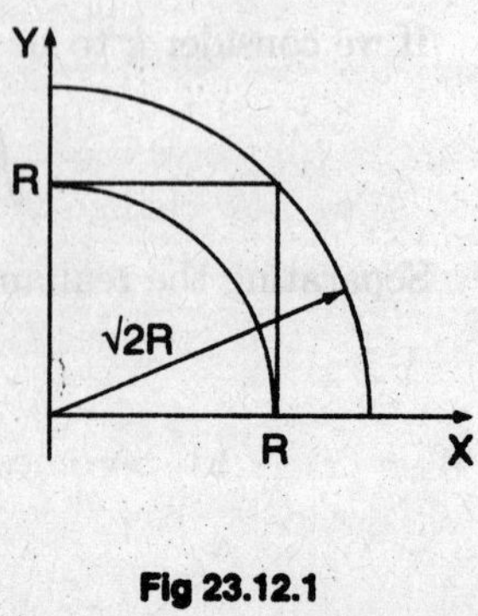

Fig 23.12.1

It is easy to see that this volume is greater than that of the figure of revolution over the circle of radius R and less than that over the circle of radius $\sqrt{2}R$. Hence if the integral

$$\int_0^{\pi/2}\int_0^{R} e^{-r^2} r dr d\theta$$

approaches a limit for $R=\infty$, we have

$$I_1^2 = \int_0^{\pi/2}\int_0^{\infty} e^{-r^2} r dr d\theta. \tag{23.12.9}$$

The integration with respect to θ merely multiplies by $\pi/2$, while in the integral

$$\int_0^{R} e^{-r^2} r dr = -\frac{1}{2}e^{-r^2}\Big|_0^R$$

the fact that we have an exact differential makes integration possible. Passing to the limit, we have

$$I_1^2 = \lim_{R\to\infty}\frac{\pi}{2}\int_0^R e^{-r^2} r dr = \frac{\pi}{4}\lim_{R\to\infty}\left(1-e^{-R^2}\right) = \frac{\pi}{4}. \tag{23.12.10}$$

We, therefore, have the desired result

$$I_1 = \int_0^{\infty} e^{-x^2} dx = \frac{\sqrt{\pi}}{2}. \tag{23.12.11}$$

If in this integral we make the change in variable $x=\sqrt{a}\,u$, $a>0$, then we obtain

$$I_1 = \int_0^{\infty} e^{-au^2}\sqrt{a}du = \frac{\sqrt{\pi}}{2} \quad \text{or,} \quad \int_0^{\infty} e^{-au^2} du = \frac{1}{2}\sqrt{\frac{\pi}{a}}, \quad a>0 \tag{23.12.12}$$

To illustrate a slightly different case, let us consider the integral

$$I = \int_0^{\infty} e^{-x^2+a^2/x^2} dx$$

This integral may be differentiated with respect to a to obtain

$$\frac{dI}{da} = +2\int_0^{\infty} \frac{a}{x^2} e^{-x^2+a^2/x^2} dx \tag{23.12.13}$$

If we now change the variable of integration by putting

$$x = \frac{a}{y} \quad \text{so that} \quad dx = -\frac{a dy}{y^2} = -\frac{x^2}{a}dy,$$

then (23.12.13) becomes

$$\frac{dI}{da} = -2\int_0^{\infty} e^{-y^2+a^2/y^2} dy = -2I. \tag{23.12.14}$$

This is a linear differential equation with constant coefficient for I. Its general solution is $I = Ce^{-2a}$ where C is an arbitrary constant to be determined. Putting $a = 0$, I reduces to

$$\int_0^\infty e^{-x^2} dx = \frac{\sqrt{\pi}}{2}.$$

Therefore, $C = \frac{\sqrt{\pi}}{2}$. Hence we have finally

$$\int_0^\infty e^{-x^2+a^2/x^2} dx = \frac{\sqrt{\pi}}{2} e^{-2a}, \quad a > 0 \tag{23.12.15}$$

The above examples illustrate typical procedures by which certain definite integrals may be evaluated.

23.13 Leibnitz's Rule

If $F(x, \alpha)$ and $\frac{\partial F(x,\alpha)}{\partial x}$ be continuous functions of x and α, then it can be shown, as given below, that

$$\frac{d}{d\alpha}\left[\int_a^b F(x,\alpha)dx\right] = \int_a^b \frac{\partial F(x,\alpha)}{\partial \alpha} dx \tag{23.13.1}$$

a, b being independent of α.

The above result is due to the German mathematician G.W. Leibnitz.

Proof. Let $\int_a^b F(x,\alpha)dx = f(\alpha)$. $\therefore\ f(\alpha+\delta\alpha) = \int_a^b F(x, \alpha+\delta\alpha)dx$

$$\therefore \quad \frac{f(\alpha+\delta\alpha) - f(\alpha)}{\delta\alpha} = \int_a^b \frac{F(x,\alpha+\delta\alpha) - F(x,\alpha)}{\delta\alpha} dx$$

$$= \int_a^b \frac{F(x,\alpha+\theta\delta\alpha)}{\delta\alpha} dx, 0 < \theta < 1 \text{ (by Mean Value Theorem)}$$

$$\therefore \quad \lim_{\delta\alpha\to 0} \frac{f(\alpha+\delta\alpha) - f(\alpha)}{\delta\alpha} = \int_a^b \frac{\partial F(x,\alpha)}{\partial\alpha} dx$$

$$\text{or, } \frac{d}{d\alpha}[f(\alpha)] = \int_a^b \frac{\partial F(x,\alpha)}{\partial\alpha} dx \quad \text{or, } \frac{d}{d\alpha}\left[\int_a^b F(x,\alpha)dx\right] = \int_a^b \frac{\partial F(x,\alpha)}{\partial\alpha} dx$$

Illustrative Examples

1. Evaluate $\int_0^\pi \log(1 + a\cos\alpha)dx$, using the method of differentiation under the sign of integration.

[B.U. (Hons.) 1981]

Ans. Let

$$f(a) = \int_0^\pi \log(1 + a\cos\alpha)dx.$$

$$\therefore \quad f'(a) = \frac{d}{da} f(a) = \int_0^{\pi} \frac{\partial}{\partial a}[\log(1 + a\cos\alpha)]dx = \int_0^{\pi} \frac{\cos\alpha}{1 + a\cos\alpha} dx$$

$$= \frac{1}{a}\int_0^{\pi} \frac{1 + a\cos\alpha - 1}{1 + a\cos\alpha} dx = \frac{1}{a}\int_0^{\pi} dx - \frac{1}{a}\int_0^{\pi} \frac{1}{1 + a\cos\alpha} dx$$

$$= \frac{\pi}{a} - \frac{1}{a}\frac{2}{\sqrt{1-a^2}}\left[\tan^{-1}\left(\sqrt{\frac{1-a}{1+a}}\tan\frac{x}{2}\right)\right]_0^{\pi}$$

$$= \frac{\pi}{a} - \frac{\pi}{a\sqrt{1-a^2}} = \frac{\pi(\sqrt{1-a^2} - 1)}{a\sqrt{1-a^2}}$$

$$\therefore \quad f(a) = \pi \int \frac{\sqrt{1-a^2} - 1}{a\sqrt{1-a^2}} da + C.$$

Now, putting $a = \sin\theta$, we have $da = \cos\theta d\theta$. With these substitutions, the above expression reduces to $f(a) = \pi\log(1 + \sqrt{1-a^2}) + C$. Again, putting $a = 0$, we have $C = -\pi\log 2$, so that $f(a) = \pi\log\frac{1}{2}(1 + \sqrt{1-a^2})$.

2. Evaluate the integral $\int_0^{\infty} e^{-x}\frac{\sin\alpha x}{x} dx$, using the method of differentiation under the sign of integration. **[B.U. (Hons.) 2000]**

Ans. Let $f(\alpha) = \int_0^{\infty} e^{-x}\frac{\sin\alpha x}{x} dx$, so that we may write

$$f'(\alpha) = \frac{d[f(\alpha)]}{d\alpha} = \int_0^{\infty} \frac{\partial}{\partial\alpha}\left[e^{-x}\frac{\sin\alpha x}{x}\right] dx$$

$$= \int_0^{\infty} \frac{xe^{-x}\cos\alpha x}{x} dx = \int_0^{\infty} e^{-x}\cos\alpha x dx$$

Let

$$I = \int e^{-x}\cos\alpha x dx = \frac{1}{\alpha}e^{-x}\sin\alpha x + \frac{1}{\alpha}\int e^{-x}\sin\alpha x dx \quad \text{(integrating by parts)}$$

$$= \frac{1}{\alpha}e^{-x}\sin\alpha x - \frac{1}{\alpha^2}\left[e^{-x}\cos\alpha x + \int e^{-x}\cos\alpha x dx\right] \quad \text{(again, integrating by parts)}$$

$$\therefore \quad \left(1 + \frac{1}{\alpha^2}\right)I = \frac{1}{\alpha}e^{-x}\sin\alpha x - \frac{1}{\alpha^2}e^{-x}\cos\alpha x$$

$$\therefore \quad I = \frac{1}{\alpha^2 + 1}\left[\alpha e^{-x}\sin\alpha x - e^{-x}\cos\alpha x\right], \quad \text{so that we have}$$

$$f'(\alpha) = \frac{1}{\alpha^2 + 1}\left[\alpha e^{-x}\sin\alpha x - e^{-x}\cos\alpha x\right]_0^{\infty} = \frac{1}{\alpha^2 + 1}$$

$$\therefore \quad f(\alpha) = \tan^{-1}\alpha + C; \text{ now putting } \alpha = 0, \text{ we have } C = 0$$

$$\therefore \quad \int_0^{\infty} e^{-x}\frac{\sin\alpha x}{x} dx = \tan^{-1}\alpha.$$

23.14 Double Integrals

In case of a single variable x, we define the definite integral as

$$\int_a^b F(x)dx = \lim_{h \to 0} h \sum_{r=1}^{n} F(a+rh), \quad h = (b-a)/n \tag{23.14.1}$$

i.e., we can express the definite integral as the limit of a sum.

If the sub-intervals are unequal and $n \to \infty$, we have

$$\int_a^b F(x)dx = \lim_{\substack{n\to\infty \\ \delta r \to 0}} \sum_{r=1}^{n} F(x_r)\delta r \tag{23.14.2}$$

In the case of double variables x and y for a function $F(x, y)$, we are to divide the total region, i.e., area A in n elementary areas $\delta A_r, r = 1, 2, \cdots, n$.

Here, also we are to consider the sum as

$$\sum_{r=1}^{n} F(x_r, y_r)\delta A_r. \tag{23.14.3}$$

Now, if the number of elementary areas is made infinitely large, i.e., as $n \to \infty, \delta A_r \to 0 (r = 1, 2, \cdots, n)$, then the limiting value (provided it exists) of the sum (23.14.3) is defined as the double integral over the defined region and is denoted by

$$\iint_A F(x, y)dA.$$

So

$$\iint_A F(x, y)dA = \lim_{n\to\infty} \sum_{r=1}^{n} F(x_r, y_r)\delta A_r.$$

This elementary area dA can be chosen according to the system of co-ordinates as discussed below.

(1) **Cartesian co-ordinates:** Let the co-ordinates of P be (x, y) and those of Q be $(x + \delta x, y + \delta y)$. So the sides of the elementary area are δx and δy.

$$\therefore \quad \text{the elementary area} = \delta A_r = \delta x_r \delta y_r.$$

$$\therefore \quad dA = \lim_{\substack{\delta x\to 0 \\ \delta y \to 0}} \delta x_r \delta y_r = dxdy$$

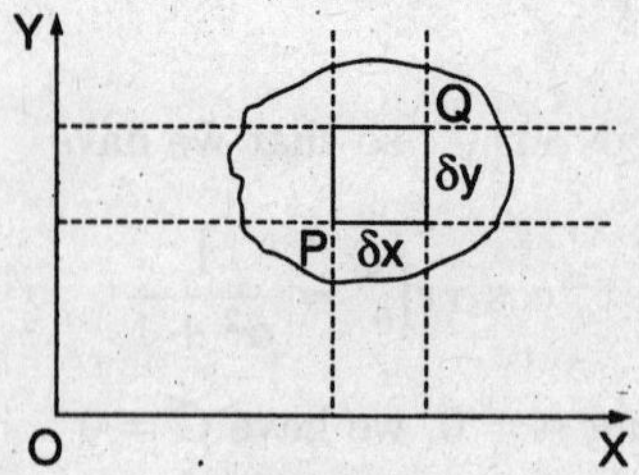

Fig 23.14.1

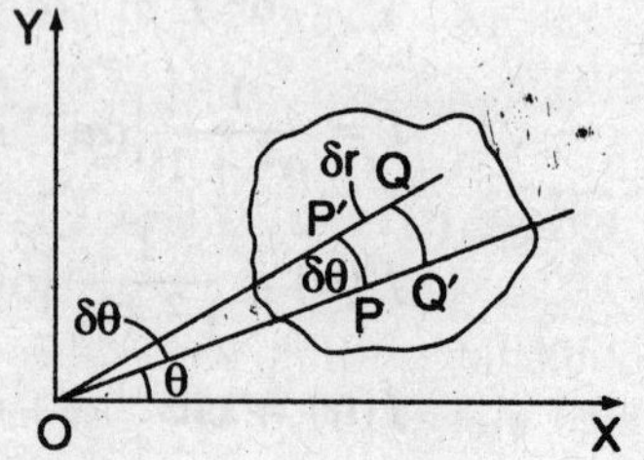

Fig 23.14.2

(2) **Polar co-ordinates:** Let the polar co-ordinates of P and Q be (r, θ) and $(r+\delta r, \theta+\delta\theta)$ respectively. Here $\angle POP' = \delta\theta$, $OP = OP' = r$.

$$\therefore \quad \text{arc } PP' = r\delta\theta \quad \text{and} \quad P'Q = \delta r$$

$$\therefore \quad \text{the elementary area} = \delta A = r\delta\theta \cdot \delta r = r\delta r\delta\theta$$

$$\therefore \quad dA = \lim_{\substack{\delta r \to 0 \\ \delta\theta \to 0}} r\delta r\delta\theta = r\,dr\,d\theta$$

Illustrative Examples

1. Evaluate $\displaystyle\int_{x=0}^{2}\int_{y=0}^{x} x^2(x+y)\,dx\,dy$.

Ans.

$$\int_{x=0}^{2}\int_{y=0}^{x} x^2(x+y)\,dx\,dy = \int_{x=0}^{2}\left[\int_{y=0}^{x}(x^3+x^2y)\,dy\right]dx = \int_{x=0}^{2}\left[x^3y+\frac{x^2y^2}{2}\right]_{y=0}^{x} dx$$

$$= \int_{x=0}^{2}\left(x^4+\frac{x^4}{2}\right)dx = \frac{3}{2}\int_0^2 x^4\,dx = \frac{3}{2}\left(\frac{x^5}{5}\right)_0^2 = \frac{3}{2}\cdot\frac{32}{5} = \frac{48}{5}.$$

2. Find by double integration the area inside the circle $r = a\sin\theta$ and the cardioide $r = a(1-\cos\theta)$.

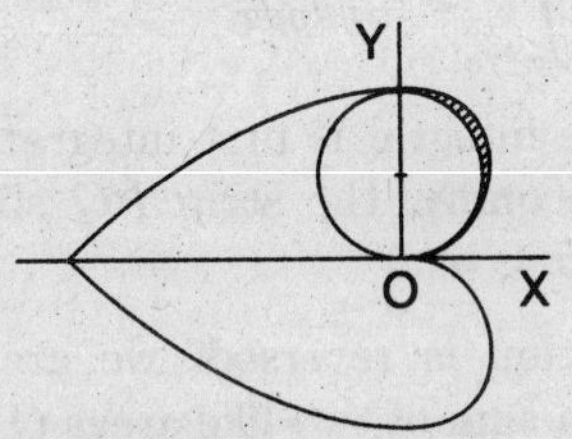

Ans. Here the points of intersections of the circle $r = a\sin\theta$ and the cardioide $r = a(1-\cos\theta)$ are $(0, 0)$ and $\left(a, \frac{\pi}{2}\right)$

$\therefore$ the required area in between the circle and the cardioide

$$= \int_{\theta=0}^{\pi/2}\int_{r=a(1-\cos\theta)}^{a\sin\theta} r\,dr\,d\theta = \int_{\theta=0}^{\pi/2}\left[\int_{r=a(1-\cos\theta)}^{a\sin\theta} r\,dr\right]d\theta = \frac{1}{2}\int_{\theta=0}^{\pi/2}\left[r^2\right]_{a(1-\cos\theta)}^{a\sin\theta} d\theta$$

$$= \frac{a^2}{2}\int_0^{\pi/2}\left[\sin^2\theta-(1-\cos\theta)^2\right]d\theta = \frac{a^2}{2}\int_0^{\pi/2}(2\cos\theta-\cos 2\theta-1)\,d\theta.$$

$$= \frac{a^2}{2}\left[2\sin\theta-\frac{1}{2}\sin 2\theta-\theta\right]_0^{\pi/2} = a^2\left(1-\frac{\pi}{4}\right).$$

23.15 Change of the Order of Integration

In case of double integration, sometimes it becomes most effective if the order of integration be changed, which makes the purpose of integration easier. It is our common experience that if we change the order, the corresponding limits of the variables are to be changed. The following examples will illustrate the above statement.

Illustrative Examples

1. Change the order of integration in the integral

$$\int_{x=0}^{1}\int_{y=x^2}^{\sqrt{x}} F(x,y)dydx.$$ **[B.U. (Hons.) 1994]**

Ans. In the given integral, the variable y should be given first preference. So the strip AB parallel to Y-axis with thickness dx is considered, and is finally made to move from $x = 0$ to $x = 1$ to generate the area.

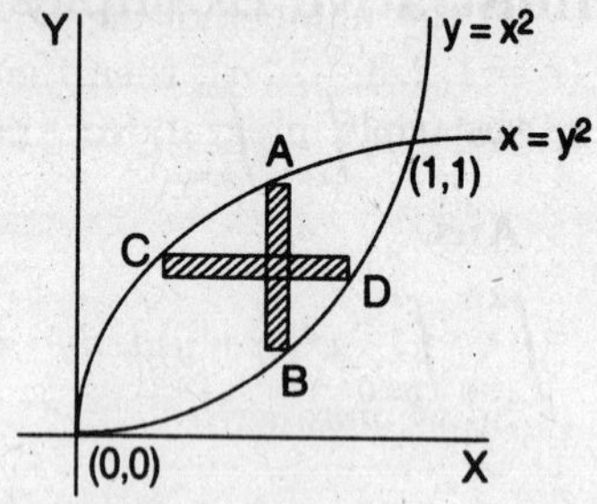

While changing the order, let us consider the strip CD parallel to X-axis with thickness dy. As C is on $x = y^2$ and D is on $x = \sqrt{y}$, and finally y changes from 0 to 1, we may write the above integral as

$$\int_{y=0}^{1}\int_{x=y^2}^{x=\sqrt{y}} F(x,y)dxdy.$$

2. Evaluate the following integral by changing the order of integration :

$$\int_{0}^{a}\int_{x^2/a}^{(2a-x)} xydydx.$$ **[B.U. (Hons.) 2000]**

Ans. In the given order, the integral is first integrated along the vertical strip PQ from $y = x^2/a$ to $y = (2a - x)$. Secondly, the strip PQ slides parallel to Y-axis from $x = 0$ to $x = a$.

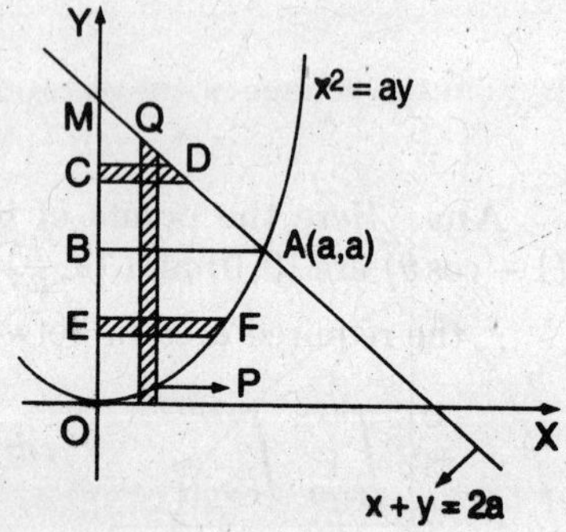

If now the order of integration is reversed, we are to consider the total area OAM as a sum of two like areas OAB and BAM.

On area OAB, we are to consider the strip EF parallel to X-axis, and the integration is to be carried out w.r.t. x from $x = 0$ to $x = \sqrt{ay}$ as EF slides from $y = 0$ to $y = a$. Again on area BAM, we are to consider the strip CD and the integration is to be carried out w.r.t. x from $x = 0$ to $x = 2a - y$ as the strip finally slides from $y = a$ to $y = 2a$.

$$I = \int_{y=0}^{a}\left[\int_{x=0}^{\sqrt{ay}} xydx\right]dy + \int_{y=a}^{2a}\left[\int_{x=0}^{(2a-y)} xydx\right]dy$$

$$= \int_{y=0}^{a}\left[\frac{x^2y}{2}\right]_0^{\sqrt{ay}} dy + \int_{y=a}^{2a}\left[\frac{x^2y}{2}\right]_0^{(2a-y)} dy$$

$$= \frac{1}{2}\int_{y=0}^{a} ay^2dy + \frac{1}{2}\int_{y=a}^{2a}(2a-y)^2ydy$$

$$= \frac{a}{2}\left[\frac{y^3}{3}\right]_0^a + \frac{1}{2}\left[2a^2y^2 - \frac{4ay^3}{3} + \frac{y^4}{4}\right]_a^{2a} = \frac{3}{8}a^4.$$

23.16 Triple Integrals

In the case of triple variables x, y and z for a function $F(x,y,z)$ we are to divide the total space, i.e., volume V into n elementary volumes $\delta V_r, r = 1, 2, \cdots, n$. Here, also we are to consider the sum as

$$\sum_{r=1}^{n} F(x_r, y_r, z_r)\delta V_r. \tag{23.16.1}$$

If the number of elementary volumes is made infinitely large, i.e., as $n \to \infty$, $\delta V_r \to 0$ $(r = 1, 2, 3, \cdots, n)$, then the limiting value (provided it exists) of the sum (23.16.1) is defined as the triple integral over the defined space and is denoted by

$$\iiint_V F(x,y,z)dV.$$

So we may write

$$\iiint_V F(x,y,z)dV = \lim_{n\to\infty} \sum_{r=1}^{n} F(x_r, y_r, z_r)\Delta V_r.$$

The elementary volume dV can be chosen according to the system of co-ordinates.

Illustrative Examples

1. Find the volume of the ellipsoid $\frac{x^2}{a^2} + \frac{y^2}{b^2} + \frac{z^2}{c^2} = 1$ by triple integration.

[B.U. (Hons.) 1996]

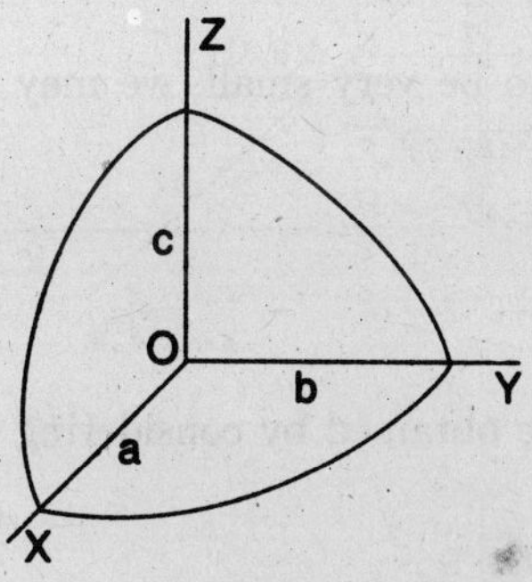

Ans. Equation to the ellipsoid is $\frac{x^2}{a^2} + \frac{y^2}{b^2} + \frac{z^2}{c^2} = 1$.
At $z = 0$, i.e., in the X-Y plane, equation to the ellipse is

$$\frac{x^2}{a^2} + \frac{y^2}{b^2} = 1 \quad \therefore \quad y = \pm b\sqrt{1 - \frac{x^2}{a^2}}.$$

$\therefore$ the volume in the first octant

$$= \int_{x=0}^{a} \int_{y=0}^{b\sqrt{1-\frac{x^2}{a^2}}} \int_{z=0}^{c\sqrt{1-\frac{x^2}{a^2}-\frac{y^2}{b^2}}} dzdydx$$

$$= c\int_{x=0}^{a} \int_{y=0}^{b\sqrt{1-\frac{x^2}{a^2}}} \sqrt{1 - \frac{x^2}{a^2} - \frac{y^2}{b^2}}\,dydx$$

$$= c\int_{x=0}^{a} \int_{y=0}^{b\sqrt{1-\frac{x^2}{a^2}}} \sqrt{\alpha^2 - \frac{y^2}{b^2}}\,dydx \text{ where } \alpha^2 = \left(1 - \frac{x^2}{a^2}\right)$$

$$= \frac{c}{b}\int_{x=0}^{a}\int_{y=0}^{b\sqrt{1-\frac{x^2}{a^2}}} \sqrt{b^2\alpha^2 - y^2}\,dy\,dx$$

$$= \frac{c}{b}\int_{x=0}^{a}\left[\frac{y\sqrt{b^2\alpha^2-y^2}}{2} + \frac{b^2\alpha^2}{2}\sin^{-1}\frac{y}{b\alpha}\right]_0^{b\alpha} dx \quad \text{(integrating by parts)}$$

$$= \frac{c}{b}\int_{x=0}^{a}\frac{b^2\alpha^2}{2}\frac{\pi}{2}dx = \frac{\pi bc}{4}\int_0^a\left(1-\frac{x^2}{a^2}\right)dx = \frac{\pi bc}{4}\left[x-\frac{x^3}{3a^2}\right]_0^a = \frac{\pi abc}{6}$$

$$\therefore \text{ total volume of the ellipsoid} = 8\cdot\frac{\pi abc}{6} = \frac{4\pi abc}{3}.$$

23.17 Errors and Approximations

Very often it happens that the result of a physical quantity calculated from the measured values of the different experimental data is approximate, i.e., the measured value differs from the actual value—due to some errors in the measurements. In these cases we may apply the notion of partial derivatives.

Let us consider a continuous function $F(x, y)$. Now, if we allow very small increments δx and δy in the variables x and y respectively, then denoting the corresponding change in $F(x, y)$ as δF, we may write $\delta F = F(x + \delta x, y + \delta y) - F(x, y)$.

Now, by Taylor's series expansion of $F(x + \delta x, y + \delta y)$, we have

$$F(x+\delta x, y+\delta y) = F(x,y) + \delta x\frac{\partial F}{\partial x} + \delta y\frac{\partial F}{\partial y} + \text{higher order terms.}$$

Neglecting higher order terms, since δx and δy are considered to be very small, we may write

$$\delta F = \delta x\frac{\partial F}{\partial x} + \delta y\frac{\partial F}{\partial y}$$

which may be considered as approximate solution for δF.

Thus an almost approximate effect of the errors in $F(x, y)$ can be obtained by considering δx and δy as errors in x and y respectively.

Illustrative Examples

1. The area of a triangle is calculated by using the formula $\frac{1}{2}bc\sin A$. If the lengths b and c are liable to an error of $\frac{1}{4}$% in measurement and the angle A measured as 60° is liable to an error of 1 minute of angle, find the greatest possible percentage error in the value of the area. **[Calicut 1984]**

Ans. Let S be the area of the triangle.

$$\therefore \quad S = \frac{1}{2}bc\sin A. \tag{1}$$

If δS be the error in S, then

$$\delta S = \frac{1}{2}\left[\delta b\frac{\partial S}{\partial b} + \delta c\frac{\partial S}{\partial c} + \delta A\frac{\partial S}{\partial A}\right]$$

$$= \frac{1}{2}[\delta b\cdot c\sin A + \delta c\cdot b\sin A + \delta A\cdot bc\cos A] \qquad (2)$$

Here $A = 60°, \delta b = \frac{b}{400}, \delta c = \frac{c}{400}, \delta A = \frac{1}{60}$.

$\therefore$ from (1), $S = \frac{\sqrt{3}}{4}bc$ and from (2), we have

$$\delta S = \frac{1}{2}\left[\frac{bc}{400}\frac{\sqrt{3}}{2} + \frac{bc}{400}\frac{\sqrt{3}}{2} + \frac{bc}{120}\right]$$

$$= \frac{bc}{80}\left[\frac{\sqrt{3}}{10} + \frac{1}{3}\right] = \frac{3\sqrt{3}+10}{30}\cdot\frac{bc}{80}$$

$$\therefore \text{ percentage error} = \frac{\delta S}{S} \times 100\% = \frac{3\sqrt{3}+10}{2400}\cdot\frac{4}{\sqrt{3}} \times 100 = \frac{3\sqrt{3}+10}{6\sqrt{3}} = 1.462.$$

2. If z varies directly as the square of x and inversely as the cube of y, and the possible errors in measuring x and y are 1% and 0.5% respectively, find the magnitude of error in z. Given that when x and y are 30 and 25 respectively, $z = 36/625$.

Ans. $z \propto \frac{x^2}{y^3}$ or, $z = k\frac{x^2}{y^3}$. Now, for $x = 30, y = 25, z = \frac{36}{625}$ we have $k = 1$. So $z = \frac{x^2}{y^3}$ or, $\log z = 2\log x - 3\log y$. $\therefore$ $\frac{\delta z}{z} = 2\frac{\delta x}{x} - 3\frac{\delta y}{y}$.

Now, from the given data

$$\frac{\delta x}{x} = \frac{1}{100} \quad \text{and} \quad \frac{\delta y}{y} = \frac{5}{1000} = \frac{1}{200}$$

$$\therefore \quad \frac{\delta z}{z} = \frac{2}{100} - \frac{3}{200} = \frac{1}{200}, \text{ i.e., } \delta z = \frac{1}{200}z, \text{ i.e., } 0.5\% \text{ of } z.$$

Exercises

1. **Find the first and the second order partial derivatives of the function $\tan^{-1}(x/y)$.**
2. **Find F_x, F_y, F_{xx}, F_{xy}, F_{yx} and F_{yy}, where $F(x,y) = ax^2 + 2hxy + by^2$.**

   [**Ans.** See worked out examples.]

3. **If $F(x,y) = xy(x^2 - y^2)/(x^2 + y^2)$, for $(x,y) \neq (0,0)$ and $F(0,0) = 0$, show that $F_{xy}(0,0) \neq F_{yx}(0,0)$.**

   [**Ans.** See worked out examples.]

   Hence write down the conditions for equality of F_{xy} and F_{yx}.

4. **Show that, if $u = \ln(x^2 + y^2) + \tan^{-1}(y/x)$, then $\frac{\partial^2 u}{\partial x^2} + \frac{\partial^2 u}{\partial y^2} = 0$.**

   [**Ans.** See worked out examples.]

5. Show that, if $u = \tan(y + ax) + \sqrt{y - ax}$, then $\dfrac{\partial^2 u}{\partial x^2} = a^2 \dfrac{\partial^2 u}{\partial y^2}$.

6. If F is a function of x and y, and $x + y = 2e^{\theta} \cos\phi, x - y = 2je^{\theta} \sin\phi$, where $j = \sqrt{-1}$, prove that

$$\frac{\partial^2 F}{\partial \theta^2} + \frac{\partial^2 F}{\partial \phi^2} = 4xy\left(\frac{\partial^2 F}{\partial x \partial y}\right).$$

7. Calculate the error in z where $z = x^2 y^3$ and the errors in x and y are Δx and Δy respectively.

[**Ans.** See worked out examples]

8. The rate of flow V of a liquid through a capillary tube of radius r and length l at a pressure difference P between its ends is given by $V = \dfrac{\pi P r^4}{8\eta l}$, where η is the coefficient of viscosity of the liquid. In an experiment, the errors in the measurements of P, r, l and V are 1%, 1.5%, 0.5% and 2% respectively. Evaluate the error in the measurement of η. **[B.U. (Hons.) '95]**

[**Ans.** See worked out examples.]

9. The density of the material of a uniform circular cylinder is determined by measuring its mass m, length l, and diameter d. Calculate the density (in kg-m^{-3}) and its error from the following data:

$m = 47.36 \pm 0.01$ g; $l = 15.28 \pm 0.05$ mm, $d = 21.37 \pm 0.04$ mm. **[B.U. (Hons.) 2001]**

[**Ans.** See worked out examples.]

10. The volume (V) of a cylinder is given by $V = \pi r^2 l$, where r is the radius, and l the length. If Δr and Δl be the errors in measuring radius and length of the cylinder, find the approximate error made in the volume.

[**Ans.** Proceed exactly as in (7) and (8).]

11. If T be the time period and l, the length of a simple pendulum, then $T = 2\pi\sqrt{l/g}$, where g = the acceleration due to gravity. Find the error in computing 'g' of a certain place, where $T = 1.13 \pm 0.01$ sec; $l = 199.7 \pm 0.25$ cm, $\pi = 22/7$.

[**Ans.** Proceed exactly as in (9).]

12. If $u = F\left(\dfrac{x}{y}, \dfrac{y}{z}, \dfrac{z}{x}\right)$, prove that $x\dfrac{\partial u}{\partial x} + y\dfrac{\partial u}{\partial y} + z\dfrac{\partial u}{\partial z} = 0$.

[**Ans.** See worked out examples.]

13. If $u = F(x - y, y - z, z - x)$, prove that $\dfrac{\partial u}{\partial x} + \dfrac{\partial u}{\partial y} + \dfrac{\partial u}{\partial z} = 0$.

14. Transform the Laplacian operator into polar form.

15. Use Taylor's series expansion, for a function of two variables, to expand the function $\sin(xy)$ in powers of $(x - 1)$ and $y - \pi/2)$ up to second degree terms.

[**Ans.** See worked out examples.]

16. Use Taylor's series expansion, to expand the function $e^x \log(1 + y)$ in powers of x and y up to 3rd degree terms.

[**Ans.** Proceed as in (15).]

17. The sides of a rectangular parallelopiped are 2.01 cms, 1.96 cms and 0.97 cm. Find the approximate length of the greatest diagonal.

[**Ans.** See worked out examples.]

18. The angle of elevation of the top of a tower is 60° from a point on the ground level at a distance 25 metres apart from the foot of the tower. If the errors in measuring the elevation and the angle are 5 cms and 0.01 degree respectively, find the error in computing the height of the tower.

[**Ans.** Proceed as in (7) and (8).]

19. Examine, for extreme values, of the function $F(x, y) = x^3 + y^2 - 3xy$.

[**Ans.** See worked out examples.]

20. Find the maxima and minima values of the function $F(x, y) = x^2y^2 - 5x^2 - 8xy - 5y^2$.

[**Ans.** See worked out examples.]

21. Investigate the maxima and minima, if there be any, of the function

$$F(x, y) = x^3 + 3x^2 + 4xy + y^2.$$

[**Ans.** Proceed as in (19) and (20).]

22. Examine the nature of extreme values of the function

$$F(x, y) = x^4 + y^4 - 6(x^2 + y^2) + 8xy.$$

[**Ans.** Proceed as in (19) and (20).]

23. Using the method of Lagrange multiplier, find the maximum of $F = 4xyz$, subject to the constraint $x^2 + y^2 + z^2 - a^2 = 0$. **[B.U. (Hons.) '97]**

[**Ans.** See worked out examples.]

24. Using the method of Lagrange multiplier, find the area of the largest rectangle which can be inscribed in a semi-circle of radius R with one of the larger sides of the rectangle coinciding with the diameter. **[B.U. (Hons.) '96]**

[**Ans.** See worked out examples.]

25. Find the height of a right circular cylinder with greatest volume that can be inscribed in a sphere of radius R. **[B.U. (Hons.) '93]**

[**Ans.** See worked out examples.]

26. Find the maximum value of $V = xyz$ subject to the condition that $\frac{x^2}{a^2} + \frac{y^2}{b^2} + \frac{z^2}{c^2} = 1$.

[**Ans.** Proceed as in (23).]

What is the geometric interpretation of this problem?

27. A rectangular box open at the top is to be constructed having a volume 108 cubic feet. Find the dimensions of the box requiring least material for its construction. **[B.U. (Hons.) 1995]**

[**Ans.** Proceed as in (23), (24) and (25).]

28. Find the critical points of the following functions and test for minima and maxima:

(i) $z = 1 + x^2 + y^2$, (ii) $z = 2x^2 - xy - 3y^2 - 3x + 7y$. **[B.U. (Hons.) '94]**

[**Ans.** Proceed as in (19) and (20).]

29. Divide 24 into three parts such that the continued product of the first, the square of the second and the cube of the third may be a maximum.

[**Ans.** Proceed as in (23), (24) and (25).]

30. Test the following function for maximum and minimum, $z = x^2 - xy + y^2 + 3x - 2y + 1$. Find also the maximum or minimum value. **[B.U. (Hons.) '92]**

[**Ans.** See worked out examples.]

31. Evaluate $\int_0^{\pi} \log(1 + a\cos x)dx$, using the method of differentiation under the sign of integration. **[B.U. (Hons.) '81]**

[**Ans.** See worked out examples.]

32. Evaluate the integral $\int_0^{\infty} e^{-x} \frac{\sin \alpha x}{x} dx$. **[B.U. (Hons.) 2000]**

[**Ans.** See worked out examples.]

33. Evaluate $\int_0^1 \frac{x^\alpha - 1}{\log x} dx, \alpha \geq 0$. **[B.U. (Hons.) '98]**

[**Ans.** See worked out examples.]

34. Show that $\int_0^{\infty} e^{-ax} \cos bx dx = \frac{a}{a^2 + b^2}$. Using this result, evaluate the integral

$$\int_0^{\infty} e^{-ax} \sin bx x dx.$$

[**Ans.** See Art. 23.12.]

35. Using Leibnitz's rule for differentiation under the integral sign, find

$$I(\alpha) = \int_0^{\infty} \frac{e^{-\alpha x} \sin x}{x} dx$$

and hence evaluate the integral

$$\int_0^{\infty} \frac{\sin x}{x} dx.$$ **[B.U. (Hons.) '92]**

[**Ans.** See worked out examples.]

36. Find the value of the integral $\int_0^{\infty} \frac{e^{-x} - e^{-\lambda x}}{x} dx, \lambda > 0$. **[B.U. (Hons.) '95]**

37. Evaluate $\int_{x=0}^{2} \int_{y=0}^{x} x^2(x + y)dxdy$.

[**Ans.** See worked out examples.]

38. Find by double integration the area inside the circle $x = a\sin\theta$ and the cardioide

$$r = a(1 - \cos\theta).$$

[**Ans.** See worked out examples.]

39. Evaluate the following integral by changing the order of integration, $\int_0^a \int_{x^2/a}^{(2a-x)} xydydx$.

[**Ans.** See worked out examples.]

40. Evaluate the integral by changing the order of integration, $\int_0^a \int_{a-\sqrt{a^2-y^2}}^{a+\sqrt{a^2-y^2}} xdydx$.

[**Ans.** See worked out examples.]

41. Evaluate the integral

$$\int_{-a}^{a}\int_{-\frac{b}{a}\sqrt{a^2-x^2}}^{\frac{b}{a}\sqrt{a^2-x^2}}(x+y)^2dxdy.$$

Show that the integral remains unaltered if the order of integration is reversed.

[**Ans.** See worked out examples.]

42. Integrate after reversing the order $\int_0^{2a}\int_0^{\sqrt{2ax-x^2}} x^2dydx.$

[**Ans.** See worked out examples.]

43. Find the volume bounded by the cylinder $x^2+y^2=1$ and the planes

$y+z=4$ and $z=0$. **[B.U. (Hons.) '96]**

[**Ans.** 4π.]

44. Find the volume of the ellipsoid $\frac{x^2}{a^2}+\frac{y^2}{b^2}+\frac{z^2}{c^2}=1$ by triple integration.

[B.U. (Hons. Old) '96]

45. Show that the equation $\frac{dI(a)}{da} = -2I(a)$ is satisfied by $I(a) = \int_0^{\infty} e^{-x^2-(a/x)^2}dx$; given $\int_0^{\infty} e^{-x^2}dx = \frac{1}{2}\sqrt{\pi}$, show that $I(a) = \frac{1}{2}\sqrt{\pi}e^{-2a}$. **[B.U. (Hons.) '99]**

Ans. $\frac{d}{da}[I(a)] = \int_0^{\infty}\frac{\partial}{\partial a}\left[e^{-x^2-(\frac{a}{x})^2}\right]dx = -2\int_0^{\infty}\frac{a}{x^2}e^{-x^2-(a/x)^2}dx.$

Let $\frac{a}{x}=y$, then for $x=0$, $y=\infty$, and for $x=\infty, y=0$ and $-\frac{a}{x^2}dx=dy$.

$$\therefore \quad \frac{d}{da}[I(a)] = -2\int_0^{\infty} e^{-y^2-(a/y)^2}dy = -2I(a)$$

$$\therefore \quad d[I(a)]+2I(a)da=0 \quad \therefore \quad I(a)=Ce^{-2a};\ \text{now for } a=0,\ I(0)=C$$

$$\therefore \quad C=I(0)=\int_0^{\infty}e^{-x^2}dx=\frac{\sqrt{\pi}}{2} \quad \therefore \quad I(a)=\frac{\sqrt{\pi}}{2}e^{-2a}$$

46. The area of a triangle is calculated by using the formula $\frac{1}{2}bc\sin A$. If the lengths b and c are liable to an error of $\frac{1}{4}$% in measurement and the angle A measured as $60°$ is liable to an error of 1 minute of angle, find the greatest possible percentage error in the value of the area.

[**Ans.** See worked out problems.]

47. If z varies directly as the square of x and inversely as the cube of y, and the possible errors in measuring x and y are 1% and 0.5% respectively, find the magnitude of error in z; given that when x and y are 30 and 25 respectively, $z=36/625$.

[**Ans.** See worked out problems.]

48. Write down the Taylor series expansion of a function $f(x,y)$ of two variables x and y about the values x_0, y_0. **[C.U. (Hons.) 2005]**

(See Arts. 23.4 and 23.4.1)

49. (a) Given a function $f(x)$, assume that the Taylor series for $f(x)$ is

$$f(x) = a_0 + a_1(x-a) + a_2(x-a)^2 + a_3(x-a)^3 + \cdots + a_n(x-a)^n + \cdots$$

Calculate $f(a), f'(a), f''(a) \cdots f^n(a)$, where $f^n(x)$ is the nth derivative of $f(x)$ at a.

(b) Write the Taylor series of $f(x)$ about $x = a$.

(c) Hence define an analytic function in terms of expansion in Taylor series.

[C.U. (Hons.) 2006]

Ans. See Q.26 of the Exercises of Chapter 21.

Chapter 24

Statistics and Probability

24.1 Introduction

In many branches of physics, processes, which are not known with exactness, are encountered frequently. In the study of phenomena of this type, the applications of the methods of statistics and the theory of probability are imperative. The theory of probability is of basic importance in quantum mechanics, kinetic theory, statistical mechanics, particle physics, etc. Statistical methods along with the laws of probability will be needed in any problem dealing with large numbers of particles or variables, where it is impossible or impractical to have complete information, such as the shot effect in vacuum tubes, radioactive decay, turbulence in hydrodynamics, some electrical network problems, information theory, etc. Also, since physical measurements are always subject to error, the theory of probability is needed in the theory of errors.

The word *probably* is frequently used in our everyday life. For instance, we very often say—*it will probably rain today* or *we will probably win this game*, etc. Such statements always imply a state of partial ignorance about the outcomes of some event. We never say *probably* about something whose outcome we know. The theory of probability tries to express more precisely just what our state of ignorance is. For example, if we have hundred balls in a bag and draw ninety-nine balls out of the bag and if all these ninety nine balls happen to be red in colour, then we say that probably the remaining ball is also red in colour. Though it is always possible that the remaining ball is of some other colour, yet it has the likelihood of being of the same colour. Thus, if we draw the conclusion that the remaining ball is also red in colour, this conclusion is not based upon certainty, but we use to say that our conclusion is based on probability.

In mathematics, a numerical measure is given to the probability. For example, in tossing a coin, either the head or the tail will turn up. We cannot say with certainty whether the head or the tail will turn up. The chances of both the things to happen (either the head or the tail up) are equal if the number of tosses made is large. Mathematically we express it by saying that the probability of getting a head in a large number of tosses of the coin is 1/2 and similarly for a tail. We mean by this that there are two possible outcomes of the experiments and that we have no reason to expect one outcome more than the other; therefore, we assign equal probabilities to the two possible outcomes.

We take another example in which we consider two persons A and B each tossing a coin and looking at his own coin, but not at each other's. They want to find the probability that both coins show heads. We suppose that A sees his coin showing head; he then says that the probability that both coins show head is one (i.e., hundred percent), because he knows that his coin shows head. On the other hand we suppose that B sees his coin showing tail; then he says that the probability of both heads is zero. None of them is correct. It we now suppose that a third person looks at both coins and says that at least one is head, then there are three possibilities HH, HT, TH—all of which are equally likely and therefore, we say that the probability of HH is 1/3. Without this information of at least one head, there are four possibilities HH, HT, TH, TT—all of which are equally likely, and we say in that case that the probability of HH is 1/4.

The last example shows that the answer to a probability question depends on the state of knowledge (or ignorance) of the person giving the answer.

24.2 Definitions of Probability: A Priori Probability or Mathematical Probability

Probability may be defined in two ways—*a priori probability* and *a posteriori probability.*

A Priori Probability or Mathematical Probability: If there are n *equally likely, mutually exclusive* and *collectively exhaustive* outcomes of an experiment, and n_s of them are favourable to an event A, then the mathematical or *a priori* probability of the event A is

$$P(A) = \frac{\text{Number of outcomes favourable to } A}{\text{Total number of outcomes}} = \frac{n_s}{n}$$

Here n is the number of equally likely outcomes of an experiment and n_s is the number of successes of the experiment. If now we *predict* that the ratio $P(A) = n_s/n$ is such that $\lim_{n\to\infty} P(A) = p_s$, then p_s is called the *a priori probability* or the prediction that the experiment will succeed.

In the above definition, the words *equally likely* mean equally probable, i.e., no happening is biased or partially bound to occur. Since we have no information which makes us expect one result more than another, so we assume the same probability for each one of the set of outcomes. The words *mutually exclusive* rule out the probability of two simultaneous happenings in a trial—for example in tossing a coin, *head* and *tail* cannot fall together but falling of one excludes the other. The words *collectively exhaustive* assure the happening of an event either in favour or against and rule out the possibility of happening of neither (in favour or against) in any trial, i.e., we must consider all possibilities.

The above definition of probability is sometimes expressed as the odds in favour of A are p to $(q-p)$ or the odds against A are $(q-p)$ to p (writing $n_s = p$ and $n = q$). More precisely, if the odds in favour of an event are m to n (or n to m for odds against the event), then the probability of happening of the event is $\frac{m}{m+n}$ or the probability of not happening is $\frac{n}{m+n}$. For example, the odds in favour of drawing a white ball from a bag containing 3 white and 4 red balls is 3 to 4 and hence the probability of drawing a white ball from the bag is $\frac{3}{3+4} = \frac{3}{7}$, and similarly the probability of drawing a red ball is $\frac{4}{3+4} = \frac{4}{7}$.

24.2.1 A Posteriori Probability or the Statistical Definition of Probability

If an experiment is performed N times and the number of times the experiment succeeds is N_s and if the ratio $P(N) = \frac{N_s}{N}$ approaches the limit $\lim\limits_{N\to\infty} P(N) = \lim\limits_{N\to\infty} \frac{N_s}{N} = P_s$, then we define the probability of the success of the experiment as P_s. Therefore, we can define the a posteriori probability as follows:

If in a large number of trials performed under the same conditions, the ratio of the number of success of an experiment to the total number of trials is unique and finite under the limit that the number of trials tends to infinity, then this ratio is a mesure of probability of the success.

In this case, the probability P_s of success is measured by actually performing the experiment, and is not predicted; therefore, P_s is called the a *posteriori probability* that the experiment will be successful.

Of course, the two types of probability should always give the same answer. If the a posteriori probability comes out to be different from the a priori probability, it will be concluded that an error has been made in the analysis. In the following discussions, no distinction will be made between the two types of probability.

From the above definitions of probability, the following conclusions can be made:

Corollary 1. If p_s be the probability of success of an experiment, then $(1 - p_s) = p_f$ is the probability of its failure. Obviously $p_s + p_f = 1$.

Similarly, if the probability of happening of an event is p, then the probability of not happening is $(1 - p) = q$, such that $p + q = 1$.

For example, if in throwing two dice together, we want to find the probability of throwing at least one ace in a single throw, the problem may be considered as follows—Since there are six faces in a die, numbered $1, 2, 3, 4, 5$ and 6, there are 5 ways of throwing for each die, so that there is no ace. In throwing two dice together, the number of ways excluding aces is $5 \times 5 = 25$. Therefore, the probability of not throwing any ace is $25/36$, since the total number of ways is $6 \times 6 = 36$. Thus the probability of throwing at least one ace is $1 - \frac{25}{36} = \frac{11}{36}$.

Corollary 2. The value of the probability P is a number beween 0 and 1 (both the limits being included), i.e., $0 < P < 1$.

Corollary 3. The probability of a certain event is one or $P(E) = 1$, where $P(E)$ is the probability of a certain event, i.e., the event which occurs in every observation.

Corollary 4. The probability of an impossible event is zero or $P(0) = 0$. By the words *an impossible event*, we mean an event which cannot occur in any observation. The following illustrative examples will give us a clear idea of the definition of probability:

Illustrative Examples:

1. If a pair of dice is thrown, what is the probability that a sum of 8 shows?

Ans. The first die can fall in 6 ways, for each of which the second die can fall in 6 ways. The total number of ways, i.e., events is $6 \times 6 = 36$—all of which are equally likely.

Now a sum of 8 can be obtained in the following 5 ways—$2 + 6$, $6 + 2$, $3 + 5$, $5 + 3$, $4 + 4$. The first event of them states that one die shows 2 and the second shows 6 and so on. Thus out of a total 36 events, 5 events are favourable to get a sum of 8.

2. In a single cast with two dice, what is the probability of throwing (i) two aces, (ii) doublets, (iii) five-six, (i.e., one die turns up five and the other six)?

Ans. As, in example 1, there are 36 equally likely cases—which are made up of the 6 throws $(1,1),(2,2),\cdots,(6,6)$, together with the 30 throws $(1,2),(2,1),\cdots,(5,6),(6,5)$.

Now we may count $(5,6),(6,5)$ as the same throw, and call it five-six.

The favourable event of throwing 2 aces is one, i.e., $(1,1)$, so that the probability of throwing two aces $= 1/36$.

Again there are the following 6 favourable events for throwing doublets $(1,1)$, $(2,2)$, $(3,3)$, $(4,4)$, $(5,5)$ and $(6,6)$, so that the probability of throwing doublets $= 6/36 = 1/6$.

Similarly, the probability of throwing five-six $= 2/36 = 1/18$.

3. In a well-shuffled deck of cards, what is the probability that the top four cards are, respectively ace, two, three and four of hearts?

Ans. The first card may be any one of 52; for each determination of that card, there remain 51 possibilities for the next and so on. Hence the total number of cases or events is 52.51.50.49 respectively for the top four cards as given in the question.

Out of all these cases, the required arrangement of four cards is obtained once. This means that only one event is favourable. Hence the desired probability $= \frac{1}{52.51.50.49}$.

In order to generalise the above case, we consider r things into r numbered spaces from a stock of n distinct things; then any particular arrangement of the objects is called a *permutation* of n things taking r at a time. The total number of such permutations is

$$\begin{aligned} {}^nP_r &= n\cdot(n-1)\cdot(n-2)\cdot\cdots\cdot(n-r+1) \\ &= \frac{n\cdot(n-1)\cdot(n-2)\cdot\cdots\cdot(n-r+1)\cdot(n-r)\cdots 3\cdot 2\cdot 1}{(n-r)\cdot(n-r-1)\cdot\cdots\cdot 3\cdot 2\cdot 1} \\ &= \frac{n!}{(n-r)!} \end{aligned}$$

4. If a hand of 4 cards is drawn from a deck of cards, what is the probability that the hand consists of ace, two, three and four of hearts?

Ans. Here the order is not relevant. Let c be the number of ways in which four cards are taken out of 52 cards, without taking into consideration the order of the cards. But each hand of 4 cards admits of ${}^4P_4$ different orderings of its members. So the number of distinct 4-card hands when the order is relevant is $c\,{}^4P_4$.

On the other hand, the number of 4-card hands when the order is counted is ${}^{52}P_4$, so that we may write $c\,{}^4P_4 = {}^{52}P_4$ or, $c = \frac{{}^{52}P_4}{{}^4P_4} = \frac{52!}{48!4!}$.

In general when r things are taken from the stock of n things, the number of ways of doing so is called *combination* of n things taken r at a time—denoted by ${}^nC_r$.

So ${}^nC_r = \frac{n!}{r!(n-r)!} = \frac{{}^nP_r}{r!}$ or, ${}^nP_r = r!\,{}^nC_r$.

5. What is the probability that a leap-year selected at random contains 53 Sundays?

Ans. A leap-year consists of 52 weeks and 2 more days, i.e., a total of 366 days. The remaining two days may form any of the following 7 sets—Sunday and Monday, Monday and Tuesday, Tuesday and Wednesday, Wednesday and Thursday, Thursday and Friday, Friday and Saturday, Saturday and Sunday.

Out of the above 7 sets, only two (first and the last) contain Sunday. Thus the required probability is 2/7.

24.3 Fundamental Laws of Probability

Let $P(A)$ be the probability of an event A occurring when a certain experiment is performed. By the definition of probability, $P(A)$ is a fraction that lies between 0 and 1. We also have, by definition, that if $P(A) = 1$, the event A is certain to happen. However, if $P(A) = 0$, the event A is certain *not* to happen.

Probability of Compound Events: When two or more simple events occur in connection with each other, their joint occurrence is called a *compound event.* For example, if in drawing two cards in succession, we get a king on first draw (A) followed by an ace on the second draw (B), then the compound event AB happens, and the probability of the compound event is known as the *compound probability.*

So to illustrate more complex possibilities, let us consider an experiment with N equally likely outcomes that involve two events A and B. Here we shall use the following notations:

n_1 = the number of outcomes in which A only occurs;

n_2 = the number of outcomes in which B only occurs;

n_3 = the number of outcomes in which both A and B occur;

n_4 = the number of outcomes in which neither A nor B occurs.

so that $N = n_1 + n_2 + n_3 + n_4$.

Now, by the use of the above notations, we may define the following probabilities:

$$P(A) = \frac{n_1 + n_3}{N} = \text{the probability of } A \text{ occurring,}$$

$$P(B) = \frac{n_2 + n_3}{N} = \text{the probability of } B \text{ occurring,}$$

$$P(A+B) = \frac{n_1 + n_2 + n_3}{N} = \text{the probability of either } A \text{ or } B \text{ (or both) occurring,}$$

$$P(AB) = \frac{n_3}{N} = \text{the probability of the joint occurrence of } A \text{ and } B,$$

$$P\left(\frac{A}{B}\right) = \frac{n_3}{n_2 + n_3} = \text{the conditional probability of the event } A, \text{ given that the event } B \text{ has occurred,}$$

$$P\left(\frac{B}{A}\right) = \frac{n_3}{n_1 + n_3} = \text{the conditional probability of the event } B, \text{ given that the event } A \text{ has occurred.}$$

From the above definitions, we see that

$$P(A+B) = \frac{n_1 + n_2 + n_3}{N} = \frac{n_1 + n_3}{N} + \frac{n_2 + n_3}{N} - \frac{n_3}{N}$$

$$\text{or,} \quad P(A+B) = P(A) + P(B) - P(AB) \tag{1}$$

which is the *additive law of probability.*

We also have, from the above definitions,

$$P(AB) = \frac{n_3}{N} = \frac{n_2+n_3}{N}\,\frac{n_3}{n_2+n_3} = \frac{n_1+n_3}{N}\,\frac{n_3}{n_1+n_3}$$

$$\text{or,}\quad P(AB) = P(B)P\left(\frac{A}{B}\right) = P(A)P\left(\frac{B}{A}\right) \qquad (2)$$

which is the *multiplicative law of probability.*

24.3.1 Mutually Exclusive Events

If A and B are *mutually exclusive* events so that the occurrence of one precludes the occurrence of the other, then the probability of their joint occurrence is zero and we have

$$P(AB) = 0. \qquad (3)$$

and for this case the additive law of probability (24.3.1) reduces to

$$P(A+B) = P(A) + P(B) = P(A \text{ or } B). \qquad (4)$$

24.3.2 Statistically Independent Events

If the events A and B are *statistically independent*, then the probability of the occurrence of A does not depend on the occurrence of B and vice versa and we may then write

$$P\left(\frac{B}{A}\right) = P(B) \quad \text{and} \quad P\left(\frac{A}{B}\right) = P(A) \qquad (5)$$

and in this case, the multiplicative law of probability (24.3.2) reduces to

$$P(AB) = P(A)P(B) \qquad (6)$$

Also, if the events A and B are statistically independent, then the additive law (24.3.1) becomes

$$P(A+B) = P(A) + P(B) - P(A)P(B). \qquad (7)$$

Illustrative Examples

1. Calculate the probability of drawing a white and a black ball in succession from an urn containing 30 black balls and 20 white balls.

Ans. The probability of drawing a white ball $P(A) = \frac{20}{50}$. After a white ball is drawn, the probability of drawing a black ball is $P\left(\frac{B}{A}\right) = \frac{30}{49}$, so that we may write

$$P(AB) = P(A)P(B/A) = \frac{20}{50}.\frac{30}{49} = \frac{12}{49}.$$

2. Find the probability of drawing 2 aces in succession from a pack of 52 cards.

Ans. The probability of drawing an ace in the first trial is $P(A) = \frac{4}{52}$. When one ace has been drawn in the first trial, the probability of drawing another ace in the second trial is $P(B/A) = \frac{3}{51}$.

So the desired probability $P(AB) = P(A)P(B/A) = \frac{4}{52} \cdot \frac{3}{51} = \frac{1}{221}$.

3. Compute the probability that when one card is drawn from each of two decks, at least one will be an ace.

Ans. In this case, the events A and B are the drawings of an ace out of a deck of 52 cards. Therefore, $P(A) = \frac{4}{52} = \frac{1}{13}$ and also $P(B) = \frac{1}{13}$.

Since the events A and B are statistically independent, we have from (24.3.7) above, $P(A+B) = P(A) + P(B) - P(A)P(B) = \frac{1}{13} + \frac{1}{13} - \frac{1}{13} \cdot \frac{1}{13} = \frac{25}{169}$.

4. In a given race, the odds in favour of horses x, y, z are 1 to 2, 1 to 3 and 1 to 4 respectively. Assuming that a dead-heat is impossible, find the chance that one of them will win the race.

Ans. The probability that the horse x wins $= \frac{1}{1+2} = \frac{1}{3} = p_1$ (say)

The probability that the horse y wins $= \frac{1}{1+3} = \frac{1}{4} = p_2$ (say)

The probability that the horse z wins $= \frac{1}{1+4} = \frac{1}{5} = p_3$ (say)

Since the three events are mutually exclusive, we have the probability that one of them wins the race, given by $p_1 + p_2 + p_3 = \frac{1}{3} + \frac{1}{4} + \frac{1}{5} = \frac{47}{60}$.

5. One bag contains 3 white balls and 2 black balls; another contains 5 white and 3 black balls. If a bag is chosen at random and a ball drawn from it, what is the chance, i.e., the probability that it is white?

Ans. The probability that the first bag is chosen is $\frac{1}{2}$, and the probability of drawing a white ball from it is $\frac{3}{5}$. Hence the probability of choosing the first bag and drawing a white ball from it is $\frac{1}{2} \cdot \frac{3}{5} = \frac{3}{10}$ (both being independent events).

Similarly, the probability of choosing the second bag and drawing a white ball from it is $\frac{1}{2} \cdot \frac{5}{8} = \frac{5}{16}$.

Hence the chance of drawing a white ball from a bag chosen at random is $\frac{3}{10} + \frac{5}{16} = \frac{49}{80}$.

6. The probability that x solves a problem is p_1 and the probability that y solves it is p_2. What is the probability that the problem will be solved?

Ans. Here x and y work independently. By the theorem of compound probability, we know that the probability that both of them solve the problem is $p_1 p_2$.

So the probability that at least one of them solves the problem is $p_1 + p_2 - p_1 p_2$ from (24.3.7).

24.4 Relations among Conditional Probabilities : Bayes's Theorem

Some interesting relations among conditional probabilities may be deduced as follows. From (24.3.2), we have

$$P\left(\frac{B}{A}\right) = \frac{P(AB)}{P(A)} \quad \text{and} \quad P(AB) = P(B)P\left(\frac{A}{B}\right).$$

Substituting the second result into the first one, we may write

$$P\left(\frac{B}{A}\right) = \frac{P(B)}{P(A)} P\left(\frac{A}{B}\right) \tag{1}$$

Replacing B by C in (24.4.1), we obtain

$$P\left(\frac{C}{A}\right) = \frac{P(C)}{P(A)} P\left(\frac{A}{C}\right) \tag{2}$$

Dividing (24.4.1) by (24.4.2), we have

$$\frac{P(B/A)}{P(C/A)} = \frac{P(B)P(A/B)}{P(C)P(A/C)} \tag{3}$$

This result is a form of Bayes's theorem in the theory of probability.

24.5 Repeated Trials

Sometimes the probability of an event cannot be determined by counting favourable cases and the total number of cases. In many such cases, the probability can be determined by repeating the trials.

We consider a series of n trials, say $T_1, T_2, \cdots, T_n$. Each trial is capable of producing one of two possible outcomes—either a success or a failure. Let us suppose that success is shown by the letter S and failure by F.

Let the event occur m times. The relative frequency is m/n.

If the event has constant probability p of success, the probability of m successes in n independent trials may be computed as follows:

We assume that (i) the probability of success in each particular trial of the series is the same. This probability is denoted by p and its complementary by $q = 1 - p$; (ii) the occurrence of success in any particular trial T_i is completely independent of that in all other trials of the series.

The number of failures is $(n - m)$. A sequence of m successes and $(n - m)$ failures is represented by a sequence of m letters S and $(n - m)$ letters F, i.e., $SSFFSS \cdots SF$.

Since the trials are independent, the probability of any one of such sequence is $ppqqpp \cdots pq = p^m q^{n-m}$, where $q = 1 - p =$ the probability of failure of any trial.

To obtain the number of favourable sequences, we observe that a sequence is determined as soon as the positions of the m letters S are fixed. The m places for these letters S can be chosen from the n places in ${}^nC_m$ ways. Hence the required probability is $p^m q^{n-m} + p^m q^{n-m} + \cdots$ up to ${}^nC_m$ terms $= {}^nC_m p^m q^{n-m}$.

If we replace m by x, then the probability of x successes in n independent trials is given by

$$B(x) = {}^nC_x p^x q^{n-x} = \frac{n!}{x!(n-x)!} p^x q^{n-x} \tag{1}$$

which is the $(x + 1)$th term in the binomial expansion of $(p + q)^n$. Here the function $B(x)$ is called the *binomial frequency function* and the statement that $B(x)$ gives the probability of x successes in n independent trials, is called the *Binomial law of probability.*

x is a single-valued function called the *binomial variable.* It represents the number of successes in n completely independent trials in which the probability of success in each particular trial is constant. If we substitute $(x-1)$ in place of x, we have

$$B(x-1) = {}^nC_{x-1}p^{x-1}q^{n-x+1} = \frac{n!}{(x-1)!(n-x+1)!}p^{x-1}q^{n-x+1}. \quad (2)$$

Dividing (24.5.1) by (24.5.2), we may write, $\dfrac{B(x)}{B(x-1)} = \dfrac{n+1-x}{x}\cdot\dfrac{p}{q} = Q_x$ (say).

Now $Q_x > 1$ for all values of x with the possible exception of $Q_n = 1$. So each probability $B(x)$ is greater than the preceding probability $B(x-1)$ for $x = 1, 2, \cdots, (n-1)$.

If $Q_n > 1$, then $B(n-1) < B(n)$ and there is just one most probable value of x, namely, $x_0 = n$. If $Q_n = 1$, then $B(n-1) = B(n)$ and there are two most probable values namely $x_0 = n$ and $x_0 = n-1$. Let us now take the following examples to illustrate the method.

Illustrative Examples

1. Calculate the probability that an event happens at least m times in n trials.

Ans. If the event happens at least m times, then it must happen exactly n times or exactly $(n-1)$ times or exactly $(n-2)$ times $\cdots$ or exactly m times. So the required probability P is given by

$$\begin{aligned} P &= {}^nC_n p^n q^{n-n} + {}^nC_1 p^{n-1}q + {}^nC_2 p^{n-2}q^2 + \cdots + {}^nC_{n-m}p^m q^{n-m} \\ &= p^n + {}^nC_1 p^{n-1}q + {}^nC_2 p^{n-2}q^2 + \cdots + {}^nC_{n-m}p^m q^{n-m}. \end{aligned}$$

2. If a die is thrown 5 times, what is the chance that an ace turns up (i) exactly 3 times, (ii) at least 3 times?

Ans. (i) The probability of getting an ace in tossing a die is $\frac{1}{6} = p$ (say). We want to calculate the probability of 3 successes in 5 independent trials and this is given by

$${}^5C_3 p^3(1-p)^2 = {}^5C_3\left(\frac{1}{6}\right)^3\left(\frac{5}{6}\right)^2 = \frac{250}{6^5}.$$

(ii) Using the result of Ex.1 above, the required probability is given by

$$\begin{aligned} & p^5 + {}^5C_1 p^4(1-p) + {}^5C_2 p^3(1-p)^2 \\ &= \left(\frac{1}{6}\right)^5 + {}^5C_1\left(\frac{1}{6}\right)^4\left(\frac{5}{6}\right) + {}^5C_2\left(\frac{1}{6}\right)^3\left(\frac{5}{6}\right)^2 = \frac{23}{648}. \end{aligned}$$

24.6 Sample Space

We know that the equally likely cases associated with the definition of probability represent the possible outcomes of an experiment. For example, if a pair of dice is tossed, then 36 cases correspond to 36 possible outcomes. A set of all possible outcomes is called a *sample space* and the points of a sample space are called *events.* Any sample space may have only those events which are mutually exclusive.

The outcomes of an experiment are represented by means of a *random variable* (discussed in the next article). Thus the sample space of a variable is the set of the points corresponding to the random variable. The sample space may be finite or infinite. Finite sample space is one which has only finite number of points. In such a space if we suppose that the point events have respective probabilities $p_1, p_2, \cdots, p_n$, then we may write

$$p_1 + p_2 + p_3 + \cdots + p_n = P(E) = 1.$$

If only first m sample points are favourable to an event A, then the probability of A is $P(A) = p_1 + p_2 + \cdots + p_m$.

The sample space is said to be *uniform* if each point of the sample space has the same probability. If the total number of points of a uniform sample space is n, then the probability of each point of this sample space is $1/n$. Then the total probability of the sample space is

$$P = \frac{1}{n} + \frac{1}{n} + \cdots + \frac{1}{n} \quad \text{(up to } n \text{ terms)} = 1. \tag{1}$$

If only first m sample points are favourable to an event A, then $P(A) = \frac{1}{n} + \frac{1}{n} + \cdots$ to m terms $= \frac{m}{n}$. If the points of a sample space correspond to different probabilities, then it is said to be *non-uniform*.

As an example of a non-uniform sample space, let us consider the number of heads in the tossing of four coins. The sample space is composed of the following events with their respective probabilities. Total number of cases = 16.

Events	Probabilities
No head	${}^4C_0/16 = \frac{1}{16} = p_1$
One head	${}^4C_1/16 = \frac{4}{16} = p_2$
Two heads	${}^4C_2/16 = \frac{6}{16} = p_3$
Three heads	${}^4C_3/16 = \frac{4}{16} = p_4$
Four heads	${}^4C_4/16 = \frac{1}{16} = p_5$

$\therefore$ the probability of the whole sample space $= p_1 + p_2 + p_3 + p_4 + p_5 = 1$.

Other Illustrative Examples

1. A coin is tossed 4 times. Construct a uniform sample space for this experiment and calculate the probability of at least 1 head in the tosses.

Ans. If we denote head by H and tail by T, then we have the following possibilities—(i) all heads, (ii) all tails, (iii) 3 heads and 1 tail, (iv) 3 tails and 1 head, (v) 2 heads and 2 tails. The sample space may therefore be constructed as follows:

Events	H	T	H	H	T	H	T	H	H	T	H	T	T	T	H	T
	H	H	T	H	T	T	H	H	T	H	H	T	T	H	T	T
	H	H	H	T	H	T	T	H	H	H	T	T	H	T	T	T
	H	H	H	H	H	H	H	T	T	T	T	H	T	T	T	T
Probability	$\frac{1}{16}$	$\frac{1}{16}$	$\frac{1}{16}$	$\frac{1}{16}$	$\frac{1}{16}$	$\frac{1}{16}$	$\frac{1}{16}$	$\frac{1}{16}$	$\frac{1}{16}$	$\frac{1}{16}$	$\frac{1}{16}$	$\frac{1}{16}$	$\frac{1}{16}$	$\frac{1}{16}$	$\frac{1}{16}$	$\frac{1}{16}$

Since at least one head occurs in all the first 15 cases, the required probability $= \frac{15}{16}$.

2. Four coins are tossed. A reliable witness tells us that there are at least as many heads as tails. What is the most probable number of events and what is its probability?

Ans. There are 16 cases in all as discussed in the preceding Ex.1. But according to the reliable witness, events $HTTT, THTT, TTHT, TTTH$ and $TTTT$ do not happen, and hence there are the remaining 11 trials which are given below:

	H	T	H	H	T	H	T	H	H	T	H
Events	H	H	T	H	T	T	H	H	T	H	H
	H	H	H	T	H	T	T	H	H	H	T
	H	H	H	H	H	H	H	T	T	T	T
Probability	$\frac{1}{11}$	$\frac{1}{11}$	$\frac{1}{11}$	$\frac{1}{11}$	$\frac{1}{11}$	$\frac{1}{11}$	$\frac{1}{11}$	$\frac{1}{11}$	$\frac{1}{11}$	$\frac{1}{11}$	$\frac{1}{11}$

From the above table it is evident that the event, where two heads and two tails occur simultaneously, happens 6 times—which is the maximum number of times. Hence this is the most probable event, and its probability is $\frac{6}{11}$.

24.7 Random Variables

A process is said to be *random*, if it is impossible to predict the final state from the initial state. If a certain group of events is described in a statistical way, it is treated as a random phenomenon. Random phenomena are characterised by the fact that they are governed (or conditioned) by so many interacting causes that it is practically impossible to analyse them accurately enough to make exact predictions.

Associated with a random process, there may be certain numerical-valued variables which themselves have random character. These variables are called *random variables*. For example, if X denotes a number obtained in tossing a die, then X is a variable, which assumes the values $1, 2, 3, 4, 5, 6$ corresponding to six events—1 show, 2 shows and so on. The random variable may also be defined as a quantity which can be determined quantitatively, and which, in different observations of the same category, can assume different values. Thus a random variable is a numerical-valued function defined in a sample space. Symbolically we may write

$$X(e_i) = x_i, i = 1, 2, 3, \cdots, n,$$

where e_i are the events of the sample space and x_i are the values of the random variable X. *Random variables* have the following *properties*:

(1) If X be a random variable, then whatever be the number t, there exists the probability, say

$$p_x(t) = P(X = t). \tag{1}$$

that X will assume the value t.

This probability is defined for all real numbers t. In other words, it is a single-valued function defined over the sample space, the points of which are shown by all real numbers t. The function $p_x(t) = P(X = t)$ defined over such a sample space, having points shown with real numbers t and representing the probability that the random variable X will assume the value t, is called the *frequency function* of the random variable X.

(2) If X be a random variable, then whatever be the real number t, there exists the probability, say

$$\phi_x(t) = P(-\infty < X \leq t) = P(-\infty, t) \tag{2}$$

that the value assumed by X will not exceed t.

The function $\phi_x(t) = P(-\infty < X \leq t)$ defined over the sample space, the points of which are shown by the real numbers t, and representing the probability that the random variable X will not exceed t, is called the *distribution function* of the random variable X. The distribution function $\phi_x(t)$ of the random variables encountered in practice will belong to one of the two following types—

(i) $\phi_x(t)$ is piece-wise constant;

(ii) $\phi_x(t)$ is continuous and piece-wise differentiable with a continuous derivative.

In the above cases, we speak of discontinuous and continuous distribution functions respectively.

(3) If a random variable X assumes the values $u_1 < u_2 < \cdots < u_n$ and no others, then the frequency function $p_x(t)$ of the variable X is equal to zero for all values of t different from $u_1, u_2, \cdots, u_n$ and

$$p_x(u_k) = P(X = u_k) \quad \text{for} \quad k = 1, 2, 3, \cdots, n \tag{3}$$

(4) If the random variable X assumes the values $u_1 < u_2 < \cdots < u_n$ and no others, then the sum of the values of the frequency function $p_x(t)$ at the points $t = u_1, u_2, \cdots, u_n$ is equal to unity, i.e.,

$$p_x(u_1) + p_x(u_2) + \cdots + p_x(u_n) = 1. \tag{4}$$

Proof. Since $X = u_1$, $X = u_2$, $\cdots$, $X = u_n$ represent all the points of the sample space, we may write $P[(X = u_1) + (X = u_2) + \cdots + (X = u_n)]$ as the probability of the whole sample space, so that we have

$$P\left[\sum_{i=1}^{n}(X = u_i)\right] = 1.$$

But according to the law of total probability, we have

$$P\left[\sum_{i=1}^{n}(X = u_i)\right] = \sum_{i=1}^{n} P(X = u_i) = \sum_{i=1}^{n} p_x(u_i) = 1 \quad \text{(Proved)}$$

If $u_1, u_2, u_3, \cdots, u_k, \cdots, u_n$ are all the different values of a random variable X and if u_k has the probability that, whatever may be any other possible value u_i of X, the probability $P(X = u_k)$ is at least as large as the probability $P(X = u_i)$

$$\text{and } p(X = u_k) \geq P(X = u_i) \text{ for } i = 1, 2, \cdots, n, \tag{5}$$

then u_k is called the *most probable value* of **X**.

24.8 The Statistical Expectation Value

Let (e_i) be a sample space of n events e_i with associated probabilities p_i. Let X be a random variable defined on the sample space (e_i) and let it assume the value x_i at the ith sample point, so that $X(e_i) = x_i$. Then the *statistical expectation* or the expected value $E(X)$ is defined as

$$E(X) = p_1x_1 + p_2x_2 + p_3x_3 + \cdots + p_nx_n = \sum_{i=1}^{n} p_ix_i. \tag{1}$$

If the sample space be uniform, i.e., every event of the sample space has a constant probability p, then $p_1 = p_2 = \cdots = p_n = p = \frac{1}{n}$, so that we may write

$$E(X) = \frac{1}{n}(x_1 + x_2 + \cdots + x_n) = \frac{1}{n}\sum_{i=1}^{n} x_i. \tag{2}$$

Thus for a uniform sample space, the expectation or the expected value of a random variable is the arithmetic mean of all the possible values of the random variable. Even in a non-uniform sample space, if a sufficiently large number of observations of the variable X be made, the mean of these observations will almost certainly be very close to $E(X)$. In this way $E(X)$ represents the average value attained by X in the long run.

24.8.1 Expectation values of the sum and product of two random variables

Let x_i be r distinct values of a random variable X and let y_j be s distinct values of another random variable Y. The sum $X + Y$ is a random variable which has the values $x_i + y_j$, where $X = x_i$ and $Y = y_j$. Thus $X + Y$ is defined on the sample space whose points consist of rs events:

$$\left.\begin{matrix} X = x_i \\ Y = y_j \end{matrix}\right\} \quad \begin{matrix} \text{with } i = 1, 2, \cdots, r \\ \text{and } j = 1, 2, \cdots, s \end{matrix}$$

The product XY of two random variables is similarly defined and given by $XY = x_iy_j$.

Theorem I. The expectation value of the sum of two random variables is equal to the sum of their expectation values or

$$E(X + Y) = E(X) + E(Y) \tag{3}$$

Proof. We suppose that p_{ij} is the probability of the compound event $\begin{bmatrix} X = x_i \\ Y = y_j \end{bmatrix}$. Then from the definition of expectation value, we have

$$E(X + Y) = \sum p_{ij}(x_i + y_j) = \sum_i x_i \left(\sum_j p_{ij}\right) + \sum_j y_j \left(\sum_i p_{ij}\right).$$

Here $\sum_j p_{ij}$ represents the probability for

$$\begin{pmatrix} X = x_i \\ Y = y_1 \end{pmatrix}, \quad \begin{pmatrix} X = x_i \\ Y = y_2 \end{pmatrix}, \quad \begin{pmatrix} X = x_i \\ Y = y_3 \end{pmatrix}, \quad \cdots, \quad \begin{pmatrix} X = x_i \\ Y = y_j \end{pmatrix},$$

$X = x_i$ being common to all these results.

Hence $\sum_j p_{ij}$ represents the probability for $X = x_i$. Let this probability for $X = x_i$ be p_i, then $\sum_j p_{ij} = p_i$; similarly, $\sum_i p_{ij} = p_j$, where p_j is the probability that $y = y_j$.

So $E(X+Y) = \sum_i x_i p_i + \sum_j y_j p_j = E(X) + E(Y)$. (Proved)

Theorem II. The expectation value of the product of two independent random variables is the product of their individual expectation values.

Proof. The product XY of two random variables is a random variable which is equal to $x_i y_j$, where $X = x_i, Y = y_j$.

If p_i be the probability that $X = x_i$ and q_j be the probability that $Y = y_j$, then assuming independence, $p_i q_j$ stands for the probability that simultaneously $X = x_i$, $Y = y_j$.

Hence

$$E(XY) = \sum_i \sum_j p_i q_j x_i y_j = \left(\sum_i p_i x_i\right)\left(\sum_j q_j y_j\right) = E(X)E(Y) \tag{4}$$

For a sample space $\sum_{i=1}^{n} p_i = p_1 + p_2 + p_3 + \cdots + p_n = 1$.

So

$$\begin{aligned} E(x) &= p_1 x_1 + p_2 x_2 + p_3 x_3 + \cdots + p_n x_n \\ &= \frac{p_1 x_1 + p_2 x_2 + \cdots + p_n x_n}{p_1 + p_2 + \cdots + p_n} = \frac{\sum_{i=1}^{n} p_i x_i}{\sum_{i=1}^{n} p_i} \end{aligned} \tag{5}$$

Thus the expectation value may also be interpreted as the centre of mass. We now consider the following examples to illustrate the above theorems.

Illustrative Examples

1. Find the expected number of heads when n coins are tossed.

Ans. We assume that the random variable X shows the number of heads. Then $X_i = 1$, if ith coin shows head and $X_i = 0$ if ith coin shows tail. So for any coin (say ith coin) $E(X_i) = \frac{1}{2}\cdot 1 + \frac{1}{2}\cdot 0 = \frac{1}{2}$ (since probability of getting head is $\frac{1}{2}$ and that of tail is also $\frac{1}{2}$).

The total number of heads is $m = X_1 + X_2 + \cdots + X_n$. So the expectation value of getting head is given by

$$\begin{aligned} E(m) &= E(X_1 + X_2 + \cdots + X_n) \\ &= E(X_1) + E(X_2) + \cdots + E(X_n) = \frac{1}{2} + \frac{1}{2} + \cdots \text{ up to } n \text{ terms} \\ &= n/2. \end{aligned}$$

2. From an urn containing 'a' white and 'b' black balls, one ball is drawn at random and set aside. What is the expected number of white balls left in the urn?

Ans. Let x be the number of white balls left. If a white ball is drawn, $x_1 = a - 1$; when a black ball is drawn, then $x_2 = a$.

Hence $E(x) = E(X_1 + X_2) = E(X_1) + E(X_2)$.

Now $E(X_1) = x_1 p_1$ and $E(X_2) = x_2 p_2$, where p_1 is the probability of drawing a white ball from the urn and p_2 is that of drawing a black ball from the urn.

So $p_1 = \frac{a}{a+b}$ and $p_2 = \frac{b}{a+b}$.

Hence $E(x) = E(X_1) + E(X_2) = x_1 p_1 + x_2 p_2 = (a-1)\frac{a}{a+b} + a\frac{b}{a+b} = a - \frac{a}{a+b}$.

24.9 Statistical Distributions

When a series of measurements are made on a set of objects, the set of objects being measured is called a *population*. A population may be, for example, the animals living in a farm, the groups of people living in a certain city, the automobiles produced by a factory in a year. the set of results of a physical or chemical experiment, etc.

If a *single quantity* is measured for each member of a population, this quantity is called a *variate*. Variates are usually denoted by symbols such as x or y. The possible observed values of a variate x will be a finite sequence or a set of numbers $x_1, x_2, x_3, \cdots, x_n$. The numbers x_k, for example, may be the heights of a group of n individuals or the prices of a certain brand of automobile in a dealer's showroom.

24.9.1 Frequency

In a given population, the number of times a particular value x_r of a variate is observed is called the *frequency* f_r of x_r. For example, an automobile dealer may have nine cars in his showroom; three cars are priced at \$3,500 each, four at \$4,000 each, and two at \$2,000 each. In this case we have x_1 = \$3,500, $f_1 = 3$, x_2 = \$4,000, $f_2 = 4$, and x_3 = \$2,000, $f_3 = 2$. The total number of automobiles in the population is N, where $N = f_1 + f_2 + f_3 = 3 + 4 + 2 = 9$.

The total value S of the automobiles is

$$S = f_1x_1 + f_2x_2 + f_3x_3 = 3(3,500) + 4(4,000) + 2(2,000) = 30,500$$

The average or mean price $\bar{x}$ is

$$\bar{x} = \frac{1}{N}(f_1x_1 + f_2x_2 + f_3x_3) = \$3,388 = \frac{1}{N}\sum_{r=1}^{3} f_r x_r, \quad N = f_1 + f_2 + f_3. \tag{1}$$

24.9.2 The Mean of a Distribution

The most elementary statistical quantity defined by a distribution is the arithmetic mean $\bar{x}$ of the variate x, usually known as the *mean*. This is the sum of the values of x for each member of the population divided by N. Therefore,

$$\bar{x} = \frac{1}{N}\sum_{r=1}^{k} f_r x_r, \quad k = \text{number of frequencies} \tag{2}$$

Let us suppose that a is a typical value of the variate x, and let us define

$$d_r = x_r - a, \ r = 1, 2, 3, \cdots, N.$$

It is thus evident that d_r is the deviation of x_r from the typical value a. The deviation d has a distribution f_r over the values d_r. The mean $\bar{d}$ of the different values of d for this distribution is

$$\bar{d} = \frac{1}{N}\sum_{r=1}^{k} f_r d_r \tag{3}$$

$\bar{d}$ is simply related to $\bar{x}$; this can be seen by writing (24.9.3) in the form:

$$\bar{d} = \frac{1}{N}\sum_{r=1}^{k} f_r(x_r - a) = \bar{x} - a. \tag{4}$$

Therefore, we have

$$\bar{x} = \bar{d} + a. \tag{5}$$

The following relation is sometimes useful:

$$d_r - \bar{d} = (x_r - a) + (a - \bar{x}) = x_r - \bar{x}, \quad r = 1, 2, 3, \cdots, N. \tag{6}$$

Equation (24.9.5) is useful in computing $\bar{x}$ quickly, provided that we choose a to be a number very close to the mean.

24.10 Second Moments and Standard Deviation

In statistical investigations, it is of interest to know whether or not a distribution is closely grouped around its mean. One of the most useful quantities for estimating the spread of a distribution is the *standard deviation* σ. Before we define σ it is instructive to note that $\bar{d}$ as given by (24.9.4) is the *first moment* of a distribution about a given value a of the variate. We note from (24.9.5) that if $a = \bar{x}$, then the first moment $\bar{d} = 0$.

24.10.1 The Second Moment

The *second moment* μ_2' of a distribution about a given value a of the variate is the average over the population of the *squares* of the differences $d_r = x_r - a$ between a and the values of the variate x_r. It is given by

$$\mu_2' = \frac{1}{N}\sum_{r=1}^{k} f_r d_r^2 = \frac{1}{N}\sum_{r=1}^{k} f_r(x_r - a)^2 \tag{1}$$

It is apparent that if most of the measured values of x for the population are approximately equal to a, μ_2' will be small, while if a differs considerably from the observed values of x, then μ_2' will be large since all the terms in (24.10.1) are positive. If we now take $a = \bar{x}$ in (24.10.1), that is, if we take the second moment about the *mean* $\bar{x}$ of the population, then this second moment μ_2 is called the *variance*. The variance is therefore given by

$$\mu_2 = \frac{1}{N}\sum_{r=1}^{k} f_r(x_r - \bar{x})^2. \tag{2}$$

The variance gives an estimate of the spread of values of x about their mean. The variance is not directly comparable to $\bar{x}$ since it is of the dimension of x^2.

24.10.2 The Standard Deviation

The positive square root of μ_2 is the ***standard deviation*** σ; since it is of the same dimension as x, σ gives us an estimate of the *spread* of the values of x which can be compared directly with $\bar{x}$. By definition, we have, therefore

$$\sigma^2 = \mu_2 = \frac{1}{N}\sum_{r=1}^{k} f_r(x_r - \bar{x})^2 = \frac{1}{N}\sum_{r=1}^{k} f_r(d_r - \bar{d})^2, \tag{3}$$

where we have made use of (24.9.6).

If (24.10.3) is expanded, we obtain the following result:

$$\begin{aligned}\sigma^2 &= \frac{1}{N}\left(\sum_{r=1}^{k} f_r d_r^2 - 2\bar{d}\sum_{r=1}^{k} f_r d_r + \bar{d}^2\sum_{r=1}^{k} f_r\right)\\ &= \frac{1}{N}(N\mu_2' - 2N\bar{d}^2 + N\bar{d}^2)\\ &= \frac{1}{N}(N\mu_2' - N\bar{d}^2) = \mu_2' - \bar{d}^2. \end{aligned} \tag{4}$$

It is thus evident that the variance μ_2 and the second moment μ_2' about any value a are related by

$$\sigma^2 = \mu_2 = \mu_2' - \bar{d}^2 \quad \text{[from (24.10.4)]}. \tag{5}$$

In order to calculate the standard deviation σ, it is usually easier to calculate the second moment μ_2' about a simple value a and then use (24.10.3) instead of using (24.10.3) directly.

Since σ^2 is the mean value of the squares of the deviations of the variate from its average value $\bar{x}$, σ (the *root mean square*) gives us an overall measure of the deviation of the set from the average $\bar{x}$ without reference to sign. There are two other features of the population which are sometimes found useful. We suppose that a frequency diagram has been constructed in which the ordinates represent the number of readings lying in successive intervals. The interval in which the ordinate attains its maximum, clearly corresponds to the most frequent or *most fashionable* value of x among the set. This value is called the *mode*; in general, it is not identical with the average or mean, but it will be if the frequency curve is symmetric about the mean value. A frequency curve may have more than one mode, but we are here concerned only with cases in which a single mode exists.

We may also arrange our data in ascending order of magnitude and divide them into two sections halfway, so that as many measurements lie above this division as below it. This position is called the *median*, and is such that it is useful in analysing the data in many experimental observations.

24.11 Continuous Distributions

In many problems encountered in the mathematical theory of probability, the possible values of a variate x are not discrete but lie in a continuous range. For example, the heights of adults normally take values in the continuous range from 4.5 to 6.5 ft. Even though observed heights are classified as taking one of a finite discrete set of values, any theoretical probability law should refer to the continuous range. If, for example, a law of probability predicts a distribution of heights at intervals of $\frac{1}{2}$ in., it will not be appropriate and adequate for comparison with a set of measurements made to the nearest centimeter.

A law of probability for a *continuous* variate is therefore expressed in terms of a *probability function* $\phi(x)$, such that $\phi(x)dx$ is the probability of a trial giving a result in the infinitesimal range $[x-(dx/2), x+(dx/2)]$ for all values of x. The probability $P(x_1, x_2)$ of the result lying between values x_1 and x_2 is then

$$P(x_1, x_2) = \int_{x_1}^{x_2} \phi(x)dx \tag{1}$$

Since the total probability is unity, we have

$$\int \phi(x)dx = 1, \tag{2}$$

where the integral is taken over the complete range of the variate x.

24.12 Expectation, Moments and Standard Deviation

The results of articles (24.9) and (24.10) apply equally well to probabilities. The mean of a probability distribution for a variate x is called the *expectation* $E(x)$ and is defined by

$$E(x) = \frac{p_1x_1 + p_2x_2 + \cdots + p_nx_n}{p_1 + p_2 + \cdots + p_n} \tag{1}$$

Since the sum of the probabilities is unity, we have

$$p_1 + p_2 + p_3 + \cdots + p_n = 1. \tag{2}$$

Therefore, the expectation $E(x)$ as given by (24.12.1) takes the form

$$E(x) = p_1x_1 + p_2x_2 + \cdots + p_nx_n = [x] \tag{3}$$

The expectation of a probability distribution of a *continuous variate* is defined by the equation

$$E(x) = \int x\phi(x)dx = [x], \tag{4}$$

where the integration takes place over the whole range of x.

The *second moment* μ_2' about a of a *discrete* probability distribution is defined by the equation

$$\mu_2' = \sum_{r=1}^{n} p_r(x_r - a)^2 \tag{5}$$

The *second moment* about a of a *continuous* probability distribution is defined by the equation

$$\mu_2' = \int \phi(x)(x-a)^2 dx = [(x-a)^2] \tag{6}$$

The *variance* μ_2 is given by (24.12.5) or (24.12.6) with $a = [x]$. The *standard deviation* σ is given by

$$\sigma^2 = \mu_2 = [(x - [x])^2] \tag{7}$$

The standard deviation σ and the second moment $\mu_2'(a)$ are related in the same manner as in (24.10.5) by the equation

$$\sigma^2 = \mu_2' - [d]^2, \tag{8}$$

where

$$d = x - a$$

so that

$$[d] = [x] - a \tag{9}$$

It is interesting to note that when $a = [x]$, then (24.12.9) becomes

$$[d] = [x] - [x] = 0. \tag{10}$$

As an example of the above definitions, we consider the following illustration. Let us suppose that cards numbered $1, 2, 3, \cdots, n$ are put in a box and thoroughly shuffled; then one card is drawn at random. If we assume that each card is equally likely to be drawn, so that the law of probability of drawing the cards is

$$p_r = \frac{1}{n}, \quad r = 1, 2, 3, \cdots, n, \tag{11}$$

then the expectation $E(x)$ of the number x on the card is

$$E(x) = [x] = \frac{1 + 2 + 3 + \cdots + n}{n} = \frac{n+1}{2}. \tag{12}$$

The second moment about $x = 0$ of the probability distribution of x is, by (24.12.5),

$$\mu_2' = [x^2] = n^{-1}(1^2 + 2^2 + \cdots + n^2) = \frac{1}{6}(n+1)(2n+1) \tag{13}$$

The variance is given by (24.12.7) in the form

$$\sigma^2 = \mu_2 = [x^2] - [x]^2 = \frac{(n+1)(2n+1)}{6} - \frac{(n+1)^2}{4} = \frac{1}{12}(n^2 - 1) \tag{14}$$

{from (24.12.12) and (24.12.13)}

It can be seen that for large values of n we have

$$\left.\begin{aligned} [x] &= \frac{n}{2} \\ \sigma &= \frac{n}{2(3)^{1/2}} \end{aligned}\right\} \quad (n \text{ large}) \tag{15}$$

It is to be expected that the values of $[x]$ and σ will become proportional to n when n is large.

As an example of a continuous distribution, we consider a variate x which is equally likely to take any value in the range $(0, X)$. In such a case the probability function $\phi(x)$ must equal a constant ϕ_0 in this range and zero elsewhere. In order to satisfy (24.11.2), we must have

$$\int_0^X \phi_0 dx = 1 \quad \text{or,} \quad \phi(x) = \phi_0 = \frac{1}{X} \tag{16}$$

The expectation $E(x)$ of x is, by (24.12.4),

$$E(x) = [x] = X^{-1}\int_0^X x\,dx = \frac{X}{2} \tag{17}$$

The variance is given by (24.12.6) with $a = [x] = X/2$, so that

$$\mu_2 = \int_0^X X^{-1}\left(x - \frac{X}{2}\right)^2 dx = \frac{X^2}{12} = \sigma^2 \tag{18}$$

Therefore the standard deviation is $\sigma = X/2(3)^{1/2}$.

24.12.1 An Important Statistical Inequality—Chebyshev's Inequality

Let us consider a variate x with probability function $f(x)$. Let μ be the mean value and σ the standard deviation of x. We shall prove that if we select any number a, the probability that x differs from its mean value μ by more than a is less than σ^2/a^2. This means that x is unlikely to differ from μ by more than a few standard deviations, and that, therefore, the standard deviation σ gives an estimate of the *spread* of the distribution. For example, if $a = 2\sigma$, we find that the probability for x to differ from μ by more than 2σ is less than $\sigma^2/(2\sigma)^2 = \frac{1}{4}$.

Proof. By the definition of the standard deviation σ, we have

$$\sigma^2 = \sum_{x=0}^{n}(x-\mu)^2 f(x). \tag{19}$$

If we now sum over values of x for which x exceeds $(\mu + a) = x_0$, we get *less* than σ^2, so that we may write

$$\sigma^2 > \sum_{x=x_0}^{n}(x-\mu)^2 f(x) \tag{20}$$

If we replace $(x - \mu)$ by a in (24.12.20), the sum is further decreased and we have

$$\sigma^2 > \sum_{x_0}^{n} a^2 f(x) = a^2 \sum_{x_0}^{n} f(x) \tag{21}$$

We may write (24.12.21) in the form

$$\sum_{x=x_0}^{n} f(x) < \frac{\sigma^2}{a^2} \quad \text{(Chebyshev's inequality)} \tag{22}$$

However, $f(x)$ is the sum of the probabilities of values of x which differ from μ by more than a, so that $x > (\mu + a)$ and (24.12.22) says that this probability is less than σ^2/a^2. This result is known as *Chebyshev's inequality.*

24.13 The Binomial Distribution

Let us suppose that a trial can have only two possible results, one called a *success* with probability p, and the other called a *failure* with a probability $q = (1-p)$. If we make n independent trials of this type in succession, then the probability of a *particular sequence* of r successes and $(n-r)$ failures is $p^r q^{n-r}$. However, the number of sequences with r successes and $(n-r)$ failures is

$$C(n,r) = {}^nC_r = \frac{n!}{[r!(n-r)!]} = \text{the binomial coefficient}$$

Therefore, the probability of there being exactly r *successes* in n trials is

$$P(r) = {}^nC_r p^r q^{n-r} = {}^nC_r p^r (1-p)^{n-r} \tag{1}$$

It will be noticed that $P(r)$ is the term containing p^r in the binomial expansion of $(p+q)^n$. For this reason the probability distribution (24.13.1) is known as the *binomial distribution.*

Characteristics of Binomial Distribution

1. Its general form depends on the parameters p, q and n.
2. The probability that there are r successes in n independent trials is given by ${}^nC_r p^r q^{n-r}$ in one set and hence in all the N sets, this is given by $N \cdot {}^nC_r p^r q^{n-r}$.
3. It is chiefly applied when the population being sampled is infinite, so that p remains unchanged by sampling.
4. It can be applied to finite populations also if they are not too small.
5. It is used under the conditions
   (a) The variable is discrete.
   (b) A dichotomy (i.e. the process of classification of collected individuals into two classes according to whether they do or do not possess a particular attribute) exists.
   (c) Statistical independence is assumed.
   (d) The exponent (power) n is finite and small.
   (e) For symmetrical distribution $p = q$ and for asymmetrical one $p \neq q$.
   (f) For calculation of *constants of the Binomial distribution,* we are to take an arbitrary origin at 0 (zero) successes so that the successive derivations are $0, 1, 2, 3, \ldots, n$.

Illustrative Examples

1. If the probability that a man aged 60 will live up to 70 is $p = 0.65$, what is the probability that 8 out of 10 men now aged 60 will live up to 70?

This is exactly the problem proposed above, with $p = 0.65, r = 8$ and $n = 10$. The answer is, therefore, [using (24.13.1)] given by

$$P(8) = {}^{10}C_8(0.65)^8(0.35)^2 = \left[\frac{10!}{8!2!}\right](0.65)^8(0.35)^2 = 45(0.65)^8(0.35)^2 = 0.176.$$

2. What is the probability of throwing an ace exactly three times in four trials with a single die?

In this case we have $n = 4$, $r = 3$, and since there is one chance in six of throwing an ace on a single trial, we have $p = \frac{1}{6}$. Therefore, we have from (24.13.1) $P(3) = {}^4C_3\left(\frac{1}{6}\right)^3 \frac{5}{6} = \frac{5}{324}$.

3. What is the probability of throwing a deuce exactly three times in three trials?

In this case $n = 3, r = 3, p = \frac{1}{6}$; therefore, we have the result from (24.13.1) $P(3) = {}^3C_3\left(\frac{1}{6}\right)^3\left(\frac{5}{6}\right)^0 = \left(\frac{1}{6}\right)^3 = \frac{1}{216}$.

24.13.1 The Mean of the Binomial Distribution

The expectation or the mean number of successes $E(x) = \bar{x}$ is given by equation (24.12.1) in the form

$$\bar{x} = \frac{\sum_{r=0}^{n} rp_r}{\sum_{r=0}^{n} p_r}, \quad \text{where } p_r = {}^nC_r p^r q^{n-r}. \tag{2}$$

Here we have

$$\begin{aligned}\sum_{r=0}^{n} p_r &= \sum_{r=0}^{n} {}^nC_r p^r q^{n-r} = (p+q)^n \\ &= [p + (1-p)]^n = 1^n = 1 \quad \text{since } q = 1-p.\end{aligned} \tag{3}$$

If we denote the partial derivative with respect to p by D_p, we note that

$$rp^r = pD_p p^r. \tag{4}$$

Since the denominator of (24.13.2) is unity, we have

$$\bar{x} = \sum_{r=0}^{n} {}^nC_r r p^r q^{n-r}. \tag{5}$$

The series in (24.13.5) is summed by treating p and q as *independent* variables. If we now use the result (24.13.4), we may write (24.13.5) in the form

$$\bar{x} = pD_p \sum_{r=0}^{n} {}^nC_r p^r q^{n-r} = pD_p(p+q)^n = np(p+q)^{n-1}. \tag{6}$$

We now substitute $q = 1-p$ in (24.13.6) and obtain the result

$$\bar{x} = np[p + (1-p)]^{n-1} = np. \tag{7}$$

Therefore the mean number of successes for the binomial distribution is $\bar{x} = np$.

24.13.2 The Standard Deviation of the Binomial Distribution

We may compute the variance of the binomial distribution by the use of Eq.(24.13.8). The second moment μ_2' about $x = 0$ of the distribution of the number of successes can be found by a method similar to that used in computing the first moment above and may be shown to be

$$\mu_2' = np[1 + p(n-1)] \tag{8}$$

As a consequence of (24.12.8) we have

$$\sigma^2 = \mu_2' - \bar{x}^2 = np[1 + p(n-1)] - (np)^2 = npq \quad [q = 1 - p] \tag{9}$$

Therefore, the *standard deviation* of the *total* number of successes in n trials is

$$\sigma = (npq)^{1/2}. \tag{10}$$

Illustrative Examples

1. Show the results of throwing 12 dice, 4096 times, a throw of 4, 5 or 6 being called a success. Find the expected frequencies and compare the actual mean with those of the expected distribution. Calculate the standard deviation.

Ans. Probability of success of fall 4, 5, or 6 $= \frac{3}{6} = \frac{1}{2} = p$ (say)

$$\therefore \quad q = 1 - p = 1/2$$

Binomial distribution is $4096 \cdot \left(\frac{1}{2} + \frac{1}{2}\right)^{12}$

Frequency of 0 success $= 4096 \times \left(\frac{1}{2}\right)^{12} = 1$

" " 1 " $= 4096\,{}^{12}C_1 q^{11} p = 12 \times \frac{4096}{2^{12}} = 12$

" " 2 " $= 4096\,{}^{12}C_2 q^{10} p^2 = 66$

" " 3 " $= 4096\,{}^{12}C_3 q^9 p^3 = 220$

" " 4 " $= 4096\,{}^{12}C_4 q^8 p^4 = 495$ etc.

Now expected mean $= np = 12 \times \frac{1}{2} = 6$

$$\sigma^2 = npq = 12 \times \frac{1}{2} \times \frac{1}{2} = 3$$

$\therefore$ Standard deviation, $\sigma = \sqrt{3} = 1.732$.

24.14 The Poisson Distribution

The Poisson distribution is derived as the limit of the binomial distribution when the number of trials n is very large and the probability of success p is very small. For the binomial distribution we have seen that the mean number of successes is given by

$$\bar{x} = np = m, \quad \text{(say)}.$$

We now investigate the limiting form taken by the binomial distribution as the number of trials n tends to infinity and the probability p tends to zero in such a manner that

$$\lim_{\substack{n\to\infty \\ p\to 0}} (np) = m = \text{a finite constant} \quad \text{or,} \quad p = \frac{m}{n}. \tag{1}$$

If we now substitute $p = m/n$ in the expression (24.13.1) for the binomial distribution, we obtain the equation

$$P(r) = n(n-1)(n-2)\cdots(n-r+1)\left(\frac{m}{n}\right)^r \frac{(1-m/n)^{n-r}}{r!} \tag{2}$$

This expression may be written in the form

$$P(r) = \frac{m^r}{r!}\left(1-\frac{1}{n}\right)\left(1-\frac{2}{n}\right)\cdots\left(1-\frac{(r-1)}{n}\right)\left(1-\frac{m}{n}\right)^n\left(1-\frac{m}{n}\right)^{-r} \tag{3}$$

Now, for any fixed value of r we have, as n tends to infinity,

$$\lim_{n\to\infty}\left(1-\frac{1}{n}\right)\left(1-\frac{2}{n}\right)\cdots\left(1-\frac{(r-1)}{n}\right) = 1,$$

$$\lim_{n\to\infty}\left(1-\frac{m}{n}\right)^{-r} = 1,$$

and $$\lim_{n\to\infty}\left(1-\frac{m}{n}\right)^{n} = e^{-m}.$$

We therefore see that the limiting form of (24.14.3), as n tends to infinity, is

$$P(r) = \frac{m^r e^{-m}}{r!} \tag{4}$$

This is known as the *Poisson distribution.*

We note that the sum of the probabilities

$$\sum_{r=0}^{\infty} P(r) = e^{-m}\sum_{r=0}^{\infty}\frac{m^r}{r!} = e^{-m}e^{m} = 1. \tag{5}$$

The following are some examples of Poisson variates:

(i) The number of defective screws per box of 100 screws.

(ii) The number of bacterial colonies in a given culture per unit area of microscopic slab.

(iii) The number of cars passing through a certain street in time 't'.

(iv) The number of suicides or deaths by heart attack in time 't'.

(v) The number of radio-active atoms decaying in time t.

Mode of the Poisson distribution is that value of r for which $\frac{e^{-m}m^r}{r!}$ is greater than the term that precedes it and the term that follows it

$$\text{i.e.} \quad \frac{e^{-m}m^{r-1}}{(r-1)!} \le \frac{e^{-m}m^r}{r!} \ge \frac{e^{-m}m^{r+1}}{(r+1)!}$$

whence we have $r \le m$ and $(r+1) \ge m$ i.e. $(m-1) \le r < m$.

Thus if m is an integer, then there are two modes $(m-1)$ and m, while if m is not an integer, the mode lies between $(m-1)$ and m.

For calculation of *constants of the Poisson distribution,* we are to take an arbitrary origin as in the case of calculation of constants of Binomial distribution.

24.14.1 The Mean of the Poisson Distribution

The mean $\bar{x}$ of the Poisson distribution may be computed by the equation

$$\bar{x} = \sum_{r=0}^{\infty} \frac{rm^r e^{-m}}{r!} = me^{-m} \sum_{r=0}^{\infty} \frac{m^{r-1}}{(r-1)!} = me^{-m}e^m = m. \tag{6}$$

24.14.2 The Standard Deviation of the Poisson Distribution

The standard deviation of the Poisson distribution may be obtained by taking the limiting value of the standard deviation for the binomial distribution as we let p tend to zero. Equation (24.13.9) may be written in the form

$$\sigma^2 = np(1-p) \tag{7}$$

We therefore have

$$\lim_{p \to 0} \sigma^2 = np = m. \tag{8}$$

Therefore,

$$\sigma = m^{\frac{1}{2}} = (np)^{\frac{1}{2}} \tag{9}$$

We thus see that the standard deviation of the Poisson distribution is $m^{\frac{1}{2}}$.

24.15 The Normal or Gaussian Distribution

A most important probability distribution that arises as a limit of the binomial distribution is a continuous distribution known as the *normal* or *gaussian* distribution, with the probability function

$$\phi(x) = \frac{(2\pi)^{-\frac{1}{2}}}{\sigma} e^{-(x-\mu)^2/2\sigma^2} \quad (-\infty < x < \infty). \tag{1}$$

[**N.B.** To understand how the present standard form of normal distribution i.e. (24.15.1) has arrived at, we are to consider that the actual normal distribution $y_x = y_0 e^{-x^2/2\sigma^2}$ has been derived as a limiting case of binomial distribution when $p = q$ or when $p \ne q$ but $p \simeq q$. Here y_x is the frequency of $(np+x)$ successes or of x successes if the origin is transferred to np successes in place of zero successes. The curve $y = y_0 e^{(-x^2/2\sigma^2)}$ called the normal curve Fig. 24.15.1(a) is symmetrical about the point $x = 0$, where the ordinate has its maximum value.

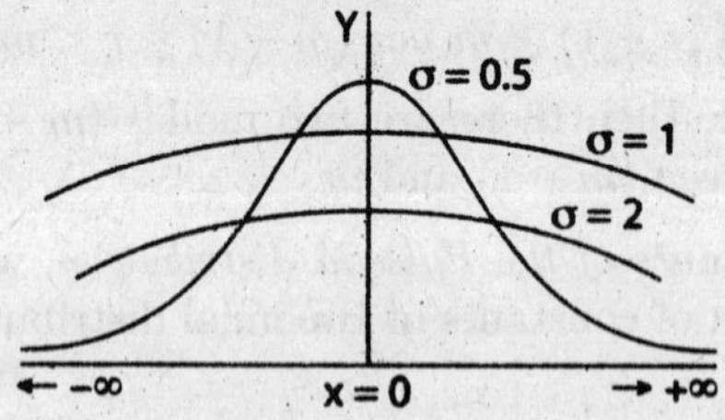

Fig 24.15.1(a)

To make the curve $y = y_0 e^{-x^2/2\sigma^2}$ conform to the standard form of normal probability curve, the value of y_0 is determined in such a way that the total frequency may be 1 i.e. the area of the normal curve is unity, and based on this consideration y_0 is calculated as $\frac{1}{\sigma\sqrt{2\pi}}$, so that the *standard form of the normal curve* is

$$y = \frac{(2\pi)^{-1/2}}{\sigma} e^{-x^2/2\sigma^2}$$

which is the normal distribution.

In arriving at the above form of the normal curve, we have taken the mean at the origin, but if however we take another point as the origin such that the excess of the mean over the value chosen as the arbitrary origin is μ, the form of the standard normal distribution becomes

$$y = \frac{(2\pi)^{-1/2}}{\sigma} e^{-(x-\mu)^2/2\sigma^2}$$ [$(x-\mu)$ being the excess of the mean over the arbitrary origin]

which represents the present day standard form of the normal distribution curve with origin shifted to $(\mu, 0)$.

Properties of the normal curve

$$y = \frac{(2\pi)^{-1/2}}{\sigma} e^{-x^2/2\sigma^2} = y_0 e^{-x^2/2\sigma^2} \quad \text{[Fig 24.15.1(a)]}$$

1. The curve is symmetrical (with one single peak) about the y-axis. The mean, median and mode coincide at the origin.

2. The curve is specified if mean (origin of x) and standard derivation are given. The value of y_0 can be calculated from the fact that the area of the curve must be equal to the total number of observations.

3. y decreases rapidly as x increases numerically. The curve extends to infinity (asymptotically) on either side of the origin.

4. The points of inflexion of the normal curve are the points where the curvature changes its direction and are given by

$$x = \pm\sigma, \quad y = \frac{(2\pi)^{-1/2}}{\sigma} e^{-1/2}$$

which can be obtained by putting $\frac{d^2y}{dx^2} = 0$ and verifying that at these points of the curve $\frac{d^3y}{dx^3} \neq 0$. Thus the points of inflexion of the normal curve are equidistant from the mean.]

We shall show that μ is the *mean* of the distribution and σ is the *standard deviation* of the normal or gaussian distribution. If the function $\phi(x)$ is to be a probability density function, it is implied that

$$A = \int_{-\infty}^{+\infty} \phi(x)dx = 1 \tag{2}$$

In order to verify that (24.15.2) is true, let us compute

$$A = \int_{-\infty}^{+\infty} \frac{(2\pi)^{-\frac{1}{2}}}{\sigma} \int_{-\infty}^{+\infty} e^{-(x-\mu)^2/2\sigma^2} dx \tag{3}$$

In order to simplify (24.15.3), we make the substitution $y = \frac{x-\mu}{\sigma}$ in the integral (24.15.3) and find that it is transformed into

$$A = (2\pi)^{-\frac{1}{2}} \int_{-\infty}^{+\infty} e^{-y^2/2} dy \tag{4}$$

From a table of definite integrals we obtain the result

$$\int_{-\infty}^{+\infty} e^{-y^2/2} dy = (2\pi)^{\frac{1}{2}} \tag{5}$$

If we now substitute (24.15.5) in (24.15.4), we find that $A = 1$.

24.15.1 The Mean Value of the Normal Distribution

We may compute the mean $\bar{x}$ of the normal distribution by the use of Eq.(24.12.4), which in this case takes the form

$$\bar{x} = \int_{-\infty}^{+\infty} x\phi(x)dx = \frac{(2\pi)^{-1/2}}{\sigma} \int_{-\infty}^{+\infty} xe^{-(x-\mu)^2/2\sigma^2} dx \tag{6}$$

To simplify (24.15.6), let us introduce the following change in variable:

$$y = \frac{(x-\mu)}{\sigma} \quad \text{or,} \quad dx = \sigma dy \quad \text{and} \quad x = (\sigma y + \mu) \tag{7}$$

With this change in variable, (24.15.6) becomes, on integration by parts,

$$\bar{x} = (2\pi)^{-\frac{1}{2}}\sigma \int_{-\infty}^{+\infty} ye^{-y^2/2}dy + \mu(2\pi)^{-\frac{1}{2}} \int_{-\infty}^{+\infty} e^{-y^2/2}dy. \tag{8}$$

From a table of definite integrals, we have the results

$$\int_{-\infty}^{+\infty} ye^{-y^2/2}dy = 0 \quad \text{and} \quad \int_{-\infty}^{+\infty} e^{-y^2/2}dy = (2\pi)^{1/2}. \tag{9}$$

If we substitute the results (24.15.9) in (24.15.8), we obtain

$$\bar{x} = \mu \tag{10}$$

We have therefore verified that μ is the mean or expected value of the distribution function $\phi(x)$ of (24.15.1).

24.15.2 The Standard Deviation of the Normal Distribution

The variance $\mu_2 = \sigma^2$ of the normal distribution may be obtained by the use of (24.12.6) by putting $a = \mu$ and thus obtaining

$$\mu_2 = \int_{-\infty}^{+\infty} (x-\mu)^2 \phi(x) dx. \tag{11}$$

If we now substitute $\phi(x)$ as given by (24.15.1) into (24.15.11) and make use of the change in variable $(x - \mu) = \sigma y$, we obtain

$$\mu_2 = (2\pi)^{-1/2}\sigma^2 \int_{-\infty}^{+\infty} y^2 e^{-y^2/2} dy = \sigma^2 (2\pi)^{-1/2} I, \tag{12}$$

where the integral

$$I = \int_{-\infty}^{+\infty} y(ye^{-y^2/2}) dy$$

may be integrated by parts to obtain

$$I = -ye^{-y^2/2}\Big|_{-\infty}^{+\infty} + \int_{-\infty}^{+\infty} e^{-y^2/2} dy = (2\pi)^{1/2} \tag{13}$$

If the above value for I is substituted into (24.15.12), we obtain

$$\mu_2 = (2\pi)^{-1/2}(2\pi)^{1/2}\sigma^2 = \sigma^2 \tag{14}$$

Therefore, σ correctly represents the standard deviation of the normal or gaussian distribution (Fig. 24.15.1).

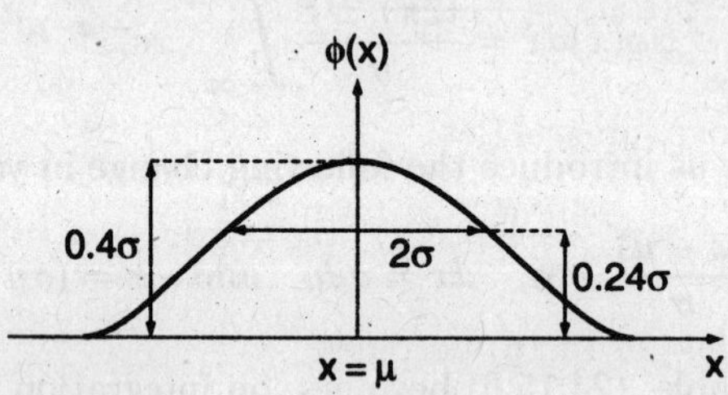

Fig 24.15.1

The *maximum* value of $\phi(x)$ is located at $x = \mu$; $\phi(\mu)$ is approximately $\frac{2}{5}\sigma$. When $x = \mu \pm \sigma$, the value of $\phi(x)$ is slightly less than $\frac{1}{4}\sigma$. It is easy to show that the points $x = \mu \pm \sigma$ are the points of *inflexion* of the normal distribution curve $\phi(x)$ obtained by differentiating (24.15.1) and putting $\frac{d^2\phi(x)}{dx^2} = 0$ and verifying that $\frac{d^3\phi(x)}{dx^3} \neq 0$ for these points. It is of great interest to know the probability that the deviation $y = x - \mu$ from the mean is within certain limits. The probability that y lies between the limits σt_1 and σt_2 is

$$\frac{(2\pi)^{-1/2}}{\sigma} \int_{\sigma t_1}^{\sigma t_2} e^{-y^2/2\sigma^2} dy = \int_{t_1}^{t_2} (2\pi)^{-1/2} e^{-t^2/2} dt \tag{15}$$

By putting $t_1 = -a$ and $t_2 = a$ in (24.15.15) and using a table of definite integrals, it can be shown that the probability that the deviation y exceeds 2σ is less than $\frac{1}{20}$ and the probability that y exceeds 3σ is of the order of 0.003. In statistics, an observation which is improbable is said to be *significant.* For a normally distributed variate, it is usual to regard a single observation with $|x| > \mu+2\sigma$ as significant and one with $|x| > \mu+3\sigma$ as *highly significant.*

The normal or gaussian distribution may be derived as a limiting form of the binomial distribution when n becomes very large and p remains constant.

The normal or gaussian distribution (24.15.1) is sometimes also known as the *normal error law,* although some statisticians object that it is not in fact a normal occurrence in nature because it is only valid as a limiting case for exceptionally large numbers. Sufficiently large numbers are, however, available in problems of applied physics, such as the kinetic theory of gases and fluctuations in the magnitude of an electric current, and so the gaussian distribution seems more real to physicists and engineers than to social and biological scientists.

The application of the gaussian distribution to the theory of errors is made reasonable by the fact that any process of random sampling tends to produce a gaussian distribution of sample values, even if the whole population from which the samples are drawn does not have a normal distribution. Hence it is usual to assume that the gaussian distribution will be a useful approximation to the distribution of errors of observation, provided that all causes of systematic error can be excluded.

24.15.3 The Area under Some Part of the Gaussian Distribution Function

The area under some part of the gaussian distribution curve is frequently needed. That is, we must evaluate the integral of $\phi(x)$ between finite values of x. For example, let it be required to evaluate the integral I, where

$$I = \int_{x_1}^{x_2} \phi(x)dx = \frac{(2\pi)^{-1/2}}{\sigma} \int_{x_1}^{x_2} e^{-(x-\mu)^2/2\sigma^2} dx. \tag{16}$$

In order to simplify the above integral, let

$$v = \frac{x-\mu}{\sigma}, \quad \text{or,} \quad dx = \sigma dv.$$

With this change in variable, (24.15.16) becomes

$$I = (2\pi)^{-1/2} \int_{v_1}^{v_2} e^{-v^2/2} dv, \tag{17}$$

where

$$v_1 = \frac{x_1-\mu}{\sigma}, \quad v_2 = \frac{x_2-\mu}{\sigma} \tag{18}$$

Now the function $F(z)$ given by

$$F(z) = (2\pi)^{-1/2} \int_{-\infty}^{z} e^{-y^2/2} dy \tag{19}$$

has been *extensively tabulated.* It is easy to see that the integral is given by

$$I = F(v_2) - F(v_1) \tag{20}$$

in terms of the function $F(z)$.

We may treat the significance of the area under the curve from a different angle.

The total area of a normal curve being treated as unity, the probability corresponding to any interval in the range of the variate is measured by the area under the curve within that interval. Hence if μ be the mean of the normal distribution, then P, the probability from μ to any value x of the variate is given by

$$P = \frac{(2\pi)^{-1/2}}{\sigma} \int_{\mu}^{x} e^{-(x-\mu)^2/2\sigma^2} dx = (2\pi)^{-1/2} \int_0^z e^{-z^2/2} dz \quad \text{putting } \frac{x-\mu}{\sigma} = z \text{ i.e. } dx = \sigma dz.$$

This value of P is known as the ***probability integral*** or ***error function.***

Actual calculation shows that

(i) $P[(\mu - \sigma) < x < \mu + \sigma] = 0.6826$—showing that 68.26% of the items in the normal distribution fall between the range $\pm\sigma$ of the mean;

(ii) $P[(\mu - 2\sigma) < x < (\mu + 2\sigma)] = 0.9544$—showing that 95.44% of the items in the normal distribution fall within the range $\pm 2\sigma$ of the mean;

(iii) $P[(\mu - 3\sigma) < x < (\mu + 3\sigma)] = 0.9972$—showing that 99.72% of the items in the normal distribution fall within the range $\pm 3\sigma$ of the mean.

Conclusively a normal curve can be used to find (i) the number of cases at any given distance from the mean; (ii) the number of cases lying within certain range of values in the distribution; (iii) the probability that a case selected at random lies above or below a given point.

24.16 Distribution of a Sum of Normal Variates

If two variates x and y are distributed normally, their sum $u = x + y$ is also distributed normally. To prove this, we shall refer each distribution to its mean and assume that the distribution functions of $v = x - \bar{x}$ and $w = y - \bar{y}$ are respectively

$$\phi_1(v) = \frac{(2\pi)^{-1/2}}{\sigma_1} e^{-v^2/2\sigma_1^2} \tag{1}$$

$$\text{and} \quad \phi_2(w) = \frac{(2\pi)^{-1/2}}{\sigma_2} e^{-w^2/2\sigma_2^2} \tag{2}$$

The probability that $(v + w)$ takes a particular value z is found by integrating over the probabilities that v and w take particular values whose sum is z. This gives

$$\int \phi_1(v)\phi_2(z - v) dv = \frac{1}{2\pi\sigma_1\sigma_2} e^{-z^2/2(\sigma_1^2+\sigma_2^2)}$$

$$\times \int_{-\infty}^{+\infty} e^{-(v - z\sigma_1^2)^2/2\sigma_1^2+\sigma_2^2(\sigma_1^2+\sigma_2^2)} dv = F(z) \tag{3}$$

If we now put

$$t = \frac{(\sigma_1^2 + \sigma_2^2)^{1/2}(v - z\sigma_1^2)}{\sigma_1\sigma_2(\sigma_1^2 + \sigma_2^2)},$$

the integral in (24.16.3) becomes

$$\int_{-\infty}^{+\infty} \frac{e^{-t^2/2}\sigma_1\sigma_2 dt}{(\sigma_1^2+\sigma_2^2)^{1/2}} = \frac{(2\pi)^{1/2}\sigma_1\sigma_2}{(\sigma_1^2+\sigma_2^2)^{1/2}}. \tag{4}$$

Hence the distribution (24.16.3) is

$$F(z) = \frac{(2\pi)^{-1/2}}{(\sigma_1^2+\sigma_2^2)^{1/2}} e^{-z^2/2(\sigma_1^2+\sigma_2^2)} \tag{5}$$

We therefore see that $z = v + w$ is normally distributed about the mean value zero, with standard deviation given by

$$\sigma^2 = \sigma_1^2 + \sigma_2^2 \tag{6}$$

The distribution of $u = x + y$, therefore, has the expectation $\bar{u} = \bar{x} + \bar{y}$, with a standard deviation given by (24.16.6).

24.17 Applications to Experimental Measurements

Let $x_1, x_2, \cdots, x_n$ be a set of n measurements. The x_i, $i = 1, 2, \cdots, n$, quantities may be the n number of determinations of the velocity of light in a certain medium. If we take z as a typical value of the measurements x_i, then d_i is the deviation of the measurement x_i from z, where

$$d_i = z - x_i, \quad i = 1, 2, 3, \cdots, n \tag{1}$$

Let y represent the sum of the squares of the deviations of the measurements x_i from the typical value z, so that

$$y = \sum_{i=1}^{n} d_i^2 = \sum_{i=1}^{n} (z - x_i)^2 \tag{2}$$

We now wish to choose the typical value z so that the sum of the squares of the deviations y is a minimum. We therefore minimise (24.17.2) in the usual way by computing the derivative, so that

$$\frac{dy}{dz} = \sum_{i=1}^{n} 2(z - x_i) \tag{3}$$

On setting the derivative dy/dz equal to zero and solving for z, we find

$$z = \bar{x} = \sum_{i=1}^{n} \frac{x_i}{n} = \text{the arithmetic mean of } x_i, \quad i = 1, 2, 3, \cdots, n \tag{4}$$

That is, we find that the arithmetic mean of the measurements x_i is the typical value that minimizes y. It will be noted that $\bar{x}$ gives the *best* typical value of the x_i measurements in the least-square sense.

The *minimum* value $y_{\min}$ of y is obtained by substituting $z = \bar{x}$ in (24.17.2) and thus obtaining

$$y_{\min} = \sum_{i=1}^{n} (\bar{x} - x_i)^2 \tag{5}$$

If we divide (24.17.5) by n, we obtain the *variance* σ^2 of the set of measurements x_i, so that

$$\sigma^2 = \frac{1}{n}\sum_{i=1}^{n}(\bar{x} - x_i)^2 = \text{the variance of } x_i \tag{6}$$

The square root of (24.17.6) is the *standard deviation* σ of the set of measurements x_i, so that

$$\sigma = \left[\sum_{i=1}^{n}\frac{(\bar{x} - x_i)^2}{n}\right]^{1/2} = \text{the standard deviation of } x_i. \tag{7}$$

24.18 The Standard Deviation of the Mean

We now turn to a very important question that arises in the theory of statistics: *What is the relation between the standard deviation of the individual measurements and the standard deviation of the mean of a set of measurements?*

In order to answer this question, we will assume that we have a parent distribution of errors which follow the *Gauss distribution.* (The justification for assuming that the distribution is gaussian is that in practice errors do follow the gaussian law.) We suppose first that we take N measurements or a sample of the parent distribution. We then take *another* sample or set of measurements. If we now compute the mean and standard deviation of both sets, these quantities will in general be different. That is, the mean and variance of a *sample* of N observations are *not* in general equal to the mean and variance of the *parent* distribution.

The process to be followed is to take M sets of N measurements, each with its own mean and standard deviation, and then compute the standard deviation of the means. The standard deviation of the means provides an indication of the reliability of any one of the means.

To facilitate the calculation let the following notations be introduced:

$$\begin{aligned}
\sigma &= \text{the standard deviation of the individual measurements,}\\
\sigma_m &= \text{the standard deviation of the means (of the various samples),}\\
x_{si} &= \text{measurement } i \text{ in the set } s,\\
\bar{x}_s &= \text{the mean of the set } s,\\
\bar{X} &= \text{the mean of all the measurements,}\\
d_{si} &= \text{the deviation of } x_{si}, x_{si} - \bar{X} = d_{si},\\
D_s &= \bar{x}_s - \bar{X} = \text{the deviation of the mean } \bar{x}_s.
\end{aligned}$$

Since we are taking M sets of measurements with N measurements in each set, there will be $n = MN$ measurements in the total. The variance of the individual measurements is given by

$$\sigma^2 = \sum_{s=1}^{M}\sum_{i=1}^{N}\frac{d_{si}^2}{MN} \tag{1}$$

The variance of the means is given by

$$\sigma_m^2 = \sum_{s=1}^{M}\frac{D_s^2}{M} \tag{2}$$

The deviations D_s of the means can be expressed in terms of the deviations d_{si} of the individual observations as follows:

$$D_s = \bar{x}_s - \bar{X} = \frac{1}{N}\sum_{i=1}^{N} x_{si} - \bar{X}. \tag{3}$$

We also have

$$d_{si} = x_{si} - \bar{X} \tag{4}$$

If we now substitute the square of (24.18.4) into (24.18.1), we obtain

$$\sigma^2 = \sum_{s=1}^{M}\sum_{i=1}^{N} \frac{(x_{si} - \bar{X})^2}{MN} \tag{5}$$

The substitution of the square of (24.18.3) into (24.18.2) gives the result

$$\sigma_m^2 = \frac{1}{M}\sum_{s=1}^{M}\left(\frac{1}{N}\sum_{i=1}^{N} x_{si} - \bar{X}\right)^2 \tag{6}$$

After some simplifications, it can be seen that (24.18.6) may be written in the following form:

$$\sigma_m^2 = \sum_{s=1}^{M}\sum_{i=1}^{N} \frac{(x_{si} - \bar{X})^2}{MN^2} \tag{7}$$

If we now compare (24.18.5) with (24.18.7), we see that

$$\sigma_m^2 = \frac{\sigma^2}{N} \tag{8}$$

or therefore

$$\sigma_m = \frac{\sigma}{N^{1/2}} \tag{9}$$

Therefore, the variance of the mean of a set of N measurements is simply the variance of the individual measurements divided by the number of measurements.

It may be mentioned that in order to reduce (24.18.6) to the form (24.18.7) it is necessary to assume that the cross-product terms are negligibly small. This is true for the gaussian distribution, and since experimental measurements so often obey the gaussian distribution, formula (24.18.8) is a very useful one.

24.19 Stirling's Approximation for the Factorial

In many phases of the mathematical theory of probability, the factorial of large numbers occurs. A classical result given by Stirling is of paramount importance in the computation of the factorial of large numbers. The formula given by Stirling may be stated in the following form:

$$\lim_{n\to\infty} n! = (2\pi)^{1/2} n^{n+\frac{1}{2}} e^{-n}. \tag{1}$$

The percentage error of this formula for finite values of n is of the order of $100/12n$ percent.

In addition to being a very useful method for the computation of factorial numbers, Stirling's formula is very useful in a theoretical sense. Its use in the computation of the factorial of large numbers may be seen by considering the computation of 10!. Instead of multiplying together a large number of integers, we have merely to calculate the factorial by the use of the expression (24.19.1), by means of a table of logarithms, and this involves far fewer operations. For example, for $n = 10$, we obtain the value 3,598,696 by Stirling's expression, while the exact value of 10! obtained by direct multiplication is 3,628,800. The percentage error obtained by the use of Stirling's formula is of the order of 5/6 percent too low.

For proof of expression (24.19.1) any standard book on Statistics and Probability may be consulted.

Illustrative Examples (Worked Out)

1. Starting from Poisson distribution, prove that $\sum_{x=0}^{\infty} P(x) = 1$.

We know

$$P(x) = \frac{e^{-m}m^x}{x!}$$

$$\therefore \sum_{x=0}^{\infty} P(x) = \sum_{x=0}^{\infty} e^{-m}\frac{m^x}{x!} = e^{-m}\sum_{x=0}^{\infty}\frac{m^x}{x!}$$

$$= e^{-m}\left[1 + \frac{m}{1!} + \frac{m^2}{2!} + \frac{m^3}{3!} + \cdots\right] \quad \text{putting } x = 0, 1, 2, 3, \text{ etc.}$$

$$= e^{-m}e^{m} = 1.$$

2. If points x_i are distributed at random on the X-axis in such a fashion that the following assumptions are valid:

(a) If $P(\Delta x)$ is the probability of two or more points in an interval of length Δx, then $\frac{P(\Delta x)}{\Delta x} \to 0$ as $\Delta x \to 0$;

(b) If $P_1(\Delta x)$ is the probability of 1 point in an interval of length Δx, then $\frac{P_1(\Delta x)}{\Delta x} \to k$ (a constant) as $\Delta x \to 0$.

Prove that in this case the probability $P_n(x)$ of n points in an interval of length x satisfies Poisson distribution $P_n(x) = e^{-kx}(kx)^n/n!$.

Let us consider an interval $(0, x + \Delta x)$ of length $x + \Delta x$. We can have n points in this interval in three mutually exclusive ways. Either there are n points in x and none in Δx, or there are $(n-1)$ in x and 1 in Δx, or there are fewer than $(n-1)$ in x and at least 2 in Δx. The probability of this last alternative may be written as $\varepsilon\Delta x$, where $\varepsilon \to 0$ as $\Delta x \to 0$ in view of assumption (a).

Now by the total and compound probability laws,

$$P_n(x + \Delta x) = P_n(x)P_0(\Delta x) + P_{n-1}(x)P_1(\Delta x) + \varepsilon\Delta x.$$

Subtracting $P_n(x)$ from both sides of the above equation and dividing by Δx, we have

$$\frac{P_n(x+\Delta x) - P_n(x)}{\Delta x} = P_n(x)\frac{P_0(\Delta x) - 1}{\Delta x} + P_{n-1}(x)\frac{P_1(\Delta x)}{\Delta x} + \varepsilon \qquad (1)$$

For the interval Δx, there are only these three possibilities and the corresponding probabilities are $P_0(\Delta x), P_1(\Delta x)$ and $P(\Delta x)$ respectively.

Hence $P_0(\Delta x) + P_1(\Delta x) + P(\Delta x) = 1$.

$$\therefore \quad \frac{P_0(\Delta x) - 1}{\Delta x} = -\frac{P_1(\Delta x)}{\Delta x} - \frac{P(\Delta x)}{\Delta x} \tag{2}$$

If $\Delta x \to 0$, then $\frac{P(\Delta x)}{\Delta x} \to 0$ and $\frac{P_1(\Delta x)}{\Delta x} \to k$.

So

$$\lim_{\Delta x \to 0} \frac{P_n(x + \Delta x) - P_n(x)}{\Delta x} = -kP_n(x) + kP_{n-1}(x)$$

since $\varepsilon \to 0$ as $\Delta x \to 0$.

$$\therefore \quad \frac{d}{dx} P_n(x) = -kP_n(x) + kP_{n-1}(x) \tag{3}$$

For $n = 0, P_{n-1}(x) = P_{-1}(x) = 0$, so that $\frac{d}{dx} P_0(x) = -kP_0(x)$ or, $P_0(x) = ce^{-kx}$

But an interval of zero length is sure to have no point.

So $P_0(x) = 1$ for $x = 0$, then $c = 1$ and

$$P_0(x) = e^{-kx} \tag{4}$$

Similarly, for $n = 1$,

$$\frac{d}{dx} P_1(x) = -kP_1(x) + kP_0(x) = -kP_1(x) + ke^{-kx}$$

This gives $P_1(x) = e^{-kx} \frac{kx}{1!}$.

Similarly if we put $n = 2$, then we have $P_2(x) = e^{-kx} \frac{(kx)^2}{2!}$ and so on.

$$\therefore \quad P_n(x) = e^{-kx} \frac{(kx)^n}{n!}$$

3. What is the probability that the ace of spades will be drawn from a deck of cards at least once in 104 consecutive trials?

Here $p = 1/52$. The probability of failure to draw the ace $P(0)$ is given by

$$P(0) = \frac{(104/52)^0}{0!} e^{-104/52} = e^{-2} = 0.133.$$

Hence the probability that at least one ace of spade is drawn $= 1 - e^{-2} = 1 - 0.133 = 0.867$.

4. Show that the constant k in Poisson distribution represents the expected number of points in a unit interval.

From the result of example 2 above, the probability of n points in a unit interval is

$$P_n(1) = e^{-k} \frac{k^n}{n!},$$

so that the expected number is given by

$$\sum_{n=1}^{\infty} \frac{e^{-k}k^n}{n!} n = e^{-k} \cdot k \sum_{n=1}^{\infty} \frac{k^{n-1}}{(n-1)!} = e^{-k} \cdot ke^k = k.$$

5. A variable x is said to be uniformly distributed on (a, b), if $f(x)$ is constant on (a, b) and zero outside (a, b). Find $f(x)$ in this case.

If the constant value of $f(x)$ for $a < x < b$ be c, then

$$\int_{-\infty}^{\infty} f(x)dx = \int_a^b c dx = c(b-a) = 1$$

$$\therefore \quad c = \frac{1}{(b-a)} = f(x) \text{ for } a < x < b \text{ and } f(x) = 0 \text{ elsewhere.}$$

6. A stick of length a is broken at random into two pieces. Find the distribution function $F(l)$ for the length l of the shorter piece. From this find the probability density $f(L)$ for the length L of the longer piece.

Here $0 \le l \le \frac{a}{2}$ in every case. For any point t between 0 and $\frac{a}{2}$, we have $l < t$, if and only if, x is on one of the intervals $(0, t)$ or (t, a) as shown in the attached figure. The probability for this is $2t/a$, since x is uniformly distributed.

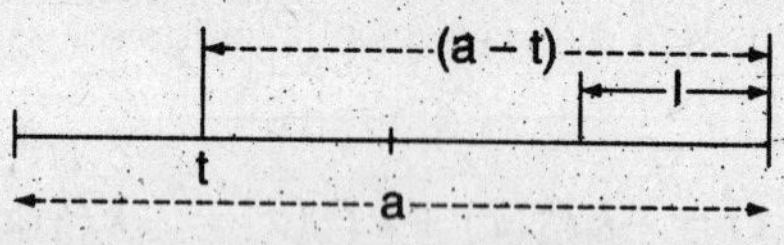

So, we may write, $F(l) = \frac{2l}{a}$ for $0 \le l \le \frac{a}{2}$, $F(l) = 0$ for $l < 0$ and $F(l) = 1$ for $l > \frac{a}{2}$.

Again, length of the longer piece $L = a - l$. Thus $L < t$, if and only if, $l > a - t$; from the result just obtained above, the probability is

$$1 - P_r[l \le (a-t)] = 1 - \frac{2(a-t)}{a} = \begin{cases} f(L) & \text{for } \frac{a}{2} < L < a \\ 0 & \text{for } L < \frac{a}{2} \\ 1 & \text{for } L > a \end{cases}$$

This gives the distribution for L. By differentiation, the density is found to be $f(L) = \frac{2}{a}$ for $\frac{a}{2} < L < a$, and $f(L) = 0$ elsewhere.

7. A coin is tossed 3 times. Construct the uniform sample space and find (i) the probability of at least two tails in succession, (ii) the probability that two consecutive tosses fall the same.

The uniform sample space for this problem contains 8 points as shown below, and we assign the probability 1/8 to each.

$$\mathrm{H\overset{\circ}{H}H} \quad \mathrm{HTH} \quad \mathrm{H\overset{\circ}{H}T} \quad \mathrm{H\overset{\circ}{T}\overset{*}{T}} \quad \mathrm{T\overset{*}{T}\overset{\circ}{T}} \quad \mathrm{THT} \quad \mathrm{T\overset{\circ}{H}H} \quad \mathrm{T\overset{*}{T}\overset{\circ}{H}}$$

By actual counting we see that there are 3 cases (shown by the mark $*$) in which two tails in succession are obtained. Events being mutually exclusive, the required probability is

$$\frac{1}{8} + \frac{1}{8} + \frac{1}{8} = \frac{3}{8}.$$

Now for two consecutive tosses falling the same, there are 6 favourable cases (shown by the mark ∘), so that the required probability in the second case is $6 \cdot \frac{1}{8} = \frac{3}{4}$.

8. By using the law of compound probability, find (a) the probability of all heads in three tossings of a coin, (b) the probability of at least one tail in 10 tosses of a coin.

(a) Probability of head in each toss is 1/2. The tosses are independent and therefore according to the law of compound probability, the required probability is $p = \frac{1}{2} \cdot \frac{1}{2} \cdot \frac{1}{2} = \frac{1}{8}$.

(b) The event of at least 1 tail in 10 tosses corresponds to the rest of the sample space except the *all heads* points. Since the sum of the probabilities of all the sample points is 1, the required probability is $p = 1 - p_1$, where p_1 is the probability of all heads in 10 tosses of a coin, and may be calculated by using the law of compound probability. Since the tosses are independent, we have $p_1 = \left(\frac{1}{2}\right)^{10}$ and hence $p = 1 - \left(\frac{1}{2}\right)^{10}$.

9. A loaded die has probabilities $\frac{1}{21}, \frac{2}{21}, \frac{3}{21}, \frac{4}{21}, \frac{5}{21}, \frac{6}{21}$ of throwing $1, 2, 3, 4, 5, 6$. (a) What is the probability of throwing two 3's in succession? (b) What is the probability of throwing a 4 the first time and not a 4 the second time?

(a) Since the throws are independent, the required probability (according to the law of compound probability) is $p = \left(\frac{3}{21}\right)^2 = \frac{1}{49}$.

(b) The probability of *not a* 4 is, to be sure, the sum of other 5 possibilities, but it is easier to find it as $1 - \frac{4}{21} = \frac{17}{21}$, since the sum of the probabilities of all possible cases is 1. Then the required probability is $p = \frac{4}{21} \cdot \frac{17}{21}$.

10. How many times must we throw a die loaded as in example 9, to have the probability greater than 1/2 of getting an ace?

The probability of getting an ace at any throw is 1/21—so that the probability of not getting an ace is $1 - \frac{1}{21} = \frac{20}{21}$. Then the probability of not getting an ace in any of n successive throws is $(\frac{20}{21})^n$. We want this probability to be less than 1/2, for then the probability of getting an ace on at least one throw is greater than 1/2. In order to solve $\left(\frac{20}{21}\right)^n < \frac{1}{2}$, we take logarithms on both sides of this inequality; we have then

$$\log\left(\frac{20}{21}\right)^n < \log\frac{1}{2} \quad \text{or,} \quad n\log\frac{20}{21} < -\log 2$$

$$\text{or,} \quad n\log\frac{21}{20} > \log 2 \quad \text{or,} \quad n > \frac{\log 2}{\log\frac{21}{20}} = \frac{0.3010}{0.0210} = 14.5.$$

So we must toss the die 15 times.

11. Find the probability of distributing 15 balls in 6 boxes, such that the number of balls in various boxes is given as

No. of Balls	3	1	4	2	3	2
Box No.	1	2	3	4	5	6

We first consider in how many ways we can select 3 balls out of 15, for the first box. Obviously, this number is ${}^{15}C_3$. Here we have not taken into consideration the order of the balls in the box.

We have now 12 balls left, out of which we are to select 1 for box 2—we can do this in ${}^{12}C_1$ ways. We can then select 4 balls for box 3 from the remaining 11 balls in ${}^{11}C_4$ ways, 2 balls for box 4 in ${}^7C_2$ ways, 3 balls for box 5 in ${}^5C_3$ ways, and finally 2 balls for box 6 in ${}^2C_2$ ways. Thus the total number of ways of putting the required number of balls into the numbered boxes is

$$ {}^{15}C_3\cdot{}^{12}C_1\cdot{}^{11}C_4\cdot{}^7C_2\cdot{}^5C_3\cdot{}^2C_2 = \frac{15!}{3!12!}\cdot\frac{12!}{1!11!}\cdot\frac{11!}{4!7!}\cdot\frac{7!}{2!5!}\cdot\frac{5!}{3!2!}\cdot\frac{2!}{2!0!}\cdot = \frac{15!}{3!4!2!3!2!} \qquad (1) $$

We now find out the probability of this particular distribution. We assume that the balls are distributed at random, into the boxes;—by this we mean that a ball has the same probability (namely, 1/6) of being put into any one box as into any other box. We can put the first ball into any one of the 6 boxes, the second ball into any one of the 6 boxes and so on. Thus the total number of ways of distributing 15 balls into 6 boxes is $6\cdot 6\cdot 6\cdots 6 = 6^{15}$ and here we are assuming that these distributions are equally likely. Then the probability that when 15 balls are distributed at random into 6 boxes, such that there will be 3 balls in box 1, 1 in box 2, etc., is

$$ \frac{\text{number of favourable cases}}{\text{total number of cases}} = \frac{15!}{3!4!2!3!2!} \div 6^{15} \quad \text{from (1).} $$

12. Find the probability that the first two chairs are vacant, when four persons are sitting in six chairs.

If we number the chairs from 1 to 6 and allow each person to choose his chair by tossing a die, we may have two or more persons, choosing the same chair. The result 6^4 (as in example 11) then does not apply to this problem. However, we may consider a uniform sample space of 6^4 points and select from it the points corresponding to the restriction in the problem (i.e., one person or none per chair). The new sample space contains ${}^6C_4 4!$ points (i.e., the number of ways of selecting the 4 chairs multiplied by the number of ways of then arranging 4 persons in 4 chairs). Since these points were equally probable in the original (uniform) sample space, we still consider them equally probable. Number of sample points corresponding to the given event (4 persons are seated and the first 2 chairs are vacant) is 4! which is the number of ways of arranging 4 persons in the last 4 chairs. Thus the required probability is $4!/{}^6C_4\cdot 4! = 1/{}^6C_4$.

13. Find the probability that in 100 tosses of a coin, we get between 45 and 55 heads.

In this case $n = 100$ and $np = 100\cdot\frac{1}{2} = 50$, since the probability of getting a head in one toss is $p = 1/2$. It is given that $45 \le x \le 55$. Moreover $\sqrt{np(1-p)} = 5$ and hence $\frac{x-np}{\sqrt{np(1-p)}} = \frac{x-50}{5}$ lies between -1 and 1. So we want the probability that $-1 \le \frac{x-np}{\sqrt{np(1-p)}} \le 1$.

Now, from the gaussian or normal probability density function given by

$$ \phi(t) = \frac{1}{\sqrt{2\pi}}\int_0^t e^{-x^2/2}dx, $$

we may write

$$ \begin{aligned} \phi(1) - \phi(-1) &= \frac{1}{\sqrt{2\pi}}\int_{-1}^{1} e^{-x^2/2}dx \\ &= \frac{2}{\sqrt{2\pi}}\int_0^1 e^{-x^2/2}dx = 2.(0.34) = 0.68 \quad \text{(from Tables)} \end{aligned} $$

14. The number of particles emitted each minute by a radioactive source is recorded for a period of 10 hours; a total of 1800 counts is registered. During how many 1-minute intervals should we expect to observe no particles, exactly one particle, etc.?

The average number of counts per minute is $1800/10 \times 60 = 3$ counts/min $= m$. Then applying Poisson distribution, we have the probability of x counts per minute given by

$$P(x) = \frac{m^x e^{-m}}{x!} = \frac{3^x e^{-3}}{x!}.$$

Thus $P(0) = e^{-3} = 0.05$; then we should expect to observe no particles in about 5% of the 600 1-minute intervals, that is during 30 1-minute intervals. Similarly we can calculate the expected number of 1-minute intervals during which 1, 2, etc. particles will be observed.

15. Out of $(2n+1)$ tickets consequently numbered, three are drawn at random. Find the chance that the numbers on them are in arithmetic progression.

Let the first selected number be 1; then the possible groups may be written as

$$(1,2,3), (1,3,5), (1,4,7), \cdots, [1, (n+1), (2n+1)]$$

in n possible ways. Similar grouping will show that when the lowest numbers selected are $1, 2, 3, \cdots (2n-2), (2n-1)$, the number of favourable ways will be respectively $n, n-1, n-1, n-2, n-2, \cdots, 1, 1$, so that the total number of favourable ways will be

$$2[1+2+3+\cdots+(n-1)+n] - n = \frac{2n(n+1)}{2} - n = n^2.$$

Again the total number of ways $= {}^{2n+1}C_3 = \frac{(2n+1)\cdot 2n\cdot(2n-1)}{3!}$.

$\therefore$ the required chance $= \frac{6n^2}{2n(2n+1)(2n-1)} = 3n/4n^2 - 1$.

16. Find the probability of atmost 5 defective fuses to be found in a box of 200 fuses, if experience shows that 2% of such fuses are defective.

The probability of a fuse to be defective is $\frac{2}{100} = \frac{1}{50} = p$, so that $np = 200 \cdot \frac{1}{50} = 4 = m$. Then from Poission distribution formula, the probabilities of $0, 1, 2, 3, 4, 5$ defective fuses are

$$e^{-4}, \frac{e^{-4}\cdot 4}{1!}, \frac{e^{-4}\cdot 4^2}{2!}, \frac{e^{-4}\cdot 4^3}{3!}, \frac{e^{-4}\cdot 4^4}{4!}, \frac{e^{-4}\cdot 4^5}{5!} \quad \text{respectively.}$$

Therefore the probability of 5 or less than 5 fuses to be defective is

$$e^{-4}\left[1+4+\frac{4^2}{2!}+\frac{4^3}{3!}+\frac{4^4}{4!}+\frac{4^5}{5!}\right]$$

$$= 0.0183\left[1+4+8+\frac{32}{3}+\frac{32}{3}+\frac{128}{15}\right] = 0.785 \text{ (approx.)}$$

Exercises

1. What is the probability that number 1 appears at least once in n throws of a dice? **[Ans.** $1-(\frac{5}{6})^n$.]

2. A hunter finds that on the average he kills once in three shots. He fires three times at a duck; on the assumption that his a priori probability of killing is $\frac{1}{3}$, what is the probability that he kills the duck? **[Ans.** 19/27.]

3. Prove that if in any continuously varying process, a certain characteristic is present to the extent of one in T units, then the probability that the characteristic does not occur in a sample of t units is $e^{-t/T}$.

   [**Hints.** Consider the Poisson distribution.]

4. It is known that 100 litres of water have been polluted with 10^6 bacteria. If 1 cc of water is drawn off, what is the probability that the sample is not polluted? [**Ans.** $e^{-10} = 4.5 \times 10^{-5}$.]

5. An air-craft company carries on the average P passengers—M miles for every passenger killed. What is the probability of a passenger completing a journey of m miles in safety? [**Ans.** $\sim e^{-m/MP}$.]

6. Prove that $\frac{P(B/A)}{P(C/A)} = \frac{P(B)P(A/B)}{P(C)P(A/C)}$, where the symbols denote usual meanings.

7. When one card is drawn from each of two decks, find the probability that at least one of them is an ace. [**Ans.** 25/169.]

8. If the probability that a man aged 60 will live up to 70 is $p = 0.65$, what is the probability that 8 out of 10 men now aged 60 will live up to 70? [**Ans.** 0.176.]

9. What is the probability of throwing an ace exactly three times in four trials with a single die? [**Ans.** 5/324.]

10. If during the week-end road traffic, 100 cars per hour pass along a certain road, each taking 1 minute to cover it, find the probability that at any instant no car will be on the road. [**Ans.** $e^{-60/36}$.]

11. Let it be supposed that in the manufacture of fire-crackers, there is a certain defect that causes only three-fourths of them to explode when lighted. What is the probability that if we choose five of these fire-crackers, they will explode? [**Ans.** $(\frac{3}{4})^5 = 0.237$.]

12. It has been observed in human reproduction that twins occur approximately once in 100 births. If the number of babies in a birth are assumed to follow a Poisson distribution, calculate the probability of the birth of quintuplets. What is the probability of the birth of octuplets?

13. On the average, how many times must a die be thrown until one gets a 6?

    [**Hints.** Use the binomial distribution.] [**Ans.** 6.]

14. If 1,000 people each select a number at random between 1 and 500, what is the probability that 3 people will select the number 29? [**Ans.** 0.17.]

15. Three coins are tossed; what is the probability that two are heads and one tail? What is the probability that first two are heads and the third tail? [**Ans.** 3/8, 1/8.]

16. Find the probabilities that

    (a) in 100 tosses of a coin, exactly 50 are heads; [**Ans.** 0.08.]

    [**Hints.** Use binomial distribution.]

    (b) in 720 tosses of a die, 120 are aces; [**Ans.** 0.04.]

    (c) in 720 tosses of a die, between 100 and 140 are aces. [**Ans.** 0.96.]

17. Prove that $P(A+B+C) = P(A) + P(B) + P(C) - P(AB) - P(AC) - P(BC) + P(ABC)$.

18. In a family of 5 children, what is the probability that there are 2 boys and 3 girls? Also find the probability that the two oldest are boys and the other three girls. [**Ans.** 5/16, 1/32.]

19. Suppose you have 5 exams during the 5 days of the examination week. Find the probability that on a given day you have no exams, just 1 exam, 2 exams, 3 exams. [**Ans.** 0.37, 0.37, 0.18, 0.06.]

20. A bread salesman sells on the average 20 cakes on a round of his route. What is the probability that he sells an even number of cakes? [**Ans.** 0.50.]

[**Hints.** Assume the Poisson distribution.]

21. If a stick is broken into two at random, what is the average length of the smaller piece?

[**Ans.** $l/4$.]

22. Shuffle an ordinary deck of 52 playing cards containing 4 aces and then turn up the cards from the top until the first ace appears. On the average, how many cards are required to produce the first ace? [**Ans.** 9.6 cards.]

23. A coin is tossed 10,000 times; the results are 5176 heads and 4824 tails. Is this a reasonable result for a symmetric coin or is it a fairly conclusive evidence that the coin is defective?

[**Hints.** Calculate the total probability for more than 5176 heads in 10,000 tosses. To do this, a gaussian distribution may be assumed.]

24. If a set of measurements is distributed in accordance with a gaussian distribution, find the probability that any single measurement will fall between $m - \frac{\sigma}{2}$ and $m + \frac{\sigma}{2}$.

25. The *probable error* of a distribution is defined as the error such that the probability of occurrence of an error whose absolute value is less than this value by $\frac{1}{2}$. Determine the probable error for the gaussian distribution and express it as a multiple of σ.

26. An object undergoes a simple harmonic motion with amplitude A and angular frequency ω according to the equation $x = A \sin \omega t$, where x represents the displacement of the object from equilibrium. Calculate the mean and standard deviation of the position and of the speed of the object.

27. Among a large number of eggs, 1 per cent were found to be rotten. In a dozen eggs, what is the probability that none is rotten?

28. It is shown in the kinetic theory of gases that the probability for the distance, a molecule travels between successive collisions, to be between x and $x + dx$, is proportional to $e^{-x/\lambda} dx$, where λ is a constant. Show that the average distance between successive collisions—called the *mean free path*—is λ.

29. Show that the probability that a variate x differs from its mean value μ by more than any number a is less than σ^2/a^2, where σ is the standard deviation of x. (This is Chebyshev's inequality.)

30. If two variates x and y are distributed normally, prove that their sum $x + y$ is also distributed normally.

31. If n grains of wheat are scattered in a haphazard manner over a surface of S units of area, show that the probability that A units of area will contain R grains of wheat is

$$P = \left(\frac{An}{S}\right)^R \frac{e^{-An/S}}{R!}.$$

[**Hints.** ndS/S represents the infinitely small probability that the small space dS contains a grain of wheat. If the selected space be A units of area, we may suppose each dS to be a trial; the number of trials will therefore be A/dS. Hence we must substitute An/S for np in the Poisson distribution $P = (np)^r e^{-np}/r!$.]

32. A popular lady receives an average of four telephone calls a day. What is the probability that on a given day she will receive no telephone calls? Just one call? Exactly four calls?

33. The Poisson distribution is used mainly for *small* values of $\mu = np$. For large values of μ, the Poisson distribution (as well as the binomial) is fairly well approximated by the normal or gaussian distribution. Show that

$$\lim_{\mu\to\text{large}} \frac{\mu^x e^{-\mu}}{x!} \approx \frac{e^{-(x-\mu)^2/2\mu}}{(2\pi\mu)^{1/2}} \quad (\mu \text{ large})$$

[Hints. The peak of the graph of the Poisson distribution is at $x = \mu$; note that the approximating normal curve is shifted to have its centre at $x = \mu$. The equation given above gives a good approximation in the central region (around $x = \mu$), where the probability is large.]

34. Starting from Poisson distribution, prove that $\sum_{x=0}^{\infty} P(x) = 1$.

35. What is the probability that the ace of spades will be drawn from a deck of cards at least once in 104 consecutive trials? **[Ans.** 0.867—use Poisson distribution.]

36. The probability that a nail chosen at random from the output of a certain machine will be defective is 0.0025. What is the probability that a bag of 1000 nails made by the machine will have at most 3 defective nails? **[Ans.** 0.758.]

37. Find the probability that the first two chairs are vacant when four people are sitting in six chairs. **[Ans.** $1/{}^6C_4$.]

38. The number of particles emitted each minute by a radioactive source is recorded for a period of 10 hours; a total of 1800 counts is registered. During how many 1-minute intervals, should we expect to observe no particles, exactly one particle, etc.? (Use Poisson distribution.)

39. Find the probability of atmost 5 defective fuses to be found in a box of 200 fuses, if experience shows that 2% of such fuses are defective. **[Ans.** 0.785.]

[Hints. Use Poisson distribution.]

40. Write down the expression for standard normal (Gaussian) distribution function. Calculate the mean and variance for this distribution. (See Arts. 24.15, 24.15.1 and 24.15.2).

[C.U. (Hons.) 2005]

41. (a) A particular experiment performed gives either of two results r_1 and r_2, occurring with probabilities p_1 and p_2 respectively. The experiment is performed N times. Calculate the probability that for the above performance, the result r_1, will occur n times. What is the probability function called? (See Art. 24.13) **[C.U. (Hons.) 2006]**

(b) The Gaussian probability distribution is given by

$$P_G(x, \mu, \sigma) = \frac{1}{\sqrt{2\pi}\sigma} \exp\left[-\frac{1}{2}\left(\frac{x-\mu}{\sigma}\right)^2\right] \; -\infty < x < \infty$$

i. Sketch the graph of P_G vs. x and show that it has two points of inflexion at $x = \mu \pm \sigma$.

[Ans. For points of inflexion, $\dfrac{d^2 P_G}{dx^2} = 0$.

Now

$$\frac{dP_G}{dx} = -\frac{1}{\sigma\sqrt{2\pi}}\frac{(x-\mu)}{\sigma^2}e^{-\frac{1}{2}\left(\frac{x-\mu}{\sigma}\right)^2}$$

$$\therefore \quad \frac{d^2P_G}{dx^2} = \frac{1}{\sigma^3\sqrt{2\pi}}(x-\mu)\frac{(x-\mu)}{\sigma^2}e^{-\frac{1}{2}\left(\frac{x-\mu}{\sigma}\right)^2} - \frac{1}{\sigma^3\sqrt{2\pi}}e^{-\frac{1}{2}\left(\frac{x-\mu}{\sigma}\right)^2}$$

$$= \frac{1}{\sigma^3\sqrt{2\pi}}e^{-\frac{1}{2}\left(\frac{x-\mu}{\sigma}\right)^2}\left[\left(\frac{x-\mu}{\sigma}\right)^2 - 1\right] = 0$$

whence we have $\left(\frac{x-\mu}{\sigma}\right)^2 - 1 = 0$ if $\frac{d^2P_G}{dx^2} = 0$ or $(x-\mu)^2 = \sigma^2$ or $x - \mu = \pm\sigma$ or $x = \mu \pm \sigma$]

ii. **What is the area under the curve? What does it signify? (See Art. 24.15.2, 24.15.3).**

(c) A random variable x has a probability function $f(x)$. Define variance (dispersion) and standard deviation of x. See Arts. 24.10.1 and 24.10.2. [C.U. (Hons.) 2006]

(d) A radio-active substance decays with a fairly long half-life time. We observe such a system for an interval of time which is very small compared to the half-life time. Show that the probability of getting n counts during an interval of time 't' is given by

$$P_n(t) = \frac{(\mu t)^n}{n!}e^{-\mu t}$$

where μ = probability of one particle emitted per unit time. What is the identity of the above probability distribution? (See Art. 24.14) [C.U. (Hons.) 2006]

42. (a) Calculate the mean and variance for standard normal distribution. (See Arts. 24.15.1 and 24.15.2) [C.U. (Hons.) 2007]

(b) Write down the expression for standard normal distribution function. Calculate standard deviation for this distribution. (See Art. 24.15) [C.U. (Hons.) 2007]

43. (a) One card is drawn from a well-shuffled ordinary deck of 52 cards. Then another card is drawn from the remaining 51 cards. Find the probability that they are both aces. (See worked out examples).

(b) The weather record shows that on the average, Kolkata has 5 norwesters in 30 days of April. What is the binomial probability that a particular April has at least 4 norwesters? (See worked out examples) [C.U. (Hons.) 2008]

Appendix A

Vectors and Tensors

A.1 Introduction

A linear vector function $\vec{A}$ is a constant scalar multiple of another vector $\vec{B}$ and may be represented by

$$\vec{A} = a\vec{B} \tag{1}$$

where a is a constant. Using rectangular cartesian co-ordinates, if

$$\left.\begin{aligned} \vec{A} &= \hat{\imath}A_1 + \hat{\jmath}A_2 + \hat{k}A_3 \\ \text{and } \vec{B} &= \hat{\imath}B_1 + \hat{\jmath}B_2 + \hat{k}B_3 \end{aligned}\right\} \tag{2}$$

then the more general type of linear vector requires

$$A_1 = a_1 B_1, \quad A_2 = a_2 B_2 \quad \text{and} \quad A_3 = a_3 B_3.$$

Here we have used the suffixes, 1, 2, 3 for x, y, z. The vector $\vec{A}$ is different from $\vec{B}$ not only in magnitude, but also in direction. For the most general case, in which each component of $\vec{A}$ is a linear function of all the three components of $\vec{B}$, the relations between the components in rectangular co-ordinates may be written as

$$\left.\begin{aligned} A_1 &= a_{11}B_1 + a_{12}B_2 + a_{13}B_3 \\ A_2 &= a_{21}B_1 + a_{22}B_2 + a_{23}B_3 \\ A_3 &= a_{31}B_1 + a_{32}B_2 + a_{33}B_3 \end{aligned}\right\} \tag{3}$$

The above expression may be written in a reduced form given by

$$A_i = \sum_j a_{ij} B \tag{4}$$

The nine coefficients a_{ij} characterise the transformation of the components of one vector into the components of the other vector. The scheme of coefficients a_{ij} has an independent meaning if the correspondence of vector $\vec{A}$ with vector $\vec{B}$ is independent of a particular co-ordinate system in which the vectors are resolved into components. In this case, the coefficients a_{ij} are called the components of a *tensor.* A *vector,* on the other hand, is a physically directed quantity in space, the components of which depend on the choice of axes.

A.2 Symmetric and Antisymmetric Tensors

The equations (A.1.3) may also be written in the following form known as the matrix form:

$$\begin{bmatrix} A_1 \\ A_2 \\ A_3 \end{bmatrix} = \begin{bmatrix} a_{11} & a_{12} & a_{13} \\ a_{21} & a_{22} & a_{23} \\ a_{31} & a_{32} & a_{33} \end{bmatrix} \begin{bmatrix} B_1 \\ B_2 \\ B_3 \end{bmatrix}. \tag{1}$$

Here $a_{11}, a_{12}, \cdots, a_{33}$ are called the components of a tensor of rank two. The tensor is given by

$$a_{ij} = \begin{bmatrix} a_{11} & a_{12} & a_{13} \\ a_{21} & a_{22} & a_{23} \\ a_{31} & a_{32} & a_{33} \end{bmatrix} \tag{2}$$

A tensor of which $a_{ij} = a_{ji}$ for all i and j, is called a symmetric tensor. A tensor is said to be antisymmetric, if for all i and j, $a_{ij} = -a_{ji}$. The diagonal elements of such a tensor, i.e., a_{11}, a_{22} and a_{33} vanish.

A.3 Scalars, Vectors and Tensors

A vector may be called a tensor of rank one since it is connected with one axis and a scalar may be called a tensor of zero rank as it has no connection with any axis. A component of a tensor of rank two is connected with two axes of which one belongs to each vector. A component of a tensor of rank three and four, on the other hand, is connected with three and four axes, and so on.

A.4 The Origin and Concept of Tensor Calculus

Tensor Calculus—also called Absolute Differential Calculus—is a generalisation of the Differential Geometry of Gauss and Riemann, and systematically developed by Levi-Civita in his book under the title *Absolute Differential Calculus*. Einstein found it to be a most suitable tool for his General Relativity. The reason is that a physicist requires a formulation of the laws of physics which remain the same (i.e., *invariant*) when we pass on from one observer to any other. One of the main objects of Tensor Calculus is the study of the change of an expression, when we pass on *from one system of co-ordinates to any other.*

The laws of Newtonian mechanics do not possess this invariance. In fact they are formulated by making use of inertial frames, in which the acceleration of a particle, not subjected to any force will vanish, if it is appreciated w.r.t. the fixed axes. Again, in a rotating frame of reference, the acceleration of the particle does not vanish, but it moves as if it were under the action of a *fictitious force*—the centrifugal force. Einstein wanted a formulation of the laws of gravitation, where this *distinction between real and fictitious forces disappear.*

Again, if the rectangular co-ordinates of a point $P(x, y)$ change to (x', y') in a rotation of amplitude θ, about the origin O, then $x = x'\cos\theta - y'\sin\theta$, $y = x'\sin\theta + y'\cos\theta$, so that $x^2 + y^2 = (x'\cos\theta - y'\sin\theta)^2 + (x'\sin\theta + y'\cos\theta)^2 = x'^2 + y'^2$.

It is interesting to note that both sides of this equation express the same *geometric quantity*—the square of the distance OP.

The above example may be used to introduce the fundamental idea of the Theory of Relativity. The above two systems of axes Oxy and $Ox'y'$ may be considered as *two observers*, who are looking at a *geometrical object—the segment* OP. The length of this segment has a value independent of the particular observer. This is the reason why we must have $x^2+y^2 = x'^2+y'^2$ = an invariant.

Einstein has generalised this idea into *the Principle of Relativity—the mathematical formulation of the laws of gravitation and electricity must be the same, whatever be the system of reference.*

Clearly, this does not apply to the Laws of Newtonian Mechanics—which are true only if the accelerations are evaluated w.r.t. the fixed axes. But we know that on account of Michelson's experiment, the laws of Newton are only an *approximation*—valid only if the velocities of the particles are *small* compared to the velocity of light. With the aid of Tensor Calculus, Einstein gave the exact formulation, which does satisfy the principle of relativity. A similar difficulty arises in electro-magnetic field theory, and is overcome by giving a tensorial form to the laws of Maxwell.

A.5 Definitions—Contravariant and Covariant vectors

Contravariant and Covariant vectors: We have seen above that tensors are important in many areas of physics, including general relativity and electro-dynamics. Scalars and vectors are special cases of tensors. A quantity that does not change under rotations of the coordinate system in three dimensional space i.e. *an invariant* is called *a scalar*. A scalar is specified by one real number and is *a tensor of rank zero*. A quantity whose components transform under rotations like those of the distance of a point from a chosen origin is called *a vector*. The transformation of the components of the vector under *a rotation of the coordinates* preserves the vector as a geometric entity (such as an arrow in space), independent of the orientation of the reference frame. In three-dimensional space, a vector is specified by 3 = 3 real numbers, for example, its cartesian components, and is a *tensor of rank one*. A *tensor of rank n* has 3^n components which transform in a definite way*.

There is a possible ambiguity in the transformation law of vector

$$A'_i = \sum_j a_{ij} A_j \tag{1}$$

in which a_{ij} is the cosine of the angle between x'_i-axis and the x_j-axis.

If we start with a differential distance vector $d\vec{r}$, then, taking dx'_i to be a function of the unprimed variables,

$$dx'_i = \sum_j \frac{\partial x'_i}{\partial x_j} dx_j \tag{2}$$

by partial differentiation. If we set

$$a_{ij} = \frac{\partial x'_i}{\partial x_j}, \tag{3}$$

*In N-dimensional space a tensor of rank n has N^n components.

eqs. (A.5.1) and (A.5.2) are consistent. Any set of quantities A_j transforming according to

$$A'_i = \sum_j \frac{\partial x'_i}{\partial x_j} A_j \tag{4}$$

is defined as *a contravariant vector*.

However, we are already familiar with a slightly different type of vector transformation. The gradient of a scalar, $\vec{\nabla}\phi$, defined by

$$\vec{\nabla}\phi = \hat{x}\frac{\partial \phi}{\partial x_1} + \hat{y}\frac{\partial \phi}{\partial x_2} + \hat{z}\frac{\partial \phi}{\partial x_3} \tag{5}$$

(using x_1, x_2, x_3 for x, y, z), transforms as

$$\frac{\partial \phi'}{\partial x'_i} = \sum_j \frac{\partial \phi}{\partial x_j}\frac{\partial x_j}{\partial x'_i}, \tag{6}$$

using $\phi = \phi(x, y, z) = \phi(x', y', z') = \phi', \phi$ defined as a scalar quantity. It is to be noted that this differs from Eq. (A.5.4) in that we have $\partial x_j/\partial x'_i$ instead of $\partial x'_i/\partial x_j$. Equation (A.5.6) is taken as the definition of *a covariant vector* with the gradient as the prototype.

In cartesian coordinates

$$\frac{\partial x_j}{\partial x'_i} = \frac{\partial x'_i}{\partial x_j} = a_{ij}, \tag{7}$$

and there is no difference between contravariant and covariant transformations. In other systems eq. (A.5.7) in general does not apply.

Appendix B

Mechanics and Central Force-field Problems

B.1 Transformation of Co-ordinates through matrix operation—Rotation of co-ordinates in two dimensions

Transformation of physical quantities expressed in one co-ordinate system to another can be obtained through transformation matrix as discussed below. We consider rotation of a co-ordinate frame in two dimensions as shown in the figure 1. Any point P whose co-ordinates w.r.t axes OX and OY are (x, y) has co-ordinates (x', y') w.r.t the co-ordinate frame with axes OX' and OY' obtained by rotation of OX and OY through an angle θ. We may then write, with reference to the figure 1.

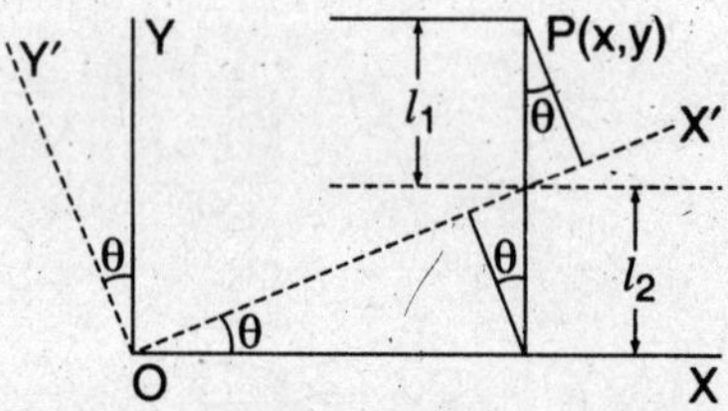

Fig 1

$$
\begin{aligned}
x' &= x\cos\theta + l_2\sin\theta + l_1\sin\theta \\
&= x\cos\theta + (l_1 + l_2)\sin\theta \\
&= x\cos\theta + y\sin\theta.
\end{aligned}
$$

Similarly

$$y' = y\cos\theta - x\sin\theta = -x\sin\theta + y\cos\theta.$$

The above two relations, can be equivalently, written in matrix notation as

$$\begin{bmatrix} x' \\ y' \end{bmatrix} = \begin{bmatrix} \cos\theta & \sin\theta \\ -\sin\theta & \cos\theta \end{bmatrix} \begin{bmatrix} x \\ y \end{bmatrix} = [T] \begin{bmatrix} x \\ y \end{bmatrix} \tag{1}$$

Equation (B.1.1) gives us the transformation matrix $[T]$—also known as the *rotational* matrix—relating the fixed co-ordinates with rotating co-ordinates. Inverting the rotation matrix, x and y can be expressed in terms of x' and y'. Writing $x = x_1$ and $y = x_2$, the above transformation equations become

$$
\begin{aligned}
x_1' &= x_1\cos\theta + x_2\sin\theta \quad \text{or} \quad x_1' = a_{11}x_1 + a_{12}x_2 \\
\text{and} \quad x_2' &= -x_1\sin\theta + x_2\cos\theta \quad \text{or} \quad x_2' = -a_{21}x_1 + a_{22}x_2
\end{aligned}
$$

which can be equivalently expressed in matrix notation as

$$\begin{bmatrix} x_1' \\ x_2' \end{bmatrix} = \begin{bmatrix} a_{11} & a_{12} \\ -a_{21} & a_{22} \end{bmatrix} \begin{bmatrix} x_1 \\ x_2 \end{bmatrix} \tag{2}$$

where $a_{11} = \cos(x_1', x_1)$, $a_{12} = \cos(x_1', x_2)$, $a_{21} = \cos(x_2', x_1)$, $a_{22} = \cos(x_2', x_2)$. In general $a_{ij} = \cos(x_i', x_j)$ and from figure 1, we have $\cos(x_1', x_1) = \cos\theta = a_{11}$, $\cos(x_1', x_2) = \cos\left(\frac{\pi}{2} - \theta\right) = \sin\theta = a_{12}$, $\cos(x_2', x_1) = \cos\left(\frac{\pi}{2} + \theta\right) = -\sin\theta = a_{21}$; $\cos(x_2', x_2) = \cos\theta = a_{22}$. In case of three dimensions, $(x_1, x_2, x_3) \to (x_1', x_2', x_3')$ and this transformation is represented by

$$\begin{bmatrix} x_1' \\ x_2' \\ x_3' \end{bmatrix} = \begin{bmatrix} a_{11} & a_{12} & a_{13} \\ a_{21} & a_{22} & a_{23} \\ a_{31} & a_{32} & a_{33} \end{bmatrix} \begin{bmatrix} x_1 \\ x_2 \\ x_3 \end{bmatrix} \tag{3}$$

where $a_{ij} = \cos(x_i', x_j)$.

If, however, rotation takes place about Z-axis through an angle of $\frac{\pi}{2}$, as represented by the attached figure 2, then we may write

$$\begin{aligned} a_{11} &= \cos(x_1', x_1) = \cos\tfrac{\pi}{2} = 0, & a_{12} &= \cos(x_1', x_2) = \cos 0^\circ = 1, & a_{13} &= \cos(x_1', x_3) = \cos\tfrac{\pi}{2} = 0 \\ a_{21} &= \cos(x_2', x_1) = \cos\pi = -1, & a_{22} &= \cos(x_2', x_2) = \cos\tfrac{\pi}{2} = 0, & a_{23} &= \cos(x_2', x_3) = \cos\tfrac{\pi}{2} = 0 \\ a_{31} &= \cos(x_3', x_1) = \cos\tfrac{\pi}{2} = 0, & a_{32} &= \cos(x_3', x_2) = \cos\tfrac{\pi}{2} = 0, & a_{33} &= \cos(x_3', x_3) = \cos 0^\circ = 1 \end{aligned}$$

so that the transformation or rotation matrix is written as

$$[R] = \begin{bmatrix} a_{11} & a_{12} & a_{13} \\ a_{21} & a_{22} & a_{23} \\ a_{31} & a_{32} & a_{33} \end{bmatrix} = \begin{bmatrix} 0 & 1 & 0 \\ -1 & 0 & 0 \\ 0 & 0 & 1 \end{bmatrix} \tag{4}$$

$\therefore$ from (B.1.3) we have

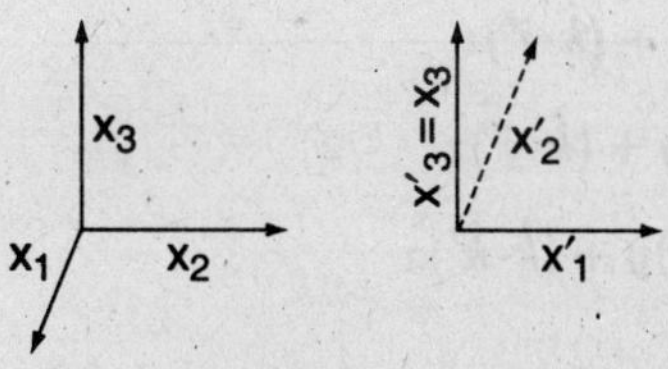

Fig 2

$$\begin{bmatrix} x_1' \\ x_2' \\ x_3' \end{bmatrix} = \begin{bmatrix} 0 & 1 & 0 \\ -1 & 0 & 0 \\ 0 & 0 & 1 \end{bmatrix} \begin{bmatrix} x_1 \\ x_2 \\ x_3 \end{bmatrix} = \begin{bmatrix} x_2 \\ -x_1 \\ x_3 \end{bmatrix}$$

so that we may write $x_1' = x_2$, $x_2' = -x_1$, $x_3' = x_3$.

If the rotation about Z-axis is through an angle θ, as shown in figure 3 such that $a_{11} = \cos\theta$, $a_{12} = \cos(90 - \theta) = \sin\theta$, $a_{13} = 0$, $a_{21} = \cos\left(\frac{\pi}{2} + \theta\right) = -\sin\theta$, $a_{22} = \cos\theta$, $a_{23} = 0$, $a_{31} = 0$, $a_{32} = 0$, $a_{33} = 1$, we may write the rotation matrix as

$$[R] = \begin{bmatrix} \cos\theta & \sin\theta & 0 \\ -\sin\theta & \cos\theta & 0 \\ 0 & 0 & 1 \end{bmatrix} \tag{5}$$

For arbitrary rotation of the co-ordinate axes, we consider a fixed space-set of axes defined by unit vectors $\hat{i}, \hat{j}, \hat{k}$ and a body-set of axes defined by unit vectors $\hat{i}', \hat{j}', \hat{k}'$ as shown in figure 4;

we have then $\vec{r} = \hat{\imath}x + \hat{\jmath}y + \hat{k}z = \hat{\imath}'x' + \hat{\jmath}'y' + \hat{k}'z'$, whence

$$
\begin{aligned}
x &= \vec{r}\cdot\hat{\imath} = (\hat{\imath}'x' + \hat{\jmath}'y' + \hat{k}'z')\cdot\hat{\imath} = (\hat{\imath}'\cdot\hat{\imath})x' + (\hat{\jmath}'\cdot\hat{\imath})y' + (\hat{k}'\cdot\hat{\imath})z' \\
y &= \vec{r}\cdot\hat{\jmath} = (\hat{\imath}'x' + \hat{\jmath}'y' + \hat{k}'z')\cdot\hat{\jmath} = (\hat{\imath}'\cdot\hat{\jmath})x' + (\hat{\jmath}'\cdot\hat{\jmath})y' + (\hat{k}'\cdot\hat{\jmath})z' \\
\text{and } z &= \vec{r}\cdot\hat{k} = (\hat{\imath}'x' + \hat{\jmath}'y' + \hat{k}'z')\cdot\hat{k} = (\hat{\imath}'\cdot\hat{k})x' + (\hat{\jmath}'\cdot\hat{k})y' + (\hat{k}'\cdot\hat{k})z'.
\end{aligned}
$$

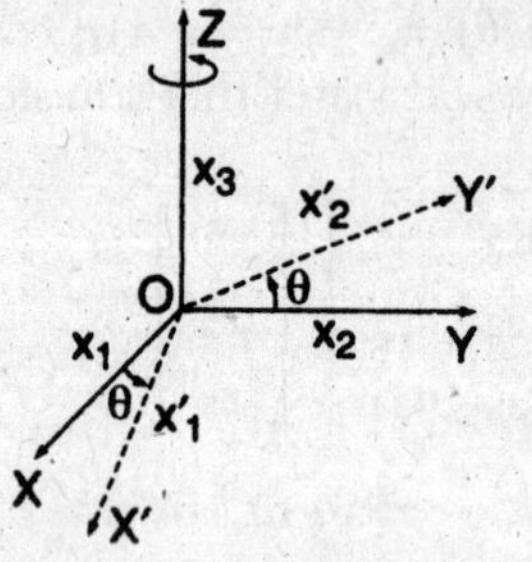

Fig 3

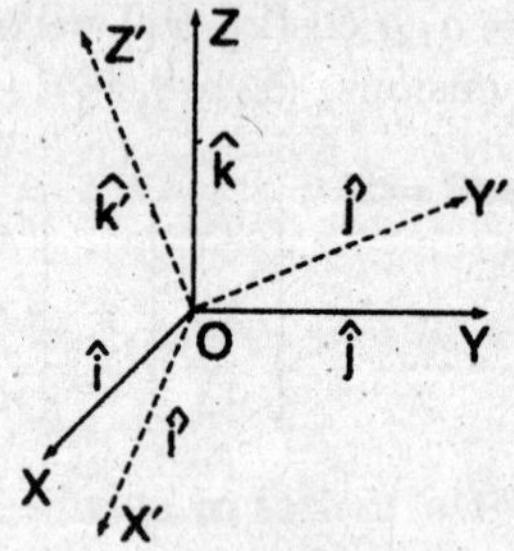

Fig 4

The above set of equations can equivalently be written in the matrix notation, as

$$
\begin{bmatrix} x \\ y \\ z \end{bmatrix} = \begin{bmatrix} (\hat{\imath}'\cdot\hat{\imath}) & (\hat{\jmath}'\cdot\hat{\imath}) & (\hat{k}'\cdot\hat{\imath}) \\ (\hat{\imath}'\cdot\hat{\jmath}) & (\hat{\jmath}'\cdot\hat{\jmath}) & (\hat{k}'\cdot\hat{\jmath}) \\ (\hat{\imath}'\cdot\hat{k}) & (\hat{\jmath}'\cdot\hat{k}) & (\hat{k}'\cdot\hat{k}) \end{bmatrix} \begin{bmatrix} x' \\ y' \\ z' \end{bmatrix} = [R] \begin{bmatrix} x' \\ y' \\ z' \end{bmatrix} \tag{6}
$$

where $[R]$ is the rotation matrix, and is given by $[R] = \begin{bmatrix} a_{11} & a_{12} & a_{13} \\ a_{21} & a_{22} & a_{23} \\ a_{31} & a_{32} & a_{33} \end{bmatrix}$

Interchanging $\hat{\imath}$, $\hat{\jmath}$, $\hat{k}$ with $\hat{\imath}'$, $\hat{\jmath}'$, $\hat{k}'$, we may have

$$
\begin{aligned}
x' &= \vec{r}\cdot\hat{\imath}' = (\hat{\imath}x + \hat{\jmath}y + \hat{k}z)\cdot\hat{\imath}' = (\hat{\imath}\cdot\hat{\imath}')x + (\hat{\jmath}\cdot\hat{\imath}')y + (\hat{k}\cdot\hat{\imath}')z \\
y' &= \vec{r}\cdot\hat{\jmath}' = (\hat{\imath}x + \hat{\jmath}y + \hat{k}z)\cdot\hat{\jmath}' = (\hat{\imath}\cdot\hat{\jmath}')x + (\hat{\jmath}\cdot\hat{\jmath}')y + (\hat{k}\cdot\hat{\jmath}')z \\
z' &= \vec{r}\cdot\hat{k}' = (\hat{\imath}x + \hat{\jmath}y + \hat{k}z)\cdot\hat{k}' = (\hat{\imath}\cdot\hat{k}')x + (\hat{\jmath}\cdot\hat{k}')y + (\hat{k}\cdot\hat{k}')z
\end{aligned}
$$

The above set of equations can similarly be written as

$$
\begin{bmatrix} x' \\ y' \\ z' \end{bmatrix} = \begin{bmatrix} (\hat{\imath}\cdot\hat{\imath}') & (\hat{\jmath}\cdot\hat{\imath}') & (\hat{k}\cdot\hat{\imath}') \\ (\hat{\imath}\cdot\hat{\jmath}') & (\hat{\jmath}\cdot\hat{\jmath}') & (\hat{k}\cdot\hat{\jmath}') \\ (\hat{\imath}\cdot\hat{k}') & (\hat{\jmath}\cdot\hat{k}') & (\hat{k}\cdot\hat{k}') \end{bmatrix} \begin{bmatrix} x \\ y \\ z \end{bmatrix} = \begin{bmatrix} a_{11} & a_{12} & a_{13} \\ a_{21} & a_{22} & a_{23} \\ a_{31} & a_{32} & a_{33} \end{bmatrix} \begin{bmatrix} x \\ y \\ z \end{bmatrix}
$$

where

$$
[R] = \begin{bmatrix} a_{11} & a_{12} & a_{13} \\ a_{21} & a_{22} & a_{23} \\ a_{31} & a_{32} & a_{33} \end{bmatrix}
$$

is the rotation matrix.

B.2 Calculation of the centre of mass (C.M) in some important cases

Case 1 : C.M of a uniform semi-circular wire of radius a

From symmetry, as shown in the attached figure, we see that the C.M should be on the Y-axis, so that $\bar{x} = 0$. If ρ be the mass per unit length of the wire, then an elementary arc (on the wire) between θ and $\theta + d\theta$, as shown in the figure, has length $dl = a d\theta$, so that we may write

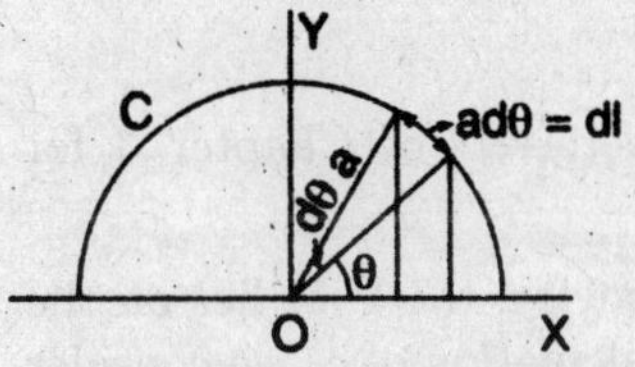

$$\bar{y} = \frac{\int_c y dl \rho}{\int_c dl \rho} = \frac{\int_0^{\pi} (a \sin\theta) a d\theta}{\int_0^{\pi} a d\theta} = \frac{2a^2}{\pi a} = \frac{2a}{\pi}$$

Case 2 : C.M of a uniform semi-circular region of radius a

Method 1 : Using rectangular co-ordinates. Considering the region as shown in the attached figure (a), we see that, if ρ be the mass per unit area, assumed constant, the co-ordinates of the C.M are given by

$$\bar{x} = \frac{\int_R x\rho dA}{\int_R \rho dA} = \frac{\iint_R x dx dy}{\iint_R dx dy} = \frac{\int_{-a}^{+a} x dx \int_0^{\sqrt{a^2-x^2}} dy}{\int_{-a}^{+a} dx \int_0^{\sqrt{a^2-x^2}} dy} = 0$$

[$\because$ the equation of the circle C is $x^2 + y^2 = a^2$, we have $y = \sqrt{a^2 - x^2}$]

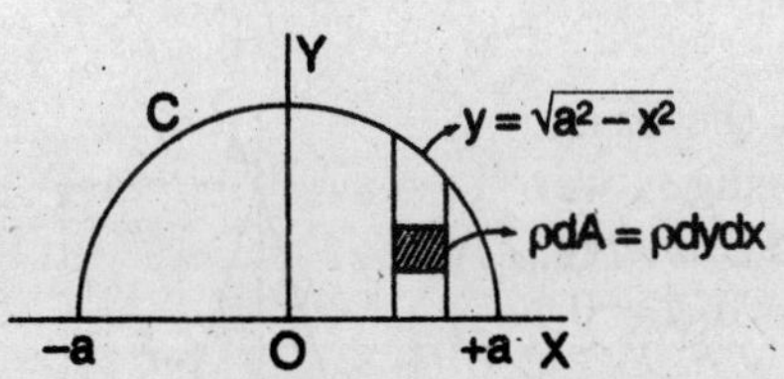

Fig (a)

$$\bar{y} = \frac{\int_R y\rho dA}{\int_R \rho dA} = \frac{\iint_R y dx dy}{\iint_R dx dy} = \frac{\int_{-a}^{+a} dx \int_0^{\sqrt{a^2-x^2}} y dy}{\int_{-a}^{+a} dx \int_0^{\sqrt{a^2-x^2}} dy}$$

$$= \frac{2a^3/3}{\pi a^2/2} = \frac{4a}{3\pi}$$

It is to be noted that, since by symmetry the C.M is on the Y-axis, we can readily write $\bar{x} = 0$. Again the denominator for $\bar{y}$ simply represents the semi-circular area which is $\frac{\pi a^2}{2}$ and as such there is no necessity for its evaluation by integration.

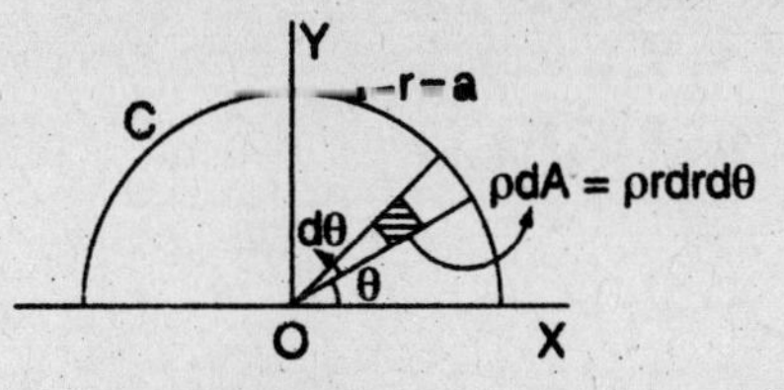

Fig (b)

Method 2 : Using plane polar co-ordinates. Considering a semi-circular region of radius a as shown in the attached figure (b), we see that, by symmetry, the C.M lies on the Y-axis, so that $\bar{x} = 0$. Since $y = r\sin\theta$ and an area dA [indicated between r and $r + dr$ and between θ and $\theta + d\theta$ in the figure (b)] $= r dr d\theta$ in polar co-ordinates, we can write

$$\bar{y} = \frac{\int_R y\rho dA}{\int_R \rho dA} = \frac{\int_{\theta=0}^{\pi}\int_{r=0}^{a}(r\sin\theta)rdrd\theta}{\int_{\theta=0}^{\pi}\int_{r=0}^{a} rdrd\theta} \quad \left[\begin{array}{l}\because \text{the equation of the circle} \\ \text{is } x^2+y^2=a^2 \text{ or} \\ r^2\cos^2\theta + r^2\sin^2\theta = a^2 \\ \text{or } r=a\end{array}\right]$$

$$= \frac{2a^2/3}{\pi a^2/2} = \frac{4a}{3\pi}$$

Case 3 : C.M of a uniform solid hemisphere of radius a

Method 1 : Using rectangular co-ordinates. See "Exercises" of Chapter 3 for calculation

Method 2 : Using spherical polar co-ordinates. By symmetry, the C.M lies on the Z-axis as shown in the attached figure, so that $\bar{x}=0$, $\bar{y}=0$. For calculation of $\bar{z}$, we consider an elementary volume dV defined between r and $r+dr$, ϕ and $\phi+d\phi$, θ and $\theta+d\theta$, so that $dV = r^2\sin\theta d\theta d\phi dr$

$$\therefore \bar{z} = \frac{\int_V z\rho dV}{\int_V \rho dV} = \frac{\iiint r\cos\theta \times r^2\sin\theta d\theta d\phi dr}{\iiint r^2\sin\theta d\theta d\phi dr} \quad \left[\begin{array}{l}\rho = \text{mass per} \\ \text{unit volume}\end{array}\right]$$

$$= \frac{\int_0^a r^2 dr\int_0^{\pi/2}\sin\theta\cos\theta d\theta\int_0^{2\pi}d\phi}{\int_0^a r^2 dr\int_0^{\pi/2}\sin\theta d\theta\int_0^{2\pi}d\phi} = \frac{\frac{a^4}{4}\times\frac{1}{2}\times 2\pi}{\frac{a^3}{3}\times 1\times 2\pi} = \frac{3}{8}a$$

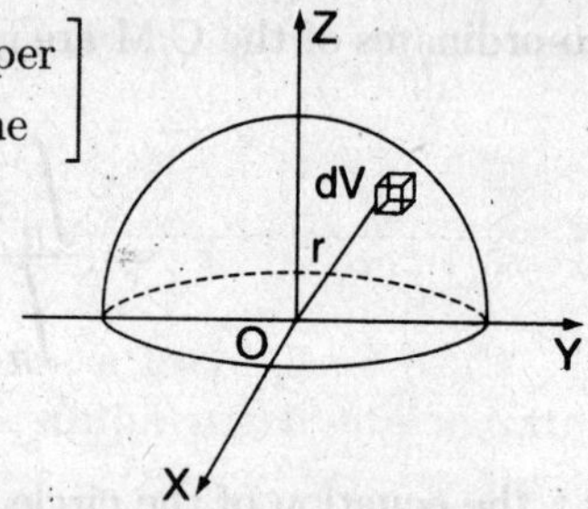

Case 4 : C.M of a uniform right circular solid cone of height h and radius a of the base

Considering a section between heights z and $z+dz$ from the base, (as shown in the attached figure), the radius of the section being r, we see that the volume enclosed is $dV = \pi r^2 dz$. Again since, by symmetry the C.M will be on the line OA along the Z-axis, we have $\bar{x}=0, \bar{y}=0$, so that

$$\bar{z} = \frac{\int_0^h \pi r^2 dz\rho z}{\int_0^h \pi r^2 dz\rho}, \quad (\rho = \text{mass per unit volume})$$

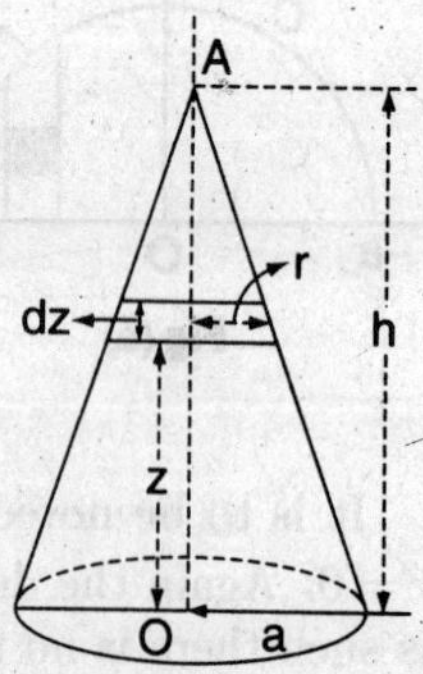

Now from the geometry of the figure, we have $\frac{r}{a} = \frac{h-z}{h}$ or $r = \frac{h-z}{h}a$, so that the numerator of the above expression for the co-ordinate of C.M

$$= \pi\int_0^h a^2\frac{(h-z)^2}{h^2}zdz = \frac{\pi a^2}{h^2}\int_0^h(h^2z - 2hz^2 + z^3)dz$$

$$= \frac{\pi a^2}{h^2}\left(h^2\frac{h^2}{2} - 2h\frac{h^3}{3} + \frac{h^4}{4}\right) = \frac{\pi a^2h^2}{12};$$

similarly the denominator

$$= \pi \int_0^h a^2 \frac{(h-z)^2}{h^2} dz = \frac{\pi a^2}{h^2} \int_0^h (h^2 - 2hz + z^2) dz = \frac{1}{3}\pi a^2 h$$

$$\therefore \quad \bar{z} = \frac{\pi a^2 h^2/12}{\pi a^2 h/3} = \frac{h}{4}$$

i.e., the C.M will be at a distance of $\frac{3}{4}h$ from the vertex.

Case 5. C.M of a solid of constant density consisting of *a* right circular cone of radius *a* of the base and height *h*, surmounted by a hemisphere of radius *a*.

By symmetry, the C.M should lie on the axis of symmetry, i.e., the Z-axis, so that $\bar{x} = 0$, $\bar{y} = 0$. The cone has mass $M_1 = \frac{1}{3}\pi a^2 h\rho$, having the C.M, C_1 at a height $\bar{z}_1 = \frac{3}{4}h$ w.r.t O. The hemisphere has mass $M_2 = \frac{2}{3}\pi a^3 \rho$, having its C.M, C_2 at a height $\bar{z}_2 = \left(h + \frac{3}{8}a\right)$ w.r.t O. Therefore the co-ordinate $\bar{z}$ of the C.M, C of the solid w.r.t O is given by (as shown in the attached figure)

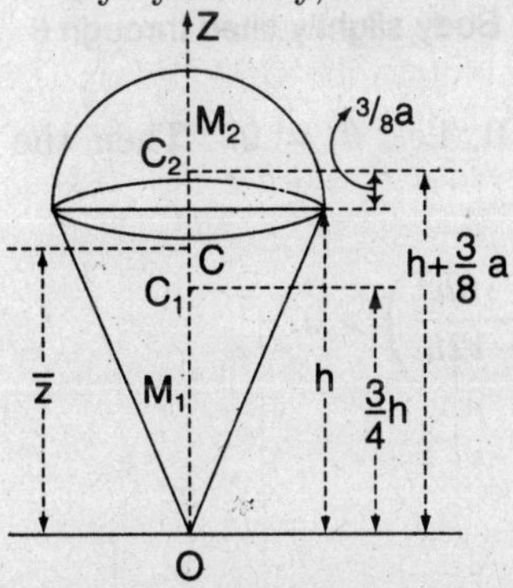

$$\bar{z} = \frac{M_1\bar{z}_1 + M_2\bar{z}_2}{M_1 + M_2} = \frac{\left(\frac{1}{3}\pi a^2 h\rho \times \frac{3}{4}h\right) + \left[\frac{2}{3}\pi a^3\rho \times \left(h + \frac{3}{8}a\right)\right]}{\left(\frac{1}{3}\pi a^2 h\rho + \frac{2}{3}\pi a^3 \rho\right)} \; (\rho = \text{mass/vol.})$$

$$= \frac{\frac{3}{4}h^2 + 2ah + \frac{3}{4}a^2}{h + 2a} = \frac{\frac{3}{4}(h^2 + a^2) + 2ah}{h + 2a}$$

Case 6. C.M of a solid of constant density consisting of a cylinder of radius *a* and height *h* surmounted by a hemisphere of radius *a*, as shown in the attached figure.

From symmetry, the C.M of the body should lie on the axis OZ of symmetry, so that $\bar{x} = 0$, $\bar{y} = 0$. Let $\bar{z}$ be the co-ordinate of the C.M, C of the body, w.r.t the base. The hemisphere has a mass $M_1 = \frac{2}{3}\pi a^3 \rho$ having its C.M, C_1 at a distance $\bar{z}_1 = \left(h + \frac{3}{8}a\right)$ from the base. Again the cylinder has a mass $M_2 = \pi a^2 h\rho$, having its C.M, C_2 at a distance $\bar{z}_2 = \frac{h}{2}$ from the base (ρ = mass per unit volume).

Therefore the co-ordinate of the C.M, C of the body, w.r.t the base, is given by

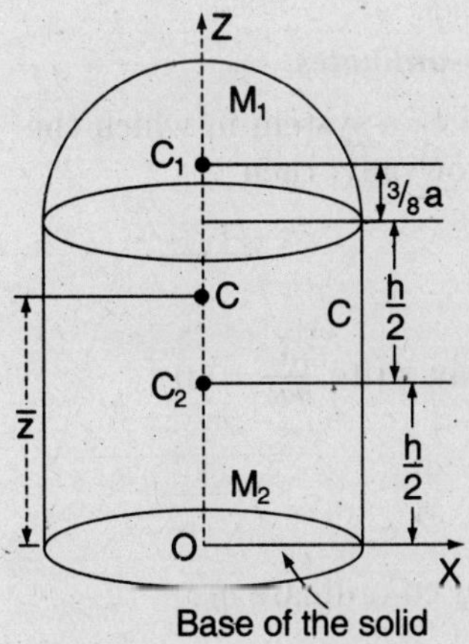

$$\bar{z} = \frac{M_1\bar{z}_1 + M_2\bar{z}_2}{M_1 + M_2} = \frac{\left(\frac{2}{3}\pi a^3 \rho\right)\left(h + \frac{3}{8}a\right) + \left(\pi a^2 h\rho\right)\left(\frac{h}{2}\right)}{\frac{2}{3}\pi a^3\rho + \pi a^2 h\rho}$$

$$= \frac{\frac{h^2}{2} + \frac{2ah}{3} + \frac{a^2}{4}}{\left(h + \frac{2}{3}a\right)}$$

$$= \frac{3a^2 + 8ah + 6h^2}{8a + 12h} \tag{1}$$

N.B. If the solid body under consideration is placed in an inverted position with its axis slightly displaced through an angle θ as shown in the attached figure, then we can show in the following way that the solid body will be in *stable equilibrium* on the horizontal plane, if and only if $a/h > \sqrt{2}$.

From (B.2.1), the C.M, C, in the inverted position, is at a distance CB (from the centre B of the hemisphere) given by

$$h - \frac{3a^2 + 8ah + 6h^2}{8a + 12h} = \frac{6h^2 - 3a^2}{8a + 12h}$$

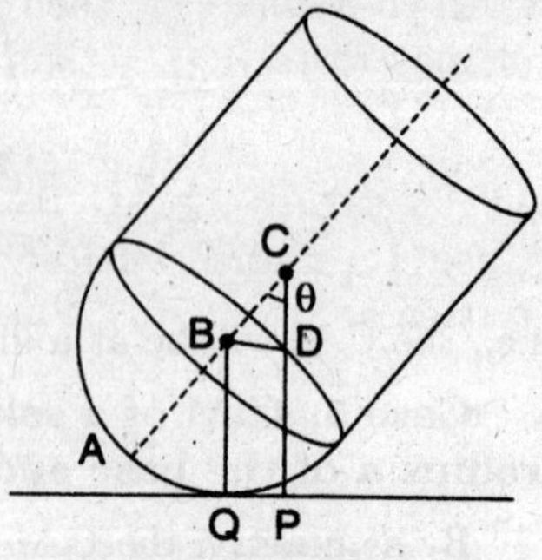

Body slightly tilted through θ

The distance of the C.M, C above the plane is then

$$CP = CD + DP = CB\cos\theta + BQ = \frac{6h^2 - 3a^2}{8a + 12h}\cos\theta + a,$$

so that the potential energy (or potential) of the system is

$$U = Mg\left(\frac{6h^2 - 3a^2}{8a + 12h}\cos\theta + a\right)$$

Equilibrium occurs where $\frac{\partial U}{\partial \theta} = 0$ or $Mg\left(\frac{3a^2 - 6h^2}{8a + 12h}\right)\sin\theta = 0$, i.e., $\theta = 0$. Then the *equilibrium* will be *stable* if

$$\left.\frac{\partial^2 U}{\partial \theta^2}\right|_{\theta=0} = Mg\left(\frac{3a^2 - 6h^2}{8a + 12h}\right)\cos\theta\bigg|_{\theta=0} = Mg\left(\frac{3a^2 - 6h^2}{8a + 12h}\right) > 0,$$

when U is *minimum*, i.e., $3a^2 - 6h^2 > 0$ or $a > \sqrt{2}h$ or $a/h > \sqrt{2}$.

Again the *equilibrium* will be *unstable* if

$$\left.\frac{\partial^2 U}{\partial \theta^2}\right|_{\theta=0} = Mg\left(\frac{3a^2 - 6h^2}{8a + 12h}\right) < 0,$$

when U is *maximum*, i.e., $3a^2 - 6h^2 < 0$ or $a/h < \sqrt{2}$.

B.3 Lagrangian formulation of conservation theorems

For the purpose, we first define *generalised momentum* and *cyclic co-ordinates*.

Generalised momentum: We consider a conservative system i.e. a system in which the forces are derivable from a potential function V dependent on position only; then

$$\frac{\partial L}{\partial \dot{q}_k} = \frac{\partial (T - V)}{\partial \dot{q}_k} = \frac{\partial T}{\partial \dot{q}_k} - \frac{\partial V}{\partial \dot{q}_k} = \frac{\partial T}{\partial \dot{q}_k}$$

(since potential V depends only on position, we may write $\frac{\partial V}{\partial \dot{q}_k} = 0$)

$$= \frac{\partial}{\partial \dot{q}_k}\sum_k \frac{1}{2} m_k \dot{q}_k^2 = m_k \dot{q}_k = p_k$$

$=$ Generalised momentum associated with generalised co-ordinate q_k.

Thus the generalised momentum corresponding to the generalised co-ordinate q_k is defined as the quantity $\frac{\partial L}{\partial \dot{q}_k}$ and is represented by p_k, i.e.,

$$p_k = \frac{\partial L}{\partial \dot{q}_k}. \tag{1}$$

Cyclic or ignorable co-ordinates: If the Lagrangian of a system is not a function of a given co-ordinate q_k, then that co-ordinate q_k is said to be *cyclic* or *ignorable.* In such a case, we may write

$$\frac{\partial L}{\partial q_k} = 0. \tag{2}$$

Case 1. Conservation theorem for generalised momentum: Lagrange's equation is written as

$$\frac{d}{dt}\left(\frac{\partial L}{\partial \dot{q}_k}\right) - \frac{\partial L}{\partial q_k} = 0. \tag{3}$$

If the generalised co-ordinate q_k is cyclic or ignorable, then from (B.3.2) we have $\dfrac{\partial L}{\partial q_k} = 0$, so that equation (B.3.3) reduces to

$$\frac{d}{dt}\left(\frac{\partial L}{\partial \dot{q}_k}\right) = 0 \quad \text{or} \quad \frac{\partial L}{\partial \dot{q}_k} = p_k = \text{constant; [using (B.3.1)]}$$

This gives the conservation theorem for generalised momentum which states that *the generalised momentum corresponding to a cyclic or ignorable co-ordinate is conserved.*

Case 2. Conservation theorem for energy: We consider a conservative system i.e. a system in which the forces are derivable from a potential function V dependent on position only. The Lagrangian L is given by

$$L = L(q_1, q_2 \cdots q_k, \cdots, \dot{q}_1, \dot{q}_2, \cdots, \dot{q}_k, \cdots, t)$$

We now assume that L does not depend upon time explicity. Then

$$L = L(q_1, q_2, \cdots, q_k, \cdots, \dot{q}_1, \dot{q}_2, \cdots, \dot{q}_k, \cdots).$$

The total time derivative of L is given by

$$\begin{aligned} \frac{dL}{dt} &= \frac{\partial L}{\partial q_1}\dot{q}_1 + \frac{\partial L}{\partial q_2}\dot{q}_2 + \cdots + \frac{\partial L}{\partial q_k}\dot{q}_k + \cdots + \frac{\partial L}{\partial \dot{q}_1}\ddot{q}_1 + \frac{\partial L}{\partial \dot{q}_2}\ddot{q}_2 + \cdots + \frac{\partial L}{\partial \dot{q}_k}\ddot{q}_k + \cdots \\ &= \sum_k \frac{\partial L}{\partial q_k}\dot{q}_k + \sum_k \frac{\partial L}{\partial \dot{q}_k}\ddot{q}_k. \end{aligned} \tag{4}$$

Now according to Lagrange's equation

$$\frac{d}{dt}\left(\frac{\partial L}{\partial \dot{q}_k}\right) - \frac{\partial L}{\partial q_k} = 0 \quad \text{or} \quad \frac{\partial L}{\partial q_k} = \frac{d}{dt}\left(\frac{\partial L}{\partial \dot{q}_k}\right),$$

so that with this substitution, equation (B.3.4) reduces to

$$\frac{dL}{dt} = \sum_k \frac{d}{dt}\left(\frac{\partial L}{\partial \dot{q}_k}\right)\dot{q}_k + \sum_k \frac{\partial L}{\partial \dot{q}_k}\ddot{q}_k = \sum_k \frac{d}{dt}\left(\dot{q}_k \frac{\partial L}{\partial \dot{q}_k}\right)$$

$$\text{or} \quad \frac{d}{dt}\left(L - \sum_k \dot{q}_k \frac{\partial L}{\partial \dot{q}_k}\right) = 0$$

which shows that

$$L - \sum_k \dot{q}_k \frac{\partial L}{\partial \dot{q}_k} = \text{consant.} \tag{5}$$

But

$$\frac{\partial L}{\partial \dot{q}_k} = \frac{\partial}{\partial \dot{q}_k}(T - V) = \frac{\partial T}{\partial \dot{q}_k} \quad \left(\text{since } \frac{\partial V}{\partial \dot{q}_k} = 0 \text{ for a conservative system}\right)$$

$$= \frac{\partial}{\partial \dot{q}_k}\left(\sum_k \frac{1}{2} m_k \dot{q}_k^2\right) = m_k \dot{q}_k,$$

so that with substitution of this result in (B.3.5) we obtain

$$L - \sum_k m_k \dot{q}_k^2 = \text{constant}$$

or $\quad L - 2T = \text{constant} \quad \text{or} \quad T - V - 2T = \text{constant}$

or $\quad -(T + V) = \text{constant}. \quad \therefore \quad T + V = \text{Total energy} = \text{constant}$

which gives the conservation theorem for energy which states that *if the Lagrangian function does not contain time explicitly, the total energy of the conservative system is conserved.*

Case 3. Conservation theorem for Linear Momentum: We consider a conservative system, in which the potential energy V is dependent on position only and the kinetic energy is independent of position, so that

$$\frac{\partial V}{\partial \dot{q}_k} = 0 \quad \text{and} \quad \frac{\partial T}{\partial q_k} = 0. \tag{6}$$

Lagrange's equation for such a system may be written as

$$\frac{d}{dt}\left(\frac{\partial L}{\partial \dot{q}_k}\right) - \frac{\partial L}{\partial q_k} = 0 \quad \text{or} \quad \frac{d}{dt}\left(\frac{\partial L}{\partial \dot{q}_k}\right) + \frac{\partial V}{\partial q_k} = 0; \quad \text{[using (B.3.6)]}$$

So

$$\dot{p}_k = \frac{d}{dt}\left(\frac{\partial L}{\partial \dot{q}_k}\right) = -\frac{\partial V}{\partial q_k} = Q_k. \tag{7}$$

Now if we can show that the generalised force Q_k represents the component of total force along the direction of translation of the system and the generalised momentum p_k is the component of total linear momentum along the same direction, then equation (B.3.7) will represent the equation of motion for total linear momentum.

The generalised force Q_k is given by

$$Q_k = \sum_i \vec{F}_i \cdot \frac{\partial \vec{r}_i}{\partial q_k}.$$

If $\hat{n}$ is the unit vector along the direction of translation of the system, then $\delta \vec{r}_i = \hat{n} \delta q_k$ i.e. $\dfrac{\partial \vec{r}_i}{\partial q_k} = \hat{n}$, so that we may write

$$Q_k = \sum_i \vec{F}_i \cdot \hat{n} = \hat{n} \cdot \vec{F} \tag{8}$$

which represents the component of total force along the direction of $\hat{n}$ (i.e. translation).

Now the kinetic energy $T = \frac{1}{2}\sum_i m_i \dot{r}_i^2$.

So the generalised momentum

$$p_k = \frac{\partial L}{\partial \dot{q}_k} = \frac{\partial T}{\partial \dot{q}_k} = \sum_i m_i \dot{\vec{r}}_i \cdot \frac{\partial \dot{\vec{r}}_i}{\partial \dot{q}_k}$$

$$= \sum_i m_i \vec{v}_i \cdot \frac{\partial \vec{r}_i}{\partial q_k} \quad \left[\because \frac{\partial \dot{\vec{r}}_i}{\partial \dot{q}_k} = \frac{\partial \vec{r}_i}{\partial q_k} \quad \text{(see deduction of Lagrange's equation in chapter 3)}\right]$$

$$= \sum_i m_i \vec{v}_i \cdot \hat{n} = \hat{n} \cdot \sum_i m_i \vec{v}_i \qquad (9)$$

which show that p_k represents the component of total linear momentum along the direction of translation.

So we can say that the equation (B.3.7) i.e. $\dot{p}_k = Q_k$ represents the equation of motion for total linear momentum.

If $Q_k = 0$, $\dot{p}_k = 0$, i.e., p_k = constant.

This gives the conservation theorem for linear momentum, and states that *if a given component of the total applied force vanishes, the corresponding component of the linear momentum is conserved.*

Case 4. Conservation theorem for Angular Momentum: We consider a conservative system in which the potential energy V is dependent on position only and the kinetic energy is indepenent of position, so that

$$\frac{\partial V}{\partial \dot{q}_k} = 0 \quad \text{and} \quad \frac{\partial T}{\partial q_k} = 0 \qquad (10)$$

Lagrange's equation for a such a system may be written as

$$\frac{d}{dt}\left(\frac{\partial L}{\partial \dot{q}_k}\right) - \frac{\partial L}{\partial q_k} = 0 \quad \text{or} \quad \frac{d}{dt}\left(\frac{\partial L}{\partial \dot{q}_k}\right) + \frac{\partial V}{\partial q_k} = 0; \quad \text{[using (B.3.10)]}$$

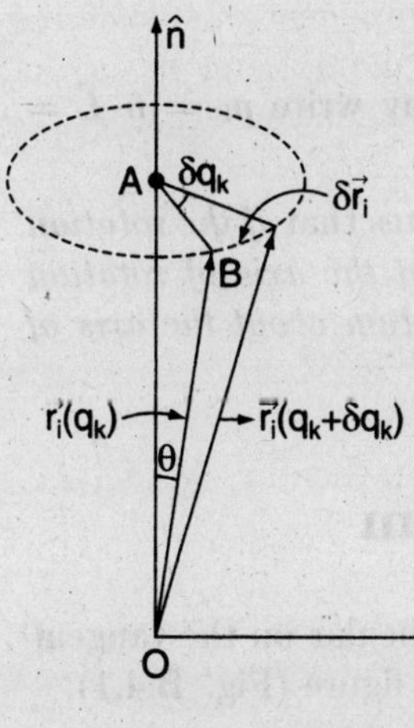

So

$$\dot{p}_k = \frac{d}{dt}\left(\frac{\partial L}{\partial \dot{q}_k}\right) = -\frac{\partial V}{\partial q_k} = Q_k. \qquad (11)$$

Now if we can show that with rotation co-ordinate q_k, the generalised force Q_k represents the component of the total applied torque about the axis of rotation and the generalised momentum p_k represents the component of the total angular momentum about the same axis, then equation (B.3.11) will represent the equation of motion for the total angular momentum.

Now the generalised force Q_k is given by

$$Q_k = \sum_i \vec{F}_i \cdot \frac{\partial \vec{r}_i}{\partial q_k}.$$

The magnitude of the change in position co-ordinate $\vec{r}_i(q_k)$ due to a change in rotation q_k is given by

$$|\delta\vec{r}_i| = AB\delta q_k = r_i \sin\theta\delta q_k \quad \text{or} \quad \left|\frac{\partial\vec{r}_i}{\partial q_k}\right| = r_i \sin\theta. \tag{12}$$

If $\hat{n}$ is the unit vector along the axis of rotation, equation (B.3.12) may be written as

$$\frac{\partial\vec{r}_i}{\partial q_k} = \hat{n} \times \vec{r}_i. \tag{13}$$

So the generalised force can be expressed as

$$Q_k = \sum_i \vec{F}_i \cdot \hat{n} \times \vec{r}_i = \sum_i \hat{n} \cdot \left(\vec{r}_i \times \vec{F}_i\right) = \sum_i \hat{n} \cdot \vec{N}_i = \hat{n} \cdot \sum_i \vec{N}_i = \hat{n} \cdot \vec{N} \tag{14}$$

where $\vec{N} = \sum_i \vec{N}_i$ is the total torque, so that from equation (B.3.14) we may conclude that Q_k is the component of the total torque about the axis of rotation.

Again

$$\begin{aligned} p_k &= \frac{\partial L}{\partial \dot{q}_k} = \frac{\partial T}{\partial \dot{q}_k} = \sum_i m_i \vec{v}_i \cdot \frac{\partial \vec{r}_i}{\partial q_k} \quad \text{[from equation (B.3.9)]} \\ &= \sum_i m_i \vec{v}_i \cdot \hat{n} \times \vec{r}_i \quad \text{[using equation (B.3.13)]} \\ &= \sum_i \hat{n} \cdot \vec{r}_i \times m_i \vec{v}_i = \hat{n} \cdot \sum_i \vec{L}_i = \hat{n} \cdot \vec{L}, \end{aligned} \tag{15}$$

where $\vec{L} = \sum_i \vec{L}_i$ is the total angular momentum. Equation (B.3.15) therefore shows that p_k is the component of the total angular momentum about the axis of rotation, so that we can conclude that equation (B.3.11) i.e. $\dot{p}_k = Q_k$ represents the equation of motion for the total angular momentum of the system about an axis.

If however q_k is a cyclic co-ordinate, then

$$\hat{n} \cdot \vec{N} = Q_k = -\frac{\partial V}{\partial q_k} = -\frac{\partial L}{\partial q_k} = 0.$$

Hence from equation (B.3.11), it follows that $\dot{p}_k = 0$ whence we may write $p_k = \hat{n} \cdot \vec{L} =$ constant [using (B.3.15)].

This gives the conservation theorem for angular momentum, which states that *if the rotation co-ordinate q_k is cyclic i.e. if the component of the applied torque about the axis of rotation vanishes, then the corresponding component of the total angular momentum about the axis of rotation is conserved.*

B.4 Equation of Central Orbit in Pedal Form

Let the position vector to any point P of the orbit be $\vec{r}$ and the perpendicular on the tangent at P be of length p. The angles θ and ϕ are as indicated in the attached figure (Fig. B.4.1).

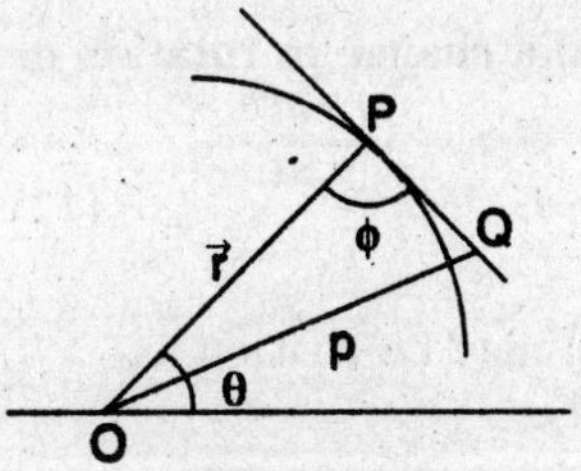

Fig B.4.1

Then, $\tan\phi = r\frac{d\theta}{dr}, p = r\sin\phi$.

$$\therefore \quad \frac{1}{p^2} = \frac{1}{r^2}\operatorname{cosec}^2\phi = \frac{1}{r^2}(1+\cot^2\phi) = \frac{1}{r^2} + \frac{1}{r^4}\left(\frac{dr}{d\theta}\right)^2$$

$$\text{or,} \quad \frac{1}{p^2} = u^2 + u^4 \cdot \frac{1}{u^4}\left(\frac{du}{d\theta}\right)^2 = u^2 + \left(\frac{du}{d\theta}\right)^2, \quad \because r = \frac{1}{u}$$

Differentiating w.r.t. u, we get,

$$-\frac{2}{p^3}\frac{dp}{du} = 2u + 2\frac{du}{d\theta}\cdot\frac{d^2u}{d\theta^2}\cdot\frac{d\theta}{du} = 2u + 2\frac{d^2u}{d\theta^2}$$

$$\text{or,} \quad -\frac{1}{p^3}\frac{dp}{du} = u + \frac{d^2u}{d\theta^2} = \frac{f}{h^2u^2}, \quad f \text{ is the acceleration}$$

$$\text{or,} \quad -\frac{h^2}{p^3}\frac{dp}{dr}\cdot\frac{dr}{du} = \frac{f}{u^2} = fr^2$$

$$\text{or,} \quad \frac{h^2}{p^3}\frac{dp}{dr} = f = \frac{\mu}{r^2}, \quad \mu = +GM,$$

where G is the universal constant of gravitation.

This equation is the pedal equation of a central orbit.

Now,

$$mvp = H \quad \text{or,} \quad vp = \frac{H}{m} = h \text{ (constant)}$$

$$\therefore \quad v^2 = \frac{h^2}{p^2} = h^2\left[u^2 + \left(\frac{du}{d\theta}\right)^2\right]$$

Integrating the orbit equation w.r.t. r,

$$h^2\int \frac{1}{p^3}\frac{dp}{dr}dr = \int \frac{\mu}{r^2}dr$$

$$\text{or,} \quad -\frac{h^2}{2p^2} = -\frac{\mu}{r} + A \quad \text{or,} \quad \frac{h^2}{p^2} = \frac{2\mu}{r} + A_1$$

$$\therefore \quad v^2 = \frac{h^2}{p^2} = \frac{2\mu}{r} + A_1.$$

The equations of an ellipse, a parabola and a hyperbola are respectively given by,

$$\frac{b^2}{p^2} = \frac{2a}{r} \quad 1 \quad (ellipse)$$

$$\frac{p^2}{r} = \text{constant} \quad (parabola)$$

$$\frac{b^2}{p^2} = \frac{2a}{r} + 1 \quad (hyperbola)$$

So the planetary orbit is,

$$\text{ellipse, if } \quad \frac{h^2}{b^2} = \frac{\mu}{a} = \frac{A_1}{(-1)}$$

$$\text{parabola, if } \quad \frac{h^2}{2\mu} = \text{constant}$$

$$\text{hyperbola, if } \quad \frac{h^2}{b^2} = \frac{\mu}{A} = \frac{A_1}{1}.$$

B.4.1 Vectorial Treatment for the Central Orbit

We have the equation of motion of a particle in the gravitational field given by

$$\frac{d^2\vec{r}}{dt^2} = -\frac{\mu}{r^2}\hat{r}, \quad \mu = GM. \tag{1}$$

Also, we know that $\vec{r} \times m\vec{V} = \vec{H}$, or, $\vec{r} \times \vec{V} = \frac{\vec{H}}{m} = \vec{h}$

$$\therefore \quad \vec{h} = \vec{r} \times \vec{V} = r\hat{r}\left(\dot{r}\hat{r} + r\dot{\theta}\hat{\theta}\right) = r^2\dot{\theta}\hat{k}, \tag{2}$$

($\hat{\theta}$ being the cross-radial unit vector), where $\hat{k}$ is a unit vector perpendicular to the plane of motion.

Taking crossproduct of the equation of motion (B.4.1) with equation (B.4.2) we get,

$$\frac{1}{\mu}\left(\ddot{\vec{r}} \times \vec{h}\right) = -\frac{\hat{r}}{r^2} \times r^2\dot{\theta}\hat{k} = -\dot{\theta}\left(\hat{r} \times \hat{k}\right) = \dot{\theta}\left(\hat{k} \times \hat{r}\right) = \hat{\theta}\dot{\theta}.$$

Now, we know that,

$$\left(\frac{d\hat{r}}{dt} = \frac{d\hat{r}}{d\theta} \cdot \frac{d\theta}{dt} = \hat{\theta}\dot{\theta}.\right.$$

Therefore,

$$\frac{1}{\mu}\left(\ddot{\vec{r}} \times \vec{h}\right) = \frac{d\hat{r}}{dt}. \tag{3}$$

Integrating this vector equation w.r.t. time get,

$$\frac{1}{\mu}\left(\vec{V} \times \vec{h}\right) = \hat{r} + e\hat{c}, \tag{4}$$

where $\hat{c}$ is a unit vector, $e\hat{c}$ is the constant of integration—a vector in the plane of motion because $\vec{V} \times \vec{h}$ and $\hat{r}$ all lie in the plane of motion. Taking the dot product of the last equation with $\vec{r}$,

$$\vec{r} \cdot \frac{1}{\mu}\left(\vec{V} \times \vec{h}\right) = \vec{r} \cdot \hat{r} + e\hat{c}.\vec{r}$$

$$\text{or,} \quad \frac{1}{\mu}\left(\vec{r} \times \vec{V}\right) \cdot \vec{h} = r + er\cos\theta$$

[see attached figure B.4.2]

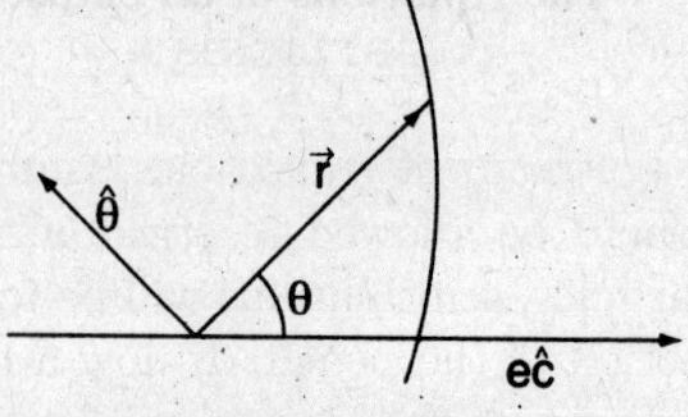

Fig B.4.2

$$\text{or,} \quad \frac{1}{\mu}\vec{h} \cdot \vec{h} = r(1 + e\cos\theta). \quad \therefore \quad r = \frac{h^2/\mu}{1 + e\cos\theta}.$$

This gives the equation of a conic. Now, we are to show that the conic is an ellipse. We will follow the same argument as given earlier that the velocity of escape is always greater than the velocity of the planet since otherwise the planet will move out of the solar system.

From the vector equation (B.4.4) we may write

$$\frac{1}{\mu}\left(\vec{V}\times\vec{h}\right)\cdot\frac{1}{\mu}\left(\vec{V}\times\vec{h}\right) = (\hat{r}+e\hat{c})\cdot(\hat{r}+e\hat{c})$$

$$\text{or,}\quad \frac{1}{\mu^2}\left|\vec{V}\times\vec{h}\right|^2 = 1+2e\hat{r}\cdot\hat{c}+e^2 = 1+2e\cos\theta+e^2$$

$$\text{or,}\quad \frac{1}{\mu^2}v^2h^2\sin^2 90^\circ = 1+2e\cos\theta+e^2$$

$$\text{or,}\quad v^2 = \frac{\mu^2}{h^2}\left(\frac{2h^2}{\mu r}+e^2-1\right) = \frac{2\mu}{r}+\left(\frac{\mu^2}{h^2}\right)(e^2-1). \qquad (5)$$

Now, V_e = velocity of escape $= \sqrt{\frac{2\mu}{r}}$.

Hence, equation (B.4.5) gives,

$$v^2 = V_e^2 + \frac{\mu^2}{h^2}(e^2-1). \qquad (6)$$

Now we already know that $V_e > v$ always for a moving planet.

$\therefore$ equation (B.4.6) shows that $e^2-1<0$, or, $e<1$ proving that the path followed is an ellipse.

In case of planetary motion, we have shown that the orbit of the planet bound to the sun by an inverse square law of force is elliptical with the sun at one of the foci.

We will now consider a few other important cases of central orbit problem.

B.4.2 Motion of an electron in an atom

An electron is bound to the nucleus of atomic number Z by an inverse square law given by

$$\vec{F} = -\frac{Ze^2}{4\pi\epsilon_0 r^2}\hat{r} = -\frac{k}{r^2}\hat{r}, \quad k = Ze^2/4\pi\epsilon_0.$$

So the orbit in this case is also elliptical.

B.4.3 Scattering in a central force field—Rutherford's scattering of α-particles

If a stream of α-particles, limited by means of suitable diaphragms to a narrow cylindrical pencil, be allowed to strike a zinc-sulphide screen placed at right angles to the path of the particles, scintillations will be found to occur over a well-dfined circular area equal to the cross-section of the pencil. If now a thin film of matter, such as gold or silver foil, is inter-posed in the path of the rays, it will be found that they pass quite readily thrown the foil, but the area over which the scintillations occur becomes larger and loses its definite boundary

—indicating that some of the particles have been deflected from their original directions. *This spreading out of the stream of α-particles on passing through thin layers of matter,* solid or gaseous, *is called "Rutherford scattering"*

—since the abovementioned experiments were conducted mainly by Rutherford along with his colleagues Geiger and Marsden. Also it was Rutherford who gave an adequate theory of such scattering, and ultimately established the nuclear atom model named after him.

An α-particle is identified with He nucleus having 2 units of positive charge. Scattering of α-particle by matter is treated in classical mechanics as a central force field problem. In considering this problem, we will not consider the wave aspects of the α-particle. We will assume the nucleus and the particle as point charges (neglecting the dimensions of the particles) and lastly, the nucleus being heavy is supposed to be at rest.

Let us have the fixed nucleus at the point N. The positively charged particle projected in a certain direction will be deviated from the straight path under the repulsive force which increases enormously as the particle approaches within a certain distance from the nucleus. The closest distance of approach between the nucleus and the positively charged particle is known as the *impact parameter, p.* Since at the point of the closest approach to the nucleus, the positively charged probing particle experiences the strongest repulsion, it flies off such that at long distance on its retreating path, the path becomes straight line. The initial direction of motion of the particle and straight line of retreat produced backwards will be inclined at some angle ϕ called the *scattering angle.*

Let O be the point of intersection of the two asymptotes to the approaching and retreating branches respectively of path followed by the particle, the asymptotes being the initial and final directions of motion (as shown in the Fig. B.4.3).

Let us first solve for the orbit followed by the charged particle carrying charge, say $Z'e$ (in case of α-particle, this is $2e$) and moving under the repulsive force of the nucleus of charge Ze, where Z = atomic number of the scatterer. Then, we can write the equation of motion of the particle as, $\vec{F} = m_e\ddot{\vec{r}} = \frac{Z'Ze^2}{4\pi\epsilon_0 r^2}\hat{r}$, where $\vec{r}$ is the position vector of the particle w.r.t. the nucleus N supposed to be at rest at any instant of time t, $\hat{r}$ being the unit vector along $\vec{r}$ and ϵ_0 is the electrical permittivity of free space. If we write $k = \frac{Z'Ze^2}{4\pi\epsilon_0}$, then,

$$m_e\ddot{\vec{r}} = \frac{k}{r^2}\hat{r}$$

which shows that the problem may be treated as a central force field problem.

Now, we know that,

$$\vec{r} \times m_e\vec{V} = \vec{H}, \quad \text{or,} \quad m_e\left(\vec{r} \times \vec{V}\right) = \vec{H}$$

$$\text{or,} \quad \vec{H} = m_e\left[r\hat{r} \times \left(\dot{r}\hat{r} + r\dot{\theta}\hat{\theta}\right)\right] = m_e r^2\dot{\theta}\left(\hat{r} \times \hat{\theta}\right) = m_e r^2\dot{\theta}\hat{k}$$

$$\therefore \quad m_e r^2\dot{\theta} = H = \text{constant.}$$

Now the radial acceleration is given by

$$\ddot{r} - r\dot{\theta}^2 = \frac{k}{m_e r^2}.$$

Putting $r = \frac{1}{u}$, we may write,

$$\frac{dr}{dt} = -\frac{1}{u^2}\frac{du}{d\theta}\frac{d\theta}{dt} = -\frac{1}{u^2}\cdot u^2\cdot\frac{H}{m_e}\frac{du}{d\theta} = -\frac{H}{m_e}\frac{du}{d\theta}$$

$$\therefore\ \frac{d^2r}{dt^2} = -\frac{H^2}{m_e^2}\cdot u^2\cdot\frac{d^2u}{d\theta^2},$$

so that the differential equation of the orbit is written as

$$-\frac{H^2}{m_e^2}u^2\frac{d^2u}{d\theta^2} - \frac{H^2}{m_e^2}u^3 = \frac{k}{m_e}u^2, \quad \text{or,} \quad \frac{d^2u}{d\theta^2} + u = -\frac{m_e k}{H^2}.$$

Solving we get,

$$u = \frac{1}{r} = -\frac{m_e k}{H^2} + A\cos(\theta - \theta_0) = -\frac{m_e k}{H^2}\left[1 - \frac{AH^2}{m_e k}\cos(\theta - \theta_0)\right],$$

where A and θ_0 are arbitrary constants. The constant θ_0 just determines the orientation of the orbit in the (r, θ) plane. We can choose the X-axis such that $\theta_0 = 0$ and the equation of the orbit becomes

$$u = \frac{1}{r} = -\frac{m_e k}{H^2}\left[1 - \frac{AH^2}{m_e k}\cos\theta\right]$$

$$\therefore\ \varepsilon = \text{the eccentricity of the orbit} = \frac{AH^2}{m_e k}.$$

Now total energy of the particle is given by

$$E = \frac{1}{2}m_e\left(\dot{r}^2 + r^2\dot{\theta}^2\right) \quad \text{or,} \quad E = \frac{1}{2}\frac{H^2}{m_e}\left(A^2 - \frac{m_e^2 k^2}{H^4}\right).$$

$$\therefore\ A = \frac{m_e k}{H^2}\sqrt{1 + \frac{2EH^2}{m_e k^2}}. \quad \therefore\ \varepsilon = \sqrt{1 + \frac{2EH^2}{m_e k^2}}.$$

In the present problem, applying the principle of conservation of energy, we obtain

$$\frac{1}{2}m_e v_i^2 = \frac{1}{2}m_e\left(\dot{r}^2 + r^2\dot{\theta}^2\right) + \frac{k}{r} = E\ (\text{Total}),$$

where, v_i is the velocity of the particle initially when the particle did not come under the central force field effectively.

$$\therefore\ \varepsilon = \sqrt{1 + \frac{v_i^2 H^2}{k^2}} = \sqrt{1 + \frac{(4\pi\epsilon_0)^2 v_i^2 H^2}{(Z'Ze)^2}} > 1,$$

hence, the path in this case is hyperbolic.

Now, scattering cross-section in a given direction is defined as,

$$\sigma(\omega)d\omega = \frac{\text{number of particles scattered into the solid angle per unit time}}{\text{incident intensity}}.$$

The solid angle defined between ϕ and $\phi + d\phi$ is, $d\Omega_\phi = 2\pi \sin\phi d\phi$ (Fig B.4.4).

The number of particles scattered into solid angle $d\Omega_\phi$ lying between ϕ and $\phi + d\phi$ must be equal to the number of incident particles having impact parameter between p and $p + dp$ (Fig B.4.4).

$\therefore$ $2\pi I \sin\phi d\phi = -2\pi p dp I$, where I is the intensity of the incident particles; $-$ve sign is used because an increase in p will cause a decrease in ϕ.

$$\therefore \quad \sigma = -\frac{pdp}{\sin\phi d\phi}.$$

Now, applying conservation of energy and momentum, $\frac{1}{2}m_e v_i^2$ = total energy at any point on the orbit = E, where, v_i = initial velocity.

$\therefore$ $m_e v_i = \sqrt{2m_e E}$ and $m_e v_i p = m_e r\dot{\theta}\cdot r = m_e r^2\dot{\theta} = H$.

We have shown that,

$$\varepsilon = \sqrt{1 + \frac{2EH^2}{m_e k^2}} = \sqrt{1 + \frac{2E}{m_e k^2}\cdot m_e^2 v_i^2 p^2} = \sqrt{1 + \frac{2Ep^2}{m_e k^2}\cdot 2m_e E} = \sqrt{1 + \left(\frac{2Ep}{k}\right)^2}.$$

Let the angle between the two asymptotes be equal to 2ψ. Then angle of scattering ϕ is given by $\psi = \frac{\pi - \phi}{2}$.

Now when $r \to \infty$, $\theta \to \psi$, the angle between the X-axis and the asymptote of the hyperbola.

Again we have already deduced that

$$\frac{1}{r} = -\frac{m_e k}{H^2}\left[1 - \frac{AH^2}{m_e k}\cos\theta\right].$$

So, we obtain

$$0 = -1 + \frac{AH^2}{m_e k}\cos\psi, \quad \text{or,} \quad \varepsilon = \frac{1}{\cos\psi}, \quad \text{a property of the hyperbola.}$$

We have, then,

$$\cos\psi = \cos\left(\frac{\pi}{2} - \frac{\phi}{2}\right) = \sin\frac{\phi}{2} = \frac{1}{\varepsilon}$$

$$\therefore \quad \cot^2\frac{\phi}{2} = \text{cosec}^2\frac{\phi}{2} - 1 = \varepsilon^2 - 1 = \left(\frac{8\pi\epsilon_0 Ep}{ZZ'e^2}\right)^2$$

$$\text{or,} \quad \cot\frac{\phi}{2} = \frac{8\pi\epsilon_0 Ep}{ZZ'e^2}, \quad \text{or,} \quad p = \frac{ZZ'e^2}{8\pi\epsilon_0 E}\cot\frac{\phi}{2}$$

$$\therefore \quad dp = -\frac{ZZ'e^2}{8\pi\epsilon_0 E}\cdot\frac{1}{2\sin^2\frac{\phi}{2}}d\phi = -\frac{ZZ'e^2}{16\pi\epsilon_0 E}\frac{1}{\sin^2\frac{\phi}{2}}d\phi$$

[Putting $E = \frac{1}{2}m_e v_i^2$ and $Z' = 2$, we may write

$$\cot\frac{\phi}{2} = \frac{2\pi\epsilon_0 m_e v_i^2 p}{Ze^2} \quad \text{and} \quad p = \frac{Ze^2}{2\pi\epsilon_0 m_e v_i^2}\cot\frac{\phi}{2}]$$

so that, the scattering cross-section becomes,

$$\sigma = -\frac{pdp}{\sin\phi d\phi} = \frac{ZZ'e^2}{8\pi\epsilon_0 E}\cot\frac{\phi}{2}\cdot\frac{1}{\sin\phi d\phi}\cdot\frac{ZZ'e^2}{16\pi\epsilon_0 E}\frac{1}{\sin^2\frac{\phi}{2}}d\phi = \frac{1}{4}\left(\frac{ZZ'e^2}{8\pi\epsilon_0 E}\right)^2\frac{1}{\sin^4\frac{\phi}{2}}.$$

Now, for α-particle, $Z' = 2$,

$$\therefore \quad \sigma = \left(\frac{Ze^2}{8\pi\epsilon_0 E}\right)^2\cdot\frac{1}{\sin^4\frac{\phi}{2}} = \left(\frac{Ze^2}{4\pi\epsilon_0 m_e v_i^2}\right)^2\cdot\frac{1}{\sin^4\frac{\phi}{2}}, \quad (7)$$

which is known as *Rutherford's scattering formula for an α-particle.* Here σ gives the number of α-particles deviated per unit solid angle, and also denotes the probability of deviation per unit solid angle.

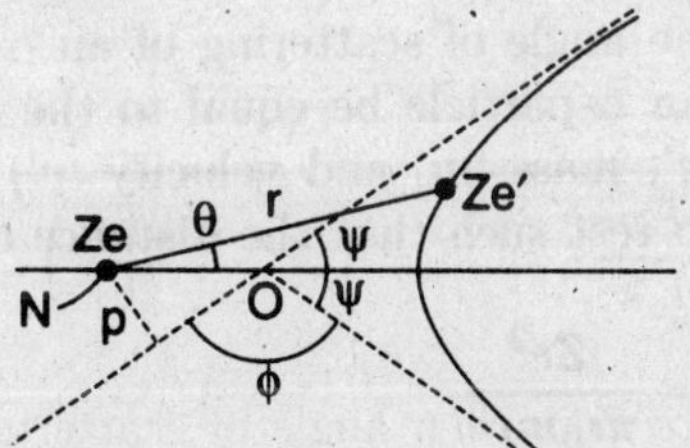

Fig B.4.3

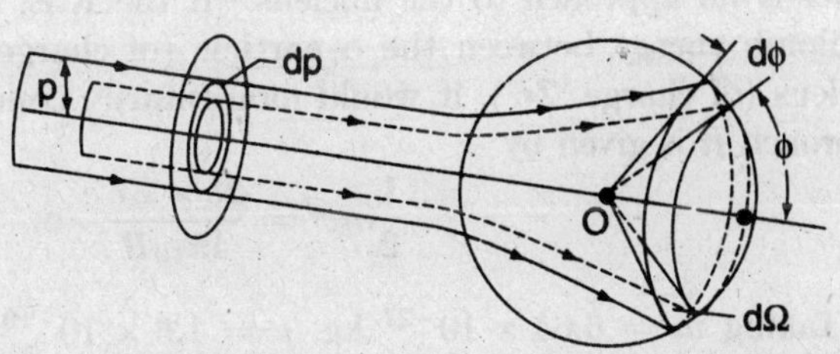

Fig B.4.4

The above formula (B.4.7) however is subject to a *correction for the motion imparted to the scattering nucleus.* This correction is appreciable only in light nuclei.

It follows from (B.4.7) that if N_0 mono-energetic α-particles be incident normally per unit area on a thin foil of thickness t containing n atoms (and hence n nuclei) per unit volume, the number N of scintillations produced by α-particles falling normally on unit area of the screen along ϕ is

$$N = N_0 nt\cdot\sigma = N_0 nt\left(\frac{Ze^2}{4\pi\epsilon_0 m_e v_i^2}\right)^2\frac{1}{\sin^4\frac{\phi}{2}}$$

which is yet another form of Rutherford's formula, and states that if Rutherford's conception of the nuclear atom model is correct, the amount of scattering, i.e., the number of α-particles N striking unit area of a screen at a given distance from the scatterer and perpendicular to the direction of motion of the scattered particles should be proportional to (i) $1/\sin^4\frac{\phi}{2}$, (ii) the thickness t of the material of the scatterer, (iii) the square of the nuclear charge $(\text{Ze})^2$ and (iv) v_i^{-4} i.e. inversely to the square of the K.E.

The above relationships connecting N with (Ze), m_e, v_i, ϕ and t were experimentally tested by Geiger and Marsden by counting the scattered particles and the agreement between the formula and the experimental results is, is general, highly accurate.

Important conclusions from the study of the experimental results: Most of the α-particles were scattered through small angles, that is, they passed almost straight through the metallic foil. This indicates that *most of the space in an atom is empty.* A very small fraction of the particles however was deviated through large angles, very very few by even nearly 180°. The phenomenon of *large angle scattering* was indeed startling, for calculation on the basis of existing Thomson model showed that the probability of scattering through angles as large as 90° was vanishingly small. It was so surprising that Rutherford himself stated: "It is just as surprising as if a gunner fired a shell at a single piece of paper and for some reason or other the projectile bounced back again." Rutherford proposed that large angle scattering suggests that α-particles must have suffered much larger Coulomb force than what the distributed positive charge of Thomson model can bring forth. He concluded that *within the atom there was a small massive core, carrying positive charge* and called it the *nucleus*. The electrons were assumed to be extra-nuclear that is, they exist outside the nucleus. Rutherford's model is thus known as *nuclear atom model*. To explain the mechanical stability of the atom, Rutherford assumed the electrons to revolve round the nucleus like planets revolving round the sum. The electrostatic attraction between the nucleus and the extra-nuclear electrons provides the necessary centripetal force.

Nuclear size: The size of the nucleus was first estimated from Rutherford's α-ray scattering by various atoms of different scatterers. The larger the angle of scattering of an α-particle, closer is its approach to the nucleus. If the K.E. of an α-particle be equal to the repulsive Coulomb energy between the α-particle (of charge '$2e$', mass 'm' and velocity 'v') and the nucleus (of charge 'Ze'), it would momentarily come to rest such that the distance of closest approach R is given by

$$\frac{1}{2}mv^2 = \frac{2e \times Ze}{4\pi\epsilon_0 R} \quad \text{or} \quad R = \frac{Ze^2}{\pi\epsilon_0 mv^2}$$

Taking $m = 6.64 \times 10^{-27}$ kg, $e = 1.6 \times 10^{-19}$ C and $v = 10^7$ m/s, the value of R for $Z = 20$ is calculated as $\sim 1.5 \times 10^{-14}$ m (with standard value of ϵ_0)—which is in fairly accurate agreement with the value obtained from different methods.

Again experimental observations indicate that the majority of atomic nuclei are spherical or nearly so in shape. The radius R of various nuclei has been theoretically calculated and is approximately given by by $R = r_0 A^{1/3}$ where A is the mass number and r_0, a constant called *nuclear radius parameter*. The value of r_0 ranges from $(1.1 - 1.5) \times 10^{-15}$ m and can be evaluated by a number of different methods.

Appendix C

Stess and Strain Tensors

Many crystals have the property that if they are stessed along one axis, electric charges are found to appear along other axes of the crystal. Thus the crystal is electrically polarised and this phenomenon is called Piezo electric effect. The relation between electric polarisation $\vec{P}$ and the stress S can be put in the form

$$\vec{P}_i = \sum_{j,k} d_{ijk} S_{jk}.$$

Here $\vec{P}_i$ is a vector and S_{jk} is a tensor of rank two and d_{ijk} is a tensor of rank three. d_{ijk} is called Piezo electric coefficient or Piezo electric tensor. Each coefficient is connected with one axis of $\vec{P}$ and two axes of S. Stress and strain are both tensors of rank two and they are related by the equation

$$S_{ij} = \sum_{k,l} C_{ijkl} e_{kl}.$$

Here S_{ij} and e_{kl} are the stress and the strain tensors and C_{ijkl} is the coefficient which relates the two tensors. C_{ijkl} is a tensor of rank four and we call it the tensor of elasticity. It is to be noted that the stress and the strain tensors are also symmetrical.

Appendix D

Introduction to Fourier Transforms

D.1 Integral Transforms

The operations of differentiation and integration, when applied to a differentiable and integrable function $F(t)$ transform it to the functions $F'(t)$ and $f(x)$ respectively, such that

$$D\{F(t)\} = F'(t), \quad \text{(for differentiation)}$$

$$\text{and} \quad I\{F(t)\} = \int_0^x F(t)dt = f(x), \quad \text{(for integration)}$$

These equations may be considered as transformations—transforming the function $F(t)$ to its *images* or *transforms*—$F'(t)$ in the first case, and $f(x)$ in the second case. Any general transformation expressed as $T\{F(t)\}$ of the function $F(t)$ is said to be linear, if

$$T\{k_1F_1(t) + k_2F_2(t)\} = k_1T\{F_1(t)\} + k_2T\{F_2(t)\}, \tag{1}$$

for every pair of functions $F_1(t)$ and $F_2(t)$, and for every pair of constants k_1 and k_2, i.e., if the transformation is linear, the transform of a linear combination of two functions is equal to the same linear combination of their transforms. It is evident that the transforms $F'(t)$ and $f(x)$ are linear. A general linear integral transformation of the function $F(t)$ is represented by the following equation

$$f(s) = T\{F(t)\} = \int_a^b F(t)k(s,t)dt, \tag{2}$$

where $f(s)$ is called the *integral transform* of $F(t)$ with respect to the *kernel* $k(s,t)$—which denotes some prescribed function of the variable t and a parameter s.

Integral transforms are used in a variety of applications—for example—in solving ordinary differential equations or partial differential equations, and in the evaluation of contour integrals, etc. There are a few different kinds of integral transforms with different names depending on what kernel $k(s,t)$ we use and what the range of integration is. For example, we have the following useful transforms of the function $F(t)$ for the kernels $e^{-st}, tJ_n(st), t^{s-1}$ and e^{-jst} respectively:

$$f(s) = \int_0^\infty F(t)e^{-st}dt \cdots\cdots \text{ Laplace transform} \tag{3}$$

$$f(s) = \int_0^\infty F(t)tJ_n(st)dt \cdots\cdots \text{ Hankel transform (Fourier-Bessel transform)} \tag{4}$$

$$f(s) = \int_0^\infty F(t)t^{s-1}dt \cdots\cdots \text{ Mellin transform} \tag{5}$$

$$f(s) = \int_0^\infty F(t)e^{-jst}dt \cdots\cdots \text{ Fourier transform} \tag{6}$$

These transforms are useful in mathematical analysis and in physical problems, although the possible types are unlimited. We have actually used the Mellin transform in the definition of gamma function for $f(t) = e^{-t}$ in chapter 17 ɪne Hankel transform given by (D.1.4), which may also be considered as the Fourier transform for a Bessel function expansion, represents a limiting case of a Fourier-Bessel series defined in chapter 17. In this chapter, we shall discuss only the Fourier transform and its important applications.

D.2 The Fourier Transform

Methods of Fourier transforms are used for solving the boundary value problems, for evaluating different types of integrals, for evaluation of contour integrations, and for representing the functions in convenient forms. A general Fourier transform is defined by equation (D.1.6) above—which is based on the kernel e^{-jst} and its real and imaginary parts. The Fourier transform thus appears frequently in the mathematical theory of communication engineering, wave propagation and other branches of applied mathematics—specially in the studies of waves and of the extraction of information from waves, particularly when information regarding phase is involved. The general inverse Fourier transform also may be derived from equation (D.1.6).

We may now proceed with the formal definitions of the Fourier transform and its inverse as follows:

If $f(x)$ be a function in ordinary space defined in the domain $(-\infty, \infty)$ and be piece-wise continuous in each finite partial interval and integrable in $(-\infty, \infty)$, then,

$$F[f(x)] = \frac{1}{\sqrt{2\pi}} \int_{-\infty}^{\infty} f(x)e^{+jkx}dx = g(k)$$

is known as the Fourier transform in any space (say k-space) and its inverse gives the function in x-space and is written as

$$F^{-1}\{g(k)\} = \frac{1}{\sqrt{2\pi}} \int_{-\infty}^{\infty} g(k)e^{-jkx}dk = f(x).$$

Sometimes $F[f(x)]$ is also written as $f(k)$.

Fourier *sine* and *cosine* Transforms: The cosine and sine transform pairs are defined as

$$g_c(k) = \sqrt{\frac{2}{\pi}}\int_0^\infty f(x)\cos kx dx, \quad f(x) = \sqrt{\frac{2}{\pi}}\int_0^\infty g_c(k)\cos kx dk,$$

$$g_s(k) = \sqrt{\frac{2}{\pi}}\int_0^\infty f(x)\sin kx dx, \quad f(x) = \sqrt{\frac{2}{\pi}}\int_0^\infty g_s(k)\sin kx dk.$$

The Fourier transform method is a very useful method for solving differential equations involving variables defined in the whole domain ($-\infty$ to ∞).

D.3 Some Important Properties of the Fourier Transform

(a) Linearity property. If $f_1(k)$ and $f_2(k)$ be the Fourier transforms of two functions $f_1(x)$ and $f_2(x)$ respectively, then $F[af_1(x) + bf_2(x)] = af_1(k) + bf_2(k)$, a and b being constants.

Proof. Here we have

$$F[f_1(x)] = f_1(k) = \frac{1}{\sqrt{2\pi}}\int_{-\infty}^{\infty} f_1(x)e^{jkx}dx$$

$$\text{and} \quad F[f_2(x)] = f_2(k) = \frac{1}{\sqrt{2\pi}}\int_{-\infty}^{\infty} f_2(x)e^{jkx}dx,$$

so that we may write

$$\begin{aligned} F[af_1(x) + bf_2(x)] &= \frac{1}{\sqrt{2\pi}}\int_{-\infty}^{\infty}[af_1(x) + bf_2(x)]e^{jkx}dx \\ &= \frac{1}{\sqrt{2\pi}}\int_{-\infty}^{\infty} af_1(x)e^{jkx}dx + \frac{1}{\sqrt{2\pi}}\int_{-\infty}^{\infty} bf_2(x)e^{jkx}dx \\ &= af_1(k) + bf_2(k) \end{aligned}$$

(b) Change of scale property. If $f(k)$ be the Fourier transform of $f(x)$, then $F[f(ax)] = \frac{1}{a}f(k/a)$, a being a constant.

Proof. Here we have

$$f(k) = F[f(x)] = \frac{1}{\sqrt{2\pi}}\int_{-\infty}^{\infty} f(x)e^{jkx}dx; \quad \therefore \quad F[f(ax)] = \frac{1}{\sqrt{2\pi}}\int_{-\infty}^{\infty} f(ax)e^{jkx}dx$$

We put $ax = p$, then $adx = dp$ or, $dx = \frac{1}{a}dp$

$$\begin{aligned} \therefore \quad F[f(ax)] &= \frac{1}{\sqrt{2\pi}}\int_{-\infty}^{\infty}\frac{1}{a}f(p)e^{j(k/a)p}dp = \frac{1}{a}\frac{1}{\sqrt{2\pi}}\int_{-\infty}^{\infty} f(p)e^{jk'p}dp \\ &= \frac{1}{a}F[f(p)] = \frac{1}{a}f(k') = \frac{1}{a}f(k/a). \end{aligned}$$

(c) Shifting property. If $f(k)$ be the complex Fourier transform of the function $f(x)$, then the Fourier transform of $f(x-a)$ is given by, $F[f(x-a)] = e^{jka} f(k)$.

Proof. Here we have

$$f(k) = F[f(x)] = \frac{1}{\sqrt{2\pi}} \int_{-\infty}^{\infty} e^{jkx} f(x) dx; \quad \therefore \quad F[f(x-a)] = \frac{1}{\sqrt{2\pi}} \int_{-\infty}^{\infty} e^{jkx} f(x-a) dx.$$

Let us put $x - a = p$, then $dx = dp$.

$$\therefore \frac{1}{\sqrt{2\pi}} \int_{-\infty}^{\infty} e^{jkx} f(x-a) dx = \frac{1}{\sqrt{2\pi}} \int_{-\infty}^{\infty} f(p) e^{jk(p+a)} dp = e^{jka} \frac{1}{\sqrt{2\pi}} \int_{-\infty}^{\infty} f(p) e^{jkp} dp$$

$$= e^{jka} f(k).$$

(d) If $f(k)$ be the complex Fourier transform of $f(x)$, then the Fourier transform of

$$f(x) \cos bx \text{ is } \tfrac{1}{2}[f(k-b) + f(k+b)].$$

Proof. Here we have

$$f(k) = F[f(x)] = \frac{1}{\sqrt{2\pi}} \int_{-\infty}^{\infty} f(x) e^{jkx} dx,$$

so that we may write

$$F[f(x) \cos bx] = \frac{1}{\sqrt{2\pi}} \int_{-\infty}^{\infty} f(x) \cos bx e^{jkx} dx = \frac{1}{\sqrt{2\pi}} \left[\int_{-\infty}^{\infty} f(x) \cdot \frac{e^{jbx} + e^{-jbx}}{2} \cdot e^{jkx} dx \right]$$

$$= \frac{1}{\sqrt{2\pi}} \left[\frac{1}{2} \int_{-\infty}^{\infty} f(x) e^{j(k+b)x} dx + \frac{1}{2} \int_{-\infty}^{\infty} f(x) e^{j(k-b)x} dx \right]$$

$$= \frac{1}{2}[f(k+b) + f(k-b)]$$

(e) Convolution property. If $f(k)$ and $g(k)$ be the Fourier transforms of $f(x)$ and $g(x)$ respectively, then the Fourier transform of the convolution of $f(x)$ and $g(x)$ is the product of their Fourier transforms.

The convolution of $f(x)$ and $g(x)$ is usually denoted as f^*g and defined as

$$f^*g = \frac{1}{\sqrt{2\pi}} \int_{-\infty}^{\infty} f(u) g(x-u) du.$$

Proof. Here we have

$$f(k) = \frac{1}{\sqrt{2\pi}} \int_{-\infty}^{\infty} f(x) e^{jkx} dx, \qquad g(k) = \frac{1}{\sqrt{2\pi}} \int_{-\infty}^{\infty} g(x) e^{jkx} dx,$$

so that we may write

$$F[f^*g] = F\left[\frac{1}{\sqrt{2\pi}} \int_{-\infty}^{\infty} f(u) g(x-u) du \right] = \frac{1}{\sqrt{2\pi}} \int_{-\infty}^{\infty} \left[\frac{1}{\sqrt{2\pi}} \int_{-\infty}^{\infty} f(u) g(x-u) du \right] e^{jkx} dx$$

$$= \frac{1}{2\pi} \int_{-\infty}^{\infty} f(u) \left[\int_{-\infty}^{\infty} g(x-u) e^{jkx} dx \right] du = \frac{1}{\sqrt{2\pi}} \int_{-\infty}^{\infty} f(u) [e^{jku} g(k)] du$$

$$= g(k) \cdot \frac{1}{\sqrt{2\pi}} \int_{-\infty}^{\infty} f(u) e^{jku} du = g(k) f(k)$$

(f) Fourier transform of the derivative. If $f(k)$ is the Fourier transform of $f(x)$, then the Fourier transform of $f'(x)$, the first derivative of $f(x)$, is $-jkf(k)$.

Proof. Here we have

$$f(k) = F[f(x)] = \frac{1}{\sqrt{2\pi}}\int_{-\infty}^{\infty} f(x)e^{jkx}\,dx,$$

$$\begin{aligned}
\therefore\ F[f'(x)] &= \frac{1}{\sqrt{2\pi}}\int_{-\infty}^{\infty} f'(x)e^{jkx}\,dx \\
&= \frac{1}{\sqrt{2\pi}}\int_{-\infty}^{\infty} \lim_{h\to 0}\frac{f(x+h)-f(x)}{h}\cdot e^{jkx}\,dx \\
&= \frac{1}{\sqrt{2\pi}}\lim_{h\to 0}\int_{-\infty}^{\infty}\frac{f(x+h)}{h}\cdot e^{jkx}\,dx - \frac{1}{\sqrt{2\pi}}\lim_{h\to 0}\int_{-\infty}^{\infty}\frac{f(x)}{h}e^{jkx}\,dx \\
&= \lim_{h\to 0}\frac{1}{\sqrt{2\pi}}\int_{-\infty}^{\infty}\frac{f(x+h)}{h}e^{jk(x+h)}\cdot d(x+h)\cdot e^{-jkh} - \lim_{h\to 0}\frac{1}{\sqrt{2\pi}}\int_{-\infty}^{\infty}\frac{f(x)}{h}e^{jkx}\,dx \\
&= \lim_{h\to 0}\frac{f(k)\cdot e^{-jkh}}{h} - \lim_{h\to 0}\frac{f(k)}{h} = f(k)\lim_{h\to 0}\left[\frac{e^{-jkh}}{h}-\frac{1}{h}\right] = f(k)\lim_{h\to 0}\left[\frac{e^{-jkh}-1}{h}\right] \\
&= f(k)\lim_{h\to 0}\left[\frac{1-jkh+\frac{1}{2!}(jkh)^2-\cdots-1}{h}\right]
\end{aligned}$$

(**N.B.** We can use L' Hospital's theorem to evaluate the above limit.)

$$= f(k)(-jk) = -jkf(k)$$

For the nth derivative, the Fourier transform is equal to $(-jk)^n$ times $f(k)$ if $f(k)$ is the Fourier transform of $f(x)$ and if the first $(n-1)$ derivatives of $f(x)$ vanish as $x\to\pm\infty$, i.e., $F[f^n(x)] = (-jk)^n f(k)$.

D.3.1 Some Important Properties of Sine and Cosine Fourier Transforms

(i) If $f_s(k)$ be the sine transform of $f(x)$, then Fourier transform of $f(ax)$ is $\frac{1}{a}f_s(k/a)$. Similar property is also possessed by cosine transform.

(ii) If $f_s(k)$ and $f_c(k)$ be the sine and cosine transforms of $f(x)$, then

$$\begin{aligned}
F_s[f(x)\cos bx] &= \frac{1}{2}[f_s(k+b)+f_s(k-b)] \\
F_c[f(x)\sin bx] &= \frac{1}{2}[f_s(k+b)-f_s(k-b)] \\
F_s[f(x)\cos bx] &= \frac{1}{2}[f_c(k-b)-f_c(k+b)]
\end{aligned}$$

Illustrative Examples

1. (a) Find the Fourier transform of $f(x)$ defined by

$$f(x) = \begin{cases} 1, & -a \le x \le a \\ 0, & |x| > a \end{cases}$$

Ans. Here we have

$$\begin{aligned} f(k) = F[f(x)] &= \frac{1}{\sqrt{2\pi}} \int_{-\infty}^{\infty} f(x) e^{jkx} dx \\ &= \frac{1}{\sqrt{2\pi}} \int_{-\infty}^{-a} f(x) e^{jkx} dx + \frac{1}{\sqrt{2\pi}} \int_{-a}^{a} f(x) e^{jkx} dx + \frac{1}{\sqrt{2\pi}} \int_{a}^{\infty} f(x) e^{jkx} dx \\ &= \frac{1}{\sqrt{2\pi}} \int_{-a}^{a} f(x) e^{jkx} dx = \frac{1}{\sqrt{2\pi}} \int_{-a}^{a} e^{jkx} dx = \frac{1}{\sqrt{2\pi}} \cdot \frac{1}{jk} e^{jkx} \Big|_{-a}^{a} \\ &= \frac{1}{\sqrt{2\pi}} \cdot \frac{1}{jk} (e^{jka} - e^{-jka}) = \frac{1}{\sqrt{2\pi}} \cdot \frac{2 \sin ka}{k}, k \neq 0. \end{aligned}$$

For $k = 0$, $f(k) = \dfrac{1}{\sqrt{2\pi}} \cdot 2a \cdot \dfrac{\sin ka}{ka} = \dfrac{2a}{\sqrt{2\pi}}$.

(b) Find the Fourier transform of $e^{-x^2/2}$.

Ans. Here we may write

$$\begin{aligned} F[e^{-x^2/2}] &= \frac{1}{\sqrt{2\pi}} \int_{-\infty}^{\infty} e^{-x^2/2} \cdot e^{jkx} dx = \frac{1}{\sqrt{2\pi}} \int_{-\infty}^{\infty} e^{-\frac{1}{2}(x-jk)^2} \cdot e^{-k^2/2} dx \\ &= \frac{e^{-k^2/2}}{\sqrt{2\pi}} \int_{-\infty}^{\infty} e^{-\frac{1}{2}(x-jk)^2} dx \end{aligned}$$

Let us put $\frac{1}{\sqrt{2}}(x - jk) = y$, then $dx = \sqrt{2} dy$

$$\therefore \quad F[e^{-x^2/2}] = \frac{e^{-k^2/2}}{\sqrt{2\pi}} \int_{-\infty}^{\infty} e^{-y^2} \cdot \sqrt{2} dy = \frac{e^{-k^2/2}}{\sqrt{\pi}} \cdot \sqrt{\pi} = e^{-k^2/2}.$$

2. Find the Fourier transform of $f(x)$ defined by

$$f(x) = \begin{cases} e^{jwx}, & a < x < b \\ 0, & x < a, x > b \end{cases}$$

Ans. Here we have

$$
\begin{aligned}
F[f(x)] &= f(k) = \frac{1}{\sqrt{2\pi}}\int_{-\infty}^{\infty} f(x)e^{jkx}dx \\
&= \frac{1}{\sqrt{2\pi}}\int_{-\infty}^{a} f(x)e^{jkx}dx + \frac{1}{\sqrt{2\pi}}\int_{a}^{b} f(x)e^{jkx}dx + \frac{1}{\sqrt{2\pi}}\int_{b}^{\infty} f(x)e^{jkx}dx \\
&= \frac{1}{\sqrt{2\pi}}\int_{a}^{b} f(x)e^{jkx}dx = \frac{1}{\sqrt{2\pi}}\int_{a}^{b} e^{jwx}\cdot e^{jkx}dx \\
&= \frac{1}{\sqrt{2\pi}}\int_{a}^{b} e^{j(w+k)x}dx = \frac{1}{\sqrt{2\pi}}\cdot\frac{1}{j(w+k)}e^{j(w+k)x}\Big|_a^b \\
&= \frac{1}{\sqrt{2\pi}}\cdot\frac{1}{j(w+k)}\left[e^{j(w+k)b} - e^{j(w+k)a}\right] \\
&= \frac{j}{\sqrt{2\pi}}\cdot\frac{1}{(w+k)}\left[e^{j(w+k)a} - e^{-j(w+k)b}\right].
\end{aligned}
$$

3. Find the Fourier sine transform of $f(x)$, where $f(x) = 1/x$.

Ans. $F_s[f(x)] = f_s(k) = \sqrt{\frac{2}{\pi}}\int_0^{\infty} f(x)\sin kx dx = \sqrt{\frac{2}{\pi}}\int_0^{\infty}\frac{1}{x}\sin kx dx.$

Now, $\int_0^{\infty} e^{-\alpha x}\sin kx dx = \frac{k}{\alpha^2+k^2}.$

Integrating w.r.t. α between α_1 and α_2, we obtain

$$\int_0^{\infty}\left[\int_{\alpha_2}^{\alpha_1} e^{-\alpha x}d\alpha\right]\sin kx dx = \int_{\alpha_2}^{\alpha_1}\frac{k}{(\alpha^2+k^2)}d\alpha$$

or, $$\int_0^{\infty}\frac{e^{-\alpha_1 x} - e^{-\alpha_2 x}}{x}\sin kx dx = \tan^{-1}\frac{\alpha_2}{k} - \tan^{-1}\frac{\alpha_1}{k}$$

As $\alpha_1 \to 0$ and $\alpha_2 \to \infty$, we obtain,

$$\int_0^{\infty}\frac{\sin kx}{x}dx = \frac{\pi}{2}.$$

$$\therefore\quad f_s(k) = \sqrt{\frac{2}{\pi}}\int_0^{\infty}\frac{\sin kx}{x}dx = \sqrt{\frac{2}{\pi}}\times\frac{\pi}{2} = \sqrt{\frac{\pi}{2}}.$$

4. Find the Fourier cosine transform of $f(x) = \frac{1}{1+x^2}$.

Ans. Here we have

$$F_c[f(x)] = f_c(k) = \sqrt{\frac{2}{\pi}}\int_0^{\infty} f(x)\cos kx dx = \sqrt{\frac{2}{\pi}}\int_0^{\infty}\frac{1}{1+x^2}\cos kx dx.$$

Differentiating both sides w.r.t. k, we get,

$$\begin{aligned}\frac{d}{dk}[f_c(k)] &= -\sqrt{\frac{2}{\pi}}\int_0^\infty \frac{x\sin kx}{1+x^2}dx = -\sqrt{\frac{2}{\pi}}\int_0^\infty \frac{(1+x^2-1)\sin kx}{x(1+x^2)}dx \\ &= -\sqrt{\frac{2}{\pi}}\left[\int_0^\infty \frac{\sin kx}{x}dx - \int_0^\infty \frac{1}{x(1+x^2)}\sin kx dx\right] \\ &= -\sqrt{\frac{2}{\pi}}\times\frac{\pi}{2}+\sqrt{\frac{2}{\pi}}\int_0^\infty \frac{\sin kx}{x(1+x^2)}dx. \quad (1)\end{aligned}$$

Differentiating again w.r.t. k, we get,

$$\frac{d^2}{dk^2}[f_c(k)] = \sqrt{\frac{2}{\pi}}\int_0^\infty \frac{x\cos kx}{x(1+x^2)}dx = f_c(k).$$

On solving we have the solution given by

$$f_c(k) = c_1e^k + c_2e^{-k}. \quad (2)$$

Now, when $k = 0$, from

$$f_c(k) = \sqrt{\frac{2}{\pi}}\int_0^\infty \frac{\cos kx}{1+x^2}dx,$$

we may write

$$f_c(k)\Big|_{k=0} = \sqrt{\frac{2}{\pi}}\int_0^\infty \frac{1}{1+x^2}dx = \sqrt{\frac{2}{\pi}}\tan^{-1}x\Big|_0^\infty = \sqrt{\frac{2}{\pi}}\times\frac{\pi}{2} = \sqrt{\frac{\pi}{2}}.$$

$$\therefore \quad c_1 + c_2 = \sqrt{\frac{\pi}{2}} \quad \text{[from (2)]}$$

Again,

$$\frac{d}{dk}f_c(k) = -\sqrt{\frac{\pi}{2}} \text{ at } k = 0. \quad \text{[from (1)]}$$

$$\therefore \quad c_1 - c_2 = -\sqrt{\frac{\pi}{2}}. \quad \text{[from (2)]}$$

Solving for c_1 and c_2 we get, $c_1 = 0$, $c_2 = \sqrt{\frac{\pi}{2}}$.

$$\therefore \quad f_c(k) = \sqrt{\frac{\pi}{2}}e^{-k}.$$

5. Find the Fourier cosine transform of $f(x) = e^{-x^2}$.

Ans. Here we have

$$\begin{aligned}F_c[f(x)] &= f_c(k) = \sqrt{\frac{2}{\pi}}\int_0^\infty f(x)\cos kx dx \\ &= \sqrt{\frac{2}{\pi}}\int_0^\infty e^{-x^2}\cos kx dx = I \quad \text{(say)}\end{aligned}$$

Differentiating w.r.t. k, we obtain,

$$\frac{dI}{dk} = \sqrt{\frac{2}{\pi}}\int_0^\infty -xe^{-x^2}\sin kx dx = \frac{1}{2}\cdot\sqrt{\frac{2}{\pi}}\int_0^\infty \left(-2xe^{-x^2}\right)\sin kx dx$$

$$= \frac{1}{2}\cdot\sqrt{\frac{2}{\pi}}\left[\int_0^\infty d\left(e^{-x^2}\right)\sin kx\right]$$

$$= \frac{1}{2}\sqrt{\frac{2}{\pi}}\left[\sin kx\, e^{-x^2}\Big|_0^\infty - \int_0^\infty k\cos kx\, e^{-x^2}dx\right] \quad \text{(on integration by parts)}$$

$$= -\frac{1}{2}\sqrt{\frac{2}{\pi}}k\int_0^\infty e^{-x^2}\cos kx dx = -\frac{k}{2}I.$$

$$\therefore \quad \frac{dI}{I} = -\frac{k}{2}dk \quad \text{or,} \quad \ln I = -\frac{k^2}{4} + \ln A \quad \text{or,} \quad I = Ae^{-k^2/4}. \tag{1}$$

Now,

$$I = \sqrt{\frac{2}{\pi}}\int_0^\infty e^{-x^2}\cos kx dx; \quad \text{when } k = 0,$$

$$I = \sqrt{\frac{2}{\pi}}\int_0^\infty e^{-x^2}dx = \sqrt{\frac{2}{\pi}}\times\frac{\sqrt{\pi}}{2} = \frac{1}{\sqrt{2}}$$

$$\therefore A = \frac{1}{\sqrt{2}} \quad \text{[from (1)]}.$$

Hence, $f_c(k) = I = \frac{1}{\sqrt{2}}e^{-k^2/4}$.

D.4 Applications of Fourier Transforms

(a) Evaluation of some important integrals

Evaluate the integrals: (i) $\int_0^\infty \frac{\cos kx}{a^2+k^2}dk$ and (ii) $\int_0^\infty \frac{k\sin kx}{a^2+k^2}dk$.

Let

$$I_1 = \sqrt{\frac{2}{\pi}}\int_0^\infty e^{-ax}\cos kx dx \quad \text{and} \quad I_2 = \sqrt{\frac{2}{\pi}}\int_0^\infty e^{-ax}\sin kx dx.$$

Now,

$$I_1 = \sqrt{\frac{2}{\pi}}\int_0^\infty e^{-ax}\cos kx dx = \sqrt{\frac{2}{\pi}}\left[-\frac{1}{a}\cos kx e^{-ax}\Big|_0^\infty + \int_0^\infty k\sin kx\times\left(-\frac{1}{a}\right)e^{-ax}dx\right]$$

(on integration by parts)

$$= \left(\sqrt{\frac{2}{\pi}}\cdot\frac{1}{a} - \frac{k}{a}I_2\right).$$

Similarly, $I_2 = \frac{k}{a} I_1$.
Solving, we obtain

$$I_1 = \sqrt{\frac{2}{\pi}} \frac{a}{a^2 + k^2}, \quad \text{and} \quad I_2 = \sqrt{\frac{2}{\pi}} \frac{k}{a^2 + k^2}.$$

If we take $f(x) = e^{-ax}$, then,

$$I_1 = F_c[f(x)] = f_c(k) = \sqrt{\frac{2}{\pi}} \frac{a}{a^2 + k^2} = \sqrt{\frac{2}{\pi}} \int_0^\infty e^{-ax} \cos kx dx$$

$$\text{and} \quad I_2 = F_s[f(x)] = f_s(k) = \sqrt{\frac{2}{\pi}} \frac{k}{a^2 + k^2} = \sqrt{\frac{2}{\pi}} \int_0^\infty e^{-ax} \sin kx dx$$

By inversion we get,

$$e^{-ax} = \sqrt{\frac{2}{\pi}} \int_0^\infty f_c(k) \cos kx dk = \frac{2}{\pi} \int_0^\infty \frac{a}{k^2 + a^2} \cos kx dk$$

$$\text{or,} \quad \int_0^\infty \frac{a}{a^2 + k^2} \cos kx dk = \frac{\pi}{2} e^{-ax}, \quad \text{or,} \quad \int_0^\infty \frac{\cos kx}{a^2 + k^2} dk = \frac{\pi}{2a} e^{-ax}.$$

Similarly,

$$\int_0^\infty \frac{k}{a^2 + k^2} \sin kx dk = \frac{\pi}{2} e^{-ax}.$$

(b) Parseval's relation and evaluation of integral $\int_0^\infty \frac{\sin^2 ax}{x^2} dx$.

The Parseval's relation or identity, as it is sometimes called, is defined as follows.

(i) If $f(k)$ be the Fourier transform of $f(x)$, then $\int_{-\infty}^{\infty} |f(x)|^2 dx = \int_{-\infty}^{\infty} |f(k)^2| dk$,

(ii) $\int_{-\infty}^{\infty} f(x) g^*(x) dx = \int_{-\infty}^{\infty} f(k) g^*(k) dk$,

where $g^*(x)$ is the complex conjugate of $g(x)$.

Illustrative Examples

1. Evaluate $\int_0^\infty \frac{\sin^2 ax}{x^2} dx$.

Ans. Let

$$f(x) = \begin{cases} 1, & |x| < a \\ 0, & |x| > a \end{cases} \qquad \therefore \quad f(k) = \frac{1}{\sqrt{2\pi}} \frac{2 \sin ka}{k}$$

(See Illustrative Ex. 1 at the end of Art. D.3.)

According to Parseval's identity,

$$\int_{-\infty}^{\infty} |f(x)|^2 dx = \int_{-\infty}^{\infty} |f(k)|^2 dk = \frac{1}{2\pi} \int_{-\infty}^{\infty} \frac{4 \sin^2 ka}{k^2} dk$$

$$\text{or,} \quad \int_{-\infty}^{\infty} \frac{\sin^2 ka}{k^2} dk = \frac{\pi}{2} \int_{-\infty}^{\infty} |f(x)|^2 dx$$

Now, the meaning of $|x|$ is given as, $|x| = \begin{cases} -x, & \text{when } x \text{ is negative} \\ x, & \text{when } x \text{ is positive} \end{cases}$

So

$$\begin{aligned} \int_{-\infty}^{\infty} |f(x)|^2 dx &= \int_{-\infty}^{-a} |f(x)|^2 dx + \int_{-a}^{a} |f(x)|^2 dx + \int_{a}^{\infty} |f(x)|^2 dx \\ &= \int_{-a}^{a} |f(x)|^2 dx = \int_{-a}^{a} 1^2 dx = 2a. \end{aligned}$$

$$\therefore \int_{-\infty}^{\infty} \frac{\sin^2 ka}{k^2} dk = \frac{\pi}{2} \times 2a = \pi a, \text{ or, } 2\int_{0}^{\infty} \frac{\sin^2 ka}{k^2} dk = \pi a, \text{ or, } \int_{0}^{\infty} \frac{\sin^2 ax}{x^2} dx = \frac{\pi a}{2}.$$

2. Evaluate $\displaystyle\int_0^{\infty} \frac{\cos \lambda x}{\lambda^2 + 1} d\lambda.$

Ans. This is similar to $\displaystyle\int_0^{\infty} \frac{\cos kx}{a^2 + k^2} dk$; here $\lambda = k$ and $a = 1$.

$$\therefore \int_0^{\infty} \frac{\cos \lambda x}{1 + \lambda^2} d\lambda = \frac{\pi}{2} e^{-x}.$$

D.5 Application of Fourier Transforms to Solve Differential Equations

(a) Solution of Ordinary Differential Equation

We consider the problem of damped harmonic oscillator to be solved by Fourier transform method.

The equation of motion of such an oscillator is given by

$$\frac{d^2 x}{dt^2} + 2b\frac{dx}{dt} + w_0^2 x = f(t), \tag{D.5.1}$$

where $f(t)$ is any function such that

$$f(t) = \frac{1}{\sqrt{2\pi}} \int_{-\infty}^{\infty} f(w) e^{-jwt} dw \quad \text{and} \quad f(w) = F[f(t)] = \frac{1}{\sqrt{2\pi}} \int_{-\infty}^{\infty} f(t) e^{jwt} dt$$

Taking Fourier transform of both sides of the above equation, we obtain,

$$F\left[\frac{d^2x}{dt^2}\right] + 2bF\left[\frac{dx}{dt}\right] + w_0^2 F[x] = F[f(t)]$$

or, $$\frac{1}{\sqrt{2\pi}}\int_{-\infty}^{\infty}\frac{d^2x}{dt^2}e^{jwt}dt + \frac{2b}{\sqrt{2\pi}}\int_{-\infty}^{\infty}\frac{dx}{dt}e^{jwt}dt + \frac{w_0^2}{\sqrt{2\pi}}\int_{-\infty}^{\infty}xe^{jwt}dt$$

$$= \frac{1}{\sqrt{2\pi}}\int_{-\infty}^{\infty} f(t)e^{jwt}dt$$

or, $$(+jw)^2 \times (w) + 2b(+jw)^1 \times (w) + w_0^2 \times (w) = f(w)$$

or, $$(-w^2 + j2bw + w_0^2)X = f(w) \quad \text{or,} \quad X = \frac{f(w)}{w_0^2 - w^2 + 2jbw}$$

By inversion we get,

$$x(t) = \frac{1}{\sqrt{2\pi}}\int_{-\infty}^{\infty}\frac{f(w)e^{-jwt}}{w_0^2 - w^2 + 2jbw}dw.$$

(b) Solution of partial differential equations

(i) Heat conduction equation. $\frac{\partial u}{\partial t} = h^2\frac{\partial^2 u}{\partial x^2}$, given $u(x,0) = f(x)$:

Let us define the following Fourier transforms:

$$u(k,t) = \frac{1}{\sqrt{2\pi}}\int_{-\infty}^{\infty} u(x,t)e^{jkx}dx \tag{D.5.2}$$

and $$f(k) = \frac{1}{\sqrt{2\pi}}\int_{-\infty}^{\infty} f(x)e^{jkx}dx \tag{D.5.3}$$

Taking Fourier transform of both sides of the heat conduction equation, we get,

$$h^2\cdot\frac{1}{\sqrt{2\pi}}\int_{-\infty}^{\infty}\frac{\partial^2 u}{\partial x^2}e^{jkx}du = \frac{1}{\sqrt{2\pi}}\int_{-\infty}^{\infty}\frac{\partial u}{\partial t}e^{jkx}dx$$

or, $$h^2(+jk)^2u(k,t) = \frac{1}{\sqrt{2\pi}}\frac{\partial}{\partial t}\int_{-\infty}^{\infty} ue^{jkx}dx = \frac{\partial}{\partial t}u(k,t)$$

or, $$-h^2k^2u(k,t) = \frac{\partial}{\partial t}u(k,t) \quad \text{or,} \quad \frac{du}{u} = -h^2k^2dt$$

or, $$\ln u(k,t) = -h^2k^2t + \ln A \quad \text{or,} \quad u(k,t) = Ae^{-h^2k^2t}.$$

$$\therefore \quad u(k,0) = A$$

Again from eqn. (D.5.2), we get

$$u(k,0) = \frac{1}{\sqrt{2\pi}}\int_{-\infty}^{\infty} u(x,0)e^{jx}dx = \frac{1}{\sqrt{2\pi}}\int_{-\infty}^{\infty} f(x)e^{jx}dx = f(k).$$

$$\therefore \quad A = f(k). \qquad \therefore \quad u(k,t) = f(k)e^{-h^2k^2t}.$$

By inversion, we get,

$$u(x,t) = \frac{1}{\sqrt{2\pi}}\int_{-\infty}^{\infty} f(k)e^{-h^2k^2t}e^{-jkx}dk = \frac{1}{\sqrt{2\pi}}\int_{-\infty}^{\infty}\left(f(k)e^{-h^2k^2t}\right)e^{-jkx}dk$$

$$= \frac{1}{\sqrt{2\pi}}\int_{-\infty}^{\infty}[f(k)g(k)]e^{-jkx}dk$$

By convolution theorem

$$\frac{1}{\sqrt{2\pi}}\int_{-\infty}^{\infty} f(k)g(k)e^{jkx}dk = \frac{1}{\sqrt{2\pi}}\int_{-\infty}^{\infty} f(\xi)g(x-\xi)d\xi,$$

where $g(x-\xi)$ is the Fourier transform inverse of $e^{-h^2k^2t}$, i.e.,

$$g(x) = \frac{1}{\sqrt{2\pi}}\int_{-\infty}^{\infty} e^{-h^2k^2t}\cdot e^{-jkx}dk.$$

Let us put $u^2 = 2k^2h^2t$, then $udu = 2h^2tkdk$

$$\text{or,}\quad dk = \frac{udu}{2h^2tk} = \frac{\sqrt{2}kht^{1/2}du}{2h^2tk} = \frac{1}{\sqrt{2}}\frac{du}{h\sqrt{t}} = \frac{du}{h\sqrt{2t}}.$$

$$\therefore\ g(x) = \frac{1}{\sqrt{2\pi}}\int_{-\infty}^{\infty} e^{-u^2/2}e^{-jkx}dk = \frac{1}{\sqrt{2\pi}}\int_{-\infty}^{\infty} e^{-u^2/2}\cdot e^{-j(u/h\sqrt{2t})x}\cdot\frac{du}{h\sqrt{2t}}$$

$$= \frac{1}{h\sqrt{4\pi t}}\int_{-\infty}^{\infty} e^{-u^2/2}\cdot e^{-jk'u}du = \frac{1}{h\sqrt{2t}}\cdot\frac{1}{\sqrt{2\pi}}\int_{-\infty}^{\infty} e^{-u^2/2}\cdot e^{-jk'u}du,\ \left(k' = \frac{x}{h\sqrt{2t}}\right)$$

$$= \frac{1}{h\sqrt{2t}}e^{-k'^2/2} = \frac{1}{h\sqrt{2t}}e^{-x^2/4h^2t}.$$

$$\therefore\quad u(x,t) = \frac{1}{\sqrt{2\pi}}\int_{-\infty}^{\infty} f(\xi)\cdot\frac{1}{h\sqrt{2t}}e^{-(x-\xi)^2/4h^2t}d\xi$$

$$= \frac{1}{2h\sqrt{\pi t}}\int_{-\infty}^{\infty} f(\xi)e^{-(x-\xi)^2/4h^2t}d\xi$$

(ii) Wave equation: We consider the vibration of a string fixed at both ends, satisfying the wave equation

$$\frac{\partial^2 u}{\partial t^2} = c^2\frac{\partial^2 u}{\partial x^2},$$

under the condition that at $t = 0$, the string is at rest but it has a displacement $f(x)$ for $-\infty < x < \infty$.

Taking Fourier transform of both sides of the wave equation we can write

$$\frac{1}{\sqrt{2\pi}}\int_{-\infty}^{\infty}\left(\frac{\partial^2 u}{\partial t^2}\right)e^{jkx}dx = \frac{c^2}{\sqrt{2\pi}}\int_{-\infty}^{\infty}\left(\frac{\partial^2 u}{\partial x^2}\right)e^{jkx}dx$$

$$\frac{d^2u(k,t)}{dt^2} = c^2(+jk)^2u(k,t) = -c^2k^2u(k,t).$$

The general solution to the above eqn. is given by

$$u(k,t) = c_1 e^{jckt} + c_2 e^{-jckt}. \tag{D.5.4}$$

Initially the string is at rest, so $\frac{\partial u}{\partial t} = 0$ at $t = 0$.

$$\therefore \quad F\left[\frac{\partial u}{\partial t}\right] = 0 \text{ or, } \frac{1}{\sqrt{2\pi}}\int_{-\infty}^{\infty}\left(\frac{\partial u}{\partial t}\right)e^{jkx}dx = 0$$

$$\text{or, } \frac{1}{\sqrt{2\pi}}\frac{\partial}{\partial t}\int_{-\infty}^{\infty} ue^{jkx}dx = 0$$

$$\text{or, } \frac{\partial}{\partial t}u(k,0) = 0$$

Differentiating eqn. (D.5.4) w.r.t. t, we get,

$$\frac{\partial u}{\partial t} = jck(c_1e^{jckt} - c_2e^{-jckt}). \quad \therefore \quad \left.\frac{\partial u}{\partial t}\right|_{t=0} = 0 = (c_1 - c_2).$$

Also, at $t = 0$

$$u(x,0) = f(x).$$

$$\therefore \quad u(k,0) = \frac{1}{\sqrt{2\pi}}\int_{-\infty}^{\infty} f(x)e^{jkx}dx = c_1 + c_2, \quad \text{[from (D.5.4)]}$$

$$\therefore \quad c_1 - c_2 = 0 \quad \text{and} \quad c_1 + c_2 = u(k,0)$$

Solving we get

$$c_1 = \frac{u}{2}(k,0), \quad c_2 = \frac{u}{2}(k,0) \qquad \therefore \quad u(k,t) = \frac{u}{2}(k,0)\left(e^{jckt} + e^{-jckt}\right).$$

Inverting the above eqn., we get

$$\begin{aligned}
u(x,t) &= \frac{1}{\sqrt{2\pi}}\int_{-\infty}^{\infty} u(k,t)e^{-jkx}dk \\
&= \frac{1}{\sqrt{2\pi}}\left[\int_{-\infty}^{\infty}\frac{u}{2}(k,0)\left(e^{jckt} + e^{-jckt}\right)e^{-jkx}dk\right] \\
&= \frac{1}{\sqrt{2\pi}}\left[\frac{1}{2}\int_{-\infty}^{\infty} u(k,0)e^{jckt}\cdot e^{-jkx}dk + \frac{1}{2}\int_{-\infty}^{\infty} u(k,0)e^{-jckt}\cdot e^{-jkx}dk\right] \\
&= \frac{1}{\sqrt{2\pi}}\left[\frac{1}{2}\int_{-\infty}^{\infty} u(k,0)e^{jk(ct-x)}dk + \frac{1}{2}\int_{-\infty}^{\infty} u(k,0)e^{-jk(ct+x)}dk\right] \\
&= \frac{1}{2}\left[\frac{1}{\sqrt{2\pi}}\int_{-\infty}^{\infty} u(k,0)e^{-jk\sigma}dk + \frac{1}{\sqrt{2\pi}}\int_{-\infty}^{\infty} u(k,0)e^{-jk\sigma'}dk\right] \\
&= \frac{1}{2}[f(x-ct) + f(x+ct)] \quad \{\sigma = x - ct \text{ and } \sigma' = x + ct\} \\
&= \frac{1}{2}f(x+ct) + \frac{1}{2}f(x-ct)
\end{aligned} \tag{D.5.5}$$

(iii) Solution of Poisson's equation: Let us take Poisson's equation represented as

$$\nabla^2 u(\bar{r}) = -\frac{\rho(\bar{r})}{\varepsilon_0},$$

where $u(\bar{r})$is the potential function and $\rho(\bar{r})$ is the charge density.

Let us write in three dimensions, the Fourier transform of $u(\bar{r})$ as $u(\bar{k})$, so that

$$u(\bar{k}) = \frac{1}{(2\pi)^{3/2}} \int_{-\infty}^{\infty} u(\bar{r}) e^{j\bar{k}\cdot\bar{r}} d^3r, \quad \text{where } d^3r = dxdydz.$$

The inverse transform is given by

$$u(\bar{r}) = \frac{1}{(2\pi)^{3/2}} \int_{-\infty}^{\infty} u(\bar{k}) e^{-j\bar{k}\cdot\bar{r}} d^3k,$$

where d^3k is the elementary volume in k-space.

Taking Fourier transform of both sides of the above eqn., we get

$$-k^2 u(\bar{k}) = -\frac{\rho(\bar{k})}{\varepsilon_0} \tag{D.5.6}$$

$$\text{or,} \quad u(\bar{k}) = \frac{\rho(\bar{k})}{\varepsilon_0 k^2}, \text{ where } \rho(\bar{k}) = \frac{1}{(2\pi)^{3/2}} \int_{-\infty}^{\infty} \rho(\bar{r}') e^{j\bar{k}\cdot\bar{r}} d^3r'$$

So Fourier inversion of the above eqn. gives,

$$\begin{aligned} u(\bar{r}) &= \frac{1}{\varepsilon_0 (2\pi)^{3/2}} \int_{-\infty}^{\infty} \frac{\rho(\bar{k})}{k^2} e^{-j\bar{k}\cdot\bar{r}} d^3k \\ &= \frac{1}{(2\pi)^{3/2}\varepsilon_0} \int_{-\infty}^{\infty} \left[\frac{1}{(2\pi)^{3/2}} \int_{-\infty}^{\infty} \frac{\rho(\bar{r}')}{k^2} e^{j\bar{k}\cdot\bar{r}} d^3r' \right] e^{-j\bar{k}\cdot\bar{r}} d^3k \\ &= \frac{1}{(2\pi)^3 \epsilon_0} \int_{-\infty}^{\infty} \int_{-\infty}^{\infty} \rho(\bar{r}') \frac{e^{-\bar{k}\cdot(\bar{r}-\bar{r}')}}{k^2} d^3k d^3r'. \end{aligned}$$

Now, in the expression

$$\int_{-\infty}^{\infty} \frac{e^{-j\bar{k}\cdot(\bar{r}-\bar{r}')}}{k^2} d^3k,$$

d^3k is the elementary volume in k-space and if spherical polar co-ordinates are chosen in k-space, then $d^3k = k^2 \sin\theta d\theta d\phi dk$.

$$\begin{aligned} \therefore \quad \int_{-\infty}^{\infty} \frac{e^{-j\bar{k}\cdot(\bar{r}-\bar{r}')}}{k^2} d^3k &= \int_0^{\infty} \int_0^{\pi} \int_0^{2\pi} \frac{e^{-jk|\bar{r}-\bar{r}'|\cos\theta}}{k^2} k^2 \sin\theta d\theta d\phi dk \\ &= \int_0^{\infty} \left[\int_0^{\pi} \left(e^{-jk|\bar{r}-\bar{r}'|\cos\theta} \sin\theta d\theta \right) \right] dk \times 2\pi \\ &= 2\pi \int_0^{\infty} \left[\int_0^{\pi} e^{-jk|\bar{r}-\bar{r}'|\cos\theta} \sin\theta d\theta \right] dk \end{aligned} \tag{D.5.7}$$

Now to consider the integral

$$\int_0^\pi e^{-jk|\bar{r}-\bar{r}'|\cos\theta}\sin\theta d\theta,$$

let us put $-jk|\bar{r}-\bar{r}'|\cos\theta = x$; then $jk|\bar{r}-\bar{r}'|\sin\theta d\theta = dx$.

$$\therefore \quad \int e^{-jk|\bar{r}-\bar{r}'|\cos\theta}\sin\theta d\theta = \int e^x \frac{dx}{jk|\bar{r}-\bar{r}'|} = \frac{e^x}{jk|\bar{r}-\bar{r}'|}$$

$$\therefore \quad \int_0^\pi e^{-jk|\bar{r}-\bar{r}'|\cos\theta}\sin\theta d\theta = \left.\frac{e^{-jk|\bar{r}-\bar{r}'|\cos\theta}}{jk|\bar{r}-\bar{r}'|}\right|_0^\pi = \frac{e^{jk|\bar{r}-\bar{r}'|}-e^{-jk|\bar{r}-\bar{r}'|}}{jk|\bar{r}-\bar{r}'|}$$

So the integral in (D.5.7) can be written as

$$\int_0^\infty \frac{e^{+jk|\bar{r}-\bar{r}'|}-e^{-jk|\bar{r}-\bar{r}'|}}{jk|\bar{r}-\bar{r}'|}dk = 2\int_0^\infty \frac{\sin k|\bar{r}-\bar{r}'|}{k|\bar{r}-\bar{r}'|}dk$$

$$\therefore \quad \int_{-\infty}^\infty \frac{e^{-j\bar{k}.|\bar{r}-\bar{r}'|}}{k^2}d^3k = 4\pi\int_0^\infty \frac{\sin k|\bar{r}-\bar{r}'|}{k|\bar{r}-\bar{r}'|}dk = \frac{4\pi}{|\bar{r}-\bar{r}'|}\int_0^\infty \frac{\sin k|\bar{r}-\bar{r}'|}{k}dk$$

$$= \frac{4\pi}{|\bar{r}-\bar{r}'|}\cdot\frac{\pi}{2} = \frac{2\pi^2}{|\bar{r}-\bar{r}'|}$$

$$\therefore \quad u(\bar{r}) = \frac{1}{(2\pi)^3\varepsilon_0}\int_{-\infty}^\infty \rho(\bar{r}')d^3r'\cdot\frac{2\pi^2}{|\bar{r}-\bar{r}'|} = \frac{1}{4\pi\varepsilon_0}\int_{-\infty}^\infty \frac{\rho(\bar{r}')}{|\bar{r}-\bar{r}'|}d^3r' \quad \text{(D.5.8)}$$

The solution as given by equation (D.5.8) is in accordance with the definition of potential. $\rho(\bar{r}')$ is the charge density at $P'(\bar{r}')$; then $\rho(\bar{r}')d^3r'$ defines the charge associated with the volume element d^3r'; hence the potential due to this charge at the point $P(\bar{r})$ distant $|\bar{r}-\bar{r}'|$ with respect to $P'(\bar{r}')$ is given by

$$\frac{1}{4\pi\epsilon_0}\frac{\rho(\bar{r}')d^3r'}{|\bar{r}-\bar{r}'|}.$$

Hence the total potential at P is given by

$$u(r) = \frac{1}{4\pi\epsilon_0}\int_{-\infty}^\infty \frac{\rho(\bar{r}')d^3r'}{|\bar{r}-\bar{r}'|}.$$

D.6 The Dirac-delta Function—its important properties—its Fourier Transform and some Simple illustrations

(a) The Dirac-delta function is used to get a precise notation for dealing with quantities involving a certain kind of infinity, i.e., the quantities having infinitely high, infinitely thin spike—as in the description of an impulsive force or charge density for a point charge, or short range forces like nuclear forces. It is of interest to note that the Dirac-delta function is not a function (or, rather an improper function), but is simply a symbol $\delta(x)$. However, it may be treated as a function, but for certain clearly defined processes. It is a generalisation of the

Kronečker-delta function to the case of a continuous variable. The Kronečker delta δ_{ij} may be defined for the discrete variables i and j in the following way:

$$\delta_{ij} = \begin{cases} 0, & \text{if } i \neq j \\ 1, & \text{if } i = j \end{cases} \tag{D.6.1}$$

Alternatively it may be defined as

$$f(j) = \sum_{i=1}^{\infty} \delta_{ij} f(i), \tag{D.6.2}$$

where i and j may take on the discrete values such as $1, 2, 3, \cdots$.

Actually the Dirac delta function denotes something physical which gets concentrated into a point in space or an instant of time like a point charge or an impulse of a force. Let us suppose that we start with a charge q which gets distributed along X-axis with linear charge density $\rho(x)$, i.e.,

$$\int_{-\infty}^{\infty} \rho(x)dx = q. \tag{D.6.3}$$

Now, if this charge is concentrated into a smaller and smaller region around $x = x'$, then it is evident that $\rho(x)$ will rise in this region with its peak at x'. Finally, in the limit of the charge concentrated at x', $\rho(x)$ will be having a sufficiently large value (infinity at x') and $\rho(x)$ will be zero except at $x = x'$. This limit of $\rho(x)$ is denoted as $q\delta(x - x')$ and will define the δ-function in the follwing way:

$$\int_{-\infty}^{\infty} \rho(x)dx = \int_{-\infty}^{\infty} q\delta(x - x')dx = q \quad \text{at } x = x'$$

$$\text{or,} \quad \int_{-\infty}^{\infty} \delta(x - x')dx = 1 \quad (\text{at } x = x') \tag{D.6.4}$$

$$\text{and} \quad q\delta(x - x') = 0 \quad \text{for } x \neq x' \quad \text{or,} \quad \delta(x - x') = 0 \quad \text{for } x \neq x' \tag{D.6.5}$$

Equations (D.6.4) and (D.6.5) are the mathematical definitions of the δ-function. As the value of δ-function at $x = x'$ cannot be stated, it means that this function is not a function in the usual mathematical sense (i.e., it is an improper function). However, we may define one-dimensional Dirac-delta function $\delta(x - x')$ as a generalised *singular function* which vanishes everywhere except at $x = x'$, and at $x = x'$, it is so large that it may be obtained only through the integral (D.6.4).

(b) For a continuous function $f(x)$, the *most useful property of the δ-function* is the substitution property which may be written as

$$\int_{-\infty}^{\infty} f(x)\delta(x - x')dx = f(x') \tag{D.6.6}$$

If the integral is zero except at $x = x'$, it is possible to replace $f(x)$ by $f(x')$ everywhere and can then be taken outside the integral, yielding

$$\text{L.H.S.} = f(x') \int_{-\infty}^{\infty} \delta(x - x')dx = f(x') \text{ from equation (D.6.4).}$$

When we put $x' = 0$, we obtain

$$f(0) = \int_{-\infty}^{\infty} \delta(x) f(x) dx \qquad \text{(D.6.7)}$$

When $f(x)$ is an odd function, then the curve obtained by plotting $f(x)$ versus x passes through the origin and thus $f(0) = 0$ and so the integral will be zero only provided the integral is a function of x. But $f(x)$ is considered to be odd. Therefore, $\delta(x)$ must be even, i.e.,

$$\left.\begin{aligned} \delta(-x) &= \delta(x) \\ \delta(x - x') &= \delta(x' - x) \end{aligned}\right\} \qquad \text{(D.6.8)}$$

(c) Representation of Dirac-delta function in different forms: When the Fourier integral formula written as

$$f(x') = \frac{1}{2\pi} \int_{-\infty}^{\infty} dk \int_{-\infty}^{\infty} f(x) e^{jk(x-x')} dx$$

is compared with equation (D.6.6), the following expression for δ-function may be obtained

$$\delta(x - x') = \frac{1}{2\pi} \int_{-\infty}^{\infty} e^{jk(x-x')} dk. \qquad \text{(D.6.9)}$$

Again it is possible to extend the definition of the delta function to three dimensions by using (D.6.9). Putting $x' = 0$ in (D.6.9) we have

$$\delta(x) = \frac{1}{2\pi} \int_{-\infty}^{\infty} e^{jkx} dk.$$

But

$$\delta(\vec{r}) = \frac{1}{(2\pi)^3} \int_{-\infty}^{\infty} \int_{-\infty}^{\infty} \int_{-\infty}^{\infty} e^{[j(k_x x + k_y y + k_z z)]} d^3 k = \frac{1}{(2\pi)^3} \int_{-\infty}^{\infty} e^{(j\vec{k}\cdot\vec{r})} d^3 k.$$

$$\therefore \quad \delta(\vec{r}) = \frac{1}{(2\pi)^3} \int_{-\infty}^{\infty} e^{(j\vec{k}\cdot\vec{r})} d^3 k. \qquad \text{(D.6.10)}$$

It is also possible to represent δ-function in another way as the limit of a sequence of analytic functions. One such representation is

$$\delta(x) = \lim_{L\to\infty} \frac{\sin Lx}{\pi x}. \qquad \text{(D.6.11)}$$

It can be proved that eqns.(D.6.4) and (D.6.5) are obtained from (D.6.11). As the function $\frac{\sin Lx}{\pi x}$ is having the value $\frac{L}{\pi}$ at $x = 0$, it means that this represents an oscillating function with period $\frac{2\pi}{L}$, with continually decreasing amplitude. The integral of this function will be equal to the area under the curve in which the main contribution is done by a triangle of height $\frac{L}{\pi}$ and base equal to $\frac{2\pi}{L}$, i.e.,

$$\text{Area} = \lim_{L\to\infty} \int_{-\infty}^{\infty} \frac{\sin Lx}{\pi x} dx = \lim_{L\to\infty} \frac{1}{2} \cdot \frac{L}{\pi} \cdot \frac{2\pi}{L} = 1.$$

Hence the function in eqn.(D.6.11) has the following properties:

(i) it possesses a unit integral independent of the value of L;

(ii) it becomes infinitely large at $x = 0$;

(iii) it possesses an amplitude which becomes zero elsewhere.

Thus, $\lim_{L\to\infty} \frac{\sin Lx}{\pi x}$ has all the properties of a δ-function.

Let us now consider the integral $\int_{-\infty}^{\infty} e^{jkx} dk$. It can also be written as

$$\begin{aligned} \lim_{L\to\infty} \int_{-L}^{L} e^{jkx} dk &= \lim_{L\to\infty} \frac{e^{jLx} - e^{-jLx}}{jx} = 2\pi \lim_{L\to\infty} \frac{e^{jLx} - e^{-jLx}}{2j(\pi x)} \\ &= 2\pi \lim_{L\to\infty} \frac{\sin Lx}{\pi x} = 2\pi\delta(x) \\ \text{or,} \quad \delta(x) &= \frac{1}{2\pi} \int_{-\infty}^{\infty} e^{jkx} dk \end{aligned} \tag{D.6.12}$$

Separating the real and imaginary parts, we see that

$$\frac{1}{2\pi} \int_{-\infty}^{\infty} \cos kx dk = \delta(x) \tag{D.6.13}$$

$$\text{and} \quad \frac{1}{2\pi} \int_{-\infty}^{\infty} \sin kx dk = 0 \tag{D.6.14}$$

Eqn.(D.6.13) is the most commonly used explicit expression for $\delta(x)$.

(d) Derivative of δ-function. The derivative of δ-function w.r.t. x can also be defined. It is denoted as $\delta'(x)$. One of the representations of $\delta'(x)$ is

$$\delta'(x) = \frac{1}{\pi} \lim_{L\to\infty} \left[\frac{L\cos Lx}{x} - \frac{\sin Lx}{x^2} \right] \tag{D.6.15}$$

The evaluation of integrals containing $\delta'(x)$ is done by integrating by parts and using $\delta(x) = 0$ for $x \neq 0$ (definition of δ-function). Thus we have

$$\int \delta'(x) f(x) dx = -f'(0). \tag{D.6.16}$$

Derivative $\delta'(x)$ satisfies the equation

$$x\delta'(x) = -\delta(x), \tag{D.6.17}$$

which follows from the definition (D.6.15); since $\delta(x)$ is an even function of x, its derivative $\delta'(x)$ is an odd function.

(e) The Fourier transform $g(k)$ in any space (say k-space) of the Dirac-delta function $\delta(x - x')$ is written as

$$F[f(x)] = g(k) = \frac{1}{\sqrt{2\pi}} \int_{-\infty}^{\infty} \delta(x - x') e^{jkx} dx \tag{D.6.18}$$

and its inversion gives the δ-function in x-space and is written as

$$F^{-1}\{g(k)\} = \delta(x - x') = \frac{1}{\sqrt{2\pi}} \int_{-\infty}^{\infty} g(k)e^{-jkx}dk. \tag{D.6.19}$$

Equations (D.6.18) and (D.6.19) represent a Fourier transform pair. From the knowledge of either function, it becomes possible to find the other. In the study of Quantum Mechanics, the Fourier transform (technique) of δ-function plays a very important role in the transformation from the co-ordinate space to the momentum space.

(f) We shall now give a number of equations satisfied by the δ-function. The meaning of these equations is that the left and right hand sides, when used as a multiplying factor under an integral sign lead to the same result. These equations are

$$\delta(-x) = \delta(x) \tag{D.6.20}$$

$$x\delta(x) = 0 \tag{D.6.21}$$

$$\delta(ax) = \frac{1}{|a|}\delta(x) \tag{D.6.22}$$

$$f(x)\delta(x - a) = \delta(a)\delta(x - a) \tag{D.6.23}$$

$$\int \delta(a - x)\delta(x - b)dx = \delta(a - b) \tag{D.6.24}$$

$$\delta(x^2 - a^2) = \frac{\delta(x - a) + \delta(x + a)}{2|a|}, \tag{D.6.25}$$

where the eqn.(D.6.20) is trivial, since $\delta(x)$ is an even function of x. Eqn.(D.6.21) is also obvious, since for any continuous function $f(x)$, we have

$$\int f(x)x\delta(x)dx = 0 \quad [\text{using } \delta(x) = 0 \text{ for } x \neq 0]$$

which shows that $x\delta(x)$, as a factor in an integrand, is equivalent to zero. Similarly, equations (D.6.22) and (D.6.25) may be verified. In order to verify eqn.(D.6.24), let us consider a continuous function $f(a)$ of a. Then

$$\begin{aligned}\int f(a)da \int \delta(a - x)dx\delta(x - b) &= \int \delta(x - b)dx \int f(a)da\delta(a - x) \\ &= \int \delta(x - b)dxf(x) = \int f(a)da\delta(a - b)\end{aligned}$$

which shows that $\int \delta(a-x)\delta(x-b)dx$ and $\delta(a-b)$ are equivalent. Eqn.(D.6.23) directly follows from

$$\int_{-\infty}^{\infty} f(x)\delta(x - a)dx = f(a)$$

which is obtained from (D.6.6) by putting $x' = a$.

(g) Some simple illustrations of δ-function: (i) Equation (D.6.21) shows that if we divide both sides of an equation by a variable x, which can assume the value zero, we should add on to one side an arbitrary multiple of $\delta(x)$, i.e., from an equation $a = b$, we cannot conclude that $\frac{a}{x} = \frac{b}{x}$, but we can only infer that

$$\frac{a}{x} = \frac{b}{x} + c\delta(x), \tag{D.6.26}$$

where c is unknown.

To illustrate this, let us consider the differentiation of $\log x$ for which the usual formula

$$\frac{d}{dx}(\log x) = \frac{1}{x} \tag{D.6.27}$$

requires the consideration for the neighbourhood of $x = 0$. In order to make this function well defined in the neighbourhood of $x = 0$, we must impose an extra condition, such that its integral from $-\varepsilon$ to ε vanishes, where ε is an infinitesimally small positive number. With this condition, the integral of r.h.s. of (D.6.27) vanishes, but that of l.h.s. is equal to $\log(-1)$, so that (D.6.27) is not a correct expression. It can be corrected by keeping in mind that if we take principal values, $\log x$ has a pure imaginary term $j\pi$ for negative values of x. This pure imaginary term vanishes discontinuously, as x passes through the value zero. This purely imaginary term gives, $-j\pi\delta(x)$ on differentiation. Thus equation (D.6.27) must be corrected to

$$\frac{d}{dx}(\log x) = \frac{1}{x} - j\pi\delta(x) \tag{D.6.28}$$

Similarly, the integrals involving singular function

$$\xi(x) = \lim_{t\to\infty} \frac{1 - e^{-jxt}}{jx} \tag{D.6.29}$$

can be evaluated by bearing in mind that

$$\xi(x) = \pi\delta(x) - jP\frac{1}{x}, \tag{D.6.30}$$

where the notation P indicates that we need to take the principal value of the integral.

(ii) The representation of δ-function is used very often in terms of a complete ortho-normal system of functions. For the function $\psi_n(x)$, which corresponds to a discrete spectrum, the ortho-normality is described as

$$\delta(x - x') = \sum_{n=1} \psi_n^*(x)\psi_n(x'), \tag{D.6.31}$$

while if the function $\psi_F(x)$ corresponds to a continuous spectrum, we have

$$\delta(x - x') = \int \psi_F^*(x)\psi_F(x')dF. \tag{D.6.32}$$

Exercises

1. Consider the function

$$\begin{aligned} f(x) &= 1 \quad -1 < x < 1 \\ &= 0 \quad |x| > 1 \end{aligned}$$

Find the Fourier transform of $f(x)$. (See worked out examples). **[C.U. (Hons.) 2006]**

2. Define Fourier transform of a function. Obtain the Fourier transform of a function

$$f(t) = \exp(-\alpha t^2) \quad \text{i.e.} \quad e^{-\alpha t^2} \quad (\alpha = \text{constant})$$

Given

$$\int_{-\infty}^{+\infty} \exp(-\alpha x^2 - \beta x)dx = \sqrt{\frac{\pi}{\alpha}} e^{\beta^2/4\alpha}$$

[C.U. (Hons.) 2007]

(See worked out examples)

3. What is the Fourier transform of $\delta(x - a)$ where a is constant? **[C.U. (Hons.) 2008]**

(See Appendix D.6)

4. Write down the expression for the Fourier transform of a function $f(x)$ defined in the domain $-\infty < x < \infty$. Calculate the Fourier transform when $f(x)$ is the Dirac delta function $\delta(x)$.

[C.U. (Hons.) 2005]

(See Appendix D.2 and D.6)

Bibliography

B.D. Gupta, Mathematical Physics.

B.K. Dutta, Introduction of Partial Differential Equations.

B.S. Agarwal, Mechanics.

Besant and Ramsay. A Treatise of Hydro-mechanics.

Bitsadze and Kalinichenko. A Collection of Problems on the Equations of Mathematical Physics.

C.J. Smith, General Properties of Matter.

Chakarabarti and Kundu, A Textbook of Mathematical Physics.

Champion and Davy, General Properties of Matter.

Charlie Harper, Mathematical Physics.

D.P. Roy Chowdhuri, Mathematical Physics.

D.S. Mathur, Elements of Properties of Matter.

Debaprasanna Biswas, Mathematical Methods in Physics.

Edser, A Textbook of General Physics.

Feynman, Lecturers on Physics—Vol. I.

Frank Ayres, Theory and Problems of Differential Equations (Schaum's outline series).

Frank Ayres. Theory and Problems of Matrices (Schaum's online series).

G.R. Chatwal and S.K. Anand, Quantum Mechanics (Edited by M. Arora and S. Puri).

Goldstein, Classical Mechanics.

Grechko, Sugakov and others, Problems in Theoretical Physics (Mr Publishers, Moscow).

Gupta and Satyaprakash, Classical Mechanics.

Gupta, Kumar and Sharma, Classical Mechanics.

Gupta, Yadav and Malik, Mathematical Physics.

Halliday and Resnick. Physics (Vol. I and Vol. II).

Huges and Brighton. Theory and Problems of Fluid Dynamics (Schaum's outline series).

Irodov. Problems in General Physics (Mir Publishers, Moscow).

Joos, Theoretical Physics.

Joshi, Matrices and Tensors in Physics.

K. Bhattacharyya, General Properties of Matter (for advanced students).

K.C. Kar, Introduction to Theoretical Physics—Vol. I.

Kalyan Kumar Mukherjee, Probability and Statistics.

Kaufmann, Fluid Mechanics.

Kittel, Knight and Ruderman, Berkeley Physics Course—Vol. I, Mechanics.

Kompaneyets, Theoretical Physics—Vol. I.

Loney, Dynamics of a Particle.

Min Chen, Physics Problems with Solutions (Berkeley).

Newman and Searle, The General Properties of Matter.

P.K. Chakrabarti, Mechanics and General Properties of Matter.

Parimal Mukhopadhyay, Theory of Probability.

Pipes and Harvill, Applied Mathematics for Engineering and Physicist.

Poynting and Thomson, General Properties of Matter.

Prescot, General Properties of Matter.

R.K. Ghosh and K.C. Maity, An Introduction to Analysis: Differential Calculus (Part I).

R.K. Ghosh and K.C. Maity, An Introduction to Analysis: Differential Calculus (Part II).

R.K. Ghosh and K.C. Maity, An Introduction to Analysis: Integral Calculus.

Rajput and Yog Prakash, Mathematical Physics.

Ramsay, Theory of Newtonian Attraction.

Rev. Fr. F. Goreux, Introduction of Tensor Calculus.

Roy Choudhuri and Maity, Mechanics and General Properties of Matter.

S. Banerjee, Mathematical Analysis—An Insight into Functions of Several Variables.

S.N. Goswami, Advanced Problems and Solutions in Physics.

Satya Prakash and C.K. Singh, Quantum Mechanics.

Spiegel, Theory and Problems of Theoretical Mechanics (Schaum's outline series).

Spiegel, Theory and Problems of Vector Analysis (Schaum's outline series).

Startling, General Properties of Matter.

Symon, Mechanics.

Synge and Griffith, Principles of Mechanics.

Takwale and Puranik, Introduction to Classical Mechanics.

Timo Shienko, Theory of Elasticity.

Index

A

Abbe Nollet's experiment, 779
Acceleration due to gravity, 345
—different methods for determination, 379
—effect of altitude, depth, latitude, 375–78
—measurement at sea, 396
—variation, 375
Acceleration—in rectangular co-ordinates, 73
—centrifugal, 187–88
—coriolis, 187
—cylindrical polar co-ordinates, 77–81
—radial and transverse in polar co-ordinates, 74–77, 84–85
—spherical polar co-ordinates, 81–84
—tangential and normal components, 73–74
Activity, 87
Adam N.K., measurement of angle of contact, 626–27
Adjoint matrix, 1022
Air correction of compound pendulum, 367–70
Amplitude, 297
Angle of contact, 623–24
—measurement of, 626–28
Angular momentum, 89, 94, 178-79, 237, 270, 452, 516
Application of partial derivatives
—thermodynamical problems, 984–85
Applications of Bernoulli's theorem, 767–69, 771–73
Applications of Fourier transforms
—Evaluation of some important integrals, 1294–95
—Parseval's relation and evaluation of integrals, 1295–96
—solution of ordinary differential equations, 1296–97
—solution of partial differential equations, 1297–1301
Applications of linear equations to problems in physics, 907–20
Applications of the matrix method, 1029–48
—diagonalisation of a matrix, 1036
—process of diagonalisation, 1042–43
—solution of problems involving matrix algebra, 1029–33
—solution of a set of linear equations, 1033
—solution of the problem of linear inversion, 1034–36
—technique of diagonalisation, 1037–39
Arc (*finite*) correction of pendulum, 350, 366
Archimedes' principle, 755–56
Areal velocity, 89, 458
Artificial satellites, 464
—condition to put into orbit, 465
—geo-stationary orbit, 467
—orbital speed, period, 465–67
Associated Legendre's differential equation, 939
—associated Legendre's polynomial, 939
—solution, 939
Astronomical units, 402
Attraction and Potential (*gravitational*)
—homogeneous cylinder, 433–35
—spherical bodies, 411-23
graphical variation, 420
—thin uniform straight rod, 435, 437
—uniform circular disc, 429–31
—uniform ring, 431
—uniform rod bent into a circular arc, 436
Atwood's machine, 379
Average velocity, 799
Avogadro's number from Brownian motion, 823–28
Avogadro's law from kinetic theory, 796
Axial modulus, 538, 546
Axial vector, 38

B

Balance, Common, Poynting's method for determination of G, 495–97
Bar pendulum, 360
Beam, bending, 558
Bending moment, Internal, 558–59
—diagram, 584
Berkeley and Hartley's apparatus, 782
Bernoulli's differential equation, 890
Bernoulli's equation, 760–61

—gas, 771–73
—irrotational flow of liquid, 763–64
—rotational flow of liquid, 764–65
Bernoulli's theorem, 761–66, 773–75
Bessel's computed time-period, 361
Bessel's differential equation, 926
—Bessel's function $J_n(x)$, 927–28
—generating function for $J_n(x)$, 929
—recurrence formula for $J_n(x)$, 930–31
—solution, 926–29
Bi-filar suspension, 381–84
Body oscillating on a concave surface, 384
Borda's pendulum, 356
Born's method
—determination of mean free path, 812–13
Bouguer's experiment, 490
Boys' experiment, 493–95
Brown R.C., 692
Brownian movement, 822
—gases, 828–29
—horizontal distribution—Einstein's equation, 825–28
—Perrin's experiment to determine Avogadro's number 823–24, 826, 828
—rotational, 829
—vertical distribution of Brownian particles, 823–24
Bubbles
—excess pressure due to surface tension, 634–35
Bulk modulus, 537
Buoyancy correction of compound pendulum, 367–69

C

Cantilever, 559–61
—reciprocal theorem, 565
—transverse vibration, 582
Capillarity, 630
—capillary tube of insufficient length, 631
—elevation or depression, 630
—energy changes associated with capillary rise, 632–33
—rise of a liquid in a capillary tube, 630–31
—rise of a liquid in a conical capillary tube, 635
Capillary tube
—shape of liquid meniscus, 628–29
Cavendish's experiment, 492
Cayley-Hamilton's theorem, 1048
Cenco rotary oil pump, 851
Central forces, 88
—conservative character of the field, 90
—conservation of energy, 91–92
—planer orbit, 89
—KE, PE, 91–93
Centre of gravity, 176
Centre of mass, 174–75
—calculation in some important cases, 1269–72
—motion of, 176–77
—oscillation, 358
—suspension, 358
—two body problem, 177–78
Charle's law from kinetic theory, 796
Circular disc
—gravitational potential at a point on the axis, 429
Circular motion, 328, 334, 336
Coefficient of viscosity, 705–06
Cofactor matrix, 1025
Cohesion, 616, 624, 629, 630
Collision, 200–10
—CM frame, 205
—elastic and inelastic, 201
—linear and angular momentum conservation-ballistic pendulum, 208
—loss of energy, 203, 207
—maximum energy transfer, 204
—Newton's collision rule, 200–01
Column matrix, 1014
Complementary function, 895–97
Complex conjugate matrix, 1023
Complex valued functions
—analytic function, 1107
—Argand diagram, 1103–04
—methods of finding the residue at a finite pole, 1119–20
—Cauchy-Riemann conditions, 1108–09
—complex numbers, 1102
—continuity, 1105
—differentiability, 1106
—domain, 1104–05
—essential or non-removable singularity, 1115
—evaluation of residues, 1117–18
—necessary conditions for a function to be analytic, 1108
—sufficient conditions for a function to be analytic, 1109–10
—functions of complex variable, 1104
—geometrical representation of the imaginary number j, 1103
—graphical representation of complex numbers, 1103
—isolated singularity, 1113
—Laplace's equation, 1112

—limit point of a set, 1104
—neighbourhood of a point, 1104
—non-esseitial or removable singularity, 1114
—path of a complex variable, 1105
—polar form of Cauchy-Riemann conditions, 1111–12
—pole of order of m, 1115
—residue at infinity, 1118
—set of points, 1104
—simple pole, 1115
—singularities of an analytic function, 1113
—the point at infinity, 1116
—zeros of an analytic function, 1112–43
Composition of simple harmonic motions, 325
—different cases, 325–31, 336
—graphical treatment, 331–35
Compound pendulum, 357
—condition for minimum time-period, 359
—different corrections, 366–74
Conductance of a tube, 848
Conical pendulum, 355
Conservative forces, 88
Constraints, 112
—holonomic and non-holonomic, 113
—scleronomic, rhenomic, 113
Contravariant and Covariant vectors, 1264–65
Coriolis force, 189–90
Cornu's method, 601
Corrections to Poiseuille's formula, 712–13
Couette's method, 732–33
Critical velocity, 707
Cross-product of two vectors, 16
Curl of a vector, 28
Curvature
—cantilever beam, 560
—knife-edge correction of a compound pendulum, 371
Curvilinear co-ordinates, 38–39
—expressions for gradient, divergence, curl and Laplacian, 40–41

D

D'Alembert's principle, 110, 757
Dalton's law from kinetic theory, 796
Damped motion, 313
—large, low and critical damping, 313–14
—log decrement, 320
—power dissipation, 315
—quality factor, 316
Degrees of freedom and ratio of specific heats, 806–07
Degrees of freedom of a mechanical system, 112
Demerits of a simple pendulum, 356
Density of the earth, 488
Dependence of viscosity on temperature, 730–31
Depression of a beam
—different cases, 566–71
Determination of coefficient of diffusion, 788–91
Determination of elastic constants, 596–603
—Young's modulus
different methods, 596–98
—Rigidity modulus,
dynamical, 599
statical, 598
—Poisson's ratio
different methods, 601–03
—Searle's method for Y and n, 599–600
Determination of G
—Airy's mine experiment, 488
—Laboratory methods
—Boys' modification, 493–95
—Cavendish's experiment, 492–93
—Heyl's method, 497–99
—Poynting's method, 495–97
—mountain experiments
Bouguer's experiment, 490–91
Maskelyne's experiment, 491
Determination of osmotic pressure, 781–82
Determination of surface tension
—Balance method, 676
—capillary (Laboratory) rise method, 675
—drop-weight method, 684–86
—Jaeger's method, 678–79
—modern methods, 691–92
—Quincke's method, 677–78
—Rayleigh's jet method, 686–87
—Rayleigh's ripple method, 688–90
—Sugden's method, 679–82
—U-tube arrangement, 683
Determination of viscosity of gases, 738–43
—Rankine's method, 741–42
—Searle's method, 739–40
—U-tube method, 738–39
Determination of viscosity of liquid, 732–37
—Couette's method, 732–33
—Poiseuille's method, 732
—rotational viscometer, 735–36
—Searle's viscometer, 735
—Stokes' method, 737
—Viscometer for comparison of viscosities, 733–34
Diagonal matrix, 1021
Diameter of a molecule
—mean free path phenomena, 807–08
—Van der Waal's constant, 836–37
Differential equation for S.H.M., 299

—solution, 300
Differential equations
—classification, 886
—exact and inexact, 891–93
—linear, 888–89
—second order equations, 893–900
—determination of complementary function and particular integral, 895–907
—method of undetermined coefficients, 898–900
—solution by other methods 900–07
Diffusion, 786
—determination of the coefficient, 788–91
—differential equation, 787
Dimensions, 2–9
—principle of homogeneity, 3
Dirac-delta function
—definition, 1301–02
—properties, 1302–04
—representation in different forms, 1303
—Fourier transform of Dirac-delta function, 1304–05
—Equations satisfied by Dirac-delta function, 1305
—some simple illustrations, 1305–06
Dirichlet's conditions, 859
Divergence of a vector, 25, 26
Dot product of two vectors, 15
Downward flow of a liquid through a vertical narrow tube, 713–15
Drop
—sessile, 649–50, 652–54
Drop-weight method, 684–86
Dynamical method
—modulus of rigidity, 599
Dynamics
—rigid body, 237
—system of particles, 174

E
Eccentric spheres
—intensity, 428–29
—potential, 428
Effect of pressure on viscosity of liquids and gases, 731
Efflux of gases, 772–73
Einstein's Theory of Brownian motion, 827–28
Elastic constants, 536–38
—fatigue, 541
—limit, 540
—properties of matter, 540
—property, 532
—relations between. 544–49
Elasticities of liquids and gases, 557
—isothermal and adiabatic, 557
Electrical oscillation, 306–07
Ellipsoid,
—inertia of, 263
—strain, 535
Elliptic disc
—moment of inertia, 252
Elliptic orbit
—central force-field motion, 452, 457
Energy—conservation of, 91–92
—kinetic of rigid body, 237, 269
—potential, 92
—strain, 588–91
—surface, 616–17
Eötvös' law, 673
Equation of continuity, 759
Equilibrium of a fluid
—condition, 754–55
Equilibrium
stable and unstable, 119–20
Equipotential surface, 410–11
Equivalent simple pendulum, 357–58
Euler's equation of motion
—rigid body, 515–17
—fluid dynamics, 757
Evaluation of the constants (Fourier), 859–60
—Riemann Zeta function by Fourier expansion, 863–64
Ewing's extensometer, 597
Expansion of a periodic function
—by Fourier's series, 868–69
—extension of the range, 869
—half-range series, 870
Experimental arrangement
—illustration of Bernoulli's theorem, 766
Experimental determination of viscosity
—gases, 738–43
—liquids, 732–37

F
Falling bodies, 11
Ferguson's method
—determination of surface tension. 693
Fick's law, 787
Fictitious force, 183, 186, 188
Field
—gravitational, 407
Films
—shape of, 636–40
—surface films, 616–17
Flexural rigidity, 559
Floating needle

—condition of equilibrium, 654–55
Flow of liquid through an annular space, 711–12
Flow of liquid through an inclined narrow tube, 715
Flow of liquid through capillaries in series and parallel, 716–17
Fluid dynamics, 753, 756
Fluid statics, 753, 755
Flux of gravitational intensity, 446
Fly-wheel—determination of moment of inertia, 286–87
Force and the law of inertia, 10
Force-free rotation of a symmetrical rigid body, 518–19
Force
—body, 753
—between two parallel plates due to surface tension, 659–60
—to pull a plate from liquid surface, 660–61
Forced vibration, 316–18
—resonance, 317
—sharpness of, 318
Formation of drops and fog, 664–66
Fourier equation for heat conduction
—one-dimensional, 986
—solution, 994–95
—three-dimensional, 986
Fourier representation of even and odd functions, 862–63
Fourier's theorem, 858
—analysis and synthesis with examples, 865–68
—importance, 871
Fourier series, 858–61
—advantages, 870
Fourier transform
—definition, 1287
—Some important properties, 1288–90
Frames of reference
—accelerated, 184–85
—inertial and non-inertial, 182
—rotating, 186–87
Fugitive elasticity, 706
Functions of several real variables
—chain rule for Jacobians, 1154
—conditions for continuity, differentiability and existence of partial differentials, 1140–43
—continuity for a function of two variables, 1138–39
—directional derivative, partial derivative, differentiability and differential, 1140–43
—Euler's theorem and its converse, 1151–52
—existence theorem of implicit function 1165–66
—function of three variables, 1149–50
—functional relationship, 1162–63
—functions of more than one variable, 1132–33
—higher order partial derivatives, 1143–46
—homogeneous function, 1151
—implicit function, 1165
—Jacobian of implicit functions, 1155–56
—Jacobians, 1154
—limit of function of two variables, 1134
—necessary and sufficient condition for function relationship, 1163
—real valued function, 1132
—repeated or iterated limits, 1137–38
—simultaneous limit or double limit, 1134–36
—some important formulae involving Jacobian, 1159–62
—theorem of Schwarz, 1147
—theorem of Young, 1147

G
Gaede's diffusion pump, 851–52
—rotary oil pump, 850
Galilean transformation
—invariance of Newton's laws, 183–84
Gamma Function, 923
Gauss's divergence theorem, 27
Gauss's theorem in gravitation, 446–48
Geissler pump, 849–50
General theory of bending, 571–72
—depression of beam in different cases, 572–82
Generalised co-ordinates, 114
—generalised momentum, 115
—generalised velocity, 115
Geophysical prospecting, 394
Glass
—determination of Poisson's ratio, 601–02
Gradient of a vector, 24
Graphical treatment
—linear motion, 11
—S.H.M. for different cases, 331–35
Gravitation
—attraction in the field, 407
—calculation in differnt cases, 416–24, 428–37
—Laplace's and Poisson's equations for potential, 444–45
—Newton's law, 401
—potential in the field, 408–09

—calculation in different cases, 411–15, 428–37
Gravity gradient, 393
Gravity meter, 387
—Balliden, 387–88
—Eötvös' balance, 389–93
—Gulf, 388
—Quartz, 387
Gravity Survey, 386
Green's theorem, 31
Gyration
—radius of, 238
Gyroscope, 515
Gyroscopic applications, 530–31
—gyroscopic compass, 530
—pendulum gyro-compass, 530
Gyroscopic motion, 514

H

Half-range Series, 870
Hamilton's action integral, 61
Hamilton's equations, 117–18
Hamilton's principle, 61
Harkins' method,
—determination of surface tension, 685
Harmonic vibration (*Damped*), 313–16
Hermite's differential equation, 940
—orthogonal property, 946
—polynomials, generating function, 943–45
—recurrence formulae for $H_n(x)$, 945–46
—solution, 940–43
Hermitian conjugate, 1024
Hermitian matrix, 1024
Heyl's method
—determination of G, 497–99
Homogeneous strain, 534
Hooke's Law, 536
Hoop, circular
—potential and attraction, 431
Horizontal directive tendency, 394
Horizontal flat ring
—force to detach from a liquid surface, 662

I

Identity matrix, 1021
Impulsive force, 85
Inclined
—plane, body rolling down, 285–86, 380
—solid surface, rise of liquid, 647–50
India rubber
—Poisson's ratio for, 602–03
Inertia, 9
—moment of, 237–38
—calculation in different cases, 242–62, 276–81
—experimental determination, 284–87
—physical significance, 238–39
Infinite Sequences
—convergent, divergent, oscillatory, 1066
—definition, 1065
—monotone, 1066
Infinite Series, 1067
—absolute and conditional convergence, 1080–81
—alternating series, 1079–80
—Cauchy's Condensation Test, 1078–79
—Cauchy's Logarithmic Test, 1078
—Cauchy's Ratio Test, 1073–77
—Cauchy's Root Test, 1077
—Cauchy's Integral Test, 1072–73
—comparison test for convergence and divergence, 1071–72
—convergent, 1067
—D'Alembort's Ratio Test, 1073
—divergent, oscillatory, 1068
—general principle of convergence, 1070
—geometric, 1068
—interval of convergence, 1081–82
—nature, 1069–70
—power series, 1081
—Raabe's Test, 1077–78
—theorems regarding power series, 1082–87
—use of series for computation of functions, 1087–88
Intensity of the field, gravitational, 407
—calculation in different cases, 416–24, 428–37
Inter-diffusion of solids, 791
Inter-facial surface, shape of, 646–47
Internal bending moment, 558–59
Integral transforms, 1286–87
Intrinsic pressure, 670–71
—surface energy, 616–17
Invariable pendulum, 386
Inverse of a matrix, 1026
—properties, 1027–28
Ionisaton gauge, 854–55
Irrotational vector, 36
Isothermal elasticity, 557

J

Jaeger's method, 678–79
Jessop
—angle of contact, 627
Jurin's law,
—capillary rise, 631

K

Kater's pendulum, 360
—Bessel's computed time period, 361
Kepler's laws, 449
—derivation from **Newton**'s law and vice-versa, 449–60
Kinematics, 72
Kinetic energy
—rotational motion of rigid body, 237, 269
Kinetic theory of gases, 792
Kinetic theory of matter, 792
Knife-edge correction
—compound pendulum, 371
Knudsen's gauge, 855–56
König's method
—measurement of angle of turning of the ends of a beam, 597–98
Kronečker delta function δ_{ij}, 272, 1302
Kronečker's formula, 880

L

Lagrange's equation, 116–17
Lagrangian formulation of conservation theorems, 1272–76
Laminar body
—moment of inertia of, 242–59
—parallel axes theorem, 240–41
—perpendicular axes theorem, 239
Langmuir pump, 852
Laplace's equation in three dimensions
—Cartesian co-ordinates for potential, 966–68
—solution by the method of separation of variables, 969–70
—cylindrical co-ordinates, 970–71
—solution by the method of separation of variables, 971–74
—Bessel's function $J_n(x)$, 972
—cylindrical harmonic, 973
—generating function for $J_n(x)$, 929
—spherical polar co-ordinates, 974–76
—solution, surface as well as zonal harmonic, 976–79
—Legendre's polynomial $P_n(x)$, 934–35, 978–79
—generating function for $P_n(x)$, 935–36
—orthogonal property, 938
Laplace's equation in two dimensions
—problem of steady flow of heat, 979–81
Lateral strain, 538, 545
Law of equipartition of energy, 804
Laws of matrix algebra, 1016–19
Legendre's differential equation, 932
—Associated Legendre's differential equation, 939, 978
—solution, 932–35, 939, 978–79
—Associated Legendre's polynomial, 939, 978
Linear equations and matrix notation, 1013–14
Limit, elastic, 540
Limitations of the method of dimensions, 6
Line integral, 28
Linear differential equations
—first order, 888–91
—second order, 893–900
Linear motion, accelerated, 11
Lines of force-gravitational, 411
Lissajous' figures
—methods for producing, 338–39
—uses of, 337
Longitudinal stress, 537
—strain, 537

M

Maskelyne's experiment
—determination of G, 491
Mass and weight, 9
Mass of the earth, 488
—of the sun, 476–77
Matrix addition, 1015
Matrix subtraction, 1016
Matrix multiplication, 1017–18
Matrix division, 1026
Maxwell's distribution law of molecular velocities, 797–99
—energy distribution, 802
—experimental verification, 803–04
—temperature dependence and reduced form, 803
Mcleod gauge, 853–54
Mean density of the earth, 488
Mean free path phenomena, 807–13
—Born's method for determination of, 812–13
—Clausius's mean free path, 808
—Maxwell's free path, 811–12
—probability of no collision and the law of distribution of free paths, 809–10
Mean square velocity, 794, 800
Measurement of osmotic pressure, 781–82
Measurement of variation of viscosity of a liquid with temperature, 733
Measurement of viscosity of a liquid by Searle's viscometer, 735–36
—by Stokes' method, 737
Mine experiment, Airy's, 488
Minor and rank of a matrix, 1029

Modulus of elasticity, 536
Molecular pump, 851
Momental ellipsoid, 262–63
Most probable velocity, 800
Motion of a body falling through a viscous medium, 725–26
—terminal velocity, 726
Motion of a body in tunnel, 424–25
Motion of projectiles, 196–99
Mountain experiment for determination of G
—Bouguer's experiment, 490
—Maskelyne's experiment, 491
Moving Frames of reference, 184–85

N

Neumann's triangle for interfacial surface tensions, 625
Neutral,
—axis, 558
—surface, 558
Newton's
—equation of motion in moving frames of reference, 184–85
—law of gravitation, 401
—its universality, 402
—and Einstein's theory of relativity, 403
—laws of motion, 10
—their invariance in Galilean transformation, 184–85
Non-Newtonian liquids, 743
Non-singular matrix, 1014
Normal co-ordinates, 125
Null matrix, 1016

O

Orthogonal matrix, 1025
—properties, 1025–26
Origin and concept of Tensor calculus, 1263–64
Oscillating body on a concave surface, 384
Osmosis, 779
Osmotic pressure
—association or dissociation of molecules, 785
—boiling point, 784
—freezing point, 785
—vapour pressure, 783–84

P

Parallel axes theorem, 240–41
—plates vertical and close to each other, rise of liquid in between, 658–59
—plates (vertical), rise of liquid in between, 656–58
—plates, separated by a thin layer of liquid, force, 659–60
Partial Differentiation, 1174
—change of the order of integration, 1209–10
—change of variables, 1186
—continuity, 1174–75
—derivatives, 1174
—differential of implicit functions, 1189–90
—differentiation of a definite integral, 1200–01
—differentiation of composite functions—chain rule, 1184–85, 1191
—double integrals, 1208–09
—errors and approximation, 1178–80, 1212–13
—evaluation of certain important definite integrals, 1203–06
—Integration under the integral sign—multiple integrations, 1202–03
—Lagrange's method of undetermined multipliers, 1196–1200
—Leibnitz's rule, 1206
—maxima and minima, 1192–96
—one equation, more than two variables, 1190
—order of differentiation, 1175
—second and higher order derivatives, 1185–86
—symbolic form of Taylor's Theorem, 1182–83
—Taylor's theorem for functions of two variables, 1180–81
—total differential, 1177–78, 1186–88
—triple integrals, 1211–12
Particular integral, 895, 898
Pascal's law for hydro-statics, 755
Path integral, 28
Pedal form of equation for a central orbit, 1276–78
Pendulum, bar, 360
—compound, 357–59
—conical, 355
—Foucault's, 191–94
—Kater's, 360
—second's, 348
—simple, 345
—torsional, 303, 599
Periodic motion, 297
Perpendicular axes theorem, 239–40
Pfeffer's method, 781
Pirani gauge, 854
Pitot tube, 769
Planetary bodies, motion of, 449–60
Plastic body, 532

Plateau's spherule, 640
—instability of films, 636–40
Plate
—force to pull from a liquid surface, 660–61
Poise, 706
Poiseuille's formula for
—flow of gases, 728–29
—flow of liquids, 708–11
Poisson's equation for potential, 445
Poisson's ratio, 538
Polar vector, 38
Potential energy
—self-attracting system, 441–43
—surface energy, 616–17
Potential in a field—gravitational, 408–09
—calculation in different cases, 411-15, 428–37
Power series solution of differential equations, 920–23
—Frobenius' method, 924–25, 960–64
—singularity test, 925–26, 1113–16
Power, 87
Poynting's method
—determination of G, 495–97
Precessional torque, 514
—precession of earth's axis of rotation, 519
Pressure exerted by a perfect gas, 792–95
—deduction of
Avogadro's law, 796
Boyle's law, 795
Charles' law, 796
Dalton's law, 796
—expression, 794
Pressure gauges, 853–56
Pressure on a curved membrane
—due to surface tension, 634
—excess pressure in different cases, 634–35
Principle of virtual work, 111
Principal axes, 263, 265, 276–81
—axes of strain, 535
—moments of inertia, 263, 265, 276–81
Production and measurement of low pressure, 847
Pseudo force, 183, 186

Q

Quincke's method, 677–78
—surface tension of a sessile drop, 649–50, 652–54

R

Radial velocity and acceleration, 74–77, 84–85
Radius of gyration, 238
Rankine's experiment for viscosity of a gas, 741–43
Rayleigh's jet method,
—determination of surface tension, 686–87
Reciprocal matrix, 1022
Reciprocal vectors, 36
Rectangular matrix, 1014
Recurrence formulae
—Bessel's function $J_n(x)$, 930–31
—Hermite's polynomial $H_n(x)$, 945–46
—Legendre's polynomial $P_n(x)$, 940
Relation between osmotic pressure and other properties of a solution, 783–85
Relations between elastic constants, 544–49
Repsold's pendulum, 369
—elimination of corrections of compound pendulum, 369
Reynold's number, 707–08
Rigid body
—angular momentum, 237, 270
—K.E., 237, 269
Rigidity modulus, 537–38
Ring—moment of inertia, 244–45
—gravitational potential and attraction, 431
Ripple method
—determination of surface tension, 688–92
Rise of a liquid along
—an inclined solid surface, 647–50
—the side of a plane plate dipped vertically, 650–52
Rodrigues' formula for Legendre's polynomials, 937–38
Rods under thrust
—Struts, 585–87
Rotary oil pumps, 850–51
Rotating cylinder method
—determination of viscosity, 722–23, 735–36
Rotating Frames, 186–94
—centrifugal force, 188
—coriolis force, 189–91
—Foucault's pendulum, 191–94
Rotational Brownian movement, 829
Routh's rule, 283
Row matrix, 1014

S

Saw-tooth wave, 866-68
Scalar matrix, 1021
Scalar product, triple, 15, 18
Scalars, vectors and tensors, 1263
Scattering in a central force field—
—Rutherford's scattering of α-particles, 1279–84
Schiehallion
—determination of G, 491

Searle's method
—determination of Y and n, 599–600
Seconds pendulum, 348
Sessile drop, 649–50, 652–54
—Quincke's method, 677–78
Shape of an interfacial surface, 646–47
Shape of films, 636
—stability and instability, 636–40
Shear, 537–38
—equivalent to a compression and an equal extension at right angles, 541
Shearing force diagram, 584
Shell spherical
—gravitational potential, 411-12, 415
—gravitational attraction, 416–17, 419
Simple harmonic motion, 297–98
—differential equation of motion, 299
—graphical representation, 302
—K.E., P.E. and total energy, 301
—period, amplitude, phase, 298
—solution, 300
—various examples, 302–13
Simple pendulum, 345
—differential equation, 346–47
—finite arc correction, anharmonic oscillator, 350–52
—gain or loss of time of a pendulum clock, 348–49
—laws of, 347–48
—with movable support, 353–54
Skew matrix, 1023
Small oscillations, 121–33
—coupled oscillation, 128–29
—double pendulum, 126–28
—general problem, 122–25
—linear triatomic molecule, 129–33
—one degree of freedom, 125
—two degrees of freedom, 126
Singular matrix, 1014
Solenoidal vector, 36
Some practical aspects of surface tension, 693–94
Speed of a pump, 847–48
Spiral springs
—flat, 591–94
—non-flat, 594–96
Spraying of a liquid into droplets-work done, 619
—inverse phenomenon, 619–20
Sprengel pump, 849
Square wave, 865–66
Square matrix, 1014
Stability of films, 636–39
Statical method
—modulus of rigidity, 598
Statistics and Probability, 1219
—a posteriori probability, 1221
—a priori probability, 1220
—application to experimental measurements, 1249
—area under some part of gaussian distribution curve, 1247–48
—Bayes's theorem, 1225–26
—binomial distribution, 1239
—characteristics of, 1239
—binomial frequency function, 1226
—binomial law of probability, 1226
—Chebyshev's inequality, 1238
—continuous distribution, 1235–36
—distribution of a sum of normal variates, 1248–49
—expectation values of the sum and product of two random variables, 1231–32
—expectation, moments and standard deviation, 1236–37
—frequency, 1233
—fundamental laws of probability, 1223–24
—gaussian distribution, 1243–44
—mean of a distribution, 1233–34
—mean of binomial distribution, 1240
—mean of gaussian distribution, 1245
—mean of poisson distribution, 1243
—mutually exclusive events, 1224
—poisson distribution, its mode and examples of poisson variates, 1241–42
—properties of random variable, 1229–30
—random variables, 1229
—relations among conditional probabilities, 1225–26
—repeated trials, 1226
—sample space and events, 1227–28
—second moment, 1234
—standard deviation of binomial distribution, 1241
—standard deviation of poisson distribution, 1243
—standard deviation of gaussian, i.e., normal distribution, 1246
—properties of the normal distribution curve, 1244, 1246–47
—standard deviation of the mean, 1250–51
—standard deviation, 1235
—statistical distributions, 1233
—statistical expectation value, 1231
—statistically independent events, 1224
—Stirling's approximation for the factorial, 1251–52
Stoke's theorem

—body falling through a viscous medium, 723–25
—vectors, 30
Stokes' method—determination of viscosity of a transparent viscous liquid, 737
Strain, 532
—energy, 588–91
calculation in different cases
bending of a beam, 589–90
stretching of a wire, 588–89
twisting of rod, 591
—homogeneous, 534
—principal axes of, 535
—principal, 536
Streamline motion, 707
—experimental demonstration, 708
Stress and strain tensors, 1285
Stress, 532–33
—normal and tangential, 533
Sub matrix, 1014
Sugden's method, 679–82
Surface energy and surface tension, 616–18
Surface tension and vapour pressure, 662–64
Surface tension,
—boiling, 666–67
—estimation of intrinsic pressure (K), 670–72
—estimation of the range of molecular action, 670
—evaporation, 664–66
—explanation from Laplace's theory of capillary action, 667–72
Surface tension
—determination by various methods, 675–93
Symmetric and anti-symmetric matrix, 1023
Symmetric and antisymmetric tensors, 1263
Symmetrical bodies
—calculation of moments, products and ellipsoids of inertia, 276–82
System of particles
—angular momentum, 178–79
—conservation of energy, 182
—dynamics of, 174
—K.E., P.E., 180–81
—linear momentum, 178
—torques, internal and external, 179–80
System of variable mass
—rocket motion, 470–72

T

Temperature
—variation of elasticity, 603–04
—variation of surface tension, 673–74
—variation of viscosity, 730–31
Tensor of inertia, 271–73
Theory of osmotic pressure, 783
Theory of semi-permeability, 783
Thick spherical shell
—calculation of gravitational potential and attraction, 415, 419
Thin spherical shell
—calculation of gravitational potential and attraction, 411–12, 416–17
Time integral of force, 85
Top motion
—steady precession, 520–22
Top symmetrical
—general motion, nutation and precession, 523–26
—Eulerian angles, 527
—in a gravitational field, 527–29
Toque on a cylinder immersed in a rotating fluid, 722–23
Torque
—and moment of inertia, 238
—external and internal, 179–80
Torricelli's theorem
—velocity of efflux of a liquid, 768–69
Torsion of a cylinder or wire, 552
—torsional rigidity, 553–54
Trace of a matrix, 1028
Transformation of co-ordinates through matrix operation, 1266–68
Translational Brownian moment in liquids—Einstein's equation, 825–28
—gases, 828–29
Transport phenomena, 814
—diffusion from kinetic theory, 817–19
—general treatment, 819–21
—thermal conductivity from kinetic theory, 816–17
—viscosity from kinetic theory, 814–16
Transpose of a matrix, 1019–20
Turbulent motion, 707
—experimental demonstration, 708

U

Uniform shell bounded by non-intersecting eccentric spheres, 428
—calculation of gravitational potential and intensity, 428–29
Uniform solid sphere
—calculation of gravitational potential and attraction, 413–14, 418
Unit matrix, 1021
Unitary matrix, 1024

Units, 1
—different systems, 1, 2
Unit vectors—rectangular, 14, 15

V

Van der Waal's equation of state, 830–31
Van't Hoff's equation and the laws of osmotic pressure, 780–81
Variation in G, 500
Variation in the direction of g with time
—horizontal pendulum, 395–96
Variation of elasticity with temperature, 603
Variation of surface tension with temperature, 673–74
Variation of viscosity with pressure
—gases, 731
—liquids, 731
Variation of viscosity with temperature
—gases, 731
—730–31
Vector Analysis, 13
Vectors and Tensors, 1262
Vector product, triple, 16, 19
Vectors, tensors and scalars, 1263
—contravariant and covariant vectors, 1264–65
Vector treatment for central orbit, 1278–79
Velocity of escape, 463
Venturimeter, 767–68
Vibration (transverse) of a cantilever, 582
Vibratory motion, 297
Virial theorem, 831–33
—deduction of perfect gas equation, 833
—deduction of Van der Waal's equation of state, 833–37
Virtual work
—principle of 111
Viscosity,
—of high viscous liquids, 723–26
Volume strain in terms of principal strains, 536

W

Water jet pump, 849
Wave equation
—one dimensional Cartesian co-ordinates, 986
—solution—D'Alembert's method, 986–87
—solution—method of separation of variables, 987–90
—two dimensional Cartesian co-ordinates, 985
—solution—transverse vibration of a rectangular membrane, 990–92
—polar co-ordinates, 993
—solution-vibrations of a circular membrane, 993–94
—three dimensional Cartesian co-ordinates, 985
—solution, 996–97
—cylindrical co-ordinates, 999
—solution, 999–1000
—spherical co-ordinates, solution, 998–99
Waves
—capillary, 688–90
—gravity 688
Work, 87
Woulfe's bottle, 678–79

Y

Yield point, 541
Yielding of support
—compound pendulum, 372–74
Young's modulus, 536–37

Questions of C.U. Physics Honours First Paper and Second Paper (New Syllabus) 2009—relevant to different chapters of this book—with answers

First Paper—Unit IA

1. (a) Let A be a square finite dimensional matrix with real entries such that $AA^T = I$, where A^T denotes the transpose of A. Show that $A^T A = 1$. (*Ans.* See Art. 19.15.1)

(b) Determine the sum and the product of the eigen values of the matrix $\begin{pmatrix} a & b \\ c & d \end{pmatrix}$.

(*Ans.* See Q.42(a) of the Exercises of Chapter 19)

(c) The random variable x_1 follows a Gaussian distribution with mean μ and standard deviation σ_1. A second random variable x_2 also follows a Gaussian distribution with same mean μ but standard deviation σ_2, which is greater than σ_1. Roughly sketch the two probability density functions.

(*Ans.* See Fig. 24.15.1 and then add—The curve with σ_2 as standard deviation will be flatter than that with σ_1 as standard deviation ($\because \sigma_2 > \sigma_1$))

(d) Find whether $d\phi$ is an exact differential where $d\phi = (x^2 - y)\, dx + x\, dy$.

(*Ans.* See worked-out examples on page 892 of Chapter 17)

(e) Show that $\bar{\nabla}\phi$ is a vector perpendicular to the surface $\phi(x, y, z) =$ constant.

(*Ans.* See worked-out examples on page 48 of Chapter 2)

(f) Keeping the mass same, if the earth is quenched and its radius reduced to half its present value, how will the *length of the day* be changed?

(*Ans.* Gravitational force on a body of mass m at a distance r from the centre of the earth of mass M is $\frac{GMm}{r^2}$. Equating this to mg, the weight of the body, g being the acceleration due to gravity, we have $g = \frac{GM}{r^2} = \frac{GM}{R^2}$, (on the surface of the earth putting $r = R$, the radius of the earth) $\therefore g \propto \frac{1}{R^2}$, i.e. if the earth shrinks, g on the surface increases provided $M =$ constant. So, if the radius is halved, the value of g on the surface increases four times, and as a result time period of revolution, i.e. the length of the day will also be changed to a value given by $T = 2\pi\sqrt{\frac{R/2}{4g}} = 2\pi\sqrt{\frac{R}{8g}} = \pi\sqrt{\frac{R}{2g}}$)

(g) The potentials within two homogeneous spherical shells of same surface density are in the ratio 1 : 2. Calculate the ratio of their radii.

(*Ans.* See worked-out example at the end of Art. 7.11.1 of Chapter 7.)

(h) A hollow shaft is stronger than a solid one of same mass. Justify.

(*Ans.* See Art. 9.13.1 of Chapter 9)

(i) When some wax is rubbed into a cloth it becomes waterpoof. Explain.

(*Ans.* The surface tension of water prevents its passing through the pores in a piece of cloth. Touching the inside of the piece of cloth having spherical water drops rolling outside over the cloth, will break the surface of water and the piece of cloth begins to leak. If, however, some wax is rubbed into the piece of cloth, inter-facial surface tension at the interface of wax and cloth becomes prominent and prevents its passing through the pores totally, and as such the piece of cloth appears to be waterproof.)

2. (a) Let $a =$ constant > 0. A function is defined as follows

$$f(x) = \begin{cases} \frac{1}{2a} & \text{for } |x| < a \\ 0 & \text{for } |x| > a \end{cases}$$

Calculate the Fourier transform of $f(x)$ and state its limiting value as $a \to 0$.

(*Ans.* See worked-out examples of Appendix D)

(b) In rectangular coordinates the position vector is $\vec{r} = x\hat{i} + y\hat{j} + z\hat{k}$. Express the velocity $\vec{V}$ in cylindrical coordinates. (*Ans.* See Art. 3.7 of Chapter 3)

(c) State Gauss's divergence theorem for a vector field $\vec{A} = \vec{A}(x, y, z)$ having continuous first derivatives. (*Ans.* See Art. 2.18 of Chapter 2)

3. (a) Prove that the series $\frac{1}{1.2} + \frac{1}{2.3} + \frac{1}{3.4} + \cdots$ is convergent and deduce its sum.

[*Hints.* Here $u_n = \frac{1}{n \cdot (n+1)}$ and $u_{n+1} = \frac{1}{(n+1) \cdot (n+2)}$ and then proceed in the usual way as shown in different worked-out examples of Chapter 20]

(b) Evaluate the double integral

$$\iint \left(1 + \sqrt{x^2 + y^2}\right) dx\, dy$$

where R is the region bounded by the circle $x^2 + y^2 = 1$.

(*Ans.* Presence of $\sqrt{x^2 + y^2}$ in the integrand suggests the use of polar coordinates $x = r\cos\theta$, $y = r\sin\theta$ whereby $\sqrt{x^2 + y^2} = r$ and the double integral changes to

$$\iint_R (1 + r) \cdot r\, dr\, d\theta = \iint_R r\, dr\, d\theta + \iint_R r^2\, dr\, d\theta = 4\left[\int_0^{\pi/2} d\theta \int_0^1 r\, dr + \int_0^{\pi/2} d\theta \int_0^1 r^2\, dr\right]$$

with the help of symmetry and remembering the circle $x^2 + y^2 = 1$ to have radius $= 1$)

(c) Find the eigenvalues and eigenvectors of the matrix

$$\begin{pmatrix} 1 & -2 \\ -2 & -2 \end{pmatrix}$$

(*Ans.* See worked-out examples of Chapter 19)

4. (a) The generating function of the Legendre polynomials is

$$G(x, t) = (1 - 2xt + t^2)^{-1/2} = \sum_{n=0}^{\infty} t^n P_n(x), \quad |t| < 1$$

Show that if

$$(x - t)\frac{\partial G}{\partial x} = t\frac{\partial G}{\partial t}$$

for all t, then

$$x\frac{dP_l(x)}{dx} - \frac{dP_{l-1}(x)}{dx} = lP_l(x)$$

(*Ans.* Given

$$(1 - 2xt + t^2)^{-1/2} = \sum_{n=0}^{\infty} t^n P_n(x) \quad [\text{for } |t| < 1] \tag{1}$$

Differentiating (1) w.r.t., we have

$$-\frac{1}{2}(1 - 2xt + t^2)^{-3/2} \cdot (-2x + 2t) = \sum_{0}^{\infty} nt^{n-1} P_n(x)$$

$$\text{or} \quad (x - t)(1 - 2xt + t^2)^{-3/2} = \sum_{0}^{\infty} nt^{n-1} P_n(x) \tag{2}$$

Again differentiating (1) w.r.t. x, we obtain

$$t(1-2xt+t^2)^{-3/2} = \sum_0^\infty t^n P_n'(x) \quad \left[\text{where } P_n'(x) = \frac{d}{dx}P_n(x)\right] \tag{3}$$

Now on dividing (2) by (3), we have

$$\frac{(x-t)}{t} = \frac{\sum_0^\infty nt^{n-1}P_n(x)}{\sum_0^\infty t^n P_n'(x)} \quad \text{or} \quad (x-t)\sum_0^\infty t^n P_n'(x) = t\sum_0^\infty nt^{n-1}P_n(x)$$

On equating the coefficients of t^n from both sides, we may write

$$xP_n'(x) - P_{n-1}'(x) = nP_n(x)$$

which is the desired result.

(b) Solve by the method of separation of variables

$$3\frac{\partial u}{\partial x} + 2\frac{\partial u}{\partial y} = 0; \quad u(x,0) = 4e^{-x}$$

(*Ans.* See Q.32 of the Exercises of Chapter 18)

(c) Two balls are simultaneously drawn at random from a box containing 6 red balls, 4 white balls and 5 blue balls. Determine the probability that at least one among them is red.

(*Ans.* See Art. 24.13 of Chapter 24)

5. (a) Solve the differential equation $yy'' + y'^2 + 4 = 0$ for $x = 1 \rightarrow y = 3,\ y' = 0$.

(*Ans.* See Q.54 and Q.55 of the Exercises of Chapter 17)

(b) Consider the two dimensional Laplace equation

$$\frac{\partial^2 u}{\partial x^2} + \frac{\partial^2 u}{\partial y^2} = 0$$

Let $P = x + \lambda y$ and assume $u(x,y) = f(P)$ to be a solution of this partial differential equation. Show that $\lambda^2 + 1 = 0$.

(*Ans.* See Art. 18.5 and the worked-out examples of Chapters 17 and 18)

(c) Let $V(x,y) = \tan^{-1}\left(\frac{y}{x}\right)$. Calculate $\bar{A} = \bar{\nabla}V$ and evaluate $\oint_c \bar{A}\cdot d\bar{V}$, where c is the unit circle with centre at the origin.

(*Ans.* $\vec{A}$ to be calculated from

$$\vec{A} = \left(\hat{\imath}\frac{\partial}{\partial x} + \hat{\jmath}\frac{\partial}{\partial y} + \hat{k}\frac{\partial}{\partial z}\right)V;$$

then find $d\vec{V}$ from

$$\left(\hat{\imath}\frac{\partial V}{\partial x}dx + \hat{\imath}\frac{\partial V}{\partial y}dy + \hat{k}\frac{\partial \vec{V}}{\partial z}dz\right) \quad \text{where } V(x,y) = \tan^{-1}\left(\frac{y}{x}\right),$$

and then calculate $\oint_c \vec{A}\cdot d\vec{V}$, c being the unit circle with centre at the origin.)

6. (a) A particle of unit mass moves according to the equation $\bar{r} = \hat{\imath}(2+3t^2) + \hat{\jmath}5t^2 + \hat{k}t$; find the force $\vec{F}$ acting on it, torque $\vec{N}$ and angular momentum $\vec{L}$ about the origin. Hence verify $\vec{N} = \frac{d\vec{L}}{dt}$. (*Ans.* See worked-out examples at the end of Art. 3.17A of Chapter 3)

(b) Two binary stars of masses 10^{21} kg and 2×10^{21} kg are 10^6 km apart. They rotate about their centre of mass with angular velocity ω. Find ω. Given $G = 6.67 \times 10^{-11}$ Nm^2 kg^{-2}.

(*Ans.* See worked-out example no. 10 on page 484 and also worked-out example 5 at the end of Art. 7.25.3 of Chapter 7)

7. (a) A rigid body is rotating with angular velocity $\vec{\omega}$ about an axis through O, the origin and having direction cosines l, m, n.

Find an expression for the kinetic energy of rotation. How is the expression modified for principal axes system? (*Ans.* See Art. 4.10 of Chapter 4)

(b) A flat circular disc of mass 0.05 kg and diameter 0.02 m rolls on its edge on a smooth horizontal surface with a velocity of 0.05 m/s. Calculate its total kinetic energy.

(*Ans.* See worked-out examples given in different cases of Art. 4.7 of Chapter 4)

8. (a) Prove that the potential due to a circular disc of uniform surface density σ and radius a at a point on its circumference is $-4G\sigma a$.

(*Ans.* See worked-out example 1 at the end of Art. 7.17.3 of Chapter 7)

(b) If the gravitational force between two bodies had been inversely proportional to the fourth power of the distance between them, then find out the escape velocity at the surface of the earth, given $g = 10\,\text{m/sec}^2$ and radius of the earth $= 6400 \times 10^3$ m.

(*Ans.* See worked-out example 4 at the end of Art. 7.25.3 and also the Exercises of Chapter 7)

(c) Find the scalar potential for the force field given by $\vec{F} = -r\vec{r}$.

(*Ans.* See worked-out examples at the end of Art. 3.17A and in the Exercises of Chapter 3)

9. (a) Show that within elastic limit, the depression at a point P in a horizontal beam due to force F applied at a point Q is the same as the depression at Q due to the same force F applied at P. (*Ans.* See Art. 9.19 of Chapter 9)

(b) When some wax is rubbed on a cloth it becomes waterproof. Explain.

(*Ans.* See answer to Q.1(i) above)

(c) A spherical soap bubble of radius 1 cm is formed inside another of radius 2 cm. Find the radius of a single soap bubble whose internal pressure is the same as that of the smaller bubble.

(*Ans.* See worked out examples at the end of Art. 10.10 and also the Exercises of Chapter 10.)

10. (a) Set up Euler's equation in hydrodynamics for an incompressible fluid.

(*Ans.* See Art. 12.5 of Chapter 12)

(b) From the above equation deduce Bernoulli's theorem for steady streamline flow.

(*Ans.* See Art. 12.7 of Chapter 12)

(c) State Stoke's law in connection with the viscous drag experienced by a moving spherical body. Hence find the terminal velocity of a rain drop with radius 0.001 cm falling through air. [η is air $= 1.8 \times 10^{-4}$ erg unit]

(*Ans.* See Art. 11.13 and 11.13.1 of Chapter 11)

11. (a) The displacement of a particle of mass m executing under-damped simple harmonic motion is given by

$$x(t) = Ae^{-bt}\cos(w't - \phi)$$

where $\omega' = \sqrt{\omega_0^2 - b^2}$, $\omega_0 =$ natural frequency and $b =$ damping factor.

Calculate A and ϕ subject to the initial conditions at $t_0 = 0$, $x(t) = 0$ and $\dot{x}(t) = v_0$. Hence show that in this situation, the energy of the oscillator is

$$E(t) = \frac{1}{2}mv_0^2 \cdot e^{-2bt}$$

(*Ans.* See worked-out examples given in the Exercises of Chapter 5)

(b) Show that in forced vibration

$$\frac{\text{average kinetic energy}}{\text{average potential energy}} = \frac{\omega^2}{\omega_0^2}$$

where ω_0 is the natural frequency of the oscillator and ω is the frequency of the forcing system. (*Ans.* See worked-out examples at the end of Art. 5.16 of Chapter 5)

(c) A particle vibrates harmonically at a frequency 0.25 Hz. At the initial moment it is in an equilibrium position moving at a speed of 10 cm/sec. Write down the equation of motion.

(*Ans.* General equation of motion of a particle vibrating harmonically is given by

$$y = a \sin \frac{2\pi}{\lambda}(ct - x) \tag{1}$$

(a = amplitude, c = velocity). In this case at $x = 0$, $c = n\lambda$ or $n = \frac{c}{\lambda}$ (n = frequency)—whence λ can be found.

$\therefore y = a \sin 2\pi nt$ (put $n = 0.25$ Hz) is the equation at $x = 0$ and putting the values of c and λ in (1) we can write down the general equation.

Second Paper—Unit IIA

1. (a) What do you mean by self-diffusion? Define the coefficient of self-diffusion.

(*Ans.* See Arts. 13.7 and 13.8)

(b) Check that $df = x^2y\,dx + 3x^3\,dy$ is a perfect differential.

$\left(\textit{Ans.}\text{ Find } \frac{\partial f}{\partial x} \text{ and } \frac{\partial^2 f}{\partial x \partial y}\text{; again find } \frac{\partial f}{\partial y} \text{ and } \frac{\partial^2 f}{\partial y \partial x} \text{ and show that } \frac{\partial^2 f}{\partial x \partial y} = \frac{\partial^2 f}{\partial y \partial x} = 2x + 9x^2\right)$

(c) An Iodine molecule can be modelled as two hard spheres connected by an elastic spring. What is the average energy of one mole of Iodine at absolute temperature T? (*Ans.* See worked-out examples given in the Exercises of Chapter 14)

(f) Assuming that the gas molecules are non-interacting and only can have kinetic energy, write down the Maxwell's speed distribution formula.

(*Ans.* Due to collisions, there will be a non-uniform distribution of velocities or speeds of the molecules in a given volume of any gas. When however a steady state corresponding to a particular temperature will be attained, according to the laws of probability, there will always be a fixed number of molecules per c.c., moving between specified velocity range c and $c + dc$. This number is given by Maxwell's velocity or speed distribution formula and is given by

$$dN_c = 4\pi N\left(\frac{m}{2\pi kT}\right)^{3/2} e^{-mc^2/2kT} c^2 dc$$

where the symbols have usual meanings and significance.)

2. Maxwells' energy distribution law is given by

$$n(\epsilon)\,d\epsilon = A\sqrt{\epsilon}\,e^{-\frac{\epsilon}{kT}}$$

(a) Calculate the overall normalisation, A, of this distribution, for N numbers of total gas molecules present at a temperature T.

One can use: $\int_{-\infty}^{\infty} e^{-\beta x^2}\, dx = \sqrt{\frac{\pi}{\beta}}$;

$$\int_{-\infty}^{\infty} x^2 e^{-\beta x^2}\, dx = -\frac{d}{d\beta}\int_{-\infty}^{\infty} e^{-\beta x^2}\, dx$$

(*Ans.* For calculation of A, See Art. 2.9 of Gupta-Roy's Thermal Physics, and Art. 14.8 of our Book.)

(b) Hence calculate the most probable speed $\tilde{c}_r$ at a temperature T.

(*Ans.* See Art. 2.10 of Gupta-Roy's Thermal Physics and Art. 14.6.3 and 14.8 of our Book)

(c) At any temperature T, if we re-scale the speed of all the gas molecules, by $\tilde{c}_r$, show that the speed distribution becomes independent of temperature.

(*Ans.* See Art. 14.9.1 of Chapter 14)

3. (a) i. What do you mean by the *mean free path* of gas molecules?

(*Ans.* The molecules of any gas move at random and because of their finite size, in course of their movements they collide with each other. Each collision would result in a change of velocity (magnitude) and also direction. It may so happen that a molecule encounters a collision within a very short distance, then comes to another collision after traversing a longer distance. So the path of a molecule amidst the molecules in chaos consists of zig-zags, in which the length of each segment of the path is not the same. If we could follow the path of the molecule, the path would be as shown in the figure given below. The path described between any two successive collisions is called *free path* of the molecule. Since these paths vary in length, it is convenient to use the term *mean free path*—which is the average distance traversed between successive collisions. This is obtained by adding up the different free paths and then dividing the sum by the number of free paths. The mean free path of a molecule is usually denoted by λ.)

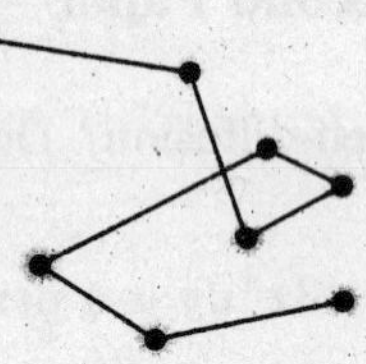

ii. Estimate the mean free path $\bar{\lambda}$ of atomic Helium at STP. Given, mass of a He atom is 5×10^{-27} kg. Density of He at STP is 1.78×10^{-4} kg.m^{-3}. One can assume He-atoms as hard spheres with radius 0.1 nm.

(*Ans.* To find n, use $mn = \rho$ and then apply the formula (14.13.4))

(b) i. Colloidal particles each of mass m are suspended in air under the force of gravitation. Show that the number density of the particles decreases exponentially with the height. One can neglect the density of air w.r.t. the density of the particles.

(*Ans.* Relation (14.20.2) to be deduced with the help of the *Alternative method* given in Art. 14.20 of Chapter 14.)

ii. Suppose, these particles are positively charged and an electric field is applied along the vertical, in a direction opposite to the gravitational force. How will the above distribution be modified?

(*Ans.* Since the electric field is applied along the vertical in a direction opposite to the force of gravity, the number density of the particles should increase with height—thereby modifying the above distribution to some extent.)

4. (a) Two monatomic gases have atomic weights A_1 and A_2. Show that the ratio of their thermal conductivities is given by

$$\frac{K_1}{K_2} = \frac{\eta_1 A_1}{\eta_2 A_2}$$

η_1 and η_2 are their coefficients of viscosity. (*Ans.* See Q.17 of the Exercises of Chapter 14)

(b) i. Using kinetic theory of gas, find out an expression of the coefficient of self diffusion D, for an ideal gas in which concentration gradient exists only along one direction. (*Ans.* See Art. 14.17.3 or 14.18 of Chapter 14)

ii. Show that $D \propto P^{-1}T^{1.5}$, P and T being the pressure and temperature of the gas.

(*Ans.* From $D = \frac{1}{3}\lambda\bar{c}$ and $\lambda = \frac{1}{\sqrt{2}\pi\sigma^2 n}$, we may write $D = \frac{\bar{c}}{3\sqrt{2}\pi\sigma^2 n}$. From this relation, we can readily deduce the dependence of the diffusion coefficient D on temperature and pressure. From the above relation, we may write $D \propto \frac{1}{n} \propto \frac{T}{p}$ ($\because p = nkT$ or $n \propto \frac{p}{T}$, symbols having usual meanings). Again $D \propto \bar{c} \propto \sqrt{T}$ ($\because \bar{c} = \sqrt{\frac{8kT}{\pi m}}$, symbols having usual meanings). So $D \propto p^{-1}T^{1.5}$. The observed results from the experimental measurements on diffusion show that D indeed varies inversely as pressure p, but the variation of D with temperature is not as $T^{1.5}$. The power of T lies between 1.75 and 2. This more rapid increase of D with temperature is due to the attractive forces between the molecules.)